Table 7–14

Summary of Fatigue Equations $\mathbf{S}_e = \mathbf{k}_a k_b \mathbf{k}_c \mathbf{k}_d \mathbf{k}_e \boldsymbol{\phi}_{0.30} \bar{S}_{ut}$ in SI Units for Steels*

Quantity	Relation	Table or Equation
Ultimate strength	$\mathbf{S}_{ut} = 3.41(1, 0.041)\mathbf{H}_B$	Eq. (5–23)
Fatigue ratio	$\boldsymbol{\phi}_{0.30} = 0.506\mathbf{LN}(1, 0.138)$, $\bar{S}_{ut} \le 1460$ MPa	
Endurance limit		
Bending	$\mathbf{S}'_e = \boldsymbol{\phi}_{0.30}\bar{S}_{ut} = 0.506\bar{S}_{ut}\mathbf{LN}(1, 0.138)$, $\bar{S}_{ut} \le 1460$ MPa	
Axial	$(\mathbf{S}'_e)_{ax} = 0.724\bar{S}_{ut}^{-0.0778}\bar{S}_{ut}\mathbf{LN}(1, 0.264)$	Above Eq. (7–20)
Torsion	$\mathbf{S}'_{se} = 0.130\bar{S}_{ut}^{0.125}\bar{S}_{ut}\mathbf{LN}(1, 0.263)$	Above Eq. (7–20)
Surface factor		
Ground	$\mathbf{k}_a = 1.58\bar{S}_{ut}^{-0.086}\mathbf{LN}(1, 0.120)$	Table 7–5
Machined, CR	$\mathbf{k}_a = 4.45\bar{S}_{ut}^{-0.265}\mathbf{LN}(1, 0.058)$	Table 7–5
Hot-rolled	$\mathbf{k}_a = 58.1\bar{S}_{ut}^{-0.719}\mathbf{LN}(1, 0.110)$	Table 7–5
As-forged	$\mathbf{k}_a = 271\bar{S}_{ut}^{-0.995}\mathbf{LN}(1, 0.045)$	Table 7–5
Size	$k_b = (d_e/7.62)^{-0.107} = 1.24d_e^{-0.107}$	Eq. (7–10)
Loading factor		
Bending	$\mathbf{k}_c = \mathbf{LN}(1, 0)$	Eq. (7–20)
Axial	$\mathbf{k}_c = 1.43\bar{S}_{ut}^{-0.0778}\mathbf{LN}(1, 0.125)$	Eq. (7–21)
Torsion	$\mathbf{k}_c = 0.258\bar{S}_{ut}^{0.125}\mathbf{LN}(1, 0.125)$	Eq. (7–22)
Temperature factor	$\mathbf{k}_d$ (as applicable)	Eqs. (7–23), (7–24)
Miscellaneous factor	$\mathbf{k}_e$ (as applicable)	
Stress-concentration		
Infinite life	$\mathbf{K}_6 = (\mathbf{K}_f)_{10^6} = \dfrac{K_t\mathbf{LN}(1, C_{Kf})}{1 + \dfrac{2}{\sqrt{r}}\dfrac{K_t - 1}{K_t}\sqrt{a}}$	Table 7–12
Finite life		Eq. (7–27)
	$\bar{K}_3 = (\bar{K}'_f)_{10^3} = 1 + \left[(\bar{K}_f)_{10^6} - 1\right]\left[-0.18 + 0.624(10^{-3})\bar{S}_{ut} - 0.948(10^{-7})\bar{S}_{ut}^2\right]$	
	$\bar{K}_N = \dfrac{\bar{K}_3^2}{\bar{K}_6} N^{(-1/3)\log(\bar{K}_3/\bar{K}_6)}$	Eq. (7–34)

* S'_e and S_{ut} in MPa, d and d_e in mm.

Mechanical Engineering Design

McGRAW-HILL SERIES IN MECHANICAL ENGINEERING

Anderson:	*Computational Fluid Dynamics: The Basics with Applications*
Anderson:	*Modern Compressible Flow: With Historical Perspective*
Arora:	*Introduction to Optimum Design*
Borman and Ragland:	*Combustion Engineering*
Cengel and Boles:	*Thermodynamics: An Engineering Approach*
Cengel and Turner:	*Fundamentals of Thermal-Fluid Sciences*
Cengel:	*Heat Transfer: A Practical Approach*
Cengel:	*Introduction to Thermodynamics & Heat Transfer*
Culp:	*Principles of Energy Conversion*
Dieter:	*Engineering Design: A Materials & Processing Approach*
Doebelin:	*Engineering Experimentation: Planning, Execution, Reporting*
Driels:	*Linear Control Systems Engineering*
Edwards and McKee:	*Fundamentals of Mechanical Component Design*
Gibson:	*Principles of Composite Material Mechanics*
Hamrock:	*Fundamentals of Fluid Film Lubrication*
Hamrock/Schmid/Jacobson:	*Fundamentals of Machine Elements*
Heywood:	*Internal Combustion Engine Fundamentals*
Histand and Alciatore:	*Introduction to Mechatronics and Measurement Systems*
Holman:	*Experimental Methods for Engineers*
Jaluria:	*Design and Optimization of Thermal Systems*
Kays and Crawford:	*Convective Heat and Mass Transfer*
Kelly:	*Fundamentals of Mechanical Vibrations*
Martin:	*Kinematics and Dynamics of Machines*
Mattingly:	*Elements of Gas Turbine Propulsion*
Modest:	*Radiative Heat Transfer*
Norton:	*Design of Machinery*
Oosthuizen and Carscallen:	*Compressible Fluid Flow*
Oosthuizen and Naylor:	*Introduction to Convective Heat Transfer Analysis*
Reddy:	*An Introduction to Finite Element Method*
Rosenberg and Karnopp:	*Introduction to Physical Systems Dynamics*
Schlichting:	*Boundary-Layer Theory*
Shames:	*Mechanics of Fluids*
Shigley and Mischke:	*Mechanical Engineering Design*
Shigley and Uicker:	*Theory of Machines and Mechanisms*
Stoecker:	*Design of Thermal Systems*
Stoecker and Jones:	*Refrigeration and Air Conditioning*
Turns:	*An Introduction to Combustion: Concepts and Applications*
Ullman:	*The Mechanical Design Process*
Wark:	*Advanced Thermodynamics for Engineers*
Wark and Richards:	*Thermodynamics*
White:	*Viscous Fluid Flow*
Zeid:	*CAD/CAM Theory and Practice*

Mechanical Engineering Design

Sixth Edition

Joseph E. Shigley
Professor Emeritus, The University of Michigan

Charles R. Mischke
Professor of Mechanical Engineering Emeritus, Iowa State University

Boston Burr Ridge, IL Dubuque, IA Madison, WI New York San Francisco
St. Louis Bangkok Bogotá Caracas Lisbon London Madrid Mexico City
Milan New Delhi Seoul Singapore Sydney Taipei Toronto

McGraw-Hill Higher Education

A Division of The ***McGraw-Hill*** *Companies*

MECHANICAL ENGINEERING DESIGN

Published by McGraw-Hill, an imprint of The McGraw-Hill Companies, Inc. 1221 Avenue of the Americas, New York, NY 10020.

This book is printed on acid-free paper.

2 3 4 5 6 7 8 9 0 VNH/VNH 0 9 8 7 6 5 4 3 2 1

ISBN 0-07-365939-8

President: *Kevin T. Kane*
Senior sponsoring editor: *Jonathan Plant*
Senior developmental editor: *Kelley Butcher*
Marketing manager: *John Wannemacher*
Project manager: *Jim Labeots*
Production supervisor: *Heather Burbridge*
Coordinator freelance design: *Gino Cieslik*
Supplement coordinator: *Matthew Perry*
Media technology producer: *Phillip Meek*
Cover design: *Gino Cieslik*
Interior design: *Joel Davies, Z Graphics*
Cover illustration: *Tony Stone Images*
Compositor: *Techsetters, Inc.*
Typeface: *10/12 Times*
Printer: *Von Hoffmann Press, Inc.*

Library of Congress Cataloging-in-Publication Data

Shigley, Joseph Edward.
Mechanical engineering design / Joseph Edward Shigley, Charles R. Mischke.–6th ed.
p. cm.
ISBN 0-07-365939-8 (alk. paper)
1. Machine design. I. Mischke, Charles R. II. Title.

TJ230 .S5 2001
621.8'15–dc21

00-055039

www.mhhe.com

Dedication

To the over 3500 of my students who asked many good questions of me, learned to ask them of themselves, to answer them, and then went on to make their Alma Maters proud.

Charles R. Mischke

About the Authors

Charles R. Mischke has held positions on the faculty of the University of Kansas, was professor and chairman of mechanical engineering at Pratt Institute in New York, and professor of mechanical engineering at Iowa State University. He was named Alcoa Foundation Professor in 1974, received the Ralph R. Teeter Award of the Society of Automotive Engineers in 1977, received Iowa State's Outstanding Teaching Award in 1980, named Fellow of the American Society of Mechanical Engineers in 1986, received the Association of American Publishers Award in 1987, the American Society of Mechanical Engineers Machine Design Award in 1990, the Iowa Legislature Teaching Excellence Award in 1991, the Ralph Coats Roe Award of the American Society for Engineering Education in 1991, Life Fellow of A.S.M.E., 1993, and the Centennial Certificate of A.S.E.E. in 1993.

In addition to many research papers he authored *Elements of Mechanical Analysis*, 1963, *Introduction to Computer-Aided Design*, 1968, *Mathematical Modelbuilding*, 1980, eight Mechanical Designers Notebooks, 1990, coauthored *Mechanical Engineering Design*, 5th ed., 1989, and was Coeditor-in-Chief of the *Standard Handbook of Machine Design*, 1986, 1996.

He was awarded the BSME and MME degrees of Cornell University, and the Ph.D. of the University of Wisconsin. He is a licensed Professional Engineer in Iowa and Kansas. He is a member of the Iowa Railroad Historical Society, and a diesel locomotive engineer on the B&SV Railroad.

Joseph E. Shigley (deceased May 1994)
He was Professor Emeritus at the University of Michigan, Fellow in the American Society of Mechanical Engineers, received the Worcester Reed Warner medal in 1977, and their Machine Design Award in 1985. He was the author of eight books, including *Theory of Machines and Mechanisms* (with John J. Uicker, Jr.), and *Applied Mechanics of Materials*. He was Coeditor-in-Chief of the *Standard Handbook of Machine Design*. He began *Machine Design* as sole author in 1956, and it evolved into *Mechanical Engineering Design*, setting the model for such textbooks.

He was awarded the B.S.M.E. and B.S.E.E. degrees of Purdue University, and the M.S. of the University of Michigan.

Joseph Edward Shigley indeed made a difference.

Contents in Brief

Preface to the Student xvii
Preface to the Sixth Edition xix

Part 1 Basics 1

1 **Introduction 3**
2 **Addressing Uncertainty 47**
3 **Stress 93**
4 **Deflection and Stiffness 175**

Part 2 Failure Prevention 253

5 **Materials 255**
6 **Failures Resulting from Static Loading 315**
7 **Failure Resulting from Variable Loading 359**

Part 3 Design of Mechanical Elements 443

8 **Screws, Fasteners, and the Design of Nonpermanent Joints 445**
9 **Welding, Brazing, Bonding, and the Design of Permanent Joints 527**
10 **Mechanical Springs 589**
11 **Rolling-Contact Bearings 689**
12 **Lubrication and Journal Bearings 739**
13 **Gearings—General 831**
14 **Spur and Helical Gears 897**
15 **Bevel and Worm Gears 951**
16 **Clutches, Brakes, Couplings, and Flywheels 991**
17 **Flexible Mechanical Elements 1049**
18 **Shafts and Axles 1107**

Appendixes

A **Statistical Relations** **1153**

B **Regression Relations** **1161**

C **Propagation of Error Relations** **1163**

D **Simulation** **1165**

E **Useful Tables** **1169**

F **Answers to Selected Problems** **1231**

Index 1237

Contents

Preface to the Student xvii
Preface to the Sixth Edition xix

Part 1 Basics 1

1 Introduction 3

1–1 Design 5
1–2 Mechanical Engineering Design 7
1–3 Your Path to Competence 12
1–4 Technology Can Be Fragile 12
1–5 Interaction between Design Process Elements 14
1–6 Codes and Standards 17
1–7 Economics 18
1–8 Safety and Product Liability 20
1–9 The Adequacy Assessment 20
1–10 Uncertainty 22
1–11 Stress and Strength 26
1–12 Design Factors and Factors of Safety 29
1–13 Reliability 30
1–14 Numbers, Units, and Preferred Units 31
Problems 37

2 Addressing Uncertainty 47

2–1 Questions Come with the Territory 49
2–2 Estimating Statistical Parameters 50
2–3 Probability Density Function and Cumulative Distribution Function 53
2–4 Linear Regression 55
2–5 Propagation of Error 58
2–6 Simulation 61
2–7 Design Factor and Factor of Safety 63
2–8 Limits and Fits 68
2–9 Dimensions and Tolerances 71
2–10 Summary 77
Problems 79

3 Stress 93

3–1 Stress Components 94
3–2 Mohr Circles 96
3–3 Triaxial Stress 100
3–4 Uniformly Distributed Stress 102
3–5 Elastic Strain 103
3–6 Stress-Strain Relations 104
3–7 Equilibrium 104
3–8 Shear and Moment 106
3–9 Singularity Functions 108
3–10 Normal Stress in Flexure 111
3–11 Beams with Asymmetrical Sections 118
3–12 Shear Stresses in Beams 118
3–13 Shear Stresses in Rectangular-Section Beams 121
3–14 Torsion 123
3–15 Stress Concentration 129
3–16 Stresses in Cylinders 132
3–17 Rotating Rings 135
3–18 Press and Shrink Fits 135
3–19 Temperature Effects 137
3–20 Curved Members in Flexure 138
3–21 Contact Stress 144
3–22 Propagation of Error 149
3–23 Summary 154
Problems 154

4 Deflection and Stiffness 175

4–1 Spring Rates 176
4–2 Tension, Compression, and Torsion 177
4–3 Deflection Due to Bending 178
4–4 Finding Deflection by Integration 180
4–5 Finding Deflection by the Area-Moment Method 187
4–6 Finding Deflection by the Use of Singularity Functions 190

- 4-7 Strain Energy 193
- 4-8 Castigliano's Theorem 195
- 4-9 Statistically Indeterminate Problems 198
- 4-10 Deflection of Curved Members 200
- 4-11 Compression Members—General 204
- 4-12 Long Columns with Central Loading 206
- 4-13 Intermediate-Length Columns with Central Loading 210
- 4-14 Columns with Eccentric Loading 210
- 4-15 Struts, or Short Compression Members 214
- 4-16 An Application: Round-Bar Clamps 216
- 4-17 Deflection of Energy-Dissipative Assemblies 220
- 4-18 Shock and Impact 229
- 4-19 Suddenly-Applied Loading 230
- 4-20 Propagation of Error 233
- Problems 237

Part 2 Failure Prevention 253

5 Materials 255

- 5-1 Static Strength 256
- 5-2 Plastic Deformation 261
- 5-3 Strength and Cold Work 265
- 5-4 Hardness 268
- 5-5 Impact Properties 269
- 5-6 Temperature Effects 271
- 5-7 Numbering Systems 272
- 5-8 Sand Casting 274
- 5-9 Shell Molding 274
- 5-10 Investment Casting 275
- 5-11 Powder-Metallurgy Process 275
- 5-12 Hot-Working Processes 275
- 5-13 Cold-Working Processes 276
- 5-14 The Heat Treatment of Steel 277
- 5-15 Alloy Steels 279
- 5-16 Corrosion-Resistant Steels 280
- 5-17 Casting Materials 281
- 5-18 Nonferrous Metals 283
- 5-19 Plastics 285
- 5-20 Notch Sensitivity 287
- 5-21 Introduction to Fracture Mechanics 288
- 5-22 Stress-Corrosion Cracking 303
- 5-23 Quantitative Estimation of Properties of Cold-Worked Metals 303
- 5-24 Quantitative Estimation of Properties of Heat-Treated Steels 307
- Problems 308

6 Failures Resulting from Static Loading 315

- 6-1 Static Strength 316
- 6-2 Stress Concentration 319
- 6-3 Hypotheses of Failure 322
- 6-4 Ductile Materials: Maximum-Shear-Stress (Tresca or Guest) Hypothesis 324
- 6-5 Ductile Materials: Strain-Energy Hypotheses 326
- 6-6 Ductile Materials: Internal-Friction Hypothesis 332
- 6-7 Criticism of Hypotheses by Data in Ductile Materials 334
- 6-8 Brittle Materials: Maximum-Normal-Stress (Rankine) Hypothesis 335
- 6-9 Brittle Materials: Modifications of the Mohr Hypothesis 337
- 6-10 The Criticism of Hypotheses by Data in Brittle Materials 341
- 6-11 What Our Failure Models Tell Us 342
- 6-12 Interference—General 343
- 6-13 Static or Quasi-Static Loading on a Shaft 347
- Problems 352

7 Failure Resulting from Variable Loading 359

- 7-1 Introduction to Fatigue in Metals 360
- 7-2 Strain-Life Relationships 361
- 7-3 Stress-Life Relationships 367
- 7-4 The Endurance Limit 369
- 7-5 Fatigue Strength 372
- 7-6 Endurance-Limit Modifying Factors 374
- 7-7 Stress Concentration and Notch Sensitivity 383

7–8 Applying What We Have Learned about Endurance Limit and Endurance Strength 387
7–9 The Distributions 395
7–10 Characterizing Fluctuating Stresses 396
7–11 Failure Loci under Variable Stresses 398
7–12 Torsional Fatigue Strength under Pulsating Stresses 408
7–13 Combinations of Loading Modes 408
7–14 Stochastic Failure Loci under Fluctuating Stresses 411
7–15 Cumulative Fatigue Damage 414
7–16 The Fracture-Mechanics Approach 421
7–17 Surface Fatigue Strength 423
7–18 The Designer's Fatigue Diagram 429
7–19 An Important Design Decision: The Design Factor in Fatigue 431
Problems 436
Summary of Parts 1 and 2 441

Part 3 Design of Mechanical Elements 443

8 Screws, Fasteners, and the Design of Nonpermanent Joints 445

8–1 Thread Standards and Definitions 446
8–2 The Mechanics of Power Screws 450
8–3 Threaded Fasteners 457
8–4 Joints—Fastener Stiffness 458
8–5 Joints—Member Stiffness 461
8–6 Bolt Strength 466
8–7 Tension Joints—The External Load 470
8–8 Relating Bolt Torque to Bolt Tension 471
8–9 Statistically Loaded Tension Joint—Preload 477
8–10 Gasketed Joints 483
8–11 Tension Joints—Dynamic Loading 484
8–12 Adequacy Assessment, Specification Set, Decision Set, and Design 492
8–13 Shear Joints 498
8–14 Setscrews 504
8–15 Pins and Keys 504
Problems 513

9 Welding, Brazing, Bonding, and the Design of Permanent Joints 527

9–1 Welding Symbols 528
9–2 Butt and Fillet Welds 530
9–3 Stresses in Welded Joints in Torsion 535
9–4 Stresses in Welded Joints in Bending 540
9–5 The Strength of Welded Joints 542
9–6 Specification Set, Adequacy Assessment, and Decision Set 544
9–7 Static Loading 549
9–8 Fatigue Loading 554
9–9 Resistance Welding 557
9–10 Bolted and Riveted Joints Loaded in Shear 558
9–11 Adhesive Bonding and Design Considerations 562
Problems 579

10 Mechanical Springs 589

10–1 Stresses in Helical Springs 590
10–2 The Curvature Effect 591
10–3 Deflection of Helical Springs 592
10–4 Extension Springs 592
10–5 Compression Springs 595
10–6 Stability 596
10–7 Spring Materials 598
10–8 Helical Compression Springs for Static Service 609
10–9 Critical Frequency of Helical Springs 620
10–10 Fatigue Loading 622
10–11 Helical Compression Springs for Dynamic Service 625
10–12 Design of a Helical Compression Spring for Dynamic Service 629
10–13 Design of Extension Springs 637
10–14 Designing Helical Coil Torsion Springs 664

10–15 Belleville Springs 678
10–16 Miscellaneous Springs 678
10–17 Summary 680
Problems 683

11 Rolling-Contact Bearings 689

11–1 Bearing Types 690
11–2 Bearing Life 693
11–3 Bearing Load–Life Trade-off at Constant Reliability 694
11–4 Bearing Survival: The Reliability–Life Trade-off 696
11–5 Load–Life–Reliability Trade-off 697
11–6 Combined Radial and Thrust Loading 699
11–7 Variable Loading 704
11–8 Selection of Ball and Cylindrical Roller Bearings 709
11–9 Selection of Tapered Roller Bearings 714
11–10 Adequacy Assessment for Selected Rolling-Contact Bearings 724
11–11 Lubrication 728
11–12 Mounting and Enclosure 729
Problems 732

12 Lubrication and Journal Bearings 739

12–1 Types of Lubrication 741
12–2 Viscosity 741
12–3 Petroff's Equation 744
12–4 Stable Lubrication 750
12–5 Thick-Film Lubrication 751
12–6 Hydrodynamic Theory 752
12–7 Design Considerations 757
12–8 The Relations of the Variables 759
12–9 Steady-State Conditions in Self-Contained Bearings 722
12–10 Clearance 781
12–11 Pressure-Fed Bearings 792
12–12 Loads and Materials 803
12–13 Bearing Types 805
12–14 Thrust Bearings 806
12–15 Boundary-Lubricated Bearings 807
Problems 823

13 Gearing—General 831

13–1 Types of Gears 832
13–2 Nomenclature 833
13–3 Tooth Systems 835
13–4 Conjugate Action 837
13–5 Involute Properties 838
13–6 Fundamentals 839
13–7 Contact Ratio 844
13–8 Interference 845
13–9 The Forming of Gear Teeth 848
13–10 Straight Bevel Gears 850
13–11 Parallel Helical Gears 851
13–12 Worm Gears 855
13–13 Gear Trains 856
13–14 Force Analysis—Spur Gearing 860
13–15 Force Analysis—Bevel Gearing 863
13–16 Force Analysis—Helical Gearing 866
13–17 Force Analysis—Worm Gearing 869
13–18 Gear Ratios and Numbers of Teeth 874
13–19 Gear-Shaft Speeds and Bearings 878
Problems 883

14 Spur and Helical Gears 897

14–1 The Lewis Bending Equation 898
14–2 Surface Durability 907
14–3 AGMA Stress Equations 909
14–4 AGMA Strength Equations 910
14–5 Geometry Factors I and J (Z_I and Y_J) 915
14–6 Elastic Coefficient C_p (Z_E) 920
14–7 Dynamic Factor K'_V 920
14–8 Overload Factor K_O 922
14–9 Surface Condition Factors C_f and Z_R 922
14–10 Size Factor K_s 923
14–11 Load-Distribution Factor K_m or K_H 923
14–12 Hardness-Ratio Factor C_H 924
14–13 Load Cycles Factors Y_N and Z_N 926
14–14 Reliability Factors K_R and Y_Z 927
14–15 Temperature Factors K_T and Y_θ 928
14–16 Rim-Thickness Factor K_B 928

14–17 Safety Factors S_F and S_H 929
14–18 Analysis 929
14–19 An Adequacy Assessment of a Gear Mesh 940
14–20 Design of a Gear Mesh 942
Problems 947

15 Bevel and Worm Gears 951
15–1 Bevel Gearing—General 952
15–2 Bevel-Gear Stresses and Strengths 954
15–3 AGMA Equation Factors 957
15–4 Straight-Bevel Gear Analysis 969
15–5 Design of a Straight-Bevel Gear Mesh 972
15–6 Worm Gearing—AGMA Equation 974
15–7 Worm-Gear Analysis 978
15–8 Designing a Worm-Gear Mesh 980
15–9 Buckingham Wear Load 985
Problems 986

16 Clutches, Brakes, Couplings, and Flywheels 991
16–1 Rudiments of Brake Analysis 993
16–2 Internal Expanding Rim Clutches and Brakes 999
16–3 External Contracting Rim Clutches and Brakes 1008
16–4 Band-Type Clutches and Brakes 1011
16–5 Friction-Contact Axial Clutches 1013
16–6 Disk Brakes 1016
16–7 Cone Clutches and Brakes 1022
16–8 Self-Locking Tapers and Torque Capacity 1024
16–9 Energy Considerations 1026
16–10 Temperature Rise 1027
16–11 Friction Materials 1031
16–12 Miscellaneous Clutches and Couplings 1032
16–13 Flywheels 1034
16–14 Adequacy Assessment for Clutches and Brakes 1039
Problems 1040

17 Flexible Mechanical Elements 1049
17–1 Belts 1050
17–2 Flat- and Round-Belt Drives 1053
17–3 V Belts 1069
17–4 Timing Belts 1077
17–5 Roller Chain 1079
17–6 Wire Rope 1088
17–7 Flexible Shafts 1097
Problems 1098

18 Shafts and Axles 1107
18–1 Introduction 1108
18–2 Sufficing Geometric Constraints 1111
18–3 Sufficing Strength Constraints 1120
18–4 The Adequacy Assessment 1128
18–5 Shaft Materials 1134
18–6 Hollow Shafts 1135
18–7 Critical Speeds 1135
18–8 Shaft Design 1141
18–9 Computer Considerations 1142
Problems 1146

Appendixes
A Statistical Relations 1153
B Linear Regression 1161
C Propagation of Error Relations 1163
D Simulation 1165
E Useful Tables 1169
F Solutions to Selected Problems 1231

Index 1237

Preface to the Student

Design is the essential task of engineering. Engineers who do not design, support the designer. Whether you design or not, you must understand it, and have knowledge of the designer's needs in order to effectively help. The course that uses this text will start you down this path.

Design is learned with and under a master designer. There are talents and bodies of knowledge that a designer needs.

- Knowledge, ability, talent to generate ideas, possibilities.
- Knowledge of how to evaluate ideas and sort wheat from chaff.
- Knowledge of the structure of design, and how to tailor the process to the task at hand.

These bodies of knowledge are largely independent of the necessary knowledge about how the universe behaves. Design can be learned, and it is the role of this course to begin the process. Industrial experience and your mentors will continue the process.

In this book you will stand on a large amount of prerequisite information. We will bring it together as it applies to machinery, extend it further, and add to your knowledge about the universe. This is essential business to be sure, but it does not address the design process. As we move along in the first two parts of the book, some design ideas will be folded in. Just because they are initially in the minority doesn't mean they are unimportant. Then, as we begin Part 3, we can do "little designs" as we learn about individual machine elements. As we go further we can and will do more.

If you, your instructor, and the authors all do our parts, you will be largely unaware of the folding together of knowledge of the universe and knowledge about design. It's akin to the background music in a dramatic movie. If it is well done, the response to the question, "How did you like the music?", can be another query, "What music?"

Of course, machinery design is not an emotional, but an intellectual experience, and you must stay attentive. You will find it very interesting.

Charles R. Mischke
Ames, Iowa

Preface to the Sixth Edition

Course and Prerequisites

This book has been written for engineering students who are beginning a course of study in mechanical engineering design. Such students will have acquired a set of engineering tools consisting, essentially, of mathematics, computer languages, and the ability to use the English language to express themselves in the spoken and written forms. Mechanical design involves a great deal of geometry, too; therefore, another useful tool is the ability to sketch and draw the various configurations that arise. Students will also have studied physics and a number of basic engineering sciences, including engineering mechanics, mechanics of materials, materials science, manufacturing as well as heat, mass, and momentum transport. These, tools and sciences, constitute the foundation for the practice of engineering, and so, at this stage of undergraduate education, it is appropriate to introduce some professional aspects of engineering. These professional studies should integrate and use the tools and sciences to the accomplishment of an engineering objective. The pressures upon the undergraduate curricula today require that we do this in the most efficient manner. Most engineering educators are agreed that mechanical design integrates and uses a greater number of the tools and disciplines than any other professional study. Mechanical design is also the very core of other professional and design types of studies in mechanical engineering. Thus studies in mechanical design seem to be the most effective method of starting the student in the practice of mechanical engineering.

One of the reasons for preparing a new edition now is the recent emphasis on the creative aspect of design. In the early 1950s a committee on evaluation of engineering education of the American Society for Engineering Education stated:

> "Training for the creative and practical phases of economic design, involving analysis, synthesis, development, and engineering research, is the most distinctive feature of professional engineering education."
>
> "The technical goal of engineering education is preparation for performance of the functions of analysis and design, or of the functions of construction, production, or operation with full knowledge of analysis and design of the structure, machine or process involved."

Though these goals were stated nearly 50 years ago, they are valid today. Ways must be found to involve the engineering student in genuine design experiences.

Approach

The approach of this book is to suggest and present short design problems or situations to illustrate the decision-making process without demanding an inordinate amount of the student's precious time. Good short design projects are certainly needed in the professional design studies. These are most effective when they are created from the instructor's own professional background and presented with enthusiasm and thoroughness which this background allows. With such an approach new and updated projects can always be devised to meet current needs and ideas.

Additional major reasons for publishing this new edition include

- Addressing uncertainty with more quantitative assurance.
- Reliability goals need to be met.
- Methodology now exists for the estimate of the chance of failure.
- Design and analysis methodologies need to be intertwined.
- There is a need for an increased number of examples, and many new problems.
- The fading of mainframe computers, and the increasing use of personal computers, networked, makes an impact on engineering use.

Uncertainty in Engineering Design

Uncertainty plagues the designer. He or she needs the ability to express it quantitatively, so that decisions can be sound. The traditional deterministic approach to design simply cannot cope with the problem of uncertainty. It is not that one must learn an entirely new way of approaching problems. The deterministic approach is not to be lost. The traditional methodology simply propagates the mean values through the problem. One has to learn how to propagate the variability through the problem. The Accrediting Board for Engineering and Technology is requiring student ability to apply statistics to engineering problems. The basic statistical knowledge is either scattered among several engineering courses, or concentrated in a formal statistics course. Fortunately we often only have to organize a little statistics in order to find the mean design factor which will permit the attainment of the reliability goal. After that, the problem can be treated deterministically. We also learn how poorly we know some basic information, and we have to accommodate to that vagueness. Product liability and quality control concerns can be addressed as needed.

If one addresses uncertainty quantitatively, then in and among that information are the tools for expressing reliability. The methodology now exists for quantitative estimation of the chance of failure. There are data to allow this in steels. Since reliability is one minus the probability of failure, we can quantitatively estimate the chance of survival. This opens the door to designing to a reliability specification.

Design Content

Some engineers look forward to receiving a design task, and the blank pieces of paper that go with it. Other engineers do not look forward to either. Part of the dichotomy during the educational process is attributable to design and analysis being separated. Design content at the end of a course gets less attention. To ensure that both design and analysis get consideration, they are blended in this edition. As a guide to both student and instructor, end-of-chapter problems are identified as to kind. There are 775 problems, half of which are new.

Examples

A noticeable change is an increase in the number of examples; first for the improvement of clarity, second, to allow illustration of new material, and third, to address the subtleness of the additional design content. The increased emphasis on design requires development of a designer's viewpoint early in the book, and many threads have to be carried throughout all the chapters. There are many more examples, which are revisited in later examples, and in end-of-chapter problems. Examples in earlier chapter are revisited.

Curriculum time allotments require that this be done without increasing the course time. Many ideas and experiences have to be interwoven with the subject-matter to allow sufficient repetition for learning and retention. A viewpoint develops, which increases confidence and reduces apprehension in the face of the unknown.

Computer Use

Access to mainframe computers continues, but, increasingly, personal computers are available to students, and many handheld calculators are miniature computers that go everywhere with the student. All this makes computing a powerful tool to be reached for when needed. Programming is no longer a "big deal." Today's student can do the same in less time, or more in the same time. Simulations can be conducted and parametric studies allow broader and more global views than heretofore. A book such as this doesn't have to provide programs for everything, although there are a few. Just listing algebraic and logical steps in algorithmic form is enough to get the student "off and running." The book takes advantage of this for the learning process. Computer-oriented problems are marked with an icon.

One of the roles of a book such as this is to do the necessary chores for the student and the instructor in a methodical and sequenced way, freeing the instructor to be more of a coach, and enabling the instructor to specifically accomplish syllabus objectives. The book reduces the instructor's need to "cover" material. The instructor is freer to mold and build a framework for knowledge and approach, and to answer the many questions that arise as students feel their way. There are highlighted Case Studies in chapters for which newly-acquired background makes the student receptive, and provides an opportunity for the instructor to make coaching points while asking pointed questions. Adequacy assessment is the designer's primal skill. Problems involving this skill are marked with an analysis icon, and the words adequacy assessment are added in the margin, or in the problem statement.

Organization and Content

The book is still in three parts. Part 1 is basic and includes a more comprehensive introduction, definitions, statistical considerations, stress analysis, deflection, and stiffness estimation. The student will be prepared on most of these topics, but not always. The basic material may need some presentation, if only to serve as a review of basic reference material, always in the hands of the student, and presented in the symbolism and terminology of the subsequent parts of the book. Readers will notice a little history has been added. Some material useful in mentoring has been added for both the student and the instructor to "mull over."

Part 2 addresses failure prevention. It makes use of and integrates the fundamentals of Part 1 toward the goal of analyzing and designing mechanical elements to achieve satisfactory levels of preserving function, safety, reliability, competitiveness, usability, manufacturability, and marketability.

Part 3 examines specific mechanical elements such as fasteners, weldments, adhesives, springs, bearings, gears, clutches, brakes, shafts, belts, chains, and so on, and specifically addresses analysis, selection, and design. The new material in Parts 1 and 2 has greatly enhanced the development and presentations in Part 3. It gives the instructor wider opportunities to augment, enrich, and pursue goals outside the scope of the book.

Chapter 1, *Introduction,* contains a fuller treatment of design ideas (and history) and raises some important considerations early. The task of assuring that a product is functional, safe, reliable, competitive, usable, manufacturable, and marketable is presented immediately, along with the design imperative. The subsequent engineering tasks are presented. The different goals of science (to explain what is, and why) and engineering (to create what never was) and the differing skills are noted. The necessary talent in both cases is identified. Suitability, feasibility, and acceptability tests are presented. Specification sets, decision sets, adequacy assessments, and Skill 1 are initially defined. Figures of merit, optimization, synthesis, and Skill 2 are also discussed. Since the reader has embarked on a path to competence, mileposts (kilometer sticks?) are cited along with the changing *modi operandi,* so that the student is aware of the nature of the game being played at the outset. The fragility of technology is noted, the importance of the designer's notebook is mentioned, and the role of the computer is identified.

Uncertainty is ubiquitous, and continually plagues the designer, and its historical treatment is traced. Insightful contributions to bending are noted, and the engineer's debt to Professor Irving P. Church of Cornell University is acknowledged. Factor of safety ideas are identified with more precision than is the usual case. Useful distinctions among the seven kinds of numbers engineers use are made. The adequacy assessment is introduced in broad description for its continual expansion and development throughout the book.

Chapter 2, *Addressing Uncertainty,* has been completely rewritten to support ABET's increased emphasis on statistical applications in design, and to demonstrate the rational treatment of uncertainty. Additional information has been included in Appendices A, B, C, and D.

Chapter 3, *Stress,* has had material added on open and closed thin-walled sections, the Smith–Liu equations for contact stresses in the presence of surface shearing traction, as well as propagation of error.

Chapter 4, *Deflection and Stiffness,* now includes snubbers, round-bar clamps, and concludes with propagation of error examples illustrating the four cases of Chapter 3 applied to relations between load and distortion.

Chapter 5, *Materials,* has had information added which shows the limitation of single-number tables of material properties. There is additional text on plastic deformation, quantitative estimation of properties of cold-worked metals (method of Datsko), and some direction on where to find quantitative treatments of properties of heat-treated steels.

Chapter 6, *Failures Resulting from Static Loading,* addresses assessment of static strength, distinguishes between hypotheses of failure and theories of failure, and addresses criticism of hypotheses in ductile and brittle materials, then considers what our theories of failure tell us, with a number of examples. Static and quasi-static shaft stress analysis concludes the chapter.

Chapter 7, *Failures Resulting from Variable Loading,* addresses fatigue. New information is presented about the tensile strength correlation method, and what it has to say about fatigue ratio and its variability. Included are improved estimations of a and b of $S_f = aN^b$ when the true stress at fracture σ'_f (available from the static tensile test) is known, or can be well-estimated from $\bar{S}_{ut}$. Marin endurance limit modification factors are presented for both stochastic and deterministic approaches. More information is given on relating fatigue stress-concentration factor $\mathbf{K}_f$ to notch-sensitivity factor $\mathbf{q}$. The modified Neuber equation for stress concentration factor $\mathbf{K}_f$ (after Heywood) is given and used because its statistical basis is more extensive than the notch-sensitivity q (after Peterson). A section replete with examples is presented applying what was learned about endurance limit and endurance strength. After characterization of fluctuating stresses, fatigue failure loci are identified. Those cannot be rejected statistically

(Gerber-parabolic and ASME-elliptic) are used principally. Those that can be rejected (Goodman and Soderberg) are included briefly for completeness, so that we can follow the work of others, if necessary. Cumulative fatigue damage includes a modified rain-flow technique. Surface fatigue information has been augmented. The designer's fatigue diagram is presented, and a shaft analysis for fatigue completes the chapter.

Chapter 8, *Screw Fasteners and the Design of Nonpermanent Joints,* addresses threaded fasteners principally, with some attention to pins and miscellany. Addition material is presented on identifying threaded and unthreaded length within a joint fastener in order to improve the bolt stiffness estimate. Because of the reader's improved stochastic understanding, initial tightening can be treated statistically, and preload given a more complete treatment because of the insights available. More material on gaskets is presented. Specification set and decision set ideas are further developed. An example of a pin failure is included.

Chapter 9, *Welding, Brazing, Bonding, and the Design of Permanent Joints,* treats weldments principally, with more material relating to the difference between an analytical, and a throat-stress approach to sizing of welds. More weld-fatigue information is included. Adhesive bonding presentation has been expanded.

Chapter 10, *Mechanical Springs,* has been revised to take advantage of what the topic can teach, and the student can learn, about design. Care has been taken to distinguish between stress-concentration factors and stress-augmentation factors. Additional material on moduli of elasticity has been incorporated. Fundamental zero-max test information is the starting point for a fatigue failure locus on the designer's fatigue diagram. Such a test is carried out on actual springs, and so includes surface, size, and loading effects. However the test cannot be run to zero stress, so it is run from a low stress to a maximum. Just the act of reducing the data to a zero-max basis requires adopting a failure locus type (Goodman, Wahl, Sines, Gerber or ASME-elliptic). Published data rarely identifies the hypothesis, and a fundamental point is in doubt. Differing zero-max properties for the same material from different investigators underscore the problem.

The chapter presents analyses, adequacy assessments, specification sets and decision sets, and gives examples for helical-coil compression springs, helical-coil extensions springs, and helical-coil torsion springs.

Chapter 11, *Rolling Contact Bearings,* recognizes that L_{10} ratings, which are the basis for many manufacturer's catalogs, are too low for machinery use. Consequently, the load-life-reliability tradeoff relations of rolling-contact bearings are a central focus. Distinctions are drawn between stepwise-constant and continuously varying loads, and the methodology for treating them is included. Variable loading on rolling-contact bearings is addressed in such a way that the Miner rule drops out of a linear-damage hypothesis. The section on selection of tapered-roller bearings has been expanded. The question of misalignment is included.

Chapter 12, *Lubrication and Journal Bearings,* begins with Petroff's equation, an early quantitative model of a journal bearing. From this model the Sommerfeld number was (inappropriately) defined. Since that definition persists, the nature of the deviation is shown. The introduction to hydrodynamic theory of the previous edition is presented. In design considerations Trumpler's criteria for journal bearings is presented and used. Raimondi and Boyd charts are combined with Trumpler's criteria. Bearing temperature rise is given more attention than previously, and more examples are provided. Adequacy assessments and decision sets are explored.

The above approach is applied to pressure-fed bearings as well. With these fundamentals in mind, the problem of journal and bushing tolerance is discussed. The result of this is the realization that all journal-and-bushing assemblies are different bearings with different properties. The design-window viewpoint recognizes this.

Boundary lubrication and wear are examined in more detail than in the previous edition. Design procedures are presented, with example.

Chapter 13, *Gearing—General,* is an introduction to four of the principal types of gearing. Interference in spur and helical gearsets is treated more quantitatively than before. Forces and moments are the primary emphasis. Additional attention is given to tooth-count and mechanical efficiency. A planetary gear train is examined to identify the appropriate Sommerfeld numbers.

Chapter 14, *Spur and Helical Gears,* has been modified to reflect the current ANSI/AGMA standards, including velocity factor K_V which is now the reciprocal of K_V heretofore. Surface durability, bending fatigue and factors of safety are explored. Analysis "maps" are provided as well as adequacy assessments of gear meshes. The design of spur and helical gear meshes is laid out in detail, with examples.

Chapter 15, *Bevel and Worm Gears,* begins with ANSI/AGMA standards, and provides analysis "maps" for bending and pitting resistance of straight-bevel gears. Worm gearing begins with an ANSI/AGMA standard for cylindrical worms. An example of a worm-gear mesh is provided. The design of a worm-gear mesh, beginning with a decision set is included. The Buckingham wear-load equation is presented, which allows for analysis and design of cylindrical-worm gearing with other than the hard steel worm and the bronze wheel material combinations.

Chapter 16, *Clutches, Brakes, Couplings, and Flywheels,* has additional material. Brakes are introduced with the relationships surrounding a door stop to identify the essential ideas. The internal brake shoe pressure distribution is derived from fundamentals, and applied. External brake shoes follow. Band-type clutches and brakes are explored. Clutch and brake relationships are summarized dimensionlessly, and there are valuable things to be learned from doing this. Frictional-contact axial clutches are included. Caliper brakes are examined quantitatively. Cone clutches and brakes are closely related to axial clutches and brakes.

Self-locking tapers are examined and their relationship to clutches and brakes explored. Temperature-rise in clutches and brakes is treated quantitatively.

Chapter 17, *Flexible Mechanical Elements,* now provides more insight into flat belt theory, and examines the importance of initial tension. Attention is given to ways to achieve initial tension and to sustain it. The decision set for specifying a flexible flat belt is identified. Metal flat belts are introduced, with example, and durability is examined. V belts are presented along with the importance of initial tension. Attention is given to life expectancy. Adequacy assessments, specification set and decision sets are identified, and the similarity to flat belts noted. Roller-chain rating basis is presented in more detail. Design life other than 15 000 h is quantitatively studied. The presentation of wire rope includes adequacy assessment, specification set and decision set, and an example of a minehoist is provided to show the relationship between nominal wire rope diameter, number of supporting ropes and fatigue factor of safety.

Chapter 18, *Shafts and Axles,* begins with a plan for addressing the task of shaft design, sufficing static and fatigue constraints, sufficing geometric constraints, with examples. In the process distortion-energy—Gerber and distortion-energy—elliptic equations are emphasized. The adequacy assessment for strength for both stochastic and deterministic approaches is illustrated by example. The preparation of the student to compare and contrast the methods, and, what they can and cannot accomplish is at hand. This is addressed. Shaft material and hollow shafts have brief presentations. Critical speeds have now been added. With the foregoing background an approach to shaft design is described. Computer considerations are identified. Programming Task No. 3 that completes the chapter is an important opportunity for a student to develop an insight.

The Appendix has had some additions. Three of the charts for geometric stress-concentration factors K_t and K_{ts} have been replaced by the results of finite element studies. Two additional tables giving stochastic parameters of strength distribution are included, along with two tables of dimensions of American Standard Plain Washers, the gamma function, the correlation coefficient r, the t-statistic, and several helical spring specification forms.

Supplements

This edition is supported by a number of supplements, made available to adopters through the publisher.

Solutions Manual—an instructor's manual which contains solutions to most end-of-chapter nondesign problems.

PowerPoint® Slides—approximately 200 slides of important figures and tables from the text are provided in PowerPoint format for use in lectures. These files are available on the book website.

Website—a website for the book has been established at www.mhhe.com/engcs/mech/shigley. This site contains information about the text, text updates and errata, PowerPoint slides, and the password-protected solutions manual.

Charles R. Mischke
Ames, Iowa

Acknowledgements

This edition has been influenced by changes in engineering education, suggestions by colleagues, users, and the publisher's reviewers and checkers. Their comments and suggestions were sound and constructive, and many of them have been incorporated. The most comprehensive critiques were offered by the following.

Reviewers:

Nels S. Christopherson, *Michigan Technological University, Houghton, MI*
J. Darrell Gibson, *Rose-Hulman Institute of Technology, Terre Haute, IN*
Vladimir Glozman, *California State Polytechnic University, Pomona, CA*
E. William Jones, *Mississippi State University, Mississippi State, MS*
Frank M. Kelso, *University of Minnesota, Minneapolis, MN*
Steven Y. Liang, *Georgia Institute of Technology, Atlanta, GA*
Clarence Maday, *North Carolina State University, Raleigh, NC*
Richard H. Messier, *University of Maine, Orono, ME*
Robert Paasch, *Oregon State University, Corvallis, OR*
Gordon R. Pencock, *Purdue University, West Lafayette, IN*
Richard A. Scott, *The University of Michgan, Ann Arbor, MI*
Steve Yugartis, *Clarkson University, Potsdam, NY*

Checkers:

Gregory V. Aloe, *The University of Michigan, Ann Arbor, MI*
Suresh Ananthsuresh, *The University of Pennsylvania, Philadelphia, PA*
Nels S. Christopherson, *Michigan Technological University, Houghton, MI*
Michael Colonna, *Cornell University, Ithaca, NY*
Richard E. Dippery, *Kettering University, Flint, MI*
William S. Larsen, *Iowa State University, Ames, IA*
Gordon Pennock, *Purdue University, West Lafayette, IN*
Albert J. Shih, *North Carolina State University, Raleigh, NC*
Richard Stanley, *Kettering University, Flint, MI*

I am grateful to all who assisted for their attention, and in helping sustain focus.

Special thanks are due to Dr. David A. Dillard, Director of the Center for Adhesive and Sealant Science, Virginia Polytechnic Institute and State University, Blacksburg, Virginia, for his assistance in the preparation of Sec. 9-11, and to Bonding Systems Division, 3M, Saint Paul, Minnesota for technical support.

The ultimate responsibility for design decisions rests with the engineer in charge of the project. Only he or she can judge if the conditions surrounding the application are congruent with the presentations here, in papers and references, as well as other literary sources. In view of the large number of considerations that enter into any design, it is impossible for the authors to assume any responsibility for the manner in which material presented here is used in a design.

Care has been taken to avoid error. The authors will appreciate being informed of any errors discovered, so that they may be eliminated in subsequent printings.

Charles R. Mischke
Ames, Iowa

List of Symbols

This is a list of common symbols used in machine design and in this book. Specialized use in a subject-matter area often attracts fore and post subscripts and superscripts. To make the table brief enough to be useful the symbol kernels are listed. See Table 14–1, pp. 899–900 for spur and helical gearing symbols, and Table 15–1, pp. 955–956 for bevel-gear symbols.

Symbol	Meaning
A	Area, coefficient
$\mathbf{A}$	Area variate
a	Distance, regression constant
$\hat{a}$	Regression constant estimate
$\mathbf{a}$	Distance variate
B	Coefficient
Bhn	Brinell hardness
$\mathbf{B}$	Variate
b	Distance, Weibull shape parameter, range number, regression constant, width
$\hat{b}$	Regression constant estimate
$\mathbf{b}$	Distance variate
C	Basic load rating, bolted-joint constant, center distance, coefficient of variation, column end condition, correction factor, specific heat capacity, spring index
c	Distance, viscous damping, velocity coefficient
CDF	Cumulative distribution function
COV	Coefficient of variation
$\mathbf{c}$	Distance variate
D	Helix diameter
d	Diameter, distance
E	Modulus of elasticity, energy, error
e	Distance, eccentricity, efficiency, Naperian logarithmic base
F	Force, fundamental dimension force
f	Coefficient of friction, frequency, function
fom	Figure of merit
fos	Factor of safety
G	Torsional modulus of elasticity
g	Acceleration due to gravity, function
H	Heat, power
H_B	Brinell hardness
HRC	Rockwell C-scale hardness
h	Distance, film thickness

I Integral, linear impulse, mass moment of inertia, second moment of area
i Index
$\mathbf{i}$ Unit vector in x-direction
J Mechanical equivalent of heat, polar second moment of area, geometry factor
$\mathbf{j}$ Unit vector in the y-direction
K Service factor, stress-concentration factor, stress-augmentation factor, torque coefficient
k Marin endurance limit modifying factor, spring rate
$\mathbf{k}$ k variate, unit vector in the z-direction
L Length, life, fundamental dimension length
~LN Lognormally distributed
l Length
ℓ Length
M Fundamental dimension mass, moment
$\mathbf{M}$ Moment vector, moment variate
m Mass, slope, strain-strengthening exponent
N Normal force, number, rotational speed
~N Normally-distributed
n Load factor, rotational speed, safety factor
n_d Design factor
P Force, pressure
PDF Probability density function
p Pitch, pressure, probability
Q First moment of area, imaginary force, volume
$\dot{Q}$ Heat transfer rate
q Distributed load, notch sensitivity
R Radius, reaction force, reliability, Rockwell hardness, stress ratio
$\mathbf{R}$ Vector reaction force
r Correlation coefficient, radius
$\mathbf{r}$ Distance vector
S Sommerfeld number, strength
$\mathbf{S}$ S variate
s Distance, sample standard deviation, stress
T Temperature, tolerance, torque, fundamental dimension time
$\mathbf{T}$ Torque vector, torque variate
t Distance, Student's t-statistic, time, tolerance
U Strain energy
~U Uniformly distributed
u Strain energy per unit volume
V Linear velocity, shear force
v Linear velocity

W	Cold-work factor, load, weight
w	Distance, gap, load intensity
$\mathbf{w}$	Vector distance
X	Coordinate, truncated number
x	Coordinate, true value of a number, Weibull parameter
$\mathbf{x}$	x variate
Y	Coordinate
y	Coordinate, deflection
$\mathbf{y}$	y variate
Z	Coordinate, section modulus, viscosity
z	Standard deviation of the unit normal distribution
$\mathbf{z}$	Variate of z
α	Coefficient, coefficient of linear thermal expansion, end-condition for springs, thread angle
β	Bearing angle, coefficient
Δ	Change, deflection
δ	Deviation, elongation
ϵ	Eccentricity ratio, engineering (normal) strain
$\boldsymbol{\epsilon}$	Normal distribution with a mean of 0 and a standard deviation of s
ε	True or logarithmic strain
Γ	Gamma function
γ	Pitch angle, shear strain, specific weight
λ	Slenderness ratio for springs
$\boldsymbol{\lambda}$	Unit lognormal with a mean of 1 and a standard deviation equal to COV
μ	Absolute viscosity, population mean
ν	Poisson ratio
ω	Angular velocity, circular frequency
ϕ	Angle, wave length
ψ	Slope integral
ρ	Radius of curvature
σ	Normal stress
σ'	Von Mises stress
$\boldsymbol{\sigma}$	Normal stress variate
$\hat{\sigma}$	Standard deviation
τ	Shear stress
$\boldsymbol{\tau}$	Shear stress variate
θ	Angle, Weibull characteristic parameter
¢	Cost per unit weight
$	Cost

PART 1 Basics

1 Introduction

1-1 Design **5**
1-2 Mechanical Engineering Design **7**
1-3 Your Path to Competence **12**
1-4 Technology Can Be Fragile **12**
1-5 Interaction between Design Process Elements **14**
1-6 Codes and Standards **17**
1-7 Economics **18**
1-8 Safety and Product Liability **20**
1-9 The Adequacy Assessment **20**
1-10 Uncertainty **22**
1-11 Stress and Strength **26**
1-12 Design Factors and Factors of Safety **29**
1-13 Reliability **30**
1-14 Numbers, Units, and Preferred Units **31**

This chapter introduces a number of ideas, many of which are new to you either in context or substance. Mechanical design is a complex undertaking, requiring many skills. A vocabulary that allows large relationships to be subdivided into a series of simple tasks is needed. The complexity of the subject suggests a sequence in which ideas are introduced and revisited. Toward this end, we consider a number of topics briefly in this chapter to serve as an orientation. Later these topics are developed in detail, their bones fleshed out, and they will become part of you.

We first address the nature of design in general, then mechanical engineering design in particular. Learning and mastering is an ongoing process as you finish your formal education and throughout your career. Technology can be fragile, and the interaction between elements in the design process is iterative. There are roles to be played by codes and standards, ever-present economics, safety, and considerations of product liability. In these matters this book identifies the cast of characters, and your instructor provides the perspective to aid in the achievement of course goals.

Then the focus of the chapter narrows, and the primal skill of the designer—adequacy assessment—is considered. Matters of uncertainty have been ever-present in engineering design, and methods have been evolved which draw on growing knowledge. Attention is given to stress and strength and to the distinction between design factor and factor of safety. Reliability is considered, as are numbers, measurement, units, and preferred units. All this is prologue to the organization of the rest of the book. The book is a study of the decision-making processes with which mechanical engineers plan for the physical realization of machines, devices, and systems. These processes are common to all disciplines in the field of engineering design—not just to mechanical engineering design. But, since our subject *is* mechanical engineering design, we will use mechanical engineering as the vehicle for understanding these decision-making processes and for applying them to practical situations in which we can see them pay off.

The book consists of three parts. Part 1 begins by explaining some differences between design and analysis and introducing some fundamental notions and approaches to design. It continues with a chapter on statistical methods. Statistical ideas, up to the ability to quantify a design factor to achieve a reliability goal, are important to the study of the rest of the book. Next there is a review of stress analysis and an introduction to stiffness and deflection analysis.

Part 2, on failure prevention, consists of one chapter on materials and two chapters on the prevention of failure of mechanical parts. The chapter on materials—Chap. 5—is included so that definitions and analyses relating to mechanical properties used in design can be presented in one place, at the same time they are related to a variety of materials. Why machine parts fail and how they can be designed to prevent failure are difficult questions, and so we take two chapters to answer them, one on preventing failure due to static loads, and the other on preventing failure due to fatigue and dynamic loads.

In Part 3, the material of Parts 1 and 2 is applied to the analysis, selection, and design of specific mechanical elements, such as fasteners, weldments, springs, rolling contact bearings, film bearings, gears, belts, chains, and wire ropes. The design of shafts ties together the influences of several of the preceding elements.

As we proceed through the book, we will need the ability to handle few, many, or an infinite number of alternatives—and not lose track. Our vocabulary will facilitate keeping track of decisions identifying the number of decisions, and discerning their impacts on the number of independent variables.

Engineers say a product design must be functional, safe, reliable, competitive, usable, manufacturable, and marketable. In this usage these words are meant to convey the following:

- *Functional*: The product must perform to fill its intended need and customer expectation.
- *Safe*: The product is not hazardous to the user, bystanders, or surrounding property. Hazards which cannot be "designed out" are eliminated by guarding (a protective enclosure); if that is not possible, appropriate directions or warnings are provided.
- *Reliable*: Reliability is the conditional probability, at a given confidence level, that the product will perform its intended function satisfactorily or without failure at a given age.
- *Competitive*: The product is a contender in its market.
- *Usable*: The product is "user-friendly," accommodating to human size, strength, posture, reach, force, power, and control.
- *Manufacturable*: The product has been reduced to a "minimum" number of parts, suited to mass production, with dimensions, distortion, and strength under control.
- *Marketable*: The product can be bought, and service (repair) is available.

1–1 Design

To design is either to formulate a plan for the satisfaction of a specified need or to solve a problem. If the plan results in the creation of something having a physical reality, then the product must be functional, safe reliable, competitive, usable, manufacturable, and marketable. To remind us that designs are constrained, and have to exhibit qualities known at the outset, a *design imperative* can be expressed as follows:

Design
(subject to certain problem-solving constraints)
a component, system, or process
that will perform a specified task
(subject to certain solution constraints)
optimally.

The parenthetical expressions refer to qualifications placed on the design. The solution methodology is constrained by what the designer knows, or can do; the solution, in addition to being functional, safe, reliable, competitive, usable, manufacturable, and marketable, must also be legal and conform to applicable codes and standards.

It is important that the designer begin by identifying exactly how he or she will recognize a satisfactory alternative, and how to distinguish between two satisfactory alternatives in order to identify the better. From this kernel, optimization strategies can be formed or selected. Then, the following tasks unfold:

- Invent alternative solutions.
- Through analysis and test, simulate and predict the performance of each alternative, retain satisfactory alternatives, and discard unsatisfactory ones.
- Choose the best satisfactory alternative discovered as an approximation to optimality.
- Implement the design.

The outputs of the design process are decisions concerning components and their connectivity, geometry, "forming" processes, thermomechanical treatments, and associated tolerances. All this is laid out in the plans and specifications. In Fig. 1–1, the "black box" representing the design process is a light bulb, a reminder to the reader that without the creative-inventive component, useful results are rare.

Figure 1–1

The input to, and the output from, the design process when the design is a physical reality.

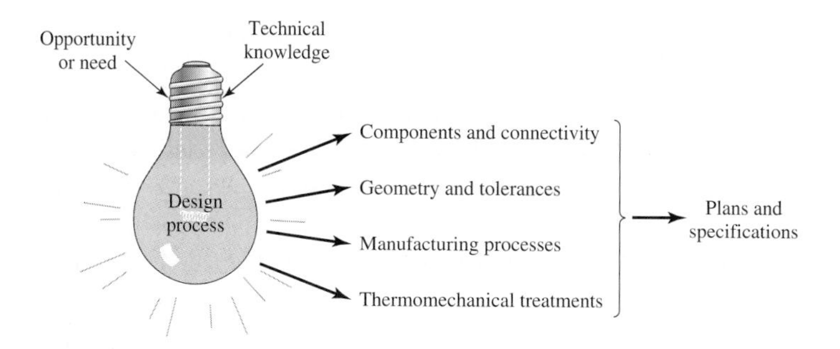

The need in Fig. 1–1 may be well-defined, such as "We need a solar-powered car," or the nature of the problem may be unclear, as in "We need a safer automobile." Often the designer's initial task is to clearly define (identify) the need, including preferences of engineering managers, marketers, and customers, before doing anything else. The needs, or problems, include opportunities, and they may be multiple.

The characterization of a design task as a design *problem* can introduce the idea that, as a problem, it has a solution. This may not be so. The design space may be empty. Some situations may simply have to be endured. To relieve the absence of solutions, some constraint(s) may have to be renegotiated in order to admit solutions. Then again, even when solutions are possible, the designer may not be creative enough, inventive enough, to conceive of them. This admits to the design problem the necessity of individual talent or skill in this area.

There may be more than one solution, and distinguishing among them to choose the best may require the ability to handle a large number of solutions without being overwhelmed. Solutions, if they exist, can be characterized as satisfactory, some better than others, some clearly good, and one, the best by some criterion. Solutions can have a time dependency, for what is acceptable today may not be so tomorrow, and vice versa.

Design is an innovative and iterative process. It is also a decision-making process. Decisions sometimes have to be made with too little information, occasionally with just the right amount of information, or with a surfeit of partially contradictory information. A man with a watch knows what time it is; with two watches, he is never sure. Decisions are sometimes made tentatively, reserving the right to adjust as more becomes known. The point is that the engineering designer has to be personally comfortable with a decision-making, problem-solving role. It should be a satisfying and welcomed activity. If it is not, there can be personal ramifications (such as stress) that can interfere, even threatening the designer's health.

Design is a communication-intensive activity in which both words and pictures are used, and written and oral forms are employed. Engineers have to communicate effectively, and persuade people who know more than they do, or less than they do. These are important skills, and an engineer's ability to function depends on them.

Designs are subject to problem-solving constraints. A designer can only apply methods he or she knows and understands. Corporations performing design tasks know and apply more methods than an individual designer does, and the problem-solving constraints are less severe. A corporation can tap the knowledge of consultants and reduce constraints even more. Eventually, the limit of what is known by humankind is reached (the state of the art). Thus time, money, and corporate collective knowledge are common problem-solving constraints imposed on a design.

A designer's personal resources of creativeness, communicative ability, and problem-solving skill are intertwined with knowledge of technology and first principles. Engineering tools (such as mathematics, statistics, computer, graphics, and language) are combined to produce a plan, which, when carried out, produces a product that is functional, safe, reliable, competitive, usable, manufacturable, and marketable, regardless of who builds it or who uses it.

1–2 Mechanical Engineering Design

Mechanical engineers are associated with the production and processing of energy, providing the means of production, the tools of transportation, and the techniques of automation. The skill and knowledge base are extensive. Among the disciplinary bases are mechanics of solids, fluids, mass and momentum transport, manufacturing processes, electrical and information theory. Mechanical engineering design involves all the disciplines of mechanical engineering.

Problems resist compartmentation. A simple journal bearing involves fluid flow, heat transfer, friction, energy transport, material selection, thermomechanical treatments, statistical descriptions, and so on. A building is environmentally controlled. The heating, ventilation, and air-conditioning considerations are sufficiently specialized that some speak of *heating, ventilating, and air-conditioning design* as if it is separate and distinct from mechanical engineering design. Similarly, *internal-combustion engine design, turbo-machinery design,* and *jet-engine design* are sometimes considered discrete entities. The leading adjectival string of words preceding the word *design* is merely a product-descriptive aid to the communication process. There are phrases such as *machine design, machine-element design, machine-component design, systems design, and fluid-power design*. All of these phrases are somewhat more focused examples of mechanical engineering design. They all draw on the same bodies of knowledge, are similarly organized, and require similar skills.

In the academic world, with its clustering of knowledge into efficient learning groups, we encounter subjects, courses, disciplines, and fields. Curricula consist of sequences of courses. The arrangement of courses present the opportunity to study machine elements and machines earlier than the last semester. Thus machine design often represents the student's first serious design experience with a substantial knowledge base. Some, but not many machine elements can be understood without a complete thermofluid base, but before you know it, we are into mechanical engineering design.

Science explains what *is,* engineering creates what *never was*. Mathematics is neither science nor engineering. Physics and chemistry are science, but not engineering. As suggested in Fig. 1–2, it takes one kind of talent to be a scientist and a different talent to create what never was. Engineers and scientists know something of each other's work, but only in rare cases are both talents developed in an individual. It takes talent and ability to create and innovate, talent to be a consistently successful problem solver and decision maker, and talent to be an effective communicator. Preparation, you see, is the developing and polishing of talent, whatever the endeavor.

Design situations previously encountered in your curriculum drew on the very small information base available, and the idea was to briefly present the design side of engineering. The prerequisite base of this book, however, is formidable. The amount of relevant detail is now sufficiently large that we must employ formalisms which organize, permit insights, and allow reduction in clutter. The individual talent and creativeness needed now are larger than before. If your talents are not developing as rapidly as you would like, explore with your instructor ways of enhancing your development and realizing your potential. There are good innovative engineers who are not great analysts, and fine analysts who cannot innovate. The world needs both, and they work together.

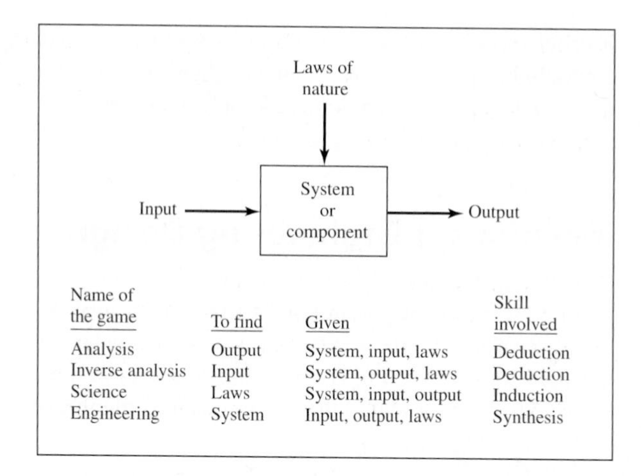

Figure 1–2

The name(s) of the game(s). Note distinctions between analysis, science, and engineering and the significant skills involved.

Viewpoint

Viewpoint to an engineer is to be able to be in a position to have a commanding view of relevant things. This view has to be communicated to others. Viewpoint is not just an important thing; viewpoint is everything![1] As engineers "talk to themselves," then, to the extent that humans think with words, words and their meanings can be helpful or hurtful. If we do not have the words, or do not know the words, key thoughts often do not penetrate our stream of consciousness. Engineers also use geometric thinking and drawings because verbal language is inadequate to the task.

Since words are very important in engineering communication, we should be aware of the debasement of words, their meaning, and representations of reality brought about by our culture. Children and many adults have difficulty distinguishing between fiction and reality. Consider television commercials, which are almost entirely false. The repairman is not a repairman but a hired impersonator, espousing a provided text, crafted to persuade rather than inform, and delivered on a sham set. Language can be used to inform, or it can be used to persuade. Language used to inform places all relevant factors before the listener, so that a person is in a position to think and to bring that information to bear on a solution to a problem. Language used to persuade employs selected facts which support a bias, and omission is a standard tactic.

Rational Decision-Making

Designers have to make decisions, few or many, some apriori, some in concert. We need a bookkeeping system to track where we are, a useful vocabulary, a shorthand of "pegs" upon which to hang kindred ideas.

Peg 1: Suitability, Feasibility, Acceptability

When the military establishment recognized the importance of clear thinking and rational decision making for its general officers, it sought engineers' advice on how to proceed. The military reasoned that engineers commit enormous resources to large projects, with no possibility of starting over. The advice offered can be distilled to the following:

1. A general technical lecture of the same title was given to the attendees of the 10th Biennial Conference on Reliability, Stress Analysis and Failure Prevention, the 14th Biennial Conference on Mechanical Vibration and Noise, the 5th International Conference on Design Theory, and the 19th Design Automation Conference on Design Theory, at the ASME Design Technical Conference, Albuquerque, September 1993, Even 88 (GL2), Charles R. Mischke.

- A contemplated action is *suitable* if its adoption will indeed accomplish the intended purpose.
- A contemplated action is *feasible* if the action can be carried out with the knowledge, personnel, money, and material at hand, or if it can be assembled in time.
- A contemplated action is *acceptable* if the probable results are worth the anticipated costs.

Peg 2: Satisfactory Alternative

If a contemplated action is suitable, feasible, and acceptable, it becomes a *satisfactory alternative*, and it is set aside to be compared with other satisfactory alternatives. If one can compare two satisfactory alternatives and choose the better, an *optimization strategy* can be crafted or selected to deal with a large number of satisfactory alternatives.

Peg 3: Specification Set

A *specification set* is the ensemble of drawings, text, bills of materials, and directions that constitutes the decision record in a form that enables the builder or user to realize function safely, reliably, competitively, and usably, having been fabricated and serviced to the customer's satisfaction.

The specification set for a helical coil compression spring for static service can be examined in Appendix E–37. Not all blanks need be filled in. Such a set uniquely defines a spring and allows the springmaker to create it. What the specification set does not reveal is which of its elements are addressing, singly or in combination, matters of function, safety, reliability, and competitiveness. A designer needs an alternative (equivalent) set that shows each necessary decision, allows bifurcation into a priori and design decisions, allows tagging to show which quality (function, safety, reliability, or competitiveness) is addressed, and reveals the dimensionality of the problem.

Peg 4: Decision Set

A *decision set* is a list of decisions required to establish the specification set. The decision set is equivalent to the specification set. Either may be deduced from the other on the basis of convenience to clear thinking. The decision set is expressed in terms of the designer's thinking parameters, and it easily focuses on function, safety, reliability, and so on. For example, the *specification set* for a helical coil spring for static service can be displayed as follows:

- Material and condition.
- End treatment.
- Coil inner or outer diameter and tolerance.
- Total turns and tolerance.
- Free length and tolerance.
- Wire size and tolerance.

Note that these items are required by the springmaker to replicate the unique spring ordered. It is not clear how, or if, a coil diameter specification addresses function, safety, and so on, nor does the springmaker care. However, the designer does and therefore organizes the equivalent *decision set* as follows:

- Material and condition.
- End treatment.
- Force F_1 and end contraction y_1 or F_1 and length L_1 (function).

- Work over a rod: d_{rod} (function).
- Fractional overrun to coil closure $\xi : \xi = 0.15$ (safety, reliability, and spring linearity [robustness of the mathematical model]).
- Wire diameter d (competitiveness through optimality).

Note that these items are used by the designer to identify decisions, what they address, and the dimensionality of the problem. Tolerances can be expressed as a function of median values and can be quantified in a subsequent step. In Chap. 10 you will learn how to prove the equivalence between the decision set and the specification set for a helical coil spring.

Composing a decision set to be properly revealing and useful is a skill developed through knowledge and practice. There is some duplication between entries in the preceding specification set and decision set, but observe the explicit appearance of thinking parameters. The load F_1 that this spring must exhibit at end contraction y_1 (or at spring length L_1) addresses function. Working over a rod is necessary to function. The fractional overrun-to-closure ξ set at 0.15 addresses safety and reliability, protecting the spring closed solid, intentionally or inadvertently, and preserving the linearity of the spring so the mathematical model remains congruent to nature.

Peg 5: A Priori Decisions vs. Design Variables

The first five decisions in the decision set can be made a priori (they are called *a priori decisions*). The last decision, that of wire size d, is called a *design variable* before we make the decision and the *design decision* after we have made the decision. It is through this variable that the designer attends to issues of preserving function, safety, and reliability, specifically using it to address competitiveness through optimality. In this case, knowledge that there is one independent variable influences selection of the methodology used to establish the wire size d.

Peg 6: The Adequacy Assessment (Skill 1)

An *adequacy assessment* consists of the cerebral, empirical, and related mathematical modeling steps the designer takes to ensure that a given specification set is satisfactory (suitable, feasible, and acceptable, remember?). An adequacy assessment is the primal skill of the designer. It is how he or she recognizes a satisfactory specification set or alternatively, a corresponding decision set. It is so important that it is called *skill* 1. Much of a first course in design of machine elements focuses on building and refining this skill in many applications. Its centrality is seen in Fig. 1–3.

Peg 7: Figure of Merit

If, in the coil spring example, the designer finds several wire sizes that pass the adequacy assessment, he or she uses a *figure of merit* to help identify the best. It is not much of an exaggeration to say that springs "sell by the pound." The volume of material used to form the spring is an index to cost. It is a robust figure of merit, abbreviated f.o.m. or, simply fom. Quantitatively, it can be expressed in the case of the helical coil compression spring as

$$\text{fom} = -\frac{\pi d^2 N_t D}{4}$$

where d is the wire diameter, N_t is the total turns, and D is the mean coil diameter. The minus sign makes increasing merit attend smaller volume. A figure of merit is a number whose magnitude is a monotonic index to the merit, or desirability, of the spring. A figure

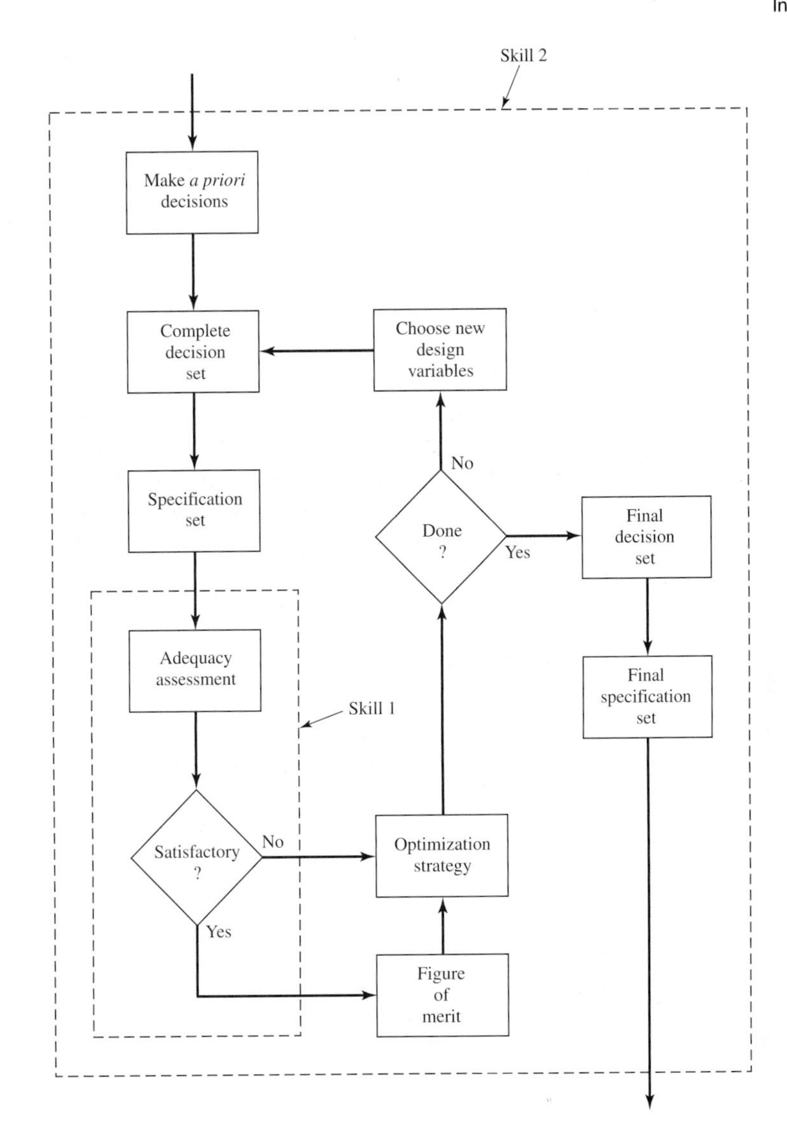

Figure 1–3

A logic flowchart of designers' skills 1 and 2. Note that the analysis skill 1 is embedded in the synthesis skill 2. Depending on how the optimization strategy is formulated, skill 2 can amount to antianalysis, much in the same sense that the integral in calculus can be viewed as an antiderivative.

of merit permits rapid choice among several satisfactory designs. Should a large number of satisfactory designs exist, an optimization strategy is used to identify the best without exhaustive examination.

Peg 8: A Skill of Synthesis (Skill 2)

The skill of synthesis involves an optimization strategy, a figure of merit, and skill 1. A flowchart of skills 1 and 2 is seen in Fig. 1–3. Note that skill 1 is embedded in skill 2, and it needs to be mastered first.

There is an irony here which should not escape us. In a way we have been building a checklist, but a checklist with an unusual objective. The objective of most checklists is remind people to perform tasks without overlooking a thinking situation and thus render a credible performance. The checklist here is for the purpose of stimulating and facilitating original thinking, thus enabling engineers with a design task to teach themselves what they need to know about the structure and connectivity of the elements in the task at hand.

We have introduced a number of pegs to suggest the importance of *viewpoint* as an essential ingredient in the design process. They will be developed in detail later in the book. It is hoped that this overview will let the reader begin to appreciate that the design process is an ensemble of innovative talent, knowledge, facile viewpoint, and accumulated experience that is brought to bear on the task of converting a need into a successful product design.

1–3 Your Path to Competence

You've been in school most of your life. Schools answer almost all needs with courses, and it is easy to form the habit of expecting everything to be explained to you in a course format. It is also easy to assume that your habits and skills of learning up to now can persist and be successful. It is recommended that you read Dreyfus[2] and discover that along the path from novice to expert performer instruction will cease, and self-instruction will replace it. Imitations of prior experience simply will not do; in fact, they can become counterproductive. Follow Dreyfus as he moves from *novice* to *beginner*, to *competent performer*, to *proficient performer*, and to *expert performer*, and see how learning and performing changes.

1–4 Technology Can Be Fragile

We need to be aware that technology can be fragile, and we must be on our guard. Of concern to us is engineering technology and computer technology. We begin with a few engineering examples.

The practitioners are gone. During World War II the United States developed high-speed, high-performance aircraft (called pursuit ships) using, for the most part, multibank radial engines to drive the propellers. After the war the U.S. government stored jigs, dies, and plans, as well as some new aircraft in case of future need. If we wanted to, we could duplicate the aircraft, but we would be hard-put to improve the engines because the engineers with the know-how by now are dead.

The technology did not transfer. An aerospace company contracted with a large city to build streetcars (light-rail vehicles). In the long run the company could not provide properly performing vehicles and forfeited a large sum of money. The engineers could not do the job well enough.

The practitioners were scattered. An interurban electric railway ordered three vest-pocket streamlined trains capable of reaching 80 miles per hour for high-speed limited service on its principal routes. Although guaranteed to operate anywhere on the system, the trains could not. The same builder in 1941 delivered to another railroad two high-speed streamlined electric trains that were a complete success. The builder was certainly in command of the technology during the 1939–1941 construction period. But a war intervened, personnel changed, and the company could not do what it had done before.

Time passes, and engineers should leave lucid tracks so that the art is not lost to those who follow. Students in particular should understand that while a homework problem takes a page or two of computation and comment, problems on the job can take dozens or hundreds of pages. The detail itself is enormous. Remember, you or someone else will be extracting the key results (numbers). It is important that *no* mistakes are made in doing this. Rationales, decisions, and adequacy assessments should be recorded as

---

2. H. L. Dreyfus, "What Computers Still Can't Do," *The Key Reporter*, vol. 59, no. 2, Winter 1993/1994, pp. 4–9; and *What Computers Still Can't Do*, MIT Press, Cambridge, 1992.

they occur in a *designer's notebook*. In this course you have the opportunity to begin, practice, and develop a style. Your instructor will give you the guidelines.

Now we turn to some examples from computer technology.

A surfeit of computer tools. There are now many computer tools to aid the designer in completing tasks. There is spatial (geometric) information, number-crunching, and blends of these. There are many programs—Matlab, MathCad, Excel, Quattro-Pro, EES, Maple, TKsolver, ANSYS, I-DEAS, Pro/E, and AutoCAD, to name a few. Your instructor is the best source of information concerning which programs may be available to you, with recommendations as to which are useful for particular tasks.

Turnkey computer programs may contain errors that do not surface until someone goes down a virgin or erroneous pathway in the program. Use programs new to you skeptically. Always do some benchmark testing. Give the program problems whose answers you already know. Humans write programs, and they make errors in coding, logic, and use theory outside a program's domain of applicability.

Often wrong, but never in doubt. A case occurred in which equations were correct but important logic was not coded. A plane-stress program was provided, accepting the normal stresses and shear stresses on orthogonal planes, and providing the principal stresses and the maximum shear stress. Sometimes in a plane-stress problem, the maximum shear stress occurs out of the plane, and the answer from the program was simply wrong from time to time.

There may be problems to solve that require you to write code. In this circumstance it is useful to keep in mind what a computer does well.

- It can remember data and programs.
- It can calculate.
- It can branch conditionally or unconditionally. Branching based on truth or falseness is akin to decision making.
- It can iterate, do a repetitive task a fixed or appropriate number of times.
- It can read and write both alphabetic and numerical information.
- It can draw, sometimes fast enough to animate in real time.
- It can pause and wait for external decisions or thoughtful input.
- It does not tire.

It is also useful to keep in mind what humans can do.

- They can understand the problem.
- They can judge what is important or unimportant.
- They can plan strategies and modify them in the light of experience.
- They can weigh intangibles.
- They can be skeptical, suspicious, and unconvinced.
- They can program computers!

Routinely the designer should delegate to the computer that which the computer does well and reserve for humans those things they do well. Interactive programming allows you to do this. Don't forget to document your program, because after a period of time you will have forgotten the details. You are really documenting for your later use, and you will be glad you took the time.

1–5 Interaction between Design Process Elements

The total design process is of interest to us in this chapter. How does it begin? Does the engineer simply sit down at his or her desk with a blank sheet of paper and jot down some ideas? What happens next? What factors influence or control the decisions which have to be made? Finally, how does this design process end?

The complete process, from start to finish, is often outlined as in Fig. 1–4. The process begins with a recognition of a need and a decision to do something about it. After many iterations, the process ends with the presentation of the plans for satisfying the need. In the next several sections, we shall examine these steps in the design process in detail.

Recognition and Identification

Sometimes, but not always, design begins when an engineer recognizes a need and decides to do something about it. *Recognition of the need* and phrasing the need often constitute a highly creative act, because the need may be only a vague discontent, a feeling of uneasiness, or a sensing that something is not right. The need is often not evident at all; recognition is usually triggered by a particular adverse circumstance or a set of random circumstances which arise almost simultaneously. For example, the need to do something about a food-packaging machine may be indicated by the noise level, by the variation in package weight, and by slight but perceptible variations in the quality of the packaging or wrap.

It is evident that a sensitive person, one who is easily disturbed by things, is more likely to recognize a need—and also more likely to do something about it. And for this reason sensitive people are more creative. A need is easily recognized after someone else has stated it. Thus the need in many countries for cleaner air and water, for more parking facilities in the cities, for better public transportation systems, and for faster traffic flow has become quite evident.

There is a distinct difference between the statement of the need and the identification of the problem which follows this statement (Fig. 1–4). The problem is more specific. If the need is for cleaner air, the problem might be that of reducing the dust discharge from power-plant stacks, of reducing the quantity of irritants from automotive exhausts, or of quickly extinguishing forest fires.

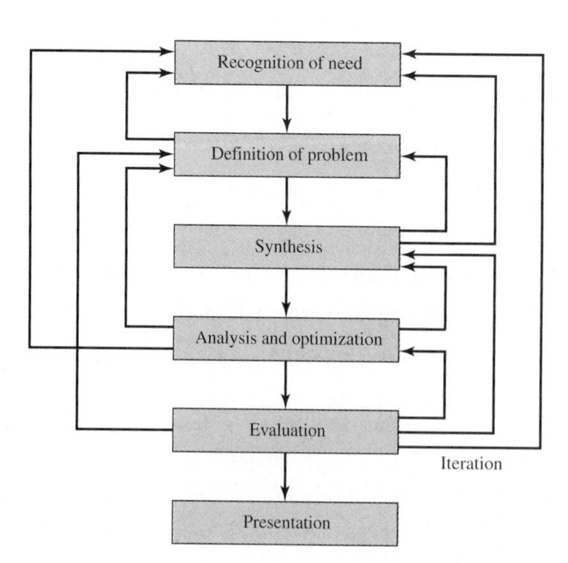

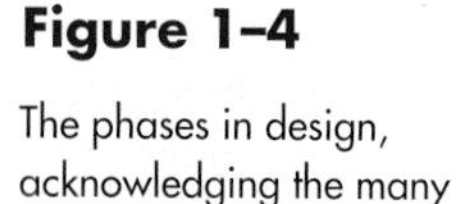

Figure 1–4

The phases in design, acknowledging the many feedbacks and iterations.

Definition of the problem must include all the specifications for the thing that is to be designed. The specifications are the input and output quantities, the characteristics and dimensions of the space the thing must occupy, and all the limitations on these quantities. We can regard the thing to be designed as something in a black box. In this case we must specify the inputs and outputs of the box, together with their characteristics and limitations. The specifications define the cost, the number to be manufactured, the expected life, the range, the operating temperature, and the reliability. Obvious items in the specifications are the speeds, feeds, temperature limitations, maximum range, expected variations in the variables, and dimensional and weight limitations.

There are many implied specifications which result either from the designer's particular environment or from the nature of the problem itself. The manufacturing processes which are available, together with the facilities of a certain plant, constitute restrictions on a designer's freedom, and hence are a part of the implied specifications. It may be that a small plant, for instance, does not own cold-working machinery. Knowing this, the designer selects other metal-processing methods which can be performed in the plant. The labor skills available and the competitive situation also constitute implied constraints. Anything which limits the designer's freedom of choice is a constraint. Many materials and sizes are listed in supplier's catalogs, for instance, but these are not all easily available and shortages frequently occur. Furthermore, inventory economics requires that a manufacturer stock a minimum number of materials and sizes.

The synthesis of a scheme connecting possible system elements is sometimes called the *invention of the concept*. This is the first step in the synthesis task. As the fleshing out of the scheme progresses, analyses must be performed to assess whether the system performance is satisfactory or better and, if satisfactory, just how well it will perform. This is an analysis task. System schemes that do not survive analysis are revised, improved, or discarded. Those with potential are optimized to determine the best performance of which the scheme is capable. Competing schemes are compared so that the path leading to the most competitive product can be chosen. Figure 1–4 shows that *synthesis* and *analysis and optimization* are intimately and iteratively related. Synthesis draws heavily on talent. In this iteration the specification set is formed.

We have noted, and we shall do so again and again, that design is an iterative process in which we proceed through several steps, evaluate the results, and then return to an earlier phase of the procedure. Thus we may synthesize several components of a system, analyze and optimize them, and return to synthesis to see what effect this has on the remaining parts of the system. Both analysis and optimization require that we construct or devise abstract models of the system which will admit some form of mathematical analysis. We call these models *mathematical models*. In creating them it is our hope that we can find one which will simulate the real physical system very well.

As indicated in Fig. 1–4, *evaluation* is a significant phase of the total design process. Evaluation is the final proof of a successful design and usually involves the testing of a prototype in the laboratory. Here we wish to discover if the design really satisfies the need or needs. Is it reliable? Will it compete successfully with similar products? Is it economical to manufacture and to use? Is it easily maintained and adjusted? Can a profit be made from its sale or use? How likely is it to result in product-liability lawsuits? And is insurance easily and cheaply obtained? Is it likely that recalls will be needed to replace defective parts or systems?

Communicating the design to others is the final, vital presentation step in the design process. Undoubtedly, many great designs, inventions, and creative works have been lost to posterity simply because the originators were unable or unwilling to explain their accomplishments to others. Presentation is a selling job. The engineer, when presenting a new solution to administrative, management, or supervisory persons, is attempting to sell or to prove to them that this solution is a better one. Unless this can be done successfully, the time and effort spent on obtaining the solution have been largely wasted. When designers sell a new idea, they also sell themselves. If they are repeatedly successful in selling ideas, designs, and new solutions to management, they begin to receive salary increases and promotions; in fact, this is how anyone succeeds in his or her profession.

Basically, there are only three means of communication. These are the *written*, the *oral*, and the *graphical* forms. Therefore the successful engineer will be technically competent and versatile in *all three forms* of communication. An otherwise competent person who lacks ability in any one of these forms is severely handicapped. If ability in all three forms is lacking, no one will ever know how competent that person is! The three forms of communication—writing, speaking, and drawing—are skills, that is, abilities which can be developed or acquired by any reasonably intelligent person. Skills are acquired only by practice—hour after monotonous hour of it. Musicians, athletes, surgeons, typists, writers, dancers, aerialists, and artists, for example, are skillful because of the number of hours, days, weeks, months, and years they have practiced. Nothing worthwhile in life can be achieved without work, often tedious, dull, and monotonous, and lots of it; and engineering is no exception.

Ability in writing can be acquired by writing letters, reports, memos, papers, and articles. It does not matter whether or not the articles are published—the practice is the important thing. Ability in speaking can be obtained by participating in fraternal, civic, church, and professional activities. This participation provides abundant opportunities for practice in speaking. To acquire drawing ability, pencil sketching should be employed to illustrate every idea possible. The written or spoken word often requires study for comprehension, but pictures are readily understood and should be used freely.

The competent engineer should not be afraid of the possibility of not succeeding in a presentation. In fact, occasional failure should be expected, because failure or criticism seems to accompany every really creative idea. There is a great deal to be learned from a failure, and the greatest gains are obtained by those willing to risk defeat. In the final analysis, the real failure would lie in deciding not to make the presentation at all.

The purpose of this section is to note the importance of *presentation* as the final step in the design process. No matter whether you are planning a presentation to your teacher or to your employer, you should communicate thoroughly and clearly, for this is the payoff. Helpful information on report writing, public speaking, and sketching or drafting is available from countless sources, and you should take advantage of these aids.

Design Considerations

Sometimes the strength required of an element in a system is an important factor in the determination of the geometry and the dimensions of the element. In such a situation we say that *strength* is an important design consideration. When we use the expression *design consideration,* we are referring to some characteristic which influences the design of the element or, perhaps, the entire system. Usually quite a number of such characteristics must be considered in a given design situation. Many of the important ones are as follows:

1 Strength/stress
2 Distortion/deflection/stiffness
3 Wear
4 Corrosion
5 Safety
6 Reliability
7 Friction
8 Usability
9 Utility
10 Cost
11 Processing
12 Weight

13	Life	**20**	Surface
14	Noise	**21**	Lubrication
15	Styling	**22**	Marketability
16	Shape	**23**	Maintenance
17	Size	**24**	Volume
18	Control	**25**	Liability
19	Thermal properties	**26**	Scrapping/recycling

Some of these have to do directly with the dimensions, the material, the processing, and the joining of the elements of the system. Other considerations affect the configuration of the total system. We shall be giving our attention to these factors and other considerations throughout the book.

In this book you will be faced with a great many design situations in which engineering fundamentals must be applied, usually in a mathematical approach, to resolve the problem or problems. This is completely correct and appropriate in an academic atmosphere, where the need is actually to utilize these fundamentals in the resolution of professional problems. To keep the correct perspective, however, it should be observed that in many design situations the important design considerations are such that no calculations or experiments are necessary in order to define an element or system. Students, especially, are often confounded when they run into situations in which it is virtually impossible to make a single calculation and yet an important design decision must be made. These are not extraordinary occurrences at all; they happen every day. Suppose that it is desirable from a sales standpoint—for example, in medical laboratory machinery—to create an impression of great strength and durability. Thicker parts assembled with larger-than-usual oversize bolts can be used to create a rugged-looking machine. Sometimes machines and their parts are designed purely from the standpoint of styling and nothing else. These points are made here so that you will not be misled into believing that there is a rational mathematical approach to every design decision.

1–6 Codes and Standards

Once upon a time there were no standards for bolts, nuts, and screw threads. One manufacturer would produce, say, $\frac{1}{2}$-inch bolts with 9 threads per inch; another used 12. Some fasteners had left-handed threads and sometimes the thread profiles differed. It wasn't unusual in the early days of the automobile to see a mechanic lay out the fasteners in a line as they were disassembled in order to avoid mixing them during reassembly. This lack of standards and uniformity was costly and inefficient for a great variety of reasons. It is no wonder that a person, disgusted with his or her inability to find a replacement for a damaged fastener, sometimes used baling wire to fasten parts together.

A *standard* is a set of specifications for parts, materials, or processes intended to achieve uniformity, efficiency, and a specified quality. One of the important purposes of a standard is to place a limit on the number of items in the specifications so as to provide a reasonable inventory of tooling, sizes, shapes, and varieties.

A *code* is a set of specifications for the analysis, design, manufacture, and construction of something. The purpose of a code is to achieve a specified degree of safety, efficiency, and performance or quality. It is important to observe that safety codes *do not* imply *absolute safety.* In fact, absolute safety is impossible to obtain. Sometimes the unexpected event really does happen. Designing a building to withstand a 120 mi/h wind does not mean that the designer thinks a 140 mi/h wind is impossible; it simply means that he or she thinks it is highly improbable.

All of the organizations and societies listed below have established specifications for standards and safety or design codes. The name of the organization provides a clue to the nature of the standard or code. Some of the standards and codes, as well as addresses, can be obtained in most technical libraries. The organizations of interest to mechanical engineers are

Aluminum Association (AA)
American Gear Manufacturers Association (AGMA)
American Institute of Steel Construction (AISC)
American Iron and Steel Institute (AISI)
American National Standards Institute (ANSI)[3]
American Society for Metals (ASM)
American Society of Mechanical Engineers (ASME)
American Society of Testing and Materials (ASTM)
American Welding Society (AWS)
Anti-Friction Bearing Manufacturers Association (AFBMA)
British Standards Institution (BSI)
Industrial Fasteners Institute (IFI)
Institution of Mechanical Engineers (I. Mech. E.)
International Bureau of Weights and Measures (BIPM)
International Standards Organization (ISO)
National Institute for Standards and Technology (NIST)[4]
Society of Automotive Engineers (SAE)

1–7 Economics

The consideration of cost plays such an important role in the design decision process that we could easily spend as much time in studying the cost factor as in the study of the entire subject of design. Here we introduce only a few general approaches and simple rules.

First, observe that nothing can be said in an absolute sense concerning costs. Materials and labor usually show an increasing cost from year to year. But the costs of processing the materials can be expected to exhibit a decreasing trend because of the use of automated machine tools and robots. The cost of manufacturing a single product will vary from city to city and from one plant to another because of overhead, labor, taxes, and freight differentials and the inevitable slight manufacturing variations.

Standard Sizes

The use of standard or stock sizes is a first principle of cost reduction. An engineer who specifies an AISI 1020 bar of hot-rolled steel 53 mm square, called a hot-rolled square, has added cost to the product, provided a bar 50 or 60 mm square, both of which are preferred sizes, would do equally well. The 53-mm size can be obtained by special order or by rolling or machining a 60-mm square, but these approaches add cost to the product. To ensure that standard or preferred sizes are specified, the designer must have access to stock lists of the materials he or she employs. These are available in libraries or can be obtained directly from the suppliers.

---

3. In 1966 the American Standards Association (ASA) changed its name to the United States of America Standards Institute (USAS). Then, in 1969, the name was again changed, to American National Standards Institute, as shown above and as it is today. This means that you may occasionally find ANSI standards designated as ASA or USAS.

4. Former National Bureau of Standards (NBS).

A further word of caution regarding the selection of preferred sizes is necessary. Although a great many sizes are usually listed in catalogs, they are not all readily available. Some sizes are used so infrequently that they are not stocked. A rush order for such sizes may mean more expense and delay. Thus you should also have access to a list such as those in Table E–17 for preferred inch and millimeter sizes.

There are many purchased parts, such as motors, pumps, bearings, and fasteners, which are specified by designers. In the case of these, too, you should make a special effort to specify parts that are readily available. Parts that are made and sold in large quantities usually cost somewhat less than the odd sizes. The cost of rolling bearings, for example, depends more upon the quantity of production by the bearing manufacturer than upon the size of the bearing.

Large Tolerances

Among the effects of design specifications on costs, those of tolerances are perhaps most significant. Tolerances in design influence the producibility of the end product in many ways; close tolerances may necessitate additional steps in processing or even render a part completely impractical to produce economically. Tolerances cover dimensional variation and surface-roughness range and also the variation in mechanical properties resulting from heat treatment and other processing operations.

Since parts having large tolerances can often be produced by machines with higher production rates, labor costs will be smaller than if skilled workers were required. Also, fewer such parts will be rejected in the inspection process, and they are usually easier to assemble.

Breakeven Points

Sometimes it happens that, when two or more design approaches are compared for cost, the choice between the two depends upon a set of conditions such as the quantity of production, the speed of the assembly lines, or some other condition. There then occurs a point corresponding to equal cost which is called the *breakeven point.*

As an example, consider a situation in which a certain part can be manufactured at the rate of 25 parts per hour on an automatic screw machine or 10 parts per hour on a hand screw machine. Let us suppose, too, that the setup time for the automatic is 3 h and that the labor cost for either machine is $20 per hour, including overhead. Figure 1–5 is a graph of cost versus production by the two methods. The breakeven point corresponds to

Figure 1–5

A breakeven point.

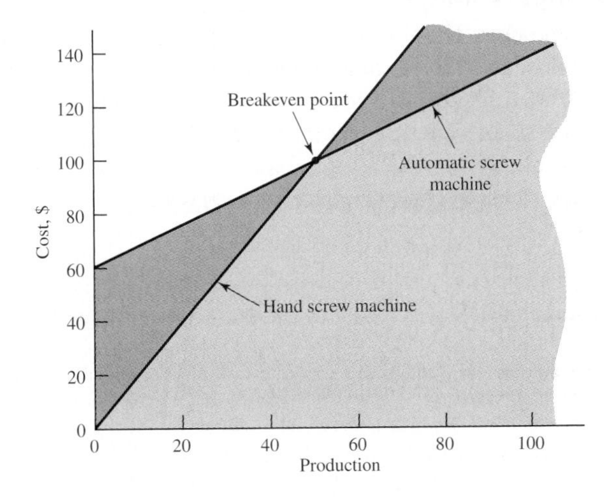

50 parts. If the desired production is greater than 50 parts, the automatic machine should be used.

Cost Estimates

There are many ways of obtaining relative cost figures so that two or more designs can be roughly compared. A certain amount of judgment may be required in some instances. For example, we can compare the relative value of two automobiles by comparing the dollar cost per pound of weight. Another way to compare the cost of one design with another is simply to count the number of parts. The design having the smaller number of parts is likely to cost less. Many other cost estimators can be used, depending upon the application, such as area, volume, horsepower, torque, capacity, speed, and various performance ratios.

1–8 Safety and Product Liability

The *strict liability* concept of product liability generally prevails in the United States. This concept states that the manufacturer of an article is liable for any damage or harm that results because of a defect. And it doesn't matter whether the manufacturer knew about the defect, or even could have known about it. For example, suppose an article was manufactured, say, 10 years ago. And suppose at that time the article could not have been considered defective on the basis of all technological knowledge then available. Ten years later, according to the concept of strict liability, the manufacturer is still liable. Thus, under this concept, the plaintiff only needs to prove that the article was defective and that the defect caused some damage or harm. Negligence of the manufacturer need not be proved.

One of the touchy subjects that sometimes comes up in the practice of engineering is what to do if you detect something that you consider poor engineering. If possible, of course, you should attempt to correct it or run sufficient tests to prove that your fears are groundless. If neither of these approaches is feasible, then another approach is to place a memo in the design file and to keep a copy of the memo at home in case the original is "lost." While this approach may protect you, it may also result in your being passed over for promotion, or even in your being discharged. In the long run, if you really feel strongly about the fact that the engineering is poor, and you cannot live with it, then you should transfer or seek a new position.

The best approaches to the prevention of product liability are good engineering in both analysis and design, quality control, and comprehensive testing procedures. Advertising managers often make glowing promises in the warranties and sales literature for a product. These statements should be reviewed carefully by the engineering staff to eliminate excessive promises and to insert adequate warnings and instructions for use.

Several books, all by one author,[5] can be helpful at this point and you may want to consult them from time to time later.

1–9 The Adequacy Assessment

An adequacy assessment was defined in Sec. 1–2, together with other concepts useful in mechanical engineering design. An adequacy assessment is such a useful and important idea that it is also referred to as *skill* 1, the primal skill of the designer. An adequacy

---

5. L. J. Kamm, *Successful Engineering: A Guide to Achieving Your Career Goals*, McGraw-Hill, New York, 1989; *Real World Engineering: A Guide to Achieving Career Success*, IEEE Press, New York, 1991; *Understanding Electro-Mechanical Engineering: An Introduction to Mechatronics*, IEEE Press, New York, 1996; and *Designing Cost-Efficient Mechanisms*, Society of Automotive Engineers, Warrentown, Pa., 1993.

assessment begins with the specification set of a simple design or with the entire design description (assembly drawings, detail drawings, and the bill of materials) when the design is complex.

When confronted with a finished design (all decisions made), the designer must ask whether all constraints have been sufficed in a manner that either equals or exceeds expectations or in a less-than expected fashion. The phrase "constraints have been sufficed" is an often-used verbal abbreviation. Additionally, some criterion is used to examine for merit, and optimal merit is often sought or approximated. An adequacy assessment is a judgment as to whether constraints have been sufficed and optimality realized.

An adequacy assessment is a process by which information is gathered and used in deciding whether constraints and optimality received appropriate attention. It can vary in some detail(s) from case to case, so that it is always familiar, yet somewhat different. It consists of various empirical and theoretical steps and mathematical modeling, ideas that an engineer uses to decide if an extant design (and its parts) is functional, safe, reliable, and so on. An adequacy assessment is composed of simple ideas. One becomes familiar with the process by performing some parts of it and analyzing increasingly complex situations, thus gaining proficiency.

Begin by asking if each of the 26 design considerations listed in Sec. 1–5 is adequately met, presuming all 26 apply. In considering item 1, strength (and stress), the following questions are useful:

- Which strength (ultimate, tensile, yield, fatigue, creep, and so on)?
- Which stress (tensile, shear, principal, von Mises, and so on)?
- Which pair of stress and strength is most revealing of functional integrity, and which location or locations are critical?
- How much of a disparity between stress and strength is "enough"?

The second characteristic is reliability, which can involve the following questions:

- Which kinds of loss of function are present?
- What is the probability of failure in each mode?
- Is the reliability goal met or exceeded?

Additional items in the list of considerations then are addressed.

An effective adequacy assessment identifies the questions to be asked, elicits quantitative estimates of the answers, includes mathematical modeling, and considers compatibility with nature and the marketplace. As with any intellectual skill, proficiency is the result of beginning simply with examples, drill problems in applications, then exercises which demand specific assessments in situations new to the student. It is important to develop a feeling for the relation of nature and the marketplace. Proficiency here is important, for the congruence of these forces is the basis for a design algorithm to be used later. It involves knowledge, judgment, proficiency, and consistency.

The adequacy assessment draws from the analysis portions of prior course work, with no respect for the compartmentalization of courses. What you choose to use depends on your goals. And the course which this book supports is no exception. It builds on many disciplinary roots in the reader's prior educational experience. Early problems are simpler, and it is easy to see what is important and to draft a pencil-and-paper adequacy assessment. As you proceed through the book and gain experience, it is useful to revisit earlier problems to see if you would still do them the same way. Your instructor is key to handling the many questions and details that surround skill 1. Consult your instructor as you learn to use skill 1 and apply it to situations developed in this book, or if you are heading in a different direction as necessitated by the course objectives.

Without function, a design is lost. Predictably, function is the first attribute examined. Consider a turnbuckle used to create an adjustable level of tension in a rodlike part. Its function is to maintain (say) a 5000-lb tension rod. Loss of function in this case occurs when

- The rod thread is permanently distorted.
- The turnbuckle thread is permanently distorted.
- Permanent distortion occurs in the turnbuckle center, either under load or during adjustment.

A paper-and-pencil adequacy assessment checks for yielding or applying more than maximum load at all of these places. A test fixture can be built to hydraulically apply a load, and prototype parts can be checked for permanent distortion. If analysis techniques are robust, then pencil-and-paper assessment may suffice. Tests can be done whenever there is doubt.

A pencil-and-paper assessment for safety on a shear key (a kind of mechanical fuse) in a power train would estimate the failing load (torque) and check to see if it is high enough to transmit the working torque, but not high enough to endanger other elements in the transmission line. It would also make sure no debris is thrown to endanger operators or bystanders. The corresponding experimental assessment would measure these torques.

A pencil-and-paper assessment for reliability under rated loading of a device that includes a rolling-contact bearing would start with an estimate of the reliability under the condition of radial and thrust loading, rotational speed, and time that the bearing should survive. The reliability estimate would be compared to the reliability goal. Such calculations are ultimately based on extensive testing carried out by the suppliers of the bearing and expressed in the AFBMA standards and equations. An experimental check, which would involve running prototypes of the machine of which the bearing is a part, is an element of an acceptance test during design development. Field service data on observed failures as reported from service records of a dealership are part of continuous monitoring by a manufacturer.

A pencil-and-paper adequacy assessment for competitiveness focuses on the cost of manufacture, shipping, and service organization costs, along with comparisons with competitive products. An optimal design is a difficult product with which to compete. It is important, however, that the figure of merit used is the one to which the marketplace is sensitive. Sales records constitute an after-the-fact competitiveness evaluation, since they are not available during the design process.

A pencil-and-paper adequacy assessment for manufacturability includes experiential opinion as to whether things such as section geometry can be maintained, tolerances sustained, or necessary heat treatments routinely delivered. Drawings are examined to check for accessibility of items such as boltheads and nuts by standard wrenches, availability of room to swing the wrench, and space in which to position, hold, and assemble such fastener parts by automatic tooling during assembly by the manufacturer or by hand during servicing or maintenance.

1–10 Uncertainty

Uncertainties in machinery design abound. Among these are uncertainties concerning

- Composition of material and the effect of variation on properties.
- Variations in properties from place to place within a bar of stock.
- Effect of processing locally, or nearby, on properties.

- Effect of nearby assemblies such as weldments and shrink fits on stress conditions.
- Effect of thermomechanical treatment on properties.
- Intensity and distribution of loading.
- Validity of mathematical models used to represent reality.
- Intensity of stress concentrations.
- Influence of time on strength and geometry.
- Effect of corrosion.
- Effect of wear.
- Uncertainty as to the length of any list of uncertainties.

Engineers must accommodate uncertainty. Many machine elements were known to engineers in ancient times (third century B.C.), but the modern forms might be unrecognizable to them. Innovations in materials and manufacturing techniques, insightful theory, computational advantages, and the inventive talent of some of our more recent ancestors have led to enormous changes.

Uncertainty always accompanies change. Material properties, load variability, fabrication fidelity, and validity of mathematical models used are among concerns to designers. The methods used by engineers to address such concerns evolved as engineering developed. We next take a brief look at the history of this field. The first two methods discussed predate the concept of stress, which was introduced by Cauchy in 1882.

The Roman Method

The first known method to address mechanical uncertainty dates to Macedonian times, several centuries B.C. The method was to replicate a proven, durable design. Toward this end encyclopedias of architectural and engineering works were compiled, such as the compendium by Vitruvius (first century B.C.). And going back to the time of Aristoteles (384–322 B.C.), the Greeks knew how to determine moments and reactions in cantilevered and simply supported beams, and the influence of width and depth of rectangular cross-section beams on load-carrying capacity.

The Factor-of-Safety Method of Philon

The second method, of which the earliest record is that of Philon of Byzantium (third century B.C.), was to separate the loss-of-function load and the impressed load by employing the ratio n_d defined as

$$n_d = \frac{\text{loss-of-function load}}{\text{impressed load}} \tag{1–1}$$

where n_d is the *design factor*.[6] An *allowable load* can be found from

$$\text{Allowable load} = \frac{\text{loss-of-function load}}{n_d} \tag{1–2}$$

The Permissible-Stress Method

After the concept of stress was formulated, a permissible-stress viewpoint was advanced. The permissible stress is chosen as a fraction of a significant material property, strength.

6. Interestingly, the design factor formulated was to protect the engineer in a military application in which the cube root of a ratio had to be extracted, and no method was known to do this. The magnitude of n_d was 1.1, a number recognizable to the astronauts of today.

The fraction is chosen on the basis of the aggregate experience of engineers with successful designs. This method has the advantage of providing advice regarding the fraction of a simply measured property. Whereas the emphasis in the first two methods was on loading, here it has shifted to a property of a material. The twentieth century saw its most rapid development.

This method is used in civil engineering (structures) and is embraced by AISC. One also sees it in weldment technology.

The Allowable Stress by Design Factor Method

The techniques of analysis have matured faster than those of synthesis in the last 150 years. As a result, much synthesis was accomplished by anti-analysis—that is, by visualizing a solution then analyzing it to see if it was satisfactory. If an engineer began with a factor of safety, proceeded with the method of Philon, and made all the necessary decisions, the resulting factor of safety increased due to rounding. For example, 5.3 necessary bolts became 6, or necessary bolt diameters of 0.9 in became 1 in due to standardization, which results in mass-produced fasteners in discrete nominal sizes. Thus, the factor of safety increased during the design process, and the same words couldn't be used to describe different things.

Engineers began to draw a distinction between goal (design factor) and realization (factor of safety). The allowable stress is defined as

$$\sigma_{\text{all}} = \frac{\text{strength}}{n_d^m} \tag{1–3}$$

where m is the exponent of load in the load-to-stress equation, and n_d is the design factor. We follow the accepted practice of using the Greek sigma (σ) to denote stress. Commonly the exponent m is unity, because in the simpler loadings (tension, compression, shear, bending, torsion) stresses induced are directly and linearly proportional to load. Many engineers think of σ_{all} as (strength)/n_d. There is nothing amiss in doing this *as long as one is alert to the encounter with nonlinear load-to-stress equations*, as occurs in contact stresses in rolling-contact bearings and in gear teeth. The design factor is still chosen by the aggregate experience of engineers and corporations.

The Stochastic Design Factor Method

The increased attention focused by the engineering community in the twentieth century on aviation and space, control systems, and the behavior of aggregates of machines as systems led to quantifying *reliability*. Attention was focused not only on reliability of a system as measured from experience with the system, but as an a priori and realizable goal in the design of the system and its components.

Uncertainty in stress and strength can be quantified. For stress linearly proportional to load, the design factor is defined as

$$\mathbf{n}_d = \frac{\mathbf{S}}{\boldsymbol{\sigma}} \tag{1–4}$$

where $\mathbf{n}_d$, $\mathbf{S}$, and $\boldsymbol{\sigma}$ are random variables with statistical parameters, such as mean and standard deviation, and exhibiting distributions. The role of $\bar{n}_d$ of $\mathbf{n}_d$ is to sufficiently displace the mean stress $\bar{\sigma}$ of $\boldsymbol{\sigma}$ from the mean strength of $\mathbf{S}$ to realize or exceed the reliability goal. The procedure is to find $\bar{n}_d$ from stochastic considerations, then use ordinary algebra to accomplish the following:

- Propagate *means* through the systemic equations.

- Propagate *variances* through the same relations using propagation of error techniques, which employ real-number algebra.
- Identify the distributions.

We will discuss this procedure in Chapter 2. The advantage of the method is that the variability in the strength **S** and in the load **P**, which are quantitative rather than qualitative, and the reliability goal are used to establish $\bar{n}_d$. Then

$$\bar{\sigma}_{\text{all}} = \frac{\bar{S}}{\bar{n}_d} \tag{1–5}$$

The Stochastic Method

The stochastic method does not use a design factor at all; rather, attention is focused on the probability of survival (reliability). The computer is used for the integrations and simulations involved.

Summary

Table 1–1 offers advice based on experience through 1948. This is presented to give some perspective to the foregoing discussion. Notice in the table the use of imprecise words having little meaning. The intent of Visodic, who originated the categorizations in the table, was to communicate approaches to uncertainty before statistical methods had been developed. Corporations have design manuals, which have been fine-tuned by in-house engineering supervisors and reflect corporate experience, for the guidance of product designers.

Engineers have been considering uncertainty since the beginning of engineering and have met with varying degrees of success. Advantages we have over our predecessors include a reservoir of experience that has been reported and codified, experimental prowess, and insightful theory. These enable us to teach ourselves what we need to know, for all the answers are not yet in.

As an example, reflect on the conceptual roadblocks met by many gifted people. Consider bending, which confounded Galileo (1564–1642). Hooke (1660) correctly contributed a "theory of springs." Mariotte (1686) thought fibers of a beam compressed and extended. Assuming half compressed and half extended, he suspected a neutral surface in the "middle." John Bernoulli (1774) thought plane sections remained plane before and after bending, although he did not carry this idea further. Euler (1725) found the elastic curve without understanding what was happening internally.

Table 1–1

Factors of Safety and Circumstances in Which They Are Used

Source: The germ of this table is due to J. P. Visodic, "Design Stress Factors," *Proc. ASME*, vol. 55, May 1948.

	Knowledge of Loads	Knowledge of Stress	Knowledge of Environment	Knowledge of Material	Factor of Safety
1	Accurate determination	Accurate determination	Controllable conditions	Well-known	1.25–1.5
2	Readily determined	Readily determined	Reasonably constant	Well-known	1.5–2.0
3	Determinable	Determinable	Ordinary	Average	2.0–2.5
4	Average	Average	Average	Less-tried, or brittle	2.5–3.0
5	Average	Average	Average	Untried	3.0–4.0
6	Uncertain	Uncertain	Uncertain	Better-known	3.0–4.0

Notes: For repeated loads fos is applied to endurance strength. For impact loads use items 3–6, but include an impact factor. For brittle materials when ultimate strength is used, use items 1–6 doubled. If higher factors seem desirable, refine your analysis and testing.

In the mid-1700s testing machines began to appear, and the concept of strain was added. Coulomb (1773) advanced a strain argument to locate the neutral surface. Young (1807) did not have a clear idea of where the neutral surface was, but he wrote so much about it ("neutral line," "neutral axis") that the terms survive to this day. Navier (1821) wrote the equations of equilibrium and used a "force intensity." Cauchy (1822) combined the notions of stress and strain with the work of Hooke, Young, and Poisson to create the two-parameter theory of elasticity. By 1827 he was applying it to practical problems, such as bending. The understanding of bending took over 200 years, with many people contributing.

The genie was finally out of the bottle when Professor I. P. Church of Cornell University invented the *free-body diagram* (1887), a concept which allowed people lacking genius to use knowledge, concepts, and first principles to construct mathematical models with embedded reality, which in turn creates insights that teach us more and suggest directions for further study.

In this book we use the six preceding methods for addressing uncertainty—the Roman method, the factor-of-safety method, the permissible-stress method, the allowable-stress method, the stochastic design factor method, and the stochastic method. We also use some variations to compare and contrast the methods to become familiar with them. In following the work of others, it is helpful to be familiar with the methods they employ, even if the method would not be our method of choice in the circumstances.

1–11 Stress and Strength

Sections 1–5 listed 26 of the more common foci of attention during the design process. The survival of many products depends on how the designer adjusts load-induced stress to be less than the strength at a location of interest. The designer must allow the strength to exceed the stress by a sufficient margin so that despite the uncertainties, failure is rare.

In focusing on the stress–strength comparison at a critical (controlling) location, we often look for "strength in the geometry and condition of use." Strengths are stresses at which something of interest occurs, such as the proportional limit, 0.2 percent-offset yielding, or fracture. In many cases, such events represent the stress level at which loss of function occurs.

Permissible-stress level is established by agreement upon the appropriate fraction of relevant strength that is a satisfactory working stress level. An allowable stress is found by dividing the relevant strength by the design factor. Although these are algebraically similar, the fraction may include no load information, whereas the design factor can. This distinction can be lost in language used by designers. Finer distinctions among terms—*allowable stress*, *permissible stress*, or simply *allowables* or *permissibles*, or, with loss of precision, simply *allowables*—can get blurred, and if the distinction is important, careful listening and interrogation are in order.

Strength is a *property* of a material or of a mechanical element. The strength of an element depends upon the choice, the treatment, and the processing of the material. Consider, for example, a shipment of 1000 springs. We can associate a strength S_i with the ith spring. When this spring is incorporated into a machine, external forces are applied that result in stresses in the spring, the magnitudes of which depend upon its geometry and are independent of the material and its processing. If the spring is removed from the machine unharmed, the stress due to the external forces will drop to zero, its value before assembly. But the strength S_i remains as one of the properties of the spring. Remember, then, that *strength is an inherent property of a part*, a property built into the part because of the use of a particular material and process.

Various metalworking and heat-treating processes, such as forging, rolling, and cold forming, cause variations in the strength from point to point throughout a part. The spring cited above is quite likely to have a strength on the outside of the coils different from its strength on the inside because the spring has been formed by a cold winding process and the two sides may not have been deformed by the same amount. Remember, too, therefore, that a strength value given for a part may apply to only a particular point or set of points on the part.

In this book we shall use the capital letter S to denote strength, as above, with appropriate markings and subscripts to denote the kind of strength. Thus, S_s is a shear strength, S_y a yield strength, S_u an ultimate strength, and $\bar{S}$ a mean strength obtained by sampling test data.

In accordance with accepted practice, we shall employ the Greek letters σ (sigma) and τ (tau) to designate normal and shear stresses, respectively. Again, various markings and subscripts will indicate some special characteristic. For example, σ_1 is a principal stress, σ_y a stress component in the y direction, and σ_r a stress component in the radial direction.

One of the basic problems in dealing with stress and strength is how to relate the two in order to develop a safe, economical, and efficient design. In studying this problem it will be useful to examine this relationship as it applies in a particular branch of engineering design. The American Institute of Steel Construction (AISC), founded in 1921, is a nonprofit society whose objectives are to improve and advance the use of fabricated structural steel. To accomplish these objectives, the AISC publishes manuals, textbooks, specifications, and technical booklets. The best known and most widely used is the *Manual of Steel Construction*,[7] which holds a highly respected position in engineering literature. Important standards included in this manual are the Specification for the Design, Fabrication, and Erection of Structural Steel for Buildings and the Code of Standard Practice for Steel Buildings and Bridges.

Table 1–2 lists minimum yield strengths S_y and minimum tensile strengths S_u for certain ASTM steels as given in the AISC specifications. In this table, S_y represents either the minimum yield point, if the material has one, or the minimum yield strength.

The ANSI-ASTM standard B483-78 defines *minimum strength* as follows:

> **A2.1** Standard mechanical property limits for the respective size ranges are based on an analysis of data from standard production material and are established at a level [at] which at least 99 percent of the population of values obtained from all standard material in the size range meets the established value.

Many readers may prefer to qualify *minimum* by always using the adjective *ASTM*, as in *ASTM minimum strength*. The unqualified word *minimum* may be misleading, since there

Table 1–2

Specified Minimum Strengths of Certain ASTM Steels

Steel Type	ASTM No.	S_y, kpsi	S_u, kpsi	Size, in, up to
Carbon	A36	36	58	8
Carbon	A529	42	60	$\frac{1}{2}$
Low alloy	A572	42	60	6
Low alloy	A572	50	65	2
Stainless	A588	50	70	4
Alloy Q&T	A514	100	110	$2\frac{1}{2}$

7. Now in its eighth edition; new editions are published periodically.

is a chance that up to 1 percent of the materials involved may have a strength smaller than the ASTM minimum.

To ensure that materials having these minimum strengths are actually used in the construction, the AISC specification states that:

> Certified mill test reports or certified reports of tests made by the fabricator or a testing laboratory in accordance with ASTM A6 or A568, as applicable, and the governing specification shall constitute sufficient evidence of conformity with one of the ASTM standards. Additionally, the fabricator shall, if requested, provide an affidavit stating that the structural steel furnished meets the requirements of the grade specified.

The AISC is developing the permissible-stress method and the notion of design factor will not be involved. Note also that the variability in S is not described in any way, and reliability is not mentioned.

Finding reduced values constitutes the first part of the AISC procedure. Following AISC vocabulary, let us designate allowable normal stress σ_{all} and allowable shear stress as τ_{all}. Then the relationship between allowable stresses and specified minimum strengths using the AISC code is specified as:

$$\begin{array}{ll} \text{TENSION} & 0.45S_y \le \sigma_{\text{all}} \le 0.60S_y \\ \text{SHEAR} & \tau_{\text{all}} = 0.40S_y \\ \text{BENDING} & 0.60S_y \le \sigma_{\text{all}} \le 0.75S_y \\ \text{BEARING} & \sigma_{\text{all}} = 0.90S_y \end{array} \tag{1–6}$$

The next part of the AISC code deals with determining the loads or forces that are used to obtain the stresses. The procedure can be presented succinctly by the equation

$$F = \sum W_d + \sum W_l + \sum K F_l + F_w + \sum F_{\text{misc}} \tag{1–7}$$

where F is the force to be used in the appropriate stress equation; see Chap. 3, for example. The components of this force are defined as follows.

The term ΣW_d is the sum of the *dead loads*. These consist of the weight of the steelwork, the materials fastened to it, and the parts supported by it.

The term ΣW_l is the sum of all the stationary or static *live loads*. This includes the weight of equipment, occupants, fixtures, and the snow load if specified by an applicable code.

The force or the resultant of forces due to equipment that may cause impact or dynamic loading is also considered to be a live load and is represented by the term F_l. This factor is to be multiplied by a *service factor* K obtained from Table 1–3.

The term F_w in Eq. (1–7) is the wind load on the structure. Appropriate guidelines for this may be specified by local or regional codes.

The term ΣF_{misc} must be included in some localities to account for the effects of earthquakes, hurricanes, or other extraordinary regional conditions.

The final step in the AISC procedure is to select dimensions of the member to be sized such that the design stress computed from the force F does not exceed the allowable stress as given by Eq. (1–6); in other words, to select dimensions or geometry such that

$$\sigma \le \sigma_{\text{all}} \qquad \text{or} \qquad \tau \le \tau_{\text{all}} \tag{1–8}$$

where σ and τ may be called the design values of the normal and shear stresses, respectively. This presumes a linear proportionality between load and stress.

In summary, Eq. (1–6) defines allowable stresses as the specified minimum strengths reduced by multiplication factors varying from 40 to 90 percent to ensure safety. Equation

Table 1–3

AISC Service Factors for Use in Eq. (1–7)

For supports of elevators	$K = 2$
For cab-operated traveling-crane support girders and their connections	$K = 1.25$
For pendant-operated traveling-crane support girders and their connections	$K = 1.10$
For supports of light machinery, shaft- or motor-driven	$K \geq 1.20$
For supports of reciprocating machinery or power-driven units	$K \geq 1.50$
For hangers supporting floors and balconies	$K = 1.33$

(1–7) accounts for all possible loads, and the service factors in Table 1–3 provide an additional degree of safety for dynamic loading. This is an implementation of the method of permissible stress with the strength characterized at the 99-percentile level.

1–12 Design Factor and Factor of Safety

The AISC method for relating stress and strength is also used in some other specialized design areas. However, it is not a general approach, since it addresses only specific materials and loadings. In mechanical engineering design there are categories of situations confronting the engineer.

- The product is made in large quantities, is valuable, or is dangerous, justifying elaborate testing of materials, components, and prototypes in the field.
- The product is made in sufficient quantities to justify a modest material test program, perhaps as small as ultimate tensile tests.
- The product is made in such small quantities that no testing of materials is performed at all.

The latter two situations are the more challenging. There is also the faulty product inquiry: Why? How can it be remedied?

A general approach to the allowable-load—loss-of-function load problem is the *design factor method*, used in one form or another since Philon, and sometimes called the classical method of design. The fundamental equation is Eq. (1–2):

$$\text{Allowable load} = \frac{\text{loss-of-function load}}{n_d}$$

where n_d is called the *design factor*.[8] For a given loss-of-function load, doubling the design factor halves the permissible load. To be most useful, this property is made to persist regardless of the linearity, or lack thereof, of stress with load. The adequacy assessment in the design-factor methods consists in part of estimating the *factor of safety* of the completed design. Factor of safety n has the same definition as the design factor, but it differs numerically due to the rounding (usually up) caused by using standard sizes and off-the-shelf components.

In assessing the factor of safety of an element, say a gear tooth that can fail in bending fatigue or surface fatigue, we note the tooth has a factor of safety guarding against bending fatigue and another factor of safety guarding against surface fatigue. If these factors of safety are numerically 1.5 and 1.3, then surface fatigue will occur before bending fatigue (1.3 vs. 1.5), and a 30 percent increase in power (torque) will risk this mode of failure.

8. See D. W. Lapedes, *Dictionary of Scientific and Technical Terms*, McGraw-Hill, New York, 1978, under *factor of safety*.

When Hooke discovered elastic behavior in metal springs, the insightful concept of stress was still 162 years away. The concept of stress presented the opportunity to express design factor in terms of stress. With the invention of testing machines,[9] where particular modes of failure could be induced, design factors came to be expressed in terms of a stress and a relevant strength, and expressions such as

$$n_d = \frac{\sigma\,(\text{loss-of-function})}{\sigma\,(\text{allowable})} = \frac{\text{strength}}{\text{stress}} \tag{1–9}$$

were possible. For modes of failure where stresses were not linearly proportional to loads, the form changes to

$$n_d = \left(\frac{\text{strength}}{\text{stress}}\right)^3 \quad \text{for sphere in contact fatigue} \tag{1–10}$$

$$n_d = \left(\frac{\text{strength}}{\text{stress}}\right)^2 \quad \text{for cylinders in contact fatigue} \tag{1–11}$$

reflecting the nonlinear relationship between stress and load. In Eqs. (1–9) through (1–11) strengths can be minimums, yields, tensiles, fatigues, and shears, as well as some others. Of course the stress used must correspond in type and units to the strength. Also, the stress and strength must be at the same location.

Equation (1–6) was used to account for any uncertainties regarding the actual strength of a member. Equation (1–7) was used to account for any uncertainties as to actual loads. The design factor in Eq. (1–9) is used to account for *both* uncertainties. The same is true of Eqs. (1–10) and (1–11), as well as Eqs. (1–4) and (1–5).

1–13 Reliability

In these days of greatly increasing numbers of liability lawsuits and the need to conform to regulations issued by governmental agencies such as EPA and OSHA, it is very important for the designer and the manufacturer to know the reliability of their product. The *reliability method* of design is one in which we learn or determine the distribution of stresses and the distribution of strengths and then relate these two in order to achieve an acceptable success rate.

The statistical measure of the probability that a mechanical element will not fail in use is called the *reliability* of that element. The reliability R can be expressed by a number having the range

$$0 \le R < 1 \tag{1–12}$$

A reliability of $R = 0.90$ means that there is a 90 percent chance that the part will perform its proper function without failure. The failure of 6 parts out of every 1000 manufactured might be considered an acceptable failure rate for a certain class of products. This represents a reliability of

$$R = 1 - \frac{6}{1000} = 0.994$$

or 99.4 percent.

9. Musschenbrock (1729), Perronet (1768), then Gauthey, Rondelet, Girard, and Williams contributed. The first to employ the hydraulic press and the balanced beam was Lagerhjelm in 1827. The feature that is important is the controlling of the mode of failure during measurement and the introduction of the notion of material strength, verified by test, and using the concept of stress due to Cauchy (1822).

In the *reliability method of design*, the designer's task is to make a judicious selection of materials, processes, and geometry (size) so as to achieve a reliability goal. Thus, if the objective reliability is to be 99.4 percent, as above, what combination of materials, processing, and dimensions is needed to meet this goal?

Analyses which lead to an assessment of reliability address uncertainties, or their estimates, in parameters which describe the situation. Stochastic variables such as stress, strength, load, or size are described in terms of their means, standard deviations, and distributions. If bearing balls are produced by a manufacturing process in which a diameter distribution is created, we can say upon choosing a ball that there is uncertainty as to size. If we wish to consider weight or moment of inertia in rolling, this size uncertainty can be considered to be *propagated* to our knowledge of weight or inertia. There are ways of estimating the statistical parameters describing weight and inertia from those describing size and density. These methods are variously called *propagation of error*, *propagation of uncertainty*, or *propagation of dispersion*. These methods are integral parts of analysis or synthesis tasks when probability of failure is involved.

The stochastic design factor method and the stochastic method approach to design are relatively new, considering the scope of machine design history. We have records of the historical behavior of ensembles of materials and their statistical descriptions. Using them the designer can proceed toward a specified reliability goal. In doing this the geometric sizes of parts will be larger than in the case where the designer has specific stochastic information on the material finally selected.

The statistical approach offers the opportunity to consider reliability goals and other ideas that the methods of antiquity could not. If the loading to which a tractor transmission will be subjected in a field is uncertain, then instrumented drawbars in field studies yield information that is better understood in stochastic rather than deterministic interpretation. Statistics allows the quantification and explanation of variation. Nature is full of individual variability with long-term regularity. Stability is not uniqueness in magnitude, but in a pattern of variation, a mix of systematic and random effects. Statistics allows these to be separated; it is about variation and its measurement, the ways models describe data and data criticize models, and about the sensitive use of data to illuminate the dark places. The design-factor method of Philon will be given statistical assistance.

1–14 Numbers, Units, and Preferred Units

Appendix A begins by defining *approximate numbers, approximation-error numbers, incomplete numbers, significant numbers,* and *rounded numbers,* all being real numbers. Incomplete and rounded numbers serve our real number needs. A *random number* **x** is denoted in boldface. It has statistical parameters (mean and standard deviation, usually) and a distribution. The standard deviation is an incomplete number, possibly rounded. The standard deviation relates to the precision of the mean, *not* the displayed digits of the mean. Confidence bounds are also incomplete numbers representing ends of a confidence interval.

Units

In the symbolic equation of Newton's second law

$$F = MLT^{-2} \tag{1–13}$$

F stands for force, M for mass, L for length, and T for time. Units chosen for any three of these quantities are called *base units.* The first three having been chosen, the fourth unit is called a *derived unit.* When force, length, and time are chosen as base units, the mass is the derived unit and the system that results is called a *gravitational system of units.* When mass, length, and time are chosen as base units, force is the derived unit and the system that results is called an *absolute system of units.*

In the English-speaking countries, the *U.S. customary foot-pound-second system* (fps) and the *inch-pound-second system* (ips) are the two standard gravitational systems most used by engineers. In the fps system the unit of mass is

$$M = \frac{FT^2}{L} = \frac{\text{(pound-force)(second)}^2}{\text{foot}} = \text{lbf} \cdot \text{s}^2/\text{ft} = \text{slug} \tag{1–14}$$

Thus, length, time, and force are the three base units in the fps gravitational system.

The unit of time in the fps system is the second, abbreviated s. The unit of force in the fps system is the pound, more properly the *pound-force.* We shall seldom abbreviate this unit as lbf; the abbreviation lb is permissible, since we shall be dealing only with the U.S. customary gravitational system. In some branches of engineering it is useful to represent 1000 lb as a kilopound and to abbreviate it as kip. Many writers add the letter "s" to kip to obtain the plural, but to be consistent with the practice of using only singular forms for units we shall not do so here. Thus, 1 kip and 3 kip are used to designate 1000 and 3000 lb, respectively. Finally, we note in Eq. (1–14) that the derived unit of mass in the fps gravitational system is the lb · s^2/ft, called a *slug*; there is no abbreviation for slug.

The unit of mass in the ips gravitational system is

$$M = \frac{FT^2}{L} = \frac{\text{(pound-force)(second)}^2}{\text{inch}} = \text{lb} \cdot \text{s}^2/\text{in} \tag{1–15}$$

The units lbf · s^2/in have no official name, unfortunately for students. Professor R. L. Norton of Worchester Polytechnic Institute suggests calling this unit the *blob,* abbreviated *bl,* an anagram for the abbreviation of pound force. The blob is larger than the slug by a factor of 12.

The International System of Units (SI) is an absolute system. The base units are the meter, the kilogram (for mass), and the second. The unit of force is derived using Newton's second law and is called the *newton* to distinguish it from the kilogram, which, as indicated, is the unit of mass. The units constituting the newton (N) are

$$F = \frac{ML}{T^2} = \frac{\text{(kilogram)(meter)}}{\text{(second)}^2} = \text{kg} \cdot \text{m/s}^2 = \text{N} \tag{1–16}$$

The weight of an object is the force exerted upon it by gravity. Designating the weight as W and the acceleration due to gravity as g, we have

$$W = mg \tag{1–17}$$

In the fps system, standard gravity is $g = 32.1740$ ft/s^2. For most cases this is rounded off to 32.2. Thus the weight of a mass of 1 slug in the fps system is

$$W = mg = (1\ \text{slug})(32.2\ \text{ft/s}^2) = 32.2\ \text{lb}$$

In the ips system, standard gravity is 386.088 or about 386 in/s^2. Thus, in this system, a unit mass weighs

$$W = (1\ \text{lb} \cdot \text{s}^2/\text{in})(386\ \text{in/s}^2) = 386\ \text{lb}$$

With SI units, standard gravity is 9.806 or about 9.80 m/s^2. Thus, the weight of a 1-kg mass is

$$W = (1\ \text{kg})(9.80\ \text{m/s}^2) = 9.80\ \text{N}$$

In view of the fact that weight is the force of gravity acting upon a mass, the following quotation is pertinent:

> The greater advantage of SI units is that there is one, and only one unit for each physical quantity—the meter for length, the kilogram for mass, the newton for force, the second for time, etc. To be consistent with this unique feature, it follows that a given unit or word should not be used as an accepted technical name for two physical quantities. However, for generations the term "weight" has been used in both technical and nontechnical fields to mean either the force of gravity acting upon a body or the mass of the body itself. The reason for this double use of the term "weight" for two different physical quantities—force and mass—is attributed to the dual use of the pound units in our present customary gravitational system in which we often use weight to mean both force and mass.[10]

The seven SI base units, with their symbols, are shown in Table 1–4. These are dimensionally independent. Lowercase letters are used for the symbols unless they are derived from a proper name; then a capital is used for the first letter of the symbol. Note that the unit of mass uses the prefix kilo; this is the only base unit having a prefix.

Table 1–4 shows that the SI unit of temperature is the kelvin. The Celsius temperature scale (once called centigrade) is not a part of SI, but a difference of one degree on the Celsius scale equals one kelvin.

A second class of SI units comprises the derived units, many of which have special names. Table 1–5 is a list of those we shall find most useful in our work in this book.

The radian (symbol rad) is a supplemental unit in SI for a plane angle.

A series of names and symbols to form multiples and submultiples of SI units has been established to provide an alternative to the writing of powers of 10. Table E–1 includes these prefixes and symbols.

Rules for Use of SI Units

The International Bureau of Weights and Measures (BIPM), the international standardizing agency for SI, has established certain rules and recommendations for the use of SI. These are intended to eliminate differences which occur among various countries of the world in scientific and technical practices.

Number Groups

Numbers having four or more digits are placed in groups of three and separated by a space instead of a comma. However, the space may be omitted for the special case of numbers having four digits. A period is used as a decimal point. These recommendations avoid the confusion caused by certain European countries in which a comma is used as

Table 1–4

SI Base Units

Quantity	Name	Symbol
Length	meter	m
Mass	kilogram	kg
Time	second	s
Electric current	ampere	A
Thermodynamic temperature	kelvin	K
Amount of matter	mole	mol
Luminous intensity	candela	cd

10. From "S.I., The Weight/Mass Controversy," *Mech. Eng,* vol. 99, no. 9, September 1977, p. 40, and vol. 101, no. 3, March 1979, p. 42.

Table 1–5

Examples of SI Derived Units*

Quantity	Unit	SI Symbol	Formula
Acceleration	meter per second squared		$m \cdot s^{-2}$
Angular acceleration	radian per second squared		$rad \cdot s^{-2}$
Angular velocity	radian per second		$rad \cdot s^{-1}$
Area	square meter		m^2
Circular frequency	radian per second	ω	$rad \cdot s^{-1}$
Density	kilogram per cubic meter		$kg \cdot m^{-3}$
Energy	joule	J	$N \cdot m$
Force	newton	N	$kg \cdot m \cdot s^{-2}$
Force couple	newton-meter		$N \cdot m$
Frequency	hertz	Hz	s^{-1}
Power	watt	W	$J \cdot s^{-1}$
Pressure	pascal	Pa	$N \cdot m^{-2}$
Quantity of heat	joule	J	$N \cdot m$
Speed (rotary)	revolution per second		s^{-1}
Stress	pascal	Pa	$N \cdot m^{-2}$
Torque	newton-meter		$N \cdot m$
Velocity	meter per second		$m \cdot s^{-1}$
Volume	cubic meter		m^3
Work	joule	J	$N \cdot m$

*In this book, negative exponents are seldom used; thus, circular frequency, for example, would be expressed in rad/s.

a decimal point, and by the English use of a centered period. Examples of correct and incorrect usage are as follows:

1924 or 1 924 but not 1,924
0.1924 or 0.192 4 but not 0.192,4
192 423.618 50 but not 192,423.61850

The decimal point should always be preceded by a zero for numbers less than unity.

Use of Prefixes

The multiple and submultiple prefixes in steps of 1000 only are recommended (Table E–1). This means that length can be expressed in mm, m, or km, but not in cm, unless a valid reason exists.

When SI units are raised to a power, the prefixes are also raised to the same power. This means that the km^2 is defined thus:

$$1\ km^2 = (1000\ m)^2 = (1000)^2\ m^2 = 10^6\ m^2$$

Similarly, the mm^2 is defined as follows:

$$(0.001\ m)^2 = (0.001)^2\ m^2 = 10^{-6}\ m^2$$

When raising prefixed units to a power, it is permissible, though not often convenient, to use the nonpreferred prefixes such as cm^2 or dm^3.

Except for the kilogram, which is a base unit, prefixes should not be used in the denominators of derived units. Thus the meganewton per square meter, MN/m^2, is satisfactory, but the newton per square millimeter, N/mm^2, is not to be used. Note that this recommendation avoids a proliferation of derived units.

Double prefixes should not be used. Thus, instead of millimillimeters (mmm), use micrometers (μm).

Preferred Units

As a general rule, it is both convenient and good practice to select prefixes such that the number strings will contain no more than four digits to the left of the decimal point. By applying this rule, we find that some of the preferred units are kilopounds per square inch (kpsi) and megapascals (MPa) for stress, pounds per square inch (psi) and kilopascals (kPa) for air or hydraulic pressure, pounds (lb) and kilonewtons (kN) for force, and inches to the fourth power (in^4) and centimeters to the fourth power (cm^4) for second moment of area.

Table E–1 is of particular value when you are using units with mixed prefixes in an equation. Suppose we wish to solve the deflection equation (see Chap. 4)

$$y = \frac{64Fl^3}{3\pi d^4 E}$$

where $F = 1.30$ kN, $l = 300$ mm, $d = 2.5$ cm, and $E = 207$ GPa. It is convenient to show the solution in two parts, the first containing the numbers, and the second containing the prefixes. Thus

$$y = \frac{64(1.30)(300)^3}{3\pi(2.5)^4(207)} \frac{(\text{kilo})(\text{milli})^3}{(\text{centi})^4(\text{giga})}$$

Now compute the numerical value of the first part and substitute the prefix values in the second. This gives

$$y = [29.48(10)^3]\left[\frac{10^3(10^{-3})^3}{(10^{-2})^4(10^9)}\right] = 29.48(10)^{-4}\ \text{m}$$
$$= 2.948\ \text{mm}$$

In many cases, similar equations are used so often that we remember the prefixes to use in order to obtain a result having a convenient prefix. Thus, in the above equation, if we use F in kN, l and d in mm, and E in GPa, the result y is in mm. Try it. Tables E–3 and E–4 will also be of help to you.

CASE STUDY 1

A Henhouse Heater

It was a severe winter, and forecasts predicted it would worsen. The oil-fired henhouse heaters a farmer used to maintain a desired temperature level each night were barely holding their own. The nearest electrical source, 800 ft away at the barn, provided 120-V 60-Hz single-phase alternating current. He had on hand about 4000 ft on no. 10 solid copper wire with a resistance of approximately $0.1\ \Omega/100$ ft. His objective was to get more heat to the henhouse.

His solution was to run the wire to the henhouse, hang coils of wire on the walls, and run a return, all in one piece. His daughter was home during the December semester break. He asked her if she thought it would work. The sophomore mechanical engineering student did the following.

Need: Augment present heaters with electrical energy.
Function: Deliver the energy to the henhouse.

Plan A

The circuit consists of the impedance of the leads and the impedance of the wire in the henhouse, Z_1 and Z_2, respectively.

$$Z_1 = \frac{2(0.1)800}{100} = 1.6\ \Omega$$

$$Z_2 = \frac{0.1(2400)}{100} = 2.4\ \Omega$$

The current I through the circuit is

$$I = \frac{E}{Z} = \frac{120}{1.6 + 2.4} = 30\ \text{A}$$

The power dissipated in the henhouse is $I^2 Z_2$; therefore,

$$P = I^2 Z_2 = \left(\frac{E}{Z_1 + Z_2}\right)^2 Z_2 = \left(\frac{120}{1.6 + 2.4}\right)^2 2.4 = 2160\ \text{W}$$

Plan B

Can one do better with these resources? If the figure of merit M is the power dissipated in the henhouse, then

$$M = \left(\frac{E}{Z_1 + Z_2}\right)^2 Z_2$$

The stationary-point maximum is found by differentiating M with respect to Z_2 equating to zero, to find that $Z_1 = Z_2$. By shortening the wire to 3200 ft (in other words, shorting out 800 ft of the henhouse coil), then the power would be

$$P = \left(\frac{E}{Z_1 + Z_2}\right)^2 Z_2 = \left(\frac{120}{1.6 + 1.6}\right)^2 1.6 = 2250\ \text{W}$$

$$I = \frac{E}{Z_1 + Z_2} = \frac{120}{1.6 + 1.6} = 37.5\ \text{A}$$

Plan C

Another plan would double the conductors to and from the barn, cutting line losses as shown in Fig. CS–1. In this case the power dissipated is

$$P = \left(\frac{120}{0.8 + 0.8}\right)^2 0.8 = 4500\ \text{W}$$

and henhouse current is

$$I_2 = \frac{E}{Z} = \frac{120}{1.6} = 75\ \text{A}$$

Resolution

The daughter's response to her father is: "Your solution (plan A) will release power in the henhouse at the rate of 2160 W with a line current of 30 A. This is within the room temperature capability of the insulation. Plan B would involve shorting out 800 ft of the henhouse wire to maximize the heat release at 2250 W with a current of 37.5 A. Plan C would involve doubling the leads from the barn, leaving 800 ft of wire coiled in the henhouse, boosting the heat release to 4500 W, with 75 A in the henhouse wire and 37.5 A in the lead wires. The insulation on the henhouse wires would be damaged, and should not be operated unattended even if fused. Plan C represents optimal performance for a double-wire scheme. All three schemes involve modest heat release and have the virtue of simplicity. As emergency measures they can 'buy time,' to take some data as well as plan other combustion-based measures."

Can you identify the innovative contributions of the father? Of the daughter? Can you contribute?

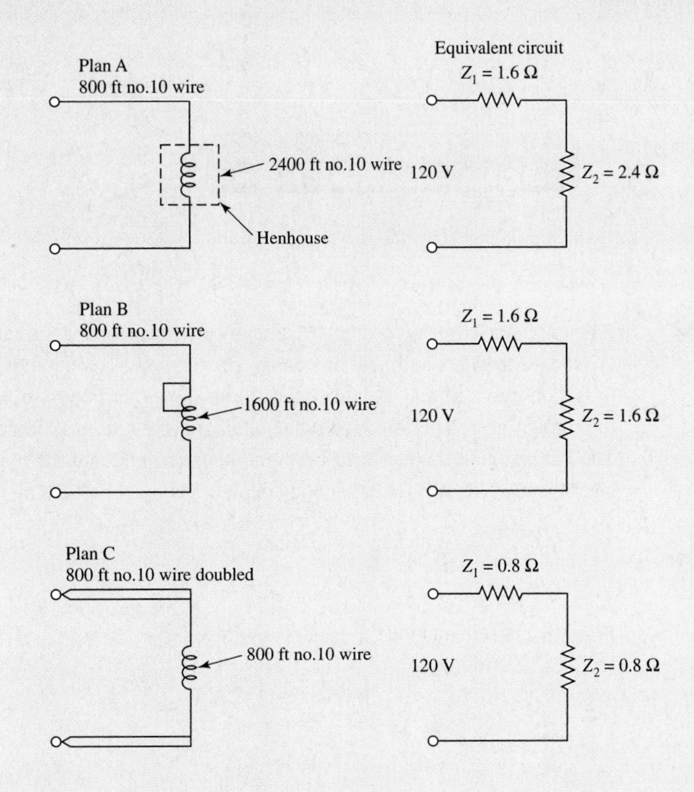

Figure CS–1

Plans A, B, and C for an emergency electric henhouse heater using 4000 ft of no. 10 solid copper wire.

PROBLEMS

Among the ideas identified in this chapter was Church's contribution to you as a reader and potential user of the elements of machinery design, either as a student or as an engineer. His insightful conceptual contribution allowed that someone who was not a genius could practice engineering through the *understanding* of the freebody, that is, being self-taught about the situation. Up to this point in your education freebody diagrams have been largely used as a mean of getting numerical answers to problems, or following the logic of others. Now they will increasingly assume their role in teaching the practitioner. Your experience thus far has shown you that some freebody diagrams are easy to identify, and others take a considerable lapse of calendar time for the insightful ensemble to occur to you. The ideas of system, surroundings, cause-effect-extent models of the influence of the surroundings on system, and the use of first principles are ingredients of a primordial soup from which an understanding is gleaned. The first few exercises of this chapter are for seeking understanding through the self-instructional role of the freebody diagram. Other exercises will coax responses different from those you may be used to. All this is to get flowing the juices that will be useful as the course unfolds.

1–1 A "loose" pin in a hole can be used as a sliding kinematic pair. If a force F is applied to the pin as shown in the figure at an angle to the axis of the pin (and the potential direction of motion), will the pin move? If not, why not? If so, what is the acceleration (or force) causing motion? How can one tell which is the case? Does reversing the sense of the force F change any of the answers?

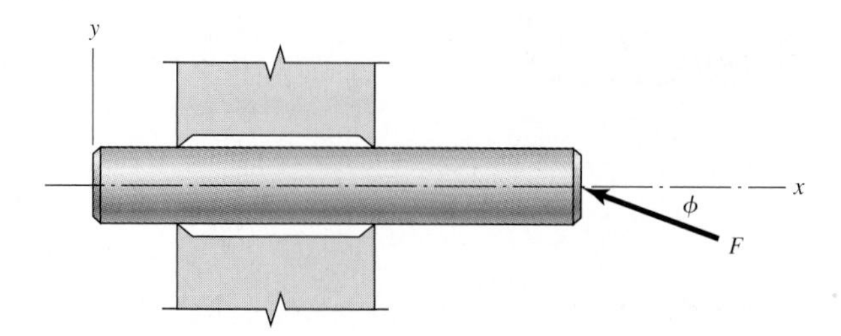

Problem 1–1

ANALYSIS

1–2 A detent (a lessening of tension, specifically a mechanism that initiates, allows, or locks movement) is a device to index and hold movement between parts. Some move in response to a force, others in response to a torque. A jaw-clutch detent shown in the figure has straightflanked teeth similar to a crown gear. The effective radius is r, and the retaining force is F, as shown in the figure. The flank angle θ is measured between the horizontal and the normal force vector, and f is the coefficient of friction. The free body depicts a contact particle on top of the jaw. Show that

$$F = N(\sin\theta - f\cos\theta)$$

$$T = Fr\frac{1 + f\tan\theta}{\tan\theta - f} = KFr$$

Is the detent self-locking?

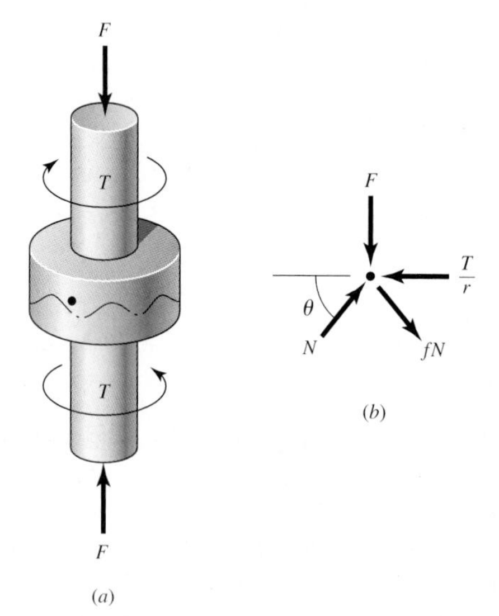

Problem 1–2
A jaw-clutch detent.

ANALYSIS

1–3 In Prob. 1–2 the axial force F is supplied by a helical compression spring. The compressive force is F_0 when the jaws are fully engaged. If x is the location of a particle from its lowest possible position, and k is the constant spring rate, then $F = F_0 + kx$.

(*a*) At what torque T_1 does axial motion begin?

(*b*) What torque T_2 allows indexing to occur?

ANALYSIS

1–4 A jaw-clutch detent as in Prob. 1–2 has a spring force $F = 10 + 2.5x$ lb, an effective jaw circle radius of $r = 2$ in, a tooth height of 0.2 in, a tooth flank angle $\theta = 60°$, and a coefficient of friction $f = 0.25$. Estimate the initial and final values of spring force F, normal force N, constant K, and the torque T.

ANALYSIS

1–5 Observe the effect of varying the tooth angle θ in Prob. 1–4 for initial motion with θ varying between 90° and critical angle θ_{cr}.

θ	N	K	T
90°			
80°			
⋮			
θ_{cr}			

ANALYSIS

1-6 Highway tunnel traffic (two parallel lanes in the same direction) experience indicates the average spacing between vehicles increases with speed. Data from a New York tunnel show that between 15 and 35 mi/h, the space x between vehicles (in miles) is $x = 0.324/(42.1 - v)$ where v is the vehicle's speed in miles per hour.

(*a*) Ignoring the length of individual vehicles, what speed will give the tunnel the largest volume in vehicles per hour?

(*b*) Does including the length of the vehicles cut the tunnel capacity prediction significantly?

(*c*) Does the optimal speed change much?

ANALYSIS

1-7 Problem 1–6 raises some interesting points. The vehicle-spacing function is bounded, that is, it exists from a lower range number, 15 mi/h, and an upper range number, 35 mi/h. The number of vehicles per hour Q is likewise a bounded function. We are seeking a global maximum of Q. Such an extreme can occur at

- An interior stationary point, a place where the slope is zero and continuous first derivatives simultaneously vanish.
- An interior point where discontinuities in the function Q or its first derivative appear.
- At a boundary between feasible and infeasible regions (at a boundary or at the intersection of boundaries).

What methods should be used to find the maximum if we know it occurs at an interior stationary point? If we don't know where the extreme will occur, what method or methods will work?

DESIGN

1-8 The engineering designer must create (invent) the concept and connectivity of the elements that constitute the design, and not lose sight of the need to develop the idea(s) with optimality in mind. A useful figure of merit can be cost, which can be related to the amount of material used (volume or weight). When you think about it, the weight is a function of the geometry and density. When the design is fleshed out, finding the weight is a straightforward, sometime tedious task. The figure depicts a simple bracket frame which has supports that project from a wall column. The bracket supports a chain-fall hoist. Pinned joints are used to avoid bending. The cost of a link can be *approximated* by $\$ = ¢Al\gamma$, where ¢ is the cost of the link per unit weight, A is the cross-sectional area of the prismatic link, l is the pin-to-pin link length, and γ is the specific weight. To be sure, this is approximate because no decisions have been made concerning the geometric form of the links or their fittings. By investigating cost now in this approximate way, one can detect whether a particular set of proportions of the bracket (indexed by angle θ) is advantageous. Is there a preferable angle θ? Show that the figure of merit can be expressed as

$$\text{fom} = \frac{\gamma ¢ W l_2}{S}\left(\frac{1+\cos^2\theta}{\sin\theta\cos\theta}\right)$$

where W is the weight of the hoist and load, and S is the allowable tensile or compressive load in the link material (no column action). Is the decision on angle θ independent of others? What is the desirable angle θ corresponding to least cost?

Problem 1–8
(a) A chain-hoist bracket frame.
(b) Free body of pin.

DESIGN **1–9** A conceptual variation to the bracket problem (Prob. 1–8) is depicted in the figure. The load W is to be carried a distance x from the bracket supports. If the cost of the two links is expressed as $\$ = -(\gamma_1 ¢_1 A_1 l_1 + \gamma_2 ¢_2 A_2 l_2)$, are there any advantageous angles α and β? Show that

$$\text{fom} = -\frac{\gamma ¢ W x}{S}\left[\left(\frac{\cos\beta}{\cos\alpha} + \frac{\cos\alpha}{\cos\beta}\right)\frac{1}{\sin(\alpha+\beta)}\right]$$

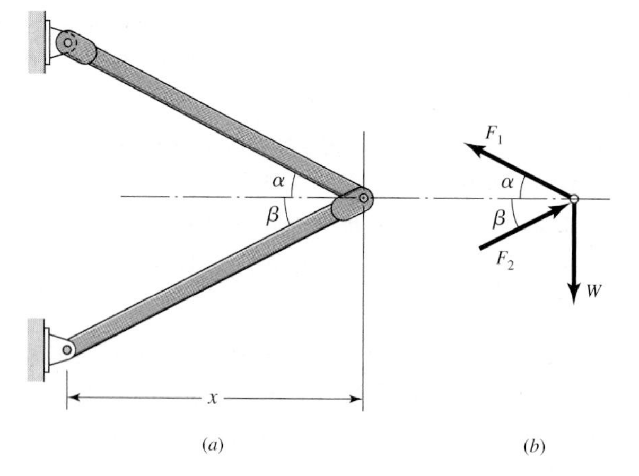

Problem 1–9
(a) A chain-hoist bracket frame.
(b) Free body of pin.

1–10 Engineering around 350 B.C. consisted largely of duplication of designs that worked earlier with only marginal changes. Aristoteles (384–322 B.C.) in *Problems of Machines* showed that he could calculate moments and reactions in cantilevered and simply supported beams and knew that too much moment led to collapse. The moment leading to collapse was doubled if a rectangular cross-section width was doubled, and it was quadrupled if the depth was doubled. In a simply supported beam, a load that could be safely placed at the center could be placed anywhere, and the largest moment occurred under a central load. By looking at extant successful beams, new beams could be proportioned. How large a load could be placed on a 4 × 4 in cross-section fir beam 160 in long, simply supported, if an existing cantilever of fir 2 × 2 × 12 in has carried a load of 155 lbf at the end for years? You are to consider yourself an apprentice to Aristoteles. For your convenience the units are not drachmas and cubits, and you need not perform your calculations using Roman numerals. However, since the equal sign was first used in *Whetstone of Wit* by Robert Record in 1557, resist the urge to use it. Also, the free-body diagram was not invented until 1887. The concept of stress, pioneered by Cauchy in 1822, is also unavailable to you. Grading policies then were an eye for an eye, so an engineer whose work injured someone suffered the same fate. As to fatal injuries, don't ask!

1–11 When one knows the true values x_1 and x_2 and has approximations X_1 and X_2 at hand, one can see where errors may arise. By viewing error as something to be added to an approximation to attain a true value, it follows that the error e_i is related to X_i and x_i as $x_i = X_i + e_i$.

(*a*) Show that the error in a sum $X_1 + X_2$ is

$$x_1 + x_2 = X_1 + X_2 + (e_1 + e_2)$$

(*b*) Show that the error in a difference $X_1 - X_2$ is

$$x_1 - x_2 = X_1 - X_2 + (e_1 - e_2)$$

(*c*) Show that the error in a product $X_1 X_2$ is

$$x_1 x_2 = X_1 X_2 \left(1 + \frac{e_1}{X_1} + \frac{e_2}{X_2}\right)$$

(*d*) Show that in a quotient X_1/X_2 the error is

$$\frac{x_1}{x_2} = \frac{X_1}{X_2}\left(1 + \frac{e_1}{X_1} - \frac{e_2}{X_2}\right)$$

1–12 Use the true values $x_1 = \sqrt{5}$ and $x_2 = \sqrt{6}$.

(*a*) Demonstrate the correctness of the error equation from Prob. 1–11 for addition if three correct digits are used for X_1 and X_2.

(*b*) Demonstrate the correctness of the error equation using three-digit significant numbers for X_1 and X_2.

ANALYSIS

1–13 With the true values $x_1 = \sqrt{12}$ and $x_2 = \sqrt{11}$, investigate the difference $X_1 - X_2$ using three-digit truncated approximate numbers for X_1 and X_2.

1–14 The true value of $x_1 = \sqrt{2}$ and $x_2 = \sqrt{3}$.

(*a*) Using three-digit truncated approximation for X_1 and X_2, demonstrate the validity of the product error relation of Prob. 1–11.

(*b*) For three-digit significant approximations for X_1 and X_2, demonstrate the validity of the product error relation.

1–15 The true value of $x_1 = \sqrt{2}$ and $x_2 = \sqrt{3}$. Using truncated three-digit approximations for X_1 and X_2, demonstrate the validity of the quotient error relation of Prob. 1–11.

1–16 When one does not know the true values a worst-case scenario is used. Three-digit truncated approximate numbers, $X_1 = 2.23$ and $X_2 = 2.44$, are to be added.

(*a*) What are the range numbers on the sum $X_1 + X_2$ and how many correct digits appear in the $X_1 + X_2$ sum?

(*b*) For X_1 and X_2 use the significant number approximations 2.24 and 2.45 and find the range numbers on $X_1 + X_2$ and the number of correct digits in the sum.

1–17 The errors in approximate numbers do not have to be of similar magnitude. If $X_1 = 3160$ and $X_2 = 3.16$, each having three correct digits, how many correct digits appear in the sum?

1–18 The numbers $X_1 = 3.46$ and $X_2 = 3.31$ are known to be truncated approximate numbers. What are the range numbers for the difference $X_1 - X_2$?

1–19 The numbers $X_1 = 1.41$ and $X_2 = 1.73$ are truncated approximate numbers. What are the range numbers for the product $X_1 X_2$ and the number of correct digits?

ANALYSIS

1–20 The numbers $X_1 = 1.41$ and $X_2 = 1.73$ are known to be approximate three-digit numbers. What are the range numbers for the quotient X_1/X_2 and the number of correct digits in the quotient?

ANALYSIS

1–21 Convert the following to appropriate SI units:

(*a*) A stress of 20 000 psi
(*b*) A force of 350 lb
(*c*) A moment of 1200 lb · in
(*d*) An area of 2.4 in^2
(*e*) A second moment of area of 17.4 in^4
(*f*) An area of 3.6 mi^2
(*g*) A modulus of elasticity of 21 Mpsi
(*h*) A speed of 45 mi/h
(*i*) A volume of 60 in^3

ANALYSIS

1–22 Convert the following to appropriate SI units:

(*a*) A length of 60 in
(*b*) A stress of 90 kpsi
(*c*) A pressure of 160 psi
(*d*) A section modulus of 11.2 in^3
(*e*) A unit weight of 2.61 lb/ft
(*f*) A deflection of 0.002 in
(*g*) A velocity of 1200 ft/min
(*h*) A unit strain of 0.0021 in/in
(*i*) A volume of 8 gal (U.S.)

ANALYSIS

1–23 Generally, final design results are rounded to or fixed to three digits because the given data cannot justify a greater display. In addition, prefixes should be selected so as to limit number strings to no more than four digits to the left of the decimal point. Using these rules, as well as those for the choice of prefixes, solve the following relations:

(*a*) $\sigma = M/Z$, where $M = 200$ N · m and $Z = 15.3\ \text{cm}^3$.
(*b*) $\sigma = F/A$, where $F = 42$ kN and $A = 6\ \text{cm}^2$.
(*c*) $y = Fl^3/3EI$, where $F = 1200$ N, $l = 800$ mm, $E = 207$ GPa, and $I = 6.4\ \text{cm}^4$.
(*d*) $\theta = Tl/GJ$, where $J = \pi d^4/32$, $T = 1100$ N · m, $l = 250$ mm, $G = 79.3$ GPa, and $d = 25$ mm. Convert results to degrees of angle.

ANALYSIS

1–24 Repeat Prob. 1–23 for the following:

(*a*) $\sigma = F/wt$, where $F = 600$ N, $w = 20$ mm, and $t = 6$ mm
(*b*) $I = bh^3/12$, where $b = 8$ mm and $h = 24$ mm
(*c*) $I = \pi d^4/64$, where $d = 32$ mm
(*d*) $\tau = 16T/\pi d^3$, where $T = 16$ N · m and $d = 25$ mm

ANALYSIS

1–25 Repeat Prob. 1–23 for:

(*a*) $\tau = F/A$, where $A = \pi d^2/4$, $F = 120$ kN, and $d = 20$ mm
(*b*) $\sigma = 32Fa/\pi d^3$, where $F = 800$ N, $a = 800$ mm, and $d = 32$ mm
(*c*) $Z = (\pi/32d)(d^4 - d_i^4)$ for $d = 36$ mm and $d_i = 26$ mm
(*d*) $G = 79.3$ GPa, $D = 19.2$ mm, and $N = 32$ (a dimensionless number)

ANALYSIS

1–26 The size of a certain hollow shaft is found to be governed by the relation $e = 0.5 - [d/(d^4 - d_i^4)]$, where d and d_i are the outside (OD) and inside (ID) diameters, respectively, and e is an error number. Find suitable values for d and d_i to the nearest $\frac{1}{8}$ in by requiring that e be a small positive number. This exercise says something about mathematical continuity in engineering calculations. Can you express the caution?

1-27 Numerical integration is often forced upon the designer. Simpson's rule is of the form

$$\int_a^b f(x)\,dx = \frac{h}{3}(y_0 + 4y_1 + 2y_2 + 4y_3 + \cdots + 4y_{n-1} + y_n)$$

where the y's are evenly spaced function evaluations of $f(x)$ in the interval $[a, b]$ and the constant h is the interval between function evaluations. The number of intervals n should be an even number, and there are advantages to making it an even number divisible by 4. If this is done, the error in the integration can be estimated by integrating again using every other ordinate, calling this integration $I_{n/2}$ and estimating the numerical integration error to be

$$E = \frac{I_n - I_{n/2}}{15}$$

We express the integral as $I_n + E$ in applying Richardson's correction to improve Simpson's value. Apply Simpson's rule to the integral

$$I = \int_0^1 x^5\,dx$$

using but four intervals. Apply Richardson's correction and show that the result is exact.

1-28 Use your experience with Prob. 1–27 and others as the basis for a computer program to perform a Simpson's rule integration. Such a program will be useful to you in the future.

1-29 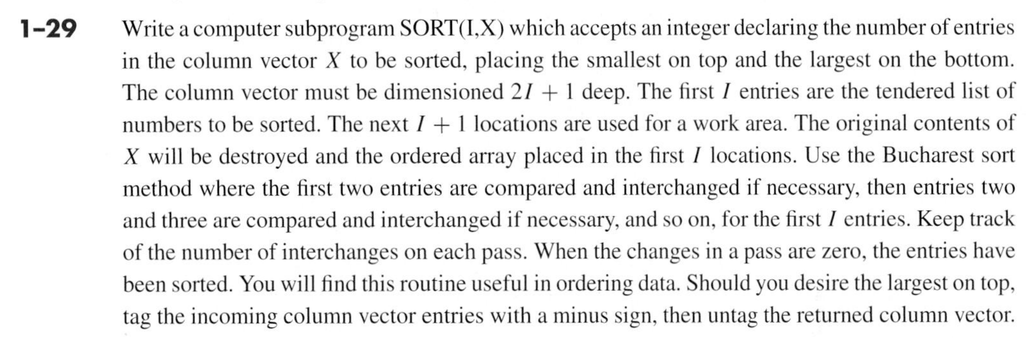
Write a computer subprogram SORT(I,X) which accepts an integer declaring the number of entries in the column vector X to be sorted, placing the smallest on top and the largest on the bottom. The column vector must be dimensioned $2I + 1$ deep. The first I entries are the tendered list of numbers to be sorted. The next $I + 1$ locations are used for a work area. The original contents of X will be destroyed and the ordered array placed in the first I locations. Use the Bucharest sort method where the first two entries are compared and interchanged if necessary, then entries two and three are compared and interchanged if necessary, and so on, for the first I entries. Keep track of the number of interchanges on each pass. When the changes in a pass are zero, the entries have been sorted. You will find this routine useful in ordering data. Should you desire the largest on top, tag the incoming column vector entries with a minus sign, then untag the returned column vector.

1-30 Write a computer subprogram for finding the real roots to a one-independent-variable function with assured convergence. If the user is confronted with finding a root of $f(x) = 0$, begin by solving for x algebraically or simply add x to both sides of the equation $f(x) = 0$, thus, $x = x + f(x) = F(x)$. Use an iteration equation of the form

$$x_{i+1} = [(1 - k)x + kF(x)]_i$$

This is convergent in the neighborhood of the root if k is selected as $k = 1/[1 - F'(x)]$. If $F(x)$ is not easily differentiable, simply evaluate $F'(x)$ by finite differences. Convergence is accelerated if k is adjusted with every iteration. The user supplies a function subprogram named F, which provides $F(x)$ for the subroutine. Don't forget the declaration EXTERNAL F since F is a name in the argument list and not a variable name. One suitable argument list is SSUB(F,MONITOR,DIFF,GUESS,ROOT). The integer MONITOR can turn off a convergence monitor by being declared 0 or turn it on by being declared 1. DIFF is the final successive difference in subsequent estimations of the root at termination. GUESS is an initial estimate of the location of the root. ROOT is the final estimate.

1-31 Write a computer subprogram CDF(z) which will give the area under the probability density function of the unit normal $N(0, 1)$ between $-\infty$ and z, as tabulated in Table E–10. The method was suggested on p. 204 of the fourth edition of this book. The method has an error of less than $7.5(10^{-8})$.

$$P(z) = 1 - Q(z) \qquad z \geq 0 \qquad \text{and} \qquad P(z) = Q(z) \qquad z < 0$$

where

$$Q(z) = f(z)(b_1 t + b_2 t^2 + b_3 t^3 + b_4 t^4 + b_5 t^5)$$

in which

$$f(z) = (1/\sqrt{2\pi})\exp(-z^2/2), \quad t = 1/(1 + p|z|)$$
$$p = 0.231\ 641\ 9, \qquad b_1 = 0.319\ 381\ 530$$
$$b_2 = -0.356\ 563\ 782, \qquad b_3 = 1.781\ 477\ 937$$
$$b_4 = -1.821\ 255\ 978, \quad b_5 = 1.330\ 274\ 429$$

Check your program against Table E–10.

1–32 Write a computer subprogram ZEE(CDF) which converts a cumulative distribution function value to the standardized z variable of the unit normal $N(0, 1)$. The method was suggested on p. 206 of the fourth edition of this book. The error is less than $4.4(10^{-4})$.

$$y = t - \frac{c_0 + c_1 t + c_2 t^2}{1 + d_1 t + d_2 t^2 + d_3 t^3}$$

where

$$t = \begin{cases} \sqrt{\ln(1/F^2)} & 0 < F \leq 0.5 \\ \sqrt{\ln(1/(1-F)^2} & 0.5 < F < 1 \end{cases}$$

$$c_0 = 2.515\ 517, \quad c_1 = 0.802\ 853$$

$$c_2 = 0.010\ 328, \quad d_1 = 1.432\ 788$$

$$d_2 = 0.189\ 269, \quad d_3 = 0.001\ 308$$

$$z = \begin{cases} y & 0 < F \leq 0.5 \\ -y & 0.5 < F < 1 \end{cases}$$

Check your program against Table E-10.

1–33 A detent using a ball plunger, as is depicted in the figure, fully develops friction force fN under N and fP under P. The friction force under spring force F is not fully developed; it is denoted (fF). Show that force summation in the vertical and horizontal directions and summation of moments about the ball center lead to the equations

$$F + fP = N(\sin\theta - f\cos\theta)$$
$$-P + (fF) + N\cos\theta + fN\sin\theta = 0$$
$$(fF) = fN - fP$$

Eliminating (fF) in the second equation using the third equation gives

$$P = \frac{N}{1+f}(\cos\theta + f\sin\theta + f)$$

Dividing the first equation by the fourth equation gives

$$P = F\frac{\cos\theta + f(1 + \sin\theta)}{\sin\theta + f(\sin\theta - 2\cos\theta) - f^2(1 + \sin\theta + \cos\theta)}$$

and from the first equation

$$N = \frac{F + fP}{\sin\theta - f\cos\theta}$$

Can the detent jam? Under what conditions?

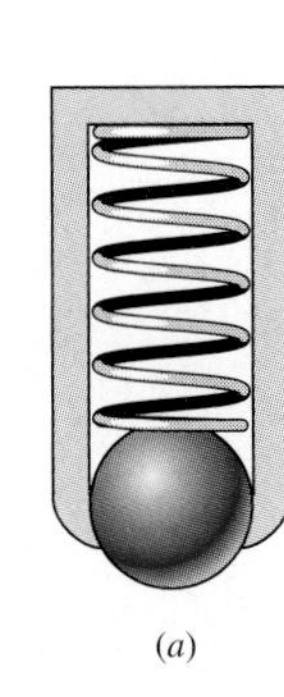
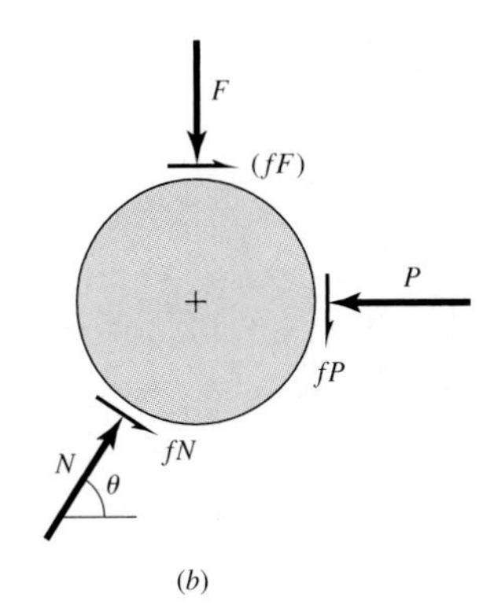

Problem 1–33
(*a*) A ball-plunger detent. (*b*) Free body of ball.

1–34 A ball plunger operates over a conical "dimple" with an apex angle of 90° to form a detent. If the coefficient of friction of dry steel on dry steel is 0.25 and the plunger spring force is 10 lbf, draw a free body of the ball and find P, fP, N, fN, and (fF), annotate with numerical values of the forces, and check your answers numerically. See the figure.

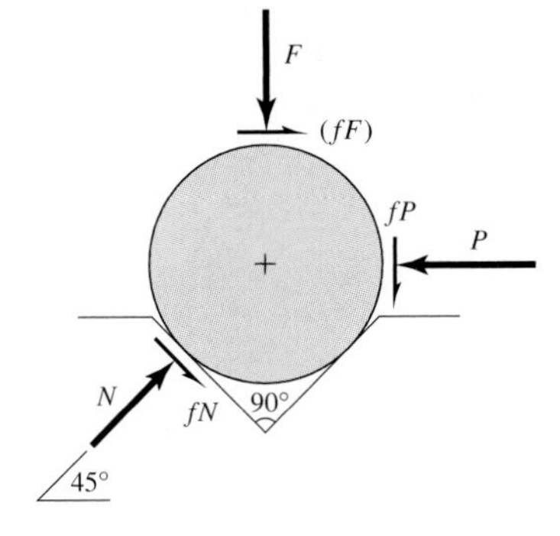

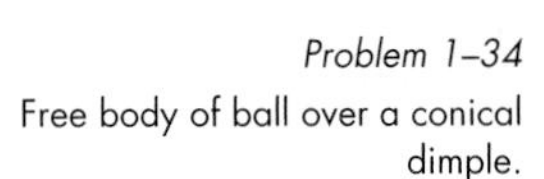
Problem 1–34
Free body of ball over a conical dimple.

1–35 Ball button inserts to create "dimples" for detents are commercially available with a conical apex angle of 90°. For the insert shown in the figure, for a spring force of 45 N and a coefficient of friction of 0.25, find P, fP, N, fN, and (fF), annotate the free body with numerical results, and check your results.

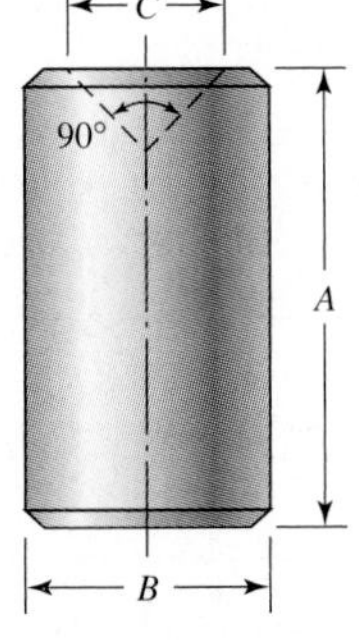

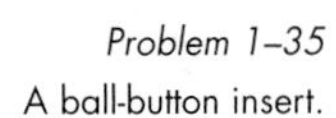
Problem 1–35
A ball-button insert.

1–36 A 5/8 in threaded-spring ball plunger has the dimensions shown in the figure. The initial end force is 7 lbf and the final end force is 50 lbf. The dimensions are $B = 0.3750$ in, $C = 0.984$ in, and $D = 0.096$ in. Create a numerical expression for the spring force $F = F_0 + ky$.

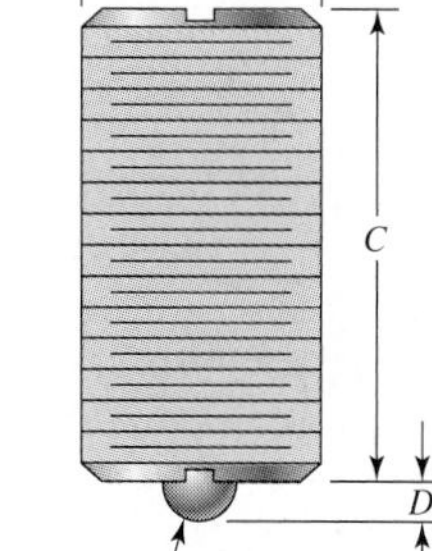

Problem 1–36
A threaded commercial spring-ball assembly.

1–37 A commercial ball button "dimple" has the dimensions shown in the preceding figure (Problem 1-35) with $A = 5/8$ in, $B = 0.5472 \begin{smallmatrix} +0.0002 \\ -0.0000 \end{smallmatrix}$ in, $C = 0.330$ in. What is the largest diameter ball that should be used with this button?

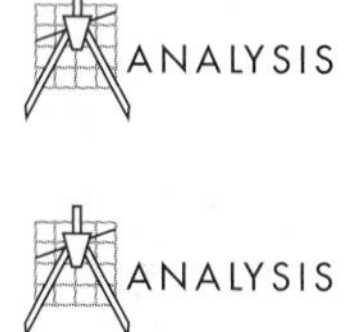

1–38 From your experience with Probs. 1-33 through 1-37 and knowing what you do about free bodies, prepare an adequacy assessment, function portion, for a ball plunger–ball button detent.

2 Addressing Uncertainty

2–1	Questions Come with the Territory **49**
2–2	Estimating Statistical Parameters **50**
2–3	Probability Density Function and Cumulative Distribution Function **53**
2–4	Linear Regression **55**
2–5	Propagation of Error **58**
2–6	Simulation **61**
2–7	Design Factor and Factor of Safety **63**
2–8	Limits and Fits **68**
2–9	Dimensions and Tolerances **71**
2–10	Summary **77**

As noted in Sec. 1–5, uncertainties abound and require quantitative treatment. The algebra of real numbers, by itself, is not well-suited to describing behavior in the presence of variation. This has been apparent for some time as engineers struggled with such notions as minimum value, minimum guaranteed value, and safety as the absence of failure. Despite these not-quite-right concepts, credible engineering work was accomplished, because any discrepancies between notions and performance were resolved by turning to nature as the final arbiter.

It is now clear that consistencies in nature are stable, not in magnitude, but in the *pattern of variation*. Evidence gathered from nature by measurement is a mixture of systematic and random effects. It is the role of statistics to separate these and, through the sensitive use of data, to illuminate the dark places.

Some students start this book after completing a formal course in statistics; others have merely encountered statistics in their engineering courses. Appendixes A, B, C, and D briefly summarize relevant statistical ideas in a notation consistent with this book for handy reference. Now would be a good time to acquaint yourself with the material there. In this chapter we draw on a subset that is needed to answer important questions.

CASE STUDY 2a

Why Engineers Use Statistics

Unless the truth of nature has been carefully built into the equations that we use, there is no knowledge of nature or its consequences in the results. Accuracy in arithmetic is not equivalent to accuracy of knowledge.

Statistics is concerned with the development and application of methods and techniques for collecting, analyzing, and interpreting quantitative data in such a way that the reliability of conclusions based on the data may be evaluated objectively in terms of probability statements. Investigation of probability began with tavern games of chance, drawing colored balls from urns, and wagering on outcomes. Mid-eighteenth-century studies in probability led to quantitative treatment of measurement errors and to "laws of error."

Statistical *decision theory* gained prominence during World War II, as a means of judging the best possible course in the face of uncertainty. Even before this, descriptive data were amassed in agriculture and biology, and pressure built for good answers to a basic question: "What does this mean?" How does one discern the underlying truth masked by all this randomness?

When astronaut Walter M. Schirra was asked what he thought about during the final part of the launch countdown—during which he had little to do—Schirra responded that he could not help but reflect on the fact that all the equipment under him was supplied by the lowest bidder! Schirra was thinking about *reliability*. So had the engineers who designed the equipment. Aviation and the space program led to many advances.

Statistical methods have developed in response to the special needs of various fields: agriculture, biology, medicine, natural science, engineering, manufacturing, economics, business, and government. Methods for working with (necessarily) small samples had been conceived and developed since the 1920s. Competition demands not just lower cost, but equal or better *quality*. *Measurement* is an important part of any field. Just what is the anatomy of a measurement? Statistical *design of experiments* shows great leverage in the amount of information obtained from experimental work as well as accuracy. *Arrangements* can balance batch and equipment effects. *Curve-fitting* and *regression* are important. *Random numbers* and *simulation* can solve difficult problems. Every specification for a product implies a corresponding test procedure to detect whether the specification is met. *Sampling* is a perennial topic for all who deal with materials. *Quality control* and *quality assurance* are essential in mass production.

Operations research views "systems" as a whole. It originated in a nonquantitative way in the Royal Air Force during World War II. The British were losing too many bombers.

Someone arrived at an air base and requested the use of a lightly damaged plane, which would be temporarily painted. Using repair records of bombers of the same type, he carefully plotted on the surface the location, size, and shape of all shrapnel damage. After much staring and contemplation of "holes" and their distribution, it struck the observer that he was looking *not* at damage inflicted by Germans on British bombers, but rather at damage only on planes that returned to base. By shifting attention to damage-free zones, inspection of the interior revealed an unprotected fuel pump. Once again, viewpoint is important.

In the 1960s two engineers went through test data accumulated in the preceding 20 years by several corporations, reexamining past data in the light of current statistical methods. They found that fully 75 percent was not statistically significant. This means that conclusions drawn from the data, and some of the decisions based on the data, may not have been valid at all.

We view the world through a cloudy crystal ball in search of what may be true. There are consistencies in nature, but the closer we look the more aware we become of individual variability and long-term regularity. Variation is omnipresent. Stability is not uniqueness of magnitude, but a stable pattern of variation, often a mix of systematic and random effects. Statistics allows these to be separated and, by the sensitive use of data, to illuminate the dark places. In this way we discover the probability that something is true. Statistics is the tool that helps us build knowledge. This is why engineers use statistics.

2–1 Questions Come with the Territory

Engineers incessantly ask questions, hoping to find answers that will help in the design process. Designers consult books, papers, reports, databases, their own experience, and the experience of others, as well as asking nature directly. The designer's first resort is research: is the response of nature already known and recorded somewhere, or is it incorporated as an integral part of theory? Experiments consume much time and money, so to keep costs down, asking nature is often the last resort. The following examples illustrate the process.

1. How high must a doorway be to allow 95 percent of male humans to pass through without striking their heads on the lintel? The answer, from an ergonomics resource, is that human males have a 95 percentile height of 73.46 in (186.6 cm). Allowing 1 in for heels and 2 in for a hat, $73.42 + 1 + 2 = 76.46$ in. Notice the mix of laboratory measurements on *many* men, and some questionable opinion.
2. Experience with 930 heats of 1035 steel gives the ultimate tensile strength mean as 86.0 kpsi with a standard deviation of 4.94 kpsi. What mean tensile strength of a heat is exceeded by 99.9 percent of the heats? The answer uses data from Fig. 2–1*a* and a presumption of a normal distribution. From Table E–10, $z = -3.090$ and

$$x_{0.999} = \bar{x} + z\sigma_x = 86.0 + (-3.09)4.94 = 70.7 \text{ kpsi}$$

Consider that Fig. 2–1*a* reports *no* observations less than 74.5 kpsi. We know, however, that experience with many more heats will produce observations less than 74.5 kpsi. Indeed, continuous distribution curves are fitted to histographic evidence because the parent population is distributed as a continuous probability density function (PDF) curve, and the histogram is a sampling event. Statistics addresses the relation between populations and samples drawn from them. Arguments center more on which distribution should be fitted than on whether a distribution should be fitted. How can this be done?

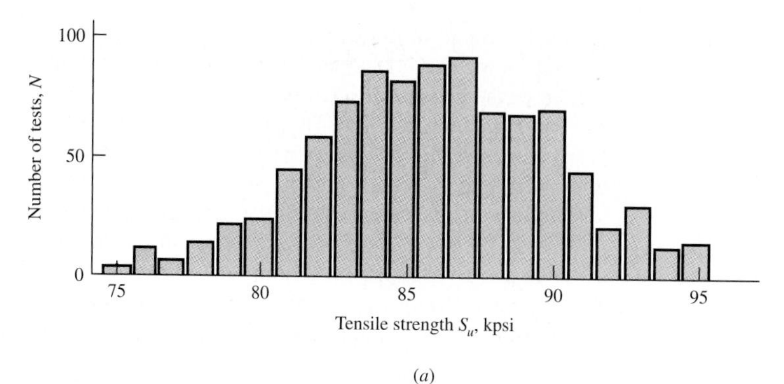

(*a*)

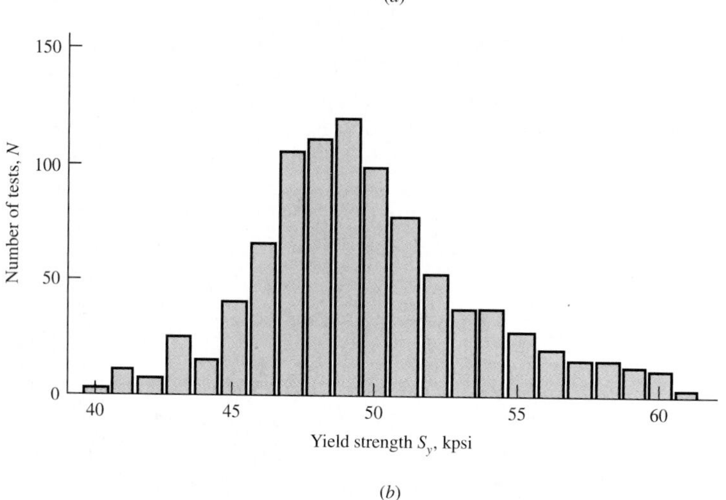

(*b*)

Figure 2–1

Histogram of tensile properties of hot-rolled 1035 steel, as-rolled condition. These tests were made from round bars, varying in diameter from 1 to 9 in. (*a*) Tensile strength distribution from 930 heats: $\bar{S}_{ut} = 86.0$ kpsi, $\sigma_{Sut} = 4.94$ kpsi. (*b*) Yield strength histogram from 899 heats: $S_y = 49.5$ kpsi, $\sigma_{Sy} = 5.36$. (By permission, *Metals Handbook*, vol. 1, 8th ed., American Society for Metals, Metals Park, Ohio, 1961, p. 64.)

3 When two or more parts are to be assembled, there are reasons to have enough clearance so that the assembly operation can be performed from randomly selected, independently mass-produced parts. Questions of tolerances on the parts and their relationships loom. The chance of interference (if interference is disabling) must be evaluated.

These three questions are very different in detail, but they have a common relation to uncertainty. Quantitative responses are needed. The proper tools and their application are considered next.

2–2 Estimating Statistical Parameters

We begin with some examples.

EXAMPLE 2–1

Five tons of 2-in round rod of 1030 hot-rolled steel has been received for workpiece stock. Nine standard-geometry tensile test specimens have been machined from random locations in various rods. In the test report, the ultimate tensile strength was given in kpsi. In ascending order (not necessary), these are displayed in Table 2–1. Find the mean $\bar{x}$, the standard deviation s, and the coefficient of variation C from the sample, such that these are best estimates of the parent population (the stock your plant will convert to product).

Table 2–1

Data Worksheet from Nine Tensile Test Specimens Taken from a Shipment of 1030 Hot-Rolled Steel Barstock

S_{ut}, kpsi x	x^2
62.8	3 943.84
64.4	4 147.36
65.8	4 329.64
66.3	4 395.69
68.1	4 637.61
69.1	4 774.81
69.8	4 872.04
71.5	5 112.25
74.0	5 476.00
Σ 611.8	41 689.24

Solution Examine Eqs. (A–3) and (A–4).[1]

$$\bar{x} = \frac{1}{N}\sum_{i=1}^{9} x_i \tag{A–3}$$

and with similar summations

$$s = \sqrt{\frac{\sum x_i^2 - (\sum x_i)^2/N}{N-1}} \tag{A–4}$$

It is computationally efficient to generate $\sum x$ and $\sum x^2$ before evaluating $\bar{x}$ and s. This has been done in Table 2–1.

$$\bar{x} = \frac{1}{9}(611.8) = 67.9\underline{7} = 67.98 \text{ kpsi}$$

$$s = \sqrt{\frac{41\,689.24 - 611.8^2/9}{9-1}} = 3.543\,225\,226 = 3.543 \text{ kpsi}$$

From Eq. (B–1),

$$C = \frac{s}{\bar{x}} = \frac{3.543\,225\,226}{67.977\,777\,77} = 0.052\,123\,287 = 0.0521$$

The data are treated as given as *incomplete numbers*, considered exact, and carried to the full computational digit complement of the hand-held calculator used. Then for economy of display, results are presented as *rounded numbers* (see Appendix B). All three statistics are estimates of the parent population statistical parameters. Note that these results are independent of the distribution.

Multiple data entries may be identical or may be grouped in histographic form to suggest a distributional shape. If the original data are lost to the designer, the grouped data can still be reduced, although with some loss in computational precision.

1. Equation numbers beginning with a letter are to be found in the appendix of the same letter.

EXAMPLE 2–2

Table 2–2

Grouped Data of Ultimate Tensile Strength from Nine Tensile Test Specimens from a Shipment of 1030 Hot-Rolled Steel Barstock

Class Midpoint x, kpsi	Class Frequency f	Extension fx	Extension fx^2
63.5	2	127	8 064.50
66.5	2	133	8 844.50
69.5	3	208.5	14 480.75
72.5	2	145	10 513.50
Σ	9	613.5	41 912.25

The data in Ex. 2–1 have come to the designer in the histographic form of Table 2–2. The data have been extended to provide $\sum fx$ and $\sum fx^2$.

Solution

From Eq. (A–5),

$$\bar{x} = \frac{1}{N}\sum_{i=1}^{4} f_i x_i = \frac{1}{9}(613.5) = 68.1\underline{6} = 68.17 \text{ kpsi}$$

From Eq. (A–6),

$$s = \sqrt{\frac{41\,912.25 - 613.5^2/9}{9-1}} = 3.391\,164\,992 = 3.39 \text{ kpsi}$$

From Eq. (B–1),

$$C = \frac{s}{\bar{x}} = \frac{3.391\,167\,992}{68.166\,666\,6} = 0.04975$$

Note the small changes in $\bar{x}$, s, and C due to small changes in the summation terms.

The descriptive statistics developed, whether from ungrouped or grouped data, describe the ultimate tensile strength $\mathbf{S}_{ut}$ of the material from which we will form parts. Such description is not possible with a single number. In fact, sometimes two or three numbers plus identification or, at least, a robust approximation of the distribution are needed. As you look at the data in Ex. 2–1, consider the answers to these questions:

- Can we characterize the ultimate tensile strength by the mean, $\bar{S}_{ut}$?
- Can we take the lowest ultimate tensile strength of 62.8 kpsi as a minimum? If we do, we will encounter some lesser ultimate strengths, because some of 100 specimens will be lower.
- Can we find the distribution of the ultimate tensile strength of the 1030 stock in Ex. 2–1? Yes, but it will take more specimens and require plotting on coordinates that rectify the data string.

Such plotting may require some regression ideas which contribute their own statistical insight. There will be more on this point later.

2–3 Probability Density Function and Cumulative Distribution Function

Superposing a probability density function curve on histographic data is a visual way to get an idea of the goodness-of-fit.

EXAMPLE 2–3 One thousand specimens of 1020 steel were tested to rupture and the ultimate tensile strengths were reported as grouped data in Table 2–3. From Eq. (A–5),

$$\bar{x} = \frac{63\,625}{1000} = 63.625 \text{ kpsi}$$

From Eq. (A–6),

$$s = \sqrt{\frac{4\,054\,864 - 63\,625^2/1000}{1000 - 1}} = 2.594\,245 = 2.594 \text{ kpsi}$$

$$C = \frac{s}{\bar{x}} = \frac{2.594\,245}{63.625} = 0.040\,773 = 0.0408$$

From Eq. (A–12) the probability density function for a normal distribution with a mean of 63.625 and a standard deviation of 2.594 245 is

$$f(x) = \frac{1}{2.594\,245\sqrt{2\pi}} \exp\left[-\frac{1}{2}\left(\frac{x - 63.625}{2.594\,245}\right)^2\right]$$

Table 2–3

Worksheet for Ex. 2–3

Class Midpoint, kpsi	Frequency f_i	Extension $x_i f_i$	Extension $x_i^2 f_i$	Observed PDF $f_i/(Nw)$*	Normal Density $f(x)$	Lognormal Density $g(x)$
56.5	2	113.0	6 384.5	0.002	0.0035	0.0026
57.5	18	1 035.0	59 512.5	0.018	0.0095	0.0082
58.5	23	1 345.5	78 711.75	0.023	0.0218	0.0209
59.5	31	1 844.5	109 747.75	0.031	0.0434	0.0440
60.5	83	5 021.5	303 800.75	0.083	0.0744	0.0773
61.5	109	6 703.5	412 265.25	0.109	0.110	0.1143
62.5	138	8 625.0	539 062.5	0.138	0.140	0.1434
63.5	151	9 588.5	608 869.75	0.151	0.1536	0.1539
64.5	139	8 965.5	578 274.75	0.139	0.1453	0.1424
65.5	130	8 515.0	577 732.5	0.130	0.1184	0.1142
66.5	82	5 453.0	362 624.5	0.082	0.0832	0.0800
67.5	49	3 307.5	223 256.25	0.049	0.0504	0.0493
68.5	28	1 918.0	131 382.0	0.028	0.0263	0.0268
69.5	11	764.5	53 132.75	0.011	0.0118	0.0129
70.5	4	282.0	19 881.0	0.004	0.0046	0.0056
71.5	2	143.0	10 224.5	0.002	0.0015	0.0022
Σ	1 000	63 625	4 054 864	1.000		

*N = sample size = 1000; w = width of class interval = 1 kpsi

For example, $f(63.625) = 0.1538$. The probability density $f(x)$ is evaluated at class midpoints to form the column of normal density in Table 2–3.

EXAMPLE 2–4 Continue Ex. 2–3, but fit a lognormal density function.

Solution From Eqs. (A–17) and (A–18),

$$\mu_y = \ln \mu_x - \ln \sqrt{1 + C_x^2} = \ln\ 63.625 - \tfrac{1}{2}\ln(1 + 0.040\ 773^2) = 4.1522$$

$$\sigma_y = \sqrt{\ln(1 + C_x^2)} = \sqrt{\ln(1 + 0.040\ 773^2} = 0.0408$$

The probability density of a lognormal distribution is given in Eq. (A–16) as

$$g(x) = \frac{1}{x\ 0.0408\ \sqrt{2\pi}} \exp\left[-\frac{1}{2}\left(\frac{\ln x - 4.1522}{0.0408}\right)^2\right]$$

For example, $g(63.625) = 0.1537$. This lognormal density has been added to Table 2–3. Plot the lognormal PDF superposed on the histogram of Ex. 2–3 along with the normal density. As seen in Fig. 2–2, both normal and lognormal densities fit well. Formal quantitative goodness-of-fit tests include the chi-square and the Kolmogorov-Smirnov tests. Cumulative data can also be plotted on coordinates that rectify the candidate's distribution cumulative distribution function (CDF) to inspect the data string for a visual goodness-of-fit presentation, Chi-square tests, PDFs, and Kolmogorov-Smirnov tests CDFs. (See any statistics textbook.) Regression techniques can be used to capture the best straight line through the data.

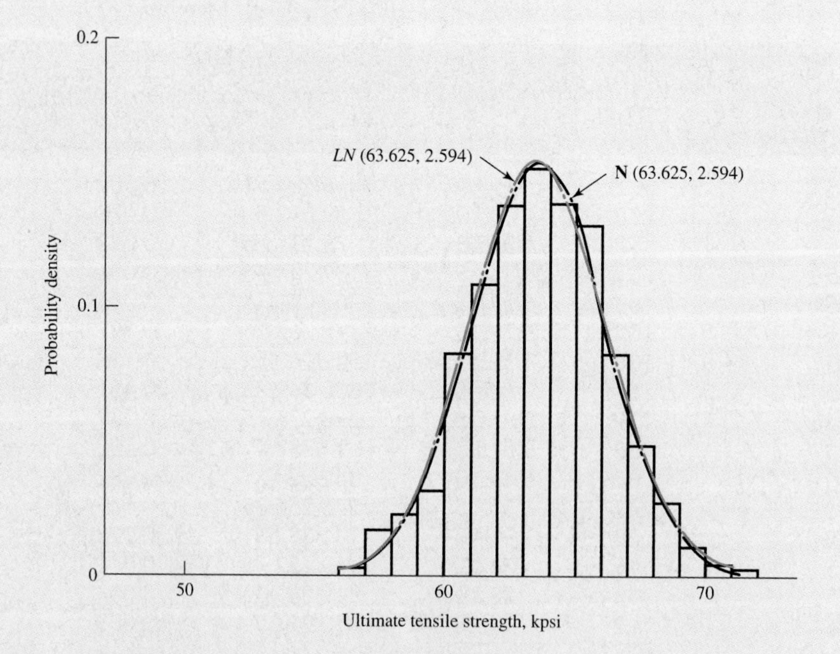

Figure 2–2

Histogram for Ex. 2–3 and Ex. 2–4 with normal and lognormal probability density functions superposed.

2–4 Linear Regression

The least-square method of fitting a straight line to a "straight" data string has valuable statistical overtones, as described in Appendix C. Our first example involves finding the regression parameters $\hat{a}$ and $\hat{b}$, intercept and slope, of the regression line, and the correlation coefficient $\hat{r}$.

The uniform distribution arises, among other places in manufacturing, when a part dimension is mass-produced in an automatic operation and the dimension gradually changes through tool wear and increased tool forces between setups. If N is the part sequence or processing number, and N_f is the sequence number of the last-produced part before another setup, then the dimension x graphs linearly when plotted against the sequence number N. If the last proof part made during setup by the technician has a dimension x_i, and the final part produced has the dimension x_f, the magnitude of the dimension x_j can be expressed as

$$x_j = x_i + (x_f - x_i)\frac{N}{N_f} = x_i + (x_f - x_i)F_j \tag{2–1}$$

since N/N_f is a good approximation to the CDF. Solving for F_j gives

$$F_j = \frac{x_j - x_i}{x_f - x_i} \tag{2–2}$$

Compare this equation with the middle form of Eq. (A–24).

EXAMPLE 2–5 An automatic screw machine produces a right circular cylindrical surface of nominal $\frac{1}{4}$-in diameter. The last setup proof part had a diameter of 0.2482 in. After 600 production parts, the final diameter was 0.2516 in. The following data were reported during the production run:

N	100	200	300	400	500
y, in	0.2489	0.2495	0.2500	0.2506	0.2511

We wish to show that the data are uniform random. One way is to plot data on coordinates on which the uniform distribution CDF is a straight line. Plotting the data, establishing the regression line, and examining the correlation coefficient $\hat{r}$ is one way to use sparse data to show that the distribution is (or is not) uniform.

Solution Interchanging x and y in Eq. (B–6) leaves $\hat{r}$ unaffected. Regressing x on y or y on x will not affect $\hat{r}$. The equation we want to use is

$$y = \hat{a} + \hat{b}F = \hat{a} + \hat{b}x$$

Table 2–4 is the worksheet for this problem. From Eq. (B–4),

$$\hat{a} = \frac{1.527\,777\,776(1.2501) - 2.500(0.625\,966\,665)}{5(1.527\,777\,776) - 2.5^2} = 0.248\,370\,003$$

From Eq. (B–5),

$$\hat{b} = \frac{5(0.625\,996\,995) - 2.5(1.2501)}{5(1.527\,777\,776) - 2.5^2} = 0.003\,299\,994$$

From Eq. (B–6), obtaining s_x and s_y from a hand-held calculator, we write

$$\hat{r} = \frac{\hat{b}s_x}{s_y} = \frac{0.003\,299\,994(0.263\,417\,919)}{0.000\,870\,057} = 0.9997$$

Table 2–4

Worksheet for Ex. 2–5

	Diameter y, in	Frequency x	x^2	xy
	0.2489	0.167	0.027 777 777	0.041 483 333
	0.2495	0.333	0.111 111 111	0.083 166 667
	0.2500	0.500	0.250 000 000	0.125 000 000
	0.2506	0.667	0.444 444 444	0.167 066 667
	0.2511	0.833	0.694 444 444	0.209 250 250
Σ	1.2501	2.500	1.527 777 776	0.625 966 665

At the 0.01 significance level (0.99 confidence level) for four degrees of freedom ν and two parameters (a and b) from Table E–35,

$$r(0.01, 4, 2) = 0.917$$

Since $\hat{r}$ is greater than $r(0.01, 4, 2)$, that is, $0.9997 > 0.917$, the fit is significant and we accept the distribution as uniform. The regression equation is $\hat{a} + \hat{b}F = \hat{a} + \hat{b}N/N_f$, so we write

$$y = 0.246\ 370\ 003 + \frac{0.003\ 299\ 994}{600}N$$

We can examine how well the regression line reflects the data. See Fig. 2–3 also.

	N	Diameter y, in
Proof setup	0	0.248 37
	100	0.248 92
	200	0.249 47
	300	0.250 02
	400	0.250 57
	500	0.251 12
Last part	600	0.251 67

Our principal purpose was to show that the data were uniform (or not). Note that if the 600 pieces are thoroughly mixed, withdrawal from bins is from a uniform distribution. If the parts were binned in order of processing, they are not random but ordered. If the parts

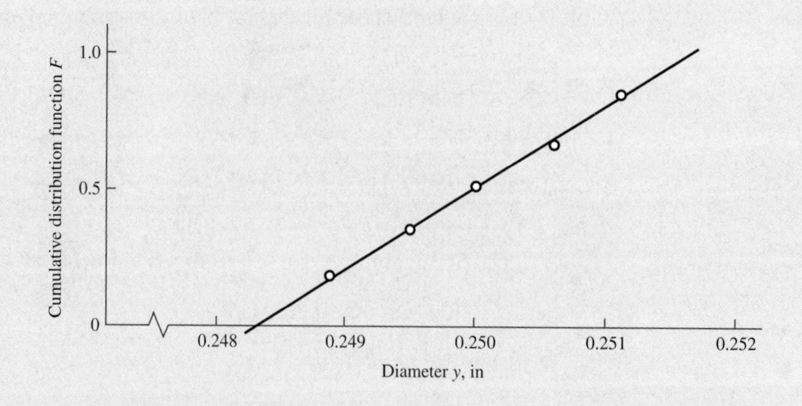

Figure 2–3

Rectified plot of CDF for the data of Ex. 2–5, showing the regression line $y = a + bF$.

were large control valve globe bodies which are removed from the automatic machinery by forklift, palletized, and moved to storage, they may never be mixed, and the statistics do not apply.

Regression procedures allow confidence statements on the parameters $\hat{a}$ and $\hat{b}$, as seen in the next example.

EXAMPLE 2–6 A specimen of a medium carbon steel was tested in tension. With the extensometer in place, this single specimen was loaded then unloaded, to see if the extensometer reading returned to the no-load reading, then the next higher load was applied. The loads and extensometer elongations were reduced to stress σ and ϵ, producing the following data:

σ, psi	5033	10 068	15 104	20 143	35 267
ϵ	0.000 20	0.000 30	0.000 50	0.000 65	0.001 15

Find the mean Young's modulus and the confidence at the 0.95 level. Since the extensometer seems to have an initial reading at no-load, use a $y = a + bx$ regression.

Solution From Table 2–5, $\bar{x} = 0.002\ 80/5 = 0.000\ 56$, $\bar{y} = 85\ 615/5 = 17\ 123$. A regression line always contains the data centroid. From Eq. (B–4),

$$\hat{a} = \frac{0.000\ 002\ 125(85\ 615) - 0.002\ 80(65.229)}{5(0.000\ 002\ 125) - 0.0028^2} = -254.694\ 794 \text{ psi}$$

From Eq. (B–5),

$$\hat{b} = \frac{5(65.229 - 0.0028(85\ 615)}{5(0.000\ 002\ 125) - 0.0028^2} = 31\ 031\ 597.85 \text{ psi}$$

From Eq. (B–6), obtaining s_x and s_y from a hand-held calculator,

$$\hat{r} = \frac{\hat{b}s_x}{s_y} = \frac{31\ 031\ 597.85(3\ 162\ 163\ 10^{-4})}{11\ 601.11} = 0.998\ 164\ 674 = 0.998$$

From Table E–35, $r(0.01, 4, 2) = 0.917$. Since $\hat{r} > r(0.01, 4, 2)$, the fit is significant. The regression line slop $\hat{b}$ is the mean Young's modulus. From Eq. (B–7), the scatter about the regression line is measured by the standard deviation $s_{y \cdot x}$ and is equal to

$$s_{y \cdot x} = \sqrt{\frac{\sum y^2 - \hat{a}\sum y - \hat{b}\sum xy}{n-2}}$$

$$= \sqrt{\frac{2\ 004\ 328\ 267 - (-254.69)85\ 615 - 31\ 031\ 297.85(65.229)}{5-2}}$$

$$= 811.060\ 006 \text{ psi}$$

From Eq. (B–8), the standard deviation of b is

$$s_b = \frac{s_{y \cdot x}}{\sqrt{\sum(x - \bar{x})^2}} = \frac{811.060\ 006}{\sqrt{0.000\ 000\ 558}} = 1\ 085\ 765.11 \text{ psi}$$

Table 2–5

Worksheet for Ex. 2–6

	y σ, psi	x ϵ	x^2	xy	y^2	$(x-\bar{x})^2$
	5 033	0.000 20	0.000 000 040	1.006 600	25 330 089	0.000 000 130
	10 068	0.000 30	0.000 000 090	3.020 400	101 364 624	0.000 000 069
	15 104	0.000 50	0.000 000 250	7.552 000	228 130 816	0.000 000 004
	20 143	0.000 65	0.000 000 423	13.092 950	405 740 449	0.000 000 008
	35 261	0.001 15	0.000 001 323	40.557 050	1 243 761 289	0.000 000 348
Σ	85 615	0.002 80	0.000 002 125	65.229 000	2 004 328 267	0.000 000 556

Note: $\bar{y} = 85\ 615/5 = 17\ 123$ psi, $\bar{x} = 0.002\ 80/5 = 0.000\ 56$.

From Table E–36, $t(0.05, 3) = 3.182$. From Eq. (B–9),

$$b = \hat{b} \pm t(1-\beta, n-2) = 31\ 031\ 597.85 \pm 3.182(1\ 085\ 765.11)$$
$$= 31\ 031\ 597 \pm 2\ 458\ 366 = 27\ 573\ 232,\ 34\ 489\ 964 \text{ psi}$$

This tells us that we are 95 percent confident the Young modulus lies between 27 573 232 and 34 489 964 psi. This is what the statistics of regression states. The deterministic result is that $E = 31\ 031\ 598$ psi. Statistics has a way of telling us what we know, which is much less than we want to know. We could improve matters by greatly increasing the number of points, but that costs time and money. Trying to get by without needed information, however, can be worse. See Fig. 2–4 for regression plot.

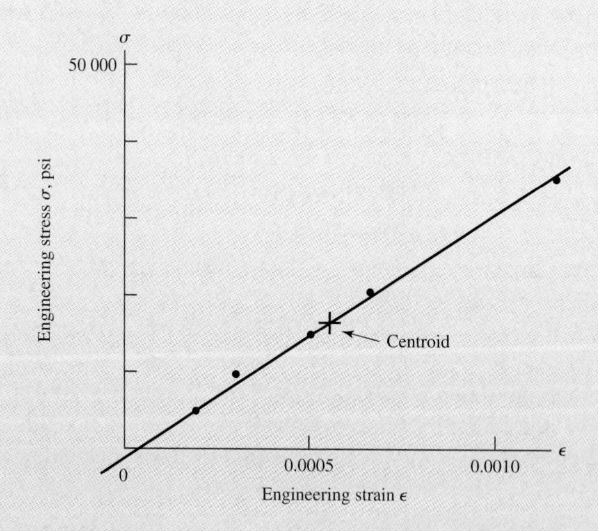

Figure 2–4

The data from Ex. 2–6 is plotted. The regression line passes through the data centroid and among the data points, minimizing the squared deviations.

2–5 Propagation of Error

When working with random variables, it is necessary to propagate error (variability) through systemic equations. Appendix C offers fundamental ideas, and Table 2–6 summarizes some of them in convenient form. Equations (C-7) and (C-8) are associated with the partial derivative estimation method and are stated here for convenience. If ϕ is a function of $\mathbf{x}_1, \mathbf{x}_2, \ldots$, then the mean and standard deviation of ϕ can be estimated from

$$\mu_\phi = \phi(x_1, x_2, \ldots, x_n)_\mu = \phi(\bar{x}_1, \bar{x}_2, \ldots, \bar{x}_n) \tag{C-7}$$

$$\sigma_\phi = \left[\sum_{i=1}^{n}\left(\frac{\partial \phi}{\partial x_i}\right)_\mu^2 \sigma_{xi}^z\right]^{1/2} \tag{C-8}$$

where quantities are evaluated using mean values. These equations are very important in what they suggest about computations involving stochastic variables:

- The estimate of the mean of a function relationship comes from substituting the mean values of the variates.
- Equation (C-8) says the variance σ_ϕ^2 of ϕ is simply the weighted variances of the constitutive variances, the weighting factors being the squares of the partial derivatives evaluated at the means.

Table 2–6

Means and Standard Deviations for Simple Algebraic Operations on Independent (Uncorrelated) Random Variables

Function	Mean	Standard Deviation
a	a	0
$\mathbf{x}$	μ_x	σ_x
$\mathbf{x}+a$	$\mu_x + a$	σ_x
$a\mathbf{x}$	$a\mu_x$	$a\sigma_x$
$\mathbf{x}+\mathbf{y}$	$\mu_x + \mu_y$	$\left(\sigma_x^2+\sigma_y^2\right)^{1/2}$
$\mathbf{x}-\mathbf{y}$	$\mu_x - \mu_y$	$\left(\sigma_x^2+\sigma_y^2\right)^{1/2}$
$\mathbf{xy}$	$\mu_x\mu_y$	$\mu_{xy}\left(C_x^2+C_y^2+C_x^2C_y^2\right)^{1/2}$
$\mathbf{x}/\mathbf{y}$	μ_x/μ_y	$\mu_{x/y}\left[\left(C_x^2+C_y^2\right)/\left(1+C_y^2\right)\right]^{1/2}$
$\mathbf{x}^n$	$\mu_x^n\left[1+\frac{n(n-1)}{2}C_x^2\right]$	$\lvert n\rvert\mu_x^nC_x\left[1+\frac{(n-1)^2}{4}C_x^2\right]$
$1/\mathbf{x}$	$\frac{1}{\mu_x}\left(1+C_x^2\right)$	$\frac{C_x}{\mu_x}\left(1+C_x^2\right)$
$1/\mathbf{x}^2$	$\frac{1}{\mu_x^2}\left(1+3C_x^2\right)$	$\frac{2C_x}{\mu_x^2}\left(1+\frac{9}{4}C_x^2\right)$
$1/\mathbf{x}^3$	$\frac{1}{\mu_x^3}\left(1+6C_x^2\right)$	$\frac{3C_x}{\mu_x^3}\left(1+4C_x^2\right)$
$1/\mathbf{x}^4$	$\frac{1}{\mu_x^4}\left(1+10C_x^2\right)$	$\frac{4C_x}{\mu_x^4}\left(1+\frac{25}{4}C_x^2\right)$
$\sqrt{\mathbf{x}}$	$\sqrt{\mu_x}\left(1-\frac{1}{8}C_x^2\right)$	$\frac{\sqrt{\mu_x}}{2}C_x\left(1+\frac{1}{16}C_x^2\right)$
$\mathbf{x}^2$	$\mu_x^2\left(1+C_x^2\right)$	$2\mu_x^2C_x\left(1+\frac{1}{4}C_x^2\right)$
$\mathbf{x}^3$	$\mu_x^3\left(1+3C_x^2\right)$	$3\mu_x^3C_x\left(1+C_x^2\right)$
$\mathbf{x}^4$	$\mu_x^4\left(1+6C_x^2\right)$	$4\mu_x^4C_x\left(1+\frac{9}{4}C_x^2\right)$

Note: The coefficient of variation of variate $\mathbf{x}$ is $C_x = \sigma_x/\mu_x$. For small COVs their square is small compared to unity, so the first term in the powers of $\mathbf{x}$ expressions are excellent approximations, and these are used in the partial derivative estimation method of Sec. 2-5. For correlated products and quotients see Charles R. Mischke, *Mathematical Model Building*, 2nd rev. ed., Iowa State University Press, Ames, 1980, app. C.

These computations use *deterministic mathematics*. The necessary derivatives can be found by formal differentiation or computer-assisted finite differencing. It is reassuring to know that engineers' previous experience with deterministic problems is useful provided mean values are used. Stochastic problems require the additional effort of propagating variances through the same relationships. Equations (C-7) and (C-8) are approximations. If you find they are not robust, Eqs. (C–9) and (C–10) give the second terms of the Taylor series.

In summary,

- A random variable or a function of random variables can be characterized by statistical parameters, often the means and standard deviations, and a distributional function, recognized or simulated, or a robust approximation that is goodness-of-fit tested.
- Ordinary deterministic algebra, using means of variates, is used to estimate means, and standard deviations are used to estimate standard deviations of functions of variates.
- The distribution of a function of random variables can sometimes be identified from closure theorems of statistics.
- Computer simulation techniques can address problems of unrecognized distributions.

EXAMPLE 2–7

If 12 random selections are made from a uniform distribution $U[0, 1]$ and the real number 6 is subtracted from the sum of the 12 selections, what are the mean and standard deviation of the result?

Solution

Note the square brackets in the presentation $U[0, 1]$. Square brackets in this context denote the enclosed parameters are *other* than the mean and standard deviation. In this case the range numbers a and b are used. There are no values of x less than a or greater than b. The sum ϕ is defined by the problem statement as

$$\boldsymbol{\phi} = \mathbf{x}_1 + \mathbf{x}_2 + \cdots + \mathbf{x}_{12} - 6$$

From Eqs. (A–25) and (A–26), the mean and variance are given as

$$\mu_x = \frac{a+b}{2} = \frac{0+1}{2} = \frac{1}{2}$$

$$\sigma_x^2 = \frac{(b-a)^2}{12} = \frac{(1-0)^2}{12} = \frac{1}{12}$$

From Table 2–1, the mean of a sum is the sum of the means:

$$\mu_\phi = \mu_1 + \mu_2 + \cdots + \mu_{12} - 6 = \left[\frac{1}{2} + \frac{1}{2} + \cdots + \frac{1}{2}\right] - 6 = 0$$

and, for independent random variables

$$\sigma_\phi = \left[\left(\frac{1}{12} + \frac{1}{12} + \cdots + \frac{1}{12}\right) + 0\right]^{1/2} = 1$$

The central limit theorem of statistics states that the sum of random variables asymptotically approaches normal. With 12 variates, one cannot statistically reject a null hypothesis of normality, thus

$$\boldsymbol{\phi} = \mathbf{N}(\mu_\phi, \sigma_\phi) = \mathbf{N}(0, 1)$$

Computer machinery manufacturers supply a machine-specific pseudo random number generator $U[0, 1]$. Such a program is a building block for other random number distribution generators, built by software. Program these steps in the language of your choice:

1. Opening declaration, name, argument list, NORMAL(μ_x, σ_x, x).
2. Set a running sum to zero (SUM = 0).
3. Obtain 12 random numbers, $u_1, u_2, \ldots$, and in turn add them to SUM in the form SUM = SUM + u_i.
4. End the repetitive loop. SUM now contains $u_1 + u_2 + \cdots + u_{12}$.
5. Perform the calculation $x = \mu_x + (\text{SUM} - 6.)\sigma_x$.
6. Return to calling program a single normally distributed random number x.

Notice that in this example you found the mean, standard deviation, and the distribution of a stochastic function. You also discovered how to build a software normal random number generator.

2–6 Simulation

Appendix E presents some basic ideas useful in Monte Carlo simulation and the process of generating random numbers from several common distributions (uniform, normal, lognormal, and Weibull). In the Monte Carlo technique the random nature of each variate is defined either by its cumulative distribution function or its probability density function. Such a function can be represented analytically, histographically, or by tabulated data lookup and interpolation. The individual random number generators create, on call, a single instance of each variate. If $\boldsymbol{\phi} = \phi(\mathbf{x}_1, \mathbf{x}_2, \ldots, \mathbf{x}_n)$, then each of the n variables has a random number instance substituted in the function to generate a single instance of $\boldsymbol{\phi}$. Then statistics on $\boldsymbol{\phi}$ are assembled from N trials of $\boldsymbol{\phi}$.

For instance, if the weight **W** of a cylindrical part of specific weight γ, diameter **d**, and length, then

$$\mathbf{W} = \frac{\gamma \pi \mathbf{d}^2 \mathbf{l}}{4}$$

If

$$\mathbf{l} \sim \mathbf{N}(\mu_l, \sigma_l)$$
$$\mathbf{d} \sim \mathbf{U}[a, b]$$
$$\boldsymbol{\gamma} \sim \mathbf{LN}(\mu_\gamma, \sigma_\gamma)$$

then a thousand trials create running sums $\sum W$ and $\sum W^2$ and estimates of the mean and standard deviation of **W** obtained. Order statistics provides a way of identifying the distribution of **W**. The observations of **W** are ordered smallest to largest and are plotted against a CDF of $\bar{F}_i = i/(n+1)$ or $\tilde{F}_i = (i-0.3)/(n+0.4)$ on coordinates that will rectify particular CDFs. In this way a distribution can be recognized or a robust approximation identified. The existence of the computer for number crunching and graphing makes the method tractable.

Probabilities of failure can be assessed by computing random instances of stress and strength, deflection and permissible deflection, wear and permissible wear, corrosion and permissible corrosion, or life and required life.

As problems increase in complexity, computer time increases. Ways of increasing efficiency are always sought by those with big problems. Note the part of Appendix D which addresses the relation between number of trials and number of correct digits on the results.

CASE STUDY 2b Two-Engine vs. Four-Engine Aircraft

The reliability of satisfactory performance of aircraft engines depends on the duration of flight. Given engine reliability q, does an airplane of two engines have a higher or lower probability of a successful flight than a four-engined plane? A successful flight is one that is completed with no more than half the engines down. For a two-engine aircraft, this means completion on one engine; for a four-engine craft, with two engines down, even if they are on the same side.

A priori probability determinations consist in large measure of counting the possible satisfactory outcomes, counting all the possible outcomes, then dividing one by the other to find the probability of success. It is an accounting procedure that may be simple or involved. If q is the probability of engine function until the end of flight, and p is the probability of an engine failure before the end of flight ($p + q = 1$), then for a two-engine aircraft shown in Fig. CS–2*a*, the probability of no failure is q^2, the probability of one failure is $2qp$, and the probability of a successful flight P_2 is

$$P_2 = q^2 + 2qp = q^2 + 2q(1 - q) = q(2 - q)$$

From the information in Fig. CS–2*b*, in a four-engine plane the probability of no failures is q^4. There are four ways to have a single-engine failure, so the probability of one engine failure in flight is $4q^3p$. There are six ways to have two engine failures, so the probability of two engine failures is $6q^2p^2$. The probability of a successful flight P_4 is

$$P_4 = q^4 + 4q^3p + 6q^2p^2 = q^4 + 4q^3(1 - q) + 6q^2(1 - q)^2$$
$$= q^2(6 - 8q + 3q^2)$$

The difference $P_4 - P_2$ is a number which, if positive, tells us that a four-engine craft has a higher probability of successfully completing a flight. The difference can be expressed as

$$P_4 - P_2 = q^2(6 - 8q + 3q^2) - q(2 - q)$$
$$= q(3q^3 - 8q^2 + 7q - 2)$$

The zero places (roots) of $P_4 - P_2$ are 0, 2/3, 1, 1. We confirm this by forming a table.

The answer to the basic question is that if aircraft engine reliabilities are 2/3 or higher, the four-engine plane is more likely to have a successful flight. The surprise is that if engine reliabilities are lower than $0.6\underline{6}$, then the two-engine craft is more likely to have a successful flight. If this seems contrary to you, note that you have had no experience with low engine reliabilities, and engine reliabilities have been greater than $0.6\underline{6}$ for a long, long time. The problem may be too little experience with stochastic processes. It is time to fix that.

q	P_4	P_2	$P_4 - P_2$
0.00	0.000	0.000	0.000
0.10	0.052	0.190	−0.138
0.20	0.181	0.360	−0.179
0.30	0.318	0.510	−0.162
0.40	0.525	0.640	−0.115
0.50	0.688	0.750	−0.063
0.60	0.821	0.840	−0.019
$0.6\underline{6}$	0.899	0.899	0
0.70	0.916	0.910	0.006
0.80	0.973	0.960	0.013
0.90	0.996	0.990	0.006
0.99	0.956	0.940	0.0496
1	1	1	0

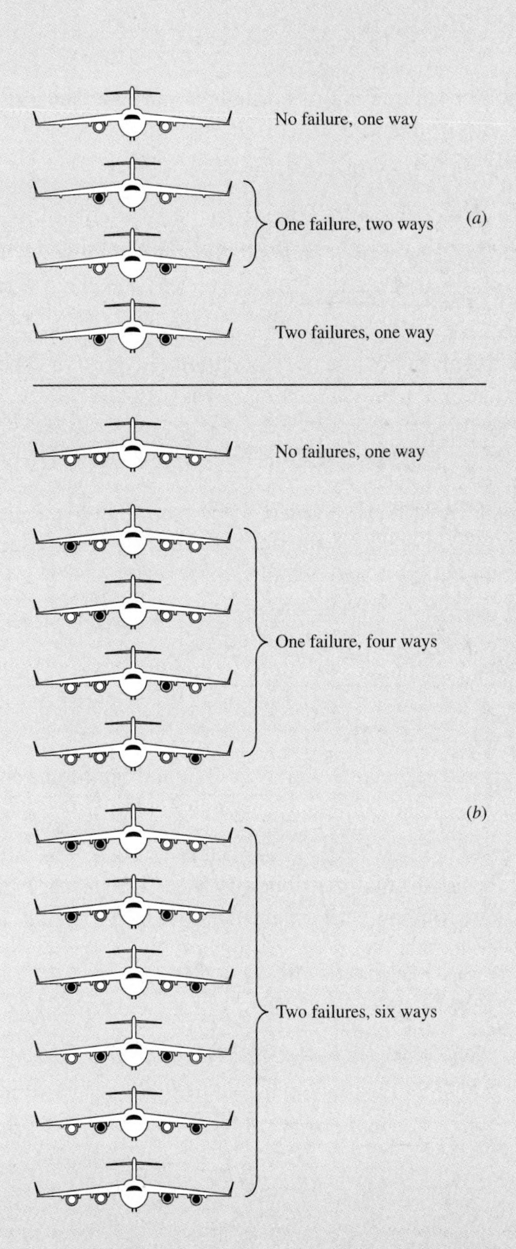

Figure CS–2

(*a*) Engine failure modes on a two-engine aircraft. (*b*) Engine failure modes on a four-engine aircraft.

2–7 Design Factor and Factor of Safety

Reliability is the probability that equipment such as parts, assemblies, components, machines, and systems of machines will perform the intended function satisfactorily, or without failure. In many machinery applications, failure is the result of excessive stress, distortion, wear, or corrosion. We next consider some typical cases involving stress-strength interference.

Normal–Normal Case

Consider the normal distributions, $\mathbf{S} \sim \mathbf{N}(\mu_S, \sigma_S)$ and $\sigma \sim \mathbf{N}(\mu_\sigma, \sigma_\sigma)$, and define a *stress margin* $\mathbf{m}$ such that $\mathbf{m} = \mathbf{S} - \sigma$. The margin $\mathbf{m}$ is normally distributed because the addition or subtraction of normals is normal. Thus $\mathbf{m} \sim \mathbf{N}(\mu_m, \sigma_m)$. Reliability is the probability p that $m > 0$.

$$R = p(S > \sigma) = p(S - \sigma > 0) = p(m > 0) \tag{2-3}$$

To find the chance that $m > 0$ we form the z variable of $\mathbf{m}$ and substitute $m = 0$. Noting that $\mu_m = \mu_S - \mu_\sigma$ and $\sigma_m = (\sigma_S^2 + \sigma_\sigma^2)^{1/2}$, we write

$$z = \frac{m - \mu_m}{\sigma_m} = \frac{0 - \mu_m}{\sigma_m} = -\frac{\mu_m}{\sigma_m} = -\frac{\mu_S - \mu_\sigma}{(\sigma_S^2 + \sigma_\sigma^2)^{1/2}} \tag{2-4}$$

Equation (2-4) is called the *normal coupling equation*. The reliability associated with z is given by

$$R = \int_x^\infty \frac{1}{\sqrt{2\pi}} \exp\left(-\frac{u^2}{2}\right) du = 1 - F = 1 - \Phi(z) \tag{2-5}$$

The body of Table E–10 gives R when $z > 0$ and $(1 - R = F)$ when $z \leq 0$. Noting that $\bar{n} = \mu_S/\mu_\sigma$, square both sides of Eq. (2–4), introduce C_S and C_σ, and solve the resulting quadratic for n to obtain

$$\bar{n} = \frac{1 \pm \sqrt{1 - (1 - z^2 C_S^2)(1 - z^2 C_\sigma^2)}}{1 - z^2 C_S^2} \tag{2-6}$$

The plus sign is associated with $R > 0.5$, and the minus sign with $R < 0.5$.

Lognormal–Lognormal Case

Consider the lognormal distributions $\mathbf{S} \sim \mathbf{LN}(\mu_S, \sigma_S)$ and $\boldsymbol{\sigma} \sim \mathbf{LN}(\mu_\sigma, \sigma_\sigma)$. If we interrupt their companion normals using Eqs. (A–17) and (A–18), we obtain

$$\left.\begin{aligned} \mu_{\ln S} &= \ln \mu_S - \ln \sqrt{1 + C_S^2} \\ \sigma_{\ln S} &= \sqrt{\ln\left(1 + C_S^2\right)} \end{aligned}\right\} \quad (strength)$$

and

$$\left.\begin{aligned} \mu_{\ln \sigma} &= \ln \mu_\sigma - \ln \sqrt{1 + C_\sigma^2} \\ \sigma_{\ln \sigma} &= \sqrt{\ln\left(1 + C_\sigma^2\right)} \end{aligned}\right\} \quad (stress)$$

Using Eq. (2–4) for interfering normal distributions,

$$z = -\frac{\mu_{\ln S} - \mu_{\ln \sigma}}{\left(\sigma_{\ln S}^2 + \sigma_{\ln \sigma}^2\right)^{1/2}} = -\frac{\ln\left(\dfrac{\mu_S}{\mu_\sigma}\sqrt{\dfrac{1 + C_\sigma^2}{1 + C_S^2}}\right)}{\sqrt{\ln\left(1 + C_S^2\right)\left(1 + C_\sigma^2\right)}} \tag{2-7}$$

The reliability R is expressed by Eq. (2–5). The design factor $\mathbf{n}$ is the random variable that is the quotient of $\mathbf{S}/\boldsymbol{\sigma}$. The quotient of lognormals is lognormal, so pursuing the z variable of the lognormal $\mathbf{n}$, we note

$$\mu_n = \frac{\mu_S}{\mu_\sigma}, \qquad C_n = \sqrt{\frac{C_S^2 + C_\sigma^2}{1 + C_\sigma^2}}, \qquad \sigma_n = C_n \mu_n$$

The companion normal to $\mathbf{n} \sim \mathbf{LN}(\mu_n, \sigma_n)$, using Eqs. (A–17) and (A–18), has a mean and standard deviation of

$$\mu_y = \ln \mu_n - \ln \sqrt{1 + C_n^2}, \qquad \sigma_y = \sqrt{\ln(1 + C_n^2)}$$

The z variable for the companion normal y distribution is

$$z = \frac{y - \mu_y}{\sigma_y}$$

Failure will occur when the stress is greater than the strength, when $n < 1$, or when $y < 0$.

$$z = \frac{0 - \mu_y}{\sigma_y} = -\frac{\mu_y}{\sigma_y} = -\frac{\ln \mu_n - \ln \sqrt{1 + C_n^2}}{\sqrt{\ln(1 + C_n^2)}} \doteq -\frac{\ln\left(\mu_n/\sqrt{1 + C_n^2}\right)}{\sqrt{\ln(1 + C_n^2)}} \qquad (2\text{–}8)$$

Solving for μ_n gives

$$\mu_n = \bar{n}_d = \exp\left[-z\sqrt{\ln(1 + C_n^2)} + \ln\sqrt{1 + C_n^2}\right] \doteq \exp\left[C_n\left(-z + \frac{C_n}{2}\right)\right] \qquad (2\text{–}9)$$

Equations (2-6) and (2-9) are remarkable for several reasons:

- They relate design factor $\bar{n}$ to the reliability goal (through z) and the coefficients of variation of strength and stress.
- They are *not* functions of the means of stress and strength.
- They estimate the design factor necessary to achieve the reliability goal before decisions involving means are made. The C_S depends slightly on the particular material. The C_σ has the coefficient of variation (COV) of the load, and that is generally given. This will be developed in Chapters 6 and 7.

The coupling equations (2-4) and (2-7) are useful in adequacy assessments.

EXAMPLE 2–8 A round cold-drawn 1018 steel rod has an 0.2 percent yield strength $\mathbf{S}_y \sim \mathbf{N}(78.4, 5.90)$ kpsi and is to be subjected to a static load of $\mathbf{P} \sim \mathbf{N}(50, 4.1)$ kip. What value of the design factor $\bar{n}$ corresponds to a reliability of 0.999 against yielding ($z = -3.09$)?

Solution $C_S = 5.90/78.4 = 0.0753$, and

$$\sigma = \frac{\mathbf{P}}{A} = \frac{4\mathbf{P}}{\pi d^2}$$

Since the COV of the diameter is an order of magnitude less than the COV of the load or strength, the diameter is treated deterministically:

$$C_\sigma = C_P = \frac{4.1}{50} = 0.082$$

From Eq. (2–6),

$$\bar{n} = \frac{1 + \sqrt{1 - (1 - 3.09^2 0.0753^2)(1 - 3.09^2 0.082^2)}}{1 - 3.09^2 0.0753^2} = 1.4156$$

The diameter is found deterministically:

$$d = \sqrt{\frac{4\bar{P}}{\pi \bar{S}_y/\bar{n}}} = \sqrt{\frac{4(50\,000)}{\pi(78\,400)/1.42}} = 1.074 \text{ in}$$

Check $\mathbf{S}_y \sim \mathbf{N}(78.4, 5.90)$ kpsi, $\mathbf{P} \sim \mathbf{N}(50, 4.1)$ kip, and $d = 1.074$ in. Then

$$A = \frac{\pi d^2}{4} = \frac{\pi(1.074^2)}{4} = 0.9059 \text{ in}^2$$

$$\bar{\sigma} = \frac{\bar{P}}{A} = \frac{4(50\,000)}{0.9059} = 55\,191 \text{ psi}$$

$$C_P = C_\sigma = \frac{4.1}{50} = 0.082$$

$$\hat{\sigma}_\sigma = C_\sigma \bar{\sigma} = 0.082(55\,191) = 4526 \text{ psi}$$

$$\hat{\sigma}_S = 5.90 \text{ kpsi}$$

From Eq. (2–4)

$$z = -\frac{78.4 - 55.191}{(5.90^2 + 4.526^2)^{1/2}} = -3.12$$

From Appendix Table E–10, $R = \Phi(-3.12) = 0.9991$.

EXAMPLE 2–9 Rework Ex. 2–8 with lognormally distributed stress and strength.

Solution $C_S = 5.90/78.4 = 0.0753$, and $C_\sigma = C_P = 4.1/50 = 0.082$. Then

$$\boldsymbol{\sigma} = \frac{\mathbf{P}}{A} = \frac{4\mathbf{P}}{\pi d^2}$$

$$C_n = \sqrt{\frac{C_S^2 + C_\sigma^2}{1 + C_\sigma^2}} = \sqrt{\frac{0.0753^2 + 0.082^2}{1 + 0.082^2}} = 0.1110$$

From Table E–10, $z = -3.09$. From Eq. (2–9),

$$\bar{n}_d = \exp\left[-(-3.09)\sqrt{\ln(1 + 0.111^2)} + \ln\sqrt{1 + 0.111^2}\right] = 1.416$$

$$d = \sqrt{\frac{4\bar{P}}{\pi \bar{S}_y/\bar{n}}} = \sqrt{\frac{4(50\,000)}{\pi(78\,400)/1.416}} = 1.0723 \text{ in}$$

Check $\mathbf{S}_y \sim \mathbf{LN}(78.4, 5.90)$, $\mathbf{P} \sim \mathbf{LN}(50, 4.1)$ kip. Then

$$A = \frac{\pi d^2}{4} = \frac{\pi(1.0723^2)}{4} = 0.9031$$

$$\bar{\sigma} = \frac{\bar{P}}{A} = \frac{50\,000}{0.9031} = 55\,367 \text{ psi}$$

$$C_\sigma = C_P = \frac{4.1}{50} = 0.082$$

$$\hat{\sigma}_\sigma = C_\sigma \mu_\sigma = 0.082(55\,367) = 4540 \text{ psi}$$

From Eq. (2–7),

$$z = -\frac{\ln\left(\dfrac{78.4}{55.367}\sqrt{\dfrac{1 + 0.082^2}{1 + 0.0753^2}}\right)}{\sqrt{\ln(1 + 0.0753^2)(1 + 0.082^2)}} = -3.1399$$

Appendix Table E–10 gives $R = 0.99914$. See Fig. 2–5 which locates $\bar{n}_d$ as μ_n.

Figure 2–5

For Ex. 2–9 the lognormal PDF of design factor *n* and the companion normal PDF of *y*. The probability of failure is shown by the shaded area under the lower tails in both distributions

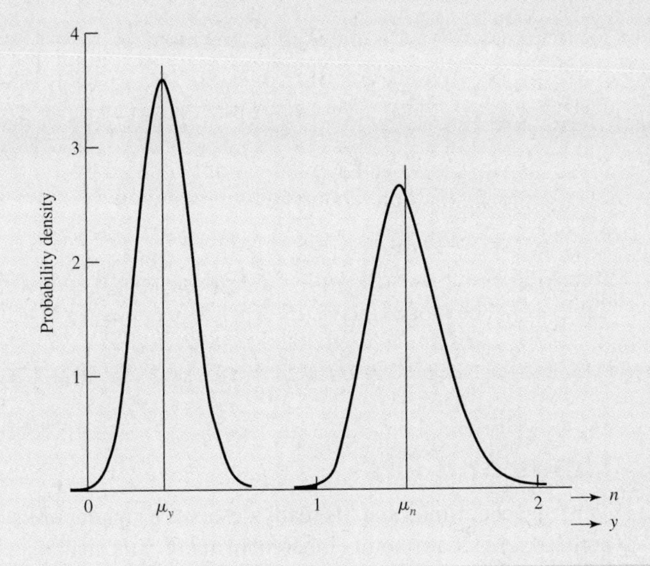

EXAMPLE 2–10 Plot the probability density functions of the design factor **n** and its companion normal **y** for Ex. 2–9.

From Ex. 2–9, $\bar{n} = 1.416$ and $\sigma_n = 0.110$; therefore, $\mathbf{n}_d \sim$ **LN**(1.416, 0.1110). The mean and standard deviation of the companion normal **y** are

$$\mu_y = \ln 1.416 - \ln \sqrt{1 + 0.111^2} = 0.342\,880$$

$$\sigma_y = \sqrt{\ln(1 + 0.111^2)} = 0.1107$$

From Eq. (A–12) for **y**, the PDF is expressed as

$$f(y) = \frac{1}{0.1107\sqrt{2\pi}} \exp\left[-\frac{1}{2}\left(\frac{y - 0.342\,880}{0.1107}\right)^2\right]$$

Table 2–7

Worksheet to Prepare Plotting Coordinates for PDFs of **y** and **n** in Ex. 2–10

y	*f*(*y*)	*n* = exp(*y*)	*g*(*n*) = *f*(*y*)/*n* = *f*(*y*)/exp(*y*)
0	0.030	1.000	0.030
0.05	0.109	1.051	0.104
0.10	0.325	1.105	0.294
0.15	0.790	1.162	0.680
0.20	1.567	1.221	1.283
0.25	2.535	1.284	1.974
0.30	3.343	1.350	2.477
0.35	3.596	1.419	2.534
0.40	3.155	1.492	2.115
0.45	2.256	1.568	1.439
0.50	1.316	1.649	0.798
0.55	0.626	1.733	0.361
0.60	0.243	1.822	0.133
0.65	0.077	1.916	0.040
0.70	0.020	2.014	0.010
0.75	0.004	2.117	0.002

From Eq. (A–16) for **n**, the PDF is expressed as

$$g(n) = \frac{1}{n(0.1107)\sqrt{2\pi}} \exp\left[-\frac{1}{2}\left(\frac{\ln n - 0.342\,880}{0.1107}\right)^2\right]$$

Since $y = \ln n$, $g(n)$ and $f(y)$ are simply related by

$$g(n) = \frac{f(y)}{n} = \frac{f(y)}{\exp(y)}$$

It may be easier to work with $f(y)$, $n = \exp(y)$, and $g(n) = f(y)/n$, so we prepare Table 2–7. The plot of the PDFs of **y** and **n** appear in Fig. 2–5.

2–8 Limits and Fits

The subject of limits and fits really deserves a chapter of its own. The subject is included here because the variability inherent in many of the fit classes is so useful in demonstrating the practical application of the statistical ideas presented.

The designer is free to adopt any geometry of fit for shafts and holes that will ensure the intended function. There is sufficient accumulated experience with commonly recurring situations to make standards useful. There are two standards for limits and fits in the United States, one based on inch units and the other based on metric units.[2] These differ in nomenclature, definitions, and organization. No point would be served by separately studying each of the two systems. The metric version is the newer of the two and is well organized, and so here we present only the metric version but include a set of inch conversions to enable the same system to be used with either system of units.

In using the standard, *capital letters always refer to the hole; lowercase letters are used for the shaft.*

The definitions illustrated in Fig. 2–6 are explained as follows:

- *Basic size* is the size to which limits or deviations are assigned and is the same for both members of the fit.
- *Deviation* is the algebraic difference between a size and the corresponding basic size.
- *Upper deviation* is the algebraic difference between the maximum limit and the corresponding basic size.
- *Lower deviation* is the algebraic difference between the minimum limit and the corresponding basic size.
- *Fundamental deviation* is either the upper or the lower deviation, depending on which is closer to the basic size.
- *Tolerance* is the difference between the maximum and minimum size limits of a part.
- *International tolerance grade* numbers IT designate groups of tolerances such that the tolerances for a particular IT number have the same relative level of accuracy but vary depending on the basic size.
- *Hole basis* represents a system of fits corresponding to a basic hole size. The fundamental deviation is H.
- *Shaft basis* represents a system of fits corresponding to a basic shaft size. The fundamental deviation is h. The shaft-basis system is not included here.

2. *Preferred Metric Limits and Fits for Cylindrical Parts,* ANSI B4.1-1967. *Preferred Metric Limits and Fits,* ANSI B4.2-1978.

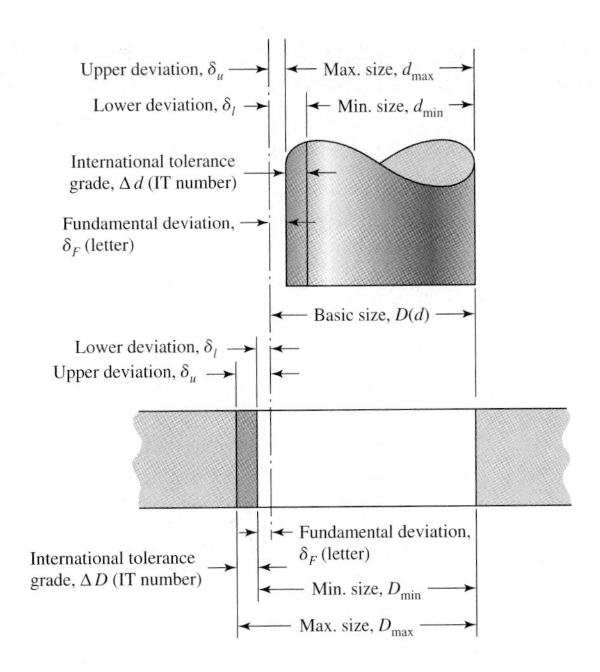

Figure 2–6

Definitions applied to a cylindrical fit.

The magnitude of the tolerance zone is the variation in part size and is the same for both the internal and the external dimensions. The tolerance zones are specified in international tolerance grade numbers, called IT numbers. The smaller grade numbers specify a smaller tolerance zone. These range from IT0 to IT16, but only grades IT6 to IT11 are needed for the preferred fits. These are listed in Tables E–11 to E–13 for basic sizes up to 16 in or 400 mm.

The standard uses *tolerance position letters,* with capital letters for internal dimensions (holes) and lowercase letters for external dimensions (shafts). As shown in Fig. 2–6, the fundamental deviation locates the tolerance zone relative to the basic size.

Table 2–8 shows how the letters are combined with the tolerance grades to establish a preferred fit. The ISO symbol for the hole for a sliding fit with a basic size of 32 mm is 32H7. Inch units are not a part of the standard. However, the designation ($1\frac{3}{8}$ in)H7 includes the same information and is recommended for use here. In both cases, the capital letter H establishes the fundamental deviation and the number 7 defines a tolerance grade of IT7.

For the sliding fit, the corresponding shaft dimensions are defined by the symbol 32g6 [($1\frac{3}{8}$ in)g6].

The fundamental deviations for shafts are given in Tables E–11 and E–13. For letter codes c, d, f, g, and h,

Upper deviation = fundamental deviation
Lower deviation = upper deviation − tolerance grade

For letter codes k, n, p, s, and u, the deviations for shafts are

Lower deviation = fundamental deviation
Upper deviation = lower deviation + tolerance grade

The lower deviation H (for holes) is zero. For these, the upper deviation equals the tolerance grade.

As shown in Fig. 2–6, we use the following notation:

D = basic size of hole
d = basic size of shaft
δ_u = upper deviation

Table 2–8

Descriptions of Preferred Fits Using the Basic Hole System

Source: Preferred Metric Limits and Fits, ANSI B4.2-1978. See also BS 4500.

Type of Fit	Description	Symbol
Clearance	*Loose running* fit: for wide commercial tolerances or allowances on external members	H11/c11
	Free running fit: not for use where accuracy is essential, but good for large temperature variations, high running speeds, or heavy journal pressures	H9/d9
	Close running fit: for running on accurate machines and for accurate location at moderate speeds and journal pressures	H8/f7
	Sliding fit: where parts are not intended to run freely, but must move and turn freely and locate accurately	H7/g6
	Locational clearance fit: provides snug fit for location of stationary parts, but can be freely assembled and disassembled	H7/h6
Transition	*Locational transition* fit for accurate location, a compromise between clearance and interference	H7/k6
	Locational transition fit for more accurate location where greater interference is permissible	H7/n6
Interference	*Locational interference* fit: for parts requiring rigidity and alignment with prime accuracy of location but without special bore pressure requirements	H7/p6
	Medium drive fit: for ordinary steel parts or shrink fits on light sections, the tightest fit usable with cast iron	H7/s6
	Force fit: suitable for parts which can be highly stressed or for shrink fits where the heavy pressing forces required are impractical	H7/u6

δ_l = lower deviation
δ_F = fundamental deviation
ΔD = tolerance grade for hole
Δd = tolerance grade for shaft

Note that these quantities are all deterministic. Thus, for the hole,

$$D_{\max} = D + \Delta D \qquad D_{\min} = D \tag{2–10}$$

For shafts with clearance fits c, d, f, g, and h,

$$d_{\max} = d + \delta_F \qquad d_{\min} = d + \delta_F - \Delta d \tag{2–11}$$

For shafts with interference fits k, n, p, s, and u,

$$d_{\min} = d + \delta_F \qquad d_{\max} = d + \delta_F + \Delta d \tag{2–12}$$

EXAMPLE 2–11 Find the shaft and hole dimensions for a loose running fit with a 34-mm basic size.

Solution From Table 2–6, the ISO symbol is 34H11/c11. From Table E–11, we find that tolerance grade IT11 is 0.160 mm. The symbol 34H11/c11 therefore says that $\Delta D = \Delta d = 0.160$ mm. Using Eq. (2–10) for the hole, we get

Answer $D_{\max} = D + \Delta D = 34 + 0.160 = 34.160$ mm

Answer $D_{min} = D = 34.000$ mm

The shaft is designated as a 34c11 shaft. From Table E–12, the fundamental deviation is $\delta_F = -0.120$ mm. Using Eq. (2–11), we get for the shaft dimensions

Answer $d_{max} = d + \delta_F = 34 + (-0.120) = 33.880$ mm

Answer $d_{min} = d + \delta_F - \Delta d = 34 + (-0.120) - 0.160 = 33.720$ mm

EXAMPLE 2–12 Find the hole and shaft limits for a medium drive fit using a basic hole size of 2 in.

Solution The symbol for the fit, from Table 2–6, in inch units is (2 in)H7/s6. For the hole, we use Table E–13 and find the IT7 grade to be $\Delta D = 0.0010$ in. Thus, from Eq. (2–10),

Answer $D_{max} = D + \Delta D = 2 + 0.0010 = 2.0010$ in

Answer $D_{min} = D = 2.0000$ in

The IT6 tolerance for the shaft is $\Delta d = 0.0006$ in. Also, from Table E–14, the fundamental deviation is $\delta_F = 0.0017$ in. Using Eq. (2–12), we get for the shaft that

Answer $d_{min} = d + \delta_F = 2 + 0.0017 = 2.0017$ in

Answer $d_{max} = d + \delta_F + \Delta d = 2 + 0.0017 + 0.0006 = 2.0023$ in

2–9 Dimensions and Tolerances

The following terms are used generally in dimensioning:

- *Nominal size.* The size we use in speaking of an element. For example, we may specify a $\frac{1}{2}$-in bolt or a $1\frac{1}{2}$-in pipe. Either the theoretical size or the actual measured size may be quite different. The bolt, say, may actually measure 0.492 in. And the theoretical size of a $1\frac{1}{2}$-in pipe is 1.900 in for the outside diameter.
- *Basic size.* The exact theoretical size. Limiting dimensions in either the plus or the minus direction begin from the basic dimension.
- *Limits.* The stated maximum and minimum dimensions.
- *Tolerance.* The difference between the two limits.
- *Bilateral tolerance.* The variation in both directions from the basic dimension. That is, the basic size is between the two limits; for example, 1.005 ± 0.002 in. The two parts of the tolerance need not be equal.
- *Unilateral tolerance.* The basic dimension is taken as one of the limits, and variation is permitted in only one direction; for example,

$$1.005 \begin{array}{l} +0.004 \\ -0.000 \end{array} \text{ in}$$

- *Natural tolerance.* A tolerance equal to plus and minus three standard deviations from the mean. For a normal distribution, this ensures that 99.73 percent of production is within the tolerance limits.

- *Clearance.* A general term which refers to the mating of cylindrical parts such as a bolt and a hole. The word *clearance* is used only when the internal member is smaller than the external member. The *diametral clearance* is the measured difference in the two diameters. The *radial clearance* is the difference in the two radii.
- *Interference.* The opposite of clearance, for mating cylindrical parts in which the internal member is larger than the external member.
- *Allowance.* The minimum stated clearance or the maximum stated interference for mating parts.

Variations in dimensions of parts in a production process may occur purely by chance as well as for specific reasons. For example, the operation temperature of a machine tool changes during start-up, and this may have an effect on the dimensions of parts produced during the first 30 min or so of running time. When all such variations in part dimensions have been eliminated, the production process is said to be in *statistical control.* Under these conditions all variations in dimensions occur at random.

When several parts are assembled the gap (or interference) depends on the dimensions and tolerances of the individual parts. Consider the array of parallel vectors in Fig. 2–7. The x's are directed to the right, and the y's are directed to the left. As parallel vectors they add and subtract algebraically. The bilateral tolerance is t_i on the mean $\bar{x}_i$, and the bilateral tolerance is t_j on the mean $\bar{y}_j$, all positive numbers. The gap which is not closed is called w. From Fig. 2–7,

$$(x_1 + x_3 + \cdots) - (y_2 + y_4 + \cdots) - w = 0$$

which can be succinctly written as

$$w = \sum x_i - \sum y_j \tag{2–13}$$

The largest gap $w_{\max}$ occurs when the right-tending vectors are the largest possible and the left-tending vectors are the smallest possible. We write

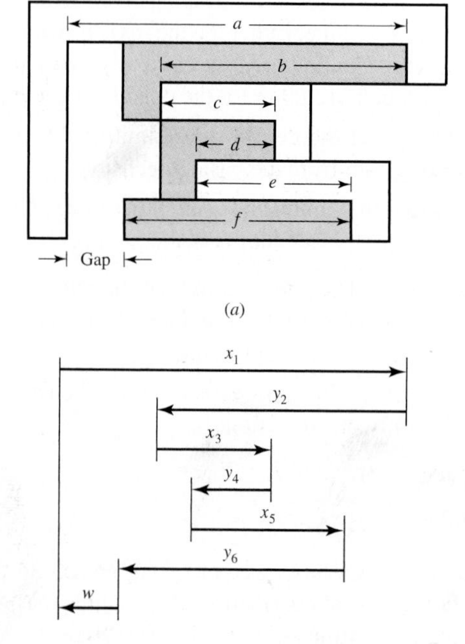

Figure 2–7

An assembly considered as a group of displacement vectors. (*a*) Six L-shaped blocks *A*, *B*, *C*, *D*, *E*, and *F* are assembled as shown. The dimensions *a*, *b*, *c*, *d*, *e*, and *f* are toleranced, making them random variables. (*b*) Corresponding vector diagram, showing right-tending displacement as *x* and left-tending displacement as *y* and *w*.

$$w_{max} = \sum(x_i + t_i) - \sum(y_j - t_j) = \sum \bar{x}_i - \sum \bar{y}_j + \sum_{all} t \tag{2–14}$$

Similarly for the smallest gap, w_{min}, we can write

$$w_{min} = \sum(x_i - t_i) - \sum(y_j + t_j) = \sum \bar{x}_i - \sum \bar{y}_j - \sum_{all} t \tag{2–15}$$

The midrange (mean) value $\bar{w} = (w_{max} + w_{min})/2$. Substituting Eqs. (2–14) and (2–15) into the expression for w and simplifying, we obtain

$$\bar{w} = \sum \bar{x}_i - \sum \bar{y}_j \tag{2–16}$$

The bilateral tolerance on the gap is $t_w = (w_{max} - w_{min})/2$. Substituting Eqs. (2–14) and (2–15) into the expression for t_w and simplifying, we obtain

$$t_w = \sum_{all} t \tag{2–17}$$

Equation (2-17) is the basis for the expression "the stacking of tolerances" used in describing the condition of the gap. All constituent tolerances *add* to give the tolerance on the gap. If the gap instance w is negative, there is interference. Equations (2-16) and (2-17) are the iron rules of tolerance. Since *all* observations of w fall between $\bar{w} - t_w$ and $\bar{w} + t_w$ and none are outside these range numbers, Eqs. (2–16) and (2–17) represent an *absolute tolerance system.*

Gap dimensions near the gap limits are rare events, requiring simultaneous observations near the bounds of individual dimensions, all on a particular side; this suggests a small-risk statistical tolerance system. The operative equation is (2-13) written for random variables:

$$\mathbf{w} = \sum \mathbf{x}_i - \sum \mathbf{y}_j \tag{2–18}$$

The mean gap $\bar{w}$ is the algebraic sum of the constituent means:

$$\bar{w} = \sum \bar{x}_i - \sum \bar{y}_j \tag{2–19}$$

which agrees with Eq. (2–16). The variance in **w** is the positive sum of all the (uncorrelated) constituent variances:

$$\sigma_w^2 = \sum \sigma_i^2 + \sum \sigma_j^2 = \sum_{all} \sigma^2$$

It follows that the standard deviation of the gap **w** is

$$\sigma_w = \sqrt{\sum_{all} \sigma^2} \tag{2–20}$$

Equations (2-19) and (2-20) are the operative equations for the *statistical tolerance system.* A common formulation of Eq. (2–20) is

$$t_w = \sqrt{\sum_{all} t^2}$$

which requires a presumption of normality in the individual dimensions, a rare occurrence. To find the distribution of **w** and/or the probability of observing values of $w > w_{crit}$ or, alternatively, $w < w_{crit}$ requires a computer simulation in most cases. See Appendix D for information on Monte Carlo simulation and the determination of the number of trials needed and of the accuracy of the estimate of probability.

The following example will be enriched using some of the ideas from Appendix D. The example shows the discrepancy between the predictions of probability using a normal assumption and a Monte Carlo simulation.

EXAMPLE 2–13

A shouldered screw contains three hollow right circular cylindrical parts on the screw before a nut is tightened against the shoulder. To sustain the function, the gap w must equal or exceed 0.003 in. The assembly depicted in Fig. 2–8 has dimensions and tolerances as follows:

$a = 1.750 \pm 0.003$ in $\qquad b = 0.750 \pm 0.001$ in
$c = 0.120 \pm 0.005$ in $\qquad d = 0.875 \pm 0.001$ in

All parts except the part with the dimension d are supplied by vendors. The part containing the dimension d is made in-house.

(a) Estimate the mean and tolerance on the gap **w**.
(b) What value of $\bar{d}$ will assure $w > 0.003$ in?
(c) Assume there is a reason for $\bar{d}$ to be greater than 0.867. What is the chance that $w \geq 0.003$ in?

Solution

(a) We use the preceding equations to find

$$\bar{w} = 1.750 - 0.750 - 0.120 - 0.875 = 0.005 \text{ in} \tag{2–16}$$

$$t_w = 0.003 + 0.001 + 0.005 + 0.001 = 0.010 \text{ in} \tag{2–17}$$

Then

$$w_{\max} = \bar{w} + t_w = 0.005 + 0.010 = 0.015 \text{ in} \tag{2–14}$$

$$w_{\min} = \bar{w} - t_w = 0.005 - 0.010 = -0.005 \text{ in} \tag{2–15}$$

There is both clearance and interference.

(b) If $w_{\min}$ is to be 0.003 in, then

$$\bar{w} = w_{\min} + t_w = 0.003 + 0.010 = 0.013 \text{ in}$$

Solve Eq. (2–19) for $\bar{d}$:

$$\bar{d} = \bar{a} - \bar{b} - \bar{c} - \bar{w} = 1.750 - 0.750 - 0.120 - 0.013 = 0.867 \text{ in}$$

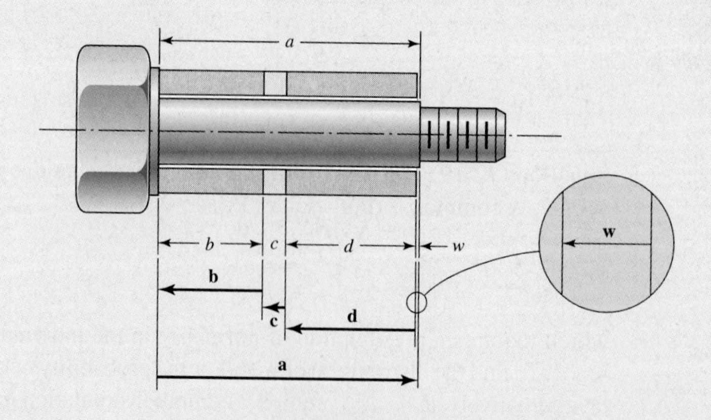

Figure 2–8

An assembly of three cylindrical sleeves of lengths a, b, and c on a shoulder bolt shank of length a. The gap **w** is of interest.

(*c*) To answer this question, we note that **a**, **b**, **c**, and **d** are uniform variates. The standard deviations are

$$\sigma_a = \frac{0.003}{\sqrt{3}}, \qquad \sigma_b = \frac{0.001}{\sqrt{3}}, \qquad \sigma_c = \frac{0.005}{\sqrt{3}}, \qquad \sigma_d = \frac{0.001}{\sqrt{3}}$$

From Eq. (2–20),

$$\sigma_w = \left[\left(\frac{0.003}{\sqrt{3}}\right)^2 + \left(\frac{0.001}{\sqrt{3}}\right)^2 + \left(\frac{0.005}{\sqrt{3}}\right)^2 + \left(\frac{0.001}{\sqrt{3}}\right)^2\right]^{1/2} = 0.003\,464 \text{ in}$$

We can describe the gap. It is $\mathbf{w} = \mathbf{a} - \mathbf{b} - \mathbf{c} - \mathbf{d}$. We even know its mean and standard deviation. What we do not know is its distribution, for it is the result of four added uniform variates. We will use a Monte Carlo simulation and describe the CDF of the lower tail. Also, to learn more we will build a table and run nine simulations. Table 2–9 is developed in the following way:

- Choose a mean dimension $\bar{d}$.
- Find w, t_w, $w_{\max}$, $w_{\min}$.
- Run a Monte Carlo simulation to find the probability of failure (that $w \geq 0.003$ in) corresponding to $\bar{d}$.
- Note the number of significant figures m from Eq. (D–7):

$$m = \log \sqrt{\frac{N}{p(1-p)4z\beta^2}}$$

- Truncate the Monte Carlo simulation results p_f to reflect the number of significant digits.
- Compare the simulation p_f with that predicted by a normal model for **w** to determine whether the sum of uniform variates (four) is close enough to normal for the model to be robust:

$$z = \frac{0.003 - \bar{w}}{0.003\,464} \qquad (p_f \text{ from Table E–10 or computer equivalent})$$

Table 2–9

A Tabulation Prepared for Ex. 2–13 Comparing the Results of Monte Carlo Simulations and a Normal Model

Part d Mean Size $\bar{d}$, in	Gap w Max. w_{max}, in	Gap w Min. w_{min}, in	Failure Probability p_f	Significant Digits Obtained m	Significant Failure Probability p_f	z	Normal Model p_f
0.875	0.015	−0.005	0.3023	2.7	0.30	−0.577	0.281 8
0.874	0.016	−0.004	0.2180	2.8	0.21	−0.866	0.193 2
0.873	0.017	−0.003	0.1389	2.9	0.13	−1.115	0.124 0
0.872	0.018	−0.002	0.0803	3.0	0.08	−1.443	0.074 6
0.871	0.019	−0.001	0.0389	3.2	0.038	−1.732	0.041 6
0.870	0.020	0.000	0.0138	3.3	0.013	−2.021	0.021 7
0.869	0.021	0.001	0.0029	3.7	0.002	−2.309	0.010 046
0.868	0.022	0.002	0.000 19	4.3	0.0001	−2.598	0.004 687
0.8675	0.0225	0.0025	0.000 026	4.6	0.0000	−2.742	0.003 05
0.867	0.023	0.003	0		0	−2.887	0.002 0

For example, the line for $\bar{d} = 0.869$ in is developed as follows:

$$\bar{w} = 1.750 - 0.750 - 0.120 - 0.869 = 0.011 \text{ in}$$

$$t_w = 0.003 + 0.001 + 0.005 + 0.001 = 0.010 \text{ in} \qquad \text{(unchanged)}$$

$$w_{\max} = 0.011 + 0.010 = 0.021 \text{ in}$$

$$w_{\min} = 0.011 - 0.010 = 0.001 \qquad \text{(which is less than 0.003 in)}$$

$$\sigma_w = 0.003\,464 \text{ in} \qquad \text{(unchanged)}$$

Simulation to 10^6 trials gives $p_f = 0.0029$. The number of significant figures for $\beta = 0.95 (z_\beta = 1.96)$ is

$$m = \log \sqrt{\frac{10^6}{0.0029(1 - 0.0029)4(1.96)^2}} = 3.676$$

We can take 0.002 as correct digits and 0.003 as a significant number:

$$z = \frac{0.003 - 0.011}{0.003\,464} = -2.307$$

and p_f under a normal model is 0.010 46, which is approximately fivefold larger than 0.002. We have learned that the normal model is not robust. We can see what is happening by examining Table 2–9. Our specific problem can be answered from Table 2–9 as $p_f = 0.013$ at the 0.95 confidence level.

Some of the following may yet have to be done:

- If the probability estimate of 1 in 500 is acceptable for the probability of an unacceptable assembly ($w \geq 0.003$ in), we have nothing more to do and the Monte Carlo simulation was satisfactory.
- If the probability of failure needs to be known better than 0.002, then more trials are necessary. At 10^6 trials, as here, the uniform random number generator may be approaching its limit (it can start repeating). It will take time and effort to find a better random number generator.
- A statistics book may be useful, particularly if it explains how the piecewise polynomials that constitute the PDF of **w** can be found from the problem data. This also will take time and effort.

CASE STUDY 2c Birthdays

There are n persons in an auditorium. (*a*) What is the probability that at least two persons have the same birthday, that is, have their birthday on the same day and month of the year? Make an estimate right now for the case where $n = 20$. (*b*) What is the smallest number n such that the probability is one-half or better that at least two have the same birthday? (*c*) What is the probability that in an auditorium of 100 persons, at least two have the same birthday?

(*a*) Exclude February 29 and use a 365-day year. There are 365^n possibilities for the birthdays of n people. For the situation that no two people have the same birthday, the probability p, given n is

n	Probability p
2	$\dfrac{364}{365}$
3	$\dfrac{364}{365}\cdot\dfrac{363}{365}$
⋮	
n	$\dfrac{364}{365}\cdot\dfrac{363}{365}\cdots\dfrac{366-n}{365}$

The probability that at least two birthdays are the same is the ones complement of p, namely, $1 - p$, therefore

$$P = 1 - p = 1 - \frac{364}{365}\cdot\frac{363}{365}\cdots\frac{366-n}{365} = 1 - \prod_{i=2}^{n}\frac{366-i}{365}$$

(*b*) Write a computer program to output P for a given n. Form the following table.

n	P	n	P
2	0.002 74	20	0.411 44
3	0.008 20	23	0.507 30
4	0.016 36	30	0.706 32
5	0.027 14	50	0.970 37
10	0.116 95	100	0.999 999 7

(*c*) From the table in part *b*, $P = 0.999\,999\,7$. Consider one lecturer's experience: "I will now recite the days of the year, and as I reach your birthday raise your hand momentarily. January 1, 2, 3. . . ." Two simultaneous hands usually went up in January, occasionally later. Once the experiment resulted in no coincidence of birthdays. After the laughter subsided, the question was modified for a repetition: "Let's do it again with your mother's birthday." The group did not get out of January.

This is the nature of probability realized over a large number of trials, and not the outcome of a single experiment. *Probability gives you the odds, not the outcome*. Another lesson is that intuition based on little awareness, and biased by a lifetime of deterministic thinking, is a poor guide. But with a few fundamentals, logic, and a computer for number-crunching, one can prevail.

The statistics present the situation as it is, and you should appreciate that quality no matter how painful.

Statistical problems and methods are the bane of too many people. Statistics tells you, all too often, that you need to collect more data, that the confidence you place on a result is less than you hoped for, and that improvement is going to cost time and money. The statistics present the situation as it is, and you should appreciate that quality no matter how painful!

2–10 Summary

This chapter is an excursion into the consideration of uncertainty. Appendixes A, Statistical Relations, B, Regression Relations, C, Propagation of Error Relations, and D, Simulation, offer some essential ideas; we have used a few here to obtain answers to problems facing the designer.

Some variability encountered is intrinsic to the phenomenon itself, and some is introduced by the way in which the observer measures. All is relevant to the task of separating the mix of the underlying systematic and the obscuring random components. We view the systematic backbone through a haze or fog of randomness. Our crystal ball is cloudy, yet we have need of what is partly obscured.

The ability to use what you know when it is needed is essential. As your body of experience grows, you will become receptive to viewpoints that are new to you, which will enable you to refine the stochastic description of reality. You will be teaching yourself more, learning less from courses and workshops and more from digging in the literature or from personal investigations.

A note of caution as you reflect on this chapter. Descriptive statistics infers things about a population from examination of samples. It is crucial that the sample be not only unbiased, but also drawn from the population you intend to describe. Example 2-6 involved a measurement of Young's modulus. The sample was observations from a *single specimen*. The results, which could be improved by taking 100 observations on this single specimen, will say nothing about another specimen from the other end of the rod, from another rod in the same bundle, or from a future shipment from a supplier from the same heat of steel. Concerning this larger population we have but a single observation. In reporting your work, and in order to understand the work of others, you must insist on identification of the applicable population. Engineers are not used to this.

The importance of the number of observations on the quality of what we discover from the observations must be appreciated. When the number of points in a regression n is quadrupled, then $\sum x$, $\sum y$, $\sum xy$, $\sum x^2$, and $\sum y^2$ are approximately quadrupled, but s_a and s_b are approximately halved. In the limit as n approaches infinity, $\hat{a} \to a$, $\hat{b} \to b$, $\hat{r} \to r$, $s_a \to 0$, $s_b \to 0$, $s_{y \cdot x} \to \sigma_{y \cdot x}$, and the standard deviation of a future observation $s_{yp} \to \sigma_{y \cdot x}$.

As we shall discover in Chaps. 6 and 7, Eqs. (2–6) and (2–9) will become valuable to us. Table 2–10 gives the transformations in ordinate x and abscissa F that rectify the data strings for several distributions. The right-hand column gives expressions for the mean μ and σ in terms of the least squares regression constants a and b. See Prob. 2–4.

Table 2–10

Transformations to Rectify Data Strings in CDF-Assignment Plots

Distribution	Ordinate x	Abscissa F	Mean CDF Relations
Uniform	1	1	$\mu_x = a + b/2$
			$\sigma_x = b/(2\sqrt{3})$
Normal	1	$\Phi^{-1}(F)$	$\mu_x = a$
			$\sigma_x = b$
Lognormal	$\ln x$	$\Phi^{-1}(F)$	$\mu_x = \exp(a + b^2/2)$
			$\sigma_x = [\exp(2a + 2b^2) - \exp(2a + b^2)]^{1/2}$
Weibull, $2p$	$\ln x$	$\ln \ln[1/(1 - F)]$	$\mu_x = \exp(a)\Gamma(1 + b)$
			$\sigma_x = \exp(a)[\Gamma(1 + 2b) - \Gamma^2(1 + b)]^{1/2}$
Exponential, $2p$	x	$\ln[1/(1 - F)]$	$\mu_x = a + b$
			$\sigma_x = b$

PROBLEMS

All problems are analysis problems.

2–1 At a constant amplitude, completely reversed bending stress level, the cycles-to-failure experience with 69 specimens of 5160H steel from 1.25-in hexagonal bar stock was as tabulated follows:

L	60	70	80	90	100	110	120	130	140	150	160	170	180	190	200	210
f	2	1	3	5	8	12	6	10	8	5	2	3	2	1	0	1

where L is the life in thousands of cycles, and f is the class frequency of failures.

(*a*) Construct a histogram with class frequency f as ordinate.

(*b*) Construct a histogram with ordinate $f/(N\,\Delta x)$.

(*c*) Estimate the mean and standard deviation of the life for the population from which the sample was drawn.

2–2 Apply Sturges's rule (Eq. A–7 in Appendix A) to the data of Prob. 2–1 to shape the histogram into its most revealing form.

(*a*) Construct a histogram with class frequency f as ordinate.

(*b*) Construct a histogram with $f/(N\,\Delta x)$ as ordinate.

(*c*) Estimate the mean and standard deviation of the life for the population from which the sample was drawn.

(*d*) Compare results with those of Prob. 2–1.

2–3 Experience with cycles-to-failure at constant-amplitude bending stress levels is that the life L is lognormally distributed. For the data of Prob. 2–2 find the probability density function $f(x)$ and superpose the plot on the $f/(N\,\Delta x)$ plot of Prob. 2–2.

2–4 For Prob. 2–3 qualitatively look at the goodness of fit of the lognormal by plotting ln L as ordinate against the z variable corresponding to the cumulative distribution function F as estimated from the data in Prob. 2–2.

(*a*) Example the straightness of the data string; the straighter the string, the better the lognormal representation. Table 2–10 suggests the coordinate transforms that rectify a lognormal.

(*b*) If life L is normally distributed, plotting L as ordinate against the abscissa of part *a* will rectify a normally distributed data string. Compare the quality of the two attempts at rectification. There are quantitative goodness-of-fit criteria such as the chi-squared and the Kolmogorov-Smirnoff methods which are tabular and computer implementable, but seeing a graph is very instructive.

2–5 Determinations of the ultimate tensile strength S_{ut} of stainless steel sheet (17-7PH, condition TH 1050), in sizes from 0.016 to 0.062 in, in 197 tests combined into seven classes were

S_{ut}, kpsi	174	182	190	198	206	214	220
Frequency, f	6	9	44	67	53	12	6

where f is the class frequency. Find the mean and standard deviation and compare a normal and a lognormal fit.

2–6 A total of 58 AISI 1018 cold-drawn steel bars were tested to determine the 0.2 percent offset yield strength S_y. The results were:

S_y, kpsi	64	68	72	76	80	84	88	92
f	2	6	6	9	19	10	4	2

where S_y is the class midpoint and f is the class frequency.

(*a*) Estimate the mean and standard deviation of S_y, its PDF, assuming a normal distribution.

(*b*) Compare the fraction with a tensile yield strength less than or equal to 72 kpsi, both from the histogram and the normal distribution fit. Which is the better estimate?

2–7 The base 10 logarithm of 55 cycles-to-failure observations on specimens subjected to a constant stress level in fatigue have been classified as follows:

y	5.625	5.875	6.125	6.375	6.625	6.875	7.125	7.375	7.625	7.875	8.125
f	1	0	0	3	3	6	14	15	10	2	1

Here y is the class midpoint and f is the class frequency.

(*a*) Estimate the mean and standard deviation of the population from which the sample was taken and establish the normal PDF.

(*b*) Plot the histogram and superpose the predicted class frequency from the normal fit.

(*c*) Since $y = \ln x$, if you are satisfied with a normal distribution fit to the data, then it follows that you are satisfied with $\mathbf{x}$ being lognormal. From your work find the mean and standard deviation of $\mathbf{x}$ and find the expression for the probability density function of $\mathbf{x}$.

2–8 A source of stochastic data for common engineering materials is a histogram published in connection with examination of some material property. From such a histogram, class frequencies can be scaled, this information can be used as in Probs. 2–5 and 2–6, and stochastic information can be acquired. From your experience with these problems you can appreciate the steps necessary to find the parameters (usually the mean and standard deviation) and to find a robust distributional fit (often lognormal or normal). The key judgmental element is satisfaction or dissatisfaction with the goodness of fit. All this suggests that most of the effort consists of chores to prepare the engineer for the judgmental step. Write an interactive computer program to assist you in tasks such as those associated with Probs. 2–5 and 2–6.

2–9 Be sure to note the magnitudes of the coefficients of variation of the properties $\mathbf{S}_{ut}$, $\mathbf{S}_y$, and $\mathbf{L}$ in Probs. 2–4, 2–5, and 2-6. The usual case is that $\mathbf{S}_{ut}$ is the tightest and $\mathbf{S}_y$ is more variable (in part because of the methodology of the test for the tensile yield strength). Finite fatigue life has the most impressive coefficient of variation. The companion fatigue endurance strength likewise has a large coefficient of variation. As variation gets larger, making predictions is a daunting task, and stochastic methods are increasingly necessary.

2–10 A $\frac{1}{2}$-in nominal diameter round is formed in an automatic screw machine operation which is initially set to produce a 0.5000-in diameter and is reset when tool wear produces diameters in excess of 0.5008 in. The stream of parts is thoroughly mixed and produces a uniform distribution of diameters.

(*a*) Estimate the mean and standard deviation of the large batch of parts from setup to reset.

(*b*) Find the expressions for the PDF and CDF of the population.

(*c*) If, by inspection, the diameters less than 0.5002 in are removed, what are the new PDF and CDF as well as the mean and standard deviation of the diameters of the survivors of the inspection?

2–11 The only detail drawing of a machine part has a dimension smudged beyond legibility. The round in question was created in an automatic screw machine and 1000 parts are in stock. A random sample of 50 parts gave a mean dimension of $\bar{d} = 0.6241$ and a standard deviation of $s = 0.000\,581$. Toleranced dimensions elsewhere are given in integral thousandths of an inch. Estimate the missing information on the drawing.

2–12 (a) The CDF of the variate **x** is $F(x) = 0.555x - 33$, where x is in millimeters. Find the PDF, the mean, the standard deviation, and the range numbers of the distribution.

(b) In the expression $\boldsymbol{\sigma} = \mathbf{F}/\mathbf{A}$, the force $\mathbf{F} \sim \mathbf{LN}(3600, 300)$ lbf and the area is $\mathbf{A} \sim LN(0.112, 0.001)$ in^2. Estimate the mean, standard deviation, coefficient of variation, and distribution of $\boldsymbol{\sigma}$.

2–13 A regression model $y = a_0 + a_1x + a_2x^2$ is desired to fit a set of n points. Multiplying the regression model by 1 and summing for all points, multiplying by x and summing for all points, and multiplying by x^2 and summing for all points, we obtain the equations (called the normal equations)

$$\sum y = na_0 + a_1 \sum x + a_2 \sum x^2$$

$$\sum xy = a_0 \sum x + a_1 \sum x^2 + a_2 \sum x^3$$

$$\sum x^2 y = a_0 \sum x^2 + a_1 \sum x^3 + a_2 \sum x^4$$

which can be solved simultaneously for a_0, a_1, and a_2.

(a) Solve the equations.

(b) Write an interactive computer program to solicit the data and provide a_0, a_1, and a_2.

2–14 A regression model of the form $y = a_1x + a_2x^2$ is desired. From the normal equations

$$\sum y = -a_1 \sum x + a_2 \sum x^2$$

$$\sum xy = a_1 \sum x^2 + a_2 \sum x^3$$

show that

$$a_1 = \frac{\sum y \sum x^3 - \sum xy \sum x^2}{\sum x \sum x^3 - (\sum x^2)^2} \qquad \text{and} \qquad a_2 = \frac{\sum x \sum xy - y \sum x^2}{\sum x \sum x^3 - (\sum x^2)^2}$$

For the data set

x	0.0	0.2	0.4	0.6	0.8	1.0
y	0.01	0.15	0.25	0.25	0.17	−0.01

find the regression equation, plot the data, and write the equation for examination. Write an interactive computer program to solicit the data and, report a_1 and a_2, print out the data and the regression ordinate at the data abscissas for plotting convenience.

2–15 R. W. Landgraf reported the following axial (push–pull) endurance strengths for steels of differing ultimate strengths:

S_u	S'_e	S_u	S'_e	S_u	S'_e
65	29.5	325	114	280	96
60	30	238	109	295	99
82	45	130	67	120	48
64	48	207	87	180	84
101	51	205	96	213	75
119	50	225	99	242	106
195	78	325	117	134	60
210	87	355	122	145	64
230	105	225	87	227	116
265	105				

(a) Plot the data with S'_e as ordinate and S_u as abscissa.

(b) Using the $y = a + bx$ linear regression model, find the regression line and plot.

(c) Using the $y = a_1 x + a_2 x^2$ regression model, find the regression curve and plot. Compare the regression lines.

2–16 In fatigue studies a parabola of the Gerber type

$$\frac{\sigma_a}{S_e} + \left(\frac{\sigma_m}{S_{ut}}\right)^2 = 1$$

is useful. Solved for σ_a the preceding equation becomes

$$\sigma_a = S_e - \frac{S_e}{S_{ut}^2}\sigma_m^2$$

This implies a regression model of the form $y = a_0 + a_2 x^2$. Show that the normal equations are

$$\sum y = n a_0 + a_2 \sum x^2$$

$$\sum xy = a_0 \sum x + a_2 \sum x^3$$

and that

$$a_0 = \frac{\sum x^3 \sum y - \sum x^2 \sum xy}{n \sum x^3 - \sum x \sum x^2} \quad \text{and} \quad a_2 = \frac{n \sum xy - \sum x \sum y}{n \sum x^3 - \sum x \sum x^2}$$

Plot the data

x	20	40	60	80
y	19	17	13	7

superposed on a plot of the regression line.

2–17 The most frequently applied least-squares regression is the linear $y = a + bx$ because the straight line is useful and many curves can be rectified. Recall the drill:

$$\hat{a} = \frac{\sum x^2 \sum y - \sum x \sum xy}{n \sum x^2 - (\sum x)^2}, \qquad \hat{b} = \frac{n \sum xy - \sum x \sum y}{n \sum x^2 - (\sum x)^2}$$

$$s_{y \cdot x} = \sqrt{\frac{\sum \epsilon^2}{n-2}} = \sqrt{\frac{\sum y^2 - \hat{a}\sum y - \hat{b}\sum xy}{n-2}}$$

$$s_b = \frac{s_{y \cdot x}}{\sqrt{\sum (x - \bar{x})^2}}, \qquad r = \frac{\hat{b} s_x}{s_y}$$

$$s_a = s_{y \cdot x}\left[\frac{1}{n} + \frac{x^2}{\sum (x - \bar{x})^2}\right]^{1/2}, \qquad s_{yp} = s_{y \cdot x}\left[1 + \frac{1}{n} + \frac{(x_i - \bar{x})^2}{\sum (x - \bar{x})^2}\right]^{1/2}$$

Write an interactive computer program to solicit the data and report the foregoing information.

2–18 It is worthwhile to repeat Prob. 2–17 for the linear $y = bx$ model for the same reasons. Recall the drill:

$$\hat{b} = \frac{\sum xy}{\sum x^2}, \qquad \hat{r} = \frac{\hat{b} s_x}{s_y}$$

$$s_{y \cdot x} = \sqrt{\frac{\sum y^2 - b^2 \sum x^2}{n-1}}$$

$$s_b = \frac{s_{y \cdot x}}{\sqrt{\sum x^2}}, \qquad s_{yp} = s_{y \cdot x}\left[1 + \frac{1}{n} + \frac{(x_i - \bar{x})^2}{\sum (x - \bar{x})^2}\right]^{1/2}$$

Write an interactive computer program to solicit the data and report the foregoing information.

2–19 Use your program which resulted from Prob. 2–17 to answer questions concerning data collected on a single helical coil extension spring with an initial extension F_i and a spring rate k suspected of being related by the equation $F = F_i + kx$ where x is the deflection beyond initial. The data are

x, in	0.2	0.4	0.6	0.8	1.0	2.0
F, lbf	7.1	10.3	12.1	13.8	16.2	25.2

(*a*) Estimate the mean and standard deviation of the initial tension F_i.

(*b*) Estimate the mean and standard deviation of the spring rate k.

(*c*) Our knowledge of F_i and k is not very good. How can the confidence intervals on these parameters be improved?

2–20 The form $Y = \alpha X^b \boldsymbol{\lambda}$ where $\boldsymbol{\lambda}$ is lognormal is often transformed into the regression model $y = a + bx + \xi$ where $\xi \sim N(0, s_{y \cdot x})$ by taking logarithms of the data. A programmer cannot anticipate whether $s_{y \cdot x}$ will be large enough to require the complete expression for $\boldsymbol{\lambda}$:

$$\boldsymbol{\lambda} = \exp(s_{y \cdot x}^2/2) LN\left(1, \frac{\sqrt{\exp(2s_{y \cdot x}^2) - \exp(s_{y \cdot x}^2)}}{\exp(s_{y \cdot x}^2/2)}\right)$$

or whether the approximation $\boldsymbol{\lambda} \sim LN(1 + s_{y \cdot x}^2/2, s_{y \cdot x})$, or the even simpler approximation, $\boldsymbol{\lambda} \sim LN(1, s_{y \cdot x})$, is appropriate. The programmer can made the code present all three for user convenience. If you wish to add this feature to the program written in Prob. 2–17, do so at this time.

2–21 In the expression for uniaxial strain $\boldsymbol{\epsilon} = \boldsymbol{\delta}/\mathbf{l}$, the elongation is specified as $\boldsymbol{\delta} \sim (0.0015, 0.000\,092)$ in and the length as $\mathbf{l} \sim (2.0000, 0.0081)$ in. What are the mean, the standard deviation, and the coefficient of variation of the corresponding strain $\boldsymbol{\epsilon}$.

2–22 In Hooke's law for uniaxial stress, $\boldsymbol{\sigma} = \boldsymbol{\epsilon}\mathbf{E}$, the strain is given as $\boldsymbol{\epsilon} \sim (0.0005, 0.000\,034)$ and Young's modulus as $\mathbf{E} \sim (29.5, 0.885)$ Mpsi. Find the mean, the standard deviation, and the coefficient of variation of the corresponding stress $\boldsymbol{\sigma}$ in psi.

2–23 The stretch of a uniform rod in tension is given by the formula $\delta = Fl/AE$. Suppose the terms in this equation are random variables and have parameters as follows:

$\mathbf{F} \sim (14.7, 1.3)$ kip $\mathbf{A} \sim (0.226, 0.003)$ in^2
$\mathbf{l} \sim (1.5, 0.004)$ in $\mathbf{E} \sim (29.5, 0.885)$ Mpsi

Estimate the mean, the standard deviation, and the coefficient of variation of the corresponding elongation $\boldsymbol{\delta}$ in inches.

2–24 The maximum bending stress in a round bar in flexure occurs in the outer surface and is given by the equation $\sigma = 32M/\pi d^3$. If the moment is specified as $\mathbf{M} \sim (15000, 1350)$ lb·in and the diameter is $\mathbf{d} \sim (2.00, 0.005)$ in, find the mean, the standard deviation, and the coefficient of variation of the corresponding stress $\boldsymbol{\sigma}$ in psi.

2–25 When a production process is wider than the tolerance interval, inspection rejects a low-end scrap fraction α with $x < x_1$ and an upper-end scrap fraction β with dimensions $x > x_2$. The surviving population has a new density function $g(x)$ related to the original $f(x)$ by a multiplier a. This is because any two observations x_i and x_j will have the same relative probability of occurrence as before. Show that

$$a = \frac{1}{F(x_2) - F(x_1)} = \frac{1}{1 - (\alpha + \beta)}$$

and

$$g(x) = \begin{cases} \dfrac{f(x)}{F(x_2) - F(x_1)} = \dfrac{f(x)}{1 - (\alpha + \beta)} & x_1 \le x \le x_2 \\ 0 & \text{otherwise} \end{cases}$$

2–26 An automatic screw machine produces a run of parts with $\mathbf{d} \sim U[0.748, 0.751]$ in because it was not reset when the diameters reached 0.750 in. The square brackets contain range numbers.

(*a*) Estimate the mean, standard deviation, and PDF of the original production run if the parts are thoroughly mixed.

(*b*) Using the results of Prob. 2–25, find the new mean, standard deviation, and PDF. Superpose the PDF plots and compare.

2–27 Using the results of Prob. 2–25, show that the mean of a truncated distribution is

$$\hat{\mu}_{x,t} = \int_{x_1}^{x_2} \frac{x f(x)\, dx}{F(x_2) - F(x_1)}$$

and that the variance is

$$\hat{\sigma}^2_{x,t} = \int_{x_1}^{x_2} \frac{x^2 f(x)\, dx}{F(x_1) - F(x_2)} - \hat{\mu}^2_{x,t}$$

and that the new CDF is

$$G(x_t) = \int_{x_1}^{x_t} \frac{f(x)\, dx}{F(x_2) - F(x_1)}$$

2–28 A springmaker is supplying helical coil springs meeting the requirement for a spring rate k of 10 ± 1 lb/in. The test program of the springmaker shows that the distribution of spring rate is well approximated by a normal distribution. The experience with inspection has shown that 8.1 percent are scrapped with $k < 9$ and 5.5 percent are scrapped with $k > 11$. Estimate the probability density function.

2–29 The lives of parts are often expressed as the number of cycles of operation that a specified percentage of a population will exceed before experiencing failure. The symbol B is used to designate this definition of life. Thus we can speak of B10 life as the number of cycles to failure exceeded by 90 percent of a population of parts. Using the mean and standard deviation for the data of Prob. 2–1, a normal distribution model, estimate the corresponding B10 life.

2–30 Fit a normal distribution to the histogram of Prob. 2–2. Superpose the probability density function on the $f/(N\Delta x)$ histographic plot.

2–31 For Prob. 2–5, plot the histogram with $f/(N\Delta x)$ as ordinate and superpose a normal distribution density function on the histographic plot.

2–32 For Prob. 2–6 plot the histogram with $f/(N\Delta x)$ as ordinate and superpose a normal distribution probability density function on the histographic plot.

2–33 A 1018 cold-drawn steel has a 0.2 percent tensile yield strength $\mathbf{S}_y \sim N(78.4, 5.90)$ kpsi. A round rod in tension is subjected to a load $\mathbf{P} \sim N(40, 8.5)$ kip.

(*a*) If rod diameter d is 1.000 in, what is the probability that a random static tensile load P from $\mathbf{P}$ imposed on the shank with a 0.2 percent tensile load S_y from $\mathbf{S}_y$ will not yield?

(*b*) If the reliability goal is 0.995, what diameter d of the rod can achieve it?

2–34 A hot-rolled 1035 steel has a 0.2 percent tensile yield strength $\mathbf{S}_y \sim LN(49.6, 3.81)$ kpsi. A round rod in tension is subjected to a load $\mathbf{P} \sim LN(30, 5.1)$ kip.

(*a*) If the rod diameter d is 1.000 in, what is the probability that a random static tensile load P from $\mathbf{P}$ on a shank with a 0.2 percent yield strength S_y from $\mathbf{S}_y$ will not yield?

(*b*) If the reliability goal is 0.95, what diameter rod can achieve it?

2–35 In Prob. 2–28, what is the new probability density function when 8.1 percent low-end and 5.5 percent high-end scrap are generated in a normally distributed population? The results of Prob. 2–25 will be useful.

2–36 Three blocks A, B, and C and a grooved block D have dimensions a, b, c, and d as follows:

$a = 1.000 \pm 0.001$ in $b = 2.000 \pm 0.003$ in
$c = 3.000 \pm 0.005$ in $d = 6.020 \pm 0.006$ in

The blocks are assembled as shown in the figure.

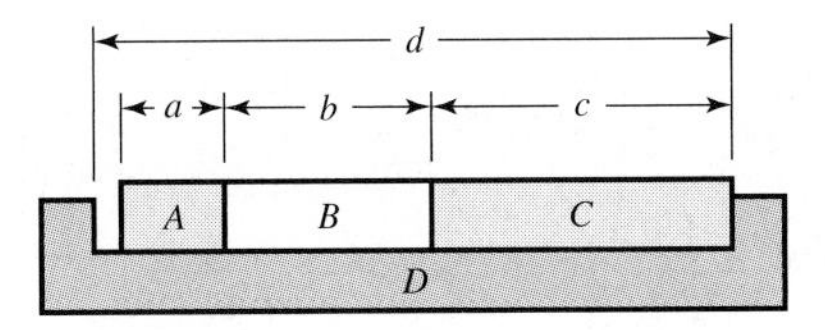

Problem 2–36

(*a*) Using the absolute tolerance system, determine the nominal gap $\bar{w}$ and its bilateral tolerance amplitude t_w.

(*b*) Using the statistical tolerance system, determine the nominal gap $\bar{w}$ and its bilateral tolerance amplitude t_w.

2–37 If $x = a \pm \Delta a$, $y = b \pm \Delta b$, and $z = c \pm \Delta c$, show that for the volume V of a rectangular parallelepiped, $V = xyz$,

$$\frac{\Delta V}{\bar{V}} = \frac{\Delta a}{\bar{a}} + \frac{\Delta b}{\bar{b}} + \frac{\Delta c}{\bar{c}}$$

Use this result to place range numbers (absolute tolerance bounds) on the volume of a rectangular parallelepiped with the dimensions

$a = 1.250 \pm 0.001$ in $b = 1.875 \pm 0.002$ in $c = 2.750 \pm 0.003$ in

2–38 A pivot in a linkage has a pin as depicted in the figure whose dimension $a \pm t_a$ is to be established. The thickness of the link clevis is 1.000 ± 0.002 in. The designer has concluded that a gap of between 0.004 and 0.14 in will satisfactorily sustain the function of the linkage pivot.

(*a*) Determine the dimension a and its tolerance using the absolute tolerance method.

(*b*) Establish the dimension a and its tolerance using the statistical tolerance method.

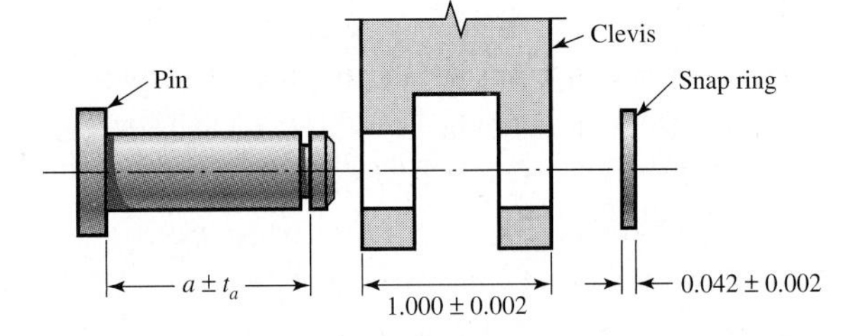

Problem 2–38
Dimensions in inches.

2–39 A guide pin is required to align the assembly of a two-part fixture. The nominal size of the pin is 15 mm. Make the dimensional decisions for a 15-mm basic size locational clearance fit.

2–40 An interference fit of a cast-iron hub of a gear on a steel shaft is required. Make the dimensional decisions for a 45-mm basic size medium drive fit.

2–41 A pin is required for forming a linkage pivot. Find the dimensions required for a 50-mm basic size pin and clevis with a sliding fit.

2–42 A journal bearing and bushing need to be described. The nominal size is 1 in. What dimensions are needed for a 1-in basic size with a close running fit if this is a lightly loaded journal and bushing assembly?

2–43 A circular cross-section O ring has the dimensions shown in the figure. In particular, a No. 240 O ring has an inside diameter D_i and a cross-section diameter W of

$$D_i = 3.734 \pm 0.028 \text{ in} \qquad W = 0.139 \pm 0.004 \text{ in}$$

Using the absolute tolerance system, estimate the mean outside diameter $\bar{D}_o$ and its bilateral tolerance.

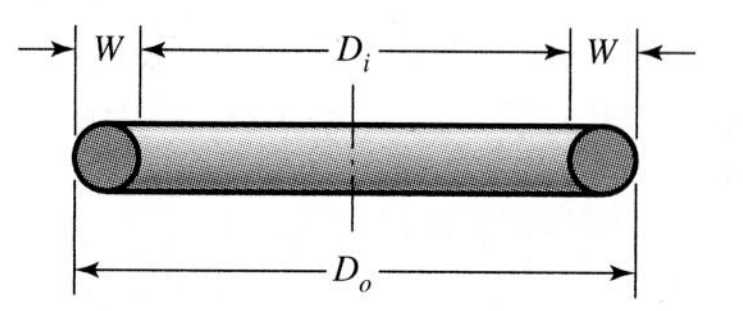

Problem 2–43

2–44 A No. 370 O ring has the dimensions

$$D_i = 208.92 \pm 1.30 \text{ mm} \qquad W = 5.33 \pm 0.13 \text{ mm}$$

Using the absolute tolerance method, estimate the mean outside diameter $\bar{D}_o$ and its bilateral tolerance.

2–45 Estimate the mean outside diameter $\bar{D}_o$ and its bilateral tolerance of the No. 240 O ring of Prob. 2–43 if **W** is independent of $\mathbf{D}_i$ and the statistical tolerancing method is used.

2–46 Find the outside diameter $\bar{D}_o$ and its bilateral tolerance for the No. 370 O ring of Prob. 2–44 if **W** is independent of $\mathbf{D}_i$ and the statistical tolerancing method is used.

2–47 The gland for an O ring under internal pressure is shown in the figure. For a No. 240 O ring, the recommended gland dimensions are

$$G = 0.185 \pm 0.005 \text{ in} \qquad F = 0.106 \pm 0.003 \text{ in}$$

One manufacturer of O rings recommends the gland outer diameter

$$Y_{\text{max}} = D_o \qquad Y_{\text{min}} = \max[0.99\bar{D}_o, \bar{D}_o - 0.060] \text{ in}$$

The outside diameter of a No. 240 O ring is 4.012 ± 0.036 in and the cross-section diameter is 0.139 ± 0.004 in.

(*a*) When the end plate is bolted into position, all O rings are compressed. What is the minimum compression (squeeze) of a No. 240 O ring in inches?

(*b*) When the O ring is placed in the gland prior to the assembly of the end plate, is the ring free or under compression?

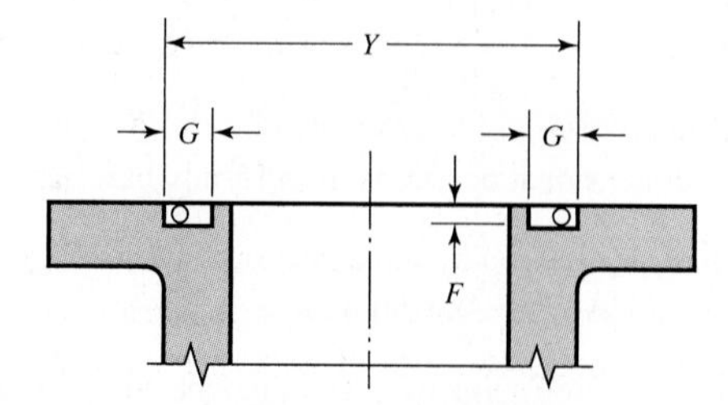

Problems 2–47 and 2–48

2–48 The gland for an O ring under internal pressure is shown in the figure. For a No. 370 O ring, the recommended gland dimensions are

$$F = 4.32 \pm 0.13 \text{ mm} \qquad G = 7.24 \pm 0.13 \text{ mm}$$

One manufacturer of O rings recommends that the gland's largest diameter be

$$Y_{\max} = D_o \qquad Y_{\min} = \max[0.99D_o, D_o - 1.52] \text{ mm}$$

The outside diameter of a No. 370 O ring is 219.58 ± 0.34 mm, and the cross-sectional diameter is 5.33 ± 0.13 mm.

(*a*) When the end plate is bolted in position, the O ring is compressed. What is the minimum compression (squeeze) of a No. 370 O ring in millimeters?

(*b*) When an O ring is placed in the gland prior to assembly of the end plate, is the ring free or under compression?

2–49 Two flanges of a bolted joint compress a soft gasket in a controlled manner so that the "squeeze" is between 0.020 in and 0.040 in. Too little squeeze results in leakage and too much in gasket "flow." As shown in the figure, a shouldered cap screw is used to control the gasket compression, and the critical factor in sustaining the function of the joint is the length a of the cap screw. If the dimensions of the members are as shown and gasket production thickness is 0.120 ± 0.005 in, find the length $\bar{a}$ and the tolerance t_a using the absolute tolerance system and the statistical tolerance system.

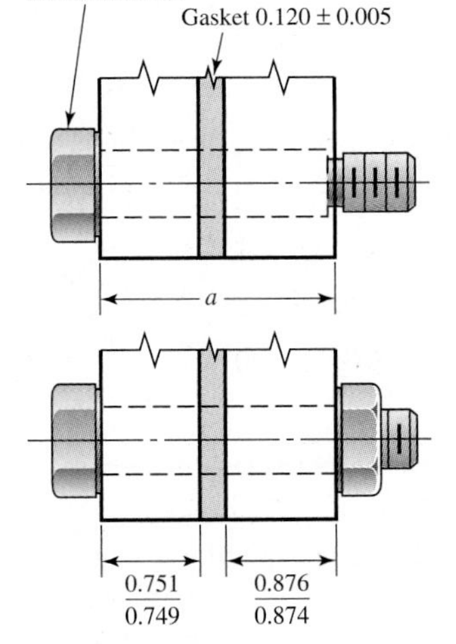

Problem 2–49

2–50 For Prob. 2–49, conduct a computer simulation with the dimensions

$$a = 1.715 \pm 0.003 \qquad b = 0.750 \pm 0.001$$
$$c = 0.120 \pm 0.005 \qquad d = 0.875 \pm 0.001$$

all in inches, to check the compatibility of the above dimensions with a gasket squeeze in the range 0.020 to 0.040 in. To do this, simulate a gap w from the necessary relation $w = a - b - c - d$ where

$$\mathbf{a} \sim U[1.712, 1.718] \qquad \mathbf{b} \sim U[0.749, 0.751]$$
$$\mathbf{c} \sim U[0.115, 0.125] \qquad \mathbf{d} \sim U[0.874, 0.876]$$

and check the largest and smallest values of w in 1000 evaluations of w. Be sure to recognize the independence of a, b, c, and d by using a different random number for each dimensional simulation that forms a single instance of w. At the same time, histographic information can be assembled so that the general shape of the distribution of w may be seen. Would you expect it to be a uniform random distribution? Why or why not?

2–51 The tensile 0.2 percent offset yield strength of AISI 1137 cold-drawn steel rounds up to 1 inch in diameter from 2 mills and 25 heats is reported histographically as follows:

S_y	93	95	97	99	101	103	105	107	109	111
f	19	25	38	17	12	10	5	4	4	2

where S_y is the class midpoint in kpsi and f is the number in each class. Presuming the distribution is normal, what is the yield strength exceeded by 99 percent of the population? How can you check on the goodness-of-fit of a normal distribution to the data?

2–52 Repeat Prob. 2–51, presuming the distribution is lognormal. What is the yield strength exceeded by 99 percent of the population? Check the goodness of fit by plotting the data on appropriate coordinates so as to rectify the ordered data. Compare the normal fit of Prob. 2–51 with the lognormal fit by superposing the PDFs and the histographic PDF.

2–53 When we wish to use a triangular distribution to represent range data, and have no basis for suspecting skewness, we use a symmetrical (isosceles) triangular distribution. Show that

Left	Right
$f(x) = \dfrac{2h}{x_3 - x_1}(-x_1 + x)$	$f(x) = \dfrac{2h}{x_3 - x_1}(x_3 + x)$
$F(x) = 2\left(\dfrac{x - x_1}{x_3 - x_1}\right)^2$	$F(x) = 1 - 2\left(\dfrac{x_3 - x}{x_3 - x_1}\right)^2$
$x = x_1 + (x_3 - x_1)\sqrt{\dfrac{F}{2}}$	$x = x_3 + (x_3 - x_1)\sqrt{\dfrac{1 - F}{2}}$
$E(x) = \dfrac{x_1 + x_3}{2}$	
$\sigma_x = \dfrac{X_3 - x_1}{2\sqrt{6}}$	

2–54 A 1046 steel, water-quenched and tempered for 2 h at 1210°F, has a mean tensile strength of 105 kpsi and a yield mean strength of 82 kpsi. Test data from endurance strength testing at 10^4-cycle life gives $(\mathbf{S}'_{fe})_{10^4} \sim W[79,\ 86.2,\ 2.60]$ kpsi. What are the mean, standard deviation, and coefficient of variation of $(\mathbf{S}'_{fe})_{10^4}$?

2–55 An ASTM grade 40 cast iron has the following result from testing for ultimate tensile strength: $\mathbf{S}_{ut} \sim W[27.7, 46.2, 4.38]$ kpsi. Find the mean and standard deviation of $\mathbf{S}_{ut}$, and estimate the chance that the ultimate strength is less than 40 kpsi.

2–56 A cold-drawn 301SS stainless steel has an ultimate tensile strength given by $\mathbf{S}_{ut} \sim W[151.9,$ $193.6, 8.00]$ kpsi. Find the mean and standard deviation.

2–57 A 100-70-04 nodular iron has tensile and yield strengths described by

$$\mathbf{S}_{ut} \sim W[47.6, 125.6, 11.84] \text{ kpsi}$$
$$\mathbf{S}_y \sim W[64.1, 81.0, 3.77] \text{ kpsi}$$

What is the chance that S_{ut} is less than 100 kpsi? What is the chance that S_y is less than 70 kpsi?

2–58 A 1038 heat-treated steel bolt in finished form provided the material from which a tensile test specimen was made. The testing of many such bolts led to the description $\mathbf{S}_{ut} \sim W[122.3, 134.6, 3.64]$ kpsi. What is the probability that the bolts meet the SAE grade 5 requirement of a minimum tensile strength of 120 kpsi? What is the probability that the bolts meet the SAE grade 7 requirement of a minimum tensile strength of 133 kpsi?

2–59 A 5160H steel was tested in fatigue and the distribution of cycles to failure at constant stress level was found to be $\mathbf{n} \sim W[36.9, 133.6, 2.66]$ in 10^3 cycles. Plot the PDF of n and the PDF of the lognormal distribution having the same mean and standard deviation. What is the B10 life (see Prob. 2–29) predicted by both distributions?

2–60 A material was tested at steady fully reversed loading to determine the number of cycles to failure using 100 specimens. The results were

$(10^{-5})L$	3.05	3.55	4.05	4.55	5.05	5.55	6.05	6.55	7.05	7.55	8.05	8.55	9.05	9.55	10.05
f	3	7	11	16	21	13	13	6	2	0	4	3	0	0	1

where L is the life in cycles and f is the number in each class. Convert to CDF, plot on coordinates that will rectify the ordered data string, draw the eyeball best-fit line, and find the lognormal parameters ($\hat{\mu}$ and $\hat{\sigma}$). Plot the PDF and histographic PDF for comparison.

2–61 The ultimate tensile strength of an AISI 1117 cold-drawn steel is Weibullian, with $\mathbf{S}_u \sim W[70.3, 84.4, 2.01]$. What are the mean, the standard deviation, and the coefficient of variation?

2–62 A 60-45-15 nodular iron has a 0.2 percent yield strength S_y with a mean of 49.0 kpsi, a standard deviation of 4.2 kpsi, and a guaranteed yield strength of 33.8 kpsi. What are the Weibull parameters θ and b?

2–63 A 35018 malleable iron has a 0.2 percent offset yield strength given by the Weibull distribution $\mathbf{S}_y \sim W[34.7, 39.0, 2.93]$ kpsi. What are the mean, the standard deviation, and the coefficient of variation? Is the distribution left- or right-skewed?

2–64 The results of testing many heats of AISI 1020 steel included 1000 separate determinations of the ultimate tensile strength. Using the data in columns 1 and 2 of Table 2–3, fit a normal and a lognormal distribution and plot the PDFs of all three distributions with the histographic PDF superposed for comparison. Which fit seems best?

2–65 The histographic results of steady load tests on 237 rolling-contact bearings are:

L	1	2	3	4	5	6	7	8	9	10	11	12
f	11	22	38	57	31	19	15	12	11	9	7	5

where L is the life in millions of revolutions and f is the number of failures. Fit a lognormal distribution to these data and plot the PDF with the histographic PDF superposed. From the lognormal distribution, estimate the life at which 10 percent of the bearings under this steady loading will have failed.

2–66 The histographic results of a steady load test on one type of rolling-contact bearing were given in Prob. 2–65. Fit a Weibull distribution to these data and plot the PDF with the histographic PDF superposed. Estimate the life at which 90 percent of the bearings under this steady loading will still be in operation.

2–67 When your objective is to identify the statistical parameters, mean and standard deviation, of a column vector of observations $\{x_n\}$ it can be useful to

- Order the data smallest to largest and assign order numbers i.
- Use the unbiased median CDF order statistic estimator $F_i = (i - 0.3)/(n + 0.4)$.
- Plot data on transformed coordinates of x and F based on Table 2–4.
- Draw an eyeball-best line and find its equation.
- Find $\bar{x}$ and σ_x from equations found in Table 2–4.

As a simple example consider data from a uniform population $\{x_n\} = \{1, 2, 3\}$. The associated median CDF assignments are $F_1 = 0.206$, $F_2 = 0.500$, $F_3 = 0.794$. Table 2–4 indicates the transformation are unity on x and F. Plot x as ordinate and F as abscissa, draw in the best line, recover its equation, $x = a + bF$, and find $\bar{x}$ and σ_x. Note the bonus of having the CDF function and confirmation of the distribution type.

2–68 From the data of Ex. 2–3, find the distribution parameters using the order statistic plotting approach. Compare answers.

2–69 Repeat Prob. 2–2, but use the order statistic approach to help you find the mean and standard deviation, expecting a lognormal population parent.

2–70 Solve Prob. 2–5 using the order statistic plotting approach and compare with your previous solution.

2–71 Solve Prob. 2–7 using the order statistic plotting approach and compare with your previous solution.

2–72 A shipment of 700 helical coil springs has been received. A random sample of 21 was tested for spring rate k in pounds per inch, and the ordered results were:

7.58	7.60	7.71	7.77	7.77	7.78	7.79
7.85	7.90	7.92	7.96	7.98	7.99	7.99
8.07	8.08	8.14	8.18	8.20	8.30	8.31

Plot the CDF on coordinates that will rectify a normal and a lognormal distribution, that is, k versus z for normal, and $\ln k$ versus z for lognormal. Since the sample is small, use the CDF estimator Eq. (4–33) $F_i = (i - 0.3)/(n + 0.4)$, where i is the order number and n is the sample size. Draw the eyeball best-fit lines in each case, and recover the equation of the CDF. Choose the best fit and estimate the value of k that will be exceeded by 95 percent of the shipment.

2–73 In a binomial distribution histogram the relation between the number of classes k and the sum of class frequencies N is

k	1	2	3	4	5	$\cdots$	k
N	1	2	4	8	16	$\cdots$	2^{k-1}

A binomial distribution with a large population approaches a normal distribution. Also, the number of observations in the first and last class interval is 1. In order that the first and last class interval is occupied by an observation in a normal distribution, it is desirable that the number of classes k be chosen from N according to

$$N = 2^{k-1}$$

Show that Sturges's rule, Eq. (A–7), follows.

2–74 It would be useful to develop an expression for the coefficient of variation for a three-parameter Weibull distribution from the parameters x_0, θ, b. Take Eq. (A–32) for σ_S and divide by Eq. (A–31) written in the form $\bar{S} - x_0 = (\theta - x_0)\Gamma(1 + 1/b)$, express σ_S as $C_S\bar{S}$, and show that

$$C_S = \frac{\bar{S} - x_0}{\bar{S}}\sqrt{\frac{\Gamma(1 + 2/b)}{\Gamma^2(1 + 1/b)} - 1}$$

Plot the logarithm of the radical against $\ln b$ and find a fit of the form b^m to substitute for the radical. Would you agree to

$$C_S = \frac{\bar{S} - x_0}{\bar{S}} b^{-0.926}$$

In what range of b would such an equation be useful?

2–75 Problem 2–74 presented an expression for the coefficient of variation of a Weibull-distributed variable from the parameters x_0, θ, and b. Find the coefficient of variation of the ultimate tensile strength of a 1018 steel with $x_0 = 30.8$ kpsi, $\theta = 90.1$ kpsi, and $b = 12$. Check the result with Table E–25.

2–76 For the ultimate tensile strength of a 2024 T6 aluminum with a Weibull distribution parameter set of $x_0 = 55.9$ kpsi, $\theta = 68.1$ kpsi, and $b = 9.26$, find the coefficient of variation using the equation of Prob. 2–74, and check the result with Table A–25.

2–77 Problem 2–53 showed that for a symmetric triangular distribution $\mu = (x_1 + x_3)/2$ and $\sigma = (x_3 - x_1)/(2\sqrt{6})$, where x_1 and x_3 are the smallest and largest observations, respectively. If a qualification on a table with a single tabular entry for a property is ± 5 percent, develop an expression for estimating the standard deviation and the coefficient of variation of the tabulated property.

3 Stress

3–1 Stress Components **94**

3–2 Mohr Circles **96**

3–3 Triaxial Stress **100**

3–4 Uniformly Distributed Stresses **102**

3–5 Elastic Strain **103**

3–6 Stress-Strain Relations **104**

3–7 Equilibrium **104**

3–8 Shear and Moment **106**

3–9 Singularity Functions **108**

3–10 Normal Stresses in Flexure **111**

3–11 Beams with Asymmetrical Sections **118**

3–12 Shear Stresses in Beams **118**

3–13 Shear Stresses in Rectangular-Section Beams **121**

3–14 Torsion **123**

3–15 Stress Concentration **129**

3–16 Stresses in Cylinders **132**

3–17 Rotating Rings **135**

3–18 Press and Shrink Fits **135**

3–19 Temperature Effects **137**

3–20 Curved Members in Flexure **138**

3–21 Contact Stresses **144**

3–22 Propagation of Error **149**

3–23 Summary **154**

Figure 3–1

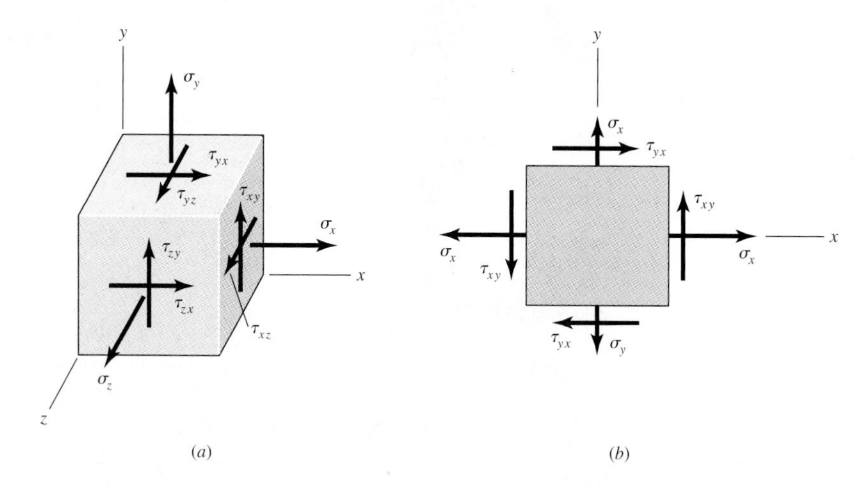

Before we consider the details of stresses at a point and the relationships among the components of stress, it is important to note the overall relationship between loading and components of stress at a point. The word *loading* here is generic, and it is a synonym for an external influence such as a force, torque, or shearing traction. Stresses at a point in an elastic homogeneous body are proportional to load to some power, say, P^m, usually. In simple tension, compression shear, and bending, the exponent m equals unity. For contact stresses in cylinders $m = \frac{1}{2}$, and in spheres $m = \frac{1}{3}$. Regardless, the stresses at a point change in concert as the load changes. If $m = 1$ and the load doubles, *all* stress components are doubled. If $m = \frac{1}{2}$ and the load is doubled, *all* stress components are larger by a factor $\sqrt{2}$. This is an example of direct $\rho = 1$ correlation. If a load is stochastic, so are the many components of stress, exhibiting the same type of distribution as that of the load, with the same coefficient of variation. If α is deterministic, it is possible to develop the subject matter of this chapter without regard to stochastics because of this correlation. For the simpler cases, probabilistic statements can be added as an afterthought.

Through control of geometry and material, including thermomechanical treatment, the designer guards the integrity of the design, that is, makes it functional, safe, reliable, usable, manufacturable, and marketable. Loss of function, when parts yield, rupture, fatigue, creep, and so on, must be anticipated. The most useful insight into the problem begins with understanding of the influence of external loading on stresses within the body.

Stress and strain were first understood rather late in the long history of mechanics. In 1660 Robert Hooke had some perceptive notions concerning bodies (of steel, iron, and brass) which exhibited elastic deformation under load. Thomas Young's work, about 1807, and his modulus, improved the understanding of materials so that by 1820, notions of stress were forming. The powerful perceptions that followed help us today. In 1822 Cauchy formalized the notion of stress and laid the foundation of what today is called *theory of elasticity*. Others have used this concept of stress to simplify the description of stress at a point. Nine quantitative statements about stress are needed to uniquely describe stress at a point in an elastic isotropic homogeneous body, as seen in Fig. 3-1*a*. Can this number be reduced?

3–1 Stress Components

Figure 3–1*a* is a general three-dimensional stress element, showing three normal stresses $\sigma_x, \sigma_y, \sigma_z$, all positive, and six shear stresses τ_{xy}, τ_{yx}, τ_{yz}, τ_{zy}, τ_{zx}, and τ_{xz}, also all positive. The element is in equilibrium, hence

$$\tau_{xy} = \tau_{yx}, \qquad \tau_{yz} = \tau_{zy}, \qquad \tau_{zx} = \tau_{xz} \tag{3–1}$$

Outwardly directed normal stresses are called *tension* or *tensile stresses*, and they are considered positive. Shear stresses on a positive face of an element are positive if they act in the positive direction of a reference axis, as is the case in Fig. 3–1*a*. The first subscript of a shear-stress component is the coordinate normal to the element face. The shear-stress component is parallel to the axis of the second subscript. Since the element is shown in static equilibrium, the negative faces will have shear stresses acting in the opposite direction; these are also considered positive.

We have just described the classical convention for stress. In this book we use a sign convention different from the classical one in order to obtain agreement in direction in the measurement of angles and shear stresses. Figure 3–1*b* illustrates a state of *biaxial*, or *plane*, stress. The two normal stresses are shown in the positive direction. Shear stresses will be taken as positive when they are in the clockwise (cw) sense. Thus in Fig. 3–1*b*, τ_{yx} is cw and positive; τ_{xy} is counterclockwise (ccw) and negative. Equations (3–1) reduce the number of stress components from *nine* to *six*. Further progress has been made.

- Mohr circles for plane stress consist of three circles for triaxial stress. When the three principal stresses are known, others can be deduced. This results in another dramatic reduction of *six* components to *three*.
- Investigations of stresses on octahedral surfaces (there are eight such surfaces) associated with the three principal stress directions establish description by *two* components, the octahedral normal stresses and the octahedral shear stress, another reduction in complexity.
- The distortion energy per unit volume at a point is a function of a *single* stress component called the *von Mises normal stress*. For materials that lose function by yielding or fatiguing in response to distortion-energy intensity level, the von Mises stress is the *single* descriptor of the stress condition at a point.
- Failure hypotheses and theories use one or more of these one-, two-, and three-component descriptions of stress at a point to explain the onset of failure in a class, or classes, of materials. You will notice in this chapter a movement toward simplifications which are helpful in understanding the stress condition at a point.

Professor Gaetano Lanza of MIT in 1885 in *Applied Mechanics* used the "method of sections" in trusses, but the "free-body diagram" as we now know it was championed by Professor Irving Church of Cornell *Mechanics of Engineering* in 1887. It was late in the development of mechanics that Professor Church made it possible for engineers to teach themselves what they needed to know, rigorously. "Draw the free-body diagram" is an admonition, that is key to your becoming an engineer.

A *principle of superposition* is useful in that stresses (at a point on a given plane) due to different external loads may be computed separately and added algebraically, provided the stresses do not exceed the proportional limit of the material, and in the absence of buckling. If there is a pattern to analysis problems involving stresses, it resembles the following:

1. Identify the external loads, and check by equilibrium equations and free-body diagrams to ensure that you have identified all of the loads and their senses.
2. Use more free-body diagrams to find the internal loading.
3. Find the corresponding stresses.
4. Use Mohr circles to identify the three principal stresses, the three maximum shear stresses, and stresses on planes of interest.
5. Compare the kinds of stresses that cause failure with the corresponding strengths to assess the possibility of failure.

In the historical development that follows, the results were considered to be true—that is, exact (and deterministic)—by their discoverers. In the light of what was presented in Chap. 2, their results can be considered to be relationships among means (strictly, medians). The additional information needed for variability description begins in Sec. 3–2.

3–2 Mohr Circles

Suppose the element of Fig. 3–1*b* is cut by an oblique plane at angle ϕ to the y axis as shown in Fig. 3–2. This section is concerned with the stresses σ and τ which act upon this oblique plane. By summing the forces caused by all the stress components to zero, the stresses σ and τ are found to be

$$\sigma = \frac{\sigma_x + \sigma_y}{2} + \frac{\sigma_x - \sigma_y}{2} \cos 2\phi + \tau_{xy} \sin 2\phi \tag{3–2}$$

$$\tau = -\frac{\sigma_x - \sigma_y}{2} \sin 2\phi + \tau_{xy} \cos 2\phi \tag{3–3}$$

Differentiating the first equation with respect to ϕ and setting the result equal to zero gives

$$\tan 2\phi = \frac{2\tau_{xy}}{\sigma_x - \sigma_y} \tag{3–4}$$

Equation (3–4) defines two particular values for the angle 2ϕ, one of which defines the maximum normal stress σ_1 and the other, the minimum normal stress σ_2. These two stresses are called the *principal stresses*, and their corresponding directions, the *principal directions*. The angle ϕ between the principal directions is 90°.

In a similar manner, we differentiate Eq. (3–3), set the result equal to zero, and obtain

$$\tan 2\phi = -\frac{\sigma_x - \sigma_y}{2\tau_{xy}} \tag{3–5}$$

Equation (3–5) defines the two values of 2ϕ at which the shear stress τ reaches an extreme value.

It is interesting to note that Eq. (3–4) can be written in the form

$$2\tau_{xy} \cos 2\phi = (\sigma_x - \sigma_y) \sin 2\phi$$

or

$$\sin 2\phi = \frac{2\tau_{xy} \cos 2\phi}{\sigma_x - \sigma_y} \tag{a}$$

Figure 3–2

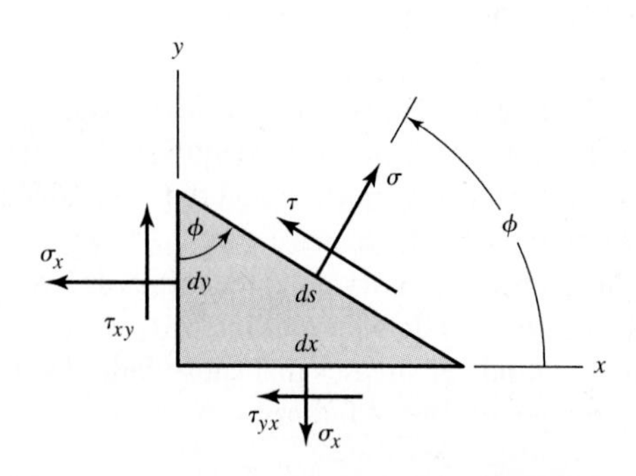

Now substitute Eq. (a) for sin 2ϕ in Eq. (3–3). We obtain

$$\tau = -\frac{\sigma_x - \sigma_y}{2}\frac{2\tau_{xy}\cos 2\phi}{\sigma_x - \sigma_y} + \tau_{xy}\cos 2\phi = 0 \tag{3–6}$$

Equation (3–6) states that the shear stress associated with both principal directions is zero.

Solving Eq. (3–5) for sin 2ϕ, in a similar manner, and substituting the result in Eq. (3–2) yields

$$\sigma = \frac{\sigma_x + \sigma_y}{2} \tag{3–7}$$

Equation (3–7) tells us that the two normal stresses associated with the directions of the two maximum shear stresses are equal.

Formulas for the two principal stresses can be obtained by substituting the angle 2ϕ from Eq. (3–4) in Eq. (3–2). The result is

$$\sigma_1, \sigma_2 = \frac{\sigma_x + \sigma_y}{2} \pm \sqrt{\left(\frac{\sigma_x - \sigma_y}{2}\right)^2 + \tau_{xy}^2} \tag{3–8}$$

In a similar manner the two extreme-value shear stresses are found to be

$$\tau_1, \tau_2 = \pm\sqrt{\left(\frac{\sigma_x - \sigma_y}{2}\right)^2 + \tau_{xy}^2} \tag{3–9}$$

Your particular attention is called to the fact that an extreme value of the shear stress may not be the same as the maximum value. See Sec. 3–3.

A graphical method for expressing the relations developed in this section, called a *Mohr circle diagram*, is a very effective means of visualizing the stress state at a point and keeping track of the directions of the various components associated with plane stress. In Fig. 3–3 we create a coordinate system with normal stresses plotted along the abscissa and

Figure 3–3

Mohr circle diagram.

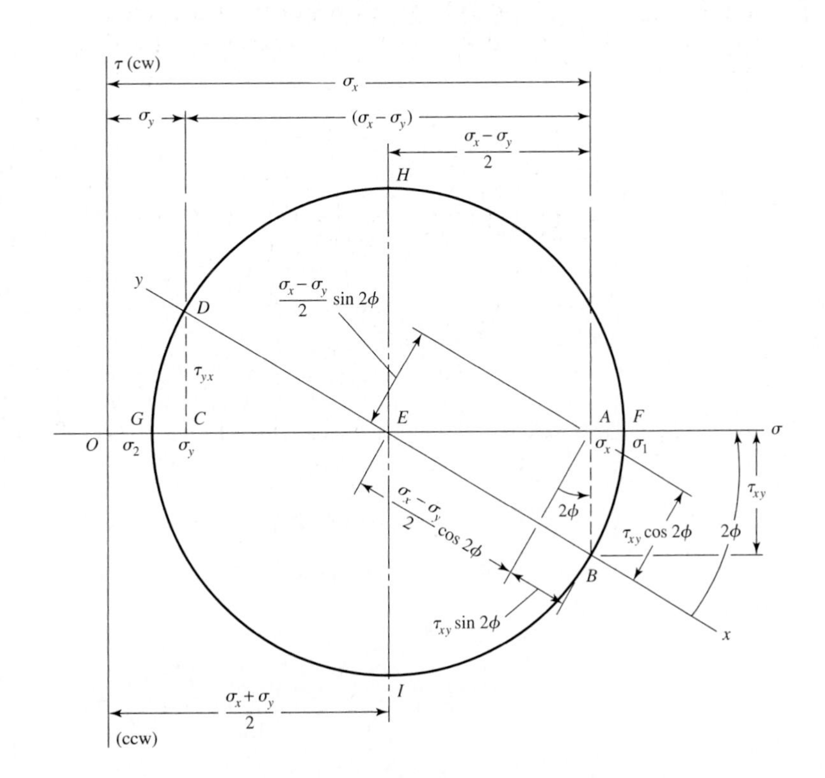

shear stresses plotted as the ordinates. On the abscissa, tensile (positive) normal stresses are plotted to the right of the origin O and compressive (negative) normal stresses to the left. On the ordinate, positive (cw) shear stresses are plotted up; counterclockwise (ccw) shear stresses are negative and are plotted down.

Using the stress state of Fig. 3–1b, we plot the Mohr circle diagram (Fig. 3–3) by laying off σ_x as OA, τ_{xy} as AB, σ_y as OC, and τ_{yx} as CD. The line DEB is the diameter of the Mohr circle with center at E on the σ axis. Point B represents the stress coordinates σ_x, τ_{xy} on the x faces, and point D the stress coordinates σ_y, τ_{yx} on the y faces. Thus EB corresponds to the x axis and ED to the y axis. The angle 2ϕ, measured counterclockwise from EB to ED, is 180°, which corresponds to $\phi = 90°$, measured counterclockwise from x to y, on the stress element of Fig. 3–1b. The maximum principal normal stress σ_1 occurs at F, and the minimum principal normal stress σ_2 at G. The two extreme-value shear stresses, one clockwise and one counterclockwise, occur at I and H, respectively. You should demonstrate for yourself that the geometry of Fig. 3–3 satisfies all the relations developed in this section.

Though Eqs. (3–4) and (3–8) can be solved directly for the principal stresses and directions, a semigraphical approach is easier and quicker and offers fewer opportunities for error. This method is illustrated by the following example.

EXAMPLE 3–1

A stress element has $\sigma_x = 80$ MPa and $\tau_{xy} = 50$ MPa cw.[1]

(a) Using a Mohr circle, find the principal stresses and directions, and show these on a stress element correctly aligned with respect to the xy coordinates. Draw another stress element to show τ_1 and τ_2, find the corresponding normal stresses, and label the drawing completely.

(b) Repeat part a using an algebraic approach.

Solution

(a) We shall construct a Mohr circle corresponding to the given data and then read the results directly from the diagram. You can construct this diagram with compass and scales and find the required information with the aid of scales and protractor, if you choose to do so. Such a complete graphical approach is perfectly satisfactory and sufficiently accurate for most purposes.

In the semigraphical approach used here, we first make an approximate freehand sketch of the Mohr circle and then use the geometry of the figure to obtain the desired information.

Draw the σ and τ axes first (Fig. 3–4a) and locate $\sigma_x = 80$ MPa along the σ axis. Then, from σ_x, locate $\tau_{xy} = 50$ MPa in the cw direction of the τ axis to establish point A. Corresponding to $\sigma_y = 0$, locate $\tau_{yx} = 50$ MPa in the ccw direction along the τ axis to obtain point D. The line AD forms the diameter of the required circle, which can now be drawn. The intersection of the circle with the σ axis defines σ_1 and σ_2 as shown. The x axis is the line CA; the y axis is the line CD. Now, noting the triangle ABC, indicate on the sketch the length of the legs AB and BC as 50 and 40 MPa, respectively. The length of the hypotenuse AC is

$$\tau_1 = \sqrt{(50)^2 + (40)^2} = 64.0 \text{ MPa}$$

and this should be labeled on the sketch too. Since intersection C is 40 MPa from the

1. Any stress components such as σ_y and τ_{zx} that are not given in a problem are always taken as zero.

Figure 3–4

All stresses in MPa.

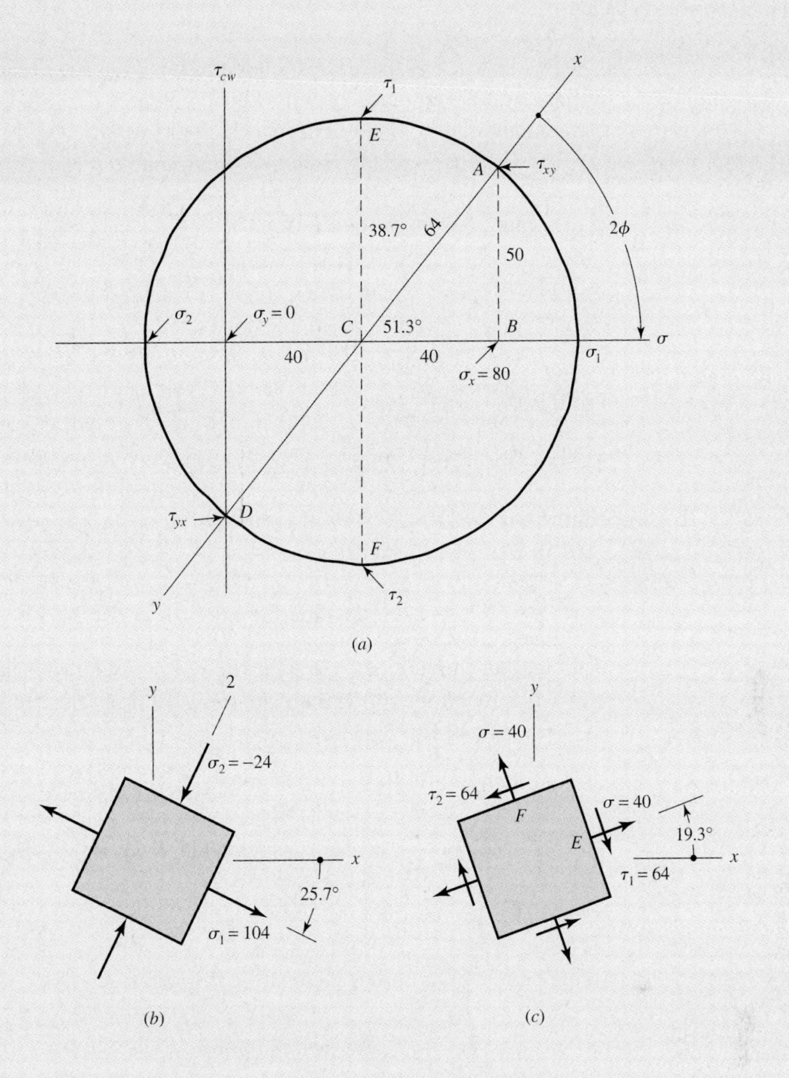

origin, the principal stresses are now found to be

$$\sigma_1 = 40 + 64 = 104 \text{ MPa} \qquad \sigma_2 = 40 - 64 = -24 \text{ MPa}$$

The angle 2ϕ from the x axis cw to σ_1 is

$$2\phi = \tan^{-1} \frac{50}{40} = 51.3°$$

If your calculator has rectangular-to-polar conversion, you should always use it to obtain τ_1 and 2ϕ.

To draw the principal stress element (Fig. 3–4*b*), sketch the x and y axes parallel to the original axes. The angle ϕ on the stress element must be measured in the *same* direction as is the angle 2ϕ on the Mohr circle. Thus, from x measure 25.7° (half of 51.3°) clockwise to locate the σ_1 axis. The stress element can now be completed and labeled as shown.

The two maximum shear stresses occur at points E and F in Fig. 3–4*a*. The two normal stresses corresponding to these shear stresses are each 40 MPa, as indicated. Point E is 38.7° ccw from point A on the Mohr circle. Therefore, in Fig. 3–4*c*, draw a stress element oriented 19.3° (half of 38.7°) ccw from x. The element should then be labeled with magnitudes and directions as shown.

In constructing these stress elements it is important to indicate the x and y directions of the original reference system. This completes the link between the original machine element and the orientation of its principal stresses.

(*b*) An algebraic approach is programmable. From Eq. (3–4),

$$\phi = \frac{1}{2}\tan^{-1}\left(\frac{2\tau_{xy}}{\sigma_x - \sigma_y}\right) = \frac{1}{2}\tan^{-1}\left[\frac{2(50)}{80}\right] = 25.7^\circ$$

From Eq. (3–2),

$$\sigma = \frac{80+0}{2} + \frac{80-0}{2}\cos[2(25.7^\circ)] + 50\sin[2(25.7^\circ)] = 104.03 \text{ MPa}$$

From Eq. (3–3),

$$\tau = -\frac{80-0}{2}\sin[2(25.7^\circ)] + 50\cos[2(25.7^\circ)] = 0 \text{ MPa}$$

confirming that 104.03 MPa is a principal stress. Adding 180° to 2ϕ gives $2\phi = 51.34^\circ + 180^\circ = 231.34^\circ$. From Eq. (3–2),

$$\sigma = \frac{80+0}{2} + \frac{80-0}{2}\cos(231.34^\circ) + 50\sin(231.34^\circ) = -24.03 \text{ MPa}$$

Anticipating the advice following Eq. (3–11), we order the principal stresses, 104, 0, −24 MPa, mentally noting there are *three* Mohr circles, and we find the maximum shear stress τ_{max} from Eq. (3–11),

$$\tau_{max} = \tau_{1/3} = \frac{\sigma_1 - \sigma_3}{2} = \frac{104 - (24)}{2} = 64.0 \text{ MPa}$$

The plane of maximum shear is located at, from Eq. (3–5),

$$2\phi = \tan^{-1}\left[-\frac{80-0}{2(50)}\right] = -38.66^\circ$$

so the plane of maximum shear stress is located $-\frac{38.66}{2}$ or -19.3° (ccw) from the x axis.

3–3 Triaxial Stress

As in the case of plane or biaxial stress, a particular orientation of the stress element occurs in space for which all shear-stress components are zero. When an element has this particular orientation, the normals to the faces correspond to the principal directions, and the normal stresses associated with these faces are the principal stresses. Since there are six faces, there are three principal directions and three principal stresses σ_1, σ_2, and σ_3.

In our studies of plane stress we were able to specify any stress state σ_x, σ_y, and τ_{xy} and find the principal stresses and principal directions. But six components of stress are required to specify a general state of stress in three dimensions, and the problem of determining the principal stresses and directions is much more difficult. It turns out that it is rarely necessary in design, and so we shall not investigate the problem in this book. The process involves finding the three roots to the cubic equation

$$\begin{aligned}\sigma^3 - (\sigma_x + \sigma_y + \sigma_z)\sigma^2 &+ (\sigma_x\sigma_y + \sigma_x\sigma_z + \sigma_y\sigma_z - \tau_{xy}^2 - \tau_{yz}^2 - \tau_{zx}^2)\sigma \\ &- (\sigma_x\sigma_y\sigma_z + 2\tau_{xy}\tau_{yz}\tau_{zx} - \sigma_x\tau_{yz}^2 - \sigma_y\tau_{zx}^2 - \sigma_z\tau_{xy}^2) = 0\end{aligned} \tag{3–10}$$

In plotting Mohr circles for triaxial stress, the principal normal stresses are ordered so that $\sigma_1 \geq \sigma_2 \geq \sigma_3$. Then the result appears as in Fig. 3–5*a*. The stress coordinates σ_N, τ_N for any arbitrarily located plane will always lie within the shaded area.

Figure 3–5*a* also shows the three *principal shear stresses* $\tau_{1/2}$, $\tau_{2/3}$, and $\tau_{1/3}$.[2] Each of these occurs on the two planes, one of which is shown in Fig. 3–5*b*. The figure shows that the principal shear stresses are given by the equations

$$\tau_{1/2} = \frac{\sigma_1 - \sigma_2}{2} \qquad \tau_{2/3} = \frac{\sigma_2 - \sigma_3}{2} \qquad \tau_{1/3} = \frac{\sigma_1 - \sigma_3}{2} \tag{3-11}$$

Of course, $\tau_{max} = \tau_{1/3}$ when the normal principal stresses are ordered $(\sigma_1 > \sigma_2 > \sigma_3)$, so always order your principal stresses. Do this in any computer code you generate and you'll always generate τ_{max}.

Octahedral Stresses

Now visualize a principal stress element having the stresses σ_1, σ_2, and σ_3, as shown in Fig. 3–6. Cut the stress element by a plane that forms equal angles with each of the three principal stresses as shown by plane *ABC*. This is called an *octahedral plane*. Note that the solid cut from the stress element retains one of the original corners. Since there are eight of these corners in all, there are a total of eight such planes.

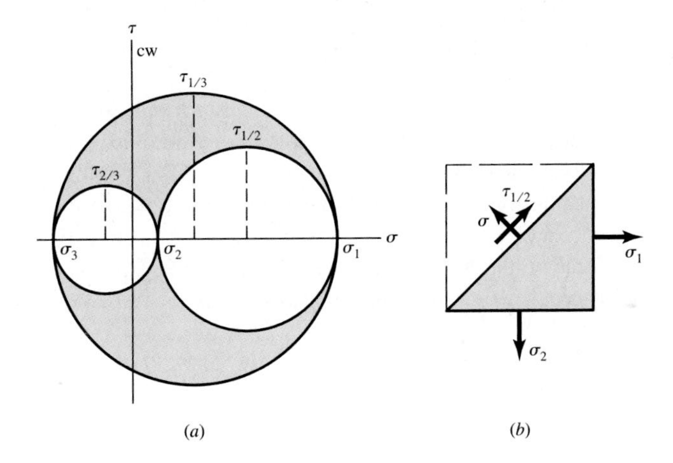

Figure 3–5

Mohr circles for triaxial stress.

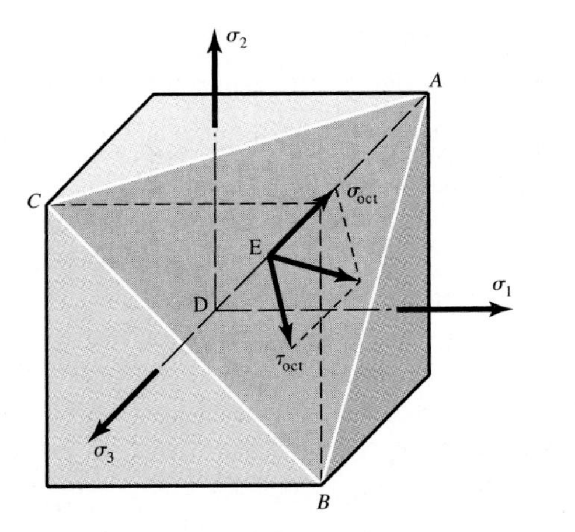

Figure 3–6

One of eight octrahedral planes.

2. Note the difference between this notation and that for a shear stress, say, τ_{xy}. The use of the shilling mark is not accepted practice, but it is used here to emphasize the distinction.

Figure 3–6 may be regarded as a free-body diagram when each of the stress components shown is multiplied by the area over which it acts to obtain the corresponding force. It is possible, then, to sum these forces to zero in each of the three coordinate directions. When this is done, it is found that a force, called the *octahedral force*, exists on plane ABC. When this force is divided by the area of face ABC, it can be resolved into two components. One of these is normal to plane ABC and so is called the *octahedral normal stress*. The other is in plane ABC and is called the *octahedral shear stress*.[3] The magnitudes of these stress components are

$$\begin{aligned}\tau_{\text{oct}} &= \tfrac{2}{3}(\tau_{1/2}^2 + \tau_{2/3}^2 + \tau_{1/3}^2)^{1/2} \\ &= \tfrac{1}{3}[(\sigma_1 - \sigma_2)^2 + (\sigma_2 - \sigma_3)^2 + (\sigma_3 - \sigma_1)^2]^{1/2} \\ &= \tfrac{1}{3}[(\sigma_x - \sigma_y)^2 + (\sigma_y - \sigma_z)^2 + (\sigma_z - \sigma_x)^2 + 6(\tau_{xy}^2 + \tau_{xz}^2 + \tau_{yz}^2)]^{1/2} \end{aligned} \tag{3–12}$$

$$\sigma_{\text{oct}} = \tfrac{1}{3}(\sigma_1 + \sigma_2 + \sigma_3) = \tfrac{1}{3}(\sigma_x + \sigma_y + \sigma_z) \tag{3–13}$$

The importance of these equations is that any three-dimensional state of stress can be represented by a single pair of stress components. In later chapters we shall find this to be a very powerful tool in design analysis made even simpler. See Sec. 6–7.

3–4 Uniformly Distributed Stresses

The assumption of a uniform distribution of stress is frequently made in design. The result is then often called *pure tension*, *pure compression*, or *pure shear*, depending upon how the external load is applied to the body under study. The word *simple* is sometimes used instead of *pure* to indicate that there are no other complicating effects. The tension rod is typical. Here a tension load F is applied through pins at the ends of the bar. The assumption of uniform stress means that if we cut the bar at a section remote from the ends and remove one piece, we can replace its effect by applying a uniformly distributed force of magnitude σA to the cut end. So the stress σ is said to be uniformly distributed. It is calculated from the equation

$$\sigma = \frac{F}{A} \tag{3–14}$$

This assumption of uniform stress distribution requires that:

- The bar be straight and of a homogeneous material
- The line of action of the force contains the centroid of the section
- The section be taken remote from the ends and from any discontinuity or abrupt change in cross section

The same equation and assumptions hold for simple compression. A slender bar in compression may, however, fail by buckling, and this possibility must be eliminated from consideration before Eq. (3–14) is used.[4]

Use of the equation

$$\tau = \frac{F}{A} \tag{3–15}$$

3. For a derivation see A. Higdon, E. H. Olsen, W. B. Stiles, J. A. Weese and W. F. Riley, *Mechanics of Materials*, 4th ed., John Wiley & Sons, 1985, pp. 493–495.

4. See Sec. 4–11.

for a body, say, a bolt, in shear assumes a uniform stress distribution too. It is very difficult in practice to obtain a uniform distribution of shear stress. The equation is included because occasions do arise in which this assumption is utilized.

3–5 Elastic Strain

When a straight bar is subjected to a tensile load, the bar becomes longer. The additional amount is called *stretch* or *elongation*. The elongation per unit length of the bar is called *unit elongation* or *strain*. The expression for strain is

$$\epsilon = \frac{\delta}{l} \tag{3–16}$$

where δ is the total elongation of the bar within the length l.

Shear strain γ is the change in a right angle of a stress element subjected to pure shear.

Elasticity is that property of a material which enables it to regain its original shape and dimensions when the load is removed. Hooke's law states that, within certain limits, the stress in a material is proportional to the strain which produced it. An elastic material does not necessarily obey Hooke's law, since it is possible for some materials to regain their original shape without the limiting condition that stress be proportional to strain. On the other hand, a material which obeys Hooke's law is elastic. For the condition in which stress is proportional to strain, we can write the relations

$$\sigma = E\epsilon \qquad \tau = G\gamma \tag{3–17}$$

where E and G are the constants of proportionality. Since the strains are dimensionless numbers, the units of E and G are the same as the units of stress. The constant E is called the *modulus of elasticity*. The constant G is called the *shear modulus of elasticity* or, sometimes, the *modulus of rigidity*. Both E and G, however, are numbers which are indicative of the stiffness or rigidity of the materials. These two constants represent fundamental properties.

By substituting $\sigma = F/A$ and $\epsilon = \delta/l$ into Eq. (3–17) and rearranging, we obtain the equation for the total deformation of a bar loaded in axial tension or compression:

$$\delta = \frac{Fl}{AE} \tag{a}$$

Experiments demonstrate that when a material is placed in tension, there exists not only an axial strain, but also a lateral strain. Poisson demonstrated that these two strains were proportional to each other within the range of Hooke's law. This constant is expressed as

$$\nu = -\frac{\text{lateral strain}}{\text{axial strain}} \tag{3–18}$$

and is known as *Poisson's ratio*. These same relations apply for compression, except that a lateral expansion takes place instead.

The three elastic constants are related to each other as follows:

$$E = 2G(1 + \nu) \tag{3–19}$$

Table 3–1

Elastic Stress-Strain Relations

Type of Stress	Principal Strains	Principal Stresses
Uniaxial	$\epsilon_1 = \dfrac{\sigma_1}{E}$	$\sigma_1 = E\epsilon_1$
	$\epsilon_2 = -\nu\epsilon_1$	$\sigma_2 = 0$
	$\epsilon_3 = -\nu\epsilon_1$	$\sigma_3 = 0$
Biaxial	$\epsilon_1 = \dfrac{\sigma_1}{E} - \dfrac{\nu\sigma_2}{E}$	$\sigma_1 = \dfrac{E(\epsilon_1 + \nu\epsilon_2)}{1-\nu^2}$
	$\epsilon_2 = \dfrac{\sigma_2}{E} - \dfrac{\nu\sigma_1}{E}$	$\sigma_2 = \dfrac{E(\epsilon_2 + \nu\epsilon_1)}{1-\nu^2}$
	$\epsilon_3 = \dfrac{\nu\sigma_1}{E} - \dfrac{\nu\sigma_2}{E}$	$\sigma_3 = 0$
Triaxial	$\epsilon_1 = \dfrac{\sigma_1}{E} - \dfrac{\nu\sigma_2}{E} - \dfrac{\nu\sigma_3}{E}$	$\sigma_1 = \dfrac{E\epsilon_1(1-\nu) + \nu E(\epsilon_2 + \epsilon_3)}{1 - \nu - 2\nu^2}$
	$\epsilon_2 = \dfrac{\sigma_2}{E} - \dfrac{\nu\sigma_1}{E} - \dfrac{\nu\sigma_3}{E}$	$\sigma_2 = \dfrac{E\epsilon_2(1-\nu) + \nu E(\epsilon_1 + \epsilon_3)}{1 - \nu - 2\nu^2}$
	$\epsilon_3 = \dfrac{\sigma_3}{E} - \dfrac{\nu\sigma_1}{E} - \dfrac{\nu\sigma_2}{E}$	$\sigma_3 = \dfrac{E\epsilon_3(1-\nu) + \nu E(\epsilon_1 + \epsilon_2)}{1 - \nu - 2\nu^2}$

Note: Plane stress situation requires one principal stress to be zero. Plane strain situation requires one principal strain associated with one principal stress to be zero.

3–6 Stress-Strain Relations

There are many experimental techniques which can be used to measure strain. Thus, if the relationship between stress and strain is known, the stress state at a point can be calculated after the state of strain has been measured. We define the *principal strains* as the strains in the direction of the principal stresses. It is true that the shear strains are zero, just as the shear stresses are zero, on the faces of an element aligned in the principal directions. Table 3-1 contains the relations for all three types of stress. See Table E-5 for values of Poisson's ratio ν.

3–7 Equilibrium

The law of particle motion states that *any force F acting on a particle of mass will produce an acceleration of the particle*. If we assume that all members to be studied are motionless or, at most, have a constant velocity, then every particle has zero acceleration. Applying the law of particle motion gives

$$\mathbf{F}_1 + \mathbf{F}_2 + \mathbf{F}_3 + \cdots + \mathbf{F}_i = \sum \mathbf{F} = \mathbf{0} \qquad (a)$$

where $\Sigma\mathbf{F}$ is the vector sum of all the forces acting on the particle.

Whenever Eq. (*a*) holds, the forces acting on the particle are said to be *balanced* and the particle is said to be in *equilibrium*. The phrase *static equilibrium* is also used to imply that the particle is *at rest*.

The word *system* will be used to denote any part or portion of a machine or structure, including all of it if desired, that we may wish to study. A system, under this definition, may consist of a particle, several particles, a part of a rigid body or an entire rigid body, or even several rigid bodies.

An *internal force* or *internal moment* is an action of one part of a system on another part of the same system. An *external force* or an *external moment* is an action that is applied to the system from the outside.

Equation (a) expresses the conditions for the equilibrium of a single particle. But what about the equilibrium of a system containing many particles? We can apply Eq. (a) to each particle in the system. We shall select one of these, say the jth one. Let $\mathbf{F}_e$ be the sum of the external forces and $\mathbf{F}_i$ be the sum of the internal forces. Then for the jth particle, Eq. (a) becomes

$$\sum \mathbf{F}_j = \mathbf{F}_e + \mathbf{F}_i = \mathbf{0} \tag{b}$$

If there are n particles in the system and if we add them all together, we get

$$\sum_1^n \mathbf{F}_e + \sum_1^n \mathbf{F}_i = \mathbf{0} \tag{c}$$

Now Newton's third law, called the law of action and reaction, states that *when two particles react, a pair of interacting forces come into existence, that these forces have the same magnitude, have opposite senses, and act along the straight line common to the two particles*. This means that, in the system under study, all the internal forces acting between particles occur in pairs. Each force of the pair, according to the third law, is equal in magnitude and opposite in direction to its mate. Therefore the second term in Eq. (c) is zero, and we have left that

$$\sum_1^n F_e = \mathbf{0} \tag{3–20}$$

which states that *the sum of all the external force vectors acting upon a system in equilibrium is zero*.

A similar procedure can be used to demonstrate that *the sum of all the external moment vectors acting upon a system in equilibrium is zero*. Or, in mathematical form,

$$\sum_1^n M_e = \mathbf{0} \tag{3–21}$$

The statements accompanying Eqs. (3–20) and (3–21) can be reversed; thus it is also true that *if these two equations are simultaneously satisfied*, then the system is in *static equilibrium*.

Free-Body Diagrams

In examining the problem of analyzing the behavior, performance, or efficiency of a complex structure or device, such as a bridge, typewriter, or tractor, the beginner is faced with a bewildering array of complicated parts and geometries. Fortunately, the problem is not as difficult as it seems. One of the most powerful analytical techniques of mechanics is that of isolating or freeing a portion of a system in our imagination in order to study the behavior of one of its segments. When the segment is isolated, the original effect of the system on the segment is replaced by the interacting forces and moments. Figure 3–7 is a symbolic illustration of the process. Let Fig. 3–7a be a total system, such as a bridge. Then we might decide to analyze just one part or segment of the bridge, such as a beam, or even several members joined together. We remove this segment, as in

Figure 3–7

The isolation of a subsystem.

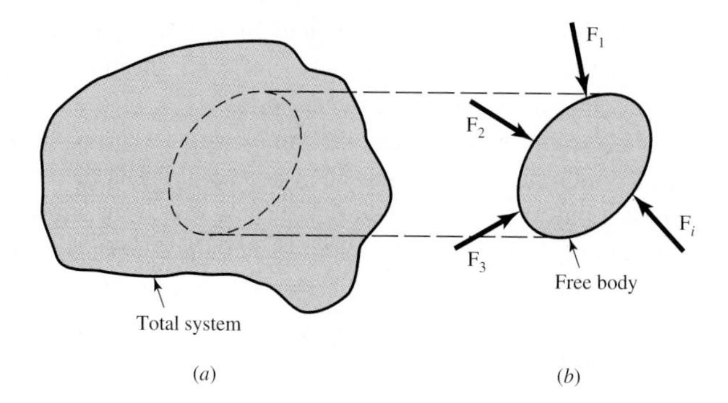

Fig. 3–7*b*, and replace the effect of the whole system on the segment by various forces and moments that would necessarily act at the interfaces of the segment and the system. Although these forces may be internal effects upon the whole system, they are external effects when applied to the segment. These interface forces are represented symbolically by the force vectors $\mathbf{F}_1$, $\mathbf{F}_2$, $\mathbf{F}_3$, and $\mathbf{F}_i$ in Fig. 3–7*b*. The isolated subsystem that results, together with all forces and moments due to any external effects and the reactions with the main system, is called a *free-body diagram.*

We can greatly simplify the analysis of a very complex structure or machine by successively isolating each element and studying and analyzing it by the use of free-body diagrams. When all the members have been treated in this manner, the knowledge can be assembled to yield information concerning the behavior of the total system. Thus, free-body diagramming is essentially a means of breaking a complicated problem into manageable segments, analyzing these simple problems, and then, usually, putting the information together again.

Using free-body diagrams for force analysis serves the following important purposes:

- The diagram establishes the directions of reference axes, provides a place to record the dimensions of the subsystem and the magnitudes and directions of the known forces, and helps in assuming the directions of unknown forces.
- The diagram simplifies your thinking because it provides a place to store one thought while proceeding to the next.
- The diagram provides a means of communicating your thoughts clearly and unambiguously to other people.
- Careful and complete construction of the diagram clarifies fuzzy thinking by bringing out various points that are not always apparent in the statement or in the geometry of the total problem. Thus, the diagram aids in understanding all facets of the problem.
- The diagram helps in the planning of a logical attack on the problem and in setting up the mathematical relations.
- The diagram helps in recording progress in the solution and in illustrating the methods used.
- The diagram allows others to follow your reasoning, showing *all* forces.

3–8 Shear and Moment

Figure 3–8*a shows a beam supported by* reactions R_1 and R_2 and loaded by the concentrated forces F_1, F_2, and F_3. The direction chosen for the y axis is the clue to the sign

Figure 3–8

Free-body diagram of simply-supported beam with V and M shown in positive directions.

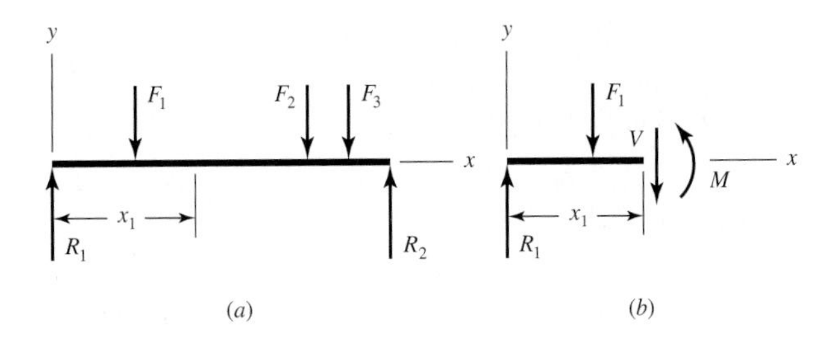

convention for the forces. F_1, F_2, and F_3 are negative because they act in the negative y direction; R_1 and R_2 are positive.

If the beam is cut at some section located at $x = x_1$ and the left-hand portion is removed as a free body, an *internal shear force* V and *bending moment* M must act on the cut surface to ensure equilibrium. The shear force is obtained by summing the forces to the left of the cut section. The bending moment is the sum of the moments of the forces to the left of the section taken about an axis through the section. Shear force and bending moment are related by the equation

$$V = \frac{dM}{dx} \tag{3–22}$$

Sometimes the bending is caused by a distributed load. Then, the relation between shear force and bending moment may be written

$$\frac{dV}{dx} = \frac{d^2M}{dx^2} = -w \tag{3–23}$$

where w is a downward-acting load of w units of force per unit length.

The sign conventions used for bending moment and shear force in this book are shown in Fig. 3–9.

The loading w of Eq. (3–23) is uniformly distributed. A more general distribution can be defined by the equation

$$q = \lim_{\Delta x \to 0} \frac{\Delta F}{\Delta x}$$

where q is called the *load intensity;* thus $q = -w$.

Equations (3–22) and (3–23) reveal additional relations if they are integrated. Thus, if we integrate between, say, x_A and x_B, we obtain

$$\int_{V_A}^{V_B} dV = \int_{x_A}^{x_B} q\,dx = V_B - V_A \tag{3–24}$$

Figure 3–9

Sign conventions for bending and shear.

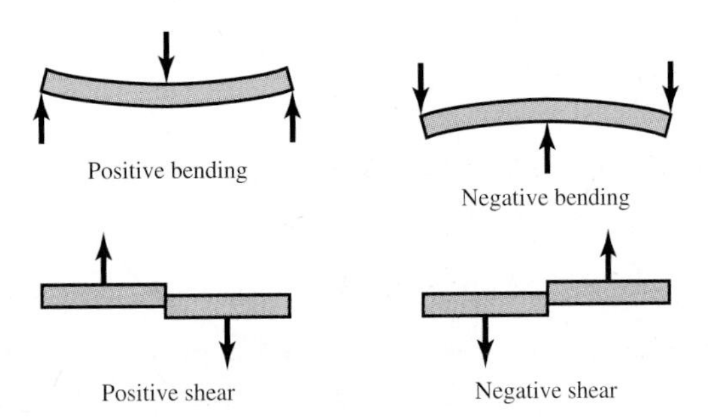

which states that *the change in shear force from A to B is equal to the area of the loading diagram between x_A and x_B.*

In a similar manner,

$$\int_{M_A}^{M_B} dM = \int_{x_A}^{x_B} V\,dx = M_B - M_A \tag{3-25}$$

which states that *the change in moment from A to B is equal to the area of the shear-force diagram between x_A and x_B.*

3–9 Singularity Functions

The five singularity functions defined in Table 3–2 constitute a useful and easy means of integrating across discontinuities. By their use, general expressions for shear force and bending moment in beams can be written when the beam is loaded by concentrated

Table 3–2

Singularity and Macaulay Functions

Function	Graph of $f_n(x)$	Meaning
Concentrated moment (unit doublet)	$<x-a>^{-2}$	$\langle x-a\rangle^{-2} = 0 \quad x \neq a$ $\int_{-\infty}^{x} \langle x-a\rangle^{-2}\,dx = \langle x-a\rangle^{-1}$ $\langle x-a\rangle^{-2} = \pm\infty \quad x = a$
Concentrated force (unit impulse)	$<x-a>^{-1}$	$\langle x-a\rangle^{-1} = 0 \quad x \neq a$ $\int_{-\infty}^{x} \langle x-a\rangle^{-1}\,dx = \langle x-a\rangle^{0}$ $\langle x-a\rangle^{-1} = +\infty \quad x = a$
Unit step	$<x-a>^{0}$	$\langle x-a\rangle^{0} = \begin{cases} 0 & x < a \\ 1 & x \geq a \end{cases}$ $\int_{-\infty}^{x} \langle x-a\rangle^{0}\,dx = \langle x-a\rangle^{1}$
Ramp	$<x-a>^{1}$	$\langle x-a\rangle^{1} = \begin{cases} 0 & x < a \\ x-a & x \geq a \end{cases}$ $\int_{-\infty}^{x} \langle x-a\rangle^{1}\,dx = \frac{\langle x-a\rangle^{2}}{2}$
Parabolic	$<x-a>^{2}$	$\langle x-a\rangle^{2} = \begin{cases} 0 & x < a \\ (x-a)^2 & x \geq a \end{cases}$ $\int_{-\infty}^{x} \langle x-a\rangle^{2}\,dx = \frac{\langle x-a\rangle^{3}}{3}$

moments or forces. As shown in the table, the concentrated moment and force functions are zero for all values of x not equal to a. The functions are undefined for values of $x = a$. Note that the unit step, ramp, and parabolic functions are zero only for values of x that are less than a. The integration properties shown in the table constitute a part of the mathematical definition too. The examples which follow show how these functions are used.

EXAMPLE 3–2 Derive expressions for the loading, shear-force, and bending-moment diagrams for the beam of Fig. 3–10.

Solution Using Table 3–2 and $q(x)$ for the loading function, we find

$$q = R_1\langle x\rangle^{-1} - F_1\langle x - a_1\rangle^{-1} - F_2\langle x - a_2\rangle^{-1} + R_2\langle x - l\rangle^{-1} \tag{1}$$

Next, we use Eq. (3–24) to get the shear force. Note that $V = 0$ at $x = -\infty$.

$$V = \int_{-\infty}^{x} q\,dx = R_1\langle x\rangle^{0} - F_1\langle x - a_1\rangle^{0} - F_2\langle x - a_2\rangle^{0} + R_2\langle x - l\rangle^{0} \tag{2}$$

A second integration, in accordance with Eq. (3–25), yields

$$M = \int_{-\infty}^{x} V\,dx = R_1\langle x\rangle^{1} - F_1\langle x - a_1\rangle^{1} - F_2\langle x - a_2\rangle^{1} + R_2\langle x - l\rangle^{1} \tag{3}$$

The reactions R_1 and R_2 can be found by taking a summation of moments and forces as usual, or they can be found by noting that the shear force and bending moment must be zero everywhere except in the region $0 \le x \le l$. This means that Eq. (2) should give $V = 0$ at x slightly larger than l. Thus

$$R_1 - F_1 - F_2 + R_2 = 0 \tag{4}$$

Since the bending moment should also be zero in the same region, we have, from Eq. (3),

$$R_1 l - F_1(l - a_1) - F_2(l - a_2) = 0 \tag{5}$$

Equations (4) and (5) can now be solved for the reactions R_1 and R_2.

Figure 3–10

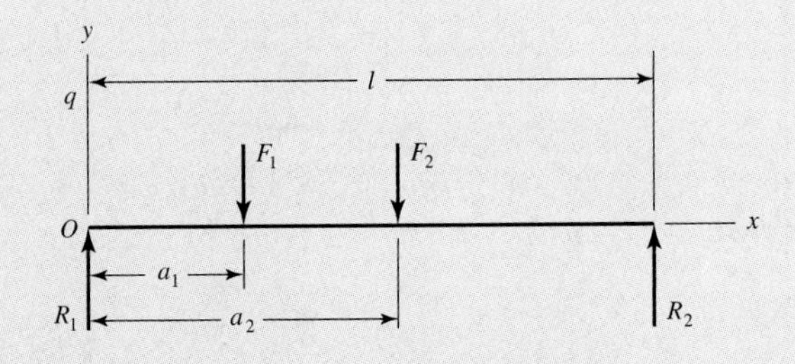

EXAMPLE 3–3 Figure 3–11a shows the loading diagram for a beam cantilevered at O and having a uniform load w acting on the portion $a \leq x \leq l$. Derive the shear-force and moment relations. M_1 and R_1 are the support reactions.

Solution Following the procedure of Example 3–2, we find the loading function to be

$$q = -M_1\langle x\rangle^{-2} + R_1\langle x\rangle^{-1} w\langle x - a\rangle^0 \tag{1}$$

Then integrating successively gives

$$V = \int_{-\infty}^{x} q\,dx = -M_1\langle x\rangle^{-1} + R_1\langle x\rangle^0 - w\langle x - a\rangle^1 \tag{2}$$

$$M = \int_{-\infty}^{x} V\,dx = -M_1\langle x\rangle^0 + R_1\langle x\rangle^1 - \frac{w}{2}\langle x - a\rangle^2 \tag{3}$$

The reactions are found by making x slightly larger than l because both V and M are zero in this region. Equation (2) will then give

$$-M_1(0) + R_1 - w(l - a) = 0 \tag{4}$$

which can be solved for R_1. From Eq. (3) we get

$$-M_1 + R_1 l - \frac{w}{2}(l - a)^2 = 0 \tag{5}$$

which can be solved for M_1. Figure 3–11b and c shows the shear-force and bending-moment diagrams.

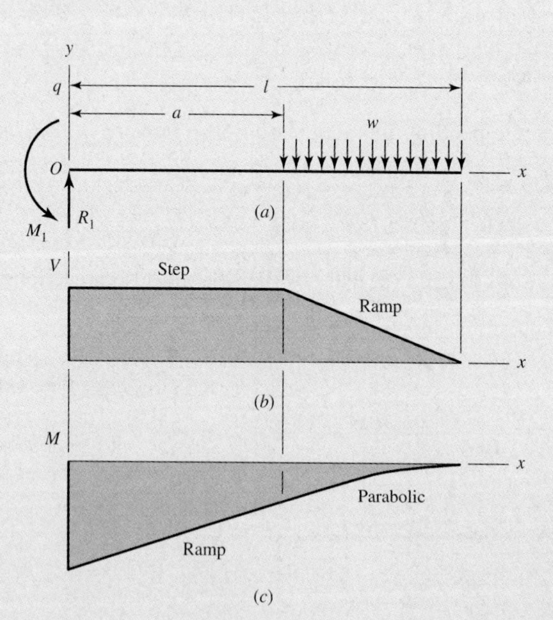

Figure 3–11

(a) Loading diagram for a beam cantilevered at O; (b) shear-force diagram; (c) bending-moment diagram.

CASE STUDY 3a Things Are Not Always What They Appear To Be

It was a cloudy spring night and the 100-car grain train was moving across gently rolling countryside along a gently rising and falling track grade, reflecting the economics of a rural branch line, as the slack was running in and out. Suddenly, the brakes went into emergency and, in the fail-safe fashion designed into railroad air brakes, both sections of the parted train came to rest without colliding. Investigation by the brakeman revealed that a coupler knuckle had dropped to the ground and the pin was missing. A spare knuckle pin was obtained from the "treasure chest" and with appropriate body and verbal coaxing of the conductor and the brakeman, the coupler knuckle was retrieved and reinstalled, then the spare pin was dropped in and cotter-pinned in place. A coupler knuckle pin is an approximately $1\frac{3}{8}$-in-diameter headed pin nearly 14 in long. The old pin could not be located in the dark. The brakeman said, "We sheared off a coupler pin." Do you think that's what happened?

You know enough to formulate an opinion. Look at the figure for Prob. 3–32. The mathematical model seen in part *c* of the figure can be shown (it is left to the student to do this) to give a bending moment at the first load of

$$M = \frac{F}{2}\left(\frac{a}{2} + \frac{b}{4}\right)$$

The bending stress at the first load is

$$\sigma = \frac{Mc}{I} = \frac{32M}{\pi d^3} = \frac{32F}{2\pi d^3}\left(\frac{a}{2} + \frac{b}{4}\right)$$

The largest shear stress in the pivot plane is

$$\tau = K\frac{V}{A} = \frac{4}{3}\frac{F/2}{\pi d^2/4} = \frac{8F}{3\pi d^2}$$

The ratio of bending stress to shear stress $\frac{\sigma}{\tau}$ is expressed as

$$\frac{\sigma}{\tau} = \frac{32F}{2\pi d^3}\left(\frac{a}{2} + \frac{b}{4}\right)\frac{3\pi d^2}{8F} = \frac{6}{d}\left(\frac{a}{2} + \frac{b}{4}\right)$$

For the conditions of Prob. 3–32 the ratio of $\frac{\sigma}{\tau}$ is approximately 5, meaning the normal stress due to bending exceeds the transverse shear stress about five fold.

The next day the two parts of the pin were located. The plane of failure was near the upper pivot face. The failure was a *classic bending fatigue fracture*. The bottom of the pin fell clear and the top part, which had to jump several inches up, was found on the ground about 10 ft to the rear of the bottom fragment.

We have yet to investigate the relative strengths in bending and shear, and we shall return to this case later. We have gone far enough to recognize that bending is the threat. With pins, do not forget bending. The brakeman was inaccurate but solved the problem satisfactorily.

3–10 Normal Stresses in Flexure

In deriving the relations for the normal bending stresses in beams, we make the following assumptions:

1 The beam is subjected to pure bending; this means that the shear force is zero, and that no torsion or axial loads are present.

2 The material is isotropic and homogeneous.
3 The material obeys Hooke's law.
4 The beam is initially straight with a cross section that is constant throughout the beam length.
5 The beam has an axis of symmetry in the plane of bending.
6 The proportions of the beam are such that it would fail by bending rather than by crushing, wrinkling, or sidewise buckling.
7 Cross sections of the beam remain plane during bending.

In Fig. 3–12*a* we visualize a portion of a beam acted upon by the positive bending moment M. The y axis is the axis of symmetry. The x axis is coincident with the *neutral axis* of the section, and the xz plane, which contains the neutral axes of all cross sections, is called the *neutral plane*. Elements of the beam coincident with this plane have zero strain. The location of the neutral axis with respect to the cross section has not yet been defined.

Application of the positive moment will cause the upper surface of the beam to bend downward, and the neutral axis will then be curved, as shown in Fig. 3–12b. Because of the curvature, a section AB originally parallel to CD, since the beam was straight, will rotate through the angle $d\phi$ to $A'B'$. Since AB and $A'B'$ are both straight lines, we have utilized the assumption that plane sections remain plane during bending. If we now specify the radius of curvature of the neutral axis as ρ, the length of a differential element of the neutral axis as ds, and the angle subtended by the two adjacent sides CD and $A'B'$ as $d\phi$, then, from the definition of curvature, we have

$$\frac{1}{\rho} = \frac{d\phi}{ds} \tag{a}$$

As shown in Fig. 3–12*b*, the elongation of a "fiber" at distance y from the neutral axis is

$$dx = y\, d\phi \tag{b}$$

The strain is the elongation divided by the original length, or

$$\epsilon = -\frac{dx}{ds} \tag{c}$$

Figure 3–12

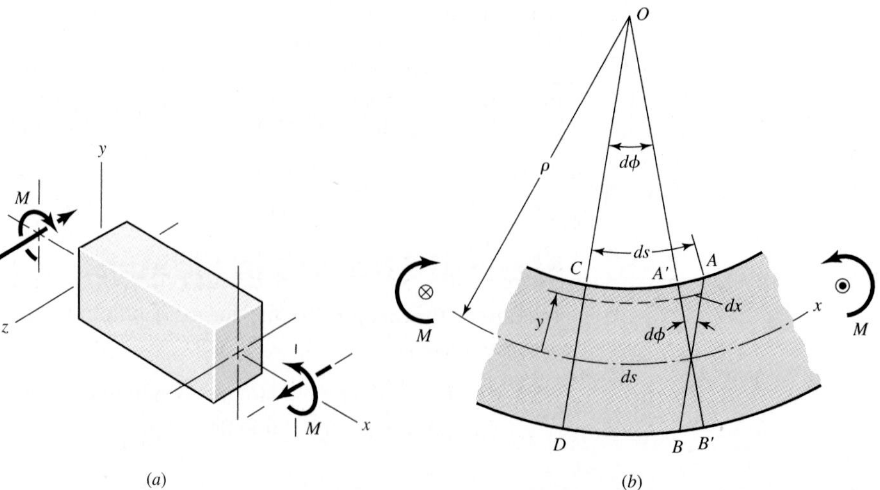

where the negative sign indicates compression. Solving Eqs. (a), (b), and (c) simultaneously gives

$$\epsilon = -\frac{y}{\rho} \tag{d}$$

Thus the strain is proportional to the distance y from the neutral axis. Now, since $\sigma = E\epsilon$, we have for the stress

$$\sigma = -\frac{Ey}{\rho} \tag{e}$$

We are now dealing with pure bending, which means that there are no axial forces acting on the beam. We can state this in mathematical form by summing all the horizontal forces acting on the cross section and equating this sum to zero. The force acting on an element of area dA is $\sigma\, dA$; therefore

$$\int \sigma\, dA = -\frac{E}{\rho}\int y\, dA = 0 \tag{f}$$

Equation (f) defines the location of the neutral axis. The moment of the area about the neutral axis is zero, and hence the neutral axis passes through the centroid of the cross-sectional area.

Next we observe that equilibrium requires that the internal bending moment created by the stress σ be the same as the external moment M. In other words,

$$M = \int y\sigma\, dA = \frac{E}{\rho}\int y^2\, dA \tag{g}$$

The second integral in Eq. (g) is the *second moment of area* about the z axis. This is

$$I = \int y^2\, dA \tag{3–26}$$

If we next solve Eqs. (g) and (3–26) and rearrange them, we have

$$\frac{1}{\rho} = \frac{M}{EI} \tag{3–27}$$

This is an important equation in the determination of the deflection of beams, and we shall employ it in Chap. 4. Finally, we eliminate ρ from Eqs. (e) and (3–27) and obtain

$$\sigma = -\frac{My}{I} \tag{3–28}$$

Equation (3–28) states that the bending stress σ is directly proportional to the distance y from the neutral axis and the bending moment M, as shown in Fig. 3–13. It is customary to designate $c = y_{\max}$, to omit the negative sign, and to write

$$\sigma = \frac{Mc}{I} \tag{3–29}$$

Figure 3–13

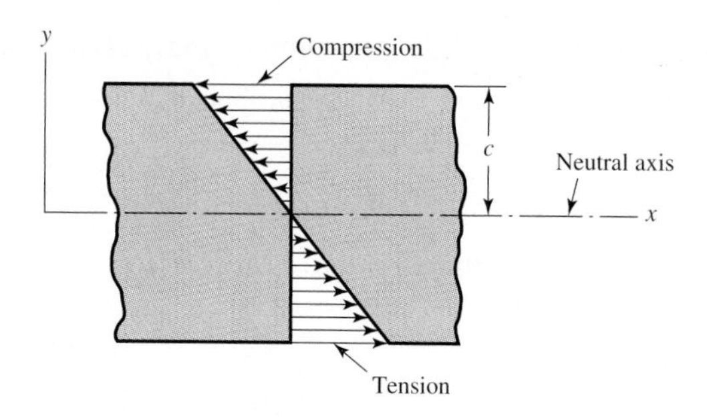

where it is understood that Eq. (3–29) gives the maximum stress. Then tensile or compressive maximum stresses are determined by inspection when the sense of the moment is known.

Equation (3–29) is often written in the two alternative forms

$$\sigma = \frac{M}{I/c} \qquad \sigma = \frac{M}{Z} \tag{3–30}$$

where $Z = I/c$ is called the *section modulus*.

EXAMPLE 3–4

A beam having a T section with the dimensions shown in Fig. 3–14 is subjected to a bending moment of 1600 N · m. Locate the neutral axis and find the maximum tensile and compressive bending stresses. The sign of the moment produces tension at the top surface.

Solution

The area of the composite section is $A = 1956\ \text{mm}^2$. Now divide the T section into two rectangles, numbered 1 and 2, and sum the moments of these areas about the top edge. We then have

$$1956c_1 = 12(75)(6) + 12(88)(56)$$

and hence $c_1 = 33.0$ mm. Therefore $c_2 = 100 - 33.0 = 67.0$ mm.

Next we calculate the second moment of area of each rectangle about its own centroidal axis. Using Table E–18, we find for the top rectangle

$$I_1 = \frac{bh^3}{12} = \frac{7.5(1.2)^3}{12} = 1.08\ \text{cm}^4$$

Figure 3–14

Dimensions in millimeters.

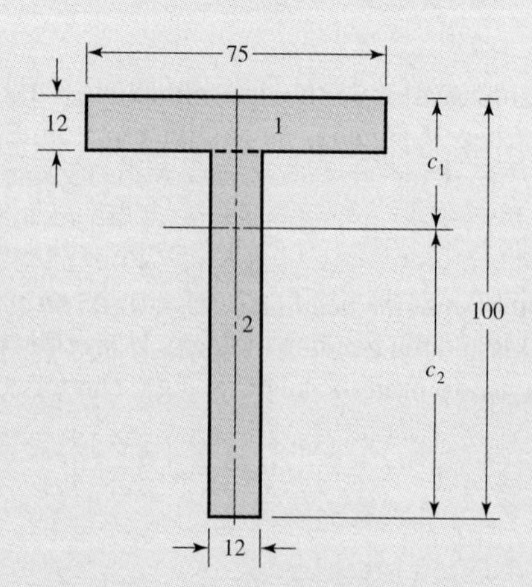

As indicated in Sec. 1–14, it is permissible to use nonpreferred prefixes when raising prefixed units to a power. In this case the use of cm in this equation instead of mm or m yields the shortest number string and, hence, is preferred. For the bottom rectangle, we have

$$I_2 = \frac{bh^3}{12} = \frac{1.2(8.8)^3}{12} = 68.15 \text{ cm}^4$$

We now employ the parallel-axis theorem to obtain the second moment of area of the composite figure about its own centroidal axis. This theorem states

$$I_x = I_{cg} + Ad^2$$

where I_{cg} is the second moment of area about its own centroidal axis and I_x is the second moment of area about any parallel axis a distance d removed. For the top rectangle, the distance is

$$d_1 = 33.0 - 6 = 27.0 \text{ mm}$$

and for the bottom rectangle,

$$d_2 = 67.0 - 44 = 23.0 \text{ mm}$$

Using the parallel-axis theorem twice, we now find that

$$\begin{aligned} I &= [1.08 + 9(2.70)^2] + [68.15 + 10.56(2.30)^2] \\ &= 190.7 \text{ cm}^4 \end{aligned}$$

Finally, then, the maximum tensile stress, which occurs at the top surface, is found to be

$$\sigma = \frac{Mc_1}{I} = \frac{1600(3.30)}{190.7} = 27.69 \text{ MPa}$$

Note that we have used Table E–3 in solving this equation. Similarly, the maximum compressive stress at the lower surface is found to be

Answer

$$\sigma = -\frac{Mc_2}{I} = -\frac{1600(6.70)}{190.7} = -56.21 \text{ MPa}$$

EXAMPLE 3–5 Determine the diameter for the solid round shaft, 18 in long, shown in Fig. 3–15*a*. The shaft is supported by self-aligning bearings at the ends. Mounted upon the shaft are a V-belt sheave, which contributes a radial load of 400 lb to the shaft, and a gear, which contributes a radial load of 150 lb. The two loads are in the same plane and have the same directions. The bending stress is not to exceed 10 kpsi.

Solution

As indicated in Chap. 1, certain assumptions are necessary. In this problem we decide on the following conditions:

- The weight of the shaft is neglected.

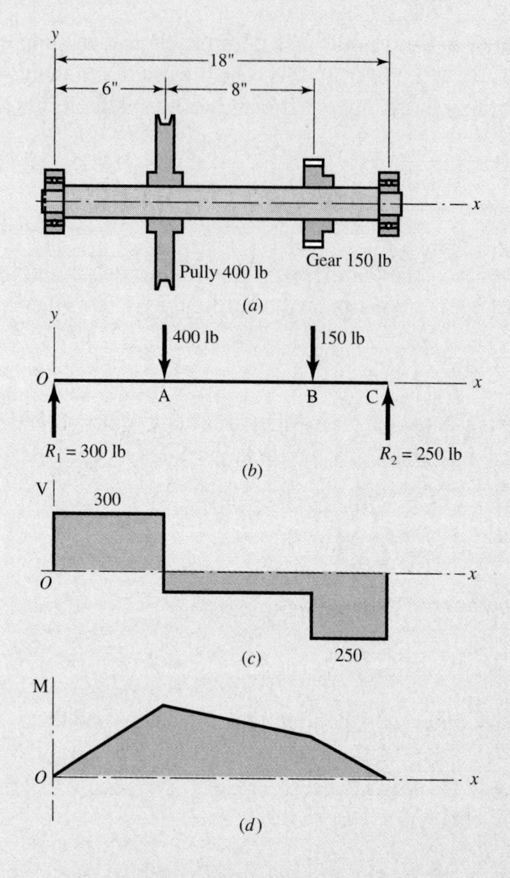

Figure 3–15

(a) Shaft drawing, dimensions in inches; (b) loading diagram; (c) shear-force diagram; (d) bending-moment diagram.

- Since the bearings are self-aligning, the shaft is assumed to be simply supported and the loads and bearing reactions to be concentrated.
- The normal bending stress is assumed to govern the design at the location of the maximum bending moment.

The assumed loading diagram is shown in Fig. 3–15*b*. Using Eq. (3–21) by taking moments about C gives

$$\Sigma M_C = -18R_1 + 12(400) + 4(150) = 0$$
$$R_1 = 300 \text{ lb}$$

To get R_2, we can sum forces using Eq. (3–20), or we can use Eq. (3–21) again for point O. With either method, we find

$$R_2 = 250 \text{ lb}$$

This statics problem can also be solved by vectors. We define the unit vector triad **ijk** in association with the x, y, and z axes, respectively. Then position vectors from the origin O to loading points A, B, and C are

$$\mathbf{r}_A = 6\mathbf{i} \qquad \mathbf{r}_B = 14\mathbf{i} \qquad \mathbf{r}_C = 18\mathbf{i}$$

Also, the force vectors are

$$\mathbf{R}_1 = R_1\mathbf{j} \qquad \mathbf{F}_A = -400\mathbf{j} \qquad \mathbf{F}_B = -150\mathbf{j} \qquad \mathbf{R}_2 = R_2\mathbf{j}$$

Then taking moments about O yields the vector equation

$$\mathbf{r}_A \times \mathbf{F}_A + \mathbf{r}_B \times \mathbf{F}_B + \mathbf{r}_C \times \mathbf{R}_2 = \mathbf{0} \tag{1}$$

The terms in Eq. (1) are called *vector cross products*. The cross product of the two vectors

$$\mathbf{A} = x_A\mathbf{i} + y_A\mathbf{j} + z_A\mathbf{k} \qquad \mathbf{B} = x_B\mathbf{i} + y_B\mathbf{j} + z_B\mathbf{k}$$

is

$$\mathbf{A} \times \mathbf{B} = (y_A z_B - z_A y_B)\mathbf{i} + (z_A x_B - x_A z_B)\mathbf{j} + (x_A y_B - y_A x_B)\mathbf{k} \tag{3-31}$$

But this relation is more conveniently viewed as the determinant

$$\mathbf{A} \times \mathbf{B} = \begin{vmatrix} \mathbf{i} & \mathbf{j} & \mathbf{k} \\ x_A & y_A & z_A \\ x_B & y_B & z_B \end{vmatrix} \tag{3-32}$$

Thus we compute the terms in Eq. (1) as follows:

$$\mathbf{r}_A \times \mathbf{F}_A = \begin{vmatrix} \mathbf{i} & \mathbf{j} & \mathbf{k} \\ 6 & 0 & 0 \\ 0 & -400 & 0 \end{vmatrix} = -2400\mathbf{k}$$

$$\mathbf{r}_B \times \mathbf{F}_B = \begin{vmatrix} \mathbf{i} & \mathbf{j} & \mathbf{k} \\ 14 & 0 & 0 \\ 0 & -150 & 0 \end{vmatrix} = -2100\mathbf{k}$$

$$\mathbf{r}_C \times \mathbf{R}_2 = \begin{vmatrix} \mathbf{i} & \mathbf{j} & \mathbf{k} \\ 18 & 0 & 0 \\ 0 & R_2 & 0 \end{vmatrix} = 18R_2\mathbf{k}$$

Substituting these three terms into Eq. (1) and solving the resulting algebraic equation gives $R_2 = 250$ lb. Hence $\mathbf{R}_2 = 250\mathbf{j}$. Next, we write

$$\mathbf{R}_1 = -\mathbf{F}_A - \mathbf{F}_B - \mathbf{R}_2 = -(-400\mathbf{j}) - (-150\mathbf{j}) - 250\mathbf{j} = 300\mathbf{j} \text{ lb}$$

Though the vector approach does seem more laborious, it is easy to program and is especially useful for three-dimensional problems.

The next step is to draw the shear-force and bending-moment diagrams (Fig. 3–15*c* and *d*). The maximum bending moment is found to be

$$M = 300(6) = 1800 \text{ lb} \cdot \text{in}$$

The section modulus is

$$\frac{I}{c} = \frac{\pi d^3}{32} = 0.0982d^3 \tag{2}$$

Then, using Eq. (3-30),

$$\sigma = \frac{M}{I/c} = \frac{1800}{0.0982d^3}$$

Substituting $\sigma = 10\,000$ psi and solving for d yields

$$d = \sqrt[3]{\frac{1800}{0.0982(10000)}} = 1.22 \text{ in}$$

Therefore we select $d = 1\frac{1}{4}$ in for the shaft diameter.

3–11 Beams with Asymmetrical Sections

The relations developed in Sec. 3–10 can also be applied to beams having asymmetrical sections, provided that the plane of bending coincides with one of the two principal axes of the section. We have found that the stress at a distance y from the neutral axis is

$$\sigma = -\frac{Ey}{\rho} \tag{a}$$

Therefore, the force on the element of area dA in Fig. 3–16 is

$$dF = \sigma\, dA = -\frac{Ey}{\rho}\, dA$$

Taking moments of this force about the y axis and integrating across the section gives

$$M_y = \int z\, dF = \int \sigma z\, dA = -\frac{E}{\rho}\int yz\, dA \tag{b}$$

We recognize that the last integral in Eq. (*b*) is the product of inertia I_{yz}. If the bending moment on the beam is in the plane of one of the principal axes, then

$$I_{yz} = \int yz\, dA = 0 \tag{c}$$

With this restriction, the relations developed in Sec. 3–10 hold for any cross-sectional shape. Of course, this means that the designer has a special responsibility to ensure that the bending loads do, in fact, come onto the beam in a principal plane!

3–12 Shear Stresses in Beams

Most beams have *both* shear forces and bending moments present. It is only occasionally that we encounter beams subjected to pure bending, that is to say, beams having zero shear force. And yet, the flexure formula was developed using the assumption of pure bending. As a matter of fact, the reason for assuming pure bending was simply to eliminate the

Figure 3–16

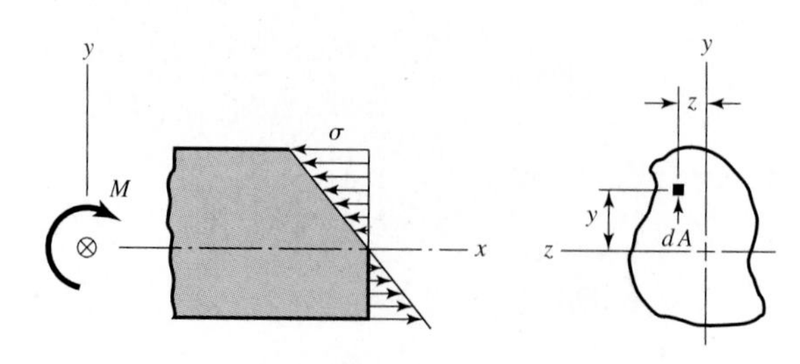

complicating effects of shear force in the development. For engineering purposes, the flexure formula is valid no matter whether a shear force is present or not. For this reason, we shall utilize the same normal bending-stress distribution [Eqs. (3–28) and (3–29)] when shear forces are present too.

In Fig. 3–17, we show a beam of constant cross section subjected to a shear force V and a bending moment M. The direction of the bending moment is easier to visualize by associating the hollow vector with your right hand. The hollow vector points in the negative z direction. If you will place the thumb of your right hand in the negative z direction, then your fingers, when bent, will indicate the direction of the moment M. By Eq. (3–22), the relationship of V to M is

$$V = \frac{dM}{dx} \tag{a}$$

At some point along the beam, we cut a transverse section of length dx down to a distance y_1 above the neutral axis, as illustrated. We remove this section to study the forces which act. Because a shear force is present, the bending moment is changing as we move along the x axis. Thus, we can designate the bending moment as M on the near side of the section and as $M + dM$ on the far side. The moment M produces a normal stress σ, and the moment $M + dM$, a normal stress $\sigma + d\sigma$, as shown. These normal stresses produce normal forces on the vertical faces of the element, the compressive force on the far side being greater than on the near side. The resultant of these two would cause the section to tend to slide in the $-x$ direction, and so this resultant must be balanced by a shear force acting in the $+x$ direction on the bottom of the section. This shear force results in a shear stress τ, as shown. Thus, there are three resultant forces acting on the element: F_N, due to σ, acts on the near face; F_F, due to $\sigma + d\sigma$, acts on the far face; and F_B, due to τ, acts on the bottom face. Let us evaluate these forces.

For the near face, select an element of area dA. The stress acting on this area is σ, and so the force is the stress times the area, or $\sigma\, dA$. The force acting on the entire near

Figure 3–17

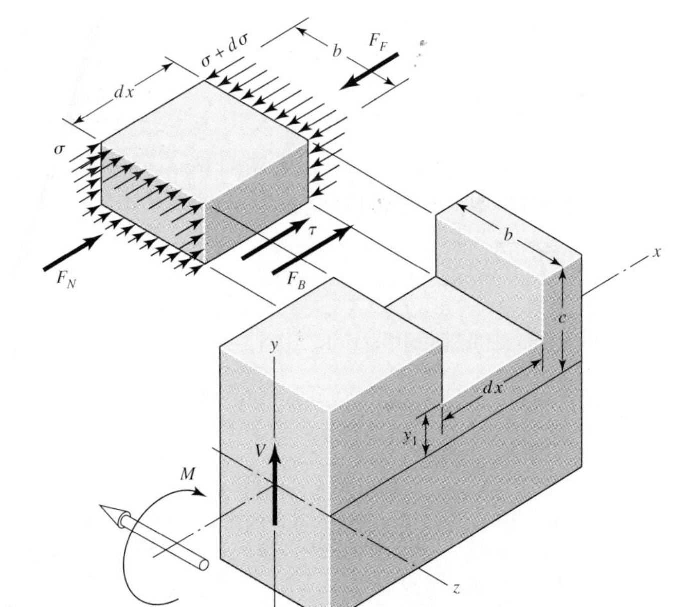

face is the sum of all the $\sigma\, dA$'s, or

$$F_N = \int_{y_1}^{c} \sigma\, dA \tag{b}$$

where the limits indicate that we integrate from the bottom $y = y_1$ to the top $y = c$. Using $\sigma = My/I$ [Eq. (3-28)], Eq. (*b*) becomes

$$F_N = \frac{M}{I} \int_{y_1}^{c} y\, dA \tag{c}$$

The force on the far face is found in a similar manner. It is

$$F_F = \int_{y_1}^{c} (\sigma + d\sigma)\, dA = \frac{M + dM}{I} \int_{y_1}^{c} y\, dA \tag{d}$$

The force on the bottom face is the shear stress τ times the area of the bottom face. Since this area is $b\, dx$, we have

$$F_B = \tau_b\, dx \tag{e}$$

Summing these three forces in the x direction gives

$$\Sigma F_x = +F_N - F_F + F_B = 0 \tag{f}$$

If we substitute Eqs. (*c*) and (*d*) for F_N and F_F and solve the result for F_B, we get

$$F_B = F_F - F_N = \frac{M + dM}{I} \int_{y_1}^{c} y\, dA - \frac{M}{I} \int_{y_1}^{c} y\, dA = \frac{dM}{I} \int_{y_1}^{c} y\, dA \tag{g}$$

Next, using Eq. (*e*) for F_B and solving for the shear stress gives

$$\tau = \frac{dM}{dx} \frac{1}{Ib} \int_{y_1}^{c} y = dA \tag{h}$$

Then, by the use of Eq. (*a*), we finally get the shear-stress formula as

$$\tau = \frac{V}{Ib} \int_{y_1}^{c} y\, dA \tag{3–33}$$

In this equation, the integral is the first moment of the area of the vertical face about the neutral axis. This moment is usually designated Q. Thus,

$$Q = \int_{y_1}^{c} y\, dA \tag{3–34}$$

With this final simplification, Eq. (3-33) may be written as

$$\boxed{\tau = \frac{VQ}{Ib}} \tag{3–35}$$

In using this equation, note that b is the width of the section at the particular distance y_1 from the neutral axis. Also, I is the second moment of area of the entire section about the neutral axis.

3–13 Shear Stresses in Rectangular-Section Beams

The purpose of this section is to show how the equations of the preceding section are used to find the shear-stress distribution in a beam having a rectangular cross section. Figure 3–18 shows a portion of a beam subjected to a shear force V and a bending moment M. As a result of the bending moment, a normal stress σ is developed on a cross section such as A-A, which is in compression above the neutral axis and in tension below. To investigate the shear stress at a distance y_1 above the neutral axis, we select an element of area dA at a distance y above the neutral axis. Then, $dA = b\,dy$, and so Eq. (3–34) becomes

$$Q = \int_{y_1}^{c} y\,dA = b\int_{y_1}^{c} y\,dy = \frac{by^2}{2}\bigg|_{y_1}^{c} = \frac{b}{2}(c^2 - y_1^2) \tag{a}$$

Substituting this value for Q into Eq. (3-35) gives

$$\tau = \frac{V}{2I}(c^2 - y_1^2) \tag{3–36}$$

This is the general equation for shear stress in a rectangular beam. To learn something about it, let us make some substitutions. From Table E–18, we learn that the second moment of area for a rectangular section is $I = bh^3/12$; substituting $h = 2c$ and $A = bh = 2bc$ gives

$$I = \frac{Ac^2}{3} \tag{b}$$

If we now use this value of I for Eq. (3–36) and rearrange, we get

$$\tau = \frac{3V}{2A}\left(1 - = \frac{y_1^2}{c^2}\right) \tag{3–37}$$

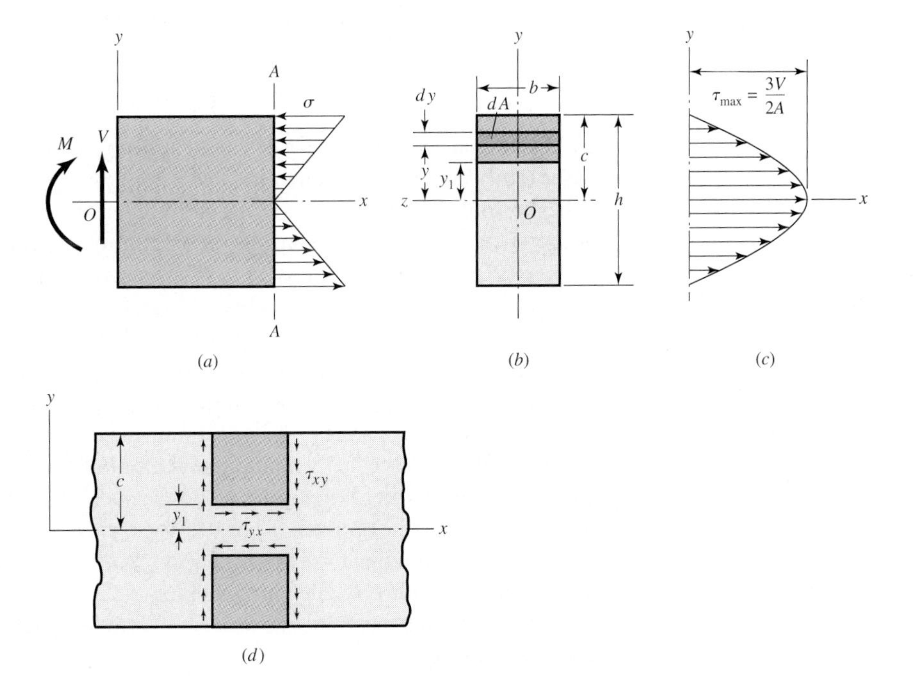

Figure 3–18

Note in part d that τ_{xy} denote the vertical shear, and that τ_{yx} denotes the horizontal shear.

Table 3–3

Variation of Shear Stress $\tau = C(V/A)$

Distance y_1	0	0.2c	0.4c	0.6c	0.8c	c
Factor C	1.50	1.44	1.26	0.96	0.54	0

Table 3–4

Formulas for Maximum Shear Stress Due to Bending

Beam Shape	Formula	Beam Shape	Formula
Rectangular	$\tau_{max} = \frac{3V}{2A}$	Hollow round	$\tau_{max} = \frac{2V}{A}$
Circular	$\tau_{max} = \frac{4V}{3A}$	Structural (Web)	$\tau_{max} = \frac{V}{A_{web}}$

Now let us substitute various values of y_1, beginning with $y_1 = 0$ and ending with $y_1 = c$. The results are displayed in Table 3–3. We note that the maximum shear stress exists with $y_1 = 0$, which is at the neutral axis. Thus

$$\tau_{max} = \frac{3V}{2A} \tag{3–38}$$

for a rectangular section. As we move away from the neutral axis, the shear stress decreases until it is zero at the outer surface where $y_1 = c$. This is a parabolic distribution and is shown in Fig. 3–18*c*. It is particularly interesting and significant here to observe that the shear stress is maximum at the neutral axis, where the normal stress due to bending is zero, and that the shear stress is zero at the outer surfaces, where the bending stress is a maximum. Horizontal shear stress is always accompanied by vertical shear stress of the same magnitude, and so the distribution can be diagrammed as shown in Fig. 3–18*d*. Figure 3–18*c* shows that the shear τ_{xy} on the vertical surfaces varies with y. We are almost always interested in the horizontal shear, τ_{yx} in Fig. 2–18*d*, which is nearly uniform with constant y. The maximum horizontal shear occurs where the vertical shear is largest. This is usually at the neutral axis but may not be if the width b is smaller somewhere else. Furthermore, if the section is such that b can be minimized on a plane not horizontal, then the horizontal shear stress occurs on an inclined plane. For example, with tubing, the horizontal shear stress occurs on a radial plane and the corresponding "vertical shear" is not vertical, but tangential.

Formulas for the maximum flexural shear stress for the most commonly used shapes are listed in Table 3–4.

3–14 Torsion

Any moment vector that is collinear with an axis of a mechanical element is called a *torque vector*, because the moment causes the element to be twisted about that axis. A bar subjected to such a moment is also said to be in *torsion*.

As shown in Fig. 3–19, the torque T applied to a bar can be designated by drawing arrows on the surface of the bar to indicate direction or by drawing torque-vector arrows along the axes of twist of the bar. Torque vectors are the hollow arrows shown on the x axis in Fig. 3–19. Note that they conform to the right-hand rule for vectors.

The *angle of twist* for a solid round bar is

$$\theta = \frac{Tl}{GJ} \tag{3–39}$$

where T = torque
l = length
G = modulus of rigidity
J = polar second moment of area

For a solid round bar, the shear stress is zero at the center and maximum at the surface. The distribution is proportional to the radius ρ and is

$$\tau = \frac{T\rho}{J} \tag{3–40}$$

Designating r as the radius to the outer surface, we have

$$\tau_{\text{max}} = \frac{Tr}{J} \tag{3–41}$$

The assumptions used in the analysis are:

- The bar is acted upon by a pure torque, and the sections under consideration are remote from the point of application of the load and from a change in diameter.
- Adjacent cross sections originally plane and parallel remain plane and parallel after twisting, and any radial line remains straight.
- The material obeys Hooke's law.

Figure 3–19

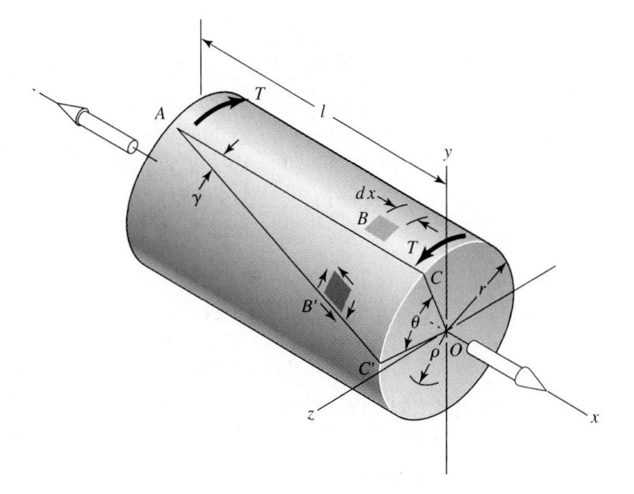

Equation (3–41) applies only to circular sections. For a solid round section,

$$J = \frac{\pi d^4}{32} \tag{3–42}$$

where d is the diameter of the bar. For a hollow round section,

$$J = \frac{\pi}{32}(d_o^4 - d_i^4) \tag{3–43}$$

where the subscripts o and i refer to the outside and inside diameters, respectively (OD and ID).

In using Eq. (3–41) it is often necessary to obtain the torque T from a consideration of the power and speed of a rotating shaft. For convenience, three forms of this relation are

$$H = \frac{FV}{33\,000} = \frac{2\pi Tn}{33\,000(12)} = \frac{\pi Tn}{198\,000} \doteq \frac{Tn}{63\,025} \tag{3–44}$$

where H = power, hp
T = torque, lb · in
n = shaft speed, r/min
F = force, lb
V = velocity, ft/min

When SI units are used, the equation is

$$H = T\omega \tag{3–45}$$

where H = power, W
T = torque, N · m
ω = angular velocity, rad/s

The torque T corresponding to the power in watts is given approximately by

$$T = 9.55\frac{H}{n} \tag{3–46}$$

where n is in revolutions per minute.

There are some applications in machinery for noncircular cross-section members and shafts where a regular polygonal cross section is useful in transmitting torque to a gear or pulley that can have an axial change in position. Because no key or keyway is needed, the possibility of a list key is avoided. Saint Venant (1855) showed that the maximum shearing stress in a rectangular $b \times c$ section bar occurs in the middle of the *longest* side b and is of the magnitude

$$\tau_{\max} = \frac{T}{\alpha bc^2} \doteq \frac{T}{bc^2}\left(3 + \frac{1.8}{b/c}\right) \tag{3–47}$$

where b is the longer side, c the shorter side, and α a factor which is a function of the ratio b/c as shown in the table[5] below. The angle of twist is of the form $\theta = Tl/C$, where C is the torsional rigidity of the shaft. For a circular shaft $C = GJ$ and for a

5. S. Timoshenko, *Strength of Materials*, Part I, 3rd ed., D. Van Nostrand Company, New York, 1955, p. 290.

rectangular shaft $C = \beta bc^3 G$, where β is a function of b/c, as shown in the table. For a rectangular section shaft,

$$\theta = \frac{Tl}{\beta bc^3 G} \tag{3-48}$$

b/c	1.00	1.50	1.75	2.00	2.50	3.00	4.00	6.00	8.00	10	∞
α	0.208	0.231	0.239	0.246	0.258	0.267	0.282	0.299	0.307	0.313	0.333
β	0.141	0.196	0.214	0.228	0.249	0.263	0.281	0.299	0.307	0.313	0.333

In Eqs. (3-47) and (3-48) b and c are the width (long side) and thickness (short side) of the bar, respectively. They cannot be interchanged. Equation (3-47) is also approximately valid for equal-sided angles; these can be considered as two rectangles, each of which is capable of carrying half the torque.[6]

EXAMPLE 3–6

Figure 3-20 shows a crank loaded by a force $F = 300$ lb which causes twisting and bending of a $\frac{3}{4}$-in-diameter shaft fixed to a support at the origin of the reference system. In actuality, the support may be an inertia which we wish to rotate, but for the purposes of a stress analysis we can consider this as a statics problem.

(*a*) Draw separate free-body diagrams of the shaft AB and the arm BC, and compute the values of all forces, moments, and torques which act. Label the directions of the coordinate axes on these diagrams.

(*b*) Compute the maxima of the torsional stress and the bending stress in the arm BC and indicate where these act.

(*c*) Locate a stress element on the top surface of the shaft at A, and calculate all the stress components which act upon this element.

Figure 3–20

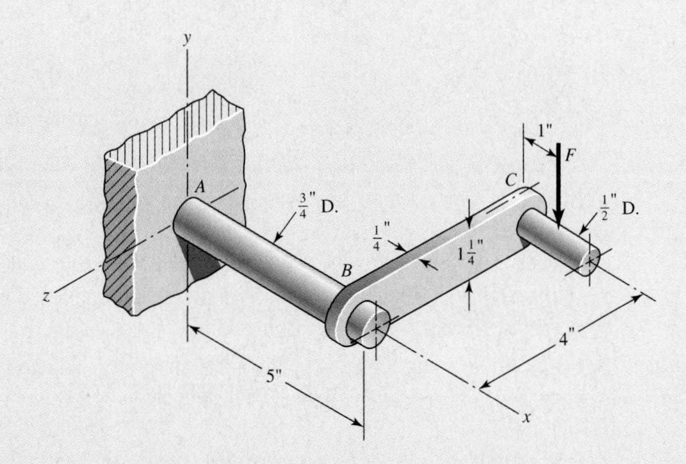

6. For other sections see W. C. Young, *Roark's Formulas for Stress and Strain*, 6th ed., McGraw-Hill, New York, 1989.

Solution (*a*) The two free-body diagrams are shown in Fig. 3–21. The results are

AT C: $\mathbf{F} = -300\mathbf{j}$ lb, $\mathbf{T} = -300\mathbf{k}$ lb · in

AT END B OF ARM BC: $\mathbf{F} = 300\mathbf{j}$ lb, $\mathbf{M} = 1200\mathbf{i}$ lb · in, $\mathbf{T} = 300\mathbf{k}$ lb · in

AT END B OF SHAFT AB: $\mathbf{F} = -300\mathbf{j}$ lb, $\mathbf{T} = -1200\mathbf{i}$ lb · in, $\mathbf{M} = -300\mathbf{k}$ lb · in

AT A: $\mathbf{F} = 300\mathbf{j}$ lb, $\mathbf{M} = 1800\mathbf{k}$ lb · in, $\mathbf{T} = 1200\mathbf{i}$ lb · in

(*b*) For arm BC, the bending stress will reach a maximum near the shaft at B. If we assume this is 1200 lb · in, then the bending stress for a rectangular section will be

Answer

$$\sigma = \frac{M}{I/c} = \frac{6M}{bh^2} = \frac{6(1200)}{0.25(1.25)^2} = 18\,400 \text{ psi}$$

Of course, this is not exactly correct, because at B the moment is actually being transferred into the shaft, probably through a weldment.

For the torsional stress, use Eq. (3–47). Thus

Answer

$$\tau_{\max} = \frac{T}{bc^2}\left(3 + \frac{1.8}{b/c}\right) = \frac{300}{1.25(0.25^2)}\left(3 + \frac{1.8}{1.25/0.25}\right) = 12\,900 \text{ psi}$$

This stress occurs at the middle of the $1\frac{1}{4}$-in side.

(*c*) For a stress element at A, the bending stress is

Answer

$$\sigma_x = \frac{M}{I/c} = \frac{32M}{\pi d^3} = \frac{32(1800)}{\pi(0.75)^3} = 43\,400 \text{ psi}$$

The torsional stress is

Answer

$$\tau_{xz} = \frac{T}{J/c} = \frac{16T}{\pi d^3} = \frac{16(1200)}{\pi(0.75)^3} = 14\,500 \text{ psi}$$

Figure 3–21

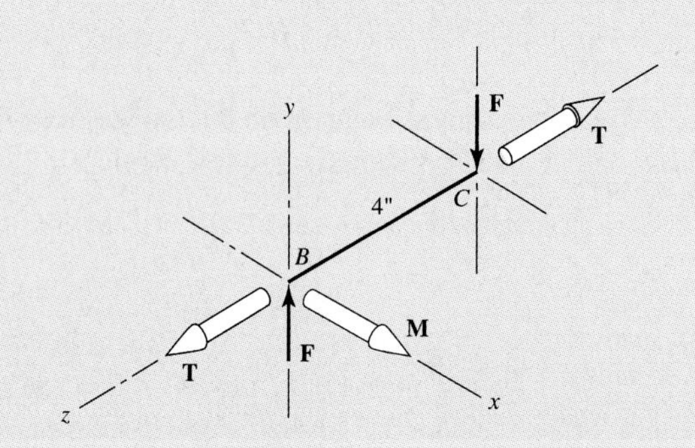

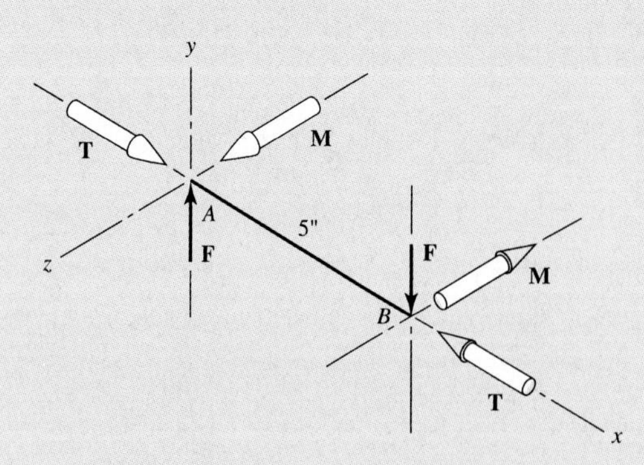

Figure 3–22

The depicted cross-section is elliptical, but the section need not be symmetrical.

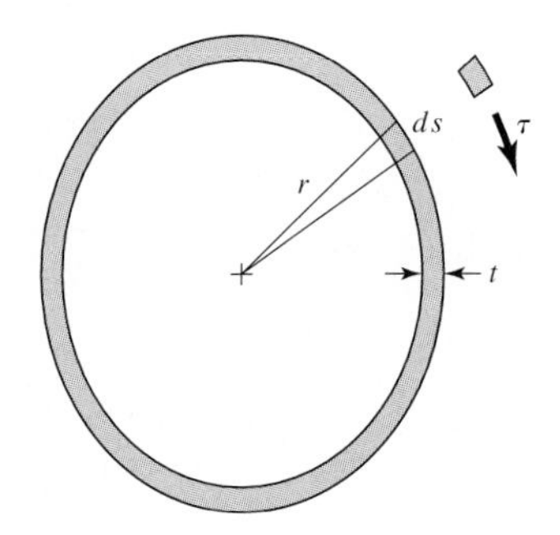

Closed Thin-Walled Tubes

In closed thin-walled tubes, it can be argued from membrane analogy theory that the product of shear stress times thickness of the wall τt is constant, and that the shear stress τ is inversely proportional to the wall thickness t. The total torque T on a tube such as depicted in Fig. 3–22 is given by

$$T = \int \tau t r\, ds = (\tau t) \int r\, ds = \tau t(2A) = 2At\tau$$

where A is the area enclosed by the section median line. Solving for τ gives

$$\tau = \frac{T}{2At} \tag{a}$$

The angular twist per unit of length of the tube θ_1 is given by

$$\theta_1 = \frac{TL}{4GA^2t} \tag{b}$$

where L is the perimeter of the section median line. For constant wall thickness t, Eqs. (a) and (b) can be combined in a useful way:

$$G\theta_1 t = \frac{TL}{4A^2} = \frac{Lt\tau}{2A} \tag{3–49}$$

These equations presume the buckling of the tube is prevented by ribs, stiffeners, bulkheads, and so on, and that the stresses are below the proportional limit.

EXAMPLE 3–7 A welded steel tube is 40 in long, has a $\frac{1}{8}$-in wall thickness, and a 2.5-in by 3.6-in rectangular cross section as shown in Fig. 3–23. Assume an allowable shear stress of 11 500 psi and a shear modulus of $11.5(10^6)$ psi.

(a) Estimate the allowable torque T.

(b) Estimate the angle of twist due to the torque.

Solution (a) Within the section median line, the area enclosed is

$$A = (2.5 - 0.125)(3.6 - 0.125) = 8.253 \text{ in}^2$$

and the length of the median perimeter is

$$L = 2[(2.5 - 0.125) + (3.6 - 0.125)] = 11.70 \text{ in}$$

Figure 3–23

A rectangular steel tube produced by welding.

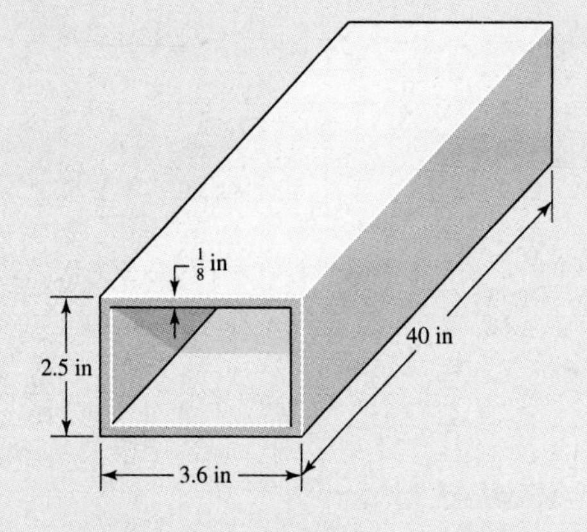

From Eq. (a) the torque T is

$$T = 2At\tau = 2(8.253)0.125(11\,500) = 23\,728 \text{ in} \cdot \text{lb}$$

(b) The angle of twist θ from Eq. (3–49) is

$$\theta = \theta_1 \ell = \frac{\tau L \ell}{2AG} = \frac{11\,500(11.70)40}{2(8.253)11.5(10^6)} = 0.0284 \text{ rad} = 1.63°$$

EXAMPLE 3–8

Compare the shear stress on a right circular cylindrical tube with an outside diameter of 1 in and an inside diameter of 0.9 in, predicted by Eq. (3–41), to that estimated by Eq. (3–49).

Solution

From Eq. (3–41),

$$\tau_{\max} = \frac{Tr}{J} = \frac{Tr}{(\pi/32)(d_o^4 - d_i^4)} = \frac{T(0.5)}{(\pi/32)(1^4 - 0.9^4)} = 14.809T$$

From Eq. (3.49),

$$\tau = \frac{T}{2At} = \frac{T}{2(\pi 0.95^2/4)0.05} = 14.108T$$

Taking Eq. (3–41) as correct, the proportional error in the thin-wall estimate is $(14.809 - 14.108)/14.809 = 0.047$.

Open Thin-Walled Sections

When the median wall line is not closed, it is said to be *open*. Figure 3–24 presents some examples. Open sections in torsion, where the wall is thin, have relations derived from

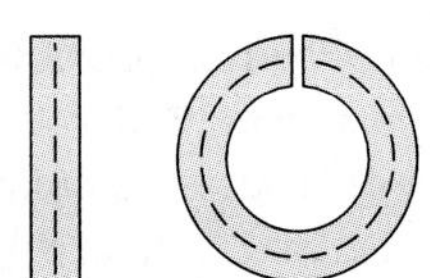
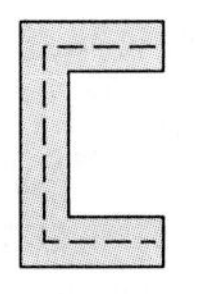
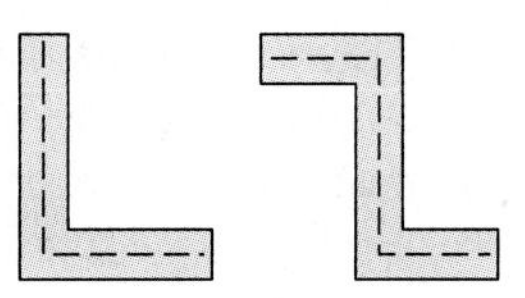

Figure 3–24

Some open thin-wall sections.

the membrane analysis as follows:

$$\tau = G\theta_1 c = \frac{3T}{Lc^2} \tag{3–50}$$

where τ is the shear stress, G is the shear modulus, θ_1 is the angle of twist per unit length, T is torque, and L is the length of the median line. The wall thickness is designated c (rather than t) to remind you that you are in open sections. By studying the table that follows Eq. (3–48) you will discover that membrane theory presumes $b/c \rightarrow \infty$.

EXAMPLE 3–9 A 12-in-long strip of steel is 1/16 in thick and 1 in wide, as shown in Fig. 3–25. If the allowable shear stress is 11 500 psi and the shear modulus is 11.5 (10⁶) psi, find the torque corresponding to the allowable shear stress and the angle of twist, in degrees, (*a*) using Eq. (3–50) and (*b*) using Eqs. (3–47) and (3–48).

Solution (*a*) The length of the median line is 1 in. From Eq. (3–50),

$$T = \frac{Lc^2\tau}{3} = \frac{(1)(1/16)^2 11\,500}{3} = 14.97 \text{ in} \cdot \text{lb}$$

$$\theta = \theta_1 \ell = \frac{3T\ell}{Lc^3G} = \frac{3(14.97)12}{(1)(1/16)^3 11.5(10^6)} = 0.192 \text{ rad} = 11°$$

A torsional spring rate k_t can be expressed as T/θ:

$$k_t = 14.97/01.92 = 77.97 \text{ in} \cdot \text{lb/rad}$$

(*b*) From Eq. (3–47),

$$T = \frac{\tau_{\max} bc^2}{3 + 1.8/(b/c)} = \frac{11\,500(1)(0.0625)^2}{3 + 1.8/(1/0.0625)} = 14.44 \text{ in} \cdot \text{lb}$$

From Eq. (3-48),

$$\theta = \frac{T\ell}{\beta bc^3 G} = \frac{T\ell}{0.319(1)0.0625^3(11.5)10^6} = 0.1935 \text{ rad } 11.1°$$

$$k_t = 14.44/0.1935 = 74.63 \text{ in} \cdot \text{lb/rad}$$

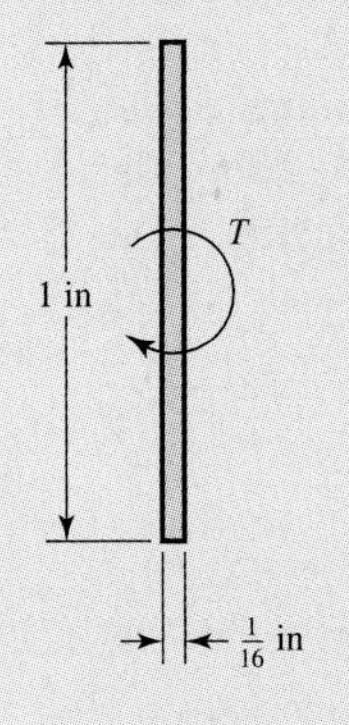

Figure 3–25

The cross-section of a thin strip of steel which is subjected to a torsional moment *T*.

3–15 Stress Concentration

In the development of the basic stress equations for tension, compression, bending, and torsion, it was assumed that no irregularities occurred in the member under consideration. But it is quite difficult to design a machine without permitting some changes in the cross sections of the members. Rotating shafts must have shoulders designed on them so that the bearings can be properly seated and so that they will take thrust loads; and the shafts must have key slots machined into them for securing pulleys and gears. A bolt has a head on one end and screw threads on the other end, both of which account for abrupt changes in the cross section. Other parts require holes, oil grooves, and notches of various kinds. Any discontinuity in a machine part alters the stress distribution in the neighborhood of the discontinuity so that the elementary stress equations no longer describe the state of stress in the part. Such discontinuities are called *stress raisers*, and the regions in which they occur are called areas of *stress concentration*.

The distribution of elastic stress across a section of a member may be uniform as in a bar in tension, linear as a beam in bending, or even rapid and curvaceous as in a sharply curved beam. Stress concentration comes from some irregularity not inherent in the member, such as tool marks, holes, notches, grooves, or threads. The *nominal stress* is said to exist if the member is free of the stress raiser. This definition is not always honored, so check the definition on the stress-concentration chart or table you are using.

To help in understanding this effect, examine Fig. 3–26. Note that the stress trajectories are uniform everywhere except in the vicinity of the hole. But at the hole these lines of force must bend to get around. Stress concentration is a highly localized effect. The stress on the tension plate is highest at the edge of the hole on plane A-A; this stress drops rapidly as points are examined farther from the hole edge and soon becomes uniform again.

A *theoretical*, or *geometric*, *stress-concentration factor* K_t or K_{ts} is used to relate the actual maximum stress at the discontinuity to the nominal stress. The factors are defined by the equations

$$K_t = \frac{\sigma_{\max}}{\sigma_0} \qquad K_{ts} = \frac{\tau_{\max}}{\tau_0} \tag{3–51}$$

where K_t is used for normal stresses and K_{ts} for shear stresses. The nominal stress σ_0 or τ_0 is more difficult to define. Generally, it is the stress calculated by using the elementary stress equations and the net area, or net cross section. But sometimes the gross cross section is used instead, and so it is always wise to check before calculating the maximum stress.

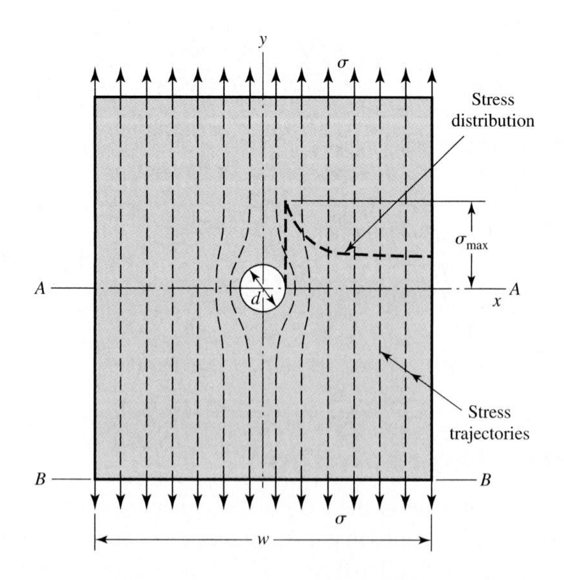

Figure 3–26

Stress distribution near a hole in a plate loaded in tension. The tensile stress on a section B-B, remote from the hole is $\sigma = F/A$, where $A = wt$ and t is the plate thickness. On a section at A-A, through the hole, the area $A_0 = (w - d)t$ and the normal stress is $\sigma_0 = F/A_0$. Note the difference between the nominal stress and the stress at a section remote from the discontinuity.

The subscript t in K_t means that this stress-concentration factor depends for its value only on the *geometry* of the part. That is, the particular material used has no effect on the value of K_t. This is why it is called a *theoretical* stress-concentration factor.

The analysis of geometric shapes to determine stress-concentration factors is a difficult problem, and not many solutions can be found. One such solution is that of an infinite plate containing an elliptical hole loaded in uniform tension. The result is

$$K_t = 1 + \frac{2b}{a} \tag{3–52}$$

where, after replacing the hole in Fig. 3–26 with an ellipse, b is the half-width, a is the half-height, and $w = \infty$. Thus, for a circular hole, $b = a$ and $K_t = 3$.

Note that Eq. (3–52) can be applied to a transverse crack, where $b \gg a$, or to a longitudinal crack ($b \ll a$).

Most stress-concentration factors are found by using experimental techniques.[7] Though the finite-element method has been used, the fact that the elements are indeed finite prevents finding the true maximum stress. Experimental approaches generally used include photoelasticity, grid methods, brittle-coating methods, and electrical strain-gauge methods. Of course, the grid and strain-gauge methods both suffer from the same drawback as the finite-element method.

Stress-concentration factors for a variety of geometries may be found in Tables E–15 and E–16.

In *static loading* stress-concentration factors are applied as follows. In ductile ($\epsilon_f \geq 0.05$) materials the stress-concentration factor *is not* usually applied to predict the critical stress, because plastic strain in the region of the stress has a strengthening effect. In *brittle materials* ($\epsilon_f < 0.05$) the geometric stress-concentration factor K_t *is* applied to the nominal stress before comparing with strength. Gray cast iron has so many inherent stress raisers that the stress raisers introduced by the designer have only a modest (but additive) effect.

CASE STUDY 3b Be Alert to Viewpoint

On a "spade" rod end (or lug) a load is transferred through a pin to a rectangular-cross-section rod or strap. The theoretical or geometric stress-concentration factor (SCF) for this geometry is known as follows, based on the net area $A = (w - d)t$ as shown in Fig. CS-3b.

d/w	0.15	0.20	0.25	0.30	0.35	0.40	0.45	0.50
K_t	7.4	5.4	4.6	3.7	3.2	2.8	2.6	2.45

As presented in the table K_t is a decreasing monotone. This rod end is similar to the square-ended lug depicted in Table E-15-12.

$$\sigma_{\max} = K_t \sigma_0 \tag{a}$$

$$\sigma_{\max} = \frac{K_t F}{A} = K_t \frac{F}{(w - d)t} \tag{b}$$

7. The best source book is R. E. Peterson, *Stress Concentration Factors*, Wiley, New York, 1974.

It is insightful to base the SCF on the *unnotched* area.

$$\sigma_{\max} = K'_t \frac{F}{wt} \tag{c}$$

By equating Eqs. (*b*) and (*c*) and solving for K'_t we obtain

$$K'_t = \frac{wt}{F}\frac{F}{(w-d)t} = \frac{K_t}{1-d/w} \tag{d}$$

A regression curve-fit for the data in the above table in the form $K_t = a(d/w)^b$ with the aid of a ln ln transform gives the result $a = \exp(0.204\,521\,2) = 1.227$, $b = -0.935$, and $r = -0.997$. Thus

$$K_t = 1.227\left(\frac{d}{w}\right)^{-0.935} \tag{e}$$

which is a decreasing monotone (and unexciting). However,

$$K'_t = \frac{1.227}{1-d/w}\left(\frac{d}{w}\right)^{-0.935} \tag{f}$$

Form another table from Eq. (*f*):

d/w	0.15	0.20	0.25	0.30	0.35	0.40	0.45	0.50	0.55	0.60
K'_t	8.507	6.907	5.980	5.403	5.038	4.817	4.707	4.692	4.769	4.946

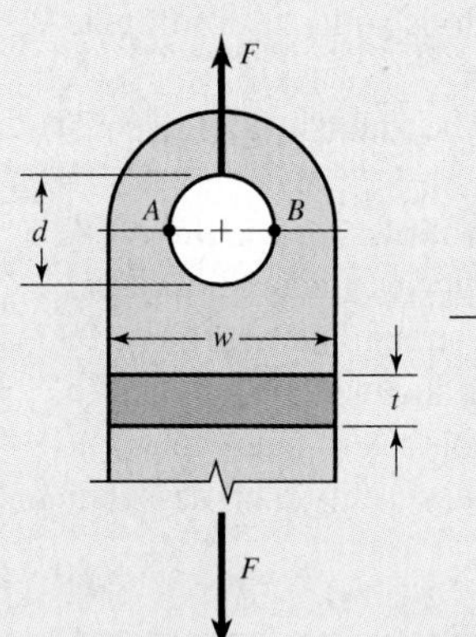

Figure CS–3b

A round-ended lug end to a rectangular cross-section rod. The maximum tensile stress in the lug occurs at locations *A* and *B*. The net area $A = (w - b)t$ is used in the definition of K_t, but there is an advantage to using the total area wt.

which shows a stationary-point minimum for K'_t. This can be found by differentiating Eq. (*f*) with respect to d/w and setting it equal to zero:

$$\frac{dK'_t}{d(d/w)} = \frac{(1-d/w)ab(d/w)^{b-1} - a(d/w)^b}{[1-(d/w)]^2} = 0$$

from which

$$\left(\frac{d}{w}\right)^* = \frac{b}{b-1} = \frac{-0.935}{-0.935-1} = 0.483$$

with a corresponding K'_t of 4.687. Knowing the section $w \times t$ lets the designer specify the strongest lug immediately by specifying a pin diameter of $0.483w$ (or, as a rule of thumb, of half the width). The theoretical SCF data in its original form, or a plot based on the data using net area, would not suggest this. The right viewpoint can suggest valuable insights.

3–16 Stresses in Cylinders

Cylindrical pressure vessels, hydraulic cylinders, gun barrels, and pipes carrying fluids at high pressures develop both radial and tangential stresses with values that are dependent upon the radius of the element under consideration. In determining the radial stress σ_r and the tangential stress σ_t, we make use of the assumption that the longitudinal elongation is constant around the circumference of the cylinder. In other words, a right section of the cylinder remains plane after stressing.

Referring to Fig. 3–27, we designate the inside radius of the cylinder by r_i, the outside radius by r_o, the internal pressure by p_i, and the external pressure by p_o. Then it can be shown that tangential and radial stresses exist whose magnitudes are

$$\sigma_t = \frac{p_i r_i^2 - p_o r_o^2 - r_i^2 r_o^2 (p_o - p_i)/r^2}{r_o^2 - r_i^2}$$
$$\sigma_r = \frac{p_i r_i^2 - p_o r_o^2 + r_i^2 r_o^2 (p_o - p_i)/r^2}{r_o^2 - r_i^2} \qquad (3–53)$$

As usual, positive values indicate tension and negative values, compression.

The special case of $p_o = 0$ gives

$$\sigma_t = \frac{r_i^2 p_i}{r_o^2 - r_i^2}\left(1 + \frac{r_o^2}{r^2}\right)$$
$$\sigma_r = \frac{r_i^2 p_i}{r_o^2 - r_i^2}\left(1 - \frac{r_o^2}{r^2}\right) \qquad (3–54)$$

The equations of set (3–54) are plotted in Fig. 3–28 to show the distribution of stresses over the wall thickness.

It should be realized that longitudinal stresses exist when the end reactions to the internal pressure are taken by the pressure vessel itself. This stress is found to be

$$\sigma_l = \frac{p_i r_i^2}{r_o^2 - r_i^2} \qquad (3–55)$$

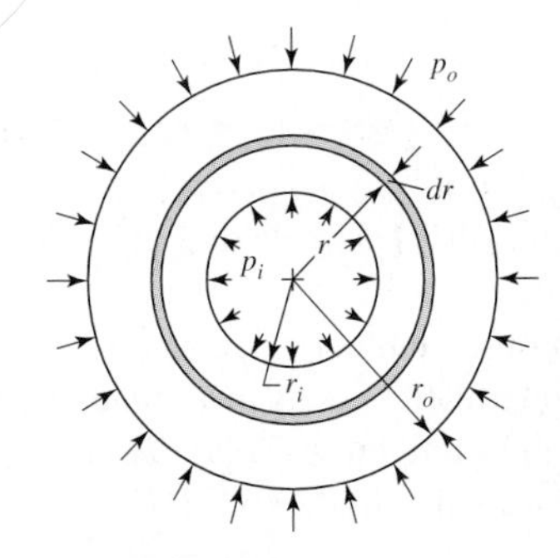

Figure 3–27

A cylinder subjected to both internal and external pressure.

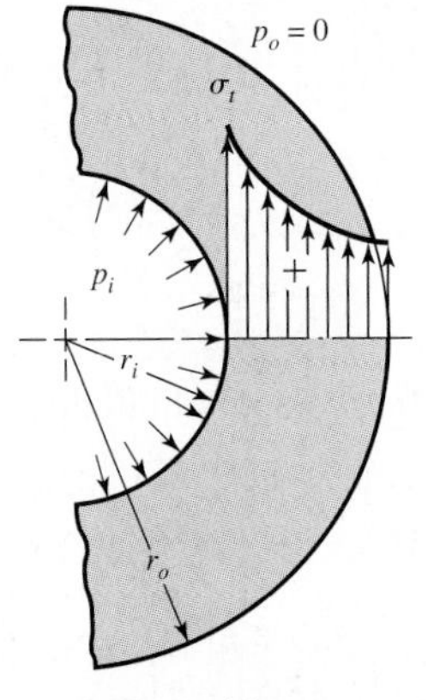

(a) Tangential stress distribution

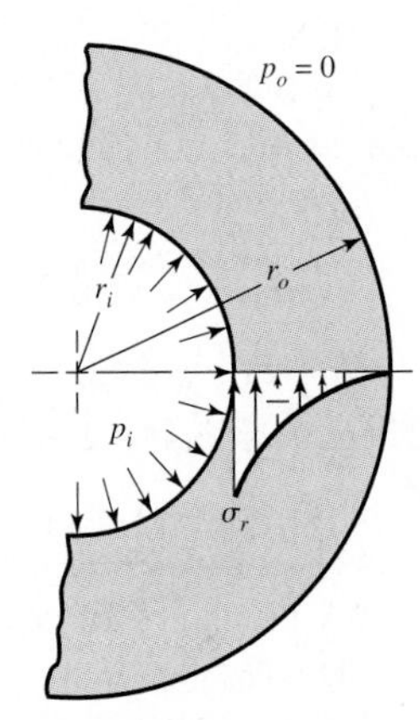

(b) Radial stress distribution

Figure 3–28

Distribution of stresses in a thick-walled cylinder subjected to internal pressure.

We further note that Eqs. (3–53), (3–54), and (3–55) apply only to sections taken a significant distance from the ends and away from any areas of stress concentration.

Thin-Walled Vessels

When the wall thickness of a cylindrical pressure vessel is about one-twentieth, or less, of its radius, the radial stress which results from pressurizing the vessel is quite small compared with the tangential stress. Under these conditions the tangential stress can be obtained as follows: Let an internal pressure p be exerted on the wall of a cylinder of thickness t and inside diameter d_i. The force tending to separate two halves of a unit length of the cylinder is pd_i. This force is resisted by the tangential stress, also called the *hoop stress*, acting uniformly over the stressed area. We then have $pd_i = 2t\sigma_t$, or

$$\sigma_{t,\mathrm{av}} = \frac{pd_i}{2t} \tag{3–56}$$

This equation gives the *average* tangential stress and is valid regardless of the wall thickness. For a thin-walled vessel an approximation to the maximum tangential stress is

$$\sigma_{t,\max} = \frac{p(d_i + t)}{2t} \tag{3–57}$$

where $d_i + t$ is the average diameter.

In a closed cylinder, the longitudinal stress σ_l exists because of the pressure upon the ends of the vessel. If we assume this stress is also distributed uniformly over the wall thickness, we can easily find it to be

$$\sigma_l = \frac{pd_i}{4t} \tag{3–58}$$

EXAMPLE 3–10

An aluminum-alloy pressure vessel is made of tubing having an outside diameter of 8 in and a wall thickness of $\frac{1}{4}$ in.

(*a*) What pressure can the cylinder carry if the permissible tangential stress is 12 kpsi and the theory for thin-walled vessels is assumed to apply?

(*b*) On the basis of the pressure found in part (*a*), compute all of the stress components using the theory for thick-walled cylinders.

Solution

(*a*) Here $d_i = 8 - 2(0.25) = 7.5$ in, $r_i = 7.5/2 = 3.75$ in, and $r_o = 8/2 = 4$ in. Then $t/r_i = 0.25/3.75 = 0.067$. Since this ratio is greater than $\frac{1}{20}$, the theory for thin-walled vessels may not yield safe results.

We first solve Eq. (3–57) to obtain the allowable pressure. This gives

Answer

$$p = \frac{2t\sigma_{t,\max}}{d_i + t} = \frac{2(0.25)(12)(10)^3}{7.5 + 0.25} = 774 \text{ psi}$$

Then, from Eq. (3–58), we find the average longitudinal stress to be

$$\sigma_l = \frac{pd_i}{4t} = \frac{774(7.5)}{4(0.25)} = 5800 \text{ psi}$$

(*b*) The maximum tangential stress will occur at the inside radius, and so we use $r = r_i$ in the first equation of the pair (3–54). This gives

Answer $$\sigma_t = p_i \frac{r_o^2 + r_i^2}{r_o^2 - r_i^2} = 774 \frac{4^2 + 3.75^2}{4^2 - 3.75^2} = 12\,000 \text{ psi}$$

Similarly, the maximum radial stress is found, from the second equation of the pair (3–54), to be

Answer $$\sigma_r = -p_i = -774 \text{ psi}$$

Equation (3–55) gives the longitudinal stress as

Answer $$\sigma_l = \frac{p_i r_i^2}{r_o^2 - r_i^2} = \frac{774(3.75)^2}{4^2 - 3.75^2} = 5620 \text{ psi}$$

These three stresses σ_t, σ_r, and σ_l are principal stresses, since there is no shear. Note that there is no significant difference in the tangential stresses in parts (*a*) and (*b*), and so the thin-wall theory can be considered satisfactory.

3–17 Rotating Rings

Many rotating elements, such as flywheels and blowers, can be simplified to a rotating ring to determine the stresses. When this is done it is found that the same tangential and radial stresses exist as in the theory for thick-walled cylinders except that they are caused by inertial forces acting on all the particles of the ring. The tangential and radial stresses so found are subject to the following restrictions:

- The outside radius of the ring, or disk, is large compared with the thickness $r_o \geq 10t$.
- The thickness of the ring or disk is constant.
- The stresses are constant over the thickness.

The stresses are

$$\sigma_t = \rho\omega^2 \left(\frac{3+\nu}{8}\right)\left(r_i^2 + r_o^2 + \frac{r_i^2 r_o^2}{r^2} - \frac{1+3\nu}{3+\nu} r^2\right)$$
$$\sigma_r = \rho\omega^2 \left(\frac{3+\nu}{8}\right)\left(r_i^2 + r_o^2 - \frac{r_i^2 r_o^2}{r^2} - r^2\right) \tag{3–59}$$

where r is the radius to the stress element under consideration, ρ is the mass density, and ω is the angular velocity of the ring in radians per second. For a rotating disk, use $r_i = 0$ in these equations.

3–18 Press and Shrink Fits

When two cylindrical parts are assembled by shrinking or press-fitting one part upon another, a contact pressure is created between the two parts. The stresses resulting from this pressure may easily be determined with the equations of the preceding sections.

Figure 3–29*b* shows two cylindrical members which have been assembled with a shrink fit. A contact pressure p exists between the members at the transition radius R, causing radial stresses $\sigma_r = -p$ in each member at the contacting surfaces. From Sec. 3–16, we find the tangential stress at the transition radius of the inner member to be

$$\sigma_{it}(\text{at } R) = -p \frac{R^2 + r_i^2}{R^2 - r_i^2} \tag{3–60}$$

Figure 3–29

Notation for press and shrink fits. (*a*) Unassembled parts; (*b*) after assembly.

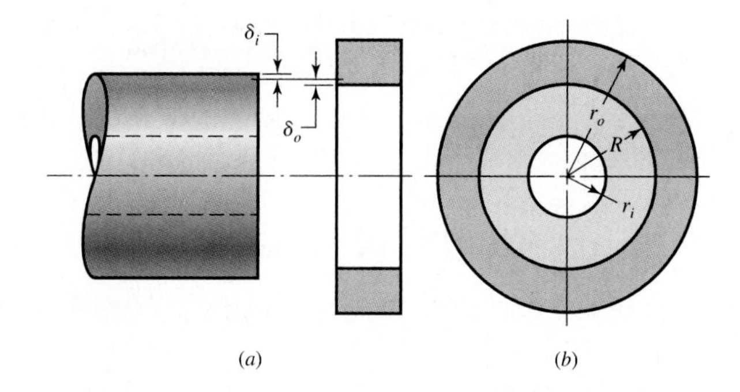

In the same manner, the tangential stress at the inner surface of the outer member is found to be

$$\sigma_{ot}(\text{at } R) = p\frac{r_o^2 + R^2}{r_o^2 - R^2} \tag{3–61}$$

These equations cannot be solved until the contact pressure is known. In obtaining a shrink fit, the radius of the male member is made larger than the radius of the female member. The difference in these dimensions is called the *radial interference* and is the radial deformation which the two members must experience. Since these dimensions are usually known, the deformation should be introduced in order to evaluate the stresses. As shown in Fig. 3–29*a*, δ_i and δ_o symbolize the changes in the radii of the inner and outer members, respectively. The total radial interference is, therefore,

$$\delta = |\delta_i| + |\delta_o| \tag{a}$$

The tangential strain at the transition radius of the outer cylinder is measured by the change in circumference, and is

$$\epsilon_{ot} = \frac{2\pi(R+\delta_o) - 2\pi R}{2\pi R} = \frac{\delta_o}{R} \tag{b}$$

and so $\delta_o = R\epsilon_{ot}$. But, since

$$\epsilon_{ot} = \frac{\sigma_{ot}}{E_o} - \frac{\nu_o \sigma_{or}}{E_o} \tag{c}$$

then, from Eqs. (3-60) and (3-61), we have

$$\delta_o = \frac{pR}{E_o}\left(\frac{r_o^2 + R^2}{r_o^2 - R^2} + \nu_o\right) \tag{d}$$

This is the change in radius of the outer member. In a similar manner, the change in radius of the inner member is found to be

$$\delta_i = -\frac{pR}{E_i}\left(\frac{R^2 + r_i^2}{R^2 - r_i^2} - \nu_i\right) \tag{e}$$

Then, from Eq. (*a*), we have for the total deformation

$$\delta = \frac{pR}{E_o}\left(\frac{r_o^2 + R^2}{r_o^2 - R^2} + \nu_o\right) + \frac{pR}{E_i}\left(\frac{R^2 + r_i^2}{R^2 - r_i^2} - \nu_i\right) \tag{3–62}$$

This equation can be solved for the pressure p when the radial interference δ is given. If the two members are of the same material, $E_o = E_i = E$, $\nu_o = \nu_i$, and the relation simplifies to

$$p = \frac{E\delta}{R}\left[\frac{(r_o^2 - R^2)(R^2 - r_i^2)}{2R^2(r_o^2 - r_i^2)}\right] \tag{3–63}$$

The value of the interface pressure p from either Eq. (3–62) or Eq. (3–63) can now be used to obtain the stress state at the specified radius in either cylinder.

Assumptions

In addition to the assumptions both stated and implied by the development, it is necessary to assume that both members have the same length. In the case of a hub which has been press-fitted to a shaft, this assumption would not be true, and there would be an increased pressure at each end of the hub. It is customary to allow for this condition by the employment of a stress-concentration factor. The value of this factor depends upon the contact pressure and the design of the female member, but its theoretical value is seldom greater than 2.

3–19 Temperature Effects

When the temperature of an unrestrained body is uniformly increased, the body expands, and the normal strain is

$$\epsilon_x = \epsilon_y = \epsilon_z = \alpha(\Delta T) \tag{3–64}$$

where α is the coefficient of thermal expansion and ΔT is the temperature change, in degrees. In this action the body experiences a simple volume increase with the components of shear strain all zero.

If a straight bar is restrained at the ends so as to prevent lengthwise expansion and then is subjected to a uniform increase in temperature, a compressive stress will develop because of the axial constraint. The stress is

$$\sigma = \epsilon E = \alpha(\Delta T)E \tag{3–65}$$

In a similar manner, if a uniform flat plate is restrained at the edges and also subjected to a uniform temperature rise, the compressive stress developed is given by the equation

$$\sigma = \frac{\alpha(\Delta T)E}{1 - \nu} \tag{3–66}$$

The stresses expressed by Eqs. (3–65) and (3–66) are called *thermal stresses*. They arise because of a temperature change in a clamped or restrained member. Such stresses, for example, occur during welding, since parts to be welded must be clamped before welding. Table 3–5 lists approximate values of the coefficients of thermal expansion.

A *thermal stress* is one which arises because of the existence of a *temperature gradient* in a member. Figure 3-30 is an example. Shown are the stress distributions within a slab of infinite dimensions during heating and cooling. During cooling, the maximum stress is the surface tension. During heating, the external surfaces are hot and tend to expand but are restrained by the cooler center. This causes compression in the surface and tension in the center as shown.

Table 3–5

Coefficients of Thermal Expansion (Linear Mean Coefficients for the Temperature Range 0–100°C)

Material	Celsius Scale	Fahrenheit Scale
Aluminum	$23.9(10)^{-6}$	$13.3(10)^{-6}$
Brass, cast	$18.7(10)^{-6}$	$10.4(10)^{-6}$
Carbon steel	$10.8(10)^{-6}$	$6.0(10)^{-6}$
Cast iron	$10.6(10)^{-6}$	$5.9(10)^{-6}$
Magnesium	$25.2(10)^{-6}$	$14.0(10)^{-6}$
Nickel steel	$13.1(10)^{-6}$	$7.3(10)^{-6}$
Stainless steel	$17.3(10)^{-6}$	$9.6(10)^{-6}$
Tungsten	$4.3(10)^{-6}$	$2.4(10)^{-6}$

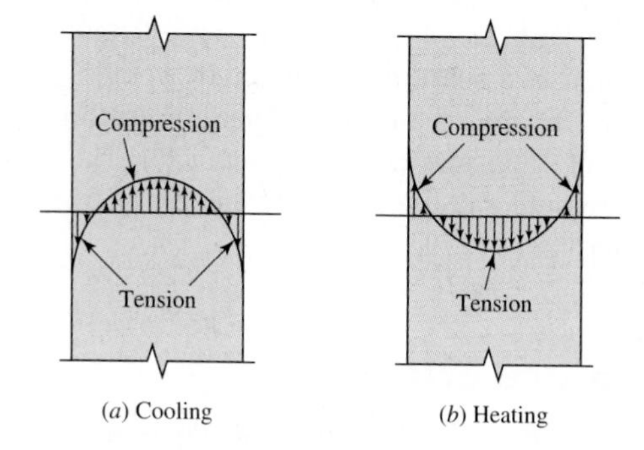

Figure 3–30

Thermal stresses in an infinite slab during heating and cooling.

3–20 Curved Members in Flexure

The distribution of stress in a curved flexural member is determined by using the following assumptions:

- The cross section has an axis of symmetry in a plane along the length of the beam.
- Plane cross sections remain plane after bending.
- The modulus of elasticity is the same in tension as in compression.

We shall find that the neutral axis and the centroidal axis of a curved beam, unlike a straight beam, are not coincident and also that the stress does not vary linearly from the neutral axis. The notation shown in Fig. 3–31 is defined as follows:

r_o = radius of outer fiber
r_i = radius of inner fiber
h = depth of section
c_o = distance from neutral axis to outer fiber
c_i = distance from neutral axis to inner fiber
r_n = radius of neutral axis
R = radius of centroidal axis
e = distance from centroidal axis to neutral axis

Figure 3–31 shows that the neutral and centroidal axes are not coincident.[8] It turns out that the location of the neutral axis with respect to the center of curvature O is given by

8. For a complete development of the relations in this section, see Joseph E. Shigley, *Mechanical Engineering Design*, First Metric Edition, McGraw-Hill, New York, 1986, pp. 72–75.

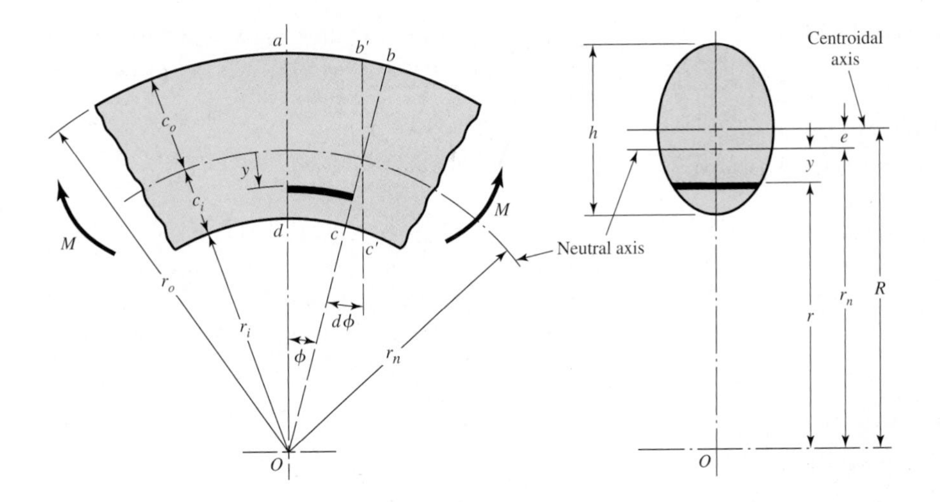

Figure 3–31

Note that y is positive in the direction toward point O.

the equation

$$r_n = \frac{A}{\int \frac{dA}{r}} \tag{3–67}$$

The stress distribution can be found by balancing the external applied moment against the internal resisting moment. The result is found to be

$$\sigma = \frac{My}{Ae(r_n - y)} \tag{3–68}$$

where M is positive in the direction shown in Fig. 3–31. Equation (3–67) shows that the stress distribution is hyperbolic. The critical stresses occur at the inner and outer surfaces and are

$$\sigma_i = \frac{Mc_i}{Aer_i} \qquad \sigma_o = -\frac{Mc_o}{Aer_o} \tag{3–69}$$

These equations are valid for pure bending. In the usual and more general case, such as a crane hook, the U frame of a press, or the frame of a clamp, the bending moment is due to forces acting to one side of the cross section under consideration. In this case the bending moment is computed about the *centroidal axis*, not the neutral axis. Also, an additional axial tensile or compressive stress must be added to the bending stresses given by Eqs. (3–68) and (3–69) to obtain the resultant stresses acting on the section.

EXAMPLE 3–11 Plot the distribution of stresses across section *A-A* of the crane hook shown in Fig. 3–32*a*. The cross section is rectangular, with $b = 0.75$ in and $h = 4$ in, and the load is $F = 5000$ lb.

Solution Since $A = bh$, we have $dA = b\,dr$ and, from Eq. (3-67),

$$r_n = \frac{A}{\int \frac{dA}{r}} = \frac{bh}{\int_{r_i}^{r_o} \frac{b}{r}\,dr} = \frac{h}{\ln \frac{r_o}{r_i}} \tag{1}$$

Figure 3–32

(*a*) Plan view of crane hook; (*b*) cross section and notation; (*c*) resulting stress distribution. There is no stress concentration.

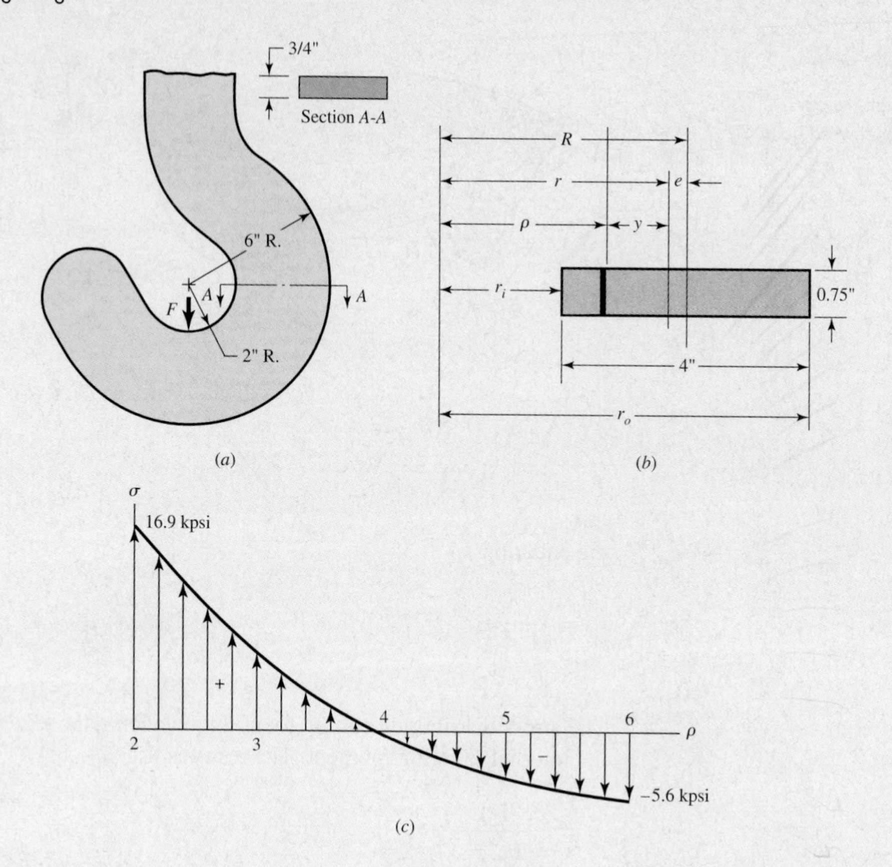

From Fig. 3–32*b*, we see that $r_i = 2$ in, $r_o = 6$ in, $R = 4$ in, and $A = 3$ in^2. Thus, from Eq. (1),

$$r_n = \frac{h}{\ln (r_o/r_i)} = \frac{4}{\ln \frac{6}{2}} = 3.641 \text{ in}$$

and so the eccentricity is $e = R - r_n = 4 - 3.641 = 0.359$ in. The moment M is positive and is $M = FR = 5000(4) = 20\ 000$ lb · in. Adding the axial component of stress to Eq. (3–68) gives

$$\sigma = \frac{F}{A} + \frac{My}{Ae(r_n - y)} = \frac{5000}{3} + \frac{(20\ 000)(3.641 - r)}{3(0.359)r} \tag{2}$$

Substituting values of r from 2 to 6 in results in the stress distribution shown in Fig. 3–32*c*. The stresses at the inner and outer radii are found to be 16.9 and −5.6 kpsi, respectively, as shown.

Sections most frequently encountered in the stress analysis of curved beams are shown in Fig. 3–33. Formulas for the rectangular section were developed in Example 3-11, but they are repeated here for convenience:

$$R = r_i + \frac{h}{2} \tag{3–70}$$

Figure 3–33

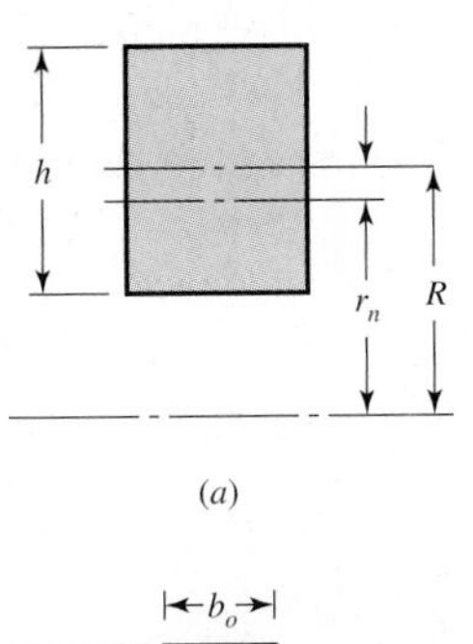

(a)

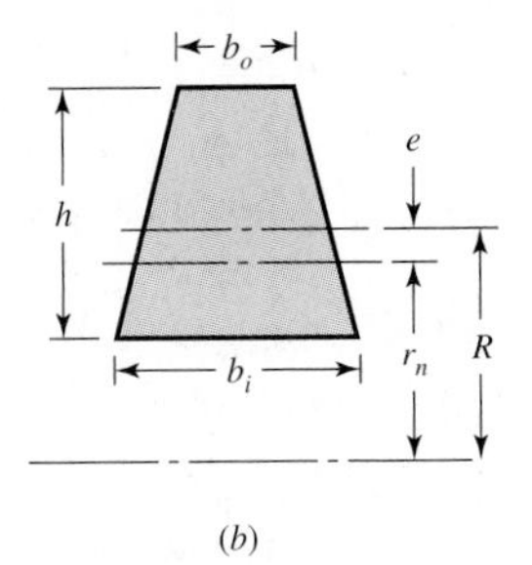

(b)

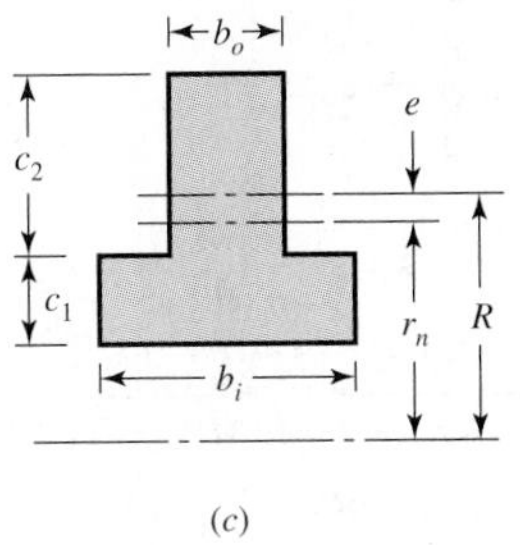

(c)

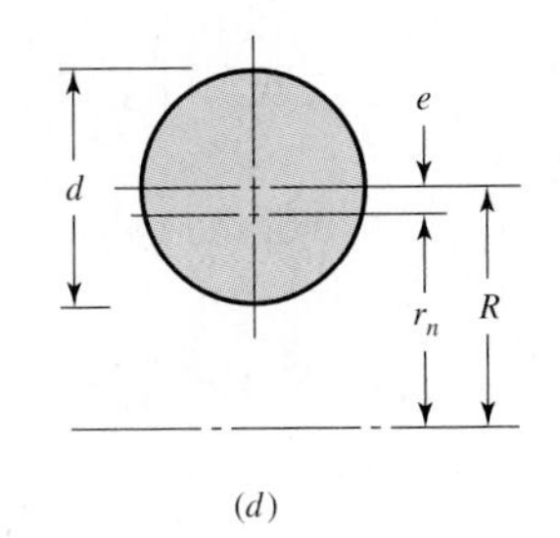

(d)

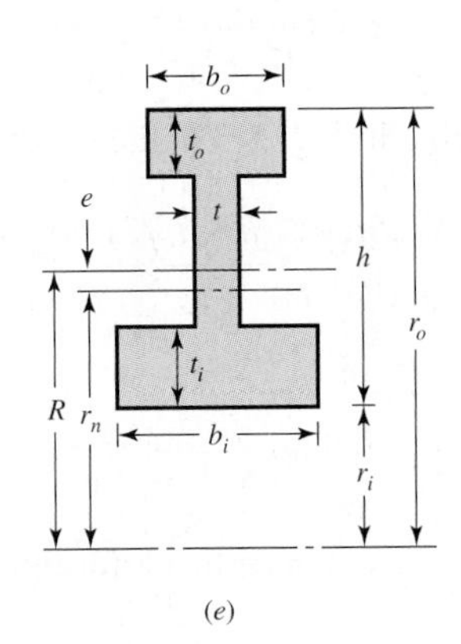

(e)

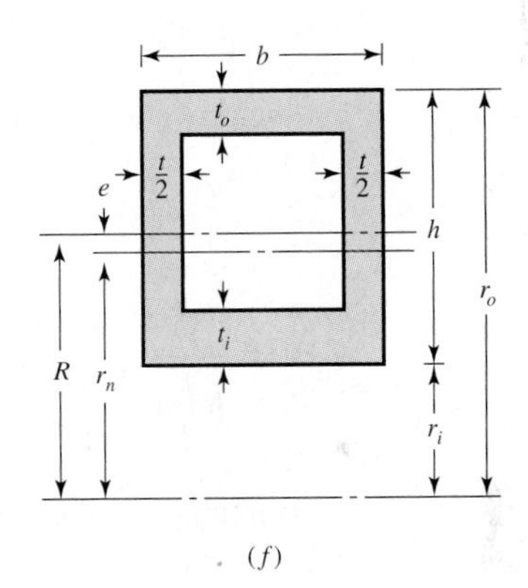

(f)

$$r_n = \frac{h}{\ln (r_o/r_i)} \tag{3–71}$$

$$e \doteq \frac{h^2}{12R} \tag{3–72}$$

For the trapezoidal section in Fig. 3–33*b*, the formulas are

$$R = r_i + \frac{h}{3}\frac{b_i + 2b_o}{b_i + b_o} \tag{3–73}$$

$$r_n = \frac{A}{b_o - b_i + [(b_i r_o - b_o r_i)/h]\ln(r_o/r_i)} \tag{3–74}$$

For the T section in Fig. 3–33*c*, we have

$$R = r_i + \frac{b_i c_1^2 + 2b_o c_1 c_2 + b_o c_2^2}{2(b_o c_2 + b_i c_1)} \tag{3–75}$$

$$r_n = \frac{b_i c_1 + b_o c_2}{b_i \ln[(r_i + c_1)/r_i)] + b_o \ln[r_o/(r_i + c_1)]} \tag{3–76}$$

The equations for the solid round section of Fig. 3–33*d* are

$$R = r_i + \frac{d}{2} \tag{3–77}$$

$$r_n = \frac{d^2}{4(2R - \sqrt{4R^2 - d^2})} \tag{3–78}$$

$$e = \bar{r}\left[\frac{1}{16}\left(\frac{d}{\bar{r}}\right)^2 + \frac{1}{128}\left(\frac{d}{\bar{r}}\right)^4\right] \doteq \frac{d^2}{16\bar{r}} \tag{3–79}$$

For the I shape in Fig. 3–33*e*, we have

$$R = r_i + \frac{\frac{1}{2}h^2 t + \frac{1}{2}t_i^2(b_i - t) + t_o(b_o - t)(h - t_o/2)}{t_i(b_i - t) + t_o(b_o - t) + ht} \tag{3–80}$$

$$r_n = \frac{t_i(b_i - t) + t_o(b_o - t) + ht_o}{b_i \ln \dfrac{r_i + t}{r_i} + t \ln \dfrac{r_o - t_o}{r_i + t_i} + b_o \ln \dfrac{r_o}{r_o - t_o}} \tag{3–81}$$

Finally, for the rectangular tubing in Fig. 3–33*f*, the results are

$$R = r_i + \frac{\frac{1}{2}h^2 t + \frac{1}{2}t_i^2(b - t) + t_o(b - t)(h - t_o/2)}{ht + (b - t)(t_i + t_o)} \tag{3–82}$$

$$r_n = \frac{(b - t)(t_i + t_o) + ht}{b\left(\ln \dfrac{r_i + t_i}{r_i} + t \ln \dfrac{r_o}{r_o + t_o}\right) + t \ln \dfrac{r_o - t_o}{r_i + t_i}} \tag{3–83}$$

Formulas for other sections can be obtained by performing the integration indicated by Eq. (3–67).

Many cases arise in which numerical integration must be used. These may occur because

- A digital computer is being used.
- It is not possible to integrate the function by any other means.
- The function to be integrated is described only by data.

A method of integration by Simpson's rule consists of defining equally spaced ordinates in the integration interval. Then parabolic curves are assumed to pass through each contiguous set of three ordinates. Using the notation of Fig. 3–34, the area under the curve AB, by Simpson's rule, is

$$I \doteq \frac{H}{3}(Y_0 + 4Y_1 + 2Y_2 + 4Y_3 + 2Y_4 + \cdots + 4Y_{N-1} + Y_N)$$

$$\doteq \frac{H}{3}(Y_0 + Y_N + 4\,\Sigma Y_{\text{odd}} + 2\,\Sigma Y_{\text{even}}) \tag{3–84}$$

where H is the width of the interval and is

$$H = \frac{X_N - X_0}{N} \tag{3–85}$$

The terms ΣY_{odd} and ΣY_{even} are the sums, respectively, of the odd-numbered and even-numbered subscripted ordinates. Equation (3–84) then gives Simpson's approximation to the equation

$$I = \int_{X_0}^{X_N} F(X)\, dX \tag{3–86}$$

Fortunately, Eq. (3–84) gives good accuracy. The result can be checked using *Richardson's error estimate*.[9] This is obtained by performing the integration twice, once with all the ordinates, and again with every other ordinate. Designating the first integration by I_1 and the second by I_2, Richardson's error is

$$E \doteq \frac{I_1 - I_2}{15} \tag{3–87}$$

The sign of the result *is significant*. Once E has been obtained from Eq. (3–87), a better estimate of the integral is

$$I \doteq I_1 + E \tag{3–88}$$

For the analysis of a curved beam of any arbitrary cross section, divide the cross section into an even number of strips of thickness Δr and length b_I, where b_I is the length of the Ith strip. Then the equations to be solved are

$$A = \int_{r_i}^{r_o} b\, dr \tag{3–89}$$

$$R = \int_{r_i}^{r_o} \frac{br\, dr}{A} \tag{3–90}$$

$$r_n + \frac{A}{\displaystyle\int_{r_i}^{r_o} \frac{b\, dr}{r}} \tag{3–91}$$

$$e = R - r_n \tag{3–92}$$

Numerical integration methods are easy to program; see Fig. 3–35 for a simplified flow diagram.

Figure 3–34

Notation for integration by Simpson's first rule. Note that N is an even number.

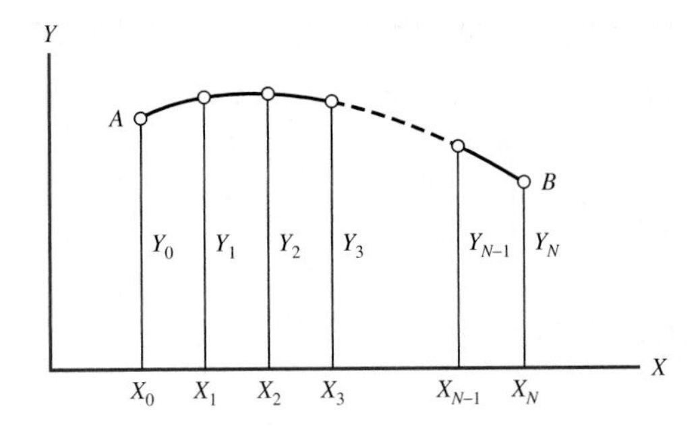

9. See B. Carnahan, H. A. Luther, and J. O. Wilkes, *Applied Numerical Analysis*, Wiley, New York, 1969, p. 79.

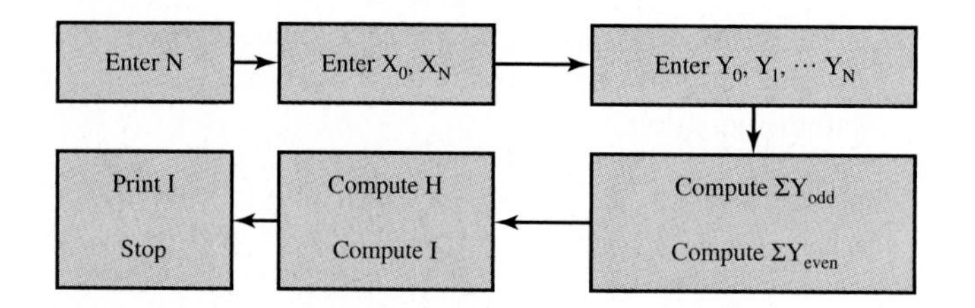

Figure 3–35

The sequence of steps in a computer implementation of Simpson's first rule of numerical integration.

3–21 Contact Stresses

When two bodies having curved surfaces are pressed together, point or line contact changes to area contact, and the stresses developed in the two bodies are three-dimensional. Contact-stress problems arise in the contact of a wheel and a rail, in automotive valve cams and tappets, in mating gear teeth, and in the action of rolling bearings. Typical failures are seen as cracks, pits, or flaking in the surface material.

The most general case of contact stress occurs when each contacting body has a double radius of curvature; that is, when the radius in the plane of rolling is different from the radius in a perpendicular plane, both planes taken through the axis of the contacting force. Here we shall consider only the two special cases of contacting spheres and contacting cylinders.[10] The results presented here are due to Hertz and so are frequently known as *Hertzian stresses*.

When two solid spheres of diameters d_1 and d_2 are pressed together with a force F, a circular area of contact of radius a is obtained. Specifying E_1, ν_1 and E_2, ν_2 as the respective elastic constants of the two spheres, the radius a is given by the equation

$$a = \sqrt[3]{\frac{3F}{8} \frac{(1-\nu_1^2)/E_1 + (1-\nu_2^2)/E_2}{1/d_1 + 1/d_2}} \tag{3-93}$$

The pressure within each sphere has a semielliptical distribution, as shown in Fig. 3–36. The maximum pressure occurs at the center of the contact area and is

$$p_{\max} = \frac{3F}{2\pi a^2} \tag{3-94}$$

Equations (3–93) and (3–94) are perfectly general and also apply to the contact of a sphere and a plane surface or of a sphere and an internal spherical surface. For a plane surface, use $d = \infty$. For an internal surface, the diameter is expressed as a negative quantity.

The maximum stresses occur on the z axis, and these are principal stresses. Their values are

$$\sigma_x = \sigma_y = -p_{\max}\left[\left(1 - \frac{z}{a}\tan^{-1}\frac{1}{z/a}\right)(1+\nu) - \frac{1}{2\left(1+\dfrac{z^2}{a^2}\right)}\right] \tag{3-95}$$

$$\sigma_z = \frac{-p_{\max}}{1 + \dfrac{z^2}{a^2}} \tag{3-96}$$

These equations are valid for either sphere, but the value used for Poisson's ratio must

10. A good explanation of the general case may be found in Arthur P. Boresi, Omar M. Sidebottom, Fred B. Seely, and James O. Smith, *Advanced Mechanics of Materials*, 3d ed., Wiley, New York, 1978, pp. 581–627.

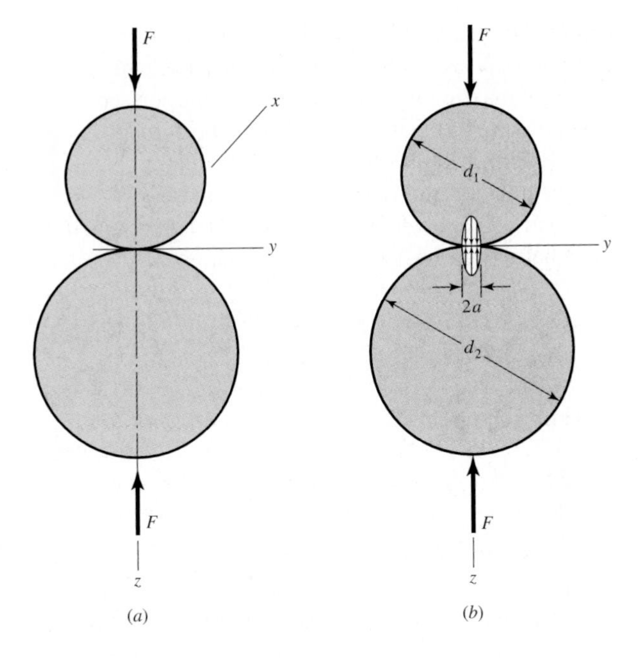

Figure 3–36

(*a*) Two spheres held in contact by force F; (*b*) contact stress has an elliptical distribution across contact zone diameter $2a$.

correspond with the sphere under consideration. The equations are even more complicated when stress states off the z axis are to be determined, because here the x and y coordinates must also be included. But these are not required for design purposes, because the maxima occur on the z axis.

The Mohr circles for the stress state described by Eqs. (3–95) and (3–96) are a point and two coincident circles. Since $\sigma_x = \sigma_y$, we have $\tau_{xy} = 0$ and

$$\tau_{xz} = \tau_{yz} = \frac{\sigma_x - \sigma_z}{2} = \frac{\sigma_y - \sigma_z}{2} \tag{3–97}$$

Figure 3–37 is a plot of Eqs. (3–95) and (3–96) for a distance of $3a$ below the surface. Note that the shear stress reaches a maximum value slightly below the surface. It is the

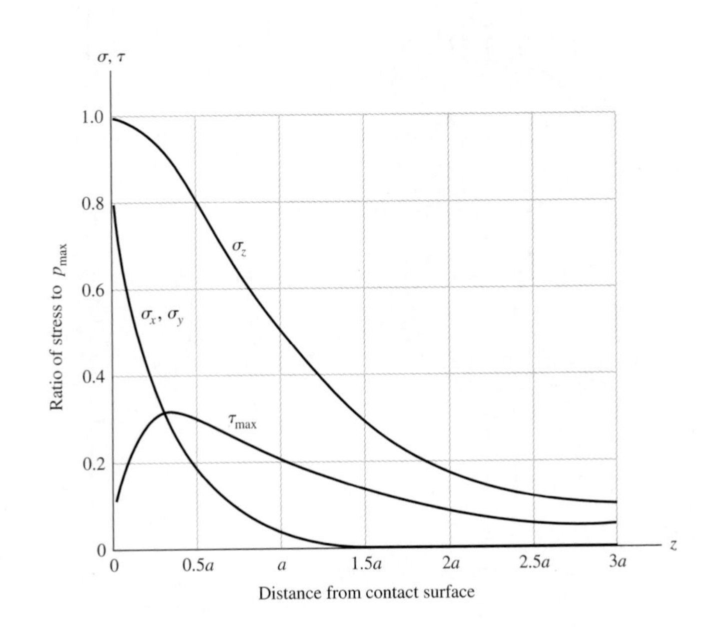

Figure 3–37

Magnitude of the stress components below the surface as a function of the maximum pressure of contacting spheres. Note that the maximum shear stress is slightly below the surface and is approximately $0.3p_{max}$. The chart is based on a Poisson ratio of 0.30. Note that the normal stresses are all compressive stresses.

opinion of many authorities that this maximum shear stress is responsible for the surface fatigue failure of contacting elements. The explanation is that a crack originates at the point of maximum shear stress below the surface and progresses to the surface and that the pressure of the lubricant wedges the chip loose.

Figure 3–38 illustrates a similar situation in which the contacting elements are two cylinders of length l and diameters d_1 and d_2. As shown in Fig. 3–38b, the area of contact is a narrow rectangle of width $2b$ and length l, and the pressure distribution is elliptical. The half-width b is given by the equation

$$b = \sqrt{\frac{2F}{\pi l} \frac{(1-\nu_1^2)/E_1 + (1-\nu_2^2)/E_2}{1/d_1 + 1/d_2}} \tag{3–98}$$

The maximum pressure is

$$p_{\max} = \frac{2F}{\pi b l} \tag{3–99}$$

Equations (3–98) and (3–99) apply to a cylinder and a plane surface, such as a rail, by making $d = \infty$ for the plane surface. The equations also apply to the contact of a cylinder and an internal cylindrical surface; in this case d is made negative.

The stress state on the z axis is given by the equations

$$\sigma_x = -2\nu p_{\max}\left(\sqrt{1+\frac{z^2}{b^2}} - \frac{z}{b}\right) \tag{3–100}$$

$$\sigma_y = -p_{\max}\left[\left(2 - \frac{1}{1+\frac{z^2}{b^2}}\right)\sqrt{1+\frac{z^2}{b^2}} - 2\frac{z}{b}\right] \tag{3–101}$$

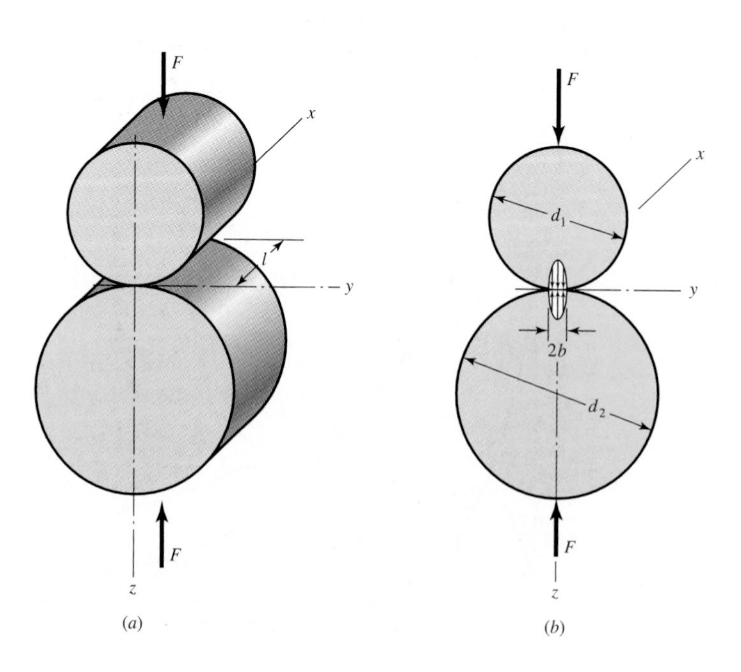

Figure 3–38

(a) Two right circular cylinders held in contact by forces F uniformly distributed along cylinder length l. (b) Contact stress has an elliptical distribution across the contact zone width $2b$.

$$\sigma_z = \frac{-p_{\max}}{\sqrt{1+z^2/b^2}} \tag{3–102}$$

These three equations are plotted in Fig. 3–39 up to a distance of $3b$ below the surface. Though τ_{zy} is not the largest of the three shear stresses for all values of z/b, it is a maximum at about $z/b = 0.75$ and is larger at that point than either of the other two shear stresses for any value of z/b.

Figure 3–40 shows the variation of the octahedral shear stress below the surface for contacting spheres and contacting cylinders. This stress is sometimes used, instead of the maximum shear stress, to define failure. The octahedral shear stress and the maximum shear stress reach their highest values on the curves at the same depth z, but the location of this point is very sensitive to the value of Poisson's ratio.

Hertz (1881) provided the preceding mathematical models of the stress field when the contact zone is free of shear stress. Another important contact stress case is *line-of-contact* with friction providing the shearing stress on the contact zone. Such shearing stresses are small with cams and rollers, but in cams with flatfaced followers, wheel-rail contact, and gear teeth, the stresses are elevated above the Hertzian field. Investigations

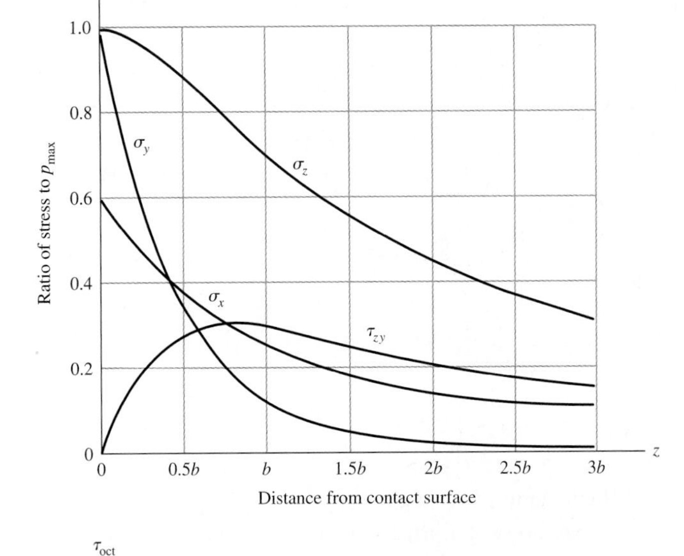

Figure 3–39

Magnitude of the stress components below the surface as a function of the maximum pressure for contacting cylinders. Shear stress τ_{xy} becomes the largest of the three shear stresses at about $z/b = 0.75$. Its maximum value is $0.30p_{max}$. The chart is based on a Poisson ratio of 0.30. Can you tell which two principal stresses are used to determine τ_{max} when $z/b = 0.75$? Note that all normal stresses are compressive stresses.

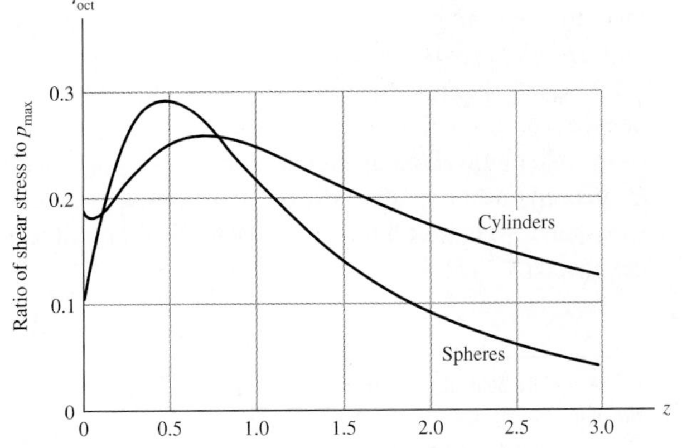

Figure 3–40

Plot of the octahedral shear stress for distances z/a and z/b below the surface for contacting spheres and cylinders. Based on $\nu = 0.30$.

of the effect on the stress field due to normal and shear stresses in the contact zone was begun theoretically by Lundberg (1939), and continued by Mindlin (1949), Smith-Liu (1949), and Poritsky (1949) independently. The Smith-Liu equations[11] for contacting cylinders are

$$b = \sqrt{\frac{2F}{\pi l}\frac{(1-\nu_1^2)/E_1 + (1-\nu_2^2)/E_2}{1/d_1 + 1/d_2}}$$

$$M = \sqrt{(b+y)^2 + z^2} \qquad N = \sqrt{(b-y)^2 + z^2}$$

$$\phi_1 = \frac{\pi(M+N)}{MN\sqrt{2MN + 2y^2 + 2z^2 - 2b^2}}$$

$$\phi_2 = \frac{\pi(M-N)}{MN\sqrt{2MN + 2y^2 + 2z^2 - 2b^2}}$$

$$\Delta' = \frac{1}{1/(2R)_1 + 1/(2R)_2}\left(\frac{1-\nu_1^2}{E_1} + \frac{1-\nu_2^2}{E_2}\right)$$

$$\sigma_x = -\frac{2\nu b}{\pi\Delta'}\left\{z\left(\frac{b^2+y^2+z^2}{b}\phi_1 - \frac{\pi}{b} - 2y\phi_2\right) + f\left[(y^2 - b^2 - z^2)\phi_2 + \frac{\pi y}{b} + (b^2 - y^2 - z^2)\frac{y}{b}\phi_1\right]\right\} \tag{3-103}$$

$$\sigma_y = -\frac{b}{\pi\Delta'}\left\{z\left(\frac{b^2+2z^2+2y^2}{b}\phi_1 - \frac{2\pi}{b} - 3y\phi_2\right) + f\left[(2y^2 - 2b^2 - 3z^2)\phi_2 + \frac{2\pi y}{b} + 2(b^2 - y^2 - z^2)\frac{y}{b}\phi_1\right]\right\} \tag{3-104}$$

$$\sigma_z = -\frac{b}{\pi\Delta'}[z(b\phi_1 - y\phi_2 + fz\phi_2)] \tag{3-105}$$

$$\tau_{yz} = -\frac{b}{\pi\Delta'}\left\{z^2\phi_2 + f\left[(b^2 + 2y^2 + 2z^2)\frac{z}{b}\phi_1 - 2\pi\frac{z}{b} - 3yz\phi_2\right]\right\} \tag{3-106}$$

In the Smith-Liu case, axes y and z of Fig. 3–38a are no longer principal axes, and σ_y and σ_z are no longer principal stresses. A glance at these equations will convince the reader that although this was reported in 1953,[12] applications grew slowly as computing aids improved. A way to study these equations is to program them to create a table running $-1 \le y/b \le 1$ across the top and $0 \le y/b \le 1$ down. Fill the body of the table with one of $\sigma_x, \sigma_y, \sigma_z, \tau_{xy}, \sigma_1, \sigma_2, \sigma_3, \tau_{12}, \tau_{13}, \tau_{23}, \tau_{max}$, and σ' (von Mises), normalized by dividing by the maximum Hertzian contact zone stress, p_{max}. Failure theory investigators examine (1) maximum compressive stress, (2) maximum tensile stress, (3) largest maximum shear stress, and (4) largest alternating shear-stress amplitude. See the problems at the end of the chapter.

11. A widely held library reference is G. A. Castleberry, "Analyzing Contact Stresses More Accurately," *Machine Design*, April 12, 1984, pp. 92–97.

12. J. O. Smith and Chang Keng Liu, "Stresses Due to Tangential and Normal Loads on an Elastic Solid with Application to Some Contact Stress Problems," *Journal of Applied Mechanics*, March 1953.

3–22 Propagation of Error

It is useful to consider the following specific situations arising when relating load to stress accounting for variability:

1. The simple case modeled by $\boldsymbol{\sigma} = \alpha\mathbf{F}$ where the proportionality constant α is deterministic, correlation is clear, and the distribution of $\mathbf{F}$ is propagated to the stress $\boldsymbol{\sigma}$.
2. The case modeled by $\boldsymbol{\sigma} = \boldsymbol{\alpha}\mathbf{F}$ where the proportionality factor $\boldsymbol{\alpha}$ is stochastically independent of load $\mathbf{F}$ and the distribution of stress $\boldsymbol{\sigma}$ may be identifiable by closure theorems.
3. The case where $\boldsymbol{\alpha}$ of $\boldsymbol{\sigma} = \boldsymbol{\alpha}\mathbf{F}$ is sufficiently complex that simulation is the only way to quantify the CDF curve of the stress $\boldsymbol{\sigma}$.
4. This case involves nonlinear relationships of load to stress.

Case 1

When the stress or component of stress is proportional to the load $\mathbf{F}$ through a deterministic constant α, or is robustly so, we can say the correlation coefficient ρ is unity between the load $\mathbf{F}$ and the stress $\boldsymbol{\sigma}$, and we can say ρ is unity in the correlation of stress components with each other. All the stresses have the same distribution type and the same coefficient of variation. If the distribution type is generically denoted as ψ as in $\mathbf{F} \sim \psi(\mu_F, \varsigma_F)$, then we can say

$$\boldsymbol{\sigma} = \alpha\mathbf{F}$$

and the mean and standard deviation of $\boldsymbol{\sigma}$ are given by Table 2-1 as

$$\mu_\sigma = \alpha\mu_F \tag{3–107}$$

$$\sigma_\sigma = \alpha\sigma_F = \alpha\mu_F\frac{\sigma_F}{\mu_F} = \mu_\sigma C_F = \alpha\mu_F C_F \tag{3–108}$$

where $C_F = \sigma_F/\mu_F$. Note that $C_\sigma = \sigma_\sigma/\mu_\sigma = \alpha\mu_F C_F/(\alpha\mu_F) = C_F$.

EXAMPLE 3–12 A 25-mm circular-cross-section rod is subjected to a static axial tensile force F of $\mathbf{F} \sim N(50, 4.56)$ kN.
(*a*) Describe the principal stress $\boldsymbol{\sigma}_1$.
(*b*) Using the nomenclature of Fig. 3–2, describe the oblique plane stresses $\boldsymbol{\sigma}$ and $\boldsymbol{\tau}$.
(*c*) Describe the maximum shear stress $\boldsymbol{\tau}_{max}$.

Solution (*a*) The mean principal stresses are

$$\bar{\sigma}_1 = \frac{\bar{F}}{A} = \frac{50(10^3)}{\pi(25^2)/4} = 101.9 \text{ MPa}, \qquad \bar{\sigma}_2 = 0, \qquad \bar{\sigma}_3 = 0$$

Since the coefficient of variation of the force is $C_F = 4.56/50 = 0.0912$, using Eq. (3–108),

$$\sigma_{\sigma 1} = \mu_{\sigma 1} C_F \qquad \text{and} \quad \boldsymbol{\sigma}_1 \sim N(101.9, 9.29) \text{ MPa}$$
$$\text{or} \quad \boldsymbol{\sigma}_1 \sim 101.9N(1, 0.0912) \text{ MPa}$$

(*b*) With $\sigma_x = 101.9$ MPa, $\sigma_y = 0$, $\tau_{xy} = 0$, Eqs. (3–2) and (3–3) become, respectively,

$$\bar{\sigma} = \frac{101.9 + 0}{2} + \frac{101.9}{2} \cos 2\theta + 0 = 50.95(1 + \cos 2\theta) \text{ MPa}$$

$$\bar{\tau} = -\frac{101.9 - 0}{2} \sin 2\theta = -50.95 \sin 2\theta \text{ MPa}$$

Therefore,

$$\boldsymbol{\sigma} \sim 50.95(1 + \cos 2\theta)\mathbf{N}(1, 0.0912) \text{ MPa}$$
$$\bar{\tau} \sim 50.95 \sin 2\theta\ \mathbf{N}(1, 0.0912) \text{ MPa}$$

(*c*) From Eq. (3–9), $\bar{\tau}_{max}$, $\hat{\sigma}_{\tau max}$, and $\boldsymbol{\tau}_{max}$ become

$$\bar{\tau}_{max} = \sqrt{\left(\frac{101.9 - 0}{2}\right)^2 + 0^2} = 50.95 \text{ MPa}$$

$$\hat{\sigma}_{\tau_{max}} = 50.95(0.0912) = 4.65 \text{ MPa}$$

$$\boldsymbol{\tau}_{max} \sim \mathbf{N}(50.95, 4.65) \text{ MPa} \quad \text{or} \quad \boldsymbol{\tau}_{max} \sim 50.95(1, 0.0912) \text{ MPa}$$

The situation where α is robustly deterministic can be examined by considering the cylinder diameter of Ex. 3-12 when $\mathbf{d} \sim \mathbf{U}(25, 0.1)$ mm. In considering the quotient $4\mathbf{F}/(\pi \mathbf{d}^2)$, the smallness of the coefficient of variation of C_d of $0.1/25 = 0.004$ lets the variability of **F** dominate; then α is considered to be robustly deterministic, and the distribution of $\boldsymbol{\sigma}$ is robustly normal. The precision in machinery dimensions because of their close geometric tolerances makes this kind of robustness common.

Case 2

When the proportionality factor $\boldsymbol{\alpha}$ is not deterministic, or robustly deterministic, then propagation of error is used to find the mean and standard deviation of $\boldsymbol{\sigma}$.

EXAMPLE 3–13

A nominal 25-mm diameter rod, $\mathbf{d} \sim \mathbf{LN}(25, 1)$ mm, is axially loaded with a static tensile force F of $\mathbf{F} \sim \mathbf{LN}(50, 4.56)$ kN.
(*a*) Describe the principal stress $\boldsymbol{\sigma}_1$.
(*b*) Describe the oblique plane stresses $\boldsymbol{\sigma}$ and $\boldsymbol{\tau}$ using the nomenclature of Fig. 3–2.
(*c*) Describe the maximum shear stress $\boldsymbol{\tau}_{max}$.

Solution (*a*)

$$\bar{\sigma}_1 = \frac{\bar{F}}{\bar{A}} = \frac{50(10^3)}{\pi 25^2/4} = 101.9 \text{ MPa}, \qquad \bar{\sigma}_2 = 0, \qquad \bar{\sigma}_3 = 0$$

Since the coefficient of variation of force is $C_F = 4.56/50 = 0.0912$, and of the diameter is $C_d = 1/25 = 0.04$, then from Table 2–6 the coefficient of variation of $\mathbf{d}^2$ is $C_{d^2} = 2C_d = 2(0.04) = 0.08$. For the quotient $\mathbf{F}/\mathbf{d}^2$, the coefficient of variation of $\boldsymbol{\sigma}_1$

is

$$C_\sigma = \sqrt{\frac{C_{F^2} + C_{d^2}^2}{1 + C_{d^2}}} = \sqrt{\frac{0.0912^2 + 0.08^2}{1 + 0.08^2}} = 0.121$$

Thus

$$\sigma_{\sigma_1} = \sigma_1 C_{\sigma_1} = 101.9(0.121) = 12.3 \text{ MPa}$$
$$\boldsymbol{\sigma}_1 \sim LN(101.9, 12.3) \text{ MPa} \quad \text{or} \quad \boldsymbol{\sigma}_1 \sim 101.9LN(1, 0.121) \text{ MPa}$$

The distribution of $\boldsymbol{\sigma}_1$ is lognormal from the closure theorem of lognormals.
(*b*) Now $\bar{\sigma}$ and $\bar{\tau}$ are the same as in Ex. 3–9,

$$\bar{\sigma} = 50.95(1 + \cos 2\theta) \text{ MPa}, \qquad \bar{\tau} = -50.95 \sin 2\theta \text{ MPa}$$

and

$$\boldsymbol{\sigma} \sim 50.95(1 + \cos 2\theta)LN(1, 0.121) \text{ MPa}$$
$$\boldsymbol{\tau} \sim -50.95 \sin 2\theta LN(1, 0.121) \text{ MPa}$$

(*c*) Similarly, from Eq. (3–9), $\bar{\tau}_{max} = 50.95$ MPa as before and $\sigma_{\tau_{max}} = 50.95(0.121) = 6.16$ MPa

$$\boldsymbol{\tau}_{max} \sim LN(-50.95, 6.16) \text{ MPa} \quad \text{or} \quad \boldsymbol{\tau}_{max} \sim -50.95 \ln(1, 0.121) \text{ MPa}$$

Note that the means of $\boldsymbol{\sigma}_1$, $\boldsymbol{\sigma}$, $\boldsymbol{\tau}$, and $\boldsymbol{\tau}_{max}$ have not changed but the dispersion of all has increased.

Case 3

Simulation is necessary where there is no clue as to the distribution propagation.

EXAMPLE 3–14

A cylindrical rod with a diameter d from $\mathbf{d} \sim U[24, 26]$ mm is subjected to a static tensile force F of $\mathbf{F} \sim N(50, 4.56)$ kN. Describe $\boldsymbol{\sigma}_1$.

Solution

A computer program was written to generate by simulation the parameters of $\boldsymbol{\sigma}_1$ and to check on those of **d** and **F**. With 10^5 trials the data for the following table were collected.

	Mean	Standard Deviation	COV
Stress $\boldsymbol{\sigma}_1$	101.982 1	10.438 32	0.102 354 5
Diameter **d**	25.000 43	0.584 510 6	0.023 392 0
Force **F**	49.981 77	4.557 829	0.091 189 8

The standard deviation of diameter **d** is

$$\sigma_d = \frac{26 - 24}{2\sqrt{3}} = 0.577\,350$$

thus

$$\mathbf{d} \sim U(25, 0.577) \text{ mm}$$

The coefficient of variation of **d** is $C_d = 0.577\,350/25 = 0.023\,094$. The coefficient of variation of $\mathbf{d}^2$ is, from Table 2–6,

$$C_{d^2} = 2C_d = 2(0.023\,094) = 0.046\,118\,8$$

The coefficient of variation of $4F/(\pi d^2)$ is, from Table 2–6,

$$C_{\sigma_1} = \sqrt{\frac{C_F^2 + C_{d^2}^2}{1 + C_{d^2}^2}} = \sqrt{\frac{0.0912^2 + 0.0462^2}{1 + 0.0462^2}} = 0.1022$$

which checks the table. The problem now is to create an approximation of the CDF of $\boldsymbol{\sigma}_1$, or its PDF, or either, in just a locality. We choose to generate some of the lower tail of the CDF of $\boldsymbol{\sigma}_1$. The program already written can be amended to find the number of instances of σ_1 in a specified interval of stress. This allows a histogram in PDF style to be generated, class interval by class interval, or by making the lower bound very small, instances of F to be generated at the specified upper bound, thereby generating points on a $F\sigma_1$ locus. For 10^5 trials the number of observations below a specific value of σ_1 was generated, producing the following table.

σ_1	Number up to σ_1	CDF of σ_1
69	2	0.0002
70	4	0.0004
71	7	0.0007
72	9	0.0009
73	8	0.0008
74	23	0.0023
75	19	0.0019
76	47	0.0047
77	67	0.0067
78	82	0.0082
79	111	0.0111
80	130	0.130

The CDF shows some scatter. The value of CDF from the table unsmoothed is 0.0067 for $\sigma_1 = 77$ MPa. If we hoped that the principal stress was robustly normal, we might note that

$$\boldsymbol{\sigma}_1 \sim N(101.9, 10.4) \text{ MPa} \qquad \text{(assumption)}$$

$$C_{\sigma_1} = \frac{10.4}{101.9} = 0.102$$

$$z = \frac{\sigma_1 - \bar{\sigma}_1}{\hat{\sigma}_{\sigma_1}} = \frac{77 - 101.9}{10.4} = -2.39$$

$$p = 1 - \Phi(z) = 1 - \Phi(-2.39) = 1 - 0.9916 = 0.0084$$

A plot of the $F\sigma_1$ locus shows there are only a few scattered points, and a least-squared polynomial will not change the $\sigma_1 = 77$ MPa value much, so the normal assumption overestimates the CDF in this locality. Simulation will give us a good approximation to the actual CDF.

Case 4

Contact stresses are examples of stresses that have a nonlinear relationship to load. Consider the Hertzian stress $\mathbf{p}_{\max}$ of a sphere on a sphere carrying load $\mathbf{F}$. Substituting Eq. (3–88) into Eq. (3–89) gives

$$p_{\max} = \frac{3F^{1/3}}{2\pi}\left[\frac{3}{8}\frac{(1-\nu_1^2)/E_1 + (1-\nu_2^2)/E_2}{1/d_1 + 1/d_2}\right]^{-2/3}$$

Considering this a Case 1, and construing the form to be $\mathbf{p}_{\max} = \alpha\mathbf{F}^{1/3}$, we write

$$\mu_p = \alpha\mu_F^{1/3}$$

$$\sigma_p = \frac{\alpha\mu_F^{1/3}C_F}{3}$$

$$C_p = \frac{\sigma_p}{\mu_p} = \frac{\frac{1}{3}\alpha\mu_F^{1/3}C_F}{\alpha\mu_F^{1/3}} = \frac{C_F}{3}$$

and

$$\mathbf{p}_{\max} \sim \frac{3\bar{F}^{1/3}}{2\pi}\left[\frac{3}{8}\frac{(1-\nu_1^2)/E_1 + (1-\nu_2^2)/E_2}{1/d_1 + 1/d_2}\right]^{-2/3}\psi\left(1, \frac{C_F}{3}\right) \tag{3–109}$$

where ψ is the distribution type of load $\mathbf{F}$.

Consider the Hertzian stress $\mathbf{p}_{\max}$ of contacting cylinders carrying load $\boldsymbol{F}$. Substituting Eq. (3–93) into Eq. (3–94) gives

$$p_{\max} = \frac{2\sqrt{F}}{\pi l}\left[\frac{2}{\pi l}\frac{(1-\nu_1^2)/E_1 + (1-\nu_2^2)/E_2}{1/d_1 + 1/d_2}\right]^{-1/2}$$

which is of the form $\mathbf{p}_{\max} = \beta\sqrt{\mathbf{F}}$. We write for Case 1 treatment

$$\mu_p = \beta\mu_F^{1/2}$$

$$\sigma_p = \frac{\beta\mu_F^{1/2}C_F}{2}$$

$$C_p = \frac{\sigma_p}{\mu_p} = \frac{\frac{1}{2}\beta\mu_F^{1/2}C_p}{\beta\mu_F^{1/2}} = \frac{C_F}{2}$$

thus

$$\mathbf{p}_{\max} \sim \frac{2\sqrt{\bar{F}}}{\pi l}\left[\frac{2}{\pi l}\frac{(1-\nu_1^2)/E_1 + (1-\nu_2^2)/E_2}{1/d_1 + 1/d_2}\right]^{-1/2}\psi\left(1, \frac{C_F}{2}\right) \tag{3–110}$$

where ψ is the distribution type of load $\mathbf{F}$.

If ν_1, ν_2, E_1, and E_2 are stochastic, then a Case 3 simulation is the only way to proceed, given the algebraic complexity.

3–23 Summary

The ability to quantify the stress condition at a critical location in a machine element is an important skill of the engineer. Why? Whether the member fails or not is assessed by comparing the (damaging) stress at a critical location with the corresponding material strength at this location. This chapter has addressed the description of stress.

Stresses can be estimated with great precision where the geometry is sufficiently simple that theory easily provides the necessary quantitative relationships. In other cases, approximations are used. There are numerical approximations such as finite element analysis (FEA) whose results tend to converge on the true values. There are experimental measurements, strain-gauging, for example, allowing *inference* of stresses from the measured strain conditions. Whatever the method(s), the goal is a robust description of the stress condition at a critical location.

The nature of research results and understanding in any field is that the longer we work on it, the more involved things seem to be, and new approaches are sought to help with the complications. As newer schemes are introduced, engineers, hungry for the improvement the new approach *promises*, begins to use the approach. Optimism usually recedes, as further experience adds concerns. Tasks that promised to extend the capabilities of the nonexpert eventually show that expertise is not optional.

In stress analysis, the computer can be helpful if the necessary equations are available. Spreadsheet analysis can quickly reduce complicated calculations for parametric studies, easily handling "what if" questions relating trade-offs (e.g., less of a costly material or more of a cheaper material). It can even give insight into optimization opportunities.

When the necessary equations are not available, then methods such as FEA are attractive, but cautions are in order. Even when you have access to a powerful FEA code, you should be near an expert while you are learning. There are nagging questions of convergence at discontinuities. Elastic analysis is much easier than elastic-plastic analysis. The results are no better than the modeling of reality that was used to formulate the problem.

PROBLEMS

3–1[13] For each of the stress states listed below, draw a Mohr circle diagram properly labeled, find the principal normal and shear stresses, and determine the angle from the x axis to σ_1. Draw a stress element as in Fig. 3–5b and label all details.

(*a*) $\sigma_x = 12, \sigma_y = 6, \tau_{xy} = 4$ cw
(*b*) $\sigma_x = 16, \sigma_y = 9, \tau_{xy} = 5$ ccw
(*c*) $\sigma_x = 10, \sigma_y = 24, \tau_{xy} = 6$ ccw
(*d*) $\sigma_x = 9, \sigma_y = 19, \tau_{xy} = 8$ cw

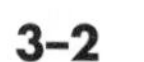

3–2 Repeat Prob. 3–1 for:

(*a*) $\sigma_x = -4, \sigma_y = 12, \tau_{xy} = 7$ ccw
(*b*) $\sigma_x = 6, \sigma_y = -5, \tau_{xy} = 8$ ccw
(*c*) $\sigma_x = -8, \sigma_y = 7, \tau_{xy} = 6$ cw
(*d*) $\sigma_x = 9, \sigma_y = -6, \tau_{xy} = 3$ cw

---

13. Stress components are always assumed to be zero unless given. Thus, in this example, $\sigma_z = 0$ for all four cases.

ANALYSIS

3–3 Repeat Prob. 3–1 for:

(*a*) $\sigma_x = 20, \sigma_y = -10, \tau_{xy} = 8$ cw
(*b*) $\sigma_x = 30, \sigma_y = -10, \tau_{xy} = 10$ ccw
(*c*) $\sigma_x = -10, \sigma_y = 18, \tau_{xy} = 9$ cw
(*d*) $\sigma_x = -12, \sigma_y = 22, \tau_{xy} = 12$ cw

ANALYSIS

3–4[14] For each of the stress states listed below, find all three principal normal and shear stresses, and the octahedral shear stress. Draw a complete Mohr three-circle diagram and label all points of interest.

(*a*) $\sigma_x = 10, \sigma_y = -4$
(*b*) $\sigma_x = 10, \tau_{xy} = 4$ ccw
(*c*) $\sigma_x = -2, \sigma_y = -8, \tau_{xy} = 4$ cw
(*d*) $\sigma_x = 10, \sigma_y = -30, \tau_{xy} = 10$ ccw

3–5 Repeat Prob. 3-4 for:

(*a*) $\sigma_x = -80, \sigma_y = -30, \tau_{xy} = 20$ cw
(*b*) $\sigma_x = 30, \sigma_y = -60, \tau_{xy} = 30$ cw
(*c*) $\sigma_x = 40, \sigma_z = -30, \tau_{xy} = 20$ ccw
(*d*) $\sigma_x = 50, \sigma_z = -20, \tau_{xy} = 30$ cw

3–6 A $\frac{1}{2}$-in-diameter steel tension rod is 72 in long and carries a load of 2000 lb. Find the tensile stress, the total deformation, the unit strains, and the change in the rod diameter.

3–7 Twin diagonal aluminum alloy tension rods 15 mm in diameter are used in a rectangular frame to prevent collapse. The rods can safely support a tensile stress of 135 MPa. If the rods are initially 3 m in length, how much must they be stretched to develop this stress?

3–8 Electrical strain gauges were applied to a notched specimen to determine the stresses in the notch. The results were $\epsilon_x = 0.0021$ and $\epsilon_y = -0.00067$. Find σ_x, σ_y, and the octahedral shear stress if the material is carbon steel.

3–9 The symbol W is used in the various figure parts to specify the weight of an element. If not given, assume the parts are weightless. For each figure part sketch a free-body diagram of each element including the frame. Try to get the forces in the proper directions, but do not compute magnitudes.

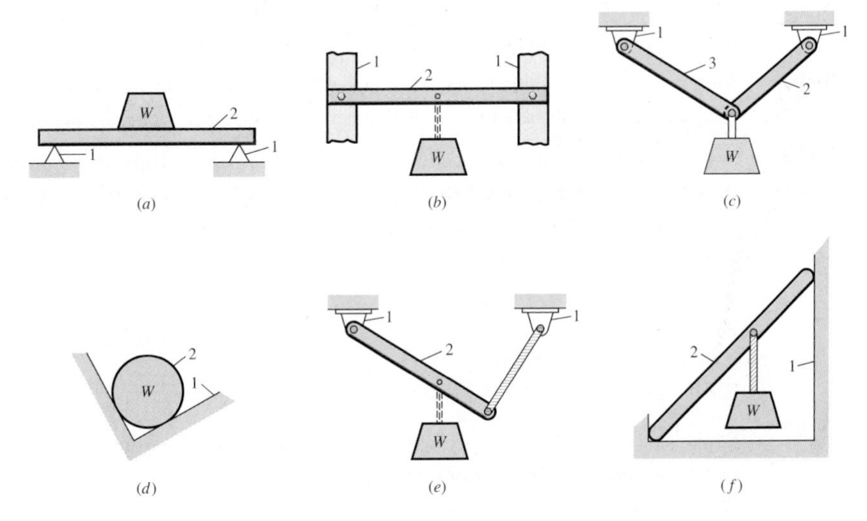

Problem 3–9

14. Stress components are always assumed to be zero unless given. Thus, in this example, $\sigma_z = 0$ for all four cases.

ANALYSIS

3–10 Using the figure part selected by your instructor, sketch a free-body diagram of each element in the figure. Compute the magnitude and direction of each force using an algebraic or vector method, as specified.

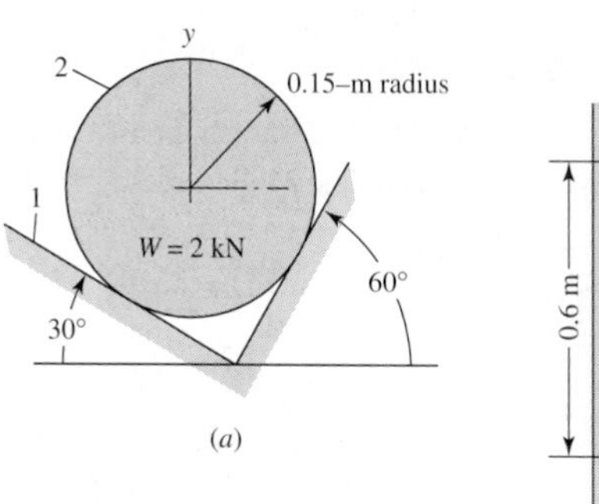

(*a*)

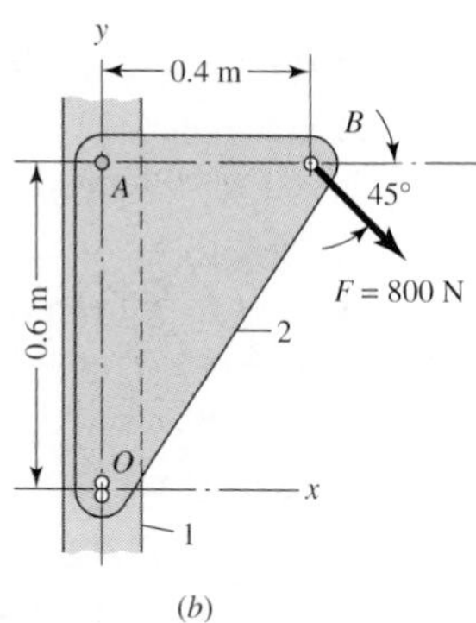

(*b*)

Problem 3–10

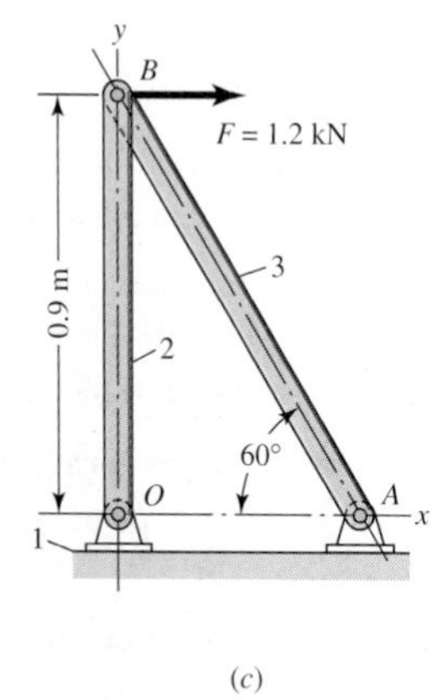

(*c*)

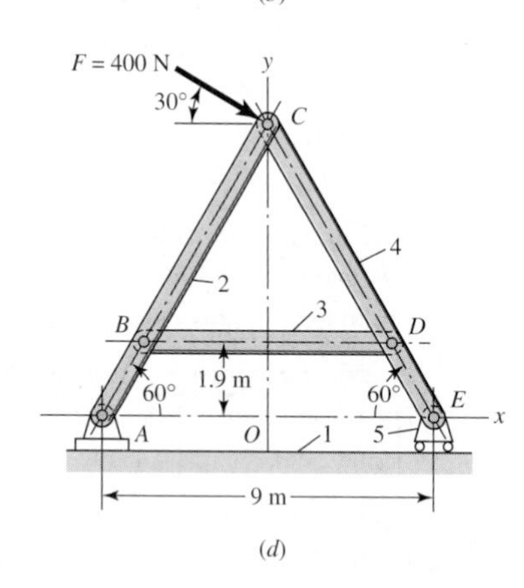

(*d*)

ANALYSIS

3–11 Find the reactions at the supports and plot the shear-force and bending-moment diagrams for each of the beams shown in the figure on p. 157. Label the diagrams properly.

3–12 Select a beam from Table E-9 and find general expressions for the loading, shear-force, bending-moment, and support reactions. Use the method specified by your instructor.

DESIGN

3–13 A beam carrying a uniform load is simply supported with the supports set back a distance a from the ends as shown in the figure. The bending moment at x can be found from summing moments to zero at section x:

$$\Sigma M = M_x + \frac{1}{2}w(a+x)^2 \frac{1}{2}wlx = 0$$

or

$$M_x = \frac{w}{2}[lx - (a+x)^2]$$

where w is the loading intensity in lb/in. The designer wishes to minimize the necessary weight of the supporting beam by choosing a setback a resulting in the smallest possible maximum bending stress.

(*a*) If the beam is configured with $a = 2.25$ in, $l = 10$ in, and $w = 100$ lb/in, find the magnitude of the severest bending moment in the beam.

(*b*) Since the configuration in part (*a*) is not optimal, find the optimal setback which will result in the lightest-weight beam.

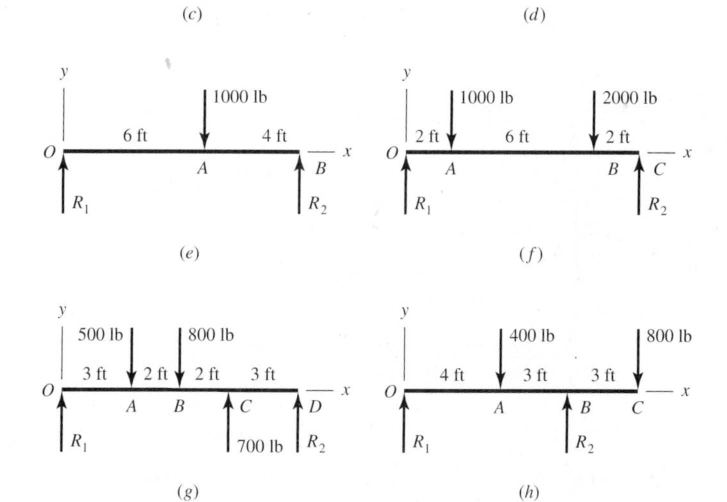

Problem 3–11

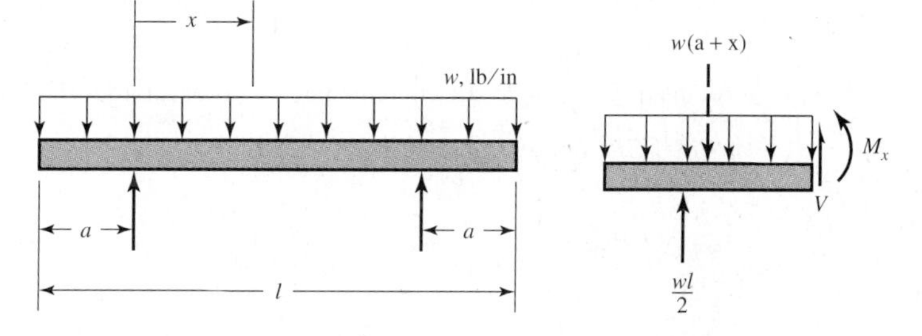

Problem 3–13

3–14 An artist wishes to construct a mobile using pendants, string, and span wire with eyelets as shown in the figure on p. 158.

(*a*) At what positions w, x, y, and z should the suspension strings be attached to the span wires?
(*b*) Is the mobile stable? If so, justify; if not, suggest a remedy.

3–15 Two steel thin-wall tubes of axial length z are to be compared. The first is of square cross section, side length b, and wall thickness t. The second is a round of diameter b and wall thickness t. The largest allowable shear stress is τ_{all} and is to be the same in both cases. How does the angle of twist per unit length compare in each case?

3–16 Begin with a 1-in-square thin-wall steel tube, wall thickness $t = 0.05$ in, length 40 in, then introduce corner radii of inside radii r_i, with allowable shear stress ψ_{all} of 11 500 psi, shear

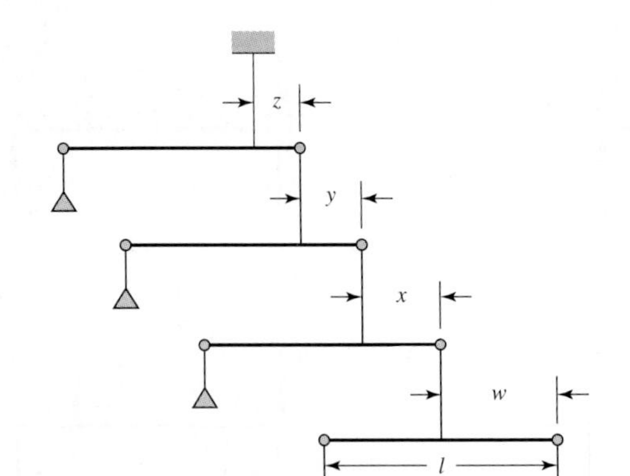

Problem 3–14

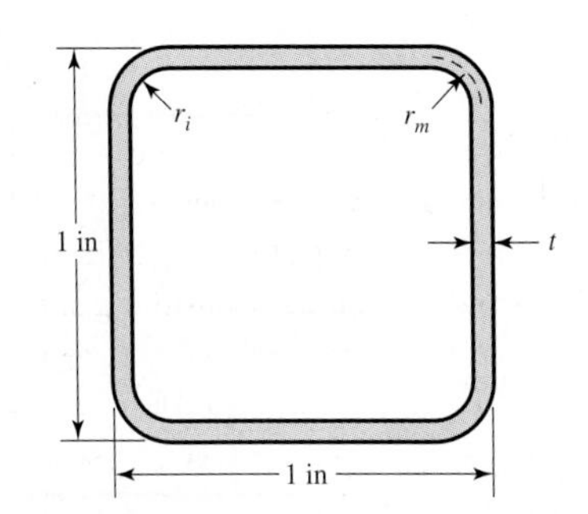

Problem 3–16

modulus of $11.5(10^6)$ psi; now form a table. Use a column of inside corner radii in the range $0 \leq r_i \leq 0.45$ in. Useful columns include median line radius r_m, periphery of the median line L, area enclosed by median curve, torque T, and the angular twist θ. The cross section will vary from square to circular round. A computer program will reduce the calculation effort. Study the table. What have you learned?

3–17 An unequal leg angle shown in the figure carries a torque T. Show that

$$T = \frac{G\theta_1}{3} \sum L_i c_i^3$$

$$\tau_{\max} = G\theta c_{\max}$$

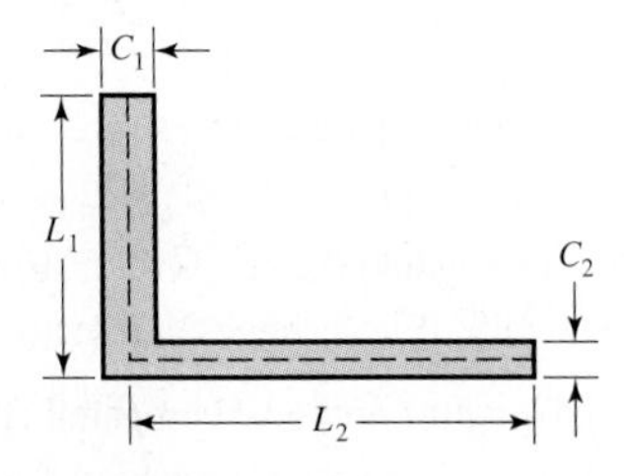

Problem 3–17

ANALYSIS

3–18 In Prob. 3-17 the angle has one leg thickness $\frac{1}{16}$ in and the other $\frac{3}{32}$ in, with both leg lengths $\frac{1}{2}$ in. The allowable shear stress is $\tau_{\text{all}} = 11\ 500$ psi for this steel angle.

(a) Find the torque carried by each leg, and the largest shear stress therein.

(b) Find the angle of twist of the angle.

3–19 Find the torque, angular twist, and torsional spring rate for the two thin rectangular-cross-section strips place together as shown in the figure. Compare these with a single strip of cross section 1 in by $\frac{1}{8}$ in, 12 in long.

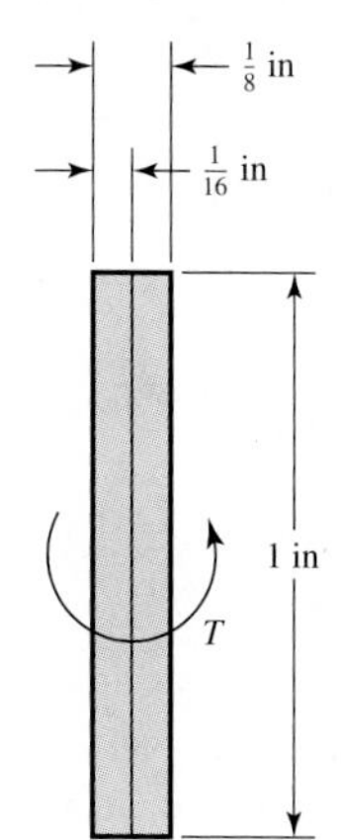

Problem 3–19

3–20 The Roman method for addressing uncertainty in design was to build a copy of a design that was satisfactory and had proven durable. Although the early Romans did not have the intellectual tools to deal with scaling size up or down, you do. Consider a simply supported, rectangular-cross-section beam with a concentrated load F, as depicted in the figure.

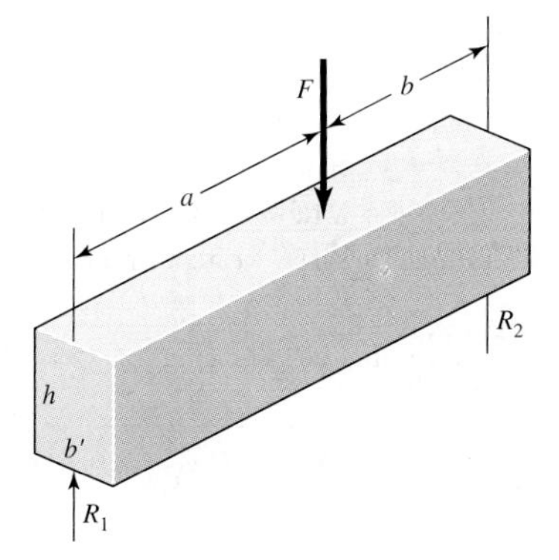

Problem 3–20

(*a*) Show that the stress-to-load equation is

$$F = \frac{\sigma b' h^2}{6ab}$$

(*b*) Subscript every parameter with m (for model) and divide into the above equation. Introduce a scale factor, $s = a_m/a$, for example. Since the Roman method was to not “lean on” the material any more than the proven design, set $\sigma_m/\sigma = 1$. Express F_m in terms of the scale factors and F, and comment on what you have learned.

3–21 Using our experience with concentrated loading on a simple beam, consider a uniformly loaded simple beam (Table E-9-7).

(*a*) Show that the stress-to-load equation for a rectangular-cross-section beam is given by

$$W = \frac{2}{3} \frac{\sigma b' h^2}{l}$$

where $W = w/l$.

(*b*) Subscript every parameter with m (for model) and divide the model equation into the prototype equation. Introduce the scale factor s as in Prob. 3–20, setting $\sigma_m/\sigma = 1$. Express W_m and w_m in terms of the scale factor, and comment on what you have learned.

ANALYSIS

3–22 Problem 3–21 produced an up-scale model load of $W_m = s^2 W$. As structural weight becomes significant in a beam, it is useful to decompose W into a productive load W_p and a beam weight W_b.

(*a*) Noting that beam weight $W_{bm} = s^3 W_b$, show that

$$W_{pm} = s^2[W_p - W_b(s - 1)]$$

(*b*) We can see that as s^2 increases from 1, W_{pm} does not grow as $s^2 W_p$, but less. Identify the scale factor at which W_{pm} is maximum.

(*c*) Identify the scale factor at which $W_{pm} = 0$.

(*d*) For $W_p = 1000$ lb, $W_b = 200$ lb, plot W_{bm} versus s.

ANALYSIS

3–23 The Chicago North Shore & Milwaukee Railroad was an electric railway running between the cities in its corporate title. It had passenger cars as shown in the figure which weighed 104.4 kip, had 32 ft, 8 in truck centers, 7 ft wheelbase trucks, and a coupled length of 55 ft, $3\frac{1}{4}$ in. Consider the case of a single car on a 100-ft-long, simply supported deck plate girder bridge.

(*a*) What was the largest bending moment in the bridge?

(*b*) Where on the bridge was the moment located?

(*c*) What was the position of the car on the bridge?

(*d*) Under which axle is the bending moment?

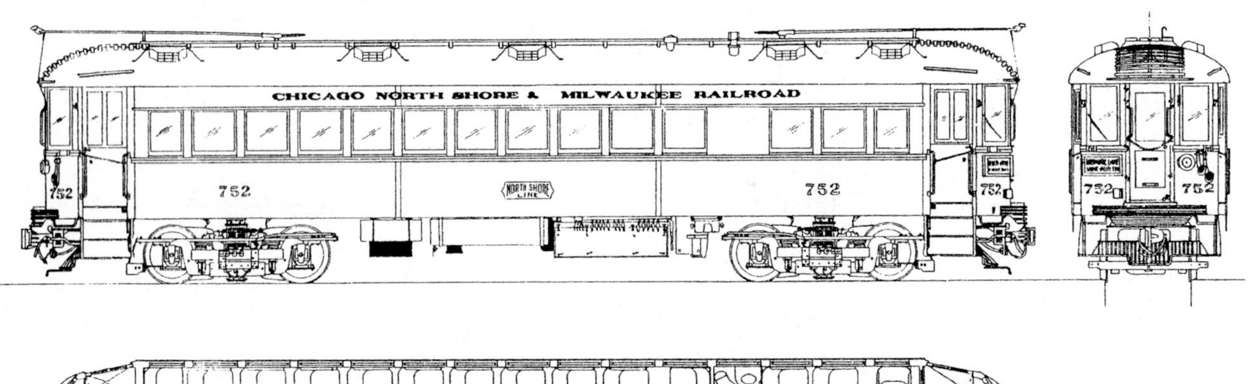

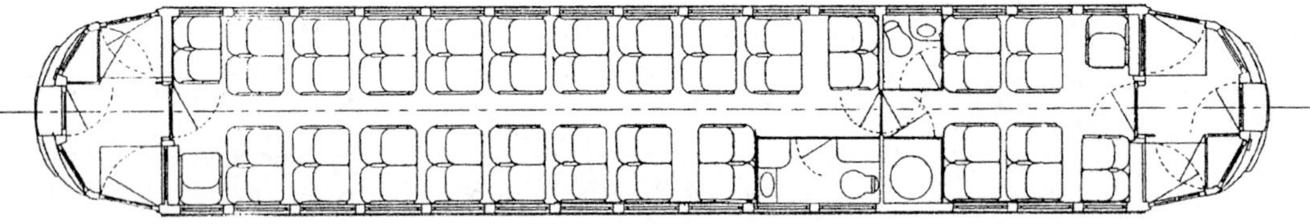

Drawing: LF, RGB

Scale in feet

0 10 15

Coaches 752–776

AS ORIGINALLY BUILT

Problem 3–23

3–24 For each section illustrated, find the second moment of area, the location of the neutral axis, and the distances from the neutral axis to the top and bottom surfaces. Suppose a positive bending moment of 10 kip · in is applied; find the resulting stresses at the top and bottom surfaces and at every abrupt change in cross section.

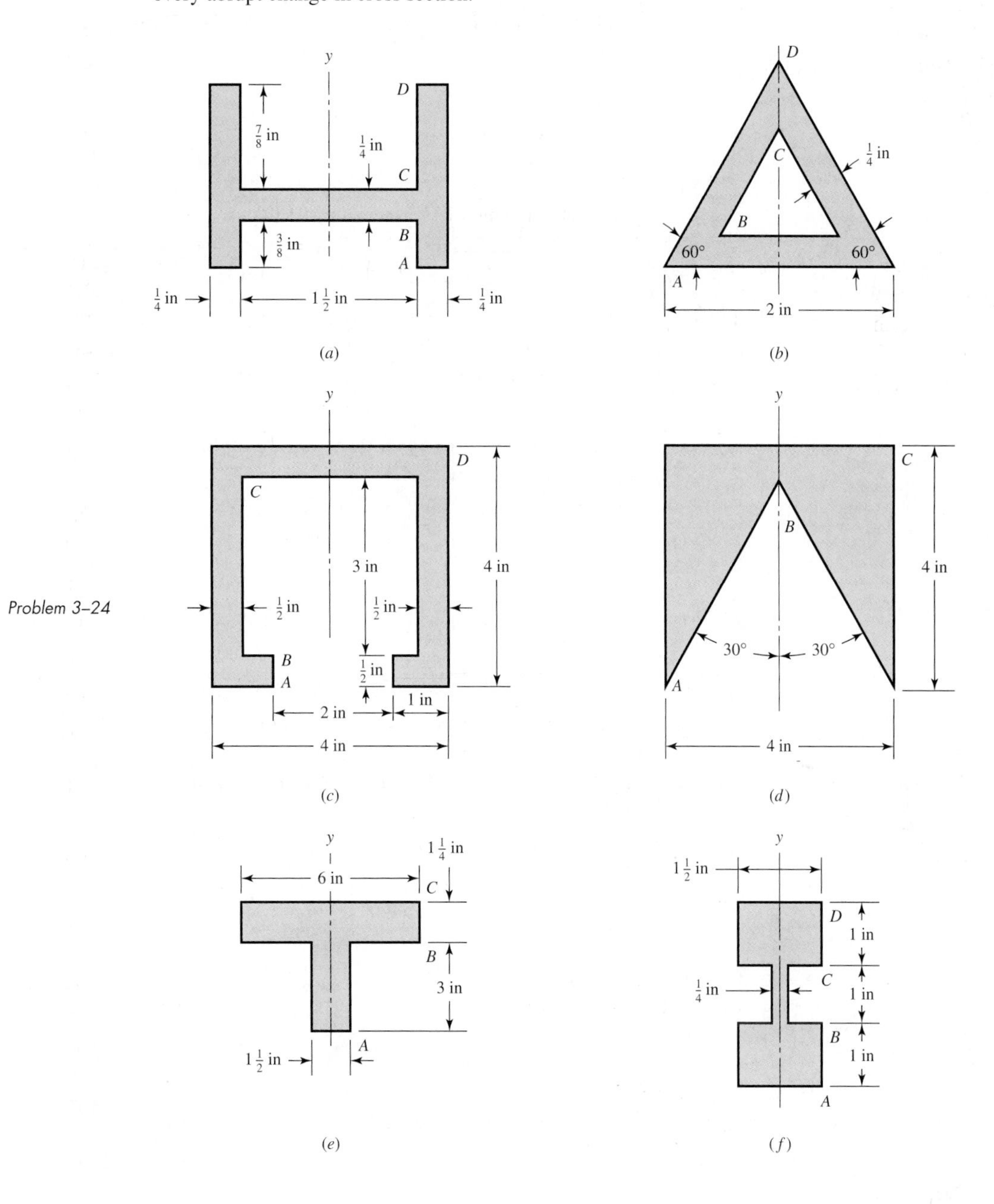

Problem 3–24

3–25 Find the x and y coordinates of the center of curvature corresponding to the place where the beam is bent the most, for each beam shown in the figure on p. 162. The beams are both made of Douglas fir (see Table A-5) and have rectangular sections.

3–26 For each beam illustrated in the figure on p. 162, find the locations and magnitudes of the maximum tensile bending stress and the maximum shear stress.

Problem 3–25

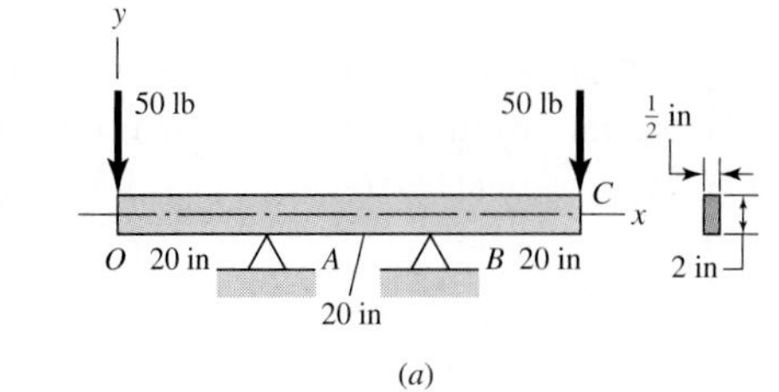

(a)

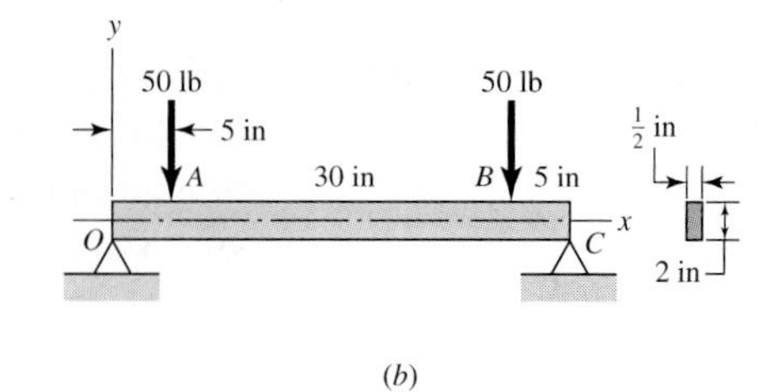

(b)

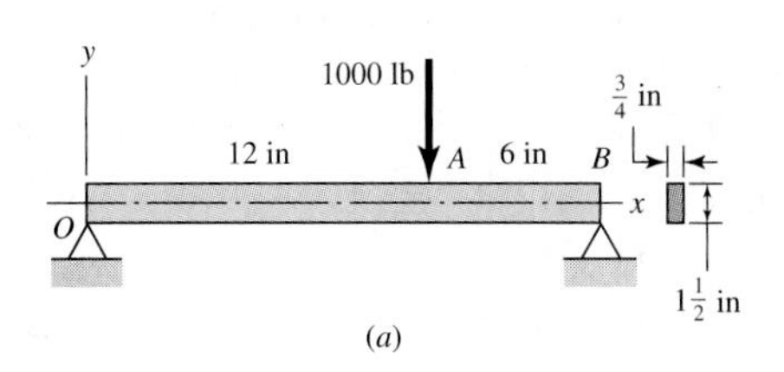

(a)

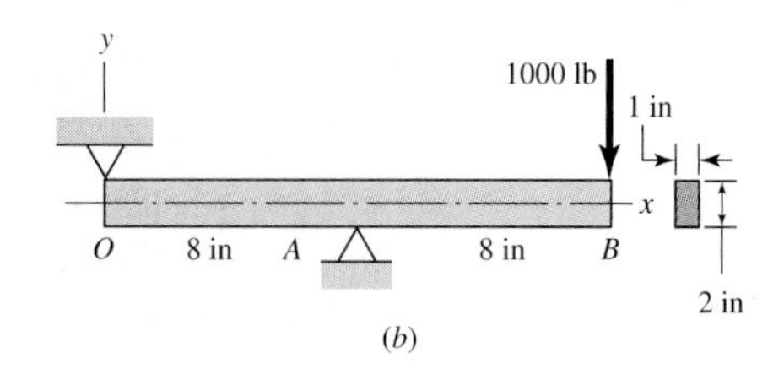

(b)

Problem 3–26

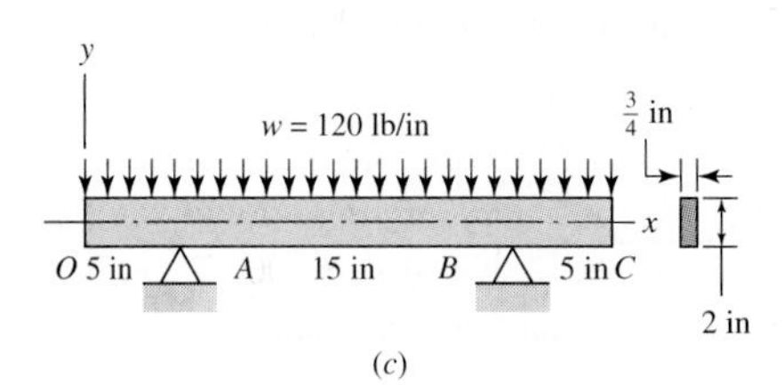

(c)

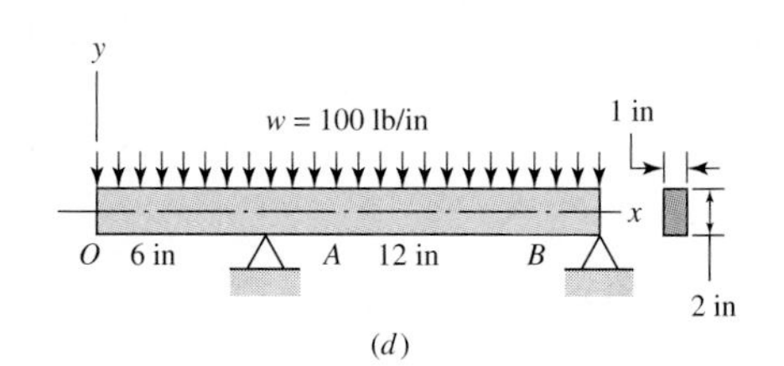

(d)

3–27 The figure illustrates a number of beam sections. Use an allowable stress of 1.2 kpsi for wood and 12 kpsi for steel and find the maximum safe uniformly distributed load that each beam can carry if the given lengths are between simple supports.

(*a*) Wood joist $1\frac{1}{2}$ by $9\frac{1}{2}$ in and 12 ft long

(*b*) Steel tube, 2 in OD by $\frac{3}{8}$-in wall thickness, 48 in long

(*c*) Hollow steel tube 3 by 2 in, outside dimensions, formed from $\frac{3}{16}$-in material and welded, 48 in long

(*d*) Steel angles $3 \times 3 \times \frac{1}{4}$ in and 72 in long

(*e*) A 5.4-lb, 4-in steel channel, 72 in long

(*f*) A 4-in × 1-in steel bar, 72 in long

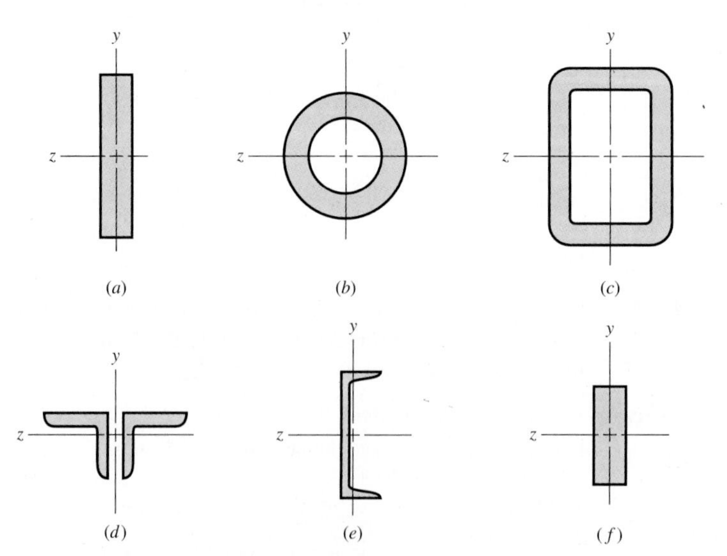

Problem 3–27

ANALYSIS

3–28 Using a maximum allowable shear stress of 8000 psi, find the shaft diameter needed to transmit 50 hp when

(*a*) The shaft speed is 2000 r/min.

(*b*) The shaft speed is 200 r/min.

ANALYSIS

3–29 A 15-mm-diameter steel bar is to be used as a torsion spring. If the torsional stress in the bar is not to exceed 110 MPa when one end is twisted through an angle of 30°, what must be the length of the bar?

ANALYSIS

3–30 A 70-mm-diameter solid steel shaft, used as a torque transmitter, is replaced with a 70-mm hollow shaft having a 6-mm wall thickness. If both materials have the same strength, what is the percentage reduction in torque transmission? What is the percentage reduction in shaft weight?

ANALYSIS

3–31 A hollow steel shaft is to transmit 5400 N · m of torque and is to be sized so that the torsional stress does not exceed 150 MPa.

(*a*) If the inside diameter is three-fourths of the outside diameter, what size shaft should be used? Use preferred sizes.

(*b*) What is the stress on the inside of the shaft when full torque is applied?

ANALYSIS

3–32 A pin in a knuckle joint carrying a tensile load F deflects somewhat on account of this loading, making the distribution of reaction and load as shown in part *b* of the figure. The usual designer's assumption of loading is shown in part *c*; others sometimes choose the loading shown in part *d*. If $a = 0.5$ in, $b = 0.75$ in, $d = 0.5$ in, and $F = 1000$ lb, estimate the maximum bending stress and the maximum shear stress for each approximation.

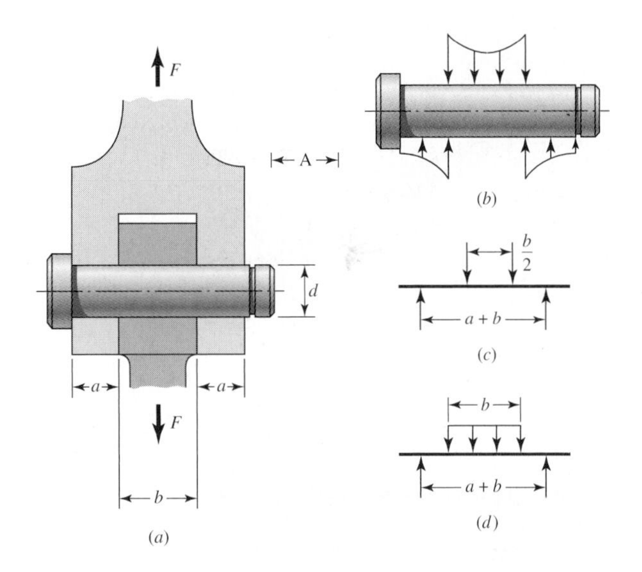

Problem 3–32

DESIGN

3–33 The figure on p. 164 illustrates two gears which are to be mounted on a shaft on the x axis. The gear loads are $F_1 = 2$ kip and $F_2 = 1.1$ kip. If one bearing is located at A, should the second bearing be located between the gears, or outboard, say, near B? Possible solutions will involve designing for minimum bending moment, or for equal bearing reactions.

DESIGN

3–34 The figure on p. 164 illustrates a pin tightly fitted into a hole of a substantial member. A usual analysis is that of assuming concentrated reactions R and M at distance l from F. Suppose the reaction is distributed along distance a. Is the resulting moment reaction larger or smaller than the concentrated reaction? What is the loading intensity q? What do you think of using the usual assumption?

DESIGN

3–35 The figure on p. 164 shows an endless-belt conveyor drive roll. The roll has a diameter of 6 in and is driven at 5 r/min by a geared-motor source rated at 1 hp. Determine a suitable shaft diameter d_C for an allowable torsional stress of 14 kpsi.

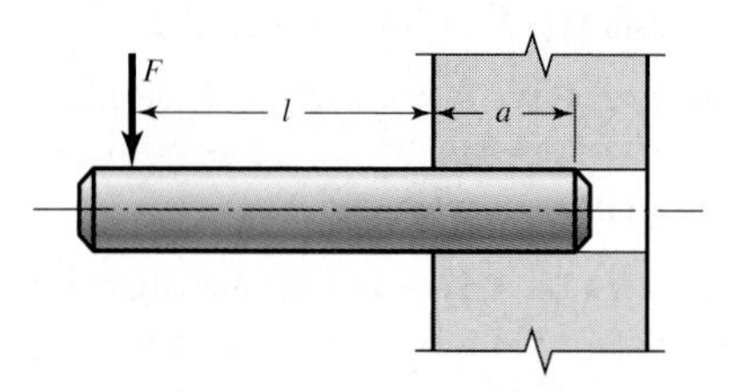

Problem 3–33

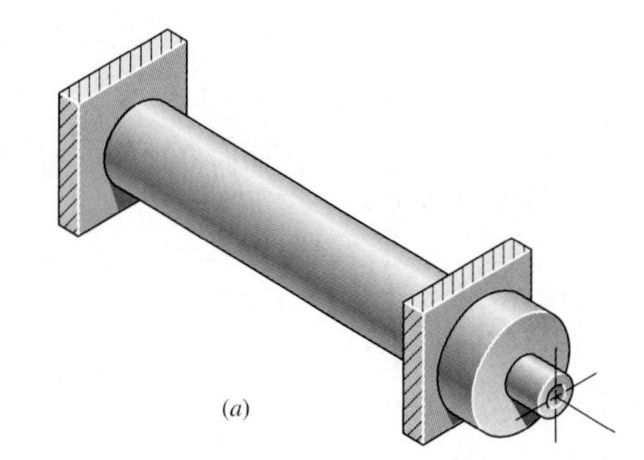

Problem 3–34

Problem 3–35

(a)

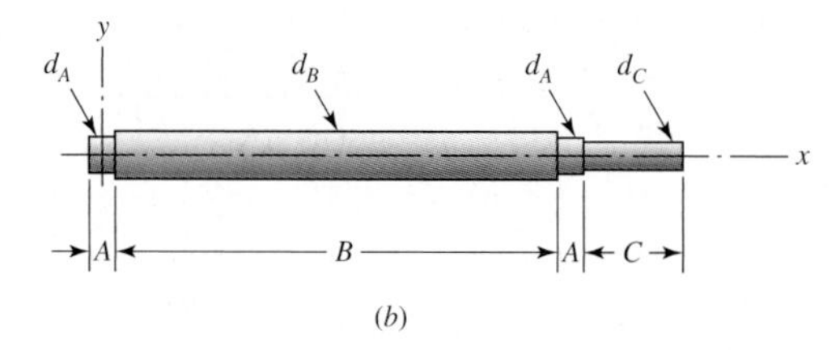

(b)

(a) What would be the stress in the shaft you have sized if the motor starting torque is twice the running torque?

(b) Is bending stress likely to be a problem? What is the effect of different roll lengths B on bending?

3–36 The conveyer drive roll in the figure for Prob. 3-35 is 150 mm in diameter and is driven at 8 rev/min by a geared-motor source rated at 1 kW. Find a suitable shaft diameter d_C based on an allowable torsional stress of 75 MPa.

3–37 If the tension-loaded plate of Fig. 3-26 is infinitely wide, then the stress state anywhere in the plate can be described using polar coordinates as

$$\sigma_r = \frac{\sigma}{2}\left(1 - \frac{d^2}{4r^2}\right) - \frac{\sigma}{2}\left(1 - \frac{d^2}{4r^2}\right)\left(1 - \frac{3d^2}{4r^2}\right)\cos 2\theta$$

$$\sigma_\theta = \frac{\sigma}{2}\left(1 + \frac{d^2}{4r^2}\right) + \frac{\sigma}{2}\left(1 + \frac{3d^4}{16r^4}\right)\cos 2\theta$$

$$\tau_{r\theta} = \frac{\sigma}{2}\left(1 - \frac{d^2}{4r^2}\right)\left(1 + \frac{3d^2}{4r^2}\right)\sin 2\theta$$

for the radial, tangential, and shear components, respectively. Here r is the distance to the point of interest and θ is measured positive from the x axis.

(*a*) Find the stress components at the top and side of the hole for $r = d/2$.

(*b*) Plot a graph of the stress distribution, similar to that of Fig. 3-26, out to $r = 20$ mm for $d = 10$ mm.

3–38 Develop the formulas for the maximum radial and tangential stresses in a thick-walled cylinder due to internal pressure only.

3–39 Repeat Prob. 3-38 where the cylinder is subject to external pressure only. At what radii do the maximum stresses occur?

3–40 Develop the stress relations for a thin-walled spherical pressure vessel.

3–41 A pressure cylinder has a diameter of 6 in and has a $\frac{1}{4}$-in wall thickness. What pressure can this vessel carry if the maximum shear stress is not to exceed 4000 psi?

3–42 A pressure vessel has an outside diameter of 240 mm and a wall thickness of 10 mm. If the internal pressure is 2400 kPa, what is the maximum shear stress in the vessel walls?

3–43 An AISI 1020 cold-drawn steel tube has an ID of $1\frac{1}{4}$ in and an OD of $1\frac{3}{4}$ in. What maximum external pressure can this tube take if the largest principal normal stress is not to exceed 80 percent of the minimum yield strength of the material?

3–44 An AISI 1020 cold-drawn steel tube has an ID of $1\frac{3}{4}$ in and an OD of $2\frac{3}{8}$ in. What maximum internal pressure can this tube take if the largest principal normal stress is not to exceed 80 percent of the minimum yield strength of the material?

3–45 Find the maximum shear stress in a 10-in circular saw if it runs idle at 7200 r/min. The saw is 14-gauge (0.0747 in) and is used on a $\frac{3}{4}$-in arbor. The thickness is uniform. What is the maximum radial component of stress?

3–46 The maximum recommended speed for a 300-mm-diameter abrasive grinding wheel is 2069 r/min. Assume that the material is isotropic; use a bore of 25 mm, $\nu = 0.24$, and a mass density of 3320 kg/m^3; and find the maximum tensile stress at this speed.

3–47 An abrasive cutoff wheel has a diameter of 6 in, is $\frac{1}{16}$ in thick, and has a 1-in bore. It weighs 6 oz and is designed to run at 10 000 rev/min. If the material is isotropic and $\nu = 0.20$, find the octahedral shear stress at the design speed.

3–48 to 3–53 The table on p. 166 lists the maximum and minimum hole and shaft dimensions for a variety of standard press and shrink fits. The materials are both hot-rolled steel. Find the maximum and minimum values of the radial interference and the interface pressure. Use a collar diameter of 80 mm for the metric sizes and 3 in for those in inch units.

3–54 An engineer wishes to determine the shearing strength of a certain epoxy cement. The problem is to devise a test specimen such that the joint is subject to pure shear. The joint shown in the figure, in which two bars are offset at an angle θ so as to keep the loading force F centroidal with the

Problem Number	Fit Designation*	Basic Size	Hole D_{max}	Hole D_{min}	Shaft d_{max}	Shaft d_{min}
3-48	40H7/p6	40 mm	40.025	40.000	40.042	40.026
3-49	(1.5 in)H7/p6	1.5 in	1.5010	1.5000	1.5016	1.5010
3-50	40H7/s6	40 mm	40.025	40.000	40.059	40.043
3-51	(1.5 in)H7/s6	1.5 in	1.5010	1.5000	1.5023	1.5017
3-52	40H7/u6	40 mm	40.025	40.000	40.076	40.060
3-53	(1.5 in)H7/u6	1.5 in	1.5010	1.5000	1.5030	1.5024

* *Notes:* See Table 4–5 for description of fits.

straight shanks, seems to accomplish this purpose. Using the contact area A and designating S_{su} as the *ultimate shearing strength*, the engineer obtains

$$S_{su} = \frac{F}{A}\cos\theta$$

The engineer's supervisor, in reviewing the test results, says the expression should be

$$S_{su} = \frac{F}{A}\left(1 + \frac{1}{4}\tan^2\theta\right)^{1/2}\cos\theta$$

Resolve the discrepancy. What is your position?

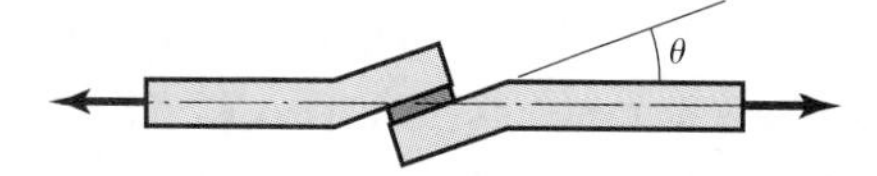

Problem 3–54

DESIGN

3–55 Using the results of Sec. 3-19, design a compact thermostat for use at room temperatures.

3–56 For a given cross section, a uniformly loaded beam of length l is simply supported as shown in the figure. If the supports are set back a short distance, the largest transverse bending moment is reduced. If the largest transverse bending moment can be minimized, the smallest beam section can be used. Write a computer program to discover the optimal setback a. Submit a listing of the programs you have written, the input and output of the production run, and the analysis on which your programming is based.

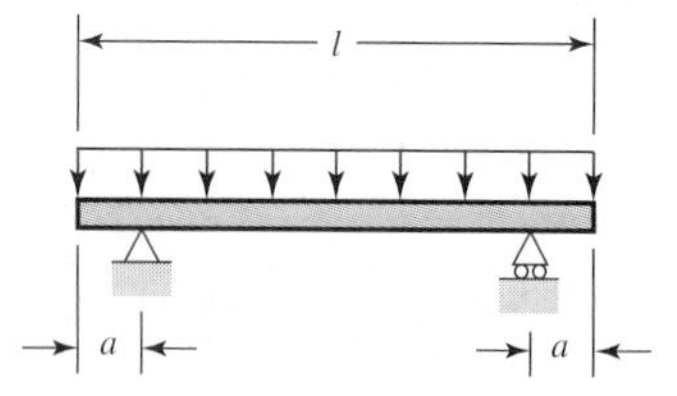

Problem 3–56

3–57 A rotary lawn-mower blade rotates at 3000 r/min. The blade has a uniform cross section $\frac{1}{4}$ in thick by $1\frac{1}{4}$ in wide, and has a $\frac{1}{2}$-in-diameter hole in the center as shown in the figure. Estimate the nominal tensile stress at the central section due to rotation.

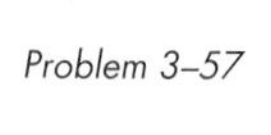

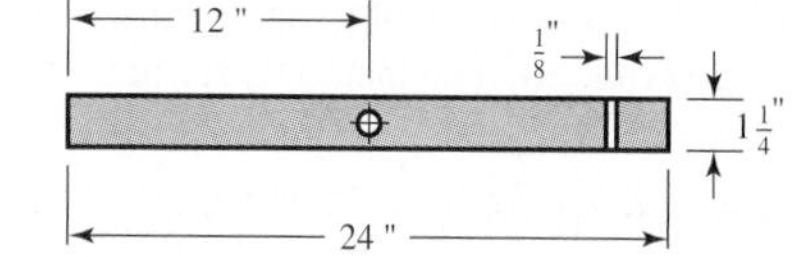

Problem 3–57

3–58 A utility hook was formed from a 1-in-diameter round rod into the geometry shown in the figure. What are the stresses at the inner and outer surfaces at section A-A if the load F is 1000 lb?

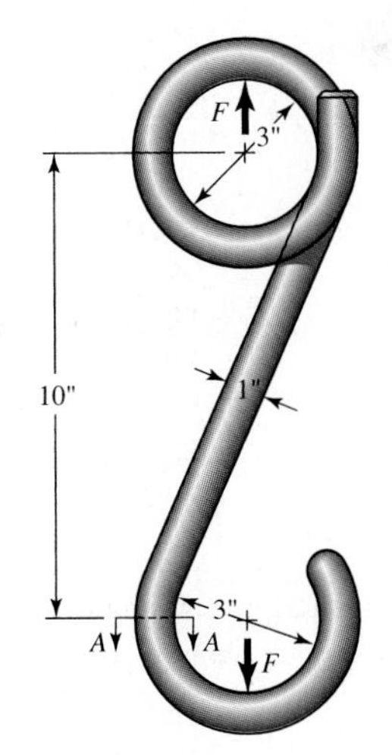

Problem 3–58

3–59 The steel eyebolt shown in the figure is loaded with a force F of 100 lb. The bolt is formed of $\frac{1}{4}$-in-diameter wire to a $\frac{3}{8}$-in radius in the eye and at the shank. Estimate the stresses at the inner and outer surfaces at sections A-A and B-B.

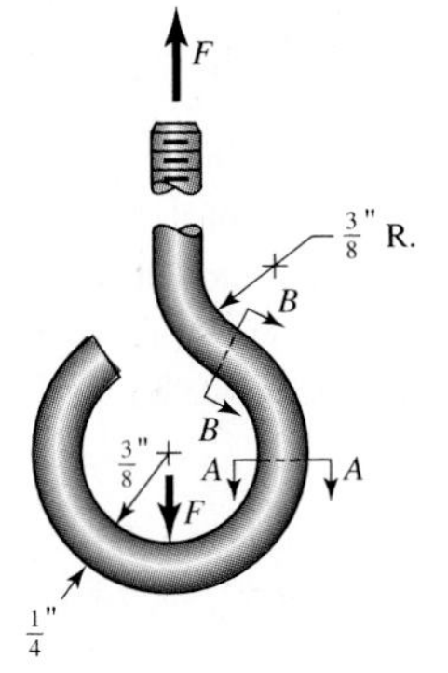

Problem 3–59

3–60 Shown in the figure is a 12-gauge (0.1094-in) by $\frac{3}{4}$-in latching spring which supports a load of $F = 3$ lb. The inside radius of the bend is $\frac{1}{8}$ in. Estimate the stresses at the inner and outer surfaces at the critical section.

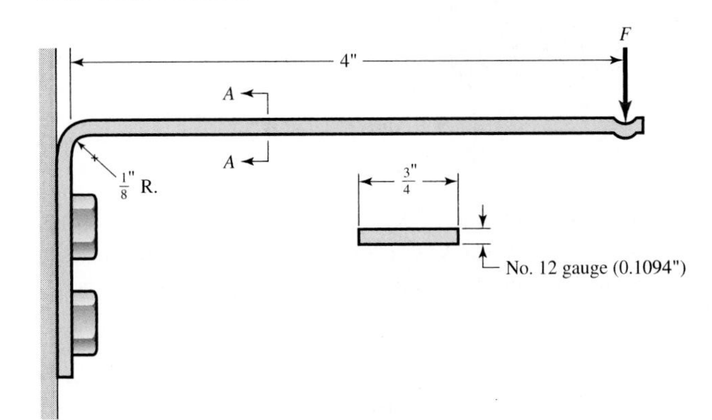

Problem 3–60

3–61 A cast-iron bell-crank lever which is depicted in the figure is acted upon by forces F_1 of 250 lb and F_2 of 333 lb. The section A-A at the central pivot has a curved inner surface with a radius of $r_i = 1$ in. Estimate the stresses at the inner and outer surfaces of the curved portion of the lever.

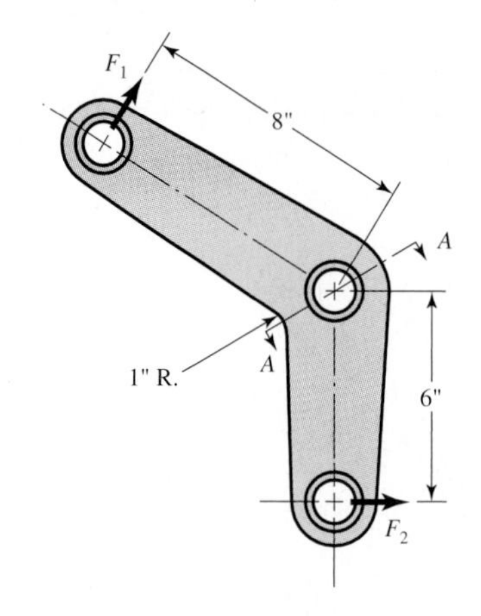

Problem 3–61

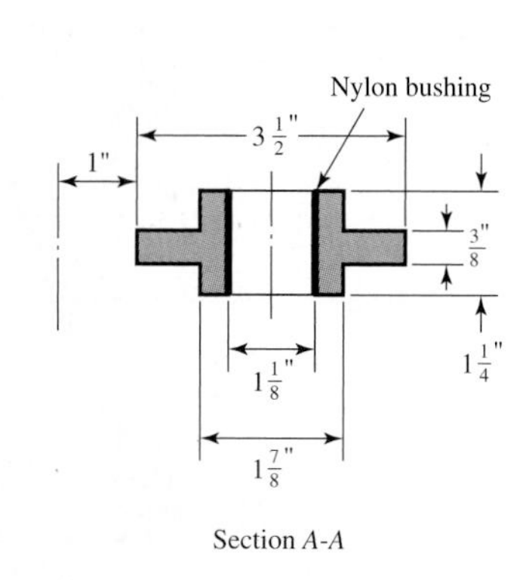

Section A-A

ANALYSIS

3–62 Find the stress at the inner and outer surfaces at section A-A of the crane hook shown in Fig. 3-32 and Example 3-10, using Simpson's rule and numerical integration with ordinates at $r = 2$, 3, 4, 5, and 6 in. What is the error in percent in the distance between the centroidal radius R and the neutral surface radius r_n? What can be done in the integration to reduce this error?

ANALYSIS

3–63 The crane hook depicted in Fig. 3-32 has a 1-in-diameter hole in the center of the critical section. For a load of 5 kip, estimate the bending stresses at the inner and outer surfaces at the critical section.

ANALYSIS

3–64 A 20-kip load is carried by the crane hook shown in the figure. The cross section of the hook uses two concave flanks. The width of the cross section is given by $b = 2/r$, where r is the radius from the center. The inside radius r_i is 2 in, and the outside radius $r_o = 6$ in. Find the stresses at the inner and outer surfaces at the critical section by (a) exact integration and (b) Simpson's rule, using ordinates placed at $r = 2$, 3, 4, 5, and 6 in.

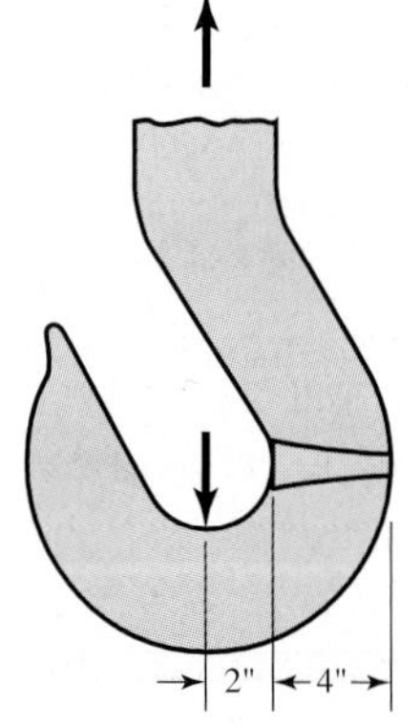

Problem 3–64

3–65 An offset tensile link is shaped to clear an obstruction with a geometry as shown in the figure. The cross section at the critical location is elliptical, with a major axis of 4 in and a minor axis of 2 in. For a load of 20 kip, estimate the stresses at the inner and outer surfaces of the critical section.

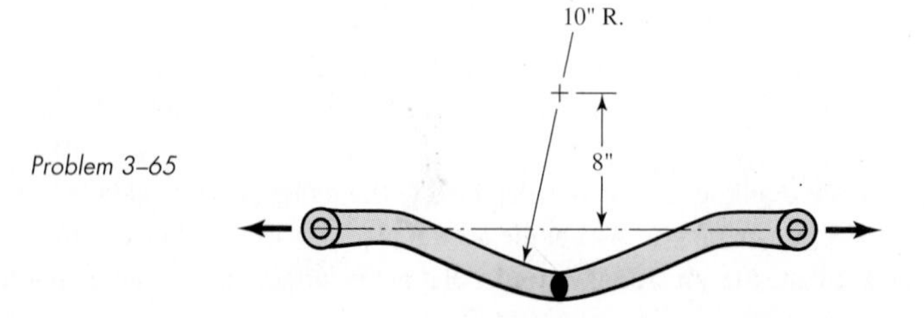

Problem 3–65

3–66 A cast-steel C frame as shown in the figure has a rectangular cross section of 1 in by 1.6 in, with a 0.4-in-radius semicircular notch on both sides which forms midflank fluting as shown. Estimate A, R, r_n, and e, and for a load of 3000 lb, estimate the inner and outer surface stresses at the throat C.

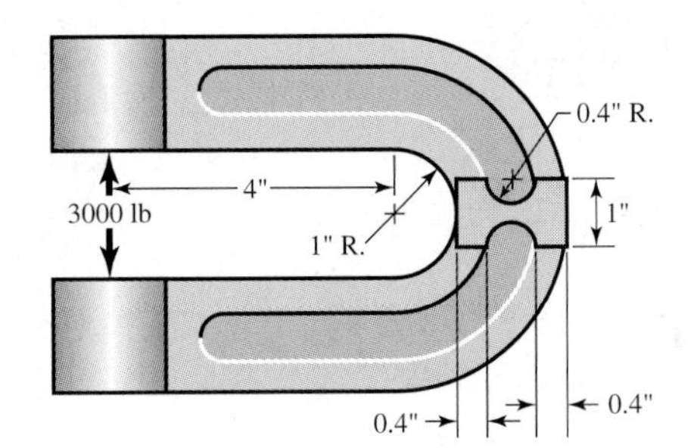

Problem 3–66

3–67 Two carbon steel balls, each 25 mm in diameter, are pressed together by a force F. In terms of the force F, find the maximum values of the principal stress, the shear stress, and the octahedral shear stress, in MPa.

3–68 One of the balls in Prob. 3-67 is replaced by a flat carbon steel plate. If $F = 18$ N, at what depth does the maximum octahedral shear stress occur?

3–69 An aluminum alloy roller with diameter 1 in and length 2 in rolls on the inside of a cast-iron ring having an inside radius of 4 in, which is 2 in thick. Find the maximum contact force F that can be used if the octahedral shear stress is not to exceed 4000 psi.

3–70 The figure shows a hip prosthesis consisting of a stem which is cemented into a reamed cavity in the femur. The cup is cemented and fastened to the hip with bone screws. Shown are porous layers of titanium into which bone tissue will grow to form a longer-lasting bond than that afforded by cement alone. The bearing surfaces are a plastic cup and a titanium femoral head. The lip shown in the figures bears against the cutoff end of the femur to transfer the load to the leg from the hip. Walking will induce several million stress fluctuations per year for an average person, so there is danger that the prosthesis will loosen the cement bonds or that metal cracks may occur because of the many repetitions of stress. Prostheses like this are made in many different sizes. Typical dimensions are ball diameter 50 mm, stem diameter 15 mm, stem length 155 mm, offset 38 mm, and neck length 39 mm. Develop an outline to follow in making a complete stress analysis of this prosthesis. Describe the material properties needed, the equations required, and how the loading is to be defined.

3–71 Based on torsion what is the relative increase in cost of material when choosing a square shaft over a round shaft of the same material for a high-production part?

3–72 Consider a Case 1 application of Eq. (3-30) of Sec. 3-10, which amounts to the relationship between the means of $\bar{\sigma} = \bar{M}/(I/c)$. The coefficient of variation of I/c reflects dimensional variation which is negligible. If the distribution of **M** is expressed as $\mathbf{M} \sim \psi(\mu_M, \sigma_M)$, show that the bending stress $\boldsymbol{\sigma}$ can be described as $\boldsymbol{\sigma} \sim \psi(c\mu_M/I,\ c\sigma_M/I)$ or $\boldsymbol{\sigma} \sim c\mu_M/I\psi(1, C_M)$. If in Ex. 3-4 $\mathbf{M} \sim N(1600, 144)$ N·m, describe the bending stress $\boldsymbol{\sigma}$.

3–73 In Sec. 3-13, Eq. (3-38) can be viewed as $\tau_{\max} = 3\mathbf{V}/(2A)$ in Case 1 circumstances. Show that $\tau_{\max}$ can be expressed as $\tau_{\max} \sim 3\mu_V/(2A)\psi(1, C_V)$.

3–74 Equation (3-41) of Sec. 3-14 can be viewed as $\boldsymbol{\tau}_{\max} = \mathbf{T}r/J$ in a Case 1 situation. Show that $\tau_{\max}$ can be expressed as $\boldsymbol{\tau}_{\max} \sim r\mu_T/J\psi(1, C_T)$.

3–75 In Sec. 3–15, Eqs. (3–51) can be viewed as

$$\boldsymbol{\sigma}_t = \frac{r_i^2 \mathbf{p}_i}{r_0^2 - r_i^2}\left(1 + \frac{r_0^2}{r^2}\right) \qquad \boldsymbol{\sigma}_r = \frac{r_i^2 \mathbf{p}_i}{r_0^2 - r_i^2}\left(1 - \frac{r_0^2}{r^2}\right)$$

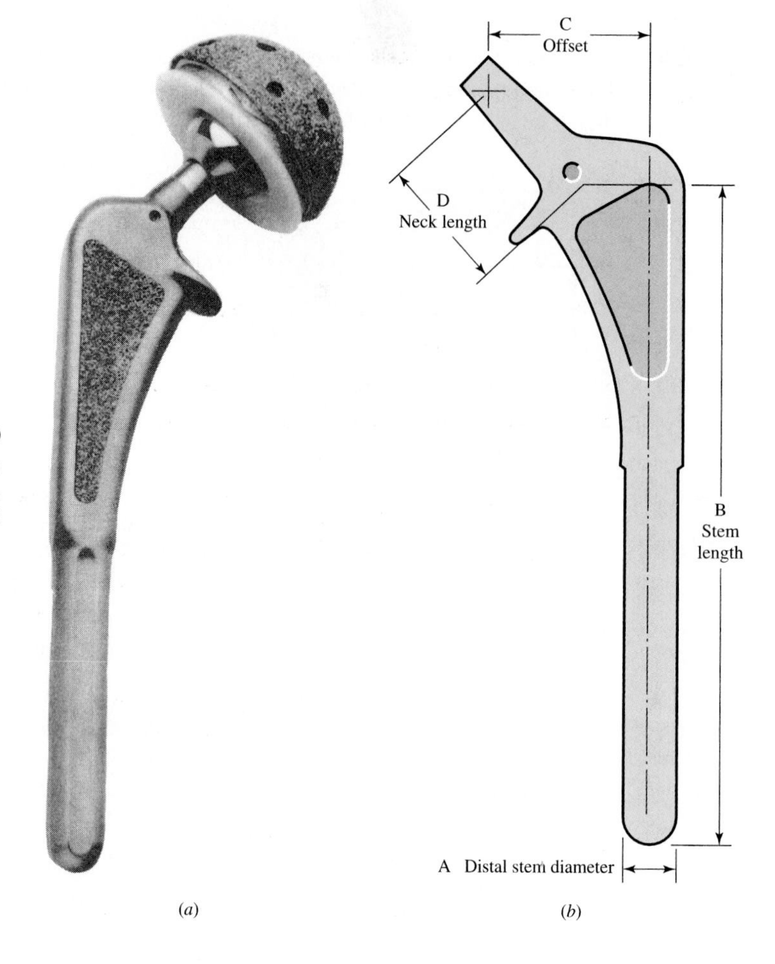

Problem 3–70
Porous hip prosthesis. (Photograph and drawing courtesy of Zimmer, Inc., Warsaw, Indiana.)

in a Case 1 circumstance. Show that $\boldsymbol{\sigma}_t$ and $\boldsymbol{\sigma}_r$ can be expressed as

$$\boldsymbol{\sigma}_t \sim \frac{r_i^2 \mu_{pi}}{r_0^2 - r_i^2}\left(1 + \frac{r_0^2}{r^2}\right)\psi(1, C_{pi}) \qquad \boldsymbol{\sigma}_r \sim \frac{r_i^2 \mu_{pi}}{r_0^2 - r_i^2}\left(1 - \frac{r_0^2}{r^2}\right)\psi(1, C_{pi})$$

ANALYSIS

3–76 In Sec. 3-18 the dominant variable is the interference $\boldsymbol{\delta} \sim \psi(\mu_\delta, \sigma_\delta)$, Eq. (3-63) in a Case 1 circumstance. Show that the stochastic pressure $\mathbf{p}$ is given by

$$\mathbf{p} \sim \frac{E\mu_\delta}{R}\left[\frac{(r_0^2 - R^2)(R^2 - r_i^2)}{2R^2(r_0^2 - r_i^2)}\right]\psi(1, C_\varsigma)$$

ANALYSIS

3–77 In Sec. 3-20, Eq. (3-69) for a crane hook can be viewed as a Case 1 circumstance in which

$$\boldsymbol{\sigma}_i = \frac{\mathbf{M}r_i}{Aer_i} + \frac{\mathbf{F}}{A} = \frac{\mathbf{F}}{A}\left(\frac{\bar{r}c_i}{er_i} + 1\right)$$

$$\boldsymbol{\sigma}_0 = \frac{\mathbf{M}c_0}{Aer_0} + \frac{\mathbf{F}}{A} = \frac{\mathbf{F}}{A}\left(\frac{\bar{r}c_0}{er_0} - 1\right)$$

If $\mathbf{F} \sim \psi(\mu_F, \sigma_F)$, show that

$$\boldsymbol{\sigma}_i \sim \frac{\mu_F}{A}\left(\frac{\bar{r}c_i}{er_i} + 1\right)\psi(1, C_F)$$

$$\boldsymbol{\sigma}_0 \sim \frac{\mu_F}{A}\left(\frac{\bar{r}c_0}{er_0} - 1\right)\psi(1, C_F)$$

ANALYSIS

3–78 Compare the maximum shear stress in a square cross section shaft of side s and a circular-cross-section shaft of diameter d for the same torsional moment T. Which has the higher stress and by what multiple is it higher?

ANALYSIS

3–79 Compare the torsional deflection θ of a square-cross-section shaft of side s and length l with a circular-cross-section shaft of diameter d of the same length for a torsional moment T. Which has the greater deflection θ and by what multiple is it higher?

ANALYSIS

3–80 For the same cross-sectional area $A = s^2 = \pi d^2/4$, for a square cross-sectional area shaft and a circular cross-sectional area shaft, which has the higher maximum shear stress, and by what multiple is it higher?

ANALYSIS

3–81 For the same cross-sectional area $A = s^2 = \pi d^2/4$, for a square cross-sectional area shaft and a circular cross-sectional area shaft, both of length l, which has the greater angular twist θ, and by what multiple is it greater?

ANALYSIS

3–82 W. C. Young, in *Roark's Formulas for Stress and Strain*, considers a rectangular cross section $2a$ by $2b$, where $2a$ is the longer side, and the relation for the maximum shear stress in the center of the longest side as a function of b/a is given by

$$\tau_{\max} = \frac{3T}{8ab^2}\left[1 + 0.6095\frac{b}{a} + 0.8865\left(\frac{b}{a}\right)^2 - 1.8023\left(\frac{b}{a}\right)^3 + 0.9100\left(\frac{b}{a}\right)^4\right]$$

Compare the first two terms of the polynomial expression in b/a with Eq. (3–47). If you find a discrepancy, comment about the reason(s). Equation (3–47) has the form

$$\tau_{\max} = \frac{T}{bc^2}\frac{1}{\alpha}$$

From the tabulation of α versus b/c, plot $1/\alpha$ as ordinate and $1/(b/c)$ as abscissa. If the data string is linear or nearly so, find a_0 and a_1 of

$$\frac{1}{\alpha} = a_0 + a_1\frac{1}{b/c}$$

by linear regression. Compare the results with Eq. (3–47).

ANALYSIS

3–83 to 3–86 The table below gives data concerning the shrink fit of two cylinders of differing materials and dimensional specification. Elastic constants for different materials may be found in Table E-5. Identify the radial interference δ, then find the interference pressure p, the tangential normal stress on both sides of the fit surface, and the radial displacements δ_i and δ_0. If dimensional tolerances are given at fit surfaces, repeat the problem for the highest and lowest stress levels.

Problem Number	Inner Cylinder			Outer Cylinder		
	Material	d_i	d_0	Material	D_i	D_0
3-83	Steel	0	1.002	Steel	1.000	2.00
3-84	Steel	0	1.002	Cast iron	1.000	2.00
3-85	Steel	0	1.002/1.003	Steel	1.000/1.001	2.00
3-86	Steel	0	2.005/2.003	Aluminum	2.000/2.002	4.00

ANALYSIS

3–87 Consider a simply supported beam of rectangular cross section of constant width b and variable depth h, so proportioned that the stress at the surface due to bending is constant, when subjected to a load F at a distance a from the left support and a distance b from the right support. Show that the depth h at location x is given by

$$h = \sqrt{\frac{b}{l}\frac{x}{l}}\sqrt{\frac{6nFl}{S_y}} \qquad 0 \le x \le a$$

where the parameter are defined by Table E-9-6, S_y is the tensile yield strength, and n is the design factor.

ANALYSIS

3–88 Consider a simply supported static beam of circular cross section of diameter d, so proportioned by varying the diameter that the stress at the surface due to bending is constant, when subjected to a steady load F located at a distance a from the left support and a distance b from the right support. Show that the diameter d at a location x is given by

$$d = \left(\frac{b}{l}\frac{x}{l}\right)^{1/3}\left(\frac{32nFl}{\pi S_y}\right)^{1/3} \qquad 0 \le x \le a$$

ANALYSIS

3–89 Continuously variable cross-sectional geometries at different locations along the span, as suggested by Probs. 3-87 and 3-88, are occasionally used. Physically large beams, as are found in traveling gantry bridge cranes, are built up of steel plate, and structural shapes can show continuous variation in the depth of the section. At the supports, essential structural detail is incompatible with "zero depth," so approximations are used. This opens another line of thought. Consider a simply supported shaft with journal diameters d_1, shoulder diameters d_2 (to locate bearings), and a central diameter of size d_3. If d_3 is chosen to keep the largest bending stress at a permissible level, S_y/n, and the shaft is loaded as in Table E-9-6, then, ignoring stress concentrations, how far from the left bearing x_{23} should diameter d_2 persist until diameter d_3 starts?

3–90 In Prob. 3-89 the introduction of the step from d_2 to d_3 at x_{23} introduces the same bending stress at the step that exists under the load. As we saw, x_{23} was in a definite place when d_2 and d_3 are known. Suppose the shoulder diameter is not uniquely dictated by the bearing but can exist in a range without impairing function. Then for every possible d_2 there is a companion x_{23} which meets the allowable stress. Which choice of D_2, x_{23} results in the largest weight-saving on the shaft?

3–91 A stationary simply supported axle made of A-36 steel ($S_y = 36$ kpsi) carries a static load of 10 000 lbf as shown in the figure. The first preliminary design is a nearly constant-diameter shaft supported on 2-in-diameter journals. Assume the result of Prob. 3-90 describes an optimal step defined by $d_2^* = \sqrt{0.6}d_3$ and the distance from the left bearing is $x_{23}^* = a(0.60)^{1.5}$, and a design factor of $n = 2$ was used to find the diameter $d_3 = 2.94$ in.

(*a*) Find d_2 and x_{23} for both sides of the shaft.

(*b*) Find, for the left side of the shaft, the decrement in shaft volume from

$$\Delta V \doteq \frac{\pi}{4}(d_3^2 - d_2^2)x_{23}$$

(*c*) Increment d_2 by ± 0.2 in and show that d_2^* and x_{23}^* produce a maximum volumetric saving from the preliminary design, to the left of the load.

(*d*) Integrate the expression in the statement of Prob. 3–88 to find the volume of the shaft to the left of the load, to see the departure from true optimality.

(*e*) Sketch the geometry of the revised shaft with decisions made for both sides of the load.

3–92 Force fits of a shaft and gear are assembled in an air-operated arbor press. An estimate of assembly force and torque capacity of the fit is needed. Assume the coefficient of friction is f, the fit interface pressure is p, the nominal shaft or hole radius is R, and the axial length of the gear bore is l.

(*a*) Show that the estimate of the axial force is $F_{ax} = 2\pi f Rlp$.

(*b*) Show the estimate of the torque capacity of the fit is $T = 2\pi f R^2 lp$.

3–93 Simplify Eqs. (3–95), (3–96), and (3–97) by setting $z = 0$ and finding σ_x/p_{max}, σ_y/p_{max}, σ_z/p_{max}, and τ_{zy}/p_{max} and, for cast iron, check the ordinate intercepts of the four loci in Fig. 3–37.

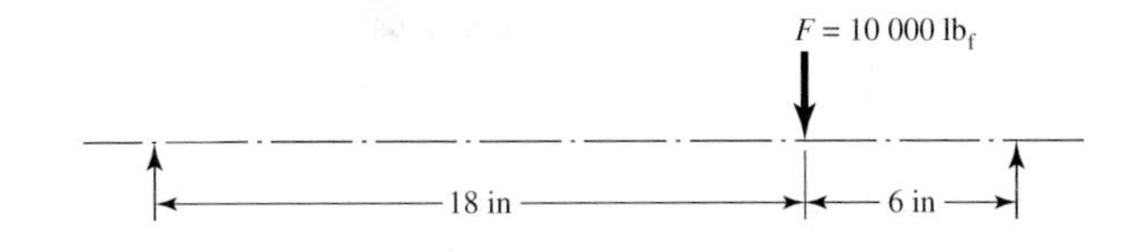

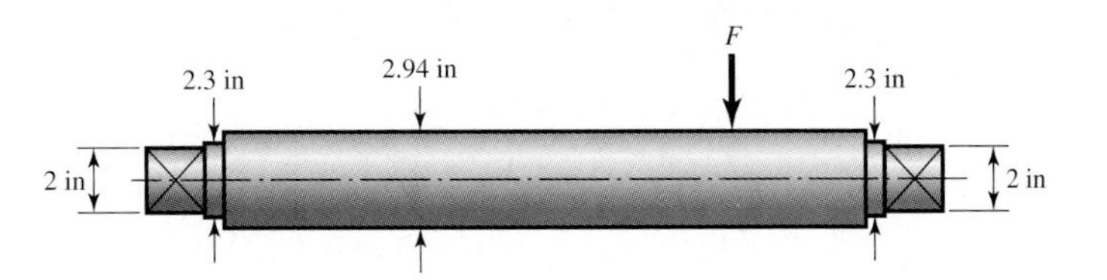

Problem 3–91

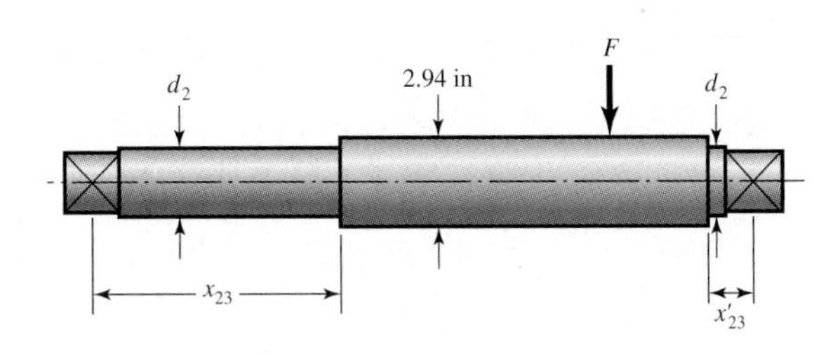

3–94 A 6-in-diameter cast-iron wheel, 2 in wide, rolls on a flat steel surface carrying a 800 lbf load.

(*a*) Find the Hertzian stresses σ_x, σ_y, σ_z, and τ_{yz}.

(*b*) What happens to the stresses at a point *A* that is 0.010 in below the wheel rim surface during a revolution?

3–95 Write an interactive computer program that will evaluate Eqs. (3-98) to (3-101) and fill out a table $0 \le y/b \le 1$ across the columns and $0 \le z/b \le 1$ for the lines, with the body of the table with any stress of interest: σ_x, σ_y, σ_z, σ_{yz}, σ_1, σ_2, σ_3, σ', τ_{max}.

3–96 Use the program of Prob. 3-95 to solve Prob. 3-94.

3–97 Use the program of Prob. 3-95 to solve Prob. 3-94 with a coefficient of friction $f = 0.10$. Notice the changes.

3–98 Explore other problems of interest with the program written in Prob. 3-95.

3–99 If, as in Prob. 3-91, a shaft with a concentrated load contains an approximation to optimality with a step on both sides of the static load, such that the stress at the now three critical locations is the same, use the binomial theorem to show

$$1 = (R + p)^3 = \underset{\text{no failure location}}{R^3} + \underset{\text{one failure location}}{3R^2 p} + \underset{\text{two failure location}}{3Rp^2} + \underset{\text{three failure location}}{p^3}$$

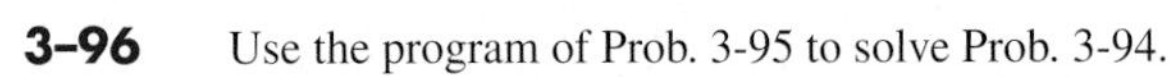

(*a*) If the reliability goal for the uniform shaft was $R = 0.99$, show that the probability of no failures is 0.970 299, the probability of a one-location failure is 0.029 403, the probability of a two-location failure is 0.000 297, and the probability of a three-location failure is 0.000 001.

(*b*) A designer wants a reliability of $R = 0.99$. He or she chooses to incorporate the stepped optimality approximation. To what reliability goal must the designer aspire in sizing diameter d_3?

3–100 Design (make the geometric decisions at the critical locations) an axle for quasi-static service to carry a load $F \sim 10\ 000 LN(1, 0.05)$ lbf in the circumstances of Prob. 3–91 using a 1018 CD steel, $\mathbf{S}_y \sim 78.4LN(1, 0.0753)$ kpsi using the method of Case 2 of Sec. 3-22. The shaft reliability goal is 0.999.

(*a*) For one critical location (under the load) make the diameter decision for the mostly uniform-diameter shaft.

(*b*) For the first approximation to minimizing the shaft weight, make the central-diameter decision and the diameter and shoulder location decisions.

4 Deflection and Stiffness

4–1 Spring Rates **176**

4–2 Tension, Compression, and Torsion **177**

4–3 Deflection Due to Bending **178**

4–4 Finding Deflection by Integration **180**

4–5 Finding Deflection by the Area-Moment Method **187**

4–6 Finding Deflections by the Use of Singularity Functions **190**

4–7 Strain Energy **193**

4–8 Castigliano's Theorem **195**

4–9 Statically Indeterminate Problems **198**

4–10 Deflection of Curved Members **200**

4–11 Compression Members—General **204**

4–12 Long Columns with Central Loading **206**

4–13 Intermediate-Length Columns with Central Loading **210**

4–14 Columns with Eccentric Loading **210**

4–15 Struts, or Short Compression Members **214**

4–16 An Application: Round-Bar Clamps **216**

4–17 Deflection of Energy-Dissipative Assemblies **220**

4–18 Shock and Impact **229**

4–19 Suddenly-Applied Loading **230**

4–20 Propagation of Error **233**

A body is said to be *rigid* if it exhibits no change in size or shape under the influences of forces or couples. All real bodies deform under load, either elastically or plastically. Classification of a real body as rigid is an idealization. A body can be sufficiently insensitive to deformation that a presumption of rigidity does not affect an analysis enough to warrant a nonrigid treatment. Rather than treatment based on an *assumption* of rigidity, the *decision* to ignore deformation represents such a small deviation from actuality that a rigid treatment is robust. If the body deformation later proves to be not negligible, then declaring rigidity was a poor decision, not a poor assumption. Strictly, *flexibility* is the ability of a body to distort by bending—that is, it is but one form of distortion—but the term is often used interchangeably with distortion. Defining the choice as bilateral—rigid *or* flexible—leaves unanswered other possibilities. A wire rope is flexible, but in tension it can be robustly rigid and it distorts enormously under attempts at compressive loading. The same body can be both rigid and nonrigid.

Deflection analysis enters into design situations in many ways. A snap ring, or retaining ring, must be flexible enough to be bent without permanent deformation and assembled with other parts; and then it must be rigid enough to hold the assembled parts together. In a transmission, the gears must be supported by a rigid shaft. If the shaft bends too much, that is, if it is too flexible, the teeth will not mesh properly, and the result will be excessive impact, noise, wear, and early failure. In rolling sheet or strip steel to prescribed thicknesses, the rolls must be crowned, that is, curved, so that the finished product will be of uniform thickness. Thus, to design the rolls it is necessary to know exactly how much they will bend when a sheet of steel is rolled between them. Sometimes mechanical elements must be designed to have a particular force-deflection characteristic. The suspension system of an automobile, for example, must be designed within a very narrow range to achieve an optimum bouncing frequency for all conditions of vehicle loading, because the human body is comfortable only within a limited range of frequencies.

The size of a load-bearing component is often determined on deflections, rather than limits on stress.

This chapter considers distortion of single bodies due to geometry (shape) and loading, then, briefly, the behavior of ensembles of bodies.

4–1 Spring Rates

Elasticity is that property of a material which enables it to regain its original configuration after having been deformed. A *spring* is a mechanical element which exerts a force when deformed. Figure 4–1*a* shows a straight beam of length l simply supported at the ends

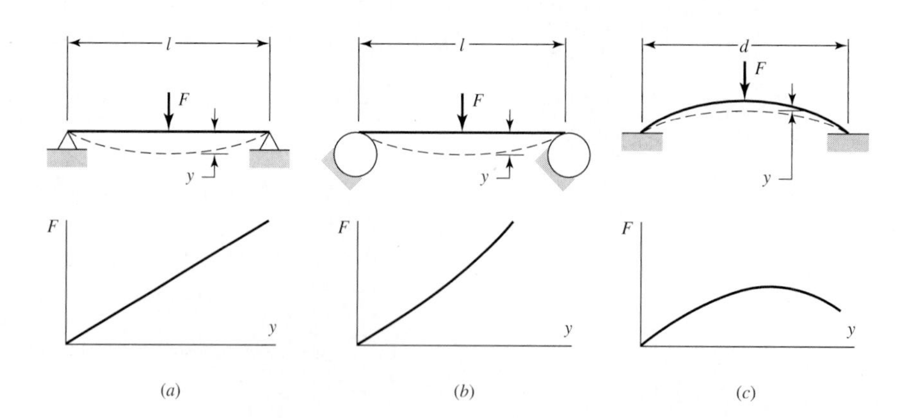

Figure 4–1

(*a*) A linear spring; (*b*) a stiffening spring; (*c*) a softening spring.

and loaded by the transverse force F. The deflection y is linearly related to the force, as long as the elastic limit of the material is not exceeded, as indicated by the graph. This beam can be described as a *linear spring.*

In Fig. 4–1b a straight beam is supported on two cylinders such that the length between supports decreases as the beam is deflected by the force F. A larger force is required to deflect a short beam than a long one, and hence the more this beam is deflected, the stiffer it becomes. Also, the force is not linearly related to the deflection, and hence this beam can be described as a *nonlinear stiffening spring.*

Figure 4–1c is a dish-shaped round disk. The force necessary to flatten the disk increases at first and then decreases as the disk approaches a flat configuration, as shown by the graph. Any mechanical element having such a characteristic is called a *nonlinear softening spring.*

If we designate the general relationship between force and deflection by the equation

$$F = F(y) \tag{a}$$

then *spring rate* is defined as

$$k(y) = \lim_{\Delta y \to 0} \frac{\Delta F}{\Delta y} = \frac{dF}{dy} \tag{4–1}$$

where y must be measured in the direction of F and at the point of application of F. Most of the force-deflection problems encountered in this book are linear, as in Fig. 4–1a. For these, k is a constant, also called the *spring constant;* consequently Eq. (4–1) is written

$$k = \frac{F}{y} \tag{4–2}$$

We might note that Eqs. (4–1) and (4–2) are quite general and apply equally well for torques and moments, provided angular measurements are used for y. For linear displacements, the units of k are often pounds per inch or newtons per meter, and for angular displacements, pound-inches per radian or newton-meters per radian.

4–2 Tension, Compression, and Torsion

The relation for the total extension or deformation of a uniform bar has already been developed in Sec. 3–5, Eq. (a). It is repeated here for convenience:

$$\delta = \frac{Fl}{AE} \tag{4–3}$$

This equation does not apply to a long bar loaded in compression if there is a possibility of buckling. Using Eqs. (4–2) and (4–3), we see that the spring constant of an axially loaded bar is

$$k = \frac{AE}{l} \tag{4–4}$$

The angular deflection of a uniform round bar subjected to a twisting moment T was given in Eq. (3–39), and is

$$\theta = \frac{Tl}{GJ} \tag{4–5}$$

where θ is in radians. If we multiply Eq. (4–5) by $180/\pi$ and substitute $J = \pi d^4/32$ for a solid round bar, we obtain

$$\theta = \frac{583.6Tl}{Gd^4} \tag{4–6}$$

where θ is in degrees.

Equation (4–5) can be rearranged to give the torsional spring rate as

$$k = \frac{T}{\theta} = \frac{GJ}{l} \tag{4–7}$$

When the word *simple* is used to describe the loading, the meaning is that no other load is present and that no geometric complexities are present. Thus, a bar loaded in *simple tension* is a uniform bar acted upon by a tensile load directed along the centroidal axis, and acted upon by no other loads.

4–3 Deflection Due to Bending

Beams deflect a great deal more than axially loaded members, and the problem of bending probably occurs more often than any other loading problem in design. Shafts, axles, cranks, levers, springs, brackets, and wheels, as well as many other elements, must often be treated as beams in the design and analysis of mechanical structures and systems. The subject of bending, however, is one which you should have studied as preparation for reading this book. It is for this reason that we include here only a brief review to establish the nomenclature and conventions to be used throughout this book.

In Sec. 3–10 we developed the relation for the curvature of a beam subjected to a bending moment M [Eq. (3–27)]. The relation is

$$\frac{1}{\rho} = \frac{M}{EI} \tag{4–8}$$

where ρ is the radius of curvature. From studies in mathematics we also learn that the curvature of a plane curve is given by the equation

$$\frac{1}{\rho} = \frac{d^2y/dx^2}{\left[1 + (dy/dx)^2\right]^{3/2}} \tag{4–9}$$

where the interpretation here is that y is the deflection of the beam at any point x along its length. The slope of the beam at any point x is

$$\theta = \frac{dy}{dx} \tag{a}$$

For many problems in bending, the slope is very small, and for these the denominator of Eq. (4–9) can be taken as unity. Equation (4–8) can then be written

$$\frac{M}{EI} = \frac{d^2y}{dx^2} \tag{b}$$

Noting Eqs. (3–22) and (3–23) and successively differentiating Eq. (*b*) yields

$$\frac{V}{EI} = \frac{d^3y}{dx^3} \tag{c}$$

$$\frac{q}{EI} = \frac{d^4y}{dx^4} \tag{d}$$

It is convenient to display these relations in a group as follows:

$$\frac{q}{EI} = \frac{d^4 y}{dx^4} \tag{4–10}$$

$$\frac{V}{EI} = \frac{d^3 y}{dx^3} \tag{4–11}$$

$$\frac{M}{EI} = \frac{d^2 y}{dx^2} \tag{4–12}$$

$$\theta = \frac{dy}{dx} \tag{4–13}$$

$$y = f(x) \tag{4–14}$$

The nomenclature and conventions are illustrated by the beam of Fig. 4–2. Here, a beam of length $l = 20$ in is loaded by the uniform load $w = 80$ lb per inch of beam length. The x axis is positive to the right, and the y axis positive upward. All quantities—loading, shear, moment, slope, and deflection—have the same sense as y; they are positive if upward, negative if downward.

The values of the quantities at the ends of the beam, where $x = 0$ and $x = l$, are called their *boundary values*. For this reason the beam problem is often called a *boundary-value problem*. The reactions $R_1 = R_2 = +800$ lb and the shear forces $V_0 = +800$ lb and $V_l = -800$ lb are easily computed using the methods of Chap. 3. The bending moment is zero at each end because the beam is simply supported. Note that the beam-deflection curve must have a negative slope at the left boundary and a positive slope at the right boundary. This is easily seen by examining the deflection of Fig. 4–2. The magnitude

Figure 4–2

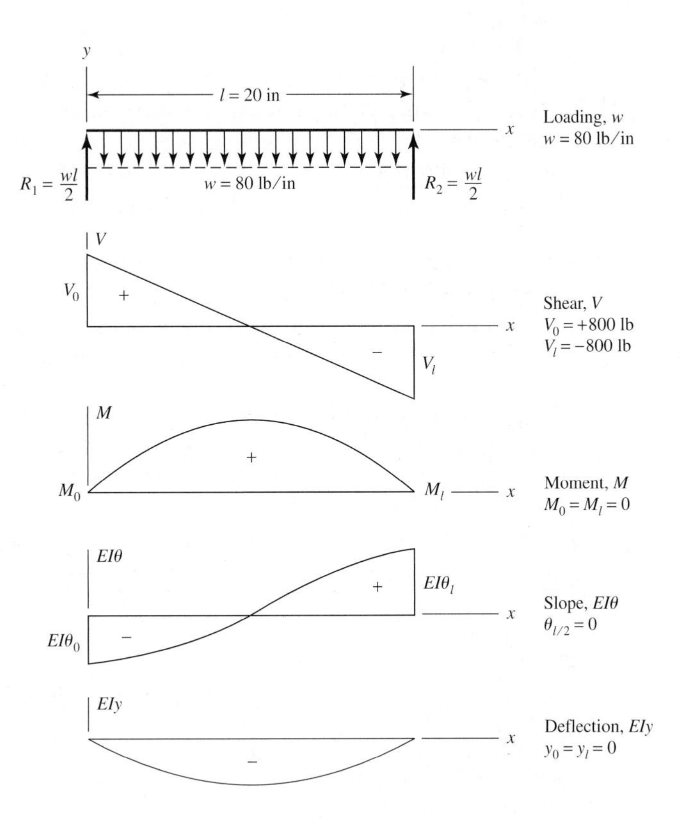

of the slope at the boundaries is, as yet, unknown; because of symmetry, however, the slope is known to be zero at the center of the beam. We note, finally, that the deflection is zero at each end.

4–4 Finding Deflection by Integration

Equations (4–10) through (4–14) are the basis for relating the intensity of loading q, vertical shear V, bending moment M, slope of the neutral surface θ, and the transverse deflection y. Beams have intensities of loading that range from q = constant (uniform loading), variable intensity $q(x)$, to Dirac delta functions (concentrated loads).

The intensity of loading usually consists of piecewise contiguous zones, the expressions for which are integrated through Eqs. (4–10) to (4–14) with varying degrees of difficulty. Another approach is to represent the deflection $y(x)$ as a Fourier series, which is capable of representing single-valued functions with a finite number of finite discontinuities, then differentiating through Eqs. (4–14) to (4–10), and stopping at some level where the Fourier coefficients can be evaluated. A complication is the piecewise continuous nature of some beams (shafts) which are stepped-diameter bodies.

All of the above constitute, in one form or another, formal integration methods, which, with properly selected problems, result in solutions for q, V, M, θ, and y. These solutions may be

1 Closed-form, or
2 Represented by infinite series, which amount to closed form if the series are rapidly convergent, or
3 Approximations obtained by evaluating the first or the first and second terms.

The series solutions can be made equivalent in convenience to the closed-form solution by the use of a computer. Roark's[1] formulas are committed to CD-ROM and can be used on a personal computer.

To understand bending it is useful to keep in mind shear and bending and, in particular, the bending moment diagram, which can be viewed as a representation of how the beam or shaft "feels" the influence of external loading. This is also halfway along the integration path represented by Eqs. (4–10) to (4–14). After an engineer experiences many bending problems, the following observations can be made:

- Moment diagrams are usually piecewise-linear; upon integration, they present piecewise-quadratic slope.
- There are discontinuities in the ordinate on moment diagrams when point couples are present. These can be accommodated at the beginning or end of a "piece."

A piecewise-linear moment diagram segment can be integrated by trapezoidal rule to produce an *exact* result. The next integration can be accomplished *exactly* by Simpson's rule. Furthermore, piecewise treatment lends itself to a tabular approach, which can be implemented with or without use of a computer.[2] Such a method begins with the moment diagram (for understanding). It gives deflection and slope only at the stations, whose the location is under the user's control. For the method to work, stations are needed at three points:

1. Warren C. Young, *Roark's Formulas for Stress and Strain*, 6th ed., McGraw-Hill, New York, 1989.

2. See Charles R. Mischke, "An Exact Numerical Method for Determining the Bending Deflections and Slopes of Stepped Shafts," in *Advances in Reliability and Stress Analysis*, Proceedings of the Winter Annual Meeting of ASME, San Francisco, December 1978, pp. 101–115.

1 Where loads and point couples are applied.
2 Where reactions occur.
3 Where beam or shaft geometry changes.

The sum of these is the minimum number of stations. If there are additional places where deflection and slope are needed, these are added to the minimum number of stations.

To accomplish this, important relationships have to be derived. We place the origin of the xy-coordinate system at the left reaction, the left end of the beam, or elsewhere. Using Eq. (4–12) we can integrate from 0 to some location x along the beam as follows:

$$\int_0^x \frac{d^2y}{dx^2}dx = \int_0^x \frac{M}{EI}dx$$

The left side becomes

$$\left.\frac{dy}{dx}\right|_x - \left.\frac{dy}{dx}\right|_0 = \int_0^x \frac{M}{EI}dx$$

The first term is what we mean by dy/dx, so we drop the subscript. The second term is the slope at the origin, a constant we shall call C_1. Rewriting, we have

$$\frac{dy}{dx} = \int_0^x \frac{M}{EI}dx + C_1 \tag{4–15}$$

Integrating Eq. (4–15) term by term from 0 to x produces

$$\int_0^x \frac{dy}{dx}dx = \int_0^x \int_0^x \left(\frac{M}{EI}dx\right)dx + \int_0^x C_1 dx$$

which becomes

$$y = \int_0^x \int_0^x \left(\frac{M}{EI}dx\right)dx + C_1 x + C_2 \tag{4–16}$$

Equations (4–15) and (4–16) are called the *prediction equations*. They use integrals which we will generate in a table. We write Eq. (4–16) twice, with $x = x_a$ and then with $x = x_b$, where x_a and x_b are the locations of the reactions. Solving such equations simultaneously for C_1 and C_2 gives

$$C_1 = \frac{\int_0^{x_a}\int_0^{x_a}\left(\frac{M}{EI}dx\right)dx - \int_0^{x_b}\int_0^{x_b}\left(\frac{M}{EI}dx\right)dx}{x_a - x_b} \tag{4–17}$$

$$C_2 = \frac{x_b\int_0^{x_a}\int_0^{x_a}\left(\frac{M}{EI}dx\right)dx - x_a\int_0^{x_b}\int_0^{x_b}\left(\frac{M}{EI}dx\right)dx}{x_a - x_b} \tag{4–18}$$

To evaluate C_1 and C_2 of the prediction equations our table must generate values of the double integrals. To perform the integrations we need to use the trapezoidal rule and Simpson's rule.[3] These rules can be expressed as follows:

$$I_{i+1} = I_i + \frac{1}{2}\left[\left(\frac{M}{EI}\right)_{i+1} + \left(\frac{M}{EI}\right)_i\right](x_{i+1} - x_i) \tag{4–19}$$

3. Brice Carnaham, H. A. Luther, and J. O. Wilkes, *Applied Numerical Methods*, Wiley, New York, 1969, p. 71, Eq. (2.8), and p. 73, Eq. (2.17), adapted.

Table 4–1

Tabular Worksheet to Allow Deflections and Slopes of Simply Supported Beams to Be Found

(1) i	(2) x	(3) M	(4) d	(5) $M/(EI)$	(6) $\int_0^x M/(EI)dx$	(7) $\int_0^x\int_0^x M/(EI)\,dx\,dx$	(8) y	(9) dy/dx
1								
2								
3								

$$I_{i+2} = I_i + \frac{1}{6}\left[\left(\frac{M}{EI}\right)_{i+2} + 4\left(\frac{M}{EI}\right)_{i+1} + \left(\frac{M}{EI}\right)_i\right](x_{i+2} - x_i) \tag{4–20}$$

The subscript i can be regarded as the line number of the table, as in Table 4–1. To carry out Simpson's rule we need intermediate lines placed midway between stations. A simple exercise will show how all these ideas fit together.

EXAMPLE 4–1

A 1-in-diameter steel shaft is simply supported between bearings spaced 10 in apart and carries a concentrated load of 4167 lb placed 6 in from the left bearing as shown in Fig. 4–3*a*. Using the minimum number of stations, find the exact shaft deflection under load.

Solution

The mandatory stations are at the left reaction, the load, and the right reaction, for a total of three. Verify the shear and moment diagrams as shown in Figs. 4–3*b* and 4–3*c*. Begin a table on a worksheet such as is shown in Table 4–2, with the following columns:

1. Station number
2. Abscissa x
3. Bending moment M (dual entry)
4. Diameter d (dual entry)
5. $M/(EI)$ (dual entry)
6. Single integral of $M/(EI)$
7. Double integral of $M/(EI)$
8. Transverse deflection y
9. Slope dy/dx

Now enter station number 1, 2, and 3 on alternate lines in column 1. Enter x as 0, 3, 6, 8, and 10 in column 2. Enter M including intermediate stations in column 3. Enter d as 1 in all places in column 4. Enter zeros in line 1, columns 5, 6, and 7. Note that $E = 30(10^6)$ psi, and second area moment $I = \pi d^4/64 = 0.0491$ in^4. On line 2,

Figure 4–3

A 1-in-diameter steel shaft simply supported in bearings, with station numbers circled. (*a*) Shaft geometry; (*b*) vertical shear diagram; (*c*) bending-moment diagram.

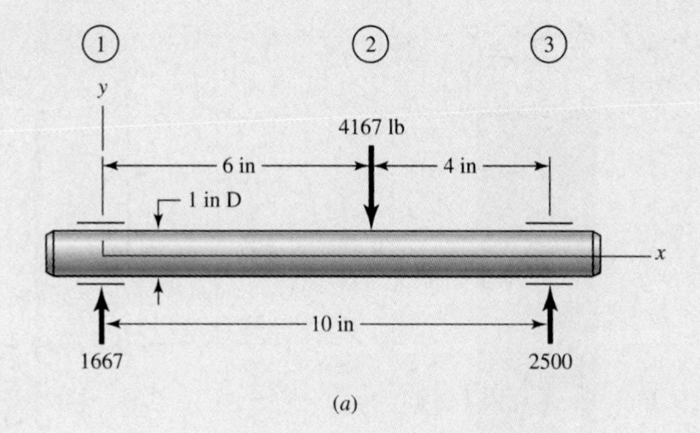

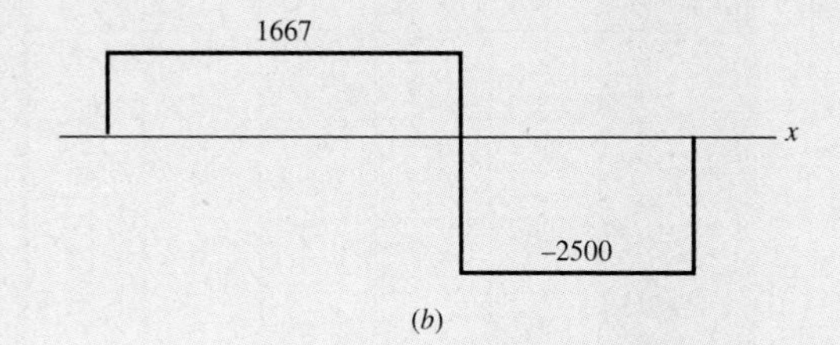

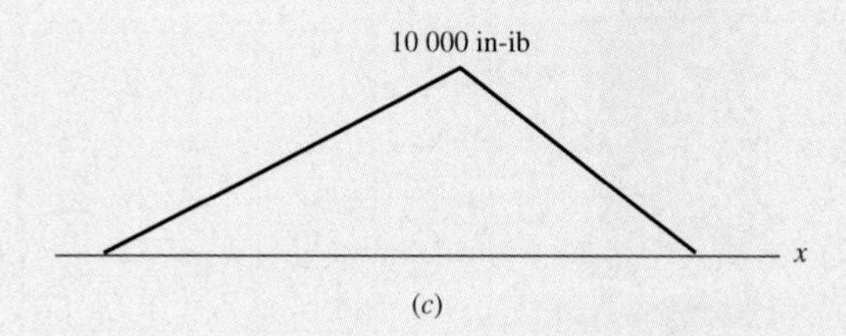

$$\frac{M}{EI} = \frac{5000}{30\left(10^{6}\right)0.0491} = 3395\left(10^{-6}\right) \qquad \text{(both entries)}$$

On line 3 (station 2),

$$\frac{M}{EI} = \frac{10\,000}{30\left(10^{6}\right)0.0491} = 6791\left(10^{-6}\right) \qquad \text{(both entries)}$$

Complete column 5. The first integration by trapezoidal rule is next. We use the entries in column 5 that flank the line separating lines 1 and 2. Noting that the first integral in column 6 is 0, we use Eq. (4–17),

$$I = 0 + \tfrac{1}{2}\left[3395\left(10^{-6}\right) + 0\right](3 - 0) = 5093\left(10^{-6}\right)$$

which we enter on the second line of the table, column 6. On the third line we write the value of the integral so far as

$$I = 5093\left(10^{-6}\right) + \tfrac{1}{2}\left[6791\left(10^{-6}\right) + 3395\left(10^{-6}\right)\right](6 - 3) = 20\,372\left(10^{-6}\right)$$

Similarly, we add 30 558(10^{-6}) and 33 953(10^{-6}) to finish column 6. We now use column 6 to produce the second integration by Simpson's rule. For column 7, station 2, using Eq. (4–18),

$$I = 0 + \tfrac{1}{6}\left[20\,372\left(10^{-6}\right) + 4(5093)10^{-6} + 0\right](6 - 0) = 40\,744\left(10^{-6}\right)$$

Table 4–2

Worksheet for Ex. 4–1

(1) Station Number	(2) x, in	(3) M, in · lb	(4) Diameter d, in	(5) $M/(EI)$	(6) $\int_0^x M/(EI)\,dx$	(7) $\int_0^x \int_0^x M/(EI)\,dx\,dx$	(8) y	(9) dy/dx
1	0	0 0	1 1	0 0	0	0	0	−0.015 845
	3	5 000 5 000	1 1	$3395(10^{-6})$ $3395(10^{-6})$	$5\,093(10^{-6})$			
2	6	10 000 10 000	1 1	$6791(10^{-6})$ $6791(10^{-6})$	$20\,372(10^{-6})$	$40\,744(10^{-6})$	$543\,26(10^{-6})$	0.004 527
	8	5 000 5 000	1 1	$3395(10^{-6})$ $3395(10^{-6})$	$30\,558(10^{-6})$			
3	10	0 0	1 1	0 0	$33\,953(10^{-6})$	$158\,449(10^{-6})$	0	0.018 108

Notes: The shaded triangle in upper-left corner of station number boxes denote bearing location. $C_1 = -0.015\,845$, $C_2 = 0$.

which we enter on the third line, station 2, column 7. For column 7, line 5, station 3,

$$I = 40\,744\left(10^{-6}\right) + \tfrac{1}{6}\left[33\,953\left(10^{-6}\right) + 4(30\,558)10^{-6} + 20\,372\left(10^{-6}\right)\right](10-6)$$
$$= 158\,449\left(10^{-6}\right)$$

which we enter. Having completed column 7, we are now able to find C_1 and C_2. Since the bearings are at stations 1 and 3,

$$C_1 = \frac{0 - (-158\,449)10^{-6}}{0 - 10} = -0.015\,845$$

$$C_2 = \frac{10(0) - 0(158\,449)10^{-6}}{0 - 10} = 0$$

We use the prediction equation, Eq. (4–16):

$$y = \int_0^x \int_0^x \left(\frac{M}{EI}dx\right) dx + (-0.015\,845)x + 0$$

At $x = 0$, $y = 0$. At $x = 6$,

$$y = 40\,744\left(10^{-6}\right) - 0.015\,845(6) = 543\,26\left(10^{-6}\right)$$

and at $x = 10$,

$$y = 158\,449\left(10^{-6}\right) - 0.015\,845(10) = 0$$

We enter these in column 8. The prediction equation for the slope is Eq. (4–15):

$$\frac{dy}{dx} = \int_0^x \frac{M}{EI}dx + (-0.015\,845) = \int_0^x \frac{M}{EI}dx - 0.015\,845$$

At station 1 (line 1), $x = 0$ and

$$\frac{dy}{dx} = 0 - 0.015\,845 = -0.015\,845$$

At station 2 (line 3), $x = 6$ and

$$\frac{dy}{dx} = 20\,372\left(10^{-6}\right) - 0.015\,845 = 0.004\,527$$

At station 3 (line 5), $x = 10$ and

$$\frac{dy}{dx} = 33\,953\left(10^{-6}\right) - 0.015\,845 = 0.018\,108$$

all of which we enter in column 9, completing the table.

The exactness of the deflection y at $x = 6$ can be demonstrated by using Table E–9–6:

$$y = \frac{Fbx}{6EIl}\left(x^2 + b^2 - l^2\right) = \frac{4167(4)6}{6(30)10^6(0.0491)10}\left(6^2 + 4^2 - 10^2\right)$$
$$= -0.054\,315 \text{ in}$$

The numerical "fuzz" is attributable to rounding of the second area moment.

If there is a limitation on slope at the bearings of the order of 0.001 (cylindrical roller bearings cease to be simple-support at this relative slope), do we have to form a new table? No. We note that an increased second area moment decreases the slope as follows:

$$I_{\text{new}} = I_{\text{old}} \frac{|dy/dx|}{0.001/n_d}$$

where n_d is the design factor. Substituting $\pi d^4/64$ for I, we obtain

$$d_{\text{new}} = d_{\text{old}} \left| \frac{n_d(dy/dx)}{0.001} \right|^{1/4} \tag{4–21}$$

In this case for $n_d = 1$,

$$d_{\text{new}} = (1) \left| \frac{(1)18\,108\,(10^{-6})}{0.001} \right|^{1/4} = 2.063 \text{ in}$$

If a limitation on deflection under the load exists, then, similarly,

$$d_{\text{new}} = d_{\text{old}} \left| \frac{n_d y_{\text{old}}}{y_{\text{all}}} \right|^{1/4} \tag{4–22}$$

Table 4–3 shows the influence of placing the origin at the extreme left of the shaft and adding two more stations. Intermediate stations are not shown. Note that C_1 is unchanged, but C_2 has changed from 0 to 0.015 845.

Complications in geometry, such as a taper on a shaft, introduce approximations which are ameliorated by increasing the number of stations along the taper. Table 4–4 is an abbreviated Table 4–2, showing the effect of having 11 stations.

When a beam or shaft has loads out of plane, two tables can be constructed for orthogonal planes. In-space deflections and slope come from Pythagorean combinations of components. The two tables also allow an end view of the shaft centerline in space. This is instructive in gear-shaft problems.

Table 4–3

Short Table for Ex. 4–1 with Five Stations Showing Effects of Placing Coordinate Origin at Extreme Left of Shaft

i	x	M	d	y	dy/dx
1	0	0 0	1 1	0.158E−1	−0.158E−1
2	1	0 0	1 1	0	−0.158E−1
3	7	10 000 10 000	1 1	−0.543E−1	0.453E−2
4	11	0 0	1 1	0	0.181E−1
5	12	0 0	1 1	0.181E−1	0.181E−1

Note: $C_1 = -0.015\,845$, $C_2 = 0.015\,845$.

Table 4–4

Short Table for Ex. 4–1 Using Eleven Stations

i	x	M	d	y	dy/dx
1	0	0	1	0	−0.158E−1
2	1	1 667	1	−0.157E−1	−0.153E−1
3	2	3 333	1	−0.302E−1	−0.136E−1
4	3	5 000	1	−0.424E−1	−0.108E−1
5	4	6 667	1	−0.513E−1	−0.679E−2
6	5	8 333	1	−0.556E−1	−0.170E−2
7	6	10 000	1	−0.543E−1	0.453E−2
8	7	7 500	1	−0.467E−1	0.105E−1
9	8	5 000	1	−0.340E−1	0.147E−1
10	9	2 500	1	−0.178E−1	0.173E−1
11	10	0	1	0	0.181E−1

Note: $C_1 = -0.015\,845$, $C_2 = 0$.

The method is powerful because it is integration, amenable to either human or computer analysis, and in most cases exact. The way in which to proceed is to do enough problems without a computer so that you can program a computer code. It can be as simple as a spreadsheet or it can be a formal computer code. Resist the temptation to code a computer to create the moments and reactions from the geometry and loading; do not forgo the opportunity to understand the problem and to gain proficiency. You can let the computer generate the intermediate station entries by interpolation, saving you work.

4–5 Finding Deflection by the Area-Moment Method

In Sec. 3–8, we learned methods for deriving the moment diagram directly from the loading diagram by employing the principles of statics. We saw in Sec. 4–3 that the moment diagram can also be obtained from the loading diagram by integrating twice. But the deflection diagram can be obtained from the moment diagram by integrating twice also. Thus it should be possible to derive the deflection diagram from the moment diagram by applying the principles of statics.

This method of deflection determination is called the *area-moment method.* It can be stated as follows: *The vertical distance between any point A on a deflection curve and a tangent through point B on the curve is the moment with respect to A of the area of the moment diagram between A and B divided by the stiffness EI.* To use this statement, first find the area of the parts of the moment diagram or, if preferred, the M/EI diagram. Then, second, multiply the areas by their centroidal distances from the axis of moments. Areas and centroidal distances for typical portions of moment diagrams are shown in Fig. 4–4.

Figure 4–4

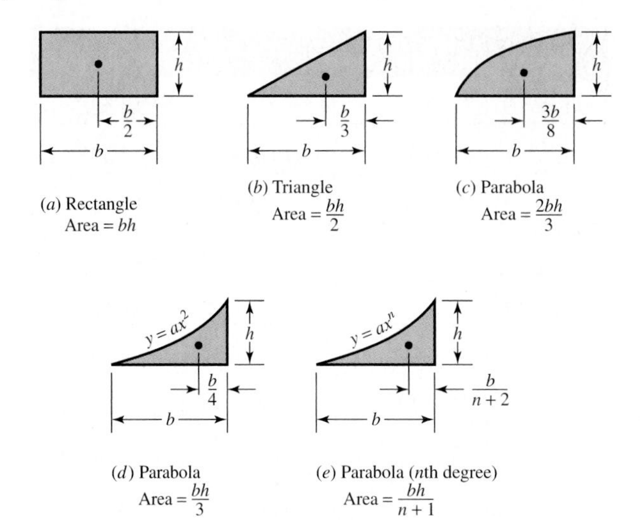

(*a*) Rectangle Area = *bh*

(*b*) Triangle Area = $\frac{bh}{2}$

(*c*) Parabola Area = $\frac{2bh}{3}$

(*d*) Parabola Area = $\frac{bh}{3}$

(*e*) Parabola (*n*th degree) Area = $\frac{bh}{n+1}$

EXAMPLE 4–2 Figure 4–5*a* shows a $1\frac{1}{4}$-in-diameter steel shaft upon which are mounted two gears. If the shaft bends excessively, the gears will mesh improperly and an early failure can be expected. In this example the gear forces, shown in Fig. 4–5*b*, are assumed to be in the same plane and of magnitude $F_1 = 120$ lb and $F_2 = 90$ lb. Find the shaft deflection at each gear. The loading diagram indicates that the bearings are self-aligning.

Solution This problem can be solved by employing the area-moment method only once, but we shall apply it twice, to illustrate the method of superposition.

The second moment of area is $I = \pi d^4/64 = \pi(1.25)^4/64 = 0.120\text{ in}^4$. Therefore $EI = 30(10)^6(0.120) = 3.6(10)^6\text{ lb}\cdot\text{in}^2$. Our first step is to calculate the deflections due only to the action of F_1. In Fig. 4–6 the loading and moment diagrams have been constructed. By statics, the two reactions are found, and then the moments at A and B. An exaggerated deflection curve is drawn on the loading diagram, and a tangent is constructed at the left-hand reaction. Then, by the area-moment method, the distance $C'C$ is equal to the moment of the area of the moment diagram about C, divided by EI.

Figure 4–5

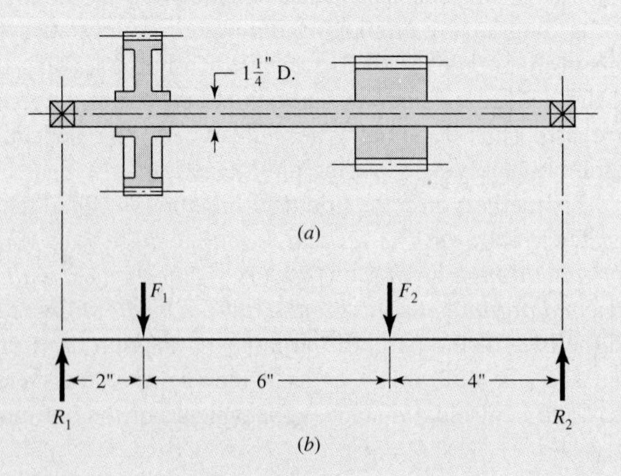

Figure 4–6

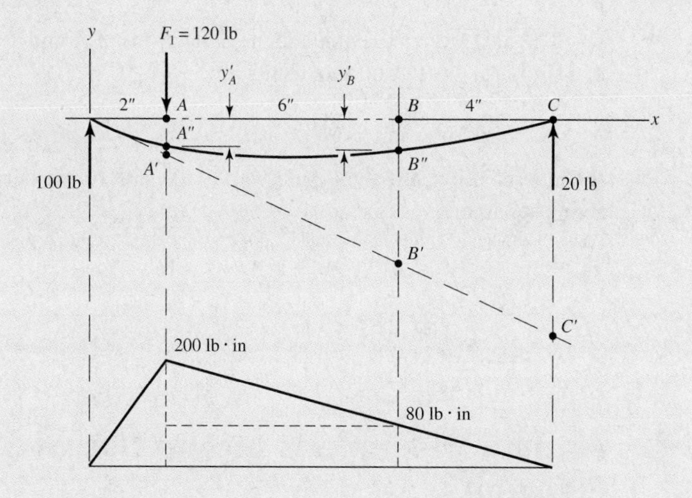

Thus

$$EI(C'C) = \overset{\text{area}}{[(200/2)(2)]}\ \overset{\text{arm}}{\left(10\tfrac{2}{3}\right)} + \overset{\text{area}}{[(200/2)(10)]}\ \overset{\text{arm}}{\left(6\tfrac{2}{3}\right)} = 8800$$

Thus

$$C'C = \frac{8800}{3.6(10)^6} = 2.44(10)^{-3} \text{ in}$$

In this approach, the areas may be either positive or negative, but the moment arms are *always* positive. The distance $C'C$ is a positive quantity, measured from C' to C. However, it is easier to keep track of the signs in this approach using carefully made sketches of the moment and deflection diagrams. By similar triangles we also find

$$A'A = 4.074(10)^{-4} \text{ in} \qquad B'B = 1.630(10)^{-3} \text{ in}$$

Next, we find the distance $A'A''$ by taking moments about A:

$$EI(A'A'') = \overset{\text{area}}{[(200/2)(2)]}\ \overset{\text{arm}}{\left(\tfrac{2}{3}\right)} = 133$$

$$A'A'' = \frac{133}{3.6(10)^6} = 3.704(10)^{-5} \text{ in}$$

The distance $B'B''$ is found by taking the moments of two triangular areas and one rectangular area about B:

$$EI(B'B'') = \overset{\text{area}}{[(200/2)(2)]}\ \overset{\text{arm}}{\left(6\tfrac{2}{3}\right)} + \overset{\text{area}}{[(80)(6)]}\ \overset{\text{arm}}{(3)} + \overset{\text{area}}{[(120/2)(6)]}\ \overset{\text{arm}}{(4)} = 4213$$

$$B'B'' = \frac{4213}{3.6(10)^6} = 1.17(10)^{-5} \text{ in}$$

The deflection at A due only to F_1 is now found to be

$$y'_A = A'A'' - A'A = 3.704(10)^{-4} \text{ in}$$

In a similar manner,

$$y'_B = B'B'' - B'B = -4.60(10)^{-4} \text{ in}$$

The next step is to calculate the deflections at A and B due only to F_2. This is done in a similar manner, and the results are

$$y''_A = -3.45(10)^{-4} \text{ in} \qquad y''_B = -7.11(10)^{-4} \text{ in}$$

By superposition, the total deflection is the sum of the deflections caused by each load acting separately. Hence

$$y_A = -7.15(10)^{-4} \text{ in} \qquad y_B = -11.71(10)^{-4} \text{ in}$$

4–6 Finding Deflections by the Use of Singularity Functions

The following examples are illustrations of the use of singularity functions.

EXAMPLE 4–3

As a first example, we choose the beam of Table E–9–6, which is a simply supported beam having a concentrated load F not in the center. Writing Eq. (4–10) for this loading gives

$$EI\frac{d^4y}{dx^4} = q = -F\langle x-a\rangle^{-1} \qquad 0 < x < l \tag{1}$$

Note that the reactions R_1 and R_2 do not appear in this equation, as they did in Chap. 3, because of the range chosen for x. If we now integrate from 0 to x—not $-\infty$ to x—according to Eq. (4–11), we get

$$EI\frac{d^3y}{dx^3} = V = -F\langle x-a\rangle^{0} + C_1 \tag{2}$$

Using Eq. (4–12) this time, we integrate again and obtain

$$EI\frac{d^2y}{dx^2} = M = -F\langle x-a\rangle^{1} + C_1x + C_2 \tag{3}$$

At $x = 0$, we have $M = 0$, and Eq. (3) gives $C_2 = 0$. At $x = l$, we again have $M = 0$, and Eq. (3) gives

$$C_1 = \frac{F(l-a)}{l} = \frac{Fb}{l}$$

Substituting these terms for C_1 and C_2 in Eq. (3) gives

$$EI\frac{d^2y}{dx^2} = M = \frac{Fbx}{l} - F\langle s-a\rangle^{1} \tag{4}$$

Note that we could have obtained this equation by summing moments about a section a distance x from the origin. Next, integrate Eq. (4) twice in accordance with Eqs. (4–13) and (4–14). This yields

$$EI\frac{dy}{dx} = EI\theta = \frac{Fbx^2}{2l} - \frac{F\langle x-a\rangle^2}{2} + C_3 \tag{5}$$

$$EIy = \frac{Fbx^3}{6l} - \frac{F\langle x-a\rangle^3}{6} + C_3x + C_4 \tag{6}$$

The constants of integration C_3 and C_4 are evaluated using the two boundary conditions $y = 0$ at $x = 0$ and $y = 0$ at $x = l$. The first condition, substituted in Eq. (6), gives $C_4 = 0$. The second condition, substituted in Eq. (6), yields

$$0 = \frac{Fbl^2}{6} - \frac{Fb^3}{6} + C_3 l$$

whence

$$C_3 = -\frac{Fb}{6l}\left(l^2 - b^2\right)$$

Upon substituting these results for C_3 and C_4 in Eq. (6), we obtain the deflection relation as

$$y = \frac{F}{6EIl}\left[bx\left(x^2 + b^2 - l^2\right) - l\langle x-a\rangle^3\right] \tag{7}$$

Compare Eq. (7) with the two deflection equations in Table E–9–6, and note that the use of singularity functions enables us to express the entire relation with a single equation.

EXAMPLE 4–4 Find the deflection relation, the maximum deflection, and the reactions for the statically indeterminate beam shown in Fig. 4–7*a* using singularity functions.

Solution The loading diagram and approximate deflection curve are shown in Fig. 4–7*b*. Based upon the range $0 < x < l$, the loading equation is

$$q = R_2\langle x-a\rangle^{-1} - w\langle x-a\rangle^0 \tag{1}$$

Figure 4–7

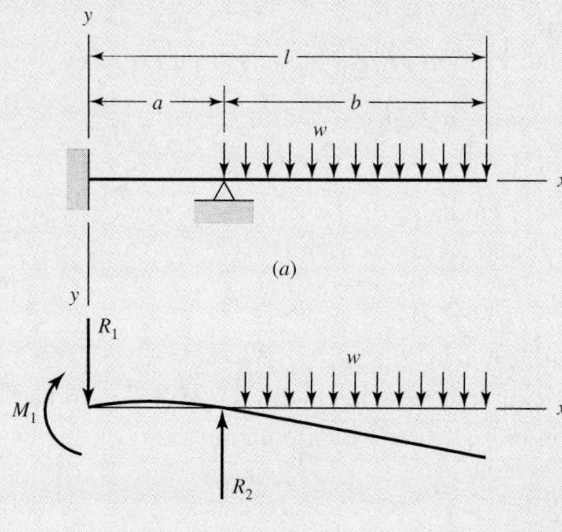

We now integrate this equation four times in accordance with Eqs. (4–10) to (4–14). The results are

$$V = R_2\langle x-a\rangle^0 - w\langle x-a\rangle^1 + C_1 \tag{2}$$

$$M = R_2\langle x-a\rangle^1 - \frac{w}{2}\langle x-a\rangle^2 + C_1x + C_2 \tag{3}$$

$$EI\theta = \frac{R_2}{2}\langle x-a\rangle^2 - \frac{w}{6}\langle x-a\rangle^3 + \frac{C_1}{2}x^2 + C_2x + C_3 \tag{4}$$

$$EIy = \frac{R_2}{6}\langle x-a\rangle^3 - \frac{w}{24}\langle x-a\rangle^4 + \frac{C_1}{6}x^3 + \frac{C_2}{2}x^2 + C_3x + C_4 \tag{5}$$

As in the previous example, the constants are evaluated by applying appropriate boundary conditions. First we note from Fig. 4–7b that both $EI\theta = 0$ and $EIy = 0$ at $x = 0$. This gives $C_3 = 0$ and $C_4 = 0$. Next we observe that at $x = 0$ the shear force is the same as the reaction. So Eq. (2) gives $V(0) = R_1 = C_1$. Since the deflection must be zero at the reaction R_2, where $x = a$, we have from Eq. (5) that

$$\frac{C_1}{6}a^3 + \frac{C_2}{2}a^2 = 0 \qquad \text{or} \qquad C_1\frac{a}{3} + C_2 = 0 \tag{6}$$

Also the moment must be zero at the overhanging end where $x = l$; Eq. (3) then gives

$$R_2(l-a) - \frac{w}{2}(l-a)^2 + C_1l + C_2 = 0$$

Simplifying and noting that $l - a = b$ gives

$$C_1a + C_2 = -\frac{wb^2}{2} \tag{7}$$

Equations (6) and (7) are now solved simultaneously for C_1 and C_2. The results are

$$C_1 = -R_1 = -\frac{3wb^2}{4a} \qquad C_2 = \frac{wb^2}{4}$$

With R_1 known, we can sum the forces in the y direction to zero to get R_2. Thus

$$R_2 = +R_1 + wb = \frac{3wb^2}{4a} + wb = \frac{wb}{4a}(4a+3b)$$

The moment reaction M_1 is obtained from Eq. (3) with $x = 0$. This gives

$$M(0) = M_1 = C_2 = \frac{wb^2}{4}$$

The complete equation of the deflection curve is obtained by substituting the known terms for R_2 and for the constants C_i in Eq. (5). The result is

$$EIy = \frac{wb}{24a}(4a+3b)\langle x-a\rangle^3 - \frac{w}{24}\langle x-a\rangle^4 - \frac{wb^2x^3}{8a} + \frac{wb^2x^2}{8} \tag{8}$$

The maximum deflection occurs at the free end of the beam, at $x = l$. By making this substitution in Eq. (8) and manipulating the resulting expression, one finally obtains

$$y_{\max} = \frac{wb^3l}{8EI} \tag{9}$$

Examination of the deflection curve of Fig. 4–7*b* reveals that the curve will have a zero slope at some point between R_1 and R_2. By replacing the constants in Eq. (4), setting $\theta = 0$, and solving the result for x we readily find that the slope is zero at $x = 2a/3$. The corresponding deflection at this point can be obtained by using the value of x in Eq. (8). The result is

$$y(x = 2a/3) = \frac{wa^2b^2}{54EI} \tag{10}$$

4–7 Strain Energy

The external work done on an elastic member in deforming it is transformed into *strain*, or *potential, energy*. If the member is deformed a distance y, this energy is equal to the product of the average force and the deflection, or

$$U = \frac{F}{2}y = \frac{F^2}{2k} \tag{a}$$

This equation is general in the sense that the force F also means torque, or moment, provided, of course, that consistent units are used for k. By substituting appropriate expressions for k, strain-energy formulas for various simple loadings may be obtained. For tension and compression and for torsion, for example, we employ Eqs. (4–4) and (4–7) and obtain

$$U = \frac{F^2 l}{2AE} \quad \text{tension and compression} \tag{4–23}$$

$$U = \frac{T^2 l}{2GJ} \quad \text{torsion} \tag{4–24}$$

To obtain an expression for the strain energy due to direct shear, consider the element with one side fixed in Fig. 4–8*a*. The force F places the element in pure shear, and the work done is $U = F\delta/2$. Since the shear strain is $\gamma = \delta/l = \tau/G = F/AG$, we have

$$U = \frac{F^2 l}{2AG} \quad \text{direct shear} \tag{4–25}$$

The strain energy stored in a beam or lever by bending may be obtained by referring to Fig. 4–8*b*. Here AB is a section of the elastic curve of length ds having a radius of

Figure 4–8

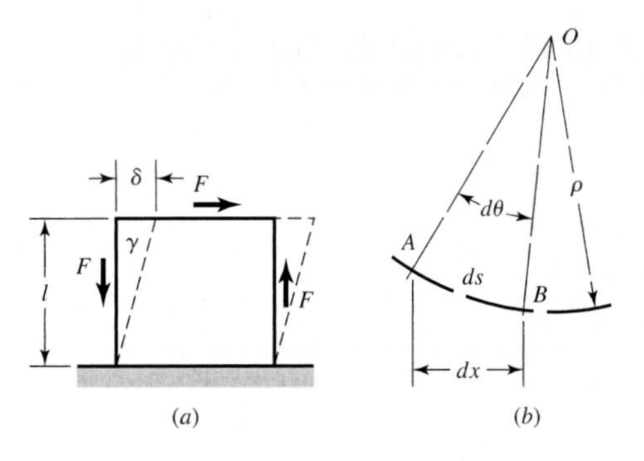

curvature ρ. The strain energy stored in this element of the beam is $dU = (M/2)d\theta$. Since $\rho\, d\theta = ds$, we have

$$dU = \frac{M\, ds}{2\rho} \tag{b}$$

We can eliminate ρ by using Eq. (4–8). Thus

$$dU = \frac{M^2\, ds}{2EI} \tag{c}$$

For small deflections, $ds \approx dx$. Then, for the entire beam

$$U = \int \frac{M^2\, dx}{2EI} \qquad \text{bending} \tag{4–26}$$

Sometimes the strain energy stored in a unit volume is a useful quantity. By dividing Eqs. (4–23) to (4–25) by the total volume lA, we obtain

$$u = \begin{cases} \dfrac{\sigma^2}{2E} & \text{tension and compression} \\ \dfrac{\tau^2}{2G} & \text{direct shear} \\ \dfrac{\tau^2_{\max}}{4G} & \text{torsion} \end{cases} \tag{4–27}$$

It is interesting to note, from Eq. (4–27), that the development of a high stress in a material with a low modulus of elasticity, or rigidity, will result in the greatest amount of energy storage.

Equation (4–26) is exact only when a beam is subject to pure bending. Even when shear is present, Eq. (4–26) continues to give quite good results, except for very short beams. The strain energy due to shear loading of a beam is a complicated problem. An approximate solution can be obtained by using Eq. (4–25) with a correction factor whose value depends upon the shape of the cross section. If we use C for the correction factor and V for the shear force, then the strain energy due to shear in bending is the integral of Eq. (4–25), or

$$U = \int \frac{CV^2\, dx}{2AG} \qquad \text{bending shear} \tag{4–28}$$

Values of the factor C are listed in Table 4–5.

Table 4–5

Strain-Energy Correction Factors for Shear

Source: Arthur P. Boresi, Omar M. Sidebottom, Fred B. Seely, and James O. Smith, *Advanced Mechanics of Materials*, 3d ed., Wiley, New York, 1978, p. 173.

Beam Cross-Sectional Shape	Factor C
Rectangular	1.50
Circular	1.33
Tubular, round	2.00
Box sections†	1.00
Structural sections†	1.00

†Use area of web only.

EXAMPLE 4–5 Find the strain energy due to shear in a rectangular cross-section beam, simply supported, and having a uniformly distributed load.

Solution Using Appendix Table E–9–7, we find the shear force to be

$$V = \frac{wl}{2} - wx$$

Substituting into Eq. (4–28), with $C = 1.5$, gives

Answer
$$U = \frac{1.5}{2AG}\int_0^l \left(\frac{wl}{2} - wx\right)^2 dx = \frac{3w^2l^3}{48AG}$$

EXAMPLE 4–6 A cantilever has a concentrated load F at the end, as shown in Fig. 4–9. Find the strain energy in the beam by neglecting shear.

Figure 4–9

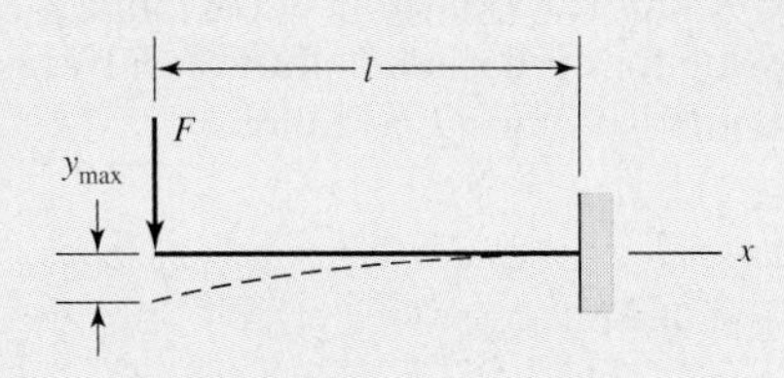

Solution At any point x along the beam, the moment is $M = -Fx$. Substituting this value of M into Eq. (4–26), we find

Answer
$$U = \int_0^l \frac{F^2x^2\,dx}{2EI} = \frac{F^2l^3}{6EI}$$

4–8 Castigliano's Theorem

A most unusual, powerful, and often surprisingly simple approach to deflection analysis is afforded by an energy method called Castigliano's theorem. It is a unique way of analyzing deflections and is even useful for finding the reactions of indeterminate structures. Castigliano's theorem states that *when forces act on elastic systems subject to small displacements, the displacement corresponding to any force, collinear with the force, is equal to the partial derivative of the total strain energy with respect to that force.* The terms *force* and *displacement* in this statement are broadly interpreted to apply equally to moments and angular displacements. Mathematically, the theorem of Castigliano is

$$\delta_i = \frac{\partial U}{\partial F_i} \tag{4–29}$$

where δ_i is the displacement of the point of application of the force F_i in the direction of F_i.

As an example, apply Castigliano's theorem using Eqs. (4–23) and (4–24) to get the axial and torsional deflections. The results are

$$\delta = \frac{\partial}{\partial F}\left(\frac{F^2 l}{2AE}\right) = \frac{Fl}{AE} \tag{a}$$

$$\theta = \frac{\partial}{\partial T}\left(\frac{T^2 l}{2GJ}\right) = \frac{Tl}{GJ} \tag{b}$$

Compare Eqs. (*a*) and (*b*) with Eqs. (4–3) and (4–5). In Example 4–6, the strain energy for a cantilever having a concentrated end load was found. According to Castigliano's theorem, the deflection at the end of the beam is

$$y = \frac{\partial U}{\partial F} = \frac{\partial}{\partial F}\left(\frac{F^2 l^3}{6EI}\right) = \frac{Fl^3}{3EI}$$

which checks with Table E–9–1.

Castigliano's theorem can be used to find the deflection at a point even though no force or moment acts there. The procedure is:

1. Set up the equation for the total strain energy U by including the energy due to a fictitious force or moment Q_i acting at the point whose deflection is to be found.
2. Find an expression for the desired deflection δ_i by taking the derivative of the total strain energy with respect to Q_i as follows:

$$\delta_i = \frac{\partial U}{\partial Q_i} \tag{4–30}$$

3. Since Q_i is a fictitious force, solve the expression obtained in step 2 by setting Q_i equal to zero.

EXAMPLE 4–7 The cantilever of Example 4–6 is a carbon steel bar 10 in long with a 1-in diameter and is loaded by a force $F = 100$ lb.

(*a*) Find the maximum deflection using Castigliano's theorem, including that due to shear.

(*b*) What error is introduced if shear is neglected?

Solution (*a*) From Eq. (4–28) and Example 4–5 data, the total strain energy is

$$U = \frac{F^2 l^3}{6EI} + \int_0^l \frac{CV^2\,dx}{2AG} \tag{1}$$

For the cantilever, $V = F$. Also, $C = 1.33$, from Table 4–5. Performing the integration and substituting these values in Eq. (1) gives, for the total strain energy,

$$U = \frac{F^2 l^3}{6EI} + \frac{1.33F^2 l}{2AG} \tag{2}$$

Then, according to the theorem, the deflection of the end is

$$y = \frac{\partial U}{\partial F} = \frac{Fl^3}{3EI} + \frac{1.33Fl}{AG} \tag{3}$$

We also find that

$$I = \frac{\pi d^4}{64} = \frac{\pi (1)^4}{64} = 0.0491 \text{ in}^4$$

$$A = \frac{\pi d^2}{4} = \frac{\pi (1)^2}{4} = 0.7854 \text{ in}^2$$

Substituting these values, together with $F = 100$ lb, $l = 10$ in, $E = 30$ Mpsi, and $G = 11.5$ Mpsi, in Eq. (3) gives

Answer $y = 0.022\,629 + 0.000\,147 = 0.022\,776$ in

Note that the result is positive because it is in the *same* direction as the force F.

(*b*) The error in neglecting shear is found to be about 0.65 percent.

EXAMPLE 4–8 Find the maximum deflection of a cantilever of length l loaded by a concentrated force F in the middle. Neglect shear.

Solution The problem is illustrated in Fig. 4–10, where the force F acts in the center and a fictitious force Q has been placed at the end where the deflection is desired.

The moments are

$$M_{AB} = F\left(x - \frac{l}{2}\right) + Q(x - l) \tag{1}$$

$$M_{BC} = Q(x - l) \tag{2}$$

The total strain energy is

$$U = \int_0^{l/2} \frac{M_{AB}^2\, dx}{2EI} + \int_{l/2}^{l} \frac{M_{BC}^2\, dx}{2EI} \tag{3}$$

Then, by Castigliano's theorem, the deflection is

$$y = \frac{\partial U}{\partial Q} = \frac{1}{2EI}\left[\int_0^{l/2} 2M_{AB}\left(\frac{\partial M_{AB}}{\partial Q}\right) dx + \int_{l/2}^{l} 2M_{BC}\left(\frac{\partial M_{BC}}{\partial Q}\right) dx\right] \tag{4}$$

Next, we have

$$\frac{\partial M_{AB}}{\partial Q} = \frac{\partial M_{BC}}{\partial Q} = x - l \tag{5}$$

Making the appropriate substitutions in Eq. (4) yields

$$y = \frac{1}{EI}\left\{\int_0^{l/2}\left[F\left(x - \frac{l}{2}\right) + Q(x - l)\right](x - l)dx + \int_{l/2}^{l}[Q(x - l)](x - l)dx\right\} \tag{6}$$

Since Q is fictitious, we substitute $Q = 0$ in Eq. (6) to get

Figure 4–10

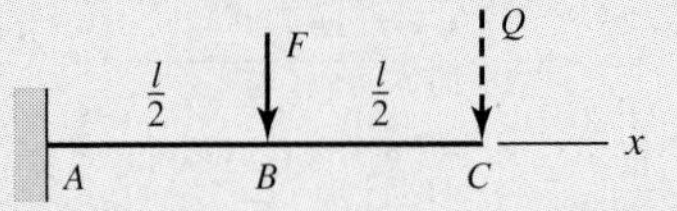

Answer $$y = \frac{F}{EI}\int_0^{l/2}\left(x - \frac{l}{2}\right)(x - l)\,dx = \frac{5Fl^3}{48EI}$$

4–9 Statically Indeterminate Problems

A system in which the laws of statics are not sufficient to determine all the unknown forces or moments is said to be *statically indeterminate.* Problems of which this is true are solved by writing the appropriate equations of static equilibrium and additional equations pertaining to the deformation of the part. In all, the number of equations must equal the number of unknowns.

A simple example of a statically indeterminate problem is furnished by the nested helical spring in Fig. 4–11*a*. When this assembly is loaded by the compressive force F, it deforms through the distance δ. What is the compressive force in each spring?

Only one equation of static equilibrium can be written. It is

$$\Sigma F = F - F_1 - F_2 = 0 \tag{a}$$

which simply says that the total force F is resisted by a force F_1 in spring 1 plus the force F_2 in spring 2. Since there are two unknowns and only one equation, the system is statically indeterminate.

To write another equation, note the deformation relation in Fig. 4–11*b*. The two springs have the same deformation. Thus, we obtain the second equation as

$$\delta_1 = \delta_2 = \delta \tag{b}$$

Figure 4–11

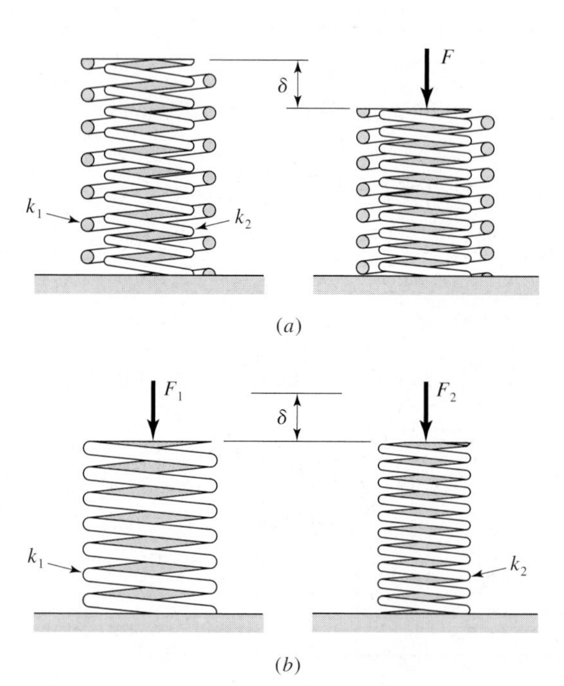

If we now substitute Eq. (4–2) in Eq. (b), we have

$$\frac{F_1}{k_1} = \frac{F_2}{k_2} \tag{c}$$

Now we solve Eq. (c) for F_1 and substitute the result in Eq. (a). This gives

$$F - \frac{k_1}{k_2}F_2 - F_2 = 0 \quad \text{or} \quad F_2 = \frac{k_2 F}{k_1 + k_2} \tag{d}$$

This completes the solution, because with F_2 known, F_1 can be found from Eq. (c).

A more elegant approach is afforded by Castigliano's theorem. The method is based on the fact that the deflection at the point of and in the direction of any reaction is zero. Thus the derivative of the total strain energy with respect to the reaction should also be zero. The procedure is:

1 Choose the redundant reaction; this is usually a force or a moment.

2 Write and solve the equations of static equilibrium for the remaining reactions in terms of the redundant reaction.

3 Write the equation for the total strain energy U.

4 Find an expression for the redundant reaction by taking the derivative of the total strain energy with respect to the reaction, say R, as follows:

$$\frac{\partial U}{\partial R} = 0 \tag{4–31}$$

5 Solve the resulting expression for the reaction R.

EXAMPLE 4–9

The indeterminate beam of Appendix Table E–9–11 is reproduced in Fig. 4–12. Determine the reactions using Castigliano's theorem to resolve the indeterminacy.

Solution

The reactions are shown in Fig. 4–12b. We choose M_1 as the redundant one. Applying the rule of static equilibrium yields the two equations

$$\Sigma F = R_1 - F + R_2 = 0$$

$$\Sigma M_O = M_1 - \frac{Fl}{2} + R_2 l = 0 \tag{1}$$

Since M_1 is selected as the indeterminate reaction, we next solve equation set (1) for R_1 and R_2 in terms of M_1. Thus

$$R_1 = \frac{F}{2} + \frac{M_1}{l} \qquad R_2 = \frac{F}{2} - \frac{M_1}{l} \tag{2}$$

Figure 4–12

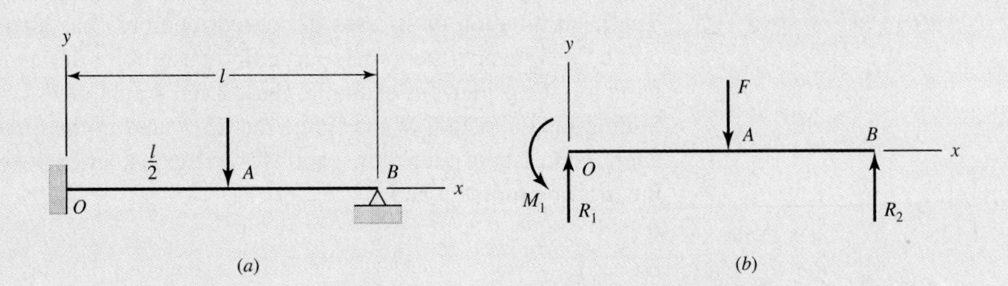

Neglecting shear, the total strain energy, from Eq. (4–26), is

$$U = \int_0^{l/2} \frac{M_{OA}^2\, dx}{2EI} + \int_{l/2}^{l} \frac{M_{AB}^2\, dx}{2EI} \tag{3}$$

The moments from O to A and from A to B are

$$M_{OA} = -M_1 + R_1 x = M_1\left(\frac{x}{l} - 1\right) + \frac{Fx}{2}$$
$$M_{AB} = -M_1 + R_1 x - F\left(x - \frac{l}{2}\right) = M_1\left(\frac{x}{l} - 1\right) + \frac{F}{2}(l - x) \tag{4}$$

The reaction M_1 is now found by taking the derivative of Eq. (3) and setting the result equal to zero. Thus

$$\frac{\partial U}{\partial M_1} = \frac{1}{2EI}\left[\int_0^{l/2} 2M_{OA}\left(\frac{\partial M_{OA}}{\partial M_1}\right) dx + \int_{l/2}^{l} 2M_{AB}\left(\frac{\partial M_{AB}}{\partial M_1}\right) dx\right] \tag{5}$$

Taking the derivatives of the moments of equation set (4) yields

$$\frac{\partial M_{OA}}{\partial M_1} = \frac{\partial M_{AB}}{\partial M_1} = \frac{x}{l} - 1 \tag{6}$$

We next substitute Eqs. (4) and (6) in Eq. (5), perform the indicated multiplication, and cancel terms. We then have

$$\frac{\partial U}{\partial M_1} = \int_0^{l/2}\left[M_1\left(\frac{x}{l} - 1\right)^2 + \frac{Fx}{2}\left(\frac{x}{l}\right)\right] dx$$
$$+ \int_{l/2}^{l}\left[M_1\left(\frac{x}{l} - 1\right)^2 + \frac{F}{2}(l - x)\left(\frac{x}{l}\right)\right] dx = 0 \tag{7}$$

After Eq. (7) is integrated and the limits substituted, we obtain

$$\frac{\partial U}{\partial M_1} = \frac{7M_1 l}{24} - \frac{Fl^2}{24} + \frac{M_1 l}{24} - \frac{Fl^2}{48} = 0 \tag{8}$$

Answer The redundant reaction is now found to be $M_1 = 3Fl/16$. This value of M_1 can now be substituted in Eq. (2) to obtain R_1 and R_2 in terms of the force F.

4–10 Deflection of Curved Members

Machine frames, springs, clips, fasteners, and the like frequently occur as curved shapes. The determination of stresses in curved members has already been illustrated in Sec. 3–20. Castigliano's theorem is particularly useful for the analysis of deflections in curved parts too. Consider, for example, the curved frame of Fig. 4–13a. We are interested in finding the deflection of the frame due to F and in the direction of F. The total strain energy consists of four terms, and we shall consider each separately. The first is due to the bending moment and is

$$U_1 = \int \frac{M^2\, d\theta}{2AeE} \tag{4–32}$$

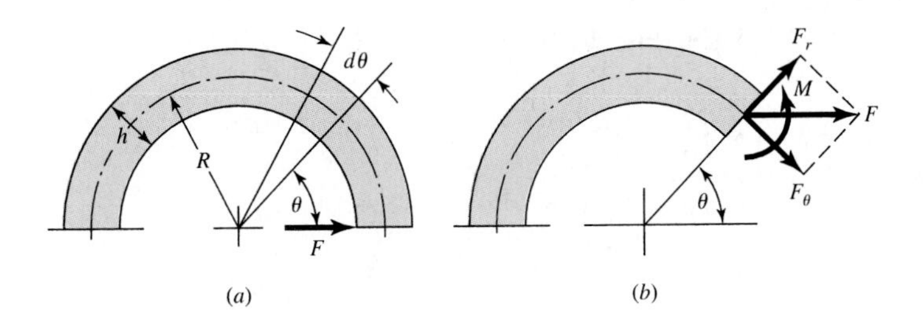

Figure 4–13

(a) Curve bar loaded by force F. R = radius to centroidal axis of section; h = section thickness. (b) Diagram showing forces acting on section taken at angle θ. $F_r = V$ = shear component of F; F_θ is component of F normal to section; M is moment caused by force F.

In this equation, the eccentricity e is

$$e = R - r_n \tag{4–33}$$

where r_n is the radius of the neutral axis as defined in Sec. 3–20 and shown in Fig. 3–31.

An approximate result can be obtained using the equation

$$U_1 \approx \int \frac{M^2 R\, d\theta}{2EI} \qquad \frac{R}{h} > 10 \tag{4–34}$$

which is obtained directly from Eq. (4–26). Note the limitation on the use of Eq. (4–34).

The strain energy component due to the normal force F_θ consists of two parts, one of which is axial and analogous to Eq. (4–23). This part is

$$U_2 = \int \frac{F_\theta^2 R\, d\theta}{2AE} \tag{4–35}$$

The force F_θ also produces a moment, which opposes the moment M in Fig. 4–13*b*. The resulting strain energy will be subtractive and is

$$U_3 = -\int \frac{M F_\theta\, d\theta}{AE} \tag{4–36}$$

The negative sign of Eq. (4–36) can be appreciated by referring to both parts of Fig. 4–13. Note that the moment M tends to decrease the angle $d\theta$. On the other hand, the moment due to F_θ tends to increase $d\theta$. Thus U_3 is negative. If F_θ had been acting in the opposite direction, then both M and F_θ would tend to decrease the angle $d\theta$.

The fourth and last term is the shear energy due to F_r. Adapting Eq. (4–28) gives

$$U_4 = \int \frac{C F_r^2 R\, d\theta}{2AG} \tag{4–37}$$

where C is the correction factor of Table 4–5.

Combining the four terms gives for the total strain energy

$$U = \int \frac{M^2\, d\theta}{2AeE} + \int \frac{F_\theta^2 R\, d\theta}{2AE} - \int \frac{M F_\theta\, d\theta}{AE} + \int \frac{C F_r^2 R\, d\theta}{2AG} \tag{a}$$

The deflection produced by the force F can now be found. It is

$$\delta = \frac{\partial U}{\partial F} = \int_0^\pi \frac{M}{AeE}\left(\frac{\partial M}{\partial F}\right) d\theta + \int_0^\pi \frac{F_\theta R}{AE}\left(\frac{\partial F_\theta}{\partial F}\right) d\theta$$
$$- \int_0^\pi \frac{1}{AE}\frac{\partial (M F_\theta)}{\partial F} d\theta + \int_0^\pi \frac{C F_r R}{AG}\left(\frac{\partial F_r}{\partial F}\right) d\theta \tag{b}$$

Using Fig. 4–13b, we find

$$M = FR\sin\theta \qquad \frac{\partial M}{\partial F} = R\sin\theta$$

$$F_\theta = F\sin\theta \qquad \frac{\partial F_\theta}{\partial F} = \sin\theta$$

$$MF_\theta = F^2R\sin^2\theta \qquad \frac{\partial MF_\theta}{\partial F} = 2FR\sin^2\theta$$

$$F_r = F\cos\theta \qquad \frac{\partial F_r}{\partial F} = \cos\theta$$

Substituting all these into Eq. (b) and factoring yields

$$\delta = \frac{FR^2}{AeE}\int_0^\pi \sin^2\theta\, d\theta + \frac{FR}{AE}\int_0^\pi \sin^2\theta\, d\theta - \frac{2FR}{AE}\int_0^\pi \sin^2\theta\, d\theta + \frac{CFR}{AG}\int_0^\pi \cos^2\theta\, d\theta$$

$$= \frac{\pi FR^2}{2AeE} + \frac{\pi FR}{2AE} - \frac{\pi FR}{AE} + \frac{\pi CFR}{2AG} = \frac{\pi FR^2}{2AeE} - \frac{\pi FR}{2AE} + \frac{\pi CFR}{2AG} \tag{4–38}$$

Because the first term contains the square of the radius, the second two terms will be small if the frame has a large radius. Also, if $R/h > 10$, Eq. (4–34) can be used. An approximate result then turns out to be

$$\delta \approx \frac{\pi FR^3}{2EI} \tag{4–39}$$

The determination of the deflection of a curved member loaded by forces at right angles to the plane of the member is more difficult, but the method is the same.[4] We shall include here only one of the more useful solutions to such a problem, though the methods for all are similar. Figure 4–14 shows a cantilevered ring segment having a span angle ϕ. The strain energy is obtained from the equation

$$U = \int_0^\phi \frac{M^2R\, d\theta}{2EI} + \int_0^\phi \frac{T^2R\, d\theta}{2GJ} \tag{4–40}$$

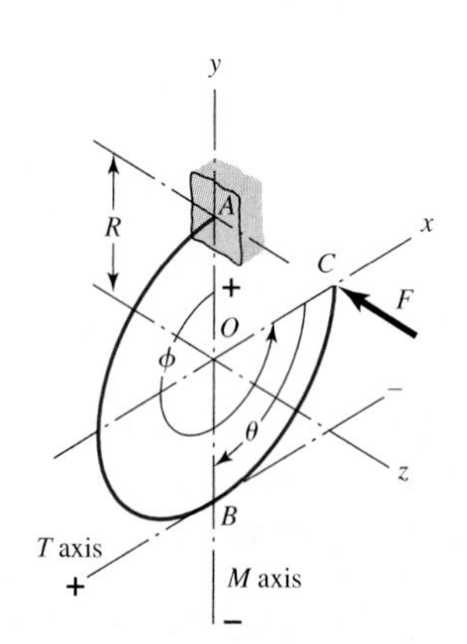

Figure 4–14

Ring ABC in the xy plane subject to force F parallel to the z axis. Corresponding to a ring segment CB at angle θ from the point of application of F, the moment axis is a line BO and the torque axis is a line in the xy plane tangent to the ring at B. Note the positive directions of the T and M axes.

The moments and torques acting on a section at B, due to the force F, are

$$M = FR\sin\theta \qquad T = FR(1-\cos\theta)$$

The deflection δ of the ring segment at F and in the direction of F is then found to be

$$\delta = \frac{\partial U}{\partial F} = \frac{FR^3}{2}\left(\frac{\alpha}{EI} + \frac{\beta}{GJ}\right) \tag{4–41}$$

where the coefficients α and β are dependent on the span angle ϕ and are defined as follows:

$$\alpha = \phi - \sin\phi\cos\phi$$
$$\beta = 3\phi - 4\sin\phi + \sin\phi\cos\phi \tag{4–42}$$

4. For more solutions than are included here, see Joseph E. Shigley, "Curved Beams and Rings," chap. 16 in Joseph E. Shigley and Charles R. Mischke (eds.), *Standard Handbook of Machine Design*, McGraw-Hill, New York, 1986.

CASE STUDY 4 Deflection in a Variable-Cross-Section Punch-Press Frame

The general result expressed in Eq. (4–38),

$$\delta = \frac{\pi F R^2}{2AeE} - \frac{\pi F R}{2AE} + \frac{\pi C F R}{2AG} \tag{4–38}$$

is useful in sections that are uniform and in which the centroidal locus is circular. This made integration tractable and general, and insightful results are obtained. The bending moment is largest where the material is farthest from the load axis. Strengthening requires a larger second area moment I. A variable-depth cross section is attractive, but it makes the integration to a closed form impossible. However, if you are seeking results, numerical integration with computer assistance is satisfactory.

Consider the steel C-frame depicted in Fig. CS4–1 in which the centroidal radius is 32 in, the cross section at the ends is 2 in × 2 in, and the depth varies sinusoidally with an amplitude of 2 in. The load is 1000 lb. It follows that $C = 1.5$, $G = 14.5(10^6)$ psi, $E = 30(10^6)$ psi. The outer and inner radii are

$$R_{\text{out}} = R + \frac{d}{2} + B \sin\theta$$

$$R_{\text{in}} = R - \frac{d}{2} - B \sin\theta$$

$$h = R_{\text{out}} - R_{\text{in}}$$

$$A = bh$$

$$r_n = \frac{h}{\ln[(R + h/2)/(R - h/2)]}$$

$$e = R - r_n$$

Note that

$$M = FR\sin\theta \qquad \partial M/\partial F = R\sin\theta$$

$$F_\theta = F\sin\theta \qquad \partial F_\theta/\partial F = \sin\theta$$

$$MF_\theta = F^2 R\sin^2\theta \qquad \partial MF_\theta/\partial F = 2FR\sin^2\theta$$

$$F_r = F\cos\theta \qquad \partial F_r/\partial F = \cos\theta$$

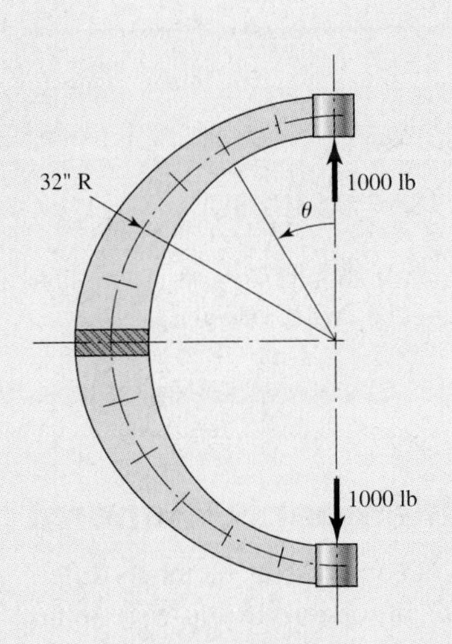

Figure CS4–1

(*a*) A steel punch press has a C-frame with a varying-depth rectangular cross section depicted. The cross section varies sinusoidally from 2 × 2 in at $\theta = 0°$ to 2 × 6 in at $\theta = 90°$, and back to 2 × 2 in at $\theta = 180°$. Of immediate interest to the designer is the deflection in the load axis direction under the load.

Substituting in Eq. (b) above Eq. (4–38) gives kernels k_1, k_2, k_3, and k_4:

$$k_1 = \frac{FR^2}{AeE}\sin^2\theta \qquad k_2 = \frac{FR}{AE}\sin^2\theta$$

$$k_3 = -\frac{2FR}{AE}\sin^2\theta \qquad k_4 = \frac{CFR}{AG}\cos^2\theta$$

noting that A and e are functions of θ. The grand kernel is $k = k_1 + k_2 + k_3 + k_4$. This may be programmed for a Simpson's rule integration using four applications of the rule involving 13 ordinates. The total deflection is given by

$$\delta = \frac{15°(\pi/180°)}{3}\sum$$

Write such a program. Benchmark it with a problem in which the sinusoidal amplitude is $B = 0$, and with $R/h \geq 10$. Then check it with Eq. (4–39),

$$\delta \doteq \frac{\pi FR^3}{2EI}$$

When all is satisfactory, solve the initial problem. The Simpson's rule table should look as follows:

Ordinate	Multiplier	Extension
8.275 862E-4	1	8.275 862E-4
1.617 215E-2	4	6.468 859E-2
2.585 165E-2	2	5.150 330E-2
2.917 900E-2	4	0.116 716
3.007 690E-2	2	6.015 379E-2
3.020 023E-2	4	0.120 801
3.018 103E-2	2	6.036 206E-2
3.020 023E-2	4	0.120 801
3.007 248E-2	2	6.014 497E-2
2.917 896E-2	4	0.116 716
2.585 165E-2	2	5.170 331E-2
1.618 460E-2	4	6.473 839E-2
8.275 862E-4	1	8.275 862E-4
	Σ =	0.890 183

The value of the integral is

$$\delta = \frac{15°(\pi/180°)}{3}0.890\ 183 = 0.077\ 68 \text{ in}$$

This programming experience should suggest to you how to handle noncircular centroidal loci with other than rectangular cross sections.

4–11 Compression Members—General

The analysis and design of compression members differs significantly from that of members loaded in tension or in torsion. If you were to take a long rod or pole, such as a meterstick, and apply gradually increasing forces at each end, nothing would happen at first, but then the stick would bend (buckle), and finally bend so much as to fracture. Try it. The other extreme would occur if you were to saw off, say, a 5-mm length of the meterstick and perform the same experiment on the short piece. You would then observe that the failure exhibits itself as a mashing of the specimen, that is, a simple compressive failure. For these reasons it is convenient to classify compression members according to their length and according to whether the loading is central or eccentric. The term *column* is applied to all such members except those in which failure would be by simple or pure compression. Columns can be categorized then as:

1. Long columns with central loading
2. Intermediate-length columns with central loading
3. Columns with eccentric loading
4. Struts or short columns with eccentric loading

Classifying columns as above makes it possible to develop methods of analysis and design specific to each category. Furthermore, these methods will also reveal whether or not you have selected the category appropriate to your particular problem. The four sections which follow correspond, respectively, to the four categories of columns listed above.

S. Timoshenko[5] offered a telling demonstration of the role of stability of members in compression. Consider a straight *rigid* bar carrying a compressive load P as shown in Fig. 4–15. The straight, concentrically loaded bar is held vertical by a pair of springs with a combined spring rate of k lbf/in. For a light load any perturbation creating a displacement of the bar to the right or left causes the spring to restore the bar to the vertical position (stable equilibrium). At some high load the spring is unable to correct, and the displacement increases (unstable equilibrium). At some intermediate load equilibrium is neutral. For a small displacement θ the load P is lowered a distance $l(1-\cos\theta)$. The loss of potential energy of the load, E_p, is

$$E_p = Pl(1-\cos\theta) = Pl\left[1-\left(1-\frac{\theta^2}{2}+\frac{\theta^4}{4!}-\cdots\right)\right] \doteq \frac{Pl\theta^2}{2}$$

Figure 4–15

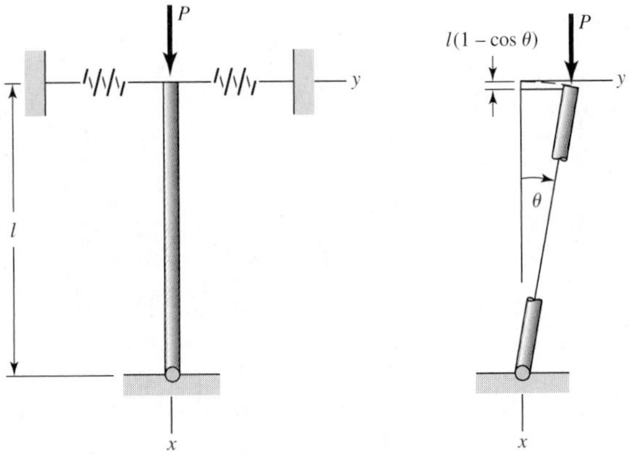

P y l x $l(1-\cos\theta)$ θ

5. S. Timoshenko, *Theory of Elastic Stability*, McGraw-Hill, New York, 1936, pp. 77–78.

using the first two terms of the cosine series. The increase in strain energy of the springs, E_{strain}, is

$$E_{\text{strain}} = \frac{k(l\theta)^2}{2}$$

The strain energy of the bar is zero if rigid and small if not. For neutral equilibrium $E_{\text{strain}} = E_p$, or

$$\frac{k(l\theta)^2}{2} = \frac{Pl\theta^2}{2}$$

from which

$$P_{\text{crit}} = kl \tag{4–43}$$

The strength of the column material does not control. The stresses in the column are not involved in Eq. (4–43). The failure to carry the load is due to a condition of elastic instability. The years of experience which engineers accumulate guarding against and monitoring stress-controlled failures dulls alertness to instability. General experience with "heft," proportions, and what works and does not work is not a reliable guide in column action. Such failure can be a surprise.

4–12 Long Columns with Central Loading

The relationship between the critical load and the column material and geometry is developed with reference to Fig. 4–16*a*. We assume a bar of length l loaded by a force P acting along the centroidal axis on rounded or pinned ends. The figure shows that the bar is bent in the positive y direction. This requires a negative moment, and hence

$$M = -Py \tag{a}$$

If the bar should happen to bend in the negative y direction, a positive moment would result, and so $M = -Py$, as before. Using Eq. (4–12), we write

$$\frac{d^2y}{dx^2} = -\frac{P}{EI}y \tag{b}$$

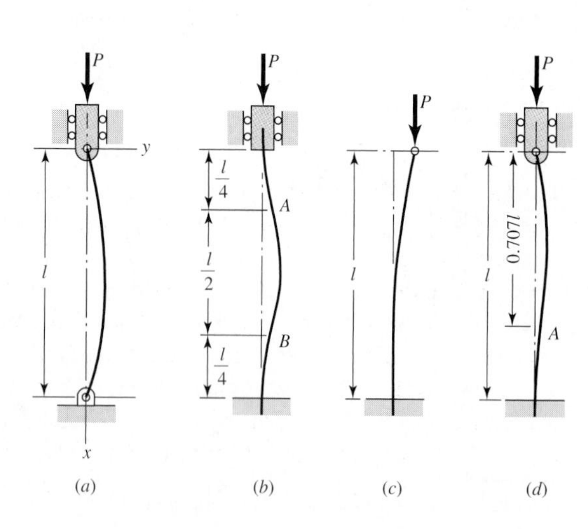

Figure 4–16

(*a*) Both ends rounded or pivoted; (*b*) both ends fixed; (*c*) one end free, one end fixed; (*d*) one end rounded and pivoted, and one end fixed.

or

$$\frac{d^2y}{dx^2} + \frac{P}{EI}y = 0 \tag{4–44}$$

This resembles the well-known differential equation for simple harmonic motion. The solution is

$$y = A \sin \sqrt{\frac{P}{EI}}x + B \cos \sqrt{\frac{P}{EI}}x \tag{c}$$

where A and B are constants of integration and must be determined from the boundary conditions of the problem. We evaluate them using the conditions that $y = 0$ at $x = 0$ and at $x = l$. This gives $B = 0$, and

$$0 = A \sin \sqrt{\frac{P}{EI}}l \tag{d}$$

The trivial solution of no buckling occurs with $A = 0$. However, if $A \neq 0$, then

$$\sin \sqrt{\frac{P}{EI}}l = 0 \tag{e}$$

Equation (*e*) is satisfied by $\sqrt{P/EI}\ l = n\pi$, where $n = 1, 2, 3, \ldots$. Solving for P when $n = 1$ gives the first critical load

$$P_{cr} = \frac{\pi^2 EI}{l^2} \tag{4–45}$$

which is called the *Euler column formula;* it applies only to rounded-end columns. If we substitute these results back in Eq. (*c*), we get the equation of the deflection curve as

$$y = A \sin \frac{\pi x}{l} \tag{f}$$

which indicates that the deflection curve is a half-wave sine. We are interested only in the minimum critical load, which occurs with $n = 1$. However, though it is not of any importance here, values of n greater than 1 result in deflection curves which cross the axis at points of inflection and are multiples of half-wave sines.

Using the relation $I = Ak^2$, where A is the area and k the radius of gyration, enables us to rearrange Eq. (4–45) into the more convenient form

$$\frac{P_{cr}}{A} = \frac{\pi^2 E}{(l/k)^2} \tag{4–46}$$

where l/k is called the *slenderness ratio.* This ratio, rather than the actual column length, will be used in classifying columns according to length categories.

The quantity P_{cr}/A in Eq. (4–46) is the *critical unit load.* It is the load per unit area necessary to place the column in a condition of *unstable equilibrium.* In this state any small crookedness of the member, or slight movement of the support or load, will cause the column to collapse. The unit load has the same units as strength, but this is the strength of a specific column, not of the column material. Doubling the length of a member, for example, will have a drastic effect on the value of P_{cr}/A but no effect at all on, say, the yield strength S_y of the column material itself.

Equation (4–46) shows that the critical unit load depends only upon the modulus of elasticity and the slenderness ratio. Thus a column obeying the Euler formula made of high-strength alloy steel is no stronger than one made of low-carbon steel, since E is the same for both.

The critical loads for columns with different end conditions can be obtained by solving the differential equation or by comparison. Figure 4–16*b* shows a column with both ends fixed. The inflection points are at A and B, a distance $l/4$ from the ends. The distance AB is the same curve as a rounded-end column. Substituting the length $l/2$ for l in Eq. (4–45), we obtain

$$P_{cr} = \frac{\pi^2 EI}{(l/2)^2} = \frac{4\pi^2 EI}{l^2} \tag{4–47}$$

In Fig. 4–16*c* is shown a column with one end free and one end fixed. This curve is equivalent to half the curve for columns with rounded ends, so that if a length of $2l$ is substituted in Eq. (4–45), the critical load becomes

$$P_{cr} = \frac{\pi^2 EI}{(2l)^2} = \frac{\pi^2 EI}{4l^2} \tag{4–48}$$

A column with one end fixed and one end rounded, as in Fig. 4–16*d*, occurs frequently. The inflection point is at A, a distance of $0.707l$ from the rounded end. Therefore

$$P_{cr} = \frac{\pi^2 EI}{(0.707l)^2} = \frac{2\pi^2 EI}{l^2} \tag{4–49}$$

We can account for these various end conditions by writing the Euler equation in the two following forms:

$$P_{cr} = \frac{C\pi^2 EI}{l^2}, \qquad \frac{P_{cr}}{A} = \frac{C\pi^2 E}{(l/k)^2} \tag{4–50}$$

Here, the factor C is called the *end-condition constant*, and it may have any one of the theoretical values $\frac{1}{4}$, 1, 2, and 4, depending upon the manner in which the load is applied. In practice it is difficult, if not impossible, to fix the column ends so that the factor $C = 2$ or $C = 4$ would apply. Even if the ends are welded, some deflection will occur. Because of this, some designers never use a value of C greater than unity. However, if liberal factors of safety are employed, and if the column load is accurately known, then a value of C not exceeding 1.2 for both ends fixed, or for one end rounded and one end fixed, is not unreasonable, since it supposes only partial fixation. Of course, the value $C = \frac{1}{4}$ must always be used for a column having one end fixed and one end free. These recommendations are summarized in Table 4–6.

Table 4–6

End-Condition Constants for Euler Columns [to Be Used with Eq. (4–50)]

	End-Condition Constant C		
Column End Conditions	**Theoretical Value**	**Conservative Value**	**Recommended Value***
Fixed-free	$\frac{1}{4}$	$\frac{1}{4}$	$\frac{1}{4}$
Rounded-rounded	1	1	1
Fixed-rounded	2	1	1.2
Fixed-fixed	4	1	1.2

*To be used only with liberal factors of safety when the column load is accurately known.

When Eq. (4–50) is solved for various values of the unit load P_{cr}/A in terms of the slenderness ratio l/k, we obtain the curve PQR shown in Fig. 4–17. Since the yield strength of the material has the same units as the unit load, the horizontal line through S_y and Q has been added to the figure. This would appear to make the figure cover the entire range of compression problems from the shortest to the longest compression member. Thus it would appear that any compression member having an l/k value less than $(l/k)_Q$ should be treated as a pure compression member while all others are to be treated as Euler columns. Unfortunately, this is not true.

In the actual design of a member which functions as a column, the designer will be aware of the end conditions shown in Fig. 4–16. He or she will endeavor to configure the ends, using bolts, welds, or pins, for example, so as to achieve the required ideal end condition. In spite of these precautions, the result, following manufacture, is likely to contain defects such as initial crookedness or load eccentricities. The existence of such defects and the methods of accounting for them will usually involve a factor-of-safety approach or a stochastic analysis. These methods work well for long columns and for simple compression members. However, tests show numerous failures for columns with slenderness ratios below and in the vicinity of point Q, as shown in the shaded area in Fig. 4–17. These have been reported as occurring even when near-perfect geometric specimens were used in the testing procedure.

A column failure is always sudden, total, and unexpected, and hence dangerous. There is no advance warning. A beam will bend and give visual warning that it is overloaded; but not so for a column. For this reason neither simple compression methods nor the Euler column equation should be used when the slenderness ratio is near $(l/k)_Q$. Then what should we do? The usual approach is to choose some point T on the Euler curve of Fig. 4–17. If the slenderness ratio is specified as $(l/k)_1$ corresponding to point T, then use the Euler equation only when the actual slenderness ratio is greater than $(l/k)_1$. Otherwise, use one of the methods in the sections which follow. See Examples 4–10 and 4–11.

Most designers select point T such that $P_{cr}/A = S_y/2$. Using Eq. (4–50), we find the corresponding value of $(l/k)_1$ to be

$$\left(\frac{l}{k}\right)_1 = \left(\frac{2\pi^2 CE}{S_y}\right)^{1/2} \tag{4–51}$$

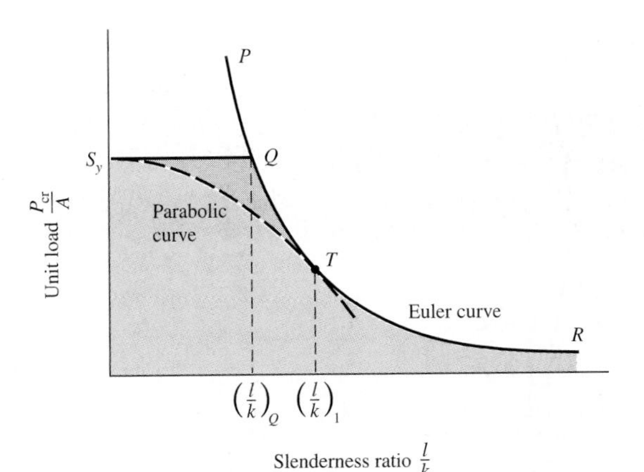

Figure 4–17

Euler curve plotted using Eq. (4–50) with $C = 1$.

4–13 Intermediate-Length Columns with Central Loading

Over the years there have been a number of column formulas proposed and used for the range of l/k values for which the Euler formula is not suitable. Many of these are based on the use of a single material; others, on a so-called safe unit load rather than the critical value. Most of these formulas are based on the use of a linear relationship between the slenderness ratio and the unit load. The *parabolic* or *J. B. Johnson formula* now seems to be the preferred one among designers in the machine, automotive, aircraft, and structural-steel construction fields.

The general form of the parabolic formula is

$$\frac{P_{cr}}{A} = a - b\left(\frac{l}{k}\right)^2 \tag{4–52}$$

where a and b are constants that are evaluated by fitting a parabola to the Euler curve of Fig. 4–17 as shown by the dashed line ending at T. If the parabola is begun at S_y, then $a = S_y$. If point T is selected as previously noted, then Eq. (4–51) gives the value of $(l/k)_1$ and the constant b is found to be

$$b = \left(\frac{S_y}{2\pi}\right)^2 \frac{1}{CE} \tag{a}$$

Upon substituting the known values of a and b into Eq. (4–52), we obtain, for the parabolic equation,

$$\frac{P_{cr}}{A} = S_y - \left(\frac{S_y}{2\pi}\frac{l}{k}\right)^2 \frac{1}{CE} \qquad \frac{l}{k} \le \left(\frac{l}{k}\right)_1 \tag{4–53}$$

4–14 Columns with Eccentric Loading

We have noted before that deviations from an ideal column, such as load eccentricities or crookedness, are likely to occur during manufacture and assembly. Though these deviations are often quite small, it is still convenient to have a method of dealing with them. Frequently, too, problems occur in which load eccentricities are unavoidable.

Figure 4–18*a* shows a column in which the line of action of the column forces is separated from the centroidal axis of the column by the eccentricity e. This problem is

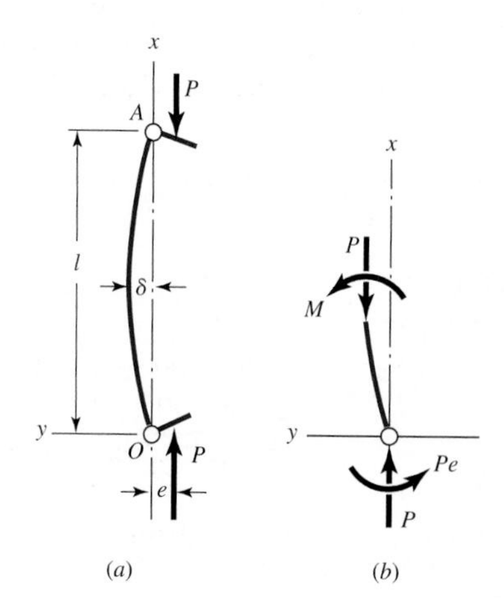

Figure 4–18

Notation for an eccentrically loaded column.

developed using the free-body diagram of Fig. 4–18b and summing the moments to zero about the origin O. This gives

$$\Sigma M_O = M + Pe + Py = 0 \tag{a}$$

Using Eq. (4–12) and rearranging produces the differential equation

$$\frac{d^2y}{dx^2} + \frac{P}{EI}y = -\frac{Pe}{EI} \tag{b}$$

Equation (b) is solved in a manner quite similar to others we have solved in this chapter. The boundary conditions are found to be

$$x = 0 \qquad y = 0$$

$$x = \frac{l}{2} \qquad \frac{dy}{dx} = 0$$

By substituting $x = l/2$ in the resulting solution, we find the deflection at midspan and the maximum bending moment to be

$$\delta = e\left[\sec\left(\frac{l}{2}\sqrt{\frac{P}{EI}}\right) - 1\right] \tag{4–54}$$

$$M_{\max} = -P(e + \delta) = -Pe\sec\left(\frac{l}{2}\sqrt{\frac{P}{EI}}\right) \tag{4–55}$$

The maximum compressive stress at midspan is found by superposing the axial component and the bending component. This gives

$$\sigma_c = \frac{P}{A} - \frac{Mc}{I} = \frac{P}{A} - \frac{Mc}{Ak^2} \tag{c}$$

Substituting $M_{\max}$ from Eq. (4–55) yields

$$\sigma_c = \frac{P}{A}\left[1 + \frac{ec}{k^2}\sec\left(\frac{l}{2k}\sqrt{\frac{P}{EA}}\right)\right] \tag{4–56}$$

Note that the length l occurs only in the secant function. Since the secant of an acute angle is a number greater than unity, the length amplifies the eccentricity through the secant term. By imposing the yield strength S_{yc} as the maximum value of σ_c, we can write Eq. (4–56) in the form

$$\frac{P}{A} = \frac{S_{yc}}{1 + (ec/k^2)\sec[(l/2k)\sqrt{P/AE}]} \tag{4–57}$$

This is called the *secant column formula.* The term ec/k^2 is called the *eccentricity ratio.* Figure 4–19 is a plot of Eq. (4–57) for a steel having a compressive (and tensile) yield strength of 40 kpsi. Note how the P/A contours asymptotically approach the Euler curve as l/k increases.

Equation (4–57) cannot be solved explicitly for the load P. Design charts, in the fashion of Fig. 4–19, can be prepared for a single material if much column design is

Figure 4–19

Comparison of secant and Euler equations.

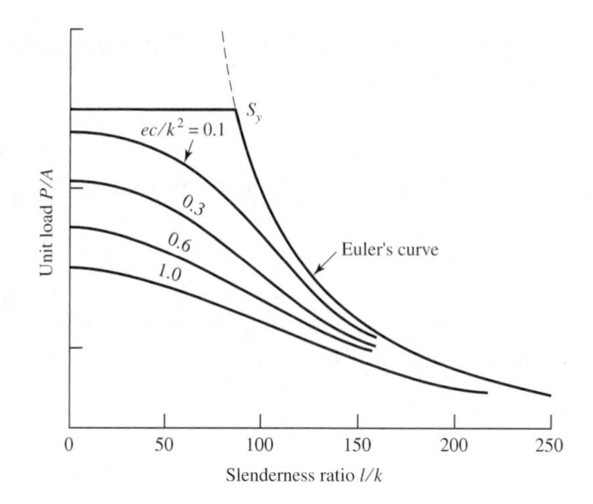

to be done. Otherwise, a root-finding technique using numerical methods must be used. With these, successive substitution with assured convergence can be applied.[6]

EXAMPLE 4–10

Develop specific Euler equations for the sizes of columns having

(*a*) Round cross sections

(*b*) Rectangular cross sections

Solution

(*a*) Using $A = \pi d^2/4$ and $k = d/4$ with Eq. (4–50) gives

Answer

$$d = \left(\frac{64 P_{\text{cr}} l^2}{\pi^3 C E}\right)^{1/4} \tag{4–58}$$

(*b*) For the rectangular column, we specify a height h and a width b with the restriction that $h \le b$. If the end conditions are the *same* for buckling in both directions, then buckling will occur in the direction of the least thickness. Therefore

$$I = \frac{bh^3}{12} \qquad A = bh \qquad k^2 = \frac{h^2}{12}$$

Substituting these in Eq. (4–50) gives

Answer

$$b = \frac{12 P_{\text{cr}} l^2}{\pi^2 C E h^3} \tag{4–59}$$

Note very particularly, though, that rectangular columns do not generally have the same end conditions in both directions.

6. See Charles R. Mischke, "Computational Considerations in Design," Chap 5 in Joseph E. Shigley and Charles R. Mischke (eds.), *Standard Handbook of Machine Design*, 2 ed., McGraw-Hill, New York, 1996, p. 5.13.

EXAMPLE 4–11 Specify the diameter of a round column 1.5 m long which is to carry a maximum load estimated to be 22 kN. Use a design factor $n_d = 4$ and consider the ends as pinned (rounded). The column material selected has a minimum yield strength of 500 MPa and a modulus of elasticity of 207 GPa.

Solution We shall design the column for a critical load of

$$P_{cr} = n_d P = 4(22) = 88 \text{ kN}$$

Then, using Eq. (4–58) with $C = 1$ gives

$$d = \left(\frac{64P_{cr}l^2}{\pi^3 CE}\right)^{1/4} = \left[\frac{64(88)(1.5)^3}{\pi^3(1)(207)}\right]^{1/4} \left(\frac{10^3}{10^9}\right)^{1/4} (10^3) = 37.48 \text{ mm}$$

Table E–17 shows that the preferred size is 40 mm. The slenderness ratio for this size is

$$\frac{l}{k} = \frac{l}{d/4} = \frac{1.5\left(10^3\right)}{40/4} = 150$$

To be sure that this is an Euler column, we use Eq. (4–51) and obtain

$$\left(\frac{l}{k}\right)_1 = \left(\frac{2\pi^2 CE}{S_y}\right)^{1/2} = \left[\frac{2\pi^2(1)(207)}{500}\right]^{1/2} \left(\frac{10^9}{10^6}\right)^{1/2} = 90.4$$

which indicates that it is indeed an Euler column. So select

Answer $d = 40$ mm

EXAMPLE 4–12 Repeat Example 4–10 for J. B. Johnson columns.

Solution (*a*) For round columns,

Answer

$$d = 2\left(\frac{P_{cr}}{\pi S_y} + \frac{S_y l^2}{\pi^2 CE}\right)^{1/2} \tag{4–60}$$

(*b*) For a rectangular section with dimensions $h \leq b$, we find

Answer

$$b = \frac{P_{cr}}{hS_y\left(1 - \frac{3l^2 S_y}{\pi^2 CEh^2}\right)} \qquad h \leq b \tag{4–61}$$

EXAMPLE 4–13 Choose a set of dimensions for a rectangular link which is to carry a maximum compressive load of 5000 lb. The material selected has a minimum yield strength of 75 kpsi and a modulus of elasticity $E = 30$ Mpsi. Use a design factor of 4 and an end condition

Table 4–7

Table Generated to Solve Ex. 4–12

h	b	A	l/k	Type	Eq. No.
0.375	3.46	1.298	139	Euler	(4–57)
0.500	1.46	0.730	104	Euler	(4–57)
0.625	0.76	0.475	83	Johnson	(4–58)
0.5625	1.03	0.579	92	Euler	(4–58)

constant $C = 1$ for buckling on the weakest direction, and design for (*a*) a length of 15 in, and (*b*) a length of 8 in with a minimum thickness of $\frac{1}{2}$ in.

Solution (*a*) Using Eq. (4–51), we find the limiting slenderness ratio to be

$$\left(\frac{l}{k}\right)_1 = \left(\frac{2\pi^2 CE}{S_y}\right)^{1/2} = \left[\frac{2\pi^2(1)(30)\,(10^6)}{75(10)^3}\right]^{1/2} = 88.8$$

By using $P_{cr} = n_d P = 4(5000) = 20\ 000$ lb, Eqs. (4–59) and (4–61) are solved, using various values of h, to form Table 4–7. The table shows that a cross section of $\frac{5}{8}$ by $\frac{3}{4}$ in, which is marginally suitable, gives the least area.

(*b*) An approach similar to that above is used with $l = 8$ in. All trial computations are found to be in the J. B. Johnson region of l/k values. A minimum area occurs when the section is a near square. Thus a cross section of $\frac{1}{2}$ by $\frac{3}{4}$ in is found to be suitable and safe.

4–15 Struts, or Short Compression Members

A short bar loaded in pure compression by a force P acting along the centroidal axis will shorten in accordance with Hooke's law, until the stress reaches the elastic limit of the material. At this point, permanent set is introduced and usefulness as a machine member may be at an end. If the force P is increased still more, the material either becomes "barrel-like" or fractures. When there is eccentricity in the loading, the elastic limit is encountered at small loads.

A *strut* is a *short compression member* such as the one shown in Fig. 4–20. The compressive stress in the x direction at point D in an intermediate section is the sum of a simple component P/A and a flexural component My/I; that is,

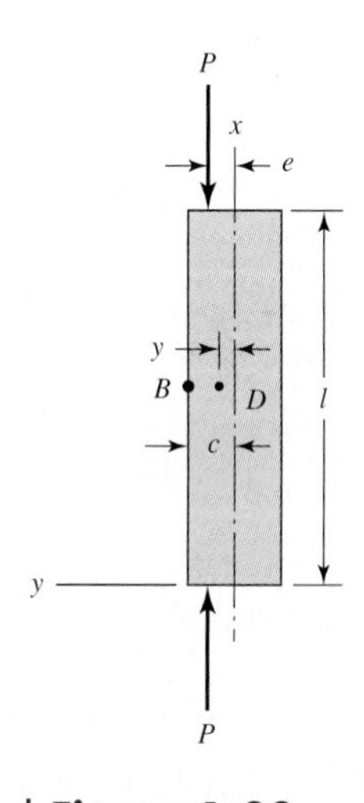

Figure 4–20

Eccentrically loaded strut.

$$\sigma_c = \frac{P}{A} + \frac{My}{I} = \frac{P}{A} + \frac{PeyA}{IA} = \frac{P}{A}\left(1 + \frac{ey}{k^2}\right) \tag{4–62}$$

where $k = (I/A)^{1/2}$ and is the radius of gyration, y is the coordinate of point D, and e is the eccentricity of loading. The y coordinate of a line parallel to the x axis along which the normal stress is zero is found by setting Eq. (4–62) equal to zero and solving for y. This gives

$$y = -\frac{k^2}{e} \tag{a}$$

As the eccentricity is increased, the line of zero stress moves toward the cross-section centroid. As e is decreased, the line moves far from the section, and the entire section has a compressive normal stress. The largest compressive stress occurs at point B in

Fig. 4–20, where $y = c$. Equation (4–62) then becomes

$$\sigma_c = \frac{P}{A}\left(1 + \frac{ec}{k^2}\right) \tag{4–63}$$

Note that the length of the strut does not appear in Eq. (4–63). In order to use the equation for design or analysis, we ought, therefore, to know the range of lengths for which the equation is valid. In other words, how long is a short member?

The difference between the secant formula and Eq. (4–63) is that the secant equation, unlike Eq. (4–63), accounts for an increased bending moment due to bending deflection. Thus the secant equation shows the eccentricity to be magnified by the bending deflection. This difference between the two formulas suggests that one way of differentiating between a "secant column" and a strut, or short compression member, is to say that in a strut, the effect of bending deflection must be limited to a certain small percentage of the eccentricity. If we decide that the limiting percentage is to be 1 percent of e, then the limiting slenderness ratio turns out to be

$$\left(\frac{l}{k}\right)_2 = 0.282\left(\frac{AE}{P_{\text{cr}}}\right)^{1/2} \tag{4–64}$$

(see Prob. 4–53). This equation then gives the limiting slenderness ratio for using Eq. (4–63). If the actual slenderness ratio is greater than $(l/k)_2$, then use the secant formula; otherwise, use Eq. (4–63).

EXAMPLE 4–14

Figure 4–21*a* shows a workpiece clamped to a milling machine table by a bolt tightened to a tension of 2000 lb. The clamp contact is offset from the centroidal axis of the strut by a distance $e = 0.10$ in, as shown in part *b* of the figure. The strut, or block, is steel, 1 in square and 4 in long, as shown. Determine the maximum compressive stress in the block.

Solution

First we find $A = bh = 1(1) = 1\ \text{in}^2$, $I = bh^3/12 = 1(1)^3/12 = 0.0833\ \text{in}^4$, $k^2 = I/A = 0.0833/1 = 0.0833\ \text{in}^2$, and $l/k = 4/(0.0833)^{1/2} = 13.9$. Equation (4–64) gives the limiting slenderness ratio as

$$\left(\frac{l}{k}\right)_2 = 0.282\left(\frac{AE}{P_{\text{cr}}}\right)^{1/2} = 0.282\left[\frac{1(30)\,(10^6)}{1000}\right]^{1/2} = 48.8$$

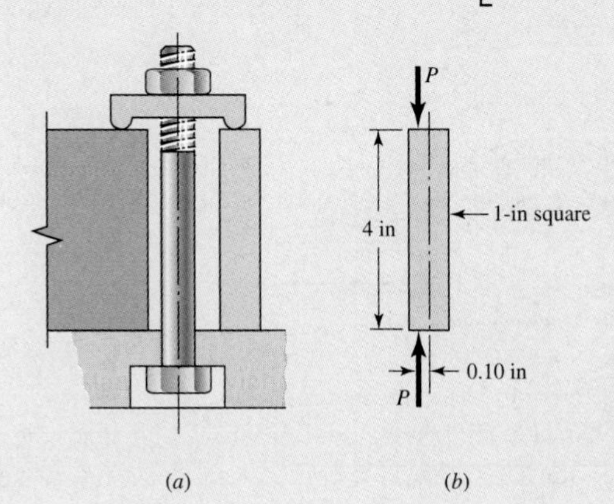

Figure 4–21
A strut that is part of a workpiece clamping assembly.

Thus the block could be as long as

$$l = 48.8k = 48.8(0.0833)^{1/2} = 14.1 \text{ in}$$

before it need be treated using the secant formula. So Eq. (4–63) applies and the maximum compressive stress is

Answer $$\sigma_c = \frac{P}{A}\left(1 + \frac{ec}{k^2}\right) = \frac{1000}{1}\left[1 + \frac{0.1(0.5)}{0.0833}\right] = 1600 \text{ psi}$$

4–16 An Application: Round-Bar Clamps

Clamps to hold round bars and resist torsion are often made of cast iron as depicted in Fig. 4–22. The cap screw closes the running fit gap and clamps the bar. We want to know how much of the cap screw preload F_i is devoted to closing the gap and how much is devoted to clamping. Proportions are such that the curved bar is thick and of rectangular cross section.

In estimating stresses, the distance between the section centroidal axis and the neutral axis $R - r_n$ is the eccentricity e. This distance plays a role in both stress and deflection analysis. For a rectangular cross section, Timoshenko's approximation, Eq. (3–72), $e \doteq h^2/(12R)$, is useful. For deflection analysis the upper half of the clamp is treated as a thick curved bar. In Fig. 4–23c the screw tension P induces a bending moment M, a shearing force V, and a hoop tension N. From statics, in the notation of Sec. 4–10, we can write

$$M = P[a + R(1 - \cos\theta)] \qquad \frac{\partial M}{\partial P} = [a + R(1 - \cos\theta)]$$

$$N = P\cos\theta \qquad \frac{\partial N}{\partial P} = \cos\theta$$

$$V = P\sin\theta \qquad \frac{\partial V}{\partial P} = \sin\theta$$

$$MN = P^2\cos\theta[a + R(1 - \cos\theta)] \qquad \frac{\partial(MN)}{\partial P} = 2P\cos\theta[a + R(1 - \cos\theta)]$$

The deflection at point A, δ, in the direction of P will be found using Castigliano's theorem. Beginning with Eq. (a) following Eq. (4–37),

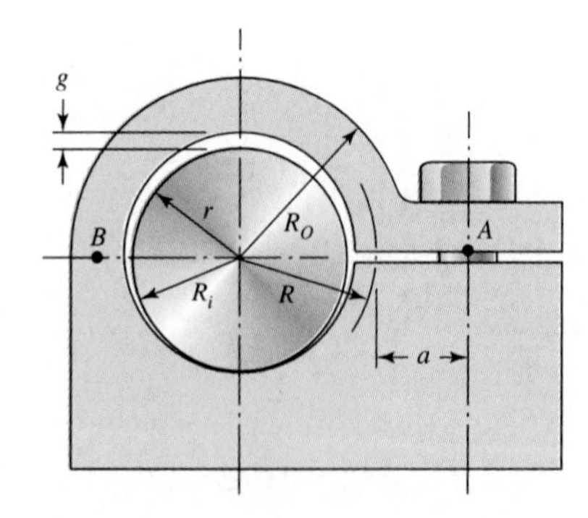

Figure 4–22

A round-bar clamp.

Figure 4–23

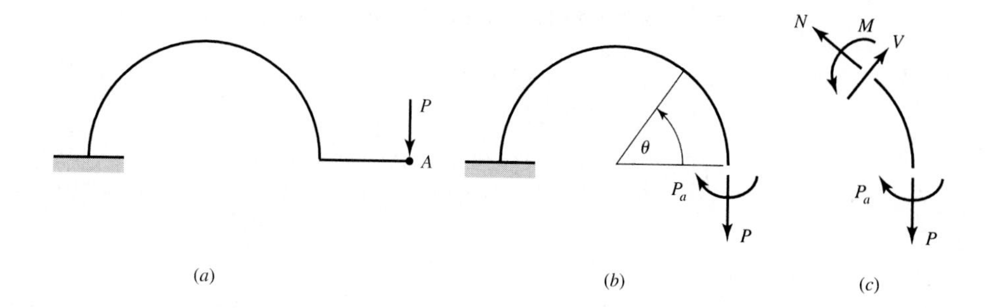

(a) (b) (c)

$$U = \int \frac{M^2 d\theta}{2AEe} + \int \frac{N^2 R\, d\theta}{2AE} - \int \frac{MN\, d\theta}{AE} + \int \frac{KV^2 R\, d\theta}{2AG} \tag{4–65}$$

The strain energy U_1 due to moment is related to the portion of the deflection δ_1 at point A by

$$\begin{aligned}\delta_1 &= \frac{\partial U_1}{\partial P} = \int_0^\pi \frac{M}{AEe}\frac{\partial M}{\partial P} d\theta = \int_0^\pi \frac{P[a + R(1 - \cos\theta)][a + R(1 - \cos\theta)]}{AEe} d\theta \\ &= \frac{\pi P}{AEe}\left[(a + R)^2 + \frac{R^2}{2}\right]\end{aligned}$$

The deflection δ_2 due to strain energy U_2 because of hoop tension N is

$$\delta_2 = \frac{\partial U_2}{\partial P} = \int_0^\pi \frac{N}{AE}\frac{\partial N}{\partial P} R\, d\theta = \int_0^\pi \frac{(P\cos\theta)\cos\theta R\, d\theta}{AE} = \frac{\pi PR}{2AE}$$

The deflection δ_3 due to strain energy due to the interaction of M and N is

$$\begin{aligned}\delta_3 &= \frac{\partial U_3}{\partial P} = \int_0^\pi \frac{1}{AE}\frac{\partial(MN)}{\partial P} d\theta = \int_0^\pi \frac{1}{AE} 2P\cos\theta[a + R(1 - \cos\theta)]d\theta \\ &= \frac{\pi PR}{AE}\end{aligned}$$

The deflection δ_4 due to the strain energy U_4 due to shear V is

$$\delta_4 = \frac{\partial U_4}{\partial P} = \int_0^\pi \frac{KVR}{AG}\frac{\partial V}{\partial P} d\theta = \int_0^\pi \frac{K(P\sin\theta)R(\sin\theta)\, d\theta}{AG} = \frac{KPR}{2AG}$$

The total deflection δ at point A due to P in the direction of P is given by

$$\begin{aligned}\delta &= \delta_1 + \delta_2 + \delta_3 + \delta_4 \\ &= \frac{\pi P}{AEe}\left[(a + R)^2 + \frac{R^2}{2}\right] + \frac{\pi PR}{2AE} - \frac{\pi PR}{AE} + \frac{\pi KPR}{2AG} \\ &= \frac{\pi P}{AE}\left[\frac{(a + R)^2 + R^2/2}{e} + \frac{R}{2}\left(\frac{KE}{G} - 1\right)\right]\end{aligned} \tag{4–66}$$

For a rectangular cast-iron grade-20 section with $E = 14(10^6)$ psi, $G = 5.6(10^6)$ psi, and $K = 3/2$,

$$\frac{E}{G} = \frac{14(10^6)}{5.6(10^6)} = 2.50 \qquad \frac{KE}{G} = 1.5(2.50) = 3.75$$

substituting into Eq. (4–66) and simplifying,

$$\delta = \frac{\pi P}{AE}\left[\frac{(a + R^2) + R^2/2}{e} + 1.375R\right] \tag{4–67}$$

With the pivot of the upper bar treated as in the neighborhood of point B in Fig. 4–22, the following proportion involving the gap g can be written:

$$\frac{2R + a}{\delta} = \frac{R}{g}$$

and

$$\delta = g\left(2 + \frac{a}{R}\right) \tag{4–68}$$

Solving Eq. (4–67) for the bolt tension P and substituting from Eq. (4–68), we have

$$P = \frac{AE}{\pi} g\left(2 + \frac{a}{R}\right)\left[\frac{(a + R)^2 + R^2/2}{e} + 1.375R\right]^{-1} \tag{4–69}$$

By using the simplifications depicted in Fig. 4–24, the clamping force F_c is found to be

$$F_c = (F_i - P)\left(2 + \frac{a}{R}\right) \tag{4–70}$$

The resistance to axial movement of the bar is the force F_a:

$$F_a = 2fF_c = 2f(F_i - P)\left(2 + \frac{a}{R}\right) \tag{4–71}$$

The torque required to rotate the bar is

$$T = 2fR_iF_c = 2fR_i(F_i - P)\left(2 + \frac{a}{R}\right) \tag{4–72}$$

Figure 4–24

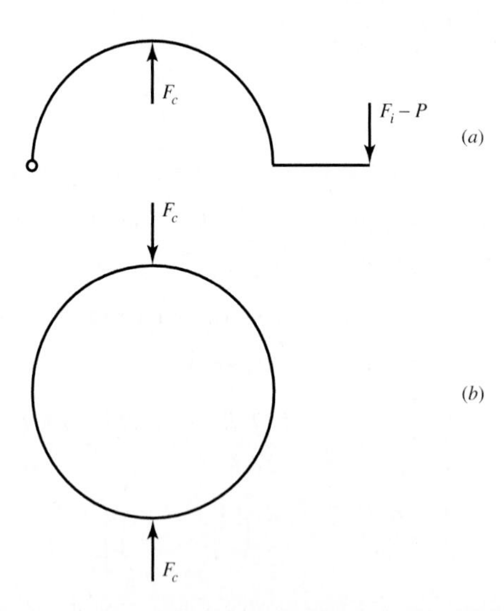

EXAMPLE 4–15 A nominal 4-in diameter bar used in a clamp, as is depicted in Fig. 4–22, has a sliding fit in a round-bar clamp. The interfacing diameters are toleranced as follows.

Bore	Bar
4.0014 in	3.9995 in
4.0000 in	3.9986 in

The other relevant information is $h = 1.1875$ in, $R_0 = 3.1875$ in, $R_i = 2$ in, $a = 1.5$ in, $f = 0.3$, and $f_i = 1000$ lb. The maximum gap g_{max} is $4.0014 - 3.9986 = 0.0028$ in. Find the screw tension P, the total deflection δ at point A due to P, the clamping force F_c, the resistance to axial movement F_a, and the torque required to move the bar. Also find the bending stresses σ_i and σ_o.

Solution

Applying Eqs. (3–70), (3–71), (4–68), and (4–69) results in

$$R = R_i + \frac{h}{2} = 2 + \frac{1.1875}{2} = 2.593\,750 \text{ in}$$

$$r_n = \frac{h}{\ln(R_o/R_i)} = \frac{1.1875}{\ln(3.1875/2)} = 2.547\,793 \text{ in}$$

$$e = R - r_n = 2.593\,750 - 2.547\,793 = 0.045\,957 \text{ in}$$

$$\delta = 0.0028\left(2 + \frac{1.5}{2.593\,750}\right) = 0.007\,22 \text{ in}$$

$$P = \frac{1.1875(1)14(10^6)}{\pi}0.007\,22\left[\frac{(1.5 + 2.594)^2 + 2.594^2/2}{0.045\,957} + 1.375(2.594)\right]^{-1}$$

Applying Eqs. (4–70), (4–71), and (4–72) results in

$$F_c = (1000 - 86.6)\left(2 + \frac{1.5}{2.594}\right) = 2355 \text{ lb}$$

$$F_a = 2(0.3)2355 = 1413 \text{ lb}$$

$$T = 2(0.3)2(2355) = 2826 \text{ in} \cdot \text{lb}$$

The bending stresses at point B stop increasing as the gap g closes, but deflection at point A continues. The only further increase is in the hoop tension N due to further increments in screw tension. The moment at point B is estimated to be

$$M|_{\theta=\pi} = P(a + 2R) = 86.6[1.5 + 2(2.594)] = 579.2 \text{ in} \cdot \text{lb}$$

$$c_i = r_n - R_i = 2.547\,793 - 2 = 0.548 \text{ in}$$

$$c_o = R_o - r_n = 3.1875 - 2.547\,793 = 0.640 \text{ in}$$

The bending stress at point B when the gap g closes, including compression, is

$$\sigma_i = -\frac{Mc_i}{AeR_i} - \frac{P}{A} = -\frac{579.2(0.548)}{1.1875(0.46)2} = \frac{86.6}{1.1875} = -2978 \text{ psi}$$

$$\sigma_o = \frac{Mc_o}{AeR_o} - \frac{P}{A} = \frac{579.2(0.640)}{1.1875(0.046)3.1875} - \frac{86.6}{1.1875} = 2056 \text{ psi}$$

4–17 Deflection of Energy-Dissipative Assemblies

Previous sections considered distortions that were resisted internally by nondestructive energy-exchange phenomena. The body absorbed energy under load and stored it as strain energy; when the load was removed, the member did work on the load as it elastically restored itself to its load-free geometry. These distortions were described by spatial differential equations (deflection a function of the independent variable x), rendered algebraic in Sec. 4–4, for example.

Deflection is of interest when arresting a moving body. The force-time and the force-displacement relations are important to carrying out the arrest in the space available, and the arrested body must be able to survive the decelerations. We still need differential equations to describe the time-dependent deflection. The equations add time as an independent variable. Examples include energy-absorbing automobile bumpers, which absorb energy, crush, and are replaced, and energy-absorbing front ends, which crush and are replaced, both being single-arrest schemes. Of particular interest here are energy-absorbing, viscous-damped devices that create fluid turbulence absorbing the energy, which is transferred as heat to the environment, and do no damage to the internal parts of the device. The device can be reset to repeat the arrest.

Design engineers do not normally design such devices; rather they select satisfactory devices from vendors' information. The devices must be understood, however, to ensure a proper choice is made. Sometimes this is referred to as *design by selection*.

Numerous types of viscously damped hydraulic cylinders are used to absorb energy in stopping a moving body. The arrested body may be as small as a part on an assembly-line conveyor, a traversing tool, an indexing fixture, or as large as a gantry crane overrunning its electrical limit switch and still under power. Devices manufactured specifically for such purposes are known by names such as shock absorbers, snubbers, dampers, buffers, dashpots, decelerators, or isolators. Their goal is energy absorption with tolerable forces, impulses, decelerations, and distance to stop (strokes). If the arresting force is repetitive, then the thermal dissipation capability of the arrestor must be considered.

A rough description of a snubber is a piston acting on a confined fluid, pumped through orifices that vary in number, size, and location along the stroke. The orifices can be "programmed" to exhibit a decreasing, steady, or increasing arresting force (and attendant deceleration). It is useful to analyze a single-orifice viscous snubber (similar to a vibration damper) to understand its characteristics and shortcomings. A manufactured-for-the-purpose snubber is depicted in Fig. 4–25. A simple viscous damper is shown in Fig. 4–26, with a free-body diagram of the weight to be arrested at partial stroke. Newton's law expressed for this case is

$$\Sigma F = m\ddot{x} = (w/g)\ddot{x} = -c\dot{x}$$

When rearranged, this is the differential equation

$$\frac{w}{g}\ddot{x} + c\dot{x} = 0 \tag{4–73}$$

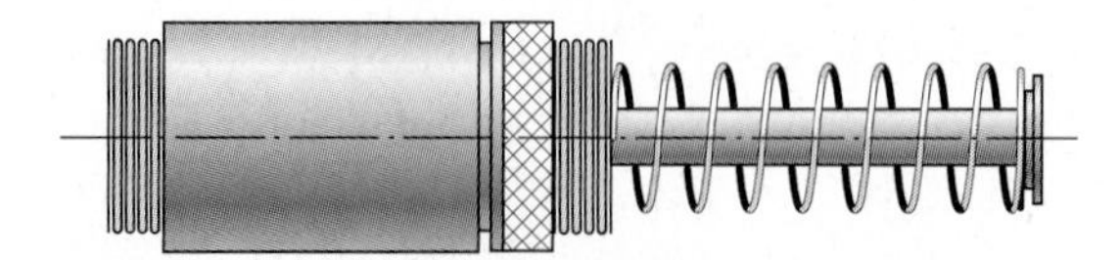

Figure 4–25

A stylized view of a snubber.

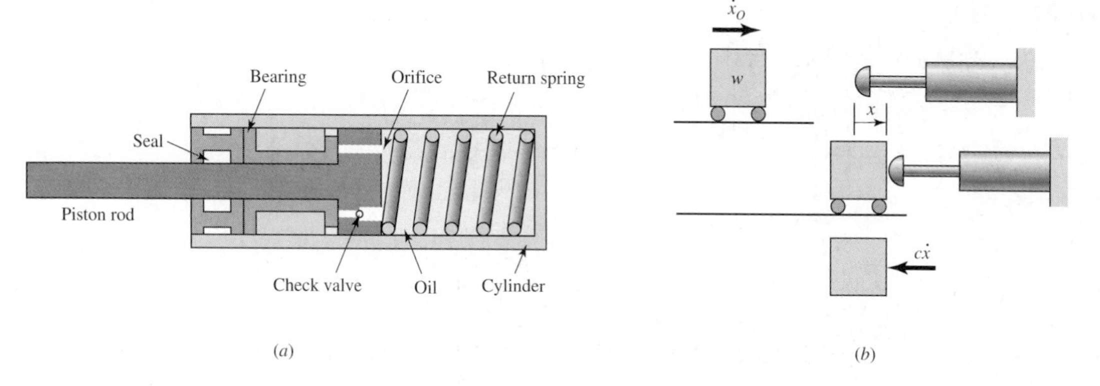

Figure 4–26

Schematic of a single-orifice snubber.

where c is the viscous coefficient. The initial conditions

$$\dot{x}|_{t=0} = \dot{x}_0 \qquad x|_{t=0} = 0$$

give rise to the solution of Eq. (4–73), which expresses the displacement x as a function of time t:

$$x = \frac{w\dot{x}_0}{cg}\left[1 - \exp\left(-\frac{cgt}{w}\right)\right] \tag{4–74}$$

The velocity x is found by differentiation Eq. (4–74) with respect to time, giving

$$\dot{x} = \frac{dx}{dt} = \dot{x}_0 \exp\left(-\frac{cgt}{w}\right) \tag{4–75}$$

The decelerating force F is given by

$$F = c\dot{x} = c\dot{x}_0 \exp\left(-\frac{cgt}{w}\right) \tag{4–76}$$

Eliminating the exponential in Eqs. (4–74) and (4–76) expresses the arresting force as a function of the displacement x, as

$$F = c\dot{x}_0 - \frac{c^2 gx}{w} \tag{4–77}$$

From Eq. (4–74), the distance to stop x_s when $t \to \infty$ is

$$\boxed{x_s = \frac{w\dot{x}_0}{cg}} \tag{4–78}$$

Figure 4–27 is a graph of the linear Eq. (4–77) showing F as a linearly decreasing function of x. This means the deceleration force F deteriorates as the stroke progresses. If the largest deceleration force is acceptable, why not sustain it and shorten the arresting stroke? Before examining an improved snubber, let us check our work so far.

Figure 4–27

Graph of arresting force F as a function of displacement x.

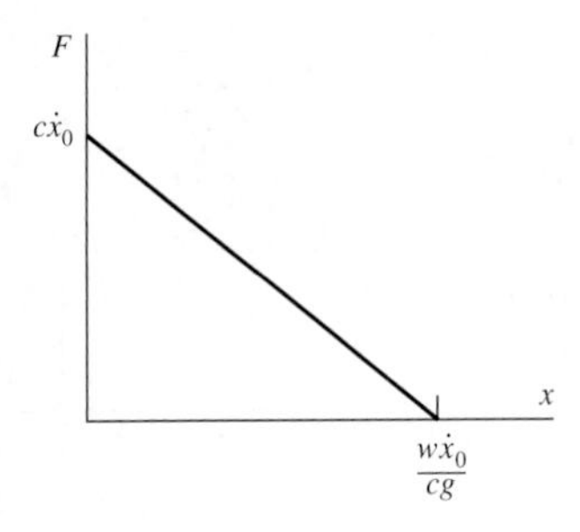

Examining Fig. 4–27 we note that the energy absorbed E_k is the area under the curve:

$$E_k = \frac{1}{2}(c\dot{x}_0)\frac{w\dot{x}_0}{cg} = \frac{1}{2}\frac{w\dot{x}_0^2}{g}$$

It is equal to the kinetic energy of the body before impact. The linear impulse I is

$$I = \int_0^{\infty} F\,dt = (c\dot{x}_0)\int_0^{\infty} \exp\left(-\frac{cgt}{w}\right)dt = \frac{w\dot{x}_0}{g}$$

which is equal to the change in momentum of the body brought to rest. It is interesting to note that although it takes infinite time to stop, the arresting distance x_s is finite. With this check, we proceed.

Figure 4–28 depicts a multiorifice which can cut the arresting distance about one-half. By arranging orifices of varying size, number, and spacing the force can be made more nearly constant. Figures 4–28*b* and 4–28*c* are ideal and actual force traces. For the ideal we write

$$\ddot{x} = -\ddot{x}_0$$

From the initial conditions we find

$$\dot{x} = -\ddot{x}_0 t + \dot{x}_0 \tag{4–79}$$

$$x = -\frac{\ddot{x}_0 t^2}{2} + \dot{x}_0 t \tag{4–80}$$

Figure 4–28

Schematic of a multiple-orifice snubber.

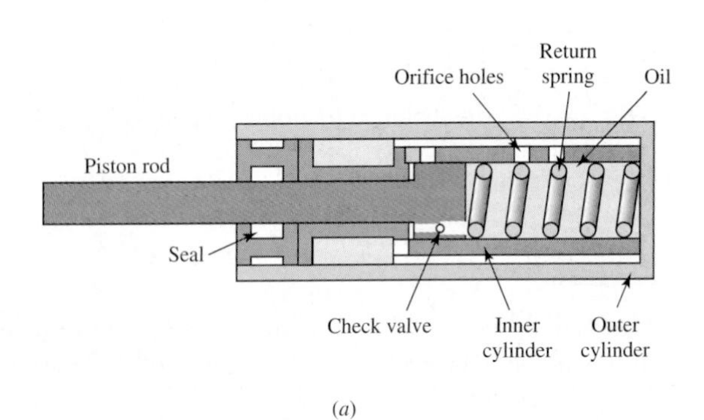

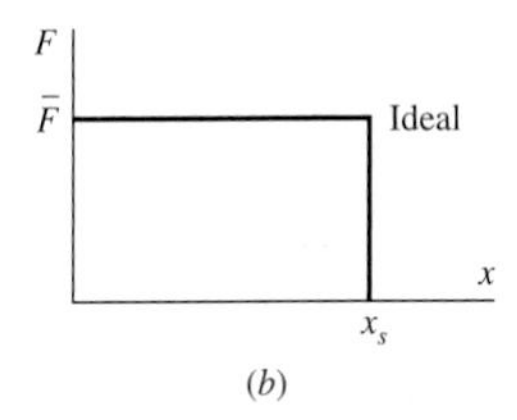

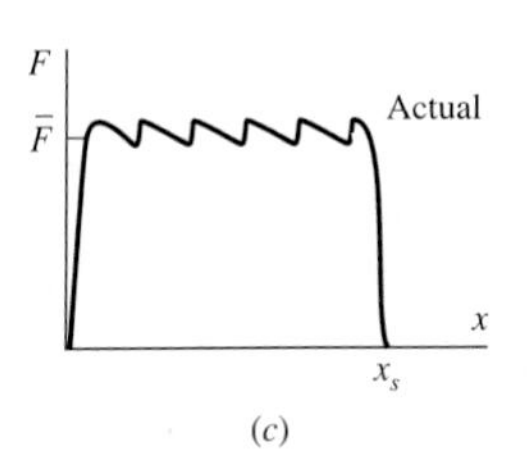

Elimination of the time t between Eqs. (4–79) and (4–80) gives the arresting distance x_s as a function of initial speed and the deceleration:

$$x_s = \frac{\dot{x}_0^2}{2\ddot{x}_0} \tag{4–81}$$

which indeed cuts the stopping distance in half. To check again we note that $F = ma = w\ddot{x}_0/g$ and the time to stop is $t = \dot{x}_0/\ddot{x}_0$, and we write

$$E_k = \left(\frac{w}{g}\ddot{x}_0\right)\left(\frac{\dot{x}_0^2}{2\ddot{x}_0}\right) = \frac{w\dot{x}_0^2}{2g}$$

$$I = Ft = \frac{w\ddot{x}_0}{g}\left(\frac{\dot{x}_0}{\ddot{x}_0}\right) = \frac{w}{g}\dot{x}_0$$

Some commercial snubbers have an orifice adjustment which allows the deceleration to be adjusted to an initial speed, so that the full stroke x_s required does not exceed the device's physical stroke limit.

It is possible to have the force increase as the stroke progresses, so three kinds of devices are available:

- Constant-orifice snubber (force declines with stroke progress)
- Conventional snubber (constant force)
- Progressive snubber (increasing force with stroke progress)

Their force-stroke curves resemble those in Fig. 4–29. Snubbers usually are not designed by the product engineer but selected from off-the-shelf offerings of vendors. The following steps are used in the selection process:

1. Identify the weight or mass to be stopped.
2. Identify the velocity of impact $\dot{x}_0$.
3. Identify forces(s) doing additional work on the snubber during arrest.
4. Identify repetitive cycle frequency.

There is a discrepancy between the ideal constant force and the actual largest force. Because of the gross "trapezoidal" shape of Fig. 4–28*c*, the largest actual force has to be higher than the ideal in order to have the same area under the curves in Figs. 4–28*b* and 4–28*c*. In one vendor's line the maximum actual force is 18 percent larger than ideal, so

$$F_{\max} = F_{\text{ideal}}\left(\frac{F_{\max}}{F_{\text{ideal}}}\right) = \frac{w\dot{x}_0^2}{2gx_s}\left(\frac{1}{0.85}\right)$$
$$= \frac{(1/2)(w\dot{x}_0^2/g)}{0.85x_s}$$

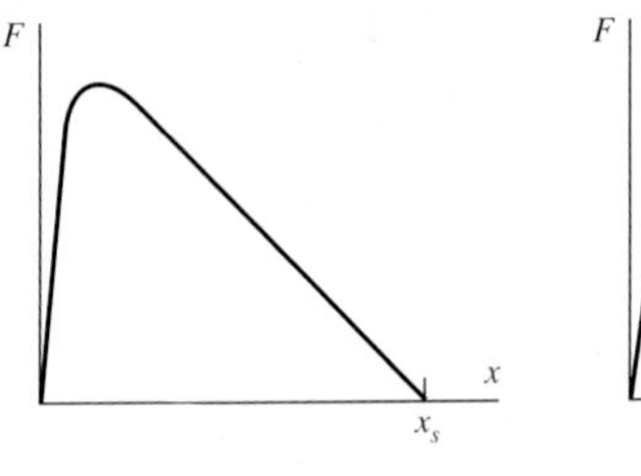

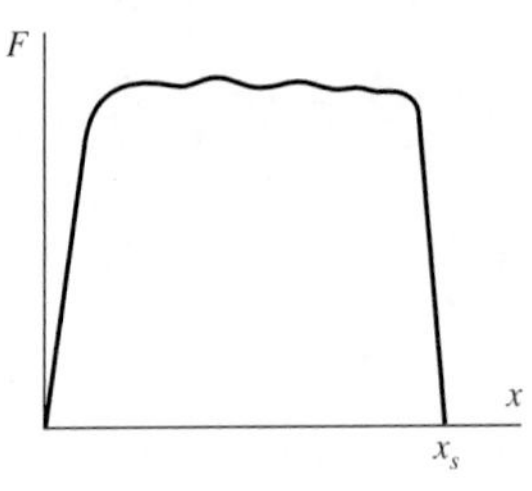

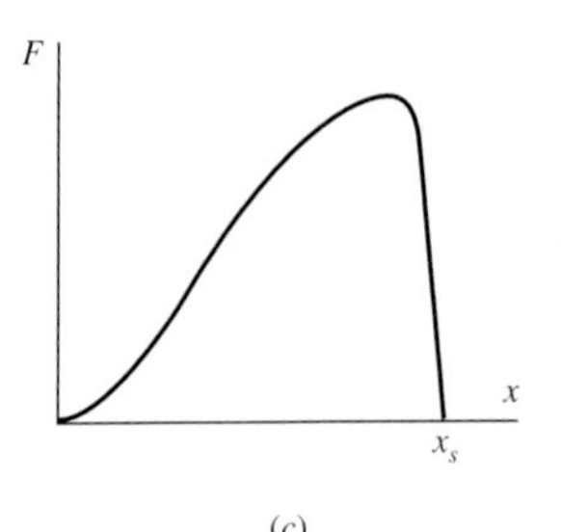

Figure 4–29

Stroke types: (*a*) declining force; (*b*) steady force; (*c*) increasing force.

Always check this point with the vendor's engineering catalog. Also remember that this force has to be acceptable to both the arrested body and the snubber.

EXAMPLE 4–16 A 3000 lbf weight drops 20 in to a snubber's ready position as shown in Fig. 4–30. Select a satisfactory snubber from those listed in Table 4–8 for 60 cycle/h service.

Solution Try a 6-in stroke snubber.

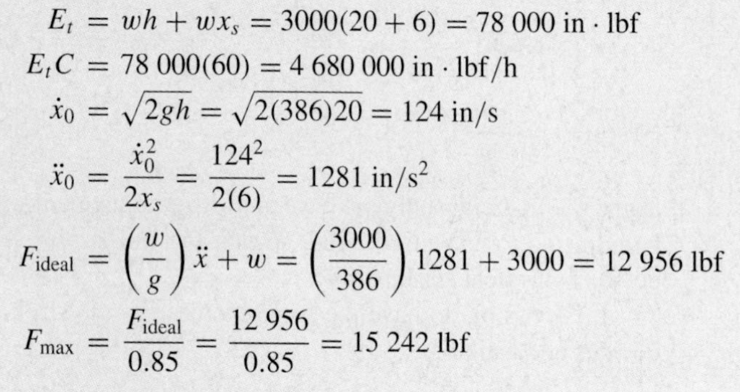

$$E_t = wh + wx_s = 3000(20 + 6) = 78\,000 \text{ in} \cdot \text{lbf}$$

$$E_t C = 78\,000(60) = 4\,680\,000 \text{ in} \cdot \text{lbf/h}$$

$$\dot{x}_0 = \sqrt{2gh} = \sqrt{2(386)20} = 124 \text{ in/s}$$

$$\ddot{x}_0 = \frac{\dot{x}_0^2}{2x_s} = \frac{124^2}{2(6)} = 1281 \text{ in/s}^2$$

$$F_{\text{ideal}} = \left(\frac{w}{g}\right)\ddot{x} + w = \left(\frac{3000}{386}\right)1281 + 3000 = 12\,956 \text{ lbf}$$

$$F_{\max} = \frac{F_{\text{ideal}}}{0.85} = \frac{12\,956}{0.85} = 15\,242 \text{ lbf}$$

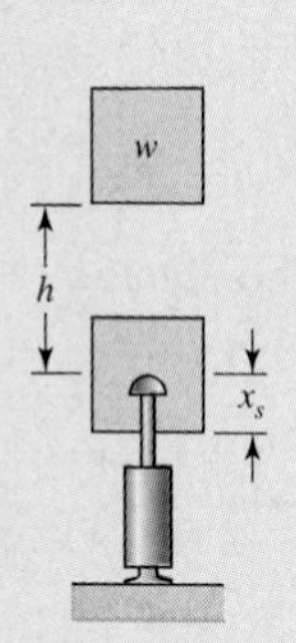

Figure 4–30

Choose the 4.0 M × 6 snubber. Now perform an adequacy assessment.

Maximum in·lbf/cycle = 102 000 (78 000 required)
Maximum in lbf/h = $18(10^6)$(4 680 000 in lbf/h required)
Maximum shock force = 25 000 lbf (15 242 lbf required)
Work against reset spring 40(6) + (30/2)6 = 330 in·lbf (negligible)
Weight 44 lbf (no information)

Snubbers can arrest rotating bodies, too. The kinematic equivalent of a rigid body, as depicted in Fig. 4–31*a*, rotating about a fixed point, consists of two concentrated weights (or masses), W_1 placed at the center of rotation and W_2 placed at the center of percussion as shown in Fig. 4–31*b*. Writing force and moment equations and solving simultaneously for W_1 and W_2 gives

$$W_1 = W\left(1 - \frac{c}{k}\right) \tag{a}$$

$$W_2 = \frac{Wc}{k} \tag{b}$$

where W is the weight of the body, c is the distance to the centroid, and k is the radius of gyration (distance to the center of percussion). The polar moment of inertia is

$$I_p = \frac{W_2}{g}k^2 \tag{c}$$

A uniform rod (or plate) pivoted about its end, of length l, has $c = l/2$ and $k = l/\sqrt{3}$, so

$$W_1 = W\left(1 - \frac{l}{2}\frac{\sqrt{3}}{l}\right) = W\left(1 - \frac{\sqrt{3}}{2}\right) = 0.134W \tag{d}$$

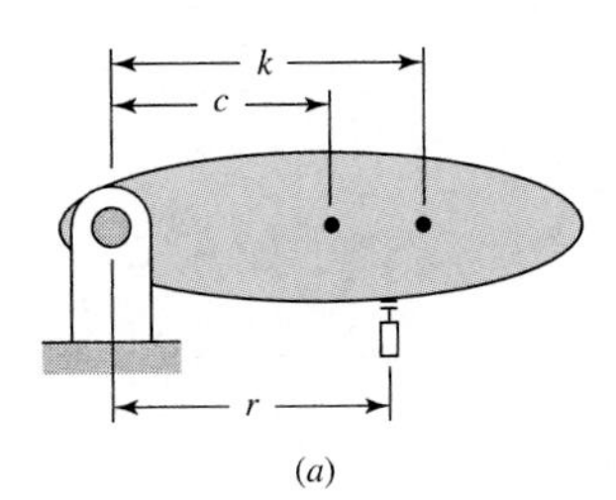

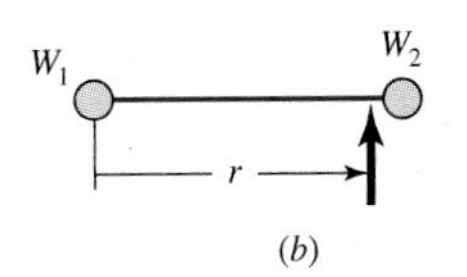

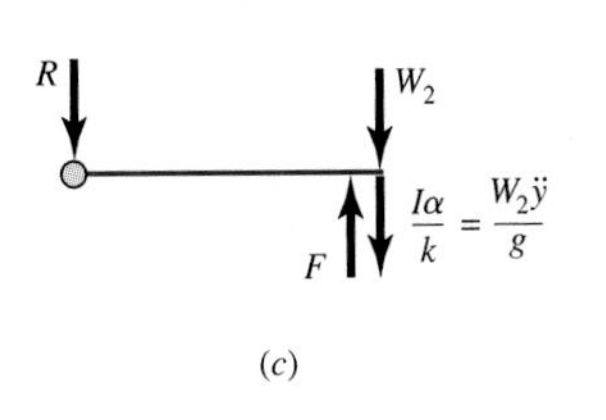

Figure 4–31

(*a*) Snubber arresting a rotating body; (*b*) equivalent system; (*c*) free body.

$$W_2 = W\frac{l}{2}\frac{\sqrt{3}}{l} = \frac{\sqrt{3}}{2}W = 0.866W \tag{e}$$

If the snubber is placed at the radius of gyration, there is no reaction at the pivot. Often this location is not available, so the reaction force must be found, as its repeatedly applied nature should be examined for static and fatigue implications. Placing d'Alembert's inertia force on the equivalent body in Fig. 4–31*c*, summing moments, and solving for snubber force F gives

$$F = \frac{I}{r}\left(\alpha + \frac{g}{k}\right) \tag{f}$$

Summing forces and solving for the reaction force R gives

$$R = F - \frac{I\alpha}{k} - W_2 \tag{g}$$

The snubber has a force-stroke signature involving a largest force $f_{\max} = F/0.85$. The deceleration responds to this, so $\alpha_{\max} = \alpha/0.85$, and the maximum reaction force is given by

$$R = \frac{1}{0.85}\left(F - \frac{I\alpha}{k}\right) - W_2 \tag{h}$$

EXAMPLE 4–17

A rotating body falls from a vertical position to a horizontal position onto a snubber placed at radius r from the pivot. Select a suitable snubber from Table 4–8.

(*a*) A concentrated weight of 1540 lbf is located 24 in from the pivot of a rod of negligible weight. The snubber is located with $r = 24$ in as shown in Fig. 4–32*a*.

(*b*) The conditions are the same as in part *a* except the snubber is positioned at 20 in from the pivot as shown in Fig. 4–32*b*.

(*c*) A steel trap door weighing 1778 lbf and having a length of 41.6 in is pivoted at its end with the snubber located at 20 in from the pivot as shown in Fig. 4–29*c*.

Solution

(*a*) We note that $c = 24$ in, $r = 24$ in, $W_1 = W(1-1) = 0$, $W_2 = W(1/1) = W$. The loss in potential energy from the vertical position to the horizontal position is equal to the kinetic energy of the weight W_2 in the horizontal position:

$$(\text{PE})_{\text{loss}} = Wh = 1540(24) = 36\,960 = \frac{W}{2g}\dot{y}_0^2$$

$$\dot{y}_0 = \sqrt{\frac{36\,960(386)2}{1540}} = 136 \text{ in/s}$$

For a trial stroke $y_s = 4$ in,

$$\ddot{y}_0 = \frac{\dot{y}_0^2}{2y_s} = \frac{136^2}{2(4)} = 2312 \text{ in/s}^2$$

$$m\ddot{y}_0 = \frac{1540}{386}(2312) = 9224 \text{ lbf}$$

Table 4–8

Example of Tabular Engineering Data Presented by One Manufacturer of Hydraulic Shock Absorbers for Two of Many Models

Models OEM & AOEM						Nominal Coil Spring Force			
Model	Bore Size, in	Stroke, in	Max. In-Lb per Cycle	Max. In-Lb per Hour	Max. Shock Force, lb	Extended, lb	Compressed, lb	Max. Propelling Force, lb	Model Weight, lb
1.5M × 1		1	1 800	1 120 000		10	15		2.0
1.5M × 2	$\frac{3}{4}$	2	3 600	1 475 000	2 500	7	15	650	2.2
1.5M × 3		3	5 400	1 775 000		7	20		2.6
2.0M × 2		2	10 000	2 400 000		36	52		7.5
2.0M × 4	$1\frac{1}{8}$	4	20 000	3 200 000	7 500	18	52	1 500	8.8
2.0M × 6		6	30 000	3 730 000		18	100		11.0
3.0M × 2		2	20 000	3 290 000		25	45		15.5
3.0M × 3.5	$1\frac{1}{2}$	3.5	35 000	5 770 000	15 000	25	45	2 700	20
3.0M × 5		5	50 000	8 260 000		16	45		24
3.0M × 6.5		6.5	65 000	10 750 000		27	75		30
4.0M × 2		2	34 000	13 300 000		50	65		33
4.0M × 4		4	68 000	16 000 000		35	65		40
4.0M × 6	2	6	102 000	18 600 000	25 000	30	70	4 800	44
4.0M × 8		8	136 000	21 300 000		40	80		66
4.0M × 10		10	170 000	24 000 000		30	80		73

Notes: 1. Air/Oil (AOEM) Models—Maximum n/lb per hour in 20% higher than the standard OEM.
2. All shock absorbers will function satisfactorily at 5% of their maximum rated energy per cycle. If less than 5%, a smaller model should be specified.
3. Provide a POSITIVE STOP to prevent internal bottoming of the shock absorber.
4. Provide a POSITIVE STOP to pervent bottoming of Clevis Mount shock absorbers in extension.

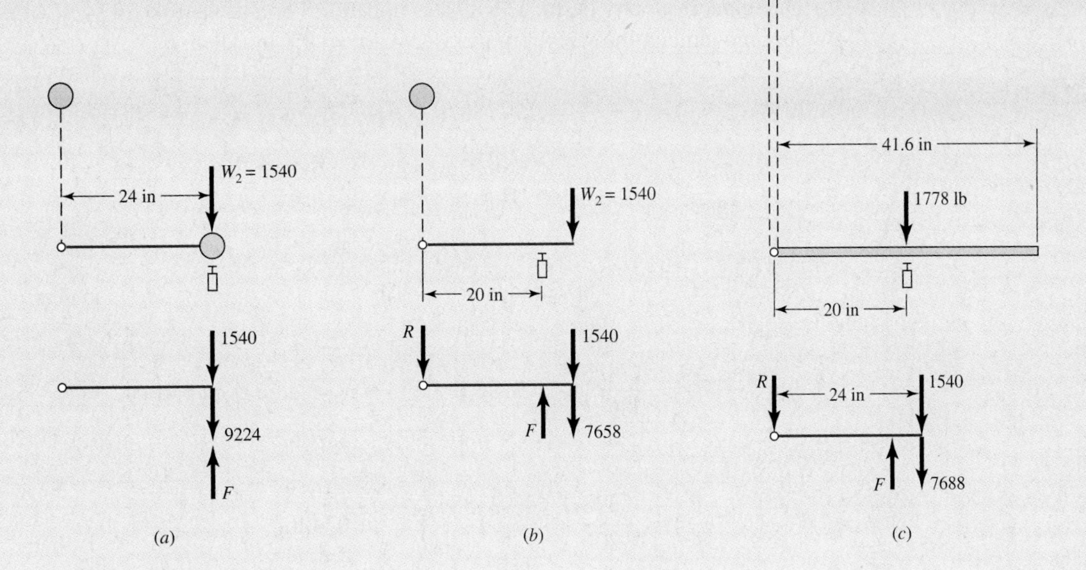

Figure 4–32

(*a*) A pivoted concentrated weight, associated free body showing weight 1540 lb, reversed D'Alembert force 9224 lb, and arresting force *F*. (*b*) The snubber is placed 20 in from the pivot. (*c*) A steel trap door with a length of 41.6 in, and associated free body.

$$\alpha = \frac{\ddot{y}}{k} = \frac{2312}{24} = 96.3 \text{ rad/s}^2$$

$$I = m_2 k^2 = \frac{1540}{386} 24^2 = 2298 \text{ in} \cdot \text{lbf} \cdot \text{s}^2$$

From Eq. (f),

$$F = \frac{T}{r}\left(\alpha + \frac{g}{k}\right) = \frac{2298}{24}\left(96.3 + \frac{386}{24}\right) = 10\,761 \text{ lbf}$$

From Eq. (g),

$$R = F - \frac{I\alpha}{k} - W_2 = 10\,761 - \frac{2298(96.3)}{24} - 1540 = 0$$

Then

$$F_{\text{max}} = \frac{F}{0.85} = \frac{10\,761}{0.85} = 12\,660 \text{ lbf}$$

$$R_{\text{max}} = 0$$

$$E_{\text{total}} = (\text{PE})_{\text{loss}} + W_2 y_s = 36\,960 + 1540(4) = 43\,120 \text{ in} \cdot \text{lbf}$$

Select a 4.0M × 4 snubber from Table 4–8.

Check

in · lbf/cycle = E/∼= 68 000 (43 120 required)
Shock force = 25 000 lbf (12 669 lbf required)
Maximum propelling force = 4800 lbf (1540 lbf required)

(b) The impact velocity is still 136 in/s, as in part a. For a trial stroke of $y_s = 4$ in of the snubber movement, the weight moves $4(24/20) = 4.8$ in. Then

$$\ddot{y}_0 = \frac{\dot{y}_0^2}{2(4.8)} = \frac{136^2}{2(4.8)} = 1927 \text{ in/s}$$

$$m_2\ddot{y}_0 = \frac{1540}{386}(1927) = 7688 \text{ lbf}$$

$$\alpha = \frac{\ddot{y}_0}{k} = \frac{1927}{24} = 80.3 \text{ rad/s}^2$$

$$I = \frac{W_2}{g}k^2 = \frac{1540}{386}24^2 = 2298 \text{ in} \cdot \text{lbf} \cdot \text{s}^2 \qquad \text{(same as in part } a\text{)}$$

From Eq. (f),

$$F = \frac{I}{r}\left(\alpha + \frac{g}{k}\right) = \frac{2298}{20}\left(80.3 + \frac{386}{24}\right) = 11\,074 \text{ lbf}$$

From Eq. (g),

$$R = F - \frac{I\alpha}{k} - W_2 = 11\,074 - \frac{2298(80.3)}{24} - 1540 = 1845 \text{ lbf}$$

Then

$$F_{\text{max}} = \frac{1845}{0.85} = 13\,029 \text{ lbf}$$

$$R_{\text{max}} = \frac{1}{0.85}\left(F - \frac{I\alpha}{k}\right) - W_2 = \frac{1}{0.85}\left[11\,074 - \frac{2298(80.3)}{24}\right] - 1540 = 2443 \text{ lbf}$$

The influence of the repeatedly applied nature of the 2443-lbf force on the pivot needs to be examined.

$$E_{\text{total}} = 36\,960 + 1540(4.8) = 44\,352 \text{ lbf}$$

Select a 4.04M × 4 snubber from Table 4–8.

Check

in · lbf/cycle = E/~= 68 000 in · lbf (44 352 in · lbf required)
Shock force = 25 000 lbf (13 029 lbf required)
Maximum propelling force 4800 lbf (1540 lbf required)

(c) First consider kinetic equivalency. From Eqs. (d), and (e) successively

$$W_1 = 0.134(1778) = 238 \text{ lbf}$$

$$W_2 = 0.866(1778) = 1540 \text{ lbf}$$

$$k = \frac{l}{\sqrt{3}} = \frac{41.6}{\sqrt{3}} = 24 \text{ in}$$

From this point on the solution is the same as in part b.

4–18 Shock and Impact

Impact refers to the collision of two masses with initial relative velocity. In some cases it is desirable to achieve a known impact in design; for example, this is the case in the design of coining, stamping, and forming presses. In other cases, impact occurs because of excessive deflections, or because of clearances between parts, and in these cases it is desirable to minimize the effects. The rattling of mating gear teeth in their tooth spaces is an impact problem caused by shaft deflection and the clearance between the teeth. This impact causes gear noise and fatigue failure of the tooth surfaces. The clearance space between a cam and follower or between a journal and its bearing may result in crossover impact and also cause excessive noise and rapid fatigue failure.

Shock is a more general term which is used to describe any suddenly applied force or disturbance. Thus the study of shock includes impact as a special case. There are two general approaches to the study of shock, depending upon whether only statics is used in the analysis or both statics and dynamics are used. Lifetimes have been spent investigating shock and impact phenomena. For reasons of space, the material presented here is intended only to indicate the scope of the subject and to give a basic understanding of what is involved.

Figure 4–33 represents a highly simplified mathematical model of an automobile in collision with a rigid obstruction. Here m_1 is the lumped mass of the engine. The displacement, velocity, and acceleration are described by the coordinate x_1 and its time derivatives. The lumped mass of the vehicle less the engine is denoted by m_2, and its motion by the coordinate x_2 and its derivatives. Springs k_1, k_2, and k_3 represent the linear and nonlinear stiffnesses of the various structural elements which compose the vehicle. Friction can and should be included, but is not shown in this model. The determination of the spring rates for such a complex structure will almost certainly have to be performed experimentally. Once these values—the k's, m's, and frictional coefficients—are obtained, a set of nonlinear differential equations can be written and a computer solution obtained for any impact velocity.

Figure 4–34 is another impact model. Here mass m_1 has an initial velocity v and is just coming into contact with spring k_1. The part or structure to be analyzed is represented by mass m_2 and spring k_2. The problem facing the designer is to find the maximum deflection of m_2 and the maximum force exerted by k_2 against m_2. In the analysis it doesn't matter whether k_1 is fastened to m_1 or to m_2, since we are interested only in a solution up to the point in time for which x_2 reaches a maximum. That is, the solution for the rebound isn't needed. The differential equations are not difficult to derive. They are

$$
\begin{aligned}
m_1\ddot{x}_1 + k_1(x_1 - x_2) &= 0 \\
m_2\ddot{x}_2 + k_2x_2 - k_1(x_1 - x_2) &= 0
\end{aligned}
\tag{4–82}
$$

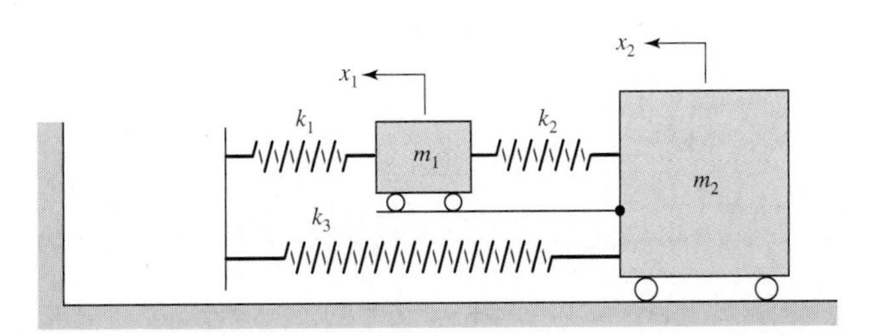

Figure 4–33

Two-degree-of-freedom mathematical model of an automobile in collision with a rigid obstruction.

Figure 4–34

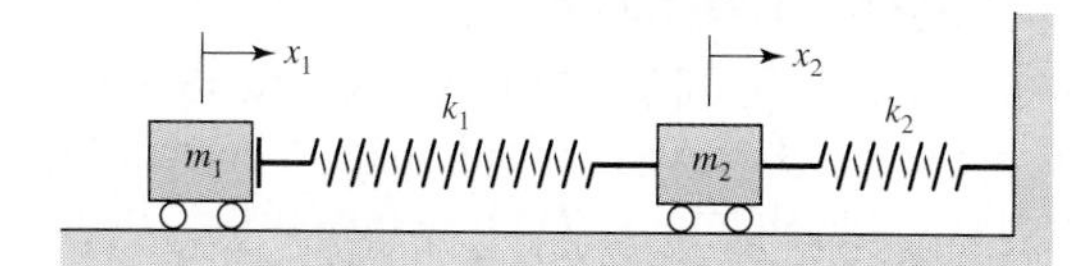

Equation pair (4–82) is rather awkward to use in obtaining a general analytical solution. But, if the values of the m's and k's are specified, you can probably find an integration routine in your computer-program library for use in obtaining a solution.

4–19 Suddenly-Applied Loading

A simple case of impact is illustrated in Fig. 4–35*a*. Here a weight W, moving at a constant velocity v on a frictionless surface, strikes a cantilever of stiffness EI and length l. We want to find the maximum deflection and the maximum bending moment in the beam due to the impact.

Figure 4–35*b* shows an abstract model of the system. Using Table E–9–1, we find the spring rate to be $k = F/y = 3EI/l^3$. We choose to count the time at the instant the weight strikes the spring. Thus at $t = 0$ the motion is started with $y = 0$ and $\dot{y} = v$. The differential equation is

$$\frac{W}{g}\ddot{y} = -ky \tag{a}$$

where the spring force ky is negative because it is opposite to the deflection y. The solution to this equation is well known, and is

$$y = A\cos\left(\frac{kg}{W}\right)^{1/2} t + B\sin\left(\frac{kg}{W}\right)^{1/2} t \tag{b}$$

The velocity is

$$\dot{y} = -A\left(\frac{kg}{W}\right)^{1/2}\sin\left(\frac{kg}{W}\right)^{1/2} t + B\left(\frac{kg}{W}\right)^{1/2}\cos\left(\frac{kg}{W}\right)^{1/2} t \tag{c}$$

The constants A and B are evaluated using the starting conditions in Eqs. (*b*) and (*c*). These are

$$A = 0 \qquad B = \frac{v}{(kg/W)^{1/2}}$$

Figure 4–35

(*a*) Collision of a weight with a cantilever beam; (*b*) modeling the cantilever beam as a spring with rate *k*.

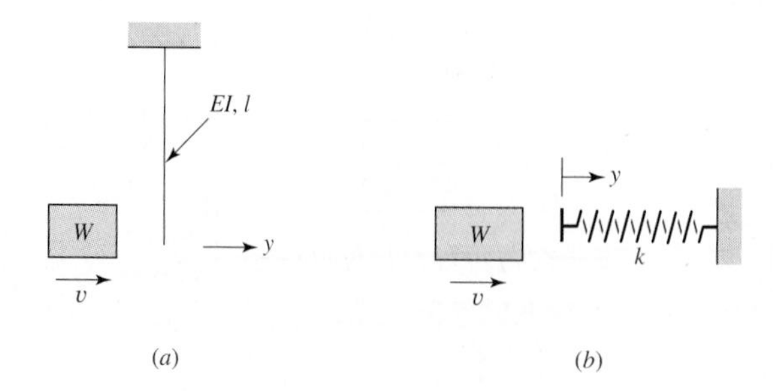

Substituting these back in Eq. (*b*) gives the solution as

$$y = \frac{v}{(kg/W)^{1/2}} \sin \left(\frac{kg}{W} \right)^{1/2} t \tag{4–83}$$

Of course, this solution is valid only as long as the weight remains in contact with the beam. The maximum deflection is

$$y_{\max} = \frac{v}{(kg/W)^{1/2}} = v \left(\frac{Wl^3}{3EIg} \right)^{1/2} \tag{4–84}$$

Also, the maximum bending moment is the product of the spring force and the beam length. The result is

$$M_{\max} = kly_{\max} = v \left(\frac{3EIW}{gl} \right)^{1/2} \tag{4–85}$$

As a second example, consider the weight W, in Fig. 4–36*a*, falling a distance h and impacting some structure or member whose spring rate is k. We choose the origin of the coordinate y corresponding to the position of the weight when time t is zero, as before. Two free-body diagrams, shown in Fig. 4–36*b* and *c*, are necessary—one when $y \leq h$, and another when $y > h$—to account for the spring force. For each of these free-body diagrams we can write Newton's law by stating that the inertia force $(W/g)\ddot{y}$ is equal to the sum of the external forces acting on the weight. We then have

$$\begin{aligned} \frac{W}{g}\ddot{y} &= W & y \leq h \\ \frac{W}{g}\ddot{y} &= -k(y-h) + W & y > h \end{aligned} \tag{d}$$

We must also include in the mathematical statement of the problem the knowledge that the weight is released with zero initial velocity. Equation pair (*d*) constitutes a set of *piecewise differential equations.* Each equation is linear, but each applies only for a certain range of y. The solution to the set is valid for all values of t, but, as before, we are interested in values of y only up until the time that the spring or structure reaches its maximum deflection.

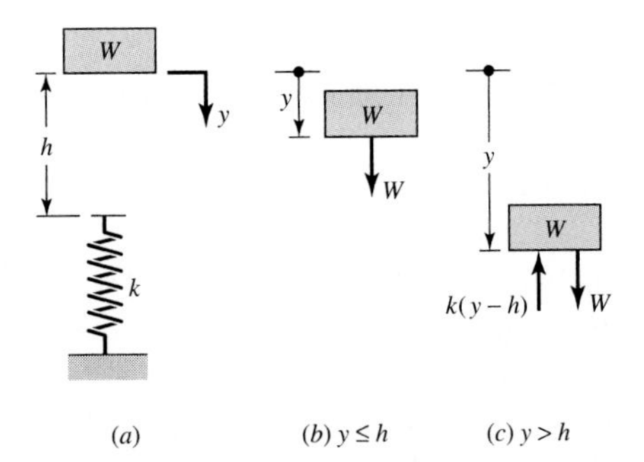

Figure 4–36

(*a*) A weight free to fall a distance h to free end of a spring. (*b*) Free body of weight during fall. (*c*) Free body of weight during arrest.

The solution to the first equation in the set is

$$y = \frac{gt^2}{2} \qquad y \le h \tag{4–86}$$

and you can verify this by direct substitution. Equation (4–86) is no longer valid after $y = h$; call this time t_1. Then

$$t_1 = \left(\frac{2h}{g}\right)^{1/2} \tag{e}$$

Differentiating Eq. (4–86) to get the velocity gives

$$\dot{y} = gt \qquad y \le h \tag{f}$$

and so the velocity of the weight at $t = t_1$ is

$$\dot{y}_1 = gt_1 = (2gh)^{1/2} \tag{g}$$

Having moved from $y = 0$ to $y = h$, we then need to solve the second equation of the set (*d*). It is convenient to define a new time $t' = t - t_1$. Thus $t' = 0$ at the instant the weight strikes the spring. Applying your knowledge of differential equations, you should find the solution to be

$$y = A\cos\left(\frac{kg}{W}\right)^{1/2} t' + B\sin\left(\frac{kg}{W}\right)^{1/2} t' + h + \frac{W}{k} \qquad y > h \tag{h}$$

If you cannot derive Eq. (*h*), you should at least prove to yourself that it is a solution by substituting it and its second derivative in the second equation of set (*d*) to show that the equality in Eq. (*h*) is satisfied identically.

The constants A and B are evaluated as shown in the previous solution. When these are substituted back, the result can be transformed to

$$y = \left[\left(\frac{W}{k}\right)^2 + \frac{2Wh}{k}\right]^{1/2} \cos\left[\left(\frac{kg}{W}\right)t' - \phi\right] + h + \frac{W}{k} \qquad y > h \tag{4–87}$$

where ϕ, though of no interest here, is given by

$$\phi = \frac{\pi}{2} + \tan^{-1}\left(\frac{w}{2kh}\right)^{1/2} \tag{i}$$

The maximum deflection of the spring (structure) occurs when the cosine term in Eq. (4–87) is unity. We designate this as δ and, after rearranging, find it to be

$$\delta = y - h = \frac{W}{k} + \frac{W}{k}\left[1 + \left(\frac{2hk}{W}\right)\right]^{1/2} \tag{4–88}$$

The maximum force acting on the spring or structure is now found to be

$$F = k\delta = W + W\left[1 + \left(\frac{2hk}{W}\right)\right]^{1/2} \tag{4–89}$$

Note, in this equation, that if $h = 0$, then $F = 2W$. This says that when the weight is released while in contact with the spring but is not exerting any force on the spring, the largest force is double the weight.

Most systems are not as ideal as those explored here, so be wary about using these relations for nonideal systems.

4–20 Propagation of Error

The four cases discussed in Sec. 3–22 for stochastic relations between load and stress apply to relations between load and deflection, and they apply generally.

EXAMPLE 4–18

Consider a shaft as depicted in Fig. P4–19. The bearing span ℓ is 10 in, the set-in b of the load is 6 in, the load F is 800 lbf, and the slope limit for roller bearings ξ is 0.001, all deterministic. Young's modulus is 30(10⁶) psi. The shaft diameter to keep the left bearing without life penalty due to excessive slope is, from the statement of Prob. 4–19,

$$d_L = \left| \frac{32Fb(b^2 - l^2)}{3\pi EI\xi} \right|^{1/4}$$

(*a*) Considering all parameters deterministic, what is the necessary shaft diameter?
(*b*) If the load is $\mathbf{F} \sim LN(800,\ 64)$ lbf and all other parameters are robustly deterministic, describe **d** and give its bound for the 99 percent satisfactory quantile.
(*c*) If the load is $\mathbf{F} \sim LN(800,\ 64)$ lbf and $\mathbf{E} \sim LN(30\ 10^6,\ 0.9\ 10^6)$ psi and all the other parameters are robustly deterministic, describe **d** and give its bound for the 99 percent satisfactory quantile.
(*d*) If the load is $\mathbf{F} \sim LN(800,\ 64)$ lbf, $\mathbf{E} \sim LN(30\,10^6,\ 0.9\,10^6)$ psi, $\mathbf{b} \sim U[5.9,\ 6.1]$ in, and $\mathbf{l} \sim U[9.9,\ 10.1]$ in, describe **d** and give its 99 percent satisfactory quantile.

Solution (*a*) For deterministic parameters the diameter d is

Answer
$$d = \left| \frac{32(800)6(6^2 - 10^2)}{3\pi 30(10^6)10(0.001)} \right|^{1/4} = 1.365\ 509 = 1.366 \text{ in}$$

(*b*) Treat as Case 1 of Sec. 3–22.

$$\mathbf{d} = \alpha \mathbf{F}^{1/4} \qquad C_F = \frac{64}{800} = 0.08$$

$$\mu_d = \alpha \mu_F^{1/4} \qquad \sigma_d = \frac{\alpha \mu_F^{1/4} C_F}{4} \qquad C_d = \frac{C_F}{4}$$

$$\alpha = \left| \frac{32(6)(6^2 - 10^2)}{3\pi 30(10^6)10(0.001)} \right|^{1/4} = 0.256\ 757$$

$$\mu_d = 0.256\ 757(800)^{0.25} = 1.365\ 509 \text{ in}$$

$$\sigma_d = \mu_d C_d = 1.365\ 509 \left(\frac{0.08}{4} \right) = 0.027\ 310 \text{ in}$$

$$C_d = \frac{0.256\ 757}{1.365\ 509} = 0.02$$

therefore $\mathbf{d} \sim 1.366LN(1,\ 0.02)$ in. From Eq. (2–37),

$$\mu_y = \ln 1.366 - \ln \sqrt{1 + 0.02^2} = 0.311\ 687$$

From Eq. (2–38),

$$\sigma_y = \sqrt{\ln(1 + 0.02^2)} = 0.019\ 998$$

$$y = \mu_y + z\sigma_y = 0.311\ 687 + (+2.326)0.019\ 998 = 0.358\ 202$$

Answer $x = \exp(0.358\,202) = 1.430\,755 = d_{0.99} = 1.431$ in

(*c*) Treat as Case 2 of Sec. 3–22.

$$\mathbf{d} = \left| \frac{32b(b^2 - l^2)}{3\pi \mathbf{E} l \xi} \right|^{1/4} \mathbf{F}^{1/4} = \boldsymbol{\beta}\mathbf{F}^{1/4}$$

which defines $\boldsymbol{\beta}$ as

$$\boldsymbol{\beta} = \left| \frac{32(6)(6^2 - 10^2)}{3\pi(10)0.001} \right|^{1/4} \left(\frac{1}{\mathbf{E}}\right)^{1/4} = 19.002 \left(\frac{1}{\mathbf{E}}\right)^{1/4}$$

$$\mu_\beta = 19.002 \left(\frac{1}{30 \cdot 10^6}\right)^{1/4} = 0.256\,755$$

$$C_\beta = \frac{C_E}{4} = \frac{0.9/30}{4} = 0.0075$$

$$\mu_{F^{1/4}} = \mu_{F^{0.25}} = (800)^{0.25} = 5.318\,29$$

$$C_{F^{1/4}} = \frac{C_F}{4} = \frac{64/800}{4} = 0.02$$

$$\mu_d = \mu_\beta \mu_{F^{1/4}} = 0.256\,755(5.318\,29) = 1.365\,498 = 1.366 \text{ in}$$

$$C_d = \sqrt{C_\beta^2 + C_{F^{1/4}}^2} = \sqrt{0.0075^2 + 0.02^2} = 0.021\,36$$

so $\mathbf{d} \sim 1.366LN(1,\ 0.021\,36)$ in. From Eq. (2–37),

$$\mu_y = \ln 1.366 - \ln\sqrt{1 + 0.021\,36^2} = 0.311\,659$$

From Eq. (2–38),

$$\sigma_y = \sqrt{\ln(1 + 0.021\,36)^2} = 0.021\,358$$

$$y = \mu_y + z\sigma_y = 0.311\,659 + (2.326)0.021\,58 = 0.361\,338$$

Answer $x = \exp(0.361\,338) = 1.435\,248 = d_{0.99} = 1.436$ in

(*d*) Treat as Case 3 of Sec. 3–22. The equation for $\mathbf{d}$ is

$$\mathbf{d} = \left| \frac{32\mathbf{Fb}(\mathbf{b}^2 - l^2)}{3\pi \mathbf{E} l \xi} \right|^{1/4}$$

Before conducting a simulation it is instructive to examine the COVs of the variates:

$$C_F = \frac{64}{800} = 0.08 \qquad C_E = \frac{0.9}{30} = 0.03$$

$$\mu_b = \frac{5.9 + 6.1}{3} = 6 \qquad \sigma_b = \frac{6.1 - 5.9}{2\sqrt{3}} = 0.057\,735$$

$$C_b = \frac{0.057\,735}{6} = 0.009\,623$$

$$\mu_l = \frac{9.9 + 10.1}{2} = 10$$

$$\sigma_l = \frac{10.1 - 9.9}{2\sqrt{3}} = 0.057\,735$$

$$C_l = \frac{0.057\,735}{10} = 0.005\,774$$

Because C_F dominates and C_E contributes little, and C_b and C_l contribute very little, we can conclude that the part *c* solution is robust and that a simulation will increase $d_{0.99}$

only slightly. To demonstrate, a simulation of 10^5 trials gives the following tabulated results:

	Mean	Standard Deviation	COV
d	1.364 726	0.029 913	0.021 919
b	6.000 547	0.057 476	0.009 578
l	10.000 84	0.056 183	0.005 618
E	30 000 845	892 043	0.029 726
F	799.949 8	64.567 94	0.080 715

From Eq. (2–37),

$$\mu_y = \ln 1.365 - \ln\sqrt{1 + 0.021\,594^2} = 0.310\,921$$

From Eq. (2–38),

$$\sigma_y = \sqrt{\ln(1 + 0.021\,594^2)} = 0.021\,591$$

$$y = \mu_y + z\sigma_y = 0.310\,921 + (2.326)0.021\,591 = 0.361\,142$$

Answer $x = \exp(0.361\,142) = 1.434\,967 = d_{0.99} = 1.435$ in

The solution to part *a* gives a diameter of 1.366 in, which if acted upon will incur a life penalty in roughly half of the left bearings. Part *b* gives 1.431 in, part *c* gives 1.436 in, and part *d* 1.435 in. One can see that because of the Pythagorean combination of the COVs, the distribution of **F** dominates and **E** helps. Because of the nonlinearity (the 1/4 power) the COV of **F** is cut to one-fourth of its magnitude, as seen in **d**. In this example a Case 1 treatment is useful, and the rest were shown for comparison.

With the experience of Ex. 4–18 in mind, consider Ex. 4–19.

EXAMPLE 4–19 The shaft of Ex. 4–18 has been established with a uniform diameter of 1.440 ± 0.001. Estimate the deflection under the load.

Solution From Table E–9–6,

$$\delta|_{x=l-b} = \frac{Fbx}{6EIl}(x^2 + b^2 - l^2)$$

$$= \frac{32Fb(l-b)}{3\pi Ed^4 l}[(l-b)^2 + b^2 - l^2]$$

after substituting $I = \pi d^4/64$ and $x = l - b$. Now

$$\mathbf{d} \sim U[1.439, 1.441] \text{ in}$$

$$\mu_d = \frac{1.439 + 1.441}{2} = 1.440 \text{ in}$$

$$\sigma_d = \frac{1.441 - 1.439}{2\sqrt{3}} = 0.000\,577$$

$$C_d = \frac{0.000\,577}{1.440} = 0.000\,401$$

The COV of **F** is 0.08, of **E** is 0.03, of **d** is 0.0004, and of $\mathbf{d}^4$ is 4(0.0004) = 0.0016. The COV of **b** is 0.0096 and of **l** is 0.0058. Clearly, the **F** distribution dominates with a small assist from **E**.

$$\mu_\delta = \frac{32(800)6(10-6)}{3\pi 30(10^6)1.44^4(10)}[(10-6)^2 + 6^2 - 10^2] = -0.002\,426 \text{ in}$$

$$C_\delta = C_{F/E} \doteq \sqrt{C_F^2 + C_E^2} = \sqrt{0.08^2 + 0.03^2} = 0.0854$$

so δ is robustly described as $\boldsymbol{\delta} \sim -0.0024LN(1, 0.085)$ in.

The distribution of **d** in Case 2 of Ex. 4–18 exhibited a COV of 0.021 36 because of uncertainties in **F** and **E**. Once the decision was made in Ex. 4–19 to establish the diameter as 1.440 ± 0.001 in, the distribution of **d** became uniform and its COV became 0.0004 because of the manufacturing method and tolerances imposed.

In conducting a computer simulation it is important to observe a protocol in order to avoid bias. The equation for **d** in part *d* of Ex. 4–18 is

$$\mathbf{d} = \left| \frac{32\mathbf{F}\mathbf{b}(\mathbf{b}^2 - \mathbf{l}^2)}{3\pi \mathbf{E}\mathbf{l}\xi} \right|^{1/4}$$

You will notice that variates **b** and **l** appear twice. It is important to remember to

- Call your random number generator of the proper distribution to provide *single* instances of **b**, **l**, **F**, and **E**.
- Substitute only these four values to create a single instance of **d**.
- Work with these instances of **d** to generate information about **d**.

If you observe these rules, the subtle influences of correlation will be taken into account without effort on your part, and no bias will be introduced.

Suppose $\mathbf{x} \sim U[0, 1]$; then

$$\mu_x = \frac{(0+1)}{2} = 0.5$$

$$\sigma_x = \frac{1-0}{2\sqrt{3}} = 0.288\,675$$

$$C_x = \frac{0.288\,675}{0.5} = 0.577\,350$$

Consider the equation $\mathbf{w} = \mathbf{x} \cdot \mathbf{x}$ as a subject for simulation. By substituting a single instance of **x**, say u_1, to obtain a single value of $w = u_1 \cdot u_1 = u_1^2$, we really have $\mathbf{w} = \mathbf{x}^2$. From Table 2–1, recognizing that C_x^2 is *not* small compared to unity,

$$\mu_w = \mu_{x^2} = \mu_x^2(1 + C_x^2) = 0.5^2(1 + 0.577350^2) = 0.333$$

$$\sigma_w = 2\mu_x^2 C_x\left(1 + \frac{C_x^2}{4}\right) = 2(0.5^2)0.577\,350\left(1 + \frac{0.577\,350^2}{4}\right) = 0.312\,731$$

$$C_w = \frac{0.312\,731}{0.333} = 0.938\,193$$

If we substitute two different values of **x**, say u_1 and u_2, and multiply them together as $z = u_1 \cdot u_2$, they are independent. It is as if the simulation equation was $\mathbf{z} = \mathbf{x} \cdot \mathbf{y}$. From Table 2–1

$$\mu_z = \mu_x \mu_y = 0.5(0.5) = 0.25$$

$$\begin{aligned}\sigma_z &= \mu_{xy}\left(C_x^2 + C_y^2 + C_x^2 C_y^2\right)^{1/2} \\ &= 0.25[0.577\,350^2 + 0.577\,350^2 + 0.577\,350^2(0.577\,350^2)]^{1/2} \\ &= 0.220\,479\end{aligned}$$

$$C_{xy} = \frac{0.220\,479}{0.25} = 0.881\,916$$

The cases are different, and simulation results converge on different asymptotes. A simulation gave

	Mean	Sigma
x^2	0.333 003	0.297 953
xy	0.247 165	0.220 007

PROBLEMS

ANALYSIS

4–1 Structures can often be considered to be composed of a combination of tension and torsion members and beams. Each of these members can be analyzed separately to determine its force-deflection relationship and its spring rate. It is possible, then, to obtain the deflection of a structure by considering it as an assembly of springs having various series and parallel relationships.

(*a*) What is the overall spring rate of three springs in series?
(*b*) What is the overall spring rate of three springs in parallel?
(*c*) What is the overall spring rate of a single spring in series with a pair of parallel springs?

ANALYSIS

4–2 The figure shows a torsion bar OA fixed at O, supported at A, and connected to a cantilever AB. The spring rate of the torsion bar is k_T, in newton-meters per radian, and that of the cantilever is k_C, in newtons per meter. What is the overall spring rate based on the deflection y at point B?

Problem 4–2

ANALYSIS

4–3 A torsion-bar spring consists of a prismatic bar, usually of round cross section, which is twisted at one end and held fast at the other to form a stiff spring. An engineer needs a stiffer one than usual and so considers building in both ends and applying the torque somewhere in the central portion of the span, as shown in the figure. If the bar is uniform in diameter, that is, if $d = d_1 = d_2$, investigate how the allowable angle of twist, the largest torque, and the spring rate depend on the location x at which the torque is applied.

ANALYSIS

4–4 An engineer is forced by geometric considerations to apply the torque on the spring of Prob. 4–3 at the location $x = 0.2l$. For a uniform-diameter spring, this would cause the long leg of the span to be underutilized when both legs have the same diameter. If the diameter of the long leg is reduced sufficiently, the shear stress in the two legs can be made equal. How would this change affect the allowable angle of twist, the largest torque, and the spring rate?

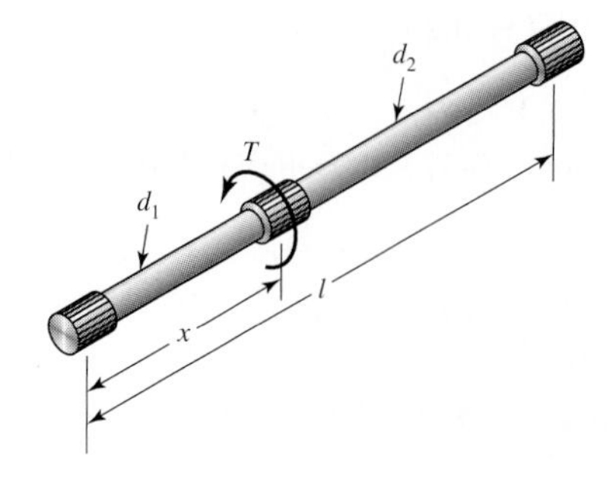

Problem 4–3

ANALYSIS

4–5 The figure shows a cantilever consisting of steel angles size $4 \times 4 \times \frac{1}{2}$ in mounted back to back. Find the deflection at B and the maximum stress in the beam.

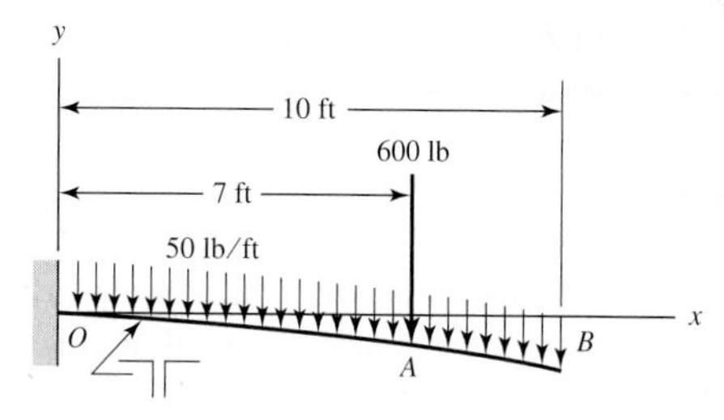

Problem 4–5

ANALYSIS

4–6 A simply supported beam loaded by two forces is shown in the figure. Select a pair of structural steel channels mounted back to back to support the loads in such a way that the deflection at midspan will not exceed $\frac{1}{16}$ in and the maximum stress will not exceed 6 kpsi.

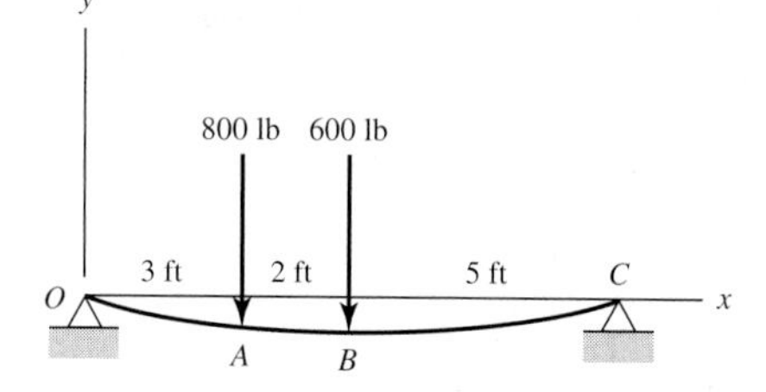

Problem 4–6

ANALYSIS

4–7 Find the deflection of the steel shaft at A in the figure. Find the deflection at midspan. By what percentage do these two values differ?

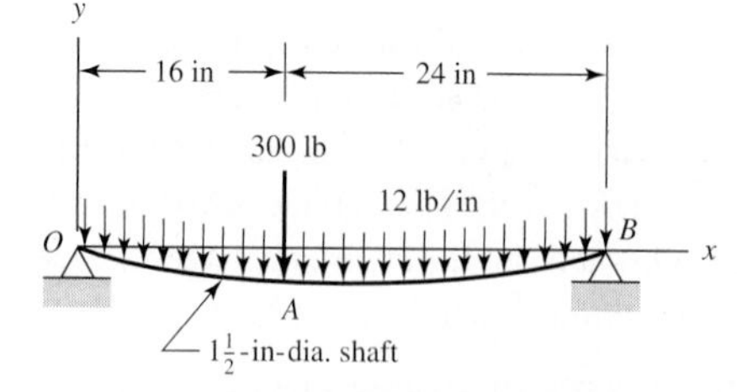

Problem 4–7

ANALYSIS

4–8 A rectangular steel bar supports the two overhanging loads shown in the figure. Find the deflection at the ends and at the center.

Problem 4–8

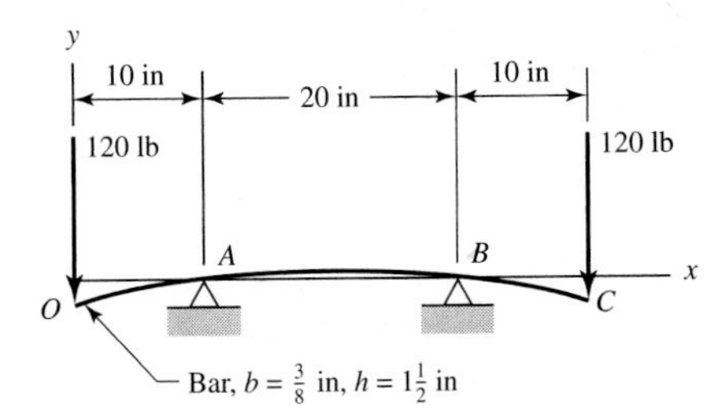

ANALYSIS

4–9 The deflection at B of the cantilever shown in the figure can be found by adding the deflection caused by the force at A acting alone to that caused by the force at B acting alone. This is called the *method of superposition;* it is applicable whenever the deflection is linearly related to the force. Using the formulas in Appendix Table E–9 and superposition, find the deflection of the cantilever at B if $I = 13\text{ in}^4$ and $E = 30$ Mpsi.

Problem 4–9

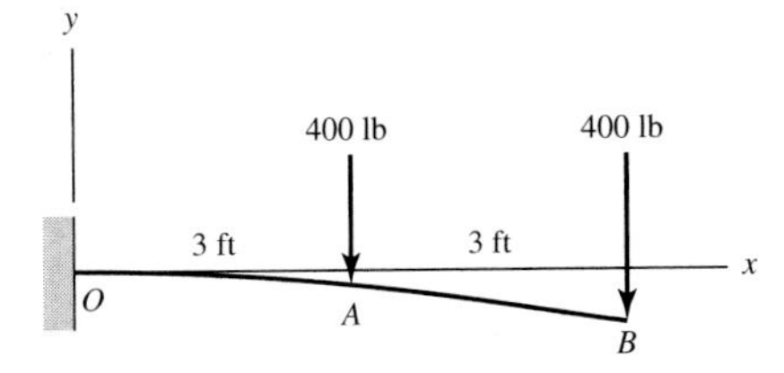

ANALYSIS

4–10 The cantilever shown in the figure is a 4-in, 5.4-lb structural-steel channel. Such sections may be designated by the symbol shown in the figure or by the capital letter C. Using the approach suggested in Prob. 4–9, find the deflection at A. Where must the load be placed to avoid twisting of the channel section?

Problem 4–10

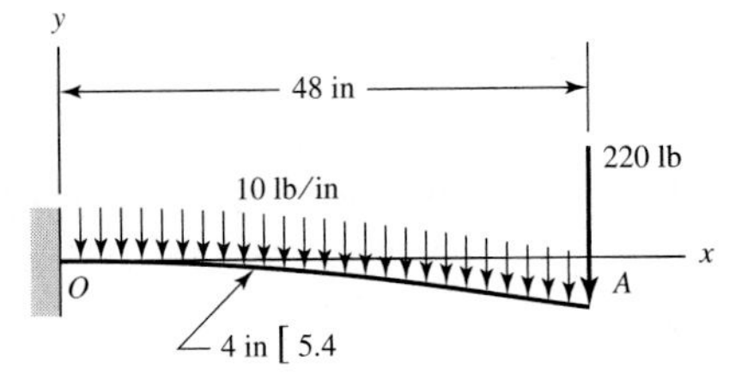

ANALYSIS

4–11 Determine the maximum deflection of the beam shown in the figure. The material is carbon steel.

Problem 4–11

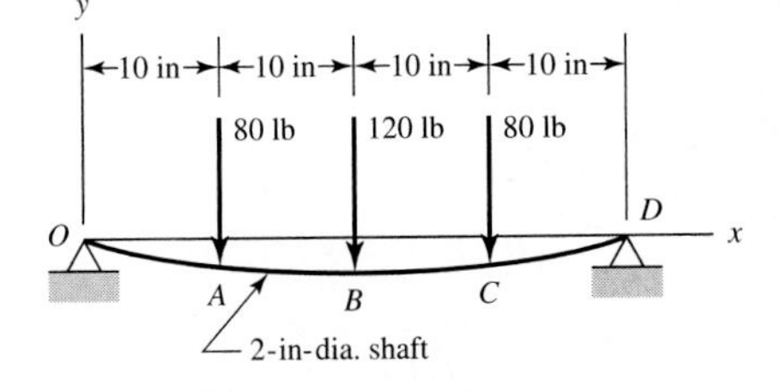

DESIGN

4–12 Illustrated is a rectangular steel bar with simple supports at the ends and loaded by a force F at the middle; the bar is to act as a spring. The ratio of the width to the thickness is to be about $b = 16h$, and the desired spring scale is 2400 lb/in.

(a) Find a set of cross-section dimensions, using preferred sizes.

(*b*) What deflection would cause a permanent set in the spring if this is estimated to occur at a normal stress of 90 kpsi?

Problem 4–12

F
A
A
4 ft
b
h
Section A–A

DESIGN

4–13 Illustrated in the figure is a $1\frac{1}{2}$-in-diameter steel countershaft which supports two pulleys. Pulley *A* delivers power to a machine causing a tension of 600 lb in the tight side of the belt and 80 lb in the loose side, as indicated. Pulley *B* receives power from a motor. The belt tensions on pulley *B* have the relation $T_1 = 0.125T_2$. Find the deflection of the shaft in the *z* direction at pulleys *A* and *B*. Assume that the bearings constitute simple supports.

Problem 4–13

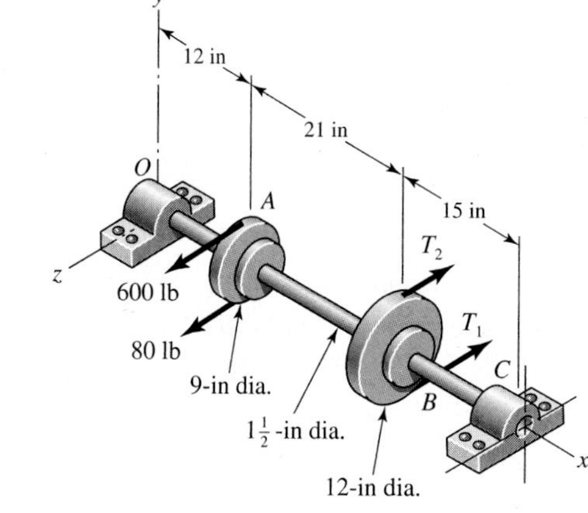

4–14 The figure shows a steel countershaft which supports two pulleys. Pulley *C* receives power from a motor producing the belt tensions shown. Pulley *A* transmits this power to another machine through the belt tensions T_1 and T_2 such that $T_1 = T_2/8$.

Problem 4–14

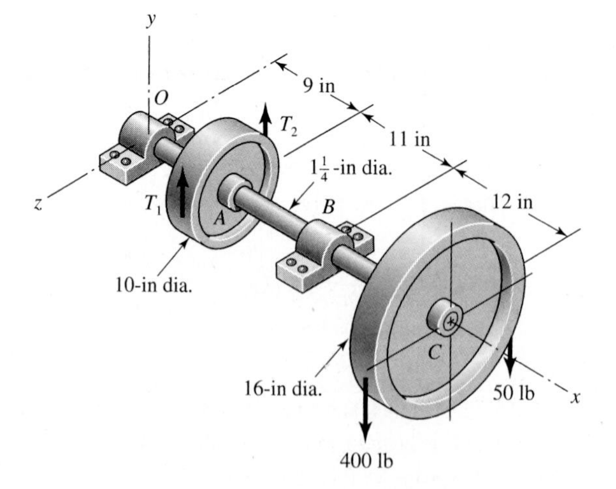

(*a*) Find the deflection of the overhanging end of the shaft, assuming simple supports at the bearings.

(*b*) If roller bearings are used, the slope of the shaft at the bearings should not exceed 0.06° for good bearing life. What shaft diameter is needed to conform to this requirement? Use $\frac{1}{8}$-in increments in any iteration you may make. What is the deflection at pulley C now?

4–15 A $\frac{5}{8}$-in-diameter plow-steel wire rope has a *rope* modulus of elasticity of $E_r = 12$ Mpsi and is used to support a 500-lb cage and a 2000-lb load in a mine hoist. If the length of the rope between the hoisting drum and the cage is 100 ft, find the stretch in the rope as a mine cart and its load are moved onto the cage platform. The load due to the mine cart is $N = 3000$ lb. The cross-sectional area of the individual rope wire in tension can be estimated from $A_m = 0.4d_r^2$, where d_r is the nominal diameter of the rope.

4–16 When a hoisting cable is long, the weight of the cable itself contributes to the elongation. If a cable has a weight of w newtons per meter, a length of l meters, and a load P attached to the free end, show that the cable elongation is

$$\delta = \frac{Pl}{AE} + \frac{wl^2}{2AE}$$

4–17 For a uniformly loaded beam supported by simple supports, the central deflection is easily found. Suppose an opportunity presented itself to lessen this deflection by moving one support toward the center by a distance a. This would lessen the largest deflection between the supports, but now, the free end would also deflect. What offset a of the right support in the figure would result in a minimum beam deflection?

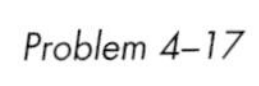
Problem 4–17

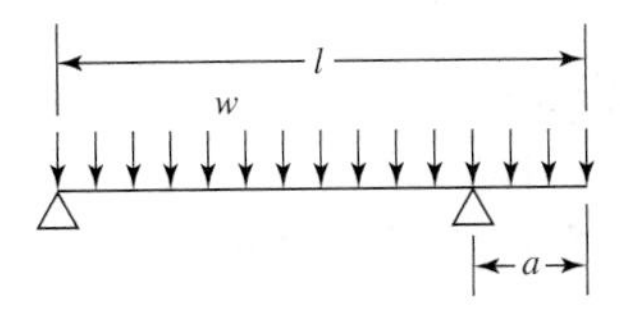

DESIGN

4–18 The structure of a diesel-electric locomotive is essentially a composite beam supporting a deck. Above the deck are mounted the diesel prime mover, generator or alternator, radiators, switch gear, and auxiliaries. Beneath the deck are found fuel and lubricant tanks, air reservoirs, and small auxiliaries. This assembly is supported at bolsters by the trucks which house the traction motors and brakes. This equipment is distributed as uniformly as possible in the span between the bolsters. In an approximate way, the loading can be viewed as uniform between the bolsters and simply supported. Because the hoods that shield the equipment from the weather have many rectangular access doors, which are mass-produced, it is important that the hood structure be level and plumb and sit on a flat deck. Aesthetics plays a role too. The center sill beam has a second moment of area of $I = 5450$ in^4, the bolsters are 36 ft apart, and the deck loading is 5000 lb/ft.

(*a*) What is the camber of the curve to which the deck will be built in order that the service-ready locomotive will have a flat deck?

(*b*) What equation would you give to locate points on the curve of part (*a*)?

4–19 The designer of a shaft usually has a slope constraint imposed by the bearings used. This limit will be denoted as ξ. If the shaft shown in the figure is to have a uniform diameter d except in the locality of the bearing mounting, it can be approximated as a uniform beam with simple supports. Show that the diameters to meet the slope constraints at the left and right bearings are, respectively,

$$d_L = \left|\frac{32Fb\,(b^2 - l^2)}{3\pi El\xi}\right|^{1/4} \qquad d_R = \left|\frac{32Fa\,(l^2 - a^2)}{3\pi El\xi}\right|^{1/4}$$

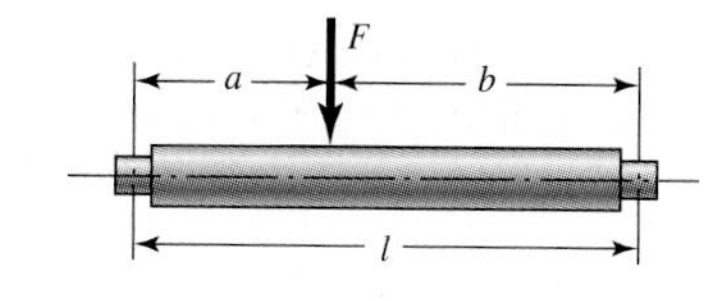

Problem 4–19

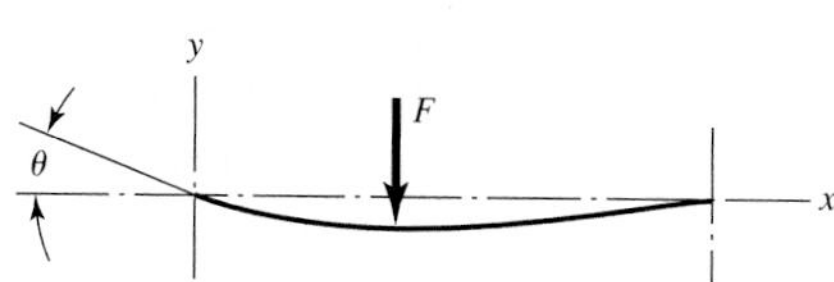

DESIGN

4–20 For the shaft shown in the figure, let $a_1 = 4$ in, $b_1 = 12$ in, $a_2 = 10$ in, $F_1 = 100$ lb, $F_2 = 300$ lb, and $E = 30$ Mpsi. The shaft is to be sized so that the maximum slope at either bearing A or bearing B does not exceed 0.001 rad. Determine a suitable diameter d.

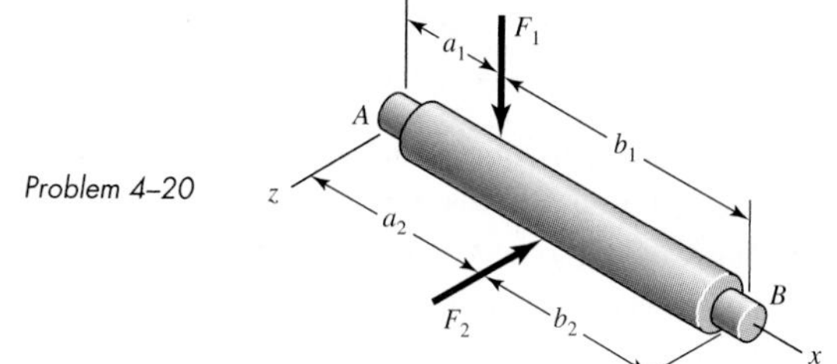

Problem 4–20

DESIGN

4–21 A shaft is to be designed so that it is supported by roller bearings. The basic geometry is shown in the figure. The allowable slope at the bearings is 0.001 mm/mm without bearing life penalty. For a design factor of 1.28, what uniform-diameter shaft will support the 3.5-kN load 100 mm from the left bearing without penalty? Use $E = 207$ GPa.

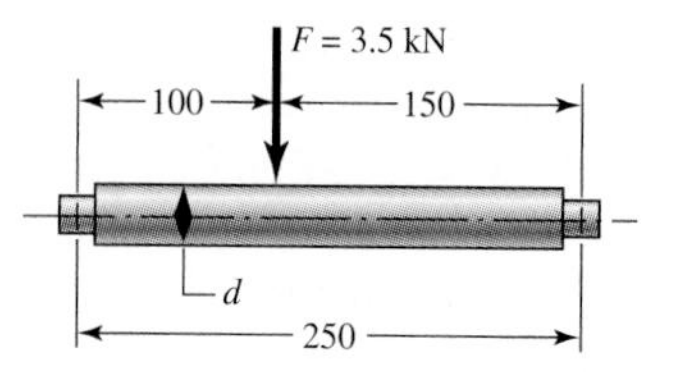

Problem 4–21
Dimensions in millimeters.

ANALYSIS

4–22 Determine the maximum deflection of the shaft of Prob. 4–21.

4–23 See Prob. 4–20 and the accompanying figure. The loads and dimensions are $F_1 = 800$ lb, $F_2 = 600$ lb, $a_1 = 4$ in, $b_1 = 6$ in, and $a_2 = 7$ in. Find the uniform shaft diameter necessary to limit the slope at the bearings to 0.001 in/in. Use a design factor of $n_d = 1.5$ and $E = 29.8$ Mpsi.

DESIGN

4–24 Shown in the figure is a uniform-diameter shaft with bearing shoulders at the ends; the shaft is subjected to a concentrated moment $M = 1200$ lb · in. The shaft is of carbon steel and has $a = 5$ in and $l = 9$ in. The slope at the ends must be limited to 0.002 rad. Find a suitable diameter d.

Problem 4–24

4–25 Rolling mills are used to reduce sheet metal cross sections and to process other sheet materials. Rolling devices should provide uniform pressure on the material passing between them. The "window" between them, shown in the figure, should be of constant height. Since the forces can be large, the rollers deflect as the material is rolled. How should a roll be crowned so that after deflection the pressure is uniform and the roller contact surfaces are parallel?

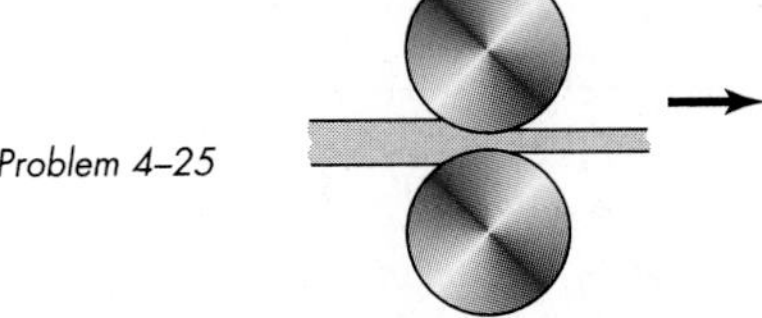

Problem 4–25

4–26 A bar in tension has a circular cross section and includes a conical portion of length l, as shown. The task is to find the spring rate of the entire bar. Equation (4–4) is useful for the outer portions of diameters d_1 and d_2, but a new relation must be derived for the tapered section. If α is the apex half-angle, as shown, show that the spring rate of the tapered portion of the shaft is

$$k = \frac{EA_1}{l}\left(1 + \frac{2l}{d_1}\tan\alpha\right)$$

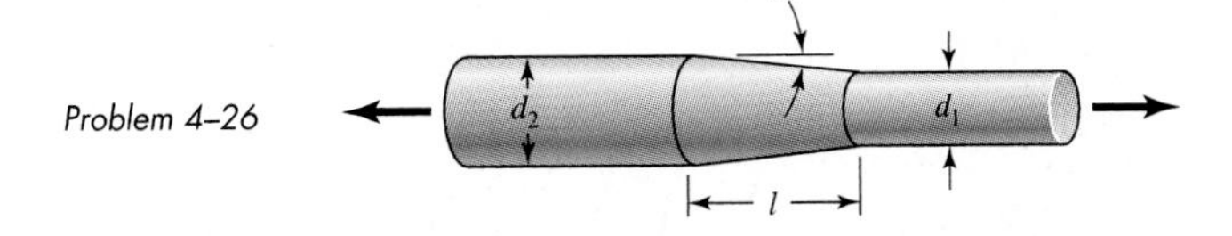

Problem 4–26

ANALYSIS

4–27 Find expressions for the maximum values of the spring force and deflection y of the impact system shown in the figure. Can you think of a realistic application for this model?

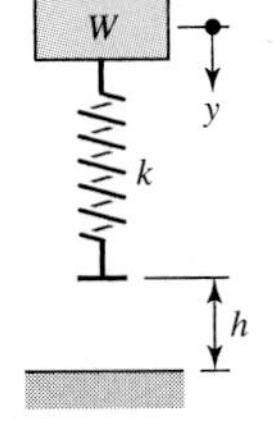

Problem 4–27

4–28 As shown in the figure, the weight W_1 strikes W_2 from a height h. Find the maximum values of the spring force and the deflection of W_2. Name an actual system for which this model might be used.

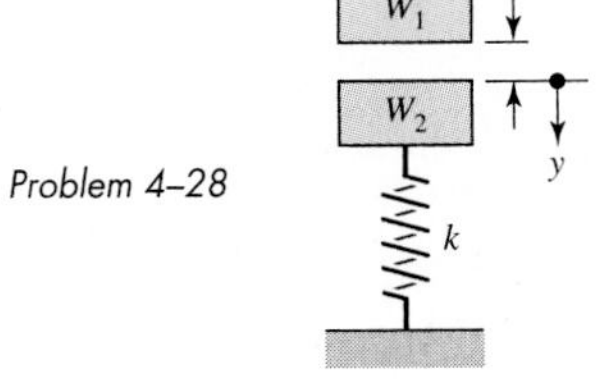

Problem 4–28

4–29 Part a of the figure shows a weight W mounted between two springs. If the free end of spring k_1 is suddenly displaced through the distance $x = a$, as shown in part b, what would be the maximum displacement y of the weight?

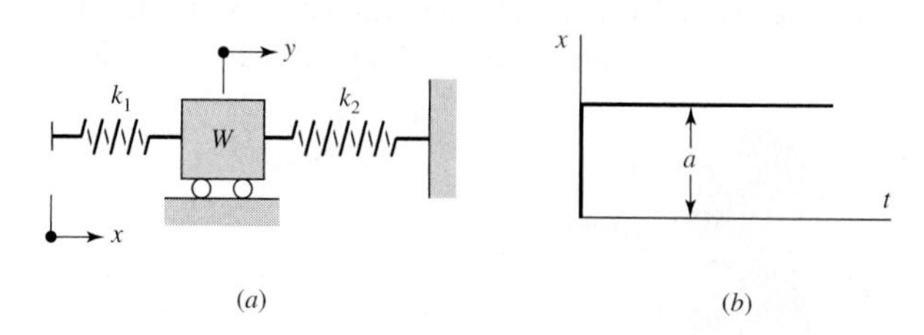

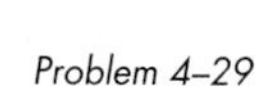
Problem 4–29

ANALYSIS

4–30 In a uniform-cross-section beam with simple supports at the ends loaded by a single concentrated load, the location of the maximum deflection will never exceed $x = l/\sqrt{3} = 0.577l$ regardless of the location of the load along the beam. The importance of this is that you can always get a quick estimate of y_{max} by using $x = l/2$. Prove this statement.

ANALYSIS

4–31 Use Castigliano's theorem to verify the maximum deflection for the uniformly loaded beam of Appendix Table E–9–7. Neglect shear.

ANALYSIS

4–32 The rectangular member OAB, shown in the figure, is held horizontal by the round hooked bar AC. The modulus of elasticity of both parts is 10 Mpsi. Use Castigliano's theorem to find the deflection at B due to a force $F = 80$ lb.

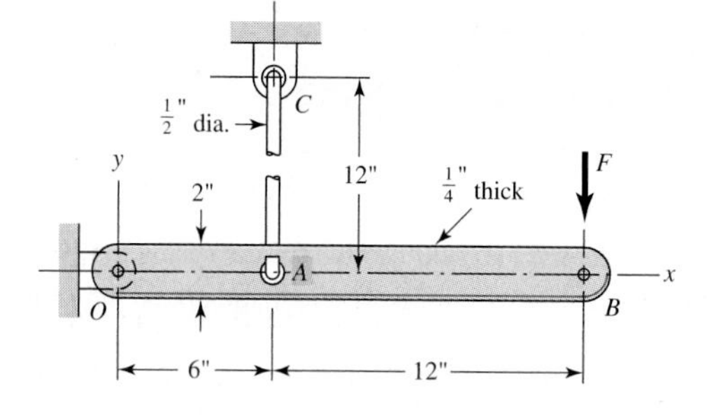

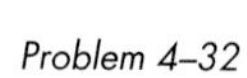
Problem 4–32

ANALYSIS

4–33 The figure illustrates a torsion-bar spring OA having a diameter $d = 12$ mm. The actuating cantilever AB also has $d = 12$ mm. Both parts are of carbon steel. Use Castigliano's theorem and find the spring rate k corresponding to a force F acting at B.

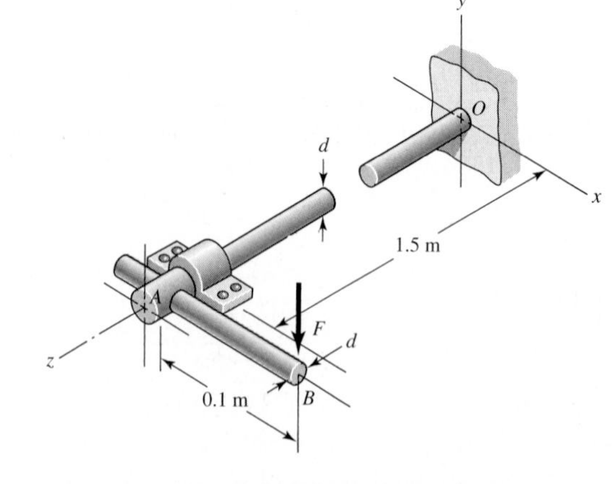

Problem 4–33

4–34 A cable is made using a 16-gauge (0.0625-in) steel wire and three strands of 12-gauge (0.0801-in) copper wire. Find the stress in each wire if the cable is subjected to a tension of 250 lb.

ANALYSIS

4–35 The figure shows a pressure cylinder of diameter 4 in which uses six SAE grade 5 bolts having a grip of 12 in. These bolts have a proof strength (see Chap. 8) of 85 kpsi for this size of bolt. Suppose the bolts are tightened to 90 percent of this strength in accordance with some recommendations.

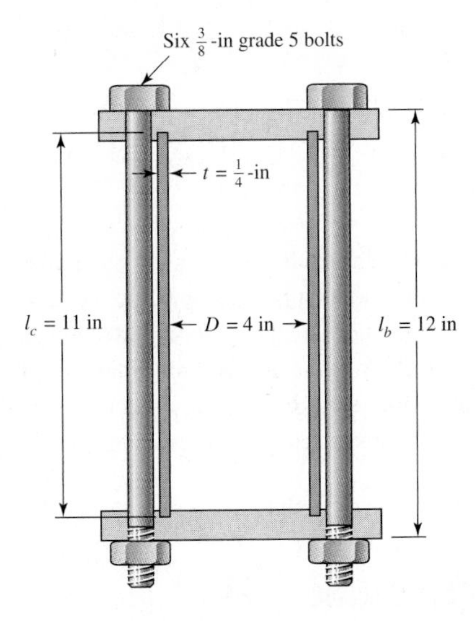

Problem 4–35

(*a*) Find the tensile stress in the bolts and the compressive stress in the cylinder walls.

(*b*) Repeat part (*a*), but assume now that a fluid under a pressure of 600 psi is introduced into the cylinder.

ANALYSIS

4–36 A torsion bar of length L consists of a round core of stiffness $(GJ)_c$ and a shell of stiffness $(GJ)_s$. If a torque T is applied to this composite bar, what percentage of the total torque is carried by the shell?

ANALYSIS

4–37 The figure shows a $\frac{3}{8}$- by $1\frac{1}{2}$-in rectangular steel bar welded to fixed supports at each end. The bar is axially loaded by the forces $F_A = 10$ kip and $F_B = 5$ kip acting on pins at A and B. Assuming that the bar will not buckle laterally, find the reactions at the fixed supports.

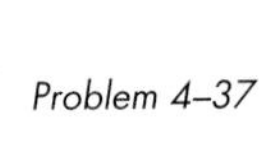
Problem 4–37

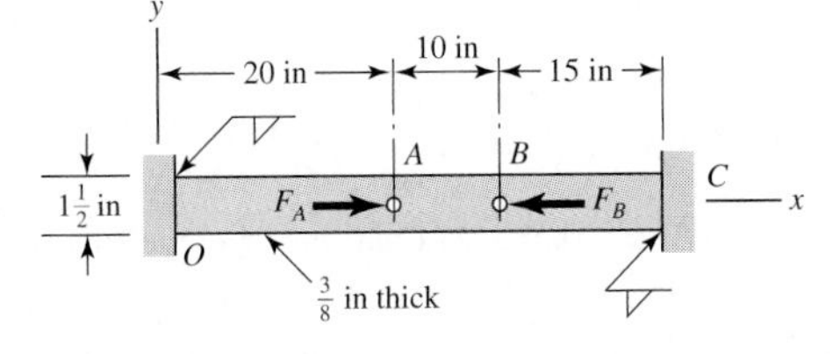

ANALYSIS

4–38 A rectangular aluminum bar $\frac{1}{2}$ in thick and 2 in wide is welded to fixed supports at the ends, and the bar supports a load $W = 800$ lb, acting through a pin as shown. Find the reactions at the supports.

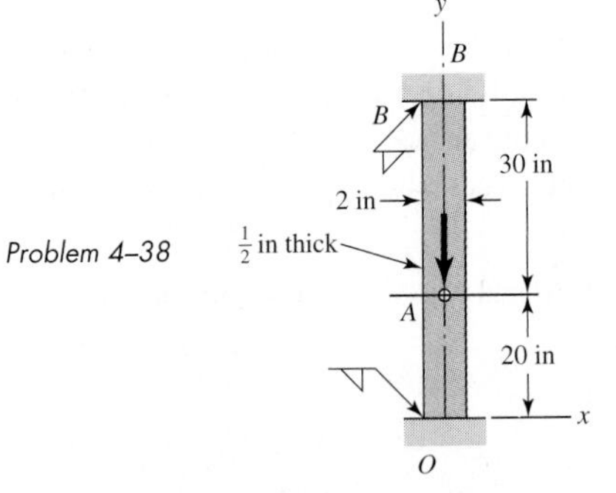

Problem 4–38

4–39 The steel shaft shown in the figure is subjected to a torque T applied at point A. Find the torque reactions at O and B.

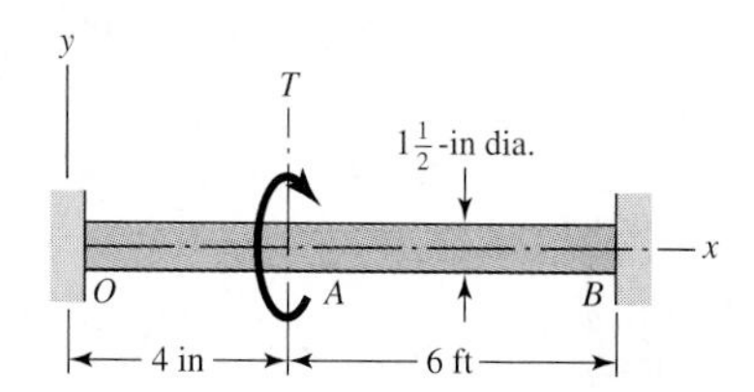

Problem 4–39

ANALYSIS

4–40 In testing the wear life of gear teeth, the gears are assembled using a pretorsion. In this way, a large torque can exist even though the power input to the tester is small. The arrangement shown in the figure uses this principle. Note the symbol used to indicate the location of the shaft bearings used in the figure. Gears A, B, and C are assembled first, and then gear C is held fixed. Gear D is assembled and meshed with gear C by twisting it through an angle of 4° to provide the pretorsion. Find the maximum shear stress in each shaft resulting from this preload.

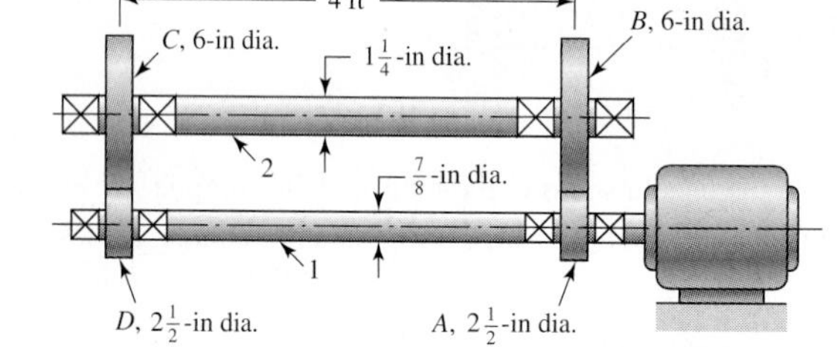

Problem 4–40

4–41 Examine the expression for the deflection of the cantilever beam, end-loaded, shown in Appendix Table E–9–1 for some intermediate point, $x = a$, as

$$y\Big|_{x=a} = \frac{F_1 a^2}{6EI}(a - 3l)$$

In Table E–9–2, for a cantilever with intermediate load, the deflection at the end is

$$y\Big|_{x=l} = \frac{F_2 a^2}{6EI}(a - 3l)$$

These expressions are remarkably similar and become identical when $F_1 = F_2 = 1$. In other words, the deflection at $x = a$ (station 1) due to a unit load at $x = l$ (station 2) is the same as the deflection at station 2 due to a unit load at station 1. Prove that this is true generally for an elastic body even when the lines of action of the loads are not parallel. This is known as a special case of *Maxwell's reciprocal theorem.* (*Hint:* Consider the potential energy of strain when the body is loaded by two forces in either order of application.)

4–42 A steel shaft of uniform 2-in diameter has a bearing span l of 23 in and an overhang of 7 in on which a coupling is to be mounted. A gear is to be attached 9 in to the right of the left bearing and will carry a radial load of 400 lb. We require an estimate of the bending deflection at the coupling. Appendix Table E–9–6 is available, but we can't be sure of how to expand the equation to predict the deflection at the coupling.

(*a*) Show how Appendix Table E–9–10 and Maxwell's theorem (see Prob. 4–41) can be used to obtain the needed estimate.

(*b*) Check your work by finding the slope at the right bearing and extending it to the coupling location.

4–43 A thin ring is loaded by two equal and opposite forces F in part a of the figure. A free-body diagram of one quadrant is shown in part b. This is a statically indeterminate problem, because the moment M_A cannot be found by statics. We wish to find the maximum bending moment in the ring due to the forces F. Assume that the radius of the ring is large so that Eq. (4–34) can be used.

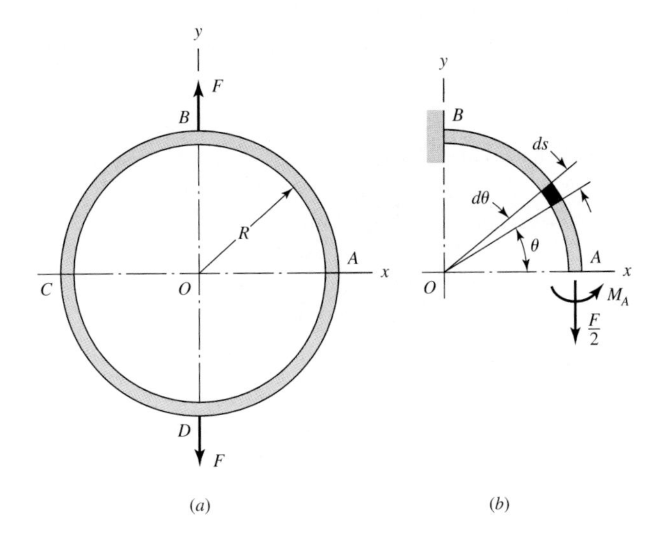

Problem 4–43

ANALYSIS

4–44 Find the decrease in the diameter of the ring of Prob. 4–43 due to the forces F and along the y axis.

4–45 A cast-iron piston ring has a mean diameter of 81 mm, a radial height $h = 6$ mm, and a thickness $b = 4$ mm. The ring is assembled using an expansion tool which separates the split ends a distance δ by applying a force F as shown. Use Castigliano's theorem and determine the deflection δ as a function of F. Use $E = 131$ GPa and assume Eq. (4–34) applies.

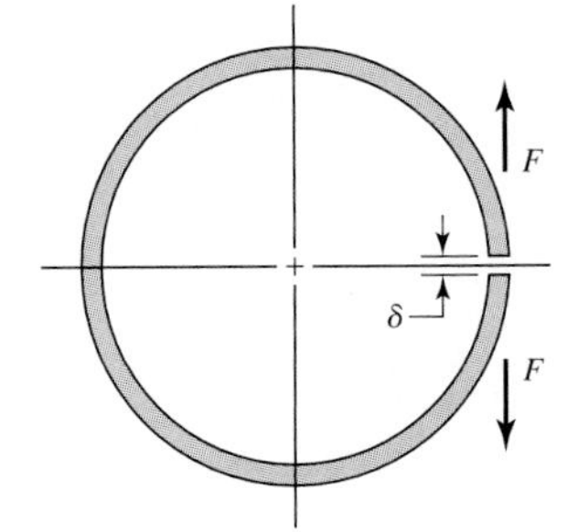

Problem 4–45

4–46 A round tubular column has outside and inside diameters of D and d, respectively, and a diametral ratio of $K = d/D$. Show that buckling will occur when the outside diameter is

$$D = \left[\frac{64 P_{cr} l^2}{\pi^3 C E \left(1 - K^4\right)}\right]^{1/4}$$

4–47 For the conditions of Prob. 4–46, show that buckling according to the parabolic formula will occur when the outside diameter is

$$D = 2\left[\frac{P_{cr}}{\pi S_y (1-K^2)} + \frac{S_y l^2}{\pi^2 C E (1+K^2)}\right]^{1/2}$$

ANALYSIS

4–48 Link 2, shown in the figure, is 1 in wide, has $\frac{1}{2}$-in-diameter bearings at the ends, and is cut from low-carbon steel bar stock having a minimum yield strength of 24 kpsi. The end-condition constants are $C = 1$ and $C = 1.2$ for buckling in and out of the plane of the drawing, respectively.

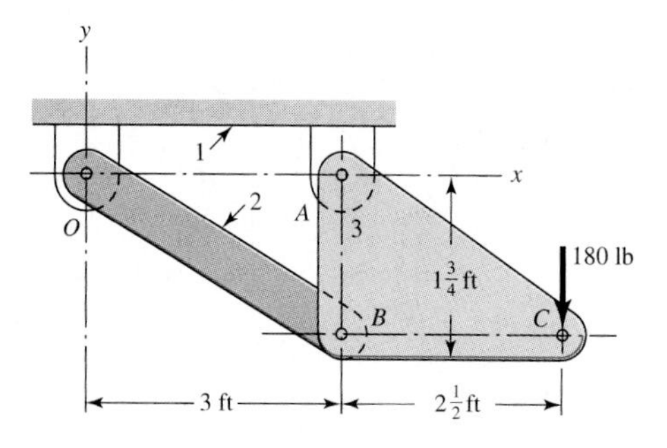

Problem 4–48

(*a*) Using a design factor $n_d = 5$, find a suitable thickness for the link.

(*b*) Are the bearing stresses at O and B of any significance?

DESIGN

4–49 Link 3, shown schematically in the figure, acts as a brace to support the 1.2-kN load. For buckling in the plane of the figure, the link may be regarded as pinned at both ends. For out-of-plane buckling, the ends are fixed. Select a suitable material and a method of manufacture, such as forging, casting, stamping, or machining, for casual applications of the brace in oil-field machinery. Specify the dimensions of the cross section as well as the ends so as to obtain a strong, safe, well-made, and economical brace.

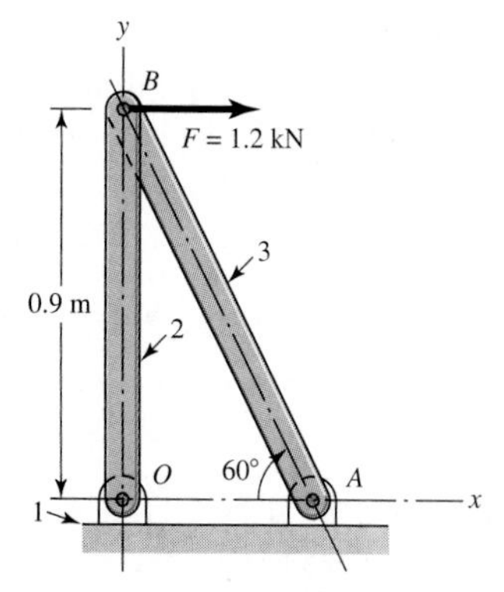

Problem 4–49

4–50 The hydraulic cylinder shown in the figure has a 3-in bore and is to operate at a pressure of 800 psi. With the clevis mount shown, the piston rod should be sized as a column with both ends rounded for any plane of buckling. The rod is to be made of forged AISI 1030 steel without further heat treatment.

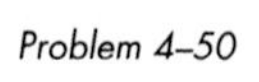
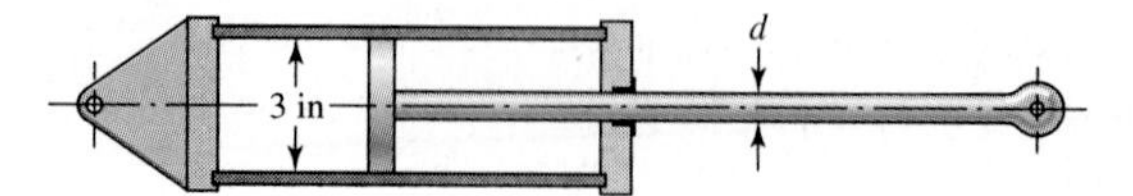

Problem 4–50

(a) Use a design factor $n_d = 3$ and select a preferred size for the rod diameter if the column length is 60 in.

(b) Repeat part (a) but for a column length of 18 in.

(c) What factor of safety actually results for each of the cases above?

4–51 The figure shows a schematic drawing of a vehicular jack which is to be designed to support a maximum weight of 400 kg based on the use of a design factor $n_d = 2.50$. The opposite-handed threads on the two ends of the screw are cut to allow the link angle θ to vary from 15 to 70°. The links are to be machined from AISI 1020 hot-rolled steel bars with a minimum yield strength of 380 MPa. Each of the four links is to consist of two bars, one on each side of the central bearings. The bars are to be 300 mm long and have a bar width of 25 mm. The pinned ends are to be designed to secure an end-condition constant of at least $C = 1.4$ for out-of-plane buckling. Find a suitable preferred thickness and the resulting factor of safety for this thickness.

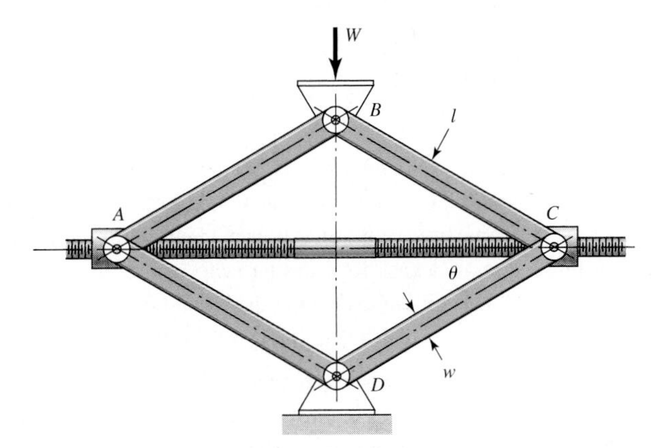

Problem 4–51

4–52 If drawn, a figure for this problem would resemble that for Prob. 4–35. A strut that is a hollow right circular cylinder has an inside diameter of 4 in and a wall thickness of $\frac{3}{8}$ in and is compressed between two circular end plates held by four bolts equally spaced on a bolt circle of 5.68-in diameter. All four bolts are hand-tightened, and then bolt A is tightened to a tension of 2000 lb and bolt C, diagonally opposite, is tightened to a tension of 10 000 lb. The strut axis of symmetry is coincident with the center of the bolt circles. Find the maximum compressive load, the eccentricity of loading, and the largest compressive stress in the strut.

4–53 The secant column equation incorporates the bending deflection, which has the effect of magnifying the eccentricity e at midspan. What is the greatest value of the slenderness ratio for which this influence will not exceed 1 percent of e?

4–54 Design link CD of the hand-operated toggle press shown in the figure. Specify the cross-section dimensions, the bearing size and rod-end dimensions, the material, and the method of processing.

4–55 The method of Sec. 4–6 is so powerful that it should be programmed for your use. Use it to build a table such as Table 4–1. Prepare a program and test it on problems that you can solve analytically. The program will be particularly useful for shafts with rolling contact bearings where there is an active constraint on a shaft slope at a bearing location.

4–56 When the program in Prob. 4–55 is working, there is a tendency to adjust shaft diameters to meet a slope constraint in a hit or miss manner because the computing is easy. The prospect of performing such iteration with a hand-held calculator, paper, and pencil is not attractive. Show that the new diameter at every shaft step is given by

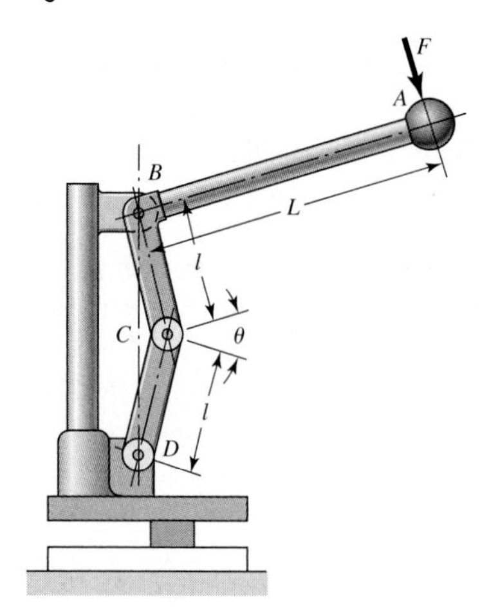

Problem 4–54
$L = 12$ in, $l = 4$ in, $\theta_{min} = 0°$.

$$(d_{\text{new}}/d_{\text{old}}) = \left|\frac{n\, dy/ds}{\xi}\right|^{1/4}$$

where n is the design factor and ξ is the upper bound of bearing slope that does not incur a life penalty in the bearing. Apply this equation at both bearings and choose the larger $d_{\text{new}}/d_{\text{old}}$ and multiple *all* diameters by this ratio. In fact, after your program of Prob. 4–55 displays the table for your inspection, it can ask for a design factor and ξ, make all the adjustments in diameters, and present the table again. In examining the second table you will find, in the slope column, a slope at one of the bearings which is $|\xi/n|$, and the other will be less.

4–57 A tabular presentation from a computer program analyzing a constant 1-in-diameter shaft shows a slope of −0.0158 rad at the left bearing location and 0.0181 rad at the right bearing location. What diameter shaft will meet a bearing slope upper limit of $\xi = 0.001$ rad with a design factor of $n = 1$? What is the shaft diameter with $n = 2$?

4–58 A shaft has four diameters: 0.750 in, 0.875 in, 1.125 in, and 0.750 in. A tabular display by the program of Prob. 4–55 shows a slope of 0.010 284 at the left bearing location and a slope of 0.007 712 at the right bearing location. For a slope limit at a bearing of 0.001 rad without a life penalty and a design factor of $n = 2$, show that the diameters are 1.597 in, 1.863 in, 2.396 in, and 1.597 in.

4–59 The result in Prob. 4–19 is so useful that it deserves an interactive computer program implementing it. The program can give the diameter of a uniform-diameter shaft for the loading that is bearing-friendly. This diameter identifies the constraint-meeting "heft" of the shaft needed. As a designer creates steps in the shaft for the assembly of gears and pulleys, and providing shoulders for locating, he or she knows that reducing diameter (in steps) near the ends has to be compensated by increasing diameters in the central region. The program creates an informed starting point and a useful perspective. Write and test such a program.

4–60 With your program from Prob. 4–59, solve a problem such as Prob. 4–20 by analyzing the horizontal plane situation and the vertical plane situation. Note that the slopes at any location θ_H and θ_V combine vectorially at a shaft station, that is, $\theta = \sqrt{\theta_H^2 + \theta_V^2}$. Change the diameter using the approach employed in Prob. 4–56. The computer program developed in Prob. 4–59 can be made three-dimensional using this idea. In fact, the conditions in the vertical plane, for example, can be made to be the superposition of a series of loads and locations in that plane. In this way a general force-loaded shaft can be analyzed for deflection and slope.

4–61 A $2 \times \frac{1}{4}$-in cold-drawn steel tubular column is 48 in long, has a 1.374 in^2 cross-sectional area, a radius of gyration of 0.625 in, distance to the outer fiber from the tube centerline of 1 in, Young's modulus of $30(10^6)$ psi, and a mean eccentricity of 0.125 in. The column is pin-ended. From the secant equation for this situation for the compressive stress σ_c, show that when eccentricity $\mathbf{e}$ is the dominant stochastic variable, the equation for $\boldsymbol{\sigma}_c$ can be written as $\boldsymbol{\sigma}_c = a(1 + b\mathbf{e})$.

(*a*) Find expressions for the mean, standard deviation, and coefficient of variation of the compressive stress $\boldsymbol{\sigma}_c$.

(*b*) For a load $P = 44\,680$ lbf, evaluate a and b.

(*c*) If $\mathbf{e} \sim \psi(0.125, 0.0125)$, express $\boldsymbol{\sigma}_c$.

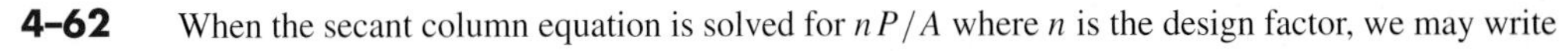

ANALYSIS

4–62 When the secant column equation is solved for nP/A where n is the design factor, we may write

$$\frac{nP}{A} = \frac{1}{1 + \dfrac{ec}{k^2}\sec\left(\dfrac{l}{2k}\sqrt{\dfrac{nP}{AE}}\right)}$$

Finding a numerical value for nP/A requires an iteration method. Use the method of Prob. 1–40. A finite-difference approach is used to evaluate $F'(x) = F'(nP/A)$. The parameter k' of Prob. 1–30 is written as

$$k' = \frac{1}{1 - \dfrac{F(1.001nP/A) - F(nP/A)}{0.001nP/A}}$$

and the iteration equation is

$$\left(\frac{nP}{A}\right)_{i+1} = \left[(1 - k')\left(\frac{nP}{A}\right) + k'F\left(\frac{nP}{A}\right)\right]_i$$

Begin with an initial estimate of $(nP/A)_0$ of 20 000 and show that the successive approximations to nP/A are

20 000
34 004
32 548
32 518
32 518

If you programmed Prob. 1–30, use your program. If not, use your hand-held calculator. Vary your estimate of $(nP/A)_0$ to demonstrate range of convergence and robustness of the strategy of the method. If one were to perform a simulation with $\mathbf{S}_y$ and $\mathbf{e}$ as stochastic variables, the foregoing procedure would be embedded in the step for creating a single instance of $\mathbf{S}_y$, $\mathbf{e}$, and the harder one, (nP/A).

ANALYSIS

4–63 When an initially straight beam sags under transverse loading, the ends contract because the neutral surface of zero strain neither extends nor contracts. The length of the deflected neutral surface is the same as the original beam length l. Consider a segment of the initially straight beam Δs. After bending, the x-direction component is shorter than Δs, namely, Δx. The contraction is $\Delta s - \Delta x$, and these summed for the entire beam gives the end contraction λ. Show that

$$\lambda = \frac{1}{2}\int_0^l \left(\frac{dy}{dx}\right)^2 dx$$

ANALYSIS

4–64 A neutral surface of a simply supported beam has the shape

$$y = -\frac{4ax}{l^2}(l - x)$$

where a is the midspan deflection magnitude. What is the end contraction from the straight original condition?

ANALYSIS

4–65 The neutral surface of a simply supported beam has the shape

$$y = a \sin \frac{\pi x}{l}$$

What is the end contraction from the straight original condition?

ANALYSIS

4–66 One manufacturer of snubbers gives the following equation for the kinetic energy of a translating body:

$$E = 0.2WV^2$$

using V ft/s, W lbf, and E in · lbf. Is this an approximation? Do you think it is robust?

PART 2

Failure Prevention

5 Materials

5–1 Static Strength 256
5–2 Plastic Deformation 261
5–3 Strength and Cold Work 265
5–4 Hardness 268
5–5 Impact Properties 269
5–6 Temperature Effects 271
5–7 Numbering Systems 272
5–8 Sand Casting 274
5–9 Shell Molding 274
5–10 Investment Casting 275
5–11 Powder-Metallurgy Process 275
5–12 Hot-Working Processes 275
5–13 Cold-Working Processes 276
5–14 The Heat Treatment of Steel 277
5–15 Alloy Steels 279
5–16 Corrosion-Resistant Steels 280
5–17 Casting Materials 281
5–18 Nonferrous Metals 283
5–19 Plastics 285
5–20 Notch Sensitivity 287
5–21 Introduction to Fracture Mechanics 288
5–22 Stress-Corrosion Cracking 303
5–23 Quantitative Estimation of Properties of Cold-Worked Metals 303
5–24 Quantitative Estimation of Properties of Heat-Treated Steels 307

In Chaps. 3 and 4, methods for estimating stresses in machine members and deflections of machine members were presented. For deflections and stability evaluations the elastic properties of the material are involved. The stress at a critical location in a machine member has to be compared with the strength at that location in the geometry and condition of use. This strength is a property found by tests and adjusted to the geometry and condition of use as necessary.

The selection of a material for a machine part or structural member is one of the decisions the designer is called upon to make. This decision is usually made before the dimensions of the part are established. After choosing the process of creating the desired geometry and the material (the two cannot be divorced), the designer can proportion the member so that loss of function can be avoided or the chance of loss of function can be held to an acceptable risk.

As important as stress and deflection are in the design of mechanical parts, the selection of a material is not always based upon these factors. Many parts have no loads on them whatever. Parts may be designed merely to fill up space or for aesthetic qualities. Members must frequently be designed to resist corrosion. Sometimes temperature effects are more important in design than stress and strain. So many other factors besides stress and strain may govern the design of parts that the designer must have the versatility that comes only with a broad background in materials and processes.

5–1 Static Strength

The standard tensile test is used to obtain a variety of characteristics and strengths that are used in design. Figure 5–1 illustrates a typical tension-test specimen and some of the dimensions that are often employed.[1] The original diameter d_0 and length of the gauge l_0, used to measure the strains, are recorded before the test is begun. The specimen is then mounted in the test machine and slowly loaded in tension while the load and strain are observed. At the conclusion of, or during, the test the results are plotted as a *stress-strain diagram* (Fig. 5–2).

Point P in Fig. 5–2 is called the *proportional limit*. This is the point at which the curve first begins to deviate from a straight line. No permanent set will be observable in the specimen if the load is removed at this point. Point E is called the *elastic limit*. Between P and E the diagram is not a perfectly straight line, even though the specimen is elastic. Thus Hooke's equation, which states that stress is proportional to strain, applies only up to the proportional limit.

During the tension test, many materials reach a point at which the strain begins to increase very rapidly without a corresponding increase in stress. This point is called the *yield point*. Not all materials have an obvious yield point. For this reason, *yield strength* S_y is often defined by an *offset method* as shown in Fig. 5–2. Such a yield strength corresponds to a definite or stated amount of permanent set, usually 0.2 or 0.5 percent of the original gauge length, although 0.01, 0.1, and 0.5 percent are sometimes used.

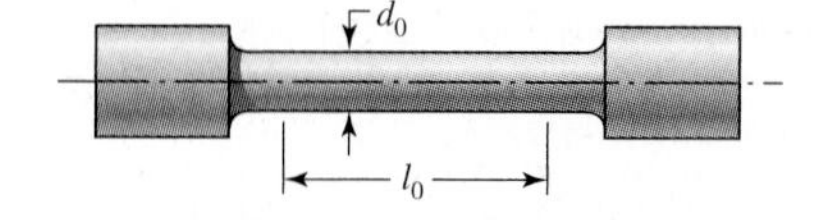

Figure 5–1

A typical tension-test specimen. Some of the standard dimensions used for d_0 are 2.5, 6.25, and 12.5 mm and 0.5 in, but other sections and sizes are in use. Common gauge lengths ℓ_0 used are 10, 25, and 50 mm and 1 and 2 in.

1. See ASTM standards E8 and E-8 m for standard dimensions.

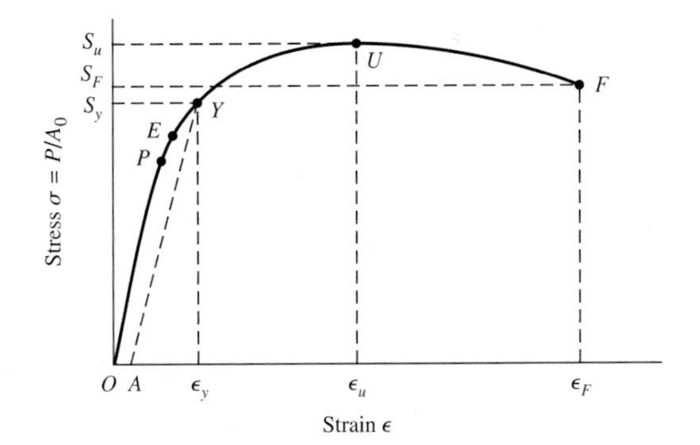

Figure 5–2

Stress-strain diagram obtained from the standard tensile test of a ductile material. P marks the proportional limit; E, the elastic limit; Y, the offset-yield strength as defined by offset strain OA; U, the maximum or ultimate strength; and F, the fracture strength.

The *ultimate*, or *tensile*, *strength* S_u or S_{ut} corresponds to point U in Fig. 5–2 and is the maximum stress reached on the stress-strain diagram.[2] Some materials exhibit a downward trend after the maximum stress is reached. These fracture at point F on the diagram in Fig. 5–2. Others, such as some of the cast irons and high-strength steels, fracture while the stress-strain locus is still rising.

There is some subtlety in ideas here, which should be pondered carefully before continuing. Figure 5–2 depicts the result of a single tension test (one specimen, now broken). It is common for engineers to consider these important stress values (at points P, E, Y, U, and F) as *properties* and to denote them as *strengths* with a special notation, uppercase S, in lieu of lowercase sigma σ, with subscripts added: S_P for proportional limit, S_y for offset yield strength, S_u for ultimate tensile strength (S_{ut} or S_{uc} if tensile or compressive sense is important).

All this is premature, since Fig. 5–2 refers to *a single specimen*. If there were 10 nominally identical specimens, each would exhibit *different* strengths. What one is observing is a consistent pattern of distribution, and, as observed in Chap. 2, the description has to be stochastic (statistical). The ideal of strength, and property, is distributional. The table below is the histographic report of 1000 tensile tests on a 1020 steel from a single heat, seeking the ultimate tensile strength S_{ut}:

Class frequency f_i	2	18	23	31	83	109	13	157	139	130	82	49	28	11	4	2
Class midpoint x_i, kpsi	56.5	57.5	58.5	59.5	60.5	61.5	62.6	63.5	64.5	65.5	66.5	67.5	68.5	69.5	70.5	71.5

Now $\Sigma x_i f_i = 63\,625$ and $\Sigma x_i^2 f_i = 4\,054\,864$, so μ_x and σ_x are

$$\mu_x = \frac{x_i f_i}{N} = \frac{63\,625}{1000} = 63.625 \text{ kpsi}$$

$$\sigma_x = \sqrt{\frac{\Sigma x_i^2 f_i - (\Sigma x_i f_i)^2/N}{N-1}} = \sqrt{\frac{4\,054\,864 - 63\,625^2/1000}{1000-1}} = 2.594 \text{ kpsi}$$

$$C_x = \frac{2.5942}{63.625} = 0.040\,773$$

For a lognormal, the companion normal as the parameters μ_y and σ_y as given by

2. Usage varies. For a long time engineers used the term *ultimate strength* hence the subscript u in S_u or S_{ut}, however in material science/metallurgy the term *tensile strength* is used.

Eqs. (2–37) and (2–38):

$$\mu_y = \ln 63.625 - \ln\sqrt{1 + 0.040\,773^2} = 4.152$$

$$\sigma_y = \sqrt{\ln(1 + 0.040\,773^2)} = 0.0408$$

The density function of the ultimate tensile strength x is given by Eq. (2–36) as

$$g(x) = \frac{1}{0.0408x\sqrt{2\pi}} \exp\left[-\frac{1}{2}\left(\frac{\ln x - 4.152}{0.0408}\right)^2\right]$$

The description of the strength S_{ut} is expressed in terms of its statistical parameters and its distribution type. In this case $\mathbf{S}_{ut} \sim LN(63.625, 2.5942)$ kpsi.

Note that the test program has described 1020 property $\mathbf{S}_{ut}$ for only one heat of one supplier. Testing is an involved and expensive process. Tables of properties are often prepared to be helpful to other persons. A stochastic quantity is described by its mean, standard deviation, and distribution type. Many tables display a single number, which is often the mean, or some percentile, such as the 99th percentile. Always read the footnotes to the table. If in a single-entry table no qualification is made, the table is not useful.

Since it is no surprise that useful descriptions of a property are stochastic in nature, engineers, when ordering property tests, should couch the instructions so the data generated are sufficient to observe the statistical parameters and to identify the distributional characteristic (or its robust characterization). The tensile test program on 1000 specimens of 1020 steel is a large one. Faced with putting something in a table of ultimate tensile strengths, if constrained to a single number, what would it be, and just how would your footnote read?

To express the strain relations for the stress-strain test depicted in Fig. 5–2, let

l_0 = original gauge length

l_i = gauge length corresponding to load P_i

A_0 = original cross-sectional area

A_i = area of smallest cross section under load P_i

We can express the engineering (nominal) strain ϵ as

$$\epsilon = \frac{l_i - l_0}{l_0} \tag{5–1}$$

and the engineering (nominal) stress as

$$\sigma = \frac{P}{A_0}$$

As noted in Sec. 1–11, *strength*, as used in this book, is a built-in property of a material, or of a mechanical element, because of the selection of a particular material or process or both. The strength of a connecting rod at the critical location in the geometry and condition of use, for example, is the same no matter whether it is already an element in an operating machine or whether it is lying on a workbench awaiting assembly with other parts.

On the other hand, stress is something that occurs in a part, usually as a result of being assembled into a machine and loaded. However, stresses may be built into a part by the processing or handling. For example, shot peening produces a compressive *stress* in the outer surface of a part, and also improves the fatigue strength of the part.

In this book we wish to be very careful in distinguishing between *strength*, designated by S, and *stress*, designated by σ or τ.

In some books, the terms *proof stress* and *yield stress* are used. The term *proof stress* corresponds to point Y in Fig. 5–2 which, in this book, is called the *offset yield strength* S_y or, sometimes, simply the *yield strength*. In most standards the offset is usually specified; usual values are 0.2, 0.5, and 1.0 percent.

The term *yield stress* as used in some books is the stress corresponding to the yield point. Some materials have both an upper and a lower yield point.

The term *true stress* is used to indicate the result obtained when any load used in the tension test is divided by the *true* or *actual* cross-sectional area of the specimen. This means that the load and the cross-sectional area must be measured simultaneously during the test. If the specimen has necked, special care must be taken to measure the area at the smallest part.

In plotting the true stress-strain diagram it is customary to use a term called *true strain* or, sometimes, *logarithmic strain*. True strain is the sum of the incremental elongations divided by the current length of the filament, or

$$\varepsilon = \int_{l_0}^{l_i} \frac{dl}{l} = \ln \frac{l_i}{l_0} \tag{5–2}$$

where l_0 is the original gauge length and l_i is the gauge length corresponding to load P_i.

The most important characteristic of a true stress-strain diagram (Fig. 5–3) is that the true stress increases all the way to fracture. Thus, as shown in Fig. 5–3, the true fracture stress σ_F is greater than the true ultimate stress σ_u. Contrast this with Fig. 5–2, where the engineering fracture strength S_F is less than the engineering ultimate strength S_u.

Bridgman has pointed out that the true stress-strain diagram of Fig. 5–3 should be corrected because of the triaxial stress state that exists in the neck of the specimen.[3] He observes that the tension is greatest on the axis and smallest on the periphery and that the stress state consists of an axial tension uniform all the way across the neck, plus a hydrostatic tension, which is zero on the periphery and increases to a maximum value on the axis.

Bridgman's correction for the true stress during necking is particularly significant. Designating σ_C as the computed true stress, σ_{ACT} as the corrected or actual stress, R as the radius of the neck (Fig. 5–4), and D as the smallest neck diameter, the equation is

$$\sigma_{ACT} = \frac{\sigma_C}{\left(1 + \dfrac{4R}{D}\right)\left[\ln\left(1 + \dfrac{D}{4R}\right)\right]} \tag{5–3}$$

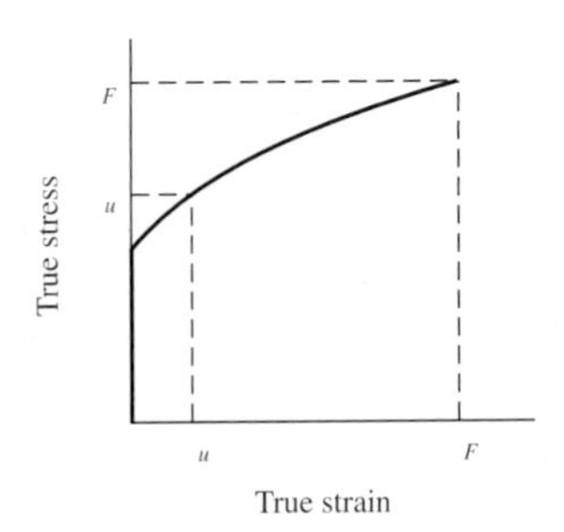

Figure 5–3

True stress–true strain diagram plotted using Cartesian coordinates.

3. P. W. Bridgman, "The Stress Distribution at the Neck of a Tension Specimen," *ASM*, vol. 32, 1944, p. 553.

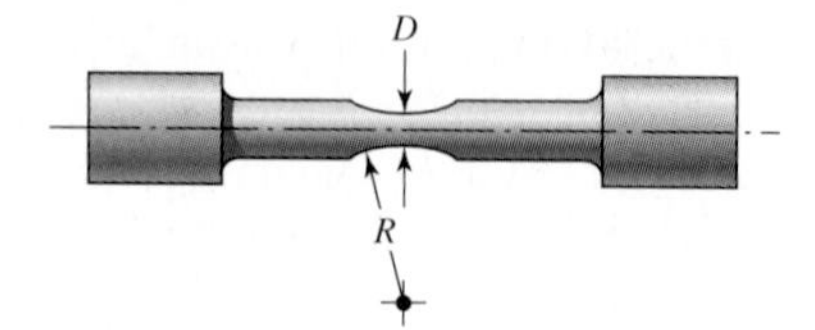

Figure 5–4

Tension specimen after necking. Radius of neck is R; diameter of smallest portion of neck is D.

When necking occurs, the engineering strain given by Eq. (5–1) will not be the same at all points within the gauge length. A more satisfactory relation can be obtained by using areas. Since the volume of material remains the same during the test, $A_0 l_0 = A_i l_i$. Consequently, $l_i = l_0(A_0/A_i)$. Substituting this value of l_i in Eq. (5–1) and canceling terms gives

$$\epsilon = \frac{A_0 - A_i}{A_i} \tag{5–4}$$

But see also Eq. (5–8).

Compression tests are more difficult to make, and the geometry of the test specimens differs from the geometry of those used in tension tests. The reason for this is that the specimen may buckle during testing or it may be difficult to get the stresses distributed evenly. Other difficulties occur because ductile materials will bulge after yielding. However, the results can be plotted on a stress-strain diagram, too, and the same strength definitions can be applied. For many materials the compressive strengths are about the same as the tensile strengths. When substantial differences occur, however, as is the case with the cast irons, the tensile and compressive strengths should be stated separately.

Torsional strengths are found by twisting bars and recording the torque and the twist angle. The results are then plotted as a *torque-twist diagram*. By using the equations in Chap. 3 for torsional stress, both the elastic limit and the *torsional yield strength* S_{sy} may be found. The maximum point on a torque-twist diagram, corresponding to point U on Fig. 5–2, is T_u. The equation

$$S_{su} = \frac{T_u r}{J} \tag{a}$$

where r is the radius of the bar and J is the polar second moment of area, defines the *modulus of rupture* for the torsion test. Note that the use of Eq. (a) implies that Hooke's law applies to this case. This is not true, because the outermost area of the bar is in a plastic state at the torque T_u. For this reason the quantity S_{su} is called the modulus of rupture. It is incorrect to call S_{su} the ultimate torsional strength.

All of the stresses and strengths defined by the stress-strain diagram of Fig. 5–2 and similar diagrams are specifically known as *engineering stresses* and *strengths* or *nominal stresses* and *strengths*. These are the values normally used in all engineering design. The adjectives *engineering* and *nominal* are used here to emphasize that the stresses are computed using the *original* or *unstressed cross-sectional area* of the specimen. In this book we shall use these modifiers only when we specifically wish to call attention to this distinction.

Sometimes the question arises, Is the compressive strength S_c a negative quantity? The answer is no. All strengths are treated as absolute quantities.

5–2 Plastic Deformation

The best current explanation of the relationships between stress and strain is by Datsko.[4] He describes the plastic region of the true stress–true strain diagram by the equation

$$\sigma = \sigma_0 \varepsilon^m \tag{5–5}$$

where σ = true stress

σ_0 = a strength coefficient, or strain-strengthening coefficient

ε = true plastic strain

m = strain-strengthening exponent

A graph of this equation is a straight line when plotted on log-log paper, as shown in Fig. 5–5. The graph contains three zones of interest: the elastic zone, on line AB, called type I behavior; the plastic zone on line Y_2C, defining type II behavior; and the intermediate zone.

According to Hooke's law, the equation of the elastic portion is

$$\sigma = E\varepsilon \tag{5–6}$$

where E is the modulus of elasticity. Taking the logarithm of both sides of Eq. (5–6) and recognizing the equation of a straight line, we get

$$\log\ \sigma = \log\ E + 1\ \log\ \varepsilon$$

and from Eq. (5–5) we obtain

$$\log\ \sigma = \log\ \sigma_0 + m\ \log\ \varepsilon$$

From this we conclude that the elastic portion of the line is the same for all materials, having a slope of unity and passing through the point ($\sigma = E, \varepsilon = 1$). We also see that the elastic portion has an exponent $m = 1$ and an intercept $\sigma_0 = E$ relative to the true stress-strain equation [Eq. (5–5)].

Figure 5–5

True stress–true strain diagram plotted on loglog paper. Since the values of ϵ are less than unity, their logarithms are negative. At point E, $\epsilon = 1$, $\log \epsilon = 0$, and the ordinate through E locates D and defines the logarithm of the constant σ_0 at F.

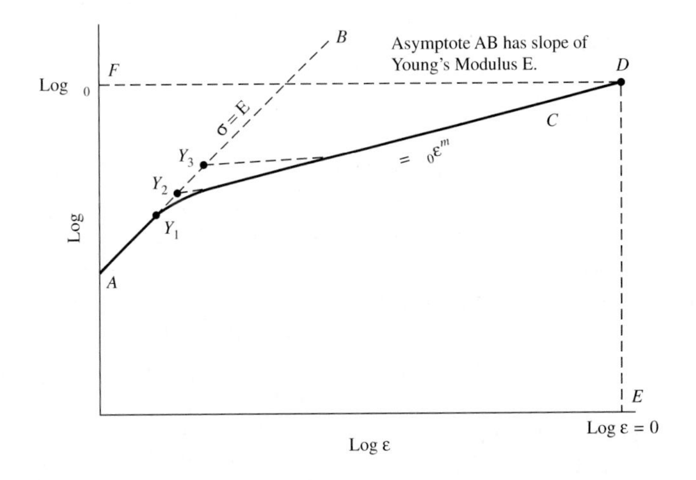

4. Joseph Datsko, "Solid Materials," Chap. 7 in Joseph E. Shigley and Charles R. Mischke (eds.), *Standard Handbook of Machine Design*, McGraw-Hill, New York, 1986. See also Joseph Datsko, "New Look at Material Strength," *Machine Design*, vol. 58, no. 3, Feb. 6, 1986, pp. 81–85.

The constant σ_0 in Eq. (5–5) is the true stress corresponding to a true strain of unity. This constant can be obtained by extending the plastic stress-strain line until it intersects an ordinate through $\varepsilon = 1$ ($\log\ \varepsilon = 0$). The height of this ordinate is $\log\ \sigma_0$, as measured parallel to the $\log\ \sigma$ axis of Fig. 5–5.

The shape of the elastic-plastic zone between the two straight lines varies from one material to another. The three possible yield points Y_1, Y_2, and Y_3 describe the various possibilities that might be observed. An extension of the plastic line would intersect the elastic line at Y_2 and describe an ideal material. Most engineering materials are said to *overyield* to Y_3, because they have a yield strength greater than the ideal value. The alloys of the steels, coppers, brasses, and nickels all have this characteristic. The point Y_1 describes what might be called *underyielding*. Only a few engineering materials have this characteristic, a fully annealed aluminum alloy being one of them.

The relationship between logarithmic strain and engineering strain can be obtained by rearranging Eq. (5–1) to read

$$\epsilon = \frac{l_i - l_0}{l_0} = \frac{l_i}{l_0} - 1 \tag{5–7}$$

Then we see that

$$\frac{l_i}{l_0} = \epsilon + 1$$

and so, from Eq. (5–2), we get

$$\boxed{\varepsilon = \ln\ (\epsilon + 1)} \tag{5–8}$$

A similar relationship can be derived between the true stress and the engineering stress. For convenience in this derivation, let σ = true stress and s = engineering stress. Then for the purpose of relating σ, s, ϵ and ε we write

$$\sigma = \frac{F_i}{A_i} \quad \text{and} \quad s = \frac{F_i}{A_0}$$

Now, since

$$A_i = A_0 \frac{l_0}{l_i}$$

then

$$\sigma = \frac{F_i}{A_0(l_0/l_i)} = s\frac{l_i}{l_0} = s(\epsilon + 1)$$

But $\epsilon + 1 = \ln^{-1} \varepsilon = \exp(\varepsilon)$. Therefore

$$\boxed{\sigma = s\ \exp(\varepsilon)} \tag{5–9}$$

Thus the engineering stress is related to the true stress through

$$\boxed{s = \sigma \exp(-\varepsilon)} \tag{5–10}$$

or, with Eq. (5–5),

$$s = \sigma_0 \varepsilon^m \exp(-\varepsilon) \tag{5–11}$$

Equations (5–7) and (5–8) are used to plot the elastic portion of the diagram. Thus, for this portion of the experiment, the data are acquired in the conventional manner, using an extensometer to obtain the elongation. The extensometer is not used for the plastic portion of the true stress-strain diagram, because the average strain is no longer useful. Its use would also expose an expensive precision instrument to damage when the specimen fractures.

The approach for the plastic region consists in measuring the area of the specimen, being particularly careful to obtain this value at the smallest cross section between the gauge points. Sometimes it is necessary to measure "diameters" of the specimen in two directions, perpendicular to each other, in case the cross section becomes oval-shaped.

From Eq. (5–4), we have

$$\frac{A_0}{A_i} = \epsilon + 1$$

Thus, from Eq. (5–8), we find the logarithmic strain in terms of areas to be

$$\varepsilon = \ln \frac{A_0}{A_i} \tag{5–12}$$

If data are to be corrected in the necking region, then the necking radius R should also be measured and the stress corrected using the Bridgman equation [Eq. (5–3)].

The exponent m represents the slope of the plastic line, as we have seen. This slope is easily obtained after the plastic line has been drawn through the points in the plastic region of the diagram. Another, and easier, method of obtaining the exponent is possible for materials having an ultimate strength greater than the nominal stress at fracture. For these materials the exponent is the same as the logarithmic strain corresponding to the ultimate strength. The proof is as follows:

$$P_i = \sigma A_i = \sigma_0 A_i (\varepsilon)^m \tag{a}$$

where we have used Eq. (5–5). Now, from Eq. (5–12), we have

$$A_i = A_0 \exp(-\varepsilon) \tag{b}$$

and so Eq. (a) becomes

$$P_i \sigma_0 A_0 \varepsilon^m \exp(-\varepsilon) \tag{c}$$

The maximum ordinate on the load-deformation diagram, or nominal stress-strain diagram, for some materials is a stationary point (exhibiting zero slope). For such materials, the derivative of the load P_i with respect to the logarithmic strain can be made to vanish:

$$\frac{dP_i}{d\varepsilon} = \sigma_0 A_0 \frac{d}{d\varepsilon}\left[\varepsilon^m \exp(-\varepsilon)\right] = \sigma_0 A_0 \left[m\varepsilon^{m-1} \exp(-\varepsilon) - \varepsilon^m \exp(-\varepsilon)\right] = 0 \tag{d}$$

Solving Eq. (d) gives $\varepsilon = m$. This corresponds to ultimate load, so

$$m = \varepsilon_u \tag{5–13}$$

Note again that this relation is valid only if the load-deformation exhibits a stationary point (a place of zero slope).

EXAMPLE 5–1

The first three columns in Table 5–1 list the results obtained from a tensile test of annealed A-40 titanium.

(a) Plot the engineering and the true stress-strain diagrams.
(b) Find the modulus of elasticity, the yield strength, and the ultimate strength.
(c) Find the plastic strain-strengthening coefficient and exponent.

Table 5–1

Results of a Tensile Test of Annealed A-40 Titanium as Reported by Datsko.* (Specimen size is $d_0 = 0.505$ in, $l_0 = 2$ in)

Observed Test Results				Analytical Results			
				Engineering		True	
Load P, kip (1)	Gauge Length l, in (2)	Diameter d, in (3)	Area A, in² (4)	Stress P/A_0, kpsi (5)	Strain ϵ, in/in (6)	Stress P/A_i, kpsi (7)	Strain ε, in/in (8)
0	2.0000	0.505	0.2003	0	0	0	0
1.00	2.0006			5.0	0.000 30	5.0	0.000 30
2.00	2.0012			10.0	0.000 60	10.0	0.000 60
3.00	2.0018			15.0	0.000 90	15.0	0.000 90
4.00	2.0024			20.0	0.001 20	20.0	0.001 20
5.00	2.0035			25.0	0.001 75	25.0	0.001 75
6.00	2.0044			30.0	0.002 20	30.0	0.002 20
7.00	2.0057			34.9	0.002 85	34.9	0.002 85
8.00	2.0070			39.9	0.003 50	39.9	0.003 49
9.00	2.0094	0.504	0.1995	44.9	0.004 70	44.9	0.004 69
10.00	2.0140			49.9	0.007 00	49.9	0.006 98
12.00		0.501	0.1971	59.9	0.016 23	60.9	0.016 00
14.00		0.493	0.1909	69.9	0.049 24	73.3	0.048 07
14.50		0.486	0.1855	72.4	0.079 78	78.2	0.076 76
14.95	2.310	0.470	0.1735	74.6	0.154 47	86.2	0.143 64
14.50		0.442	0.1534	72.4	0.305 74	94.5	0.266 77
14.00		0.425	0.1419	69.9	0.411 56	98.7	0.344 70
11.50	2.480	0.352	0.0973	57.4	1.058 58	118.2	0.722 02

*Data source: Joseph Datsko, *Materials in Design and Manufacturing*, published by the author, Ann Arbor, Mich., 1977, chap. 5.

Solution

(*a*) The engineering stress-strain diagram (Fig. 5–6) is plotted from the data in columns 5 and 6 of Table 5–1. The first 11 values in column 6 are obtained from Eq. (5–1). The remaining values in column 6 are found from Eq. (5–4). Note that the strain for the last entry in column 6 does not really mean that the test specimen has elongated 1.058 58 in/in, because necking has occurred and the result is based on a change of areas. As shown in the table, the final length is actually 2.48 in.

(*b*) Columns 7 and 8 of Table 5–1 are the computed values of the true stress and strain (Fig. 5–7). The true strain is obtained using Eq. (5–8).

The ultimate strength is $S_{ut} = 74.6$ kpsi; it is the value of the engineering stress in column 5 corresponding to the ultimate load $P = 14.95$ kip.

(*c*) Figure 5–7 shows that the plastic strain-strengthening coefficient σ_0 is found at the intersection of the stress-strain line and the ordinate corresponding to $\varepsilon = 1$.

The strain-strengthening exponent is found to be $m = 0.144$, from Eq. (5–13), and corresponds to the ultimate load $P = 14.95$ kip.

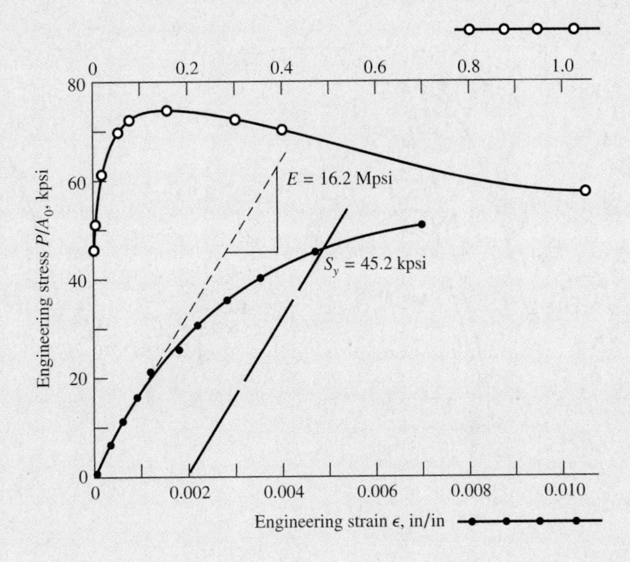

Figure 5–6

Engineering stress–engineering strain diagram from the test data on annealed A-40 titanium as listed in Table 5–1. Note that two strain scales are used in order to plot all data points.

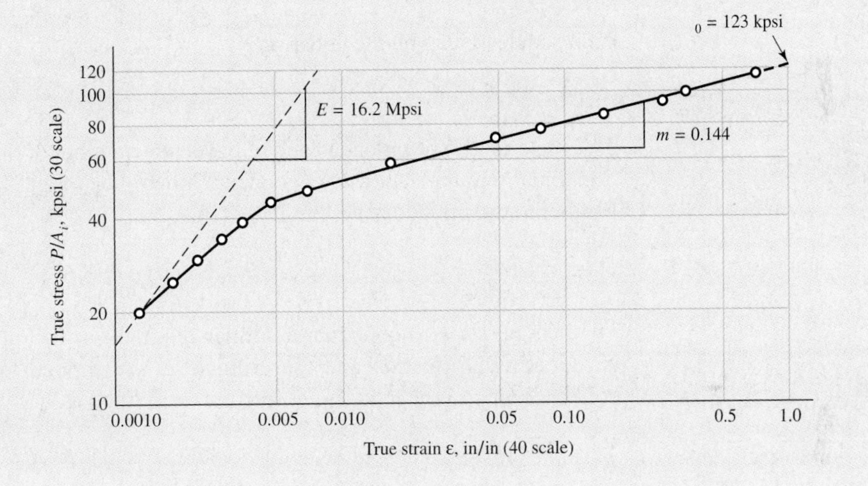

Figure 5–7

True stress–true strain diagram plotted from test data on annealed A-40 titanium in Table 5.1. The elastic modulus E is the slope of the dashed line. The strain-strengthening exponent $m = 0.144$ is the slope of the linear portion of the diagram in the plastic region. Note that the logarithmic scale for the ordinate differs from that for the abscissa. These scales were selected in the manner to create a more accurate diagram.

5–3 Strength and Cold Work

Cold working is the process of plastic straining below the recrystallization temperature in the plastic region of the stress-strain diagram. Materials can be deformed plastically by the application of heat, as in blacksmithing or hot rolling, but the resulting mechanical properties are quite different from those obtained by cold working. The purpose of this section is to explain what happens to the significant mechanical properties of a material when that material is cold-worked.

Consider the stress-strain diagram of Fig. 5–8*a*. Here a material has been stressed beyond the yield strength at Y to some point I, in the plastic region, and then the load removed. At this point the material has a permanent plastic deformation ϵ_p. If the load corresponding to point I is now reapplied, the material will be elastically deformed by the amount ϵ_e. Thus at point I the total unit strain consists of the two components ϵ_p

Figure 5–8

(*a*) Stress-strain diagram showing unloading and reloading at point *I* in the plastic region; (*b*) analogous load-deformation diagram.

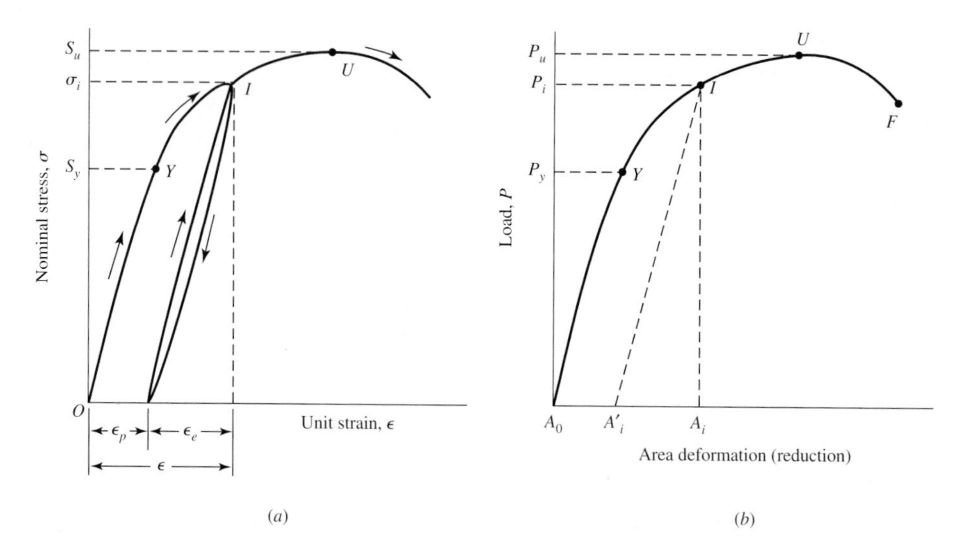

and ϵ_e and is given by the equation

$$\epsilon = \epsilon_p + \epsilon_e \tag{a}$$

This material can be unloaded and reloaded any number of times from and to point I, and it is found that the action always occurs along the straight line which is approximately parallel to the initial elastic line OY. Thus

$$\epsilon_e = \frac{\sigma_i}{E} \tag{b}$$

It is possible to construct a similar diagram, as in Fig. 5–8*b*, where the abscissa is the area deformation and the ordinate is the applied load. The *reduction in area* corresponding to the load P_f, at fracture, is defined as

$$R = \frac{A_0 - A_f}{A_0} = 1 - \frac{A_f}{A_0} \tag{5–14}$$

where A_0 is the original area. The quantity R in Eq. (5–14) is usually expressed in percent and tabulated in lists of mechanical properties as a measure of *ductility*. See Appendix Table A–20, for example. Ductility is an important property because it measures the ability of a material to absorb overloads and to be cold-worked. Thus such operations as bending, drawing, heading, and stretch forming are metal-processing operations that require ductile materials.

Figure 5-8*b* can also be used to define the quantity of cold work. Thus the *cold-work factor* W is

$$W = \frac{A_0 - A'_i}{A_0} \approx \frac{A_0 - A_i}{A_0} \tag{5–15}$$

where A'_i corresponds to the area after the load P_i has been released. The approximation in Eq. (5–15) results because of the difficulty of measuring the small diametral changes in the elastic region. If the amount of cold-work is known, then Eq. (5–15) can be solved for the area A'_i. The result is

$$A'_i = A_0(1 - W) \tag{5–16}$$

Cold working a material produces a new set of values for the strengths, as can be seen from stress-strain diagrams. If point I is to the left of point U, that is, if $P_i < P_u$, then the new yield strength is

$$S'_y = \frac{P_i}{A'_i} = \sigma_0 \varepsilon_i^m \qquad P_i \leq P_u \tag{5–17}$$

Because of the reduced area, that is, because $A'_i < A_0$, the ultimate strength also changes, and is

$$S'_u = \frac{P_u}{A'_i} \tag{c}$$

Since $P_u = S_u A_0$, we find, with Eq. (5–13), that

$$S'_u = \frac{S_u A_0}{A_0(1 - W)} = \frac{S_u}{1 - W} \qquad \varepsilon_i \leq \varepsilon_u \tag{5–18}$$

which is valid only when point I is to the left of point U.

For points to the right of U, the yield strength is approaching the ultimate strength, and, with small loss in accuracy,

$$S'_u \approx S'_y \approx \sigma_0 \varepsilon_i^m \qquad \varepsilon_i \leq \varepsilon_u \tag{5–19}$$

A little thought will reveal that a bar will have the same ultimate load in tension after being strain-strengthened in tension as it had before. The new strength is of interest to us not because the static ultimate load increases, but—since fatigue strengths are correlated with the local ultimate strengths—because the fatigue strength improves. Also the yield strength increases, giving a larger range of sustainable *elastic* loading.

EXAMPLE 5–2 An annealed AISI 1018 steel (see Table E–22) has $S_y = 32.0$ kpsi, $S_u = 49.5$ kpsi, $\sigma_f = 91.1$ kpsi, $\sigma_0 = 90$ kpsi, $m = 0.25$, and $\varepsilon_f = 1.05$ in/in. Find the new values of the strengths if the material is given 15 percent cold work.

Solution From Eq. (5–13), we find the true strain corresponding to the ultimate strength to be

$$\varepsilon_u = m = 0.25$$

The ratio A_0/A_i is, from Eq. (5–15),

$$\frac{A_0}{A_i} = \frac{1}{1 - W} = \frac{1}{1 - 0.15} = 1.176$$

The true strain corresponding to 15 percent cold work is obtained from Eq. (5–12). Thus

$$\varepsilon_i = \ln \frac{A_0}{A_i} = \ln\ 1.176 = 0.1625$$

Since $\varepsilon_i < \varepsilon_u$, Eqs. (5–17) and (5–18) apply. Therefore,

Answer $S'_y = \sigma_0 \varepsilon_i^m = 90(0.1625)^{0.25} = 57.1$ kpsi

Answer $$S'_u = \frac{S_u}{1 - W} = \frac{49.5}{1 - 0.15} = 58.2 \text{ kpsi}$$

5–4 Hardness

The resistance of a material to penetration by a pointed tool is called *hardness*. Though there are many hardness-measuring systems, we shall consider here only the two in greatest use.

Rockwell hardness tests are described by ASTM standard hardness method E–18 and measurements are quickly and easily made, they have good reproducibility, and the test machine for them is easy to use. In fact, the hardness number is read directly from a dial. Rockwell hardness scales are designated as $A, B, C, \ldots$, etc. The indenters are described as a diamond, a 1/16" diameter ball, and a diamond for scales A, B, and C respectively. And the load applied is either 60, 100, or 150 kg. Thus the Rockwell B scale, designated R_B, uses a 100-kg load and a No. 2 indenter, which is a $\frac{1}{16}$-in-diameter ball. The Rockwell C scale R_C uses a diamond cone, which is the No. 1 indenter, and a load of 150 kg. Hardness numbers so obtained are relative. Thus a hardness $R_C = 50$ has meaning only in relation to another hardness number using the same scale.

The *Brinell hardness* (note that *Brinell* is spelled with a single letter *n* and rhymes with the word *bell*) is another test in very general use. In testing, the indenting tool through which force is applied is a ball and the hardness number H_B is found as a number equal to the applied load divided by the spherical surface area of the indentation. Thus the units of H_B are the same as those of stress, though they are seldom used. Brinell hardness testing takes more time, since H_B must be computed from the test data. The primary advantage of both methods is that they are nondestructive in most cases. Both are empirically and directly related to the ultimate strength of the material tested. This means that the strength of parts could, if desired, be tested part by part during manufacture.

For 111 data pairs of carbon and low-alloy wrought steels, $200 \le H_B \le 450$,

$$\mathbf{S}_{ut} = \begin{cases} 0.495(1, 0.041)\bar{H}_B \text{ kpsi} \\ 3.41(1, 0.041)\bar{H}_B \text{ MPa} \end{cases} \qquad (S_{ut})_{0.99} = \begin{cases} 0.45\bar{H}_B \text{ kpsi} \\ 3.10\bar{H}_B \text{ MPa} \end{cases} \tag{5–20}$$

Similar relationships with higher variance for cast iron can be derived from data supplied by Krause.[5] Data from 72 tests of gray iron produced by one foundry and poured in two sizes of test bars are reported in graph form. The minimum strength, as defined by the ASTM, is found from these data to be

$$S_u = \begin{cases} 0.23H_B - 12.5 \text{ kpsi} \\ 1.58H_B - 86 \text{ MPa} \end{cases} \tag{5–21}$$

Walton[6] shows a chart from which the SAE minimum strength can be obtained. The result is

$$S_u = 0.2375H_B - 16 \text{ kpsi} \tag{5–22}$$

which is even more conservative than the values obtained from Eq. (5–21).

EXAMPLE 5–3 It is necessary to ensure that a certain part supplied by a foundry always meets or exceeds SAE grade 20 specifications for cast iron (see Table E–24). What hardness should be specified? (ASTM uses the word *class* instead of *grade*.)

---

5. D. E. Krause, "Gray Iron—A Unique Engineering Material," ASTM Special Publication 455, 1969, pp. 3–29, as reported in Charles F. Walton (ed.), *Iron Castings Handbook*, Iron Founders Society, Inc., Cleveland, 1971, pp. 204, 205.

6. Ibid.

Solution From Eq. (5–21), we have

Answer $$H_B = \frac{S_u + 12.5}{0.23} = \frac{20 + 12.5}{0.23} = 141$$

where S_u is the minimum strength. If the foundry can control the hardness within 20 points, routinely, then specify $145 < H_B < 165$. This imposes no hardship on the foundry and assures the designer that ASTM grade 20 will always be supplied at a predictable cost.

When hardness is a random variable, then, corresponding to Eqs. (5– 20),

$$\mathbf{S}_{ut} = \begin{cases} 0.495(1, 0.041)\mathbf{H}_B \text{ kpsi} \\ 3.41(1, 0.041)\mathbf{H}_B \text{ MPa} \end{cases} \tag{5–23}$$

For cast irons the equations corresponding to Eqs. (5–21) become

$$\mathbf{S}_{ut} = \begin{cases} 0.23\mathbf{H}_B - 9 + (0, 1.5) \text{ kpsi} \\ 1.58\mathbf{H}_B - 62 + (0, 10) \text{ MPa} \end{cases} \tag{5–24}$$

EXAMPLE 5–4 Brinell hardness tests made on a random sample of steel parts during processing gave observations 248, 253, 247, 244, and 246.

(*a*) Estimate $\mathbf{S}_{ut}$ in SI units and predict $(S_{ut})_{0.99}$, the ASTM strength.

(*b*) If dispersion in hardness was to be severe, how would you answer part *a*?

Solution (*a*) $\bar{H}_B = 247.6$ MPa, $\sigma_{HB} = 3.36$ MPa, $C_{HB} = 0.0136$. From Eq. (5–20),

$$\mathbf{S}_{ut} = 3.41(1, 0.041)247.6$$

Answer $$\bar{S}_{ut} = 3.41(247.6) = 844.3 \text{ MPa}$$

Answer $$\sigma_{Sut} = C_{Sut}\bar{S}_{ut} = 0.041(844.3) = 34.6 \text{ MPa}$$

Answer $$(S_{ut})_{0.99} = 3.10(247.6) = 768 \text{ MPa}$$

(*b*) If C_{HB} is large, it can indicate variation in $\mathbf{S}_{ut}$ that is due to both the fuzzy correlation multiplier 3.41 (1, 0.041) and ultimate strength itself. Expand Eq. (5–20) to let $\mathbf{H}_B$ be random:

$$\mathbf{S}_{ut} = 3.41(1, 0.041)\mathbf{H}_B = 3.41(1, 0.041)247.6(1, 0.0136)$$

Answer $$\bar{S}_{ut} = 3.41(247.6) = 844.3 \text{ MPa}$$

Answer $$C_{Sut} = [0.041^2 + 0.0136^2]^{1/2} = 0.043$$

Answer $$(S_{ut})_{0.99} = \bar{S}_{ut}(1 + z_{0.99}C_{Sut}) = 844.3[1 + (-2.326)(0.043) = 760 \text{ MPa}$$

5–5 Impact Properties

An external force applied to a structure or part is called an *impact load* if the time of application is less than one-third the lowest natural period of vibration of the part or structure. Otherwise it is called simply a static load.

The *Charpy* (commonly used) and *Izod* (rarely used) *notched-bar tests* utilize bars of specified geometries to determine brittleness and impact strength. These tests are helpful in comparing several materials and in the determination of low-temperature brittleness. In both tests the specimen is struck by a pendulum released from a fixed height, and the energy absorbed by the specimen, called the *impact value*, can be computed from the height of swing after fracture, but is read from a dial which essentially "computes" the result.

The effect of temperature on impact values is shown in Fig. 5–9 for a material showing a ductile-brittle transition. Not all materials show this transition. Notice the narrow region of critical temperatures where the impact value increases very rapidly. In the low-temperature region the fracture appears as a brittle, shattering type, whereas the appearance is a tough, tearing type above the critical-temperature region. The critical temperature seems to be dependent on both the material and the geometry of the notch. For this reason designers should not rely too heavily on the results of notched-bar tests.

The average strain rate used in obtaining the stress-strain diagram is about 0.001 in/(in · s) or less. When the strain rate is increased, as it is under impact conditions, the strengths increase, as shown in Fig. 5–10. In fact, at very high strain rates the yield strength seems to approach the ultimate strength as a limit. But note that the curves show

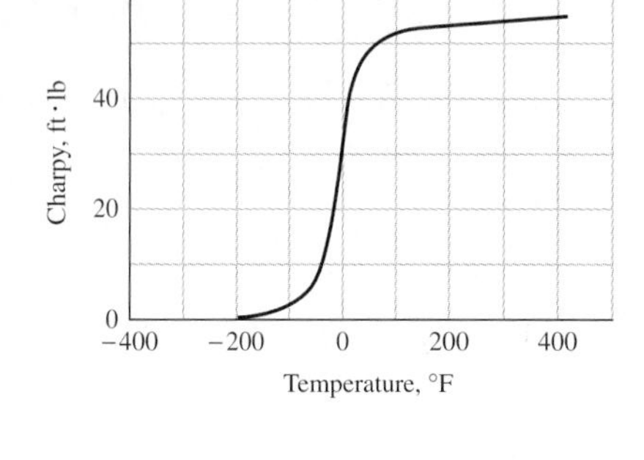

Figure 5–9

A mean trace shows the effect of temperature on impact values. The result of interest is the brittle-ductile transition temperature, often defined as the temperature at which the mean trace passes through the 15 ft-lb level. The critical temperature *is* dependent on the geometry of the notch, which is why the Charpy V-notch is closely defined.

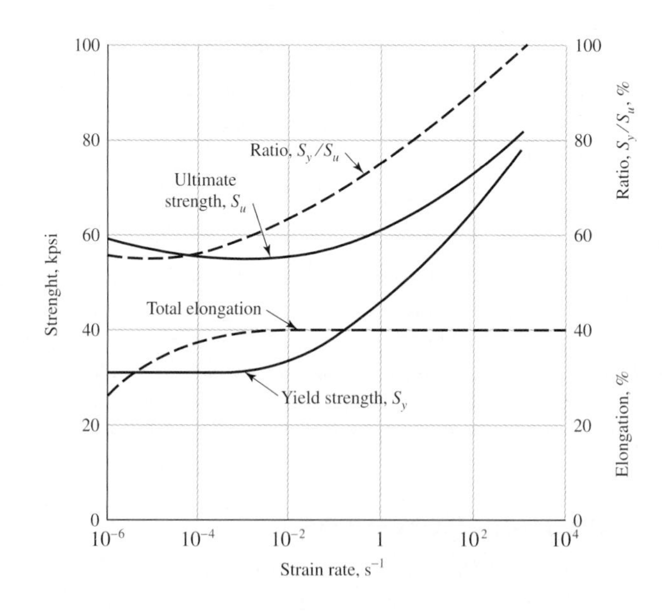

Figure 5–10

Influence of strain rate on tensile properties.

little change in the elongation. This means that the ductility remains about the same. Also, in view of the sharp increase in yield strength, a mild steel could be expected to behave elastically throughout practically its entire strength range under impact conditions.

The Charpy and Izod tests really provide toughness data under dynamic, rather than static, conditions. It may well be that impact data obtained from these tests are as dependent on the notch geometry as they are on the strain rate. For these reasons it may be better to use the concepts of notch sensitivity, fracture toughness, and fracture mechanics, discussed later in this chapter, to assess the possibility of cracking or fracture.

CASE STUDY 5 **Transition Temperature, Anyone?**

The destruction of the *Titanic* was a complex disaster that occurred on April 15, 1912, at 21 knots when the world's then largest ship collided with an iceberg, and all but 711 of the 2224 persons aboard perished. [See *Mechanical Engineering*, vol. 120, no. 8 (August 1998), pp. 54–58, for a brief review and references.] Of immediate materials concern is the role of material properties of the hull and structure. Charpy V-notch tests on samples of recovered hull plate, tested in orthogonal directions, revealed the ductile-to-brittle transition temperatures of 20°C in one direction and 30°C in the other. An A-36 structural steel has a transition temperature of −15°C. Large grain structure, cold- punched unreamed rivet holes, and the sulfur and phosphorus content, in addition to the low ductility at the freezing point of water, contributed to the breakup of the *Titanic*.

In Boston, early in 1919, a famous static tank failure by rupture drowned 21 people (and many horses) and destroyed houses with more than 2 million gallons of molasses. Some World War II tankers and Liberty Ships broke in two at their outfitting dock without cargo.

Always check the brittle-to-ductile transition temperature of parts subjected to outdoor temperatures.

5–6 Temperature Effects

Strength and ductility, or brittleness, are properties affected by the temperature of the operating environment.

The effect of temperature on the static properties of steels is typified by the strength versus temperature chart of Fig. 5–11. Note that the tensile strength changes only a small amount until a certain temperature is reached. At that point it falls off rapidly. The yield strength, however, decreases continuously as the environmental temperature is increased. There is a substantial increase in ductility, as might be expected, at the higher temperatures.

Many tests have been made of ferrous metals subjected to constant loads for long periods of time at elevated temperatures. The specimens were found to be permanently deformed during the tests, even though at times the actual stresses were less than the yield strength of the material obtained from short-time tests made at the same temperature. This continuous deformation under load is called *creep*.

One of the most useful tests to have been devised is the long-time creep test under constant load. Figure 5–12 illustrates a curve which is typical of this kind of test. The curve is obtained at a constant stated temperature. A number of tests are usually run simultaneously at different stress intensities. The curve exhibits three distinct regions. In the first stage are included both the elastic and the plastic deformation. This stage shows a decreasing creep rate which is due to the strain hardening. The second stage

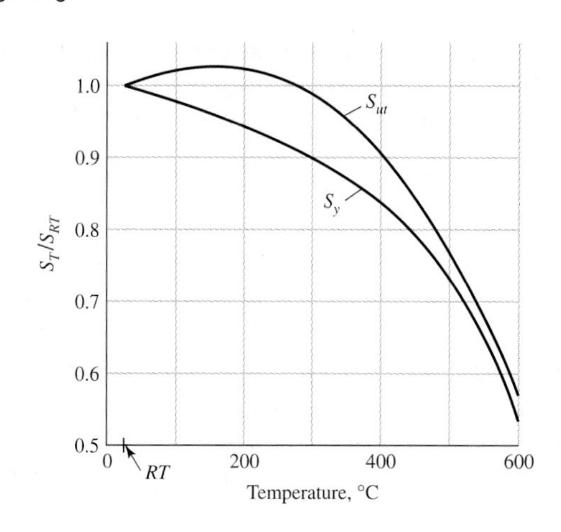

Figure 5–11

A plot of the results of 145 tests of 21 carbon and alloy steels showing the effect of operating temperature on the yield strength S_y and the ultimate strength S_{ut}. The ordinate is the ratio of the strength at the operating temperature to the strength at room temperature. The standard deviations were $0.0442 \leq \sigma_{Sy} \leq 0.152$ for S_y and $0.099 \leq \sigma_{Sut} \leq 0.11$ for S_{ut}. (*Data source*: E. A. Brandes (ed.), *Smithells Metal Reference Book*, 6th ed., Butterworth, London, 1983 pp. 22–128 to 22–131.)

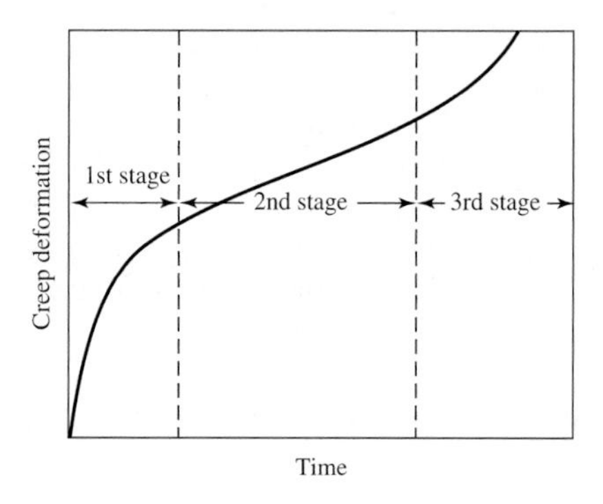

Figure 5–12

Creep-time curve.

shows a constant minimum creep rate caused by the annealing effect. In the third stage the specimen shows a considerable reduction in area, the true stress is increased, and a higher creep eventually leads to fracture.

When the operating temperatures are lower than the transition temperature (Fig. 5–9), the possibility arises that a part could fail by a brittle fracture. This subject will be discussed later in this chapter.

Of course, heat treatment, as will be shown, is used to make substantial changes in the mechanical properties of a material.

Heating due to electric and gas welding also changes the mechanical properties. Such changes may be due to clamping during the welding process, as well as heating; the resulting stresses then remain frozen in when the parts have cooled and the clamps have been removed. Hardness tests can be used to learn whether the strength has been changed by welding, but such tests will not reveal the presence of residual stresses.

5–7 Numbering Systems

The Society of Automotive Engineers (SAE) was the first to recognize the need, and to adopt a system, for the numbering of steels. Later the American Iron and Steel Institute (AISI) adopted a similar system. In 1975 the SAE published the Unified Numbering System for Metals and Alloys (UNS); this system also contains cross-reference numbers for other material specifications.[7] The UNS uses a letter prefix to designate the material, as, for example, G for the carbon and alloy steels, A for the aluminum alloys, C for the copper-base alloys, and S for the stainless or corrosion-resistant steels. For some materials, not enough agreement has as yet developed in the industry to warrant the establishment of a designation.

For the steels, the first two numbers following the letter prefix indicate the composition, excluding the carbon content. The various compositions used are as follows:

G10	Plain carbon	G46	Nickel-molybdenum
G11	Free-cutting carbon steel with more sulfur or phosphorus	G48	Nickel-molybdenum
		G50	Chromium
G13	Manganese	G51	Chromium
G23	Nickel	G52	Chromium
G25	Nickel	G61	Chromium-vanadium
G31	Nickel-chromium	G86	Chromium-nickel-molybdenum
G33	Nickel-chromium	G87	Chromium-nickel-molybdenum
G40	Molybdenum	G92	Manganese-silicon
G41	Chromium-molybdenum	G94	Nickel-chromium-molybdenum
G43	Nickel-chromium-molybdenum		

The second number pair refers to the approximate carbon content. Thus, G10400 is a plain carbon steel with a carbon content of 0.37 to 0.44 percent. The fifth number following the prefix is used for special situations. For example, the old designation AISI 52100 represents a chromium alloy with about 100 points of carbon. The UNS designation is G52986.

The UNS designations for the stainless steels, prefix S, utilize the older AISI designations for the first three numbers following the prefix. The next two numbers are reserved for special purposes. The first number of the group indicates the approximate composition. Thus 2 is a chromium-nickel-manganese steel, 3 is a chromium-nickel steel, and 4 is a chromium alloy steel. Sometimes stainless steels are referred to by their alloy content. Thus S30200 is often called an 18-8 stainless steel, meaning 18 percent chromium and 8 percent nickel.

The prefix for the aluminum group is the letter A. The first number following the prefix indicates the processing. For example, A9 is a wrought aluminum, while A0 is a casting alloy. The second number designates the main alloy group as shown in Table 5–2. The third number in the group is used to modify the original alloy or to designate the impurity limits. The last two numbers refer to other alloys used with the basic group.

Table 5–2

Aluminum Alloy Designations

Aluminum 99.00% pure and greater	Ax1xxx
Copper alloys	Ax2xxx
Manganese alloys	Ax3xxx
Silicon alloys	Ax4xxx
Magnesium alloys	Ax5xxx
Magnesium-silicon alloys	Ax6xxx
Zinc alloys	Ax7xxx

7. Many of the materials discussed in the balance of this chapter are listed in the Appendix tables. Be sure to review these.

The American Society for Testing and Materials (ASTM) numbering system for cast iron is in widespread use. This system is based on the tensile strength. Thus ASTM A18 speaks of classes; e.g., 30 cast iron has a minimum tensile strength of 30 kpsi. Note from the Appendix, however, that the *typical* tensile strength is 31 kpsi. You should be careful to designate which of the two values is used in design and problem work because of the significance of factor of safety.

5–8 Sand Casting

Sand casting is a basic low-cost process, and it lends itself to economical production in large quantities with practically no limit to the size, shape, or complexity of the part produced. Despite its apparent simplicity, sand casting is highly complex.

In sand casting, the casting is made by pouring molten metal into sand molds. A pattern, constructed of metal or wood, is used to form the cavity into which the molten metal is poured. Recesses or holes in the casting are produced by sand cores introduced into the mold. The designer should make an effort to visualize the pattern and casting in the mold. In this way the problems of core setting, pattern removal, draft, and solidification can be studied. Castings to be used as test bars of cast iron are cast separately and properties may vary.

Steel castings are the most difficult of all to produce, because steel has the highest melting temperature of all materials normally used for casting. This high temperature aggravates all casting problems.

The following rules will be found quite useful in the design of any sand casting:

1. All sections should be designed with a uniform thickness.
2. The casting should be designed so as to produce a gradual change from section to section where this is necessary.
3. Adjoining sections should be designed with generous fillets or radii.
4. A complicated part should be designed as two or more simple castings to be assembled by fasteners or by welding.

Steel, gray iron, brass, bronze, and aluminum are most often used in castings. The minimum wall thickness for any of these materials is about 5 mm; though with particular care, thinner sections can be obtained with some materials.

5–9 Shell Molding

The shell-molding process employs a heated metal pattern, usually made of cast iron, aluminum, or brass, which is placed in a shell-molding machine containing a mixture of dry sand and thermosetting resin. The hot pattern melts the plastic, which, together with the sand, forms a shell about 5 to 10 mm thick around the pattern. The shell is then baked at from 400 to 700°F for a short time while still on the pattern. It is then stripped from the pattern and placed in storage for use in casting.

In the next step the shells are assembled by clamping, bolting, or pasting; they are placed in a backup material, such as steel shot; and the molten metal is poured into the cavity. The thin shell permits the heat to be conducted away so that solidification takes place rapidly. As solidification takes place, the plastic bond is burned and the mold collapses. The permeability of the backup material allows the gases to escape and the casting to air-cool. All this aids in obtaining a fine-grain, stress-free casting.

Shell-mold castings feature a smooth surface, a draft that is quite small, and close tolerances. In general, the rules governing sand casting also apply to shell-mold casting.

5–10 Investment Casting

Investment casting uses a pattern which may be made from wax, plastic, or other material. After the mold is made, the pattern is melted out. Thus a mechanized method of casting a great many patterns is necessary. The mold material is dependent upon the melting point of the cast metal. Thus a plaster mold can be used for some materials while others would require a ceramic mold. After the pattern is melted out, the mold is baked or fired; when firing is completed, the molten metal may be poured into the hot mold and allowed to cool.

If a number of castings are to be made, then metal or permanent molds may be suitable. Such molds have the advantage that the surfaces are smooth, bright, and accurate, so that little, if any, machining is required. *Metal-mold castings* are also known as *die castings* and *centrifugal castings*.

5–11 Powder-Metallurgy Process

The powder-metallurgy process is a quantity-production process that uses powders from a single metal, several metals, or a mixture of metals and nonmetals. It consists essentially of mechanically mixing the powders, compacting them in dies at high pressures, and heating the compacted part at a temperature less than the melting point of the major ingredient. The particles are united into a single strong part similar to what would be obtained by melting the same ingredients together. The advantages are (1) the elimination of scrap or waste material, (2) the elimination of machining operations, (3) the low unit cost when mass-produced, and (4) the exact control of composition. Some of the disadvantages are (1) the high cost of dies, (2) the lower physical properties, (3) the higher cost of materials, (4) the limitations on the design, and (5) the limited range of materials which can be used. Parts commonly made by this process are oil-impregnated bearings, incandescent lamp filaments, cemented-carbide tips for tools, and permanent magnets. Some products can only be made by powder metallurgy, surgical implants for example. The structure is different from what can be obtained by melting the same ingredients.

5–12 Hot-Working Processes

By *hot working* are meant such processes as rolling, forging, hot extrusion, and hot pressing, in which the metal is heated above its recrystallation temperature.

Hot rolling is usually used to create a bar of material of a particular shape and dimensions. Figure 5–13 shows some of the various shapes which are commonly produced by the hot-rolling process. All of them are available in many different sizes as well as in different materials. The materials most available in the hot-rolled bar sizes are steel, aluminum, magnesium, and copper alloys.

Tubing may be manufactured by hot-rolling strip or plate. The edges of the strip are rolled together, creating seams which are either butt-welded or lap-welded. Seamless tubing is manufactured by roll-piercing a solid heated rod with a piercing mandrel.

Extrusion is the process by which great pressure is applied to a heated metal billet or blank, causing it to flow through a restricted orifice. This process is more common with materials of low melting point, such as aluminum, copper, magnesium, lead, tin, and zinc. Stainless steel extrusions are available on a more limited basis.

Forging is the hot working of metal by hammers, presses, or forging machines. In common with other hot-working processes, forging produces a refined grain structure which results in increased strength and ductility. Compared with castings, forgings have greater strength for the same weight. In addition, drop forgings can be made smoother

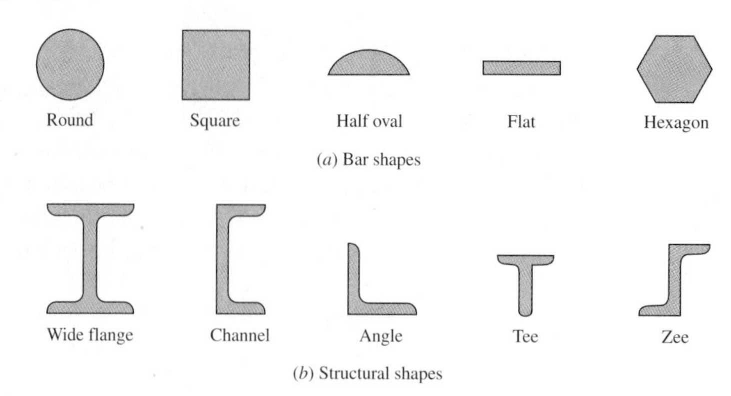

Figure 5–13

Common shapes available through hot rolling.

and more accurate than sand castings, so that less machining is necessary. However, the initial cost of the forging dies is usually greater than the cost of patterns for castings, although the greater unit strength rather than the cost is usually the deciding factor between these two processes.

5–13 Cold-Working Processes

By *cold working* is meant the forming of the metal while at a low temperature (usually room temperature). In contrast to parts produced by hot working, cold-worked parts have a bright new finish, are more accurate, and require less machining.

Cold-finished bars and shafts are produced by rolling, drawing, turning, grinding, and polishing. Of these methods, by far the largest percentage of products are made by the cold-rolling and cold-drawing processes. Cold rolling is now used mostly for the production of wide flats and sheets. Practically all cold-finished bars are made by cold drawing but even so are sometimes mistakenly called "cold-rolled bars." In the drawing process, the hot-rolled bars are first cleaned of scale and then drawn by pulling them through a die which reduces the size about $\frac{1}{32}$ to $\frac{1}{16}$ in. This process does not remove material from the bar but reduces, or "draws" down, the size. Many different shapes of hot-rolled bars may be used for cold-drawing.

Cold rolling and cold drawing have the same effect upon the mechanical properties. The cold-working process does not change the grain size but merely distorts it. Cold working results in a large increase in yield strength, an increase in ultimate strength and hardness, and a decrease in ductility. In Fig. 5–14 the properties of a cold-drawn bar are compared with those of a hot-rolled bar of the same material.

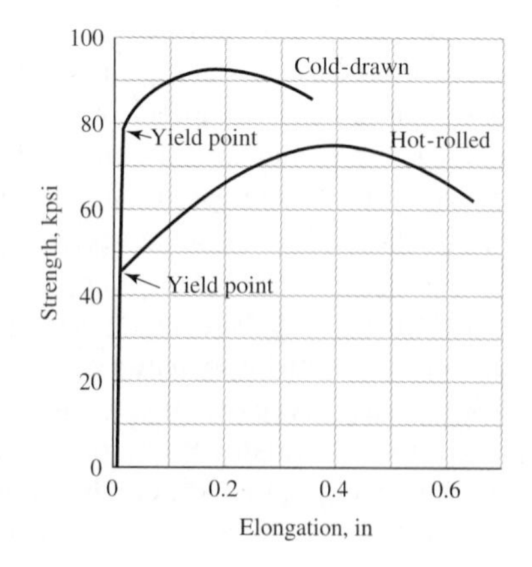

Figure 5–14

Stress-strain diagram for hot-rolled and cold-drawn AISI 1035 steel.

Heading is a cold-working process in which the metal is gathered, or upset. This operation is commonly used to make screw and rivet heads and is capable of producing a wide variety of shapes. *Roll threading* is the process of rolling threads by squeezing and rolling a blank between two serrated dies. *Spinning* is the operation of working sheet material around a rotating form into a circular shape. *Stamping* is the term used to describe punch-press operations such as *blanking, coining, forming*, and *shallow drawing*.

5–14 The Heat Treatment of Steel

Heat treatment refers to processes which interrupt or vary the transformation process described by the equilibrium diagram. Other mechanical or chemical operations are sometimes grouped under the heading of heat treatment. The common heat-treating operations are annealing, quenching, tempering, and case hardening.

Annealing

When a material is cold- or hot-worked, residual stresses are built in, and, in addition, the material usually has a higher hardness due to these working operations. These operations change the structure of the material so that it is no longer represented by the equilibrium diagram. Full-annealing and normalizing are a heating operation which permits the material to transform according to the equilibrium diagram. The material to be annealed is heated to a temperature which is approximately 100°F above the critical temperature. It is held at this temperature for a time which is sufficient for the carbon to become dissolved and diffused through the material. The object being treated is then allowed to cool slowly, usually in the furnace in which it was treated. If the transformation is complete, then it is said to have a full anneal. Annealing is used to soften a material and make it more ductile, to relieve residual stresses, and to refine the grain structure.

The term *annealing* includes the process called *normalizing*. Parts to be normalized may be heated to a slightly higher temperature than in full annealing. This produces a coarser grain structure, which is more easily machined if the material is a low-carbon steel. In the normalizing process the part is cooled in still air at room temperature. Since this cooling is more rapid than the slow cooling used in full annealing, less time is available for equilibrium, and the material is harder than fully annealed steel. Normalizing is often used as the final treating operation for steel. The cooling in still air amounts to a slow quench.

Quenching

Eutectoid steel which is fully annealed consists entirely of pearlite, which is obtained from austenite under conditions of equilibrium. A fully annealed hypoeutectoid steel would consist of pearlite plus ferrite, while hypereutectoid steel in the fully annealed condition would consist of pearlite plus cementite. The hardness of steel of a given carbon content depends upon the structure that replaces the pearlite when full annealing is not carried out.

The absence of full annealing indicates a more rapid rate of cooling. The rate of cooling is the factor which determines the hardness. A controlled cooling rate is called *quenching*. A mild quench is obtained by cooling in still air, which, as we have seen, is obtained by the normalizing process. The two most widely used media for quenching are water and oil. The oil quench is quite slow but prevents quenching cracks caused by rapid expansion of the object being treated. Quenching in water is used for carbon steels and for medium-carbon, low-alloy steels.

The effectiveness of quenching depends upon the fact that when austenite is cooled it does not transform into pearlite instantaneously but requires time to initiate and complete the process. Since the transformation ceases at about 800°F, it can be prevented by rapidly cooling the material to a lower temperature. When the material is cooled rapidly to 400°F or less, the austenite is transformed into a structure called *martensite*. Martensite is a supersaturated solid solution of carbon in ferrite and is the hardest and strongest form of steel.

If steel is rapidly cooled to a temperature between 400 and 800°F and held there for a sufficient length of time, the austenite is transformed into a material which is generally called *bainite*. Bainite is a structure which is intermediate between pearlite and martensite. Although there are several structures which can be identified between the temperatures given, depending upon the temperature used, they are collectively known as bainite. By the choice of this transformation temperature, almost any variation of structure may be obtained. These range all the way from coarse pearlite to fine martensite.

Tempering

When a steel specimen has been fully hardened it is very hard and brittle and has high residual stresses. The steel is unstable and tends to contract on aging. This tendency is increased when the specimen is subjected to externally applied loads, because the resultant stresses contribute still more to the instability. These internal stresses can be relieved by a modest heating process called *stress relieving*, or a combination of stress relieving and softening called *tempering* or *drawing*. After the specimen has been fully hardened by being quenched from above the critical temperature, it is reheated to some temperature below the critical temperature for a certain period of time and then allowed to cool in still air. The temperature to which it is reheated depends upon the composition and the degree of hardness or toughness desired.[8] This reheating operation releases the carbon held in the martensite, forming carbide crystals. The structure obtained is called *tempered martensite*. It is now essentially a superfine dispersion of iron carbide(s) in fine-grained ferrite.

The effect of heat-treating operations upon the various mechanical properties of a low alloy steel is shown graphically in Fig. 5–15.

Case Hardening

The purpose of case hardening is to produce a hard outer surface on a specimen of low-carbon steel while at the same time retaining the ductility and toughness in the core. This is done by increasing the carbon content at the surface. Either solid, liquid, or gaseous carburizing materials may be used. The process consists of introducing the part to be carburized into the carburizing material for a stated time and at a stated temperature, depending upon the depth of case desired and the composition of the part. The part may then be quenched directly from the carburization temperature and tempered, or in some cases it must undergo a double heat treatment in order to ensure that both the core and the case are in proper condition. Some of the more useful case-hardening processes are pack carburizing, gas carburizing, nitriding, cyaniding, induction hardening, and flame hardening. In the last two cases carbon is not added to the steel in question, generally a medium carbon steel, for example SAE/AISI 1144.

---

8. For the quantitative aspects of tempering in plain carbon and low-alloy steels, see Charles R. Mischke, "The Strength of Cold-Worked and Heat-Treated Steels," chap. 8 in Joseph E. Shigley and Charles R. Mischke (eds.), *Standard Handbook of Machine Design*, 2nd ed., McGraw-Hill, New York, 1996.

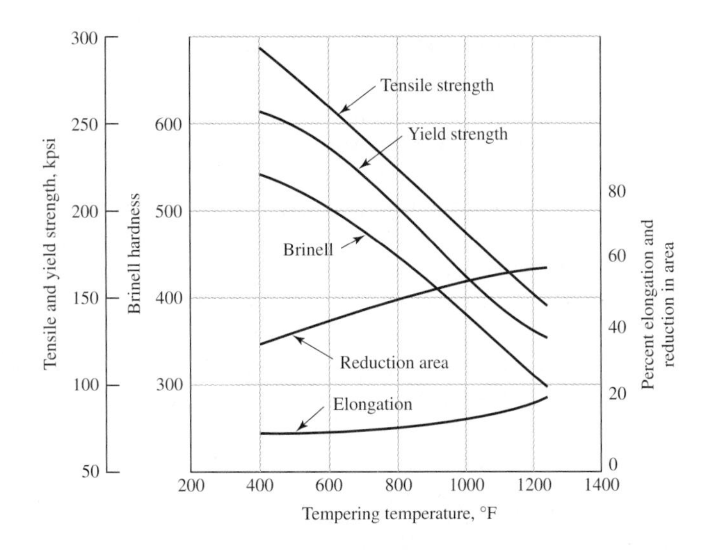

Figure 5–15

The effect of thermal-mechanical history on the mechanical properties of AISI 4340 steel. (Prepared by the International Nickel Company.)

Condition	Tensile strength, kpsi	Yield strength, kpsi	Reduction in area, %	Elongation in 2 in, %	Brinell hardness, Bhn
Normalized...	200	147	20	10	410
As rolled.......	190	144	18	9	380
Annealed......	120	99	43	18	228

5–15 Alloy Steels

Although a plain carbon steel is an alloy of iron and carbon with small amounts of manganese, silicon, sulfur, and phosphorus, the term *alloy steel* is applied when one or more elements other than carbon are introduced in sufficient quantities to modify its properties substantially. The alloy steels not only possess more desirable physical properties but also permit a greater latitude in the heat-treating process.

Chromium

The addition of chromium results in the formation of various carbides of chromium which are very hard, yet the resulting steel is more ductile than a steel of the same hardness produced by a simple increase in carbon content. Chromium also refines the grain structure so that these two combined effects result in both increased toughness and increased hardness. The addition of chromium increases the critical range of temperatures and moves the eutectoid point to the left. Chromium is thus a very useful alloying element.

Nickel

The addition of nickel to steel also causes the eutectoid point to move to the left and increases the critical range of temperatures. Nickel is soluble in ferrite and does not form carbides or oxides. This increases the strength without decreasing the ductility. Case hardening of nickel steels results in a better core than can be obtained with plain carbon steels. Chromium is frequently used in combination with nickel to obtain the toughness and ductility provided by the nickel and the wear resistance and hardness contributed by the chromium.

Manganese

Manganese is added to all steels as a deoxidizing and desulfurizing agent, but if the sulfur content is low and the manganese content is over 1 percent, the steel is classified as a manganese alloy. Manganese dissolves in the ferrite and also forms carbides. It causes the eutectoid point to move to the left and lowers the critical range of temperatures. It increases the time required for transformation so that oil quenching becomes practicable.

Silicon

Silicon is added to all steels as a deoxidizing agent. When added to very-low-carbon steels, it produces a brittle material with a low hysteresis loss and a high magnetic permeability. The principal use of silicon is with other alloying elements, such as manganese, chromium, and vanadium, to stabilize the carbides.

Molybdenum

While molybdenum is used alone in a few steels, it finds its greatest use when combined with other alloying elements, such as nickel, chromium, or both. Molybdenum forms carbides and also dissolves in ferrite to some extent, so that it adds both hardness and toughness. Molybdenum increases the critical range of temperatures and substantially lowers the transformation point. Because of this lowering of the transformation point, molybdenum is most effective in producing desirable oil-hardening and air-hardening properties. Except for carbon, it has the greatest hardening effect, and because it also contributes to a fine grain size, this results in the retention of a great deal of toughness.

Vanadium

Vanadium has a very strong tendency to form carbides; hence it is used only in small amounts. It is a strong deoxidizing agent and promotes a fine grain size. Since some vanadium is dissolved in the ferrite, it also toughens the steel. Vanadium gives a wide hardening range to steel, and the alloy can be hardened from a higher temperature. It is very difficult to soften vanadium steel by tempering; hence it is widely used in tool steels.

Tungsten

Tungsten is widely used in tool steels because the tool will maintain its hardness even at red heat. Tungsten produces a fine, dense structure and adds both toughness and hardness. Its effect is similar to that of molybdenum, except that it must be added in greater quantities.

5–16 Corrosion-Resistant Steels

Iron-base alloys containing at least 12 percent chromium are called *stainless steels*. The most important characteristic of these steels is their resistance to many, but not all, corrosive conditions. The four types available are the ferritic chromium steels, the austenitic chromium-nickel steels, and the martensitic and precipitation-hardenable stainless steels.

The ferritic chromium steels have a chromium content ranging from 12 to 27 percent. Their corrosion resistance is a function of the chromium content, so that alloys containing less than 12 percent still exhibit some corrosion resistance, although they may rust. The quench-hardenability of these steels is a function of both the chromium and the carbon content. The very-high-carbon steels have good quench-hardenability up to about 18 percent chromium, while in the lower carbon ranges it ceases at about 13 percent. If a little nickel is added, these steels retain some degree of hardenability up to 20 percent chromium. If the chromium content exceeds 18 percent, they become difficult to weld, and at the very high chromium levels the hardness becomes so great that very careful attention must be paid to the service conditions. Since chromium is expensive, the designer will choose the lowest chromium content consistent with the corrosive conditions.

The chromium-nickel stainless steels retain the austenitic structure at room temperature; hence they are not amenable to heat treatment. The strength of these steels can be greatly improved by cold working. They are not magnetic unless cold-worked. Their work-hardenability properties also cause them to be difficult to machine. All the chromium-nickel steels may be welded. They have greater corrosion-resistant properties than the plain chromium steels. When more chromium is added for greater corrosion resistance, more nickel must also be added if the austenitic properties are to be retained.

5–17 Casting Materials

Gray Cast Iron

Of all the cast materials, gray cast iron is the most widely used. This is because it has a very low cost, is easily cast in large quantities, and is easy to machine. The principal objections to the use of gray cast iron are that it is brittle and that it is weak in tension. In addition to a high carbon content (over 1.7 percent and usually greater than 2 percent), cast iron also has a high silicon content, with low percentages of sulfur, manganese, and phosphorus. The resultant alloy is composed of pearlite, ferrite, and graphite, and under certain conditions the pearlite may decompose into graphite and ferrite. The resulting product then contains all ferrite and graphite. The graphite, in the form of thin flakes distributed evenly throughout the structure, darkens it; hence, the name *gray cast iron*.

Gray cast iron is not readily welded, because it may crack, but this tendency may be reduced if the part is carefully preheated. Although the castings are generally used in the as-cast condition, a mild anneal reduces cooling stresses and improves the machinability. The tensile strength of gray cast iron varies from 100 to 400 MPa (15 to 60 kpsi), and the compressive strengths are 3 to 4 times the tensile strengths. The modulus of elasticity varies widely, with values extending all the way from 75 to 150 GPa (11 to 22 Mpsi).

Ductile and Nodular Cast Iron

Because of the lengthy heat treatment required to produce malleable cast iron, a cast iron has long been desired which would combine the ductile properties of malleable iron with the ease of casting and machining of gray iron and at the same time would possess these properties in the as-cast conditions. A process for producing such a material using magnesium-containing material seems to fulfill these requirements.

Ductile cast iron, or *nodular cast iron*, as it is sometimes called, is essentially the same as malleable cast iron, because both contain graphite in the form of spheroids. However, ductile cast iron in the as-cast condition exhibits properties very close to those of malleable iron, and if a simple 1-h anneal is given and is followed by a slow cool, it exhibits even more ductility than the malleable product. Ductile iron is made by adding MgFeSi to the melt; since magnesium boils at this temperature, it is necessary to alloy it with other elements before it is introduced.

Ductile iron has a high modulus of elasticity (172 GPa or 25 Mpsi) as compared with gray cast iron, and it is elastic in the sense that a portion of the stress-strain curve is a straight line. Gray cast iron, on the other hand, does not obey Hooke's law, because the modulus of elasticity steadily decreases with increase in stress. Like gray cast iron, however, nodular iron has a compressive strength which is higher than the tensile strength, although the difference is not as great. In forty years it has become extensively used.

White Cast Iron

If all the carbon in cast iron is in the form of cementite and pearlite, with no graphite present, the resulting structure is white and is known as *white cast iron*. This may be produced in two ways. The composition may be adjusted by keeping the carbon and silicon content low, or the gray-cast-iron composition may be cast against chills in order to promote rapid cooling. By either method a casting with large amounts of cementite is produced, and as a result the product is very brittle and hard to machine but also very resistant to wear. A chill is usually used in the production of gray-iron castings in order to provide a very hard surface within a particular area of the casting, while at the same time retaining the more desirable gray structure within the remaining portion. This produces a relatively tough casting with a wear-resistant area.

Malleable Cast Iron

If white cast iron within a certain composition range is annealed, a product called *malleable cast iron* is formed. The annealing process frees the carbon so that it is present as graphite, just as in gray cast iron but in a different form. In gray cast iron the graphite is present in a thin flake form, while in malleable cast iron it has a nodular form and is known as *temper carbon*. A good grade of malleable cast iron may have a tensile strength of over 350 MPa (50 kpsi), with an elongation of as much as 18 percent. The percentage elongation of a gray cast iron, on the other hand, is seldom over 1 percent. Because of the time required for annealing (up to 6 days for large and heavy castings), malleable iron is necessarily somewhat more expensive than gray cast iron.

Alloy Cast Irons

Nickel, chromium, and molybdenum are the most common alloying elements used in cast iron. Nickel is a general-purpose alloying element, usually added in amounts up to 5 percent. Nickel increases the strength and density, improves the wearing qualities, and raises the machinability. If the nickel content is raised to 10 to 18 percent, an austenitic structure with valuable heat- and corrosion-resistant properties results. Chromium increases the hardness and wear resistance and, when used with a chill, increases the tendency to form white iron. When chromium and nickel are both added, the hardness and strength are improved without a reduction in the machinability rating. Molybdenum added in quantities up to 1.25 percent increases the stiffness, hardness, tensile strength, and impact resistance. It is a widely used alloying element.

Cast Steels

The advantage of the casting process is that parts having complex shapes can be manufactured at costs less than fabrication by other means, such as welding. Thus the choice of steel castings is logical when the part is complex and when it must also have a high strength. The higher melting temperatures for steels do aggravate the casting problems and require closer attention to such details as core design, section thicknesses, fillets, and the progress of cooling. The same alloying elements used for the wrought steels can be used for cast steels to improve the strength and other mechanical properties. Cast-steel parts can also be heat-treated to alter the mechanical properties, and, unlike the cast irons, they can be welded.

5–18 Nonferrous Metals

Aluminum

The outstanding characteristics of aluminum and its alloys are their strength-weight ratio, their resistance to corrosion, and their high thermal and electrical conductivity. The density of aluminum is about 2770 kg/m^3 (0.10 lb/in^3), compared with 7750 kg/m^3 (0.28 lb/in^3) for steel. Pure aluminum has a tensile strength of about 90 MPa (13 kpsi), but this can be improved considerably by cold working and also by alloying with other materials. The modulus of elasticity of aluminum, as well as of its alloys, is 71 GPa (10.3 Mpsi), which means that it has about one-third the stiffness of steel.

Considering the cost and strength of aluminum and its alloys, they are among the most versatile materials from the standpoint of fabrication. Aluminum can be processed by sand casting, die casting, hot or cold working, or extruding. Its alloys can be machined, press-worked, soldered, brazed, or welded. Pure aluminum melts at 660°C (1215°F), which makes it very desirable for the production of either permanent or sand-mold castings. It is commercially available in the form of plate, bar, sheet, foil, rod, and tube and in structural and extruded shapes. Certain precautions must be taken in joining aluminum by soldering, brazing, or welding; these joining methods are not recommended for all alloys.

The corrosion resistance of the aluminum alloys depends upon the formation of a thin oxide coating. This film forms spontaneously because aluminum is inherently very reactive. Constant erosion or abrasion removes this film and allows corrosion to take place. An extra-heavy oxide film may be produced by the process called *anodizing*. In this process the specimen is made to become the anode in an electrolyte, which may be chromic acid, oxalic acid, or sulfuric acid. It is possible in this process to control the color of the resulting film very accurately.

The most useful alloying elements for aluminum are copper, silicon, manganese, magnesium, and zinc. Aluminum alloys are classified as *casting alloys* or *wrought alloys*. The casting alloys have greater percentages of alloying elements to facilitate casting, but this makes cold working difficult. Many of the casting alloys, and some of the wrought alloys, cannot be hardened by heat treatment. The alloys that are heat-treatable use an alloying element which dissolves in the aluminum. The heat treatment consists of heating the specimen to a temperature which permits the alloying element to pass into solution, then quenching so rapidly that the alloying element is not precipitated. The aging process may be accelerated by heating slightly, which results in even greater hardness and strength. One of the better-known heat-treatable alloys is duraluminum, or 2017 (4 percent Cu, 0.5 percent Mg, 0.5 percent Mn). This alloy hardens in 4 days at room temperature. Because of this rapid aging, the alloy must be stored under refrigeration after quenching and before forming, or it must be formed immediately after quenching. Other alloys (such as 5053) have been developed which age-harden much more slowly, so that only mild refrigeration is required before forming. After forming, they are artificially aged in a furnace and possess approximately the same strength and hardness as the 2024 alloys. Those alloys of aluminum which cannot be heat-treated can be hardened only by cold working. Both work hardening and the hardening produced by heat treatment may be removed by an annealing process.

Magnesium

The density of magnesium is about 1800 kg/m^3 (0.065 lb/in^3), which is two-thirds that of aluminum and one-fourth that of steel. Since it is the lightest of all commercial metals, its greatest use is in the aircraft and automotive industries, but uses are now being found for it in other applications. Although the magnesium alloys do not have great strength, because of their light weight the strength-weight ratio compares favorably with the stronger aluminum and steel alloys. Even so, magnesium alloys find their greatest use in applications where strength is not an important consideration. Magnesium will not withstand elevated temperatures; the yield point is definitely reduced when the temperature is raised to that of boiling water.

Magnesium and its alloys have a modulus of elasticity of 45 GPa (6.5 Mpsi) in tension and in compression, although some alloys are not as strong in compression as in tension. Curiously enough, cold working reduces the modulus of elasticity. A range of cast magnesium alloys are also available.

Copper-base Alloys

When copper is alloyed with zinc, it is usually called *brass*. If it is alloyed with another element, it is often called *bronze*. Sometimes the other element is specified too, as, for example, *tin bronze* or *phosphor bronze*. There are hundreds of variations in each category.

Brass with 5 to 15 Percent Zinc

The low-zinc brasses are easy to cold work, especially those with the higher zinc content. They are ductile but often hard to machine. The corrosion resistance is good. Alloys included in this group are *gilding brass* (5 percent Zn), *commercial bronze* (10 percent Zn), and *red brass* (15 percent Zn). Gilding brass is used mostly for jewelry and articles to be gold-plated; it has the same ductility as copper but greater strength, accompanied by poor machining characteristics. Commercial bronze is used for jewelry and for forgings and stampings, because of its ductility. Its machining properties are poor, but it has excellent cold-working properties. Red brass has good corrosion resistance as well as high-temperature strength. Because of this it is used a great deal in the form of tubing or piping to carry hot water in such applications as radiators or condensers.

Brass with 20 to 36 Percent Zinc

Included in the intermediate-zinc group are *low brass* (20 percent Zn), *cartridge brass* (30 percent Zn), and *yellow brass* (35 percent Zn). Since zinc is cheaper than copper, these alloys cost less than those with more copper and less zinc. They also have better machinability and slightly greater strength; this is offset, however, by poor corrosion resistance and the possibility of season cracking at points of residual stresses. Low brass is very similar to red brass and is used for articles requiring deep-drawing operations. Of the copper-zinc alloys, cartridge brass has the best combination of ductility and strength. Cartridge cases were originally manufactured entirely by cold working; the process consisted in a series of deep draws, each draw being followed by an anneal to place the material in condition for the next draw; hence, the name cartridge brass. Although the hot-working ability of yellow brass is poor, it can be used in practically any other fabricating process and is therefore employed in a large variety of products.

When small amounts of lead are added to the brasses, their machinability is greatly improved and there is some improvement in their abilities to be hot-worked. The addition of lead impairs both the cold-working and welding properties. In this group are *low-leaded brass* ($32\frac{1}{2}$ percent Zn, $\frac{1}{2}$ percent Pb), *high-leaded brass* (34 percent Zn, 2 percent Pb), and *free-cutting brass* ($35\frac{1}{2}$ percent Zn, 3 percent Pb). The low-leaded brass is not only easy to machine but has good cold-working properties. It is used for various screw-machine parts. High-leaded brass, sometimes called *engraver's brass*, is used for instrument, lock, and watch parts. Free-cutting brass is also used for screw-machine parts and has good corrosion resistance with excellent mechanical properties.

Admiralty metal (28 percent Zn) contains 1 percent tin, which imparts excellent corrosion resistance, especially to saltwater. It has good strength and ductility but only fair machining and working characteristics. Because of its corrosion resistance it is used in power-plant and chemical equipment. *Aluminum brass* (22 percent Zn) contains 2 percent aluminum and is used for the same purposes as admiralty metal, because it has nearly the same properties and characteristics. In the form of tubing or piping, it is favored over admiralty metal, because it has better resistance to erosion caused by high-velocity water.

Brass with 36 to 40 Percent Zinc

Brasses with more than 38 percent zinc are less ductile than cartridge brass and cannot be cold-worked as severely. They are frequently hot-worked and extruded. *Muntz metal* (40 percent Zn) is low in cost and mildly corrosion-resistant. *Naval brass* has the same composition as Muntz metal except for the addition of 0.75 percent tin, which contributes to the corrosion resistance.

Bronze

Silicon bronze, containing 3 percent silicon and 1 percent manganese in addition to the copper, has mechanical properties equal to those of mild steel, as well as good corrosion resistance. It can be hot- or cold-worked, machined, or welded. It is useful wherever corrosion resistance combined with strength is required.

Phosphor bronze, made with up to 11 percent tin and containing small amounts of phosphorus, is especially resistant to fatigue and corrosion. It has a high tensile strength and a high capacity to absorb energy, and it is also resistant to wear. These properties make it very useful as a spring material.

Aluminum bronze is a heat-treatable alloy containing up to 12 percent aluminum. This alloy has strength and corrosion-resistance properties which are better than brass, and in addition, its properties may be varied over a wide range by cold working, heat treating, or changing the composition. When iron is added in amounts up to 4 percent, the alloy has a high endurance limit, a high shock resistance, and excellent wear resistance.

Beryllium bronze is another heat-treatable alloy, containing about 2 percent beryllium. This alloy is very corrosion-resistant and has high strength, hardness, and resistance to wear. Although it is expensive, it is used for springs and other parts subjected to fatigue loading where corrosion resistance is required.

With slight modification most copper-based alloys are available in cast form.

5–19 Plastics

The term *thermoplastics* is used to mean any plastic that flows or is moldable when heat is applied to it; the term is sometimes applied to plastics moldable under pressure. Such plastics can be remolded when heated.

A *thermoset* is a plastic for which the polymerization process is finished in a hot molding press where the plastic is liquefied under pressure. Thermoset plastics cannot be remolded.

Table 5–3 lists some of the most widely used thermoplastics, together with some of their characteristics and the range of their properties. Table 5–4, listing some of the thermosets, is similar. These tables are presented for information only and should not be used to make a final design decision. The range of properties and characteristics that can be obtained with plastics is very great. The influence of many factors, such as cost, moldability, coefficient of friction, weathering, impact strength, and the effect of fillers

Table 5–3

The Thermoplastics *Source:* These data have been obtained from the *Machine Design Materials Reference Issue*, published by Penton/IPC, Cleveland. These reference issues are published about every 2 years and constitute an excellent source of data on a great variety of materials.

Name	S_u, kpsi	E, Mpsi	Hardness Rockwell	Elongation %	Dimensional Stability	Heat Resistance	Chemical Resistance	Processing
ABS group	2–8	0.10–0.37	60–110R	3–50	Good	*	Fair	EMST
Acetal group	8–10	0.41–0.52	80–94M	40–60	Excellent	Good	High	M
Acrylic	5–10	0.20–0.47	92–110M	3–75	High	*	Fair	EMS
Fluoroplastic group	0.50–7	. . .	50–80D	100–300	High	Excellent	Excellent	MPR†
Nylon	8–14	0.18–0.45	112–120R	10–200	Poor	Poor	Good	CEM
Phenylene oxide	7–18	0.35–0.92	115R, 106L	5–60	Excellent	Good	Fair	EFM
Polycarbonate	8–16	0.34–0.86	62–91M	10–125	Excellent	Excellent	Fair	EMS
Polyester	8–18	0.28–1.6	65–90M	1–300	Excellent	Poor	Excellent	CLMR
Polyimide	6–50	. . .	88–120M	Very low	Excellent	Excellent	Excellent†	CLMP
Polyphenylene sulfide	14–19	0.11	122R	1.0	Good	Excellent	Excellent	M
Polystyrene group	1.5–12	0.14–0.60	10–90M	0.5–60	. . .	Poor	Poor	EM
Polysulfone	10	0.36	120R	50–100	Excellent	Excellent	Excellent†	EFM
Polyvinyl chloride	1.5–7.5	0.35–0.60	65–85D	40–450	. . .	Poor	Poor	EFM

*Heat-resistant grades available.

†With exceptions.

C Coatings L Laminates R Resins E Extrusions M Moldings S Sheet F Foams P Press and sinter methods T Tubing

Table 5–4

The Thermosets *Source:* These data have been obtained from the *Machine Design Materials Reference Issue*, published by Penton/IPC, Cleveland. These reference issues are published about every 2 years and constitute an excellent source of data on a great variety of materials.

Name	S_u, kpsi	E, Mpsi	Hardness Rockwell	Elongation %	Dimensional Stability	Heat Resistance	Chemical Resistance	Processing
Alkyd	3–9	0.05–0.30	99M*	. . .	Excellent	Good	Fair	M
Allylic	4–10	. . .	105–120M	. . .	Excellent	Excellent	Excellent	CM
Amino group	5–8	0.13–0.24	110–120M	0.30–0.90	Good	Excellent*	Excellent*	LR
Epoxy	5–20	0.03–0.30*	80–120M	1–10	Excellent	Excellent	Excellent	CMR
Phenolics	5–9	0.10–0.25	70–95E	. . .	Excellent	Excellent	Good	EMR
Silicones	5–6	. . .	80–90M	. . .	. . .	Excellent	Excellent	CLMR

*With exceptions.

C Coatings L Laminates R Resins E Extrusions M Moldings S Sheet F Foams P Press and sinter methods T Tubing

and reinforcements, must be considered. Manufacturers' catalogs will be found quite helpful in making possible selections.

5–20 Notch Sensitivity

In Sec. 3–15 it was pointed out that the existence of irregularities or discontinuities, such as holes, grooves, or notches, in a part increases the theoretical stresses significantly in the immediate vicinity of the discontinuity. And Eq. (3–51) defined a stress concentration factor K_t which is used with the nominal stress to obtain the maximum resulting stress due to the irregularity or defect. It turns out that some materials are not fully sensitive to the presence of notches and hence, for these, a reduced value of K_t can be used. For these materials, the maximum stress is, in fact,

$$\sigma_{\max} = K_f \sigma_0 \qquad \text{or} \qquad \tau_{\max} = K_{fs} \tau_0 \tag{5–25}$$

where K_f is a reduced value of K_t and σ_0 is the nominal stress. The factor K_f is commonly called a *fatigue stress-concentration factor*, and hence the subscript f. So it is convenient to think of K_f as a stress-concentration factor reduced from K_t because of lessened sensitivity to notches. The resulting factor is defined by the equation

$$K_f = \frac{\text{maximum stress in notched specimen}}{\text{stress in notch-free specimen}} \tag{a}$$

Notch sensitivity q is defined by the equation

$$q = \frac{K_f - 1}{K_t - 1} \qquad \text{or} \qquad \frac{K_{fs} - 1}{K_{ts} - 1} = q_{\text{shear}} \tag{b}$$

where q is usually between zero and unity. Equation (*b*) shows that if $q = 0$, the $K_f = 1$, and the material has no sensitivity to notches at all. On the other hand, if $q = 1$, then $K_f = K_t$, and the material has full notch sensitivity. In analysis or design work, find K_t first, from the geometry of the part. Then specify the material, find q, and solve for K_f from the equation

$$K_f = 1 = q(K_t - 1) \qquad \text{or} \qquad K_{fs} = 1 + q_{\text{shear}}(K_t - 1) \tag{5–26}$$

For steels and 2024 aluminum alloys, use Fig. 5–16 to find q for bending and axial loading. For shear loading, use Fig. 5–17. In using these charts it is well to know that the

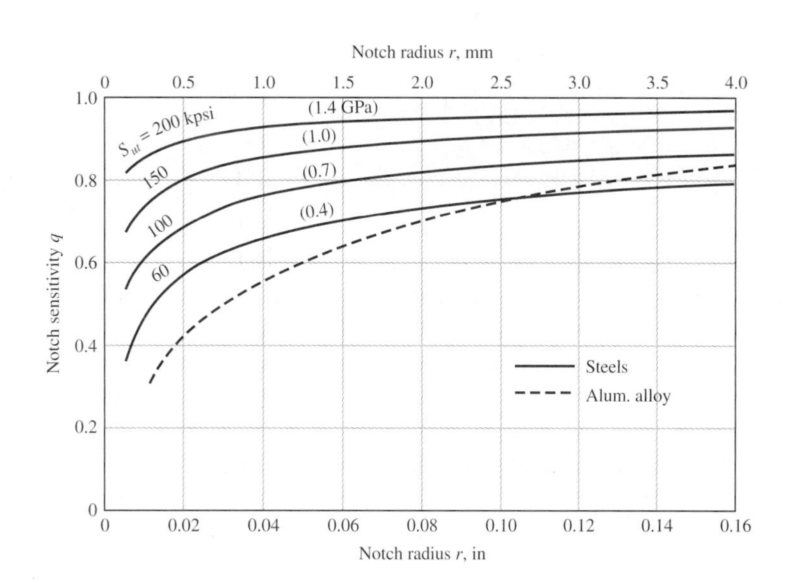

Figure 5–16

Notch-sensitivity charts for steels and UNS A92024-T wrought aluminum alloys subjected to reversed bending or reversed axial loads. For larger notch radii, use the values of q corresponding to the $r = 0.16$-in (4-mm) ordinate. [Reproduced by permission from George Sines and J. L. Waisman (eds.), *Metal Fatigue*, McGraw-Hill, New York, 1969, pp. 296, 298.]

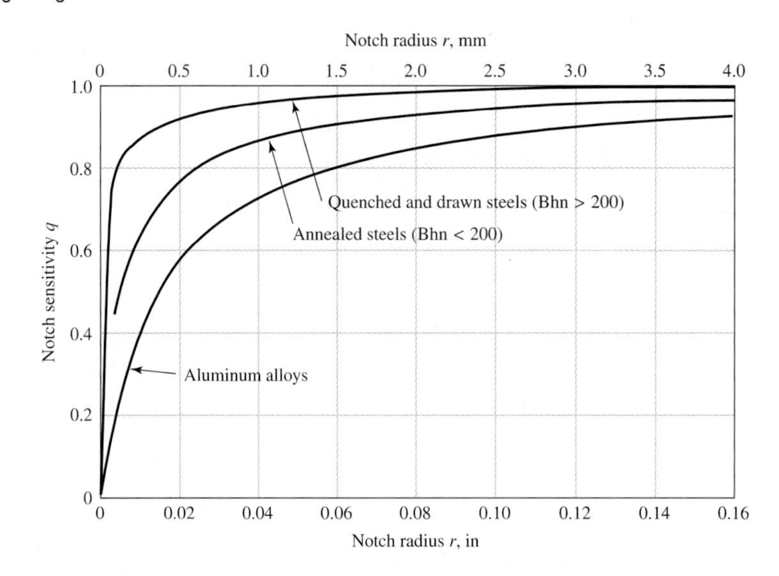

Figure 5–17

Notch-sensitivity curves for materials in reversed torsion. For larger notch radii, use the values of q corresponding to $r = 0.16$ in (4 mm).

Table 5–5

Coefficients of Variation C_{Kf} for Steels

Notch Type	Coefficient of Variation C_{Kf}
Transverse hole	0.10
Shoulder	0.11
Groove	0.15

Notes: Heywood's coefficients of variation. Notch sensitivity charts can be avoided using a modified Neuber equation. See Sec. 7–7.

actual test results from which the curves were derived exhibit a large amount of scatter. Because of this scatter it is always safe to use $K_f = K_t$ if there is any doubt about the true value of q. Also, note that q is not far from unity for large notch radii.

The notch sensitivity of the cast irons is very low, varying from 0 to about 0.20, depending upon the tensile strength. To be on the conservative side, it is recommended that the value $q = 0.20$ be used for all grades of cast iron.

Statistical Results

When the notch sensitivity q is obtained from Figs. 5–16 and 5–17, the resulting value of K_f from Eq. (5–26) may be treated as the mean value. The coefficient of variation then depends upon the type of discontinuity. Table 5–5 can be used to find values for the steels.

5–21 Introduction to Fracture Mechanics

The idea that cracks exist in parts even before service begins, and that cracks can grow during service, has led to the descriptive phrase "damage-tolerant design." The focus of this philosophy is on crack growth until it becomes critical, and the part is removed from service. The analysis tool is *linear elastic fracture mechanics* (LEFM). Inspection and maintenance are essential in the decision to retire parts before cracks reach catastrophic size. We shall now examine some of the basic ideas and vocabulary needed for the potential of the approach to be appreciated.

The use of elastic stress-concentration factors provides an indication of the average load required on a part for the onset of plastic deformation, or yielding; these factors are also useful for analysis of the loads on a part that will cause fatigue fracture. However, stress-concentration factors are limited to structures for which all dimensions are precisely known, particularly the radius of curvature in regions of high stress concentration. When there exists a crack, flaw, inclusion, or defect of unknown small radius in a part, the elastic stress-concentration factor approaches infinity as the root radius approaches zero, thus rendering the stress-concentration factor approach useless. Furthermore, even if the radius of curvature of the flaw tip is known, the high local stresses there will lead to local plastic deformation surrounded by a region of elastic deformation. Elastic stress-concentration factors are no longer valid for this situation, so analysis from the point of view of stress-concentration factors does not lead to criteria useful for design when very sharp cracks are present.

By combining analysis of the gross elastic changes in a structure or part that occur as a sharp brittle crack grows with measurements of the energy required to produce new fracture surfaces, it is possible to calculate the average stress (if no crack were present) which will cause crack growth in a part. Such calculation is possible only for parts with cracks for which the elastic analysis has been completed, and for materials that crack in a relatively brittle manner and for which the fracture energy has been carefully measured. The term *relatively brittle* is rigorously defined in the test procedures,[9] but it means, roughly, *fracture without yielding occurring throughout the fractured cross section.*

Thus glass, hard steels, strong aluminum alloys, and even low-carbon steel below the ductile-to-brittle transition temperature can be analyzed in this way. Fortunately, ductile materials blunt sharp cracks, as we have previously discovered, so that fracture occurs at average stresses of the order of the yield strength, and the designer is prepared for this condition. The middle ground of materials that lie between "relatively brittle" and "ductile" is now being actively analyzed, but exact design criteria for these materials are not yet available.

Quasi-Static Fracture

Many of us have had the experience of observing brittle fracture, whether it is the breaking of a cast-iron specimen in a tensile test or the twist fracture of a piece of blackboard chalk. It happens so rapidly that we think of it as instantaneous, that is, the cross section simply parting. Fewer of us have skated on a frozen pond in the spring, with no one near us, heard a cracking noise, and stopped to observe. The noise is due to cracking. The cracks move slowly enough for us to see them run. The phenomenon is not instantaneous, since some time is necessary to feed the crack energy from the stress field to the crack for propagation. Quantifying these things is important to understanding the phenomenon "in the small." In the large, a static crack may be stable and will not propagate. Some level of loading can render the crack unstable, and the crack propagates to fracture.

Early investigations of cracking by the *energy and compliance* method led to the definition of *fracture toughness* R as the energy to open a crack (thereby producing new area) per unit area. The usual units are kJ/m^2. In the fracture mechanics approach, a *stress-intensity factor* K has also come to be called fracture toughness. The quantitative definition of stress-intensity factor K is

$$K = \sigma\sqrt{\pi a} \qquad (a)$$

where σ is the normal stress on the gross area and a is the crack length. The usual units of K are MPa $\sqrt{\text{m}}$ or kpsi $\sqrt{\text{in}}$. In searching for data concerning fracture toughness one

9. BS 5447:1977 and ASTM E399-78.

may encounter either R or K. These are related by

$$K = \sqrt{ER} \tag{b}$$

where E is Young's modulus. To avoid confusing R with K we can speak of the "fracture toughness R" or the "fracture toughness K." If you mean fracture toughness K, why not say stress-intensity factor?

Deformation Modes and Stress-Intensity Factor

For mode I loading (see Fig. 5–20) the stress-intensity factor K_{I} characterizes the magnitude of the stresses in the vicinity of an ideally sharp crack tip in a linearly elastic and isotropic material. The stresses σ_x, σ_y, and τ_{xy} are proportional to K_{I}. For the case of Fig. 5–18, $a/b = 0$,

$$K_O = \sigma\sqrt{\pi a} \text{ MPa} \sqrt{\text{m}} \tag{5–27}$$

When $a/b \neq 0$, say $a/b = 0.5$, the stress intensity K_{I} equals

$$\begin{aligned} K_{\mathrm{I}} &= \beta\sigma\sqrt{\pi a} = \beta(\text{geometry}, a/b)\sigma\sqrt{\pi a} = (K_{\mathrm{I}}/K_0)\sigma\sqrt{\pi a} \\ &= (1.32/1)\sigma\sqrt{\pi a} = 1.32\sigma\sqrt{\pi a} \text{ MPa} \sqrt{\text{m}} \end{aligned} \tag{5–28}$$

When loading in modes II and III occur, figures similar to Figs. 5–19 to 5–26 exist.

Fracture Toughness

Through carefully controlled testing of a given material, the stress-intensity factor at which a crack will propagate to fracture is measured. This *critical stress-intensity factor* is denoted K_c. Thus, for a known magnitude of applied stress σ acting on a part with a known crack length $2a$, when the magnitude of K_{I} reaches K_c, fracture ensues. The

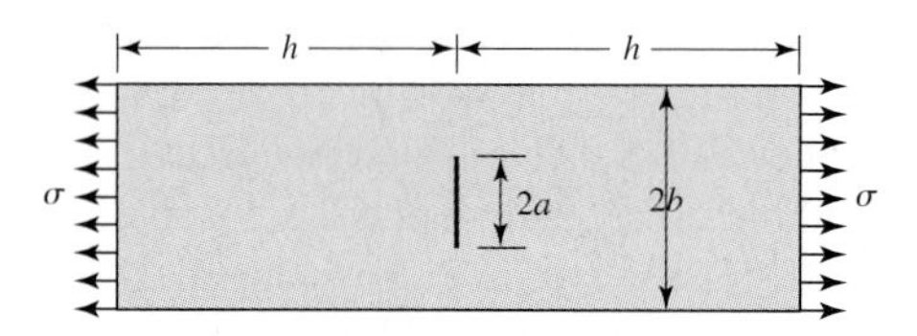

Figure 5–18

Plate length $2h$, width $2b$, containing a central crack of length $2a$; tensile stress σ acts in longitudinal direction.

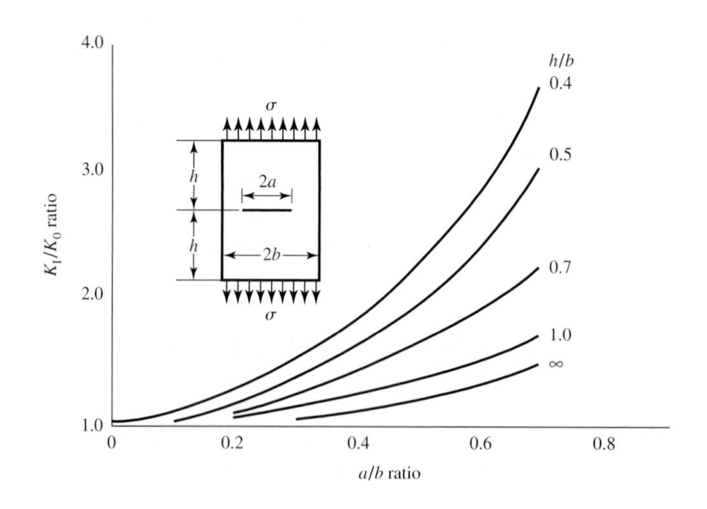

Figure 5–19

Plate containing a central crack loaded in longitudinal tension.

strength-to-stress ratio $K_c/K_{\rm I}$ can be used as a factor of safety:

$$n = \frac{S}{\sigma} = \frac{K_c}{K_{\rm I}} \tag{5–29}$$

EXAMPLE 5–5

A steel ship deck plate is 30 mm thick, 12 m wide, and 20 m long (in the tensile stress direction). It is loaded with a nominal tensile stress of 50 MPa. It is operated below its ductile-to-brittle transition temperature with K_c equal to 28.3 MPa$\sqrt{\text{m}}$. If a 65-mm-long central transverse crack is present, estimate tensile stress at which catastrophic failure will occur. Compare this stress with the yield strength of 240 MPa for this steel.

Solution

From Fig. 5–18, we see $2a = 65$ mm, $2b = 12$ m, $2h = 20$ m. Consequently, $a/b = 32.5/6(10^3) = 0.005$, $h/b = 10/6 = 1.67$. Since a/b is so small, treat the plate as robustly infinite with $a/b = 0$ and $\beta = K_{\rm I}/K_0 = 1$. The load-induced stress is given as 50 MPa, so the stress intensity factor $K_{\rm I}$ is, from Eq. (5–28),

$$K_{\rm I} = \beta\sigma\sqrt{\pi a} = (1)50\sqrt{\pi(32.5 \cdot 10^{-3})} = 16.0 \text{ MPa}\sqrt{\text{m}}$$

From Eq. (5–29),

$$n = \frac{K_c}{K_{\rm I}} = \frac{28.3}{16.0} = 1.77$$

The stress at which catastrophic failure occurs is

$$\sigma_c = \frac{K_c}{\beta\sqrt{\pi a}} = \frac{28.3}{(1)\sqrt{\pi(32.5 \cdot 10^{-3})}} = 88.6 \text{ MPa}$$

The yield strength is 240 MPa, catastrophic failure occurs at $88.6/240 = 0.37$, or at 37 percent of yield. The factor of safety in this circumstance is $K_c/K_{\rm I} = 28.3/16 = 1.77$, and not $240/50 = 4.8$.

Fracture toughness $K_{\rm Ic}$ for engineering metal lies in the range $20 \le K_{\rm Ic} \le 200$ MPa$\sqrt{\text{m}}$; for engineering polymers and ceramics, $1 \le K_{\rm Ic} \le 5$ MPa$\sqrt{\text{m}}$. For a 4340 steel, where the yield strength due to heat treatment ranges from 800 to 1600 MPa, $K_{\rm Ic}$ *decreases* from 190 to 40 MPa.

One of the first problems facing the designer is that of deciding whether the conditions exist, or not, for a brittle fracture. Low-temperature operation, that is, operation below room temperature, is a key indicator that brittle fracture is a possible failure mode. Tables of transition temperatures for various materials have not been published, possibly because of the wide variation in values, even with a single material. Thus, in many situations, laboratory testing may give the only clue to the possibility of brittle fracture.

Another key indicator of the possibility of fracture is the ratio of the yield strength to the ultimate strength. A high ratio S_y/S_u indicates there is only a small ability to absorb energy in the plastic region and hence there is a likelihood of brittle fracture.

Figure 5–20

Deformation modes: mode I is tension; modes II and III are both shear modes.

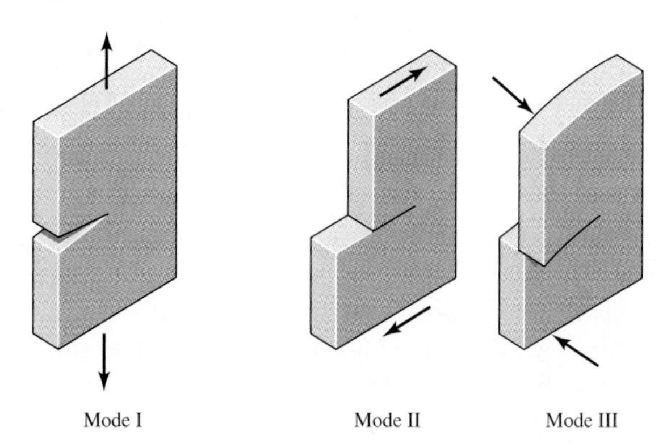

Low Charpy impact values can be used as a rough indicator of the possibility of a brittle fracture, though it is impossible to specify a transition value.[10]

The three possible ways of separating a plate are shown in Fig. 5–20. Note that modes II and III are fundamentally shear modes of fracture, but model II stresses and deformation stay within the plane of the plate. Mode III is out-of-plane shear.

Some stress analyses and fewer critical stress-intensity factor measurements have been made for modes I and III, but they are still limited in scope. The nomenclature K_{I} for stress-intensity factor and $K_{\mathrm{I}c}$ for critical stress-intensity factor under mode I conditions are in general use, so for clarity the subscript I will be added here. When analysis of K and measurements of K_c are made generally available for modes II and III, then more design can be extended to these modes. The procedure is exactly the same as for mode I analysis.

Stress-Intensity Factor Information

A substantial number of geometries for stress-intensity factors have been compiled in recent years.[11] Some of these are included here as Figs. 5–21 to 5–26. If K_{I} is needed for a configuration not included in the literature, the designer's only recourse is to carry out the complete analysis alone. A large body of literature on this subject is summarized in a form useful to the designer,[12] and typical values of $K_{\mathrm{I}c}$ are listed in Table 5–6.[13]

Note carefully in Table 5–6 the general inverse relationship between yield strength and $K_{\mathrm{I}c}$. This often leads to the choice of a material of lower yield strength and higher $K_{\mathrm{I}c}$, as is shown in the example that follows.

10. Charpy values are available for a wide range of materials. See, for example, Eric A. Brandes (ed.), *Smithell's Metals Reference Book*, 6th ed., Butterworth, London, 1983, Chap. 22.

11. H. Tada, P. C. Paris, and G. R. Irwin, *The Stress Analysis of Cracks Handbook,* Del Research, Hellertown, Pa., 1973; G. C. M. Sih, *Handbook of Stress Intensity Factors,* Lehigh University, Bethlehem, Pa., 1973; D. P. Rooke and D. J. Cartwright, *Compendium of Stress Intensity Factors,* H.M.S.O., Hillingdon Press, Uxbridge, England, 1976.

12. David K. Felbeck and Anthony G. Atkins, *Strength and Fracture of Engineering Solids,* Prentice-Hall, Englewood Cliffs, N.J., 1984; Kåre Hellan, *Introduction to Fracture Mechanics,* McGraw-Hill, New York, 1984.

13. For an extensive compilation of K_c values, see *Damage Tolerant Design Handbook,* MCIC-HB-01, Air Force Materials Laboratory, Wright-Patterson Air Force Base, Ohio, December 1972 and supplements.

Figure 5–21

Off-center crack in a plate in longitudinal tension; solid curves are for the crack tip at *A*; dashed curves are for the tip at *B*.

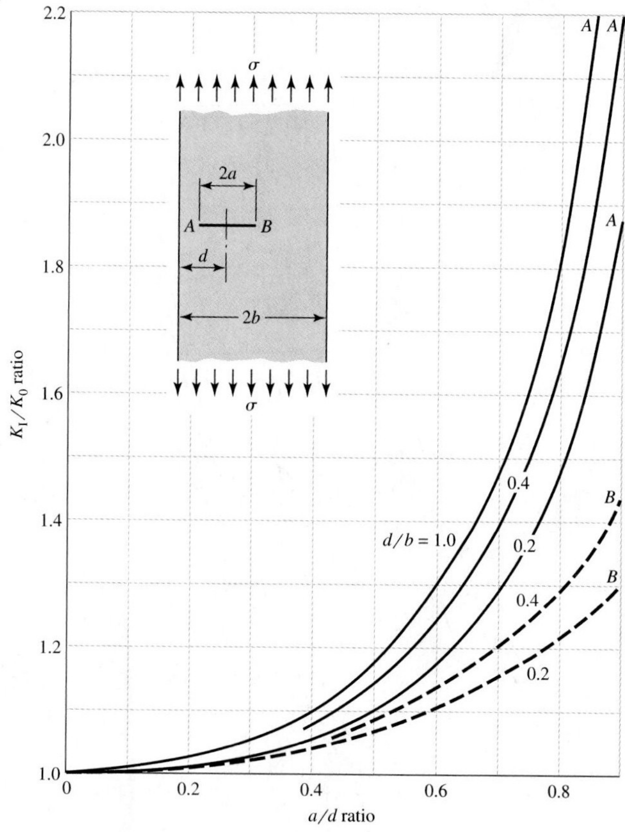

Figure 5–22

Plate loaded in longitudinal tension with a crack at the edge; for the solid curve there are no constraints to bending; the dashed curve was obtained with bending constraints added.

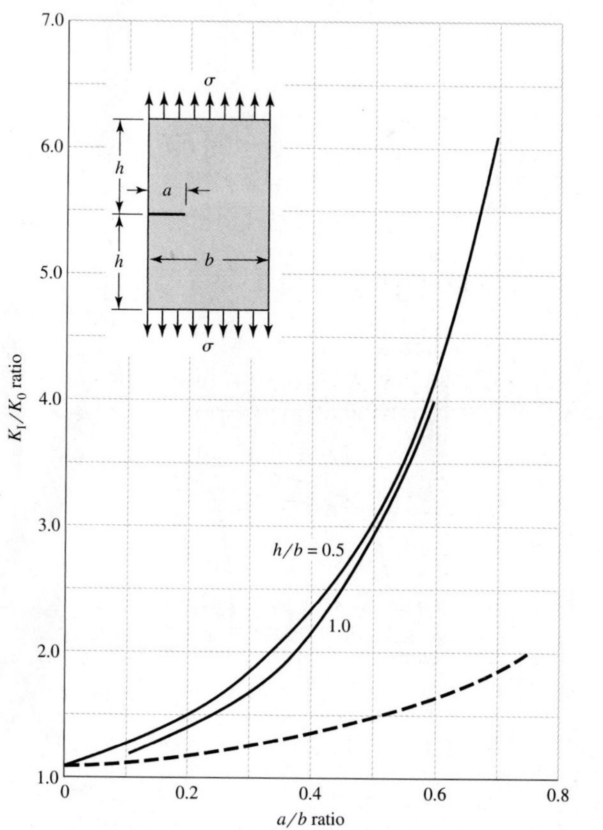

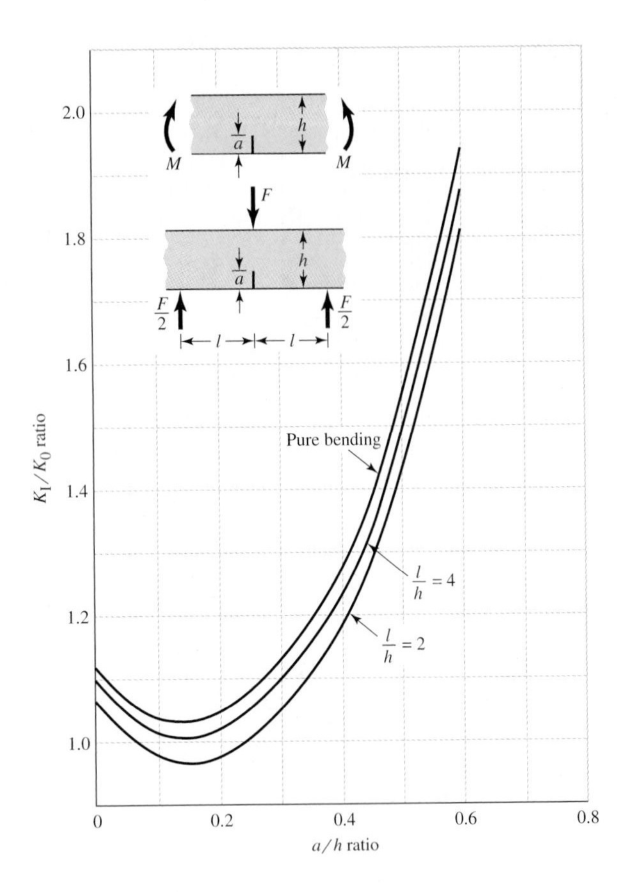

Figure 5–23

Beams of rectangular cross section having an edge crack.

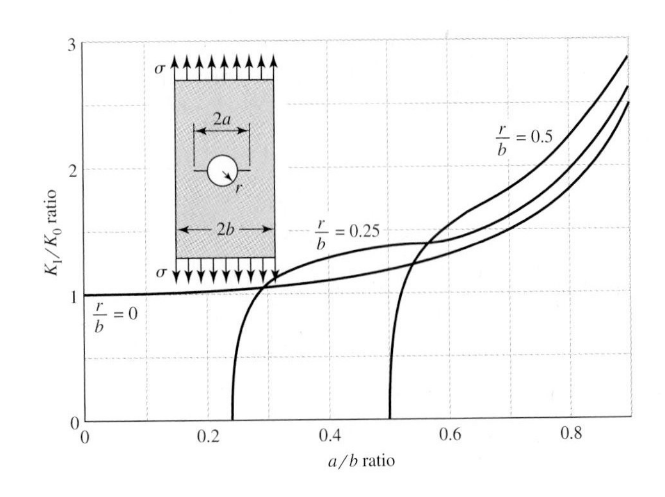

Figure 5–24

Plate in tension containing a circular hole with two cracks.

Figure 5–25

A cylinder loading in axial tension having a radial crack of depth a extending completely around the circumference of the cylinder.

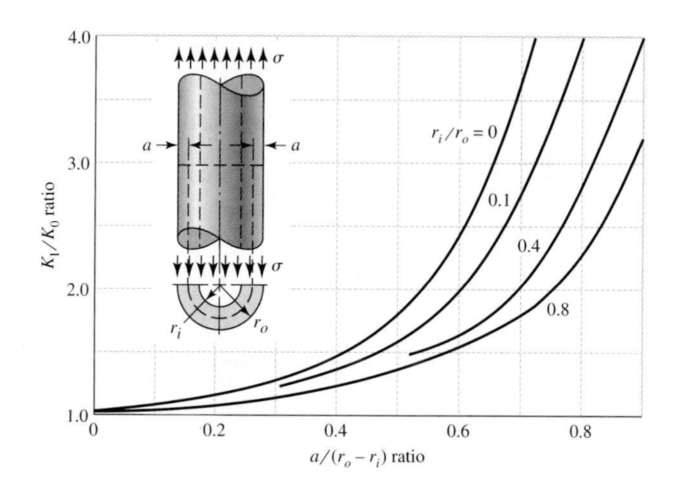

Figure 5–26

Cylinder subjected to internal pressure p, having a radial crack in the longitudinal direction of depth a. Use Eq. (3–54) for the tangential stress at $r = r_0$.

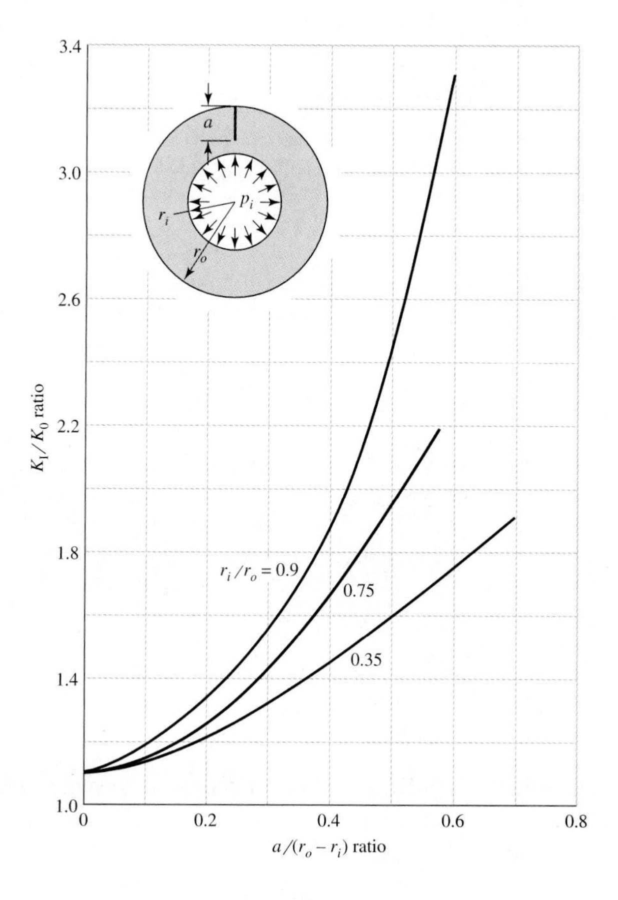

Table 5–6

Values of K_{Ic} for Some Engineering Materials

Material	K_{Ic}, MPa$\sqrt{m}$	S_y, MPa
Aluminum		
2024	26	455
7075	24	495
7178	33	490
Titanium		
Ti-6AL-4V	115	910
Ti-6AL-4V	55	1035
Steel		
4340	99	860
4340	60	1515
52100	14	2070

EXAMPLE 5–6 A plate of width 1.4 m and length 2.8 m is required to support a tensile force in the 2.8-m direction of 4.0 MN. Inspection procedures will only detect through-thickness edge cracks larger than 2.7 mm. The two Ti-6AL-4V alloys in Table 5–6 are being considered for this application, for which the safety factor must be 1.3 and minimum weight is important. Which alloy should be used?

Solution

(*a*) We elect first to estimate the thickness required to resist yielding. Since $\sigma = P/wt$, we have $t = P/w\sigma$. But

$$\sigma_{\text{all}} = \frac{S_y}{n} = \frac{910}{1.3} = 700 \text{ MPa}$$

Thus

$$t = \frac{P}{w\sigma_{\text{all}}} = \frac{4.0(10)^3}{1.4(700)} = 4.08 \text{ mm or greater}$$

where we have $S_y = 910$ MPa for the weaker titanium alloy. For the stronger alloy, we have, from Table 5–6,

$$\sigma_{\text{all}} = \frac{1035}{1.3} = 796 \text{ MPa}$$

and so the thickness is

Answer

$$t = \frac{P}{w\sigma_{\text{all}}} = \frac{4.0(10)^3}{1.4(796)} = 3.59 \text{ mm or greater}$$

(*b*) Now let us find the thickness required to prevent crack growth. Using Fig. 5–22, we have

$$\frac{h}{b} = \frac{2.8/2}{1.4} = 1 \qquad \frac{a}{b} = \frac{2.7}{1.4(10^3)} = 0.001\,93$$

Corresponding to these ratios we find from Fig. 5–22 that $\beta = K_{\text{I}}/K_0 = 1.1$, and $K_{\text{I}} = \beta\sigma\sqrt{\pi a} = 1.1$. From Table 5–6, $K_{\text{Ic}} = 115$ MPa$\sqrt{\text{m}}$ for the weaker of the two alloys:

$$n = 1.3 = \frac{K_{Ic}}{K_I} = \frac{115\sqrt{10^3}}{1.1\sigma\sqrt{\pi a}}, \qquad \sigma = \frac{K_{Ic}}{nK_I\sqrt{\pi a}}$$

Solving for σ with $n = 1$ gives the fracture stress

$$\sigma = \frac{115\sqrt{10^3}/1.3}{1.1\sqrt{\pi 2.7}} = 1135 \text{ MPa}$$

which is greater than the yield strength of 910 MPa, and so yield strength is the basis for the geometry decision. For the stronger alloy $S_y = 1035$ MPa, with $n = 1$ the fracture stress is

$$\sigma = \frac{K_{Ic}}{nK_I} = \frac{55\sqrt{10^3}/1.3}{(1)1.1\sqrt{\pi(2.7)}} = 542.9 \text{ MPa}$$

which is less than the yield strength. The thickness t is

$$t = \frac{P}{w\sigma_{all}} = \frac{4.0(10^3)}{1.4(542.9/1.3)} = 6.84 \text{ mm or greater}$$

This example shows that the fracture toughness K_{Ic} limits the geometry when the stronger alloy is used, and so a thickness of 6.84 mm or larger is required. When the weaker alloy is used the geometry is limited by the yield strength, giving a thickness of only 4.08 mm or greater. Thus the weaker alloy leads to a thinner and lighter weight choice since the failure modes differ.

Fatigue

The first phase of fatigue cracking is designated as *stage I fatigue*. Crystal slip that extends through several contiguous grains, inclusions, and surface imperfections are presumed to play a role. Since most of this is invisible to the observer, we just say that stage I involves several grains. The second phase, that of crack extension, is called *stage II fatigue*. The advance of the crack (that is, new crack area is created) does produce evidence that can be observed on micrographs from an electron microscope. The growth of the crack is orderly. Final fracture occurs during *stage III fatigue*, although fatigue is not involved. When the crack is sufficiently long that $K_I = K_{Ic}$ for the stress amplitude involved, then K_{Ic} is the critical stress intensity for the undamaged metal, and there is sudden, catastrophic failure of the remaining cross section in tensile overload. Stage III fatigue is associated with rapid acceleration of crack growth then fracture.

Crack Growth

Fatigue cracks nucleate and grow when stresses vary and there is some tension in each stress cycle. The crack nucleates at or very near a free surface or large discontinuity, as seen in Fig. 5–27. If a is the crack length, and N the cycle count, then the plot of the data is of the form of Fig. 5–28, with different loci corresponding to different stress ranges, beginning at initial crack length a_i. Notice the effect of the stress range $\Delta\sigma_{nom}$ in Fig. 5–28 is the production of longer cracks at a particular cycle count.

When the rate of crack growth per cycle, da/dN in Fig. 5–28, is plotted as shown in Fig. 5–29, the data from all three stress range levels superpose to give a sigmoidal locus. The three stages of crack development are observable, and the stage II data are linear on loglog coordinates, within the domain of LEFM validity. A group of similar

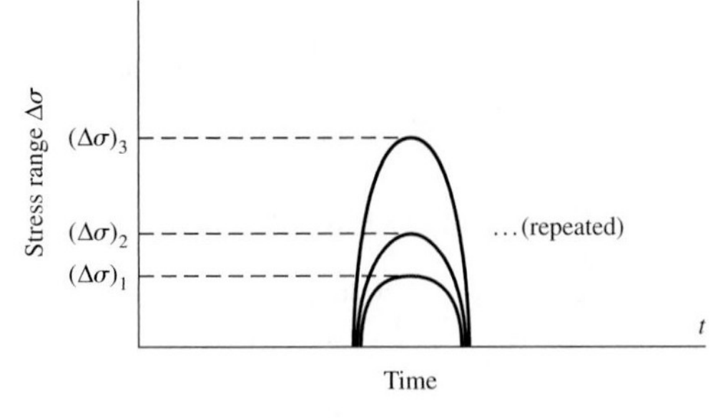

Figure 5–27

Repeatedly applied stresses producing the stress ranges $(\Delta\sigma)_1$, $(\Delta\sigma)_2$, and $(\Delta\sigma)_3$, with associated stress-intensity factor ranges $(\Delta K_I)_1$, $(\Delta K_I)_2$, and $(\Delta K_I)_3$ See Eq. (5–31).

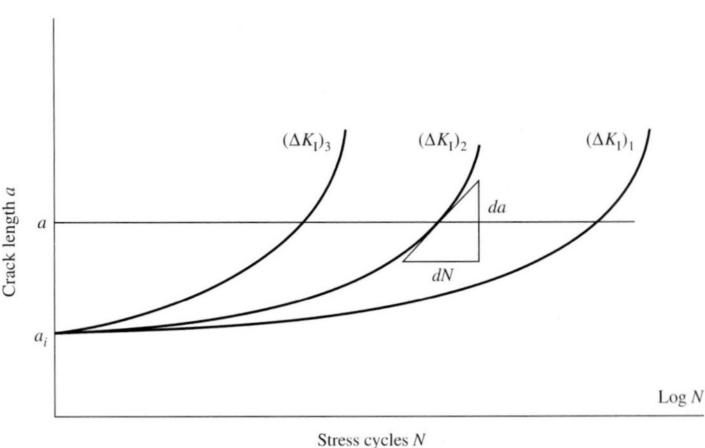

Figure 5–28

The increase in crack length a from an initial length of a_i as a function of cycle count for the three stress ranges of Fig. 5–27.

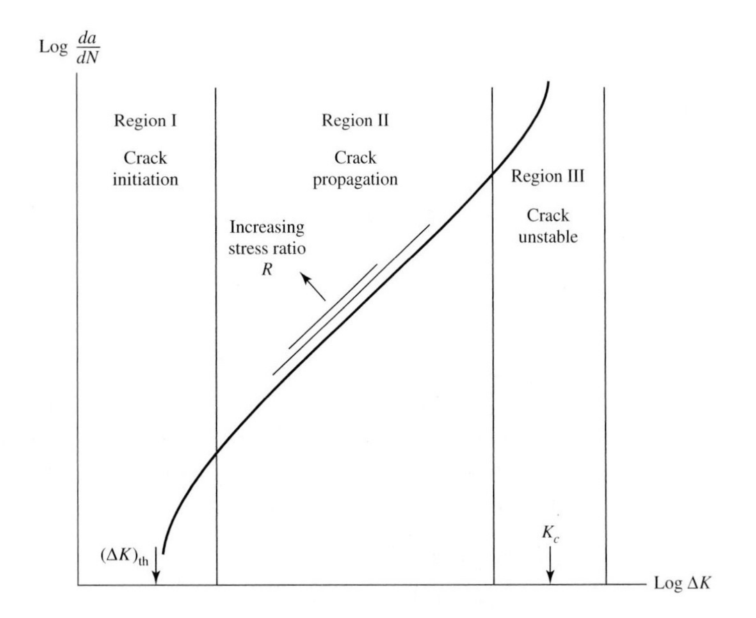

Figure 5–29

When da/dN is measured in Fig. 5–28 and plotted on loglog coordinates, the data for different stress ranges *superpose*, giving rise to a sigmoid curve as shown. $(\Delta K_I)_{th}$ is the threshold value of ΔK_I, below which a crack does not grow. From threshold to rupture an aluminum alloy will spend 85–90 percent of life in region I, 5–8 percent in region II, and 1–2 percent in region III.

curves can be generated by changing the stress ratio $R = \sigma_{\min}/\sigma_{\max}$ of the experiment. If the specimens of Fig. 5–27 are tested under repeatedly applied loading, then not only is

$$K_I = \beta\sigma\sqrt{\pi a} \tag{5–30}$$

but the change in stress-intensity factor ΔK_I is

$$\Delta K_I = \beta\sqrt{\pi a}(\sigma_{\max} - \sigma_{\min}) = \beta\sqrt{\pi a}\,\Delta\sigma \tag{5–31}$$

Life Prediction

From Fig. 5–29, the linearity of data on the transformed coordinates in region II suggests the curve fit called the *Paris equation*, which is of the form

$$\frac{da}{dN} = A(\Delta K_{\mathrm{I}})^n \tag{5–32}$$

where A and n are the Paris coefficient and Paris exponent, respectively. Life prediction is possible using the Paris equation as follows:

$$N - N_i = \frac{1}{A}\int_{a_i}^{a} \frac{da}{(\Delta K_{\mathrm{I}})} = \frac{\mathrm{I}}{A}\int_{a_i}^{a} \frac{da}{(\beta\sqrt{\pi}\,\Delta\sigma)^n a^{n/2}} \tag{5–33}$$

For austenitic stainless steel, Barsom[14] reports $n = 3.25$ and $A = 5.60(10^{-2})$ in MKS and $3.00(10^{-10})$ in ips. The prediction of cycle lives involves knowledge of initial crack length a_i, geometry factor β, and crack size where $K_{\mathrm{I}} = K_{\mathrm{I}c}$.

Equation (5–33) can be integrated to give

$$N - N_i = \frac{a^{1-n/2} - a_i^{1-n/2}}{A(\beta\sqrt{\pi}\,\Delta\sigma)^n(1 - n/2)} \tag{5–34}$$

Another prediction equation, due to Forman,[15] not only includes stress ratio R, but is also valid in both regions II and III:

$$\frac{da}{dN} = \frac{A(\Delta K)^n}{(1 - R)K_c - \Delta K} \tag{5–35}$$

where K_c is the critical stress-intensity factor. The constants A and n can be found from the data string in region II. Integration of Eq. (5–35) gives

$$N - N_i = \int_{a_i}^{a} \frac{(1 - R)K_c - \Delta K}{A(\Delta K)^n} \tag{5–36}$$

The Paris constants A and n can be found from two points on the region II locus:

$$\left(\frac{da}{dN}\right)_1 = A[(\Delta K)_1]^n$$

$$\left(\frac{da}{dN}\right)_2 = A[(\Delta K)_2]^n$$

from which A and n can be found. A quotient of the two preceding equations eliminates A to give an equation for n. The Forman constants A and n will be different:

$$\left(\frac{da}{dN}\right)_1 = \frac{A[(\Delta K)_1]^n}{(1 - R)K_c - (\Delta K)_1}$$

$$\left(\frac{da}{dN}\right)_2 = \frac{A[(\Delta K)_2]^n}{(1 - R)K_c - (\Delta K)_2}$$

Again a quotient of the two preceding equations eliminates A to give an equation for n.

---

14. J. M. Barsom and S. T. Rolfe, *Fracture and Fatigue Control in Structures*, 2nd Ed., Prentice-Hall, Englewood Cliffs, New Jersey, 1987, pp. 288–291.

15. R. G. Forman, V. E. Kearney, and R. M. Engle, "Numerical Analysis of Crack Propagation in Cycle Load Structures", *J. Bas. Engin.*, Trans. ASME, Ser. D 89, 1967, p. 459.

Table 5–7

Several Entries from NASA/FLAGRO 2.0 Materials Constants

Material	S_y	S_{ut}	K_{Ie}	K_{Ic}	A_k	B_k	C	n	p	q
A36 structural steel	44	78	100	70	0.75	0.5	$0.100(10^{-8})$	3.0	0.5	0.5
1005-1012 HR plate	25	45	100	70	0.75	0.5	$0.80(10^{-10})$	3.60	0.5	0.5
80-55-06 cast iron	58	80	65	32	0.75	0.5	$0.70(10^{-9})$	2.90	0.5	0.5
AISI 4340	155	170	190	135	0.75	0.5	$0.17(10^{-8})$	2.70	0.25	0.25
AISI 301/302 stainless steel	190	205	210	80	1.0	0.5	$0.55(10^{-8})$	2.20	0.25	0.25
7075-T6 aluminum	75	84	37	27	1.0	1.0	$0.145(10^{-6})$	2.497	0.50	1.0

Units are U.S. customary, kpsi and kpsi $\sqrt{\text{in}}$. A more extensive table can be found in Bahram Farahmand, *Fatigue and Fracture Mechanics of High Risk Parts*, Chapman and Hall, New York, 1997, Appendix A.

Equation (5–36) can be broken into two integrals and evaluated as

$$N - N_i = \frac{(1-R)K_c\left(a^{1-n/2} - a_i^{1-n/2}\right)}{A(\beta\sqrt{\pi}\Delta\sigma)^n(1-n/2)} - \frac{2\left(a^{(3-n)/2} - a_i^{(3-n)/2}\right)}{A(\beta\sqrt{\pi}\Delta\sigma)^{n-1}(3-n)} \qquad n > 2 \tag{5–37}$$

The Forman-Newman-de Koning (FNK) equation[16] applies to all three regions of crack growth,

$$\frac{da}{dN} = \frac{C(1-f)^n \Delta K^n (1 - \Delta K_{\text{th}}/\Delta K)^p}{(1-R)^n\{1 - \Delta K/[(1-R)K_c]\}^q} \tag{5–38}$$

where ΔK = change in stress-intensity factor
ΔK_{th} = threshold stress-intensity factor range
C, n = region II curve-fit constants
p, q = empirical constants
f = crack-opening function

The parameter f has some complexity in its evaluation. A source of material information is NASA/FLAGRO 2.0, a fatigue crack growth computer program. Table 5–7 lists a few material parameters. The FNK model is used in the analysis of high-risk parts in the aircraft transportation industry.

EXAMPLE 5–7

A 7075-T6 aluminum alloy centrally cracked plate (Fig. 5–19) has a thickness of $\frac{1}{8}$ in and a width of 4 in. The load is 10 000 lb, repeatedly applied. The material properties are $S_y = 76$ kpsi, $S_{ut} = 84$ kpsi, $(\Delta K_{\text{I}})_c = 28$ kpsi $\sqrt{\text{in}}$, $(\Delta K_{\text{I}})_{\text{th}} = 2$ kpsi $\sqrt{\text{in}}$. The $R = 0$ da/dN versus ΔK curve is shown in Fig. 5–30. The repeatedly applied tensile stress is 20 kpsi.

(*a*) What are the Paris equation constants A and n?

(*b*) If the initial crack size is 0.005 in, how many cycles will increase the crack to 0.3 in?

16. B. Farahmand, *Fatigue and Fracture Mechanics of High Risk Parts*, Chapman & Hall, International Thomson Publishing, New York, 1997, p. 186.

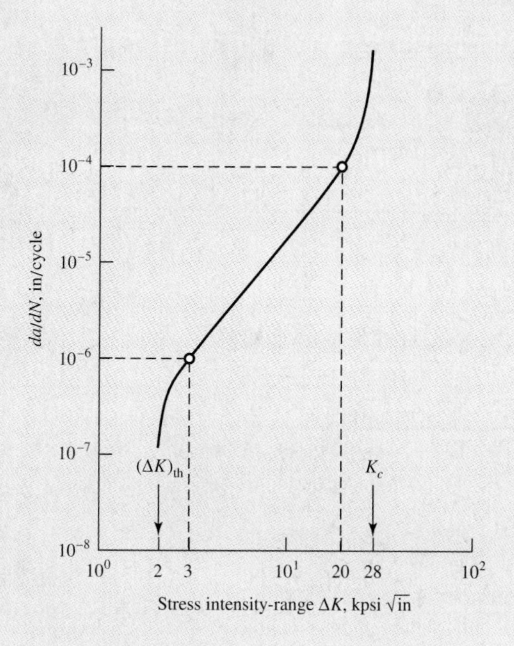

Figure 5–30

Crack growth rate curve for aluminum alloy of Ex. 5–7 and Ex. 5–8.

Solution (*a*) Select two points on Fig. 5–30, say,

$$10^{-4} = A(20)^n$$
$$10^{-6} = A(3)^n$$

from which $n = 2.43$ and $A = 0.695(10^{-7})$.

(*b*) Write Eq. (5–34) in the following form:

$$N - N_1 = \frac{a^{-0.215} - 0.005^{-0.215}}{0.695(10^{-7})(25\sqrt{\pi})^{2.43}(1 - 2.43/2)}$$

Instead of substituting 0.3 in for the final crack length a, we build a table for perspective:

a	N	ΔK
0.005	0	3.13
0.01	2 868	4.43
0.1	9 850	14.0
0.2	11 361	19.8
0.3	12 147	24.3

Note that we are in region II and the cycle count is 12 147 cycles.

EXAMPLE 5–8 Given the conditions of Ex. 5–7 and $R = 0$,

(*a*) Find the Forman constants A and n.

(*b*) Form a table as in Ex. 5–7 as far as K_c.

(*c*) Estimate the number of cycles to failure.

Solution (*a*) Select the same two points as in Ex. 5–7 to find A and n:

$$10^{-4} = \frac{A(20)^n}{28 - 20}$$

$$10^{-6} = \frac{A(3)^n}{28 - 3}$$

from which $n = 1.83$ and $A = 0.333(10^{-5})$.

(*b*) Write Eq. (5–36) as

$$N - N_i = \frac{28(a^{0.085} - 0.005^{0.085})}{0.333(10^{-5})(25\sqrt{\pi})^{1.83}(0.085)} - \frac{2(a^{0.585} - 0.005^{0.585})}{0.333(10^{-5})(25\sqrt{\pi})^{0.83}(1.17)}$$

$$\Delta K = \beta\sigma\sqrt{\pi a} = (1)25\sqrt{\pi a}$$

Form a table:

a, in	$N - N_i$ Cycles	ΔK, kpsi $\sqrt{\text{in}}$
0.005	0	3.13
0.01	3 215	4.43
0.1	12 997	14.0
0.2	14 917	19.8
0.3	15 548	24.3
0.4	15 693	28.0

(*c*) Note that the ΔK column is the same as in Ex. 5–7. Notice also that the crack length was carried to 0.4 in with a ΔK of 28 kpsi $\sqrt{\text{in}}$, which is critical. Since the Forman equation is valid in regions II and III, we can predict the crack length goes unstable at 15 693 cycles, which we could not do with the Paris equation.

Summary

Equations such as (5–33) and (5–36) permit life estimates. The FNK equation (5–38), applies to all three regions, permitting life estimates, but with complexity, for a small set of materials. The parameter f is called the *crack-opening function*. It is akin to β but it is expanded as a function of stress ratio R because of a cracking-opening phenomenon. Integration of da/dN raises the question of limits. The final value of a is the one corresponding to catastrophic failure. The initial value of a is from measurement on a part, or some kind of average over several parts, or, conservatively, the smallest detectable crack length by the measurement system used.

A damage-tolerant design method such as LEFM has appeal and drawbacks:

- Fracture mechanics constants must be available for materials and loading mode.
- Determination of initial crack length is essential.
- Monitoring crack growth by scheduled maintenance requires part accessibility and identification.
- Part retirement in time is possible.

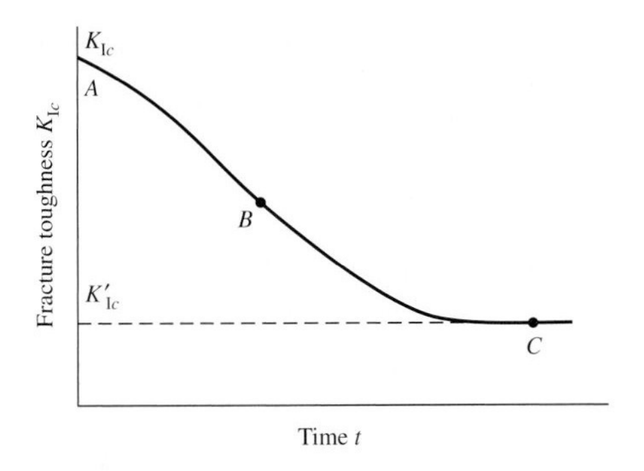

Figure 5–31

Change in fracture toughness with time.

There are costs, which are so high that the method has been applied to space and aviation equipment. Dowling[17] has a good discussion of why LEFM is not generally used. In the words of the early radio announcers, "Stay tuned."

5–22 Stress-Corrosion Cracking

Parts subjected to continuous static loads in certain corrosive environments may, over a period of time, develop serious cracks. This phenomenon is known as *stress-corrosion cracking*. Examples of such parts are door-lock springs, watch springs, lock washers, marine and bridge cables, and other highly stressed parts subjected to atmospheric or other corrosive surroundings. The stress, environment, time, and alloy structure of the part all seem to have an influence on the cracking, with each factor speeding up the influence of the others.

Stress-time tests[18] can be made on specimens in a corrosive environment in order to determine the limiting value of the fracture toughness. The curve shown in Fig. 5–31 typifies the results of many of these experiments. The tests must be run on a number of specimens, each subject to a constant but different load and each having the same size initial crack. It will then be found that the rate of crack growth depends both upon stress and upon time. When the times to fracture corresponding to the values of K_I are noted and plotted, a curve like that of Fig. 5–31 will be obtained. The limiting value of the stress-intensity factor is here designated as K'_{Ic}, corresponding to point C on the curve. Crack growth will not be obtained from stress-intensity factors less than this value, no matter how long the loaded specimen remains in the environment. Unfortunately, these tests require a great deal of time for completion, usually not less than 1000 h.[19]

5–23 Quantitative Estimation of Properties of Cold-Worked Metals

Critical or controlling locations in many machines parts are located in places where cold-work or cold-forming has occurred, incidentally or purposefully, in creating the part geometry. Plastic strains or a series of plastic strains changes yield and ultimate strengths (and associated endurance limits). There are ways to quantitatively estimate the new properties resulting from the plastic strains.

17. N. E. Dowling, *Mechanical Behavior of Materials*, Prentice Hall, Englewood Cliffs, N.J. 1993, pp. 498–501.

18. See H. O. Fuchs and R. I. Stephens, *Metal Fatigue in Engineering*, Wiley, New York, 1980, p. 218.

19. For some values of the stress-intensity factors K'_{Ic}, see *Damage Tolerant Handbook*, Metals and Ceramics Information Center, Battelle, Columbus, Ohio, 1975.

A great amount of research has been done on this subject by Joseph Datsko of the University of Michigan. Much of the methodology shown here is the result of his efforts. Any discussion of strength that is directional requires identification of

1. The kind of strength: ultimate u, yield y, fracture f, endurance e.
2. The sense of the strength: tensile t, compressive c, shear s.
3. The direction of orientation of the strength: longitudinal L, long transverse B, short transverse D, axial a, radial R, circumferential c.
4. The sense of the most recent prior strain in the axial direction of an envisioned specimen: tension t, compression c. If there is no prior strain, the subscript 0 is used.

Datsko[20] suggests a notation $(S_1)_{234}$ where the subscripts correspond to 1, 2, 3, and 4 above.

In Fig. 5–32, an axially deformed round plate and a rolled plate are depicted. A strength denoted $(S_u)_{tLC}$ would denote the engineering ultimate strength S_u in tension,

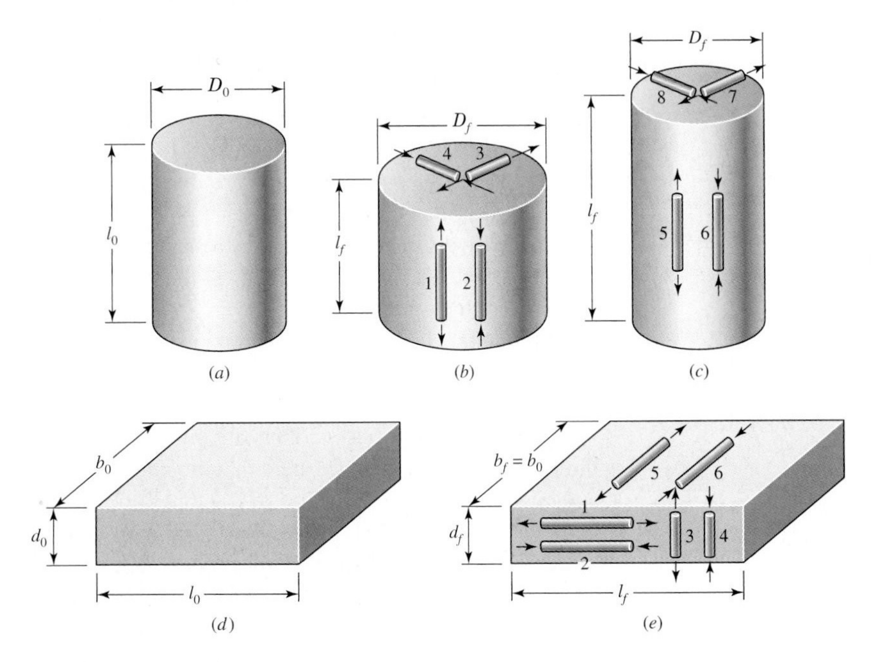

Figure 5–32

Designation of strengths in bars and plates. (*a*) Original bar before axial deformation. (*b*) Bar with permanent compressive deformation and the resulting directional strengths. (*c*) Bar with permanent tensile deformation and resulting directional strengths. (*d*) Plate prior to rolling. (*e*) Plate with permanent deformations due to rolling and the resulting directional strengths. (Adapted from Joseph Datsko, *Materials in Design and Manufacturing*, published by the author, Ann Arbor, Mich., 1977, p. 7–7.)

(*a*) Original bar before axial deformation.

	Specimen	Sense of strength	Direction in the bar	Prior strain	Designation
(*b*)	1	t	L	c	$(S)_{tLc}$
	2	c	L	c	$(S)_{cLc}$
	3	t	T	t	$(S)_{tTt}$
	4	c	T	t	$(S)_{cTt}$
(*c*)	5	t	L	t	$(S)_{tLt}$
	6	c	L	t	$(S)_{cLt}$
	7	t	T	c	$(S)_{tTc}$
	8	c	T	c	$(S)_{cTc}$

(*d*) Plate prior to rolling.

	Specimen	Sense of strength	Direction in the bar	Prior strain	Designation
(*e*)	1	t	L	t	$(S)_{tLt}$
	2	c	L	t	$(S)_{cLt}$
	3	t	D	c	$(S)_{tDc}$
	4	c	D	c	$(S)_{cDc}$
	5	t	B	0	$(S)_{tB0}$
	6	c	B	0	$(S)_{cB0}$

20. Joseph Datsko, *Materials in Design and Manufacturing*, published by the author, Ann Arbor, MI, 1977.

$(S_u)_t$, in the longitudinal direction, $(S_u)_{tL}$, after a last prior strain in the specimen direction that was compressive, $(S_u)_{tLc}$. In this notation the 0.2 percent yield strength in a tension test would be reported as $(S_y)_{tL0}$. By implication, so far in this book, a tensile test could establish S_{yt} (with the longitudinal direction L understood, and the absence of prior strain understood). By implication, we mean tensile yield unless otherwise stated, and we find S_y as a notation adequate and useful.

In the world of properties after prior plastic strains, we need Datsko's more complete notation. Datsko's research has yielded six observations that he calls "rules":

1. Strain is a bulk mechanism, exhibiting changes in strength in directions free of applied strain.
2. The maximum strain that can be imposed lies between the true strain at ultimate load ε_u and the true stain at fracture ε_f. In upsetting procedures devoid of flexure (bending), the limit is ε_f as determined in the tension test.
3. The significant strain in a sequence of strains is the largest absolute strain. In a round $\varepsilon_w = \max(|\varepsilon_r|, |\varepsilon_\theta|, |\varepsilon_x|)$. The largest absolute strain is used in the calculation of the equivalent plastic strain ε_q, which is defined for two categories of strength, ultimate and yield, and four groups of strength in Table 5–8.
4. In the case of several strains applied sequentially (say, cold-rolling then upsetting) in determining ε_{qu}, the significant strains in cycle step ε_{wi} are added in decreasing order of absolute magnitude, rather than in chronological order.
5. If the plastic strain is imposed below the material's recrystallization temperature, the ultimate strength is given by

$$
\begin{aligned}
S_u &= (S_u)_0 \exp(\varepsilon_{qu}) && \varepsilon_{qu} < m \\
&= \sigma_0(\varepsilon_{qu})^m && \varepsilon_{qu} > m
\end{aligned}
$$

Table 5–8

Strength Relations for Plastically Deformed Metals*

Source: Joseph Datsko, *Materials in Design and Manufacturing*, published by the author, Ann Arbor, Mich., 1977, and M. P. Borden, "Multidimensional Tensile Properties of Materials Subjected to Large Cyclic Strains," Ph.D. diss., University of Michigan, Ann Arbor, 1975

$(S_y)_w = \bar{\sigma}_0(\varepsilon_{qy})^m \qquad (S_u)_w = \begin{cases} (S_u)_0 \exp \varepsilon_{qu} & \varepsilon_{qu} < m \\ \bar{\sigma}_w & \varepsilon_{qu} > m \end{cases}$

Group	Strength Designation	ε_{qu}	ε_{qy}
1	$(S)_{cLc}$ $(S)_{tLt}$ $(S)_{tB0}$ $(S)_{cB0}$ $(S)_{cDc}$	$\varepsilon_{qus} = \sum_{i=1}^{n} \frac{\varepsilon_{wi}}{i}$	$\varepsilon_{qys} = \frac{\varepsilon_{qus}}{1 + 0.2\varepsilon_{qus}}$
2	$(S)_{tTt}$ $(S)_{cTc}$	$\varepsilon_{qus} = \sum_{i=1}^{n} \frac{\varepsilon_{wi}}{i}$	$\varepsilon_{qys} = \frac{\varepsilon_{qus}}{1 + 0.5\varepsilon_{qus}}$
3	$(S)_{cLt}$ $(S)_{tLc}$ $(S)_{tDc}$	$\varepsilon_{qu0} = \sum_{i=1}^{n} \frac{\varepsilon_{wi}}{i+1}$	$\varepsilon_{qy0} = \frac{\varepsilon_{qu0}}{1 + 2\varepsilon_{qu0}}$
4	$(S)_{tTc}$ $(S)_{cTt}$	$\varepsilon_{qu0} = \sum_{i=1}^{n} \frac{\varepsilon_{wi}}{i+1}$	—†

*Plastic deformation below material's recrystallization temperature.

†$(S_y)_{tTc} = (S_y)_{cTt} = 0.95(S_y)_{tTt}$ or $0.95(S_y)_{cTc}$

ε_{qus} = equivalent strain when prestrain sense is same as sense of strength

ε_{qu0} = equivalent strain when prestrain sense is opposite to sense of strength

6 The yield strength of a material whose recrystallization temperature is not exceeded is given by

$$S_y = \sigma_0(\varepsilon_{qu})^m$$

Table 5–8 summarizes the strength relations for plastically deformed metals.

EXAMPLE 5–9 A 1045 hot-rolled bar has a 0.2 percent yield strength S_y of 60 kpsi, an ultimate tensile strength of $S_u = 92.5$ kpsi, a fractional area reduction $AR = 0.44$ at fracture, and a strain-strengthening exponent of $m = 0.14$, all by tension test. The material is to be used to form an integral pinion and shaft blank, by cold-working a $2\frac{1}{4}$-in diameter to a 2-in diameter, then upsetting to $2\frac{1}{2}$ in to form the pinion blank as depicted in Fig. 5–33. Use Datsko's rules to find the ultimate strength in the direction resisting tooth bending at the root of the teeth to be cut into the blank.

Solution At ultimate load, from Eq. (*c*) of Sect. 5–2,

$$\frac{P_u}{A_0} = S_u = \sigma_0 \varepsilon^m \exp(-\varepsilon) = \sigma_0 m^m \exp(-m)$$

Solving for $= \sigma_0$ gives

$$\sigma_0 = S_u \exp(m) m^{-m} = 92.5 \exp(0.14) 0.14^{-0.14} = 140.1 \text{ kpsi}$$

The fraction reduction of area at fracture is given by

$$AR = \frac{A_0 - A_f}{A_0} = 1 - \frac{A_f}{A_0}$$

From Eq. (5-7),

$$\epsilon = \frac{l_i}{l_0} - 1 = \frac{A_0}{A_i} - 1$$

The true strain at fracture ε_f, after Eq. (5-8), is

$$\varepsilon_f = \ln(\varepsilon + 1) = \ln\left(\frac{A_0}{A_f} - 1 + 1\right) = \ln\frac{A_0}{A_f} = \ln\frac{1}{1 - AR}$$

$$= \ln\frac{1}{1 - 0.44} = 0.58$$

which is the limiting strain when deformation is free of bending (rule 2). In the first

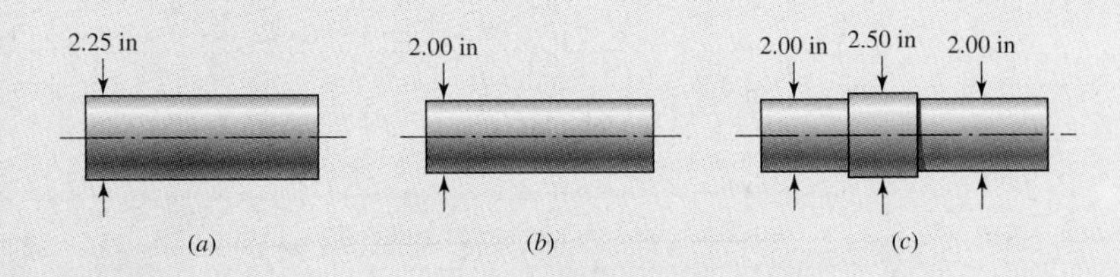

Figure 5–33

(*a*) Workpiece stock, hot-rolled, machined. (*b*) Workpiece after cold-forming. (*c*) Workpiece after upsetting to form gear blank.

deformation step (cold-rolling) the largest strain is axial, so

$$\varepsilon_1 = \left|\ln\left(\frac{D_0}{D_i}\right)^2\right| = \left|\ln\left(\frac{2.25}{2}\right)^2\right| = 0.236$$

In the second plastic deformation step (upsetting) the largest strain is axial, so

$$\varepsilon_2 = \left|\ln\left(\frac{D_i}{D_0}\right)^2\right| = \left|\ln\left(\frac{2}{2.5}\right)^2\right| = |-0.466| = 0.466$$

The significant strains are ε_{w1} and ε_{w2} (rule 4), thus $\varepsilon_{w1} = 0.466$ and $\varepsilon_{w2} = 0.236$. The group 1 strength is, from Table 5–8,

$$\varepsilon_{qu} = \sum \frac{\varepsilon_{wi}}{i} = \frac{0.466}{1} + \frac{0.236}{2} = 0.564$$

$$S_u = \sigma_0(\varepsilon_{qu})^m = 140.1(0.564)^{0.14} = 129.4 \text{ kpsi}$$

According to rule 5, $\varepsilon_{qu} > m$. The group 2 strength is the same as group 1 strength. The group 3 strength is

$$\varepsilon_{qu} = \sum \frac{\varepsilon_{wi}}{i+1} = \frac{0.466}{2} + \frac{0.236}{3} = 0.302$$

$$S_u = \sigma_0(\varepsilon_{qu})^m = 140.1(0.302)^{0.14} = 118.5 \text{ kpsi}$$

The group 4 strength is the same as group 3. The ultimate tensile strength resisting tooth bending is $(S_u)_{tTt}$, a group 3 strength which is 129.3 kpsi. The improvement in S_u due to cold work in the tooth-bending direction is about 40 percent (endurance limit improves in the same proportion).

After a virgin material undergoes plastic straining, the strengths in various directions differ. See the *Standard Handbook of Machine Design* (2nd ed., Chap. 8) for a Fortran code for implementing the method of Datsko for quantitative estimation of properties of cold-worked metals. An interactive computer program makes estimations routine and convenient.

5–24 Quantitative Estimation of Properties of Heat-Treated Steels

Courses in metallurgy (or material science) for mechanical engineers usually present the addition method of Crafts and Lamont for the prediction of heat-treated properties from the Jominy test for plain carbon steels.[21] If this has not been in the prerequisite experience of the reader, then refer to the *Standard Handbook of Machine Design*, where the addition method is covered with examples.[22] If this book is a textbook for a machine elements course, it is a good class project (many hands make light work) to study the method and report to the class.

---

21. W. Crafts and J. L. Lamont, *Hardenability and Steel Selection*, Pitman and Sons, London, 1949.

22. Charles R. Mischke, Chap. 8, in Joseph E. Shigley and Charles R. Mischke (eds.), *Standard Handbook of Machine Design*, 2nd ed., McGraw-Hill, New York, 1996, pp. 8-10–8-32.

For low-alloy steels the multiplication method of Grossman[23] and Field[24] is explained in the *Standard Handbook of Machine Design* (pp. 13-27–13-30).

Modern Steels and their Properties explains how to predict the Jominy curve by the method of Grossman and Field from a ladle analysis and grain size.[25] Bethlehem steel has developed a circular plastic slide rule which is convenient to the purpose.

PROBLEMS

ANALYSIS

5–1 A specimen of medium-carbon steel having an initial diameter of 0.503 in was tested in tension using a gauge length of 2 in. The following data were obtained for the elastic and plastic states:

Elastic State		Plastic State	
Load P, lb	Elongation, in	Load P, lb	Area A_i, in²,
1 000	0.0004	8 800	0.1984
2 000	0.0006	9 200	0.1978
3 000	0.0010	9 100	0.1963
4 000	0.0013	13 200	0.1924
7 000	0.0023	15 200	0.1875
8 400	0.0028	17 000	0.1563
8 800	0.0036	16 400	0.1307
9 200	0.0089	14 800	0.1077

Note that there is some overlap in the data. Plot the engineering or nominal stress-strain diagram using two scales for the unit strain ϵ, one from zero to about 0.02 in/in and the other from zero to maximum strain. From this diagram find the modulus of elasticity, the 0.2 percent offset yield strength, the ultimate strength, and the percent reduction in area.

ANALYSIS

5–2 Compute the true stress and the logarithmic strain using the data of Prob. 5–1 and plot the results on log-log paper. Then find the plastic strength coefficient σ_0 and the strain-strengthening exponent m. Find also the yield strength and the ultimate strength after the specimen has had 20 percent cold work.

5–3 The stress-strain data from a tensile test on a cast-iron specimen are

Engineering stress, kpsi	5	10	16	19	26	32	40	46	49	54
Engineering strain, $\epsilon \cdot 10^{-3}$ in/in	0.20	0.44	0.80	1.0	1.5	2.0	2.8	3.4	4.0	5.0

Plot the stress-strain locus and find the 0.1 percent offset yield strength, and the tangent modulus of elasticity at zero stress and at 20 kpsi.

ANALYSIS

5–4 Having completed the task of Prob. 5–3, an engineer is unsatisfied with the accuracy of the result, depending as it does on drawing tangents to unknown curves. Noting that the stress-strain locus on Cartesian coordinates looks smooth and parabolic, the engineer writes

$$\epsilon = c_0 + c_1\sigma + c_2\sigma^2$$

23. M. A. Grossman, *AIME*, February 1942.

24. J. Field, *Metals Progress*, March 1943.

25. *Modern Steels and Their Properties,* 7th ed., Handbook 2757, Bethlehem Steel, 1972, pp. 46–50.

and then notes the necessity for $c_0 = 0$ and replots the data with ϵ/σ as the ordinate and σ as the abscissa. If the plotted locus is satisfactorily straight, find the constants c_1 and c_2 by regression. From this regression equation, determine the tangent modulus at zero load and at 20 kpsi, as well as the 0.1 percent offset yield strength. How do these determinations compare with the technique of Prob. 5–3? What can you say about the error?

5–5 The true stress–true strain data from a tensile test of a polyethylene plastic are

True stress, kpsi	1.0	1.5	2.0	2.5	3.0	3.5
True strain, in/in	0.15	0.32	0.46	0.60	0.70	0.82

Use these data from the plastic range and by regression determine the strength coefficient σ_0 and the strain-strengthening exponent m. Can you say anything about the error in either of these determinations?

5–6 A straight bar of arbitrary cross section and thickness h is cold-formed to an inner radius R about an anvil as shown in the figure. Some surface at distance N having an original length L_{AB} will remain unchanged in length after bending. This length is

$$L_{AB} = L_{AB'} = \frac{\pi(R+n)}{2}$$

The lengths of the outer and inner surfaces, after bending, are

$$L_o = \frac{\pi}{2}(R+h) \qquad L_i = \frac{\pi}{2}R$$

Using Eq. (5–8), we then find the true strains to be

$$\varepsilon_o = \ln\frac{R+h}{R+N} \qquad \varepsilon_i = \ln\frac{R}{R+N}$$

Tests show that $|\varepsilon_o| = |\varepsilon_i|$. Show that

$$N = R\left[\left(1+\frac{h}{R}\right)^{1/2} - 1\right]$$

and

$$\varepsilon_o = \ln\left(1+\frac{h}{R}\right)^{1/2}$$

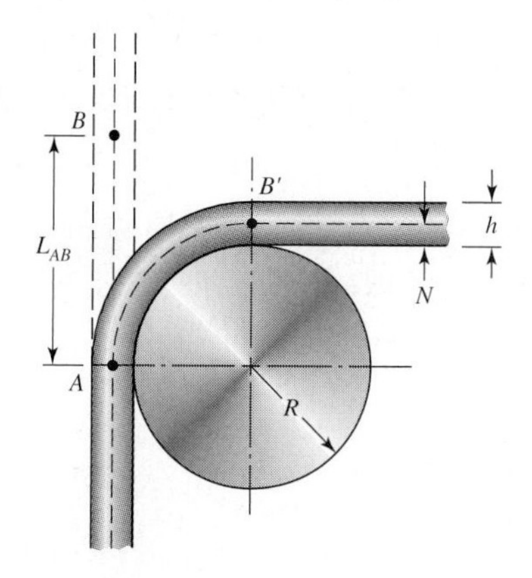

Problem 5–6

5–7 A 12-gauge (0.1094-in-thick) strap is to be cold-bent around a $\frac{1}{8}$-in-radius anvil. If the material is annealed AISI 1018 steel, use the results of Prob. 5–6 to

(*a*) Estimate the plastic strain at the surfaces.
(*b*) Estimate the new ultimate strength at the inner and outer surfaces.
(*c*) Estimate the new yield strength at the surfaces of the curve.

5–8 The strain-strengthening exponent for some 1000-series steels has been measured as follows:

Steel	1002	1008	1010	1018	1020	1045
m	0.29	0.24	0.26	0.25	0.22	0.14
	0.27	0.24	0.23			
			0.23			
			0.23			

Is there a correlation between carbon content in percent (or in points) and the strain-strengthening exponent m?

5–9 For heat-treated AISI 1045 steel, the following Brinell hardnesses and strain-strengthening exponents have been observed:

H_B	225	390	410	450	500	595
m	0.61	0.45	0.60	0.35	0.25	0.07
	1.00					

Is m a decreasing monotone with hardness? What is the regression relation?

ANALYSIS

5–10 In the torsion of a circular shaft of radius r, length l, plastically twisted through the angle θ, Datsko[26] has shown that at some angle α to the axial direction there exists a surface filament having the largest tensile strain. This angle is given by the equation

$$\alpha = \frac{1}{2}\tan^{-1}\left(\frac{2l}{r\theta}\right)$$

and the true strain by

$$\epsilon = \ln\left\{\cos\alpha\left[1 + \left(\tan\alpha + \frac{r\theta}{l}\right)^2\right]^{1/2}\right\}$$

A shaft has a diameter of 1.0 in, is 10 in long, and is twisted through an angle of 10 rad.

(*a*) Find the true strain of an axial filament.
(*b*) Find the maximum true strain at the surface.
(*c*) Determine whether a shaft of 2011-T6 aluminum alloy can be cold-worked to this extent.

5–11 A moment of decision: interpret the distribution of the 1020 tensile test data of the example of Sec. 5–1 as normal (Gaussian). Superpose the plots of the lognormal PDF and the normal PDF. Which is the better fit to the data?

5–12 A tensile test on a standard specimen is a quasi-static measuring experiment on a single specimen which gives load and corresponding elongation, as well as yield load, maximum load, ultimate load, and fractional area reduction. Organize a computation scheme that will enable you to find the

---

26. Joseph Datsko, *Materials in Design and Manufacture*, published by the author, Ann Arbor, Mich., 1977, pp. 7–19 to 7–23.

yield strength S_y, the true stress at yield σ_y, ultimate tensile strength S_{ut}, true stress at maximum load σ_u, true strain at ultimate load ε_u, true strain at fracture load ϵ_f, nominal stress at fracture S_{fracture}, true stress at fracture σ_{fracture}, strain-strengthening exponent m, strain coefficient σ_0, and modulus of elasticity E.

5–13 For the method of Ex. 5–9, with the same strain program of the example, find the four group yield strengths.

5–14 Consider executing the cold-rolling deformation step in Ex. 5–9 in two increments: reducing the diameter from $2\frac{1}{4}$ in to $2\frac{1}{8}$ in, then reducing the diameter from $2\frac{1}{8}$ in to 2 in. Would there by any advantage to doing this in the final properties?

5–15 Repeat Ex. 5–9 with the following materials, if possible.

(*a*) HR 1020 with $S_y = 42$ kpsi, $S_{ut} = 66.2$ kpsi, $\sigma_0 = 115$ kpsi, $m = 0.22$, $\varepsilon_f = 0.90$.
(*b*) HR 4340 with $S_y = 132$ kpsi, $S_{ut} = 151$ kpsi, $\sigma_0 = 210$ kpsi, $m = 0.09$, $\varepsilon_f = 0.45$.
(*c*) 17-4 PH stainless steel, 1050°F, aged with $S_y = 155$ kpsi, $S_{ut} = 185$ kpsi, $\sigma_0 = 225$ kpsi, $m = 0.05$, $\varepsilon_f = 0.90$.
(*d*) 2024T4 aluminum with $S_y = 43.0$ kpsi, $S_{ut} = 64.8$ kpsi, $\sigma_0 = 100$ kpsi, $m = 0.15$, $\varepsilon_f = 0.18$.
(*e*) 200 nickel, WQ from 1500°F with $S_y = 16.2$ kpsi, $S_{ut} = 72.1$ kpsi, $\sigma_0 = 150$ kpsi, $m = 0.375$, $\varepsilon_f = 1.805$.
(*f*) 18-8 stainless steel, annealed, $S_y = 37$ kpsi, $S_{ut} = 89.5$ kpsi, $\sigma_0 = 210$ kpsi, $m = 0.51$, $\varepsilon_f = 1.08$.

5–16 In LEFM predicting crack length a for a particular number of cycles is essential to the method. When the integral of $(da/dN)dN$ can be expressed in closed form, the problem is relatively simple, as indicated by Ex. 5–7 and Ex. 5–8. There are times when the closed form is not available and numerical integration methods must be employed. A heuristic marching method is to develop an expression for Δa for one cycle, estimate its value, and add it to the integral of the crack length so far. Write a computer program to do this and compare tables such as those found in Ex. 5–7 and Ex. 5–8 with the results of your program.

5–17 Numerical integration is often forced upon the designer. Simpson's rule is of the form

$$\int_a^b f(x)\,dx = \frac{h}{3}(y_0 + 4y_1 + 2y_2 + 4y_3 + \cdots + 4y_{n-1} + y_n)$$

where the y's are evenly spaced function evaluations of $f(x)$ in the interval $[a, b]$ and the constant h is the interval between function evaluations. The number of intervals n should be an even number, and there are advantages to making it an even number divisible by 4. If this is done, the error in the integration can be estimated by integrating again using every other ordinate, calling this integration $I_{n/2}$ and estimating the numerical integration error to be

$$E = \frac{I_n - I_{n/2}}{15}$$

We express the integral as $I_n + E$ in applying Richardson's correction to improve Simpson's value. Apply Simpson's rule to the integral

$$I = \int_0^1 x^5\,dx$$

using but four intervals. Apply Richardson's correction and show that the result is exact.

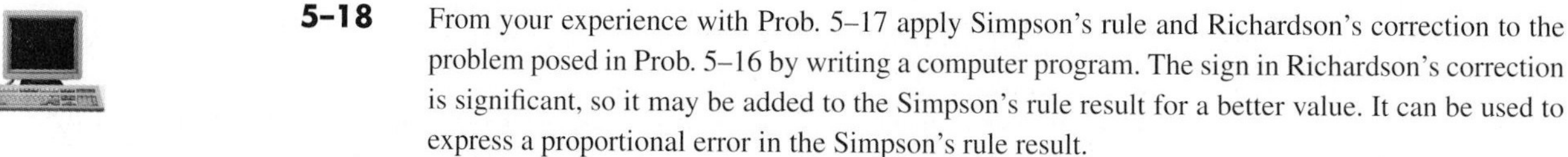

5–18 From your experience with Prob. 5–17 apply Simpson's rule and Richardson's correction to the problem posed in Prob. 5–16 by writing a computer program. The sign in Richardson's correction is significant, so it may be added to the Simpson's rule result for a better value. It can be used to express a proportional error in the Simpson's rule result.

ANALYSIS

5–19 As in Ex. 5–9, a 1045 hot-rolled steel bar is cold-worked from a diameter of $2\frac{1}{4}$ in to a diameter of 2 in. Find the new group 1 strengths S_u and S_y using the method of Datsko.

ANALYSIS

5–20 Suppose that for press capacity reasons two upsetting steps are used on Prob. 5–19 to cold-work the diameter from $2\frac{1}{4}$ in to $2\frac{1}{8}$ in, then from $2\frac{1}{8}$ in to 2 in. Find the resulting group 1 strength S_u and S_y. Compare with the results of Prob. 5–19.

Failures Resulting from Static Loading

6–1	Static Strength **316**
6–2	Stress Concentration **319**
6–3	Hypotheses of Failure **322**
6–4	Ductile Materials: Maximum-Shear-Stress (Tresca or Guest) Hypothesis **324**
6–5	Ductile Materials: Strain-Energy Hypotheses **326**
6–6	Ductile Materials: Internal-Friction Hypothesis **332**
6–7	Criticism of Hypotheses by Data in Ductile Materials **334**
6–8	Brittle Materials: Maximum-Normal-Stress (Rankine) Hypothesis **335**
6–9	Brittle Materials: Modifications of the Mohr Hypothesis **337**
6–10	The Criticism of Hypotheses by Data in Brittle Materials **341**
6–11	What Our Failure Models Tell Us **342**
6–12	Interference—General **343**
6–13	Static or Quasi-Static Loading on a Shaft **347**

In Chap. 1 we learned that *strength is a property or characteristic of a mechanical element*. This property results from the material identity, the treatment and processing incidental to creating its geometry, and the loading, and it is at the controlling or critical location. These elements must be identified and considered before we can speak about *the strength* of a part as a useful characteristic of descriptive worth. Tables of properties of engineering materials are *not* about strengths of parts. The strength of a mechanical part does not depend on that part being subjected to its intended load. In fact, this strength property is a characteristic of the element before it is assembled with other elements into a machine or system. The strength of a part is a valuable descriptor, but it exists only if we meet all the qualifications enumerated above.

In addition to considering the strength of a single part, we must be cognizant that the strengths of the mass-produced parts will all be somewhat different from the others in the collection or ensemble because of variations in dimensions, machining, forming, and composition. Descriptors of strength are necessarily stochastic in nature, involving statistical parameters (often mean and standard deviation) and distributional identification (or its robust stand-in).

A *static load* is a stationary force or couple applied to a member. To be stationary, the force or couple must be unchanging in magnitude, point or points of application, and direction. A static load can produce axial tension or compression, a shear load, a bending load, a torsional load, or any combination of these. To be considered static, the load cannot change in any manner.

In this chapter we consider the relations between strength and static loading in order to make the decisions concerning material and its treatment, fabrication, and geometry for satisfying the requirements of functionality, safety, reliability, competitiveness, usability, manufacturability, and marketability. How far we go down this list is related to the scope of the examples.

"Failure" is the first word in the chapter title. Failure can mean a part has separated into two or more pieces; has become permanently distorted, thus ruining its geometry; has had its reliability downgraded; or has had its function compromised, whatever the reason. A designer speaking of failure can mean any or all of these possibilities. In this chapter our attention is focused on the predictability of permanent distortion or separation. In strength-sensitive situations the designer must separate mean stress and mean strength at the critical location sufficiently to accomplish his or her purposes.

Figures 6–1 to 6–13 are photographs of several failed parts. The photographs exemplify the need of the designer to be well-versed in failure prevention. Toward this end we shall consider one-, two-, and three-dimensional stress states, with and without stress concentrations, for both ductile and brittle materials.

(*a*)

(*b*)

Figure 6–1

(*a*) Failure of a truck drive-shaft spline due to corrosion fatigue. Note that it was necessary to use clear tape to hold the pieces in place. (*b*) Direct end view of failure.

Figure 6–2

Fatigue failure of an automotive cooling fan due to vibrations caused by a defective water pump.

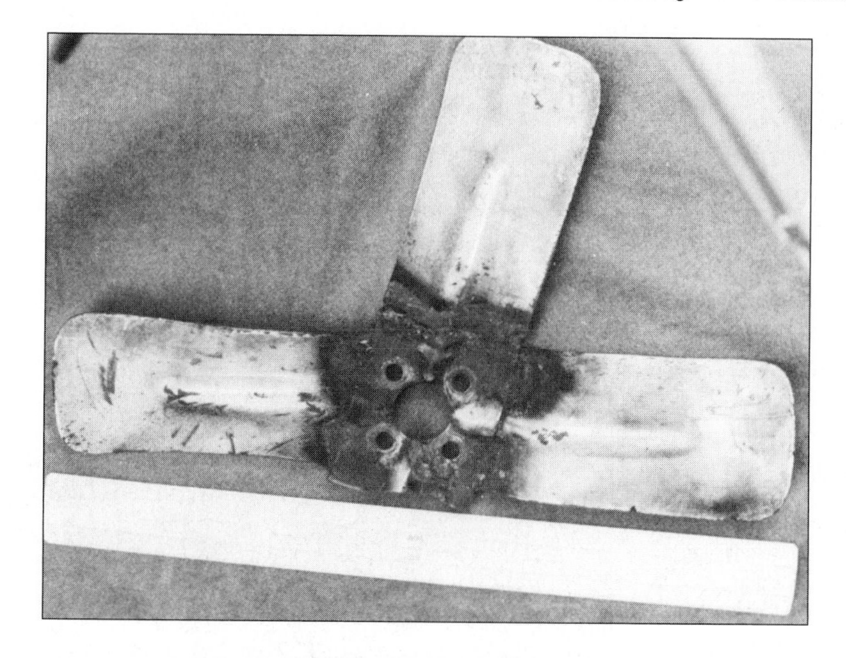

Figure 6–3

Typical failure of a stamped steel alternator bracket after about 40 000 km. The failure was probably due to residual stresses caused by the cold-forming operation. The high failure rate prompted the manufacturer to redesign the bracket as a die casting.

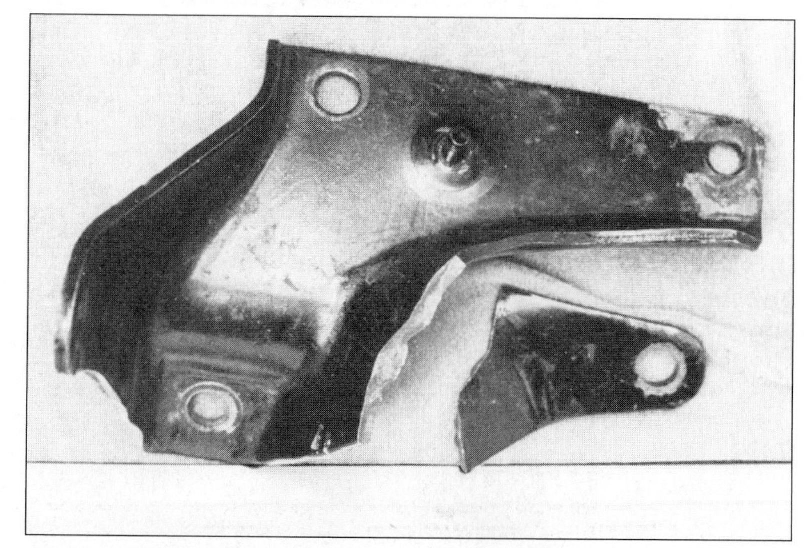

Figure 6–4

Failure of an automotive drag link. The failure occurred after about 225 000 km. Fortunately the car was in park and against a curb. Such a failure results in total disconnect of the steering wheel from the steering mechanism.

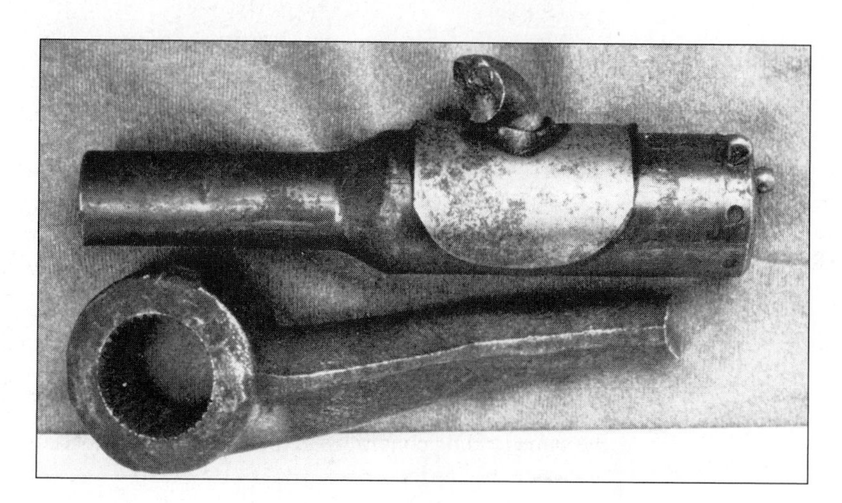

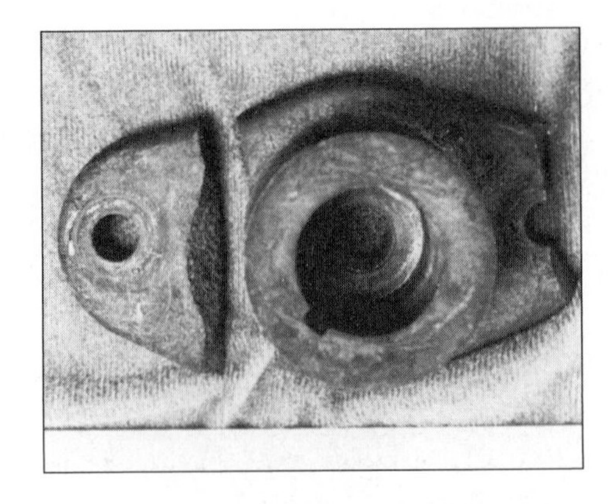

Figure 6–5

Impact failure of a lawnmower blade driver hub. The blade impacted a surveying pipe marker.

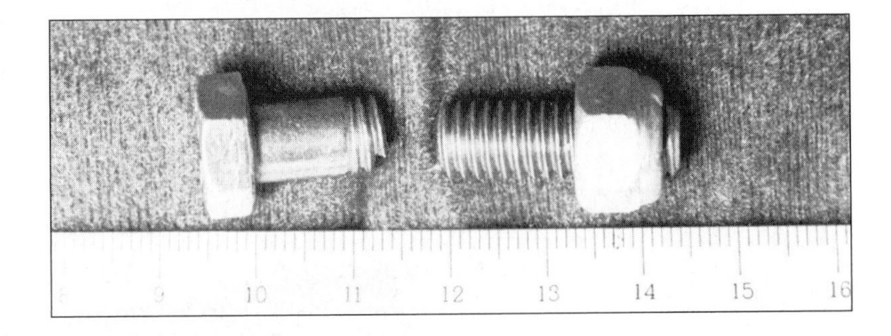

Figure 6–6

Failure of an overhead-pulley retaining bolt on a weightlifting machine. A manufacturing error caused a gap that forced the bolt to take the entire moment load.

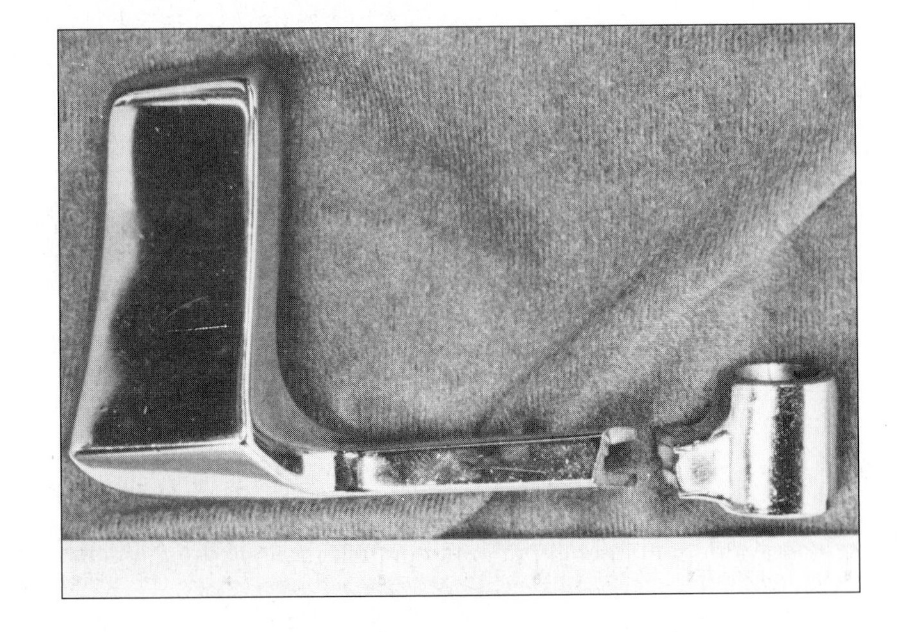

Figure 6–7

Failure of an interior die-cast car-door handle. Failure occurred about every 72 000 km. Probable causes were the electroplating material, stress concentration, the long lever arm required to operate a "sticky" door-release mechanism, and the high actuation forces.

6–1 Static Strength

Ideally, in designing any machine element, the engineer should have at his or her disposal the results of a great many strength tests of the particular material chosen. These tests should be made on specimens having the same heat treatment, surface finish, and size as the element the engineer proposes to design; and the tests should be made under exactly the same loading conditions as the part will experience in service. This means that if the

(a)

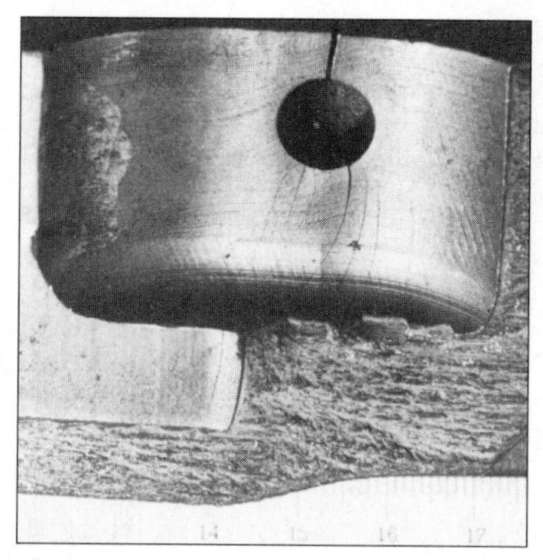
(b)

Figure 6–8

Chain test fixture that failed in one cycle. To alleviate complaints of excessive wear, the manufacturer decided to case-harden the material. (a) Two halves showing fracture; this is an excellent example of brittle fracture initiated by stress concentration. (b) Enlarged view of one portion to show cracks induced by stress concentration at the support-pin holes.

Figure 6–9

Automotive rocker-arm articulation-joint fatigue failure.

part is to experience a bending load, it should be tested with a bending load. If it is to be subjected to combined bending and torsion, it should be tested under combined bending and torsion. If it is made of heat-treated AISI 1040 steel drawn at 500°C with a ground finish, the specimens tested should be of the same material prepared in the same manner. Such tests will provide very useful and precise information. Whenever such data are available for design purposes, the engineer can be assured that he or she is doing the best possible job of engineering.

The cost of gathering such extensive data prior to design is justified if failure of the part may endanger human life or if the part is manufactured in sufficiently large quantities. Refrigerators and other appliances, for example, have very good reliabilities because the parts are made in such large quantities that they can be thoroughly tested in advance of manufacture. The cost of making these tests is very low when it is divided by the total number of parts manufactured.

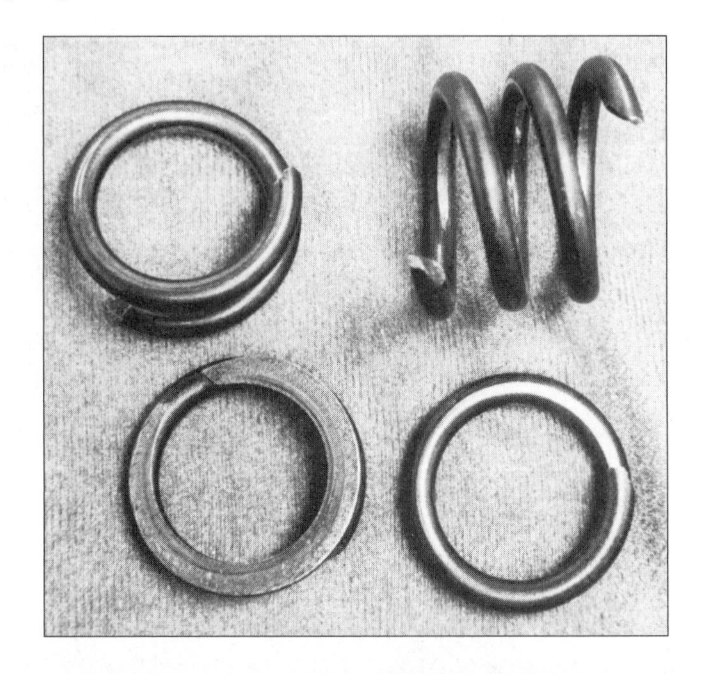

Figure 6–10

Valve-spring failure caused by spring surge in an oversped engine. The fractures exhibit the classic 45° shear failure.

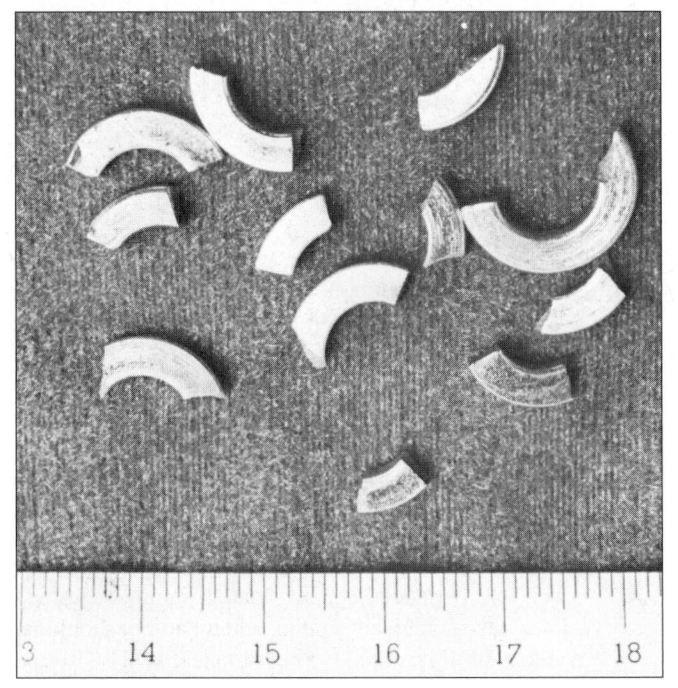

Figure 6–11

Brittle fracture of a lock washer in one-half cycle. The washer failed when it was installed.

You can now appreciate the following four design categories:

1 Failure of the part would endanger human life, or the part is made in extremely large quantities; consequently, an elaborate testing program is justified during design.

2 The part is made in large enough quantities that a moderate series of tests is feasible.

3 The part is made in such small quantities that testing is not justified at all; or the design must be completed so rapidly that there is not enough time for testing.

4 The part has already been designed, manufactured, and tested and found to be unsatisfactory. Analysis is required to understand why the part is unsatisfactory and what to do to improve it.

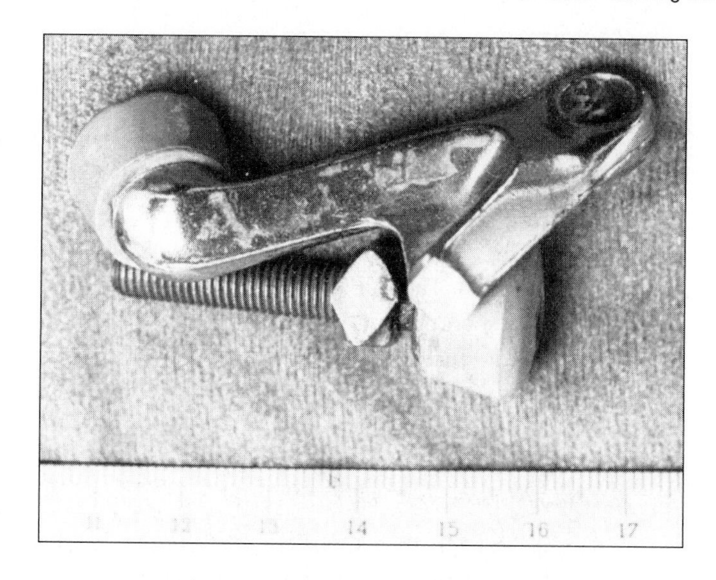

Figure 6–12

Fatigue failure of a die-cast residence door bumper. This bumper is installed on the door hinge to prevent the doorknob from impacting the wall.

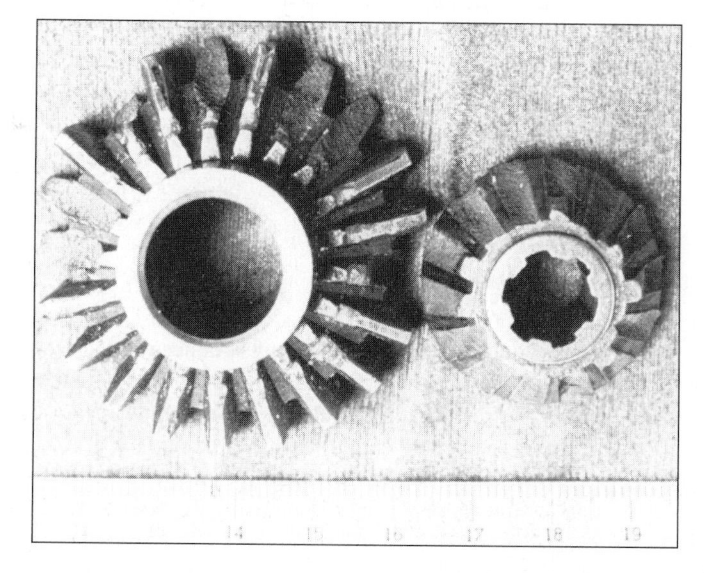

Figure 6–13

A gear failure from a $7\frac{1}{2}$-hp (5.6-kW) American-made outboard motor. The large gear has a $1\frac{7}{8}$-in (47.6-mm) outside diameter and had 21 teeth; 6 are broken. The pinion had 14 teeth; all are broken. Failure occurred when the propeller struck a steel auger placed in the lake bottom as an anchorage. The owner had replaced the shear pin with a substitute pin.

More often than not it is necessary to design using only published values of yield strength, ultimate strength, percentage reduction in area, and percentage elongation, such as those listed in Appendix E. How can one use such meager data to design against both static and dynamic loads, biaxial and triaxial stress states, high and low temperatures, and very large and very small parts? These and similar questions will be addressed in this chapter and those to follow; but think how much better it would be to have data available that duplicate the actual design situation.

6–2 Stress Concentration

Stress concentration (see Sec. 5-20) is a highly localized effect. In some instances it may be due to a surface scratch. If the material is ductile and the load static, the design load may cause yielding in the critical location in the notch. This yielding can involve strain-strengthening of the material and an increase in yield strength at the critical notch location. Since the loads are static, that part can carry them satisfactorily with no general yielding. In these cases the designer sets the geometric (theoretical) stress concentration factor K_t to unity.

The rationale can be expressed as follows. The worst-case scenario is that of an idealized non–strain-strengthening material shown in Fig. 6–14. The stress-strain locus rises linearly to the yield strength S_y, then proceeds at constant stress, which is equal to S_y. Consider a notched rectangular bar as depicted in Fig. E–15–5, where the cross-section area of the small shank is 1 in^2. If the material is ductile, with a yield point of 40 kpsi, and the theoretical stress-concentration factor (SCF) K_t is 2,

- A load of 20 kip induces a tensile stress of 20 kpsi in the shank as depicted at point A in Fig. 6–14. The SCF is $K = \sigma_{\text{max}}/\sigma_{\text{nom}} = 40/20 = 2$.
- A load of 30 kip induces a tensile stress of 30 kpsi in the shank at point B. At the critical location in the fillet the stress is 40 kpsi, and the SCF $K = \sigma_{\text{max}}/\sigma_{\text{nom}} = S_y/\sigma = 40/30 = 1.33$.
- At a load of 40 kip the induced tensile stress (point C) is 40 kpsi in the shank. At the critical location in the fillet the stress (at point E) is 40 kpsi. The SCF $K = \sigma_{\text{max}}/\sigma_{\text{nom}} = S_y/\sigma = 40/40 = 1$.

For materials that strain-strengthen, the critical location in the notch has a higher S_y. The shank area is at a stress level a little below 40 kpsi, is carrying load, and is very near its failure-by-general-yielding condition. This is the reason designers do not apply K_t in static loading of a ductile material loaded elastically, instead setting $K_t = 1$.

When using this rule for ductile materials with static loads, be careful to assure yourself that the material is not susceptible to brittle fracture (see Sec. 5–21) in the environment of use. The usual definition of geometric (theoretical) stress-concentration factor for normal stress K_t or shear stress K_{ts} is

$$\sigma_{\text{max}} = K_t \sigma_{\text{nom}} \tag{a}$$

$$\tau_{\text{max}} = K_{ts} \tau_{\text{nom}} \tag{b}$$

Since your attention is on the stress-concentration factor, and the definition of σ_{nom} or τ_{nom} is given in the graph caption or from a computer program, be sure the value of nominal stress is appropriate for the section carrying the load, and not what the graph preparer found convenient for presentation.

Brittle materials do not exhibit a plastic range. A brittle material "feels" the SCF K_t or K_{ts}, and K_t or K_{ts} is applied using Eq. (a) or (b).

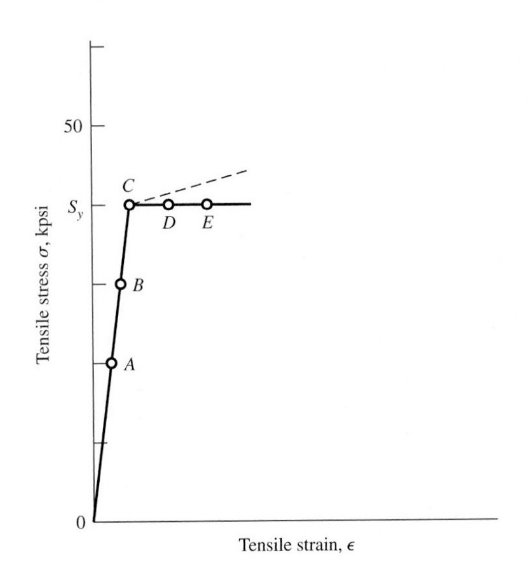

Figure 6–14

An idealized stress-strain curve. The dashed line depicts a strain-strengthening material.

An exception to this rule is a brittle material which inherently contains microdiscontinuity stress concentration, worse than the macrodiscontinuity which the designer has in mind. Sand molding introduces sand particles, air, and water vapor bubbles. The grain structure of cast iron contains graphite flakes (with little strength), which are literally cracks introduced during the solidification process. When a tensile test on a cast iron is performed, the strength reported in the literature *includes* this stress concentration. In such cases K_t or K_{ts} need not be applied.

An important source of stress-concentration factors is R. E. Peterson, who compiled them from his own work and that of others.[1] Peterson developed the style of presentation in which the stress-concentration factor K_t is multiplied by the nominal stress σ_{nom} to estimate the magnitude of the largest stress in the locality. His approximations were based on photoelastic studies of two-dimensional strips (Hartman and Levan, 1951; Wilson and White, 1973), with some limited data from three-dimensional photoelastic tests of Hartman and Levan. A contoured graph was included in the presentation of each case. Filleted shafts in tension were based on two-dimensional strips.

The tool of finite element analysis (FEA) was applied by Tipton, Sorem, and Rolovic.[2] Not only has the accuracy been improved, but a change in style of presentation is to be noted. Tipton and his colleagues employed a FEA of a shaft shoulder in axial tension. The results were presented in the form

$$\sigma_1 = K_t \sigma_{\text{nom}} \tag{c}$$

at the fillet location ϕ of the maximum principal stress $(\sigma_1)_{\max}$ in the locality [denoted σ_1 in Eq. (*a*)], and $(K_t)_{\text{tens}}$ is presented in curve-fitted, dimensionless form. For an axial tension P, $(K_t)_{\text{tens}}$ is reported as

$$(K_t)_{\text{tens}} = 0.493 + 0.48\left(\frac{D}{d}\right)^{-2.43} + \left(\frac{r}{d}\right)^{-0.48}\sqrt{\frac{3.43 - 3.41(D/d)^2 - 0.0232(D/d)^4}{1 - 8.85(D/d)^2 - 0.078(D/d)^4}} \tag{d}$$

If the situation involved a brittle material in axial loading, the maximum largest principal stress in the locality is given by

$$\sigma_1 = (K_t)_{\text{tens}}\,\sigma_{\text{nom}} = (K_t)_{\text{tens}}\,\frac{4P}{\pi d^2} \tag{e}$$

If the failure criterion is couched in terms of the largest principal stress, then one is ready to proceed. If the material is ductile, then *all* of the principal stresses (σ_1, σ_2, and σ_3) are important in formulating the von Mises (or octahedral) stress, the largest value of which occurs at another place, and one proceeds as follows:

$$\begin{aligned} \sigma_1 &= (K_t)_{\text{tens}}\sigma_{\text{nom}} \\ \sigma_2 &= \frac{1}{2}\left(\sigma_1 - \frac{\sigma_1}{|\sigma_1|}\sqrt{4\sigma'^2 - 3\sigma_1^2}\right) \\ \sigma_3 &= 0 \end{aligned} \tag{f}$$

where the von Mises stress σ' is given by

$$\sigma' = \sqrt{\sigma_1^2 - \sigma_1\sigma_2 + \sigma_2^2} \tag{g}$$

1. R. E. Peterson, "Design Factors for Stress Concentration," *Machine Design*, vol. 23, no. 2, February 1951; no. 3, March 1951; no. 5, May 1951; no. 6, June 1951; no. 7, July 1951.

2. S. M. Tipton, J. R. Sorem Jr., and R. D. Rolovic, "Updated Stress-Concentration Factors for Filleted Shafts in Bending and Tension," *Trans. ASME, Journal of Mechanical Design*, vol. 118, September 1996, pp. 321–327.

Tipton et al. defined a stress concentration factor $K_{\sigma'}$ replacing the steps implied by Eqs. (f) and Eq. (g) of the form

$$\sigma' = (K_{\sigma'})\,\sigma_{\text{nom}} \tag{h}$$

relating the von Mises stress σ' directly to the nominal stress σ_{nom}, if this is convenient to the user. For a filleted shaft in axial tension, $(K_{\sigma'})_{\text{tens}}$ was reported as

$$(K_{\sigma'})_{\text{tens}} = 0.496 + 0.472\left(\frac{D}{d}\right)^{-2.85} + \left(\frac{r}{d}\right)^{-0.48}\sqrt{\frac{2.921 - 2.945(D/d)^2 + 0.0217(D/d)^4}{1 - 9.59(D/d)^2 + 0.053(D/d)^4}} \tag{i}$$

For the filleted shaft in axial tension Tipton et al. reported a stress-concentration factor $(K_t)_{\text{tens}}$ to be applied to the nominal stress σ_{nom} to estimate the *maximum ordered principal stress*, and another, $(K_{\sigma'})_{\text{tens}}$, to be applied to the nominal stress σ_{nom} to estimate the *largest von Mises stress* in the locality. The method has produced more accurate information and *two* stress-concentration factors. They are not likely to be confused, but one must choose the appropriate one to the circumstance. Tipton et al. also reported equations for the angle ϕ into the fillet at which the largest ordered principal stress appears, and a different angle ϕ at which the largest von Mises stress appears. If this information is important, refer to the article, which also compares the results to those reported by Peterson. The stress concentrations in the filleted shaft in tension and bending reported by Peterson are low, especially at sharp fillet radii.

In the subsequent sections concerning criteria for failure, a particular kind of stress is related to a material property at incipient failure. The particular stress may be estimated from geometry and load with exact closed-form equations or approximations to them, or by utilizing stress-concentration factors. Having an idea of the origin of stresses can help in the understanding of the following sections.

6–3 Hypotheses of Failure

Section 6-1 illustrated some ways that loss of function is manifested. Events such as distortion, permanent set, cracking, and rupturing are among the ways that a machine element fails. Testing machines appeared in the 1700s, and specimens were pulled, bent, and twisted in simple loading processes. Experience in gathering these data was instrumental to the formation of the idea of strain. Cauchy linked stress, strain, and elastic constants in what we today would call theory of elasticity. Always looking for a simple, insightful "mechanism" of failure, humankind persevered. As with all things, postulations lead to predictions, and these predictions are criticized by the use of additional data. Ideas are modified, and the process continues, as does the process of understanding. The history of hypotheses of failure is a recitation of our learning experience. To cut immediately to the current level of understanding (we are *not* finished learning) would leave unexamined our path of growth.

If the failure mechanism is simple, then simple tests can give clues. Just what is simple? The tension test is uniaxial (that's simple) and elongations are largest in the axial direction, so strains can be measured and stresses inferred up to "failure." Just what is important: a critical stress, a critical strain, a critical energy? Do growing Mohr circles, as depicted in Fig. 3–5, expand until they "hit something"? If so, what? If not, what is important? In the next several sections, we shall show failure hypotheses that have helped answer some of these questions.

In subsequent sections we shall explore hypotheses of failure under static loading, which have the following outcomes:

- The onset of failure occurs when a particular type of stress reaches the level of a corresponding material strength.
- A failure locus can be plotted which separates a "safe" domain from an "unsafe" domain.
- A locus of possible stresses, called a *load line*, runs (usually) radially from the origin and intersects the failure locus. The intersection defines the onset of failure.
- A factor of safety can be identified for the deterministic case, and a probability of failure can be identified for the stochastic case.

A hypothesis of failure that explains all the data, and continues to do so as new data are assembled, can be elevated to the status of theory.

When a part is loaded so that the stress state is uniaxial, then the stress and the strength can be compared directly to determine the degree of safety, or to learn whether the part will fail. The method is simple, because there is only one value of stress and there is only one value of strength, be it yield strength, ultimate strength, shear strength, or whatever, as appropriate.

The problem becomes more complicated when the stress state is biaxial or triaxial. In such cases there are a multitude of stresses, but still only one significant strength. So how do we learn whether the part is safe or not, and, if so, how safe?

As an example consider the oldest failure hypothesis attributed to Rankine, called the maximum-normal-stress hypothesis. It states that *failure occurs whenever one of the three principal stresses equals the strength.* Suppose we order the principal stresses

$$\sigma_1 > \sigma_2 > \sigma_3 \tag{a}$$

then failure occurs whenever

$$\sigma_1 = S_t \quad \text{or} \quad \sigma_3 = -S_c \tag{b}$$

where S_t and S_c are tensile and compressive strengths, usually the yield or ultimate, respectively. Figures 6–15 and 6–16 illustrate states associated with safety (within) and failure (without). The failure locus separates these regions. Figure 6–17 for biaxial stress states uses the nonzero principal stresses σ_A and σ_B. Since $\sigma_A > \sigma_B$, only the solid-line

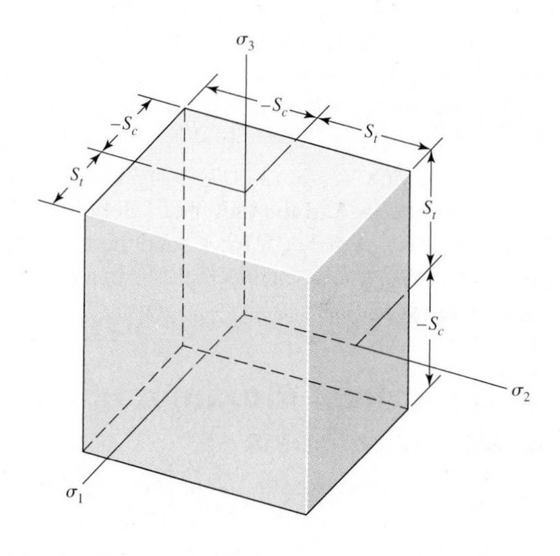

Figure 6–15

The maximum-normal-stress (MNS) hypothesis in three dimensions. The right rectangular prism encloses all safe values of any combination of stress components. The compressive strength S_c need not equal the tensile strength S_t. For the hypothesis, these strengths may be either yield or ultimate strengths. Note, too, that strengths are always positive quantities, but stresses may be either positive or negative.

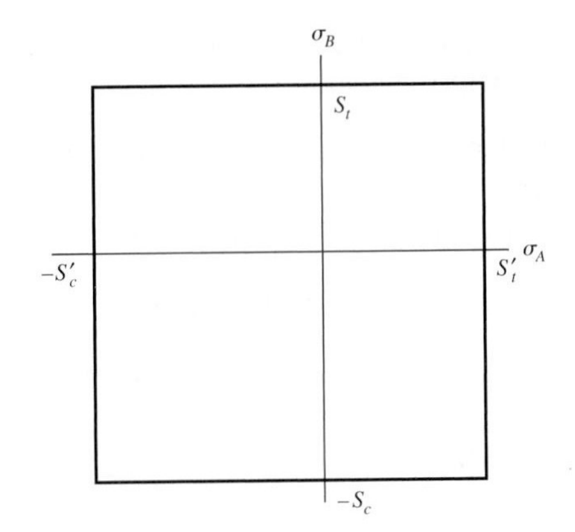

Figure 6–16

Graph of maximum-normal-stress (MNS) hypothesis of failure for biaxial stress states when $S_c > S_t$. Stress states that plot inside the failure locus are safe.

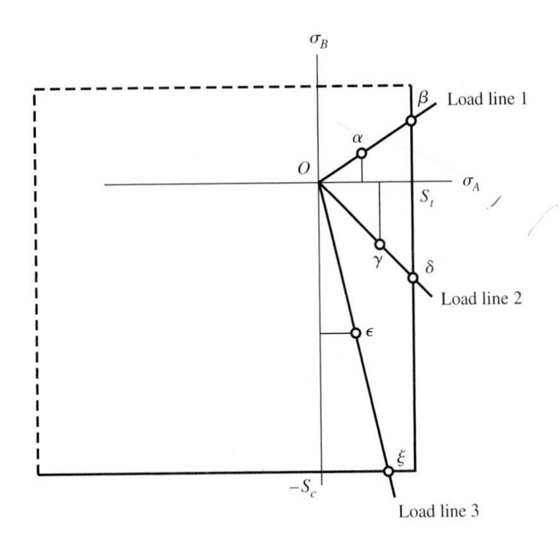

Figure 6–17

Maximum-normal-stress (MNS) failure hypothesis for biaxial stress states showing load lines.

failure locus exists. The dashed-line failure locus is not really part of the diagram. It is conventional to retain the mirror image about the diagonal because, when failure data are plotted, congested data points can be "thinned out" by plotting on the mirror image, reducing clutter and improving clarity.

The load line 1 is the locus of possible stress states for a given circumstance. Point α is the stress condition at the critical location. Point β is the corresponding strength on the failure locus. Philon's factor of safety can be expressed as $O\beta/O\alpha$ or as their projections on the σ_A axis:

$$n_1 = O\beta/O\alpha = S_t/(\sigma_A)_\alpha \tag{c}$$

The slope of the load line is $r_1 = (\sigma_B)_\alpha / (\sigma_A)_\alpha$. Load line 2 is for the case of pure torsion, so $r_2 = -1$, $\sigma_B = -\sigma_A$, and the factor of safety is $n_2 = O\delta/O\gamma = S_t/(\sigma_A)_\gamma$. The slope of load line 3 is $r_3 = (\sigma_B)_\epsilon / (\sigma_A)_\epsilon$, and the factor of safety is $n_3 = -S_c/(\sigma_B)_\epsilon$.

In this general way, ideas such as these form a part of the failure hypotheses developed in subsequent sections.

6–4 Ductile Materials: Maximum-Shear-Stress (Tresca or Guest) Hypothesis

The maximum-shear-stress hypothesis states that *yielding begins whenever the maximum shear stress in any element becomes equal to the maximum shear stress in a tension-test specimen of the same material when that specimen begins to yield.*

If we order the principal normal stresses as $\sigma_1 > \sigma_2 > \sigma_3$, then the maximum-shear-stress hypothesis predicts that yielding will occur whenever

$$\tau_{max} \geq \frac{S_y}{2} \quad \text{or} \quad \sigma_1 - \sigma_3 \geq S_y \tag{6–1}$$

Note that this hypothesis also states that the yield strength in shear is given by the equation

$$S_{sy} = 0.50S_y \tag{6–2}$$

To develop an even better understanding of this hypothesis, we repeat Eq. (3–11) for the three principal shear stresses here. These are

$$\tau_{1/2} = \frac{\sigma_1 - \sigma_2}{2} \qquad \tau_{2/3} = \frac{\sigma_2 - \sigma_3}{2} \qquad \tau_{1/3} = \frac{\sigma_1 - \sigma_3}{2} \tag{6–3}$$

The largest of $\tau_{1/2}$, $\tau_{2/3}$, and $\tau_{1/3}$ is τ_{max} of Eq. (6–1). Suppose we decompose the normal principal stresses into the components

$$\begin{aligned} \sigma_1 &= \sigma_1' + \sigma_1'' \\ \sigma_2 &= \sigma_2' + \sigma_2'' \\ \sigma_3 &= \sigma_3' + \sigma_3'' \end{aligned} \tag{a}$$

such that

$$\sigma_1'' = \sigma_2'' = \sigma_3'' \tag{b}$$

The stresses in Eq. (*b*) are called the *hydrostatic components* since they are equal. If it should happen that $\sigma_1' = \sigma_2' = \sigma_3' = 0$, then the three shear stresses, given by Eq. (6–3), would all be zero and there could be no yielding regardless of the magnitudes of the hydrostatic stresses. Thus the hydrostatic components have no effect on the size of the Mohr circle but merely serve to shift it along the normal-stress axis. It is for this reason that the yielding criterion for the general stress state can be represented by the oblique regular hexagonal cylinder of Fig. 6–18. Figure 6–19 illustrates the hypothesis for biaxial stresses.

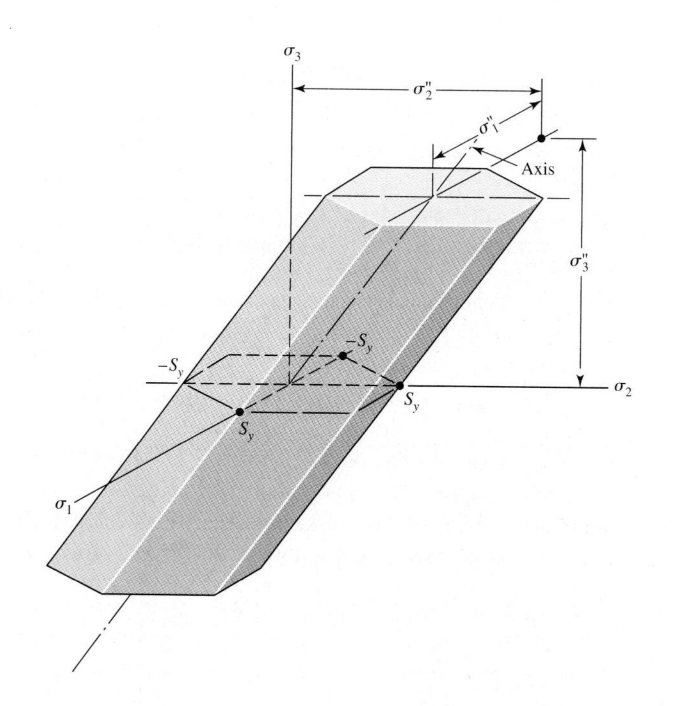

Figure 6–18

The maximum-shear-stress (MSS) hypothesis graphically represented in three dimensions. The hexagonal cylinder encloses all safe (free of yielding) components of the stress state given by σ_1, σ_2, and σ_3. The axis of the cylinder is inclined equally to each of the three principal directions and is the locus of the points described by the triad of hydrostatic components σ_1'', σ_2'', and σ_3''.

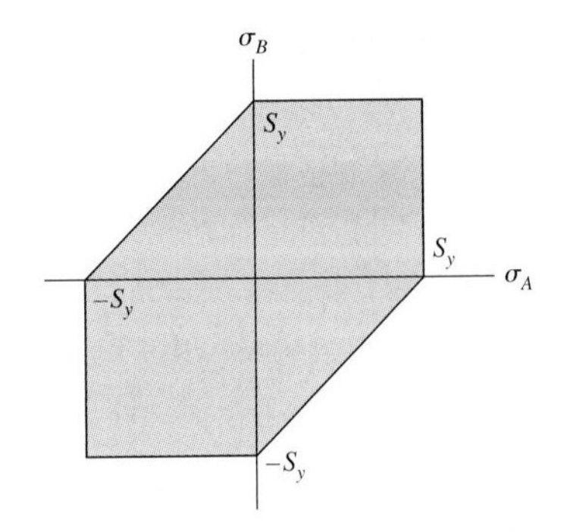

Figure 6–19

The maximum-shear-stress (MSS) hypothesis for biaxial stresses σ_A and σ_B, which are the two nonzero principal stresses.

This hypothesis predicts that hydrostatic stresses produce no yielding. It also predicts the shearing yield strength is one-half the tensile yield strength. This is about 15 percent low. It is simple to apply, and it is useful in welding technology because other concerns (geometric form) lead to even greater errors. Since it does not predict existing data, it has not been accorded the status of theory, but it finds occasional use nonetheless.

A body under huge hydrostatic stresses, well above the usual uniaxial failure stress, does not yield or rupture. When the stresses are released the body recovers elastically. Thus the hydrostatic stresses do not contribute toward failure. For this reason the maximum-shear-stress hypothesis was useful until superseded by the distortion-energy theory for ductile materials.

6–5 Ductile Materials: Strain-Energy Hypotheses

The *maximum-strain-energy hypothesis* predicts that failure by *yielding occurs when the total strain energy in a unit volume reaches or exceeds the strain energy in the same volume corresponding to the yield strength in tension or in compression.*

The strain energy stored in a unit volume when stressed uniaxially to the yield strength can be found from Eq. (4–27). Thus

$$u_S = \frac{S_y^2}{2E} \tag{a}$$

With the help of the triaxial stress-strain relations in Table 3–1, we find the total strain energy in a unit volume subjected to combined stresses to be

$$\begin{aligned} u_\sigma &= \frac{\epsilon_1\sigma_1}{2} + \frac{\epsilon_2\sigma_2}{2} + \frac{\epsilon_3\sigma_3}{2} \\ &= \frac{1}{2E}\left[\sigma_1^2 + \sigma_2^2 + \sigma_3^2 - 2\nu\left(\sigma_1\sigma_2 + \sigma_2\sigma_3 + \sigma_3\sigma_1\right)\right] \end{aligned} \tag{b}$$

Since this hypothesis is no longer favored, no graph is shown here and the biaxial stress equations are not given. You may wish, however, to determine these to satisfy your own curiosity.

The *distortion-energy hypothesis* originated because of the observation that ductile materials stressed hydrostatically exhibited yield strengths greatly in excess of the values given by the simple tension test. Therefore it was postulated that yielding was not a simple tensile or compressive phenomenon at all, but, rather, that it was related somehow to the angular distortion of the stressed element. To develop the theory, note, in Fig. 6–20*a*, the unit volume subjected to any three-dimensional stress state designated by the stresses σ_1, σ_2, and σ_3. The stress state shown in Fig. 6–20*b* is one of hydrostatic tension due to the stresses σ_{av} acting in each of the same principal directions as in Fig. 6–20*a*. The

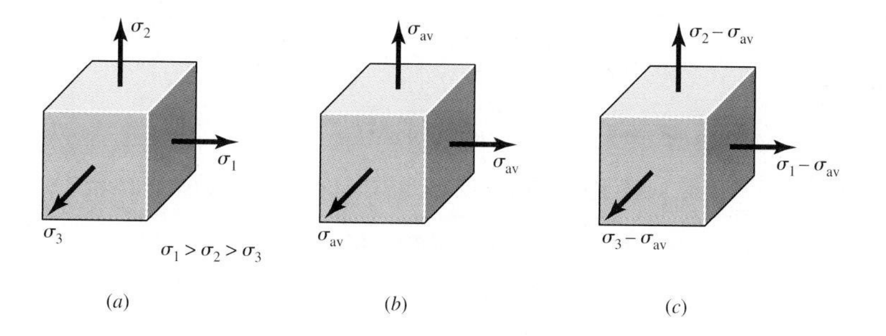

Figure 6–20

(*a*) Element with triaxial stresses; this element undergoes both volume change and angular distortion. (*b*) Element under hydrostatic tension undergoes only volume change. (*c*) Element has angular distortion without volume change.

formula for σ_{av} is

$$\sigma_{av} = \frac{\sigma_1 + \sigma_2 + \sigma_3}{3} \tag{c}$$

Thus the element in Fig. 6–20*b* undergoes pure volume change, that is, no angular distortion. If we regard σ_{av} as a component of σ_1, σ_2, and σ_3, then this component can be subtracted from them, resulting in the stress state shown in Fig. 6–20*c*. This element is subjected to pure angular distortion, that is, no volume change.

Equation (*b*) gives the total strain energy for the element of Fig. 6–20a. The strain energy for producing only volume change can be obtained by substituting σ_{av} for σ_1, σ_2, and σ_3 in Eq. (*b*). The result is

$$u_v = \frac{3\sigma_{av}^2}{2E}(1 - 2\nu) \tag{d}$$

If we now substitute the square of Eq. (*c*) in Eq. (*d*) and simplify the expression, we get

$$u_v = \frac{1 - 2\nu}{6E}\left(\sigma_1^2 + \sigma_2^2 + \sigma_3^2 + 2\sigma_1\sigma_2 + 2\sigma_2\sigma_3 + 2\sigma_3\sigma_1\right) \tag{6–4}$$

Then the distortion energy is obtained by subtracting Eq. (6–4) from Eq. (*b*). This gives

$$u_d = u_\sigma - u_v = \frac{1 + \nu}{3E}\left[\frac{(\sigma_1 - \sigma_2)^2 + (\sigma_2 - \sigma_3)^2 + (\sigma_3 - \sigma_1)^2}{2}\right] \tag{6–5}$$

Note that the distortion energy is zero if $\sigma_1 = \sigma_2 = \sigma_3$.

In words, *the distortion-energy hypothesis predicts that yielding will occur whenever the distortion energy in a unit volume equals the distortion energy in the same volume when uniaxially stressed to the yield strength.* For the simple tension test, let $\sigma_1 = \sigma'$, $\sigma_2 = \sigma_3 = 0$. The distortion energy is

$$u_d = \frac{1 + \nu}{3E}\sigma'^2 \tag{6–6}$$

Setting Eqs. (6–5) and (6–6) equal to each other gives

$$\sigma' = \left[\frac{(\sigma_1 - \sigma_2)^2 + (\sigma_2 - \sigma_3)^2 + (\sigma_3 - \sigma_1)^2}{2}\right]^{1/2} \tag{6–7}$$

Therefore yielding is predicted to occur when

$$\boxed{\sigma' \geq S_y} \tag{6–8}$$

The stress σ' should be called by a special name, because it represents the entire stress state σ_1, σ_2, and σ_3. The preferred names are the *effective stress* and the *von Mises stress*, after Dr. R. von Mises, who contributed to the hypothesis.

For the biaxial stress state, let σ_A and σ_B be the two nonzero principal stresses. Then, from Eq. (6–7), we get

$$\sigma' = \left(\sigma_A^2 - \sigma_A\sigma_B + \sigma_B^2\right)^{1/2} \tag{6–9}$$

The distortion-energy hypothesis is also called:

- The *shear-energy hypothesis*
- The *von Mises–Hencky hypothesis*
- The *octahedral-shear-stress hypothesis*

Under the name of the octahedral-shear-stress hypothesis, *failure is assumed to occur whenever the octahedral shear stress for any stress state equals or exceeds the octahedral shear stress for the simple tension-test specimen at failure.* Equation (3–12) is

$$\tau_{\text{oct}} = \frac{1}{3}\left[(\sigma_1 - \sigma_2)^2 + (\sigma_2 - \sigma_3)^2 + (\sigma_3 - \sigma_1)^2\right]^{1/2} \tag{e}$$

Using the tension test results $\sigma_1 = \sigma'$, $\sigma_2 = \sigma_3 = 0$, as before, we find from Eq. (*e*) that

$$\tau_{\text{oct}} = \frac{\sigma'}{3}(2)^{1/2} \tag{f}$$

Solving Eqs. (*e*) and (*f*) for σ' yields

$$\sigma' = \left[\frac{(\sigma_1 - \sigma_2)^2 + (\sigma_2 - \sigma_3)^2 + (\sigma_3 - \sigma_1)^2}{2}\right]^{1/2} \tag{g}$$

which is identical with Eq. (6–7).

For coordinate axes of convenience xyz the von Mises stress σ' of Eq. (*g*) can be written as

$$\sigma' = \frac{1}{\sqrt{2}}\left[\left(\sigma_x - \sigma_y\right)^2 + \left(\sigma_y - \sigma_z\right)^2 + \left(\sigma_z - \sigma_x\right)^2 + 6\left(\tau_{xy}^2 + \tau_{yz}^2 + \tau_{zx}^2\right)\right]^{1/2} \tag{h}$$

and for plane stress

$$\sigma' = \left(\sigma_x^2 - \sigma_x\sigma_y + \sigma_y^2 + 3\tau_{xy}^2\right)^{1/2} \tag{i}$$

Figure 6–21 shows the distortion-energy hypothesis for triaxial stress states. Notice that the hydrostatic components σ_1'', σ_2'', and σ_3'', as defined by Eq. (*a*) in Sec. 6–4, always lie on the axis of the cylinder no matter how far it is extended from the origin. The representation for biaxial stress states is shown in Fig. 6–22. This is a truer representation of the ellipse because of the distortion inherent in pictorial representation.[3]

The mathematical manipulation involved in the development of a hypothesis often tends to obscure the real value and usefulness of the result. Equations (6–7) and (6–8) mean that a complex stress situation can be represented by a single value. Just think about it. The von Mises stress can be used to represent the most complicated stress situation

3. Figure 6–21 was drafted using oblique pictorial drawing. The receding axis is scaled at 50 percent. In an isometric drawing, the scales of the three axes are equal. But isometric drawing could not be used in this case because the axis of the ellipse would overlap a coordinate axis.

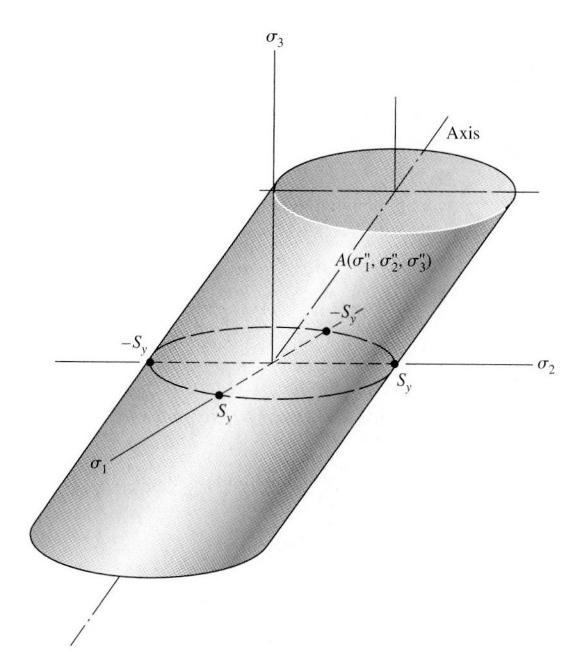

Figure 6–21

The distortion-energy (DE) theory graphically represented in three dimensions. The oblique elliptical cylinder encloses all safe values (free of yielding) of the general stress components σ_1', σ_2', and σ_3'. The axis of the cylinder is inclined equally to each of the three principal directions and is the locus of the points described by the triad of hydrostatic components σ_1'', σ_2'', and σ_3''.

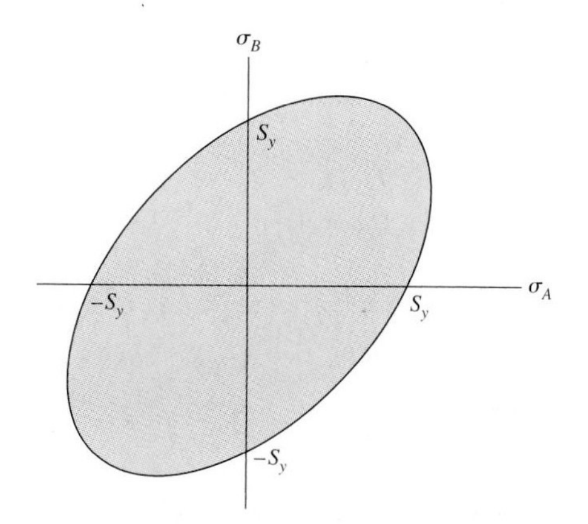

Figure 6–22

The distortion-energy (DE) theory for biaxial stress states. This is a true plot of points obtained from Eq. (6–9) with $\sigma' = S_y$.

you can think of! For example, the stress state σ_x, σ_y, σ_z, τ_{xy}, τ_{xz}, τ_{yz} can be represented by the single value σ'.

Furthermore, the distortion-energy hypothesis predicts no failure under hydrostatic stress and agrees with all the data. It has been given the status of *theory* and, accordingly, is widely used. It is a fine example of elegance, a word from the vocabulary of mathematics, which means introducing simplicity where before there had been complexity.

EXAMPLE 6–1

A hot-rolled steel has a yield strength of $S_{yt} = S_{yc} = 100$ kpsi and a true strain at fracture of $\varepsilon_f = 0.55$. Estimate the factor of safety for the following principal stress states:

(*a*) 70, 70, 0 kpsi.
(*b*) 30, 70, 0 kpsi.
(*c*) 0, 70, −30 kpsi.
(*d*) 0, −30, −70 kpsi.
(*e*) 30, 30, 30 kpsi.

Solution Since $\varepsilon_f > 0.05$ and S_{yc} and S_{yt} are equal, the distortion-energy (DE) theory applies. The maximum-shear-stress (MSS) hypothesis results will be compared to the DE results.

(*a*) The ordered principal stresses are $\sigma_1 = 70$, $\sigma_2 = 70$, $\sigma_3 = 0$ kpsi.

DE. From Eq. (6–7),

$$\sigma' = \left[\frac{(70-70)^2 + (70-0)^2 + (0-70)^2)}{2}\right]^{1/2} = 70 \text{ kpsi}$$

Answer
$$n = \frac{S_y}{\sigma'} = \frac{100}{70} = 1.43$$

MSS. Using Eq. (6–3),

$$\tau_{\max} = \tau_{1/3} = \frac{\sigma_1 - \sigma_3}{2} = \frac{70-0}{2} = 35 \text{ kpsi}$$

From Eq. (6–1),

Answer
$$n = \frac{S_y}{2\tau_{\max}} = \frac{100}{2(35)} = 1.43$$

(*b*) The ordered principal stresses are $\sigma_1 = 70$, $\sigma_2 = 30$, and $\sigma_3 = 0$ kpsi.

DE

$$\sigma' = \left[\frac{(70-30)^2 + (30-0)^2 + (0-70)^2}{2}\right]^{1/2} = 60.8 \text{ kpsi}$$

Answer
$$n = \frac{S_y}{\sigma'} = \frac{100}{60.8} = 1.64$$

MSS

$$\tau_{\max} = \frac{\sigma_1 - \sigma_3}{2} = \frac{70-0}{2} = 35 \text{ kpsi}$$

Answer
$$n = \frac{S_y}{2\tau_{\max}} = \frac{100}{2(35)} = 1.43$$

(*c*) The ordered principal stresses are $\sigma_1 = 70$, $\sigma_2 = 0$, $\sigma_3 = -30$ kpsi.

DE

$$\sigma' = \left[\frac{(70-0)^2 + [0-(-30)]^2 + (-30-70)^2}{2}\right]^{1/2} = 88.9 \text{ kpsi}$$

Answer
$$n = \frac{S_y}{\sigma'} = \frac{100}{88.9} = 1.13$$

MSS

$$\tau_{\max} = \frac{\sigma_1 - \sigma_3}{2} = \frac{70-(-30)}{2} = 50 \text{ kpsi}$$

Answer
$$n = \frac{S_y}{2\tau_{\max}} = \frac{100}{2(50)} = 1.00$$

(*d*) The ordered principal stresses are $\sigma_1 = 0$, $\sigma_2 = -30$, $\sigma_3 = -70$ kpsi.

DE

$$\sigma' = \left[\frac{[0-(-30)]^2 + [-30-(-70)]^2 + (-70-0)^2}{2}\right]^{1/2} = 60.8 \text{ kpsi}$$

Answer $$n = \frac{S_y}{\sigma'} = \frac{100}{60.8} = 1.64$$

MSS

$$\tau_{max} = \frac{\sigma_1 - \sigma_3}{2} = \frac{-70 - 0}{2} = 35 \text{ kpsi}$$

Answer $$n = \frac{S_y}{2\tau_{max}} = \frac{100}{2(35)} = 1.43$$

(*e*) The ordered principal stresses are $\sigma_1 = 30$, $\sigma_2 = 30$, $\sigma_3 = 30$ kpsi.

DE

$$\sigma' = \left[\frac{(30 - 30)^2 + (30 - 30)^2 + (30 - 30)^2}{2}\right]^{1/2} = 0 \text{ kpsi}$$

Answer $$n = \frac{S_y}{\sigma'} = \frac{100}{0} \rightarrow \infty$$

MSS

$$\tau_{max} = \frac{\sigma_1 - \sigma_3}{2} = \frac{30 - 30}{2} = 0 \text{ kpsi}$$

Answer $$n = \frac{S_y}{2\tau_{max}} = \frac{100}{2(0)} \rightarrow \infty$$

A tabular summary of the factors of safety is useful:

	(*a*)	(*b*)	(*c*)	(*d*)	(*e*)
DE	1.43	1.64	1.13	1.64	∞
MSS	1.43	1.43	1.00	1.43	∞

In these instances the factor of safety under MSS was equal to or less than under DE. Under DE the stress states under the first four parts are biaxial, and their load lines are plotted in Fig. 6–23. Note that the factors of safety can be found graphically by dividing

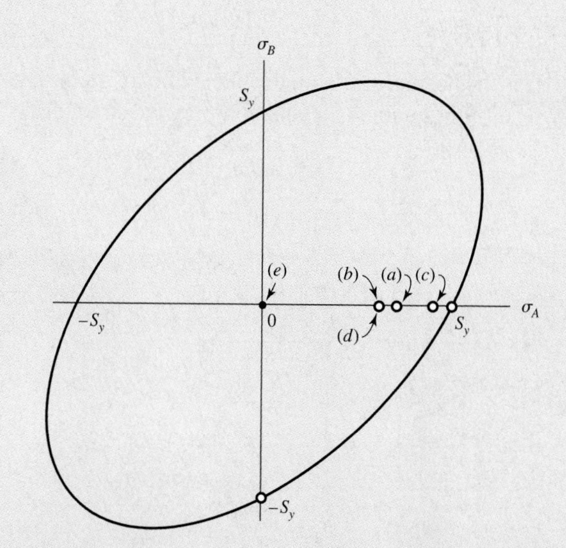

Figure 6–23

Biaxial DE diagram for Ex. 6–1. Since the distortion energy of strain is important, load lines map onto the *x* axis. The labels on the stress state characterized by the von Mises stress correspond to the parts of Ex. 6–1.

the measured radial distance from the origin to the failure locus, by the distance from the origin to the stress state. Try it.

6–6 Ductile Materials: Internal-Friction Hypothesis

The Mohr hypothesis dates to 1900, a date that is relevant to its presentation. There were no computers, just slide rules, compasses, and French curves. Graphical procedures, common then, are still useful. The idea of Mohr is based on three "simple" tests, tension, compression, and shear, to yielding if the material can yield, or to rupture. It is easier to define shear yield strength as S_{sy} than it is to test for it.

The practical difficulties aside, the hypothesis of Mohr was to use the results of tensile, compressive, and shear tests to construct three circles of Fig. 6–24 in order to define a failure envelope, depicted as straight line in the figure, above the σ axis. The failure envelope need not be straight. The argument amounted to the three Mohr circles describing the stress state in a body (Fig. 3–5) growing during loading until one of them became tangent to the failure envelope, defining failure. Was the form of the failure envelope straight, circular, or quadratic? A compass or a French curve defined the failure envelope. The Mohr hypothesis can now be stated:

> Failure is predicted to occur in a multiaxial state of stress when the larger Mohr circle associated with the state of stress at the critical location becomes tangent to, or exceeds the bounds of the failure envelope established by conditions of failure in simple tensile, compressive, and torsion tests using specimens of the same material and condition.

Suppose the material has equal tensile and compressive strengths. If the failure envelope is a straight line parallel to the σ axis of Fig. 6–24, then for a body stress state, after Fig. 3–5,

$$\sigma_1 - \sigma_3 = S_y \qquad (6\text{–}10)$$

which is the same result the maximum-shear-stress hypothesis projects. This is an indication that the failure envelope cannot be $\tau_{\max} = $ constant. A failure envelope of the form

$$\tau_{\max}^2 = a\sigma_n^2 + b$$

where σ_n is the normal stress on the plane of maximum shear stress, gives the distortion-energy result for equal tensile and compressive strengths.

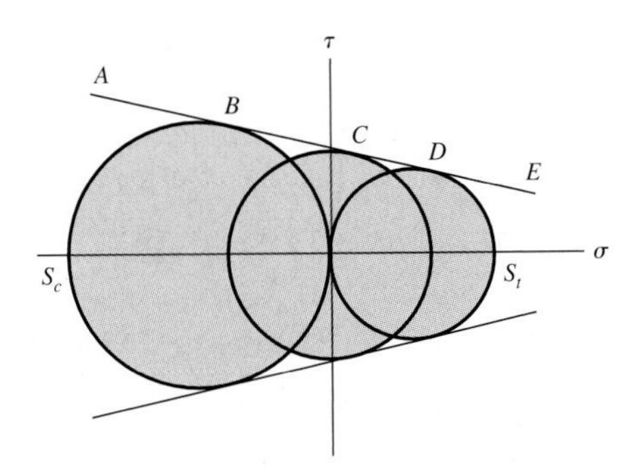

Figure 6–24

Three Mohr circles, one for the uniaxial compression test, one for the test in pure shear, and one for the uniaxial tension test, are used to define failure by the Mohr hypothesis. The strengths S_c and S_t are the compressive and tensile strengths, respectively; they can be used for yield or ultimate strength.

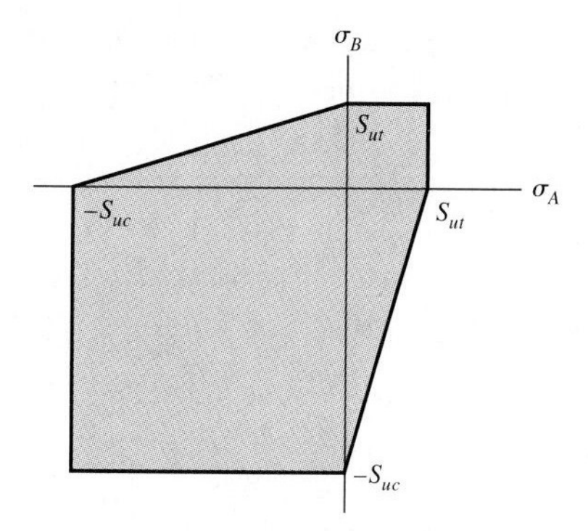

Figure 6–25

Plot of the internal-friction, or Coulomb-Mohr hypothesis of failure for biaxial stress states with $S_{uc} = 3S_{ut}$.

There is a useful spin-off of the Mohr straight-line failure envelope. Given the tensile and compressive yield strengths, the torsional yield strength can be predicted from

$$S_{sy} = \frac{S_{yt} S_{yc}}{S_{yt} + S_{yc}} \tag{6–11}$$

An internal-friction hypothesis can be based on the straight-line failure envelope line of Fig. 6–24. This still gives it maximum-shear-stress properties. It is called the *Coulomb-Mohr* hypothesis. This hypothesis is based on the assumption that line *BCD* in Fig. 6–24 is straight. Order the three principal stresses so that $\sigma_1 > \sigma_2 > \sigma_3$. Then for any stress state producing a circle tangent to line *BCD*, between points *B* and *D*, it is true that σ_1 and σ_3 have opposite signs. For this state of stress the Mohr hypothesis applies and the two stresses and the strengths are related by the equation

$$\frac{\sigma_1}{S_t} - \frac{\sigma_3}{S_c} = 1 \qquad \sigma_1 \geq 0, \sigma_3 \leq 0 \tag{6–12}$$

For biaxial stress states in which σ_1 and σ_3 have like signs, the internal-friction hypothesis is the same as the maximum-normal-stress hypothesis and failure is predicted by

$$\begin{aligned} \sigma_1 &= S_t \qquad & \sigma_1 > 0 \\ \sigma_3 &= -S_c \qquad & \sigma_3 < 0 \end{aligned} \tag{6–13}$$

Either yield strength or ultimate strength can be used with Eqs. (6–12) and (6–13). Note again that strengths are always treated as positive values. The internal-friction hypothesis is shown in Fig. 6–25 for a biaxial stress state. The nonzero stresses are σ_A and σ_B. This figure is plotted for a material, such as gray cast iron, in which $S_{uc} > S_{ut}$

EXAMPLE 6–2

A 25-mm-diameter shaft is statically torqued to 230 N·m. It is made of cast 195-T6 aluminum, with a yield strength in tension of 160 MPa and a yield strength in compression of 170 MPa. It is machined to final diameter. Estimate the factor of safety in the 25-mm shank.

Solution

Since N·m, cm, and MPa form a consistent set of dimensions, we express the diameter in cm:

$$\tau = \frac{16T}{\pi d^3} = \frac{16(230)}{\pi 2.5^3} = 75 \text{ MPa}$$

The two nonzero principal stresses are 75 and −75 MPa, making the ordered principal stresses $\sigma_1 = 75$, $\sigma_2 = 0$, and $\sigma_3 = -75$ MPa. From Eq. (6–12),

Answer
$$n = \frac{1}{\sigma_1/S_{yt} - \sigma_3/S_{yc}} = \frac{1}{75/160 - (-75)/170} = 1.10$$

Alternatively, from Eq. (6–11),

$$S_{sy} = \frac{S_{yt}S_{yc}}{S_{yt} + S_{yc}} = \frac{160(170)}{160 + 170} = 82.4 \text{ MPa}$$

$$\tau_{\max} = \frac{\sigma_1 - \sigma_3}{2} = \frac{75 - (-75)}{2} = 75 \text{ MPa}$$

Answer
$$n = \frac{S_{sy}}{\tau_{\max}} = \frac{82.4}{75} = 1.10$$

6–7 Criticism of Hypotheses by Data in Ductile Materials

The first requirement for a hypothesis to attain the status of theory is the ability to explain all the facts known at the time and to continue to do so in the future. Hypotheses are on "good behavior," which can be rescinded at any time. When there is sufficient data and all of it is deemed to be explained sufficiently well by nearly all of the scientific community involved, a hypothesis is given the status of theory.[4]

Evidence

In this section we limit ourselves to material and parts that are known to fail in a ductile manner. To help decide on appropriate and workable hypotheses of ductile material failure, Marin collected data from many sources.[5] Some of the data points for ductile materials are shown on the graph in Fig. 6–26. Marin also collected many data for copper and nickel alloys; if shown, the data points would be mingled with those already diagrammed. Figure 6–26 shows that in the fourth quadrant the distortion-energy theory is acceptable for design and analysis. You may wish to plot other hypotheses using a red or blue pencil on Fig. 6–26 to show why they are not acceptable or are used only in specific domains.

Figure 6–26 shows dimensionless data, σ_A/S_y as abscissa and σ_B/S_y as ordinate. Without resorting to statistical goodness-of-fit programs, one can see that the maximum-shear-stress hypothesis is biased "below." The distortion-energy hypothesis locus is among the data. The distortion-energy hypothesis is the best available, the most useful in predicting the onset of yielding. It has the status of theory. This evidence is for static yielding. We shall see more of distortion energy in fatigue.

For ductile materials with unequal yield strengths, S_{yt} in tension and S_{yc} in compression, the Mohr hypothesis is the best available. It can be implemented graphically with the tensile, compression, and torsional test results. Note that it takes

- Testing of the material involved in three modes.
- A French curve or circular-arc failure locus, constructed graphically.

4. Whereas a president of the United States can be chosen by a plurality and a pope by a simple majority, a theory requires near unanimity in the scientific community affected.

5. Joseph Marin was one of the pioneers in the collection, development, and dissemination of material on the failure of engineering elements. He wrote many books and papers on the subject. Here the reference used is Joseph Marin, *Engineering Materials*, Prentice Hall, Englewood Cliffs, N.J., 1952. (See pp. 156 and 157 for some data points used here.)

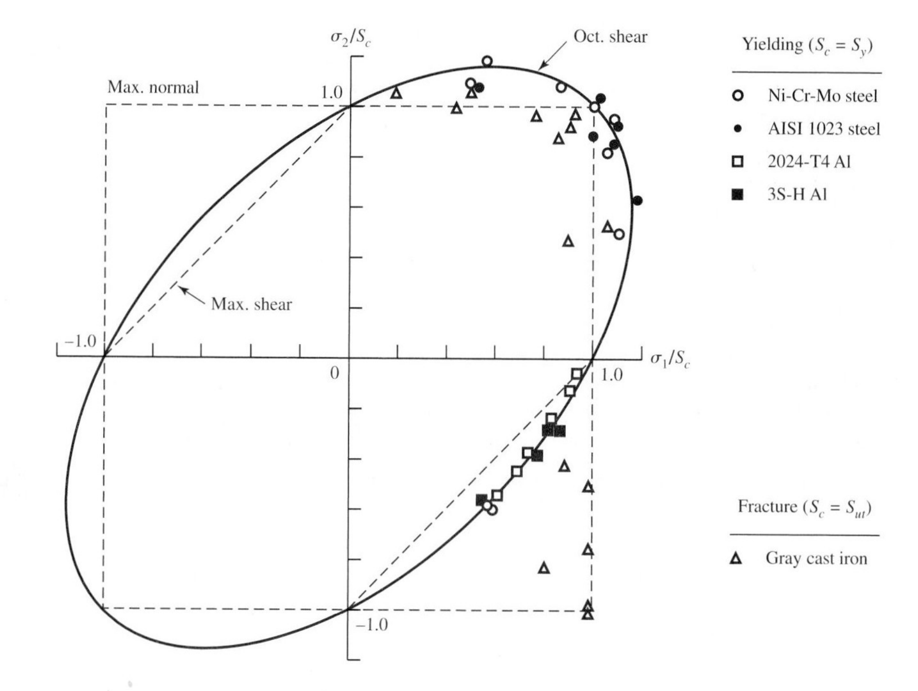

Figure 6–26

Experimental data superposed on failure hypotheses. (Reproduced from Fig. 7.11, p. 252, *Mechanical Behavior of Materials*, N. E. Dowling, Prentice Hall, Englewood Cliffs, N.J., 1993.)

- The superposing of the largest Mohr circle associated with the stress state.

The testing in torsion removes the tie to maximum-shear-stress theory. The alternative to the graphics is a computer program which can fit a polynomial of choice, or other curve, tangent to the three circles. Then the center of the largest Mohr circle of the stress state is known, find the diameter and the abscissa intercepts. In a learning situation black box computer codes do not give meaning and insight to the problem experience. Another drawback is the necessity of material testing.

The fallback position is a Coulomb-Mohr hypothesis using tensile and compressive yield strengths. It follows that

$$n\sigma_A = S_{yt} \qquad 0 \le \sigma_A \le S_{yt}, \quad 0 < \sigma_B < S_{yt} \tag{6–12}$$

$$\frac{n\sigma_A}{S_{yt}} - \frac{n\sigma_B}{S_{yc}} = 1 \qquad 0 \le \sigma_A \le S_{yt}, \quad -S_{yc} \le \sigma_B \le 0 \tag{6–13}$$

which is a variation of Coulomb-Mohr that will be developed in Sec. 6-9 for brittle materials. These recommendations are presented in Fig. 6–27. See also Fig. 6–28.

6–8 Brittle Materials: Maximum-Normal-Stress (Rankine) Hypothesis

The maximum-normal-stress (MNS) hypothesis states that *failure occurs whenever one of the three principal stresses equals or exceeds the strength.* Suppose we arrange the three principal stresses for any stress state in the ordered form

$$\sigma_1 > \sigma_2 > \sigma_3 \tag{6–14}$$

This hypothesis then predicts that failure occurs whenever

Figure 6–27

Models describe data, and data criticize models. This protocol represents what is known about static failure. Note that only one category has a theory. When hypotheses are used in their narrow constrained applicability they are thought of as "little theories," but the word "theory" should not be used *unless you recite all the constraining phrases.*

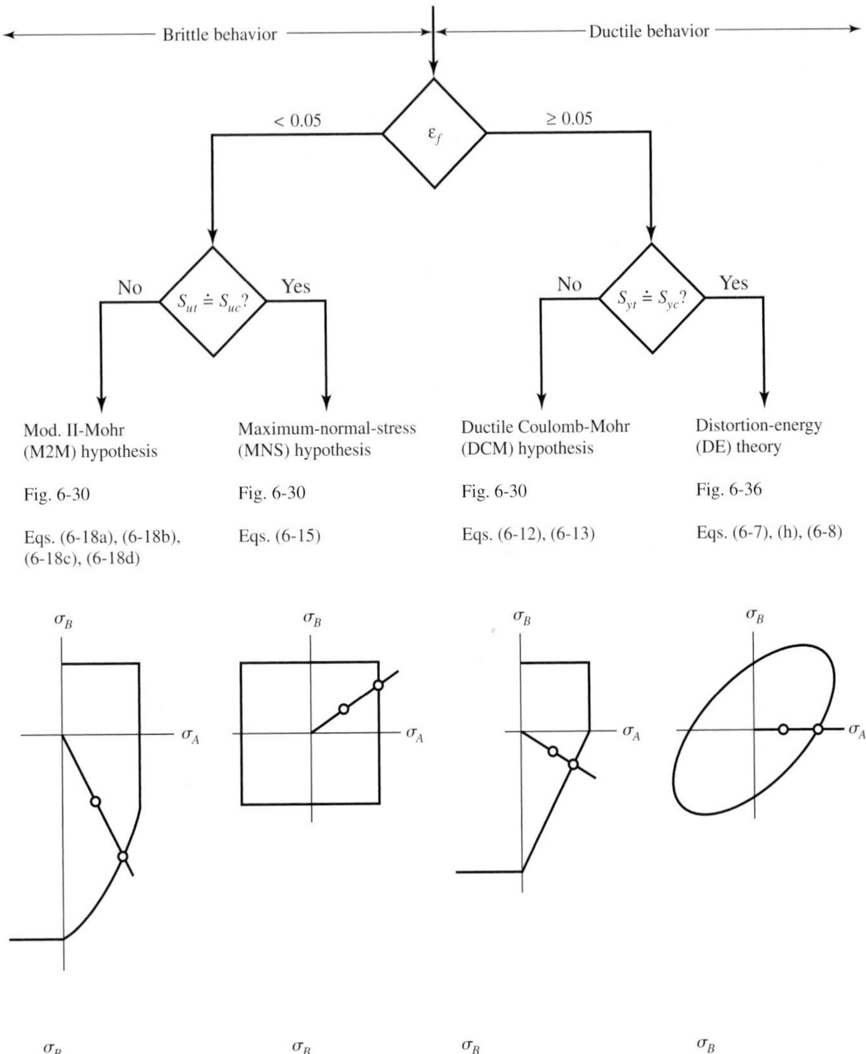

Figure 6–28

A load line, a stress state described by σ_A and σ_B, and the intersection with the failure locus described by S_A and S_B. The factor of safety n is given by S_A/σ_A, or S_B/σ_B, or by measuring $\overline{OQ}$ and $\overline{OP}$ and writing $n = \overline{OQ}/\overline{OB}$ or their projections on the axes. Scale factors on the ordinate and abscissa must agree. (*a*) Mod. II-Mohr (M2M) hypothesis; (*b*) maximum-normal-stress (MNS) hypothesis; (*c*) ductile Coulomb-Mohr (DCM) hypothesis; and (*d*) distortion-energy (DE) theory.

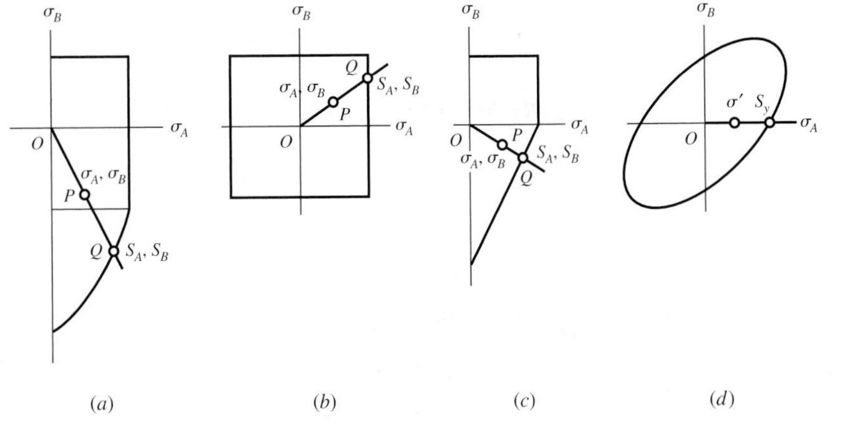

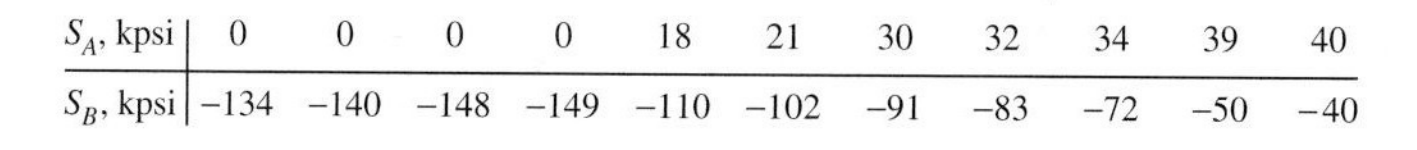

S_A, kpsi	0	0	0	0	18	21	30	32	34	39	40
S_B, kpsi	−134	−140	−148	−149	−110	−102	−91	−83	−72	−50	−40

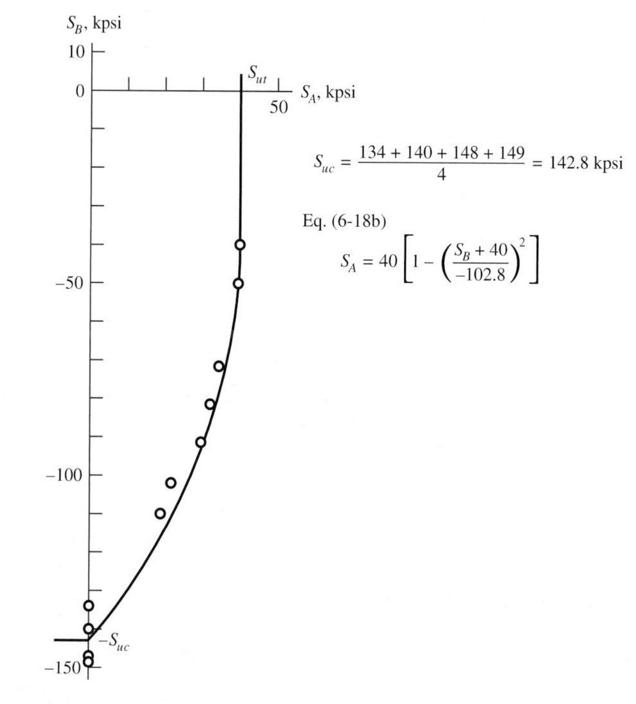

Figure 6–29

Fourth quadrant S_A S_B data for a grade 40 cast iron, compared with a mod. II-Mohr fracture locus. (Source of data: Charles F. Walton (ed.), *Iron Casting Handbook*, Iron Founders' Society, 1971, p. 215, Cleveland, Ohio.)

$$n\sigma_1 = S_t \quad \text{or} \quad n\sigma_3 = -S_c \tag{6–15}$$

where S_t and S_c are tensile and compressive strengths, usually either the yield or ultimate, respectively. Figures 6–15 and 6–16 illustrate stress states associated with safety or failure. The no-loss-of-function combinations of principal stresses lie within the prism, and the loss-of-function combinations lie outside the prism. The prism surface defines the failure surface or failure locus.

The maximum-normal-stress hypothesis fails to predict the lack of damage due to large hydrostatic stresses. There is little information in the second and third quadrant of the biaxial stress diagram. Figures 6–29 and 6–30 show first- and fourth-quadrant cast-iron data following the MNS hypothesis in the region $-S_{ut} < \sigma_2 < S_{ut}$, outside of which the MNS hypothesis becomes conservative. The MNS hypothesis is of limited use for equal-strength brittle failure in static loading; it appears in Fig. 6–27 since it falls among the data as shown in Fig. 6–30.

6–9 Brittle Materials: Modifications of the Mohr Hypothesis

When static brittle material failure data are plotted on the Coulomb-Mohr biaxial plot, the data in the lower part of the fourth quadrant fall below the failure locus (see Figs. 6–29 and 6–30). A modification was made when $\sigma_B < S_{ut}$, and the failure line was drawn as shown as a full-weight solid line in the figure. The modifications to the Coulomb-Mohr scheme are called modified-Mohr hypotheses, or mod. I-Mohr or mod. II-Mohr, or M1M or M2M for short.

Under biaxial stress conditions, using nonzero principal stresses σ_A and σ_B, the Coulomb-Mohr and modifications can be quantitatively expressed as follows.

Figure 6–30

Biaxial fracture data of gray cast iron compared with various failure criteria. (From Fig. 7.13, p. 255, *Mechanical Behavior of Materials*, N. E. Dowling, Prentice Hall, Englewood Cliffs, N.J., 1993. Data from Grassi and I. Cornet, "Fracture of Gray Cast Iron Tubes under Biaxial Stresses," *Journal of Applied Mechanics*, vol. 16, 1949, p. 178.

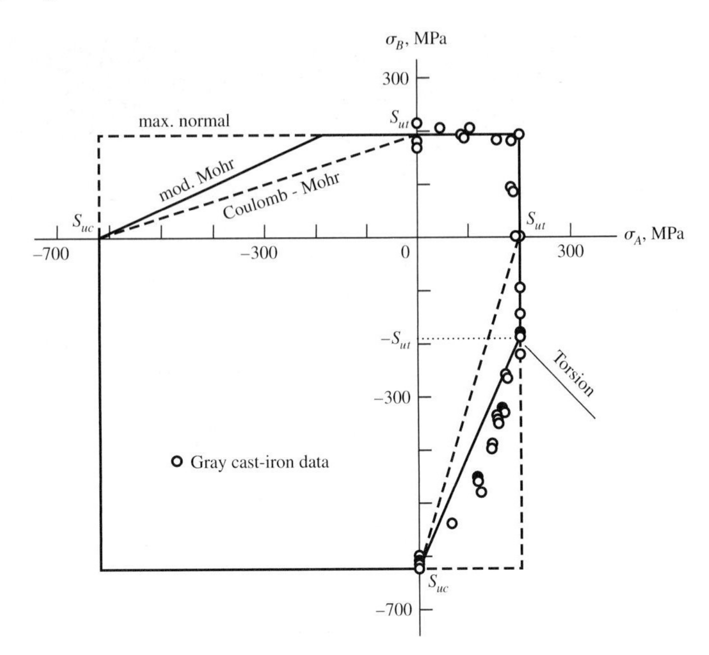

- Coulomb-Mohr:

$$\sigma_A = \frac{S_{ut}}{n} \qquad 0 < \sigma_A < S_{ut}, \quad 0 < \sigma_B < S_{ut} \tag{6–16a}$$

$$\frac{\sigma_A}{S_{ut}} - \frac{\sigma_B}{S_{uc}} = \frac{1}{n} \qquad 0 < \sigma_A < S_{ut}, \quad -S_{uc} < \sigma_B < 0 \tag{6–16b}$$

or, in terms of load line slope r,

$$S_A = n\sigma_A = \frac{S_{uc}S_{ut}}{(S_{uc} - rS_{ut})} \qquad \sigma_B = r\sigma_A, \qquad 0 < \sigma_A < S_{ut}, \quad -S_{uc} < \sigma_B < 0 \tag{6–16c}$$

- Mod. I-Mohr:

$$\sigma_A = \frac{S_{ut}}{n} \qquad 0 < \sigma_A < S_{ut}, \quad -S_{ut} < \sigma_B < S_{ut} \tag{6–17a}$$

$$\sigma_A - \frac{S_{ut}\sigma_B}{S_{uc} - S_{ut}} = \frac{S_{uc}S_{ut}}{n\,(S_{uc} - S_{ut})} \qquad 0 < \sigma_A < S_{ut}, \quad -S_{uc} < \sigma_B < -S_{ut} \tag{6–17b}$$

or, in terms of load line slope r,

$$S_A = n\sigma_A = \frac{S_{uc}S_{ut}}{S_{uc} - (1 + r)S_{ut}} \qquad \sigma_B = r\sigma_A, \quad 0 < \sigma_A < S_{ut}, \quad -S_{uc} < \sigma_B < -S_{ut} \tag{6–17c}$$

- Mod. II-Mohr:

$$\sigma_A = \frac{S_{ut}}{n} \qquad 0 \le \sigma_A \le S_{ut}, \quad -S_{ut} \le \sigma_B \le S_{ut} \tag{6–18a}$$

$$\frac{n\sigma_A}{S_{ut}} = 1 - \left(\frac{n\sigma_B + S_{ut}}{-S_{uc} + S_{ut}}\right)^2 \qquad 0 \le \sigma_A \le S_{ut}, \quad -S_{uc} \le \sigma_B \le -S_{ut} \tag{6–18b}$$

Table 6–1

Modified II-Mohr Static Failure Hypothesis* under Biaxial Stress Conditions for Brittle Materials

First Quadrant $\sigma_A \geq 0, \sigma_B \geq 0$	**Fourth Quadrant** $\sigma_A \geq 0, \sigma_B \leq 0$
$n\sigma_A = S_A = S_{ut}$	$n\sigma_A = S_A = S_{ut}, \sigma_A \geq 0, -S_{ut} \leq \sigma_B \leq 0$
	Parabolic failure locus segment $-S_{uc} \leq \sigma_B \leq -S_{ut}$ $\frac{S_A}{S_{ut}} = 1 - \left(\frac{S_B + S_{ut}}{-S_{uc} + S_{ut}}\right)^2$ $\frac{n\sigma_A}{S_{ut}} = 1 - \left(\frac{n\sigma_B + S_{ut}}{-S_{uc} + S_{ut}}\right)^2$ In terms of the load line slope $r = \sigma_B/\sigma_A$ $\Delta = -S_{uc} + S_{ut}$ $b = \frac{\Delta^2}{r^2 S_{ut}} + \frac{2S_{ut}}{r}$ $c = \frac{S_{ut}^2}{r^2} - \frac{\Delta^2}{r^2}$ $S_A = \frac{-b + \sqrt{b^2 - 4c}}{2}, S_B = rS_A, \sigma_B = r\sigma_A$
	$n\sigma_B = -S_{uc}, \sigma_A \leq 0$

* Of maximum-normal-stress, Coulomb-Mohr, mod. I-Mohr, and mod. II-Mohr, only the mod. II-Mohr hypothesis lies among the data in the first and fourth quadrants. This feature is necessary in formulating probability-of-failure estimates.

or, in terms of the load line slope $r = \sigma_B/\sigma_A$, $\Delta = -S_{uc} + S_{ut}$,

$$n\sigma_A = \frac{-b + \sqrt{b^2 - 4c}}{2} \tag{6–18c}$$

where

$$b = \frac{\Delta^2}{r^2 S_{ut}} + \frac{2S_{ut}}{r}, \quad c = \frac{S_{ut}^2}{r^2} - \frac{\Delta^2}{r^2} \quad 0 \leq \sigma_A \leq S_{ut}, \quad -S_{uc} \leq \sigma_B \leq -S_{ut} \tag{6–18d}$$

See Fig. 6–29. The mod. II-Mohr locus passes among the data.

The original Mohr hypothesis, constructed with tensile, compression, and torsion tests, with a curved failure locus is the best hypothesis we have, but the difficulty of applying it without a computer leads engineers to choose modifications, namely, Coulomb-Mohr, mod. I-Mohr, or mod. II-Mohr. Even the original Mohr is still a hypothesis, because there is a scarcity of data in the second and third quadrants. The hypotheses must be used *only* in their *verified* domains. The mod. II-Mohr fits the data best. See Table 6–1, and Fig. 6–27 for a summary concerning static loading of brittle and ductile materials and effective analysis methods.

EXAMPLE 6–3 This example illustrates the use of a failure model to estimate the strength of a mechanical element or component. The example may help distinguish among *strength of a machine part* and *strength of a material* and *strength of a part at a point.*

A static force F applied at D near the end of the 15-in lever shown in Fig. 6–31, which is quite similar to a socket wrench, results in certain stresses in the cantilevered bar $OABC$. This bar is made of cast iron, machined to dimension. The force F required to fracture this part can be regarded as the strength of the component part. If the material is ASTM grade 30 cast iron, find the force F with

(*a*) Coulomb-Mohr failure model.
(*b*) Mod. I-Mohr failure model.
(*c*) Mod. II-Mohr failure model.

Solution We assume that the lever DC is strong, and not part of the problem. Since grade 30 cast iron is a brittle material *and* cast iron, the stress-concentration factors K_t and K_{ts} are set to unity. From Table E-24, the tensile ultimate strength is 31 kpsi and the compressive ultimate strength is 109 kpsi. The stress element at A on the top surface will be subjected to a tensile bending stress and a torsional stress. This location, on the 1-in-diameter section fillet, is the weakest location, and it governs the strength of the assembly. The normal stress σ_x and the shear stress at A are given by

$$\sigma_x = K_t \frac{M}{I/c} = K_t \frac{32M}{\pi d^3} = (1)\frac{32(14F)}{\pi(1)^3} = 142.6F$$

$$\tau_{xy} = K_{ts} \frac{Tr}{J} = K_{ts} \frac{16T}{\pi d^3} = (1)\frac{16(15F)}{\pi(1)^3} = 76.4F$$

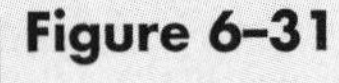
Figure 6–31

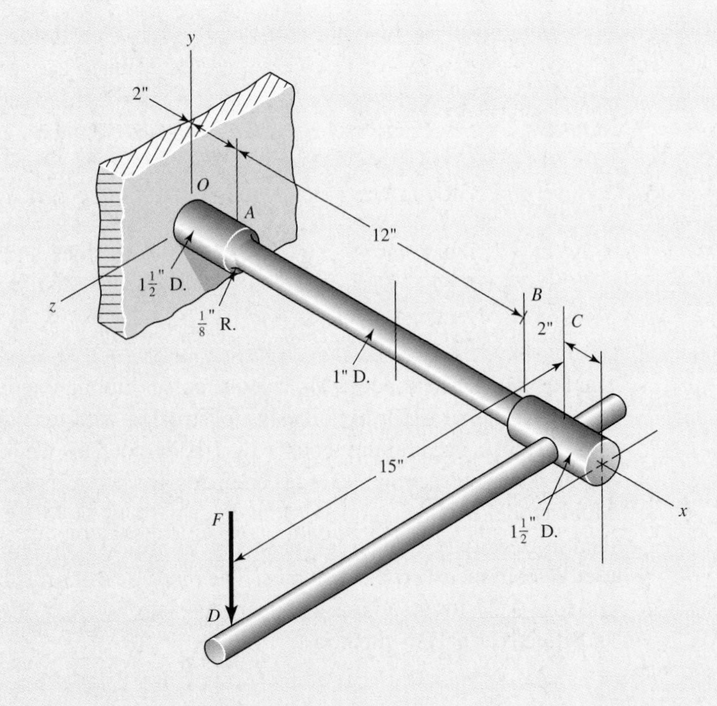

From Eq. (3–8) the nonzero principal stresses σ_A and σ_B are

$$\sigma_A, \sigma_B = \frac{142.6F + 0}{2} \pm \sqrt{\left(\frac{142F - 0}{2}\right)^2 + (76.4F)^2} = 175.6F, -33.2F$$

The slope of the load line r is $r = \sigma_B/\sigma_A = -33.2F/175.8F = -0.189$.

(*a*) From Eq. (6–16c), the strength S_A is

$$S_A = n\sigma_A = \frac{109(31)}{109 - (-0.189)31} = 29.4 \text{ kpsi}$$

$$S_B = rS_A = -0.189(29.4) = -5.56 \text{ kpsi}$$

Equating $n\sigma_A$ to 29 400 psi and solving for F gives

Answer $$F = \frac{29\,400}{175.8} = 167 \text{ lb}$$

(*b*) From Eq. (6–17a), $S_A = S_{ut} = 31$ kpsi and

Answer $$F = \frac{31\,000}{175.8} = 176 \text{ lb}$$

Answer (*c*) The solution is the same as in part *b*; $F = 176$ lb.

As one would expect from inspection of Fig. 6–30, Coulomb-Mohr is conservative.

6–10 The Criticism of Hypotheses by Data in Brittle Materials

We have identified failure or strength of brittle materials that conform to the usual meaning of the word *brittle*, relating to those materials whose true strain at fracture is 0.05 or less. We also have to be aware of normally ductile materials which for some reason may develop a brittle fracture or crack if used below the transition temperature. Figure 6–32 shows data for a nominal grade 30 cast iron taken under biaxial stress conditions, with several brittle failure hypotheses shown, superposed. We note the following:

- In the first quadrant the data appear on both sides and along the failure locus of maximum-normal-stress, Coulomb-Mohr, mod. I-Mohr, and mod. II-Mohr. All failure loci are the same, and data fit well.
- In the fourth quadrant only the mod. II-Mohr hypothesis produces a failure locus which lies in and among the data.
- In the third quadrant the points A, B, C, and D are too few to make any suggestion concerning a fracture locus.
- It is clear that unless the designer is prepared to do considerable testing, he or she should avoid second- and third-quadrant loadings on a contemplated concept. Changing the concept to stay in the first or fourth quadrant is the less costly path.

The pressure to get things done leads engineers to employ loadings which are understood quantitatively. At any point in time, know what we can and cannot do, and act accordingly.

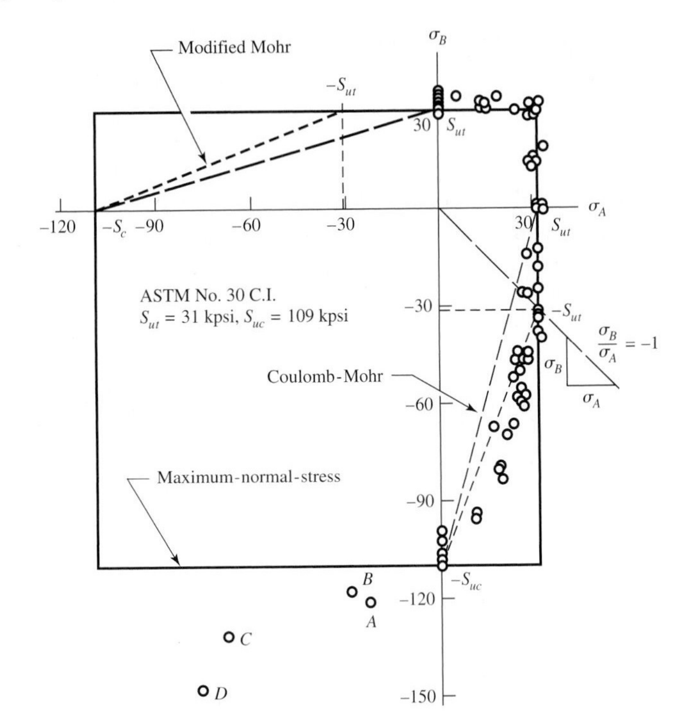

Figure 6–32

A plot of experimental data points obtained from tests on cast iron. Shown also are the graphs of three failure hypotheses of possible usefulness for brittle materials. Note points *A*, *B*, *C*, and *D*. To avoid congestion in the first quadrant, points have been plotted for $\sigma_A > \sigma_B$ as well as for the opposite sense. (Source of data: Charles F. Walton (ed.), *Iron Castings Handbook*, Iron Founders' Society, 1971, pp. 215, 216, Cleveland, Ohio.)

6–11 What Our Failure Models Tell Us

In the case of ductile materials with equal yield strengths in tension and compression, the distortion-energy theory is the explanation of the data. This means that if you accept this thesis and act upon it, your predictions will be in congruence with nature. In Sec. 6–5, Eqs. (6–7) and (6–8) are the crux of the matter, along with the following Eq. (*h*). Since the mean lay of the data was the basis of the relations, these equations are relations among means. A more powerful statement is "The probability of encountering a yielding loss of function is the probability that an instance of von Mises stress σ' exceeds an instance of a yield strength S_y."

One might approach the uncertainty, as outlined in Sec. 1–5, using Philon's method by writing

$$\sigma' = S_y/n_d \tag{6–19}$$

where the design factor comes from the aggregate experience of engineering successes and failures in the past. Another approach is the stochastic design factor method where

$$\mathbf{n}_d = \frac{\mathbf{S}_y}{\boldsymbol{\sigma}'} \tag{6–20}$$

Still another is the stochastic method.

Ponder Fig. 6–26. The designer needs to establish a failure ellipse which is defined by S_y. If this is historical all-heats information as in Fig. 2–1*b*, the COV of $\mathbf{S}_y$ for 1035-classifiable steels is $5.36/49.5 = 0.108$, and $\mathbf{S}_y \sim LN(49.5, 5.36)$ kpsi. If the designer or his or her company has not yet ordered a 1035 steel, and the part manufacture is so far in the future that the steel has not yet been smelted, this is all the knowledge that is available. The ability to locate the DE ellipse for yielding failure is clouded by the level of uncertainty in our knowledge of $\mathbf{S}_y$. If a particular heat is in stock at a steel distributor's yard, then $\mathbf{S}_y$ can be constructed by test. If delivered, the actual population from which the workpieces will be selected is available. The COV will be smaller than in Fig. 2–1*b*, but the mean can be higher or lower. The shrinking of the COV of $\mathbf{S}_y$ is important.

It is useful to view the DE ellipse as a bagel-half, squeezed to make an elliptical footprint for Figs. 6–22 and 6–26. The radial load line slices defining PDFs. We could draw nested ellipses to represent yield strength probability contours.

Similarly, our knowledge of the von Mises stress $\boldsymbol{\sigma}'$ is stochastic for reasons of load and geometry variability. The stochastic design factor $\mathbf{n}_d = \mathbf{S}_y/\boldsymbol{\sigma}'$ can be constructed from our knowledge of $\mathbf{S}_y$ and $\boldsymbol{\sigma}'$. We can often use $\bar{n}_d$, $\bar{S}_y$, and $\bar{\sigma}'$ and proceed deterministically to achieve an acceptable risk of failure (yielding).

In the case of brittle materials, MNS, Coulomb-Mohr, mod. I-Mohr, and mod. II-Mohr agree in the first quadrant of the biaxial stress situation. In the fourth quadrant Coulomb-Mohr is conservative, and the other three agree and are among the data the region $-S_{ut} < \sigma_B < 0$. In the region $-S_{uc} < \sigma_B < -S_{ut}$ mod. I-Mohr is conservative and mod. II-Mohr lies among the data. For the entirety of the first and fourth quadrants, only the mod. I-Mohr is congruent to the data.

In the case of brittle materials, Philon's method can be expressed as

$$n\sigma_A = S_A \tag{6–21}$$

where S_A is set equal to the expression for S_A in Eqs. (6–16a) or (6–16c), (6–17a) or 6–17c), or (6–18a) or (6–18c).

In addressing uncertainty and probability of failure only the mod. II-Mohr can be used for $\mathbf{S}_A$, and $\bar{S}_A$ and σ_{SA} are found by propagation of error through Eq. (6–18c) and its auxiliaries, Eqs. (6–18d). The load and geometry effects on $\boldsymbol{\sigma}_A$ are handled as with ductile materials, as

$$\mathbf{n}_d = \frac{\mathbf{S}_A}{\boldsymbol{\sigma}''} \tag{6–22}$$

or

$$\bar{n}_d = \frac{\bar{S}_A}{\bar{\sigma}'} \tag{6–23}$$

6–12 Interference—General

In the previous sections, we have employed interference theory to estimate reliability when the distributions are both normal and when they are both lognormal. Sometimes, however, it turns out that the strength has, say, a Weibull distribution while the stress is distributed lognormally. In fact, stresses are quite likely to have a lognormal distribution, because the multiplication of variates that are normally distributed produces a result that approaches lognormal. What all this means is that we must expect to encounter interference problems involving mixed distributions and we need a general method to handle the problem.

It is quite likely that we will use interference theory for problems involving distributions other than strength and stress. For this reason we employ the subscript 1 to designate the strength distribution and the subscript 2 to designate the stress distribution. Figure 6–33 shows these two distributions aligned so that a single cursor x can be used to identify points on both distributions. We can now write

$$\begin{pmatrix}\text{Probability that}\\ \text{stress is less}\\ \text{than strength}\end{pmatrix} = dp(\sigma < x) = dR = F_2(x)\,dF_1(x)$$

By substituting $1 - R_2$ for F_2 and $-dR_1$ for dF_1, we have

$$dR = -[1 - R_2(x)]\,dR_1(x)$$

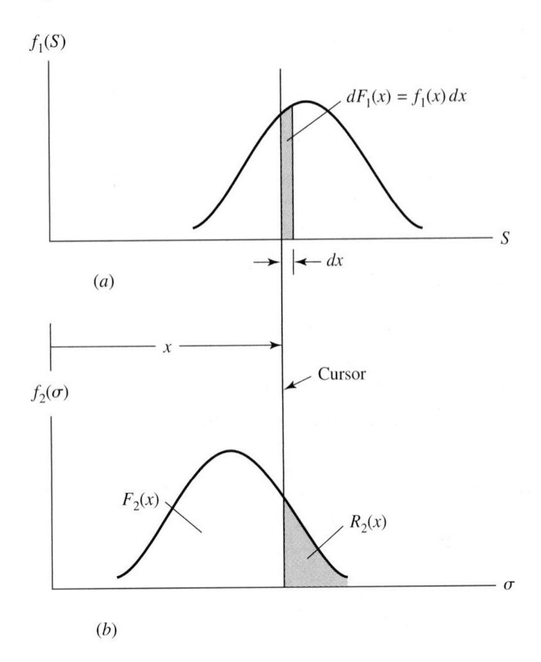

Figure 6–33

(*a*) PDF of the strength distribution; (*b*) PDF of the load-induced stress distribution.

The reliability for all possible locations of the cursor is obtained by integrating x from $-\infty$ to ∞; but this corresponds to an integration from 1 to 0 on the reliability R_1. Therefore

$$R = -\int_1^0 [1 - R_2(x)]\, dR_1(x)$$

which can be written

$$R = 1 - \int_0^1 R_2\, dR_1 \tag{6–24}$$

where

$$R_1(x) = \int_x^\infty f_1(S)\, dS \tag{6–25}$$

$$R_2(x) = \int_x^\infty f_2(\sigma)\, d\sigma \tag{6–26}$$

For the usual distributions encountered, plots of R_1 versus R_2 appear as shown in Fig. 6–34. Both of the cases shown are amenable to numerical integration and computer

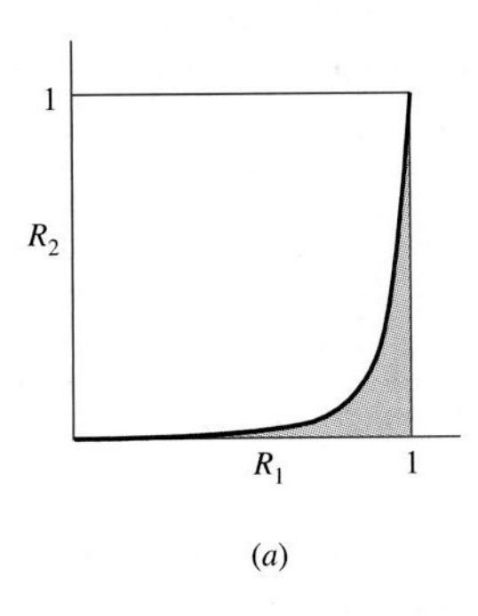

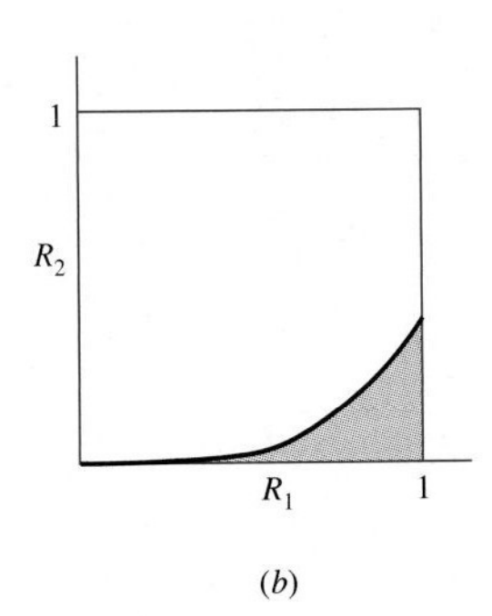

Figure 6–34

Curve shapes of the $R_1 R_2$ plot. In each case the shaded area is equal to $1 - R$ and is obtained by numerical integration. (*a*) Typical curve for asymptotic distributions; (*b*) curve shape obtained from lower truncated distributions such as the Weibull.

solution. When the reliability is high, the bulk of the integration area is under the right-hand spike of Fig. 6–34*a*. A system of increasingly finer Simpson's rule ordinate spacing is in order.

EXAMPLE 6–4 Interfere $\mathbf{S} \sim N(100, 5)$ kpsi with $\boldsymbol{\sigma} \sim N(80, 4)$ kpsi using Eq. (6-24) and Simpson's rule to find the reliability.

Solution We will find the reliability by forming a table as is shown in Table 6–2. The headings of the columns will be R_1, corresponding z_1, cursor location x, corresponding z_2, corresponding R_2, Simpson's rule multiplier, and extension. Now

$$z_1 = \frac{x - \mu_S}{\sigma_S} \qquad z_2 = \frac{x - \mu_\sigma}{\sigma_\sigma}$$

From the first equation above the cursor location is

$$x = \mu_S + z_1 \sigma_S = 100 + 5z_1$$

and

$$z_2 = \frac{x - 80}{4}$$

For the second row of Table 6–2 $R_1 = 0.1$ and

$$z_1 = 1.28 \text{ from Table E-10}$$

$$x = 100 + 5(1.28) = 106.4$$

$$z_2 = (106.4 - 80)/4 = 6.60$$

From Table E–10, $R_2 = 0.000\,000\,020\,6$, which is essentially zero.

The numerical value of the integral of Eq. (6–24) is

$$I = \frac{h}{3}\sum = \frac{0.1}{3}(1.001\,4800) = 0.033\,383$$

and R of Eq. (6–24) is

$$R = 1 - I = 1 - 0.033\,383 = 0.996\,617$$

Table 6–2

Numerical Integration by Simpson's Rule for the Interference of $\mathrm{S} \sim N(100, 5)$ kpsi with $\sigma \sim N(80, 4)$ kpsi

R_1	z_1	x	z_2	R_2	Multiplier	Extension
0	∞	∞	∞	0	1	0
0.1	1.28	106.40	6.60	0	4	0
0.2	0.84	104.20	6.05	0	2	0
0.3	0.52	102.60	5.65	0	4	0
0.4	0.25	101.25	5.31	0	2	0
0.5	0.00	100.00	5.00	0	4	0
0.6	−0.25	98.75	4.69	0	2	0
0.7	−0.52	97.40	4.35	0.000 01	4	0.000 04
0.8	−0.84	95.80	3.95	0.000 04	2	0.000 08
0.9	−1.28	93.60	3.40	0.000 34	4	0.001 36
1	−∞	−∞	−∞	1	1	1.000 00
						Σ 1.001 48

Checking this result using Eq. (2–4) gives

$$z = -\frac{\mu_S - \mu_\sigma}{\left(\sigma_S^2 + \sigma_\sigma^2\right)^{1/2}} = -\frac{100 - 80}{\left(5^2 + 4^2\right)^{1/2}} = -3.12$$

From Table E–10

$$R = 1 - \Phi(z) = 1 - \Phi(-3.12) = 0.9991$$

Observe the discrepancy in the reliability between the tabulation reliability and the result of Eq. (2–4). Why? It is integration error. The problem is that the R_1R_2 plot is different from Fig. 6–34*a*. The area is in the lower right-hand corner with the function R_2 rising rapidly. The spacing of the Simpson ordinates is too coarse. This kind of error can go undetected in a computer. This is why you should (1) draw a picture such as Fig. 6–34*a* and (2) print out a table from the computer. What is the next move?

- Form the new table upside down, so the contributions to the area appear at the top.
- Build the table down until you have captured all of the area (until the ordinate R_2 is 0.01 or less). Where does this occur?

$$R_2 = 0.1$$

$$z_2 = 2.33 \qquad z_1 = \frac{x - \mu_S}{\sigma_S}$$

- Eliminate x and solve for z_1:

$$z_1 = \frac{\mu_\sigma - \mu_S + z_2\sigma_\sigma}{\sigma_S} = \frac{80 - 100 + 2.33(4)}{5} = -2.14$$

$$R_2 = 0.9838$$

- Plan to run the table in the range $0.98 \le R_1 \le 1$. Table 6–3 is the revised table.

From Table 6–3,

$$I = \frac{h}{3}\sum = \frac{0.002}{3}(1.8005) = 0.0012$$

$$R = 1 - I = 1 - 0.0012 = 0.9988$$

which compares well to 0.9991 given by Eq. (2–4).

In writing a computer program use more computing digits, output R_1 and R_2 in tabular form so you can plot it (or have the computer plot it) to see if you have captured the area, then print out a table similar to Table 6–3 for record. You see, in problems where you do not have an equation such as Eq. (2–4) to tell you what the reliability is,

Table 6–3

Recommended Form for Numerical Integration by Simpson's Rule for Interference of $\mathbf{S} \sim N(100, 5)$ kpsi with $\sigma \sim N(80, 4)$ kpsi using Eq. (6–29)

R_1	z_1	x	z_2	R_2	Multiplier	Extension
1.000				1	1	1.0000
0.998	−2.88	85.60	1.40	0.0808	4	0.3232
0.996	−2.65	86.75	1.69	0.0455	2	0.0910
0.994	−2.51	87.45	1.86	0.0314	4	0.1256
0.992	−2.41	87.95	1.99	0.0233	2	0.0466
0.990	−2.33	88.35	2.09	0.0183	4	0.0732
0.988	−2.26	88.70	2.18	0.0146	2	0.0292
0.986	−2.20	89.00	2.25	0.0122	4	0.0480
0.984	−2.15	89.25	2.31	0.0104	2	0.0208
0.982	−2.10	89.50	2.38	0.0087	4	0.0348
0.980	−2.05	89.75	2.44	0.0073	1	0.0073
						Σ 1.8005

your numerical integration *must be robust*. Program defensively, giving yourself enough information so that you can believe the result, or take other steps.

6–13 Static or Quasi-Static Loading on a Shaft

The fundamental kinematic construct of our mechanical universe is the wheel and axle. An essential part of this revolute joint is the shaft. It is a good example of a static, quasi-static, and dynamically loaded body. Inasmuch as this chapter addresses static and quasi-static loading, application of the information developed in this chapter to shafts is useful and necessary.

The stress at an element located on the surface of a solid round shaft of diameter d subjected to bending, axial loading, and twisting is

$$\sigma_x = \frac{32M}{\pi d^3} + \frac{4F}{\pi d^2} \tag{a}$$

$$\tau_{xy} = \frac{16T}{\pi d^3} \tag{b}$$

where the axial component of the normal stress σ_x may be additive or subtractive. We observe that the three loadings M, F, and T occur at a section containing the specific surface element under scrutiny.

By use of the Mohr circle it can be shown that the two nonzero principal stresses σ_A and σ_B are

$$\sigma_A, \sigma_B = \frac{\sigma_x}{2} \pm \left[\left(\frac{\sigma_x}{2}\right)^2 + \tau_{xy}^2\right]^{1/2} \tag{6–27}$$

These principal stresses can be combined to obtain the von Mises stress σ' as

$$\sigma' = \left(\sigma_A^2 - \sigma_A\sigma_B + \sigma_B^2\right)^{1/2} = \left(\sigma_x^2 + 3\tau_{xy}^2\right)^{1/2} \tag{6–28}$$

Should the maximum-shear hypothesis be useful,

$$\tau_{\max} = \frac{\sigma_A - \sigma_B}{2} = \left[\left(\frac{\sigma_x}{2}\right)^2 + \tau_{xy}^2\right]^{1/2} \tag{6–29}$$

By substituting Eqs. (a) and (b) into Eqs. (6–27) and (6–28), we obtain

$$\sigma' = \frac{4}{\pi d^3}\left[(8M + Fd)^2 + 48T^2\right]^{1/2} \tag{6–30}$$

for von Mises stress, and using Eq. (6–29) for the maximum-shear-stress approximation, we obtain

$$\tau_{\max} = \frac{2}{\pi d^3}\left[(8M + Fd)^2 + (8T)^2\right]^{1/2} \tag{6–31}$$

Equations (6–30) and (6–31) permit an estimate of σ' or $\tau_{\max}$ when diameter d is given, or an estimate of d when an allowable value of σ' or $\tau_{\max}$ is given. For a design factor of n_d the distortion-energy theory of ductile failure gives an allowable stress of

$$\sigma'_{\text{all}} = \frac{S_y}{n_d} \tag{6–32}$$

For a design factor of n_d the maximum-shear hypothesis of ductile failure gives an allowable shear stress τ_{all} of

$$\tau_{all} = \frac{S_{sy}}{n_d} = \frac{S_y}{2n_d} \tag{6–33}$$

Static or Quasi-Static Loading of a Shaft—Bending and Torsion

Under many conditions, the axial force F in Eqs. (6–30) and (6–31) is either zero or so small that its effect may be neglected. With $F = 0$, Eqs. (6–30) and (6–31) become

$$\sigma' = \frac{16}{\pi d^3}\left(4M^2 + 3T^2\right)^{1/2} \tag{6–34}$$

$$\tau_{max} = \frac{16}{\pi d^3}\left(M^2 + T^2\right)^{1/2} \tag{6–35}$$

It is easier to solve these equations for diameter d than Eqs. (6–30) and (6–31). By substituting the allowable stresses from Eqs. (6–32) and (6–33) we find, for the von Mises theory of failure,

$$d = \left[\frac{16n}{\pi S_y}\left(4M^2 + 3T^2\right)^{1/2}\right]^{1/3} \tag{6–36}$$

$$\frac{1}{n} = \frac{16}{\pi d^3 S_y}\left(4M^2 + 3T^2\right)^{1/2} \tag{6–37}$$

and for the maximum-shear-stress approximation

$$d = \left[\frac{32n}{\pi S_y}\left(M^2 + T^2\right)^{1/2}\right]^{1/3} \tag{6–38}$$

$$\frac{1}{n} = \frac{32}{\pi d^3 S_y}\left(M^2 + T^2\right)^{1/2} \tag{6–39}$$

EXAMPLE 6–5

The integral pinion shaft shown in Fig. 6–35 is to be mounted in bearings at the locations shown and is to have a gear (not shown) mounted on the right-hand overhanging end. The loading diagram (Fig. 6–34*b*) shows the pinion force at A and the gear force at C are in the same plane. Equal and opposite torques T_A and T_B are represented as concentrated at A and C, as are the forces. The bending moment diagram of Fig. 6–35*c* shows an extreme at A and at B. The smaller diameter at B makes the location B the critical location at the center of the right-hand bearing. Since the shaft is used at intermittent emergencies, its use will not exceed 1000 revolutions at full load, so the problem can be treated as quasi-static.

(*a*) The material is a heat-treated carbon steel with a mean yield strength of 66 kpsi. In this circumstance the project engineer decides to use a design factor of 1.80. What is the smallest right-hand bearing journal diameter determined from distortion-energy theory and by maximum-shear-hypothesis approximation?

(*b*) The material has a yield strength $\mathbf{S}_y \sim LN(66.0, 5.3)$ kpsi and the applied moment $\mathbf{M} \sim LN(1925, 96)$ in · lbf and a correlated torque $\mathbf{T} \sim LN(3300, 765)$ in · lbf. If the reliability against yielding at the right-hand journal is 0.999 995, what is the smallest right-hand bearing journal diameter as estimated from distortion-energy theory and estimated from the maximum-shear-hypothesis approximation?

Figure 6–35

Left-bearing diameter is 1.000 in, left-bearing shoulder diameter is 2.000 in, dedendum diameter of gear is 3.43 in; right-hand bearing shoulder diameter is 2.000 in; the overhang shoulder is $1\frac{1}{8}$ in by $\frac{1}{4}$ in long; diameter of overhanging gear seat is 1.000 in.

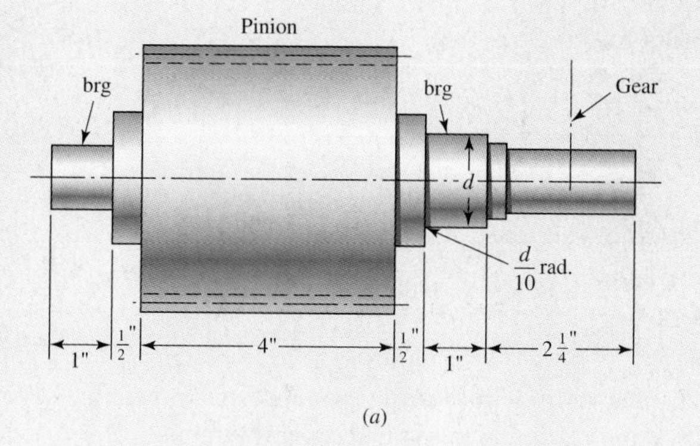

(a)

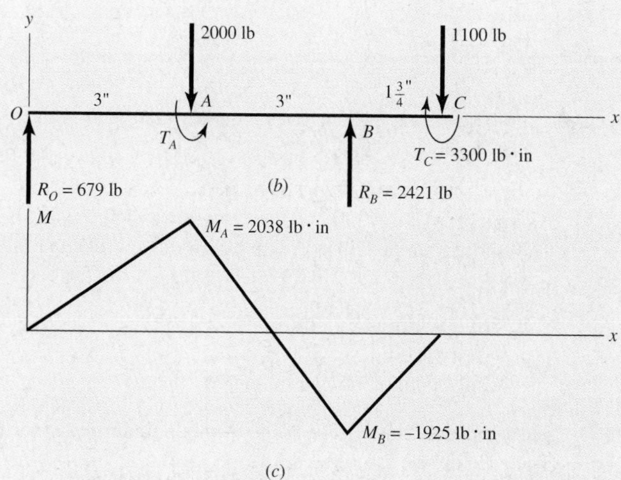

(b)

(c)

Solution (a) From Eq. (6–36),

$$d = \left\{ \frac{16(1.80)}{\pi 66\,000} \left[4(-1925)^2 + 3(3300)^2\right]^{1/2} \right\}^{1/3} = 0.986 \text{ in}$$

From Eq. (6–38),

$$d = \left\{ \frac{32(1.80)}{\pi 66\,000} \left[(-1925)^2 + 3300^2\right]^{1/2} \right\}^{1/3} = 1.02 \text{ in}$$

The diameter is dependent on the decision for the design factor.

(b) Equation (6–36) can be expressed stochastically as

$$\boldsymbol{\sigma}' = \frac{16}{\pi d^3}\left(4\mathbf{M}^2 + 3\mathbf{T}^2\right)^{1/2} = \frac{16\mathbf{M}}{\pi d^3}\left[4 + 3\left(\frac{\mathbf{T}}{\mathbf{M}}\right)^2\right]^{1/2}$$

since there is no stress-concentration factor for static loading. Inasmuch as **T** and **M** are correlated ($\rho = 1$), the quotient **T/M** is a deterministic constant, $\bar{T}/\bar{M}$. The variability in stress $\boldsymbol{\sigma}'$ is the same as in **M**. From Eq. (2–59),

$$C_n \doteq \sqrt{C_S^2 + C_P^2} = \left[\left(\frac{5.3}{66}\right)^2 + \left(\frac{96}{1925}\right)^2\right]^{1/2} = 0.0945$$

$$\bar{n} = \exp\left[C_n\left(|z| + C_n/2\right)\right] = \exp[0.0945(|-4.417| + 0.0945/2)] = 1.52$$

From Eq. (6–36),

$$d = \left\{ \frac{16(1.52)}{\pi 66\,000} \left[4(-1925)^2 + 3(3300)^2 \right]^{1/2} \right\}^{1/3} = 0.890 \text{ in}$$

and from Eq. (6–38)

$$d = \left\{ \frac{32(1.52)}{\pi 66\,000} \left[(-1925)^2 + 3300^2 \right]^{1/2} \right\}^{1/3} = 0.964 \text{ in}$$

The answer to part (*b*) utilizes stochastic knowledge of loading (**M** and **T**) and strength ($\mathbf{S}_y$) and responds to a stated reliability goal.

CASE STUDY 6 A Preliminary Axle Design

A conceptual study has produced the drawing for a flat-belt conveyor depicted in Fig. CS6–1. The immediate task is to create rough design for the pulley axle. The nominal conditions are

Belt pull:	3600 lb
Belt width:	30 in
Belt speed:	100 ft/min
Pulley diameter:	16 in

The proposed axle will be made in small quantities, since no more than 250 sales are expected. The axle drawing will be developed and filed until someone designs the pulley in detail. The axle is to have a long, trouble-free life.

Concept The inner cones of the tapered roller bearings are to have a manufacturer's recommended press fit. The axle will be clamped to the conveyor belt frame at the end. To allow for

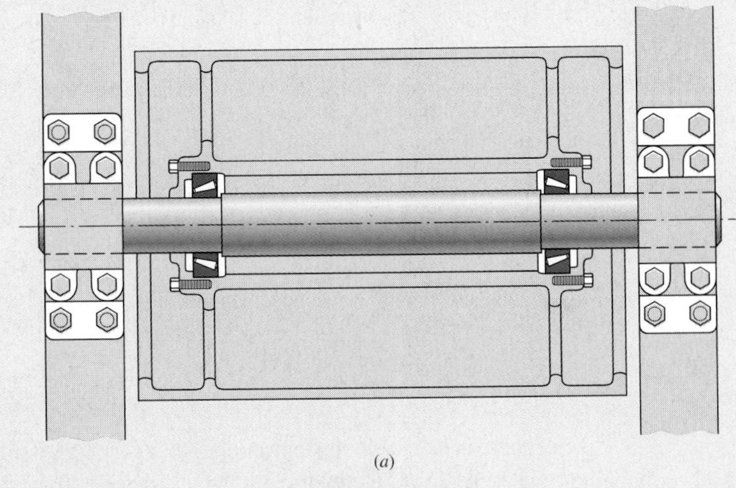

(*a*)

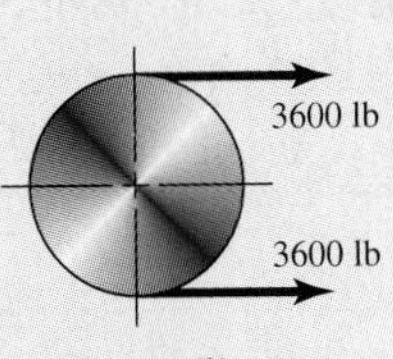

(*b*)

Figure CS6–1

(*a*) Drawing from a conceptual study for a 16-in-diameter tail-pulley assembly for a conveyor belt. The belt has a maximum nomial belt tension of 3600 lb, a 30-in wide flat belt traveling at 100 ft/min. (*b*) End view of the tail-pulley showing the axle loading. (Part *a* courtesy of the Timken Roller Bearing Company.)

variations in belt width, the pulley will be made 32 in long. A clearance of 1 in between the ends of the pulley and the structural supporting members will be provided. It is expected that the structural members will be 4-in angles. This makes the target overall length of the axle 42 in. The bearings will be placed at the approximate quarter-points of the axle, as the concept sketch, Fig. CS6–2, shows.

A plain carbon steel will be used, cold drawn to increase strength, which keeps size down. The final axle design will depend on the bearing selection. The axle is subjected to an essentially steady load. The axle is in nonrotating bending static loading. Loads and reactions are to be treated as concentrated forces. Stress concentrations at the shoulder and fillet seat fillets are neglected because of the static loading of a ductile material. On hand is 1018 cold-drawn steel from which workpiece stock will be earmarked. The yield strength is $\mathbf{S}_y \sim 78.4\mathbf{LN}(1, 0753)$ kpsi. Failure would be from general permanent set (yielding). For a total production of 250, a 10 percent chance of failure of one axle is acceptable, so the reliability goal is $R = 1 - (1/10)(1/250) = 0.9996$ ($z = -3.355$). The maximum belt tension will be a variable, even a function of mishaps in use. We will estimate the COV of the maximum load to be 0.20. Thus

$$\mathbf{2F} \sim 2(3600)\ \mathbf{LN}(1, 0.20)$$

and it follows that the COV of the design factor is

$$C_n = \sqrt{C_F^2 + C_{Sy}^2} = \sqrt{(0.20^2 + 0.0753^2)} = 0.214$$

The mean design factor $\bar{n}$ is given by

$$\bar{n} = \exp\left[-3\sqrt{\ln\left(1 + 0.214^2\right)} + \ln\sqrt{1 + 0.214^2}\right] = 2.05$$

The maximum bending moment $\mathbf{M}$ is

$$\mathbf{M} = 3600\ \mathbf{LN}(1, 0.20)(9.5)\ \text{in} \cdot \text{lb}$$

and its mean is $\bar{M} = 34\ 200$ in · lb. The strength-stress equation is

$$\bar{\sigma} = \frac{\bar{S}_y}{\bar{n}} = \frac{\bar{M}c}{I} = \frac{32\bar{M}}{\pi d^3}$$

from which the diameter d is found to be

$$d = \sqrt[3]{\frac{32nM}{\pi S_y}} = \sqrt[3]{\frac{32(2.05)34\ 200}{\pi(78\ 400)}} = 2.09\ \text{in}$$

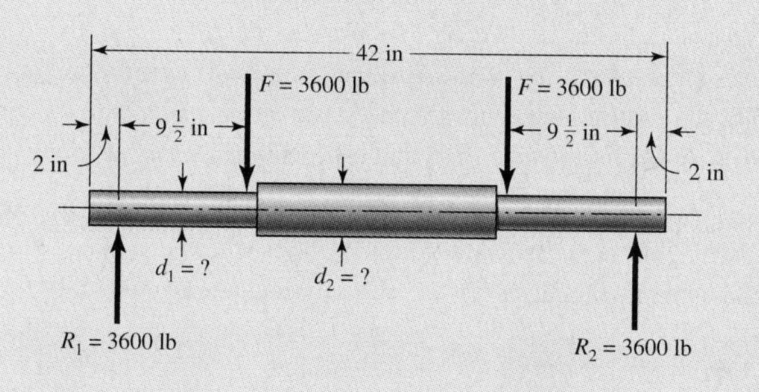

Figure CS6–2

Concept sketch showing the gross geometry of the axle before the designer refines it with workable dimensions and features.

Figure CS6–3

Refined axle concept with bearing seats for mounting the inner race (cup) of the tapered-roller bearing when selected. Shoulder fillets will be sized according to the bearing manufacturer's recommendations.

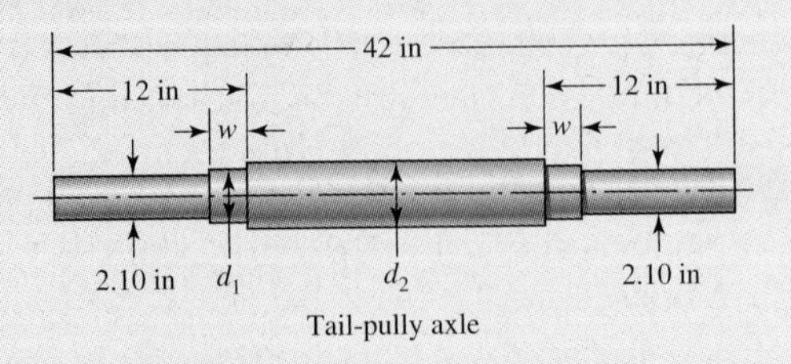

Decisions

Pulley axle material:	1018 CD steel
Pulley length:	32 in
C-to-C structural support:	38 in
Width of structural support:	4 in

The end diameters of the axle are made 2.10 in. The bearing seat diameter d_1 depends on d_1 and w of the bearing selected. The diameter d_2 will reflect the manufacturer's recommended shoulder for the bearing selected. These are summarized in Fig. CS6–3.

The task turned out to be simple, but only because of the many topics treated in this and earlier chapters.

PROBLEMS

ANALYSIS

6–1 Among the decisions which a designer must make is selection of the failure locus which is applicable to the material and its static loading. A 1020 hot-rolled steel has the following properties: $S_y = 42$ kpsi, $S_{ut} = 66.2$ kpsi, and true strain at fracture $\varepsilon_f = 0.90$. Plot the failure locus and, for the static stress states at the critical locations listed below, plot the load line and estimate the factor of safety analytically and graphically.

(*a*) $\sigma_x = 9$ kpsi, $\sigma_y = -5$ kpsi.
(*b*) $\sigma_x = 12$ kpsi, $\tau_{xy} = 3$ kpsi ccw.
(*c*) $\sigma_x = -4$ kpsi, $\sigma_y = -9$ kpsi, $\tau_{xy} = 5$ kpsi cw.
(*d*) $\sigma_x = 11$ kpsi, $\sigma_y = 4$ kpsi, $\tau_{xy} = 1$ kpsi cw.

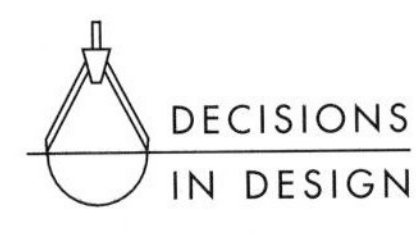

6–2 A 4142 steel Q&T at 80°F exhibits $S_{yt} = 235$ kpsi, $S_{yc} = 275$ kpsi, and $\varepsilon_f = 0.06$. Choose and plot the failure locus and, for the static stresses at the critical locations which are 10 times those in Prob. 6–1, plot the load lines and estimate the factors of safety analytically and graphically.

6–3 For grade 20 cast iron, Table E–24 gives $S_{ut} = 22$ kpsi, $S_{uc} = 83$ kpsi. Choose and plot the failure locus and, for the static loadings inducing the stresses at the critical locations of Prob. 6–1, plot the load lines and estimate the factors of safety analytically and graphically.

6–4 A cast aluminum 195-T6 has an ultimate strength in tension of $S_{ut} = 36$ kpsi and ultimate strength in compression of $S_{uc} = 35$ kpsi, and it exhibits a true strain at fracture $\varepsilon_f = 0.045$. Choose and plot the failure locus and, for the static loading inducing the stresses at the critical locations of Prob. 6–1, plot the load lines and estimate the factors of safety analytically and graphically.

6–5 An ASTM cast iron, grade 30 (see Table E–24), carries static loading resulting in the stress state listed below at the critical locations. Choose the appropriate failure locus, plot it and the load lines, and estimate the factors of safety analytically and graphically.

ANALYSIS

(*a*) $\sigma_A = 20$ kpsi, $\sigma_B = 20$ kpsi.
(*b*) $\tau_{xy} = 15$ kpsi.
(*c*) $\sigma_A = \sigma_B = -80$ kpsi.
(*d*) $\sigma_A = 20$ kpsi, $\sigma_B = -10$ kpsi.

6–6 This problem illustrates that the factor of safety for a machine element depends on the particular point selected for analysis. Here you are to compute factors of safety, based upon the distortion-energy theory, for stress elements at A and B of the member shown in the figure. This bar is made of AISI 1006 cold-drawn steel and is loaded by the forces $F = 0.55$ kN, $P = 8.0$ kN, and $T = 30$ N · m.

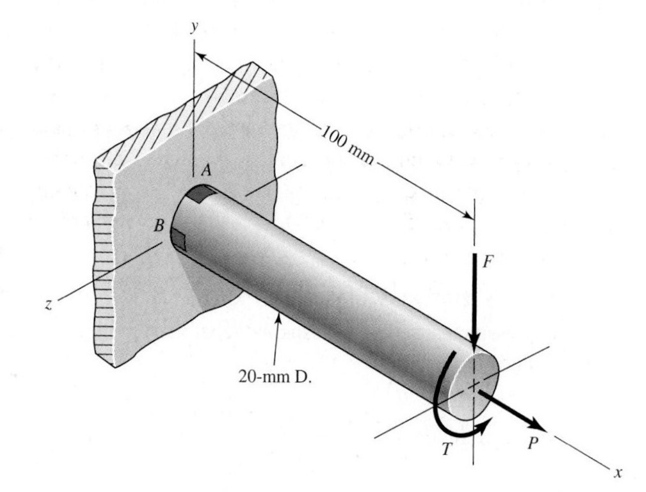

Problem 6–6

6–7 * Design the lever arm *CD* of Fig. 6–31 by specifying a suitable size and material.

6–8 The figure shows a crank loaded by a force $F = 300$ lb which causes twisting and bending of the $\frac{3}{4}$-in-diameter shaft fixed to a support at the origin of the reference system. In actuality, the support may be an inertia which we wish to rotate, but for the purposes of a strength analysis we can consider this to be a statics problem. The material of the shaft AB is hot-rolled AISI 1018 steel (Table E–20). Using the maximum-shear-stress hypothesis, find the factor of safety based on the stress at point A.

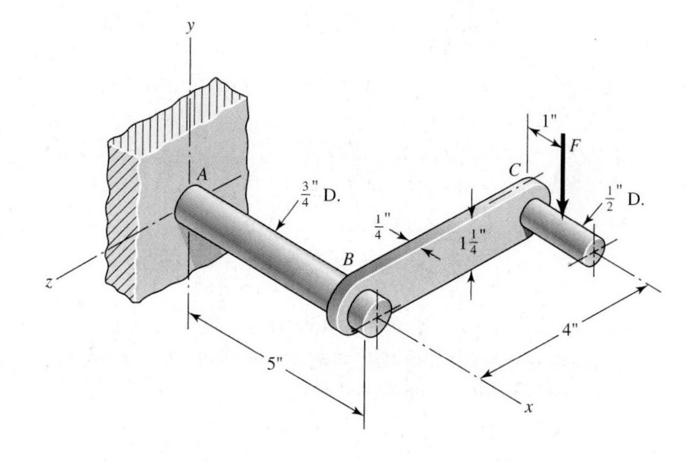

Problem 6–8

---

* The asterisk indicates a problem that may not have a unique result or a particularly challenging problem.

6–9 A spherical pressure vessel is formed of 18-gauge (0.05-in) cold-drawn AISI 1018 sheet steel. If the vessel has a diameter of 8 in, estimate the pressure necessary to initiate yielding. What is the estimated bursting pressure?

6–10 This problem illustrates that the strength of a machine part can sometimes be measured in units other than those of force or moment. For example, the maximum speed that a flywheel can reach without yielding or fracturing is a measure of its strength. In this problem you have a rotating ring made of hot-forged AISI 1020 steel; the ring has a 6-in inside diameter and a 10-in outside diameter and is 1.5 in thick. What speed in revolutions per minute would cause the ring to yield? At what radius would yielding begin? [*Note:* The maximum radial stress occurs at $r = (r_o r_i)^{1/2}$; see Eq. (3-59).]

6–11 A light pressure vessel is made of 2024-T3 aluminum alloy tubing with suitable end closures. This cylinder has a $3\frac{1}{2}$-in OD, a 0.065-in wall thickness, and $\nu = 0.334$. The purchase order specifies a minimum yield strength of 46 kpsi. What is the factor of safety if the pressure-release valve is set at 500 psi?

6–12 A cold-drawn AISI 1015 steel tube is 300 mm OD by 200 mm ID and is to be subjected to an external pressure caused by a shrink fit. What maximum pressure would cause the material of the tube to yield?

6–13 What speed would cause fracture of the ring of Prob. 6–10 if it were made of grade 30 cast iron?

6–14 The figure shows a shaft mounted in bearings at A and D and having pulleys at B and C. The forces shown acting on the pulley surfaces represent the belt pulls. The shaft is to be made of ASTM grade 25 cast iron using a design factor $n_d = 2.8$. What diameter should be used for the shaft?

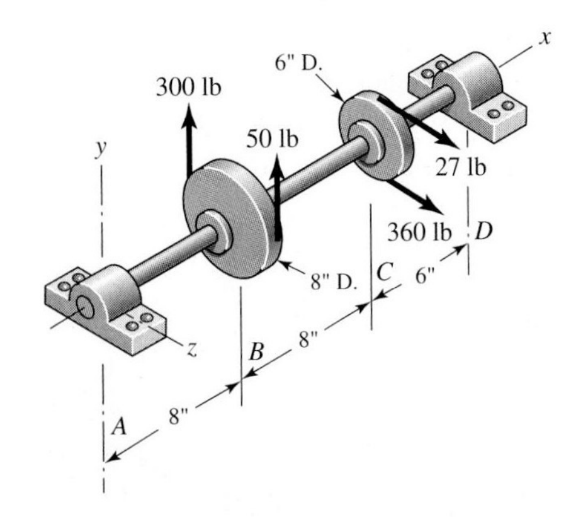

Problem 6–14

6–15 By modern standards, the shaft design of Prob. 6–14 is poor because it is so long. Suppose it is redesigned by halving the length dimensions. Using the same material and design factor as in Prob. 6–14, find the new shaft diameter.

6–16 The clevis pin shown in the figure is 12 mm in diameter and has the dimensions $a = 12$ mm and $b = 18$ mm. The pin is machined from AISI 1018 hot-rolled steel (Table E–20) and is to be loaded to no more than 4.4 kN. Determine whether or not the assumed loading of c yields a factor of safety any different from that of d. Use the maximum-shear-stress hypothesis.

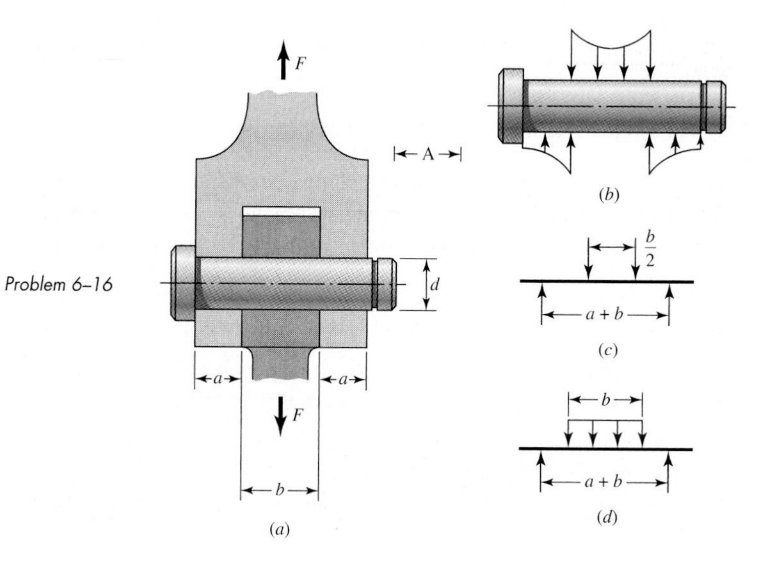

Problem 6–16

6–17 Repeat Prob. 6–16, but this time use the distortion-energy theory.

6–18 A stochastic force $F \sim N(410, 45)$ lb applied at D near the end of the 15-in lever shown in the figure results in certain stresses in the cantilevered bar $OABC$.

(*a*) Determine the mean and standard deviation of the principal stresses for a stress element on the upper surface of the 1-in section at A.

(*b*) What are the mean and standard deviation of the von Mises stresses corresponding to the stress element at A?

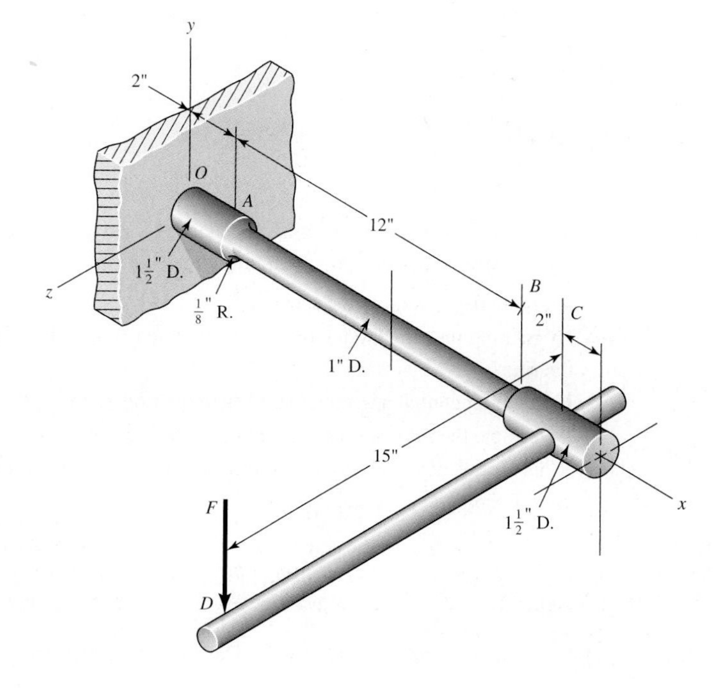

Problem 6–18

6–19 A carbon steel collar of length 1 in is to be machined to inside and outside diameters, respectively, of

$$D_i = 0.750 \pm 0.0004 \text{ in} \qquad D_o = 1.125 \pm 0.002 \text{ in}$$

This collar is to be shrink-fitted to a hollow steel shaft having inside and outside diameters, respectively, of

$$d_i = 0.375 \pm 0.002 \text{ in} \qquad d_o = 0.752 \pm 0.0004 \text{ in}$$

These tolerances are assumed to have a normal distribution, to be centered in the spread interval, and to have a total spread of ± 4 standard deviations. Determine the means and the standard deviations of the tangential stress components for both cylinders at the interface.

6–20 Suppose the collar of Prob. 6–19 has a yield strength of $\mathbf{S}_y \sim \mathbf{N}(95.5, 6.59)$ kpsi. What is the probability that the material will not yield?

ANALYSIS

6–21 The clevis pin of Prob. 6–16 is to be made of a malleable cast iron with a tensile strength of $\mathbf{S}_{ut} \sim \mathbf{N}(44.5, 4.34)$ kpsi. The load is given as $\mathbf{F} \sim \mathbf{N}(1500, 150)$ lb.

(*a*) Determine the reliability for a point at midspan and on the bottom of the pin. Note at this point that the direct shear is zero.

(*b*) Repeat part (*a*), but with a strength distribution that is Weibullian and is given as $\mathbf{S}_{ut} \sim W(27.7, 46.2, 4.32)$ kpsi.

ANALYSIS

6–22 A carbon steel tube has an outside diameter of 1 in and a wall thickness of $\frac{1}{8}$ in. The tube is to carry an internal hydraulic pressure given as $\mathbf{p} \sim N(6000, 500)$ psi. The material of the tube has a yield strength of $\mathbf{S}_y \sim N(50, 4.1)$ kpsi. Find the reliability using thin-wall theory.

ANALYSIS

6–23* A split-ring clamp-type shaft collar is shown in the figure. The collar is 2 in OD by 1 in ID by $\frac{1}{2}$ in wide. The screw is designated as $\frac{1}{4}$-28 UNF. The relation between the screw tightening torque T, the nominal screw diameter d, and the tension in the screw F_i is approximately $T = 0.2\, F_i d$. The shaft is sized to obtain a close running fit. Find the axial holding force F_x of the collar as a function of the coefficient of friction and the screw torque.

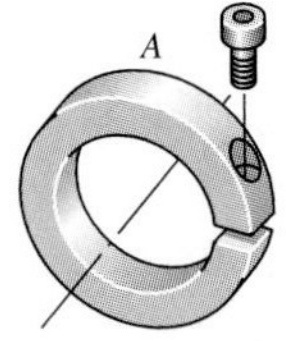

Problem 6–23

ANALYSIS

6–24* Suppose the collar of Prob. 6–23 is tightened using a screw torque of 190 lb · in. The collar material is AISI 1040 steel heat-treated to a minimum tensile yield strength of 63 kpsi.

(*a*) Estimate the tension in the screw.

(*b*) By relating the tangential stress to the hoop tension, find the internal pressure of the shaft on the ring.

(*c*) Find the tangential and radial stresses in the ring at the inner surface.

(*d*) Determine the maximum shear stress and the von Mises stress.

(*e*) What are the factors of safety based on the maximum-shear-stress hypothesis and the distortion-energy theory?

DESIGN

6–25 In Prob. 6–23, the role of the screw was to induce the hoop tension which produces the clamping. The screw should be placed so that no moment is induced in the ring. Just where should the screw be located?

ANALYSIS

6–26 A tube has another tube shrunk over it. The specifications are:

	Inner Member	Outer Member
ID	1.000 ± 0.002 in	1.999 ± 0.0004 in
OD	2.000 ± 0.0004 in	3.000 ± 0.004 in

Both tubes are made of a plain carbon steel.

(*a*) Find the shrink-fit pressure and the stresses at the fit surface.

(*b*) If the inner tube is changed to solid shafting with the same outside dimensions, find the shrink-fit pressure and the stresses at the fit surface.

6–27 Steel tubes with a Young's modulus of 207 GPa have the specifications:

	Inner Tube	Outer Tube
ID	25 ± 0.050 mm	49.98 ± 0.010 mm
OD	50 ± 0.010 mm	75 ± 0.10 mm

These are shrink-fitted together. Find the shrink-fit pressure and the von Mises stress in each body at the fit surface.

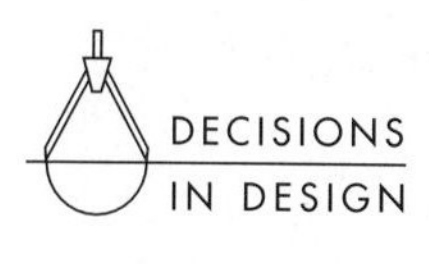

6–28 A 2-in-diameter solid steel shaft has a gear with ASTM grade 20 cast-iron hub ($E = 14.5$ Mpsi) shrink-fitted to it. The specifications for the shaft are

$$2.000 \begin{array}{l} +0.0000 \\ -0.0004 \end{array} \text{ in}$$

The hole in the hub is sized at 1.999 ± 0.0004 in and an OD of $4.00 \pm \frac{1}{32}$ in. Use the modified Mohr hypothesis and estimate the factor of safety guarding against fracture in the gear hub due to the shrink fit.

6–29 A steel shaft with a diameter of

$$1.875 \begin{array}{l} +0.0000 \\ -0.0004 \end{array} \text{ in}$$

is shrink-fitted to the hub of a pulley with a bore of 1.870 ± 0.002 in and a hub diameter of $3.25 \pm \frac{1}{32}$ in. The pulley hub, which is grade 40 cast iron, has an ultimate tensile strength of $\mathbf{S}_{ut} \sim \mathbf{N}(44.5, 4.34)$ kpsi, a compressive ultimate strength of $\mathbf{S}_{uc} \sim \mathbf{N}(140, 13.1)$ kpsi, and a modulus of elasticity $E = 17$ Mpsi. Estimate the reliability of the hub against fracture due to the shrink-fit pressure.

6–30 This problem is an exercise in using Eq. (6–24) to estimate reliability. A tensile strength is uniformly distributed in the interval $46 \le S_{ut} \le 54$ kpsi; that is, $\mathbf{S} \sim U[48, 54]$ kpsi. The load-induced stress is $\sigma \sim U[40, 48]$ kpsi. Estimate the reliability and check by drawing a figure similar to Fig. 6–34.

6–31 Solve Prob. 6–30 using the idea of stress margin, taking careful note that the distributions of **S** and σ are uniform random. The stress margin is the difference between two uniform random variables which overlap. The distribution of **m** is isosceles triangular between the range numbers of **m**. Determine the reliability and note the factor of safety.

6–32 Write a computer program to simulate conditions of Prob. 6–30 using a uniform random number generator supplied to your machine by the computer manufacturer. Generate independent random instances of the stress and strength in order to create random instances of the stress margin. Keep track of the number of instances in which $m > 0$. The reliability is this number divided by the number of trials. (Four instances of 10 000 trials estimate the reliability as 0.9718, 0.9671, 0.9682, and 0.9690. Do you agree?)

7

Failure Resulting from Variable Loading

7–1	Introduction to Fatigue in Metals **360**
7–2	Strain-Life Relationships **361**
7–3	Stress-Life Relationships **367**
7–4	The Endurance Limit **369**
7–5	Fatigue Strength **372**
7–6	Endurance-Limit Modifying Factors **374**
7–7	Stress Concentration and Notch Sensitivity **383**
7–8	Applying What We Have Learned about Endurance Limit and Endurance Strength **387**
7–9	The Distributions **395**
7–10	Characterizing Fluctuating Stresses **396**
7–11	Failure Loci under Variable Stresses **398**
7–12	Torsional Fatigue Strength under Pulsating Stresses **408**
7–13	Combinations of Loading Modes **408**
7–14	Stochastic Failure Loci under Fluctuating Stresses **411**
7–15	Cumulative Fatigue Damage **414**
7–16	The Fracture-Mechanics Approach **421**
7–17	Surface Fatigue Strength **423**
7–18	The Designer's Fatigue Diagram **429**
7–19	An Important Design Decision: The Design Factor in Fatigue **431**

In Chap. 6 we considered the analysis and design of parts subjected to static loading. The behavior of machine parts is entirely different when subjected to time-varying loading. In this chapter we shall examine how parts fail under variable loading and how to proportion them to successfully resist such conditions.

7–1 Introduction to Fatigue in Metals

In most testing of those properties of materials that relate to the stress-strain diagram, the load is applied gradually, to give sufficient time for the strain to fully develop. Furthermore, the specimen is tested to destruction, and so the stresses are applied only once. Testing of this kind is applicable, then, to what are known as *static conditions*; such conditions closely approximate the actual conditions to which many structural and machine members are subjected.

The condition frequently arises, however, in which the stresses vary or they fluctuate between levels. For example, a particular fiber on the surface of a rotating shaft subjected to the action of bending loads undergoes both tension and compression for each revolution of the shaft. If the shaft is part of an electric motor rotating at 1725 rev/min, the fiber is stressed in tension and compression 1725 times each minute. If, in addition, the shaft is also axially loaded (as it would be, for example, by a helical or worm gear), an axial component of stress is superposed upon the bending component. In this case, some stress is always present in any one fiber, but now the *level* of stress is fluctuating. These and other kinds of loading occurring in machine members produce stresses which are called *variable*, *repeated*, *alternating*, or *fluctuating* stresses.

Often, machine members are found to have failed under the action of repeated or fluctuating stresses; yet the most careful analysis reveals that the actual maximum stresses were below the ultimate strength of the material, and quite frequently even below the yield strength. The most distinguishing characteristic of these failures is that the stresses have been repeated a very large number of times. Hence the failure is called a *fatigue failure*.

A fatigue failure begins with a small crack. The initial crack is so minute that it cannot be detected by the naked eye and is even quite difficult to locate in a Magnaflux or x-ray inspection. The crack will develop at a point of discontinuity in the material, such as a change in cross section, a keyway, or a hole. Less obvious points at which fatigue failures are likely to begin are inspection or stamp marks, internal cracks, or even irregularities caused by machining. Once a crack is initiated, the stress-concentration effect becomes greater and the crack progresses more rapidly. As the stressed area decreases in size, the stress increases in magnitude until, finally, the remaining area fails suddenly. A fatigue failure, therefore, is characterized by two distinct regions (Fig. 7–1). The first of these is due to the progressive development of the crack, while the second is due to the sudden fracture. The zone of sudden fracture is very similar in appearance to the fracture of a brittle material, such as cast iron, that has failed in tension.

When machine parts fail statically, they usually develop a very large deflection, because the stress has exceeded the yield strength; and the part is replaced before fracture actually occurs. Thus many static failures give visible warning in advance. But a fatigue failure gives no warning! It is sudden and total, and hence dangerous. It is relatively simple to design against a static failure, because our knowledge is comprehensive. Fatigue is a much more complicated phenomenon, only partially understood, and the engineer seeking competence must acquire as much knowledge of the subject as possible. Anyone who lacks knowledge of fatigue can double or triple design factors and formulate a design that will not fail. However, such designs cannot compete in the marketplace. Neither can the engineers who produce them.

Figure 7–1

A fatigue failure of a $7\frac{1}{2}$-in-diameter forging at a press fit. The specimen is UNS G10450 steel, normalized and tempered, which has been subjected to rotating bending. (Courtesy of The Timken Company.)

7–2 Strain-Life Relationships

The best hypothesis yet advanced to explain the nature of fatigue failure is called by some the *strain-life hypothesis*. The hypothesis can be used to estimate fatigue strengths, but when it is so used it is necessary to compound several idealizations, and so some uncertainties will exist in the results. For this reason, the hypothesis is presented here only because of its value in explaining the nature of fatigue.

A fatigue failure almost always begins at a local discontinuity such as a notch, crack, or other area of stress concentration. When the stress at the discontinuity exceeds the elastic limit, plastic strain occurs. If a fatigue fracture is to occur, there must exist cyclic plastic strains. Thus we shall need to investigate the behavior of materials subject to cyclic deformation.

In 1910, Bairstow verified by experiment Bauschinger's theory that the elastic limits of iron and steel can be changed, either up or down, by the cyclic variations of stress.[1] In general, the elastic limits of annealed steels are likely to increase when subjected to cycles of stress reversals, while cold-drawn steels exhibit a decreasing elastic limit.

Test specimens subjected to reversed bending are not suitable for strain cycling, because of the difficulty of measuring plastic strains. Consequently, most of the research has been done using axial specimens. By using electrical transducers, it is possible to generate signals that are proportional to the stress and strain, respectively. These signals can then be displayed on an oscilloscope or plotted on an *XY* plotter. R. W. Landgraf has investigated the low-cycle fatigue behavior of a large number of very high-strength steels, and during his research he made many cyclic stress-strain plots.[2] Figure 7–2 has been constructed to show the general appearance of these plots for the first few cycles of controlled cyclic strain. In this case the strength decreases with stress repetitions, as evidenced by the fact that the reversals occur at ever-smaller stress levels. As previously noted, other materials may be strengthened, instead, by cyclic stress reversals.

Slightly different results may be obtained if the first reversal occurs in the compressive region; this is probably due to the fatigue-strengthening effect of compression.

Landgraf's paper contains a number of plots that compare the monotonic stress-strain relations in both tension and compression with the cyclic stress-strain curve.[3] Two of these have been redrawn and are shown in Fig. 7–3. The importance of these is that

1. L. Bairstow, "The Elastic Limits of Iron and Steel under Cyclic Variations of Stress," *Philosophical Transactions*, Series A, vol. 210, Royal Society of London, 1910, pp. 35–55.

2. R. W. Landgraf, *Cyclic Deformation and Fatigue Behavior of Hardened Steels*, Report no. 320, Department of Theoretical and Applied Mechanics, University of Illinois, Urbana, 1968, pp. 84–90.

3. Ibid., pp. 58–62.

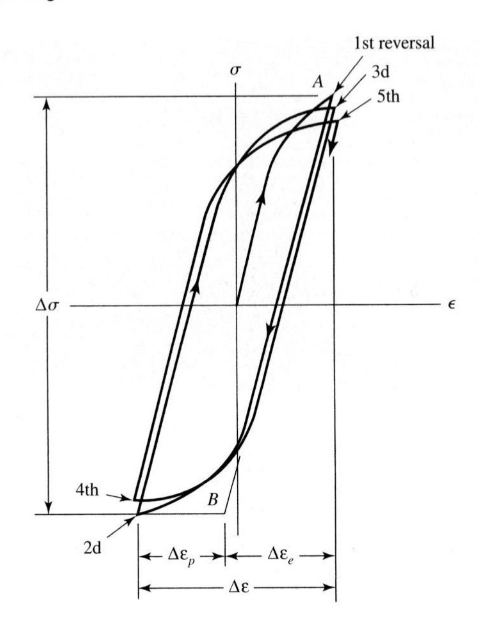

Figure 7–2

True stress–true strain hysteresis loops showing the first five stress reversals of a cyclic-softening material. The graph is slightly exaggerated for clarity. Note that the slope of the line *AB* is the modulus of elasticity *E*. The stress range is $\Delta\sigma$, $\Delta\varepsilon_p$ is the plastic-strain range, and $\Delta\varepsilon_e$ is the elastic strain range. The total-strain range is $\Delta\varepsilon = \Delta\varepsilon_p + \Delta\varepsilon_e$.

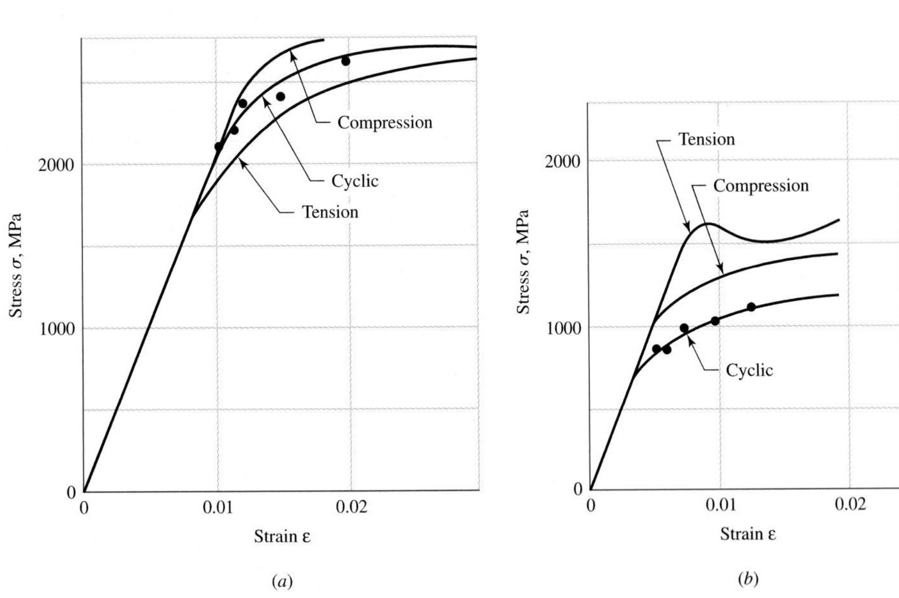

Figure 7–3

Monotonic and cyclic stress-strain results. (*a*) Ausformed H-11 steel, 660 Brinell; (*b*) SAE 4142 steel, 400 Brinell.

they emphasize the difficulty of attempting to predict the fatigue strength of a material from known values of monotonic yield or ultimate strengths in the low-cycle region.

The SAE Fatigue Design and Evaluation Steering Committee released a report in 1975 in which the life in reversals to failure is related to the strain amplitude.[4] The report contains a plot of this relationship for SAE 1020 hot-rolled steel; the graph has been reproduced as Fig. 7–4. To explain the graph, we first define the following terms:

Fatigue ductility coefficient ε'_F is the true strain corresponding to fracture in one reversal (point *A* in Fig. 7–2). The plastic-strain line begins at this point in Fig. 7–4.

4. *Technical Report on Fatigue Properties*, SAE J1099, 1975.

Figure 7–4

A log-log plot showing how the fatigue life is related to the true-strain amplitude for hot-rolled SAE 1020 steel. (Reproduced from Tech. Rep. SAE J1099 by permission.)

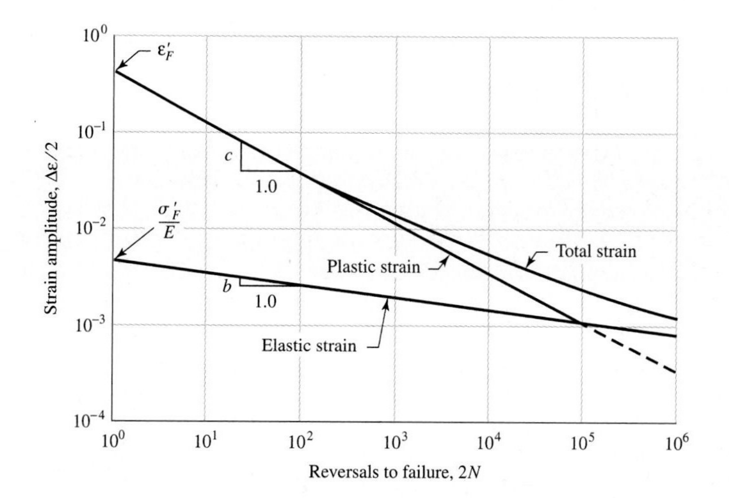

Fatigue strength coefficient σ'_F is the true stress corresponding to fracture in one reversal (point *A* in Fig. 7–2). Note in Fig. 7–4 that the elastic-strain line begins at σ'_F/E.

Fatigue ductility exponent c is the slope of the plastic-strain line in Fig. 7–4 and is the power to which the life $2N$ must be raised to be proportional to the true plastic-strain amplitude.

Fatigue strength exponent b is the slope of the elastic-strain line, and is the power to which the life $2N$ must be raised to be proportional to the true-stress amplitude.

Now, from Fig. 7–2, we see that the total strain is the sum of the elastic and plastic components. Therefore the total strain amplitude is

$$\frac{\Delta\varepsilon}{2} = \frac{\Delta\varepsilon_e}{2} + \frac{\Delta\varepsilon_p}{2} \tag{a}$$

The equation of the plastic-strain line in Fig. 7–4 is

$$\frac{\Delta\varepsilon_p}{2} = \varepsilon'_F(2N)^c \tag{7–1}$$

The equation of the elastic strain line is

$$\frac{\Delta\varepsilon_e}{2} = \frac{\sigma'_F}{E}(2N)^b \tag{7–2}$$

Therefore, from Eq. (*a*), we have for the total-strain amplitude

$$\frac{\Delta\varepsilon}{2} = \frac{\sigma'_F}{E}(2N)^b + \varepsilon'_F(2N)^c \tag{7–3}$$

which is the Manson-Coffin relationship between fatigue life and total strain.[5] Some values of the coefficients and exponents are listed in Table 7–1. Many more are included in the SAE J1099 report.

Though Eq. (7–3) is a perfectly legitimate equation for obtaining the fatigue life of a part when the strain and other cyclic characteristics are given, it appears to be of little use to the designer. The question of how to determine the total strain at the bottom of

5. J. F. Tavernelli and L. F. Coffin, Jr., "Experimental Support for Generalized Equation Predicting Low Cycle Fatigue," and S. S Manson, discussion, *Trans. ASME, J. Basic Eng.*, vol. 84, no. 4, pp. 533–537.

Table 7–1

Cyclic Properties of Some High-Strength Steels

AISI Number	Processing	Brinell Hardness H_B	Cyclic Yield Strength S'_y, kpsi	Fatigue Strength Coefficient σ'_F, kpsi	Fatigue Ductility Coefficient ε'_F	Fatigue Strength Exponent b	Fatigue Ductility Exponent c	Fatigue Strain-Hardening Exponent m
1045	Q & T 80°F	705	. . .	310	. . .	−0.065	−1.0	0.10
1045	Q & T 360°F	595	250	395	0.07	−0.055	−0.60	0.13
1045	Q & T 500°F	500	185	330	0.25	−0.08	−0.68	0.12
1045	Q & T 600°F	450	140	260	0.35	−0.07	−0.69	0.12
1045	Q & T 720°F	390	110	230	0.45	−0.074	−0.68	0.14
4142	Q & T 80°F	670	300	375	. . .	−0.075	−1.0	0.05
4142	Q & T 400°F	560	250	385	0.07	−0.076	−0.76	0.11
4142	Q & T 600°F	475	195	315	0.09	−0.081	−0.66	0.14
4142	Q & T 700°F	450	155	290	0.40	−0.080	−0.73	0.12
4142	Q & T 840°F	380	120	265	0.45	−0.080	−0.75	0.14
4142*	Q & D 550°F	475	160	300	0.20	−0.082	−0.77	0.12
4142	Q & D 650°F	450	155	305	0.60	−0.090	−0.76	0.13
4142	Q & D 800°F	400	130	275	0.50	−0.090	−0.75	0.14

* Deformed 14 percent.

Source: Data from R. W. Landgraf, *Cyclic Deformation and Fatigue Behavior of Hardened Steels*, Report no. 320, Department of Theoretical and Applied Mechanics, University of Illinois, Urbana, 1968.

a notch or discontinuity has not been answered. There are no tables or charts of strain concentration factors in the literature. It is possible that strain concentration factors will become available in research literature very soon because of the increase in the use of finite-element analysis. Moreover, finite-element analysis can of itself approximate the strains that will occur at all points in the subject structure.

Data for curves similar to those in Fig. 7-4 have from 3 to 15 points on each curve. The intercepts ε'_F and σ'_F/E and the slopes b and c are established by regression procedures. A linear regression may be robust in the data interval, but ordinate intercepts have a wider dispersion. As a measurement, intercepts can be imprecise and inaccurate. This is because the intercepts are removed from the data and are based on an extrapolation through a region with no data. These intercept numbers can help describe a robust line *in the range of the data*, but another data set can give a very different intercept. It is convenient to take a number from a table such as Table 7–1, but notice there is no mention of the number of points, their range, or regression information. A sense of the nature of these data sets is conveyed by Boller and Seeger.[6] The algebra of these strain-life equations is exponential. Computer codes are commonly used to enhance speed and provide relief from computational drudgery, although this can obscure the fact that useful constants (intercepts) can be broadly distributed and their ranges of usefulness unstated.

Equation (7–2) in the form $\sigma_A = \sigma'_F(2N)^b$ is known as the Basquin relation. At the time (1910) it was thought that the number of *stress reversals* $2N$ could expedite cumulative damage fatigue. This later proved not to be the case,[7] but the notation has become part of the literature.

6. C. Boller and T. Seeger, *Materials Data for Cyclic Loading*, vols. 1–5, Elsevier, New York, 1987.

7. See J. A. Bannantine, J. J. Comer, and J. H. Handrock, *Fundamentals of Metal Fatigue Analysis*, Prentice Hall, Englewood Cliffs, N.J., 1990.

Table 7–2

Completely-Reversed Strain-Controlled Fatigue Tests on RQC-100 Steel (First 4 Columns are Data, Last 3 are Extensions)

Source: N. E. Dowling, *Mechanical Behavior of Materials*, Prentice Hall, Englewood Cliffs, N.J., 1993, p. 628.

ε_a	σ_a, kpsi	ε_{pa}	N_f, cycles	ln σ_a	ln ε_{pa}	ln$(2N_f)$
0.020 2	91.6	0.016 95	227	4.517	−4.077	6.118
0.010 0	83.3	0.007 05	1 030	4.422	−4.955	7.630
0.004 5	73.3	0.001 93	6 450	4.295	−6.250	9.465
0.003 0	68.5	0.000 64	22 250	4.227	−7.354	10.703
0.002 3	66.0	0.000 10	110 000	4.190	−9.210	12.301

$E =$ Young's modulus $= 29\ 000$ Mpsi
$S_y = 99$ kpsi
$S_{ut} = 110$ kpsi

Dowling presents data for polished specimens on completely reversed, strain-controlled, fatigue tests on RQC-100 steel, as seen in Table 7–2. The constants σ'_f and b can be obtained from Eq. (7–2) in the form

$$\sigma_a = \sigma'_F \left(2N_f\right)^b \qquad \text{or} \qquad \ln \sigma_a = \ln \sigma'_F + b \ln \left(2N_f\right) \tag{b}$$

Associating y with $\ln \sigma_a$ and x with $\ln(2N_f)$, using Eqs. (B–4), (B–5), (B–6), (B–7), (B–8), and (B–10), a $y = a + bx$ regression gives $a = 4.480\ 354$, $b = -0.0551$, $\sigma'_F = \exp(4.480\ 354) = 126.5$ kpsi, $r = -0.986$, $s_{y\cdot x} = 0.0263$, $s_a = 0.051$, and $s_b = 0.005\ 37$.

Using Eq. (7–1) in the form

$$\frac{\Delta \varepsilon_p}{2} \varepsilon_{pa} = \varepsilon'_F \left(2N_f\right)^c \qquad \text{or} \qquad \ln \varepsilon_{pa} = \ln \varepsilon'_F + c \ln \left(2N_f\right) \tag{c}$$

and, associating y with $\ln \varepsilon_{pa}$ and x with $\ln(2N_f)$, a $y = a + bx$ regression gives $a = 1.189\ 119$ and $b = -0.818$, resulting in $\varepsilon'_F = \exp(1.189\ 119) = 3.284$, $c = b = -0.818$, $r = -0.990$, $s_{y\cdot x} = 0.331$, $s_a = 0.643$, and $s_b = 0.0677$.

The general result is

$$\varepsilon_{ea} = \frac{\sigma_a}{E} = \frac{\sigma'_F}{E} \left(2N_f\right)^b = \frac{126.5}{29\ 000} \left(2N_f\right)^{-0.0551} = 0.004\ 362 \left(2N_f\right)^{-0.0551} \tag{d}$$

$$\varepsilon_{pa} = \varepsilon'_F \left(2N_f\right)^c = 3.284 \left(2N_f\right)^{-0.818} \tag{e}$$

$$\varepsilon_a = \varepsilon_{ea} + \varepsilon_{pa} = 0.004\ 36 \left(2N_f\right)^{-0.0551} + 3.284 \left(2N_f\right)^{-0.818} \tag{f}$$

See Prob. 7–33 concerning matters of precision. We will use Basquin's relation $\sigma_a = \sigma'_F (2N_f)^b$ in Sec. 7–5 in high-cycle fatigue.

Strain-life data originate from polished specimens. The methodology for adjusting for machine part surfaces is still under development. Similarly, size effects, strength effects, and notch effects need quantitative treatment. Furthermore, the Manson-Coffin equation does not exhibit an endurance limit, posing a problem in steels, with work to be done. Nevertheless, facets of strain-life investigations have been found useful. For instance, if Eq. (*d*) is equated to Eq. (*e*), namely, setting $\varepsilon_{pa} = \varepsilon_{ea}$, it follows that

$$N_f = \frac{1}{2} \left(\frac{\sigma'_F}{E \varepsilon'_F} \right)^{1/(c-b)} \tag{g}$$

establishing the cycle count which partitions elastic-strain dominance (high-cycle fatigue) from plastic-strain dominance (low-cycle fatigue). This partition is useful in partitioning strain-life methods from stress-life methods. Note that Eq. (g) includes the Manson-Coffin constants σ'_F, ε'_F, E, b, and c. For the example above, using Eq. (g),

$$N_f = \frac{1}{2}\left(\frac{126.5}{29\,000(3.284)}\right)^{1/[-0.818-(-0.0551)]} = 2954 \text{ cycles}$$

Another approach is to use Eq. (e) with $\varepsilon_{pa} = 0.002$, the strain associated with the common S_y property, which gives

$$N_f = \frac{1}{2}\left(\frac{0.002}{\varepsilon'_F}\right)^{1/c} \tag{h}$$

which uses 0.002, ε'_F, and c of Eq. (7–3). For the example, Eq. (h) gives

$$N_f = \frac{1}{2}\left(\frac{0.002}{3.284}\right)^{1/-0.818} = 4263 \text{ cycles}$$

We are speaking of an arbitrary division between low-cycle and high-cycle fatigue. The method may depend on how much one knows.

We have been presenting the strain-life failure model with considerable detail. The model is summarized briefly in a boxed form for ready reference.

Strain-Life Fatigue Failure Model

Low-cycle life:

$$\Delta\varepsilon_p/2 = \varepsilon'_F\,(2N_f)^c$$

High-cycle life:

$$\sigma_a = \sigma'_F\,(2N_f)^b$$

Transition life N_T between low-cycle and high-cycle domains:

$$N_T = \frac{1}{2}\left(\frac{\sigma'_F}{E\varepsilon'_F}\right)^{1/(c-b)}$$

General:

$$\sigma_a = S_a = \sigma'_F\,(2N_f)^b + E\varepsilon'_F\,(2N_f)^c$$

$$\Delta\varepsilon/2 = \Delta\varepsilon_e/2 + \Delta\varepsilon_p/2 = (\sigma'_F/E)\,(2N_f)^b + \varepsilon'_F\,(2N_f)^c$$

σ'_F and ε'_F are true stress and true strain, respectively. When expressing σ as a strength S, it is customary to convert to nominal engineering strength. At high cycles there is little difference.

Engineers are always lacking some data, and approximations are useful. In the absence of σ'_F, ε'_F, b, and c, Collins[8] suggests setting parameters equal to the following

8. J. A. Collins, *Failure of Materials in Mechanical Design*, Wiley-Interscience, New York, 1981, p. 288.

static tensile test results:

$$\sigma'_F \doteq \sigma_f \qquad \text{(true stress at fracture)}$$

$$\varepsilon'_F \doteq \varepsilon_f \qquad \text{(true fracture ductility)}$$

$$c \doteq -0.6$$

$$b \doteq -\frac{\log\left(2\sigma_f/S_u\right)}{\log\left(2N_e\right)} = -0.16\log\left(2\sigma_f S_u\right) = -0.16\log\left[2m^{-m}\exp(m)\right]$$

and, for cumulative damage, the Palmgren-Miner rule is competitive with other proposals.

Another approximation, based on knowing S_{ut}, ε_f, and E, setting $b = -0.12$ and $c = -0.6$, and evaluating Eq. (7–3) for total strain amplitude is called the *universal slope method.* There are other approximations. Investigators are getting only approximate agreement. Involvement sometimes clouds investigators' perspective. Load- or stress-driven cycling gives different properties than distortion- or strain-driven cycling. Machinery has more examples of load-driven cycles than distortion-driven cycles. Machine designers pay closer attention to the former than the latter. There are things to be learned from this ongoing strain-driven work, so it is watched carefully. Many references are available.

7–3 Stress-Life Relationships

To determine the strength of materials under the action of fatigue loads, specimens are subjected to repeated or varying forces of specified magnitudes while the cycles or stress reversals are counted to destruction. The most widely used fatigue-testing device is the R. R. Moore high-speed rotating-beam machine. This machine subjects the specimen to pure bending (no transverse shear) by means of weights. The specimen, shown in Fig. 7–5, is very carefully machined and polished, with a final polishing in an axial direction to avoid circumferential scratches. Other fatigue-testing machines are available for applying fluctuating or reversed axial stresses, torsional stresses, or combined stresses to the test specimens.

To establish the fatigue strength of a material, quite a number of tests are necessary because of the statistical nature of fatigue. For the rotating-beam test, a constant bending load is applied, and the number of revolutions (stress reversals) of the beam required for failure is recorded. The first test is made at a stress which is somewhat under the ultimate strength of the material. The second test is made at a stress which is less than that used in the first. This process is continued, and the results are plotted as an *S-N* diagram (Fig. 7–6). This chart may be plotted on semilog paper or on log-log paper. In the case of ferrous metals and alloys, the graph becomes horizontal after the material has been stressed for a certain number of cycles. Plotting on log paper emphasizes the bend in the curve, which might not be apparent if the results were plotted by using Cartesian coordinates.

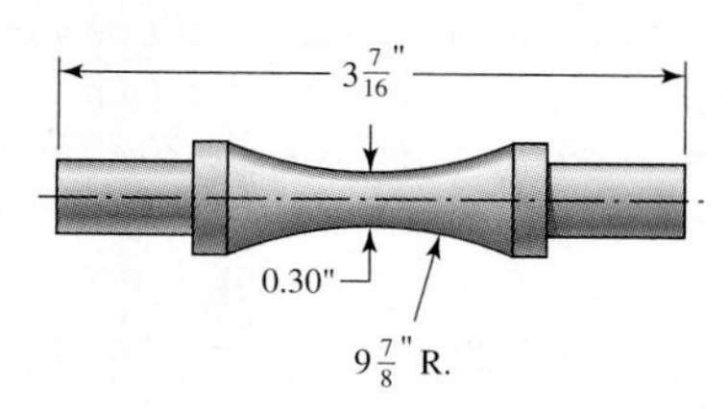

Figure 7–5

Test-specimen geometry for the R. R. Moore rotating-beam machine. The bending moment is uniform over the curved at the highest-stressed portion, a valid test of material, whereas a fracture elsewhere (not at the highest-stress level) is grounds for suspicion of material flaw.

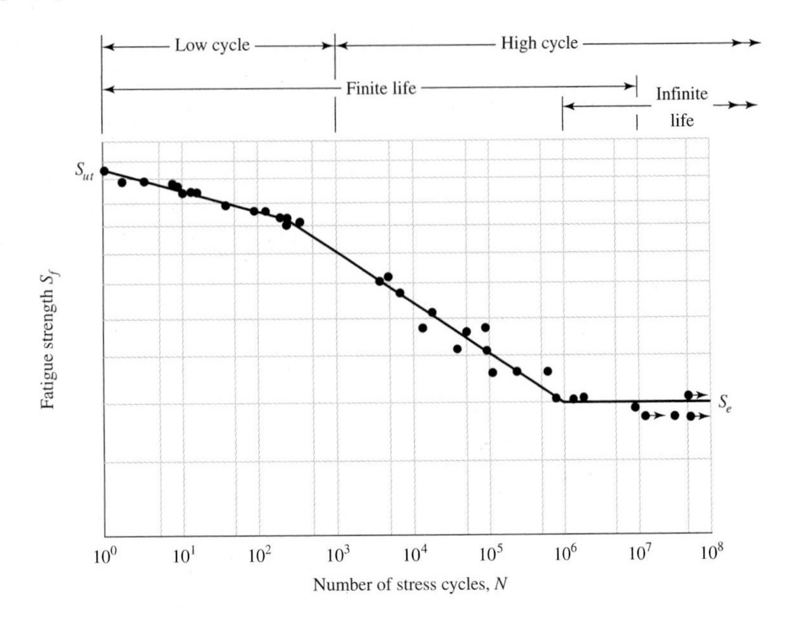

Figure 7–6

An S-N diagram plotted from the results of completely reversed axial fatigue tests. Material: UNS G41300 steel, normalized; $S_{ut} = 116$ kpsi; maximum $S_{ut} = 125$ kpsi. (Data from NACA Tech. Note 3866, December 1966.)

The ordinate of the S-N diagram is called the *fatigue strength* S_f; a statement of this strength must always be accompanied by a statement of the number of cycles N to which it corresponds.

Soon we shall learn that S-N diagrams can be determined either for a test specimen or for an actual mechanical element. Even when the material of the test specimen and that of the mechanical element are identical, there will be significant differences between the diagrams for the two.

In the case of the steels, a knee occurs in the graph, and beyond this knee failure will not occur, no matter how great the number of cycles. The strength corresponding to the knee is called the *endurance limit* S_e, or the fatigue limit. The graph of Fig. 7–6 never does become horizontal for nonferrous metals and alloys, and hence these materials do not have an endurance limit.

We note that a stress cycle ($N = 1$) constitutes a single application and removal of a load and then another application and removal of the load in the opposite direction. Thus $N = \frac{1}{2}$ means the load is applied once and then removed, which is the case with the simple tension test.

The body of knowledge available on fatigue failure from $N = 1$ to $N = 1000$ cycles is generally classified as *low-cycle fatigue*, as indicated in Fig. 7–6. *High-cycle fatigue*, then, is concerned with failure corresponding to stress cycles greater than 10^3 cycles.

We also distinguish a *finite-life region* and an *infinite-life region* in Fig. 7–6. The boundary between these regions cannot be clearly defined except for a specific material; but it lies somewhere between 10^6 and 10^7 cycles for steels, as shown in Fig. 7–6.

As noted previously, it is always good engineering practice to conduct a testing program on the materials to be employed in design and manufacture. This, in fact, is a requirement, not an option, in guarding against the possibility of a fatigue failure. *Because of this necessity for testing, it would really be unnecessary for us to proceed any further in the study of fatigue failure except for one important reason: the desire to know why fatigue failures occur so that the most effective method or methods can be used to improve fatigue strength.* Thus our primary purpose in studying fatigue is to understand why failures occur so that we can guard against them in an optimum manner. For this reason, the analytical design approaches presented in this book, or in any other book, for that matter, do not yield absolutely precise results. The results should be taken as a guide, as something which indicates what is important and what is not important in designing against fatigue failure.

The methods of fatigue-failure analysis represent a combination of engineering and science. Often science fails to provide the answers which are needed. But the airplane must still be made to fly—safely. And the automobile must be manufactured with a reliability that will ensure a long and trouble-free life and at the same time produce profits for the stockholders of the industry. Thus, while science has not yet completely explained the actual mechanism of fatigue, the engineer must still design things that will not fail. In a sense this is a classic example of the true meaning of engineering as contrasted with science. Engineers use science to solve their problems *if* the science is available. But available or not, the problem must be solved, and whatever form the solution takes under these conditions is called engineering.

7–4 The Endurance Limit

The determination of endurance limits by fatigue testing is now routine, though a lengthy procedure. Generally, stress testing is preferred to strain testing for endurance limits.

For preliminary and prototype design and for some failure analysis as well, a quick method of estimating endurance limits is needed. There are great quantities of data in the literature on the results of rotating-beam tests and simple tension tests of specimens taken from the same bar or ingot. By plotting these as in Fig. 7–7, it is possible to see whether there is any correlation between the two sets of results. The graph appears to suggest that the endurance limit ranges from about 40 to 60 percent of the tensile strength for steels up to about 200 kpsi (1400 MPa). Beginning at about $S_{ut} = 200$ kpsi (1400 MPa), the scatter appears to increase, but the trend seems to level off, as suggested by the dashed horizontal line at $S'_e = 100$ kpsi (700 MPa).

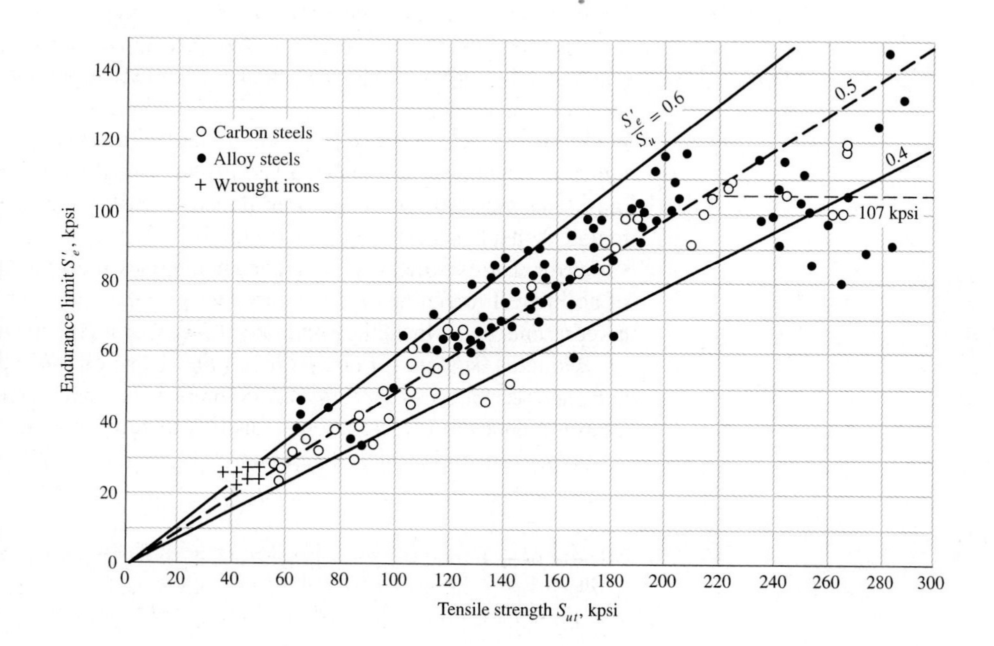

Figure 7–7

Graph of endurance limits versus tensile strengths from actual test results for a large number of wrought irons and steels. Ratios of S'_e/S_{ut} of 0.60, 0.50, and 0.40 are shown by the solid and dashed lines. Note also the horizontal dashed line for $S'_e = 107$ kpsi. Points shown having a tensile strength greater than 214 kpsi have a mean endurance limit of $S'_e = 107$ kpsi and a standard deviation of 13.5 kpsi. (Collated from data compiled by H. J. Grover, S. A. Gordon, and L. R. Jackson in *Fatigue of Metals and Structures*, Bureau of Naval Weapons Document NAVWEPS 00-25-534, 1960; and from *Fatigue Design Handbook*, SAE, 1968, p. 42.)

Table 7–3

Endurance-Limit Ratio S'_e/S_{ut} for Various Steel Microstructures

	Ferrite		Pearlite		Martensite	
	Range	Average	Range	Average	Range	Average
Carbon steel	0.57–0.63	0.60	0.38–0.41	0.40	. . .	0.25
Alloy steel	. . .	. . .	. . .	. . .	0.23–0.47	0.35

Source: Adapted from L. Sors, *Fatigue Design of Machine Components*, Pergamon Press, Oxford, England, 1971.

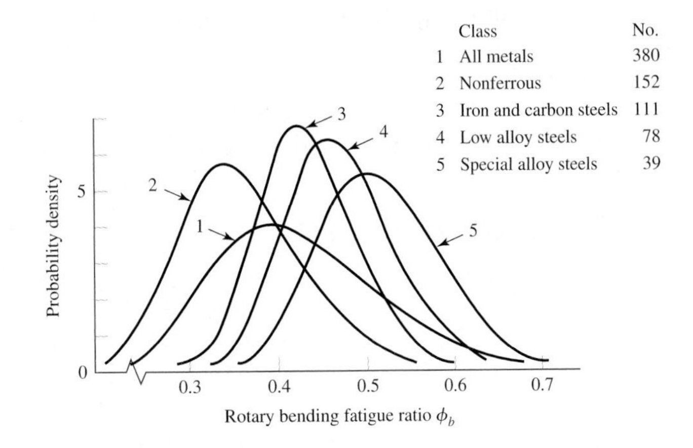

Figure 7–8

The lognormal probability density PDF of the fatigue ratio ϕ_b of Gough.

Another series of tests, this time for various microstructures, is shown in Table 7–3. In this table the endurance limits vary from about 23 to 63 percent of the tensile strength.[9]

Now, it is important to observe that the dispersion of the endurance limit is *not* due to a dispersion in the tensile strengths of the specimen, but rather that the spread occurs even when the tensile strengths of a large number of specimens remain exactly the same. Keep this in mind when choosing factors of safety.

We shall now present the *tensile strength correlation method*, a method for estimating endurance limits. The ratio $\boldsymbol{\phi} = \mathbf{S}'_e/\bar{S}_{ut}$ is called the *fatigue ratio*. For ferrous metals, most of which exhibit an endurance limit, the endurance limit is used as a numerator. For materials that do not show an endurance limit, an endurance strength at a specified number of cycles-to-failure is used and noted. Gough[10] reported the stochastic nature of the fatigue ratio $\boldsymbol{\phi}$ for several classes of metals, and this is shown in Fig. 7–8. The first item to note is that the coefficient of variation is of the order 0.10 to 0.15, and the distribution varies for classes of metals. The second item to note is that Gough's data include materials of no interest to engineers. In the absence of testing, engineers use the correlation that $\boldsymbol{\phi}$ represents to estimate the endurance limit $\mathbf{S}'_e$ from the mean ultimate strength $\bar{S}_{ut}$.

Gough's data are for ensembles of metals, some chosen for metallurgical interest, and include materials that are not commonly selected for machine parts. Mischke[11] analyzed data for 133 common steels and treatments in varying diameters in rotating bending,[12] and the result was

9. But see H. O. Fuchs and R. I. Stephens, *Metal Fatigue in Engineering*, Wiley, New York, 1980, pp. 69–71, which reports a range of 35 to 60 percent for steels having $S_{ut} < 1400$ MPa and as low as 20 percent for high-strength steels.

10. In J. A. Pope, *Metal Fatigue*, Chapman and Hall, London, 1959.

11. Charles R. Mischke, "Prediction of Stochastic Endurance Strength," *Trans. ASME, Journal of Vibration, Acoustics, Stress, and Reliability in Design*, vol. 109, no. 1, January 1987, pp. 113–122.

12. Data from H. J. Grover, S. A. Gordon, and L. R. Jackson, *Fatigue of Metals and Structures*, Bureau of Naval Weapons, Document NAVWEPS 00-2500435, 1960.

$$\phi = 0.445d^{-0.107}\mathbf{LN}(1, 0.138)$$

where d is the specimen diameter in inches and $\mathbf{LN}(1, 0.138)$ is a unit lognormal variate with a mean of 1 and a standard deviation (and coefficient of variation) of 0.138. For the standard R. R. Moore specimen,

$$\boldsymbol{\phi}_{0.30} = 0.445(0.30)^{-0.107}\mathbf{LN}(1, 0.138) = 0.506\mathbf{LN}(1, 0.138)$$

Also, 25 plain carbon and low-alloy steels with $S_{ut} > 212$ kpsi are described by

$$\mathbf{S}'_e = 107.3\mathbf{LN}(1, 0.139)\text{kpsi}$$

In summary,

$$\mathbf{S}'_e = \begin{cases} 0.506\bar{S}_{ut}\mathbf{LN}(1, 0.138) \text{ kpsi or MPa} & \bar{S}_{ut} \leq 212 \text{ kpsi (1460 MPa)} \\ 107\mathbf{LN}(1, 0.139) \text{ kpsi} & \bar{S}_{ut} > 212 \text{ kpsi (1460 MPa)} \\ 740\mathbf{LN}(1, 0.139) \text{ MPa} & \bar{S}_{ut} > 1460 \text{ MPa} \end{cases} \tag{7–4}$$

where $\bar{S}_{ut}$ is the *mean* ultimate tensile strength. The prime mark on $\mathbf{S}'_e$ refers to the rotating-beam tensile specimen itself. The unprimed symbol $\mathbf{S}_e$ is reserved for the endurance limit of a particular machine element (at the critical location and in the geometry and condition of use). These two strengths, $\mathbf{S}'_e$ and $\mathbf{S}_e$, may be quite different, as we shall soon learn.

The Eqs. (7–4) represent the state of information before an engineer has chosen a material. In choosing, the designer has made a random choice from the ensemble of possibilities, and the statistics can give the odds of disappointment. If the testing is limited to finding an estimate of the ultimate tensile strength mean $\bar{S}_{ut}$ with the chosen material, Eqs. (7–4) are directly helpful. If there is to be rotary-beam fatigue testing, then statistical information on the endurance limit is gathered and there is no need for the correlation above.

The data of Table 7–3 emphasize the difficulty of attempting to provide a single rule for deriving the endurance limit from the tensile strength. The table also shows a part of the cause of this difficulty. Steels treated to give different microstructures have different S'_e/S_{ut} ratios. It appears that the more ductile microstructures have a higher ratio. Martensite has a very brittle nature and is highly susceptible to fatigue-induced cracking; thus the ratio is low. When designs include detailed heat-treating specifications to obtain specific microstructures, it is possible to use an estimate of the endurance limit based on test data for the particular microstructure; such estimates are much more reliable and indeed should be used.

The endurance limits for various classes of cast irons, polished or machined, are given in Table E–24. Aluminum alloys do not have an endurance limit. The fatigue strengths of some aluminum alloys at $50(10^7)$ cycles of reversed stress are given in Table E–24b.

Table 7–4 compares approximate mean values of the fatigue ratio $\bar{\phi}_{0.30}$ for several classes of ferrous materials.

Table 7–4

Comparison of Approximate Values of Mean Fatigue Ratio for Some Classes of Metals

Material Class	$\bar{\phi}_{0.30}$
Wrought steels	0.50
Cast steels	0.40
Powdered steels	0.38
Gray cast iron	0.35
Malleable cast iron	0.40
Normalized nodular cast iron	0.33

7–5 Fatigue Strength

As shown in Fig. 7–6, a region of low-cycle fatigue extends from $N = 1$ to about 10^3 cycles. In this region the fatigue strength S_f is only slightly smaller than the tensile strength S_{ut}. An analytical approach has been given by Mischke[13] for both high-cycle and low-cycle regions, requiring the parameters of the Manson-Coffin equation plus the strain-strengthening exponent m. Engineers often have to work with less information.

Figure 7–6 indicates that the high-cycle fatigue domain extends from 10^3 cycles for steels to the endurance limit life N_e, which is about 10^6 to 10^7 cycles. The purpose of this section is to develop methods of approximation of the S-N diagram in the high-cycle region, when information may be as sparse as the results of a simple tension test. Experience has shown high-cycle fatigue data are rectified by a logarithmic transform to both stress and cycles-to-failure. Engineers can work with Eq. (7–2) in the following way:

$$(S_f)_{10^3 \text{ cycles}} = \sigma'_F \left(2 \cdot 10^3\right)^b = f S_{ut}$$

where f is the fraction of S_{ut} represented by $(S_f)_{10^3 \text{ cycles}}$. Solving for f gives

$$f = \frac{\sigma'_F}{S_{ut}} \left(2 \cdot 10^3\right)^b \tag{a}$$

Now $\sigma'_F = \sigma_0 \varepsilon^m$, and if this true-stress–true-strain equation is not known, the SAE approximation[14] for steels with $H_B \le 500$ may be used:

$$\sigma'_F = S_{ut} + 50 \text{ kpsi} \qquad \text{or} \quad \sigma'_F = S_{ut} + 345 \text{ MPa} \tag{b}$$

The exponent b is found from $\sigma_a = S_e = \sigma'_F(2N_e)^b$ as

$$b = -\frac{\log\left(\sigma'_F/S_e\right)}{\log\left(2N_e\right)} \tag{c}$$

Thus the equation $S_f = \sigma'_F(2N)^b$ is known. For example, if $S_{ut} = 105$ kpsi and $S_e = 52.5$ kpsi at 10^6 cycles-to-failure,

$$\sigma'_F = 105 + 50 = 155 \text{ kpsi}$$

$$b = -\frac{\log(155/52.5)}{\log\left(2 \cdot 10^6\right)} = -0.0746$$

$$f = \frac{155}{105}\left(2 \cdot 10^3\right)^{-0.0746} = 0.837$$

and

$$S_f = 155(2N)^{-0.0746} \tag{d}$$

This equation is valid from the transition cycle count Eq. (g) of Sec. 7–2 to the elastic limit cycle count N_e, approximately $10^3 \le N \le N_e$.

Empirically, the common curve fit is $S_f = aN^b$, where N is cycles-to-failure and the constants a and b are defined by the points 10^3, S_f and 10^6, S_e with $(S_f)_{10^3} = f S_{ut}$:

$$a = \frac{f^2 S_{ut}^2}{S_e} \tag{7–5}$$

13. J. E. Shigley and C. R. Mischke, *Standard Handbook of Machine Design*, 2nd ed., McGraw-Hill, New York, 1996, pp. 13.34–13.37.

14. *Fatigue Design Handbook*, vol. 4, Society of Automotive Engineers, New York, 1958, p. 27.

$$b = -\frac{1}{3}\log\left(\frac{f S_{ut}}{S_e}\right) \tag{7–6}$$

Continuing the informal example,

$$a = \frac{0.837^2 105^2}{52.5} = 147.1 \text{ kpsi}$$

$$b = -\frac{1}{3}\log\frac{0.837(105)}{52.5} = -0.0746$$

and the resulting equation is

$$S_f = 147.1N^{-0.0746} \tag{e}$$

Note that $2^{-0.0746}(155) = 147.2$, so Eqs. (*d*) and (*e*) are really the same. There are popular curve fits with f treated as a constant, but f varies with S_{ut}. The following table shows the nature of such an approximation:

S_{ut}, kpsi	60	90	120	200
f	0.93	0.86	0.82	0.77

If a completely reversed stress σ_a is given, the number of cycles-to-failure can be expressed as

$$N = \left(\frac{\sigma_a}{a}\right)^{1/b} \tag{7–7}$$

Low-cycle fatigue is often defined (see Fig. 7–6) as failure which occurs in a range of $1 \le N \le 10^3$ cycles. On a loglog plot such as Fig. 7–6 the failure locus in this range is nearly linear below 10^3 cycles. A straight line between 10^3, $f S_{ut}$ and 1, S_{ut} (transformed) is conservative, and it is given by

$$S_f \ge S_{ut} N^{(\log f)/3} \qquad 1 \le N \le 10^3 \tag{f}$$

EXAMPLE 7–1 A 1050 HR steel has a mean ultimate tensile strength of $S_{ut} = 105$ kpsi and a mean yield strength of 60 kpsi, and 0.51 RA (fractional reduction in area).

(*a*) Estimate the rotating-beam endurance limit.

(*b*) Estimate the endurance strength for a polished rotating-beam specimen corresponding to 10^4 cycles-to-failure.

(*c*) Estimate the expected life under a completely reversed stress of 55 kpsi.

Solution (*a*) From Eq. (7–4),

Answer $S'_e = 0.506(105) - 53.1$ kpsi

(*b*) From Eq. (*b*),

$$\sigma'_F = 105 + 50 = 155 \text{ kpsi}$$

From Eq. (*c*),

$$b = -\frac{\log(155/53.1)}{\log(2 \cdot 10^6)} = -0.0738$$

From Eq. (*a*),

$$f = \frac{155}{105}\left(2 \cdot 10^3\right)^{-0.0738} = 0.842$$

From Eq. (7–5),

$$a = \frac{0.842^2 105^2}{53.1} = 147.2 \text{ kpsi}$$

From Eq. (7–6),

$$b = -\frac{1}{3}\log\left[\frac{0.842(105)}{53.1}\right] = -0.0738 \qquad \text{(verification)}$$

$$S_f = 147.2N^{-0.0738}$$

Answer $$(S_f)_{10^4} = 147.2\left(10^4\right)^{-0.0738} = 74.6 \text{ kpsi}$$

(*c*) From Eq. (7–7),

Answer $$N = \left(\frac{55}{147.2}\right)^{1/-0.0738} = 621\ 290 = 0.62\left(10^6\right) \text{ cycles}$$

7–6 Endurance-Limit Modifying Factors

We have seen that the rotating-beam specimen used in the laboratory to determine endurance limits is prepared very carefully and tested under closely controlled conditions. It is unrealistic to expect the endurance limit of a mechanical or structural member to match the values obtained in the laboratory. Some differences include

- *Material:* composition, basis of failure, variability.
- *Manufacturing:* method, heat treatment, fretting corrosion, surface condition, stress concentration.
- *Environment:* corrosion, temperature, stress state, relaxation times.
- *Design:* size, shape, life, stress state, stress concentration, speed, fretting, galling.

Marin[15] identified factors which quantified the effects of surface condition, size, loading, temperature, and miscellaneous items. The question of whether to adjust the endurance limit by subtractive corrections or multiplicative corrections was resolved by an extensive statistical analysis of a 4340 (electric furnace, aircraft quality) steel, in which a correlation coefficient of 0.85 was found for the multiplicative form and 0.40 for the additive form. A Marin equation is therefore written

$$\mathbf{S}_e = \mathbf{k}_a k_b \mathbf{k}_c \mathbf{k}_d \mathbf{k}_e \mathbf{S}'_e \tag{7–8}$$

where $\mathbf{k}_a$ = surface condition modification factor
k_b = size modification factor (deterministic)
$\mathbf{k}_c$ = load modification factor
$\mathbf{k}_d$ = temperature modification factor
$\mathbf{k}_e$ = miscellaneous-effects modification factor
$\mathbf{S}'_e$ = rotary-beam endurance limit

15. Joseph Marin, *Mechanical Behavior of Engineering Materials*, Prentice-Hall, Englewood Cliffs, N.J., 1962, p. 224.

Table 7–5

Parameters in Marin Surface Condition Factor

$\mathbf{k}_a = aS_{ut}^b\,\mathbf{LN}(1, C)$				
Surface Finish	***a*** **kpsi**	***a*** **MPa**	***b***	**Coefficient of Variation, *C***
Ground*	1.34	1.58	−0.086	0.120
Machined or Cold-rolled	2.67	4.45	−0.265	0.058
Hot-rolled	14.5	56.1	−0.719	0.110
As-forged	39.8	271	−0.995	0.145

* Due to the wide scatter in ground surface data, an alternate function is $\mathbf{k}_a = 0.878\mathbf{LN}(1, 0.120)$. Note: S_{ut} in kpsi or MPa.

$\mathbf{S}_e$ = endurance limit at the critical location of a machine part in the geometry and condition of use

When endurance tests of *parts* are not available, estimations are made by applying Marin factors to the endurance limit.

Surface Factor k_a

The surface of a rotating-beam specimen is highly polished, with a final polishing in the axial direction to smooth out any circumferential scratches. The modification factor depends on the quality of the finish of the part surface and on the tensile strength. To find quantitative expressions for common finishes of machine parts (ground, machined or cold-drawn, hot-rolled, and as-forged), the coordinates of data points were recaptured from a plot of endurance limit versus ultimate tensile strength of data gathered by Lipson and Noll and reproduced by Horger.[16] The result of regression analysis by Mischke was of the form

$$\mathbf{k}_a = a\bar{S}_{ut}^b\,\mathbf{LN}(1, C) \qquad (S_{ut} \text{ in kpsi or MPa}) \tag{7–9}$$

where Table 7–5 gives values of a, b, and C for various surface conditions. The symbol $\mathbf{LN}(1, C)$ is a unit variate lognormally distributed, with a mean of 1 and a standard deviation (and coefficient of variation) of C. The mean and standard deviation of $\mathbf{k}_a$ are given by

$$\bar{k}_a = aS_{ut}^b \qquad \text{and} \qquad \sigma_{ka} = C\bar{k}_a$$

EXAMPLE 7–2

A steel has a mean ultimate strength of 520 MPa and a machined surface. Estimate $\mathbf{k}_a$.

Solution

From Table 7–5,

$$\mathbf{k}_a = 4.45(520)^{-0.265}\mathbf{LN}(1, 0.058)$$
$$\bar{k}_a = 4.45(520)^{-0.265}(1) = 0.848$$
$$\sigma_{ka} = 4.45(520)^{-0.265}(0.058) = 0.049$$

so $\mathbf{k}_a \sim \mathbf{LN}(0.848, 0.049)$. The deterministic value is simply 0.848, the mean.

16. C. J. Noll and C. Lipson, "Allowable Working Stresses," *Society for Experimental Stress Analysis*, vol. 3, no. 2, 1946, p. 29. Reproduced by O. J. Horger (ed.), *Metals Engineering Design ASME Handbook*, McGraw-Hill, New York, 1953, p. 102.

Size Factor k_b

The size factor has been evaluated using 133 sets of data points.[17] The results for bending and torsion may be expressed as

$$k_b = \begin{cases} (d/0.3)^{-0.107} = 0.879d^{-0.107} & 0.11 \le d < 2 \text{ in} \\ 0.859 - 0.021\,25d & 2 < d \le 10 \text{ in} \\ (d/7.62)^{-0.107} = 1.24d^{-0.107} & 2.79 \le d \le 51 \text{ mm} \\ 0.859 - 0.000\,837d & 51 < d \le 254 \text{ mm} \end{cases} \qquad (7\text{–}10)$$

For axial loading there is no size effect, so

$$k_b = 1 \qquad (7\text{–}11)$$

but see k_c.

One of the problems that arise in using Eq. (7–10) is what to do when a round bar in bending is not rotating, or when a noncircular cross section is used. For example, what is the size factor for a bar 6 mm thick and 40 mm wide? The approach to be used here employs an *effective dimension* d_e obtained by equating the volume of material stressed at and above 95 percent of the maximum stress to the same volume in the rotating-beam specimen.[18] It turns out that when these two volumes are equated, the lengths cancel, and so we need only consider the areas. For a rotating round section, the 95 percent stress area is the area in a ring having an outside diameter d and an inside diameter of $0.95d$. So, designating the 95 percent stress area $A_{0.95\sigma}$, we have

$$A_{0.95\sigma} = \frac{\pi}{4}\left[d^2 - (0.95d)^2\right] = 0.0766d^2 \qquad (7\text{–}12)$$

This equation is also valid for a rotating hollow round. For nonrotating solid or hollow rounds, the 95 percent stress area is twice the area outside of two parallel chords having a spacing of $0.95D$, where D is the diameter. Using an exact computation, this is

$$A_{0.95\sigma} = 0.0105D^2 \qquad (7\text{–}13)$$

when rounded. Setting Eqs. (7–12) and (7–13) equal to each other enables us to solve for the effective diameter. This gives

$$d_e = 0.370D \qquad (7\text{–}14)$$

as the effective size of a round corresponding to a nonrotating solid or hollow round.

A rectangular section of dimensions $h \times b$ has $A_{0.95\sigma} = 0.05hb$. Using the same approach as before, we have, as displayed in Table 7–6,

$$d_e = 0.808(hb)^{1/2} \qquad (7\text{–}15)$$

These sections are shown in Fig. 7–9 together with a channel and a wide-flange or I-beam section. For the channel,

$$A_{0.95\sigma} = \begin{cases} 0.05ab & \text{axis 1-1} \qquad (7\text{–}16) \\ 0.052xa + 0.1t_f(b - x) & \text{axis 2-2} \qquad (7\text{–}17) \end{cases}$$

The 95 percent stress area for the wide flange is

$$A_{0.95\sigma} = \begin{cases} 0.10at_f & \text{axis 1-1} \qquad (7\text{–}18) \\ 0.05ba \quad t_f > 0.025a & \text{axis 2-2} \qquad (7\text{–}19) \end{cases}$$

---

17. Mischke, op. cit., Table 3. Additional data modify the exponent slightly; it can be taken as −0.11.

18. See R. Kuguel, "A Relation between Theoretical Stress Concentration Factor and Fatigue Notch Factor Deduced from the Concept of Highly Stressed Volume," *Proc. ASTM*, vol. 61, 1961, pp. 732–748.

Table 7–6

Equivalent Diameter for Size Factor

Section	Equivalent Diameter, d_e
Round, rotary bending, torsion	d
Round, nonrotating bending	$0.37d$
Rectangle, nonrotating bending	$0.808(bh)^{1/2}$

Figure 7–9

(*a*) Solid round; (*b*) Rectangular section; (*c*) channel section; (*d*) wide-flange section.

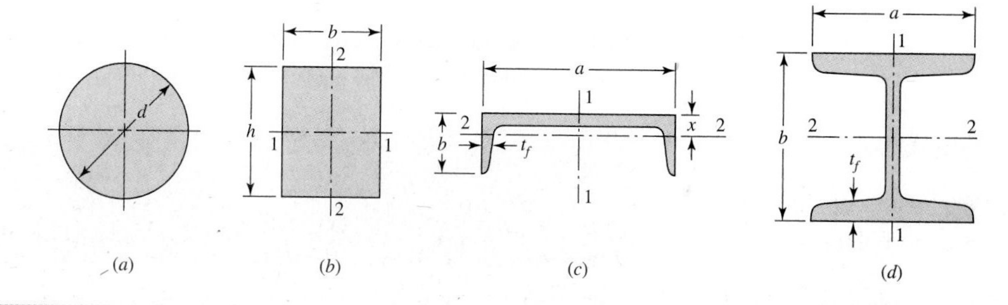

EXAMPLE 7–3 A steel shaft loaded in bending is 32 mm in diameter, abutting a filleted shoulder 38 mm in diameter. The shaft material has a mean ultimate tensile strength of 690 MPa. Estimate the Marin size factor k_b if the shaft is used in
(*a*) A rotating mode.
(*b*) A nonrotating mode.

Solution (*a*) From Eq. (7–10),

Answer
$$k_b = \left(\frac{d}{7.62}\right)^{-0.107} = \left(\frac{32}{7.62}\right)^{-0.107} = 0.858$$

(*b*) From Table 7–6,

$$d_e = 0.37\text{d} = 0.37(32) = 11.84 \text{ mm}$$

From Eq. (7–10),

Answer
$$k_b = \left(\frac{11.84}{7.62}\right)^{-0.107} = 0.954$$

Loading Factor k_c

When fatigue tests are carried out with rotating bending, axial (push-pull), and torsional loading, the endurance limits differ, as shown below, with $\bar{S}_{ut}$ in kpsi,

$$\mathbf{S}'_e = \boldsymbol{\phi}_b \bar{S}_{ut} = 0.506\bar{S}_{ut}\mathbf{LN}(1, 0.138)$$
$$\mathbf{S}'_{ax} = \boldsymbol{\phi}_{ax} \bar{S}_{ut} = 0.623\bar{S}_{ut}^{-0.0778}\bar{S}_{ut}\mathbf{LN}(1, 0.264)$$
$$\mathbf{S}'_{se} = \boldsymbol{\phi}_t \bar{S}_{ut} = 0.166\bar{S}_{ut}^{0.125}\bar{S}_{ut}\mathbf{LN}(1, 0.263)$$

If one substitutes $\phi_{0.30}\bar{S}_{ut}$ for S'_e in Eq. (7–8) *regardless* of loading, then it is clear that $\mathbf{k}_c$ is defined, with $\bar{S}_{ut}$ in kpsi, as

$$(\mathbf{k}_c)_{\text{bending}} = \frac{\boldsymbol{\phi}_{0.30}}{\boldsymbol{\phi}_{0.30}} = (1, 0) \tag{7–20}$$

$$(\mathbf{k}_c)_{\text{axial}} = \frac{\phi_{ax}}{\phi_{0.30}} = 1.23\bar{S}_{ut}^{-0.0778}\mathbf{LN}(1, 0.125) \tag{7–21}$$

$$(\mathbf{k}_c)_{\text{torsion}} = \frac{\phi_t}{\phi_{0.30}} = 0.328\bar{S}_{ut}^{0.125}\mathbf{LN}(1, 0.125) \tag{7–22}$$

There are fewer data to study for axial fatigue. Equation (7–21) was deduced from the data of Landgraf and of Grover, Gordon, and Jackson (as cited earlier).

Table 7–7

Parameters in Marin Loading Factor

$\mathbf{k}_c = \alpha\bar{\mathbf{S}}_{ut}^{\beta}$ LN(1, C)					
Mode of Loading	α **kpsi**	α **MPa**	β	**C**	**Average k_c**
Bending	1	1	0	0	1
Axial	1.23	1.43	−0.078	0.125	0.85
Torsion	0.328	0.258	0.125	0.125	0.59

Table 7–8

Average Marin Loading Factor for Axial Load

S_{ut}, kpsi	k_c^*
50	0.907
100	0.860
150	0.832
200	0.814

* Average entry 0.85.

Table 7–9

Average Marin Loading Factor for Torsional Load

S_{ut}, kpsi	k_c^*
50	0.535
100	0.583
150	0.614
200	0.636

* Average entry 0.59.

Table 7–10

Average Marin Torsional Loading Factor k_c for Several Materials

Material	Range	n	$\bar{k}_c$	σ_{kc}
Wrought steels	0.52–0.69	31	0.60	0.04
Wrought Al	0.43–0.74	13	0.55	0.09
Wrought Cu and alloy	0.41–0.67	7	0.56	0.10
Wrought Mg and alloy	0.49–0.61	2	0.54	0.12
Titanium	0.37–0.57	3	0.48	0.12
Cast iron	0.79–1.01	9	0.90	0.07
Cast Al, Mg, and alloy	0.71–0.91	5	0.85	0.09

Source: The table is an extension of P. G. Forrest, *Fatigue of Metals*, Pergamon Press, London, 1962, Table 17, p. 110, with standard deviations estimated from range and sample size using Table A–1 in J. B. Kennedy and A. M. Neville, *Basic Statistical Methods for Engineers and Scientists*, 3rd ed., Harper & Row, New York, 1986, pp. 54–55.

Torsional data are sparser, and Eq. (7–22) is deduced from data in Grover et al. Notice the mild sensitivity to strength in the axial and torsional load factor, so $\mathbf{k}_c$ in these cases is not constant. Average values are shown in the last column of Table 7–7, and as footnotes to Tables 7–8 and 7–9. Table 7–10 shows the influence of material classes on the load factor $\mathbf{k}_c$. Distortion energy theory predicts $(k_c)_{\text{torsion}} = 0.577$ for materials to which the distortion-energy theory applies.

EXAMPLE 7–4 Estimate the Marin loading factor $\mathbf{k}_c$ for a 1–in-diameter bar that is used as follows.
(*a*) In bending. It is made of steel with $\mathbf{S}_{ut} \sim 100\mathbf{LN}(1, 0.035)$ kpsi, and the designer intends to use the correlation $\mathbf{S}'_e = \boldsymbol{\phi}_{0.30}\bar{S}_{ut}$ to predict $\mathbf{S}'_e$.
(*b*) In bending, but endurance testing gave $\mathbf{S}'_e \sim 55\mathbf{LN}(1, 0.081)$ kpsi.
(*c*) In push-pull (axial) fatigue, $\mathbf{S}_{ut} \sim \mathbf{LN}(86.2, 3.92)$ kpsi, and the designer intended to use the correlation $\mathbf{S}'_e = \boldsymbol{\phi}_{0.30}\bar{S}_{ut}$.
(*d*) In torsional fatigue. The material is cast iron, and $\mathbf{S}'_e$ is known by test.

Solution (*a*) From Eq. (7–20),

Answer $\mathbf{k}_c = (1, 0)$

(*b*) Since the test is in bending and use is in bending,

Answer $\mathbf{k}_c = (1, 0)$

(*c*) From Eq. (7–21),

Answer
$$(\mathbf{k}_c)_{ax} = 1.23(86.2)^{-0.0778}\mathbf{LN}(1, 0.125)$$
$$\bar{k}_c = 1.23(86.2)^{-0.0778}(1) = 0.870$$
$$\sigma_{kc} = C\bar{k}_c = 0.125(0.870) = 0.109$$

(*d*) From Table 7–10, $k_c = 0.90$, $\sigma_{kc} = 0.07$, and

Answer $C_{kc} = \frac{0.07}{0.90} = 0.08$

Temperature Factor k_d

When operating temperatures are below room temperature, brittle fracture is a strong possibility and should be investigated first. When the operating temperatures are higher than room temperature, yielding should be investigated first because the yield strength drops off so rapidly with temperature; see Fig. 5–11. Any stress will induce creep in a material operating at high temperatures; so this factor must be considered too. Finally, it may be true that there is no fatigue limit for materials operating at high temperatures. Because of the reduced fatigue resistance, the failure process is, to some extent, dependent on time.

The limited amount of data available show that the endurance limit for steels increases slightly as the temperature rises and then begins to fall off in the 400 to 700°F range, not unlike the behavior of the tensile strength shown in Fig. 5–11. For this reason it is probably true that the endurance limit is related to tensile strength at elevated temperatures in the same manner as at room temperature.[19] It seems quite logical, therefore,

19. For more, see Table 2 of ANSI/ASME B106. 1M-1985 shaft standard, and E. A. Brandes (ed.), *Smithell's Metals Reference Book*, 6th ed., Butterworth, London, 1983, pp. 22–134 to 22–136, where endurance limits from 100 to 650°C are tabulated.

Table 7–11

Effect of Operating Temperature on the Tensile Strength of Steel.* (S_T = tensile strength at operating temperature; S_{RT} = tensile strength at room temperature; $0.099 \leq \hat{\sigma} \leq 0.110$)

Temperature, °C	S_T/S_{RT}	Temperature, °F	S_T/S_{RT}
20	1.000	70	1.000
50	1.010	100	1.008
100	1.020	200	1.020
150	1.025	300	1.024
200	1.020	400	1.018
250	1.000	500	0.995
300	0.975	600	0.963
350	0.943	700	0.927
400	0.900	800	0.872
450	0.843	900	0.797
500	0.768	1000	0.698
550	0.672	1100	0.567
600	0.549		

* *Data source*: Fig. 5–11.

to employ the same relations to predict endurance limit at elevated temperatures as are used at room temperature, at least until more comprehensive data become available. At the very least, this practice will provide a useful standard against which the performance of various materials can be compared.

Table 7–11 has been obtained from Fig. 5–11 using only the tensile-strength data. Note that the table represents 145 tests of 21 different carbon and alloy steels and that the maximum standard deviation is just 0.110. A 4th order polynomial curve fit to the data underlying Fig. 5-11 gives a mean locus, which is multiplied by $\mathbf{LN}(1, 0.11)$ to give $\mathbf{k}_d$ as

$$\mathbf{k}_d = \left[0.975 + 0.432\,(10^{-3})\,t_f - 0.115\,(10^{-5})\,t_f^2 + 0.104\,(10^{-8})\,t_f^3 - 0.595\,(10^{-12})\,t_f^4\right]\mathbf{LN}(1, 0.11) \tag{7–23}$$

where $70 \leq t_f \leq 1000°\text{F}$.

Two types of problems arise when temperature is a consideration. If the rotating-beam endurance limit is known at room temperature, then use

$$k_d = \frac{S_T}{S_{RT}} \tag{7–24}$$

from Table 7–11 and proceed as usual. If the rotating-beam endurance limit is not given, then compute it using Eq. (7–4) and the temperature-corrected tensile strength obtained using the factor from Table 7–11. Then use $k_d = 1$.

EXAMPLE 7–5

A 1035 steel has a mean ultimate tensile strength of 70 kpsi and is to be used for a part that sees 450°F in service. Estimate the Marin temperature modification factor and $(S_e')_{450°}$ if

(*a*) The room-temperature endurance limit by test is $(\mathbf{S}_e')_{70°} \sim \mathbf{LN}(39.0, 1.76)$ kpsi.

(*b*) Only the mean ultimate tensile strength at room temperature is known.

Solution (*a*) $(\mathbf{S}'_e)_{450°} = \mathbf{k}_d(\mathbf{S}'_e)_{70°}$. First, from Eq. (7–23),

$$\begin{aligned}\mathbf{k}_d &= \left[0.975 + 0.432\left(10^{-3}\right)(450) - 0.115\left(10^{-5}\right)450^2 \right.\\ &\quad \left. + \ 0.104\left(10^{-8}\right)450^3 - 0.595\left(10^{-12}\right)450^4\right]\mathbf{LN}(1, 0.11)\\ &= 0.007\mathbf{LN}(1, 0.11)\end{aligned}$$

$$\begin{aligned}\left(\mathbf{S}'_e\right)_{450°} &= \mathbf{k}_d\left(\mathbf{S}'_e\right)_{70°} = \mathbf{k}_d\mathbf{LN}(39.0, 1.76) = \mathbf{k}_d \cdot 39\mathbf{LN}(1, 0.045)\\ &= 1.007\mathbf{LN}(1, 0.11)39\mathbf{LN}(1, 0.045)\end{aligned}$$

The mean of $(\mathbf{S}'_e)_{450°}$ is given by

$$\left(\bar{S}'_e\right)_{450°} = 1.007(39.0) = 39.27 \text{ kpsi}$$

and the standard deviation is given by

$$(\sigma_{Se'})_{450°} = \sqrt{0.11^2 + 0.045^2}(39.27) = 0.119(39.27) = 4.71 \text{ kpsi}$$

and the elevated endurance limit can be written as

Answer $(\mathbf{S}'_e)_{450°} \sim \mathbf{LN}(39.27, 4.71)$ kpsi

(*b*) From Table 7–11 the interpolation entry is $(1.018 + 0.995)/2 = 1.007$. Thus

$$\begin{aligned}\left(\bar{S}_{ut}\right)_{450°} &= \frac{\left(\bar{S}_{ut}\right)_{450°}}{\left(\bar{S}_{ut}\right)_{70°}}\left(\bar{S}_{ut}\right)_{70°} = \frac{S_T}{S_{RT}}\left(\bar{S}_{ut}\right)_{70°} = 1.007\left(\bar{S}_{ut}\right)_{70°}\\ &= 1.007(70) = 70.5 \text{ kpsi}\end{aligned}$$

then

Answer $(S'_e)_{450°} = 0.506(S_{ut})_{450°} = 0.506(70.5) = 35.7$ kpsi

Part a gives the better estimate due to testing of the particular material, rather than reciting the history of steels with a tensile strength of 70 kpsi at room temperature when tested for endurance limit.

Miscellaneous-Effects Factor k_e

Though the factor k_e is intended to account for the reduction in endurance limit due to all other effects, it is really intended as a reminder that these must be accounted for, because actual values of k_e are not always available.

Residual stresses may either improve the endurance limit or affect it adversely. Generally, if the residual stress in the surface of the part is compression, the endurance limit is improved. Fatigue failures appear to be tensile failures, or at least to be caused by tensile stress, and so anything which reduces tensile stress will also reduce the possibility of a fatigue failure. Operations such as shot peening, hammering, and cold rolling build compressive stresses into the surface of the part and improve the endurance limit significantly. Of course, the material must not be worked to exhaustion.

The endurance limits of parts which are made from rolled or drawn sheets or bars, as well as parts which are forged, may be affected by the so-called *directional characteristics* of the operation. Rolled or drawn parts, for example, have an endurance limit in the transverse direction which may be 10 to 20 percent less than the endurance limit in the longitudinal direction.

Figure 7–10

The failure of a case-hardened part in bending or torsion. In this example, failure occurs in the core.

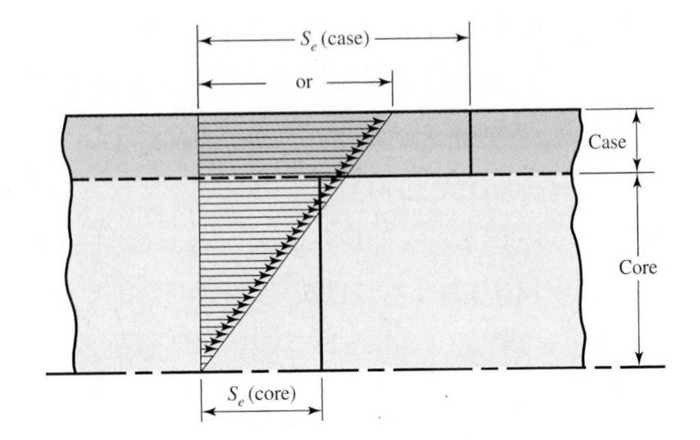

Parts which are case-hardened may fail at the surface or at the maximum core radius, depending upon the stress gradient. Figure 7–10 shows the typical triangular stress distribution of a bar under bending or torsion. Also plotted as a heavy line in this figure are the endurance limits S_e for the case and core. For this example the endurance limit of the core rules the design because the figure shows that the stress σ or τ, whichever applies, at the outer core radius, is appreciably larger than the core endurance limit.

Of course, if stress concentration is also present, the stress gradient is much steeper, and hence failure in the core is unlikely.

Corrosion

It is to be expected that parts which operate in a corrosive atmosphere will have a lowered fatigue resistance. This is, of course, true, and it is due to the roughening or pitting of the surface by the corrosive material. But the problem is not so simple as the one of finding the endurance limit of a specimen which has been corroded. The reason for this is that the corrosion and the stressing occur at the same time. Basically, this means that in time any part will fail when subjected to repeated stressing in a corrosive atmosphere. There is no fatigue limit. Thus the designer's problem is to attempt to minimize the factors that affect the fatigue life; these are:

- Mean or static stress
- Alternating stress
- Electrolyte concentration
- Dissolved oxygen in electrolyte
- Material properties and composition
- Temperature
- Cyclic frequency
- Fluid flow rate around specimen
- Local crevices

Electrolytic Plating

Metallic coatings, such as chromium plating, nickel plating, or cadmium plating, reduce the endurance limit by as much as 50 percent. In some cases the reduction by coatings has been so severe that it has been necessary to eliminate the plating process. Zinc plating does not affect the fatigue strength. Anodic oxidation of light alloys reduces bending endurance limits by as much as 39 percent but has no effect on the torsional endurance limit.

Metal Spraying

Metal spraying results in surface imperfections that can initiate cracks. Limited tests show reductions of 14 percent in the fatigue strength.

Cyclic Frequency

If, for any reason, the fatigue process becomes time-dependent, then it also becomes frequency-dependent. Under normal conditions, fatigue failure is independent of frequency. But when corrosion or high temperatures, or both, are encountered, the cyclic rate becomes important. The slower the frequency and the higher the temperature, the higher the crack propagation rate and the shorter the life at a given stress level.

Frettage Corrosion

The phenomenon of frettage corrosion is the result of microscopic motions of tightly fitting parts or structures. Bolted joints, bearing-race fits, wheel hubs, and any set of tightly fitted parts are examples. The process involves surface discoloration, pitting, and eventual fatigue. The frettage factor k_e depends upon the material of the mating pairs and ranges from 0.24 to 0.90.

7–7 Stress Concentration and Notch Sensitivity

Notch sensitivity q was defined in Eq. (b) of Sec. 5–20. The stochastic equivalent is

$$\mathbf{q} = \frac{\mathbf{K}_f - 1}{K_t - 1} \tag{7–25}$$

where K_t is the theoretical (or geometric) stress-concentration factor, a deterministic quantity. A study of lines 3 and 4 of Table 2–1 will reveal that adding a scalar to (or subtracting one from) a variate $\mathbf{x}$ will affect only the mean. Also, multiplying (or dividing) by a scalar affects both the mean and standard deviation. With this in mind, we can relate the statistical parameters of the fatigue stress-concentration factor $\mathbf{K}_f$ to those of notch sensitivity $\mathbf{q}$. It follows that

$$\mathbf{q} = \mathbf{LN}\left(\frac{\bar{K}_f - 1}{K_t - 1}, \frac{C\bar{K}_f}{K_t - 1}\right)$$

where $C = C_{Kf}$ and

$$\begin{aligned} \bar{q} &= \frac{\bar{K}_f - 1}{K_t - 1} \\ \sigma_q &= \frac{C\bar{K}_f}{K_t - 1} \\ C_q &= \frac{C\bar{K}_f}{\bar{K}_f - 1} \end{aligned} \tag{7–26}$$

The fatigue stress-concentration factor $\mathbf{K}_f$ has been investigated more in England than in the United States. Values of C_{kf} for transverse holes, shoulders, and grooves are listed in Table 7–12. Once $\mathbf{K}_f$ is described, $\mathbf{q}$ can also be quantified using the set Eqs. (7–26).

Table 7–12

Heywood's Parameters for $\sqrt{a}$ and C_{Kf} for Steels

	$\sqrt{a}$		
Feature	**$\bar{S}_{ut}$, kpsi**	**$\bar{S}_{ut}$, Mpa**	**C_{Kf}**
Transverse hole	$5/\bar{S}_{ut}$	$174/\bar{S}_{ut}$	0.10
Shoulder	$4/\bar{S}_{ut}$	$139/\bar{S}_{ut}$	0.11
Groove	$3/\bar{S}_{ut}$	$104/\bar{S}_{ut}$	0.15

The modified Neuber equation (after Heywood) gives the fatigue stress-concentration factor K_f as

$$\mathbf{K}_f = \frac{K_t \mathbf{LN}(1, C_{Kf})}{1 + \dfrac{2}{\sqrt{r}} \dfrac{K_t - 1}{K_t} \sqrt{a}} \tag{7–27}$$

where $\mathbf{K}_f$ is lognormally distributed, r is the notch radius, and $\sqrt{a}$ is a function of the mean ultimate tensile strength $\bar{S}_{ut}$. Table 7–12 gives values of $\sqrt{a}$ for steels. Equations (7–25), (7–26), and (7–27) have their parallel in shear, using q_s, K_{ts}, and K_{fs}.

The fatigue stress-concentration factor $\mathbf{K}_f$ is applied to the nominal stress σ_0 as $\mathbf{K}_f\sigma_0$ as an augmentation of stress (preferred because in combined stresses situations the value of $\mathbf{K}_f$ is different for each component) or as a strength reduction factor $\mathbf{k}_e = 1/\mathbf{K}_f$ (not preferred but used occasionally when only one component of stress exists). Both $\mathbf{K}_f$ and $\mathbf{k}_e$ are lognormally distributed with the same coefficient of variation. For stochastic methods Eq. (7–27) is preferred to Eq. (*b*) of Sec. 5–20 as it has a much larger statistical database. When σ_0 is broken into an amplitude component σ_{a0} and a steady component σ_{m0}, $\mathbf{K}_f$ is applied to both as long as no yielding occurs at the notch, which must be checked.

A handy stress-concentration factor summary is provided for easy reference.

Stress-Concentration Factor Summary

Brittle material:

Static: Set $K_t = K_t$, $K_{ts} = K_{ts}$
Dynamic: Set $K_t = K_t$, $K_{ts} = K_{ts}$

Ductile material:

Static: Set $K_t = K_{ts} = 1$
Dynamic: Set $K_t = K_t$, $K_{ts} = K_{ts}$, $K_f = f_1(K_t, q)$, $K_{fs} = f_2(K_{ts}, q_s)$

Apply to both σ_{a0} and σ_{m0} as long as no yielding occurs.

EXAMPLE 7–6 A steel shaft has a mean ultimate tensile strength of 690 MPa and a shoulder with a fillet radius of 3 mm connecting a 32-mm diameter with a 38-mm diameter.
(*a*) Estimate K_t and $\mathbf{K}_f$.
(*b*) Estimate $\mathbf{q}$.

Solution (*a*) From Table E–15–9, using $D/d = 38/32 = 1.1875$, $r/d = 3/32 = 0.093\,75$, we read the graph to find $K_t = 1.65$. In preparation of using Eq. (7–27) we use Table 7–12:

$$\sqrt{a} = \frac{4}{\bar{S}_{ut}} = \frac{4}{690/6.89} = 0.040 \qquad r = \frac{3}{25.4} = 0.118 \text{ in} \qquad C_{Kf} = 0.11$$

From Eq. (7–27),

Answer
$$\mathbf{K}_f = \frac{1.65\mathbf{LN}(1, 0.11)}{1 + \dfrac{2}{\sqrt{0.118}} \dfrac{1.65 - 1}{1.65} 0.040} = 1.51\mathbf{LN}(1, 0.11)$$

(*b*) From Eq. (7–26)

$$\bar{q} = \frac{1.51 - 1}{1.65 - 1} = 0.78$$

$$C_q = \frac{C_{Kf}\bar{K}_f}{\bar{K}_f - 1} = \frac{0.11(1.51)}{1.51 - 1} = 0.326$$

$$\sigma_q = C_q\bar{q} = 0.326(0.78) = 0.254$$

so

Answer $\mathbf{q} \sim \mathbf{LN}(0.78, 0.254)$

EXAMPLE 7–7

Conditions are the same as in Ex. 7–6. Suppose you had an experiment in which you could measure K_f part by part and find the corresponding value of q.
(*a*) What fraction of the observations of q would be larger than unity?
(*b*) Compare the value of q using Eq. (7–27) and Fig. 5–16.

Solution

(*a*) From Ex. 7–6, $\mathbf{q} \sim \mathbf{LN}(0.78, 0.254)$, $C_q = 0.326$. The companion normal to $\mathbf{q}$ has a mean and standard deviation of

$$\mu_y = \ln(0.78) - 0.326^2/2 = -0.302$$

$$\sigma_y = \sqrt{\ln\left(1 + C_x^2\right)} = \sqrt{\ln\left(1 + 0.326^2\right)} = 0.318$$

$$z = \frac{y - \mu_y}{\sigma_y} = \frac{\ln(1) - (-0.302)}{0.318} = 0.95$$

From Table E–10, $\alpha = 0.1711$, so about 17 percent of the observations of q would exceed unity. A great deal of literature on the subject has been written with attention implicitly focused on a deterministic q, really the mean of $\mathbf{q}$, and the impression can be left that the notch sensitivity q lies in the range $0 \leq q \leq 1$, whereas the above insight shows that things are different.
(*b*) Entering Fig. 5–16 with $r = 3$ mm, $S_{ut} = 0.7$ GPa gives $q = 0.84$, compared to 0.78 in Ex. 7–6. See the explanation that follows.

The discrepancy in notch sensitivities in Ex. 7–6, with 0.78, and Ex. 7–7, with 0.84, arises because Fig. 5–17 has as its basis a Neuber (rather than a modified Neuber) equation. The Neuber equation is

$$K_f = 1 + \frac{K_t - 1}{1 + \sqrt{a/r}} \tag{7–28}$$

It follows from Eq. (7–28) that the notch sensitivity is

$$q = \frac{1}{1 + \sqrt{a/r}} \tag{7–29}$$

A third-order polynomial curve fit of the data for $\sqrt{a}$, with S_{ut} in kpsi, is

$$\sqrt{a} = 0.220\,353 - 0.275\,605\left(10^{-2}\right)\bar{S}_{ut} + 0.113\,449\left(10^{-4}\right)\bar{S}_{ut}^2 - 0.247\,328\left(10^{-7}\right)\bar{S}_{ut}^3 \tag{7–30}$$

Substituting $\bar{S}_{ut} = 100$ kpsi, $\sqrt{a} = 0.055$ and

$$q = \frac{1}{1 + 0.055/\sqrt{3}} = 0.848$$

The reason Figs. 5–16 and 5–17 are included is that they are widely used and the reader should be acquainted with them. Note that neither they nor Eqs. (7–28) and (7–29) distinguish between the kinds of notch (as graphs). If you wish to use Eqs. (7–28) and (7–29) for torsion for low-alloy steels, increase the mean ultimate tensile strength by 20 kpsi in Eq. (7–30) and apply this value of $\sqrt{a}$.

Since our knowledge of dispersion in q comes from data supporting the modified Neuber equation (after Heywood), and distinction in kinds of notches can be made, we will use Eq. (7–27) in connection with fatigue stress-concentration factor $\mathbf{K}_f$ and notch sensitivity $\mathbf{q}$.

When cycles-to-failure are less than 10^6 there is evidence that the fatigue stress-concentration factor $\mathbf{K}_f$ is less. Endurance tests with notched and unnotched specimens show, as noted before, that unnotched failure data for unnotched specimens are linear on loglog paper. The failure locus is represented by $\bar{S}_f = aN^b$, and using Eqs. (7–6),

$$\bar{S}_f = \frac{f^2 \bar{S}_{ut}^2}{\bar{S}'_e} N^{(-1/3)\log(f\bar{S}_{ut}/\bar{S}'_e)} \tag{7–31}$$

which exhibits ordinate values at 10^3 and 10^6 cycles-to-failure of

$$(\bar{S}_f)_{10^3} = \frac{f^2 \bar{S}_{ut}^2}{\bar{S}'_e} 10^{3(-1/3)\log(f\bar{S}_{ut}/\bar{S}'_e)} = \frac{f^2 \bar{S}_{ut}^2}{\bar{S}'_e} \frac{\bar{S}'_e}{f\bar{S}_{ut}} = f\bar{S}_{ut}$$

$$(\bar{S}_f)_{10^6} = \frac{f^2 \bar{S}_{ut}^2}{\bar{S}'_e} 10^{6(-1/3)\log(f\bar{S}_{ut}/\bar{S}'_e)} = \frac{f^2 \bar{S}_{ut}^2}{\bar{S}'_e} \frac{\bar{S}'^2_e}{f^2\bar{S}_{ut}^2} = \bar{S}'_e$$

as it should. With a notched specimen, a notch sensitivity q'_{10^3} is defined where k'_f is $(K_f)_{10^3}$ and

$$q'_{10^3} = \frac{K'_f - 1}{K_f - 1} = -0.18 + 0.43\left(10^{-2}\right)\bar{S}_{ut} - 0.45\left(10^{-5}\right)\bar{S}_{ut}^2 \tag{7–32}$$

where $\bar{S}_{ut}$ is in kpsi, $\bar{S}_{ut} < 330$ kpsi, and K_f is the fatigue stress-concentration factor for 10^6 cycles. The ordinates to the notched failure locus are $\bar{S}'_e/(\bar{K}_f)_{10^6}$, which we will abbreviate to $\bar{S}'_e/\bar{K}_6$; and the ordinate to the fatigue locus at 10^3 cycles-to-failure is $f\bar{S}_{ut}/K'_f$, which we will abbreviate to $f\bar{S}_{ut}/K_3$. The data are still straight on a loglog plot, so with these new ordinates

$$\bar{S}'_f = a'N^{b'} = \frac{f^2\bar{S}_{ut}^2 K_6}{K_3^2\bar{S}'_e} N^{(1/3)\log(\bar{S}'_e K_3)/(K_6 f\bar{S}_{ut})} \tag{7–33}$$

which exhibits ordinates at 10^3 and 10^6 cycles-to-failure of

$$(\bar{S}'_f)_{10^3} = \frac{f^2\bar{S}_{ut}^2 K_6}{K_3^2\bar{S}'_e} 10^{3(1/3)\log(\bar{S}'_e K_3/K_6 f\bar{S}_{ut})} = \frac{f^2\bar{S}_{ut}^2 K_6}{K_3^2\bar{S}'_e} \frac{\bar{S}'_e K_3}{K_6 f\bar{S}_{ut}} = \frac{f\bar{S}_{ut}}{K_3}$$

$$(S'_f)_{10^6} = \frac{f^2\bar{S}_{ut}^2 K_6}{K_3^2\bar{S}'_e} 10^{6(1/3)\log(\bar{S}'_e K_3/K_6 f\bar{S}_{ut})} = \frac{f^2\bar{S}_{ut}^2 K_6}{K_3^2\bar{S}'_e} \frac{\bar{S}'^2_e K_3^2}{K_6^2 f^2\bar{S}_{ut}^2} = \frac{\bar{S}'_e}{K_6}$$

as they should. Now S_f/K_N is equal to $(S'_f)_N$, and using this gives

$$K_N = \frac{K_3^2}{K_6} N^{(-1/3)\log(K_3/K_6)} \tag{7–34}$$

Checking at $N = 10^3$ and 10^6 cycles-to-failure gives

$$K_{10^3} = \frac{K_3^2}{K_6} 10^{3(-1/3)\log(K_3/K_6)} = \frac{K_3^2}{K_6}\frac{K_6}{K_3} = K_3$$

$$K_{10^6} = \frac{K_3^2}{K_6} 10^{6(-1/3)\log(K_3/K_6)} = \frac{K_3^2}{K_6}\frac{K_6^2}{K_3^2} = K_6$$

as it should. An alternative form of Eq. (7–34) is given by

$$(K_f)_N = (K_f')_{10^3} \left[\frac{(K_f)_{10^6}}{(K_f')_{10^3}} \right]^{[(1/3)\log N - 1]} \tag{7–35}$$

EXAMPLE 7–8 Using the results of Ex. 7–6, find the value of $(K_f)_{10^5}$.

Solution From Ex. 7–6, $(K_f)_{10^6} = K_6 = 1.51$, $S_{ut} = 690/6.89 = 100$ kpsi. From Eq. (7-32),

$$q'_{10^3} = -0.18 + 0.43\left(10^{-2}\right)100 - 0.45\left(10^{-5}\right)100^2 = 0.205$$

$$(K_f)_{10^3} = K_3 = 1 + (K_f - 1)\,q'_{10^3} = 1 + (1.51 - 1)0.205 = 1.105$$

From Eq. (7–34),

Answer

$$K_N = \frac{1.105^2}{1.51} 10^{5(-1/3)\log(1.105/1.51)} = \frac{1.105^2}{1.51}\left(\frac{1.51}{1.105}\right)^{5/3} = 1.361$$

or Eq. (7–35),

$$(K_f)_N = 1.05\left[\frac{1.51}{1.105}\right]^{[(1/3)\log 10^5 - 1]} = 1.361$$

7–8 Applying What We Have Learned about Endurance Limit and Endurance Strength

We have examined many ideas that will enable us to quantitatively estimate endurance limits and endurance strengths. It is appropriate to summarize our tools and to demonstrate their value by applying them. Table 7–13 has been prepared to gather in one place the elements of Marin's equation for endurance limit $\mathbf{S}_e$ or endurance strength $\mathbf{S}_f$ of the critical location of a steel machine part, in the geometry and condition of use. Table 7–13 uses customary engineering units. Table 7–14 contains the same information, but the equations are expressed in SI units. The tables are handy to place open before you while you are working on a fatigue problem.

Table 7–13

Summary of Fatigue Equation $\mathbf{S}_e = \mathbf{k}_a k_b \mathbf{k}_c \mathbf{k}_d \mathbf{k}_e \phi_{0.30} \bar{S}_{ut}$ in Customary Engineering Units for Steels*

Quantity	Relation	Table or Equation
Ultimate strength	$S_{ut} = 0.495\mathbf{LN}(1, 0.041)\mathbf{H}_B$	Eq. (5–23)
Fatigue ratio	$\boldsymbol{\phi}_{0.30} = 0.506\mathbf{LN}(1, 0.138)$, $\bar{S}_{ut} \leq 212$ kpsi	Above Eq. (7–20)
Endurance limit		
Bending	$\mathbf{S}'_e = \boldsymbol{\phi}_{0.30}\bar{S}_{ut} = 0.506\bar{S}_{ut}\mathbf{LN}(1.\ 0.138)$, $\bar{S}_{ut} \leq 212$ kpsi	Above Eq. (7–20)
Axial	$(\mathbf{S}'_e)_{ax} = 0.623\bar{S}_{ut}^{-0.0778}\bar{S}_{ut}\mathbf{LN}(1, 0.264)$	Above Eq. (7–20)
Torsion	$\mathbf{S}'_{se} = 0.166\bar{S}_{ut}^{0.125}\bar{S}_{ut}\mathbf{LN}(1, 0.263)$	Above Eq. (7–20)
Surface factor		
Ground	$\mathbf{k}_a = 1.34\bar{S}_{ut}^{-0.086}\mathbf{LN}(1, 0.120)$	Table 7–5
Machined, CR	$\mathbf{k}_a = 2.67\bar{S}_{ut}^{-0.265}\mathbf{LN}(1, 0.058)$	Table 7–5
Hot-rolled	$\mathbf{k}_a = 14.5\bar{S}_{ut}^{-0.719}\mathbf{LN}(1, 0.110)$	Table 7–5
As-forged	$\mathbf{k}_a = 39.8\bar{S}_{ut}^{-0.995}\mathbf{LN}(1, 0.146)$	Table 7-5
Size	$k_b = (d_e/0.30)^{-0.107} = 0.879d_e^{-0.107}$	Eq. (7–10)
Loading factor		
Bending	$\mathbf{k}_c = \mathbf{LN}(1, 0)$	Eq. (7–20)
Axial	$\mathbf{k}_c = 1.23\bar{S}_{ut}^{-0.0778}\mathbf{LN}(1, 0.125)$	Eq. (7–21)
Torsion	$\mathbf{k}_c = 0.328\bar{S}_{ut}^{0.125}\mathbf{LN}(1, 0.125)$	Eq. (7–22)
Temperature factor	$\mathbf{k}_d$ (as applicable)	Eqs. (7–23), (7–24)
Miscellaneous factor	$\mathbf{k}_e$ (as applicable)	
Stress-concentration		
Infinite life	$K_6 = (\mathbf{K}_f)_{10^6} = \dfrac{K_t\mathbf{LN}(1, C_{Kf})}{1 + \dfrac{2}{\sqrt{r}}\dfrac{K_t - 1}{K_t}\sqrt{a}}$	Table 7–12
Finite Life	$\bar{K}_3 = (\bar{K}'_f)_{10^3} = 1 + \left[(\bar{K}_f)_{10^6} - 1\right]\left[-0.18 + 0.43(10^{-2})\bar{S}_{ut} - 0.45\left(10^{-5}\right)\bar{S}_{ut}^2\right]$	Eq. (7–27)
	$\bar{K}_N = \dfrac{\bar{K}_3^2}{\bar{K}_6}N^{(-1/3)\log(\bar{K}_3/\bar{K}_6)}$	Eq. (7–34)

*S'_e and S_{ut} in kpsi, d and d_e in inches.

EXAMPLE 7–9 A 1015 hot-rolled steel bar has been machined to a diameter of 1 in. It is to be placed in rotary bending for 70 000 cycles-to-failure in an operating environment of 550°F.
(*a*) Estimate the endurance limit and endurance strength at the surface of this bar using ASTM "minimum" properties.
(*b* Repeat part (*a*) using mean properties.

Solution (*a*) From Table E–20, $S_{ut} = 50$ kpsi at 70°F. From Table 7–11,

$$\left(\frac{S_T}{S_{RT}}\right)_{550} = \frac{0.995 + 0.963}{2} = 0.979$$

From Eq. (7–4),

$$(S_{ut})_{550} = 0.979(50) = 49.0 \text{ kpsi}$$

$$\left(S'_e\right)_{550} = \phi_{0.30} S_{ut550} = 0.506(49) = 24.8 \text{ kpsi}$$

From Table 7–5,

$$k_a = 2.67(49)^{-0.265} = 0.952$$

From Eq. (7–10),

$$k_b = (d/0.30)^{-0.107} = (1/0.30)^{-0.107} = 0.879$$

$$k_c = k_d = k_e = 1$$

From Eq. (7–8),

$$(S_e)_{550} = 0.952(0.879)24.8 = 20.8 \text{ kpsi}$$

From Sec. 7–5, Eq. (*b*),

$$\sigma'_F = 49 + 50 = 99 \text{ kpsi}$$

From Sec. 7–5, Eq. (*c*),

$$b = \frac{-\log\left(\sigma'_F/S_e\right)}{\log\left(2N_e\right)} = -\frac{\log(99/20.8)}{\log\left(2\cdot 10^6\right)} = -0.108$$

From Sec. 7–5, Eq. (*a*),

$$f = \frac{\sigma'_F}{S_{ut}}\left(2\cdot 10^3\right)^b = \frac{99}{49}\left(2\cdot 10^3\right)^{-0.108} = 0.889$$

From Eq. (7–5),

$$a = \frac{f^2 S_{ut}^2}{S_e} = \frac{0.889^2 49^2}{20.9} = 91.2 \text{ kpsi}$$

Answer From Eq. (7–6),

$$\left(S_f\right)_{550} = aN^b = 91.2(70\,000)^{-0.108} = 27.3 \text{ kpsi}$$

(*b*) From Table E–21, $\bar{S}_{ut} = 61$ kpsi at 70°F. From part (*a*), $(S_T/S_{RT})_{550} = 0.979$. From Eq. (7–4),

$$\left(\bar{S}_{ut}\right)_{550} = 0.979(61) = 59.7 \text{ kpsi}$$

$$\left(\bar{S}'_e\right)_{550} = 0.506(59.7) = 30.2 \text{ kpsi}$$

$$\bar{k}_a = 2.67(59.7)^{-0.265} = 0.903$$

$$k_b = 0.879 \qquad k_c = k_d = k_e = 1$$

as in part (*a*). Then from Eq. (7–8),

$$\left(\bar{S}'_e\right)_{550} = 0.903(0.879)30.2 = 24.0 \text{ kpsi}$$

From Sec. 7–5, Eq. (*b*),

$$\sigma'_F = 59.7 + 50 = 109.7 \text{ kpsi}$$

From Sec. 7–5, Eq. (*c*),

$$b = -\frac{\log\left(\sigma'_F/S_e\right)}{\log\left(2N_e\right)} = -\frac{\log(109.7/24)}{\log\left(2\cdot 10^6\right)} = -0.105$$

From Sec. 7–5, Eq. (a)

$$f = \frac{\sigma'_F}{S_{ut}}\left(2\cdot 10^3\right)^b = \frac{109.7}{59.7}\left(2\cdot 10^3\right)^{-0.105} = 0.827$$

From Eq. (7–5),

$$a = \frac{f^2 S_{ut}^2}{S_e} = \frac{0.827^2\ 59.7^2}{24} = 101.6\ \text{kpsi}$$

From Eq. (7–6),

Answer $(S_f)_{550} = aN^b = 101.6(70\ 000)^{0.105} = 31.5$ kpsi

Table 7–14

Summary of Fatigue Equations $\mathbf{S}_e = \mathbf{k}_a k_b \mathbf{k}_c \mathbf{k}_d \mathbf{k}_e \boldsymbol{\phi}_{0.30}\bar{S}_{ut}$ in SI Units for Steels*

Quantity	Relation	Table or Equation
Ultimate strength	$\mathbf{S}_{ut} = 3.41(1, 0.041)\mathbf{H}_B$	Eq. (5–23)
Fatigue ratio	$\boldsymbol{\phi}_{0.30} = 0.506\mathbf{LN}(1, 0.138),\ \bar{S}_{ut} \le 1460$ MPa	
Endurance limit		
Bending	$\mathbf{S}'_e = \boldsymbol{\phi}_{0.30}\bar{S}_{ut} = 0.506\bar{S}_{ut}\mathbf{LN}(1, 0.138),\ \bar{S}_{ut} \le 1460$ MPa	
Axial	$(\mathbf{S}'_e)_{ax} = 0.724\bar{S}_{ut}^{-0.0778}\bar{S}_{ut}\mathbf{LN}(1, 0.264)$	Above Eq. (7–20)
Torsion	$\mathbf{S}'_{se} = 0.130\bar{S}_{ut}^{0.125}\bar{S}_{ut}\mathbf{LN}(1, 0.263)$	Above Eq. (7–20)
Surface factor		
Ground	$\mathbf{k}_a = 1.58\bar{S}_{ut}^{-0.086}\mathbf{LN}(1, 0.120)$	Table 7–5
Machined, CR	$\mathbf{k}_a = 4.45\bar{S}_{ut}^{-0.265}\mathbf{LN}(1, 0.058)$	Table 7–5
Hot-rolled	$\mathbf{k}_a = 58.1\bar{S}_{ut}^{-0.719}\mathbf{LN}(1, 0.110)$	Table 7–5
As-forged	$\mathbf{k}_a = 271\bar{S}_{ut}^{-0.995}\mathbf{LN}(1, 0.045)$	Table 7–5
Size	$k_b = (d_e/7.62)^{-0.107} = 1.24d_e^{-0.107}$	Eq. (7–10)
Loading factor		
Bending	$\mathbf{k}_c = \mathbf{LN}(1, 0)$	Eq. (7–20)
Axial	$\mathbf{k}_c = 1.43\bar{S}_{ut}^{-0.0778}\mathbf{LN}(1, 0.125)$	Eq. (7–21)
Torsion	$\mathbf{k}_c = 0.258\bar{S}_{ut}^{0.125}\mathbf{LN}(1, 0.125)$	Eq. (7–22)
Temperature factor	$\mathbf{k}_d$ (as applicable)	Eqs. (7–23), (7–24)
Miscellaneous factor	$\mathbf{k}_e$ (as applicable)	
Stress-concentration		
Infinite life	$\mathbf{K}_6 = (\mathbf{K}_f)_{10^6} = \dfrac{K_t\mathbf{LN}(1, C_{Kf})}{1 + \dfrac{2}{\sqrt{r}}\dfrac{K_t - 1}{K_t}\sqrt{a}}$	Table 7–12
Finite life	$\bar{K}_3 = (\bar{K}'_f)_{10^3} = 1 + \left[(\bar{K}_f)_{10^6} - 1\right]\left[-0.18 + 0.624(10^{-3})\bar{S}_{ut} - 0.948(10^{-7})\bar{S}_{ut}^2\right]$	Eq. (7–27)
	$\bar{K}_N = \dfrac{\bar{K}_3^2}{\bar{K}_6}N^{(-1/3)\log(\bar{K}_3/\bar{K}_6)}$	Eq. (7–34)

* S'_e and S_{ut} in MPa, d and d_e in mm.

EXAMPLE 7–10 Figure 7–11*a* shows a rotating axle supported in ball bearings at *A* and *D* and loaded by a nonrotating force *F* of 6.8 kN. Using ASTM "minimum" strengths, estimate the life of the part.

Solution From Fig. 7–11*b* we learn that failure will probably occur at *B* rather than at *C* or at the point of maximum moment. Point *B* has a smaller cross section, a higher bending moment, and a higher stress-concentration factor than *C*, and the location of maximum moment has a larger size and no stress-concentration factor.

We shall solve the problem by first estimating the strength at point *B*, since the strength will be different elsewhere, and comparing this strength with the stress at the same point.

From Table E–20 we find $S_{ut} = 690$ MPa and $S_y = 580$ MPa. The endurance limit S'_e is estimated as

$$S'_e = 0.506(690) = 349.1 \text{ MPa}$$

From Table 7–5,

$$k_a = 4.45(690)^{-0.265} = 0.787$$

From Eq. (7–10),

$$k_b = (32/7.62)^{-0.107} = 0.858$$

To find the geometric stress-concentration factor K_t we enter Table E–15–9 with $D/d = 38/32 = 1.1875$ and $r/d = 3/32 = 0.093\ 75$ and read $K_t = 1.65$. From Table 7–12, $\sqrt{a} = 139/690 = 0.201$. From Eq. (7–27),

$$K_f = \frac{K_t}{1 + \dfrac{2}{\sqrt{r}}\dfrac{K_t - 1}{K_t}\sqrt{a}} = \frac{1.65}{1 + \dfrac{2}{\sqrt{3}}\dfrac{1.65 - 1}{1.65}0.201} = 1.51$$

This stress-concentration factor applies to 10^6 cycles or more.
Since $k_c = k_d = k_e = 1$,

$$S_e = 0.787(0.858)341.1 = 235.7 \text{ MPa}$$

The next step is to estimate the bending stress at point *B*. The bending moment is

$$M_B = R_1 x = \frac{225F}{550}250 = \frac{225(6.8)}{550}250 = 695 \text{ MPa}$$

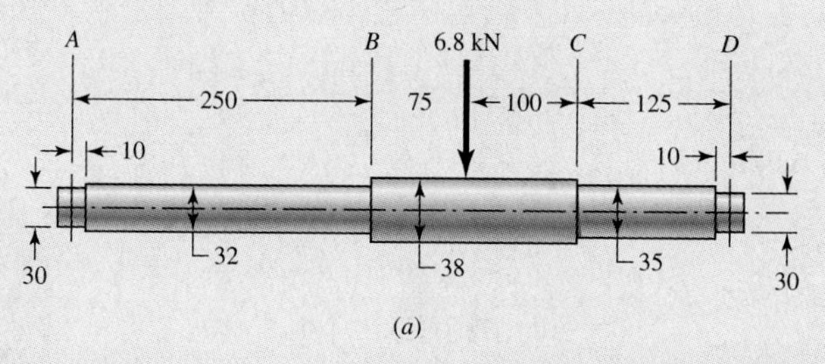

(*a*)

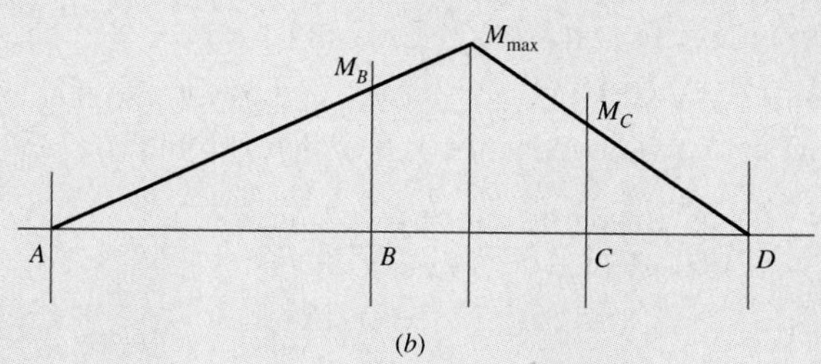

(*b*)

Figure 7–11

(*a*) Shaft drawing showing all dimensions in millimeters; all fillets 3-mm radius. The shaft rotates and the load is stationary; material is machined from AISI 1060 cold-drawn steel. (*b*) Bending-moment diagram.

The section modulus is $I/c = \pi d^3/32 = \pi 3.2^3/32 = 3.22\ \text{cm}^3$. The bending stress is, assuming infinite life,

$$\sigma = K_f \frac{M_B}{I/c} = \frac{1.51(695)}{3.22} = 325.9\ \text{MPa}$$

This stress is greater than S_e and less than S_y. It means we have both finite life and no yielding on the first cycle. The fatigue stress-concentration factor K_N is unknown because the cycles-to-failure N is unknown. Because we don't know K_N, instead of applying it to stress σ we will correct our endurance strength by reducing it. Our stress is

$$\sigma = \frac{M_B}{I/c} = \frac{695}{3.22} = 215.8\ \text{MPa}$$

From Table 7–14, $K_6 = 1.51$ and

$$K_3 = \left(K'_f\right)_{10^3} = 1 + \left[(K'_f)_{10^6} - 1\right]\left[-0.18 + 0.624\left(10^{-3}\right)690 - 0.948\left(10^{-7}\right)690^2\right]$$
$$= 1 + (1.51 - 1)0.205 = 1.105$$

From Sec. 7–5, Eq. (*b*),

$$\sigma'_F = S_{ut} + 345 = 690 + 345 = 1035\ \text{MPa}$$

From Sec. 7–5, Eq. (*c*),

$$b = -\frac{\log\left(\sigma'_F/S_e\right)}{\log\left(2N_e\right)} = -\frac{\log(1035/235.7)}{\log\left(2\cdot 10^6\right)} = -0.102$$

From Sec. 7–5, Eq. (*a*),

$$f = \frac{\sigma'_F}{S_{ut}}\left(2\cdot 10^3\right)^b = \frac{1035}{690}\left(2\cdot 10^3\right)^{-0.102} = 0.691$$

From Eq. (7–5),

$$a = \frac{f^2 S_{ut}^2}{S_e} = \frac{0.691^2 690^2}{235.7} = 964.5\ \text{MPa}$$

With $a = 964.5$ MPa and $b = -0.102$ we can describe the material corrected for surface and size with no stress concentration:

$$\left(S_f\right)_{10^6} = 964.5\left(10^6\right)^{-0.102} = 235.7\ \text{MPa}$$

$$\left(S_f\right)_{10^3} = 964.5\left(10^3\right)^{-0.102} = 476.8\ \text{MPa}$$

From Eq. (7–33),

$$a' = \frac{f^2 S_{ut}^2 K_6}{K_3^2 S_e} = a\frac{K_6}{K_3^2} = 964.5\frac{1.51}{1.105^2} = 1192.8\ \text{MPa}$$

$$b' = \frac{1}{3}\log\frac{235.7(1.105)}{1.51(0.691)690} = -0.147$$

making

$$S_f = 1192.8 N^{-0.147}$$

then

$$\left(S_f\right)_{10^6} = 1192.8\left(10^6\right)^{-0.147} = 156.5\ \text{MPa}$$

$$\left(S_f\right)_{10^3} = 1192.8\left(10^3\right)^{-0.147} = 432.1\ \text{MPa}$$

With $476.8/432.1 = 1.103 = K_3$ and $235.7/156.5 = 1.52 = K_6$ we have an indication that our strength reduction is complete. From Eq. (7–34),

$$K_N = \frac{K_3^2}{K_6} N^{(-1/3)\log(1.105/1.51)} = 0.809 N^{0.045}$$

For a stress level of $\sigma_a = 215.8$ MPa, the cycles-to-failure is

$$N = \left(\frac{\sigma_a}{1192.8}\right)^{-1/b} = \left(\frac{215.8}{1192.8}\right)^{-1/0.147} = 0.112\left(10^6\right) \text{ cycles}$$

and

Answer $K_N = 0.809(0.112\ 10^6)^{0.045} = 1.365$

Had we known K_N at the outset, the life would be predicted by

$$N = \left(\frac{\sigma_a}{a}\right)^{-1/b} = \left(\frac{1.365(215.8)}{964.5}\right)^{-1/0.102} = 0.112(10^6) \text{ cycles}$$

Example 7–10 showed a finite life problem with stress concentration, and the complication that ensues, and an efficient procedure for modifying the strength. It is also an illustration of the use of ASTM "minimum" property analysis. It is a deterministic analysis. There is no clue that the cycles-to-failure prediction has a large coefficient of variation.

EXAMPLE 7–11 The bar shown in Fig. 7–12 is machined from a cold-rolled flat having an ultimate strength of $\mathbf{S}_{ut} \sim \mathbf{LN}(87.6, 5.74)$ kpsi. The axial load shown is completely reversed. The load amplitude is $\mathbf{F}_a \sim \mathbf{LN}(1000, 120)$ lb.

(*a*) Estimate the reliability.

(*b*) Reestimate the reliability when a rotating bending endurance test shows that $\mathbf{S}'_e \sim \mathbf{LN}(40, 2)$ kpsi.

Solution (*a*) $\mathbf{S}'_e = 0.506 S_{ut}\mathbf{LN}(1, 0.138) = 0.506(87.6)\mathbf{LN}(1, 0.138)$
$= 44.2\mathbf{LN}(1, 0.138)$ kpsi

From Table 7–5,

$$\mathbf{k}_a = 2.67\bar{S}_{ut}^{-0.265}\mathbf{LN}(1, 0.058) = 2.67(87.6)^{-0.265}\mathbf{LN}(1, 0.058)$$
$$= 0.816\mathbf{LN}(1, 0.058)$$

$k_b = 1$ (axial loading)

From Table 7–7,

$$\mathbf{k}_c = 1.23\bar{S}_{ut}^{-0.078}\mathbf{LN}(1, 0.125) = 1.23(87.6)^{-0.078}\mathbf{LN}(1, 0.125)$$
$$= 0.868\mathbf{LN}(1, 0.125)$$

$$\mathbf{k}_d = \mathbf{k}_e = (1, 0)$$

Figure 7–12

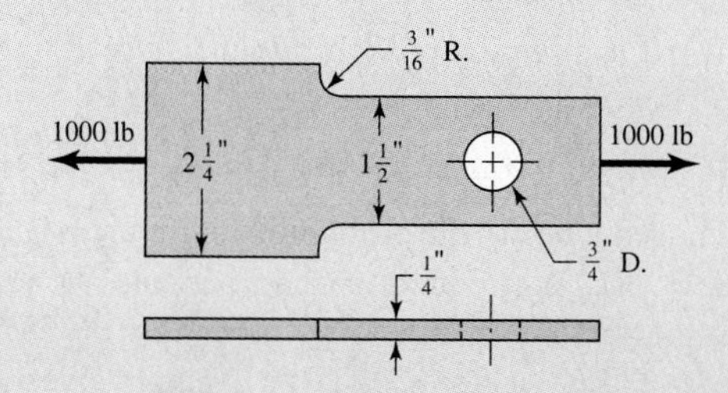

The endurance strength, from Eq. (7–8), is

$$\mathbf{S}_e = \mathbf{k}_a k_b \mathbf{k}_c \mathbf{k}_d \mathbf{k}_e \mathbf{S}'_e$$

$$\mathbf{S}_e = 0.816(1, 0.058)(1)0.868(1, 0.125)(1)(1)44.2(1, 0.138)$$

where all variates are lognormal. The parameters are

$$\bar{S}_e = 0.816(0.868)44.2 = 31.3 \text{ kpsi}$$

$$C_{Se} = \left(0.058^2 + 0.125^2 + 0.138^2\right)^{1/2} = 0.195$$

so $\mathbf{S}_e \sim 31.3\mathbf{LN}(1, 0.195)$ kpsi.

In computing the stress the section at the hole governs. Using the terminology of Table E–15–1 we find $d/w = 0.50$, therefore $K_t = 2.18$. From Table 7–12, $\sqrt{a} = 5/S_{ut} = 5/87.6 = 0.057$, $r = 0.375$ in, $C_{kf} = 0.10$. From Eq. (7–27)

$$\mathbf{K}_f = \frac{2.18\mathbf{LN}(1, 0.10)}{1 + \dfrac{2}{\sqrt{0.375}}\dfrac{2.18 - 1}{2.18}0.057} = 1.98\text{LN}(1, 0.10)$$

The stress at the hole is

$$\boldsymbol{\sigma} = \mathbf{K}_f \frac{\mathbf{F}}{A} = 1.98\mathbf{LN}(1, 0.10)\frac{1000\mathbf{LN}(1, 0.12)}{0.25(0.75)}$$

$$\bar{\sigma} = 1.98\frac{1000}{0.25(0.75)}10^{-3} = 10.56 \text{ kpsi}$$

$$C_\sigma = \left(0.10^2 + 0.12^2\right)^{1/2} = 0.156$$

so stress can be expressed as $\boldsymbol{\sigma} \sim 10.56\mathbf{LN}(1, 0.156)$ kpsi.

The endurance limit is considerably greater than the load-induced stress, indicating that finite life is not a problem. Using an approximation of Eq. (2-55),

$$z = -\frac{\ln \bar{S}/\bar{\sigma}}{\left(C_S^2 + C_\sigma^2\right)^{1/2}} = -\frac{\ln 31.3/10.56}{\left(0.195^2 + 0.156^2\right)^{1/2}} = -4.35$$

From Table E–10 the probability of failure $p_f = \Phi(-4.35) = .000\,0.007\,98$, and the reliability is

Answer $\quad R = 1 - 0.000\,007\,98 = 0.999\,992\,02$

(*b*) The rotary endurance tests are described by $\mathbf{S}'_e \sim 40\mathbf{LN}(1, 0.05)$ kpsi whose mean is *less* than the predicted mean in part *a*. The mean endurance strength $\bar{S}_e$ is

$$\bar{S}_e = 0.816(0.868)40 = 28.3 \text{ kpsi}$$

$$C_{Se} = \left(0.058^2 + 0.125^2 + 0.05^2\right)^{1/2} = 0.147$$

so the endurance strength can be expressed as $\mathbf{S}_e \sim 28.3\mathbf{LN}(1, 0.147)$ kpsi.
From Eq. (2-55),

$$z = -\frac{\ln(28.3/10.56)}{\left(0.147^2 + 0.156^2\right)^{1/2}} = -4.6$$

Using Table E–10, we see the probability of failure $P_f = \Phi(-4.69) = 0.000\,002\,11$, and

$$R = 1 - 0.000\,003\,40 = 0.999\,997\,89$$

an increase! The reduction in the probability of failure is $(0.000\,002\,11 - 0.000\,007\,98)/0.000\,007\,98 = -0.73$, a reduction of 73 percent. We are analyzing an extant design, so in part (*a*) the factor of safety was $\bar{n} = \bar{S}/\bar{\sigma} = 31.3/10.56 = 2.96$. In part (*b*) $\bar{n} = 28.3/10.56 = 2.68$, a *decrease*. This example gives you the opportunity to see the role of the design factor. Given knowledge of $\bar{S}$, C_S, $\bar{\sigma}$, C_σ, and reliability (through z), the mean factor of safety (as a design factor) separates $\bar{S}$ and $\bar{\sigma}$ so that the reliability goal is achieved. Knowing $\bar{n}$ alone *says nothing about the probability of failure*. Looking at $\bar{n} = 2.96$ and $\bar{n} = 2.68$ says nothing about the respective probabilities of failure. Rewriting the approximation to Eq. (2-55) used in this example we obtain

$$\bar{S} = \bar{\sigma} \exp\left[-z\left(C_S^2 + C_\sigma^2\right)^{1/2}\right]$$

Since $\bar{\sigma}$ and $\bar{C}_\sigma$ are fixed, by the loading, then with $z = -4.28$, the preceding equation is a trade-off between C_S and $\bar{S}$ for the same reliability. After the rotating-bending tests C_S was reduced, so

$$\bar{S}_e = 10.56 \exp\left[-(-4.35)\sqrt{0.147^2 + 0.156^2}\right] = 26.82 \text{ kpsi}$$

In other words, the reduction in C_S can also be accompanied by a reduction of $\bar{S}_e$ to as low as 27 kpsi without reducing the reliability. Since the tests did not reduce $\bar{S}_e$ that much (from 31.3 to 28.3 kpsi), the reliability was *increased*. When a mean design factor (or mean factor of safety) defined as $\bar{S}_e/\bar{\sigma}$ is said to be *silent* on matters of frequency of failures, it means that a scalar factor of safety by itself does not offer any information about probability of failure. Nevertheless, some engineers let the factor of safety speak up, and they can be wrong in their conclusions.

As revealing as Ex. 7–11 is concerning the meaning (and lack of meaning) of a design factor or factor of safety, let us remember that the rotary testing associated with part (*b*) changed *nothing* about the part, but only our knowledge about the part. The mean endurance limit was 40 kpsi all the time, and our adequacy assessment had to move with what was known.

7–9 The Distributions

We have given both deterministic and stochastic substance to the fatigue modification factors of Eq. (7–8), also called the *Marin modification factors*. The question of the distribution of the endurance strength $\mathbf{S}_e$ is answered by the central limit theorem of statistics. The individual distributional nature of the elements of the product is not vital, for the distribution of the multiple-element product approaches lognormal regardless. Since elements are (1) continuous, (2) not seriously skewed, and (3) several in number, it would be difficult to detect deviation from lognormality. The presumption of lognormality in this circumstance is said to be *robust*. If the elements of a product are lognormal themselves, then the product is lognormal *exactly* for two, three, four, or more elements. In this case the multiplicative operation is said to be *closed*. In finite-life fatigue, the fatigue strength S'_f is lognormal, because in the interval 10^3 to 10^6 cycles-to-failure, loglog transformed data are rectified and normally distributed about the regression line, and the inverse transformation makes $\mathbf{S}'_f$ lognormal.

In strength-limited design, a load-induced stress is related to a fatigue strength. When the problem involves an axial tension in a notched round rod, the load-induced stress $\boldsymbol{\sigma}$ can be expressed as

$$\boldsymbol{\sigma} = \mathbf{K}_f\sigma_0 = \mathbf{K}_f\frac{\mathbf{F}}{\mathbf{A}} = \frac{4}{\pi}\mathbf{K}_f\mathbf{F}\left(\frac{1}{\mathbf{d}}\right)^2$$

or, for bending of a notched flat bar,

$$\boldsymbol{\sigma} = \mathbf{K}_f\sigma_0 = \mathbf{K}_f\frac{6\mathbf{M}}{\mathbf{bh}^2} = 6\mathbf{K}_f\frac{\mathbf{Fx}}{\mathbf{bh}^2}$$

Note that there are three stochastic factors in the first case and five in the second. If all are lognormal, the stress $\boldsymbol{\sigma}$ is lognormal. If the factors are continuous, not seriously skewed, and several in number, the stress $\boldsymbol{\sigma}$ is robustly lognormal. If there is doubt, and the individual distributions are known, a computer simulation will comment on the robustness and also answer statistical questions. In load-induced stress $\boldsymbol{\sigma}$, when machining causes the geometric factors to have very small coefficients of variation, the load dominates, or the load and stress-concentration factor $\mathbf{K}_f$ dominate because of the controlling coefficient of variation.

7–10 Characterizing Fluctuating Stresses

Fluctuating stresses in machinery often take the form of a sinusoidal pattern because of the nature of some rotating machinery. However, other patterns, some quite irregular, do occur. It has been found that in periodic patterns exhibiting a single maximum and a single minimum of force, the shape of the wave is not important, but the peaks on both the high side (maximum) and the low side (minimum) are important. Thus $F_{\max}$ and $F_{\min}$ in a cycle of force can be used to characterize the force pattern. It is also true that ranging above and below some baseline can be equally effective in characterizing the force pattern. If the largest force is $F_{\max}$ and the smallest force is $F_{\min}$, then a steady component and an alternating component can be constructed as follows:

$$F_m = \frac{F_{\max} + F_{\min}}{2} \qquad F_a = \left| \frac{F_{\max} - F_{\min}}{2} \right|$$

where F_m is the midrange component of force, and F_a is the amplitude component of force.

Figure 7–13 illustrates some of the various stress-time traces which occur. The components of stress, some of which are shown in Fig. 7–13d, are

$\sigma_{\min}$ = minimum stress

$\sigma_{\max}$ = maximum stress

σ_a = amplitude component

σ_m = midrange component

σ_r = range of stress

σ_s = static or steady stress

The steady, or static, stress is *not* the same as the midrange stress; in fact, it may have any value between $\sigma_{\min}$ and $\sigma_{\max}$. The steady stress exists because of a fixed load or preload applied to the part, and it is usually independent of the varying portion of the load. A helical compression spring, for example, is always loaded into a space shorter than the free length of the spring. The stress created by this initial compression is called the steady, or static, component of the stress. It is not the same as the midrange stress.

We shall have occasion to apply the subscripts of these components to shear stresses as well as normal stresses.

The following relations are evident from Fig. 7–13:

$$\sigma_m = \frac{\sigma_{\max} + \sigma_{\min}}{2} \qquad\qquad (7\text{–}36)$$
$$\sigma_a = \left| \frac{\sigma_{\max} - \sigma_{\min}}{2} \right|$$

Equations (7–36) utilize symbols σ_a and σ_m as the stress components at the location under scrutiny. This means, in the absence of a notch, σ_a and σ_b are equal to the nominal stresses σ_{ao} and σ_{mo} induced by loads F_a and F_m, respectively; in the presence of a notch they are $K_f \sigma_{ao}$ and $K_f \sigma_{mo}$, respectively, as long as the material remains without plastic strain. In other words, the fatigue stress concentration factor K_f is applied to *both* components.

When the steady stress component is high enough to induce localized notch yielding, the designer has a problem. The first-cycle local yielding produces plastic strain and

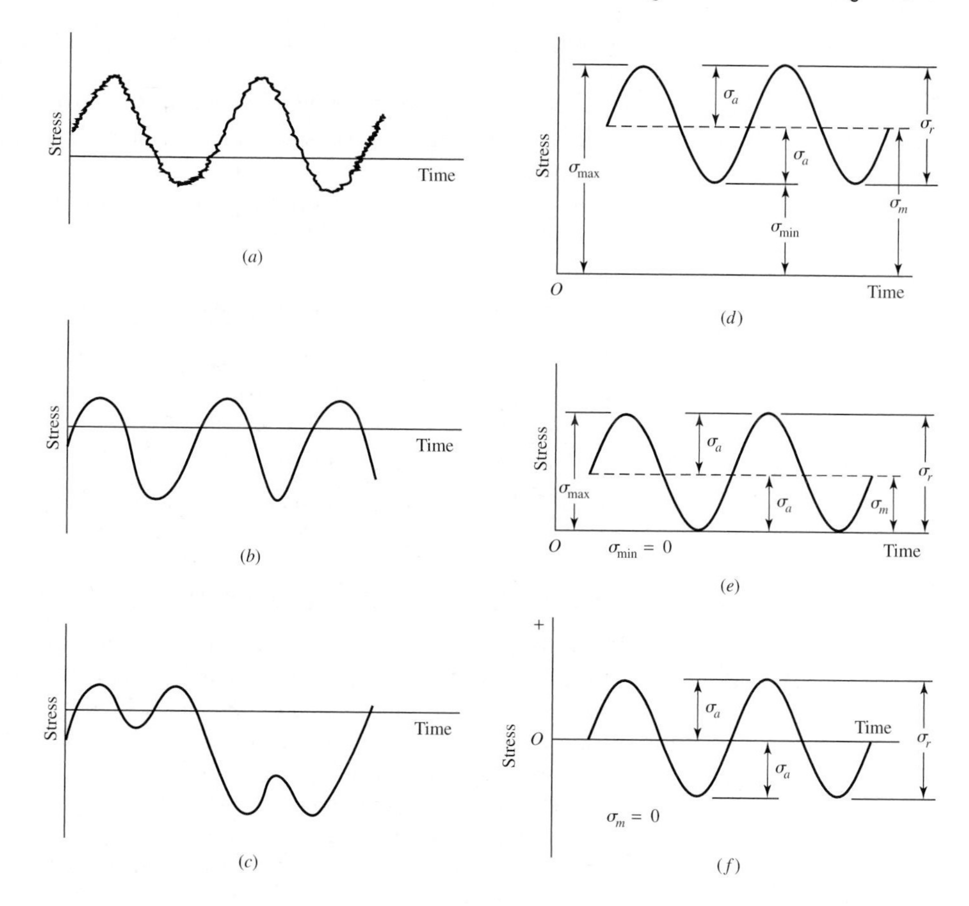

Figure 7–13

Some stress-time relations: (*a*) fluctuating stress with high-frequency ripple; (*b* and *c*) nonsinusoidal fluctuating stress; (*d*) sinusoidal fluctuating stress; (*e*) repeated stress; (*f*) repeated stress; (*f*) completely reversed sinusoidal stress.

strain-strengthening. This is occurring at the location when fatigue crack nucleation and growth are most likely. The material properties (S_y and S_{ut}) are new and difficult to quantify. The prudent engineer controls the concept, material and condition of use, and geometry so that no plastic strain occurs. There are discussions concerning possible ways of quantifying what is occurring under localized and general yielding in the presence of a notch, referred to as the *nominal mean stress* method, *residual stress* method, and the like.[20] The nominal mean stress method (set $\sigma_a = K_f \sigma_{ao}$ and $\sigma_m = \sigma_{mo}$) gives roughly comparable results to the residual stress method, but both are *approximations*.

There is the method of Dowling[21] for ductile material, which, for materials with a pronounced yield point and approximated by an elastic–perfectly plastic behavior model, quantitatively expresses the steady stress component stress-concentration factor K_{fm} as

$$\begin{aligned} K_{fm} &= K_f & K_f|\sigma_{\max,o}| &< S_y \\ K_{fm} &= \frac{S_y - K_f \sigma_{ao}}{|\sigma_{mo}|} & K_f|\sigma_{\max,o}| &> S_y \\ K_{fm} &= 0 & K_f|\sigma_{\max,o} - \sigma_{\min,o}| &> 2S_y \end{aligned} \tag{7–37}$$

20. R. C. Juvinall, *Stress, Strain, and Strength*, McGraw-Hill, New York, 1967, articles 14.9–14.12; R. C. Juvinall and K. M. Marshek, *Fundamentals of Machine Component Design*, Wiley, New York, 1991, Sec. 8.11; M. E. Dowling, *Mechanical Behavior of Materials*, Prentice Hall, Englewood Cliffs, N.J., 1993, Secs. 10.3–10.5.

21. Dowling, op. cit., p. 415.

For the purposes of this book, for ductile materials in fatigue,

- Avoid localized plastic strain at a notch. Set $\sigma_a = K_f \sigma_{a,o}$ and $\sigma_m = K_f \sigma_{mo}$.
- When plastic strain at a notch cannot be avoided, and a Dowling material is involved, use Eqs. (7–37).
- If a non-Dowling material is involved, set $\sigma_a = K_f \sigma_{ao}$, and use $K_{mf} = 1$, that is, $\sigma_m = \sigma_{mo}$.

In addition to Eq. (7–36), the *stress ratio*

$$R = \frac{\sigma_{\min}}{\sigma_{\max}} \tag{7–38}$$

and the stress ratio

$$A = \frac{\sigma_a}{\sigma_m} \tag{7–39}$$

are also defined and used in connection with fluctuating stresses.

7–11 Failure Loci under Variable Stresses

Now that we have defined the various components of stress associated with a part subjected to fluctuating stress, we want to vary both the midrange stress and the stress amplitude, or alternating component, to learn something about the fatigue resistance of parts when subjected to such situations. Three methods of plotting the results of such tests are in general use and are shown in Figs. 7–14, 7–15, and 7–16.

The early viewpoint expressed on a $\sigma_a \sigma_m$ diagram was that there existed a locus which divided safe from unsafe combinations of σ_a, σ_m. Ensuing proposals included

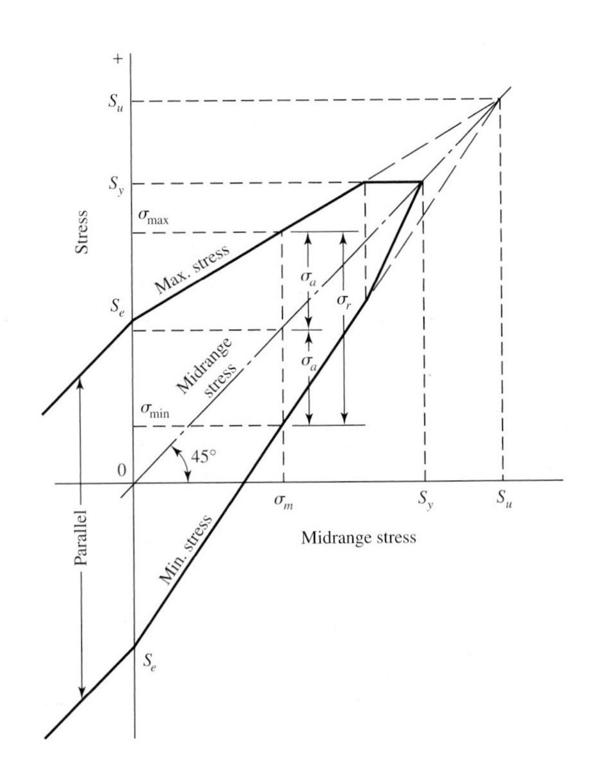

Figure 7–14

Modified Goodman diagram showing all the strengths and the limiting values of all the stress components for a particular midrange stress.

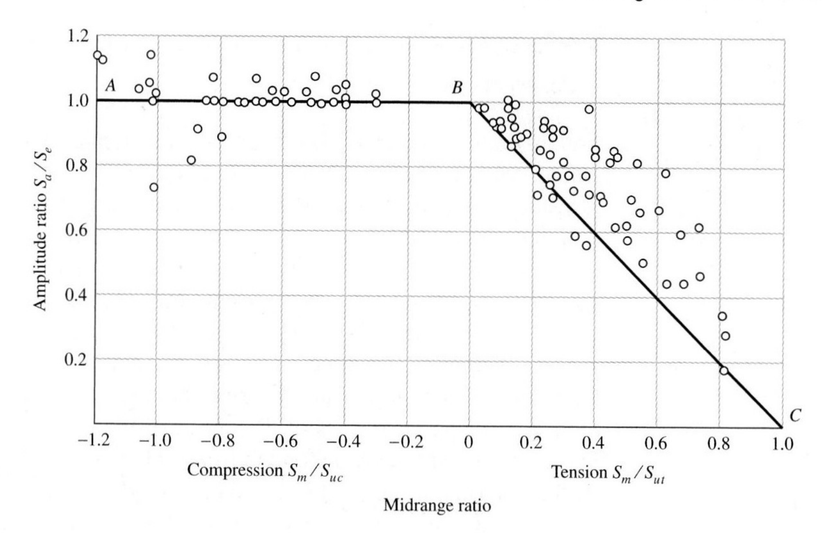

Figure 7–15

Plot of fatigue failures for midrange stresses in both tensile and compressive regions. Normalizing the data by using the ratio of steady strength component to tensile strength S_m/S_{ut}, steady strength component to compressive strength S_m/S_{uc}, and strength amplitude component to endurance limit S_m/S'_e enables a plot of experimental results for a variety of steels. [Data source: Thomas J. Dolan, "Stress Range," Sec. 6.2 in O. J. Horger (ed.), *ASME Handbook—Metals Engineering Design*, McGraw-Hill, New York, 1953.]

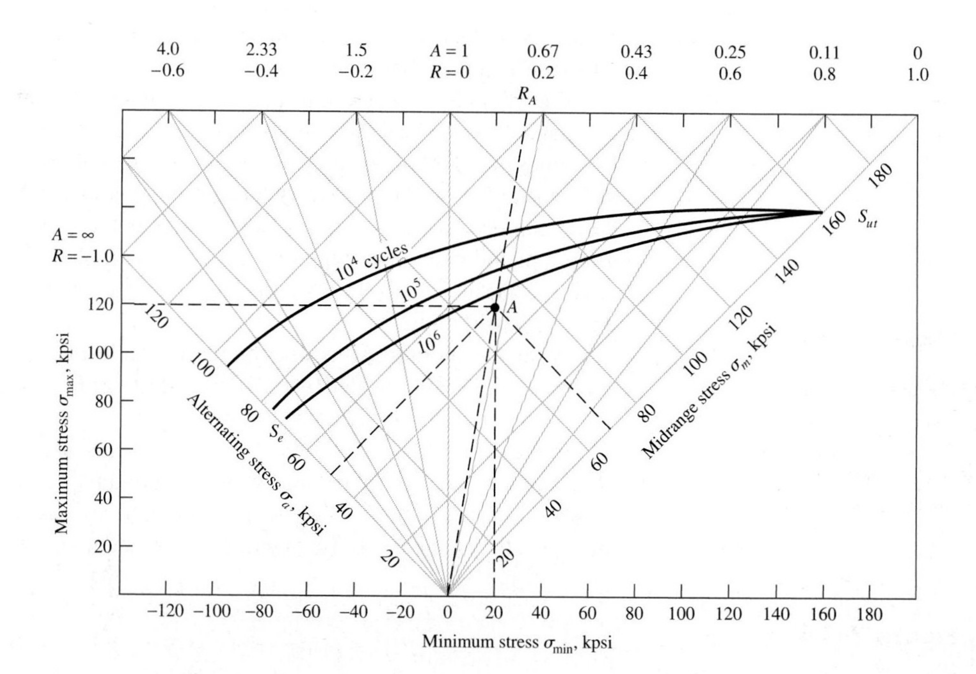

Figure 7–16

Master fatigue diagram created for AISI 4340 steel having S_{ut} = 147 kpsi. The stress components at A are σ_{min} = 20, σ_{max} = 120, σ_m = 70, and σ_a = 50, all in kpsi. (Source: H. J. Grover, *Fatigue of Aircraft Structures*, U.S. Government Printing Office, Washington, D.C., 1966, pp. 317, 322. See also J. A. Collings, *Failure of Materials in Mechanical Design*, Wiley, New York, 1981, p. 216.)

the parabola of Gerber (1874), the Goodman (1890)[22] (straight) line, and the Soderberg (1930) (straight) line. As more data were generated it became clear that a fatigue locus, rather than being a "fence," was more like a zone or band wherein the probability of failure could be estimated. We include the failure locus of Goodman because

- It is a straight line and the algebra is linear and easy.
- It is easily graphed, every time for every problem.

22. It is difficult to date Goodman's work because it went through several modifications and was never published.

- It reveals subtleties of insight into fatigue problems.
- Answers can be scaled from the diagrams as a check on the algebra.

We also caution that it is deterministic and the phenomenon is not. It is biased and we cannot quantify the bias. It is not conservative. It is a steppingstone to understanding; it is history; and to read the work of other engineers and to have meaningful oral exchanges with them, it is necessary that you understand the Goodman approach should it arise.

The *modified Goodman diagram* of Fig. 7–14 has the midrange stress plotted along the abscissa and all other components of stress plotted on the ordinate, with tension in the positive direction. The endurance limit, fatigue strength, or finite-life strength, whichever applies, is plotted on the ordinate above and below the origin. The midrange-stress line is a 45° line from the origin to the tensile strength of the part. The modified Goodman diagram consists of the lines constructed to S_e (or S_f) above and below the origin. Note that the yield strength is also plotted on both axes, because yielding would be the criterion of failure if $\sigma_{\max}$ exceeded S_y.

Another way to display test results is shown in Fig. 7–15. Here the abscissa represents the ratio of the midrange strength to the ultimate strength, with tension plotted to the right and compression to the left. The ordinate is the ratio of the alternating strength to the endurance limit. The line BC then represents the modified Goodman criterion of failure. Note that the existence of midrange stress in the compressive region has little effect on the endurance limit.

The very clever diagram of Fig. 7–16 is unique in that it displays four of the stress components as well as the two stress ratios. A curve representing the endurance limit for values of R beginning at $R = -1$ and ending with $R = 1$ begins at S_e on the σ_a axis and ends at S_{ut} on the σ_m axis. Constant-life curves for $N = 10^5$ and $N = 10^4$ cycles have been drawn too. Any stress state, such as the one at A, can be described by the minimum and maximum components, or by the midrange and alternating components. And safety is indicated whenever the point described by the stress components lies below the constant-life line.

When the midrange stress is compression, failure occurs whenever $\sigma_a = S_e$ or whenever $\sigma_{\max} = S_{yc}$, as indicated by the left-hand side of Fig. 7–15. Neither a fatigue diagram nor any other failure criteria need be developed.

In Fig. 7–17, the tensile side of Fig. 7–15 has been redrawn using strengths, instead of strength ratios, with the same modified Goodman criterion together with three additional criteria of failure. Such diagrams are often constructed for analysis and design purposes; they are easy to use and the results can be scaled off directly.

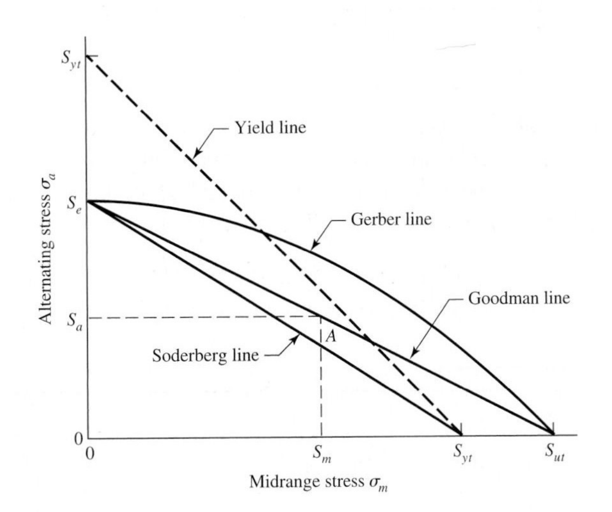

Figure 7–17

Fatigue diagram showing various criteria of failure. For each criterion, points on or "above" the respective line indicate failure. Some point A on the Goodman line, for example, gives the strength S_m as the limiting value of σ_m corresponding to the strength S_a, which, paired with σ_m, is the limiting value of σ_a.

Either the fatigue limit S_e or the finite-life strength S_f is plotted on the ordinate of Fig. 7–17. These values will have already been corrected using the Marin factors of Eq. (7–13). Note that the yield strength S_{yt} is plotted on the ordinate too. This serves as a reminder that first-cycle yielding rather than fatigue might be the criterion of failure.

The midrange-stress axis of Fig. 7–17 has the yield strength S_{yt} and the tensile strength S_{ut} plotted along it.

Four criteria of failure are diagrammed in Fig. 7–17: the Soderberg, the modified Goodman, the Gerber, and yielding. The diagram shows that only the Soderberg criterion guards against yielding, but is biased low.

The linear hypotheses of Fig. 7–17 can be placed in equation form for machine computation by writing the equation of a straight line in intercept form. This form is

$$\frac{x}{a} + \frac{y}{b} = 1 \tag{a}$$

where a and b are the x and y intercepts, respectively. This equation for the Soderberg line is

$$\frac{S_a}{S_e} + \frac{S_m}{S_{yt}} = 1 \tag{7–40}$$

Similarly, we find the modified Goodman relation to be

$$\frac{S_a}{S_e} + \frac{S_m}{S_{ut}} = 1 \tag{7–41}$$

Examination of Fig. 7–15 shows that both a parabola and an ellipse have a better opportunity to pass among the data and to permit quantification of the probability of failure. The Gerber failure locus is written as

$$\frac{S_a}{S_e} + \left(\frac{S_m}{S_{ut}}\right)^2 = 1 \tag{7–42}$$

and the distortion-energy elliptic is written as

$$\left(\frac{S_a}{S_e}\right)^2 + \left(\frac{S_m}{S_{ut}}\right)^2 = 1 \tag{7–43}$$

The *Langer* first-cycle-yielding locus is used in connection with the fatigue locus:

$$\frac{S_a}{S_{yt}} + \frac{S_m}{S_{yt}} = 1 \tag{7–44}$$

The stresses $n\sigma_a$ and $n\sigma_m$ can replace S_a and S_m, where n is the design factor or factor of safety. Then, Eq. (7–40), the Soderberg line, becomes

$$\frac{\sigma_a}{S_e} + \frac{\sigma_m}{S_y} = \frac{1}{n} \tag{7–45}$$

Equation (7–41), the modified Goodman line, becomes

$$\frac{\sigma_a}{S_e} + \frac{\sigma_m}{S_{ut}} = \frac{1}{n} \tag{7–46}$$

Equation (7–42), the Gerber locus, becomes

$$\frac{n\sigma_a}{S_e} + \left(\frac{n\sigma_m}{S_{ut}}\right)^2 = 1 \tag{7–47}$$

Equation (7–43), the DE-elliptic locus, becomes

$$\left(\frac{n\sigma_a}{S_e}\right)^2 + \left(\frac{n\sigma_m}{S_y}\right)^2 = 1 \tag{7–48}$$

We will work principally with DE-Gerber and DE-elliptic for fatigue failure loci and the Langer for first-cycle yielding. The failure loci are used in conjunction with a load line,

Table 7–15

Amplitude and Steady Coordinates of Strength and Important Intersections in First Quadrant for DE-Gerber and Langer Failure Loci

Intersecting Loci	Intersection Coordinates
$\dfrac{S_a}{S_e} + \left(\dfrac{S_m}{S_{ut}}\right)^2 = 1$	$S_a = \dfrac{r^2 S_{ut}^2}{2S_e}\left[-1 + \sqrt{1 + \left(\dfrac{2S_e}{r S_{ut}}\right)^2}\right]$
Load line $r = \dfrac{S_a}{S_m}$	$S_m = \dfrac{S_a}{r}$
$\dfrac{S_a}{S_y} + \dfrac{S_m}{S_y} = 1$	$S_a = \dfrac{r S_y}{1+r}$
Load line $r = \dfrac{S_a}{S_m}$	$S_m = \dfrac{S_y}{1+r}$
$\dfrac{S_a}{S_e} + \left(\dfrac{S_m}{S_{ut}}\right)^2 = 1$	$S_m = \dfrac{S_{ut}^2}{2S_e}\left[1 - \sqrt{1 + \left(\dfrac{2S_e}{S_{ut}}\right)^2 \left(1 - \dfrac{S_y}{S_e}\right)}\right]$
$\dfrac{S_a}{S_y} + \dfrac{S_m}{S_y} = 1$	$S_a = S_y - S_m,\ r_{\text{crit}} = S_a/S_m$

Fatigue factor of safety

$$n_f = \frac{1}{2}\left(\frac{S_{ut}}{\sigma_m}\right)^2 \frac{\sigma_a}{S_e}\left[-1 + \sqrt{1 + \left(\frac{2\sigma_m S_e}{S_{ut}\sigma_a}\right)^2}\right]$$

Table 7–16

Amplitude and Steady Coordinates of Strength and Important Intersections in First Quadrant for DE-Elliptic and Langer Failure Loci

Intersecting Loci	Intersection Coordinates
$\left(\dfrac{S_a}{S_e}\right)^2 + \left(\dfrac{S_m}{S_y}\right)^2 = 1$	$S_a = \sqrt{\dfrac{r^2 S_e^2 S_y^2}{S_e^2 + r^2 S_y^2}}$
Load line $r = S_a/S_m$	$S_m = \dfrac{S_a}{r}$
$\dfrac{S_a}{S_y} + \dfrac{S_m}{S_y} = 1$	$S_a = \dfrac{r S_y}{1+r}$
Load line $r = S_a/S_m$	$S_m = \dfrac{S_y}{1+r}$
$\left(\dfrac{S_a}{S_e}\right)^2 + \left(\dfrac{S_m}{S_y}\right)^2 = 1$	$S_a = 0,\ \dfrac{2}{S_y\left(1/S_e^2 + 1/S_y^2\right)}$
$\dfrac{S_a}{S_y} + \dfrac{S_m}{S_y} = 1$	$S_m = S_y - S_a,\ r_{\text{crit}} = S_a/S_m$

Fatigue factor of safety

$$n_f = \sqrt{\frac{1}{(\sigma_a/S_e)^2 + \left(\sigma_m/S_y\right)^2}}$$

$r = S_a/S_m = \sigma_a/\sigma_m$. Principal intersections are tabulated in Tables 7–15 and 7–16, and are depicted graphically in Fig. 7–18. Formal expressions for fatigue factor of safety are given in the lower panel of Tables 7–15 and 7–16.

Some examples will help solidify the ideas just discussed.

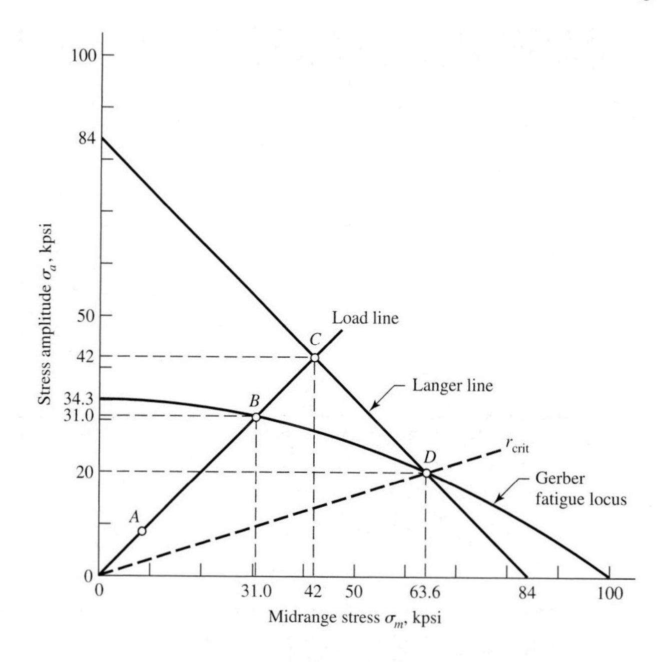

Figure 7–18

Principal points *A*, *B*, *C*, and *D* on the designer's diagram drawn for DE-Gerber, Langer, and load line.

EXAMPLE 7–12

A 1.5-in-diameter bar has been machined from an AISI 1050 cold-drawn bar. This part is to withstand a fluctuating tensile load varying from 0 to 16 kip. Because of the ends, and the fillet radius, a fatigue stress-concentration factor K_f is 1.85 for 10^6 or larger life. Find S_a and S_m and the factor of safety guarding against fatigue and first-cycle yielding for (*a*) a Gerber fatigue locus and (*b*) a DE-elliptic fatigue locus.

Solution

We begin with some preliminaries. From Table E–20, $S_{ut} = 100$ kpsi and $S_y = 84$ kpsi. Note that $F_a = F_m = 8$ kip. The Marin factors from Table 7–13 are, deterministically,

$$k_a = 2.67(100)^{-0.265} = 0.788$$

$$k_b = 1 \text{ (axial loading, see } k_c)$$

$$k_c = 1.23(100)^{-0.0778} = 0.860$$

$$k_d = k_e = 1$$

$$S_e = 0.788(1)0.860(1)(1)0.506(100) = 34.3 \text{ kpsi}$$

The nominal normal stress components σ_{ao} and σ_{mo} are

$$\sigma_{ao} = \frac{4F_a}{\pi d^2} = \frac{4(8)}{\pi 1.5^2} = 4.53 \text{ kpsi} \qquad \sigma_{mo} = \frac{4F_m}{\pi d^2} = \frac{4(8)}{\pi 1.5^2} = 4.53 \text{ kpsi}$$

Applying K_f to both components σ_{ao} and σ_{mo} constitutes a prescription of no notch yielding:

$$\sigma_a = K_f \sigma_{ao} = 1.85(4.53) = 8.38 \text{ kpsi} \quad = \sigma_m$$

The load line slope is $r = \sigma_a/\sigma_m = 1$.

(*a*) From the first panel of Table 7–15,

$$S_a = \frac{(1)^2 100^2}{2(34.3)} \left\{ -1 + \sqrt{1 + \left[\frac{2(34.3)}{(1)100}\right]^2} \right\} = 31.0 \text{ kpsi}$$

$$S_m = \frac{S_a}{r} = \frac{31.0}{1} = 31.0 \text{ kpsi}$$

In Fig. 7–18 the intersection of the load line and the Gerber locus is point B. Point A on the load line represents the stress components σ_a and σ_m. Point C represents the intersection of the load line and Eq. (7–44). Solving Eq. (7–44) simultaneously with $S_m = S_a/r$ gives

$$S_a = \frac{rS_y}{1+r} = \frac{(1)84}{1+1} = 42 \text{ kpsi} \qquad S_m = \frac{S_a}{r} = 42.0 \text{ kpsi}$$

The load line is the locus of possible stress states. As loading increases, point A will move toward point B and point C. The first encounter is with point B on the Gerber line, so the threat to the part is from fatigue. The factor of safety in fatigue n_f is

$$n_f = \frac{OB}{OA} = \frac{(S_a)_{\text{Gerber}}}{\sigma_a} = \frac{31.0}{8.38} = 3.70$$

Alternatively, from the fourth panel of Table 7–15,

$$n_f = \frac{1}{2}\left(\frac{100}{8.38}\right)^2\left(\frac{8.38}{34.3}\right)\left\{-1+\sqrt{1+\left[\frac{2(8.38)34.3}{100(8.38)}\right]^2}\right\} = 3.70$$

The factor of safety guarding against first-cycle yielding is

$$n_y = \frac{OC}{OA} = \frac{(S_a)_{\text{Langer}}}{\sigma_a} = \frac{42.0}{8.38} = 5.01$$

This confirms that there is no local yielding at the fillet. Point D represents the changeover from fatigue failure to first-cycle yielding. The coordinates of point D can be found from the simultaneous solution of Eqs. (7–42) and (7–44), as shown in the third panel of Table 7–15:

$$S_m = \frac{100^2}{2(34.3)}\left[1-\sqrt{1+\left(\frac{2(34.3)}{100}\right)^2\left(1-\frac{84}{34.3}\right)}\right] = 63.6 \text{ kpsi}$$

$$S_a = S_y - S_m = 84 - 63.6 = 20.4 \text{ kpsi}$$

The critical slope of the load line r_{crit} is

$$r_{\text{crit}} = \frac{S_a}{S_m} = \frac{20.4}{63.6} = 0.321$$

Since the load line slope is $r = 1$, and $r_{\text{crit}} < r$, there is a primary threat from fatigue.

(*b*) From panel 1 of Table 7–16, with $r = 1$, we obtain the coordinates S_a and S_m of point B in Fig. 7–19:

$$S_a = \sqrt{\frac{(1)^2 34.3^2 84^2}{34.3^2 + (1)^2 84^2}} = 31.8 \text{ kpsi} \qquad S_m = \frac{S_a}{r} = \frac{31.4}{1} = 31.8 \text{ kpsi}$$

From panel 3 of Table 7–16, the coordinates S_a and S_m of point D are

$$S_a = \frac{2}{84\left(1/34.3^2 + 1/84^2\right)} = 24.0 \text{ kpsi} \qquad S_m = S_y - S_a = 84 - 24.0 = 60 \text{ kpsi}$$

$$r_{\text{crit}} = \frac{S_a}{S_m} = \frac{24.0}{60} = 0.400$$

Since $r_{\text{crit}} < r$, the primary threat is from fatigue. The factor of safety in fatigue n_f is given by

$$n_f = \frac{S_a}{\sigma_a} = \frac{31.8}{8.38} = 3.79$$

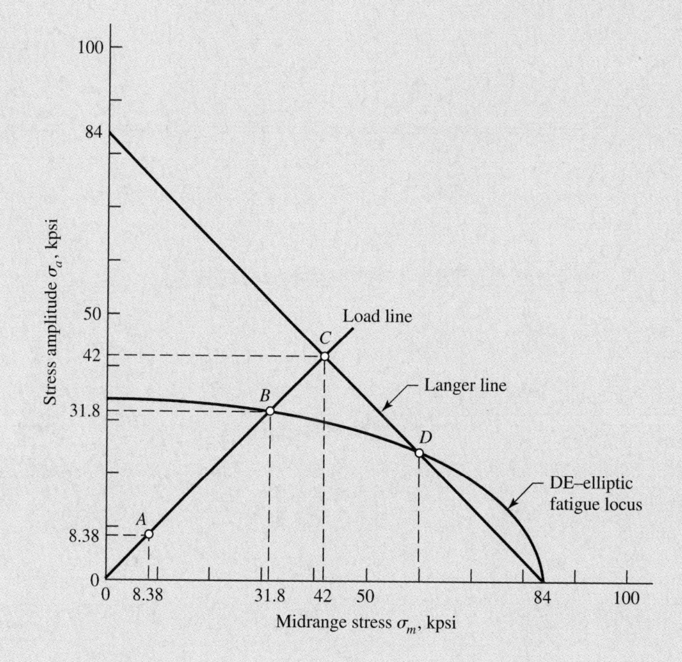

Figure 7–19

Principal points *A*, *B*, *C*, and *D* on the designer's diagram drawn for DE-elliptic, Langer, and load line.

Alternatively, from the fourth panel of Table 7–16,

$$n_f = \sqrt{\frac{1}{(8.38/34.3)^2 + (8.38/84)^2}} = 3.79$$

The factor of safety guarding against first-cycle yielding n_y is

$$n_y = \frac{(S_a)_y}{\sigma_a} = \frac{42}{8.38} = 5.01$$

The DE-Gerber and the DE-elliptic fatigue failure loci are very close to each other and are used interchangeably. The ANSI/ASME Standard B106.1M–1985 uses DE-elliptic for shafting.

EXAMPLE 7–13

A flat-leaf spring is used to retain an oscillating flat-faced follower in contact with a plate cam. The range of motion is 2 in and fixed, so the alternating component of force, bending moment, and stress is fixed, too. The spring is preloaded to adjust to various cam speeds. The preload must be increased to prevent follower float or jump. For lower speeds the preload should be decreased to obtain longer life of cam and follower surfaces. The spring is a steel cantilever 32 in long, 2 in wide, and $\frac{1}{4}$ in thick, as seen in Fig. 7–20a. The spring strengths are $S_{ut} = 150$ kpsi, $S_y = 127$ kpsi, and $S_e = 28$ kpsi fully corrected. The total cam motion is 2 in. The designer wishes to preload the spring by deflecting it 2 in for low speed and 5 in for high speed.

(*a*) Plot the Gerber-Langer failure locus and the load line.

(*b*) What are the strength factors of safety corresponding to 2 in and 5 in preload?

(*c*) What are the factors of safety based on preload deflection?

Figure 7–20

Cam follower retaining spring. (*a*) Geometry; (*b*) designer's fatigue diagram for Ex. 7-13.

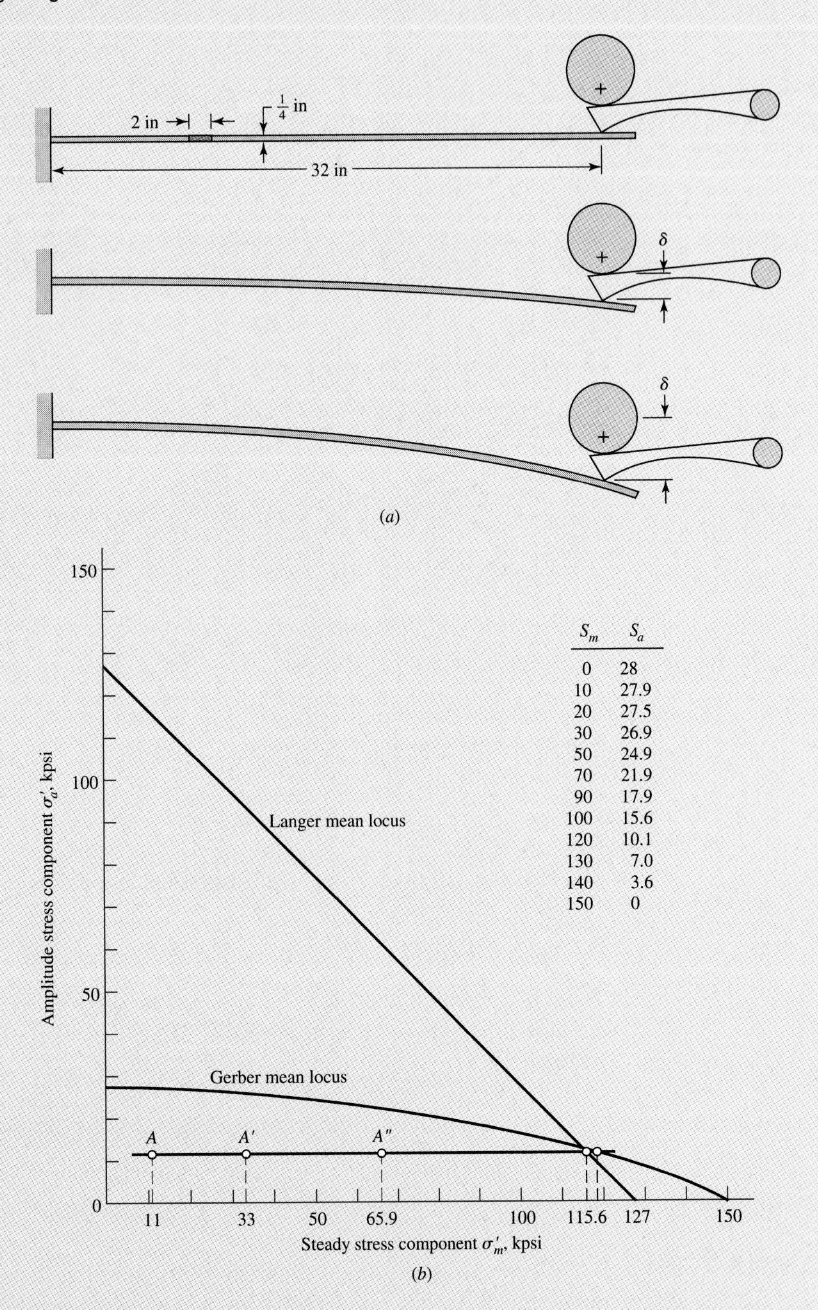

S_m	S_a
0	28
10	27.9
20	27.5
30	26.9
50	24.9
70	21.9
90	17.9
100	15.6
120	10.1
130	7.0
140	3.6
150	0

Solution We begin with preliminaries. The second area moment of the cantilever cross section is

$$I = \frac{bh^3}{12} = \frac{2(0.25)^3}{12} = 0.0026 \text{ in}^4$$

Since force F and deflection y in a cantilever are related by $F = 3EIy/\ell^3$, then stress σ and deflection y are related by

$$\sigma = \frac{Mc}{I} = \frac{32Fc}{I} = \frac{32(3EIy)}{\ell^3}\,\frac{c}{I} = \frac{96ECy}{\ell^3} = Ky$$

$$K = \frac{96\left(30 \cdot 10^6\right) 0.125}{32^3} \doteq 11 \text{ kpsi/in}$$

where F, y and σ can be subscripted max and min. Now $y_{\text{min}} = \delta$ and $y_{\text{max}} = 2 + \delta$, so

$$F_{\text{min}} = \frac{3EI\delta}{\ell^3} \qquad F_{\text{max}} = \frac{3EI(2+\delta)}{\ell^3}$$

The stress components are

$$\sigma_{\text{max}} = K(2+\delta) \qquad \sigma_a = \frac{K(2+\delta) - K\delta}{2} = K$$

$$\sigma_{\text{max}} = K\delta \qquad \sigma_m = \frac{K(2+\delta) + 2\delta}{2} = K\delta$$

For $\delta = 0$,

$$\sigma_{\text{max}} = K(2+\delta) = 11(2) = 22 \text{ kpsi} \qquad \sigma_a = (22-0)/2 = 11 \text{ kpsi}$$

$$\sigma_{\text{min}} = k\delta = 0 \text{ kpsi} \qquad \sigma_m = (22+0)/2 = 11 \text{ kpsi}$$

For $\delta = 2$,

$$\sigma_{\text{max}} = K(2+2) = 11(4) = 44 \text{ kpsi} \qquad \sigma_a = (44-22)/2 = 11 \text{ kpsi}$$

$$\sigma_{\text{min}} = K\delta = 11(2) = 22 \text{ kpsi} \qquad \sigma_m = (44+22)/2 = 33 \text{ kpsi}$$

For $\delta = 5$,

$$\sigma_{\text{max}} = K(2+\delta) = 11(2+5) = 76.9 \text{ kpsi} \qquad \sigma_a = (76.9-54.9)/2 = 11 \text{ kpsi}$$

$$\sigma_{\text{min}} = K\delta = 11(5) = 54.9 \text{ kpsi} \qquad \sigma_m = (76.9+54.9)/2 = 65.6 \text{ kpsi}$$

(*a*) A plot of the Gerber and Langer loci is shown in Fig. 7–20*b*. The three preload deflections of 0, 2, and 5 in are are shown as points A, A', and A''. Note the load line is horizontal and does not contain the origin. The intersection between the Gerber locus and the load line is found from solving Eq. (7–42) for S_m and substituting 11 kpsi for S_a:

$$S_m = S_{ut}\sqrt{1 - \frac{S_a}{S_e}} = 150\sqrt{1 - \frac{11}{28}} = 116.9 \text{ kpsi}$$

The intersection of the Langer locus and the load line is found from solving Eq. (7–44) for S_m and substituting 11 kpsi for S_a:

$$S_m = S_y - S_a = 127 - 11 = 116 \text{ kpsi}$$

The threats from fatigue and first-cycle yielding are approximately equal.
(*b*) For $\delta = 2$ in,

$$n_f = \frac{S_m}{\sigma_m} = \frac{116.9}{33} = 3.54 \qquad n_y = \frac{116}{33} = 3.52$$

and for $\delta = 5$ in,

$$n_f = \frac{116.9}{65.9} = 1.77 \qquad n_y = \frac{116}{65.9} = 1.76$$

(*c*) A factor of safety based on preload deflection involves finding the preload deflection associated with failure. Treating fatigue as the threat, Gerber's relation can be expressed as

$$\frac{11}{28} + \left(\frac{K\delta^*}{150}\right)^2 = 1$$

from which

$$\delta^* = \frac{150}{K}\sqrt{1 - \frac{11}{28}} = \frac{150}{11}\sqrt{1 - \frac{11}{28}} = 10.6 \text{ in}$$

For slower speeds the factor of safety can be defined as

Answer $$n = \frac{\text{loss-of-function preload deflection}}{\text{working preload deflection}} = \frac{\delta^*}{\delta} = \frac{10.6}{2} = 5.3$$

For higher speeds the same definition applies and the factor of safety is

Answer $$n = \frac{10.6}{5} = 2.12$$

7–12 Torsional Fatigue Strength under Pulsating Stresses

Extensive tests by Smith[23] provide some very interesting results on pulsating torsional fatigue. Smith's first result, based on 72 tests, shows that the existence of a torsional steady-stress component not more than the torsional yield strength has no effect on the torsional endurance limit, provided the material is *ductile*, *polished*, *notch-free*, and *cylindrical*.

Smith's second result applies to materials with stress concentration, notches, or surface imperfections. In this case, he finds that the torsional fatigue limit decreases monotonically with torsional steady stress. Since the great majority of parts will have surfaces that are less than perfect, this result indicates DE-Gerber, DE-elliptic, and other approximations are useful. Joerres[24] of Associated Spring-Barnes Group, confirms Smith's results and recommends the use of the modified Goodman relation for pulsating torsion. In constructing the Goodman diagram, Joerres uses

$$S_{su} = 0.67 S_{ut} \tag{7–49}$$

Also, from Chap. 6, $S_{sy} = 0.577 S_{yt}$ from distortion-energy theory, and the mean load factor $\bar{k}_c$ is given by Eq. (7–22), or 0.577 if one uses a constant. This is elaborated on in Chap. 10.

7–13 Combinations of Loading Modes

In Sec. 7–8 we learned that a load factor k_c is used to obtain the endurance limit and hence that the result is dependent on whether the loading is axial, bending, or torsion. In this section we want to answer the question, How do we proceed when the loading is a mixture of, say, axial, bending, and torsional loads? In addition to the complication introduced by the fact that a separate endurance limit is associated with each mode of loading, there may also be multiple stress-concentration factors, one also for each mode of loading. Fortunately, the answer turns out to be rather simple:

1. For the strength, use the fully corrected endurance limit for bending, S_e or $\mathbf{S}_e$.
2. Apply the appropriate fatigue stress-concentration factors to the torsional stress, the bending stress, and the axial stress components.
3. Multiply any alternating axial stress components by the factor $1/k_{c,\text{ax}}$.
4. Enter the resultant stresses into a Mohr circle analysis and find the principal stresses.
5. Using the results of step 4, find the von Mises alternating stress σ'_a or the stress variate $\boldsymbol{\sigma}'_a$.
6. Compare $\sigma/_a$ with S_a to find the factor of safety, or interfere $\boldsymbol{\sigma}'_a$ with S_a to find the reliability.

23. James O. Smith, "The Effect of Range of Stress on the Fatigue Strength of Metals," *Univ. of Ill. Eng. Exp. Sta. Bull.* 334, 1942.

24. Robert E. Joerres, "Springs," chap. 24 in Joseph E. Shigley and Charles R. Mischke, *Standard Handbook of Machine Design*, 2nd ed., McGraw-Hill, New York, 1996.

This approach is based on the assumption that all stress components are completely reversed and are always in time phase with each other. If they are not in phase but have the same frequency, the maxima can be found by expressing each component in trigonometric terms, using phase angles, and then finding the sum. If two or more stress components have differing frequencies, the problem is difficult; one solution is to assume that the two (or more) components often reach an in-phase condition, so that their magnitudes are additive.

If midrange stresses are also present, then steps 4 and 5 can be repeated for them and the resulting steady von Mises stress component σ'_m used with σ'_a in forming a DE-Gerber or DE-elliptic solution. Both the steady and amplitude components are augmented by K_f or K_{fs} stress-concentration factor. It is worth noting that the analysis outline above incorporates a size factor for axial loading that is the same for bending and torsion. When bending is present, the existence of an axial component is relatively small, so in most cases this loss of accuracy is small and always conservative.

EXAMPLE 7–14

A rotating shaft is made of 42- × 4-mm AISI 1018 cold-drawn steel tubing and has a 6-mm-diameter hole drilled transversely through it. Estimate the factor of safety guarding against fatigue and static failures for the following loading conditions:
(*a*) The shaft is subjected to a completely reversed torque of 120 N · m in phase with a completely reversed bending moment of 150 N · m.
(*b*) The shaft is subjected to a pulsating torque fluctuating from 20 to 160 N · m and a steady bending moment of 150 N · m.

Solution

Here we follow the procedure of estimating the strengths and then the stresses, followed by relating the two.

From Table E–20 we find the minimum strengths to be $S_{ut} = 440$ MPa and $S_{yt} = 370$ MPa. The endurance limit of the rotating-beam specimen is 0.506(440) = 223 MPa. The surface factor, obtained from Table 7–14, is

$$k_a = 4.45 S_{ut}^{-0.265} = 4.45(440)^{-0.265} = 0.886$$

From Table 7–14 the size factor is

$$k_b = \left(\frac{d}{7.62}\right)^{-0.107} = \left(\frac{42}{7.62}\right)^{-0.107} = 0.833$$

The remaining Marin factors are all unity, so the endurance S_e is

$$S_e = 0.886(0.833)223 = 165 \text{ MPa}$$

(*a*) Theoretical stress-concentration factors are found from Table E–16. Using $a/D = 6/42 = 0.143$ and $d/D = 34/42 = 0.810$, and using linear interpolation, we obtain $A = 0.798$ and $K_t = 2.366$ for bending; and $A = 0.89$ and $K_{ts} = 1.75$ for torsion. Thus, for bending,

$$Z_{\text{net}} = \frac{\pi A}{32D}\left(D^4 - d^4\right) = \frac{\pi(0.798)}{32(4.2)}\left[(4.2)^4 - (3.4)^4\right] = 3.31 \text{ cm}^3$$

and for torsion

$$J_{\text{net}} = \frac{\pi A}{32}\left(D^4 - d^4\right) = \frac{\pi(0.89)}{32}\left[(4.2)^4 - (3.4)^4\right] = 15.5 \text{ cm}^4$$

Next, using Figs. 5–16 and 5–17 we find the notch sensitivities to be 0.78 for bending and 0.96 for torsion. The two corresponding fatigue stress-concentration factors are obtained from Eq. (5–26) as

$$K_f = 1 + q(K_t - 1) = 1 + 0.78(2.366 - 1) = 2.07$$
$$K_{fs} = 1 + 0.96(1.75 - 1) = 1.72$$

The bending stress is now found to be

$$\sigma_x = K_f \frac{M}{Z_{\text{net}}} = 2.07 \frac{150}{3.31} = 93.8 \text{ MPa}$$

and the torsional stress is

$$\tau_{sy} = K_{fs} \frac{TD}{2J_{\text{net}}} = 1.72 \frac{120(4.2)}{2(15.5)} = 28.0 \text{ MPa}$$

The von Mises steady-stress component σ'_m is zero. The amplitude component σ'_a is given by

$$\sigma'_a = \left(\sigma_{xa}^2 + 3\tau_{xya}^2\right)^{1/2} = \left[93.8^2 + 3\left(28^2\right)\right]^{1/2} = 105.6 \text{ MPa}$$

Since $S_e = S_a$, the fatigue factor of safety n_f is

$$n_f = \frac{S_a}{\sigma'_a} = \frac{165}{105.6} = 1.56$$

The first-cycle yield factor of safety is

$$n_y = \frac{S_{yt}}{\sigma'_a} = \frac{370}{105.6} = 3.50$$

There is no localized yielding; the threat is from fatigue. See Fig. 7–21.

(*b*) This part asks us to find the factors of safety when the alternating component is due to pulsating torsion, and a steady component is due to both torsion and bending. We have $T_a = (160 - 20)/2 = 70$ N · m and $T_m = 20 + 70 = 90$ N · m. The corresponding amplitude and steady-stress components are

$$\tau_{xya} = K_{fs} \frac{T_a D}{2J_{\text{net}}} = 1.72 \frac{70(42)}{2(15.5)} = 16.3 \text{ MPa}$$

$$\tau_{xym} = K_{fs} \frac{T_m D}{2J_{\text{net}}} = 1.72 \frac{90(42)}{2(15.5)} = 21.0 \text{ MPa}$$

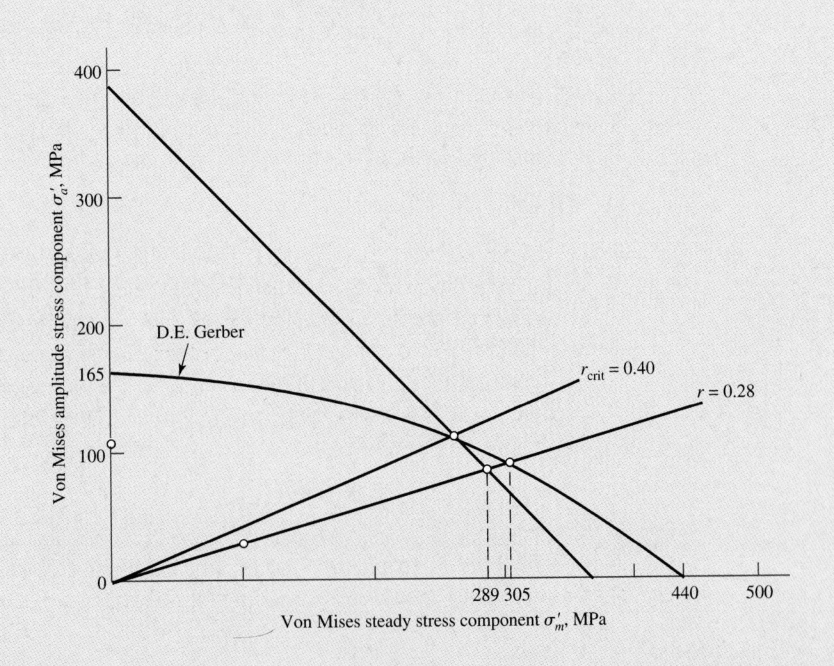

Figure 7–21

Designer's fatigue diagram for Ex. 7-14.

The steady bending stress component σ_{xm} is

$$\sigma_{xm} = K_f \frac{M_m}{Z_{\text{net}}} = 2.07 \frac{150}{3.31} = 93.8 \text{ MPa}$$

The von Mises components σ'_a and σ'_m are

$$\sigma'_a = [3(16.3)]^{1/2} = 28.2 \text{ MPa}$$

$$\sigma'_m = \left[93.8^2 + 3(21)^2\right]^{1/2} = 100.6 \text{ MPa}$$

The slope of the load line is $r = \sigma'_a/\sigma'_m = 28.2/100.6 = 0.28$. The strength amplitude component S_a and steady-strength component S_m are

$$S_a = \frac{0.28^2 440^2}{2(165)} \left\{ -1 + \sqrt{1 + \left[\frac{2(165)}{0.28(440)}\right]^2} \right\} = 85.5 \text{ MPa}$$

$$S_m = \frac{85.5}{0.28} = 305.4 \text{ MPa}$$

The fatigue factor of safety n_f is

Answer $$n_f = \frac{S_a}{\sigma'_a} = \frac{85.5}{28.2} = 3.03$$

The first-cycle yield factor of safety n_y is

Answer $$n_y = \frac{S_y}{\sigma'_a + \sigma'_m} = \frac{370}{28.2 + 100.6} = 2.87$$

There is no notch yielding. The threat is from first-cycle yielding at the notch. See the plot in Fig. 7–21.

7–14 Stochastic Failure Loci under Fluctuating Stresses

Deterministic failure loci which lie among the data are candidates for regression models. Included among these are the Gerber, ASME-elliptic, and, for brittle materials, Smith-Dolan models, which use mean values in their presentation. The Gerber parabola is

$$\frac{\bar{S}_a}{\bar{S}_e} + \left(\frac{\bar{S}_m}{\bar{S}_{ut}}\right)^2 = 1 \tag{7–50}$$

Just as the deterministic failure loci are located by endurance strength and ultimate tensile (or yield) strength, so too are stochastic failure loci located by $\mathbf{S}_e$ and by $\mathbf{S}_{ut}$ or $\mathbf{S}_y$. Figure 7–21 shows how the mean locus of Eq. (7–50) fits a parabola to form the Gerber mean locus. We also need to establish a contour located one standard deviation from the mean. Since stochastic loci are most likely to be used with a radial load line, we will develop the equation using the load line slope $\bar{S}_a/\bar{S}_m = 5$. Substituting in Eq. (7–50) $\bar{S}_m = \bar{S}_a/r$ and solving for $\bar{S}_a$ gives

$$\bar{S}_a = \frac{r^2 \bar{S}_{ut}^2}{2\bar{S}_e} \left[-1 + \sqrt{1 + \left(\frac{2\bar{S}_e}{r\bar{S}_{ut}}\right)^2} \right] \tag{7–51}$$

Because of the positive correlation between $\mathbf{S}_e$ and $\mathbf{S}_{ut}$, we increment $\bar{S}_e$ by $C_{Se}\bar{S}_e$, $\bar{S}_{ut}$ by $C_{Sut}\bar{S}_{ut}$, and $\bar{S}_a$ by $C_{Sa}\bar{S}_a$, substitute into Eq. (7–51), and solve for C_{Sa} to obtain

$$C_{Sa} = \frac{(1+C_{Sut})^2}{1+C_{Se}} \frac{\left\{-1+\sqrt{1+\left[\dfrac{2\bar{S}_e(1+C_{Se})}{r\bar{S}_{ut}(1+C_{Sut})}\right]^2}\right\}}{\left[-1+\sqrt{1+\left(\dfrac{2\bar{S}_e}{r\bar{S}_{ut}}\right)^2}\right]} - 1 \tag{7–52}$$

Equation (7–52) can be viewed as an interpolation formula for C_{Sa} which falls between C_{Se} and C_{Sut} depending on load line slope r. Note that $\mathbf{S}_a \sim \bar{S}_a\mathbf{LN}(1, C_{Sa})$.

The ASME-elliptic locus is expressed in terms of its means as

$$\left(\frac{\bar{S}_a}{\bar{S}_e}\right)^2 + \left(\frac{\bar{S}_m}{\bar{S}_{yt}}\right)^2 = 1 \tag{7–53}$$

Similarly, substituting $\bar{S}_m = \bar{S}_a/r$ into Eq. (7–53) and solving for $\bar{S}_a$ gives

$$\bar{S}_a = \frac{r\bar{S}_y\bar{S}_e}{\sqrt{r^2\bar{S}_y^2 + \bar{S}_e^2}} \tag{7–54}$$

Similarly, we increment $\bar{S}_e$ by $C_{Se}\bar{S}_e$, $\bar{S}_y$ by $C_{Sy}\bar{S}_y$, and $\bar{S}_a$ by $C_{Sa}\bar{S}_a$, substitute into Eq. (7–54), and solve for $C_{S'a}$:

$$C_{Sa} = (1+C_{Sy})(1+C_{Se})\sqrt{\frac{r^2\bar{S}_y^2+\bar{S}_e^2}{r^2\bar{S}_y^2(1+C_{Sy})^2+\bar{S}_e^2(1+C_{Se})^2}} - 1 \tag{7–55}$$

Many *brittle* materials follow a Smith-Dolan failure locus, written deterministically as

$$\frac{n\sigma_a}{S_e} = \frac{1 - n\sigma_m/S_{ut}}{1 + n\sigma_m/S_{ut}} \tag{7–56}$$

Expressed in terms of its means,

$$\frac{\bar{S}_a}{\bar{S}_e} = \frac{1 - \bar{S}_m/\bar{S}_{ut}}{1 + \bar{S}_m/\bar{S}_{ut}} \tag{7–57}$$

For a radial load line slope of r, we substitute $\bar{S}_a/r$ for $\bar{S}_m$ and solve for $\bar{S}_a$, obtaining

$$\bar{S}_a = \frac{r\bar{S}_{ut}+\bar{S}_e}{2}\left[-1+\sqrt{1+\frac{4r\bar{S}_{ut}\bar{S}_e}{(r\bar{S}_{ut}+\bar{S}_e)^2}}\right] \tag{7–58}$$

and the expression for C_{Sa} is

$$C_{Sa} = \frac{r\bar{S}_{ut}(1+C_{Sut})+\bar{S}_e(1+C_{Se})}{2\bar{S}_a} \cdot \left\{-1+\sqrt{1+\frac{4r\bar{S}_{ut}\bar{S}_e(1+C_{Se})(1+C_{Sut})}{[r\bar{S}_{ut}(1+C_{Sut})+\bar{S}_e(1+C_{Se})]^2}}\right\} - 1 \tag{7–59}$$

EXAMPLE 7–15 A rotating shaft experiences a steady torque $\mathbf{T} \sim 1360\mathbf{LN}(1, 0.05)$ in · lbf, and at a shoulder with a 1.1-in small diameter, a fatigue stress-concentration factor $\mathbf{K}_f \sim 1.50\mathbf{LN}(1, 0.11)$, $K_{fs} \sim 1.28\mathbf{LN}(1, 0.11)$, and at that location a bending moment of $\mathbf{M} \sim 1260\mathbf{LN}(1, 0.05)$ in · lbf. The material of which the shaft is machined is hot-rolled 1035 with $\mathbf{S}_{ut} \sim 86.2\mathbf{LN}(1, 0.045)$ kpsi and $\mathbf{S}_y \sim 56.0\mathbf{LN}(1, 0.077)$ kpsi. Estimate the reliability using a stochastic Gerber failure zone.

Solution Establish the endurance strength. From Table 7–13,

$$\mathbf{S}'_e = 0.506(86.2)\mathbf{LN}(1, 0.138) = 43.6\mathbf{LN}(1, 0.138) \text{ kpsi}$$

$$\mathbf{k}_a = 2.67(86.2)^{-0.265}\mathbf{LN}(1, 0.058) = 0.820\mathbf{LN}(1, 0.058)$$

$$k_b = (1.1/0.30)^{-0.107} = 0.870$$

$$\mathbf{k}_c = \mathbf{k}_d = \mathbf{k}_e = \mathbf{LN}(1, 0)$$

$$\mathbf{S}_e = 0.820\mathbf{LN}(1, 0.058)0.870(43.6)\mathbf{LN}(1, 0.138)$$

$$\bar{S}_e = 0.820(0.870)43.6 = 31.1 \text{ kpsi}$$

$$C_{Se} = \left(0.058^2 + 0.138^2\right)^{1/2} = 0.150$$

and so $S_e = 31.1\mathbf{LN}(1, 0.150)$ kpsi.

Stress:

$$\boldsymbol{\sigma}'_{xa} = \frac{32\mathbf{K}_f\mathbf{M}_a}{\pi d^3} = \frac{32(1.50)\mathbf{LN}(1, 0.11)1.26\mathbf{LN}(1, 0.05)}{\pi(1.1)^3}$$

$$\bar{\sigma}_{xa'} = \frac{32(1.50)1.26}{\pi(1.1)^3} = 14.5 \text{ kpsi}$$

$$C_{\sigma xa'} = \left(0.11^2 + 0.05^2\right)^{1/2} = 0.121$$

$$\tau_{xym} = \frac{16\mathbf{K}_{fs}\mathbf{T}_m}{\pi d^3} = \frac{16(1.28)\mathbf{LN}(1, 0.11)1.36\mathbf{LN}(1, 0.05)}{\pi(1.1)^3}$$

$$\bar{\tau}_{xym} = \frac{16(1.28)1.36}{\pi(1.1)^3} = 6.66 \text{ kpsi}$$

$$C_{\tau xym} = \left(0.11^2 + 0.05^2\right)^{1/2} = 0.121$$

$$\bar{\sigma}'_a = \left(\bar{\sigma}'^2_{xa} + 3\bar{\tau}^2_{xya}\right)^{1/2} = \left[14.5^2 + 3(0)^2\right]^{1/2} = 14.5 \text{ kpsi}$$

$$\bar{\sigma}'_m = \left(\bar{\sigma}'^2_{xm} + 3\bar{\tau}^2_{xym}\right)^{1/2} = \left[0 + 3(6.66)^2\right]^{1/2} = 11.54 \text{ kpsi}$$

$$r = \frac{\bar{\sigma}'_a}{\bar{\sigma}'_m} = \frac{14.5}{11.54} = 1.26$$

Strength: From Eqs. (7–51) and (7–52),

$$\bar{S}_a = \frac{1.26^2 86.2^2}{2(31.1)}\left\{-1 + \sqrt{1 + \left[\frac{2(31.1)}{1.26(86.2)}\right]^2}\right\} = 28.9 \text{ kpsi}$$

$$C_{Sa} = \frac{(1 + 0.045)^2}{1 + 0.150}\,\frac{-1 + \sqrt{1 + \left[\dfrac{2(31.1)(1 + 0.15)}{1.26(86.2)(1 + 0.045)}\right]^2}}{-1 + \sqrt{1 + \left[\dfrac{2(31.1)}{1.26(86.2)}\right]^2}} - 1 = 0.134$$

Reliability: Since $\mathbf{S}_a \sim \mathbf{LN}(1, 0.134)$ kpsi and $\boldsymbol{\sigma}'_a \sim 14.5\mathbf{LN}(1, 0.121)$ kpsi,

$$z = -\frac{\ln(28.9/14.5)}{\left(0.134^2 + 0.121^2\right)^{1/2}} \doteq -3.8$$

From Table E–10 the probability of failure is $p_f = 0.000\,0.72$, and the reliability is, against fatigue,

Answer $R = 1 - p_f = 1 - 0.000\,072 = 0.999\,93$

The chance of first-cycle yielding is estimated by interfering $\mathbf{S}_y$ with $\boldsymbol{\sigma}'_{\max}$. The quantity $\boldsymbol{\sigma}'_{\max}$ is formed from $\boldsymbol{\sigma}'_a + \boldsymbol{\sigma}'_m$. The mean of $\boldsymbol{\sigma}'_{\max}$ is $\bar{\sigma}'_a + \bar{\sigma}'_m = 14.5 + 11.54 = 26.04$ kpsi. The sum $\boldsymbol{\sigma}'_a + \boldsymbol{\sigma}'_m$ is correlated with $\rho = 1$. Thus the coefficient of variation of the sum is 0.121, since both COVs are 0.121, thus $C_{\sigma\,\max} = 0.121$. We interfere $\mathbf{S}_y \sim 56\mathbf{LN}(1, 0.077)$ kpsi with $\boldsymbol{\sigma}'_{\max} \sim 26.04\mathbf{LN}(1, 0.121)$ kpsi. The corresponding z variable is

$$z = -\frac{\ln(56/26.04)}{\left(0.077^2 + 0.121^2\right)^{1/2}} = -5.3$$

which represents, from Table E–10, a probability of failure of approximately 0.0^7579 of first-cycle yield in the fillet.

The probability of observing a fatigue failure exceeds the probability of a yield failure, something a deterministic analysis does not foresee and in fact could lead one to expect a yield failure should a failure occur. Look at the $\boldsymbol{\sigma}'_a\mathbf{S}_a$ interference and the $\boldsymbol{\sigma}'_{\max}\mathbf{S}_y$ interference and examine the z expressions. These control the relative probabilities. A deterministic analysis is oblivious to this and can mislead. Check your statistics text for events that are not mutually exclusive, but are independent, to quantify the probability of failure:

$$\begin{aligned} p_f &= p(\text{yield}) + p(\text{fatigue}) - p(\text{yield and fatigue}) \\ &= p(\text{yield}) + p(\text{fatigue}) - p(\text{yield})p(\text{fatigue}) \\ &= 0.579\left(10^{-7}\right) + 0.72\left(10^{-4}\right) - 0.579\left(10^{-7}\right)0.72\left(10^{-4}\right) = 0.721\left(10^{-4}\right) \end{aligned}$$

$$R = 1 - 0.721\left(10^{-4}\right) = 0.999\,928$$

against either or both modes of failure.

Examine Fig. 7–22, which depicts the results of Ex. 7–15. The problem distribution of $\mathbf{S}_e$ was compounded of historical experience with $\mathbf{S}'_e$ and the uncertainty manifestations due to features requiring Marin considerations. The Gerber "failure zone" displays this. The interference with load-induced stress predicts the risk of failure. If additional information is known (R. R. Moore testing, with or without Marin features), the stochastic Gerber can accommodate to the information. Usually, the accommodation to additional test information is movement and contraction of the failure zone. In its own way the stochastic failure model accomplishes more precisely what the deterministic models and conservative postures intend. Additionally, stochastic models can estimate the probability of failure, something a deterministic approach cannot address.

7–15 Cumulative Fatigue Damage

Instead of a single fully reversed stress history block composed of n cycles, suppose a machine part, at a critical location, is subjected to

Figure 7–22

Designer's fatigue diagram for Ex. 7–15.

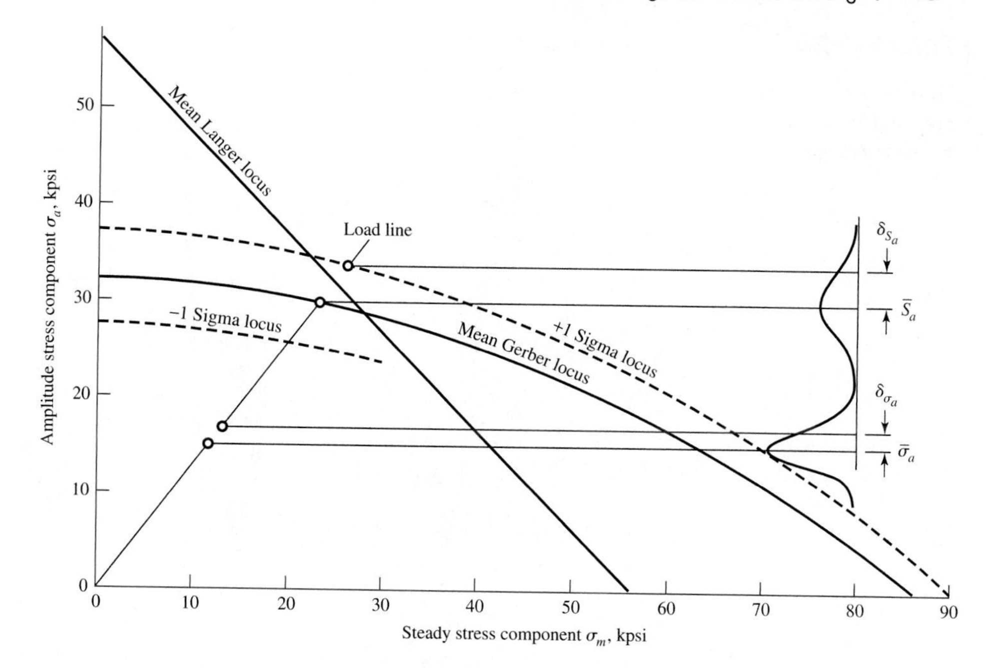

- A fully reversed stress σ_1 for n_1 cycles, σ_2 for n_2 cycles, . . . , or
- A "wiggly" time line of stress exhibiting many and different peaks and valleys.

What stresses are significant, what counts as a cycle, and what is the measure of damage incurred? Consider a fully reversed cycle with stresses varying 60, 80, 40, and 60 kpsi and a second fully reversed cycle −40, −60, −20, and −40 kpsi as depicted in Fig. 7–23*a*. First, it is clear that to impose the pattern of stress in Fig. 7–23*a* on a part it is necessary that the time trace look like the solid line plus the dashed line in Fig. 7–23*a*. Figure 7–23*b* moves the shapshot to exist beginning with 80 kpsi and ending with 80 kpsi. Acknowledging the existence of a single stress-time trace is to discover a "hidden" cycle shown as the dashed line in Fig. 7–23*b*. If there are 100 applications of the all-positive stress cycle, then 100 applications of the all-negative stress cycle, the hidden cycle is applied but once. If the all-positive stress cycle is applied alternatively with the all-negative stress cycle, the hidden cycle is applied 100 times.

To ensure that the hidden cycle is not lost, begin on the shapshot with the largest (or smallest) stress and add previous history to the right side, as was done in Fig. 7–23*b*. Characterization of a cycle takes on a max-min-same max (or min-max-same min) form. We identify the hidden cycle first by moving along the dashed-line trace in Fig. 7–23*b* identifying a cycle with an 80-kpsi max, a 60-kpsi min, and returning to 80 kpsi. Mentally deleting the used part of the trace (the dashed line) leaves a 40, 60, 40 cycle and a −40, −20, −40 cycle. Since failure loci are expressed in terms of stress amplitude component σ_a and steady component σ_m, we use Eq. (7–36) to construct the table below:

Cycle Number	σ_{max}	σ_{min}	σ_a	σ_m
1	80	−60	70	10
2	60	40	10	50
3	−20	−40	10	−30

Figure 7–23

Variable stress diagram prepared for assessing cumulative damage.

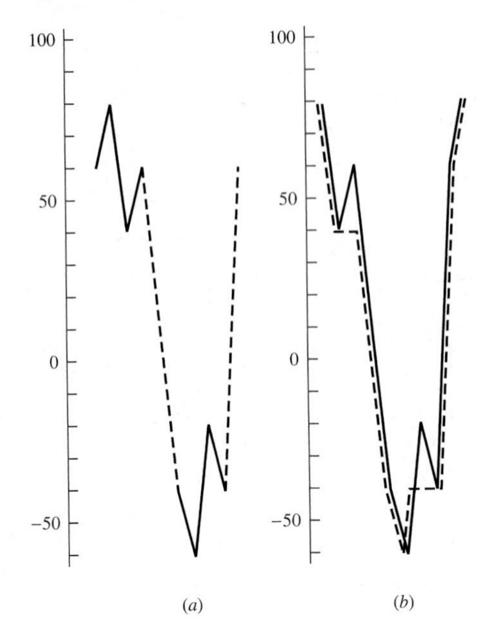

The most damaging cycle is number 1. It could have been lost.

Methods for counting cycles include:

- Number of tensile peaks to failure.
- All maxima above the wave-form mean, all minima below.
- The global maxima between crossings above the mean and the global minima between crossings below the mean.
- All positive slope crossings of levels above the mean, and all negative slope crossings of levels below the mean.
- A modification of the preceding method with only one count made between successive crossings of a level associated with each counting level.
- Each local maxi-min excursion is counted as a half-cycle, and the associated amplitude is half-range.
- The preceding method plus consideration of the local mean.
- Rain-flow counting technique.

The method used here amounts to a variation of the *rain-flow counting technique*.

The *Palmgren-Miner*[25] *cycle-ratio summation rule*, also called *Miner's rule*, is written

$$\sum \frac{n_i}{N_i} = c \tag{7–60}$$

where n_i is the number of cycles, as defined above, and N_i is the number of cycles-to-failure at stress level σ_i. The parameter c has been determined by experiment; it is usually found in the range $0.7 < c < 2.2$ with an average value near unity. The stochastic

25. A. Palmgren, "Die Lebensdauer von Kugellagern," *ZVDI*, vol. 68, pp. 339–341, 1924; M. A. Miner, "Cumulative Damage in Fatigue," *J. Appl. Mech.*, vol. 12, *Trans. ASME*, vol. 67, pp. A159–A164, 1945.

equivalent can be written as

$$\sum \frac{n_i}{\mathbf{N}_i} = \boldsymbol{\lambda} \tag{7-61}$$

where n_i are real numbers and $\mathbf{N}_i \sim LN$ and $\boldsymbol{\lambda} \sim LN$. The deterministic approach is to set $c = 1$, and the stochastic approach is to use the quantile of $\boldsymbol{\lambda}$ associated with the reliability goal.

Using the deterministic formulation as a linear damage rule we write

$$D = \sum \frac{n_i}{N_i} \tag{7-62}$$

where D is the accumulated damage. When $D = c = 1$, failure ensues.

EXAMPLE 7–16 For the loading of Fig. 7–23 on a part, the following properties at the critical location exist: $\bar{S}_{ut} = 151$ kpsi, $\sigma_0 = 210$ kpsi, $\varepsilon_f = 0.45$, $m = 0.09$, $\bar{S}_e = 67.5$ kpsi. Estimate the number of repetitions of the stress-time block in Fig. 7–23 that can be made before failure.

Solution $\sigma'_F = \sigma_0 \varepsilon^m = 210(0.45)^{0.09} = 195.4$ kpsi

From Sec. 7–7, Eq. (*c*),

$$b = \frac{\log(195.4/67.5)}{\log\left(2 \cdot 10^6\right)} = -0.0733$$

From Sec. 7–5, Eq. (*b*),

$$f = \frac{\sigma'_F}{S_{ut}}\left(2 \cdot 10^3\right)^b = \frac{195.4}{151}\left(2 \cdot 10^3\right)^{-0.0733} = 0.741$$

From Eq. (7–5),

$$a = \frac{0.741^2 151^2}{67.5} = 186 \text{ kpsi}$$

so

$$S_f = 186N^{-0.0733} \qquad N = \left(\frac{\sigma_a}{186}\right)^{1/-0.0733} = \left(\frac{\sigma_a}{186}\right)^{-13.64}$$

We prepare to add two columns to the previous table. Using a Gerber fatigue locus on the designer's fatigue diagram we can write

$$S_f = \begin{cases} \dfrac{\sigma_a}{1 - (\sigma_m/S_{ut})^2} & \sigma_m > 0 \\ S_e & \sigma_m \le 0 \end{cases}$$

Cycle 1: $r = 70/10 = 7$, and the strength amplitude is

$$S_a = \frac{7^2 151^2}{2(67.5)}\left\{-1 + \sqrt{1 + \left[\frac{2(67.5)}{7(151)}\right]^2}\right\} = 67.2 \text{ kpsi}$$

Since $\sigma_a > S_a$, that is, $70 > 67.2$, life is reduced.

$$S_f = \frac{70}{1 - (10/151)^2} = 70.3 \qquad N = \left(\frac{70.3}{186}\right)^{-13.64} = 0.58\,(10^6)$$

Cycle 2: $r = 10/50 = 0.2$, and the strength amplitude is

$$S_a = \frac{0.2^2 151^2}{2(67.5)}\left\{-1 + \sqrt{1 + \left[\frac{2(67.5)}{0.2(151)}\right]^2}\right\} = 24.2 \text{ kpsi}$$

Since $\sigma_a < S_a$, that is $10 < 24.2$, then $S_f = S_e$ and indefinite life follows.
Cycle 3: $r = 10/-30 = -0.333$, and since $\sigma_m < 0$, $S_f = S_e$, indefinite life follows.

Cycle Number	S_f, kpsi	N, cycles
1	70.3	$0.58(10^6)$
2	67.5	∞
3	67.5	∞

From Eq. (7–62) the damage per block

$$D = \frac{1}{N_i} = \frac{1}{0.58\,(10^6)} + \frac{1}{\infty} + \frac{1}{\infty} = \frac{1}{0.58\,(10^6)}$$

To illustrate the use of the Miner rule, let us choose a steel having the properties $S_{ut} = 80$ kpsi and $S'_{e,0} = 40$ kpsi, where we have used the designation $S'_{e,0}$ instead of the more usual S'_e to indicate the endurance limit of the *virgin*, or *undamaged*, *material*. The log S–log N diagram for this material is shown in Fig. 7–24 by the heavy solid line. Now apply, say, a reversed stress $\sigma_1 = 60$ kpsi for $n_1 = 3000$ cycles. Since $\sigma_1 > S'_{e,0}$, the endurance limit will be damaged, and we wish to find the new endurance limit $S'_{e,1}$ of the damaged material using the Miner rule. The figure shows that the material has a life $N_1 = 8320$ cycles, and consequently, after the application of σ_1 for 3000 cycles, there are $N_1 - n_1 = 5320$ cycles of life remaining. This locates the finite-life strength

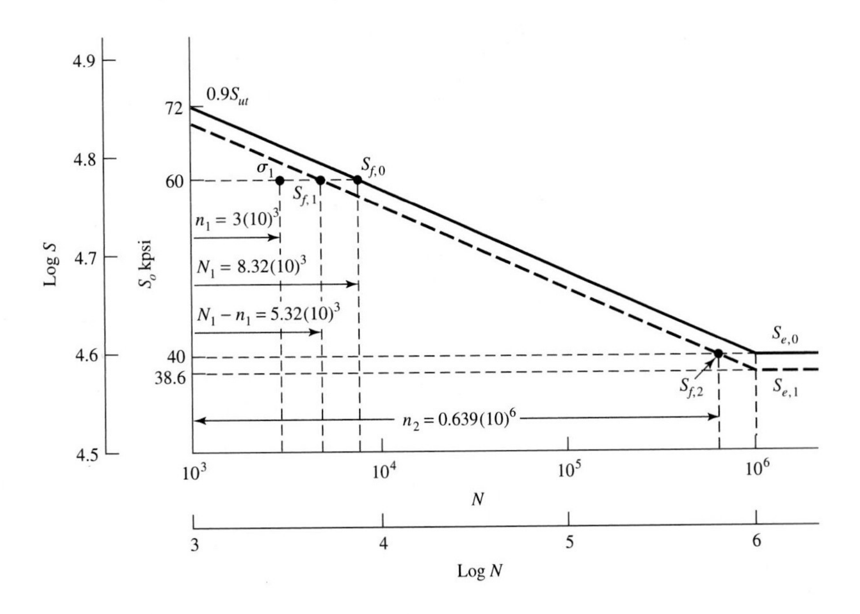

Figure 7–24

Use of the Miner rule to predict the endurance limit of a material which has been overstressed for a finite number of cycles.

$S_{f,1}$ of the damaged material, as shown in Fig. 7–24. To get a second point, we ask the question, With n_1 and N_1 given, how many cycles of stress $\sigma_2 = S'e, 0$ can be applied before the damaged material fails? This corresponds to n_2 cycles of stress reversal, and hence, from Eq. (7–60), we have

$$\frac{n_1}{N_1} + \frac{n_2}{N_2} = 1 \tag{a}$$

or

$$n_2 = \left(1 - \frac{n_1}{N_1}\right) N_2 \tag{b}$$

Then

$$n_2 = \left[1 - \frac{3(10)^3}{8.32(10)^3}\right](10)^6 = 0.639(10)^6 \text{ cycles}$$

This corresponds to the finite-life strength $S_{f,2}$ in Fig. 7–24. A line through $S_{f,1}$ and $S_{f,2}$ is the log S–log N diagram of the damaged material according to the Miner rule. The new endurance limit is $S_{e,1} = 38.6$ kpsi.

The equation of the virgin material failure locus in Fig. 7–24 in the 10^3 to 10^6 range is

$$S_f = aN^b = 129.6N^{-0.085\,318}$$

The cycles-to-failure at stress level $\sigma_1 = 60$ kpsi is

$$N_1 = \left(\frac{\sigma_1}{129.6}\right)^{1/-0.085\,318} = \left(\frac{60}{129.6}\right)^{1/-0.085\,318} = 8320$$

From Eq. (7–62) after 3000 cycles at stress level 1, the damage is

$$D = \frac{3000}{8320} = 0.3606$$

This makes

$$\frac{n_2}{N_2} = 1 - D = 1 - 0.3606 = 0.6394$$

At level 2 σ_2 is 40 kpsi, which is S_e, with an associated life of 10^6, so

$$n_2 = (1 - D)N_2 = 0.6394\left(10^6\right)$$

We could leave it at this, but a little more investigation can be helpful. We have two points on the new fatigue locus, $N_1 - n_1, \sigma_1$ and n_2, σ_2. It is useful to prove that the slope of the new line is still b. For the equation $S_f = a'N^{b'}$, where the values of a' and b' are established by two points α and β, the equation for b' is

$$b' = \frac{\log \sigma_\alpha/\alpha_\beta}{\log N_\alpha/N_\beta} \tag{a}$$

Examine the denominator of Eq. (a):

$$\log\frac{N_\alpha}{N_\beta} = \log\frac{N_1 - n_1}{(1-D)N_2} = \log\frac{1 - n_1/N_1}{(1-D)N_2/N_1} = \log\frac{1-D}{1-D}\frac{N_1}{N_2}$$

$$= \log\frac{N_1}{N_2} = \log\frac{\left(\sigma_1/a'\right)^{1/b}}{(\sigma_2/a')^{1/b}} = \log\left(\frac{\sigma_1}{\sigma_2}\right)^{1/b} = \frac{1}{b}\log\frac{\sigma_1}{\sigma_2}$$

Substituting into Eq. (a) gives

$$b' = \frac{\log(\sigma_1/\sigma_2)}{(1/b)\log(\sigma_1/\sigma_2)} = b$$

Figure 7–25

Use of the Manson method to predict the endurance limit of a material that has been overstressed for a finite number of cycles.

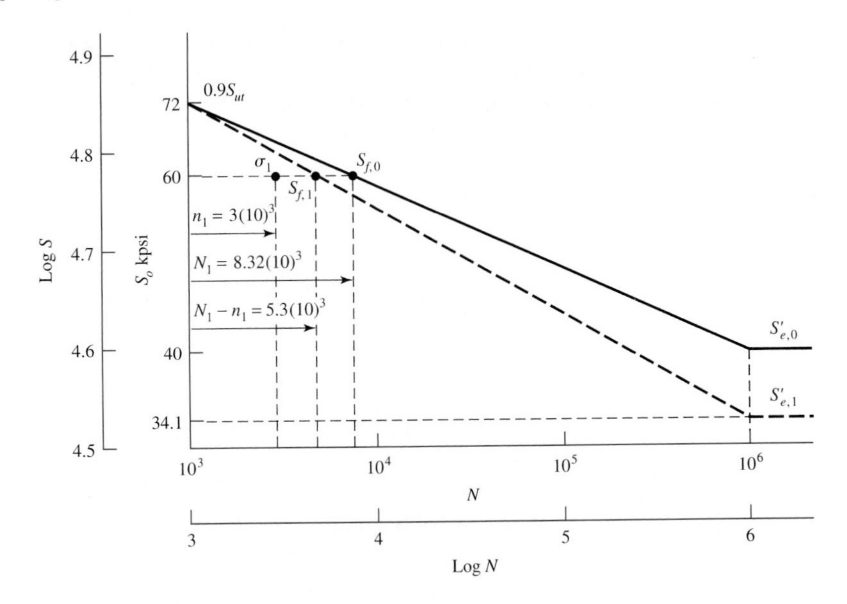

which means the damaged material line has the same slope as the virgin material line; therefore, the lines are parallel. This information can be helpful in writing a computer program for the Palmgren-Miner hypothesis.

Though the Miner rule is quite generally used, it fails in two ways to agree with experiment. First, note that this theory states that the static strength S_{ut} is damaged, that is, decreased, because of the application of σ_1; see Fig. 7–24 at $N = 10^3$ cycles. Experiments fail to verify this prediction.

The Miner rule, as given by Eq. (7–60), does not account for the order in which the stresses are applied, and hence ignores any stresses less than $S'_{e,0}$. But it can be seen in Fig. 7–24 that a stress σ_3 in the range $S'_{e,1} < \sigma_3 < S'_{e,0}$ would cause damage if applied after the endurance limit had been damaged by the application of σ_1.

Manson's[26] approach overcomes both of the deficiencies noted for the Palmgren-Miner method; historically it is a much more recent approach, and it is just as easy to use. Except for a slight change, we shall use and recommend the Manson method in this book. Manson plotted the S-log N diagram instead of a log S–log N plot as is recommended here. Manson also resorted to experiment to find the point of convergence of the S-log N lines corresponding to the static strength, instead of arbitrarily selecting the intersection of $N = 10^3$ cycles with $S = 0.9S_{ut}$ as is done here. Of course, it is always better to use experiment, but our purpose in this book has been to use the simple tension-test data to learn as much as possible about fatigue failure.

The method of Manson, as presented here, consists in having all log S–log N lines, that is, lines for both the damaged and the virgin material, converge to the same point, $0.9S_{ut}$ at 10^3 cycles. In addition, the log S–log N lines must be constructed in the same historical order in which the stresses occur.

The data from the preceding example are used for illustrative purposes. The results are shown in Fig. 7–25. Note that the strength $S_{f,1}$ corresponding to $N_1 - n_1 = 5.32(10)^3$ cycles is found in the same manner as before. Through this point and through $0.9S_{ut}$ at 10^3 cycles, draw the heavy dashed line to meet $N = 10^6$ cycles and define the endurance

26. S. S. Manson, A. J. Nachtigall, C. R. Ensign, and J. C. Fresche, "Further Investigation of a Relation for Cumulative Fatigue Damage in Bending," *Trans. ASME, J. Eng. Ind.*, ser. B, vol. 87, No. 1, pp. 25–35, February 1965.

limit $S'_{e,1}$ of the damaged material. In this case the new endurance limit is 34.1 kpsi, somewhat less than that found by the Miner method.

It is now easy to see from Fig. 7–25 that a reversed stress $\sigma = 36$ kpsi, say, would not harm the endurance limit of the virgin material, no matter how many cycles it might be applied. However, if $\sigma = 36$ kpsi should be applied *after* the material was damaged by $\sigma_1 = 60$ kpsi, then additional damage would be done.

Both these rules involve a number of computations, which are repeated every time damage is estimated. For complicated stress-time traces, this might be every cycle. Clearly a computer program is useful to perform the tasks, including scanning the trace and identifying the cycles.

Collins said it well: "In spite of all the problems cited, the Palmgren linear damage rule is frequently used because of its simplicity and the experimental fact that other more complex damage theories do not always yield a significant improvement in failure prediction reliability."[27]

7–16 The Fracture-Mechanics Approach

Here we present a method of estimating the remaining life in a specimen after discovery of a crack. The method is known as *linear elastic fracture mechanics*, and it requires the assumption that plane strain conditions prevail.[28]

Assuming an elastic isotropic material, the growth of a fatigue crack is approximated by the equation

$$\frac{da}{dN} = \frac{C\,(\Delta K - \Delta K_{\mathrm{th}})^m}{(1 - R)K_c - \Delta K} \tag{7–63}$$

where C and m are empirical constants. The stress ratio R is given by Eq. (7–38), and K_c is the critical stress-intensity factor as defined in Sec. 5–21. The quantity ΔK depends upon the stress range, and is

$$\Delta K = \sigma_r (\pi a)^{1/2} \left(\frac{K_{\mathrm{I}}}{K_0} \right) \tag{7–64}$$

where σ_r is the stress range as defined in Sec. 7-10. The ratio $K_I K_0$ corrects for the geometry of the problem and is the ordinate in Figs. 5–19 and 5–21 to 5–26.

The quantity ΔK_{th} in Eq. (7–63) is called the threshold value; crack growth is not expected for lesser values of ΔK.

We can bypass much of the complexity of Eq. (7–63) by using the power equation

$$\frac{da}{dN} = C\,(\Delta K_{\mathrm{I}})^m \tag{a}$$

Values of C and m for several classes of materials are listed in Table 7–17. By rearranging and integrating both sides of Eq. (a), we have

$$\int_{a_0}^{a_f} da = \int_0^{N_f} C\,(\Delta K_{\mathrm{I}})^m\, dN \tag{7–65}$$

Here a_0 is the initial crack length, a_f is the final crack length corresponding to failure, and N_f is the estimated number of cycles required to produce a failure. Note that the ratio

27. J. A. Collins, *Failure of Materials in Mechanical Design*, Wiley, New York, 1981, p. 243.

28. Recommended references are: Collins, op. cit.; H. O. Fuchs and R. I. Stephens, *Metal Fatigue in Engineering*, Wiley, New York, 1980; and Harold S. Reemsnyder, "Constant Amplitude Fatigue Life Assessment Models," *SAE Trans. 820688*, vol. 91, November 1983.

Table 7–17

Values of the Factor C and Exponent m in Eq. (a)

Material	Equation and Units
Ferritic-pearlitic steels	$\frac{da}{dN}(\text{m/cycle}) = 6.9\,(10^{-12})\,(\Delta K\ \text{MPa}\sqrt{\text{m}})^{3.0}$ $\frac{da}{dN}(\text{in/cycle}) = 3.6\,(10^{-10})\,(\Delta K\ \text{kpsi}\sqrt{\text{in}})^{3.0}$
Martensitic steels	$\frac{da}{dN}(\text{m/cycle}) = 1.35\,(10^{-10})\,(\Delta K\ \text{MPa}\sqrt{\text{m}})^{2.25}$ $\frac{da}{dN}(\text{in/cycle}) = 6.6\,(10^{-9})\,(\Delta K\ \text{kpsi}\sqrt{\text{in}})^{2.25}$
Austenitic stainless steels	$\frac{da}{dN}(\text{m/cycle}) = 5.6\,(10^{-12})\,(\Delta K\ \text{MPa}\sqrt{\text{m}})^{3.25}$ $\frac{da}{dN}(\text{in/cycle}) = 3.0\,(10^{-10})\,(\Delta K\ \text{kpsi}\sqrt{\text{in}})^{3.25}$

Source: Fuchs and Stephens, op. cit., pp. 85, 86.

K_{I}/K_0 may vary in the integration interval. If this should happen, then Reemsnyder[29] suggests the use of numerical integration employing the algorithm

$$\begin{aligned}\delta a_j &= C(\Delta K)_j^m(\delta N)_j \\ a_{j+1} &= a_j + \delta a_j \\ N_{j+1} &= N_j + \delta N_j \\ N_j &= \sum \delta N_j\end{aligned} \qquad (7\text{–}66)$$

Here δa_j and δN_j are increments of the crack length and the number of cycles. The procedure is to select a value of δN_j, compute ΔK using a_0, and then find the next value of a. Repeat until $a = a_f$. The procedure can be illustrated using a simple example.

EXAMPLE 7–17 The bar shown in Fig. 7–26 is subjected to a repeated moment $M_{\max} = 208$ lb · in. The bar is hot-rolled steel with $S_{ut} = 60$ kpsi and $S_y = 33$ kpsi. As shown, a nick of size 0.004 in has been discovered on the bottom of the bar. Estimate the number of cycles of life remaining.

Solution The stress range σ_r for use in Eq. (7–64) is always computed using the nominal (uncracked) area. Thus

$$\frac{I}{c} = \frac{bh^2}{6} = \frac{0.25(0.5)^2}{6} = 0.0104\ \text{in}^3$$

Figure 7–26

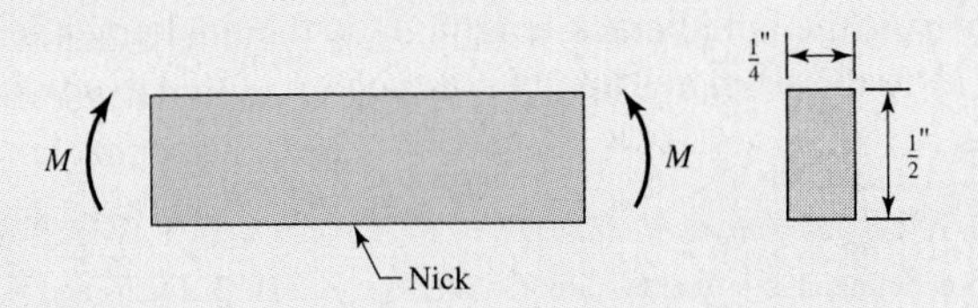

29. Op. cit.

Therefore the stress range is

$$\sigma_r = \frac{M}{I/c} = \frac{208}{0.0104}\left(10^{-3}\right) = 20 \text{ kpsi}$$

If the crack grows, it will eventually become long enough that the bar will fail by yielding. Designate this crack length a_f. The section modulus corresponding to this length is

$$\left(\frac{I}{c}\right)_f = \frac{b\left(0.5 - a_f\right)^2}{6}$$

Since $\sigma_{\max} = S_y = M/(I/c)_f$, we have $M/S_y = (I/c)_f$, or

$$\frac{208}{33\left(10^3\right)} = \frac{0.25}{6}\left(0.5 - a_f\right)^2 \tag{1}$$

The solution to this equation is $a_f = 0.111$ in.

Table 7–15 gives $C = 3.6(10^{-10})$ and $m = 3.0$ for use in Eq. (a). Referring next to Fig. 5–23, we compute the largest a/h ratio as

$$\frac{a}{h} = \frac{0.111}{0.5} = 0.222$$

Thus a/h varies from near zero to 0.222, and for this range $K_{\mathrm{I}}/K_0 = 1.06$ and is nearly a constant. We shall assume it to be so, and integrate Eq. (7–65) analytically. Thus

$$K_{\mathrm{I}} = \sigma_r(\pi a)^{1/2}\left(\frac{K_{\mathrm{I}}}{K_0}\right) = \left[20(\pi)^{1/2}(1.06)\right]^{3.0} a^{1.5}$$

Thus Eq. (7–65) becomes

$$\int_{a_0}^{a_f} \frac{da}{a^{1.5}} = \int_0^{N_f} 3.6\left(10^{-10}\right)\left[20(\pi)^{1/2}(1.06)\right]^{3.0} dN$$

Performing the indicated operations, we have

$$\left.\frac{-2}{\sqrt{a}}\right|_{0.004}^{0.111} = 1.91\left(10^{-5}\right) N\Big|_0^{N_f}$$

Answer from which we find $N_f = 1.34(10^6)$ cycles, which is the estimated remaining life.

7–17 Surface Fatigue Strength

The surface fatigue mechanism is not definitively understood. The contact-affected zone, in the absence of surface shearing tractions, entertains compressive principal stresses. Rotary fatigue has its cracks grown at or near the surface in the presence of tensile stresses which are associated with crack propagation, to catastrophic failure. There are shear stresses in the zone, which are largest just below the surface. Cracks seem to grow from this stratum until small pieces of material are expelled, leaving pits on the surface. Because engineers had to design durable machinery before the surface fatigue phenomenon was understood in detail, they had taken the posture of conducting tests, observing pits on the surface, and declaring failure at an arbitrary projected area of hole, and they related this to the Hertzian contact pressure. This compressive stress did not produce the failure directly, but whatever the failure mechanism, whatever the stress type that was instrumental in the failure, the contact stress was an *index* to its magnitude.

Buckingham[30] conducted a number of tests relating the fatigue at 10^8 cycles to endurance strength (Hertzian contact pressure). While there is evidence of an endurance limit at about $3(10^7)$ cycles for cast materials, hardened steel rollers showed no fatigue limit up to $4(10^8)$ cycles. Subsequent testing on hard steel shows no endurance limit. Hardened steel exhibits such high endurance strengths that its use in resisting surface fatigue is widespread.

Our studies thus far have dealt with the failure of a machine element by yielding, by fracture, and by fatigue. The endurance limit obtained by the rotating-beam test is frequently called the *flexural endurance limit*, because it is a test of a rotating beam. In this section we shall study a property of *mating materials* called the *surface endurance shear*. The design engineer must frequently solve problems in which two machine elements mate with one another by rolling, sliding, or a combination of rolling and sliding contact. Obvious examples of such combinations are the mating teeth of a pair of gears, a cam and follower, a wheel and rail, and a chain and sprocket. A knowledge of the surface strength of materials is necessary if the designer is to create machines having a long and satisfactory life.

When two surfaces roll or roll and slide against one another with sufficient force, a pitting failure will occur after a certain number of cycles of operation. Authorities are not in complete agreement on the exact mechanism of the pitting; although the subject is quite complicated, they do agree that the Hertz stresses, the number of cycles, the surface finish, the hardness, the degree of lubrication, and the temperature all influence the strength. In Sec. 3–21 it was learned that, when two surfaces are pressed together, a maximum shear stress is developed slightly below the contacting surface. It is postulated by some authorities that a surface fatigue failure is initiated by this maximum shear stress and then is propagated rapidly to the surface. The lubricant then enters the crack which is formed and, under pressure, eventually wedges the chip loose.

To determine the surface fatigue strength of mating materials, Buckingham designed a simple machine for testing a pair of contacting rolling surfaces in connection with his investigation of the wear of gear teeth. Buckingham and, later, Talbourdet gathered large numbers of data from many tests so that considerable design information is now available. To make the results useful for designers, Buckingham defined a *load-stress factor*, also called a *wear factor*, which is derived from the Hertz equations. Equations (3–98) and (3–99) for contacting cylinders are found to be

$$b = \sqrt{\frac{2F}{\pi l} \frac{\left(1 - \nu_1^2\right)/E_1 + \left(1 - \nu_2^2\right)/E_2}{(1/d_1) + (1/d_2)}} \tag{7–67}$$

$$p_{\max} = \frac{2F}{\pi b l} \tag{7–68}$$

where b = half width of rectangular contact area
F = contact force
l = width of cylinders
ν = Poisson's ratio
E = modulus of elasticity
d = cylinder diameter

It is more convenient to use the cylinder radius; so let $2r = d$. If we then designate the width of the cylinders as w instead of l and remove the square root sign, Eq. (7–67) becomes

30. Earle Buckingham, *Analytical Mechanics of Gears*, McGraw-Hill, New York, 1949.

$$b^2 = \frac{4F}{\pi w} \frac{\left(1 - \nu_1^2\right)/E_1 + \left(1 - \nu_2^2\right)/E_2}{1/r_1 + 1/r_2} \tag{7–69}$$

We can define a *surface endurance strength* S_C using

$$p_{\max} = \frac{2F}{\pi b w} \tag{7–70}$$

as

$$S_C = \frac{2F}{\pi b w} \tag{7–71}$$

which may also be called *contact strength*, the *contact fatigue strength*, or the *Hertzian endurance strength*. The strength is the contacting pressure which, after a specified number of cycles, will cause failure of the surface. Such failures are often called *wear* because they occur over a very long time. They should not be confused with abrasive wear, however. By substituting the value of b in Eq. (7–69) and substituting the result into Eq. (7–71), we obtain

$$\frac{F}{w}\left(\frac{1}{r_1} + \frac{1}{r_2}\right) = \pi S_c^2 \left[\frac{1 - \nu_1^2}{E_1} + \frac{1 - \nu_2^2}{E_2}\right] = K_1 \tag{7–72}$$

The left expression consists of parameters a designer may seek to control independently. The central expression consists of material properties that come with the material and condition specification. The third expression is the parameter K_1, Buckingham's load-stress factor, determined by a test fixture with values F, W, r_1, r_2 and the number of cycles associated with the first tangible evidence of fatigue. In gear studies a similar K factor is used:

$$K_g = \frac{K_1}{4} \sin\phi \tag{7–73}$$

where ϕ is the tooth pressure angle, and the term $[(1 - \nu_1^2)/E_1 + (1 - \nu_2^2)/E_2]$ is defined as $1/(\pi C_P^2)$, so that

$$S_C = C_P \sqrt{\frac{F}{w}\left(\frac{1}{r_1} + \frac{1}{r_2}\right)} \tag{7–74}$$

Buckingham and others reported K_1 for 10^8 cycles and nothing else. This gives only one point on the $S_C N$ curve. For cast metals this may be sufficient, but for wrought steels, heat-treated, some idea of the slope is useful in meeting design goals of other than 10^8 cycles.

Experiments show that K_1 versus N, K_g versus N, and S_C versus N data are rectified by loglog transformation. This suggests that

$$K_1 = \alpha_1 N^{\beta_1} \qquad K_g = a N^b \qquad S_C = \alpha N^{\beta}$$

The three exponents are given by

$$\beta_1 = \frac{\log(K_1/K_2)}{\log(N_1/N_2)} \qquad b = \frac{\log\left(K_{g1}/K_{g2}\right)}{\log(N_1/N_2)} \qquad \beta = \frac{\log(S_{C1}/S_{C2})}{\log(N_1/N_2)} \tag{7–75}$$

Data on induction-hardened steel on steel gives $(S_C)_{10^7} = 271$ kpsi and $(S_C)_{10^8} =$ 239 kpsi, so β, from Eq. (7–75), is

$$\beta = \frac{\log(271/239)}{\log\left(10^7/10^8\right)} = -0.055$$

It may be of interest that the American Gear Manufacturers Association (AGMA) uses -0.056 between $10^4 < N < 10^{10}$ if the designer has no data to the contrary beyond 10^7 cycles.

A long standing correlation in steels between S_C and H_B at 10^8 cycles is

$$(S_C)_{10^8} = \begin{cases} 0.4H_B - 10 \text{ kpsi} \\ 2.76H_B - 70 \text{ MPa} \end{cases} \tag{7–76}$$

AGMA uses

$${}_{0.99}(S_C)_{10^7} = 0.327H_B + 26 \text{ kpsi} \tag{7–77}$$

Equation (7–74) can be used in design to find an allowable surface stress by using a design factor. Since this equation is nonlinear in its stress-load transformation, the designer must decide if loss of function denotes inability to carry the load. If so, then to find the allowable stress, one divides the load F by the design factor n_d:

$$\sigma_C = C_P\sqrt{\frac{F}{wn_d}\left(\frac{1}{r_1}+\frac{1}{r_2}\right)} = \frac{C_P}{\sqrt{n_d}}\sqrt{\frac{F}{w}\left(\frac{1}{r_1}+\frac{1}{r_2}\right)} = \frac{S_C}{\sqrt{n_d}}$$

and $n_d = (S_C/\sigma_C)^2$. If the loss of function is focused on stress, then $n_d = S_C/\sigma_C$. It is recommended that an engineer

- Decide whether loss of function is failure to carry load or stress.
- Define the design factor and factor of safety accordingly.
- Announce what he or she is using and why.
- Be prepared to defend his or her position.

In this way everyone who is party to the communication knows what a design factor (or factor of safety) of 2 means and adjusts, if necessary, the judgmental perspective.

7–18 The Designer's Fatigue Diagram

It was shown in Chap. 3 that the *nine* elements that describe the state of stress at a point in a body can be reduced to *three* if principal stresses are used, to *two* if octahedral shear and octahedral normal stress are used, and to *one* if distortion energy (von Mises stress) is used. For ductile metals that are described by the distortion-energy theory, two important facts are useful to us:

- The von Mises stress is the essential determinant if the onset of yielding is to be reliably described.
- The von Mises components in variable stress problems can describe the fatigue failure locus.

Since loss of function in machine parts can be permanent distortion (yielding) as well as fatigue fracture, both of these threats can be monitored using von Mises stresses. Several variable stress diagrams have been used so far in this chapter. They are examples of partial or full use of the *designer's fatigue diagram* or its variations. The beauty of this concept is that just two components, the amplitude component σ'_a and the midrange component σ'_m, can describe the stress conditions at the intended (design) point. A *load line* shows the locus of stress conditions which can occur if the "loading" is changed. The intersection of the load line with the failure locus defined the stress (strength) conditions at failure using the components of strength S_a and S_m. The comparison of the design stress condition, described by σ'_a and σ'_m, and the failure condition, described by S_a and

S_m, allows quantitative inspection of the margin that separated the usage condition from the failure condition. All this can be *graphed on a plane*, which makes inspection and contemplation simple and allows geometric thought and reasoning to help because the situation can be visualized.

Two principal concerns of loss of function are (1) yielding and (2) fatigue fracture. The failure loci of both these threats can be mapped onto the designer's fatigue diagram. The information needed for the first-cycle yield line is the yield strength as determined by the tensile test. The yield locus, when drawn on the diagram, is a straight line with coordinate intercepts of tensile yield strength in the first quadrant and compressive yield strength in the second quadrant. Such lines are sometimes called Langer lines. Fatigue fracture loci require endurance strength and one static (yield or ultimate tensile) strength. From these the fatigue locus can be constructed as shown in Fig. 7–27*a*. If endurance testing is not performed on the material to be used in manufacture, then estimates of endurance strength are made using correlations discussed in this chapter. If the load lines are radial, as depicted in Fig. 7–27*b* where the fatigue fracture locus contour is displayed symbolically, we find the design condition, described by σ'_a, σ'_m, within a failure boundary. If the design point is one load line,

- OA or OD, the threat is from first-cycle yielding.
- OB or OC, the threat is from fatigue fracture.

The ASME-elliptic locus curves from point S_e, 0 to point 0, S_{yt}, having crossed the Langer line. To avoid drawing the full designer fatigue diagram, engineers have defined

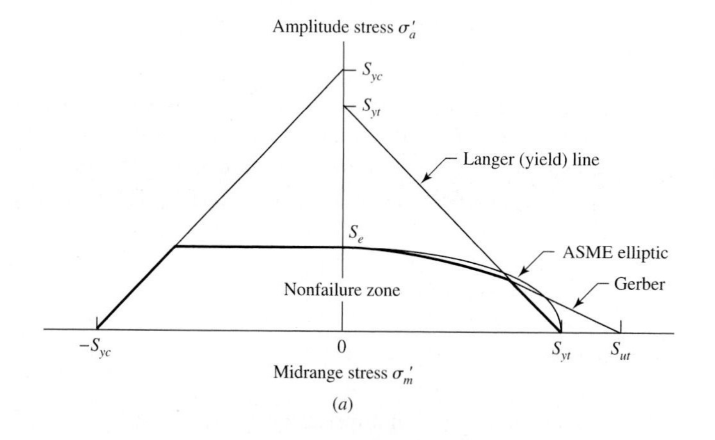

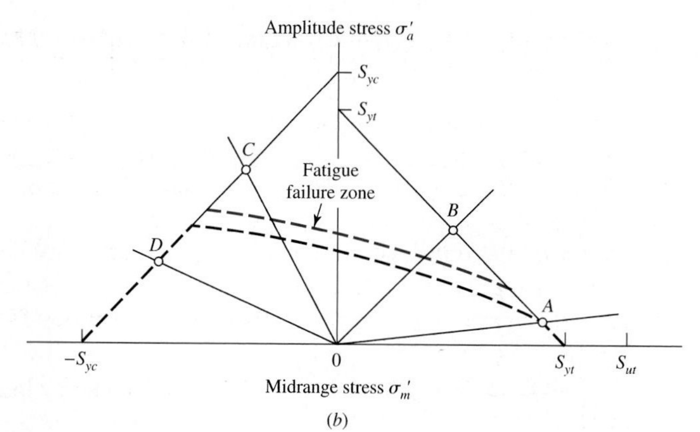

Figure 7–27

The designer's fatigue diagram displayed (*a*) by itself, and (*b*) with various load lines. It is possible to assess whether the threat is from fatigue or first-cycle yielding. The dashed fatigue locus is a zone.

the load lines that pass through the intersections of failure loci as having *critical slope*. The critical slope can be deduced from property relationships. Given the slope of a radial load line, if it is greater than the critical slope, the threat is from fatigue; if less, the threat is from first-cycle yielding.

In spite of the complexity of the ductile material failure problem, the evolution to the designer's fatigue diagram has enabled engineers to address a wide variety of problems. As simple as the diagram is (now), it is still the designer's responsibility to establish the fatigue failure locus that is appropriate to the problem at hand. In ductile material, Gerber-parabolic and ASME-elliptic are in good agreement, but a choice has to be made. The ASME *Design of Transmission Shafting* (Standard ANSI/ASME B106.1M, second printing, 1985) will be helpful in deciding.

Nonferrous metals (excluding aluminums) exhibit fatigue failure loci which are more often Goodman than Gerber. Aluminums are mostly Gerber with some exceptions. In plotting failure points, care has to be taken to delete those that appear where the yield strength was exceeded. If the points are not deleted, then coldwork has modified the properties, and the specimen tested is not the specimen made. In precision machinery, shafting which yields on the first cycle can be bent in an unpredictable way and has incurred loss of function. Such points that involve yielding should not influence decisions where yielding is supposedly excluded.

When searching out and plotting data on your designer's fatigue diagram in order to decide on an appropriate failure locus, you may have to discover how the raw data gathered by the investigator(s) were manipulated into published data points. When a rotary specimen is loaded with a bending moment and a steady torque, one has to be careful what one plots. The normal and shear stresses are

$$\sigma_x = \sigma_{xa} = \frac{32M_a}{\pi d^3} \qquad \tau_{xy} = \tau_{xym} = \frac{16T_m}{\pi d^3}$$

If the investigator takes σ_a as σ_x and draws a Mohr circle for amplitude component stresses, he or she finds the largest associated shear stress to be $\tau_a = \sigma_a/2 = 16M_a/(\pi d^3)$. Constructing a Mohr circle for midrange components using $\tau_{xy} = \tau_{xym} = 16T_m/(\pi d^3)$ as the largest shear, the associated σ_a and σ_m, following the maximum shear hypothesis, are

$$\sigma_a = 2\tau_a = 2\frac{16M_a}{\pi d^3} = \frac{32M_a}{\pi d^3} \tag{7–78}$$

$$\sigma_m = 2\tau_m = 2\frac{16T_m}{\pi d^3} = \frac{32T_m}{\pi d^3} \tag{7–79}$$

A distortion-energy interpretation gives

$$\sigma'_a = \left(\sigma_{xa}^2 + 3\tau_{xya}^2\right)^{1/2} = \left[\left(\frac{32M_a}{\pi d^3}\right)^2 + 3(0)\right]^{1/2} = \frac{32M_a}{\pi d^3} \tag{7–80}$$

$$\sigma'_m = \left(\sigma_{xm}^2 + 3\tau_{xym}^2\right)^{1/2} = \left[0 + 3\left(\frac{16\sqrt{3}T_m}{\pi d^3}\right)^2\right]^{1/2} = \sqrt{3}\frac{16T_m}{\pi d^3} \tag{7–81}$$

Compare the coordinates

$$\sigma_a = \frac{32M_a}{\pi d^3} \qquad \sigma_m = \frac{32T_m}{\pi d^3} \qquad \text{(maximum-shear interpretation)} \tag{7–82}$$

$$\sigma'_a = \frac{32M_a}{\pi d^3} \qquad \sigma'_m = \frac{16\sqrt{3}T_m}{\pi d^3} \qquad \text{(distortion-energy interpretation)} \tag{7–83}$$

The abscissas differ by the ratio $\sigma_a'/\sigma_m' = 1.155$. Before you plot fatigue data on your designer's fatigue diagram, you should ask

- Are you embracing and plotting data with a maximum shear interpretation of data on a $\sigma_a\sigma_m$ diagram?
- Are you embracing the distortion-energy theory, plotting data with that interpretation on a $\sigma_a'\sigma_m'$ diagram?
- Are you mixing modes?

The obfuscation of raw data in technical writing leads engineers to discard data of obscure lineage. The author of a technical paper can save the day (for the data) by making available to you information that was not clear, or not in the paper at all. It can be obtained by a phone call.

The fatigue diagram for a brittle material differs markedly from that of a ductile material:

- Yielding is not involved since the material may not have a yield strength.
- Characteristically, the compressive ultimate strength exceeds the ultimate tensile strength severalfold.
- First-quadrant fatigue failure locus is concave-upward, Smith-Dolan, for example, and as flat as Goodman. Brittle materials are more sensitive to midrange stress, being lowered, but compressive midrange stresses are beneficial.
- Not enough work has been done on brittle fatigue to discover insightful generalities, so we stay in the first and a bit of the second quadrant.

The most likely domain of designer use is in the range from $-S_{ut} \leq \sigma_m \leq S_{ut}$. The locus in the first quadrant is Goodman, Smith-Dolan, or something in between. The portion of the second quadrant that is used is represented by a straight line between the points $-S_{ut}$, S_{ut} and O, S_e which has the equation

$$S_a = S_e + \left(\frac{S_e}{S_{ut}} - 1\right) S_m \qquad -S_{ut} \leq S_m \leq 0 \quad \text{(for cast iron)}$$

Table E–24 gives properties of gray cast iron. The endurance limit stated is really $k_a k_b S_e'$ and only corrections K_c, d_d, and k_e need be made. The average k_c for axial and torsional loading is 0.9 (see Table 7–10 for torsional k_c). Example 7–18 might be helpful.

EXAMPLE 7–18

A grade 30 gray cast iron is subjected to a load F applied to a 1 by 3/8-in cross-section link with a 1/4-in-diameter hole drilled in the center as depicted in Fig. 7–28*a*. The surfaces are machined. In the neighborhood of the hole, what is the factor of safety guarding against failure under the following conditions:
(*a*) The load $F = 1000$ lb tensile, steady.
(*b*) The load is 1000 lb repeatedly applied.
(*c*) The load fluctuates between -1000 lb and 300 lb without column action.
Use the Smith-Dolan fatigue locus.

Solution

Some preparatory work is needed. From Table E–24, $S_{ut} = 31$ kpsi, $S_{uc} = 109$ kpsi, $k_a k_b S_e' = 14$ kpsi. Since k_c for axial loading is 0.9, then $S_e = (k_a k_b S_e')k_c = 14(0.9) = 12.6$ kpsi. Using Table E–15–1, $A = t(w - d) = 0.375(1 - 0.25) = 0.281\ \text{in}^2$, $d/w = 0.25/1 = 0.25$, and $K_t = 2.45$. The notch sensitivity for cast iron is 0.20, so

$$K_f = 1 + q(K_t - 1) = 1 + 0.20(2.45 - 1) = 1.29$$

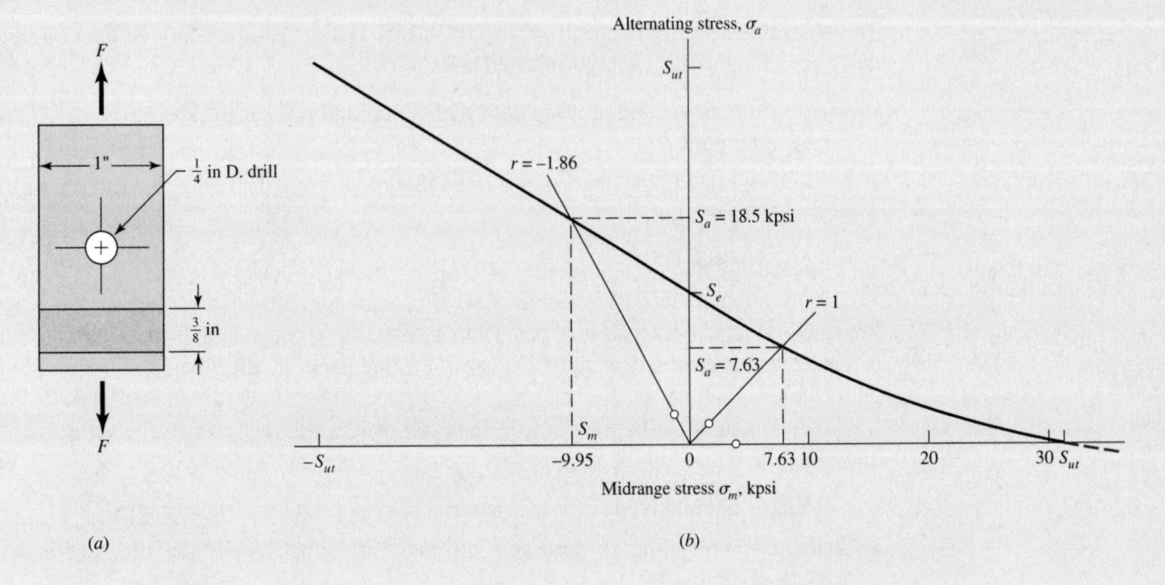

Figure 7–28

The grade 30 cast-iron part in axial fatigue with (*a*) its geometry displayed and (*b*) its designer's fatigue diagram for the circumstances of Ex. 7–18.

$$(a)\ \sigma_a = \frac{K_f F_a}{A} = \frac{1.29(0)}{0.281} = 0 \qquad \sigma_m = \frac{K_f F_m}{A} = \frac{1.29(1000)}{0.281} = 4.59 \text{ kpsi}$$

and

Answer $$n = \frac{S_{ut}}{\sigma_m} = \frac{31.0}{4.59} = 6.75$$

$$(b)\ \ F_a = \frac{F}{2} = \frac{1000}{2} = 500 \text{ lbf} \qquad F_m = F_a = 500 \text{ lbf}$$

$$\sigma_a = \sigma_m = \frac{K_f F_a}{A} = \frac{1.29(500)}{0.281} = 2.30 \text{ kpsi}$$

$$r = \frac{\sigma_a}{\sigma_m} = 1$$

From Eq. (7–58),

$$S_a = \frac{(1)31 + 12.6}{2}\left[-1 + \sqrt{1 + \frac{4(1)31(12.6)}{[(1)31 + 12.6)]^2}}\right] = 7.63 \text{ kpsi}$$

Answer $$n = \frac{S_a}{\sigma_a} = \frac{7.73}{2.30} = 3.32$$

$$(c)\ \ F_a = \frac{1}{2}|300 - (-1000)| = 650 \text{ lbf} \qquad \sigma_a = \frac{1.29(650)}{0.281} = 3.0 \text{ kpsi}$$

$$F_m = \frac{1}{2}[300 + (-1000)] = -350 \text{ lbf} \qquad \sigma_m = \frac{1.29(-350)}{0.281} = -1.61 \text{ kpsi}$$

$$r = \frac{\sigma_a}{\sigma_m} = \frac{-1.0}{1.61} = -1.86$$

Since $S_a = S_e + (S_e/S_{ut} - 1)S_m$ and $S_a = rS_m$, it follows that

$$S_a = \frac{S_e}{1 - \dfrac{1}{r}\left(\dfrac{S_e}{S_{ut}} - 1\right)} = \frac{12.6}{1 - \dfrac{1}{-1.86}\left(\dfrac{12.6}{31} - 1\right)} = 18.5 \text{ kpsi}$$

Answer $$n = \frac{S_a}{\sigma_a} = \frac{18.5}{3.0} = 6.17$$

The Fig. 7–28*b* shows the portion of the designer's fatigue diagram that was constructed.

7–19 An Important Design Decision: The Design Factor in Fatigue

The designer, in envisioning how he or she will execute the geometry of a part subject to the imposed constraints, can begin making a priori decisions without realizing the impact on the design task. Now is the time to note how these things are related to the reliability goal.

The mean value of the design factor is given by Eq. (2–9),

$$\bar{n} = \exp\left[-z\sqrt{\ln\left(+C_n^2\right)} + \ln\sqrt{1 + C_n^2}\right] \tag{2–9}$$

in which, from Table 2–6 for the quotient $\mathbf{n} = \mathbf{S}/\boldsymbol{\sigma}$,

$$C_n = \sqrt{\frac{C_S^2 + C_\sigma^2}{1 + C_\sigma^2}}$$

where C_S is the COV of the significant strength and C_σ is the COV of the significant stress at the critical location. Note that $\bar{n}$ is a function of the reliability goal (through z) and the COVs of the strength and stress. There are no means present, just measures of variability. The nature of C_S in a fatigue situation may be C_{Se} for fully reversed loading, or C_{Sa} otherwise. Also, experience shows $C_{Se} < C_{Sa} < C_{Sut}$, so C_{Se} can be used as a conservative estimate of C_{Sa}. If the loading is bending or axial, the form of σ'_a might be

$$\boldsymbol{\sigma}'_a = \mathbf{K}_f \frac{\mathbf{M}_a c}{I} \qquad \text{or} \qquad \boldsymbol{\sigma}'_a = \mathbf{K}_f \frac{\mathbf{F}}{A}$$

respectively. This makes the COV of $\boldsymbol{\sigma}'_a$, namely $C_{\sigma'_a}$, expressible as

$$C_{\sigma'_a} = \left(C_{Kf}^2 + C_F^2\right)^{1/2}$$

again, a function of variabilities. The COV of $\mathbf{S}_e$, namely C_{Se}, is

$$\begin{aligned} C_{Se} &= \left(C_{ka}^2 + C_{kc}^2 + C_{kd}^2 + C_{ke}^2 + C_\phi^2\right)^{1/2} \\ &= \left(C_{ka}^2 + C_{kc}^2 + C_{kd}^2 + C_{ke}^2 + C_{Se'}^2\right)^{1/2} \end{aligned}$$

again, a function of variabilities. An example will be useful.

EXAMPLE 7–19 A strap to be made from a cold-drawn steel strip workpiece is to carry a fully reversed axial load $\mathbf{F} \sim \mathbf{LN}(1000, 120)$ lb as shown in Fig. 7–29. Consideration of adjacent parts established the geometry as shown in the figure, except for the thickness t. Make a decision as to the magnitude of the design factor if the reliability goal is to be 0.999 95, then make a decision as to the workpiece thickness t.

Solution Let us take each a priori decision and note the consequence:

A Priori Decision	Consequence
Use 1018 CD steel	$\bar{S}_{ut} = 87.6$ kpsi, $C_{Sut} = 0.0655$
Function:	
Carry axial load	$C_F = 0.12,\ C_{kc} = 0.125$
$R \geq 0.999\ 95$	$z = -3.891$
Machined surfaces	$C_{ka} = 0.058$
Hole critical	$C_{Kf} = 0.10,\ C_{\sigma'_a} = (0.10^2 + 0.12^2)^{1/2} = 0.156$
Ambient temperature	$C_{kd} = 0$
Correlation method	$C_\phi = 0.138$
Hole drilled	$C_{Se} = (0.058^2 + 0.125^2 + 0.138^2)^{1/2} = 0.195$
	$C_n = \sqrt{\dfrac{C_{Se}^2 + C_{\sigma'_a}^2}{1 + C_{\sigma'_a}^2}} = \sqrt{\dfrac{0.195^2 + 0.156^2}{1 + 0.156^2}} = 0.2467$
	$\bar{n} = \exp\left[-(-3.891)\sqrt{\ln(1 + 0.2467^2)} + \ln\sqrt{1 + 0.2467^2}\right] = 2.65$

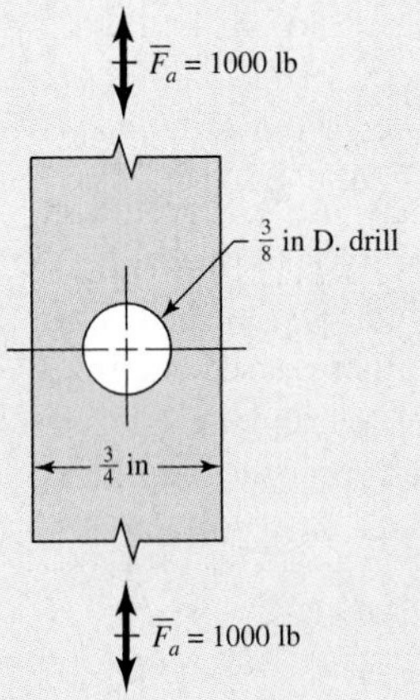

These eight a priori decisions have quantified the mean design factor as $\bar{n} = 2.65$. Proceeding deterministically hereafter we write

$$\sigma'_a = \frac{\bar{S}_e}{\bar{n}} = \bar{K}_f \frac{\bar{F}}{(w-d)t}$$

from which

$$t = \frac{\bar{K}_f \bar{n} \bar{F}}{(w-d)\bar{S}_e}$$

To evaluate the preceding equation we need $\bar{S}_e$ and $\bar{K}_f$. The Marin factors are

$$\mathbf{k}_a = 2.67\bar{S}_{ut}^{-0.265}\mathbf{LN}(1, 0.058) = 2.67(87.6)^{-0.265}\mathbf{LN}(1, 0.058)$$

$$\bar{k}_a = 0.816$$

$$k_b = 1 \text{ (see } \mathbf{k}_c)$$

$$\mathbf{k}_c = 1.23\bar{S}_{ut}^{-0.078}\mathbf{LN}(1, 0.125) = 0.868\mathbf{LN}(1, 0.125)$$

$$\bar{k}_c = 0.868$$

$$\bar{k}_d = \bar{k}_e = 1$$

$$\bar{\phi}_{0.030} = 0.506$$

and the endurance strength is

$$\bar{S}_e = 0.816(0.868)(1)(1)(1)0.506(87.6) = 31.4 \text{ kpsi}$$

The hole governs. From Table E–15–1 we find $d/w = 0.50$, therefore $K_t = 2.18$. From Table 7–12 $\sqrt{a} = 5/\bar{S}_{ut} = 5/87.6 = 0.057$, $r = 0.375$ in. From Eq. (7–27) the fatigue stress concentration factor is

$F_a = 1000$ lb

$\frac{3}{8}$ in D. drill

$\frac{3}{4}$ in

$F_a = 1000$ lb

Figure 7–29

A strap with a thickness t is subjected to a fully reversed axial load of 1000 lb. Example 7–19 considers the thickness necessary to attain a reliability of 0.999 95 against a fatigue failure.

$$\bar{K}_f = \frac{2.18}{1 + \dfrac{2}{\sqrt{0.375}} \dfrac{2.18 - 1}{2.18} 0.057} = 1.98$$

The thickness t can now be determined from

$$t \geq \frac{\bar{K}_f \bar{n} \bar{F}}{(w - d)S_e} = \frac{1.98(2.65)1000}{(0.75 - 0.375)31\,400} = 0.446 \text{ in}$$

Use 1/2-in-thick strap for the workpiece. The 1/2-in thickness attains and, in the rounding to available nominal size, exceeds the reliability goal. It is left as an exercise for the student to perform an adequacy assessment that demonstrates this.

The example demonstrates that, for a given reliability goal, the fatigue design factor which facilitates its attainment is decided by the variabilities of the situation. Furthermore, the necessary design factor is not a constant independent of the way the concept unfolds. Rather, it is a function of a number of seemingly unrelated a priori decisions which are made in giving definition to the concept. The involvement of stochastic methodology can be limited to defining the necessary design factor. In particular, in the example, the design factor is not a function of the design variable t; rather, t follows from the design factor.

CASE STUDY 7 Exploring an Interesting Idea

A company engineer has conceived a novel way to make a light duty hoisting hook by simply hot-forming it from a rod by bending to the shape in Fig. CS7-1 in a jig. This avoids casting or forging a small production run item for in-plant use. The first step was to find the stresses at the critical location due to a load F. Noting that $D = 3$ in, $D = 1$ in, $r_i = 1.5$ in, $r_o = 2.5$ in, $R = (3 + 1)/2 = 2$ in, the radius of the neutral surface r_n can be found from Eq. (3–78):

$$r_n = \frac{1^2}{4^2(2[2] - \sqrt{4[2] - 1})} = 1.968\,246 \text{ in}$$

$$e = R - r_n = 2 - 1.968\,'246 = 0.031\,754 \text{ in}$$

$$c_i = r_n - r_i = 1.968\,246 - 1.5 = 0.468\,246 \text{ in}$$

$$c_o = r_o - r_n = 2.5 - 1.968\,254 = 0.531\,754 \text{ in}$$

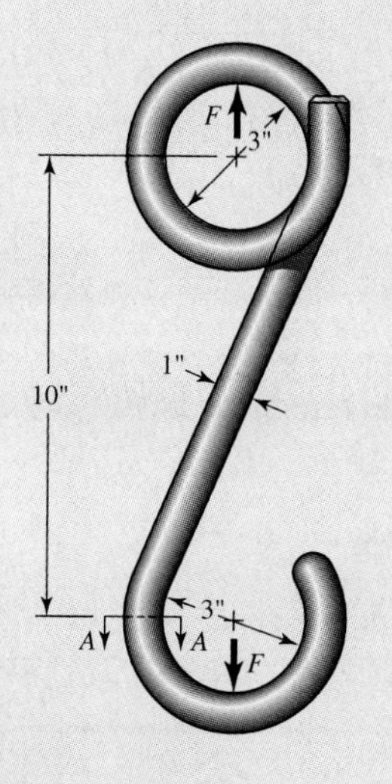

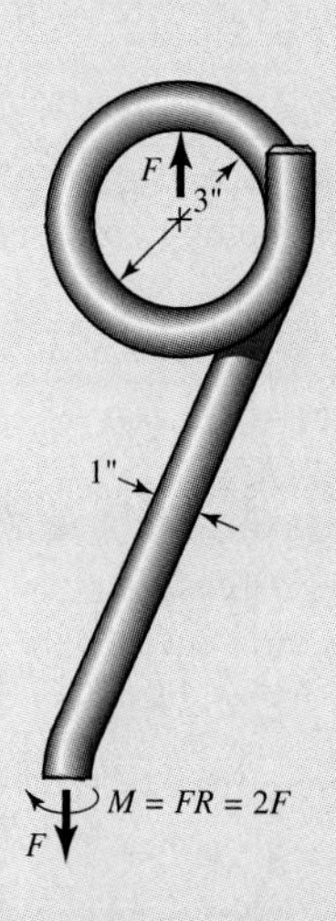

Figure CS7–1

Wire-formed hoisting hook of case study CS7.

$$A = \pi\left(1^2\right)/4 = 0.7854 \text{ in}^2$$
$$M = FR = F(2) = 2F$$

The inner surface stress σ_i is, using Eq. (3–69) in part,

$$\sigma_i = \frac{F}{A} + \frac{Mc_i}{Aer_i} = \frac{F}{0.7854} + \frac{F(2)0.468\ 246}{0.7854(0.031\ 754)1.5} = 26.31F \text{ kpsi}$$

The outer surface stress σ_o is, using Eq. (3–69) in part,

$$\sigma_o = \frac{F}{A} - \frac{Mc_o}{Aer_o} = \frac{F}{0.7854} - \frac{F(2)0.531\ 754}{0.7854(0.031\ 754)2.5} = -15.784F \text{ kpsi}$$

The decision was to use 1035 hot-rolled steel rod. From Table E–25, $\bar{S}_{ut} = 86.2$ kpsi, $C_{Sut} = 0.445$, $\bar{S}_y = 49.6$ kpsi, and $C_{Sy} = 0.0768$. For a low-use hook (less than 1000 lifts), the engineer chose to use a static (yield) analysis. The reliability goal corresponds to a probability of 0.1 of a single failure in 50 hooks, and is, since there are *two* nominally identical sections, $R = \sqrt{1 - p_f}$ is the goal.

$$R = \sqrt{1 - p_f} = \sqrt{1 - \frac{1}{50}\frac{1}{10}} = \sqrt{0.998} = 0.999 \qquad (z = -3.090)$$

The stress-strength relationship is

$$\boldsymbol{\sigma}_{\text{all}} = \frac{\mathbf{S}_y}{\mathbf{n}_y}$$

The hoisting wire or chain will carry a load cell, which prevents a hoisting tension that exceeds design load $\mathbf{F}_{\text{max}}$. To allow for miscalibration or missetting of the load cell, a COV of 0.10 on F_{max} was provided by the engineer. The COV of the design factor n_y is

$$C_n \doteq \sqrt{C_{Sy}^2 + C_{F_{\text{max}}}^2} = \sqrt{0.0768^2 + 0.10^2} = 0.126$$

From Eq. (2–9),

$$\bar{n} = \exp\left[-(-3.09)\sqrt{\ln(1 + 0.126^2)} + \ln\sqrt{1 + 0.126^2}\,\right] = 1.48$$

Proceeding deterministically from here on,

$$\bar{\sigma}_{\text{all}} = \bar{\sigma}_i = 26.31\bar{F}_{\text{max}} = \bar{S}_y/\bar{n}_y$$

from which

$$\bar{F}_{\text{max}} = \frac{S_y}{26.31\bar{n}_y} = \frac{49.6}{26.31(1.48)} = 1.27 \text{ kip}$$

For fatigue loading for indefinite life, the load is repeatedly applied. The components of load are

$$F_a = \left|\frac{F_{\text{max}} - F_{\text{min}}}{2}\right| = \frac{F_{\text{max}}}{2} \qquad F_m = \frac{F_{\text{max}} + F_{\text{min}}}{2} = \frac{F_{\text{max}}}{2}$$

and the slope of the loadline is $r = \bar{F}_a/\bar{F}_m = 1$. The expression for σ_a at the critical location is, there being no stress concentration,

$$\bar{\sigma}_a = 26.31\bar{F}_a$$

A Priori Decision	Consequence
Use 1035 HR steel	$S_{ut} = 86.2$ kpsi
HR condition	$C_{ka} = 0.110$
Correlation method	$C_\phi = 0.138$
	$C_S = \sqrt{C_{ka}^2 + C_\phi^2} = \sqrt{0.11^2 + 0.138^2} = 0.176$
	$C_n \doteq \sqrt{C_S^2 + C_F^2} = \sqrt{0.176^2 + 0.10^2} = 0.202$

The mean factor of safety in fatigue $\bar{n}_f$ is

$$\bar{n}_f = \exp\left[-(-3.09)\sqrt{\ln\left(1 + 0.202^2\right)} + \ln\sqrt{1 + 0.202^2}\,\right] = 1.89$$

Estimate $\bar{S}_e$:

$$\bar{k}_a = 14.5(86.2)^{-0.719} = 0.588$$

$$D_e = 0.37(1) = 0.37, \qquad k_b = (0.37/0.30)^{-0.107} = 0.978$$

$$\bar{k}_c = \bar{k}_d = \bar{k}_e = 1$$

$$\bar{S}_e = 0.588(0.978)0.506(86.2) = 25.08 \text{ kpsi}$$

See Fig. CS7-2.

Estimate strength amplitude: Using DE-Gerber failure locus, Table 7–15, panel 1,

$$\bar{S}_a = \frac{1^2 86.2^2}{2(25.08)}\left\{-1 + \sqrt{1 + \left[\frac{2(25.08)}{1(86.2)}\right]^2}\,\right\} = 23.25 \text{ kpsi}$$

Estimate allowable load:

$$\bar{\sigma}_a = 26.31\bar{F}_a = 26.31\frac{\bar{F}_{\max}}{2} = \frac{\bar{S}_a}{\bar{n}_f}$$

Solve for $\bar{F}_{\max}$:

$$\bar{F}_{\max} = \frac{2\bar{S}_a}{26.31\bar{n}_f} = \frac{2(23.25)}{26.31(1.89)} = 0.935 \text{ kip}$$

Contemplation

The engineer is going to write a brief report, then present the idea to his supervisor (and any other the supervisor invites). Before this happens the engineer muses . . .

- Does it make sense that the capacity of this hook is small despite the modest factors of safety?
- Could this bar be bent to the desired shape cold?
- Is the reliability goal appropriate for hoisting hook service?
- Will impending static failure give warning?
- Will impending fatigue failure give warning?

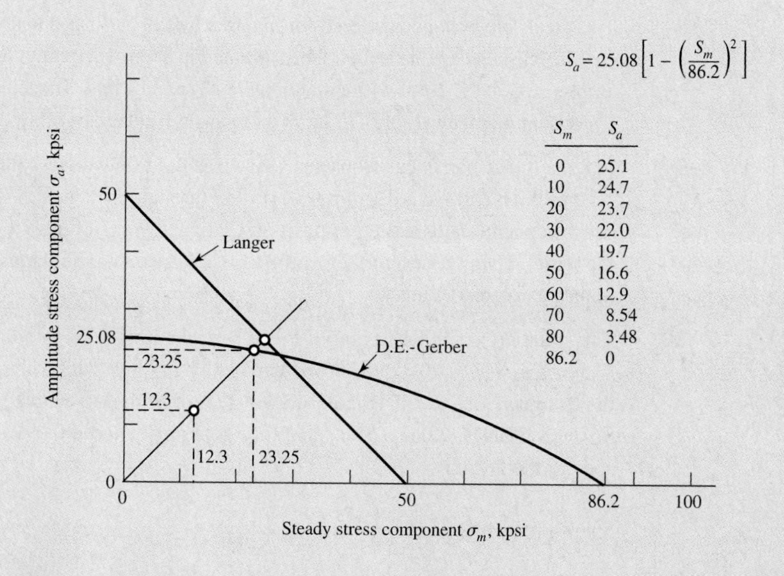

Figure CS7–2

Designer's fatigue diagram for case study CS7 using a DE-Gerber fatigue failure locus.

Why not answer these questions yourself. Also, do you have a suggestion to improve the load-carrying capability with the same silhouette? Examine the implications of your suggestion.

PROBLEMS

ANALYSIS

7–1 Estimate the fatigue strength of a rotating-beam specimen made of AISI 1020 hot-rolled steel corresponding to a life of 12.5 kilocycles of stress reversal. Also, estimate the life of the specimen corresponding to a stress amplitude of 36 kpsi. The known properties are $S_{ut} = 66.2$ kpsi, $\sigma_0 = 115$ kpsi, $m = 0.22$, and $\varepsilon_f = 0.90$.

ANALYSIS

7–2 Derive Eq. (f) following Eq. (7–7). For the specimen of Prob. 7–1, estimate the strength corresponding to 500 cycles.

ANALYSIS

7–3 For the interval $10^3 \leq N \leq 10^6$ cycles, develop an expression for the endurance strength $(S'_f)_{ax}$ in axial fatigue for the polished specimens of 4130 used to obtain Fig. 7–6. The ultimate strength is $\bar{S}_{ut} = 125$ kpsi and the endurance limit is $(\bar{S}'_e)_{ax} = 49$ kpsi.

ANALYSIS

7–4 A bar of steel has the minimum properties $S_e = 276$ MPa, $S_y = 413$ MPa, and $S_{ut} = 551$ MPa. For each of the cases below, find the factor of safety guarding against a static failure, and either the factor of safety guarding against a fatigue failure or the expected life of the part.

(a) A steady torsional stress of 103 MPa and an alternate bending stress of 172 MPa.
(b) A steady torsional stress of 138 MPa and an alternating torsional stress of 69 MPa.
(c) A steady torsional stress of 103 MPa, an alternating torsional stress of 69 MPa, and an alternating bending stress of 83 MPa.
(d) An alternating torsional stress of 207 MPa.
(e) An alternating torsional stress of 103 MPa and a steady tensile stress of 103 MPa.

7–5 A 3/16-in drill rod was heat-treated and ground. The measured hardness was found to be 490 Brinell. Estimate the endurance strength if the rod is used in rotating bending.

ANALYSIS

7–6 Solve Prob. 7–5 if the hardness of production pieces is found to be $\mathbf{H}_b = 495\mathbf{LN}(1, 0.03)$.

7–7 Estimate the endurance strength of a 32-mm-diameter rod of AISI 1035 steel having a machined finish and heat-treated to a tensile strength of 710 MPa.

ANALYSIS

7–8 Two steels are being considered for manufacture of as-forged connecting rods. One is AISI 4340 Cr-Mo-Ni steel capable of being heat-treated to a tensile strength of 260 kpsi. The other is a plain carbon steel AISI 1040 with an attainable S_{ut} of 113 kpsi. If each rod is to have a size giving an equivalent diameter d_e of 0.75 in, is there any advantage to using the alloy steel?

ANALYSIS

7–9 A rectangular bar is cut from an AISI 1018 cold-drawn steel flat. The bar is 60 mm wide by 10 mm thick and has a 12-mm hole drilled through the center as depicted in Table E–15–1. The bar is concentrically loaded in push-pull fatigue by axial forces F_a, uniformly distributed across the width. Using a design factor of $n_d = 1.8$, estimate the largest force F_a that can be applied ignoring column action.

DESIGN

7–10 The situation is similar to that of Prob. 7–9 wherein the imposed completely reversed axial load $F_a = 15LN(1, 0.20)$ kN is to be carried by the link with a thickness to be specified by you, the designer. Use the 1018 cold-drawn steel of Prob. 7–9 with $\mathbf{S}_{ut} = 440\mathbf{LN}(1, 0.30)$ MPa and $\mathbf{S}_{yt} = 370\mathbf{LN}(1, 0.061)$. The reliability goal must exceed 0.999. Using the correlation method, specify the thickness t.

7-11 Bearing reactions R_1 and R_2 are exerted on the shaft shown in the figure, which rotates at 1150 rev/min and supports a 10-kip bending force. Use a 1095 steel. Specify a diameter d using a design factor of $n_d = 1.6$ for a life of 3 min. The surfaces are machined.

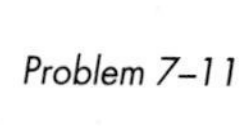
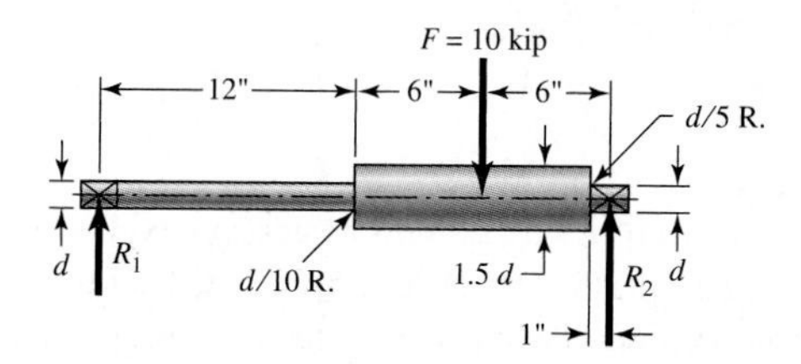

Problem 7–11

7-12 A solid round steel bar is machined to a diameter of 1.25 in. A groove 1/8 in deep with a radius of 1/8 in is cut into the bar. The material has a mean tensile strength of 110 kpsi. A completely reversed bending moment $M = 1400$ lb · in is applied. Estimate the reliability. The size factor should be based on the gross diameter. The bar rotates.

7-13 Repeat Prob. 7–12, with a completely reversed torsional moment of $T = 1400$ lb · in applied.

7-14 A 1 1/4-in-diameter hot-rolled steel bar has a 1/8-in diameter hole drilled transversely through it. The bar is nonrotating and is subject to a completely reversed bending moment of $M = 1600$ lb · in in the same plane as the axis of the transverse hole. The material has a mean tensile strength of 58 kpsi. Estimate the reliability. The size factor should be based on the gross size. Use Table E–16 for K_t.

7-15 Repeat Prob. 7–14, with the bar subject to a completely reversed torsional moment of 2400 lb · in.

7-16 The cold-drawn AISI 1018 steel bar shown in the figure is subjected to a tensile load fluctuating between 800 and 3000 lb. Estimate the factors of safety n_y and n_f using (*a*) a DE-Gerber fatigue failure locus as part of the designer's fatigue diagram, and (*b*) a DE-elliptic fatigue failure locus as part of the designer's fatigue diagram.

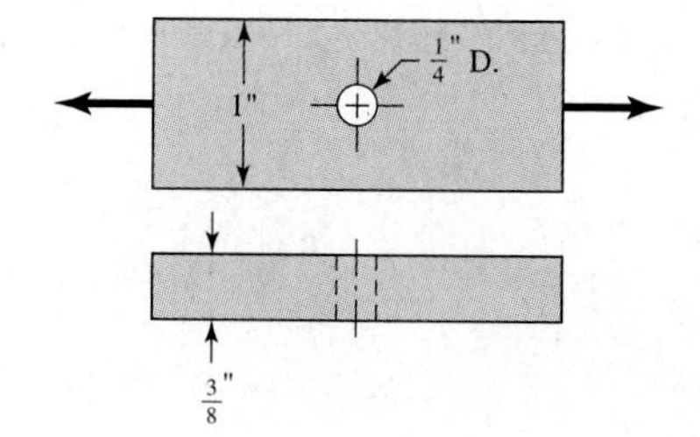

Problem 7–16

7-17 Repeat Prob. 7–16, with the load fluctuating between −800 and 3000 lb. Assume no buckling.

7-18 Repeat Prob. 7–16, with the load fluctuating between 800 and −3000 lb. Assume no buckling.

7-19 The figure shows a formed round-wire cantilever spring subjected to a varying force. The hardness tests made on 25 springs gave a minimum hardness of 380 Brinell. It is apparent from the mounting details that there is no stress concentration. A visual inspection of the springs indicates that the surface finish corresponds closely to a hot-rolled finish. What number of applications is likely to cause failure?

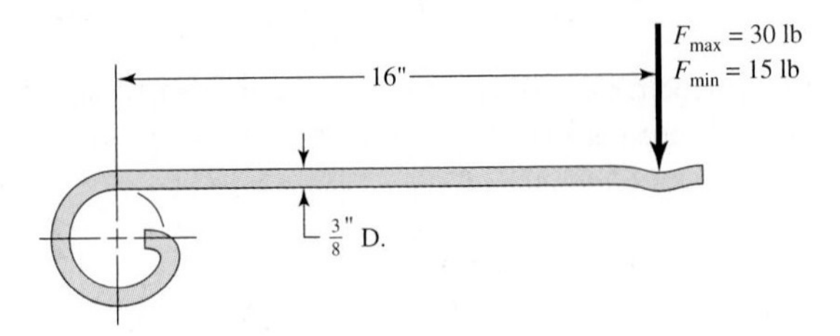

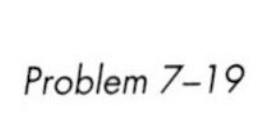
Problem 7–19

ANALYSIS

7–20 The figure is a drawing of a 3- by 18-mm latching spring. A preload is obtained during assembly by shimming under the bolts to obtain an estimated initial deflection of 2 mm. The latching operation itself requires an additional deflection of exactly 4 mm. The material is ground high-carbon steel, bent then hardened and tempered to a minimum hardness of 490 Bhn. The radius of the bend is 3 mm.

(*a*) Find the maximum and minimum latching forces.

(*b*) Is it likely the spring will fail in fatigue?

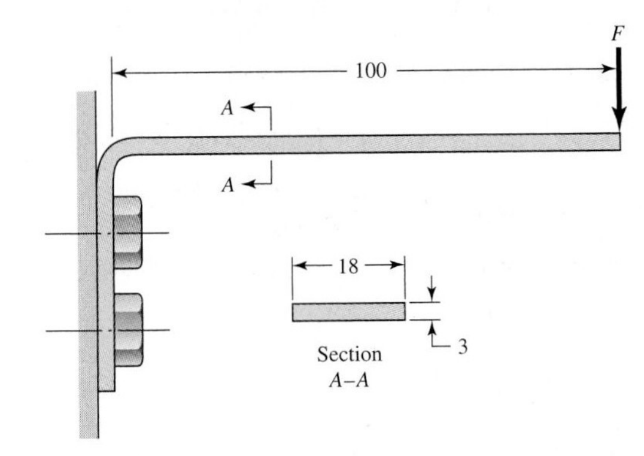

Problem 7–20
Dimensions in millimeters.

7–21 The figure shows the free-body diagram of a connecting-link portion having stress concentration at three sections. The dimensions are $r = 0.25$ in, $d = 0.75$ in, $h = 0.50$ in, $w_1 = 3.75$ in, and $w_2 = 2.5$ in. The forces F fluctuate between a tension of 4 kip and a compression of 16 kip. Neglect column action and find the least factor of safety if the material is cold-drawn AISI 1018 steel.

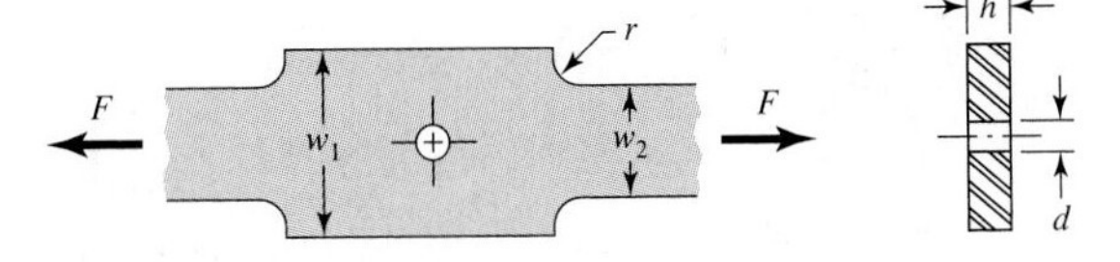

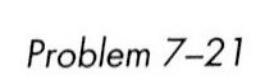
Problem 7–21

7–22 The plan view of a link is the same as in Prob. 7–21; however, the forces F are completely reversed, the reliability goal is 0.998, and the material properties are $\mathbf{S}_{ut} = 64\mathbf{LN}(1, 0.045)$ kpsi and $\mathbf{S}_y = 54\mathbf{LN}(1, 0.077)$ kpsi. Treat F_a as deterministic, and specify the thickness h.

7–23 A 1/4- by 1 1/2-in steel bar has a 3/4-in drilled hole located in the center, much as is shown in Table E–15–1. The bar is subjected to a completely reversed axial load with a deterministic load of 1200 lb. The material has a mean ultimate tensile strength of $\bar{S}_{ut} = 80$ kpsi.

(*a*) Estimate the reliability.

(*b*) Conduct a computer simulation to confirm your answer to part *a*.

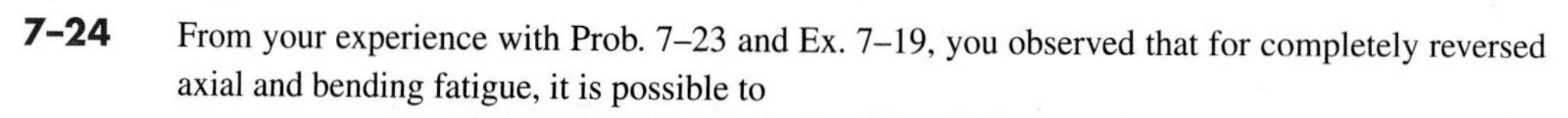

7-24 From your experience with Prob. 7–23 and Ex. 7–19, you observed that for completely reversed axial and bending fatigue, it is possible to

- Observe the COVs associated with a priori design considerations.
- Note the reliability goal.
- Find the mean design factor $\bar{n}_d$ which will permit making a geometric design decision that will attain the goal using deterministic methods in conjunction with $\bar{n}_d$.

Formulate an interactive computer program that will enable the user to find $\bar{n}_d$. While the material properties $\mathbf{S}_{ut}$, $\mathbf{S}_y$, and the load COV must be input by the user, all of the COVs associated with $\boldsymbol{\phi}_{0.30}$, $\mathbf{k}_a$, $\mathbf{k}_c$, $\mathbf{k}_d$, and $\mathbf{K}_f$ can be internal, and answers to questions will allow C_σ and C_S, as well as C_n and $\bar{n}_d$, to be calculated. Later you can add improvements. Test your program with problems you have already solved.

7-25 When using the Gerber fatigue failure locus in a stochastic problem, Eqs. (7–51) and (7–52) are useful. They are also computationally complicated. It is helpful to have a computer subroutine or procedure that performs these calculations. When writing an executive program, and it is appropriate to find S_a and C_{Sa}, a simple call to the subroutine does this with a minimum of effort. Also, once the subroutine is tested, it is always ready to perform. Write and test such a program.

7-26 Repeat Problem. 7–25 for the ASME-elliptic (DE-elliptic) fatigue failure locus, implementing Eqs. (7–54) and (7–55).

7-27 Fatigue failure locus, implementing Eqs. (7–58) and (7–59).

7-28 Write and test computer subroutines or procedures that will implement

(*a*) Table 7–5, returning a, b, C, and $\bar{k}_a$.
(*b*) Equation (7–10) using Table 7–6, returning k_b.
(*c*) Table 7–7, returning α, β, C, and $\bar{k}_c$.
(*d*) Equation (7–23), returning $\bar{k}_d$ and C_{kd}.

7-29 Write and test a computer subroutine or procedure that implements Eqs. (7–25) and (7–26), returning $\bar{q}$, σ_q, and C_q.

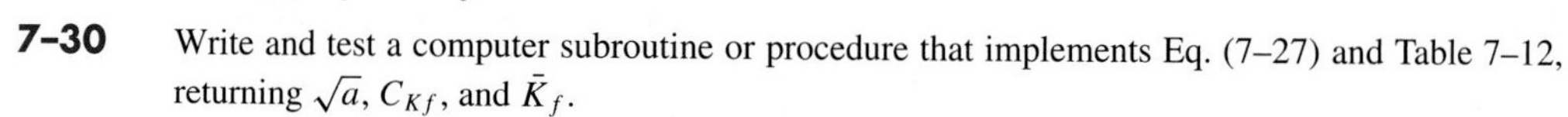

7-30 Write and test a computer subroutine or procedure that implements Eq. (7–27) and Table 7–12, returning $\sqrt{a}$, C_{Kf}, and $\bar{K}_f$.

7-31 Landgraf[31] reported data that are useful in correlating the cycle-to-failure count N_T (where elastic and plastic strains are equal) to Brinell hardness H_B. One possible correlation equation is

$$\bar{N}_T = 10^6 \exp(-0.0197\bar{H}_B) \text{ cycles}$$

For a 1045 steel with a Brinell hardness of 230 Bhn, predict the transition cycle count N_T.

7-32 Strain-life relations have been historically posed as a function of stress reversals-to-failure $2N_f$ and stress-life relations as a function of cycles-to-failure N_f. A plot such as that of Fig. 7–4 can use N_f or $2N_f$ as abscissa. Just what the ordinate intercepts are, and where they are to be found, is important to serious plotting or regression interpretation. Explore these before attempting Prob. 7–33.

7-33 Data in support of behavior as depicted in Fig. 7–4 is as follows, where ε_{ae} is the elastic strain amplitude (half-range), ε_{ap} is the plastic strain amplitude, ε_a is $\varepsilon_{ae} + \varepsilon_{ap}$ the total strain amplitude,

---

31. R. W. Landgraf, "Resistance of Metals to Cyclic Deformation," *Achievement of High Fatigue Resistance in Metals and Alloys*, ASTM Special Technical Publication 467, 1970, American Society for Testing and Materials, Philadelphia, pp. 3–36.

and N_f is the cycles-to-failure count. The material is 1045 hot-rolled steel, with a Young's modulus of 199 700 N/mm^2 and a Brinell hardness of 230 Bhn:

ε_{ae}	ε_{ap}	N_f
0.001 139	0.000 056	960 000
0.001 64	0.000 257	86 000
0.002 08	0.000 695	17 750
0.002 27	0.002 05	3 192
0.003 69	0.007 23	434

Plot on loglog coordinates $2N_f$ as abscissa, and ε_{ae} and ε_{ap} as ordinates, finding a good strain line fit for (*a*) the plastic strain amplitude curve, finding the slope c and the intercept ε'_F, and (*b*) the elastic strain amplitude curve, finding the slope b and the intercept σ'_F/E. If you use a linear regression procedure, take the opportunity to draw in the point by point (or line-as-a-whole) 0.95 confidence bounds on the line, note the confidence interval on b and c, and note the confidence interval on the intercepts. What have you learned?

7–34 Repeat Prob. 7–23, using N_f as the abscissa. You should have agreement on ε'_F, σ'_F, b, and c. If you do not (especially if you are using a linear regression procedure), return to Prob. 7–32. What have you learned?

Summary of Parts 1 and 2

The first recommendation is to reread Chap. 1. With the experience you have gathered so far, you will gain from doing it. With the meat you have added to the bare bones of the introductory chapter, it will have a greater meaning. In Sec. 1–6, there are two dozen design considerations. We have addressed item 1, the question of the strength/stress relationship in a loss-of-function for ductile and brittle materials, for steady and fatigue loading, and for finite and indefinite life. We have also started on item 2, reliability, as it applies to stress/strength relationships. In investigating the stress/strength relations, the reader should now be prepared to

- Identify the critical location(s), either by inspection, or, if not obvious, by analyzing the several candidates, and identifying the "worst case."
- Identify the significant strength at that location.
- Identify the significant stress at that location.
- Address the question of whether the disparity between stress and strength is sufficient such that function will be preserved in the face of service loading.

This preparation took a long time because an extensive set of ideas and insights had to be identified in and among your prerequisite studies, and placed in a useful context.

The question of stiffness, distortion, and deflection, item 2, and their influence on loss of function has also been addressed. The reader should now be prepared to identify

- The level of distortion that risks loss-of-function.
- The location(s) at which loss-of-function due to distortion is possible.
- The level of distortion present.
- Whether the difference is sufficient.

Some other considerations will be touched on in Part 3, and those just noted will be further developed for the application at hand. As we proceed into Part 3 our focus becomes more specific as we consider particular machine elements and their applications.

For now, the reader should feel comfortable with a kit of tools from which an adequacy assessment is devised. Skill 1 will take on additional substance as applications unfold. In addition to focus on individual elements, design/synthesis ideas will appear more often, and skill 2 will take form and grow.

PART 3

Design of Mechanical Elements

8

Screws, Fasteners, and the Design of Nonpermanent Joints

8–1	Thread Standards and Definitions **446**
8–2	The Mechanics of Power Screws **450**
8–3	Threaded Fasteners **457**
8–4	Joints—Fastener Stiffness **458**
8–5	Joints—Member Stiffness **461**
8–6	Bolt Strength **466**
8–7	Tension Joints—The External Load **470**
8–8	Relating Bolt Torque to Bolt Tension **471**
8–9	Statically Loaded Tension Joint—Preload **477**
8–10	Gasketed Joints **483**
8–11	Tension Joints—Dynamic Loading **484**
8–12	Adequacy Assessment, Specification Set, Decision Set, and Design **492**
8–13	Shear Joints **498**
8–14	Setscrews **504**
8–15	Pins and Keys **504**

To meet a need, or an opportunity, the designer often has to conceive (or invent, dream up, or adapt) a form and express the connectivity of its parts in order to achieve a goal. Much of this mental sculpting is geometric in nature, because shape often permits the achievement of function. Just consider the one-piece can opener which creates a triangular aperture in a can top. This is a single body which, because of shape, completes a machine (the lever), letting the can provide the fulcrum, allowing a human to supply the energy to reliably cut and fold metal. Geometry and geometric thinking are always involved in formulating a mechanical concept.

In earlier times, casting, with all its freedom of shape, was important to the realization of desirable forms in metal. Limitations on the shapes and sizes that can be cast, however, restrict geometries. The idea of *joints* (an old one, too)—places where separate bodies could be assembled to create geometric forms not possible with casting—allows the designer more variation in shape, therefore more latitude with form. What would be the largest bridge if bridges could not be assembled from many parts, on site? Just as the mortise-and-tenon joint, and dowel pins, helped with wood structures, so did the nut and bolt and the rivet afford metal structures more freedom of form. The idea of a joint is not new, but it can still do much for you as a designer.

One kind of permanent joint is a nonpermanent joint that is never opened. In other words, a nonpermanent joint continues a built-in option: to disassemble or not. The descriptive phrase "nonpermanent joint" is useful only if you remember not to remove it from your arsenal of permanent joints.

The helical-thread screw was an important invention. It is the basis of threaded fasteners, and an important element in nonpermanent joints.

A threat to the function of a joint is the possibility that it opens unintentionally, thereby compromising its function, which is to preserve a greater, useful geometry. Strength and stresses due to load interplay under the control of the designer. If a designer cannot ensure the integrity of a joint, then its use is not available to achieve a grander geometric form, limiting conceptual opportunities. Hence we study joints first, to keep our conceptual horizons broad.

8–1 Thread Standards and Definitions

The terminology of screw threads, illustrated in Fig. 8–1, is explained as follows:

The *pitch* is the distance between adjacent thread forms measured parallel to the thread axis. The pitch in U.S. units is the reciprocal of the number of thread forms per inch N.

The *major diameter* d is the largest diameter of a screw thread.

The *minor diameter* d_r or d_1 is the smallest diameter of a screw thread.

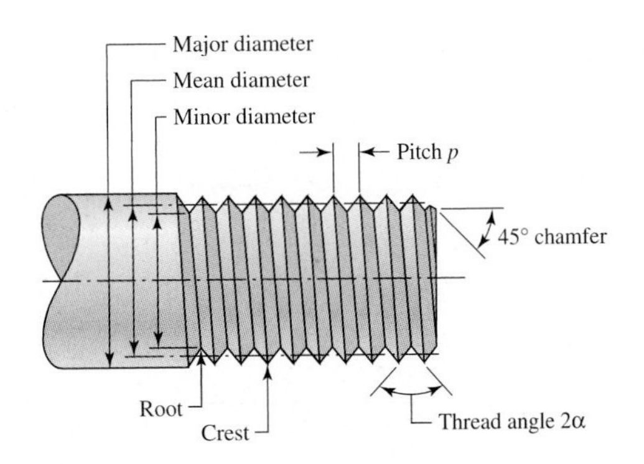

Figure 8–1

Terminology of screw threads. Sharp vee threads shown for clarity; the crests and roots are actually flattened or rounded during the forming operation.

The lead l, not shown, is the distance the nut moves parallel to the screw axis when the nut is given one turn. For a single thread, as in Fig. 8–1, the lead is the same as the pitch.

A *multiple-threaded* product is one having two or more threads cut beside each other (imagine two or more strings wound side by side around a pencil). Standardized products such as screws, bolts, and nuts all have single threads; a *double-threaded* screw has a lead equal to twice the pitch, a *triple-threaded* screw has a lead equal to 3 times the pitch, and so on.

All threads are made according to the *right-hand rule* unless otherwise noted.

The *American National (Unified)* thread standard has been approved in this country and in Great Britain for use on all standard threaded products. The thread angle is 60° and the crests of the thread may be either flat or rounded.

Figure 8–2 shows the thread geometry of the metric M and MJ profiles. The M profile replaces the inch class and is the basic ISO 68 profile with 60° symmetric threads. The MJ profile has a rounded fillet at the root of the external thread and a larger minor diameter of both the internal and external threads. This profile is especially useful where high fatigue strength is required.

Tables 8–1 and 8–2 will be useful in specifying and designing threaded parts. Note that the thread size is specified by giving the pitch p for metric sizes and by giving the number of threads per inch N for the Unified sizes. The screw sizes in Table 8–2 with diameter under $\frac{1}{4}$ in are numbered or gauge sizes. The second column in Table 8–2 shows that a No. 8 screw has a nominal major diameter of 0.1640 in.

A great many tensile tests of threaded rods have shown that an unthreaded rod having a diameter equal to the mean of the pitch diameter and minor diameter will have the same tensile strength as the threaded rod. The area of this unthreaded rod is called the tensile-stress area A_t of the threaded rod; values of A_t are listed in both tables.

Two major Unified thread series are in common use: UN and UNR. The difference between these is simply that a root radius must be used in the UNR series. Because of reduced thread stress-concentration factors, UNR series threads have improved fatigue strengths. Unified threads are specified by stating the nominal major diameter, the number of threads per inch, and the thread series, for example, $\frac{5}{8}''$-18 UNRF or 0.625″-18 UNRF.

Metric threads are specified by writing the diameter and pitch in millimeters, in that order. Thus, M12 × 1.75 is a thread having a nominal major diameter of 12 mm and a

Figure 8–2

Basic thread profile for metric M and MJ threads. $D (d)$ = basic major diameter of internal (external) thread; $D_1 (d_1)$ = basic minor diameter of internal (external) thread; $D_2 (d_2)$ = basic pitch diameter of internal (external) thread; p = pitch; $H = 0.5(3)^{1/2}p$.

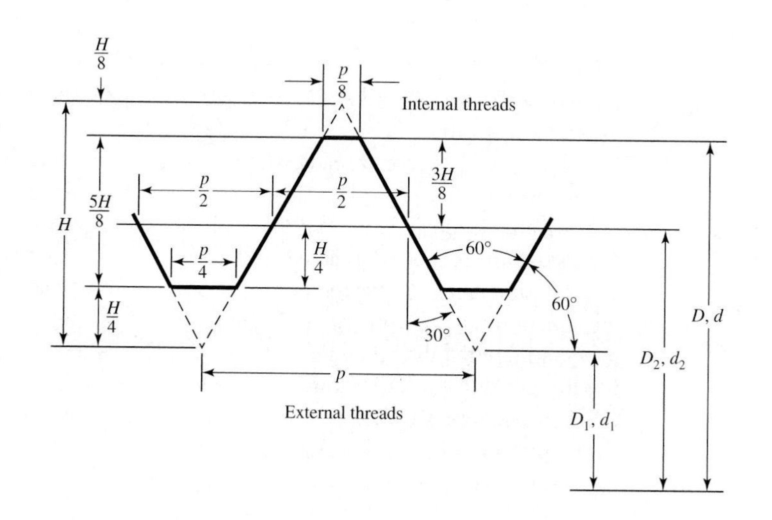

Table 8–1

Diameters and Areas of Coarse-Pitch and Fine-Pitch Metric Threads. (All Dimensions in Millimeters)*

	Coarse-Pitch Series			Fine-Pitch Series		
Nominal Major Diameter d	Pitch p	Tensile-Stress Area A_t	Minor-Diameter Area A_r	Pitch p	Tensile-Stress Area A_t	Minor-Diameter Area A_r
1.6	0.35	1.27	1.07			
2	0.40	2.07	1.79			
2.5	0.45	3.39	2.98			
3	0.5	5.03	4.47			
3.5	0.6	6.78	6.00			
4	0.7	8.78	7.75			
5	0.8	14.2	12.7			
6	1	20.1	17.9			
8	1.25	36.6	32.8	1	39.2	36.0
10	1.5	58.0	52.3	1.25	61.2	56.3
12	1.75	84.3	76.3	1.25	92.1	86.0
14	2	115	104	1.5	125	116
16	2	157	144	1.5	167	157
20	2.5	245	225	1.5	272	259
24	3	353	324	2	384	365
30	3.5	561	519	2	621	596
36	4	817	759	2	915	884
42	4.5	1120	1050	2	1260	1230
48	5	1470	1380	2	1670	1630
56	5.5	2030	1910	2	2300	2250
64	6	2680	2520	2	3030	2980
72	6	3460	3280	2	3860	3800
80	6	4340	4140	1.5	4850	4800
90	6	5590	5360	2	6100	6020
100	6	6990	6740	2	7560	7470
110				2	9180	9080

* The equations and data used to develop this table have been obtained from ANSI B1.1-1974 and B18.3.1-1978. The minor diameter was found from the equation $d_r = d - 1.226\ 869p$, and the pitch diameter from $d_m = d - 0.649\ 519p$. The mean of the pitch diameter and the minor diameter was used to compute the tensile-stress area.

pitch of 1.75 mm. Note that the letter M, which precedes the diameter, is the clue to the metric designation.

Square and Acme threads, shown in Fig. 8–3*a* and *b*, respectively, are used on screws when power is to be transmitted. Table 8–3 lists the preferred pitches for inch-series Acme threads. However, other pitches can be and often are used, since the need for a standard for such threads is not great.

Modifications are frequently made to both Acme and square threads. For instance, the square thread is sometimes modified by cutting the space between the teeth so as to have an included thread angle of 10 to 15°. This is not difficult, since these threads are usually cut with a single-point tool anyhow; the modification retains most of the high efficiency inherent in square threads and makes the cutting simpler. Acme threads are sometimes modified to a stub form by making the teeth shorter. This results in a larger minor diameter and a somewhat stronger screw.

Table 8–2

Diameters and Area of Unified Screw Threads UNC and UNF*

Size Designation	Nominal Major Diameter in	Coarse Series—UNC Threads per Inch N	Coarse Series—UNC Tensile-Stress Area A_t in²	Coarse Series—UNC Minor-Diameter Area A_r in²	Fine Series—UNF Threads per Inch N	Fine Series—UNF Tensile-Stress Area A_t in²	Fine Series—UNF Minor-Diameter Area A_r in²
0	0.0600				80	0.001 80	0.001 51
1	0.0730	64	0.002 63	0.002 18	72	0.002 78	0.002 37
2	0.0860	56	0.003 70	0.003 10	64	0.003 94	0.003 39
3	0.0990	48	0.004 87	0.004 06	56	0.005 23	0.004 51
4	0.1120	40	0.006 04	0.004 96	48	0.006 61	0.005 66
5	0.1250	40	0.007 96	0.006 72	44	0.008 80	0.007 16
6	0.1380	32	0.009 09	0.007 45	40	0.010 15	0.008 74
8	0.1640	32	0.014 0	0.011 96	36	0.014 74	0.012 85
10	0.1900	24	0.017 5	0.014 50	32	0.020 0	0.017 5
12	0.2160	24	0.024 2	0.020 6	28	0.025 8	0.022 6
$\frac{1}{4}$	0.2500	20	0.031 8	0.026 9	28	0.036 4	0.032 6
$\frac{5}{16}$	0.3125	18	0.052 4	0.045 4	24	0.058 0	0.052 4
$\frac{3}{8}$	0.3750	16	0.077 5	0.067 8	24	0.087 8	0.080 9
$\frac{7}{16}$	0.4375	14	0.106 3	0.093 3	20	0.118 7	0.109 0
$\frac{1}{2}$	0.5000	13	0.141 9	0.125 7	20	0.159 9	0.148 6
$\frac{9}{16}$	0.5625	12	0.182	0.162	18	0.203	0.189
$\frac{5}{8}$	0.6250	11	0.226	0.202	18	0.256	0.240
$\frac{3}{4}$	0.7500	10	0.334	0.302	16	0.373	0.351
$\frac{7}{8}$	0.8750	9	0.462	0.419	14	0.509	0.480
1	1.0000	8	0.606	0.551	12	0.663	0.625
$1\frac{1}{4}$	1.2500	7	0.969	0.890	12	1.073	1.024
$1\frac{1}{2}$	1.5000	6	1.405	1.294	12	1.581	1.521

* This table was compiled from ANSI B1.1-1974. The minor diameter was found from the equation $d_r = d - 1.299\ 038p$, and the pitch diameter from $d_m = d - 0.649\ 519p$. The mean of the pitch diameter and the minor diameter was used to compute the tensile-stress area.

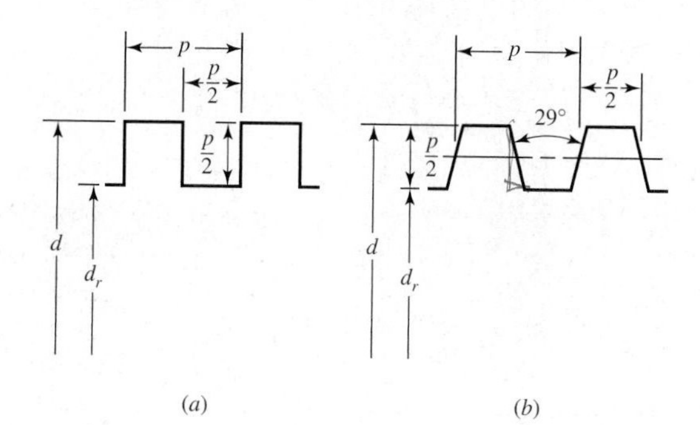

Figure 8–3

(*a*) Square thread; (*b*) Acme thread.

Table 8–3

Preferred Pitches for Acme Threads

d, in	$\frac{1}{4}$	$\frac{5}{16}$	$\frac{3}{8}$	$\frac{1}{2}$	$\frac{5}{8}$	$\frac{3}{4}$	$\frac{7}{8}$	1	$1\frac{1}{4}$	$1\frac{1}{2}$	$1\frac{3}{4}$	2	$2\frac{1}{2}$	3
p, in	$\frac{1}{16}$	$\frac{1}{14}$	$\frac{1}{12}$	$\frac{1}{10}$	$\frac{1}{8}$	$\frac{1}{6}$	$\frac{1}{6}$	$\frac{1}{5}$	$\frac{1}{5}$	$\frac{1}{4}$	$\frac{1}{4}$	$\frac{1}{4}$	$\frac{1}{3}$	$\frac{1}{2}$

8–2 The Mechanics of Power Screws

A power screw is a device used in machinery to change angular motion into linear motion, and, usually, to transmit power. Familiar applications include the lead screws of lathes, and the screws for vises, presses, and jacks.

An application of power screws to a power-driven jack is shown in Fig. 8–4. You should be able to identify the worm, the worm gear, the screw, and the nut. Is the worm gear supported by one bearing or two?

In Fig. 8–5 a square-threaded power screw with single thread having a mean diameter d_m, a pitch p, a lead angle λ, and a helix angle ψ is loaded by the axial compressive force F. We wish to find an expression for the torque required to raise this load, and another expression for the torque required to lower the load.

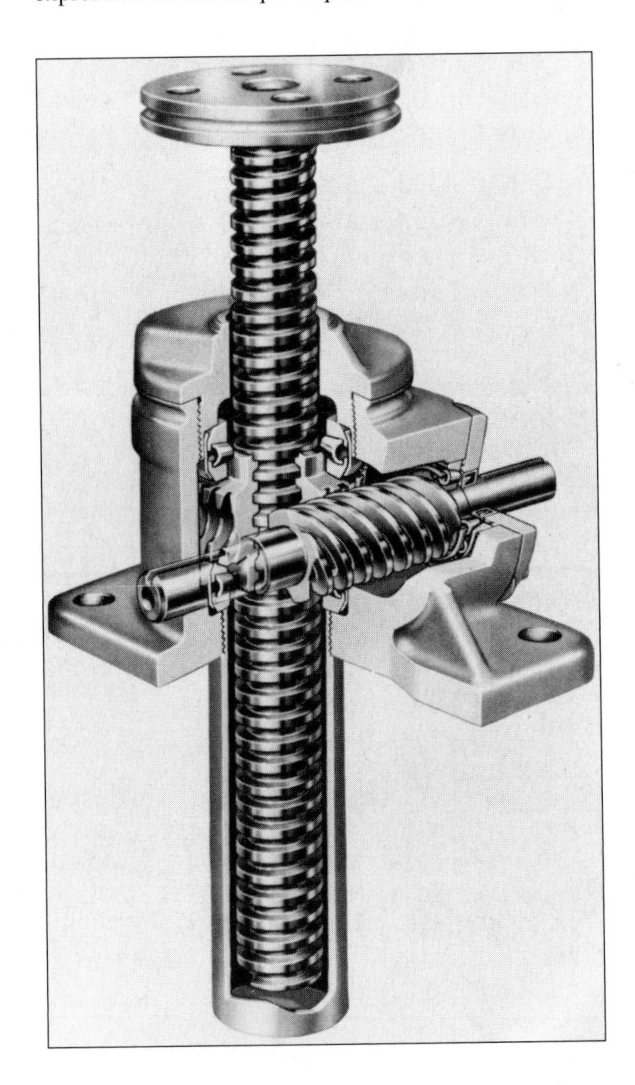

Figure 8–4

The Joyce worm-gear screw jack. (Courtesy Joyce-Dayton Corp., Dayton, Ohio.)

Figure 8–5

Portion of a power screw.

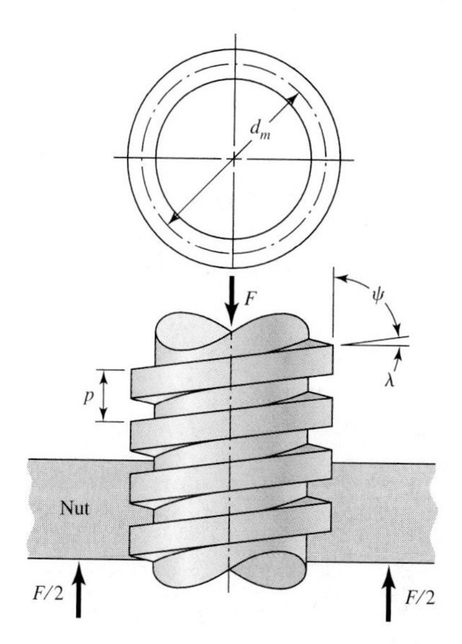

First, imagine that a single thread of the screw is unrolled or developed (Fig. 8–6) for exactly the single turn. Then one edge of the thread will form the hypotenuse of a right triangle whose base is the circumference of the mean-thread-diameter circle and whose height is the lead. The angle λ, in Figs. 8–5 and 8–6, is the lead angle of the thread. We represent the summation of all the unit axial forces acting upon the normal thread area by F. To raise the load, a force P acts to the right (Fig. 8–6*a*), and to lower the load, P acts to the left (Fig. 8–6*b*). The friction force is the product of the coefficient of friction f with the normal force N, and acts to oppose the motion. The system is in equilibrium under the action of these forces, and hence, for raising the load, we have

$$\begin{aligned} \sum F_H &= P - N\sin\lambda - fN\cos\lambda = 0 \\ \sum F_V &= F + fN\sin\lambda - N\cos\lambda = 0 \end{aligned} \qquad (a)$$

In a similar manner, for lowering the load, we have

$$\begin{aligned} \sum F_H &= -P - N\sin\lambda + fN\cos\lambda = 0 \\ \sum F_V &= F - fN\sin\lambda - N\cos\lambda = 0 \end{aligned} \qquad (b)$$

Since we are not interested in the normal force N, we eliminate it from each of these sets of equations and solve the result for P. For raising the load, this gives

$$P = \frac{F(\sin\lambda + f\cos\lambda)}{\cos\lambda - f\sin\lambda} \qquad (c)$$

Figure 8–6

Force diagrams: (*a*) lifting the load; (*b*) lowering the load.

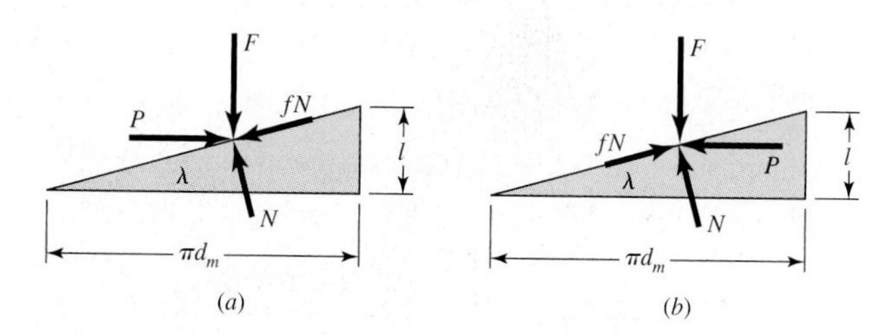

and for lowering the load,

$$P = \frac{F(f \cos \lambda - \sin \lambda)}{\cos \lambda + f \cos \lambda} \tag{d}$$

Next, divide the numerator and the denominator of these equations by $\cos \lambda$ and use the relation $\tan \lambda = l/\pi d_m$ (Fig. 8–6). We then have, respectively,

$$P = \frac{F\,[(l/\pi d_m) + f]}{1 - (fl/\pi d_m)} \tag{e}$$

$$P = \frac{F\,[f - (l/\pi d_m)]}{1 + (fl/\pi d_m)} \tag{f}$$

Finally, noting that the torque is the product of the force P and the mean radius $d_m/2$, for raising the load we can write

$$T = \frac{F d_m}{2}\left(\frac{l + \pi f d_m}{\pi d_m - fl}\right) \tag{8–1}$$

where T is the torque required for two purposes: to overcome thread friction and to raise the load.

The torque required to lower the load, from Eq. (f), is found to be

$$T = \frac{F d_m}{2}\left(\frac{\pi f d_m - l}{\pi d_m + fl}\right) \tag{8–2}$$

This is the torque required to overcome a part of the friction in lowering the load. It may turn out, in specific instances where the lead is large or the friction is low, that the load will lower itself by causing the screw to spin without any external effort. In such cases, the torque T from Eq. (8–2) will be negative or zero. When a positive torque is obtained from this equation, the screw is said to be *self-locking*. Thus the condition for self-locking is

$$\pi f d_m > l$$

Now divide both sides of this inequality by πd_m. Recognizing that $l/\pi d_m = \tan \lambda$, we get

$$f > \tan \lambda \tag{8–3}$$

This relation states that self-locking is obtained whenever the coefficient of thread friction is equal to or greater than the tangent of the thread lead angle.

An expression for efficiency is also useful in the evaluation of power screws. If we let $f = 0$ in Eq. (8–1), we obtain

$$T_0 = \frac{Fl}{2\pi} \tag{g}$$

which, since thread friction has been eliminated, is the torque required only to raise the load. The efficiency is therefore

$$e = \frac{T_0}{T} = \frac{Fl}{2\pi T} \tag{8–4}$$

The preceding equations have been developed for square threads where the normal thread loads are parallel to the axis of the screw. In the case of Acme or other threads, the normal thread load is inclined to the axis because of the thread angle 2α and the lead angle λ. Since lead angles are small, this inclination can be neglected and only the effect of the thread angle (Fig. 8–7*a*) considered. The effect of the angle α is to increase the frictional force by the wedging action of the threads. Therefore the frictional terms in Eq. (8–1) must be divided by $\cos\alpha$. For raising the load, or for tightening a screw or bolt, this yields

$$T = \frac{F d_m}{2}\left(\frac{l + \pi f d_m \sec\alpha}{\pi d_m - f l \sec\alpha}\right) \tag{8–5}$$

In using Eq. (8–5), remember that it is an approximation because the effect of the lead angle has been neglected.

For power screws, the Acme thread is not as efficient as the square thread, because of the additional friction due to the wedging action, but it is often preferred because it is easier to machine and permits the use of a split nut, which can be adjusted to take up for wear.

Usually a third component of torque must be applied in power-screw applications. When the screw is loaded axially, a thrust or collar bearing must be employed between the rotating and stationary members in order to carry the axial component. Figure 8–7*b* shows a typical thrust collar in which the load is assumed to be concentrated at the mean collar diameter d_c. If f_c is the coefficient of collar friction, the torque required is

$$T_c = \frac{F f_c d_c}{2} \tag{8–6}$$

For large collars, the torque should probably be computed in a manner similar to that employed for disk clutches.

Nominal body stresses in power screws can be related to thread parameters as follows. The nominal shear stress τ in torsion of the screw body can be expressed as

$$\tau = \frac{16T}{\pi d_r^3} \tag{8–7}$$

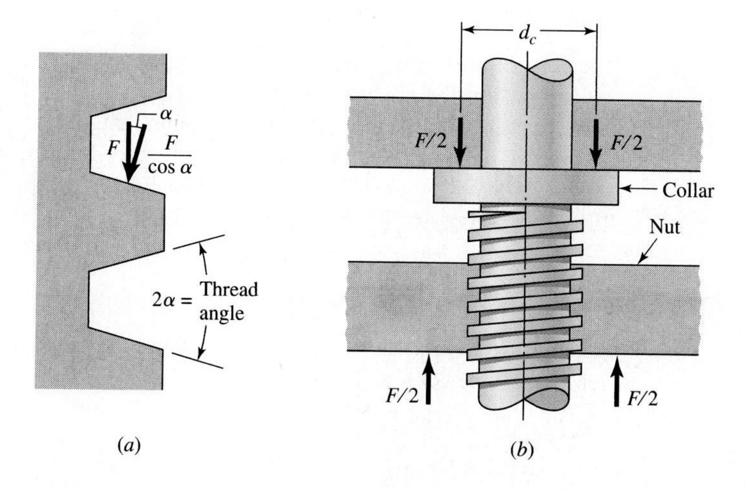

Figure 8–7

(*a*) Normal thread force is increased because of angle α; (*b*) thrust collar has frictional diameter d_c.

The axial stress σ in the body of the screw due to load F is

$$\sigma = \frac{F}{A} = \frac{4F}{\pi d_r^2} \tag{8-8}$$

in the absence of column action. For a short column the J. B. Johnson equation is

$$\left(\frac{F}{A}\right)_{\text{crit}} = S_y - \left(\frac{S_y}{2\pi}\frac{\ell}{k}\right)\frac{1}{CE} \tag{8-9}$$

after Eq. (4–53).

Nominal thread stresses in power screws can be related to thread parameters as follows. The bearing stress in Fig. 8–8, σ_B, is

$$\sigma_B = \frac{F}{\pi d_m n_t p/2} = \frac{2F}{\pi d_m n_t p} \tag{8-10}$$

where n_t is the number of engaged threads. The bending stress at the root of the thread σ_b is found from

$$\frac{I}{c} = \frac{(\pi d_r n_t)(p/2)^2}{6} = \frac{\pi}{24} d_r n_t p^2 \qquad M = \frac{Fp}{4}$$

so

$$\sigma_b = \frac{M}{I/c} = \frac{Fp}{4}\frac{24}{\pi d_r n_t p^2} = \frac{6F}{\pi d_r n_t p} \tag{8-11}$$

The transverse shear stress τ at the center of the root of the thread due to load F is

$$\tau = \frac{3V}{2A} = \frac{3}{2}\frac{F}{\pi d_r n_t p/2} = \frac{3F}{\pi d_r n_t p} \tag{8-12}$$

and at the top of the root it is zero. The von Mises stress σ' at the top of the root "plane" is found by first identifying the orthogonal normal stresses and the shear stresses. From the coordinate system of Fig. 8–8, we note

$$\sigma_x \quad \frac{6F}{\pi d_r n_t p} \qquad \tau_{xy} = \frac{16T}{\pi d_r^3}$$

$$\sigma_y = 0 \qquad \tau_{yz} = 0$$

$$\sigma_z = -\frac{4F}{\pi d_r^2} \qquad \tau_{zx} = 0$$

then use Eq. (h) of Sec. 6–7. Alternatively, one can recognize that this is a plane-stress situation, find the nonzero principal stresses, and find the von Mises stress using Eq. (6–13).

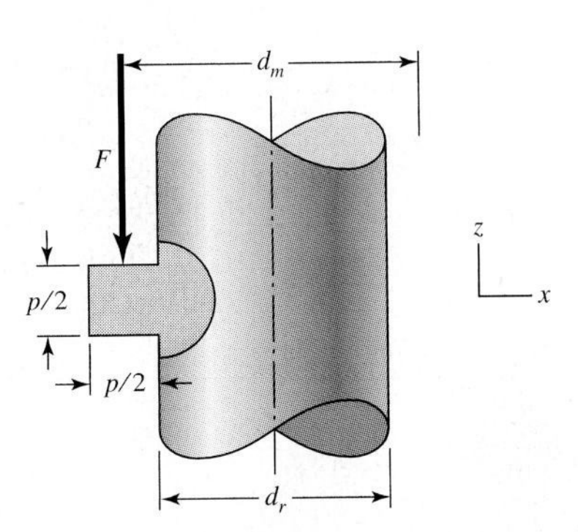

Figure 8–8

Geometry of square thread useful in finding bending and transverse shear stresses at the thread root.

The screw-thread form is complicated from an analysis viewpoint. Remember the origin of the tensile-stress area A_t which comes from experiment. A power screw lifting a load is in compression and its threat pitch is *shortened* by elastic deformation. Its engaging nut is in tension and its thread pitch is *lengthened*. The engaged threads cannot share the load equally. Some experiments show that the first engaged thread carries 0.38 of the load, the second 0.25, the third 0.18, and the seventh is free of load. In estimating thread stresses using the equations above, substituting $0.38F$ for F and setting n_t to 1 will give the largest level of stresses in the thread-nut combination.

EXAMPLE 8–1

A square-thread power screw has a major diameter of 32 mm and a pitch of 4 mm with double threads, and it is to be used in an application similar to Fig. 8–4. The given data include $f = f_c = 0.08$, $d_c = 40$ mm, and $F = 6.4$ kN per screw.
(*a*) Find the thread depth, thread width, pitch diameter, minor diameter, and lead.
(*b*) Find the torque required to raise and lower the load.
(*c*) Find the efficiency during lifting the load.
(*d*) Find the body stresses, torsional and compressive.
(*e*) Find the bearing stress.
(*f*) Find the thread stresses bending at the root, shear at the root, and von Mises stress and maximum shear stress at the same location.

Solution

From Fig. 8–3*a* the thread depth and width are the same and equal to half the pitch, or 2 mm. Also

$$d_m = d - p/2 = 32 - 4/2 = 30 \text{ mm}$$

Answer

$$d_r = d - p = 32 - 4 = 28 \text{ mm}$$

$$l = np = 2(4) = 8 \text{ mm}$$

(*b*) Using Eqs. (8–1) and (8–6), and assuming positive torque is a load-raising torque, the torque required to turn the screw against the load is

$$T = \frac{Fd_m}{2}\left(\frac{l - \pi f d_m}{\pi d_m - fl}\right) + \frac{Ff_c d_c}{2}$$

$$= \frac{6.4(30)}{2}\left[\frac{8 + (0.08)(30)}{(30) - 0.08(8)}\right] + \frac{6.4(0.08)40}{2}$$

Answer

$$= 15.94 + 10.24 = 26.18 \text{ N} \cdot \text{m}$$

Using Eqs. (8–2) and (8–6), negative torque being a load-lowering torque,

$$T = \frac{-Fd_m}{2}\left(\frac{\pi f d_m - l}{\pi d_m + fl}\right) - \frac{Ff_c d_c}{2}$$

$$= \frac{-6.4(30)}{2}\left[\frac{\pi(0.08)30 - 8}{\pi(30) + 0.08(8)}\right] - \frac{6.4(0.08)(40)}{2}$$

Answer

$$= +0.466 - 10.24 = -9.77 \text{ N} \cdot \text{m}$$

The plus sign in the first term indicates that the screw alone is not self-locking and would rotate under the action of the load except for the fact that the collar friction is present and must be overcome, too. Thus the torque required to rotate the screw "with" the load is less than is necessary to overcome collar friction alone.
(*c*) The overall efficiency in raising the load is

Answer

$$e = \frac{Fl}{2\pi T} = \frac{6.4(8)}{2\pi(26.18)} = 0.311$$

(*d*) The body shear stress τ due to torsional moment T at the outside of the screw body is

Answer
$$\tau = \frac{16T}{\pi d_r^3} = \frac{16(26.18)\left(10^3\right)}{\pi\left(28^3\right)} = 6.07 \text{ MPa}$$

The axial nominal normal stress σ is

Answer
$$\sigma = -\frac{4F}{\pi d_r^2} = -\frac{4(6.4)10^3}{\pi\left(28^3\right)} = -10.39 \text{ MPa}$$

(*e*) The bearing stress σ_B is, with one thread carrying $0.38F$,

Answer
$$\sigma_B = \frac{2(0.38F)}{\pi d_m n_t p} = \frac{2(0.38)(6.4)10^3}{\pi(30)(1)(4)} = 12.9 \text{ MPa (compressive)}$$

(*f*) The thread-root bending stress σ_b with one thread carrying $0.38F$ is

$$\sigma_b = \frac{6(0.38F)}{\pi d_r n_t p} = \frac{6(0.38)(6.4)10^3}{(28)(1)4} = 41.5 \text{ MPa}$$

The transverse shear at the extreme of the root cross section due to bending is zero. However, there is a circumferential shear stress at the extreme of the root cross section of the thread as shown in part (*d*) of 6.07 MPa. The three-dimensional stresses, after Fig. 8–8, noting the y coordinate is into the page, are

$$\sigma_x = 41.5 \text{ MPa} \qquad \tau_{xy} = 6.07 \text{ MPa}$$
$$\sigma_y = 0 \qquad \tau_{yz} = 0$$
$$\sigma_z = -10.39 \text{ MPa} \qquad \tau_{zx} = 0$$

Equation (*h*) of Sec. 6–7 can be written as

$$\sigma' = \frac{1}{\sqrt{2}}\left\{(41.5-0)^2 + [0-(-10.39)]^2 + (-10.39-41.4)^2 + 6(6.07)^2\right\}^{1/2}$$
$$= 48.7 \text{ MPa}$$

Alternatively, this can be viewed as a plane-stress situation with

$$\sigma_x = 41.5 \text{ MPa} \qquad \tau_{xy} = 6.07 \text{ MPa}$$
$$\sigma_z = -10.39 \text{ MPa} \qquad \tau_{zx} = 0$$

The principal stresses, nonzero, are

$$\sigma_1, \sigma_2 = \frac{41.5 + (-10.39)}{2} \pm \sqrt{\left[\frac{41.5-(-10.39)}{2}\right]^2 + 6.07^2}$$
$$= 15.56 \pm 26.65 = 42.21, -11.09 \text{ MPa}$$

From Eq. (6–13),

Answer
$$\sigma' = \left[42.21^2 - 42.21(-11.09) + (-11.09)^2\right]^{1/2} = 48.7 \text{ MPa}$$

Since we have the principal stresses, the maximum shear stress can be found by ordering the principal stresses as 42.21, 0, −11.09 MPa and following Eqs. (3–11), which amounts to

Answer
$$\tau_{\max} = \max\left[\frac{42.21-0}{2}, \frac{0-(-11.09)}{2}, \frac{42.21-(-11.09)}{2}\right]$$
$$= \max[21.11, 5.55, 26.65] = 26.65 \text{ MPa}$$

Table 8–4

Screw Bearing Pressure p_b

Source: H. A. Rothbart, *Mechanical Design and Systems Handbook*, 2nd ed., McGraw-Hill, New York, 1985.

Screw Material	Nut Material	Safe p_b, psi	Notes
Steel	Bronze	2500–3500	Low speed
Steel	Bronze	1600–2500	10 fpm
	Cast iron	1800–2500	8 fpm
Steel	Bronze	800–1400	20–40 fpm
	Cast iron	600–1000	20–40 fpm
Steel	Bronze	150–240	50 fpm

Table 8–5

Coefficients of Friction f for Threaded Pairs

Source: H. A. Rothbart, *Mechanical Design and Systems Handbook*, 2nd ed., McGraw-Hill, New York, 1985.

Screw Material	Nut Material			
	Steel	Bronze	Brass	Cast Iron
Steel, dry	0.15–0.25	0.15–0.23	0.15–0.19	0.15–0.25
Steel, machine oil	0.11–0.17	0.10–0.16	0.10–0.15	0.11–0.17
Bronze	0.08–0.12	0.04–0.06	—	0.06–0.09

Table 8–6

Thrust-Collar Friction Coefficients

Source: H. A. Rothbart, *Mechanical Design and Systems Handbook*, 2nd ed., McGraw-Hill, New York, 1985.

Combination	Running	Starting
Soft steel on cast iron	0.12	0.17
Hard steel on cast iron	0.09	0.15
Soft steel on bronze	0.08	0.10
Hard steel on bronze	0.06	0.08

Ham and Ryan[1] showed that the coefficient of friction in screw threads is independent of axial load, practically independent of speed, decreases with heavier lubricants, shows little variation with combinations of materials, and is best for steel on bronze. Sliding coefficients of friction in power screws are about 0.10–0.15.

Table 8–4 shows safe bearing pressures on threads, to protect the moving surfaces from abnormal wear. Table 8–5 shows the coefficients of sliding friction for common material pairs. Table 8–6 shows coefficients of starting and running friction for common material pairs.

8–3 Threaded Fasteners

As you study the sections on threaded fasteners and their use, be alert to the presence of a blend of stochastic and deterministic viewpoints. In most cases the threat is from over-proof loading of fasteners, and this is best addressed using statistical methods. The threat from fatigue is lower, and deterministic methods can be adequate.

Figure 8–9 is a drawing of a standard hexagon-head bolt. Points of stress concentration are at the fillet, at the start of the threads (runout), and at the thread-root fillet in the plane of the nut when it is present. See Table E–26 for dimensions. The diameter of the washer face is the same as the width across the flats of the hexagon. The thread length

1. Ham and Ryan, *An Experimental Investigation of the Friction of Screw-threads*, Bulletin 247, University of Illinois Experiment Station, Champaign-Urbana, IL.

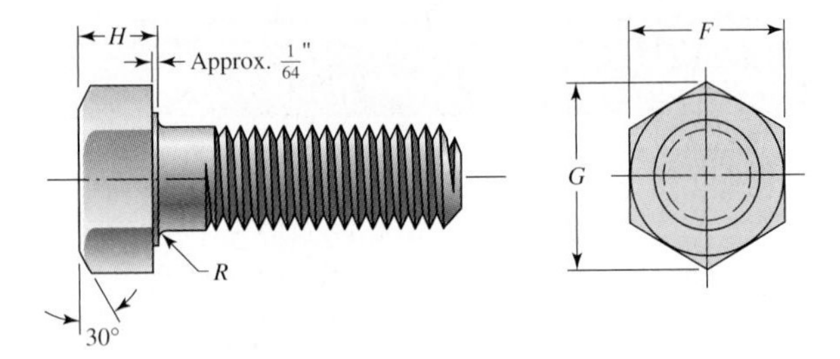

Figure 8–9

Hexagon-head bolt; note the washer face, the fillet under the head, the start of threads, and the chamfer on both ends. Bolt lengths are always measured from below the head.

of inch-series bolts, where D is the nominal diameter, is

$$L_T = \begin{cases} 2D + \frac{1}{4} \text{ in} & L \leq 6 \text{ in} \\ 2D + \frac{1}{2} \text{ in} & L > 6 \text{ in} \end{cases} \tag{8–13}$$

and for metric bolts is

$$L_T = \begin{cases} 2D + 6 & L \leq 125 \quad D \leq 48 \\ 2D + 12 & 125 < L \leq 200 \\ 2D + 25 & L > 200 \end{cases} \tag{8–14}$$

where the dimensions are in millimeters. The ideal bolt length is one in which only one or two threads project from the nut after it is tightened. Bolt holes may have burrs or sharp edges after drilling. These could bite into the fillet and increase stress concentration. Therefore, washers must always be used under the bolt head to prevent this. They should be of hardened steel and loaded onto the bolt so that the rounded edge of the stamped hole faces the washer face of the bolt. Sometimes it is necessary to use washers under the nut too.

The purpose of a bolt is to clamp two or more parts together. The clamping load stretches or elongates the bolt; the load is obtained by twisting the nut until the bolt has elongated almost to the elastic limit. If the nut does not loosen, this bolt tension remains as the preload or clamping force. When tightening, the mechanic should, if possible, hold the bolt head stationary and twist the nut; in this way the bolt shank will not feel the thread-friction torque.

The head of a hexagon-head cap screw is slightly thinner than that of a hexagon-head bolt. Dimensions of hexagon-head cap screws are listed in Table E–27. Hexagon-head cap screws are used in the same applications as bolts and also in applications in which one of the clamped members is threaded. Three other common cap-screw head styles are shown in Fig. 8–10.

A variety of machine-screw head styles are shown in Fig. 8–11. Inch-series machine screws are generally available in sizes from No. 0 to about $\frac{3}{8}$ in.

Several styles of hexagonal nuts are illustrated in Fig. 8–12; their dimensions are given in Table E–28. The material of the nut must be selected carefully to match that of the bolt. During tightening, the first thread of the nut tends to take the entire load; but yielding occurs, with some strengthening due to the cold work that takes place, and the load is eventually divided over about three nut threads. For this reason you should never reuse nuts; in fact, it can be dangerous to do so.

8–4 Joints—Fastener Stiffness

When a connection is desired which can be disassembled without destructive methods and which is strong enough to resist external tensile loads, moment loads, and shear loads,

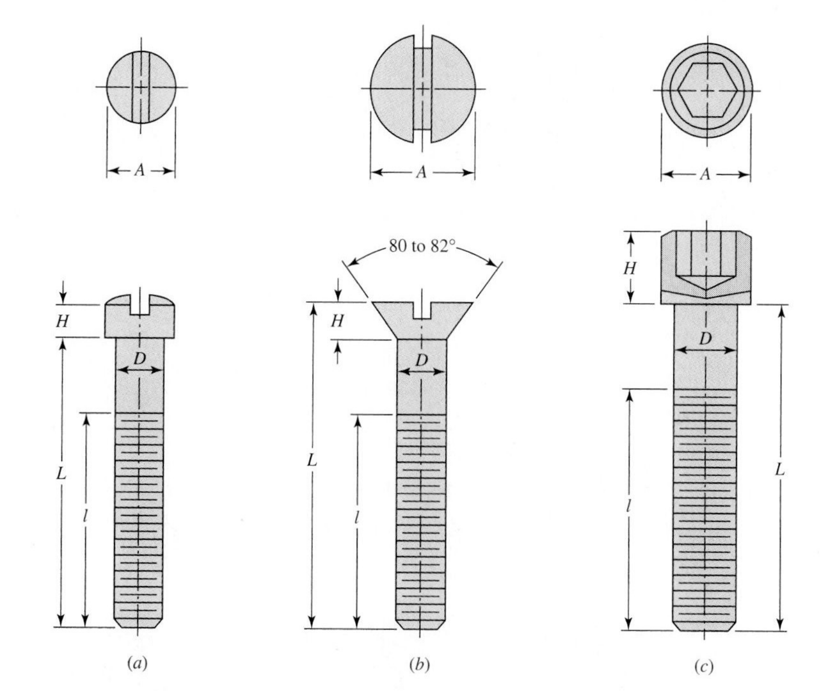

Figure 8–10

Typical cap-screw heads: (*a*) fillister head; (*b*) flat head; (*c*) hexagonal socket head. Cap screws are also manufactured with hexagonal heads similar to the one shown in Fig. 8–9, as well as a variety of other head styles. This illustration uses one of the conventional methods of representing threads.

or a combination of these, then the simple bolted joint using hardened-steel washers is a good solution. Such a joint can also be dangerous unless it is properly designed and assembled by a *trained* mechanic.

A section through a tension-loaded bolted joint is illustrated in Fig. 8–13. Notice the clearance space provided by the bolt holes. Notice, too, how the bolt threads extend into the body of the connection.

As noted previously, the purpose of the bolt is to clamp the two, or more, parts together. Twisting the nut stretches the bolt to produce the clamping force. This clamping force is called the *pre-tension* or *bolt preload.* It exists in the connection after the nut has been properly tightened no matter whether the external tensile load P is exerted or not.

Of course, since the members are being clamped together, the clamping force which produces tension in the bolt induces compression in the members.

Figure 8–14 shows another tension-loaded connection. This joint uses cap screws threaded into one of the members. An alternative approach to this problem (of not using a nut) would be to use studs. A stud is a rod threaded on both ends. The stud is screwed into the lower member first; then the top member is positioned and fastened down with hardened washers and nuts. The studs are regarded as permanent, and so the joint can be disassembled merely by removing the nut and washer. Thus the threaded part of the lower member is not damaged by reusing the threads.

The *spring rate* is a limit as expressed in Eq. (4–1). For an elastic member such as a bolt, as we learned in Eq. (4–2), it is the ratio between the force applied to the member and the deflection produced by that force. We can use Eq. (4–4) and the results of Prob. 4–1 to find the stiffness constant of a fastener in any bolted connection.

The *grip* of a connection is the total thickness of the clamped material. In Fig. 8–13 the grip is the sum of the thicknesses of both members and both washers. In Fig. 8–14 the grip is the thickness of the top member plus that of the washer.

The stiffness of the portion of a bolt or screw within the clamped zone will generally consist of two parts, that of the unthreaded shank portion and that of the threaded portion.

Figure 8–11

Types of heads used on machine screws.

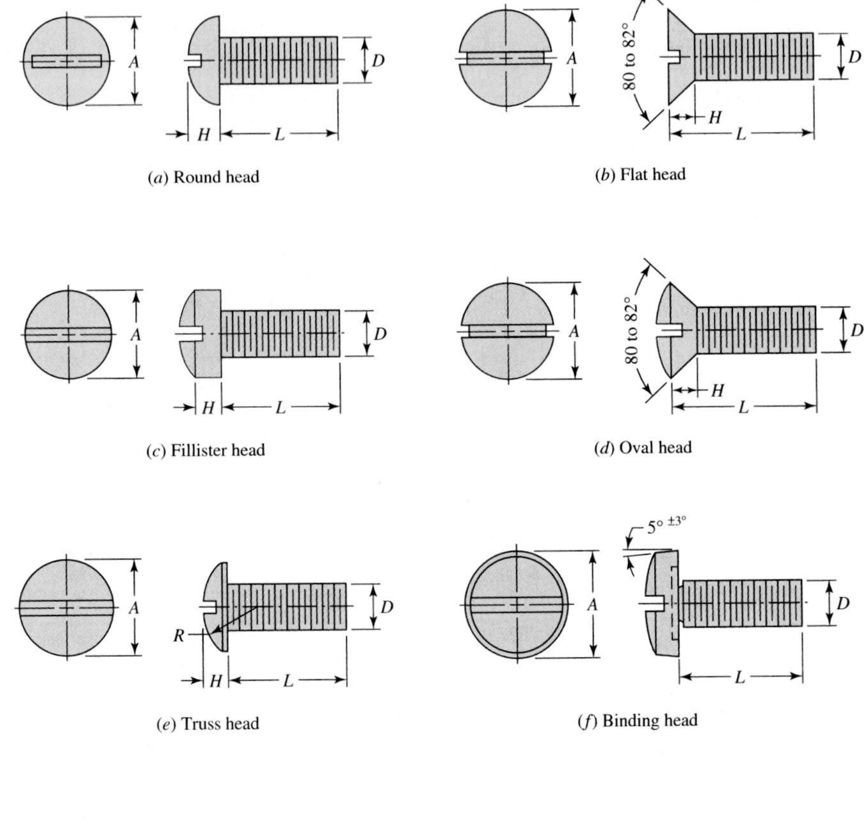

(a) Round head (b) Flat head (c) Fillister head (d) Oval head (e) Truss head (f) Binding head

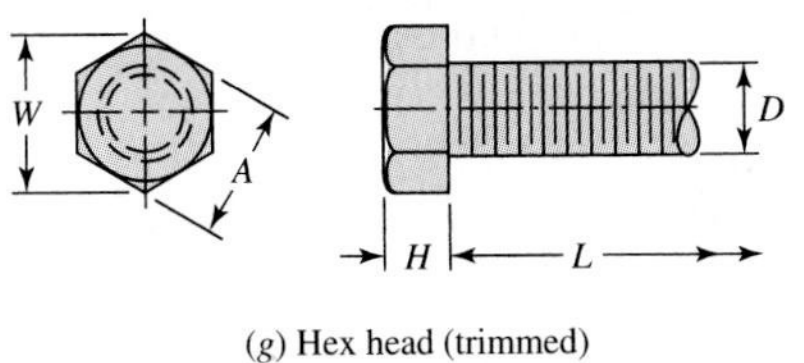

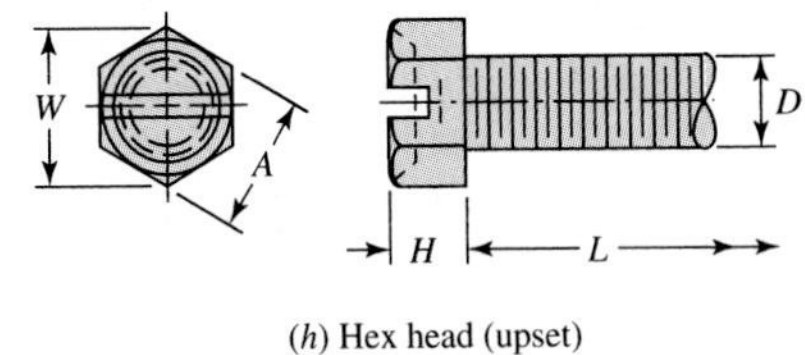

(g) Hex head (trimmed) (h) Hex head (upset)

Figure 8–12

Hexagonal nuts: (a) end view, general; (b) washer-faced regular nut; (c) regular nut chamfered on both sides; (d) jam nut with washer face; (e) jam nut chamfered on both sides.

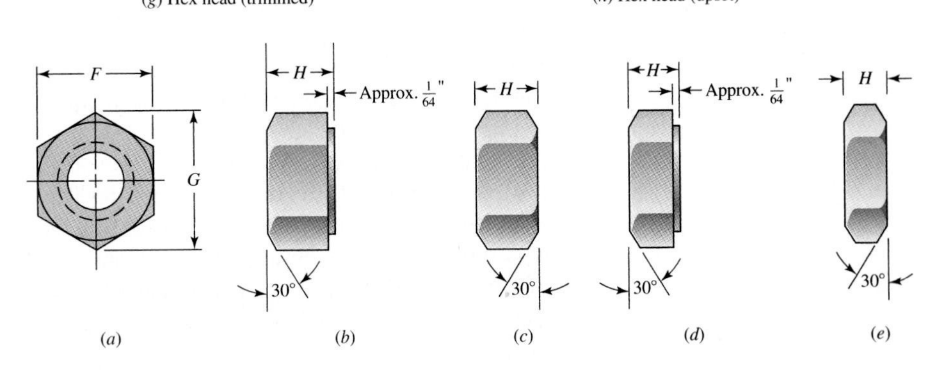

Thus the stiffness constant of the bolt is equivalent to the stiffnesses of two springs in series. Using the results of Prob. 4–1, we find

$$\frac{1}{k} = \frac{1}{k_1} + \frac{1}{k_2} \quad \text{or} \quad k = \frac{k_1 k_2}{k_1 + k_2} \tag{8-15}$$

for two springs in series. From Eq. (4–4), the spring rates of the threaded and unthreaded portions of the bolt in the clamped zone are, respectively,

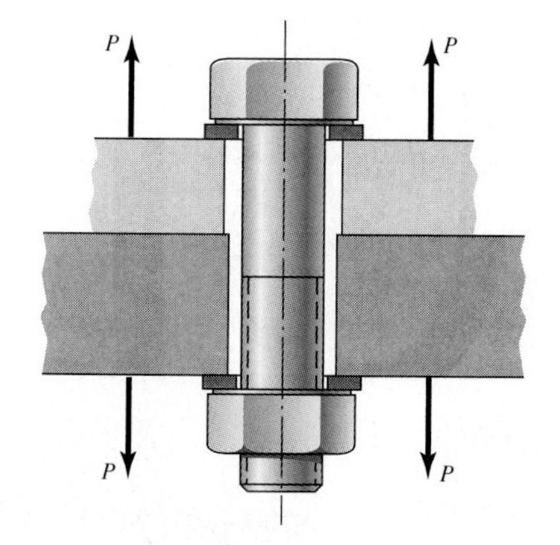

Figure 8–13

A bolted connection loaded in tension by the forces *P*. Note the use of two washers. A simplified conventional method is used here to represent the screw threads. Note how the threads extend into the body of the connection. This is usual and is desired.

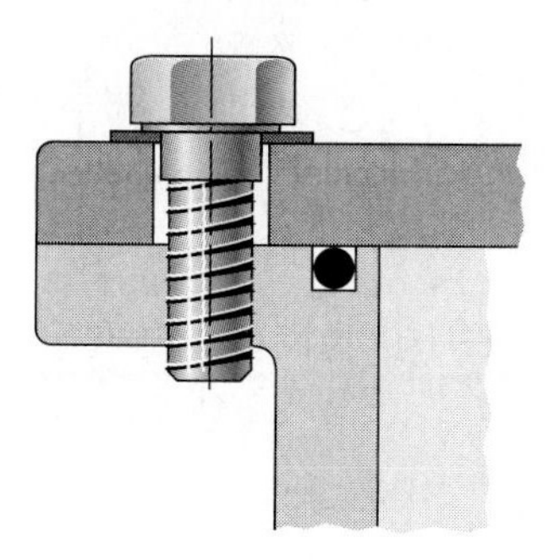

Figure 8–14

Section of cylindrical pressure vessel. Hexagon-head cap screws are used to fasten the cylinder head to the body. Note the use of an O-ring seal.

$$k_T = \frac{A_t E}{l_t} \qquad k_d = \frac{A_d E}{l_d} \tag{8–16}$$

where A_t = tensile-stress area (Tables 8–1, 8–2)
l_t = length of threaded portion of grip
A_d = major-diameter area of fastener
l_d = length of unthreaded portion in grip

Substituting these stiffnesses in Eq. (8–15) gives

$$k_b = \frac{A_d A_t E}{A_d l_t + A_t l_d} \tag{8–17}$$

where k_b is the estimated effective stiffness of the bolt or cap screw in the clamped zone. For short fasteners, the one in Fig. 8–14, for example, the unthreaded area is small and so the first of the expressions in Eq. (8–16) can be used to find k_b; and in the case of long fasteners, the threaded area is relatively small, and so the second expression in Eq. (8–16) can be used. Table 8–7 is useful.

8–5 Joints—Member Stiffness

In the previous section, we determined the stiffness of the fastener in the clamped zone. In this section, we wish to study the stiffness of the members in the clamped zone. Both of these stiffnesses must be known in order to learn what happens when the assembled connection is subjected to an external tensile loading.

There may be more than two members included in the grip of the fastener. All together these act like compressive springs in series, and hence the total spring rate of

Table 8–7

Suggested Procedure for Finding Fastener Stiffness

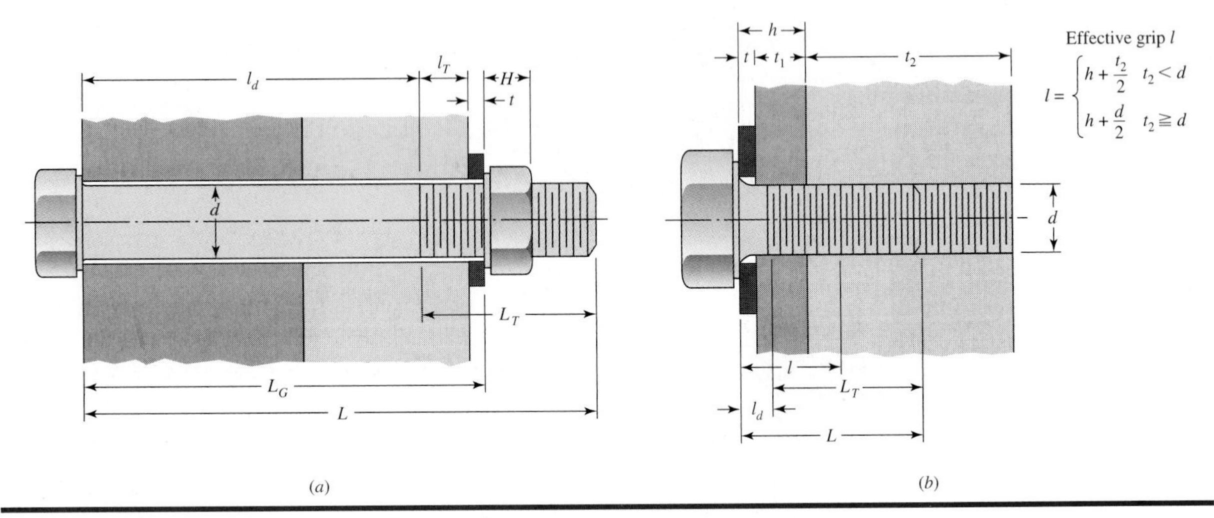

	Given fastener diameter d and pitch p or number of threads	
Grip is thickness L_G		Effective grip $= \begin{cases} h + t_2/2, & t_2 < d \\ h + d/2, & t_2 > d \end{cases}$
	Washer thickness from Table E–32 or E–33 Threaded length L_T Inch series: $L_T = \begin{cases} 2D + \frac{1}{4}\text{ in}, & L \le 6\text{ in} \\ 2D + \frac{1}{2}\text{ in}, & L > 6\text{ in} \end{cases}$ Metric series: $L_T = \begin{cases} 2D + 6\text{ mm}, & L < 125,\ D \le 48\text{ mm} \\ 2D + 12\text{ mm}, & 125 < L < 200\text{ mm} \\ 2D + 25\text{ mm}, & L > 200\text{ mm} \end{cases}$	
Fastener length: $L > L_G + H$	Round up using Table E–17*	Fastener length: $L > h + 1.5d$
Length of useful unthreaded portion: $l_d = L - L_T$ Length of threaded portion: $l_T = L_G - l_d$		Length of useful unthreaded portion: $l_d = L - L_T$ Length of useful threaded portion: $l_T = l - l_d$
	Area of unthreaded portion: $A_d = \pi d^2/4$ Area of threaded portion: $A_T = A_t$, Table 8–1 or 8–2 Fastener stiffness: $k_b = \frac{A_d A_t E}{A_d l_t + A_t l_d}$	

*Bolts and cap screws may not be available in all the preferred lengths listed in Table E–17. Large fasteners may not be available in fractional inches or in millimeter lengths ending in a nonzero digit. Check with your bolt supplier for availability.

the members is

$$\frac{1}{k_m} = \frac{1}{k_1} + \frac{1}{k_2} + \frac{1}{k_3} + \cdots + \frac{1}{k_i} \tag{8–18}$$

If one of the members is a soft gasket, its stiffness relative to the other members is usually so small that for all practical purposes the others can be neglected and only the gasket stiffness used.

If there is no gasket, the stiffness of the members is rather difficult to obtain, except by experimentation, because the compression spreads out between the bolt head and the nut and hence the area is not uniform. There are, however, some cases in which this area can be determined.

Ito[2] has used ultrasonic techniques to determine the pressure distribution at the member interface. The results show that the pressure stays high out to about 1.5 bolt radii. The pressure, however, falls off farther away from the bolt. Thus Ito suggests the use of Rotscher's pressure-cone method for stiffness calculations with a variable cone angle. This method is quite complicated, and so here we choose to use a simpler approach using a fixed cone angle.

Figure 8–15*b* illustrates the general cone geometry using a half-apex angle α. An angle $\alpha = 45°$ has been used, but Little[3] reports that this overestimates the clamping stiffness. When loading is restricted to a washer-face annulus (hardened steel, cast iron, or aluminum), the proper apex angle is smaller. Osgood[4] reports a range of $25° \le \alpha \le 33°$ for most combinations. In this book we shall use $\alpha = 30°$ except in cases in which the material is insufficient to allow the frusta to exist.

Referring now to Fig. 8–15, the elongation of an element of the cone of thickness dx subjected to a tensile force P is, from Eq. (4–3),

$$d\delta = \frac{P\,dx}{EA} \tag{a}$$

The area of the element is

$$\begin{aligned} A &= \pi\left(r_o^2 - r_i^2\right) = \pi\left[\left(x\tan\alpha + \frac{D}{2}\right)^2 - \left(\frac{d}{2}\right)^2\right] \\ &= \pi\left(x\tan\alpha + \frac{D+d}{2}\right)\left(x\tan\alpha + \frac{D-d}{2}\right) \end{aligned} \tag{b}$$

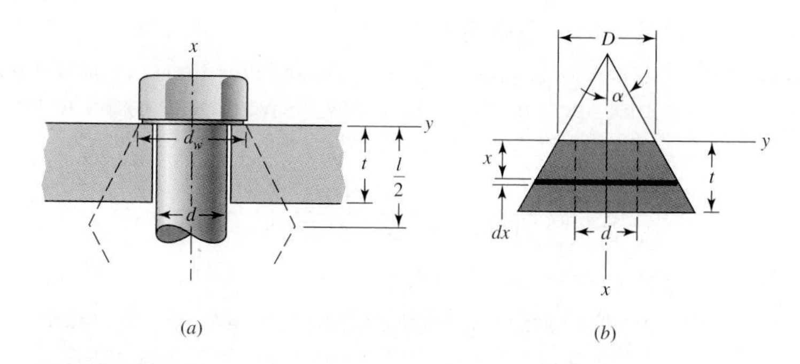

Figure 8–15

Compression of a member with the equivalent elastic properties represented by a frustum of a hollow cone.

2. Y. Ito, J. Toyoda, and S. Nagata, "Interface Pressure Distribution in a Bolt-Flange Assembly," ASME paper no. 77-WA/DE-11, 1977.

3. R. E. Little, "Bolted Joints: How Much Give?," *Machine Design*, Nov. 9, 1967.

4. C. C. Osgood, "Saving Weight on Bolted Joints," *Machine Design*, Oct. 25, 1979.

Substituting this in Eq. (a) and integrating the left side gives the elongation as

$$\delta = \frac{P}{\pi E}\int_0^t \frac{dx}{[x\tan\alpha + (D+d)/2][x\tan\alpha + (D-d)/2]} \tag{c}$$

Using a table of integrals, we find the result to be

$$\delta = \frac{P}{\pi E d\tan\alpha}\ln\frac{(2t\tan\alpha + D - d)(D+d)}{(2t\tan\alpha + D + d)(D-d)} \tag{d}$$

Thus the spring rate or stiffness of this frustum is

$$k = \frac{P}{\delta} = \frac{\pi E d\tan\alpha}{\ln\dfrac{(2t\tan\alpha + D - d)(D+d)}{(2t\tan\alpha + D - d)(D-d)}} \tag{8–19}$$

With $\alpha = 30°$, this becomes

$$k = \frac{0.577\pi E d}{\ln\dfrac{(1.15t + D - d)(D+d)}{(1.15t + D + d)(D-d)}} \tag{8–20}$$

Equation (8–20), or (8–19), must be solved separately for each frustum in the joint. Then individual stiffnesses are assembled to obtain k_m using Eq. (8–18).

If the members of the joint have the same Young's modulus E with symmetrical frusta back to back, then they act as two identical springs in series. From Eq. (8–18) we learn that $k_m = k/2$. Using the grip as $l = 2t$ and d_w as the diameter of the washer face, we find the spring rate of the members to be

$$k_m = \frac{\pi E d\tan\alpha}{2\ln\dfrac{(l\tan\alpha + d_w - d)\,(d_w + d)}{(l\tan\alpha + d_w + d)\,(d_w - d)}} \tag{8–21}$$

The diameter of the washer face is about 50 percent greater than the fastener diameter for standard hexagon-head bolts and cap screws. Thus we can simplify Eq. (8–21) by letting $d_w = 1.5d$. If we also use $\alpha = 30°$, then Eq. (8–21) can be written as

$$k_m = \frac{0.577\pi E d}{2\ln\left(5\dfrac{0.577l + 0.5d}{0.577l + 2.5d}\right)} \tag{8–22}$$

It is easy to program the numbered equations in this section, and you should do so. The time spent in programming will save many hours of formula plugging.

To see how good Eq. (8–21) is, solve it for k_m/Ed:

$$\frac{k_m}{Ed} = \frac{\pi\tan\alpha}{2\ln\left[\dfrac{(l\tan\alpha + d_w - d)\,(d_w + d)}{(l\tan\alpha + d_w + d)\,(d_w - d)}\right]}$$

Earlier in the section use of $\alpha = 30°$ was recommended for hardened steel, cast iron, or aluminum members. Wileman, Choudury, and Green[5] conducted a finite-element study

5. J. Wileman, M. Choudury, and I. Green, "Computation of Member Stiffness in Bolted Connections," *Trans. ASME, J. Mech. Design*, vol. 113, December 1991, pp. 432–437.

of this problem. The results, which are depicted in Fig. 8–16, agree with the $\alpha = 30°$ recommendation, coinciding exactly at the aspect ratio $d/l = 0.4$. Additionally, they offered an exponential curve-fit of the form

$$\frac{k_m}{Ed} = A\exp(Bd/l) \tag{8–23}$$

with constants A and B defined in Table 8–8. For standard washer faces and members of the same material, Eq. (8–23) offers a simple calculation for member stiffness k_m. For departure from these conditions, Eq. (8–20) remains the basis for the approach to the problem.

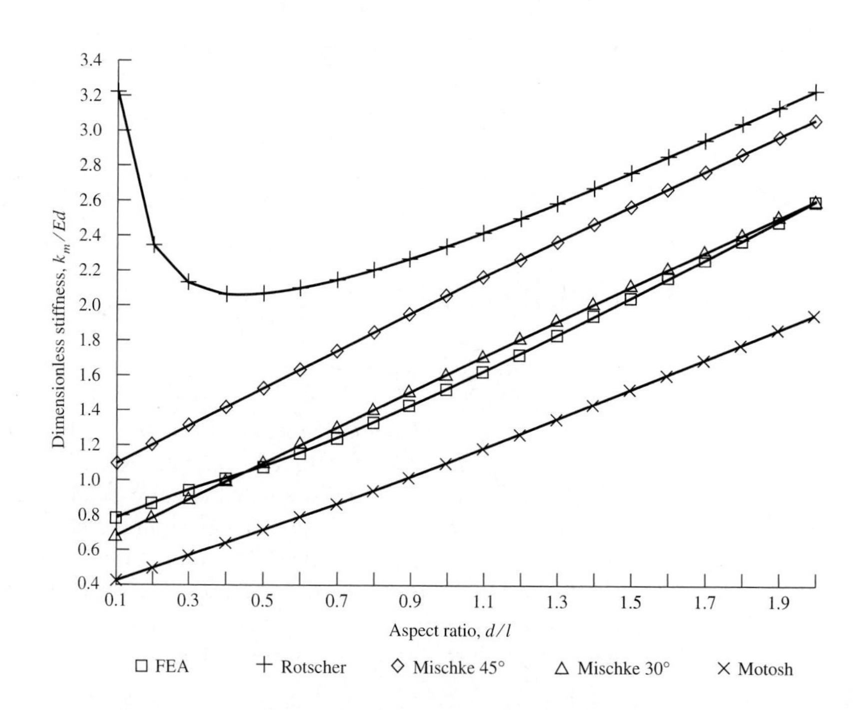

Figure 8–16

The dimensionless plot of stiffness versus aspect ratio of the members of a bolted joint, showing the relative accuracy of methods of Rotscher, Mischke, and Motosh, compared to a finite-element analysis (FEA) conducted by Wileman, Choudury, and Green. (See footnote to Table 8–8 for further information.)

Table 8–8

Stiffness Parameters of Various Member Materials

Source: J. Wileman, M. Choudury, and I. Green, "Computation of Member Stiffness in Bolted Connections," *Trans. ASME, J. Mech. Design*, vol. 113, December 1991, pp. 432–437.

Material Used	Poisson Ratio	Elastic Modulus GPa	Elastic Modulus Mpsi	A	B
Steel	0.291	207	30.0	0.787 15	0.628 73
Aluminum	0.334	71	10.3	0.796 70	0.638 16
Copper	0.326	119	17.3	0.795 68	0.635 53
Gray cast iron	0.211	100	14.5	0.778 71	0.616 16
General expression				0.789 52	0.629 14

EXAMPLE 8–2 Two 1/2-in-thick steel plates with a modulus of elasticity of $30(10^6)$ psi are clamped by washer-faced 1/2-in-diameter UNC SAE grade 5 bolts with a 0.095-in-thick washer under the nut. Find the member spring rate k_m using the method of conical frusta, and compare the result with the finite element analysis (FEA) curve-fit method of Wileman et al.

Solution The grip is $0.5 + 0.5 + 0.095 = 1.095$ in. Using Eq. (8–22) with $l = 1.095$ and $d = 0.5$ in, we write

$$k_m = \frac{0.577\pi 30\left(10^6\right)0.5}{2\ln\left[5\dfrac{0.577(1.095) + 0.5(0.5)}{0.577(1.095) + 2.5(0.5)}\right]} = 15.97\left(10^6\right)\text{ lbf/in}$$

From Table 8–8, $A = 0.787\ 15$, $B = 0.628\ 73$. Equation (8–23) gives

$$\begin{aligned} k_m &= 30\left(10^6\right)(0.5)(0.787\ 15)\exp[0.628\ 73(0.5)/1.095] \\ &= 15.73\left(10^6\right)\text{ lbf/in} \end{aligned}$$

The agreement between Eqs. (8–22) and (8–23) is good.

8–6 Bolt Strength

In the specification standards for bolts, the strength is specified by stating ASTM minimum quantities, the *minimum proof strength*, or *minimum proof load*, and the *minimum tensile strength*.

The *proof load* is the maximum load (force) that a bolt can withstand without acquiring a permanent set. The *proof strength* is the quotient of the proof load and the tensile-stress area. The proof strength thus corresponds roughly to the proportional limit and corresponds to 0.0001 in permanent set in the fastener (first measurable deviation from elastic behavior). The value of the mean proof strength, the mean tensile strength, and the corresponding standard deviations are not part of the specification codes, so it is the designer's responsibility to obtain these values, perhaps by laboratory testing, before designing to a reliability specification. Figure 8–17 shows the distribution of ultimate

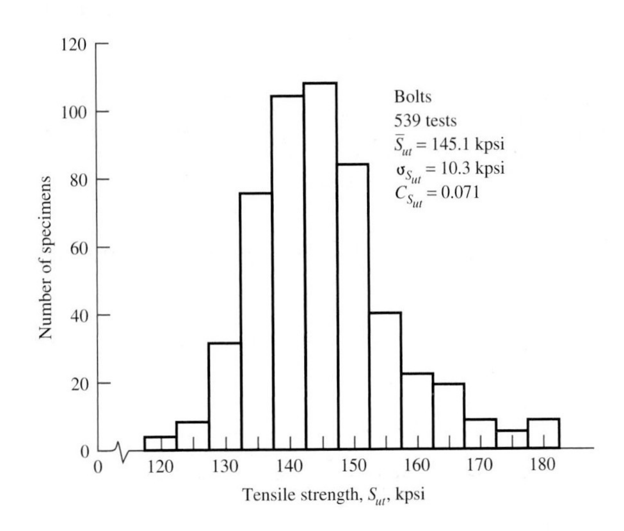

Figure 8–17

Histogram of bolt ultimate tensile strength based on 539 tests displaying a mean ultimate tensile strength $\bar{S}_{ut} = 145.1$ kpsi and a standard deviation of $\sigma_{S_{ut}} = 10.3$ kpsi.

tensile strength from a bolt production run. If the ASTM minimum strength equals or exceeds 120 kpsi, the bolts can be offered as SAE grade 5. The designer does not see this histogram. Instead, in Table 8–9, he or she sees the entry $S_{ut} = 120$ kpsi under the 1/4–1-in size in grade 5 bolts. Similarly, minimum strengths are shown in Tables 8–10 and 8–11.

The SAE specifications are found in Table 8–9. The bolt grades are numbered according to the tensile strengths, with decimals used for variations at the same strength level. Bolts and screws are available in all grades listed. Studs are available in grades 1, 2, 4, 5, 8, and 8.1. Grade 8.1 is not listed.

Table 8–9

SAE Specifications for Steel Bolts

SAE Grade No.	Size Range Inclusive, in	Minimum Proof Strength,* kpsi	Minimum Tensile Strength,* kpsi	Minimum Yield Strength,* kpsi	Material	Head Marking
1	$\frac{1}{4}$–$1\frac{1}{2}$	33	60	36	Low or medium carbon	
2	$\frac{1}{4}$–$\frac{3}{4}$	55	74	57	Low or medium carbon	
	$\frac{7}{8}$–$1\frac{1}{2}$	33	60	36		
4	$\frac{1}{4}$–$1\frac{1}{2}$	65	115	100	Medium carbon, cold-drawn	
5	$\frac{1}{4}$–1	85	120	92	Medium carbon, Q&T	
	$1\frac{1}{8}$–$1\frac{1}{2}$	74	105	81		
5.2	$\frac{1}{4}$–1	85	120	92	Low-carbon martensite, Q&T	
7	$\frac{1}{4}$–$1\frac{1}{2}$	105	133	115	Medium-carbon alloy, Q&T	
8	$\frac{1}{4}$–$1\frac{1}{2}$	120	150	130	Medium-carbon alloy, Q&T	
8.2	$\frac{1}{4}$–1	120	150	130	Low-carbon martensite, Q&T	

* Minimum strengths are strengths exceeded by 99 percent of fasteners.

Table 8–10

ASTM Specifications for Steel Bolts

ASTM Designation No.	Size Range, Inclusive, in	Minimum Proof Strength,* kpsi	Minimum Tensile Strength,* kpsi	Minimum Yield Strength,* kpsi	Material	Head Marking
A307	$\frac{1}{4}$–$1\frac{1}{2}$	33	60	36	Low carbon	
A325, type 1	$\frac{1}{2}$–1	85	120	92	Medium carbon, Q&T	A325
	$1\frac{1}{8}$–$1\frac{1}{2}$	74	105	81		
A325, type 2	$\frac{1}{2}$–1	85	120	92	Low-carbon, martensite, Q&T	A325
	$1\frac{1}{8}$–$1\frac{1}{2}$	74	105	81		
A325, type 3	$\frac{1}{2}$–1	85	120	92	Weathering steel, Q&T	A325
	$1\frac{1}{8}$–$1\frac{1}{2}$	74	105	81		
A354, grade BC					Alloy steel, Q&T	BC
A354, grade BD	$\frac{1}{4}$–4	120	150	130	Alloy steel, Q&T	
A449	$\frac{1}{4}$–1	85	120	92	Medium-carbon, Q&T	
	$1\frac{1}{8}$–$1\frac{1}{2}$	74	105	81		
	$1\frac{3}{4}$–3	55	90	58		
A490, type 1	$\frac{1}{2}$–$1\frac{1}{2}$	120	150	130	Alloy steel, Q&T	A490
A490, type 3					Weathering steel, Q&T	A490

* Minimum strengths are strengths exceeded by 99 percent of fasteners.

ASTM specifications are listed in Table 8–10. ASTM threads are shorter because ASTM deals mostly with structures; structural connections are generally loaded in shear, and the decreased thread length provides more shank area.

Specifications for metric fasteners are given in Table 8–11.

Table 8–11

Metric Mechanical-Property Classes for Steel Bolts, Screws, and Studs*

Property Class	Size Range, Inclusive	Minimum Proof Strength,† MPa	Minimum Tensile Strength,† MPa	Minimum Yield Strength,† MPa	Material	Head Marking
4.6	M5–M36	225	400	240	Low or medium carbon	4.6
4.8	M1.6–M16	310	420	340	Low or medium carbon	4.8
5.8	M5–M24	380	520	420	Low or medium carbon	5.8
8.8	M16–M36	600	830	660	Medium carbon, Q&T	8.8
9.8	M1.6–M16	650	900	720	Medium carbon, Q&T	9.8
10.9	M5–M36	830	1040	940	Low-carbon martensite, Q&T	10.9
12.9	M1.6–M36	970	1220	1100	Alloy, Q&T	12.9

* The thread length for bolts and cap screws is

$$L_T = \begin{cases} 2d + 6 & L \le 125 \\ 2d + 12 & 125 < L \le 200 \\ 2d + 25 & L > 200 \end{cases}$$

where L is the bolt length. The thread length for structural bolts is slightly shorter than given above.

† Minimum strengths are strength exceeded by 99 percent of fasteners.

It is worth noting that all specification-grade bolts made in this country bear a manufacturer's mark or logo, in addition to the grade marking, on the bolt head. Such marks confirm that the bolt meets or exceeds specifications. If such marks are missing, the bolt may be imported; for imported bolts there is no obligation to meet specifications.

Bolts in fatigue axial loading fail at the fillet under the head, at the thread runout, and at the first thread engaged in the nut. If the bolt has a standard shoulder under the head, it has a value of K_f from 2.1 to 2.3, *and* this shoulder fillet is protected from scratching or scoring by a washer. If the thread runout has a 15° or less half-cone angle, the stress is higher at the first engaged thread in the nut. Bolts are sized by examining the loading at the plane of the washer face of the nut. This is the weakest part of the bolt *if and only if* the conditions above are satisfied (washer protection of the shoulder fillet and thread runout $\leq 15°$). Inattention to this requirement has led to a record of 15 percent fastener fatigue failure under the head, 20 percent at thread runout, and 65 percent where the designer is focusing attention. It does little good to concentrate on the plane of the nut washer face if it is not the weakest location. It is akin to playing football on a soccer field. You catch the arcing ball with your hands, tuck it, and run for the touchdown only to have the referee whistle you down and hand the ball to the opposition.

Nuts are graded so that they can be mated with their corresponding grade of bolt. The purpose of the nut is to have its threads deflect to distribute the load of the bolt more evenly to the nut. The nut's properties are controlled in order to accomplish this. The grade of the nut should be the grade of the bolt.

8–7 Tension Joints—The External Load

Let us now consider what happens when an external tensile load P, as in Fig. 8–13, is applied to a bolted connection. It is to be assumed, of course, that the clamping force, which we will call the *preload* F_i, has been correctly applied by tightening the nut *before* P is applied. The nomenclature used is:

F_i = preload
P = external tensile load
P_b = portion of P taken by bolt
P_m = portion of P taken by members
$F_b = P_b + F_i$ = resultant bolt load
$F_m = P_m - F_i$ = resultant load on members
C = fraction of external load P carried by bolt
$1 - C$ = fraction of the external load P carried by members

The load P is tension, and it causes the connection to stretch, or elongate, through some distance δ. We can relate this elongation to the stiffnesses by recalling that k is the force divided by the deflection. Thus

$$\delta = \frac{P_b}{k_b} \quad \text{and} \quad \delta = \frac{P_m}{k_m} \tag{a}$$

or

$$P_b = P_m \frac{k_b}{k_m} \tag{b}$$

Since $P = P_b + P_m$, we have

$$P_b = \frac{k_b P}{k_b + k_m} \tag{c}$$

Therefore the resultant bolt load is

$$F_b = P_b + F_i = \frac{k_b P}{k_b + k_m} + F_i = CP + F_i \qquad F_m < 0 \tag{8–24}$$

Table 8–12

Computation of Bolt and Member Stiffnesses. Steel members clamped using a $\frac{1}{2}''$-13 NC steel bolt. $C = \dfrac{k_b}{k_b + k_m}$

	Stiffnesses, Mlb/in			
Bolt Grip, in	k_n	k_m	C	$1 - C$
2	2.57	12.69	0.168	0.832
3	1.79	11.33	0.136	0.864
4	1.37	10.63	0.114	0.886

and the resultant load on the connected members is

$$F_m = P_m - F_i = \frac{k_m P}{k_b + k_m} - F_i = (1 - C)P - F_i \qquad F_m < 0 \tag{8–25}$$

Of course, these results are valid only as long as some clamping load remains in the members; this is indicated by the qualifier in the equations.

Table 8–12 is included to provide some information on the relative values of the stiffnesses encountered. The grip contains only two members, both of steel, and no washers. The ratios C and $1 - C$ are the coefficients of P in Eqs. (8–24) and (8–25), respectively. They describe the proportion of the external load taken by the bolt and by the members, respectively. In all cases, the members take over 80 percent of the external load. Think how important this is when fatigue loading is present. Note also that making the grip longer causes the members to take an even greater percentage of the external load.

8–8 Relating Bolt Torque to Bolt Tension

Having learned that a high preload is very desirable in important bolted connections, we must next consider means of ensuring that the preload is actually developed when the parts are assembled.

If the overall length of the bolt can actually be measured with a micrometer when it is assembled, the bolt elongation due to the preload F_i can be computed using the formula $\delta = F_i l/(AE)$. Then the nut is simply tightened until the bolt elongates through the distance δ. This ensures that the desired preload has been attained.

The elongation of a screw cannot usually be measured, because the threaded end is often in a blind hole. It is also impractical in many cases to measure bolt elongation. In such cases the wrench torque required to develop the specified preload must be estimated. Then torque wrenching, pneumatic-impact wrenching, or the turn-of-the-nut method may be used.

The torque wrench has a built-in dial which indicates the proper torque.

With impact wrenching, the air pressure is adjusted so that the wrench stalls when the proper torque is obtained, or in some wrenches, the air automatically shuts off at the desired torque.

The turn-of-the-nut method requires that we first define the meaning of snug-tight. The *snug-tight* condition is the tightness attained by a few impacts of an impact wrench, or the full effort of a person using an ordinary wrench. When the snug-tight condition is attained, all additional turning develops useful tension in the bolt. The turn-of-the-nut method requires that you compute the fractional number of turns necessary to develop the required preload from the snug-tight condition. For example, for heavy hexagon structural bolts, the turn-of-the-nut specification states that the nut should be turned a minimum of 180° from the snug-tight condition under optimum conditions. Note that this is also about the correct rotation for the wheel nuts of a passenger car.

Although the coefficients of friction may vary widely, we can obtain a good estimate of the torque required to produce a given preload by combining Eqs. (8–5) and (8–6):

$$T = \frac{F_i d_m}{2}\left(\frac{l + \pi f d_m \sec\alpha}{\pi d_m - f l \sec\alpha}\right) + \frac{F_i f_c d_c}{2} \tag{a}$$

Since $\tan\lambda = l/\pi d_m$, we divide the numerator and denominator of the first term by πd_m and get

$$T = \frac{F_i d_m}{2}\left(\frac{\tan\lambda + f \sec\alpha}{l - f \tan\lambda \sec\alpha}\right) + \frac{F_i f_c d_c}{2} \tag{b}$$

The diameter of the washer face of a hexagonal nut is the same as the width across flats and equal to $1\frac{1}{2}$ times the nominal size. Therefore the mean collar diameter is $d_c = (d + 1.5d)/2 = 1.25d$. Equation (*b*) can now be arranged to give

$$T = \left[\left(\frac{d_m}{2d}\right)\left(\frac{\tan\lambda + f \sec\alpha}{1 - f \tan\lambda \sec\alpha}\right) + 0.625 f_c\right] F_i d \tag{c}$$

We now define a *torque coefficient* K as the term in brackets, and so

$$K = \left(\frac{d_m}{2d}\right)\left(\frac{\tan\lambda + f \sec\alpha}{1 - f \tan\lambda \sec\alpha}\right) + 0.625 f_c \tag{8–26}$$

Equation (*c*) can now be written

$$T = K F_i d \tag{8–27}$$

The coefficient of friction depends upon the surface smoothness, accuracy, and degree of lubrication. On the average, both f and f_c are about 0.15. The interesting fact about Eq. (8–26) is that $K \approx 0.20$ for $f = f_c = 0.15$ no matter what size bolts are employed and no matter whether the threads are coarse or fine.

Blake and Kurtz have published results of numerous tests of the torquing of bolts.[6] By subjecting their data to a statistical analysis, we can learn something about the distribution of the torque coefficients and the resulting preload. Blake and Kurtz determined the preload in quantities of unlubricated and lubricated bolts of size $\frac{1}{2}''$-20 UNF when torqued to 800 lb · in. This corresponds roughly to an M12 × 1.25 bolt torqued to 90 N · m. The statistical analyses of these two groups of bolts, converted to SI units, are displayed in Tables 8–13 and 8–14.

We first note that both groups have about the same mean preload, 34 kN. The unlubricated bolts have a standard deviation of 4.9 kN and a COV of about 0.15. The lubricated bolts have a standard deviation of 3 kN and a COV of about 0.9.

Table 8–13

Distribution of Preload F_i for 20 Tests of Unlubricated Bolts Torqued to 90 N · m

23.6,	27.6,	28.0,	29.4,	30.3,	30.7,	32.9,	33.8,	33.8,	33.8,
34.7,	35.6,	35.6,	37.4,	37.8,	37.8,	39.2,	40.0,	40.5,	42.7

* Mean value $\bar{F}_i = 34.3$ kN. Standard deviation, $\hat{\sigma} = 4.91$ kN.

6. J. C. Blake and H. J. Kurtz, "The Uncertainties of Measuring Fastener Preload," *Machine Design*, vol. 37, Sept. 30, 1965, pp. 128–131.

Table 8–14

Distribution of Preload F_i for 10 Tests of Lubricated Bolts Torqued to 90 N · m

30.3,	32.5,	32.5,	32.9,	32.9,	33.8,	34.3,	34.7,	37.4,	40.5

* Mean value $\bar{F}_i = 34.18$ kN. Standard deviation, $\hat{\sigma} = 2.88$ kN.

The means obtained from the two samples are nearly identical, approximately 34 kN; using Eq. (8–27), we find, for both samples, $K = 0.208$.

Bowman Distribution, a large manufacturer of fasteners, recommends the values shown in Table 8–15. In this book we shall use these values and use $K = 0.2$ when the bolt condition is not stated.

We shall now take a closer look at the nature of the tension-induced bolt stresses. Equation (8–24) can be written without joint load P by equating P to zero, to obtain

$$\mathbf{F}_b = \mathbf{F}_i = \boldsymbol{\sigma} A_t$$

where $\mathbf{F}_b$, $\mathbf{F}_i$, and $\boldsymbol{\sigma}$ are random variables. The variability in the bolt stress $\boldsymbol{\sigma}$ is transmitted to the initial tension $\mathbf{F}_b$. The coefficient of variation of the stress is equal to that in $\mathbf{F}_i$, that is, $C_\sigma = C_{Fi}$. While it is best that all fasteners are to be replaced when a bolted joint is disassembled, there are always these tempting circumstances:

- The replacement fasteners have not arrived or are elsewhere.
- The old fasteners, when cleaned in solvent, look good.
- In the press of haste, who will know?

The prudent designer protects against these circumstances because the reassembled joint it *different* owing to the permanent set in the fasteners, for which the torque-tension relationship is now unknown. Furthermore, for the same torque, the initial tension will be less than the designer had specified.

The integrity of the design has moved to another person's hands. To protect against this development, the prudent designer can specify a tightening torque that will provide an acceptably low probability of causing permanent thread distortion in bolt or nut. Such a probability assessment requires a stress-strength interference. The bolt preload tightening torque is given by

$$T = KdF_i = Kd\xi_1 S_{0.99} A_t \quad \text{(customary engineering units)}$$
$$T = Kd\xi_1 S_{0.99} A_t \left(10^{-3}\right) \quad \text{(SI units)} \qquad (8\text{–}28)$$

In SI units, d is mm, $S_{0.99}$ is MPa, A_t is mm^2, and T is N · m. The parameter ξ_1 is the fraction of proof stress felt by the bolt when tightened, $\xi_1 = \sigma_i / S$. The parameter ξ_2 is

Table 8–15

Torque Factors K for Use with Eq. (8–27)

Bolt Condition	K
Nonplated, black finish	0.30
Zinc-plated	0.20
Lubricated	0.18
Cadmium-plated	0.16
With Bowman Anti-Seize	0.12
With Bowman-Grip nuts	0.09

the fraction of proof stress felt by the bolt under external load. The fraction ξ_1 is selected so that acceptable small risk of over-proof loading exists when the load imposes ξ_2. This protects the integrity of the joint so that the design is not compromised by the reuse of fasteners.

We can find ξ_1 and ξ_2 as follows. In Fig. 8–18*a* we see the distribution of the initial tightening stress $\boldsymbol{\sigma}$. Displaced slightly to the right is the distribution of the bolt stress while load P is carried by the joint. To the right is the distribution of the proof strength $\mathbf{S}_p$. The proof strength is the sum of the bolt stress $\bar{\sigma}_b = \xi_2 \bar{S}_p$ and the product of the z variable, and the standard deviation written as

$$S_p = \xi_2 S_p + z_2 \xi_1 S_p C_F$$

where C_F is the COV of the induced initial tension. The proof strengths S_p cancel and $1 = \xi_2 + z_2 \xi_1 C_F$. Noting the bolt equation $\boldsymbol{\sigma}_b = CP/A_t + \boldsymbol{\sigma}_i$ can be expressed as $\xi_1 = \xi_2 - CP/(S_p A_t)$, we substitute for ξ_1 to obtain

$$1 = \xi_2 + z_2 C_F \left(\xi_2 - \frac{CP}{S_p A_t} \right)$$

from which

$$\xi_2 = \frac{1 + z_2 CPC_F/(S_p A_t)}{1 + z_2 C_F} \tag{8–29}$$

and we find ξ_1 from

$$\xi_1 = \xi_2 - \frac{CP}{S_p A_t} \tag{8–30}$$

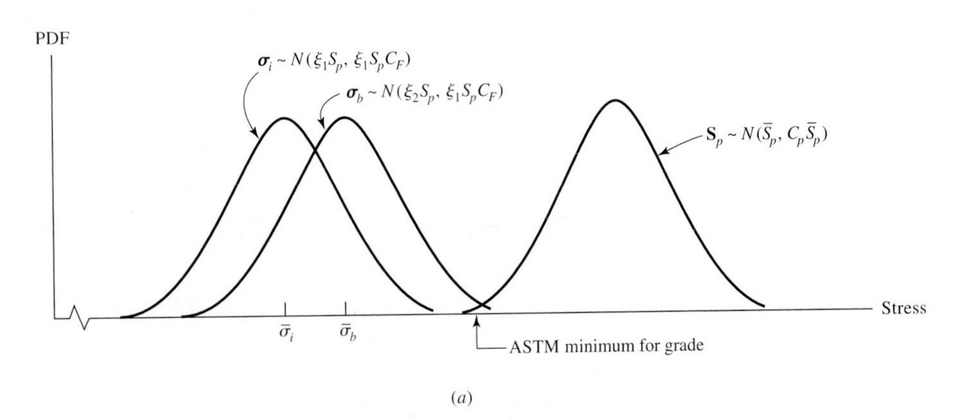

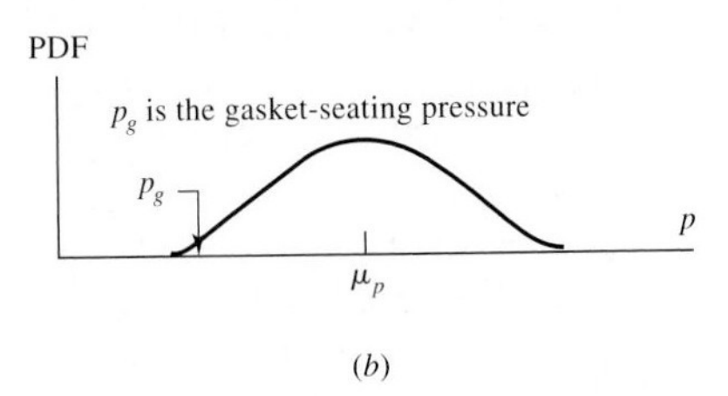

Figure 8–18

(*a*) Three distributions: preload stress σ_i, loaded bolt stress σ_b, and proof strength S_p, in an externally-loaded bolted joint. (*b*) PDF of gasket pressure p in a gasketed bolted joint containing a fluid.

Solving the equation preceding Eq. (8–29) for z_2 gives

$$z_2 = \frac{1 - \xi_2}{\xi_1 C_F} \tag{8–31}$$

Substituting Eq. (8–30) to eliminate ξ_2 gives

$$z_2 = \frac{1}{C_F}\left(\frac{1 - CP/(S_p A_t)}{\xi_1} - 1\right) \tag{8–32}$$

We will now carry out the interference between the $\boldsymbol{\sigma}_b$ distribution and the $\mathbf{S}_p$ distribution shown in Fig. 8–18. The ASTM minimum strength is a value of strength exceeded by 99 percent of the population. We can write, noting $|z_{0.99}| = 2.326$,

$$_{0.99}S_p = \bar{S}_p - 2.326\,\sigma_{Sp} = \bar{S}_p - 2.326 C_{Sp}\bar{S}_p = \bar{S}_p\left(1 - 2.326 C_{Sp}\right)$$

To reduce the complexity of the notation, we will drop the subscript P until the manipulation is complete. The means and the standard deviations of the strength distribution $\mathbf{S}$ and the bolt stress distribution $\boldsymbol{\sigma}_b$ can be expressed as follows, from Fig. 8–18:

$$\bar{S} = \frac{S_{0.99}}{1 - 2.326 C_S} \qquad \sigma_S = C_S \bar{S} = \frac{C_S S_{0.99}}{1 - 2.326 C_S}$$

$$\bar{\sigma}_b = \xi_2 S_{0.99} \qquad \sigma_{\sigma_b} = C_F \bar{\sigma}_b = C_F \xi_1 S_{0.99}$$

The normal-normal coupling equation, Eq. (2–52), is used to find the probability of over-proof stressing:

$$z_2 = -\frac{\mu_S - \mu_\sigma}{\left(\sigma_S^2 + \sigma_\sigma^2\right)^{1/2}}$$

$$z_2 = -\frac{\dfrac{S_{0.99}}{1 - 2.326 C_S} - \xi_2 S_{0.99}}{\left[\left(\dfrac{C_S S_{0.99}}{1 - 2.326 C_S}\right)^2 + (\xi_1 S_{0.99} C_F)^2\right]^{1/2}}$$

$$z_2 = -\frac{\dfrac{1}{1 - 2.326 C_S} - \xi_2}{\left[\left(\dfrac{C_S}{1 - 2.326 C_S}\right)^2 + (\xi_1 C_F)^2\right]^{1/2}} \tag{8–33}$$

Equation (8–33) is remarkable in that the population means $\bar{\sigma}_b$ and $\bar{S}$ are absent, and no strength or stress percentile is present. Only the coefficients of variation of the proof strength C_{Sp} and the initial tension C_{Fi} appear. When $C_{Sp} = 0$, Eq. (8–33) simplifies to Eq. (8–31).

The situation is that fastener manufacturers know by test the S_p distribution of their bolts. Manufacturers must assure themselves that 99 percent exceed the grade-specified minimum. This information is not generally available to the designer. The designer can instigate testing to find this distribution, but it takes a high volume of joint manufacture to recover costs. The designer models the situation by concentrating the S_p distribution as a point at the grade-minimum S_p and uses Eqs. (8–29) to (8–32) knowing that they are conservative. The probability of over-proof stressing a bolt is less than given by Eqs. (8–32) and (8–33).

From $\mathbf{F}_b = CP + \mathbf{F}_i$ we can write

$$\boldsymbol{\sigma}_2 A_t = CP + \boldsymbol{\sigma}_1 A_t$$

$$\boldsymbol{\sigma}_2 = \frac{CP}{A_t} + \boldsymbol{\sigma}_1$$

The induced tensile stress $\boldsymbol{\sigma}_1$ augmented by the load P to $\boldsymbol{\sigma}_2$ both have the same standard deviation, namely, $\hat{\sigma}_{\sigma_1} = \hat{\sigma}_{\sigma_2} = C_{Fi}\bar{\sigma}_1$. Because the means are different, the coefficient of variation of the $\boldsymbol{\sigma}_2$ distribution is less, due to the increased mean, namely

$$C_{\text{loaded}} = \frac{\hat{\sigma}_{\sigma_2}}{\bar{\sigma}_2} = \frac{\bar{\sigma}_1 C_{Fi}}{\bar{\sigma}_2} \tag{8–34}$$

If one is interfering $\boldsymbol{\sigma}_1$ with $\mathbf{S}_p$, the coefficient of variation of the stress is C_{Fi}. If one is interfering $\boldsymbol{\sigma}_2$ with $\mathbf{S}_p$, the coefficient of variation of the stress is $C_{Fi}\bar{\sigma}_1/\bar{\sigma}_2$.

EXAMPLE 8–3

A 3/4″-16 UNF × 2 1/2″ SAE grade 5 bolt is 1 of 10 2 1/2″ bolts subjected to a load P of 6 kip in a tension joint. The initial tension is $F_i = 25$ kip. The joint constant C is 0.320. The threads are lubricated.

(*a*) Find the proof stress fractions ξ_1 and ξ_2.
(*b*) Find the bolt stresses σ_1 and σ_2.
(*c*) Use Eq. (8–31) with ξ_2 to find z and probability of failure p_f.
(*d*) Find ξ_1 and ξ_2 using Eqs. (8–29) and (8–30).
(*e*) What is the probability that one or more bolts will be over-proof loaded?
(*f*) What is the torque-wrench setting?
(*g*) Solve part (*c*) using Eq. (8–33) if $C_{Sp} = 0.10$, and compare.

Solution

(*a*) The fraction of proof stress ξ_1 attained by torquing the bolt is

Answer
$$\xi_1 = \frac{F_i}{S_p A_t} = \frac{25\ 000}{85\ 000(0.373)} = 0.789$$

Answer
$$\xi_2 = \xi_1 + \frac{CP}{S_p A_t} = 0.789 + \frac{0.320(6000)}{85\ 000(0.373)} = 0.850$$

Answer
$$(b)\ \sigma_i = \frac{F_i}{A_t} = \frac{25\ 000}{0.373} = 67\ 024 \text{ psi}$$

Answer
$$\sigma_b = \sigma_i + \frac{CP}{A_t} = 67\ 024 + \frac{0.320(6000)}{0.373} = 72\ 171 \text{ psi}$$

(*c*) Using Eq. (8–31),

Answer
$$z_2 = \frac{1-\xi_2}{\xi_1 C_{Fi}} = \frac{1-0.850}{0.789(0.09)} = 2.11$$

Answer From Table E–10, $p_f = 0.0174$, which means $p_f < 0.0174$.

(*d*) Using Eq. (8–29),

Answer
$$\xi_2 = \frac{1 + 2.11 + \dfrac{0.320(6000)0.09}{85\ 000(0.373)}}{1 + 2.11(0.09)} = 0.850$$

Answer Using Eq. (8–30),

$$\xi_1 = 0.850 - \frac{0.320(6000)}{85\,000(0.373)} = 0.790$$

(*e*) $p_f = 0.017$

$$R_j = R_b^N = (1 - 0.017)^{10} = 0.842$$

Answer $p_{fj} = 1 - R_j = 1 - 0.842 = 0.158$

Answer (*f*) $T = KdF_i = Kd\xi_1 S_p A_t = 0.20(0.75)0.789(85\,000)0.373 = 3752$ in · lbf

(*g*) $$z_2 = -\frac{\dfrac{1}{1 - 2.326(0.10)} - 0.850}{\left\{\left[\dfrac{0.10}{1 - 2.326(0.10)}\right]^2 + [0.09(0.790)]^2\right\}^{1/2}} = -2.79$$

Answer From Table E–10, $p_f = 0.002\,64$, as compared to 0.0174, demonstrating the conservatism of Eq. (8–31).

The experience with Ex. 8–3 suggests the following. The mean bolt load corresponding to ASTM minimum proof strength S_p is $S_p A_t$. The mean bolt load corresponding to the torquing operation and the external load on the joint is $\xi_2 S_p A_t$. The mean factor of safety guarding against permanent set in the fastener is

$$n_p = \frac{S_p A_t}{\xi_2 S_p A_t} = \frac{1}{\xi_2} \tag{8–35}$$

In Ex. 8–3 this was

$$n_p = \frac{1}{0.850} = 1.18$$

This factor of safety represents the displacement of the mean bolt stress from the ASTM minimum proof strength as a ratio S_p/σ_b.

Bolted joints containing a fluid under pressure are tested hydrostatically at some multiple m of the fluid pressure to ensure freedom from leakage. This brings us to the subject of preload of the next section.

8–9 Statically Loaded Tension Joint—Preload

Let us write Eq. (8–24) for a bolted joint carrying fluid under pressure

$$F_b = \frac{k_b n(mP)}{k_b + k_m} + F_i = Cn(mP) + F_i \tag{8–36}$$

where m is a multiplier which represents the ratio of hydrostatic test pressure to operating pressure, n is the design factor, and C is the joint constant $k_b/(k_b + k_m)$. Equation (8–25) can be written as

$$F_m = (1 - C)n(mP) - F_i \tag{8–37}$$

where n represents a design factor.

We consider the bolt first. The tensile stress in the bolt can be found by dividing Eq. (8–36) by the tensile-stress area A_t of the bolt:

$$\frac{F_b}{A_t} = \sigma_b = \frac{Cn(mP)}{A_t} + \frac{F_i}{A_t} \tag{a}$$

Solving Eq. (*a*) for load design factor n and substituting $\xi_1 S_p A_t$ for F_i gives

$$n = \frac{\sigma_b A_t - F_i}{C(mP)} = \frac{\sigma_b A_t - \xi_1 S_p A_t}{C(mP)} \tag{b}$$

The deterministic view is that the loss-of-function stress [the numerator of Eq. (*b*)] occurs when $\sigma_b = S_p$. The bolt stress will not be S_p because ξ_2 is related to ξ_1, and not arbitrary. Substitution of S_p for σ_b introduces *reserve capacity* with respect to the load capacity in the joint. Setting bolt stress to S_p gives

$$n = \frac{S_p A_t}{C(mp)}(1 - \xi_1) \tag{8–38}$$

The bolt stress is $\xi_2 S_p$ and we substitute $\xi_2 S_p$ in Eq. (*b*). The design factor is $n = 1$ because ξ_1 and ξ_2 were chosen to be at the stochastic limit in accordance with the probability of failure goal chosen to accommodate fastener reuse:

$$n = 1 = \frac{\xi_2 S_p A_t - \xi_1 S_p A_t}{C(mP)} = \frac{S_p A_t}{C(mP)}(\xi_2 - \xi_1)$$

From this,

$$\frac{S_p A_t}{C(mP)} = \frac{1}{\xi_2 - \xi_1}$$

Substituting this into Eq. (8–38) gives

$$n = \frac{1 - \xi_1}{\xi_2 - \xi_1} \tag{8–39}$$

which expressed the load factor of safety (in this form) in terms of ξ_1 and ξ_2. In Ex. 8–3, $\xi_1 = 0.788\ 519$, to greater precision, $\xi_2 = 0.849\ 077$ and $\xi_2 - \xi_1 = 0.060\ 558$. Using Eq. (8–39),

$$n = \frac{1 - 0.788\ 519}{0.060\ 558} = 3.49$$

This means the load mP needs to be increased 3.49-fold to bring σ_b to S_p.

Another threat to joint function is the possibility that the external load on the joint may open the joint. Such a separation causes the entire external load to be placed on the fastener, loss of structural rigidity, and leaking of fluid if present. Let P_0 be the value of the external load that will cause joint separation. At separation the member force $F_m = 0$ in Eq. (8–37), so we write

$$F_i = (1 - C)P_0 = (1 - C)n(mP)$$

where n is the design factor, P is the applied load per bolt, and m is the hydrostatic test pressure multiple. The design factor (as factor of safety) is $n = P_0/(mP)$. Solving for

n in the above equation, and denoting it as n_0 the factor of safety against joint opening, then

$$n_0 = \frac{F_i}{(1-C)mP} = \frac{\xi_1 S_p A_t}{(1-C)mP} \tag{8–40}$$

Equation (8–40) shows that preload fraction of S_p, namely ξ_1, controls the safety margin represented by n_0, which is an external load P multiple.

Figure 8–19 is the stress-strain diagram of a good-quality bolt material. Notice that there is no clearly defined yield point and that the diagram progresses smoothly up to fracture, which corresponds to the tensile strength. This means that no matter how much preload is given the bolt, it will retain its load-carrying capacity. This is what keeps the bolt tight and determines the joint strength. The pre-tension is the "muscle" of the joint, and its magnitude is determined by the bolt strength. If the full bolt strength is not used in developing the pre-tension, then money is wasted and the joint is weaker.

Good-quality bolts can be preloaded into the plastic range to develop more strength. Some of the bolt torque used in tightening produces torsion, which increases the principal tensile stress. However, this torsion is held only by the friction of the bolt head and nut; in time it relaxes and lowers the bolt tension slightly. Thus, as a rule, a bolt will either fracture during tightening, or not at all.

Above all, do not rely too much on wrench torque; it is not a good indicator of preload. Actual bolt elongation should be used whenever possible—especially with fatigue loading. In fact, if high reliability is a requirement of the design, then preload should always be determined by bolt elongation.

RB&W recommendations for preload are 60 kpsi for SAE grade 5 bolts for nonpermanent connections, and that A325 bolts (equivalent to SAE grade 5) used in structural applications be tightened to proof load or beyond (85 kpsi up to a diameter of 1 in).[7] Bowman[8] recommends a preload of 75 percent of proof load, which is about the same as the RB&W recommendations for reused bolts. In view of these guidelines, it is

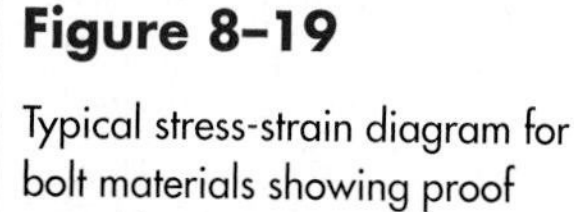

Figure 8–19

Typical stress-strain diagram for bolt materials showing proof strength S_p, yield strength S_y, and ultimate tensile strength S_{ut}.

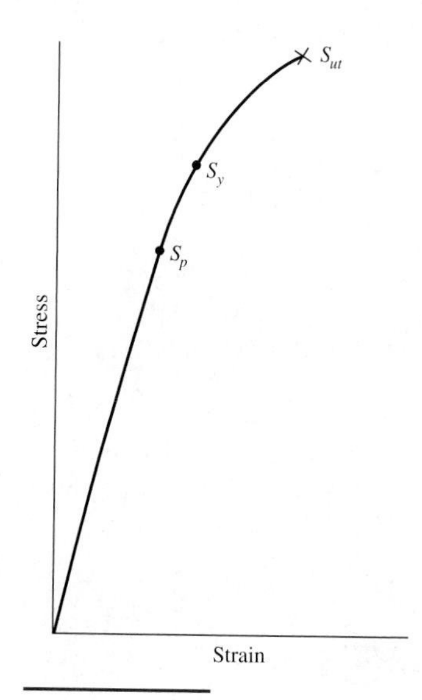

7. Russell, Burdsall & Ward Inc., *Helpful Hints for Fastener Design and Application*, Mentor, Ohio, 1965, p. 42.

8. Bowman Distribution–Barnes Group, *Fastener Facts*, Cleveland, 1985, p. 90.

recommended for both static and fatigue loading that the following be used for preload:

$$F_i = \begin{cases} 0.75F_p & \text{for nonpermanent connections, reused fasteners} \\ 0.90F_p & \text{for permanent connections} \end{cases} \tag{8-41}$$

where F_p is the proof load, obtained from the equation

$$F_p = A_t S_p \tag{8-42}$$

Here S_p is the proof strength obtained from Tables 8–9 to 8–11. For other materials, an approximate value is $S_p = 0.85S_y$. Be very careful not to use a soft material in a threaded fastener. For high-strength steel bolts used as structural steel connectors, if advanced tightening methods are used, tighten to yield.

You can see that the RB&W recommendations on preload are in line with what we have encountered in this chapter. The purposes of development were to give the reader the perspective to appreciate Eqs. (8–41) and a methodology with which to handle cases more specifically than the recommendations.

An adequacy assessment for a statically loaded joint develops and scrutinizes the

- Joint constant C.
- Preload F_i.
- Fractions ξ_1 and ξ_2.
- Various factors of safety.
- Estimates of probability of failure.

EXAMPLE 8–4

Figure 8–20 is a cross section from a grade 25 cast-iron pressure cylinder with ten 5/8″-11 UNC × 2 1/4″ SAE grade 5 lubricated bolts torqued to 1700 in · lbf. The external tension force of the joint is 18 kip. Given that a hydrostatic test is conducted to twice the operating pressure, perform an adequacy assessment for static loading.

Solution

For SAE grade 5, $S_p = 85$ kpsi, $A_t = 0.226$ in^2, $A_d = 0.3068$ in^2, $P = 18\,000/10 = 1800$ lbf/bolt, $l_T = 0.75$ in, $l_d = 0.75$ in, $K = 0.20$, $l = 1.5$ in, $C_{Fi} = 0.09$, $E_m = 12$ Mpsi, $m = 2$. From Eq. (8–17),

$$k_b = \frac{0.3068(0.226)30\left(10^6\right)}{0.3068(0.75) + 0.226(0.75)} = 5.21 \text{ Mlb/in}$$

Figure 8–20

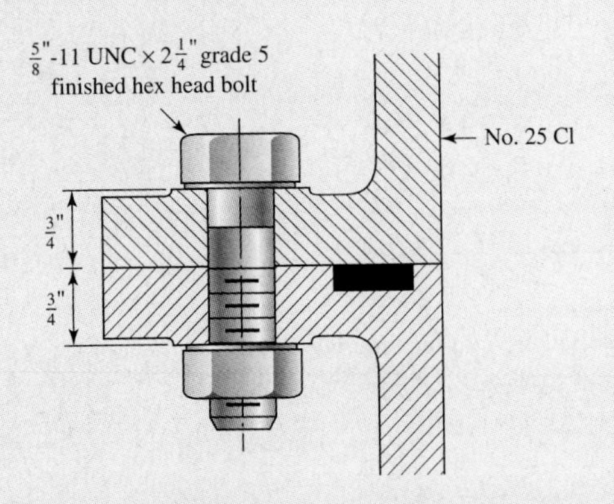

From Eq. (8–22),

$$k_m = \frac{0.577(12)10^6(0.625)}{2\ln\left[5\dfrac{0.577(1.5)+0.5(0.625)}{0.577(1.5)+2.5(0.625)}\right]} = 7.67 \text{ Mlbf/in}$$

From Eq. (8–36),

$$C = \frac{5.21}{5.21+7.67} = 0.404$$

From Eq. (8–28),

$$\xi_1 = \frac{T}{KdS_pA_t} = \frac{1700}{0.20(0.625)85\,000(0.226)} = 0.707\,965$$

$$F_i = \xi_1 S_p A_t = 0.708(85\,000)0.226 = 13\,600 \text{ lbf}$$

Under hydrostatic test:

$$\xi_2 = \xi_1 + \frac{CmP}{S_pA_t} = 0.707\,965 + \frac{0.404(2)1800}{85\,000(0.226)} = 0.783\,675$$

From Eq. (8–39),

$$n = \frac{1-0.707\,965}{0.783\,675-0.707\,965} = 3.86$$

or, alternatively, from Eq. (8–38),

$$n = \frac{85\,000(0.226)(1-0.784)}{0.404(2)1800} = 3.86$$

From Eq. (8–40),

$$n_0 = \frac{0.708(85\,000)0.226}{(1-0.404)2(1800)} = 6.34$$

$$n_p = \frac{1}{\xi_2} = \frac{1}{0.784} = 1.28$$

From Eq. (8–32),

$$z_2 = \frac{1}{0.09}\left[\frac{1-\dfrac{0.404(2)1800}{85\,000(0.226)}}{0.708} - 1\right] = 3.39$$

From Table E–10,

$$R = 1 - \Phi(z) = 1 - 0.000\,39 = 0.999\,64$$

$$R_j = 0.999\,64^{10} = 0.9964 \qquad p_{fj} = 1 - 0.9964 = 0.0036$$

Under operating load:

$$\xi_1 = 0.708$$

$$\xi_2 = \xi_1 + \frac{CmP}{S_pA_t} = 0.708 + \frac{0.404(1)1800}{85\,000(0.226)} = 0.7459$$

From Eq. (8–39),

$$n = \frac{1 - 0.708}{0.7459 - 0.708} = 7.71$$

From Eq. (8–40),

$$n_0 = \frac{0.708(85\,000)0.226}{(1 - 0.404)(1)1800} = 12.68$$

$$n_P = \frac{1}{\xi_2} = \frac{1}{0.7459} = 1.34$$

From Eq. (8–32),

$$z_2 = \frac{1}{0.09}\left[\frac{1 - \dfrac{0.404(1)1800}{85\,000(0.226)}}{0.708} - 1\right] \doteq 4.0$$

From Table E–10,

$$R = 1 - \Phi(z) = 1 - 0.000\,031\,7 = 0.999\,968\,3$$
$$R_j = 0.999\,968\,3^{10} = 0.999\,683 \qquad p_f = 0.000\,317$$

Some key results are tabulated below:

Parameter	Hydrostatic Test at Twice Operating Pressure, *m* = 2	At Operating Pressure, *m* = 1
n	3.86	7.71
n_0	6.34	12.68
n_P	1.28	1.34
p_f	0.003 6	0.000 317

- Notice that the factor of safety n is inversely proportional to the load.
- Notice that the factor of safety n_0 is inversely proportional to the load.
- Notice that the factor of safety n_P is not inversely proportional to the load.
- Notice that the probability of over-proof stressing is not inversely proportional to load.

The world is nonlinear. In a pencil-and-paper adequacy assessment such as this, one looks at "everything," focusing particularly on merit-sensitive numbers such as those in the table above. Remember elastic bodies behaving elastically can have properties not linearly or proportionally related to load. Whether or not this joint is satisfactory depends on over-proof stressing and how the extant probability of failure relates to the designer's goal, here unstated.

This is the beginning of your development of designer's skill 1. It is the ability to take an existing design and decide whether it is satisfactory or not.

8–10 Gasketed Joints

If a full gasket is present in the joint, the gasket pressure p is found by dividing the force on the members by the area of the gasket per bolt, A_g/N:

$$p = -\frac{F_m}{A_g/N}$$

where N is the number of bolts in the joint. From Eq. (8–37), if the joint is unloaded,

$$F_m = -F_i = -\frac{pA_g}{N} = -\sigma A_t$$

where σ is the bolt stress. The gasket pressure and the bolt stress are random variables, so we write

$$\mathbf{p} = \frac{NA_t}{A_g}\boldsymbol{\sigma}$$

The mean of $\mathbf{p}$ is

$$\mu_p = \frac{NA_t\bar{\sigma}}{A_g} = \frac{NA_t}{A_g}\xi_1 S_P$$

The standard deviation of $\mathbf{p}$ is

$$\sigma_p = \frac{NA_t}{A_g}C_{Fi}\xi_1 S_P$$

The designer wants the chance (expressed in terms of z) that the gasket pressure exceed the gasket-seating pressure p_g to be quantitatively expressible. After inspecting Fig. 8–18b we may state

$$\mu_p = p_g + |z|\sigma_p = p_g + |z|C_{Fi}\mu_p$$

Solving for μ_p gives

$$\mu_p = \frac{p_g}{1 - |z|C_{Fi}} = \psi p_g \tag{a}$$

where ψ is given by

$$\psi = \frac{1}{1 - |z|C_{Fi}} = \frac{\xi_1}{2\xi_1 - 1}$$

which is related to ξ_1 through its definition. This means that when the proof-stress fraction ξ_1 is decided, so also is the value of μ_p as well as ψ. Solving for $|z|$ from Eq. (a) gives

$$|z| = \frac{1}{C_{Fi}}\left(1 - \frac{p_g}{\mu_p}\right) = \frac{1}{C_{Fi}}\left(1 - \frac{p_g A_g}{\xi_1 S_p A_t N}\right)$$

which expresses the change of gasket-seating (through $|z|$) at initial tightening of the joint. When the load is applied, from Eq. (8–37),

$$F_m = (1 - C)n_0(mP) - F_i$$

$$-\frac{\psi p_g A_t}{N} = (1 - C)n(mP) - \xi_1 S_p A_t$$

$$\psi = \frac{N}{p_g A_t}\left[\xi_1 S_p A_t - (1 - C)n(mP)\right] \tag{8–43}$$

This value of ψ is less than that prevailing after tightening. Checking for gasket pressure above gasket-seating pressure is part of a sufficiency check made during adequacy assessment.

In full-gasketed joints uniformity of pressure on the gasket is important. To maintain adequate uniformity of pressure adjacent bolts should not be placed more than six nominal diameters apart on the bolt circle. To maintain wrench clearance bolts should be placed at least three diameters apart. A rough rule for bolt spacing around a bolt circle is

$$3 \leq \frac{\pi D_b}{Nd} \leq 6 \tag{8–44}$$

where D_b is the diameter of the bolt circle and N is the number of bolts.

8–11 Tension Joints—Dynamic Loading

Tension-loaded bolted joints subjected to fatigue action can be analyzed directly by the methods of Chap. 7. Table 8–16 lists average fatigue-strength-reduction factors for the fillet under the bolt head and also at the beginning of the threads on the bolt shank. These are already corrected for notch sensitivity and for surface finish. Designers should be aware that situations may arise in which it would be advisable to investigate these factors more closely, since they are only average values. In fact, Peterson[9] observes that the distribution of typical bolt failures is about 15 percent under the head, 20 percent at the end of the thread, and 65 percent in the thread at the nut face.

Use of rolled threads is the predominant method of thread-forming in screw fasteners, where Table 8–16 applies. In thread-rolling the amount of cold work and strain-strengthening is unknown to the designer; therefore, fully corrected (including K_f) axial endurance strength is reported in Table 8–17. For cut threads, the methods of Chap. 7 are useful. Anticipate that the endurance strengths will be considerably lower.

Most of the time, the type of fatigue loading encountered in the analysis of bolted joints is one in which the externally applied load fluctuates between zero and some maximum force P. This would be the situation in a pressure cylinder, for example, where a pressure either exists or does not exist.

If we have a ductile material in fatigue which fits the elastic-perfectly plastic model of Dowling, Eqs. (7–31) are available to estimate K_{fm}. In the case of bolts in fatigue we have strong materials that have a $\sigma\epsilon$ curve which is a steeply increasing monotone up to rupture with no yield point, requiring an offset-defined yield strength and not fitting Dowling's model. Fatigue crack nucleation and growth comes from actual stress intensity, so $K_f\sigma_{a0}$ is σ_a. The root fillet notch is sharp and the stress in the root-circle plane is nominal except for small plastic flow (which strain-strengthens) on the first

Table 8–16

Fatigue Stress-Concentration Factors K_f for Threaded Elements

SAE Grade	Metric Grade	Rolled Threads	Cut Threads	Fillet
0 to 2	3.6 to 5.8	2.2	2.8	2.1
4 to 8	6.6 to 10.9	3.0	3.8	2.3

9. R. E. Peterson, *Stress Concentration Factors*, Wiley, New York, 1974, p. 253.

Table 8–17

Fully Corrected Endurance Strengths for Bolts and Screws with Rolled Threads*

Grade or Class	Size Range	Endurance Strength
SAE 5	$\frac{1}{4}$–1 in	18.6 kpsi
	$1\frac{1}{8}$–$1\frac{1}{2}$ in	16.3 kpsi
SAE 7	$\frac{1}{4}$–$1\frac{1}{2}$ in	20.6 kpsi
SAE 8	$\frac{1}{4}$–$1\frac{1}{2}$ in	23.2 kpsi
ISO 8.8	M16–M36	129 MPa
ISO 9.8	M1.6–M16	140 MPa
ISO 10.9	M5–M36	162 MPa
ISO 12.9	M1.6–M36	190 MPa

* Repeatedly-applied, axial loading, fully corrected.

cycle if the stress is high enough in the fillet. This is tantamount to the use of $K_{fm} = 1$. Interestingly, Dowling's model gives a similar result.

In the bolt we want the joint to be safe from fatigue and preload just short to fastener permanent set. Gerber's failure locus is

$$\frac{S_a}{S_e} + \left(\frac{S_m}{S_{ut}}\right)^2 = \frac{K_f n\sigma_{a0}}{S_e} + \left(\frac{n\sigma_{m0}}{S_{ut}}\right)^2 = \frac{n\sigma_{a0}}{S_e/K_f} + \left(\frac{n\sigma_{m0}}{S_{ut}}\right)^2 = 1$$

Note that the factor of safety is *unaffected* by K_f being applied to σ_{a0} or the endurance strength S_e being divided by K_f. It is numerically of no consequence. The latter approach can be viewed as using a *fully corrected* endurance strength. We will use nominal stresses in equations. All the familiar coordinate intercepts in the designer's fatigue diagram will be present (except S_e, which is fully corrected).

A common application of the bolted tension joint is devices such as a pressure cylinder in which the external load varies from a lower extreme of $P = 0$ to an upper extreme of P, per bolt. This is a repeatedly applied loading situation. The amplitude and steady components of load, P_a and P_m, are both half the largest load P, or $P/2$. From the bolt equation expression of stress as

$$\sigma_b = \frac{CP}{A_t} + \frac{F_i}{A_t}$$

we can write expressions for $(\sigma_b)_{\min}$ and $(\sigma_b)_{\max}$ as follows:

$$\sigma_a = \frac{1}{2}(\sigma_{\max} - \sigma_{\min}) = \frac{1}{2}\left(\frac{CP}{A_t} + \frac{F_i}{A_t} - \frac{F_i}{A_t}\right) = \frac{CP}{2A_t} \tag{8–45}$$

$$\sigma_m = \frac{1}{2}(\sigma_{\max} + \sigma_{\min}) = +\frac{1}{2}\left(\frac{CP}{A_t} + \frac{F_i}{A_t} + \frac{F_i}{A_t}\right) = \frac{CP}{2A_t} + \frac{F_i}{A_t} \tag{8–46}$$

On the designer's fatigue diagram the load line is

$$\sigma_a = \sigma_m - \sigma_i \tag{8–47}$$

The next problem is to find the strength components S_a and S_m of the fatigue failure locus. (See Fig. 8–21.) These depend on the failure locus:

Figure 8–21

Designer's fatigue diagram showing a Goodman failure locus and how a load line is used to define failure and safety in preloaded bolted joints in fatigue. Point *B* represents nonfailure; point *C*, failure.

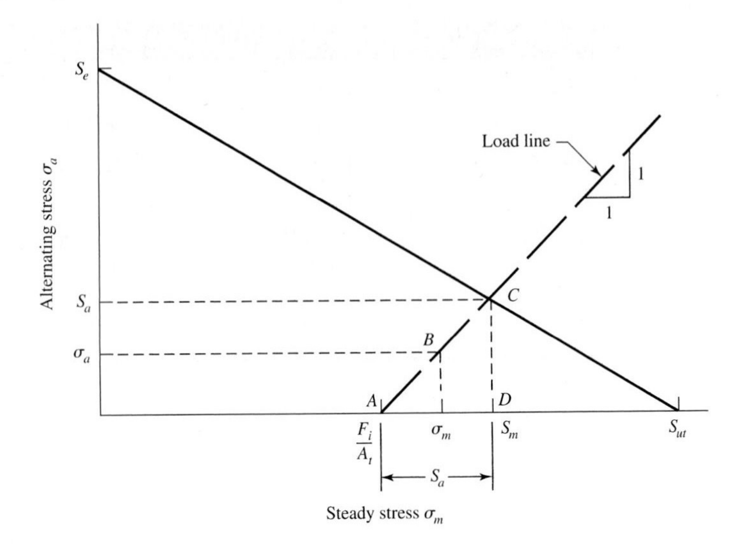

Goodman:

$$\frac{S_a}{S_e} + \frac{S_m}{S_{ut}} = 1 \tag{8–48}$$

Gerber:

$$\frac{S_a}{S_e} + \left(\frac{S_m}{S_{ut}}\right)^2 = 1 \tag{8–49}$$

ASME-elliptic:

$$\left(\frac{S_a}{S_e}\right)^2 + \left(\frac{S_m}{S_p}\right)^2 = 1 \tag{8–50}$$

For simultaneous solution between Eq. (8–47) as $S_a = S_m - \sigma_i$, and each of Eqs. (8–48), (8–49), and (8–50) gives

DE-Goodman:

$$S_m = \frac{S_{ut}\,(S_e + \sigma_i)}{S_e + S_{ut}} \tag{8–51}$$

$$S_a = S_m - \sigma_i \tag{8–52}$$

DE-Gerber:

$$S_m = \frac{S_{ut}^2}{2S_e}\left[-1 + \sqrt{1 + \frac{4S_e}{S_{ut}^2}\,(S_e + \sigma_i)}\right] \tag{8–53}$$

$$S_a = S_m - \sigma_i$$

DE-elliptic:

$$S_a = \frac{1}{S_P^2\left(\dfrac{1}{S_e^2} + \dfrac{1}{S_P^2}\right)}\left\{-\sigma_i + \sqrt{\sigma_i^2 + \left[1 - \left(\frac{\sigma_i}{S_P}\right)^2\right]\left(\frac{1}{S_e^2} + \frac{1}{S_P^2}\right)S_P^4}\right\} \tag{8–54}$$

$$S_m = S_a + \sigma_i$$

Examination of Eqs. (8–48) to (8–54) shows parametric equations which evaluate the coordinates to loci of interest. Factors of safety are given by

$$n_f = \frac{S_a}{\sigma_a} = \frac{S_m - \sigma_i}{\sigma_a} \tag{8–55}$$

$$n_P = \frac{1}{\xi_2} \tag{8–56}$$

$$n_{f0} = \left.\frac{S_a}{\sigma_a}\right|_{\substack{\sigma_i=0 \\ C=1}} = \frac{S_a|_{\sigma_i=0}}{\sigma_a|_{C=1}} \tag{8–57}$$

Preload is beneficial for resisting fatigue when n_f/n_{f0} is greater than unity. For Goodman, $n_f/n_{f0} \geq 1$ puts an upper bound on σ_i and ξ_1 of

$$\sigma_i \leq (1 - C)S_{ut} \qquad \xi_1 \leq (1 - C)S_{ut}/S_p \tag{8–58}$$

These parametric equations for the coordinates of loci on the designer's fatigue diagram may be computationally efficient, but global understanding is not available by inspection. The fatigue failure loci of Gerber and ASME-elliptic are regression lines that cannot be easily rejected statistically. For all that virtue, they are second order, and back-substitutions become complicated. However, we will use Goodman first to give limited insight, but use the others for analysis. The Goodman line equation in conjunction with other bolted-joint relations will indicate direction but will be myopic as to how far.

Substituting $n_f\sigma_a$ and $n_f\sigma_m$ for S_a and S_m, respectively, in Eq. (8–48) and solving for n_f gives

$$n_f = \frac{S_{ut}A_t - F_i}{(CP/2)\,(1 + S_{ut}/S_e)} \tag{8–59}$$

when preload F_i is present. With no preload (loosened joint), $C = 1$, $F_i = 0$, and Eq. (8–50) becomes

$$n_{f0} = \frac{S_{ut}A_t}{(P/2)\,(1 + S_{ut}/S_e)} \tag{8–60}$$

It is clear from inspection of Eq. (8–59) that increasing preload *reduces* the fatigue resistance of the joint. At this point we need to recall our priorities:

- A joint exists to provide a body whose shape helps fulfill a function.
- The body must stay intact, not lose its rigidity. To do that, the bolts must not fatigue or loosen, and the joint must not open.

Why do bolts loosen? They are friction devices, and vibration as well as some other subtle effects allow the fasteners to lose tension over time. How does one fight loosening? The tighter the preload the better. A rule of thumb is that preloads of 60 percent of proof load rarely loosen. If more is better, how much more? Well, not enough to create reused fasteners as a future threat. Alternatively, fastener-locking schemes can be employed.

EXAMPLE 8–5

Figure 8–22 shows a connection using 5/8″-11 UNC SAE 5 cap screws with a $\frac{1}{16}$-in-thick hardened steel washer. The joint constant C is 0.280. The preload is 75 percent of proof load. The hydrostatic test multiple is $m = 2$. Find the factors of safety using the Goodman criterion of fatigue. The external load is 5 kip/screw. Fully corrected S_e is 18.6 kpsi.

Figure 8–22

Pressure-cone frusta member model for a cap screw. For this model the significant sizes are
$l = \begin{cases} h + t_2/2 & t_2 < d \\ h + d/2 & t_2 \geq d \end{cases}$
$D_1 = d_w + l \tan\alpha = 1.5d + 0.577l$
$D_2 = d_w = 1.5d$
where l = effective grip. The solutions are for $\alpha = 30°$ and $d_w = 1.5d$.

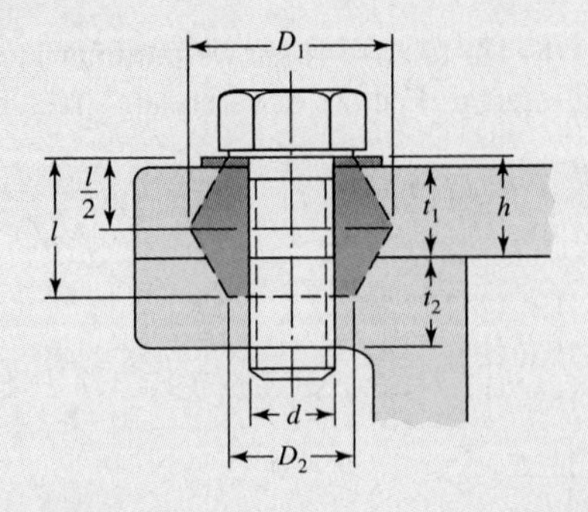

Solution Parameters that are independent of failure theory are

$$F_i = \xi_1 S_p A_t = 0.75(85)0.226 = 14.4 \text{ kip}$$

$$\sigma_i = \frac{F_i}{A_t} = \frac{14.4}{0.226} = 63.72 \text{ kpsi}$$

With preload:

From Eq. (8–45),

$$\sigma_a = \frac{CP}{2A_t} = \frac{0.280(5)}{2(0.226)} = 3.10 \text{ kpsi}$$

From Eq. (8–46),

$$\sigma_m = \sigma_a + \sigma_i = 3.10 + 63.72 = 66.82 \text{ kpsi}$$

From Eq. (8–33),

$$\xi_2 = \xi_1 + \frac{CP}{S_P A_t} = 0.75 + \frac{0.280(5)}{85(0.226)} = 0.822\,879$$

Without preload:

From Eq. (8–45),

$$\sigma_a = \frac{P}{2A_t} = \frac{5}{2(0.226)} = 11.06 \text{ kpsi}$$

From Eq. (8–46),

$$\sigma_m = \sigma_a = 11.06 \text{ kpsi}$$

Parameters that are specific to failure theory (see point c, Fig. 8–23) are, from Eq. (8–51),

$$S_m = \frac{120(18.6 + 63.72)}{18.6 + 120} = 71.3 \text{ kpsi}$$

From Eq. (8–52),

$$S_a = 71.3 - 63.72 = 7.55 \text{ kpsi}$$

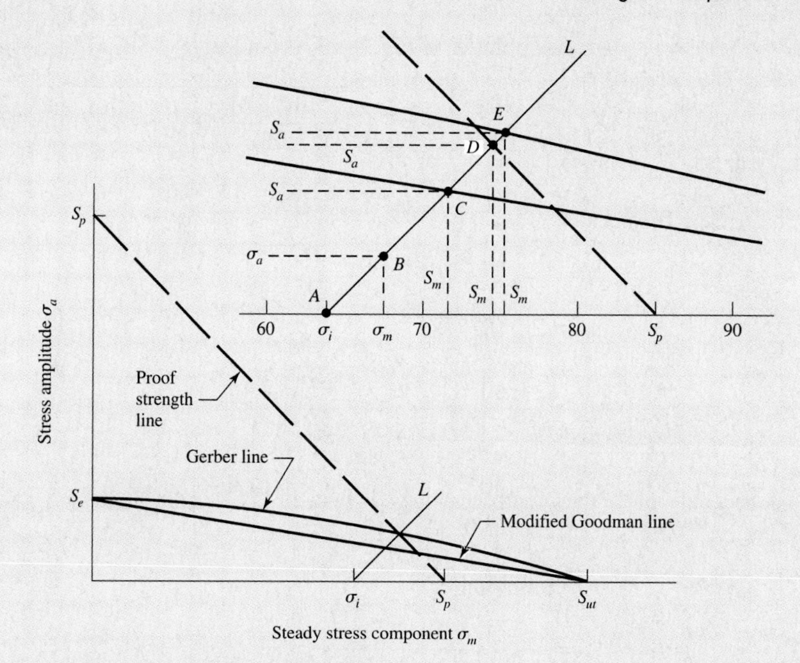

Figure 8–23

Designer's fatigue diagram for preloaded bolts, drawn to scale showing the modified Goodman locus, the Gerber locus, and the Langer proof-strength locus, with an exploded view of the area of interest. The strengths used are $S_p = 85$ kpsi, $S_e = 18.6$ kpsi, and $S_{ut} = 120$ kpsi. The coordinates are A, $\sigma_i = 63.72$ kpsi; B, $\sigma_a = 3.10$ kpsi, $\sigma_m = 66.82$ kpsi; C, $S_a = 7.55$ kpsi, $S_m = 71.29$ kpsi; D, $S_a = 10.64$ kpsi, $S_m = 74.36$ kpsi; E, $S_a = 11.32$ kpsi, $S_m = 75.04$ kpsi.

Answer
$$n_f = \frac{S_a}{\sigma_a} = \frac{7.55}{3.10} = 2.44$$

Alternatively, from Eq. (8–59),

$$n_f = \frac{120(0.226) - 14.4}{\dfrac{0.280(5)}{2}\left(1 + \dfrac{120}{18.6}\right)} = 2.44$$

From Eq. (8–35),

Answer
$$n_P = \frac{1}{\xi_2} = \frac{1}{0.823} = 1.22$$

From Eq. (8–57),

Answer
$$n_{f0} = \frac{S_{ut} S_e}{S_e + S_{ut}} \frac{2A_t}{P} = \frac{120(18.6)}{18.6 + 120} \frac{2(0.226)}{5} = 1.46$$

It is clear that the presence of a preload was beneficial in this case since it raised an open joint fatigue factor of safety from 1.46 to 2.44. If $\xi_1 < 0.75$, n_f is even greater than 2.44.

EXAMPLE 8–6 (*a*) Apply Gerber fatigue failure loci to the problem of Ex. 8–5 and note differences.
(*b*) Repeat part (*a*) using ASME-elliptic failure loci.

Solution (*a*) See point E in Fig. 8–23. From Eq. (8–53),

$$S_m = \frac{120^2}{2(18.6)}\left[-1+\sqrt{1+\frac{4(18.6)}{120^2}(18.6+63.72)}\right] = 75.05 \text{ kpsi}$$

$$S_a = S_m - \sigma_i = 75.05 - 63.72 = 11.3 \text{ kpsi}$$

$$n_f = \frac{S_a}{\sigma_a} = \frac{11.3}{3.10} = 3.65$$

From Eq. (8–57),

$$n_{f0} = \frac{S_{ut}^2}{2S_e}\left[-1+\sqrt{1+\frac{4S_e^2}{S_{ut}^2}}\right]\frac{2A_t}{P}$$

$$= \frac{120^2}{2(18.6)}\left[-1+\sqrt{1+\frac{4(18.6)^2}{120^2}}\right]\frac{2(0.226)}{5} = 1.64$$

(*b*) From Eq. (8–54),

$$S_a = \frac{1}{85^2\left(\frac{1}{18.6^2}+\frac{1}{85^2}\right)}$$

$$\cdot\left\{-63.72+\sqrt{63.72^2+\left[1-\left(\frac{63.72}{85}\right)^2\right]\left(\frac{1}{18.6^2}+\frac{1}{85^2}\right)85^4}\right\}$$

$$= 9.45 \text{ kpsi}$$

$$S_m = 9.45 + 63.72 = 73.2 \text{ kpsi}$$

$$n_f = \frac{S_a}{\sigma_a} = \frac{9.45}{3.10} = 3.05$$

From Eq. (8–57),

$$n_{f0} = \frac{1}{\sqrt{\frac{1}{S_e^2}+\frac{1}{S_p^2}}}\frac{2A_t}{P} = \frac{1}{\sqrt{\frac{1}{18.6^2}+\frac{1}{85^2}}}\frac{2(0.226)}{5} = 1.64$$

EXAMPLE 8–7 For the conditions of Ex. 8–5 and Ex. 8–6 prepare a table of values for factor of safety in fatigue n_f for Goodman, Gerber, and ASME-elliptic fatigue loci, comparing values over the range of preload fraction $0 \le \xi_1 \le 1$.

Solution *For Goodman:*

$$S_m = \frac{120\,(18.6+85\xi_1)}{18.6+120} \qquad S_a = S_m - 85\xi_1$$

$$n_f = \frac{S_a}{3.10}$$

For Gerber:

$$S_m = \frac{120^2}{2(18.6)}\left[-1 + \sqrt{1 = \frac{4(18.6)}{120^2}\,(18.6 + 85\xi_1)}\right]$$

$$S_a = S_m - 85\xi_1 \qquad n_f = S_a/3.10$$

For ASME-elliptic:

$$S_a = \frac{1}{85^2\left(\dfrac{1}{18.6^2} + \dfrac{1}{85^2}\right)}\left[-85\xi_1 + \sqrt{(85\xi_1)^2 + \left(1 - \xi_1^2\right)\left(\frac{1}{18.6^2} + \frac{1}{85^2}\right)85^4}\right]$$

$$n_f = \frac{S_a}{3.10}$$

Using the values of n_{f0} from Ex. 8–5 and Ex. 8–6, the following table is prepared:

	n_f			
ξ_1	**Goodman**	**Gerber**	**ASME-Elliptic**	n_p
0	5.19 (1.46)	5.86 (1.64)	5.87 (1.64)	13.72
0.1	4.83	5.72	5.71	5.78
0.2	4.46	5.52	5.50	3.66
0.3	4.09	5.28	5.23	2.68
0.4	3.72	4.49	4.90	2.11
0.5	3.36	4.65	4.49	1.75
0.6	2.99	4.28	4.00	1.49
0.7	2.62	3.87	3.40	1.29
0.8	2.25	3.34	2.66	1.15
0.9	1.88	2.95	1.67	1.03
1.0	1.51	2.43	0	0.93

Figure 8–24 is a plot of the values in the table. Notice how the shape of each n_f curve mimics the shape of the fatigue locus from which it was derived. The terminal point of each n_f curve represents the preload fraction ξ_1 at which the factor of safety has decreased to the open-joint value. While it is true that for most of the fractional preload range the

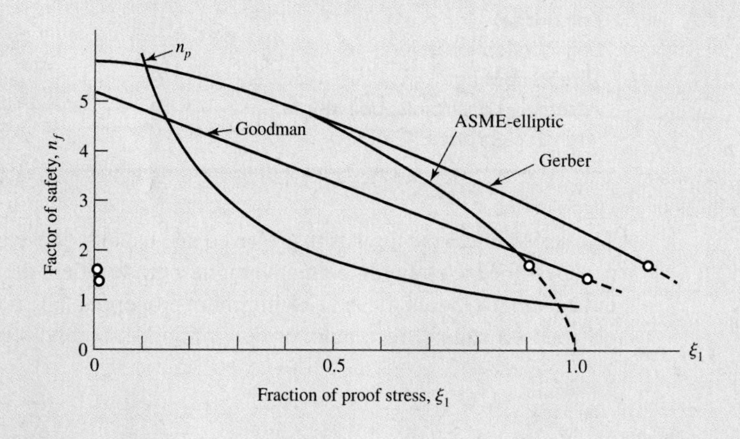

Figure 8–24

A plot of n_f for conditions of Ex. 8–5 and Ex. 8–6 for Goodman, Gerber, and ASME-elliptic failure loci, terminating at n_{f0}, and showing the range in which the preload fraction ξ_i is beneficial. Note how the shapes of n_f mimic that of the related loci.

fatigue factor of safety is higher than the open joint (and preload can be claimed to be effective), notice also that increasing the preload fraction ξ_1 *monotonically decreases* the fatigue resistance of the joint. The fact that the safer increment is downward should be helpful when you are confronted with the need to round a torque wrench setting.

8–12 Adequacy Assessment, Specification Set, Decision Set, and Design

The adequacy assessment for a bolted tension joint consists of examining the factors of safety for fatigue n_f, open-joint fatigue n_{f0}, for over proof stressing n_p, or the reliabilities related to these, or a mix of factors of safety and reliabilities. Included in the adequacy assessment are a series of inequalities such as

$R_j \geq 0.99$

$N \geq 6 \quad \text{(even?)}$

$n_f \geq 1.9$

$$3 \leq \frac{\pi D_b}{Nd} \leq 6$$

The specification set for a bolted tension joint consists of

- Fastener size (diameter and length)
- Washers
- Thread series (fine, coarse, or other)
- Bolt grade
- Number of fasteners in pattern
- Bolt location pattern
- Thread coating (dry or lubricated or plated)
- Torque if applicable

The decision set is similar but arranged as follows:

A Priori Decisions	Design decisions
Washers, if used, what type Thread series Bolt grade Bolt location pattern Thread coating Assembly constraints; bolt length, wrench clearance, . . .	Diameter d (washer thickness is coupled) Number of fasteners N

The series of coarse threads is preferred and usually chosen. The grade 5 fastener is the cheapest for the strength, being the highest grade where the vendor is free to cold-work or heat-treat to attain the ASTM minimum properties. The bolt location pattern is usually obvious—a square, rectangle, or bolt circle. Assembly is usually lubricated. All these decisions can be made a priori. The design problem has two independent design variables, which are not continuous but discrete. The diameters are from a standard series, and the number of bolts is limited to integers. This establishes not only the dimensionality but the character of the task. Since any function of the design variables is not continuous, the idea of derivatives is not useful in the problem. It is common to handle such problems in a tabular ways, in a computer DO-loop treatment, and hugging some margin in pursuing optimality.

The following steps can be used in carrying out the design process for a joint with a bolt circle:

1. Choose d, a trial diameter.
2. $N_{\text{low}} = \frac{\pi D_b}{6d}$, $N_{\text{high}} = \frac{\pi D_b}{3d}$; choose N, if $\left(\frac{\pi D_b}{Nd}.lt.3..or.\frac{\pi D_b}{Nd}.gt.6.\right)$ go to 1.
3. Having d and N, a design is established. Evaluate C, the joint constant.
4. $z = \Phi^{-1}(R^{1/n})$, $\xi_1 = \frac{1-CP/(S_pA_t)}{1+zC_{Fi}}$, $\xi_2 = \xi_1 + \frac{CP}{S_pA_t}$.
5. If $(\xi_1.lt.0.6)$ go to 1.
6. $T = Kd\xi_1 S_p A_t$, $F_i = \xi_1 S_p A_t$, $\sigma_i = \xi_1 S_p$.
7. S_e from Table 8–17, or otherwise.
8. Find σ_a, σ_m with and without preload.
9. Find S_a, S_m from failure locus.
10. Find n_f, n_{f0}, n_p, p_f.
11. If (n_f .lt. n_f (goal)) go to 1.
12. Estimate figure of merit and record it.
13. Go to 1.

Figure of merit for bolted and screwed joints consider

- Cost of the fastener set.
- Cost of holes, finishing or tapping.
- Cost of assembly.

The difficulty in formulating these in the abstract is that (1) fasteners are mass-produced in great quantities, so the cost per pound is small; (2) holes are rarely created in comparable numbers, so the cost of holes in manufacturing is not small; and (3) the labor and tool costs for joint assembly are plant- and volume-specific. Notwithstanding the difficulty of having a viable figure of merit before the reader is on the job, we will use a surrogate figure of merit representing the cost of fasteners plus holes, for instructional purpose, as displayed below in pidgin Fortran:

if (d.LT.t$_1$) $C_1 = t_1/d + 1$
if (d.GE.t$_1$) $C_1 = d/t_1$
if (d.LT.t$_2$) $C_2 = t_2/d + 1$
if (d.GE.t$_2$) $C_2 = d/t_2$
if (ISAE.LE.5)fom= $-(0.50 + 0.05\text{ISAE} + C_1^{1.5} + C_2^{1.5})$
if (ISAE.gt.5)fom= $-(0.03(\text{ISAE})^2 + C_1^{1.5} + C_2^{1.5})$
fom=fom $\times N(0.785)d^2(t_1 + t_2)$

where t_1 and t_2 are member thicknesses, d is nominal fastener diameter and ISAE is the variable name for grade. The units of the figure of merit are cost of fasteners and holes. Grade 5 fasteners are the most economical in strength-per-dollar terms, since grade 5 is the highest grade in which heat treatment is not essential to meet the minimums for the grade. Suppliers must sustain the ASTM minimums, but in any way they elect (through composition, cold-working, or heat treatment).

In using the 13-step process, it is useful to note that the a priori decisions and the design variables are discrete (not continuous) so that the natural programming form is a set of nested DO-loops with the bottom two (in this case) being the design variables. The program makes it easy for the designer to make decisions, adjust, and readjust until optimality is achieved. All the number crunching is done by machine; decisions are made by the designer.

EXAMPLE 8–8

A steel hanger depicted in Fig. 8–25 is one of two to be used to suspend a transformer from overhead steel beams. Each bracket is to support 9000 lbf and employ a static design factor of 3. Size the joint using UNC cap screws, allowing for reuse of fasteners at 0.01 level.

Solution

A priori decisions: (1) no washer, (2) UNC cap screws, (3) grade 5 fasteners, (4) two screws per bracket, (5) lubricate with machine oil, and (6) $N = 2$. Design variable: trial $d = 1/2$ in, establishing *a* design.

Estimate C: $A_t = 0.1419$ in^2. From Table 8–7, $l = 1/2 + 0.5/2 = 0.75$ in, $L_T = 2(0.5) + 0.25 = 1.25$ in, $L \geq 0.5 + 1.5(0.5) = 1.25$. $L = 1.25$ in, all threads. Then

$$k_b = \frac{AE}{l} = \frac{0.1419(30)10^6}{0.75} = 5.68\left(10^6\right)\ \text{lbf/in}$$

$$k_m = \frac{0.577\pi(30)10^6 0.5}{2\ln\left[5\dfrac{0.577(0.75)+0.5(0.5)}{0.577(0.75)+2.5(0.5)}\right]} = 19.2\left(10^6\right)\ \text{lbf/in}$$

$$C = 5.68/(5.68 + 19.2) = 0.228$$

Estimate z, ξ_1, ξ_2:

$$z = \Phi^{-1}\left(R^{1/2}\right) = \Phi^{-1}\left(0.99^{1/2}\right) = \Phi^{-1}(0.995) = 2.58$$

$$\xi_1 = \frac{1 - \dfrac{CP}{S_p A_t}}{1 + zC_{Fi}} = \frac{1 - \dfrac{0.228(4500)}{85\,000(0.1419)}}{1 + 2.58(0.09)} = 0.743$$

$$\xi_2 = 0.743 + \frac{0.228(4500)}{85\,000(0.1419)} = 0.828$$

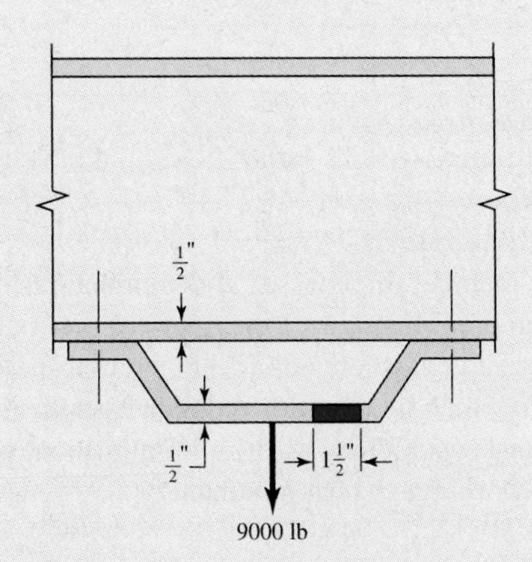

Figure 8–25

The steel hanger for Ex. 8–8.

Estimate torque:

$$T = Kd\xi_1 S_p A_t = 0.18(0.5)0.743(85\,000)0.1419 = 807 \text{ in} \cdot \text{lbf}$$

Factors of safety and probabilities of failure:

$$n = \frac{1-\xi_1}{\xi_2-\xi_1} = \frac{1-0.743}{0.828-0.743} = 3.02$$

$$n_p = 1/\xi_2 = 1/0.828 = 1.21$$

If this design is satisfactory, estimate the fatigue of merit (fom):

$$C_1 = d/t_1 = 0.5/0.5 = 1$$
$$C_2 = d/t_2 = 0.5/0.5 = 1$$
$$\text{fom} = -\left(0.50 + 0.05(5) + 1^{1.5} + 1^{1.5}\right) = -2.75$$
$$\text{fom} = (-2.75)2(0.785)0.5^2(0.5+0.5) = -1.08 \text{ dollars}$$

Note corresponding d, the figure of merit, and any other attributes of interest, and examine the next design ($d = 0.625$ in). When all satisfactory designs have been discovered, identify the design with the highest figure of merit. Check one last time for adequacy, then implement your design.

Study the structure of Ex. 8–8. Note that in this case the specification set and the decision set are identical. Look at Fig. 1–2 and identify how this small design problem fits the structure of skill 2. It is the responsibility of the designer to invent the concept and its connectivity, formulate the specification set and decision set, identify an adequacy assessment processor, then formulate a skill 2, together with a meaningful figure of merit to give guidance during the design process.

In Ex. 8–8 the number of fasteners is given by the statement of the problem. For bolted joints with a bolt circle on which fasteners are placed, the number of fasteners and their nominal diameters are both design variables. They are both discrete (N integer, d standard) and notions of slope or gradient are not useful. The view of the situation in this case is to lay out a plane of squares with the coordinates d and N as depicted below:

	N				
d	**6**	**7**	**8**	**9**	**10**
0.25	X	X	X	X	X
0.375	X	X	X	X	5.9
0.500	X	X	5.5	4.9	4.4
0.5625	X	5.6	4.9	4.3	3.9
0.750	4.9	4.2	3.7	3.3	X
0.875	4.2	3.6	3.1	X	X
1.000	3.7	3.1	X	X	X

The first task is to identify a feasible domain in which satisfactory designs lie. This is done by placing in the squares (temporarily) the bolt spacing along the bolt circle, in diameters, that is, $3 \le \pi D_b/(Nd) \le 6$. This has been done in the table above. The unsatisfactory designs are marked with an X, and we will not look there. The satisfactory designs (from a bolt-spacing viewpoint) lie in a band extending from the lower left toward the upper right in the table. The figure of merit is such that the most competitive designs are to found along a row where the bolt diameter is about one-half of the grip. We also have to understand that some now-feasible designs will be eliminated by the fatigue factor of safety falling short of the desired value. This does not affect the figure of merit; it just eliminates its presence in the grid.

When executing such a design manually, begin a first trial with a grade 5 bolt with a diameter of half the grip, with the smallest number of bolts permitted. If not successful, move to the next square to the right. Once you break into a feasible domain in which n_f exceeds $(n_f)_{\text{goal}}$, maneuver to find the design with the best figure of merit.

If you have a computer, you can set up a program to accept the combination d, N and conduct an adequacy assessment. Among the evaluated elements is the fatigue factor of safety, and if it exceeds the designer's goal, that is, $n_f \geq (n_f)_{\text{goal}}$, print the figure of merit for the design in the proper box; otherwise print a zero. This can be accomplished by putting your adequacy assessment program within a double DO-loop and printing the table line-by-line. Then, by inspection, you can pick out the best design. Furthermore, if you arrange the program, following your display and your decision, to print out the entirety of the adequacy assessment after you type in your d, N decision, you have documentation.

EXAMPLE 8–9

A steel 4-in (inside diameter) pipe has a 9-in-diameter spool flange $\frac{1}{2}$ in thick. A $\frac{1}{2}$-in-thick steel blanking flange is to be bolted to the pipe flange to hold 2000 psi air with a hard joint using an O-ring seal. The bolt circle is to be 7 in in diameter. The hydrostatic test multiple is $m = 1.5$. The designer wants the conventional fatigue factor of safety to be $n_f \geq 3$ using a Goodman fatigue locus. The reliability of the joint in the event of a reused fastener set is to be 0.99. The designer wants to use special $\frac{1}{16}$-in-thick hardened steel washers under the heads and nuts. Design the joint.

Solution

The bolt spacing on the bolt circle is $\pi D_b/Nd = \pi(7)/(Nd) \doteq 22/(Nd)$. The first pass at identifying the feasible domain is to make a table of the bolt spacings in diameters in the range 3 to 6:

	N			
d	**6**	**8**	**10**	**12**
0.375	X	X	5.9	4.9
0.500	X	5.5	4.4	3.9
0.5625	X	4.9	3.9	3.3
0.625	5.9	4.4	3.5	X
0.75	4.9	3.7	X	X

The maximum load $P = \pi d_i^2 p/4 = \pi(4^2)2000/4 = 25\ 133$ lbf or $25\ 133/N$ lbf/bolt. For the first trial set $d = 0.5625$ in, half-grip diameter, with the least number of bolts, $N = 8$:

$$\text{Grip} = l = w_1 + t_1 + w_2 + t_2 = 0.0625 + 0.5 + 0.0625 + 0.5 = 1.125 \text{ in}$$

Bolt length, following Table 8–7:

$$L_G = l = 1.125 \text{ in} \qquad L_T = 2(0.5625) + 0.25 = 1.375 \text{ in}$$

$$L \geq L_G + H = 1.125 + 31/64 = 1.61 \Rightarrow 1.75 \text{ in} \qquad \text{(Table E–17)}$$

$$l_d = 1.75 - 1.375 = 0.375 \text{ in}$$

$$l_T = L_G - l_d = 1.125 - 0.375 = 0.75 \text{ in}$$

$$A_d = \pi d^2/4 = \pi\left(0.5625^2\right)/4 = 0.249 \text{ in}^2$$

$$A_t = 0.182 \text{ in}^2$$

$$k_b = \frac{0.249(0.182)30\left(10^6\right)}{0.249(0.75) + 0.182(0.375)} = 5.33\left(10^6\right) \text{ lbf/in}$$

$$k_m = \frac{0.577\pi 0.5625(30)10^6}{2\ln\left[5\dfrac{0.577(1.125)+0.5(0.5625)}{0.577(1.125)+2.5(0.5625)}\right]} = 18.7\left(10^6\right) \text{ lbf/in}$$

$$C = 5.33/(5.33 + 18.7) = 0.222$$

$$z = \Phi^{-1}\left(0.99^{1/8}\right) = \Phi^{-1}(0.9987) \doteq 3.0$$

$$P = 25\,133/8 = 3142 \text{ lbf/bolt}$$

$$\xi_1 = \frac{1 - \dfrac{0.222(3142)}{85\,000(0.182)}}{1 + 3.0(0.09)} = 0.752$$

$$F_i = \xi_1 S_p A_t = 0.752(85\,000)0.182 = 11\,633 \text{ lbf}$$

$$\sigma_i = F_i/A_t = 11\,633/0.182 = 63\,920 \text{ psi}$$

$$\sigma_a = CP/(2A_t) = 0.222(3142)/[2(0.182)] = 1916 \text{ psi}$$

$$\sigma_m = \sigma_a + \sigma_i = 1916 + 63\,920 = 65\,836$$

$$S_m = \frac{S_{ut}\left(S_e + \sigma_i\right)}{S_e + S_{ut}} = \frac{120(18.6 + 63\,920)}{18.6 + 120} = 71.45 \text{ kpsi}$$

$$S_a = S_m - \sigma_i = 71.45 - 67.92 = 7.53 \text{ kpsi}$$

$$n_f = S_a/\sigma_a = 7.53/1.916 = 3.91$$

$$\text{fom} = -\$6.24$$

Having found feasible space ($n_f \geq 3$), we can try one box above and below, and one to the right. We can make our decision and look at all of the elements in the adequacy assessment.

A view of our table with the figures of merit inserted appears below:

	N			
d	**6**	**8**	**10**	**12**
0.375	X	X	−$8.70	−$10.44
0.500	X	−$4.32	−$5.40	−$6.48
0.5625	X	−$6.24	−$7.79	−$9.35
0.625	−$6.53	−$6.70	−$10.87	−$13.05
0.75	−$11.72	−$15.64	X	X

The best design is eight $\frac{1}{2}''$-13 $\times 1.5''$ UNC bolts, SAE grade 5.

8–13 Shear Joints

Joints can and should be loaded in shear so that the fasteners see no additional stress beyond the initial tightening. The shear loading is resisted in two principal ways:

- The shear load is carried by friction between the members and ensured by the clamping action of the bolts or cap screws. Should the friction be insufficient, the shear load is carried by only two of the fasteners in the pattern. This occurs because errors in hole size and placement preclude a uniform sharing of the shear load. The analysis problem involves identifying the two fasteners that represent the worst case.
- The shear load is carried by dowel pins in reamed holes, placed in both parts while clamped together to ensure alignment. The dowels will carry the shear load. Pins are often tapered to allow firm setting and easy removal if necessary.

In days when driving hot rivets was a common structural joining method, the driving ensured that the rivets filled every hole completely and, upon cooling, provided a clamping preload. This kind of rivet pattern can share a shear load.

Integral to the analysis of a shear joint is locating the center of relative motion between the two members. In Fig. 8–26 let A_1 to A_5 be the respective cross-sectional areas of a group of five pins or hot-driven rivets. Bolts as usually used cannot meet the underlying assumption that the shearing force on each of the five areas is proportional to the distance from the relative pivot point. Under this assumption the pivot point lies at the centroid of the cross-sectional area pattern of the pin or rivets. Using statics, we learn that the centroid G is located by the coordinates $\bar{x}$ and $\bar{y}$, where x_1 and y_i are the distances to the ith area center:

$$\bar{x} = \frac{A_1x_1 + A_2x_2 + A_3x_3 + A_4x_4 + A_5x_5}{A_1 + A_2 + A_3 + A_4 + A_5} = \frac{\sum_1^n A_i x_i}{\sum_1^n A_i}$$
$$\bar{y} = \frac{A_1y_1 + A_2y_2 + A_3y_3 + A_4y_4 + A_5y_5}{A_1 + A_2 + A_3 + A_4 + A_5} = \frac{\sum_1^n A_i y_i}{\sum_1^n A_i} \tag{8–61}$$

In many instances the centroid can be located by symmetry. Note that the $\sum$ key on your programmable hand-held calculator can be useful here.

An example of eccentric loading of fasteners is shown in Fig. 8–27. This is a portion of a machine frame containing a beam A subjected to the action of a bending load. In this case, the beam is fastened to vertical members at the ends with specially prepared load-sharing bolts. You will recognize the schematic representation in Fig. 8–27b as an indeterminate beam with both ends fixed and with the moment reaction M and the shear reaction V at the ends.

Figure 8–26

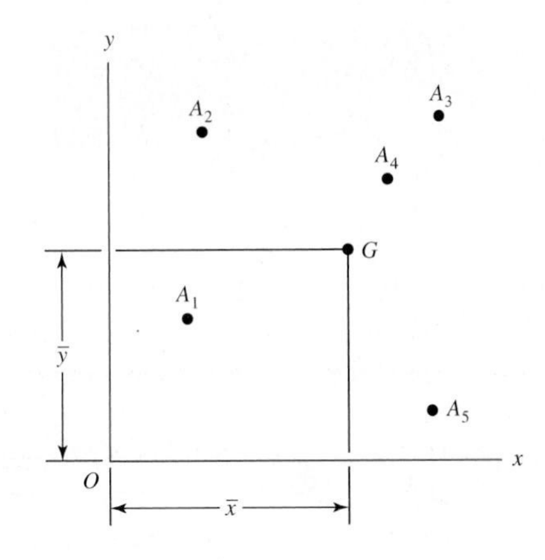

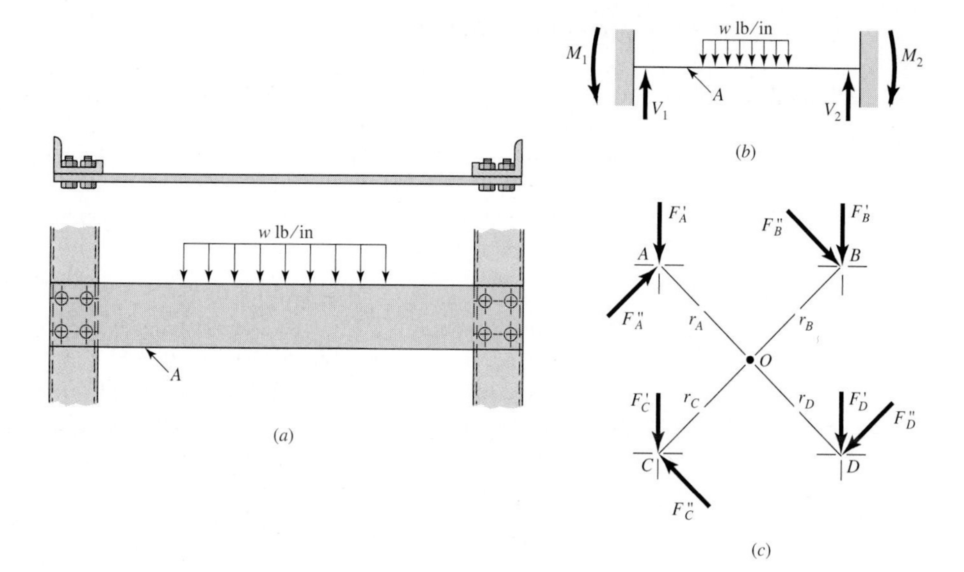

Figure 8–27

(*a*) Beam bolted at both ends with distributed load; (*b*) free-body diagram of beam; (*c*) enlarged view of bolt group showing primary and secondary shear forces.

For convenience, the centers of the bolts at one end of the beam are drawn to a larger scale in Fig. 8–27*c*. Point O represents the centroid of the group, and it is assumed in this example that all the bolts are of the same diameter. The total load taken by each bolt will be calculated in three steps. In the first step the shear V is divided equally among the bolts so that each bolt takes $F' = V/n$, where n refers to the number of bolts in the group and the force F' is called the *direct load*, or *primary shear*.

It is noted that an equal distribution of the direct load to the bolts assumes an absolutely rigid member. The arrangement of the bolts or the shape and size of the members sometimes justify the use of another assumption as to the division of the load. The direct loads F' are shown as vectors on the loading diagram (Fig. 8–27*c*).

The *moment load*, or *secondary shear*, is the additional load on each bolt due to the moment M. If r_A, r_B, r_C, etc., are the radial distances from the centroid to the center of each bolt, the moment and moment load are related as follows:

$$M = F''_A r_A + F''_B r_B + F''_C r_C + \cdots \tag{a}$$

where F'' is the moment load. The force taken by each bolt depends upon its radial distance from the centroid; that is, the bolt farthest from the centroid takes the greatest load, while the nearest bolt takes the smallest. We can therefore write

$$\frac{F''_A}{r_A} = \frac{F''_B}{r_B} = \frac{F''_C}{r_C} \tag{b}$$

Solving Eqs. (*a*) and (*b*) simultaneously, we obtain

$$F''_n = \frac{M r_n}{r_A^2 + r_B^2 + r_C^2 + \cdots} \tag{8–62}$$

where the subscript n refers to the particular bolt whose load is to be found. These moment loads are also shown as vectors on the loading diagram.

In the third step the direct and moment loads are added vectorially to obtain the resultant load on each bolt. Since all the bolts or rivets are usually the same size, only that bolt having the maximum load need be considered. When the maximum load is found, the strength may be determined, using the various methods already described.

EXAMPLE 8–10 Shown in Fig. 8–28 is a 15- by 200-mm rectangular steel bar to be cantilevered to a 250-mm steel channel using two pins, located at E and F, and four bolts. On the basis of a steady external load of 16 kN, find the shear loads in the *pins* should be the clamping action fail.

Solution The point O, the centroid of the pin cross-sectional areas, is found by symmetry. If a free-body diagram of the beam is considered, the shear reaction V would pass through the centroid O, and the moment reaction would be about O. The reactions are $V = 16$ kN, and the moment is $M = 16(425) = 6800$ N · m. The primary shear force F' on each pin is equal to $V/2 = 16/2 = 8$ kN. The secondary shear forces F'' are also equal (and opposite), and from Eq. (8–62),

$$F'' = \frac{Mr}{r^2} = \frac{M}{2r} = \frac{6800}{2(75)} = 45.33 \text{ kN}$$

making $F_E = 45.33 - 8 = 37.33$ kN (downward) and $F_F = 45.33 + 8 = 53.33$ kN (upward). Since the pin forces differ, the designer may wish to work the pin material equally and consider increasing the size of A_F to $A_F = A_E 53.33/37.33 = 1.43 A_E$. It is instructive to pursue this as it represents an instance of free bodies teaching us what we need to know. From Eq. (8–61), the distance to the centroid $\bar{x}$ is

$$\bar{x} = \frac{A_E(0) + A_F(150)}{A_E + A_F} = \frac{0 + 1.43(150)A_E}{A_E + 1.43A_E} = \frac{1.43(150)}{1 + 1.43} = 88.3 \text{ mm}$$

The new moment reaction about the new pivot point O' is $M = 16(61.7 + 350) = 6588$ N · m and the shear reaction is still 16 kN. The total reactions at E and F can be found by taking moments about O' and summing forces vertically. The equations are

$$88.3F_E + 61.7F_F = 6588$$
$$-F_E + F_F = 16$$

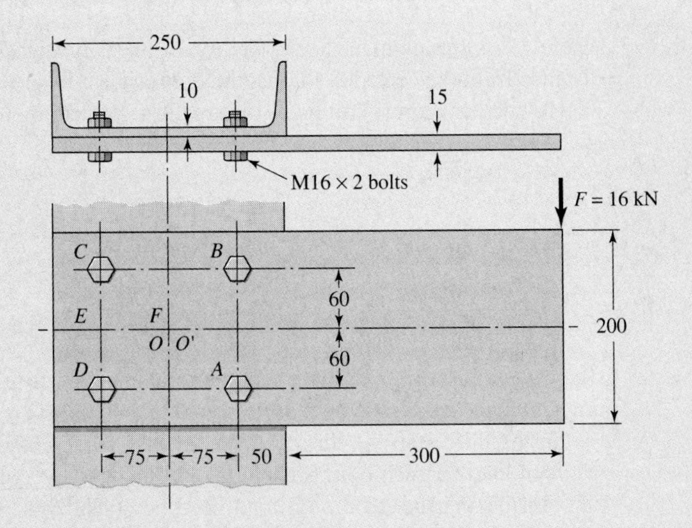

Figure 8–28

Dimensions in millimeters.

from which the simultaneous solution for F_E and F_F gives

$$F_E = 37.33 \text{ kN}$$
$$F_F = 53.3 \text{ kN}$$

The reactions have not changed. (Why?) The shear stress on the pin cross sections are

$$\tau_E = \frac{F_E}{A_E} = \frac{37.3}{A_E}$$

$$\tau_F = \frac{F_F}{A_F} = \frac{F_F}{1.43A_E} = \frac{53.3}{1.43A_E} = \frac{37.3}{A_E}$$

which are equal, the goal is achieved.

In Ex. 8–10 the role of the four bolts is to hold the joint together. A preload of 0.6 proof load is about the lower limit; about 0.75 of proof load would be better. The bolts see no shear load, and the pins carry that. The bolts have not yet been sized.

If the pins are omitted in Fig. 8–28 and the four bolt shanks are tightly fitted in reamed holes, one may proceed as in Ex. 8–11.

EXAMPLE 8–11

The situation is similar to that of Ex. 8–10 except that there are no pins and the bolts are tightly fitted. For a $F = 16$ kN load find

(*a*) The resultant load on each bolt
(*b*) The maximum load on each bolt
(*c*) The maximum bearing stress
(*d*) The critical bending stress in the bar

Solution

(*a*) Point O, the centroid of the bolt group in Fig. 8–28, is found by symmetry. If a free-body diagram of the beam were constructed, the shear reaction V would pass through O and the moment reactions M would be about O. These reactions are

$$V = 16 \text{ kN} \qquad M = 16(425) = 6800 \text{ N} \cdot \text{m}$$

In Fig. 8–29, the bolt group has been drawn to a larger scale and the reactions are shown. The distance from the centroid to the center of each bolt is

$$r = \sqrt{(60)^2 + (75)^2} = 96.0 \text{ mm}$$

The primary shear load per bolt is

$$F' = \frac{V}{n} = \frac{16}{4} = 4 \text{ kN}$$

Since the secondary shear forces are equal, Eq. (8–62) becomes

$$F'' = \frac{Mr}{4r^2} = \frac{M}{4r} = \frac{6800}{4(96.0)} = 17.7 \text{ kN}$$

The primary and secondary shear forces are plotted to scale in Fig. 8–29 and the resultants obtained by using the parallelogram rule. The magnitudes are found by measurement (or analysis) to be

Figure 8–29

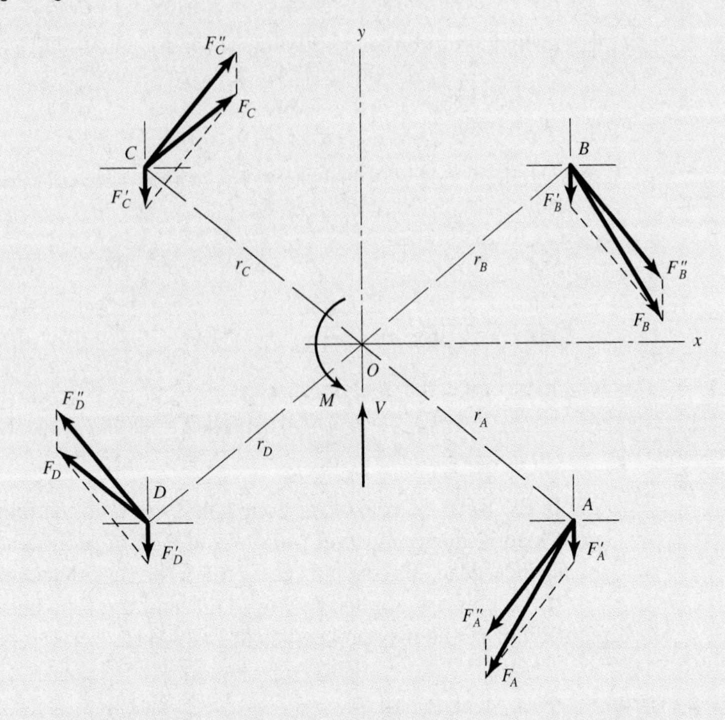

Answer $F_A = F_B = 21.0$ kN

Answer $F_C = F_D = 13.8$ kN

(*b*) Bolts *A* and *B* are critical because they carry the largest shear load. Does this shear act on the threaded portion of the bolt, or on the unthreaded portion? The bolt length will be 25 mm plus the height of the nut plus about 2 mm for a washer. Table E–28 gives the nut height as 14.8 mm. This adds up to a length of 41.8 mm, and so a bolt 46 mm long will be needed. From Table 8–11 we compute the thread length as $L_T = 22$ mm. Thus the unthreaded portion of the bolt is $45 - 22 = 23$ mm long. This exceeds the 15 mm for the plate in Fig. 8–28, and so the bolt will tend to shear across its major diameter. Therefore the shear-stress area is $A_s = \pi d^2/4 = \pi(16)^2/4 = 201$ mm^2, and so the shear stress is

Answer $$\tau = \frac{F}{A_s} = \frac{21.0(10)^3}{201} = 104 \text{ MPa}$$

(*c*) The channel is thinner than the bar, and so the largest bearing stress is due to the pressing of the bolt against the channel web. The bearing area is $A_b = td = 10(16) = 160$ mm^2. Thus the bearing stress is

Answer $$\sigma = \frac{F}{A_b} = -\frac{21.0(10)^3}{160} = -131 \text{ MPa}$$

(*d*) The critical bending stress in the bar is assumed to occur in a section parallel to the *y* axis and through bolts *A* and *B*. At this section the bending moment is

$$M = 16(300 + 50) = 5600 \text{ N} \cdot \text{m}$$

The second moment of area through this section is obtained by the use of the transfer formula, as follows:

$$I = I_{\text{bar}} - 2\left(I_{\text{holes}} + d^2 A\right)$$

$$= \frac{15(200)^3}{12} - 2\left[\frac{15(16)^3}{12} + (60)^2(15)(16)\right] = 8.26(10)^6 \text{ mm}^4$$

Then

Answer $$\sigma = \frac{Mc}{I} = \frac{5600(100)}{8.26(10)^6}(10)^3 = 67.8 \text{ MPa}$$

Should two bolts carry the entire shear load—the worst case—which pair would it be? A little thought indicates the bolts located at C and D due to the leverage of F. In this case $F'_C = 8$ kN (down), $F'_D = 8$ kN (down), and $F''_C = M/120 = 16(500)/120 = 66.67$ kN (to the right) and $F''_D = 66.67$ (to the left). The resultant loads on bolts at C and D have the magnitude

$$F_C = F_D = \sqrt{8^2 + 66.67^2} = 67.1 \text{ kN}$$

Following part (b),

$$\tau = \frac{F}{A_s} = \frac{67.1\left(10^3\right)}{201} = 333.8 \text{ MPa}$$

This increase is more than threefold that given by the assumption that all bolts share the load equally. In part (c) the bearing stress becomes $-131(67.1)/21.0 = -419$ MPa. One had to prepare for the worst. The object of Ex. 8–11 is to show the analysis procedure if a rectangular array of pins, rivets, or bolts prepared with care carry a shear load, and to indicate the danger with bolts when they do not.

The mechanics of a shear-loaded bolted joint is relatively simple. The assumptions, however, are difficult because they are in doubt. Should ordinary bolts carry a shear load? Then failing to do so, the bolts have to be made oversize. What is the best way to proceed? If a designer facing a situation akin to Ex. 8–11 should choose to resist the shear by friction, the question of geometry of the contacting surface, the distribution of pressure, and the necessary integration with point symmetry of secondary shear distribution coupled with cartesian boundaries prompts the designer to create a controlled geometry that is tractable to analysis. One thing that could be done is to insert washers on the bolts between the channel and the strap (between the members). This allows an analysis as in Ex. 8–11 to find the shearing traction to be resisted at each bolt. The largest traction occurs at bolts A and B when the force is $F_A = F_B = 21.0$ kN. For a steel-on-steel coefficient of friction $f = 0.25$, the bolt preload must be at least $21.0/f = 21.0/0.25 = 84$ kN. For a design factor of $n_d = 2$, the initial bolt tension F_i must be 2(84) = 168 kN. For a metric grade of 8.8, with $\xi_1 = 0.75$ and $S_P = 600$ MPa, from Table 8–11,

$$F_i = \xi_1 S_P A_t$$

$$A_t = \frac{F_i}{\xi_1 S_p} = \frac{168\left(10^3\right)}{0.75(600)} = 373 \text{ mm}^2$$

From Table 8–1, select an M30 × 3.5 grade 8.8 with $A_t = 561$ mm^2. The tightening torque to be specified is, after Eq. (8–28),

$$T = \xi_1 K S_P A_t d\left(10^{-3}\right) = 0.75(0.20)600(561)30\left(10^{-3}\right) = 1515 \doteq 1550 \text{ N} \cdot \text{m}$$

8–14 Setscrews

Unlike bolts and cap screws, which depend on tension to develop a clamping force, the setscrew depends on compression to develop the clamping force. The resistance to axial motion of the collar or hub relative to the shaft is called *holding power*. This holding power, which is really a force resistance, is due to frictional resistance of the contacting portions of the collar and shaft as well as any slight penetration of the setscrew into the shaft.

Figure 8–30 shows the point types available with socket setscrews. These are also manufactured with screwdriver slots and with square heads.

Table 8–18 lists values of the seating torque and the corresponding holding power for inch-series setscrews. The values listed apply to both axial holding power, for resisting thrust, and the tangential holding power, for resisting torsion. Typical factors of safety are 1.5 to 2.0 for static loads and 4 to 8 for various dynamic loads.

Setscrews should have a length of about half of the shaft diameter. Note that this practice also provides a rough rule for the radial thickness of a hub or collar.

8–15 Pins and Keys

Pins

Pins preserve alignment, and under force-fit conditions they retain parts relative to each other (locate). Frequent removal and reinsertion wears pins. In such a case a tapered pin is used. The intent is to subject a pin to a transverse shear stress. When the pin is interference-fitted to its hold there is a compressive stress, like a two-dimensional liquid. When the pin hold permits clearance, bending occurs in addition to the transverse shear.

The standard for pins is the American National Standard Hardened and Ground Dowel Pin (ANSI B18.8.2-1978), which comes in $\frac{1}{16}$, $\frac{3}{32}$, $\frac{1}{8}$, $\frac{3}{16}$, $\frac{1}{4}$, $\frac{5}{16}$, $\frac{3}{8}$, $\frac{7}{16}$, $\frac{1}{2}$, $\frac{5}{8}$, $\frac{3}{4}$, $\frac{7}{8}$, and 1 in diameters. The standard requires a minimum shearing yield strength of $S_{sy} = 130$ kpsi. With Production Dowel Pins the shearing yield strength minimum is $S_{sy} = 102$ kpsi. Unhardened Dowel Pins have $S_{sy} = 64$ kpsi. Often the hole (member) into which a pin is inserted is a structural A36 steel with a minimum tensile yield strength of $S_y = 36$ kpsi. In such a case the member limits the pin load.

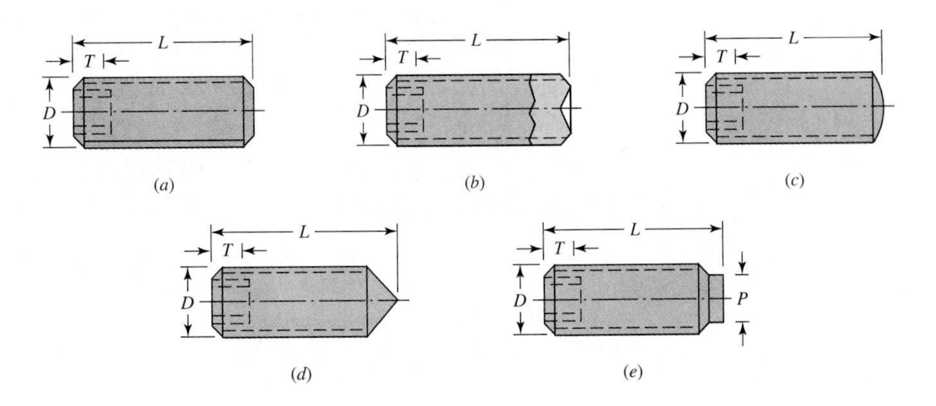

Figure 8–30

Socket setscrews: (*a*) flat point; (*b*) cup point; (*c*) oval point; (*d*) cone point; (*e*) half-dog point.

Table 8–18

Typical Holding Power (Force) for Socket Setscrews*

Source: Unbrako Division, SPS Technologies, Jenkintown, Pa.

Size, in	Seating Torque, lb · in	Holding Power, lb
#0	1.0	50
#1	1.8	65
#2	1.8	85
#3	5	120
#4	5	160
#5	10	200
#6	10	250
#8	20	385
#10	36	540
$\frac{1}{4}$	87	1000
$\frac{5}{16}$	165	1500
$\frac{3}{8}$	290	2000
$\frac{7}{16}$	430	2500
$\frac{1}{2}$	620	3000
$\frac{9}{16}$	620	3500
$\frac{5}{8}$	1325	4000
$\frac{3}{4}$	2400	5000
$\frac{7}{8}$	5200	6000
1	7200	7000

* Based on allow-steel screw against steel shaft, class 3A coarse or fine threads in class 2B holes, and cup-point socket setscrews.

A force-fitted pin subjected to transverse shear is in a plane stress condition, where

$$\sigma_x = -p \qquad \tau_{xy} = 0$$
$$\sigma_y = -p \qquad \tau_{yz} = 0$$
$$\sigma_z = 0 \qquad \tau_{xz} = \tau_{xz}$$

where τ_{xz} is responding to the transverse shearing force V. Substituting in the von Mises Eq. (h) of Sec. 6–7 gives

$$\sigma' = \frac{1}{\sqrt{2}}\left[\left(\sigma_x - \sigma_y\right)^2 + \sigma_y^2 + \sigma_x^2 + 6\tau_{xz}^2\right]^{1/2}$$
$$= \left(\sigma_x^2 + \sigma_y^2 - \sigma_x\sigma_y + 6\tau_{xz}^2\right)^{1/2}$$
$$\sigma' = \left[(-p)^2 + (-p)^2 - (-p)(-p) + 6\tau_{xz}^2\right]^{1/2}$$
$$= \left(p^2 + 3\tau_{xz}^2\right)^{1/2} = \frac{S_y}{n_d} = \frac{\sqrt{3}S_{sy}}{n_d}$$

Solving for the allowable shear stress gives

$$(\tau_{xz})_{\text{all}} = \frac{1}{\sqrt{3}}\sqrt{3\left(\frac{S_{sy}}{n_d}\right)^2 - p^2} = \frac{1}{\sqrt{3}}\sqrt{\left(\frac{S_y}{n_d}\right)^2 - p^2} \tag{8–63}$$

The magnitude p of the interference fit pressure is found from Eq. (3–60) with $r_i = 0$, $r_0 \to \infty$, which results in

$$p = \frac{E\delta}{2R} \tag{8–64}$$

where R is the nominal pin size. The largest interference pressure exists when the maximum pin radius and the minimum hole radius are assembled:

$$\delta = (r_{\text{pin}})_{\max} - (r_{\text{hole}})_{\min} \tag{8–65}$$

A 1-in hardened and ground dowel pin has a diameter of $\begin{smallmatrix}1.0003\\1.0001\end{smallmatrix}$ in and the reamed hole has a diameter of $\begin{smallmatrix}1.0000\\0.9995\end{smallmatrix}$ in. The radial interference δ is, from Eq. (8–65),

$$(\delta)_{\max} = \frac{1.003 - 0.9995}{2} = 0.0004 \text{ in}$$

and the interference pressure is, from Eq. (8–64),

$$p = \frac{30\left(10^6\right)0.0004}{2(1.0002/2)} = 12\,000 \text{ psi}$$

The allowable shear stress $(\tau_{xz})_{\text{all}}$ is, for a design factor $n_d = 2$, from Eq. (8–63),

$$(\tau_{xz})_{\text{all}} = \frac{1}{\sqrt{3}}\sqrt{3\left(\frac{130}{2}\right)^2 - 12^2} = 64.6 \text{ kpsi}$$

The influence of the interference pressure is small. Had we neglected it, $(\tau_{xz})_{\text{all}}$ would be 65 kpsi. For heavy force-fits the pressure is significant in its influence on the allowable shear stress on the pin due to transverse shear load.

The member, from Eqs. (3–51) for σ_t and σ_r, sees a compressive pressure of p in the neighborhood of the interface, and Eq. (8–63) applies. Members are often structural steels (A36) with S_y of 32 or 36 (two grades). In such cases

$$(\tau_{xz})_{\text{all}} = \frac{1}{\sqrt{3}}\sqrt{\left(\frac{S_y}{n_d}\right)^2 - p^2} = \frac{1}{\sqrt{3}}\sqrt{\left(\frac{36}{2}\right)^2 - 12^2} = 7.75 \text{ kpsi}$$

Keys

Keys and pins are used on shafts to secure rotating elements, such as gears, pulleys, or other wheels. Keys are used to enable the transmission of torque from the shaft to the shaft-supported element. Pins are used for axial positioning and for the transfer of torque or thrust or both.

Figure 8–31 shows a variety of keys and pins. Pins are useful when the principal loading is shear and when both torsion and thrust are present. Taper pins are sized according to the diameter at the large end. Some of the most useful sizes of these are listed in Table 8–19. The diameter at the small end is

$$d = D - 0.0208L \tag{8–66}$$

where d = diameter at small end, in
D = diameter at large end, in
L = length, in

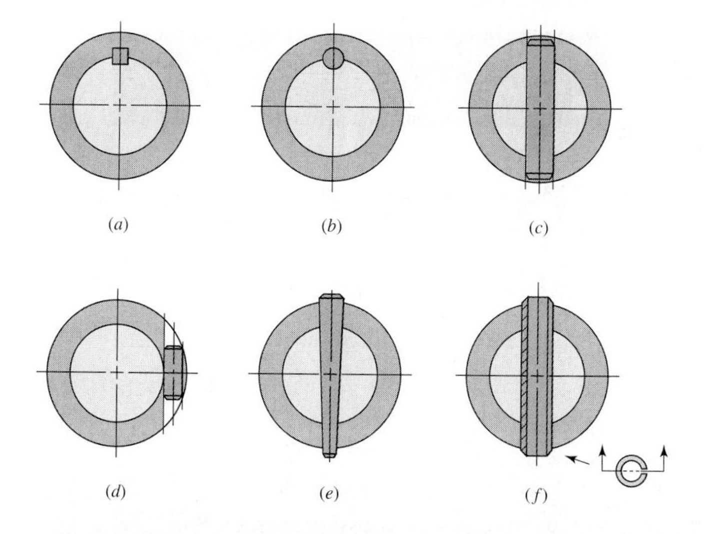

Figure 8–31

(*a*) Square key; (*b*) round key; (*c* and *d*) round pins; (*e*) taper pin; (*f*) split tubular spring pin. The pins in parts (*e*) and (*f*) are shown longer than necessary, to illustrate the chamfer on the ends, but their lengths should be kept smaller than the hub diameters to prevent injuries due to projections on rotating parts.

Table 8–19

Dimensions at Large End of Some Standard Taper Pins—Inch Series

	Commercial		Precision	
Size	Maximum	Minimum	Maximum	Minimum
4/0	0.1103	0.1083	0.1100	0.1090
2/0	0.1423	0.1403	0.1420	0.1410
0	0.1573	0.1553	0.1570	0.1560
2	0.1943	0.1923	0.1940	0.1930
4	0.2513	0.2493	0.2510	0.2500
6	0.3423	0.3403	0.3420	0.3410
8	0.4933	0.4913	0.4930	0.4920

For less important applications, a dowel pin or a drive pin can be used. A large variety of these are listed in manufacturers' catalogs.[10]

The square key, shown in Fig. 8–31*a*, is also available in rectangular sizes. Standard sizes of these, together with the range of applicable shaft diameters, are listed in Table 8–20. The length of the key is based on the hub length and the torsional load to be transferred.

The gib-head key, in Fig. 8–32, is tapered so that, when firmly driven, it acts to prevent relative axial motion. This also gives the advantage that the hub position can be adjusted for the best axial location. The head makes removal possible without access to the other end, but the projection may be hazardous.

Also shown in Fig. 8–32*a* is the sled-runner keyway. This has less stress concentration than an end-milled keyway.

The Woodruff key, shown in Fig. 8–32*b*, is of general usefulness, especially when a wheel is to be positioned against a shaft shoulder, since the keyslot need not be machined into the shoulder stress-concentration region. The use of the Woodruff key also yields better concentricity after assembly of the wheel and shaft. This is especially important at high speeds, as, for example, with a turbine wheel and shaft. Dimensions for some standard Woodruff key sizes can be found in Table 8–21, and Table 8–22 gives the shaft diameters for which the different keyseat widths are suitable.

10. See also Joseph E. Shigley, "Unthreaded Fasteners," Chap. 22 in Joseph E. Shigley and Charles R. Mischke (eds.), *Standard Handbook of Machine Design*, 2nd ed., McGraw-Hill, New York, 1996.

Table 8–20

Inch Dimensions for Some Standard Square- and Rectangular-Key Applications

Source: Joseph E. Shigley, "Unthreaded Fasteners," chap. 22 in Joseph E. Shigley and Charles R. Mischke (eds.), *Standard Handbook of Machine Design*, 2nd ed., McGraw-Hill, New York, 1996.

Shaft Diameter		Key Size		Keyway Depth
Over	To (Incl.)	w	h	
$\frac{5}{16}$	$\frac{7}{16}$	$\frac{3}{32}$	$\frac{3}{32}$	$\frac{3}{64}$
$\frac{7}{16}$	$\frac{9}{16}$	$\frac{1}{8}$	$\frac{3}{32}$	$\frac{3}{64}$
		$\frac{1}{8}$	$\frac{1}{8}$	$\frac{1}{16}$
$\frac{9}{16}$	$\frac{7}{8}$	$\frac{3}{16}$	$\frac{1}{8}$	$\frac{1}{16}$
		$\frac{3}{16}$	$\frac{3}{16}$	$\frac{3}{32}$
$\frac{7}{8}$	$1\frac{1}{4}$	$\frac{1}{4}$	$\frac{3}{16}$	$\frac{3}{32}$
		$\frac{1}{4}$	$\frac{1}{4}$	$\frac{1}{8}$
$1\frac{1}{4}$	$1\frac{3}{8}$	$\frac{5}{16}$	$\frac{1}{4}$	$\frac{1}{8}$
		$\frac{5}{16}$	$\frac{5}{16}$	$\frac{5}{32}$
$1\frac{3}{8}$	$1\frac{3}{4}$	$\frac{3}{8}$	$\frac{1}{4}$	$\frac{1}{8}$
		$\frac{3}{8}$	$\frac{3}{8}$	$\frac{3}{16}$
$1\frac{3}{4}$	$2\frac{1}{4}$	$\frac{1}{2}$	$\frac{3}{8}$	$\frac{3}{16}$
		$\frac{1}{2}$	$\frac{1}{2}$	$\frac{1}{4}$
$2\frac{1}{4}$	$2\frac{3}{4}$	$\frac{5}{8}$	$\frac{7}{16}$	$\frac{7}{32}$
		$\frac{5}{8}$	$\frac{5}{8}$	$\frac{5}{16}$
$2\frac{3}{4}$	$3\frac{1}{4}$	$\frac{3}{4}$	$\frac{1}{2}$	$\frac{1}{4}$
		$\frac{3}{4}$	$\frac{3}{4}$	$\frac{3}{8}$

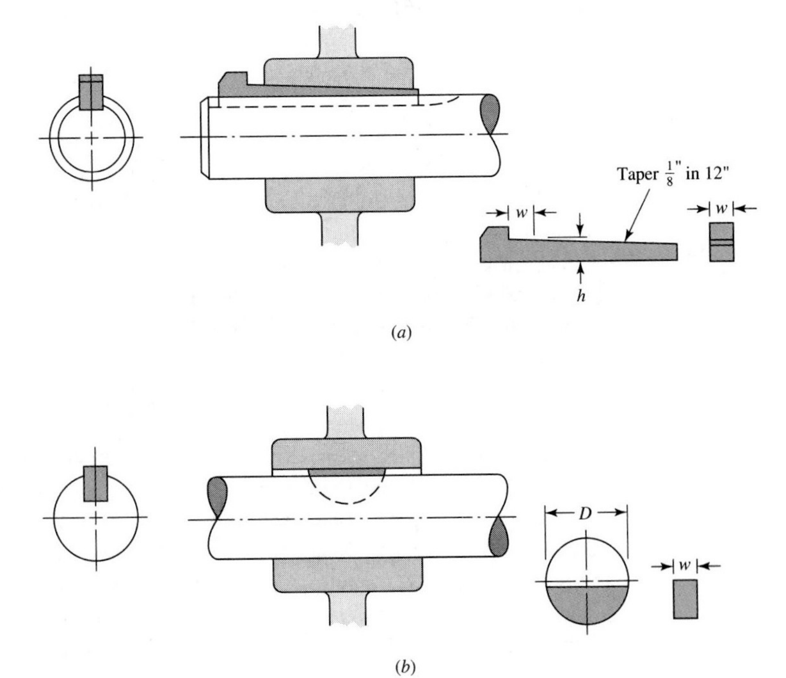

Figure 8–32

(*a*) Gib-head key; (*b*) Woodruff key.

Table 8–21

Dimensions of Woodruff Keys—Inch Series

Key Size		Height	Offset	Keyseat Depth	
w	D	b	e	Shaft	Hub
$\frac{1}{16}$	$\frac{1}{4}$	0.109	$\frac{1}{64}$	0.0728	0.0372
$\frac{1}{16}$	$\frac{3}{8}$	0.172	$\frac{1}{64}$	0.1358	0.0372
$\frac{3}{32}$	$\frac{3}{8}$	0.172	$\frac{1}{64}$	0.1202	0.0529
$\frac{3}{32}$	$\frac{1}{2}$	0.203	$\frac{3}{64}$	0.1511	0.0529
$\frac{3}{32}$	$\frac{5}{8}$	0.250	$\frac{1}{16}$	0.1981	0.0529
$\frac{1}{8}$	$\frac{1}{2}$	0.203	$\frac{3}{64}$	0.1355	0.0685
$\frac{1}{8}$	$\frac{5}{8}$	0.250	$\frac{1}{16}$	0.1825	0.0685
$\frac{1}{8}$	$\frac{3}{4}$	0.313	$\frac{1}{16}$	0.2455	0.0685
$\frac{5}{32}$	$\frac{5}{8}$	0.250	$\frac{1}{16}$	0.1669	0.0841
$\frac{5}{32}$	$\frac{3}{4}$	0.313	$\frac{1}{16}$	0.2299	0.0841
$\frac{5}{32}$	$\frac{7}{8}$	0.375	$\frac{1}{16}$	0.2919	0.0841
$\frac{3}{16}$	$\frac{3}{4}$	0.313	$\frac{1}{16}$	0.2143	0.0997
$\frac{3}{16}$	$\frac{7}{8}$	0.375	$\frac{1}{16}$	0.2763	0.0997
$\frac{3}{16}$	1	0.438	$\frac{1}{16}$	0.3393	0.0997
$\frac{1}{4}$	$\frac{7}{8}$	0.375	$\frac{1}{16}$	0.2450	0.1310
$\frac{1}{4}$	1	0.438	$\frac{1}{16}$	0.3080	0.1310
$\frac{1}{4}$	$1\frac{1}{4}$	0.547	$\frac{5}{64}$	0.4170	0.1310
$\frac{5}{16}$	1	0.438	$\frac{1}{16}$	0.2768	0.1622
$\frac{5}{16}$	$1\frac{1}{4}$	0.547	$\frac{5}{64}$	0.3858	0.1622
$\frac{5}{16}$	$1\frac{1}{2}$	0.641	$\frac{7}{64}$	0.4798	0.1622
$\frac{3}{8}$	$1\frac{1}{4}$	0.547	$\frac{5}{64}$	0.3545	0.1935
$\frac{3}{8}$	$1\frac{1}{2}$	0.641	$\frac{7}{64}$	0.4485	0.1935

Table 8–22

Sizes of Woodruff Keys Suitable for Various Shaft Diameters

Keyseat Width, in	Shaft Diameter, in	
	From	To (inclusive)
$\frac{1}{16}$	$\frac{5}{16}$	$\frac{1}{2}$
$\frac{3}{32}$	$\frac{3}{8}$	$\frac{7}{8}$
$\frac{1}{8}$	$\frac{3}{8}$	$1\frac{1}{2}$
$\frac{5}{32}$	$\frac{1}{2}$	$1\frac{5}{8}$
$\frac{3}{16}$	$\frac{9}{16}$	2
$\frac{1}{4}$	$\frac{11}{16}$	$2\frac{1}{4}$
$\frac{5}{16}$	$\frac{3}{4}$	$2\frac{3}{8}$
$\frac{3}{8}$	1	$2\frac{5}{8}$

Figure 8–33

Typical uses for retaining rings. (*a*) External ring and (*b*) its application; (*c*) internal ring and (*d*) its application.

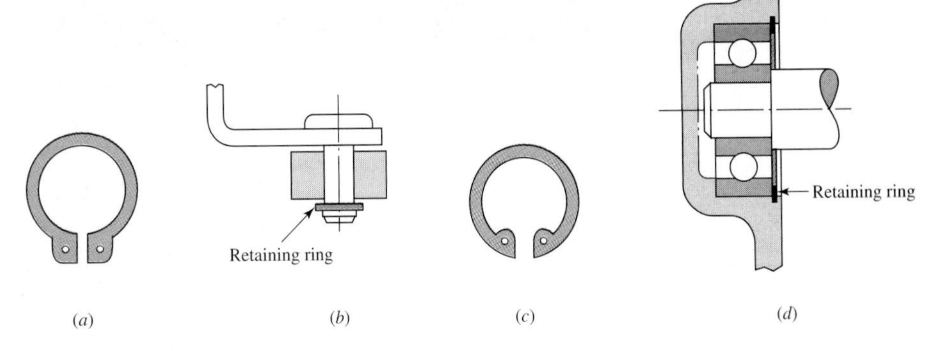

Stress-concentration factors for keyways depend for their values upon the fillet radius at the bottom and ends of the keyway, according to Peterson.[11] For fillets cut by standard milling-machine cutters, Peterson's charts give $K_t = 2.14$ for bending and $K_{ts} = 2.62$ for torsion. These are for end-milled keyseats, which permit definite key positioning longitudinally. In bending, the sled-runner seat will have less stress concentration. The values of K_t for the sled-runner keyway are based on the full values of I and J for shafting; that is, the second moments are not reduced by the slot dimensions. Other results reported are 1.79 for an end-milled keyway and 1.38 for the sled-runner type.

The use of shrink fits eliminates the use of keys to transmit torque. These were discussed in Sec. 2–8, and corresponding data are listed in Tables E–11 to E–14. Peterson reports $K_f = 2.0$ for bending at each end of a press-fitted collar. Tests seem to indicate that the interface pressure has little effect on K_t. However, fretting corrosion introduces a complication.

A retaining ring is frequently used instead of a shaft shoulder or a sleeve to axially position a component on a shaft or in a housing bore. As shown in Fig. 8–33, a groove is cut in the shaft or bore to receive the spring retainer. The tapered design of both the internal and external rings ensures uniform pressure against the bottom of the groove. For sizes, dimensions, and ratings, the manufacturers' catalogs should be consulted.

EXAMPLE 8–12

A UNS G10350 steel shaft, heat-treated to a minimum yield strength of 75 kpsi, has a diameter of $1\frac{7}{16}$ in. The shaft rotates at 600 r/min and transmits 40 hp through a gear. Select an appropriate key for the gear.

Solution

A $\frac{3}{8}$-in square key is selected, UNS G10200 cold-drawn steel being used. The design will be based on a yield strength of 65 kpsi. A factor of safety of 2.80 will be employed in the absence of exact information about the nature of the load.

The torque is obtained from the horsepower equation

$$T = \frac{63\,025H}{n} = \frac{(63\,025)(40)}{600} = 4200 \text{ lb} \cdot \text{in}$$

Referring to Fig. 8–34, the force F at the surface of the shaft is

$$F = \frac{T}{r} = \frac{4200}{0.719} = 5850 \text{ lb}$$

By the distortion-energy theory, the shear strength is

11. R. E. Peterson, *Stress Concentration Factors*, Wiley, New York, 1974, pp. 245, 266, 267.

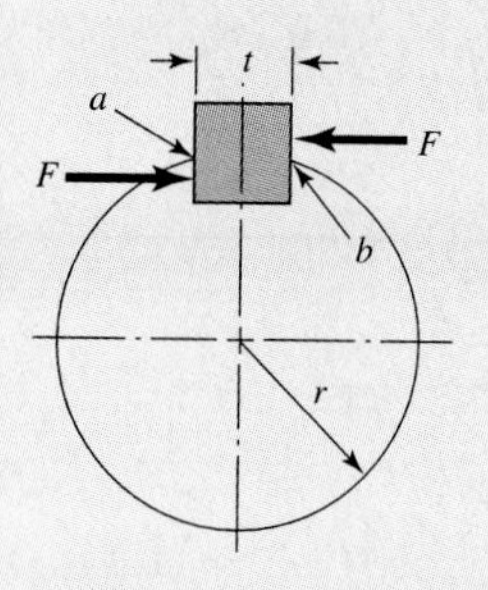

Figure 8–34

$$S_{sy} = 0.577S_y = (0.577)(65) = 37.5 \text{ kpsi}$$

Failure by shear across the area ab will create a stress of $\tau = F/tl$. Substituting the strength divided by the factor of safety for τ gives

$$\frac{S_{sy}}{n} = \frac{F}{tl} \quad \text{or} \quad \frac{37.5(10)^3}{2.80} = \frac{5850}{0.375l}$$

or $l = 1.16$ in. To resist crushing, the area of one-half the face of the key is used:

$$\frac{S_y}{n} = \frac{F}{tl/2} \quad \text{or} \quad \frac{65(10)^3}{2.80} = \frac{5850}{0.375l/2}$$

and $l = 1.34$ in. The hub length of a gear is usually greater than the shaft diameter, for stability. If the key, in this example, is made equal in length to the hub, it would therefore have ample strength, since it would probably be $1\frac{7}{16}$ in or longer.

Loose Pins

In some applications a pin is not a tight fit in its holes but is loose enough to allow movement. One such example is the knuckle pin for a railroad car or locomotive coupler as seen in Fig. 8–35. The pin is loose enough to allow unimpeded rotation of the knuckle even in the presence of dirt and rust. Problem 3–32 presented an exercise in the bending and shear stresses present in a loose-fitting clevis pin, suggesting two methods of analysis. Using a similar model, the effective beam span length, from Fig. 8–35, is

$$l = \frac{2.25}{2} + 2.25 + 3.75 + 2.25 + \frac{2.25}{2} = 10.5 \text{ in}$$

The distance from the left reaction to the left concentrated flexural loading is 2.25 in. The largest bending moment in the beam is estimated as

$$M_{\max} = \frac{F}{2}(2.25) = 1.125F$$

The associated bending stress σ and the von Mises stress σ_b' are

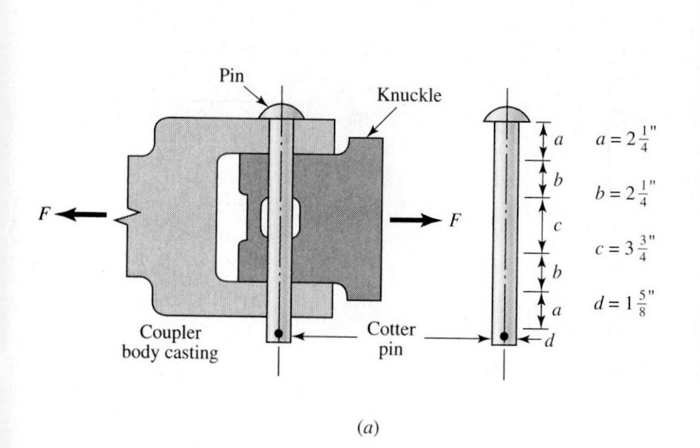

(*a*)

(*b*)

Figure 8–35

National AAR E freight car coupler and some geometric detail.

$$\sigma_b' = \sigma = \frac{Mc}{I} = \frac{32M}{\pi d^3} = \frac{32(1.125F)}{\pi\left(1.625^3\right)} = 2.67F$$

The shearing stress due to double shear at the shearing separation planes (at the top and bottom surfaces of the knuckle) is

$$\tau_{\max} = \frac{1}{2}\frac{4V}{3A} = \frac{1}{2}\frac{4}{3}\frac{V}{\pi d^2/4} = \frac{1}{2}\frac{4}{3}\frac{F/2}{\pi\left(1.625^2\right)/4} = 0.161F$$

The associated von Mises stress is $\sigma_s' = \sqrt{3}\tau_{\max} = 0.279F$. The ratio is

$$\frac{\sigma_b'}{\sigma_s'} = \frac{2.67}{0.279} = 9.6$$

Freight cars, including the coupling system, are designed to tolerate a large load F. The history of the buffeting force F is not known, but if the pin fails in bending fatigue rather than shear fatigue, the location and character of the failure reveal where the stress is the highest. In this service the pin failed in bending fatigue at a section about $3\frac{1}{2}$ in below the head of the pin.[12] Clearly this is not a shear separation plane. Our preceding analysis indicates a 9.6-fold higher von Mises stress due to bending rather than shear. Always investigate bending with loose pins. Furthermore, a bolt in a shear joint is loose in its hole to allow easy joint assembly. If bolts are allowed to carry the shear load on their body, they are in a similar situation to the knuckle pin. Bolts are kept from "seeing" such a loading by friction due to clamping provided by the initial tightening, or by providing tight pins to carry the shear load to protect the joint if friction fails to carry the load.

CASE STUDY 8 Learning While Doing

It has been said that engineering is a methodology for teaching oneself what one needs to know. There is sufficient truth in this statement to merit further examination.

The phone rings, and an engineer receives a new assignment: "Pull the print for angle bracket blank 6847 and design a fastener set for a 1250-lb static load 8 in from the vertical support." The engineer obtains the print for a jig-welded ribbed angle bracket, which is depicted in Fig. CS8-1. With this assignment comes many choices: number of fasteners, pattern, fastener type, size, series, and grade. In writing the moment equation, the engineer questions the location of point A, the pivot point. Finding the location x may require a finite-element analysis—too costly in time and money. The engineer examines the magnitude of forces P_1 and P_2 at various values of x including $x = 0$ and $x = 3$ in. From what he learns he decides to force x to be 3 in by placing a washer *between* the bracket and the support at each bolt.

He decides upon four identical bolts in rectangular array; four midjoint washers to control the pivot-point location, protecting the joint design in the case of fastener reuse; SAE grade 5 bolts and nuts for greatest strength to the dollar; $\frac{1}{2}$-in fasteners to start, since smaller fasteners can be stripped or broken by a person with an ordinary wrench; and, since it is a single machine and one joint, a probability of 1 in 100 of over-proof loading one or more bolts.

His design consisted of four 1/2″-13 × 1.75″ SAE grade 5, washer-faced finished bolts and nuts torqued to 940 in · lb, four 1/2-in washers, ID of 0.562 in, OD of 1.5 in, 0.109 in thick. Note that this is an OD larger than that of an American Standard plain washer. The joint is to be assembled lubricated, torqued to 940 in · lb.

12. The lower part of the pin dropped to the roadbed, the upper part of the pin bounced out, the knuckle pulled out, and train parted, the emergency brakes automatically engaged, and the crew substituted a new pin. The two parts of the pin were found by a track-walker within 15 feet of each other. The parts were unseen by the train crew, since the incident occurred at night and the pieces were under the train, to the back.

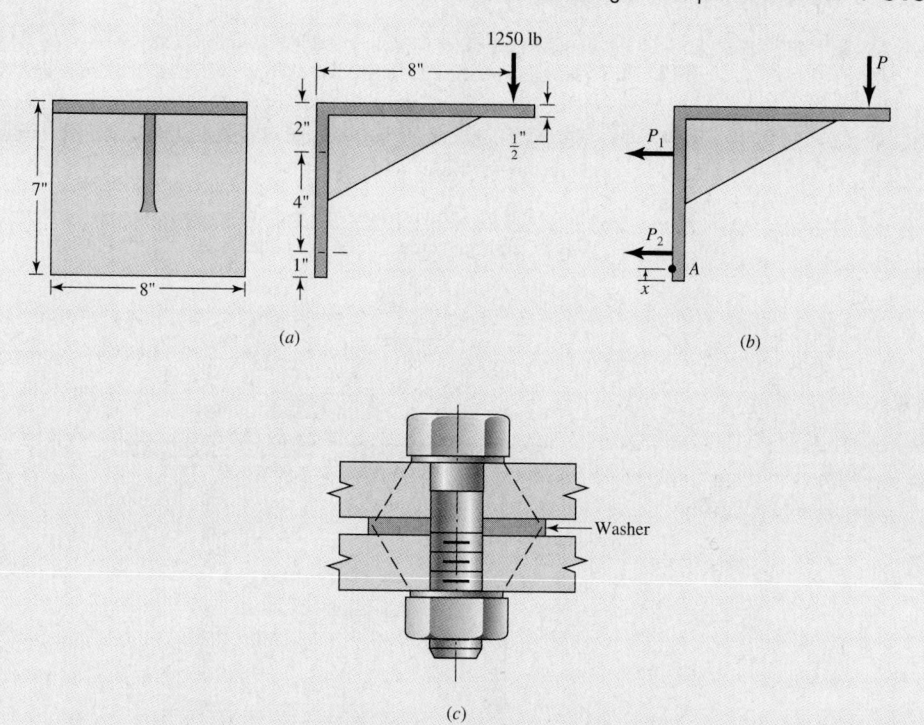

Figure CS8–1

(*a*) Drawing 6487 for a welded angle bracket blank with weld detail omitted; (*b*) planning and learning sketch; (*c*) sketch of bolt-member-washer detail.

An adequacy assessment in this case can be k_b from Eq. (8–17), k_m using Eq. (8–22) or (8–23), double-frusta equatorial diameter from $L_G \tan 30° + 1.5d$, F_i using Eq. (8–21), ξ_1 from $F_i/(S_p A_t)$, ξ_2 using $CP/(S_p A_t) + \xi_1$, σ_i from F_i/A_t, σ_b using $\sigma_i + CP/A_t$, F_b from Eq. (8–24), F_m using Eq. (8–28), n from Eq. (8–31), n_0 using Eq. (8–40), n_p from Eq. (8–35), z_2 using Eq. (8–31), $p_f = 1 - \Phi(z_2)$, $R_b = 1 - p_f$, $R_j = R_b^{1/N_b}$, and $p_{fj} = 1 - R_j$. Carry the vertical shear by friction clamping, and check if it can be carried directly by two (why two?) bolts.

After completing the adequacy assessment, discussion of the many aspects of this case will be fruitful. Was the approach to optimality robust? Are the factors of safety n and n_0 sufficiently large to convince you the threat is from over-proof loading (n_p smaller than n and n_0)? Just what were the innovative actions of the designer? What would you have done?

PROBLEMS

8–1 A power screw is 25 mm in diameter and has a thread pitch of 5 mm.

(*a*) Find the thread depth, the thread width, the mean and root diameters, and the lead, provided square threads are used.

(*b*) Repeat part (*a*) for Acme threads.

8–2 Using the information in the footnote of Table 8–1, show that the tensile-stress area is

$$A_t = \frac{\pi}{4}(d - 0.938\,194p)^2$$

8–3 Show that for zero collar friction the efficiency of a square-thread screw is given by the equation

$$e = \tan\lambda \frac{1 - f\tan\lambda}{\tan\lambda + f}$$

Plot a curve of the efficiency for lead angles up to 45°. Use $f = 0.08$.

8–4 A single-threaded 25-mm power screw is 25 mm in diameter with a pitch of 5 mm. A vertical load on the screw reaches a maximum of 6 kN. The coefficients of friction are 0.05 for the collar and 0.08 for the threads. The frictional diameter of the collar is 40 mm. Find the overall efficiency and the torque to "raise" and "lower" the load.

8–5 The machine shown in the figure can be used for a tension test but not for a compression test. Why? Can both screws have the same hand?

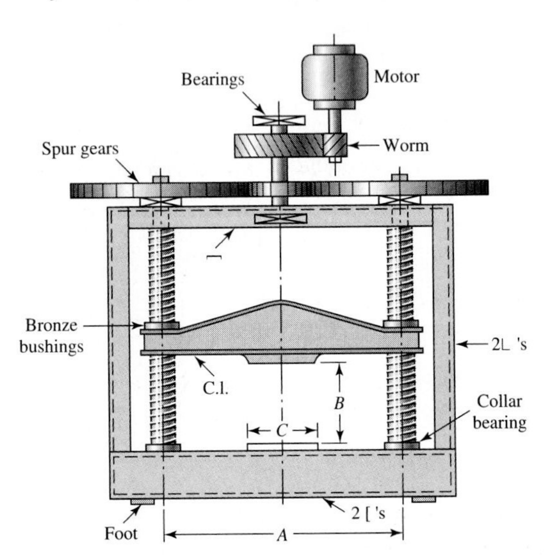

Problem 8–5

8–6 The press shown for Prob. 8–5 has a rated load of 5000 lb. The twin screws have Acme threads, a diameter of 3 in, and a pitch of $\frac{1}{2}$ in. Coefficients of friction are 0.05 for the threads and 0.06 for the collar bearings. Collar diameters are 5 in. The gears have an efficiency of 95 percent and a speed ratio of 75:1. A slip clutch, on the motor shaft, prevents overloading. The full-load motor speed is 1720 rev/min.

(*a*) When the motor is turned on, how fast will the press head move?

(*b*) What should be the horsepower rating of the motor?

8–7 A screw clamp similar to the one shown in the figure has a handle with diameter $\frac{3}{16}$ in made of cold-drawn AISI 1006 steel. The overall length is 3 in. The screw is $\frac{7}{16}''$-14 UNC and is $5\frac{3}{4}$ in long, overall. Distance A is 2 in. The clamp will accommodate parts up to $4\frac{3}{16}$ in high.

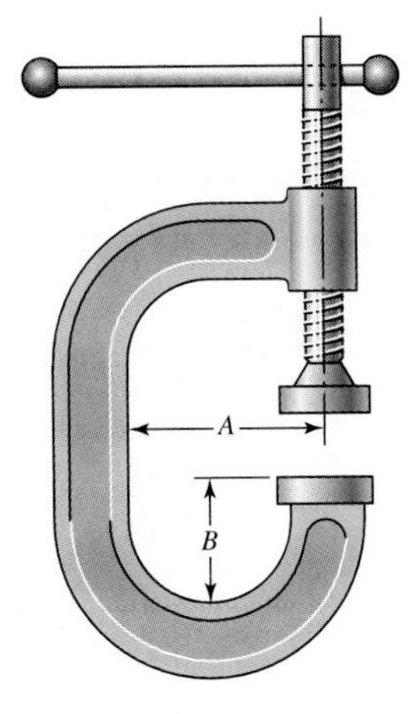

Problem 8–7

(*a*) What screw torque will cause the handle to bend permanently?

(*b*) What clamping force will the answer to part (*a*) cause if the collar friction is neglected and if the thread friction is 0.075?

(*c*) What clamping force will cause the screw to buckle?

(*d*) Are there any other stresses or possible failures to be checked?

8–8 The C clamp shown in the figure for Prob. 8–7 uses a $\frac{5}{8}''$-6 Acme thread. The frictional coefficients are 0.15 for the threads and for the collar. The collar, which in this case is the anvil striker's swivel joint, has a friction diameter of $\frac{7}{16}$ in. Calculations are to be based on a maximum force of 6 lb applied to the handle at a radius of $2\frac{3}{4}$ in from the screw centerline. Find the clamping force.

8–9 Find the power required to drive a 40-mm power screw having double square threads with a pitch of 6 mm. The nut is to move at a velocity of 48 mm/s and move a load of $F = 10$ kN. The frictional coefficients are 0.10 for the threads and 0.15 for the collar. The frictional diameter of the collar is 60 mm.

8–10 A single square-thread power screw has an input power of 3 kW at a speed of 1 rev/s. The screw has a diameter of 36 mm and a pitch of 6 mm. The frictional coefficients are 0.14 for the threads and 0.09 for the collar, with a collar friction radius of 45 mm. Find the axial resisting load F and the combined efficiency of the screw and collar.

8–11 A bolted joint is to have a grip consisting of two $\frac{1}{2}$-in steel plates and one $\frac{1}{2}$-in American Standard plain washer to fit under the head of the $\frac{1}{2}''$-13 × 1.75″ UNC hex-head bolt.

(*a*) What is the length of the thread L_T for this diameter inch-series bolt?

(*b*) What is the length of the grip L_G?

(*c*) What is the height H of the nut?

(*d*) Is the bolt long enough? If not, round to the next larger preferred length (Table E–17).

(*e*) What is the length of the shank and threaded portions of the bolt within the grip? These lengths are needed in order to estimate the bolt spring rate k_b.

8–12 A bolted joint is to have a grip consisting of two 14-mm steel plates and one 14R metric plain washer to fit under the head of the M14 × 2 hex-head bolt, 50 mm long.

(*a*) What is the length of the thread L_T for this diameter metric coarse-pitch series bolt?

(*b*) What is the length of the grip L_G?

(*c*) What is the height H of the nut?

(*d*) Is the bolt long enough? If not, round to the next larger preferred length (Table E–17).

(*e*) What is the length of the shank and the threaded portions of the bolt within the grip? These lengths are needed in order to estimate bolt spring rate k_b.

8–13 A blanking disk 0.875 in thick is to be fastened to a spool whose flange is 1 in thick, using eight $\frac{1}{2}''$-13 × 1.75″ hex-head cap screws.

(*a*) What is the length of threads L_T for this cap screw?

(*b*) What is the effective length of the grip L'_G?

(*c*) Is the length of this cap screw sufficient? If not, round up.

(*d*) Find the shank length l_d and the useful thread length l_T within the grip. These lengths are needed for the estimate of the fastener spring rate k_b.

8–14 A blanking disk is 20 mm thick and is to be fastened to a spool whose flange is 25 mm thick, using eight M12 × 40 hex-head metric cap screws.

(*a*) What is the length of the threads L_T for this fastener?

(*b*) What is the effective grip length L'_G?

(*c*) Is the length of this fastener sufficient? If not, round to the next preferred length.

(*d*) Find the shank length l_d and the useful threaded length in the grip l_T. These lengths are needed in order to estimate the fastener spring rate k_b.

8–15 A 3/4″-16 UNF series SAE grade 5 bolt has a 3/4-in ID tube 13 in long clamped between washer faces of bolt and nut, by turning the nut snug and adding 1/3 of a turn. The tube OD is the washer-face diameter $d_w = 1.5d = 1.5(0.75) = 1.125$ in = OD.

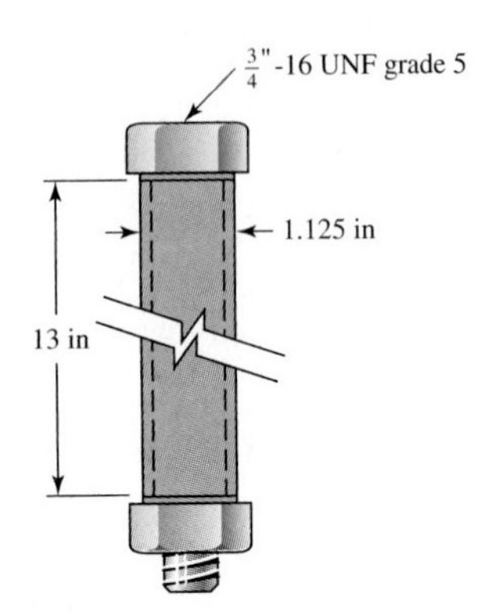

Problem 8–15

(*a*) What is the spring rate of the bolt and the tube, if the tube is made of steel? What is the joint constant C?

(*b*) When the 1/3 turn-of-nut is applied, what is the initial tension F_i in the bolt?

(*c*) What is the bolt tension at opening if additional tension is applied to the bolt external to the joint?

8–16 From your experience with Prob. 8–15, generalize your solution to develop a turn-of-nut equation

$$t = \frac{\theta}{360^\circ} = NF_i\left[\frac{1}{k_b} + \frac{1}{k_m}\right] = \frac{NF_i}{k_j}$$

where t = turn of the nut from snug tight
θ = twist of the nut in degrees
N = number of thread/in ($1/p$ where p is pitch)
F_i = initial preload
k_b, k_m, k_j = spring rates of the bolt, members, and joint, respectively

Use this equation to find the relation between torque-wrench setting T and turn-of-nut t. ("Snug tight" means the joint has been tightened to perhaps half the intended preload to flatten asperities on the washer faces and the members. Then the nut is loosened and retightened finger tight, and the nut is rotated the number of degrees indicated by the equation. Properly done, the result is competitive with torque-wrenching.)

8–17 RB&W[13] recommends turn-of-nut from snug fit to preload as follows: 1/3 turn for bolt grips of 1–4 diameters, 1/2 turn for bolt grips 4–8 diameters, and 2/3 turn for grips of 8–12 diameters. These recommendations are for structural steel fabrication (permanent joints), producing preloads of 100 percent of proof strength and beyond. Machinery fabricators with fatigue loadings and possible joint disassembly have much smaller turns-of-nut. The RB&W recommendation enters the nonlinear plastic deformation zone.

13. Russell, Burdsall & Ward, Inc., Metal Forming Specialists, Mentor, Ohio.

Problem 8–17
Turn-of-nut method

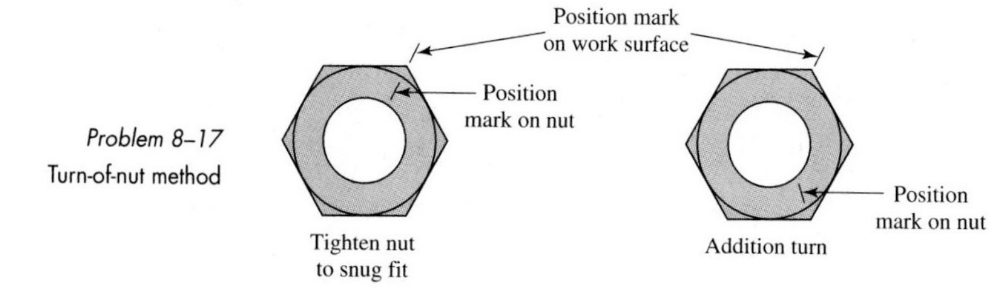

(*a*) Use the results of Prob. 8–16 and the torque from Ex. 8–4 to find the turn-of-nut in degrees for $\xi_1 = 0.75$ and $\xi_1 = 0.90$ for the cast in member joint.

(*b*) Apply the results of Prob. 8–16 and the torque from Ex. 8–5 to find the turn-of-nut in degrees for ξ_1 of 0.75 and 0.90 for the steel members.

ANALYSIS

8–18 Take Eq. (8–22) and express $k_m/(Ed)$ as a function of l/d, then compare with Eq. (8–23) for $d/l = 0.5$.

ANALYSIS

8–19 A joint has the same geometry as Ex. 8–4, but the lower member is steel. Use Eq. (8–23) to find the spring rate of the members in the grip.

ADEQUACY ASSESSMENT

8–20 A cylinder head and the cylinder depicted in the figure are made of steel and have the dimensions $A = 5$ in, $B = 9$ in, $C = 13$ in, $D = 0.5$ in, and $E = 0.5$ in. There are ten 1/2″-13 × 1.5″ grade 5 bolts torqued to 800 in · lb using lubricated bolts. The pressure load in use varies from 0 to 18 000 lb. The hydrostatic multiple is 2. There are no washers on the bolts. An adequacy assessment consists of two general tasks: a static analysis of the 2(18 000)-lb hydrostatic test and of the 0 to 18 000-lb dynamic loading. Perform the first task using a fully corrected endurance strength of 18 600 psi.

Problem 8–20
Cylinder head is steel; cylinder is grade 30 cast iron.

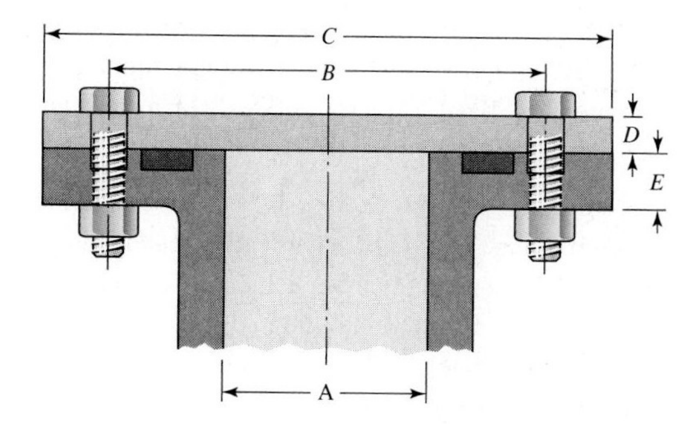

ADEQUACY ASSESSMENT

8–21 Continue Prob. 8–20 and perform an adequacy assessment for the dynamically loaded joint, using a fully corrected endurance strength of 18 600 psi.

DESIGN

8–22 Take the conditions of geometry and loading from Prob. 8–20 and design a fastener set for the joint using UNC-series grade 5 bolts and nuts, choosing the cheapest assembly. When the design is finalized, conduct an adequacy assessment.

ANALYSIS

8–23 The computer can be very helpful to the engineer. In matters of analysis it can take the drudgery out of calculations and improve accuracy. In synthesis, good programming is a matter of organizing decisions that must be made, soliciting them while displaying enough information, accepting them, and doing the number crunching. In either case, one cannot program what one does not understand. Understanding comes from experience with problems executed manually. It is useful to program the protocol of Table 8–7 because it is so easy to make a mistake in longhand. Focusing on the fastener, recognize two situations: (1) the fastener has been chosen, its diameter and length are known, and the designer needs to know all the pertinent dimensions, including the effective grip of a cap-screw joint and whether the length is adequate; and (2) the fastener diameter, nut, and washers are chosen, and the designer has to make the length decision, after which documentation of pertinent dimensions is in order. Code the protocol of Table 8–7, bearing in mind that you may wish to embed some of it in a larger program.

ANALYSIS

8–24 For a static loaded tension joint carrying an external load P, write a computer program to help you with an adequacy assessment using the fastener specifications as they affect performance. You will want to scrutinize the joint constant C, preload F_i, the fractions ξ_1 and ξ_2, and the various deterministic factors of safety, and estimate the probability of over-proof loading one or more bolts. Begin simply—write the code for just inch or metric fasteners—then add features as you exercise your program. When all is satisfactory, you can accommodate the other series to determine whether it worth blending the two.

DESIGN

8–25 For a static tension joint carrying an external load P, write a computer program that will allow you to synthesize the fastener specification for the joint. This will be done in an iterative manner. The a priori decisions will likely be washers, series, grade, and pattern; the design decisions will likely be fastener diameter and, possibly, number of fasteners. It is useful to focus on a single fastener. The structure of the program is likely to be a totem pole of iterative loops, beginning with the a priori decisions, with the design variable(s) being at the bottom. The discussion in Sec. 8–12 might be helpful.

ANALYSIS

8–26 Your experience with Prob. 8–25 will suggest that as you get into the feasible domain, the attention changes from attainment of satisfaction to pursuit of an optimal design. This may interest you in making the result of Prob. 8–24 more convenient to the task. Develop the two tables of Ex. 8–9 to provide an overview showing where to look from the first tabulation. This makes it easy to find the top of column entries in the second table.

ANALYSIS

8–27 Figure P8–20 illustrates the connection of a cylinder head to a pressure vessel using 10 bolts and a confined-gasket seal. The effective sealing diameter is 150 mm. Other dimensions are: $A = 100$, $B = 200$, $C = 300$, $D = 20$, and $E = 25$, all in millimeters. The cylinder is used to store gas at a static pressure of 6 MPa. ISO class 8.8 bolts with a diameter of 12 mm have been selected. This provides an acceptable bolt spacing. What load factor n results from this selection?

ANALYSIS

8–28 We wish to alter the figure for Prob. 8–27 by decreasing the inside diameter of the seal to the diameter $A = 100$ mm. This makes an effective sealing diameter of 120 mm. Then, by using cap screws instead of bolts, the bolt circle diameter B can be reduced as well as the outside diameter C. If the same bolt spacing and the same edge distance are used, then eight 12-mm cap screws can be used on a bolt circle with $B = 160$ mm and an outside diameter of 260 mm, a substantial savings. With these dimensions and all other data the same as in Prob. 8–27, find the load factor.

ANALYSIS

8–29 In the figure for Prob. 8–27, the bolts have a diameter of $\frac{1}{2}$ in and the cover plate is steel, with $D = \frac{1}{2}$ in. The cylinder is cast iron, with $E = \frac{5}{8}$ in and a modulus of elasticity of 18 Mpsi. The $\frac{1}{2}$-in SAE washer to be used under the nut has OD $= 1.062$ in and is 0.095 in thick. Find the stiffnesses of the bolt and the members and the joint constant C.

8–30 The same as Prob. 8–29, except that $\frac{1}{2}$-in cap screws are used with washers.

ANALYSIS

8–31 In addition to the data of Prob. 8–29, the dimensions of the cylinder are $A = 3.5$ in and an effective seal diameter of 4.25 in. The internal static pressure is 1500 psi. The outside diameter of the head is $C = 8$ in. The diameter of the bolt circle is 6 in, and so a bolt spacing in the range of 3 to 5 bolt diameters would require from 8 to 13 bolts. Select ten SAE grade 5 bolts and find the resulting load factor n.

ANALYSIS

8–32 A $\frac{3}{8}$-in class 5 cap screw and steel washer are used to secure a cast-iron frame of a machine having a blind threaded hole. The washer is 0.065 in thick. The cap has a modulus of elasticity of 14 Mpsi and is $\frac{1}{4}$ in thick. The screw is 1 in long. The material in the frame has a modulus of elasticity of 14 Mpsi. Find the stiffnesses k_b and k_m of the bolt and members.

8–33 Bolts distributed about a bolt circle are often called upon to resist an external bending moment as shown in the figure. The external moment is 12 kip · in and the bolt circle has a diameter of 8 in. The neutral axis for bending is a diameter of the bolt circle. What needs to be determined is the most severe external load seen by a bolt in the assembly.

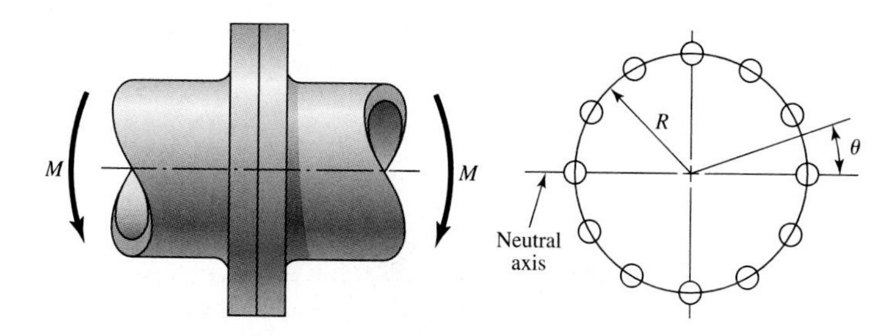

Problem 8–33
Bolted connection subjected to bending.

(a) View the effect of the bolts as placing a line load around the bolt circle whose intensity F_b', in pounds per inch, varies linearly with the distance from the neutral axis according to the relation $F_b' = F_{b,\max}' R \sin\theta$. The load on any particular bolt can be viewed as the effect of the line load over the arc associated with the bolt. For example, there are 12 bolts shown in the figure. Thus each bolt load is assumed to be distributed on a 30° arc of the bolt circle. Under these conditions, what is the largest bolt load?

(b) View the largest load as the intensity $F_{b,\max}'$ multiplied by the arc length associated with each bolt and find the largest bolt load.

(c) Express the load on any bolt as $F = F_{\max} \sin\theta$, sum the moments due to all the bolts, and estimate the largest bolt load. Compare the results of these three approaches to decide how to attack such problems in the future.

ANALYSIS

8–34 A $\frac{3}{8}''$-24 UNF SAE grade 5 bolt is lubricated and then tightened to a torque-wrench specification of 450 lb · in. This corresponds to a preload of 90 percent of proof strength. [See Eq. (8–41).] Data in the footnote of Table 8–14 show that the coefficient of variation of the preload is $C_{Fi} = 2.88/34.18 = 0.0843$.

(a) Find the mean and the standard deviation of the bolt tension.

(b) What is the probability that the proof strength will be exceeded?

ANALYSIS

8–35 The figure shows a cast-iron bearing block which is to be bolted to a steel ceiling joist and is to support a gravity load. Bolts used are M20 ISO 8.8 with coarse threads and with 3.4-mm-thick steel washers under the bolt head and nut. The joist flanges are 16 mm in thickness, and the dimension A, shown in the figure, is 20 mm. The modulus of elasticity of the bearing is 135 GPa.

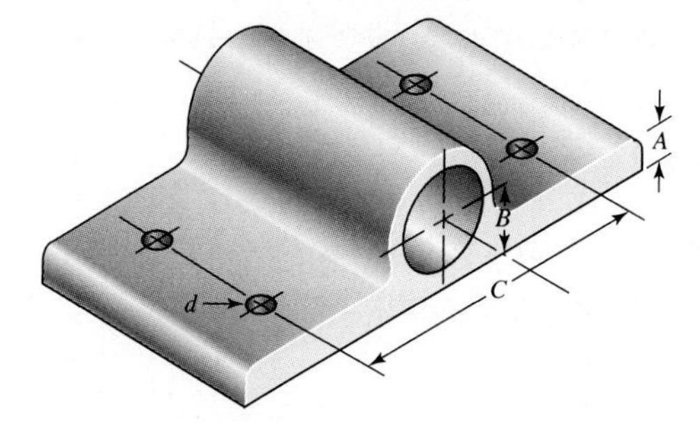

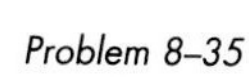
Problem 8–35

(a) Find the wrench torque required if the fasteners are lubricated during assembly and the joint is to be permanent.

(*b*) Estimate values of the least and the greatest preloads taken by the four bolts. Use $C_F = 0.09$.

8–36 The upside-down A frame shown in the figure is to be bolted to steel beams on the ceiling of a machine room using ISO grade 8.8 bolts. This frame is to support the 40-kN radial load as illustrated. The total bolt grip is 48 mm, which includes the thickness of the steel beam, the A-frame feet, and the steel washers used. The bolts are size M20 × 2.5.

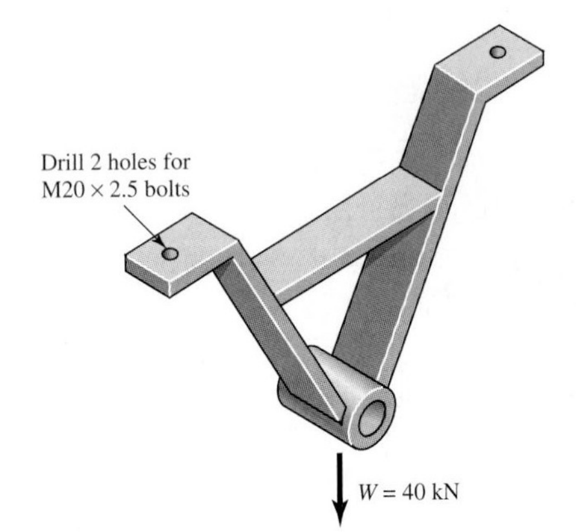

Problem 8–36

(*a*) What tightening torque should be used if the connection is permanent and the fasteners are lubricated?

(*b*) What portion of the external load is taken by the bolts? By the members?

8–37 In the figure for Prob. 8–20, let $A = 0.9$ m, $B = 1$ m, $C = 1.10$ m, $D = 20$ mm, and $E = 25$ mm. The cylinder is made of ASTM No. 35 cast iron ($E = 96$ GPa), and the head, of low-carbon steel. There are thirty-six M10 × 1.5 ISO 10.9 bolts tightened to 75 percent of proof load. During use, the cylinder pressure fluctuates between 0 and 550 kPa. Using the Gerber relation, find the factor of safety guarding against a fatigue failure of a bolt.

8–38 A 1-in-diameter hot-rolled AISI 1144 steel rod is hot-formed into an eyebolt similar to that shown in the figure for Prob. 3–59, with a 2-in-diameter eye. The threads are 1″-12 UNF and are die-cut.

(*a*) For a repeatedly applied load collinear with the thread axis, is fatigue failure more likely in the thread or in the eye?

(*b*) What can be done to strengthen the bolt at the weaker location?

(*c*) If the factor of safety guarding against a fatigue failure is $n_f = 2$, what repeatedly applied load can be applied to the eye?

8–39 The section of the sealed joint shown in the figure is loaded by a repeated force $P = 6$ kip. The members have $E = 16$ Mpsi. All bolts have been carefully preloaded to $F_i = 25$ kip each.

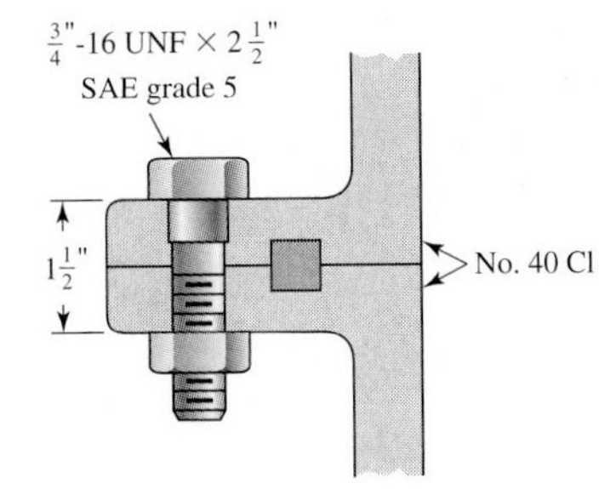

Problem 8–39

(*a*) If hardened-steel washers 0.134 in thick are to be used under the head and nut, what length of bolts should be used?

(*b*) Find k_b, k_m, and C.

(*c*) Using the Gerber failure locus, find the factor of safety guarding against a fatigue failure.

(*d*) Find the load factor guarding against over-proof loading.

ANALYSIS

8–40 Suppose the welded steel bracket shown in the figure is bolted underneath a structural-steel ceiling beam to support a fluctuating tensile load imposed on it by a pin and yoke. The bolts are $\frac{1}{2}$ in, coarse threads, SAE grade 5, tightened to recommended preload. The stiffnesses have already been computed and are $k_b = 4.94$ Mlb/in and $k_m = 15.97$ Mlb/in.

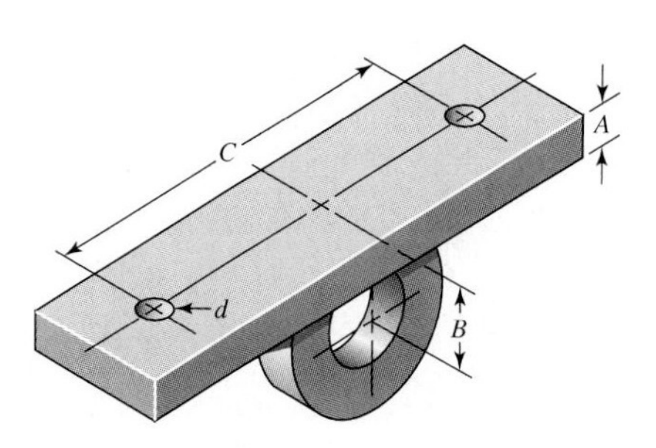

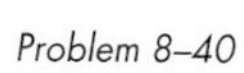
Problem 8–40

(*a*) Assuming that the bolts, rather than the welds, govern the strength of this design, determine the safe repeated load P that can be imposed on this assembly using the Gerber criterion and a fatigue design factor of 2.

(*b*) Compute the load factors based on the load found in part (*a*).

ANALYSIS

8–41 Using the Gerber fatigue locus and a fatigue-design factor of 2, determine the external repeated load P that a $1\frac{1}{4}$-in SAE grade 5 coarse-thread bolt can take compared with that for a fine-thread bolt. The joint constants are $C = 0.30$ for coarse- and 0.32 for fine-thread bolts.

ANALYSIS

8–42 An M30 × 3.5 ISO 8.8 bolt is used in a joint at recommended preload, and the joint is subject to a repeated tensile fatigue load of $P = 80$ kN per bolt. The joint constant is $C = 0.33$. Find the load factors and the factor of safety guarding against a fatigue failure based on the Gerber fatigue locus.

ANALYSIS

8–43 The figure shows a fluid-pressure linear actuator (hydraulic cylinder) in which $D = 4$ in, $t = \frac{3}{8}$ in, $L = 12$ in, and $w = \frac{3}{4}$ in. Both brackets as well as the cylinder are of steel. The actuator has been designed for a working pressure of 2000 psi. Six $\frac{3}{8}$-in SAE grade 5 coarse-thread bolts are used, tightened to 75 percent of proof load.

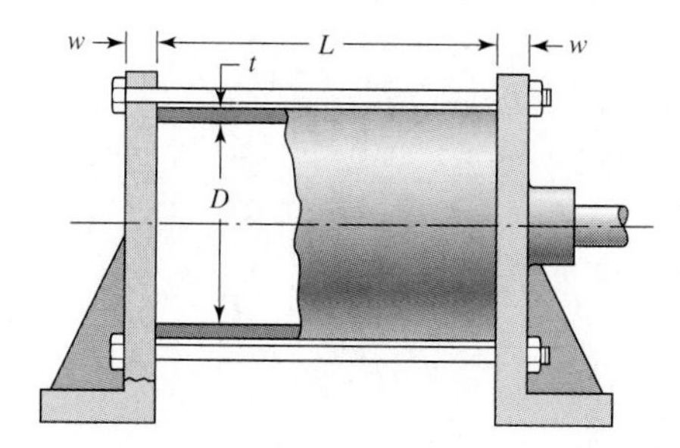

Problem 8–43

(*a*) Find the stiffnesses of the bolts and members, assuming that the entire cylinder is compressed uniformly and that the end brackets are perfectly rigid.

(*b*) Using the Gerber fatigue locus, find the factor of safety guarding against a fatigue failure.

(*c*) What pressure would be required to cause total joint separation?

ANALYSIS

8–44 The figure shows a bolted lap joint that uses SAE grade 8 bolts. The members are made of cold-drawn AISI 1040 steel. Find the safe tensile shear load F that can be applied to this connection if the following factors of safety are specified: shear of bolts 3, bearing on bolts 2, bearing on members 2.5, and tension of members 3.

Problem 8–44

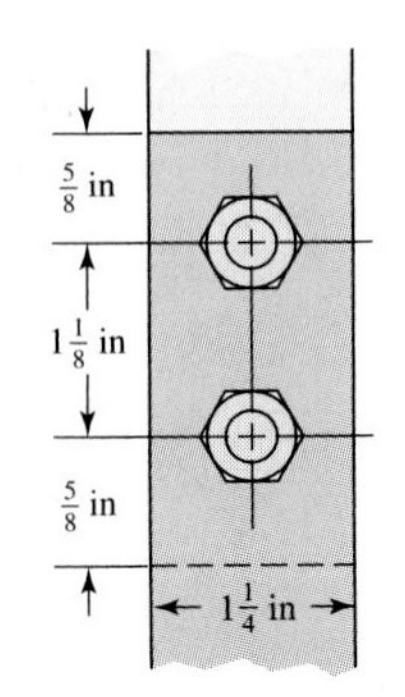

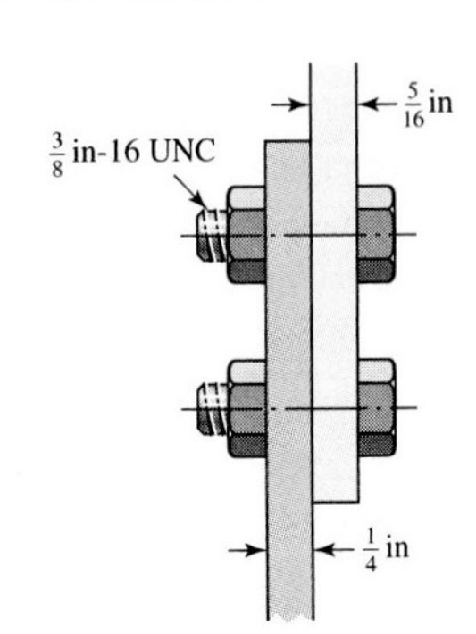

ANALYSIS

8–45 The bolted connection shown in the figure uses SAE grade 5 bolts. The members are hot-rolled AISI 1018 steel. A tensile shear load $F = 4000$ lb is applied to the connection. Find the factor of safety for all possible modes of failure.

Problem 8–45

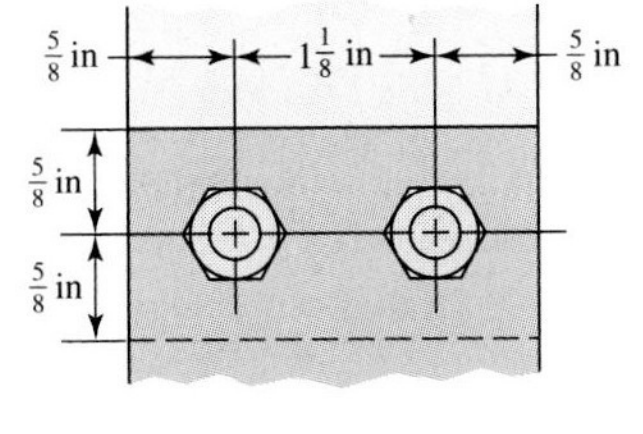

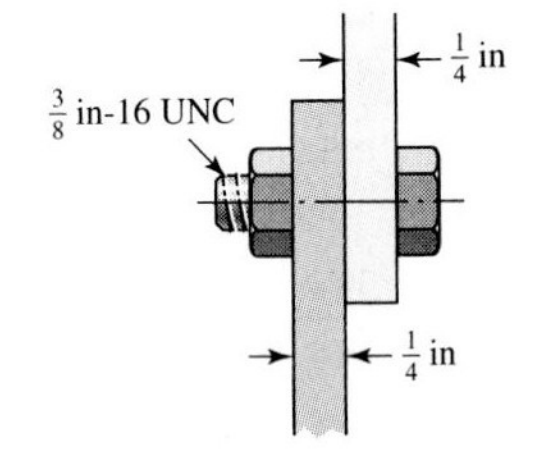

8–46 A bolted lap joint using SAE grade 5 bolts and members made of cold-drawn SAE 1040 steel is shown in the figure. Find the tensile shear load F that can be applied to this connection if the following factors of safety are specified: shear of bolts 1.8, bearing on bolts 2.2, bearing on members 2.4, and tension of members 2.6.

Problem 8–46

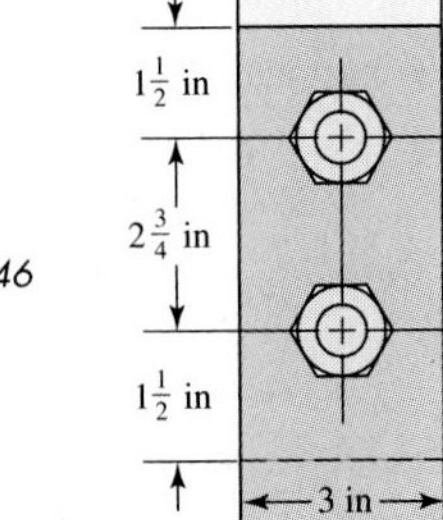

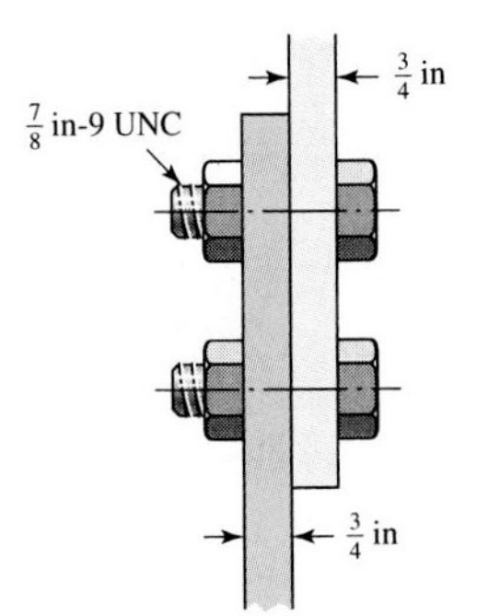

8–47 The bolted connection shown in the figure is subjected to a tensile shear load of 20 kip. The bolts are SAE grade 5 and the material is cold-drawn AISI 1015 steel. Find the factor of safety of the connection for all possible modes of failure.

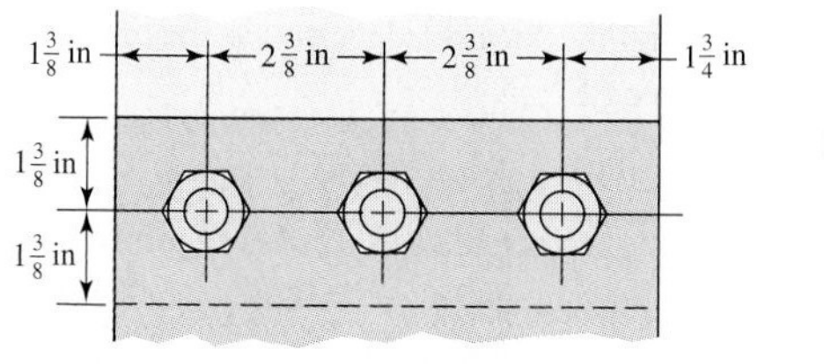

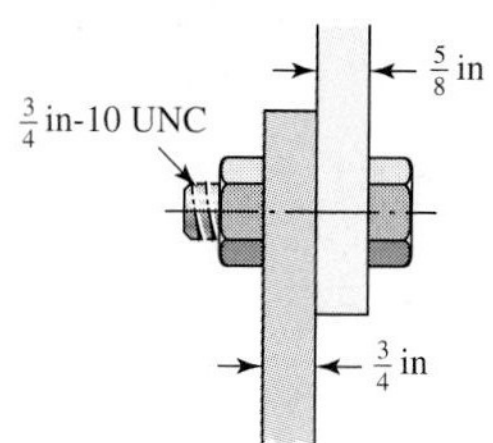

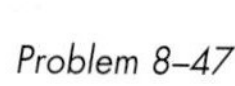
Problem 8–47

8–48 The figure shows a connection which employs three SAE grade 5 bolts. The tensile shear load on the joint is 5400 lb. The members are cold-drawn bars of AISI 1020 steel. Find the factor of safety for each possible mode of failure.

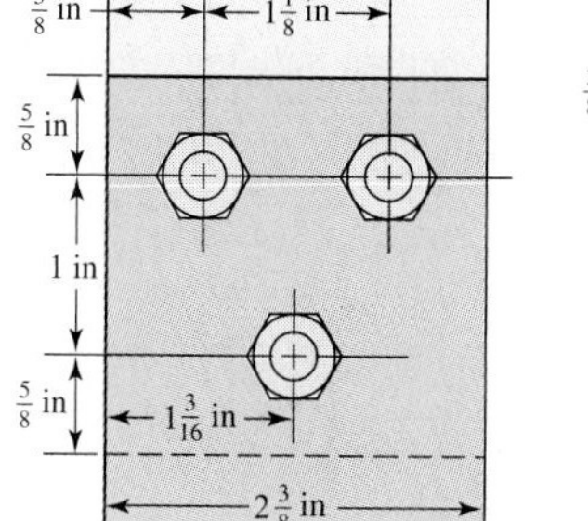

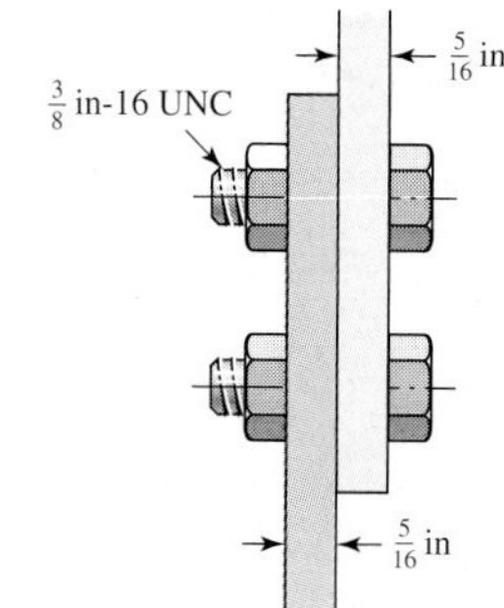

Problem 8–48

8–49 A beam is made up by bolting together two cold-drawn bars of AISI 1018 steel as a lap joint, as shown in the figure. The bolts used are ISO 5.8. Ignoring any twisting, determine the factor of safety of the connection.

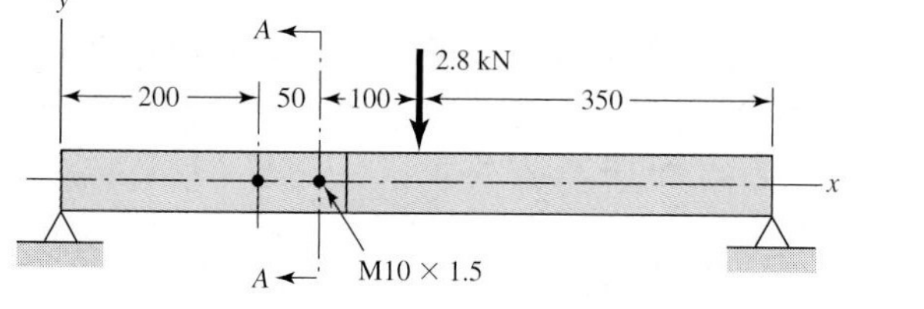

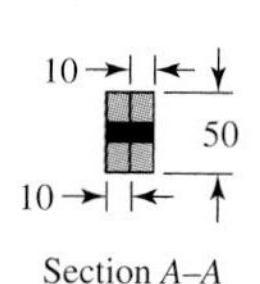

Problem 8–49
Dimensions in millimeters.

8–50 Standard design practice, as exhibited by the solutions to Probs. 8–44 to 8–48, is to assume that the bolts, or rivets, share the shear equally. For many situations, such an assumption may lead to an unsafe design. Consider the yoke bracket of Prob. 8–40, for example. Suppose this bracket is bolted to a wide-flange column with the centerline through the two bolts in the vertical direction. A vertical load through the yoke-pin hole at distance B from the column flange would place a shear load on the bolts as well as a tensile load. The tensile load comes about because the bracket tends to pry itself about the bottom corner, much like a claw hammer, exerting a large tensile load on the upper bolt. In addition, it is almost certain that both the spacing of the bolt holes and their diameters will be slightly different on the column flange from what they are on the yoke bracket. Thus, unless yielding occurs, only one of the bolts will take the shear load. The designer has no way of knowing which bolt this will be.

In this problem the bracket is 8 in long, $A = \frac{1}{2}$ in, $B = 3$ in, $C = 6$ in, and the column flange is $\frac{1}{2}$ in thick. The bolts are $\frac{1}{2}$-in UNC SAE 5. Steel washers 0.095 in thick are used under the nuts. The nuts are tightened to 75 percent of proof load. The vertical yoke-pin load is 3000 lb.

If the upper bolt takes all the shear load as well as the tensile load, how closely does the bolt stress approach the proof strength?

ANALYSIS

8–51 The bearing of Prob. 8–35 is bolted to a vertical surface and supports a horizontal shaft. The bolts used have coarse threads and are M20 ISO 5.8. The joint constant is $C = 0.30$, and the dimensions are $A = 20$ mm, $B = 50$ mm, and $C = 160$ mm. The bearing base is 240 mm long. The bearing load is 12 kN. If the bolts are tightened to 75 percent of proof load, will the bolt stress exceed the proof strength? Use worst-case loading, as discussed in Prob. 8–50.

ANALYSIS

8–52 A split-ring clamp-type shaft collar such as is described in Prob. 6–23 must resist an axial load of 1000 lb. Using a design factor of $n = 3$ and a coefficient of friction of 0.12, specify an SAE Grade 5 cap screw using fine threads. What wrench torque should be used if a lubricated screw is used?

ANALYSIS

8–53 A vertical channel 152 × 76 (see Table E–7) has a cantilever bolted to it as shown. The channel is hot-rolled AISI 1006 steel. The bar is of hot-rolled AISI 1015 steel. The bolts are M12 × 1.75 ISO 5.8. For a design factor of 2.8, find the safe force F that can be applied to the cantilever.

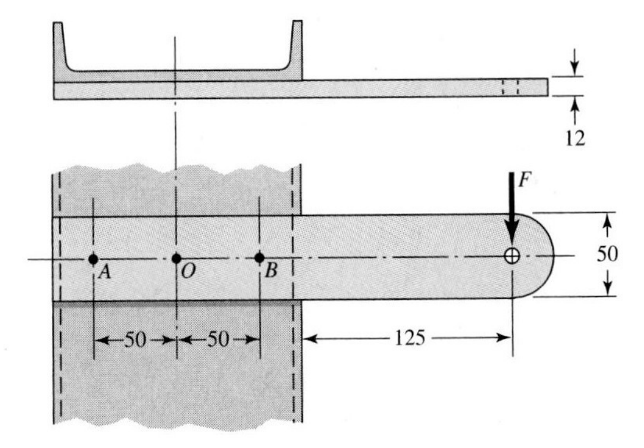

Problem 8–53
Dimensions in millimeters.

ANALYSIS

8–54 Find the total shear load on each of the three bolts for the connection shown in the figure and compute the significant bolt shear stress and bearing stress. Find the second moment of area of the 8-mm plate on a section through the three bolt holes, and find the maximum bending stress in the plate.

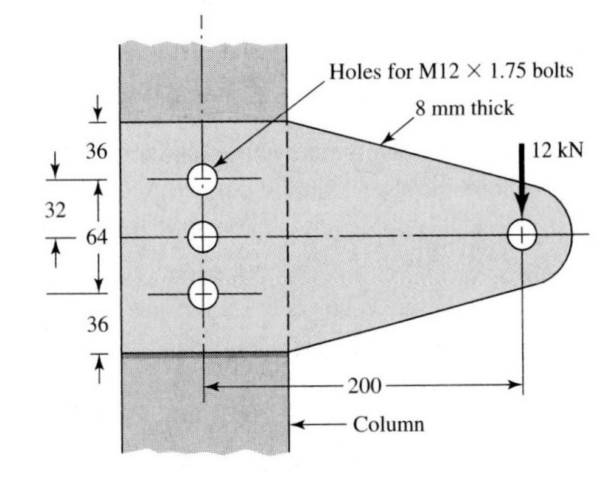

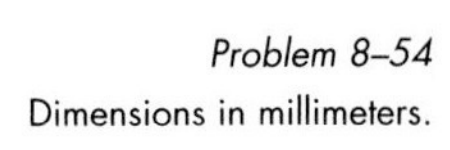
Problem 8–54
Dimensions in millimeters.

ANALYSIS

8–55 A $\frac{3}{8}$- × 2-in AISI 1018 cold-drawn steel bar is cantilevered to support a static load of 300 lb as illustrated. The bar is secured to the support using two $\frac{1}{2}''$-13 UNC SAE 5 bolts. Find the factor of safety for the following modes of failure: shear of bolt, bearing on bolt, bearing on member, and strength of member.

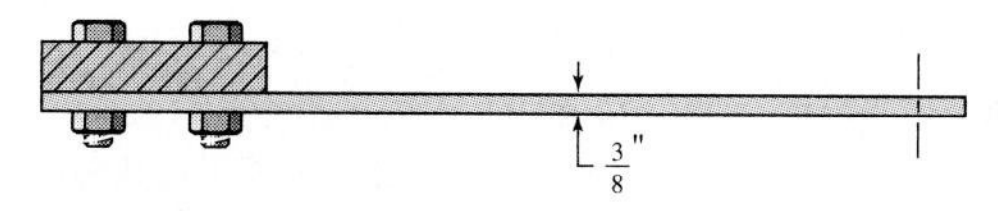

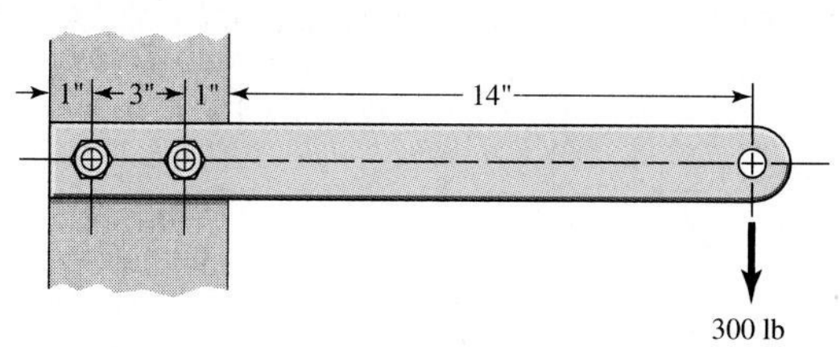

Problem 8–55

8–56 The figure shows a welded fitting which has been tentatively designed to be bolted to a channel so as to transfer the 2500-lb load into the channel. The channel is made of hot-rolled low-carbon steel having a minimum yield strength of 46 kpsi; the two fitting plates are of hot-rolled stock having a minimum S_y of 45.5 kpsi. The fitting is to be bolted using six standard SAE grade 2 bolts. Check the strength of the design by computing the factor of safety for all possible modes of failure.

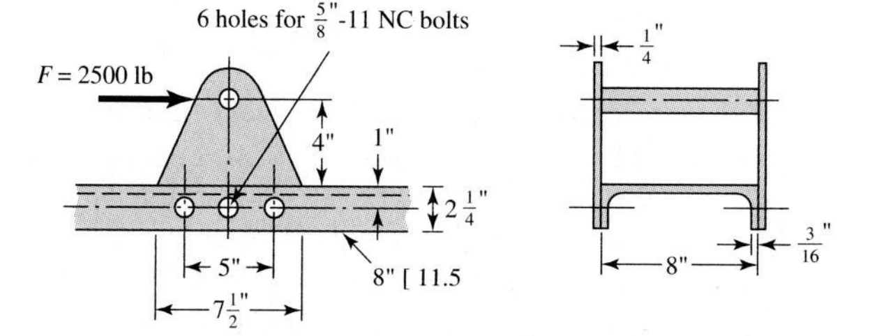

Problem 8–56

8–57 Tests with carefully torqued bolts lubricated with machine oil have shown the initial tension predicted by Eq. (8–27) to be approximately normally distributed with a coefficient of variation of 0.09. What fraction of proof load should be the target if a designer wishes to run a risk of 1 chance in 100 of tightening a bolt over the recommended proof load? Apply this result to a $\frac{5}{8}''$-11 UNC grade 5 bolt tightening torque.

ANALYSIS

8–58 Use the decision of Example 8–4 and estimate the probability of tightening any of the bolts too much during a maximum load of nP if the bolts are torqued to 60 percent of recommended value.

8–59 What should be the torque-wrench setting for the bolts of Example 8–4 if the chance of getting too much proof load is set at 1 in 100? The lubricant is machine oil.

DESIGN

8–60 The task is to design a joint similar to that of Ex. 8–9 with the following exception: use the distortion-energy Gerber fatigue failure locus.

(*a*) Prepare a table showing the feasible and infeasible domains, by placing the bolt spacing in diameters as entries.

(*b*) Prepare another table showing the figure of merit, eliminating the designs where bolt spacing is unsatisfactory.

(*c*) How does your design compare with Ex. 8–9 results?

(*d*) Declare your specifications and perform an adequacy assessment for your design.

8–61 Write a computer program for the adequacy assessment for a joint such as that of Prob. 8–60, carrying a load varying from 0 to P. The program should accept the fastener specifications. You can build in the option of using Goodman, Gerber, and DE-elliptic fatigue failure loci and a Wileman et al. member stiffness. The statements of Probs. 8–20 and 8–21 may contain hints. Limit yourself to inch- or metric-series fasteners.

DESIGN

8–62 A cantilever is to be attached to the flat side of a 6-in, 13.0 lb/in channel used as a column. The cantilever is to carry a load as shown in the figure. To a designer the choice of a bolt array is usually an a priori decision. Such decisions are made from a background of knowledge of the effectiveness of various patterns.

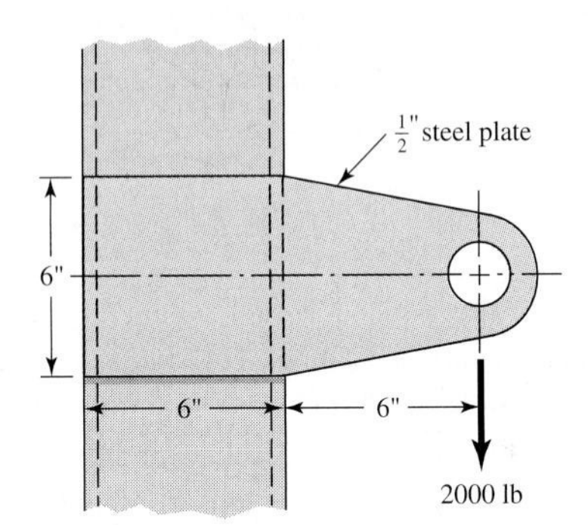

Problem 8–62

(*a*) If two fasteners are used, should the array be arranged vertically, horizontally, or diagonally? How would you decide?

(*b*) If three fasteners are used, should a linear or triangular array be used? For a triangular array, what should be the orientation of the triangle? How would you decide?

8–63 Using your experience with Prob. 8–62, specify a bolt pattern for Prob. 8–62, and size the bolts.

DESIGN

8–64 Unstated in the presentations of many of the problems is the designer's objective.

(*a*) Should the designer seek to carry primary and secondary shear on the joint members by friction, avoiding any shear on the bolts? If so, how can the designer ascertain whether the joint is doing this? How is a quantitative estimate of the factor of safety made?

(*b*) If there is some reason (vibration, inept assembly) that the bolts are too loose for friction to carry all the load by friction, do the bolts share the shear load? For what situation should the engineer design?

(*c*) Is there merit to providing two pins at the joint to carry the loading, aligning the members, and using the fasteners (how many?) to hold the joint together.

Welding, Brazing, Bonding, and the Design of Permanent Joints

9–1	Welding Symbols **528**
9–2	Butt and Fillet Welds **530**
9–3	Stresses in Welded Joints in Torsion **535**
9–4	Stresses in Welded Joints in Bending **540**
9–5	The Strength of Welded Joints **542**
9–6	Specification Set, Adequacy Assessment, and Decision Set **544**
9–7	Static Loading **549**
9–8	Fatigue Loading **554**
9–9	Resistance Welding **557**
9–10	Bolted and Riveted Joints Loaded in Shear **558**
9–11	Adhesive Bonding and Design Considerations **562**

Form can more readily pursue function by joining processes such as welding, brazing, soldering, cementing, and gluing—processes that are used extensively in manufacturing today. Whenever parts have to be assembled or fabricated, there is usually good cause for considering one of these processes in preliminary design work. Particularly when sections to be joined are thin, one of these methods may lead to significant savings. The elimination of individual fasteners, with their holes, and assembly costs is an important factor. Also, some of the methods allow rapid machine assembly, furthering their attractiveness.

Riveted permanent joints were common as the means of fastening rolled steel shapes to one another to form a permanent joint. The childhood fascination of seeing a cherry-red hot rivet thrown with tongs across a building skeleton to be unerringly caught by a person with a conical bucket, to be hammered pneumatically into its final shape, is all but gone. Two developments relegated riveting to lesser prominence. The first was the development of high-strength steel bolts whose preload could be controlled. The second was the improvement of welding, competing both in cost and in latitude of possible form.

9–1 Welding Symbols

A weldment is fabricated by welding together a collection of metal shapes, cut to particular configurations. During welding, the several parts are held securely together, often by clamping or jigging. The welds must be precisely specified on working drawings, and this is done using the welding symbol, shown in Fig. 9–1, as standardized by the American Welding Society (AWS). The arrow of this symbol points to the joint to be welded. The body of the symbol contains as many of the following elements as are deemed necessary:

- Reference line
- Arrow
- Basic weld symbols as in Fig. 9–2
- Dimensions and other data
- Supplementary symbols
- Finish symbols

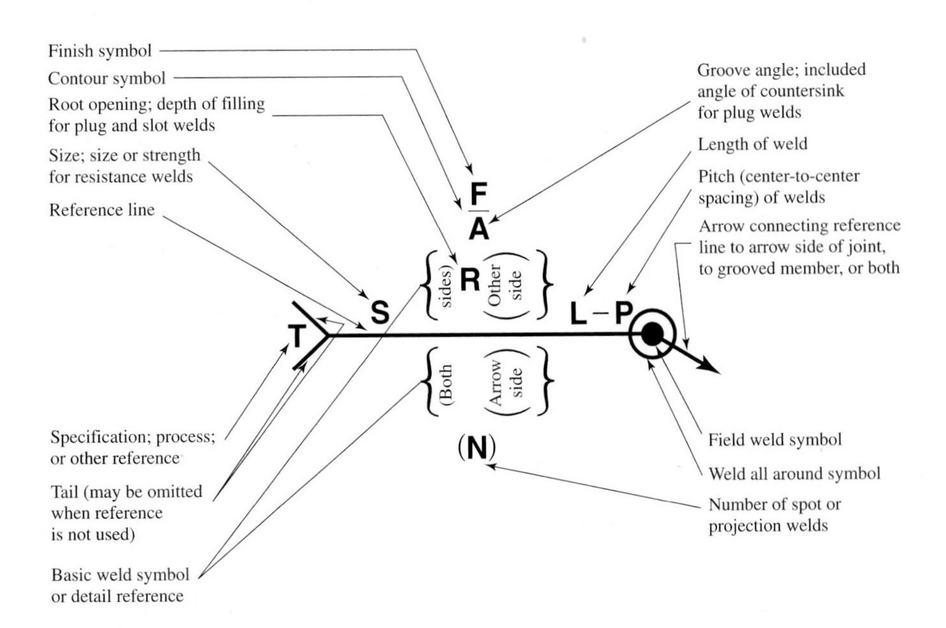

Figure 9–1

The AWS standard welding symbol showing the location of the symbol elements.

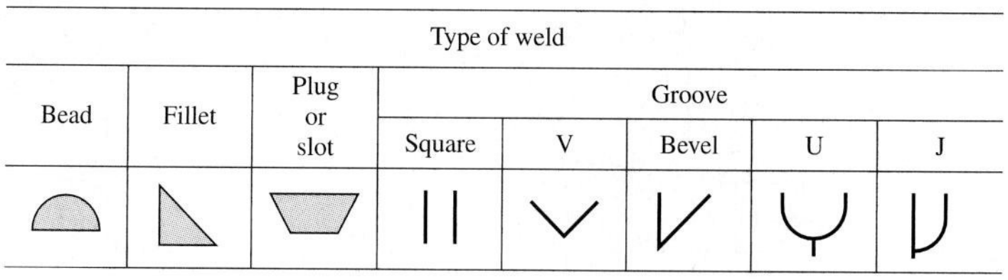

Figure 9–2

Arc- and gas-weld symbols.

Type of weld							
Bead	Fillet	Plug or slot	Groove				
			Square	V	Bevel	U	J

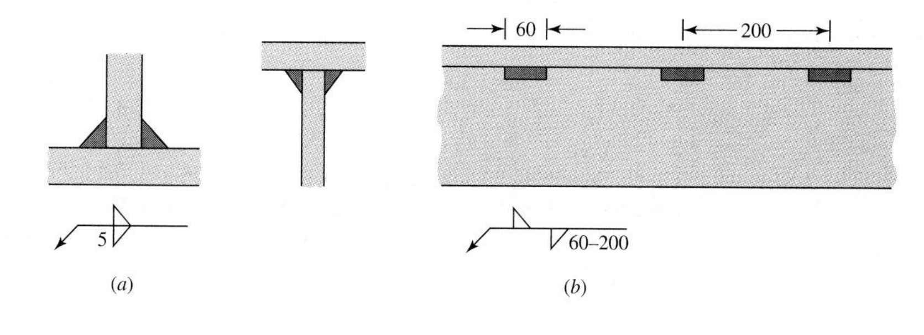

Figure 9–3

Fillet welds. (*a*) The number indicates the leg size; the arrow should point only to one weld when both sides are the same. (*b*) The symbol indicates that the welds are intermittent and staggered 60 mm along on 200-mm centers.

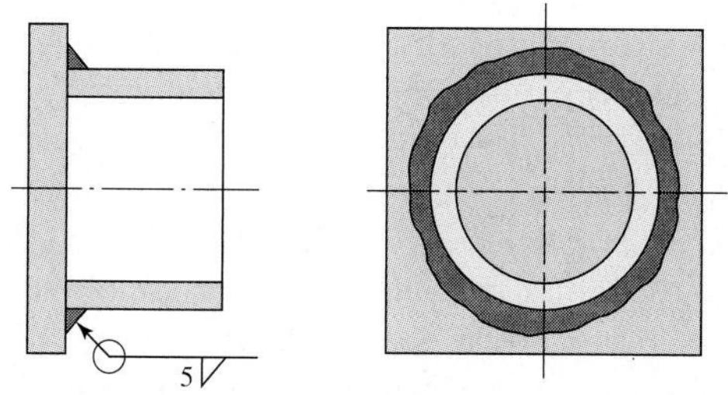

Figure 9–4

The circle on the weld symbol indicates that the welding is to go all around.

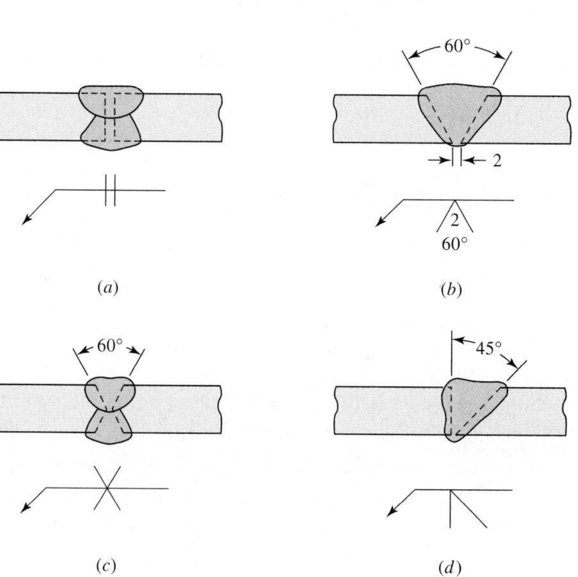

Figure 9–5

Butt or groove welds: (*a*) square butt-welded on both sides; (*b*) single V with 60° bevel and root opening of 2 mm; (*c*) double V; (*d*) single bevel.

- Tail
- Specification or process

The *arrow side* of a joint is the line, side, area, or near member to which the arrow points. The side opposite the arrow side is the *other side*.

Figures 9–3 to 9–6 illustrate the types of welds used most frequently by designers. For general machine elements most welds are fillet welds, though butt welds are used a great deal in designing pressure vessels. Of course, the parts to be joined must be

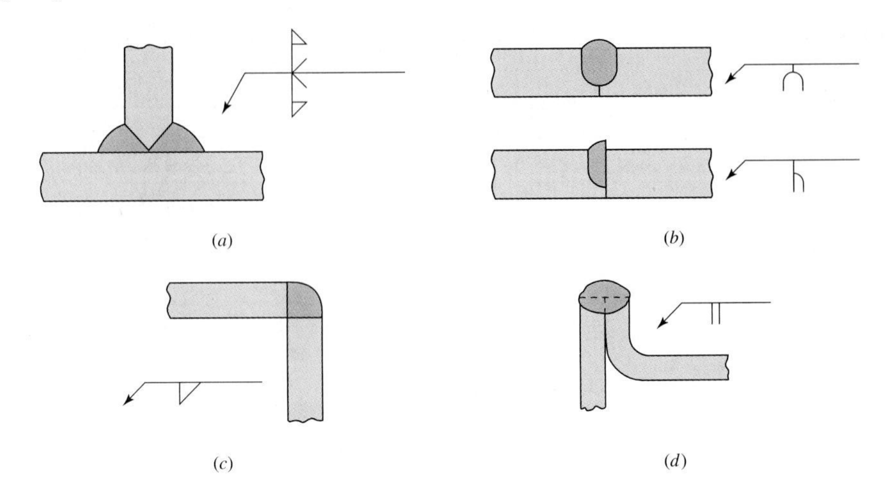

Figure 9–6

Special groove welds: (*a*) T joint for thick plates; (*b*) U and J welds for thick plates; (*c*) corner weld (may also have a bead weld on inside for greater strength but should not be used for heavy loads); (*d*) edge weld for sheet metal and light loads.

arranged so that there is sufficient clearance for the welding operation. If unusual joints are required because of insufficient clearance or because of the section shape, the design may be a poor one and the designer should begin again and endeavor to synthesize another solution.

Since heat is used in the welding operation, there are metallurgical changes in the parent metal in the vicinity of the weld. Also, residual stresses may be introduced because of clamping or holding or, sometimes, because of the order of welding. Usually these residual stresses are not severe enough to cause concern; in some cases a light heat treatment after welding has been found helpful in relieving them. When the parts to be welded are thick, a preheating will also be of benefit. If the reliability of the component is to be quite high, a testing program should be established to learn what changes or additions to the operations are necessary to ensure the best quality.

9–2 Butt and Fillet Welds

Figure 9–7 shows a single V-groove weld loaded by the tensile force F. For either tension or compression loading, the average normal stress is

$$\sigma = \frac{F}{hl} \tag{9-1}$$

where h is the weld throat and l is the length of the weld, as shown in the figure. Note that the value of h does not include the reinforcement. The reinforcement can be desirable, but it varies somewhat and does produce stress concentration at point A in the figure. If fatigue loads exist, it is good practice to grind or machine off the reinforcement.

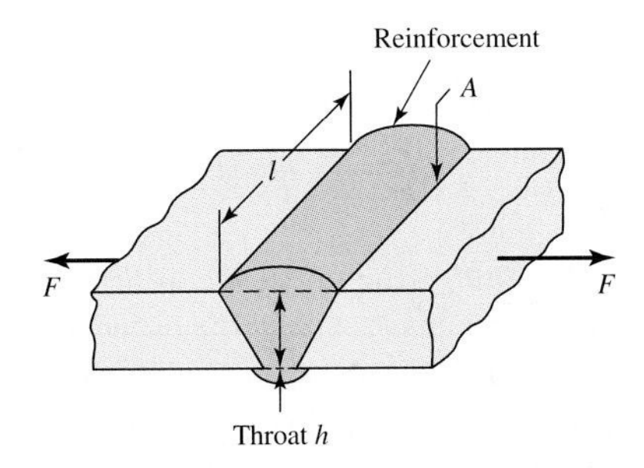

Figure 9–7

A typical butt joint.

The average stress in a butt weld due to shear loading is

$$\tau = \frac{F}{hl} \tag{9–2}$$

Figure 9–8 illustrates a typical transverse fillet weld. In Fig. 9–9 a portion of the welded joint has been isolated from Fig. 9–8 as a free body. At angle θ the forces on each weldment consist of a normal force F_n and a shear force F_s. Summing forces in the x and y directions gives

$$\sum F_x = 0 = -F_s \cos\theta + F_n \sin\theta \tag{a}$$

$$\sum F_y = 0 = 2F - 2F_s \sin\theta - 2F_n \cos\theta \tag{b}$$

Solving Eqs. (a) and (b) simultaneously for F_n gives

$$F_n = F_s \frac{\cos\theta}{\sin\theta} \tag{c}$$

Backsubstituting Eq. (c) into Eq. (a) gives

$$F - F_s \sin\theta - F_s \frac{\cos^2\theta}{\sin\theta} = 0$$

from which

$$F_n = F \cos\theta \tag{d}$$

The throat length t at θ in terms of the leg size h is

$$t = \frac{h}{\cos\theta + \sin\theta} \tag{e}$$

The nominal stresses at the angle θ in the weldment, τ and σ, are

$$\tau = \frac{F_s}{A} = \frac{F \sin\theta(\cos\theta + \sin\theta)}{hl} = \frac{F}{hl}\left(\sin\theta \cos\theta + \sin^2\theta\right) \tag{f}$$

Figure 9–8

A transverse fillet weld.

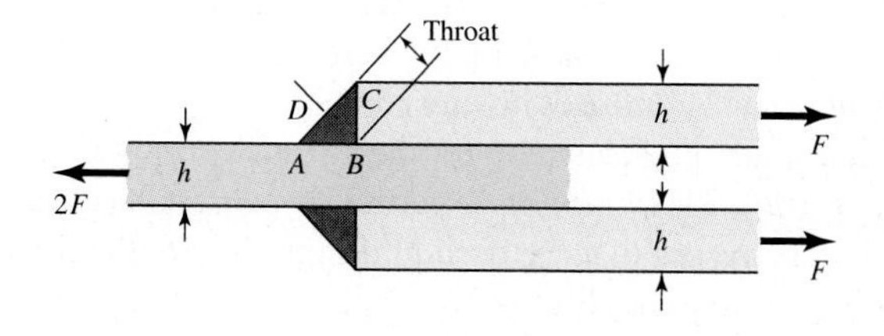

Figure 9–9

Free body from Fig. 9–8.

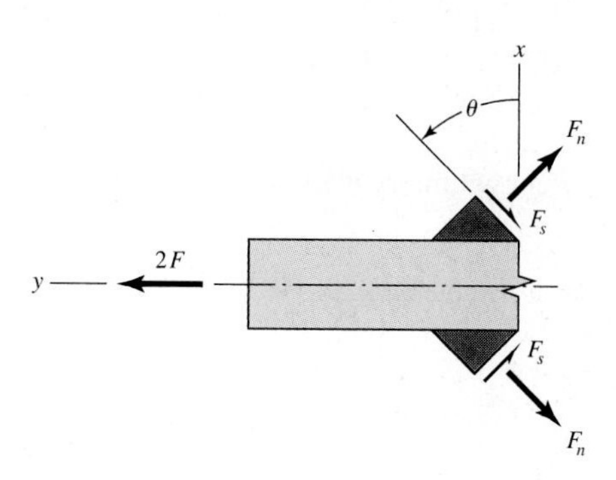

Figure 9–10

Stress distribution in fillet welds: (*a*) stress distribution on the legs as reported by Norris; (*b*) distribution of principal stresses and maximum shear stress as reported by Salakian.

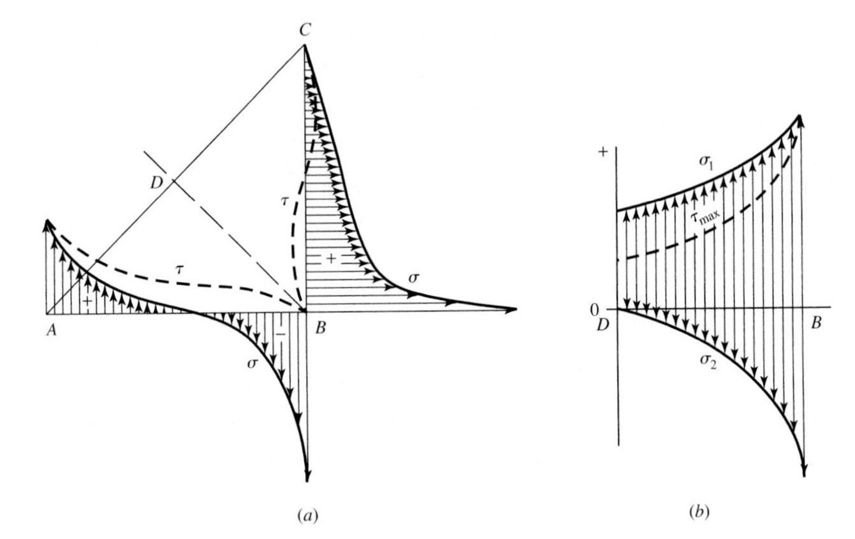

$$\sigma = \frac{F_n}{A} = \frac{F\cos\theta(\cos\theta + \sin\theta)}{hl} = \frac{F}{hl}\left(\cos^2\theta + \sin\theta\cos\theta\right) \tag{g}$$

The von Mises stress σ' at angle θ is

$$\sigma' = \left(\sigma^2 + 3\tau^2\right)^{1/2} = \frac{F}{hl}\left[\left(\cos^2\theta + \sin\theta\cos\theta\right)^2 + 3\left(\sin^2\theta + \sin\theta\cos\theta\right)^2\right]^{1/2} \tag{h}$$

The largest von Mises stress occurs at $\theta = 62.5°$ with a value of $\sigma' = 2.16F/(hl)$. The corresponding values of τ and σ are $\tau = 1.196F/(hl)$ and $\sigma = 0.623F/(hl)$.

The maximum shear stress can be found by differentiating Eq. (f) with respect to θ and equating to zero. The stationary point occurs at $\theta = 67.5°$ with a corresponding $\tau_{\max} = 1.207F/(hl)$ and $\sigma = 0.5F/(hl)$.

There are some experimental and analytical results that are helpful in evaluating Eqs. (f) through (h) and consequences. A model of the transverse fillet weld of Fig. 9–8 is easily constructed for photoelastic purposes and has the advantage of a balanced loading condition. Norris constructed such a model and reported the stress distribution along the sides AB and BC of the weld.[1] An approximate graph of the results he obtained is shown as Fig. 9–10*a*. Note that stress concentration exists at A and B on the horizontal leg and at B on the vertical leg. Norris states that he could not determine the stresses at A and B with any certainty.

Salakian[2] presents data for the stress distribution across the throat of a fillet weld (Fig. 9–10*b*). This graph is of particular interest because we have just learned that it is the throat stresses that are used in design. Again, the figure shows stress concentration at point B. Note that Fig. 9–10*a* applies either to the weld metal or to the parent metal, and that Fig. 9–10*b* applies only to the weld metal. Figure 9–11 shows a parallel fillet weld. Table 9–1 gives K_{fs} for fillet welds.

Equations (a) through (h) and their consequences seem familiar, and we can become comfortable with them. The net result of photoelastic and finite-element analysis of transverse fillet weld geometry is more like that shown in Fig. 9–10 than those given by

1. C. H. Norris, "Photoelastic Investigation of Stress Distribution in Transverse Fillet Welds," *Welding J.*, vol. 24, 1945, p. 557s.

2. A. G. Salakian and G. E. Claussen, "Stress Distribution in Fillet Welds: A Review of the Literature," *Welding J.*, vol. 16, May 1937, pp. 1–24.

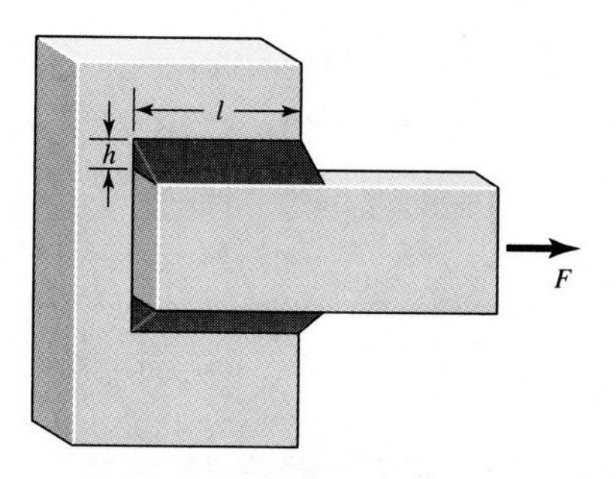

Figure 9–11

A parallel fillet weld.

Table 9–1

Transverse and Parallel Fillet Welds*

Kind of Loading	Throat-Shear Stress (Welding Code Method)
1. Transverse fillet weld 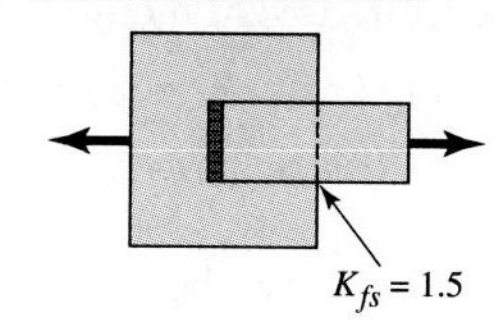 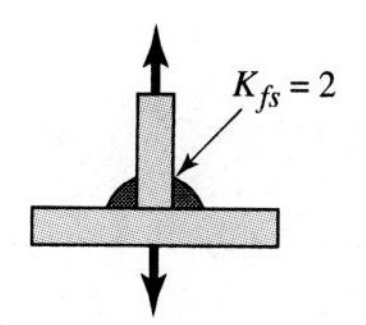	$\tau = \dfrac{\sqrt{2}F}{h\Delta x}$
2. Parallel fillet weld 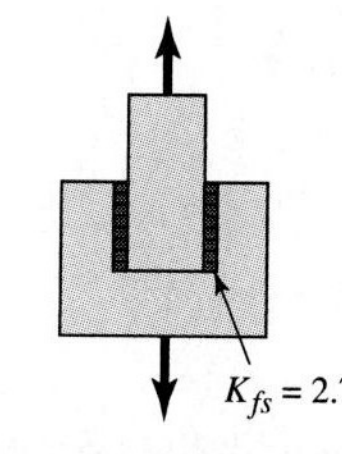	$\tau = \dfrac{\sqrt{2}F}{h\Delta x}$

*The approach to fillet analysis and design provides for carrying the external loading as a pure shear stress on the throat area.

mechanics of materials or elasticity methods. The most important concept here is that we have *no analytical approach that predicts the extant stresses*. The geometry of the fillet is crude by machinery standards, and even if it were ideal, the macrogeometry is too abrupt and complex for our methods. There are also subtle bending stresses due to eccentricities. Still, in the absence of robust analysis, weldments must be specified and the resulting joints must be safe. Absent finite element analysis tools the approach has been to use a simple *and conservative* model, verified by testing as conservative. The approach has been to

- Consider the external loading to be carried by shear forces on the throat area of the weld. By ignoring the normal stress on the throat, the shearing stresses are inflated sufficiently to render the model conservative.

- Use distortion energy for significant stresses.
- Circumscribe typical cases by code.

By using this model the basis for weld analysis or design employs

$$\tau = \frac{F}{0.707hl} = \frac{1.414F}{hl} \tag{9–3}$$

Under circumstances of combined loading we

- Examine primary shear stresses due to external forces.
- Examine secondary shear stresses due to torsional and bending moments.
- Estimate the strength(s) of the parent metal(s).
- Estimate the strength of deposited weld metal.
- Estimate permissible load(s) for parent metal(s).
- Estimate permissible load for deposited weld metal.

In fatigue the Gerber failure locus is the best, but you will find the Goodman locus in common use. The specification set and decision set can be as follows:

Specification Set	Decision Set
A Priori Decisions	
Pattern of weld bead	Pattern of weld bead
Electrode	Electrode
Type of weld	Type of weld
Length of weld, ℓ	Length of weld, ℓ
Decision Variable	
Leg size, h	Leg size, h

With this view, consider Fig. 9–9 with the normal forces neglected and θ set to 45°:

$$\Sigma F_x = -F_s \cos 45^\circ + F_s \cos 45^\circ = 0$$

$$\Sigma F_y = 2F - 2F_s \sin 45^\circ = 0$$

from which

$$F_s = \frac{F}{\sin 45^\circ}$$

$$\tau = \frac{F_s}{A} = \frac{F}{hl \sin 45^\circ} = \frac{1.414F}{hl} \tag{9–4}$$

which agrees with Eq. (9–3). Compare this with Eq. (f), with normal stress existing:

$$\tau = \frac{F}{hl}\left(\sin 45^\circ \cos 45^\circ + \sin^2 45^\circ\right) = \frac{F}{hl}$$

The shear stress on the throat in Eq. (9–4) is inflated in the ratio 1.414/1 by neglecting the normal stress on the throat. Also note that the shear stress $1.414F/(hl)$ is larger than the maximum shear stress of $1.207F/(hl)$. The shear stress corresponding to the largest von Mises stress is $2.16(0.577)F/(hl) = 1.246F/(hl)$, which is less than $1.414F/(hl)$. This demonstrates the *inflation* of the shear stress beyond maxima as a result of neglecting normal stress. One must rely on the extensive testing behind welding codes and practice for assurance of the *conservatism* of the method.

9–3 Stresses in Welded Joints in Torsion

Figure 9–12 illustrates a cantilever of length l welded to a column by two fillet welds. The reaction at the support of a cantilever always consists of a shear force V and a moment M. The shear force produces a *primary shear* in the welds of magnitude

$$\tau' = \frac{V}{A} \tag{9–5}$$

where A is the throat area of all the welds.

The moment at the support produces *secondary shear* or *torsion* of the welds, and this stress is given by the equation

$$\tau'' = \frac{Mr}{J} \tag{9–6}$$

where r is the distance from the centroid of the weld group to the point in the weld of interest and J is the second polar moment of area of the weld group about the centroid of the group. When the sizes of the welds are known, these equations can be solved and the results combined to obtain the maximum shear stress. Note that r is usually the farthest distance from the centroid of the weld group.

Figure 9–13 shows two welds in a group. The rectangles represent the throat areas of the welds. Weld 1 has a throat width $b_1 = 0.707h_1$; and weld 2 has a throat width $d_2 = 0.707h_2$. Note that h_1 and h_2 are the respective weld sizes. The throat area of both welds together is

$$A = A_1 + A_2 = b_1 d_1 + b_2 d_2 \tag{a}$$

This is the area that is to be used in Eq. (9–5).

The x axis in Fig. 9–13 passes through the centroid G_1 of weld 1. The second moment of area about this axis is

$$I_x = \frac{b_1 d_1^3}{12}$$

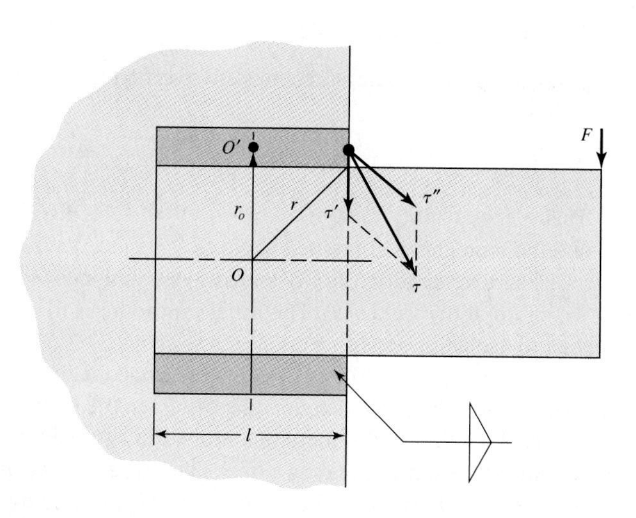

Figure 9–12

This is a *moment connection;* such a connection produces *torsion* in the welds.

Figure 9–13

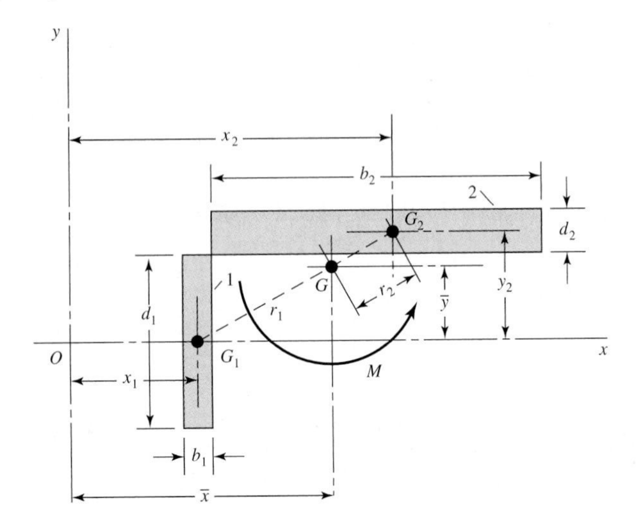

Similarly, the second moment of area about an axis through G_1 parallel to the y axis is

$$I_y = \frac{d_1 b_1^3}{12}$$

Thus the second polar moment of area of weld 1 about its own centroid is

$$J_{G1} = I_x + I_y = \frac{b_1 d_1^3}{12} + \frac{d_1 b_1^3}{12} \tag{b}$$

In a similar manner, the second polar moment of area of weld 2 about its centroid is

$$J_{G2} = \frac{b_2 d_2^3}{12} + \frac{d_2 b_2^3}{12} \tag{c}$$

The centroid G of the weld group is located at

$$\bar{x} = \frac{A_1 x_1 + A_2 x_2}{A} \qquad \bar{y} = \frac{A_1 y_1 + A_2 y_2}{A}$$

Using Fig. 9–13 again, we see that the distances r_1 and r_2 from G_1 and G_2 to G, respectively, are

$$r = \left[(\bar{x} - x_1)^2 + \bar{y}^2\right]^{1/2} \qquad r_2 = \left[(y_2 - \bar{y})^2 + (x_2 - \bar{x})^2\right]^{1/2}$$

Now, using the parallel-axis theorem, we find the second polar moment of area of the weld group to be

$$J = \left(J_{G1} + A_1 r_1^2\right) + \left(J_{G2} + A_2 r_2^2\right) \tag{d}$$

This is the quantity to be used in Eq. (9–6). The distance r must be measured from G and the moment M computed about G.

The reverse procedure is that in which the allowable shear stress is given and we wish to find the weld size. The usual procedure is to estimate a probable weld size and then to use iteration.

Observe in Eq. (b) that the second term contains the quantity b_1^3, which is the cube of the weld width, and that the quantity d_2^3 in the first term of Eq. (c) is also the cube of the weld width. Both of these quantities can be set equal to unity. This leads to the idea of treating each fillet weld as a line. The resulting second moment of area is then a *unit second polar moment of area*. The advantage of treating the weld size as a line is that the value of J_u is the same regardless of the weld size. Since the throat width of a fillet weld is $0.707h$, the relationship between J and the unit value is

$$J = 0.707h J_u \tag{9–7}$$

in which J_u is found by conventional methods for an area having unit width. The transfer formula for J_u must be employed when the welds occur in groups, as in Fig. 9–12. Table 9–2 lists the throat areas and the unit second polar moments of area for the most common

Table 9–2

Torsional Properties of Fillet Welds*

Weld	Throat Area	Location of G	Unit Second Polar Moment of Area
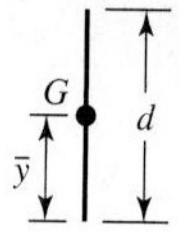 G, d, $\bar{y}$	$A = 0.707hd$	$\bar{x} = 0$ $\bar{y} = d/2$	$J_u = d^3/12$
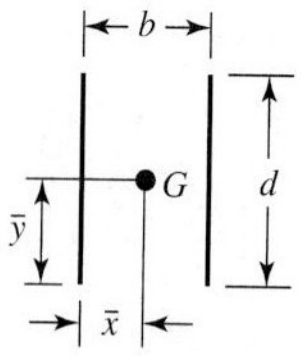 b, G, d, $\bar{y}$, $\bar{x}$	$A = 1.414hd$	$\bar{x} = d/2$ $\bar{y} = d/2$	$J_u = \dfrac{d(3b^2 + d^2)}{6}$
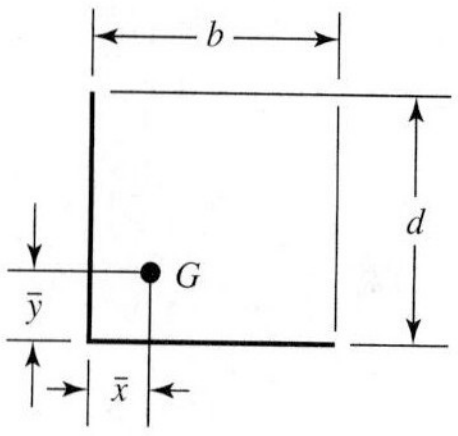 b, d, G, $\bar{y}$, $\bar{x}$	$A = 0.707h(2b + d)$	$\bar{x} = \dfrac{b^2}{2(b + d)}$ $\bar{y} = \dfrac{d^2}{2(b + d)}$	$J_u = \dfrac{(b + d)^4 - 6b^2d^2}{12(b + d)}$
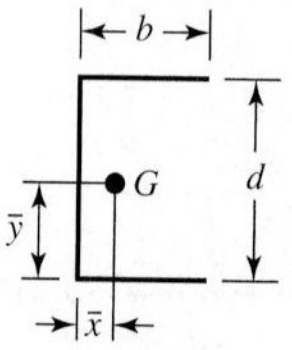 b, G, d, $\bar{y}$, $\bar{x}$	$A = 0.707h(2b + d)$	$\bar{x} = \dfrac{b^2}{2b + d}$ $\bar{y} = d/2$	$J_u = \dfrac{8b^3 + 6bd^2 + d^3}{12} - \dfrac{b^4}{2b + d}$
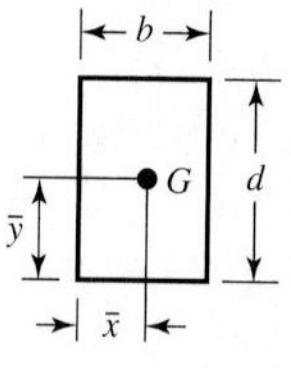 b, G, d, $\bar{y}$, $\bar{x}$	$A = 1.414h(b + d)$	$\bar{x} = b/2$ $\bar{y} = d/2$	$J_u = \dfrac{(b + d)^3}{6}$
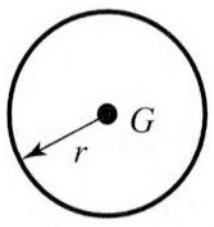 G, r	$A = 1.414\pi hr$		$J_u = 2\pi r^3$

*G is centroid of weld group; h is weld size; plane of torque couple is in the plane of the paper; all welds are of unit width.

fillet welds encountered. The example that follows is typical of the calculations normally made.

EXAMPLE 9–1

A 50-kN load is transferred from a welded fitting into a 200-mm steel channel as illustrated in Fig. 9–14. Estimate the maximum stress in the weld.

Solution[3]

(*a*) Label the ends and corners of each weld by letter. Sometimes it is desirable to label each weld of a set by number. See Fig. 9–15.

(*b*) Estimate the primary shear stress τ'. As shown in Fig. 9–14, each plate is welded to the channel by means of three 6-mm fillet welds. Figure 9–15 shows that we have divided the load in half and are considering only a single plate. From case 4 of Table 9–2 we find the throat area as

$$A = 0.707(6)\,[2(56) + 190] = 1280 \text{ mm}^2$$

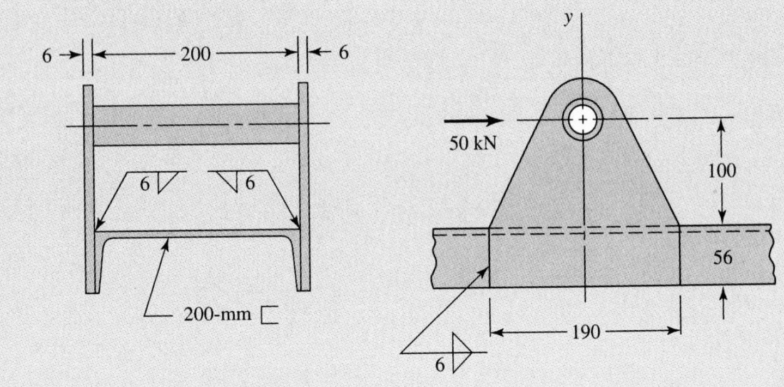

Figure 9–14

Dimensions in millimeters.

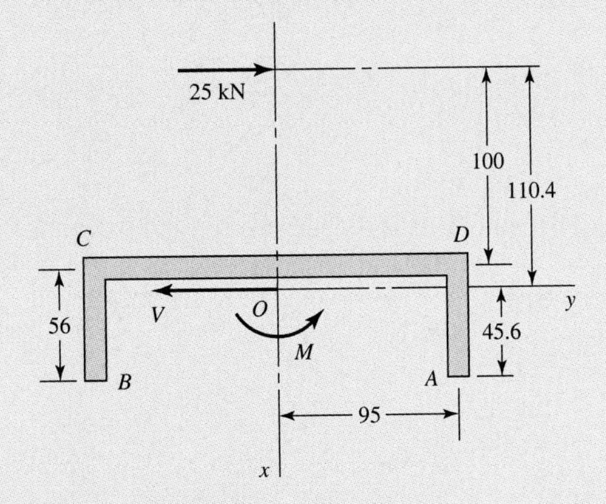

Figure 9–15

Diagram showing the weld geometry; all dimensions in millimeters. Note that *V* and *M* represent loads applied by the welds *to the plate*.

3. We are indebted to Professor George Piotrowski of the University of Florida for the detailed steps, presented here, of his method of weld analysis. J.E.S., C.R.M.

Then the primary shear stress is

$$\tau' = \frac{V}{A} = \frac{25(10)^3}{1280} = 19.5 \text{ MPa}$$

(*c*) Draw the τ' stress, to scale, at each lettered corner or end. See Fig. 9–16.
(*d*) Locate the centroid of the weld pattern. Using case 4 of Table 9–2, we find

$$\bar{x} = \frac{(56)^2}{2(56) + 190} = 10.4 \text{ mm}$$

This is shown as point O on Figs. 9–15 and 9–16.
(*e*) Find the distances r_i (see Fig. 9–16):

$$r_A = r_B = \left[(190/2)^2 + (56 - 10.4)^2\right]^{1/2} = 105 \text{ mm}$$

$$r_C = r_D = \left[(190/2)^2 + (10.4)^2\right]^{1/2} = 95.6 \text{ mm}$$

These distances can also be scaled from the drawing.
(*f*) Find J. Using case 4 of Table 9–2 again, we get

$$J = 0.707(6)\left[\frac{8(56)^3 + 6(56)(190)^2 + (190)^3}{12} - \frac{(56)^4}{2(56) + 190}\right]$$
$$= 7.07(10)^6 \text{ mm}^4$$

(*g*) Find M:

$$M = Fl = 25(100 + 10.4) = 2760 \text{ N} \cdot \text{m}$$

(*h*) Estimate the secondary shear stresses τ'' at each lettered end or corner:

$$\tau''_A = T''_B = \frac{Mr}{J} = \frac{2760(10)^3(105)}{7.07(10)^6} = 41.0 \text{ MPa}$$

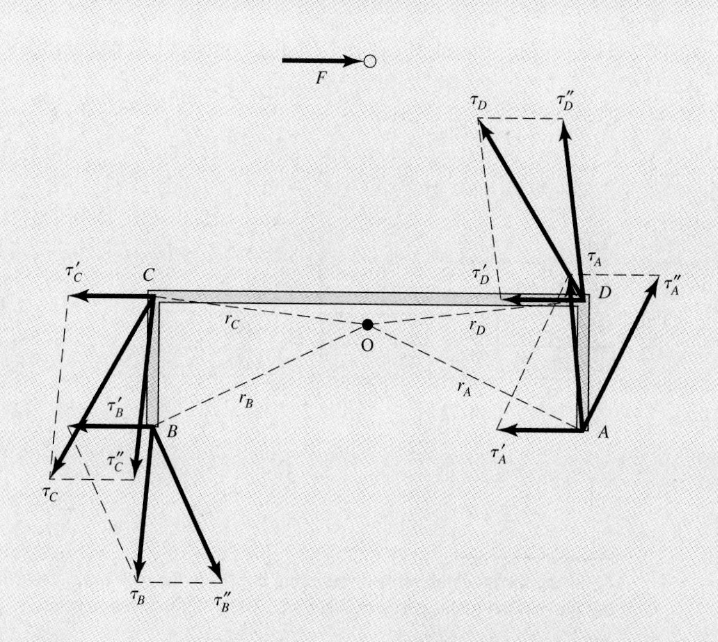

Figure 9–16

Free-body diagram of one of the side plates.

$$\tau''_C = \tau''_D = \frac{2760(10)^3(95.6)}{7.07(10)^6} = 37.3 \text{ MPa}$$

(*i*) Draw the τ'' stress, to scale, at each corner and end. See Fig. 9–16. Note that this is a free-body diagram of one of the side plates, and therefore the τ' and τ'' stresses represent what the channel is doing to the plate (through the welds) to hold the plate in equilibrium.
(*j*) At each letter, combine the two stress components as vectors. This gives

$$\tau_A = \tau_B = 37 \text{ MPa}$$

$$\tau_C = \tau_D = 44 \text{ MPa}$$

(*k*) Identify the most highly stressed point:

Answer $\tau_{\max} = \tau_C = \tau_D = 44 \text{ MPa}$

9–4 Stresses in Welded Joints in Bending

Figure 9–17*a* shows a cantilever welded to a support by fillet welds at top and bottom. A free-body diagram of the beam would show a shear-force reaction V and a moment reaction M. The shear force produces a primary shear in the welds of magnitude

$$\tau' = \frac{V}{A} \tag{a}$$

where A is the total throat area.

The moment M induces a throat shear stress component of 0.707τ in the welds.[4] Treating the two welds of Fig. 9–17*b* as lines we find the unit second moment of area to be

$$I_u = \frac{bd^2}{2} \tag{b}$$

The second moment of area I based on weld throat area is

$$I = 0.707hI_u = 0.707h\frac{bd^2}{2} \tag{c}$$

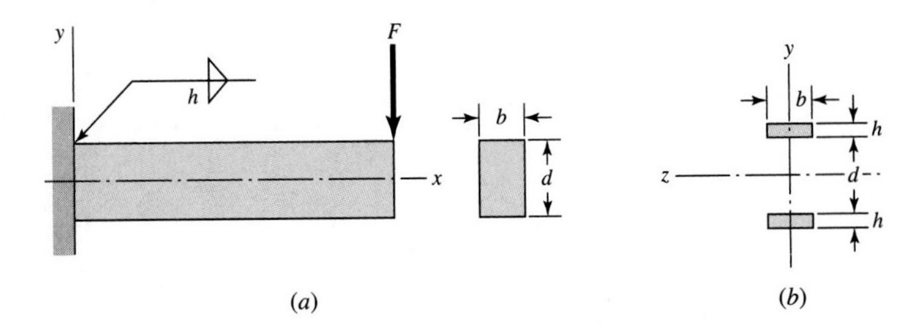

Figure 9–17
A rectangular cross-section cantilever welded to a support at the top and bottom edges.

4. Using the model described preceding Eq. (9–3) the moment is carried by components of the shear stress 0.707τ parallel to the x-axis of Fig. 9–17. The y-components cancel.

The nominal throat sheat stress is now found to be

$$\tau = \frac{Mc}{I} = \frac{Md/2}{0.707hbd^2/2} = \frac{1.414M}{bdh} \qquad (d)$$

The model gives the coefficient of 1.414, in contrast to the predictions of Sec. 9–2 of 1.197 from distortion energy, or 1.207 from maximum shear. The conservatism of the model's 1.414 is not that it is simply larger than either 1.196 or 1.207, but the tests carried out to validate the model show that it is large enough.

The second moment of area in Eq. (*d*) is based on the distance *d* between the two welds. If this moment is found by treating the two welds as having rectangular footprints, the distance between the weld throat centroids is approximately $(d + h)$. This would produce a slightly larger second moment of area, and result in a smaller level of stress. This method of treating welds as a line does not interfere with the conservatism of the model. It also makes Table 9–3 possible with all the conveniences that ensue.

Table 9–3

Bending Properties of Fillet Welds*

Weld	Throat Area	Location of G	Unit Second Moment of Area
G, d, $\bar{y}$	$A = 0.707hd$	$\bar{x} = 0$ $\bar{y} = d/2$	$I_u = \frac{d^3}{12}$
b, G, d, $\bar{x}$, $\bar{y}$	$A = 1.414hd$	$\bar{x} = b/2$ $\bar{y} = d/2$	$I_u = \frac{d^3}{6}$
b, G, d, $\bar{x}$, $\bar{y}$	$A = 1.414hb$	$\bar{x} = b/2$ $\bar{y} = d/2$	$I_u = \frac{bd^2}{2}$
b, G, d, $\bar{x}$, $\bar{y}$	$A = 0.707h(2b + d)$	$\bar{x} = \frac{b^2}{2b + d}$ $\bar{y} = d/2$	$I_u = \frac{d^2}{12}(6b + d)$

Table 9–3

Continued

Weld	Throat Area	Location of G	Unit Second Moment of Area
	$A = 0.707h(b + 2d)$	$\bar{x} = b/2$ $\bar{y} = \dfrac{d^2}{b + 2d}$	$I_u = \dfrac{2d^3}{3} - 2d^2\bar{y} + (b + 2d)\bar{y}^2$
	$A = 1.414h(b + d)$	$\bar{x} = b/2$ $\bar{y} = d/2$	$I_u = \dfrac{d^2}{6}(3b + d)$
	$A = 0.707h(b + 2d)$	$\bar{x} = b/2$ $\bar{y} = \dfrac{d^2}{b + 2d}$	$I_u = \dfrac{2d^3}{3} - 2d^2\bar{y} + (b + 2d)\bar{y}^2$
	$A = 1.414h(b + d)$	$\bar{x} = b/2$ $\bar{y} = d/2$	$I_u = \dfrac{d^2}{6}(3b + d)$
	$A = 1.414\pi hr$		$I_u = \pi r^3$

*I_u, unit second moment of area, is taken about a horizontal axis through G, the centroid of the weld group, h is weld size; the plane of the bending couple is normal to the plane of the paper and parallel to the y-axis; all welds are of the same size.

9–5 The Strength of Welded Joints

The matching of the electrode properties with those of the parent metal is usually not so important as speed, operator appeal, and the appearance of the completed joint. The properties of electrodes vary considerably, but Table 9–4 lists the minimum properties for some electrode classes.

It is preferable, in designing welded components, to select a steel that will result in a fast, economical weld even though this may require a sacrifice of other qualities such as machinability. Under the proper conditions, all steels can be welded, but best results

Table 9–4

Minimum Weld-Metal Properties

AWS Electrode Number*	Tensile Strength kpsi (MPa)	Yield Strength, kpsi (MPa)	Percent Elongation
E60xx	62 (427)	50 (345)	17–25
E70xx	70 (482)	57 (393)	22
E80xx	80 (551)	67 (462)	19
E90xx	90 (620)	77 (531)	14–17
E100xx	100 (689)	87 (600)	13–16
E120xx	120 (827)	107 (737)	14

*The American Welding Society (AWS) specification code numbering system for electrodes. This system uses an E prefixed to a four- or five-digit numbering system in which the first two or three digits designate the approximate tensile strength. The last digit includes variables in the welding technique, such as current supply. The next-to-last digit indicates the welding position, as, for example, flat, or vertical, or overhead. The complete set of specifications may be obtained from the AWS upon request.

will be obtained if steels having a UNS specification between G10140 and G10230 are chosen. All these steels have a tensile strength in the hot-rolled condition in the range of 60 to 70 kpsi.

The designer can choose factors of safety or permissible working stresses with more confidence if he or she is aware of the values of those used by others. One of the best standards to use is the American Institute of Steel Construction (AISC) code for building construction.[5] The permissible stresses are now based on the yield strength of the material instead of the ultimate strength, and the code permits the use of a variety of ASTM structural steels having yield strengths varying from 33 to 50 kpsi. Provided the loading is the same, the code permits the same stress in the weld metal as in the parent metal. For these ASTM steels, $S_y = 0.5S_u$. Table 9–5 lists the formulas specified by the code for calculating these permissible stresses for various loading conditions. The factors of safety implied by this code are easily calculated. For tension, $n = 1/0.60 = 1.67$. For shear, $n = 0.577/0.40 = 1.44$, using the distortion-energy theory as the criterion of failure.

It is important to observe that the electrode material is often the strongest material present. If a bar of AISI 1010 steel is welded to one of 1018 steel, the weld metal is actually a mixture of the electrode material and the 1010 and 1018 steels. Furthermore, a welded cold-drawn bar has its cold-drawn properties replaced with the hot-rolled properties in the vicinity of the weld. Finally, remembering that the weld metal is usually the strongest, do check the stresses in the parent metals.

Table 9–5

Stresses Permitted by the AISC Code for Weld Metal

Type of Loading	Type of Weld	Permissible Stress	n^*
Tension	Butt	$0.60S_y$	1.67
Bearing	Butt	$0.90S_y$	1.11
Bending	Butt	$0.60–0.66S_y$	1.52–1.67
Simple compression	Butt	$0.60S_y$	1.67
Shear	Butt or fillet	$0.30S_{ut}^{\dagger}$	

*The factor of safety n has been computed using the distortion-energy theory.
†Shear stress on base metal will not exceed $0.40S_y$ of base metal.

5. For a copy, write the AISC, 400 N. Michigan Ave., Chicago, IL 60611.

Table 9–6

Fatigue Stress-Concentration Factors, K_{fs}

Type of Weld	K_{fs}
Reinforced butt weld	1.2
Toe of transverse fillet weld	1.5
End of parallel fillet weld	2.7
T-butt joint with sharp corners	2.0

The AISC code, as well as the AWS code, for bridges includes permissible stresses when fatigue loading is present. The designer will have no difficulty in using these codes, but their empirical nature tends to obscure the fact that they have been established by means of the same knowledge of fatigue failure already discussed in Chap. 7. Of course, for structures covered by these codes, the actual stresses *cannot* exceed the permissible stresses; otherwise the designer is legally liable. But in general, codes tend to conceal the actual margin of safety involved.

The fatigue stress-concentration factors listed in Tables 9–1 and 9–6 are suggested for use. These factors should be used for the parent metal as well as for the weld metal. Table 9–7 gives steady-load information and minimum fillet sizes.

9–6 Specification Set, Adequacy Assessment, and Decision Set

Specification Set

Adequacy assessment begins with the establishment of the specification set used to decide whether the welded joint is satisfactory. The specification set for a welded joint includes

- Pattern of weldment (distribution including b and d)
- Electrode identification
- Type of weld
- Length of weld bead l
- Leg size h

The pattern may be described as a configuration such as depicted in Table 9–2, 9–3, or 9–8. It can be described on a drawing by one or more standard welding symbols such as those described in Figs. 9–1 to 9–6. Electrode identification can be made using the American Welding Society specification code numbering system for electrodes. The type of weld—bead, fillet, plug or slot, or groove—can be included in the welding symbol or stated elsewhere. The length of the weld can be implied by the welding symbol (this side only, both sides, pitch, or weld-all-around) or stated otherwise. The leg size can be stated in the weld symbol or elsewhere.

Adequacy Assessment

Method of Philon

When the factor of safety n equals or exceeds the design factor n_d—that is, when $n \geq n_d$—the design is satisfactory with respect to strength. Historically, minimum values of yield strength S_y, ultimate tensile strength S_{ut}, and ultimate shear strength estimated by $S_{su} = 0.67 S_{ut}$ are used mainly because they are available for electrode materials and for many structural steels. Mean values of properties are available to designers only

Table 9–7

Allowable Steady Loads and Minimum Fillet Weld Sizes

Schedule A: Allowable Load for Various Sizes of Fillet Welds

	Strength Level of Weld Metal (EXX)						
	60*	70*	80	90*	100	110*	120
	Allowable shear stress on throat, ksi (1000 psi) of fillet weld or partial penetration groove weld						
$\tau =$	18.0	21.0	24.0	27.0	30.0	33.0	36.0
	Allowable Unit Force on Fillet Weld, kip/linear in						
†$f =$	$12.73h$	$14.85h$	$16.97h$	$19.09h$	$21.21h$	$23.33h$	$25.45h$
Leg Size h, in	Allowable Unit Force for Various Sizes of Fillet Welds kip/linear in						
1	12.73	14.85	16.97	19.09	21.21	23.33	25.45
7/8	11.14	12.99	14.85	16.70	18.57	20.41	22.27
3/4	9.55	11.14	12.73	14.32	15.92	17.50	19.09
5/8	7.96	9.28	10.61	11.93	13.27	14.58	15.91
1/2	6.37	7.42	8.48	9.54	10.61	11.67	12.73
7/16	5.57	6.50	7.42	8.35	9.28	10.21	11.14
3/8	4.77	5.57	6.36	7.16	7.95	8.75	9.54
5/16	3.98	4.64	5.30	5.97	6.63	7.29	7.95
1/4	3.18	3.71	4.24	4.77	5.30	5.83	6.36
3/16	2.39	2.78	3.18	3.58	3.98	4.38	4.77
1/8	1.59	1.86	2.12	2.39	2.65	2.92	3.18
1/16	0.795	0.930	1.06	1.19	1.33	1.46	1.59

*Fillet welds actually tested by the joint AISC-AWS Task Committee.

†$f = 0.707h\, \tau_{\text{all}}$.

Schedule B: Minimum Fillet Weld Size, h

Material Thickness of Thicker Part Joined, in		Weld Size, in
*To 1/4 incl.		1/8
Over 1/4	To 1/2	3/16
Over 1/2	To 3/4	1/4
†Over 3/4	To 1 1/2	5/16
Over 1 1/2	To 2 1/4	3/8
Over 2 1/4	To 6	1/2
Over 6		5/8

Not to exceed the thickness of the thinner part.
†For minimum fillet weld size, schedule does not go above 5/16 in fillet weld for every 3/4 in material.
*Minimum size for bridge application does not go below 3/16 in.

Source: Adapted from Omer W. Blodgett (ed.), *Stress Allowables Affect Weldment Design*, D412, The James F. Lincoln Arc Welding Foundation, Cleveland, May 1991, p. 3.

Table 9–8

AISC Fatigue Allowables *Source:* Adapted from *AISC Manual,* Appendix K, 9th ed., 1989 in Omer W. Blodgett (ed.), *Stress Allowables Affect Weldment Design,* D412, The James F. Lincoln Arc Welding Foundation, Cleveland, May 1991, p. 4, 5.

Schedule A: Welded Joint Configurations 1–26 Showing Categories A–F

① (1)

Base metal — no attachments — rolled or clean surfaces

(A)

② (15)

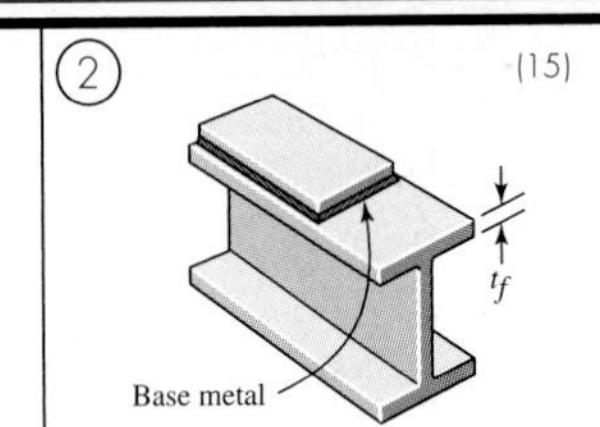

Base metal at ends of partial length welded cover plates narrower than the flange — square or tapered ends — with or without welds across the ends.

$t_f \le 0.8$ in. (E)
$t_f > 0.8$ in. (E')

③ (25)

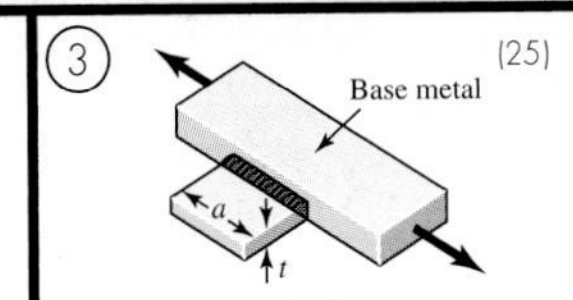

Longitudinal loading
Base metal — full penetration groove weld
Weld termination ground smooth
Weld reinforcement not removed
Not necessarily equal thickness

$2 < a < 12$ b or 4 in. (D)
$a > 12$ b or 4 in. when $b \le 1$ in. (E)
$a > 12$ b or 4 in. when $b > 1$ in. (E')

④ (19)

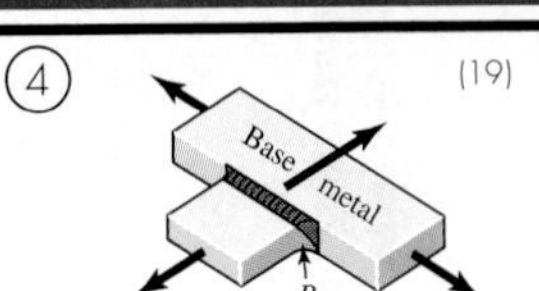

Base metal — full penetration groove weld.
Weld termination ground smooth.
Longitudinal load.
If also transverse load, weld must be inspected by radiography or ultrasound.

$R > 24$ in. (B)
24 in. $> R >$ 6 in. (C)
6 in. $> R >$ 2 in. (D)
2 in. $> R$ (E)

⑨ (2)

(B)

Base metal — built up plates or shapes — connected by continuous complete penetration groove welds or fillet welds — without attachments. Note: don't use this as a fatigue allowable for the fillet weld to transfer a load — this would be (F).

⑩ (6)

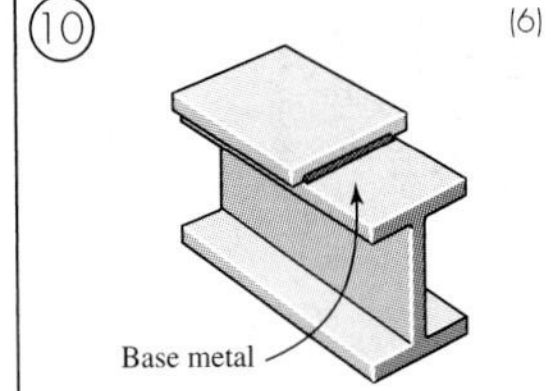

Base metal at ends of partial length welded cover plates wider than the flange.

Weld across end. $t_f \le 0.8$ in. (E)
$t_f > 0.8$ in. (E')

No weld across end (E').

⑪ (26)

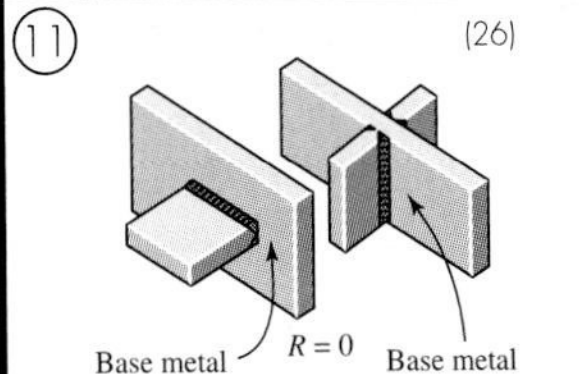

Base metal at detail attached by fillet welds.
Longitudinal loading

$a < 2$ in. (C)
2 in. $< a < 12$ b (D)
$a > 12$ b or 4 in. when $b \le 1$ in. (E)
$a > 12$ b or 4 in. when $b > 1$ in. (E')

⑫ (27)

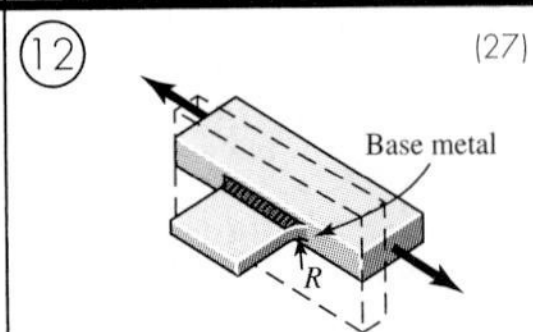

Base metal attached by partial penetration groove welds.
Longitudinal loading.
Weld termination ground smooth
Weld reinforcement left on
Using a transition radius of
$R > 2$ in. (D)
$R \le 2$ in. (E)

⑮ (3)

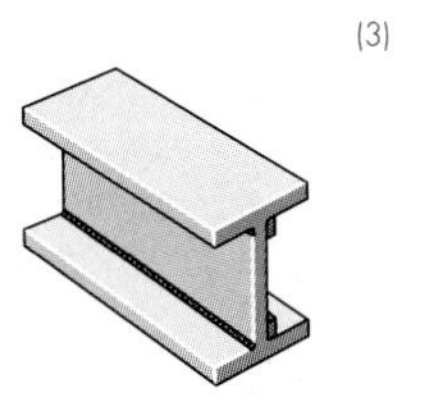

Base metal — built up plates or shapes — connected by continuous complete penetration groove welds with backing bars not removed — or partial penetration groove welds.

(B')

⑯ (4)

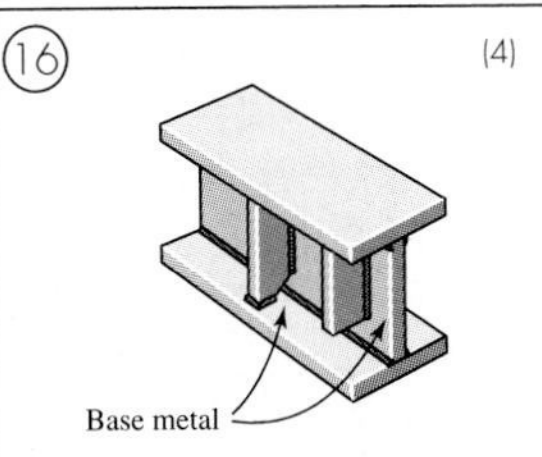

Base metal — at end of welds on transverse stiffeners to web and flanges of girder.

(C)

⑰ (26)

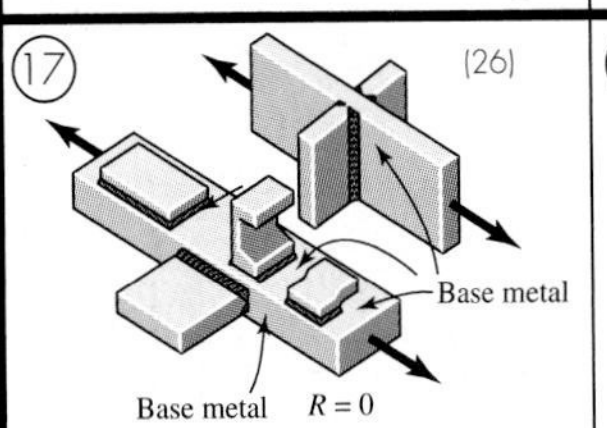

Base metal at detail attached by fillet welds.
Longitudinal loading

$a < 2$ in. (C)
2 in. $< a < 12$ b (D)
$a > 12$ b or 4 in. when $b \le 1$ in. (E)
$a > 12$ b or 4 in. when $b > 1$ in. (E')

⑱ (27)

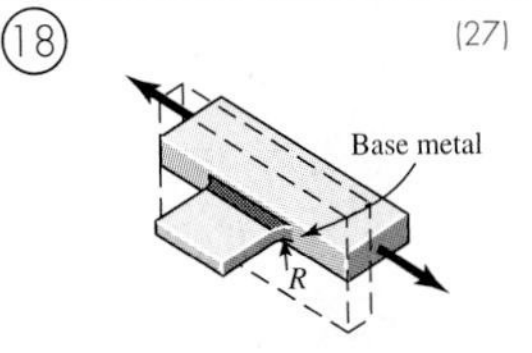

Base metal attached by fillet welds
Longitudinal loading
Weld termination ground smooth
Using a transition radius of
$R > 2$ in. (D)
$R \le 2$ in. (E)

㉑ (7) (8)

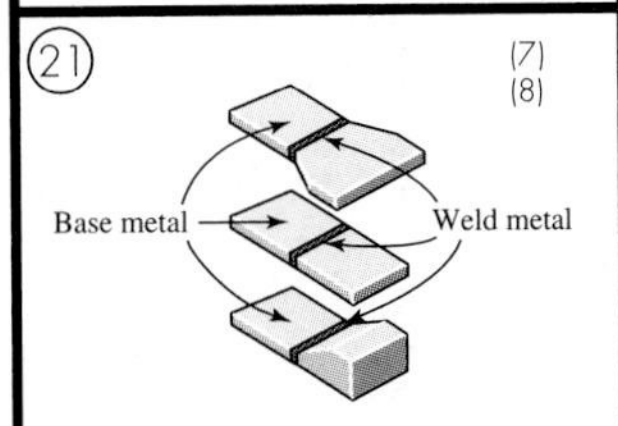

Base metal and weld metal at full penetration groove welds — changes in thickness or width not to exceed a slope of 1 in. 2½ (22°). Ground flush and inspected by radiography or ultrasound (B).

For A514 steel (B')

㉒ (9)

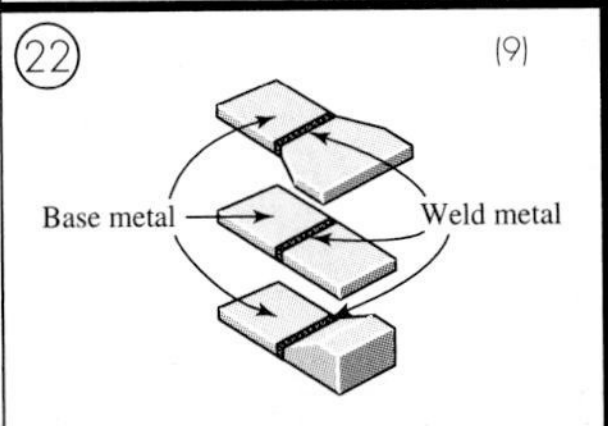

Base metal and weld metal at full penetration groove welds — changes in thickness or width not to exceed a slope of 1 in. 2½ (22°). Weld reinforcement not removed. Inspected by radiography or ultrasound.

(C)

㉓ (12)

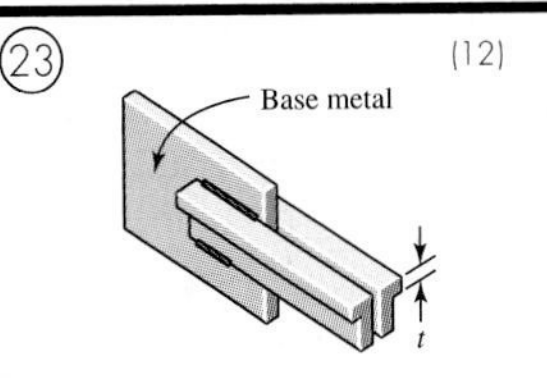

Base metal at junction of axially loaded members with fillet welded end connections. Welds balanced about C.G. of member.

$t_f \le 0.8$ in. (E)
$t_f > 0.8$ in. (E')

㉔ (13)

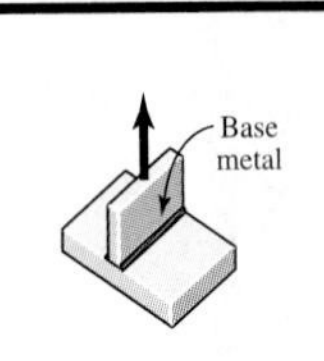

Base metal connected with transverse fillet welds.

$t_f \le 1/2$ in. (C)
$t_f > 1/2$ in. (C)

Table 9–8

Continued

Schedule A: Welded Joint Configurations 1–26 Showing Categories A–F

(5) (20) (21)

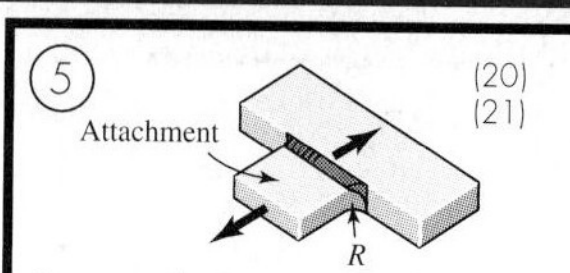

Transverse loading
Attachment — full penetration groove weld.
Weld termination ground smooth.
Equal thickness

	Weld reinforcement removed	Weld reinforcement not removed
R > 24 in.	B	C
24 in. > R > 6 in.	C	C
6 in. > R > 2 in.	D	D
2 in. > R	E	E

(6) (22) (23)

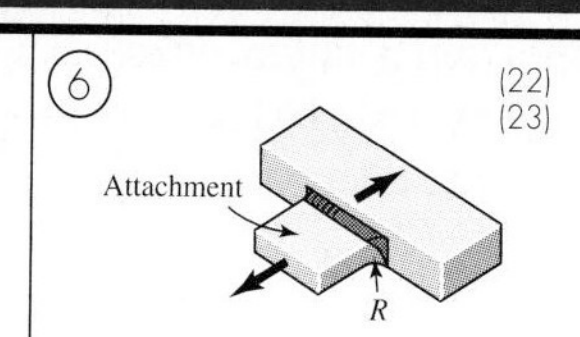

Transverse loading
Attachment — full penetration groove weld.
Weld termination ground smooth.
Unequal thickness
Weld reinforcement removed
R > 2 in. (D)
R ≤ 2 in. (E)

Weld reinforcement not removed
any R (E)

(7) (24)

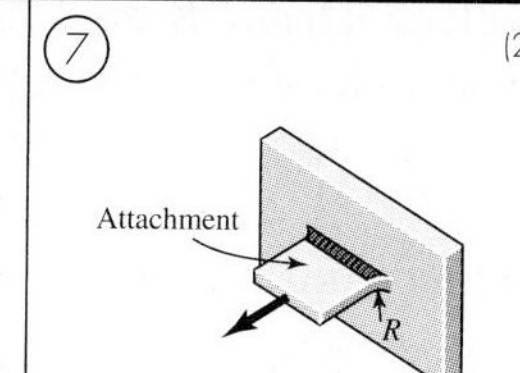

Transverse loading
Attachment — full penetration groove weld.
Weld termination ground smooth
Weld reinforcement not removed
R > 6 in. (C)
6 in. > R > 2 in. (D)
R ≤ 2 in. (E)

(8) (28)

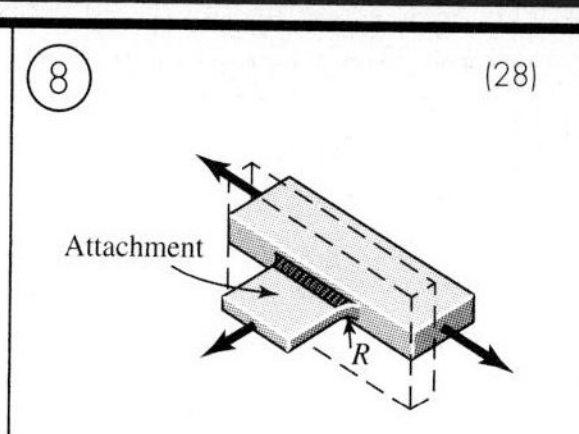

Attachment —Transverse load
Fillet welds
Weld termination ground smooth
Main material longitudinal loaded
Using a transition radius of
R > 2 in. (D)
R ≤ 2 in. (E)

(13) (29) (30)

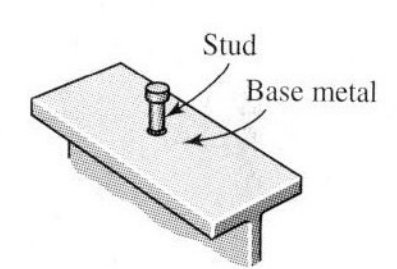

Base metal at stud — type shear connector — fillet weld attachment or automatic end weld. (C)

Shear stress on nominal area of stud-type shear connectors. (F)

(14) (15)

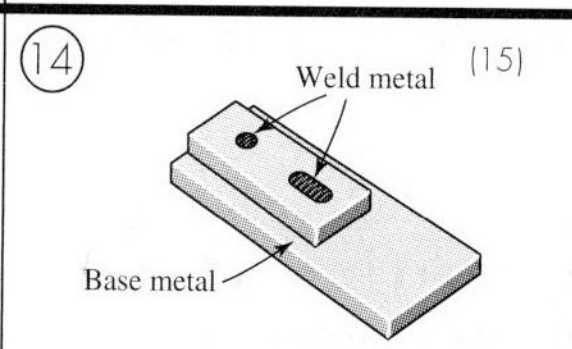

Base metal at plug or slot welds.
(E)

Shear on plug or shot welds.
(F)

(19) (11)

Base metal

Base metal

Base metal at intermittent fillet welds.

(E)

(20) (16) (17) (18)

Base metal at *gross section* of high strength bolted *slip-critical* connections
— No out of plane bending. (B)

Base metal at *net section* of fully tensioned high strength, bolted *bearing* connections. (B)

Base metal at net section of other mechanical fastened joints. (D)

(25) (14)

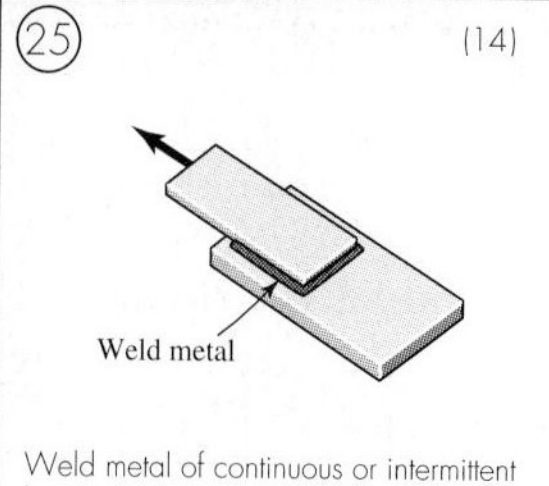

Weld metal of continuous or intermittent longitudinal or transverse fillet welds.

(F)

(26) (10)

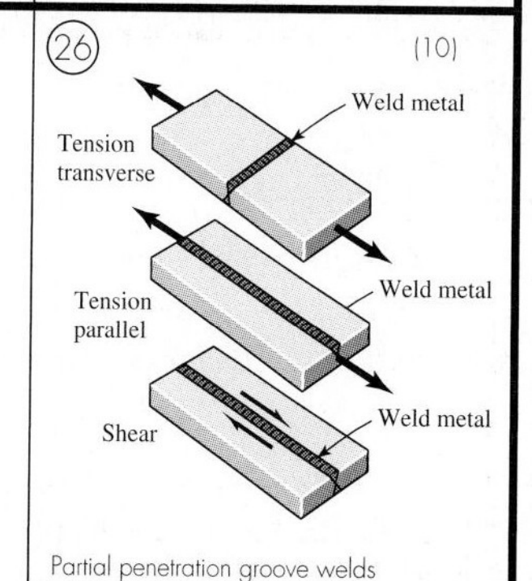

Partial penetration groove welds

Schedule B: AISC Allowable Fatigue Stress (σ_{sr} and τ_{sr}), kpsi

Category (from Table A-K4.2)	Allowable Stress Range, Ksi: 20,000 to 100,000 ~	100,000 to 500,000 ~	500,000 to 2×10^6 ~	Over 2×10^6 ~
A	63	37	24	24
B	49	29	18	16
B'	39	23	15	12
C	35	21	13	10[a]
D	28	16	10	7
E	22	13	8	5
E'	16	9	6	3
F	15	12	9	8

[a] Flexural stress range of 12 ksi permitted at toe of stiffener welds on flanges.

allowable fatigue stress $\boxed{\sigma_{max} = \dfrac{\sigma_{sr}}{1-K}}$ for normal stress σ

$\boxed{\tau_{max} = \dfrac{\tau_{sr}}{1-K}}$ for shear stress τ

but shall not exceed steady allowables

σ_{max} or τ_{max} = maximum allowable fatigue stress
σ_{sr} or τ_{sr} = allowable range of stress from table

$$K = \frac{\sigma_{min}}{\sigma_{max}} = \frac{M_{min}}{M_{max}} = \frac{F_{min}}{F_{max}} = \frac{\tau_{min}}{\tau_{max}} = \frac{V_{min}}{V_{max}}$$

S = shear
T = tension
R = reversal
M = stress in metal
W = stress in weld

$$\sigma_{sr} = \sigma_{sr}^c \left(\frac{0.71 - 0.65\frac{2a}{t_p} + 0.79\frac{w}{t_p}}{1.10\, t_p^{1/6}} \right)$$

for fillet welds $\frac{2a}{t_p} = 1.0$

$$\sigma_{sr} = \sigma_{sr}^c \left(\frac{0.06 + 0.79\frac{w}{t_p}}{1.10\, t_p^{1/6}} \right)$$

but $\sigma_{xr} \le \sigma_{xr}^c$

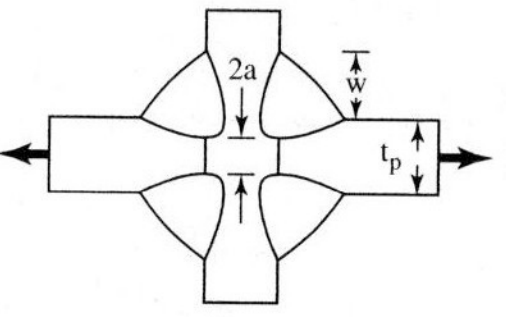

σ_{sr} = stress range for this condition
σ_{sr}^c = stress range for category (C)
w = leg size of fillet weld
2a = lack of penetration into joint
t_p = plate thickness

by test. Factor-of-safety definitions are "bent" (minimum strength divided by $\theta = 45°$ weld throat mathematical model shear-stress predictions at the critical location) and have meaning only relative to each other. The design factor n_d is selected by the designer (or corporate design manual) in light of this and thus is experiential, personal, or corporate. In fatigue situations the DE-Goodman failure locus has a history of common use. In fatigue situations, electrode properties can be taken into account. See Table 9–4. Member adequacy under loading is part of the adequacy assessment.

Welding Code Method

In static loading the allowable shear stress τ_{all} is compared to the extant shear stress that is predicted by the $\theta = 45°$ mathematical model at the critical location. If $\tau < \tau_{\text{all}}$, the joint is considered satisfactory for strength of the weld. At the equality $\tau = \tau_{\text{all}}$ there is an unannounced factor of safety contained in the code. That factor of safety, too, is experiential from a national performance base. It has been adjusted over the years as more satisfactory performance has been observed, and adjustments have been incorporated as changes in the code. Table 9–4 gives strength properties for several electrode materials. Table 9–7 is a guide to fillet size.

In fatigue problems, the variable-stress descriptive components are stress range σ_r and K factor, $\sigma_{\min}/\sigma_{\max}$, rather than stress amplitude and steady (midrange) stress components. Table 9–8 defines the K factor in five convenient ways. By context, the stresses are free of fatigue stress-concentration factors (where applicable). Useful relations between shear-stress amplitude τ_a, steady stress τ_m, maximum shear stress $\tau_{\max}$, and minimum shear stress $\tau_{\min}$ are

$$\tau_{\max} = \tau_a + \tau_m \qquad \tau_{\min} = -(\tau_a - \tau_m)$$

or

$$\sigma_{\max} = \sigma_a + \sigma_m \qquad \sigma_{\min} = -(\sigma_a - \sigma_m)$$

broadening the definition of K factor to

$$K = \frac{\sigma_{\min}}{\sigma_{\max}} = \frac{M_{\min}}{M_{\max}} = \frac{F_{\min}}{F_{\max}} = \frac{V_{\min}}{V_{\max}} = \frac{\tau_{\min}}{\tau_{\max}} = -\frac{\sigma_a - \sigma_m}{\sigma_a + \sigma_m} = -\frac{\tau_a - \tau_m}{\tau_a + \tau_m} \tag{9–8}$$

Schedule B in Table 9–8 lists categories A through F and displays allowable stress range σ_{sr} or τ_{sr} for four life ranges. The allowable maximum stress is given by

$$(\sigma_{\max})_{\text{all}} = \frac{\sigma_{sr}}{1 - K} \qquad \text{or} \qquad (\tau_{\max})_{\text{all}} = \frac{\tau_{sr}}{1 - K} \tag{9–9}$$

This allowable maximum stress is compared to the extant maximum stress at the critical location:

$$\sigma_{\max} \le (\sigma_{\max})_{\text{all}} \qquad \text{or} \qquad \tau_{\max} \le (\tau_{\max})_{\text{all}} \tag{9–10}$$

When this inequality is met the design is satisfactory as to weld fatigue strength. The code does not respond to electrode identity, or properties of the base metal. The fatigue locus is probably DE-Goodman and contains an unannounced factor of safety.

Decision Set

The designer uses the decision set during the synthesis process. Since it is often identical to the specification set, or only slightly different, the path from the decision set of the final design to the corresponding specification set is easily found. The decision set can be

- Pattern of weldment
- Electrode identification
- Type of weld
- Length of weld

} a priori decisions

- Let size h — design variable

or a small variation of it. For design decisions to be made it is necessary to have a robust figure of merit. Often it is the negative of the cost of the weld, including fixtures, jigging, material, labor, and overhead costs. These are usually not available in an instructional situation, so our surrogate will be the negative of the volume of material laid down.

Welding codes focus on steel-framed structures and weldments. They are not static documents. Updates include new processes, higher strength levels, higher allowed stresses, and additional steels. They recognize developments and experience in welded steel design. Codes represent living documents and should be at hand whenever design work is done. The codes include drawings such as those in Table 9–8, as well as much explanatory information and qualifying text, which is not reproduced in this book.

In many structural applications the electrode metal is the strongest metal appearing in the joint. Member properties often control the strength of the weldment. In allowed fatigues, the electrode metal is not even mentioned. Welding codes do influence the design of machinery, machine tools, machine elements, and agricultural equipment.

In machinery applications, member materials are often chosen for properties that are not structural and are not addressed by the codes. The mechanical designer needs a methodology independent of the welding codes and should be prepared to use or not use the codes, as the situation requires. In this book we analyze and design weldments either way.

9–7 Static Loading

Some examples of statically loaded joints are useful in comparing and contrasting the conventional method of analysis and the welding code methodology.

EXAMPLE 9–2 A 1/2-in by 2-in rectangular-cross-section 1015 bar carries a static load of 16.5 kip. It is welded to a gusset plate with a 3/8-in fillet weld 2 in long on both sides with an E70XX electrode as depicted in Fig. 9–18. Use the welding code method.

(a) Is the weld metal strength satisfactory?

(b) Is the attachment strength satisfactory?

Figure 9–18

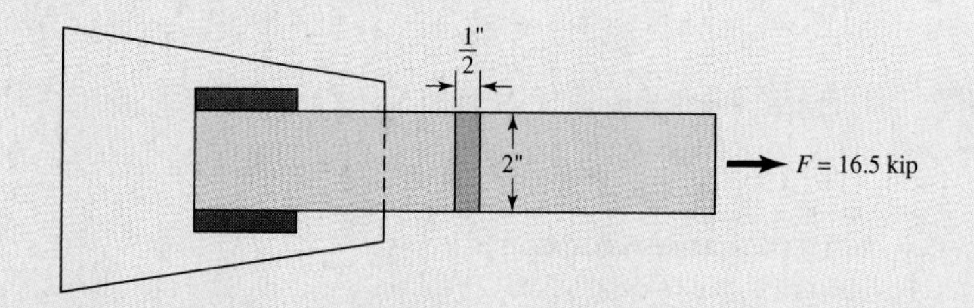

Solution (*a*) From Table 9–7, allowable force per unit length for a 3/8-in E70 electrode metal is 5.57 kip/in of weldment; thus

$$F = 5.57l = 5.57(4) = 22.28 \text{ kip}$$

Since 22.28 > 16.5 kip, weld metal strength is satisfactory.

(*b*) Check shear in attachment adjacent to the welds. From Table 9–5 and Table E–20, from which $(S_y)_a = 27.5$ kpsi,

$$\tau_{\text{all}} = 0.4\left(S_y\right)_a = 0.4(27.5) = 11 \text{ kpsi}$$

The shear stress τ on the base metal adjacent to the weld is

$$\tau = \frac{F}{2hl} = \frac{16.5}{2(0.375)2} = 11 \text{ kpsi}$$

Since $\tau_{\text{all}} \geq \tau$, that is, $11 \geq 11$, attachment is satsifactory near the weld beads. The tensile stress in the shank of the attachment σ is

$$\sigma = \frac{F}{tl} = \frac{16.5}{(1/2)2} = 16.5 \text{ kpsi}$$

The allowable tensile stress σ_{all}, from Table 9–5, is 0.6 and y preserving welding code safety level,

$$\sigma_{\text{all}} = 0.6\left(S_y\right)_a = 0.6(27.5) = 16.5 \text{ kpsi}$$

Since $\sigma_{\text{all}} \geq \sigma$, that is, $16.5 \geq 16.5$, shank tensile stress is satisfactory.

EXAMPLE 9–3

A specially rolled A36 structural steel section for the attachment has a cross section as shown in Fig. 9–19 and has a yield strength of 36 kpsi and an ultimate tensile strength of 58 kpsi. It is statically loaded through the attachment centroid by a load of $F = 48$ kip. Unsymmetrical weld tracks can compensate for eccentricity such that there is no moment to be resisted by the welds. Specify the weld track lengths l_1 and l_2 for a 5/16-in fillet weld using an E70XX electrode. This is part of a design problem in which the design variables include weld lengths and the fillet leg size.

Solution The y coordinate of the section centroid of the attachment is

$$\bar{y} = \frac{\sum y_i A_i}{\sum A_i} = \frac{1(0.75)2 + 3(2)0.375}{0.75(2) + 2(0.375)} = 1.67 \text{ in}$$

Figure 9–19

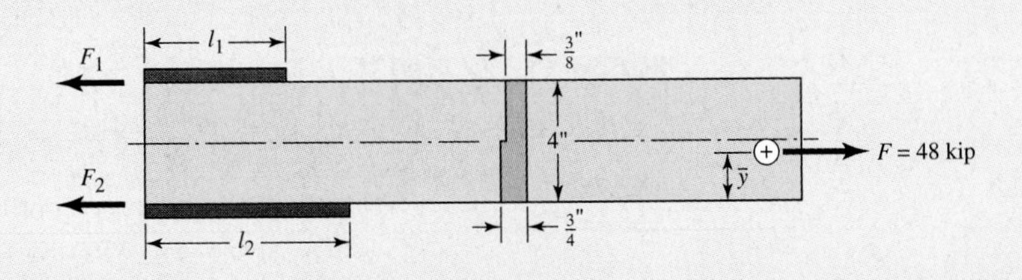

Summing moments about point B to zero gives

$$\sum M_B = 0 = -F_1 d + F\bar{y}$$

from which

$$F_1 = \frac{F\bar{y}}{d} \qquad F_2 = F - F_1$$

It follows that

$$F_1 = \frac{48(1.67)}{4} = 20.0 \text{ kip}$$

$$F_2 = 48 - 20.0 = 28.0 \text{ kip}$$

The weld throat areas have to be in the ratio $28/20 = 1.4$, that is, $l_2 = 1.4l_1$. The weld length design variables are coupled by this relation, so l_1 is the weld length design variable. The other design variable is the fillet weld leg size h, which has been decided by the problem statement. From Table 9–5 the allowable shear stress on the throat τ_{all} is

$$\tau_{\text{all}} = 0.3(70) = 21 \text{ kpsi}$$

The shear stress τ on the 45° throat is

$$\tau = \frac{F}{2(0.707)h\,(l_1 + l_2)} = \frac{F}{2(0.707)h\,(l_1 + 1.4l_1)}$$

$$= \frac{F}{2(0.707)h\,(2.4l_1)} = \tau_{\text{all}} = 21 \text{ kpsi}$$

from which the weld length l_1 is

$$l_1 = \frac{48}{21(2)0.707(0.3125)2.4} = 2.2 \text{ in}$$

and

$$l_2 = 1.4l_1 = 1.4(2.2) = 3.08 \text{ in}$$

These are the weld-bead lengths required by weld metal strength. The attachment shear stress allowable in the base metal, from Table 9–5, is

$$\tau_{\text{all}} = 0.4\left(S_y\right)_b = 0.4(36) = 14.4 \text{ kpsi}$$

The shear stress τ in the base metal adjacent to the weld is

$$\tau = \frac{F}{h\,(l_1 + l_2)} = \frac{F}{h\,(l_1 + 1.4l_1)} = \frac{F}{h\,(2.4l_1)} = \tau_{\text{all}} = 14.4 \text{ kpsi}$$

from which

$$l_1 = \frac{F}{14.4h(2.4)} = \frac{48}{14.4(0.3125)2.4} = 4.44 \text{ in}$$

$$l_2 = 1.4l_1 = 1.4(4.44) = 6.22 \text{ in}$$

These are the weld-bead lengths required by base metal (attachment) strength. The base metal controls the weld lengths. For the allowable tensile stress σ_{all} in the shank of the attachment, the AISC allowable for tension members is $0.6S_y$, therefore,

$$\sigma_{\text{all}} = 0.6\left(S_y\right)_b = 0.6(36) = 21.6 \text{ kpsi}$$

The nominal tensile stress σ is *uniform* across the attachment cross section because of the load application at the centroid. The stress σ is

$$\sigma = \frac{F}{A} = \frac{48}{0.75(2) + 2(0.375)} = 21.3 \text{ kpsi}$$

Since $\sigma_{\text{all}} \geq \sigma$, that is, $21.6 \geq 21.3$, the shank section is satisfactory. Note the attachment is almost at its load-carrying capacity. With l_1 set to a nominal 4 1/2 in, l_2 should be $1.4(4.5) = 6.30$ in.

Decision

Set $l_1 - 4\ 1/2$ in, $l_2 = 6\ 1/4$ in. The small magnitude of the departure from $l_2/l_1 = 1.4$ is not serious. The joint is essentially moment-free.

EXAMPLE 9–4

Perform an adequacy assessment of the statically loaded welded cantilevel carrying 500 lb depicted in Fig. 9–20. The cantilever is made of AISI 1018 HR steel and welded with a 3/8-in fillet weld as shown in the figure. An E6010 electrode was used, and the design factor was 3.0.
(*a*) Use the conventional method (of Philon) for the weld metal.
(*b*) Use the conventional method for the attachment (cantilever) metal.
(*c*) Use a welding code for the weld metal.

Solution

(*a*) From Table 9–4, $S_y = 50$ kpsi, $S_{ut} = 62$ kpsi. From Table 9–3, second pattern, $b = 0.375$ in, $d = 2$ in, so

$$A = 1.414hd = 1.414(0.375)2 = 1.06 \text{ in}^2$$
$$I_u = d^3/6 = 2^3/6 = 1.33 \text{ in}^3$$
$$I = 0.707hI_u = 0.707(0.375)1.33 = 0.353 \text{ in}^4$$

Primary shear:

$$\tau' = \frac{500\left(10^{-3}\right)}{2(0.707)0.375(2)} = 0.472 \text{ kpsi}$$

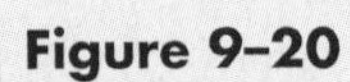
Figure 9–20

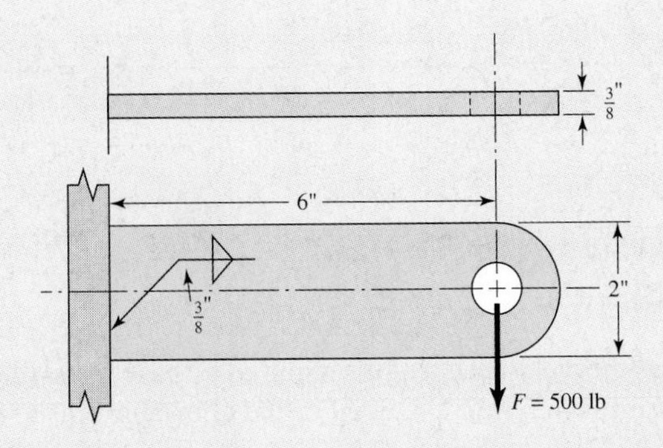

Secondary shear:

$$\tau'' = \frac{Mr}{I} = \frac{0.500(6)(1)}{0.353} = 8.50 \text{ kpsi}$$

The shear magnitude τ is the Pythagorean combination

$$\tau = \left(\tau'^2 + \tau''^2\right)^{1/2} = \left(0.472^2 + 8.50^2\right)^{1/2} = 8.51 \text{ kpsi}$$

The factor of safety based on a minimum strength is

Answer $$n = \frac{S_{sy}}{\tau} = \frac{0.577(50)}{8.51} = 3.39$$

Since $n \geq n_d$, that is, $3.39 \geq 3.0$, the weld metal has satisfactory strength.

(*b*) From Table E–20, minimum strengths are $S_{ut} = 58$ kpsi and $S_y = 32$ kpsi. Then

$$\sigma = \frac{M}{I/c} = \frac{M}{bd^2/6} = \frac{500\left(10^{-3}\right)6}{0.375\left(2^2\right)/6} = 12 \text{ kpsi}$$

$$n = \frac{S_y}{\sigma} = \frac{32}{12} = 2.67$$

Since $n < n_d$, that is, $2.67 < 3.0$, the joint is unsatisfactory as to the attachment strength.

(*c*) From part (*a*), $\tau = 8.51$ kpsi. For an E6010 electrode Table 9–7 give the allowable shear stress τ_{all} as 18 kpsi. Since $\tau < \tau_{\text{all}}$, that is, $8.51 < 18$, the weld is satisfactory. Since the code already has a design factor of $0.577(50)/18 = 1.6$ included at the equality, the corresponding factor of safety to part (*a*) is

Answer $$n = 1.6\frac{18}{8.51} = 3.39$$

which is consistent.

EXAMPLE 9–5 A 3/8-in-thick 2010 steel triangular attachment as depicted in Fig. 9–21*a* is to be welded to a 1/2-in 1015 plate in order to carry a 35-kip static load. Design the weldment using the welding code method.

Solution The decision set consists of the following a priori decisions:

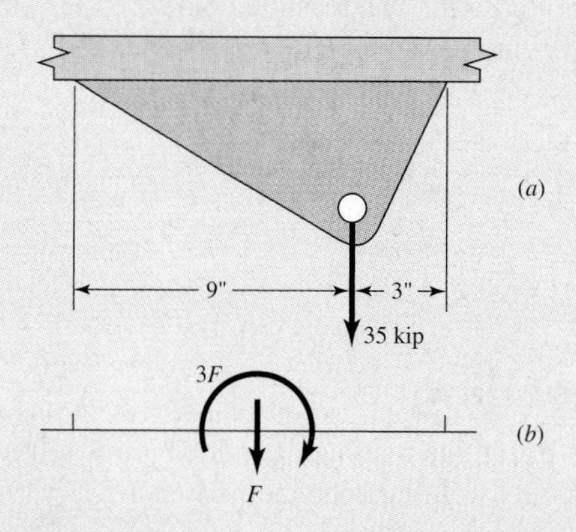

Figure 9–21

Triangular attachment to be welded to horizontal support. (*a*) Essential geometry; (*b*) static equivalent loading on the centroid of the weld-bead pattern.

- Pattern: Use a complete 12-in-long fillet weld on both sides of the attachment, that is, $b = 3/8$ in, $d = 12$ in, as shown in the second pattern of Table 9–3.
- Electrode: E6010.
- Type: Ordinary fillet weld.
- Length: $l = 12$ in on each of two fillets.

The design variable will be the leg size h. The figure of merit will be the negative of the material volume:

$$\text{fom} = -h^2(12)/2 = -6h^2$$

From Table 9–7, the inequality constraints on h are $0.25 \le h \le 0.375$ in with h confined to integer sixteenths. Additionally, unsatisfactory leg sizes as dictated by the code stress requirements further cut into feasible space and are not yet known. However, finding the first (least) satisfactory thickness locates the locals of the largest figure of merit. Alternatively, we could try leg thickness 1/4, 5/16, and 3/8 in, respectively, adopting the first satisfactory one. We choose the first alternative. From Table 9–3,

$$A = 2(0.707)12h = 17h \text{ in}^2$$
$$I_u = d^3/6 = 12^3/6 = 288 \text{ in}^3$$
$$I = 0.707 I_u h = 0.707(288)h = 203.6h$$

From the equivalent loading shown in Fig. 9–19*b*, the primary shear is

$$\tau' = \frac{F}{A} = \frac{35\,000}{17h} = \frac{2059}{h} \text{ psi}$$

The secondary shear is

$$\tau'' = \frac{Mc}{I} = \frac{3(35\,000)6}{203.6h} = \frac{3094}{h} \text{ psi}$$

The total shear is additive with the critical location at the right corner:

$$\tau = \tau' + \tau'' = \frac{2059}{h} + \frac{3094}{h} = \frac{5153}{h} \text{ psi}$$

From Table 9–7, for E6010 electrodes, $\tau_{\text{all}} = 18$ kpsi:

$$\tau = \tau_{\text{all}} = \frac{5153}{h} = 18\,000$$

from which $h = 0.286$ in. The feasible domain on leg size is reduced to $0.286 \le h \le 0.375$. The closest integer sixteenth larger than 0.286 in is 5/16, which has the highest figure of merit.

Answer $h = 5/16$ in.

9–8 Fatigue Loading

Some examples of fatigue loading of welded joints follow. They compare and contrast the conventional and welding code methods.

EXAMPLE 9–6 The 1018 steel strap of Fig. 9–22 has a 1000-lb, completely reversed load applied. Perform that portion of an adequacy assessment addressing fatigue strength of the weldment for infinite life if the design factor n_d is 3.
(*a*) Use the conventional factor-of-safety method.
(*b*) Use AISC fatigue allowable.

Solution From Table E–20 for the 1018 attachment metal the strengths are $S_{ut} = 58$ kpsi and $S_y = 32$ kpsi. For the E6010 electrode, $S_{ut} = 62$ kpsi and $S_y = 50$ kpsi. The fatigue stress-concentration factor $K_{fs} = 2.7$ occurs where the weld and attachment metals mingle during welding. The analysis is based on the weaker of the two materials.
(*a*) From Table 7–5, $k_a = 39.8(58)^{-0.995} = 0.700$. The shear stress is distributed uniformly over the throat:

$$A = 2(0.707)0.375(2) = 1.061 \text{ in}^2$$

For a uniform shear stress on the throat, $k_b = 1$.
From Table 7–7,

$$k_c = 0.328(58)^{0.125} = 0.545 \qquad k_d = k_e = 1$$

From Eqs. (7–4) and (7–8),

$$S_{se} = 0.700(1)0.545(1)(1)0.506(58) = 11.2 \text{ kpsi}$$

$$K_{fs} = 2.7 \qquad F_a = 1000 \text{ lb} \qquad F_m = 0$$

Only primary shear is present:

$$\tau'_a = \frac{K_{fs}F_a}{A} = \frac{2.7(1000)}{1.061} = 2545 \text{ psi} \qquad \tau'_m = 0 \text{ psi}$$

Figure 9–22

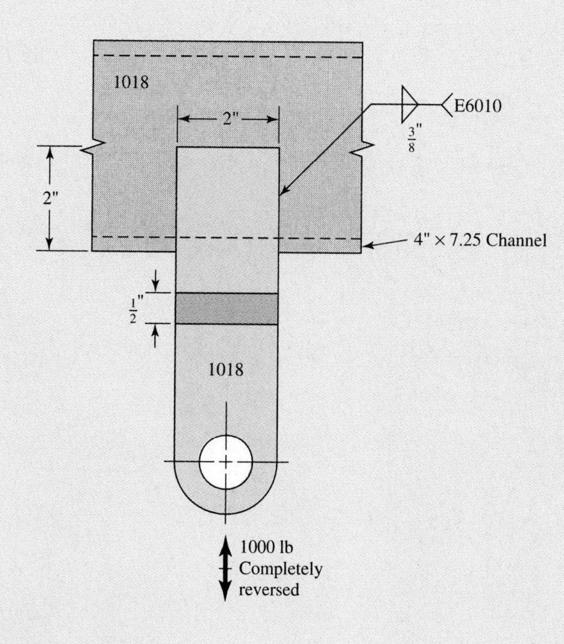

In the absence of a midrange component, the fatigue factor of safety n_f is given by

Answer $$n_f = \frac{S_{se}}{\tau'_a} = \frac{11\,200}{2545} = 4.40$$

Since $n_f > n_d$, that is, $4.40 > 3$, the weldment is satisfactory in fatigue.
(*b*) The nominal shear stress amplitude $\tau_a = 2545/2.7 = 943$ psi. Then

$$K = -\frac{\tau_a - \tau_m}{\tau_a + \tau_m} = -\frac{943 - 0}{943 + 0} = -1$$

From Table 9–8, item 25, category F, the allowable shear-stress range is 8 kpsi for indefinite life, so

$$(\tau_{max})_{all} = \frac{8}{1 - K} = \frac{8}{1 - (-1)} = 4 \text{ kpsi}$$

Answer The nominal $\tau_{max} = \tau_a + \tau_m = 943 + 0 = 943$ psi. Since $\tau_{max} = 0.943$ kpsi is less than allowable 4 kpsi, that is, $0.943 < 4$, the weldment is satisfactory in fatigue.

EXAMPLE 9–7 The 1018 steel strap of Fig. 9–23 has a repeatedly applied load of 2000 lb ($F_a = F_m = 1000$ lb). Perform that part of an adequacy assessment addressing the fatigue strength of the weldment.
(*a*) Use the conventional factor-of-safety method.
(*b*) Use the AISC fatigue allowables.

Solution (*a*) From Table 7–5, $k_a = 39.8(58)^{-0.995} = 0.700$.

$$A = 2(0.707)0.375(2) = 1.061 \text{ in}^2$$

For uniform shear stress on the throat $k_b = 1$.
From Table 7–7, $k_c = 0.328(58)^{0.125} = 0.545$. From Eqs. (7–4) and (7–8),

$$S_{se} = 0.700(1)0.545(1)(1)0.506(58) = 11.2 \text{ kpsi}$$

From Table 9–1, $K_{fs} = 2$. Only primary shear is present:

$$\tau'_a = \frac{K_{fs}F_a}{A} = \frac{2(1000)}{1.061} = 1885 \text{ psi} \qquad \tau'_m = 1885 \text{ psi}$$

Figure 9–23

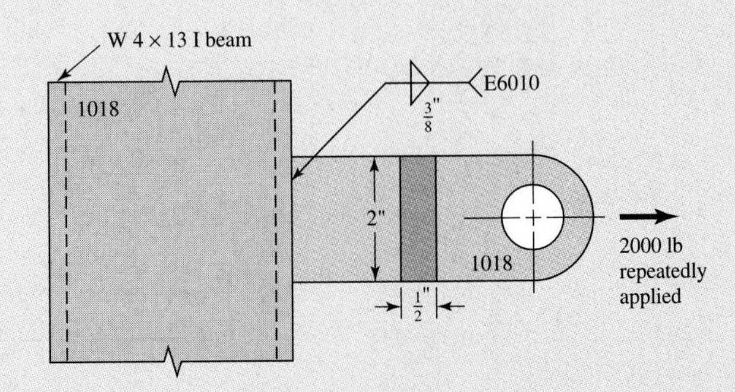

From Table 7–15, for shear stresses, the DE-Gerber fatigue failure locus, noting $S_{su} \doteq 0.67S_{ut}$, is

$$n_f = \frac{1}{2}\left(\frac{0.67S_{ut}}{\tau_m}\right)^2 \frac{\tau_a}{S_{se}}\left[-1 + \sqrt{1 + \left(\frac{2\tau_m S_{se}}{0.67S_{ut}\tau_a}\right)^2}\right]$$

$$n_f = \frac{1}{2}\left[\frac{0.67(58)}{1.885}\right]^2 \frac{1.885}{11.2}\left\{-1 + \sqrt{1 + \left[\frac{2(1.885)11.2}{0.67(58)1.885}\right]^2}\right\} = 5.52$$

Answer Since $n_f > n_d$, that is $5.52 > 3$, the weldment is satisfactory in fatigue.

(*b*) The nominal shear stresses are $\tau_a = \tau_m = 1885/2 = 942.5$ psi. Then

$$\tau_{\max} = \tau_a + \tau_m = 942.5 + 942.5 = 1885 \text{ psi}$$

$$\tau_{\min} = -(\tau_a - \tau_m) = -(942.5 - 942.3) = 0$$

$$\text{Thus } R = K = \frac{\tau_{\min}}{\tau_{\max}} = \frac{0}{942.5} = 0$$

From Table 9–8, item 24, $t = 1/2$ in, the nominal allowable stress range, line C, is 10 kpsi, thus

$$(\tau_{\max})_{\text{all}} = \frac{\tau_r}{1-K} = \frac{10}{1-0} = 10 \text{ kpsi}$$

Answer Since $(\tau_{\max})_{\text{all}} > \tau_{\max}$, that is, $10 > 1.885$ kpsi, the weldment is satisfactory in fatigue.

9–9 Resistance Welding

The heating and consequent welding that occur when an electric current is passed through several parts that are pressed together is called *resistance welding*. *Spot welding* and *seam welding* are forms of resistance welding most often used. The advantages of resistance welding over other forms are the speed, the accurate regulation of time and heat, the uniformity of the weld, and the mechanical properties which result. In addition the process is easy to automate, and filler metal and fluxes are not needed.

The spot- and seam-welding processes are illustrated schematically in Fig. 9–24. Seam welding is actually a series of overlapping spot welds, since the current is applied in pulses as the work moves between the rotating electrodes.

Failure of a resistance weld occurs either by shearing of the weld or by tearing of the metal around the weld. Because of the possibility of tearing, it is good practice to avoid loading a resistance-welded joint in tension. Thus, for the most part, design so that the spot or seam is loaded in pure shear. The shear stress is then simply the load divided by the area of the spot. Because the thinner sheet of the pair being welded may tear, the strength of spot welds is often specified by stating the load per spot based on the thickness of the thinnest sheet. Such strengths are best obtained by experiment.

Somewhat larger factors of safety should be used when parts are fastened by spot welding rather than by bolts or rivets, to account for the metallurgical changes in the materials due to the welding.

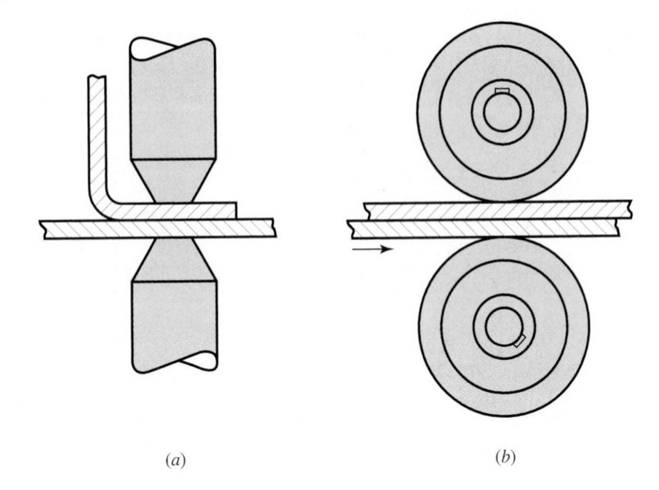

Figure 9–24

(*a*) Spot welding; (*b*) seam welding.

9–10 Bolted and Riveted Joints Loaded in Shear[6]

Riveted and bolted joints loaded in shear are treated exactly alike in design and analysis.

In Fig. 9–25*a* is shown a riveted connection loaded in shear. Let us now study the various means by which this connection might fail.

Figure 9–25*b* shows a failure by bending of the rivet or of the riveted members. The bending moment is approximately $M = Ft/2$, where F is the shearing force and t is the grip of the rivet, that is, the total thickness of the connected parts. The bending stress in the members or in the rivet is, neglecting stress concentration,

$$\sigma = \frac{M}{I/c} \tag{9–11}$$

where I/c is the section modulus for the weakest member or for the rivet or rivets, depending upon which stress is to be found. The calculation of the bending stress in this manner is an assumption, because we do not know exactly how the load is distributed to the rivet or the relative deformations of the rivet and the members. Although this equation can be used to determine the bending stress, it is seldom used in design; instead its effect is compensated for by an increase in the factor of safety.

In Fig. 9–25*c* failure of the rivet by pure shear is shown; the stress in the rivet is

$$\tau = \frac{F}{A} \tag{9–12}$$

where A is the cross-sectional area of all the rivets in the group. It may be noted that it is standard practice in structural design to use the nominal diameter of the rivet rather than the diameter of the hole, even though a hot-driven rivet expands and nearly fills up the hole.

---

6. The design of bolted and riveted connections for boilers, bridges, buildings, and other structures in which danger to human life is involved is strictly governed by various construction codes. When designing these structures, the engineer should refer to the *American Institute of Steel Construction Handbook*, the American Railway Engineering Association specifications, or the Boiler Construction Code of the American Society of Mechanical Engineers.

Figure 9–25

Modes of failure in shear loading of a bolted or riveted connection: (*a*) shear loading; (*b*) bending of rivet; (*c*) shear of rivet; (*d*) tensile failure of members; (*e*) bearing of rivet on members or bearing of members on rivet; (*f*) shear tear-out; (*g*) tensile tear-out.

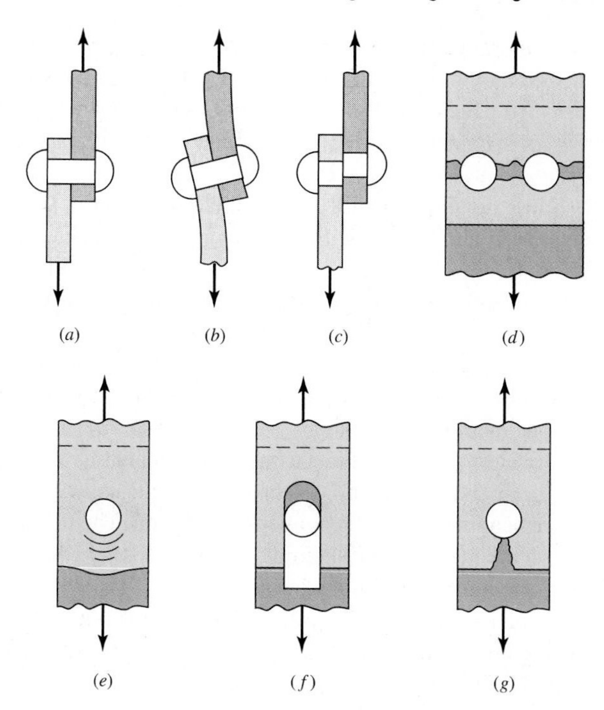

Rupture of one of the connected membes or plates by pure tension is illustrated in Fig. 9–25*d*. The tensile stress is

$$\sigma = \frac{F}{A} \tag{9–13}$$

where A is the net area of the plate, that is, the area reduced by an amount equal to the area of all the rivet holes. For brittle materials and static loads and for either ductile or brittle materials loaded in fatigue, the stress-concentration effects must be included. It is true that the use of a bolt with an initial preload and, sometimes, a rivet will place the area around the hole in compression and thus tend to nullify the effects of stress concentration, but unless definite steps are taken to ensure that the preload does not relax, it is on the conservative side to design as if the full stress-concentration effect were present. The stress-concentration effects are not considered in structural design, because the loads are static and the materials ductile.

In calculating the area for Eq. (9–13), the designer should, of course, use the combination or rivet or bolt holes which gives the smallest area.

Figure 9–25*e* illustrates a failure by crushing of the rivet or plate. Calculation of this stress, which is usually called a *bearing stress*, is complicated by the distribution of the load on the cylindrical surface of the rivet. The exact values of the forces acting upon the rivet are unknown, and so it is customary to assume that the components of these forces are uniformly distributed over the projected contact area of the rivet. This gives for the stress

$$\sigma = \frac{F}{A} \tag{9–14}$$

where the projected area for a single rivet is $A = td$. Here, t is the thickness of the thinnest plate and d is the rivet or bolt diameter.

Edge shearing, or tearing, of the margin is shown in Fig. 9–25f and g, respectively. In structural practice this failure is avoided by spacing the rivets at least 1 1/2 diameters away from the edge. Bolted connections usually are spaced an even greater distance than this for satisfactory appearance, and hence this type of failure may usually be neglected in an adequacy assessment.

In structural design it is customary to select in advance the number of rivets, their diameters, and their spacing. The strength is then determined for each method of failure. If the strength is found to be unsatisfactory, a change is made in the diameter, spacing, or the number of rivets used to increase the strength. The mode of operation is to complete the decision set and perform an adequacy assessment; if the design is satisfactory, the next step is to estimate the figure of merit and continue to improve the figure of merit if possible. It is not usual instructional practice to consider the combined effects of the various methods of failure.

In nonpermanent joints using bolts, upon disassembly the fasteners are replaced, but the members, and other elements such as splice plates or lug plates, are reused. They are permanent members of a nonpermanent joint. The focus in Chap. 8 was on the fasteners, not on the members of the joint. In this chapter we have the opportunity to examine the permanent members of the joint.

In a rivet joint, the rivets all share the load in shear, bearing in the rivet, bearing in the member, and shear in the rivet. Other failures are participated in by only some of the joint. In a bolted joint, shear is taken by clamping friction, and bearing does not exist. When bolt preload is lost, one bolt begins to carry the shear and bearing until yielding slowly brings other fasteners in to share the shear and bearing. Finally, all participate, and this is the basis of most bolted-joint analysis if loss of bolt preload is complete. The usual analysis involves

- Bearing in the bolt (all bolts participate)
- Bearing in members (all holes participate)
- Shear of bolt (all bolts participate eventually)
- Distinguishing between thread and shank shear
- Edge shearing and tearing of member (edge bolts participate)
- Tensile yielding of member across bolt holes
- Checking member capacity

EXAMPLE 9–8 Two 1-in by 4-in 1018 cold-rolled steel bars are butt-spliced with two 1/2-in by 4-in 1018 cold-rolled splice plates using four 3/4"-16 UNF grade 5 bolts as depicted in Fig. 9–26. For a design factor of $n_d = 1.5$ estimate the static load F that can be carried if the bolts lose preload.

Solution From Table E–20, minimum strengths of $S_y = 54$ kpsi and $S_{ut} = 64$ kpsi are found for the members, and from Table 8–9 minimum strengths of $S_p = 85$ kpsi and $S_{ut} = 120$ kpsi for the bolts are found.

Bearing in bolts, all bolts loaded:

$$\sigma = \frac{F}{4td} = \frac{S_p}{n_d}$$

$$F = \frac{4tdS_p}{n_d} = \frac{4(1)(3/4)85}{1.5} = 170 \text{ kip}$$

Figure 9–26

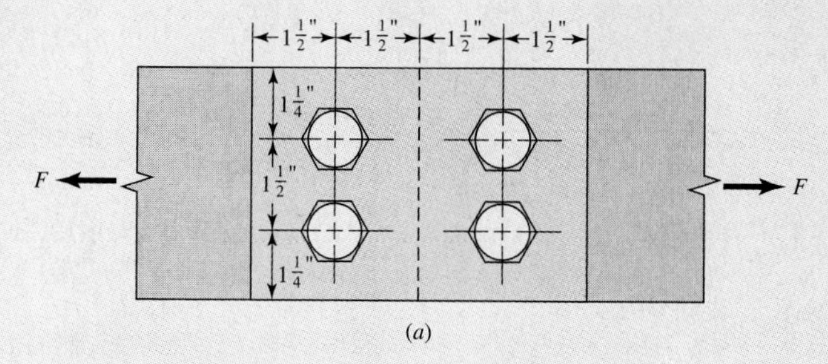

(a)

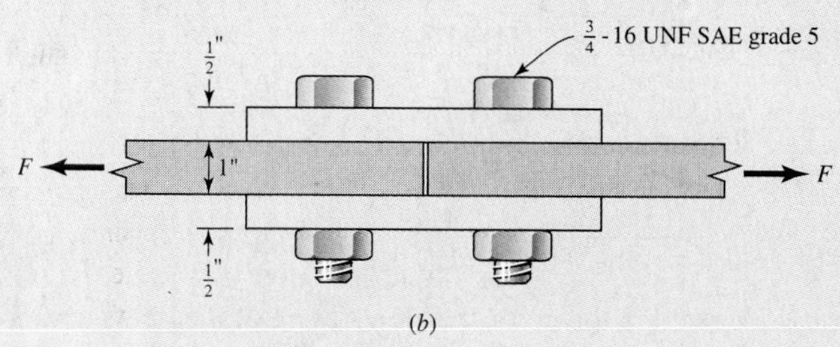

(b)

Bearing in members, all bolts active:

$$\sigma = \frac{F}{4td} = \frac{(S_y)_{\text{mem}}}{n_d}$$

$$F = \frac{4td\,(S_y)_{\text{mem}}}{n_d} = \frac{4(1)(3/4)54}{1.5} = 108 \text{ kip}$$

Shear of bolt, all bolts active: If the bolt threads do not extend into the shear planes for four shanks:

$$\tau = \frac{F}{4\pi d^2/4} = 0.577\frac{S_p}{n_d}$$

$$F = 0.577\pi d^2 \frac{S_p}{n_d} = 0.577\pi(0.75)^2 \frac{85}{1.5} = 57.8 \text{ kip}$$

If the bolt threads extend into both shear planes, for four roots:

$$\tau = \frac{F}{4A_r} = 0.577\frac{S_p}{n_d}$$

$$F = \frac{0.577(4)A_r S_p}{n_d} = \frac{0.577(4)0.351(85)}{1.5} = 45.9 \text{ kip}$$

Edge shearing of member at two margin bolts: From Fig. 9–27,

$$\tau = \frac{F}{4at} = \frac{0.577\,(S_y)_{\text{mem}}}{n_d}$$

$$F = \frac{4at0.577\,(S_y)_{\text{mem}}}{n_d} = \frac{4(1.125)(1)0.577(54)}{1.5} = 93.5 \text{ kip}$$

Tensile yielding of members across bolt holes:

$$\sigma = \frac{F}{[4 - 2(3/4)]t} = \frac{(S_y)_{\text{mem}}}{n_d}$$

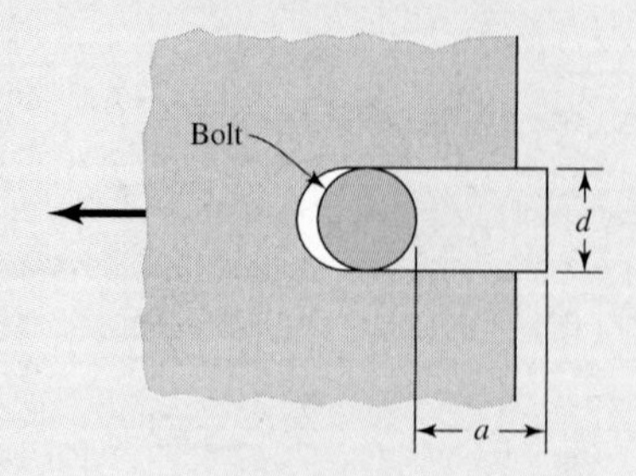

Figure 9–27

Edge shearing of member.

$$F = \frac{[4 - 2(3/4)]t\,(S_y)_{\text{mem}}}{n_d} = \frac{[4 - 2(3/4)](1)54}{1.5} = 90 \text{ kip}$$

Member yield:

$$F = \frac{dt\,(S_y)_{\text{mem}}}{n_d} = \frac{4(1)54}{1.5} = 144 \text{ kip}$$

Answer Neglecting the bolts, tensile yielding of members across the bolt holes limits the load to 90 kip. The joing efficiency is $90/144 = 0.63$.

9–11 Adhesive Bonding and Design Considerations[7]

The use of polymeric adhesives to join components for structural, semistructural, and nonstructural applications has expanded greatly in recent years as a result of the unique advantages adhesives may offer for certain assembly processes and the development of new adhesives with improved robustness and environmental acceptability. The increasing complexity of modern assembled structures and the diverse types of materials used have led to many joining applications that would not be possible with more conventional joining techniques. Adhesives are also being used either in conjunction with or to replace mechanical fasteners and welds. Reduced weight, sealing capabilities, and reduced part count and assembly time, as well as improved fatigue and corrosion resistance, all combine to provide the designer with opportunities for customized assembly. In 1998, for example, adhesives were a $20 billion industry with 24 trillion pounds of adhesives produced and sold. Figure 9–28 illustrates the numerous places where adhesives are used on a modern automobile. Indeed, the fabrication of many modern vehicles, devices, and structures are dependent on adhesives.

In well-designed joints and with proper processing procedures, use of adhesives can result in significant reductions in weight. Eliminating mechanical fasteners eliminates the weight of the fasteners, and also may permit the use of thinner gage materials because stress concentrations associated with the holes are eliminated. The capability of polymeric adhesives to dissipate energy can significantly reduce noise, vibration, and harshness (NVH), crucial in modern automobile performance. Adhesives can be used to assemble heat-sensitive materials or components that might be damaged by drilling holes for mechanical fasteners. They can be used to join dissimilar materials or thin-gage stock which cannot be joined through other means.

7. This section was prepared with the assistance of Professor David A. Dillard, Professor of Engineering Science and Mechanics and Director of the Center for Adhesive and Sealant Science, Virginia Polytechnic Institute and State University, Blacksburg, Virginia, and with the encouragement and technical support of the Bonding Systems Division of 3M, Saint Paul, Minnesota.

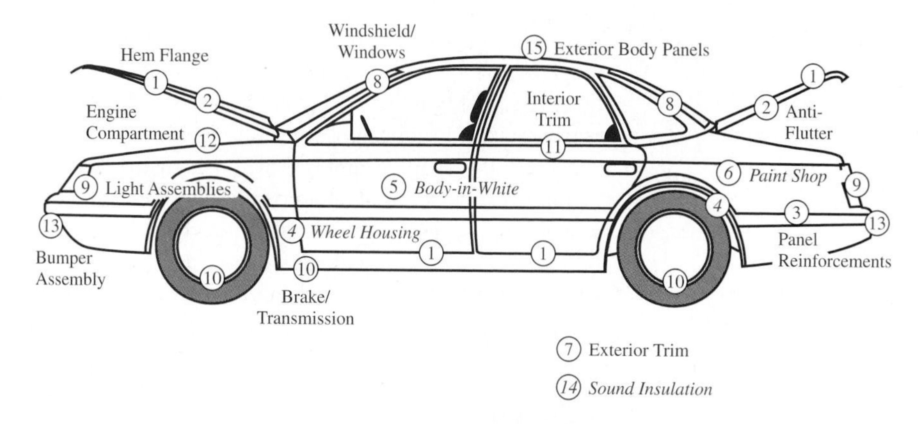

Figure 9–28

Diagram of an automobile body showing at least 15 locations at which adhesives and sealants could be used or are being used. Particular note should be made of the windshield (8), which is considered a load-bearing structure in modern automobiles and is adhesively bonded. Also attention should be paid to hem flange bonding (1), at which adhesives are used to bond and seal. Adhesives are used to bond friction surfaces in brakes and clutches (10). Antiflutter adhesive bonding (2) helps control deformation of hood and trunk lids under wind shear. Thread-sealing adhesives are used in engine applications (12). *(A. V. Pocius, Adhesion and Adhesives Technology, Hanser/Gardner Publications, Cincinnati, Ohio, 1997, reproduced by permission of the publisher.)*

In spite of a number of attractive features associated with the use of adhesives either alone or in combination with traditional fastening methods, design methods have not been widely incorporated into engineering design courses. In part this arises because of some of the complications associated with the various types of joints and the difficulty in establishing design procedures that can be universally applied. Furthermore, the use of adhesive bonding often requires skills beyond what a single individual educated in a given discipline may possess. Surface preparation, polymer processing, and mechanics are all important concepts for the design team to understand. Because adhesive bonding is so important to the design and construction of efficient structures of the future, this section is intended to provide a basic introduction for the designer considering the use of adhesives.

Mechanisms of Adhesion

Adhesives are substances that are used to join two or more components together through attractive forces acting across the interfaces. The components being joined are commonly referred to as *adherends* or *substrates*. To function as an adhesive, a polymer must be able to take on the characteristics of both a liquid and a solid at different times in its use. Bonding takes place when the liquidlike adhesive flows and wets the surfaces of interest, providing intimate contact between the adhesive and the substrate. The adhesive must then take on solidlike properties in order to be capable of sustaining loads over the service life. Adhesives may solidify by curing or cross-linking in the case of thermosets, by cooling and possibly by crystallizing (as in the case of hot melts), by drying in the case of solventborne solutions, or by drying and coalescing in the case of waterborne latex adhesives.[8] The ability to properly wet a given surface depends on two aspects: a positive spreading coefficient, a thermodynamic quantity related to the surface energies of the materials being brought into contact; and sufficient time to wet the substrate.

8. An apparent exception to this two-stage process is pressure-sensitive adhesives that remain relatively soft. Under light pressure conditions, the soft adhesive is able to flow to fill surface asperities, wetting the substrate. Once it has completely wet the adherend, a significant amount of energy is required to remove the adhesive. This is especially true for relatively rapid withdrawal, where the adhesive molecules do not have sufficient time to move, making the adhesive act like a solid.

Increasing pressure, temperature, or time of bonding may accelerate the latter kinetic process. If the adhesive and adherend surfaces are incompatible thermodynamically, however, poor bonding will result because of the difficulty of achieving intimate contact between adhesive and substrate.

Over the years, a number of mechanisms have been used to explain adhesion in its many forms. Current understanding of the matter is that secondary bonding between atoms in close proximity is able to account for much of adhesion we observe. These bonds are referred to as dispersion or van der Waals' forces, and they account for thermodynamic work of adhesion values of several tens of mJ/m^2.[9] Primary chemical bonds can exist in limited situations, frequently involving bonds to silicon or silane coupling agents. Molecular interdiffusion can occur between two miscible polymers; a good example of this is rubber cement. In some situations, electrical double layers can develop across the bond plane, leading to electrostatic attraction.[10] In certain situations, enhanced bond strength can result from mechanically roughening the surface of substrates through abrasion or grit blasting. While commonly thought to enhance mechanical interlocking of the adhesive within the crevices and pores of the surface, mechanical roughening can remove weakly bound surface layers and increase the available bond area as well.

The attractive forces between the atoms of the adhesive and the substrate give rise to very small energies, known as the thermodynamic work of adhesion, which are often measured in tens of mJ/m^2. On the other hand, when real adhesive joints are mechanically broken, fracture energies of several kJ/m^2 are often measured. The latter energies, referred to as practical adhesion, are several orders of magnitude larger than the thermodynamic surface energies. This significant multiplication of energies is associated with the dissipation of energy in the adhesive and sometimes the adherends through viscoelastic or plastic deformation processes. These practical adhesion energies are quite dependent on rate of loading and temperature, and they often depend on adhesive and adherend thickness as well. Ideally, fracture energies should be measured in such a way that plastic deformation of the adherends does not occur, as it can significantly overestimate the adhesive toughness. The large values of practical adhesion energies for high-quality modern adhesives provide significant resistance to debond growth. This is one reason that properly designed adhesive joints utilizing tough adhesives can be more durable under sustained fatigue loading than joints assembled with mechanical fasteners. Even pressure-sensitive adhesives can dissipate large amounts of energy through the viscoelastic deformation processes associated with debonding.

Types of Adhesive

There are numerous adhesive types for various applications. They may be classified in a variety of ways depending on their chemistries (e.g., epoxies, polyurethanes, polyimides), their form (e.g., paste, liquid, film, pellets, tape), their type (e.g., hot melt, reactive hot melt, thermosetting, pressure sensitive, contact), or their load-carrying capability (structural, semistructural, or nonstructural).

Structural adhesives are relatively strong adhesives that are normally used well below their glass transition temperature; common examples include epoxies and certain acrylics. Such adhesives can carry significant stresses, and they lend themselves to structural applications. For many engineering applications, semistructural applications (where failure would be less critical) and nonstructural applications (of headliners, etc.,

9. For comparison, the surface tension of water is 72 mJ/m^2.

10. To see an example of this, rapidly peel a pressure-sensitive adhesive tape from glass in a dark room to which your eyes have become accustomed. You will likely see small blue sparks associated with electrical breakdown of this charged double layer.

for aesthetic purposes) are also of significant interest to the design engineer, providing cost-effective means required for assembly of finished products. These include *contact adhesives*, where a solution or emulsion containing an elastomeric adhesive is coated onto both adherends, the solvent is allowed to evaporate, and then the two adherends are brought into contact. Examples include rubber cement and adhesives used to bond laminates to countertops. *Pressure-sensitive adhesives* are very low modulus elastomers which deform easily under small pressures, permitting them to wet surfaces. When the substrate and adhesive are brought into intimate contact, van der Waals' forces are sufficient to maintain the contact and provide relatively durable bonds. Pressure-sensitive adhesives are normally purchased as tapes or labels for nonstructural applications, although there are also double-sided foam tapes that can be used in semistructural applications. As the name implies, *hot melts* become liquid when heated, wetting the surfaces and then cooling into a solid polymer. These materials are increasingly used in a wide array of engineering applications using more sophisticated versions of the glue guns in popular use. *Anaerobic adhesives* cure within narrow spaces deprived of oxygen; such materials have been widely used in mechanical engineering applications to lock bolts or bearings in place. Cure in other adhesives may be induced by exposure to ultraviolet light or electron beams, or it may be catalyzed by certain materials that are ubiquitous on many surfaces, such as water.

Adhesives of various chemistries are available in many different forms as well. For structural applications, adhesives are available as pastes, liquids, films, and supported films. The latter are supported on loose-knit or mat scrim cloth to improve the handling properties and also to offer some measure of thickness control. Many of these adhesives produce little or no outgassing when cured, significantly reducing the likelihood of voids within the adhesive. It is important that these adhesives be kept dry, as sorbed moisture can create significant void problems. Thermosetting structural adhesives are normally available in two-part forms that are mixed through carefully controlled stoichiometry into a product that cures within the desired time. One-part forms are also available in which the resin and hardener (cross-linking agent) are already mixed together. These one-part forms must be kept at sufficiently low temperature that the reaction does not occur prematurely, sometimes utilizing latent cross-linking agents that are not active at low temperatures. One-part thermosetting adhesives often have limited *shelf life*, and often must be stored at low temperatures, but they offer very high performance capabilities. *Pot life* refers to the time after a two-part adhesive is mixed during which it is workable and will still make a satisfactory bond. Materials with too short of a pot life will harden too fast and do not give the workers sufficient time to assemble the product. An excessively long pot life may delay the cure time and slow the assembly process.

Adhesives may be applied in a variety of ways depending on their original form. Adhesives may be spread on a surface manually or dispensed using a variety of sophisticated nozzles and robotic equipment. Maintaining adherend cleanliness, providing proper jigs and fixturing during cure, and providing adequate cure conditions may all be important considerations for certain types of adhesives.

One of the most important properties of any polymer is the *glass transition temperature*, which refers to the temperature vicinity in which the amorphous portion of the polymer transitions from a hard, glassy material to soft, rubbery material. Although specific temperatures are often quoted for the glass transition temperature, it is important to remember that this transition temperature is a rate-dependent process.[11] For thermoset-

11. For example, Silly Putty at room temperature will readily flow when pulled slowly, will bounce like a rubber ball when dropped on the floor, or can shatter in a brittle fashion when struck with a hammer.

Table 9–9

Mechanical Performance of Various Types of Adhesives *Source:* A. V. Pocius, *Adhesion and Adhesives Technology,* Hanser, New York, 1997, p. 262.

Adhesive Chemistry or Type	Room Temperature Lap-Shear Strength, MPa (psi)		Peel Strength per Unit Width, kN/m (lb/in)	
Pressure-sensitive	0.01–0.07	(2–10)	0.18–0.88	(1–5)
Starch-based	0.07–0.7	(10–100)	0.18–0.88	(1–5)
Cellosics	0.35–3.5	(50–500)	0.18–1.8	(1–10)
Rubber-based	0.35–3.5	(50–500)	1.8–7	(10–40)
Formulated hot melt	0.35–4.8	(50–700)	0.88–3.5	(5–20)
Synthetically designed hot melt	0.7–6.9	(100–1000)	0.88–3.5	(5–20)
PVAc emulsion (white glue)	1.4–6.9	(200–1000)	0.88–1.8	(5–10)
Cyanoacrylate	6.9–13.8	(1000–2000)	0.18–3.5	(1–20)
Protein-based	6.9–13.8	(1000–2000)	0.18–1.8	(1–10)
Anaerobic acrylic	6.9–13.8	(1000-2000)	0.18–1.8	(1–10)
Urethane	6.9–17.2	(1000–2500)	1.8–8.8	(10–50)
Rubber-modified acrylic	13.8–24.1	(2000–3500)	1.8–8.8	(10–50)
Modified phenolic	13.8–27.6	(2000–4000)	3.6–7	(20–40)
Unmodified epoxy	10.3–27.6	(1500–4000)	0.35–1.8	(2–10)
Bis-maleimide	13.8–27.6	(2000–4000)	0.18–3.5	(1–20)
Polyimide	13.8–27.6	(2000–4000)	0.18–0.88	(1–5)
Rubber-modified epoxy	20.7–41.4	(3000–6000)	4.4–14	(25–80)

ting structural adhesives, the glass transition temperature should normally be 50°C higher than the expected service temperature.[12] Unless there are significant exotherms associated with the cure process, the glass transition temperature of an adhesive seldom exceeds the cure temperature. High-performance structural bonds often require an elevated temperature cure to provide a sufficiently high glass transition temperature in a reasonable cure time. One concern with such conditions, however, is the residual stresses that may develop when an assembled joint is cooled from the cure temperature to the service conditions.

For less structural applications, many additional types of adhesive are available. See Table 9–9 for some typical adhesive properties.

Two Basic Stress Distributions

Although adequate design procedures have been developed for specific industries, most notably the aircraft industry, generalized design procedures have not been widely available for general engineering applications. Within the framework of this brief section, industry-specific design procedures cannot be conveyed, but we do seek to present some general principles that will give insights into the design of bonded joints. These concepts will prove useful in designing certain generic classes of adhesive bonds. Two principles will be addressed, the shear-lag concept and the beam on elastic foundation concept, which provide fundamental insights into the stress distributions in a host of bonded joints, as well as many other structures the design engineer may face. For practical joint

12. The glass transition temperature of epoxies and other adhesives can be significantly reduced by moisture absorption, a factor that should be considered when designing for humid applications.

design, numerical analysis of the joint details using the finite-element method provides information not available through closed-form solutions.

Shear Lag

Good design practice normally requires that adhesive joints be constructed in such a manner that the adhesive carries the load in shear rather than tension. Bonds are typically much stronger when loaded in shear rather than in tension across the bond plate. Lap-shear joints represent an important family of joints, both for test specimens to evaluate adhesive properties and for actual incorporation into practical designs. Generic types of lap joints that commonly arise are illustrated in Fig. 9–29.

The simplest analysis of lap joints suggests the applied load is uniformly distributed over the bond area. Lap joint test results, such as those obtained following the ASTM D1002 for single-lap joints, report the "apparent shear strength" as the breaking load divided by the bond area. Although this simple analysis can be adequate for stiff adherends bonded with a soft adhesive over a relatively short bond length, significant peaks in shear stress occur except for the most flexible adhesives. In an effort to point out the problems associates with such practice, ASTM D4896 outlines some of the concerns associated with taking this simplistic view of stresses within lap joints.

In 1938, O. Volkersen presented an analysis of the lap joint, known as the *shear-lag model*. It provides valuable insights into the shear-stress distributions in a host of lap joints. Bending induced in the single-lap joint due to eccentricity significantly complicates the analysis, so here we will consider a symmetric double-lap joint to illustrate the

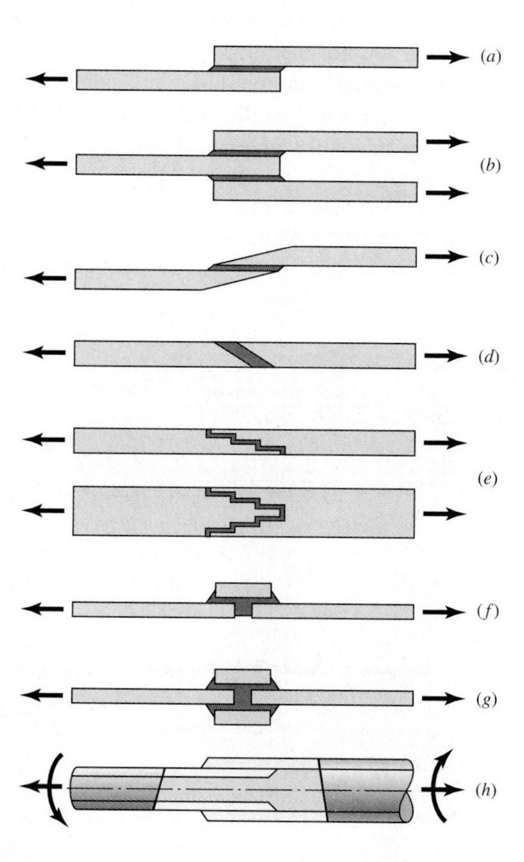

Figure 9–29

Common types of lap joints used in mechanical design: (*a*) single lap; (*b*) double lap; (*c*) scarf; (*d*) bevel; (*e*) step; (*f*) butt strap; (*g*) double butt strap; (*h*) tubular lap. (Adapted from R. D. Adams, J. Comyn, and W. C. Wake, *Structural Adhesive Joints in Engineering*, 2nd ed., Chapman and Hall, New York, 1997.

principles. The shear-stress distribution is given by

$$\tau(x) = \frac{P\omega}{4b\sinh(\omega l/2)}\cosh(\omega x) + \left[\frac{P\omega}{4b\cosh(\omega l/2)}\left(\frac{2E_o t_o - E_i t_i}{2E_o t_o + E_i t_i}\right) + \frac{(\alpha_i - \alpha_o)\,\Delta T\omega}{(1/E_o t_o + 2/E_i t_i)\cosh(\omega l/2)}\right]\sinh(\omega x) \tag{9–15}$$

where

$$\omega = \sqrt{\frac{G}{h}\left(\frac{1}{E_o t_o} + \frac{2}{E_i t_i}\right)}$$

and E_o and E_i are the moduli of the outer and inner adherends, respectively, t_o and t_i are the thicknesses of the outer and inner adherends, α_o and α_i are the coefficients of thermal expansion for the outer and inner adherends, h is the thickness of the adhesive layer, G is the shear modulus of the adhesive, and l is the length of the bond. P is the force applied to the inner adherend, and b is the width of the bond. The reciprocal of ω has units of length and is the characteristic shear-lag distance, a measure of how quickly (in spatial distance) the load is transferred from one adherend to the other.

This analysis assumes that no shear deformation occurs within the adherends and that no axial stresses are carried within the adhesive. These assumptions are approximately satisfied for many practical adhesion problems, although they become suspect with anisotropic adherends such as laminated composite or wood adherends that may have relatively small shear moduli. Because this model gives only the shear stresses (assumed to be constant throughout the thickness of the bond), the normal stresses associated with the coefficient of thermal expansion (CTE) mismatch between the adhesive and the adherends are not included. This model does not predict peel stresses, which are smaller in magnitude in double-lap joints than in single-lap joints, where significant adherend bending can occur, as illustrated in Fig. 9–30. Nonetheless, peel stresses can lead to failure even in double-lap joints. Critical joints should be analyzed using numeri-

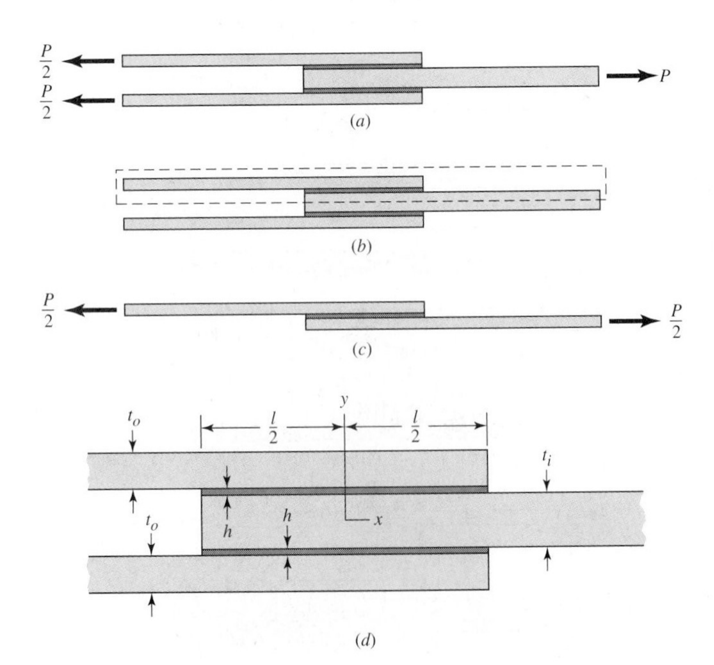

Figure 9–30

Double-lap joint analysis. (*a*) The extant double-lap joint, which is free of external moment. (*b*) The system chosen as the basis of analysis. Loading symmetry keeps the lower boundary straight. (*c*) Free-body of the system. (*d*) Nomenclature used in analysis. (Adapted with permission of D. A. Dillard, Virginia Polytechnic Institute and State University, Blacksburg, Virginia.)

cal procedures to better understand the complex stress states that can occur. Detrimental tensile peel stresses are present to some extent in most practical joints; minimization of these stresses is critical in the design process.

Several significant points become clear as we consider the shear-stress state predicted by the shear-lag model. Thicker and softer (in comparison to the adherend stiffness) adhesive layers tend to reduce the peaks in the shear stress, whether for mechanical or thermal loading. As with any design problem, however, compromises must be considered. If the adhesive layer is made too thick, bond quality often suffers because of adhesive flow and voids, becoming impractical at larger thicknesses. Furthermore, the softer adhesives may lack the strength and stiffness retention over time, resulting in systems that cannot satisfactorily carry the loads over long periods of time. They may also result in excessive deformations, a particular problem when close dimensional tolerances are to be maintained. Common adhesive applications typically call for adhesive layer thicknesses of 0.1–0.5 mm for many structural bonds, although depending on the application and tolerances of the mating parts, both thicker and thinner bonds are sometimes encountered.

Although design considerations for single-lap joints are beyond the scope of this chapter, one should note that the load eccentricity is an important aspect in the stress state of single-lap joints. Adherend bending can result in shear stresses that may be as much as double those given for the double-lap configuration (for a given total bond area). In addition, peel stresses can be quite large and often account for joint failure. Finally, plastic bending of the adherends can lead to high strains, which less ductile adhesives cannot withstand, leading to bond failure as well. Bending stresses in the adherends at the end of the overlap can be four times greater than the average stress within the adherend; thus they must be considered in the design. Figure 9–31 shows the shear and peel stresses present in a typical single-lap joint that corresponds to the ASTM D1002 test specimen. Note that the shear stresses are significantly larger than predicted by the Volkersen analysis, a result of the increased adhesive strains associated with adherend bending.

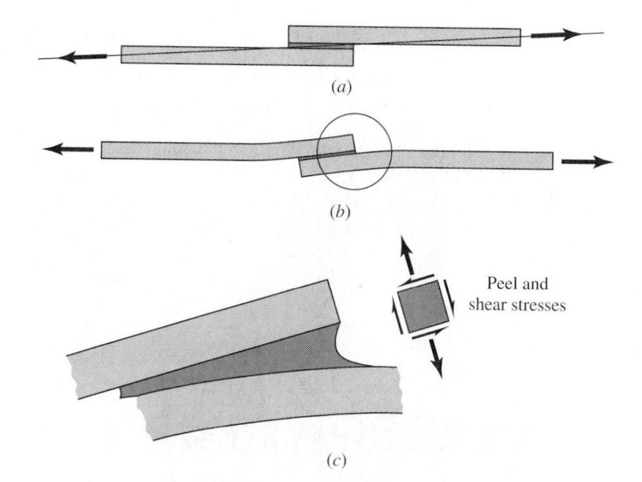

Figure 9–31

Stresses within a single-lap joint. (*a*) Lap-joint tensile forces have a line of action that is not initially parallel to the adherend sides. (*b*) As the load increases the adherends and bond bend. (*c*) In the locality of the end of an adherend peel and shear stresses appear, and the peel stresses often induce joint failure. (*d*) The seminal Goland and Reissner stress predictions (*J. Appl. Mech.*, vol. 77, 1944) are shown. Note that the predicted shear-stress maximum is higher than that predicted by the Volkersen shear-lag model because of adherend bending.

Figure 9–31

Continued

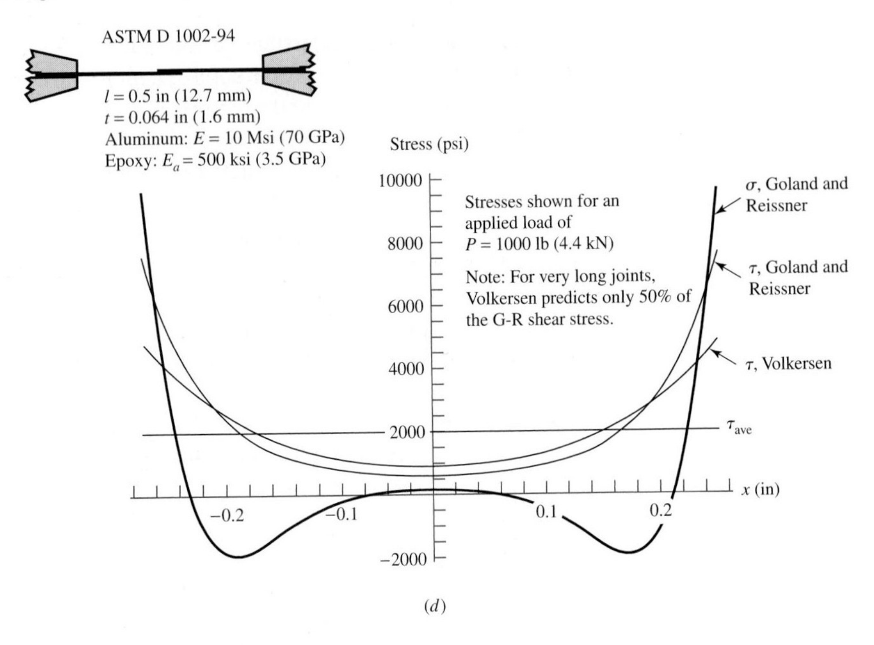

(*d*)

Beam on Elastic Foundation

In 1867, Emil Winkler presented his analysis of a beam on an elastic foundation, a key concept relevant in many mechanics problems ranging from rails deflecting under a train to the peeling of a pressure-sensitive adhesive tape from a rigid substrate. In considering adhesion problems, we assume that the adherend acts as a simple beam resting on the adhesive, which acts as an elastic foundation, represented by a continuous series of axial springs as seen in Fig. 9–32. The foundation exerts a stress directly proportional to the deflection. If a lateral force and a moment are applied to the end of a relatively long beam[13] supported on such a foundation, the resulting peel stress distribution is given by

$$\sigma(x) = \frac{E_a}{2hEI\beta^3} e^{-\beta x}\{F\cos(\beta x) + M\beta[\cos(\beta x) - \sin(\beta x)]\} \tag{9–16}$$

where E_a is the modulus of the adhesive, h is the thickness of the adhesive, F is the applied force, M is the applied moment, E is the modulus of the adherend beam, I is the second moment of area of the adherend beam, w is the width of the bond, and

$$\beta = \sqrt[4]{\frac{E_a w}{4EIh}}$$

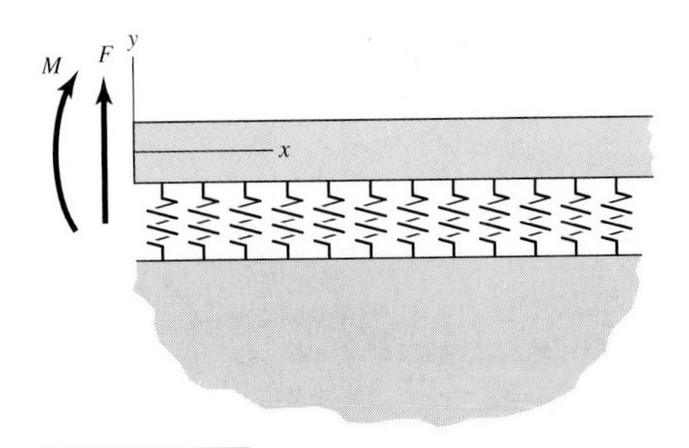

Figure 9–32

A semi-infinite beam on an elastic foundation is subjected to an applied moment and a force at the left end. The foundation is illustrated by a discrete set of springs, although the analysis and applications to adhesive layers are for continuous support.

13. If the beam is shorter than $5/\beta$, these equations are no longer valid.

The reciprocal of β also has units of length and is a measure of the distance over which the stresses are distributed. As β becomes smaller, the stresses are distributed over wider areas, effectively reducing the peak stresses. An important feature of the beam on elastic foundation model is that the overall stress distribution from the applied moment produces no net force. The compressive and tensile zones counteract one another so that no net force is present, although they do constitute a couple (see Fig. 9–33). Although the areas under these respective portions of the curve are equal, the peak of the region at the end of the bond is five times larger in magnitude than the inner peak. Since adhesives are considerably weaker in tension than in compression, having the outer (and larger magnitude) peak be tensile is of greater concern. This is especially the case since the environment has access to the outermost region and may increase the likelihood for debonding, especially over time.

Stress Concentrations in Bonded Joints

Stress concentrations associated with holes are well known. When forces acting on rivets or bolts load these holes, the resulting stress concentrations can become even larger. Eliminating rivet and bolt holes through the appropriate use of adhesives obviously reduces the stress concentrations associated with mechanical fasteners. In some cases, this allows for the use of thinner gage material, thereby contributing to weight savings. It would be erroneous, however, to conclude that there are no stress concentrations associated with bonded joints. Indeed, we have already seen for the shear-lag model that stresses are augmented at bond terminations. These peak stresses can be considerably larger than the average stress.

In bonded joints, additional stress concentrations can arise because of the abrupt angles and changes in material properties. A simple lap joint, for example, would have reentrant angles that might look similar to the fillets seen in Fig. E–15–4, which is for a homogeneous material. In fact, sharp corners can lead to severe stress concentrations, which are particularly detrimental when using brittle adhesives. For adhesives that are able to yield and plastically deform, however, these severe stress concentrations are greatly mitigated. Although strains may remain significant, the magnitudes of the stresses are tolerable. Again, localized ductility in the adhesive is a critical feature in producing

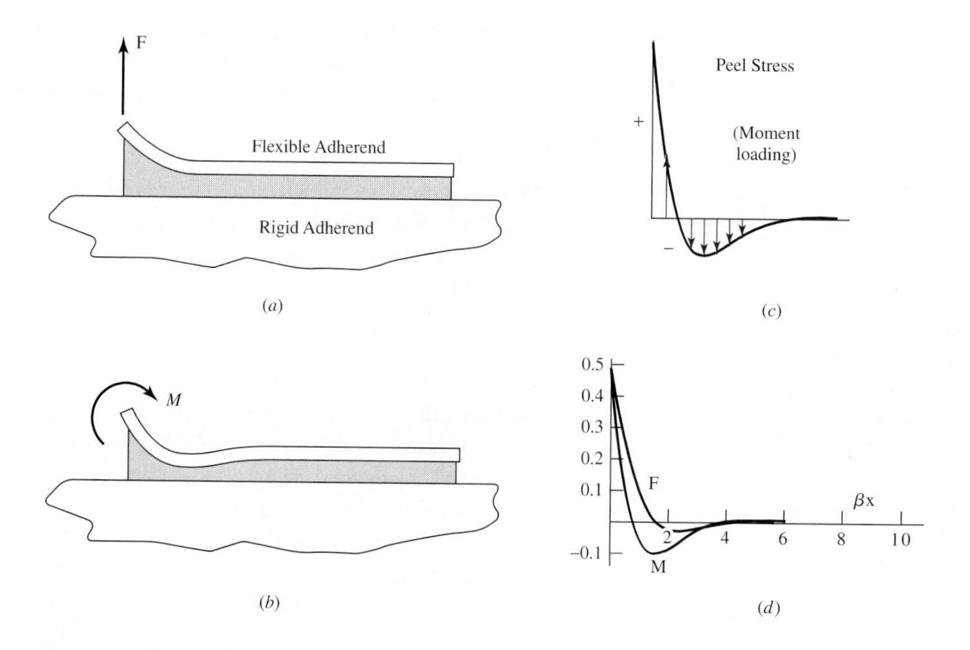

Figure 9–33

Deflected shapes for a beam on an elastic foundation for (a) an applied force acting alone or (b) an applied moment acting alone. (c) The peel stress distribution for the applied moment case. (d) A plot of x as abscissa and dimensionless normal stress $\sigma/[E_a F/(hEI\beta^3)]$ or $\sigma/[E_a M/(hEI\beta^2)]$ in the case of force F or moment M, respectively.

a "forgiving" design. Nonetheless, details of the bond terminations such as amount of adhesive spew and sharpness of adherend corners can have a measurable effect on bond strength.

Residual Stresses in Adhesive Bonds

In many bonded joints, the residual stresses associated with the mismatch in coefficient of thermal expansion between the adhesive and adherends or between the two adherends may be significant. This is especially important when adhesives are cured at elevated temperatures for adhesives with higher glass transition temperatures, which are more resistant to creep, and elevated temperature service conditions. For relatively stiff adherends made of the same material, the biaxial in-plane stress within the bondline may be estimated by

$$\sigma_0 = \frac{E_a}{1 - \nu_a} (\alpha - \alpha_a)\, \Delta T \tag{9–17}$$

where E_a is the modulus of the adhesive, ν_a is the Poisson's ratio of the adhesive, α_a is the CTE of the adhesive, α is the CTE for the substrates, and ΔT is the temperature change from the stress-free temperature (approximately the glass transition temperature). In addition to these normal stresses, significant shear stresses can also develop near the edges of the bond. These shear stresses decay rapidly as one moves away from the edges, or from cracks, defects, holidays, and so on, which may result from the fabrication process or develop in service. For laminated structures made with different adherends, these residual shear stresses can be even more significant, as predicted by Eq. (9–17). The residual stresses within the adherends can also induce considerable warpage as the structure cools from the cure condition. Such possibilities should be included in the design process.

On the Use of Fracture Mechanics with Adhesive Bonds

Design based on a strength analysis typically assumes that the bond is well-formed and does not contain voids. For many situations, designing from a fracture standpoint may be more appropriate. Whereas a strength analysis should indicate whether a failure would occur in a properly bonded joint, fracture mechanics examines whether a flaw in a bond will propagate. Imperfections in adherends, surface preparation, cleanliness, or adhesive curing often lead to initial debonds or voids within the bond. In service, debonds can be initiated by excess loads, fatigue, or environmental degradation. A rational approach to adhesive joint design should ensure that the joint is strong enough to prevent failure, but also tough enough to arrest a growing debond in the event that failure is initiated.

To use a fracture mechanics approach to design with adhesive bonds, the designer needs the critical energy release rate for the adhesive. (For dissimilar materials, the energy release rate approach is more convenient than the stress intensity factor approach to fracture mechanics.) If the available energy release rate for a given debond reaches the critical energy release rate, failure will occur spontaneously. This critical energy release rate may be reduced significantly by environmental exposure, so the allowable design factors should be determined in conditions that accurately reflect the service environment. Debonds can propagate under fatigue conditions as well; finite life structures can be designed. Because of the relatively high exponents involved, designing below the threshold value of the critical energy release rate is recommended for long service life.

Fracture mechanics involves an additional length scale which is often the length of the debond, although with compliant adhesives it can be the thickness of the adhesive layer. Proper application of fracture mechanics requires some quantification of the size of the debonds present within the bond. Nondestructive evaluation (NDE) is an important feature in determining the acceptability of critical bonds. A variety of nondestructive evaluation techniques have routinely been used to characterize bonds in aircraft and spacecraft applications. As the capabilities of these techniques improve, and as adhesives find more widespread use in structural applications, NDE has become a routine part of the production line for the automotive industry as well.

Designing with Adhesives

Use of adhesives opens up new design opportunities. Adhesives may reduce the need for access holes (as required for spot welds or mechanical fasteners), resulting in stiffer and stronger structures. The use of adhesives in conjunction with spot welding results in a process known as *weld-bonding*. The spot welds serve to fixture the bond until it is cured. Once cured, the adhesive greatly increases the stiffness and damping characteristics of the automobile and also seals the joint from salt-water intrusion. The resulting auto body has improved noise, vibration, and harshness (NVH) characteristics, preventing squeaks over many years.

Well-designed joints should be as strong as the adherends themselves. Ideally, bonded structures should fail outside the bonded region. If failures occur within the adhesive layer, people often prefer cohesion failures within the adhesive layer rather than adhesion failures between the adhesive and substrate, indicating the strength of the adhesive is limited by the properties of the polymer rather than its ability to bond to the substrates of interest. Factors of safety have not been standardized for adhesives, although common usage dictates maximum stresses of 10–20 percent of the strength of the adhesives. Even these significant factors of safety can be insufficient when combinations of environment, temperature, and time produce degradation in the adhesion. An understanding of potential degradation mechanisms is essential if long-term durability is a concern.

Two key concepts should be kept in mind when designing with adhesives. First, adhesive bonds should be seen as *material systems*. Although tests on the adhesive can be useful, much of the evaluation should focus on the material system, consisting of the adhesive, primers, surface preparation methods, and the adherends to be bonded. Omitting some of these components from the testing phase of the evaluation process can lead to failures in identifying potential degradation mechanisms. A second point of critical importance, is that every adhesive bond should be seen as a *structure*, with inherent complexities in stress distribution. Even the simplest test methods result in nonuniform stresses, which should be taken into account in determining allowable stresses and in the design process. Some additional basic guidelines that should be used in adhesive joint design include:

- Design to place bondline in shear, not peel. Beware of peel stresses focused at bond terminations. When necessary, reduce peel stresses through tapering the adherend ends, increasing bond area where peel stresses occur, or utilizing rivets at bond terminations where peel stresses can initiate failures.
- Where possible, use adhesives with adequate ductility. The ability of an adhesive to yield reduces the stress concentrations associated with the ends of joints and increases the toughness to resist debond propagation.
- Recognize environmental limitations of adhesives and surface preparation methods. Exposure to water, solvents, and other diluents can significantly degrade adhesive performance in some situations, through displacing the adhesive from the surface or degrading the polymer. Certain adhesives may be susceptible to environmental stress

cracking in the presence of certain solvents. Exposure to ultraviolet light can also degrade adhesives.

- Design in a way that permits or facilitates inspections of bonds where possible. A missing rivet or bolt is often easy to detect, but debonds or unsatisfactory adhesive bonds are not readily apparent.
- Understand the role of nondestructive testing in quality control of the production process and in service-life monitoring for critical structures.
- Allow for sufficient bond area so that the joint can tolerate some debonding before going critical. This increases the likelihood that debonds can be detected. Having some regions of the overall bond at relatively low stress levels can significantly improve durability and reliability.
- Where possible, bond to multiple surfaces to offer support to loads in any direction. Bonding an attachment to a single surface can place peel stresses on the bond, whereas bonding to several adjacent planes tends to permit arbitrary loads to be carried predominantly in shear.
- Incorporate adhesive bond requirements early in the design process to ensure that adhesives can be used effectively. Simply replacing welds and mechanical fasteners with adhesives may not lead to successful designs.
- Where possible, use robust adhesives and surface preparation techniques that show little sensitivity to processing variations and delays, ensuring that reproducible bonds can routinely be made.
- Remember to test the *material system*, and remember that all bonded joints are *structures* with complex stress distributions for even the simplest geometries.
- Unless disassembly is desired, design for adherend failures where possible.
- Take care in the design process to ensure that cure and fixturing will be satisfactory. Satisfactory curing of adhesives often requires adequate time at the cure temperature and the ability to hold the adherends in place until the cure is complete.
- When designing with rate-dependent materials, consider effective strain rate in adhesive. A debond propagating rapidly in an adhesive layer can represent quite high relative strain rates.
- Remember rate and temperature dependence of all polymeric materials. Adhesives tested under laboratory conditions may appear more ductile than when tested under impact or low-temperature conditions. They may also exhibit more creep when loaded for long time periods at higher temperatures.

Figure 9–34 presents examples of improvements in adhesive bonding.

Cautions for Adhesive Use

In spite of numerous advantages which adhesive joints enjoy over conventional mechanical fasteners and welds, the designer must address several concerns when selecting appropriate joining methods for a given application. Process robustness is a concern with adhesive bonds. Seemingly minor changes in surface pretreatment, application method, or curing conditions can actually lead to significant reductions in performance. Chemists, material scientists, or process engineers working closely with the designer can help avoid such problems.

All adhesives have certain environmental limitations that must be avoided, even for very brief exposure. Depending on the type, adhesives may soften flow, melt, and lose strength when raised to elevated temperatures. At sufficiently low temperatures, or when loaded rapidly, adhesives fail in a brittle fashion. All polymeric materials are viscoelastic, exhibiting time-dependent properties such as creep, relaxation, and damping. These are

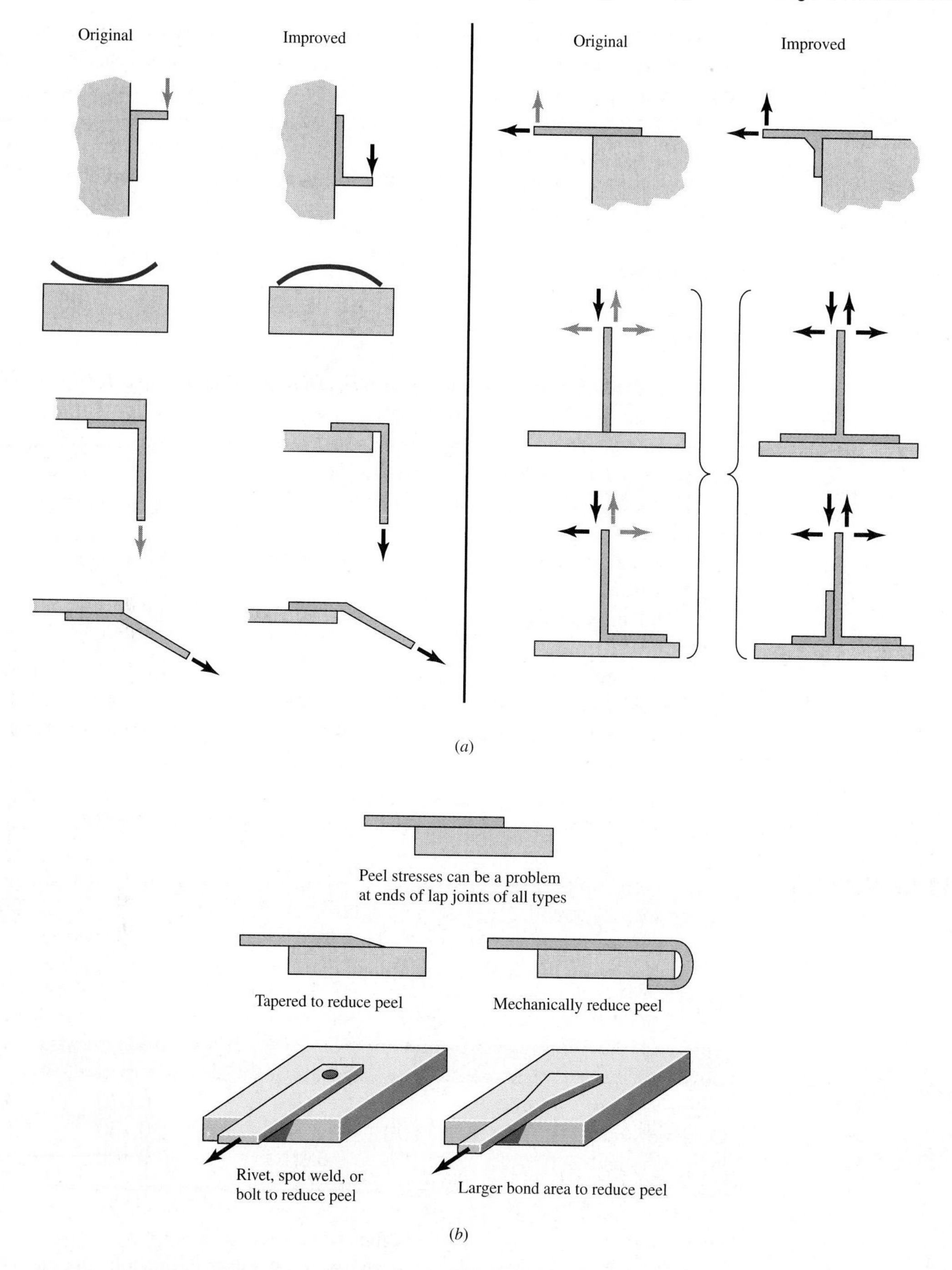

Figure 9–34

Design practices which improve adhesive bonding. (*a*) Gray load vectors are to be avoided as resulting strength is poor. (*b*) Means to reduce peel stresses in lap-type joints.

strongly dependent on temperature and rate of loading, the interdependence of which is strongly related (see time-temperature superposition principle in other texts). Adhesive properties can also be strongly affected by moisture, solvents, and other diluents.

As with any design process, adhesive selection often involves compromises: ductile adhesives are more forgiving, but they may lack adequate creep resistance over time. Failure to adequately prepare the surfaces to be bonded may significantly reduce bond performance or durability. Finally, bonding steel plates with steel bolts may require knowledge of only a single material type. Use of adhesives should involve an interdisciplinary understanding of the adherends, surfaces, and adhesives and how they interact over the service life of the bonded joint.

References

A number of good references are available for analyzing and designing adhesive bonds, including the following:

G. P. Anderson, S. J. Bennett, and K. L. DeVries, *Analysis and Testing of Adhesive Bonds*, Academic Press, New York, 1977.

R. D. Adams, J. Comyn, and W. C. Wake, *Structural Adhesive Joints in Engineering*, 2nd ed., Chapman and Hall, New York, 1997.

H. F. Brinson (ed.), *Engineered Materials Handbook, vol. 3: Adhesives and Sealants*, ASM International, Metals Park, Ohio, 1990.

A. J. Kinloch, *Adhesion and Adhesives: Science and Technology*, Chapman and Hall, New York, 1987.

A. J. Kinloch (ed.), *Durability of Structural Adhesives*, Applied Science Publishers, New York, 1983.

W. A. Lees, *Adhesives in Engineering Design*, Springer-Verlag, New York, 1984.

F. L. Matthews, *Joining Fibre-Reinforced Plastics*, Elsevier, New York, 1986.

A. V. Pocius, *Adhesion and Adhesives Technology: An Introduction*, Hanser, New York, 1997.

EXAMPLE 9–9

A double-lap joint depicted in Fig. 9–30 consists of aluminum outer adherends and an inner steel adherend. The assembly is cured at 250°F and is stress-free at 200°F. The completed bond is subjected to an axial load of 2000 lb at a service temperature of 70°F. The width b is 1 in, the length of the bond l is 1 in. Additional information is tabulated below:

	G, psi	E, psi	α, in/(in · °F)	Thickness, in
Adhesive	$0.2(10^6)$		$55(10^{-6})$	0.020
Outer adherend		$10(10^6)$	$13.3(10^{-6})$	0.150
Inner adherend		$30(10^6)$	$6.0(10^{-6})$	0.100

Sketch a plot of the shear stress as a function of the length of the bond due to (*a*) thermal stress, (*b*) load-induced stress, and (*c*) the sum of stresses in *a* and *b*; and (*d*) find where the largest shear stress is maximum.

Solution In Eq. (9–15) the parameter ω is given by

$$\omega = \sqrt{\frac{G}{h}\left(\frac{1}{E_o t_o} + \frac{2}{E_i t_i}\right)}$$

$$= \sqrt{\frac{0.2\,(10^6)}{0.020}\left[\frac{1}{10\,(10^6)\,0.15} + \frac{2}{30\,(10^6)\,0.10}\right]} = 3.65\ \text{in}^{-1}$$

(*a*) $\alpha_i - \alpha_o = 6(10^{-6}) - 13.3(10^{-6}) = 0.000\,007\,3$, $\Delta T = 70 - 200 = -130°$ F,

$$\tau_{th}(x) = \frac{(\alpha_i - \alpha_o)\,\Delta T\,\omega\sinh(\omega x)}{(1/E_o t_o + 2/E_i t_i)\cosh(\omega l/2)}$$

$$\tau_{\text{th}}(x) = \frac{-0.000\,007\,3(-130)3.65\sinh(3.65x)}{\left[\dfrac{1}{10\,(10^6)\,0.150} + \dfrac{2}{30\,(10^6)\,0.100}\right]\cosh\left[\dfrac{3.65(1)}{2}\right]}$$

$$= 816.4\sinh(3.65x)$$

$$\tau_{\text{th}}(0) = 0\ \text{psi}$$

$$\tau_{\text{th}}(0.5) = 816.4(3.0208) = 2466\ \text{psi}$$

$$\tau_{\text{th}}(-0.5) = 816.4(-3.0208) = -2466\ \text{psi}$$

(*b*) The bond is "balanced" ($E_o t_o = E_i t_i/2$), so the load-induced stress is given by

$$\tau_i(x) = \frac{P\omega\cosh(\omega x)}{4b\sinh(\omega l/2)} = \frac{2000(3.65)\cosh(3.65x)}{4(1)3.0208} = 604.1\cosh(3.65x)$$

$$\tau_i(0) = 604.1(1) = 604\ \text{psi}$$

$$\tau_i(0.5) = 604.1(3.1820) = 1922\ \text{psi}$$

$$\tau_i(-0.5) = 604.1(3.1820) = 1922\ \text{psi}$$

(*c*) Total stress table:

	$\tau(-0.5)$	$\tau(0)$	$\tau(0.5)$
Thermal only	−2466	0	2466
Load-induced only	1922	604	1922
Combined	−544	604	4388

(*d*) The maximum shear stress predicted by the shear-lag model will always occur at the ends. See the plot in Fig. 9–35. Since the residual stresses are always present, significant

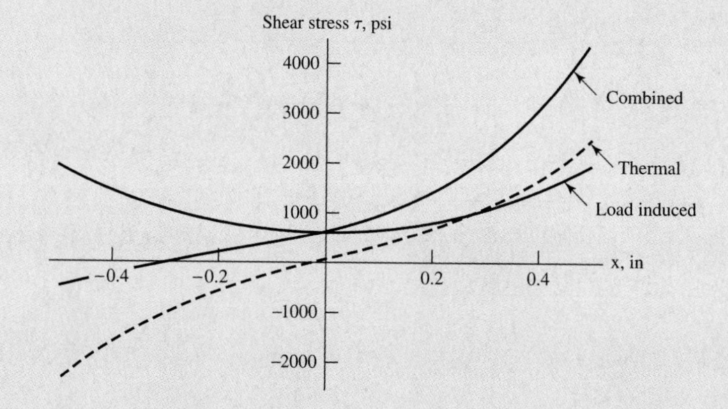

Figure 9–35

Plot for Ex. 9–9.

shear stresses already exist prior to application of the load. The large stresses present for the combined-load case could result in local yielding of a ductile adhesive or failure of a more brittle one. The significance of the thermal stresses serves as a caution against joining dissimilar adherends when large temperature changes are involved.

EXAMPLE 9–10 A double-sided acrylic foam tape is used to bond a flat molding strip to a curved section of an automobile panel. Recognize that the moment M required to produce the same curvature in the molding strip is $M = EI/\rho$ where E is Young's modulus, I is the second area moment and ρ is the radius of curvature of the neutral surface. One can represent this situation by the equivalent case of a strip subjected to an external moment as shown in Fig. 9–36. Young's modulus for the molding strip is $0.1(10^6)$ psi, and E_a for the adhesive is 100 psi. Foam tapes such as this are widely used for attaching automotive molding strips.

(*a*) Find I for the molding strip.

(*b*) Find β for Eq. (9–16).

(*c*) Find an expression for σ at $x = 0$ and the minimum radius ρ for a maximum peel stress of 5 psi.

(*d*) Discuss why soft, relatively thick elastomeric foam tapes are advantageous for minimizing stress in such applications.

Solution (*a*) Estimate I for the molding strip:

$$I = \frac{bh^3}{12} = \frac{(1)0.2^3}{12} = 0.000\,667\ \text{in}^4$$

(*b*) Estimate β from Eq. (9–16):

$$\beta = \sqrt[4]{\frac{E_a w}{4EIh}} = \sqrt[4]{\frac{100(1)}{4\left[(0.1\,(10^6)\,0.000\,667(0.125)\right]}} = 1.32\ \text{in}^{-1}$$

(*c*) Maximum tensile stress will occur at the end, so we want to set $x = 0$ ($F = 0$ also). From Eq. (9–16),

$$\sigma(x) = \frac{E_a}{2hEI\beta^3}\exp(-\beta x)M\beta(\cos\beta x - \sin\beta x)$$

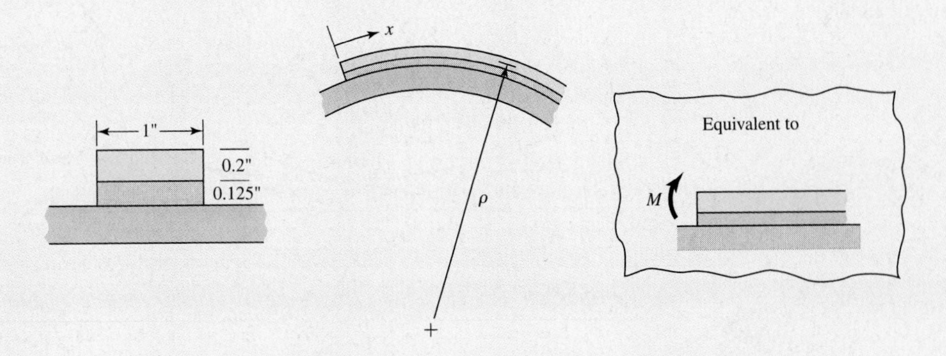

Figure 9–36

Bonding of an initially straight molding strip to a curve body panel of an automobile.

Substituting $M = EI/\rho$ and $x = 0$ gives

$$\sigma(0) = \frac{E_a}{2hEI\beta^3}\exp(0)\frac{EI}{\rho}\beta = \frac{E_a}{2\rho h\beta^2}$$

Solving for ρ gives

$$\rho = \frac{E_a}{2\sigma(0)h\beta^2} = \frac{100}{2(5)0.125(1.32)^2} = 45.9 \text{ in}$$

(d) The normal stress equation at $x = 0$ reduces to

$$\sigma(0) = \frac{E_a}{2\rho h\beta^2}$$

Now σ can be reduced by reducing E_a and increasing h, which is to say soft (small E_a) and thick (large h) reduces the magnitude of the maximum tensile stress $\sigma(0)$.

PROBLEMS

ANALYSIS

9–1 The figure shows a horizontal steel bar 3/8 in thick loaded in steady tension and welded to a vertical support. Find the load F that will cause a shear stress of 20 kpsi in the throats of the welds.

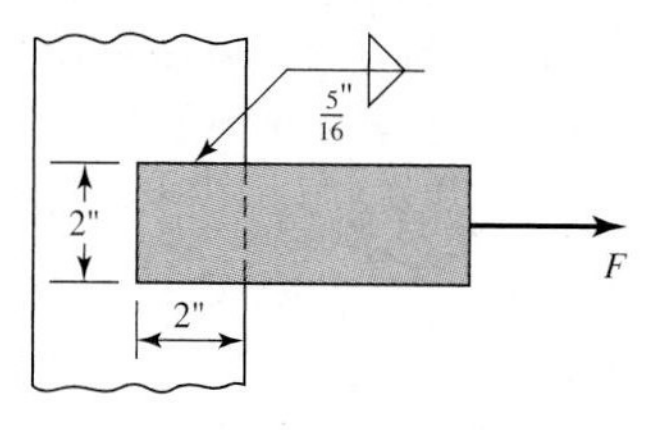

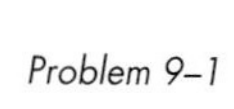
Problem 9–1

9–2 For the weldment of Prob. 9–1 the electrode specified is E7010. For the electrode metal, what is the allowable load on the weldment?

ANALYSIS

9–3 The members being joined in Prob. 9–1 are cold-rolled 1018 for the bar and hot-rolled 1018 for the vertical support. What load on the weldment is allowable because member metal is incorporated into the welds?

ANALYSIS

9–4 A 5/16-in steel bar is welded to a vertical support as shown in the figure. What is the shear stress in the throat of the welds if the force F is 32 kip?

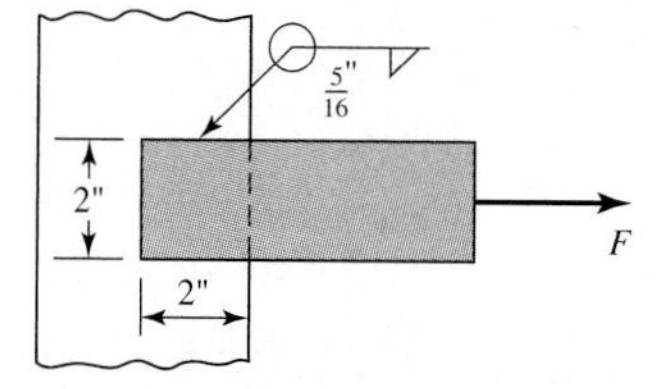

Problem 9–4

9–5 A 3/4-in-thick steel bar, to be used as a beam, is welded to a vertical support using two fillet welds as illustrated.

(a) Find the safe bending force F if the permissible shear stress in the welds is 20 kpsi.

(*b*) In part *a* you found a simple expression for F in terms of the allowable shear stress. Find the allowable load if the electrode is E7010, the bar is hot-rolled 1020, and the support is hot-rolled 1015.

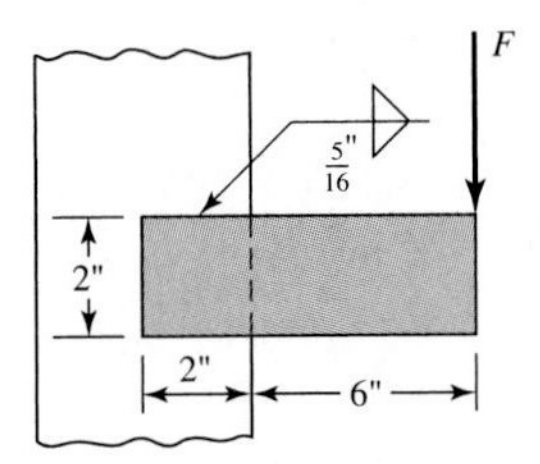

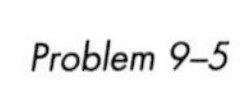
Problem 9–5

ANALYSIS

9–6 The figure shows a weldment just like that of Prob. 9–5 except that there are four welds instead of two. Show that the weldment is twice as strong as that of Prob. 9–5.

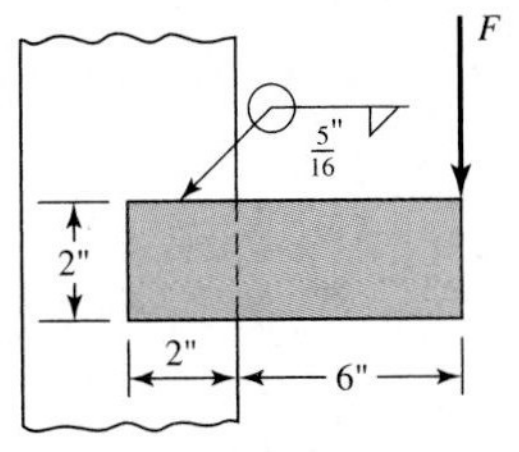

Problem 9–6

ANALYSIS

9–7 The weldment shown in the figure is subjected to an alternating force F. The hot-rolled steel bar is 10 mm thick and is of AISI 1010 steel. The vertical support is likewise of 1010 steel. The electrode is 6010. Estimate the fatigue load F the bar will carry if three 6-mm fillet welds are used. Use a conventional fatigue analysis.

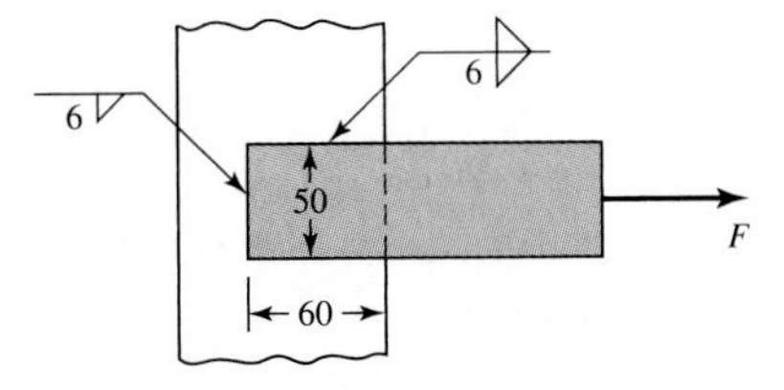

Problem 9–7

ANALYSIS

9–8 Repeat Prob. 9–7 using a welding code analysis for comparison.

9–9 The permissible shear stress for the weldment illustrated is 140 MPa. Estimate the bending load stress that will cause this stress in the weldment throat.

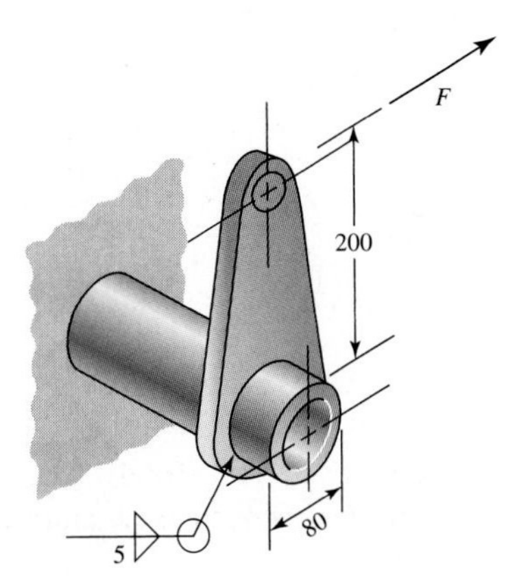

Problem 9–9

9–10 In the design weldments in torsion it is helpful to have a hierarchical perception of the relative efficiency of common patterns. For example, the weld-bead patterns shown in Table 9–2 can be ranked for desirability. Assume the space available is an $a \times a$ square. Use a formal figure of merit that is directly proportional to J and inversely proportional to the volume of weld metal laid down:

$$\text{fom} = \frac{J}{\text{vol}} = \frac{0.707hJ_u}{(h^2/2)\,l} = 1.414\frac{J_u}{hl}$$

A tactical figure of merit could omit the constant, that is, $\text{fom}' = J_u/(hl)$. Rank the six patterns of Table 9–2 from most to least efficient.

9–11 In your ordering of weld metal distribution patterns in Table 9–2 in response to Prob. 9–10, you found the second line (parallel beads) and the fifth line (beads in a square) tied for first place as the most efficient distribution of metal. In Probs. 9–5 and 9–6 you estimated the loads that would induce the maximum allowable stress of 20 kpsi at the critical location. You also found the square pattern would carry about twice the load. In Prob. 9–10 the tactical figure of merit fom′ is related to the formal figure of merit by $\text{fom} = \text{fom}'(1.414)$.

(*a*) Find the formal figure of merit from Prob. 9–10 results.

(*b*) Compare with the formal figure of merit from Probs. 9–5 and 9–6.

9–12 The space available for a weld-bead pattern subject to bending is $a \times a$. Place the patterns of Table 9–3 in hierarchical order of efficiency of weld metal placement to resist bending. A formal figure of merit can be directly proportion to I and inversely proportional to the volume of weld metal laid down:

$$\text{fom} = \frac{I}{\text{vol}} = \frac{0.707hI_u}{(h^2/2)\,l} = 1.414\frac{I_u}{hl}$$

The tactical figure of merit can omit the constant 1.414, that is, $\text{fom}' = I_u/(hl)$. Omit the patterns intended for T-beams and I-beams. Rank the remaining seven.

DESIGN

9–13 Among the possible forms of weldment problems are the following:

- The attachment and the member(s) exist and only the weld specifications need to be decided.
- The members exist, but both the attachment and the weldment must be designed.
- The attachment, member(s), and weldment must be designed.

What follows is a design task of the first category. The attachment shown in the figure is made of 1018 HR steel 1/2 in thick. The static force is 25 kip. The member is 4 in wide, such as that shown in Prob. 9–4. Specify the weldment (give the pattern, electrode number, type of weld, length of weld, and leg size).

Problem 9–13

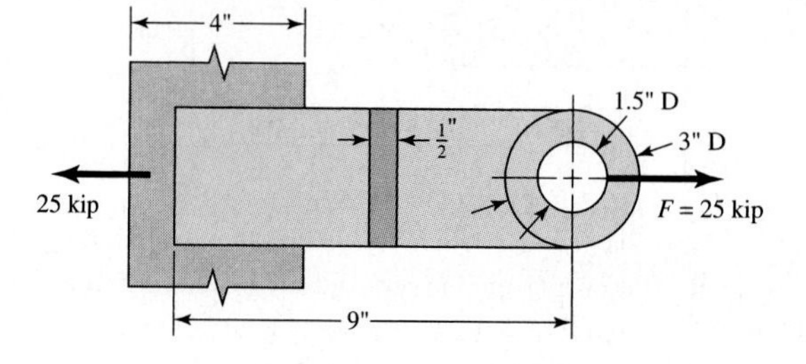

9–14 The attachment of Prob. 9–13 carries a bending load of 3 kip. The clearance a is to be 6 in. The load is a static 3000 lb. Specify the weldment (give the pattern, electrode number, type of weld, length of weld, and leg size). See figure on page 582.

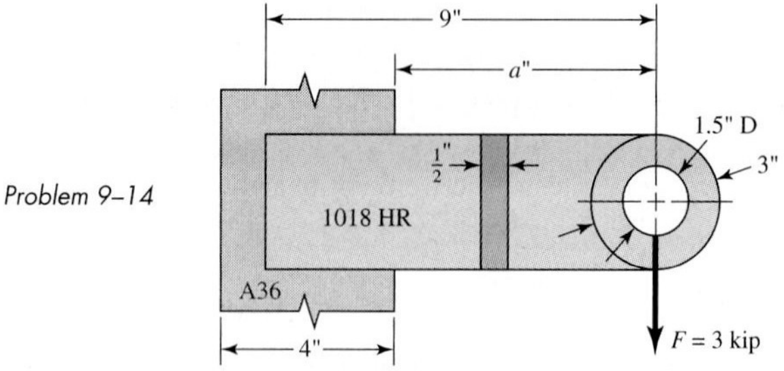

Problem 9–14

9–15 The attachment in Prob. 9–14 has not had its length determined. The static force is 3000 lb; the clearance a is to be 6 in. The member is 4 in wide. Specify the weldment (give the pattern, electrode number, type of weld, length of bead, and leg size). Specify the attachment length.

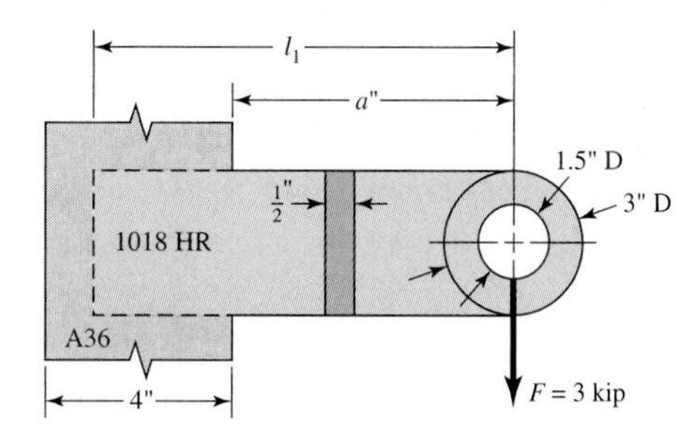

Problem 9–15

9–16 A vertical column of A36 structural steel ($S_y = 36$ kpsi, $S_{ut} = 58$–80 kpsi) is 10 in wide. An attachment has been designed to the point shown in the figure. The static load of 20 000 lb is applied, and the clearange a of 6.25 in has to be equaled or exceeded. The attachment is 1018 hot-rolled steel, to be made from 1/2-in plate with weld-on bosses when all dimensions are known. Specify the weldment (give the pattern, electrode number, type of weld, length of weld bead, and leg size). Specify also the length l_1 for the attachment.

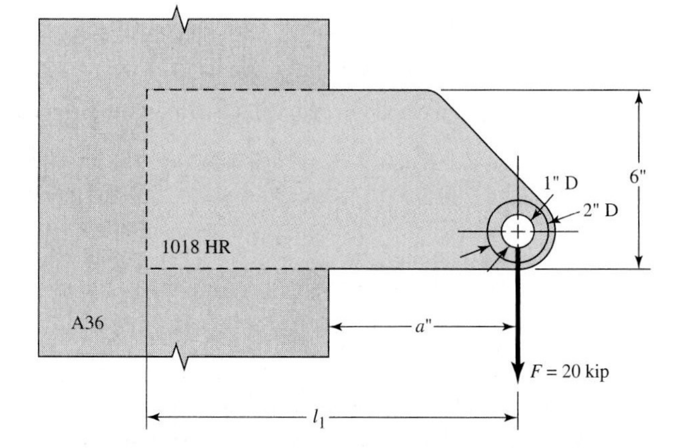

Problem 9–16

9–17 Problem 9–16 has a rectangular weldment area available for the weld beads. We wish to place weld metal as far away from the centroid as possible. Choose a rectangular weld-all-around pattern with $b = 10$ in and $d = 6$ in. For the same loads and materials as in Prob. 9–16 find the necessary leg size h and the figure of merit.

(a) Compare these with those of Prob. 9–16.

(b) Tabulate h and fom using some other widths and note any trends you see.

ANALYSIS

9–18 Write a computer program to assist with a task such as that of Prob. 9–16 with a rectangular weld-bead pattern for a torsional shear joint. In doing so solicit the force F, the clearance a, and the largest allowable shear stress. Then, as part of an iterative loop, solicit the dimensions b and d of the rectangle. These can be your design variables. Output all the parameters after the leg size has been determined by computation. In effect this will be your adequacy assessment when you stop iterating. Include the figure of merit $J_u/(hl)$ in; the output. The fom and the leg size h with available width will give you a useful insight into the nature of this class of welds. Use your program to verify your solutions to Probs. 9–16 and 9–17.

9–19 Fillet welds in joints resisting bending are interesting in that they can be simpler than those resisting torsion. The algebraic protocol can obscure some useful insights. From Prob. 9–12 you learned that your objective is to place weld metal as far away from the weld-bead centroid as you can, but in an orientation of being distributed parallel to the x axis. Furthermore, placement on the top and bottom of the built-in end of a cantilever with rectangular cross section results in parallel weld beads, each element of which is in the ideal position. Study the full weld bead and the interrupted weld-bead pattern. Consider the case of the figure: $F = 10\,000$ lb, $a = 10$ in, $b = 8$ in, $d = 8$ in, and let b_1 be your design variable. Now let $b = 8$ in, $b_1 = 2$ in, and $d = 8$ in. Study the two cases. What do you notice about τ''_x, τ'_y, and τ_{max}? Why is this happening?

9–20 For a rectangular weld-bead track resisting bending, develop the necessary equations to treat cases of vertical welds, horizontal welds, and weld-all-around patterns with depth d and width b and allowing central gaps in parallel beads of length b_1 and d_1. Do this by superposition of parallel tracks, vertical tracks subtracting out the gaps. Then put the two together for a rectangular weld bead with central gaps of length b_1 and d_1. Show that the results are

$$A = 1.414\,(b - b_1 + d - d_1)\,h$$

$$I_u = \frac{(b - b_1)\,d^2}{2} + \frac{d_3 - d_1^3}{6}$$

$$I = 0.707 h I_u$$

$$l = 2\,(b - b_1) + 2\,(d - d_1)$$

$$\text{fom} = \frac{I_u}{hl}$$

9–21 Write a computer program based on Prob. 9–20 protocol. Solicit the largest allowable shear stress, the force F, and the clearance a, as well as the dimensions b and d. Begin an iterative loop by soliciting b_1 and d_1. Either or both of these can be your design variables. Program to find the leg size corresponding to a shear-stress level at the maximum allowable at a corner. Output all your parameters including the figure of merit. Use the program to check any previous problems to which it is applicable. Play with it in a "what if" mode and learn from the trends in your parameters.

9–22 When comparing two different weldment patterns it is useful to observe the resistance to bending or torsion and the volume of weld metal deposited. Measure of effectiveness, defined as second moment of area divided by weld-metal volume, is useful. If a 6-in by 8-in section of a cantilever carries a static 10 000-lb bending load 10 in from the weldment plane, with an allowable shear stress of 12 800 psi realized, compare horizontal weldments with vertical weldments. The horizontal beads are to be 6 in long and the vertical beads, 8 in long.

9–23 The following is given. The weldment zone is a 10-in-diameter circle. The member is in static bending with a cross section $b \times d$. The intent is a weld-all-around bead. If the bending force is 10 000 lb, its moment arm is 10 in from the plane of the weldment, and the attained permissible stress is 12 800 psi, determine the desirable cross section of the cantilever member.

9–24 A torque $T = 20(10^3)$ lb · in is applied to the weldment shown. Estimate the maximum shear stress in the weld throat.

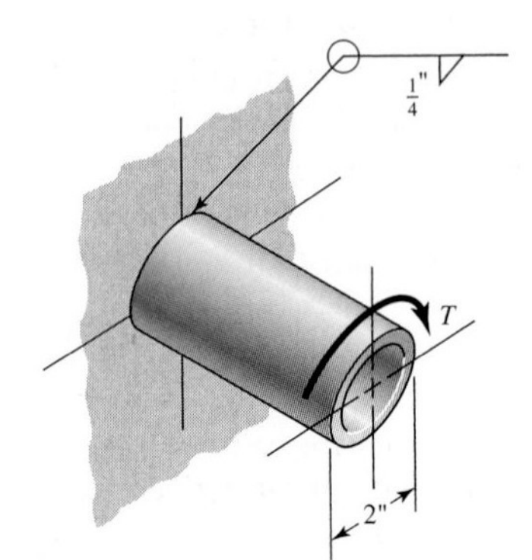

Problem 9–24

9–25 Find the maximum shear stress in the throat of the weld metal in the figure.

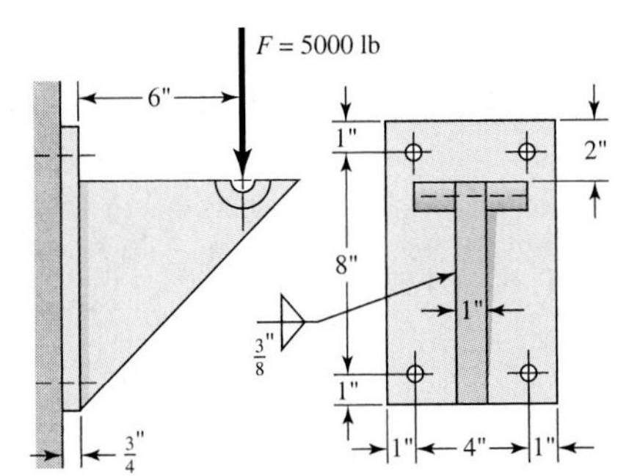

Problem 9–25

9–26 The figure shows a welded steel bracket loaded by a static force F. Estimate the factor of safety if the allowable shear stress in the weld throat is 120 MPa.

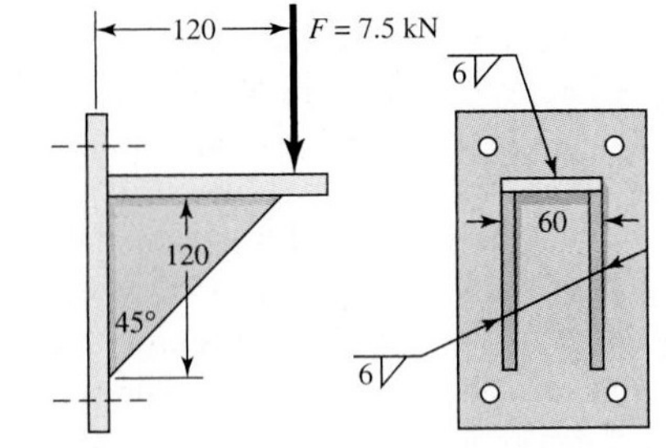

Problem 9–26

9–27 The figure on page 585 shows a formed sheet-steel bracket. Instead of securing it to the support with machine screws, welding has been proposed. If the combined stress in the weld metal is limited to 900 psi, estimate the total load W the bracket will support.

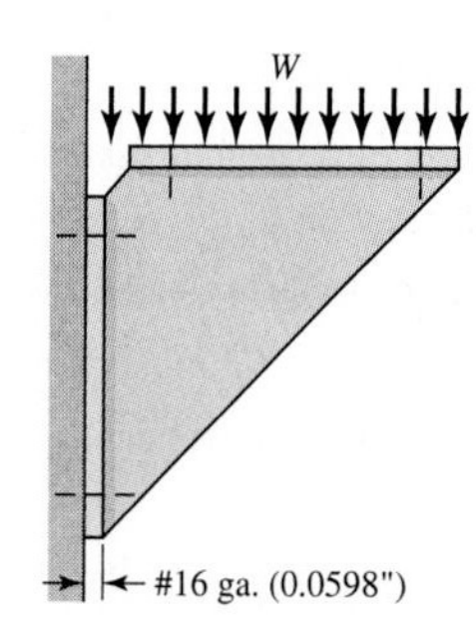

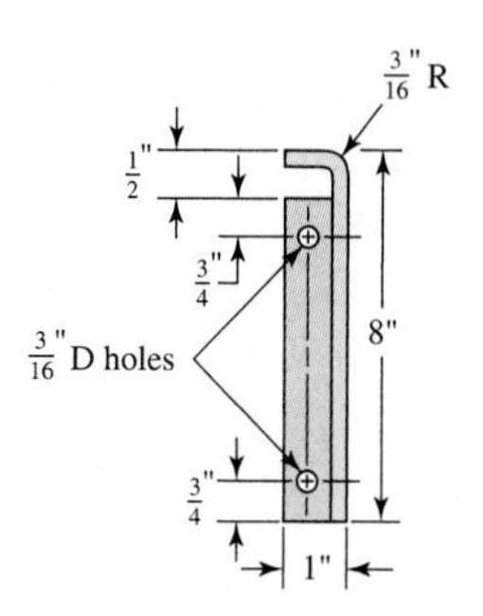

Problem 9–27
Structural support is A26 structural steel, bracket is 1020 press cold-formed steel. The weld electrode is 6010.

9–28 Without bracing, a machinist can exert only about 100 lb on a wrench or tool handle. The lever shown in the figure has $t = 1/2$ in and $w = 2$ in. We wish to specify the fillet-weld size to secure the lever to the tubular part at A. Both parts are of steel, and the shear stress in the weld throat should not exceed 3000 psi. Find a safe weld size.

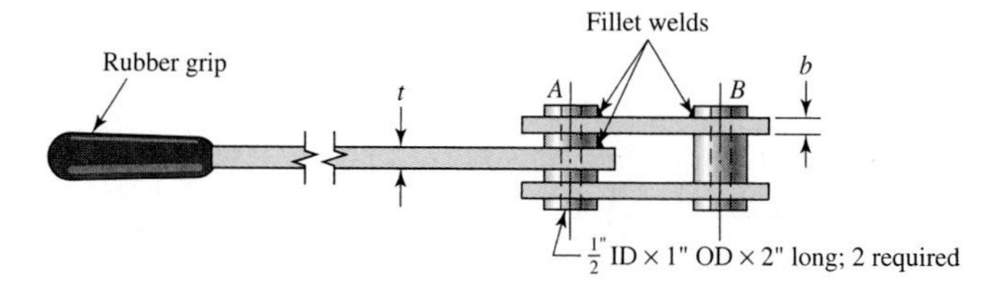

Problem 9–28

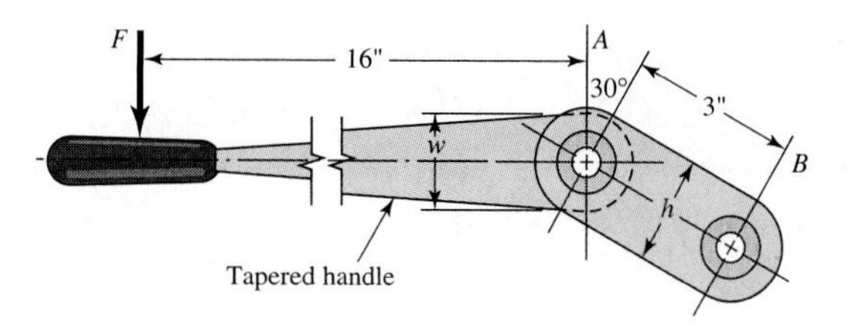

9–29 Estimate the safe static load F for the weldment shown in the figure if an E6010 electrode is used and the design factor is to be 2. Use conventional analysis.

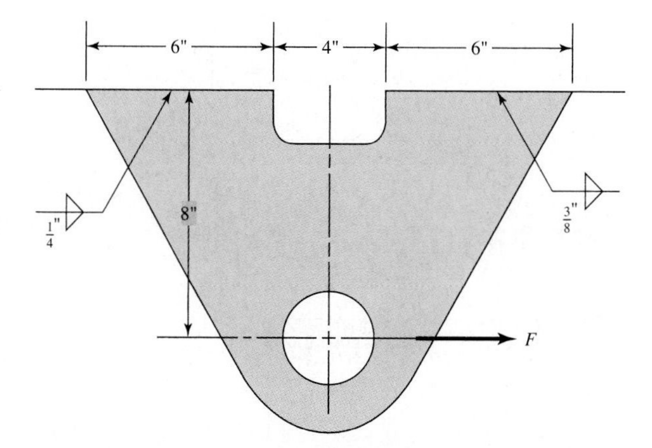

Problem 9–29

9–30 Brackets, such as the one depicted in the figure on page 586, are used in mooring small watercraft. Failure of such brackets is usually caused by bearing pressure of the mooring clip against the side of the hole. Our purpose here is to get an idea of the static and dynamic margins of safety involved. We use a bracket 1/4 in thick made of hot-rolled 1018 steel. We then assume wave action on the boat will create force F no greater than 1200 lb.

(*a*) Identify the moment M which produces a shear stress on the throat resisting bending action with a "tension" at A and "compression" at C.

(*b*) Find the force component F_y which produces a shear stress at the throat resisting a "tension" throughout the weld.

(*c*) Find the force component F_x that produces an in-line shear throughout the weld.

(*d*) Find A, I_u, and I using Table 9–3, in part.

(*e*) Find the shear stress τ_1 at A due to F_y and M, the shear stress τ_2 due to F_x, and combine to find τ.

(*f*) Find the factor of safety guarding against shear yielding in the weldment.

(*g*) Find the factor of safety guarding against a static failure in the parent metal at the weld.

(*h*) Find the factor of safety guarding against a fatigue failure in the weld metal using a Gerber failure locus.

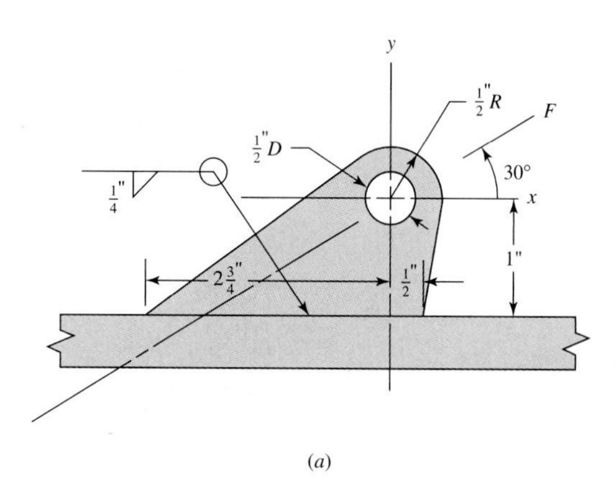

Problem 9–30
Small water-craft mooring bracket.

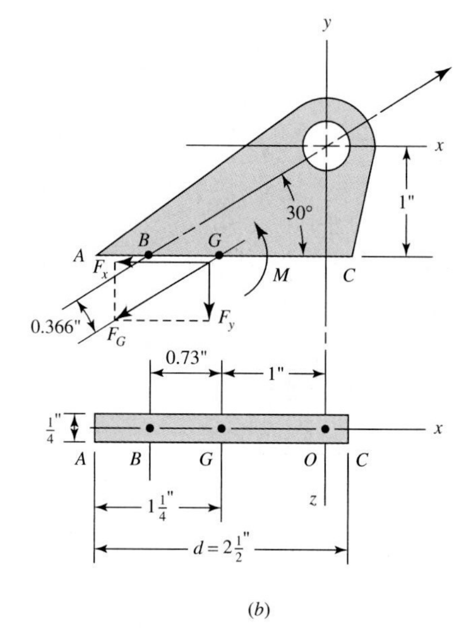

9–31 Compare your solution to Prob. 9–30 to the welding code method of analysis.

9–32 For the sake of perspective it is always useful to look at the matter of scale. Double all dimensions in Prob. 9–5 and find the allowable load. By what factor has it increased? First make a guess, then carry out the computation. Would you expect the same ratio if the load had been variable?

9–33 Hardware stores often sell plastic hooks that can be mounted on walls with pressure-sensitive adhesive foam tape. Suppose the maximum allowable design stress in *tension* for the tape is 2 psi. The back plate has a flexural rigidity EI of 100 lb·in^2, the modulus of elasticity E_a of the adhesive is 60 psi, and consider the wall rigid in comparison. What is the largest load the hook could hold? How would this capacity be changed if the hook were to be inverted as depicted in part b of the figure? Explain the difference in these two configurations and indicate which one you would buy. (*Hint*: Because the adhesive layer is so soft, the shear stress is carried over a large area. The tensile stresses, however, will be focused at the end where the moment is applied, so the beam on an elastic foundation analysis is quite realistic for this geometry.)

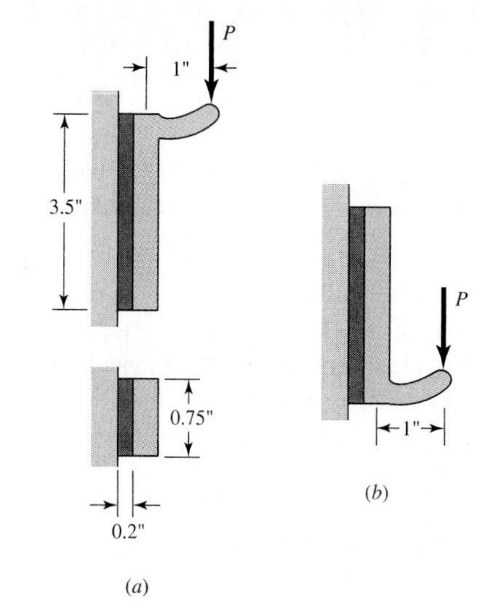

Problem 9–33

9–34 Estimate the magnitude of the residual tensile-stress state that would develop in an epoxy adhesive bonding two relatively thick aluminum plates. The stress-free temperature is 150°C, and the service temperature is −35°C. The moduli of elasticity of the aluminum plates is 70 GPa and the adhesive is 3 GPa. The linear thermal coefficients of thermal expansion are 23.9(10^{-6}) and 60(10^{-6}) mm/(mm °C) for the plates and adhesive respectively. Poisson's ratio for the adhesive is 0.35.

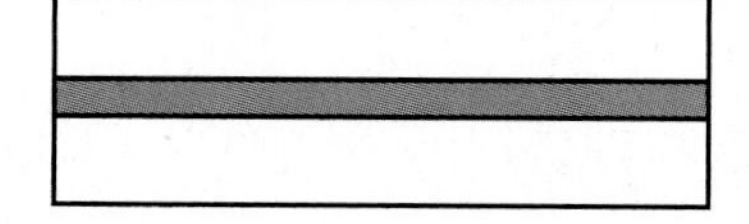

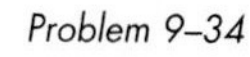
Problem 9–34

9–35 For a balanced double-lap joint cured at room temperature, Volkersen's equation simplifies to

$$\tau(x) = \frac{P\omega \cosh(\omega x)}{4b \sinh(\omega l/2)} = A_1 \cosh(\omega x)$$

(a) Show that the average stress $\bar{\tau}$ is $P/(2bl)$.

(b) Show that the largest shear stress is $P\omega/[4b \tanh(\omega l/2]$.

(*c*) Define a stress-augmentation factor K such that

$$\tau(l/2) = K\bar{\tau}$$

and it follows that

$$K = \frac{P\omega}{4b\tanh(\omega l/2)}\frac{2bl}{P} = \frac{\omega l/2}{\tanh(\omega l/2)} = \frac{\omega l}{2}\frac{\exp(\omega l/2)+\exp(-\omega l/2)}{\exp(\omega l/2)-\exp(-\omega l/2)}$$

9–36 In Ex. 9–10 the molding strip was bonded on the convex side of the panel. How would the analysis change if the flat molding strip were applied to a concave surface instead? Repeat the analysis for this situation.

9–37 Program the shear-lag solution for the shear-stress state into your computer using Eq. (9–15). Determine the maximum shear stress for each of the following scenarios:

Part	E_a, psi	t_o, in	t_i, in	E_o, psi	E_i, psi	h, in
a	$0.2(10^6)$	0.125	0.250	$30(10^6)$	$30(10^6)$	0.005
b	$0.2(10^6)$	0.125	0.250	$30(10^6)$	$30(10^6)$	0.015
c	$0.2(10^6)$	0.125	0.125	$30(10^6)$	$30(10^6)$	0.005
d	$0.2(10^6)$	0.125	0.250	$30(10^6)$	$10(10^6)$	0.005

Provide plots of the actual stress distributions predicted by this analysis. You may omit thermal stresses from the calculations, assuming that the service temperature is similar to the stress-free temperature. If the allowable shear stress is 800 psi and the load to be carried is 300 lb, estimate the respective factors of safety for each geometry. Let $l = 1.25$ in and $b = 1$ in.

10 Mechanical Springs

10–1 Stresses in Helical Springs **590**
10–2 The Curvature Effect **591**
10–3 Deflection of Helical Springs **592**
10–4 Extension Springs **592**
10–5 Compression Springs **595**
10–6 Stability **596**
10–7 Spring Materials **598**
10–8 Helical Compression Springs for Static Service **609**
10–9 Critical Frequency of Helical Springs **620**
10–10 Fatigue Loading **622**
10–11 Helical Compression Springs for Dynamic Service **625**
10–12 Design of a Helical Compression Spring for Dynamic Service **629**
10–13 Design of Extension Springs **637**
10–14 Designing Helical Coil Torsion Springs **664**
10–15 Belleville Springs **678**
10–16 Miscellaneous Springs **678**
10–17 Summary **680**

When a designer wants rigidity, negligible deflection is an acceptable approximation as long as it does not compromise function. Flexibility is sometimes needed and is often provided by metal bodies with cleverly controlled geometry. These bodies can exhibit flexibility to the degree the designer seeks. Such flexibility can be linear or nonlinear in relating deflection to load. These devices allow controlled application of force or torque; the storing and release of energy can be another purpose. Flexibility allows temporary distortion for access and the immediate restoration of function. Because of machinery's value to designers, springs have been intensively studied; moreover, they are mass-produced (and therefore low cost), and ingenious configurations have been found for a variety of desired properties. In this chapter we will discuss the more frequently used types of springs, their necessary parametric relationships, their adequacy assessment, and their design.

In general, springs may be classified as wire springs, flat springs, or special-shaped springs, and there are variations within these divisions. Wire springs include helical springs of round or square wire, made to resist tensile, compressive, and torsional loads. Flat springs include cantilever and elliptical types, wound motor- or clock-type power springs, and flat spring washers, usually called Belleville springs.

10–1 Stresses in Helical Springs

Figure 10–1a shows a round-wire helical compression spring loaded by the axial force F. We designate D as the *helix spring diameter* and d as the *wire diameter*. Now imagine that the spring is cut at some point (Fig. 10–1b), a portion of it removed, and the effect of the removed portion replaced by the internal forces. Then, as shown in the figure, the cut portion would exert a direct shear force F and a torsion T on the remaining part of the spring.

To visualize the torsion, picture a coiled garden hose. Now pull one end of the hose in a straight line perpendicular to the plane of the coil. As each turn of hose is pulled off the coil, the hose twists or turns about its own axis. The flexing of a helical spring creates a torsion in the wire in a similar manner.

The maximum stress in the wire may be computed by superposition of Eqs. (3–15) and (3–41). The result is

$$\tau_{\max} = \pm\frac{Tr}{J} + \frac{F}{A} \tag{a}$$

where the term Tr/J is the torsion formula and F/A is the direct (not flexural) shear

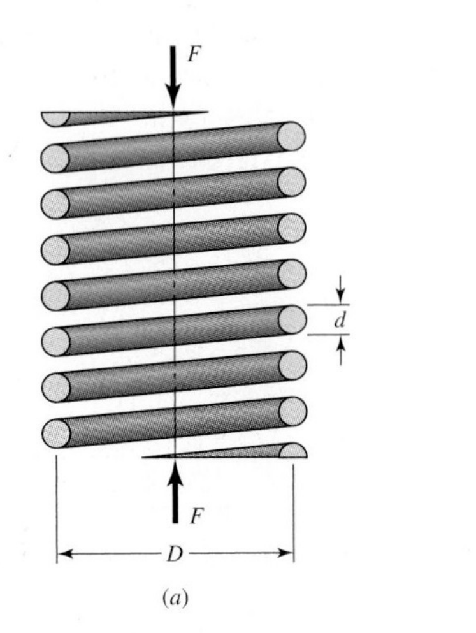

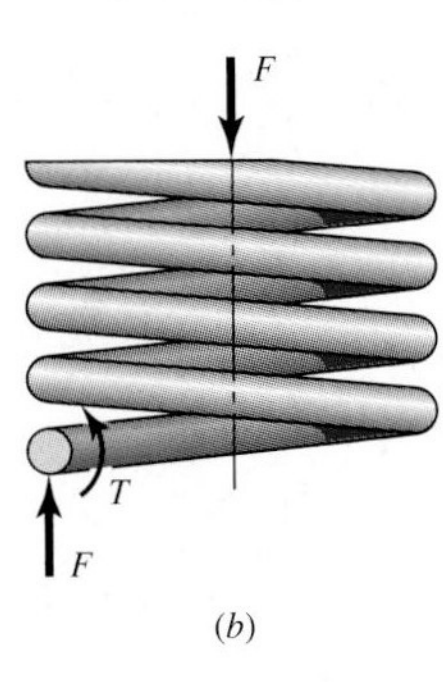

Figure 10–1

(a) Axially loaded helical spring; (b) free-body diagram showing that the wire is subjected to a direct shear and a torsional shear.

stress. Replacing the terms by $T = FD/2$, $r = d/2$, $J = \pi d^4/32$, and $A = \pi d^2/4$ gives

$$\tau = \frac{8FD}{\pi d^3} + \frac{4F}{\pi d^2} \tag{10–1}$$

In this equation the subscript indicating maximum shear stress has been omitted as unnecessary. The positive signs of Eq. (*a*) have been retained, and hence Eq. (10–1) gives the shear stress at the *inside* fiber of the spring.

Now we define the *spring index*

$$C = \frac{D}{d} \tag{10–2}$$

as a measure of coil curvature. With this relation, Eq. (10–1) can be rearranged to give

$$\tau = K_s \frac{8FD}{\pi d^3} \tag{10–3}$$

where K_s is a *shear-stress augmentation factor* and is defined by the equation

$$K_s = \frac{2C + 1}{2C} \tag{10–4}$$

For most springs, C will range from about 6 to 12. Equation (10–3) is quite general and applies for both static and dynamic loads.

The use of square or rectangular wire is not recommended for springs unless space limitations make it necessary. Springs of special wire shapes are not made in large quantities, as are those of round wire; they have not had the benefit of refining development and hence may not be as strong as springs made from round wire. When space is severely limited, the use of nested round-wire springs should always be considered. They may have an economical advantage over the special-section springs, as well as a strength advantage.

10–2 The Curvature Effect

An effect occurs at the inside surface of a helical spring. The curvature of the wire increases the stress on the inside of the spring but decreases it only slightly on the outside. This curvature stress is important only in fatigue because the loads are lower and there is no opportunity for localized yielding. For static loading, these stresses can be neglected because of strain-strengthening with the first application of load.

Unfortunately, it is necessary to find the curvature factor in a roundabout way. The reason for this is that the published equations include the effect of the direct shear stress too. Suppose K_s in Eq. (10–3) is replaced by another K factor which corrects for both curvature and direct shear. Then this factor is given by either of the equations

$$K_W = \frac{4C - 1}{4C - 4} + \frac{0.615}{C} \tag{10–5}$$

$$K_B = \frac{4C + 2}{4C - 3} \tag{10–6}$$

The first of these is called the Wahl factor, and the second, the Bergsträsser factor.[1] Since

1. Cyril Samónov, "Some Aspects of Design of Helical Compression Springs," *Int. Symp. Design and Synthesis*, Tokyo, 1984.

the results of these two equations differ by less than 1 percent, Eq. (10–6) is preferred. The curvature correction factor can now be obtained by canceling out the effect of the direct shear. Thus, using Eq. (10–6) with Eq. (10–4), the curvature correction factor is found to be

$$K_c = \frac{K_B}{K_s} = \frac{2C(4C+2)}{(4C-3)(2C+1)} \tag{10–7}$$

Now, K_s, K_B or K_W, and K_c are simply stress augmentation factors applied multiplicatively to Tr/J at the critical location to estimate a particular stress. There is *no* stress concentration factor. In this book we will use $\tau = K_B(8FD)/(\pi d^3)$ to predict the largest shear stress.

10–3 Deflection of Helical Springs

The deflection-force relations are quite easily obtained using Castigliano's theorem. The total strain energy for a helical spring is composed of a torsional component and a shear component. From Eqs. (4–24) and (4–25), the strain energy is

$$U = \frac{T^2 l}{2GJ} + \frac{F^2 l}{2AG} \tag{a}$$

Substituting $T = FD/2$, $l = \pi DN$, $J = \pi d^4/32$, and $A = \pi d^2/4$ results in

$$U = \frac{4F^2 D^3 N}{d^4 G} + \frac{F^2 DN}{d^2 G} \tag{b}$$

where $N = N_a$ = number of active coils. Then, using Castigliano's theorem to find total deflection y,

$$y = \frac{\partial U}{\partial F} = \frac{8FD^3 N}{d^4 G} + \frac{4FDN}{d^2 G} \tag{c}$$

Since $C = D/d$, Eq. (*c*) can be rearranged to yield

$$y = \frac{8FD^3 N}{d^4 G}\left(1 + \frac{1}{2C^2}\right) \approx \frac{8FD^3 N}{d^4 G} \tag{10–8}$$

The spring rate is $k = F/y$, and so

$$k = \frac{d^4 G}{8D^3 N} \tag{10–9}$$

10–4 Extension Springs

Extension springs necessarily must have some means of transferring the load from the support to the body of the spring. Although this can be done with a threaded plug or a swivel hook, both of these add to the cost of the finished product, and so one of the methods shown in Fig. 10–2 is usually employed. In designing a spring with a hook end, the stress-augmentation effect must be considered.

In Fig. 10–3*a* and *b* is shown a much-used method of designing the end. Stresses due to the sharp bend make it impossible to design the hook as strong as the spring. Tests as well as analyses show that the stress-augmentation factors for points A and B of Fig. 10–3 are

$$(K)_A = \frac{4C_1^2 - C_1 - 1}{4C_1(C_1 - 1)}, \quad C_1 = \frac{2r_1}{d} \tag{10–10a}$$

Figure 10–2

Types of ends used on extension springs. (Courtesy of Associated Spring Corporation.)

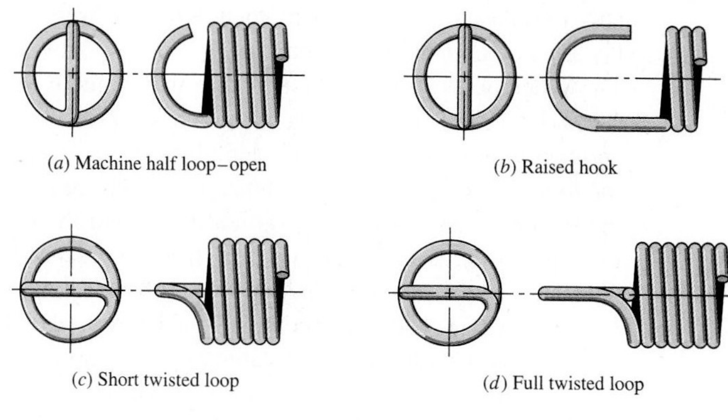

Figure 10–3

Ends for extension springs. (*a*) Usual design; stress at *A* is due to combined axial force and bending moment. (*b*) Side view of part *a*; stress is mostly torsion at *B*. (*c*) Improved design; stress at *A* is due to combined axial force and bending moment. (*d*) Side view of part *c*; stress at *B* is mostly torsion.

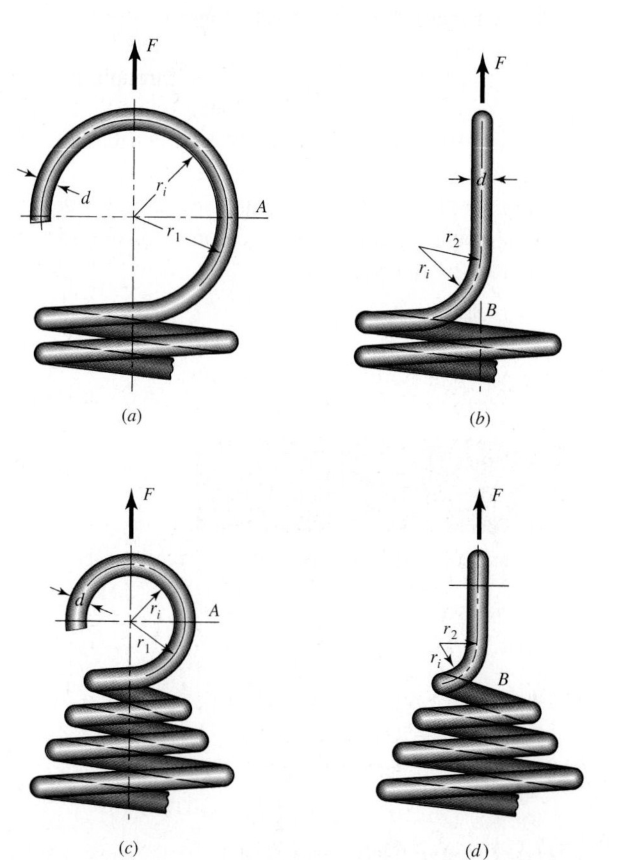

$$(K)_B = \frac{4C_2 - 1}{4C_2 - 4}, \quad C_2 = \frac{2r_2}{d} \tag{10–10b}$$

which holds for bending stress and occurs when the hook is offset, and for torsional stress. Figure 10-3*c* and *d* shows an improved design due to a reduced coil diameter, not to elimination of stress augmentation. The reduced coil diameter results in a lower stress because of the shorter moment arm. No stress-augmentation factor is needed for the axial component of the load.

When extension springs are made with coils in contact with one another, they are said to be *close-wound*. Spring manufacturers prefer some initial tension in close-wound springs in order to hold the free length more accurately. Table 10–1 gives the range of torsional stress due to the pre-tension as preferred by spring manufacturers. Stresses below the given range cause difficulty in controlling the free length. The customary tolerance on preload is ±10 percent, though lesser values can be obtained. The initial tension is created in the winding process by twisting the wire as it is wound onto the mandrel. When the spring is completed and removed from the mandrel, the initial tension is locked in because the spring cannot get any shorter.

The direction of the stresses can be visualized by reference to Fig. 10–4. In Fig. 10–4*a*, block *A* simulates the effect of the stacked coils, and the free length of the spring is the length L_0 with no external force applied. In Fig. 10–4*b*, an external force *F* has been applied, causing the spring to elongate through the distance *y*. Note, very particularly, that the stresses in the spring are in the *same direction* in Fig. 10–4*a* and *b*.

Figure 10–4*c* shows the relation between the external force and the spring elongation. Here we see that *F* must exceed the initial tension F_i before a deflection *y* is experienced.

The free length L_0 of an extension spring is equal to the body length plus 2 times the hook distance, and is measured on the inside surface of the hooks. The body length L_B is given by the equation

$$L_B = d(N_a + 1) \tag{10–11}$$

where N_a is the number of active coils.

Table 10–1

Preferred Range of Torsional Stresses Due to Initial Tension for Steel Helical Extension Springs

Source: Associated Spring–Barnes Group, *Design Handbook*, Bristol, Conn., 1987, p. 50.

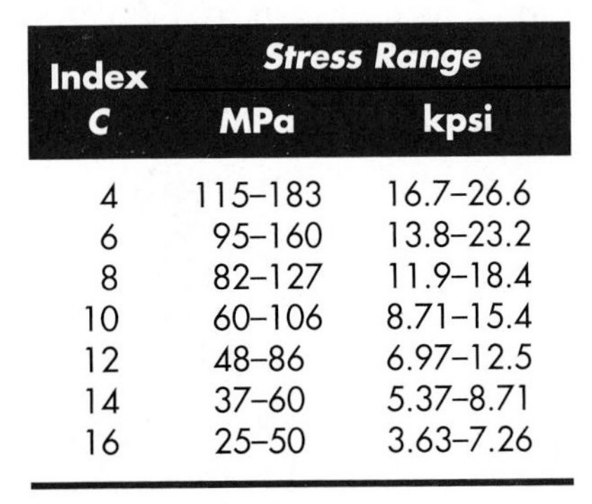

Index	Stress Range	
C	MPa	kpsi
4	115–183	16.7–26.6
6	95–160	13.8–23.2
8	82–127	11.9–18.4
10	60–106	8.71–15.4
12	48–86	6.97–12.5
14	37–60	5.37–8.71
16	25–50	3.63–7.26

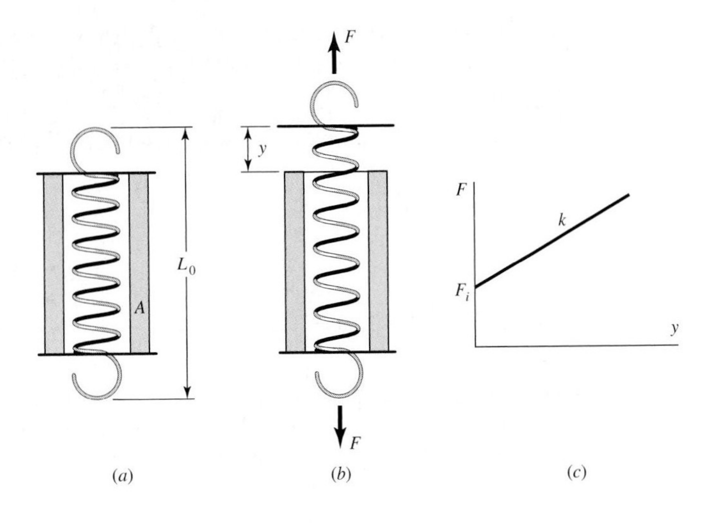

Figure 10–4

Simulation of an extension spring with initial coil body tension. (*a*) No external force; spring compresses block *A* with initial force F_i. The free length is L_0. (*b*) Spring extended a distance *y* by external force *F*. (*c*) Force-deflection locus.

10–5 Compression Springs

The four types of ends generally used for compression springs are illustrated in Fig. 10–5. A spring with *plain ends* has a noninterrupted helicoid; the ends are the same as if a long spring had been cut into sections. A spring with plain ends that are *squared* or *closed* is obtained by deforming the ends to a zero-degree helix angle. Springs should always be both squared and ground for important applications, because a better transfer of the load is obtained.

Table 10–2 shows how the type of end used affects the number of coils and the spring length.[2] Note that the digits 0, 1, 2, and 3 appearing in Table 10–2 are often used without question. *Some of these need closer scrutiny as they may not be integers.* This depends on how a springmaker forms the ends. Forys[3] pointed out that squared and ground ends give a solid length L_s of

$$L_s = (N_t - a)d$$

where a varies, with an average of 0.75, so the entry dN_t in Table 10–2 may be overstated. The way to check these variations is to take springs from a particular springmaker, close them solid, and measure the solid height. Another way is to look at the spring and count the wire diameters in the solid stack. It is useful to define Q as the number of "dead" turns, $Q = N_t - N_a$, and Q' as the quantity to add to N_a in finding the solid height. For

Figure 10–5

Types of ends for compression springs: (*a*) both ends plain; (*b*) both ends squared; (*c*) both ends squared and ground; (*d*) both ends plain and ground.

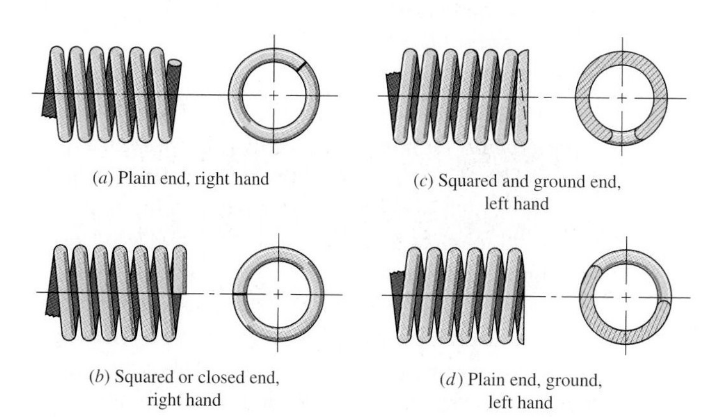

Table 10–2

Formulas for Compression-Spring Dimensions. (N_a = Number of Active Coils)

Source: Associated Spring–Barnes Group, *Design Handbook*, Bristol, Conn., 1987, p. 32.

	Type of Spring Ends			
Term	**Plain**	**Plain and Ground**	**Squared or Closed**	**Squared and Ground**
End coils, N_e	0	1	2	2
Total coils, N_t	N_a	$N_a + 1$	$N_a + 2$	$N_a + 2$
Free length, L_0	$pN_a + d$	$p(N_a + 1)$	$pN_a + 3d$	$pN_a + 2d$
Solid length, L_s	$d(N_t + 1)$	dN_t	$d(N_t + 1)$	dN_t
Pitch, p	$(L_0 - d)/N_a$	$L_0/(N_a + 1)$	$(L_0 - 3d)/N_a$	$(L_0 - 2d)/N_a$

2. For a thorough discussion and development of these relations, see Cyril Samónov, "Computer-Aided Design of Helical Compression Springs," ASME paper No. 80-DET-69, 1980.

3. Edward L. Forys, "Accurate Spring Heights," *Machine Design*, vol. 56, no. 2, January 26, 1984.

example, for squared and ground ends, the solid height is

$$L_s = (N_t - 1)d = (N_a + Q - 1)d = (N_a + Q')d$$

With $Q = 2$,

$$Q' = Q - 1 = 2 - 1 = 1$$

Therefore,

$$L_s = (N_a + 1)d \tag{10–12}$$

which varies somewhat from Forys's estimate. Spotts[4] points out the average value of Q for squared and ground ends is 1.75; for plain ends 0.5; and for plain ends, ground, 1. For the purposes of this book the following table is offered:

	Q	Q'
Plain ends	0*	1
Plain, ground ends	1	1
Squared or closed ends	2	3
Squared and ground ends	2†	1‡

* Some tests suggest 0.5 (Spotts).
† Some tests suggest 1.75 (Spotts).
‡ Some tests suggest 1.25 (Forys).

It all depends on how your springmaker prepares the ends. Find out! Get samples and test, and have your computer programs solicit the input of Q and Q'.

The length of wire used to wind a spring is $L \doteq \pi(N_a + Q)D$ and the volume of wire used to *make* the spring is $V \doteq \pi^2 d^2(N_a + Q)D/4$, which is often a component of a figure of merit.

Set removal or *presetting* is a process used in the manufacture of compression springs to induce useful residual stresses. It is done by making the spring longer than needed and then compressing it to its solid height. This operation *sets* the spring to the required final free length and, since the torsional yield strength has been exceeded, induces residual stresses opposite in direction to those induced in service. Springs to be preset should be designed so that 10 to 30 percent of the initial free length is removed during the operation. If the stress at the solid height is greater than 1.3 times the torsional yield strength, distortion may occur. If this stress is much less than 1.1 times, it is difficult to control the resulting free length.

Set removal increases the strength of the spring and so is especially useful when the spring is used for energy-storage purposes. However, set removal should not be used when springs are subject to fatigue.

10–6 Stability

In Chap. 4 we learned that a column will buckle when the load becomes too large. Similarly, compression coil springs will buckle when the deflection becomes too large.

4. M. F. Spotts, *Design of Machine Elements*, 6th ed., Prentice Hall, Englewood Cliffs, N.J., 1985, p. 232.

Table 10–3

End-Condition Constants α for Helical Compression Springs*

End Condition	Constant α
Spring supported between flat parallel surfaces (fixed ends)	0.5
One end supported by flat surface perpendicular to spring axis (fixed); other end pivoted (hinged)	0.707
Both ends pivoted (hinged)	1
One end clamped; other end free	2

*Ends supported by flat surfaces must be squared and ground.

The critical deflection is given by the equation

$$y_{cr} = L_0 C_1' \left[1 - \left(1 - \frac{C_2'}{\lambda_{eff}^2} \right)^{1/2} \right] \tag{10–13}$$

where y_{cr} is the deflection corresponding to the onset of instability. Samónov[5] states that this equation is cited by Wahl[6] and verified experimentally by Haringx.[7] The quantity λ_{eff} in Eq. (10–13) is the *effective slenderness ratio* and is given by the equation

$$\lambda_{eff} = \frac{\alpha L_0}{D} \tag{10–14}$$

C_1' and C_2' are the elastic constants and are defined by the equations

$$C_1' = \frac{E}{2(E - G)}$$

$$C_2' = \frac{2\pi^2(E - G)}{2G + E}$$

Equation (10-14) contains the *end-condition constant* α. This depends upon how the ends of the spring are supported. Table 10–3 gives values of α for usual end conditions. Note how closely these resemble the end conditions for columns.

Absolute stability occurs when, in Eq. (10–13), the term C_2'/λ_{eff}^2 is less than unity. This means that the condition for absolute stability is that

$$L_0 < \frac{\pi D}{\alpha} \left[\frac{2(E - G)}{2G + E} \right]^{1/2} \tag{10–15}$$

For steels, this turns out to be

$$L_0 < 2.63 \frac{D}{\alpha} \tag{10–16}$$

For squared and ground ends $\alpha = 0.5$ and $L_0 \le 5.26D$.

5. "Computer-Aided Design."

6. A. M. Wahl, *Mechanical Springs*, 2d ed., McGraw-Hill, New York, 1963.

7. J. A. Haringx, "On Highly Compressible Helical Springs and Rubber Rods and Their Application for Vibration-Free Mountings," I and II, *Philips Res. Rep.*, vol. 3, December 1948, pp. 401–449, and vol. 4, February 1949, pp. 49–80.

10–7 Spring Materials

Springs are manufactured either by hot- or cold-working processes, depending upon the size of the material, the spring index, and the properties desired. In general, prehardened wire should not be used if $D/d < 4$ or if $d > \frac{1}{4}$ in. Winding of the spring induces residual stresses through bending, but these are normal to the direction of the torsional working stresses in a coil spring. Quite frequently in spring manufacture, they are relieved, after winding, by a mild thermal treatment.

A great variety of spring materials are available to the designer, including plain carbon steels, alloy steels, and corrosion-resisting steels, as well as nonferrous materials such as phosphor bronze, spring brass, beryllium copper, and various nickel alloys. Descriptions of the most commonly used steels will be found in Table 10–4. The UNS steels listed in the Appendix should be used in designing hot-worked, heavy-coil springs, as well as flat springs, leaf springs, and torsion bars.

Spring materials may be compared by an examination of their tensile strengths; these vary so much with wire size that they cannot be specified until the wire size is known. The material and its processing also, of course, have an effect on tensile strength. It turns out that the graph of tensile strength versus wire diameter is almost a straight line for some materials when plotted on log-log paper. Writing the equation of this line as

$$S_{ut} = \frac{A}{d^m} \tag{10–17}$$

furnishes a good means of estimating minimum tensile strengths when the intercept A and the slope m of the line are known. Values of these constants have been worked out from recent data and are given for strengths in units of kpsi and MPa in Table 10–5. In Eq. (10–17) when d is measured in mm, then A is in MPa · mm^m and when d is measured in in, then A is in kpsi·in^m.

Equation (10–17) is valid only for the materials listed in Table 10–5. Figure 10–6 has been prepared to use in obtaining the strength of stainless steel wire (type 302) and of hard phosphor bronze wire. Note that phosphor bronze is very nearly a straight line on the semilog plot.

Although the torsional yield strength is needed to design the spring and to analyze the performance, spring materials customarily are tested only for tensile strength—perhaps because it is such an easy and economical test to make. A very rough estimate of the torsional yield strength can be obtained by assuming that the tensile yield strength is between 60 and 90 percent of the tensile strength. Then the distortion-energy theory can be employed to obtain the torsional yield strength. This approach results in the range

$$0.35S_{ut} \le S_{sy} \le 0.52S_{ut} \tag{10–18}$$

for steels. Tables 10–6 through 10–11 present useful spring information.

A helical compression spring with an original free length L_0' is closed solid and returns to a reduced value L_0. If the free length differs, the spring has undergone permanent set. With this permanent set, the material of the periphery of the spring wire, at places where stress is highest, undergoes cold work and attendant strain-strengthening. The former yield strength has been increased selectively, regions of higher strain-strengthening more, regions of lesser strain-strengthening not as much. Wherever the stress exceeded the prior material yield strength, the yield strength is improved. This process of inducing permanent set is used by springmakers to meet a free-length specification or to strengthen the spring. The springmaker calls the process *set removal*.

Table 10–4

High-Carbon and Alloy Spring Steels

Source: By permission from Harold C. R. Carlson, "Selection and Application of Spring Materials," *Mech. Eng.*, vol. 78, 1956, pp. 331–334.

Name of Material	Similar Specifications	Description
Music wire, 0.80–0.95*C*	UNS G10850 AISI 1085 ASTM A228-51	This is the best, toughest, and most widely used of all spring materials for small springs. It has the highest tensile strength and can withstand higher stresses under repeated loading than any other spring material. Available in diameters 0.12 to 3 mm (0.005 to 0.125 in). Do not use above 120°C (250°F) or at subzero temperatures
Oil-tempered wire, 0.60–0.70*C*	UNS G10650 AISI 1065 ASTM 229-41	This general-purpose spring steel is used for many types of coil springs where the cost of music wire is prohibitive and in sizes larger than available in music wire. Not for shock or impact loading. Available in diameters 3 to 12 mm (0.125 to 0.5000 in), but larger and smaller sizes may be obtained. Not for use above 180°C (350°F) or at subzero temperatures
Hard-drawn wire, 0.60–0.70*C*	UNS G10660 AISI 1066 ASTM A227-47	This is the cheapest general-purpose spring steel and should be used only where life, accuracy, and deflection are not too important. Available in diameters 0.8 to 12 mm (0.031 to 0.500 in). Not for use above 120°C (250°F) or at subzero temperatures
Chrome vanadium	UNS G61500 AISI 6150 ASTM 231-41	This is the most popular alloy spring steel for conditions involving higher stresses than can be used with the high-carbon steels and for use where fatigue resistance and long endurance are needed. Also good for shock and impact loads. Widely used for aircraft-engine valve springs and for temperatures to 220°C (425°F). Available in annealed or pretempered sizes 0.8 to 12 mm (0.031 to 0.500 in) in diameter
Chrome silicon	UNS G92540 AISI 9254	This alloy is an excellent material for highly stressed springs that require long life and are subjected to shock loading. Rockwell hardnesses of C50 to C53 are quite common, and the material may be used up to 250°C (475°F). Available from 0.8 to 12 mm (0.031 to 0.500 in) in diameter

Table 10–5

Constants A and m of $S_{ut} = A/d^m$ for Estimating Minimum Tensile Strength of Common Spring Wires

Source: Associated Spring–Barnes Group, *Design Handbook,* p. 19. Bristol, Conn., 1987. The graphical information has been curvefitted, relative costs of p. 20 added.

Material	ASTM No.	Exponent m	Diameter, in	A, kpsi·in^m	Diameter, mm	A, MPa·mm^m	Relative Cost of wire
Music wire*	A228	0.145	0.004–0.256	201	0.10–6.5	2211	2.6
OQ&T wire†	A229	0.187	0.020–0.500	147	0.5–12.7	1855	1.3
Hard-drawn wire‡	A227	0.190	0.028–0.500	140	0.7–12.7	1783	1.0
Chrome-vanadium wire§	A232	0.168	0.032–0.437	169	0.8–11.1	2005	3.1
Chrome-silicon wire‖	A401	0.108	0.063–0.375	202	1.6–9.5	1974	4.0
302 Stainless wire#	A313	0.146	0.013–0.10	169	0.3–2.5	1867	7.6–11
		0.263	0.10–0.20	128	2.5–5	2065	
		0.478	0.20–0.40	90	5–10	2911	
Phosphor-bronze wire**	B159	0	0.004–0.022	145	0.1–0.6	1000	8.0
		0.028	0.022–0.075	121	0.6–2	913	
		0.064	0.075–0.30	110	2–7.5	932	

*Surface is smooth, free of defects, and has a bright, lustrous finish.

†Has a slight heat-treating scale which must be removed before plating.

‡Surface is smooth and bright with no visible marks.

§Aircraft-quality tempered wire, can also be obtained annealed.

‖Tempered to Rockwell C49, but may be obtained untempered.

#Type 302 stainless steel.

**Temper CA510.

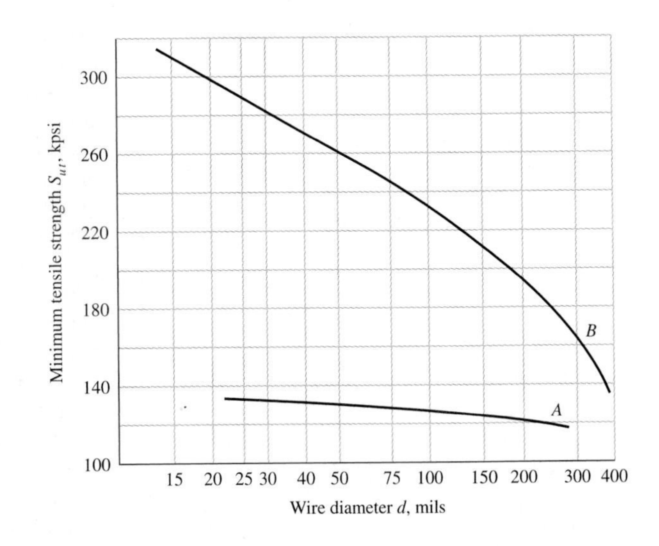

Figure 10–6

Minimum tensile strengths; A, hard phosphor-bronze wire, $E = 15$ Mpsi, $G = 6.5$ Mpsi; B, ASTM A313 stainless steel wire (type 302), $E = 28$ Mpsi, $G = 10$ Mpsi.

Table 10–6

Preferred Diameters for Spring Steel Wire

Source: Associated Spring–Barnes Group, *Design Handbook,* Bristol, Conn., 1987, p. 20.

Metric Sizes, mm			English Sizes, in.	
First Preference	Second Preference	Third Preference	First Preference	Second Preference
0.10			0.004	
	0.11		0.005	
0.12			0.006	
	0.14		0.008	
0.16				0.009
	0.18		0.010	
0.20				0.011
	0.22		0.012	
0.25				0.013
	0.28		0.014	
0.30				0.015
	0.35		0.016	
0.40				0.017
	0.45		0.018	
0.50				0.019
	0.55		0.020	
0.60				0.021
	0.65		0.022	
	0.70		0.024	
0.80			0.026	
	0.90		0.028	
1.0			0.030	
	1.1			0.031
1.2				0.033
		1.3	0.035	
	1.4		0.038	
1.6				0.040
	1.8		0.042	
2.0			0.045	
		2.1		0.047
	2.2		0.048	
		2.4	0.051	
2.5			0.055	
		2.6	0.059	
	2.8		0.063	
3.0			0.067	
		3.2	0.072	
	3.5		0.076	
		3.8	0.081	
4.0			0.085	
		4.2	0.092	
	4.5		0.098	
		4.8		0.102
5.0			0.105	
	5.5		0.112	
6.0				0.120
	6.5		0.125	
	7.0			0.130
		7.5	0.135	

Table 10–6

Continued

Metric Sizes, mm			English Sizes, in.	
First Preference	Second Preference	Third Preference	First Preference	Second Preference
8.0				0.140
		8.5	0.148	
	9.0			0.156
		9.5	0.162	
10.0				0.170
	11.0		0.177	
12.0			0.192	
	13.0			0.200
14.0			0.207	
	15.0			0.218
			0.225	
16.0			0.250	
				0.262
			0.281	
				0.306
			0.312	
			0.343	
			0.362	
			0.375	
			0.406	
			0.437	
			0.469	
			0.500	

*Steel wire size specifications are usually by either music wire gage or steel wire gage (Washburn and Moen). See Table E-25. Sizes displayed above are diameters in order to give perspective at a glance.

Table 10–7

Ranking of Relative Costs of Common Spring Wires

Source: Associated Spring–Barnes Group, *Design Handbook*, Bristol, Conn., 1987, p. 20.

		Relative Cost of 2 mm (0.079 in) Diameter	
Wire	Specification	Mill Quantities	Warehouse Lots
Patented and cold drawn	ASTM A227	1.0	1.0
Oil tempered	ASTM A229	1.3	1.3
Music	ASTM A228	2.6	1.4
Carbon valve spring	ASTM A230	3.1	1.9
Chrome-silicon valve	ASTM A401	4.0	3.9
Stainless steel (Type 302)	ASTM A313 (302)	7.6	4.7
Phosphor-bronze	ASTM	8.0	6.7
Stainless steel (Type 631) (17–7 PH)	ASTM A 313 (631)	11	8.7
Beryllium copper	ASTM B197	27	17
Inconel alloy X–750		44	31

Table 10–8

Standard Tolerances for Spring Wire

Source: Associated Spring–Barnes Group, *Design Handbook,* Bristol, Conn., 1987, p. 20.

Diameter		Tolerance		Maximum Out-of-Roundness	
mm	in	mm	in	mm	in
0.51–0.71	0.020–0.028	± 0.010	0.0004	0.010	0.0004
0.71–2.00	0.028–0.078	± 0.015	0.0006	0.015	0.0006
2.00–3.00	0.078–0.118	± 0.020	0.0008	0.020	0.0008
3.00–6.00	0.118–0.240	± 0.030	0.00118	0.030	0.0012
6.00–9.00	0.240–0.354	± 0.050	0.00197	0.050	0.002
9.50–16.00	0.375–0.625	± 0.070	0.00276	0.070	0.0028

Most spring wires can be purchased to tighter tolerances. Music wire and most nonferrous materials are regularly made to closer tolerances.

Table 10–9

Free-Length Tolerances of Squared and Ground Helical Compression Springs

Source: Associated Spring–Barnes Group, *Design Handbook,* Bristol, Conn., 1987, p. 42.

Number of Active Coils per mm (in)	Tolerances, ± mm/mm (in/in) of Free Length						
	Spring Index (D/d)						
	4	6	8	10	12	14	16
0.02 (0.5)	0.010	0.011	0.012	0.013	0.015	0.016	0.016
0.04 (1)	0.011	0.013	0.015	0.016	0.017	0.018	0.019
0.08 (2)	0.013	0.015	0.017	0.019	0.020	0.022	0.023
0.2 (4)	0.016	0.018	0.021	0.023	0.024	0.026	0.027
0.3 (8)	0.019	0.022	0.024	0.026	0.028	0.030	0.032
0.5 (12)	0.021	0.024	0.027	0.030	0.032	0.034	0.036
0.6 (16)	0.022	0.026	0.029	0.032	0.034	0.036	0.038
0.8 (20)	0.023	0.027	0.031	0.034	0.036	0.038	0.040

For springs less than 12.7 mm (0.500 in) long, use the tolerances for 12.7 mm (0.500 in). For closed ends not ground, multiply above values by 1.7.

For wires listed in Table 10–12 (page 606), the maximum allowable shear stress in a spring can be seen in column 3. Music wire and hard-drawn steel spring wire have a low end of range $S_{sy} = 0.45S_{ut}$. Valve spring wire, Cr-Va, Cr-Si, and other (not shown) hardened and tempered carbon and low-alloy steel wires as a group have $S_{sy} \geq 0.50S_{ut}$. Many nonferrous materials (not shown) as a group have $S_{sy} \geq 0.35S_{ut}$. In view of

Table 10–10

Coil Diameter Tolerances of Helical Compression and Extension Springs

Source: Associated Spring–Barnes Group, *Design Handbook*, Bristol, Conn., 1987, p. 43.

Wire Diameter, mm (in)	Tolerances, ± mm (in)						
	Spring Index (D/d)						
	4	6	8	10	12	14	16
0.38 (0.015)	0.05 (0.002)	0.05 (0.002)	0.08 (0.003)	0.10 (0.004)	0.13 (0.005)	0.15 (0.006)	0.18 (0.007)
0.58 (0.023)	0.05 (0.002)	0.08 (0.003)	0.10 (0.004)	0.15 (0.006)	0.18 (0.007)	0.20 (0.008)	0.25 (0.010)
0.89 (0.035)	0.05 (0.002)	0.10 (0.004)	0.15 (0.006)	0.18 (0.007)	0.23 (0.009)	0.28 (0.011)	0.33 (0.013)
1.30 (0.051)	0.08 (0.003)	0.13 (0.005)	0.18 (0.007)	0.25 (0.010)	0.30 (0.012)	0.38 (0.015)	0.43 (0.017)
1.93 (0.076)	0.10 (0.004)	0.18 (0.007)	0.25 (0.010)	0.33 (0.013)	0.41 (0.016)	0.48 (0.019)	0.53 (0.021)
2.90 (0.114)	0.15 (0.006)	0.23 (0.009)	0.33 (0.013)	0.46 (0.018)	0.53 (0.021)	0.64 (0.025)	0.74 (0.029)
4.34 (0.171)	0.20 (0.008)	0.30 (0.012)	0.43 (0.017)	0.58 (0.023)	0.71 (0.028)	0.84 (0.033)	0.97 (0.038)
6.35 (0.250)	0.28 (0.011)	0.38 (0.015)	0.53 (0.021)	0.71 (0.028)	0.90 (0.035)	1.07 (0.042)	1.24 (0.049)
9.53 (0.375)	0.41 (0.016)	0.51 (0.020)	0.66 (0.026)	0.94 (0.037)	1.17 (0.046)	1.37 (0.054)	1.63 (0.064)
12.70 (0.500)	0.53 (0.021)	0.76 (0.030)	1.02 (0.040)	1.57 (0.062)	2.03 (0.080)	2.54 (0.100)	3.18 (0.125)

this Joerres[8] uses the maximum allowable torsional stress for static application shown in Table 10–13 (page 606). For specific materials for which you have torsional yield information use this table as a guide. Joerres provides set-removal information in Table 10–13, that $S_{sy} \geq 0.65S_{ut}$ increases strength through cold work, but at the cost of an additional operation by the springmaker. Sometimes the additional operation can be done by the manufacturer during assembly. Some correlations with carbon steel springs show that the tensile yield strength of spring wire in torsion can be estimated from $0.75S_{ut}$. The corresponding estimate of the yield strength in shear based on distortion energy theory is $S_{sy} = 0.577(0.75)S_{ut} = 0.433S_{ut} \doteq 0.45S_{ut}$. Samónov discusses the problem of allowable stress and shows that

$$S_{sy} = \tau_{\text{all}} = 0.56S_{ut}$$

for high-tensile spring steels which is close to the value given by Joerres for hardened alloy steels. He points out that this value of allowable stress is specified by Draft Standard 2089 of the German Federal Republic when Eq. (10–3) is used without stress-augmentation factor.

8. Robert E. Joerres, "Springs," Chap. 24 in Joseph E. Shigley and Charles R. Mischke (eds.), *Standard Handbook of Machine Design*, 2nd ed., McGraw-Hill, New York, 1996.

Table 10–11

Load Tolerances of Helical Compression Springs *Source:* Associated Spring–Barnes Group, *Design Handbook*, Bristol, Conn., 1987, p. 44.

Load Tolerance, ± mm (in)	Tolerances, ± % of Load. Start with Tolerance from Table 10–9 Multiplied by L_F — Deflection from Free Length to Load, mm (in)														
	1.27 (0.050)	2.54 (0.100)	3.81 (0.150)	5.08 (0.200)	6.35 (0.250)	7.62 (0.300)	10.2 (0.400)	12.7 (0.500)	19.1 (0.750)	25.4 (1.00)	38.1 (1.50)	50.8 (2.00)	76.2 (3.00)	102 (4.00)	152 (6.00)
0.13 (0.005)	12.0	7.0	6.0	5.0											
0.25 (0.010)		12.0	8.5	7.0	6.5	5.5	5.0								
0.51 (0.020)		22.0	15.5	12.0	10.0	8.5	7.0	6.0	5.0						
0.76 (0.030)			22.0	17.0	14.0	12.0	9.5	8.0	6.0	5.0					
1.0 (0.040)				22.0	18.0	15.5	12.0	10.0	7.5	6.0	5.0				
1.3 (0.050)					22.0	19.0	14.5	12.0	9.0	7.0	5.5				
1.5 (0.060)					25.0	22.0	17.0	14.0	10.0	8.0	6.0	5.0			
1.8 (0.070)						25.0	19.5	16.0	11.0	9.0	6.5	5.5			
2.0 (0.080)							22.0	18.0	12.5	10.0	7.5	6.0	5.0		
2.3 (0.090)							25.0	20.0	14.0	11.0	8.0	6.0	5.0		
2.5 (0.100)								22.0	15.5	12.0	8.5	7.0	5.5		
5.1 (0.200)										22.0	15.5	12.0	8.5	7.0	5.5
7.6 (0.300)											22.0	17.0	12.0	9.5	7.0
10.2 (0.400)												21.0	15.0	12.0	8.5
12.7 (0.500)												25.0	18.5	14.5	10.5

First load test at not less than 15% of available deflection.
Final load test at not more than 85% of available deflection.

Table 10–12

Mechanical Properties of Some Spring Wires

Material	Elastic Limit, Percent of S_{ut} Tension	Torsion	Diameter d, in	E Mpsi	E GPa	G Mpsi	G GPa
Music wire A228	65–75	45–60	<0.032	29.5	203.4	12.0	82.7
			0.033–0.063	29.0	200	11.85	81.7
			0.064–0.125	28.5	196.5	11.75	81.0
			>0.125	28.0	193	11.6	80.0
HD spring A227	60–70	45–55	<0.032	28.8	198.6	11.7	80.7
			0.033–0.063	28.7	197.9	11.6	80.0
			0.064–0.125	28.6	197.2	11.5	79.3
			>0.125	28.5	196.5	11.4	78.6
Oil tempered A239	85–90	45–50		28.5	196.5	11.2	77.2
Valve spring A230	85–90	50–60		29.5	203.4	11.2	77.2
Chrome vanadium A231	88–93	65–75		29.5	203.4	11.2	77.2
A232	88–93			29.5	203.4	11.2	77.2
Chrome silicon A401	85–93	65–75		29.5	203.4	11.2	77.2
Stainless steel							
A313*	65–75	45–55		28	193	10	69.0
17-7PH	75–80	55–60		29.5	208.4	11	75.8
414	65–70	42–55		29	200	11.2	77.2
420	65–75	45–55		29	200	11.2	77.2
431	72–76	50–55		30	206	11.5	79.3
Phosphor bronze B159	75–80	45–50		15	103.4	6	41.4
Beryllium copper B197	70	50		17	117.2	6.5	44.8
	75	50–55		19	131	7.3	50.3
Inconel alloy X-750	65–70	40–45		31	213.7	11.2	77.2

*Also includes 302, 304, and 316.

Note: See Table 10–13 for allowable torsional stress design values.

Table 10–13

Maximum Allowable Torsional Stresses for Helical Compression Springs in Static Applications

Source: Robert E. Joerres, "Springs," Chap. 24 in Joseph E. Shigley and Charles R. Mischke (eds.), *Standard Handbook of Machine Design,* 2nd ed., McGraw-Hill, New York, 1996.

Material	Maximum Percent of Tensile Strength: Before Set Removed (includes K_W or K_B)	After Set Removed (includes K_s)
Music wire and cold-drawn carbon steel	45	60–70
Hardened and tempered carbon and low-alloy steel	50	65–75
Austenitic stainless steels	35	55–65
Nonferrous alloys	35	55–65

EXAMPLE 10–1 A helical compression spring is made of No. 16 (0.037) music wire. The outside diameter of the spring is $\frac{7}{16}$ in. The ends are squared and there are $12\frac{1}{2}$ total turns.

(*a*) Estimate the torsional yield strength of the wire.

(*b*) Estimate the static load corresponding to the yield strength.

(*c*) Estimate the scale of the spring.

(*d*) Estimate the deflection that would be caused by the load in part (*b*).

(*e*) Estimate the solid length of the spring.

(*f*) What length should the spring be to ensure that when it is compressed solid and then released, there will be no permanent change in the free length?

(*g*) Given the length found in part (*f*), is buckling a possibility?

(*h*) What is the pitch of the body coil?

Solution (*a*) Using Eq. (10–17) and Table 10–5, we find $A = 201$ kpsi and $m = 0.145$ Therefore,

$$S_{ut} = \frac{A}{d^m} = \frac{201}{0.037^{0.145}} = 324 \text{ kpsi}$$

Then, from Table 10–13,

Answer $S_{sy} = 0.45 S_{ut} = 0.45(324) = 146$ kpsi

(*b*) The mean spring diameter is $D = \frac{7}{16} - 0.037 = 0.400$ in, and so the spring index is $C = \frac{0.400}{0.037} = 10.8$. Then, from Eq. (10–6),

$$K_B = \frac{4C + 2}{4C - 3} = \frac{4(10.8) + 2}{4(10.8) - 3} = 1.124$$

Now rearrange Eq. (10–3) and solve for F_s, using the torsional yield strength instead of shear stress, giving

Answer
$$F_s = \frac{\pi d^3 S_{sy}}{8 K_B D} = \frac{\pi(146)10^3(0.037)^3}{8(1.124)0.400} = 6.46 \text{ lbf}$$

(*c*) From Table 10–2, $N_a = 12.5 - 2 = 10.5$ turns. Using $G = 11.85$ Mpsi, the scale of the spring is found to be, from Eq. (10–9),

Answer
$$k = \frac{Gd^4}{8D^3 N_a} = \frac{11.85(10^6)0.037^4}{8(0.400)^3 10.5} = 4.13 \text{ lbf/in}$$

Answer (*d*) $y_s = \dfrac{F_s}{k} = \dfrac{6.46}{4.13} = 1.56$ in.

(*e*) From Table 10–2 and the table following,

Answer $L_s = (N_a + Q')d = (10.5 + 3)0.037 = 0.500$ in

Answer (*f*) $L_0 = y_s + L_s = 1.56 + 0.500 = 2.06$ in.

(*g*) $2.63\dfrac{D}{\alpha} = 2.63\dfrac{0.400}{0.5} = 2.104$

Mathematically, a free length of 2.06 in is less than 2.104 in, and buckling is unlikely. However, the forming of the ends will control how close α is to 0.5. This has to be investigated and an inside rod or exterior tube or hole may be needed.

Answer (*h*) $p = \dfrac{L_0 - 3d}{N_a} = \dfrac{2.06 - 3(0.037)}{10.5} = 0.186$ in

Our ability to predict (estimate) useful attributes of springs in the manner of Ex. 10–1 is related to discreteness of wire size (Table 10–8), free length (Table 10–9), coil diameter (Table 10–10), load (Table 10–11), number of active turns, and variability of material properties.

Spring wire drawing is a friction/wear process and, as we saw in Chap. 2, gives rise to a uniform distribution provided the wire segments used to wind springs are thoroughly mixed as springs. The stochastic description of wire diameter $\mathbf{d}$ is

$$\mathbf{d} \sim U(d, \sigma_d) \qquad \text{or} \qquad \mathbf{d} \sim \bar{d}U(1, C_d)$$

From Eq. (2–27) the standard deviation of a uniform distribution is $(b-a)/(2\sqrt{3})$. For the spring of diameter $d = 0.037$ in of Ex. 10–1, the bilateral tolerance $t = \pm 0.0006$ in, making estimations of standard deviation σ_d and the coefficient of variation C_d, using Table 10–7,

$$\sigma_d = \frac{0.0006}{\sqrt{3}} = 0.000\,346 \text{ in}$$

$$C_d = \frac{0.000\,346}{0.037} = 0.009\,362$$

Table 10–8 is used to estimate the statistical parameters of the coil helix diameter $\mathbf{D}$, which can be expressed as

$$\mathbf{D} \sim N(D, \sigma_D) \qquad \text{or} \qquad \mathbf{D} \sim \bar{D}N(1, C_d)$$

The entries in Table 10–8 are "natural tolerances" representing $\pm 3\sigma$ of a normal distribution. For the spring of Ex. 10–1, $t = \pm 0.008$ in. The standard deviation σ_D and the coefficient of variation C_D of the helix diameter are estimated to be

$$\sigma_D = \frac{t}{3} = \frac{0.008}{3} = 0.002\,667 \text{ in}$$

Since the mean helix diameter is $D = \frac{7}{16} - 0.037 = 0.401$ in,

$$C_D = \frac{0.002\,667}{0.401} = 0.006\,551$$

The tolerances on free length of a helical compression spring are displayed in Table 10–9:

$$\mathbf{L}_0 \sim \mathbf{N}(L_0, \sigma_{L0}) \qquad \text{or} \qquad L_0 \sim \bar{L}_0\mathbf{N}(1, C_{L0})$$

The ratio $N_a/L_0 = 10.5/2.2 = 4.77$. The tolerance t is approximately $t \doteq \pm 0.024$ $(4.77) = \pm 0.1145$ in, so

$$\sigma_{L0} = \frac{0.1145}{3} = 0.0382$$

$$C_{L0} = \frac{0.0382}{2.2} = 0.017\,364$$

The coefficient of variation of the torsional modulus $\mathbf{G}$ is approximately 0.01. The tolerance on total turns T, and therefore on active turns N_a, is $\pm\frac{1}{4}$ turn (check with your springmaker):

$$\sigma_{N_a} = \frac{0.25}{3} = 0.08\underline{3} \text{ turn}$$

$$C_{N_a} = \frac{0.08\underline{3}}{10.5} = 0.007\,937$$

Now consider the spring rate given by Eq. (10–9) and used in Ex. 10–1. That computation is really an estimate of the mean of **k**. Consider the coefficient of variation of **k** (Table 2–1 is helpful):

$$\mathbf{k} = \mathbf{d}^4 \cdot \mathbf{G} \cdot \frac{1}{8} \cdot \frac{1}{\mathbf{D}^3} \cdot \frac{1}{\mathbf{N}_a}$$

$$C_{d^4} = 4C_d = 4(0.009\,362) = 0.0374$$

$$C_{1/D^3} = 3C_D = 3(0.006\,651) = 0.019\,953$$

$$C_{1/N_a} = C_{N_a} = 0.007\,937 \qquad C_G = 0.01$$

and the coefficient of variation of **k** is Pythagorean, that is,

$$C_k = (0.0374^2 + 0.019\,953^2 + 0.007\,937^2 + 0.01^2)^{1/2} = 0.04$$

and the standard deviation of **k** is 4.01(0.0442), or 0.178. The "natural spread" of the spring rate is $k = 4.01 \pm 3(0.178)$, or 4.01 ± 0.534. This makes the spread $3.48 < k < 4.54$ lbf/in. This represents poor control of spring rate, something to be kept in mind.

Consider another important point. Example 10–1 used music wire. The torsional modulus G is a function of wire diameter so, from Table 10–12,

d, in	G, psi
Up to 0.032	12 000 000
0.033–0.063	11 850 000
0.064–0.125	11 750 000
Over 0.125	11 600 000

Had the example used $11.5(10^6)$ psi for torsional modulus, it would allow this postproblem observation. The best estimate of $\bar{G}$ is 11 850 000 psi. Coursework in mechanics of materials builds the habit of using $30(10^6)$ for Young's modulus for steel (although this is actually a bit high) and $11.5(10^6)$ psi for torsional modulus, so that we rarely look it up. Is it hard to find?[9] It almost seem as though we believe that if we use a familiar value for G, and check one another's work, we can agree on the correctness of our incorrect mathematical drill. As Pogo said, "I have met the enemy and they is us."

There are larger statistical spreads in engineering work than imprecise knowledge of the torsional modulus G (mean and standard deviation). Find what's important and deal with that first. Do as much more as you can. The danger is forgetting there is imprecision and failing to act accordingly when it is important.

10–8 Helical Compression Springs for Static Service

The preferred range of spring index is $4 \le C \le 12$, with the lower indexes being more difficult to form (because of the danger of surface cracking) and springs with higher indexes tending to tangle often enough to require individual packing. This can be the first item of the adequacy assessment. The range of active turns is $3 \le N_a \le 15$. To keep linearity when a spring is about to close, it is necessary to avoid the gradual touching of coils (due to nonperfect pitch). The overrun-to-closure is controlled. A helical coil

9. Try H. Carlson, *Spring Designer's Handbook*, Marcel Dekker, New York, 1978.

spring characteristic is ideally linear. Practically, it is nearly so, but not everywhere. The spring force is not reproducible up to point A in Fig. 10–7. After a linear segment of the force–deflection curve, nonlinear behavior begins at B as the number of active turns diminishes as coils begin to touch. The designer confines the spring's operating point to the central 75 percent of the locus between no load, $y = 0$, and soliding, $y = y_s$. From the geometry of Fig. 10–8,

$$L_0 = L_1 + y_1 = L_s + y_s \tag{10–19}$$

$$L_1 - L_s = y_s - y_1 = \xi y_1 \tag{10–20}$$

where ξ is defined as the fractional overrun of y_1 to closure, y_s. It follows that

$$y_s = (1 + \xi) y_1 \tag{10–21}$$

To use all of the linear 75 percent of the spring characteristic, y is $\frac{7}{8}$ of y_s:

$$y_s = (1 + \xi) y_1 = (1 + \xi) \left(\frac{7}{8}\right) y_s$$

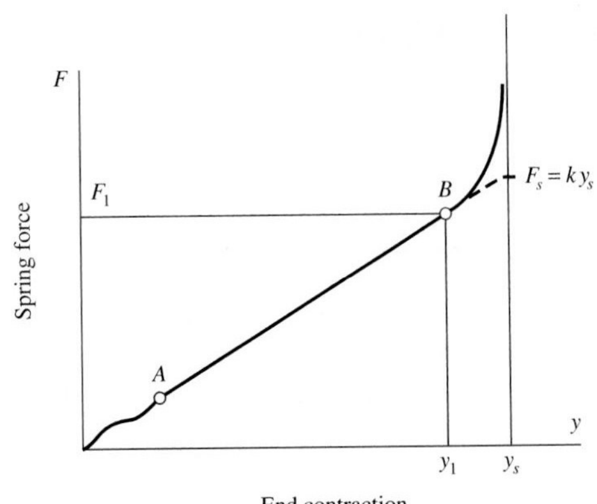

Figure 10–7

AB is the usable linear portion of the spring's force-end contraction locus. Point F_s, y_s lies on an extrapolation of the linear locus. The operating point for a static-service spring is best placed at or slightly below point *B*.

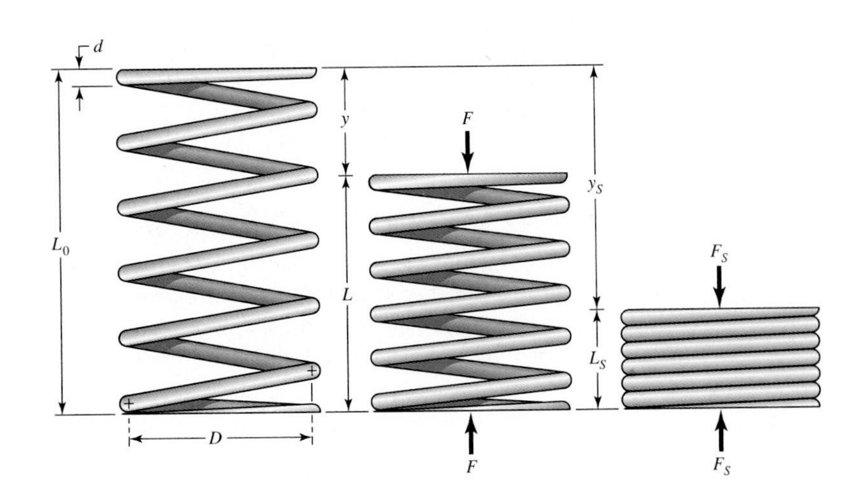

Figure 10–8

The geometry of a helical compression spring with squared and ground ends. The wire diameter is *d*, the load is *F*, the free length is L_0, the loaded length is *L*, and the end contraction is *y*. At soliding load F_s the end contraction is y_s and the spring length is L_s.

From the outer equality

$$\xi = \frac{1}{7} = 0.143 \doteq 0.15 \tag{10–22}$$

In keeping springs no longer than they have to be, it is prudent to place F_1, y_1 at B in Fig. 10–7, which is to set $\xi = 0.15$. It is useful to place $\xi \geq 0.15$ in your adequacy assessment. Spring nomenclature is defined in Fig. 10–8. Note that

$$\xi = \frac{y_s}{y_1} - 1 = \frac{L_0 - L_s}{L_0 - L_1} - 1 \tag{10–23}$$

When a spring is closed solid during assembly, or while someone plays with it, we want no yielding to occur whether set has been removed or not. If there is yielding, then we have a *different* spring than was intended. To avoid this we use a design factor $n_s = S_{sy}/\tau$. We set n_s equal to or greater than 1.2. The shearing yield strength is set according to Table 10-13, the column of choice depending on whether or not set is removed. Prevention of yielding when the spring is compressed solid is a design practice which produces a spring which is inherently safe in regard to yielding due to overload.

The adequacy assessment should check for the first surging frequency and keep well away from that of the source of excitation. Designers commonly keep the first surging frequency 20 times any exciting frequency. Buckling should be checked for with a fivefold margin on critical load for other than squared and ground ends on flat plates. In inequality from the adequacy assessment can be written as

$$4 \leq C \leq 12 \tag{10–24}$$

$$3 \leq N_a \leq 15 \tag{10–25}$$

$$\xi \geq 0.15 \tag{10–26}$$

$$n_s \geq 1.2 \tag{10–27}$$

$$S_{sy} = \begin{cases} 0.45S_{ut} & \text{music wire and cold-drawn carbon steels} \\ 0.50S_{ut} & \text{OQ\&T plain carbon and low-alloy steels} \\ 0.35S_{ut} & \text{stainless austenitic steel and nonferrous alloys} \\ 0.65S_{ut} & \text{all the above, but with set removal} \end{cases} \tag{10–28}$$

$$\tau_s = \begin{cases} K_B \dfrac{8F_s D}{\pi d^3} & \text{(as wound)} \\ K_s \dfrac{8F_s D}{\pi d^3} & \text{(set removed)} \end{cases} \tag{10–29}$$

$$\frac{S_{sy}}{\tau_s} \geq (n_s)_d \tag{10–30}$$

$$L_0 \leq (L_0)_{\text{cr}} = \frac{2.63D}{\alpha} \tag{10–31}$$

Excitation frequencies rarely disturb static springs. Critical frequency estimation will be addressed with dynamic service springs. We add to the preceding core of adequacy assessment all the inequalities which are geometric constraints:

- Free length L_0 $(L_0)_{\text{small}} \le L_0 \le (L_0)_{\text{large}}$
- Solid height L_s $(L_s)_{\text{small}} \le L_s \le (L_s)_{\text{large}}$
- Inside diameter ID $(\text{ID})_{\text{small}} \le \text{ID} \le (\text{ID})_{\text{large}}$
- Outside diameter OD $(\text{OD})_{\text{small}} \le \text{OD} \le (\text{OD})_{\text{large}}$
- Additional constraints peculiar to the application

Specification Set The specification set can be

- Material and condition
- End treatment
- As-wound or set removed
- Coil ID or OD and tolerance
- Total turns and tolerance
- Free length and tolerance
- Wire size and tolerance

Tolerance can be tighter than ordinary commercial tolerances, but at additional cost.

If load deflection must be controlled closer than would result from the specification set above, then a different specification set may be used:

- Material and condition
- End treatment
- As-wound or set removed
- Coil ID or OD and tolerance
- F_1 at L_1 (or F_1 at y_1) and tolerance on F_1
- Total turns and tolerance
- Wire size and tolerance

Decision Set The decision set can be

- Material and condition
- End treatment
- Robust linearity: $\xi = 0.15$
- As-wound or set removed
- Compressed solid safety factor: $(n_s)_d = 1.2$
- Function: F_1 at y_1 (or F_1 and L_1)
- Wire size d

The first six are a priori decisions and the wire size a design variable or variables. Tolerances can be assigned later, which simplifies the problem.

Figure of Merit The figure of merit can be the cost of the wire from which the spring is wound:

$$\text{fom} = -(\text{relative material cost})\frac{\gamma\pi^2 d^2 N_t D}{4} \tag{10–32}$$

For steels, the specific weight γ can be omitted.

Optimization Strategy Make the a priori decisions, with hard-drawn steel wire the first choice (relative material cost is 1.0). Choose a wire size d. With all decisions made, generate a column of parameters: d, D, C, OD or ID, N_a, L_s, L_0, $(L_0)_{cr}$, L_1, n_s, and fom. By incrementing wire sizes available, we can scan the rectangular table and apply the adequacy assessment by inspection. After wire sizes are eliminated, choose the spring design with the highest figure of merit. We will have found the optimal design despite the presence of a discrete design variable d and aggregation of equality and inequality constraints. The column vector of information can be generated using the flow chart displayed in Fig. 10–9. It is general enough to accommodate to the situations of as-wound and set-removed springs, operating over a rod, in a hole free of rod or hole. In as-wound springs the controlling equation must be solved for spring index by successive substitution. The equation is convergent for this purpose, gaining one correct digit each iteration every repetition after the first. Efficient Fortran coding for this operation can be

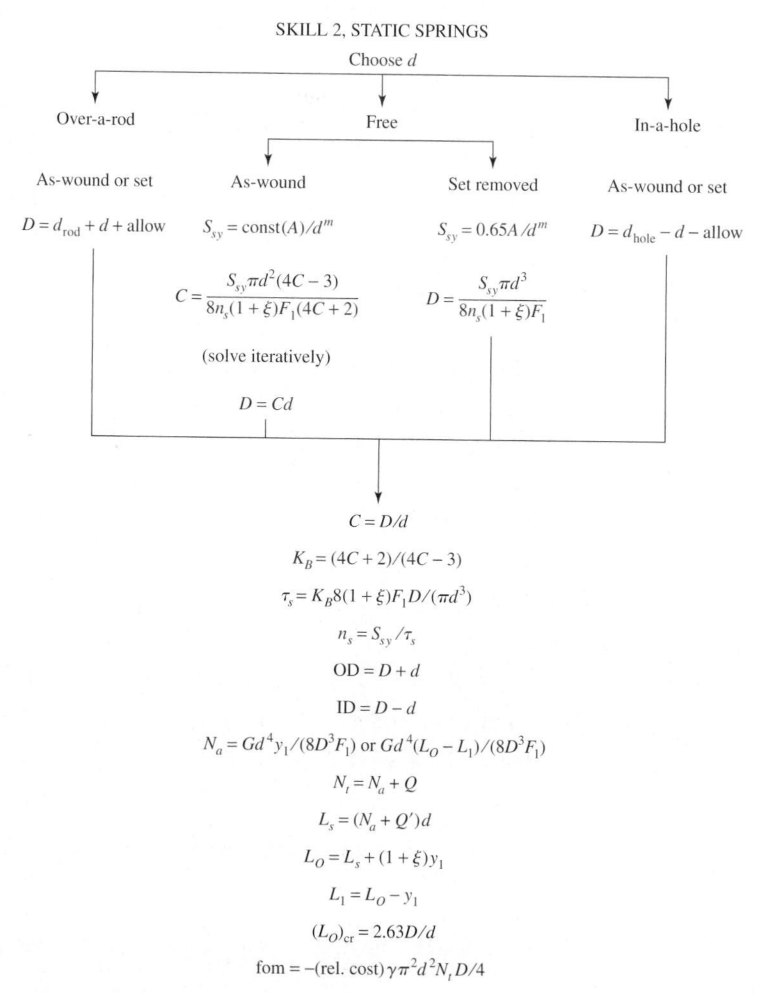

Figure 10–9

Skill 2 for a helical coil compression spring.

```
.
.
.
C=8
Cold=C
do 10 i=1, 100
Cnew=Ssy+pi*d**2*(4.*C-3.)/ns/8./(1.+xi)/F1/(4.*C+2.)
if(abs(Cold-Cnew).le.0.00001)go to 11
C=Cnew
Cold=Cnew
10 continue
    .
    .
    .
```

EXAMPLE 10–2 A music wire helical compression spring is needed to support a 20-lb load after being compressed 2 in. Because of assembly considerations the solid height cannot exceed 1 in and the free length cannot be more than 4 in. Design the spring.

Solution The a priori decisions are

- Music wire, A228, $A = 201\ 000$, $m = 0.145$; from Table 10–5, $E = 28.5$ Mpsi, $G = 11.75$ Mpsi (expect $d > 0.064$ in), Table 10–12
- Ends squared and ground: $Q = 2$, $Q' = 1$
- Function: $F_1 = 20$ lb, $y_1 = 2$ in
- Safety: use design factor at solid height of $(n_s)_d = 1.2$
- Robust linearlity: $\xi = 0.15$
- Use as-wound spring (cheaper), $S_{sy} = 0.45 S_{ut}$ per Table 10–13
- Decision variable: $d = 0.080$ in, music wire gage #30, Table E-25. Using Fig. 10–9,

$$S_{sy} = 0.45 \frac{201\ 000}{0.080^{0.145}} = 130\ 455 \text{ psi}$$

Use force at solid height of $F_s = (1 + \xi)F_1$ with stress of $\tau_s = S'_{sy}/n_s$, or Fig. 10–9

$$C = \frac{130\ 455\pi 0.080^2(4C - 3)}{8(1.2)(1 + 0.15)20(4C + 2)}$$

Begin with $C = 8$, iterate, $C = 10.53348$, say 10.53,

$$D = Cd = 10.53(0.080) = 0.843$$

$$K_s = \frac{2(10.53) + 1}{2(10.53)} = 1.047$$

$$K_B = \frac{4(10.53) + 2}{4(10.53) - 2} = 1.128$$

$$\tau_s = 1.128 \frac{8(1 + 0.15)20(0.843)}{\pi(0.080)^3} = 108\ 776 \text{ psi}$$

$$n_s = \frac{130\,445}{108\,776} = 1.2$$

$$\text{OD} = 0.843 + 0.080 = 0.943 \text{ in}$$

$$N_a = \frac{0.080^4(11.75)10^6(2)}{8(0.843)^3 20} = 10.05 \text{ turns}$$

$$N_t = 10.05 + 2 = 12.05 \text{ total turns}$$

$$L_s = (10.05 + 1)0.080 = 0.884 \text{ in}$$

$$L_0 = 0.884 + (1 + 0.15)2 = 3.184 \text{ in}$$

$$L_1 = L_0 - y_1 = 3.184 - 2 = 1.184 \text{ in}$$

$$(L)_{\text{cr}} = 2.63(0.843/0.5) = 4.43 \text{ in}$$

$$\text{fom} = -2.6\pi^2(0.080)^2 12.05(0.843)/4 = -0.417$$

Repeat the above for other wire diameters and form a table:

d:	**0.059**	**0.067**	**0.071**	**0.075**	**0.080**	**0.085**	**0.090**	**0.095**
D	0.313	0.479	0.578	0.688	0.843	1.017	1.211	1.427
C	5.297	7.153	8.143	9.178	10.53	11.96	13.46	15.02
OD	0.372	0.546	0.649	0.763	0.923	1.102	1.301	1.522
N_a	58.3	26.9	19.3	14.2	10.1	7.3	5.4	4.1
L_s	3.500	1.869	1.442	1.144	0.844	0.705	0.578	0.486
L_0	5.800	4.169	3.742	3.444	3.184	3.005	2.878	2.786
$(L_0)_{cr}$	1.64	2.52	3.04	3.62	4.43	5.35	6.37	7.51
n_s	1.2	1.2	1.2	1.2	1.2	1.2	1.2	1.2
fom	−0.421	−0.399	−0.398	−0.404	−0.417	−0.438	−0.467	−0.505

Now examine the table and perform the adequacy assessment. The constraint $3 \le N_a \le 15$ rules out wire diameters less than 0.080 in. The spring index constraint $4 \le C \le 12$ rules out diameters larger than 0.085 in. The $L_s > 1$ constraint rules out diameters less than 0.080 in. The $L_0 > 4$ constraint rules out diameters less than 0.071 in. The buckling criterion rules out free lengths longer than $(L_0)_{\text{cr}}$, which rules out diameters less than 0.075 in. The factor of safety n_s is exactly 1.20 because the mathematics forced it. Had the spring been in a hole or over a rod, the helix diameter would be chosen without reference to $(n_s)_d$. The result is that there are only two springs in the feasible domain, one with a wire diameter of 0.080 in and the other with a wire diameter of 0.085. The figure of merit decides and the decision is the design with 0.080 in wire diameter. It is useful to plot the figure of merit as ordinate and the wire diameter as abscissa. Mark the feasible designs with hollow points. Note that the curve does not exist; only the discrete points do. There is no stationary point maximum because there is no such thing as curve slope. Even if there were, the maximum is elsewhere, and we found it. This plot is shown in Fig. 10–10. A specification sheet checklist is shown in Table E–37 in the Appendix.

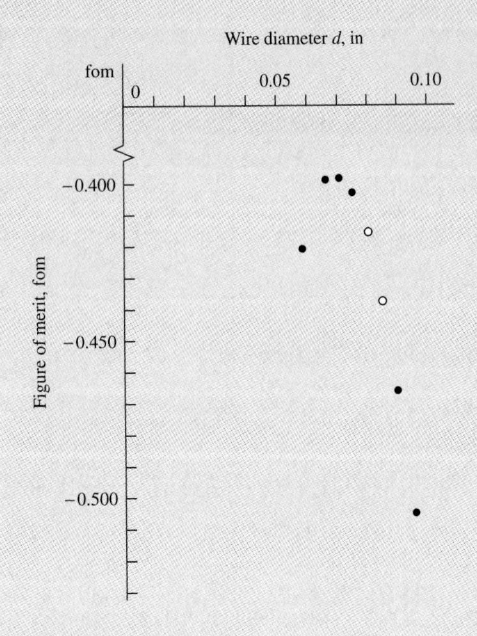

Figure 10–10

The figure of merit plotted against wire diameter d for eight wire sizes. The hollow plotting points denote the only two satisfactory springs.

Having designed a spring, will we have it made to our specifications? Not necessarily. There are vendors who stock literally thousands of music wire compression springs. By browsing their catalogs, we will usually find several that are close. Maximum deflection and maximum load are listed in the display of characteristics. Check to see if this allows soliding without damage. Often it does not. Spring rates may only be close. At the very least this situation allows a small number of springs to be ordered "off the shelf" for testing. The decision often hinges on the economics of special order versus the acceptability of a close fit.

EXAMPLE 10–3

Indexing is used in machine operations when a circular part being manufactured must be divided into a certain number of segments. Figure 10–11 shows a portion of an indexing fixture used to successively position a part for the operation. When the knob is pulled up, part 6, which holds the workpiece, is rotated to the next position and locked in place by releasing the index pin. In this example we wish to design the spring to exert a force of about 3 lb and to fit in the space defined in the figure caption.

Solution

Since the fixture is not a high-production item, a stock spring will be selected. These are available in music wire. In one catalog there are 76 stock springs available having an outside diameter of 0.480 in and designed to work in a $\frac{1}{2}$-in hole. These are made in seven different wire sizes, ranging from 0.038 up to 0.063 in, and in free lengths from $\frac{1}{2}$ to $2\frac{1}{2}$ in, depending upon the wire size.

Since the pull knob must be raised $\frac{3}{4}$ in for indexing and the space for the spring is $1\frac{3}{8}$ in long when the pin is down, the solid length cannot be more than $\frac{5}{8}$ in.

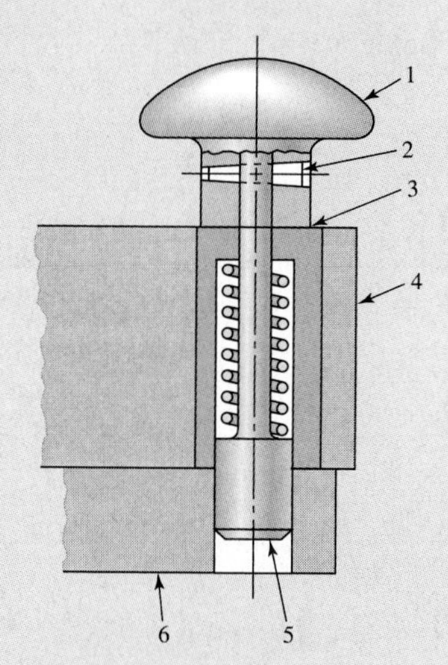

Figure 10–11

Part 1, pull knob; part 2, tapered retaining pin; part 3, hardened bushing with press fit; part 4, body of fixture; part 5, indexing pin; part 6, workpiece holder. Space of the spring is $\frac{5}{8}$ in OD, $\frac{1}{4}$ in ID, and $1\frac{3}{8}$ in long, with the pin down as shown. The pull knob must be raised $\frac{3}{4}$ in to permit indexing.

Let us begin by selecting a spring having an outside diameter of 0.480 in, a wire size of 0.051 in, a free length of $1\frac{3}{4}$ in, $11\frac{1}{2}$ total turns, and plain ends. Then $m = 0.145$ and $A = 201$ kpsi for music wire. Then

$$S_{sy} = 0.45\frac{A}{d^m} = 0.45\frac{201}{0.051^{0.145}} = 139.3 \text{ kpsi}$$

With plain ends $Q = 0$, $Q' = 1$, so the number of active turns is

$$N_a = N_t - Q = 11.5 - 0 = 11.5 \text{ turns}$$

The helix diameter is $D = D_0 - d = 0.480 - 0.051 = 0.429$ in. From Eq. (10–9) the spring rate is, using $G = 11.6(10^6)$ psi from Table 10–12,

$$k = \frac{d^4 G}{8D^3 N_a} = \frac{0.051^4(11.6)10^6}{8(0.429)^3 11.5} = 10.8 \text{ lb/in}$$

The solid height L_s is

$$L_s = (N_a + Q')d = (11.5 + 1)0.051 = 0.638 \text{ in}$$

The spring force when the pin is down, F_{min}, is

$$F_{\text{min}} = ky_{\text{min}} = 10.8(1.75 - 1.375) = 4.05 \text{ lb}$$

When the spring is compressed solid, the spring force F_s is

$$F_s = ky_s = k(L_0 - L_s) = 10.8(1.75 - 0.638) = 12 \text{ lb}$$

Since the spring index is $C = \frac{D}{d} = \frac{0.429}{0.051} = 8.41$,

$$K_B = \frac{4C + 2}{4C - 3} = \frac{4(8.41) + 2}{4(8.41) - 3} = 1.163$$

and for the as-wound spring, the shear stress at soliding is

$$\tau_s = K_B \frac{8F_s D}{\pi d^3} = 1.163 \frac{8(12)0.429}{\pi(0.051)^3} = 114\,934 \text{ psi}$$

The factor of safety at soliding is

$$n_s = \frac{S_{sy}}{\tau_s} = \frac{139.3}{114.9} = 1.21$$

Since n_s is adequate but L_s is larger than $\frac{5}{8}$ in, we must investigate other springs with a smaller wire size. After several investigations another spring has possibilities. It is as-wound music wire, $d = 0.045$ in, 20 gage (see Table E-25) $D_O = 0.480$, $N_t = 11.5$ turns, $L_O = 1.75$ in. S_{sy} is still 139.3 kpsi, and

$$D = D_O - d = 0.480 - 0.045 = 0.435 \text{ in}$$

$$N_a = N_t - Q = 11.5 - 0 = 11.5 \text{ turns}$$

$$k = \frac{0.051^4(11.6)10^6}{8(0.435)^3 11.5} = 6.28 \text{ lb/in}$$

$$L_s = (N_a + Q')d = (11.5 + 1)0.045 = 0.563 \text{ in}$$

$$F_{\min} = ky_{\min} = 6.28(1.75 - 1.375) = 2.36 \text{ lb}$$

$$F_s = 6.28(1.75 - 0.563) = 7.45 \text{ lb}$$

$$C = \frac{D}{d} = \frac{0.435}{0.045} = 9.67$$

$$K_B = \frac{4(9.67) + 2}{4(9.67) - 3} = 1.140$$

$$\tau_s = 1.140 \frac{8(7.45)0.435}{\pi (0.045)^3} = 103\ 253 \text{ psi}$$

$$n_s = \frac{S_{sy}}{\tau_s} = \frac{139.3}{103.2} = 1.35$$

Now $n_s > 1.2$, buckling is not possible as the coils are guarded by the hole surface, and the solid length is less than $\frac{5}{8}$ in, so this spring is selected. By using a stock spring, economy of scale is available.

Before leaving static springs, perhaps some perspective on the compressed solid design factor $(n_s)_d$ for as-wound carbon steel and music wire springs would be helpful. We can write n_s as

$$n_s = \frac{S_{sy}}{\tau_s} = \frac{0.45A}{d^m} \frac{\pi d^3}{K_B 8 F_s D}$$

Substituting the force to closure

$$F_s = ky_s = \frac{d^4 G}{8D^3 N_a}[L_O - (N_a + Q')d]$$

into the equation for n_s gives

$$n_s = \frac{0.45 A \pi D^2 N_a}{K_B d^{1+m} G[L_O - (N_a + Q')d]} \tag{10-33}$$

Because of tolerances on d, D, N_a, and L_0 there is a range in n_s. The worst stacking of tolerances occurs when all deviations from midrange are on one side and of largest value. We can write

$$\Delta n_s = \left|\frac{\partial n_s}{\partial d}\Delta d\right| + \left|\frac{\partial n_s}{\partial D}\Delta D\right| + \left|\frac{\partial n_s}{\partial N_a}\Delta N_a\right| + \left|\frac{\partial n_s}{\partial L_0}\Delta L_0\right| \tag{10–34}$$

For a case where $A = 186\,000$ psi, $m = 0.163$, $\bar{d} = 0.045$ in, $\bar{D} = 0.435$ in, $\bar{N}_a = 11.5$ turns, $\bar{L}_0 = 0.563$ in, $G = 11.6(10^6)$ psi, the midrange value of $\bar{n}_s$ is 1.342 360. By programming Eq. (10–33) on a hand-held programmable calculator so to allow entering values of d, D, N_a, and L_0, we can numerically evaluate the partial derivatives as follows:

$$\frac{\partial n_s}{\partial d} \doteq \frac{n_s(\bar{d}+\Delta d) - n_s(\bar{d})}{\Delta d} = \frac{1.322\,402 - 1.342\,360}{0.001} = -19.96$$

$$\frac{\partial n_s}{\partial D} \doteq \frac{n_s(\bar{D}+\Delta D) - n_s(\bar{D})}{\Delta D} = \frac{1.348\,539 - 1.342\,360}{0.001} = 6.179$$

$$\frac{\partial n_s}{\partial N_a} = \frac{n_s(\bar{N}_a+\Delta N_a) - n_s(\bar{N}_a)}{\Delta N_a} = \frac{1.342\,528 - 1.342\,360}{0.001} = 0.168$$

$$\frac{\partial n_s}{\partial L_0} = \frac{n_s(\bar{L}_0+\Delta L_0) - n_s(\bar{L}_0)}{\Delta L_0} = \frac{1.341\,231 - 1.342\,360}{0.001} = -1.129$$

If the bilateral tolerances assigned were

$$d = 0.045 \pm 0.0006 \text{ in} \qquad \text{(Table 10–8)}$$
$$D = 0.435 \pm 0.009 \text{ in} \qquad \text{(Table 10–9)}$$
$$N_a = 11.5 \pm 0.25 \text{ turns}$$
$$L_0 = 1.75 \pm 0.044 \text{ in} \qquad \text{(Table 10–10)}$$

then

$$\begin{aligned}\Delta n_s &= |-19.96(0.0006)| + |16.179(0.009)| + |0.168(0.25)| \\ &\quad + |-1.129(0.044)| \\ &= 0.011\,976 + 0.055\,611 + 0.042 + 0.497 = 0.16\end{aligned}$$

Since $(n_s)_{\min} = \bar{n}_s - \Delta n_s$, we can write

$$\bar{n}_s = (n_s)_{\min} + \Delta n_s$$

The smallest acceptable value of $(n_s)_{\min}$ is 1, so

$$\bar{n}_s = 1 + \Delta n_s = 1 + 0.16 = 1.16$$

so far.

The excursions from a midrange value of n_s are bounded by the tolerances and the inspection procedures associated with spring manufacture. The distribution of wire size $\mathbf{d}$ is uniform if the springs are mixed. The distributions of $\mathbf{D}$, $\mathbf{N}_a$, and $\mathbf{L}_0$ are probably truncated or untruncated lognormal. The distribution of $\mathbf{n}_s$ is nearly lognormal by the central limit theorem.

The ultimate tensile strengths in spring materials are usually reported as "minimum" values. The percentile ranges for S_{sy} as a function of S_{ut} in Table 10–12 are kept as "minimums" by always using the lower bound, so the values of S_{sy} are "minimums." The values of E and G in Table 10–12 are mean values. The introduction of a minimum S_{sy} in Eq. (10–33) for n_s does not reduce n_s. The actual value can increase n_s, but we will not notice it in our estimate of n_s, so the variability increases (but does not decrease) n_s. This is conservative. The presence of G in Eq. (10–33) for n_s can reduce n_s as G varies. Since there is no inspection for G, its distribution is not truncated. An adjustment can be made. From Eq. (10–33) the product $n_s G$ is a constant for purposes of finding the partial derivative $\frac{\partial_{n_s}}{\partial G}$. We write

$$\psi = n_s G = 1.342\,360(11.6)10^6 = 15.544(10^6)$$

so

$$n_s = \frac{\psi}{G} = \frac{15.544(10^6)}{G}$$

$$\frac{\partial n_s}{\partial G} = -\frac{15.544(10^6)}{G^2} = \frac{15.544(10^6)}{[11.6(10^6)]^2} = 1.155\,172(10^{-7})$$

The mean and standard deviation of $\mathbf{n}_s$ are

$$\bar{n}_s = \frac{\phi}{\bar{G}} = \frac{15.544(10^6)}{11.6(10^6)} = 1.342\,360$$

$$\sigma_{n_s} = \left[\left(\frac{\partial n_s}{\partial G}\right)^2 \sigma_G^2\right]^{1/2} = \left|\frac{\partial n_s}{\partial G}\sigma_G\right| = \frac{\partial n_s}{\partial_G} C_G \bar{n}_s$$

$$\sigma_{n_s} = 1.155\,172(10^{-7})0.01[11.6(10^6)] = 0.0134$$

where 0.01 is the coefficient of variation of the shear modulus $\mathbf{G}$. The three-sigma range is

$$(\Delta n_s)_G = 3\sigma_{ns} = 3(0.0134) = 0.0402$$

The contribution to n_s by G is added to the 0.16 so that the value of $(n_s)d$ can be written as

$$(n_s)_d = 1 + (0.16 + 0.04) = 1.2 \tag{10–35}$$

Another reason for investigating $(n_s)_d$ is to show which tolerances are the most influential in causing variations in n_s and how to find them. The wire tolerance is the least influential. The others are comparable, and the influence of the torsional modulus uncertainty is of the same order.

10–9 Critical Frequency of Helical Springs

If a wave is created by a disturbance at one end of a swimming pool, this wave will travel down the length of the pool, be reflected back at the far end, and continue in this back-and-forth motion until it is finally damped out. The same effect occurs in helical springs, and it is called *spring surge*. If one end of a compression spring is held against a flat surface and the other end is disturbed, a compression wave is created that travels back and forth from one end to the other exactly like the swimming-pool wave.

Spring manufacturers have taken slow-motion movies of automotive valve-spring surge. These pictures show a very violent surging, with the spring actually jumping out of contact with the end plates. Figure 10–12 is a photograph of such a failure.

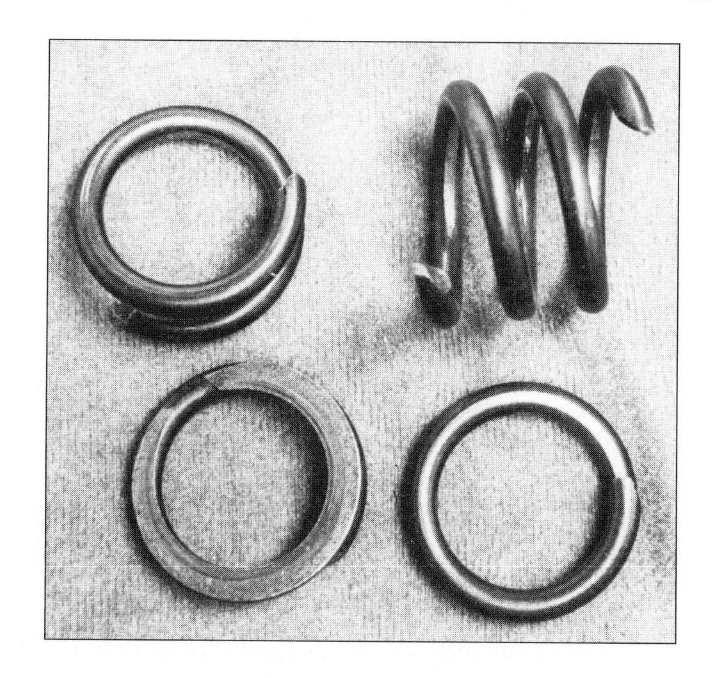

Figure 10–12

Valve-spring failure in an overrevved engine. Fracture is along the 45° line of maximum principal stress associated with pure torsional loading.

When helical springs are used in applications requiring a rapid reciprocating motion, the designer must be certain that the physical dimensions of the spring are not such as to create a natural vibratory frequency close to the frequency of the applied force; otherwise, resonance may occur, resulting in damaging stresses, since the internal damping of spring materials is quite low.

The governing equation for a spring placed between two flat and parallel plates is the wave equation

$$\frac{\partial^2 u}{\partial y^2} = \frac{W}{kgl^2}\frac{\partial^2 u}{\partial t^2} \tag{10–36}$$

where
k = spring rate
g = acceleration due to gravity
l = length of spring between plates
W = weight of spring
y = coordinate along length of spring
u = motion of any particle at distance y

The solutions to this equation are obtained using well-known methods. Here we are interested only in the natural frequencies; in radians per second, these are

$$\omega = m\pi\sqrt{\frac{kg}{W}}$$

where the fundamental frequency is found for $m = 1$, the second harmonic for $m = 2$, and so on. We are usually interested in the frequency in cycles per second; since $\omega = 2\pi f$, we have, for the fundamental frequency,

$$f = \frac{1}{2}\sqrt{\frac{kg}{W}} \tag{10–37}$$

Wolford and Smith[10] show that the frequency is

$$f = \frac{1}{4}\sqrt{\frac{kg}{W}} \tag{10–38}$$

where the spring has one end against a flat plate and the other end free. They also point out that Eq. (10–37) applies when one end is against a flat plate and the other end is driven with a sine-wave motion.

The weight of the active part of a helical spring is

$$W = AL\gamma = \frac{\pi d^2}{4}(\pi D N_a)(\gamma) = \frac{\pi^2 d^2 D N_a \gamma}{4} \tag{10–39}$$

where γ is the specific weight.

The fundamental critical frequency should be from 15 to 20 times the frequency of the force or motion of the spring in order to avoid resonance with the harmonics. If the frequency is not high enough, the spring should be redesigned to increase k or decrease W.

10–10 Fatigue Loading

Fatigue information bearing on spring behavior comes in several forms. Research into the effect of alternating torsion on straight spring wire develops information concerning endurance strength of materials in straight wires. The testing of springs in a zero-max repeatedly applied loading (or nearly so) is another kind of report. At this point it is useful to recall that the only thing that *may be* true is the data; everything else is a human construct of varying utility. A fatigue-failure locus passes through two cardinal points, often ordinal intercepts for convenience in the form of the mathematical model. If zero-max endurance strength is used, together with the ultimate shearing strength, the mathematical form is more complicated. The failure curve can be extrapolated to find the *constructive* ordinate intercept (completely reversed torsional endurance strength). Remember that we have not measured it but have created a constructive intercept whose only virtue is that it gives us a failure locus in convenient form. However, we have to embrace a particular locus type (e.g., Gerber, ASME-elliptic, Goodman) in order to find the intercept. For each locus the constructive intercept is different. In the analysis drill, finding the constructive intercept is simply part of the procedure.

The most insightful presentation of data concerning spring wires is that there is no difference among springs wound with wires ranging in size $0.020 \le d \le 0.207$ in in the flat part of the SN curve. Further, in wire sizes $d \le 0.375$ in there is no difference in the flat part of the SN curve of carbon and alloy steels. Unpeened springs were tested from a minimum torsional stress of 20 kpsi to a maximum of 90 kpsi and peened springs in the range 20 kpsi to 135 kpsi. Table 10–14 shows the corresponding strength components for infinite life. For example, for a spring with $S_{su} = 211.5$ kpsi the constructive Gerber ordinate intercept is

$$S_{se} = \frac{S_{sa}}{1 - \left(\dfrac{S_{sm}}{S_{su}}\right)^2} = \frac{35}{1 - \left(\dfrac{55}{211.5}\right)^2} = 37.5 \text{ kpsi} \qquad \text{(unpeened)}$$

For a Goodman failure locus the intercept would be 47.3 kpsi. Each possible wire size would change these numbers. The invariants are given in Table 10–14.

10. J. C. Wolford and G. M. Smith, "Surge of Helical Springs," *Mech. Eng. News*, vol. 13, no. 1, February 1976, pp. 4–9.

Table 10–14

Zimmerli Endurance Strength Components S_{sa} and S^*_{sm} Corresponding to Data from Carbon Steels, Music Wire, and Alloy Steels Corrected for Surface Condition and Size†

Component	Unpeened	Peened
S_{sa}		
kpsi	35	57.5
MPa	241	398
S_{sm}		
kpsi	55	77.5
MPa	379	534

*From F. P. Zimmerli, "Human Failures in Spring Applications," *The Mainspring*, no. 17, Associated Spring Corporation, Bristol, Conn., August–September 1957. This remarkable invariance was reported as a result of spring fatigue tests with torsional stresses 20 000 psi minimum to 90 000 psi maximum on unpeened springs and 20 000 psi minimum to 135 000 psi maximum on peened springs.

†At room temperature, noncorrosive environment, and in the absence of surging.

An extended study[11] of available literature regarding torsional fatigue found that for polished, notch-free, cylindrical specimens subjected to torsional shear stress, the maximum alternating stress that may be imposed without causing failure is *constant* and independent of the mean stress in the cycle provided that the maximum stress range does not equal or exceed the torsional yield strength of the metal. With notches and abrupt section changes this consistency is not found. Springs are free of notches and surfaces are often very smooth. This failure locus is known as the Sines failure criterion in torsional fatigue.

Wahl recognized a curved failure locus (akin to the Gerber), and when zero-max fatigue data were available he made one cardinal point on the stress ratio line $R = 0$ (or load line $r = 1$) at which the maximum shear stress $\tau_{max} = 2\tau_a = 2\tau_m = (\tau_{max})_{all} = \tau_{max}\,|_{r=1} = 2S_{sr}$. The torsional endurance strength S_{sr} is $(\tau_{max})_{all}/2$. He chose as the other cardinal point the abscissa intercept S_{sy}. From this Wahl derived

$$\frac{S_{sm} - S_{sa}}{S_{sy}} + \frac{2S_{sa}}{S_{sr}} = 1 \tag{10–40}$$

or, using a fatigue factor of safety,

$$\frac{\tau_m - \tau_a}{S_{sy}} + \frac{2\tau_a}{S_{sr}} = \frac{1}{n_f} \tag{10–41}$$

The key properties are S_{sy} and S_{sr}. The equation is simple, linear, and "conservative." For the circumstances when zero-max data were not available but fatigue data were available at known τ'_{max} and τ'_{min}, he wrote

$$\frac{\tau_m}{S_{sy}} + \frac{\tau_a}{(\tau'_{max} - \tau'_{min})/2}\left[1 - \frac{(\tau'_{max} + \tau'_{min})/2}{S_{sy}}\right] = \frac{1}{n_f} \tag{10–42}$$

11. Oscar J. Horger (ed.), *Metals Engineering: Design Handbook*, McGraw-Hill, New York, 1953, p. 84.

EXAMPLE 10–4 For an unpeened music wire spring with a 0.039-in wire diameter ($S_{ut} = 321.7$ kpsi, $S_{su} = 215.6$ kpsi, $S_{sy} = 144.8$ kpsi), find, using Zimmerli's unpeened data, the failure loci of (*a*) Gerber, (*b*) Sines, (*c*) Goodman, and (*d*) Wahl. Plot them on the designer's fatigue diagram.

Solution (*a*) For Gerber, the constructive ordinate intercept is

$$S_{se} = \frac{S_{sa}}{1 - (S_{sm}/S_{su})^2} = \frac{35}{1 - (55/215.6)^2} = 37.4 \text{ kpsi}$$

and

Answer

$$\frac{n_f \tau_a}{37.4} + \left(\frac{n_f \tau_m}{215.6}\right)^2 = 1$$

(*b*) For Sines, $S_{se} = S_{sa} = 35$ kpsi and

Answer

$$\frac{n_f \tau_a}{35} = 1$$

(*c*) For Goodman, the constructive ordinate intercept is

$$S_{se} = \frac{S_{sa}}{1 - (S_{sm}/S_{su})} = \frac{35}{1 - (55/215.6)} = 47.0 \text{ kpsi}$$

and

Answer

$$\frac{\tau_a}{47.0} + \frac{\tau_m}{215.6} = \frac{1}{n_f}$$

(*d*) For Wahl, $\tau'_{\max} = 90$ kpsi and $\tau'_{\min} = 20$ kpsi; thus

$$\frac{1}{n_f} = \frac{\tau_m}{144.8} + \frac{\tau_a}{(90 - 20)/2}\left[1 - \frac{(90 + 20)/2}{144.8}\right]$$

or

Answer

$$\frac{\tau_a}{56.4} + \frac{\tau_m}{144.8} = \frac{1}{n_f}$$

The answers to parts (*a*), (*b*), (*c*), and (*d*) are plotted in superposition in Fig. 10–13.

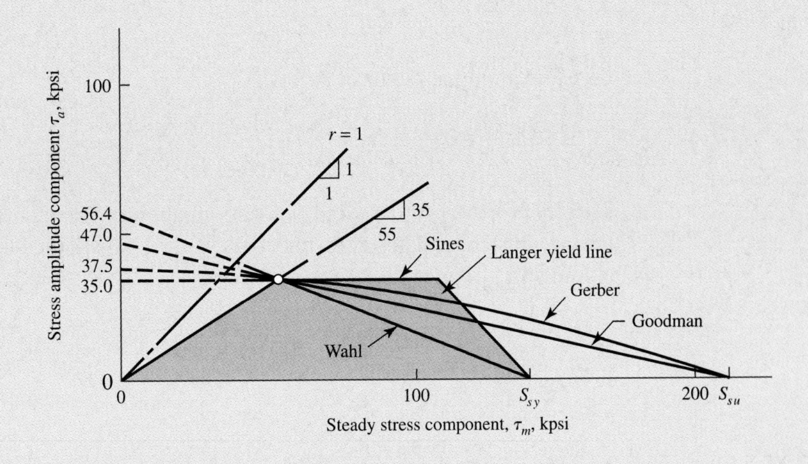

Figure 10–13

Superposition of the fatigue-failure loci of Sines, Gerber, Goodman, and Wahl for Zimmerli's unpeened spring fatigue data point ($\tau_a = 35$ kpsi, $\tau_m = 55$ kpsi). The shaded area shows the possible locations of spring operating points.

In constructing certain failure loci on the designers' torsional fatigue diagram, the torsional modulus of rupture S_{su} is needed. We shall continue to employ Eq. (7–43), which is

$$S_{su} = 0.67 S_{ut} \tag{10–43}$$

10–11 Helical Compression Springs for Dynamic Service

Springs are made to be used, and consequently they are almost always subject to fatigue loading. In many instances the number of cycles of required life may be small, say, several thousand for a padlock spring or a toggle-switch spring. But the valve spring of an automotive engine must sustain millions of cycles of operation without failure; so it must be designed for infinite life.

In the case of shafts and many other machine members, fatigue loading in the form of completely reversed stresses is quite ordinary. Helical springs, on the other hand, are never used as both compression and extension springs. In fact, they are usually assembled with a preload so that the working load is additional. Thus the stress-time diagram of Fig. 7–12*d* expresses the usual condition for helical springs. The worst condition, then, would occur when there is no preload, that is, when $\tau_{\min} = 0$.

Now, we define

$$F_a = \left| \frac{F_{\max} - F_{\min}}{2} \right| \tag{10–44a}$$

$$F_m = \frac{F_{\max} + F_{\min}}{2} \tag{10–44b}$$

where the subscripts have the same meaning as those of Fig. 7–12*d* when applied to the axial spring force F. Then the shear stress amplitude is

$$\tau_a = K_B \frac{8 F_a D}{\pi d^3} \tag{10–45}$$

where K_B is the Bergsträsser factor, obtained from Eq. (10–6), and corrects for both direct shear and the curvature effect. As noted in Sec. 10–2, the Wahl factor K_W can be used instead, if desired.

The midrange shear stress is given by the equation

$$\tau_m = K_B \frac{8 F_m D}{\pi d^3} \tag{10–46}$$

Specification Set To improve the fatigue strength of dynamically loaded springs, shot peening can be used. It can increase the torsional fatigue strength 20 percent or more. Shot size is about $\frac{1}{64}$ in, so spring coil wire diameter and pitch must allow for complete coverage of the spring surface. The specification set for a dynamic service spring has the additional entry for peening or not. The specification set can be

- Material and condition
- Peened or unpeened
- End treatment
- As-wound or set removed
- Total coils and tolerance
- OD or ID and tolerance
- Free length and tolerance
- Wire size and tolerance

If loading is important in its precision, the specification set can be

- Material and condition
- Peened or unpeened
- End treatment
- As-wound or set removed
- Coil OD or ID and tolerance
- F_1 at L_1, F_2 at L_2 (or at y_1 and y_2) and tolerance on F_1, F_2
- Total turns and tolerance
- Wire size and tolerance

Since this specification set requires spring testing, the springs cost more due to discard of scrap.

Decision Set The decision set can be

- Material and condition
- Peened or unpeened
- End treatment
- Robust linearity: $\xi = 0.15$
- As-wound or set removed
- Fatigue safe: n_f
- Function: F_1, y_1, F_2, y_2 (or F_1, L_1, F_2, L_2)

} a priori decisions

- Wire diameter d } design variable(s)

Tolerance can be assigned later, which partitions a larger problem in two smaller problems addressed in sequence. The figure of merit can be

$$\text{fom} = -(\text{relative material cost})\,\frac{\gamma\pi^2 d^2 N_t D}{4} \tag{10–47}$$

Optimization strategy encompasses a priori decisions, with cheapest springs attempted first. This dictates hard-drawn spring wire, unpeened, squared and ground ends in as-wound condition. For robust linearity ξ is set to 0.15. We set n_f but not n_s, because we cannot set both. Thus $n_s \geq 1.2$ is placed in the adequacy assessment, since an inequality is a non-decision. We choose a wire size that makes the design variable a definite value and find d, D, OD, ID, C, N_a, L_s, L_0, $(L_0)_{\text{cr}}$, n_f, n_s, and fom. By incrementing available wire sizes we can form a rectangular table as we did with static service springs, perform the adequacy assessment by scanning the table, removing from consideration springs with violated constraints, then from the satisfactory springs choose the design with the highest figure of merit.

An analysis problem will provide some experience with the fatigue problem, which will help us plan the execution of the design problem.

EXAMPLE 10–5

An as-wound helical compression spring, made of music wire, has a wire size of 0.092 in, an outside coil diameter of $\frac{9}{16}$ in, a free length of $4\frac{3}{8}$ in, 21 active coils, and both ends squared and ground. The spring is unpeened. This spring is to be assembled with a preload of 5 lbf and will operate with a maximum load of 35 lbf during use.

(*a*) Estimate the factor of safety guarding against fatigue failure using a torsional Gerber fatigue failure locus, Zimmerli data.

(*b*) Repeat part (*a*) using the Sines torsional fatigue locus (steady stress component has no effect), Zimmerli data.

(*c*) Repeat using a torsional Goodman failure locus, Zimmerli data.

(*d*) Repeat using a Wahl failure locus, Zimmerli data.

(*e*) Estimate the critical frequency of the spring.

Solution The helic diameter is $D = 0.5625 - 0.092 = 0.4705$ in. The spring index is $C = \frac{D}{d} = \frac{0.4705}{0.092} = 5.11$. Then

$$K_B = \frac{4C+2}{4C-3} = \frac{4(5.11)+2}{4(5.11)-3} = 1.287$$

Using Eqs. (10–44),

$$F_a = \frac{35-5}{2} = 15 \text{ lbf} \qquad F_m = \frac{35+5}{2} = 20 \text{ lbf}$$

The alternating shear-stress component is found from Eq. (10–45) to be

$$\tau_a = K_B \frac{8F_a D}{\pi d^3} = 1.287 = \frac{8(15)0.4705}{\pi(0.092)^3}(10^{-3}) = 29.7 \text{ kpsi}$$

Equation (10–46) gives the midrange shear-stress component

$$\tau_m = K_B \frac{8F_m D}{\pi d^3} = 1.287 \frac{8(20)0.4705}{\pi(0.092)^3}(10^{-3}) = 39.6 \text{ kpsi}$$

From Table 10–5 we find $A = 201$ and $m = 0.145$. The ultimate tensile strength is estimated from

$$S_{ut} = \frac{A}{d^m} = \frac{201}{0.092^{0.145}} = 284.1 \text{ kpsi}$$

Also the shearing ultimate strength is estimated from

$$S_{su} = 0.67 S_{ut} = 0.67(284.1) = 190.3 \text{ kpsi}$$

and

$$S_{sy} = 0.45 S_{ut} = 0.45(284.1) = 127.8 \text{ kpsi}$$

The load-line slope $r = \frac{\tau_a}{\tau_m} = \frac{29.7}{39.6} = 0.75$.

(*a*) The Gerber constructive ordinate intercept for Zimmerli data is

$$S_{se} = \frac{S_{sa}}{1-(S_{sm}/S_{su})^2} = \frac{35}{1-(55/190.3)} = 38.2 \text{ kpsi}$$

The amplitude component of strength S_{sa} is

$$S_{sa} = \frac{0.75^2 190.3^2}{2(38.2)}\left\{-1+\sqrt{1+\left[\frac{2(38.2)}{0.75(190.3)}\right]^2}\right\} = 35.8 \text{ kpsi}$$

and the fatigue factor of safety n_f is given by

Answer

$$n_f = \frac{S_{sa}}{\tau_a} = \frac{35.8}{29.7} = 1.21$$

(*b*) The Sines failure locus for Zimmerli data is $n_f \frac{\tau_a}{35} = 1$, from which

Answer $$n_f = \frac{35}{\tau_a} = \frac{35}{29.7} = 1.18$$

(*c*) The constructive ordinate intercept S_{se} for the Goodman failure locus for the Zimmerli data is

$$S_{se} = \frac{S_{sa}}{1 - (S_{sm}/S_{su})} = \frac{35}{1 - (55/190.3)} = 49.2 \text{ kpsi}$$

The amplitude component of the strength S_{sa} for the Goodman locus is

$$S_{sa} = \frac{1}{1/S_{se} + 1/(rS_{su})} = \frac{1}{1/49.2 + 1/[0.75(190.3)]} = 36.6 \text{ kpsi}$$

The fatigue factor of safety is given by

Answer $$n_f = \frac{S_{sa}}{\tau_a} = \frac{36.6}{29.7} = 1.23$$

(*d*) The Wahl equation for Zimmerli data can be written

$$\frac{1}{n_f} = \frac{\tau_m}{127.8} + \frac{\tau_a}{(90-20)/2}\left[1 - \frac{(90+20)/2}{127.8}\right] = \frac{\tau_m}{127.8} + \frac{\tau_a}{61.4}$$

Substituting S_{sa} for $n_f\tau_a$, $S_{sm} = n_f\tau_m = S_{sa}/r$ gives

$$\frac{S_{sa}}{61.4} + \frac{S_{sa}}{127.8} = 1$$

from which

$$S_{sa} = \frac{1}{1/61.4 + 1/(127.8r)} = \frac{1}{1/61.4 + 1/[127.8(0.75)]} = 37.4 \text{ kpsi}$$

The fatigue factor of safety is given by

Answer $$n_f = \frac{S_{sa}}{\tau_a} = \frac{37.4}{29.7} = 1.26$$

(*e*) Using Eq. (10–9) we estimate the spring rate as

$$k = \frac{d^4G}{8D^3N_a} = \frac{0.092^4[11.75(10^6)]}{8(0.4705)^3 21} = 48.1 \text{ lbf/in}$$

From Eq. (10–39) we estimate the spring weight as

$$W = \frac{\pi^2(0.092^2)0.4705(21)0.284}{4} = 0.0586 \text{ lbf}$$

and from Eq. (10–37) the frequency of the fundamental wave is

Answer $$f_n = \frac{1}{2}\left[\frac{48.1(386)}{0.0586}\right]^{1/2} = 281 \text{ Hz}$$

If the operating or exciting frequency is much more than $\frac{281}{20} = 14.1$ Hz, the spring may have to be redesigned.

We used four approaches to estimating the fatigue factor of safety in Ex. 10–5. The results were 1.21 (Gerber), 1.18 (Sines), 1.23 (Goodman), and 1.26 (Wahl). These are in the order of smallest to largest, as one would expect from Fig. 10–13 for a 0.75 load line.

Some evaluation is in order. The fatigue locus that is a regression candidate, going in and among the data, is the Gerber fatigue-failure locus. Sines has its authority rooted in straight scratch-free wire, which is not commercially commonplace. Goodman and Wahl are low when compared to data. We will use Gerber, but be prepared to use Sines, Goodman, or Wahl when we are trying to follow the work of others.

10–12 Design of a Helical Compression Spring for Dynamic Service

Let us begin with the statement of a problem. In order to compare a static spring to a dynamic spring, we shall design the spring in Ex. 10–2 for dynamic service at 5 Hz.

EXAMPLE 10–6

A music wire helical compression spring with infinite life is needed to resist a dynamic load that varies from 5 to 20 lb while the end deflection varies from $\frac{1}{2}$ to 2 in. Because of assembly considerations the solid height cannot exceed 1 in, and the free length cannot be more than 4 in. The springmaker has the following dies in stock: 0.069, 0.071, 0.080, 0.085, 0.090, 0.095, 0.105, and 0.112 in.

Solution

The a priori decisions are:

- Material and condition: $A = 201$, $m = 0.145$, $G = 11.75(10^6)$ psi; relative cost is 2.6
- Surface treatment: unpeened
- End treatment: squared and ground, $Q = 2$, $Q' = 1$
- Robust linearity: $\xi_1 = 0.15$
- Set: use in as-wound condition
- Fatigue-safe: $n_f = 1.5$ using the Sines-Zimmerli fatigue-failure locus
- Function: $F_{\min} = 5$ lbf, $F_{\max} = 20$ lbf, $y_{\min} = 0.5$ in, $y_{\max} = 2$ in, spring operates free (no rod or hole)
- Decision variable: wire size d

The figure of merit will be the volume of wire to wind the spring, Eq. (10–32). The optimization strategy will be to set wire size d to the die sizes in turn, build a table, perform the adequacy assessment by inspecting the table, and choose the satisfactory spring with the highest figure of merit.

Solution

Set $d = 0.112$ in. Then

$$F_a = \frac{20 - 5}{2} = 7.5 \text{ lb} \qquad F_m = \frac{20 + 2}{2} = 12.5 \text{ lb}$$

$$k = \frac{F_{\max}}{y_{\max}} = \frac{20}{2} = 10 \text{ lb/in}$$

$$S_{ut} = \frac{201}{0.112^{0.145}} = 276.1 \text{ kpsi} \qquad S_{su} = 0.67(276.1) = 185.0 \text{ kpsi}$$

$$S_{sy} = 0.45(276.1) = 124.2 \text{ kpsi}$$

From Table 10–15 for Sines-Zimmerli $(S_{se})_{\text{constr}} = 35$ kpsi. From $n_f = S_{sa}/\tau_a = (\pi d^3 S_{sa})/(8K_B F_a D)$. Set $D/d = C$ and solve for C:

Table 10–15

Key Relationships* for Various Fatigue-Failure Loci for Compression Springs

	Zimmerli Basis	Zero-Max Basis (unpeened)
Sines	$(S_{se})_{\text{constr}} = S_{sa} = \begin{cases} 35 \text{ kpsi} & \text{(unpeened)} \\ 57.5 \text{ kpsi} & \text{(peened)} \end{cases}$ $\quad 0 \le S_{sm} \le S_{sy} - S_{sa}$	$(S_{se})_{\text{constr}} = \dfrac{S_{sr}}{2} = S_{sa} \quad 0 \le S_{sm} \le S_{sy} - S_{sa}$
Gerber	$(S_{se})_{\text{constr}} = \dfrac{35}{1-(55/S_{su})^2}$ (unpeened) $= \dfrac{57.5}{1-(77.5/S_{su})^2}$ (peened) $S_{sa} = \dfrac{r^2 S_{su}^2}{2S_{se}}\left[-1+\sqrt{1+\left(\dfrac{2}{r}\dfrac{S_{se}}{S_{su}}\right)^2}\right]$	$(S_{se})_{\text{constr}} = \dfrac{S_{sr}/2}{1-[S_{sr}/(2S_{su})]^2}$ $S_{sa} = \dfrac{r^2 S_{su}^2}{2S_{se}}\left[-1+\sqrt{1+\left(\dfrac{2}{r}\dfrac{S_{se}}{S_{su}}\right)^2}\right]$
Goodman	$(S_{se})_{\text{constr}} = \dfrac{35}{1-55/S_{su}}$ (unpeened) $= \dfrac{57.5}{1-77.5/S_{su}}$ (peened) $S_{sa} = \dfrac{1}{1/S_{se}+1/(rS_{su})}$	$(S_{se})_{\text{constr}} = \dfrac{S_{sr}}{2-S_{sr}/S_{su}}$ $S_{sa} = \dfrac{1}{1/S_{se}+1/(rS_{su})}$
Wahl	$(S_{se})_{\text{constr}} = \dfrac{35}{1-55/S_{sy}}$ (unpeened) $= \dfrac{57.5}{1-77.5/S_{sy}}$ (peened) $S_{sa} = \dfrac{1}{1/(rS_{sy})+(1-55/S_{sy})/35}$ (unpeened) $S_{sa} = \dfrac{1}{1/(rS_{sy})+(1-77.5/S_{sy})/57.5}$ (peened)	$(S_{se})_{\text{constr}} = \dfrac{S_{sr}}{2-S_{sr}/S_{sy}}$ $S_{sa} = \dfrac{1}{1/(rS_{sy})+2[1-S_{sr}/(2S_{sy})]/S_{sr}}$

*Fatigue factor of safety $n_f = S_{sa}/\tau_a$ applies to all eight loci. The loadline of slope r passes through the origin for all S_{sa} equations.

$$C = \frac{\pi d^2 S_{sa}}{8 n_f K_B F_a}$$

Next solve by iteration. Trial $C = 10$:

$$K_B \frac{4C+2}{4C-3} = \frac{4(10)+2}{4(10)-3} = 1.135$$

$$C = \frac{(0.112)^2 35\,000}{8(1.5)1.135(7.5)} = 13.500\,936$$

Subsequent substitutions give 13.937 140, 14.000 316, 14.001 262, 14.004 621, 14.004 654, 14.004 657. Note the gain of approximately one additional correct digit from each iteration.

$$D = Cd = 14.005(0.112) = 1.569 \text{ in}$$

$$F_s = (1 + 0.15)20 = 23 \text{ lb}$$

$$N_a = \frac{Gd^4 y_{\max}}{8D^3 F_{\max}} = \frac{11.75(10^6)0.112^4(2)}{8(1.569)^3 20} = 5.98 \text{ turns}$$

$$N_t = N_a + Q = 5.98 + 2 = 7.98 \text{ turns}$$

$$L_s = (N_a + Q')d = (5.98 + 1)0.112 = 0.782 \text{ in}$$

$$L_0 = L_s + \frac{F_s}{k} = 0.782 + \frac{23}{10} = 3.082 \text{ in}$$

$$\text{ID} = 1.569 - 0.112 = 1.457 \text{ in}$$

$$\text{OD} = 1.569 + 0.112 = 1.681 \text{ in}$$

$$y_s = L_0 - L_s = 3.082 - 0.782 = 2.30 \text{ in}$$

$$L_1 = L_0 - \frac{F_{\max}}{k} = 3.082 - \frac{23}{10} = 1.082 \text{ in}$$

$$L_2 = L_0 - \frac{F_{\min}}{k} = 3.082 - \frac{5}{10} = 2.582 \text{ in}$$

$$(L_0)_{cr} < \frac{2.63D}{\alpha} = 2.63\frac{(1.569)}{0.5} = 8.253 \text{ in}$$

$$K_B = \frac{4(14.005) + 2}{4(14.005) - 3} = 1.094$$

$$W = \frac{\pi^2 d^2 D N_a \gamma}{4} = \frac{\pi^2 0.112^2(1.569)0.284}{4} = 0.0825 \text{ lb}$$

$$f_n = 0.5\sqrt{\frac{386k}{W}} = 0.5\sqrt{\frac{386(10)}{0.0825}} = 108 \text{ Hz}$$

$$\tau_a = K_B \frac{8F_a D}{\pi d^3} = 1.094\frac{8(7.5)1.569}{\pi 0.112^3} = 23\,304 \text{ psi}$$

$$\tau_m = \tau_a \frac{F_m}{F_a} = 23\,304\frac{12.5}{7.5} = 38\,900 \text{ psi}$$

$$\tau_s = \tau_a \frac{F_s}{F_a} = 23\,304\frac{23}{7.5} = 71\,466 \text{ psi}$$

$$n_f = \frac{S_{sa}}{\tau_a} = \frac{35\,000}{23\,304} = 1.5$$

$$n_s = \frac{S_{sy}}{\tau_s} = \frac{124\,200}{71\,466} = 1.74$$

$$\text{fom} = \frac{-2.6\pi^2 0.112^2(5.98 + 2)1.569}{4} = -1.01$$

Inspection of the foregoing development shows all conditions are met. This design is satisfactory. Are there better ones? Repeat the process for the other available wire sizes and develop the following table:

d:	0.069	0.071	0.080	0.085	0.090	0.095	0.105	0.112
D	0.297	0.332	0.512	0.631	0.767	0.919	1.274	1.569
ID	0.228	0.261	0.432	0.547	0.677	0.824	1.169	1.457
OD	0.366	0.403	0.592	0.717	0.857	1.014	1.379	1.681
C	4.33	4.67	6.40	7.44	8.53	9.67	12.13	14.00
N_a	127.2	102	44.8	30.5	21.3	15.4	8.63	5.99
L_s	8.847	7.343	3.660	2.665	2.010	1.560	1.01	0.783
L_0	11.15	9.643	5.590	4.965	4.310	3.860	3.311	3.083
$(L_0)_{cr}$	1.562	1.74	2.964	3.325	4.036	4.833	6.703	8.25
n_f	1.50	1.50	1.50	1.50	1.50	1.50	1.50	1.50
n_s	1.86	1.85	1.83	1.81	1.79	1.78	1.75	1.74
f_n	87.5	89.7	96.9	99.7	101.9	103.8	106	108
fom	-1.17	-1.12	-0.983	-0.948	-0.930	-0.927	-0.958	-1.01

The problem-specific inequality constraints are

$$L_s \le 1 \text{ in}$$
$$L_0 \le 4 \text{ in}$$
$$f_n \ge 5(20) = 100 \text{ Hz}$$

The general constraints are

$$3 \le N_a \le 15$$
$$4 \le C \le 16$$
$$(L_0)_{cr} > L_0$$

The tight constraint is L_s, which limits the smallest diameter of wire to 0.112 in. See the plot in Fig. 10–14.

A specification set can be entered on a form shown in Table E–37 of the appendix.

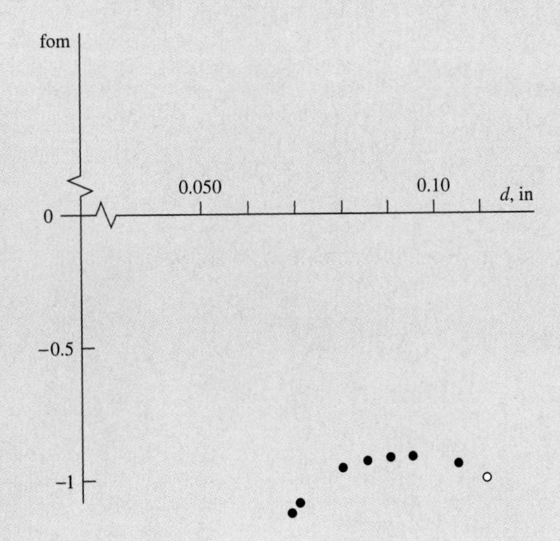

Figure 10–14

The figure of merit in Ex. 10–6 plotted against wire size for eight designs. The hollow plotting point symbols denote one satisfactory design and show that the best spring has a wire diameter of 0.112 in.

It is common among springmakers to speak of *light service*, *average service*, and *severe service*. This refers, in an approximate way, to the cycles-to-failure of springs. A required life of 10^3 to 10^4 cycles is called light service. Average service is 10^5 to 10^6 cycles. Severe service is a requirement for greater life, including infinite life. See Fig. 10–15.

When a designer places a fatigue locus on the designer's fatigue diagram, the question of how many cycles-to-failure it corresponds to arises. All coordinates on a strength locus have the same cycle life. By specifying the cycles-to-failure at one point on the locus, we have described the entire locus life. See Fig. 10–16. Zero-max testing (or reporting), or any other described testing, will give information. Table 10–18 gives such information. For example, zero-max endurance strength for music wire is included as

$$S_{sr} = \frac{50}{d^{0.154}}$$

For a 0.039-in wire we know $S_{ut} = \frac{201}{0.039^{0.145}} = 321.7$ kpsi and $S_{su} = 0.67S_{ut} = 0.67(321.7) = 215.6$ kpsi. Now

$$S_{sr} = \frac{50}{0.039^{0.154}} = 82.4 \text{ kpsi}$$

This means the shear-stress range is 0 to 82.4 kpsi and it results in infinite life. This occurs at a place on the fatigue locus where $r = 1$. From Table 10–15 the Gerber fatigue locus has a constructive endurance strength $(S_{se})_{\text{constr}}$ of

$$(S_{se})_{\text{constr}} = \frac{S_{sr}/2}{1 - [S_{sr}/(2S_{su})]^2} = \frac{82.4/2}{1 - [82.4/(2 \cdot 215.6]^2} = 42.8 \text{ kpsi}$$

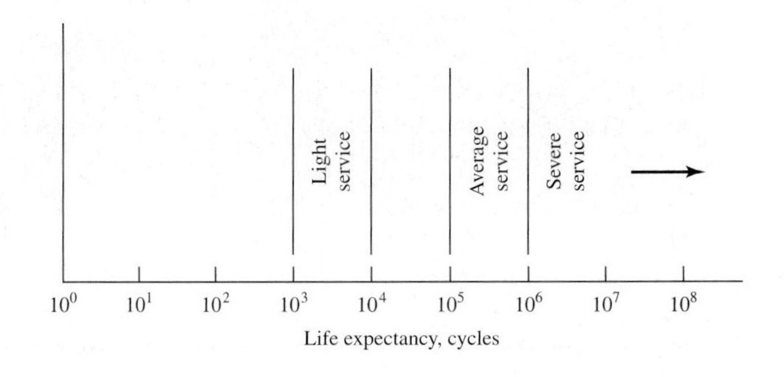

Figure 10–15

Vocabulary used by springmakers in describing life expectancy of springs in service.

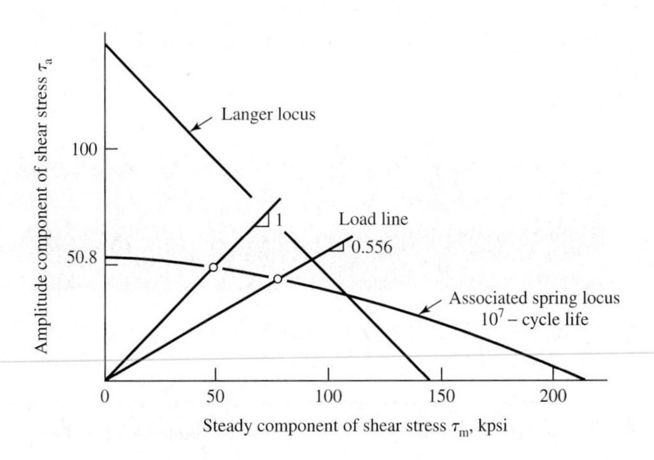

Figure 10–16

The 10^7-cycle life Gerber fatigue-failure locus for Ex. 10–7. The operating point $\tau_a = 43.5$ kpsi, $\tau_m = 78.3$ kpsi is not on the Gerber locus, but slightly below it. The $r = 1$ stresses are $\tau_a = 47.7$ kpsi, $\tau_m = 47.7$ kpsi, and that point is below the Gerber locus.

The fatigue locus can be expressed as

$$\frac{S_{sa}}{42.8} + \left(\frac{S_{sm}}{215.6}\right)^2 = 1$$

The fraction $\frac{S_{sr}}{S_{ut}} = \frac{82.4}{321.7} = 0.26$, which in Table 10–16 indicates a life exceeding 10^7 cycles. From Table 10–17,

$$N_f = \left(\frac{\tau_{\max}/S_{ut}}{0.568}\right)^{-1/0.0396} = \left(\frac{82.4/321.7}{0.568}\right)^{-1/0.0396} = 3.3(10^8) \text{ cycles}$$

which is an extrapolation and should be taken as infinite life.

If one wishes to establish a strength locus corresponding to 10^5 cycles-to-failure,

$$S_{sr} = 0.36 S_{ut} = 0.36(321.7) = 115.8 \text{ kpsi}$$

and

$$(S_{se})_{\text{constr}} = \frac{S_{sr}/2}{1 - [S_{sr}/(2S_{su})]^2} = \frac{115.8/2}{1 - [115.8/(2 \cdot 215.6)]^2} = 62.4 \text{ kpsi}$$

The locus can be described by

$$\frac{S_{sa}}{62.4} + \left(\frac{S_{sm}}{215.6}\right)^2 = 1$$

Table 10–16

Maximum Allowable Torsional Stress in Zero-Max Loading for Round-Wire Helical Coil Compression Springs

Source: "Springs," Chap. 24 in Joseph E. Shigley and Charles R. Mischke (eds.), *Standard Handbook of Machine Design*, 2nd ed., McGraw-Hill, New York, 1996.

Fatigue Life	$(\tau_{max})_{all}/S_{ut}$			
	ASTM 228, Austenitic Stainless Steel and Nonferrous		Oil-Tempered Steels A230, A232	
N_f	Unpeened	Peened	Unpeened	Peened
10^5	0.36	0.42	0.42	0.49
10^6	0.33	0.39	0.48	0.47
10^7	0.30	0.36	0.38	0.46

Table 10–17

Cycles-to-Failure N_f Estimate for Round-Wire Helical Compression Coil Springs*

Music Wire, Austenitic Stainless Steel and Nonferrous		Oil-Tempered Steels A230, A232	
Unpeened	Peened	Unpeened	Peened
$\left(\frac{\tau_{\max}/S_{ut}}{0.568}\right)^{-1/0.0396}$	$\left(\frac{\tau_{\max}/S_{ut}}{0.618}\right)^{-1/0.0355}$	$\left(\frac{\tau_{\max}/S_{ut}}{0.540}\right)^{-1/0.0217}$	$\left(\frac{\tau_{\max}/S_{ut}}{0.576}\right)^{-1/0.0142}$

Least-squares curve-fit to data in Table 10–14 which obviates logarithmic interpolation.

Table 10–18

Zero-Max Endurance Strength of Compression Springs Undergoing Repeatedly Applied Loading. Equation form: $S_{sr} = B/d^{m_1}$.

Source: Virgil Moring Faires, *Design of Machine Elements*, 4th ed., Macmillan, New York, 1965, Table AT–17, pp. 590–591. The table is more comprehensive than appears above.

Material	Diameter d		Exponent	Constant B	
	in	mm	m_1	kpsi	MPa
Hard-drawn A227	$0.15 \le d \le 0.625$	$3.8 \le d \le 15.9$	0.34	30	621
Oil-tempered A229	$0.041 \le d \le 0.15$	$1 \le d \le 3.8$	0.10	47	448
Music wire A228	$0.018 \le d \le 0.18$	$0.46 \le d \le 4.6$	0.154	50	567
Carbon steel A230	$0.093 \le d \le 0.25$	$2.36 \le d \le 6.35$	0.15	49	548
Cr-V steel A232	$0.028 \le d \le 0.5$	$0.71 \le d \le 12.7$	0.15	49	548
Cr-Si A401	$0.032 \le d \le 0.375$	$0.81 \le d \le 9.53$	0.15	49	548
Cr-Ni (stainless) A313	$0.01 \le d \le 0.375$	$0.25 \le d \le 9.53$	0.17	30	358
Be-Cu*	$0.10 \le d \le 0.5$	$2.29 \le d \le 12.7$	0.20	35	461
P-bronze†	$0.09 \le d \le 0.5$	$2.29 \le d \le 12.7$	0.20	15.3	201

* $S_{ut} = 56$ kpsi (386 MPa) maximum.

† $S_{ut} = 28$ kpsi (193 MPa) maximum.

The explicit solution for S_{sa} for this Gerber locus is

$$S_{sa} = \frac{r^2 S_{su}^2}{2S_{se}}\left[-1 + \sqrt{1 + \left(\frac{2S_{se}}{rS_{su}}\right)^2}\,\right]$$

For $r = 1$,

$$S_{sa} = \frac{215.6^2}{2(62.4)}\left\{-1 + \sqrt{1 + \left[\frac{2(62.4)}{215.6}\right]^2}\,\right\} = 57.9 \text{ kpsi}$$

The stress range $\tau_{\max}$ is

$$\tau_{\max} = 2S_{sr} = 2(57.9) = 115.8 \text{ kpsi}$$

Using Table 10–16 to check the cycle-to-failure life gives

$$N_f = \left(\frac{115.8/321.7}{0.568}\right)^{-1/0.0396} \doteq 10^5$$

Spring Life

The designer places a 10^7 cycles-to-failure (infinite life) locus on the designer's fatigue diagram. Associated Spring places one point on the locus. Table 10–16, for unpeened springs, places the maximum allowable shear stress $\tau_{\max}$ at $0.30S_{ut}$ when $\tau_{\min}$ is zero. This we will call S_{sr} (shear strength, repeatedly applied), which is equal to $2\tau_a$, to $2\tau_m$; thus the plotted point is $0.15S_{ut}, 0.15S_{ut}$. A Gerber fatigue-failure locus is placed through this point and the abscissa intercept S_{su}. From this the ordinate intercept S_{se} is found. The next step is to pass a fatigue locus through the operating point and the abscissa intercept S_{su}. On this second locus we find the coordinates at $r = 1$. Table 10–17 can be used to find the cycles-to-failure N_f. Table 10–18 gives zero-max endurance strengths, repeatedly applied loading. An example will be useful.

EXAMPLE 10–7 Consider an unpeened music wire helical coil compression spring formed with 0.039-in wire with squared and ground ends, and with the following dimensions and characteristics. The helix diameter D is 0.312 in, the free length L_0 is 0.807 in, the solid height L_s is 0.315 in, intended to be flexed between $L_{\min} = 0.394$ in and $L_{\max} = 0.689$ in, and wound with 8 total turns. Estimate the fatigue cycles to failure N_f using a Gerber failure locus using Associated Spring data.

Solution Begin with the preliminaries. From Table 10–5,

$$S_{ut} = \frac{201}{0.039^{0.145}} = 321.7 \text{ kpsi}$$

From Eq. (10–43),

$$S_{su} = 0.67S_{ut} = 0.67(321.7) = 215.5 \text{ kpsi}$$

From Table 10–13,

$$S_{sy} = 0.45(321.7) = 144.8 \text{ kpsi}$$

Now

$$C = \frac{D}{d} = \frac{0.312}{0.039} = 8.0$$

$$k = \frac{11.85(10^6)0.039^4}{8(0.312)^3 6} = 18.8 \text{ lbf/in}$$

$$K_B = \frac{4(8)+2}{4(8)-3} = 1.172$$

$$F_{\min} = (L_0 - L_{\min})k = (0.807 - 0.689)18.8 = 2.22 \text{ lb}$$

$$F_{\max} = (L_0 - L_{\max})k = (0.807 - 0.334)18.8 = 7.76 \text{ lb}$$

$$F_s = (L_0 - L_s)k = (0.807 - 0.313)18.8 = 9.25 \text{ lb}$$

$$\tau_{\min} = K_B \frac{8F_{\min}D}{\pi d^3} = 1.172 \frac{8(2.22)0.312}{\pi 0.039^3} = 34.8 \text{ kpsi}$$

$$\tau_{\max} = K_B \frac{8F_{\max}D}{\pi d^3} = 1.172 \frac{8(7.76)0.312}{\pi 0.039^3} = 121.8 \text{ kpsi}$$

$$\tau_s = K_B \frac{8F_s D}{\pi d^3} = 1.172 \frac{8(9.25)0.312}{\pi 0.039^3} = 145.2 \text{ kpsi}$$

$$\tau_a = \frac{121.8 - 34.8}{2} = 43.5 \text{ kpsi}$$

$$\tau_m = \frac{121.8 + 34.8}{2} = 78.3 \text{ kpsi}$$

$$r = \frac{\tau_a}{\tau_m} = \frac{43.5}{78.3} = 0.556$$

The Associated Spring value for S_{sr} from Table 10–16 for 10^7 cycle life is

$$S_{sr} = 0.30S_{ut} = 0.30(312.7) = 96.51 \text{ kpsi}$$

The Gerber fatigue locus ordinate intercept S_{se} of Fig. 10–16 is

$$S_{se} = \frac{S_{sr}/2}{1 - \left(\dfrac{S_{sr}/2}{S_{su}}\right)^2} = \frac{96.51/2}{1 - \left(\dfrac{96.51/2}{215.5}\right)^2} = 50.8 \text{ kpsi}$$

allowing us to write the equation for the 10^7 Gerber fatigue locus as

$$\frac{S_{sa}}{50.8} + \left(\frac{S_{sm}}{215.5}\right)^2 = 1$$

All points on this locus have a 10^7-cycle life. The operating point is $\tau_a = 43.5$ kpsi, $\tau_m = 78.3$ kpsi. A Gerber parabola through the operating point and abscissa intercept S_{su} has an ordinate intercept of

$$S_{se} = \frac{\tau_a}{1 - (\tau_m/S_{su})^2} = \frac{43.5}{1 - (78.3/215.5)^2} = 50.1 \text{ kpsi}$$

Since $50.8 > 50.1$, life exceeds 10^7 cycles; also, this locus is slightly below the Gerber strength locus, thus the operating point is below the strength locus. At $r = 1$,

$$S_{sa} = \frac{215.5^2}{2(50.1)}\left\{-1 + \sqrt{1 + \left[\frac{2(50.1)}{215.5}\right]^2}\right\} = 47.7 \text{ kpsi } = S_{sm}$$

These are the strength components for the number of cycles of life extant. The stress range is $S_{sa} + S_{sm} = 47.7 + 47.7 = 95.4$ kpsi. From Table 10–17,

$$N_f = \left(\frac{95.4/312.7}{0.568}\right)^{-1/0.0398} = 0.61(10^7) \text{ cycles-to-failure}$$

which is regarded as indefinite life.

10–13 Design of Extension Springs

As shown in Sec. 10–4 extension springs carry tensile loading, they have end loops or hooks, and the body of the spring is wound with an initial tension. The load-deflection curve is depicted in Fig. 10–17a. The initial tension F_i is inferred from a linear backwards extrapolation of the linear characteristic, which does not begin until all body coils have completely separated. The equation of the operating characteristic is

$$F = F_i + ky \tag{10–48}$$

where k is the spring rate and y is the extension beyond the free length L_0. All dimensions in Fig. 10–17b are to the insides of the loops.

The amount of initial tension that a springmaker can routinely incorporate is as shown in Fig. 10–17c. The preferred range can be expressed in terms of the *uncorrected* torsional stress τ_i as

$$\tau_i = \frac{33\,500}{\exp(0.105C)} \pm 1000\left(4 - \frac{C-3}{6.5}\right) \text{ psi} \tag{10–49}$$

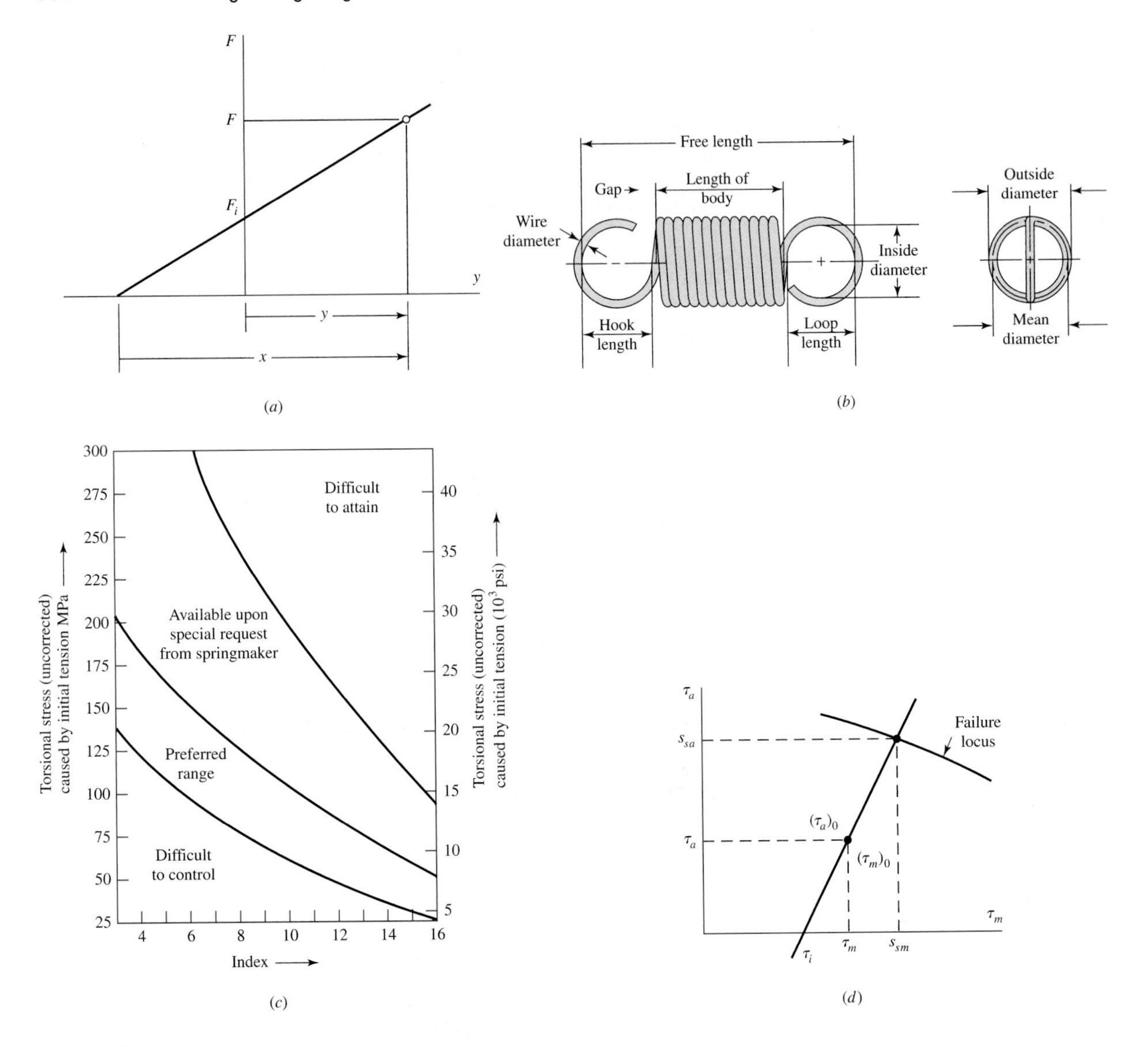

Figure 10–17

(*a*) Geometry of the force F and extension y locus of an extension spring; (*b*) geometry of the extension spring; (*c*) torsional stresses due to initial tension as a function of spring index C in helical extension springs; and (*d*) the designer's fatigue-failure diagram for an extension spring coil body showing the load line and the fatigue-failure locus.

where C is the spring index. The free length L_0 of a spring measured inside the end loops or hooks can be expressed as

$$L_0 = 2(D - d) + (N_b + 1)d = (2C - 1 + N_b)d \tag{10–50}$$

where D is the helix diameter and N_b is the number of body coils. In fatigue and yielding considerations the initial shear stress τ_i must be taken into account in a manner similar to the preload in a bolted joint. With ordinary twisted endloops as shown in Fig. 10–17*b* the equivalent number of active helical turns N_a with the same spring rate as the extension spring is

$$N_a = N_b + \frac{G}{E} \tag{10–51}$$

where G and E are the torsional and tensile moduli of elasticity.

Extension springs differ in detail and approach from compression springs. Figure 10–17a is a force-deflection diagram for an extension spring. In the x-coordinate system, the force at any extension from zero-force length is x and it follows that

$$F = kx \tag{a}$$

where k is the familiar spring rate. This is a convenient equation since the force F is proportional to x. Using the y coordinate,

$$F = F_i + ky \tag{b}$$

When $y = 0$ the spring force F is F_i, the initial tension manufactured into the spring. The transformation between the x coordinate and the y coordinate is

$$x = y + \frac{F_i}{k} \tag{c}$$

As a check, substitute Eq. (c) into Eq. (a):

$$F = kx = k\left(y + \frac{F_i}{k}\right) = ky + F_i$$

which is a form of Eq. (b). The torsional stress in the spring is given by

$$\tau = \frac{8K_B F D}{\pi d^3} = \frac{8K_B(kx)D}{\pi d^3} \tag{d}$$

When $x = F_i/k$, $\tau = \tau_i$ and Eq. (d) becomes

$$\tau_i = \frac{8K_B(kF_i/k)D}{\pi d^3} = \frac{8K_B F_i D}{\pi d^3} \tag{e}$$

At some place $y = y_1$ (or $x - x_1$) and τ_1 becomes, from Eq. (d),

$$\tau_1 = \frac{8K_B(kx_1)D}{\pi d^3} = \frac{8K_B[k(y_1 + F_i/k)]D}{\pi d^3} = \frac{8K_B(ky_1 + F_i)D}{\pi d^3}$$

From Eq. (b) $ky_1 = F - F_i$ and τ_1 becomes

$$\begin{aligned}\tau_1 &= \frac{8K_B(F_1 - F_i + F_i)D}{\pi d^3} = \frac{8K_B(F_1 - F_i)D}{\pi d^3} + \frac{8K_B F_i D}{\pi d^3} \\ &= \frac{8K_B(F_1 - F_i)D}{\pi d^3} + \tau_i\end{aligned} \tag{f}$$

If τ_i is included, the other component of shear stress must be formed with $(F_1 - F_i)$. It also follows that

$$\tau_1 = \frac{8K_B F_1 D}{\pi d^3} \tag{g}$$

without regard to τ_i. It is useful to remember that the force-induced torsional shear stress τ_1 is directly proportional to the load F_1 whether you are using the x- or y-coordinate equations.

On the designer's fatigue diagram we convert F_{max} and F_{min} (which are loads) to steady and amplitude components of load, F_m and F_a. Although they have units of load, F_a and F_m are not really loads, because they are not applied to the spring. The point that characterizes the unique state of loading, $(\tau_a)_0$ and $(\tau_m)_0$, is plotted on the designer's fatigue diagram, Fig. 10–17*d*. The initial torsional stress τ_i is shown on the abscissa. The load line connecting them is straight. It can be expressed as

$$\tau_a = a + b\tau_m \tag{h}$$

The derivative of τ_a with respect to τ_m gives

$$\frac{d\tau_a}{d\tau_m} = b$$

so Eq. (h) becomes

$$0 = a + b\tau_i$$

from which $a = -b\tau_i$. Backsubstituting into Eq. (h) gives

$$\tau_a = -b\tau_i + b\tau_m = b(\tau_m - \tau_i) \tag{i}$$

The slope of the load line in Fig. 10–17*d* can be expressed as

$$b = \frac{(\tau_a)_0}{(\tau_m)_0 - \tau_i} = \frac{(F_a)_0}{(F_m)_0 - F_i} \tag{j}$$

since torsional stresses are proportional to loading. Equation (i) becomes

$$\tau_a = \frac{F_a}{F_m - F_i}(\tau_m - \tau_i) \tag{10–52}$$

dropping the operating point notation $(\ )_0$, and carrying it as understood. Solving Eq. (10–52) for τ_m gives

$$\tau_m = \frac{F_m - F_i}{F_a}\tau_a + \tau_i \tag{10–53}$$

In the end hooks *all* loads from zero to F_1 are felt by the end coils and resisted by stresses proportional to load. Here $F_i = 0$, as does τ_i, and Eq. (10–53) becomes

$$\tau_m = \frac{F_m}{F_a}\tau_a \tag{10–54}$$

Coil Body

For the fatigue factor of safety $(n_f)_{body}$, we first find τ_i, τ_a, and τ_m:

$$\tau_i = \frac{8K_B F_i D}{\pi d^3} \qquad \tau_a = \frac{8K_B F_a D}{\pi d^3} \qquad \tau_m = \frac{F_m - F_i}{F_a}\tau_a + \tau_i$$

From Table 10–16 or 10–18 we find S_{sr}, then S_{se}:

$$S_{se} = \frac{S_{sr}/2}{1 - \left(\dfrac{S_{sr}/2}{S_{su}}\right)^2} \tag{10–55}$$

Take the Gerber fatigue locus

$$\frac{n_f\tau_a}{S_{se}} + \left(\frac{n_f\tau_m}{S_{su}}\right)^2 = 1$$

and solve the quadratic for n_f:

$$(n_f)_{\text{body}} = \frac{1}{2}\frac{\tau_a}{S_{se}}\left(\frac{S_{su}}{\tau_m}\right)^2\left[-1+\sqrt{1+\left(2\frac{\tau_m}{S_{su}}\frac{S_{se}}{\tau_a}\right)^2}\right] \tag{10–56}$$

For the yield factor of safety in the coil body, we take the Langer line $S_{sa} + S_{sm} = S_{sy}$ and Eq. (10–53) written as

$$S_{sm} = \frac{F_m - F_i}{F_a} S_{sa} + \tau_i$$

and solve these two equations simultaneously:

$$S_{sa} + S_{sm} = S_{sy}$$

$$\frac{F_m - F_i}{F_a} S_{sa} - S_{sm} = -\tau_i$$

The variables S_{sa} and S_{sm} separate by addition, and we obtain

$$\left(1 + \frac{F_m - F_i}{F_a}\right) S_{sa} = S_{sy} - \tau_i$$

and S_{sa} for the load line–yield line intersection is designated $(S_{sa})_y$ and is equal to

$$(S_{sa})_y = \frac{S_{sy} - \tau_i}{1 + \dfrac{F_m - F_i}{F_a}} \tag{10–57}$$

The yield factor of safety is expressed as

$$(n_y)_{\text{body}} = \frac{(S_{sa})_y}{\tau_a} \tag{10–58}$$

End-Hook Bending

At end-coil location at point A, the stress-augmentation factor $(K)_A$ is well-approximated by

$$(K)_A = \frac{4C_1^2 - C_1 - 1}{4C_1(C_1 - 1)} \qquad C_1 = \frac{2r_1}{d}$$

The bending stress amplitude component σ_a is given by

$$\sigma_a = F_a\left[\frac{16(K)_A D}{\pi d^3} + \frac{4}{\pi d^2}\right] \tag{10–59}$$

and following Eq. (10–54) the steady component σ_m is given by

$$\sigma_m = \sigma_a \frac{F_m}{F_a} \tag{10–60}$$

We note that $r = \frac{F_a}{F_m}$. We take S_r from Table 10–22, in bending, and write

$$S_e = \frac{S_r/2}{1 - \left(\dfrac{S_r/2}{S_{ut}}\right)^2} \tag{10–61}$$

Equation (10–56), expressed for bending, will find our factor of safety $(n_f)_A$:

$$(n_f)_A = \frac{1}{2}\frac{\sigma_a}{S_e}\left(\frac{S_{ut}}{\sigma_m}\right)^2\left[-1+\sqrt{1+\left(2\frac{\sigma_m}{S_{ut}}\frac{S_e}{\sigma_a}\right)^2}\right] \tag{10–62}$$

To obtain the factor of safety in yielding, the Langer line $S_a + S_m = S_y$ is combined with the bending equivalent of Eq. (10–54):

$$S_m = \frac{F_m}{F_a}S_a \tag{10–63}$$

and solved simultaneously:

$$S_a + S_m = S_y$$

$$\frac{F_m}{F_a}S_a - S_m = 0$$

The variables S_a and S_m separate by addition and

$$\left(1+\frac{F_m}{F_a}\right)S_a = S_y$$

The amplitude component of the yield strength $(S_a)_y$ is expressed as

$$(S_a)_y = \frac{S_y}{1+F_m/F_a} \tag{10–64}$$

and the factor of safety $(n_y)_A$ is

$$(n_y)_A = \frac{(S_a)_y}{(\sigma_a)_A} \tag{10–65}$$

End-Hook Torsion

At location B in the end loop the stress-augmentation factor $(K)_B$ is well-approximated by

$$(K)_B = \frac{4C_2-1}{4C_2-4} \qquad C_2 = \frac{2r_2}{d}$$

The shear-stress components are, using Eq. (10–54),

$$(\tau_a)_B = \frac{8(K)_B F_a D}{\pi d^3} \qquad (\tau_m)_B = (\tau_a)_B\frac{F_m}{F_a} \tag{10–66}$$

From Table 10–22, "In Torsion, End" column, find S_{sr}, and from Eq. (10–55)

$$S_{se} = \frac{S_{sr}/2}{1-\left(\dfrac{S_{sr}/2}{S_{su}}\right)^2} \tag{10–67}$$

The factor of safety $(n_f)_B$ is given by

$$(n_f)_B = \frac{1}{2}\frac{\tau_a}{S_{se}}\left(\frac{S_{su}}{\tau_m}\right)^2\left[-1+\sqrt{1+\left(2\frac{\tau_m}{S_{su}}\frac{S_{se}}{\tau_a}\right)^2}\right] \tag{10–68}$$

The factor of safety in yielding $(n_y)_B$ is found from Eq. (10–64) expressed for shear:

$$(S_{sa})_y = \frac{S_{sy}}{1+F_m/F_a} \tag{10–69}$$

and then

$$(n_y)_B = \frac{(S_{sa})_y}{(\tau_a)_B} \tag{10–70}$$

Tolerances

Wire size and OD tolerances are the same as for helical compression springs. Tables 10–19 through 10–23 give information that differs from information for compression springs. An example of a specification sheet for extension springs can be found in Table E–37.

Table 10–19

Commercial Free-Length Tolerances for Helical Extension Springs with Initial Tension

Source: Associated Spring–Barnes Group, *Design Handbook*, Bristol, Conn., 1987, p. 53.

Spring Free Length (inside hooks)		Tolerance	
mm	in	± mm	in
Up to 12.7	0.500	0.51	0.020
Over 12.7–25.4	0.500–1.00	0.76	0.030
Over 25.4–50.8	1.00–2.00	1.0	0.040
Over 50.8–102	2.00–4.00	1.5	0.060
Over 102–203	4.00–8.00	2.4	0.093
Over 203–406	8.00–16.0	4.0	0.156
Over 406–610	16.0–24.0	5.5	0.218

Table 10–20

Tolerances on Angular Relationship of Extension Spring Ends *Source:* Associated Spring–Barnes Group, *Design Handbook*, Bristol, Conn., 1987, p. 53.

Index	4	5	6	7	8	9	10	12	14	16
Angular Tolerance per Coil, ± Degrees	0.75	0.9	1.1	1.3	1.5	1.7	1.9	2.3	2.6	3

For example, tolerance for a 10-coil spring with an index of 8 is $10 \times \pm 1.5 = \pm 15°$.

If angular tolerance is greater than $\pm$ 45°, or if closer tolerances than indicated must be held, consult with Associated Spring.

Table 10–21

Maximum Allowable Stresses (K_W or K_B corrected) for Helical Extension Springs in Static Applications

Source: Associated Spring–Barnes Group, *Design Handbook*, Bristol, Conn., 1987, p. 52.

	Percent of Tensile Strength		
	In Torsion		In Bending
Materials	Body	End	End
Patented, cold-drawn or hardened and tempered carbon and low-alloy steels	45–50	40	75
Austenitic stainless steel and nonferrous alloys	35	30	55

This information is based on the following conditions: set not removed and low temperature heat treatment applied.

For springs that require high initial tension, use the same percent of tensile strength as for end.

Table 10–22

Maximum Allowable Stresses for ASTM A228 and Type 302 Stainless Steel Helical Extension Springs in Cyclic Applications

Source: Associated Spring–Barnes Group, *Design Handbook*, Bristol, Conn., 1987, p. 52.

Number of Cycles	Percent of Tensile Strength: In Torsion, Body	In Torsion, End	In Bending, End
10^5	36	34	51
10^6	33	30	47
10^7	30	28	45

This information is based on the following conditions: not shot-peened, no surging and ambient environment with a low temperature heat treatment applied. Stress ratio = 0.

Table 10–23

Load Tolerances for Helical Extension Springs *Source:* Associated Spring–Barnes Group, *Design Handbook*, Bristol, Conn., 1987, p. 54.

Index D/d	Body Length Divided by Deflection L_B/f	Tolerances: ± Percent of Specified Load (Closer tolerances may require additional operations.) Wire Diameter, mm (in)										
		0.38 (0.015)	0.56 (0.022)	0.81 (0.032)	1.0 (0.041)	1.6 (0.062)	2.3 (0.092)	3.2 (0.125)	4.7 (0.187)	6.4 (0.250)	4.5 (0.375)	11.1 (0.437)
4	12	20.0	18.5	17.6	16.9	16.2	15.5	15.0	14.3	13.8	13.0	12.6
	8	18.5	17.5	16.7	15.8	15.0	14.5	14.0	13.2	12.5	11.5	11.0
	6	16.8	16.1	15.5	14.7	13.8	13.2	12.7	11.8	11.2	9.9	9.4
	4.5	15.0	14.7	14.1	13.5	12.6	12.0	11.5	10.3	9.7	8.4	7.9
	2.5	13.1	12.4	12.1	11.8	10.6	10.0	9.1	8.5	8.0	6.8	6.2
	1.5	10.2	9.9	9.3	8.9	8.0	7.5	7.0	6.5	6.1	5.3	4.8
	0.5	6.2	5.4	4.8	4.6	4.3	4.1	4.0	3.8	3.6	3.3	3.2
5	12	17.8	16.5	15.7	15.5	14.8	14.1	13.5	12.8	12.3	12.0	11.5
	8	16.8	15.7	14.9	14.3	13.5	13.0	12.5	11.7	11.2	10.6	10.1
	6	15.8	14.8	13.8	13.2	12.3	11.8	11.4	10.7	10.0	9.3	8.7
	4.5	14.2	13.1	12.3	11.7	11.1	10.6	10.2	9.6	8.8	8.0	7.4
	2.5	12.3	11.3	10.8	10.0	9.6	9.0	8.7	8.1	7.5	6.5	6.1
	1.5	10.0	9.3	8.9	8.4	8.0	7.7	7.2	6.5	6.2	5.0	4.5
	0.5	6.2	5.4	4.9	4.6	4.4	4.2	4.1	3.9	3.7	3.4	3.3
6	12	17.0	15.5	14.6	14.1	13.5	13.1	12.7	12.0	11.5	11.2	10.7
	8	16.2	14.7	13.9	13.4	12.6	12.2	11.7	11.0	10.5	10.0	9.5
	6	15.2	14.0	12.9	12.3	11.6	10.9	10.7	10.0	9.4	8.8	8.3
	4.5	13.7	12.4	11.5	11.0	10.5	10.0	9.6	9.0	8.3	7.6	7.1
	2.5	11.9	10.8	10.2	9.8	9.4	9.0	8.5	7.9	7.2	6.2	6.0
	1.5	9.9	9.0	8.3	7.7	7.3	7.0	6.7	6.4	6.0	4.9	4.7
	0.5	6.3	5.5	4.9	4.7	4.5	4.3	4.1	4.0	3.7	3.5	3.4
8	12	15.8	14.3	13.1	13.0	12.1	12.0	11.5	10.8	10.2	10.0	9.5
	8	15.0	13.7	12.5	12.1	11.4	11.0	10.6	10.1	9.4	9.0	8.6
	6	14.2	13.0	11.7	11.2	10.6	10.0	9.7	9.3	8.6	8.1	7.6
	4.5	12.8	11.7	10.7	10.1	9.7	9.0	8.7	8.3	7.8	7.2	6.6
	2.5	11.2	10.2	9.5	8.8	8.3	7.9	7.7	7.4	6.9	6.1	5.6
	1.5	9.5	8.6	7.8	7.1	6.9	6.7	6.5	6.2	5.8	4.9	4.5
	0.5	6.3	5.6	5.0	4.8	4.5	4.4	4.2	4.1	3.9	3.6	3.5

Table 10–23

Continued

Index D/d	Body Length Divided by Deflection L_B/f	*Tolerances: ± Percent of Specified Load (Closer tolerances may require additional operations.)* *Wire Diameter, mm (in)*										
		0.38 (0.015)	0.56 (0.022)	0.81 (0.032)	1.0 (0.041)	1.6 (0.062)	2.3 (0.092)	3.2 (0.125)	4.7 (0.187)	6.4 (0.250)	4.5 (0.375)	11.1 (0.437)
10	12	14.8	13.3	12.0	11.9	11.1	10.9	10.5	9.9	9.3	9.2	8.8
	8	14.2	12.8	11.6	11.2	10.5	10.2	9.7	9.2	8.6	8.3	8.0
	6	13.4	12.1	10.8	10.5	9.8	9.3	8.9	8.6	8.0	7.6	7.2
	4.5	12.3	10.8	10.0	9.5	9.0	8.5	8.1	7.8	7.3	6.8	6.4
	2.5	10.8	9.6	9.0	8.4	8.0	7.7	7.3	7.0	6.5	5.9	5.5
	1.5	9.2	8.3	7.5	6.9	6.7	6.5	6.3	6.0	5.6	5.0	4.6
	0.5	6.4	5.7	5.1	4.9	4.7	4.5	4.3	4.2	4.0	3.8	3.7
12	12	14.0	12.3	11.1	10.8	10.1	9.8	9.5	9.0	8.5	8.2	7.9
	8	13.2	11.8	10.7	10.2	9.6	9.3	8.9	8.4	7.9	7.5	7.2
	6	12.6	11.2	10.2	9.7	9.0	8.5	8.2	7.9	7.4	6.9	6.4
	4.5	11.7	10.2	9.4	9.0	8.4	8.0	7.6	7.2	6.8	6.3	5.8
	2.5	10.5	9.2	8.5	8.0	7.8	7.4	7.0	6.6	6.1	5.6	5.2
	1.5	8.9	8.0	7.2	6.8	6.5	6.3	6.1	5.7	5.4	4.8	4.5
	0.5	6.5	5.8	5.3	5.1	4.9	4.7	4.5	4.3	4.2	4.0	3.3
14	12	13.1	11.3	10.2	9.7	9.1	8.8	8.4	8.1	7.6	7.2	7.0
	8	12.4	10.9	9.8	9.2	8.7	8.3	8.0	7.6	7.2	6.8	6.4
	6	11.8	10.4	9.3	8.8	8.3	7.7	7.5	7.2	6.8	6.3	5.9
	4.5	11.1	9.7	8.7	8.2	7.8	7.2	7.0	6.7	6.3	5.8	5.4
	2.5	10.1	8.8	8.1	7.6	7.1	6.7	6.5	6.2	5.7	5.2	5.0
	1.5	8.6	7.7	7.0	6.7	6.3	6.0	5.8	5.5	5.2	4.7	4.5
	0.5	6.6	5.9	5.4	5.2	5.0	4.8	4.6	4.4	4.3	4.2	4.0
16	12	12.3	10.3	9.2	8.6	8.1	7.7	7.4	7.2	6.8	6.3	6.1
	8	11.7	10.0	8.9	8.3	7.8	7.4	7.2	6.8	6.5	6.0	5.7
	6	11.0	9.6	8.5	8.0	7.5	7.1	6.9	6.5	6.2	5.7	5.4
	4.5	10.5	9.1	8.1	7.5	7.2	6.8	6.5	6.2	5.8	5.3	5.1
	2.5	9.7	8.4	7.6	7.0	6.7	6.3	6.1	5.7	5.4	4.9	4.7
	1.5	8.3	7.4	6.6	6.2	6.0	5.8	5.6	5.3	5.1	4.6	4.4
	0.5	6.7	5.9	5.5	5.3	5.1	5.0	4.8	4.6	4.5	4.3	4.1

EXAMPLE 10–8 A hard-drawn wire extension spring has a wire diameter of 0.035 in, an outside coil diameter of 0.248 in, generous hook radii, and an initial tension of 1.19 lbf; the number of body turns is 12.17. Find the factors of safety under a static 5.25-lbf load.

Solution

$$D = \text{OD} - d = 0.248 - 0.035 = 0.213 \text{ in}$$

$$r_1 = \frac{D}{2} = \frac{0.213}{2} = 0.106 \text{ in}$$

$$C = \frac{D}{d} = \frac{0.213}{0.035} = 6.086$$

$$K_B = \frac{4(6.086) + 2}{4(6.086) - 3} = 1.234 \text{ in}$$

$$N_a = 12.17 + 11.5/30 = 12.55 \text{ turns}$$

$$N_b = N_a - G/E = 12.55 - 11.5/30 = 12.17 \text{ turns}$$

$$k = \frac{Gd^4}{8D^3N_a} = \frac{11.5(10^6)0.035^4}{8(0.213)^3 12.55} = 17.78 \text{ lbf/in}$$

$$L_0 = 2(D - d) + (N_b + 1)_d = 2(0.213 - 0.035) + (12.17 + 1)0.035$$
$$= 0.817 \text{ in}$$

$$y_1 = \frac{F_1 - F_i}{k} = \frac{5.25 - 1.19}{17.78} = 0.228 \text{ in}$$

$$L_1 = L_0 + y_1 = 0.817 + 0.228 = 1.045 \text{ in}$$

The uncorrected initial stress is

$$(\tau_i)_{\text{uncorr}} = \frac{8F_iD}{\pi d^3} = \frac{8(1.19)0.213}{\pi(0.035)^3} = 15.1 \text{ kpsi}$$

$$(\tau_i)_{\max} = \frac{33\,500}{\exp[0.105(6.086)]} + 1000\left(4 - \frac{6.086 - 3}{6.5}\right)$$
$$= 17\,681 + 3525 = 21.2 \text{ kpsi}$$

$$(\tau_i)_{\min} = 17\,681 - 3525 = 14.2 \text{ kpsi}$$

The initial tension is in the preferred range.

$$\tau_1 = \frac{8K_BF_1D}{\pi d^3} = \frac{8(1.234)5.25(0.213)}{\pi(0.035)^3} = 82.0 \text{ kpsi}$$

$$S_{ut} = \frac{A}{d^m} = \frac{137}{0.035^{0.201}} = 268.8 \text{ kpsi}$$

$$S_y = 0.75S_{ut} = 0.75(268.8) = 201.6 \text{ kpsi}$$

$$S_{sy} = 0.45S_{ut} = 0.45(268.8) = 121 \text{ kpsi}$$

$$n_s = \frac{S_{sy}}{\tau_1} = \frac{121}{82} = 1.48$$

The situation in the end-hook bending at A is

$$C_1 = \frac{2r_1}{D} = \frac{D}{d} = C = 6.086$$

$$(K)_A = \frac{4C_1^2 - C_1 - 1}{4C_1(C_1 - 1)} = \frac{4(6.086^2) - 6.086 - 1}{4(6.086)(6.086 - 1)} = 1.14$$

$$(\sigma_1)_A = 5.25\left[\frac{16(1.14)0.213}{\pi(0.035)^3} + \frac{4}{\pi(0.035)^2}\right] = 156.9 \text{ kpsi}$$

Answer $$(n_y)_A = \frac{S_y}{(\sigma_1)_A} = \frac{201.6}{156.9} = 1.28$$

The situation in the end-hook torsion at B is

$$C_2 = \frac{2r_2}{d} = \frac{D - d}{d} = \frac{0.213 - 0.035}{0.035} = 5.086$$

$$(K)_B = \frac{4C_2 - 1}{4C_2 - 4} = \frac{4(5.086) - 1}{4(5.086) - 4} = 1.18$$

$$(\tau_1)_B = \frac{8(1.18)5.25(0.213)}{\pi 0.035^3} = 78.4 \text{ kpsi}$$

Answer $$(n_y)_B = \frac{S_{sy}}{(\tau_1)_B} = \frac{121}{78.4} = 1.54$$

EXAMPLE 10–9 The helical coil extension spring of Ex. 10–8 is subjected to a dynamic loading from 1.5 to 5 lbf. Estimate the factors of safety using a Gerber failure locus for (*a*) coil fatigue, (*b*) coil yielding, (*c*) end-hook bending fatigue at point *A* of Fig. 10–3*a*, and (*d*) end-hook torsional fatigue at point *B* of Fig. 10–3*b*.

Solution A number of quantities are the same as in Ex. 10–8: $d = 0.035$ in, $D = 0.213$ in, $r_1 = 0.106$ in, $C = 6.086$, $K_B = 1.234$, $(K)_A = 1.14$, $(K)_B = 1.18$, $N_b = 12.17$ turns, $L_0 = 0.817$ in, $k = 17.78$ lbf/in, $F_i = 1.19$ lbf, $\tau_i = 18.6$ kpsi, and it is in the preferred (makeable) range. Then

$$F_a = (F_{\max} - F_{\min})/2 = (5 - 1.5)/2 = 1.75 \text{ lbf}$$

$$F_m = (5 + 1.5)/2 = 3.25 \text{ lbf}$$

The strengths from Ex. 10–8 include $S_{ut} = 268.8$ kpsi, $S_y = 201.6$ kpsi, and $S_{sy} = 121$ kpsi. The ultimate shear strength is estimated as

$$S_{su} = 0.67 S_{ut} = 0.67(268.8) = 180.1 \text{ kpsi}$$

(*a*) Body-coil fatigue:

$$\tau_a = \frac{8 K_B F_a D}{\pi d^3} = \frac{8(1.234)1.75(0.213)}{\pi 0.035^3} = 27.3 \text{ kpsi}$$

$$\tau_m = \frac{F_m - F_i}{F_a}\tau_a + \tau_i = \frac{3.25 - 1.19}{1.75}(27.3) + 18.6 = 50.8 \text{ kpsi}$$

From Table 10–22 for 10^7 cycle life, $S_{sr} = 0.30 S_{ut} = 0.30(268.8) = 80.6$ kpsi. From Eq. (10–55),

$$S_{se} = \frac{80.6/2}{1 - \left(\dfrac{80.6/2}{180.1}\right)^2} = 42.4 \text{ kpsi}$$

Answer $$(n_f)_{\text{body}} = \frac{1}{2}\frac{27.3}{42.4}\left(\frac{180.1}{58.8}\right)^2\left[-1 + \sqrt{1 + \left(2\frac{50.8}{180.1}\frac{42.4}{27.3}\right)^2}\,\right] = 1.33$$

(*b*) Coil yielding: From Eq. (10–55),

$$(S_{sa})_y = \frac{121 - 18.6}{1 + \dfrac{3.25 - 1.19}{1.75}} = 47.0 \text{ kpsi}$$

and from Eq. (10–58),

Answer
$$(n_y)_{\text{body}} = \frac{47.0}{27.3} = 1.72$$

(*c*) End-hook bending fatigue: Using Eqs. (10–59) and (10–60),

$$\sigma_a = 1.75\left[\frac{16(1.14)0.213}{\pi 0.035^3} + \frac{4}{\pi 0.035^2}\right] = 52.3 \text{ kpsi}$$

$$\sigma_m = 52.3\frac{3.25}{1.75} = 97.1 \text{ kpsi}$$

From Table 10–22, in bending, end column $S_r = 0.45S_{ut} = 0.45(268.8) = 121$ kpsi. Eq. (10–61) gives

$$S_e = \frac{121/2}{1 - \left(\dfrac{121/2}{268.8}\right)^2} = 63.7 \text{ kpsi}$$

From Eq. (10–62),

Answer
$$(n_f)_A = \frac{1}{2}\frac{52.3}{63.7}\left(\frac{268.8}{97.1}\right)^2\left[-1 + \sqrt{1 + \left(2\frac{97.1}{268.8}\frac{63.7}{52.3}\right)^2}\right] = 1.03$$

(*d*) End-hook torsional fatigue: From Eqs. (10–66),

$$(\tau_a)_B = \frac{8(1.18)1.75(0.213)}{\pi 0.035^3} = 26.1 \text{ kpsi}$$

$$(\tau_m)_B = 26.1\frac{3.25}{1.75} = 48.5 \text{ kpsi}$$

From Table 10–22, in torsion, end column find $S_{sr} = 0.28S_{ut} = 0.28(268.8) = 75.3$ kpsi. From Eq. (10–67),

$$S_{se} = \frac{75.3/2}{1 - \left(\dfrac{75.3/2}{180.1}\right)^2} = 39.4 \text{ kpsi}$$

From Eq. (10–68),

Answer
$$(n_f)_B = \frac{1}{2}\frac{26.1}{39.4}\left(\frac{180.1}{48.5}\right)^2\left[-1 + \sqrt{1 + \left(2\frac{48.5}{180.1}\frac{39.4}{26.1}\right)^2}\right] = 1.32$$

The analyses in Ex. 10–8 and Ex. 10–9 show how extension springs differ from compression springs. The end hooks are usually the weakest part, with bending usually controlling. We should also appreciate that a fatigue failure separates the extension spring under load. Flying fragments, lost load, and machine shutdown are threats to personal safety as well as machine function. For these reasons higher design factors are used in extension-spring design than in the design of compression springs.

Examples 10–8 and 10–9, which were analyses, set the stage for consideration of the specification set, and the corresponding decision set, for extension springs. The simplest specification set for an extension spring occurs when commercial load tolerances are acceptable:

Static Service Specification Set	Corresponding Decision Set
Material and condition	Material and condition
End-loop treatment	End-loop treatment
Bend radius	Bend radius
Body turns	Static-safe
Coil OD	Function: F_1 at y_1
Free length	Wire diameter } design variables
Wire size	Initial tension } design variables

Let us use this decision set in designing a static service extension spring.

EXAMPLE 10–10

Design a full-loop end extension spring with the broadest possible loop radius to statically carry a 5.25-lb load with an extension of 0.228 in. The free length cannot exceed $1\frac{1}{4}$ in, the body turns must be between 3 and 15, and the outside diameter cannot exceed $\frac{1}{2}$ in. Use hard-drawn spring wire.

Solution

The a priori decisions are:

- Material: HD spring wire, $A = 137$, $m = 0.201$, $G = 11.85(10^6)$ psi, $E = 29.0(10^6)$ psi
- End-loop treatment: Broadest possible: use coil turn as loop
- Bend radius: Largest possible radius, $r = D/2$
- Static-safe: Use design factor $n_y = 1.5$
- Function: $F_1 = 5.25$ lbf at $y_1 = 0.228$ in
- Design decision 1: Set $d = 0.035$ in, then

$$S_{ut} = \frac{137}{0.035^{0.201}} = 268.8 \text{ kpsi}$$

$$S_y = 0.75(268.8) = 201.6 \text{ kpsi}$$

$$S_{sy} = 0.45(268.8) = 121.0 \text{ kpsi}$$

The weakest point in the spring is end-loop bending at point A. Set

$$n_y = \frac{S_y}{\sigma_A} = \frac{S_y}{F_1\left[\dfrac{16(K)_A D}{\pi d^3} + \dfrac{4}{\pi d^2}\right]}$$

Incorporating $C_1 = D/d = C$ and $(K)_A = (4C^2 - C - 1)/[4C(C-1)]$,

$$C = \frac{\pi d^2 S_y}{4 n_y F_1 \left\{4\left[\dfrac{4C^2 - C - 1}{4C(C-1)}\right] + \dfrac{1}{C}\right\}}$$

which must be solved iteratively. Beginning with $C = 10$, successive values of the spring index are 5.570, 5.135, 5.054, 5.038, and 5.034. Thus

$$D = 5.034(0.035) = 0.176 \text{ in}$$

$$\text{OD} = 0.176 + 0.035 = 0.211 \text{ in}$$

$$K_B = \frac{4(5.034) + 2}{4(5.034) - 3} = 1.292$$

The midrange value of the initial shear stress τ_i (uncorrected) is a function of the spring index:

$$(\tau_i)_{\text{uncorr}} = 33\,500 \exp(-0.105C) = 33\,500 \exp[-0.105(4.631)]$$
$$= 20.6 \text{ kpsi}$$

The associated tolerance is

$$\text{tol} = 1000\left(4 - \frac{C - 3}{6.5}\right) = 1000\left(4 - \frac{4.631 - 3}{6.5}\right) = 3.75 \text{ kpsi}$$

The largest value of initial tension that can be manufactured is

$$(F_i)_{\max} = \frac{[(\tau_i)_{\text{uncorr}} + \text{tol}]\pi d^3}{8D} = \frac{(20.6 + 3.75)\pi(0.035)^3}{8(0.162)} = 2.53 \text{ lb}$$

$$(F_i)_{\min} = \frac{(20.6 - 3.75)\pi(0.035)^3}{8(0.162)} = 1.75 \text{ lbf}$$

For a given wire size, increasing F_i increases spring rate k, increases body turns N_b, and increases free length L_0. We will place the initial tension at the bottom of the makeable range.

- Design decision 2: Initial tension, $F_i = 1.75$ lb, then

$$k = \frac{F_1 - F_i}{y_1} = \frac{5.25 - 1.75}{0.228} = 15.35 \text{ lb/in}$$

$$N_a = \frac{Gd^4}{8D^3k} = \frac{11.85(10^6)0.035^4}{8(0.176)^3 15.35} = 26.56 \text{ turns}$$

$$N_b = 26.56 - \frac{11.85}{29.0} = 26.15 \text{ turns}$$

$$L_0 = 2(0.176 - 0.035) + (26.15 + 1)0.035 = 1.232 \text{ in}$$

$$L_1 = 1.232 + 0.228 = 1.460 \text{ in}$$

$$\tau_1 = \frac{8(1.292)5.25(0.162)}{\pi(0.035)^3} = 70.9 \text{ kpsi}$$

$$n_s = \frac{S_{sy}}{\tau_1} = \frac{121.0}{70.9} = 1.71$$

Bending at A should exhibit a factor of safety of 1.5, but we will check anyway:

$$(K)_A = \frac{4C^2 - C - 1}{4C(C - 1)} = \frac{4(5.034)^2 - 5.034 - 1}{4(5.034)(5.034 - 1)} = 1.174$$

$$\sigma_A = 5.75\left[\frac{16(1.174)0.176}{\pi 0.035^3} + \frac{4}{\pi 0.035^2}\right] = 143.3 \text{ kpsi}$$

$$(n_y)_A = \frac{S_y}{\sigma_A} = \frac{201.6}{134.3} = 1.50$$

Table 10–24

A Tabulation of the Results of Ex. 10–10 with Initial Tension Set to the Minimum Value for Manufacturability

d: F_i:	0.035 1.75	0.038 1.49	0.042 1.21	0.045 1.04	0.048 0.89	0.051 0.76	0.055 0.61
D	0.176	0.229	0.314	0.387	0.501	0.565	0.706
OD	0.211	0.267	0.356	0.432	0.550	0.616	0.761
C	5.034	6.040	7.469	8.61	10.22	11.07	12.84
k	15.35	16.49	17.72	18.46	19.12	19.68	20.35
N_a	26.50	15.50	8.43	5.66	3.55	2.83	1.89
N_b	26.09	15.09	8.02	5.25	3.14	2.42	1.48
L_0	1.230	0.994	0.922	0.966	1.107	1.201	1.439
L_1	1.458	1.222	1.150	1.194	1.335	1.429	1.667
$(n_y)_b$	1.71	1.72	1.74	1.75	1.75	1.76	1.76
$(n_y)_A$	1.50	1.50	1.50	1.50	1.50	1.50	1.50
$(n_y)_B$	1.77	1.79	1.81	1.82	1.82	1.82	1.82
fom	−0.0150	−0.0140	−0.0137	−0.0140	−0.0153	−0.0160	−0.0184

as it should be. Investigate the torsion at B:

$$2r_2 = D - d$$

$$C_2 = \frac{2r_2}{d} = \frac{(D-d)}{d} = C - 1 = 5.034 - 1 = 4.034$$

$$(K)_B = \frac{4C_2 - 1}{4C_2 - 4} = \frac{4(4.034) - 1}{4(4.034) - 4} = 1.247$$

$$\tau_B = \frac{8(1.247)5.25(0.176)}{\pi 0.035^3} = 68.4 \text{ kpsi}$$

$$(n_y)_B = \frac{S_{sy}}{\tau_B} = \frac{121.0}{68.4} = 1.77$$

Looking back on our work we note that body turns are too large, as is the free length. After repeating the process for wire sizes 0.038, 0.042, 0.045, 0.048, 0.051, and 0.055 in, we construct Table 10–24. Perform the adequacy assessment by inspecting the table. Note that the wire sized 0.042 and 0.045 in results in satisfactory designs. Figure 10–18 (see page 652) is a plot of the figure of merit. There are two satisfactory springs. The figure of merit is maximized by the spring with the wire size of 0.042 in. Notice from the table that the free length goes through a minimum in the range $0.035 \leq d \leq 0.055$-in in wire size. There was the possibility that no free length under $1\frac{1}{4}$ in was possible. In such a case try a stronger material (OQ&T steel, at a 30 percent increase in cost). A spring specification checklist is shown in Table E–37 in the Appendix. Tables 10–24, 10–25, and 10–26 (see pages 652 and 653) have useful information relevant to the ensuing discussion.

Now is the time to allow viewpoint to teach us something about the structure of this design problem. We declared two decision variables, but in reality we reduced them to one by something we did. The expectation that increasing the initial tension F_i increases the free length L_0, and that we were watching a free-length constraint, led us to exercise design decision 2 in deciding to use the initial tension corresponding to minimum tension makeable spring. This is giving up a free-choice option. Even if this is sound engineering guiding "where to look," it is still loss of a degree of freedom.

Figure 10–18

The figure of merit as a function of (*a*) wire size *d*, and (*b*) initial tension F_i. The hollow dot plotting symbol denotes the two satisfactory designs.

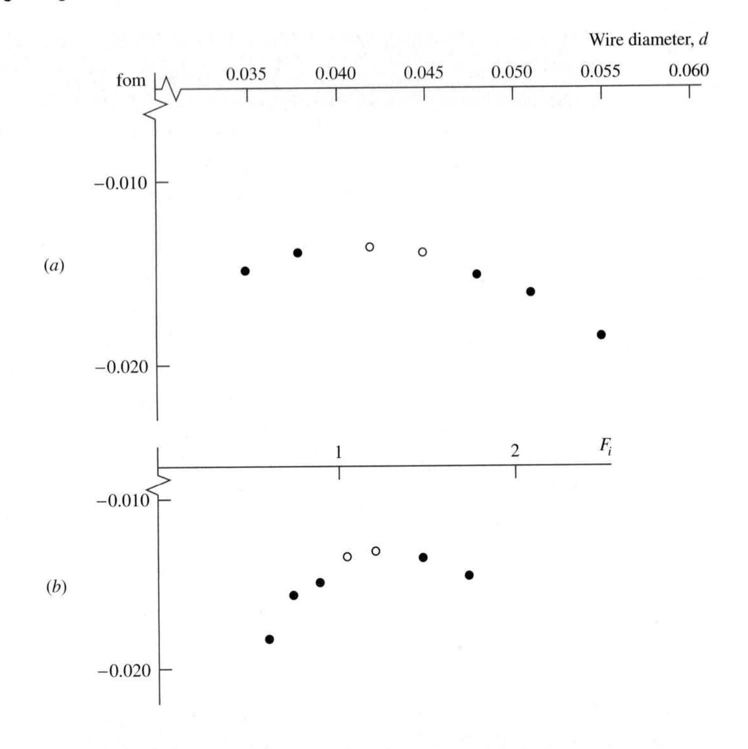

Table 10–25

A Tabulation of the Results of Ex. 10–10 with Initial Tension Set to the Mid-Range Value for Manufacturability

d: F_i:	0.035 1.90	0.038 1.67	0.042 1.42	0.045 1.25	0.048 1.10	0.051 0.97	0.055 0.80
D	0.176	0.229	0.314	0.387	0.471	0.565	0.706
OD	0.211	0.267	0.356	0.432	0.519	0.616	0.761
C	5.033	6.039	7.469	8.610	9.810	11.07	12.84
k	14.69	15.70	16.80	17.54	18.20	18.77	19.52
N_a	27.68	16.28	8.89	5.95	4.14	2.97	1.97
N_b	27.27	15.87	8.48	5.54	3.73	2.56	1.56
L_0	1.272	1.024	0.942	0.979	1.073	1.209	1.443
L_1	1.499	1.252	1.169	1.207	1.301	1.437	1.671
$(n_y)_b$	1.705	1.721	1.737	1.745	1.752	1.758	1.764
$(n_y)_A$	1.500	1.500	1.500	1.500	1.500	1.500	1.500
$(n_y)_B$	1.765	1.794	1.811	1.820	1.819	1.820	1.820
fom	−0.0156	−0.0146	−0.0143	−0.0146	−0.0153	−0.0165	−0.0188

Let us consider this design problem two more times, first choosing the midrange initial tension and building another table, then repeating with the maximum initial tension and a third table. Now we plot the figure-of-merit surface above the d–F_i plane. Rather than draw the three-dimensional picture of the merit surfaces, we will plot points representing the various designs on the plane of the paper as Fig. 10–19, using the plotted points as the decimal points for displaying the figure-of-merit magnitudes. We can sketch in contours of constant merit and visualize the search problem. From an arbitrary initial point, following the gradient "uphill," we would move perpendicular to the merit contour lines until we encountered the $(F_i)_{\min}$ constraint, then we would move along the constraint until encountering the optimal merit. Bear in mind that the wire size is discrete, but initial tension is continuous. A constrained optimization program that treats

Table 10–26

A Tabulation of the Results of Ex. 10–10 with Initial Tension Set to Maximum Value for Manufacturability

d: F_i:	0.035 2.24	0.038 2.00	0.042 1.72	0.045 1.54	0.048 1.37	0.051 1.22	0.055 1.03
D	0.176	0.229	0.314	0.387	0.471	0.565	0.706
OD	0.211	0.267	0.356	0.432	0.519	0.616	0.671
C	5.03	6.04	7.47	8.61	9.81	11.07	12.84
k	13.20	14.25	15.48	16.27	17.02	17.67	18.51
N_a	31.81	17.93	9.64	6.42	4.43	3.15	2.08
N_b	30.40	17.52	9.23	6.01	4.02	2.74	1.67
L_0	1.381	1.087	0.973	1.000	1.087	1.218	1.449
L_1	1.069	1.315	1.201	1.228	1.315	1.446	1.677
$(n_y)_b$	1.705	1.721	1.737	1.745	1.752	1.758	1.764
$(n_y)_A$	1.500	1.500	1.500	1.500	1.500	1.500	1.500
$(n_y)_B$	1.765	1.794	1.811	1.817	1.819	1.820	1.820
fom	−0.0173	−0.0160	−0.0153	−0.0155	−0.0161	−0.0172	−0.0193

both decision variables as continuous would stop at an optimum that corresponded to an unavailable wire diameter.

Notice the domain of satisfactory designs in Fig. 10–19. The domain was identified by adequacy assessment, not by noting points interior to other constraints (which we have not drawn). There are some 0.048-in wire size springs that are satisfactory, something not discovered by the first pass of Ex. 10–10. The cost of the $d = 0.048$, $F_i = 1.37$ lb spring is $(0.0161 - 0.0137)/0.0137$, or about 17.5 percent greater than the optimal spring ($d = 0.042$ in, $F_i = 1.21$ lb design). We probably would not choose the undiscovered spring, but it is instructive to realize that it is there. Had a constraint been $4 \leq N_b \leq 5$ turns, the first method would have found no springs, although in fact there were some 0.048-in wire diameter springs that are satisfactory. It is important to realize the mathematical reason: we had one decision variable brought about by using the engineering expectation of where the best chance of finding satisfactory designs is located. A one-decision-variable problem structure looks at fewer designs than a two-decision-variable problem structure. If you give up the opportunity for two decision variables in the way you structure the problem, do this in a way that avoids excluding the high-merit satisfactory designs. We shall discuss this point further after Ex. 10–11.

Until now we have considered cases with one decision variable. In extension springs we have encountered two-decision variables. If it takes p points to satisfactorily reveal the nature of the figure-of-merit variation (with all other decision variables held constant), it will take p^2 points to show the role of the second of two variables. This is an exponential increase in effort. Note, therefore, that

- The computer becomes more attractive in carrying out the necessary calculation chores as dimensionality (the number of decision variables) increases.
- Optimization theory and programs become more attractive in finding the best design. In extension springs we have one discrete decision variable present (wire diameter), which complicates procedures.
- In learning and mastering the fundamentals, it is useful to do the chores in order to see all the intermediate parameters and their variations. Later, some of the drill can be delegated to the computer. Your intimate knowledge will help you program and trouble-shoot the inevitable snares.

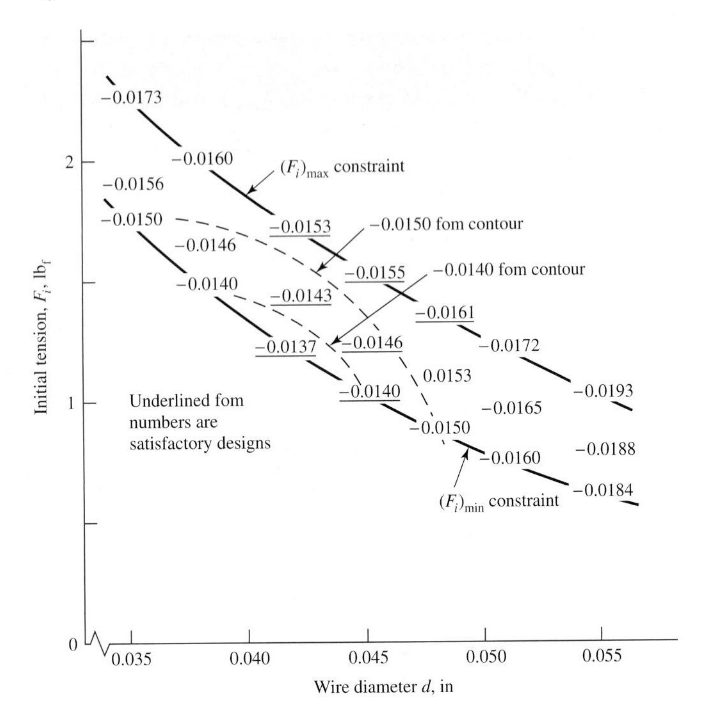

Figure 10–19

Values of the figure of merit plotted on the dF_i plane. The decimal points are the plotting positions, and underlined figure-of-merit entries denote satisfactory designs. Equal-merit contours can be visualized. There are seven satisfactory designs; the optimal occurs at wire size d of 0.042 in, initial tension of 1.21 lb, with a corresponding figure of merit of −0.0137, located against the $(F_i)_{min}$ constraint. Initial tension F_i is a continuous variable, wire size d is discrete, so there are an infinite number of satisfactory designs. Knowing that the optimum lies against the constraint allows the decision variable to be set to $(F_i)_{min}$ as an a priori decision. In effect this reduces the design problem to one decision variable d, greatly simplifying the problem.

The specification set for dynamic service is very nearly the same as for static service except for the cycle-life comment. This is because shot-peening, while beneficial in fatigue, cannot be accomplished since the pretensioned coils shield much of the coil surface from the shot. The corresponding decision set is modified to express the function as F_{max}, F_{min} with a change in length of Δy with a cycle life of N cycles. A fatigue-safe decision replaces a static-safe decision:

Dynamic Service Specification Set	Corresponding Decision Set
Material and condition N-cycle life	Material and condition
End-loop treatment	End-loop treatment
Side-bend radius	Side-bend radius
Body turns	Fatigue-safe
Coil OD	Function: F_{max}, F_{min} at Δy, N-cycle life
Free length	Wire diameter } design variables
Wire size	Initial tension }

The design plan of approach is to attend to the Achilles heel, which is bending fatigue at point A in the end hook. This is a bending situation and the failure locus can be Gerber, which is regression-worthy, or Goodman, which is not. The endurance strength in bending is given for 10^5, 10^6, or 10^7 cycle life in Table 10–22, and it is expressed as a fraction of the ultimate tensile strength. For 10^7 cycles,

$$(\sigma_{\max})_{\text{all}} = 0.45 S_{ut}$$

$$(S_e)_{\text{constr}} = \frac{(\sigma_{\max})_{\text{all}}/2}{1 - [(\sigma_{max})_{\text{all}}/(2S_{ut})]^2}$$

allowing the Gerber fatigue failure locus to be expressed as

$$\frac{S_a}{S_e} + \left(\frac{S_m}{S_{ut}}\right)^2 = 1$$

The strength amplitude S_a can be expressed as

$$S_a = \frac{r^2 S_{ut}^2}{2S_e}\left[-1 + \sqrt{1 + \left(\frac{2S_e}{rS_{ut}}\right)^2}\,\right]$$

where $r = F_a/F_m$. Now

$$F_a = \frac{F_{\max} - F_{\min}}{2} \qquad F_m = \frac{F_{\max} + F_{\min}}{2}$$

and the fatigue factor of safety at point A, $(n_f)_A$, is

$$(n_f)_A = \frac{S_a}{\sigma_A} = \frac{S_a}{F_a\left[\dfrac{16(K)_A D}{\pi d^3} + \dfrac{4}{\pi d^2}\right]}$$

which can be expressed as

$$C = \frac{\pi d^2 S_a}{4n_f F_a\left\{4\left[\dfrac{4C^2 - C - 1}{4C(C-1)}\right] + \dfrac{1}{C}\right\}}$$

which has to be solved for spring index C iteratively. Then

$$D = Cd$$

$$\text{OD} = D + d$$

$$K_B = \frac{4C + 2}{4C - 3}$$

The midrange value of the initial coil tension τ_i (uncorrected) and its tolerance are

$$(\tau_i)_{\text{uncorr}} = 33\,500 \exp(-0.105C)$$

$$\text{tol} = 1000\left(4 - \frac{C - 3}{6.5}\right)$$

and the largest and smallest initial tension that can be incorporated are

$$(F_i)_{\max} = \frac{[(\tau_i)_{\text{uncorr}} + \text{tol}]\pi d^3}{8D}$$

$$(F_i)_{\min} = \frac{[(\tau_i)_{\text{uncorr}} - \text{tol}]\pi d^3}{8D}$$

Since initial tension will be a design variable, we will place it at the lower end of the makeable-spring range in order to keep the spring length small, if this is desirable. The spring rate is established by

$$k = \frac{F_{max} - F_{min}}{\Delta y}$$

Since the spring rate k can also be expressed as $(F_{max} - F_i)/y_{max}$, we can write

$$y_{max} = \frac{(F_{max} - F_i)\Delta y}{F_{max} - F_{min}} \qquad y_{min} = \frac{(F_{min} - F_i)\Delta y}{F_{max} - F_{min}}$$

then

$$N_a = \frac{Gd^4}{8D^3k}$$

$$N_b = N_a - \frac{G}{E}$$

$$L_0 = (2C - 1 + N_b)d$$

$$L_{max} = L_0 + y_{max}$$

$$L_{min} = L_0 + y_{min}$$

The shear-stress amplitude component τ_a and midrange component τ_m are given by

$$\tau_a = \frac{8K_B F_a D}{\pi d^3} \qquad \tau_i = \frac{8K_B F_i D}{\pi d^3} \qquad \tau_m = \frac{8K_B F_m D}{\pi d^3} = \frac{F_m - F_i}{F_a}\tau_a + \tau_i$$

The coil-body failure locus can also be Gerber but expressed in shear terms. From Table 10–22, for 10^7 cycles

$$(\tau_{max})_{all} = 0.30 S_{ut}$$

$$(S_{se})_{constr} = \frac{(\tau_{max})_{all}/2}{1 - [(\tau_{max})_{all}/2/S_{su}]^2}$$

and the Gerber fatigue failure locus for body coils can be expressed as

$$S_{sa} = r\left(\frac{rS_{su}^2}{2S_{se}} + \tau_i\right)\left[-1 + \sqrt{1 + \frac{4(S_{su}^2 + \tau_i^2)}{\left(r\frac{S_{su}^2}{S_{se}} + 2\tau_i\right)^2}}\,\right]$$

where $r = F_a/F_m$, or Eq. (10–56) can be used directly. Then

$$(n_f)_{body} = \frac{S_{sa}}{\tau_a}$$

From Eq. (10–56),

$$(S_{sa})_y = \frac{S_{sy} - \tau_i}{1 + F_m/F_a}$$

and from Eq. (10–57)

$$(n_y)_{body} = \frac{(S_{sy})_y}{\tau_a}$$

Now we check the ends. At point A,

$$(K)_A = \frac{4C^2 - C - 1}{4C(C-1)}$$

$$(\sigma_a)_A = F_a\left[\frac{16(K)_A D}{\pi d^3} + \frac{4}{\pi d^2}\right] \qquad (\sigma_m)_A = (\sigma_a)_A \frac{F_m}{F_a}$$

$$(n_f)_A = \frac{S_a}{(\sigma_a)_A}$$

$$(n_y)_A = \frac{S_y}{(\sigma_a)_A + (\sigma_m)_A}$$

At point B, $C_2 = C - 1$,

$$(K)_B = \frac{4C_2 - 1}{4C_2 - 4} = \frac{4C - 5}{4C - 8}$$

$$(\tau_a)_B = \frac{8(K)_B F_a D}{\pi d^3} \qquad (\tau_m)_B = (\tau_a)_B \frac{F_m}{F_a}$$

For a Gerber fatigue failure locus,

$$(n_f)_B = \frac{1}{2}\frac{\tau_a}{S_{se}}\left(\frac{S_{su}}{\tau_m}\right)^2\left[-1 + \sqrt{1 + \left(2\frac{\tau_m}{S_{su}}\frac{S_{se}}{\tau_a}\right)^2}\right]$$

$$(n_y)_B = \frac{S_{sy}}{(\tau_a)_B + (\tau_m)_B}$$

EXAMPLE 10–11 Design an infinite-life helical extension spring with full end loops and generous loop-bend radius for a minimum load of 1.5 lb and a maximum load of 5 lb, with an accompanying stretch between the loads of 0.197 in. Use a Gerber fatigue locus. The outside diameter of the body coil should not exceed 9/16 in. The maximum length cannot exceed $1\frac{1}{2}$ in, and the spring should open at least 0.010 for minimum load.

Solution

- Material and condition: Hard-drawn spring wire, $A = 137$ kpsi, $m = 0.201$, $G = 11.85(10^6)$ psi, $E = 29(10^6)$ psi
- End-loop treatment: Full coil loop
- Side-bend radius: Generous
- Fatigue-safe: $(n_f)_d = 1.5$
- Function: $F_{\min} = 1.5$ lb, $F_{\max} = 5$ lb, $\Delta y = 0.197$ in
- Decision variable: Wire size d
- Decision variable: Initial tension F_i

Choose a trial diameter $d = 0.048$ in. Then

$$S_{ut} = 137/0.048^{0.201} = 252.2 \text{ kpsi}$$

$$S_y = 0.75 S_{ut} = 0.75(252.2) = 189.2 \text{ kpsi}$$

$$S_{su} = 0.67 S_{ut} = 0.67(252.2) = 169.0 \text{ kpsi}$$

$$S_{sy} = 0.45 S_{ut} = 0.45(252.2) = 113.5 \text{ kpsi}$$

$$F_a = (F_{\max} - F_{\min})/2 = (5 - 1.5)/2 = 1.75 \text{ lbf}$$

$$F_m = (F_{\max} + F_{\min})/2 = (5 + 1.5)/2 = 3.25 \text{ lbf}$$

In the end hoops $r = F_a/F_m = 1.75/3.25 = 0.538$ and load line is radial. The weakest point is in the end hooks at A. The wire is in bending. From Table 10–22 the zero-max bending stress maximum is

$$(\sigma_{\max})_{\text{all}} = S_r = 0.45 S_{ut} = 0.45(252.2) = 113.5 \text{ kpsi}$$

The Gerber ordinate intercept on the designer's fatigue diagram is

$$(S_{se})_A = \frac{S_r/2}{1 - \left(\frac{S_r/2}{S_{ut}}\right)^2} = \frac{113.5/2}{1 - \left(\dfrac{113.5/2}{252.2}\right)^2} = 59.8 \text{ kpsi}$$

The Gerber locus can be expressed as

$$\frac{S_a}{59.8} + \left(\frac{S_m}{252.2}\right)^2 = 1$$

The load line runs through the origin. The strength amplitude S_a is given by Eq. (7–47):

$$S_a = \frac{0.538^2 252.2^2}{2(59.8)} \left\{ -1 + \sqrt{1 + \left[\frac{2(59.8)}{0.538(252.2)}\right]^2} \right\} = 51.3 \text{ kpsi}$$

The spring index is given by rearranging the equation $(n_f)_A = (S_a)_A/(\sigma_a)_A$ as follows:

$$C = \frac{\pi d^2 S_a}{4 n_f F_a \left\{ 4\left[\dfrac{4C^2 - C - 1}{4C(C-1)}\right] + \dfrac{1}{C}\right\}} = \frac{\pi (0.048)^2 51.3}{4(1.5)1.75 \left\{ 4\left[\dfrac{4C^2 - C - 1}{4C(C-1)}\right] + \dfrac{1}{C}\right\}}$$

and this equation is solved by successive substitution giving $C = 7.76$. From this

$$D = Cd = 7.76(0.048) = 0.372 \text{ in}$$

$$\text{OD} = D + d = 0.372 + 0.048 = 0.420 \text{ in}$$

$$K_B = \frac{4(7.76) + 2}{4(7.76) - 3} = 1.178$$

From Eq. (10–49) the uncorrected initial torsional stress is

$$(\tau_i)_{\text{uncorr}} = 33\,500 \exp[-0.105(7.76)] = 14.8 \text{ kpsi}$$

$$\text{tol} = 1000\left[4 - \frac{7.76 - 3}{6.5}\right] = 3.27 \text{ kpsi}$$

Equation (10–3) can be arranged without a K-factor to express the initial tension F_i:

$$(F_i)_{\max} = \frac{(14.8 + 3.27)0.048^3 10^3}{8(0.372)} = 2.11 \text{ lbf}$$

$$(F_i)_{\min} = \frac{(14.8 - 3.27)0.048^3 10^3}{8(0.372)} = 1.348 \text{ lbf}$$

We must choose an initial tension F_i to be less than $F_{\min}$. *Decision:* $F_i = 1.35$ lb. Now

$$k = \frac{F_{\max} - F_{\min}}{\Delta y} = \frac{5 - 1.5}{0.197} = 17.77 \text{ lb/in}$$

$$y_{\max} = \frac{(F_{\max} - F_i)\Delta y}{F_{\max} - F_{\min}} = \frac{(5 - 1.35)0.197}{5 - 1.5} = 0.205 \text{ in}$$

$$y_{\min} = \frac{(F_{\min} - F_i)\Delta y}{F_{\max} - F_{\min}} = \frac{(1.5 - 1.35)0.197}{5 - 1.5} = 0.008 \text{ in}$$

The number of active turns can be found from Eq. (10–9):

$$N_a = \frac{Gd^4}{8D^3k} = \frac{11.85(10^6)0.048^4}{8(0.372^3)17.77} = 8.60 \text{ turns}$$

Rearranging Eq. (10–51) gives the number of body turns:

$$N_b = 8.60 - \frac{G}{E} = 8.60 - \frac{11.85}{29} = 8.19 \text{ turns}$$

From Eq. (10–50),

$$L_0 = [2(C - 1) + N_b]d = [2(7.76 - 1) + 8.19]0.048 = 1.042 \text{ in}$$

$$L_{\max} = 1.042 + 0.205 = 1.247 \text{ in}$$

$$L_{\min} = 1.042 + 0.008 = 1.050 \text{ in}$$

End hooks: Bending at A:

$$(K)_A = \frac{4(7.76)^2 - 7.76 - 1}{4(7.76)(7.76 - 1)} = 1.106$$

The bending stress components are found from Eqs. (10–59) and (10– 60):

$$(\sigma_a)_A = 1.75\left[\frac{16(1.106)0.372}{\pi 0.048^3} + \frac{4}{\pi 0.048^2}\right] = 34.1 \text{ kpsi}$$

$$(\sigma_m)_A = 34.1\frac{3.25}{1.75} = 63.3 \text{ kpsi}$$

The fatigue factor of safety can be found from

$$(n_f)_A = \frac{S_a}{(\sigma_a)_A} = \frac{51.3}{34.1} = 1.5 \qquad \text{(as it should)}$$

The yield factor of safety is found from

$$(n_y)_A = \frac{S_y}{(\sigma_a)_A + (\sigma_m)_A} = \frac{189.2}{34.1 + 63.3} = 1.94$$

End hooks: Torsion at point B:

$$(K)_B = \frac{4(7.76) - 5}{4(7.76) - 8} = 1.130$$

From Eq. (10–66) the shear-stress components are

$$(\tau_a)_B = \frac{8(1.13)1.75(0.372)}{\pi 0.048^3} = 16.9 \text{ kpsi}$$

$$(\tau_m)_B = \frac{16.9(3.25)}{1.75} = 31.3 \text{ kpsi}$$

The zero-maximum shear stress from Table 10–22 is

$$(\tau_{\max})_{\text{all}} = S_{sr} = 0.28S_{ut} = 0.28(252.2) = 70.6 \text{ kpsi}$$

and the corresponding endurance limit is

$$(S_{se})_B = \frac{70.6/2}{1 - \left(\dfrac{70.6/2}{169}\right)^2} = 36.8 \text{ kpsi}$$

The factors of safety, using Eq. (10–68) for $(n_f)_B$, are

$$(n_f)_B = \frac{1}{2}\frac{16.9}{36.9}\left(\frac{169}{31.3}\right)^2\left[-1 + \sqrt{1 + \left(2\frac{31.3}{169}\frac{36.9}{16.9}\right)^2}\,\right] = 1.91$$

$$(n_y)_B = \frac{S_{sy}}{(\tau_a)_B + (\tau_m)_B} = \frac{113.5}{16.9 + 31.3} = 2.36$$

Coil body: The stress components are

$$\tau_i = \frac{8(1.178)1.35(0.372)}{\pi 0.048^3} = 13.6 \text{ kpsi}$$

$$\tau_a = \frac{8(1.178)1.75(0.372)}{\pi 0.048^3} = 17.7 \text{ kpsi}$$

$$\tau_m = \frac{F_m - F_i}{F_a}\tau_a + \tau_i = \frac{3.25 - 1.35}{1.75}17.7 + 13.6 = 32.8 \text{ kpsi}$$

The zero-max allowable shear stress, from Table 10–22, is

$$(\tau_{\max})_{\text{all}} = S_{sr} = 0.30S_{ut} = 0.30(252.2) = 75.7 \text{ kpsi}$$

and the corresponding Gerber ordinate intercept S_{se} is

$$S_{se} = \frac{75.7/2}{1 - \left(\dfrac{75.7/2}{169}\right)^2} = 39.8 \text{ kpsi}$$

The fatigue factor of safety, from Eq. (10–68), is

$$(n_f)_{\text{body}} = \frac{1}{2}\frac{17.7}{39.8}\left(\frac{169}{32.8}\right)^2\left[-1 + \sqrt{1 + \left(2\frac{32.8}{169}\frac{39.8}{17.7}\right)^2}\,\right] = 1.93$$

From Eq. (10–57),

$$(S_{sa})_y = \frac{S_{sy} - \tau_i}{1 + \dfrac{F_m - F_i}{F_a}} = \frac{113.5 - 13.6}{1 + \dfrac{3.25 - 1.35}{1.75}} = 47.9 \text{ kpsi}$$

and from Eq. (10–55),

$$(n_y)_{\text{body}} = \frac{(S_{sa})_y}{(\tau_a)_{\text{body}}} = \frac{47.9}{17.7} = 2.71$$

The figure of merit is

$$\text{fom} = -(\text{cost})\frac{\pi^2 d^2 (N_b + 2) D}{4} = -(1)\frac{\pi^2 0.048^2 (8.19 + 2) 0.372}{4}$$
$$= -0.0216$$

Clearly the end-loop bending controls, as anticipated. We will build a table.

Examination of Table 10–27 reveals some interesting facts. The spring rate is the same for all springs. The designer had no choice since the function statement defined the spring rate. The 0.042-in- and 0.045-in-diameter springs are not makeable since the initial tension F_i is out of the range that can be manufactured. One can see this early in the analysis, coding the design with $F_i = F_{\text{min}}$ producing $y_{\text{max}} = 0$, which reminds us of this fact in Table 10–27. The requirement that the coils be open ($y_{\text{min}} \geq 0.010$ in) eliminates the 0.048 and smaller wire diameter springs. The maximum length held in 1.5 in or less eliminates the 0.055-in wire diameter and higher. The only satisfactory spring has a wire diameter of 0.051 in.

Figure 10–20 is the designer's fatigue diagrams. Notice that there are *three* Gerber failure loci. This occurs because there are different material *conditions* present at each critical location. The coil was wound (one amount of cold work). The end loops were

Table 10–27

A Tabulation of the Results of Ex. 10–11 with Initial Tension Set to the Minimum Value for Manufacturability

d: **F_i:**	**0.042** **1.5**	**0.045** **1.5**	**0.048** **1.35**	**0.051** **1.2**	**0.055** **1.0**
D	0.245	0.305	0.372	0.448	0.563
OD	0.287	0.350	0.420	0.499	0.618
C	5.845	6.779	7.760	8.789	10.232
k	17.77	17.77	17.77	17.77	17.77
N_a	17.54	12.04	8.56	6.26	4.28
N_b	17.13	11.63	8.15	5.88	3.87
L_0	1.168	1.089	1.088	1.144	1.283
L_{max}	1.365	1.268	1.294	1.358	1.509
L_{min}	1.168	1.089	1.097	1.161	1.312
y_{max}	0.197	0.197	0.205	0.214	0.225
y_{min}	0	0	0.008	0.017	0.028
$(n_y)_{\text{body}}$	2.74	2.76	2.71	2.65	2.58
$(n_f)_{\text{body}}$	1.91	1.92	1.93	1.94	1.95
$(n_y)_A$	1.94	1.94	1.94	1.94	1.94
$(n_f)_A$	1.50	1.50	1.50	1.50	1.50
$(n_y)_B$	2.31	2.33	2.34	2.35	2.35
$(n_f)_B$	1.88	1.89	1.90	1.91	1.91
fom	−0.0204	−0.0208	−0.0215	−0.0226	−0.0247

Figure 10–20

Designer's fatigue diagram (*a*) for shear stress due to torsion and (*b*) for bending stress at hook location *A*.

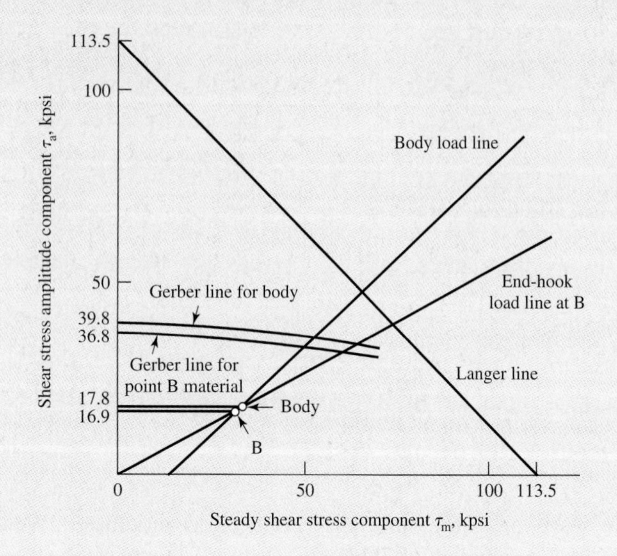

(*a*)

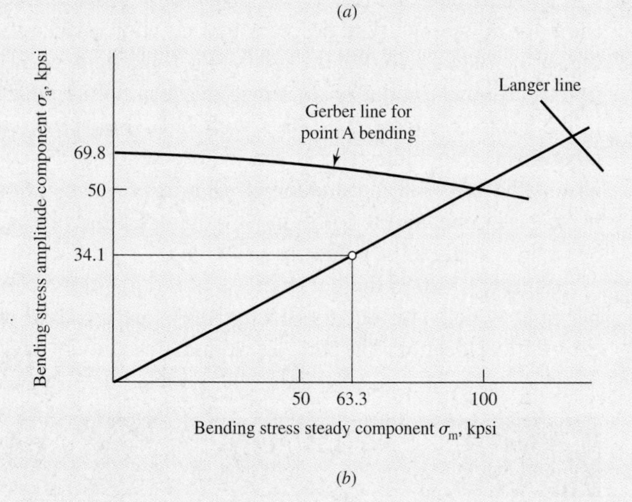

(*b*)

formed with two different bends (two additional amounts of cold work). Also, there are two kinds of stress (shear and bending). Naturally there are three different zero-max fatigue strengths. See the entries in Table 10–22 for body, torsion at end, and bending at end. Thus we recognize three "operating" points. See also Fig. 10–21 for fom traces.

Again, as was indicated after Ex. 10–10, we think we have two decision variables, but really there is but one, due to setting the initial tension to the minimum value for manufacturability. To see if we have lost any satisfactory designs because we used a single decision variable, we repeat the solution to Ex. 10–11 using midrange initial tension. The result is Table 10–28. We note several things. The columns in Table 10–28, including the figure of merit, are identical to those in Table 10–27 except for the entries for L_{max}, L_{min}, y_{max}, and y_{min}. One can see from scanning the y_{min} row in Table 10–28 that we have encountered a wasteland because initial tensions are larger and consequently y_{min} has gotten smaller—in fact, it is negative. The satisfactory solutions are concentrated along the minimum initial tension constraint.

A spring specification checklist is found in Table E–37 in the Appendix.

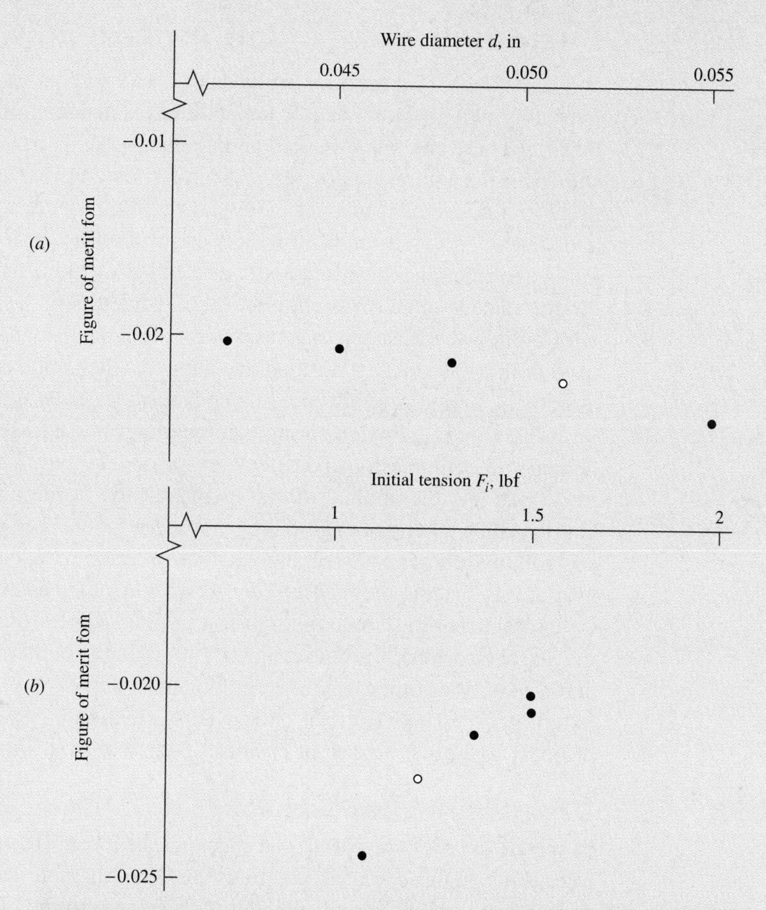

Figure 10–21

Figure-of-merit trace (*a*) in the *d* versus fom plane; (*b*) in the F_i versus fom plane. The hollow plotting point symbol denotes a satisfactory design.

Table 10–28

A Tabulation of the Results of Ex. 10–11 with Initial Tension Set to Midrange Value for Manufacturability

d: F_i:	0.042 2.15	0.045 1.95	0.048 1.73	0.051 1.55	0.055 1.33
D	0.254	0.305	0.372	0.448	0.562
OD	0.287	0.350	0.420	0.499	0.618
C	5.84	6.78	7.76	8.79	10.23
k	17.77	17.77	17.77	17.77	17.77
N_a	17.54	12.04	8.56	6.27	4.23
N_b	17.13	11.63	8.51	5.86	3.87
L_0	1.168	1.089	1.088	1.144	1.283
L_{max}	1.329	1.260	1.272	1.338	1.490
L_{min}	1.132	1.063	1.075	1.141	1.293
y_{max}	0.160	0.172	0.184	0.194	0.207
y_{min}	−0.036	−0.025	−0.013	−0.003	−0.010
$(n_y)_{body}$	3.139	3.023	2.905	2.820	2.724
$(n_f)_{body}$	1.911	1.924	1.934	1.942	1.951
$(n_y)_A$	1.937	1.937	1.937	1.937	1.937
$(n_f)_A$	1.500	1.500	1.500	1.500	1.500
$(n_y)_B$	2.313	2.332	2.341	2.347	2.350
$(n_f)_B$	1.879	1.894	1.903	1.907	1.910
fom	−0.0204	−0.0208	−0.0215	−0.0226	−0.0247

10–14 Designing Helical Coil Torsion Springs

When a helical coil spring is subjected to end torsion, it is called a *torsion spring*. It is usually close-wound, as is a helical coil extension spring, but with negligible initial tension. There are single-bodied and double-bodies types as depicted in Fig. 10–22. As shown in the figure, torsion springs have ends configured to apply torsion to the coil body in a convenient manner, with short hook, hinged straight offset, straight torsion, and special ends. The ends ultimately connect a force at a distance from the coil axis to apply a torque. The most frequently encountered (and least expensive) end is the straight torsion end. If intercoil friction is to be avoided completely, the spring can be wound with a pitch that just separates the body coils. Helical coil torsion springs are usually used with a rod or arbor for reactive support when ends cannot be built in, to maintain alignment, and to provide buckling resistance if necessary.

The wire in a torsion spring is in bending, in contrast to the torsion encountered in helical coil compression and extension springs. The springs are designed to wind tighter in service. As the applied torque increases, the inside diameter of the coil decreases. Care must be taken not to grab the pin, rod, or arbor. The bending mode in the coil might seem to invite square- or rectangular-cross-section wire, but cost, range of materials, and availability discourage its use. The treatment of the ends is dictated by application.

Torsion springs are familiar in clothespins, window shades, animal traps, where they may be seen around the house, and out-of-sight in counterbalance mechanisms, ratchets, and a variety of other machine components. There are many stock springs that can be bough off-the-shelf from a vendor. This selection can add economy of scale to small projects, avoiding the cost of custom design and small-run manufacture.

Describing the End Location

In specifying a torsion spring, the ends must be located relative to each other. Commercial tolerances on these relative positions are listed in Table 10–29. The simplest scheme for expressing the location of one end with respect to the other is in terms of an angle β defining the partial turn present in the coil body as $N_p = \beta/360°$ as shown in Fig. 10–23.

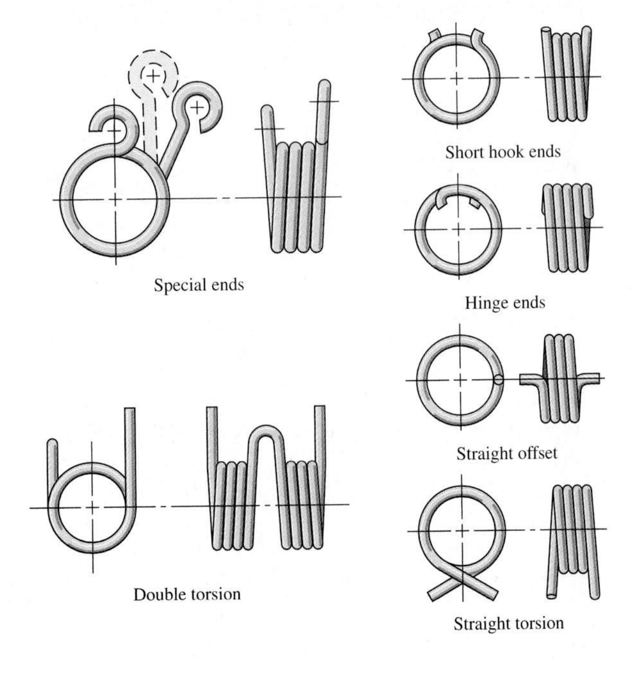

Figure 10–22

Torsion springs. (Courtesy of Associated Spring Corporation.)

Table 10–29

End Position Tolerances for Helical Coil Torsion Springs (for D/d Ratios up to and Including 16)

Source: Associated Spring–Barnes Group, *Design Handbook*, Bristol, Conn., 1987, p. 65.

Total Coils	Tolerance: ± Degrees*
Up to 3	8
Over 3–10	10
Over 10–20	15
Over 20–30	20
Over 30	25

*Closer tolerances available on request.

Figure 10–23

Angle β defines the angular location of the torsion spring ends. The partial turn present is $N_p = \beta/360°$.

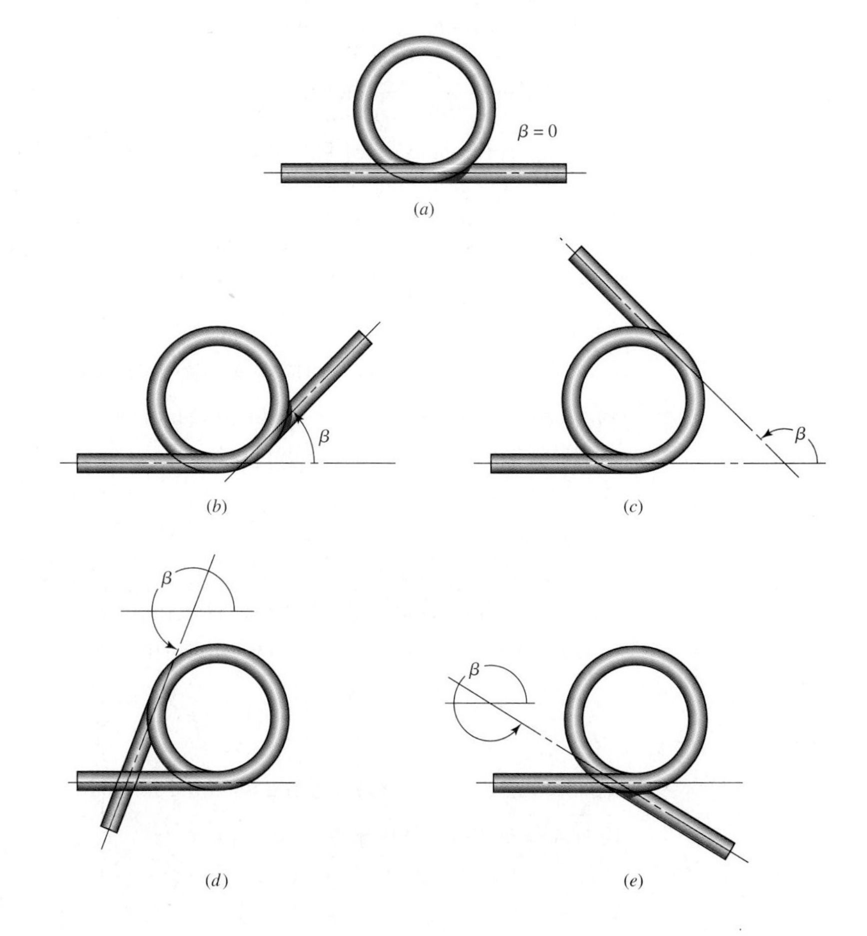

For analysis purposes the nomenclature of Fig. 10–24 can be used. Communication with a springmaker is often in terms of the back-angle α. The spring rate k' is expressed in units of torque/revolution (in·lbf/r or N·mm/r) and moment is proportional to angle θ' expressed in turns rather than radians. The spring rate can be expressed as

$$k' = \frac{M_1}{\theta'_1} = \frac{M_2}{\theta'_2} = \frac{M_2 - M_1}{\theta'_2 - \theta'_1} \tag{10–71}$$

where the moment M can be expressed as $F\ell$ or Fr.

Figure 10–24

The free-end location angle is β. The rotational coordinate θ is proportional to the product *Fl*. Its back angle is α. For all positions of the moving end $\theta + \alpha = \Sigma =$ constant.

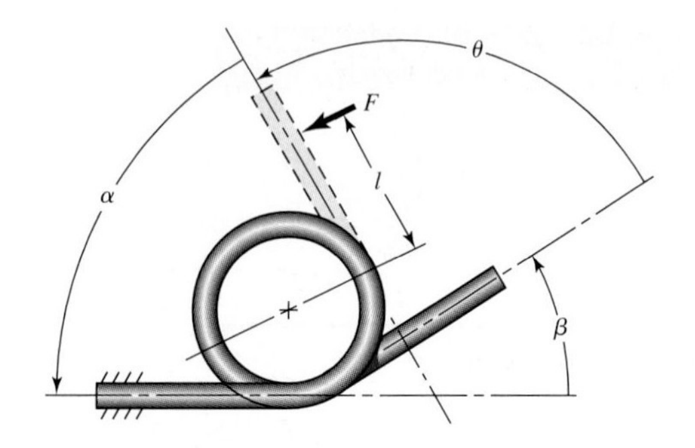

The number of body turns N_b is the number of turns in the free spring body by count. The body-turn count is related to the initial position angle β by

$$N_b = \text{integer} + \frac{\beta}{360°} = \text{integer} + N_p$$

where N_p is the number of partial turns. The above equation means that N_b takes on noninteger, discrete values such as 5.3, 6.3, 7.3, . . . , with successive differences of 1 as possibilities in designing a specific spring. This consideration will be discussed later.

The angle subtended by the end deflection of a cantilever, when viewed from the built-in ends, is y/l rad. From Table E-9-1

$$\theta_e = \frac{y}{l} = \frac{Fl^2}{3EI} = \frac{Fl^2}{3E(\pi d^4/64)} = \frac{64Ml}{3\pi d^4 E} \tag{10–72}$$

For a straight torsion end spring, end corrections such as Eq. (10–72) must be added to the body-coil deflection. The strain energy in bending is, from Eq. (4–26),

$$U = \int \frac{M^2\,dx}{2EI}$$

For a torsion spring $M = Fl = Fr$, and integration must be accomplished over the length of the body-coil wire. The force F will deflect through a distance $r\theta$ where θ is the angular deflection of the coil body. Applying Castigliano's theorem,

$$r\theta = \frac{\partial U}{\partial F} = \int_0^{\pi DN_b} \frac{\partial}{\partial F}\left(\frac{F^2r^2dx}{2EI}\right) = \int_0^{\pi DN_b} \frac{Fr^2\,dx}{EI}\Bigg)$$

Substituting $I = \pi d^4/64$ for round wire and solving for θ gives

$$\theta = \frac{64FrDN_b}{d^4E} = \frac{64MDN_b}{d^4E}$$

The total angular deflection in radians is $\theta + 2\theta_e = \theta_t$, or

$$\theta_t = \frac{64MDN_b}{d^4E} + \frac{64Ml_1}{3\pi d^4E} + \frac{64Ml_2}{3\pi d^4E} = \frac{64MD}{d^4E}\left(N_b + \frac{l_1 + l_2}{3\pi D}\right) \tag{10–73a}$$

Defining the number of end-equivalent turns as N_e, we write

$$N_e = \frac{l_1 + l_2}{3\pi D} \tag{10–74}$$

The equivalent number of active turns N_a is expressed as

$$N_a = N_b + N_e \tag{10–75}$$

The spring rate k in torque per radian is

$$k = \frac{Fr}{\theta_t} = \frac{M}{\theta_t} = \frac{d^4 E}{64 D N_a} \tag{10–76}$$

The spring rate may also be expressed as torque per turn. The expression for this is obtained by multiplying Eq. (10–76) by 2π rad/turn. Thus spring rate k' (units torque/turn) is

$$k' = \frac{2\pi d^4 E}{64 D N_a}$$

Tests show that the effect of friction between the coils and arbor are such that the constant $64/(2\pi) = 10.2$ should be increased to 10.8. The equation above, as well as those for the spring rate of the coils k'_c and the ends k'_e, becomes

$$k'_s = \frac{d^4 E}{10.8 D N_a} \qquad k'_c = \frac{d^4 E}{10.8 D N_b} \qquad k'_e = \frac{3\pi d^4 E}{10.8(l_1 + l_2)} \tag{10–77}$$

where the units of the k' are torque per turn. Equations (10–77) will give better results. Also Eq. (10–73a) becomes

$$\theta'_t = \frac{10.8 M D}{d^4 E}\left(N_b + \frac{l_1 + l_2}{3\pi D}\right) \tag{10–73b}$$

Torsion springs are frequently used over a round bar or pin. When the load is applied to a torsion spring, the spring winds up, causing a decrease in the inside diameter of the coil body. It is necessary to ensure that the inside diameter of the coil never becomes equal to or less than the diameter of the pin, in which case loss of spring function would ensue. The helix diameter of the coil D' becomes

$$D' = \frac{N_b D}{N_b + \theta'_c} \tag{10–78}$$

The new inside diameter $D'_i = D' - d$ makes the diametral clearance between the body coil and the pin Δ equal to

$$\Delta = D' - d - D_a = \frac{N_b D}{N_b + \theta'_c} - d - D_a \tag{10–79a}$$

Equation (10–79a) solved for N_b is

$$N_b = \frac{\theta'_c(\Delta + d + D_a)}{D - \Delta - d - D_a} \tag{10–79b}$$

which gives the number of body turns corresponding to a specified diametral clearance of the arbor. This angle may not be in agreement with the necessary partial-turn remainder. Thus the diametral clearance may be exceeded but not equaled.

Static Strength

First column entries in Table 10–19 can be divided by 0.577 (from distortion-energy theory) to give

$$S_y = \begin{cases} 0.78 S_{ut} & \text{music wire and cold-drawn carbon steels} \\ 0.87 S_{ut} & \text{OQ\&T carbon and low-alloy steels} \\ 0.61 S_{ut} & \text{austenitic stainless steel and nonferrous alloys} \end{cases} \tag{10–80}$$

Table 10–30

Maximum Recommended Bending Stresses for Helical Torsion Springs in Static Applications

Source: Associated Spring–Barnes Group, *Design Handbook*, Bristol, Conn., 1987, p. 63.

Material	Percent of Tensile Strength	
	Stress-Relieved* (K_B Corrected)	With Favorable Residual Stress† (No Correction Factor)
Patented, and cold drawn	80	100
Hardened and tempered carbon and low-alloy steels	85	100
Austenitic stainless steels and nonferrous alloys	60	80

*Also for springs without residual stresses.
†Springs that have not been stress-relieved and which have bodies and ends loaded in a direction that decreases the radius of curvature.

Compare the foregoing values with Table 10–30, which gives 0.80, 0.85, and 0.60, respectively.

Dynamic Strength

Since the spring wire is in bending, the Sines equation or Sines-Zimmerli modification is not applicable. The Sines model is in the presence of pure torsion, and Zimmerli's results were for compression springs (wire in pure torsion). We will use zero-max bending stress values provided by Associated Spring in Table 10–31. We will use the Gerber fatigue-failure locus incorporating the Associated Spring zero-max fatigue strength S_r:

$$(S_e)_{\text{constr}} = \frac{S_r/2}{1 - \left(\dfrac{S_r/2}{S_{ut}}\right)^2} \tag{10–81}$$

The value of S_r (and S_e) has been corrected for size, surface condition, and type of loading, but not for temperature or miscellaneous effects. The Gerber fatigue locus is now defined. The strength-amplitude component is given as

$$S_a = \frac{r^2 S_{ut}^2}{2S_e}\left[-1 + \sqrt{1 + \left(\frac{2S_e}{r\,S_{ut}}\right)^2}\,\right] \tag{10–82}$$

where the slope of the load line is $r = M_a/M_m$. The load line is radial through the origin of the designer's fatigue diagram. The factor of safety guarding against fatigue failure is

$$n_f = \frac{S_a}{\sigma_a} \tag{10–83}$$

Alternatively, we can find n_f directly using Eqs. (10–82) and (10–83):

$$n_f = \frac{1}{2}\frac{\sigma_a}{S_e}\left(\frac{S_{ut}}{\sigma_m}^2\right)\left[-1 + \sqrt{1 + \left(2\frac{\sigma_m}{S_{ut}}\frac{S_e}{\sigma_a}\right)^2}\,\right] \tag{10–84}$$

Table 10–31

Maximum Recommended Bending Stresses (K_B Corrected) for Helical Torsion Springs in Cyclic Applications as Percent of S_{ut}

Source: Associated Spring–Barnes Group, *Design Handbook*, Bristol, Conn., 1987, p. 63.

Fatigue Life, cycles	ASTM A228 and Type 302 Stainless Steel		ASTM A230 and A232	
	Not Shot-Peened	Shot-Peened*	Not Shot-Peened	Shot-Peened*
10^5	53	62	55	64
10^6	50	60	53	62

This information is based on the following conditions: no surging, springs are in the "as-stress-relieved" condition.
*Not always possible.

Bending Stress

A torsion spring has bending moment induced in the coils, rather than torsion. This means that residual stresses built in during winding are in the same direction but of opposite sign to the working stresses which occur during use. The strain-strengthening locks in residual stresses opposing working stresses *provided* the load is always applied in the winding sense. Torsion springs can operate at bending stresses exceeding the yield strength of the wire from which it was wound.

The bending stress can be obtained from curved-beam theory expressed in the form

$$\sigma = K\frac{Mc}{I}$$

where K is a stress-augmentation factor. The value of K depends on the shape of the wire cross section and whether the stress sought is at the inner or outer fiber. Wahl analytically determined the values of K to be, for round wire,

$$K_i = \frac{4C^2 - C - 1}{4C(C-1)} \qquad K_o = \frac{4C^2 + C - 1}{4C(C+1)} \tag{10–85}$$

where C is the spring index and the subscripts i and o refer to the inner and outer fibers, respectively. In view of the fact that K_o is always less than unity, we shall use K_i to estimate the stresses. When the bending moment is $M = Fr$ and the section modulus $I/c = d^3/32$, we express the bending equation as

$$\sigma = K_i\frac{32Fr}{\pi d^3} \tag{10–86}$$

which gives the bending stress for a round-wire torsion spring.

The adequacy assessment for a helical coil torsion spring consists of checking the inequality constraints on spring index, body turns, arbor diametral clearance, spring length, and outside diameter. Additionally, the factors of safety n_y and n_f are checked. If buckling or surging is possible, these are also included. Table 10–32 lists commercial tolerances for coil spring diameters.

The specification set for a static service helical coil torsion spring is usually

- Material and condition
- End treatment
- Body turns and tolerance
- M_1 at α_1 (or θ_1) and tolerance

Table 10–32

Commercial Tolerances for Torsion Spring Coil Diameters

Source: Associated Spring–Barnes Group, *Design Handbook*, Bristol, Conn., 1987, p. 65.

Wire Diameter mm (in)	Tolerances, ± mm (in)						
	Spring Index (D/d)						
	4	6	8	10	12	14	16
0.38 (0.015)	0.05 (0.002)	0.05 (0.002)	0.05 (0.002)	0.05 (0.002)	0.08 (0.003)	0.08 (0.003)	0.10 (0.004)
0.58 (0.023)	0.05 (0.002)	0.05 (0.002)	0.05 (0.002)	0.08 (0.003)	0.10 (0.004)	0.13 (0.005)	0.15 (0.006)
0.89 (0.035)	0.05 (0.002)	0.05 (0.002)	0.08 (0.003)	0.10 (0.004)	0.15 (0.006)	0.18 (0.007)	0.23 (0.009)
1.30 (0.051)	0.05 (0.002)	0.08 (0.003)	0.13 (0.005)	0.18 (0.007)	0.20 (0.008)	0.25 (0.010)	0.31 (0.012)
1.93 (0.076)	0.08 (0.003)	0.13 (0.005)	0.18 (0.007)	0.23 (0.009)	0.31 (0.012)	0.38 (0.015)	0.46 (0.018)
2.90 (0.114)	0.10 (0.004)	0.18 (0.007)	0.25 (0.010)	0.33 (0.013)	0.46 (0.018)	0.56 (0.022)	0.71 (0.028)
4.37 (0.172)	0.15 (0.006)	0.25 (0.010)	0.33 (0.013)	0.51 (0.020)	0.69 (0.027)	0.86 (0.034)	1.07 (0.042)
6.35 (0.250)	0.20 (0.008)	0.36 (0.014)	0.56 (0.022)	0.76 (0.030)	1.02 (0.040)	1.27 (0.050)	1.52 (0.060)

- Outside diameter and tolerance
- Wire size and tolerance

The specification set for a dynamic service spring is usually

- Material and condition
- End treatment
- Body turns and tolerance
- M_1 at α_a (or θ_1) and tolerance; M_2 at α_2 (or θ_2) and tolerance
- Outside coil diameter and tolerance
- Wire size and tolerance

Consider the analysis of the stock spring in the following example.

EXAMPLE 10–12

A stock spring is shown in Fig. 10–25. It is made from 0.072-in-diameter music wire and has $4\frac{1}{4}$ body turns with straight torsion ends. It works over a pin of 0.400 in diameter. The coil outside diameter is $\frac{19}{32}$ in.

(a) Find the maximum operating torque and corresponding rotation for static loading.

(b) Estimate the inside coil diameter and pin diametral clearance when the spring is subjected to the torque in part (*a*).

(c) Estimate the fatigue factor of safety n_f if the applied moment varies between $M_{\min} =$ 1 to $M_{\max} = 5$ in·lb.

Solution

(*a*) For music wire, from Table 10–5 we find that $A = 201$ kpsi and $m = 0.145$. Therefore,

$$S_{ut} = \frac{A}{d^m} = \frac{201}{(0.072)^{0.145}} = 294.4 \text{ kpsi}$$

Figure 10–25

Angles α, β, and θ are measured between the straight-end centerline translated to the coil axis. Coil OD is 19/32 in.

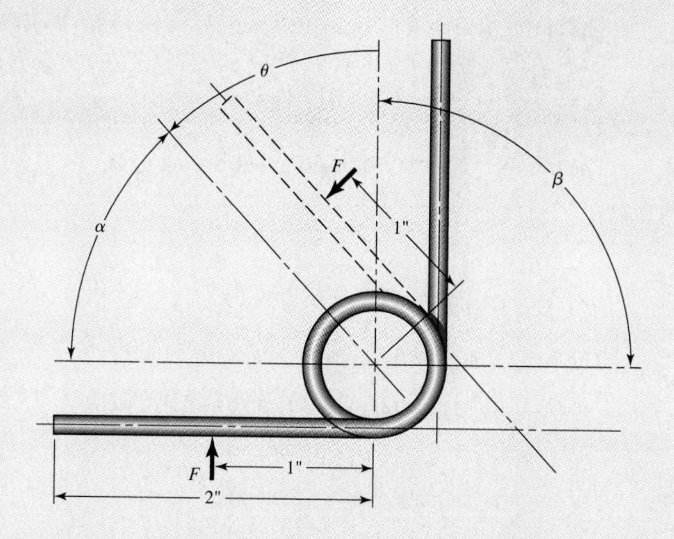

Using Eq. (10–80),

$$S_y = 0.78S_{ut} = 0.78(294.4) = 229.6 \text{ kpsi}$$

The coil helix diameter is $D = 19/32 - 0.072 = 0.522$ in. The spring index $C = D/d = 0.522/0.072 = 7.247$. The bending stress augmentation factor K_i, from Eq. (10–85), is

$$K_i = \frac{4(7.247)^2 - 7.247 - 1}{4(7.247)(7.247 - 1)} = 1.115$$

Now rearrange Eq. (10–86), substitute S_y for σ, and solve for the maximum torque Fr to obtain

$$(Fr)_{\max} = \frac{\pi d^3 S_y}{32K_i} = \frac{\pi(0.072)^3 229\,600}{32(1.115)} = 7.546 \text{ in}\cdot\text{lb}$$

Note that no factor of safety has been used. Next, from Eq. (10–77) we estimate the spring rate k'_c of the body coil as

$$k'_c = \frac{d^4 E}{10.8DN_b} = \frac{0.072^4(28.5)10^6}{10.8(0.522)4.25} = 31.97 \text{ in}\cdot\text{lb/turn}$$

The number of turns of the coil body θ'_c is

$$\theta'_c = \frac{(Fr)_{\max}}{k'_c} = \frac{7.546}{31.97} = 0.236 \text{ turn}$$

Answer $$(\theta'_c)_{\deg} = 0.236(360°) = 85.0°$$

The equivalent number of turns of the ends N_e, from Eq. (10–74), is

$$N_e = \frac{l_1 + l_2}{3\pi D} = \frac{1+1}{3\pi 0.522} = 0.407 \text{ turns}$$

The active number of turns N_a, from Eq. (10–75), is

$$N_a = N_b + N_e = 4.25 + 0.407 = 4.657 \text{ turns}$$

The spring rate of the complete spring k'_s, from Eq. (10–77), is

$$k'_s = \frac{0.072^4(28.5)10^6}{10.8(0.522)4.657} = 29.17 \text{ in·lb/turn}$$

The number of turns of the complete spring θ'_s is

$$\theta'_s = \frac{M}{k'_s} = \frac{7.546}{29.17} = 0.259 \text{ turn}$$

Answer $$(\theta'_s)_{\text{deg}} = 0.259(360°) = 93.13°$$

The spring rate for the ends k'_e, from Eq. (10–77), is

$$k'_e = \frac{3\pi d^4 E}{10.8(l_1 + l_2)} = \frac{3\pi(0.072)^4 28.5(10^6)}{10.8(1+1)} = 334.2 \text{ in·lb/turn}$$

The angular displacement of the ends θ'_e is

$$\theta'_e = \frac{M}{k'_e} = \frac{7.546}{334.2} = 0.0226 \text{ turn}$$

We can check these results by the sum of the reciprocal relationship between the spring rates k'_s, k'_c, and k'_e:

$$\frac{1}{k'_s} = \frac{1}{29.17} = \frac{1}{k'_c} + \frac{1}{k'_e} = \frac{1}{31.97} + \frac{1}{334.2} = 0.0343 = \frac{1}{29.18}$$

which is close enough.

(*b*) With no load, the helix diameter of the spring is 0.522 in. From Eq. (10–78),

$$D' = \frac{N_b D}{N_b + \theta'_c} = \frac{4.25(0.522)}{4.25 + 0.236} = 0.495 \text{ in}$$

The diametral clearance between the inside of the spring coil and the pin at load is

Answer $$\Delta = D' - d - D_a = 0.495 - 0.072 - 0.400 = 0.023 \text{ in}$$

The length of the spring unloaded is

$$(L)_{\text{unloaded}} = (N_b + 1)d = (4.25 + 1)0.072 = 0.378 \text{ in}$$

The loaded length is

$$(L)_{\text{loaded}} = (N_b + \theta'_c + 1)d = (4.25 + 0.236 + 1)0.072 = 0.395 \text{ in}$$

Yes, loaded is longer.

(*c*) Now

$$M_a = (M_{\text{max}} - M_{\text{min}})/2 = (5 - 1)/2 = 2 \text{ in·lb}$$

$$M_m = (M_{\text{max}} + M_{\text{min}})/2 = (5 + 1)/2 = 3 \text{ in·lb}$$

$$r = \frac{M_a}{M_m} = \frac{2}{3} = 0.667$$

$$\sigma_a = K_i \frac{32 M_a}{\pi d^3} = 1.115 \frac{32(2)}{\pi 0.072^3} = 60\,857 \text{ psi}$$

$$\sigma_m = \frac{M_m}{M_a}\sigma_a = \frac{3}{2}(60\,857) = 91\,285 \text{ psi}$$

From Table 10–31, $S_r = 0.50S_{ut} = 0.50(294.4) = 147.2$ kpsi. Then

$$(S_e)_{\text{constr}} = \frac{147.2/2}{1 - \left(\dfrac{147.2/2}{294.4}\right)^2} = 78.51 \text{ kpsi}$$

The Gerber fatigue locus can be expressed as

$$\frac{S_a}{78.51} + \left(\frac{S_m}{294.4}\right)^2 = 1$$

The amplitude component of the strength S_a, from Eq. (10–82), is

$$S_a = \frac{0.667^2 294.4^2}{2(78.51)}\left[-1 + \sqrt{1 + \left(\frac{2}{0.667}\frac{78.51}{294.4}\right)^2}\right] = 68.86 \text{ kpsi}$$

The fatigue factor of safety is

Answer $$n_f = \frac{S_a}{\sigma_a} = \frac{68.86}{60.857} = 1.13$$

Our next example will be a design task. For this we need to develop a decision set. The similarity between the specification sets for static and dynamic service springs encourages us to consider a decision set that will serve either. The presence of a solid coil usually precludes shot peening, so the number of decisions is the same whether static or dynamic. Consider

- Material and condition
- End treatment
- Arbor diameter
- Service-safe: n_y or $n_f = (n)_d$
- Function: M_{min} at α_1 (or θ_1); M_{max} at α_2 (or θ_2)
- Wire size

For static springs set $M_{\text{min}} = 0$. For dynamic springs we could find the spring index which allows the spring to exactly meet (n_y) or $(n_f)_d$. However, fidelity to arbor size is not controllable and few satisfactory springs can be found. Owing to the discreteness of the wire sizes, sometimes only one and sometimes none can be found. There are satisfactory solutions, but the method ignores all those with factors of safety not exactly $(n_f)_d$. Replacing the criterion with $n_f \geq (n_f)_d$ removes a decision from the decision set. The check is displaced to the adequacy assessment. In its place we might substitute N_b as a decision variable. Let us see what this might entail.

EXAMPLE 10–13 Design a straight-end torsion spring to hold a cabinet door closed. The spring should be preloaded to a torque of 0.49 in·lbf when the door is closed with $\alpha_1 = 90°$. When the door is opened 120° the maximum torque is to be 0.98 in·lbf. The moment arms at the ends are to be 0.748 in. The maximum spring length is to be 1.000 in. The pin over which the spring will operate has a diameter of 0.200 in. The spring life is to be 10^6 cycles (infinite). The material is to be music wire, $n_f \geq 1.1$, $n_y \geq 1.2$.

Solution The orientation of the spring ends is unknown. Take

$$\frac{M_1}{M_2} = \frac{\theta_1}{\theta_2} = \frac{\theta_1}{\theta_1 + 120°}$$

from which

$$\theta_1 = \frac{120M_1}{M_2 - M_1} = \frac{120(0.49)}{0.98 - 0.49} = 120°$$

The unloaded spring will look as depicted in Fig. 10–26. The invariant Σ is $\theta_1 + \alpha_1 = 120° + 90° = 210°$; $\alpha_2 = \Sigma - \theta_2 = 210° - 240° = -30°$ (or $+330°$); β is the partial-turn location angle. Then

$$N_p = \frac{\alpha_2}{360°} = \frac{330°}{360°} = +0.91\underline{6} \text{ turn} \qquad \left(\text{or } N_p = \frac{\beta}{360°} = \frac{-30°}{360°} = -0.08\underline{3} \text{ turn}\right)$$

The following table summarizes the angles:

Position	θ	α	Σ	β
0	0	210°	210°	−30°
1	120°	90°	210°	−30°
2	240°	−30°	210°	−30°

Notice that the constant Σ from $\theta + \alpha = \Sigma$ is an invariant, as is the partial-turns location angle β. The foregoing relations are for rotation of the coil body. Let us now proceed, beginning with the following decisions:

- Material and composition: music wire
- End treatment: straight torsion ends, 1 in long
- Arbor diameter: 0.200 in, $\Delta \geq 0.010$ in
- Function: $M_{min} = 0.49$ in·lb at $\theta_1 = 120°$; $M_{max} = 0.98$ in·lb at $\theta_2 = 240°$
- Decision variables: wire size d; body turns N_b

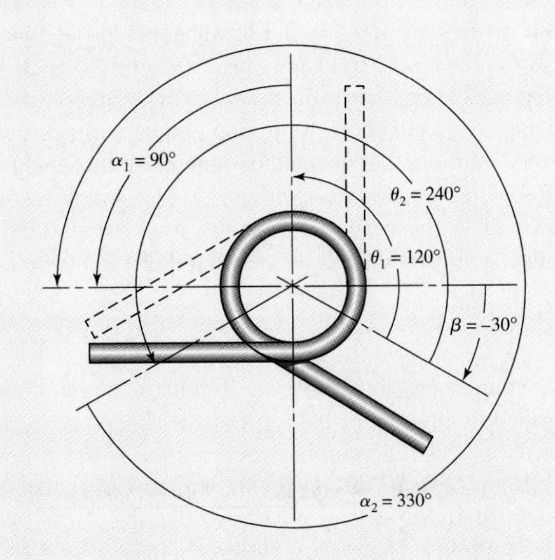

Figure 10–26

The straight end-torsion spring of Ex. 10–13. The invariant $\Sigma = \theta + \alpha$ expressed for position 1 is $\Sigma = \theta_1 + \alpha_1 = 120° + 90° = 210°$. The angle β is $180° - \Sigma = 180° - 210° = -30°$ (or $540° - 210° = 330°$).

The spring rate of the coil body k'_c is

$$k'_c = \frac{M_{\text{max}} - M_{\text{min}}}{\theta_{\text{max}} - \theta_{\text{min}}} = \frac{0.98 - 0.49}{2/3 - 1/3} = 1.47 \text{ in·lb/turn}$$

(For a static service spring one could set $M_{\text{min}} = 0$, $\theta_1 = 0$.) The amplitude and steady components of the load moments are

$$M_a = (M_{\text{max}} - M_{\text{min}})/2 = (0.98 - 0.49)/2 = 0.245 \text{ in·lb}$$

$$M_m = (M_{\text{max}} + M_{\text{min}})/2 = (0.98 + 0.49)/2 = 0.735 \text{ in·lb}$$

$$r = M_a/M_m = 0.245/0.735 = 0.333$$

The first design decision: $d = 0.042$ in. It follows, from Eq. (10–17), that

$$S_{ut} = 201\,000/0.042^{0.145} = 318\,291 \text{ psi}$$

from Eq. (10–80),

$$S_y = 0.78S_{ut} = 0.78(318\,291) = 248\,267 \text{ psi}$$

from Table 10–31,

$$S_r = 0.50S_{ut} = 0.50(318\,291) = 159\,146 \text{ psi}$$

$$S_e = \frac{159.1/2}{1 - \left(\dfrac{159.1/2}{318.3}\right)^2} = 84.85 \text{ kpsi}$$

$$S_a = \frac{0.333^2 318.3^2}{2(84.85)}\left[-1 + \sqrt{1 + \left(\frac{2}{0.333}\frac{84.85}{318.3}\right)^2}\,\right] = 58.77 \text{ kpsi}$$

The next problem is caused by the nature of the body-coil turns N_b, which is really an integer with N_p added, or an integer with 0.083 subtracted, or another integer plus $0.91\underline{6}$ turns. The analytic expression for N_b is given by

$$N_b = \frac{d^4E}{10.8k'_c}\frac{1}{D} - \frac{l_1 + l_2}{3\pi}\frac{1}{D} = \frac{1}{D}\left(\frac{d^4E}{10.8k'_c} - \frac{l_1 + l_2}{3\pi}\right) = \frac{A_1}{D}$$

which defines A_1. Now N_b has to be adjusted to an integer less $0.08\underline{3}$ or plus $0.91\underline{6}$. This has an effect of k'_c. If $N_b = A_1/D_{\text{min}}$, then the adjustment N'_b is

$$N'_b = \text{int}(N_b) + N_p = \text{int}(A_1/D) + N_p$$

The smallest possible value of helix diameter is $D = D_a + d + \Delta = 0.200 + 0.042 + 0.010 = 0.252$ in. From the definition of A_1,

$$A_1 = \frac{0.042^4(29)10^6}{10.8(1.47)} - \frac{0.746 + 0.746}{3\pi} = 5.526$$

$$N'_b = \text{int}(5.526/0.252) - 0.08\underline{3} = 20.917$$

The corresponding value of D is $5.526/20.917 = 0.264$ in. Had this come out to be 0.252 in, the reduction in clearance from the arbor *under load* would violate Δ.

Our second design decision is $N_b = 20.917$ turns. (The corresponding helix diameter is 0.264 in.) The spring index is

$$C = d/D = 0.264/0.042 = 6.29$$
$$\text{OD} = D + d = 0.264 + 0.042 = 0.306 \text{ in}$$
$$\text{ID} = D - d = 0.264 - 0.042 = 0.222 \text{ in}$$

$$K_i = \frac{4C^2 - C - 1}{4C(C-1)} = \frac{4(6.29)^2 - 6.29 - 1}{4(6.29)(6.29 - 1)} = 1.134$$

$$N_e = \frac{l_1 + l_2}{3\pi D} = \frac{0.746 + 0.746}{3\pi(0.264)} = 0.600$$

$$N_a = N_b + N_e = 20.264 + 0.600 = 21.517 \text{ turns}$$

The coil-body spring rate k_c' is

$$k_c' = \frac{d^4 E}{10.8 D N_b} = \frac{0.042^4(29)10^6}{10.8(0.264)(20.917)} = 1.513 \text{ in·lb/turn}$$

The torsion spring rate k_s' is

$$k_s' = \frac{d_4^4 E}{10.8 D N_a} = \frac{0.042^4(29)10^6}{10.8(0.264)21.517} = 1.471 \text{ in·lb/turn}$$

The end spring rate k_e' is

$$k_e' = \frac{3\pi d^4 E}{10.8(l_1 + l_2)} = \frac{3\pi(0.042^4)29(10^6)}{10.8(0.746 + 0.746)} = 52.8 \text{ in·lb/turn}$$

$$\theta_c = \frac{M_2}{k_c'} = \frac{0.98}{1.513} = 0.684 \text{ turn}$$

$$\theta_s = \frac{M_2}{k_s'} = \frac{0.98}{1.471} = 0.666 \text{ turn, or } 240°$$

$$\theta_e = \frac{M_2}{k_e'} = \frac{0.98}{52.8} = 0.0186 \text{ turn, or } 6.7°$$

From Eq. (10–79*a*),

$$\Delta = \frac{N_b D}{N_b + \theta_c'} - d - D_a = \frac{20.917(0.264)}{20.917 + 0.648} - 0.042 - 0.200 = 0.014 \text{ in}$$

The axial length of the coil is

$$L = (N_b + \theta_c + 1)d = (20.917 + 0.648 + 1)0.042 = 0.948 \text{ in}$$

The stress amplitude σ_a is

$$\sigma_a = K_i \frac{32 M_a}{\pi d^3} = 1.134 \frac{32(0.245)}{\pi(0.042)^3} = 38\ 197 \text{ psi}$$

and the fatigue factor of safety n_f is

$$n_f = \frac{S_a}{\sigma_a} = \frac{58.77}{38.2} = 1.54$$

From the Langer line,

$$(S_a)_y = \frac{r S_y}{1 + r} = \frac{0.33\underline{3}(248\ 267)}{1 + 0.33\underline{3}} = 62\ 020 \text{ psi}$$

Table 10–33

The Results of the Search in Ex. 10–13

d:	0.035	0.038	0.042	0.045
N_b:	9.919	13.932	20.917	27.877
C	7.429	6.895	6.286	5.844
D	0.260	0.262	0.264	0.263
OD	0.295	0.300	0.306	0.308
ID	0.225	0.224	0.222	0.218
Δ	0.0095	0.0125	0.0141	0.0120
L	0.404	0.591	0.948	1.331
n_f	0.933	1.170	1.539	1.855
n_y	0.985	1.235	1.624	1.958
fom	−0.0253	−0.0397	−0.0697	−0.104

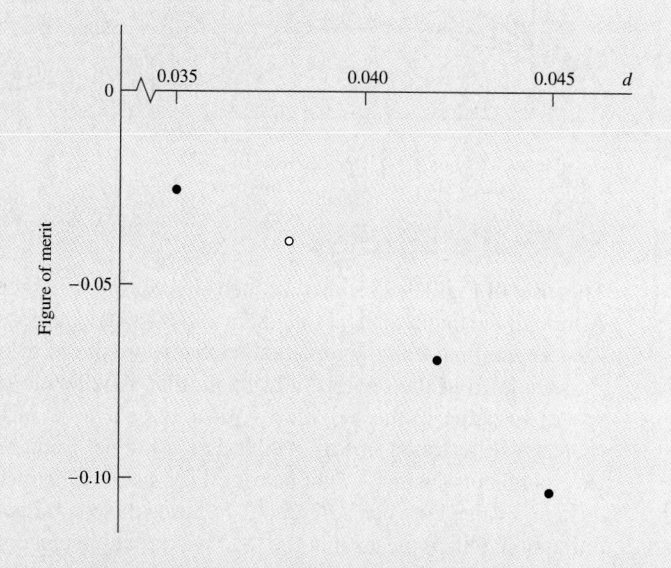

Figure 10–27

The d versus fom trace of the merit surface for the springs of Ex. 10–13. Solid plotting symbols denote unsatisfactory designs.

and the factor of safety guarding against yielding n_y is

$$n_y = \frac{(S_a)_y}{\sigma_a} = \frac{62\ 020}{38\ 197} = 1.62$$

The figure of merit is given by

$$\text{fom} = -\$\left[\frac{\pi^2 d^2 N_b D}{4} + \frac{\pi d^2 (L_1 + L_2)}{4}\right]$$

$$= -2.6\left[\frac{\pi^2 0.042^2 20.917(0.264)}{4} + \frac{\pi 0.042^2(1+1)}{4}\right] = -0.0696$$

This design meets the constraints on n_y, n_f, C, L, and Δ; however, the number of turns is greater than 15. The design is unsatisfactory and we must search further. Table 10–33 presents the results of searching additional wire diameters. Inspection of the table shows that the inequality constraints on n_y, n_f, and N_b eliminate all except the 0.038-in wire diameter. A spring specification checklist is shown in Table E–37 in the Appendix. Figure 10–27 shows the trace of the merit surface on the fom versus d plane. Table 10–32 gives commercial tolerances for torsion spring coil diameters.

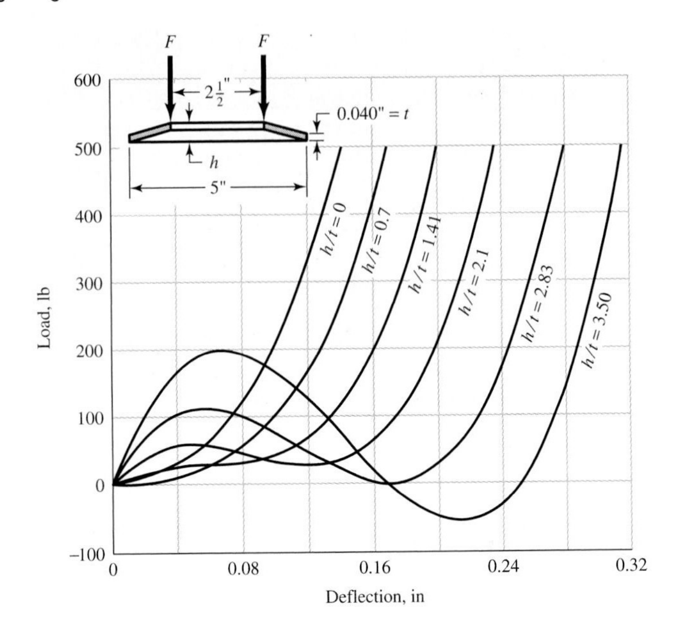

Figure 10–28

Load-deflection curves for Belleville springs. (Courtesy of Associated Spring Corporation.)

10–15 Belleville Springs

The inset of Fig. 10–28 shows a coned-disk spring, commonly called a *Belleville spring*. Although the mathematical treatment is beyond the purposes of this book, you should at least be familiar with the remarkable characteristics of these springs.

Aside from the obvious advantage that a Belleville spring occupies only a small space, variation in the h/t ratio will produce a wide variety of load-deflection curve shapes, as illustrated in Fig. 10–28. For example, using an h/t ratio of 2.83 or larger gives an S curve which might be useful for snap-acting mechanisms. A reduction of the ratio to a value between 1.41 and 2.1 causes the central portion of the curve to become horizontal, which means that the load is constant over a considerable deflection range.

A higher load for a given deflection may be obtained by nesting, that is, by stacking the springs in parallel. On the other hand, stacking in series provides a larger deflection for the same load, but in this case there is danger of instability.

10–16 Miscellaneous Springs

The extension spring shown in Fig. 10–29 is made of slightly curved strip steel, not flat, so that the force required to uncoil it remains constant; thus it is called a *constant-force spring*. This is equivalent to a zero spring rate. Such springs can also be manufactured having either a positive or a negative spring rate.

A *volute spring* is a wide, thin strip, or "flat," of steel wound on the flat so that the coils fit inside one another. Since the coils do not stack, the solid height of the spring is the width of the strip. A variable-spring scale, in a compression volute spring, is obtained by permitting the coils to contact the support. Thus, as the deflection increases, the number of active coils decreases. The volute spring, shown in Fig. 10–30*a*, has another important advantage which cannot be obtained with round-wire springs: if the coils are wound so as to contact or slide on one another during action, the sliding friction will serve to damp out vibrations or other unwanted transient disturbances.

A *conical spring*, as the name implies, is a coil spring wound in the shape of a cone. Most conical springs are compression springs and are wound with round wire. But a

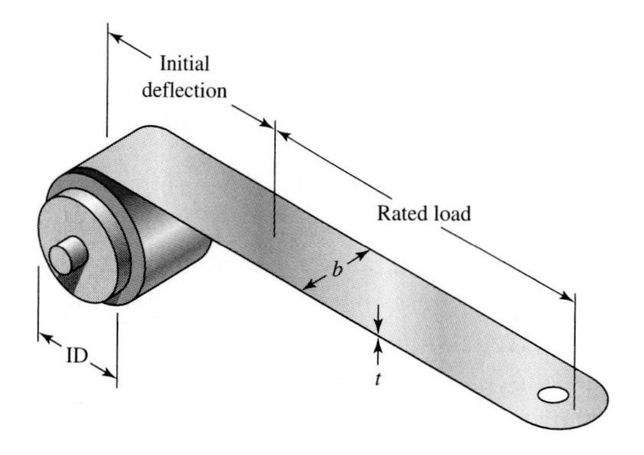

Figure 10–29

Constant-force spring. (Courtesy of Vulcan Spring & Mfg Co., Huntingdon Valley, Pa.)

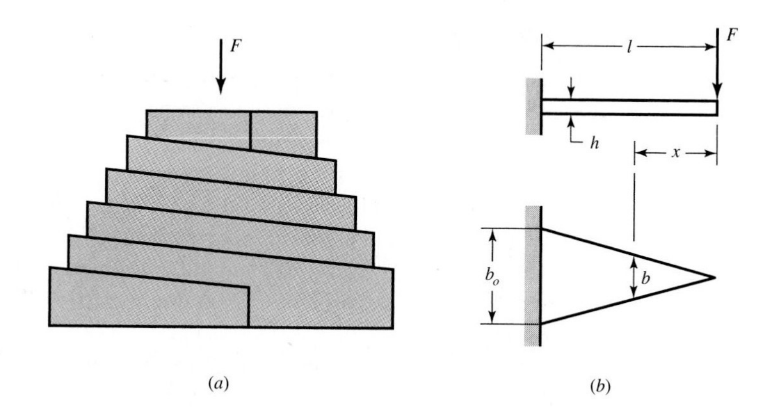

Figure 10–30

(*a*) A volute spring; (*b*) a flat triangular spring.

volute spring is a conical spring too. Probably the principal advantage of this type of spring is that it can be wound so that the solid height is only a single wire diameter.

Flat stock is used for a great variety of springs, such as clock springs, power springs, torsion springs, cantilever springs, and hair springs; frequently it is specially shaped to create certain spring actions for fuse clips, relay springs, spring washers, snap rings, and retainers.

In designing many springs of flat stock or strip material, it is often economical and of value to proportion the material so as to obtain a constant stress throughout the spring material. A uniform-section cantilever spring has a stress

$$\sigma = \frac{M}{I/c} = \frac{Fx}{I/c} \tag{a}$$

which is proportional to the distance x if I/c is a constant. But there is no reason why I/c need be a constant. For example, one might design such a spring as that shown in Fig. 10-30*b*, in which the thickness h is constant but the width b is permitted to vary. Since, for a rectangular section, $I/c = bh^2/6$, we have, from Eq. (*a*),

$$\frac{bh^2}{6} = \frac{Fx}{\sigma}$$

or

$$b = \frac{6Fx}{h^2\sigma} \tag{b}$$

Since b is linearly related to x, the width b_σ at the base of the spring is

$$b_\sigma = \frac{6Fl}{h^2\sigma} \tag{10–87}$$

But the deflection of this triangular flat spring is more difficult to obtain, because the second moment of area is now a variable. Probably the quickest solution could be obtained by using singularity functions or the method of numerical integration.

The methods of stress and deflection analysis illustrated in previous sections of this chapter have served to illustrate that springs may be analyzed and designed by using the fundamentals discussed in the earlier chapters of this book. This is also true for most of the miscellaneous springs mentioned in this section, and you should now experience no difficulty in reading and understanding the literature of such springs.

10–17 Summary

In this chapter we have considered helical coil springs in considerable detail in order to show the importance of viewpoint in approaching engineering problems or analysis or design. The idea that there exists a specification set is both useful and important. It is true that a specification set is not unique. A specification set when load precision must be tighter than commercial tolerance would provide is different from the case when it need not be tighter. The engineer has the task of identifying the specification set applicable to the problem at hand. The reward lies in the contribution to insight and all the advantages that ensue. Springmakers provide a general-purpose form as a guide, and examples have been provided.

The decision set is even more important. In a way it is the specification set in another form. It reveals the structure of the design problem, including its dimensionality. It even allows the opportunity for the designer to control the dimensionality. The insight it offers is invaluable. Considering the detail attendant with springs, can you envision attempting to execute a spring design without a decision set?

As problems become computationally heavy, and a programmable calculator or a computer is used, can you imagine how difficult creating a successful program would be without the understanding that the decision set provides? Even if the engineer has a programming assistant, the engineer has to lay out the form algebraically and check the results of the program. This is very difficult without the understanding that the decision set provides.

CASE STUDY 10a When Conventional Wisdom Is Partially Wise

Conventional wisdom, based on Eq. (10–9), indicates that the coil pitch p to which a helical compression spring is wound does not affect the spring rate k, although it will affect the shear stress at closure. A young engineer was convinced that if a coil spring was wound to *two* different pitches, the spring rate would be affected. Could this be so?

Let us use the spring of Ex. 10–1. The spring has a wire diameter d of 0.037 in and an outer coil diameter OD of 7/16 in. The helix diameter D is 0.400 in. The ends are squared with $12\frac{1}{2}$ total turns, so $Q = 2$ and $Q' = 3$, and a uniform pitch of 0.186 in makes the spring solid-safe. Let's wind the spring from the same wire with $5\frac{1}{4}$ coils having a pitch of 0.1 in, and $5\frac{1}{4}$ coils with a pitch of 0.186 in (keeping the solid-safe property). The geometry of the spring is shown in Fig. CS10a–1. The space occupied by $5\frac{1}{4}$ turns of 0.1-in pitch is $0.1(5\frac{1}{4}) = 0.525$ in. The space occupied by $5\frac{1}{4}$ turns of 0.186-in pitch is $0.186(5\frac{1}{4}) = 0.977$ in. The dead coils occupy a length of $Q'd = 3(0.037) = 0.111$ in, or 0.0555 in at each extreme of the spring. A *partition plane* divides the two windings, now viewed as two springs, and it is

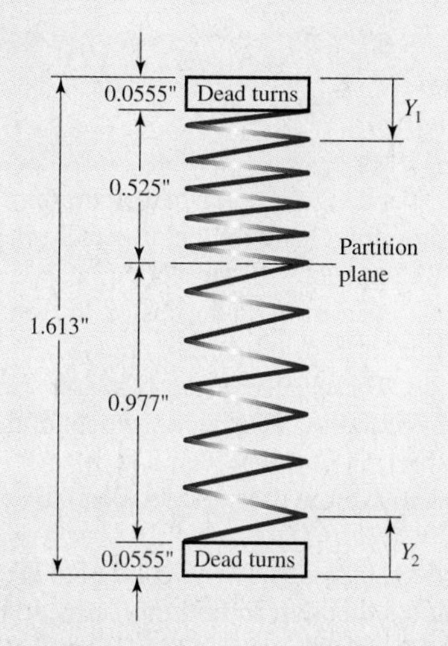

Figure CS10a–1

Schematic of a dual-pitch helical compression spring.

upon this plane, which moves with respect to the ground, that the observer will stand to note the end contractions Y_1 and Y_2. The force F is the same in both springs. The spring rate of the fine-pitch and coarse-pitch springs, without soliding, is 2(4.13) = 8.26 lb/in. Now

$$F = kY_1 = kY_2 \qquad \therefore\ Y_1 = Y_2$$

We also define $y = Y_1 + Y_2$, which is the end contraction of one end of the total spring with respect to the other. When the fine-pitch spring closes, then Y_1 is frozen at $0.525 - 0.037(5\frac{1}{4}) = 0.333$ in, but Y_2 continues to grow. The coarse-pitch spring closes at $Y_2 = 0.977 - 0.037(5\frac{1}{4}) = 0.783$ in. Form a table:

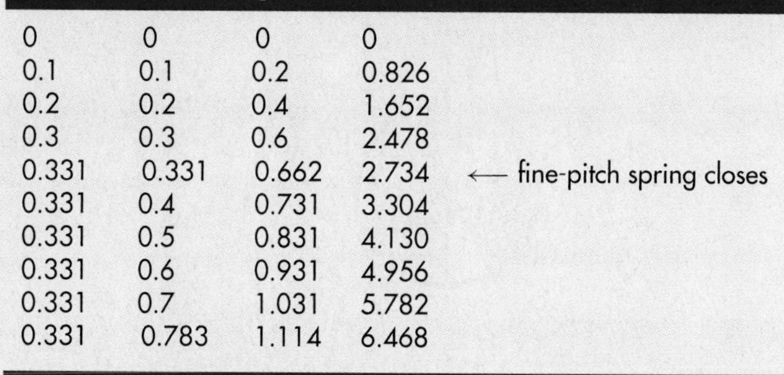

Y_1, in	Y_2, in	y, in	F, lb	
0	0	0	0	
0.1	0.1	0.2	0.826	
0.2	0.2	0.4	1.652	
0.3	0.3	0.6	2.478	
0.331	0.331	0.662	2.734	← fine-pitch spring closes
0.331	0.4	0.731	3.304	
0.331	0.5	0.831	4.130	
0.331	0.6	0.931	4.956	
0.331	0.7	1.031	5.782	
0.331	0.783	1.114	6.468	

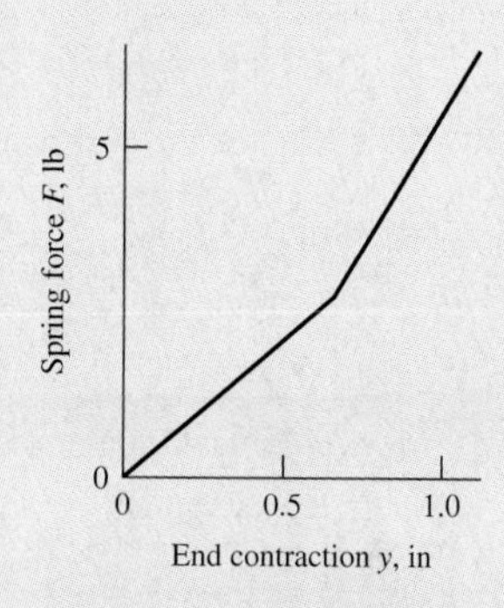

Figure CS10a–2

Force-end contraction characteristic of the dual-pitch wound helical compression spring.

A plot of y versus F from the table is shown in Fig. CS10a–2. Also from the table

$$k = \frac{F_1 - F_2}{y_1 - y_2} = \frac{1.652 - 0.826}{0.4 - 0.2} = 4.13 \text{ lb/in}$$

$$k = \frac{4.956 - 4.130}{0.931 - 0.831} = 8.26 \text{ lb/in}$$

A dual-pitch spring winding can create a dual-spring rate compression spring. Note that the spring rate of both halves is 8.26 lb/in. The closing of the fine-pitch spring *cut the number of active turns from $10\frac{1}{2}$ to $5\frac{1}{4}$ to change the spring rate*. Clever! A uniform-pitch spring does not close sequentially in the operating range, so its spring rate is constant, in this case 4.13 lb/in. The idea of a constant-varying pitch allows the spring to close coils continuously, producing a continuously varying spring rate.

CASE STUDY 10b An Innovative Design

We are all familiar with the wooden, spring-loaded clothespin, which has a torsional spring for pivot, as a source of clamping force, and to hold together pin halves by clever bending of the spring ends. When plastics came along, control of complex geometry could be accomplished in the die work. The opportunity to save on the cost of the torsional spring was presented, and the circular-ring spring made its appearance. Figure CS10b–1 shows such a clothespin with a 3:1 mechanical advantage. The beauty of the concept is the simplicity of the round-wire circular ring. The slot at A accommodates the movement of the wire when the clamp jaws are fully spread. Since a circular ring of round wire had distinct cost advantages over a torsional coil spring with formed ends, explore the parameters relating the force F in Fig. CS10b–2 to the gap size.

One could use Eqs. (4–32), (4–35), (4–36), and (4–37) to obtain the strain energy in the spring, then use Castigliano's theorem to find the deflection. Because the wire has a small diameter with respect to the ring diameter, we will use Eq. (4–34) since it dominates the energy content and the approximation is good. From Fig. CS10b–2,

$$M = FR(1 - \cos\theta) \qquad \frac{\partial M}{\partial F} = R(1 - \cos\theta)$$

$$F_\theta = F\cos\theta \qquad F_r = F\sin\theta$$

Using Eq. (4–34),

$$U = \int \frac{M^2 R\, d\theta}{2EI}$$

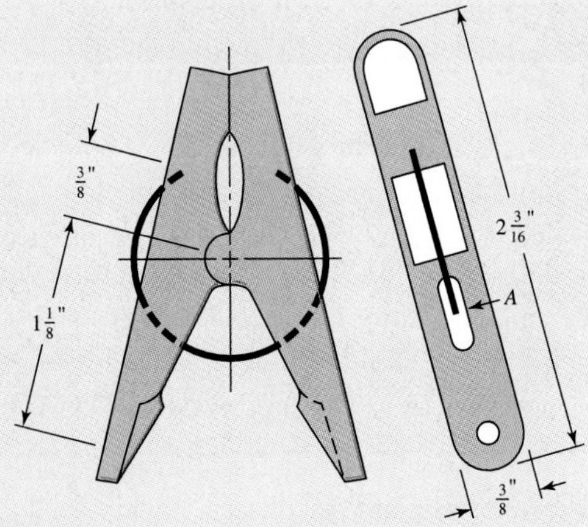

Figure CS10b–1

A plastic clothespin with a ring spring.

Figure CS10b–2

(*a*) A gapped circular ring of radius *R*. (*b*) A free-body diagram of the ring segment subtending the angle θ.

then

$$\delta = \frac{\partial U}{\partial F} = \int_0^{\varnothing} \frac{2MR}{2EI}\left(\frac{\partial M}{\partial F}\right) d\theta = \int_0^{\varnothing} \frac{FR^3(1-\cos\theta)^2\, d\theta}{EI}$$

$$= \frac{FR^3}{EI}\left[\int_0^{\varnothing} d\theta - 2\int_0^{\varnothing} \cos\theta\, d\theta + \int_0^{\varnothing} \cos^2\theta\, d\theta\right]$$

$$= \frac{FR^3}{EI}\left[\varnothing - 2\sin\varnothing + \frac{\varnothing}{2} + \frac{\sin 2\varnothing}{4}\right]$$

When $\varnothing = 2\pi$ then

$$\delta = \frac{3\pi FR^3}{EI}$$

With $I = \pi d^4/64$, and $R = D/2$, then the spring rate k is

$$k = \frac{F}{\delta} = \frac{d^4 E}{24D^3}$$

For instance, if $d = 0.060$ in and $d = 1$ in,

$$k = \frac{0.060^4 30(10^6)}{24(1)^3} = 16.2 \text{ lb/in}$$

For an initial load the arc length corresponding to $\Delta\theta$ of 30° is $s = r\Delta\theta = 0.5(30)\pi/180 =$ 0.262 in, with a corresponding preload $F_i = ks = 16.2(0.262) = 4.24$ lb. The effort P it takes to open the jaws is $P = 4.24/3 = 1.41$ lb. For sizing purposes the moment at section B is

$$M = FR(1-\cos\pi) = 2FR \qquad F_\theta = F\cos\pi = -F \qquad \text{(tensile)}$$

and the largest normal stress is

$$\sigma = \frac{M}{I/c} + \frac{|F|}{A} = F\left[\frac{64R}{\pi d^3} + \frac{4}{\pi d^2}\right]$$

Castigliano's theorem (1875) predates these clothespins. Such innovations come about because of the confluence of time, place, inventiveness, and the presence of a prepared mind. One cannot control time and place, but one can work to ensure that inventiveness and a prepared mind are ready the instant that opportunity knocks.

PROBLEMS

ANALYSIS

10–1 Make a two-view drawing or a good freehand sketch of a helical compression spring closed to its solid height and having a wire diameter of $\frac{1}{2}$ in, outside diameter of 4 in, and one active coil. The spring is to have plain ends. Make another drawing of the same spring with ends plain and ground.

ANALYSIS

10–2 It is instructive to examine the question of the units of the parameter A of Eq. (10–17). Show that for U.S. customary units

$$\dim(A) = \text{kpsi} \cdot \text{in}^m$$

and for SI units

$$\dim(A_1) = \text{MPa} \cdot \text{mm}^m$$

which make the dimensions of both A and A_1 different from every material to which Eq. (10–17) applies. Also show that the conversion from A to A_1 is given by

$$A_1 = 6.894\,757(25.40)^m A$$

ANALYSIS

10–3 A helical compression spring is wound using 0.105-in-diameter music wire. The spring has an outside diameter of 1.225 in with plain ground ends, and 12 total coils.

(*a*) What should the free length be to ensure that when the spring is compressed solid torsional stress does not exceed the yield strength, that is, that it is solid-safe?
(*b*) What force is needed to compress this spring to closure?
(*c*) Estimate the spring rate.
(*d*) Is there a possibility that the spring might buckle in service?

ANALYSIS

10–4 The spring in Prob. 10–3 is to be used with a static load of 30 lb. Apply the adequacy assessment represented by Eqs. (10–24) through (10–31) and associated commentary.

ANALYSIS

10–5 A helical compression spring is made of hard-drawn spring steel wire 2 mm in diameter and has an outside diameter of 22 mm. The ends are plain and ground, and there are $8\frac{1}{2}$ total coils.

(*a*) The spring is wound to a free length which is the largest possible with a solid-safe property. Find this free length.
(*b*) What is the pitch of this spring?
(*c*) What force is needed to compress the spring to its solid length?
(*d*) Estimate the spring rate.
(*e*) Will the spring buckle in service?

10–6 The spring of Prob. 10-5 is to be used with a static load of 75 N. Apply the adequacy assessment represented by Eqs. (10–14) through (10–31) and associated commentary.

10–7 to 10–17 Listed below are seven springs described in customary units and six springs described in SI units. Investigate these squared-and-ground-ended helical compression springs to see if they are solid-safe. If not, what is the largest free length to which they can be wound?

Problem Number	d, in	OD, in	L_0, in	N_t	Material
10–7	0.006	0.036	0.63	40	A228 music wire
10–8	0.012	0.120	0.81	15.1	B159 phosphor bronze
10–9	0.040	0.240	0.75	10.4	A313 stainless steel
10–10	0.135	2.0	2.94	5.25	A227 hard-drawn steel
10–11	0.144	1.0	3.75	13.0	A229 OQ&T steel
10–12	0.192	3.0	9.0	8.0	A232 chrome vanadium
	d, mm	OD, mm	L_0, mm	N_t	Material
10–13	0.2	0.91	15.9	40	A313 stainless steel
10–14	1.0	6.10	19.1	10.4	A228 music wire
10–15	3.4	50.8	74.6	5.25	A229 OQ&T spring steel
10–16	3.7	25.4	95.3	13.0	B159 phosphor bronze
10–17	4.3	76.2	228.6	8.0	A232 chrome vanadium

10–18 A static service music wire helical compression spring is needed to support a 20-lb load after being compressed 2 in. The solid height of the spring cannot exceed $1\frac{1}{2}$ in. The free length must not exceed 4 in. The static factor of safety must equal or exceed 1.2. For robust linearity use a fractional overrun to closure ξ of 0.15. There are two springs to be designed.

(*a*) The spring must operate over a $\frac{3}{4}$-in rod. A 0.050-in diametral clearance allowance should be adequate to avoid interference between the rod and the spring due to out-of-round coils. Design the spring.

(*b*) The spring must operate in a 1-in-diameter hole. A 0.050-in diametral clearance allowance should be adequate to avoid interference between the spring and the hole due to swelling of the spring diameter as the spring is compressed and out-of-round coils. Design the spring.

10–19 Not all springs are made in a conventional way. Consider the special steel spring in the illustration.

(*a*) Find the pitch, solid height, and number of active turns.

(*b*) Find the spring rate.

(*c*) Find the force F_s required to close the spring solid.

(*d*) Find the shear stress in the spring due to the force F_s.

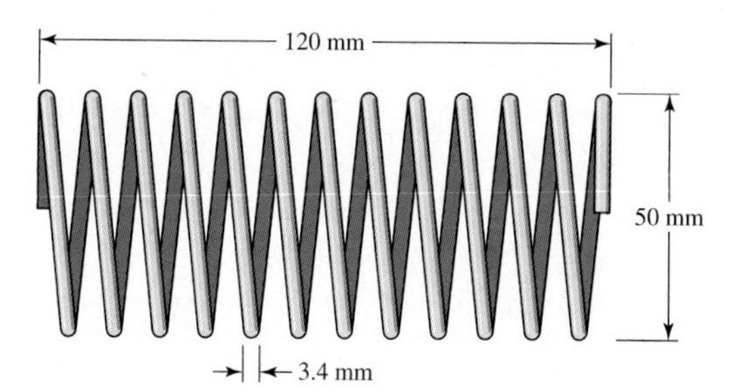

Problem 10–19

10–20 A holding fixture for a workpiece $1\frac{1}{2}$ in thick at clamp locations is being designed. The detail of one of the clamps is shown in the figure. A spring is required to drive the clamp upward while removing or inserting a workpiece. A clamping force of 10 lb is satisfactory. The base plate is $\frac{5}{8}$ in thick. The clamp screw has a $\frac{7}{16}''$-20 UNF thread. It is useful to have the free length L_0 short enough so that the clamp screw can compress the spring upon fixture reassembly during inspection and service, say $L_0 \leq 1.5 + \frac{3}{8}$ in. The spring cannot close solid at a length greater than $1\frac{1}{4}$ in. The safety factor at soliding should be $n_s \geq 1.2$, and at service load $n_1 \geq 1.5$. Design a suitable helical coil compression spring for this fixture. Specify commercial tolerances to the spring, and prepare the spring specification sheet for the springmaker.

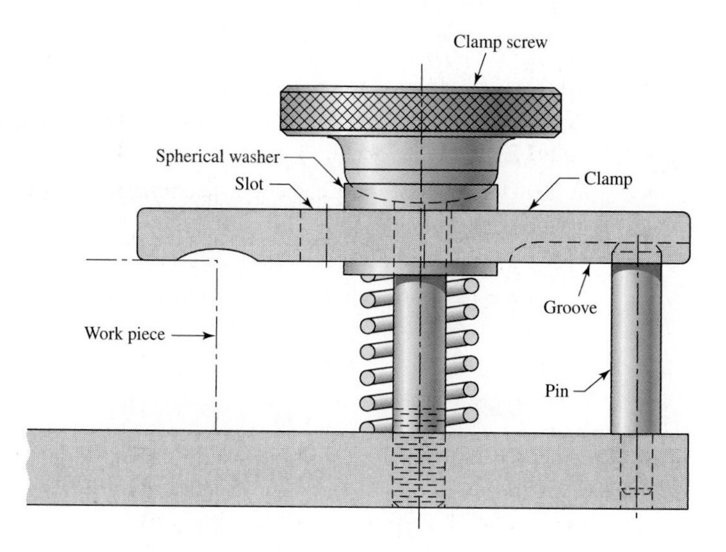

Problem 10–20
Clamping fixture.

10–21 Your instructor will provide you with a stock spring supplier's catalog, or pages reproduced from it. Accomplish the task of Prob. 10–20 by selecting an available stock spring. (This is design by *selection.*)

10–22 A helical coil compression spring is needed for food service machinery. The load varies from a minimum of 4 lb to a maximum of 18 lb. The spring lengths vary from a minimum of 1 in to a maximum of 2.5 in. The outside diameter of the spring cannot exceed $2\frac{1}{2}$ in. The springmaker has available suitable dies for drawing 0.080-, 0.0915-, 0.1055-, and 0.1205-in-diameter wire. Using a fatigue design factor n_f of 1.5, and a Gerber-Zimmerli fatigue-failure locus, design a suitable spring.

10–23 Suppose the spring you specified as the result of the design task of Prob. 10–20 was

Hard-drawn spring wire ID: 0.488 ± 0.009 in
As-wound condition N_t: $13 \pm \frac{1}{4}$ turns
Ends: squared-and-ground Free length: 1.858 ± 0.039 in
Wire: 0.081 ± 0.0008 in

(*a*) Approximate the variability of force $F_1 + \Delta F_1$ at length L_1 by a worst-case stacking of tolerances; $F = d^4 G(L_0 - L)1)/(8D^3 N_a)$.

(*b*) Investigate the variation in $\mathbf{F}_1$ at L_1 by a propagation-of-error approach on the equation $\mathbf{F} = \mathbf{d}^4\mathbf{G}(\mathbf{L}_0 - L_1)/(8\mathbf{D}^3\mathbf{N}_a)$ using $\mathbf{d} \sim U$ and the others normally distributed.

(*c*) Perform a computer simulation which will describe $\mathbf{F}$ using $\bar{F}$, σ_F, and C_F to confirm the results from part *b*.

(*d*) Use Table 10–11 to find the tolerances springmakers can routinely provide on F if specified for control.

10–24 Design the spring of Ex. 10–6 using a Gerber fatigue-failure locus, assign commercial tolerances, and fill out the spring specification sheet for the springmaker.

10–25 A hard-drawn spring steel extension spring is to be designed to carry a static load of 18 lb with an extension of $\frac{1}{2}$ in using a design factor of $n_y = 1.5$. Use full-coil end hooks with the fullest bend radius of $r = \frac{D}{2}$. The free length must be less than 3 in, and the body turns must be fewer than 30. Integer and half-integer body turns allow end hooks to be placed in the same plane. This adds extra cost and is done only when necessary. Complete the spring specification sheet and determine commercial tolerances.

10–26 The extension spring shown in the figure has full-twisted loop ends. The material is AISI 1065 OQ&T wire. The spring has 84 coils and is close-wound with a preload of 16 lb.

(*a*) Find the closed length of the spring.

(*b*) Find the torsional stress in the spring corresponding to the preload.

(*c*) Estimate the spring rate.

(*d*) What load would cause permanent deformation?

(*e*) What is the spring deflection corresponding to the load found in part *d*?

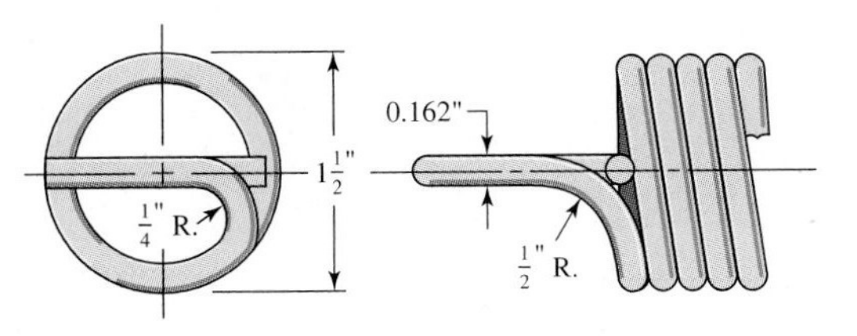

Problem 10–26

10–27 Design an infinite-life helical coil extension spring with full end loops and generous loop-bend radii for a minimum load of 9 lb and a maximum load of 18 lb, with an accompanying stretch of $\frac{1}{4}$ in. The spring is for food-service equipment and must be stainless steel. The outside diameter of the coil cannot exceed 1 in, and the free length cannot exceed $2\frac{1}{2}$ in. Using a fatigue design factor of $n_f = 2$, complete the design, identify the commercial tolerances, and prepare the spring specification sheet for the springmaker.

10–28 The figure shows a finger exerciser used by law-enforcement officers and athletes to strengthen their grip. It is formed by winding cold-drawn steel wire around a mandrel to obtain $2\frac{1}{2}$ turns when the grip is in the closed position. After winding, the wire is cut to leave the two legs as handles. The plastic handles are then molded on, the grip is squeezed together, and a wire clip is placed around the legs to obtain initial "tension" and to space the handles for the best initial gripping position. The clip is formed like a figure 8 to prevent it from coming off. The wire material is hard-drawn, 60-point carbon steel, plated. When the grip is in the closed position, the stress in the spring should not exceed the permissible stress.

(*a*) Determine the configuration of the spring before the grip is assembled.

(*b*) Find the force necessary to close the grip.

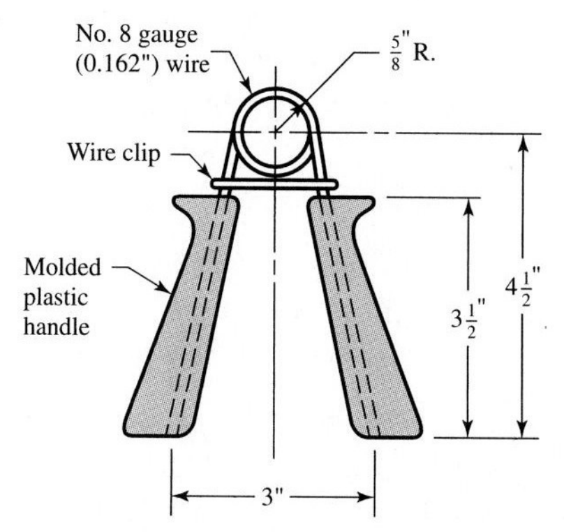

Problem 10–28

10–29 The rat trap shown in the figure uses two opposite-image torsion springs. The wire has a diameter of 0.081 in, and the outside diameter of the spring in the position shown is $\frac{1}{2}$ in. Each spring has 11 turns. Use of a fish scale revealed a force of about 8 lb is needed to set the trap.

(*a*) Find the probabable configuration of the spring prior to assembly.

(*b*) Find the maximum stress in the spring when the trap is set.

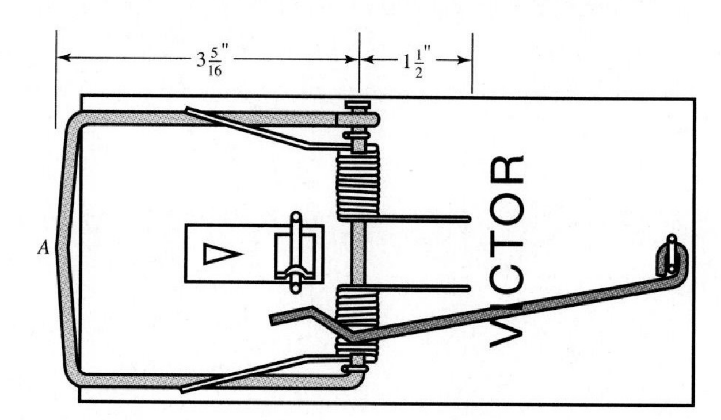

Problem 10–29

10–30 Using the experience gained with Prob. 10–22, write a computer program that would help in the design of helical coil compression springs.

10–31 Using the experience gained with Prob. 10–27, write a computer program that would help in the design of a helical coil extension spring.

11 Rolling-Contact Bearings

11-1	Bearing Types **690**
11-2	Bearing Life **693**
11-3	Bearing Load–Life Trade-Off at Constant Reliability **694**
11-4	Bearing Survival: The Reliability–Life Trade-Off **696**
11-5	Load–Life–Reliability Trade-Off **697**
11-6	Combined Radial and Thrust Loading **699**
11-7	Variable Loading **704**
11-8	Selection of Ball and Cylindrical Roller Bearings **709**
11-9	Selection of Tapered Roller Bearings **714**
11-10	Adequacy Assessment for Selected Rolling-Contact Bearings **724**
11-11	Lubrication **728**
11-12	Mounting and Enclosure **729**

The terms *rolling-contact bearing*, *antifriction bearing*, and *rolling bearing* are all used to describe that class of bearing in which the main load is transferred through elements in rolling contact rather than in sliding contact. In a rolling bearing the starting friction is about twice the running friction, but still it is negligible in comparison with the starting friction of a sleeve bearing. Load, speed, and the operating viscosity of the lubricant do affect the frictional characteristics of a rolling bearing. It is probably a mistake to describe a rolling bearing as "antifriction," but the term is used generally throughout the industry.

From the mechanical designer's standpoint, the study of antifriction bearings differs in several respects when compared with the study of other topics because the bearings they specify have already been designed. The specialist in antifriction-bearing design is confronted with the problem of designing a group of elements which compose a rolling bearing; these elements must be designed to fit into a space whose dimensions are specified; they must be designed to receive a load having certain characteristics; and finally, these elements must be designed to have a satisfactory life when operated under the specified conditions. Bearing specialists must therefore consider such matters as fatigue loading, friction, heat, corrosion resistance, kinematic problems, material properties, lubrication, machining tolerances, assembly, use, and cost. From a consideration of all these factors, bearing specialists arrive at a compromise which, in their judgment, is a good solution to the problem as stated.

We begin with an overview of bearing types, then we note that bearing life cannot be described in deterministic form. We introduce the invariant, the statistical distribution of life, which is robustly Weibullian. There are some useful deterministic equations addressing load–life trade-off at constant reliability, and we introduce the catalog rating at rating life.

The reliability–life trade-off involves Weibullian statistics. The load–life–reliability trade-off combines statistical and deterministic relationships. In this way we arrive at Eq. (11–5), which gives the designer the way to move from desired load and life to catalog rating in *one* equation.

Ball bearings resist thrust, and a unit of thrust does different damage per revolution than a unit of radial load, so we must find the equivalent pure radial load that does the same damage as the extant radial and thrust load. Next variable loading, stepwise and continuous, is approached, and the equivalent pure radial load doing the same damage is quantified. Oscillatory loading is mentioned.

With this preparation we have the tools to consider the selection of ball and cylindrical roller bearings. The question of misalignment is quantitatively approached.

Tapered roller bearings have some complications which our experience so far contributes to understanding.

Having the tools to find the proper catalog ratings, we make decisions (selections) and because we round up (or sometimes down), we must perform an adequacy assessment, and the extant reliability is quantified, as seen in Eq. (11–22). Lubrication and mounting conclude our introduction. Vendors' manuals should be consulted for specific details relating to bearings of their manufacture.

11–1 Bearing Types

Bearings are manufactured to take pure radial loads, pure thrust loads, or a combination of the two kinds of loads. The nomenclature of a ball bearing is illustrated in Fig. 11–1, which also shows the four essential parts of a bearing. These are the outer ring, the inner ring, the balls or rolling elements, and the separator. In low-priced bearings, the separator is sometimes omitted, but it has the important function of separating the elements so that rubbing contact will not occur.

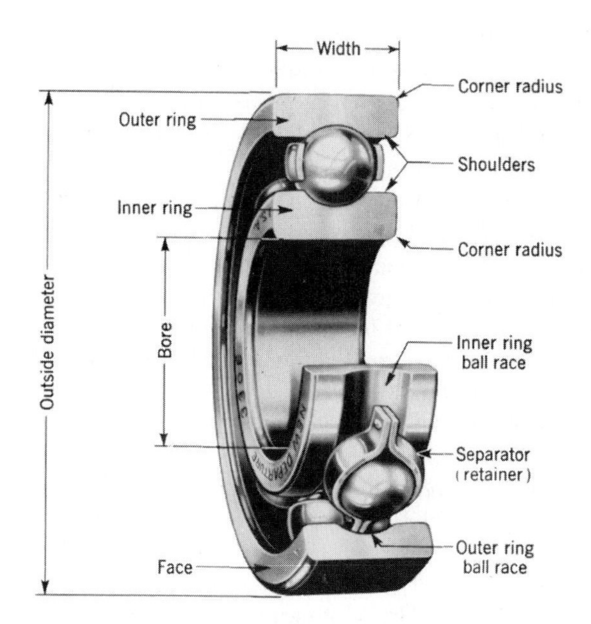

Figure 11–1

Nomenclature of a ball bearing. *(Courtesy of New Departure–Hyatt Division, General Motors Corporation.)*

In this section we include a selection from the many types of standardized bearings which are manufactured. Most bearing manufacturers provide engineering manuals and brochures containing lavish descriptions of the various types available. In the small space available here, only a meager outline of some of the most common types can be given. So you should include a survey of bearing manufacturers' literature in your studies of this section.

Some of the various types of standardized bearings which are manufactured are shown in Fig. 11–2. The single-row deep-groove bearing will take radial load as well as some thrust load. The balls are inserted into the grooves by moving the inner ring to an eccentric position. The balls are separated after loading, and the separator is then

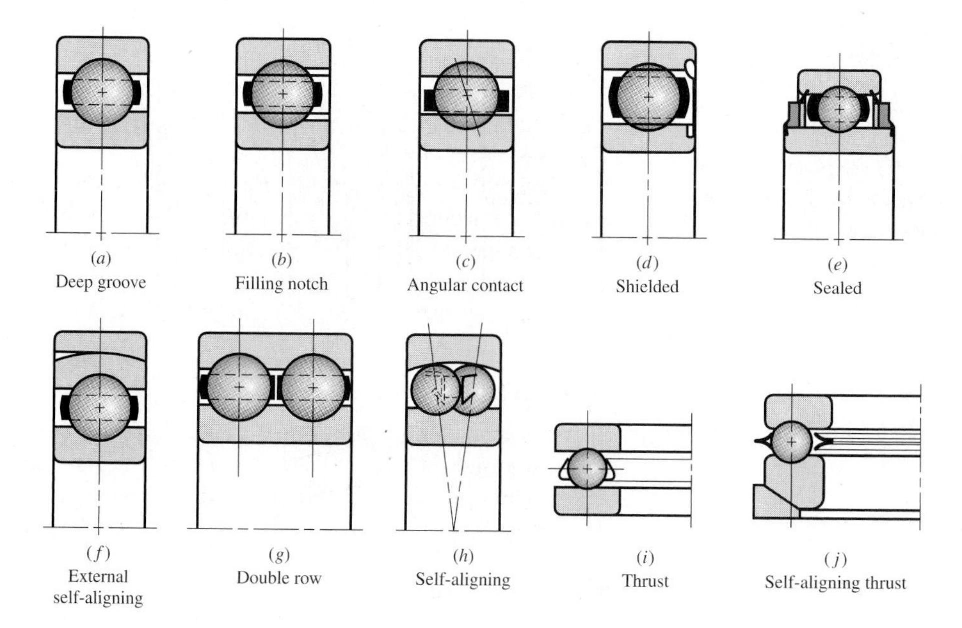

Figure 11–2

Various types of ball bearings.

inserted. The use of a filling notch (Fig. 11-2*b*) in the inner and outer rings enables a greater number of balls to be inserted, thus increasing the load capacity. The thrust capacity is decreased, however, because of the bumping of the balls against the edge of the notch when thrust loads are present. The angular-contact bearing (Fig. 11–2*c*) provides a greater thrust capacity.

All these bearings may be obtained with shields on one or both sides. The shields are not a complete closure but do offer a measure of protection against dirt. A variety of bearings are manufactured with seals on one or both sides. When the seals are on both sides, the bearings are lubricated at the factory. Although a sealed bearing is supposed to be lubricated for life, a method of relubrication is sometimes provided.

Single-row bearings will withstand a small amount of shaft misalignment of deflection, but where this is severe, self-aligning bearings may be used. Double-row bearings are made in a variety of types and sizes to carry heavier radial and thrust loads. Sometimes two single-row bearings are used together for the same reason, although a double-row bearing will generally require fewer parts and occupy less space. The one-way ball thrust bearings (Fig. 11–2*i*) are made in many types and sizes.

Some of the large variety of standard roller bearings available are illustrated in Fig. 11–3. Straight roller bearings (Fig. 11–3*a*) will carry a greater load than ball bearings of the same size because of the greater contact area. However, they have the disadvantage of requiring almost perfect geometry of the raceways and rollers. A slight misalignment will cause the rollers to skew and get out of line. For this reason, the retainer must be heavy. Straight roller bearings will not, of course, take thrust loads.

Helical rollers are made by winding rectangular material into rollers, after which they are hardened and ground. Because of the inherent flexibility, they will take considerable misalignment. If necessary, the shaft and housing can be used for raceways instead of separate inner and outer races. This is especially important if radial space is limited.

The spherical-roller thrust bearing (Fig. 11–3*b*) is useful where heavy loads and misalignment occur. The spherical elements have the advantage of increasing their contact area as the load is increased.

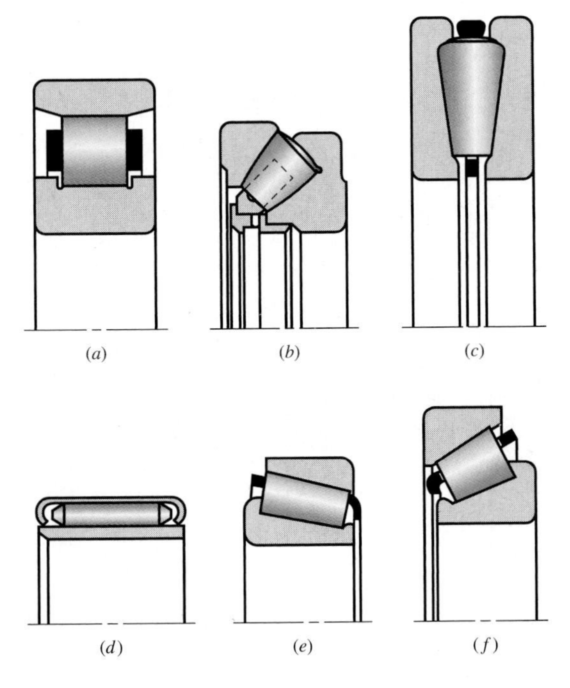

Figure 11–3

Types of roller bearings: (*a*) straight roller; (*b*) spherical roller, thrust; (*c*) tapered roller, thrust; (*d*) needle; (*e*) tapered roller; (*f*) steep-angle tapered roller. *(Courtesy of The Timken Company.)*

Needle bearings (Fig. 11–3*d*) are very useful where radial space is limited. They have a high load capacity when separators are used, but may be obtained without separators. They are furnished both with and without races.

Tapered roller bearings (Fig. 11–3*e*, *f*) combine the advantages of ball and straight roller bearings, since they can take either radial or thrust loads or any combination of the two, and in addition, they have the high load-carrying capacity of straight roller bearings. The tapered roller bearing is designed so that all elements in the roller surface and the raceways intersect at a common point on the bearing axis.

The bearings described here represent only a small portion of the many available for selection. Many special-purpose bearings are manufactured, and bearings are also made for particular classes of machinery. Typical of these are:

- Instrument bearings, which are high-precision and are available in stainless steel and high-temperature materials
- Nonprecision bearings, usually made with no separator and sometimes having split or stamped sheet-metal races
- Ball bushings, which permit either rotation or sliding motion or both
- Bearings with flexible rollers

11–2 Bearing Life

When the ball or roller of rolling-contact bearings rolls, contact stresses occur on the inner ring, the rolling element, and on the outer ring. Because the curvature of the contacting elements in the axial direction is different from that in the radial direction, the equations for these stresses are more involved than in the Hertz or Smith-Liu equations presented in Chapter 3. If a bearing is clean and properly lubricated, is mounted and sealed against the entrance of dust and dirt, is maintained in this condition, and is operated at reasonable temperatures, then metal fatigue will be the only cause of failure. Inasmuch as metal fatigue implies many millions of stress applications successfully endured, we need a quantitative life measure. Common life measures are

- Number of revolutions of the inner ring (outer ring stationary) until the first tangible evidence of fatigue
- Number of hours of use at a standard angular speed until the first tangible evidence of fatigue

The commonly used term is *bearing life*, which is applied to either of the measures just mentioned. The second thing to realize is, as in all fatigue, life as defined above is a stochastic variable and, as such, has both a distribution and associated statistical parameters. The life measure of an individual bearing is defined as the total number of revolutions (or hours at a constant speed) of bearing operation until the failure criterion is developed. Under ideal conditions, the fatigue failure consists of spalling of the load-carrying surfaces. The Anti-Friction Bearing Manufacturer's Association (AFBMA) standard states that the failure criterion is the first evidence of fatigue. The fatigue criterion used by the Timken Company laboratories is the spalling or pitting of an area of 0.01 in^2. Timkin also observes that the useful life of the bearing may extend considerably beyond this point. This is an operational definition of fatigue failure in rolling bearings.

The *rating life* is a term sanctioned by the AFBMA and used by most manufacturers. The rating life of a group of nominally identical ball or roller bearings is defined as the number of revolutions (or hours at a constant speed) that 90 percent of a group of bearings will achieve or exceed before the failure criterion develops. The terms *minimum life*, L_{10}

life, and B_{10} *life* are also used as synonyms for rating life. The rating life is the 10th percentile location of the bearing group's revolutions-to-failure distribution.

Median life is the 50th percentile life of a group of bearings. The term *average life* has been used as a synonym for median life, contributing to confusion. When many groups of bearings are tested, the median life is between four and five times the L_{10} life.

11–3 Bearing Load–Life Trade-Off at Constant Reliability

When nominally identical groups are tested to the life-failure criterion at different loads, the data are plotted on a graph as depicted in Fig. 11–4 using a log-log transformation which rectifies the data string. To establish a single point, load F_1 and the rating life of group one $(L_{10})_1$ are the coordinates which are logarithmically transformed. The reliability associated with this point, and all other points, is 0.90. Thus we gain a glimpse of the load–life trade-off function at 0.90 reliability. Using a regression equation of the form

$$FL^{1/a} = \text{constant} \tag{11–1}$$

the result of many tests for various kinds of bearings result in

- $a = 3$ for ball bearings
- $a = 10/3$ for roller bearings (cylindrical and tapered roller)

A bearing manufacturer chooses a rated cycles value of 10^6 revolutions (or in the case of the Timken Company, $90(10^6)$ revolutions), or otherwise, as declared in the manufacturer's catalog to correspond to a basic load rating in the catalog for each bearing manufactured, as their rating life. We shall call this the *catalog load rating* and display it algebraically as C_{10}, to denote it as the 10th percentile rating life for a particular bearing in the catalog. From Eq. (11–1) we can write

$$F_1 L_1^{1/a} = F_2 L_2^{1/a} \tag{a}$$

and associate load F_1 with C_{10}, life measure L_1 with L_{10}, and write

$$C_{10} L_{10}^{1/a} = F L^{1/a}$$

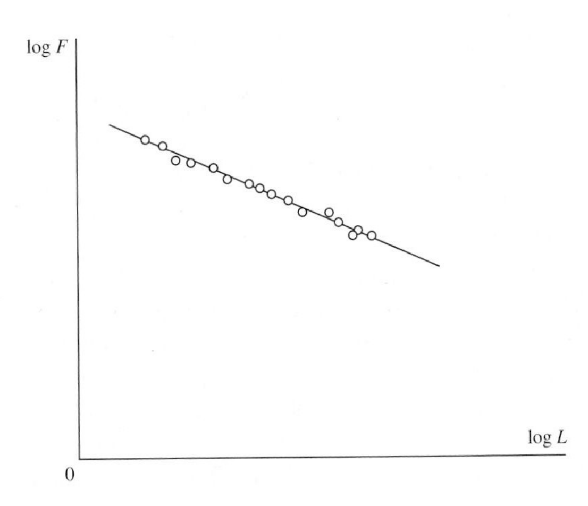

Figure 11–4

The rectification of failure experience in rolling-contact bearings by a log-log transformation, leading to an equation of the form of Eq. (11–1).

Further, we can write

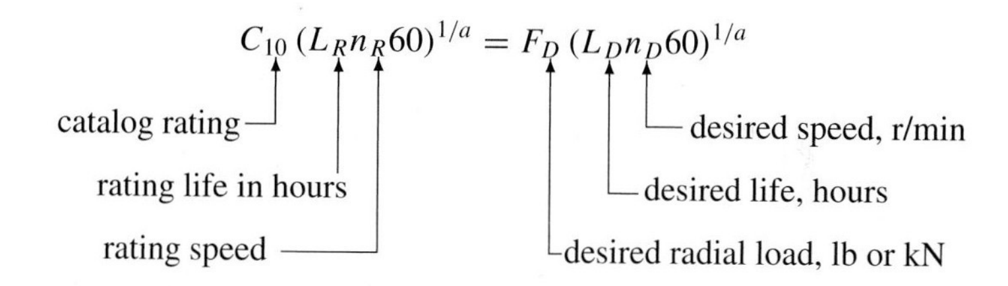

Solving for C_{10} gives

$$C_{10} = F_D \left(\frac{L_D n_D 60}{L_R n_R 60} \right)^{1/a} \tag{11–2}$$

EXAMPLE 11–1 Consider SKF, which rates its bearings for 1 million revolutions, so that L_{10} life is $60 L_R n_R = 10^6$ revolutions. The $L_R n_R 60$ product produces a familiar number. Timken, for example, uses $90(10^6)$ revolutions. If you desire a life of 5000 h at 1725 r/min with a load of 400 lbf with a reliability of 90 percent, for which catalog rating would you search in an SKF catalog?

Solution From Eq. (11–2),

$$C_{10} = F_D \left(\frac{L_D n_D 60}{L_R n_R 60} \right)^{1/a} = 400 \left[\frac{5000(1725)60}{10^6} \right]^{1/3} = 3211 \text{ lb} = 14.3 \text{ kN}$$

Now turn to Table 11–2 and Fig. 11–7 (on page 701). In searching Table 11–2 for a single-row, 02-series deep-groove ball bearing with C_{10} equaling or exceeding 14.3 kN, we find an 02-30 mm bearing has a value of $C_{10} = 19.5$ kN. The quotient $(L_D n_D 60)/(L_R n_R 60) = x_D$, where x_D is the desired (design) life expressed in multiples of rating life. The solution could have been done in two steps:

$$x_D = \frac{L_D n_D 60}{10^6} = \frac{5000(1725)60}{10^6} = 517.5 \text{ multiples of rating life}$$

Equation 11–2 can be written as

$$C_{10} = F_D\,(x_D)^{1/3} = 400(517.5)^{1/3} = 3211 \text{ lb} = 14.3 \text{ kN}$$

and the selection of the 02-30 mm deep-groove bearing would follow.

If a bearing manufacturer rates bearings at 500 h at 33 1/3 r/min with a reliability of 0.90, then $L_R n_R 60 = 500(33\ 1/3)60 = 10^6$. The tendency is to substitute 10^6 for $L_R n_R 60$ in Eq. (11–2). Although it is true that the 60s in Eq. (11-2) as displayed cancel algebraically, they are worth keeping, because at some point in your keystroke sequence on your hand-held calculator the manufacturer's magic number (10^6 or some other number) will appear to remind you of what the rating basis is and those manufacturers' catalogs to which you are limited. Of course, if you evaluate the bracketed quantity in Eq. (11–2) by alternating between numerator and denominator entries, the magic number will not appear and you will have lost an opportunity to check.

Figure 11–5

Constant reliability contours. Point A represents the catalog rating C_{10} at $x = L/L_{10} = 1$. Point B is on the target reliability locus R_D, with a load of C_{10}. Point D is a point on the desired reliability contour exhibiting the design life $x_D = L_D/L_{10}$.

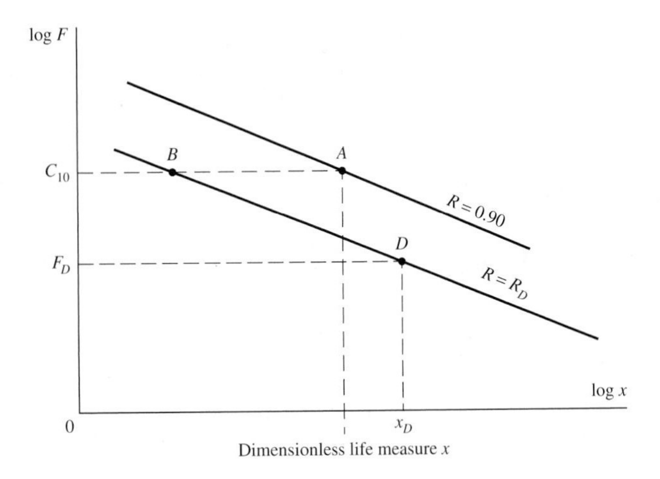

11–4 Bearing Survival: The Reliability–Life Trade-Off

At constant load, the life measure distribution is right skewed as depicted in Fig. 11–5. Candidates for a robust distributional curve fit include lognormal and Weibull. The Weibull is by far the most popular, largely due to its ability to adjust to varying amounts of skewness. If the life measure is expressed dimensionlessly as $x = L/L_{10}$, then

$$R = \exp\left[-\left(\frac{x - x_0}{\theta - x_0}\right)^b\right] \tag{11–3}$$

where R = reliability
x = life measure dimensionless variate
x_0 = guaranteed, or "minimum" value of the variate
θ = characteristic parameter corresponding to the 63.2121 percentile value of the variate
b = shape parameter which controls the skewness

Because there are three distributional parameters, x_0, θ, and b, the Weibull has a robust ability to conform to a data string. Also, in Eq. (11–3) an explicit expression for the cumulative distribution function is possible:

$$F = 1 - R = 1 - \exp\left[-\left(\frac{x - x_0}{\theta - x_0}\right)^b\right] \tag{11–4}$$

EXAMPLE 11–2

Construct the distributional properties of a 02-30 mm deep-groove ball bearing. If the Weibull parameters are $x_0 = 0.02$, $(\theta - x_0) = 4.439$, and $b = 1.483$, find the mean, median, 10th percentile life, standard deviation, and coefficient of variation.

Solution

From Eq. (A–31) the mean dimensionless life μ_x is

Answer

$$\mu_x = x_0 + (\theta - x_0)\,\Gamma\left(1 + \frac{1}{b}\right) = 0.02 + 4.439\Gamma\left(1 + \frac{1}{1.483}\right) = 4.029$$

The median dimensionless life is, after Eq. (A-33) where $R = 0.5$,

Answer $$x_{0.50} = x_0 + (\theta - x_0)\left(\ln \frac{1}{0.5}\right)^{1/b} = 0.02 + 4.439(\ln 2)^{1/1.483}$$

$$= 3.484$$

The 10th percentile value of the dimensionless life x is

Answer $$x_{0.10} = 0.02 + 4.439\left(\ln \frac{1}{0.90}\right)^{1/1.483} = 1 \qquad \text{(as it should be)}$$

The standard deviation of the dimensionless life is given by Eq. (A-32):

Answer $$\sigma_x = (\theta - x_0)\left[\Gamma\left(1 + \frac{2}{b}\right) - \Gamma^2\left(1 + \frac{1}{b}\right)\right]^{1/2}$$

$$= 4.439\left[\Gamma\left(1 + \frac{2}{1.483}\right) - \Gamma^2\left(1 + \frac{1}{1.483}\right)\right]^{1/2} = 2.568$$

The coefficient of variation of the dimensionless life is

Answer $$C_x = \frac{\sigma_x}{\mu_x} = \frac{2.568}{4.029} = 0.6374$$

11–5 Load–Life–Reliability Trade-Off

This is the designer's problem. The desired load is not the manufacturer's test load or catalog entry. The desired speed is different from the vendor's test speed, and the reliability expectation is much higher than the 0.90 accompanying the catalog entry. Figure 11–5 shows the situation. The catalog information is plotted as point A, whose coordinates are (the logs of) C_{10} and $x = L/L_{10} = 1$, a point on the 0.90 reliability contour. The design point is at D, with the coordinates (the logs of) F_D and x_D, a point that is on the $R = R_D$ reliability contour. The designer must move from point D to point A via point B as follows. Along a constant reliability contour (BD), Eq. (a) of Sec. 11–3 applies:

$$F_B x_B^{1/a} = F_D x_D^{1/a}$$

from which

$$F_B = F_D \left(\frac{x_D}{x_B}\right)^{1/a} \tag{a'}$$

Along a constant load line (AB), Eq. (11–3) applies:

$$R_D = \exp\left[-\left(\frac{x_B - x_0}{\theta - x_0}\right)^b\right]$$

Solving for x_B gives

$$x_B = x_0 + (\theta - x_0)\left(\ln \frac{1}{R_D}\right)^{1/b}$$

Now substitute this for x_B in Eq. (a') to obtain

$$F_B = F_D \left(\frac{x_D}{x_B}\right)^{1/a} = F_D \left[\frac{x_D}{x_0 + (\theta - x_0)(\ln 1/R_D)^{1/b}}\right]^{1/a}$$

However, $F_B = C_{10}$, so

$$C_{10} = F_D \left[\frac{x_D}{x_0 + (\theta - x_0)(\ln 1/R_D)^{1/b}}\right]^{1/a} \tag{11–5}$$

As useful as Eq. (11–5) is, one's attention to keystrokes and their sequence on a handheld calculator is rapt, and, as a result, the most common error is keying in the inappropriate logarithm. We have the rare opportunity to make Eq. (11–5) student-proof. Note that

$$\ln \frac{1}{R_D} = \ln \frac{1}{1 - p_f} = \ln\left(1 + p_f + \cdots\right) \doteq p_f = 1 - R_D$$

Equation (11–5) can be written as

$$C_{10} = F_D \left[\frac{x_D}{x_0 + (\theta - x_0)(1 - R_D)^{1/b}}\right]^{1/a} \qquad R \geq 0.90 \tag{11–6}$$

Loads are often nonsteady, so that the desired load is multiplied by an application factor (AF). The steady load (AF)F_D does the same damage as the variable load F_D does to the rolling surfaces. This point will be elaborated later.

EXAMPLE 11–3

The design load on a ball bearing is 413 lb and an application factor of 1.2 is appropriate. The speed of the shaft is to be 300 r/min, the life to be 30 kh with a reliability of 0.99. What is the C_{10} catalog entry to be sought (or exceeded) when searching for a deep-groove bearing in a manufacturer's catalog based on 10^6 revolutions for rating life? The Weibull parameters are $x_0 = 0.02$, $(\theta - x_0) = 4.439$, and $b = 1.483$.

Solution

$$x_D = \frac{L}{L_{10}} = \frac{60 L_D n_D}{60 L_R n_R} = \frac{60(30\,000)300}{10^6} = 540$$

The design life is 540 times the L_{10} life. From Eq. (11–6),

Answer

$$C_{10} = (1.2)(413)\left[\frac{540}{0.02 + 4.439(1 - 0.99)^{1/1.483}}\right]^{1/3} = 6696 \text{ lb}$$

We have learned to identify the catalog basic load rating corresponding to a steady radial load F_D, a desired life L_D, and a speed n_D.

Shafts generally have two bearings. Often these bearings are different. If the bearing reliability of the shaft with its pair of bearings is to be R, then R is related to the individual bearing reliabilities R_A and R_B by

$$R = R_A R_B$$

First, we observe that if the product $R_A R_B$ equals R, then, in general, R_A and R_B are both greater than R. Since the failure of either or both of the bearings results in the shutdown of the shaft, then A or B or both can create a failure. Second, in sizing bearings one can begin by making R_A and R_B equal to the square root of the reliability goal, $\sqrt{R}$. In Ex. 11–3, if the bearing was one of a pair, the reliability goal would be $\sqrt{0.99}$, or 0.995. The bearings selected are discrete in their reliability property in your problem, so the selection procedure "rounds up," and the overall reliability exceeds the goal R. Third, it may be possible, if $R_A > \sqrt{R}$, to round down on B yet have the product $R_A R_B$ still exceed the goal R.

11–6 Combined Radial and Thrust Loading

A ball bearing is capable of resisting a radial loading. It is also capable of resisting a thrust loading. Furthermore, these can be combined. If $V = 1$ when the inner ring rotates, and $V = 1.2$ when the outer ring rotates, F_a is the axial load, F_r is the radial load, and F_e is the equivalent radial load that does the same damage as the combined radial and thrust load, then two dimensionless groups can be formed: $F_e/(VF_r)$ and $F_a/(VF_r)$. When these two dimensionless groups are plotted as in Fig. 11–6, the data fall in a gentle curve that is well-approximated by two straight-line segments. The abscissa e is defined by the intersection of the two lines. The equations for the two lines shown in Fig. 11–6 are

$$\frac{F_e}{VF_r} = 1 \qquad \text{when} \quad \frac{F_a}{VF_r} \le e \tag{11–7}$$

$$\frac{F_e}{VF_r} = X + Y\frac{F_a}{VF_r} \qquad \text{when} \quad \frac{F_a}{VF_r} > e \tag{11–8}$$

It is common to express Eqs. (11–7) and (11–8) as a single equation,

$$F_e = X_i V F_r + Y_i F_a \tag{11–9}$$

where $i = 1$ when $F_a/(VF_r) \le e$ and $i = 2$ when $F_a/(VF_r) > e$. Table 11–1 lists values of X_1, Y_1, X_2, and Y_2 as a function of e, which in turn is a function of F_a/C_0, where C_0 is the bearing static load catalog rating.

In using these equations, the rotation factor V is to correct for the various rotating-ring conditions. For a rotating inner ring, $V = 1$. For a rotating outer ring, $V = 1.2$. The factor of 1.2 for outer-ring rotation is simply an acknowledgment that the fatigue life is reduced under these conditions. Self-aligning bearings are an exception; they have $V = 1$ for rotation of either ring.

The X and Y factors in Eqs. (11–7) and (11–8) depend upon the geometry of the bearing, including the number of balls and the ball diameter. The AFBMA recommendations are based on the ratio of the thrust component F_a to the *basic static load rating* C_0 and a variable reference value e. The static load rating C_0 is tabulated, along with the basic dynamic load rating C, in many of the bearing manufacturers' publications; see Table 11–2, for example.

Since straight or cylindrical roller bearings will take no axial load, or very little, the Y factor is always zero.

The AFBMA has established standard boundary dimensions for bearings which define the bearing bore, the outside diameter (OD), the width, and the fillet sizes on the shaft and housing shoulders. The basic plan covers all ball and straight roller bearings in the metric sizes. The plan is quite flexible in that, for a given bore, there are an assortment of widths and outside diameters. Furthermore, the outside diameters selected are such that, for a particular outside diameter, one can usually find a variety of bearings having different bores and widths.

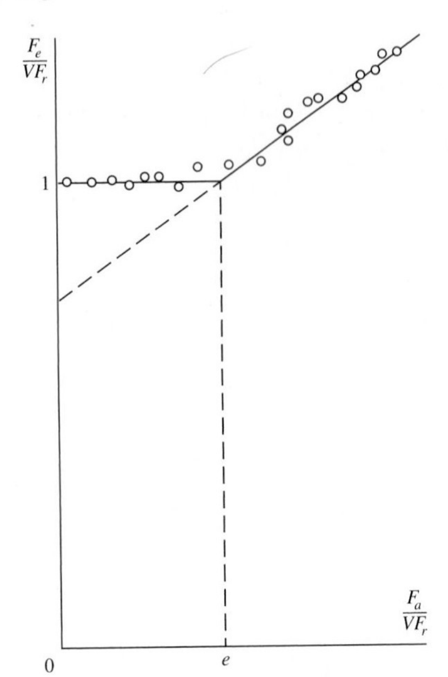

Figure 11–6

The relationship of dimensionless group $F_e/(VF_r)$ and $F_a/(VF_r)$ and the straight-line segments representing the data.

Table 11–1

Equivalent Radial Load Factors for Ball Bearings

		$F_a/(VF_r) \le e$		$F_a/(VF_r) > e$	
F_a/C_0	e	X_1	Y_1	X_2	Y_2
0.014*	0.19	1.00	0	0.56	2.30
0.021	0.21	1.00	0	0.56	2.15
0.028	0.22	1.00	0	0.56	1.99
0.042	0.24	1.00	0	0.56	1.85
0.056	0.26	1.00	0	0.56	1.71
0.070	0.27	1.00	0	0.56	1.63
0.084	0.28	1.00	0	0.56	1.55
0.110	0.30	1.00	0	0.56	1.45
0.17	0.34	1.00	0	0.56	1.31
0.28	0.38	1.00	0	0.56	1.15
0.42	0.42	1.00	0	0.56	1.04
0.56	0.44	1.00	0	0.56	1.00

*Use 0.014 if $F_a/C_0 < 0.014$.

This basic AFBMA plan is illustrated in Fig. 11–7. The bearings are identified by a two-digit number called the *dimension-series code*. The first number in the code is from the *width series*, 0, 1, 2, 3, 4, 5, and 6. The second number is from the *diameter series* (outside), 8, 9, 0, 1, 2, 3, and 4. Figure 11–7 shows the variety of bearings which may be obtained with a particular bore. Since the dimension-series code does not reveal the dimensions directly, it is necessary to resort to tabulations. The 02 series are used here as an example of what is available. See Table 11–2.

The housing and shaft shoulder diameters listed in the tables should be used whenever possible to secure adequate support for the bearing and to resist the maximum thrust loads (Fig. 11–8). Table 11–3 lists the dimensions and load ratings of some straight roller bearings.

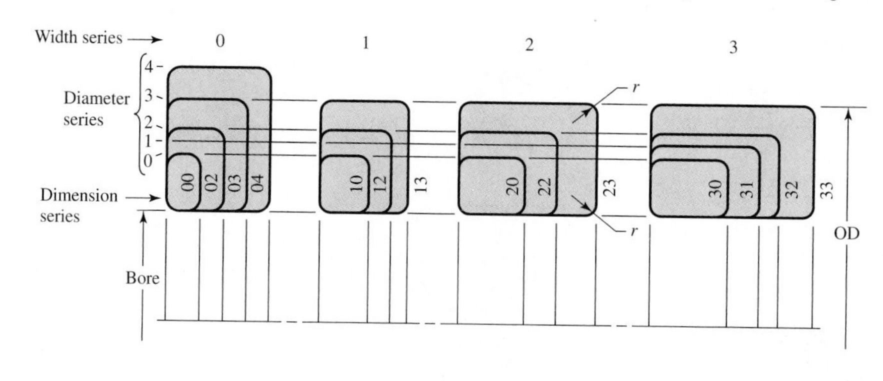

Figure 11–7

The basic AFBMA plan for boundary dimensions. These apply to ball bearings, straight roller bearings, and spherical roller bearings, but not to inch-series ball bearings or tapered roller bearings. The contour of the corner is not specified. It may be rounded or chamfered, but it must be small enough to clear the fillet radius specified in the standards.

Table 11–2

Dimensions and Load Ratings for Single-Row 02-Series Deep-Groove and Angular-Contact Bearings

Bore, mm	OD, mm	Width, mm	Fillet Radius, mm	Shoulder Diameter, mm		Load Ratings, kN			
						Deep Groove		Angular Contact	
				d_S	d_H	C	C_0	C	C_0
10	30	9	0.6	12.5	27	5.07	2.24	4.94	2.12
12	32	10	0.6	14.5	28	6.89	3.10	7.02	3.05
15	35	11	0.6	17.5	31	7.80	3.55	8.06	3.65
17	40	12	0.6	19.5	34	9.56	4.50	9.95	4.75
20	47	14	1.0	25	41	12.7	6.20	13.3	6.55
25	52	15	1.0	30	47	14.0	6.95	14.8	7.65
30	62	16	1.0	35	55	19.5	10.0	20.3	11.0
35	72	17	1.0	41	65	25.5	13.7	27.0	15.0
40	80	18	1.0	46	72	30.7	16.6	31.9	18.6
45	85	19	1.0	52	77	33.2	18.6	35.8	21.2
50	90	20	1.0	56	82	35.1	19.6	37.7	22.8
55	100	21	1.5	63	90	43.6	25.0	46.2	28.5
60	110	22	1.5	70	99	47.5	28.0	55.9	35.5
65	120	23	1.5	74	109	55.9	34.0	63.7	41.5
70	125	24	1.5	79	114	61.8	37.5	68.9	45.5
75	130	25	1.5	86	119	66.3	40.5	71.5	49.0
80	140	26	2.0	93	127	70.2	45.0	80.6	55.0
85	150	28	2.0	99	136	83.2	53.0	90.4	63.0
90	160	30	2.0	104	146	95.6	62.0	106	73.5
95	170	32	2.0	110	156	108	69.5	121	85.0

To assist the designer in the selection of bearings, most of the manufacturers' handbooks contain data on bearing life for many classes of machinery, as well as information on load-application factors. Such information has been accumulated the hard way, that is, by experience, and the beginner designer should utilize this information until he or she gains enough experience to know when deviations are possible. Table 11–4 contains recommendations on bearing life for some classes of machinery. The load-application factors in Table 11–5 serve the same purpose as factors of safety; use them to increase the equivalent load before selecting a bearing.

Figure 11–8

Shaft and housing shoulder diameters d_S and d_H should be adequate to ensure good bearing support.

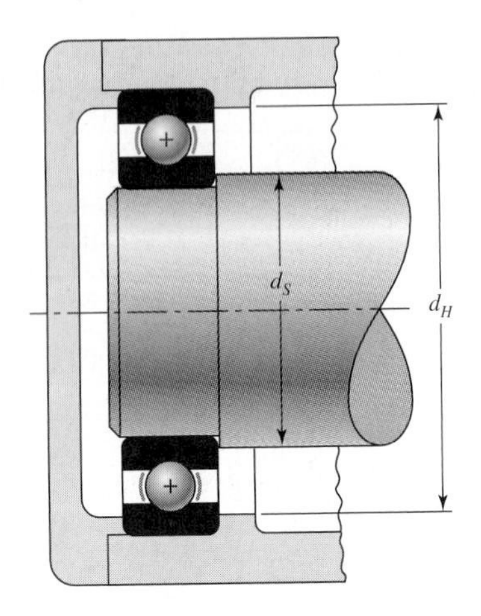

Table 11–3

Dimensions and Basic Load Ratings for Cylindrical Roller Bearings

Bore, mm	02-Series OD, mm	02-Series Width, mm	02-Series Load Rating, kN C_{10}	02-Series Load Rating, kN C_0	03-Series OD, mm	03-Series Width, mm	03-Series Load Rating, kN C_{10}	03-Series Load Rating, kN C_0
25	52	15	16.8	8.8	62	17	28.6	15.0
30	62	16	22.4	12.0	72	19	36.9	20.0
35	72	17	31.9	17.6	80	21	44.6	27.1
40	80	18	41.8	24.0	90	23	56.1	32.5
45	85	19	44.0	25.5	100	25	72.1	45.4
50	90	20	45.7	27.5	110	27	88.0	52.0
55	100	21	56.1	34.0	120	29	102	67.2
60	110	22	64.4	43.1	130	31	123	76.5
65	120	23	76.5	51.2	140	33	138	85.0
70	125	24	79.2	51.2	150	35	151	102
75	130	25	93.1	63.2	160	37	183	125
80	140	26	106	69.4	170	39	190	125
85	150	28	119	78.3	180	41	212	149
90	160	30	142	100	190	43	242	160
95	170	32	165	112	200	45	264	189
100	180	34	183	125	215	47	303	220
110	200	38	229	167	240	50	391	304
120	215	40	260	183	260	55	457	340
130	230	40	270	193	280	58	539	408
140	250	42	319	240	300	62	682	454
150	270	45	446	260	320	65	781	502

The static load rating is given in bearing catalog tables. It comes from the equations

$$C_0 = M n_b d_b^2 \qquad \text{(ball bearings)}$$

and

$$C_0 = M n_r \ell_c d \qquad \text{(roller bearings)}$$

Table 11–4

Bearing-Life Recommendations for Various Classes of Machinery

Type of Application	Life, kh
Instruments and apparatus for infrequent use	Up to 0.5
Aircraft engines	0.5–2
Machines for short or intermittent operation where service interruption is of minor importance	4–8
Machines for intermittent service where reliable operation is of great importance	8–14
Machines for 8-h service which are not always fully utilized	14–20
Machines for 8-h service which are fully utilized	20–30
Machines for continuous 24-h service	50–60
Machines for continuous 24-h service where reliability is of extreme importance	100–200

Table 11–5

Load-Application Factors

Type of Application	Load Factor
Precision gearing	1.0–1.1
Commercial gearing	1.1–1.3
Applications with poor bearing seals	1.2
Machinery with no impact	1.0–1.2
Machinery with light impact	1.2–1.5
Machinery with moderate impact	1.5–3.0

where C_0 = bearing static load rating, lb (kN)
n_b = number of balls
n_r = number of rollers
d_b = diameter of balls, in (mm)
d = diameter of rollers, in (mm)
ℓ_c = length of contact line, in (mm)

and M takes on the values of which the following table is representative:

M	in and lb	mm and kN
Radial ball	$1.78(10)^3$	$5.11(10)^3$
Ball thrust	$7.10(10)^3$	$20.4(10)^3$
Radial roller	$3.13(10)^3$	$8.99(10)^3$
Roller thrust	$14.2(10)^3$	$40.7(10)^3$

EXAMPLE 11–4

An SKF 6210 angular-contact ball bearing has an axial load F_a of 400 lb applied, a radial load F_r of 500 lb applied with the outer ring stationary. The basic static load rating C_0 is 4450 lb and the basic load rating C_{10} is 7900 lb. Estimate the L_{10} life at a speed of 720 r/min.

Solution

$V = 1$ and $F_a/C_0 = 400/4450 = 0.09$. Interpolate for e in Table 11–1:

F_a/C_0	e	
0.084	0.28	
0.09	e	from which $e = 0.285$
0.11	0.30	

Interpolate for Y_2 since $F_a/(VF_r) = 400/[(1)500] = 0.8$, which is greater than e:

F_a/C_0	Y_2	
0.084	1.55	
0.09	Y_2	from which $Y_2 = 1.527$
0.11	1.45	

From Eq. (11–9),

$$F_e = X_2 V F_r + Y_2 F_a = 0.56(1)500 + 1.527(400) = 890.8 \text{ lb}$$

Solving Eq. (11–2) for L_{10} gives

Answer $$L_{10} = \frac{60 L_R n_R}{60 n_D}\left(\frac{C_{10}}{F_e}\right)^3 = \frac{10^6}{60(720)}\left(\frac{7900}{890.8}\right)^3 = 16\,146 \text{ h}$$

We now know how to combine a steady radial load and a steady thrust load into an equivalent steady radial load F_e that inflicts the same damage per revolution as the radial–thrust combination.

11–7 Variable Loading

Bearing loads are frequently variable and occur in some identifiable patterns:

- Piecewise constant loading in a cyclic pattern
- Continuously variable loading in a repeatable cyclic pattern
- Random variation

Equation (11–1) can be written as

$$F^a L = \text{constant} = K \tag{a}$$

Note that F may already be an equivalent steady radial load for a radial–thrust load combination. Figure 11–9 is a plot of F^a as ordinate and L as abscissa and the locus of Eq. (a). If a load level of F_1 is selected and run to the failure criterion, then the area under the F_1–L_1 trace is numerically equal to K. The same is true for a load level F_2, that is, the area under the F_2–L_2 trace is numerically equal to K. The linear damage hypothesis says that in the case of load level F_1, the area from $L = 0$ to $L = L_A$ does damage measured by $F_1^a L_A = D$.

Figure 11–9

Plot of F^a as ordinate and L as abscissa showing the F^aL = constant locus. The linear damage hypothesis says that in the case of load F_1, the area under the curve from $L = 0$ to $L = L_A$ is a measure of the damage $D = F_1^a L_A$. The complete damage to failure is measured by $C_{10}^a L_B$.

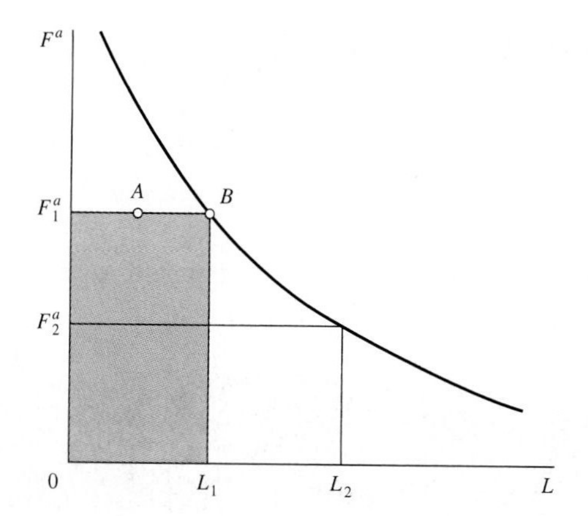

Consider the piecewise continuous cycle depicted in Fig. 11–10. The damage done by loads F_1, F_2, and F_3 is

$$D = F_1^a \ell_1 + F_2^a \ell_2 + F_3^a \ell_3 \tag{b}$$

The equivalent steady load F_{eq} when run for $\ell_1 + \ell_2 + \ell_3$ revolutions does the same damage D. Thus

$$D = F_e^a (\ell_1 + \ell_2 + \ell_3) \tag{c}$$

Equating Eqs. (b) and (c), and solving for F_{eq}, we get

$$F_{eq} = \left[\frac{F_1^a \ell_1 + F_2^a \ell_2 + f_3^a \ell_3}{\ell_1 + \ell_2 + \ell_3}\right]^{1/a} = \left[\sum f_i F_i^a\right]^{1/a} \tag{11-10}$$

where f_i is the fraction of revolution run up under load F_i. Since ℓ_i can be expressed as $n_i t_i$ where n_i is the rotational speed at load F_i and t_i is the duration of that speed, then it follows that

$$F_{eq} = \left[\frac{\sum n_i t_i F_{ei}^a}{\sum n_i t_i}\right]^{1/a} \tag{11-11}$$

The loads F_i may already be equivalent loads F_{ei}. The character of the individual loads can change, so an application factor (AF) can be prefixed to F_{ei} as $[(\text{AF})_i F_{ei}]^a$; then

Figure 11–10

A three-part piecewise-continuous periodic loading cycle involving loads F_1, F_2, and F_3. F_{eq} is the equivalent steady load inflicting the same damage when run for $\ell_1 + \ell_2 + \ell_3$ revolutions, doing the same damage D per period.

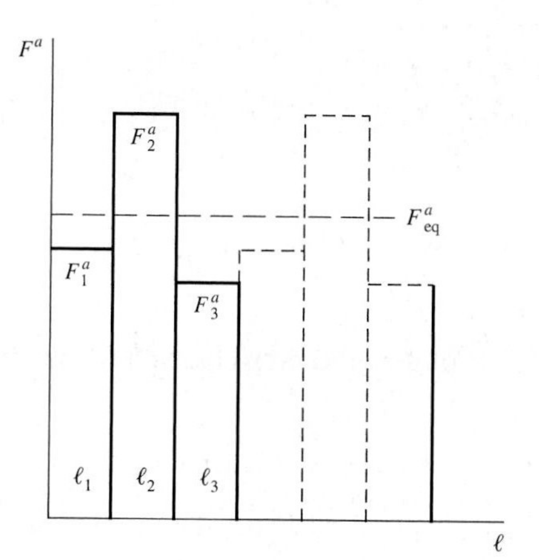

Eq. (11–10) can be written

$$F_{\text{eq}} = \left\{ \sum f_i \, [(\text{AF})_i F_{ei}]^a \right\}^{1/a} \qquad L_{\text{eq}} = \frac{K}{F_{\text{eq}}^a} \tag{11–12}$$

EXAMPLE 11–5 A ball bearing is run at four piecewise continuous steady loads as shown in the following table:

(1) Time Fraction	(2) Speed r/min	(3) Product, Column (1) × (2)	(4) Turns Fraction (3)/∑(3)	(5) F_{ri} lb	(6) F_{ai} lb	(7) F_{ei} lb	(8) $(\text{AF})_i$	(9) $(\text{AF})_i F_{ei}$ lb
0.1	2000	200	0.077	600	300	794	1.10	873
0.1	3000	300	0.115	300	300	626	1.25	795
0.3	3000	900	0.346	750	300	878	1.10	966
0.5	2400	1200	0.462	375	300	668	1.25	835
		2600	1.000					

Columns 1 and 2 are multiplied to obtain column 3. The column 3 entry is divided by 2600 to give column 4. Columns 5, 6, and 7 are the radial, axial, and equivalent loads respectively. Column 8 is the appropriate application factor. Column 9 is the product of columns 7 and 8.

Solution From Eq. (11–10), the equivalent radial load F_e is

Answer
$$F_e = \left[0.077(873)^3 + 0.115(795)^3 + 0.346(966)^3 + 0.462(835)^3\right]^{1/3} = 884 \text{ lb}$$

Sometime the question after several levels of loading is: How much life is left if the next level of stress is held until failure? Failure occurs under the linear damage hypothesis when the damage D equals the constant $K = F^a L$. Taking the first form of Eq. (11–10), we write

$$F_{\text{eq}}^a L_{\text{eq}} = F_{e1}^a \ell_1 + F_{e2}^a \ell_2 + F_{e3}^a \ell_3$$

and note that

$$K = F_{e1}^a L_1 = F_{e2}^a L_2 = F_{e3}^a L_3$$

and K also equals

$$K = F_{e1}^a \ell_1 + F_{e2}^a \ell_2 + F_{e3}^a \ell_3 = \frac{K}{L_1}\ell_1 + \frac{K}{L_2}\ell_2 + \frac{K}{L_3}\ell_3 = K \sum \frac{\ell_i}{L_i}$$

From the outer parts of the preceding equation we obtain

$$\sum \frac{\ell_i}{L_i} = 1 \tag{11–13}$$

This equation was advanced by Palmgren in 1924, and again by Miner in 1945. See Eq. (7–56).

Figure 11–11

A continuous load variation of a cyclic nature whose period is ϕ.

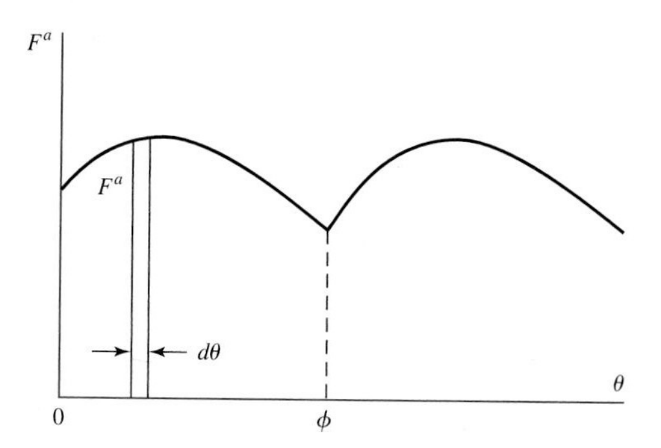

The second kind of load variation mentioned is continuous, periodic variation, depicted by Fig. 11–11. The differential damage done by F^a during rotation through the angle $d\theta$ is

$$dD = F^a d\theta$$

An example of this would be a cam whose bearings rotate with the cam through the angle $d\theta$. The total damage during a complete cam rotation is given by

$$D = \int dD = \int_0^{\phi} F^a d\theta = F_{\text{eq}}\phi$$

from which, solving for the equivalent load, we obtain

$$F_{\text{eq}} = \left[\frac{1}{\phi}\int_0^{\phi} F^a d\theta\right]^{1/a} \qquad L_{\text{eq}} = \frac{K}{F_{\text{eq}}^a} \tag{11–14}$$

The value of ϕ is often 2π, although other values occur. Simpson's first rule of numerical integration is often useful to carry out the indicated integration, particularly when a is not an integer and trigonometric functions are involved. We have now learned how to find the steady equivalent load that does the same damage as a continuously varying cyclic load.

EXAMPLE 11–6

The operation of a particular rotary pump involves a power demand of $P = \bar{P} + A' \sin\theta$ where $\bar{P}$ is the average power. The bearings feel the same variation as $F = \bar{F} + A\sin\theta$. Develop an application factor (AF) for this application for ball bearings.

Solution

From Eq. (11–14),

$$F_{\text{eq}} = (\text{AF})F = \left(\frac{1}{2\pi}\int_0^{2\pi} F^a d\theta\right)^{1/a} = \left(\frac{1}{2\pi}\int_0^{2\pi} (\bar{F} + A\sin\theta)^3\right)^{1/3}$$

$$= \left[\frac{1}{2\pi}\left(\int_0^{2\pi} \bar{F}^3 d\theta + 3\bar{F}^2 A\int_0^{2\pi} \sin\theta\, d\theta + 3\bar{F}A^2\int_0^{2\pi} \sin^2\theta\, d\theta + A^3\int_0^{2\pi} \sin^3\theta\, d\theta\right)\right]^{1/3}$$

$$F_{\text{eq}} = \left[\frac{1}{2\pi}(2\pi\bar{F}^3 + 0 + 3\pi\bar{F}A^2 + 0)\right]^{1/3} = \bar{F}\left[1 + \frac{3}{2}\left(\frac{A}{\bar{F}}\right)^2\right]^{1/3}$$

or

$$(\text{AF}) = \left[1 + \frac{3}{2}\left(\frac{A}{\bar{F}}\right)^2\right]^{1/3}$$

We can present the result in tabular form:

$A/\bar{F}$	(AF)
0	1
0.2	1.02
0.4	1.07
0.6	1.15
0.8	1.25
1.0	1.36

In dynamic loading such as in Ex. 11–6 the damage inflicted by an excursion from the mean is cumulative. The damage from a departure above the mean is more than a similar departure below the mean. The equivalent load determination needs a time trace (or a revolution trace) to estimate an application (or service) factor. In Ex. 11–6 the trace was known analytically. In some applications such as pulling a plow, the load variation is said to be "random." Field tests record these variations and the modified rainflow technique of Chap. 7 is used to estimate an application factor. Bearing suppliers have experience with tests as well as observing successful and unsuccessful applications. They share that experience in their engineering catalogs and by extending an invitation to telephone or e-mail one of their bearing application engineers. Table 11–6 is an example of a more extended application factor table than Table 11–5.

Table 11–6

Ball Bearing Load-Application Factors Categorized by Word Description of Machinery Type

Application factor 1.0–1.1	
Smooth, shock free	
Application factor 1.1–1.3	
Agricultural	Pinions
Tool room	Cranes
Foundry cranes	Blowers
Cultivators	Ball mills
Machine tools	Power, water station
Ventilating fans	Mine pumps
Wood processing	Paper pulp machinery
Reduction gears	
Application factor 1.3–1.5	
Construction	Lift truck
Earth moving	Rolling mills
Heavy-duty cranes	Shakes
Lumbering, mining	Crushers

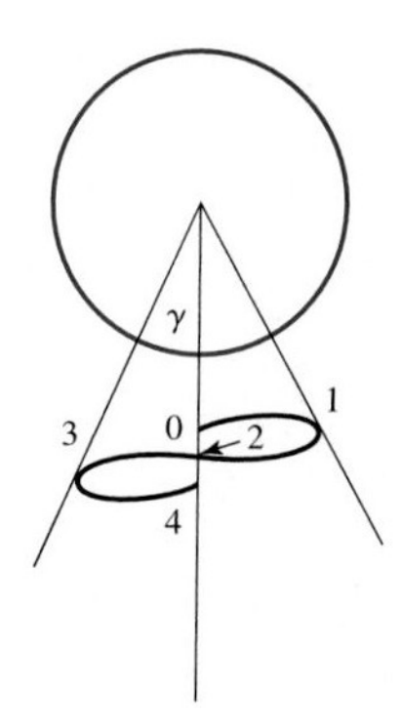

Figure 11–12

The complete oscillation angle of a bearing is 4γ. The amplitude is γ.

Rolling-contact bearings that are used to support shafts entertaining an oscillatory movement are exposed to very localized surface fatigue. The circular tracks around the inner and outer rings have very localized use. Furthermore, a limited number of balls or rollers carry the load and their peripheries may be only partially used. Additionally, lubrication is very important.

SKF defines the complete oscillatory angle as 4γ and its amplitude as γ as shown in Fig. 11–12. The relationship between the L_{10} rotary life and the L_{10} life for oscillation is

$$(L_{10})_{\text{osc}} = \frac{180}{2\gamma}(L_{10})_{\text{rot}} \tag{11–15}$$

where γ is the oscillation amplitude. The speed of oscillation is measured by the frequency of oscillation n_{osc} expressed in cycles/min, and the equivalent rotational speed n is given by

$$n = \frac{2\gamma}{180}n_{\text{osc}} \tag{11–16}$$

Rolling-contact bearings have some tolerance to misalignment between the centerline of the inner race and the centerline of the outer race without life penalty. Some quantitative information is given in Sec. 11–8.

11–8 Selection of Ball and Cylindrical Roller Bearings

We have enough information concerning the loading of rolling-contact ball and roller bearings to develop the steady equivalent radial load that will do as much damage to the bearing as the extant loading. Now let's put it to work.

EXAMPLE 11–7

The second shaft on a parallel-shaft 25-hp foundry crane speed reducer has bearing reactions as shown in Fig. 11–13. A ball bearing is to be selected for location C to accept the thrust, and a cylindrical roller bearing is to be utilized at location D. The life goal of the speed reducer is 10 kh, with a reliability for the ensemble of all four bearings to equal or exceed 0.96. The application factor is to be 1.2.
(*a*) Select the roller bearing for location D.
(*b*) Select the ball bearing (angular contact) for location C.

Solution

The radial load at D is $\sqrt{106.7^2 + 297.5^2} = 316.1$ lb, and the radial load at C is $\sqrt{356.7^2 + 297.5^2} = 464.5$ lb. The individual bearing reliabilities, if equal, must be at least $\sqrt[4]{0.96} = 0.98985 \doteq 0.99$.
(*a*) At D the desired life is, for both bearings,

$$x_D = \frac{L}{L_{10}} = \frac{60L_D n_D}{60L_R n_R} = \frac{60(10\,000)656}{10^6} = 393.6$$

From Eq. (11–6), for the roller bearing at D,

$$C_{10} = 1.2(316.1)\left[\frac{393.6}{0.02 + 4.439(1 - 0.99)^{1/1.483}}\right]^{3/10} = 3593 \text{ lb}$$

or 16 kN.

Figure 11–13

Forces in pounds applied to the second shaft of the helical gear speed reducer of Ex. 11–7.

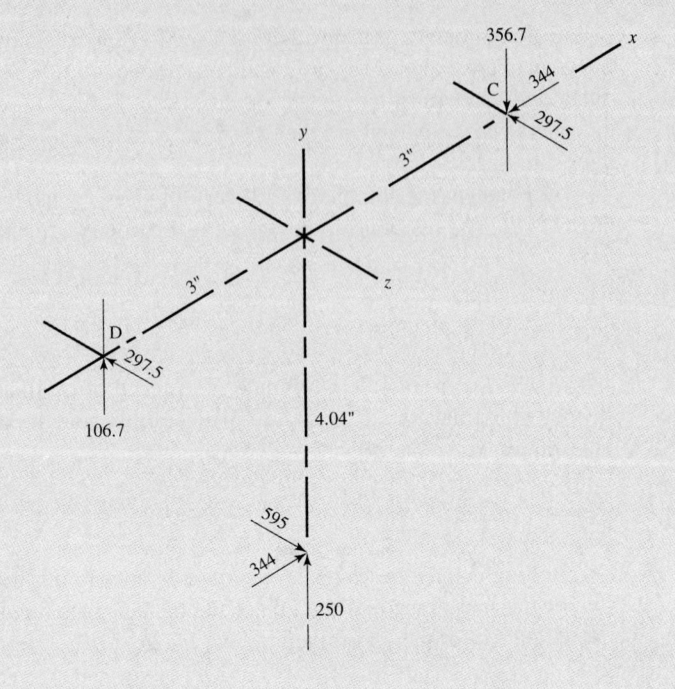

Answer The absence of a thrust component makes the selection procedure simple. Choose a 02-25 mm series, or a 03-25 mm series cylindrical roller bearing from Table 11–3.

(*b*) The ball bearing at C involves a thrust component. This selection procedure contains an iterative process:

1. Choose Y_2 from Table 11–1.
2. Find C_{10}.
3. Tentatively identify a suitable bearing from Table 11–2, note C_0.
4. Using F_a/C_0 enter Table 11–1 to obtain a new value of Y_2.
5. Find C_{10}.
6. If the same bearing is obtained, stop.
7. If not, take next bearing and go to step 4.

As a first approximation, take the middle entry from Table 11–1:

$$X_2 = 0.056 \qquad Y_2 = 1.63.$$

From Eq. (11–7),

$$\frac{F_e}{VF_r} = 0.56 + 1.63\frac{344}{(1)464.5} = 1.77$$

$$F_e = 1.77VF_r = 1.77(1)464.5 = 822.2 \text{ lb} \quad \text{or } 3.65 \text{ kN}$$

From Eq. (11–6),

$$C_{10} = 1.2(3.65)\left[\frac{393.6}{0.02 + 4.439(1 - 0.99)^{1/1.483}}\right]^{1/3} = 53.3 \text{ kN}$$

From Table 11–2, angular-contact bearing 02–60 mm has $C'_{10} = 55.9$ kN. C_0 is 35.5 kN. Step 4 becomes

$$\frac{F_a}{C_0} = \frac{344(4.45)10^{-3}}{35.5} = 0.043$$

which makes e from Table 11–1 approximately 0.24. Now $F_a/[VF_r] = 344/[(1)464.5] = 0.74$, which is greater than 0.24, so we find Y_2 by interpolation:

F_a/C_0	Y_2
0.042	1.85
0.043	Y_2
0.056	1.71

$$\frac{0.043 - 0.042}{0.056 - 0.042} = \frac{Y_2 - 1.85}{1.71 - 1.85}$$

$$Y_2 = 1.84$$

From Eq. (11–8),

$$\frac{F_e}{VF_r} = 0.56 + 1.84\frac{344}{464.5} = 1.93$$

$$F_e = 1.93VF_r = 1.93(1)464.5 = 897 \text{ lb} \quad \text{or } 3.99 \text{ kN}$$

The prior calculation for C_{10} changes only in F_e, so

$$C_{10} = \frac{3.99}{3.65}53.3 = 58.3 \text{ kN}$$

From Table 11–2 an angular contact bearing 02-65 mm has $C'_{10} = 63.7$ kN and C'_0 of 41.5. Again,

$$\frac{F_a}{C_0} = \frac{344(4.45)10^{-3}}{41.5} = 0.0369$$

making e approximately 0.23. Now $F_a/(VF_r) = 344/[1(464.5)] = 0.74$, which is greater than 0.23. We find Y_2 again by interpolation:

F_a/C_0	Y_2
0.028	1.99
0.0369	Y_2
0.042	1.85

$$\frac{0.0369 - 0.028}{0.042 - 0.028} = \frac{Y_2 - 1.99}{1.85 - 1.99}$$

$$Y_2 = 1.90$$

From Eq. (11–8),

$$\frac{F_e}{VF_r} = 0.56 + 1.90\frac{344}{464.5} = 1.967$$

$$F_e = 1.967VF_r = 1.967(1)464.5 = 913.7 \text{ lb} \quad \text{or } 4.066 \text{ kN}$$

The prior calculation for C_{10} changes only in F_e, so

$$C_{10} = \frac{4.066}{3.65}53.3 = 59.37 \text{ kN}$$

Answer From Table 11–2 an angular-contact 02-65 mm is still selected, so the iteration is complete.

In Ex. 11-7 we examined the procedure of selecting a ball or roller bearing. If the relative slope between the inner and outer race centerlines exceeds some small amount ξ_0, there is a life penalty. In 1971 Bamberger[1] reported a study on life adjustment factors for ball and roller bearings. Bearings are presumed free of moment in the plane of shaft deflection. When such a moment occurs the shaft changes from simply supported to elastically built-in ends, a bearing life penalty ensues, as compared to the moment-free life information supplied in bearing catalogs. When the inner race and outer race axes are congruent, then moved until the beginning of a resisting moment, the angle of intersection of the inner and outer race centerlines has grown to ξ_0 rad. See Fig. 11–14. In cylindrical and tapered roller bearings ξ_0 is approximately 0.001 rad. For spherical ball bearings ξ_0 is about 0.0087 rad. Deep-groove bearings lie in the range $0.0035 \le \xi_0 \le 0.0047$ rad. One must always check specific vendors' catalogs.

Common design practice is to stay within these "angles of play" and avoid any life penalty. This also preserves the robustness of simply supported beam theory and stress estimations based upon it. Avoiding the life penalty places a burden on the designer to size shafts to ensure the slope angle at a bearing is less than ξ_0 for that bearing. In Ex. 11–7 the presumption was that the shaft would be designed to accommodate to this. There are times when the slope angle cannot be made small enough to avoid a life penalty. There is a zone in Fig. 11–15 in which penalty occurs, bounded by those bearings in which the internal geometric tolerances are favorable to longer life, also bounded on the other side by those bearings whose internal geometric tolerances are adverse to long life. It should not be assumed that one manufacturer's product is poor and another's good; there simply is a range. From the designer's viewpoint, a worst-case scenario is a possible posture. A probabilistic view is another. For small production runs, the worst-case approach is appealing. For mass production, a stochastic view is useful.

We can define a mean fractional reduction in life factor as $\bar{f}_r$. From the Bamberger et al. study for cylindrical roller bearings we can write

$$\bar{f}_r = \begin{cases} 1 & \xi \le \xi_0 \\ 1.1 - 200\,(\xi - \xi_0) & \xi > \xi_0 \end{cases} \tag{11–17}$$

where ξ is the misalignment angle between the centerlines of the inner ring and outer

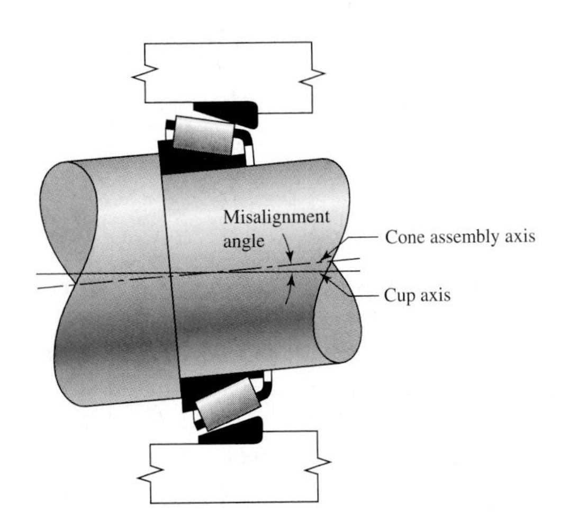

Figure 11–14

The definition of misalignment angle ξ_0.

1. E. V. Bamberger et al., *Life Adjustment Factors for Ball and Roller Bearings: An Engineering Design Guide,* Americal Society of Mechanical Engineers, New York, 1971.

Figure 11–15

(*a*) Shaded zone is the data region with linear descriptors resulting from regression fits. (*b*) In avoiding life penalty set $f_r = 1$, then $\bar{\xi} = 0.0015$ rad, $\sigma_\xi = 0.000\,17$ rad, ϕ is the shaft journal slope, $C_\phi \doteq (C_P^2 + C_E^2)^{1/2}$. Interference will estimate the change of a life penalty.

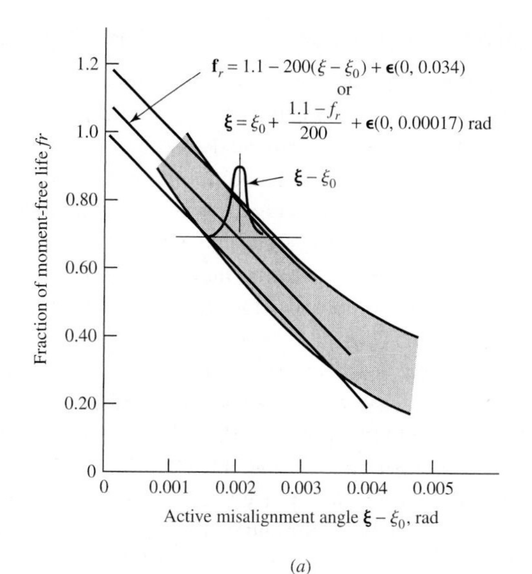

(*a*)

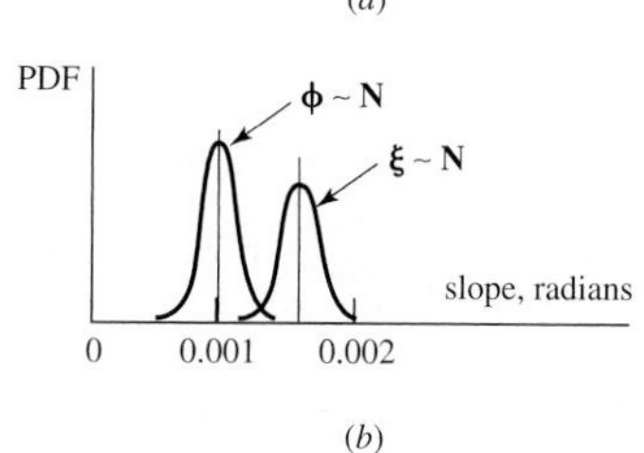

(*b*)

ring, and ξ_0 is the angle of play at which moment begins to build. For example, for a cylindrical roller bearing in which $\xi = 0.001$ rad and $\xi_0 = 0.001$ rad, then $\bar{f}_r = 1$. For the case where $\xi = 0.002$ rad and $\xi_0 = 0.001$ rad, then $\bar{f}_r = 0.90$. The mean life penalty fractional life reduction factor $\bar{f}_r$ can be accommodated in Eq. (11–15) as follows:

$$C_{10} = (\mathrm{AF})F_D \left[\frac{x_D/\bar{f}_r}{x_0 + (\theta - x_0)(1 - R_D)^{1/b}} \right]^{1/a} \tag{11–18}$$

In other words, the desired dimensionless life in rating-life multiples x_D is inflated by dividing by $\bar{f}_r$ to increase the basic load rating (catalog rating). The amount of inflation in the catalog rating C_{10} can be seen by factoring $\bar{f}_r$ out of the brackets of Eq. (11–18). For example, for $\bar{f}_r = 0.80$,

$$\left(\bar{f}_r\right)^{-1/a} = \bar{f}_r^{-3/10} = 0.80^{-3/10} = 1.069$$

or about a 7 percent increase in the catalog basic load rating required. The data underlying Eq. (11–17) were not plotted in the Bamberger paper. For Dareing and Radzimovsky[2] the reduction in life in roller bearings due to misalignment is addressed by modeling the rollers on an elastic foundation. The model requires knowledge of the number and geometry of the rollers. Under their assumption the fractional life reduction is several times as severe as given in Eq. (11–17). Should misalignment be an unavoidable problem, it is recommended that you communicate with an applications engineer of the bearing manufacturer for guidance concerning their bearings.

2. D. W. Dareing and E. I. Radzimovsky, "Misaligned Roller Bearings," *Machine Design*, 13 February 1964, pp. 175–179.

11–9 Selection of Tapered Roller Bearings

Tapered roller bearings have a number of features that make them complicated. As we address the differences between tapered roller and ball and cylindrical roller bearings, note that the underlying fundamentals are the same, but that there are differences in detail. Moreover, bearing and cup combinations are not necessarily priced in proportion to capacity. Any catalog displays a mix of high-production, low-production, and successful special-order designs. Bearing suppliers have computer programs that will take your problem descriptions, give intermediate adequacy assessment information, and list a number of satisfactory cup-and-cone combinations in order of decreasing cost. Company sales offices provide access to comprehensive engineering services to help designers select and apply their bearings. At a large original equipment manufacturer's plant there may be a resident bearing company representative.

Take a few minutes to go to your department's design library and look at a bearing supplier's engineering catalog, such as The Timken Company's *Bearing Selection Handbook—Revised* (1986). There is a log of engineering information and detail, based on long and successful experience. All we can do here is introduce the vocabulary, show congruence to fundamentals that were learned earlier, offer examples, and develop confidence. Finally, problems should reinforce the learning experience.

Form

The four components of a tapered roller bearing assembly are the

- Cone (inner ring)
- Cup (outer ring)
- Tapered rollers
- Cage (spacer-retainer)

The assembled bearing consists of two separable parts: (1) the cone assembly: the cone, the rollers, and the cage; and (2) the cup. Bearings can be made as single-row, two-row, four-row, and thrust-bearing assemblies. Additionally, auxiliary components such as spacers and closures can be used.

A tapered roller bearing can carry both radial and thrust (axial) loads, or any combination of the two. However, even when an external thrust load is not present, the radial load will induce a thrust reaction within the bearing because of the taper. To avoid the separation of the races and the rollers, this thrust must be resisted by an equal and opposite force. One way of generating this force is to always use at least two tapered roller bearings on a shaft. Two bearings can be mounted with the cone backs facing each other, in a configuration called *direct mounting*, or with the cone fronts facing each other, in what is called *indirect mounting*. Figure 11–16 shows the nomenclature of a tapered roller bearing, and the point G through which radial and axial components of load act.

A radial load will induce a thrust reaction. The load zone includes about half the rollers and subtends an angle of approximately 180°, which is referred to as the *load zone*. Using the symbol $F_{a(180)}$ for the induced thrust load from a radial load with a 180° load zone, Timken provides the equation

$$F_{a(180)} = \frac{0.47 F_r}{K} \tag{11–19}$$

where the k factor is geometry-specific, coming from the relationship

$$K = 0.389 \cot \alpha$$

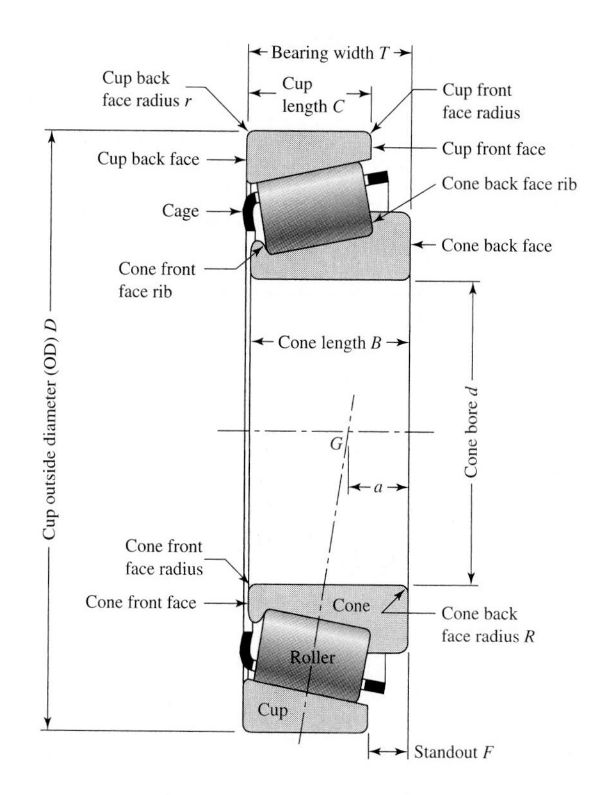

Figure 11–16

Nomenclature of a tapered roller bearing. Point G is the location of the effective load center; use this point to estimate the radial bearing load. *(Courtesy of The Timken Company.)*

where α is half the included cup angle. The K factor is the ratio of the radial load rating to the thrust load rating. The K factor can be first approximated with 1.5 for a radial bearing and 0.75 for a steep angle bearing in the preliminary selection process. After a possible bearing is identified, the exact value of K for each bearing can be found in the *Bearing Selection Handbook—Revised* (1986) in the case of Timken bearings.

Notation

The catalog rating C corresponding to 90 percent reliability was denoted C_{10} earlier in the chapter, the subscript 10 denoting 10 percent failure level. Timken denoted its catalog ratings as C_{90}, the subscript 90 standing for "at 90 million revolution." The failure fraction is still 10 percent (90 percent reliability). This should produce no difficulties since Timken's catalog ratings for radial and thrust loads display neither C_{90} nor $C_{a(90)}$ at the head of the columns. See Fig. 11–17, which is a reproduction of two Timken catalog pages.

Location of Reactions

Figure 11–18 shows a pair of tapered roller bearings mounted directly and indirectly with the bearing reaction locations A_0 and B_0 shown for the shaft. For the shaft as a beam, the span is a_e, the effective spread. It is through points A_0 and B_0 that the radial loads act perpendicular to the shaft axis, and the thrust loads act along the shaft axis. The geometric spread a_g for the direct mounting is greater than for the indirect mounting. With indirect mounting the bearings are closer together compared to the direct mounting; however, the system stability is the same (a_e is the same in both cases). Thus direct and indirect mounting involve space and compactness needed or desired, but with the same system stability.

SINGLE-ROW STRAIGHT BORE

bore	outside diameter	width	rating at 500 rpm for 3000 hours L_{10}: one-row radial	rating at 500 rpm for 3000 hours L_{10}: thrust	fac-tor	eff. load center	part numbers: cone	part numbers: cup	cone: max shaft fillet radius	cone: width	cone: backing shoulder diameters		cup: max hous-ing fillet radius	cup: width	cup: backing shoulder diameters	
d	D	T	N lbf	N lbf	K	a②			R①	B	d_b	d_a	r①	C	D_b	D_a
25.000 0.9843	52.000 2.0472	16.250 0.6398	8190 1840	5260 1180	1.56	−3.6 −0.14	◆30205	◆30205	1.0 0.04	15.000 0.5906	30.5 1.20	29.0 1.14	1.0 0.04	13.000 0.5118	46.0 1.81	48.5 1.91
25.000 0.9843	52.000 2.0472	19.250 0.7579	9520 2140	9510 2140	1.00	−3.0 −0.12	◆32205-B	◆32205-B	1.0 0.04	18.000 0.7087	34.0 1.34	31.0 1.22	1.0 0.04	15.000 0.5906	43.5 1.71	49.5 1.95
25.000 0.9843	52.000 2.0472	22.000 0.8661	13200 2980	7960 1790	1.66	−7.6 −0.30	◆33205	◆33205	1.0 0.04	22.000 0.8661	34.0 1.34	30.5 1.20	1.0 0.04	18.000 0.7087	44.5 1.75	49.0 1.93
25.000 0.9843	62.000 2.4409	18.250 0.7185	13000 2930	6680 1500	1.95	−5.1 −0.20	◆30305	◆30305	1.5 0.06	17.000 0.6693	32.5 1.28	30.0 1.18	1.5 0.06	15.000 0.5906	55.0 2.17	57.0 2.24
25.000 0.9843	62.000 2.4409	25.250 0.9941	17400 3910	8930 2010	1.95	−9.7 −0.38	◆32305	◆32305	1.5 0.06	24.000 0.9449	35.0 1.38	31.5 1.24	1.5 0.06	20.000 0.7874	54.0 2.13	57.0 2.24
25.159 0.9905	50.005 1.9687	13.495 0.5313	6990 1570	4810 1080	1.45	−2.8 −0.11	07096	07196	1.5 0.06	14.260 0.5614	31.5 1.24	29.5 1.16	1.0 0.04	9.525 0.3750	44.5 1.75	47.0 1.85
25.400 1.0000	50.005 1.9687	13.495 0.5313	6990 1570	4810 1080	1.45	−2.8 −0.11	07100	07196	1.0 0.04	14.260 0.5614	30.5 1.20	29.5 1.16	1.0 0.04	9.525 0.3750	44.5 1.75	47.0 1.85
25.400 1.0000	50.005 1.9687	13.495 0.5313	6990 1570	4810 1080	1.45	−2.8 −0.11	07100-S	07196	1.5 0.06	14.260 0.5614	31.5 1.24	29.5 1.16	1.0 0.04	9.525 0.3750	44.5 1.75	47.0 1.85
25.400 1.0000	50.292 1.9800	14.224 0.5600	7210 1620	4620 1040	1.56	−3.3 −0.13	L44642	L44610	3.5 0.14	14.732 0.5800	36.0 1.42	29.5 1.16	1.3 0.05	10.668 0.4200	44.5 1.75	47.0 1.85
25.400 1.0000	50.292 1.9800	14.224 0.5600	7210 1620	4620 1040	1.56	−3.3 −0.13	L44643	L44610	1.3 0.05	14.732 0.5800	31.5 1.24	29.5 1.16	1.3 0.05	10.668 0.4200	44.5 1.75	47.0 1.85
25.400 1.0000	51.994 2.0470	15.011 0.5910	6990 1570	4810 1080	1.45	−2.8 −0.11	07100	07204	1.0 0.04	14.260 0.5614	30.5 1.20	29.5 1.16	1.3 0.05	12.700 0.5000	45.0 1.77	48.0 1.89
25.400 1.0000	56.896 2.2400	19.368 0.7625	10900 2450	5740 1290	1.90	−6.9 −0.27	1780	1729	0.8 0.03	19.837 0.7810	30.5 1.20	30.0 1.18	1.3 0.05	15.875 0.6250	49.0 1.93	51.0 2.01
25.400 1.0000	57.150 2.2500	19.431 0.7650	11700 2620	10900 2450	1.07	−3.0 −0.12	M84548	M84510	1.5 0.06	19.431 0.7650	36.0 1.42	33.0 1.30	1.5 0.06	14.732 0.5800	48.5 1.91	54.0 2.13
25.400 1.0000	58.738 2.3125	19.050 0.7500	11600 2610	6560 1470	1.77	−5.8 −0.23	1986	1932	1.3 0.05	19.355 0.7620	32.5 1.28	30.5 1.20	1.3 0.05	15.080 0.5937	52.0 2.05	54.0 2.13
25.400 1.0000	59.530 2.3437	23.368 0.9200	13900 3140	13000 2930	1.07	−5.1 −0.20	M84249	M84210	0.8 0.03	23.114 0.9100	36.0 1.42	32.5 1.27	1.5 0.06	18.288 0.7200	49.5 1.95	56.0 2.20
25.400 1.0000	60.325 2.3750	19.842 0.7812	11000 2480	6550 1470	1.69	−5.1 −0.20	15578	15523	1.3 0.05	17.462 0.6875	32.5 1.28	30.5 1.20	1.5 0.06	15.875 0.6250	51.0 2.01	54.0 2.13
25.400 1.0000	61.912 2.4375	19.050 0.7500	12100 2730	7280 1640	1.67	−5.8 −0.23	15101	15243	0.8 0.03	20.638 0.8125	32.5 1.28	31.5 1.24	2.0 0.08	14.288 0.5625	54.0 2.13	58.0 2.28
25.400 1.0000	62.000 2.4409	19.050 0.7500	12100 2730	7280 1640	1.67	−5.8 −0.23	15100	15245	3.5 0.14	20.638 0.8125	38.0 1.50	31.5 1.24	1.3 0.05	14.288 0.5625	55.0 2.17	58.0 2.28
25.400 1.0000	62.000 2.4409	19.050 0.7500	12100 2730	7280 1640	1.67	−5.8 −0.23	15101	15245	0.8 0.03	20.638 0.8125	32.5 1.28	31.5 1.24	1.3 0.05	14.288 0.5625	55.0 2.17	58.0 2.28
25.400 1.0000	62.000 2.4409	19.050 0.7500	12100 2730	7280 1640	1.67	−5.8 −0.23	15102	15245	1.5 0.06	20.638 0.8125	34.0 1.34	31.5 1.24	1.3 0.05	14.288 0.5625	55.0 2.17	58.0 2.28
25.400 1.0000	62.000 2.4409	20.638 0.8125	12100 2730	7280 1640	1.67	−5.8 −0.23	15101	15244	0.8 0.03	20.638 0.8125	32.5 1.28	31.5 1.24	1.3 0.05	15.875 0.6250	55.0 2.17	58.0 2.28
25.400 1.0000	63.500 2.5000	20.638 0.8125	12100 2730	7280 1640	1.67	−5.8 −0.23	15101	15250	0.8 0.03	20.638 0.8125	32.5 1.28	31.5 1.24	1.3 0.05	15.875 0.6250	56.0 2.20	59.0 2.32
25.400 1.0000	63.500 2.5000	20.638 0.8125	12100 2730	7280 1640	1.67	−5.8 −0.23	15101	15250X	0.8 0.03	20.638 0.8125	32.5 1.28	31.5 1.24	1.5 0.06	15.875 0.6250	55.0 2.17	59.0 2.32
25.400 1.0000	64.292 2.5312	21.433 0.8438	14500 3250	13500 3040	1.07	−3.3 −0.13	M86643	M86610	1.5 0.06	21.433 0.8438	38.0 1.50	36.5 1.44	1.5 0.06	16.670 0.6563	54.0 2.13	61.0 2.40

Figure 11–17

Catalog entry of single-row straight-bore Timken roller bearings, in part. *(Courtesy of The Timken Company.)*

SINGLE-ROW STRAIGHT BORE

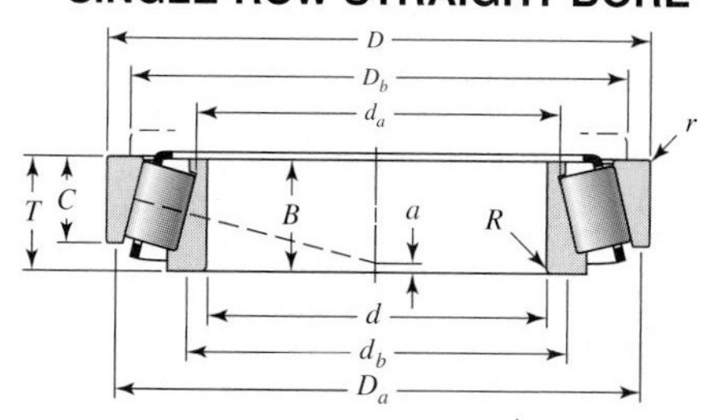

bore	outside diameter	width	rating at 500 rpm for 3000 hours L_{10} one-row radial	rating at 500 rpm for 3000 hours L_{10} thrust	factor	eff. load center	part numbers cone	part numbers cup	cone: max shaft fillet radius	cone: width	cone: backing shoulder diameters		cup: max housing fillet radius	cup: width	cup: backing shoulder diameters	
d	D	T	N lbf	N lbf	K	a②			R①	B	d_b	d_a	r①	C	D_b	D_a
25.400 1.0000	65.088 2.5625	22.225 0.8750	13100 2950	16400 3690	0.80	−2.3 −0.09	23100	23256	1.5 0.06	21.463 0.8450	39.0 1.54	34.5 1.36	1.5 0.06	15.875 0.6250	53.0 2.09	63.0 2.48
25.400 1.0000	66.421 2.6150	23.812 0.9375	18400 4140	8000 1800	2.30	−9.4 −0.37	2687	2631	1.3 0.05	25.433 1.0013	33.5 1.32	31.5 1.24	1.3 0.05	19.050 0.7500	58.0 2.28	60.0 2.36
25.400 1.0000	68.262 2.6875	22.225 0.8750	15300 3440	10900 2450	1.40	−5.1 −0.20	02473	02420	0.8 0.03	22.225 0.8750	34.5 1.36	33.5 1.32	1.5 0.06	17.462 0.6875	59.0 2.32	63.0 2.48
25.400 1.0000	72.233 2.8438	25.400 1.0000	18400 4140	17200 3870	1.07	−4.6 −0.18	HM88630	HM88610	0.8 0.03	25.400 1.0000	39.5 1.56	39.5 1.56	2.3 0.09	19.842 0.7812	60.0 2.36	69.0 2.72
25.400 1.0000	72.626 2.8593	30.162 1.1875	22700 5110	13000 2910	1.76	−10.2 −0.40	3189	3120	0.8 0.03	29.997 1.1810	35.5 1.40	35.0 1.38	3.3 0.13	23.812 0.9375	61.0 2.40	67.0 2.64
26.157 1.0298	62.000 2.4409	19.050 0.7500	12100 2730	7280 1640	1.67	−5.8 −0.23	15103	15245	0.8 0.03	20.638 0.8125	33.0 1.30	32.5 1.28	1.3 0.05	14.288 0.5625	55.0 2.17	58.0 2.28
26.162 1.0300	63.100 2.4843	23.812 0.9375	18400 4140	8000 1800	2.30	−9.4 −0.37	2682	2630	1.5 0.06	25.433 1.0013	34.5 1.36	32.0 1.26	0.8 0.03	19.050 0.7500	57.0 2.24	59.0 2.32
26.162 1.0300	66.421 2.6150	23.812 0.9375	18400 4140	8000 1800	2.30	−9.4 −0.37	2682	2631	1.5 0.06	25.433 1.0013	34.5 1.36	32.0 1.26	1.3 0.05	19.050 0.7500	58.0 2.28	60.0 2.36
26.975 1.0620	58.738 2.3125	19.050 0.7500	11600 2610	6560 1470	1.77	−5.8 −0.23	1987	1932	0.8 0.03	19.355 0.7620	32.5 1.28	31.5 1.24	1.3 0.05	15.080 0.5937	52.0 2.05	54.0 2.13
† 26.988 † 1.0625	50.292 1.9800	14.224 0.5600	7210 1620	4620 1040	1.56	−3.3 −0.13	L44649	L44610	3.5 0.14	14.732 0.5800	37.5 1.48	31.0 1.22	1.3 0.05	10.668 0.4200	44.5 1.75	47.0 1.85
† 26.988 † 1.0625	60.325 2.3750	19.842 0.7812	11000 2480	6550 1470	1.69	−5.1 −0.20	15580	15523	3.5 0.14	17.462 0.6875	38.5 1.52	32.0 1.26	1.5 0.06	15.875 0.6250	51.0 2.01	54.0 2.13
† 26.988 † 1.0625	62.000 2.4409	19.050 0.7500	12100 2730	7280 1640	1.67	−5.8 −0.23	15106	15245	0.8 0.03	20.638 0.8125	33.5 1.32	33.0 1.30	1.3 0.05	14.288 0.5625	55.0 2.17	58.0 2.28
† 26.988 † 1.0625	66.421 2.6150	23.812 0.9375	18400 4140	8000 1800	2.30	−9.4 −0.37	2688	2631	1.5 0.06	25.433 1.0013	35.0 1.38	33.0 1.30	1.3 0.05	19.050 0.7500	58.0 2.28	60.0 2.36
28.575 1.1250	56.896 2.2400	19.845 0.7813	11600 2610	6560 1470	1.77	−5.8 −0.23	1985	1930	0.8 0.03	19.355 0.7620	34.0 1.34	33.5 1.32	0.8 0.03	15.875 0.6250	51.0 2.01	54.0 2.11
28.575 1.1250	57.150 2.2500	17.462 0.6875	11000 2480	6550 1470	1.69	−5.1 −0.20	15590	15520	3.5 0.14	17.462 0.6875	39.5 1.56	33.5 1.32	1.5 0.06	13.495 0.5313	51.0 2.01	53.0 2.09
28.575 1.1250	58.738 2.3125	19.050 0.7500	11600 2610	6560 1470	1.77	−5.8 −0.23	1985	1932	0.8 0.03	19.355 0.7620	34.0 1.34	33.5 1.32	1.3 0.05	15.080 0.5937	52.0 2.05	54.0 2.13
28.575 1.1250	58.738 2.3125	19.050 0.7500	11600 2610	6560 1470	1.77	−5.8 −0.23	1988	1932	3.5 0.14	19.355 0.7620	39.5 1.56	33.5 1.32	1.3 0.05	15.080 0.5937	52.0 2.05	54.0 2.13
28.575 1.1250	60.325 2.3750	19.842 0.7812	11000 2480	6550 1470	1.69	−5.1 −0.20	15590	15523	3.5 0.14	17.462 0.6875	39.5 1.56	33.5 1.32	1.5 0.06	15.875 0.6250	51.0 2.01	54.0 2.13
28.575 1.1250	60.325 2.3750	19.845 0.7813	11600 2610	6560 1470	1.77	−5.8 −0.23	1985	1931	0.5 0.03	19.355 0.7620	34.0 1.34	33.5 1.32	1.3 0.05	15.875 0.6250	52.0 2.05	55.0 2.17

① These maximum fillet radii will be cleared by the bearing corners.

② Minus value indicates center is inside cone backface.

† For standard class **ONLY**, the maximum metric size is a whole millimetre value.

* For "J" part tolerances—see metric tolerances, page 73, and fitting practice, page 65.

◆ ISO cone and cup combinations are designated with a common part number and should be purchased as an assembly. For ISO bearing tolerances—see metric tolerances, page 73, and fitting practice, page 65.

Figure 11–17

(Continued)

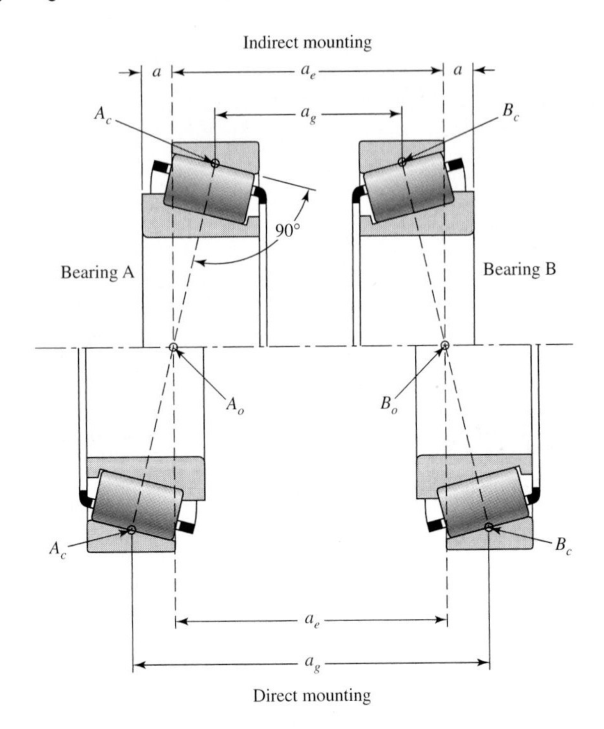

Figure 11–18

Comparison of mounting stability between indirect and direct mountings. *(Courtesy of The Timken Company.)*

Load–Life–Reliability Trade-Off

Recall Eq. (11–6). It can be written as

$$C_{10} = F_D \left(\frac{x_D}{x_B} \right)^{1/a}$$

Timken uses a two-parameter Weibull model with $\theta = 4.48$ and $b = 3/2$. Solving for x_D above gives

$$x_D = x_0 + (\theta - x_0)\,(1 - R_D)^{1/b} \left(\frac{C_{10}}{F_D} \right)^a = \left[4.48\,(1 - R_D)^{2/3}\right] \left(\frac{C_{10}}{F_D} \right)^{10/3}$$

Now x_D is desired life in multiples of rating life. Since $x_D = L_D/90(10^6)$, we obtain

$$L_D = 4.48\,(1 - R_D)^{2/3} \left(\frac{C_{10}}{F_D} \right)^{10/3} 90\left(10^6\right) \qquad \text{(in revolutions)}$$

Timken writes this equation as

$$L_D = a_1 a_2 a_3 a_4 \left(\frac{C_{10}}{F_D} \right)^{10/3} 90(60^6) \tag{11–20}$$

a_1: $4.48\,(1 - R_D)^{2/3}$

a_2: bearing material

$a_3 = a_{3k} a_{3\ell} a_{3m}$
- a_{3k}: load zone
- $a_{3\ell}$: Lubricant $a_{3\ell} = f_T f_v$
- a_{3m}: alignment

$a_4 = 1$ (spall size is 0.01 in^2)

Temperature factor f_T can be found in Fig. 11–19, and viscosity factor f_v can be found in Fig. 11–20.

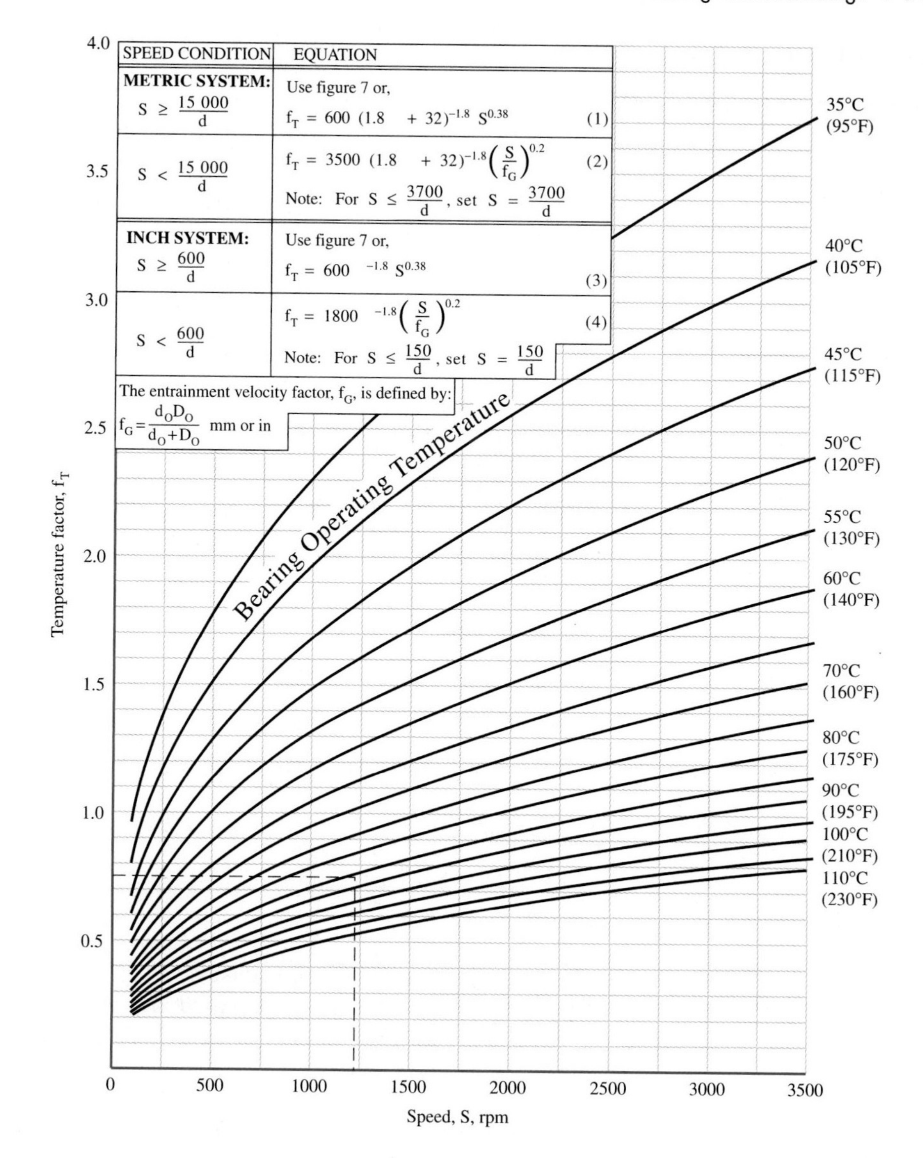

Figure 11–19

Temperature factor f_T as a function of speed and bearing operating temperature. For speed S less than 15 000/d use equation shown in inset when d is bearing bore in millimeters (less than 600/d when bearing bore is in inches). *(Courtesy of The Timken Company.)*

For the usual case, solving the preceding equation for C_{10} gives

$$C_{10} = (\text{AF})P\left[\frac{L_D}{f_T f_v 4.48\,(1 - R_D)^{2/3}\,90\left(10^6\right)}\right]^{3/10} \qquad (L_D \text{ in revolutions}) \qquad (11\text{–}21a)$$

The load P is the dynamic equivalent load of the combination F_r and F_a of Sec. 11–6. The particular values of X and Y are given in Table 11–7, in the various expressions for the radial equivalent load P in the right-hand column.

For very slow rotative speeds, say less than 10 r/min, fatigue is not the cause of failure. Timken recommends that the dynamic equivalent load rating be increased fourfold for such an application, that is,

$$4C_{10} = f_A P$$

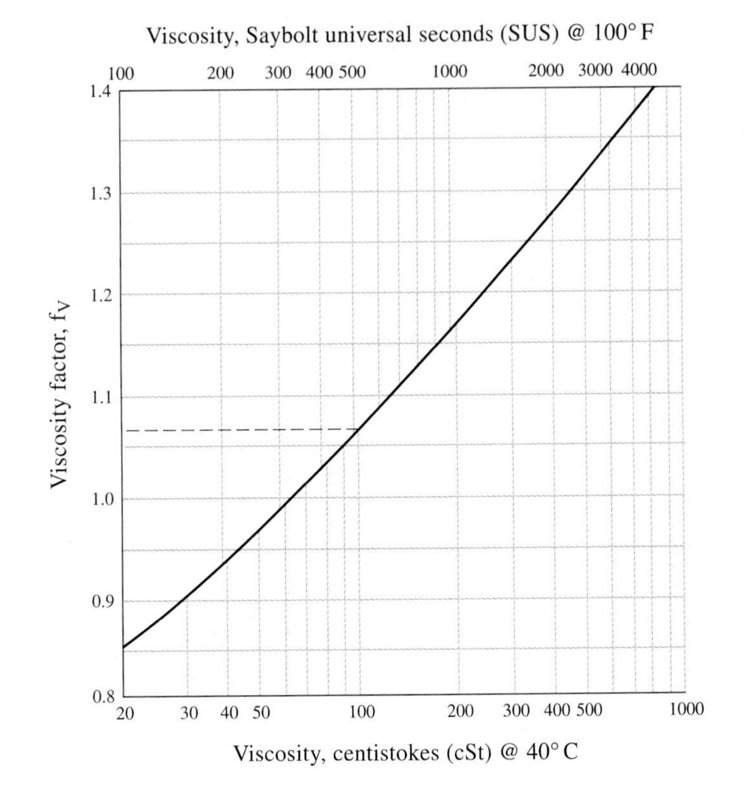

Figure 11–20

Viscosity factor f_V as a function of oil viscosity. (This graph applies to petroleum oil with a viscosity index of approximately 90.) *(Courtesy of The Timken Company.)*

or

$$C_{10} = \frac{f_A P}{4} \tag{11–21b}$$

For completely static (nonrotative) loads values of the static radial load rating C_0 and the static thrust load rating C_{0a} are based on a maximum contact stress of 4000 MPa (580 kpsi) at the center of the contact and a 180° load zone. Values for a specific bearing can be obtained from a Timken Company sales engineer.

Table 11–7

Dynamic Equivalent Radial Load Equations for P *Source:* Courtesy of The Timken Company.

Two-Row Mounting, Fixed or Floating (With No External Thrust, F_{ae} = 0) Similar Bearing Series

For two-row similar bearing series with no external thrust, $F_{ae} = 0$, the dynamic equivalent radial load, P, equals F_{rAB} or F_{rC}. Since F_{rAB} or F_{rC} is the radial load on the two-row assembly, the two-row basic dynamic radial load rating, $C_{90(2)}$, is to be used to calculate bearing life.

OPTIONAL APPROACH FOR DETERMINING DYNAMIC EQUIVALENT RADIAL LOADS

The following is a general approach to determining the dynamic equivalent radial loads and therefore is more suitable for programmable calculators and computer programming. Here a factor "m" has to be defined as $+1$ for direct mounted single-row or two-row bearings or -1 for indirect mounted bearings. Also a sign convention is necessary for the external thrust F_{ae} as follows:

a. In case of external thrust *applied to the shaft* (typical rotating cone application), F_{ae} to the right is positive; to the left is negative.

b. When external thrust is *applied to the housing* (typical rotating cup application) F_{ae} to the right is negative; to the left is positive.

Table 11–7

Continued

1. Single-Row Mounting

Design	Thrust Condition	Thrust Load	Dynamic Equivalent Radial Load
Indirect mounting (m = −1)	$\frac{0.47F_{rA}}{K_A} \le \frac{0.47F_{rB}}{K_B} - m\,F_{ae}$	$F_{aA} = \frac{0.47F_{rB}}{K_B} - m\,F_{ae}$	$P_A = 0.4F_{rA} + K_A\,F_{aA}$
		$F_{aB} = \frac{0.47F_{rB}}{K_B}$	$P_B = F_{rB}$
Direct mounting (m = 1)	$\frac{0.47F_{rA}}{K_A} > \frac{0.47F_{rB}}{K_B} - m\,F_{ae}$	$F_{aA} = \frac{0.47F_{rA}}{K_A}$	$P_A = F_{rA}$
		$F_{aB} = \frac{0.47F_{rAB}}{K_A} + m\,F_{ae}$	$P_B = 0.4F_{rB} + K_B\,F_{aB}$

Note: If $P_A < F_{rA}$, use $P_A = F_{rA}$ or if $P_B < F_{rB}$, use $P_B = F_{rB}$

2. Two Row Mounting - Fix Bearing With External Thrust, F_{ae}

(Similar or Dissimilar Series)

Design	Thrust Condition	Dynamic Equivalent Radial Load
Fixed bearing Indirect mounting (m = −1)	$F_{ae} \le \frac{0.6\,F_{rAB}}{K^*}$	$P_A = \frac{K_A}{K_A + K_B}(F_{rAB} - 1.67\,m\,K_B\,F_{ae})$
		$P_B = \frac{K_B}{K_A + K_B}(F_{rAB} + 1.67\,m\,K_A\,F_{ae})$
Fixed bearing Direct mounting (m = 1)	$F_{ae} > \frac{0.6\,F_{rAB}}{K^*}$	$P_A = 0.4\,F_{rAB} - m\,K_A\,F_{ae}$
		$P_B = 0.4F_{rAB} + m\,K_B\,F_{ae}$

*If "$m\,F_{ae}$" is positive, $K = K_B$; If "$m\,F_{ae}$" is negative, $K = K_A$

Note: F_{rAB} is the radial load on the two-row assembly. The single-row basic dynamic radial load rating, C_{90}, is to be applied when calculating life based on above equations.

EXAMPLE 11–8 The shaft depicted in Fig. 11–21*a* carries a helical gear with a tangential force of 3980 N, a separating force of 1770 N, and a thrust force of 1690 N at the pitch cylinder with directions shown. The pitch cylinder of the gear has a diameter of 200 mm. The shaft runs at a speed of 1050 r/min, and the span (effective spread) between the direct-mount bearings is 150 mm. The design life is to be 5000 h and an application factor of 1 is appropriate. The lubricant will be ISO VG 68 (68 cSt @ 40°C) oil with an estimated operating temperature of 55°C. If the reliability of the bearing set is to be 0.99, select suitable single-row tapered roller Timken bearings.

Figure 11–21

Essential geometry of helical gear and shaft. Length dimensions in mm, loads in N, couple in N·mm. (*a*) Sketch (not to scale) showing thrust and separating and tangential forces. (*b*) Forces in *xz* plane. (*c*) Forces in *xy* plane.

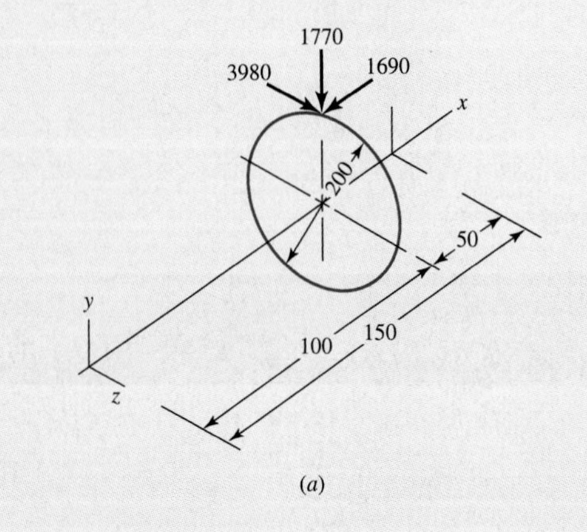

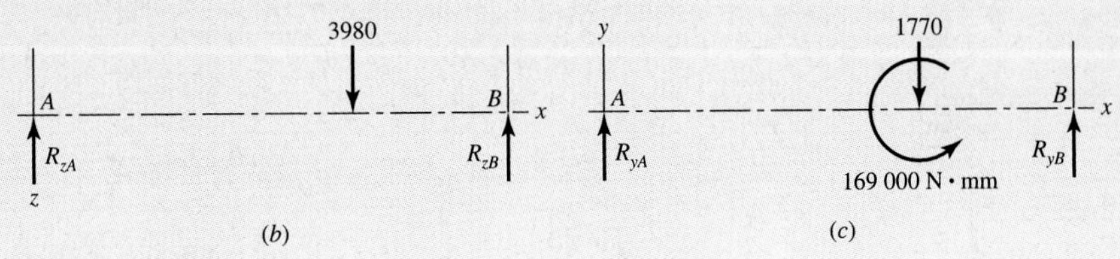

Solution The reactions in the xz plane from Fig. 11–21*b* are

$$R_{zA} = \frac{3980(50)}{150} = 1327 \text{ N}$$

$$R_{zB} = \frac{3980(100)}{150} = 2653 \text{ N}$$

The reactions in the xy plane from Fig. 11–21*c* are

$$R_{yA} = \frac{1770(50)}{150} + \frac{169\,000}{150} = 1717 \text{ N}$$

$$R_{yB} = \frac{1770(100)}{150} - \frac{169\,000}{150} = 53.3 \text{ N}$$

The radial loads F_{rA} and F_{rB} are Pythagorean combinations of R_{yA} and R_{zA}, and R_{yB} and R_{zB}, respectively:

$$F_{rA} = \left(R_{zA}^2 + R_{yA}^2\right)^{1/2} = \left(1327^2 + 1717^2\right)^{1/2} = 2170 \text{ N}$$

$$F_{rB} = \left(R_{zB}^2 + R_{yB}^2\right)^{1/2} = \left(2653^2 + 53.3^2\right)^{1/2} = 2654 \text{ N}$$

Trial 1: We will use $K_A = K_B = 1.5$ to start. From Table 11–7, noting that $m = +1$ for direct mounting and F_{ae} to the right is positive,

$$\frac{0.47F_{rA}}{K_A} <?> \frac{0.47F_{rB}}{K_B} - mF_{ae}$$

$$\frac{0.47(2170)}{1.5} <?> \left[\frac{0.47(2654)}{1.5} - (+1)(-2690)\right]$$

$$680 < 2521$$

We use the upper line of Table 11–7 to find the thrust loads:

$$F_{aA} = \frac{0.47F_{rB}}{K_B} - mF_{ae} = \frac{0.47(2654)}{1.5} - (+1)(-1690) = 2522 \text{ N}$$

$$F_{aB}\frac{0.47F_{rB}}{K_B} = \frac{0.47(2654)}{1.5} = 832 \text{ N}$$

The dynamic equivalent loads P_A and P_B are

$$P_A = 0.4F_{rA} + K_A F_{aA} = 0.4(2170) + 1.5(2522) = 4651 \text{ N}$$
$$P_B = F_{rB} = 2654 \text{ N}$$

From Fig. 11–19 for 1050 r/min, $f_T = 1.31$. From Fig.11–20, $f_v = 1.01$. For use in Eq. (11–20), $a_{3\ell} = f_T f_v = 1.31(1.01) = 1.32$. The catalog basic load rating corresponding to the load–life–reliability goals is given by Eq. (11–21*a*). Estimate R_D as $\sqrt{0.99} = 0.995$ for each bearing. For bearing A, from Eq. (11–21*a*) the catalog entry C_{10} should equal or exceed

$$C_{10} = (1)(4651)\left[\frac{5000(1050)60}{1.32(4.48)(1-0.995)^{2/3}90\left(10^6\right)}\right]^{3/10} = 11\,481 \text{ N}$$

From Fig. 11–17 tentatively select type TS 15100 cone and 15245 cup, which will work: $K_A = 1.67$, $C'_{10} = 12\,100$ N.

For bearing B, from Eq. (11–21*a*), the catalog entry C_{10} should equal or exceed

$$C_{10} = (1)2654\left[\frac{5000(1050)60)}{1.32(4.48)(1-0.995)^{2/3}90\left(10^6\right)}\right]^{3/10} = 6542 \text{ N}$$

Tentatively select the bearing identical to bearing A, which will work: $K_B = 1.67$, $C'_{10} = 12\,100$ N.

Trial 2: Use $K_A = K_B = 1.67$ from tentative bearing selection. The sense of the previous inequality $680 < 2521$ is still the same, so the same equations apply:

$$F_{aA} = \frac{0.47F_{rB}}{K_B} - mF_{ae} = \frac{0.47(2654)}{1.67} - (+1)(-1690) = 2437 \text{ N}$$
$$F_{aB} = \frac{0.47F_{rB}}{K_B} = \frac{0.47(2654)}{1.67} = 747 \text{ N}$$
$$P_A = 0.4F_{rA} + K_A F_{aA} = 0.4(2170) + 1.67(2437) = 4938 \text{ N}$$
$$P_B = F_{rB} = 2654 \text{ N}$$

For bearing A, from Eq. (11–21*a*) the corrected catalog entry C_{10} should equal or exceed

$$C_{10} = (1)(4938)\left[\frac{5000(1050)60}{1.32(4.48)(1-0.995)^{2/3}90\left(10^6\right)}\right]^{3/10} = 12\,167 \text{ N}$$

Although this catalog entry exceeds slightly the tentative selection for bearing A, we will keep it since the reliability of bearing B exceeds 0.995. In the next section we will quantitatively show that the combined reliability of bearing A and B will exceed the reliability goal of 0.99.

For bearing B, $P_B = F_{rB} = 2654$ N. From Eq. (11–21*a*),

$$C_{10} = (1)2654\left[\frac{5000(1050)60}{1.32(4.48)(1-0.995)^{2/3}90\left(10^6\right)}\right]^{3/10} = 6542 \text{ N}$$

Select cone and cup 15100 and 15245, respectively, for both bearing A and B. Note from

Fig. 11–17 the effective load center is located at $a = -5.8$ mm, that is, 5.8 mm into the cup from the back. Thus the shoulder-to-shoulder dimension should be $150 - 2(5.8) = 138.4$ mm. Note, also, the calculation for the second bearing C_{10} contains the same bracketed expression as for the first. For example, on the first trial C_{10} for bearing A is 11 464 N. C_{10} for bearing B can be easily calculated by

$$(C_{10})_B = \frac{(C_{10})_A}{P_A} P_B = \frac{11\,464}{4651} 2654 = 6542 \text{ N}$$

The computational effort can be simplified only after this is understood, and not until then.

11–10 Adequacy Assessment for Selected Rolling-Contact Bearings

In textbooks machine elements typically are treated singly. This can lead the reader to the presumption that an adequacy assessment involves only that element, in this case a rolling-contact bearing. The immediately adjacent elements (the shaft journal and the housing bore) have immediate influence on the performance. Other elements, further removed (gears producing the bearing load), also have influence. Just as some say, "If you pull on something in the environment, you find that it is attached to everything else." This should be intuitively obvious to those involved with machinery. How, then, can one check shaft attributes that aren't mentioned in a problem statement? Possibly, because the bearing hasn't been designed yet (in fine detail). All this points out the necessary iterative nature of designing, say, a speed reducer. If power, speed, and reduction are stipulated, then gear sets can be roughed in, their sizes, geometry, and location estimated, shaft forces and moments identified, bearings tentatively selected, seals identified; the bulk is beginning to make itself evident, the housing and lubricating scheme as well as the cooling considerations become clearer, shaft overhangs and coupling accommodations appear. It is time to iterate, now addressing each element again, knowing much more about all of the others. When you have completed the necessary iterations, you will know what you need for the adequacy assessment for the bearings. In the meantime you do as much of the adequacy assessment as you can, avoiding bad selections, even if tentative. Always keep in mind that you eventually have to do it all in order to pronounce your completed design satisfactory.

An outline of an adequacy assessment for a rolling contact bearing includes, at a minimum,

- Extant reliability for the load imposed and life expected
- Shouldering on shaft and housing satisfactory
- Journal finish, diameter and tolerance compatible
- Housing finish, diameter and tolerance compatible
- Lubricant type according to manufacturer's recommendations, lubricant paths and volume supplied to keep operating temperature satisfactory
- Preloads, if required are supplied

Since we are focusing on rolling-contact bearings, we can address extant reliability quantitatively, as well as shouldering. Other quantitative treatment will have to wait until the materials for shaft and housing, surface quality, and diameters and tolerances are known.

Extant Reliability

Equation (11–5) can be solved for the reliability R_D in which C'_{10} appears, it being the *basic load rating of the selected bearing:*

$$R = \exp\left(-\left\{\frac{\left[x_D - x_0\left(\frac{C'_{10}}{f_A F_D}\right)^a\right]}{(\theta - x_0)\left(\frac{C'_{10}}{f_A F_D}\right)^a}\right\}^b\right) \tag{11–22}$$

Equation (11–6) can likewise be solved for R_D:

$$R \doteq 1 - \left\{\frac{\left[x_D - x_0\left(\frac{C'_{10}}{f_A F_D}\right)^a\right]}{(\theta - x_0)\left(\frac{C'_{10}}{f_A F_D}\right)^a}\right\} \qquad R \geq 0.90 \tag{11–23}$$

EXAMPLE 11–9

In Ex. 11–3 the catalog basic load rating for 99 percent reliability at $x_D = 540$ rating lives as $C'_{10} = 6702$ lb $= 29.8$ kN, which corresponds to a 02-40 mm deep-groove ball bearing. If the bore in the application had to be 70 mm or larger (selecting a 02-70 mm deep-groove bearing), what is the extant reliability?

Solution

A 02-70 mm deep-groove ball bearing has, from Table 11–2, $C'_{10} = 61.8$ kN $=$ 13 888 lbf. Using Eq. (11–23),

$$R = 1 - \left[\frac{540 - 0.02\left(\frac{13\,888}{1.2(413)}\right)^3}{(4.439)\left(\frac{13\,888}{1.2(413)}\right)^3}\right]^{1.483} = 0.999\,963$$

Using Eq. (11–22),

$$R = \exp\left(-\left[\frac{540 - 0.02\left(\frac{13\,888}{1.2(413)}\right)^3}{(4.439)\left(\frac{13\,888}{1.2(413)}\right)^3}\right]^{1.483}\right) = 0.999\,963$$

Both equations give the same result, which is much higher than 0.99, which was to be expected.

In tapered roller bearings, or other bearing using a two-parameter Weibull distribution, Eq. (11–22) becomes, for $\theta = 4.48$, $b = 3/2$,

$$R = \exp\left\{-\left[\frac{x_D}{\theta\left(C'_{10}/\left[f_A F_D\right]\right)^a}\right]^b\right\}$$

$$= \exp\left\{-\left[\frac{x_D}{4.48 f_T f_v \left(C'_{10}/\left[f_A F_D\right]\right)^{10/3}}\right]^{3/2}\right\} \tag{11–24}$$

and Eq. (11–23) becomes

$$R \doteq 1 - \left[\frac{x_D}{\theta\left(C'_{10}/\left[f_A F_D\right]\right)^a}\right]^b = 1 - \left[\frac{x_D}{f_T f_v 4.48\left(C'_{10}/\left[f_A F_D\right]\right)^{10/3}}\right]^{3/2} \tag{11–25}$$

EXAMPLE 11–10 In Ex. 11–8 bearings A and B (cone 15100 and cup 15245) have $C'_{10} = 12\ 100$ N. What is the extant reliability of the pair of bearings A and B?

Solution The desired life x_D was $5000(1050)60/[90(10^6)] = 3.5$ rating lives. Using Eq. (11–25) for bearing A gives

$$R \doteq 1 - \left[\frac{3.5}{4.48(1.32)(12\ 100/[(1)(4938)])^{10/3}}\right]^{3/2} = 0.994\ 846$$

which is less than 0.995, as expected. Using Eq. (11–25) for bearing B gives

$$R \doteq 1 - \left[\frac{3.5}{4.48(1.32)(12\ 100/[(1)(2654)])^{10/3}}\right]^{3/2} = 0.999\ 769$$

Answer The reliability of the bearing pair is

$$R = R_A R_B = 0.994\ 846(0.999\ 769) = 0.994\ 616$$

which is greater than the overall reliability goal of 0.99. When two bearings are made identical for simplicity, or reducing the number of spares, or other stipulation, and the loading is not the same, both can be made smaller and still meet a reliability goal. If the loading is disparate, then the more heavily loaded bearing can be chosen for a reliability goal just slightly larger than the overall goal.

An additional example is useful to show what happens in cases of pure thrust loading.

EXAMPLE 11–11 Consider a constrained housing as depicted in Fig. 11–22 with two direct-mount tapered roller bearings resisting an external thrust F_{ae} of 8000 N. The shaft speed is 950 r/min, the desired life is 10 000 h, the expected shaft diameter is approximately 1 in. The lubricant is ISO VG 150 (150 cSt @ 40°C) oil with an estimated bearing operating temperature of 80°C. The reliability goal is 0.95. The application factor is appropriately $f_A = 1$.

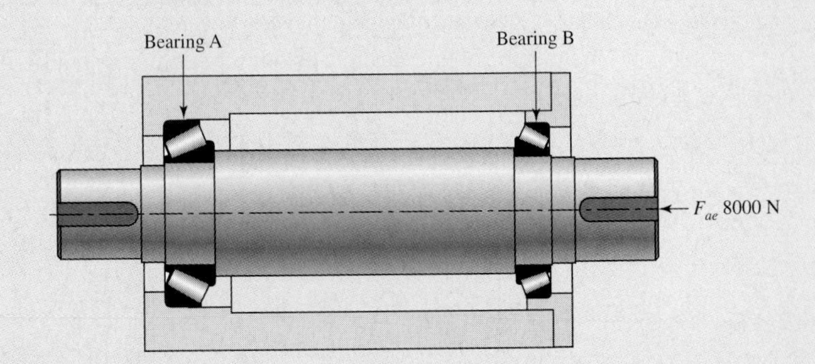

Figure 11–22
The constrained housing of Ex. 11–11.

(*a*) Choose a suitable tapered roller bearing for A.
(*b*) Choose a suitable tapered roller bearing for B.
(*c*) Find the extant reliabilities R_A, R_B, and R.

Solution (*a*) The bearing reactions at A are

$$F_{rA} = F_{rB} = 0$$

$$F_{aA} = F_{ae} = 8000 \text{ N}$$

Since bearing B is unloaded, $R = R_A = 0.95$. From Table 11–7,

$$\frac{0.47F_{rA}}{K_A} < ? > \frac{0.47F_{rB}}{K_B} - mF_{ae}$$

Noting that F_{ae} to the right is positive (Table 11–7) and $m = +1$,

$$\frac{0(0)}{K_A} < ? > \left[\frac{0.47(0)}{K_B} - (+1)(-8000)\right]$$

$$0 < 8000 \text{ N}$$

The top row in Table 11–7 applies, so

$$F_{aA} = \frac{0.47(0)}{K_B} - (+1)(-8000) = 8000 \text{ N}$$

$$F_{aB} = \frac{0.47(0)}{K_B} = 0$$

If we set $K_A = 1$, we can find C_{10} in the thrust column and avoid iteration:

$$P_A = 0.4F_{rA} + K_A F_{aA} = 0.4(0) + (1)8000 = 8000 \text{ N}$$
$$P_B = F_{rB} = 0$$

From Fig. 11–19, $f_T = 0.76$. From Fig. 11–20, $f_v = 1.12$, making $a_{3\ell} = 0.76(1.12) = 0.85$. Then

$$x_D = \frac{10\,000(950)60}{90\left(10^6\right)} = 6.3\underline{3} \text{ rating lives}$$

$$C_{10} = (1)8000\left[\frac{6.3\underline{3}}{0.85(4.48)(1-0.95)^{2/3}}\right]^{3/10} = 16\,965 \text{ N}$$

Answer Figure 11–17 presents one possibility in the 1-in bore (25.4 mm) size: cone, HM88630, cup HM88610 with $(C'_{10})_a = 17\,200$ N.

Answer (*b*) Bearing B experiences no load, and the cheapest bearing of this bore size will do, including a ball or roller bearing.
(*c*) Equation (11–25) predicts the extant reliability of bearing A as

Answer
$$R_A \doteq 1 - \left[\frac{6.3\underline{3}}{0.85(4.48)(17\,200/8000)^{10/3}}\right]^{3/2} = 0.953$$

which is greater than 0.95, as one would expect. For bearing B,

Answer
$$F_D = P_B = 0$$

$$R_B \doteq 1 - \left[\frac{6.33}{0.85(4.48)(17\,200/0)^{10/3}}\right]^{3/2} = 1 - 0 = 1$$

as one would expect. The combined reliability of bearings A and B as a pair is

Answer $R = R_A R_B = 0.953(1) = 0.953$

which is greater than the reliability goal of 0.95, as one would expect.

Matters of Fit

Table 11-2 (and Fig. 11–8), which shows the rating of single-row, 02-series, deep-groove and angular-contact ball bearings, includes shoulder diameters recommended for the shaft seat of the inner ring and the shoulder diameter of the outer ring, denoted d_S and d_H, respectively. The shaft shoulder can be greater than d_S but not enough to obstruct the annulus. It is important to maintain concentricity and perpendicularity with the shaft centerline, and to that end the shoulder diameter should equal or exceed d_S. The housing shoulder diameter d_H is to be equal to or less than d_H to maintain concentricity and perpendicularity with the housing bore axis. Neither the shaft shoulder nor the housing shoulder features should allow interference with the free movement of lubricant through the bearing annulus.

In a tapered roller bearing (Fig. 11–10), the cup housing shoulder diameter should be equal to or less than D_b. The shaft shoulder for the cone should be equal to or greater than d_b. Additionally, free lubricant flow is not to be impeded by obstructing any of the annulus. In splash lubrication, common in speed reducers, the lubricant is thrown to the housing cover (ceiling) and is directed in its draining by ribs to a bearing. In direct mounting, a tapered roller bearing pumps oil from outboard to inboard. An oil passageway to the outboard side of the bearing needs to be provided. The oil returns to the sump as a consequence of bearing pump action. With an indirect mount, the oil is directed to the inboard annulus, the bearing pumping it to the outboard side. An oil passage from the outboard side to the sump has to be provided.

11–11 Lubrication

The contacting surfaces in rolling bearings have a relative motion that is both rolling and sliding, and so it is difficult to understand exactly what happens. If the relative velocity of the sliding surfaces is high enough, then the lubricant action is hydrodynamic (see Chap. 12). *Elastohydrodynamic lubrication* (EHD) is the phenomenon that occurs when a lubricant is introduced between surfaces that are in pure rolling contact. The contact of gear teeth and that found in rolling bearings and in cam-and-follower surfaces are typical examples. When a lubricant is trapped between two surfaces in rolling contact, a tremendous increase in the pressure within the lubricant film occurs. But viscosity is exponentially related to pressure, and so a very large increase in viscosity occurs in the lubricant that is trapped between the surfaces. Leibensperger[3] observes that the change in viscosity in and out of contact pressure is equivalent to the difference between cold asphalt and light sewing machine oil.

3. R. L. Leibensperger, "When Selecting a Bearing," *Machine Design,* vol. 47, no. 8, April 3, 1975, pp. 142–147.

The purposes of an antifriction-bearing lubricant may be summarized as follows:

1 To provide a film of lubricant between the sliding and rolling surfaces
2 To help distribute and dissipate heat
3 To prevent corrosion of the bearing surfaces
4 To protect the parts from the entrance of foreign matter

Either oil or grease may be employed as a lubricant. The following rules may help in deciding between them.

Use Grease When	Use Oil When
1. The temperature is not over 200°F. 2. The speed is low. 3. Unusual protection is required from the entrance of foreign matter. 4. Simple bearing enclosures are desired. 5. Operation for long periods without attention is desired.	1. Speeds are high. 2. Temperatures are high. 3. Oiltight seals are readily employed. 4. Bearing type is not suitable for grease lubrication. 5. The bearing is lubricated from a central supply which is also used for other machine parts.

11–12 Mounting and Enclosure

There are so many methods of mounting antifriction bearings that each new design is a real challenge to the ingenuity of the designer. The housing bore and shaft outside diameter must be held to very close limits, which of course is expensive. There are usually one or more counterboring operations, several facing operations and drilling, tapping, and threading operations, all of which must be performed on the shaft, housing, or cover plate. Each of these operations contributes to the cost of production, so that the designer, in ferreting out a trouble-free and low-cost mounting, is faced with a difficult and important problem. The various bearing manufacturers' handbooks give many mounting details in almost every design area. In a text of this nature, however, it is possible to give only the barest details.

The most frequently encountered mounting problem is that which requires one bearing at each end of a shaft. Such a design might use one ball bearing at each end, one tapered roller bearing at each end, or a ball bearing at one end and a straight roller bearing at the other. One of the bearings usually has the added function of positioning or axially locating the shaft. Figure 11–23 shows a very common solution to this problem. The inner rings are backed up against the shaft shoulders and are held in position by round nuts threaded onto the shaft. The outer ring of the left-hand bearing is backed up against a housing shoulder and is held in position by a device which is not shown. The outer ring of the right-hand bearing floats in the housing.

There are many variations possible on the method shown in Fig. 11–23. For example, the function of the shaft shoulder may be performed by retaining rings, by the hub of a gear or pulley, or by spacing tubes or rings. The round nuts may be replaced by retaining rings or by washers locked in position by screws, cotters, or taper pins. The housing shoulder may be replaced by a retaining ring; the outer ring of the bearing may be grooved for a retaining ring, or a flanged outer ring may be used. The force against the outer ring of the left-hand bearing is usually applied by the cover plate, but if no thrust is present, the ring may be held in place by retaining rings.

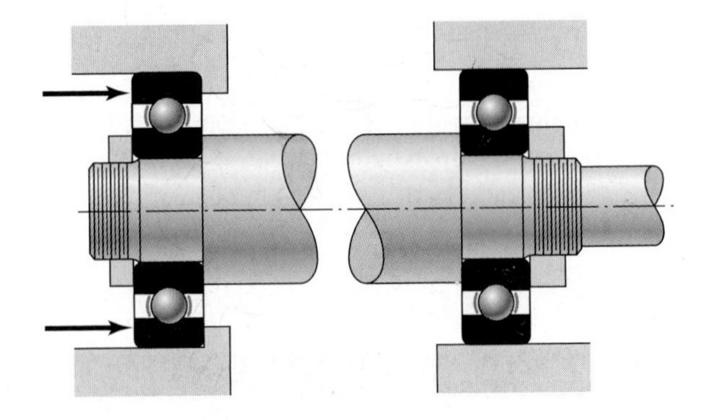

Figure 11–23

A common bearing mounting.

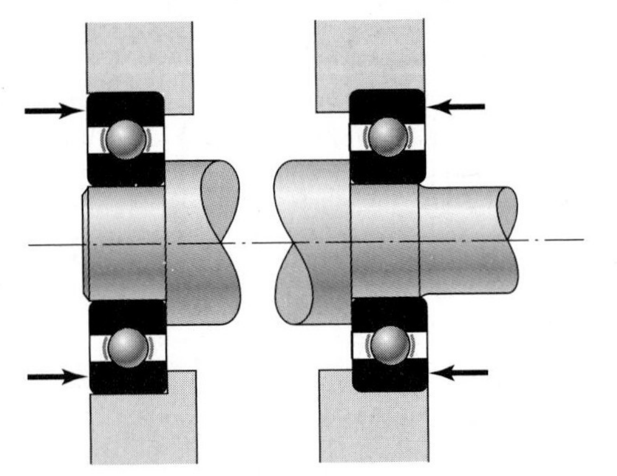

Figure 11–24

An alternative bearing mounting.

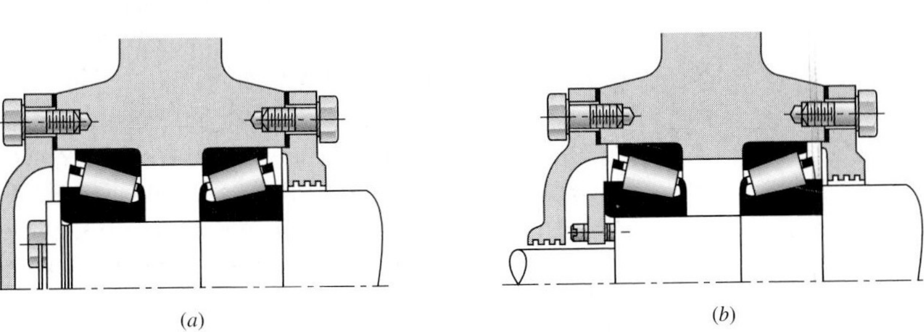

Figure 11–25

Two-bearing mountings. *(Courtesy of The Timken Company.)*

Figure 11–24 shows an alternative method of mounting in which the inner races are backed up against the shaft shoulders as before but no retaining devices are required. With this method the outer races are completely retained. This eliminates the grooves or threads, which cause stress concentration on the overhanging end, but it requires accurate dimensions in an axial direction or the employment of adjusting means. This method has the disadvantage that if the distance between the bearings is great, the temperature rise during operation may expand the shaft enough to wreck the bearings.

It is frequently necessary to use two or more bearings at one end of a shaft. For example, two bearings could be used to obtain additional rigidity or increased load capacity or to cantilever a shaft. Several two-bearing mountings are shown in Fig. 11–25. These may be used with tapered roller bearings, as shown, or with ball bearings. In either case it should be noted that the effect of the mounting is to preload the bearings in an axial direction.

Figure 11–26 shows another two-bearing mounting. Note the use of washers against the cone backs.

When maximum stiffness and resistance to shaft misalignment is desired, pairs of angular-contact ball bearings (Fig. 11–2) are often used in an arrangement called *duplex-*

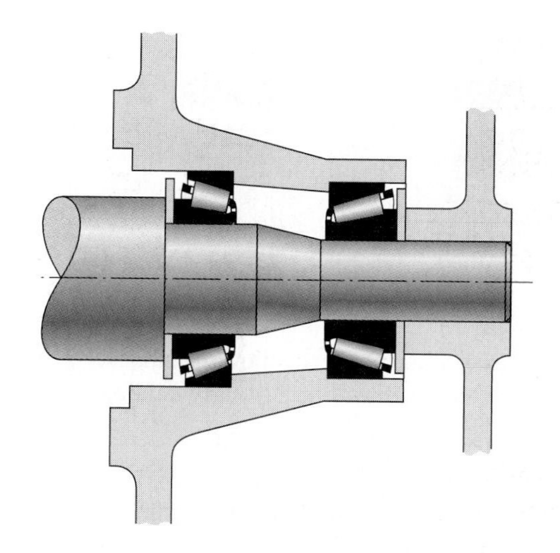

Figure 11–26

Mounting for a washing-machine spindle. (Courtesy of The Timken Company.)

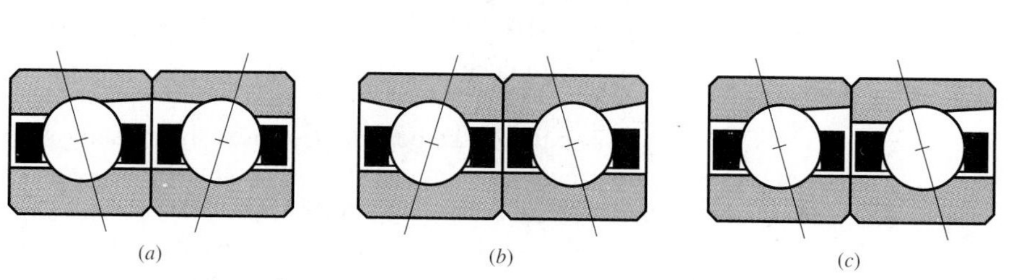

Figure 11–27

Arrangements of angular ball bearings. (*a*) DF mounting; (*b*) DB mounting; (*c*) DT mounting. *(Courtesy of The Timken Company.)*

ing. Bearings manufactured for duplex mounting have their rings ground with an offset, so that when a pair of bearings is tightly clamped together, a preload is automatically established. As shown in Fig. 11–27, three mounting arrangements are used. The face-to-face mounting, called DF, will take heavy radial loads and thrust loads from either direction. The DB mounting (back to back) has the greatest aligning stiffness and is also good for heavy radial loads and thrust loads from either direction. The tandem arrangement, called the DT mounting, is used where the thrust is always in the same direction; since the two bearings have their thrust functions in the same direction, a preload, if required, must be obtained in some other manner.

Bearings are usually mounted with the rotating ring a press fit, whether it be the inner or outer ring. The stationary ring is then mounted with a push fit. This permits the stationary ring to creep in its mounting slightly, bringing new portions of the ring into the load-bearing zone to equalize wear.

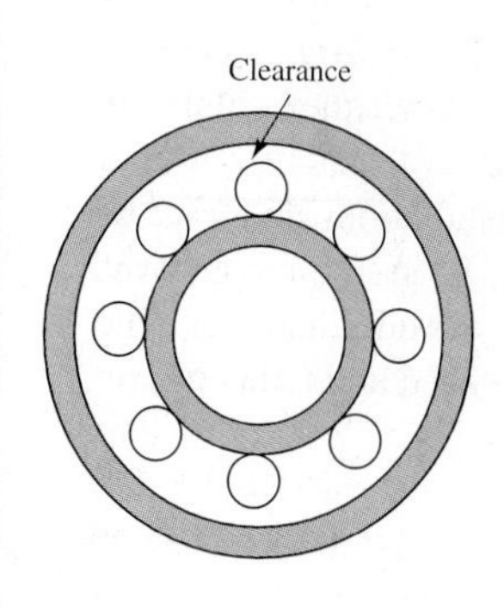

Figure 11–28

Clearance in an off-the-shelf bearing, exaggerated for clarity.

Preloading

The object of preloading is to remove the internal clearance usually found in bearings, to increase the fatigue life, and to decrease the shaft slope at the bearing. Figure 11–28 shows a typical bearing in which the clearance is exaggerated for clarity.

Preloading of straight roller bearings may be obtained by:

1. Mounting the bearing on a tapered shaft or sleeve to expand the inner ring
2. Using an interference fit for the outer ring
3. Purchasing a bearing with the outer ring preshrunk over the rollers

Ball bearings are usually preloaded by the axial load built in during assembly. However, the bearings of Fig. 11–27*a* and *b* are preloaded in assembly because of the differences in widths of the inner and outer rings.

Figure 11–29

Typical sealing methods. *(Courtesy of New Departure–Hyatt Division, General Motors Corporation.)*

(*a*) Felt seal

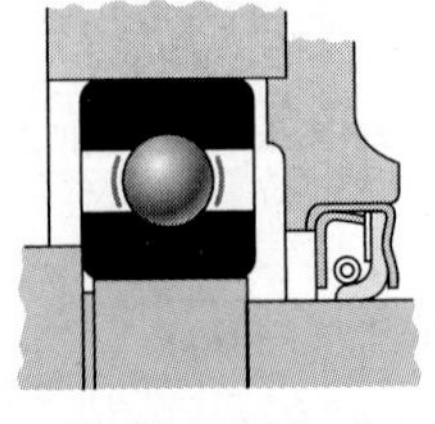

(*b*) Commercial seal

(*c*) Labyrinth seal

It is always good practice to follow manufacturers' recommendations in determining preload, since too much will lead to early failure.

Alignment

Based on the general experience with rolling bearings as expressed in manufacturers' catalogs, the permissible misalignment in cylindrical and tapered roller bearings is limited to 0.001 rad. For spherical ball bearings, the misalignment should not exceed 0.0087 rad. But for deep-groove ball bearings, the allowable range of misalignment is 0.0035 to 0.0047 rad.

The life of the bearing decreases significantly when the misalignment exceeds the allowable limits. Figure 11-15 shows that there is about a 20 percent loss in life for every 0.001 rad of neutral-axis slope beyond 0.001 rad.

Additional protection against misalignment is obtained by providing the full shoulders (see Fig. 11-8) recommended by the manufacturer. Also, if there is any misalignment at all, it is good practice to provide a safety factor of around 2 to account for possible increases during assembly.

Enclosures

To exclude dirt and foreign matter and to retain the lubricant, the bearing mountings must include a seal. The three principal methods of sealings are the felt seal, the commercial seal, and the labyrinth seal (Fig. 11-29).

Felt seals may be used with grease lubrication when the speeds are low. The rubbing surfaces should have a high polish. Felt seals should be protected from dirt by placing them in machined grooves or by using metal stampings as shields.

The *commercial seal* is an assembly consisting of the rubbing element and, generally, a spring backing, which are retained in a sheet-metal jacket. These seals are usually made by press-fitting them into a counterbored hole in the bearing cover. Since they obtain the sealing action by rubbing, they should not be used for high speeds.

The *labyrinth seal* is especially effective for high-speed installations and may be used with either oil or grease. It is sometimes used with flingers. At least three grooves should be used, and they may be cut on either the bore or the outside diameter. The clearance may vary from 0.010 to 0.040 in, depending upon the speed and temperature.

PROBLEMS

Since each bearing manufacturer makes individual decisions with respect to materials, treatments, and manufacturing processes, manufacturers' experiences with bearing life distribution differ. In solving the following problems, we will use the experience of two manufacturers as tabulated on page 733:

Manufacturer	Rating Life, revolutions	Weibull Parameters Rating Lives		
		x_0	θ	b
1	$90(10^6)$	0	4.48	1.5
2	$1(10^6)$	0.02	4.459	1.483

Tables 11–2 and 11–3 are based on manufacturer 2.

DESIGN

11–1 A certain application requires a ball bearing with the inner ring rotating, with a design life of 30 000 h at a speed of 300 r/min. The radial load is 1.898 kN and an application factor of 1.2 is appropriate. The reliability goal is 0.90. Find the multiple of rating life required, x_D, and the catalog rating C_{10} with which to enter a bearing table. Choose an 02-series deep-groove ball bearing from Table 11–2, and estimate the extant reliability in use.

DESIGN

11–2 An angular-contact, inner ring rotating, 02-series ball bearing is required for an application in which the life requirement is 50 000 h at 480 r/min. The design radial load is 610 lb. The application factor is 1.4. The reliability goal is 0.90. Find the multiple of rating life x_D required and the catalog rating C_{10} with which to enter Table 11–2. Choose a bearing and estimate the extant reliability in service.

DESIGN

11–3 The other bearing on the shaft of Prob. 11–2 is to be a 03-series cylindrical roller bearing with inner ring rotating. For a 1650-lb radial load, find the catalog rating C_{10} with which to enter Table 11–3. The reliability goal is 0.90. Choose a bearing and estimate its reliability in use.

ANALYSIS

11–4 Problems 11–2 and 11–3 raise the question of the reliability of the bearing pair on the shaft. Since the combined reliabilities R is $R_1 R_2$, what is the reliability of the two bearings (probability that either or both will not fail) as a result of your decisions in Probs. 11–2 and 11–3? What does this mean in setting reliability goals for each of the bearings of the pair on the shaft?

DESIGN

11–5 Combine Probs. 11–2 and 11–3 for an overall reliability of $R = 0.90$. Reconsider your selections, and meet this overall reliability goal.

ANALYSIS

11–6 Take Ex. 11–2 for an 02-30 mm deep-groove ball bearing, made by manufacturer 2, and show that life at rated load has the following descriptors: $\bar{L} = 4.029(10^6)$ r, $\tilde{L} = 3.484(10^6)$ r, $L_{10} = 10^6$ r, $\sigma_L = 2.568(10^6)$ r, and $C_L = 0.6374$. Plot the probability density function $f(L/L_{10})$ as a function of L/L_{10}. To plot $f(L)$ versus L, note that $f(L/L_{10})$ is divided by L_{10}.

11–7 Write a computer program to generate the plotting positions of Prob. 11–6. Make it general enough so that you are not confined to manufacturer 2. Check your program against Prob. 11–6.

ANALYSIS

11–8 An 02-series ball bearing is to be selected to carry a radial load of 8 kN and a thrust load of 4 kN. The desired life L_D is to be 5000 h with an inner-ring rotation rate of 900 r/min. What is the basic load rating that should be used in selecting a bearing for a reliability goal of 0.90?

ANALYSIS

11–9 The bearing of Prob. 11–8 is to be sized to have a reliability of 0.96. What basic load rating should be used in selecting the bearing?

11–10 A straight (cylindrical) roller bearing is subjected to a radial load of 12 kN. The life is to be 4000 h at a speed of 750 r/min and exhibit a reliability of 0.90. What basic load rating should be used in selecting the bearing from a catalog of manufacturer 2?

DESIGN

11–11 Shown in the figure is a gear-driven squeeze roll which mates with an idler roll, below. The roll is designed to exert a normal force of 30 lb/in of roll length and a pull of 24 lb/in on the material being processed. The roll speed is 300 r/min, and a design life of 30 000 h is desired. Use an application factor of 1.2, and select a pair of angular-contact 02-series ball bearings to be mounted at 0 and A. Use the same size bearings at both locations and a combined reliability of at least 0.92.

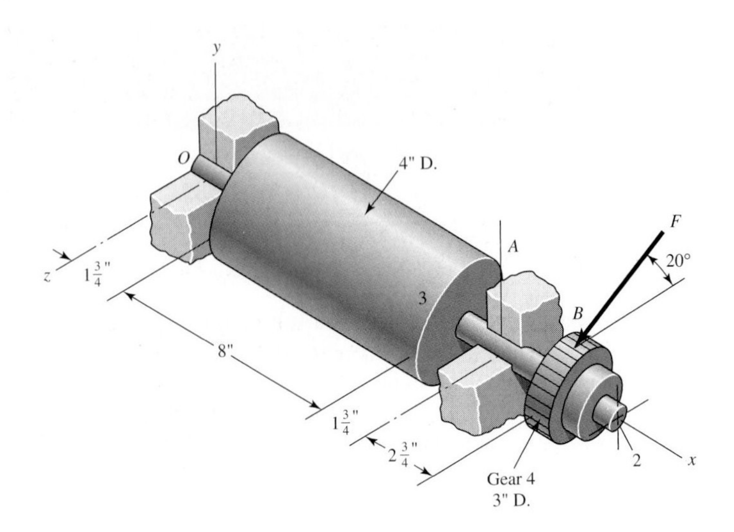

Problem 11–11
Idler roll is below powered roll.

11–12 The figure shown is a geared countershaft with an overhanging pinion at C. Select an angular-contact ball bearing for mounting at O and a straight roller bearing for mounting at B. The force on gear A is $F_A = 600$ lb, and the shaft is to run at a speed of 480 r/min. Solution of the statics problem gives force of bearings against the shaft at O as $\mathbf{R}_O = -388\mathbf{j} + 471\mathbf{k}$ lb, and at B as $\mathbf{R}_B = 317\mathbf{j} + 1620\mathbf{k}$ lb. Specify the bearings required, using an application factor of 1.4, a desired life of 50 000 h, and a combined reliability goal of 0.90.

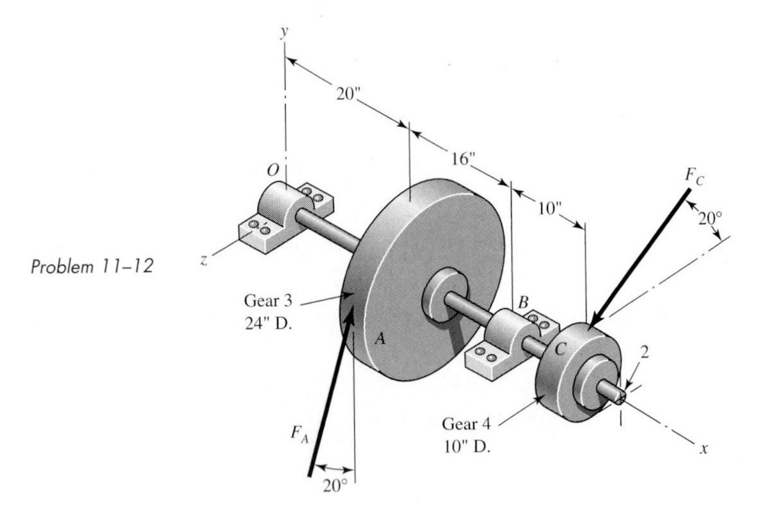

Problem 11–12

11–13 The figure is a schematic drawing of a countershaft that supports two V-belt pulleys. The countershaft runs at 1200 r/min and the bearings are to have a life of 60 kh at a combined reliability of 0.999. The belt tension on the loose side of pulley A is 15 percent of the tension on the tight side. Select bearings for use at O and E, each to have a 25-mm bore, using an application factor of unity.

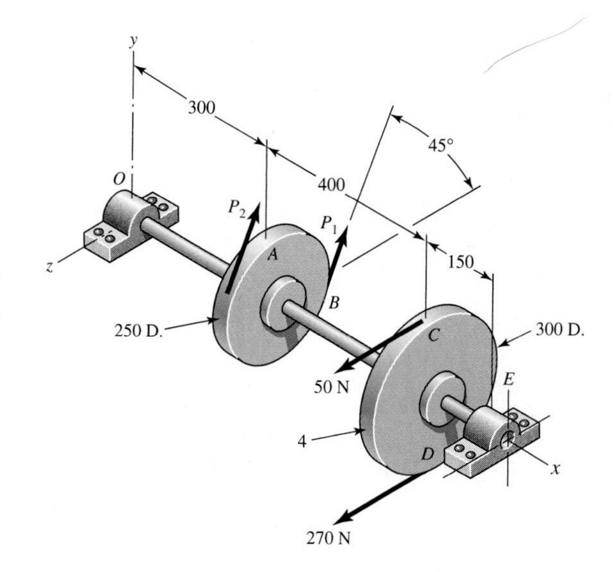

Problem 11–13
Dimensions in millimeters.

ANALYSIS

11–14 The bearing lubricant (513 SUS @ 100°F) operating point is 135°F. A countershaft is supported by two tapered roller bearings using an indirect mounting. The radial bearing loads are 1120 lb for the left-hand bearing and 2190 for the right-hand bearing. The shaft rotates at 400 r/min and is to have a desired life of 40 kh. Use an application factor of 1.4 and a combined reliability goal of 0.90. Using an initial $K = 1.5$, find the required radial rating for each bearing.

DESIGN

11–15 A gear-reduction unit uses the countershaft depicted in the figure. Find the two bearing reactions. The bearings are to be angular-contact ball bearings, having a desired life of 40 kh when used at 200 r/min. Use 1.2 for the application factor and a reliability goal for the bearing pair of 0.95. Select the bearings.

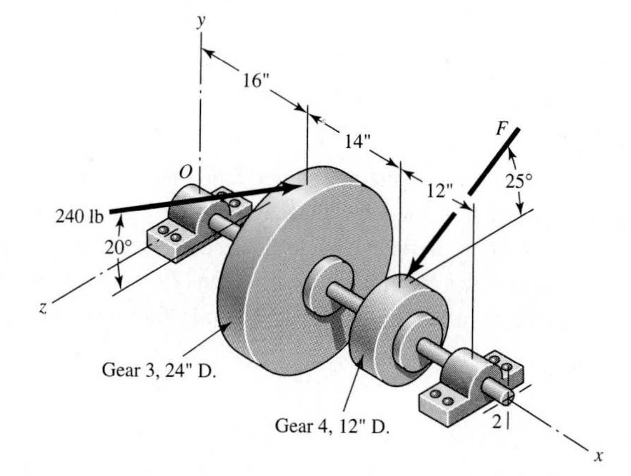

Problem 11–15

DESIGN

11–16 The worm shaft shown in part *a* of the figure transmits 1.35 hp at 600 r/min. A static force analysis gave the results shown in part *b* of the figure. Bearing *A* is to be an angular-contact ball bearing mounted to take the 555-lb thrust load. The bearing at *B* is to take only the radial load, so a straight roller bearing will be employed. Use an application factor of 1.3, a desired life of 25 kh, and a reliability goal, combined, of 0.99. Specify each bearing.

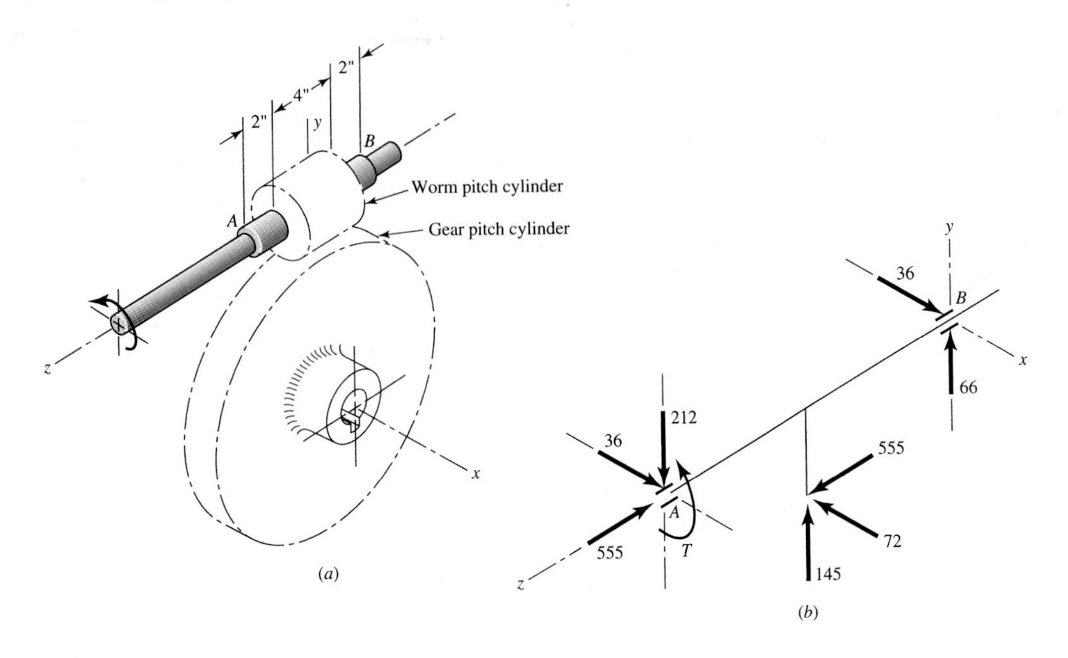

Problem 11–16
(a) Worm and worm gear; (b) force analysis of worm shaft, forces in pounds.

ANALYSIS

11–17 Take the information concerning rating life and Weibull parameters of the two manufacturers shown in the table preceding Prob. 11–1. Find mean, standard deviation, median, and mode of L/L_{10}, then on a plot of probability density with an abscissa of L/L_{10} indicate the mean, median, and mode. The mode can easily check your work.

ANALYSIS

11–18 In bearings tested at 2000 r/min with a steady radial load of 18 kN, a set of bearings showed a B_{10} life of 115 h and a B_{80} life of 600 h. The basic load rating of this bearing is 39.6 kN. Estimate the Weibull shape factor b and the characteristic life θ for a two-parameter model. This manufacturer rates ball bearings at 1 million revolutions.

DESIGN

11–19 A 16-tooth pinion drives the double-reduction spur-gear train in the figure. All gears have 25° pressure angles. The pinion rotates ccw at 1200 r/min and transmits 50 hp to the gear train. The shaft has not yet been designed, but the free bodies have been generated. The shaft speeds are 1200 r/min, 240 r/min, and 80 r/min. A bearing study is commencing with a 10-kh life and a gearbox bearing ensemble reliability of 0.99. An application factor of 1.2 is appropriate. Specify the six bearings.

ANALYSIS

11–20 Different bearing metallurgy affects bearing life. A manufacturer reports that a particular heat treatment increases bearing life at least threefold. A bearing identical to that of Prob. 11–18 except for the heat treatment, loaded to 18 kN and run at 2000 r/min, revealed a B_{10} life of 360 h and a B_{80} life of 2000 h. Do you agree with the manufacturer's assertion concerning increased life?

ANALYSIS

11–21 Estimate the remaining life in revolutions of an 02-30 mm angular-contact ball bearing already subjected to 200 000 revolutions with a radial load of 18 kN, if it is now to be subjected to a change in load to 30 kN.

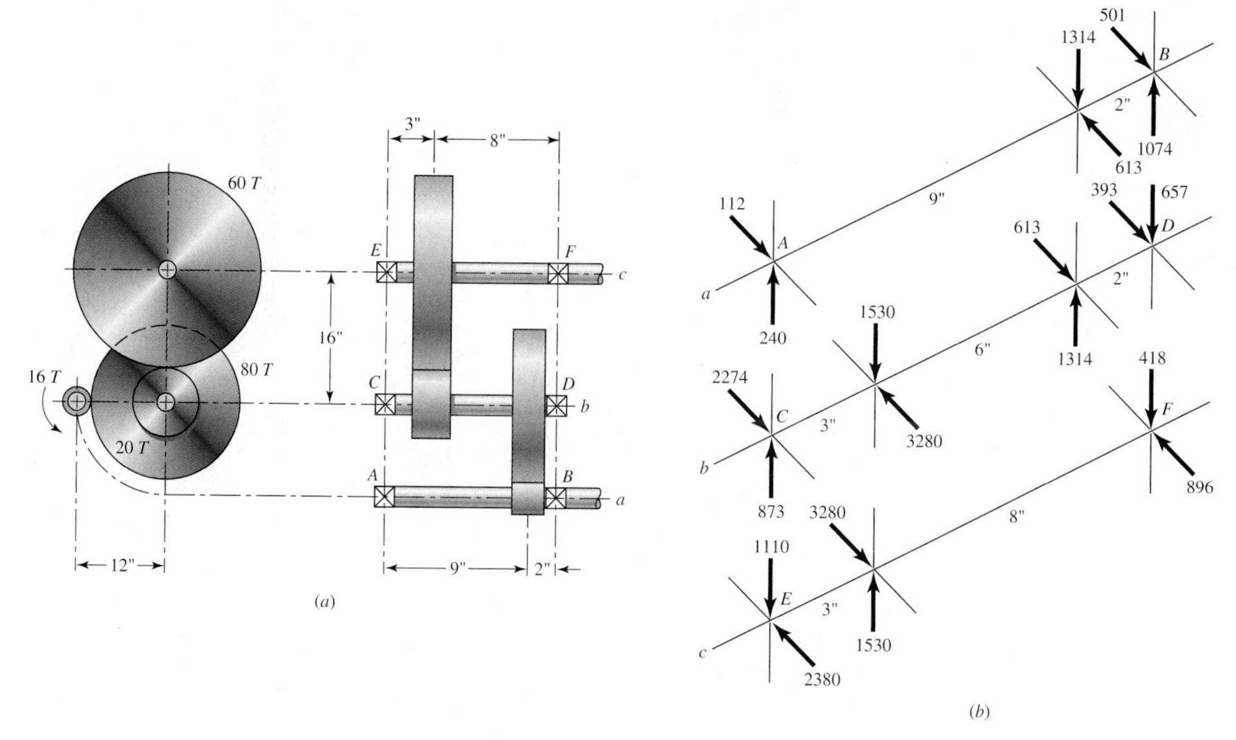

Problem 11–19
(a) Drive detail; (b) force analysis on shafts. Forces in pounds.

11–22 The same 02-30 angular-contact ball bearing as in the preceding problem is to be subjected to a two-step loading cycle of 4 min with a loading of 18 kN, and one of 6 min with a loading of 30 kN. This cycle is to be repeated until failure. Estimate the total life in revolutions, hours, and loading cycles.

11–23 The expression $F^a L$ = constant can be written using $x = L/L_{10}$, and it can be expressed as $F^a x = K$ or $\log F = (1/a)\log K - (1/a)\log x$. This is a straight line on a log-log plot, and it is the basis of Fig. 11–5. For the geometric insight provided, produce Fig. 11–5 to scale using Ex. 11–3, and

For point D: find $F_D = 1.2(413) = 495.6$ lb, $\log F_D$, x_D, $\log x_D$, K_D
For point B: find x_B, $\log x_B$, F_B, $\log F_B$, K_B
For point A: find $F_A = F_B = C_{10}$, $\log F_A$, K_{10}

and plot to scale. On this plot, also show the line containing C'_{10}, the basic load rating, of the *selected* bearing. Note how the reliability of BD locus (which does not move) changes according to Eq. (11–22).

11–24 In light of your experience with ball and cylindrical roller bearing selection, formulate for that class of bearings the (*a*) specification set, (*b*) decision set, and (*c*) adequacy assessment. Your instructor may give you additional guidance pertaining to the execution of this assignment. Having this consolidated in your notes will be useful for future reference.

12

Lubrication and Journal Bearings

12-1	Types of Lubrication **741**
12-2	Viscosity **741**
12-3	Petroff's Equation **744**
12-4	Stable Lubrication **750**
12-5	Thick-Film Lubrication **751**
12-6	Hydrodynamic Theory **752**
12-7	Design Considerations **757**
12-8	The Relations of the Variables **759**
12-9	Steady-State Conditions in Self-Contained Bearings **772**
12-10	Clearance **781**
12-11	Pressure-Fed Bearings **792**
12-12	Loads and Materials **803**
12-13	Bearing Types **805**
12-14	Thrust Bearings **806**
12-15	Boundary-Lubricated Bearings **807**

The wheel and axle, a cornerstone of our civilization, represents a conceptual breakthrough of the first magnitude. Inherent in the wheel-and-axle idea is a journal–bushing pair which kinematicians call a *revolute* pair. Its purposes are (1) to allow relative rotary motion and (2) to control (constrain) the nature of that motion. Mankind has been on a continuing search for lubricants—substances to help make the necessary relative sliding motion freer (ameliorate friction).

When faced with a new phenomenon, people have learned to use the method c. 1914 of Edgar Buckingham to reduce the dimensionality of the problem and the intensity (cost) of the experimental effort necessary to achieve a quantitative description, even when the cause–effect–extent mechanism is not known. This involves dimensionless parameter clusters called *pi terms*. When cause–effect–extent mechanisms, such as Newton's viscous effect are quantitatively understood, this knowledge can be combined with first-principle description of the physical world to permit quantitative understanding of the new phenomenon.

This textbook draws and then builds on much prerequisite material. A course in fluid mechanics rarely has the time to include bearing theory. Some of this is included here in order to let the reader observe how a body of knowledge evolved and how it is presented.

The object of lubrication is to reduce friction, wear, and heating of machine parts which move relative to each other. A lubricant is any substance which, when inserted between the moving surfaces, accomplishes these purposes. In a sleeve bearing, a shaft, or *journal,* rotates or oscillates within a sleeve, or *bushing,* and the relative motion is sliding. In an antifriction bearing, the main relative motion is rolling. A follower may either roll or slide on the cam. Gear teeth mate with each other by a combination of rolling and sliding. Pistons slide within their cylinders. All these applications require lubrication to reduce friction, wear, and heating.

The field of application for journal bearings is immense. The crankshaft and connecting-rod bearings of an automotive engine must operate for thousands of miles at high temperatures and under varying load conditions. The journal bearings used in the steam turbines of power-generating stations are said to have reliabilities approaching 100 percent. At the other extreme there are thousands of applications in which the loads are light and the service relatively unimportant; a simple, easily installed bearing is required, using little or no lubrication. In such cases an antifriction bearing might be a poor answer because of the cost, the elaborate enclosures, the close tolerances, the radial space required, the high speeds, or the increased inertial effects. Instead, a nylon bearing requiring no lubrication, a powder-metallurgy bearing with the lubrication "built in," or a bronze bearing with ring oiling, wick-feeding, or solid-lubricant film or grease lubrication might be a very satisfactory solution. Recent metallurgy developments in bearing materials, combined with increased knowledge of the lubrication process, now make it possible to design journal bearings with satisfactory lives and very good reliabilities.

Much of the material we have studied thus far in this book has been based on fundamental engineering studies, such as statics, dynamics, the mechanics of solids, metal processing, mathematics, and metallurgy. In the study of lubrication and journal bearings, additional fundamental studies, such as chemistry, fluid mechanics, thermodynamics, and heat transfer, must be utilized in developing the material. While we shall not utilize all of them in the material to be included here, you can now begin to appreciate better how the study of mechanical engineering design is really an integration of most of your previous studies and a directing of this total background toward the resolution of a single objective.

12–1 Types of Lubrication

Five distinct forms of lubrication may be identified:

1 Hydrodynamic
2 Hydrostatic
3 Elastohydrodynamic
4 Boundary
5 Solid-film

Hydrodynamic lubrication means that the load-carrying surfaces of the bearing are separated by a relatively thick film of lubricant, so as to prevent metal-to-metal contact, and that the stability thus obtained can be explained by the laws of fluid mechanics. Hydrodynamic lubrication does not depend upon the introduction of the lubricant under pressure, though that may occur; but it does require the existence of an adequate supply at all times. The film pressure is created by the moving surface itself pulling the lubricant into a wedge-shaped zone at a velocity sufficiently high to create the pressure necessary to separate the surfaces against the load on the bearing. Hydrodynamic lubrication is also called *full-film,* or *fluid, lubrication.*

Hydrostatic lubrication is obtained by introducing the lubricant, which is sometimes air or water, into the load-bearing area at a pressure high enough to separate the surfaces with a relatively thick film of lubricant. So, unlike hydrodynamic lubrication, this kind of lubrication does not require motion of one surface relative to another. We shall not deal with hydrostatic lubrication in this book, but the subject should be considered in designing bearings where the velocities are small or zero and where the frictional resistance is to be an absolute minimum.

Elastohydrodynamic lubrication is the phenomenon that occurs when a lubricant is introduced between surfaces which are in rolling contact, such as mating gears or rolling bearings. The mathematical explanation requires the Hertzian theory of contact stress and fluid mechanics.

Insufficient surface area, a drop in the velocity of the moving surface, a lessening in the quantity of lubricant delivered to a bearing, an increase in the bearing load, or an increase in lubricant temperature resulting in a decrease in viscosity—any one of these—may prevent the buildup of a film thick enough for full-film lubrication. When this happens, the highest asperities may be separated by lubricant films only several molecular dimensions in thickness. This is called *boundary lubrication.* The change from hydrodynamic to boundary lubrication is not at all a sudden or abrupt one. It is probable that a mixed hydrodynamic- and boundary-type lubrication occurs first, and as the surfaces move closer together, the boundary-type lubrication becomes predominant. The viscosity of the lubricant is not of as much importance with boundary lubrication as is the chemical composition.

When bearings must be operated at extreme temperatures, a *solid-film lubricant* such as graphite or molybdenum disulfide must be used because the ordinary mineral oils are not satisfactory. Much research is currently being carried out in an effort, too, to find composite bearing materials with low wear rates as well as small frictional coefficients.

12–2 Viscosity

In Fig. 12–1 let a plate A be moving with a velocity U on a film of lubricant of thickness h. We imagine the film as composed of a series of horizontal layers and the force F causing these layers to deform or slide on one another just like a deck of cards. The layers in contact with the moving plate are assumed to have a velocity U; those in

Figure 12–1

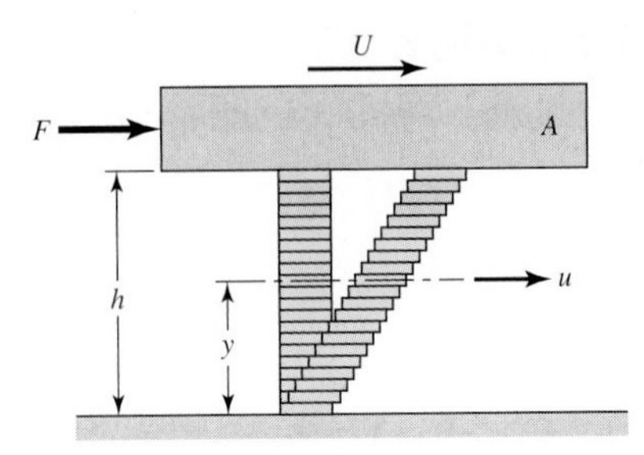

contact with the stationary surface are assumed to have a zero velocity. Intermediate layers have velocities which depend upon their distances y from the stationary surface. Newton's viscous effect states that the shear stress in the fluid is proportional to the rate of change of velocity with respect to y. Thus

$$\tau = \frac{F}{A} = \mu \frac{du}{dy} \tag{12–1}$$

where μ is the constant of proportionality and defines *absolute viscosity,* also called *dynamic viscosity*. The derivative du/dy is the rate of change of velocity with distance and may be called the rate of shear, or the velocity gradient. The viscosity μ is thus a measure of the internal frictional resistance of the fluid. If the assumption is made that the rate of shear is a constant, then $du/dy = U/h$, and from Eq. (12–1),

$$\tau = \frac{F}{A} = \mu \frac{U}{h} \tag{12–2}$$

The unit of viscosity in the ips system is seen to be the pound-force-second per square inch; this is the same as stress or pressure multiplied by time. The ips unit is called the *reyn,* in honor of Sir Osborne Reynolds.

The absolute viscosity is measured by the pascal-second (Pa · s) in SI; this is the same as a Newton-second per square meter. The conversion from ips units to SI is the same as for stress. For example, multiply the absolute viscosity in reyns by 6890 to convert to units of Pa · s.

The American Society of Mechanical Engineers (ASME) has published a list of cgs units which are not to be used in ASME documents.[1] This list results from a recommendation by the International Committee of Weights and Measures (CIPM) that the use of cgs units with special names be discouraged. Included in this list is a unit of force called the *dyne* (dyn), a unit of dynamic viscosity called the *poise* (P), and a unit of kinematic viscosity called the *stoke* (St). All of these units have been, and still are, used extensively in lubrication studies.

The poise is the cgs unit of dynamic or absolute viscosity, and its unit is the dyne-second per square centimeter (dyn · s/cm^2). It has been customary to use the centipoise (cP) in analysis, because its value is more convenient. When the viscosity is expressed in centipoises, it is designated by Z. The conversion from cgs units to SI and ips units is as follows:

$$\mu(\text{Pa} \cdot \text{s}) = (10)^{-3} Z \text{ (cP)}$$

$$\mu(\text{reyn}) = \frac{Z \text{ (cP)}}{6.89(10)^6}$$

$$\mu(\text{mPa} \cdot \text{s}) = 689\, \mu' \text{ } (\mu\text{reyn})$$

1. *ASME Orientation and Guide for Use of Metric Units,* 2d ed., American Society of Mechanical Engineers, 1972, p. 13.

The ASTM standard method for determining viscosity uses an instrument called the Saybolt Universal Viscosimeter. The method consists of measuring the time in seconds for 60 ml of lubricant at a specified temperature to run through a tube 17.6 mm in diameter and 12.25 mm long. The result is called the *kinematic viscosity,* and in the past the unit of the square centimeter per second has been used. One square centimeter per second is defined as a *stoke*. By the use of the *Hagen-Poiseuille law,* the kinematic viscosity based upon seconds Saybolt, also called *Saybolt Universal viscosity* (SUV) in seconds, is

$$Z_k = \left(0.22t - \frac{180}{t}\right) \tag{12–3}$$

where Z_k is in centistokes (cSt) and t is the number of seconds Saybolt.

In SI, the kinematic viscosity ν has the unit of the square meter per second ($\mathrm{m^2/s}$), and the conversion is

$$\nu(\mathrm{m^2/s}) = 10^{-6} Z_k\ (\mathrm{cSt})$$

Thus, Eq. (12–3) becomes

$$\nu = \left(0.22t - \frac{180}{t}\right)(10^{-6}) \tag{12–4}$$

To convert to dynamic viscosity, we multiply ν by the density in SI units. Designating the density as ρ with the unit of the kilogram per cubic meter, we have

$$\mu = \rho\left(0.22t - \frac{180}{t}\right)(10^{-6}) \tag{12–5}$$

where μ is in pascal-seconds.

Figure 12–2 shows the absolute viscosity in the ips system of a number of fluids often used for lubrication purposes and their variation with temperature.

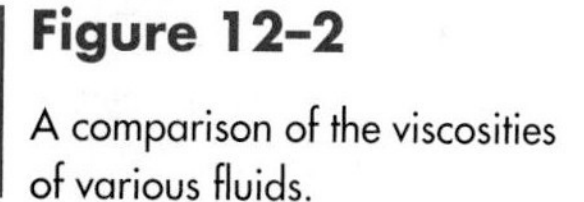

Figure 12–2

A comparison of the viscosities of various fluids.

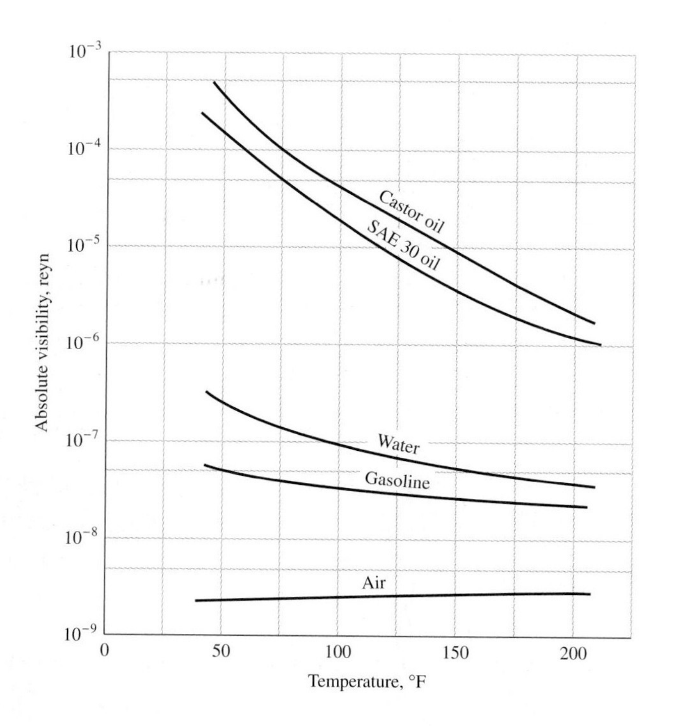

12–3 Petroff's Equation

The phenomenon of bearing friction was first explained by Petroff using the assumption that the shaft is concentric. Though we shall seldom make use of Petroff's method of analysis in the material to follow, it is important because it defines groups of dimensionless parameters and because the coefficient of friction predicted by this law turns out to be quite good even when the shaft is not concentric.

Let us now consider a vertical shaft rotating in a guide bearing. It is assumed that the bearing carries a very small load, that the clearance space c is completely filled with oil, and that leakage is negligible (Fig. 12–3). We denote the radius of the shaft by r, the radial clearance by c, and the length of the bearing by l, all dimensions being in inches. If the shaft rotates at N r/s, then its surface velocity is $U = 2\pi r N$ in/s. Since the shearing stress in the lubricant is equal to the velocity gradient times the viscosity, from Eq. (12–2) we have

$$\tau = \mu \frac{U}{h} = \frac{2\pi r \mu N}{c} \tag{a}$$

where the radial clearance c has been substituted for the distance h. The force required to shear the film is the stress times the area. The torque is the force times the lever arm r. Thus

$$T = (\tau A)(r) = \left(\frac{2\pi r \mu N}{c}\right)(2\pi r l)(r) = \frac{4\pi^2 r^3 l \mu N}{c} \tag{b}$$

If we now designate a small force on the bearing by W, in pounds-force, then the pressure P, in pounds-force per square inch of projected area, is $P = W/2rl$. The frictional force is fW, where f is the coefficient of friction, and so the frictional torque is

$$T = fWr = (f)(2rlP)(r) = 2r^2 flP \tag{c}$$

Substituting the value of the torque from Eq. (c) in Eq. (b) and solving for the coefficient of friction, we find

$$f = 2\pi^2 \frac{\mu N}{P} \frac{r}{c} \tag{12–6}$$

Equation (12–6) is called *Petroff's equation* and was first published in 1883. The two quantities $\mu N/P$ and r/c are very important parameters in lubrication. Substitution of the appropriate dimensions in each parameter will show that they are dimensionless.

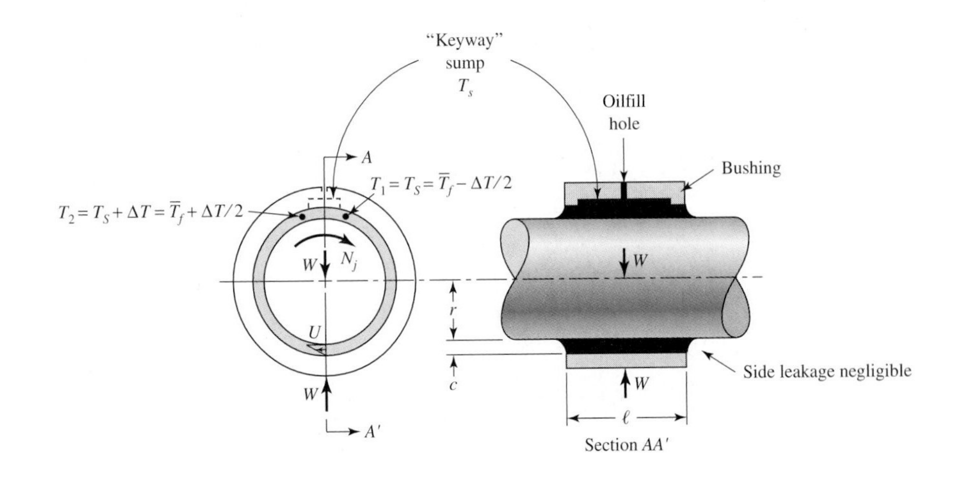

Figure 12–3

Petroff's lightly loaded journal bearing consisting of a shaft journal and a bushing with an axial-groove internal lubricant reservoir. The linear velocity gradient is shown in the end view. The clearance c is several thousandths of an inch and is grossly exaggerated for presentation purposes.

The *bearing characteristic number,* or the *Sommerfeld number,* is defined by the equation

$$S = \left(\frac{r}{c}\right)^2 \frac{\mu N}{P} \tag{12–7}$$

where S = bearing characteristic number
r = journal radius, in
c = radial clearance, in
μ = absolute viscosity, reyn
N = significant speed, r/s
P = load per unit of projected bearing area, psi

The Sommerfeld number is very important in lubrication analysis because it contains many of the parameters which specified by the designer. Note also that it is dimensionless. The quantity r/c is called the *radial clearance ratio*. If we multiply both sides of Eq. (12–6) by this ratio, we obtain the interesting relation

$$f\frac{r}{c} = 2\pi^2 \frac{\mu N}{P}\left(\frac{r}{c}\right)^2 = 2\pi^2 S \tag{12–8}$$

The dimensionless friction variable fr/c is a function of the Sommerfeld number. Rarely is the journal angular speed N in Eq. (12–8) subscripted. It should be. We will use N_j from now on. See case study CS12a in this chapter which details the reason why. Now we examine Petroff's equation.

First, there are three dimensionless groups present:

$$f\frac{r}{c} \qquad \frac{\mu N_j}{P} \qquad \frac{r}{c}$$

As knowledge about journal bearings increased it became clear that the product $(r/c)^2(\mu N_j/P)$ was an independent variable of which many other parameters of interest were a function. The designer chooses the geometry r and c which affects P through $W/(2rl)$. The rotational speed N_j is usually a given, and the choice of a lubricant affects the viscosity μ. The dimensionless friction variable fr/c is a function of $(r/c)^2(\mu N_j/P)$ as shown in Eq. (12–8). Petroff's equation applies only to lightly loaded journal bearings. Additionally it contains the key independent variable later known as the Sommerfeld number.

Petroff's equation deserves a closer look. It consists of a shaft *journal,* the *bushing,* a lubricant *sump,* and a frame, which is sometimes called the *housing*. The ensemble is called a *bearing*. In Fig. 12–3 the end view shows the lubricant sump to be an integral feature of the bushing, shaped much as the top half of a keyway slot. Essential to the working of a natural-circulation bearing is the fact that the lubricant "wets" the journal and bushing surfaces, and because of the relative motion and the adherence of the lubricant, a velocity gradient is established as depicted in Fig. 12–1 and in section AA' of Fig. 12–3. Petroff assumed a concentricity of the journal axis with the bushing axis, making the radial clearance c the same all around the annulus. This means that the lubricant velocity gradient is linear, and it is identical at every section. The shear stress and forces are predictable.

Because of the shearing action there is a uniformly distributed energy release in the lubricant which heats the lubricant as it works its way around. The temperature is uniform in the radial direction but increases from the sump temperature T_s by an amount ΔT during the lubricant pass. The exiting lubricant mixes with the sump contents, being cooled to sump temperature. The lubricant in the sump is cooled because the bushing and housing metal is at a nearly uniform lower temperature due to heat losses by convection and radiation to the surroundings at ambient temperature T_∞. In the usual configurations of such bearings, the bushing and housing metal temperature is approximately midway between the average film temperature $\bar{T}_f = T_s + \Delta T/2$ and the ambient temperature T_∞. See Sec. 12–12. The mass flow rate $\dot{m}$ past plane aa' can be expressed as the annular cross-sectional lc area times the mass density of the lubricant ρ times the average velocity $U/2$, or

$$\dot{m} = lc\rho U/2 = lc\rho\pi r N_j \tag{d}$$

where U is the journal peripheral speed and N_j is the journal angular speed in r/s. The heat loss rate $\dot{Q}_{\text{loss}}$ by convection and radiation to the surroundings from the bearing housing lateral surface area A_0, using the overall combined coefficient U_0, and bearing metal surface temperature T_b is given by

$$\dot{Q}_{\text{loss}} = U_0 A_0 (T_b - T_\infty) = \frac{U_0 A_0(\bar{T}_f - T_\infty)}{2} \tag{e}$$

The heat generation rate $\dot{Q}_{\text{gen}}$ by viscous friction is warming the lubricant during its pass:

$$\dot{Q}_{\text{gen}} = \dot{m} C_p \Delta T = lc\rho\pi r N_j C_p \Delta T \tag{f}$$

The heat generation rate $\dot{Q}_{\text{gen}}$ is also equal (at steady state) to the work rate of the torque T:

$$\dot{Q}_{\text{gen}} = \frac{2545 T N_j}{1050} = \frac{2545 f W r N_j}{1050} = \frac{2545}{1050}\frac{fr}{c}\frac{c}{r} r W N_j$$

$$= 2\pi^2 \frac{2545}{1050}\left(\frac{r}{c}\right)^2 \frac{\mu N_j (2rl)}{W}\frac{c}{r} r W N_j \tag{g}$$

If Eqs. (e) and (g) are set equal and solved for T_f, one obtains

$$T_f = T_\infty + \frac{8\pi^2 2545}{1050}\frac{\mu N_j^2 l r^3}{U_0 A_0 c} \tag{h}$$

The temperature rise ΔT during a lubricant pass can be obtained by equating Eqs. (e) and (f):

$$\Delta T = \frac{\dot{Q}_{\text{gen}}}{\dot{m} C_p} = \frac{U_0 A_0 (\bar{T}_f - T_\infty)/2}{3600 lc\rho\pi r N_j C_p} \tag{i}$$

The conversion constants 2545 Btu/hp-h, 1050 (in · lbf)(r/s)/hp, and 3600 s/h come from ips unit use.

EXAMPLE 12–1 Consider a natural-feed pillow-block bearing as is depicted in Fig. 12–3 operating at 900 r/min in a 70°F environment. The lateral area of the bearing housing is 40 in^2. The lubricant is SAE grade 20. The l/D ratio is 1, U_0 is 2.7 Btu/(h · ft^2 · °F). The journal radius is 1 in and the radial clearance is 0.001 in. For a Petroff bearing model:

(a) Express torque T, $\dot{Q}_{\text{gen}}$, $\dot{Q}_{\text{loss}}$, $\bar{T}_f$, and ΔT in terms of viscosity.

(b) Find $\bar{T}_f$, ΔT, T_s, $T_{\max}$, and T_b for a constant viscosity of 0.98 μreyn.

(c) Estimate $\bar{T}_f$, ΔT, T_s, T_{max}, and T_b for temperature-variable viscosity for SAE grade 20 using Fig. 12–11. The radial load $W = 10$ lb, specific weight of the lubricant is 0.0311 lbf/in^3, and the specific heat capacity $C_p = 0.42$ Btu/(lbf · °F).

Solution (*a*) From Eq. (*b*),

Answer
$$T = \frac{4\pi^2(1)^3(2)}{0.001}\frac{\mu'(15)}{10^6} = 1.184\mu' \text{ in} \cdot \text{lbf (Note: } \mu' \text{ in } \mu\text{reyn)}$$

From Eq. (*g*),

Answer
$$\dot{Q}_{gen} = 2\pi^2\frac{2545}{1050}\left(\frac{1}{0.001}\right)^2\frac{\mu(15)(2)(1)(2)}{10^6(10)}\frac{0.001}{1}(1)10(15) = 43.06\mu' \text{ Btu/h}$$

From Eq. (*e*),

Answer
$$\dot{Q}_{loss} = \frac{2.7(40)}{2(144)}(\bar{T}_f - 70) = 0.375(T_f - 70) \text{ Btu/h}$$

From Eq. (*h*),

Answer
$$\bar{T}_f = 70 + \frac{8\pi^2 2545}{1050}\frac{\mu'}{10^6}\frac{15^2(2)(1)^3}{2.7(40/144)(0.001)} = 70 + 114.8\mu'°\text{F}$$

From Eq. (*i*),

Answer
$$\Delta T = \frac{2.7(40/144)(\bar{T}_f - 70)/2}{3600(2)0.001(0.0311)\pi(1)(15)(0.42)} = 0.0846(\bar{T}_f - 70)°\text{F}$$

The lubricant leaves the sump at temperature T_s and makes one circuit of the annulus, exiting into the sump at temperature $T_s + \Delta T$, to be mixed and cooled back to sump temperature with heat transfer to the metal surrounding the sump at temperature T_b.

(*b*) Given $\mu' = 0.98$ μreyn, then from part (*a*),

$$T = 1.184(0.98) = 1.16 \text{ in} \cdot \text{lbf}$$

$$\dot{Q}_{gen} = 43.06(0.98) = 42.2 \text{ Btu/h}$$

Answer $\bar{T}_f = 70 + 114.8(0.98) = 182.5°\text{F}$

$\dot{Q}_{loss} = 0.375(182.5 - 70) = 42.2$ Btu/h

Answer $\Delta T = (182.5 - 70)0.0846 = 9.52°\text{F}$

Answer $T_s = \bar{T}_f - \Delta T/2 = 182.5 - 9.52/2 = 177.7°\text{F}$

Answer $T_{max} = \bar{T}_f + \Delta T/2 = 182.5 + 9.52/2 = 187.3°\text{F}$

Answer $T_b = (\bar{T}_f + 70)/2 = (182.5 + 70)/2 = 126.3°\text{F}$

(*c*) Since the viscosity varies with temperature, that function is given in Fig. 12–11. We will check for steady state by seeing if $\dot{Q}_{gen} = \dot{Q}_{loss}$ or, alternatively, if trial $\bar{T}_f$ equals the average temperature in the film. The plan is to form a table whose columns are

- Trial $\bar{T}_f$
- Estimate μ' from Fig. 12–11 in μreyns
- Estimate $\dot{Q}_{gen}$
- Estimate T_{av}

When the first and last column agree within one degree we can use the equations of part (a). The viscosity μ' is read from Fig. 12–11.

Trial $\bar{T}_f$	μ'	$\dot{Q}_{gen}$	T_{av}
200	0.98	42.2	182.5
191	1.15	49.5	202
196	1.05	45.2	190.5
194	1.08	46.5	194

The first row of the table is what was obtained in part (b). The equality of 194 to $\bar{T}_f$ and T_{av} shows we have $\bar{T}_f$ and μ'. Calculations of the sort of part (b) give $T = 128$ in · lbf, $\dot{Q}_{gen} = 46.5$ Btu/h,

Answer $\bar{T}_f = 194$°F, $\Delta T = 10.5$°F, $T_s = 189$°F, $T_{max} = 199$°F, and $T_b = 132$°F. From Eq. (12–6) with $P = 10/4 = 2.5$ psi,

$$f = 2\pi^2 \frac{1.08}{10^6} \frac{15}{2.5} \frac{1}{0.001} = 0.128$$

and the predicted torque is $T = fWr = 0.128(10)(1) = 1.28$ in · lbf. Equation (12–8) predicts that the slope of the curves in Fig. 12–17 is $2\pi^2$ when the Sommerfeld number is large (light loading). Can you verify this? This example, although long, gives you some perspective on the nature of hydrodynamic film bearings.

It has become traditional to denote the product $(r/c)^2(\mu N/P)$ as the Sommerfeld number to honor an early contributor to hydrodynamic film theory. A preoccupation with journal bearings where the angular speed of the bushing and the load vector are zero has led to the definition

$$S = \left(\frac{r}{c}\right)^2 \left(\frac{\mu N_j}{P}\right) \tag{j}$$

It has been discovered that the angular speed N which is significant to hydrodynamic film bearing performance[2] is

$$N = N_j + N_b - 2N_W \tag{k}$$

where $N_j =$ journal angular speed, r/s

$N_b =$ bushing angular speed, r/s

$N_W =$ load vector angular speed, r/s

The common practice is to present bearing performance parameters in dimensionless form with the traditional Sommerfeld number as the abscissa. When encountering a bearing of a more general type enter the chart with

$$\mathcal{S} = \left(\frac{r}{c}\right)^2 \left[\frac{\mu(N_j + N_b - 2N_W)}{P}\right] = S\left(1 + \frac{N_b}{N_j} - 2\frac{N_W}{N_j}\right) \tag{l}$$

2. Paul Robert Trumpler, *Design of Film Bearings*, Macmillan, New York, 1966, pp. 103–119.

EXAMPLE 12–2

The traditional Sommerfeld number is 2 for a bearing in which the load rotation is 100 r/min, the journal rotation is 600 r/min, and the bushing rotation is 100 r/min, all counterclockwise. At which abscissa should one enter Fig. 12–17, for example?

Solution

Because of the dimensionless form of the final term of Eq. (*l*) units need not be corrected to revolutions per second. Then

$$\mathcal{S} = S\left(1 + \frac{N_b}{N_j} - 2\frac{N_W}{N_j}\right) = 2\left(1 + \frac{100}{600} - 2\frac{100}{600}\right) = 1.67$$

CASE STUDY 12a

The General Entry Parameter for Raimondi and Boyd Charts

Many insights followed from the dimensionless groupings which have come to be known as the Sommerfeld number,

$$S = \left(\frac{r}{c}\right)^2 \frac{\mu LDN}{W}$$

where N was presumed to be the angular speed of the journal. Since journals were resisting gravity loads with fixed-in-earth bushings, awareness of the necessity to comment on the angular speed of the bushing and load vector was not widespread. We have come to understand that the proper form of N in the Sommerfeld number should be

$$N = N_j + N_b - 2Nf$$

where N_j, N_b, and N_f are the angular speed of the journal, bushing, and load vector, respectively. A vector algebra derivation due to Dr. Kwan Yu Chen is found in Trumpler.[3] Clearly the journal speed N_j does not determine the bearing performance by itself, even when N_b equals zero. A more comprehensive definition of the Sommerfeld number S is

$$\mathcal{S} = \left(\frac{r}{c}\right)^2 \frac{\mu LDN}{W} = \left(\frac{r}{c}\right)^2 \frac{\mu LDN_j}{W}\left|\left(1 + \frac{N_b}{N_j} - 2\frac{N_f}{N_j}\right)\right|$$

$$\mathcal{S} = S\left|1 + \frac{N_b}{N_j} - 2\frac{N_f}{N_j}\right|$$

It is with this presentation of the Sommerfeld number that you should enter the Raimondi and Boyd charts. It can save you grief.

If $N_f = 0$ or $N_f = N_j$, $\mathcal{S} = S$. The attitude n is the same but the location of h_0 is 90° ahead of the load, or 90° behind it. The absolute value signs signal this condition. If $N_b = 0$ and $N_f = N_j/2$, then $\mathcal{S} = 0$ and a load cannot be carried on a fluid film. If $N_f = 0$ and $N_b = N_j$, $\mathcal{S} = 2S$ and the load-carrying capability of the bearing is half of what you expect from S. Figure CS12a–1 shows several situations. Visualize them physically, and ponder how you might explain them to someone who had not encountered a bearing other than a gravity-loaded, stationary-bushing bearing. Stone and Underwood[4] have shown by experiment that journal speed is not the speed descriptor.

3. Paul Robert Trumpler, *Design of the Film Bearings,* Macmillan, New York, 1966, pp. 105–119.

4. J. M. Stone and F. A. Underwood, "Load Carrying Capacity of Journal Bearings," *SAE Q. Trans.,* vol. 1, 1947, p. 56.

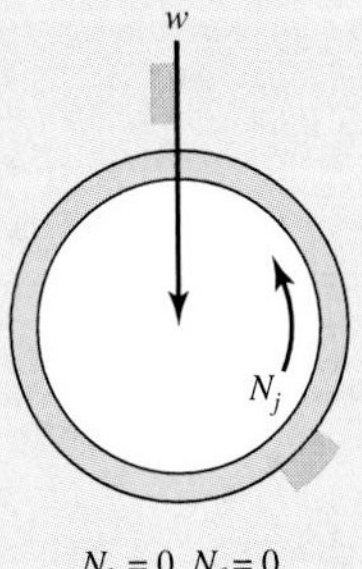

$N_b = 0, N_f = 0$

$\mathcal{S} = |S\,(1 + 0 - 2(0))| = S$

(a)

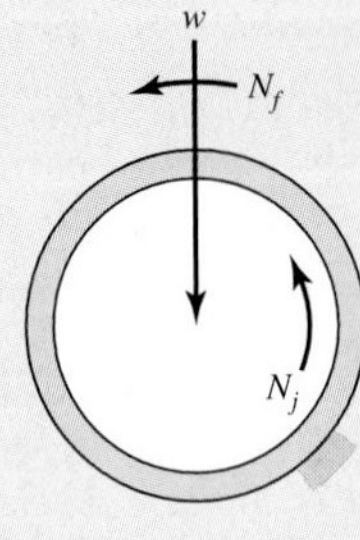

$N_b = 0, N_f = N_j$

$\mathcal{S} = |S\,(1 + 0 - 2(1))| = S$

(b)

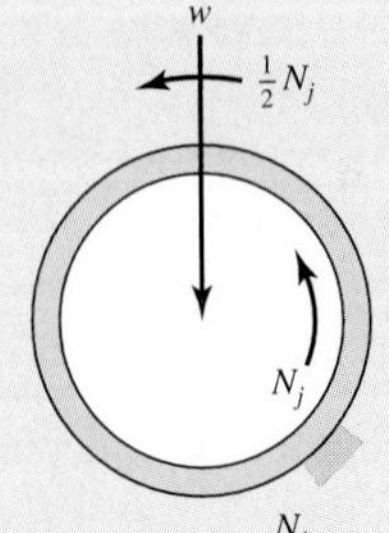

$N_b = 0, N_f = \dfrac{N_j}{2}$

$\mathcal{S} = |S\,(1 + 0 - 2(\tfrac{1}{2}))| = 0$

(c)

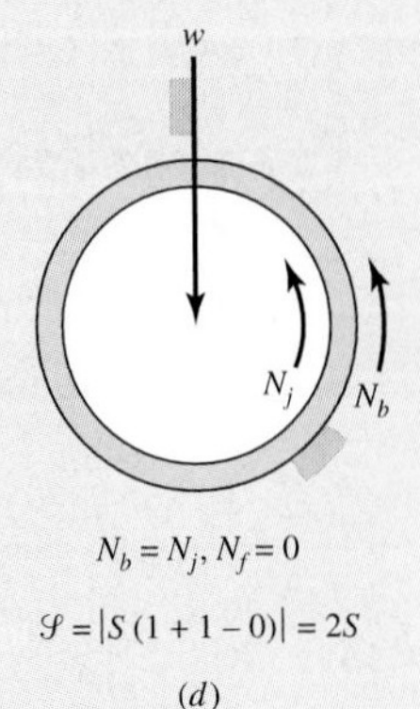

$N_b = N_j, N_f = 0$

$\mathcal{S} = |S\,(1 + 1 - 0)| = 2S$

(d)

Figure CS12a–1

How the Sommerfeld number varies. (*a*) Common bearing case. (*b*) Load vector moves at the same speed as the journal. (*c*) Load vector moves at half journal speed, no load can be carried. (*d*) Journal and bushing move at same speed, load vector stationary, capacity halved.

In retrospect, we were premature in defining the Sommerfeld number in its usual form. There is now confusion, so bear in mind:

- To save yourself some wrong answers, begin a Sommerfeld number calculation with "$\mathcal{S} = S|(1 + N_b/N_j - 2N_f/N_j)|$," where $S = (r/c)^2 \mu L D N_j / W$. This will remind you to be careful.
- Use $\mathcal{S}$ and S to enter Raimondi and Boyd charts.
- To continue to honor Sommerfeld, as is richly deserved, call $\mathcal{S}$ your *chart variable*.

12–4 Stable Lubrication

Petroff's bearing model in the form of Eq. (12–6) predicts

$$f = 2\pi^2 \left(\frac{\mu N_j}{P}\right)\left(\frac{r}{c}\right)$$

for a given r and c, and that f is proportional to $\mu N_j / P$, that is, a straight line from the origin in the first quadrant. On the coordinates of Fig. 12–4 the locus to the right of point C is an example. Petroff's model presumes thick-film lubrication, that is, no metal-to-metal contact, the surfaces being completely separated by a lubricant film. Conditions under which the thick film would fail were reported in 1932.

The difference between boundary and hydrodynamic lubrication can be explained by reference to Fig. 12–4. This plot of the change in the coefficient of friction versus the bearing characteristic $\mu N/P$ was obtained by the McKee brothers in an actual test of friction.[5] The plot is important because it defines stability of lubrication and helps us to understand hydrodynamic and boundary, or thin-film, lubrication.

The McKee abscissa was ZN/P (centipoise × r/min/psi) and abscissa B in Fig. 12–4 was 30. The corresponding $\mu N_j / P$ (reyn × r/s/psi) is $0.33\,(10^{-6})$. Designers keep

5. S. A. McKee and T. R. McKee, "Journal Bearing Friction in the Region of Thin Film Lubrication," *SAE J.*, vol. 31, 1932, pp. (T)371–377.

Figure 12–4

The variation of the coefficient of friction f with $\mu N/P$.

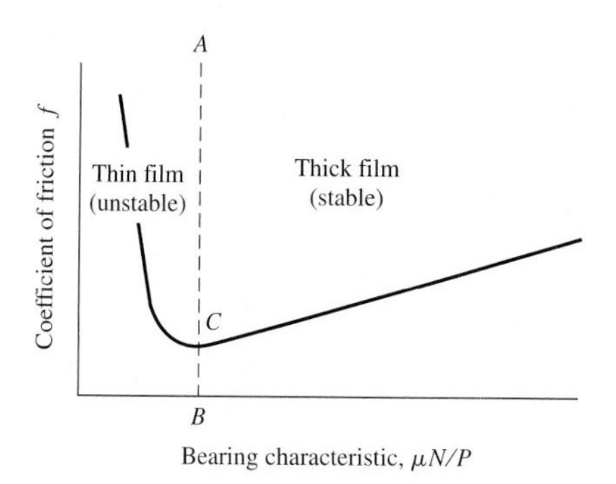

$\mu N_j/P \geq 17(10^{-6})$, which corresponds to $ZN/P \geq 150$. A design constraint to keep thick film lubrication is to be sure that

$$\frac{\mu N_j}{P} \geq 17(10^{-6}) \tag{a}$$

Suppose we are operating to the right of line BA and something happens, say, an increase in lubricant temperature. This results in a lower viscosity and hence a smaller value of $\mu N/P$. The coefficient of friction decreases, not as much heat is generated in shearing the lubricant, and consequently the lubricant temperature drops. Thus the region to the right of line BA defines *stable lubrication* because variations are self-correcting.

To the left of line BA, a decrease in viscosity would increase the friction. A temperature rise would ensue, and the viscosity would be reduced still more. The result would be compounded. Thus the region to the left of line BA represents *unstable lubrication*.

It is also helpful to see that a small viscosity, and hence a small $\mu N/P$, means that the lubricant film is very thin and that there will be a greater possibility of some metal-to-metal contact, and hence of more friction. Thus, point C represents what is probably the beginning of metal-to-metal contact as $\mu N/P$ becomes smaller.

12–5 Thick-Film Lubrication

Let us now examine the formation of a lubricant film in a journal bearing. Figure 12–5*a* shows a journal which is just beginning to rotate in a clockwise direction. Under starting conditions, the bearing will be dry, or at least partly dry, and hence the journal will climb or roll up the right side of the bearing as shown in Fig. 12–5*a*. Under the conditions of a dry bearing, equilibrium will be obtained when the friction force is balanced by the tangential component of the bearing load.

Now suppose a lubricant is introduced into the top of the bearing as shown in Fig. 12–5*b*. The action of the rotating journal is to pump the lubricant around the bearing in a

Figure 12–5

Formation of a film.

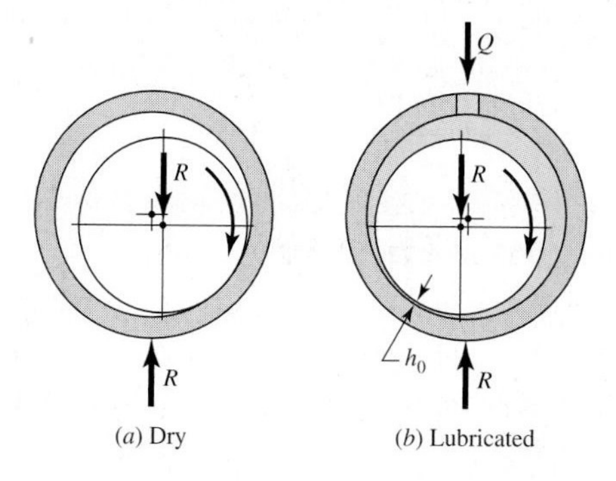

Figure 12–6

Nomenclature of a journal bearing.

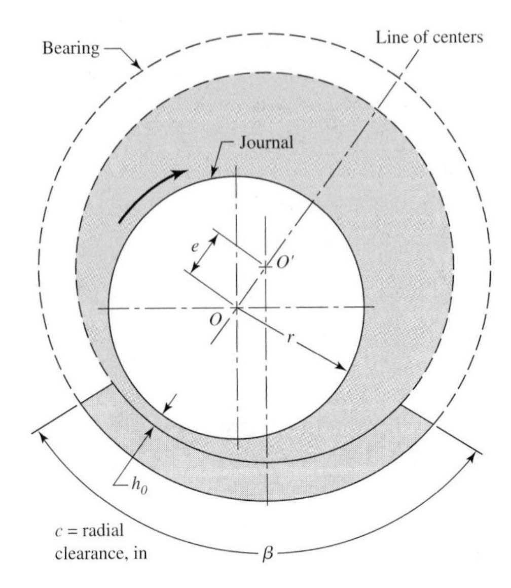

clockwise direction. The lubricant is pumped into a wedge-shaped space and forces the journal over to the other side. A *minimum film thickness* h_0 occurs, not at the bottom of the journal, but displaced clockwise from the bottom as in Fig. 12–5*b*. This is explained by the fact that a film pressure in the converging half of the film reaches a maximum somewhere to the left of the bearing center.

Figure 12–5 shows how to decide whether the journal, under hydrodynamic lubrication, is eccentrically located on the right or on the left side of the bearing. Visualize the journal beginning to rotate. Find the side of the bearing upon which the journal tends to roll. Then, if the lubrication is hydrodynamic, mentally place the journal on the opposite side.

The nomenclature of a journal bearing is shown in Fig. 12–6. The dimension c is the *radial clearance* and is the difference in the radii of the bushing and journal. In Fig. 12–6 the center of the journal is at O and the center of the bearing at O'. The distance between these centers is the *eccentricity* and is denoted by e. The *minimum film thickness* is designated by h_0, and it occurs at the line of centers. The film thickness at any other point is designated by h. We also define an *eccentricity ratio* ϵ as

$$\epsilon = \frac{e}{c}$$

The bearing shown in the figure is known as a *partial bearing*. If the radius of the bushing is the same as the radius of the journal, it is known as a *fitted bearing*. If the bushing encloses the journal, as indicated by the dashed lines, it becomes a *full bearing*. The angle β describes the angular length of a partial bearing. For example, a 120° partial bearing has the angle β equal to 120°.

12–6 Hydrodynamic Theory

The present theory of hydrodynamic lubrication originated in the laboratory of Beauchamp Tower in the early 1880s in England. Tower had been employed to study the friction in railroad journal bearings and learn the best methods of lubricating them. It was an accident or error, during the course of this investigation, that prompted Tower to look at the problem in more detail and that resulted in a discovery that eventually led to the development of the theory.

Figure 12–7 is a schematic drawing of the journal bearing which Tower investigated. It is a partial bearing, having a diameter of 4 in, a length of 6 in, and a bearing arc of 157°, and having bath-type lubrication, as shown. The coefficients of friction obtained by Tower in his investigations on this bearing were quite low, which is not now surprising. After testing this bearing, Tower later drilled a $\frac{1}{2}$-in-diameter lubricator hole through the top. But when the apparatus was set in motion, oil flowed out of this hole. In an effort to prevent this, a cork stopper was used, but this popped out, and so it was necessary to drive a wooden plug into the hole. When the wooden plug was pushed out too, Tower, at this point, undoubtedly realized that he was on the verge of discovery. A pressure gauge connected to the hole indicated a pressure of more than twice the unit bearing load. Finally, he investigated the bearing film pressures in detail throughout the bearing width and length and reported a distribution similar to that of Fig. 12–8.[6]

The results obtained by Tower had such regularity that Osborne Reynolds concluded that there must be a definite equation relating the friction, the pressure, and the velocity. The present mathematical theory of lubrication is based upon Reynolds' work following

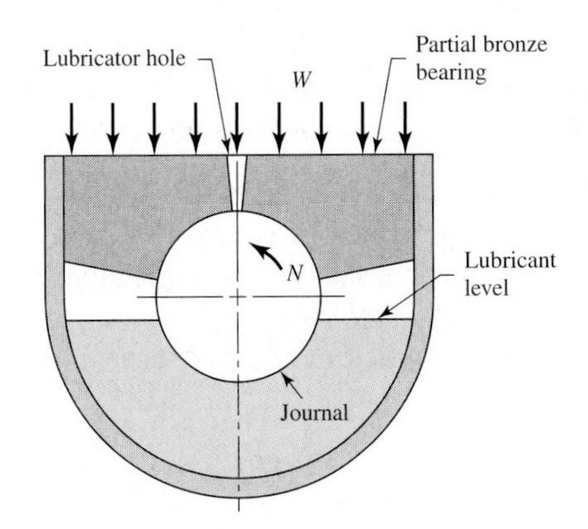

Figure 12–7

Schematic representation of the partial bearing used by Tower.

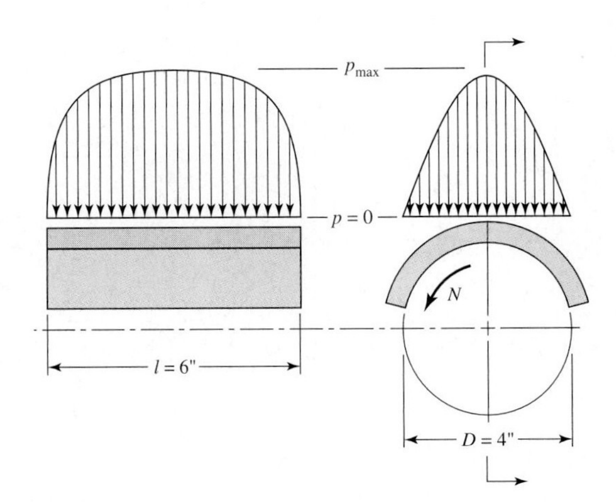

Figure 12–8

Approximate pressure-distribution curves obtained by Tower.

6. Beauchamp Tower, "First Report on Friction Experiments," *Proc. Inst. Mech. Eng.,* November 1883, pp. 632–666; "Second Report," ibid., 1885, pp. 58–70; "Third Report," ibid., 1888, pp. 173–205; "Fourth Report," ibid., 1891, pp. 111–140.

the experiment by Tower.[7] The original differential equation, developed by Reynolds, was used by him to explain Tower's results. The solution is a challenging problem which has interested many investigators ever since then, and it is still the starting point for lubrication studies.

Reynolds pictured the lubricant as adhering to both surfaces and being pulled by the moving surface into a narrowing, wedge-shaped space so as to create a fluid or film pressure of sufficient intensity to support the bearing load. One of the important simplifying assumptions resulted from Reynolds' realization that the fluid films were so thin in comparison with the bearing radius that the curvature could be neglected. This enabled him to replace the curved partial bearing with a flat bearing, called a *plane slider bearing*. Other assumptions made were:

1 The lubricant obeys Newton's viscous effect, Eq. (12–1).
2 The forces due to the inertia of the lubricant are neglected.
3 The lubricant is assumed to be incompressible.
4 The viscosity is assumed to be constant throughout the film.
5 The pressure does not vary in the axial direction.

Figure 12–9*a* shows a journal rotating in the clockwise direction supported by a film of lubricant of variable thickness h on a partial bearing which is fixed. We specify that the journal has a constant surface velocity U. Using Reynolds' assumption that curvature can be neglected, we fix a right-handed xyz reference system to the stationary bearing. We now make the following additional assumptions:

6 The bushing and journal extend infinitely in the z direction; this means there can be no lubricant flow in the z direction.
7 The film pressure is constant in the y direction. Thus the pressure depends only on the coordinate x.
8 The velocity of any particle of lubricant in the film depends only on the coordinates x and y.

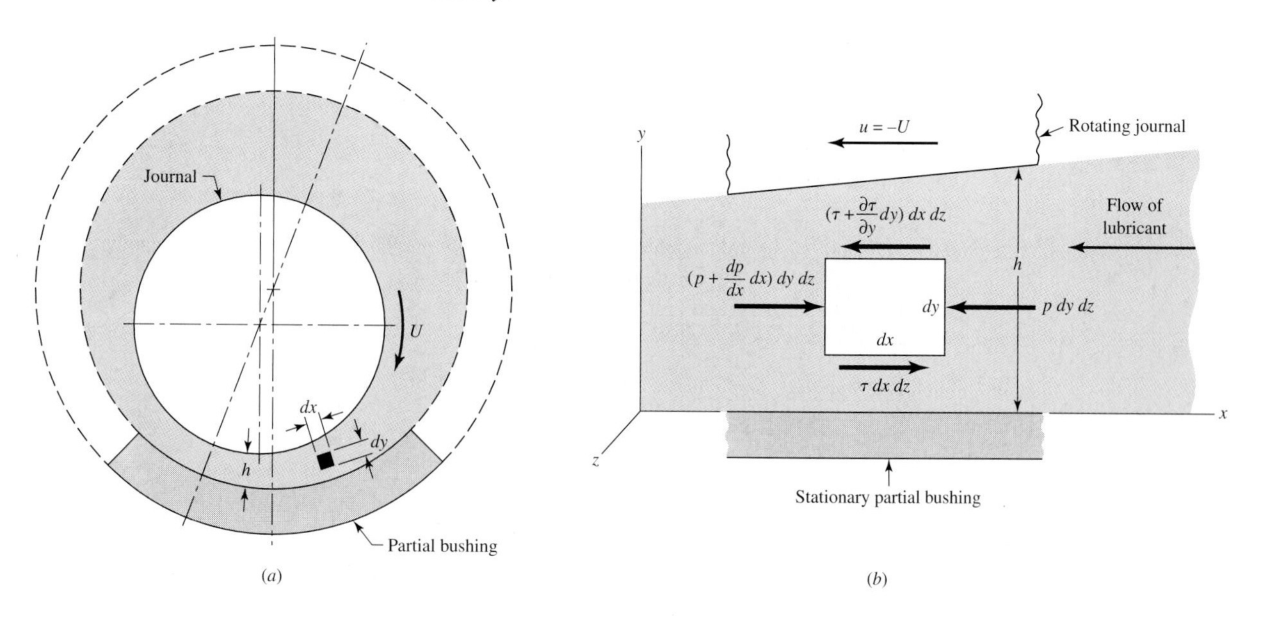

Figure 12–9

7. Osborne Reynolds, "Theory of Lubrication, Part I," *Phil. Trans. Roy. Soc. London,* 1886.

We now select an element of lubricant in the film (Fig. 12–9a) of dimensions dx, dy, and dz, and compute the forces which act on the sides of this element. As shown in Fig. 12–9b, normal forces, due to the pressure, act upon the right and left sides of the element, and shear forces, due to the viscosity and to the velocity, act upon the top and bottom sides. Summing the forces gives

$$\sum F = \left(p + \frac{dp}{dx}\,dx\right) dy\,dz + \tau\,dx\,dz - \left(\tau + \frac{\partial \tau}{\partial y}\,dy\right) dx\,dz - p\,dy\,dz = 0 \qquad (a)$$

This reduces to

$$\frac{dp}{dx} = \frac{\partial \tau}{\partial y} \qquad (b)$$

From Eq. (12–1), we have

$$\tau = \mu \frac{\partial u}{\partial y} \qquad (c)$$

where the partial derivative is used because the velocity u depends upon both x and y. Substituting Eq. (c) in Eq. (b), we obtain

$$\frac{dp}{dx} = \mu \frac{\partial^2 u}{\partial y^2} \qquad (d)$$

Holding x constant, we now integrate this expression twice with respect to y. This gives

$$\frac{\partial u}{\partial y} = \frac{1}{\mu}\frac{dp}{dx} y + C_1$$

$$u = \frac{1}{2\mu}\frac{dp}{dx} y^2 + C_1 y + C_2 \qquad (e)$$

Note that the act of holding x constant means that C_1 and C_2 can be functions of x. We now assume that there is no slip between the lubricant and the boundary surfaces. This gives two sets of boundary conditions for evaluating the constants C_1 and C_2:

$$\begin{array}{ll} y = 0 & y = h \\ u = 0 & u = -U \end{array} \qquad (f)$$

Notice, in the second condition, that h is a function of x. Substituting these conditions in Eq. (e) and solving for the constants gives

$$C_1 = -\frac{U}{h} - \frac{h}{2\mu}\frac{dp}{dx} \qquad C_2 = 0$$

or

$$u = \frac{1}{2\mu}\frac{dp}{dx}(y^2 - hy) - \frac{U}{h} y \qquad (12\text{–}9)$$

This equation gives the velocity distribution of the lubricant in the film as a function of the coordinate y and the pressure gradient dp/dx. The equation shows that the velocity distribution across the film (from $y = 0$ to $y = h$) is obtained by superposing a parabolic distribution (the first term) onto a linear distribution (the second term). Figure 12–10 shows the superposition of these two terms to obtain the velocity for particular values of x and dp/dx. In general, the parabolic term may be additive or subtractive to the linear term, depending upon the sign of the pressure gradient. When the pressure is maximum,

Figure 12–10

Velocity of the lubricant.

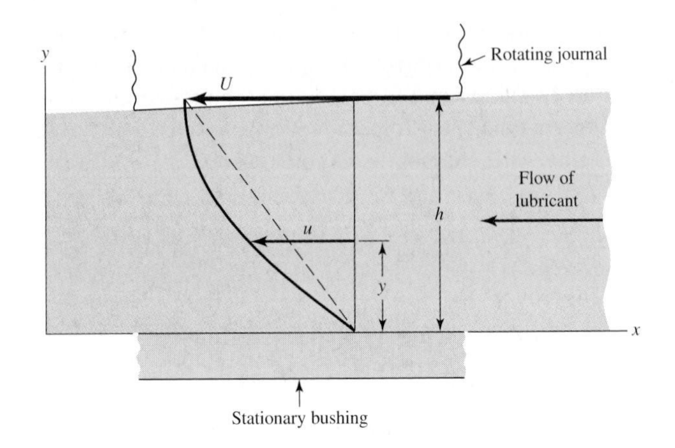

$dp/dx = 0$ and the velocity is

$$u = -\frac{U}{h}y \tag{g}$$

which is a linear relation.

We next define Q as the volume of lubricant flowing in the x direction per unit time. By using a width of unity in the z direction, the volume may be obtained by the expression

$$Q = \int_0^h u\, dy \tag{h}$$

Substituting the value of u from Eq. (12–9) and integrating gives

$$Q = -\frac{Uh}{2} - \frac{h^3}{12\mu}\frac{dp}{dx} \tag{i}$$

The next step uses the assumption of an incompressible lubricant and states that the flow is the same for any cross section. Thus

$$\frac{dQ}{dx} = 0$$

From Eq. (i),

$$\frac{dQ}{dx} = -\frac{U}{2}\frac{dh}{dx} - \frac{d}{dx}\left(\frac{h^3}{12\mu}\frac{dp}{dx}\right) = 0$$

or

$$\frac{d}{dx}\left(\frac{h^3}{\mu}\frac{dp}{dx}\right) = -6U\frac{dh}{dx} \tag{12–10}$$

which is the classical Reynolds equation for one-dimensional flow. It neglects side leakage, that is, flow in the z direction. A similar development is used when side leakage is not neglected. The resulting equation is

$$\frac{\partial}{\partial x}\left(\frac{h^3}{\mu}\frac{\partial p}{\partial x}\right) - \frac{\partial}{\partial z}\left(\frac{h^3}{\mu}\frac{\partial p}{\partial z}\right) = -6U\frac{\partial h}{\partial x} \tag{12–11}$$

There is no general solution to Eq. (12–11); approximate solutions have been obtained by using electrical analogies, mathematical summations, relaxation methods, and numerical and graphical methods. One of the important solutions is due to Sommerfeld[8] and may be expressed in the form

$$\frac{r}{c}f = \phi\left[\left(\frac{r}{c}\right)^2 \frac{\mu N}{P}\right] \tag{12–12}$$

where ϕ indicates a functional relationship. Sommerfeld found the functions for half-bearings and full bearings by using the assumption of no side leakage.

12–7 Design Considerations

We may distinguish between two groups of variables in the design of sliding bearings. In the first group are those whose values either are given or are under the control of the designer. These are:

1. The viscosity μ
2. The load per unit of projected bearing area, P
3. The speed N
4. The bearing dimensions r, c, β, and l

Of these four variables, the designer usually has no control over the speed, because it is specified by the overall design of the machine. Sometimes the viscosity is specified in advance, as, for example, when the oil is stored in a sump and is used for lubricating and cooling a variety of bearings. The remaining variables, and sometimes the viscosity, may be controlled by the designer and are therefore the *decisions* he or she makes. In other words, when these four decisions are made, the design is complete.

In the second group are the dependent variables. The designer cannot control these except indirectly by changing one or more of the first group. These are:

1. The coefficient of friction f
2. The temperature rise ΔT
3. The flow of oil Q
4. The minimum film thickness h_0

This group of variables tells us how well the bearing is performing, and hence we may regard them as *performance factors*. Certain limitations on their values must be imposed by the designer to ensure satisfactory performance. These limitations are specified by the characteristics of the bearing materials and of the lubricant. The fundamental problem in bearing design, therefore, is to define satisfactory limits for the second group of variables and then to decide upon values for the first group such that these limitations are not exceeded.

Trumpler's Design Criteria for Journal Bearings

Because the bearing assembly creates the lubricant pressure to carry a load, it reacts to loading by changing its eccentricity, which reduces the minimum film thickness h_0 until the load is carried. What is the limit of smallness of h_0? Close examination reveals that the moving adjacent surfaces of the journal and bushing are not smooth but a series of asperities which pass one another, separated by a lubricant film. In starting a bearing under

8. A. Sommerfeld, "Zur Hydrodynamischen Theorie der Schmiermittel-Reibung" ("On the Hydrodynamic Theory of Lubrication"), *Z. Math. Physik*, vol. 50, 1904, pp. 97–155.

load from rest there is metal-to-metal contact and surface asperities are broken off, free to move and circulate with the oil. Unless a filter is provided, this debris accumulates. Such particles have to be free to tumble at the section containing the minimum film thickness without snagging in a togglelike configuration, creating additional damage and debris. Trumpler, an accomplished bearing designer, provides a throat of at least 200 μin to pass particles from ground surfaces.[9] He also provides for the influence of size (tolerances tend to increase with size) by stipulating

$$h_0 \geq 0.0002 + 0.000\,04D \text{ in} \tag{a}$$

where D is the journal diameter in inches.

A lubricant is a mixture of hydrocarbons which reacts to increasing temperature by vaporizing the lighter components, leaving behind the heavier. This process (bearings have lots of time) slowly increases the viscosity of the remaining lubricant, which increases heat generation rate and elevates lubricant temperatures. This sets the stage for future failure. For light oils, Trumpler limits the maximum film temperature $T_{\max}$ to

$$T_{\max} \leq 250°\text{F} \tag{b}$$

Some oils can operate at slightly higher temperatures. Always check with the lubricant manufacturer.

A journal bearing often consists of a ground steel journal working against a softer, usually nonferrous, bushing. In starting under load there is metal-to-metal contact, abrasion, and the generation of wear particles, which, over time, can change the geometry of the bushing. The starting load divided by the projected area is limited to

$$\frac{W_{st}}{lD} \leq 300 \text{ psi} \tag{c}$$

If the load on a journal bearing is suddenly increased, the increase in film temperature in the annulus is immediate. Since ground vibration due to passing trucks, trains, and earth tremors is often present, Trumpler used a design factor of 2 or more on the running load, but not on the starting load of Eq. (*c*):

$$n_d \geq 2 \tag{d}$$

As you work problems where the radial clearance c is varied, you will note that with increasing c, h_0 first increases then decreases. If the design is such that the derivative $\partial h_0/\partial c > 0$, then wear from repeated starting can tend to increase h_0. In other words, there is a tendency for the bearing to improve its ability to pass wear particles with use. Incorporating this feature when possible may be attractive. To reiterate,

$$\frac{\partial h_0}{\partial c} > 0 \tag{e}$$

Many of Trumpler's designs are operating today, long after his consulting career was over; clearly they constitute good advice to the beginning designer.

The road to understanding solid and fluid friction has been long and not without error. Leonardo DaVinci (c. 1515) investigated dry friction by attaching a line (rope) to a ship's cable (anchor chain) laid out along a dusty road. He counted the number of sailors pulling on the line to move it along the road at a steady speed. He then stoppered the chain (tied it into an irregular body) and repeated the experiment. The number of sailors was the same. He concluded that the tangential friction force is independent of

9. Op. cit., pp. 192–194.

the area of contact, and the familiar expression $F \leq \mu N$ was developed. This was not universally true, as a railroad locomotive manufacturer was to discover after World War II. Amontons (1699) stated that the coefficient of friction was the same for all metal-on-metal combinations, and it was 0.3.[10] It was not until Coulomb (1781) that a distinction between static and dynamic friction was drawn.

Industrial necessity during the Industrial Revolution drove Towers (1883) to discover the pressure rise in a bearing due to viscous pumping and Petroff (1883) to discover Eq. (12–6). Reynolds (1886) developed the differential equations [Eqs. (12–10) and (12–11)] that quantitatively describe film lubrication. Sommerfeld (1904) solved the differential equation for the journal bearing with no side leakage. Kingsbury over the years 1912 to 1952 developed film thrust bearings for which he used an electrical analog circuit to solve Reynolds' differential equation for side leakage in thrust bearings. It was not until 1952, that Raimondi and Boyd used the digital computer to solve Reynolds' differential equation for journal bearings. Their solutions were presented in dimensionless form; they are the bases for many figures in this chapter. After all this effort by the investigators mentioned above, and many others, we still do not have general analytical solutions. Our quantitative phenomenological understanding is in these figures (and others) and we use them. It is possible to make curve fits to Raimondi and Boyd loci which result in computer subroutines to assist in the necessary iterative calculation effort associated with film bearing analysis and design.

Film bearings are quantitatively understood in terms of pi terms, dimensionless clusters of parameters, many of which are a function of the Sommerfeld number. Such terms include

$$\frac{fr}{c} \qquad \frac{h_0}{c} \qquad \phi \qquad \frac{Q}{rcNl} \qquad \frac{Q_s}{Q} \qquad \frac{P}{p_{\max}} \qquad \theta_P \qquad \frac{C_p \Delta T}{P}$$

which are discussed in the next section.[11]

12–8 The Relations of the Variables

Before proceeding to the problem of design, it is necessary to establish the relationships between the variables. Albert A. Raimondi and John Boyd, of Westinghouse Research Laboratories, used an iteration technique to solve Reynolds' equation on the digital computer.[12] This is the first time such extensive data have been available for use by designers, and consequently we shall employ them in this book.[13]

The Raimondi and Boyd papers were published in three parts and contain 45 detailed charts and 6 tables of numerical information. In all three parts, charts are used to define the variables for length-diameter (l/d) ratios of 1:4, 1:2, and 1 and for beta angles of 60 to 360°. Under certain conditions the solution to the Reynolds equation gives negative pressures in the diverging portion of the oil film. Since a lubricant cannot usually support

10. Amontons was a favorite among students of physics for simplifying problem solving, and among instructors for simplifying grading.

11. To appreciate this subject, see Charles R. Mischke, *Mathematical Model Building,* Iowa State University Press, Ames, 1980, pp. 139–164, or Joseph E. Shigley and Charles R. Mischke, "Minimizing Engineering Effort," in *Standard Handbook of Machine Design,* 2nd ed., McGraw-Hill, New York, 1996, Chap. 11, pp. 11.1–11.9.

12. A. A. Raimondi and John Boyd, "A Solution for the Finite Journal Bearing and Its Application to Analysis and Design, Parts I, II, and III," *Trans. ASLE,* vol. 1, no. 1, in *Lubrication Science and Technology,* Pergamon, New York, 1958, pp. 159–209.

13. See also the earlier companion paper, John Boyd and Albert A. Raimondi, "Applying Bearing Theory to the Analysis and Design of Journal Bearings, Part I and II," *J. Appl. Mechanics,* vol. 73, 1951, pp. 298–316.

a tensile stress, Part III of the Raimondi-Boyd papers assumes that the oil film is ruptured when the film pressure becomes zero. Part III also contains data for the infinitely long bearing; since it has no ends, this means that there is no side leakage. The charts appearing in this book are from Part III of the papers, and are for full journal bearings ($\beta = 360°$) only. Space does not permit the inclusion of charts for partial bearings. This means that you must refer to the charts in the original papers when beta angles of less than 360° are desired. The notation is very nearly the same as in this book, and so no problems should arise.

Viscosity Charts (Figs. 12–11 to 12–13)

One of the most important assumptions made in the Raimondi-Boyd analysis is that *viscosity of the lubricant is constant as it passes through the bearing*. But since work is done on the lubricant during this flow, the temperature of the oil is higher when it leaves the loading zone than it was on entry. And the viscosity charts clearly indicate that the viscosity drops off significantly with a rise in temperature. Since the analysis is based on a constant viscosity, our problem now is to determine the value of viscosity to be used in the analysis.

Some of the lubricant that enters the bearing emerges as a side flow, which carries away some of the heat. The balance of the lubricant flows through the load-bearing zone and carries away the balance of the heat generated. In determining the viscosity to be used we shall employ a temperature that is the average of the inlet and outlet temperatures, or

$$T_{av} = T_1 + \frac{\Delta T}{2} \tag{12–13}$$

where T_1 is the inlet temperature and ΔT is the temperature rise of the lubricant from inlet to outlet. Of course, the viscosity used in the analysis must correspond to T_{av}.

Viscosity varies considerably with temperature in a nonlinear fashion. The ordinates in Figs. 12–11 to 12–13 are not logarithmic, as the decades are of differing vertical length. These graphs represent the temperature versus viscosity functions for common grades of lubricating oils in both customary engineering and SI units. We have the temperature versus viscosity function only in graphical form, unless curve fits are developed. See Table 12–1.

One of the objectives of lubrication analysis is to determine the oil outlet temperature when the oil and its inlet temperature are specified. This is a trial-and-error type of problem.

To illustrate, suppose we have decided to use SAE 30 oil in an application in which the oil inlet temperature is $T_1 = 180°\text{F}$. We begin by estimating that the temperature rise will be $\Delta T = 30°\text{F}$. Then, from Eq. (12–13),

$$T_{av} = T_1 + \frac{\Delta T}{2} = 180 = \frac{30}{2} = 195°\text{F}$$

From Fig. 12–11 we follow the SAE 30 line and find that $\mu = 1.40\ \mu\text{reyn}$ at 195°F. So we use this viscosity (in an analysis to be explained in detail in due time) and find that the temperature rise is actually $\Delta T = 54°\text{F}$. Thus Eq. (12–13) gives

$$T_{av} = 180 + \frac{54}{2} = 207°\text{F}$$

This corresponds to point A on Fig. 12–11, which is above the SAE 30 line and indicates that the viscosity used in the analysis was too high.

Figure 12–11

Viscosity–temperature chart in U.S. customary units. *(Raimondi and Boyd.)*

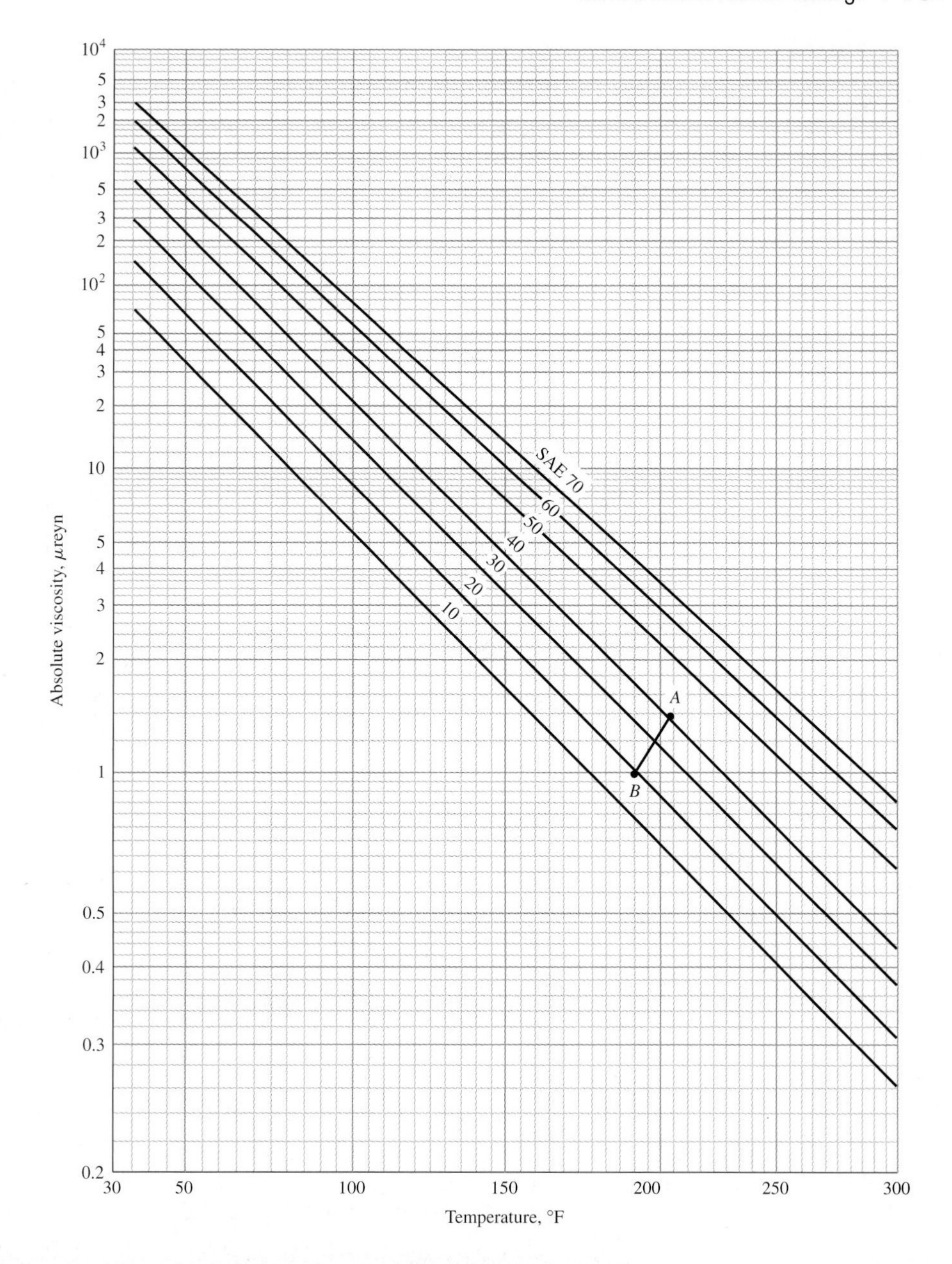

For a second guess, try $\mu = 1.00\ \mu$reyn. Again we run through an analysis and this time find that $\Delta T = 30°$F. This gives an average temperature of

$$T_{av} = 180 + \frac{30}{2} = 195°\text{F}$$

and locates point B on Fig. 12–11.

If points A and B are fairly close to each other and on opposite sides of the SAE 30 line, a straight line can be drawn between them with the intersection locating the correct values of viscosity and average temperature to be used in the analysis. For this illustration, we see from the viscosity chart that they are $T_{av} = 203°$F and $\mu = 1.26\ \mu$reyn.

Figure 12–12

Viscosity–temperature chart in SI units. *(Adapted from Fig. 12–11.)*

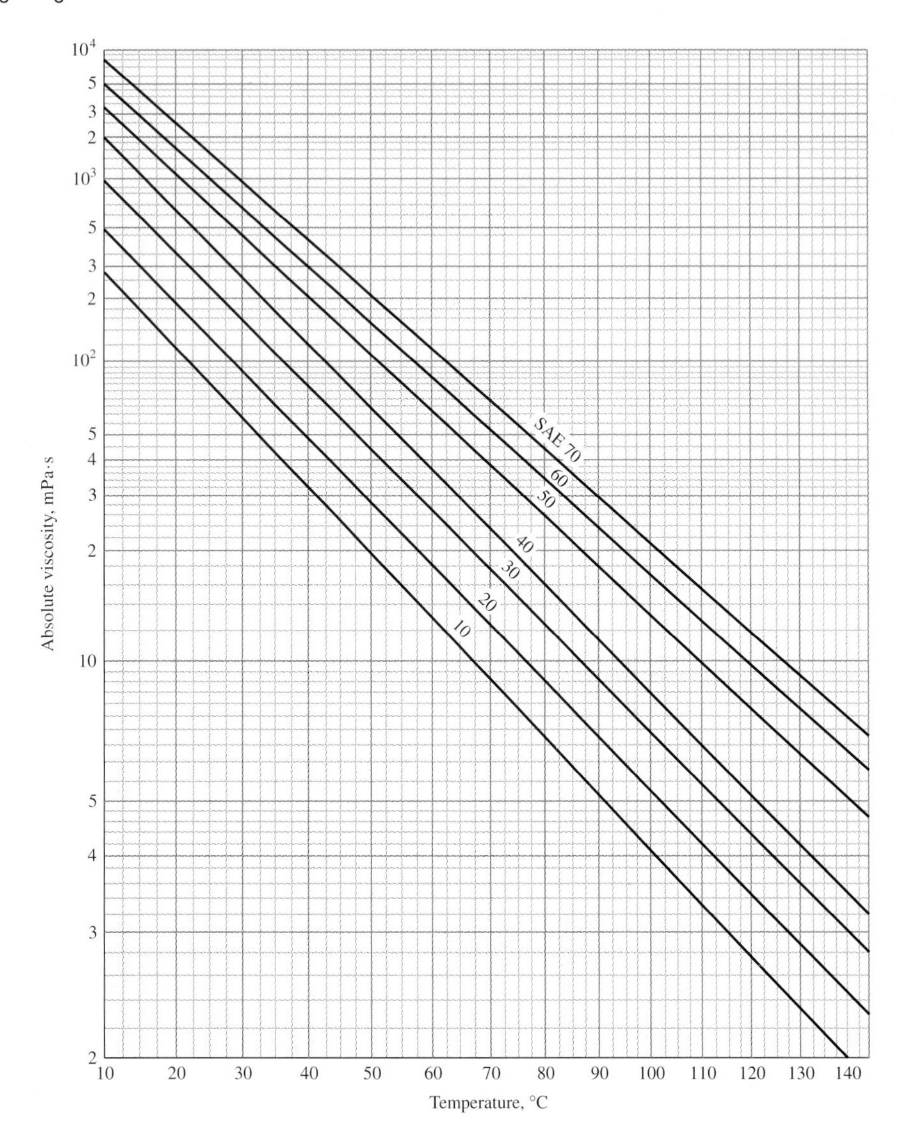

Table 12–1

Curve Fits* to Approximate the Viscosity versus Temperature Functions for SAE Grades 10–60

Source: A. S. Seireg and S. Dandage, "Empirical Design Procedure for the Thermodynamic Behavior of Journal Bearings," *J. Lubrication Technology*, vol. 104, April 1982, pp. 135–148.

Oil Grade, SAE	Viscosity μ_0, reyn	Constant b, °F
10	$0.0158(10^{-6})$	1157.5
20	$0.0136(10^{-6})$	1271.6
30	$0.0141(10^{-6})$	1360.0
40	$0.0121(10^{-6})$	1474.4
50	$0.0170(10^{-6})$	1509.6
60	$0.0187(10^{-6})$	1564.0

*$\mu = \mu_0 \exp[b/(T + 95)]$, T in °F.

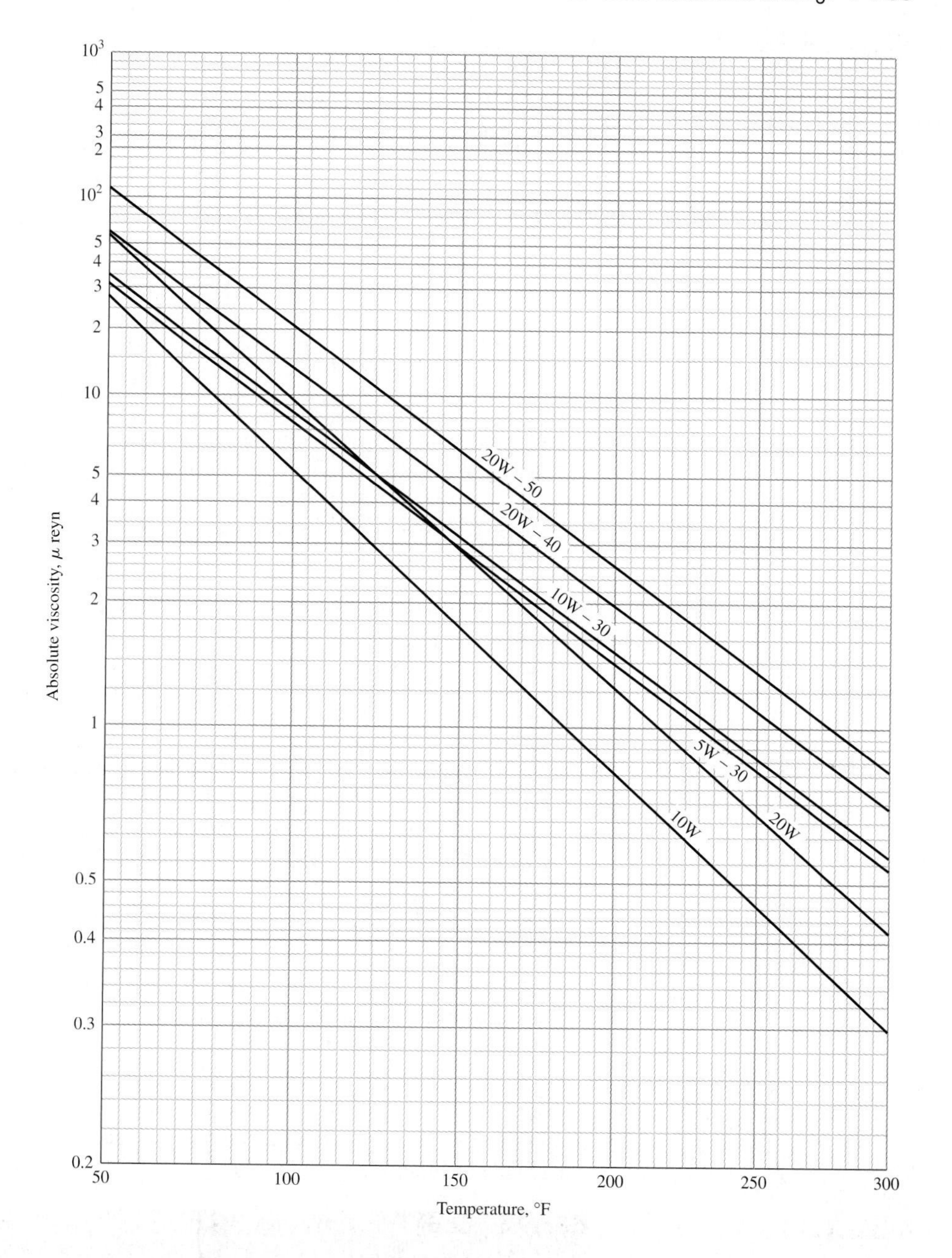

Figure 12–13

Chart for multiviscosity lubricants. This chart was derived from known viscosities at two points, 100 and 210°F, and the results are believed to be correct for other temperatures.

Minimum Film Thickness (Figs. 12–14 and 12–15)

EXAMPLE 12–3 As an example of an analysis using Raimondi and Boyd charts, consider a bearing with the following parameters known:

$$\mu = 4\ \mu\text{reyns}$$
$$N_j = 30\ \text{r/s}$$
$$W = 500\ \text{lb (bearing load)}$$

$$r = 0.75 \text{ in}$$
$$c = 0.0015 \text{ in}$$
$$l = 1.5 \text{ in}$$

The nominal bearing pressure (in projected area of the journal) is

$$P = \frac{W}{2rl} = \frac{500}{2(0.75)1.5} = 222 \text{ psi}$$

The Sommerfeld number is, from Eq. (12–7),

$$S = \left(\frac{r}{c}\right)^2 \left(\frac{\mu N_j}{P}\right) = \left(\frac{0.75}{0.0015}\right)^2 \left[\frac{4(10^{-6})30}{222}\right] = 0.135$$

Also, $l/d = 1.50/(2 \cdot 0.75) = 1$. Entering Fig. 12–14 with $S = 0.135$ and $l/d = 1$ gives $h_0/c = 0.42$ and $\epsilon = 0.58$. The quantity h_0/c is called the *minimum film thickness variable*. Since $c = 0.0015$ in, the minimum film thickness h_0 is

$$h_0 = 0.42(0.0015) = 0.000\,63 \text{ in}$$

This is shown in Fig. 12–16. We can find the angular location ϕ of the minimum film thickness from the chart of Fig. 12–15. Entering with $S = 0.135$ and $l/d = 1$ gives $\phi = 53°$.

The eccentricity ratio is $\epsilon = e/c = 0.58$. This means the eccentricity e is

$$e = 0.58(0.0015) = 0.000\,87 \text{ in}$$

This is also shown in Fig. 12–16. Note that if the journal is centered in the bushing, $e = 0$ and $h_0 = c$, corresponding to a very light (zero) load. Since $e = 0$, $\epsilon = 0$. As the load is increased the journal displaces downward; the limiting position is reached when $h_0 = 0$ and $e = c$, that is, when the journal touches the bushing. For this condition the eccentricity ratio is unity. Since $h_0 = c - e$, dividing both sides by c, we have

$$\frac{h_0}{c} = 1 - \epsilon$$

Design optima are sometimes *maximum load*, which is a load- carrying characteristic of the bearing, and sometimes *minimum parasitic power loss* or *minimum coefficient of friction*. Dashed lines appear on Fig. 12–14 for maximum load and minimum coefficient of friction, so you can easily favor one of maximum load or minimum coefficient of friction, but not both. The zone between the two dashed-line contours might be considered a desirable location for a design point.

Coefficient of Friction (Fig. 12–17)

The friction chart, Fig. 12–17, has the *friction variable* $(r/c)f$ plotted against Sommerfeld number S with contours for various values of the l/d ratio. We enter Fig. 12–17 with $S = 0.135$ and $l/d = 1$ and find $(r/c)f = 3.50$. The coefficient of friction f is

$$f = 3.50c/r = 3.50(0.0015/0.75) = 0.007$$

The friction torque on the journal is

$$T = fWr = 0.007(500)0.75 = 2.61 \text{ in} \cdot \text{lbf}$$

The power loss in horsepower is

$$(hp)_{\text{loss}} = \frac{TN_j}{1050} = \frac{2.62(30)}{1050} = 0.0748 \text{ hp}$$

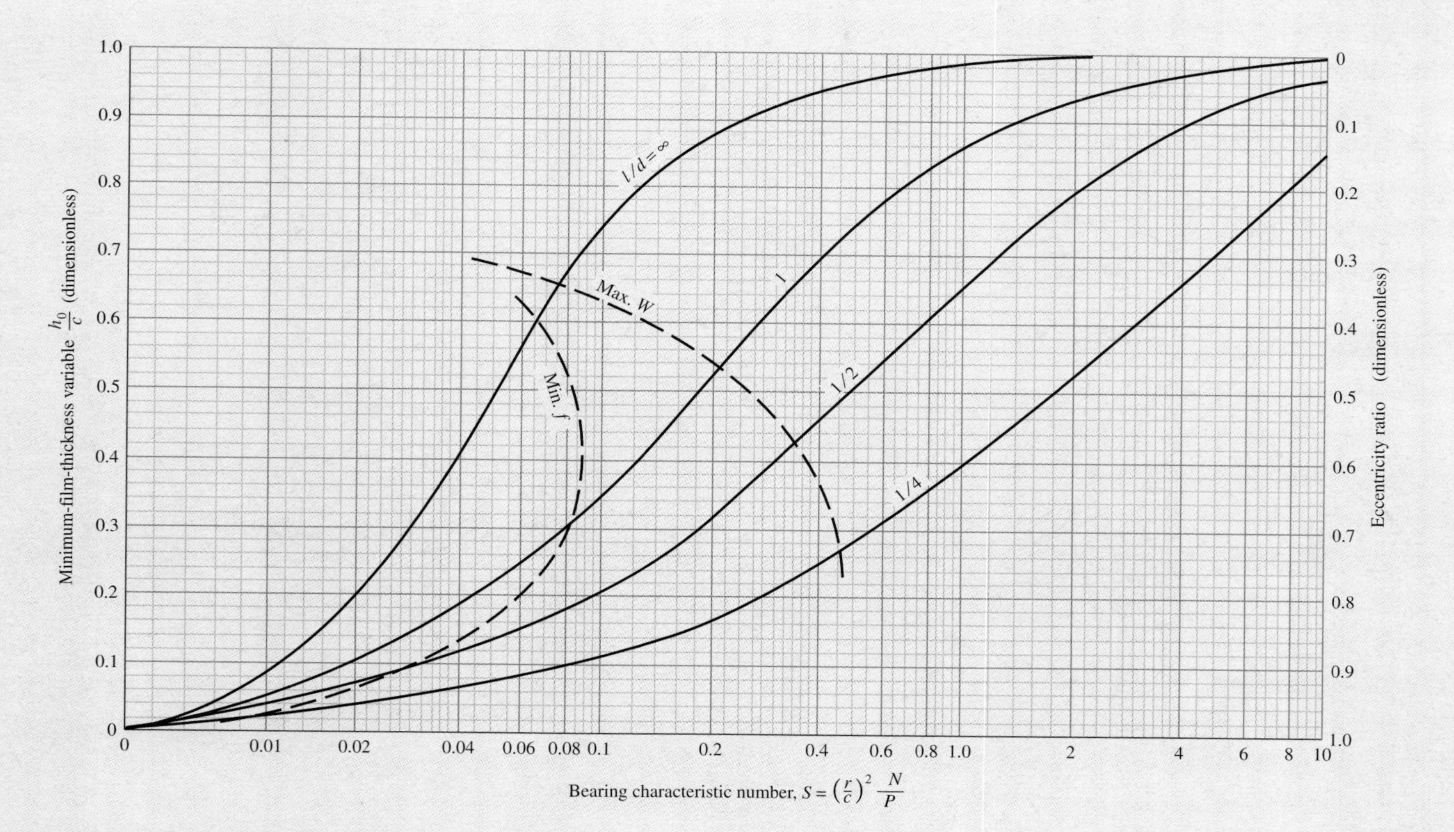

Figure 12–14

Chart for minimum film-thickness variable and eccentricity ratio. The left boundary of the zone defines the optimal h_0 for minimum friction; the right boundary is optimum h_0 for load. *(Raimondi and Boyd.)*

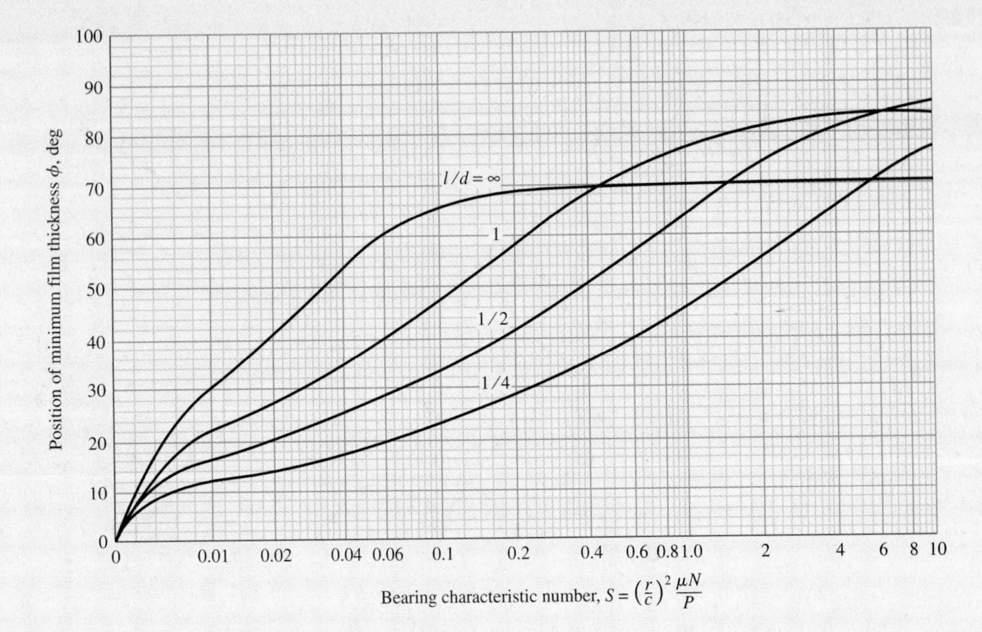

Figure 12–15

Chart for determining the position of the minimum film thickness h_0. For location of the origin, see Fig. 12–16. *(Raimondi and Boyd.)*

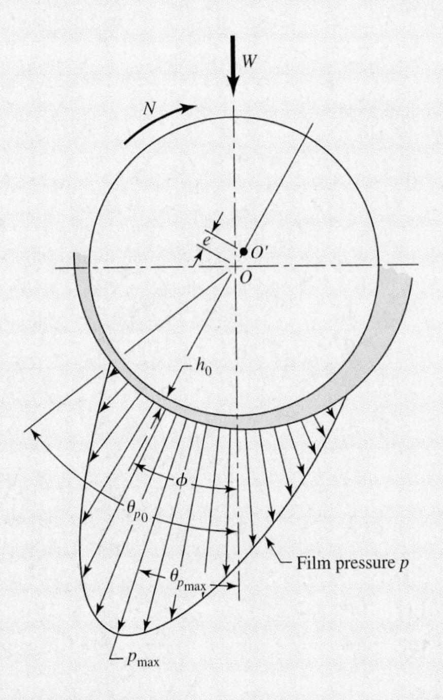

Figure 12–16

Polar diagram of the film–pressure distribution showing the notation used. *(Raimondi and Boyd.)*

or, expressed in Btu/s,

$$H = \frac{2\pi T N_j}{778(12)} = \frac{2\pi(2.62)30}{778(12)} = 0.0529 \text{ Btu/s}$$

Lubricant Flow (Figs. 12–18 and 12–19)

To estimate the lubricant flow, enter Fig. 12–18 with $S = 0.135$ and $l/d = 1$ to obtain $Q/(rcN_j l) = 4.28$. The total volumetric flow rate is

$$Q = 4.28 rcN_j l = 4.28(0.75)0.0015(30)1.5 = 0.216 \text{ in}^3/\text{s}$$

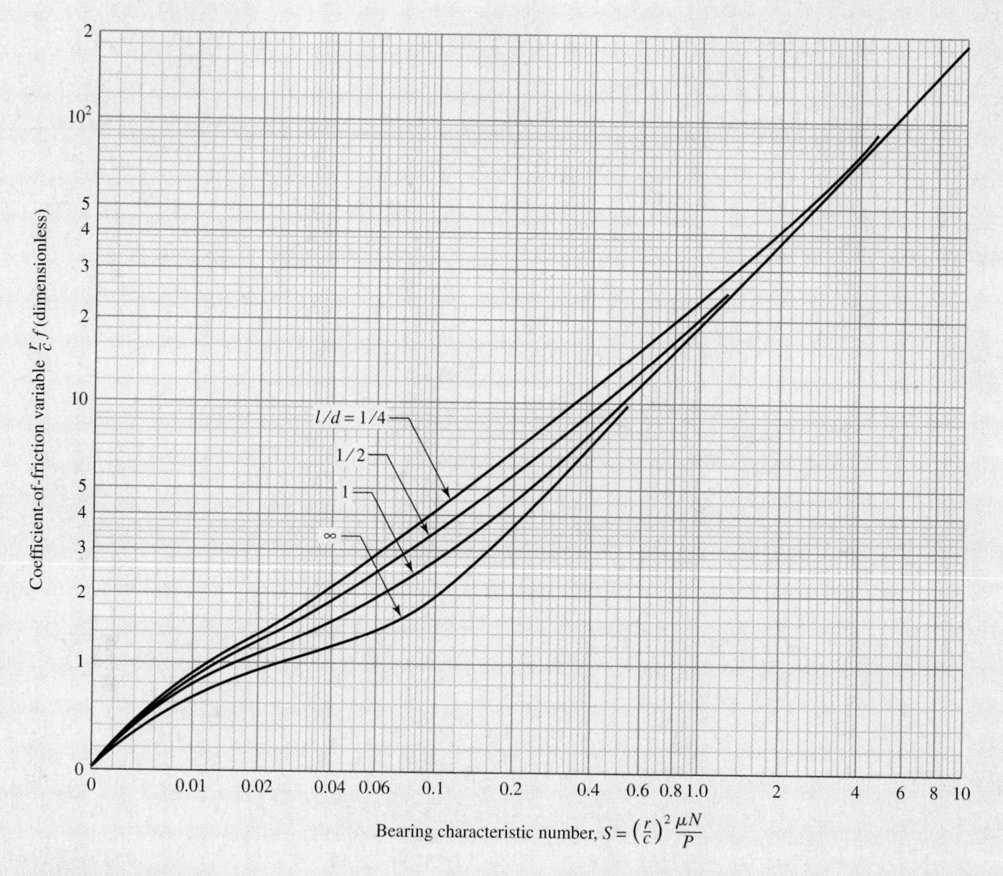

Figure 12–17

Chart for coefficient-of-friction variable; note that Petroff's equation is the asymptote. *(Raimondi and Boyd.)*

From Fig. 12–19 we find the *flow ratio* $Q_s/Q = 0.655$ and Q_s is

$$Q_s = 0.655Q = 0.655(0.216) = 0.142 \text{ in}^3/\text{s}$$

The side leakage Q_s is from the lower part of the bearing, where the internal pressure is above atmospheric pressure. The leakage forms a fillet at the journal–bushing external junction, and it is carried by journal motion to the top of the bushing, where the internal pressure is below atmospheric pressure and the gap is much larger, to be "sucked in" and returned to the lubricant sump. That portion of side leakage which leaks away from the bearing has to be made up by adding oil to the bearing sump periodically by maintenance personnel.

Film Pressure (Figs. 12–20 and 12–21)

The maximum pressure developed in the film can be estimated by finding the pressure ratio P/p_{max} from the chart in Fig. 12–20. Entering this chart with $S = 0.135$ and $l/d = 1$, we find $P/p_{max} = 0.42$. The maximum pressure P_{max} is therefore

$$p_{max} = \frac{P}{0.42} = \frac{222}{0.42} = 529 \text{ psi}$$

The location at which the largest pressure occurs is depicted in Fig. 12–16. With $S = 0.135$ and $l/d = 1$, from Fig. 12–21, $\theta_{p_{max}} = 18.5°$ and the terminating position θ_{p_0} is 75°.

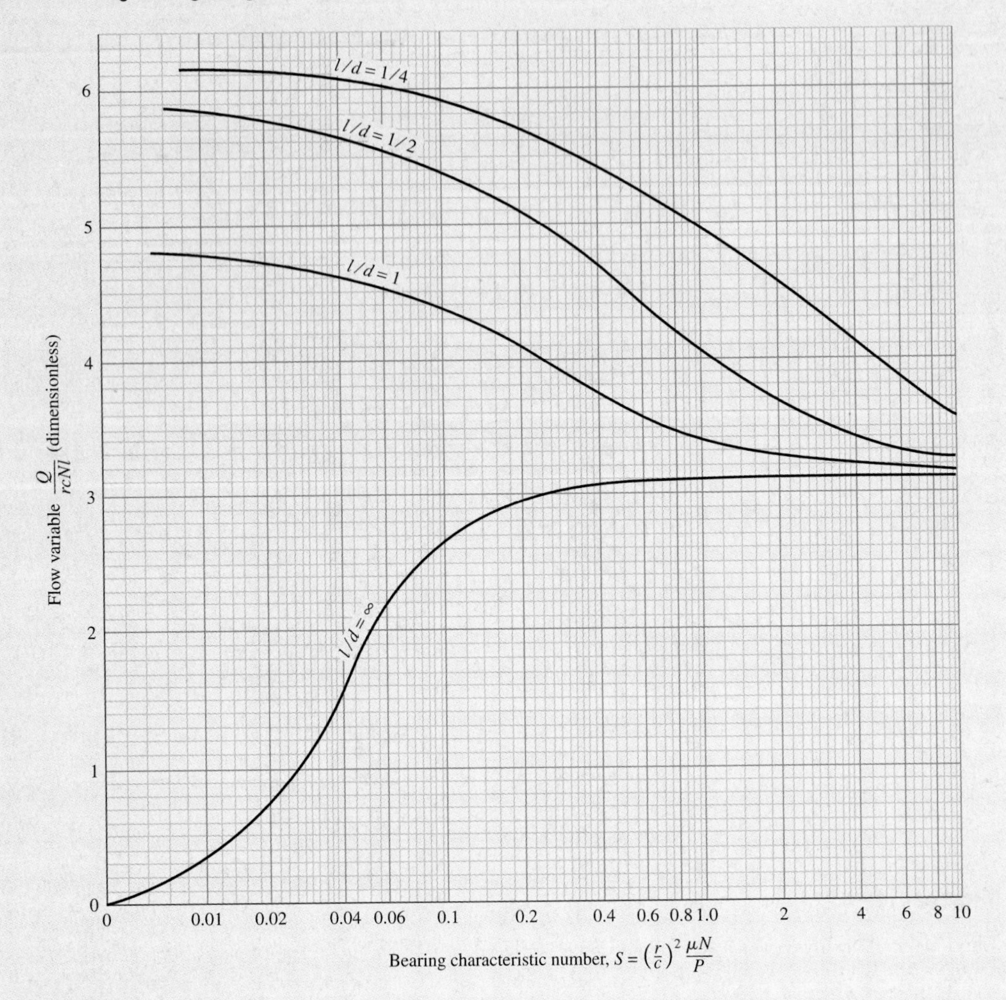

Figure 12–18

Chart for flow variable. *(Raimondi and Boyd.)*

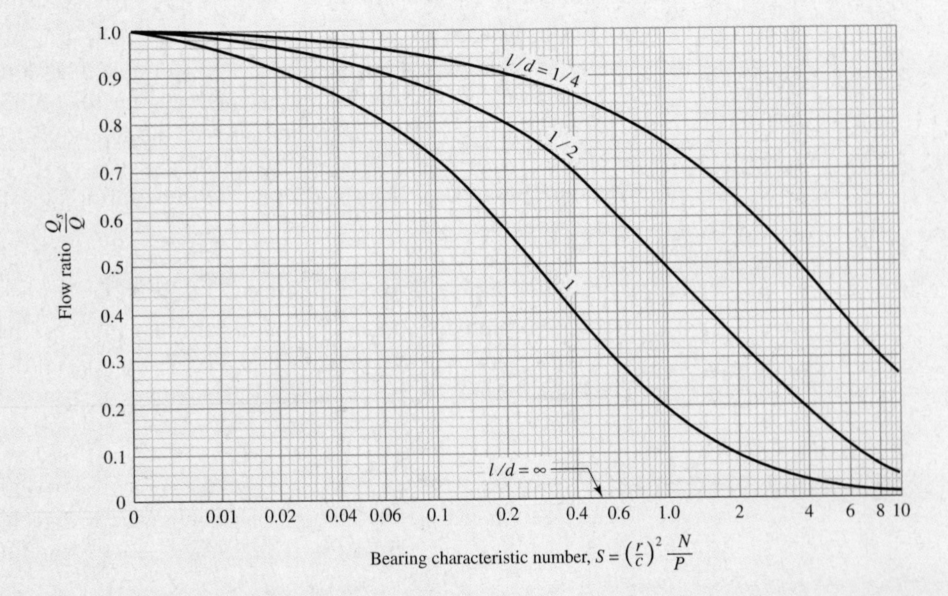

Figure 12–19

Chart for determining the ratio of side flow to total flow. *(Raimondi and Boyd).*

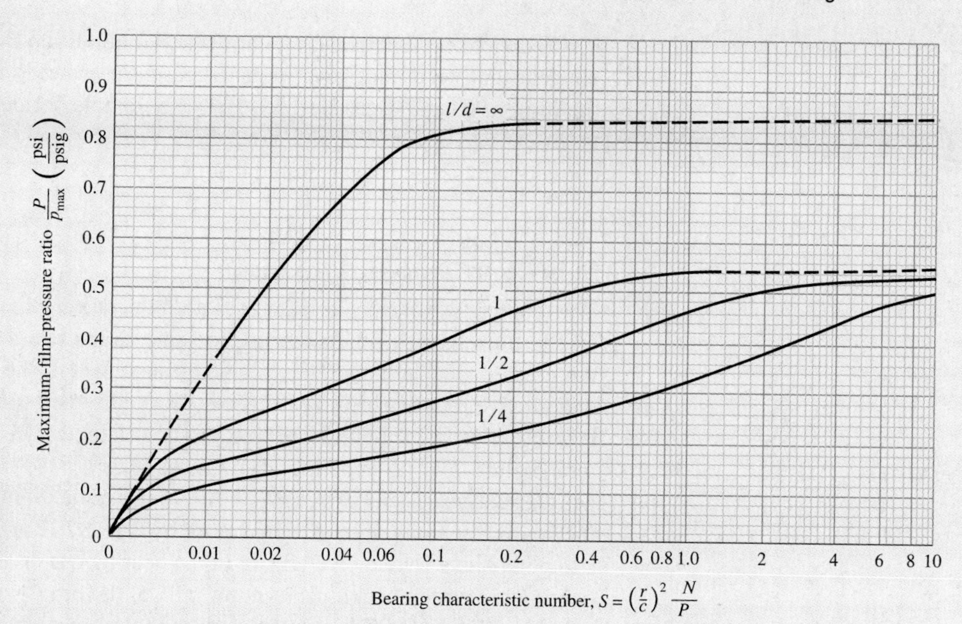

Figure 12–20

Chart for determining the maximum film pressure. *(Raimondi and Boyd.)*

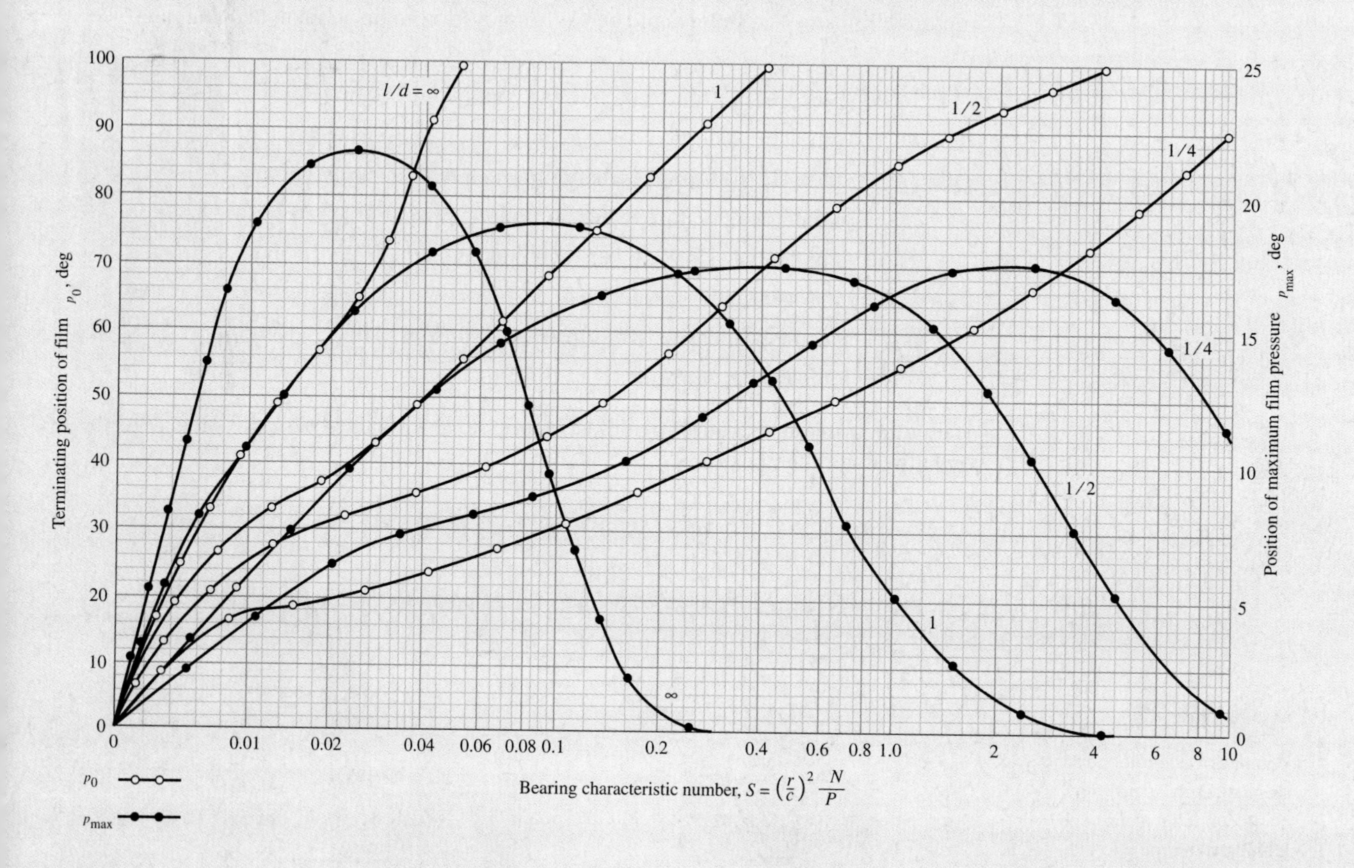

Figure 12–21

Chart for finding the terminating position of the lubricant film and the position of maximum film pressure. *(Raimondi and Boyd.)*

Example 12–3 demonstrates how the Raimondi and Boyd charts are used. It should be clear that we do not have journal–bearing parametric relations as equations, but in the form of charts. Moreover, Ex. 12–3 was simple because the steady-state equivalent viscosity was given. We will now show how the average film temperature (and the corresponding viscosity) is found from energy considerations.

Lubricant Temperature Rise

The temperature of the lubricant rises until the rate at which work is done by the journal on the film through fluid shear is the same as the rate at which heat is transferred to the greater surroundings. The specific arrangement of the bearing plumbing affects the quantitative relationships. See Fig. 12–22. A lubricant sump (internal or external to the bearing housing) supplies lubricant at sump temperature T_s to the bearing annulus at temperature $T_s = T_i$. The lubricant passes once around the bushing and is delivered at a higher lubricant temperature $T_i + \Delta T$ to the sump. Some of the lubricant leaks out of the bearing at a mixing-cup temperature of $T_i + \Delta T/2$ and is returned to the sump. The sump may be a keywaylike groove in the bearing cap or a larger chamber up to half the bearing circumference. It can occupy "all" of the bearing cap of a split bearing. In such a bearing the side leakage occurs from the lower portion and is sucked back in, into the ruptured film arc. The sump could be well-removed from the journal–bushing interface. If

Q = volumetric oil-flow rate into the bearing, in^3/s

Q_s = volumetric side-flow leakage rate out of the bearing and to the sump, in^3/s

$Q - Q_s$ = volumetric oil-flow discharge from annulus to sump, in^3/s

T_i = oil inlet temperature (equal to sump temperature T_s), °F

ΔT = temperature rise in oil between inlet and outlet, °F

ρ = lubricant density, lbm/in^3

C_P = specific heat capacity of lubricant, Btu/(lbm · °F)

J = Joulean heat equivalent, in · lbf/Btu

Using the sump as a control region, we can write an enthalpy balance. Using T_i as the enthalpy datum temperature,

$$\dot{Q}_{\text{loss}} = \rho C_p Q_s \Delta T/2 + \rho C_P (Q - Q_s)\Delta T = \rho C_P Q \Delta T\left(1 - \frac{1}{2}\frac{Q_s}{Q}\right)$$

The thermal energy loss at steady state $\dot{Q}_{\text{loss}}$ is equal to the rate the journal does work

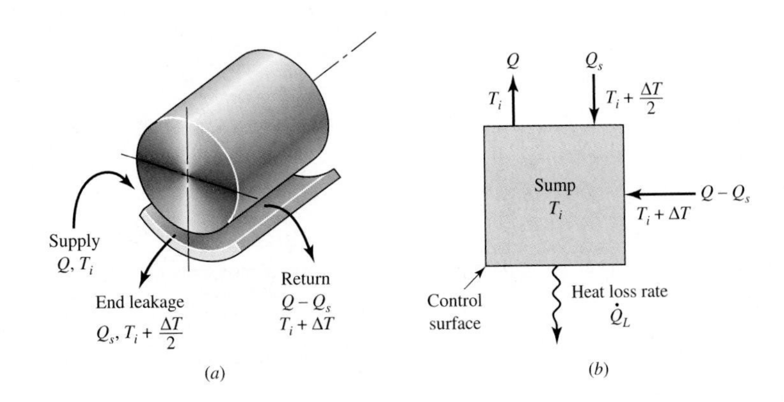

Figure 12–22

Schematic of a journal bearing with an external sump with cooling; lubricant makes one pass before returning to the sump.

on the film:

$$\dot{Q}_{\text{loss}} = \dot{W}k = \frac{2\pi T N_j}{J} = \frac{2\pi f W r N_j}{J} = \frac{2\pi W N_j c}{J}\frac{rf}{c} = \frac{4\pi P r l N_j c}{J}\frac{rf}{c}$$

Rearranging,

$$\frac{J\rho C_P \Delta T}{4\pi P} = \frac{rf/c}{(1 - \frac{1}{2}Q_s/Q)(Q/(rcN_j l)}$$

For common petroleum lubricants $\rho = 0.0311$ lbm/in^3, $C_P = 0.42$ Btu/(lbm · °F), and $J = 778(12) = 9336$ in · lbf/Btu; therefore

$$\frac{J\rho C_P \Delta t}{4\pi P} = \frac{9336(0.0311)0.42\Delta T}{4\pi P} = 9.70\frac{\Delta T}{P}$$

thus

$$\frac{9.70\Delta T_F}{P_{\text{psi}}} = \frac{rf/c}{(1 - \frac{1}{2}Q_s/Q)(Q/rcN_j l)} \tag{12–14}$$

where ΔT_F is the temperature rise in °F and P_{psi} is the bearing pressure in psi. The right side of Eq. (12–14) can be evaluated from Figs. 12–17, 12–18, and 12–19 for various Sommerfeld numbers and l/d ratios to give Fig. 12–23. It is easy to show that the left side of Eq. (12–14) can be expressed as $0.120\Delta T_C/P_{\text{MPa}}$ where ΔT_C is expressed in °C and the pressure P_{MPa} is expressed in MPa. The ordinate in Fig. 12–23 is either $0.970\Delta T_F/P_{\text{psi}}$ or $0.120\,\Delta T_C/P_{\text{MPa}}$, which is not surprising since both are dimensionless in proper units and *identical in magnitude*. Since solutions to bearing problems involve iteration and reading many graphs can introduce errors, Fig. 12–23 reduces three graphs to one, a step in the proper direction.

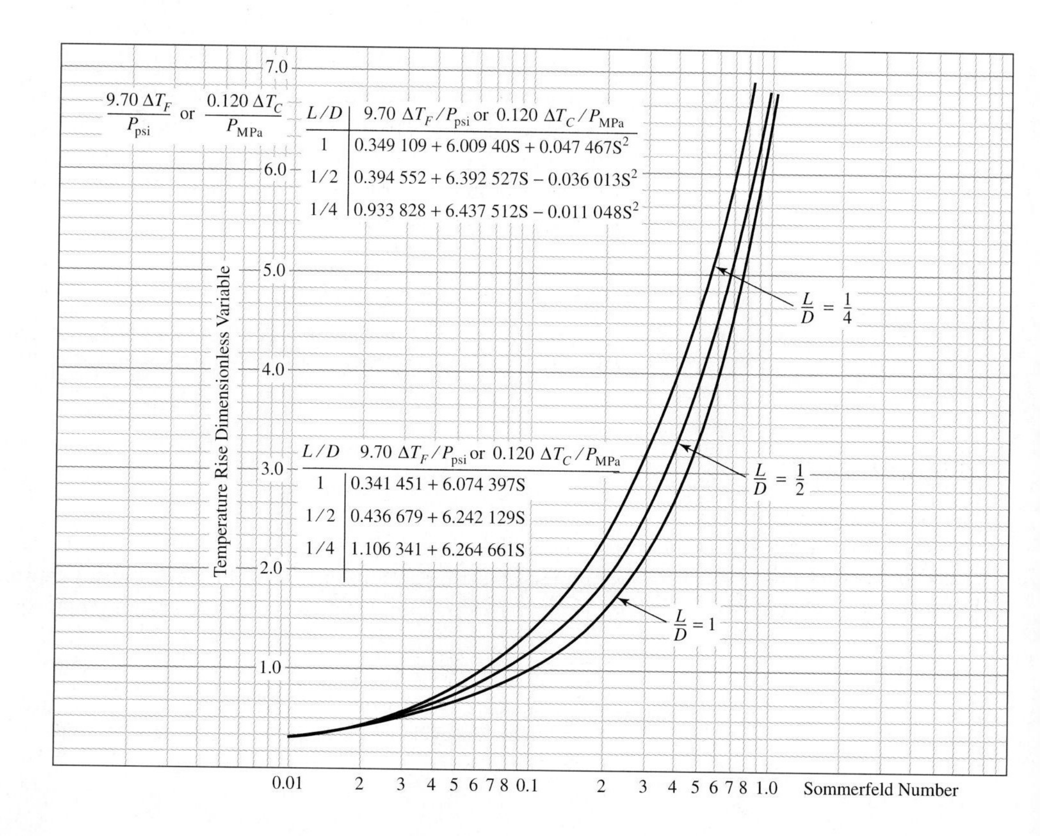

Figure 12–23

Figures 12–17, 12–18, and 12–19 combined to reduce iterative table look-up. *(Source: Chart based on work of Raimondi and Boyd boundary condition (2), i.e., no negative lubricant pressure developed. Chart is for full journal bearing using single lubricant pass, side flow emerges with temperature rise ΔT/2, thru flow emerges with temperature rise Δ T, and entire flow is supplied at datum sump temperature.)*

Interpolation

According to Raimondi and Boyd, interpolation of the chart data for other l/d ratios can be done by using the equation

$$y = \frac{1}{(l/d)^3}\left[-\frac{1}{8}\left(1-\frac{l}{d}\right)\left(1-2\frac{l}{d}\right)\left(1-4\frac{l}{d}\right)y_\infty + \frac{1}{3}\left(1-2\frac{l}{d}\right)\left(1-4\frac{l}{d}\right)y_1 - \frac{1}{4}\left(1-\frac{l}{d}\right)\left(1-4\frac{l}{d}\right)y_{1/2} + \frac{1}{24}\left(1-\frac{l}{d}\right)\left(1-2\frac{l}{d}\right)y_{1/4}\right] \qquad (12\text{–}15)$$

where y is the desired variable within the interval $\infty > l/d > \frac{1}{4}$ and y_∞, y_1, $y_{1/2}$, and $y_{1/4}$ are the variables corresponding to l/d ratios of ∞, 1, $\frac{1}{2}$, and $\frac{1}{4}$, respectively.

12–9 Steady-State Conditions in Self-Contained Bearings

The case in which the lubricant carries away all of the enthalpy increase from the journal–bushing pair has already been discussed. Bearings in which the warm lubricant stays within the bearing housing will now be addressed. These bearings are called *self-contained* bearings because the lubricant sump is within the bearing housing and the lubricant is cooled within the housing. These bearings are described as *pillow-block* or *pedestal* bearings. They find use on fans, blowers, pumps, and motors, for example. Integral to design considerations for these bearings is dissipating heat from the bearing housing to the surroundings at the same rate that enthalpy is being generated within the fluid film.

In a self-contained bearing the sump can be positioned as a keywaylike cavity in the bushing, the ends of the cavity not penetrating the end planes of the bushing. Film oil exits the annulus at about one-half of the relative peripheral speeds of the journal and bushing and slowly tumbles the sump lubricant, mixing with the sump contents. Since the film in the top "half" of the cap has cavitated, it contributes essentially nothing to the support of the load, but it does contribute friction. Bearing caps are in use in which the "keyway" sump is expanded peripherally to encompass the top half of the bearing. This reduces friction for the same load, but the included angle β of the bearing has been reduced to 180°. Charts for this case were included in the Raimondi and Boyd paper.

The heat given up by the bearing housing may be estimated from the equation

$$\dot{Q}_{\text{loss}} = UA(T_b - T_\infty) \qquad (12\text{–}16)$$

where $\dot{Q}_{\text{loss}}$ = heat dissipated, Btu/h

U = combined overall coefficient of radiation and convection heat transfer, Btu/(h · ft^2 · °F)

A = surface area of bearing housing, ft^2

T_b = surface temperature of the housing, °F

T_∞ = ambient temperature, °F

The overall coefficient U depends on the material, surface coating, geometry, even the roughness, the temperature difference between the housing and surrounding objects, and air velocity. After Karelitz,[14] and others, in ordinary industrial environments, the overall

14. G. B. Karelitz, "Heat Dissipation in Self-Contained Bearings," Trans. ASME, Vol. 64, 1942, p. 463; D. C. Lemmon and E. R. Booser, "Bearing Oil-Ring Performance," Trans. ASME, J. Bas. Engin., Vol. 88, 1960, p. 327.

coefficient U can be treated as a constant. Some representative values are

$$U = \begin{cases} 2\ \text{Btu/(h} \cdot \text{ft}^2 \cdot {}^\circ\text{F)} & \text{for still air} \\ 2.7\ \text{Btu/(h} \cdot \text{ft}^2 \cdot {}^\circ\text{F)} & \text{for shaft-stirred air} \\ 5.9\ \text{Btu/(h} \cdot \text{ft}^2 \cdot {}^\circ\text{F)} & \text{for air moving at 500 ft/min} \end{cases}$$

An expression similar to Eq. (12–16) can be written for the temperature difference $T_f - T_b$ between the lubricant film and the housing surface. This is possible because the bushing and housing are metal and very nearly isothermal. If one defines $\bar{T}_f$ as the *average* film temperature (halfway between the lubricant inlet temperature T_s and the outlet temperature $T_s + \Delta T$), then the following proportionality has been observed between $\bar{T}_f - T_b$ and the difference between the housing surface temperature and the ambient temperature, $T_b - T_\infty$:

$$\bar{T}_f - T_b = \alpha(T_b - T_\infty) \tag{a}$$

where $\bar{T}_f$ is the average film temperature and α is a constant depending on the lubrication scheme and the bearing housing geometry. Equation (*a*) may be used to estimate the bearing housing temperature. Table 12–2 provides some guidance concerning suitable values of α. The work of Karelitz allows the broadening of the application of the charts of Raimondi and Boyd, to be applied to a variety of bearings beyond the natural circulation pillow-block bearing. In other words, the experience with pillow-block bearings that produced the estimate $T_b = (\bar{T}_f - T_\infty)/2$ can be greatly expanded in generality of the Eqs. (*e*) through (*i*) of Sec. 12–3.

Solving Eq. (*a*) for T_b and substituting into Eq. (12–16) gives the bearing heat loss rate to the surroundings as

$$\dot{Q}_{\text{loss}} = \frac{U_0 A_0}{1+\alpha}(T_f - T_\infty) \tag{12–17a}$$

$$T_b = \frac{T_f + \alpha T_\infty}{1+\alpha} \tag{12–17b}$$

In beginning a steady-state analysis the average film temperature is unknown, hence the viscosity of the lubricant in a self-contained bearing is unknown. Finding the equilibrium temperatures is an iterative process wherein a trial average film temperature (and the corresponding viscosity) is used to compare the heat generation rate and the heat loss rate. An adjustment is made to bring these two heat rates into agreement. This can be done on paper with a tabular array to help adjust $\bar{T}_f$ to achieve equality between heat generation and loss rates. A root-finding algorithm can be used. Even a simple one can be programmed for a digital computer.

Karelitz's parameter α can be introduced into Eq. (*e*) of Sec. 12–3 to give

$$\dot{Q}_{\text{loss}} = \frac{U_0 A_0}{1+\alpha}(T_f - T_\infty)$$

Table 12–2

Lubrication System	Conditions	Range of α
Oil ring	Moving air	1–2
	Still air	$\frac{1}{2}$–1
Oil bath	Moving air	$\frac{1}{2}$–1
	Still air	$\frac{1}{5}$–$\frac{2}{5}$

Equation (h) of Sec. 12–3 can now read

$$\bar{T}_f = T_\infty + \frac{4(1+\alpha)\pi^2 2545}{1050}\frac{\mu N_j^2 l r^3}{U_0 A_0 c}$$

Equation (i) of Sec. 12–3 can be expressed as

$$\Delta T = \frac{U_0 A_0}{1+\alpha}\frac{T_f - T_\infty}{3600 l c \rho \pi r N_j C_P}$$

and the more general expression for the housing temperature Eq. (12–17b)

$$T_b = \frac{\bar{T}_f + \alpha T_\infty}{1+\alpha}$$

applies. Other lettered equations of Sec. 12–3 need not be modified. You can recognize the affected equations by the presence of the $1+\alpha$ term.

Analysis of a Self-Contained Natural Circulation Bearing

It is useful to present an example of the analysis of a self-contained bearing before considering either adequacy assessment or design.

EXAMPLE 12–4

Consider a pillow-block bearing with a keyway sump, whose journal rotates at 900 r/min in shaft-stirred air at 70°F with $\alpha = 1$. The lateral area of the bearing is 40 in^2. The lubricant is SAE grade 20 oil. The gravity radial load is 100 lb and the L/D ratio is unity. The bearing has a journal diameter of 2.000 + 0.000/−0.002 in, a bushing bore of 2.002 + 0.004/−0.000 in.

(a) For a minimum clearance assembly estimate the steady-state temperatures as well as the minimum film thickness and coefficient of friction.

(b) Repeat part a for a radial load of 200 lb.

Solution

(a) The least radial clearance c is $c_{\min}$, and equal to

$$c = c_{\min} = \frac{2.002 - 2.000}{2} = 0.001 \text{ in}$$

$$P = \frac{W}{lD} = \frac{100}{(2)2} = 25 \text{ psi}$$

$$S = \left(\frac{r}{c}\right)^2 \frac{\mu N_j}{P} = \left(\frac{1}{0.001}\right)^2 \frac{\mu'(15)}{10^6(25)} = 0.6\,\mu'$$

The friction horsepower is $(hp)_f$, found as follows:

$$(hp)_f = \frac{f W r N_j}{1050} = \frac{W N_j c}{1050}\frac{fr}{c} = \frac{100(15)0.001}{1050}\frac{fr}{c} = 0.001\,429\frac{fr}{c} \text{ hp}$$

The heat generation rate $\dot{Q}_{\text{gen}}$ is

$$\dot{Q}_{\text{gen}} = 2545(hp)_f = 2545(0.001\,429)fr/c = 3.634 fr/c$$

The rate of heat loss to the environment $\dot{Q}_{\text{loss}}$ is

$$\dot{Q}_{\text{loss}} = \frac{UA}{\alpha+1}(\bar{T}_f - 70) = \frac{2.7(40)}{144(1+1)}(\bar{T}_f - 70) = 0.375(\bar{T}_f - 70) \text{ Btu/h}$$

The temperature rise ΔT_f is

$$\Delta T_F = \frac{9.70\Delta T_F}{P_{\text{psi}}}\frac{P_{\text{psi}}}{9.70} = \left(\frac{9.70\Delta T_F}{P_{\text{psi}}}\right)\frac{25}{9.70} = 2.58\left(\frac{9.70\Delta T_F}{P_{\text{psi}}}\right)$$

Build a table as follows for trial values of $\bar{T}_f$ of 190 and 195°F:

Trial $\bar{T}_f$	μ'	S	fr/c	$\dot{Q}_{gen}$	$\dot{Q}_{loss}$
190	1.15	0.69	13.6	49.6	45.0
195	1.03	0.62	12.2	44.6	46.9

The temperature at which $\dot{Q}_{\text{gen}} = \dot{Q}_{\text{loss}} = 46.3$ Btu/h is 193.4°F. Rounding $\bar{T}_f$ to 193°F we find $\mu' = 1.08$ and $S = 0.6(1.08) = 0.65$. From Fig. 12–23, $9.70\Delta T/P = 4.25$ and

$$\Delta T = 2.58(4.25) = 10.97°\text{F}$$

$$T_i = T_s = \bar{T}_f - \Delta T/2 = 193 - 11.2 = 187.5°\text{F}$$

$$T_{\max} = T_i + \Delta T = 187.5 + 11 = 198.5°\text{F}$$

$$T_b = \frac{\bar{T}_f + \alpha T_\infty}{1+\alpha} = \frac{193 + (1)70}{1+1} = 131.5°\text{F}$$

The minimum film thickness from Fig. 12–14 is

$$h_0 = \frac{h_0}{c}c = 0.79(0.001) = 0.000\,79 \text{ in}$$

The coefficient of friction from Fig. 12–17 is

$$f = \frac{fr}{c}\frac{c}{r} = 12.79\frac{0.001}{1} = 0.012\,79$$

The parasitic friction torque T is

$$T = fWr = 0.012\,79(100)(1) = 1.28 \text{ in} \cdot \text{lbf}$$

(*b*) $P = 200/(2 \cdot 2) = 50$ psi, $S = 0.6\mu'(25/60) = 0.3\mu'$, and

$$(hp)_f = 2(0.001\,429\,fr/c) = 0.002\,86 fr/c$$

The heat generation rate is

$$\dot{Q}_{\text{gen}} = 2545(hp)_f = 2545(0.002\,86)fr/c = 7.27fr/c$$

The heat loss rate $\dot{Q}_{\text{loss}}$ is $0.375(\bar{T}_f - 70)$, unchanged. The temperature rise ΔT_F is

$$\Delta T_F = \frac{9.70\Delta T_F}{P_{\text{psi}}}\frac{P_{\text{psi}}}{9.70} = \frac{9.70\Delta T_F}{P_{\text{psi}}}\frac{50}{9.70} = 5.15\frac{9.70\Delta T_F}{P_{\text{psi}}}$$

Build a table using trial average film temperatures of 190 and 200°F:

Trial $\bar{T}_f$	μ'	S	fr/c	$\dot{Q}_{gen}$	$\dot{Q}_{loss}$
190	1.15	0.35	7.5	54.5	45.0
200	0.98	0.29	6.6	48.0	48.8

from which $\dot{Q}_{gen} = \dot{Q}_{loss} = 48.5$ at $\bar{T}_f = 196.3$°F. Rounding to 196°F, $\mu' = 1.01$, $S = 0.3(1.01) = 0.30$. From Fig. 12–23, $(9.70\Delta T_F/P_{psi}) = 2.2$, so

$$\Delta T_F = 5.15(2.2) = 11.3°\text{F}$$

$$T_i = T_s = \bar{T}_f - \Delta T_F/2 = 196 - 11.3/2 = 190.4°\text{F}$$

$$T_{max} = T_i + \Delta T_F = 190.4 + 11.3 = 201.7°\text{F}$$

$$T_b = \frac{\bar{T}_f + \alpha T_\infty}{1+\alpha} = \frac{196 + (1)70}{1+1} = 133°\text{F}$$

The minimum film thickness from Fig. 12–14 is

$$h_0 = \frac{h_0}{c}c = 0.62(0.001) = 0.000\,62 \text{ in}$$

The coefficient of friction from Fig. 12–17 is

$$f = \frac{fr}{c}\frac{c}{r} = 6.0\frac{0.001}{1} = 0.006$$

The parasitic friction torque T is

$$T = fWr = 0.0006(200)(1) = 1.20 \text{ in} \cdot \text{lbf}$$

Example 12–4 illustrates a highly nonlinear problem. Let's summarize some results:

W, lbf	$\bar{T}_f$, °F	ΔT_F, °F	T_i, °F	T_{max}, °F	T_b, °F	h_0, in	T, in · lbf	f
100	193	11.0	187.5	198.5	131.5	0.000 79	1.28	0.0126
200	196	11.3	190.4	201.7	133	0.000 62	1.20	0.006

When the load is doubled,

- The average film temperature $\bar{T}_f$ increases about 3°F.
- The temperature rise through the film pass increases a fraction of a degree.
- The sump temperature T_s and inlet temperature T_i increase about 3°F.
- The maximum temperature T_{max} increased about 1.5°F.
- The minimum film thickness was reduced about 20 percent.
- The coefficient of friction was cut about in half.
- The parasitic friction torque was reduced only slightly.

If the 200-lbf load was just satisfactory, then a 100-lbf load operates with a large load margin (double), but there is hardly any temperature margin. The frictional behavior is instructive, is it not?

Adequacy Assessment

Analyses provide quantitative estimations of problem parameters, but in itself, an analysis does not determine the status—satisfactory or not—of a project. Veteran bearing designer

Trumpler[15] used the criterion of a design factor $n_d \geq 2$ when one or more of the following constraints were tight:

$$h_0 \geq 0.0002 + 0.000\,04D \text{ in} \tag{12–18}$$

$$T_{\text{max}} \leq 250°\text{F} \tag{12–19}$$

$$W_{st}/(lD) \leq 300 \text{ psi} \tag{12–20}$$

Let us review Ex. 12–4 in the light of the adequacy assessment of Trumpler. If the load to be carried is 100 lbf, then the limiting minimum film thickness h_0 is

$$h_0 = 0.0002 + 0.000\,04D = 0.0002 + 0.000\,04(2) = 0.000\,280 \text{ in}$$

whereas the extant minimum film thickness at 200 lbf load is 0.000 62, and the constraint is loose. The limiting lubricant temperature is 250°F and the extant T_{max} at 200 lbf load is 201.7°F, and the constraint is loose. The starting load with no design factor is 100 lbf, so

$$\frac{W_{st}}{lD} = \frac{100}{2(2)} = 25 \text{ psi}$$

and the starting pressure constraint is loose. Since all constraints are loose, the bearing is satisfactory (and underloaded). We do not know the factor of safety, but it is more than 2. To find it, one would have to extend Ex. 12–4 by choosing loads of 300, 400, . . . , until one of the constraints became tight. Then that limiting load divided by the intended load of 100 lbf would reveal the factor of safety.

The result of repeating Ex. 12–4 for loads from 300 to 1400 lbf is tabulated below with the arrows locating Trumpler constraint violations:

W, lbf	h_0, in	T_{max}, °F	P_{st}, psi	Torque T, in · lbf
300	0.000 516	206.7	75	1.35
400	0.000 420	210.7	100	1.39
500	0.000 354	214.7	125	1.43
600	0.000 306	218.8	150	1.47
	→			
700	0.000 265	222.8	175	1.50
800	0.000 231	226.8	200	1.54
900	0.000 207	230.6	225	1.58
1000	0.000 187	234.3	250	1.61
1100	0.000 169	237.8	275	1.65
1200	0.000 153	241.3	→300	1.68
1300	0.000 140	244.5	325	1.69
1400	0.000 125	247.6	350	1.71
		→		

Given the usual impulse to keep the bearing as small as possible, for a desired load the first constraint encountered is film thickness. This observation can be the basis for finding the load corresponding to a tight film constraint. It is useful to examine Fig. 12–24.

15. Op. cit., pp. 192–194.

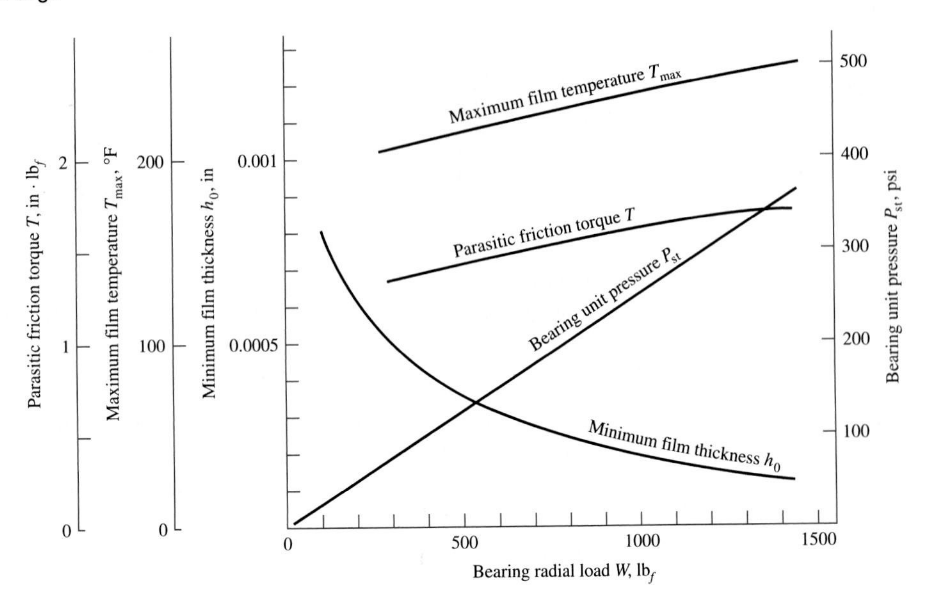

Figure 12–24

Plot of the results of repeating Ex. 12–4 with higher loads.

EXAMPLE 12–5 In Ex. 12–4 the bearing encountered the first constraint of Trumpler at a load of $W = 650$ lbf from inspection of the table above. Taking Trumpler's film thickness constraint as tight, find the load at which it occurs, and using half this load as allowable (design factor of 2), show the conditions at allowable load.

Solution With $c = 0.001$ in, $r = 1$ in, and $D = 2$ in,

$$h_0 = 0.0002 + 0.000\,04D = 0.0002 + 0.000\,04(2) = 0.000\,280 \text{ in}$$

Now

$$\frac{h_0}{c} = \frac{0.000\,280}{0.001} = 0.280$$

From Fig. 12–14 the Sommerfeld number is $S = 0.074$. From Fig. 12–17 $fr/c = 2.2$. From Fig. 12–23, $9.70\Delta T/P = 0.80$, or alternatively, from the curve fit,

$$\frac{9.70\Delta T}{P} = 0.349\,109 + 6.009\,40S + 0.047\,467S^2$$
$$= 0.349\,109 + 6.009\,40(0.074) + 0.047\,467(0.074)^2 = 0.794$$

The rate of heat generation is

$$\dot{Q}_{\text{gen}} = \frac{2545}{1050} WN_j c\frac{fr}{c} = \frac{2545}{1050} W(15)0.001\,(2.2) = 0.080W \text{ Btu/h}$$

The rate of heat loss is

$$\dot{Q}_{\text{loss}} = \frac{UA}{\alpha + 1}(\bar{T}_f - 70) = \frac{2.7(40/144)}{1+1}(\bar{T}_f - 70) = 0.375(\bar{T}_f - 70)$$
$$= 0.080W \text{ Btu/h}$$

from which

$$\bar{T}_f = \frac{0.080W}{0.375} + 70 = 0.213W + 70°\text{F}$$

Using $9.70\Delta T/P = 0.794$,

$$\Delta T = \frac{0.794W}{4(9.70)} = 0.020W\,°\text{F}$$

Introducing W into the Sommerfeld number and solving for W gives

$$W = \frac{r^2}{c^2}\frac{4\mu N_j}{S} = \frac{1^2}{0.001^2}\frac{4\mu'}{10^6}\frac{15}{0.074} = 810.8\mu'\ \text{lbf}$$

Substituting the expression for W above into the T_f equation gives

$$\bar{T}_f = 0.213(810.8\mu') + 70 = 172.7\mu' + 70 \qquad (a)$$

In this equation $\bar{T}_f$ on the left must equal the $\bar{T}_f$ used to evaluate Eq. (*a*). We will use the Seirig curve fit of Table 12–1:

$$\mu' = 0.0136\exp[1271.6/(\bar{T}_f + 95)]\mu\text{reyn} \qquad (b)$$

and construct a table:

Trial $\bar{T}_f$	Eq. (*b*)	$\bar{T}_f$, Eq. (*a*)	
200	1.013	244.9	
210	0.8794	221.9	
212	0.856	217.8	
214	0.833	213.9	(close enough)

Now $W = 810.8\mu' = 810.8(0.833) = 675$ lbf. The allowable load (after Trumpler) is $675/2 = 388$ lbf. The conditions existing at $W = 675$ lbf are $\bar{T}_f = 214$°F and

$$\Delta T = 0.020W = 0.020(675) = 13.5°\text{F}$$

$$T_i = T_s = \bar{T}_f - \Delta T/2 = 214 - 13.5/2 = 207.3°\text{F}$$

$$T_{\text{max}} = T_i + \Delta T = 207.3 + 13.5 = 220.8°\text{F}$$

$$T_b = \frac{\bar{T}_f + \alpha T_\infty}{1+\alpha} = \frac{214 + (1)70}{1+1} = 142°\text{F}$$

We remember that h_0 is 0.000 280 in and $P_{st} = 675/[(2)2] = 168.75$ psi.

Now for the conditions at $W = 338$ lbf. $P = 338/[(2)2] = 84.5$ psi. The Sommerfeld number is

$$S = \left(\frac{1}{0.001}\right)^2 \frac{\mu'}{10^6}\frac{15}{84.5} = 0.1775\mu'$$

$$\dot{Q}_{\text{gen}} = \frac{2545}{1050}WN_jc\frac{fr}{c} = \frac{2545(338)15(0.001)}{1050}\frac{fr}{c} = 12.29\frac{fr}{c}$$

$$\dot{Q}_{\text{loss}} = \frac{UA}{\alpha+1}(\bar{T}_f - 70) = \frac{2.7(40/144)}{1+1}(\bar{T}_f - 70) = 0.375(\bar{T}_f - 70)$$

$$\Delta T = \frac{P}{9.70}\left(\frac{9.70\Delta T}{P}\right) = \frac{84.5}{9.70}\left[0.349\,109 + 6.009\,40(0.1775) + 0.047\,467(0.1775)^2\right] = 12.35°\text{F}$$

We form a table to equate $\dot{Q}_{\text{gen}}$ and $\dot{Q}_{\text{loss}}$:

Trial $\bar{T}_f$	μ'	S	fr/c	$\dot{Q}_{gen}$	$\dot{Q}_{loss}$	
200	1.013	0.180	4	49.16	48.75	
201	0.998	0.177	4	49.16	49.13	(close enough)

Thus

$$\bar{T}_f = 201°\text{F}$$
$$T_i = T_s = T_f - \Delta T/2 = 201 - 12.35/2 = 194.8°\text{F}$$
$$T_{\max} = T_i + \Delta T = 201 + 12.35 = 207.2°\text{F}$$
$$T_b = \frac{T_f + \alpha T_\infty}{1+\alpha} = \frac{201 + (1)70}{1+1} = 135°\text{F}$$
$$h_0/c = 0.48 \text{ since } S = 0.177, \text{ and } h_0 = 0.48(0.001) = 0.000\,480 \text{ in}$$
$$P_{st} = 84.5 \text{ psi}$$
$$T_{\max} = 207.2°\text{F}$$
$$\text{fos} = 675/388 = 2$$

as compared with $h_0 = 0.000\,28$ in, $T_{\max} = 221°$F, and $P_{st} = 169$ psi for the $W = 675$ lbf load.

We have in Ex. 12–5 shown how to estimate the load at which Trumpler's film-thickness constraint becomes tight, halve the load to ensure that the design factor is 2, and find the conditions of the bearing confirming that all constraints (h_0, $T_{\max}$, and P_{st}) are loose. The bearing is satisfactory.

For a single bearing where the clearance is known the steps of an analysis that is part of an adequacy assessment are

- Find the limiting film thickness h_0 following Trumpler.
- Find the Sommerfeld number from h_0/c, Fig. 12–14.
- Find fr/c from Fig. 12–17.
- Find $9.70\Delta T/P$ from either Fig. 12–20 or the curve-fit equation.
- From $\dot{Q}_{gen} = \dot{Q}_{loss}$ find relationship between $\bar{T}_f$ and load W.
- From $9.70\Delta T/P$ relation find the equation relating ΔT and W.
- From the Sommerfeld number find the relationship between load and viscosity.
- By successive trials find the steady-state average film temperature $\bar{T}_f$.
- From $\bar{T}_f$ find load W where the film-thickness constraint is tight.
- Find the allowable load from $W/n_d = W/2$.
- Find the allowable P.
- Relate S to viscosity μ'.
- Form the table

Trial $\bar{T}_f$	μ'	S	fr/c	$\dot{Q}_{gen}$	$\dot{Q}_{loss}$
⋮	⋮	⋮	⋮	⋮	⋮

- Iterate until $\dot{Q}_{\text{gen}} = \dot{Q}_{\text{loss}}$, then analyze bearing for h_0, P_{st}, $T_{\max}$, and n.

We are not quite finished since the clearance in a design exists in a zone due to tolerances, and the designer has the problem that *all* bearings assembled from bushings and journal that are within tolerance must be satisfactory.

12–10 Clearance

In designing a journal bearing for thick-film lubrication, the engineer must select the grade of oil to be used, together with suitable values for P, N, r, c, and l. A poor selection of these or inadequate control of them during manufacture or in use may result in a film that is too thin, so that the oil flow is insufficient, causing the bearing to overheat and, eventually, fail. Furthermore, the radial clearance c is difficult to hold accurate during manufacture, and it may increase because of wear. What is the effect of an entire range of radial clearances, expected in manufacture, and what will happen to the bearing performance if c increases because of wear? Most of these questions can be answered and the design optimized by plotting curves of the performance as functions of the quantities over which the designer has control.

Figure 12–25 shows the results obtained when the performance of a particular bearing is calculated for a whole range of radial clearances and is plotted with clearance as the independent variable. The bearing used for this graph is the one with SAE 20 oil at an inlet temperature of 100°F. The graph shows that if the clearance is too tight, the temperature will be too high and the minimum film thickness too low. High temperatures may cause the bearing to fail by fatigue. If the oil film is too thin, dirt particles may be unable to pass without scoring or may embed themselves in the bearing. In either event, there will be excessive wear and friction, resulting in high temperatures and possible seizing.

To investigate the problem in more detail, Table 12–3 was prepared using the two types of preferred running fits which seem to be most useful for journal-bearing design (see Table 2–8). The results shown in Table 12–3 were obtained using Eqs. (2–10) and (2–11) of Sec. 2–8. Notice that there is a slight overlap, but the range of clearances for the free-running fit is about twice that of the close-running fit.

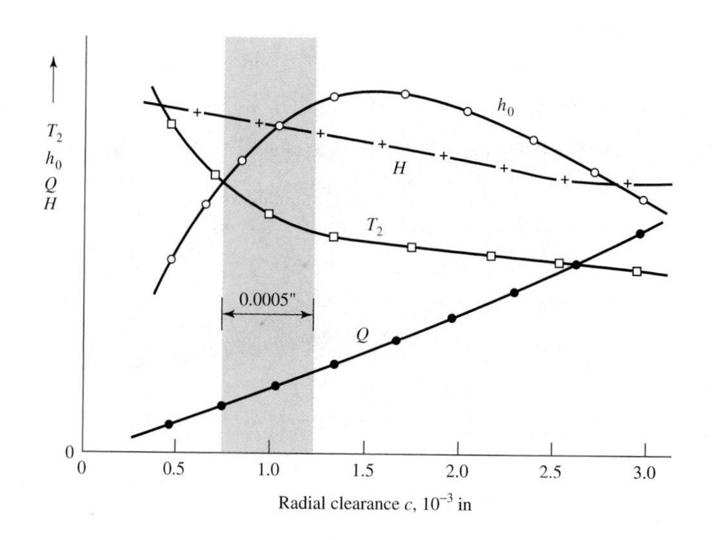

Figure 12–25

A plot of some performance characteristics of the bearing of Ex. 12–1 for radial clearances of 0.0005 to 0.003 in. The bearing outlet temperature is designated T_2. New bearings should be designed for the shaded zone, because wear will move the operating point to the right.

Table 12–3

Maximum, Minimum, and Average Clearances for 1.5-in-Diameter Journal Bearings Based on Type of Fit

Type of Fit	Symbol	Clearance c, in Maximum	Average	Minimum
Close-running	H8/f7	0.001 75	0.001 125	0.000 5
Free-running	H9/d9	0.003 95	0.002 75	0.001 55

The six clearances of Table 12–3 were used in a computer program to obtain the numerical results shown in Table 12–4. These conform to the results of Fig. 12–25, too. Both the table and the figure show that a tight clearance results in a high temperature. Figure 12–26 can be used to estimate an upper temperature limit when the characteristics of the application are known.

It would seem that a large clearance will permit the dirt particles to pass through and also will permit a large flow of oil, as indicated in Table 12–4. This lowers the temperature and increases the life of the bearing. However, if the clearance becomes too large, the bearing becomes noisy and the minimum film thickness begins to decrease again.

In between these two limitations there exist a rather large range of clearances that will result in satisfactory bearing performance.

When both the production tolerance and the future wear on the bearing are considered, it is seen, from Fig. 12–25, that the best compromise is a clearance range slightly

Table 12–4

Performance of 1.5-in-Diameter Journal Bearing with Various Clearances. (SAE 20 Lubricant, $T_1 = 100°$F, $N = 30$ r/s, $W = 500$ lb, $L = 1.5$ in)

c, in	T_2, °F	h_0, in	f	Q, in^3/s,	H, Btu/s
0.000 5	226	0.000 38	0.011 3	0.061	0.086
0.001 125	142	0.000 65	0.009 0	0.153	0.068
0.001 55	133	0.000 77	0.008 7	0.218	0.066
0.001 75	128	0.000 76	0.008 4	0.252	0.064
0.002 75	118	0.000 73	0.007 9	0.419	0.060
0.003 95	113	0.000 69	0.007 7	0.617	0.059

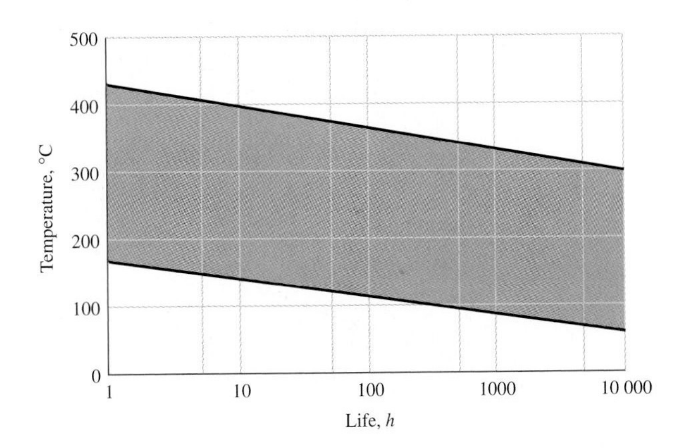

Figure 12–26

Temperature limits for mineral oils. The lower limit is for oils containing antioxidants and applies when oxygen supply is unlimited. The upper limit applies when insignificant oxygen is present. The life in the shaded zone depends on the amount of oxygen and catalysts present. *[Source: M. J. Neale (ed.), Tribology Handbook, Section B1, Newnes-Butterworth, London, 1975.]*

to the left of the top of the minimum-film-thickness curve. In this way, future wear will move the operating point to the right and increase the film thickness and decrease the operating temperature.

Journal and Bushing Tolerances and Radial Clearance

A journal bearing assembled by random selection of a journal and a bushing, each often having a uniform distribution of diameters, produces a bounded distribution of radial clearance. For example, if the journal diameter has a distribution $U[0.9985, 1.000]$ in, and the bushing diameter has a distribution $U[1.005, 1.008]$ in, then the smallest difference in diameters is $1.005 - 1.000 = 0.005$ in, and the largest difference in diameters is $1.008 - 0.9985 = 0.0095$ in. This makes the smallest radial clearance $c_{\min} = 0.005/2 = 0.0025$ in, and the largest radial clearance $c_{\max} = 0.0095/2 = 0.004\,75$ in. The radial clearance lies in the range $0.0025 \le c \le 0.004\,75$ in *for certain*. The difference between two bounded distributions is a bounded distribution. An absolute-tolerance approach is quite workable.

If a journal and a bushing are unilaterally toleranced as shown in Fig. 12–27, then the smallest radial clearance $c_{\min}$ is given by

$$c_{\min} = \frac{B - 0 - (D - 0)}{2} = \frac{B - D}{2} \tag{12–21}$$

The largest radial clearance $c_{\max}$ is given by

$$c_{\max} = \frac{B + b - (D - d)}{2} = \frac{B - D}{2} + \frac{b + d}{2} \tag{12–22}$$

The distribution of **c** is symmetrical. The mean radial clearance $\bar{c}$ is

$$\bar{c} = \frac{c_{\min} + c_{\max}}{2} = \frac{\dfrac{B - D}{2} + \dfrac{B + D}{2} + \dfrac{b + d}{2}}{2} = \frac{B - D}{2} + \frac{b + d}{4} \tag{12–23}$$

The radial clearance range can be found from Eqs. (12–21) and (12–22) to be

$$\Delta c = c_{\max} - c_{\min} = \frac{B - D}{2} + \frac{b + d}{2} - \frac{B - D}{2} = \frac{b + d}{2} \tag{12–24}$$

Vendor-supplied bushings usually have B in nominal sizes and the unilateral tolerance b is stepwise-increasing, as in $b = \text{INT}(1 + B)/1000$. The mathematical operator INT means the integer part of $(1 + B)$. With in-house production of the journals, the unilateral tolerance d is stepwise-increasing, as in $d = \text{INT}(3 + 2D)/10\,000$. In tabular form the tolerance b and d might be displayed as

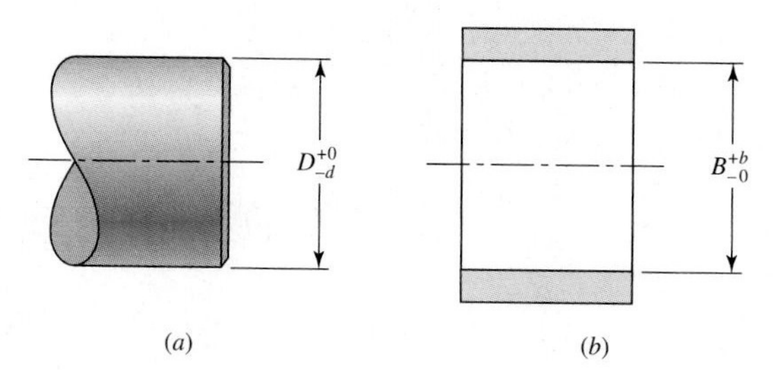

Figure 12–27

(*a*) The unilateral dimensioning of a journal diameter, plus nothing and minus *d*. The nominal diameter is really a maximum diameter. (*b*) The unilateral dimensioning of a bushing bore, plus *b* and minus nothing. The nominal bushing bore diameter *B* is really the minimum bore diameter.

B, in (including and larger)	b, in	D, in (including and larger)	d, in
0	0.001	0	0.0003
1	0.002	0.5	0.0004
2	0.003	1.0	0.0005
3	0.004	1.5	0.0006
		2.0	0.0007

Of the four parameters, D, d, B, and b, two are generally decided a priori, b and d, and D and B can be taken as decision variables. However, $r_{max} = D/2$ and c_{min} are usually taken as decision variables since film bearing theory is expressed in terms of journal radius r and clearance c. A point r_{max}, c_{min} in the rc plane has displayed adjacent to it the zone of tolerance-permitted assemblies as depicted in Fig. 12–28. With b and d as a priori decisions, from the descriptor location r_{max}, c_{min}, one can recover

$$D = 2r$$

$$B = D + 2c_{min} \qquad \text{from Eq. (12–21)}$$

$$c_{max} = \frac{B + D}{2} + \frac{b + d}{2} \qquad \text{from Eq. (12–22)}$$

The rectangular zone of possible assemblies is sometimes called the *design window*. The design window is locally of fixed geometry, but large movements in the rc plane involve stepwise changes in b or d or both. Bear in mind the design window is a "rubber window." The feasible domain for self-contained journal bearings is depicted in Fig. 12–29. The figure of merit might be the minimization of parasitic friction torque whose gradient lines are shown as arrows in Fig. 12–29. The P_{st} constraint can be further to the right and potentially active. The search for optimal location of the design window

Figure 12–28

The specifications D^{+0}_{-d} and B^{+b}_{-0} result in an ensemble of assemblies whose rc coordinates fall within the shaded parallelogram shown. The circumscribed rectangle is known as the design window. See Case Study 12b.

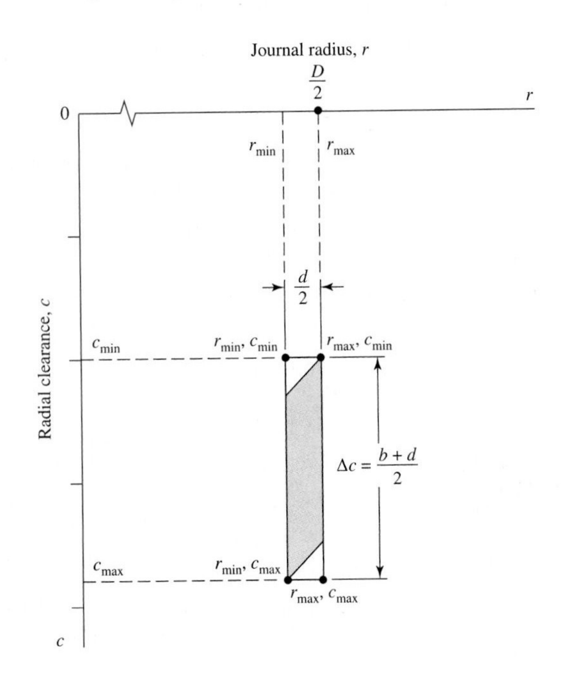

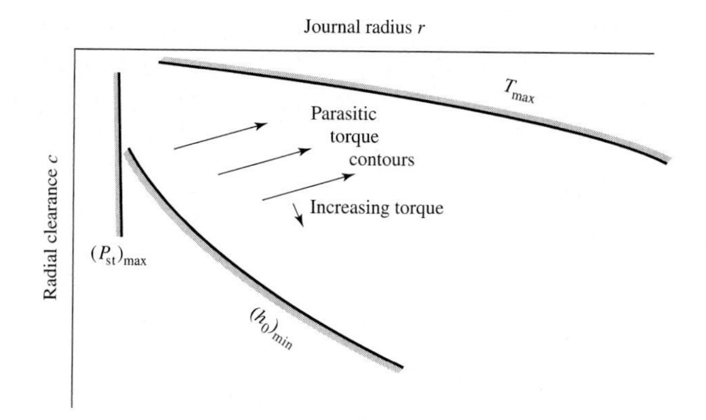

Figure 12–29

The design plane for self-contained natural-circulation journal bearings showing Trumpler's constraints and the parasitic friction torque contours in the feasible domain. The design window is to be placed in the upper left-hand region of the feasible domain, touching one or more constraints.

involves moving to the upper left in Fig. 12–29. One might locate with the lower left-hand corner of the design window touching the film-thickness constraint, or the upper right-hand corner touching the $T_{\max}$ constraint. It could have the left-hand edge touching the P_{st} constraint.

CASE STUDY 12b The Origin of the Design Window in Film Bearings

Journals and bushings are defined geometrically with unilateral tolerances as depicted in Fig. 12–27. The minimum and maximum radial clearance are expressible as

$$c_{\min} = \frac{B - D}{2} \qquad c_{\max} = \frac{B - D}{2} + \frac{b + d}{2}$$

Usually bushings are vendor-supplied with nominal-sized bushing bore diameters B and associated tolerance b. Journals are usually manufactured in-house as part of a shaft. Journals are usually ground to control size and finish. The tolerance d on the journal is usually smaller than that of the associated b of the bushing. The diameter D is used to create the bearing of choice. The design variables are usually journal radius r and radial clearance c, creating an rc plane such as the design space as depicted in Fig. 12–28. Film theory is usually expressed in terms of r and c.

A design decision is represented by r, c. The journal diameter D is $2r$, and the bushing bore diameter B is $D + 2c_{\min}$. The definition of radial clearance is $c = r_B - r$, where r_B is the radius of a particular bushing bore, assembled, and r is $D/2$ of a particular journal diameter, assembled. We rearrange the definition to read

$$r + c = r_B$$

Due to the tolerance b on B, $r + c$ lies in range $B/2 \le r_B \le (B + b)/2$, creating a pair of inequalities:

$$r + c \ge B/2 \qquad \text{or} \qquad r \ge B/2 - c$$
$$r + c \le (B + b)/2 \qquad \text{or} \qquad r \le (B + b)/2 - c$$

The upper equation graphs on the rc plane as a linear locus with ordinate and abscissa intercepts of $B/2$ and a slope of -1. The lower equation graphs as a linear locus with ordinate and abscissa intercepts of $(B + b)/2$ and a slope of -1. This is shown in Fig. CS12b–1. The clearance range is $c_{\min} \le c \le c_{\max}$. These are also plotted on Fig. CS12b–1. The shaded zone contains all the bearings resulting from the random (or otherwise) assembly of bushings and journals. Now rotate the entire graph of Fig. CS12b–1 90° counterclockwise, to agree with the presentation of Fig. 12–28. For the bearing $B = 2.000$ in, $b = 0.003$ in, $D = 1.996$ in,

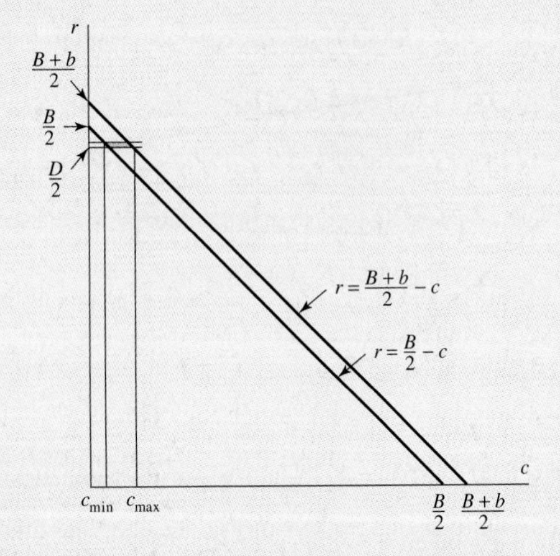

Figure CS12b–1

Bounding equations $r = B/2 - c$, $r = (B+b)/2 - c$, $c = c_{min}$, and $c = c_{max}$ shown on rc coordinates. These equations define a parallelogram containing all the assemblies.

and $d = 0.0006$ in, Fig. CS12b–2 results. All bearings in the ensemble are to be found in the parallelogram. Constraints encountered are usually at the upper right-hand corner of the parallelogram, along the left edge, or at the lower right-hand corner. By circumscribing the parallelogram with a rectangle, and designating the rectangle as the *design window*, no harm has been done, as the parallelogram and the rectangle touch constraints at common locations.

With every change in B, b, D, and d, the design window moves and sometimes changes in size. The window analogy takes on a "rubber" aspect with respect to size. This window, albeit rubbery, is the zone in which the designer sees the ensemble of bearings that result from an r, c decision, which translates into B, b, D, and d, and into a design window. Without the design window concept, the desinger only has the applicable equations and inequalities. Keeping all of them satisfied and in mind would be difficult. Programming a computer to help with the chores requires rigorous understanding[16] if the results are to be congruent with nature.

This is how a parallelogram encapsulating all assemblies in the ensemble became defined; how circumscribing the parallelogram with a rectangle did no harm; how a rectangle through which the designer can view consequences came to be called a window; and how a complex

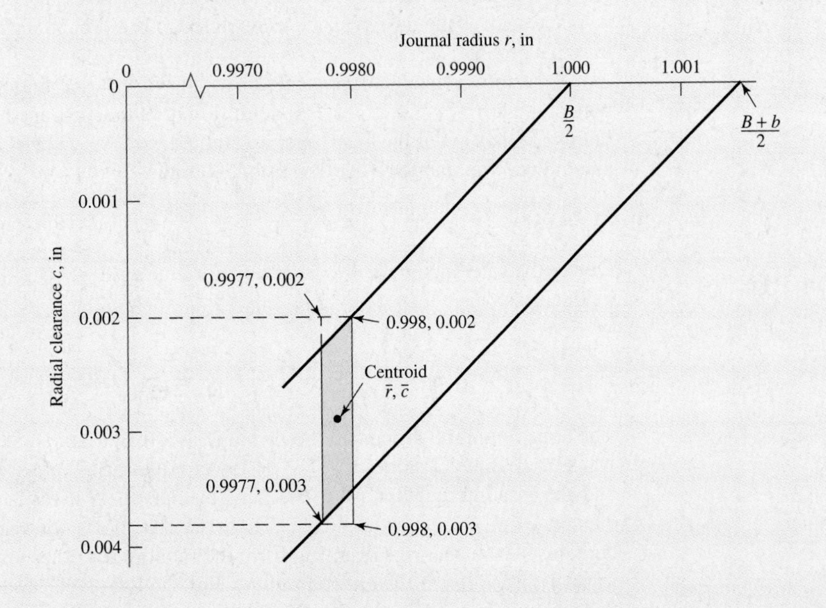

Figure CS12b–2

The parallelogram contains all members of the ensemble. Circumscribing this with a rectangle still contains all members of the ensemble. The rectangle is the design window.

16. This is true only if the programmer wants the results to agree with reality.

design problem, reduced to moving a "rubber" window into the most advantageous position as prompted by a figure-of-merit, became part of the designer's kit of tools.

Viewpoint is important. It can focus our attention on the essentials while improving simplicity.

EXAMPLE 12–6 A self-contained bearing is operating in a 70°F environment with a housing area of 40 in^2 in shaft-stirred air with $\alpha = 1$. In-house journal tolerances are $d = \text{INT}(3+2d)/10\,000$ and vendor-supplied bronze bushings have $b = \text{INT}(1+B)/1000$ in, with B available in nominal sizes in $\frac{1}{8}$-in increments. The gravity load is to be 200 lb and carried at 900 r/min. Specify a bearing for $l/D = 1$. The lubricant is SAE 20.

Solution The solution will be found by trials of bushings.

Decision Since hand assembly of journal and bushing is contemplated, set $c_{\min} = 0.001$ in.

Decision $B = 2.000$ in, then

$$D = B - 2C_{\min} = 2.000 - 2(0.001) = 1.998 \text{ in}$$

$$r = D/2 = 1.998/2 = 0.999 \text{ in}$$

$$b = \text{INT}(1+B)/1000 = \text{INT}(1+2.000)/1000 = 0.003 \text{ in}$$

$$d = \text{INT}(3+2D)/10\,000 = \text{INT}[3+2(1.998)]/10\,000 = 0.006 \text{ in}$$

Since

$$\Delta c = \frac{b+d}{2} = \frac{0.003+0.0006}{2} = 0.0018 \text{ in}$$

$$c_{\max} = c_{\min} + \Delta c = 0.001 + 0.0018 = 0.0028 \text{ in}$$

We will find the load at which the design window just touches the film-thickness constraint in the lower left-hand corner. From Eq. (12–18)

$$h_0 = 0.0002 + 0.00004(1.998) = 0.000\,280 \text{ in}$$

$$\frac{h_0}{c_{\max}} = \frac{0.000\,280}{0.0028} = 0.10$$

From Fig. 12–14, $S = 0.02$, and from Fig. 12–17, $fr/c = 1.1$. Also

$$\frac{9.70\Delta T}{P} = 0.349 + 6(0.02) + 0.047(0.02)^2 = 0.469$$

$$\dot{Q}_{\text{gen}} = \frac{2545}{1050} W N_j c \frac{fr}{c} = \frac{2545}{1050} W(15)0.0028(1.1) = 0.112W \text{ Btu/h}$$

$$\dot{Q}_{\text{loss}} = \frac{UA}{\alpha+1}(\bar{T}_f - T_\infty) = \frac{2.7(40/144)}{1+1}(\bar{T}_f - 70) = 0.375(\bar{T}_f - 70) \text{ Btu/h}$$

$$= \dot{Q}_{\text{gen}} = 0.112W$$

from which

$$\bar{T}_f = \frac{0.112W}{0.375} + 70 = 0.299W + 70°\text{F}$$

$$\Delta T = \frac{P}{9.70}\left(\frac{9.70\Delta T}{P}\right) = \frac{W/4}{9.70}0.469 = 0.012W°\text{F}$$

From the Sommerfeld number definition

$$W = \frac{r^2}{c^2}\frac{4\mu N_j}{S} = \frac{0.9987^2}{0.0028^2}\frac{4\mu' 15}{10^6 0.02} = 382\mu' \text{ lbf}$$

Substituting in the $\bar{T}_f$ equation above gives

$$\bar{T}_f = 0.299(382\mu') + 70 = 114.2\mu' + 70°\text{F} \tag{a}$$

Using the curve fit in Table 12–1,

$$\mu' = 0.0136 \exp[1271.6/(\bar{T}_f + 95)] \tag{b}$$

we find the steady-state temperature by tabulation:

Trial $\bar{T}_f$	μ', Eq. (*b*)	$\bar{T}_f$, Eq. (*a*)	
200	1.013	187.7	
193	1.125	198.5	
195	1.091	194.6	(close enough)

From the above table we declare $\bar{T}_f = 195°\text{F}$ and $\mu' = 1.091$. It follows that the corresponding load is

$$W = 382(1.091) = 417 \text{ lbf}$$

The Trumpler allowable load is $417/2 = 209$ lbf. Next we check the maximum film temperature T_{max} at the upper right corner of the design window, $c = c_{min} = 0.001$in and $r = r_{max} = 0.999$ in:

$$S = \frac{0.999^2}{0.001^2}\frac{\mu'}{10^6}\frac{15}{417/4} = 0.144\mu'$$

$$\dot{Q}_{gen} = \frac{2545}{1050}(417)15(0.001)\frac{fr}{c} = 15.16\frac{fr}{c} \text{ Btu/h}$$

$$\dot{Q}_{loss} = 0.375(T_f - 70) \text{ Btu/h}$$

We find the steady-state average film temperature by tabulation:

Trial T_f	μ'	S	fr/c	$\dot{Q}_{gen}$	$\dot{Q}_{loss}$	
200	1.013	0.146	3.3	50.03	48.75	
210	0.879	0.127	3.0	45.48	52.50	
202	0.984	0.142	3.2	48.51	49.50	(close enough)

We declare $\bar{T}_f = 202°\text{F}$, $S = 0.144$. Then

$$\Delta T = \frac{P}{9.70}\left(\frac{9.70\Delta T}{P}\right) = \frac{417/4}{9.70}[0.349 + 6(0.144) + 0.047(0.144)^2] = 13.05°\text{F}$$

$$T_{\max} = \bar{T}_f + \frac{\Delta T}{2} = 202 + \frac{13.05}{2} = 208.5°\text{F} \qquad \text{(loose constraint)}$$

The bearing with the specification

$$D = 1.998^{+0.0000}_{-0.0006} \text{ in} \qquad B = 2.000^{+0.003}_{-0.000} \text{ in}$$

when carrying a 417-lbf load has a tight film-thickness constraint. The maximum film-thickness constraint is loose and the starting pressure $417/[2(2)] = 104.3$ psi < 300 psi and therefore loose. A load of 200 lbf will have all constraints loose and the factor of safety is $n = 417/200 = 2.085$, which is greater than 2. This is a satisfactory bearing. Others should be examined.

As a matter of perspective the design window has been given nine "lights." Within each light is listed h_0, $T_{\max}$, P_{st}, and $T_{\text{parasitic}}$. Units are in °F, psi, and in · lb respectively,

	r		
c	0.9987	0.99885	0.999
0.0010	0.000 610 203.1 52.4 1.316	0.000610 203.1 52.4 1.316	0.000 611 203.2 52.4 1.316
0.0019	0.000 677 183.5 52.4 1.143	0.000 678 183.5 52.4 1.143	0.000 678 183.5 52.4 1.143
0.0028	0.000 624 177.8 52.4 1.0937	0.000 624 177.8 52.4 1.0939	0.000 624 177.9 52.4 1.0941

It is useful to examine the design window. The left-to-right direction at a radial clearance c shows very little change. In the up-and-down direction along a journal radius r, there are changes in maximum film temperature and parasitic friction torque. To characterize the friction torque of the ensemble one would use the value in the center light of $T = 1.143$ in · lb.

You will be moving back-and-forth between the journal–bushing description and the design window description. The protocols of transformation are displayed below to avoid errors caused by faulty memory of details. See Case Study 12b for how and why the design window is useful.

Journal–Bushing Geometry → Design Window Geometry

Given: B, b, D, d; then

$$r_{max} = \frac{D}{2} \qquad \Delta c = \frac{b+d}{2}$$

$$\Delta r = d/z \qquad c_{min} = \frac{B-D}{2}$$

$$r_{min} = r_{max} - \Delta r \qquad c_{max} = c_{min} + \Delta c$$

$$\bar{r} = \frac{r_{max} + r_{min}}{2} \qquad \bar{c} = \frac{c_{max} + c_{min}}{2}$$

Established:

Corner coordinates r_{max}, c_{min}; r_{min}, c_{min}; r_{min}, c_{max}; r_{max}, c_{max} or Δr by Δc located by $\bar{r}$, $\bar{c}$, or r_{max}, c_{min}

Design Window Geometry → Journal–Bushing Geometry

Given: $r_{max}, r_{min}, c_{max}, c_{min}$	*Given:* $\bar{r}, \bar{c}, \Delta r, \Delta c$	*Given:* $r_{max}, c_{min}, \Delta r, \Delta c$
$D = 2r_{max}$	$d = 2\Delta r$	$D = 2r_{max}$
$B = D + 2r_{min}$	$b = 2\Delta c - d$	$d = 2\Delta r$
$\Delta r = r_{max} - r_{min}$	$c_{max} = \bar{c} + \Delta c/2$	$B = D + 2c_{min}$
$d = 2\Delta r$	$c_{min} = \bar{c} - \Delta c/2$	$b = 2\Delta c - d$
$\Delta c = c_{max} - c_{min}$	$r_{max} = \bar{r} + \Delta r/2$	
$b = 2\Delta c - d$	$r_{min} = \bar{r} - \Delta r/2$	
	$D = 2r_{max}$	
	$B = D + 2c_{min}$	

Established: B, b, D, d

Computer Considerations

Thus far in this chapter we have focused on a step-by-step development of understanding of a self-contained bearing. We have acknowledged the c. 1914 contribution of Edgar Buckingham (using dimensionless groups to reduce the number of parameters) in helping to make complexity tractable. The work of Reynolds in developing the descriptive differential equation, of Petroff and Sommerfeld in developing some early solutions, and the digital computer in the hands of Raimondi and Boyd in solving the finite-length journal bearing with end leakage—all have been presented to enhance understanding of how design may be accomplished when our knowledge is contained in a number of graphs. Our objective is the pursuit of understanding. The exercises showed the relationships of parameters, the iteration of dimensionless plots mixed with equations, all of the necessary algebra, and the level of effort involved. This can be formidable. Journal bearings *are* a formidable subject; to understand them, one simply does the work.

Your guide is your level of understanding. If you design one bearing a year, pencil, paper, and work are the way to go. If you have a substantial number of designs to complete each year, your methodology will have to include the computer.

- Each Raimondi and Boyd chart can be made into a subroutine, accepting Sommerfeld number and l/d and returning the appropriate pi term. This involves curve-fitting, interpolation, and error-messaging.

- An executive program can be written to perform an analysis given all the a priori decisions and the rc coordinates of *a* design decision. The result can be a presentation of

  h_0
  T_{max}
  P
  Torque

  In the days of teletypes and Decwriters and large output paper, an array of analyses 11 rows by 11 columns produced a rectangular display of 121 analyses of 121 bearings. These could be produced several sheets wide and several sheets deep. With scissors and tape, the rc plane could be displayed; contours of h_0, T_{max}, P, and torque could be sketched in with crayon. Then with this overall view in mind, a design window could be moved about to find the desirable location. Returning to the computer, the desirable locale could be enlarged to find the best location. All this was done at a bearing load of $W = 2W_{design}$. Then the load was changed to W_{design} and the window as at the end of Ex. 12–6 was explored.
- The small personal computer screen severely reduces the viewing area to about 5 × 5, and unless your computer has a print screen option, building the big map is not possible. Computer graphics, if available, may be useful. Much has been learned by students exploring the terrain as window after window unfolded, and a store of experience may be acquired after much viewing of rc space.
- The view of a broad area of the rc plane might involve $50 \times 50 = 2500$ analyses. This simply cannot be done without computer assistance.
- Understanding of the problem has a human basis. What you program is based on your understanding. Output can be manually verified, and your perspective on the size of numbers will be helpful. There will be programming mistakes, theory mistakes, and iteration termination mistakes. One company, pleased with its early use of computers in gear design, suddenly realized that none of the engineers in the gear department had designed a gearset by hand! The firm introduced a requirement that once every nine months each gear designer had to design a gearset by hand, then confirm it with a run on the computer. The programming was upgraded, broadened—and even corrected.

A Decision Set for Self-Contained Full-Journal Bearings

A decision set suitable for use with self-contained full-journal bearing can be

A priori decisions:
- Length-to-diameter ratio l/D
- Lubricant grade
- Plumbing connectivity
- Material and surface roughness of journal and bushing
- Function: load and speed

Design variables:
- Journal radius r
- Radial clearance c

The design plane is the rc plane and its constraints are diagrammed in Fig. 12–29. After a priori decisions are made, the length and width of the design window can be established. The design window is moved to the most advantageous location in feasible space.

The approach in Ex. 12–6 was to place the design window in a "good" location, with the lower left-hand corner of the design window against the film-thickness constraint, and then find the corresponding load, cut the load in half, and examine the upper left-hand corner of the design window for T_{max} constraint looseness. The starting characteristic pressure P was checked for looseness. Since the bearing examined could meet Trumpler's criteria at the design load, the design was accepted. Had the allowable load been less than the required 200 lbf, the next larger size bearing would be examined using the method of Ex. 12–6.

When Ex. 12–6 is viewed as a design problem, useful perspective is added. The P_{st} constraint can be written as

$$P_{st} = \frac{W_{st}}{lD} = \frac{W_{st}}{(l/D)D^2} \leq 300 \text{ psi}$$

$$D \geq \sqrt{\frac{W_{st}}{300(l/D)}} = \sqrt{\frac{200}{300(1)}} = 0.816 \text{ in}$$

Since the bushings come in $\frac{1}{8}$-in increments a table can be made:

B	b	d ≥
0.875	0.001	0.0004
1.000	0.002	0.0004
1.125	0.002	0.0005
1.25	0.002	0.0005
1.375	0.002	0.0005
1.50	0.002	0.0006
1.625	0.002	0.0006
1.750	0.002	0.0006
1.875	0.002	0.0006
2.000	0.003	0.0006
2.125	0.003	0.0007
⋮	⋮	⋮

All the bearings in the table have a loose P_{st} constraint. Instead of an infinite number of bearings, there are several. The smallest bearings in the table will not be able to support the load. Remember the approach in Ex. 12–6 was to select an rc combination that might be satisfactory. To get the design window in the upper left of Fig. 12–29, c_{min} was made as small as consistent with assembly considerations. A bearing with $B = 2.000$ in was the initial trial. It could carry 417 lbf hitting the film-thickness constraint, and so could carry 209 lbf with a factor of safety of 2. Had the 2-in bore been unable to carry 200 lbf, the next larger possible bearing with a 2.125-in bore would be examined. So, instead of an intractable design problem, a satisfactory design can be executed by hand without undue difficulty.

12–11 Pressure-Fed Bearings

With our experience so far with self-contained natural-circulation bearings we observe that their load-carrying capacity is not impressive. Several hundred pounds of radial load required about a 2-in-diameter journal. The factor limiting better performance is the heat-dissipation capability of the bearing. A first thought of a way to increase heat dissipation is to cool the sump with an external fluid such as water. The high-temperature problem is in the film where the heat is generated but cooling is not possible in the film until later. This does not protect against exceeding the maximum allowable temperature of the lubricant. A second alternative is to reduce the *temperature rise* in the film by dramatically increasing the rate of lubricant flow. The lubricant itself is reducing the temperature rise. A water-cooled sump may still be in the picture. To increase lubricant flow, an external pump must be used with lubricant supplied at pressures of tens of pounds per square inch gage. Because the lubricant is supplied to the bearing under pressure, such bearings are called *pressure-fed bearings*.

To force a greater flow through the bearing and thus obtain an increased cooling effect, a common practice is to use a circumferential groove at the center of the bearing, with an oil-supply hole located opposite the load-bearing zone. Such a bearing is shown in Fig. 12–30. The effect of the groove is to create two half-bearings, each having a smaller l/d ratio than the original. The groove divides the pressure-distribution curve into two lobes and reduces the minimum film thickness, but it has wide acceptance among lubrication engineers because such bearings carry more load without overheating.

To set up a method of solution for oil flow, we shall assume a groove ample enough that the pressure drop in the groove itself is small. Initially we will neglect eccentricity and then apply a correction factor for this condition. The oil flow, then, is the amount which flows out of the two halves of the bearing in the direction of the concentric shaft. If we neglect the rotation of the shaft, we obtain the force situation shown in Fig. 12–31. Here we designate the supply pressure by p_s and the pressure at any point by p. Laminar flow is assumed, and we are interested in the static equilibrium of an element of width dx, thickness $2y$, and unit depth. Note particularly that the origin of the reference system has been chosen at the midpoint of the clearance space. The pressure is $p + dp$ on the left face and p on the right face, and the upper and lower surfaces are acted upon by the shear stresses τ. The equilibrium equation is

$$2y(p + dp) - 2yp - 2\tau\, dx = 0 \tag{a}$$

Expanding and canceling terms, we find that

$$\tau = y\frac{dp}{dx} \tag{b}$$

Newton's equation for viscous flow [Eq. (12–1)] is

$$\tau = \mu\frac{du}{dy}$$

However, in this case we have taken τ in a negative direction. Also, du/dy is negative because u decreases as y increases. We therefore write Newton's equation in the form

$$-\tau = \mu\left(-\frac{du}{dy}\right) \tag{c}$$

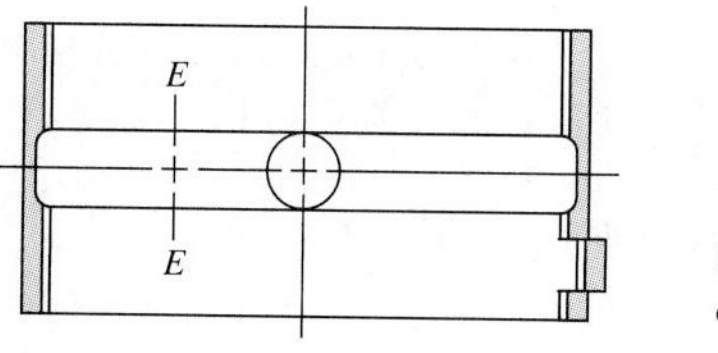

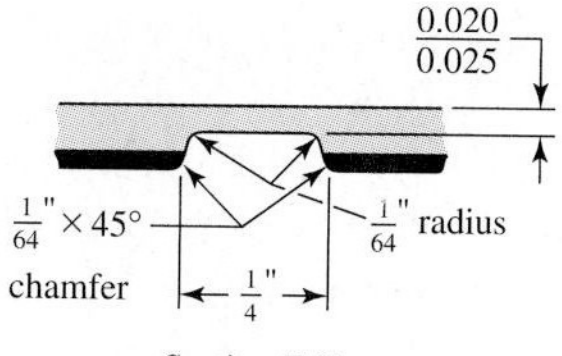

Figure 12–30

Centrally located full annular groove. *(Courtesy of the Cleveland Graphite Bronze Company, Division of Clevite Corporation.)*

Figure 12–31

Flow of lubricant from a pressure-fed bearing having a central annular groove.

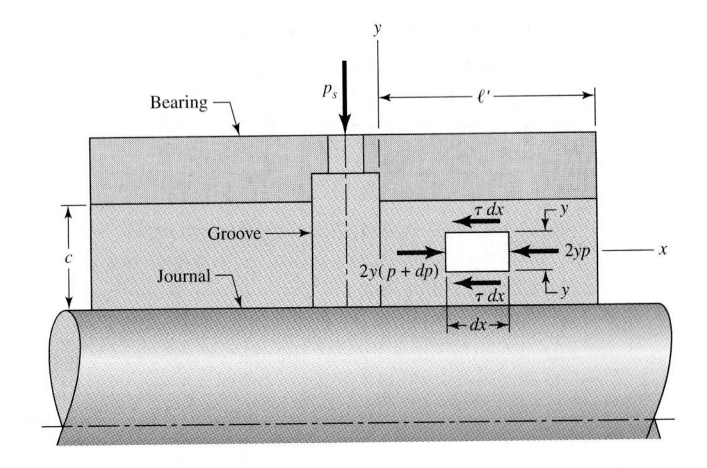

Now eliminating τ from Eqs. (b) and (c) gives

$$\frac{du}{dy} = \frac{1}{\mu}\frac{dp}{dx}y \tag{d}$$

Treating dp/dx as a constant and integrating with respect to y gives

$$u = \frac{1}{2\mu}\frac{dp}{dx}y^2 + C_1 \tag{e}$$

At the boundaries, where $y = \pm c/2$, the velocity u is zero. Using one of these conditions in Eq. (e) gives

$$0 = \frac{1}{2\mu}\frac{dp}{dx}\left(\frac{c}{2}\right)^2 + C_1$$

or

$$C_1 = -\frac{c^2}{8\mu}\frac{dp}{dx}$$

Substituting this constant in Eq. (e) yields

$$u = \frac{1}{8\mu}\frac{dp}{dx}(4y^2 - c^2) \tag{f}$$

Next let us assume that the oil pressure varies linearly from the center to the end of the bearing, as shown in Fig. 12-32. Since the equation of a straight line may be written

$$p = Ax + B$$

with $p = p_s$ at $x = 0$ and $p = 0$ at $x = l'$, substituting these end conditions gives

$$A = \frac{p_s}{l'} \qquad B = p_s$$

or

$$p = -\frac{p_s}{l'}x + p_s \tag{g}$$

and therefore

$$\frac{dp}{dx} = -\frac{p_s}{l'} \tag{h}$$

Figure 12–32

A linear oil–pressure distribution is assumed.

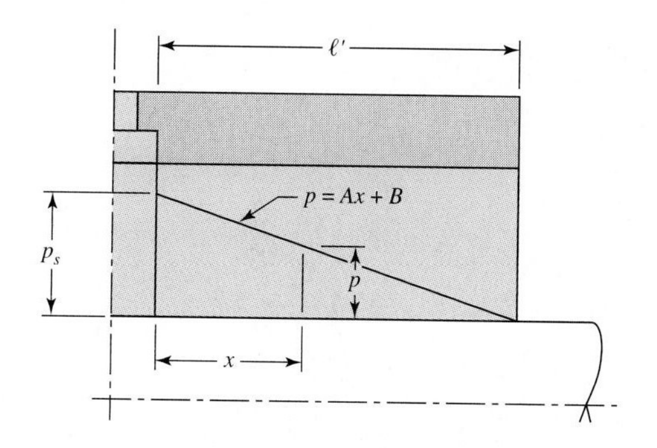

We can now substitute Eq. (h) in Eq. (f) to get the relationship between the oil velocity and the coordinate y:

$$u = \frac{p_s}{8\mu l'}(c^2 - 4y^2) \tag{12–25}$$

Figure 12–33 shows a graph of this relation fitted into the clearance space c so that you can see how the velocity of the lubricant varies from the journal surface to the bearing surface. The distribution is parabolic, as shown, with the maximum velocity occurring at the center, where $y = 0$. The magnitude is, from Eq. (12–25),

$$u_{\text{max}} = \frac{p_s c^2}{8\mu l'} \tag{i}$$

The average ordinate of a parabola is two-thirds of the maximum; if we also generalize by substituting h for c in Eq. (i), then the average velocity at any angular position θ (Fig. 12–34) is

$$u_{\text{av}} = \frac{2}{3}\,\frac{p_s h^2}{8\mu l'} = \frac{p_s}{12\mu l'}(c - e\cos\theta)^2 \tag{j}$$

We still have a little further to go in this analysis; so be patient. Now that we have an expression for the lubricant velocity, we can compute the amount of lubricant that flows out both ends; the elemental side flow at any position θ (Fig. 12–34) is

$$dQ_s = 2u_{\text{av}}\,dA = 2u_{\text{av}}(rh\,d\theta) \tag{k}$$

where dA is the elemental area. Substituting u_{av} from Eq. (j) and h from Fig. 12–34 gives

$$dQ_s = \frac{p_s r}{6\mu l'}(c - e\cos\theta)3d\theta \tag{l}$$

Figure 12–33

Parabolic distribution of the lubricant velocity.

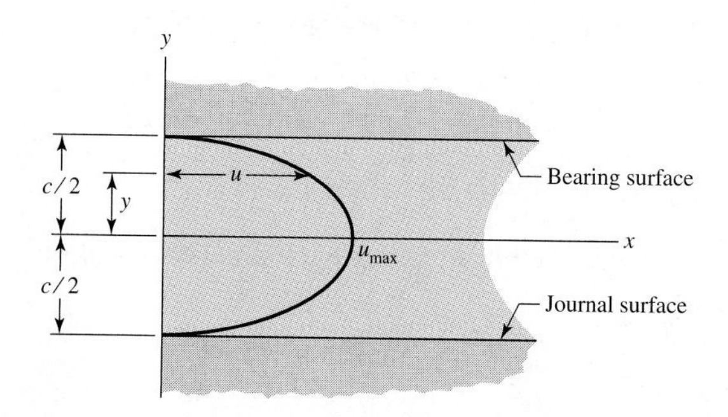

Figure 12–34

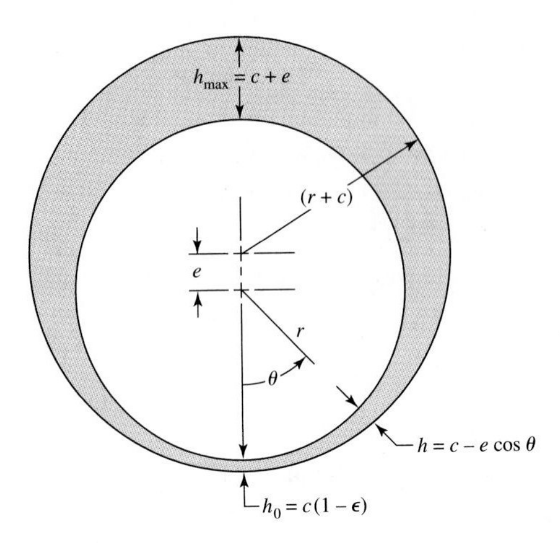

Integrating around the bearing gives the total side flow as

$$Q_s = \int dQ_s = \frac{p_s r}{6\mu l'} \int_0^{2\pi} (c - e\cos\theta)^3 d\theta$$

$$Q_s = \frac{\pi p_s r c^3}{3\mu l'}(1 + 1.5\epsilon^2) \qquad (12\text{–}26)$$

In analyzing the performance of pressure-fed bearings, the bearing length should be taken as l', as defined in Fig. 12–31. The characteristic pressure in each of the two bearings which constitute the pressure-fed bearing assembly P is given by

$$P = \frac{W/2}{2rl'} = \frac{W}{4rl'} \qquad (12\text{–}27)$$

The charts for flow variable and flow ratio (Figs. 12–18 and 12–19) do not apply to pressure-fed bearings. Also, the maximum film pressure given by Fig. 12–20 must be increased by the oil supply pressure p_s to obtain the total film pressure.

Since the oil flow has been increased by forced feed, Eq. (12–23) will give a temperature rise that is too high because of the side flow carries away all the heat generated. Likewise, Eq. (12–14) does not apply. The plumbing in a pressure-fed bearing is depicted schematically in Fig. 12–35. The oil leaves the sump at the externally maintained temperature T_s at the volumetric rate Q_s. The enthalpy balance on the sump as a thermodynamic control region, using as the enthalpy datum the temperature T_s, involves the net enthalpy rate $\dot{H}_{net}$ of the fluid streams,

$$\dot{H}_{net} = \rho C_p(Q_s/2)\Delta T + \rho C_p(Q_s/2)\Delta T - \rho C_p Q_s(0) = \rho C_p Q_s \Delta T$$

and the rate of heat transfer to the surroundings $\dot{Q}_{loss}$. Since the work done by the matter in the control region upon the surroundings is zero,

$$\dot{Q}_{loss} = \dot{H}_{net} = \rho C_p Q_s \Delta T$$

At steady state the rate at which the journal does friction work on the fluid film is

$$\dot{W}k = \frac{2\pi T N_j}{J} = \frac{2\pi f W r N_j}{J} = \frac{2\pi}{J} W N_j c \frac{fr}{c}$$

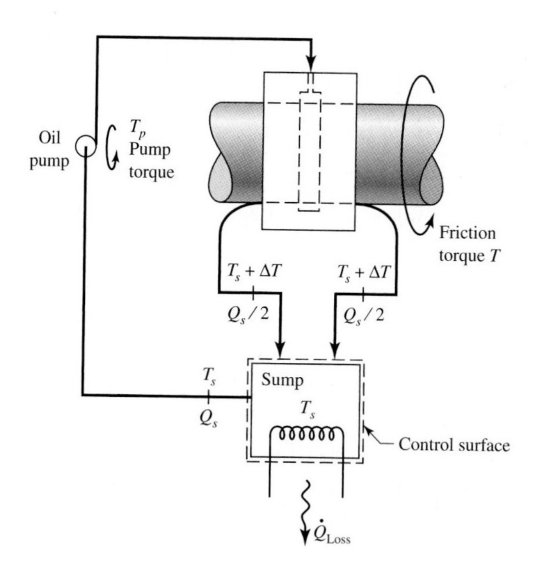

Figure 12–35

Pressure-fed centrally located full annular-groove journal bearing with external, coiled lubricant sump.

which is equal to the heat loss rate of the sump. Equating these and solving for ΔT gives

$$\Delta T = \frac{2\pi W N_j c}{J\rho C_p Q_s} \frac{fr}{c}$$

Summarizing, the temperature rise ΔT is

- Dictated by the first law of thermodynamics
- Thus $T_s + \Delta T/2$ is the mixing cup temperature of the film
- And $T_s + \Delta T$ is the average temperature of the effluxing lubricant

Thus

$$T_s = T_1$$
$$\bar{T}_f = T_1 + \Delta T/2$$
$$T_{\max} = T_1 + \Delta T$$

Pressure-fed bearings with external cooling have a known sump temperature from the outset. Natural-circulation bearings elevate the housing temperature until the heat generation and heat loss rates are equal.

Substituting Eq. (12–26) for Q_s in the equation for ΔT above gives

$$\Delta T = \frac{2\pi}{J\rho C_p} W N_j c \frac{fr}{c} \frac{3\mu l'}{(1+1.5\epsilon^2)p_s r c^3}$$

The Sommerfeld number may be expressed as

$$S = \left(\frac{r}{c}\right)^2 \frac{\mu N_j}{P} = \left(\frac{r}{c}\right)^2 \frac{4rl'\mu N_j}{W}$$

Solving for $\mu N_j l'$ in the Sommerfeld expression; substituting in the ΔT expression; and using $J = 9336$ lb · in/Btu, $\rho = 0.0311$ lbf/in^3, and $C_p = 0.42$ Btu/(lbf · °F), we find

$$\Delta T_F = \frac{6(fr/c)SW^2}{4JC_p p_s r^4} \frac{1}{(1+1.5\epsilon^2)} = \frac{0.0123(fr/c)SW^2}{(1+1.5\epsilon^2)p_s r^4} \tag{12–28}$$

where ΔT_F is in °F. The corresponding equation in SI units uses the bearing load W in kN, lubricant supply pressure p_s in kPa, and the journal radius r in mm:

$$\Delta T_C = \frac{978(10^6)}{1 + 1.5\epsilon^2} \frac{(fr/c)SW^2}{p_s r^4} \tag{12–29}$$

An analysis example of a pressure-fed bearing will be useful.

EXAMPLE 12–7

A circumferential-groove pressure-fed bearing is lubricated with SAE grade 20 oil supplied at a gage pressure of 30 psi. The journal diameter D is 1.750 in, with a unilateral tolerance of −0.002 in. The central circumferential bushing has a diameter B of 1.753 in, with a unilateral tolerance of +0.004 in. The l'/d ratio of the two "half-bearings" that constitute the complete pressure-fed bearing is 1/2. The journal angular speed is 300 r/min, or 50 r/s, and the radial steady load is 900 lb. The external sump is maintained at 120°F as long as the necessary heat transfer does not exceed 800 Btu/h. The load is due to gravity, so W_{st} is also 900 lb.

(a) Find the steady-state average film temperature.

(b) Compare h_0, $T_{\max}$, and P_{st} with the Trumpler criteria.

(c) Estimate the volumetric side flow Q_s, the heat loss rate $\dot{Q}_{\text{loss}}$, and the parasitic friction torque.

Solution

(a)

$$r = \frac{D}{2} = \frac{1.750}{2} = 0.875 \text{ in}$$

$$c_{\min} = \frac{B - D}{2} = \frac{1.753 - 1.750}{2} = 0.0015 \text{ in}$$

$$P = \frac{W}{4rl'} = \frac{900}{4(0.875)0.875} = 294 \text{ psi}$$

The Sommerfeld number S can be expressed as

$$S = \left(\frac{r}{c}\right)^2 \frac{\mu N_j}{P} = \frac{0.875^2}{0.0015^2} \frac{\mu'}{(10^6)} \frac{50}{294} = 0.0579\mu'$$

We will use a tabulation method to find the average film temperature. The first trial average film temperature $\bar{T}_f$ will be 170°F. Using the Seireg curve fit of Table 12–1, we obtain

$$\mu' = 0.0136 \exp[1271.6/(170 + 95)] = 1.650\ \mu\text{reyn}$$

$$S = 0.0579\mu' = 0.0579(1.650) = 0.0955$$

From Fig. (12–17), $fr/c = 3.3$, and from Fig. (12–14), $\epsilon = 0.80$. From Eq. (12–28),

$$\Delta T_f = \frac{0.0123(3.3)0.0955(900^2)}{[1 + 1.5(0.80)^2]30(0.875^4)} = 91.1°\text{F}$$

$$T_{\text{av}} = T_s + \frac{\Delta T}{2} = 120 + \frac{91.1}{2} = 165.6°\text{F}$$

We form a table, adding a second line with $\bar{T}_f = 168.5$°F:

Trial $\bar{T}_f$	μ'	S	fr/c	ϵ	ΔT_F	T_{av}
170	1.65	0.0955	3.3	0.800	91.1	165.6
168.5	1.693	0.0980	3.39	0.792	97.1	168.5

If the iteration had not closed, one could plot trial $\bar{T}_f$ against resulting T_{av} and draw a straight line between them, the intersection with a $\bar{T}_f = T_{av}$ line defining the new trial $\bar{T}_f$. The result of this tabulation is $\bar{T}_f = 168.5$, $\Delta T_F = 97.1°F$, and $T_{max} = 120+97.1 = 217.1°F$

(*b*) From Fig. (12–16), and using $\epsilon = \frac{e}{c}$,

$$h_0 = \left(\frac{h_0}{c}\right)c = (1-\epsilon)c = (1-0.792)0.0015 = 0.000\,300 \text{ in}$$

The required four Trumpler criteria are

$$h_0 \geq 0.0002 + 0.000\,04(1.750) = 0.000\,270 \text{ in} \qquad \text{(OK)}$$

$$T_{max} = T_s + \Delta T = 120 + 97.1 = 217.1°\text{F} \qquad \text{(OK)}$$

$$P_{st} = \frac{W_{st}}{4rl'} = \frac{900}{4(0.875)0.875} = 294 \text{ psi} \qquad \text{(OK)}$$

The factor of safety on the load is approximately unity. (Not OK.)

(*c*) From Eq. (12–26),

$$Q_s = \frac{\pi(30)0.875(0.0015)^3}{3(1.693)10^{-6}0.875}[1 + 1.5(0.800)^2] = 0.123 \text{ in}^3/\text{s}$$

$$\dot{Q}_{loss} = \rho C_p Q_s \Delta T = 0.0311(0.42)0.122(97.1) = 0.154 \text{ Btu/s}$$

or 557 Btu/h or 0.219 hp. The parasitic friction torque T is

$$T = fWr = \frac{fr}{c}Wc = 3.3(900)0.0015 = 4.46 \text{ in} \cdot \text{lb}$$

A Decision Set for Pressure-Fed, Circumferential-Groove Journal Bearings

A decision set for use with central annular-groove pressure-fed journal bearings can be

A priori decisions:

- Length-to-diameter ratio l'/D
- Lubricant grade
- Lubricant supply pressure p_s
- Lubricant sump temperature T_s
- Materials and surface roughness
- Function: load and speed; W, N_j
- Design factor n_d

Design variables:

- Journal radius r
- Radial clearance c

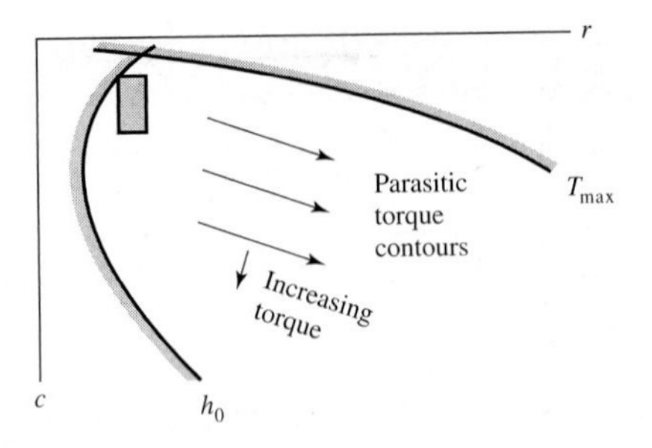

Figure 12–36

Constraints on the *rc* plane of a pressure-fed, centrally located full annular-groove journal bearing. The parasitic friction torque contours can be used as a figure of merit. The design window shown is in a very desirable location.

The rc plane and its constraints are sketched in Fig. 12–36. In pressure-fed bearings the film-thickness constraint is recurving. The desirable location for the design window is with the upper left-hand corner against the minimum film-thickness constraint. Because bushings come in nominal-size bores, the design window cannot occupy just any position in the rc plane. The value of $c_{\min}$ is controlled by the minimum film-thickness constraint. An example would be helpful in illustrating the procedure.

EXAMPLE 12–8

A pressure-fed circumferential-groove journal bearing is to be design to carry a 900-lb radial load at 3000 r/min using SAE grade 20 lubricant at a supply pressure of 30 psi. The sump is to be maintained at 120°F and the heat loss rate is less than 2000 Btu/h. The load is generated by a gear mesh, so the starting load W_{st} is comparable to W. The bushing bores are nominal sizes available in $\frac{1}{8}$-in increments. The associated vendor-supplied unilateral tolerance is $b = \text{INT}(1 + B)/1000$. The in-house manufactured unilateral journal tolerance is $\text{INT}(3 + 2D)/10\,000$. Use Trumpler's design criteria. Establish a satisfactory bearing geometry.

Solution

Our first task is to see if a nominal 2-in-diameter central-groove bushing can carry an 1800-lb load and be placed as shown in Fig. 12–36 with the design window's upper left-hand corner against the minimum film-thickness constraint. A table is prepared to keep the geometry organized:

$c_{\min}$	D	Δc	$c_{\max}$	$D - d$	$r_{\min}$	$r_{\max}$
0.001	1.998	0.0018	0.0028	1.9974	0.9987	0.9990
0.0015	1.997	0.0018	0.0033	1.9964	0.9982	0.9985
0.002	1.996	0.0018	0.0038	1.9954	0.9977	0.9980
0.0025	1.995	0.0018	0.0043	1.9944	0.9972	0.9975
0.003	1.994	0.0018	0.0048	1.9934	0.9967	0.9970

It also defines design windows. The third row is generated as follows. Set $B = 2.000$ in and $c_{\min} = c = 0.002$ in, then

$$D = B - 2c_{\min} = 2.000 - 2(0.002) = 1.996 \text{ in}$$

$$b = \text{INT}(1 + 2)/1000 = 0.003 \text{ in}$$

$$d = \text{INT}(3 + 2)[1.996])/10\,000 = 0.0006 \text{ in}$$

$$\Delta c = (b + d)/2 = (0.003 + 0.0006)/2 = 0.0018 \text{ in}$$

$$c_{\max} = c_{\min} + \Delta c = 0.002 + 0.0018 = 0.0038 \text{ in}$$

$$D - d = 1.996 - 0.0006 = 1.9954 \text{ in}$$

$$r_{\text{min}} = (D - d)/2 = 1.9954/2 = 0.9977 \text{ in}$$

$$r_{\text{max}} = \frac{D}{2} = 1.996/2 = 0.998 \text{ in}$$

Choose $l'/D = 0.5$, $D = 2r_{\text{min}}$, $l' = 0.5[2(0.9977)] = 0.9977$ in. Next for $c = 0.002$ in, $n_d W = 2(900) = 1800$ lb, find minimum film thickness h_0:

$$P = \frac{n_d W}{4rl'} = \frac{1800}{4(0.9977)0.9977} = 452.1 \text{ psi}$$

$$S = \frac{0.9977^2}{0.002^2} \frac{\mu'}{(10^6)} \frac{50}{452.1} = 0.0275\mu'$$

We iterate to find the average film temperature $\bar{T}_f$. We use the following equations and figures in sequence:

$$\mu' = 0.0136 \exp\left(\frac{1271.6}{\bar{T}_f + 95}\right)$$

$$S = 0.0275\mu'$$

fr/c from Fig. 12–17

ϵ from Fig. 12–14

$$\Delta T_F = \frac{0.0123(fr/c)S(n_d W)^2}{(1 + 1.5\epsilon^2)p_s r^4}$$

$$T_{\text{av}} = T_s + \frac{\Delta T}{2}$$

Fill out the table below. If first trial $\bar{T}_f = 150$°F, then $T_{\text{av}} = 178.14$:

$\bar{T}_f$	T_{av}
150	178.14
160	160.06
160.25	160.25

Knowing the correct $\bar{T}_f = 160.25$°F, you can assemble the following information:

c	$\bar{T}_f$	μ	S	ΔT	T_{max}	fr/c	h_0/c	ϵ	Q_s	$\dot{Q}_{loss}$
0.002	160.25	$1.982(10^{-6})$	0.0546	80.5	200.5	2.296	0.1490	0.851	0.2646	1001.40

Then

$$h_0 = c(1 - \epsilon) = 0.002(1 - 0.851) = 0.000\,298 \text{ in}$$

$$f = \frac{(fr/c)c}{r} = \frac{2.296(0.002)}{0.9977} = 0.0046$$

Trumpler:

$$h_0 = 0.0002 + 0.000\,04(1.996) = 0.000\,280 \text{ in} \qquad \text{(OK)}$$
$$T_{max} = 200.5°\text{F} \qquad \text{(OK)}$$
$$P_{st} = 225.9 \text{ psi} \qquad \text{(OK)}$$
$$n \geq 2 \qquad \text{(OK)}$$

Next try $c = 0.0015$. The iteration for $\bar{T}_f$ is

$\bar{T}_f$	T_{av}
180	179.19
170	200.89
179.7	179.7

The associated information is

c_{min}	$\bar{T}_f$	μ	S	ΔT	T_{max}	fr/c	h_0/c	ϵ	Q_s	$\dot{Q}_{loss}$
0.0015	179.7	1.393(10⁻⁶)	0.0682	119.45	239.45	2.655	0.171	0.829	0.157	868.5

$$h_0 = c(1 - \epsilon) = 0.0.0015(1 - 0.829) = 0.000\,257 \text{ in}$$

$$f = \frac{(fr/c)c}{r} = \frac{2.655(0.0015)}{0.9962} = 0.004\,00$$

Trumpler:

$$h_0 \geq 0.0002 + 0.000\,04(1.997) = 0.000\,280 \text{ in} \qquad \text{(unsatisfactory)}$$
$$T_{max} = 239.45°\text{F} \qquad \text{(OK)}$$
$$P_{st} = 225.7 \text{ psi} \qquad \text{(OK)}$$
$$n \geq 2 \qquad \text{(OK)}$$

To get an idea of the lay of the h_0 constraint, we can check the lower left-hand corner of the design window. Set $c = 0.0033$ in and iterate to find $\bar{T}_f$:

$\bar{T}_f$	T_{av}
160	127.46
150	130.46
140	135.12
137	137.01

and develop the following information:

c_{max}	$\bar{T}_f$	μ	S	ΔT	T_{max}	fr/c	h_0/c	ϵ	$\dot{Q}_{loss}$
0.0033	137.01	3.265(10⁻⁶)	0.0331	34.03	154.03	1.687	0.1077	0.892	1213.9

$h_0 = (h_0/c)c = 0.1077(0.0033) = 0.000\,355$ in

Trumpler:

$$h_0 \geq 0.0002 + 0.000\,04(1.997) = 0.000\,280 \text{ in} \qquad \text{(OK)}$$

$$T_{max} = 154.03°\text{F} \qquad \text{(OK)}$$

$$P_{st} = 225.9 \text{ psi} \qquad \text{(OK)}$$

$$n \geq 2 \qquad \text{(OK)}$$

The looseness of the h_0 constraint means that the constraint is moving away from the design window as depicted in Fig. 12–36.

We can decide the bearing geometry:

$$D = 1.996 \begin{matrix} +0.0 \\ -0.0006 \end{matrix} \text{ in} \qquad \text{(in-house manufacture)}$$

$$B = 2.000 \begin{matrix} +0.003 \\ -0.0 \end{matrix} \text{ in} \qquad \text{(vendor-supplied)}$$

The work done on the oil during delivery is $p_s Q_s$ in · lb/s or $p_s Q_s/12$ ft · lb/s. The rate of this work in units of horsepower is $p_s Q_s/[12(550)]$. Considering this work to be done by a torque T_P gives a value of

$$T_P = \frac{63\,025\text{hp}}{60N_j} = \frac{63\,025 p_s Q_s}{60N_j(12)550(\text{eff})} = \frac{0.159 p_s Q_s}{N_j(\text{eff})}$$

including mechanical efficiency of the mechanism. The equivalent parasitic torque is the sum of the fluid friction torque and the pumping torque:

$$T = \frac{fr}{c} n_d W c + \frac{0.159 p_s Q_s}{N_j(\text{eff})}$$

The negative of this torque can be used as a figure of merit. For the design above,

$$T = 2.296(2)900(0.002) + \frac{0.159(30)0.2046}{50(0.50)}$$

$$= 8.27 + 0.039 = 8.31 \text{ in} \cdot \text{lb}$$

This gives some perspective on the relative magnitudes of the fluid friction torque and the lubricant pumping torque. The dissipated power is

$$hp' = \frac{60TN_j}{63\,025} = \frac{60(8.31)50}{63\,025} = 0.396 \text{ hp}$$

The next size bushing to try is $B = 1.875$ in, with detail similar to that above. With a figure of merit, you should have no difficulty choosing among other alternatives.

12–12 Loads and Materials

Some help in choosing unit loads and bearing materials is afforded by Tables 12–5 and 12–6. Since the diameter and length of a bearing depend upon the unit load, these tables will help the designer to establish a starting point in the design.

The length-diameter ratio l/d of a bearing depends upon whether it is expected to run under thin-film-lubrication conditions. A long bearing (large l/d ratio) reduces the

Table 12–5

Range of Unit Loads in Current Use for Sleeve Bearings

Application	Unit Load psi	Unit Load MPa
Diesel engines:		
Main bearings	900–1700	6–12
Crankpin	1150–2300	8–15
Wristpin	2000–2300	14–15
Electric motors	120–250	0.8–1.5
Steam turbines	120–250	0.8–1.5
Gear reducers	120–250	0.8–1.5
Automotive engines:		
Main bearings	600–750	4–5
Crankpin	1700–2300	10–15
Air compressors:		
Main bearings	140–280	1–2
Crankpin	280–500	2–4
Centrifugal pumps	100–180	0.6–1.2

Table 12–6

Some Characteristics of Bearing Alloys

Alloy Name	Thickness, in	SAE Number	Clearance Ratio r/c	Load Capacity	Corrosion Resistance
Tin-base babbitt	0.022	12	600–1000	1.0	Excellent
Lead-base babbitt	0.022	15	600–1000	1.2	Very good
Tin-base babbitt	0.004	12	600–1000	1.5	Excellent
Lead-base babbitt	0.004	15	600–1000	1.5	Very good
Leaded bronze	Solid	792	500–1000	3.3	Very good
Copper-lead	0.022	480	500–1000	1.9	Good
Aluminum alloy	Solid		400–500	3.0	Excellent
Silver plus overlay	0.013	17P	600–1000	4.1	Excellent
Cadmium (1.5% Ni)	0.022	18	400–500	1.3	Good
Trimetal 88*				4.1	Excellent
Trimetal 77†				4.1	Very good

*This is a 0.008-in layer of copper-lead on a steel back plus 0.001 in of tin-base babbitt.

†This is a 0.013-in layer of copper-lead on a steel back plus 0.001 in of lead-base babbitt.

coefficient of friction and the side flow of oil and therefore is desirable where thin-film or boundary-value lubrication is present. On the other hand, where forced-feed or positive lubrication is present, the l/d ratio should be relatively small. The short bearing length results in a greater flow of oil out of the ends, thus keeping the bearing cooler. Current practice is to use an l/d ratio of about unity, in general, and then to increase this ratio if thin-film lubrication is likely to occur and to decrease it for thick-film lubrication or high temperatures. If shaft deflection is likely to be severe, a short bearing should be used to prevent metal-to-metal contact at the ends of the bearings.

You should always consider the use of a partial bearing if high temperatures are a problem, because relieving the non-load-bearing area of a bearing can very substantially reduce the heat generated.

The two conflicting requirements of a good bearing material are that it must have a satisfactory compressive and fatigue strength to resist the externally applied loads and that it must be soft and have a low melting point and a low modulus of elasticity.

The second set of requirements is necessary to permit the material to wear or break in, since the material can then conform to slight irregularities and absorb and release foreign particles. The resistance to wear and the coefficient of friction are also important because all bearings must operate, at least for part of the time, with thin-film lubrication.

Additional considerations in the selection of a good bearing material are its ability to resist corrosion and, of course, the cost of producing the bearing. Some of the commonly used materials are listed in Table 12–6, together with their composition and characteristics.

Bearing life can be increased very substantially by depositing a layer of babbitt, or other white metal, in thicknesses from 0.001 to 0.014 in over steel backup material. In fact, a copper-lead on steel to provide strength, with a babbitt overlay to provide surface and corrosion characteristics, makes an excellent bearing.

Small bushings and thrust collars are often expected to run with thin-film lubrication. When this is the case, improvements over a solid bearing material can be made to add significantly to the life. A powder-metallurgy bushing is porous and permits the oil to penetrate into the bushing material. Sometimes such a bushing may be enclosed by oil-soaked material to provide additional storage space. Bearings are frequently ball-indented to provide small basins for the storage of lubricant while the journal is at rest. This supplies some lubrication during starting. Another method of reducing friction is to indent the bearing wall and to fill the indentations with graphite.

With all these tentative decisions made, a lubricant can be selected and the hydrodynamic analysis made as already presented. The values of the various performance parameters, if plotted as in Fig. 12–25, for example, will then indicate whether a satisfactory design has been achieved or additional iterations are necessary.

12–13 Bearing Types

A bearing may be as simple as a hole machined into a cast-iron machine member. It may still be simple yet require detailed design procedures, as, for example, the two-piece grooved pressure-fed connecting-rod bearing in an automotive engine. Or it may be as elaborate as the large water-cooled, ring-oiled bearings with built-in reservoirs used on heavy machinery.

Figure 12–37 shows two types of bushings. The solid bushing is made by casting, by drawing and machining, or by using a powder-metallurgy process. The lined bushing is usually a split type. In one method of manufacture the molten lining material is cast continuously on thin strip steel. The babbitted strip is then processed through presses, shavers, and broaches, resulting in a lined bushing. Any type of grooving may be cut into the bushings. Bushings are assembled as a press fit and finished by boring, reaming, or burnishing.

Flanged and straight two-piece bearings are shown in Fig. 12–38. These are available in many sizes in both thick- and thin-wall types, with or without lining material. A locking lug positions the bearing and effectively prevents axial or rotational movement of the bearing in the housing.

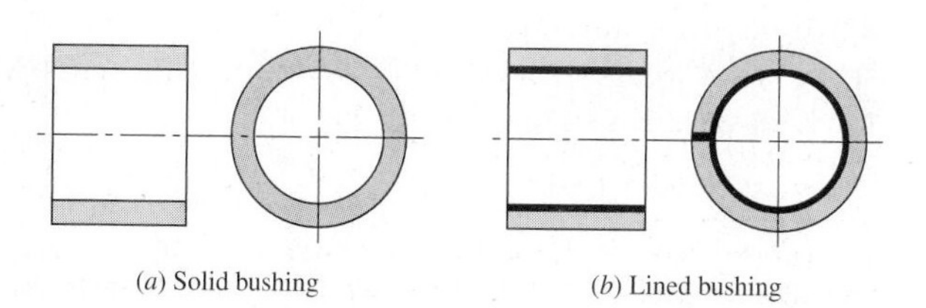

Figure 12–37

Sleeve bushings.

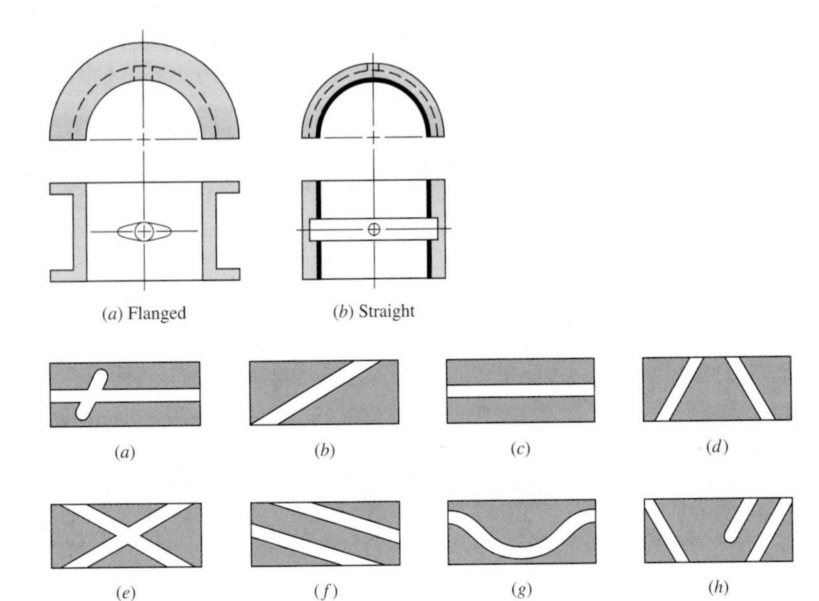

Figure 12–38
Two-piece bushings.

Figure 12–39
Developed views of typical groove patterns. *(Courtesy of the Cleveland Graphite Bronze Company, Division of Clevite Corporation.)*

Some typical groove patterns are shown in Fig. 12–39. In general, the lubricant may be brought in from the end of the bushing, through the shaft, or through the bushing. The flow may be intermittent or continuous. The preferred practice is to bring the oil in at the center of the bushing so that it will flow out both ends, thus increasing the flow and cooling action.

12–14 Thrust Bearings

This chapter is devoted to the study of the mechanics of lubrication and its application to the design and analysis of journal bearings. The design and analysis of thrust bearings is an important application of lubrication theory, too. A detailed study of thrust bearings is not included here, because it would not contribute anything significantly different and because of space limitations. Having studied this chapter, you should experience no difficulty in reading the literature on thrust bearings and applying that knowledge to actual design situations.[17]

Figure 12–40 shows a fixed-pad thrust bearing consisting essentially of a runner sliding over a fixed pad. The lubricant is brought into the radial grooves and pumped into the wedge-shaped space by the motion of the runner. Full-film, or hydrodynamic, lubrication is obtained if the speed of the runner is continuous and sufficiently high, if the lubricant has the correct viscosity, and if it is supplied in sufficient quantity. Figure 12–41 provides a picture of the pressure distribution under conditions of full-film lubrication.

We should note that bearings are frequently made with a flange, as shown in Fig. 12–42. The flange positions the bearing in the housing and also takes a thrust load. Even when it is grooved, however, and has adequate lubrication, such an arrangement is not a hydrodynamically lubricated thrust bearing. The reason for this is that the clearance space is not wedge-shaped but has a uniform thickness. Similar reasoning would apply to various designs of thrust washers.

17. Harry C. Rippel, *Cast Bronze Thrust Bearing Design Manual,* International Copper Research Association, Inc., 825 Third Ave., New York, NY 10022, 1967. CBBI, 14600 Detroit Ave., Cleveland, OH, 44107, 1967.

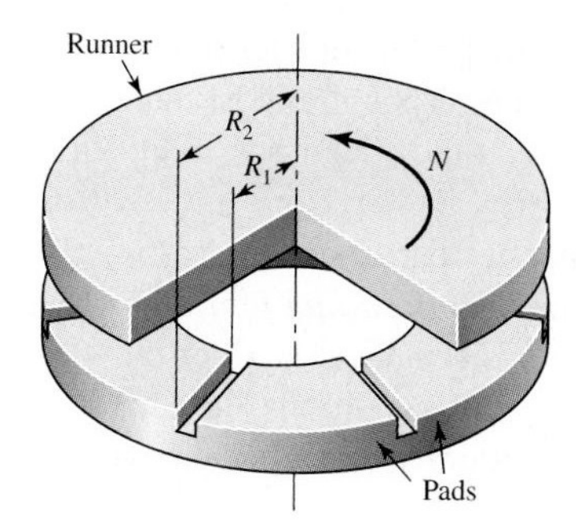

Figure 12–40

Fixed-pad thrust bearing. *(Courtesy of The Westinghouse Corporation Research Laboratories.)*

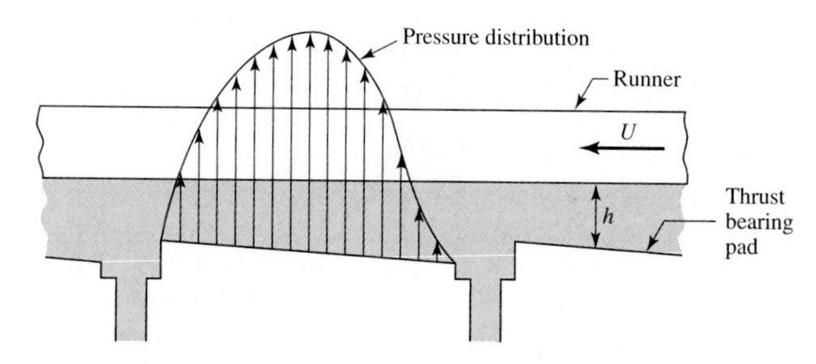

Figure 12–41

Pressure distribution of lubricant in a thrust bearing. *(Courtesy of Copper Research Corporation.)*

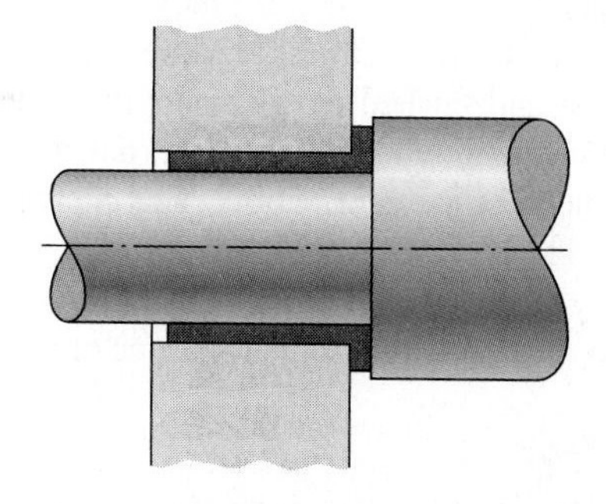

Figure 12–42

Flanged sleeve bearing takes both radial and thrust loads.

12–15 Boundary-Lubricated Bearings

When two surfaces slide relative to each other with only a partial lubricant film between them, *boundary lubrication* is said to exist. Boundary- or thin-film lubrication occurs in hydrodynamically lubricated bearings when they are starting or stopping, when the load increases, when the supply of lubricant decreases, or whenever other operating changes happen to occur. There are, of course, a very large number of cases in design in which boundary-lubricated bearings must be used because of the type of application or the competitive situation.

The coefficient of friction for boundary-lubricated surfaces may be greatly decreased by the use of animal or vegetable oils mixed with the mineral oil or grease. Fatty acids, such as stearic acid, palmitic acid, or oleic acid, or several of these, which occur in animal and vegetable fats, are called *oiliness agents*. These acids appear to reduce friction, either because of their strong affinity for certain metallic surfaces or because they form a soap film which binds itself to the metallic surfaces by a chemical reaction. Thus the fatty-acid molecules bind themselves to the journal and bearing surfaces with such great strength that the metallic asperities of the rubbing metals do not weld or shear.

Fatty acids will break down at temperatures of 250°F or more, causing increased friction and wear in thin-film-lubricated bearings. In such cases the *extreme-pressure,* or EP, lubricants may be mixed with the fatty-acid lubricant. These are composed of chemicals such as chlorinated esters or tricresyl phosphate, which form an organic film between the rubbing surfaces. Though the EP lubricants make it possible to operate at higher temperatures, there is the added possibility of excessive chemical corrosion of the sliding surfaces.

When a bearing operates partly under hydrodynamic conditions and partly under dry or thin-film conditions, a *mixed-film lubrication* exists. If the lubricant is supplied by hand oiling, by drop or mechanical feed, or by wick feed, for example, the bearing is operating under mixed-film conditions. In addition to occurring with a scarcity of lubricant, mixed-film conditions may be present when

- The viscosity is too low.
- The bearing speed is too low.
- The bearing is overloaded.
- The clearance is too tight.
- Journal and bearing are not properly aligned.

Relative motion between surfaces in contact in the presence of a lubricant is called *boundary lubrication*. This condition is present in hydrodynamic film bearings during starting, stopping, overloading, or lubricant deficiency. Some bearings are boundary lubricated (or dry) at all times. To signal this an adjective is placed before the word "bearing." Commonly applied adjectives (to name a few) are thin-film, boundary friction bearings, Oilite, Oiles, and bushed-pin. The applications include situations in which thick film will not develop and there are low journal speed, oscillating journal, padded slides, light loads, and lifetime lubrication. The characteristics include considerable friction, ability to tolerate expected wear without losing function, and light loading. Such bearings are limited by lubricant temperature, speed, characteristic pressure, galling, and cumulative wear. Table 12–7 gives some properties of a range of bushing materials.

Linear Sliding Wear

Our first task is to relate wear to other parameters. Consider the sliding block depicted in Fig. 12–43, moving along a plate with contact pressure P' acting over area A, in the presence of a coefficient of sliding friction f_s. The linear measure of wear w is

Table 12–7

Some Materials for Boundary-Lubricated Bearings and Their Operating Limits

Material	Maximum Load, psi	Maximum Temperature, °F	Maximum Speed, fpm	Maximum PV Value*
Cast bronze	4 500	325	1 500	50 000
Porous bronze	4 500	150	1 500	50 000
Porous iron	8 000	150	800	50 000
Phenolics	6 000	200	2 500	15 000
Nylon	1 000	200	1 000	3 000
Teflon	500	500	100	1 000
Reinforced Teflon	2 500	500	1 000	10 000
Teflon fabric	60 000	500	50	25 000
Delrin	1 000	180	1 000	3 000
Carbon-graphite	600	750	2 500	15 000
Rubber	50	150	4 000	
Wood	2 000	150	2 000	15 000

* P = load, psi; V = speed, fpm.

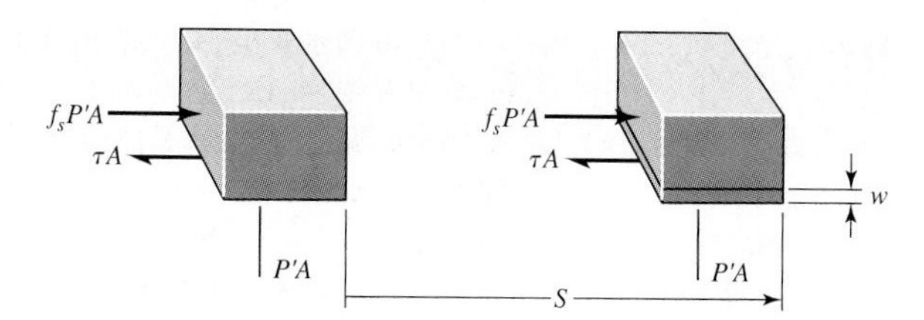

Figure 12–43

Sliding block subjected to wear.

expressed in inches or millimeters. The work done by force $f_s P'A$ during displacement S is $f_s P'AS$ or $f_s P'AVt$, where V is the sliding velocity and t is time. The material volume eroded is wA. The volume of material removed is proportional to the work done, that is

$$wA \sim f_s P'AVt$$

or

$$wA = KP'AVt$$

where K is the proportionality factor which absorbs the coefficient of sliding friction. The linear wear is expressed as

$$w = KP'Vt \tag{12–30}$$

Conventionally, P' is expressed in psi, V in ft/min, and t in hours. This makes the units of K

$$\dim[K] = \frac{\text{in}}{(\text{psi})(\text{ft/min})(\text{h})} = \frac{\text{in}^3}{\text{lbf(ft/min)h}}$$

in conventional engineering units, or

$$\dim[K_m] = \frac{\text{cm}^3(\text{min})}{\text{kg}_f(\text{m})(\text{h})}$$

The relation between K and K_m is $K_m = 111.2K$. It is useful to include a modifying factor f_1 depending on motion type, load, and speed (see Tables 12–8 and 12–9) and a factor f_2 to account for environment (temperature and cleanliness). Thus f_1 and f_2 account for departures from the laboratory conditions under which K was measured. Equation (12–30) can now be written as

$$w = f_1 f_2 KP'Vt \tag{12–31}$$

in which time is related to wear; that is, Eq. (12–31) is a time-to-wear equation. A distance-to-wear equation can be formed by noting the distance S in feet is $60Vt$, thus

$$w = \frac{f_1 f_2 KP'S}{60} \tag{12–32}$$

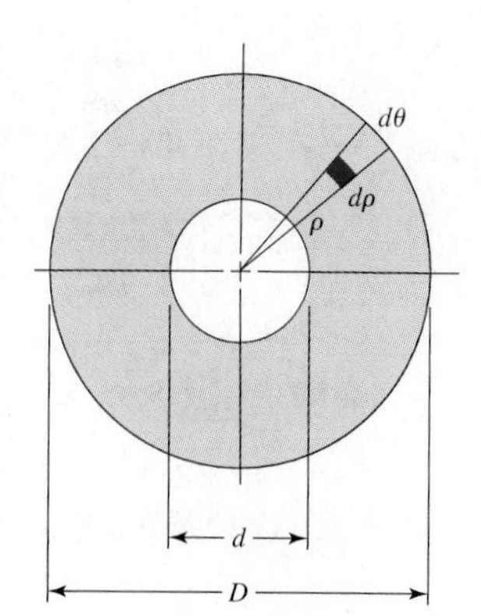

Figure 12–44

An element of area on the face of a thrust washer.

Thrust Washers

Boundary lubrication in commonly used thrust washers represents an application of Eq. (12–31). Consider the washer depicted in Fig. 12–44. Consider the steady-state wear situation, where the wear is uniform all over the annulus, making $P'\rho =$ constant. The differential normal force is $(P'\rho)\,d\rho\,d\theta$, so

$$F = (P'\rho)\int_0^{2\pi}\int_{r_i}^{r_0} d\rho d\theta = 2\pi(P'\rho)(r_0 - r_i) = \pi(P'\rho)(D - d) \tag{12–33}$$

Table 12–8

Motion-Related Factor f_1

Mode of Motion	Characteristic Pressure *P*, psi		Velocity *V*, ft/min	f_1*
Rotary	720 or less		3.3 or less	1.0
			3.3–33	1.0–1.3
			33–100	1.3–1.8
	720–3600		3.3 or less	1.5
			3.3–33	1.5–2.0
			33–100	2.0–2.7
Oscillatory	720 or less	>30°	3.3 or less	1.3
			3.3–100	1.3–2.4
		<30°	3.3 or less	2.0
			3.3–100	2.0–3.6
	720–3600	>30°	3.3 or less	2.0
			3.3–100	2.0–3.2
		<30°	3.3 or less	3.0
			3.3–100	3.0–4.8
Reciprocating	720 or less		33 or less	1.5
			33–100	1.5–3.8
	720–3600		33 or less	2.0
			33–100	2.0–7.5

* Values of f_1 based on results over an extended period of time on automotive manufacturing machinery.

Table 12–9

Environmental Factor f_2

Source: Oiles America Corp., Plymouth, Mich., 48170.

Ambient Temperature, °F	Foreign Matter	f_2
140 or lower	No	1.0
140 or lower	Yes	3.0–6.0
140–210	No	3.0–6.0
140–210	Yes	6.0–12.0

The differential friction moment $dM_0 = f_s \rho dF = f_s \rho (P'\rho) d\rho d\theta$, so

$$M_0 = f_s(P'\rho) \int_0^{2\pi} \int_{r_i}^{r_0} \rho d\rho d\theta = 2\pi f_s(P'\rho) \frac{r_0^2 - r_i^2}{2}$$

$$= \frac{\pi}{4} f_s(P'\rho)(D^2 - d^2)$$

$$= \frac{\pi}{4} f_s \frac{F}{\pi(D-d)}(D^2 - d^2) = \frac{f_s F}{4}(D + d) \tag{12–34}$$

The largest pressure P' from Eq. (12–33) is

$$P'_{\max} = \frac{2F}{\pi d(D-d)}$$

The $P'V$ product is constant:

$$P'V = \frac{F}{\pi\rho(D-d)} \frac{\pi(2\rho)(\text{rpm})}{12} = \frac{F(\text{rpm})}{6(D-d)} \tag{12–35}$$

The velocities at the inner and outer radii differ:

$$V_0 = \frac{\pi D(\text{rpm})}{12} \qquad V_i = \frac{\pi d(\text{rpm})}{12} \qquad V_{\text{max}} = \frac{\pi D(\text{rpm})}{12}$$

and the average $\bar{V}$ is often used:

$$\bar{V} = \frac{V_0 + V_i}{2} = \frac{1}{2}\left[\frac{\pi D(\text{rpm})}{12} + \frac{\pi d(\text{rpm})}{12}\right] = \frac{\pi}{24}(\text{rpm})(D+d) \tag{12–36}$$

The wear equation is, from Eq. (12–31),

$$w = f_1 f_2 K(P'V)t = \frac{f_1 f_2 K F(\text{rpm})t}{6(D-d)} \tag{12–37}$$

for thrust washers, steady-state wear.

EXAMPLE 12–9

A brass alloy thrust washer has an inside diameter of 1 in and an outside diameter of 2 in. It carries a thrust load of 1000 lb from the end of a steel shaft which rotates at 100 r/min. The wear constant K is $6(10^{-11})$ in/[(psi)(ft/min)h]. Estimate the time in hours for axial wear to reach 0.001 in. See Fig. 12–45.

Solution

$$P'_{\text{max}} = \frac{2F}{\pi d(D-d)} = \frac{2(1000)}{\pi(1)(2-1)} = 637 \text{ psi}$$

From Eq. (12–36) the average velocity V is

$$\bar{V} = \frac{\pi}{24}(\text{rpm})(D+d) = \frac{\pi}{24}(100)(2+1) = 39.3 \text{ ft/min}$$

From Eq. (12–35),

$$P'V = \frac{F(\text{rpm})}{6(D-d)} = \frac{1000(100)}{6(2-1)} = 16\,667 \text{ (psi)(ft/min)}$$

From Table 12–8, the factor f_1 is estimated by interpolation as

$\bar{V}$	f_1
33	1.3
39.3	f_1
100	1.8

$$\frac{39.3-33}{100-33} = \frac{f_1 - 1.3}{1.8-1.3}$$

from which $f_1 = 1.35$. The factor f_2 is unity. The time to allowable wear is, from Eq. (12–37),

Figure 12–45

Thrust washer in use.

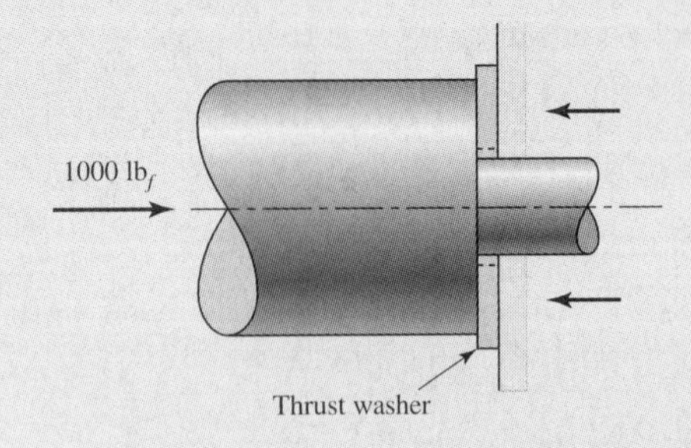

Table 12–10

Oiles 500 SP (SPBN · SPWN) Service Range and Properties

Source: Oiles America Corp., Plymouth, Michigan, 48170.

Service Range	Units	Allowable
Characteristic pressure P_{max}	psi	<3560
Velocity V_{max}	ft/min	<100
PV product	(psi)(ft/min)	<46 700
Temperature *T*	°F	<300
Properties	**Test Method, Units**	**Value**
Tensile strength	(ASTM E8) psi	>110 000
Elongation	(ASTM E8) %	>12
Compressive strength	(ASTM E9) psi	49 770
Brinell hardness	(ASTM E10) HB	>210
Coefficient of thermal expansion	(10^{-5}) °C	>1.6
Specific gravity		8.2

Answer
$$t = \frac{6w(D-d)}{f_1 f_2 K F(\text{rpm})} = \frac{6(0.001)(2-1)}{1.35(1)6(10^{-11})1000(100)} = 741 \text{ h}$$

In examining Tables 12–8, 12–9, and 12–10 we observe that the values of P'_{max}, V, and $P'V$ are all within limits, so that the value of K is valid, and the estimate of 741 h is well-founded.

Rotating-Pin Wear

For a pin diameter D, bushing length L, carrying a radial load F as depicted in Fig. 12–46, Eq. (12–31) has to be particularized for the curvilinear geometry and variable pressure. In a pin–bushing combination, initial wear is localized in a small zone at initial

Figure 12–46

Wear at the periphery of a rotating pin.

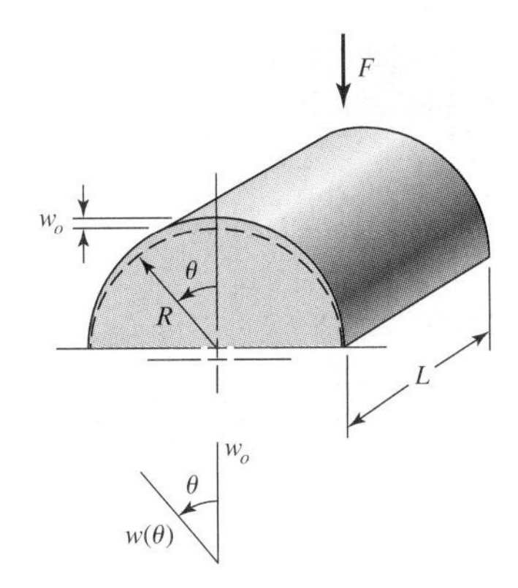

contact. Because pin and bore are similarly sized, wear quickly progresses until the pin completely invades the bushing. When the arc of contact equals 180° we observe the beginning of steady-state wear. At angle θ the rubbing scours to a depth of $w(\theta)$, which is related to the radial wear w_0 at $\theta = 0$:

$$w(\theta) = w_0 \cos\theta$$

Note that $w(\pi/2) = 0$ and there is no wear perpendicular to the radial load F direction, no scour, therefore $p(\pi/s) = 0$. Recall that in a sliding block $p = P' = F/A$. For a rotating pin in a bushing the definition of P' in Eq. (12–31) must be found. At angle θ

$$w(\theta) = Kp(\theta)Vt$$

$$w_0 \cos\theta = Kp(\theta)Vt$$

from which, using Eq. (12–31), we obtain

$$p(\theta) = \frac{w_0 \cos\theta}{KVt} = P' \cos\theta$$

The surface pressure between the pin and bushing varies approximately cosinusoidally with θ. Next we relate load F to P' in Eq. (12–31):

$$dF = p(\theta) \cos\theta RL d\theta = P'RL \cos^2\theta \, d\theta$$

Integration over θ from $-\pi/2$ to $\pi/2$ gives

$$F = \pi P'RL/2 = \pi P'DL/4$$

Consequently the definition of P' is

$$P' = \frac{4F}{\pi DL} \tag{12–38}$$

Another consideration is the general use of F/DL as the definition of the characteristic loading intensity P. Thus Eq. (12–38) becomes

$$P' = \frac{4}{\pi}\frac{F}{DL} = \frac{4}{\pi}P \tag{12–39}$$

Equations for P, V, and PV for journals and bushings are

$$P = \frac{F}{DL} \text{ psi} \tag{12–40}$$

$$V = \frac{\pi D(\text{rpm})}{12} \text{ ft/min} \tag{12–41}$$

$$PV = \frac{F}{DL}\frac{\pi D(\text{rpm})}{12} = \frac{\pi}{12}\frac{F(\text{rpm})}{L} \text{ (psi)(ft/min)} \tag{12–42}$$

Note the independence of the PV product from the journal diameter D. A time-to-wear equation is useful. Starting with Eq. (12–32),

$$w_0 = f_1 f_2 K P' V t$$

and substituting $V = \pi D(\text{rpm})/12$ and $P' = 4P/\pi$ we obtain

$$w_0 = f_1 f_2 K \frac{4}{\pi} P \frac{\pi D(\text{rpm})t}{12} = f_1 f_2 K \frac{4}{\pi}\frac{F}{DL}\frac{\pi D(\text{rpm})t}{12}$$

$$= \frac{f_1 f_2 K F(\text{rpm})t}{3L} \tag{12–43}$$

The necessary bushing length from Eq. (12–43) is

$$L = \frac{f_1 f_2 K n_d F(\text{rpm})t}{3w_0} \tag{12–44}$$

where n_d is the design factor. While Eq. (12–44) says that L is independent of diameter D, the designer is constrained as follows by P_{max} and V_{max} of the journal–bushing combination:

$$P_{\text{max}} > \frac{n_d F}{DL}$$

from which

$$D > \frac{n_d F}{L P_{\text{max}}}$$

Also, from $V_{\text{max}} > \pi D(\text{rpm})/12$,

$$D < \frac{12 V_{\text{max}}}{\pi(\text{rpm})}$$

consequently

$$\frac{n_d F}{L P_{\text{max}}} < D < \frac{12 V_{\text{max}}}{\pi(\text{rpm})} \tag{12–45}$$

EXAMPLE 12–10

An Oiles SP 500 alloy brass bushing is 1 in long with a 1 in bore and operates in a clean environment at 70°F. The allowable wear without loss of function is 0.005 in. The radial load is 700 lbf. The peripheral velocity is 33 ft/min. Estimate the number of revolutions for radial wear to be 0.005 in. See Fig. 12–47.

Solution

From Tables 12–8 and 12–9, $f_1 = 1.3$, $f_2 = 1$. From Table 12–10, $PV = 46\ 700$, $P_{\text{max}} = 3560$ psi, $V_{\text{max}} = 100$ ft/min, and Eqs. (12–40), (12–41), and (12–42) give

$$P = \frac{F}{LD} = \frac{700}{(1)(1)} = 700 \text{ psi} \qquad \text{(OK)}$$

$$V = 33 \text{ ft/min} \qquad \text{(OK)}$$

$$PV = 700(33) = 23\ 100 \text{ (psi)(ft/min)} \qquad \text{(OK)}$$

Equation (12–32) is

$$w_0 = f_1 f_2 K P' V t = f_1 f_2 K \frac{4}{\pi} \frac{F}{LD} V t$$

Figure 12–47

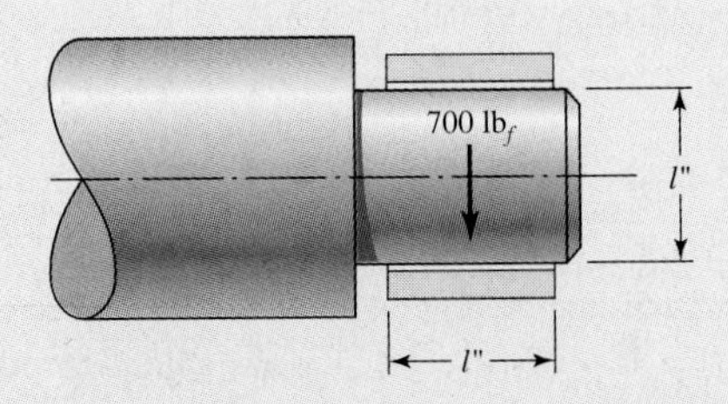

Solving for t gives

$$t = \frac{\pi L D w_0}{4 f_1 f_2 K V F} = \frac{\pi(1)(1)0.005}{4(1.3)(1)6(10^{-11})33(700)} = 2179 \text{ h} = 130\,740 \text{ min}$$

$$\text{rpm} = \frac{12V}{\pi D} = \frac{12(33)}{\pi(1)} = 126 \text{ r/min}$$

Answer $\quad \text{cycles} = 126(130\,740) = 16.5(10^6) \text{ rev}$

EXAMPLE 12–11

An Oiles 500 SP bushing carries a radial load of 100 lbf at 400 r/min. The diameter of the bore is $\frac{7}{8}$ in and the length is also $\frac{7}{8}$ in. Estimate the time to an allowable radial wear of 0.001 in. See Fig. 12–48.

Solution

From Tables 12–8 and 12–9, $f_1 = 1.3$, $f_2 = 1$. From Table 12–10, $P_{max} = 3560$ psi, $V_{max} = 100$ ft/min, and $PV = 46\,700$ (psi)(ft/min). From Eq. (12–40),

$$P = \frac{F}{LD} = \frac{100}{0.875(0.875)} = 130.6 \text{ psi} \qquad \text{(OK)}$$

From Eq. (12–41),

$$V = \frac{\pi D(\text{rpm})}{12} = \frac{\pi(0.875)400}{12} = 91.9 \text{ ft/min} \qquad \text{(OK)}$$

From Eq. (12–42),

$$PV = \frac{\pi}{12}\frac{F(\text{rpm})}{L} = \frac{\pi(100)400}{12(0.875)} = 11\,968 \text{ (psi)(ft/min)} \qquad \text{(OK)}$$

Solve Eq. (12–43) for t:

Answer

$$t = \frac{3 L w_0}{f_1 f_2 K F(\text{rpm})} = \frac{3(0.875)0.001}{1.3(1)6(10^{-11})100(400)} = 841 \text{ h}$$

Figure 12–48

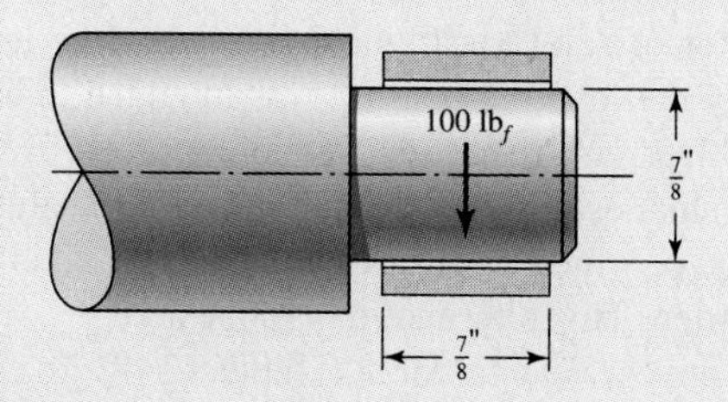

Oscillatory Journal Motion

Note that the presence of oscillatory motion affects factor f_1 as shown in Table 12–8. The distance-to-wear expression, Eq. (12–32), can incorporate Eq. (12–39) to give, using

$$p' = \frac{4}{\pi} P,$$

$$w_0 = \frac{f_1 f_2 K P' S}{60} = \frac{4 f_1 f_2 K P S}{60\pi}$$

This equation is based on a constant velocity. With oscillation present the velocity varies between 0 and V_{max}. For cosinusoidal velocity variation it can be shown that the average velocity is $(2/\pi)V_{max}$. Since Oiles data for oscillating journals is based on V_{max}, it is necessary to introduce the factor $2/\pi$ into the preceding equation:

$$w_0 = \frac{4}{\pi}\frac{2}{\pi}\frac{f_1 f_2 K P S}{60} = \frac{2}{15\pi^2} f_1 f_2 K P S \tag{12–46}$$

EXAMPLE 12–12

An Oiles SP 500 bushing 1 in in diameter and 1 in long, operating in a clean environment, oscillates through a 60° subtended angle with $V_{max} = 3.3$ ft/min and $P = 2800$ psi. See Fig. 12–49. The allowable wear is 0.005 until loss of function. Estimate the number of cycles until allowable wear occurs.

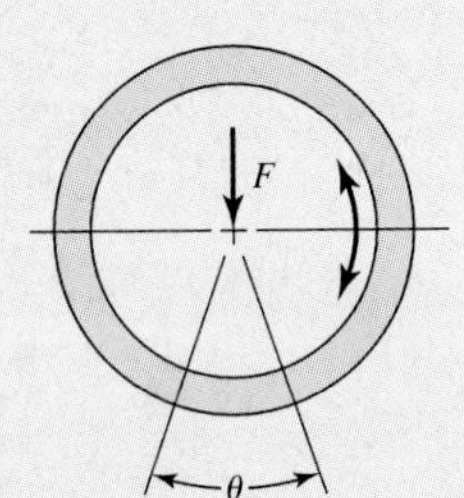

Figure 12–49

Solution

From Tables 12–8 and 12–9, $f_1 = 2$ and $f_2 = 1$. Solve Eq. (12–46) for distance S:

$$S = \frac{15\pi^2 w_0}{2 f_1 f_2 K P} = \frac{15\pi^2(0.005)}{2(2)(1)6(10^{-11})2800} = 1.10(10^6)\text{ ft} = 1.32(10^7)\text{ in}$$

The distance S' in a complete cycle is

$$S' = \frac{2(\pi D)\theta}{360} = \frac{2\pi(1)60}{360} = 1.05\text{ in}$$

Answer

$$\text{cycles} = \frac{S}{S'} = \frac{1.32(10^7)}{1.05} = 1.26(10^7)$$

Design Data

Sliding pairs are limited to pressures less than P_{max} and velocities less than V_{max}; the zones of satisfactory and unsatisfactory PV combinations are depicted in Fig. 12–50. A design factor n_d can be introduced by multiplying the load F by n_d. As in full-film bearings, the factor of safety is found by dividing the load corresponding to the first tight constraint by the load the bearing was intended to carry. Alternatively, as will be suggested in the "Design Procedure" section, a factor of safety for each constraint can be used, with the smallest of these being *the* factor of safety. Since it is not easy to have a priori knowledge of which constraint will be tight, and the simple algebra imposes no consequential burden, This alternative is suggested.

Table 12–11 gives wear factor K, Table 12–12 gives coefficients of friction, and Table 12–13 displays available bushing sizes, all for one manufacturer.

Temperature Rise

At steady state the rate at which work is done against bearing friction equals the rate at which heat is transferred from the bearing housing to the surroundings by convection and radiation. The rate of heat generation in Btu/h is given by

$$\dot{Q}_{gen} = \frac{(fF)(\pi D)(60N)}{12J} = \frac{60\pi f F D N}{12J} \tag{12–47}$$

Figure 12–50

Constraints imposed by the sliding material pair.

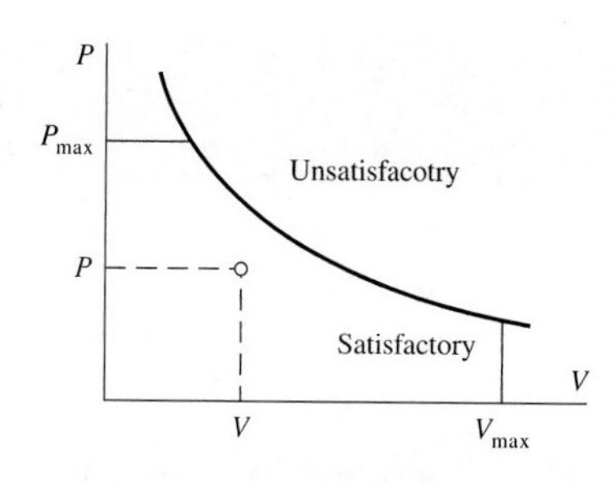

Table 12–11

Wear Factors in U.S. Customary Units*

Source: Oiles America Corp., Plymouth, Mich., 48170.

Bushing Material	Wear Factor *K*	Limiting *PV*
Oiles 800	$3(10^{-10})$	18 000
Oiles 500	$0.6(10^{-10})$	46 700
Polyactal copolymer	$50(10^{-10})$	5 000
Polyactal homopolymer	$60(10^{-10})$	3 000
66 Nylon	$200(10^{-10})$	2 000
66 Nylon + 15% PTFE	$13(10^{-10})$	7 000
+ 15% PTFE + 30% glass	$16(10^{-10})$	10 000
+ 2.5% MoS_2	$200(10^{-10})$	2 000
6 Nylon	$200(10^{-10})$	2 000
Polycarbonate + 15% PTFE	$75(10^{-10})$	7 000
Sintered bronze	$102(10^{-10})$	8 500
Phenol +25% glass fiber	$8(10^{-10})$	11 500

*dim$[K]$ = in^3 · min /(lbf · ft · h).

Table 12–12

Coefficients of Friction

Source: Oiles America Corp., Plymouth, Mich., 48170.

Type	Bearing	*f*
Placetic	Oiles 80	0.05
Composite	Drymet ST	0.03
	Toughmet	0.05
Met	Cermet M	0.05
	Oiles 2000	0.03
	Oiles 300	0.03
	Oiles 500SP	0.03

where N is journal speed in r/min and $J = 778$ ft · lbf/Btu. The rate at which heat is transferred to the surroundings is, in Btu/h,

$$\dot{Q}_{\text{loss}} = UA\Delta T = UA(T_b - T_\infty) = \frac{UA}{2}(T_f - T_\infty) \tag{12–48}$$

where A = housing surface area, ft^2

U = overall combined coefficient of heat transfer, Btu/(h · ft^2 · °F)

T_b = housing metal temperature, °F

T_f = lubricant temperature, °F

The empirical observation that T_b is about midway between T_f and T_∞ has been incor-

Table 12–13

Available Bushing Sizes of One Manufacturer*

		L													
ID	**OD**	$\frac{1}{2}$	$\frac{5}{8}$	$\frac{3}{4}$	$\frac{7}{8}$	1	$1\frac{1}{4}$	$1\frac{1}{2}$	$1\frac{3}{4}$	2	$2\frac{1}{2}$	3	$3\frac{1}{2}$	4	5
$\frac{1}{2}$	$\frac{3}{4}$	•	•	•	•	•									
$\frac{5}{8}$	$\frac{7}{8}$		•	•		•		•							
$\frac{3}{4}$	$1\frac{1}{8}$		•	•		•		•							
$\frac{7}{8}$	$1\frac{1}{4}$			•		•	•	•							
1	$1\frac{3}{8}$			•		•	•	•	•	•					
1	$1\frac{1}{2}$			•		•		•		•					
$1\frac{1}{4}$	$1\frac{5}{8}$					•	•	•	•	•					
$1\frac{1}{2}$	2					•	•	•	•	•					
$1\frac{3}{4}$	$2\frac{1}{4}$						•	•	•	•	•	•	•	•	
2	$2\frac{1}{2}$							•		•	•	•			
$2\frac{1}{4}$	$2\frac{3}{4}$							•		•	•	•			
$2\frac{1}{2}$	3							•		•		•			
$2\frac{3}{4}$	$3\frac{3}{8}$							•		•	•	•			
3	$3\frac{5}{8}$									•	•	•	•		
$3\frac{1}{2}$	$4\frac{1}{8}$									•		•		•	
4	$4\frac{3}{4}$									•		•		•	
$4\frac{1}{2}$	$5\frac{3}{8}$											•		•	•
5	6											•		•	•

*In a display such as this a manufacturer is likely to display catalog numbers where the • is displayed.

porated in Eq. (12–48). Equating Eqs. (12–47) and (12–48) gives

$$T_f = T_\infty + \frac{10\pi f F D N}{J U A} \tag{12–49}$$

One can observe that Eq. (12–49) indicates that the temperature rise $T_f - T_\infty$ is independent of length L and proportional to D. Equation (12–49) can be written

$$D \leq \frac{J U A(T_f - T_\infty)}{10\pi f F N n_d} \tag{12–50}$$

with a design factor n_d introduced by substituting $n_d F$ for F. Equation (12–50) becomes a contending upper bound for diameter D in Eq. (12–44). If the bushing is housed in a pillow block, the surface area can be roughly estimated from

$$A \doteq \frac{2\pi L D}{144} \tag{12–51}$$

EXAMPLE 12–13

Choose an Oiles 500 bushing to give a maximum wear of 0.001 in in 800 h of use with a 300 r/min journal and 50 lb radial load. Use $U = 2.7$ Btu/(h · ft² · °F), $T_{max} = 300$°F, $f = 0.03$, and a design factor $n_d = 2$.

Solution First estimate bushing length L using Eq. (12–44) with $f_1 = f_2 = 1$:

$$L = \frac{f_1 f_2 K n_d F N t}{3 w_0} = \frac{1(1)6(10^{-11})2(50)300(800)}{3(0.001)} = 0.48 \text{ in}$$

Find the bounds on D from Eq. (12–45) and Table 12–10:

$$\frac{n_d F}{L P_{\max}} \le D \le \frac{12 V_{\max}}{\pi N}$$

$$\frac{2(50)}{0.48(3560)} \le D \le \frac{12(100)}{\pi(300)}$$

$$0.059 \le D \le 1.27 \text{ in}$$

Trial 1: $D = 1$ in, $L = 1$ in.

Eq. (12.40): $$P = \frac{n_d F}{DL} = \frac{2(50)}{1(1)} = 100 \text{ psi} \quad \text{(OK)}$$

Eq. (12.41): $$V = \frac{\pi D N}{12} = \frac{\pi(1)300}{12} = 78.5 \text{ ft/min} \quad \text{(OK)}$$

$$PV = 100(78.5) = 7850 \text{ (psi)(ft/min)} \quad \text{(OK)}$$

From Table 12–8,

V	f_1	
33	1.3	
78.5	f_1	=> $f_1 = 1.64$
100	1.8	

Our second estimate of L is $0.48(1.64) = 0.787$ in. From Table 12–13, available bushing length is 1 in. Choose $D = 1$ in, too.

Answer *Check* P, V, and PV are unchanged, still OK. From Eq. (12–43),

$$w_0 = \frac{f_1 f_1 K n_d F N t}{3L} = \frac{1.64(1)6(10^{-11})2(50)300(800)}{3(1)} = 0.0008 \text{ in}$$

which is less than 0.001 in. Equation (12–50), including Eq. (12–51), gives

$$D \le \frac{J U A(T_f - T_\infty)}{10\pi f F N n_d} = \frac{778(2.7)[2\pi 1.75(1)/144](300 - 70)}{10\pi(0.03)50(300)2} = 1.30$$

The temperature constraint of D is loose, so T_f is less than 300°F.

Design Procedure

All the bullets in Table 12–13 represent available bushings from one manufacturer. In lieu of the bullets, catalog numbers could have been displayed. The specification set for a bushing could be its catalog number or, instead, a length and a diameter and a material series. The decision set could be written as

- Material combination
- Lubricant
- Function: load, speed, life, wear

} A priori decisions

- Length L
- Bore diameter D

} Design variables

Since neither L nor D is continuous, the design space will consist of many constraints defining a feasible domain. The feasible area will be populated with dots, much like Table 12–13. The figure of merit can be −(cost), which can be approximated by fom $= -DL$ in the absence of cost tables.

The adequacy assessment consists of checking constraints for looseness. Given design decisions D and L, with a priori decisions F, N, t, and $(w_0)_{\max}$, we might find, for example,

$$V = \frac{\pi DN}{12}$$

$$f_1 = \frac{V - 33}{100 - 33}(1.8 - 1.3) + 1.3$$

We will define five factors of safety as follows:

$$n_1 = \frac{LDP_{\max}}{F} \quad \text{(pressure)} \tag{12–52}$$

$$n_2 = \frac{V_{\max}}{V} = \frac{12V_{\max}}{\pi DN} \quad \text{(velocity)} \tag{12–53}$$

$$n_3 = \frac{(PV)_{\max}}{PV} = \frac{12(PV)_{\max} L}{\pi FN} \quad (PV \text{ constraint}) \tag{12–54}$$

$$n_4 = \frac{(w_0)_{\max}}{w_0} = \frac{3L(w_0)_{\max}}{f_1 f_2 KFNt} \quad \text{(wear)} \tag{12–55}$$

$$n_5 = \frac{(\Delta T)_{\max}}{\Delta T} = \frac{(\Delta T)_{\max} JUA}{10\pi fFDN} \quad \text{(temperature rise)} \tag{12–56}$$

The inequality constraints on L and D can be expressed in terms of a design factor n_d applied to the load as

$$LD > \frac{n_d F}{P_{\max}} \quad \text{(pressure)} \tag{12–57}$$

$$D < \frac{12V_{\max}}{\pi N} \quad \text{(velocity)} \tag{12–58}$$

$$L > \frac{\pi n_d FN}{12(PV)_{\max}} \quad (PV \text{ product}) \tag{12–59}$$

$$L > \frac{f_1 f_2 K n_d FNt}{3(w_0)_{max}} \quad (\text{wear; note that } f_1 \text{ is a linear function of } D) \tag{12–60}$$

$$D < \frac{JUA(\Delta T)_{\max}}{10\pi f n_d FN} \quad \text{(temperature rise)} \tag{12–61}$$

Note that in Eq. (12–61) the area A can be a function of L and D as in Eq. (12–51), which recasts Eq. (12–61) from a constraint on D to a constraint on L, in particular,

$$L > \frac{720 f n_d F N}{J U (\Delta T)_{max}} \tag{12–62}$$

It is useful to map the constraint inequalities (12–57) through (12–60) plus (12–62) onto the LD plane. Additionally, the smallest available bushing diameter D and length L can be added (see Table 12–13) as $D > 0.5$ in, $L > 0.5$ in. The purpose of the constraint mapping is to visualize the feasible domain.

EXAMPLE 12–14

An Oiles 500 bushing is to be selected which will hold wear to 0.001 in or less in 800 h. The journal speed is 300 r/min, the radial load is 30 lb. The maximum allowable lubricant temperature is 300°F. The design factor is to be $n_d \geq 2$. The bearing lateral area can be estimated from $A = 2\pi LD/144$ ft^2, $f = 0.03$.

Solution

Eq. (12–57): $\quad LD > \dfrac{n_d F}{P_{max}} = \dfrac{2(30)}{3560} = 0.017 \quad$ (pressure)

This is a rectangular hyperbola nested close to the axes.

Eq. (12–58): $\quad D < \dfrac{12 V_{max}}{\pi N} = \dfrac{12(100)}{\pi(300)} = 1.27 \text{ in} \quad$ (velocity)

Eq. (12–59): $\quad L > \dfrac{\pi n_d F N}{12(PV)_{max}} = \dfrac{\pi(2)30(300)}{12(46\,700)} = 0.10 \text{ in} \quad$ (PV constraint)

Eq. (12–60): $\quad L > \dfrac{f_1 f_2 K n_d F N t}{3(w_0)_{max}} \quad$ (wear constraint)

This constraint is a mild function of D since $V = \pi DN/12$ and

$$f_1 = \frac{V - 33}{100 - 33}(1.8 - 1.3) + 1.3 = \frac{\pi DN/12 - 33}{100 - 33}(1.8 - 1.3) + 1.3$$

To locate two points on this straight-line constraint set $D = 0$, then

$$f_1 = \frac{\pi(0)300/12 - 33}{100 - 33}(1.8 - 1.3) + 1.3 = 1.05$$

From Eq. (12–60),

$$L_{D=0} = \frac{1.05(1)6(10^{-11})2(30)300(800)}{3(0.001)} = 0.302 \text{ in}$$

and when $D = 2$ in,

$$f_1 = \frac{\pi(2)300/12 - 33}{100 - 33}(1.8 - 1.3) + 1.3 = 2.06$$

and

$$L_{D=2} = \frac{2.06}{1.05} = (0.302) = 0.592 \text{ in}$$

The wear constraint can be plotted from $L_{D=0}$ and $L_{D=2}$. Note $\Delta T_{max} = 300 - 70 = 230°F$.

$$\text{Eq. (12–62): } L > \frac{720 f n_d F N}{J U (\Delta T)_{max}} = \frac{720(0.03)2(30)300}{778(2.7)230} = 0.80 \text{ in} \quad \text{(temperature)}$$

Also

$$D \geq 0.5 \text{ in}$$
$$L \geq 0.5 \text{ in}$$

Figure 12–51 is the plot of the constraints on the design plane and the feasible domain can be identified. For a figure of merit of fom $= -DL$, the merit contours of equal merit are rectangular hyperbolas, with increasing merit in the direction "toward the origin." This makes the most desirable design point as close to the lower left-hand corner of the feasible domain as is possible with discrete bushing sizes. Now let us find the closest constraint when $D = 0.625$ in and $L = 1.00$ in:

$$\text{Eq. (12–52):} \quad n_1 = \frac{L D P_{max}}{F} = \frac{(1)0.625(3560)}{30} = 74.2 \quad \text{(pressure)}$$

$$\text{Eq. (12–53):} \quad n_2 = \frac{12 V_{max}}{\pi D N} = \frac{12(100)}{\pi(0.625)300} = 2.04 \quad \text{(velocity)}$$

$$\text{Eq. (12–54):} \quad n_3 = \frac{12(PV)_{max} L}{\pi F N} = \frac{12(46\,700)(1)}{\pi(30)300} = 19.8 \quad (PV \text{ constraint})$$

$$\text{Eq. (12–55):} \quad n_4 = \frac{3L(w_0)_{max}}{f_1 f_2 K F N t}$$

$$f_1 = \frac{\pi D N/12 - 33}{100 - 33}(1.8 - 1.3) + 1.3 = 1.42$$

$$n_4 = \frac{3(1)0.001}{1.42(1)6(10^{-11})30(300)800} = 4.89 \quad \text{(wear)}$$

$$\text{Eq. (12–56):} \quad n_5 = \frac{(\Delta T)_{max} J U A}{10\pi f F D N}$$

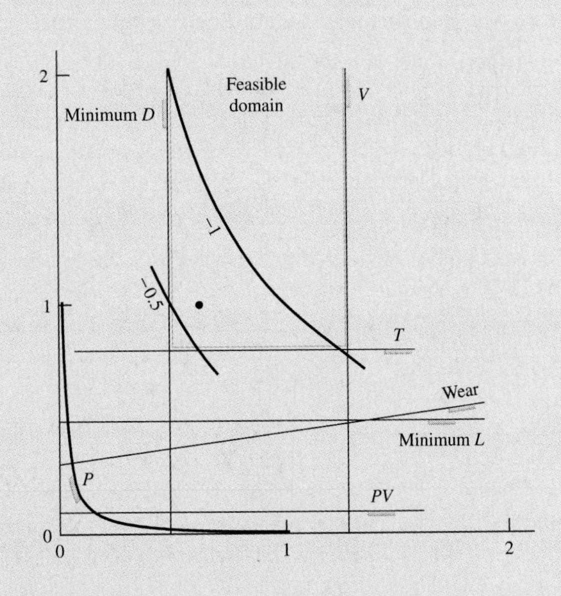

Figure 12–51

The feasible domain on the design plane for Ex. 12–14. For a figure of merit of $-DL$, contours of constant merit are rectangular hyperbolas with higher merit toward the origin. The plotted point is for $D = 0.625$ in, $L = 1.000$ in.

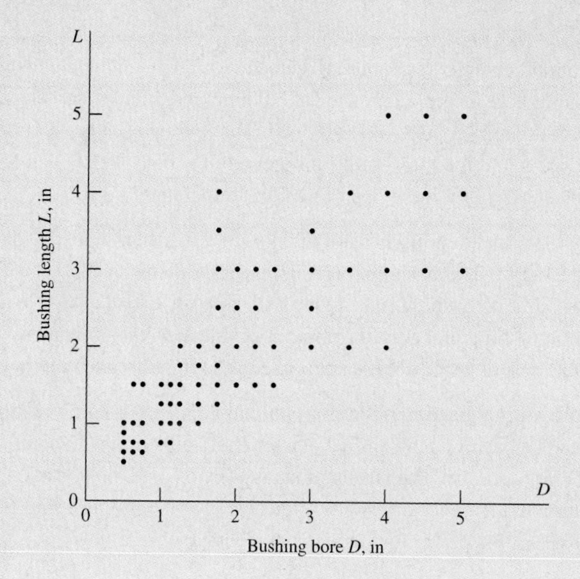

Figure 12–52

The design plane for Oiles 500 bushings with steel shafts. The plotted points identify available bushings by inside diameter.

Since $A = 2\pi DL/144 = 2\pi 0.625(1)/144 = 0.0273\ \text{ft}^2$,

$$n_5 = \frac{230(778)2.7(0.0273)}{10\pi(0.03)30(0.625)300} = 2.49 \qquad \text{(temperature)}$$

The least factor of safety is $n_2 = 2.04$. As the load is increased from 30 lb to 2.04(30) the wear constraint becomes tight. A design factor of $n_d = 2$ is realizable. Table 12–13 was designed to make the point that a manufacturer can present information in terms of common decision variables. Rotate Fig. 12–13 90° ccw and see the design plane. If only it was done to scale! And you can do it: prepare a master, and use a Xerox copy of it upon which to plot the constraints, define the feasible domain, then conveniently choose a likely candidate or candidates and use Eqs. (12–52) through (12–56) a show adequacy or inadequacy. Viewpoint is an indispensable aid to the design process. See Fig. 12–52.

PROBLEMS

ANALYSIS

12–1 A full journal bearing has a journal diameter D of 1.000 in, with a unilateral tolerance of −0.0015 in. The bushing bore has a diameter B of 1.0015 in and a unilateral tolerance of 0.003 in. The l/D ratio is unity. The load is 250 lb and the journal runs at 1100 r/min. If the average viscosity is 8 μreyn, find the minimum film thickness, the power loss, and the side flow for the minimum clearance assembly.

ANALYSIS

12–2 A full journal bearing has a journal diameter D of 1.250 in, with a unilateral tolerance of −0.001 in. The bushing bore has a diameter B of 1.252 in and a unilateral tolerance of 0.003 in. The bearing is 2.5 in long. The journal load is 400 lb and it runs at a speed of 1150 r/min. Using an average viscosity of 10 μreyn find the minimum film thickness, the maximum film pressure, and the total oil-flow rate for the minimum clearance assembly.

ANALYSIS

12–3 A journal bearing has a journal diameter D of 3.000 in, with a unilateral tolerance of −0.001 in. The bushing bore has a diameter B of 3.005 in and a unilateral tolerance of 0.004 in. The bushing is 1.5 in long. The journal speed is 600 r/min and the load is 800 lb. For both SAE 10 and SAE 40 lubricants find the minimum film thickness and the maximum film pressure for an operating temperature of 150°F for the minimum clearance assembly.

ANALYSIS

12–4 A journal bearing has a journal diameter of 3.000 in with a unilateral tolerance of −0.003 in. The bushing bore B has a diameter of 3.006 in and a unilateral tolerance of 0.004 in. The bushing is 3 in long and supports a 600-lb load. The journal speed is 750 r/min. Find the minimum oil film thickness and the maximum film pressure for both SAE 10 and SAE 20W-40 lubricants, for the tightest assembly if the operating film temperature is 140°F.

ANALYSIS

12–5 A full journal bearing has a journal with a diameter D of 2.000 in and a unilateral tolerance of −0.0012 in. The bushing has a bore with a diameter B of 2.0024 and a unilateral tolerance of 0.002 in. The bushing is 1 in long and supports a load of 600 lb at a speed of 800 r/min. Find the minimum film thickness, the power loss, and the total lubricant flow if the average film temperature is 130°F and SAE 20 lubricant is used. The tightest assembly is to be analyzed.

ANALYSIS

12–6 A full journal bearing has a shaft journal diameter of 25 mm with a unilateral tolerance of −0.01 mm. The bushing bore has a diameter B of 25.04 mm with a unilateral tolerance of 0.03 mm. The l/D ratio is unity. The bushing load is 1.25 kN, and the journal rotates at 1200 r/min. Analyze the minimum clearance assembly if the average viscosity is 50 mPa · s to find the minimum oil film thickness, the power loss, and the percentage of side flow.

ANALYSIS

12–7 A full journal bearing has a shaft journal with a diameter D of 30.00 mm and a unilateral tolerance of −0.015 mm. The bushing bore has a diameter B of 30.05 mm with a unilateral tolerance of 0.035 mm. The bushing bore is 50 mm in length. The bearing load is 2.75 kN and the journal rotates at 1120 r/min. Analyze the minimum clearance assembly and find the minimum film thickness, the coefficient of friction, and the total oil flow if the average viscosity is 60 mPa · s.

ANALYSIS

12–8 A journal bearing has a shaft diameter D of 75.00 mm with a unilateral tolerance of −0.02 mm. The bushing bore has a diameter B of 75.10 mm with a unilateral tolerance of 0.06 mm. The bushing is 36 mm long and supports a load of 2 kN. The journal speed is 720 r/min. For the minimum clearance assembly find the minimum film thickness, the heat loss rate, and the maximum lubricant pressure for SAE 20 and SAE 40 lubricants operating at an average film temperature of 60°.

ANALYSIS

12–9 A full journal bearing is 25 mm long. The shaft journal has a diameter D of 50 mm with a unilateral tolerance of −0.01 mm. The bushing bore has a diameter B of 50.05 mm with a unilateral tolerance of 0.01 mm. The load is 2000 N and the journal speed is 840 r/min. For the minimum clearance assembly find the minimum oil-film thickness, the power loss, and the side flow if the operating temperature is 55°C and SAE 30 lubricating oil is used.

ANALYSIS

12–10 A $1\frac{1}{4}$- $\times 1\frac{1}{4}$-in sleeve bearing supports a load of 700 lb and has a journal speed of 3600 r/min. An SAE 10 oil is used having an average temperature of 160°F. Using Fig. 12–14, estimate the radial clearance for minimum coefficient of friction f and for maximum load-carrying capacity W. The difference between these two clearances is called the clearance range. Is the resulting range attainable in manufacture?

ANALYSIS

12–11 A full journal bearing has a shaft diameter of 80.00 mm with a unilateral tolerance of −0.01 mm. The l/D ratio is unity. The bushing has a bore diameter D of 80.08 mm with a unilateral tolerance of 0.03 mm. The oil supply is in an axial-groove sump with a steady-state temperature of 60°C. The radial load is 3000 N. Estimate the average film temperature, the minimum film thickness, the heat loss rate, and the lubricant side-flow rate for the minimum clearance assembly, if the journal speed is 8 r/min.

ANALYSIS

12–12 A $2\frac{1}{2}$- $\times 2\frac{1}{2}$-in sleeve bearing uses 20 grade lubricant. The axial-groove sump has a steady-state temperature of 110°F. The shaft journal has a diameter D of 2.200 in with a unilateral tolerance of −0.001 in. The bushing bore has a diameter B of 2.504 in with a unilateral tolerance of 0.001 in. The journal speed is 1120 r/min and the radial load is 1200 lb. Estimate

(*a*) The magnitude and location of the minimum oil-film thickness.

(*b*) The eccentricity.

(*c*) The coefficient of friction.

(*d*) The power loss rate.

(*e*) Both the total and side oil-flow rates.

(*f*) The maximum oil-film pressure and its angular location.

(*g*) The terminating position of the oil film.

(*h*) The average temperature of the side flow.

(*i*) The oil temperature at the terminating position of the oil film.

12–13 A set of sleeve bearings has a specification of shaft journal diameter D of 1.250 in with a unilateral tolerance of -0.001 in. The bushing bore has a diameter B of 1.252 in with a unilateral tolerance of 0.003 in. The bushing is $1\frac{1}{4}$ in long. The radial load is 250 lb and the shaft rotational speed is 1750 r/min. The lubricant is SAE 10 oil and the axial-groove sump temperature at steady state T_s is 120°F. For the $c_{\min}$, c_{median}, and $c_{\max}$ assemblies analyze the bearings and observe the changes in S, ϵ, f, Q, Q_s, ΔT, $T_{\max}$, $\bar{T}_f$, hp, and H.

12–14 It is always useful to examine matters of scale, say doubling of physical dimensions. What are your expectations with respect to the parameters?

12–15 In Prob. 12–14 you found various parameters changed by 2, 2^2, and 2^3. Unfortunately the ability to hold tolerances routinely is not proportional to size, as indicated in Ex. 12–8. Repeat Ex. 12–8 by doubling the nominal sizes and change the load appropriately, then observing the effect on parameters.

12–16 The iteration process to find the average fluid film temperature T_f can be speeded up after completion of two trials. In Ex. 12–8 for $c = 0.003$ in the iteration trials were

$\bar{T}_f$	T_{av}
160	127.5
150	130.5
140	135.1
127	137.01

Plot T_{av} as ordinate, $\bar{T}_f$ as abscissa and locate the points in the table above. Additionally plot the line $T_{\text{av}} = \bar{T}_f$. Draw a straight line connecting the first two points in the table. Read the intersection between this line and the $T_{\text{av}} = \bar{T}_f$ locus. This is a recommended value of $\bar{T}_f$ for the third evaluation. Drawing a line through the new point provided by the third evaluation and the second evaluation produces another, even closer estimate. Try it.

12–17 The graphical method of Prob. 12–16 can be rendered algebraic. Two trials in search of an equilibrium temperature resulted in

	x	y
Trial	$\bar{T}_f$	T_{av}
1	145	154.13
2	160	145

(*a*) Show that a straight line defined by the two points above, using the notation x and y, intersects the locus $x = y$ $(\bar{T}_f = T_{av})$ at

$$y = x = \frac{y_1 - \dfrac{y_2 - y_1}{x_2 - x_1} x_1}{1 - \dfrac{y_2 - y_1}{x_2 - x_1}} = \frac{y_1 - x_1 \Delta}{1 - \Delta}$$

where $\Delta = (y_2 - y_1)/(x_2 - x_1)$.

(*b*) Predict the next trial $\bar{T}_f$.

12–18 The figure-of-merit development at the end of Ex. 12–8 is simple and effective. However, the design window is populated by an ensemble of bearings, all with different figures of merit. It is important that the figure of merit characterize the ensemble. If the design window is located by the coordinates of the upper right-hand corner of the design window, then it is necessary to analyze the median clearance assembly and to use its figure of merit as characteristic of the design. The average of the friction torques of the entire ensemble will agree with that of the median clearance assembly. For Ex. 12–8 analyze the minimum clearance, median clearance, and maximum clearance assemblies. Use a mechanical efficiency of unity. Prepare a three-row table and observe the changes.

12–19 In Ex. 12–8 the geometry table was begun without fanfare. How does the designer know where to begin with respect to c_{min} and B? One must be able to *start* a journal bearing before one can run it. Trumpler's third criterion gives a clue. It's a vertical line on the rc plane. Show that the smallest journal diameter $(D)_{min}$ that can be started repeatedly is

Natural Circulation	Pressure-Fed
$(D)_{min} \geq \sqrt{\dfrac{W_{st}}{300(l/D)}}$	$(D)_{min} \geq \sqrt{\dfrac{W_{st}}{2(300)(l'/D)}}$

Commonly $l/D = 1$; therefore in such cases

$$(D)_{min} \geq \sqrt{\frac{W_{st}}{300}} \qquad (D)_{min} \geq \sqrt{\frac{W_{st}}{300}}$$

12–20 Based on the outcome of Prob. 12–18, repeat Ex. 12–8 with $B = 1.75$ and 1.85. If there are satisfactory bearings, are any superior to the one in Ex. 12–8?

12–21 A bearing has the descriptors $B = 2.000$ in, $b = 0.003$, $D = 1.996$ in, and $d = 0.0006$ in. It also has an associated design window. Move the window up, decreasing the minimum clearance to $c_{min} = 0.001$ in. Draw the window on the rc plane, and describe the new bearing specifications from the window.

12–22 The original bearing in Prob. 12–21 is to have its design window moved to the left, keeping the same window depth. Do this by choosing the next available bushing whose diameter B is 1.875 in. Plot the new window on the rc plane and describe the new bearing specifications from the plotted window.

12–23 A bearing bushing and journal have the following dimensions:

$$6.0060 \begin{array}{l} +0 \\ -0.0015 \end{array} \text{ in} \qquad 6.000 \begin{array}{l} +0.0005 \\ -0 \end{array} \text{ in}$$

Convert these specifications to the following form:

$$B^{+b}_{-0} \qquad D^{+0}_{-d}$$

12–24 Consider the bearing that has the specifications $B = 2.000$, $b = 0.003$, $D = 1.996$, and $d = 0.0006$ in. The unilateral tolerances are given by

$$b = \text{INT}(1 + B)/1000 \qquad \text{and} \qquad d = \text{INT}(3 + 2D)/10\,000 \text{ in}$$

Define the design window, plot it on the rc plane, and locate it.

12–25 Consider the bearing with the specifications $B = 2.000$ in, $b = 0.003$ in, $D = 1.998$ in, and $d = 0.0006$ in. Define the design window, plot it on the rc plane, and locate it.

12–26 A design window has a width $\Delta r = 0.0003$ in and a depth $\Delta c = 0.0018$ in. Find B, b, D, and d if the window is located by

(*a*) Upper right-hand coordinates $r_{max} = 0.998$, $c_{min} = 0.002$ in.
(*b*) Centroidal coordinates $\bar{r} = 0.997\,85$ in, $\bar{c} = 0.000\,29$ in.

ANALYSIS

12–27 Consider the bearing

$$B = 2.000^{+0.003}_{-0} \qquad D = 1.996^{+0}_{-0.0006} \text{ in}$$

Define the design window, plot it on the rc plane, and locate it on the rc plane.

12–28 For the conditions of Ex. 12–8, seek an optimal bearing that minimizes the minimum parasitic friction torque for the chosen bearing.

12–29 In Ex. 12–14 find the values of the design variables D and L with the largest figure of merit.

(*a*) Prepare a machine master copy that you plotted as in Fig. 12–52.
(*b*) With a copy of the master copy, plot the constraints of Ex. 12–14, and find the optimal solution by inspection.

12–30 An interpolation equation was given by Raimondi and Boyd, and it is displayed as Eq. (12–15). This equation is a good candidate for a computer program. Write such a program for interactive use. Once ready for service it can save time and reduce errors. Another version of this program can be used with a subprogram that contains curve fits to Raimondi and Boyd charts for computer use.

12–31 If you are thinking that Raimondi and Boyd charts may be computerized, you have a choice of curve-fitting forms and techniques. Seireg and Dandage (Table 12–1) presented some curve-fitted equations for h_0/c, ϕ, fr/c, $Q/rcN_j l$, $JC_p\Delta T/P$, and P/p_{max} in the form

$$a\left(\frac{l}{D}\right)^{b_1} S^{b_2+b_3 l/D}$$

Computerizing lines of reference Table 28.15 would make a good class project, with division of labor, user-friendly ideas, and eventual class use.[18]

12–32 The lubricant pressure diagram in Fig. 12–16 shows only the positive pressure lobe of the 360° Sommerfeld bearing. If the diagram continued it could show a region of negative pressure (the journal circle is the zero-pressure contour) which carries the other "half" of the load by negative pressure. This would be the nature of the 360° Sommerfeld model. Real liquid lubricants cannot provide negative pressures of the magnitude required, and the liquid cavitates, introducing vapor bubbles. The load-carrying capacity is about half that of the Sommerfeld model. Raimondi and Boyd charts account for this in their mathematical model, as do Gumbel and Swift-Stieber models. The practical consequence is that half of the bushing can be relieved to form a much-enlarged axial-groove sump without compromising the load-carrying capacity of the bearing. The object of this

18. See Theo G. Keith Jr., Chap. 28 in Joseph E. Shigley and Charles R. Mischke (eds.), *Standard Handbook of Machine Design,* 2nd ed., McGraw-Hill, New York, 1996, pp. 28.32–28.35.

problem is to have you think about when this is appropriate, and where the bearing-cap split should (or could) be placed. If you did relieve the cap to make the enlarged sump (and reduce friction), what would you do with the value of h_0/c from Fig. 12–14? What would you do with the value of fr/c from Fig. 12–17?

DESIGN

12–33 A natural-circulation pillow-block bearing has a journal diameter D of 2.500 in with a unilateral tolerance of −0.001 in. The bushing bore diameter B is 2.504 in with a unilateral tolerance of 0.004 in. The shaft runs at an angular speed of 1120 r/min, the bearing uses SAE grade 20 oil and carries a steady load of 300 lb in shaft-stirred air at 70°F. The lateral area of the pillow-block housing is 60 in^2. Perform an adequacy assessment using minimum radial clearance for a load of 600 lb and 300 lb. Use Trumpler's criteria.

DESIGN

12–34 An eight-cylinder diesel engine has a front main bearing with a journal diameter of 3.500 in and a unilateral tolerance of −0.003 in. The bushing bore diameter is 3.505 in with a unilateral tolerance of +0.005 in. The bushing length is 2 in. The pressure-fed bearing has a central annulur groove 0.250 in wide. The SAE 30 oil comes from a sump at 120°F using a supply pressure of 50 psig. The sump's heat-dissipation capacity is 5000 Btu/h per bearing. For a minimum radial clearance, a speed of 2000 r/min, and a radial load of 4600 lb, find the average film temperature and apply Trumpler's criteria in your adequacy assessment.

DESIGN

12–35 A pressure-fed bearing has a journal diameter of 50.00 mm with a unilateral tolerance of −0.05 mm. The bushing bore diameter is 50.084 mm with a unilateral tolerance of 0.10 mm. The length of the bushing is 55 mm. Its central annular groove is 5 mm wide and is fed by SAE 30 oil is 55°C at 200 kPa supply gage pressure. The journal speed is 2880 r/min carrying a load of 10 kN. The sump can dissipate 300 watts per bearing if necessary. For minimum radial clearances perform an adequacy assessment using Trumpler's criteria.

DESIGN

12–36 Design a central annular-groove pressure-fed bearing with an l'/D ratio of 0.5, using SAE grade 20 oil, the lubricant supplied at 30 psig. The exterior oil cooler can maintain the sump temperature at 120°F for heat dissipation rates up to 1500 Btu/h. The load to be carried is 900 lb at 3000 r/min. The groove width is $\frac{1}{4}$ in. Use nominal journal diameter D as one design variable and c as the other. Use Trumpler's criteria for your adequacy assessment.

DESIGN

12–37 Repeat design problem Prob. 12–36 using the nominal bushing bore B as one decision variable and the radial clearance c as the other. Again, Trumpler's criteria to be used.

ANALYSIS

12–38 Table 12–1 gives the Seireg and Dandage curve fit for the absolute viscosity in customary U.S. engineering units. Show that in SI units of mPa · s and a temperature of $C°$ Celsius, the viscosity can be expressed as

$$\mu = 6.89(10^6)\mu_0 \exp[(b/(1.8C + 127)]$$

where μ_0 and b are from Table 12–1. If the viscosity μ_0' is expressed in μreyn, then

$$\mu = 6.89\mu_0' \exp[(b/(1.8C + 127)]$$

What is the viscosity of a grade 30 oil at 79°C?

DESIGN

12–39 For Prob. 12–36 a satisfactory design is

$$D = 2.000^{+0}_{-0.001} \text{ in} \qquad B = 2.005^{+0.003}_{-0} \text{ in}$$

Double the size of the bearing dimensions and quadruple the load to 3600 lb.

(*a*) Analyze the scaled-up bearing for median assembly.

(*b*) Compare the results of a similar analysis for the 2-in bearing, median assembly.

12–40 An Oiles 500 bushing is to be selected for use with a steel journal that will hold wear to 0.001 in or less in 800 h. The maximum allowable lubricant temperature is 250°F, in an environment in which the ambient temperature is 70°F. The air is shaft-stirred. The design factor must equal or exceed 2.5. The bearing lateral area can be estimated from $A = 2\pi DL/144$ ft². The radial load is 50 lb, and the angular speed is 500 r/min. An optimal design is desired.

DESIGN

12–41 Matters of scale provide perspective and enhance viewpoint. Consider the design result from Prob. 12–40. If the geometry of the specified bushing is doubled and the load quadrupled, how would the factors of safety [Eqs. (12–52) through (12–56)] change? Predict the ratios n_i'/n_i where n_i' is the scaled-up case.

DESIGN

12–42 The following design problem appeared in a *Sample Examination and Solutions* booklet of the National Council of Examiners for Engineering and Surveying.[19] *Situation:* A nylon plastic bearing bushing is to be selected for the shaft of an electric motor. The maximum radial load at the bearing is 50 lb. The shaft speed is 600 r/min. The bearing length is to be 1.5 times the diameter.

F = radial load, 50 lb	$P_{max} = 1000$ psi
N = rotational speed, 600 r/min	$V_{max} = 1000$ ft/min
D = inside diameter of bushing, in	$(PV)_{max} = 300$ psi(ft/min)
f = coefficient of friction, 0.03	
T = torque, $fFD/2$, in · lb	

The solution recommended a strategy of using the limits P_{max}, V_{max}, and $(PV)_{max}$ to establish constraints on D and using the smallest D to reduce friction torque.

Answer: $D = 1.75$ in, $L = 2.63$ in.

(*a*) Can you confirm the answers?
(*b*) Plot the DL diagram and place on it these three constraints, plus the fourth, $L = 1.5D$, to display the feasible line and the solution.
(*c*) In the light of your experience with Ex. 12–14, do you have any concerns?

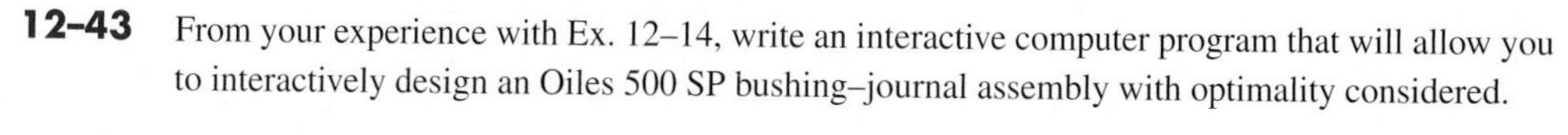

12–43 From your experience with Ex. 12–14, write an interactive computer program that will allow you to interactively design an Oiles 500 SP bushing–journal assembly with optimality considered.

19. Reported in *Mechanical Engineering,* May 1991, Vol. 113, p. 97.

13 Gears—General

13–1 Types of Gears **832**
13–2 Nomenclature **833**
13–3 Tooth Systems **835**
13–4 Conjugate Action **837**
13–5 Involute Properties **838**
13–6 Fundamentals **839**
13–7 Contact Ratio **844**
13–8 Interference **845**
13–9 The Forming of Gear Teeth **848**
13–10 Straight Bevel Gears **850**
13–11 Parallel Helical Gears **851**
13–12 Worm Gears **855**
13–13 Gear Trains **856**
13–14 Force Analysis—Spur Gearing **860**
13–15 Force Analysis—Bevel Gearing **863**
13–16 Force Analysis—Helical Gearing **866**
13–17 Force Analysis—Worm Gearing **869**
13–18 Gear Ratios and Numbers of Teeth **874**
13–19 Gear-Shaft Speeds and Bearings **878**

This chapter will address gear geometry, the kinematic relations, and the consequential forces and moments associated with the four principal types of gears. The forces and moments are transmitted to meshing gears and to the shafts and bearings to which they are attached. The torsional moment allows motion and power transmission. Other moments and forces affect the shaft and its bearings. The next two chapters will address stress, strength, safety, and reliability.

13–1 Types of Gears

Spur gears, illustrated in Fig. 13–1, have teeth parallel to the axis of rotation and are used to transmit motion from one shaft to another, parallel, shaft. Of all types, the spur gear is the simplest and, for this reason, will be used to develop the primary kinematic relationships of the tooth form.

Helical gears, shown in Fig. 13–2, have teeth inclined to the axis of rotation. Helical gears can be used for the same applications as spur gears and, when so used, are not as noisy, because of the more gradual engagement of the teeth during meshing. The inclined tooth also develops thrust loads and bending couples, which are not present with spur gearing. Sometimes helical gears are used to transmit motion between nonparallel shafts.

Bevel gears, shown in Fig. 13–3, have teeth formed on conical surfaces and are used mostly for transmitting motion between intersecting shafts. The figure actually illustrates *straight-tooth bevel gears. Spiral bevel gears* are cut so the tooth is no longer straight, but forms a circular arc. *Hypoid gears* are quite similar to spiral bevel gears except that the shafts are offset and nonintersecting.

Figure 13–1

Spur gears are used to transmit rotary motion between parallel shafts.

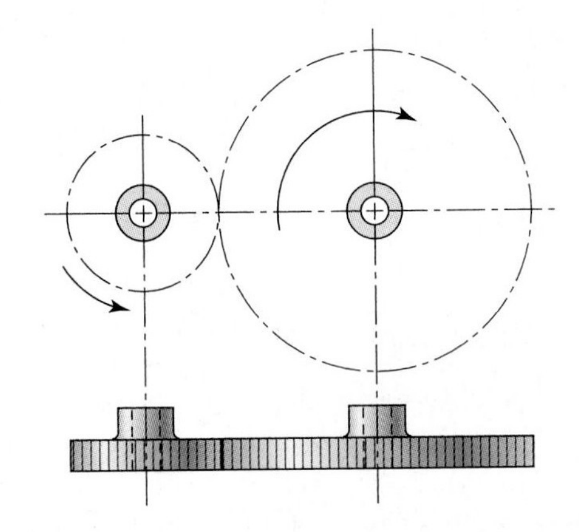

Figure 13–2

Helical gears are used to transmit motion between parallel or nonparallel shafts.

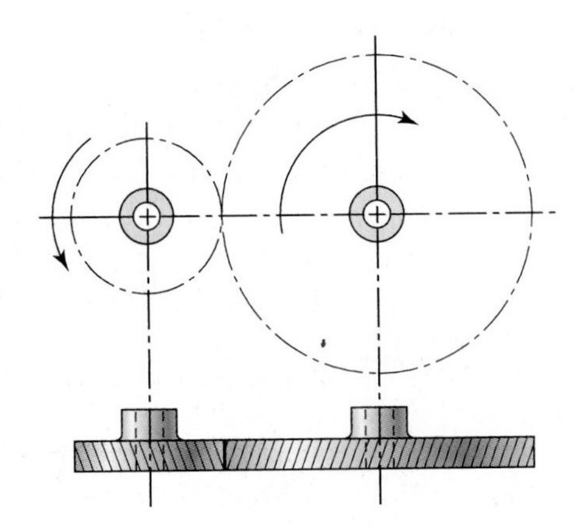

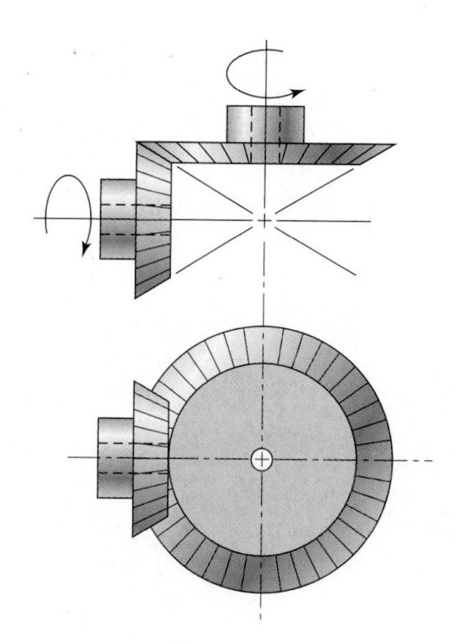

Figure 13–3

Bevel gears are used to transmit rotary motion between intersecting shafts.

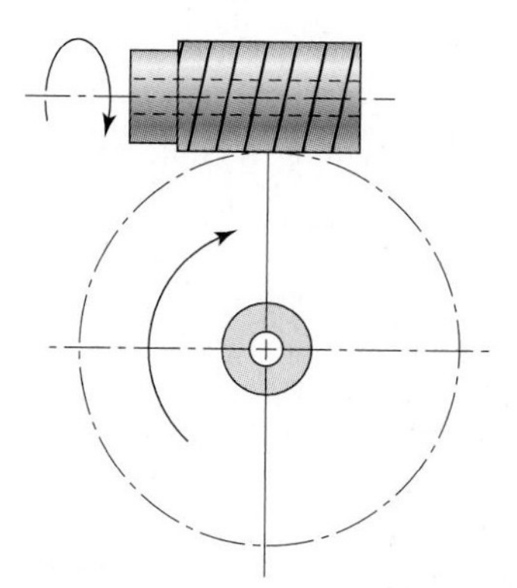

Figure 13–4

Worm gearsets are used to transmit rotary motion between nonparallel and nonintersecting shafts.

Shown in Fig. 13–4 is the fourth basic gear type, the *worm* and *worm gear.* As shown, the worm resembles a screw. The direction of rotation of the worm gear, also called the worm wheel, depends upon the direction of rotation of the worm and upon whether the worm teeth are cut right-hand or left-hand. Worm-gear sets are also made so that the teeth of one or both wrap partly around the other. Such sets are called *single-enveloping* and *double-enveloping* worm-gear sets. Worm-gear sets are mostly used when the speed ratios of the two shafts are quite high, say, 3 or more.

13–2 Nomenclature

The terminology of spur-gear teeth is illustrated in Fig. 13–5. The *pitch circle* is a theoretical circle upon which all calculations are usually based; its diameter is the *pitch diameter.* The pitch circles of a pair of mating gears are tangent to each other. A *pinion* is the smaller of two mating gears. The larger is often called the *gear.*

The *circular pitch p* is the distance, measured on the pitch circle, from a point on one tooth to a corresponding point on an adjacent tooth. Thus the circular pitch is equal to the sum of the *tooth thickness* and the *width of space.*

The *module m* is the ratio of the pitch diameter to the number of teeth. The customary unit of length used is the millimeter. The module is the index of tooth size in SI.

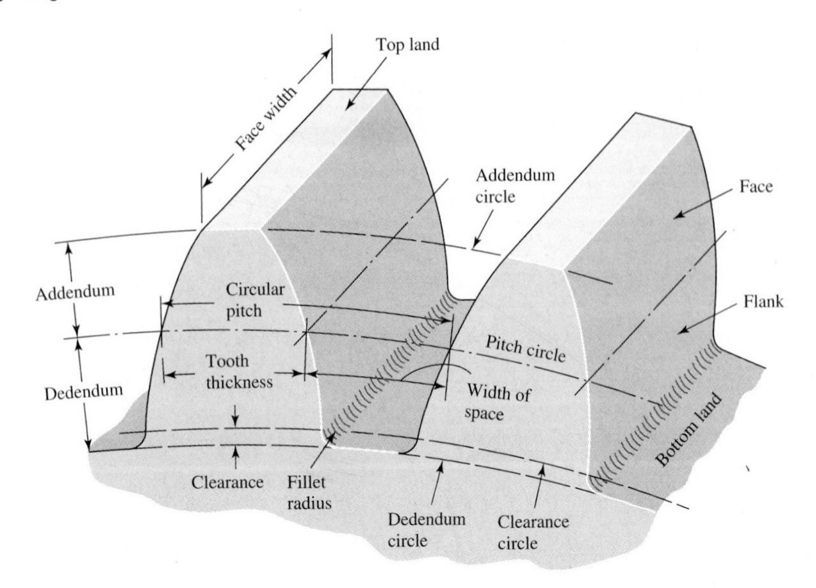

Figure 13–5

Nomenclature of spur-gear teeth.

The *diametral pitch* P is the ratio of the number of teeth on the gear to the pitch diameter. Thus, it is the reciprocal of the module. Since diametral pitch is used only with U.S. units, it is expressed as teeth per inch.

The *addendum* a is the radial distance between the *top land* and the pitch circle. The *dedendum* b is the radial distance from the *bottom land* to the pitch circle. The *whole depth* h_t is the sum of the addendum and the dedendum.

The *clearance circle* is a circle that is tangent to the addendum circle of the mating gear. The *clearance* c is the amount by which the dedendum in a given gear exceeds the addendum of its mating gear. The *backlash* is the amount by which the width of a tooth space exceeds the thickness of the engaging tooth measured on the pitch circles.

You should prove for yourself the validity of the following useful relations:

$$P = \frac{N}{d} \tag{13–1}$$

where P = diametral pitch, teeth per inch
N = number of teeth
d = pitch diameter, in

$$m = \frac{d}{N} \tag{13–2}$$

where m = module, mm
d = pitch diameter, mm

$$p = \frac{\pi d}{N} = \pi m \tag{13–3}$$

where p = circular pitch

$$pP = \pi \tag{13–4}$$

13–3 Tooth Systems[1]

A *tooth system* is a standard which specifies the relationships involving addendum, dedendum, working depth, tooth thickness, and pressure angle. The standards were originally planned to attain interchangeability of gears of all tooth numbers, but of the same pressure angle and pitch.

Table 13–1 contains the standards most used for spur gears. A $14\frac{1}{2}^\circ$ pressure angle was once used for these but is now obsolete; the resulting gears had to be comparatively larger to avoid interference problems.

Table 13–2 is particularly useful in selecting the pitch or module of a gear. Cutters are generally available for the sizes shown in this table.

Table 13–3 lists the standard tooth proportions for straight bevel gears. These sizes apply to the large end of the teeth. The nomenclature is defined in Fig. 13–21.

Standard tooth proportions for helical gears are listed in Table 13–4. Tooth proportions are based on the normal pressure angle; these angles are standardized the same as for spur gears. Though there will be exceptions, the face width of helical gears should be at least 2 times the axial pitch to obtain good helical-gear action.

Tooth forms for worm gearing have not been highly standardized, perhaps because there has been less need for it. The pressure angles used depend upon the lead angles and must be large enough to avoid undercutting of the worm-gear tooth on the side at which contact ends. A satisfactory tooth depth, which remains in about the right proportion to the lead angle, may be obtained by making the depth a proportion of the axial circular pitch. Table 13–5 summarizes what may be regarded as good practice for pressure angle and tooth depth.

Table 13–1

Standard and Commonly Used Tooth Systems for Spur Gears

Tooth System	Pressure Angle ϕ, deg	Addendum a	Dedendum b
Full depth	20	$1/P_d$ or $1m$	$1.25/P_d$ or $1.25m$
			$1.35/P_d$ or $1.35m$
	$22\frac{1}{2}$	$1/P_d$ or $1m$	$1.25/P_d$ or $1.25m$
			$1.35/P_d$ or $1.35m$
	25	$1/P_d$ or $1m$	$1.25/P_d$ or $1.25m$
			$1.35/P_d$ or $1.35m$
Stub	20	$0.8/P_d$ or $0.8m$	$1/P_d$ or $1m$

Table 13–2

Tooth Sizes in General Uses

Diametral Pitch	
Coarse	$2, 2\frac{1}{4}, 2\frac{1}{2}, 3, 4, 6, 8, 10, 12, 16$
Fine	20, 24, 32, 40, 48, 64, 80, 96, 120, 150, 200

Modules	
Preferred	1, 1.25, 1.5, 2, 2.5, 3, 4, 5, 6, 8, 10, 12, 16, 20, 25, 32, 40, 50
Next Choice	1.125, 1.375, 1.75, 2.25, 2.75, 3.5, 4.5, 5.5, 7, 9, 11, 14, 18, 22, 28, 36, 45

1. Standardized by the American Gear Manufacturers Association (AGMA). Write AGMA for a complete list of standards, because changes are made from time to time. The address is: 1500 King Street, Suite 201, Alexandria, VA 22314.

Table 13–3

Tooth Proportions for 20° Straight Bevel-Gear Teeth

Item	Formula
Working depth	$h_k = 2.0/P$
Clearance	$c = (0.188/P) + 0.002$ in
Addendum of gear	$a_G = \dfrac{0.54}{P} + \dfrac{0.460}{P(m_{90})^2}$
Gear ratio	$m_G = N_G/N_P$
Equivalent 90° ratio	$m_{90} = m_G$ when $\Sigma = 90°$
	$m_{90} = \sqrt{m_G \dfrac{\cos \gamma}{\cos \Gamma}}$ when $\Sigma \neq 90°$
Face width	$F = \dfrac{A_O}{3}$ or $F = \dfrac{10}{P}$, whichever is smaller
Minimum number of teeth	Pinion: 16, 15, 14, 13; Gear: 16, 17, 20, 30

Table 13–4

Standard Tooth Propertions for Helical Gears

Quantity*	Formula	Quantity*	Formula
Addendum	$\dfrac{1.00}{P_n}$	External gears:	
Dedendum	$\dfrac{1.25}{P_n}$	Standard center distance	$\dfrac{D+d}{2}$
Pinion pitch diameter	$\dfrac{N_P}{P_n \cos \psi}$	Gear outside diameter	$D + 2a$
Gear pitch diameter	$\dfrac{N_G}{P_n \cos \psi}$	Pinion outside diameter	$d + 2a$
Normal arc tooth thickness	$\dfrac{\pi}{P_n} - \dfrac{B_n}{2}$	Gear root diameter	$D - 2b$
Pinion base diameter	$d \cos \phi_t$	Pinion root diameter	$d - 2b$
		Internal gears:	
Gear base diameter	$D \cos \phi_t$	Center distance	$\dfrac{D-d}{2}$
Base helix angle	$\tan^{-1}(\tan \psi \cos \phi_t)$	Inside diameter	$D - 2a$
		Root diameter	$D + 2b$

*All dimensions are in inches, and angles are in degrees.

Table 13–5

Recommended Pressure Angles and Tooth Depths for Worm Gearing

Lead Angle λ, deg	Pressure Angle ϕ_n, deg	Addendum a	Dedendum b_G
0–15	$14\frac{1}{2}$	$0.3683p_x$	$0.3683p_x$
15–30	20	$0.3683p_x$	$0.3683p_x$
30–35	25	$0.2865p_x$	$0.3314p_x$
35–40	25	$0.2546p_x$	$0.2947p_x$
40–45	30	$0.2228p_x$	$0.2578p_x$

The *face width* F_G of the worm gear should be made equal to the length of a tangent to the worm pitch circle between its points of intersection with the addendum circle, as shown in Fig. 13–6.

Figure 13–6

A graphical depiction of the face width of a worm of a worm gearset.

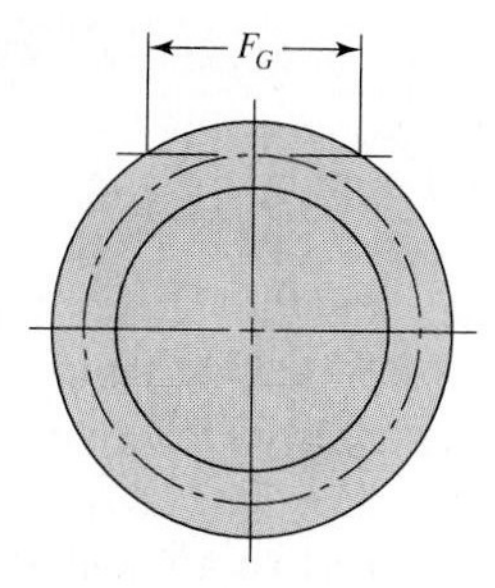

13–4 Conjugate Action

The following discussion assumes the teeth to be perfectly formed, perfectly smooth, and absolutely rigid. Such an assumption is, of course, unrealistic, because the application of forces will cause deflections.

Mating gear teeth acting against each other to produce rotary motion are similar to cams. When the tooth profiles, or cams, are designed so as to produce a constant angular-velocity ratio during meshing, these are said to have *conjugate action.* In theory, at least, it is possible arbitrarily to select any profile for one tooth and then to find a profile for the meshing tooth which will give conjugate action. One of these solutions is the *involute profile,* which, with few exceptions, is in universal use for gear teeth and is the only one with which we should be concerned.

When one curved surface pushes against another (Fig. 13–7), the point of contact occurs where the two surfaces are tangent to each other (point c), and the forces at any instant are directed along the common normal ab to the two curves. The line ab, representing the direction of action of the forces, is called the *line of action.* The line of action will intersect the line of centers O-O at some point P. The angular-velocity ratio between the two arms is inversely proportional to their radii to the point P. Circles drawn through point P from each center are called *pitch circles,* and the radius of each circle is called the *pitch radius.* Point P is called the *pitch point.*

Figure 13–7 is useful in making another observation. A pair of gears is really pairs of cams which act through a small arc and, before running off the involute contour, are replaced by another identical pair of cams. The cams can run in either direction

Figure 13–7

Cam *A* and follower *B* in contact. When the contacting surfaces are involute profiles, the ensuing conjugate action produces a constant angular-velocity ratio.

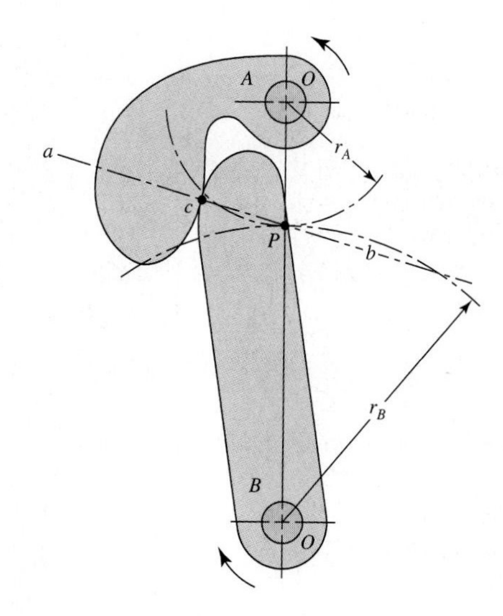

and are configured to transmit a constant angular-velocity ratio. If involute curves are used, the gears are tolerant of changes in center-to-center distance with *no* variation in constant angular-velocity ratio. Furthermore, the rack profiles are straight-flanked, making primary tooling simpler.

To transmit motion at a constant angular-velocity ratio, the pitch point must remain fixed; that is, all the lines of action for every instantaneous point of contact must pass through the same point P. In the case of the involute profile, it will be shown that all points of contact occur on the same straight line ab, that all normals to the tooth profiles at the point of contact coincide with the line ab, and, thus, that these profiles transmit uniform rotary motion.

13–5 Involute Properties

An involute curve may be generated as shown in Fig. 13–8*a*. A partial flange B is attached to the cylinder A, around which is wrapped a cord def which is held tight. Point b on the cord represents the tracing point, and as the cord is wrapped and unwrapped about the cylinder, point b will trace out the involute curve ac. The radius of the curvature of the involute varies continuously, being zero at point a and a maximum at point c. At point b the radius is equal to the distance be, since point b is instantaneously rotating about point e. Thus the generating line de is normal to the involute at all points of intersection and, at the same time, is always tangent to the cylinder A. The circle on which the involute is generated is called the *base circle*.

Let us now examine the involute profile to see how it satisfies the requirement for the transmission of uniform motion. In Fig 13–8*b*, two gear blanks with fixed centers at O_1 and O_2 are shown having base circles whose respective radii are O_1a and O_2b. We now imagine that a cord is wound clockwise around the base circle of gear 1, pulled tight between points a and b, and wound counterclockwise around the base circle of gear 2. If, now, the base circles are rotated in different directions so as to keep the cord tight, a point g on the cord will trace out the involutes cd on gear 1 and ef on gear 2. The involutes are thus generated simultaneously by the tracing point. The tracing point, therefore, represents the point of contact, while the portion of the cord ab is the generating line. The point of contact moves along the generating line; the generating line does not change position, because it is always tangent to the base circles; and since the

Figure 13–8

(*a*) Generation of an involute; (*b*) involute action.

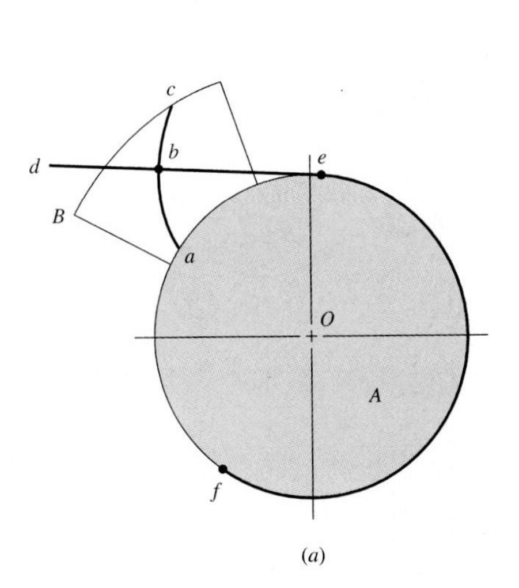

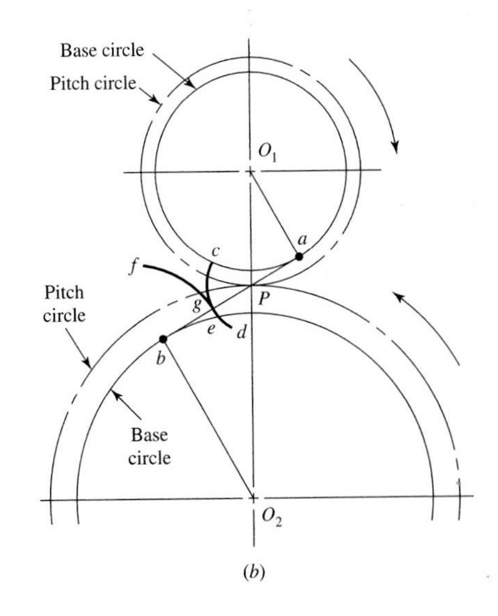

generating line is always normal to the involutes at the point of contact, the requirement for uniform motion is satisfied.

13–6 Fundamentals

Among other things, it is necessary that you actually be able to draw the teeth on a pair of meshing gears. You should understand, however, that you are not doing this for manufacturing or shop purposes. Rather, we make drawings of gear teeth to obtain an understanding of the problems involved in the meshing of the mating teeth.

First, it is necessary to learn how to construct an involute curve. As shown in Fig. 13–9, divide the base circle into a number of equal parts, and construct radial lines OA_0, OA_1, OA_2, etc. Beginning at A_1, construct perpendiculars A_1B_1, A_2B_2, A_3B_3, etc. Then along A_1B_1 lay off the distance A_1A_0, along A_2B_2 lay off twice the distance A_1A_0, etc., producing points through which the involute curve can be constructed.

To investigate the fundamentals of tooth action, let us proceed step by step through the process of constructing the teeth on a pair of gears.

When two gears are in mesh, their pitch circles roll on one another without slipping. Designate the pitch radii as r_1 and r_2 and the angular velocities as ω_1 and ω_2, respectively. Then the pitch-line velocity is

$$V = |r_1\omega_1| = |r_2\omega_2|$$

Thus the relation between the radii on the angular velocities is

$$\left|\frac{\omega_1}{\omega_2}\right| = \frac{r_2}{r_1} \tag{13–5}$$

Suppose now we wish to design a speed reducer such that the input speed is 1800 r/min and the output speed is 1200 r/min. This is a ratio of 3:2; the gear pitch diameters would be in the same ratio, for example, a 4-in pinion driving a 6-in gear. The various dimensions found in gearing are always based on the pitch circles.

We next specify that an 18-tooth pinion is to mesh with a 30-tooth gear and that the diametral pitch of the gearset is to be 2 teeth per inch. Then, from Eq. (13–1), the pitch diameters of the pinion and gear are, respectively,

$$d_1 = \frac{N_1}{P} = \frac{18}{2} = 9 \text{ in} \qquad d_2 = \frac{N_2}{P} = \frac{30}{2} = 15 \text{ in}$$

The first step in drawing teeth on a pair of mating gears is shown in Fig. 13–10. The cen-

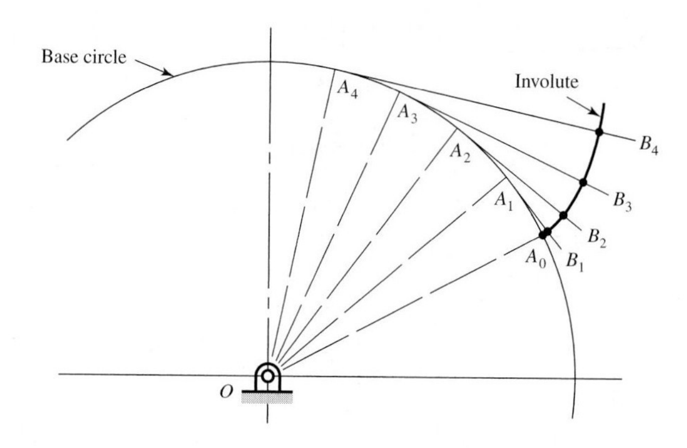

Figure 13–9

Construction of an involute curve.

Figure 13–10

Circles of a gear layout.

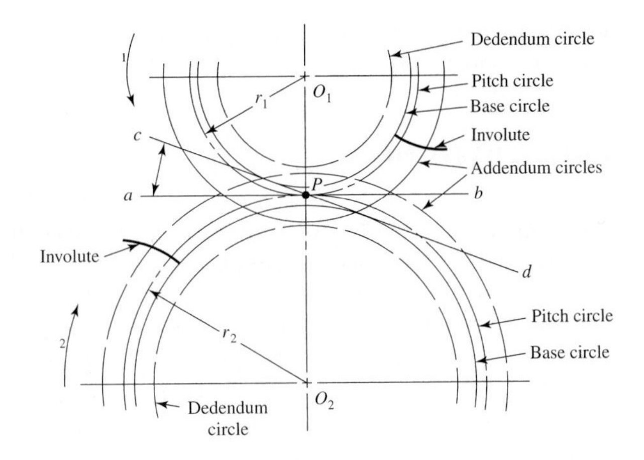

ter distance is the sum of the pitch radii, in this case 12 in. So locate the pinion and gear centers O_1 and O_2, 12 in apart. The construct the pitch circles of radii r_1 and r_2. These are tangent at P, the *pitch point.* Next draw line ab, the common tangent, through the pitch point. We now designate gear 1 as the driver, and since it is rotating counterclockwise, we draw a line cd through point P at an angle ϕ to the common tangent ab. The line cd has three names, all of which are in general use. It is called the *pressure line,* the *generating line,* and the *line of action.* It represents the direction in which the resultant force acts between the gears. The angle ϕ is called the *pressure angle,* and it usually has values of 20 or 25°, though $14\frac{1}{2}^\circ$ was once used.

Next, on each gear draw a circle tangent to the pressure line. These circles are the *base circles.* Since they are tangent to the pressure line, the pressure angle determines their size. As shown in Fig. 13–11, the radius of the base circle is

$$r_b = r\cos\phi \tag{13-6}$$

where r is the pitch radius.

Now generate an involute on each base circle as previously described and as shown in Fig. 13–10. This involute is to be used for one side of a gear tooth. It is not necessary to draw another curve in the reverse direction for the other side of the tooth, because we are going to use a template which can be turned over to obtain the other side.

The addendum and dedendum distances for standard interchangeable teeth are, as we shall learn later, $1/P$ and $1.25/P$, respectively. Therefore, for the pair of gears we

Figure 13–11

Base circle radius can be related to the pressure angle ϕ and the pitch circle radius by $r_b = r\cos\phi$.

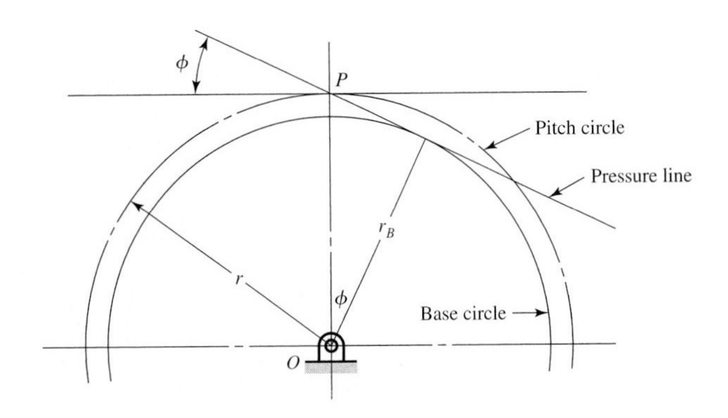

are constructing,

$$a = \frac{1}{P} = \frac{1}{2} = 0.500 \text{ in} \qquad b = \frac{1.25}{P} = \frac{1.25}{2} = 0.625 \text{ in}$$

Using these distances, draw the addendum and dedendum circles on the pinion and on the gear as shown in Fig. 13–10.

Next, using heavy drawing paper, or preferably, a sheet of 5.015- to 0.020-in clear plastic, cut a template for each involute, being careful to locate the gear centers properly with respect to each involute. Figure 13–12 is a reproduction of the template used to create some of the illustrations for this book. Note that only one side of the tooth profile is formed on the template. To get the other side, turn the template over. For some problems you might wish to construct a template for the entire tooth.

To draw a tooth, we must know the tooth thickness. From Eq. (13–4), the circular pitch is

$$p = \frac{\pi}{P} = \frac{\pi}{2} = 1.57 \text{ in}$$

Therefore, the tooth thickness is

$$t = \frac{p}{2} = \frac{1.57}{2} = 0.785 \text{ in}$$

measured on the pitch circle. Using this distance for the tooth thickness as well as the tooth space, draw as many teeth as desired, using the template, after the points have been marked on the pitch circle. In Fig. 13–13 only one tooth has been drawn on each gear. You may run into trouble in drawing these teeth if one of the base circles happens to be larger than the dedendum circle. The reason for this is that the involute begins at the base circle and is undefined below this circle. So, in drawing gear teeth, we usually draw a radial line for the profile below the base circle. The actual shape, however, will depend upon the kind of machine tool used to form the teeth in manufacture, that is, how the profile is generated.

The portion of the tooth between the clearance circle and the dedendum circle includes the fillet. In this instance the clearance is

$$c = b - a = 0.625 - 0.500 = 0.125 \text{ in}$$

The construction is finished when these fillets have been drawn.

Referring again to Fig. 13–13, the pinion with center at O_1 is the driver and turns counterclockwise. The pressure, or generating, line is the same as the cord used in Fig. 13–8*a* to generate the involute, and contact occurs along this line. The initial contact will take place when the flank of the driver comes into contact with the tip of the driven tooth. This occurs at point a in Fig. 13–13, where the addendum circle of the driven gear crosses the pressure line. If we now construct tooth profiles through point a and draw radial lines from the intersections of these profiles with the pitch circles to the gear centers, we obtain the *angle of approach* for each gear.

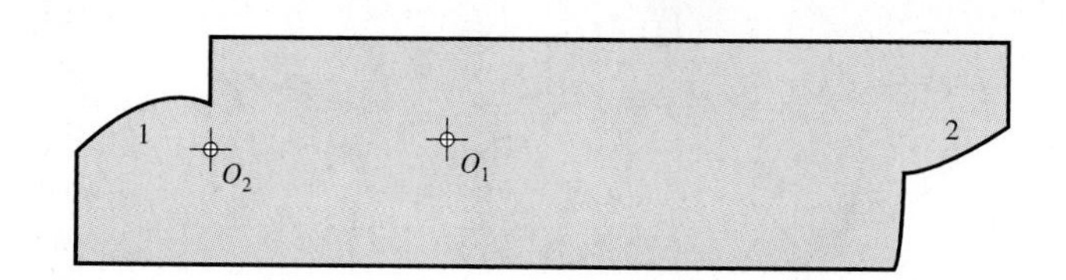

Figure 13–12

A template for drawing gear teeth.

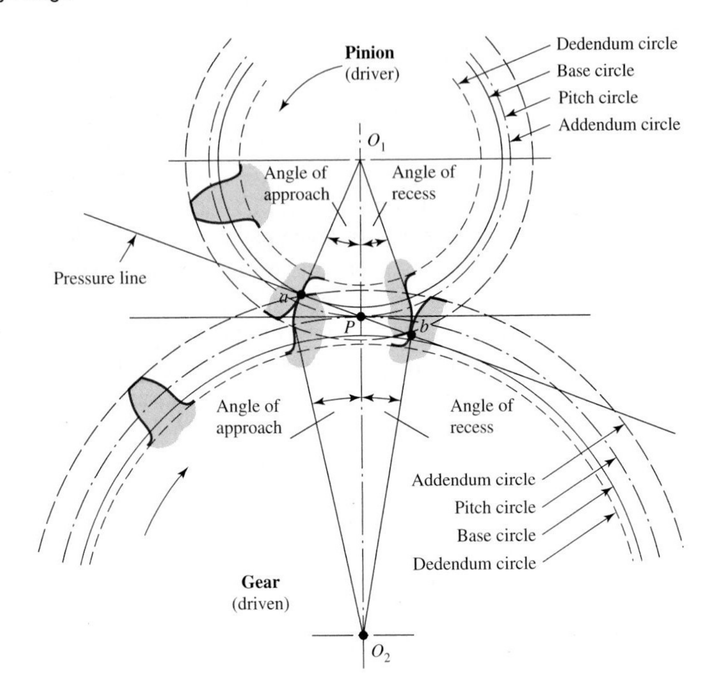

Figure 13–13

Tooth action.

As the teeth go into mesh, the point of contact will slide up the side of the driving tooth so that the tip of the driver will be in contact just before contact ends. The final point of contact will therefore be where the addendum circle of the driver crosses the pressure line. This is point b in Fig. 13–13. By drawing another set of tooth profiles through b, we obtain the *angle of recess* for each gear in a manner similar to that of finding the angles of approach. The sum of the angle of approach and the angle of recess for either gear is called the *angle of action.* The line ab is called *the line of action.*

We may imagine a *rack* as a spur gear having an infinitely large pitch diameter. Therefore, the rack has an infinite number of teeth and a base circle which is an infinite distance from the pitch point. The sides of involute teeth on a rack are straight lines making an angle to the line of centers equal to the pressure angle. Figure 13–14 shows an involute rack in mesh with a pinion. Corresponding sides on involute teeth are parallel curves; the *base pitch* is the constant and fundamental distance between them along a common normal as shown in Fig. 13–14. The base pitch is related to the circular pitch

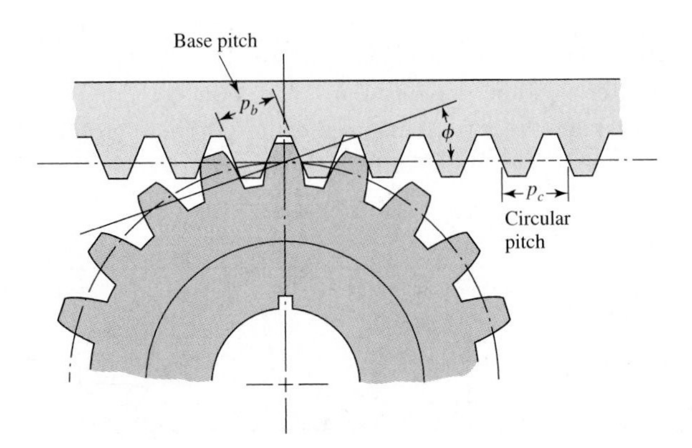

Figure 13–14

Involute-toothed pinion and rack.

Figure 13–15

Internal gear and pinion.

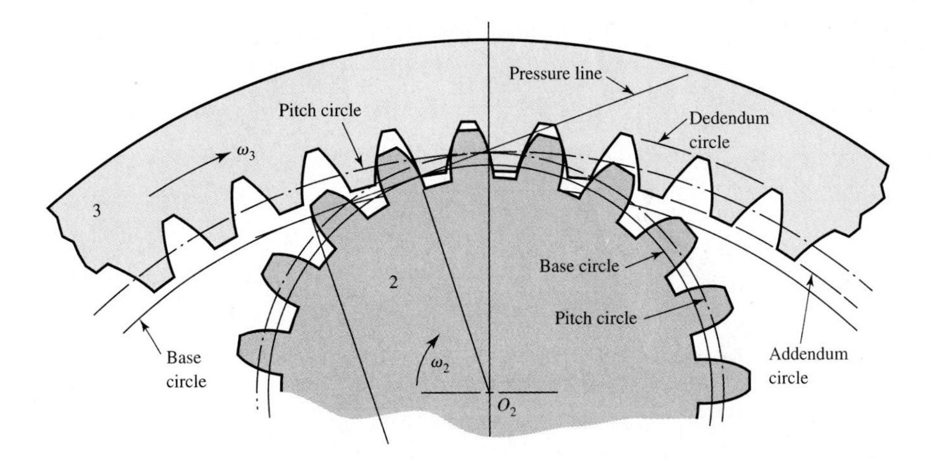

by the equation

$$p_b = p_c \cos \phi \tag{13-7}$$

where p_b is the base pitch.

Figure 13–15 shows a pinion in mesh with an *internal,* or *annular, gear.* Note that both of the gears now have their centers of rotation on the same side of the pitch point. Thus the positions of the addendum and dedendum circles with respect to the pitch circle are reversed; the addendum circle of the internal gear lies *inside* the pitch circle. Note, too, from Fig. 13–15, that the base circle of the internal gear lies inside the pitch circle near the addendum circle.

Another interesting observation concerns the fact that the operating diameters of the pitch circles of a pair of meshing gears need not be the same as the respective design pitch diameters of the gears, though this is the way they have been constructed in Fig. 13–13. If we increase the center distance, we create two new operating pitch circles having larger diameters because they must be tangent to each other at the pitch point. Thus the pitch circles of gears really do not come into existence until a pair of gears are brought into mesh.

Changing the center distance has no effect on the base circles, because these were used to generate the tooth profiles. Thus the base circle is basic to a gear. Increasing the center distance increases the pressure angle and decreases the length of the line of action, but the teeth are still conjugate, the requirement for uniform motion transmission is still satisfied, and the angular-velocity ratio has not changed.

EXAMPLE 13–1

A gearset consists of a 16-tooth pinion driving a 40-tooth gear. The diametral pitch is 2, and the addendum and dedendum are $1/P$ and $1.25/P$, respectively. The gears are cut using a pressure angle of 20°.

(*a*) Compute the circular pitch, the center distance, and the radii of the base circles.

(*b*) In mounting these gears, the center distance was incorrectly made $\frac{1}{4}$ in larger. Compute the new values of the pressure angle and the pitch-circle diameters.

Solution

Answer

(*a*) $p = \dfrac{\pi}{P} = \dfrac{\pi}{2} = 1.57$ in

The pitch diameters of the pinion and gear are, respectively,

$$d_P = \tfrac{16}{2} = 8 \text{ in} \qquad d_G = \tfrac{40}{2} = 20 \text{ in}$$

Therefore the center distance is

$$\frac{d_P + d_G}{2} = \frac{8 + 20}{2} = 14 \text{ in}$$

Since the teeth were cut on the 20° pressure angle, the base-circle radii are found to be, using $r_b = r \cos \phi$,

Answer $$r_b \text{ (pinion)} = \tfrac{8}{2} \cos 20° = 3.76 \text{ in}$$

Answer $$r_b \text{ (gear)} = \tfrac{20}{2} \cos 20° = 9.40 \text{ in}$$

(*b*) Designating d'_P and d'_G as the new pitch-circle diameters, the $\frac{1}{4}$-in increase in the center distance requires that

$$\frac{d'_P + d'_G}{2} = 14.250 \tag{1}$$

Also, the velocity ratio does not change, and hence

$$\frac{d'_P}{d'_G} = \frac{16}{40} \tag{2}$$

Solving Eqs. (1) and (2) simultaneously yields

Answer $$d'_P = 8.143 \text{ in} \qquad d'_G = 20.357 \text{ in}$$

Since $r_b = r \cos \phi$, the new pressure angle is

Answer $$\phi' = \cos^{-1} \frac{r_b \text{ (pinion)}}{d'_P/2} = \cos^{-1} \frac{3.76}{8.143/2} = 22.56°$$

13–7 Contact Ratio

The zone of action of meshing gear teeth is shown in Fig. 3–16. We recall that tooth contact begins and ends at the intersections of the two addendum circles with the pressure line. In Fig. 13–16 initial contact occurs at a and final contact at b. Tooth profiles drawn through these points intersect the pitch circle at A and B, respectively. As shown, the distance AP is called the *arc of approach* q_a, and the distance PB, the *arc of recess* q_r. The sum of these is the *arc of action* q_t.

Now, consider a situation in which the arc of action is exactly equal to the circular pitch, that is, $q_t = p$. This means that one tooth and its space will occupy the entire arc AB. In other words, when a tooth is just beginning contact at a, the previous tooth is simultaneously ending its contact at b. Therefore, during the tooth action from a to b, there will be exactly one pair of teeth in contact.

Next, consider a situation in which the arc of action is greater than the circular pitch, but not very much greater, say, $q_t \approx 1.2p$. This means that when one pair of teeth is just entering contact at a, another pair, already in contact, will not yet have reached b. Thus, for a short period of time, there will be two teeth in contact, one in the vicinity of A and

Figure 13–16

Definiton of contact ratio.

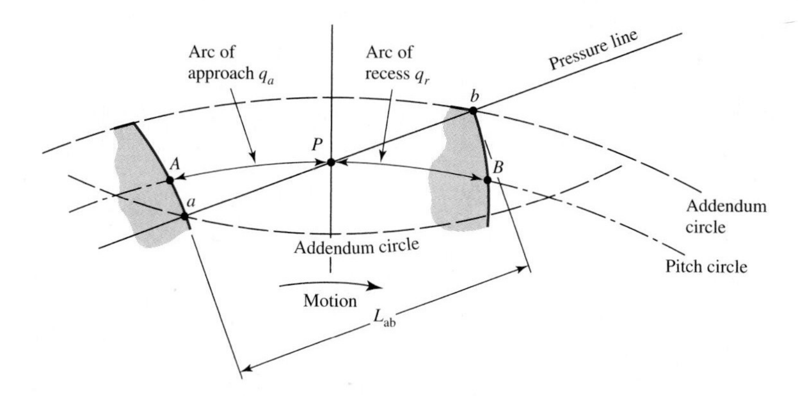

another near B. As the meshing proceeds, the pair near B must cease contact, leaving only a single pair of contacting teeth, until the procedure repeats itself.

Because of the nature of this tooth action, either one or two pairs of teeth in contact, it is convenient to define the term *contact ratio* m_c as

$$m_c = \frac{q_t}{p} \tag{13–8}$$

a number which indicates the average number of pairs of teeth in contact. Note that this ratio is also equal to the length of the path of contact divided by the base pitch. Gears should not generally be designed having contact ratios less than about 1.20, because inaccuracies in mounting might reduce the contact ratio even more, increasing the possibility of impact between the teeth as well as an increase in the noise level.

An easier way to obtain the contact ratio is to measure the line of action ab instead of the arc distance AB. Since ab in Fig. 13–16 is tangent to the base circle when extended, the base pitch p_b must be used to calculate m_c instead of the circular pitch as in Eq. (13–8). Designating the length of the line of action as L_{ab}, the contact ratio is

$$m_c = \frac{L_{ab}}{p \cos \phi} \tag{13–9}$$

in which Eq. (13–7) was used for the base pitch.

13–8 Interference

The contact of portions of tooth profiles which are not conjugate is called *interference.* Consider Fig. 13–17. Illustrated are two 16-tooth gears which have been cut using the now obsolete $14\frac{1}{2}^\circ$ pressure angle. The driver, gear 2, turns clockwise. The initial and final points of contact are designated A and B, respectively, and are located on the pressure line. Now notice that the points of tangency of the pressure line with the base circles C and D are located *inside* of points A and B. Interference is present.

The interference is explained as follows. Contact begins when the tip of the driven tooth contacts the flank of the driving tooth. In this case the flank of the driving tooth first makes contact with the driven tooth at point A, and this occurs *before* the involute portion of the driving tooth comes within range. In other words, contact is occurring below the base circle of gear 2 on the *noninvolute* portion of the flank. The actual effect is that the involute tip or face of the driven gear tends to dig out the noninvolute flank of the driver.

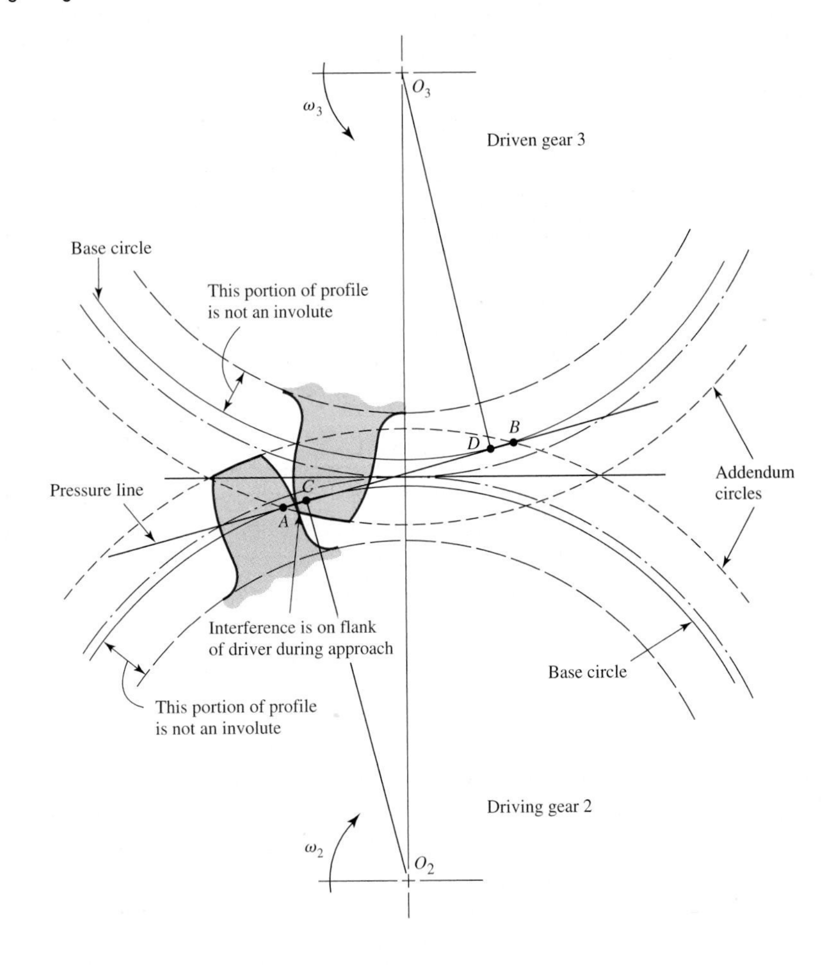

Figure 13–17

Interference in the action of gear teeth. (This is actually a rather poor figure; I drew the tooth shape using circular arcs, which is incorrect, to answer a student's question many years ago. J. E. S.)

In this example the same effect occurs again as the teeth leave contact. Contact should end at point D or before. Since it does not end until point B, the effect is for the tip of the driving tooth to dig out, or interfere with, the flank of the driven tooth.

When gear teeth are produced by a generation process, interference is automatically eliminated because the cutting tool removes the interfering portion of the flank. This effect is called *undercutting;* if undercutting is at all pronounced, the undercut tooth is considerably weakened. Thus the effect of eliminating interference by a generation process is merely to substitute another problem for the original one.

The smallest number of teeth on a spur pinion and gear[2], one-to-one gear ratio, which can exist without interference is N_P. This number of teeth for spur gears is given by

$$N_P = \frac{4k}{6\sin^2\phi}\left(1 + \sqrt{1 + 3\sin^2\phi}\right) \tag{a}$$

where $k =$ 1 for fill-depth teeth, 0.8 for stub teeth
$\phi =$ pressure angle

2. Robert Lipp, "Avoiding Tooth Interference in Gears," *Machine Design*, Vol. 54 No.1, 1982, pp. 122, 124.

For a 20° pressure angle,

$$N_P = \frac{4(1)}{6 \sin^2 20^\circ}\left(1 + \sqrt{1 + 3 \sin^2 20^\circ}\right) = 12.3 = 13 \text{ teeth}$$

Thus 13 teeth on pinion and gear are interference-free. Realize that 12.3 teeth is possible in meshing arcs, but for fully rotating gears, 13 teeth represents the least number. For a $14\frac{1}{2}^\circ$ pressure angle, $N_P = 23$ teeth, so one can appreciate why few $14\frac{1}{2}^\circ$-tooth systems are used, as the higher pressure angles can produce smaller pinion with accompanying smaller center-to-center distances.

If the mating gear has more teeth than the pinion, that is, $m_G = N_G/N_P = m$ is more than one, then the smallest number of teeth on the pinion without interference is given by

$$N_P = \frac{2k}{(1 + 2m) \sin^2 \phi}\left(m + \sqrt{m^2 + (1 + 2m) \sin^2 2\phi}\right) \tag{b}$$

If $m = 4$, $\phi = 20^\circ$,

$$N_P = \frac{2(1)}{(1 + 2[4]) \sin^2 20^\circ}\left[4 + \sqrt{4^2 + (1 + 2[4]) \sin^2 20^\circ}\right] = 15.4 = 16 \text{ teeth}$$

Thus a 16-tooth pinion will mesh with a 64-tooth gear without interference.

The smallest spur pinion that will operate with a rack without interference is

$$N_P = \frac{4(k)}{2 \sin^2 \phi} \tag{c}$$

For a 20° pressure angle full-depth tooth the smallest number of pinion teeth is

$$N_P = \frac{4(1)}{2 \sin^2 20^\circ} = 17.1 = 18 \text{ teeth}$$

The largest gear with a specified pinion that is interference-free is

$$N_G = \frac{N_P^2 \sin^2\phi - 4k^2}{4k - 2N_P \sin^2 \phi} \tag{d}$$

For a 13-tooth pinion with a pressure angle ϕ of 20°,

$$N_G = \frac{13^2 \sin^2 20^\circ - 4(1)^2}{4(1) - 2(13) \sin^2 20^\circ} = 16.45 = 16 \text{ teeth}$$

From Table 13–6, for a 13-tooth spur pinion, the maximum number of gear teeth possible without interference is 16.

Since gear-shaping tools amount to contact with a rack, and the gear-hobbing process is similar, the minimum number of teeth to prevent interference to prevent undercutting by the hobbing process is equal to the value of N_P in Table 13–6 when N_G is infinite.

The importance of the problem of teeth which have been weakened by undercutting cannot be overemphasized. Of course, interference can be eliminated by using more teeth on the pinion. However, if the pinion is to transmit a given amount of power, more teeth can be used only by increasing the pitch diameter.

Interference can also be reduced by using a larger pressure angle. This results in a smaller base circle, so that more of the tooth profile becomes involute. The demand for smaller pinions with fewer teeth thus favors the use of a 25° pressure angle even though the frictional forces and bearing loads are increased and the contact ratio decreased.

See Table 13–6 for minimum tooth numbers to avoid interference problems.

Table 13–6

Maximum Tooth Numbers on Gears to Avoid Interference. Numbers Are Based on a Normal Pressure Angle of $\phi_n = 20°$ and Full-Depth Teeth. For Spur Gears, $\psi = 0$

Source: R. Lipp, "Avoiding Interference in Gears," *Machine Design*, vol. 34, no. 1, 1982, p. 122.

Number of Pinion Teeth, N_P	Number of Gear Teeth, N_G — Helix angle, ψ, deg							
	0	5	10	15	20	25	30	35
8								12
9							12	34
10						12	26	∞
11					13	23	93	
12			12	16	24	57	∞	
13	16	17	20	27	50	1385		
14	26	27	34	53	207			
15	45	49	69	181	∞			
16	101	121	287	∞				
17	1309	∞	∞					

Note: The minimum number of teeth for the gear is N_P.

13–9 The Forming of Gear Teeth

There are a large number of ways of forming the teeth of gears, such as *sand casting, shell molding, investment casting, permanent-mold casting, die casting,* and *centrifugal casting.* Teeth can be formed by using the *powder-metallurgy process;* or, by using *extrusion,* a single bar of aluminum may be formed and then sliced into gears. Gears which carry large loads in comparison with their size are usually made of steel and are cut with either *form cutters* or *generating cutters.* In form cutting, the tooth space takes the exact form of the cutter. In generating, a tool having a shape different from the tooth profile is moved relative to the gear blank so as to obtain the proper tooth shape. One of the newest and most promising of the methods of forming teeth is called *cold forming,* or *cold rolling,* in which dies are rolled against steel blanks to form the teeth. The mechanical properties of the metal are greatly improved by the rolling process, and a high-quality generated profile is obtained at the same time.

Gear teeth may be machined by milling, shaping, or hobbing. They may be finished by shaving, burnishing, grinding, or lapping.

Milling

Gear teeth may be cut with a form milling cutter shaped to conform to the tooth space. With this method it is theoretically necessary to use a different cutter for each gear, because a gear having 25 teeth, for example, will have a different-shaped tooth space from one having, say, 24 teeth. Actually, the change in space is not too great, and it has been found that eight cutters may be used to cut with reasonable accuracy any gear in the range of 12 teeth to a rack. A separate set of cutters is, of course, required for each pitch.

Shaping

Teeth may be generated with either a pinion cutter or a rack cutter. The pinion cutter (Fig. 13–18) reciprocates along the vertical axis and is slowly fed into the gear blank to the required depth. When the pitch circles are tangent, both the cutter and the blank rotate slightly after each cutting stroke. Since each tooth of the cutter is a cutting tool, the teeth are all cut after the blank has completed one rotation. The sides of an involute rack tooth are straight. For this reason, a rack-generating tool provides an accurate method of cutting gear teeth. This is also a shaping operation and is illustrated by the drawing of Fig. 13–19. In operation, the cutter reciprocates and is first fed into the gear blank until

Figure 13–18

Generating a spur gear with a pinion cutter. *(Courtesy of Boston Gear Works, Inc.)*

Figure 13–19

Shaping teeth with a rack. (This is a drawing-board figure that I executed about 35 years ago in response to a question from a student at the University of Michigan. J.E.S.)

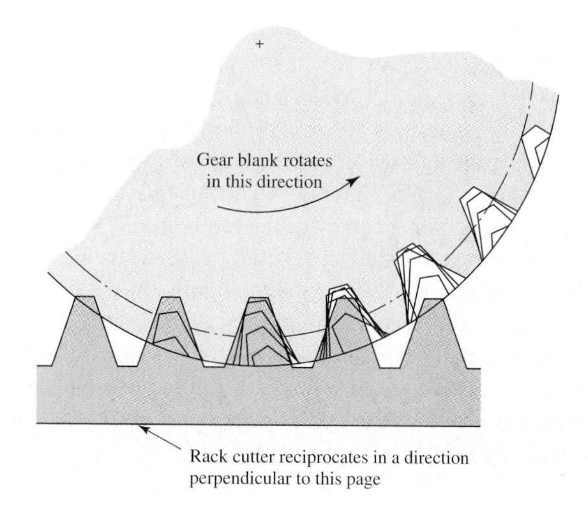

the pitch circles are tangent. Then, after each cutting stroke, the gear blank and cutter roll slightly on their pitch circles. When the blank and cutter have rolled a distance equal to the circular pitch, the cutter is returned to the starting point, and the process is continued until all the teeth have been cut.

Hobbing

The hobbing process is illustrated in Fig. 13–20. The hob is simply a cutting tool which is shaped like a worm. The teeth have straight sides, as in a rack, but the hob axis must be turned through the lead angle in order to cut spur-gear teeth. For this reason, the teeth

Figure 13–20

Hobbing a worm gear. *(Courtesy of Boston Gear Works, Inc.)*

generated by a hob have a slightly different shape from those generated by a rack cutter. Both the hob and the blank must be rotated at the proper angular-velocity ratio. The hob is then fed slowly across the face of the blank until all the teeth have been cut.

Finishing

Gears which run at high speeds and transmit large forces may be subjected to additional dynamic forces if there are errors in tooth profiles. Errors may be diminished somewhat by finishing the tooth profiles. The teeth may be finished, after cutting, by either shaving or burnishing. Several shaving machines are available which cut off a minute amount of metal, bringing the accuracy of the tooth profile within the limits of 250 μin.

Burnishing, like shaving, is used with gears which have been cut but not heat-treated. In burnishing, hardened gears with slightly oversize teeth are run in mesh with the gear until the surfaces become smooth.

Grinding and lapping are used for hardened gear teeth after heat treatment. The grinding operation employs the generating principle and produces very accurate teeth. In lapping, the teeth of the gear and lap slide axially so that the whole surface of the teeth is abraded equally.

13–10 Straight Bevel Gears

When gears are used to transmit motion between intersecting shafts, some form of bevel gear is required. A bevel gearset is shown in Fig. 13–21. Although bevel gears are usually made for a shaft angle of 90°, they may be produced for almost any angle. The teeth may be cast, milled, or generated. Only the generated teeth may be classed as accurate.

The terminology of bevel gears is illustrated in Fig. 13–21. The pitch of bevel gears is measured at the large end of the tooth, and both the circular pitch and the pitch diameter are calculated in the same manner as for spur gears. It should be noted that the clearance is uniform. The pitch angles are defined by the pitch cones meeting at the apex, as shown

Figure 13–21

Terminology of bevel gears.

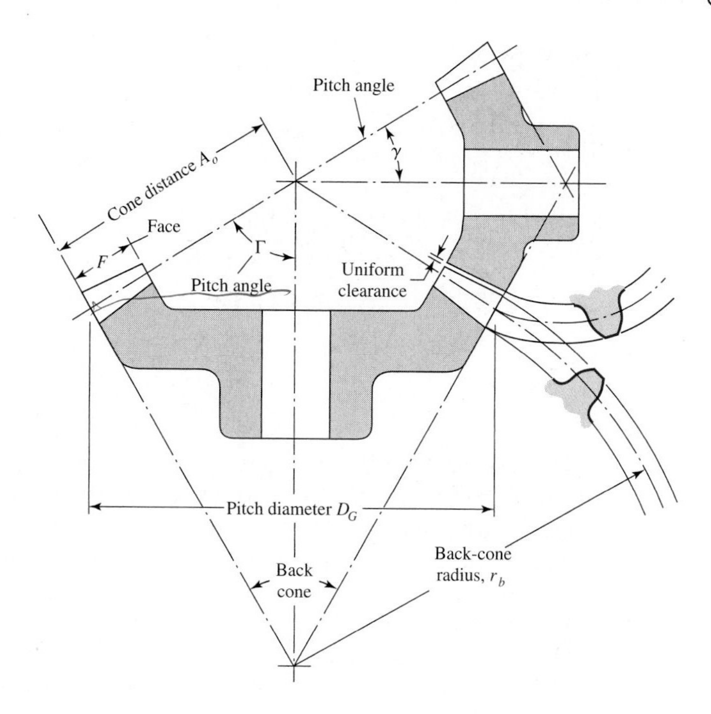

in the figure. They are related to the tooth numbers as follows:

$$\tan \gamma = \frac{N_P}{N_G} \qquad \tan \Gamma = \frac{N_G}{N_P} \tag{13–10}$$

where the subscripts P and G refer to the pinion and gear, respectively, and where γ and Γ are, respectively, the pitch angles of the pinion and gear.

Figure 13–21 shows the shape of the teeth, when projected on the back cone, is the same as in a spur gear having a radius equal to the back-cone distance r_b. This is called Tredgold's approximation. The number of teeth in this imaginary gear is

$$N' = \frac{2\pi r_b}{P} \tag{13–11}$$

where N' is the *virtual number of teeth* and p is the circular pitch measured at the large end of the teeth. Standard straight-tooth bevel gears are cut by using a 20° pressure angle, unequal addenda and dedenda, and full-depth teeth. This increases the contact ratio, avoids undercut, and increases the strength of the pinion.

13–11 Parallel Helical Gears

Helical gears, used to transmit motion between parallel shafts, are shown in Fig. 13–2. The helix angle is the same on each gear, but one gear must have a right-hand helix and the other a left-hand helix. The shape of the tooth is an involute helicoid and is illustrated in Fig. 13–22. If a piece of paper cut in the shape of a parallelogram is wrapped around a cylinder, the angular edge of the paper becomes a helix. If we unwind this paper, each point on the angular edge generates an involute curve. This surface obtained when every point on the edge generates an involute is called an *involute helicoid.*

The initial contact of spur-gear teeth is a line extending all the way across the face of the tooth. The initial contact of helical-gear teeth is a point which extends into a line

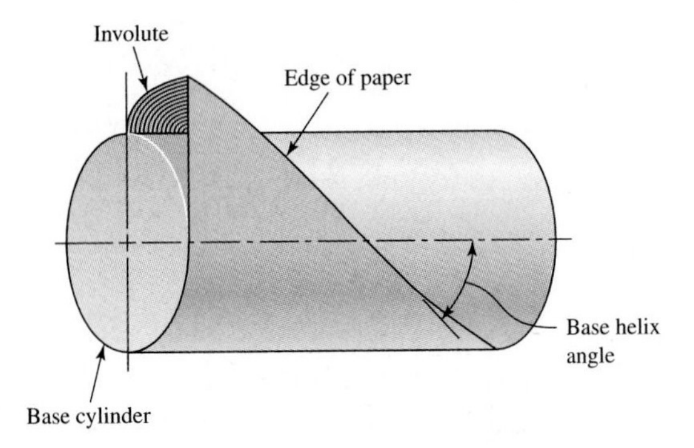

Figure 13–22

An involute helicoid.

as the teeth come into more engagement. In spur gears the line of contact is parallel to the axis of rotation; in helical gears the line is diagonal across the face of the tooth. It is this gradual engagement of the teeth and the smooth transfer of load from one tooth to another which gives helical gears the ability to transmit heavy loads at high speeds. Because of the nature of contact between helical gears, the contact ratio is of only minor importance, and it is the contact area, which is proportional to the face width of the gear, that becomes significant.

Helical gears subject the shaft bearings to both radial and thrust loads. When the thrust loads become high or are objectionable for other reasons, it may be desirable to use double helical gears. A double helical gear (herringbone) is equivalent to two helical gears of opposite hand, mounted side by side on the same shaft. They develop opposite thrust reactions and thus cancel out the thrust load.

When two or more single helical gears are mounted on the same shaft, the hand of the gears should be selected so as to produce the minimum thrust load.

Figure 13–23 represents a portion of the top view of a helical rack. Lines ab and cd are the centerlines of two adjacent helical teeth taken on the same pitch plane. The angle ψ is the *helix angle.* The distance ac is the *transverse circular pitch* p_t in the plane of rotation (usually called the *circular pitch*). The distance ae is the *normal circular pitch* p_n and is related to the transverse circular pitch as follows:

$$p_n = p_t \cos \psi \tag{13–12}$$

The distance ad is called the *axial pitch* p_x and is related by the expression

$$p_x = \frac{p_t}{\tan \psi} \tag{13–13}$$

Since $p_n P_n = \pi$, the *normal diametral pitch* is

$$P_n = \frac{P_t}{\cos \psi} \tag{13–14}$$

The pressure angle ϕ_n in the normal direction is different from the pressure angle ϕ_t in the direction of rotation, because of the angularity of the teeth. These angles are related by the equation

$$\cos \psi = \frac{\tan \phi_n}{\tan \phi_t} \tag{13–15}$$

Figure 13–24 illustrates a cylinder cut by an oblique plane ab at an angle ψ to a right section. The oblique plane cuts out an arc having a radius of curvature of R. For

Figure 13–23

Nomenclature of helical gears.

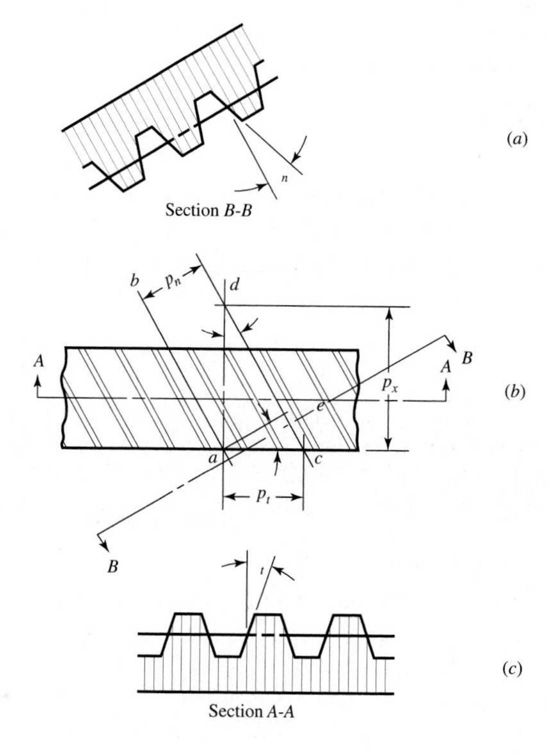

Figure 13–24

A cylinder cut by an oblique plane.

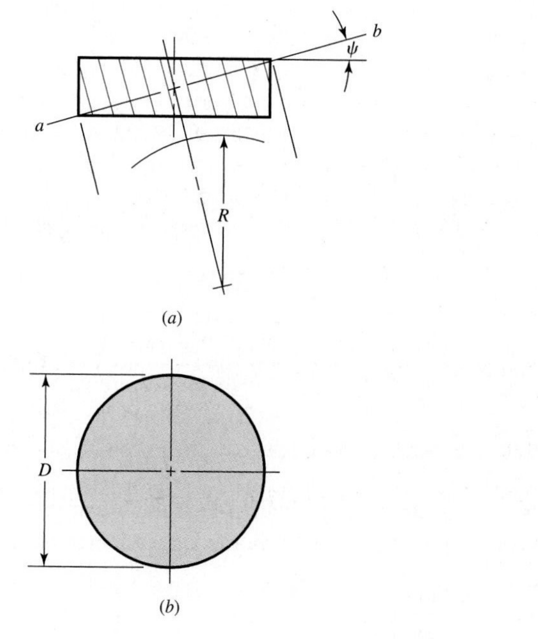

the condition that $\psi = 0$, the radius of curvature is $R = D/2$. If we imagine the angle ψ to be slowly increased from zero to 90°, we see that R begins at a value of $D/2$ and increases until, when $\psi = 90°$, $R = \infty$. The radius R is the apparent pitch radius of a helical-gear tooth when viewed in the direction of the tooth elements. A gear of the same pitch and with the radius R will have a greater number of teeth, because of the increased radius. In helical-gear terminology this is called the *virtual number of teeth.*

It can be shown by analytical geometry that the virtual number of teeth is related to the actual number by the equation

$$N' = \frac{N}{\cos^3 \psi} \tag{13–16}$$

where N' is the virtual number of teeth and N is the actual number of teeth. It is necessary to know the virtual number of teeth in design for strength and also, sometimes, in cutting helical teeth. This apparently larger radius of curvature means that few teeth may be used on helical gears, because there will be less undercutting.

EXAMPLE 13–2 A stock helical gear has a normal pressure angle of $14\frac{1}{2}°$, a helix angle of 45°, and a transverse diametral pitch of 6 teeth/in, and has 18 teeth. Find:

(*a*) The pitch diameter

(*b*) The transverse, the normal, and the axial pitches

(*c*) The normal diametral pitch

(*d*) The transverse pressure angle

Solution

Answer (*a*) $d = \frac{N}{P_t} = \frac{18}{6} = 3 \text{ in}$

Answer (*b*) $p_t = \frac{\pi}{P_t} = \frac{\pi}{6} = 0.5236 \text{ in}$

Answer $p_n = p_t \cos\psi = 0.5236 \cos 45° = 0.3702 \text{ in}$

Answer $p_x = \frac{p_t}{\tan\psi} = \frac{0.5236}{\tan 45°} = 0.5236 \text{ in}$

Answer (*c*) $P_n = \frac{P_t}{\cos\psi} = \frac{6}{\cos 45°} = 8.485 \text{ teeth/in}$

Answer (*d*) $\phi_t = \tan^{-1}\left(\frac{\tan\phi_n}{\cos\psi}\right) = \tan^{-1}\left(\frac{\tan 14.5°}{\cos 45°}\right) = 20.09°$

Just as with spur gears, helical-gear teeth can interfere. Equation (13–15) can be solved for the pressure angle ϕ_t in the tangential (rotation) direction to give

$$\phi_t = \tan^{-1}\left(\frac{\tan\phi_n}{\cos\psi}\right)$$

The smallest tooth number N_P of a helical-spur pinion that will run without interference[3] with a gear with the same number of teeth is

$$N_P = \frac{4k\cos\psi}{6\sin^2\phi_t}\left(1 + \sqrt{1 + 3\sin^2\phi_t}\right) \tag{a}$$

3. Op. cit., Robert Lipp, *Machine Design*, pp. 122, 124.

If the normal pressure angle ϕ_n is 20°, the helix angle ψ is 30°, then ϕ_t is

$$\phi_t = \tan^{-1}\left(\frac{\tan 20°}{\cos 30°}\right) = 22.80°$$

$$N_P = \frac{4(1)\cos 30°}{6\sin^2 22.80°}\left(1 + \sqrt{1 + 3\sin^2 22.80°}\right) = 8.48 = 9 \text{ teeth}$$

In Table 13–6, $\psi = 30°$ gives a minimum pinion tooth count of 9.

For a given gear ratio $m_G = N_G/N_P = m$, the smallest pinion tooth count is

$$N_P = \frac{2k\cos\psi}{(1+2m)\sin^2\phi_t}\left[m + \sqrt{m^2 + (1+2m)\sin^2\phi_t}\right] \tag{b}$$

The smallest pinion that can be run with a rack is

$$N_P = \frac{4k\cos\psi}{2\sin^2\phi_t} \tag{c}$$

For a normal pressure angle ϕ_n of 20° and a helix angle ψ of 30°, and $\phi_t = 22.80°$,

$$N_P = \frac{4(1)\cos 30°}{2\sin^2 22.80°} = 11.5 = 12 \text{ teeth}$$

Table 13–6, for $N_P = 12$, allows mesh with a rack without interference.

The largest gear with a specified pinion is given by

$$N_G = \frac{N_P^2\sin^2\phi_t - 4k^2\cos^2\psi}{4k\cos\psi - 2N_P\sin^2\phi_t} \tag{d}$$

For a nine-tooth pinion with a pressure angle ϕ_n of 20°, a helix angle ψ of 30°, and recalling that the tangential pressure angle ϕ_t is 22.80°,

$$N_G = \frac{9^2\sin^2 22.80° - 4(1)^2\cos^2 30°}{4(1)\cos 30° - 2(9)\sin^2 22.80°} = 12.03 = 12$$

Table 13–6 agrees that for these conditions N_G of 12 teeth is the largest gear-tooth count that can mesh with a 9-tooth pinion without interference.

For helical-gear teeth the number of teeth in mesh across the width of the gear will be greater than unity and a term called *face-contact ratio* is used to describe it. This increase of contact ratio, and the gradual sliding engagement of each tooth, results in quieter gears.

13–12 Worm Gears

The nomenclature of a worm gear is shown in Fig. 13–25. The worm and worm gear of a set have the same hand of helix as for crossed helical gears, but the helix angles are usually quite different. The helix angle on the worm is generally quite large, and that on the gear very small. Because of this, it is usual to specify the lead angle λ on the worm and helix angle ψ_G on the gear; the two angles are equal for a 90° shaft angle. The worm lead angle is the complement of the worm helix angle, as shown in Fig. 13–25.

In specifying the pitch of worm gearsets, it is customary to state the *axial pitch* p_x of the worm and the *transverse circular pitch* p_t, often simply called the circular pitch, of the mating gear. These are equal if the shaft angle is 90°. The pitch diameter of the gear is the diameter measured on a plane containing the worm axis, as shown in Fig. 13–25;

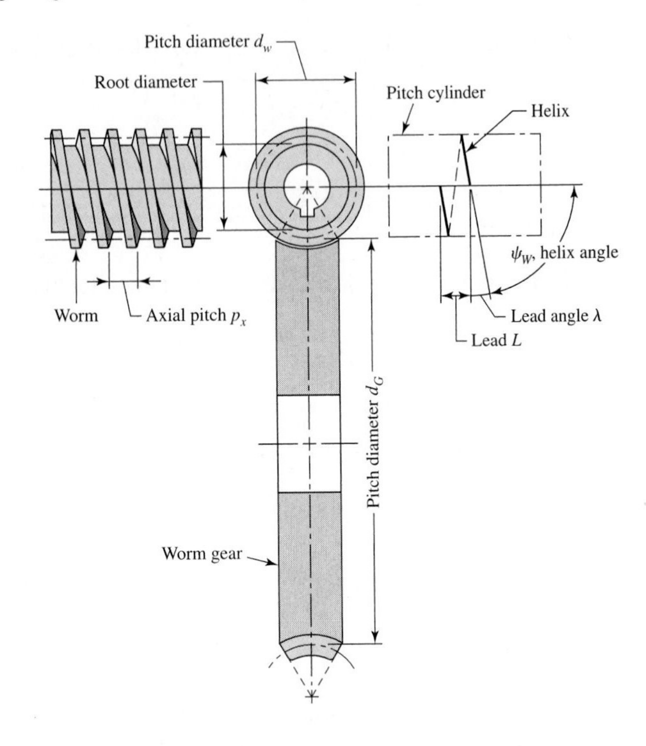

Figure 13–25

Nomenclature of a single-enveloping worm gearset.

it is the same as for spur gears and is

$$d_G = \frac{N_G p_t}{\pi} \tag{13-17}$$

Since it is not related to the number of teeth, the worm may have any pitch diameter; this diameter should, however, be the same as the pitch diameter of the hob used to cut the worm-gear teeth. Generally, the pitch diameter of the worm should be selected so as to fall into the range

$$\frac{C^{0.875}}{3.0} \le d_W \le \frac{C^{0.875}}{1.7} \tag{13-18}$$

where C is the center distance. These proportions appear to result in optimum horsepower capacity of the gearset.

The *lead* L and the *lead angle* λ of the worm have the following relations:

$$L = p_x N_W \tag{13-19}$$

$$\tan\lambda = \frac{L}{\pi d_W} \tag{13-20}$$

13–13 Gear Trains

Consider a pinion 2 driving a gear 3. The speed of the driven gear is

$$n_3 = \left|\frac{N_2}{N_3} n_2\right| = \left|\frac{d_2}{d_3} n_2\right| \tag{13-21}$$

where n = revolutions or r/min
N = number of teeth
d = pitch diameter

Equation (13–21) applies to any gearset no matter whether the gears are spur, helical, bevel, or worm. The absolute-value signs are used to permit complete freedom in choosing positive and negative directions. In the case of spur and parallel helical gears, the directions ordinarily correspond to the right-hand rule and are positive for counterclockwise rotation.

Rotational directions are somewhat more difficult to deduce for worm and crossed helical gearsets. Figure 13–26 will be of help in these situations.

The gear train shown in Fig. 13–27 is made up of five gears. The speed of gear 6 is

$$n_6 = -\frac{N_2}{N_3}\frac{N_3}{N_4}\frac{N_5}{N_6}n_2 \tag{a}$$

Hence we notice that gear 3 is an idler, that its tooth numbers cancel in Eq. (a), and hence that it affects only the direction of rotation of gear 6. We notice, furthermore, that gears 2, 3, and 5 are drivers, while 3, 4, and 6 are driven members. We define *train value* e as

$$e = \frac{\text{product of driving tooth numbers}}{\text{product of driven tooth numbers}} \tag{13–22}$$

Note that pitch diameters can be used in Eq. (13–22) as well. When Eq. (13–22) is used for spur gears, e is positive if the last gear rotates in the same sense as the first, and negative if the last rotates in the opposite sense.

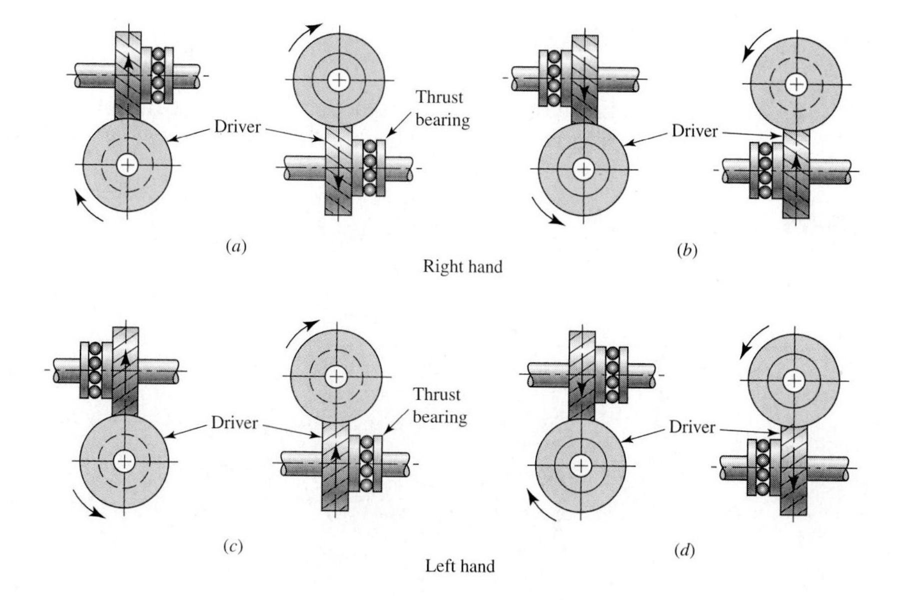

Figure 13–26

Thrust, rotation, and hand relations for crossed helical gears. Note that each pair of drawings refers to a single gearset. These relations also apply to worm gearsets. *(Courtesy of Boston Gear Works, Inc.)*

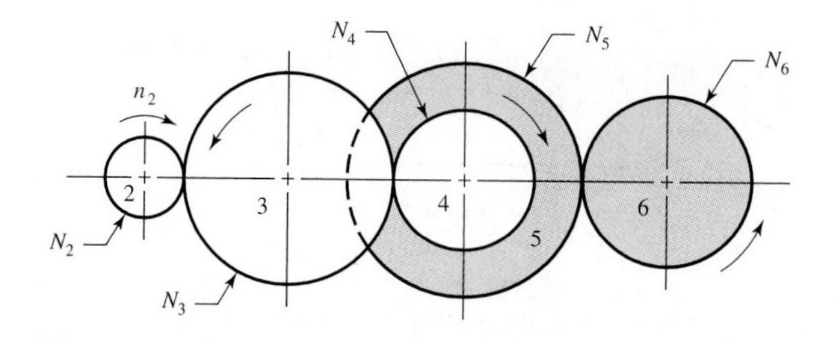

Figure 13–27

A gear train.

Now we can write

$$n_L = en_F \tag{13–23}$$

where n_L is the speed of the last gear in the train and n_F is the speed of the first.

Unusual effects can be obtained in a gear train by permitting some of the gear axes to rotate about others. Such trains are called *planetary,* or *epicyclic, gear trains.* Planetary trains always include a *sun gear,* a *planet carrier* or *arm,* and one or more *planet gears,* as shown in Fig. 13–28. Planetary gear trains are unusual mechanisms because they have two degrees of freedom; that is, for constrained motion, a planetary train must have two inputs. For example, in Fig. 13–28 these two inputs could be the motion of any two of the elements of the train. We might, in Fig. 13–28, say, specify that the sun gear rotates at 100 r/min clockwise and that the ring gear rotates at 50 r/min counterclockwise; these are the inputs. The output would be the motion of the arm. In most planetary trains one of the elements is attached to the frame and has no motion. Figure 13–29 shows a planetary train composed of a sun gear 2, an arm or carrier 3, and planet gears 4 and 5. The angular velocity of gear 2 relative to the arm in r/min is

$$n_{23} = n_2 - n_3 \tag{b}$$

Also, the velocity of gear 5 relative to the arm is

$$n_{53} = n_5 - n_3 \tag{c}$$

Dividing Eq. (c) by Eq. (b) gives

$$\frac{n_{53}}{n_{23}} = \frac{n_5 - n_3}{n_2 - n_3} \tag{d}$$

Equation (d) expresses the ratio of gear 5 to that of gear 2, and both velocities are taken relative to the arm. Now this ratio is the same and is proportional to the tooth numbers,

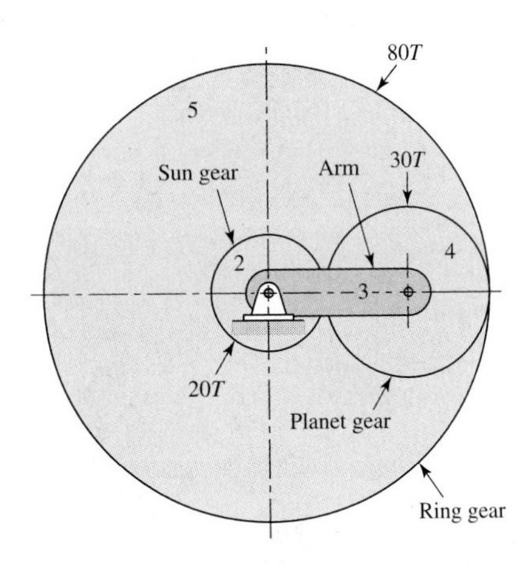

Figure 13–28

A planetary gear train.

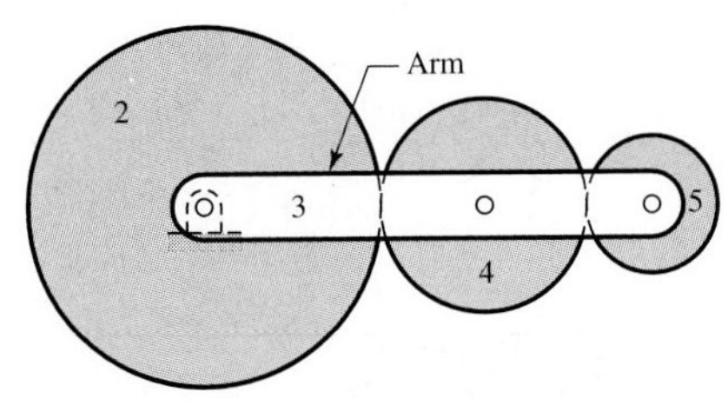

Figure 13–29

A gear train on the arm of a planetary gear train.

whether the arm is rotating or not. It is the train value. Therefore, we may write

$$e = \frac{n_5 - n_3}{n_2 - n_3} \tag{e}$$

This equation can be used to solve for the output motion of any planetary train. It is more conveniently written in the form

$$e = \frac{n_L - n_A}{n_F - n_A} \tag{13–24}$$

where n_F = r/min of first gear in planetary train
n_L = r/min of last gear in planetary train
n_A = r/min of arm

EXAMPLE 13–3 In Fig. 13–28 the sun gear is the input, and it is driven clockwise at 100 r/min. The ring gear is held stationary by being fastened to the frame. Find the r/min and direction of rotation of the arm.

Solution Designate $n_F = n_2 = -100$ r/min, and $n_L = n_5 = 0$. Unlocking gear 5 and holding the arm stationary, in our imagination, we find

$$e = -\left(\tfrac{20}{30}\right)\left(\tfrac{30}{80}\right) = -0.25$$

Substituting in Eq. (13–24),

$$-0.25 = \frac{0 - n_A}{(-100) - n_A}$$

or

Answer $n_A = -20$ r/min

To obtain the speed of gear 4, we follow the procedure outlined by Eqs. (*b*), (*c*), and (*d*). Thus

$$n_{43} = n_4 - n_3 \qquad n_{23} = n_2 - n_3$$

and so

$$\frac{n_{43}}{n_{23}} = \frac{n_4 - n_3}{n_2 - n_3} \tag{1}$$

But

$$\frac{n_{43}}{n_{23}} = -\frac{20}{30} = -\frac{2}{3} \tag{2}$$

Substituting the known values in Eq. (1) gives

$$-\frac{2}{3} = \frac{n_4 - (-20)}{(-100) - (-20)}$$

Solving gives

Answer $n_4 = 33\tfrac{1}{3}$ r/min

EXAMPLE 13–4 Figure 13–30 shows a gear train consisting of a pair of *miter gears* (same-size bevel gears) having 16 teeth each, a 4-tooth right-hand worm, and a 40-tooth worm gear. The speed of gear 2 is given as $n_2 = +200$ r/min, which corresponds to counterclockwise about the y axis. What is the speed and direction of rotation of the worm gear?

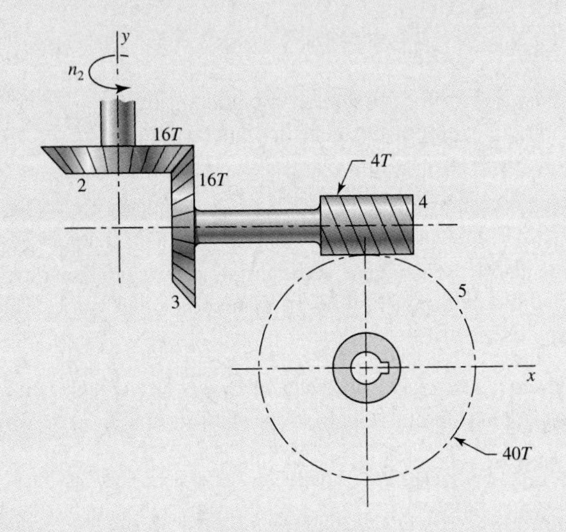

Figure 13–30

A miter gearset and a worm gearset in train.

Solution

Answer $n_5 = -\left(\frac{16}{16}\right)\left(\frac{4}{40}\right)(200) = -20$ r/min

Gear 5 rotates clockwise (negative) 20 r/min about the z axis in a right-handed coordinate system.

13–14 Force Analysis—Spur Gearing

Before beginning the force analysis of gear trains, let us agree on the notation to be used. Beginning with the numeral 1 for the frame of the machine, we shall designate the input gear as gear 2, and then number the gears successively 3, 4, etc., until we arrive at the last gear in the train. Next, there may be several shafts involved, and usually one or two gears are mounted on each shaft as well as other elements. We shall designate the shafts, using lowercase letters of the alphabet, a, b, c, etc.

With this notation we can now speak of the force exerted by gear 2 against gear 3 as F_{23}. The force of gear 2 against a shaft a is F_{2a}. We can also write F_{a2} to mean the force of a shaft a against gear 2. Unfortunately, it is also necessary to use superscripts to indicate directions. The coordinate directions will usually be indicated by the x, y, and z coordinates, and the radial and tangential directions by superscripts r and t. With this notation, F_{43}^t is the tangential component of the force of gear 4 acting against gear 3.

Figure 13–31a shows a pinion mounted on shaft a rotating clockwise at n_2 r/min and driving a gear on shaft b at n_3 r/min. The reactions between the mating teeth occur along the pressure line. In Fig. 13–31b the pinion has been separated from the gear and

Figure 13–31

Free-body diagrams of the forces and moments acting upon two gears of a simple gear train.

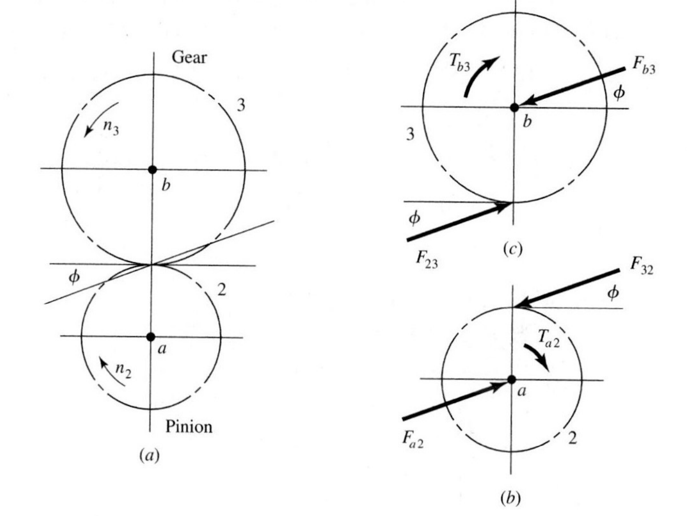

the shaft, and their effects have been replaced by forces. F_{a2} and T_{a2} are the force and torque, respectively, exerted by shaft a against pinion 2. F_{32} is the force exerted by gear 3 against the pinion. Using a similar approach, we obtain the free-body diagram of the gear shown in Fig. 13–31*c*.

In Fig. 13–32, the free-body diagram of the pinion has been redrawn and the forces have been resolved into tangential and radial components. We now define

$$W_t = F_{32}^t \tag{a}$$

as the *transmitted load.* This tangential load is really the useful component, because the radial component F_{32}^r serves no useful purpose. It does not transmit power. The applied torque and the transmitted load are seen to be related by the equation

$$T = \frac{d}{2} W_t \tag{13–25}$$

where we have used $T = T_{a2}$ and $d = d_2$ to obtain a general relation.

Figure 13–32

Resolution of gear forces.

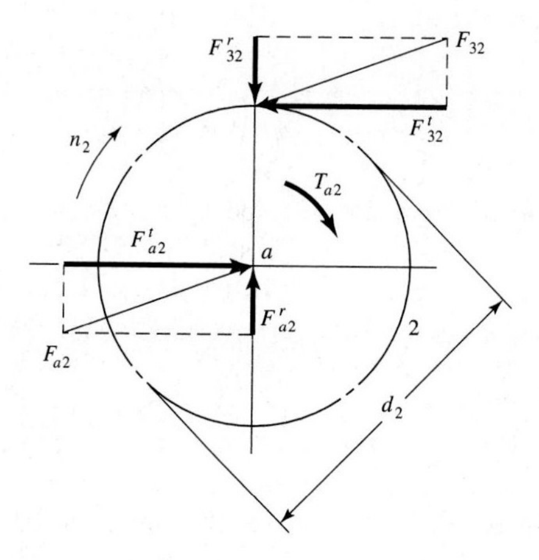

If next we designate the pitch-line velocity be V, where $V = \pi dn/12$ and is in feet per minute, the power H may be obtained from the equation

$$H = \frac{W_t V}{33\,000} \tag{13–26}$$

The corresponding equation in SI is

$$W_t = \frac{60(10)^3 H}{\pi dn} \tag{13–27}$$

where W_t = transmitted load, kN
H = power, kW
d = gear diameter, mm
n = speed, r/min

EXAMPLE 13–5 Pinion 2 in Fig. 13–33a runs at 1750 r/min and transmits 2.5 kW to idler gear 3. The teeth are cut on the 20° full-depth system and have a module of $m = 2.5$ mm. Draw a free-body diagram of gear 3 and show all the forces which act upon it.

Solution The pitch diameters of gears 2 and 3 are

$$d_2 = N_2 m = 20(2.5) = 50 \text{ mm}$$

$$d_3 = N_3 m = 50(2.5) = 125 \text{ mm}$$

From Eq. (13–27) we find the transmitted load to be

$$W_t = \frac{60(10)^3 H}{\pi d_2 n} = \frac{60(10)^3 (2.5)}{\pi(50)(1750)} = 0.546 \text{ kN}$$

Thus, the tangential force of gear 2 on gear 3 is $F_{23}^t = 0.546$ kN, as shown in Fig. 13–33b.

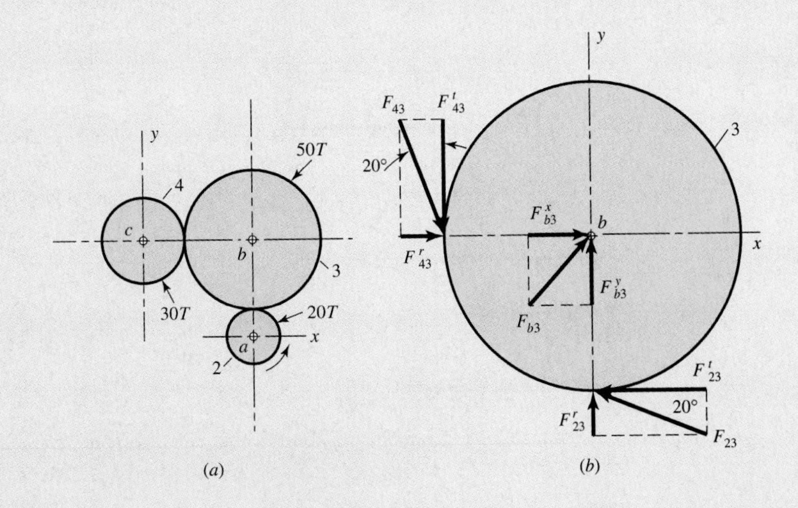

Figure 13–33

A gear train containing an idler gear. (a) The gear train. (b) Free-body of the idler gear.

Therefore

$$F^r_{23} = F^t_{23} \tan 20° = (0.546) \tan 20° = 0.199 \text{ kN}$$

and so

$$F^r_{23} = \frac{F^t_{23}}{\cos 20°} = \frac{0.546}{\cos 20°} = 0.581 \text{ kN}$$

Since gear 3 is an idler, it transmits no power (torque) to its shaft, and so the tangential reaction of gear 4 on gear 3 is also equal to W_t. Therefore

$$F^t_{43} = 0.546 \text{ kN} \qquad F^r_{43} = 0.199 \text{ kN} \qquad F_{43} = 0.581 \text{ kN}$$

and the directions are shown in Fig. 13–33*b*.

The shaft reactions in the x and y directions are

$$F^x_{b3} = -(F^t_{23} + F^r_{43}) = -(-0.546 + 0.199) = 0.347 \text{ kN}$$

$$F^y_{b3} = -(F^r_{23} + F^t_{43}) = -(0.199 - 0.546) = 0.347 \text{ kN}$$

The resultant shaft reaction is

$$F_{b3} = \sqrt{(0.347)^2 + (0.347)^2} = 0.491 \text{ kN}$$

These are shown on the figure.

13–15 Force Analysis—Bevel Gearing

In determining shaft and bearing loads for bevel-gear applications, the usual practice is to use the tangential or transmitted load which would occur if all the forces were concentrated at the midpoint of the tooth. While the actual resultant occurs somewhere between the midpoint and the large end of the tooth, there is only a small error in making this assumption. For the transmitted load, this gives

$$W_t = \frac{T}{r_{av}} \tag{13–28}$$

where T is the torque and r_{av} is the pitch radius at the midpoint of the tooth for the gear under consideration.

The forces acting at the center of the tooth are shown in Fig. 13–34. The resultant force W has three components: a tangential force W_t, a radial force W_r, and an axial force W_a. From the trigonometry of the figure,

$$W_r = W_t \tan \phi \cos \gamma \tag{13–29}$$

$$W_a = W_t \tan \phi \sin \gamma \tag{13–30}$$

The three forces W_t, W_r, and W_a are at right angles to each other and can be used to determine the bearing loads by using the methods of statics.

Figure 13–34

Bevel-gear tooth forces.

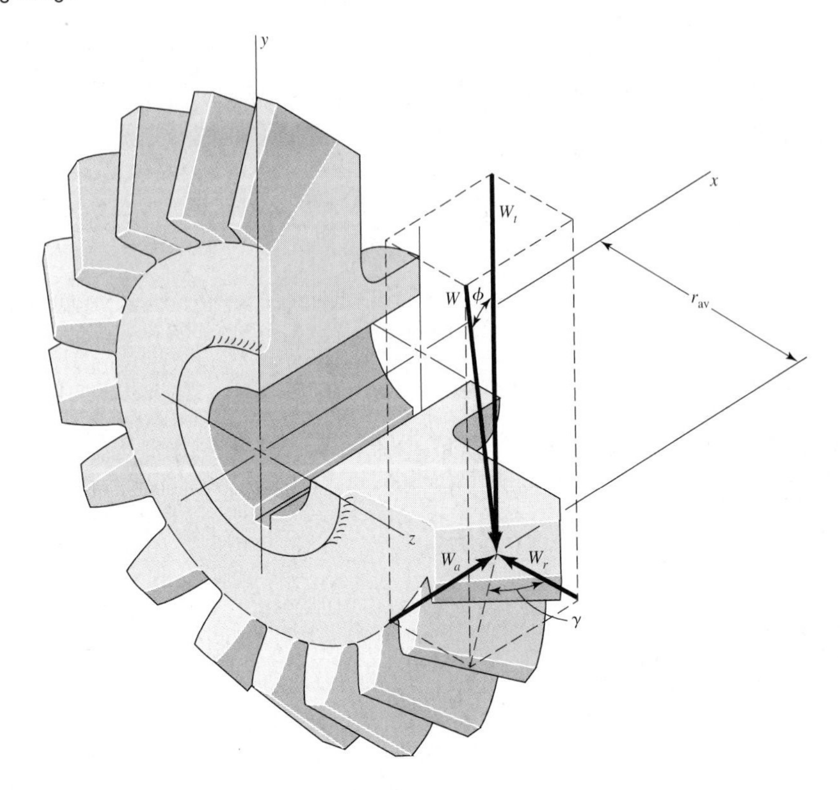

EXAMPLE 13–6 The bevel pinion in Fig. 13–35 rotates at 600 r/min in the direction shown and transmits 5 hp to the gear. The mounting distances, the location of all bearings, and the average pitch radii of the pinion and gear are shown in the figure. For simplicity, the teeth have been replaced by pitch cones. Bearings *A* and *C* should take the thrust loads. Find the bearing forces on the gearshaft.

Solution The pitch angles are

$$\gamma = \tan^{-1}\left(\tfrac{3}{9}\right) = 18.4^\circ \qquad \Gamma = \tan^{-1}\left(\tfrac{9}{3}\right) = 71.6^\circ$$

The pitch-line velocity corresponding to the average pitch radius is

$$V = \frac{2\pi r_P n}{12} = \frac{2\pi(1.293)(600)}{12} = 406 \text{ ft/min}$$

Therefore the transmitted load is

$$W_t = \frac{33\ 000H}{V} = \frac{(33\ 000)(5)}{406} = 406 \text{ lb}$$

which acts in the positive z direction, as shown in Fig. 13–36. We next have

$$W_r = W_t \tan\phi \cos\Gamma = 406 \tan 20^\circ \cos 71.6^\circ = 46.6 \text{ lb}$$

$$W_a = W_t \tan\phi \sin\Gamma = 406 \tan 20^\circ \sin 71.6^\circ = 140 \text{ lb}$$

where W_r is in the $-x$ direction and W_a is in the $-y$ direction, as illustrated in the isometric sketch of Fig. 13–36.

Figure 13–35

Bevel-gear set of Ex. 13–6.

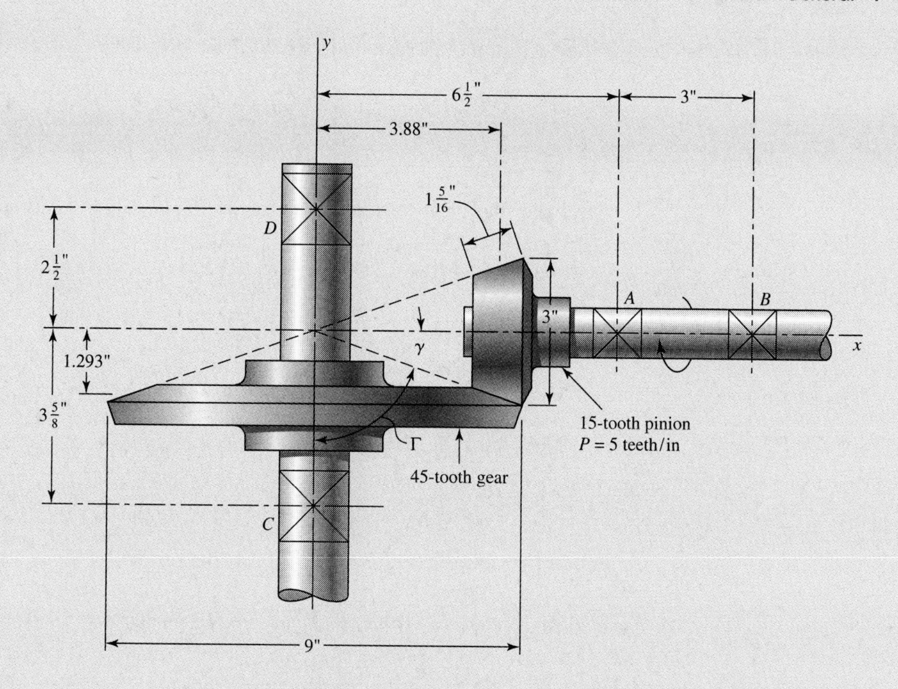

Figure 13–36

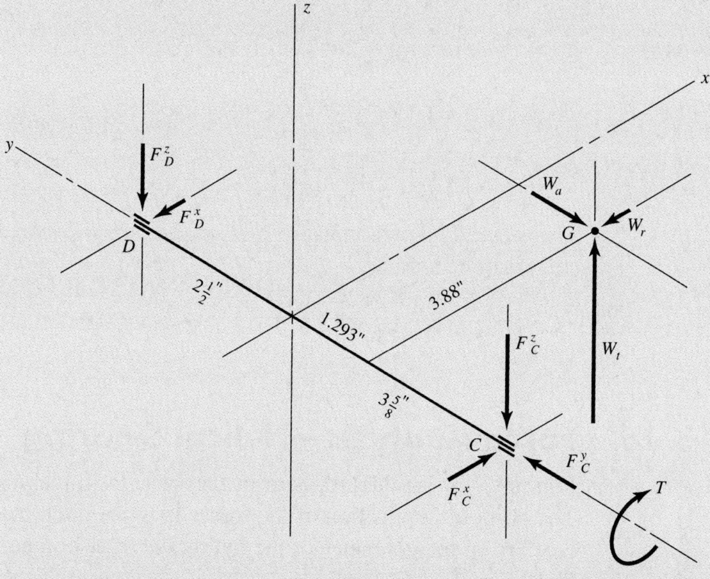

In preparing to take a sum of the moments about bearing D, define the position vector from D to G as

$$\mathbf{R}_G = 3.88\mathbf{i} - (2.5 + 1.293)\mathbf{j} = 3.88\mathbf{i} - 3.793\mathbf{j}$$

We shall also require a vector from D to C:

$$\mathbf{R}_C = -(2.5 + 3.625)\mathbf{j} = -6.125\mathbf{j}$$

Then, summing moments about D gives

$$\mathbf{R}_G \times \mathbf{W} + \mathbf{R}_C \times \mathbf{F}_C + \mathbf{T} = \mathbf{0} \tag{1}$$

When we place the details in Eq. (1), we get

$$(3.88\mathbf{i} - 3.793\mathbf{j}) \times (-46.6\mathbf{i} - 140\mathbf{j} + 406\mathbf{k}) + (6.125\mathbf{j}) \times (F_C^x\mathbf{i} + F_C^y\mathbf{j} + F_C^z\mathbf{k}) + T\mathbf{j} = \mathbf{0} \tag{2}$$

After the two cross products are taken, the equation becomes

$$(-1504\mathbf{i} - 1580\mathbf{j} - 721\mathbf{k}) + (-6.125F_C^z\mathbf{i} + 6.125F_C^x\mathbf{k}) + T\mathbf{j} = \mathbf{0}$$

from which

$$\mathbf{T} = 1580\mathbf{j} \text{ lb} \cdot \text{in} \qquad F_C^x = 118 \text{ lb} \qquad F_C^z = -246 \text{ lb} \tag{3}$$

Now sum the forces to zero. Thus

$$\mathbf{F}_D + \mathbf{F}_C + \mathbf{W} = \mathbf{0} \tag{4}$$

When the details are inserted, Eq. (4) becomes

$$(F_D^x\mathbf{i} + F_D^z\mathbf{k}) + (118\mathbf{i} + F_C^y\mathbf{j} - 246\mathbf{k}) + (-46.6\mathbf{i} - 140\mathbf{j} + 406\mathbf{k}) = \mathbf{0} \tag{5}$$

First we see that $F_C^y = 140$ lb, and so

Answer $\mathbf{F}_C = 118\mathbf{i} + 140\mathbf{j} - 246\mathbf{k}$ lb

Then, from Eq. (5),

Answer $\mathbf{F}_D = -71\mathbf{i} - 160\mathbf{k}$ lb

These are all shown in Fig. 13–36 in the proper directions. The analysis for the pinion shaft is quite similar.

13–16 Force Analysis—Helical Gearing

Figure 13–37 is a three-dimensional view of the forces acting against a helical-gear tooth. The point of application of the forces is in the pitch plane and in the center of the gear face. From the geometry of the figure, the three components of the total (normal) tooth force W are

$$\begin{aligned} W_r &= W \sin \phi_n \\ W_t &= W \cos \phi_n \cos \psi \\ W_a &= W \cos \phi_n \sin \psi \end{aligned} \tag{13–31}$$

where W = total force

W_r = radial component

W_t = tangential component; also called transmitted load

W_a = axial component; also called thrust load

Usually W_t is given and the other forces are desired. In this case, it is not difficult to

Figure 13–37

Tooth forces acting on a right-hand helical gear.

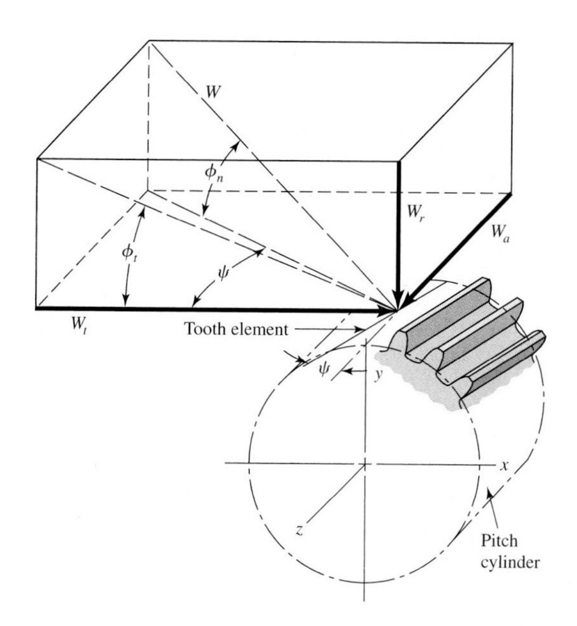

discover that

$$
\begin{aligned}
W_r &= W_t \tan \phi_t \\
W_a &= W_t \tan \psi \\
W &= \frac{W_t}{\cos \phi_n \cos \psi}
\end{aligned}
\tag{13–32}
$$

EXAMPLE 13–7 In Fig. 13–38 a 1-hp electric motor runs at 1800 r/min in the clockwise direction, as viewed from the positive x axis. Keyed to the motor shaft is an 18-tooth helical pinion having a normal pressure angle of 20°, a helix angle of 30°, and a normal diametral pitch of 12 teeth/in. The hand of the helix is shown in the figure. Make a three-dimensional sketch of the motor shaft and pinion, and show the forces acting on the pinion and the bearing reactions at A and B. The thrust should be taken out at A.

Figure 13–38

The motor and gear train of Ex. 13–7.

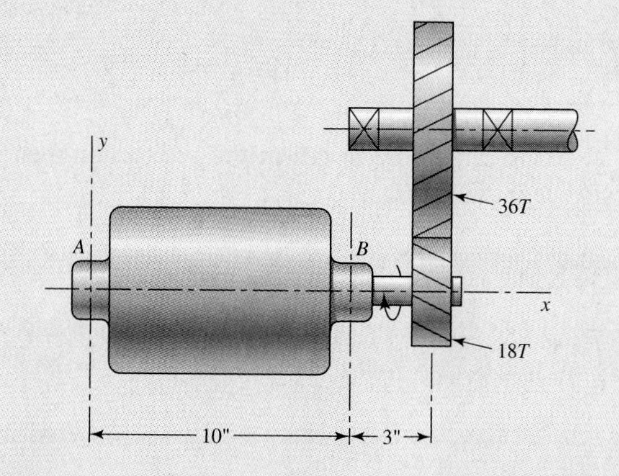

Figure 13–39

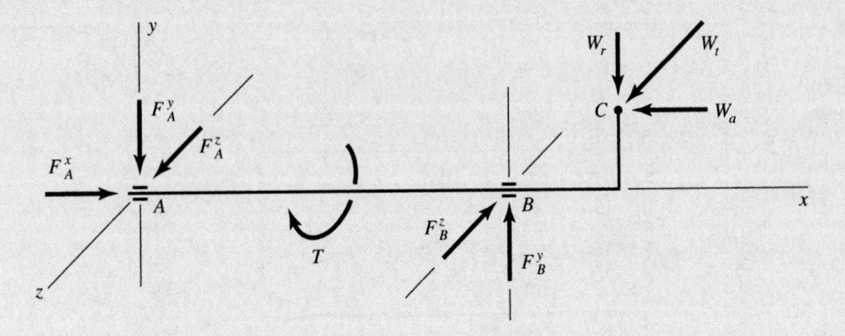

Solution From Eq. (13–15) we find

$$\phi_t = \tan^{-1}\frac{\tan\phi_n}{\cos\psi} = \tan^{-1}\frac{\tan 20^\circ}{\cos 30^\circ} = 22.8^\circ$$

Also, $P_t = P_n \cos\psi = 12 \cos 30^\circ = 10.4$ teeth/in. Therefore the pitch diameter of the pinion is $d_p = 18/10.4 = 1.73$ in. The pitch-line velocity is

$$V = \frac{\pi dn}{12} = \frac{\pi(1.73)(1800)}{12} = 815 \text{ ft/min}$$

The transmitted load is

$$W_t = \frac{33\,000H}{V} = \frac{(33\,000)(1)}{815} = 40.5 \text{ lb}$$

From Eq. (13–32) we find

$$W_r = W_t \tan\phi_t = (40.5)(0.420) = 17.0 \text{ lb}$$

$$W_a = W_t \tan\psi = (40.5)(0.577) = 23.4 \text{ lb}$$

$$W = \frac{W_t}{\cos\phi_n \cos\psi} = \frac{40.5}{(0.940)(0.866)} = 49.8 \text{ lb}$$

These three forces, W_r in the $-y$ direction, W_a in the $-x$ direction, and W_t in the $+z$ direction, are shown acting at point C in Fig. 13–39. We assume bearing reactions at A and B as shown. Then $F_a^x = W_a = 23.4$ lb. Taking moments about the z axis,

$$-(17.0)(13) + (23.4)\left(\frac{1.73}{2}\right) + 10F_B^y = 0$$

or $F_B^y = 20$ lb. Summing forces in the y direction then gives $F_A^y = 3.0$ lb. Taking moments about the y axis, next

$$10F_B^z - (40.5)(13) = 0$$

or $F_B^z = 52.7$ lb. Summing forces in the z direction and solving gives $F_A^z = 12.2$ lb. Also, the torque is $T = W_t d_p/2 = (40.5)(1.73/2) = 35 \text{ lb} \cdot \text{in}$.

EXAMPLE 13–8 Solve Example 13–7 using vectors.

Solution The force at C is

$$\mathbf{W} = -23.4\mathbf{i} - 17.1\mathbf{j} + 40.5\mathbf{k}$$

Position vectors to B and C from origin A are

$$\mathbf{R}_B = 10\mathbf{i} \qquad \mathbf{R}_C = 13\mathbf{i} + 0.865\mathbf{j}$$

Taking moments about A, we have

$$\mathbf{R}_B \times \mathbf{F}_B + \mathbf{T} + \mathbf{R}_C \times \mathbf{W} = \mathbf{0}$$

Using the directions assumed in Fig. 13–39 and substituting values gives

$$10\mathbf{i} \times (F_B^y\mathbf{j} - F_B^z\mathbf{k}) - T\mathbf{i} + (13\mathbf{i} + 0.865\mathbf{j}) \times (-23.4\mathbf{i} - 17.0\mathbf{j} + 40.5\mathbf{k}) = \mathbf{0}$$

When the cross products are formed, we get

$$(10F_B^y\mathbf{k} + 10F_B^z\mathbf{j}) - T\mathbf{i} + (35\mathbf{i} - 527\mathbf{j} - 200\mathbf{k}) = \mathbf{0}$$

whence $T = 35$ lb · in, $F_B^y = 20$ lb, and $F_B^z = 52.7$ lb.

Next, $\mathbf{F}_A = -\mathbf{F}_B - \mathbf{W}$, and so $\mathbf{F}_A = 23.4\mathbf{i} - 2.9\mathbf{j} + 12.1\mathbf{k}$ lb.

13–17 Force Analysis—Worm Gearing

If friction is neglected, then the only force exerted by the gear will be the force W, shown in Fig. 13–40, having the three orthogonal components W^x, W^y, and W^z. From the geometry of the figure, we see that

$$\begin{aligned} W^x &= W \cos\phi_n \sin\lambda \\ W^y &= W \sin\phi_n \\ W^z &= W \cos\phi_n \cos\lambda \end{aligned} \tag{13–33}$$

We now use the subscripts W and G to indicate forces acting against the worm and gear, respectively. We note that W^y is the separating, or radial, force for both the worm and the gear. The tangential force on the worm is W^x and is W^z on the gear, assuming a 90° shaft angle. The axial force on the worm is W^z, and on the gear, W^x. Since the gear forces are opposite to the worm forces, we can summarize these relations by writing

$$\begin{aligned} W_{Wt} &= -W_{Ga} = W^x \\ W_{Wr} &= -W_{Gr} = W^y \\ W_{Wa} &= -W_{Gt} = W^z \end{aligned} \tag{13–34}$$

It is helpful in using Eq. (13–33) and also Eq. (13–34) to observe that *the gear axis is parallel to the x direction and the worm axis is parallel to the z direction* and that we are employing a right-handed coordinate system.

In our study of spur-gear teeth we have learned that the motion of one tooth relative to the mating tooth is primarily a rolling motion; in fact, when contact occurs at the pitch

Figure 13–40

Drawing of the pitch cylinder of a worm, showing the forces exerted upon it by the worm gear.

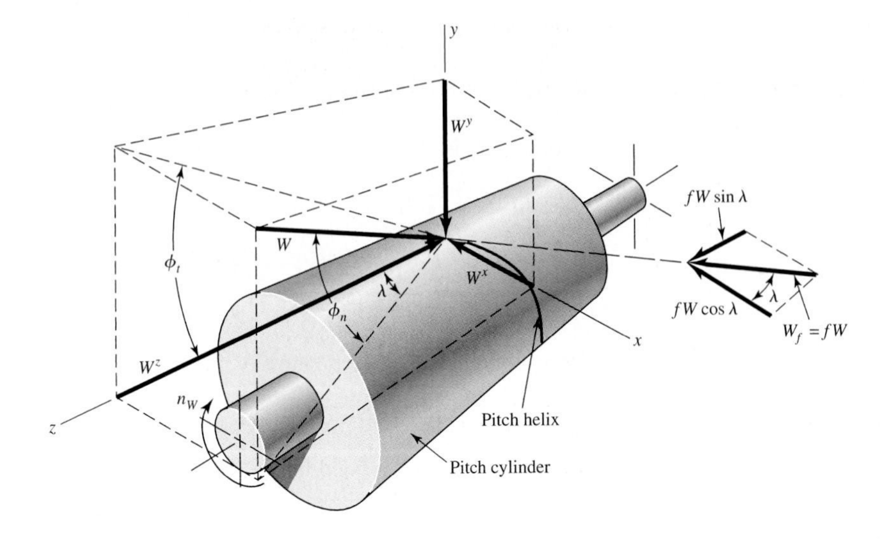

point, the motion is pure rolling. In contrast, the relative motion between worm and worm-gear teeth is pure sliding, and so we must expect that friction plays an important role in the performance of worm gearing. By introducing a coefficient of friction f, we can develop another set of relations similar to those of Eq. (13–33). In Fig. 13–40 we see that the force W acting normal to the worm-tooth profile produces a frictional force $W_f = fW$, having a component $fW \cos\lambda$ in the negative x direction and another component $fW \sin\lambda$ in the positive z direction. Equation (13–33) therefore becomes

$$\begin{aligned} W^x &= W(\cos\phi_n \sin\lambda + f\cos\lambda) \\ W^y &= W\sin\phi_n \\ W^z &= W(\cos\phi_n \cos\lambda - f\sin\lambda) \end{aligned} \tag{13–35}$$

Equation (13–34), of course, still applies.

If we substitute W^z in the third part of Eq. (13–34) and multiply both sides by f, we find the frictional force to be

$$W_f = fW = \frac{fW_{Gt}}{f\sin\lambda - \cos\phi_n \cos\lambda} \tag{13–36}$$

Another useful relation can be obtained by solving the first and third parts of Eq. (13–34) simultaneously to get a relation between the two tangential forces. The result is

$$W_{Wt} = W_{Gt} = \frac{\cos\phi_n \sin\lambda + f\cos\lambda}{f\sin\lambda - \cos\phi_n \cos\lambda} \tag{13–37}$$

Efficiency η can be defined by using the equation

$$\eta = \frac{W_{Wt}(\text{without friction})}{W_{Wt}(\text{with friction})} \tag{a}$$

Substitute Eq. (13–37) with $f = 0$ in the numerator of Eq. (a) and the same equation in the denominator. After some rearranging, you will find the efficiency to be

$$\eta = \frac{\cos\phi_n - f\tan\lambda}{\cos\phi_n + f\cot\lambda} \tag{13–38}$$

Selecting a typical value of the coefficient of friction, say $f = 0.05$, and the pressure angles shown in Table 13–7, we can use Eq. (13–38) to get some useful design infor-

Table 13–7

Efficiency of Worm Gearsets for $f = 0.05$

Helix Angle ψ, deg	Efficiency η, %
1.0	25.2
2.5	45.7
5.0	62.0
7.5	71.3
10.0	76.6
15.0	82.7
20.0	85.9
30.0	89.1

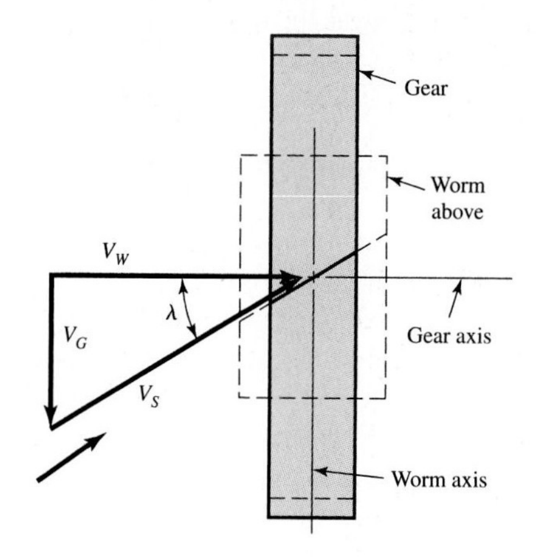

Figure 13–41

Velocity components in worm gearing.

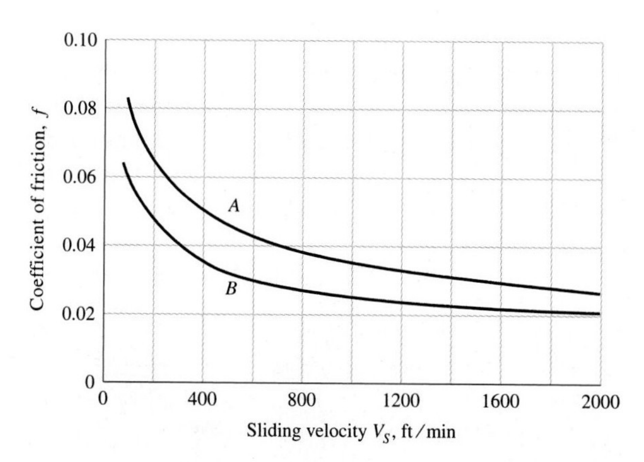

Figure 13–42

Representative values of the coefficient of friction for worm gearing. These values are based on good lubrication. Use curve *B* for high-quality materials, such as a case-hardened steel worm mating with a phosphor-bronze gear. Use curve *A* when more friction is expected, as with a cast-iron worm mating with a cast-iron worm gear.

mation. Solving this equation for helix angles from 1 to 30° gives the interesting results shown in Table 13–7.

Many experiments have shown that the coefficient of friction is dependent on the relative or sliding velocity. In Fig. 13–41, V_G is the pitch-line velocity of the gear and V_W the pitch-line velocity of the worm. Vectorially, $\mathbf{V}_W = \mathbf{V}_G + \mathbf{V}_S$; consequently,

$$V_S = \frac{V_W}{\cos\lambda} \tag{13–39}$$

Published values of the coefficient of friction vary as much as 20 percent, undoubtedly because of the differences in surface finish, materials, and lubrication. The values on the chart of Fig. 13–42 are representative and indicate the general trend.

EXAMPLE 13–9 A 2-tooth right-hand worm transmits 1 hp at 1200 r/min to a 30-tooth worm gear. The gear has a transverse diametral pitch of 6 teeth/in and a face width of 1 in. The worm has a pitch diameter of 2 in and a face width of $2\frac{1}{2}$ in. The normal pressure angle is $14\frac{1}{2}°$. The materials and quality of work needed are such that curve *B* of Fig. 13–42 should be used to obtain the coefficient of friction.

(*a*) Find the axial pitch, the center distance, the lead, and the lead angle.

(*b*) Figure 13–43 is a drawing of the worm gear oriented with respect to the coordinate system described earlier in this section; the gear is supported by bearings *A* and *B*. Find the forces exerted by the bearings against the worm-gear shaft, and the output torque.

Solution (*a*) The axial pitch is the same as the transverse circular pitch of the gear, which is

Answer
$$p_t = \frac{\pi}{P} = \frac{\pi}{6} = 0.5236 \text{ in}$$

The pitch diameter of the gear is $d_G = N_G/P = 30/6 = 5$ in. Therefore, the center distance is

Answer
$$C = \frac{d_W + d_G}{2} = \frac{2+5}{2} = 3.5 \text{ in}$$

From Eq. (13–19), the lead is

$$L = p_x N_W = (0.5236)(2) = 1.0472 \text{ in}$$

Answer Also using Eq. (13–20), find

Answer
$$\lambda = \tan^{-1}\frac{L}{\pi d_W} = \tan^{-1}\frac{1.0472}{\pi(2)} = 9.47°$$

(*b*) Using the right-hand rule for the rotation of the worm, you will see that your thumb points in the positive *z* direction. Now use the bolt-and-nut analogy (the worm is right-handed, as is the screw thread of a bolt), and turn the bolt clockwise with the right hand while preventing nut rotation with the left. The nut will move axially along the bolt toward your right hand. Therefore the surface of the gear (Fig. 13–43) in contact with

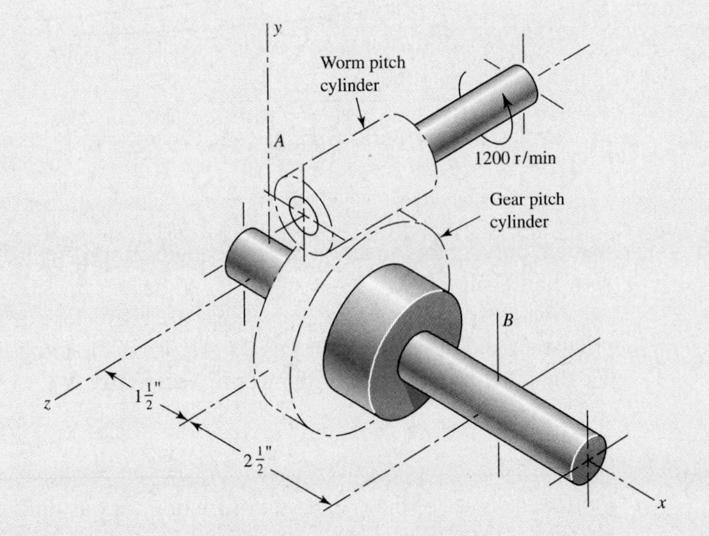

Figure 13–43
The pitch cylinders of the worm gear train of Ex. 13–9.

the worm will move in the negative z direction. Thus, the gear rotates clockwise about x, with your right thumb pointing in the negative x direction.

The pitch-line velocity of the worm is

$$V_W = \frac{\pi d_W n_W}{12} = \frac{\pi(2)(1200)}{12} = 628 \text{ ft/min}$$

The speed of the gear is $n_G = (\frac{2}{30})(1200) = 80$ r/min. Therefore the pitch-line velocity is

$$V_G = \frac{\pi d_G n_G}{12} = \frac{\pi(5)(80)}{12} = 105 \text{ ft/min}$$

Then, using Eq. (13–39), the sliding velocity V_S is found to be

$$V_S = \frac{V_W}{\cos\lambda} = \frac{628}{\cos 9.47^\circ} = 637 \text{ ft/min}$$

Getting to the forces now, we begin with the horsepower formula

$$W_{Wt} = \frac{33\,000H}{V_W} = \frac{(33\,000)(1)}{628} = 52.5 \text{ lb}$$

This force acts in the negative z direction, the same as in Fig. 13–40. Using Fig. 13–42, we find $f = 0.03$. Then, the first equation of group (13–35) gives

$$\begin{aligned} W &= \frac{W^x}{\cos\phi_n \sin\lambda + f\cos\lambda} \\ &= \frac{52.5}{\cos 14.5^\circ \sin 9.47^\circ + 0.03\cos 9.47^\circ} = 278 \text{ lb} \end{aligned}$$

Also, from Eq. (13–35),

$$\begin{aligned} W^y &= W\sin\phi_n = 278 \sin 14.5^\circ = 69.6 \text{ lb} \\ W^z &= W(\cos\phi_n \cos\lambda - f\sin\lambda) \\ &= 278(\cos 14.5^\circ \cos 9.47^\circ - 0.03 \sin 9.47^\circ) = 264 \text{ lb} \end{aligned}$$

We now identify the components acting on the gear as

$$\begin{aligned} W_{Ga} &= -W^x = 52.5 \text{ lb} \\ W_{Gr} &= -W^y = 69.6 \text{ lb} \\ W_{Gt} &= -W^z = -264 \text{ lb} \end{aligned}$$

At this point a three-dimensional line drawing should be made in order to simplify the work to follow. An isometric sketch, such as the one of Fig. 13–44, is easy to make and will help you to avoid errors. Note the y axis is vertical, while the x and z axes make angles of 30° with the horizontal. The illusion of depth is enhanced by sketching lines parallel to each of the coordinate axes through every point of interest.

We shall make B a thrust bearing in order to place the gearshaft in compression. Thus, summing forces in the x direction gives

Answer $F_B^x = -52.5$ lb

Taking moments about the z axis, we have

Answer $-(52.5)(2.5) - (69.6)(1.5) + 4F_B^y = 0 \qquad F_B^y = 58.9$ lb

Figure 13–44

An isometric sketch used in Ex. 13–9.

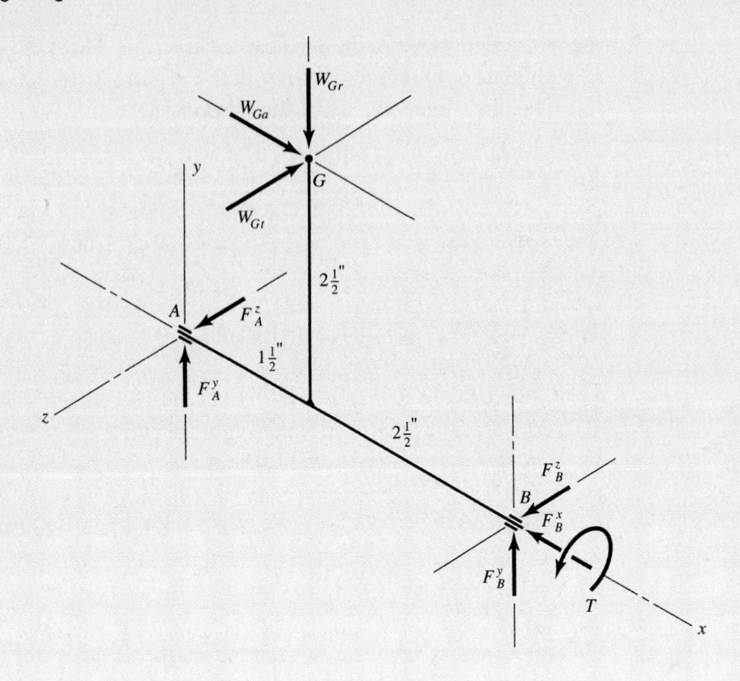

Taking moments about the y axis,

Answer $(264)(1.5) - 4F_B^z = 0 \qquad F_B^z = 99 \text{ lb}$

These three components are now inserted on the sketch as shown at B in Fig. 13–44. Summing forces in the y direction,

Answer $-69.6 + 58.9 + F_A^y = 0 \qquad F_A^y = 10.7 \text{ lb}$

Similarly, summing forces in the z direction,

Answer $-264 + 99 + F_A^z = 0 \qquad F_A^z = 165 \text{ lb}$

These two components can now be placed at A on the sketch. We still have one more equation to write. Summing moments about x,

Answer $-(264)(2.5) + T = 0 \qquad T = 660 \text{ lb} \cdot \text{in}$

It is because of the frictional loss that this output torque is less than the product of the gear ratio and the input torque.

13–18 Gear Ratios and Numbers of Teeth

Now, with the basics of gear-force analysis and gear-ratio relations in mind, we can consider closely related concerns.

Numbers of Teeth

One essential ingredient in the quantitative treatment in this chapter has been tooth counts and gear ratios. With the exception of identifying minimum pinion tooth count (size)

and maximum tooth count possible in a gear in mesh with a pinion in the avoidance of interference, not much has been said about the tooth counts themselves or associated gear ratios.

Gear ratios for completely rotating gears are made up of quotients of integers, and they are discrete. If approximations to contemplated ratios are acceptable, there are no problems, except finding the closest approximation. Figure 13–45 has an abscissa of numbers of pinion teeth and an ordinate of numbers of gear teeth. Points in the first quadrant with integer coordinates can be thought of as pins standing upright on a board. The desired gear ratio can be represented by a line (a string) through the origin with a slope of $r = (R)_{\text{desired}}$. If no pin touches the string, there is no exact representation of the desired ratio. Approximations can be found by driving another pin well beyond allowable tooth numbers with coordinates representing the desired slope. If the string is disconnected from the origin and pulled downward, the string will touch several pins. The touched pins represent possible approximations. Pulling the left end of the string horizontally will produce several more approximations.

Before the days of the digital computer, mathematical schemes (involving partial fractions and continued fractions) were developed to identify the approximations, from which the best was selected. With a computer, a nested double D0-loop can march among all the possible pins, sort the pin property $|(r - R_{\text{desired}}|$ ranked smallest to largest, and print out, for example, the first ten.

Single or Double Reduction

Another problem the designer has (even if the overall ratio is given) is in the power transmission applications. If the overall speed reduction is to be R:1, should the reduction be accomplished in one or more than one step? In two steps, what should be the reduction ratios R_1 and R_2 such that $R = R_1 R_2$?

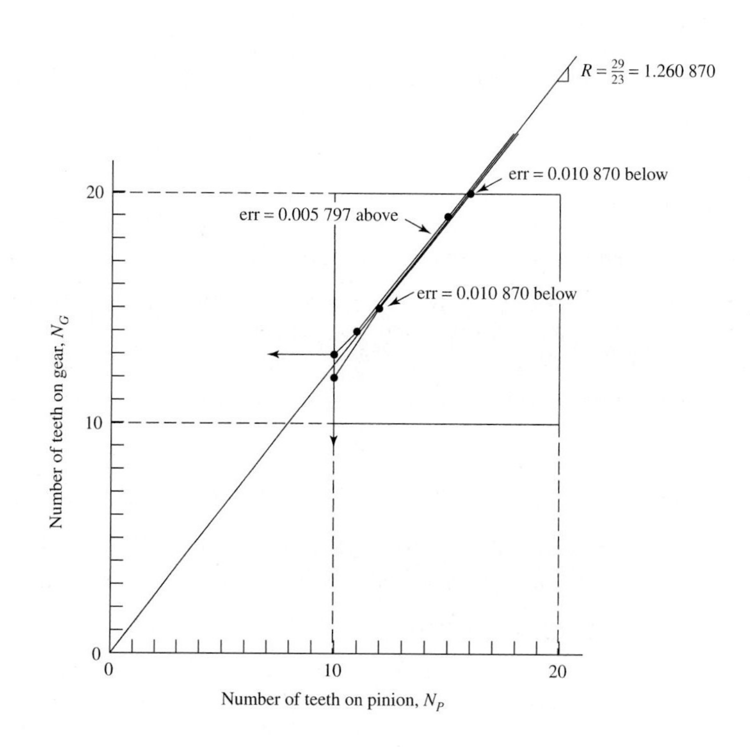

Figure 13–45

The "string method" of finding approximate gear ratios. Allowable tooth numbers: $10 \le N_P \le 20$, $10 \le N_G \le 20$, desired ratio $R = 29/23$. Upper string touches 13/10, 14/11, 19/15; lower string touches 12/10, 15/12, and 20/16. The best ratio is 19/15, to which the corresponding error is $29/23 - 19/15 = -0.005\ 797$. The computer has made this method obsolete.

One goal of a gear-box designer is to achieve a small size (compactness). From Eq. (13–1), equating the diametral pitches for first mesh,

$$P = \frac{N_1}{d_1} = \frac{N_2}{d_2}$$

and it follows that

$$d_1 = \frac{N_1}{N_2} d_2 = \frac{d_2}{R}$$

The center-to-center distance C is a measure of size. For a pair of gears in mesh with speed-reduction ratio R

$$C = \frac{d_1}{2} + \frac{d_2}{2} = \frac{d_1}{2} + \frac{d_1 R}{2} = \frac{d_1}{2}(1 + R) \tag{13–40}$$

For a double reduction the center-to-center distance C_1 for the first reduction R_1 is, from Eq. (13–40), noting the bodies are: first pinion 1, meshing gear 2, attached pinion 3, and meshing gear 4,

$$C_1 = \frac{d_1}{2}(1 + R_1)$$

For the second reduction, the center-to-center distance C_2 from Eq. (13–40), is

$$C_2 = \frac{d_3}{2}(1 + R_2)$$

The center-to-center distance C for the compound gear trains is

$$C = C_1 + C_2 = \frac{d_1}{2}(1 + R_1) + \frac{d_3}{2}(1 + R_2)$$

Using identical pinions having the same minimum number of teeth, $d_3 = d_1$, it follows that the average center-to-center distance C is

$$C = \frac{d_1}{2}(1 + R_1) + \frac{d_1}{2}(1 + R_2) = \frac{d_1}{2}(2 + R_1 + R_2) = \frac{d_1}{2}\left(2 + R_1 + \frac{R}{R_1}\right) \tag{13–41}$$

Minimizing C in Eq. (13–41) by differentiating with respect to R_1 and equating to zero to find the stationary point gives

$$\frac{dC}{dR_1} = \frac{d_1}{2}\left(0 + 1 - \frac{R}{R_1^2}\right) = 0$$

from which

$$R_1 = \sqrt{R} \qquad R_2 = \frac{R}{R_1} = \frac{R}{\sqrt{R}} = \sqrt{R} \tag{13–42}$$

Mechanical Efficiency

Since the mechanical efficiency of spur gears is high, $0.98 < e < 0.99$, equations for efficiencies are rarely displayed. Spotts gives an excellent simple approximation for the efficiency of a spur-gear mesh.[4] Wilson gives detailed equations for several kinds of

4. M. F. Spotts, *Mechanical Design Analysis*, Prentice-Hall, Englewood Cliffs, N. J., pp. 187–188.

gearing.[5] The approximation of Spotts is

$$e \doteq 1 - \pi f \left(\frac{1}{N_1} + \frac{1}{N_2} \right) \tag{13–43}$$

where f is the coefficient of sliding friction, N_1 is the pinion tooth count, and N_2 is the gear tooth count. In a double reduction

$$e = e_1 e_2 = \left[1 - \pi f \left(\frac{1}{N_1} + \frac{1}{N_2} \right) \right] \left[1 - \pi f \left(\frac{1}{N_3} + \frac{1}{N_4} \right) \right]$$

Substituting $R = R_1 R_2$ and $N_3 = N_1$ gives

$$e = 1 - \frac{\pi f}{N_1} \left(2 + \frac{1}{R_1} + \frac{R_1}{R} \right) + \frac{\pi^2 f^2}{N_1^2} \left(1 + \frac{1}{R_1} + \frac{R_1}{R} + \frac{1}{R} \right)$$

We can maximize the efficiency e by differentiating e with respect to R_1 and equating to zero to find the stationary point:

$$\frac{de}{dR_1} = 0 + \frac{\pi f}{N_1} \left(-\frac{1}{R_1^2} + \frac{1}{R} \right) + \frac{\pi^2 f^2}{N_1^2} \left(-\frac{1}{R_1^2} + \frac{1}{R} \right) = 0$$

from which $R_1 = \sqrt{R}$, $R_2 = R/\sqrt{R} = \sqrt{R}$, which is the same result as before.

Another "volume" minimization approach is to minimize the tooth sum $\sum$ given by

$$\sum = N_1 + N_2 + N_3 + N_4$$

After noting that $N_1 = N_3$, $N_2/N_1 = R_1$, $N_4/N_3 = R_2 = R/R_1$, the sum $\sum$ becomes

$$\sum = N_1 + R_1 N_1 + N_1 + R_2 N_1 = N_1(2 + R_1 + R_2) = N_1 \left(2 + R_1 + \frac{R}{R_1} \right)$$

$$\frac{d\sum}{dR_1} = N_1 \left(0 + 1 - \frac{R}{R_1^2} \right)$$

from which $R_1 = R_2 = \sqrt{R}$, as before.

It is always a good idea to "punch in some numbers" and "plot some curves." Consider a 2-in-diameter pinion and an overall reduction $R = 10$. Equation (13–40) gives a center-to-center distance C of

$$c = \frac{d_1}{2}(1 + R) = \frac{2}{2}(1 + 10) = 11 \text{ in}$$

For a double reduction Eq. (13–41) gives

$$C = \frac{d_1}{2} \left(2 + R_1 + \frac{R}{R_1} \right) = \frac{2}{2} \left(2 + R_1 + \frac{10}{R_1} \right) \tag{13–44}$$

5. Earle Buckingham, "Gears," sec. 105 in Frank F. Wilson, Editor-in-Chief, *Tool Engineers Handbook*, McGraw-Hill, New York, pp. 1703–1707.

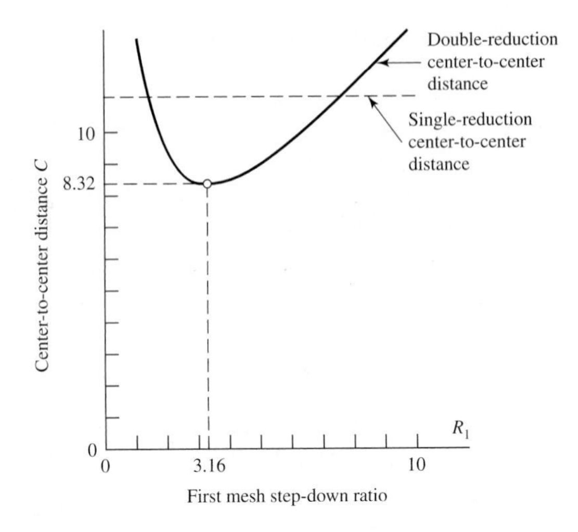

Figure 13–46

A plot of Eq. (13–44) for a 2-in-diameter pinion and an overall reduction of $R = 10$.

We make a table:

R_1	C	R_1	C
1	13.0	6	9.6
2	9.0	7	10.4
3	8.3	8	11.3
4	8.5	9	12.1
5	9.0	10	13.0

A plot as depicted in Fig. 13–46 shows that a double reduction is

- "Smaller" most of the time, but not necessarily.
- Insensitive around $\sqrt{R}$, if $\sqrt{R}$ is not available exactly, and the increase in center-to-center distance is not much.

If we choose a double reduction with $\sqrt{10}$ in each mesh for the ratio, we cannot use identical gear sets. The torque transmitted in the second reduction is larger by a factor of $\sqrt{10}$. The diametral pitch of the second pinion would have to change to provide stronger teeth. This increases the center-to-center distance in the second mesh. Also, the larger the first step in ratio, the larger the second mesh pinion and gear become. Intuitively (at this time) the optimum is displaced. With no knowledge (at this time) of gear-tooth stresses, if the same pitch is used in both meshes, the second pinion would have to be wider by a factor of $\sqrt{10}$ while not increasing the center-to-center distance. This produces a pinion that is too wide for good gear proportions and uniform distance distribution of the tooth load and the gear force.

13–19 Gear-Shaft Speeds and Bearings

Since we are familiar with thick- and thin-film bearings from Chap. 12, we can now look at how bearings in planetary gearing are affected by radial load vectors and bushing with angular velocities. An example is a good way to visualize and understand.

EXAMPLE 13–10 The 24-tooth 8-pitch 20° spur pinion depicted in Fig. 13–47 rotates clockwise at 900 r/min and transmits 3 hp to rotating machinery connected to the arm. As shown in Fig. 13–47*b* the bearing at *A* has its bushing fixed in the earth (or in body 5, the internal gear, which is grounded), and has a journal speed of n_2 r/min. Bearing *B* has its bushing fixed in the arm (body 3) and a journal speed of $n_2 - n_3$ r/min. Bearing *C* has its bushing fixed in body 3 and a journal speed of $n_2 - n_3$ r/min. The bearings appear in pairs, but since the force systems are planer we can put all the load on one bearing of the pair, and remember to divide by 2 when sizing a bearing.

(*a*) Draw a free body of the arm and each gear showing the forces and torques acting upon them.

(*b*) Find the value of N^* to be used to form the Sommerfeld number for thick-film bearings *A*, *B*, and *C*.

(*c*) For boundary-lubricated bearings find the loads and the rubbing speeds for bearings *A*, *B*, and *C*.

Solution First some preparatory steps to find the gear speeds. We note $n_2 = -900$ r/min and $n_5 = 0$. The train value e_{25} is

$$e_{25} = -\frac{24}{40}\frac{40}{104} = -\frac{24}{104}$$

From Eq. (13–24),

$$-\frac{24}{104} = e_{25} = \frac{n_5 - n_3}{n_2 - n_3} = \frac{0 - n_3}{-900 - n_3}$$

from which

$$n_3 = \frac{-900(24)/104}{1 + 24/104} = -168.75 \text{ r/min} \qquad \text{(cw)}$$

The angular velocity of body 2 with respect to the arm n_{23} is

$$n_{23} = n_2 - n_3 = -900 - (-168.75) = -731.25 \text{ r/min} \qquad \text{(cw)}$$

The angular velocity of body 4 with respect to the arm n_{43} is

$$n_{43} = -n_{23}(24/40) = -(-731.25)(24/40) = 438.75 \qquad \text{(ccw)}$$

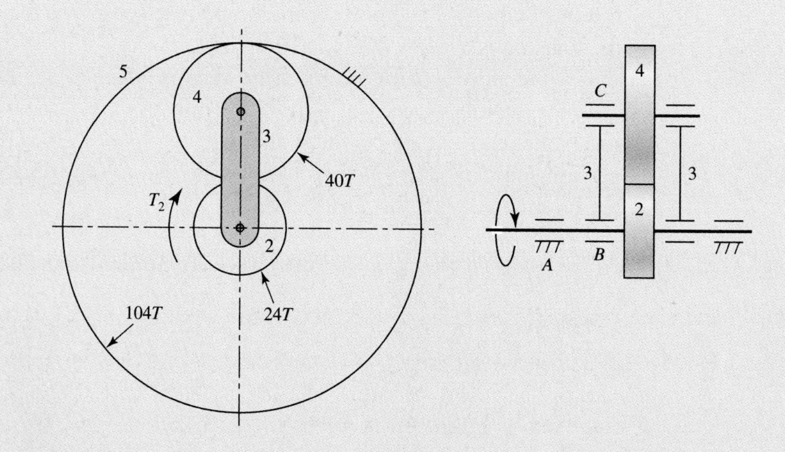

Figure 13–47

The planetary gear train of Ex. 13–10.

A table can be made:

Bearing	Bushing, fixed in	Journal Speed, r/min	Load, lb	Load magnitude, lb	Load Vector Speed, r/min
A	5 or 0	$n_2 = -900$	$\mathbf{F}_{20}$	149	−168.75
B	3	$n_{23} = -731.25$	$\mathbf{F}_{32} + \mathbf{F}_{02}$	149	−168.75
C	3	$n_{43} = 438.75$	$\mathbf{F}_{43}$	280	−168.75

(*a*) $d_2 = 24/8 = 3$ in, $d_4 = 40/8 = 5$ in, and $d_5 = 104/8 = 13$ in. The radius of the arm is $\frac{1}{2}(24 + 40)/8 = 4$ in. For the sun gear the transmitted (tangential) force is

$$W_t = \frac{33\,000(12)\text{hp}}{\pi d_2 n_2} = \frac{33\,000(12)3}{\pi(3)900} = 140 \text{ lbf}$$

The force vector of which 140 lbf is the tangential component is 140/cos 20° = 149 lbf. Figure 13–48*a* depicts the free body of the planet gear 4. $F_{54} = F_{24} = 149$ lbf. The force F_{34} is given by

$$F_{34} = 2(149 \cos 20°) = 280 \text{ lbf}$$

Figure 13–48*b* depicts the free body of the arm 3. $F_{43} = F_{34} = 280$ lbf. The force T_3 is given by

$$T_3 = 280(4) = 1120 \text{ in} \cdot \text{lbf}$$

The free body of the sun gear 2 is shown in Fig. 13–48*c*. Force F_{42} and F_{32} are placed on first, and F_{02} and T_2 are found from summation of forces and moments. The horizontal component of F_{02} is 140 lbf to the left and the angle to the horizontal is 20°. The bearing load is $F_{32} + F_{02}$. The moment T_2 is

$$T_2 = 140(3/2) = 210 \text{ in} \cdot \text{lbf}$$

Figure 13–48*d* shows the grounded ring gear. The earth also provides the support for bearing *A*. F_{20} and F_{40} are known to be 149 in · lbf in magnitude and lines of action are 20° to the horizontal. The reactive torque on the earth is 140(13/2) = 910 in · lbf. The torques on body 2 and body 3 are in opposite senses, so the net torque required to hold the housing is 1120 − 210 = 910 in · lbf.

All the forces in this problem are vectors which rotate in synchronism with the arm. All have angular speeds of −168.75 r/min. As a check on our work we compare the mechanical advantage by force analysis $T_3/T_2 = 1120/210 = 5.\underline{3}$ to the mechanical advantage by angular velocity analysis $n_2/n_3 = 900/168.75 = 5.\underline{3}$, and they agree.

(*b*) From the table above, for bearing *A*

$$N^* = N_j + N_b - 2N_F = -900 + 0 - 2(-168.75) = -562.5 \text{ r/min} \qquad \text{(ccw)}$$

For bearing *B*

$$N^* = N_j + N_b - 2N_F = n_{23} + n_3 - 2n_3 = -731.25 + (-168.75) - 2(-168.75)$$
$$= -562.5 \text{ r/min} \qquad \text{(ccw)}$$

Figure 13–48

Free bodies of principal gear train part of Fig. 13–47 for Ex. 13-10. (*a*) Free body of planet 4; (*b*) free body of arm 3; (*c*) free body of sun gear 2; and (*d*) free body of ring gear 5, which is fastened to earth 0.

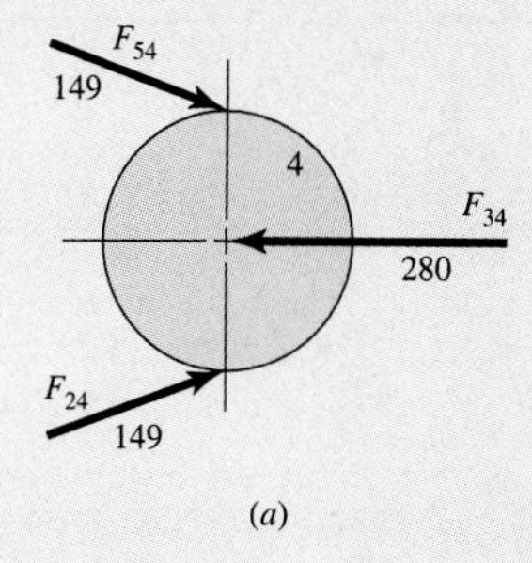

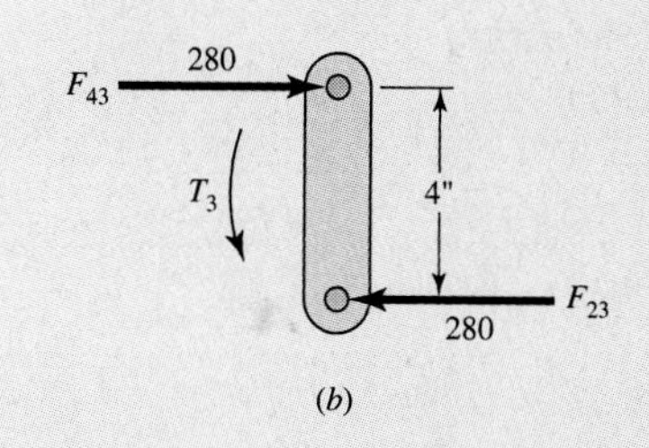

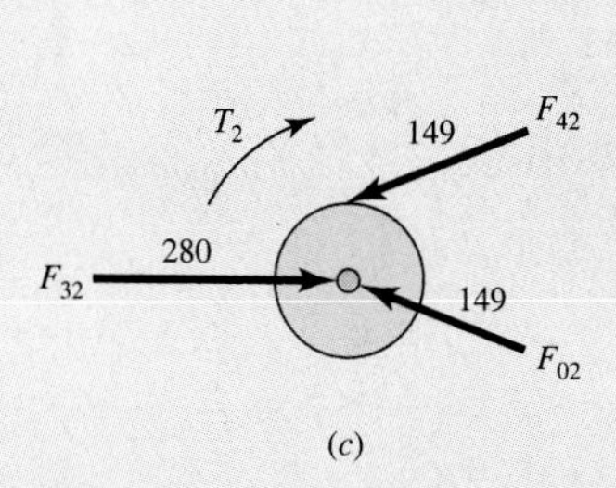

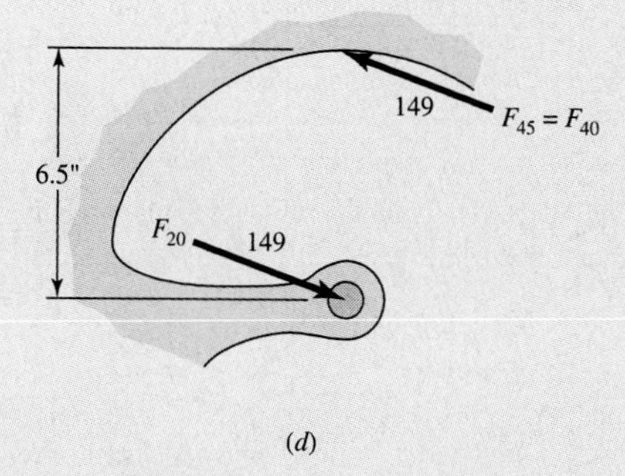

For bearing *C*

$$N^* = N_j + N_b - 2N_F = n_{43} + n_3 - 2n_3 = 438.75 + (-168.75) - 2(-168.75)$$
$$= 607.5 \text{ r/min} \qquad \text{(cw)}$$

(*c*) For boundary lubricated bearings the bearing wear responds to the relative speed between the journal and the bushing. The angular speed of the load vector converts the wear pattern from invasion of the journal into the bushing, to equal wear all around the periphery of the bushing. Time-to-wear equations for this case were not developed. For bearing *A*

$$N = N_j - N_b = n_2 - 0 = -900 \text{ r/min with load of 149 lbf}$$

For bearing *B*

$$N = N_j - N_b = n_{23} = -731.25 \text{ r/min with load of 149 lbf}$$

For bearing *C*

$$N = N_j - N_b = n_{43} = 438.75 \text{ r/min with load of 280 lbf}$$

CASE STUDY 13a They Also Serve Who Only Stand and Point

It has been suggested that the Great Wall of China was built along the climatological contour line marking the farthest consistent northward advance of the monsoons, partitioning agricultural land from grazing land. The necessities of life created different cultures. The defense of cropland to the south of the contour required in-place fortifications, hence the wall

Figure CS13–1

Photograph of the Smithsonian model of the south-pointing chariot. *(Courtesy of the Smithsonian Institution.)*

was begun with earthwork and eventually finished in stone. Food resources to the north of the contour were defended by maneuver, as grazing animals could be moved. The style of warfare depended on who was attacking.

Any northward military excursion faced the problem of direction on seemingly featureless plains, plagued by overcast skies both day and night and battle smoke screens during engagements. There were reconnaissance parties to be vectored out and returned, segmentation of forces before attacking, and messages to be sent back.

About 2600 B.C. Emperor Huang-Ti[6] navigated by a south-pointing chariot,[7] which predated the magnetic compass. A model on exhibition at the Smithsonian Institution is shown in Fig. CS13–1. Gardner and Bagci discuss the mechanism.[8] A section drawing expressed in terms of present-day artifacts is shown in Fig. CS13–2.

Our point is not the details and analysis of this two-wheel, 12-gear mechanism containing a bevel-gear differential, but do consult the references, then ponder how this device came into being. It was created by people with severe limitations on materials and executable geometric forms; they had no algebra as we know it to couple algebraic sign and sense of relative rotation. The output of this mechanism was no angular displacement, whatever the input angular displacement. The concept of zero was yet to come, as was the equal sign.

---

6. H. D. Gardner, "The Mechanism of China's South-Pointing Carriage," *Journal of the Institute of Navigation*, vol. 40, no. 1, 1993.

7. R. L. French, "Ancient Chinese South-Pointing Chariot: World's First Vehicular Navigation System," *Journal of the Institute of Navigation*, vol. 30, no. 1, 1988, p. 8.

8. Cemil Bagci, "The Elementary Theory for the Synthesis of Constant-Direction Pointing Chariots (or Rotational Neutralizers)," *Gear Technology*, November/December 1988, pp. 31–35.

Figure CS13–2

Sketch identifying the gear trains expressed in terms of bevel and spur gears. *(Courtesy of Gear Technology, Randall Publishing, Inc., Elk Grove Village, Illinois.)*

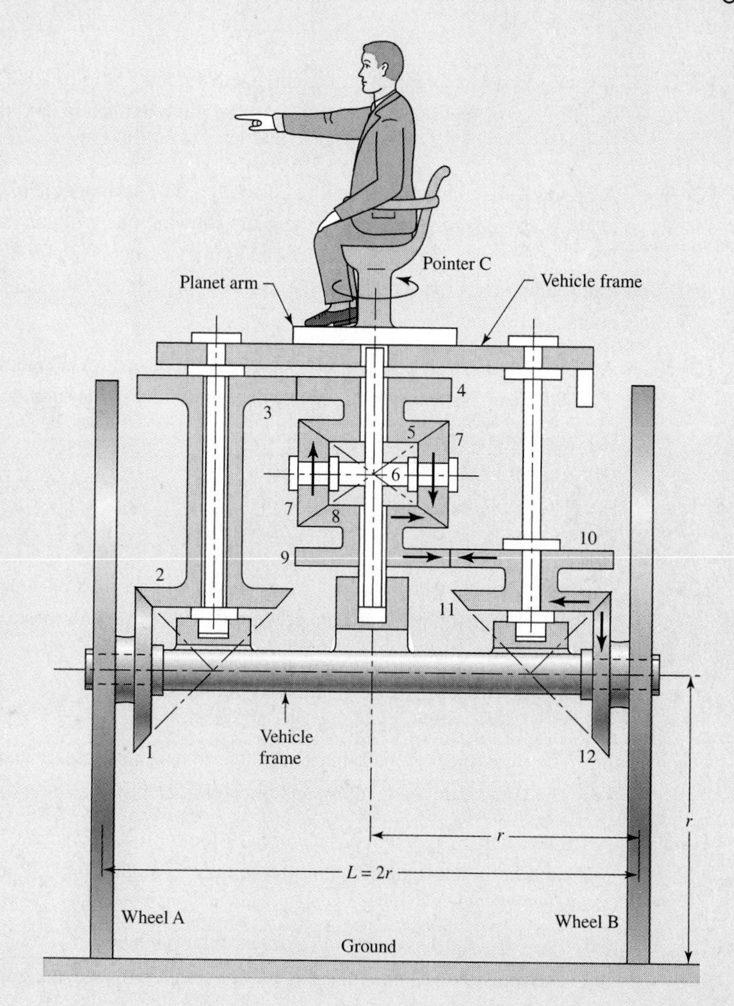

Students might consider that in the social fabric of the time partial credit was not awarded, and grades were final. The pointing figure could be placed in any direction, which would be maintained. The Chinese adopted south as their cardinal direction, whereas Europeans chose north for the same purpose.

This engineering triumph was rendered obsolete by the invention of the magnetic compass. The first written Chinese record of the magnetic compass was in A.D. 1080, and Mediterranean seafarers were using it in the twelfth century. Today, every infantrymember carries a compass and can navigate day or night.

PROBLEMS

13–1 A 17-tooth spur pinion has a diametral pitch of 8, runs at 1120 r/min, and drives a gear at a speed of 544 r/min. Find the number of teeth on the gear and the theoretical center-to-center distance.

13–2 A 15-tooth spur pinion has a module of 3 mm and runs at a speed of 1600 r/min. The driven gear has 60 teeth. Find the speed of the driven gear, the circular pitch, and the theoretical center-to-center distance.

ANALYSIS

13–3 A spur gearset has a module of 4 mm and a velocity ratio of 2.80. The pinion has 20 teeth. Find the number of teeth on the driven gear, the pitch diameters, and the theoretical center-to-center distance.

ANALYSIS

13–4 A 21-tooth spur pinion mates with a 28-tooth gear. The diametral pitch is 3 teeth/in and the pressure angle is 20°. Make a drawing of the gears showing one tooth on each gear. Find and tabulate the following results: the addendum, dedendum, clearance, circular pitch, tooth thickness, and base-circle diameters; the lengths of the arc of approach, recess, and action; and the base pitch and contact ratio.

ANALYSIS

13–5 A 20° straight-tooth bevel pinion having 14 teeth and a diametral pitch of 6 teeth/in drives a 32-tooth gear. The two shafts are at right angles and in the same plane. Find:

(*a*) The cone distance
(*b*) The pitch angles
(*c*) The pitch diameters
(*d*) The face width

ANALYSIS

13–6 A parallel helical gearset uses a 17-tooth pinion driving a 34-tooth gear. The pinion has a right-hand helix angle of 30°, a normal pressure angle of 20°, and a normal diametral pitch of 5 teeth/in. Find:

(*a*) The normal, transverse, and axial circular pitches
(*b*) The normal base circular pitch
(*c*) The transverse diametral pitch and the transverse pressure angle
(*d*) The addendum, dedendum, and pitch diameter of each gear

ANALYSIS

13–7 A parallel helical gearset consists of a 19-tooth pinion driving a 57-tooth gear. The pinion has a left-hand helix angle of 20°, a normal pressure angle of $14\frac{1}{2}°$, and a normal diametral pitch of 10-teeth/in. Find:

(*a*) The normal, transverse, and axial circular pitches
(*b*) The transverse diametral pitch and the transverse pressure angle
(*c*) The addendum, dedendum, and pitch diameter of each gear

ANALYSIS

13–8 In Ex. 13–1 find:

(*a*) The smallest pinion tooth count that will run with itself
(*b*) The smallest pinion tooth count at this ratio
(*c*) The smallest pinion that will run with a rack
(*d*) The largest gear tooth count possible with this pinion

ANALYSIS

13–9 For Ex. 13–2 find:

(*a*) The smallest pinion tooth count that will run with itself
(*b*) The smallest pinion tooth count that will run with a rack
(*c*) The largest gear tooth count possible with this pinion

ANALYSIS

13–10 The decision has been made to use $\phi_t = 20°$, $P_t = 6$ teeth/in, and $\psi = 30°$ for a 2 : 1 reduction. Choose a suitable pinion and gear tooth count to avoid interference.

ANALYSIS

13–11 Repeat Problem 13–10 with a 6 : 1 reduction.

13–12 By employing a pressure angle larger than standard, it is possible to use fewer pinion teeth, and hence obtain smaller gears without undercutting during teeth generation. If the gears are spur gears, what is the smallest possible pressure angle that can be obtained without undercutting for a 9-tooth pinion to mesh with a rack?

13–13 A parallel-shaft gearset consists of an 18-tooth helical pinion driving a 32-tooth gear. The pinion has a left-hand helix angle of 25°, a normal pressure angle of 20°, and a normal module of 3 mm. Find:

(*a*) The normal, transverse, and axial circular pitches
(*b*) The transverse module and the transverse pressure angle
(*c*) The pitch diameters of the two gears

13–14 The double-reduction helical gearset shown in the figure is driven through shaft *a* at a speed of 900 r/min. Gears 2 and 3 have a normal diametral pitch of 10 teeth/in, a 30° helix angle, and a normal pressure angle of 20°. The second pair of gears in the train, gears 4 and 5, have a normal diametral pitch of 6 teeth/in, a 25° helix angle, and a normal pressure angle of 20°. The tooth numbers are: $N_2 = 14$, $N_3 = 54$, $N_4 = 16$, $N_5 = 36$. Find:

(*a*) The directions of the thrust force exerted by each gear upon its shaft
(*b*) The speed and direction of shaft *c*
(*c*) The center distance between shafts

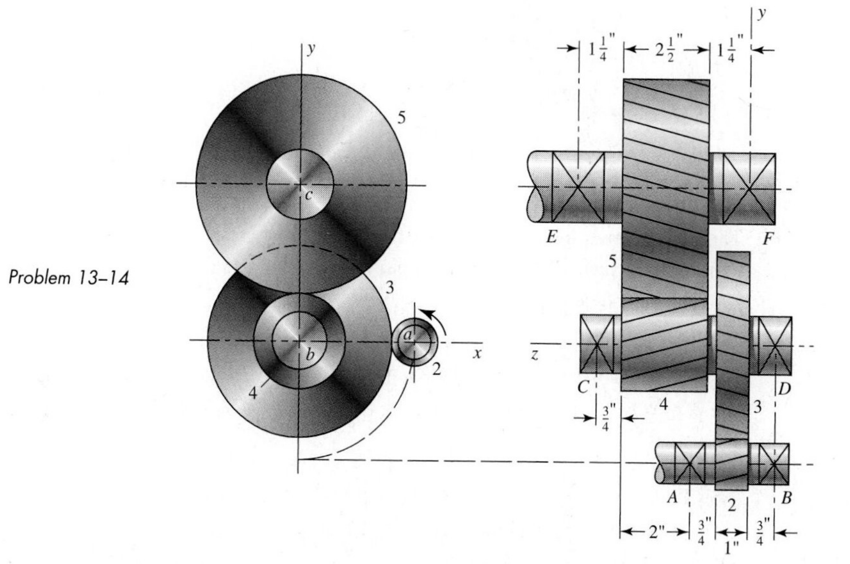

Problem 13–14

13–15 Shaft *a* in the figure rotates at 600 r/min in the direction shown. Find the speed and direction of rotation of shaft *d*.

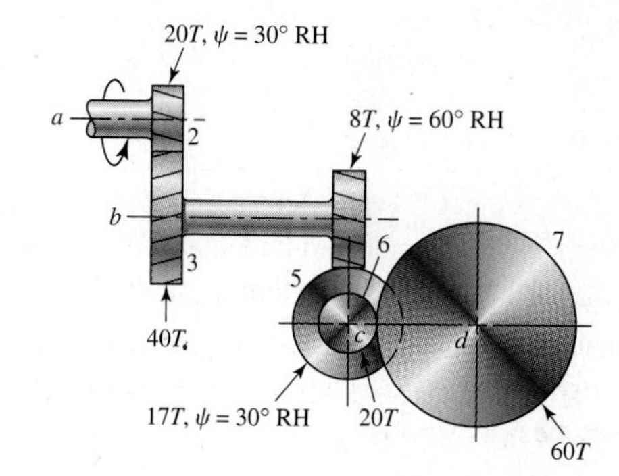

Problem 13–15

13–16 The mechanism train shown consists of an assortment of gears and pulleys to drive gear 9. Pulley 2 rotates at 1200 r/min in the direction shown. Determine the speed and direction of rotation of gear 9.

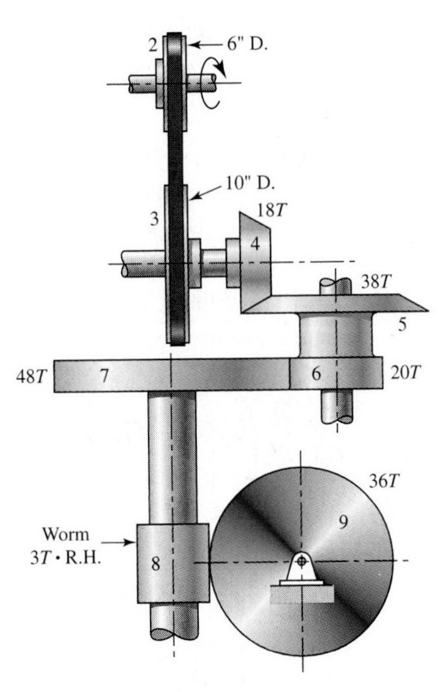

Problem 13–16

13–17 The figure shows a gear train consisting of a pair of helical gears and a pair of miter gears. The helical gears have a $17\frac{1}{2}^\circ$ normal pressure angle and a helix angle as shown. Find:

(a) The speed of shaft c

(b) The distance between shafts a and b

(c) The diameter of the miter gears

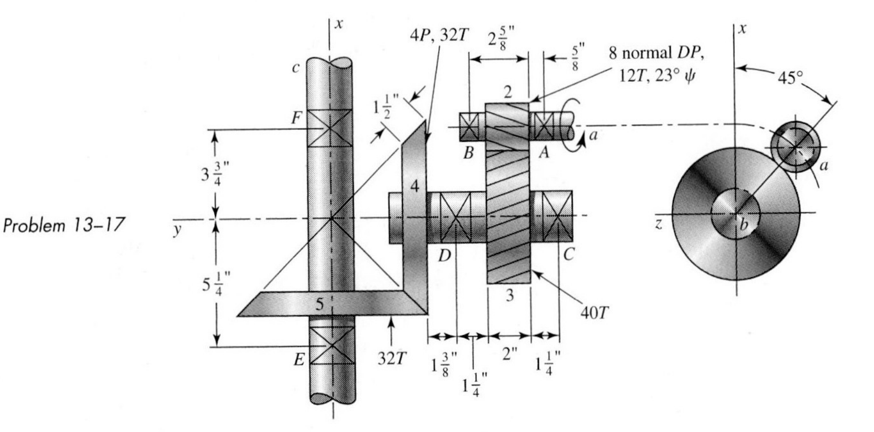

Problem 13–17

ANALYSIS

13–18 The tooth numbers for the automotive differential shown in the figure are $N_2 = 17$, $N_3 = 54$, $N_4 = 11$, $N_5 = N_6 = 16$. The drive shaft turns at 1200 r/min.

(a) What are the wheel speeds if the car is traveling in a straight line on a good road surface?

(b) Suppose the right wheel is jacked up and the left wheel resting on a good road surface. What is the speed of the right wheel?

(*c*) Suppose, with a rear-wheel drive vehicle, the auto is parked with the right wheel resting on a wet icy surface. Does the answer to part (*b*) give you any hint as to what would happen if you started the car and attempted to drive on?

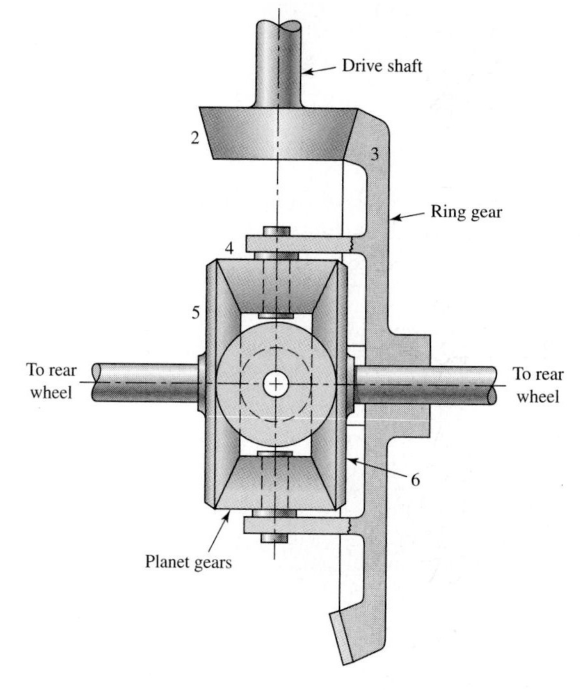

Problem 13–18

13–19 The figure illustrates an all-wheel drive concept using three differentials, one for the front axle, another for the rear, and the third connected to the drive shaft.

(*a*) Explain why this concept may allow greater acceleration.

(*b*) Suppose either the center of the rear differential, or both, can be locked for certain road conditions. Would either or both of these actions provide greater traction? Why?

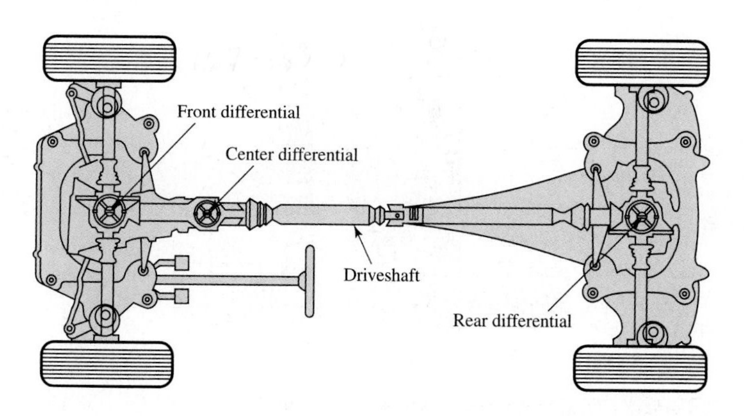

Problem 13–19
The Audi "Quattro concept," showing the three differentials which provide permanent all-wheel drive. *(By permission, Audi of America, Inc., Troy, Michigan.)*

13–20 In the reverted planetary train illustrated, find the speed and direction of rotation of the arm if gear 2 is unable to rotate and gear 6 is driven at 12 r/min in the clockwise direction. Let gear 2 be driven at 180 r/min ccw while gear 6 is held stationary. What is the speed of the arm?

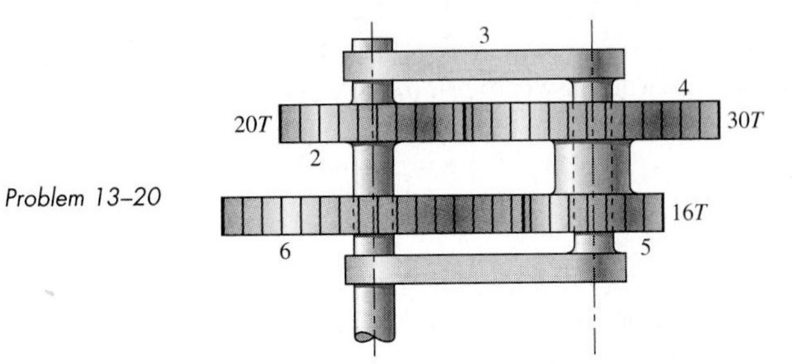

Problem 13–20

ANALYSIS

13–21 Tooth numbers for the gear train shown in the figure are $N_2 = 12$, $N_3 = 12$, and $N_4 = 12$. How many teeth must internal gear 5 have? Suppose gear 5 is fixed. What is the speed of the arm if shaft a rotates counterclockwise at 320 r/min?

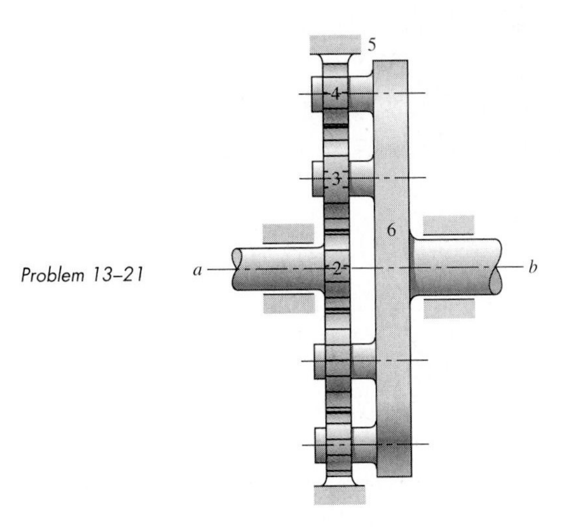

Problem 13–21

ANALYSIS

13–22 The tooth numbers for the gear train illustrated are $N_2 = 24$, $N_3 = 18$, $N_4 = 30$, $N_6 = 36$, and $N_7 = 54$. Gear 7 is fixed. If shaft b is turned through 5 revolutions, how many turns will shaft a make?

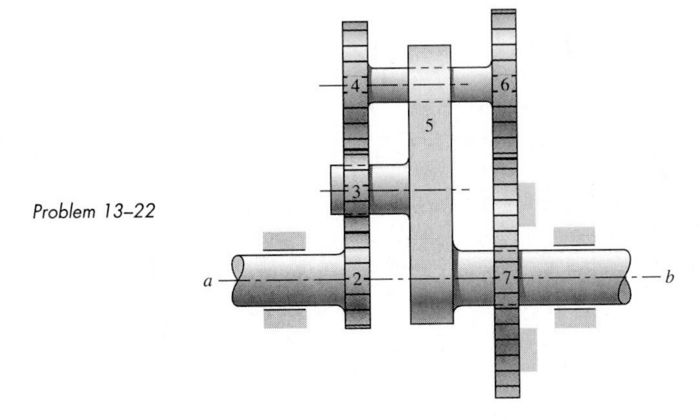

Problem 13–22

ANALYSIS

13–23 By using nonstandard gears, it is possible to mate a $99T$ with a $100T$ gear at the same distance between centers as would be required to mate a $100T$ gear with a $101T$ gear. The planetary train shown in the figure is based on this idea.

(*a*) Find the ratio of the speed of the output shaft to the speed of the input shaft.

(*b*) The housing for this planetary train is cylindrical, with the axis of the cylinder coincident with the axes of the input and output shafts. If the pitch of gears 4 and 5 is 10 teeth/in of pitch diameter, and if these gears have standard addenda, what should be the inside diameter of this housing?

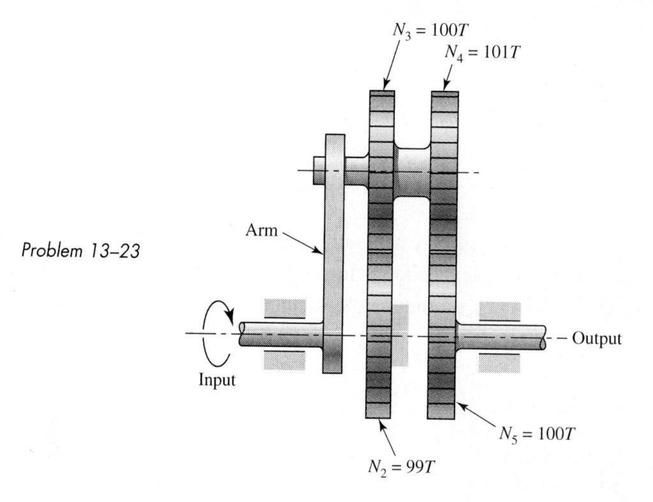

Problem 13–23

13–24 The epicyclic train shown in the figure has the arm attached to shaft a, and sun gear 2 to shaft b. Gear 5, with 111 teeth, is an internal gear and is part of the frame. The two planets, gears 3 and 4, are both fixed to the same planet shaft. If this train is used as an in-line speed reducer, which is the input shaft, a or b? Will both shafts then rotate in the same or in the opposite directions?

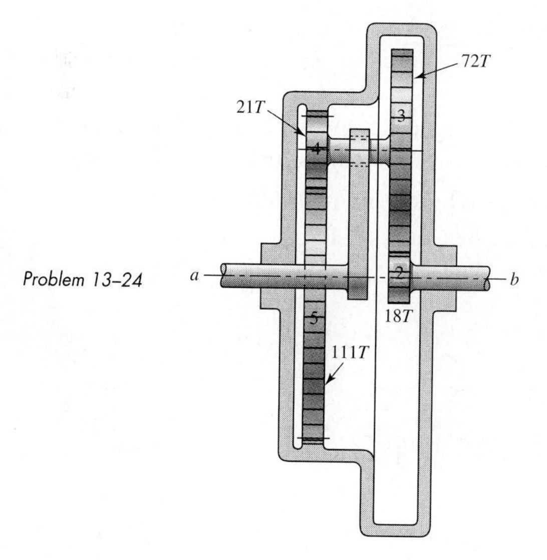

Problem 13–24

13–25 The figure shows a speed reducer in which the input shaft a is in line with output shaft b. The tooth numbers are $N_2 = 24$, $N_3 = 18$, $N_5 = 22$, and $N_6 = 64$. Find the ratio of the output speed to the input speed. Will both shafts rotate in the same direction? Note that gear 6 is a fixed internal gear.

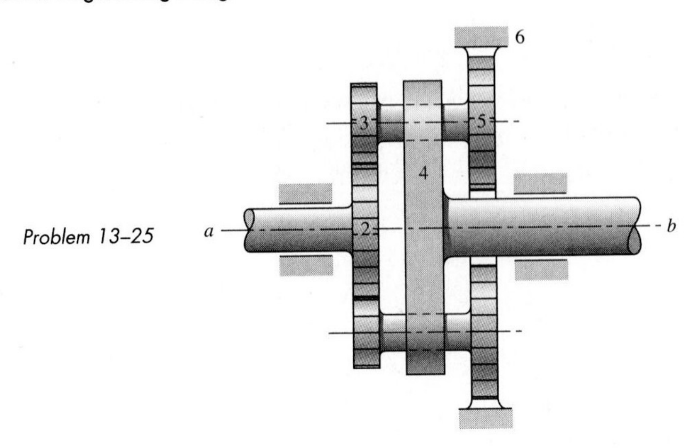

Problem 13–25

13–26 The speed reducer shown in the figure has pinion 2 fixed. The planets are gears 3 and 4, both fixed to the planet shaft. Sun gear 5 is attached to the output shaft. Input shaft a drives the arm. Find the overall speed ratio of this reducer, and the direction of rotation of the output shaft.

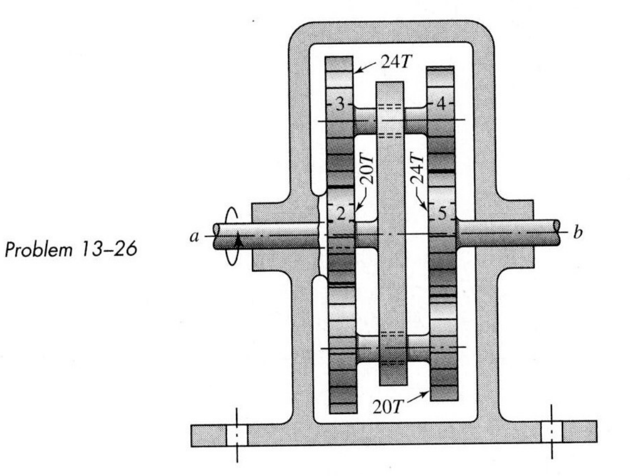

Problem 13–26

ANALYSIS

13–27 Shaft a in the figure has a power input of 75 kW at a speed of 1000 r/min in the counterclockwise direction. The gears have a module of 5 mm and a 20° pressure angle. Gear 3 is an idler.

(a) Find the force F_{3b} that gear 3 exerts against shaft b.

(b) Find the torque T_{4c} that gear 4 exerts on shaft c.

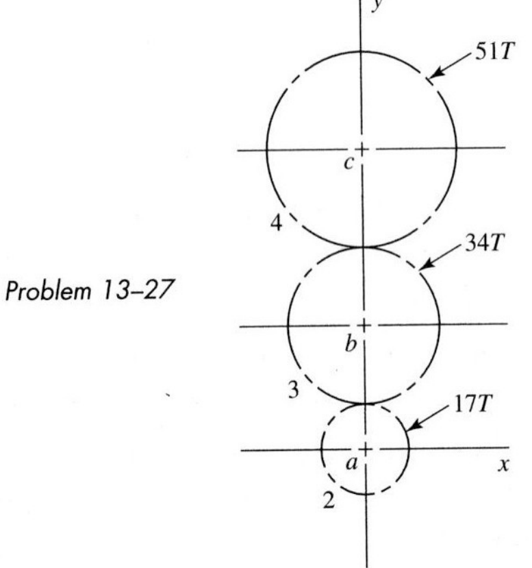

Problem 13–27

13–28 The 24T 6-pitch 20° pinion 2 shown in the figure rotates clockwise at 1000 r/min and is driven at a power of 25 hp. Gears 4, 5, and 6 have 24, 36, and 144 teeth, respectively. What torque can arm 3 deliver to its output shaft? Draw free-body diagrams of the arm and of each gear and show all forces which act upon them.

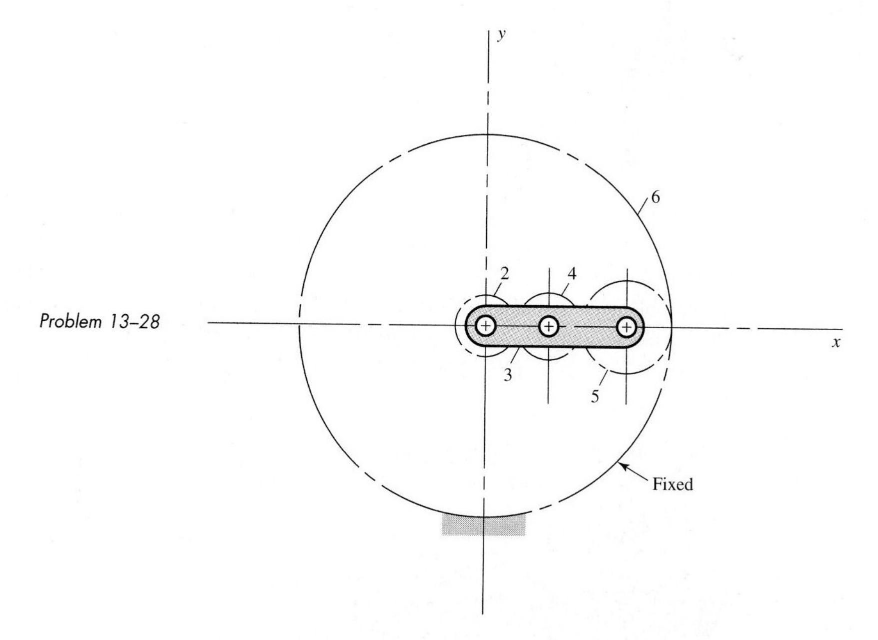

Problem 13–28

13–29 The gears shown in the figure have a diametral pitch of 2 teeth per inch and a 20° pressure angle. The pinion rotates at 1800 r/min clockwise and transmits 200 hp through the idler pair to gear 5 on shaft c. What forces do gears 3 and 4 transmit to the idler shaft?

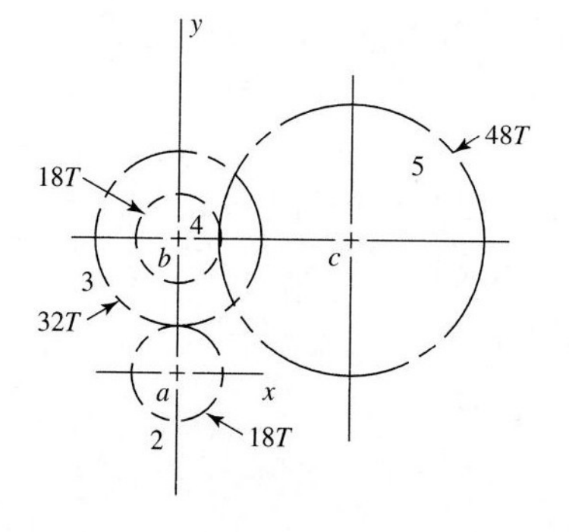

Problem 13–29

13–30 The figure shows a pair of shaft-mounted spur gears having a diametral pitch of 5 teeth/in with an 18-tooth 20° pinion driving a 45-tooth gear. The horsepower input is 32 maximum at 1800 r/min. Find the direction and magnitude of the maximum forces acting on bearings A, B, C, and D. The reliability goal for bearing ensemble is 0.999 for 30 000-h life.

(*a*) Find the direction and magnitude of the maximum forces on the bearings.

(*b*) Find the direction and magnitude of the forces on the mounting screws of the 12 × 15 × 10-in gearbox.

(*c*) Select suitable rolling-contact bearings.

(*d*) Select suitable tiedown screws or bolts.

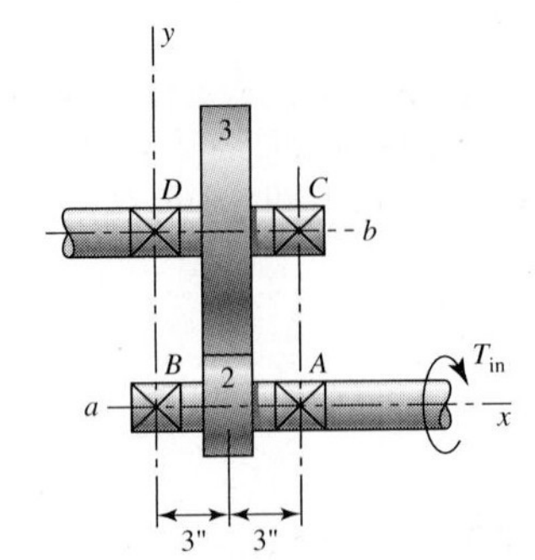

Problem 13–30

13–31 The figure shows the electric-motor frame dimensions for a 30-hp 900 r/min motor. The frame is bolted to its support using four $\frac{3}{4}$-in bolts spaced $11\frac{1}{4}$ in apart in the view shown and 14 in apart when viewed from the end of the motor. A 4-diametral pitch 20° spur pinion having 20 teeth and a face width of 2 in is keyed to the motor shaft. This pinion drives another gear whose axis is in the same xz plane. Determine the maximum shear and tensile forces on the mounting bolts based on 200 percent overload torque. Does the direction of rotation matter?

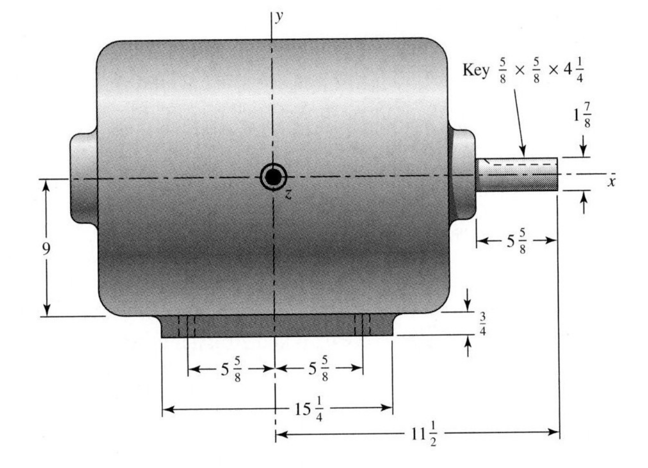

Problem 13–31
NEMA No. 364 frame; dimensions in inches. The z axis is directed out of the paper.

13–32 The figure shows a 16T 20° straight bevel pinion driving a 32T gear, and the location of the bearing centerlines. Pinion shaft a receives 2.5 hp at 240 r/min. Determine the bearing reactions at A and B if A is to take both radial and thrust loads.

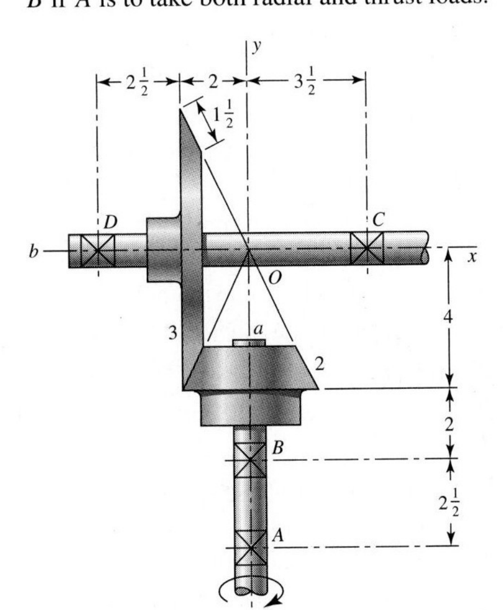

Problem 13–32
Dimensions in inches.

13–33 The figure shows a 10-diametral pitch 15-tooth 20° straight bevel pinion driving a 25-tooth gear. The transmitted load is 30 lb. Find the bearing reactions at C and D on the output shaft if D is to take both radial and thrust loads.

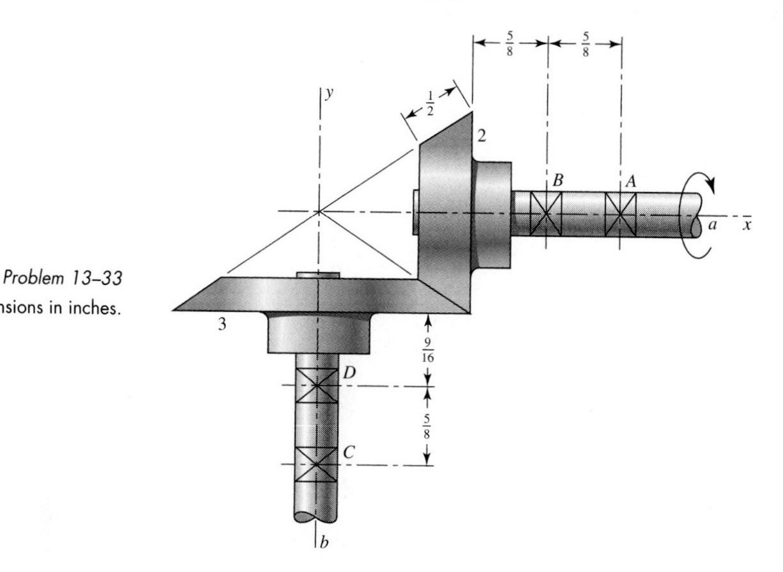

Problem 13–33
Dimensions in inches.

ANALYSIS

13–34 The gears in the two trains shown in the figure have a normal diametral pitch of 4, a normal pressure angle of 20°, and a 30° helix angle. For both gear trains the transmitted load is 800 lb. In part a the pinion rotates counterclockwise about the y axis. Find the force exerted by each gear in part a on its shaft.

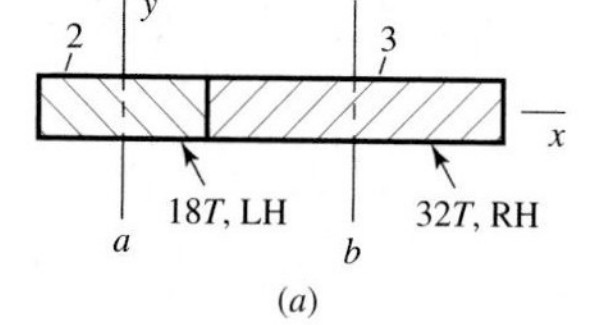

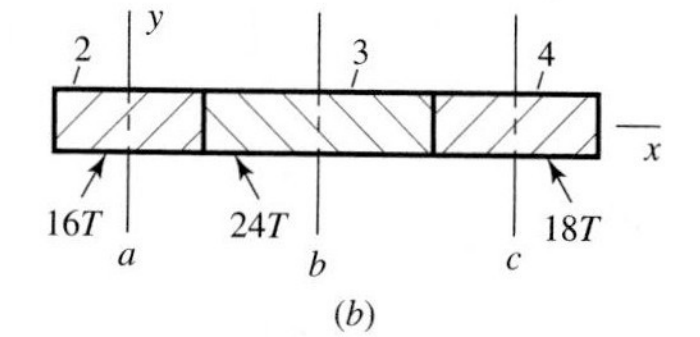

Problem 13–34

13–35 This is a continuation of Prob. 13-34. Here, you are asked to find the forces exerted by gears 2 and 3 on their shafts as shown in part b. Gear 2 rotates clockwise about the y axis. Gear 3 is an idler.

ANALYSIS

13–36 A gear train is composed of four helical gears with the three shaft axes in a single plane, as shown in the figure. The gears have a normal pressure angle of 20° and a 30° helix angle. Shaft b is an idler and the transmitted load acting on gear 3 is 500 lb. The gears on shaft b both have a normal diametral pitch of 7 teeth/in and have 54 and 14 teeth, respectively. Find the forces exerted by gears 3 and 4 on shaft b.

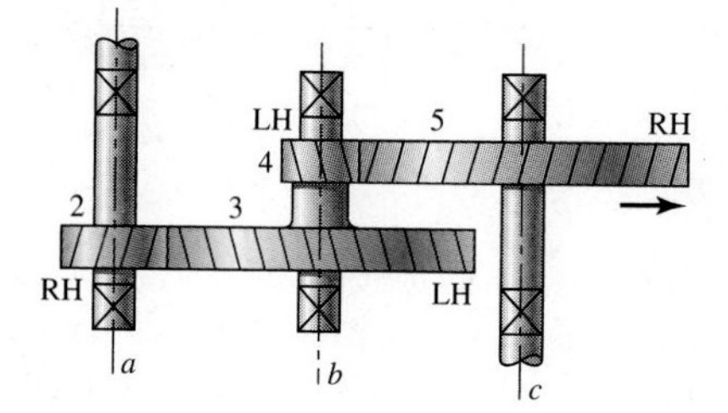

Problem 13–36

13–37 In the figure for Prob. 13-30, pinion 2 is to be a right-hand helical gear having a helix angle of 30°, a normal pressure angle of 20°, 16 teeth, and a normal diametral pitch of 6 teeth/in. A 25-hp motor drives shaft a at a speed of 1720 r/min clockwise about the x axis. Gear 3 has 42 teeth. Find the reaction exerted by bearings C and D on shaft b. One of these bearings is to take both radial and thrust loads. This bearing should be selected so as to place the shaft in compression.

ANALYSIS

13–38 Gear 2, in the figure, has 16 teeth, a 20° transverse angle, a 15° helix angle, and a normal diametral pitch of 8 teeth/in. Gear 2 drives the idler on shaft b, which has 36 teeth. The driven gear on shaft c has 28 teeth. If the driver rotates at 1720 r/min and transmits $7\frac{1}{2}$ hp, find the radial and thrust load on each shaft.

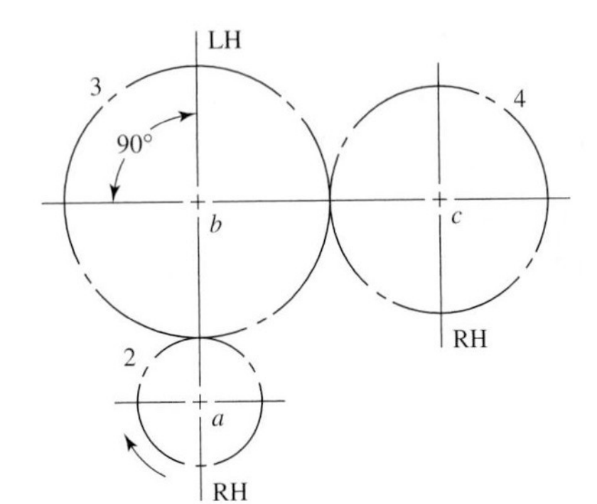

Problem 13–38

ANALYSIS

13–39 The figure shows a double-reduction helical gearset. Pinion 2 is the driver, and it receives a torque of 1200 lb · in from its shaft in the direction shown. Pinion 2 has a normal diametral pitch of 8 teeth/in, 14 teeth, and a normal pressure angle of 20° and is cut right-handed with a helix angle of 30°. The mating gear 3 on shaft b has 36 teeth. Gear 4, which is the driver for the second pair of gears in the train, has a normal diametral pitch of 5 teeth/in, 15 teeth, and a normal pressure angle of 20° and is curt left-handed with a helix angle of 15°. Mating gear 5 has 45 teeth. Find the magnitude and direction of the force exerted by the bearings C and D on shaft b if bearing C can take only radial load while bearing D is mounted to take both radial and thrust load.

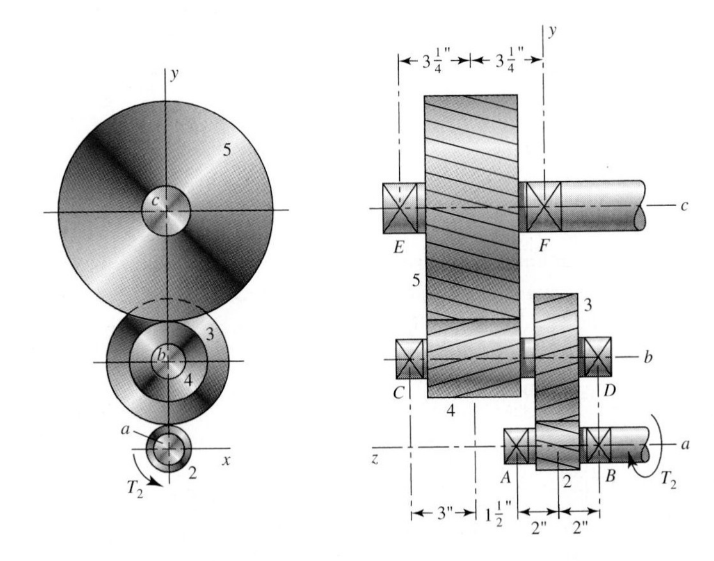

Problem 13–39

ANALYSIS

13–40 A right-hand single-tooth hardened-steel (hardness not specified) worm has a catalog rating of 2000 W at 600 r/min when meshed with a 48-tooth cast-iron gear. The axial pitch of the worm is 25 mm, the normal pressure angle is $14\frac{1}{2}°$, the pitch diameter of the worm is 100 mm, and the face widths of the worm and gear are, respectively, 100 mm and 50 mm. The figure shows bearings

A and *B* on the worm shaft symmetrically located with respect to the worm and 200 mm apart. Determine which should be the thrust bearing, and find the magnitudes and directions of the forces exerted by both bearings.

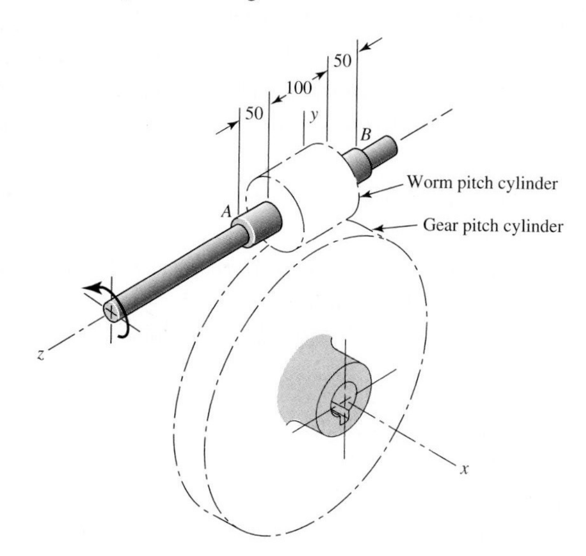

Problem 13–40
Dimensions in millimeters.

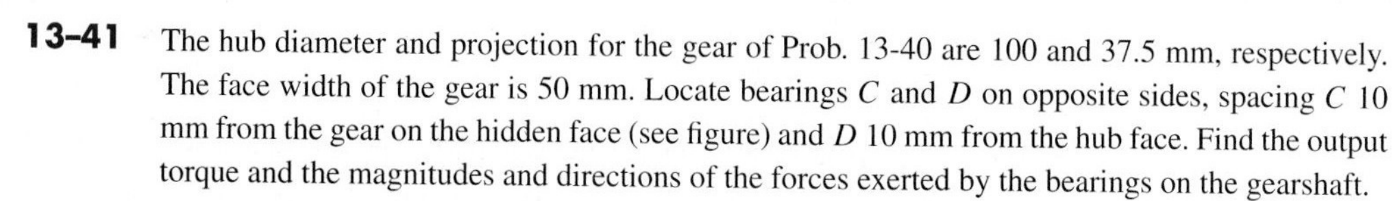

13–41 The hub diameter and projection for the gear of Prob. 13-40 are 100 and 37.5 mm, respectively. The face width of the gear is 50 mm. Locate bearings *C* and *D* on opposite sides, spacing *C* 10 mm from the gear on the hidden face (see figure) and *D* 10 mm from the hub face. Find the output torque and the magnitudes and directions of the forces exerted by the bearings on the gearshaft.

13–42 A 2-tooth left-hand worm transmits $\frac{1}{2}$ hp at 900 r/min to a 36-toth gear having a transverse diametral pitch of 10 teeth/in. The worm has a normal pressure angle of $14\frac{1}{2}^\circ$, a pitch diameter of $1\frac{1}{2}$ in, and a face width of $1\frac{1}{2}$ in. Use a coefficient of friction of 0.05 and find the force exerted by the gear on the worm and the torque input. For the same geometry as shown for Prob. 13-40, the worm velocity is clockwise about the *z* axis.

13–43 The diametral pitch of the teeth in all the spur gears of the epicyclic gear train of Prob. 13-23 is 48 teeth per inch. The pressure angle is 20°. If the 100 : 101 gear mesh is set to the proper center-to-center distance, what is the operative pressure angle of the 99 : 100 gear mesh? Show that the pressure angle ϕ' is given by the equation

$$\phi' = \cos^{-1}\left[\frac{(1+m_G)N_P\cos\phi}{2PC'}\right]$$

13–44 Write a computer program that will analyze a spur-gear or helical-mesh gear, accepting ϕ_n, ψ, P_t, N_P, and N_G; compute m_G, d_P, d_G, p_t, p_n, p_x, and ϕ_t; and give advice as to the smallest tooth count that will allow a pinion to run with itself without interference, run with its gear, and run with a rack. Also have it give the largest tooth count possible with the intended pinion.

13–45 When Prob. 13-44 is a working computer program, verify some of the numbers in Table 13–6.

13–46 Write a computer program to find a single-reduction gear-tooth pair to approximate a specified gear ratio with specified acceptable error in m_G. When your program is working, do the following:

(*a*) For $m_G = R = \sqrt{10}$, with a fractional error $\Delta R/R$ of ± 0.0005, find and specify suitable tooth numbers.

(*b*) For $R = \pi$ with fractional error of $\pm 0.000\,13$, find suitable N_P and N_G and extant error, and specify a tooth count.

(c) For $R = \sqrt{2}$ with fractional error of ± 0.0001, find possible N_P and N_G and extant error. Specify a suitable tooth count.

(d) For $R = 1.234\,567\,9$ and a fractional error of ± 0.0001, find possible extant tooth pairs and their extant error, and specify a suitable tooth count.

DESIGN

13-47 In the previous problems very little has been designed with respect to the gear meshes, gears, shafts, or bearings. It is useful to size a few bearings just to appreciate the impending bulk, as the process of fleshing out an idea develops. Your instructor may assign some design tasks associated with problems already assigned. Design is an iterative process.

ANALYSIS

13-48 A lathe headstock is spur-geared to the lead screw. Assume the lead screw has 6 threads per inch (lead $L = \frac{1}{6}$ in).

(a) Show that to cut a thread on a workpiece chucked in the lathe, the ratio of the spur gearset is NL, where N is the number of threads per inch being cut.

(b) If the number of threads per inch is 20, what is the spur-gear ratio, and which numbers of teeth will work?

(c) Suppose the same lathe is used to cut a metric thread with a pitch p of 2 mm. Is there a spur-gear ratio that will allow this to be done?

(d) Are there any approximations involved?

13-49 Within the constraints $(N_P)_{\min} \le N_P \le (N_P)_{\max}$ and $(N_G)_{\min} \le N_G \le (N_P)_{\max}$, write a computer program to investigate the error found in every quotient N_G/N_P by comparing it with the desired ratio. Then report out all the ratios that are within a specified ΔR of R, error $= R - N_j/N_i$. You can merely list the satisfactory N_P, N_G pairs with accompanying error, or rank the discovered errors, smallest to largest.

DESIGN

13-50 Use the program developed in Prob. 13-49, and specify the best gearset for the following variables:

R	ΔR	$(N_P)_{\min}$	$(N_P)_{\max}$	$(N_G)_{\min}$	$(N_G)_{\max}$
29/23	0.011	10	20	10	20
π	0.01	20	100	20	100
1.234 567 89	0.001	25	110	25	110

14 Spur and Helical Gears

14–1 The Lewis Bending Equation **898**

14–2 Surface Durability **907**

14–3 AGMA Stress Equations **909**

14–4 AGMA Strength Equations **910**

14–5 Geometry Factors I and J (Z_I and Y_J) **915**

14–6 The Elastic Coefficient C_p (Z_E) **920**

14–7 Dynamic Factor K'_V **920**

14–8 Overload Factor K_O **922**

14–9 Surface Condition Factors C_f and Z_R **922**

14–10 Size Factor K_s **923**

14–11 Load-Distribution Factor K_m or K_H **923**

14–12 Hardness-Ratio Factor C_H **924**

14–13 Load Cycles Factors Y_N and Z_N **926**

14–14 Reliability Factors K_R and Y_Z **927**

14–15 Temperature Factors K_T and Y_θ **928**

14–16 Rim-Thickness Factor K_B **928**

14–17 Safety Factors S_F and S_H **929**

14–18 Analysis **929**

14–19 An Adequacy Assessment of a Gear Mesh **940**

14–20 Design of a Gear Mesh **942**

This chapter is devoted primarily to analysis and design of spur and helical gears to resist bending failure of the teeth as well as pitting failure of tooth surfaces. Failure by bending will occur when the significant tooth stress equals or exceeds either the yield strength or the bending endurance strength. A surface failure occurs when the significant contact stress equals or exceeds the surface endurance strength. The first two sections present a little of the history of the analyses from which current methodology developed. Thus new ideas are introduced while they are still simple, and applications of fundamentals from the first part of this book is illustrated. As AGMA methodology is introduced, the reader can appreciate how these fundamentals have been adapted to a protocol that has the ability to address more complicated circumstances.

The American Gear Manufacturers Association[1] (AGMA) has for many years been the responsible authority for the dissemination of knowledge pertaining to the design and analysis of gearing. The methods this organization presents are in general use in the United States when strength and wear are primary considerations. In view of this fact it is important that the AGMA approach to the subject be presented here.

The general AGMA approach requires a great many charts and graphs—too many for a single chapter in this book. We have omitted many of these here by choosing a single pressure angle and by using only full-depth teeth. This simplification reduces the complexity but does not prevent the development of a basic understanding of the approach. Furthermore, the simplification makes possible a better development of the fundamentals and hence should constitute an ideal introduction to the use of the general AGMA method.[2] Sections 14–1 and 14–2 are elementary and serve as an examination of the foundations of the AGMA method. Table 14–1 is largely AGMA nomenclature.

14–1 The Lewis Bending Equation

Wilfred Lewis introduced an equation for estimating the bending stress in gear teeth in which the tooth form entered into the formulation. The equation, announced in 1892, still remains the basis for most gear design today.

To derive the basic Lewis equation, refer to Fig. 14–1*a*, which shows a cantilever of cross-sectional dimensions F and t, having a length l and a load W^t, uniformly distributed across the face width F. The section modulus I/c is $Ft^2/6$, and therefore the bending stress is

$$\sigma = \frac{M}{I/c} = \frac{6W^t l}{Ft^2} \tag{a}$$

Gear designers denote the components of gear-tooth forces as W_t, W_r, W_a or W^t, W^r,

1. 1500 King Street, Suite 201, Alexandria, VA 22314.

2. The standards ANSI/AGMA 2001-C95 (revised AGMA 2001-B88) and ANSI/AGMA 2101-C95 (metric edition of ANSI/AGMA 2001-C95), *Fundamental Rating Factors and Calculation Methods for Involute Spur and Helical Gear Teeth*, are used in this chapter. The use of American National Standards is completely voluntary; their existence does not in any respect preclude people, whether they have approved the standards or not, from manufacturing, marketing, purchasing, or using products, processes, or procedures not conforming to the standards.

The American National Standards Institute does not develop standards and will in no circumstances give an interpretation of any American National Standard. Requests for interpretation of these standards should be addressed to the American Gear Manufacturers Association. [Tables or other self-supporting sections may be quoted or extracted in their entirety. Credit line should read: "Extracted from ANSI/AGMA Standard 2001-C95 or 2101-C95 *Fundamental Rating Factors and Calculation Methods for Involute Spur and Helical Gear Teeth*" with the permission of the publisher, American Gear Manufacturers Association, 1500 King Street, Suite 201, Alexandria, Virginia 22314.] The foregoing is adapted in part from the ANSI foreword to these standards.

Table 14–1

Symbols, Their Names, and Locations*

Symbol	Name	Where Found
b	Net width of face of narrowest member	Eq. (14–16)
C_e	Mesh alignment correction factor	Eq. (14–35)
C_f	Surface condition factor	Eq. (14–16)
C_H	Hardness-ratio factor	Eq. (14–18)
C_{ma}	Mesh alignment factor	Eq. (14–34)
C_{mc}	Load correction factor	Eq. (14–31)
C_{mf}	Face load-distribution factor	Eq. (14–30)
C_p	Elastic coefficient	Eq. (14–13)
C_{pf}	Pinion proportion factor	Eq. (14–32)
C_{pm}	Pinion proportion modifier	Eq. (14–33)
d	Operating pitch diameter of pinion	Eq. (14–1)
d_P	Pitch diameter, pinion	Eq. (14–22)
d_G	Pitch diameter, gear	Eq. (14–22)
E	Modulus of elasticity	Eq. (14–10)
F	Net face width of narrowest member	Eq. (14–15)
f_P	Pinion surface finish	Fig. 14–13
H	Power	Fig. 14–17
H_B	Brinell hardness	Ex. 14–3
H_{BG}	Brinell hardness of gear	Sec. 14–12
H_{BP}	Brinell hardness of pinion	Sec. 14–12
hp	Horsepower	Ex. 14–1
h_t	Gear-tooth whole depth	Sec. 14–16
I	Geometry factor of pitting resistance	Eq. (14–16)
J	Geometry factor for bending strength	Eq. (14–15)
K	Contact load factor for pitting resistance	Eq. (7–72)
K_B	Rim-thickness factor	Eq. (14–40)
K_f	Fatigue stress-concentration factor	Eq. (14–9)
K_m	Load-distribution factor	Eq. (14–30)
K_O	Overload factor	Eq. (14–15)
K_R	Reliability factor	Eq. (14–17)
K_s	Size factor	Sec. 14–10
K_T	Temperature factor	Eq. (14–17)
$K_V{}'$	Dynamic factor	Eq. (14–17)
m	Metric module	Eq. (14–15)
m_B	Backup ratio	Eq. (14–34)
m_G	Gear ratio (never less than 1)	Eq. (14–22)
m_N	Load-sharing ratio	Eq. (14–21)
N	Number of stress cycles	Fig. 14–14
N_G	Number of teeth on gear	Eq. (14–22)
N_P	Number of teeth on pinion	Eq. (14–22)
n	Speed	Ex. 14–1
n_P	Pinion speed	Ex. 14–4
P	Diametral pitch	Eq. (14–2)
P_d	Diametral pitch of pinion	Eq. (14–15)
p_N	Normal base pitch	Eq. (14–24)
P_n	Normal circular pitch	Eq. (14–24)
P_x	Axial pitch	Eq. (14–19)
Q_V	Transmission accuracy level number	Eq. (14–29)
R	Reliability	Eq. (14–38)
R_a	Root-mean-squared roughness	Fig. 14–13
r_f	Tooth fillet radius	Fig. 14–1
r_G	Pitch-circle radius, gear	In standard

Table 14–1

Symbols, Their Names, and Locations*, (*continued*)

Symbol	Name	Where Found
r_P	Pitch-circle radius, pinion	In standard
r_{bP}	Pinion base-circle radius	Eq. (14–25)
r_{bG}	Gear base-circle radius	Eq. (14–25)
S_C	Buckingham surface endurance strength	Ex. 14–3
S_c	AGMA surface endurance strength	Eq. (14–18)
S_t	AGMA bending strength	Eq. (14–17)
S	Bearing span	Fig. 14–10
S_1	Pinion offset from center span	Fig. 14–10
S_F	Safety factor—bending	Eq. (14–41)
S_H	Safety factor—pitting	Eq. (14–42)
W^t or W_t[†]	Transmitted load	Fig. 14–1
Y_N	Stress cycle factor for bending strength	Fig. 14–14
Z_N	Stress cycle factor for pitting resistance	Fig. 14–15
β	Exponent	Eq. (14–44)
σ	Bending stress	Eq. (14–2)
σ_C	Contact stress from Hertzian relationships	Eq. (14–14)
σ_c	Contact stress from AGMA relationships	Eq. (14–16)
σ_{all}	Allowable bending stress	Eq. (14–17)
$\sigma_{c,all}$	Allowable contact stress, AGMA	Eq. (14–18)
ϕ	Pressure angle	Eq. (14–12)
ϕ_t	Transverse pressure angle	Eq. (14–23)
ψ	Helix angle at standard pitch diameter	Ex. 14–2

* Because ANSI/AGMA 2001-C95 introduced a significant amount of new nomenclature, this summary and references is provided for use until the reader's vocabulary has grown.

† See preference rationale following Eq. (*a*), Sec. 14–1.

Figure 14–1

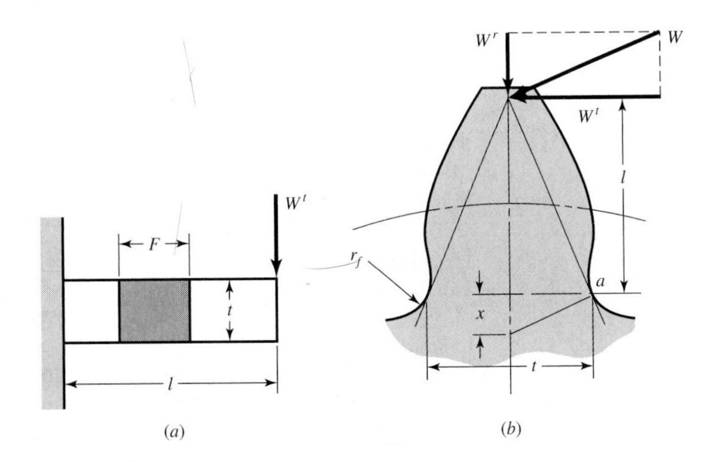

W^a interchangeably. The latter notation leaves room for post-subscripts essential to free-body diagrams. For instance, for gears 2 and 3 in mesh, W^t_{23} is the transmitted force of body 2 on body 3, and W^t_{32} is the transmitted force of body 3 on body 2. When working with double- or triple-reduction speed reducers, this notation is compact and essential to clear thinking. Since gear-force components rarely take exponents, this causes no complication. Pythagorean combinations, if necessary, can be treated with parentheses or avoided by expressing the relations trigonometrically.

Referring now to Fig. 14–1b, we assume that the maximum stress in a gear tooth occurs at point a. By similar triangles, you can write

$$\frac{t/2}{x} = \frac{l}{t/2} \quad \text{or} \quad x = \frac{t^2}{4l} \tag{b}$$

By rearranging Eq. (a),

$$\sigma = \frac{6W^t l}{Ft^2} = \frac{W^t}{F}\frac{1}{t^2/6l} = \frac{W^t}{F}\frac{1}{t^2/4l}\frac{1}{\frac{4}{6}} \tag{c}$$

If we now substitute the value of x from Eq. (b) in Eq. (c) and multiply the numerator and denominator by the circular pitch p, we find

$$\sigma = \frac{W^t p}{F\left(\frac{2}{3}\right)xp} \tag{d}$$

Letting $y = 2x/3p$, we have

$$\sigma = \frac{W^t}{Fpy} \tag{14–1}$$

This completes the development of the original Lewis equation. The factor y is called the *Lewis form factor*, and it may be obtained by a graphical layout of the gear tooth or by digital computation.

In using this equation, most engineers prefer to employ the diametral pitch in determining the stresses. This is done by substituting $P = \pi/p$ and $Y = \pi y$ in Eq. (14–1). This gives

$$\sigma = \frac{W^t P}{FY} \tag{14–2}$$

where

$$Y = \frac{2xP}{3} \tag{14–3}$$

The use of this equation for Y means that only the bending of the tooth is considered and that the compression due to the radial component of the force is neglected. Values of Y obtained from this equation are tabulated in Table 14–2.

The use of Eq. (14–3) also implies that the teeth do not share the load and that the greatest force is exerted at the tip of the tooth. But we have already learned that the contact ratio should be somewhat greater than unity, say about 1.5, to achieve a quality gearset. If, in fact, the gears are cut with sufficient accuracy, the tip-load condition is not the worst, because another pair of teeth will be in contact when this condition occurs. Examination of run-in teeth will show that the heaviest loads occur near the middle of the tooth. Therefore the maximum stress probably occurs while a single pair of teeth is carrying the full load, at a point where another pair of teeth is just on the verge of coming into contact.

Dynamic Effects

AGMA standards ANSI/AGMA 2110-C95 and 2101-C95 contain this caution:

> Dynamic factor K_V has been redefined as the reciprocal of that used in previous AGMA standards. It is now greater than 1.0. In earlier AGMA standards it was less than 1.0.

Table 14–2

Values of the Lewis Form Factor Y. (These Values Are for a Normal Pressure Angle of 20°, Full-Depth Teeth, and a Diametral Pitch of Unity in the Plane of Rotation)

Number of Teeth	Y	Number of Teeth	Y
12	0.245	28	0.353
13	0.261	30	0.359
14	0.277	34	0.371
15	0.290	38	0.384
16	0.296	43	0.397
17	0.303	50	0.409
18	0.309	60	0.422
19	0.314	75	0.435
20	0.322	100	0.447
21	0.328	150	0.460
22	0.331	300	0.472
24	0.337	400	0.480
26	0.346	Rack	0.485

For this book, published during the transition period when much information based on the earlier standard is still in use, we will use the symbol K'_V, appropriately, in all equations, with its new meaning, in order to lessen confusion.

When a pair of gears is driven at moderate or high speed and noise is generated, it is certain that dynamic effects are present. One of the earliest efforts to account for an increase in the load due to velocity employed a number of gears of the same size, material, and strength. Several of these gears were tested to destruction by meshing and loading them at zero velocity. The remaining gears were tested to destruction at various pitch-line velocities. For example, if a pair of gears failed at 500 lbf tangential load at zero velocity and at 250 lbf at velocity V_1, then a *velocity factor*, designated K'_V, of 2 was specified for the gears at velocity V_1. Then another, identical, pair of gears running at a pitch-line velocity V_1 could be assumed to have a load equal to twice the tangential or transmitted load.

In the nineteenth century, Carl G. Barth first expressed the velocity factor by the equations

$$K'_V = \frac{600 + V}{600} \qquad \text{(cast iron, cast profile)} \tag{14–4a}$$

$$K'_V = \frac{1200 + V}{1200} \qquad \text{(cut or milled profile)} \tag{14–4b}$$

where V is the pitch-line velocity in feet per minute. It is also quite probably, because of the date that the tests were made, that the tests were conducted on teeth having a cycloidal profile instead of an involute profile. Cycloidal teeth were in general use in the nineteenth century because they were easier to cast than involute teeth. Equation (14–4a) is called the *Barth equation*. The Barth equation is often modified into Eq. (14–4b), for cut or milled teeth. Later AGMA added

$$K'_V = \frac{50 + \sqrt{V}}{50} \qquad \text{(hobbed or shaped profile)} \tag{14–5a}$$

$$K'_V = \sqrt{\frac{78 + \sqrt{V}}{78}} \qquad \text{(shaved or ground profile)} \tag{14–5b}$$

In SI units, Eqs. (14–4*a*) through (14–5*b*) become

$$K'_V = \frac{3.05 + V}{3.05} \qquad \text{(cast iron, cast profile)} \tag{14–6a}$$

$$K'_V = \frac{6.1 + V}{6.1} \qquad \text{(cut or milled profile)} \tag{14–6b}$$

$$K'_V = \frac{3.56 + \sqrt{V}}{3.56} \qquad \text{(hobbed or shaped profile)} \tag{14–6c}$$

$$K'_V = \sqrt{\frac{5.56 + \sqrt{V}}{5.56}} \qquad \text{(shaved or ground profile)} \tag{14–6d}$$

where V is in meters per second (m/s).

Introducing the velocity factor into Eq. (14–2) gives

$$\sigma = \frac{K'_V W^t P}{FY} \tag{14–7}$$

The metric version of this equation is

$$\sigma = \frac{K'_V W^t}{FmY} \tag{14–8}$$

where the face width F and the module m are both in millimeters (mm). Expressing the tangential component of load W^t in newtons (N) then results in stress units of megapascals (MPa).

As a general rule, spur gears should have a face width F from three to five times the circular pitch p.

Equations (14–7) and (14–8) are important because they form the basis for the AGMA approach to the bending strength of gear teeth. They are in general use for estimating the capacity of gear drives when life and reliability are not important considerations. The equations can be useful in obtaining a preliminary estimate of gear sizes needed for various applications.

EXAMPLE 14–1

A stock spur gear is available having a diametral pitch of 8 teeth/in, a $1\frac{1}{2}$-in face, 16 teeth, and a pressure angle of 20° with full-depth teeth. The material is AISI 1020 steel in as-rolled condition. Use a design factor of $n_d = 3$ to rate the horsepower output of the gear corresponding to a speed of 1200 r/m and moderate applications.

Solution

The term *moderate applications* seems to imply that the gear can be rated using the yield strength as a criterion of failure. From Table E–20, we find $S_{ut} = 55$ kpsi and $S_y = 30$ kpsi. A design factor of 3 means that the allowable bending stress is $30/3 = 10$ kpsi. The pitch diameter is $N/P = 16/8 = 2$ in, so the pitch-line velocity is

$$V = \frac{\pi dn}{12} = \frac{\pi(2)1200}{12} = 628 \text{ ft/min}$$

The velocity factor from Eq. (14–4*b*) is found to be

$$K'_V = \frac{1200 + V}{1200} = \frac{1200 + 628}{1200} = 1.52$$

Table 14–2 gives the form factor as $Y = 0.296$ for 16 teeth. We now arrange and substitute in Eq. (14–7) as follows:

$$W^t = \frac{FY\sigma_{\text{all}}}{K'_v P} = \frac{1.5(0.296)10\,000}{1.52(8)} = 365 \text{ lbf}$$

The horsepower that can be transmitted is

Answer $$hp = \frac{W^t V}{33\,000} = \frac{365(628)}{33\,000} = 6.95 \text{ hp}$$

It is important to emphasize that this is a rough estimate, and that this approach must not be used for important applications. The example is intended to help you understand some of the fundamentals that will be involved in the AGMA approach.

EXAMPLE 14–2 Estimate the horsepower rating of the gear in the previous example based on obtaining an infinite life in bending.

Solution The rotating-beam endurance limit is estimated from

$$S'_e = 0.506 S_{ut} = 0.506(55) = 27.8 \text{ kpsi}$$

To obtain the surface finish Marin factor k_a we refer to Table 7–5 for machined surface, finding $a = 2.67$ and $b = -0.265$. Then Eq. (7–9) gives the surface finish Marin factor k_a as

$$k_a = aS_{ut}^b = 2.67(55)^{-0.265} = 0.923$$

The next step is to estimate the size factor k_b. From Table 13–2,

$$l = \frac{1}{P} + \frac{1.25}{P} = \frac{1}{8} + \frac{1.25}{8} = 0.281 \text{ in}$$

The tooth thickness t in Fig. 14–1b is given in Sec. 14–1 [Eq. (b)] as $t = (4lx)^{1/2}$ when $x = 3Y/(2P)$ from Eq. (14–3). Therefore

$$x = \frac{3Y}{2P} = \frac{3(0.296)}{2(8)} = 0.0555 \text{ in}$$

then

$$t = (4lx)^{1/2} = [4(0.281)0.0555]^{1/2} = 0.250 \text{ in}$$

We have recognized the tooth as a cantilever beam of rectangular cross section, so the equivalent rotating-beam diameter must be obtained from Eq. (7–15):

$$d_e = 0.808(hb)^{1/2} = 0.808(Ft)^{1/2} = 0.808[1.5(0.250)]^{1/2} = 0.495 \text{ in}$$

Then, Eq. (7–10) gives k_b as

$$k_b = \left(\frac{d_e}{0.30}\right)^{-0.107} = \left(\frac{0.495}{0.30}\right)^{-0.107} = 0.948$$

The load factor k_c from Eq. (7–20) is unity. The temperature factor k_d is likewise unity.

Two effects are used to evaluate the miscellaneous-effects Marin factor k_e. The first of these is the effect of one-way bending. In general, a gear tooth is subjected only to one-way bending. Exceptions include idler gears and gears used in reversing mechanisms.

For one-way bending the steady and alternating stress components are $\sigma_a = \sigma_m = \sigma/2$ where σ is the largest repeatedly applied bending stress as given in Eq. (14–7). If a material exhibited a Goodman failure locus,

$$\frac{S_a}{S'_e} + \frac{S_m}{S_{ut}} = 1$$

Since S_a and S_m are equal for one-way bending we substitute S_a for S_m and solve the preceding equation for S_a, giving

$$S_a = \frac{S'_e S_{ut}}{S'_e + S_{ut}}$$

Now replace S_a with $\sigma/2$, and in the denominator replace S'_e with $0.506S_{ut}$ to obtain

$$\sigma = \frac{2S'_e S_{ut}}{0.506S_{ut} + S_{ut}} = \frac{2S'_e}{0.506 + 1} = 1.33S'_e$$

Now $k_e = \sigma/S'_e = 1.33S'_e/S'_e = 1.33$. However, a Gerber fatigue locus gives mean values of

$$\frac{S_a}{S'_e} + \left(\frac{S_m}{S_{ut}}\right)^2 = 1$$

Setting $S_a = S_m$ and solving the quadratic in S_a gives

$$S_a = \frac{S_{ut}^2}{2S'_e}\left(-1 + \sqrt{1 + \frac{4S_e'^2}{S_{ut}^2}}\right)$$

Setting $S_a = \sigma/2$, $S_{ut} = S'_e/0.506$ gives

$$\sigma = \frac{S'_e}{0.506^2}\left[-1 + \sqrt{1 + 4(0.506)^2}\right] = 1.65S'_e$$

and $k_e = 1.54$. Since a Gerber locus runs in and among fatigue data and Goodman does not, we will use $k_e = 1.65$.

The second effect to be accounted for in using the miscellaneous-effects Marin factor k_e is stress concentration, for which we will use our fundamentals from Chap. 7. For a 20° full-depth tooth the radius of the root fillet is denoted r_f, where

$$r_f = \frac{0.300}{P} = \frac{0.300}{8} = 0.0375 \text{ in}$$

From Fig. E–15–6

$$\frac{r}{d} = \frac{r_f}{t} = \frac{0.0375}{0.250} = 0.15$$

Since $D/d = \infty$, we approximate with $D/d = 3$, giving $K_t = 1.68$. From Eq. (7–27),

$$K_f = \frac{K_t}{1 + \dfrac{2}{\sqrt{r}}\dfrac{K_t - 1}{K_t}\sqrt{a}} = \frac{1.68}{1 + \dfrac{2}{\sqrt{0.0375}}\dfrac{1.68 - 1}{1.68}\dfrac{4}{55}} = 1.29$$

The miscellaneous-effects Marin factor for stress concentration is

$$k_e = \frac{1}{K_f} = \frac{1}{1.29} = 0.775$$

The final value of k_e is the product of the two k_e factors, that is, $1.65(0.775) = 1.28$. The Marin equation for the fully corrected endurance strength is

$$\begin{aligned} S_e &= k_a k_b k_c k_d k_e S_e' \\ &= 0.923(0.948)(1)(1)1.28(27.8) = 31.1 \text{ kpsi} \end{aligned}$$

For a design factor of $n_d = 3$ applied to the load or strength is

$$\sigma_{\text{all}} = \frac{S_e}{n_d} = \frac{31.1}{3} = 10\ 400 \text{ psi}$$

The transmitted load W^t is

$$W^t = \frac{FY\sigma_{\text{all}}}{K_V' P} = \frac{1.5(0.296)10\ 400}{1.52(8)} = 380 \text{ lb}$$

and the power is

$$hp = \frac{W^t V}{33\ 000} = \frac{380(628)}{33\ 000} = 7.2 \text{ hp}$$

Again, it should be emphasized that these results should be accepted only as preliminary estimates to alert you to the nature of bending in gear teeth. Note that the repeatedly-applied allowable stress ($\sigma_{\text{all}} = 10\ 400$ psi) is slightly higher than the yield strength in Ex. 14–1 ($S_g = 10\ 000$ psi). This condition could be corrected by using a design factor somewhat larger than three.

In Ex. 14–2 our resources (Table E–15–6) did not directly address stress concentration in gear teeth. A photoelastic investigation by Dolan and Broghamer reported in 1942 constitutes a primary source of information on stress concentration.[3] Mitchiner and Mabie[4] interpret the results in term of fatigue stress-concentration factor K_f as

$$K_f = H + \left(\frac{t}{r}\right)^L \left(\frac{t}{l}\right)^M \tag{14–9}$$

where

$$\begin{aligned} H &= 0.34 - 0.458\ 366\ 2\phi \\ L &= 0.316 - 0.458\ 366\ 2\phi \\ M &= 0.290 + 0.458\ 366\ 2\phi \\ r &= \frac{(b - r_f)^2}{(d/2) + b - r_f} \end{aligned}$$

In these equations l and t are from the layout in Fig. 14–1, ϕ is the pressure angle, r_f is

3. T. J. Dolan and E. I. Broghamer, *A Photoelastic Study of the Stresses in Gear Tooth Fillets*, Bulletin 335, Univ. Ill. Exp. Sta., March 1942, See also R. E. Peterson, *Stress Concentration Factors*, Wiley, New York, 1974, pp. 250, 251.

4. R. G. Mitchiner and H. H. Mabie, "Determination of the Lewis Form Factor and the AGMA Geometry Factor J of External Spur Gear Teeth," *J. Mech. Des.*, Vol. 104, No. 1, Jan. 1982, pp. 148–158.

the fillet radius, b is the dedendum, and d is the pitch diameter. It is left as an exercise for the reader to compare K_f from Eq. (14–9) with the results of using the approximation of Fig. E–15–6 in Ex. 14–2.

14–2 Surface Durability

In this section we are interested in the failure of the surfaces of gear teeth, which is generally called *wear*. *Pitting*, as explained in Sec. 7–17, is a surface fatigue failure due to many repetitions of high contact stresses. Other surface failures are *scoring*, which is a lubrication failure, and *abrasion*, which is wear due to the presence of foreign material.

To obtain an expression for the surface-contact stress, we shall employ the Hertz theory. In Eq. (3–94) it was shown that the contact stress between two cylinders may be computed from the equation

$$p_{\max} = \frac{2F}{\pi b l} \tag{a}$$

where $p_{\max}$ = largest surface pressure

F = force pressing the two cylinders together

l = length of cylinders

and b is obtained from the equation

$$b = \left\{ \frac{2F}{\pi l} \frac{\left[\left(1-\nu_1^2\right)/E_1\right] + \left[\left(1-\nu_2^2\right)/E_2\right]}{(1/d_1) + (1/d_2)} \right\}^{1/2} \tag{14–10}$$

[Eq. (3–98)] where ν_1, ν_2, E_1, and E_2 are the elastic constants and d_1 and d_2 are the diameters, respectively, of the two contacting cylinders.

To adapt these relations to the notation used in gearing, we replace F by $W^t/\cos\phi$, d by $2r$, and l by the face width F. With these changes, we can substitute the value of b as given by Eq. (14–10) in Eq. (a). Replacing $p_{\max}$ by σ_C, the *surface compressive stress (Hertzian stress)* is found from the equation

$$\sigma_C^2 = \frac{W^t}{\pi F \cos\phi} \frac{(1/r_1) + (1/r_2)}{\left[\left(1-\nu_1^2\right)/E_1\right] + \left[\left(1-\nu_2^2\right)/E_2\right]} \tag{14–11}$$

where r_1 and r_2 are the instantaneous values of the radii of curvature on the pinion- and gear-tooth profiles, respectively, at the point of contact. By accounting for load sharing in the value of W^t used, Eq. (14–11) can be solved for the Hertzian stress for any or all points from the beginning to the end of tooth contact. Of course, pure rolling exists only at the pitch point. Elsewhere the motion is a mixture of rolling and sliding. Equation (14–11) does not account for any sliding action in the evaluation of stress.

We note that AGMA uses μ for Poisson's ratio instead of ν as is used here.

We have already noted that the first evidence of wear occurs near the pitch line. The radii of curvature of the tooth profiles at the pitch point are

$$r_1 = \frac{d_P \sin\phi}{2} \qquad r_2 = \frac{d_G \sin\phi}{2} \tag{14–12}$$

where ϕ is the pressure angle and d_P and d_G are the pitch diameters of the pinion and gear, respectively.

Note, in Eq. (14–11), that the denominator of the second group of terms contains four elastic constants, two for the pinion and two for the gear. As a simple means of combining and tabulating the results for various combinations of pinion and gear materials, AGMA defines an *elastic coefficient* C_p by the equation

$$C_p = \left[\frac{1}{\pi \left(\dfrac{1 - \nu_P^2}{E_P} + \dfrac{1 - \nu_G^2}{E_G} \right)} \right]^{1/2} \tag{14–13}$$

With this simplification, and the addition of a velocity factor K'_V, Eq. (14–11) can be written as

$$\sigma_C = -C_P \left[\frac{K'_V W^t}{F \cos\phi} \left(\frac{1}{r_1} + \frac{1}{r_2} \right) \right]^{1/2} \tag{14–14}$$

where the sign is negative because σ_C is a compressive stress.

EXAMPLE 14–3

The pinion of Examples 14–1 and 14–2 is to be mated with a 50-tooth gear manufactured of ASTM No. 50 cast iron. Using the tangential load of 380 lb, estimate the factor of safety of the drive based on the possibility of a surface fatigue failure.

Solution

From Table E–5 we find the elastic constants to be $E_P = 30$ Mpsi, $\nu_P = 0.292$, $E_G = 14.5$ Mpsi, $\nu_G = 0.211$. We substitute these in Eq. (14–13) to get the elastic coefficient as

$$C_p = \left\{ \frac{1}{\pi \left[\dfrac{1 - (0.292)^2}{30\,(10^6)} + \dfrac{1 - (0.211)^2}{14.5\,(10^6)} \right]} \right\}^{1/2} = 1817$$

From Example 14–1, the pinion pitch diameter is $d_P = 2$ in. The value for the gear is $d_G = 50/8 = 6.25$ in. Then Eq. (14–12) is used to obtain the radii of curvature at the pitch points. Thus

$$r_1 = \frac{2 \sin 20^\circ}{2} = 0.342 \text{ in} \qquad r_2 = \frac{6.25 \sin 20^\circ}{2} = 1.069 \text{ in}$$

The face width is given as $F = 1.5$ in. Use $K'_V = 1.52$ from Example 4–1. Substituting all these values in Eq. (14–14) gives the contact stress as

$$\sigma_C = -1817 \left[\frac{1.52(380)}{1.5 \cos 20^\circ} \left(\frac{1}{0.342} + \frac{1}{1.069} \right) \right]^{1/2} = -72\,259 \text{ psi}$$

The surface endurance strength of cast iron can be estimated from

$$S_C = 0.32 H_B \text{ kpsi}$$

for 10^8 cycles, where S_C is in kpsi. Table E–24 gives $H_B = 262$ for ASTM No. 50 cast iron. Therefore $S_C = 0.32(262) = 83.8$ kpsi. Contact stress is not linear with transmitted load [see Eq. (14–14)]. If the factor of safety is defined as the loss-of-function load divided by the imposed load, then the ratio of loads is the ratio of stresses squared. In other words,

$$n = \frac{\text{loss-of-function load}}{\text{imposed load}} = \frac{S_C^2}{\sigma_C^2} = \left(\frac{83.8}{72.3} \right)^2 = 1.34$$

One is free to define factor of safety as S_C/σ_C. Awkwardness comes when one compares the factor of safety in bending fatigue with the factor of safety in surface fatigue for a particular gear. Suppose the factor of safety of this gear in bending fatigue is 1.20 and the factor of safety in surface fatigue is 1.34 as above. The threat, since 1.34 is greater than 1.20, is in bending fatigue since both numbers are based on load ratios. If the factor of safety in surface fatigue is based on $S_C/\sigma_C = \sqrt{1.34} = 1.16$, then 1.20 is greater than 1.16, but the threat is not from surface fatigue. The surface fatigue factor of safety can be defined either way. One way has the burden of requiring a squared number before numbers that instinctively seem comparable can be compared.

In addition to the dynamic factor K'_V already introduced, there are transmitted load excursions, nonuniform distribution of the transmitted load over the tooth contact, and the influence of rim thickness on bending stress. Tabulated strength values can be means, ASTM minimums, or of unknown heritage. In surface fatigue there are no endurance limits. Endurance strengths have to be qualified as to corresponding cycle count, and the slope of the *S-N* curve needs to be known. In bending fatigue there is a definite change in slope of the *S-N* curve near 10^6 cycles, but some evidence indicates that an endurance limit does not exist. Gearing experience leads to cycle counts of 10^{11} or more. Evidence of diminishing endurance strengths in bending have been included in AGMA methodology.

14–3 AGMA Stress Equations

Two fundamental stress equations are used in the AGMA methodology, one for bending stress and another for pitting resistance (contact stress). In AGMA terminology, these are called *stress numbers* and are designated using a lowercase letter s instead of the Greek lower case σ we have used in this book (and shall continue to use). The fundamental equations are

$$\sigma = \begin{cases} W^t K_O K'_V K_s \dfrac{P_d}{F} \dfrac{K_m K_B}{J} \\ W^t K_O K'_V K_s \dfrac{1}{bm_t} \dfrac{K_H K_B}{Y_J} \end{cases} \tag{14–15}$$

where the upper equation uses U.S. customary units and the lower SI units. Also,

W^t is the tangential transmitted load
K_O is the overload factor
K'_V is the dynamic factor
K_s is the size factor
P_d is the transverse diametral pitch
F is the face width of the narrower member
K_m is the load-distribution factor
K_B is the rim-thickness factor
J is the geometry factor for bending strength (which includes root fillet stress-concentration factor K_f)

In the SI formulations,

b is the net face width of the narrower member
m_t is the transverse metric module
K_H is the load distribution factor

Y_J is the geometry factor for bending strength (including root fillet stress-concentration factor K_f)

Before you try to digest the meaning of all these terms in Eq. (14–15) view it as advice from AGMA concerning items the designer should consider *whether he or she follows the voluntary standard or not*. These items include issues such as

- Transmitted load magnitude
- Overload
- Dynamic augmentation of transmitted load
- Size
- Geometry: pitch and face width
- Distribution of load across the teeth
- Rim support of the tooth
- Lewis form factor and root fillet stress concentration

The fundamental equation for pitting resistance (contact stress) is

$$\sigma_c = \begin{cases} C_p\sqrt{W^t K_O K_V' K_s \dfrac{K_m}{dF}\dfrac{C_f}{I}} \\ Z_E\sqrt{W^t K_O K_V' K_s \dfrac{K_H}{d_{w1}b}\dfrac{Z_R}{Z_I}} \end{cases} \tag{14–16}$$

where the upper equation uses U.S. customary units and the lower uses SI units. Also,

C_p is an elastic coefficient, $\left(\text{lbf/in}^2\right)^{0.5}$
W^t is the transmitted tangential load
K_O is the overload factor
K_V' is the dynamic factor
K_s is the size factor
K_m is the load-distribution factor
C_f is the surface condition factor
d is the pitch diameter of the *pinion*
F is the face width of the narrowest member

In the SI formulation,

Z_E is an elastic coefficient, $\left(\text{N/mm}^2\right)^{0.5}$
K_H is the load-distribution factor
Z_R is the surface condition factor for pitting resistance
d_{w1} is the pitch diameter of the *pinion*
b is the face width of the narrowest member
Z_I is the geometry factor for pitting resistance

The evaluation of all these factors is explained in the sections that follow. The development of Eq. (14–16) is clarified in the second part of Sec. 14–5.

14–4 AGMA Strength Equations

Instead of using the term *strength,* AGMA uses data termed *allowable stress numbers* and designates these by the symbol S_a. It will be less confusing here if we continue the practice in this book of using the uppercase letter S to designate strength and the lowercase Greek letters σ and τ for stress. To make it perfectly clear we shall use the term *AGMA strength* as a replacement for the phrase *allowable stress numbers* as used by AGMA.

Following this convention, values for *AGMA bending strength*, designated here as S_t, are to be found in Figs. 14–2, 14–3, and 14–4, and in Tables 14–3 and 14–4. Since AGMA strengths are not identified with other strengths such as S_{ut}, S_e, or S_y as used elsewhere in this book, their use should be restricted to gear problems.

In the AGMA approach the strengths are modified by various factors that produce limiting values of the bending stress and the contact stress. Using the same notation as in Eq. (1–5) we shall term the resulting modifications the allowable bending stress σ_{all} and the allowable contact stress $\sigma_{c,\text{all}}$. The equation for the allowable bending stress is

$$\sigma_{\text{all}} = \begin{cases} \dfrac{S_t}{S_F}\dfrac{Y_N}{K_T K_R} \\ \dfrac{\sigma_{FP}}{S_F}\dfrac{Y_N}{Y_\theta Y_Z} \end{cases} \tag{14–17}$$

where the upper equation uses U.S. customary units and the lower equations uses SI units. Also,

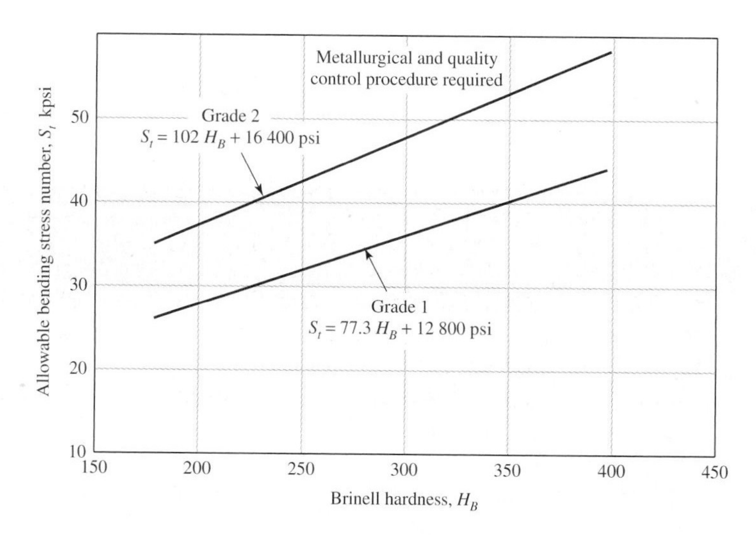

Figure 14–2

Allowable bending stress number for through-hardened steels. The SI equations are $\sigma_{FP} = 0.533H_B + 88.3$ MPa, grade 1, and $\sigma_{FP} = 0.703H_B + 113$ MPa, grade 2. (*Source: ANSI/AGMA 2001-C95 and 2101-C95.*)

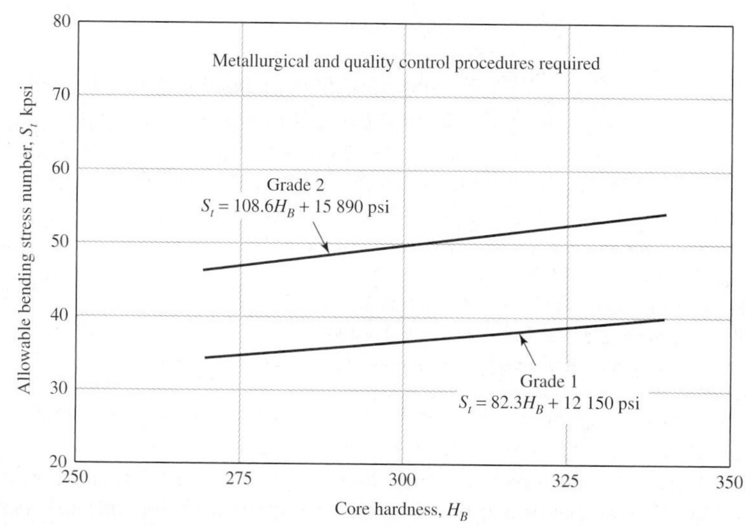

Figure 14–3

Allowable bending stress number for nitrided through-hardened steel gears (i.e., AISI 4140, 4340), S_t. The SI equations are $\sigma_{FP} = 0.568H_B + 83.8$ MPa, grade 1, and $\sigma_{FP} = 0.749H_B + 110$ MPa, grade 2. (*Source: ANSI/AGMA 2001-C95 and 2101-C95.*)

Figure 14–4

Allowable bending stress numbers for nitriding steel gears S_t. The SI equations are
$\sigma_{FP} = 0.594H_B + 87.76$ MPa Nitralloy grade 1
$\sigma_{FP} = 0.784H_B + 114.81$ MPa Nitralloy grade 2
$\sigma_{FP} = 0.7255H_B + 63.89$ MPa 2.5% chrome, grade 1
$\sigma_{FP} = 0.7255H_B + 153.63$ MPa 2.5% chrome, grade 2
$\sigma_{FP} = 0.7255H_B + 201.91$ MPa 2.5% chrome, grade 3
(Source: ANSI/AGMA 2001-C95, 2101-C95.)

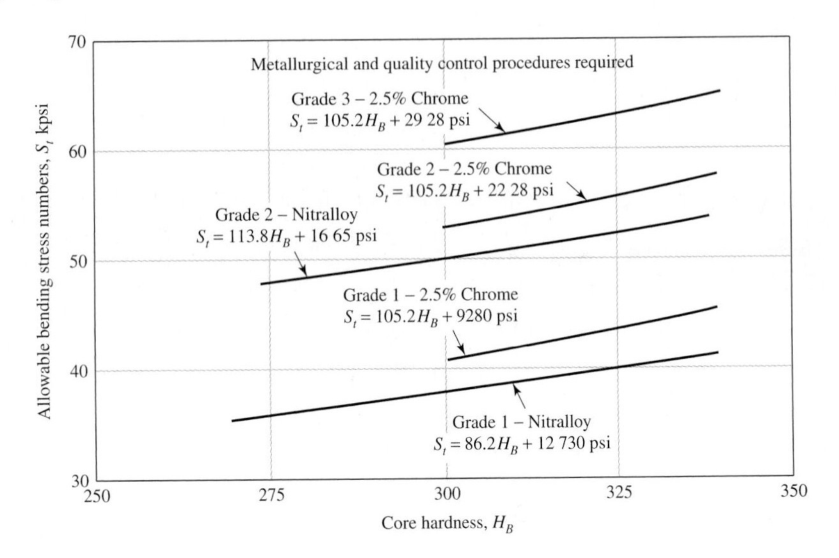

Table 14–3

Repeatedly Applied Bending Strength S_t at 10^7 Cycles and 0.99 Reliability for Steel Gears *Source: ANSI/AGMA 2001-C95.*

Material Designation	Heat Treatment	Minimum Surface Hardness[1]	Allowable Bending Stress Number S_t,[2] psi		
			Grade 1	Grade 2	Grade 3
Steel[3]	Through hardened	See Fig. 14–2	See Fig. 14–2	See Fig. 14–2	—
	Flame[4] or induction hardened[4] with type A pattern[5]	See Table 8*	45 000	55 000	—
	Flame[4] or induction hardened[4] with type B pattern[5]	See Table 8*	22 000	22 000	—
	Carburized and hardened	See Table 9*	55 000	65 000 or 70 000[6]	75 000
	Nitrided[4,7] (through-hardened steels)	83.5 HR15N	See Fig. 14–3	See Fig. 14–3	—
Nitralloy 135M, Nitralloy N, and 2.5% chrome (no aluminum)	Nitrided[4,7]	87.5 HR15N	See Fig. 14–4	See Fig. 14–4	See Fig. 14–4

Notes: See ANSI/AGMA 2001-C95 for references cited in notes 1–7.
[1]Hardness to be equivalent to that at the root diameter in the center of the tooth space and face width.
[2]See tables 7 through 10 for major metallurgical factors for each stress grade of steel gears.
[3]The steel selected must be compatable with the heat treatment process selected and hardness required.
[4]The allowable stress numbers indicated may be used with the case depths prescribed in 16.1
[5]See figure 12 for type A and type B hardness patterns.
[6]If bainite and microcracks are limited to grade 3 levels, 70,000 psi may be used.
[7]The overload capacity of nitrided gears is low. Since the shape of the effective S-N curve is flat, the sensitivity to shock should be investigated before proceeding with the design. [7]
*Tables 8 and 9 of ANSI/AGMA 2001-C95 are comprehensive tabulations of the major metallurgical factors affecting S_t and S_c of flame-hardened and induction-hardened (Table 8) and carburized and hardened (Table 9) steel gears.

Table 14–4

Repeatedly Applied Bending Strength S_t for Iron and Bronze Gears at 10^7 Cycles and 0.99 Reliability

Source: ANSI/AGMA 2001-C95.

Material	Material Designation[1]	Heat Treatment	Typical Minimum Surface Hardness[2]	Allowable Bending Stress Number, S_t,[3] psi
ASTM A48 gray cast iron	Class 20	As cast	—	5000
	Class 30	As cast	174 HB	8500
	Class 40	As cast	201 HB	13 000
ASTM A536 ductile (nodular) Iron	Grade 60–40–18	Annealed	140 HB	22 000–33 000
	Grade 80–55–06	Quenched and tempered	179 HB	22 000–33 000
	Grade 100–70–03	Quenched and tempered	229 HB	27 000–40 000
	Grade 120–90–02	Quenched and tempered	269 HB	31 000–44 000
Bronze		Sand cast	Minimum tensile strength 40 000 lb/in^2	5700
	ASTM B–148 Alloy 954	Heat treated	Minimum tensile strength 90 000 lb/in^2	23 600

Notes:

[1] See ANSI/AGMA 2004-B89, *Gear Materials and Heat Treatment Manual.*

[2] Measured hardness to be equivalent to that which would be measured at the root diameter in the center of the tooth space and face width.

[3] The lower values should be used for general design purposes. The upper values may be used when:

High quality material is used.

Section size and design allow maximum response to heat treatment.

Proper quality control is effected by adequate inspection.

Operating experience justifies their use.

S_t, σ_{FP} are allowable bending stress, lbf/in^2 (N/mm^2)

Y_N is the stress cycle factor for bending stress

K_T, Y_θ are the temperature factors

K_R, Y_Z are the reliability factors

S_F is the AGMA factor of safety, a stress ratio

The equation for the allowable contact stress $\sigma_{c,\text{all}}$ is

$$\sigma_{c,\text{all}} = \begin{cases} \dfrac{S_c}{S_H}\dfrac{Z_N C_H}{K_T K_R} \\ \dfrac{\sigma_{HP}}{S_H}\dfrac{Z_N Z_W}{Y_\theta Y_Z} \end{cases} \qquad (14\text{–}18)$$

where the upper equation is in U.S. customary units and the lower equation is in SI units, Also,

S_c, σ_{HP} are the allowable contact stress, lbf/in^2 (N/mm^2)

Z_N is the stress cycle life factor

C_H, Z_W are the hardness ratio factors for pitting resistance

K_T, Y_θ are the temperature factors

K_R, Y_Z are the reliability factors
S_H is the AGMA factor of safety, a stress ratio

The values for AGMA contact stress designated here as S_c are to be found in Fig. 14–5 and Tables 14–5, 14–6, and 14–7 (see page 916).

AGMA allowable stress numbers (strengths) for bending and contact stress are for

- Unidirectional loading
- 10 million stress cycles
- 99 percent reliability

The factors in this section, too, will be evaluated in subsequent sections.

When two-way (reversed) loading occurs, as with idler gears, AGMA recommends using 70 percent of S_t values. This is equivalent to $1/0.70 = 1.43$ as a value of k_e in Ex. 14–2. The recommendation falls between the value of $k_e = 1.33$ for a Goodman failure locus and $k_e = 1.65$ for a Gerber failure locus.

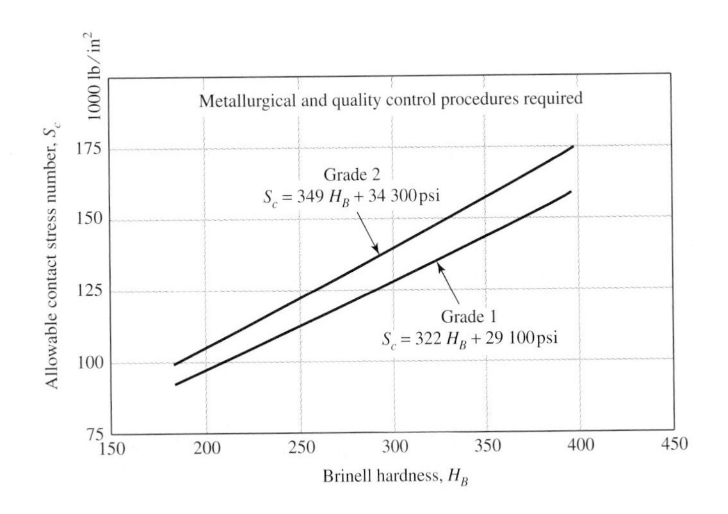

Figure 14–5

Contact-fatigue strength S_c at 10^7 cycles and 0.99 reliability for through-hardened steel gears. The SI equations are $\sigma_{HP} = 2.22H_B + 200$ MPa, grade 1, and $\sigma_{HP} = 2.41H_B + 237$ MPa, grade 2. *(Source: ANSI/AGMA 2001-C95 and 2101-C95).*

Table 14–5

Nominal Temperature Used in Nitriding and Hardnesses Obtained

Source: Darle W. Dudley, Handbook of Practical Gear Design, rev. ed., McGraw-Hill, New York, 1984.

Steel	Temperature before nitriding, °F	Nitriding, °F	Hardness, Rockwell C Scale	
			Case	Core
Nitralloy 135*	1150	975	62–65	30–35
Nitralloy 135M	1150	975	62–65	32–36
Nitralloy N	1000	975	62–65	40–44
AISI 4340	1100	975	48–53	27–35
AISI 4140	1100	975	49–54	27–35
31 Cr Mo V 9	1100	975	58–62	27–33

* Nitralloy is a trademark of the Nitralloy Corp., New York, NY.

Table 14–6

Repeatedly Applied Contact Strength S_c at 10^7 Cycles and 0.99 Reliability for Steel Gears *Source:* ANSI/AGMA 2001-C95.

Material Designation	Heat Treatment	Minimum Surface Hardness[1]	Allowable Contact Stress Number, s_c,[2] psi		
			Grade 1	Grade 2	Grade 3
Steel[3]	Through hardened[4]	See Fig. 14-5	See Fig. 14-5	See Fig. 14-5	—
	Flame[5] or induction hardened[5]	50 HRC	170 000	190 000	—
		54 HRC	175 000	195 000	—
	Carburized and hardened[5]	See Table 9*	180 000	225 000	275 000
	Nitrided[5] (through hardened steels)	83.5 HR15N	150 000	163 000	175 000
		84.5 HR15N	155 000	168 000	180 000
2.5% chrome (no aluminum)	Nitrided[5]	87.5 HR15N	155 000	172 000	189 000
Nitralloy 135M	Nitrided[5]	90.0 HR15N	170 000	183 000	195 000
Nitralloy N	Nitrided[5]	90.0 HR15N	172 000	188 000	205 000
2.5% chrome (no aluminum)	Nitrided[5]	90.0 HR15N	176 000	196 000	216 000

Notes: See ANSI/AGMA 2001-C95 for references cited in notes 1–5.

[1] Hardness to be equivalent to that at the start of active profile in the center of the face width.

[2] See Tables 7 through 10 for major metallurgical factors for each stress grade of steel gears.

[3] The steel selected must be compatible with the heat treatment process selected and hardness required.

[4] These materials must be annealed or normalized as a minimum.

[5] The allowable stress numbers indicated may be used with the case depths prescribed in 16.1.

* Table 9 of ANSI/AGMA 2001 C95 is a comprehensive tabulation of the major metallurgical factors affecting S_t and S_c of carburized and hardened steel gears.

14–5 Geometry Factors *I* and *J* (Z_I and Y_J)

We have seen how the factor Y is used in the Lewis equation to introduce the effect of tooth form into the stress equation. The AGMA factors[5] I and J are intended to accomplish the same purpose in a more involved manner.

The determination of I and J depends upon the *face-contact ratio* m_F. This is defined as

$$m_F = \frac{F}{p_x} \tag{14–19}$$

where p_x is the axial pitch and F is the face width. For spur gears, $m_F = 0$.

Low-contact-ratio (LCR) helical gears having a small helix angle or a thin face width, or both, have face-contact ratios less than unity ($m_F \leq 1$), and will not be considered here. Such gears have a noise level not too different from that for spur gears. Consequently we shall consider here only spur gears with $m_F = 0$ and conventional helical gears with $m_F > 1$.

5. A useful reference is AGMA 908-B89, *Geometry Factors for Determining Pitting Resistance and Bending Strength of Spur, Helical and Herringbone Gear Teeth.*

Table 14–7

Repeatedly Applied Contact Strength S_c 10^7 Cycles and 0.99 Reliability for Iron and Bronze Gears *Source:* ANSI/AGMA 2001-C95.

Material	Material Designation [1]	Heat Treatment	Typical Minimum Surface Hardness [2]	Allowable Contact Stress Number, S_c,[3] psi
ASTM A48 gray cast iron	Class 20	As cast	—	50 000–60 000
	Class 30	As cast	174 HB	65 000–75 000
	Class 40	As cast	201 HB	75 000–85 000
ASTM A536 ductile (nodular) iron	Grade 60–40–18	Annealed	140 HB	77 000–92 000
	Grade 80–55–06	Quenched and tempered	179 HB	77 000–92 000
	Grade 100–70–03	Quenched and tempered	229 HB	92 000–112 000
	Grade 120–90–02	Quenched and tempered	269 HB	103 000–126 000
Bronze	—	Sand cast	Minimum tensile strength 40 000 psi	30 000
	ASTM B-148 Alloy 954	Heat treated	Minimum tensile strength 90 000 psi	65 000

Notes:

[1] See ANSI/AGMA 2004-B89, *Gear Materials and Heat Treatment Manual.*

[2] Hardness to be equivalent to that at the start of active profile in the center of the face width.

[3] The lower values should be used for general design purposes. The upper values may be used when:

High-quality material is used.

Section size and design allow maximum response to heat treatment.

Proper quality control is effected by adequate inspection.

Operating experience justifies their use.

Bending-Strength Geometry Factor *J*

The AGMA factor J employs a modified value of the Lewis form factor, also denoted by Y; a *fatigue stress-concentration factor* K_f; and a tooth *load-sharing ratio* m_N. The resulting equation for J is

$$J = \frac{Y}{K_f m_N} \tag{14–20}$$

It is important to note that the form factor Y in Eq. (14–20) is *not* the Lewis factor at all. The value of Y here is obtained from a generated layout of the tooth profile in the normal plane and is based on the highest point of single-tooth contact.

The factor K_f in Eq. (14–20) is called a *stress correction factor* by AGMA. It is based on a formula deduced from a photoelastic investigation of stress concentration in gear teeth over 50 years ago.

The load-sharing ratio m_N is equal to the face width divided by the minimum total length of the lines of contact. This factor depends on the transverse contact ratio m_p, the face-contact ratio m_F, the effects of any profile modifications, and the tooth deflection. For spur gears, $m_N = 1.0$. For helical gears having a face-contact ratio $m_F > 2.0$, a conservative approximation is given by the equation

$$m_N = \frac{p_N}{0.95Z} \tag{14–21}$$

where p_N is the normal base pitch and Z is the length of the line of action in the transverse plane (distance L_{ab} in Fig. 13–16).

Use Fig. 14–6 to obtain the geometry factor J for spur gears having a 20° pressure angle and full-depth teeth. Use Figs. 14–7 and 14–8 for helical gears having a 20° normal pressure angle and face-contact ratios of $m_F = 2$ or greater. For other gears, consult the AGMA standard.

Surface-Strength Geometry Factor I and Z_I

The factor I is also called the *pitting-resistance geometry factor* by AGMA. We begin by noting that the sum of the reciprocals of Eq. (14–13) can be expressed as

$$\frac{1}{r_1} + \frac{1}{r_2} = \frac{2}{\sin \phi_t}\left(\frac{1}{d_P} + \frac{1}{d_G}\right) \tag{a}$$

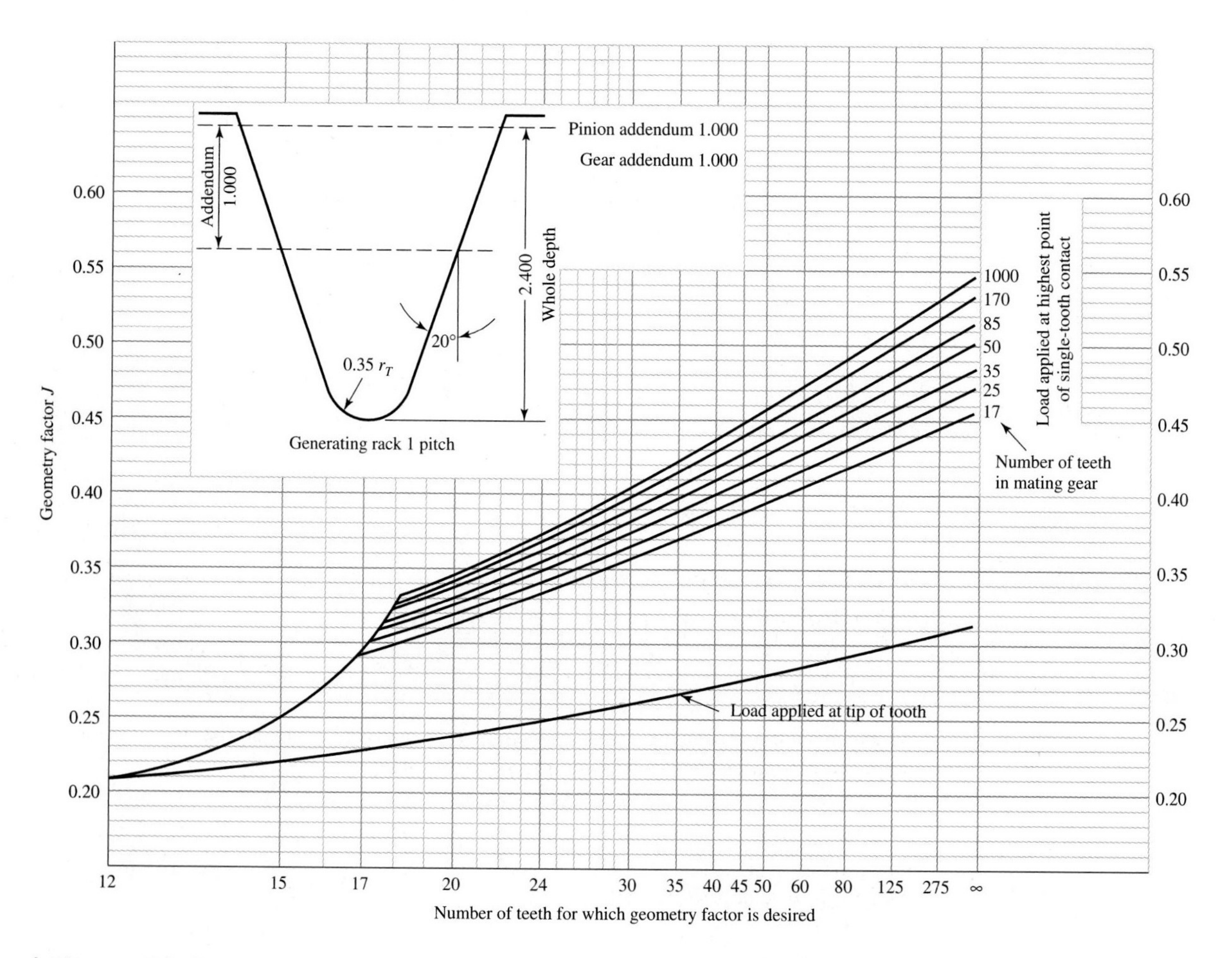

Figure 14–6

Spur-gear geometry factors J. *(Source: ANSI/AGMA 218.01.)*

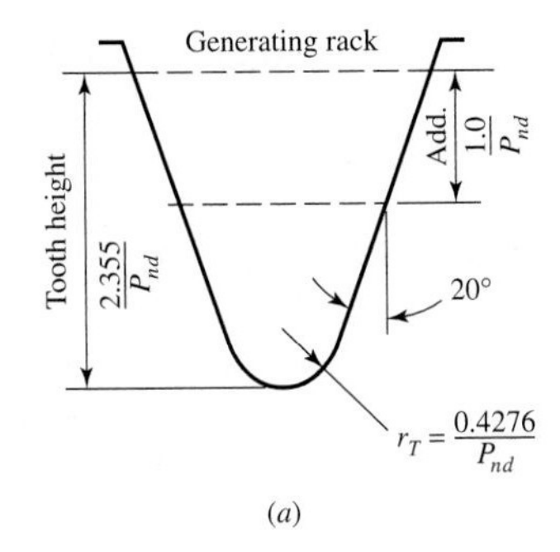

(a)

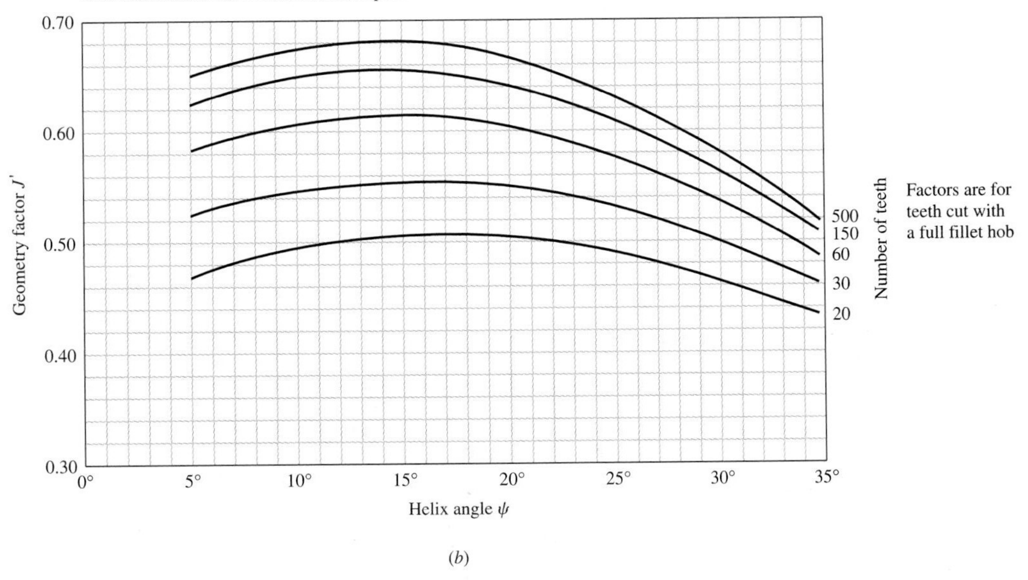

(b)

Figure 14–7

Helical-gear geometry factors J'. *(ANSI/AGMA 218.01.)*

where we have replaced ϕ by ϕ_t, the transverse pressure angle, so that the relation will apply to helical gears too. Now define *speed ratio* m_G as

$$m_G = \frac{N_G}{N_P} = \frac{d_G}{d_P} \tag{14–22}$$

Equation (*a*) can now be written

$$\frac{1}{r_1} + \frac{1}{r_2} = \frac{1}{d_P} \frac{1}{\sin \phi_t} \frac{m_G + 1}{m_G} \tag{b}$$

Now substitute Eq. (*b*) for the sum of the reciprocals in Eq. (14–14). The result is found

Figure 14–8

J′-factor multipliers for use with Fig. 14–7 to find *J*. (ANSI/AGMA 218.01.)

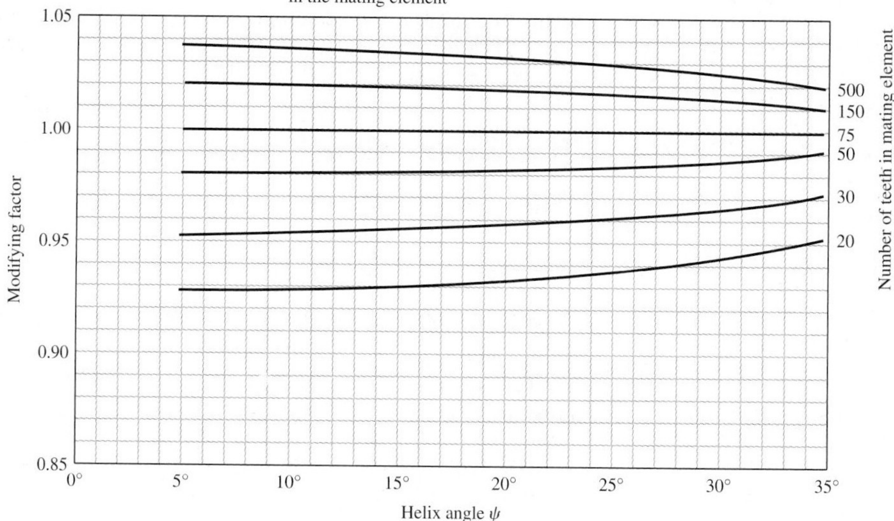

to be

$$\sigma_c = -\sigma_C = C_P \left[\frac{W^t}{d_P F} \frac{1}{\dfrac{\cos\phi_t \sin\phi_t}{2} \dfrac{m_G}{m_G+1}} \right]^{1/2} \tag{c}$$

The geometry factor I for external spur gears is the denominator of the second term in the brackets in Eq. (*c*). By adding the load-sharing ratio m_N, we obtain a factor valid for both spur and helical gears. The equation is then written as

$$I = \begin{cases} \dfrac{\cos\phi_t \sin\phi_t}{2m_N} \dfrac{m_G}{m_G+1} & \text{external gears} \\[2ex] \dfrac{\cos\phi_t \sin\phi_t}{2m_N} \dfrac{m_G}{m_G-1} & \text{internal gears} \end{cases} \tag{14–23}$$

where $m_N = 1$ for spur gears. In solving Eq. (14–21) for m_N, note that

$$p_N = p_n \cos\phi_n \tag{14–24}$$

where p_n is the normal circular pitch. If a layout of the gears is not made, the quantity Z, for use in Eq. (14–21), can be obtained from the equation

$$Z = \left[(r_P + a)^2 - r_{bP}^2\right]^{1/2} + \left[(r_G + a)^2 - r_{bG}^2\right]^{1/2} - (r_P + r_G)\sin\phi_t \tag{14–25}$$

where r_P and r_G are the pitch radii and r_{bP} and r_{bG} the base-circle radii.[6] The radius of the base circle is

$$r_b = r\cos\phi_t \tag{14–26}$$

Certain precautions must be taken in using Eq. (14–25). The tooth profiles are not con-

6. For a development, see Joseph E. Shigley and John J. Uicker Jr., *Theory of Machines and Mechanisms*, McGraw-Hill, New York, 1980, p. 262.

jugate below the base circle, and consequently, if either one or the other of the first two terms in brackets is larger than the third term, then it should be replaced by the third term. In addition, the effective outside radius is sometimes less than $r + a$, owing to removal of burrs or rounding of the tips of the teeth. When this is the case, always use the effective outside radius instead of $r + a$.

14–6 The Elastic Coefficient C_p (Z_E)

Values of C_p may be computed directly from Eq. (14–13) or obtained from Table 14–8.

14–7 Dynamic Factor K'_V

As noted earlier, dynamic factors are used to account for inaccuracies in the manufacture and meshing of gear teeth in action. *Transmission error* is defined as the departure from uniform angular velocity of the gear pair. Some of the effects which produce transmission error are:

- Inaccuracies produced in the generation of the tooth profile; these include errors in tooth spacing, profile lead, and runout
- Vibration of the tooth during meshing due to the tooth stiffness
- Magnitude of the pitch-line velocity
- Dynamic unbalance of the rotating members
- Wear and permanent deformation of contacting portions of the teeth
- Gearshaft misalignment and the linear and angular deflection of the shaft
- Tooth friction

In an attempt to gain some control over these effects, AGMA has defined a set of *quality-control numbers*.[7] These numbers define the tolerances for gears of various sizes manufactured to a specified quality class. Classes 3 to 7 will include most commercial-quality gears. Classes 8 to 12 are of precision quality. The AGMA *transmission accuracy-level number* Q_v can be taken as the same as the quality number. The following equations for the dynamic factor are based on these Q_v numbers:

$$K'_V = \begin{cases} \left(\dfrac{A+V}{A}\right)^B & V \text{ in ft/min} \\ \left(\dfrac{A+\sqrt{200V}}{A}\right)^B & V \text{ in m/s} \end{cases} \tag{14–27}$$

where

$$\begin{aligned} A &= 50 + 56(1 - B) \\ B &= 0.25\,(12 - Q_V)^{2/3} \end{aligned} \tag{14–28}$$

and the maximum velocity, representing the end point of the Q_V curve, is given by

$$(V_t)_{\max} = \begin{cases} [A + (Q_V - 3)]^2 & \text{ft/min} \\ \dfrac{[A + (Q_V - 3)]^2}{200} & \text{m/s} \end{cases} \tag{14–29}$$

7. AGMA 390.01.

Table 14–8

Elastic Coefficient C_p and Z_E *Source: AGMA 218.01*

		Gear Material and Modulus of Elasticity E_G, lb/in² (MPa)*					
Pinion Material	**Pinion Modulus of Elasticity E_p psi (MPa)***	**Steel 30×10^6 (2×10^5)**	**Malleable Iron 25×10^6 (1.7×10^5)**	**Nodular Iron 24×10^6 (1.7×10^5)**	**Cast Iron 22×10^6 (1.5×10^5)**	**Aluminum Bronze 17.5×10^6 (1.2×10^5)**	**Tin Bronze 16×10^6 (1.1×10^5)**
		$\sqrt{\text{psi}}$ ($\sqrt{\text{MPa}}$)					
Steel	30×10^6 (2×10^5)	2300 (191)	2180 (181)	2160 (179)	2100 (174)	1950 (162)	1900 (158)
Malleable iron	25×10^6 (1.7×10^5)	2180 (181)	2090 (174)	2070 (172)	2020 (168)	1900 (158)	1850 (154)
Nodular iron	24×10^6 (1.7×10^5)	2160 (179)	2070 (172)	2050 (170)	2000 (166)	1880 (156)	1830 (152)
Cast iron	22×10^6 (1.5×10^5)	2100 (174)	2020 (168)	2000 (166)	1960 (163)	1850 (154)	1800 (149)
Aluminum bronze	17.5×10^6 (1.2×10^5)	1950 (162)	1900 (158)	1880 (156)	1850 (154)	1750 (145)	1700 (141)
Tin bronze	16×10^6 (1.1×10^5)	1900 (158)	1850 (154)	1830 (152)	1800 (149)	1700 (141)	1650 (137)

Poisson's ratio = 0.30.

*When more exact values for modulus of elasticity are obtained from roller contact tests, they may be used.

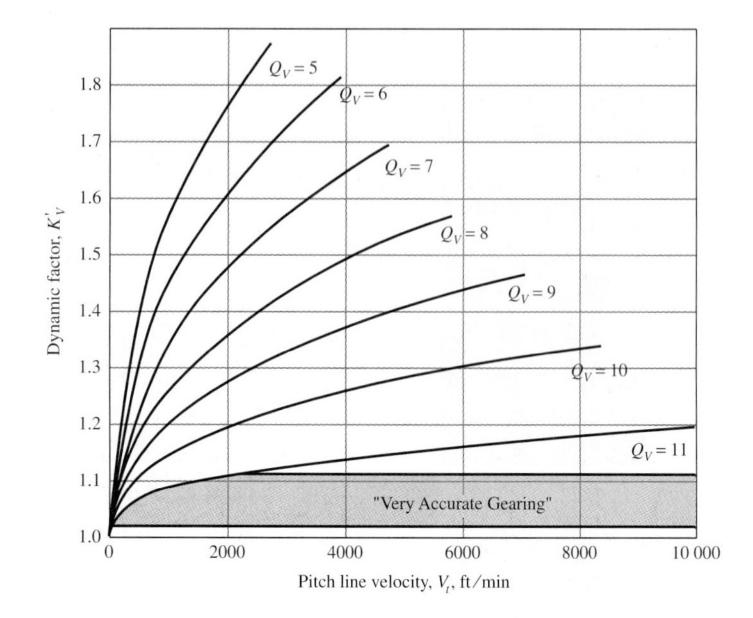

Figure 14–9

Dynamic factor K_V'. The equations to these curves are given by Eq. (14–27) and the end points by Eq. (14–29). *(ANSI/AGMA 2001-C95.)*

Figure 14–9 is a graph of K_V', the dynamic factor, as a function of pitch-line speed for graphical estimate of K_V'.

14–8 Overload Factor K_O

The overload factor K_O is intended to make allowance for all externally applied loads in excess of the nominal tangential load W^t in a particular application. Examples include variations in torque from the mean value due to firing of cylinders in an internal combustion engine or reaction to torque variations in a piston pump drive. Others call a similar factor an application factor or a service factor. These are established after considerable field experience in a particular application.[8]

14–9 Surface Condition Factors C_f and Z_R

The surface condition faction C_f or Z_R is used only in the pitting resistance equation. It depends on

- Surface finish as affected by, but not limited to, cutting, shaving, lapping, grinding, shotpeening
- Residual stress
- Plastic effects (work hardening)

Standard surface conditions for gear teeth have not yet been established. When a detrimental surface finish effect is known to exist, AGMA suggests a value of C_f greater than unity.

8. An extensive list of service factors appears in Howard B. Schwerdlin, "Couplings," Chap. 29 in Joseph E. Shigley and Charles R. Mischke (eds.), *Standard Handbook of Machine Design*, 2nd ed., McGraw-Hill, New York, 1996.

14–10 Size Factor K_s

The size factor reflects nonuniformity of material properties due to size. It depends upon

- Tooth size
- Diameter of part
- Ratio of tooth size to diameter of part
- Face width
- Area of stress pattern
- Ratio of case depth to tooth size
- Hardenability and heat treatment

Standard size factors for gear teeth have not yet been established for cases where there is a detrimental size effect. In such cases AGMA recommends a size factor greater than unity. If there is no detrimental size effect, use unity.

AGMA has identified and provided a location symbol for size factor. Also, AGMA suggests $K_s = 1$, which makes K_s a placeholder in Eqs. (14–15) and (14–16) until more information is gathered. Following the standard in this manner is a failure to bring all of your knowledge to bear. From Table 13–2, $l = 2.25/P$. The tooth thickness t in Fig. 14–6 is given in Sec. 14–1, Eq. (a), as $t = \sqrt{4lx}$ where x is $3Y/(2P)$ from Eq. (14–9). From Eq. (7–15) the equivalent diameter d_e of a rectangular section in bending is $d_e = 0.808\sqrt{Ft}$. From Eq. (7–10) $k_b = (d_e/0.30)^{-0.107}$. Noting that AGMA K_s is the reciprocal of k_b, the result of all the algebraic substitution is

$$K_s = \frac{1}{k_b} = 1.192\left(\frac{F\sqrt{Y}}{P}\right)^{0.0535} \tag{a}$$

AGMA K_s can be viewed as Lewis's geometry incorporated into the Marin size factor in fatigue. You may set AGMA $K_s = 1$, or you may elect to use the preceding Eq. (a). This is a point to discuss with your instructor. We will use Eq. (a) to remind you that you have a choice. If K_S in Eq. (a) is less than 1, use $K_S = 1$.

14–11 Load-Distribution Factor K_m or K_H

The load-distribution factor modified the stress equations to reflect nonuniform distribution of load across the line of contact. The ideal is to locate the gear "midspan" between two bearings at the zero slope place when the load is applied. However, this is not always possible. The following procedure is applicable to

- Net face width to pinion pitch diameter ratio $F/d \leq 2$
- Gear elements mounted between the bearings
- Face widths up to 40 in
- Contact, when loaded, across the full width of the narrowest member

The load-distribution factor under these conditions is given by

$$K_m = C_{mf} = 1 + C_{mc}\left(C_{pf}C_{pm} + C_{ma}C_e\right) \tag{14–30}$$

where

$$C_{mc} = \begin{cases} 1 & \text{for uncrowned teeth} \\ 0.8 & \text{for crowned teeth} \end{cases} \tag{14–31}$$

Table 14–9

Empirical Constants A, B, and C for Eq. (14–34), Face Width F in Inches*

Source: ANSI/AGMA 2001-C95.

Condition	A	B	C
Open gearing	0.247	0.0167	$-0.765(10^{-4})$
Commercial, enclosed units	0.127	0.0158	$-0.093(10^{-4})$
Precision, enclosed units	0.0675	0.0128	$-0.926(10^{-4})$
Extraprecision enclosed gear units	0.00360	0.0102	$-0.822(10^{-4})$

* See ANSI/AGMA 2101-C95, pp. 21–22, for SI formulation.

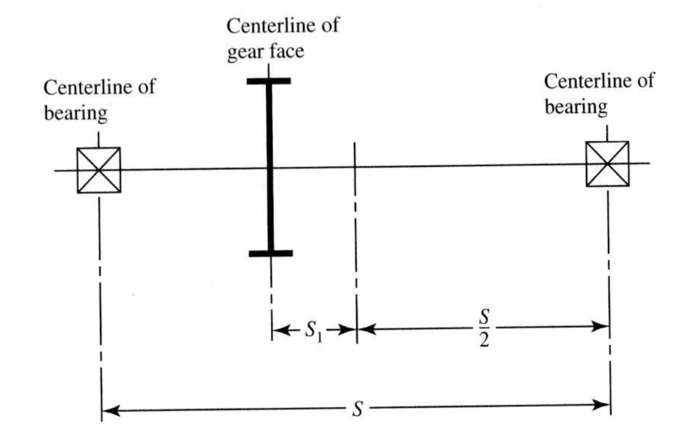

Figure 14–10

Definition of distances S and S_1 used in evaluating C_{pm}, Eq. (14–33). *(ANSI/AGMA 2001-C95.)*

$$C_{pf} = \begin{cases} \dfrac{F}{10d} - 0.025 & F \le 1 \text{ in} \\ \dfrac{F}{10d} - 0.0375 + 0.0125F & 1 < F \le 17 \text{ in} \\ \dfrac{F}{10d} - 0.1109 + 0.0207F - 0.000\,228F^2 & 17 < F \le 40 \text{ in} \end{cases} \tag{14–32}$$

Note that for values of $F/(10d) < 0.05$, use $F/(10d) = 0.05$.

$$C_{pm} = \begin{cases} 1 & \text{for straddle-mounted pinion with } S_1/S < 0.175 \\ 1.1 & \text{for straddle-mounted pinion with } S_1/S \ge 0.175 \end{cases} \tag{14–33}$$

$$C_{ma} = A + BF + CF^2 \quad \text{(see Table 14–9 for values of } A, B, \text{ and } C) \tag{14–34}$$

$$C_e = \begin{cases} 0.8 & \text{for gearing adjusted at assembly, or compatibility is improved by lapping, or both} \\ 1 & \text{for all other conditions} \end{cases} \tag{14–35}$$

See Fig. 14–10 for definitions of S and S_1 for use with Eq. (14–33), and see Fig. 14–11 for graph of C_{ma}.

14–12 Hardness-Ratio Factor C_H

The pinion generally has a smaller number of teeth than the gear and consequently is subjected to more cycles of contact stress. If both the pinion and the gear are through-hardened, then a uniform surface strength can be obtained by making the pinion harder than the gear. A similar effect can be obtained when a surface-hardened pinion is mated

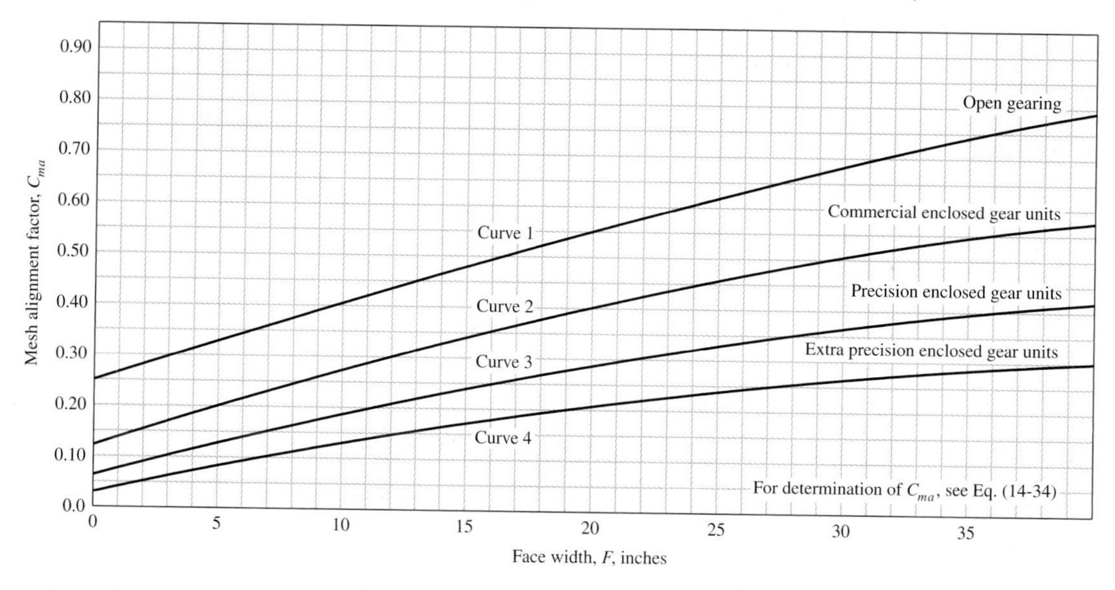

Figure 14–11

Mesh alignment factor C_{ma}. Curve-fit equations in Table 14–9. *(ANSI/AGMA 2001-C95.)*

with a through-hardened gear. The hardness-ratio factor C_H is used *only for the gear.* Its purpose is to adjust the surface strengths for this effect. The values of C_H are obtained from the equation

$$C_H = 1.0 + A'(m_G - 1.0) \tag{14–36}$$

where

$$A' = 8.98\left(10^{-3}\right)\left(\frac{H_{BP}}{H_{BG}}\right) - 8.29\left(10^{-3}\right) \quad 1.2 \le \frac{H_{BP}}{H_{BG}} \le 1.7$$

The terms H_{BP} and H_{BG} are the Brinell hardness (10-mm ball at 3000-kg load) of the pinion and gear, respectively. The term m_G is the speed ratio and is given by Eq. (14–22). See Fig. 14–12 for a graph of Eq. (14–36). For

$$\frac{H_{BP}}{H_{BG}} < 1.2, \quad A' = 0$$

$$\frac{H_{BP}}{H_{BG}} > 1.7, \quad A' = 0.006\,98$$

When surface-hardened pinions with hardnesses of 48 Rockwell C scale (Rockwell C48) or harder are run with through-hardened gears (180–400 Brinell), a work hardening occurs. The C_H factor is a function of pinion surface finish f_P and the mating gear hardness. Figure 14–13 displays the relationships:

$$C_H = 1 + B'(450 - H_{BG}) \tag{14–37}$$

where $B' = 0.000\,75 \exp[-0.0112 f_P]$ and f_P is the surface finish of the pinion expressed as root-mean-squared roughness R_a in μin.

Figure 14–12

Hardness ratio factor C_H (through-hardened steel). *(ANSI/AGMA 2001-C95.)*

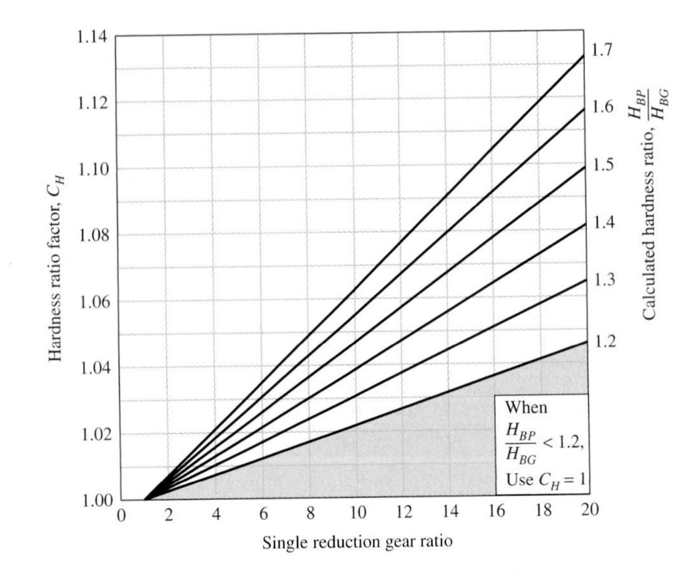

Figure 14–13

Hardness ratio factor C_H (surface-hardened steel pinion). *(ANSI/AGMA 2001-C95.)*

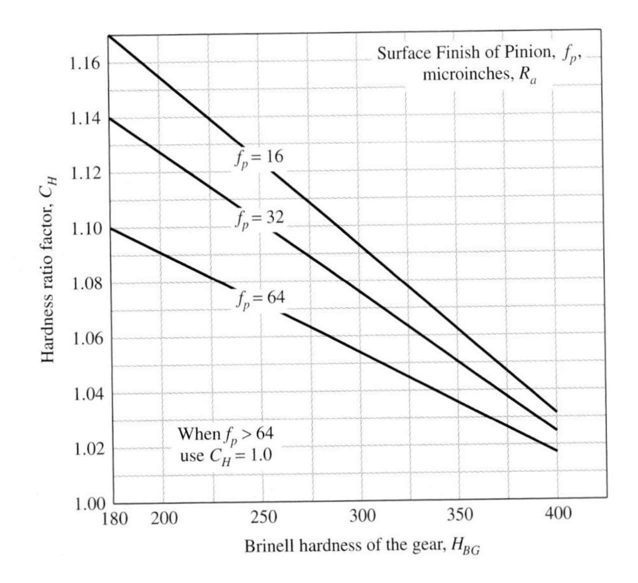

14–13 Load Cycles Factors Y_N and Z_N

The AGMA strengths as given in Figs. 14–2 through 14–4, in Tables 14–3 and 14–4 for bending fatigue, and in Fig. 14–5 and Tables 14–5 and 14–6 for contact-stress fatigue are based on 10^7 load cycles repeatedly applied. The purpose of the load cycles factors Y_N and Z_N is to modify the AGMA strength for lives other than 10^7 cycles. Values for these factors are given in Figs. 14–14 and 14–15. Note that for 10^7 cycles $Y_N = Z_N = 1$ on each graph. Note also that the equations for Y_N and Z_N change on either side of 10^7 cycles. For life goals slightly higher than 10^7 cycles, the mating gear may be experiencing fewer than 10^7 cycles and the equations for $(Y_N)_P$ and $(Y_N)_G$ can be different. The same comment applies to $(Z_N)_P$ and $(Z_N)_G$.

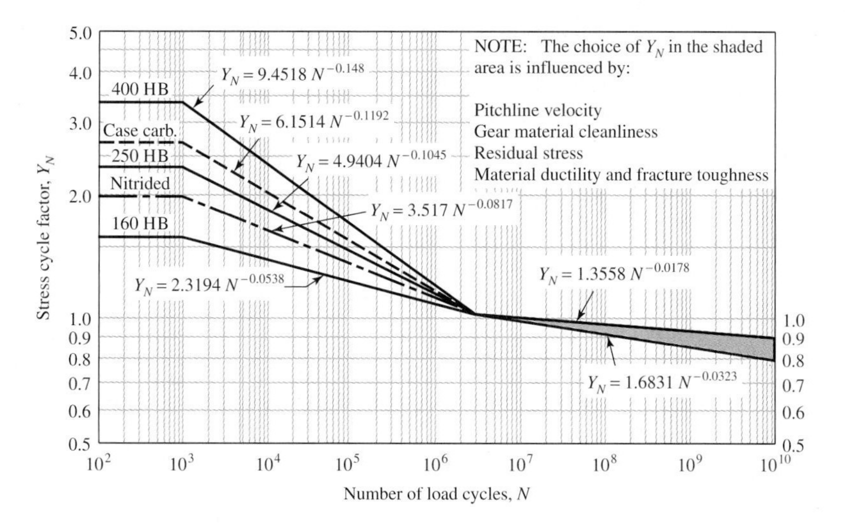

Figure 14–14

Repeatedly applied bending strength stress-cycle factor Y_N. *(ANSI/AGMA 2001-C95.)*

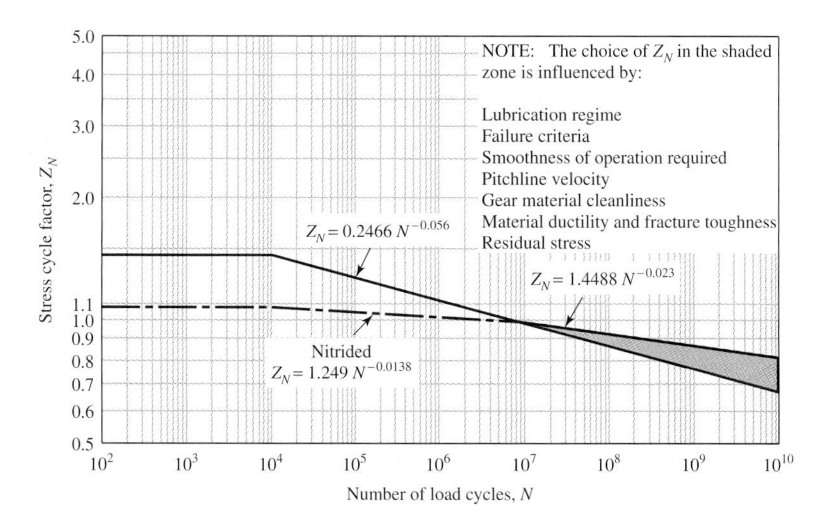

Figure 14–15

Pitting resistance stress-cycle factor Z_N. *(ANSI/AGMA 2001-C95.)*

14–14 Reliability Factors K_R and Y_Z

The reliability factor accounts for the effect of the statistical distributions of material fatigue failures. Load variation is not addressed here. The AGMA strengths S_t and S_c are based on a reliability of 99 percent. Table 14–10 is based on data developed by the U.S. Navy for bending and contact-stress fatigue failures.

The functional relationship between K_R and reliability is highly nonlinear. When interpolation is required, linear interpolation is too crude. A log transformation to each

Table 14–10

Reliability Factors K_R and Y_Z

Source: ANSI/AGMA 2001-C95.

Reliability	K_R, Y_Z
0.9999	1.50
0.999	1.25
0.99	1.00
0.90	0.85
0.50	0.70

quantity produces a linear string. A least-squares regression fit is

$$\begin{array}{ll} 0.658 - 0.0759 \ln(1 - R) & 0.5 < R < 0.99 \\ 0.50 - 0.109 \ln(1 - R) & 0.99 \le R \le 0.9999 \end{array} \qquad (14\text{–}38)$$

For cardinal values of R, take K_R from the table. Otherwise use the logarithmic interpolation afforded by Eqs. (14–38).

14–15 Temperature Factors K_T and Y_θ

For oil or gear-blank temperatures up to 250°F (120°C), use $K_T = Y_\theta = 1.0$. For higher temperatures, these factors should be greater than unity. Heat exchangers may be used to ensure that operating temperatures are considerably below this value, as is desirable for the lubricant.

14–16 Rim-Thickness Factor K_B

When the rim thickness is not sufficient to provide full support for the tooth root, the location of bending fatigue failure may be through the gear rim rather than at the tooth fillet. In such cases, the use of a stress-modifying factor K_B or (t_R) is recommended. This factor, the *rim-thickness factor* K_B, adjusts the estimated bending stress for the thin-rimmed gear. It is a function of the backup ratio m_B,

$$m_B = \frac{t_R}{h_t} \qquad (14\text{–}39)$$

where t_R = rim thickness below the tooth, in

h_t = the whole depth

The geometry is depicted in Fig. 14–16. The rim-thickness factor K_B is given by

$$K_B = \begin{cases} 1.6 \ln \dfrac{2.242}{m_B} & m_B < 1.2 \\ 1 & m_B \ge 1.2 \end{cases} \qquad (14\text{–}40)$$

Figure 14–16 also gives the value of K_B graphically. The rim-thickness factor K_B is applied in addition to the 0.70 reverse-loading factor when applicable.

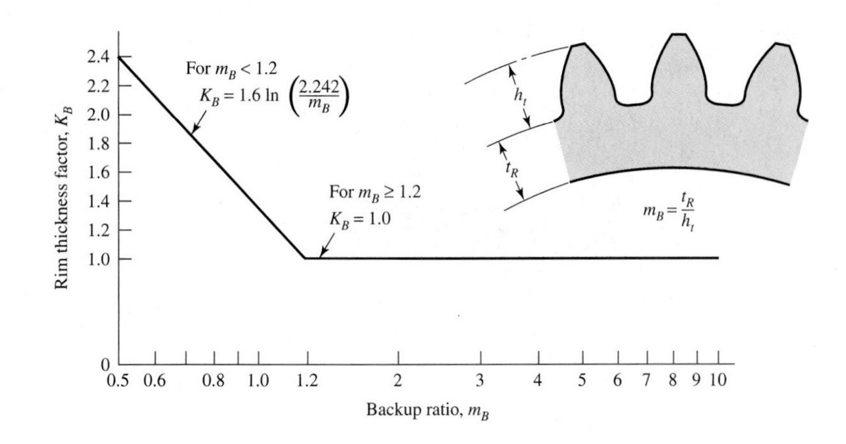

Figure 14–16

Rim thickness factor K_B. *(ANSI/AGMA 2001-C95.)*

14–17 Safety Factors S_F and S_H

The ANSI/AGMA standards 2001-C95 and 2101-C95 have reintroduced safety factor S_F guarding against bending fatigue failure and safety factor S_H guarding against pitting failure.

The definition of S_F, from Eq. (14–17), is

$$S_F = \frac{S_t Y_N/(K_T K_R)}{\sigma_{\text{imposed}}} = \frac{\text{fully corrected bending strength}}{\text{bending stress imposed}} \tag{14–41}$$

where σ_{imposed} is estimated from Eq. (14–15). It is a strength-over-stress definition in a case where the stress is linear with the transmitted load.

The definition of S_H, from Eq. (14–18), is

$$S_H = \frac{S_c Z_N C_H/(K_T K_R)}{\sigma_{\text{imposed}}} = \frac{\text{fully corrected contact strength}}{\text{contact stress imposed}} \tag{14–42}$$

when σ_{imposed} is estimated from Eq. (14–16). This, too, is a strength-over-stress definition but in a case where the stress is *not* linear with the transmitted load W^t.

While the definition of S_H does not interfere with its intended function, a caution is required when comparing S_F with S_H in an analysis in order to ascertain the nature and severity of the threat to loss of function. To render S_H linear with the transmitted load W^t it could have been defined as

$$S_H = \left(\frac{\text{fully corrected contact strength}}{\text{contact stress imposed}}\right)^2 \tag{14–43}$$

with the exponent 2 for linear or helical contact, or an exponent of 3 for crowned teeth (spherical contact). With the AGMA definition, Eq. (14–42), compare S_F with S_H^2 (or S_H^3 for crowned teeth) when trying to identify the threat to loss of function with confidence.

The role of the overload factor K_O is to include predictable excursions of load beyond W^t based on experience. A safety factor is intended to account for unquantifiable elements in addition to K_O. When designing a gear mesh, the quantity S_F becomes a design factor $(S_F)_d$ within the meanings used in this book. The quantity S_F evaluated as part of an adequacy assessment is a factor of safety. This applies equally well to the quantity S_H.

14–18 Analysis

Description of the AGMA procedure is highly detailed. The best review is a "road map" for bending fatigue and contact-stress fatigue. Figure 14–17 displays the contact-stress equation, the contact fatigue endurance strength equation, and the factor of safety S_H. Figure 14–18 identifies the AGMA bending stress equation, the endurance strength in bending equation, and the factor of safety S_F. When analyzing a gear problem, this figure is a useful reference.

The following example of a gear mesh analysis is intended to make all the details presented concerning the AGMA method more familiar.

SPUR GEAR WEAR
ANSI / AGMA 2001-C95

$$d_P = \frac{N_P}{P_d} = \frac{2C}{m_G \pm 1} \quad \begin{array}{l} + \text{External gears} \\ - \text{Internal gears} \end{array}$$

$$V = \frac{\pi d n}{12}$$

AGMA contact stress equation

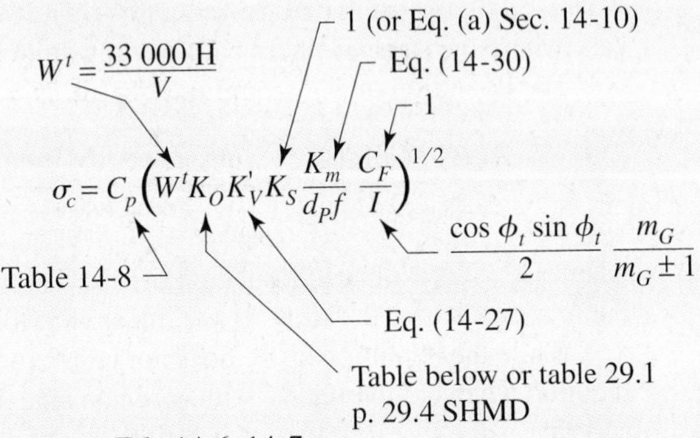

AGMA contact endurance strength

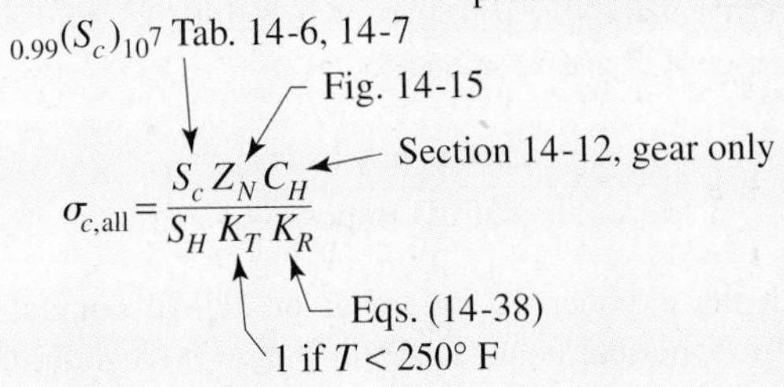

Wear factor of safety

$$S_H = \frac{S_c Z_N C_H / (K_T K_R)}{\sigma_c}$$

Remember to compare S_F with S_H^2 when deciding whether bending or wear is the threat to function. For crowned gears compare S_F with S_H^3.

Table of Overload Factors

	Driven Machine		
Power source	Uniform	Moderate shock	Heavy shock
Uniform	1.00	1.25	1.75
Light shock	1.25	1.50	2.00
Medium shock	1.50	1.75	2.25

Figure 14–17

Roadmap of AGMA wear equations, ANSI/AGMA 2001-C95.

SPUR GEAR BENDING
ANSI / AGMA 2001-C95

$$d_P = \frac{N_P}{P_d} = \frac{2C}{m_G \pm 1} \quad \begin{array}{l} + \text{External gears} \\ - \text{Internal gears} \end{array}$$

$$V = \frac{\pi d n}{12}$$

$$W^t = \frac{33\,000\ \text{H}}{V}$$

AGMA bending stress equation

$$\sigma = W^t K_O K'_V K_S \frac{P_d}{F} \frac{K_m K_B}{J}$$

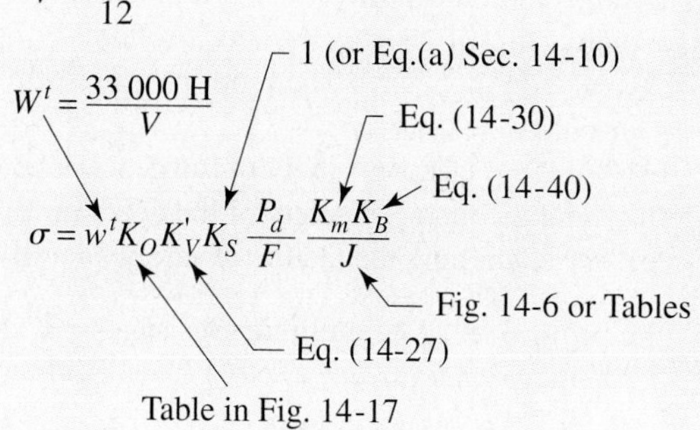

$_{0.99}(S_t)_{10^7}$ Tab. 14-3, 14-4

AGMA bending endurance strength equation

$$\sigma_{\text{all}} = \frac{S_t}{S_F} \frac{Y_N}{K_T K_R}$$

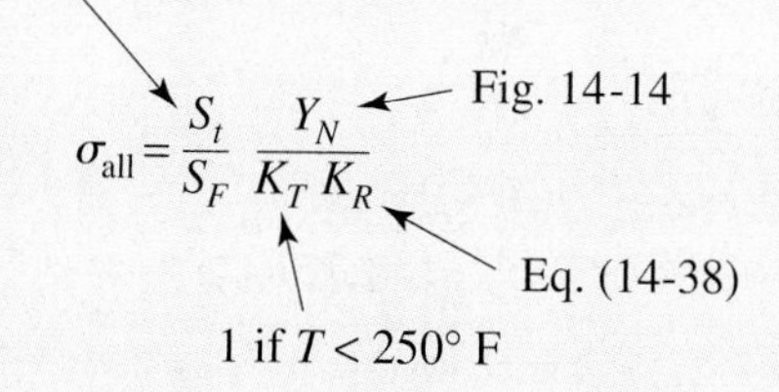

Bending factor of safety

$$S_F = \frac{S_t Y_N / (K_T K_R)}{\sigma}$$

Remember to compare S_F with S_H^2 when deciding whether bending or wear is the threat to function. For crowned gears compare S_F with S_H^3.

Table of Overload Factors

	Driven Machine		
Power source	Uniform	Moderate shock	Heavy shock
Uniform	1.00	1.25	1.75
Light shock	1.25	1.50	2.00
Medium shock	1.50	1.75	2.25

Figure 14–18

Roadmap of AGMA bending equations, ANSI/AGMA 2001-C95.

EXAMPLE 14–4 A 17-tooth 20° pressure angle spur pinion rotates at 1800 r/min and transmits 4 hp to a 52-tooth disk gear. The diametral pitch is 10, the face width 1.5 in, and the quality standard is No. 6. The gears are straddle-mounted with bearings immediately adjacent. The pinion is a grade 1 steel with a hardness of 240 Brinell tooth surface and through-hardened core. The gear is steel, through-hardened also, grade 1 material, with a Brinell hardness of 200, tooth surface and core. Poisson's ratio is 0.30, $J_P = 0.30$, $J_G = 0.40$, and Young's modulus is $30(10^6)$ psi. The loading is smooth due to motor and load. Assume a pinion life of 10^8 cycles and a reliability of 0.90, and use $Y_N = 1.3558N^{-0.0178}$, $Z_N = 1.4488N^{-0.023}$. The tooth profile is uncrowned. This is a commercial enclosed gear unit.

(a) Find the factor of safety of the gears in bending.

(b) Find the factor of safety of the gears in wear.

(c) By examining the factors of safety, identify the threat to each gear and to the mesh.

Solution We begin with some preliminaries: $K_o = 1$, $d_P = 17/10 = 1.700$ in, $d_G = 52/10 = 5.200$ in, speed ratio, $m_G = 52/17 = 3.06$, Lewis form factor $Y_P = 0.303$, $Y_G = 0.412$,

$$V = \frac{\pi d_P n_P}{12} = \frac{\pi(1.7)1800}{12} = 801 \text{ ft/min}$$

$$W^t = \frac{33\,000H}{V} = \frac{33\,000(4)}{801} = 164.8 \text{ lbf}$$

From Eq. (14–28)

$$B = 0.25(12 - 6)^{2/3} = 0.8255$$

$$A = 50 + 56(1 - 0.8255) = 59.77$$

From Eq. (14–27) velocity factor

$$K'_V = \left(\frac{59.77 + \sqrt{801}}{59.77}\right)^{0.8255} = 1.377$$

The size factors are

$$(K_s)_P = 1.192\left(\frac{1.5\sqrt{0.0303}}{10}\right)^{0.0535} = 1.043$$

$$(K_s)_G = 1.192\left(\frac{1.5\sqrt{0.412}}{10}\right)^{0.0535} = 1.0517$$

From Eq. (14–30) the load distribution factor K_m is

$$K_m = 1 + C_{mc}(C_{pf}C_{pm} + C_{ma}C_e)$$

$$C_{mc} = 1, \quad C_{pf} = 1.5/(10 \cdot 1.7) - 0.0375 + 0.0125(1.5) = 0.069$$

$C_{pm} = 1$, for commercial enclosed gear units $C_{ma} = 0.15$ from Fig. 14–11, $C_e = 1$

$$K_m = 1 + (1)[0.069(1) + 0.15(1)] = 1.219$$

The load cycle factors from Fig. 14–14 are

$$(Y_N)_P = 1.3558(10^8)^{-0.0178} = 0.977$$

$$(Y_N)_G = 1.3558(10^8/3.06)^{-0.0178} = 0.996$$

From Eq. (14–38) the reliability factor is

$$K_R = 0.657 - 0.0759 \ln(1 - 0.9) = 0.833$$

The temperature and surface condition factors are

$$K_T = 1, \quad C_f = 1$$

From Eq. (14–23) the geometry factor is

$$I = \frac{1}{2} \cos 20^\circ \sin 20^\circ \frac{3.06}{3.06 + 1} = 0.121$$

The elastic coefficient is

$$C_p = 2300\sqrt{\text{psi}}, \quad \text{Table 14–8}$$

AGMA strengths at 10^7 cycles and 0.99 reliability:

From Fig. 14–2 the allowable bending strengths are

$$(S_t)_P = 77.3(240) + 12\,800 = 31\,352 \text{ psi}$$

$$(S_t)_G = 77.3(200) + 12\,800 = 28\,260 \text{ psi}$$

From Fig. 14–5 the allowable contact strengths are

$$(S_c)_P = 322(240) + 29\,100 = 106\,380 \text{ psi}$$

$$(S_c)_G = 322(200) + 29\,100 = 93\,500 \text{ psi}$$

From Fig. 14–15 the stress cycle factors are

$$(Z_N)_P = 1.4488(10^8)^{-0.023} = 0.948$$

$$(Z_N)_G = 1.4488(10^8/3.06)^{-0.023} = 0.973$$

$$A' = 8.98(10^{-3})\frac{240}{200} - 8.29(10^{-3}) = 0.002\,49$$

$$C_H = 1 + 0.002\,49(3.06 - 1) = 1.005$$

(*a*) **Pinion tooth bending.** Substituting into Eq. (14–15) gives

$$\sigma = 164.8(1)(1.377)1.043\frac{10}{1.5}\frac{1.219(1)}{0.30} = 6412 \text{ psi}$$

Substituting into Eq. (14–17) with $S_F = 1$,

$$\sigma_{\text{all}} = \frac{31\,352(0.977)}{(1)0.833} = 36\,773 \text{ psi}$$

Answer $$S_F = \frac{36\,773}{6412} = 5.74$$

Gear tooth bending. Substituting into Eq. (14–15) gives the gear tooth bending stress

$$\sigma = 164.8(1)(1.377)1.0517\frac{10}{1.5}\frac{1.219(1)}{0.40} = 4848 \text{ psi}$$

Substituting into Eq. (14–17) with $S_F = 1$ gives

$$\sigma_{\text{all}} = \frac{282\,60(0.996)}{(1)0.833} = 33\,790 \text{ psi}$$

Answer $$S_F = \frac{33\,790}{4848} = 6.97$$

(*b*) **Pinion tooth wear.** Substituting into Eq. (14–16) gives

$$\sigma_c = 2300\left[164.8(1)1.377(1.043)\frac{1.219}{1.7(1.5)}\frac{1}{0.121}\right]^{1/2} = 70\,332 \text{ psi}$$

Substituting into Eq. (14–18) with $S_H = 1$ gives

$$\sigma_{c,\text{all}} = \frac{106\,380(0.948)}{(1)0.833} = 121\,156 \text{ psi}$$

Answer $$S_H = \frac{121\,156}{70\,332} = 1.72$$

Gear tooth wear. The Hertzian stress σ_C, changing only K_s, is 70 625 psi. Substituting in Eq. (14–18) without S_H gives

$$\sigma_{c,\text{all}} = \frac{93\,500(0.973)1.005}{(1)0.833} = 109\,760 \text{ psi}$$

Answer $$S_H = \frac{109\,760}{70\,625} = 1.55$$

(*c*) For the pinion we compare S_F with S_H^2, or 5.74 with $1.72^2 = 2.96$, so the threat in the pinion is from wear. For the gear we compare S_F with S_H^2, or 6.97 with $1.55^2 = 2.40$, so the threat in the gear is also from wear. Gear wear will probably retire the gearset.

There are perspectives to be gained from Ex. 14–4. First, the pinion is overly strong in bending compared to wear. The performance in wear can be improved by surface-hardening techniques, such as flame or induction hardening, nitriding, or carburizing and case-hardening, as well as shot-peening. This in turn permits the gearset to be made smaller. Second, in bending, the gear was stronger than the pinion, indicating that the gear core hardness could be reduced, may reduce tooth size that is, increase P and reduce diameter of the gears, or perhaps allow a cheaper material. Third, in wear, surface strength equations have the ratio $(Z_N)/K_R$. The values of $(Z_N)_P$ and $(Z_N)_G$ are affected by gear ratio m_G. The designer can control strength by specifying surface hardness. This point will be elaborated later.

Having followed a spur-gear analysis in detail in Ex. 14–4, it is timely to analyze a helical gearset under similar circumstances to observe similarities and differences.

EXAMPLE 14–5 A 17-tooth 20° normal pitch-angle helical pinion with a right-hand helix angle of 30° rotates at 1800 r/min when transmitting 4 hp to a 52-tooth helical gear. The normal diametral pitch is 10 teeth/in, the face width is 1.5 in, and the set has a quality number of 6. The gears are straddle-mounted with bearings immediately adjacent. The pinion and gear are made from a through-hardening steel with surface and core hardnesses of 240 Brinell on the pinion an surface and core hardnesses of 200 Brinell on the gear. The transmission is smooth, connecting an electric motor and a centrifugal pump. Assume a pinion life of 10^8 cycles and a reliability of 0.9 and use upper curves in Figs. 14–14 and 14–15.

(a) Find the factors of safety of the gears in bending.
(b) Find the factors of safety of the gears in wear.
(c) By examining the factors of safety identify the threat to each gear and to the mesh.

Solution Overload factor $K_O = 1$, $Y_P = 0.303$, $Y_G = 0.412$. The transverse diametral pitch is

$$P_t = P_n \cos\psi = 10\cos 30^\circ = 8.66 \text{ teeth/in}$$

Additionally, $d_P = N_P/P_t = 17/8.66 = 1.963$ in, $d_G = (52/17)1.963 = 6.004$ in, and $m_G = 52/17 = 3.06$. The pitch-line velocity is

$$V = \frac{\pi d_P n_P}{12} = \frac{\pi(1.963)1800}{12} = 925 \text{ ft/min}$$

$$W^t = \frac{33\,000(4)}{925} = 142.7 \text{ lbf}$$

We find the dynamic factor as follows:

$$B = 0.25(12-6)^{2/3} = 0.8255$$

$$A = 50 + 56(1 - 0.8255) = 59.77$$

$$K'_V = \left(\frac{59.77 + \sqrt{925}}{59.77}\right)^{0.8255} = 1.404$$

The transverse pressure angle is ϕ_t and is given by Eq. (13–15)

$$\phi_t = \tan^{-1}\left(\frac{\tan\phi_n}{\cos\psi}\right) = \tan^{-1}\left(\frac{\tan 20^\circ}{\cos 30^\circ}\right) = 22.796^\circ$$

The radii of the pinion and gear are $r_P = d_P/2 = 1.963/2 = 0.9815$ in, $r_G = 3.06(0.9815) = 3.004$ in. The addendum is $a = 1/P_n = 1/10 = 0.100$ in. The base-circle radius r_b of the pinion and gear are

$$(r_b)_P = r_P \cos\phi_t = 0.9815\cos 22.796^\circ = 0.9048 \text{ in}$$

$$(r_b)_G = r_G \cos\phi_t = 3.004\cos 22.796^\circ = 2.769 \text{ in}$$

From Eq. (14–25), the surface strength geometry factor

$$Z = \sqrt{(0.9815 + 0.1)^2 - 0.9048^2} + \sqrt{(3.004 + 0.1)^2 - 2.769^2}$$
$$-(0.9815 + 3.004)\sin 22.796^\circ$$
$$= 0.5924 + 1.4027 - 1.544\,27 = 0.4509 \text{ in}$$

Since the first two terms are less than 1.544 27, the equation for Z stands. From Eq. (14–24) the normal circular pitch P_N is

$$p_N = p_n \cos\phi_n = \frac{\pi}{P_n}\cos 20^\circ = \frac{\pi}{10}\cos 20^\circ = 0.2952 \text{ in}$$

From Eq. (14–21), the load sharing ratio

$$m_N = \frac{p_N}{0.95Z} = \frac{0.2952}{0.95(0.4509)} = 0.6891$$

Substituting in Eq. (14–23), the geometry factor I is

$$I = \frac{\sin 22.796^\circ \cos 22.796^\circ}{2(0.6891)}\,\frac{3.06}{3.06+1} = 0.195$$

From Fig. 14–6, geometry factors $J'_P = 0.45$ and $J'_G = 0.54$. Also from Fig. 14–7 the J-factor multipliers are 0.94 and 0.98, correcting J'_P and J'_G to

$$J_P = 0.45(0.94) = 0.423$$
$$J_G = 0.54(0.98) = 0.529$$

The size factors are estimated as in Eq. (a), Sec. 14–10 as

$$(K_s)_P = 1.192\left(\frac{1.5\sqrt{0.303}}{10}\right)^{0.0535} = 1.043$$

$$(K_s)_G = 1.192\left(\frac{1.5\sqrt{0.412}}{10}\right)^{0.0535} = 1.052$$

The load-distribution factor K_m is estimated from Eq. (14–32)

$$C_{pf} = \frac{1.5}{10(1.963)} - 0.0375 + 0.0125(1.5) = 0.0577$$

with $C_{mc} = 1$, $C_{pm} = 1$, $C_{ma} = 0.15$ from Fig. 14–11, and $C_e = 1$, therefore, using Eq. (14–30),

$$K_m = 1 + (1)[0.0577(1) + 0.15(1)] = 1.208$$

The stress cycle factors Y_N in bending are

$$(Y_N)_P = 1.3558(10^{-8})^{-0.0178} = 0.977$$
$$(Y_N)_G = 1.3558(10^8/3.06)^{-0.0178} = 0.996$$

The reliability factor K_R is

$$K_R = 0.0657 - 0.0759\ln(1 - 0.9) = 0.833$$

Also, $K_T = 1$ and $C_f = 1$. The AGMA strengths for 10^7 cycles and 0.99 reliability are

$$(S_t)_P = 77.3(240) + 12\,800 = 31\,352 \text{ psi}$$
$$(S_t)_G = 77.3(200) + 12\,800 = 28\,260 \text{ psi}$$
$$(S_c)_P = 332(240) + 29\,100 = 106\,380 \text{ psi}$$
$$(S_c)_G = 332(200) + 29\,100 = 93\,500 \text{ psi}$$

The stress cycle factors for pitting Z_N are

$$(Z_N)_P = 1.4488(10^8)^{-0.023} = 0.948$$
$$(Z_N)_G = 1.4488(10^8/3.06)^{-0.023} = 0.973$$

The hardness factor C_H is

$$A' = 8.98(10^{-3})\frac{240}{200} - 8.29(10^{-3}) = 0.002\,49$$

$$C_H = 1 + 0.002\,49(3.06 - 1) = 1.005$$

(a) **Pinion tooth bending.** Substituting in Eq. (14–15) using P_t gives

$$\sigma = 142.7(1)1.404(1.043)\frac{8.66}{1.5}\frac{1.208(1)}{0.423} = 3446 \text{ psi}$$

$$\sigma_{\text{all}} = \frac{31\,352(0.977)}{(1)0.833} = 36\,773 \text{ psi}$$

Answer $$(S_F)_P = \frac{36\,773}{3446} = 10.7$$

Gear tooth bending.

$$\sigma = 142.7(1)1.404(1.052)\frac{8.66}{1.5}\frac{1.208(1)}{0.529} = 2778 \text{ psi}$$

$$\sigma_{\text{all}} = \frac{28\,260(0.977)}{(1)0.833} = 33\,145 \text{ psi}$$

Answer $$(S_F)_G = \frac{33\,145}{2778} = 11.9$$

(*b*)**Pinion tooth wear.** Substituting into Eq. (14–16) gives

$$\sigma_c = 2300\left[142.7(1)1.404(1.043)\frac{1.208(1)}{1.5(1.963)0.195}\right]^{1/2} = 48\,225 \text{ psi}$$

$$\sigma_{c,\text{all}} = \frac{106\,380(0.948)}{(1)0.833} = 121\,066 \text{ psi}$$

Answer $$(S_H)_P = \frac{121\,066}{48\,225} = 2.51$$

Gear tooth wear. The contact stress σ_c, changing only K_s for the gear, namely 48 432 psi, is

$$\sigma_{c,\text{all}} = \frac{93\,500(0.973)1.005}{(1)0.833} = 109\,760 \text{ psi}$$

Answer $$(S_H)_G = \frac{109\,760}{48\,432} = 2.27$$

(*c*) For the pinion we compare S_F with S_H^2, or 10.7 with $2.51^2 = 6.30$, so the threat in the pinion is from wear. For the gear we compare S_F with S_H^2, or 11.9 with $2.77^2 = 7.7$, so the threat is also from wear in the gear. For the meshing gearset wear controls, and gear wear will probably retire the gearset.

It is worthwhile to compare Ex. 14–4 with Ex. 14–5. The spur and helical gearsets were placed in nearly identical circumstances. The helical gear teeth are of greater length because of the helix and identical face widths. The pitch diameters of the helical gears are larger. The J factors and the I factor are larger, thereby reducing stresses. The result is larger factors of safety. In the design phase the gearsets in Ex. 14–4 and Ex. 14–5 can be made smaller with control of materials and relative hardnesses.

Now that examples have given the AGMA parameters substance, it is time to examine some desirable (and necessary) relationships between material properties of spur gears in mesh. In bending, the AGMA equations are displayed side by side:

$$\sigma_P = \left(W^t K_O K_V' K_s \frac{P_d}{F}\frac{K_m K_B}{J}\right)_P \qquad \sigma_G = \left(W^t K_O K_V' K_s \frac{P_d}{F}\frac{K_m K_B}{J}\right)_G$$

$$(\sigma_{\text{all}})_P = \left(\frac{S_t Y_N}{K_T K_R}\right)_P \qquad (\sigma_{\text{all}})_G = \left(\frac{S_t Y_N}{K_T K_R}\right)_G$$

$$(S_F)_P = \left(\frac{\sigma_{\text{all}}}{\sigma}\right)_P \qquad (S_F)_G = \left(\frac{\sigma_{\text{all}}}{\sigma}\right)_G$$

Equating the factors of safety, substituting for stress and strength, canceling identical terms (K_s virtually equal or exactly equal), and solving for $(S_t)_G$ gives

$$(S_t)_G = (S_t)_P \frac{(Y_N)_P}{(Y_N)_G} \frac{J_P}{J_G} = (S_t)_P m_G^\beta \frac{J_P}{J_G} \tag{14–44}$$

Equation (14–44) shows that the gear can be less strong (lower Brinell hardness) than the pinion for the same safety factor.

EXAMPLE 14–6

A 300-Brinell 18-tooth 16-pitch 20° full-depth pinion meshes with a 64-tooth gear, both of grade 1 through-hardened steel. The form factors are $J_P = 0.32$ and $J_G = 0.41$. Using $\beta = -0.023$, what hardness can the gear have for the same factor of safety?

Solution

For through-hardening grade 1 steel the pinion strength $(S_t)_P$ is given in Fig. 14–2

$$(S_t)_P = 77.3(300) + 12\,800 = 35\,990 \text{ psi}$$

Equation (14–44) gives

$$(S_t)_G = 35\,990 \left(\frac{64}{18}\right)^{-0.023} \frac{0.32}{0.41} = 27\,282 \text{ psi}$$

Use the equation in Fig. 14–2 again.

Answer

$$(H_B)_G = \frac{27\,282 - 12\,800}{77.3} = 187 \text{ Brinell}$$

The AGMA contact-stress equations also are displayed side by side:

$$(\sigma_c)_P = C_p \left(W^t K_O K_V' K_s \frac{K_m}{d_P F} \frac{C_f}{I}\right)_P^{1/2} \qquad (\sigma_c)_G = C_p \left(W^t K_O K_V' K_s \frac{K_m}{d_P F} \frac{C_f}{I}\right)_G^{1/2}$$

$$(\sigma_{c,\text{all}})_P = \left(\frac{S_c Z_N}{K_T K_R}\right)_P \qquad (\sigma_{c,\text{all}})_G = \left(\frac{S_c Z_N C_H}{K_T K_R}\right)_G$$

$$(S_H)_P = \left(\frac{\sigma_{c,\text{all}}}{\sigma_c}\right)_P \qquad (S_H)_G = \left(\frac{\sigma_{c,\text{all}}}{\sigma_c}\right)_G$$

Equating the factors of safety, substituting stress and strength relations, canceling identical terms including K_s gives, after solving for $(S_c)_G$,

$$(S_c)_G = (S_c)_P \frac{(Z_N)_P}{(Z_N)_G} \left(\frac{1}{C_H}\right)_G = (S_C)_P m_G^\beta \left(\frac{1}{C_H}\right)_G$$

Since C_H is so close to unity, it is usually neglected; therefore

$$(S_c)_G = (S_c)_P m_G^\beta \tag{14–45}$$

EXAMPLE 14–7 For $\beta = -0.056$ for a through-hardened steel, grade 1, continue Ex. 14–6 for wear.

Solution From Fig. 14–5

$$(S_c)_P = 322(300) + 29\,100 = 125\,700 \text{ psi}$$

From Eq. (14–45),

$$(S_c)_G = (S_c)_P \left(\frac{64}{18}\right)^{-0.056} = 125\,700 \left(\frac{64}{18}\right)^{-0.056} = 117\,080 \text{ psi}$$

Answer
$$(H_B)_G = \frac{117\,080 - 29\,200}{322} = 273 \text{ Brinell}$$

slightly less than the pinion hardness of 300 Brinell.

Equations (14–44) and (14–45) apply as well to helical gears.

Since $\sigma = \sigma_{\text{all}}/n$ where n is the factor of safety (or design factor), it follows that the allowed W^t is given by

$$W^t = \frac{S_t Y_N}{n K_T K_R} \frac{F J}{K_O K_V' K_s P_d K_m K_B}$$

Also, $\sigma_c = \sigma_{c,\text{all}}/\sqrt{n}$, and it follows that

$$W_t = \frac{S_c^2 Z_N^2}{n K_T^2 K_R^2} \frac{d_P F I}{C_p^2 K_O K_V' K_s K_m C_f}$$

Since $n W^t$ is the same on a gear, say the pinion, then it follows that

$$S_{cP} = \frac{C_p}{Z_{NP}} \sqrt{\frac{S_{tP} Y_{NP} J_P K_T K_R C_f}{K_B N_P I}} \tag{14–46}$$

Equation (14–46) relates S_{cP} to S_{tP} for equal strength. Returning to Ex. 14–4, using Eq. (14–46), we have, recalling $(S_t)_P$ from Fig. 14–2 is 31 352 psi,

$$S_{cP} = \frac{2300}{0.948} \sqrt{\frac{31\,352(0.977)0.30(1)0.833(1)}{(1)17(0.121)}} = 148\,001 \text{ psi}$$

$$(H_{BP})_{\text{case}} = \frac{148\,001 - 29\,100}{322} \doteq 369 \text{ Brinell}$$

From Eq. (14–44),

$$S_{tG} = S_{tP} m G^{-0.0178} \frac{J_P}{J_G} \frac{K_{sG}}{K_{sP}} = 31\,352 \left(\frac{52}{17}\right)^{-0.0178} \frac{0.30}{0.40} \frac{1.0517}{1.043} = 23\,243 \text{ psi}$$

$$(H_{BG})_{\text{core}} = \frac{23\,243 - 12\,800}{77.3} \doteq 135 \text{ Brinell}$$

From Eq. (14–45), since Brinell hardness of gear case is slightly less than 369, $C_H = 1$

and

$$S_{cG} = S_{cP} m_G^{-0.023} \frac{1}{C_H} = 148\,001 \left(\frac{52}{17}\right)^{-0.023} (1) = 144\,244 \text{ psi}$$

$$(H_{BG})_{\text{case}} = \frac{144\,244 - 29\,100}{322} \doteq 358 \text{ Brinell}$$

The hardness schedule, following from pinion core hardness of 240 Brinell, is shown below together with the associated W^t for pinion bending, gear bending, pinion wear, and gear wear, respectively:

	Core	Case
Pinion	240	369
Gear	135	358

$$W_1^t = 944.3 \text{ lb}, \text{hp}_1 = 22.92$$
$$W_2^t = 944.1 \text{ lb}, \text{hp}_2 = 22.91$$
$$W_3^t = 945.4 \text{ lb}, \text{hp}_3 = 22.95$$
$$W_3^t = 940.1 \text{ lb}, \text{hp}_4 = 22.83$$

The rated power for a factor of safety of unity is min(22.92, 22.91, 22.95, 22.83), or 22.83 hp. We have discovered the "one-hoss shay"[9] hardness pattern corresponding to a pinion core hardness of 240 Brinell. Such a pattern is a good place to begin in a gear mesh analysis. Two very costly procedures in producing a product (or its elements) are heat treating and painting. These should be minimized. The four hardnesses of the gearset speak loudly to material and process issues.

14–19 An Adequacy Assessment of a Gear Mesh

A specification set for a gear mesh includes the following:

- Tooth system: pressure angle, addendum, dedendum, root fillet radius
- Number of teeth on pinion and gear
- Quality number (tooth-forming method, inspection method)
- Diametral pitch
- Face width
- Pinion: material, core and case hardness
- Gear: material, core and case hardness

The service descriptors include (describing function):

- Power, speed
- Overload factor
- Reliability, life
- Mounting

By now it is clear that among steel materials, the hardnesses in core and case are the focus of attention. As you know from earlier work there is a correlation between hardness and ultimate tensile strength and between endurance limit and hardness. What AGMA has done is present correlations between 99th percentile, one-way endurance

9. From the much declaimed poem "Wonderful One-Hoss Shay" by Oliver Wendell Holmes (1809–1894), also known as "The Deacon's Masterpiece"—required reading for all would-be designers.

strength in bending at the 10^7 cycle level. Similarly, AGMA has presented correlations between 99th percentile endurance strength in contact and at the 10^7 cycle level. For example, in Fig. 14–5 for grade 1 through-hardened steels the correlation relation is

$$S_c = 322 H_B + 29\,100 \text{ psi}$$

It is important that the material selected be capable of achieving the hardness specified when in production in conjunction with other processes. The production personnel want a hardness range in the specification and the designer provides it based on knowledge of their capability. If production can routinely control within a range of 20 points on the Brinell scale, then the specification is posed that way. If the designer must have 300 Brinell or more, then the specification will be a range with 300 at its lower limit.

Gear materials[10] include

- Cast irons: ASTM grades 20–60; ductile irons, malleable irons
- Untreated wrought steels with 30–40 points of carbon
- Cold-rolled steel up to 3 in for stock
- Through-hardened 1335, 3135, 4037, 4140, 5150, 8640, 8740
- Case-hardened materials:
  Cyaniding medium-carbon steels
  Flame and induction-hardening to 500 Brinell
  Carburizing to 600 Brinnell 4118, 4320, 4620, 4720, 4820, 5120, 8620
  Nitriding, Nitralloys, 4140, 4340
  Shot-peening (sometimes)

Treatment sequences include

- Blank–normalize–machine–heat-treat
- Blank–normalize–heat-treat–machine
- Blank–normalize–rough cut–heat-treat–finish cut

depending on size. Cold-rolled is usually available to about 3 in, hot-rolled to about 6 in, and casting for larger sizes.

Since a gearset has no intrinsic safety margin unless the service conditions are specified, the service conditions must be known before an adequacy assessment is attempted. For this reason function is included as the first element of the adequacy assessment.

Adequacy Assessment for Spur and Helical Gear Meshes

- Function descriptors: power, speed, overload, reliability, life, quality
- Interference
- Face width: $3\pi/P_d \leq F \leq 5\pi/P_d$
- Bending factor of safety, pinion and gear
- Wear factor of safety, pinion and gear (remember the exponent)

10. An excellent source of strength and hardenability information is Bethlehem Steel Corporation, *Modern Steels and Their Properties,* 7th ed., Handbook 2757, Bethlehem, PA, 1972.

14–20 Design of a Gear Mesh

A useful decision set for spur and helical gears includes

- Function: load, speed, reliability, life, K_O
- Unquantifiable risk: design factor n_d
- Tooth system: ϕ, ψ, addendum, dedendum, root fillet radius
- Gear ratio m_G, N_p, N_G
- Quality number Q_V

} a priori decisions

- Diametral pitch P_d
- Face width F
- Pinion material, core hardness, case hardness
- Gear material, core hardness, case hardness

} design decisions

The first item to notice is the dimensionality of the decision set. There are four design decision categories, eight different decisions if you count them separately. This is a larger number than we have encountered before. It is important to use a design strategy that is convenient in either longhand execution or computer implementation. The design decisions have been placed in order of importance (impact on the amount of work to be redone in iterations). The steps are, after the a priori decisions have been made,

- Choose a diametral pitch.
- Examine implications on face width, pitch diameters, and material properties. If not satisfactory, return to pitch decision for change.
- Choose a pinion material and examine core and case hardness requirements. If not satisfactory, return to pitch decision and tumble down, iterating until no decisions are changed.
- Choose a gear material and examine core and case hardness requirements. If not satisfactory, return to pitch decision and tumble down, iterating until no decisions are changed.

With these plan steps in mind, we can consider them in more detail.

First select a trial diametral pitch.

Pinion:

- Select a median face width for this pitch, $4\pi/P$
- Find the range of necessary ultimate strengths
- Choose a material and a core hardness
- Find face width to meet factor of safety in bending
- Choose face width
- Check factor of safety in bending

Gear:

- Find necessary companion core hardness
- Choose a material and core hardness
- Check factor of safety in bending

Pinion:

- Find necessary S_c and attendant case hardness

- Choose a case hardness
- Check factor of safety in wear

Gear:

- Find companion case hardness
- Choose a case hardness
- Check factor of safety in wear

Completing this set of steps will yield a satisfactory design. Additional designs with diametral pitches adjacent to the first satisfactory design will produce several among which to choose. A figure of merit is necessary in order to choose the best. Unfortunately, a figure of merit in gear design is complex in an academic environment because material and processing cost vary. The possibility of using a process depends on the manufacturing facility if gears are made in-house.

After examining Ex. 14–4 and Ex. 14–5 and seeing the wide range of factors of safety, one might entertain the notion of considering a "one-hoss shay" approach (all factors of safety equal).[11] In steel gears wear is usually controlling and $(S_H)_P$ and $(S_H)_G$ can be brought close to equality. The use of softer cores can bring down $(S_F)_P$ and $(S_F)_G$, but there is value in keeping them higher. A tooth broken by bending fatigue not only can destroy the gear set, but can bend shafts, damage bearings, and produce inertial stresses up- and downstream in the power train, causing damage elsewhere if the gear box locks.

EXAMPLE 14–8

Design a 4:1 spur-gear reduction for a 100-hp, three-phase squirrel-cage induction motor running at 1120 r/min. The load is smooth, providing a reliability of 0.95 at 10^9 revolutions of the pinion. Gearing space is meager. Use Nitralloy 135M, grade 1 material to keep the gear size small. The gears are heat-treated first then nitrided.

Solution

Make the a priori decisions:

- Function: 100 hp, 1120 r/min, $R = 0.95$, $N = 10^9$ cycles, $K_O = 1$
- Design factor for unquantifiable exingencies: $n_d = 2$
- Tooth system: $\phi_n = 20°$, addendum $1/P_d$, dedendum $1.25/P_d$; $r_f = 0.300/P_d$
- Tooth count: $N_P = 18$ teeth, $N_G = 72$ teeth (no interference)
- Quality number: $Q_V = 6$, use grade 1 material

Pitch: Select a trial diametral pitch: $P_d = 4$ teeth/in. The geometry factors from Fig. 14–6 are $J_P = 0.32$, $J_G = 0.415$. Preliminaries: $Y_P = 0.309$, $Y_G = 0.4324$, $d_P = 18/4 = 4.5$ in, $d_G = 72.4 = 18$ in,

$$V = \frac{\pi(4.5)1120}{12} = 1319.5 \text{ ft/min}$$

$$B = 0.25(12 - 6)^{2/3} = 0.8255$$

$$A = 50 + 56(1 - 0.8255) = 59.77$$

11. In designing gears it makes sense to define the factor of safety in wear as $(\sigma_{c,\text{all}}/\sigma_c)^2$ for uncrowned teeth, so that there is no mix-up. ANSI, in the preface to ANSI/AGMA 2001-C95 and 2101-C95, states "the use is completely voluntary. . . does not preclude anyone from using . . . procedures . . . not conforming to the standards."

$$K'_V = \left(\frac{59.77 + \sqrt{1319.5}}{59.77}\right)^{0.8255} = 1.480$$

$$W^t = \frac{33\,000(100)}{1319.5} = 2501 \text{ lbf}$$

The reliability factor from Eq. (14–38) is

$$K_R = 0.658 - 0.0759 \ln(1 - 0.95) = 0.885$$

From Fig. 14–14,

$$(Y_N)_P = 1.3558(10^9)^{-0.0178} = 0.938$$

$$(Y_N)_G = 1.3558(10^9/4)^{-0.0178} = 0.961$$

From Fig. 14–15

$$(Z_N)_P = 1.4488(10^9)^{-0.023} = 0.900$$

$$(Z_N)_G = 1.4488(10^9/4)^{-0.023} = 0.929$$

Use a midrange face width $F = 4\pi/P_d = 4\pi/4 = 3.14$ in. Then the size factor is

$$K_s = 1.192\left(\frac{3.14\sqrt{0.309}}{4}\right)^{0.0535} = 1.1403$$

$$\frac{F}{10d_P} = \frac{3.14}{10(4.5)} = 0.0698$$

so Eq. (14–32) applies. $C_{mc} = 1$, $C_{pm} = 1$, $C_{ma} = 0.175$ from Fig. 14–11, $C_e = 1$ and

$$C_{pf} = 0.0698 - 0.0375 + 0.0125(3.14) = 0.0716$$

$$K_m = 1 + (1)[0.0716(1) + 0.175(1)] = 1.247$$

Pinion tooth bending. With the above estimates of K_s and K_m from the trial dimetral pitch, we check to see if the mesh width F is controlled by bending or wear considerations. Equating Eqs. (14–15) and (14–17), substituting $n_d W^t$ for W^t, and solving for the face width $(F)_{\text{bend}}$ necessary to resist bending fatigue, we obtain

$$(F)_{\text{bend}} = n_d W^t K_O K'_V K_s P_d \frac{K_m K_B}{J_P} \frac{K_T K_R}{S_t Y_n}$$

Equating Eqs. (14–16) and (14–18), substituting $n_d W^t$ for W^t, and solving for the face width $(F)_{\text{wear}}$ necessary to resist wear fatigue, we obtain

$$(F)_{\text{wear}} = \left(\frac{C_p Z_N}{S_c K_T K_R}\right)^2 n_d W^t K_O K'_V K_s \frac{K_m C_f}{d_p I}$$

$$I = \frac{1}{2}\cos 20^\circ \sin 20^\circ \frac{4}{4+1} = 0.129$$

From Table 14–5 the hardnesses of Nitralloy 135M are Rockwell C32–36 (302–335

Brinell). Choosing a midrange hardness as attainable, using 320 Brinell. From Fig. 14–4,

$$S_t = 86.2(320) + 12\,730 = 40\,314 \text{ psi}$$

Inserting the numerical value of S_t in the equation for $(F)_b$ to estimate the face width gives

$$(F)_{\text{bend}} = 2(2501)(1)1.48(1.14)4\frac{1.247(1)(1)0.885}{0.32(40\,314)0.938} = 3.08 \text{ in}$$

From Table 14–6 for Nitralloy 135M, $S_c = 170\,000$ psi. Inserting this in the equation for $(F)_{\text{wear}}$ we find

$$(F)_{\text{wear}} = \left(\frac{2300(0.900)}{170\,000(1)0.885}\right)^2 2(2501)1(1.48)1.14\frac{1.247(1)}{4.5(0.129)} = 3.43 \text{ in}$$

Decision

Make face width 3.50 in. Correct K_s and K_m:

$$K_s = 1.192\left(\frac{3.50\sqrt{0.309}}{4}\right)^{0.0535} = 1.1469$$

$$\frac{F}{10d_P} = \frac{3.50}{10(4.5)} = 0.0777$$

$$C_{pf} = 0.0777 - 0.0375 + 0.0125(3.50) = 0.084$$

$$K_m = 1 + (1)[0.084(1) + 0.175(1)] = 1.259$$

The allowable stress in the pinion in bending is

$$\sigma_{\text{all}} = \frac{40\,314(0.938)}{(1)0.885} = 42\,728 \text{ psi}$$

The bending stress induced by W^t in bending is

$$\sigma = 2501(1)1.48(1.147)\frac{4}{3.25}\frac{1.25(1)}{0.32} = 20\,411 \text{ psi}$$

The factor of safety in bending of the pinion is

$$(S_F)_P = \frac{\sigma_{\text{all}}}{\sigma} = \frac{42\,728}{20\,411} = 2.09$$

Gear tooth bending. *Decision:* Use cast gear blank because of the 18-in pitch diameter. Use the same material, heat-treatment, and nitriding. The load-induced bending stress is in the ratio of J_P/J_G. Then

$$\sigma = 19\,088\frac{0.32}{0.415} = 14\,718 \text{ psi}$$

$$\sigma_{\text{all}} = \frac{40\,314(0.961)}{(1)0.885} = 43\,776 \text{ psi}$$

The factor of safety of the gear in bending is

$$(S_F)_G = \frac{\sigma_{\text{all}}}{\sigma} = \frac{43\,776}{14\,718} = 2.97$$

Pinion tooth wear. The case hardness of Nitralloy 135M is Rockwell C62–65, which is over 600 Brinell. We do not need the Brinell hardness of a gear specifically to find S_c. Use Table 14–6 to find $S_c = 170\,000$ psi. Then

$$\sigma_c = 2300\left[2501(1)1.48(1.147)\frac{1.25}{4.5(3.25)0.129}\right]^{1/2} = 121\,986 \text{ psi}$$

$$\sigma_{c,\text{all}} = \frac{170\,000(0.900)}{(1)0.885} = 172\,881 \text{ psi}$$

The factor of safety based on load W^t is

$$n = (S_H)_P^2 = \left(\frac{\sigma_{c,\text{all}}}{\sigma_c}\right)^2 = \left(\frac{172\,881}{121\,986}\right)^2 = 2.01$$

Gear tooth wear. The contact stress on the gear is the same as on the pinion, $\sigma_c = 117\,966$ psi, as is the wear strength $S_c = 170\,000$ psi, $C_H = 1$ since the hardnesses are the same:

$$\sigma_{c,\text{all}} = \frac{170\,000(0.929)(1)}{(1)0.885} = 178\,452 \text{ psi}$$

The factor of safety of the gear in bending based on load W^t is

$$n = (S_H)_G^2 = \left(\frac{\sigma_{c,\text{all}}}{\sigma_c}\right)^2 = \left(\frac{178\,452}{117\,866}\right)^2 = 2.29$$

Rim. Keep $m_B \geq 1.2$. The whole depth is h_t = addendum + dedendum = $1/P_d + 1.25/P_d = 2.25/P_d = 2.25/4 = 0.5625$ in. The rim thickness t_R is

$$t_R \geq m_B h_t = 1.2(0.5625) = 0.675 \text{ in}$$

In the design of the gear blank, be sure the rim thickness exceeds 0.675 in; if it does not, review and modify this mesh design.

This design example showed a satisfactory design for a four-pitch spur-gear mesh. Material could be changed, as could pitch. There are a number of other satisfactory designs, thus a figure of merit is needed to identify the best.

One can appreciate that gear design was one of the early applications of the digital computer to mechanical engineering. The program should be interactive, presenting results of calculations, pausing for a decision by the designer, and showing the consequences of the decision, with the loop back to change a decision for the better. The program can be structured in totem-pole fashion, with the most influential decision at the top, then tumbling down, decision after decision, ending with the ability to change the current decision or to begin again. Such a program would make a fine class project. Troubleshooting the coding will reinforce your knowledge, adding flexibility as well as bells and whistles in subsequent terms.

Standard gears may not be the most economical design that meets the functional requirements, because no application is standard in all respects.[12] Methods of designing custom gears are well-understood and frequently used in mobile equipment due to

12. See H. W. Van Gerpen, C. K. Reece, and J. K. Jensen, *Computer Aided Design of Custom Gears,* Van Gerpen–Reece Engineering, Cedar Falls, Iowa, 1996.

good weight-to-performance index. The required calculations including optimizations are within the capability of a personal computer.

PROBLEMS

Problems 14–1 through 14–8 are based on material presented in Secs. 14–1 and 14–2. They are couched in fundamentals learned so far, in simplified situations, so the student may gain perspective and insights concerning gear vocabulary, geometry, and levels of stress and strength. AGMA methods are for the gear designer, and thus can take more into account.

14–1 A steel spur pinion has a pitch of 6 teeth/in, 17 full-depth milled teeth, and a pressure angle of 20°. The pinion has an ultimate tensile strength at the involute surface of 116 kpsi, a Brinell hardness of 232, and a yield strength of 90 kpsi. Its shaft speed is 1120 r/min, its face width is 2 in, and its mating gear has 51 teeth. Rate the pinion for power transmission if the design factor is 2.

(*a*) Pinion bending fatigue imposes what power limitation?

(*b*) Pinion surface fatigue imposes what power limitation? The gear has identical strengths to the pinion with regard to material properties.

(*c*) Consider power limitations due to gear bending and wear.

(*d*) Rate the gearset.

14–2 A steel spur pinion and gear have a diametral pitch of 12 teeth/in, milled teeth, 17 and 30 teeth, respectively, a 20° pressure angle, and a pinion speed of 525 r/min. The tooth properties are $S_{ut} = 76$ kpsi, $S_y = 42$ kpsi and the Brinell hardness is 149. For a design factor of 2.25, a face width of $\frac{7}{8}$ in, what is the power rating of the gearset?

14–3 A milled-teeth steel pinion and gear pair have $S_{ut} = 113$ kpsi, $S_y = 86$ kpsi and a hardness at the involute surface of 262 Brinell. The diametral pitch is 3 teeth/in, the face width is 2.5 in, and the pinion speed is 870 r/min. The tooth counts are 20 and 100. For a design factor of 1.5, rate the gearset for power considering both bending and wear.

ANALYSIS

14–4 A steel pinion rotates as 1120 r/min. It has a module of 4.23 mm, a face width of 50.8 mm, and it has 17 teeth. The ultimate tensile strength at the involute is 799 MPa, yield strength of 534 MPa, and a Brinell hardness of 232. The gear has 51 teeth. For a design factor of 2, find the power rating based on pinion and gear resisting bending and wear. (This problem is identical to Prob. 14–1 cast in SI units. If you have your solution to Prob. 14–1 at hand, you can check step-by-step for snarls before proceeding too far.)

14–5 For the conditions in Prob. 14–1 substitute a grade 50 cast iron 48-tooth gear with a Brinell hardness of 262. For cast irons of grades 20–60, the surface contact strength can be estimated from $S_C = 0.32H_B$ kpsi. Rate the steel–cast iron gearset for power.

ANALYSIS

14–6 If things were as simple as Eq. (14–2) and Eq. (14–14) without K'_V, consider what would happen if all geometric sizes were doubled. What is the expected transmitted force W^t and horsepower?

14–7 A 20° full-depth steel spur pinion rotates at 1145 r/min. It has a module of 6 mm, a face width of 75 mm, and 16 milled teeth. The ultimate tensile strength at the involute is 900 MPa exhibiting a Brinell hardness of 260. The gear is steel with 30 teeth and has identical material strengths. For a design factor of 1.3 find the power rating of the gearset based on the pinion and the gear resisting bending and wear fatigue.

ANALYSIS

14–8 As a matter of size perspective, double the size of the gears in Prob. 14–1, noting that the pitch becomes 3 teeth/in. Compare these outcomes with the estimates of Prob. 14–6. Can you explain the difference?

Experience with the approximate analysis of Secs. 14–1 and 14–2, in Probs. 14–1 through 14–8, has afforded some perspective. You have seen that wear is often limiting. There are opportunities to control core and surface hardness. Pinion and gear core hardnesses can be adjusted to balance bending resistance; surface hardnesses can be adjusted to balance wear resistance and to balance wear and bending limitations. Also, although the number of cycles on gear and pinion differ, through accommodation their simultaneous wearing out is possible. AGMA gear methodology was developed with the designer in mind. These things can be addressed by designers familiar with AGMA approach. Figures 14–17 and 14–18 will become your roadmaps.

14–9 A spur gearset has 17 teeth on the pinion and 51 teeth on the gear. The pressure angle is 20° and the overload factor $K_O = 1$. The diametral pitch is 6 teeth/in and the face width is 2 in. The pinion cycle life is to be 10^8 revolutions at a reliability $R = 0.99$. The quality number is 5. The material is a through-hardened steel, grade 1, with Brinell hardnesses of 232 core and case of both gears. For a design factor of 2, rate the gearset for these conditions using the AGMA method.

14–10 In Sec. 14–10 Eq. (a) is given for K_s based on the procedure in Ex. 14–2. Derive this equation.

14–11 A speed-reducer has 20° full-depth teeth, and the single-reduction spur-gear gearset has 22 and 60 teeth. The diametral pitch is 4 teeth/in and the face width is $3\frac{1}{4}$ in. The pinion shaft speed is 1145 r/min. The life goal of 5-year 24-hour-per-day service is about $3(10^9)$ pinion revolutions. The absolute value of the pitch variation is such that the transmission accuracy level number is 6. The materials are 4340 through-hardened grade 1 steels, heat-treated to 250 Brinell, core and case, both gears. The load is moderate shock and the power is smooth. For a reliability of 0.99, rate the speed reducer for power.

14–12 The speed reducer of Prob. 14–11 is to be used for an application requiring 40 hp at 1125 r/min. Estimate the stresses of pinion bending, gear bending, pinion wear, and gear wear and the attendant AGMA factors of safety $(S_F)_P$, $(S_F)_G$, $(S_H)_P$, and $(S_H)_G$. For the reducer, what is the factor of safety for unquantifiable exingencies in W^t? What mode of failure is the most threatening?

14–13 The gearset of Prob. 14–11 needs improvement of wear capacity. Toward this end the gears are nitrided so that the grade 1 materials have hardnesses as follows: the pinion core is 250 and the pinion case hardness is 390 Brinell, and the gear core hardness is 250 core and 390 case. Estimate the power rating for the new gearset.

14–14 The gearset of Prob. 14–11 has had its gear specification changed to 9310 for carburizing and surface hardening with the result that the pinion Brinell hardnesses are 285 core and 580–600 case, and the gear hardnesses are 285 core and 580–600 case. Estimate the power rating for the new gearset.

14–15 The gearset of Prob. 14–14 is going to be upgraded in material to a quality of grade 2 9310 steel. Estimate the power rating for the new gearset.

14–16 Matters of scale always improve insight and perspective. Reduce the physical size of the gearset in Prob. 14–11 and note the result on the estimates of transmitted load W^t and power.

14–17 AGMA procedures with cast iron gear pairs differ from those with steels because life predictions are difficult; consequently $(Y_N)_P$, $(Y_N)_G$, $(Z_N)_P$, and $(Z_N)_G$ are set to unity. The consequence of this is that the fatigue strengths of the pinion and gear materials are the same. The reliability is 0.99 and the life is 10^7 revolution of the pinion ($K_R = 1$). For longer lives the reducer is derated in power. For the pinion and gear set of Prob. 14–11, use grade 40 cast iron for both gears ($H_B = 201$ Brinell). Rate the reducer for power with S_F and S_H equal to unity.

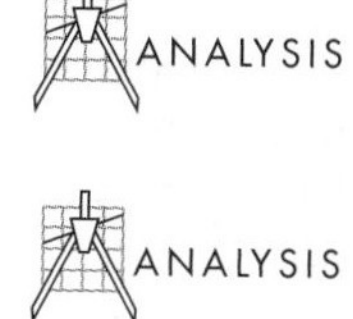
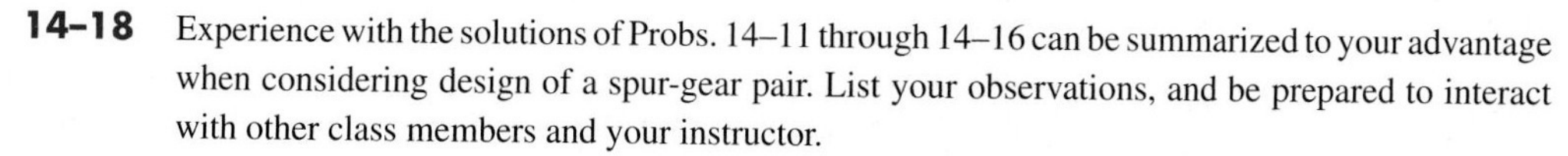

14–18 Experience with the solutions of Probs. 14–11 through 14–16 can be summarized to your advantage when considering design of a spur-gear pair. List your observations, and be prepared to interact with other class members and your instructor.

14–19 Spur-gear teeth have rolling and slipping contact (often about 8 percent slip). Spur gears tested to wear failure are reported at 10^8 cycles as Buckingham's surface fatigue load-stress factor K. This factor is related to Hertzian contact strength S_C by

$$S_C = \sqrt{\frac{1.4K}{(1/E_1 + 1/E_2)\sin\phi}}$$

where ϕ is the normal pressure angle. Cast iron grade 20 gears with $\phi = 14\frac{1}{2}^\circ$ and 20° pressure angle exhibit a minimum K of 81 and 112 psi, respectively. How does this compare with $S_C = 0.32H_B$ kpsi?

14–20 You've probably noticed that although the AGMA method is based on two equations, the details of assembling all the factors is computationally intensive. To reduce error and omissions, a computer program would be useful. Write a program to perform a power rating of an existing gearset, then use Prob. 14–11, 14–13, 14–14, 14–15, and 14–16 to test your program by comparing the results to your longhand solutions.

14–21 Equations (14–44) and (14–45) were formulated as handy approximations, and the influence of size factors K_{sP} and K_{sG} as well as hardness factor C_H were omitted. Refine Eqs. (14–44) and (14–46) to include these and apply to Ex. 14–4.

DESIGN

14–22 In Ex. 14–4 the pinion and gear had hardnesses of pinion core 240 and case 240, and the gear, core 200 and case 200 Brinell, respectively. In the same order the transmitted loads and corresponding power were, for a figure of safety of unity,

$W_1^t = 944$ lb, $\text{hp}_1 = 22.9$

$W_2^t = 1148$ lb, $\text{hp}_2 = 27.8$

$W_3^t = 489$ lb, $\text{hp}_3 = 11.9$

$W_4^t = 394$ lb, $\text{hp}_4 = 9.5$

for a reducer capacity of 9.5 hp. Using Eqs. (14–44) through (14–46) we found the hardness protocol for matching the transmitted loads, and the reducer capacity approximately doubled to 22.8 hp. The matching "one-hoss shay" hardness protocol was

	Core	Case
Pinion	244	369
Gear	135	358

As designers, to what extent can we act on this protocol?

ANALYSIS

14–23 In Ex. 14–5 use a through-hardened steel of grade 1 with Brinell hardnesses of surface and core of 240 in both gears. The limit of through-hardening is 350–400 Brinell as the limit of machine-ability is approached, and beyond 400, bending strength ebbs. Consider the manufacturer to have experience which agrees with the upper fatigue curves in Figs. 14–14 and 14–15. With factors of safety of $S_F = S_H = 1$, estimate the power capability of the mesh. What is the "one-hoss shay" hardness protocol? If the heat treatment creates a 240 Brinell core and 340 Brinell case in both gears, what is the power capacity of the gearset?

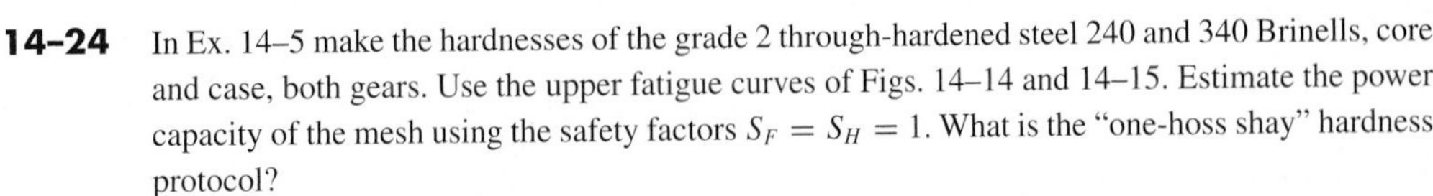

14–24 In Ex. 14–5 make the hardnesses of the grade 2 through-hardened steel 240 and 340 Brinells, core and case, both gears. Use the upper fatigue curves of Figs. 14–14 and 14–15. Estimate the power capacity of the mesh using the safety factors $S_F = S_H = 1$. What is the "one-hoss shay" hardness protocol?

14–25 In Ex. 14–5 use nitrided grade 1 steel (4140) which produces Brinell hardnesses of 250 core and 500 at the surface (case). Use the upper fatigue curves on Figs. 14–14 and 14–15. Estimate the power capacity of the mesh.

14–26 In Ex. 14–5 use carburized and case-hardened gears of grade 1. Carburizing and case-hardening can produce a 550 Brinell case. The core hardnesses are 200 Brinell. Estimate the power capacity of the mesh with factors of safety of $S_F = S_H = 1$, using the lower fatigue curves in Figs. 14–14 and 14–15.

14–27 In Ex. 14–5, use carburized and case-hardened gears of grade 2 steel. The core hardnesses are 200, and surface hardnesses are 600 Brinell. Use the lower fatigue curves of Figs. 14–14 and 14–15. Estimate the power capacity of the mesh using $S_F = S_H = 1$. Compare the power capacity with the results of Prob. 14–26.

14–28 Take the gearset of Ex. 14–5 and halve its size. Note that the normal diametral pitch will become 20 teeth/in. Estimate the transmitted load and power capacity for $S_F = S_H = 1$.

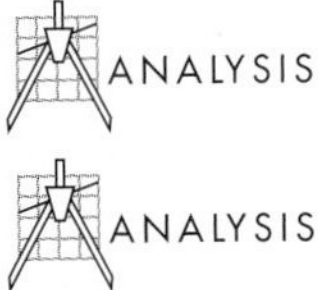

14–29 In the AGMA method K'_V addresses dynamic augmentation of the transmitted load W^t due to tooth placement error. The factor K_O addresses load excursions beyond the mean due to the nature of both driving and driven machinery. If these are known with precision, then the factors of safety S_F and S_H can be unity. The role of other values of the factors of safety is to provide a margin of capacity to accommodate unquantifiable risk with respect to load. Review Probs. 14–23 to 14–27 with a factor of safety $S_F = 2$ (and $S_H = \sqrt{2}$) and note how transmitted load W^t and power ratings compare.

14–30 Example 14–5 showed how similar a helical-gear analysis is to a spur-gear analysis. The additional parameters in the helical-gear analysis include

- Transverse diametral pitch $P_t = P_n \cos\psi$
- Transverse pressure angle $\phi_t = \tan^{-1}(\tan\phi_n / \cos\psi)$
- Pinion base-circle radius $r_o = r_P \cos\phi_t$
- Gear base-circle radius $r_b = r_G \cos\phi_t$
- Parameter Z of Eq. (14–25)
- Normal circular pitch $p_N = \frac{\pi}{p_n}\cos\phi_n$
- Load-sharing ratio $m_N = p_N/(0.95Z)$
- $I = \dfrac{\sin\phi_t \cos\phi_t}{2m_N}\dfrac{m_G}{m_G + 1}$, $m_N = 1$ for spur gears
- J'_P and J'_G from Fig. 14–7
- Multiplying factors corresponding to J'_P and J'_G from Fig. 14–8 to find J_P and J_G

Otherwise the analysis protocol is the same as for spur gears. The procedure required to complete the AGMA method is computationally intense. Write an interactive computer program that will assist in this chore for both spur and helical gearing. The program should focus on transmitted load W^t and power capacity of the mesh. The ability to "loop back" to expose a variation makes it suitable for design. For example, if the "one-hoss shay" hardness protocol is eventually displayed, one can loop back and implement it and see how it works by comparing the various ratings. All the information for an adequacy assessment will be present.

15

Bevel and Worm Gears

15–1	Bevel Gearing—General **952**
15–2	Bevel-Gear Stresses and Strengths **954**
15–3	AGMA Equation Factors **957**
15–4	Straight-Bevel Gear Analysis **969**
15–5	Design of a Straight-Bevel Gear Mesh **972**
15–6	Worm Gearing—AGMA Equation **974**
15–7	Worm-Gear Analysis **978**
15–8	Designing a Worm-Gear Mesh **980**
15–9	Buckingham Wear Load **985**

The American Gear Manufacturers Association (AGMA) has established standards for the analysis and design of the various kinds of bevel and worm gears. Chapter 14 was an introduction to the AGMA methods for spur and helical gears. AGMA has established similar methods for other types of gearing, which all follow the same general approach.

15–1 Bevel Gearing—General

Bevel gears may be classified as follows:

- Straight bevel gears
- Spiral bevel gears
- Zerol bevel gears
- Hypoid gears
- Spiroid gears

A straight bevel gear was illustrated in Fig. 13–34. These gears are usually used for pitch-line velocities up to 1000 ft/min (5 m/s) when the noise level is not an important consideration. They are available in many stock sizes and are less expensive to produce than other bevel gears, especially in small quantities.

A *spiral bevel* gear is shown in Fig. 15–1; the definition of the *spiral angle* is illustrated in Fig. 15–2. These gears are recommended for higher speeds and where the noise level is an important consideration. Spiral bevel gears are the bevel counterpart of the helical gear; it can be seen in Fig. 15–1 that the pitch surfaces and the nature of contact are the same as for straight bevel gears except for the differences brought about by the spiral-shaped teeth.

The *Zerol bevel gear* is a patented gear having curved teeth but with a zero spiral angle. The axial thrust loads permissible for Zerol bevel gears are not as large as those for the spiral bevel gear, and so they are often used instead of straight bevel gears. The Zerol bevel gear is generated using the same tool as for regular spiral bevel gears. For design purposes, use the same procedure as for straight bevel gears and then simply substitute a Zerol bevel gear.

It is frequently desirable, as in the case of automotive differential applications, to have gearing similar to bevel gears but with the shafts offset. Such gears are called

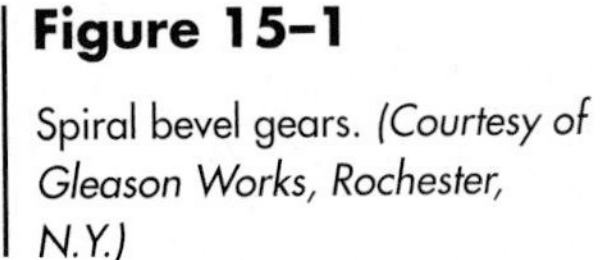

Figure 15–1

Spiral bevel gears. *(Courtesy of Gleason Works, Rochester, N.Y.)*

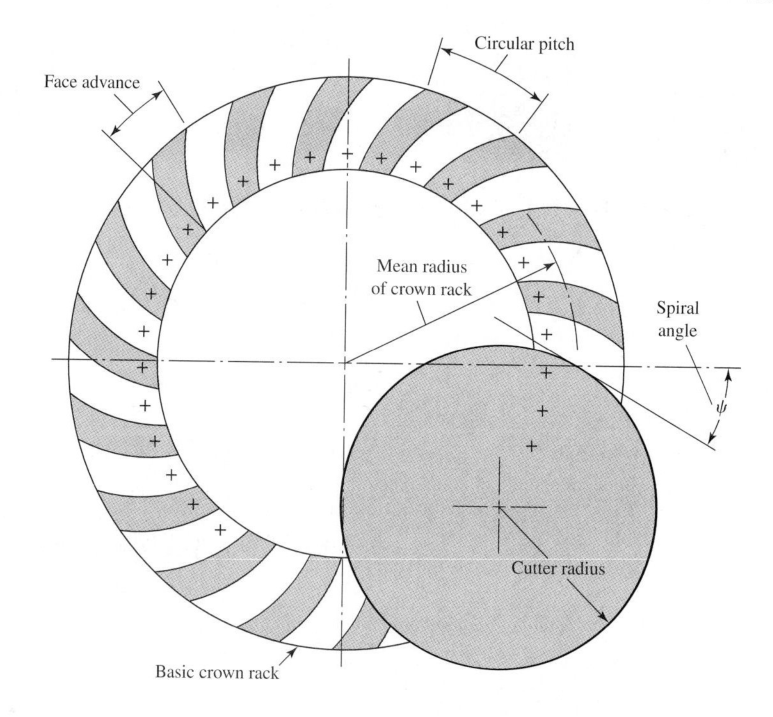

Figure 15–2

Cutting spiral-gear teeth on the basic crown rack.

hypoid gears, because their pitch surfaces are hyperboloids of revolution. The tooth action between such gears is a combination of rolling and sliding along a straight line and has much in common with that of worm gears. Figure 15–3 shows a pair of hypoid gears in mesh.

Figure 15–4 is included to assist in the classification of spiral bevel gearing. It is seen that the hypoid gear has a relatively small shaft offset. For larger offsets, the pinion begins to resemble a tapered worm and the set is then called *spiroid gearing.*

Figure 15–3

Hypoid gears. *(Courtesy of Gleason Works, Rochester, N.Y.)*

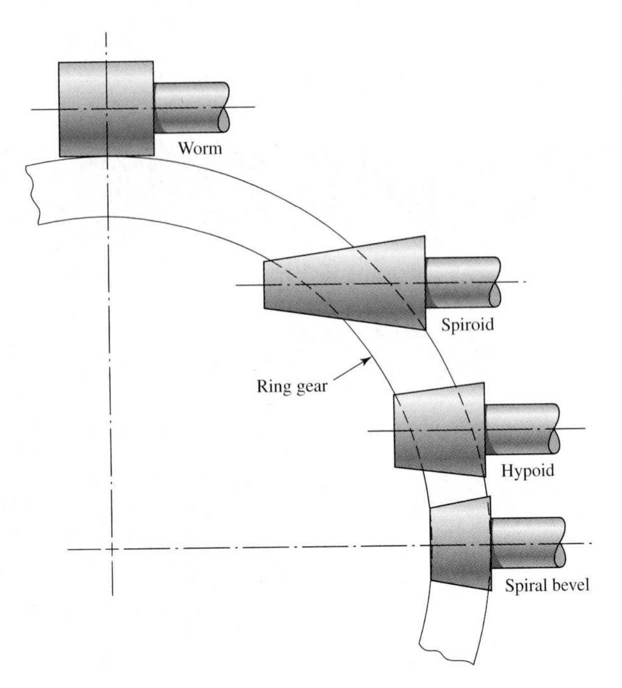

Figure 15–4

Comparison of intersecting- and offset-shaft bevel-type gearings. *(By permission, from Gear Handbook, McGraw-Hill, New York, 1962, p. 2–24.)*

15–2 Bevel-Gear Stresses and Strengths

In a typical bevel-gear mounting, Fig. 13–35, for example, one of the gears is often mounted outboard of the bearings. This means that the shaft deflections can be more pronounced and can have a greater effect on the nature of the tooth contact. Another difficulty which occurs in predicting the stress in bevel-gear teeth is the fact that the teeth are tapered. Thus, to achieve perfect line contact passing through the cone center, the teeth ought to bend more at the large end than at the small end. To obtain this condition requires that the load be proportionately greater at the large end. Because of this varying load across the face of the tooth, it is desirable to have a fairly short face width.

Because of the complexity of bevel, spiral bevel, Zerol bevel, hypoid gears, and spiroid gears, as well as the limitations of space, only a portion of the applicable standards which refer to straight-bevel gears is presented here.[1] Table 15–1 gives the symbols used in 2003-B97.

Fundamental Contact Stress Equation

$$s_c = \sigma_c = C_p \left(\frac{W^t}{F d_P I} K_o K_v' K_m C_s C_{xc} \right)^{1/2} \qquad \text{(Customary U.S. units)}$$

$$\sigma_H = Z_E \left(\frac{100 W^t}{b d Z_1} K_A K_v' K_{H\beta} Z_x Z_{xc} \right)^{1/2} \qquad \text{(SI units)} \tag{15–1}$$

We will use σ_c and σ_H for contact stress to remain consistent with our notation.

1. Figures 15–5 to 15–13 and Tables 15–1 to 15–7 have been extracted from ANSI/AGMA 2003-B97, *Rating the Pitting Resistance and Bending Strength of Generated Straight Bevel, Zerol Bevel and Spiral Bevel Gear Teeth* with the permission of the publisher, the American Gear Manufacturers Association, 1500 King Street, Suite 201, Alexandria, Virginia, 22314.

Table 15–1

Symbols Used in Gear Rating Equations, ANSI/AGMA 2003-B97 Standard. *Source:* ANSI/AGMA 2003-B97.

AGMA Symbol	ISO Symbol	Description	Units
A_m	R_m	Mean cone distance	in (mm)
A_0	R_e	Outer cone distance	in (mm)
C_H	Z_W	Hardness ratio factor for pitting resistance	
C_i	Z_i	Inertia factor for pitting resistance	
C_L	Z_{NT}	Stress cycle factor for pitting resistance	
C_p	Z_E	Elastic coefficient	$[\text{lb/in}^2]^{0.5}$ $([\text{N/mm}^2]^{0.5})$
C_R	Z_Z	Reliability factor for pitting	
C_{SF}		Service factor for pitting resistance	
C_S	Z_x	Size factor for pitting resistance	
C_{xc}	Z_{xc}	Crowning factor for pitting resistance	
D, d	d_{e2}, d_{e1}	Outer pitch diameters of gear and pinion, respectively	in (mm)
E_G, E_P	E_2, E_1	Young's modulus of elasticity for materials of gear and pinion, respectively	lb/in^2 (N/mm^2)
e	e	Base of natural (Napierian) logarithms	
F	b	Net face width	in (mm)
F_{eG}, F_{eP}	b'_2, b'_1	Effective face widths of gear and pinion, respectively	in (mm)
f_P	R_{a1}	Pinion surface roughness	μin (μm)
H_{BG}	H_{B2}	Minimum Brinell hardness number for gear material	HB
H_{BP}	H_{B1}	Minimum Brinell hardness number for pinion material	HB
h_c	$E_{ht\,\min}$	Minimum total case depth at tooth middepth	in (mm)
h_e	h'_c	Minimum effective case depth	in (mm)
$h_{e\,\lim}$	$h'_{c\,\lim}$	Suggested maximum effective case depth limit at tooth middepth	in (mm)
I	Z_I	Geometry factor for pitting resistance	
J	Y_J	Geometry factor for bending strength	
J_G, J_P	Y_{J2}, Y_{J1}	Geometry factor for bending strength for gear and pinion, respectively	
K_F	Y_F	Stress correction and concentration factor	
K_i	Y_i	Inertia factor for bending strength	
K_L	Y_{NT}	Stress cycle factor for bending strength	
K_m	$K_{H\beta}$	Load distribution factor	
K_o	K_A	Overload factor	
K_R	Y_z	Reliability factor for bending strength	
K_S	Y_X	Size factor for bending strength	
K_{SF}		Service factor for bending strength	
K_T	K_θ	Temperature factor	
K_v	K_v	Dynamic factor	
K_x	Y_β	Lengthwise curvature factor for bending strength	
	m_{et}	Outer transverse module	(mm)
	m_{mt}	Mean transverse module	(mm)
	m_{mn}	Mean normal module	(mm)
m_{NI}	ε_{NI}	Load sharing ratio, pitting	
m_{NJ}	ε_{NJ}	Load sharing ratio, bending	
N	z_2	Number of gear teeth	
N_L	n_L	Number of load cycles	
n	z_1	Number of pinion teeth	
n_P	n_1	Pinion speed	r/min
P	P	Design power through gear pair	hp (kW)
P_a	P_a	Allowable transmitted power	hp (kW)

Table 15–1

Continued

AGMA Symbol	ISO Symbol	Description	Units
P_{ac}	P_{az}	Allowable transmitted power for pitting resistance	hp (kW)
P_{acu}	P_{azu}	Allowable transmitted power for pitting resistance at unity service factor	hp (kW)
P_{at}	P_{ay}	Allowable transmitted power for bending strength	hp (kW)
P_{atu}	P_{ayu}	Allowable transmitted power for bending strength at unity service factor	hp (kW)
P_d		Outer transverse diametral pitch	in^{-1}
P_m		Mean transverse diametral pitch	in^{-1}
P_{mn}		Mean normal diametral pitch	in^{-1}
Q_v	Q_v	Transmission accuracy number	
q	q	Exponent used in formula for lengthwise curvature factor	
R, r	r_{mpt2}, r_{mpt1}	Mean transverse pitch radii for gear and pinion, respectively	in (mm)
R_t, r_t	r_{myo2}, r_{myo1}	Mean transverse radii to point of load application for gear and pinion, respectively	in (mm)
r_c	r_{c0}	Cutter radius used for producing Zerol bevel and spiral bevel gears	in (mm)
s	g_c	Length of the instantaneous line of contact between mating tooth surfaces	in (mm)
s_{ac}	$\sigma_{H\,\lim}$	Allowable contact stress number	lb/in^2 (N/mm^2)
s_{at}	$\sigma_{F\,\lim}$	Bending stress number (allowable)	lb/in^2 (N/mm^2)
s_c	σ_H	Calculated contact stress number	lb/in^2 (N/mm^2)
S_F	S_F	Bending safety factor	
S_H	S_H	Contact safety factor	
s_t	σ_F	Calculated bending stress number	lb/in^2 (N/mm^2)
s_{wc}	σ_{HP}	Permissible contact stress number	lb/in^2 (N/mm^2)
s_{wt}	σ_{FP}	Permissible bending stress number	lb/in^2 (N/mm^2)
T_P	T_1	Operating pinion torque	lb in (Nm)
T_T	θ_T	Operating gear blank temperature	°F (°C)
t_0	s_{ai}	Normal tooth top land thickness at narrowest point	in (mm)
U_c	U_c	Core hardness coefficient for nitrided gear	lb/in^2 (N/mm^2)
U_H	U_H	Hardening process factor for steel	lb/in^2 (N/mm^2)
v_t	v_{et}	Pitch-line velocity at outer pitch circle	ft/min (m/s)
Y_{KG}, Y_{KP}	Y_{K2}, Y_{K1}	Tooth form factors including stress-concentration factor for gear and pinion, respectively	
μ_G, μ_P	υ_2, υ_1	Poisson's ratio for materials of gear and pinion, respectively	
ρ_o	ρ_{yo}	Relative radius of profile curvature at point of maximum contact stress between mating tooth surfaces	in (mm)
ϕ	α_n	Normal pressure angle at pitch surface	
ϕ_t	α_{wt}	Transverse pressure angle at pitch point	
ψ	β_m	Mean spiral angle at pitch surface	
ψ_b	β_{mb}	Mean base spiral angle	

Permissible Contact Stress Number (Strength) Equation

$$s_{wc} = (\sigma_c)_{\text{all}} = \frac{s_{ac} C_L C_H}{S_H K_T C_R} \qquad \text{(Customary U.S. units)}$$

$$\sigma_{HP} = \frac{\sigma_{H\,\text{lim}} Z_{NT} Z_W}{S_H K_\theta Z_Z} \qquad \text{(SI units)} \tag{15–2}$$

Bending Stress

$$s_t = \frac{W^t}{F} P_d K_o K_v' \frac{K_s K_m}{K_x J} \qquad \text{(Customary U.S. units)}$$

$$\sigma_F = \frac{1000 W^t}{b d_{e1}} \frac{K_A K_v'}{m_{et}} \frac{Y_x K_H}{Y_\beta Y_J} \qquad \text{(SI units)} \tag{15–3}$$

Permissible Bending Stress Equation

$$s_{wt} = \frac{s_{at} K_L}{S_F K_T K_R} \qquad \text{(Customary U.S. units)}$$

$$\sigma_{FP} = \frac{\sigma_{F\,\text{lim}} Y_{NT}}{S_F K_\theta Y_z} \qquad \text{(SI units)} \tag{15–4}$$

15–3 AGMA Equation Factors

Overload Factor K_o (K_A)

The overload factor makes allowance for any externally applied loads in excess of the nominal transmitted load. Table 15–2 of Appendix A of 2003-B97 is included for your guidance.

Safety Factors S_H and S_F

The factors of safety S_H and S_F as defined in 2003-B97 are adjustments to strength, not load, and consequently cannot be used as is to assess (by comparison) whether the threat is from wear fatigue or bending fatigue. Since W^t is the same for the pinion and gear, the comparison of $\sqrt{S_H}$ to S_F allows direct comparison.

Dynamic Factor K_v'

In 2003-C87 AGMA changed the definition of K_v to its reciprocal but used the same symbol. Other standards have yet to follow this move. We will use the symbol K_v' to alert

Table 15–2

Overload Factors K_o (K_A)

Source: ANSI/AGMA 2003-B97.

Character of Prime Mover	Character of Load on Driven Machine			
	Uniform	Light Shock	Medium Shock	Heavy Shock
Uniform	1.00	1.25	1.50	1.75 or higher
Light shock	1.10	1.35	1.60	1.85 or higher
Medium shock	1.25	1.50	1.75	2.00 or higher
Heavy shock	1.50	1.75	2.00	2.25 or higher

Note: This table is for speed-decreasing drives. For speed-increasing drives, add $0.01(N/n)^2$ or $0.01(z_2/z_1)^2$ to the above factors.

you to this different definition. The symbol K'_v makes allowance for the effect of gear-tooth quality related to speed and load, and the increase in stress that follows. AGMA uses a *transmission accuracy number* Q_v to describe the precision with which tooth profiles are spaced along the pitch circle. Figure 15–5 shows graphically how pitch-line velocity and transmission accuracy number are related to the dynamic factor K'_v. Curve fits are

$$K'_v = \left(\frac{A}{A + \sqrt{v_t}}\right)^{-B} \quad \text{(Customary U.S. units)}$$

$$K'_v = \left(\frac{A}{A + \sqrt{200 v_{et}}}\right)^{-B} \quad \text{(SI units)} \tag{15–5}$$

where

$$A = 50 + 56(1 - B)$$
$$B = 0.25(12 - Q_v)^{2/3} \tag{15–6}$$

and $v_t\,(v_{et})$ is the pitch-line velocity at outside pitch diameter, expressed in ft/min (m/s):

$$v_t = \pi d_P n_P/12 \quad \text{(Customary U.S. units)}$$
$$v_{et} = 5.236(10^{-5}) d_1 n_1 \quad \text{(SI units)} \tag{15–7}$$

The maximum recommended pitch-line velocity is associated with the abscissa of the terminal points of the curve in Fig. 15–5:

$$v_{t\,\max} = [A + (Q_v - 3)]^2 \quad \text{(Customary U.S. units)}$$

$$v_{te\,\max} = \frac{[A + (Q_v - 3)]^2}{200} \quad \text{(SI units)} \tag{15–8}$$

where $v_{t\,\max}$ and $v_{et\,\max}$ are in ft/min and m/s, respectively.

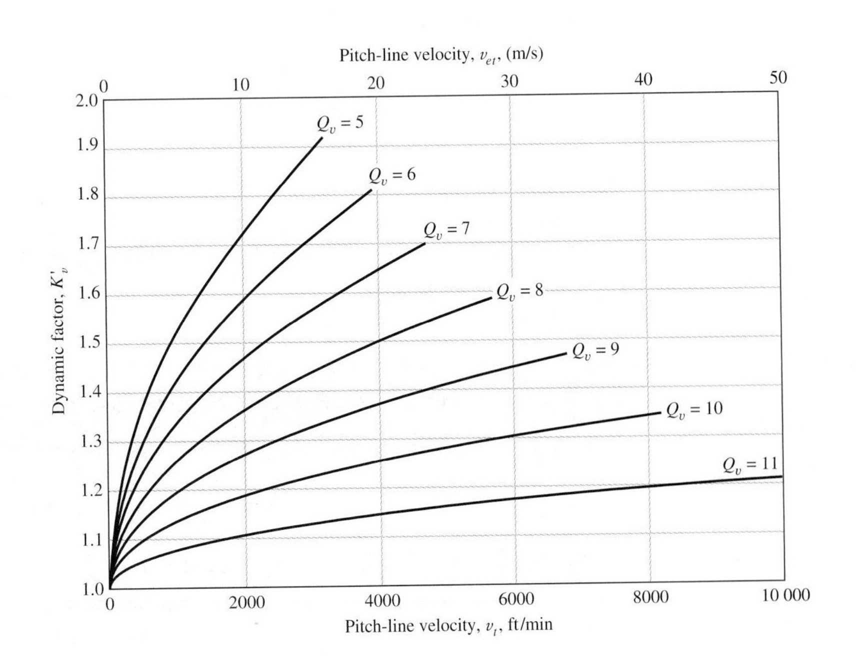

Figure 15–5

Dynamic factor K'_v. *(Source: ANSI/AGMA 2003-B97).*

Size Factor for Pitting Resistance C_s (Z_x)

$$C_s = \begin{cases} 0 & F < 0.5 \text{ in} \\ 0.125F + 0.4375 & 0.5 \leq F \leq 4.5 \text{ in} \\ 1 & F > 4.5 \text{ in} \end{cases} \qquad \text{(Customary U.S. units)}$$

$$Z_x = \begin{cases} 0.5 & b < 12.7 \text{ mm} \\ 0.004\,92b + 0.4375 & 12.7 \leq b \leq 114.3 \text{ mm} \\ 1 & b > 114.3 \text{ mm} \end{cases} \qquad \text{(SI units)} \tag{15–9}$$

Size Factor for Bending K_s (Y_x)

$$K_s = \begin{cases} 0.4867 + 0.2132/P_d & 0.5 \leq P_d \leq 16 \text{ in}^{-1} \\ 0.5 & P_d > 16 \text{ in}^{-1} \end{cases} \qquad \text{(Customary U.S. units)}$$

$$Y_x = \begin{cases} 0.5 & m_{et} < 1.6 \text{ mm} \\ 0.4867 + 0.008\,339 m_{et} & 1.6 \leq m_{et} \leq 50 \text{ mm} \end{cases} \qquad \text{(SI units)} \tag{15–10}$$

Load-Distribution Factor K_m ($K_{H\beta}$)

$$K_m = K_{mb} + 0.0036F^2 \qquad \text{(Customary U.S. units)}$$
$$K_{H\beta} = K_{mb} + 5.6(10^{-6})b^2 \qquad \text{(SI units)} \tag{15–11}$$

where

$$K_{mb} = \begin{cases} 1.00 & \text{both members straddle-mounted} \\ 1.10 & \text{one member straddle-mounted} \\ 1.25 & \text{neither member straddle-mounted} \end{cases}$$

Crowning Factor for Pitting C_{xc} (Z_{xc})

The teeth of most bevel gears are crowned in the lengthwise direction during manufacture to accommodate to the deflection of the mountings.

$$C_{xc} = Z_{xc} = \begin{cases} 1.5 & \text{properly crowned teeth} \\ 2.0 & \text{or larger noncrowned teeth} \end{cases} \tag{15–12}$$

Lengthwise Curvature Factor for Bending Strength K_x (Y_β)

For straight-bevel gears,

$$K_x = Y_\beta = 1 \tag{15–13}$$

Pitting Resistance Geometry Factor I (Z_I)

Figure 15–6 shows the geometry factor I (Z_I) for straight-bevel gears with a 20° pressure angle and 90° shaft angle. Enter the figure ordinate with the number of pinion teeth, move to the number of gear-teeth contour, and read from the abscissa.

Bending Strength Geometry Factor J (Y_J)

Figure 15–7 shows the geometry factor J for straight-bevel gears with a 20° pressure angle and 90° shaft angle.

Figure 15–6

Contact geometry factor $I(Z_I)$ for coniflex straight-bevel gears with a 20° normal pressure angle and a 90° shaft angle. *(Source: ANSI/AGMA 2003-B97.)*

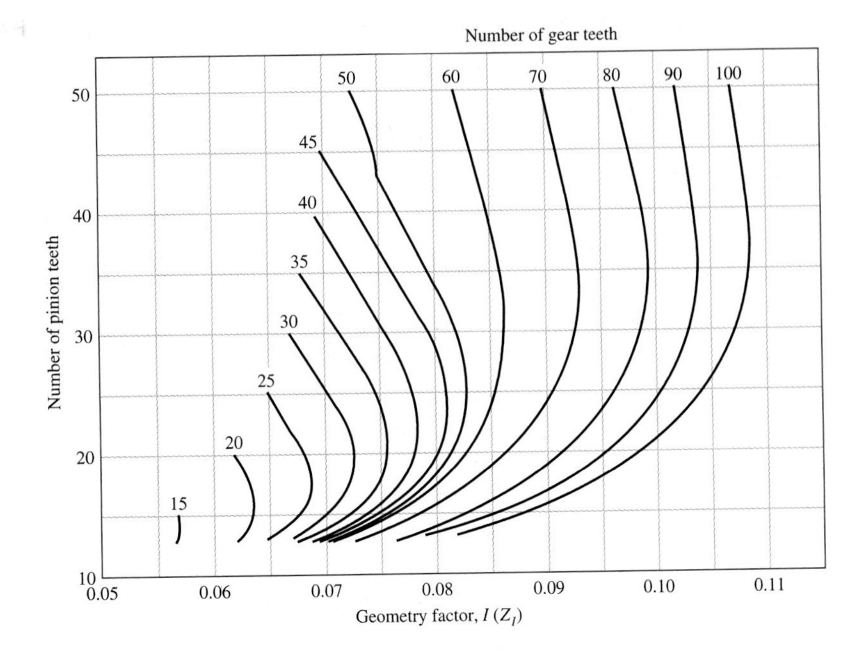

Figure 15–7

Bending factor $J(Y_J)$ for coniflex straight-bevel gears with a 20° normal pressure angle and 90° shaft angle. *(Source: ANSI/AGMA 2003-B97.)*

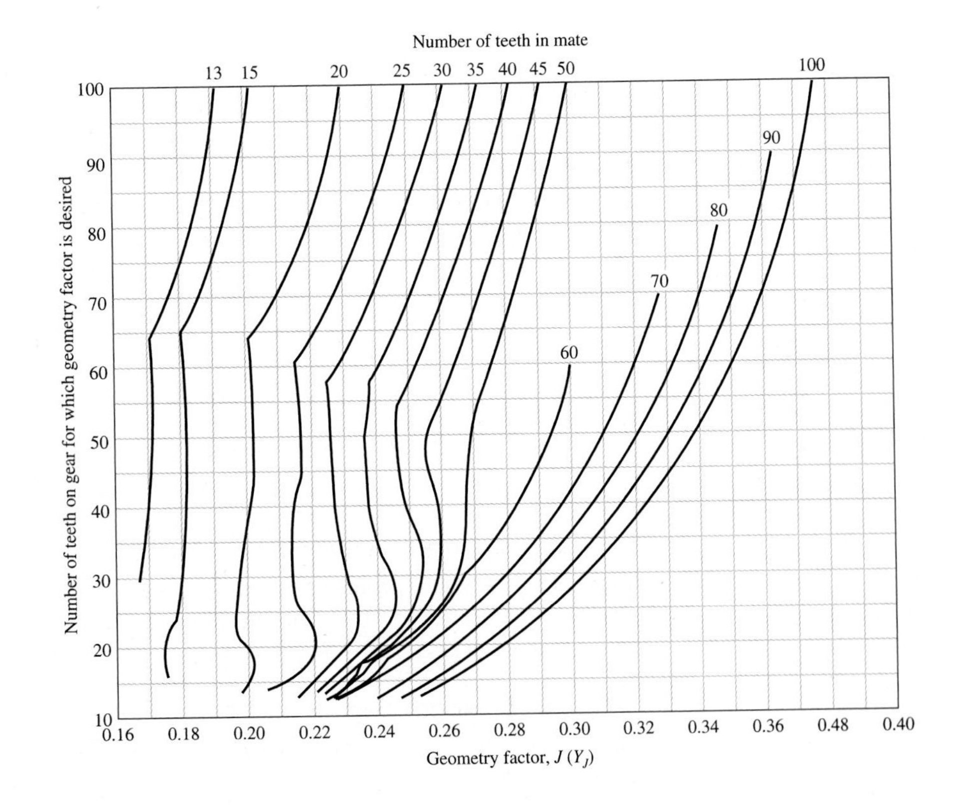

Stress-Cycle Factor for Pitting Resistance C_L (Z_{NT})

$$C_L = \begin{cases} 2 & 10^3 \le N_L < 10^4 \\ 3.4822 N_L^{-0.0602} & 10^4 \le N_L \le 10^{10} \end{cases}$$

$$Z_{NT} = \begin{cases} 2 & 10^3 \le n_L < 10^4 \\ 3.4822 n_L^{-0.0602} & 10^4 \le n_L \le 10^{10} \end{cases} \tag{15–14}$$

See Fig. 15–8 for a graphical presentation of Eqs. (15–14).

Stress-Cycle Factor for Bending Strength K_L (Y_{NT})

$$K_L = \begin{cases} 2.7 & 10^2 \le N_L < 10^3 \\ 6.1514 N_L^{-0.1182} & 10^3 \le N_L < 3(10^6) \\ 1.683 N_L^{-0.0323} & 3(10^6) \le N_L \le 10^{10} \quad \text{general} \\ 1.3558 N_L^{-0.0178} & 3(10^6) \le N_L \le 10^{10} \quad \text{critical} \end{cases}$$

$$Y_{NT} = \begin{cases} 2.7 & 10^2 \le n_L < 10^3 \\ 6.1514 n_L^{-0.1182} & 10^3 \le n_L < 3(10^6) \\ 1.683 n_L^{-0.0323} & 3(10^6) \le n_L \le 10^{10} \quad \text{general} \\ 1.3558 n_L^{-0.0323} & 3(10^6) \le n_L \le 10^{10} \quad \text{critical} \end{cases} \tag{15–15}$$

See Fig. 15–9 for a plot of Eqs. (15–15).

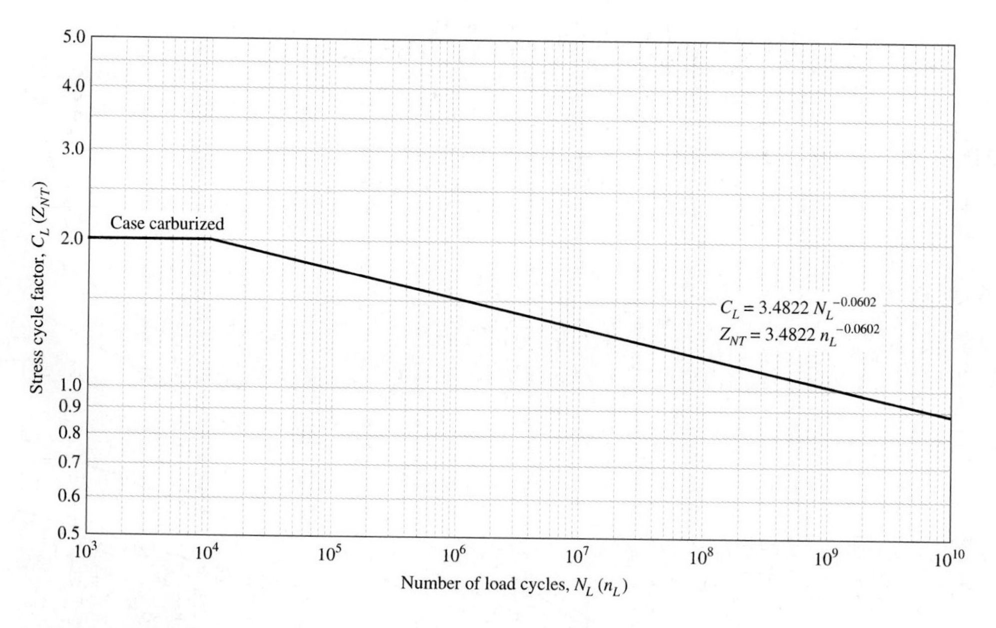

Figure 15–8

Contact stress cycle factor for pitting resistance $C_L(Z_{NT})$ for carburized case-hardened steel bevel gears. *(Source: ANSI/AGMA 2003-B97.)*

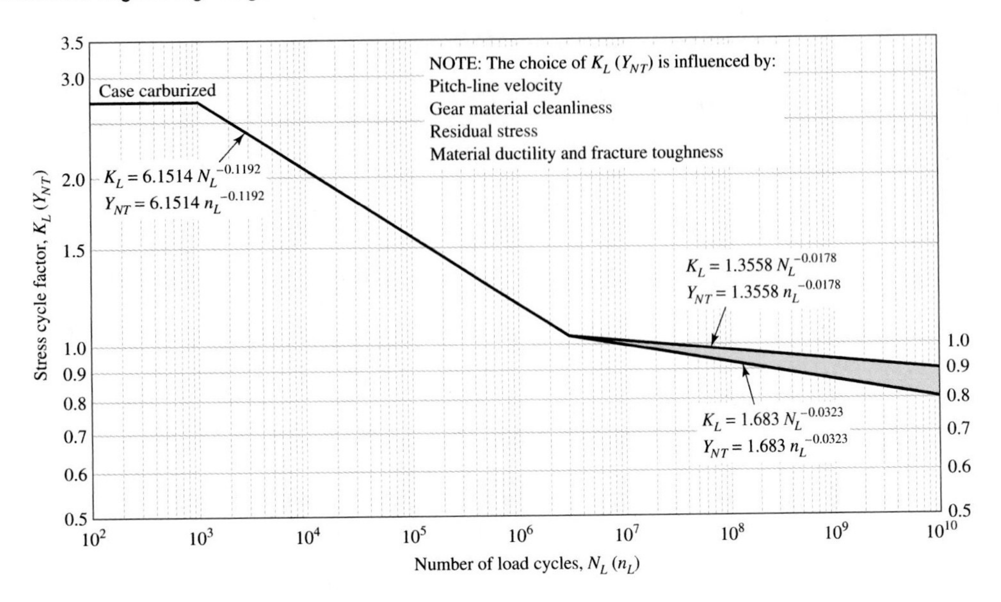

Figure 15–9

Stress cycle factor for bending strength $K_L(Y_{NT})$ for carburized case-hardened steel bevel gears. *(Source: ANSI/AGMA 2003-B97.)*

Hardness-Ratio Factor C_H (Z_W)

$$
\begin{aligned}
C_H &= 1 + B_1(N/n - 1) \qquad B_1 = 0.008\,98(H_{BP}/H_{BG}) - 0.008\,29 \\
Z_W &= 1 + B_1(z_1/z_2 - 1) \qquad B_1 = 0.008\,98(H_{B1}/H_{B2}) - 0.008\,29
\end{aligned}
\tag{15–16}
$$

The preceding equations are valid when $1.2 \le H_{BP}/H_{BG} \le 1.7$ ($1.2 \le H_{B1}/H_{B2} \le 1.7$). Figure 15–10 graphically displays Eqs. (15–16). When a surface-hardened pin-

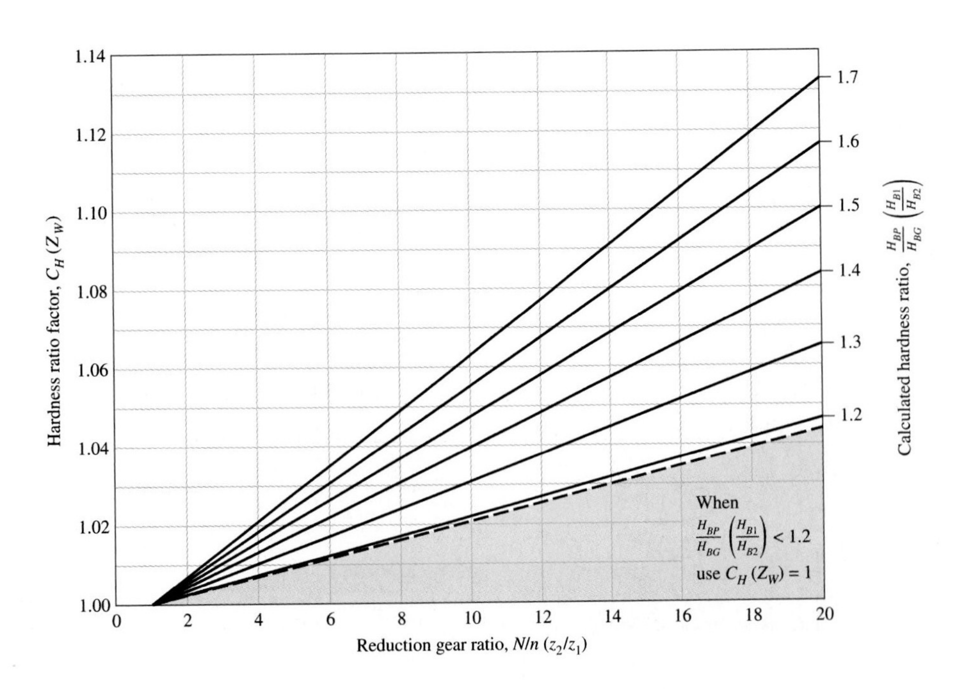

Figure 15–10

Hardness-ratio factor $C_H(Z_W)$ for through-hardened pinion and gear. *(Source: ANSI/AGMA 2003-B97.)*

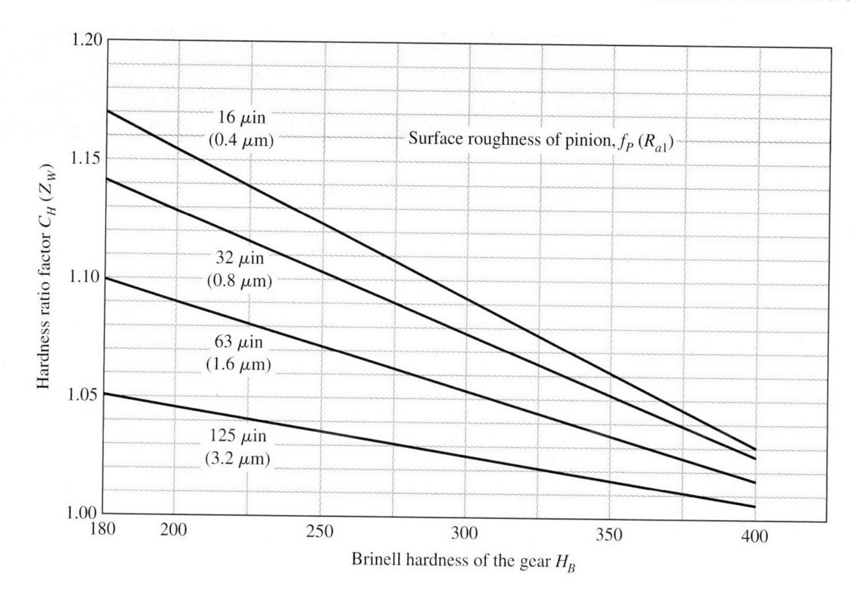

Figure 15–11

Hardness-ratio factor $C_H(Z_W)$ for surface-hardened pinions. *(Source: ANSI/AGMA 2003-B97.)*

ion (48 HRC or harder) is run with a through-hardened gear ($180 \le H_B \le 400$) a work-hardening effect occurs. The $C_H(Z_W)$ factor varies with pinion surface roughness $f_P(R_{a1})$ and the mating-gear hardness:

$$
\begin{aligned}
C_H &= 1 + B_2(450 - H_{BG}) \qquad B_2 = 0.000\,75 \exp(-0.0122 f_P) \\
Z_W &= 1 + B_2(450 - H_{B2}) \qquad B_2 = 0.000\,75 \exp(-0.52 f_P)
\end{aligned}
\tag{15–17}
$$

where $\quad f_P(R_{a1}) =$ pinion surface hardness μin (μm)

$H_{BG}(H_{B2}) =$ minimum Brinell hardness

See Fig. 15–11 for carburized steel gear pairs of approximately equal hardness $C_H = Z_W = 1$.

Temperature Factor K_T (K_θ)

$$
\begin{aligned}
K_T &= \begin{cases} 1 & 32°\text{F} \le t \le 250°\text{F} \\ (460 + t)/710 & t > 250°\text{F} \end{cases} \\
K_\theta &= \begin{cases} 1 & 0°\text{C} \le \theta \le 120°\text{F} \\ (273 + \theta)/393 & \theta > 120°\text{C} \end{cases}
\end{aligned}
\tag{15–18}
$$

Reliability Factors C_R (Z_Z) and K_R (Y_Z)

Table 15–3 displays the reliability factors. Note that $C_R = \sqrt{K_R}$ and $Z_Z = \sqrt{Y_Z}$. Logarithmic interpolation equations are

$$
Y_Z = K_R = \begin{cases} 0.50 - 0.25 \log(1 - R) & 0.99 \le R \le 0.999 \qquad (15–19) \\ 0.70 - 0.15 \log(1 - R) & 0.90 \le R < 0.99 \qquad (15–20) \end{cases}
$$

The reliability of the stress (fatigue) numbers allowable in Tables 15–4, 15–5, 15–6, and 15–7 is 0.99.

Table 15–3

Reliability Factors

Source: ANSI/AGMA 2003-B97.

Requirements of Application	Reliability Factors for Steel* C_R (Z_Z)	K_R (Y_Z)†
Fewer than one failure in 10 000	1.22	1.50
Fewer than one failure in 1000	1.12	1.25
Fewer than one failure in 100	1.00	1.00
Fewer than one failure in 10	0.92	0.85‡
Fewer than one failure in 2	0.84	0.70§

* At the present time there are insufficient data concerning the reliability of bevel gears made from other materials.

† Tooth breakage is sometimes considered a greater hazard than pitting. In such cases a greater value of $K_R(Y_Z)$ is selected for bending.

‡ At this value plastic flow might occur rather than pitting.

§ From test data extrapolation.

Table 15–4

Allowable Contact Stress Number for Steel Gears, s_{ac} $(\sigma_{H\,\mathrm{lim}})$ *Source:* ANSI/AGMA 2003-B97.

Material Designation	Heat Treatment	Minimum Surface* Hardness	Allowable Contact Stress Number, s_{ac} $(\sigma_{H\,\mathrm{lim}})$ lb/in² (N/mm²) Grade 1†	Grade 2†	Grade 3†
Steel	Through hardened‡	Fig. 15–12	Fig. 15–12	Fig. 15–12	
	Flame or induction hardened§	50 HRC	175 000 (1210)	190 000 (1310)	
	Carburized and case hardened§	2003-B97 Table 8	200 000 (1380)	225 000 (1550)	250 000 (1720)
AISI 4140	Nitrided§	84.5 HR15N		145 000 (1000)	
Nitralloy 135M	Nitrided§	90.0 HR15N		160 000 (1100)	

* Hardness to be equivalent to that at the tooth middepth in the center of the face width.

† See ANSI/AGMA 2003-B97, Tables 8 through 11, for metallurgical factors for each stress grade of steel gears.

‡ These materials must be annealed or normalized as a minumum.

§ The allowable stress numbers indicated may be used with the case depths prescribed in 21.1, ANSI/AGMA 2003-B97.

Elastic Coefficient for Pitting Resistance C_p (Z_E)

$$C_p = \sqrt{\frac{1}{\pi[(1-\nu_P^2)/E_P + (1-\nu_G^2)/E_G]}} \qquad (15\text{–}21)$$

$$Z_E = \sqrt{\frac{1}{\pi[(1-\nu_1^2)/E_1 + (1-\nu_2^2)/E_2]}}$$

Table 15–5

Allowable Contact Stress Number for Iron Gears, s_{ac} ($\sigma_{H\,\text{lim}}$) *Source:* ANSI/AGMA 2003-B97.

Material	Material Designation: ASTM	Material Designation: ISO	Heat Treatment	Typical Minimum Surface Hardness	Allowable Contact Stress Number, s_{ac} ($\sigma_{H\,\text{lim}}$) lb/in² (N/mm²)
Cast iron	ASTM A48	ISO/DR 185			
	Class 30	Grade 200	As cast	175 HB	50 000 (345)
	Class 40	Grade 300	As cast	200 HB	65 000 (450)
Ductile	ASTM A536	ISO/DIS 1083			
(nodular)	Grade 80-55-06	Grade 600-370-03	Quenched	180 HB	94 000 (650)
iron	Grade 120-90-02	Grade 800-480-02	and tempered	300 HB	135 000 (930)

Table 15–6

Allowable Bending Stress Numbers for Steel Gears, s_{at} ($\sigma_{F\,\text{lim}}$) *Source:* ANSI/AGMA 2003-B97.

Material Designation	Heat Treatment	Minimum Surface Hardness	Bending Stress Number (Allowable), s_{at} ($\sigma_{F\,\text{lim}}$) lb/in² (N/mm²)		
			Grade 1*	Grade 2*	Grade 3*
Steel	Through hardened	Fig. 15–13	Fig. 15–13	Fig. 15–13	
	Flame or induction hardened				
	Unhardened roots	50 HRC	15 000 (85)	13 500 (95)	
	Hardened roots		22 500 (154)		
	Carburized and case hardened†	2003-B97 Table 8	30 000 (205)	35 000 (240)	40 000 (275)
AISI 4140	Nitrided†,‡	84.5 HR15N		22 000 (150)	
Nitralloy 135M	Nitrided†,‡	90.0 HR15N		24 000 (165)	

* See ANSI/AGMA 2003-B97, Tables 8–11, for metallurgical factors for each stress grade of steel gears.

† The allowable stress numbers indicated may be used with the case depths prescribed in 21.1, ANSI/AGMA 2003-B97.

‡ The overload capacity of nitrided gears is low. Since the shape of the effective S-N curve is flat, the sensitivity to shock should be investigated before proceeding with the design.

where C_p = elastic coefficient, 2290 $\sqrt{\text{psi}}$ for steel

Z_E = elastic coefficient, 190 $\sqrt{\text{N/mm}^2}$ for steel

E_P and E_G = Young's moduli for pinion and gear respectively, psi

E_1 and E_2 = Young's moduli for pinion and gear respectively, N/mm²

Allowable Contact Stress

Tables 15–4 and 15–5 provide values of $s_{ac}(\sigma_H)$ for steel gears and for iron gears, respectively. Figure 15–12 graphically displays allowable stress for grade 1 and 2 materials.

Table 15–7

Allowable Bending Stress Number for Iron Gears, s_{at} ($\sigma_{F\,\lim}$) *Source:* ANSI/AGMA 2003-B97.

Material	Material Designation: ASTM	Material Designation: ISO	Heat Treatment	Typical Minimum Surface Hardness	Bending Stress Number (Allowable), s_{at} ($\sigma_{F\,\lim}$) lb/in² (N/mm²)
Cast iron	ASTM A48	ISO/DR 185			
	Class 30	Grade 200	As cast	175 HB	4500 (30)
	Class 40	Grade 300	As cast	200 HB	6500 (45)
Ductile	ASTM A536	ISO/DIS 1083			
(nodular)	Grade 80-55-06	Grade 600-370-03	Quenched	180 HB	10 000 (70)
iron	Grade 120-90-02	Grade 800-480-02	and tempered	300 HB	13 500 (95)

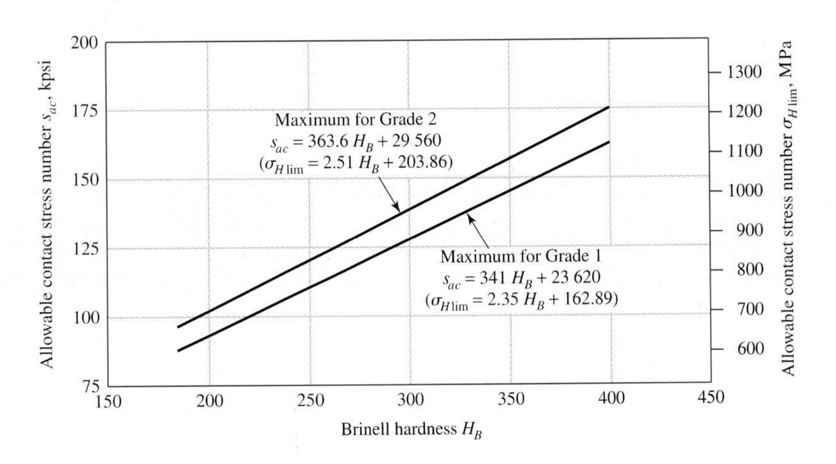

Figure 15–12

Allowable contact stress number for through-hardened steel gears, $s_{ac}(\sigma_{H\,\lim})$. *(Source: ANSI/AGMA 2003-B97.)*

The equations are

$$
\begin{aligned}
s_{ac} &= 341H_B + 23\,620 \text{ psi} && \text{grade 1}\\
\sigma_{H\ \lim} &= 2.35H_B + 162.89 \text{ MPa} && \text{grade 1}\\
s_{ac} &= 363.6H_B + 29\,560 \text{ psi} && \text{grade 2}\\
\sigma_{H\ \lim} &= 2.51H_B + 203.86 \text{ MPa} && \text{grade 2}
\end{aligned}
\qquad (15\text{–}22)
$$

Allowable Bending Stress Numbers

Tables 15–6 and 15–7 provide $s_{at}(\sigma_{F\ \lim})$ for steel gears and for iron gears, respectively. Figure 15–13 shows graphically allowable bending stress $s_{at}(\sigma_{H\ \lim})$ for through-hardened steels. The equations are

$$
\begin{aligned}
s_{at} &= 44H_B + 2100 \text{ psi} && \text{grade 1}\\
\sigma_{F\ \lim} &= 0.30H_B + 14.48 \text{ MPa} && \text{grade 1}\\
s_{at} &= 48H_B + 5980 \text{ psi} && \text{grade 2}\\
\sigma_{H\ \lim} &= 0.33H_B + 41.24 \text{ MPa} && \text{grade 2}
\end{aligned}
\qquad (15\text{–}23)
$$

Reversed Loading

AGMA recommends use of 70 percent of allowable strength in cases where tooth load is completely reversed, as in idler gears and reversing mechanisms.

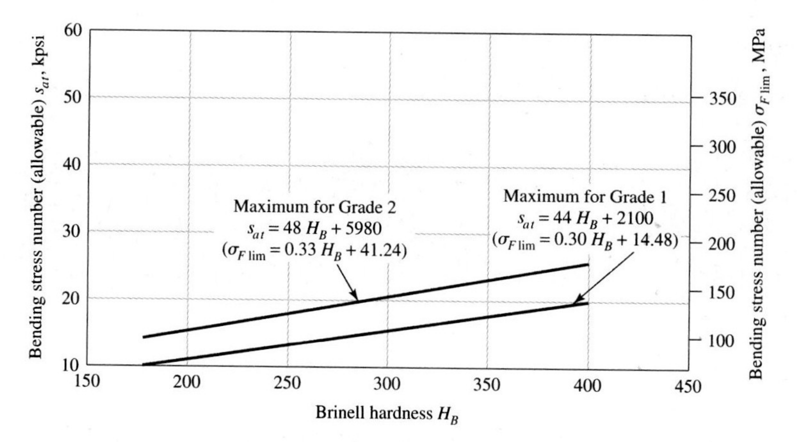

Figure 15–13

Allowable bending stress number for through-hardened steel gears, $s_{at}(\sigma_{F\,\lim})$. *(Source: ANSI/AGMA 2003-B97.)*

Summary

Figure 15–14 is a "roadmap" for straight-bevel gear wear relations using 2003-B97. Figure 15–15 is a similar guide for straight-bevel gear bending using 2003-B97.

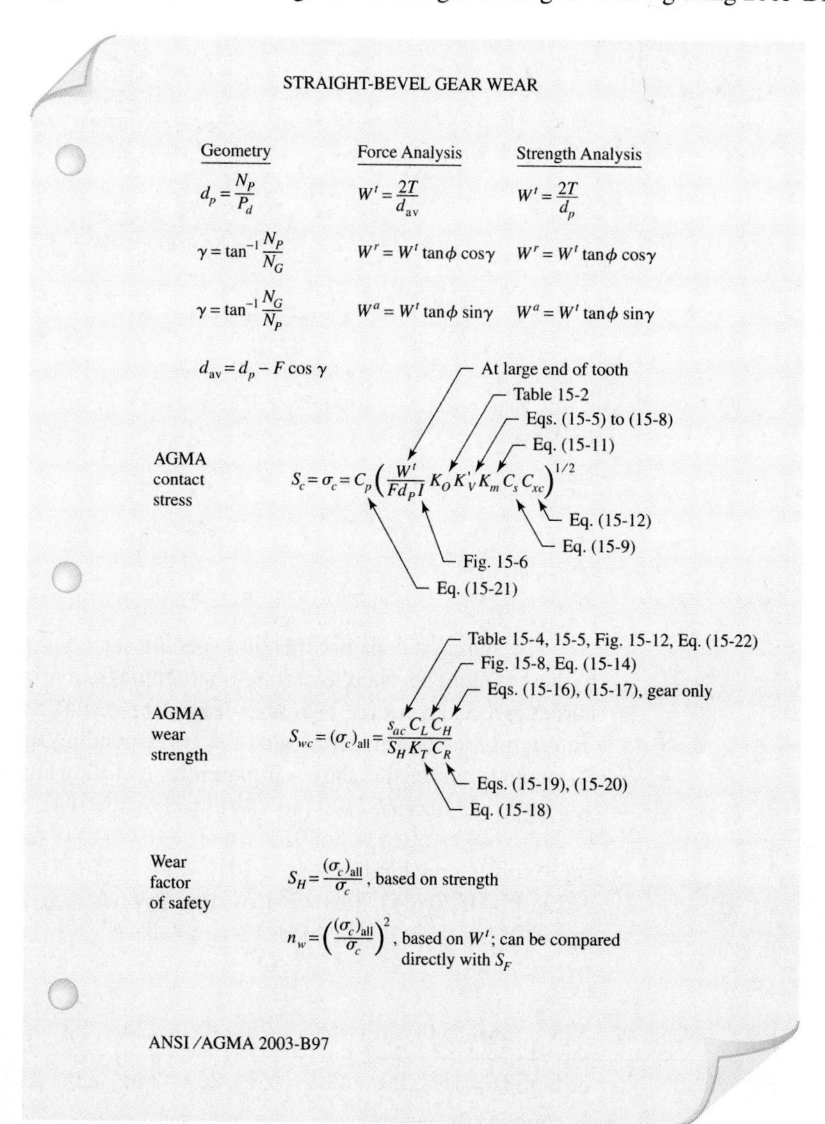

Figure 15–14

"Roadmap" summary of principal straight-bevel gear wear equations and their parameters.

Figure 15–15

"Roadmap" summary of principal straight-bevel gear bending equations and their parameters.

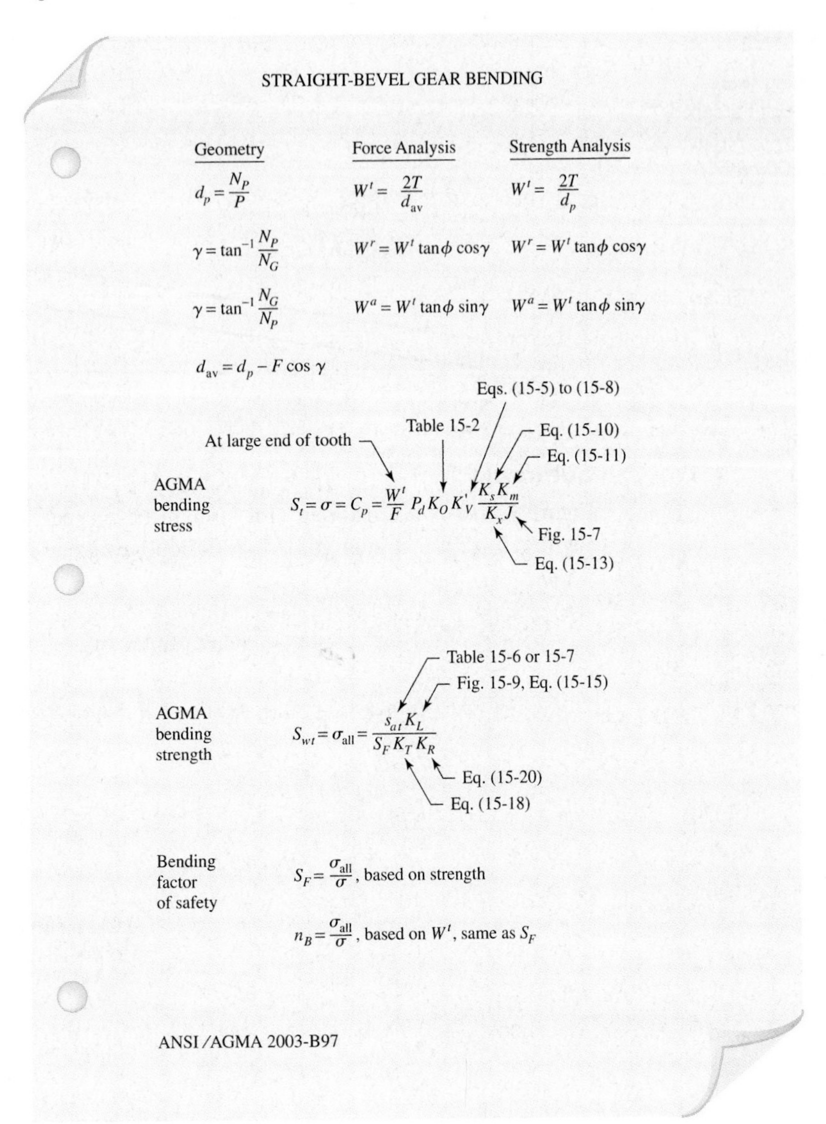

The standard does not mention specific steel but mentions the hardness attainable by heat treatments such as through-hardening, carburizing and case-hardening, flame-hardening, and nitriding. Through-hardening results depend on size (diametral pitch). Through-hardening materials and the corresponding Rockwell C-scale hardness at the 90 percent martensite shown in parentheses following include 1045 (50), 1060 (54), 1335 (46), 2340 (49), 3140 (49), 4047 (52), 4130 (44), 4140 (49), 4340 (49), 5145 (51), E52100 (60), 6150 (53), 8640 (50), and 9840 (49). For carburized case-hard materials the approximate core hardnesses are 1015 (22), 1025 (37), 1118 (33), 1320 (35), 2317 (30), 4320 (35), 4620 (35), 4820 (35), 6120 (35), 8620 (35), and E9310 (30). The conversion from HRC to H_B (300-kg load, 10-mm ball) is

HRC	42	40	38	36	34	32	30	28	26	24	22	20	18	16	14	12	10
H_B	388	375	352	331	321	301	285	269	259	248	235	223	217	207	199	192	187

Most bevel-gear sets are made from carburized case-hardened steel, and the factors incorporated in 2003-B97 largely address these high-performance gears. For through-hardening gears, 2003-B97 is silent on K_L and C_L, and Figs. 15–8 and 15–9 should prudently be considered as approximate.

15–4 Straight-Bevel Gear Analysis

EXAMPLE 15–1

A pair of identical straight-tooth miter gears listed in a catalog has a diametral pitch of 5 at the large end, 25 teeth, a 1.10-in face width, and a 20° normal pressure angle; the gears are through-hardened with a core and case hardness of 180 Brinell. The gears are intended for general industrial use. They have a quality number of $Q_v = 7$. It is likely that the application intended will require outboard mounting of the gears. Use a safety factor of 1, a 10^7 cycle life, and a 0.99 reliability.

(a) For a speed of 600 r/min find the power rating of this gearset based on AGMA bending strength.

(b) For the same conditions as in part (*a*) find the power rating of this gearset based on AGMA wear strength.

(c) For a reliability of 0.995, a gear life of 10^9 revolutions, and a safety factor of $S_F = S_H = 1.5$, find the power rating for this gearset using AGMA strengths.

Solution

We begin with

$$d_P = 25/5 = 5.000 \text{ in}$$

$$v = \pi d_P n_P/12 = \pi(5)600/12 = 785 \text{ ft/min}$$

Overload factor: uniform-uniform loading, Table 15–2, $K_o = 1.00$.
Safety factor: $S_F = 1$, $S_H = 1$.
Dynamic factor K'_v: from Eq. (15–6),

$$B = 0.25(12 - 7)^{2/3} = 0.731$$

$$A = 50 + 56(1 - 0.731) = 65.064$$

$$K'_v = \left(\frac{65.064}{65.064 + \sqrt{785}}\right)^{-0.731} = 1.299$$

From Eq. (15–8),

$$v_{t\,\max} = [65.064 + (7 - 3)]^2 = 4770 \text{ ft/min}$$

$v_t < v_{t\,\max}$, that is, $785 < 4770$ ft/min, therefore K'_v is valid. From Eq. (15–10),

$$K_s = 0.4867 + 0.2133/5 = 0.529$$

From Eq. (15–11), $K_{mb} = 1.25$ and

$$K_m = 1.25 + 0.0036(1.1)^2 = 1.254$$

From Fig. 15–6, $I = 0.065$; from Fig. 15–7, $J_P = 0.212$, $J_G = 0.212$. From Eq. (15–15),

$$K_L = 1.683(10^7)^{-0.0323} = 0.999\,96 \doteq 1$$

From Eq. (15–14),

$$C_L = 3.4822(10^7)^{-0.0602} = 1.32$$

Also $C_H = 1$, $K_T = 1$, $K_x = 1$. From Eq. (15–20),

$$K_R = 0.70 - 0.15\log(1 - 0.99) = 1 \qquad C_R = \sqrt{K_R} = \sqrt{1} = 1$$

(*a*) *Bending:* From Eq. (15–23),

$$s_{at} = 44(180) + 2100 = 10\,020 \text{ psi}$$

From Eq. (15–3),

$$s_t = \sigma = \frac{W^t}{F} P_d K_o K_v' \frac{K_s K_m}{K_x J} = \frac{W^t}{1.1}(5)(1)1.299 \frac{0.529(1.254)}{(1)0.212}$$

$$= 18.48 W^t$$

From Eq. (15–4),

$$s_{wt} = \frac{s_{at} K_L}{S_F K_T K_R} = \frac{10\,020(1)}{(1)(1)(1)} = 10\,020 \text{ psi}$$

Equating s_t and s_{wt},

$$18.48_{W^t} = 10\,020 \qquad W^t = 542.2 \text{ lb}$$

Answer $$H = \frac{W^t v}{33\,000} = \frac{542.2(785)}{33\,000} = 12.9 \text{ hp}$$

(*b*) *Wear:* From Fig. 15–12,

$$s_{ac} = 341(180) + 23\,620 = 85\,000 \text{ psi}$$

From Eq. (15–2),

$$\sigma_{c,\text{all}} = \frac{s_{ac} C_L C_H}{S_H K_T C_R}$$

Also, $C_H = 1$, $S_H = 1$, $K_T = 1$, $C_R = \sqrt{K_R} = \sqrt{1} = 1$. Substituting into the previous equation gives

$$\sigma_{c,\text{all}} = \frac{85\,000(1.32)(1)}{(1)(1)(1)} = 112\,200 \text{ psi}$$

Now C_P is 2290 from definitions following Eq. (15–21). From Eq. (15–9),

$$C_s = 0.125(1.1) + 0.4375 = 0.575 \qquad C_{xc} = 2$$

Substituting in Eq. (15–1),

$$\sigma_c = 2290 \left[\frac{W^t}{1.1(5)0.065}(1)1.299(1.254)0.575(2) \right]^{1/2} = 5242\sqrt{W^t}$$

Equating σ_c and $\sigma_{c,\text{all}}$ gives

$$5242\sqrt{W^t} = 112\,200 \qquad W^t = 458 \text{ lb}$$

$$H = \frac{458(785)}{33\,000} = 10.9 \text{ hp}$$

Rated power for the gearset is

Answer $H = \min(12.9, 10.9) = 10.9$ hp

(*c*) Life goal 10^9 cycles, $R = 0.995$, $S_F = S_H = 1.5$, and

$$K_L = 1.683(10^9)^{-0.0323} = 0.862$$

From Eq. (15–19),

$$K_R = 0.50 - 0.25\log(1 - 0.995) = 1.075 \qquad C_R = \sqrt{K_R} = \sqrt{1.075} = 1.037$$

From Eq. (15–14),

$$C_L = 3.4822(10^9)^{-0.0602} = 1$$

Bending: From Eq. (15–23) and part (*a*), $s_{at} = 10\,020$ psi. From Eq. (15–3),

$$s_t = \sigma = \frac{W^t(5)(1)1.299}{1.1}\,\frac{0.529(1.254)}{(1)0.212} = 18.48W^t$$

From Eq. (15–4),

$$s_{wt} = \frac{s_{at}K_L}{S_F K_T K_R} = \frac{10\,020(0.862)}{1.5(1)1.075} = 5356 \text{ psi}$$

Equating s_t to s_{wt} gives

$$18.48W^t = 5356 \qquad W^t = 290 \text{ lb}$$

$$H = \frac{290(785)}{33\,000} = 6.9 \text{ hp}$$

Wear: From Eq. (15–22),

$$s_{ac} = 341(180) + 23\,620 = 85\,000 \text{ psi}$$

Substituting into Eq. (15–2) gives

$$\sigma_{c,\text{all}} = \frac{s_{ac}C_L C_H}{S_H K_T C_R} = \frac{85\,000(1)(1)}{1.5(1)1.037} = 54\,644 \text{ psi}$$

Substituting into Eq. (15–1) gives

$$\sigma_c = 2290\left[\frac{W^t}{1.1(5)0.065}(1)1.299(1.254)0.575(2)\right]^{1/2} = 5242\sqrt{W^t}$$

Equating σ_c to $\sigma_{c,\text{all}}$ gives

$$\sigma_c = \sigma_{c,\text{all}} = 54\,644 = 5242\sqrt{W^t} \qquad W^t = 108.7 \text{ lb}$$

The wear power is

$$H = \frac{108.7(785)}{33\,000} = 2.58 \text{ hp}$$

Answer The mesh rated power is $H = \min(6.9, 2.58) = 2.6$ hp.

15–5 Design of a Straight-Bevel Gear Mesh

A useful decision set for straight-bevel gear design is

- Function
- Design factor
- Tooth system
- Tooth count

} A priori decisions

- Pitch and face width
- Quality number
- Gear material, core and case hardness
- Pinion material, core and case hardness

} Design variables

In bevel gears the quality number is linked to the wear strength. The J factor for the gear can be smaller than the pinion. Bending strength is not linear with face width, because added material is placed at the small end of the teeth. Consequently, face width is roughly proscribed as

$$F = \min(A_0/3,\ 10/P_d)$$

and the nonlinearity is sidestepped. Some designers use $F = 0.3A_0$ instead of $A_0/3$.

EXAMPLE 15–2 Design a straight-bevel gear mesh for shaft centerlines which intersect perpendicularly, to deliver 6.85 hp at 900 r/min with a gear ratio of 3:1, temperature of 300°F, normal pressure angle of 20°, using a design factor of 2. The load is uniform-uniform. Although the minimum number of teeth on the pinion is 13, which will mesh with 31 or more teeth without interference, use a pinion of 20 teeth. The material is to be AGMA grade 1 and the teeth are to be crowned. The reliability goal is 0.995 with a pinion life of 10^9 revolutions.

Solution First we list the a priori decisions and their immediate consequences.

Function: 6.85 hp at 900 r/min, gear ratio $m_G = 3$, 300°F environment, neither gear straddle-mounted, $K_{mb} = 1.25$, $R = 0.995$ at 10^9 revolutions of the pinion,

$$(C_L)_G = 3.4822(10^9/3)^{-0.0602} = 1.068$$

$$(C_L)_P = 3.4822(10^9)^{-0.0602} = 1$$

$$(K_L)_G = 1.683(10^9/3)^{-0.0323} = 0.8929$$

$$(KL)_P = 1.683(10^9)^{-0.0323} = 0.8618$$

$$K_R = 0.50 - 0.25\log(1 - 0.995) = 1.075$$

$$C_R = \sqrt{K_R} = \sqrt{1.075} = 1.037$$

$$K_T = C_T = (460 + 300)/710 = 1.070$$

Design factor: $n_d = 2$, $S_F = 2$, $S_H = \sqrt{2} = 1.414$.

Tooth system: crowned, straight-bevel gears, normal pressure angle 20°, $K_x = 1$, $C_{xc} = 1.5$.

Number of teeth on pinion and gear: $N_P = 20$, $N_G = 60$,

$$\gamma = \tan^{-1}(20/60) = 18.4^\circ$$
$$\Gamma = \tan^{-1}(60/20) = 71.6^\circ$$

From Fig. 15–6, $I = 0.086$; from Fig. 15–7, $J_P = 0.248$, $J_G = 0.203$. Note that $J_P > J_G$.

Decision 1: Trial diametral pitch $P_d = 6$ teeth/in,

$$K_s = 0.4867 + 0.2132/6 = 0.522$$
$$d_P = 20/6 = 3.333 \text{ in} \qquad d_G = 3.333(3) = 10.000 \text{ in}$$
$$v = \pi(3.333)900/12 = 785.3 \text{ ft/min}$$
$$W^t = 33\,000(6.85)/785.3 = 288 \text{ lb}$$
$$A_0 = (d_G/3)\sin\Gamma = (10/3)\sin 71.6^\circ = 3.163$$
$$F = \min(A_0/3,\ 10/P_d) = \min(3.163/3,\ 10/6) = \min(1.054,\ 1.667) = 1.054 \text{ in}$$

Decision 2: Make $F = 1\frac{1}{16} = 1.0625$ in. Then

$$C_s = 0.125(1.0625) + 0.4375 = 0.570$$
$$K_m = 1.25 + 0.0036(1.0625)^2 = 1.254$$

Decision 3: Make transition accuracy number $Q_v = 6$. From Eq. (15–6),

$$B = 0.25(12-6)^{2/3} = 0.8255$$
$$A = 50 + 56(1 - 0.8255) = 59.77$$

$$K_v' = \left(\frac{59.77}{59.77 + \sqrt{785.3}}\right)^{-0.8255} = 1.374$$

Decision 4: Pinion and gear material and treatment. Carburize and case-harden grade 1 ASTM 1320 to

core 21 HRC (H_B is 229 Brinell)
case 55–64 HRC (H_B is 515 Brinell minimum)

From Table 15–4, $s_{ac} = 200\,000$ psi and from Table 15–6, $s_{at} = 30\,000$ psi.

Bending, gear: The load-induced bending stress from Eq. (15–3) is

$$(s_t)_G = \frac{W^t}{F}P_d K_0 K_v' \frac{K_s K_m}{K_x J_G} = \frac{288}{1.0625}(6)(1)1.374\frac{0.522(1.254)}{(1)0.203} = 7206 \text{ psi}$$

From Eq. (15–4) the bending strength is given by

$$(s_{wt})_G = \frac{s_{at}K_L}{S_F K_T K_R} = \frac{30\,000(0.8929)}{2(1.070)1.075} = 11\,644 \text{ psi}$$

The strength exceeds the stress by a factor of 11 644/7206 = 1.62.

Bending, pinion: The load-induced bending stress can be found from

$$(s_t)_P = (s_t)_G\frac{J_G}{J_P} = 7206\frac{0.203}{0.248} = 5898 \text{ psi}$$

The strength can be found from Eq. (15–4) as

$$(s_{wt})_P = \frac{s_{at} K_L}{S_F K_T K_R} = \frac{30\,000(0.8618)}{2(1.070)1.075} = 11\,238 \text{ psi}$$

The strength exceeds the stress by a factor of 11 238/5898 = 1.91.

Wear, gear: The load-induced contact stress from Eq. (15–1) is

$$(s_c)_G = C_p \left(\frac{W^t}{F d_P I} K_O K'_v K_m C_s C_{xc} \right)^{1/2}$$

$$= 2290 \left[\frac{288}{1.0625(3.333)0.086} (1)1.374(1.254)0.570(1.5) \right]^{1/2}$$

$$= 85\,472 \text{ psi}$$

From Eq. (15–2) the contact strength is

$$(s_{wt})_G = \frac{s_{ac} C_L C_H}{S_H K_T C_R} = \frac{200\,000(1.068)(1)}{\sqrt{2}\,1.070(1.037)} = 136\,121 \text{ psi}$$

The strength exceeds the stress by the ratio 136 121/85 472 = 1.59.

Wear, pinion: The Hertzian stress is the same on the pinion and the gear, that is, $(s_c)_P = (s_c)_G = 85\,472$ psi. The contact strength is given by Eq. (15–2) as

$$(s_{wc})_P = \frac{s_{ac} C_L C_H}{S_H K_T C_R} = \frac{200\,000(1)(1)}{\sqrt{2}\,1.070(1.037)} = 127\,454 \text{ psi}$$

The ratio of strength to stress is 127 454/85 472 = 1.49.

The ratios are 1.62, 1.91, 1.59, and 1.49. Because we took $S_F = n_d = 2$, and $S_H = \sqrt{n_d} = \sqrt{2}$, we make a direct comparison of the ratios and note the threat is from pinion wear (1.49). We note that three of the ratios are comparable. Our goal would be to make changes in the design decisions that drive the ratios toward unity. The next step would be to adjust the design variables. We would choose a finer conventional pitch ($P_d = 8$), or we could tune the material properties if the changes lower costs. It is obvious that an iterative process is involved. We need a figure of merit to order the designs. A computer program clearly is desirable.

15–6 Worm Gearing—AGMA Equation

Since they are essentially nonenveloping worm gears, the *crossed helical* gears, shown in Fig. 15–16, can be considered with other worm gearing. Because the teeth of worm gears have *point contact* changing to *line contact* as the gears are used, worm gears are said to "wear in," whereas other types "wear out."

Crossed helical gears, and worm gears too, usually have a 90° shaft angle, though this need not be so. The relation between the shaft and helix angles is

$$\sum = \psi_P \pm \psi_G \tag{15–24}$$

where Σ is the shaft angle. The plus sign is used when both helix angles are of the same hand, and the minus sign when they are of opposite hand. The subscript P in Eq. (15–24) refers to the pinion (worm); the subscript W is used for this same purpose. The subscript G refers to the gear, also called *gear wheel, worm wheel,* or simply the *wheel.* Table 15–8 gives cylindrical worm dimensions common to worm and gear.

Figure 15–16

View of the pitch cylinders of a pair of crossed helical gears.

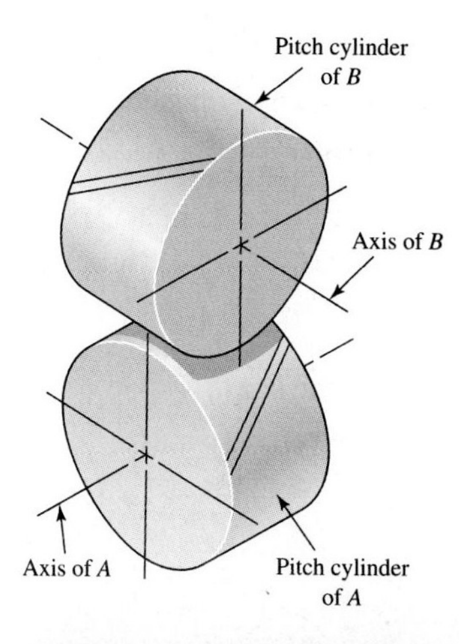

Table 15–8

Cylindrical Worm Dimensions Common to Both Worm and Gear*

Quantity	Symbol	ϕ_n 14.5° $N_W \le 2$	20° $N_W \le 2$	25° $N_W > 2$
Addendum	a	$0.3183p_x$	$0.3183p_x$	$0.286p_x$
Dedendum	b	$0.3683p_x$	$0.3683p_x$	$0.349p_x$
Whole depth	h_t	$0.6866p_x$	$0.6866p_x$	$0.635p_x$

* The table entries are for a tangential diametral pitch of the gear of $P_t = 1$.

Section 13–11 introduced worm gears, and Sec. 13–17 developed the force analysis and efficiency of worm gearing to which we will refer. Here our interest is in strength and durability. Good proportions indicate the mean worm diameter d_m falls in the range

$$\frac{C^{0.875}}{3} \le d_m \le \frac{C^{0.875}}{1.6} \tag{15–25}$$

where C is the center-to-center distance.[2] AGMA relates the tangential force on the worm-gear tooth W^t to other parameters by

$$W^t = C_s D_m^{0.8}(F_e)_G C_m C_v \tag{15–26}$$

where C_s = materials factor

D_m = mean gear diameter, in (mm)

F_e = effective face width of the gear (actual face width, but not to exceed $0.67d_m$, the mean worm diameter), in (mm)

C_m = ratio correction factor

C_v = velocity factor

The friction force W_f is given by

$$W_f = \frac{fW^T}{\cos\lambda\cos\phi_n} \tag{15–27}$$

2. ANSI/AGMA 6034-A87, March 1988, *Practice for Enclosed Cylindrical Wormgear Speed-Reducers and Gear Motors;* ANSI/AGMA 6030-C87, March 1988, *Design of Industrial Double-Enveloping Wormgears;* and ANSI/AGMA 6022-C93, Dec. 1993, *Design Manual for Cylindrical Wormgearing,* and useful references.

where f = coefficient of friction
λ = lead angle at mean worm diameter
ϕ_n = normal pressure angle

The sliding velovity V_s is

$$V_s = \frac{\pi n_w d_m}{12 \cos \lambda} \tag{15-28}$$

where n_w = rotative speed of the worm
d_m = mean worm diameter

The torque at the worm gear is

$$T_G = \frac{W^t D_m}{2} \tag{15-29}$$

where D_m is the mean gear diameter.

The parameters in Eq. (15–26) are, quantitatively,

$$C_s = 270 + 10.37C^3 \qquad C \le 3 \text{ in} \tag{15-30}$$

For sand-cast gears,

$$C_s = \begin{cases} 1000 & & d_G \le 2.5 \text{ in} \\ 1190 - 477 \log d_G & C > 3 & d_G > 2.5 \text{ in} \end{cases} \tag{15-31}$$

For chilled-cast gears,

$$C_s = \begin{cases} 1000 & C > 3 & d_G \le 8 \text{ in} \\ 1412 - 456 \log d_G & C > 3 & d_G > 8 \text{ in} \end{cases} \tag{15-32}$$

For centrifugally cast gears,

$$C_s = \begin{cases} 1000 & C > 3 & d_G \le 25 \text{ in} \\ 1251 - 180 \log d_G & C > 3 & d_G > 25 \text{ in} \end{cases} \tag{15-33}$$

The ratio correction factor C_m is given by

$$C_m = \begin{cases} 0.02\sqrt{-m_G^2 + 40m_G - 76} + 0.46 & 3 < m_G \le 20 \\ 0.0107\sqrt{-m_G^2 + 56m_G + 5145} & 20 < m_G \le 76 \\ 1.1483 - 0.006\,58m_G & m_G > 76 \end{cases} \tag{15-34}$$

The velocity factor C_v is given by

$$C_v = \begin{cases} 0.659 \exp(-0.0011 V_s) & V_s < 700 \text{ ft/min} \\ 13.31 V_s^{-0.571} & 700 \le V_s < 3000 \text{ ft/min} \\ 65.52 V_s^{-0.774} & V_s > 3000 \text{ ft/min} \end{cases} \tag{15-35}$$

AGMA reports the coefficient of friction f as

$$f = \begin{cases} 0.15 & V_s = 0 \\ 0.124 \exp(-0.074 V_s^{0.645}) & 0 < V_s \le 10 \text{ ft/min} \\ 0.103 \exp(-0.110 V_s^{0.450}) + 0.012 & V_s > 10 \text{ ft/min} \end{cases} \tag{15-36}$$

Now we examine some worm-gear mesh geometry. The addendum a and dedendum b are

$$a = \frac{p_x}{\pi} = 0.3183 p_x \tag{15–37}$$

$$b = \frac{1.157 p_x}{\pi} = 0.3683 p_x \tag{15–38}$$

The full depth h_t is

$$h_t = \begin{cases} \dfrac{2.157 p_x}{\pi} = 0.6866 p_x & p_x \geq 0.16 \text{ in} \\ \dfrac{2.200 p_x}{\pi} + 0.002 = 0.7003 p_x + 0.002 & p_x < 0.16 \text{ in} \end{cases} \tag{15–39}$$

The worm outside diameter d_0 is

$$d_0 = d + 2a \tag{15–40}$$

The worm root diameter d_r is

$$d_r = d - 2b \tag{15–41}$$

The worm-gear throat diameter D_t is

$$D_t = D + 2a \tag{15–42}$$

The worm-gear root diameter D_r is

$$D_r = D - 2b \tag{15–43}$$

The clearance c is

$$c = b - a \tag{15–44}$$

The worm face width (maximum) $(F_w)_{\text{max}}$ is

$$(F_w)_{\text{max}} = 2\sqrt{\left(\frac{D_t}{2}\right)^2 - \left(\frac{D}{2} - a\right)^2} \tag{15–45}$$

The worm-gear face width F_G is

$$F_G = \begin{cases} 2d_m/3 & p_x > 0.16 \text{ in} \\ 1.125\sqrt{(d_0 + 2c)^2 - (d_0 - 4a)^2} & p_x \leq 0.16 \text{ in} \end{cases} \tag{15–46}$$

The heat loss rate $\dot{Q}$ from the worm-gear case in ft · lb/min is

$$\dot{Q} = 33\,000(1 - e)H_{in} \text{ ft} \cdot \text{lb/min} \tag{15–47}$$

The overall coefficient U for combined convective and radiative heat transfer from the worm-gear case in ft·lb/(min · in^2 · °F) is

$$U = \begin{cases} \dfrac{n_w}{6494} + 0.13 & \text{no fan on worm shaft} \\ \dfrac{n_w}{3939} + 0.13 & \text{fan on worm shaft} \end{cases} \tag{15–48}$$

when the case lateral area A is expressed in in^2. The temperature of the oil sump t_s is given by

$$t_s = t_a + \frac{\dot{Q}}{UA} = \frac{33\,000(1-e)(H)_{\text{in}}}{UA} + t_a \tag{15–49}$$

Bypassing Eqs. (15–47), (15–48), and (15–49) one can use the AGMA recommendation for minimum lateral area A_{min} in in^2 using

$$A_{\text{min}} = 43.20^{1.7} \tag{15–50}$$

Because worm teeth are inherently much stronger than worm-gear teeth, they are not considered. The teeth in worm gears are short and thick on the edges of the face; midplane they are thinner as well as curved. Buckingham[3] adapted the Lewis equation for this case:

$$\sigma_a = \frac{W_G^t}{p_n (F_e)_G y} \tag{15–51}$$

where $p_n = p_x \cos\lambda$ and y is the Lewis form factor related to circular pitch. For $\phi_n = 14.5°$, $y = 0.100$; $\phi_n = 20°$, $y = 0.125$; $\phi_n = 25°$, $y = 0.150$; $\phi_n = 30°$, $y = 0.175$.

15–7 Worm-Gear Analysis

EXAMPLE 15–3

A single-threaded steel worm rotates at 1800 r/min, meshing with a 24-tooth worm gear transmitting 3 hp to the output shaft. The worm is 3 in in diameter and the tangential diametral pitch of the gear is 4 teeth/in. The normal pressure angle is 14.5°. The ambient temperature is 70°F. The application factor is 1.25 and the design factor was 1; gear face width is 2 in, lateral case area 600 in^2, and the gear is chill-cast bronze.

(a) Find the gear geometry.
(b) Find the gear forces and the mesh efficiency.
(c) Is the mesh sufficient to handle the loading?
(d) Estimate the lubricant sump temperature.

Solution

(*a*) $m_G = N_G/N_W = 24/1 = 24$, $d_G = N_G/P_t = 24/4 = 6.000$ in, $d_P = 3.000$ in. The axial circular pitch p_x is $p_x = \pi/P_t = \pi/4 = 0.785$ in. $C = (3+6)/2 = 4.5$ in.

Eq. (15–37) $a = p_x/\pi = 0.785/\pi = 0.250$ in

Eq. (15–38) $h = 0.3683 p_x = 0.3683(0.785) = 0.289$ in

Eq. (15–39) $h_t = 0.6866 p_x = 0.6866(0.785) = 0.539$ in

Eq. (15–40) $d_0 = 3 + 2(0.250) = 3.500$ in

Eq. (15–41) $d_r = 3 - 2(0.289) = 2.422$ in

Eq. (15–42) $D_t = 6 + 2(0.250) = 6.500$ in

Eq. (15–43) $D_r = 6 - 2(0.289) = 5.422$ in

Eq. (15–44) $c = 0.289 - 0.250 = 0.039$ in

Eq. (15–45) $(F_W)_{\max} = 2\sqrt{\left(\frac{6.5}{2}\right)^2 - \left(\frac{6}{2} - 0.250\right)^2} = 3.464$ in

3. Earle Buckingham, *Analytical Mechanics of Gears*, McGraw-Hill, New York, 1949, p. 495.

The tangential speeds of the worm V_W and gear V_G are, respectively,

$$V_W = \pi(3)1800/12 = 1413.7 \text{ ft/min} \qquad V_G = \frac{\pi(6)1800/24}{12} = 117.8 \text{ ft/min}$$

The lead of the worm is $p_x N_W = 0.785(1) = 0.785$ in. The lead angle λ is

$$\lambda = \tan^{-1}\frac{L}{\pi d_W} = \tan^{-1}\frac{0.785}{\pi(3)} = 4.764°$$

$$P_n = \frac{P_t}{\cos\lambda} = \frac{4}{\cos 4.764°} = 4.0139$$

$$p_n = \frac{\pi}{P_n} = \frac{\pi}{4.0139} = 0.7827 \text{ in}$$

The sliding velocity is

$$V_s = \frac{\pi d_W n_W}{12\cos\lambda} = \frac{\pi(3)1800}{12\cos 4.764°} = 1418.6 \text{ ft/min}$$

(*b*) The coefficient of friction from Eq. (15–36) is

$$f = 0.103\exp[-0.110(1418.6)^{0.450}] + 0.012 = 0.0179$$

The efficiency e from Eq. (13–38) is

Answer
$$e = \frac{\cos\phi_n - f\tan\lambda}{\cos\phi_n + f\cot\lambda} = \frac{\cos 14.5° - 0.0179\tan 4.764°}{\cos 14.5° + 0.0179\cot 4.764°} = 0.8172$$

The designer used $n_d = 1$, $K_a = 1.25$ and an output horsepower of $H_0 = 3$ hp. The gear tangential force component W_G^t is

Answer
$$W_G^t = \frac{33\,000 n_d H_0 K_a}{V_G e} = \frac{33\,000(1)3(1.25)}{117.8(0.818)} = 1284.2 \text{ lbf}$$

The third expression in Eqs. (13–35), solving for total force W gives

Answer
$$W = \frac{W_G^t}{\cos\phi_n\cos\lambda - f\sin\lambda} = \frac{1284.2}{\cos 14.5°\cos 4.764° - 0.0179\sin 4.764°}$$
$$= 1333 \text{ lbf}$$

The first expression in Eqs. (13–35) is

Answer
$$W_W^t = W(\cos\phi_n\sin\lambda + f\cos\lambda)$$
$$= 1333(\cos 14.5°\sin 4.764° + 0.0179\cos 4.764°) = 131 \text{ lbf}$$

(*c*)

Eq. (15–31) $\quad C_s = 1000$

Eq. (15–34) $\quad C_m = 0.0107\sqrt{-24^2 + 56(24) + 5145} = 0.823$

Eq. (15–35) $\quad C_v = 13.31(1418.6)^{-0.571} = 0.211$

Eq. (15–36) $\quad (W^t)_{\text{all}} = C_s d_G^{0.8}(F_e)_G C_m C_v$

$$= 1000(6)^{0.8}(2)0.823(0.211) = 1456 \text{ lbf}$$

Since $W_G^t < (W^t)_{\text{all}}$, the mesh will survive at least 25 000 h. The friction force W_f is given by Eq. (13–36):

$$W_f = \frac{f W_G^t}{f \sin\lambda - \cos\phi_n \cos\lambda} = \frac{0.0179(1284.2)}{0.0179 \sin 4.764° - \cos 14.5° \cos 4.764°}$$
$$= -23.9 \text{ lbf}$$

The power dissipated in frictional work H_f is given by

$$H_f = \frac{|W_f| V_s}{33\,000} = \frac{|-23.9|1418.6}{33\,000} = 1.03 \text{ hp}$$

The worm and gear powers, H_W and H_G, are given by

$$H_W = \frac{W_W^t V_W}{33\,000} = \frac{131(1413.7)}{33\,000} = 5.61 \text{ hp} \qquad H_G = \frac{W_G^t V_G}{33\,000} = \frac{1284.2(117.8)}{33\,000} = 4.58 \text{ hp}$$

Answer Gear power is satisfactory. Now,

$$P_n = P_t / \cos\lambda = 4/\cos 4.764° = 4.1039$$
$$p_n = \pi / P_n = \pi/4.0139 = 0.7827 \text{ in}$$

The bending stress in a gear tooth is given by Buckingham's adaptation of the Lewis equation, Eq. (15–51), as

$$(\sigma)_G = \frac{W_G^t}{p_n F_G y} = \frac{1284.2}{0.7827(2)(0.1)} = 8204 \text{ psi}$$

Answer Stress in gear satisfactory.

(*d*)

Eq. (15–50) $A_{\min} = 43.2C^{1.7} = 43.2(4.5)^{1.7} = 557 \text{ in}^2$

The gear case has a lateral area of 600 in^2.

Eq. (15–47) $\dot{Q} = 33\,000(1 - e)H_{\text{in}} = 33\,000(1 - 0.8172)5.61$
$= 33\,842 \text{ ft} \cdot \text{lbf/min}$

Eq. (15–48) $U = \frac{n_W}{3939} + 0.13 = \frac{1800}{3939} + 0.13 = 0.587 \text{ ft} \cdot \text{lbf/(min} \cdot \text{in}^2 \cdot °\text{F)}$

Answer Eq. (15–49) $t_s = t_a + \frac{\dot{Q}}{UA} = 70 + \frac{33\,842}{0.587(600)} = 166°\text{F}$

15–8 Designing a Worm-Gear Mesh

A usable decision set for a worm-gear mesh includes

- Function: power, speed, m_G, K_a
- Design factor: n_d
- Tooth system
- Materials and processes
- Number of threads on the worm: N_W

} A priori decisions

- Axial pitch of worm: p_x
- Pitch diameter of the worm: d_W
- Face width of gear: F_G
- Lateral area of case: A

} Design variables

Reliability information for worm gearing is not well-developed at this time. The use of AGMA Eq. (15–19) together with the factors C_s, C_m, and C_v, with an alloy steel case-hardened worm together with customary nonferrous worm-wheel materials, will result in lives in excess of 25 000 h. The worm-gear materials in the experience base are principally bronzes:

- Tin- and nickel-bronzes (chilled-casting produces hardest surfaces)
- Lead-bronze (high-speed applications)
- Aluminum- and silicon-bronze (heavy load, slow-speed application)

The factor C_s for bronze in the spectrum sand-cast, chilled-cast, and centrifugally cast increases in the same order.

Standardization of tooth systems is not as far along as it is in other types of gearing. For the designer this represents freedom of action, but acquisition of tooling for tooth-forming is more of a problem for in-house manufacturing. When using a subcontractor the designer must be aware of what the supplier is capable of providing with on-hand tooling.

Axial pitches for the worm are usually integers, and quotients of integers are common. Typical pitches are $\frac{1}{4}$, $\frac{5}{16}$, $\frac{3}{8}$, $\frac{1}{2}$, $\frac{3}{4}$, 1, $\frac{5}{4}$, $\frac{6}{4}$, $\frac{7}{4}$, and 2, but others are possible. Table 15–8 shows dimensions common to both worm gear and cylindrical worm for proportions often used. Teeth frequently are stubbed when lead angles are 30° or larger.

Worm-gear design is constrained by available tooling, space restrictions, shaft center-to-center distances, gear ratios needed, and designers' experience. ANSI/AGMA 6022-C93, *Design Manual for Cylindrical Wormgearing* offers the following guidance. Normal pressure angles are chosen from 14.5°, 17.5°, 20°, 22.5°, 25°, 27.5°, and 30°. The recommended minimum number of gear teeth is given in Table 15–9. The normal range of the number of threads on the worm is 1 through 10. Mean worm pitch diameter is usually chosen in the range

$$\frac{C^{0.875}}{3} \le d_m \le \frac{C^{0.875}}{1.6}$$

where C is the center-to-center distance. Note that midrange is $C^{0.875}/2.09$ and not $C^{0.875}/2.3$. (Why?)

Table 15–9

Minimum Number of Gear Teeth for Normal Pressure Angle ϕ_n

ϕ_n	$(N_G)_{\min}$
14.5	40
17.5	27
20	21
22.5	17
25	14
27.5	12
30	10

A design decision is the axial pitch of the worm. Since acceptable proportions are couched in terms of the center-to-center distance, which is not yet known, one chooses a trial axial pitch p_x. Having N_W and a trial worm diameter d_W,

$$N_G = m_G N_W \qquad P_t = \frac{\pi}{p_x} \qquad d_G = \frac{N_G}{P_t}$$

Then

$$(d_W)_{\text{lo}} = C^{0.875}/3 \qquad (d_W)_{\text{hi}} = C^{0.875}/1.6$$

Examine $(d_W)_{\text{lo}} \le d_w \le (d_W)_{\text{hi}}$, and refine the selection of mean worm-pitch diameter to d_{W1} if necessary. Recompute the center-to-center distance as $C_1 = (d_{W1} + d_G)/2$. There is even an opportunity to make C_1 a round number. Choose C_2 and set

$$d_{W2} = 2C_2 - d_G$$

Equations (15–37) through (15–46) apply to one usual set of proportions.

Compared to other gearing systems worm-gear meshes have a much lower mechanical efficiency. Cooling, for the benefit of the lubricant, becomes a design constraint sometimes resulting in what appears to be an oversize gear case in light of its contents. If the heat can be dissipated by natural cooling, or simply with a fan on the wormshaft, simplicity persists. Water coils within the gear case or lubricant outpumping to an external cooler is the next level of complexity. For this reason, gear-case area is a design decision.

To reduce cooling load use multiple-threaded worms. Also keep the worm pitch diameter as small as possible.

Multiple threaded worms can remove the self-locking feature of many worm-gear drives. When the worm drives the gearset the mechanical efficiency e_w is given by

$$e_w = \frac{\cos\phi_n - f\tan\lambda}{\cos\phi_n + f\cot\lambda} \tag{15–52}$$

With the gear driving the gearset the mechanical efficiency e_G is given by

$$e_G = \frac{\cos\phi_n - f\cot\lambda}{\cos\phi_n + f\tan\lambda} \tag{15–53}$$

To ensure that the worm gear will drive the worm,

$$f_{\text{stat}} < \cos\phi_n \tan\lambda \tag{15–54}$$

where values of f_{stat} can be found in ANSI/AGMA 6034-B92. To prevent the worm gear from driving the worm, refer to clause 9 of 6034-B92 for a discussion of self-locking in the static condition.

It is important to have a way to relate the tangential component of the gear force W_G^t to the tangential component of the worm force W_W^t, which includes the role of friction and the angularities of ϕ_n and λ. Refer to Eq. (13–37) solved for W_W^t:

$$W_W^t = W_G^t \frac{\cos\phi_n \sin\lambda + f\cos\lambda}{\cos\phi_n \cos\lambda - f\sin\lambda} \tag{15–55}$$

In the absence of friction

$$W_W^t = W_G^t \tan\lambda$$

The mechanical efficiency of most gearing is very high, which allows power in and power out to be used almost interchangeably. Worm gearsets have such poor efficiencies that we work with, and speak of, output power. The magnitude of the gear transmitted force W_G^t can be related to the output power H_0, the application factor K_a, and the

Table 15–10

Largest Lead Angle Associated with a Normal Pressure Angle ϕ_n for Worm Gearing

ϕ_n	Maximum Lead Angle λ_{max}
14.5°	16°
20°	25°
25°	35°
30°	45°

efficiency e by

$$W_G^t = \frac{33\,000 n_d H_0 K_a}{V_G e} \tag{15–56}$$

We use Eq. (15–55) to obtain the corresponding worm force W_W^t. It follows that

$$H_W = \frac{W_W^t V_W}{33\,000} = \frac{\pi d_W n_W W_W^t}{12(33\,000)} \text{ hp} \tag{15–57}$$

$$H_G = \frac{W_G^t V_G}{33\,000} = \frac{\pi d_G n_G W_G^t}{12(33\,000)} \text{ hp} \tag{15–58}$$

From Eq. (13–36),

$$W_f = \frac{f W_G^t}{f \sin\lambda - \cos\phi_n \cos\lambda} \tag{15–59}$$

The sliding velocity of the worm at the pitch cylinder V_s is

$$V_s = \frac{\pi d_W n_W}{12\cos\lambda} \tag{15–60}$$

and the friction power H_f is given by

$$H_f = \frac{|W_f| V_s}{33\,000} \text{ hp} \tag{15–61}$$

Table 15–10 gives the largest lead angle λ_{max} associated with normal pressure angle ϕ_n.

With these tools at our disposal it is time to design a worm-gear mesh.

EXAMPLE 15–4

Design a 10-hp 11:1 worm-gear speed-reducer mesh for a lumber mill planer feed drive for 3–10 h daily use. A 1720-r/min squirrel-cage induction motor drives the planer feed ($K_a = 1.25$), and the ambient temperature is 70°F.

Solution

Function: $H_0 = 10$ hp, $m_G = 11$, $n_W = 1720$ r/min.
Design factor: $n_d = 1.2$.
Materials and processes: case-hardened alloy steel worm, sand-cast bronze gear.
Worm threads: double, $N_W = 2$, $N_G = m_G N_W = 11(2) = 22$ Number of gear teeth acceptable according to Table 15–9.
Decision 1: Choose an axial pitch of worm $p_x = 1.5$ in. Then,

$$P_t = \pi/p_x = \pi/1.5 = 2.0944$$
$$d_G = N_G/P_t = 22/2.0944 = 10.504 \text{ in}$$

Eq. (15–37) $a = 0.3183 p_x = 0.3183(1.5) = 0.478$ in addendum

Eq. (15–38) $b = 0.3683(1.5) = 0.552$ in dedendum

Eq. (15–39) $h_t = 0.6866(1.5) = 1.030$ in

Decision 2: Choose a mean worm diameter $d_W = 2.000$ in. Then

$$C = (2.000 + 10.504)/2 = 6.252 \text{ in}$$
$$(d_W)_{\text{lo}} = 6.252^{0.875}/3 = 1.657 \text{ in}$$
$$(d_W)_{\text{hi}} = 6.252^{0.875}/1.6 = 3.112 \text{ in}$$

The range is $1.657 \le d_W \le 3.112$ in. Make $d_W = 2.500$ in. Recompute C:

$$C = (2.5 + 10.504)/2 = 6.502 \text{ in}$$

The range is now $1.715 \le d_W \le 3.216$ in, which is satisfactory. Decision: $d_W = 2.500$ in. Then

$$L = p_x N_W = 1.5(2) = 3.000 \text{ in}$$
$$\lambda = \tan^{-1}[L/(\pi d_W)] = \tan^{-1}[3/(\pi 2.5)] = 20.905° \qquad \text{lead angle OK}$$

Eq. (15–60) $$V_s = \frac{\pi d_W n_W}{12 \cos \lambda} = \frac{\pi(2.5)1720}{12 \cos 20.905°} = 1205.1 \text{ ft/min}$$

$$V_W = \frac{\pi d_W n_W}{12} = \frac{\pi(2.5)1720}{12} = 1125.7 \text{ ft/min}$$

$$V_G = \frac{\pi d_G n_G}{12} = \frac{\pi(10.504)1720/11}{12} = 430 \text{ ft/min}$$

Eq. (15–31) $C_s = 1190 - 477 \log 10.504 = 702.8$

Eq. (15–34) $C_m = 0.02\sqrt{-11^2 + 40(11) - 76} + 0.46 = 0.772$

Eq. (15–35) $C_v = 13.31(1205.1)^{-0.571} = 0.232$

Eq. (15–36) $f = 0.103 \exp[-0.11(1205.1)^{0.45}] + 0.012 = 0.0191$

Eq. (15–52) $$e_W = \frac{\cos 20° - 0.0191 \tan 20.905°}{\cos 20° + 0.0191 \cot 20.905°} = 0.942$$

(If the worm drives, $e_G = 0.939$.) To ensure nominal 10-hp output, with adjustments for K_a, n_d, and e,

Eq. (15–56) $$W_{Gt} = \frac{33\,000(1.2)10(1.25)}{430(0.942)} = 1222 \text{ lbf}$$

Eq. (15–57) $$H_W = \frac{\pi(2.5)1720(495.4)}{12(33\,000)} = 16.9 \text{ hp}$$

Eq. (15–58) $$H_G = \frac{\pi(10.504)1720/11(1222)}{12(33\,000)} = 15.93 \text{ hp}$$

Eq. (15–59) $$W_f = \frac{0.0191(1222)}{0.0191 \sin 20.904° - \cos 20° \cos 20.905°} = -26.8 \text{ lbf}$$

Eq. (15–61) $$H_f = \frac{|-26.8|1205.1}{33\,000} = 0.979 \text{ hp}$$

The required gear face width $(F_G)_{\text{req}}$ is found from Eq. (15–26).

Decision 3:

$$(F_e)_{\text{req}} = \frac{W_G^t}{C_s d_G^{0.8} C_m C_v} = \frac{1222}{702.8(10.504)^{0.8} 0.772(0.232)} = 1.479 \text{ in}$$

The available range of F_e is $(F_e)_{\text{req}} \le F_e \le 2d_W/3$ or $1.479 \le F_e \le 1.667$ in. Set $F_e = 1.5$ in.

Eq. (15–26) $W_{\text{all}}^t = 702.8(10.504)^{0.8} 1.5(0.772)0.232 = 1239.1$ lbf

This is greater than 1222 lbf. There is a little excess capacity. The force analysis stands.
Decision 4:

$$\text{Eq. (15–48)} \quad U = \frac{n_W}{6494} + 0.13 = \frac{1720}{6494} + 0.13 = 0.395 \text{ ft} \cdot \text{lbf/(min} \cdot \text{in}^2 \cdot {}^\circ\text{F)}$$

$$\text{Eq. (15–47)} \quad \dot{Q} = 33\,000(1 - e)H_W = 33\,000(1 - 0.942)16.9 = 32\,347 \text{ ft} \cdot \text{lbf/min}$$

The AGMA area is $A_{\min} = 42.3C^{1.7} = 43.2(6.502)^{1.7} = 1041.5 \text{ in}^2$. A rough estimate of the lateral area:

Vertical:	$d_W + d_G + 6 = 2.5 + 10.5 + 6 = 19$ in
Width:	$d_G + 6 = 10.5 + 6 = 16.5$ in
Thickness:	$d_W + 6 = 2.5 + 6 = 8.5$ in
Area:	$2(19)16.5 + 2(8.5)19 + 16.5(8.5) \doteq 1090 \text{ in}^2$

Expect an area of 1100 in^2. Choose: Air-cooled, no fan on worm.

$$t_s = t_a + \frac{\dot{Q}}{UA} = 70 + \frac{32\,347}{0.395(1100)} = 70 + 74.5 = 144.5^\circ\text{F}$$

Lubricant is safe with some margin for smaller area.

$$P_n = \frac{P_t}{\cos\lambda} = \frac{2.0944}{\cos 20.905^\circ} = 2.242$$

$$p_n = \frac{\pi}{P_n} = \frac{\pi}{2.242} = 1.401 \text{ in}$$

Gear bending stress, for reference,

$$\text{Eq. (15–51)} \quad \sigma = \frac{W_G^t}{p_n F_e y} = \frac{1222}{1.401(1.5)0.125} = 4652 \text{ psi}$$

The risk is from wear, which is addressed by the AGMA method that provides $(W_G^t)_{\text{all}}$.

15–9 Buckingham Wear Load

A precursor to the AGMA method was the method of Buckingham, which identified an allowable wear load in worm-gearing. Buckingham showed that the allowable gear-tooth loading for wear can be estimated from

Table 15–11

Wear Factor K_w for Worm Gearing

Source: Earle Buckingham, *Design of Worm and Spiral Gears,* Industrial Press, New York, 1981.

Material		Thread Angle ϕ_n			
Worm	**Gear**	$14\frac{1}{2}°$	**20°**	**25°**	**30°**
Hardened steel*	Chilled bronze	90	125	150	180
Hardened steel*	Bronze	60	80	100	120
Steel, 250 BHN (min.)	Bronze	36	50	60	72
High-test cast iron	Bronze	80	115	140	165
Gray iron†	Aluminum	10	12	15	18
High-test cast iron	Gray iron	90	125	150	180
High-test cast iron	Cast steel	22	31	37	45
High-test cast iron	High-test cast iron	135	185	225	270
Steel 250 BHN (min.)	Laminated phenolic	47	64	80	95
Gray iron	Laminated phenolic	70	96	120	140

* Over 500 BHN surface.

† For steel worms, multiply given values by 0.6.

$$(W_G^t)_{\text{all}} = K_w d_G F_e \tag{15–62}$$

where K_w = worm-gear load factor

d_G = gear-pitch diameter

F_e = worm-gear effective face width

Table 15–11 gives values for K_w for worm-gearsets which are a function of the material pairing and the normal pressure angle.

EXAMPLE 15–5 Estimate the allowable gear wear load $(W_G^t)_{\text{all}}$ for the gearset of Ex. 15–4 using Buckingham's wear equation.

Solution From Table 15–11 for a hardened steel worm and a bronze bear, K_w is given as 80 for $\phi_n = 20°$. Equation (15–62) gives

$$(W_G^t)_{\text{all}} = 80(10.504)1.5 = 1260 \text{ lbf}$$

which is larger than the 1239 lbf of the AGMA method. The method of Buckingham does not have refinements of the AGMA method. [Is $(W_G^t)_{\text{all}}$ linear with gear diameter?]

For material combinations not addressed by AGMA, Buckingham's method allows quantitative treatment.

PROBLEMS

15–1 An uncrowned straight-bevel pinion has 20 teeth, a diametral pitch of 6 teeth/in, and a transmission accuracy number of 6. Both the pinion and gear are made of through-hardened steel with a Brinell hardness of 300. The driven gear has 60 teeth. The gearset has a life goal of 10^9 revolutions of the pinion with a reliability of 0.999. The shaft angle is 90°; the pinion speed is 900 r/min. The face width is 1.25 in, and the normal pressure angle is 20°. The pinion is mounted outboard of its bearings, and the gear is straddle-mounted. Based on the AGMA bending strength, what is the power rating of the gearset? Use $K_0 = 1$, $S_F = 1$, and $S_H = 1$.

ANALYSIS

15–2 For the gearset and conditions of Prob. 15–1 find the power rating based on the AGMA surface durability.

15–3 An uncrowned straight-bevel pinion has 30 teeth, a diametral pitch of 6, and a transmission accuracy number of 6. The driven gear has 60 teeth. Both are made of No. 30 cast iron. The shaft angle is 90°. The face width is 1.25 in, the pinion speed is 900 r/min, and the normal pressure angle is 20°. The pinion is mounted outboard of its bearings, the bearings of the gear straddle it. What is the power rating based on AGMA bending strength? (For cast iron gearsets reliability information has not yet been developed. We say the life is greater than 10^7 revolutions; set $K_L = 1$, $C_L = 1$, $C_R = 1$, $K_R = 1$; and apply a factor of safety. Use $S_F = 2$ and $S_H = \sqrt{2}$.

ANALYSIS

15–4 For the gearset and conditions of Prob. 15–3, find the power rating based on AGMA surface durability. Using the solutions to Probs. 15–3 and 15–4, what is the power rating of the gearset?

ANALYSIS

15–5 An uncrowned straight-bevel pinion has 22 teeth, a module of 4 mm, and a transmission accuracy number of 5. The pinion and the gear are made of through-hardened steel, both having core and case hardnesses of 180 Brinell. The pinion drives the 24-tooth bevel gear. The shaft angle is 90°, the pinion speed is 1800 r/min, the face width is 25 mm, and the normal pressure angle is 20°. Both gears have an outboard mounting. Find the power rating based on AGMA pitting resistance if the life goal is 10^9 revolutions of the pinion at 0.999 reliability.

15–6 For the gearset and conditions of Prob. 15–5, find the power rating for AGMA bending strength.

15–7 In straight-bevel gearing, there are some analogs to Eqs. (14–44), (14–45), and (14–46). If we have a pinion core with a hardness of $(H_B)_{11}$ and we try the "one-hoss shay" idea for equal power ratings, the transmitted load W^t can be made equal in all four cases. It is possible to find these relations:

	Core	Case
Pinion	$(H_B)_{11}$	$(H_B)_{12}$
Gear	$(H_B)_{21}$	$(H_B)_{22}$

(*a*) For carburized case-hardened gear steel with core AGMA bending strength $(s_{at})_G$ and pinion core strength $(s_{at})_P$, show that the relationship is

$$(s_{at})_G = (s_{at})_P \frac{J_P}{J_G} m_G^{-0.0323}$$

This allows $(H_B)_{21}$ to be related to $(H_B)_{11}$.

(*b*) Show that the AGMA contact strength of the gear case $(s_{ac})_G$ can be related to the AGMA core bending strength of the pinion core $(s_{at})_P$ by

$$(s_{ac})_G = \frac{C_p}{(C_L)_G C_H} \sqrt{\frac{S_H^2}{S_F} \frac{(s_{at})_P (K_L)_P K_x J_P K_T C_s C_{xc}}{N_P I K_s}}$$

If factors of safety are applied to the transmitted load W_t, then $S_H = \sqrt{S_F}$ and S_H^2/S_F is unity. The result allows $(H_B)_{22}$ to be related to $(H_B)_{11}$.

(*c*) Show that the AGMA contact strength of the gear $(s_{ac})_G$ is related to the contact strength of the pinion $(s_{ac})_P$ by

$$(s_{ac})_P = (s_{ac})_g m_G^{0.0602} \frac{1}{C_H}$$

15–8 Refer to your solution to Probs. 15–1 and 15–2, which is to have a pinion core hardness of 300 Brinell. Use the relations from Prob. 15–7 to establish the hardness of the gear core and the case hardnesses of both gears.

15–9 Repeat Probs. 15–1 and 15–2 with the hardness protocol

	Core	Case
Pinion	300	372
Gear	352	344

which can be established by relations in Prob. 15–7, and see if the result matches transmitted loads W^t in all four cases.

15–10 A catalog of stock bevel gears lists a power rating of 5.2 hp at 1200 r/min pinion speed for a straight-bevel gearset consisting of a 20-tooth pinion driving a 40-tooth gear. This gear pair has a 20° normal pressure angle, a face width of 0.71 in, a diametral pitch of 10 teeth/in, and is through-hardened to 300 BHN. Assume the gears are for general industrial use, are generated to a transmission accuracy number of 5, and are uncrowned. Given these data, what do you think about the stated catalog power rating?

ANALYSIS

15–11 Apply the relations of Prob. 15–7 to Ex. 15–1 and find the Brinell case hardness of the gears for equal allowable load W^t in bending and wear. Check your work by reworking Ex. 15–1 to see if you are correct. How would you go about the heat treatment of the gears?

15–12 Your experience with Ex. 15–1 and problems based on it will enable you to write an interactive computer program for power rating of through-hardened steel gears. Test your understanding of bevel-gear analysis by noting the ease with which the coding develops. The hardness protocol developed in Prob. 15–7 can be incorporated at the end of your code, first to display it, then as an option to loop back and see the consequences of it.

15–13 Use your experience with Prob. 15–11 and Ex. 15–2 to design an interactive computer-aided design program for straight-steel bevel gears, implementing the ANSI/AGMA 2003-B97 standard. It will be helpful to follow the decision set in Sec. 15–5, allowing the return to earlier decisions for revision as the consequences of earlier decisions develop. When all of the design decisions appear satisfactory, add a summary akin to an adequacy assessment as part of the program.

We next go to design-of-mesh tasks in the case of cylindrical worm gearing. By "design-of-mesh" we mean that all of the decisions relating to tooth geometry, materials, and durability will be made, possibly including rim thickness, but the total worm or gear geometry, hub, web, keying, or bore, will not be addressed. The gear body might be cast iron, to which the tooth rim is attached. By now you have had sufficient gear-mesh analysis with spur, helical, and bevel gearing for the vocabulary to become part of you. It is time to begin thinking in terms of design structure.

15–14 The ANSI/AGMA method for worm gearing documents the variation of the coefficient of friction with sliding velocity for 58 HRC polished-steel worms and gear bronzes. For less-hard steel worms and bronzes as well as cast iron on cast iron, use Fig. 13–42. These graphs may be unsuitable for computer purposes. The following curve fits are offered. Figure 13–42 *A* curve:

$$f = 0.9837(10^{-1}) - 0.218\,77(10^{-3})V_s + 0.330\,22(10^{-6})V_S^2$$
$$-0.261\,075(10^{-9})V_s^3 + 0.102\,495(10^{-12})V_s^4 - 0.158\,487(10^{-16})V_s^5$$

For Fig. 13–42 *B* curve:

$$f = 0.745\,485(10^{-1}) - 0.184\,666(10^{-3})V_s + 0.311\,098\,(10^{-6})V_S^2$$
$$-0.267\,263(10^{-9})V_s^3 + 0.111\,155(10^{-12})V_s^4 - 0.177\,829(10^{-16})V_s^5$$

both for the range $100 \leq V_s \leq 2000$ ft/min. V. M. Faires[4] fit similar data as follows. For the conditions of the *A* curve:

$$f = \begin{cases} \dfrac{0.155}{V_s^{0.2}} & 3 \leq V_s \leq 70 \text{ ft/min} \\ \dfrac{0.32}{V_s^{0.36}} & 70 < V_s \leq 3000 \text{ ft/min} \end{cases}$$

and for the conditions of the *B* curve:

$$f = \begin{cases} \dfrac{0.194}{V_s^{0.2}} & 3 \leq V_s \leq 70 \text{ ft/min} \\ \dfrac{0.40}{V_s^{0.36}} & 70 < V_s \leq 3000 \text{ ft/min} \end{cases}$$

If you believe your know how a computer works, program one polynomial and one fractional power of V_s equation in separate iterative loops, and compare the CPU seconds. Before you believe what you get, check with your computation center (if you are using a time-shared system) to be sure you are recovering only your computational time. If you are on a single-user PC, check your manual.

For other than polished hard steel works on phosphor bronze, you will have to avoid AGMA friction data and use something else. Use your computer to compare the outputs, then identify the equation pairs you would want to code.

The preferred materials for worm gearing are an array of bronzes: tin-, nickel-tin-, leaded-, aluminum-, and silicon-bronzes. Cast iron, because of its durability, also finds use.

15–15 to 15–22 Design a cylindrical worm-gear mesh to connect a squirrel-cage induction motor to a liquid agitator. The motor speed is 1125 r/min, and the velocity ratio is to be 10:1. The output power requirement is 25 hp. The shaft axes are 90° to each other. An overload factor K_o (see Table 15–2) makes allowance for external dynamic excursions of load from the nominal or average load W'. For this service $K_o = 1.25$ is appropriate. Additionally, a design factor n_d of 1.1 is to be included to address other unquantifiable risks. Use the design method and the material pair in Table 15–12.

Table 15–12

Table Supporting Problems 15–15 to 15–22

Problem		Materials	
No.	Method	Worm	Gear
15–15	AGMA	Steel, HRC 58	Sand-cast bronze
15–16	AGMA	Steel, HRC 58	Chilled-cast bronze
15–17	AGMA	Steel, HRC 58	Centrifugal-cast bronze
15–18	Buckingham	Steel, 500 Bhn	Chilled-cast bronze
15–19	Buckingham	Steel, 250 Bhn	Cast bronze
15–21	Buckingham	High-test cast iron	Cast bronze
15–22	Buckingham	High-test cast iron	High-test cast iron

Since iteration in worm-gear design contains considerable detail, focus on the important items (keep your eye on the doughnut and not on the hole). The allowable transmitted load on the gear

4. V. M. Faires, *Design of Machine Elements*, 4th Ed., Macmillan, New York, 1965, p. 434.

$(W_G^t)_{\text{all}}$ is to be kept slightly higher than the design load. The parameter to monitor is C_s in the AGMA method and K_w in the Buckingham method. Reducing the difference involves reducing axial pitch p_x, bearing in mind that p_x is discrete. Further, when dealing with decision 1 (p_x), decision 2 (d_W), and decision 3 (F_W [or F_e]), there are ranges on d_W, λ, and F_W. Recall that the lead angle λ is given by

$$\lambda = \tan^{-1}\left(\frac{p_x N_W}{\pi d_W}\right)$$

and backing off from its upper-level constraint involves reducing p_x or N_W, or increasing d_W, or a combination. Sometimes revising the a priori decision N_W is helpful. Notice that in both these considerations p_x is common.

One objective of this series of design tasks is to show you that the choice of a material pair is not just a case of looking for the largest C_s or K_w. Once again, understanding derives from practice. Problems 15–15 and 15–19 offer an opportunity to compare the AGMA and Buckingham methods, as do Probs. 15–16 and 15–18. Heat transfer and gear-case area issues should be addressed in worm gearing, for they often limit the power that can be transmitted.

ANALYSIS

15–23 Perform an adequacy assessment on any of the designs presented in Probs. 15–15 to 15–22. Work with a design that you have done yourself or with a classmates design, as assigned by your instructor.

16 Clutches, Brakes, Couplings, and Flywheels

16–1 Rudiments of Brake Analysis **993**
16–2 Internal Expanding Rim Clutches and Brakes **999**
16–3 External Contracting Rim Clutches and Brakes **1008**
16–4 Band-Type Clutches and Brakes **1011**
16–5 Frictional-Contact Axial Clutches **1013**
16–6 Disk Brakes **1016**
16–7 Cone Clutches and Brakes **1022**
16–8 Self-Locking Tapers and Torque Capacity **1024**
16–9 Energy Considerations **1026**
16–10 Temperature Rise **1027**
16–11 Friction Materials **1031**
16–12 Miscellaneous Clutches and Couplings **1032**
16–13 Flywheels **1034**
16–14 Adequacy Assessment for Clutches and Brakes **1039**

This chapter is concerned with a group of elements usually associated with rotation that have in common the function of storing and/or transferring rotating energy. Because of this similarity of function, clutches, brakes, couplings, and flywheels are treated together in this book.

A simplified dynamic representation of a friction clutch or brake is shown in Fig. 16–1*a*. Two inertias, I_1 and I_2, traveling at the respective angular velocities ω_1 and ω_2, one of which may be zero in the case of brakes, are to be brought to the same speed by engaging the clutch or brake. Slippage occurs because the two elements are running at different speeds and energy is dissipated during actuation, resulting in a temperature rise. In analyzing the performance of these devices we shall be interested in:

1 The actuating force
2 The torque transmitted
3 The energy loss
4 The temperature rise

The torque transmitted is related to the actuating force, the coefficient of friction, and the geometry of the clutch or brake. This is a problem in statics which will have to be studied separately for each geometric configuration. However, temperature rise is related to energy loss and can be studied without regard to the type of brake or clutch, because the geometry of interest is that of the heat-dissipating surfaces.

The various types of devices to be studied may be classified as follows:

1 Rim types with internal expanding shoes
2 Rim types with external contracting shoes
3 Band types
4 Disk or axial types
5 Cone types
6 Miscellaneous types

A flywheel is an inertial energy-storage device. It absorbs mechanical energy by increasing its angular velocity and delivers energy by decreasing its velocity. Figure 16–1*b* is a mathematical representation of a flywheel. An input torque T_i, corresponding to a coordinate θ_i, will cause the flywheel speed to increase. And a load or output torque T_o, with coordinate θ_o, will absorb energy from the flywheel and cause it to slow down. We shall be interested in designing flywheels so as to obtain a specified amount of speed regulation.

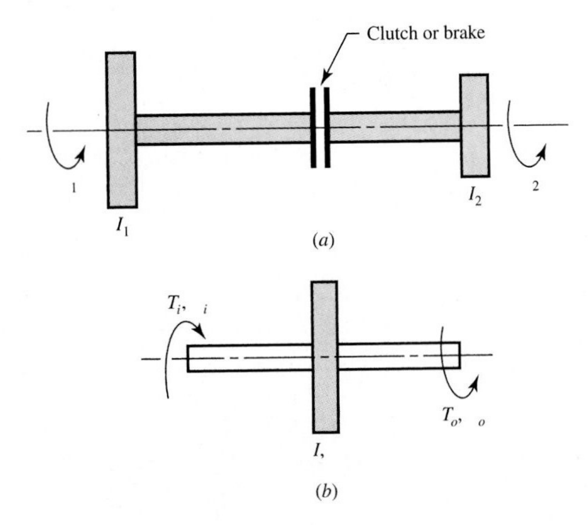

Figure 16–1

(*a*) Dynamic representation of a clutch or brake; (*b*) mathematical representation of a flywheel.

16–1 Rudiments of Brake Analysis

Many types of clutches and brakes can be analyzed by following a general procedure. The procedure entails the following tasks:

- Estimate, model, or measure the pressure distribution on the friction surfaces.
- Find a relationship between the largest pressure and the pressure at any point.
- Use the conditions of static equilibrium to find the braking force or torque and the support reactions.

Let us apply these tasks to the doorstop depicted in Fig. 16–2a. The stop is hinged at pin A. A normal pressure distribution is shown under the friction pad of magnitude $p(u)$ lbf/in^2. A similar distribution of shearing traction is on the surface and of intensity $fp(u)$ psi. The width of the pad into the paper is w_2 in. We will need certain integrals which we will now evaluate using a dummy variable of integration u:

$$w_2 \int_0^{w_1} p(u)\,du = p_{\text{av}} w_1 w_2 \tag{a}$$

$$\int_0^{w_1} p(u) u\,du = \bar{u} \int_0^{w_1} p(u)\,du = p_{\text{av}} w_1 \bar{u} \tag{b}$$

We sum the forces in the x-direction to obtain

$$\Sigma F^x = R^x \mp \int_0^{w_1} f w_2 p(u)\,du = R^x \mp f w_2 \int_0^{w_1} p(u)\,du = R^x \mp f p_{\text{av}} w_1 w_2 = 0$$

from which

$$R^x = \pm f w_2 \int_0^{w_1} p(u)\,du = \pm f p_{\text{av}} w_1 w_2 \qquad \begin{array}{l} + \text{ for leftward motion of floor} \\ - \text{ for rightward motion of the floor} \end{array} \tag{c}$$

Summing the forces in the y direction gives

$$\Sigma F^y = -F + \int_0^{w_1} p(u) w_2\,du + R^y = 0$$

from which

$$R^y = F - w_2 \int_0^{w_1} p(u)\,du = F - p_{\text{av}} w_1 w_2 \qquad \text{for either direction} \tag{d}$$

Summing moments about the pin located at A we have

$$M_A = Fb - \int_0^{w_1} w_2 p(u)(c+u)\,du \mp a f w_2 \int_0^{w_1} p(u)\,du = 0$$

A brake shoe is *self-energizing* if its moment sense helps set the brake, *self-deenergizing* if the moment resists setting the brake. Continuing,

$$F = \frac{w_2}{b}\left[\int_0^{w_1} p(u)(c+u)\,du \mp af \int_0^{w_1} p(u)\,du\right] \tag{e}$$

Can F be equal to or less than zero? Only during rightward motion of the floor when the expression in brackets in Eq. (e) is equal to or less than zero. We set the brackets to zero or less:

$$\int_0^{w_1} p(u)(c+u)\,du - af \int_0^{w_1} p(u)\,du \le 0$$

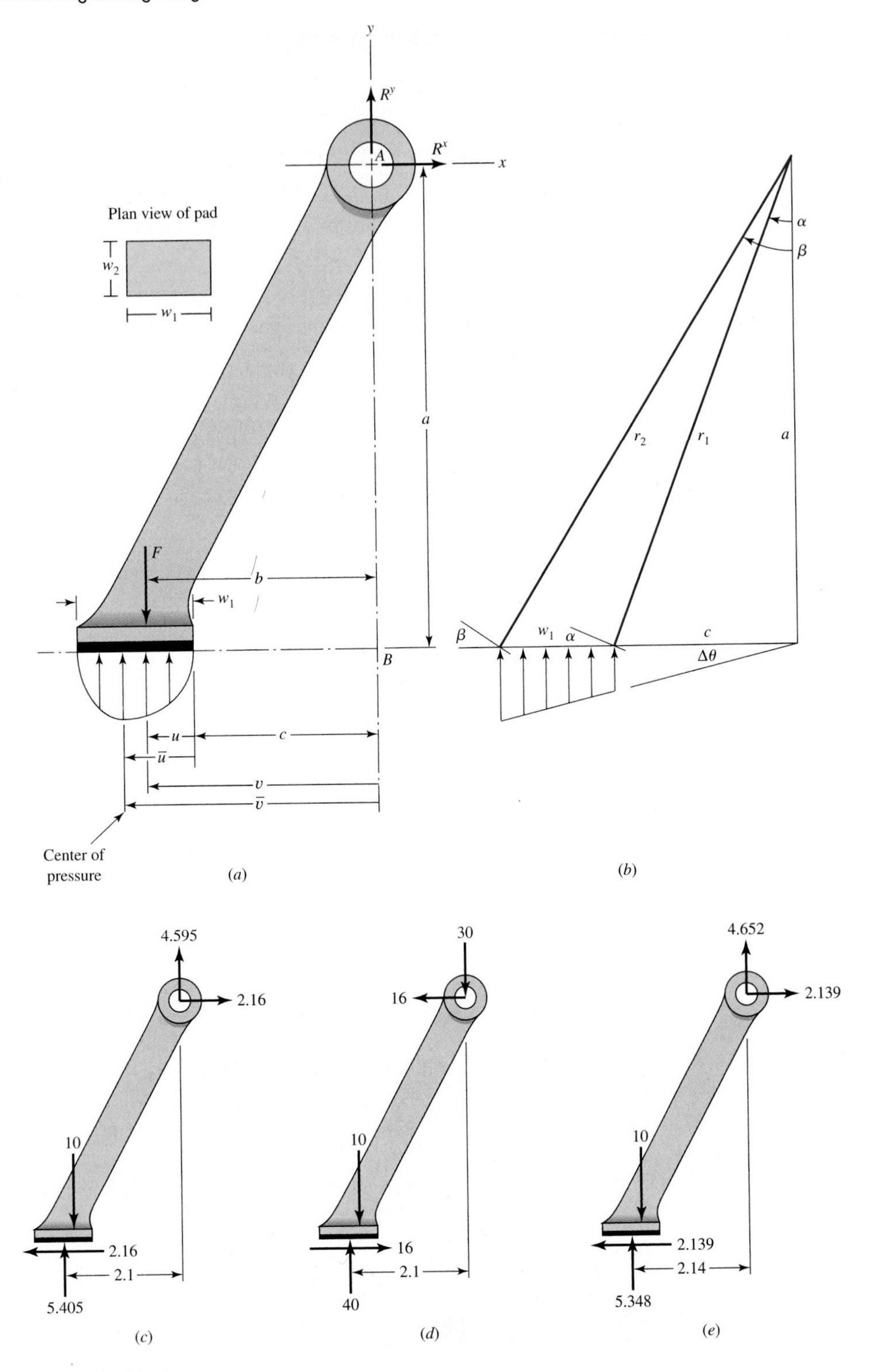

Figure 16–2

A common doorstop. (*a*) Free body of the doorstop. (*b*) Trapezoidal pressure distribution on the foot pad. (*c*) Free-body diagram for leftward movement of the floor, uniform pressure, Ex. 16–1. (*d*) Free-body diagram for rightward movement of the floor, uniform pressure, Ex. 16–1. (*e*) Free-body diagram for leftward movement of the floor, trapezoidal pressure, Ex. 16–1.

from which

$$f_{\mathrm{cr}} \geq \frac{1}{a}\frac{\displaystyle\int_0^{w_1} p(u)(c+u)\,du}{\displaystyle\int_0^{w_1} p(u)\,du} = \frac{1}{a}\frac{\displaystyle c\int_0^{w_1} p(u)\,du + \int_0^{w_1} p(u)u\,du}{\displaystyle\int_0^{w_1} p(u)\,du}$$

$$f_{\mathrm{cr}} \geq \frac{c+\bar{u}}{a} \tag{f}$$

where $\bar{u}$ is the distance of the center of pressure from the righthand edge of the pad. The conclusion that a *self-acting* or *self-locking* phenomenon is present is independent of our knowledge of the normal pressure distribution $p(u)$. Our ability to *find* the critical value of the coefficient of friction f_{cr} is dependent on our knowledge of $p(u)$, from which we derive $\bar{u}$.

We have

- Recognized self-energizing and self-deenergizing conditions without knowing the pressure distribution $p(u)$.
- Detected self-locking capability even though the distribution of the pressure is unknown.
- Found a quantitative expression for the critical coefficient of friction, which requires knowledge of the pressure distribution to evaluate. The expression also requires knowledge of the coefficient of friction extant in the designer's specification. Such a coefficient of friction can vary. Stochastic methods can be used for decision making.
- Determined that a brake shoe or pad must be self-energizing in order that it have the potential of being self-locking.

The designer brings about or precludes self-locking by control of the geometry (here the dimensions a, b, and c) and choice of friction pad material. Variability in the coefficient of friction can be treated by statistical methods. The pressure distribution must be known, robustly modeled, or measured.

An example will be helpful here.

EXAMPLE 16–1

The doorstop depicted in Fig. 16–2*a* has the following dimensions: $a = 4$ in, $b = 2$ in, $c = 1.6$ in, $w_1 = 1$ in, $w_2 = 0.75$ in, where w_2 is the depth of the pad into the plane of the paper.

(*a*) For a leftward movement of the floor, an actuating force F of 10 lbf, a coefficient of friction of 0.4, use a uniform pressure distribution p_{av}, find R^x, R^y, p_{av}, and the largest pressure p_a.

(*b*) Repeat part *a* for rightward movement of the floor.

(*c*) Model the normal pressure to be the "crush" of the pad, much as if it were composed of many small helical coil springs. Find R^x, R^y, p_{av}, and p_a for leftward movement of the floor and other conditions as in part *a*.

(*d*) For rightward movement of the floor, is the doorstop a self-acting brake?

Solution (*a*)

Eq. (*c*) $R^x = f p_{\mathrm{av}} w_1 w_2 = 0.4(1)(0.75)p_{\mathrm{av}} = 0.3p_{\mathrm{av}}$

Eq. (*d*) $R^y = F - p_{\mathrm{av}} w_1 w_2 = 10 - p_{\mathrm{av}}(1)(0.75) = 10 - 0.75p_{\mathrm{av}}$

$$\text{Eq. }(e)\ F = \frac{w_2}{b}\left[\int_0^1 p_{av}(c+u)\,du + af\int_0^1 p_{av}\,du\right]$$
$$= \frac{w_2}{b}\left(p_{av}c\int_0^1 du + p_{av}\int_0^1 u\,du + afp_{av}\int_0^1 du\right)$$
$$= \frac{w_2 p_{av}}{b}(c + 0.5 + af) = \frac{0.75}{2}[1.6 + 0.5 + 4(0.4)]p_{av}$$
$$= 1.3875p_{av}$$

Solving for p_{av} gives

$$p_{av} = \frac{F}{1.3875} = \frac{10}{1.3875} = 7.207\text{ psi}$$

We evaluate R^x and R^y as

Answer $R^x = 0.3(7.207) = 2.162$ lbf

Answer $R^y = 10 - 0.75(7.207) = 4.595$ lbf

The normal force N on the pad is $F - R^y = 10 - 4.595 = 5.405$ lbf, upward. The line of action is through the center of pressure, which is at the center of the pad. The friction force is $fN = 0.4(5.405) = 2.162$ lbf directed to the left. A check of the moments about A gives

$$\Sigma M_A = Fb - fNa - N(w_1/2 + c)$$
$$= 10(2) - 0.4(5.405)4 - 5.405(1/2 + 1.6) \doteq 0$$

Answer The maximum pressure $p_a = p_{av} = 7.207$ psi.

(*b*)

$$\text{Eq. }(c)\ R^x = -fp_{av}w_1w_2 = -0.4(1)(0.75)p_{av} = -0.3p_{av}$$
$$\text{Eq. }(d)\ R^y = F - p_{av}w_1w_2 = 10 - p_{av}(1)(0.75) = 10 - 0.75p_{av}$$
$$\text{Eq. }(e)\ F = \frac{w_2}{b}\left[\int_0^1 p_{av}(c+u)\,du + af\int_0^1 p_{av}\,du\right]$$
$$= \frac{w_2}{b}\left(p_{av}c\int_0^1 du + p_{av}\int_0^1 u\,du + afp_{av}\int_0^1 du\right)$$
$$= \frac{0.75}{2}p_{av}[1.6 + 0.5 - 4(0.4)] = 0.1875p_{av}$$

from which

$$p_{av} = \frac{F}{0.1875} = \frac{10}{0.1875} = 53.\underline{3}\text{ psi}$$

which makes

Answer $R^x = -0.3(53.\underline{3}) = -16$ lbf

Answer $R^y = 10 - 0.75(53.\underline{3}) = -30$ lbf

The normal force N on the pad is $10 + 30 = 40$ lbf upward. The friction shearing force is $fN = 0.4(40) = 16$ lbf to the right. We now check the moments about A:

$$M_A = fNa + Fb - N(c + 0.5) = 16(4) + 10(2) - 40(1.6 + 0.5) = 0$$

Note the change in average pressure from 7.207 psi in part a to 53.$\underline{3}$ psi. Also note how directions of forces have changed. The maximum pressure p_a is the same as p_{av}, which has changed from 7.207 psi to 53.$\underline{3}$ psi.

(c) From Fig. 16–2b we evaluate

$$\alpha = \tan^{-1}\frac{c}{a} = \tan^{-1}\frac{1.6}{4} = 21.80^\circ$$

$$\beta = \tan^{-1}\frac{c + w_1}{a} = \tan^{-1}\frac{1.6 + 1}{4} = 33.02^\circ$$

$$r_1 = \sqrt{c^2 + a^2} = \sqrt{1.6^2 + 4^2} = 4.308 \text{ in}$$

$$r_2 = \sqrt{(c + w_1)^2 + a^2} = \sqrt{(1.6 + 1)^2 + 4^2} = 4.771 \text{ in}$$

$$y_1 = r_1 \Delta\theta \sin\alpha = 4.308\Delta\theta \sin 21.80^\circ = 1.60\Delta\theta \text{ in}$$

$$y_2 = r_2 \Delta\theta \sin\beta = 4.771\Delta\theta \sin 33.02^\circ = 2.6\Delta\theta \text{ in}$$

The spring rate of this conceptual bed of springs will be expressed as pressure per inch of end compression, which we will designate as k. Thus ky is the pressure p at the location where the end contraction is y. We form a table:

Location	ky	p	v	u
Right edge	$1.60k\ \Delta\theta$	$1.60k\ \Delta\theta$	1.6	0
Left edge	$2.60k\ \Delta\theta$	$2.60k\ \Delta\theta$	2.6	1

We will be interested in p as a function of v and x. These linear functions can be shown to be

$$p(v) = k\Delta\theta v = C_1 v$$

$$p(u) = k\Delta\theta(1.6 + u) = C_1(1.6 + u)$$

$$\int_0^1 p(u)\,du = C_1\left(\int_0^1 1.6\,du + \int_0^1 u\,du\right) = C_1(1.6 + 0.5) = 2.1C_1$$

$$\int_0^1 p(u)u\,du = C_1\left(\int_0^1 1.6u\,du + \int_0^1 u^2\,du\right) = C_1[1.6(0.5) + 0.\underline{3}] = 1.1\underline{3}C_1$$

$$\bar{u} = \frac{1.1\underline{3}C_1}{2.1C_1} = 0.5937 \text{ in} \qquad \bar{v} = c + \bar{u} = 1.6 + 0.5397 = 2.1397 \text{ in}$$

$$p(\bar{u}) = p(0.5397) = C_1(1.6 + 0.5397) = 2.1397C_1$$

$$\text{Eq. }(c)\ R^x = f w_2 \int_0^1 p(u)\,du = f w_2 p_{\text{av}} = 0.4(0.75)p_{\text{av}} = 0.3p_{\text{av}}$$

$$\text{Eq. }(d)\ R^y = F - w_2 \int_0^1 p(u)\,du = 10 - 0.75p_{\text{av}}$$

$$\text{Eq. }(e)\ F = \frac{w_2}{b}\left[\int_0^1 p(u)c\,du + \int_0^1 p(u)u\,du + af\int_0^1 p(u)\,du\right]$$

$$= \frac{0.75}{2}[1.6p_{\text{av}} + 0.5397p_{\text{av}} + 4(0.4)p_{\text{av}}] = 1.402p_{\text{av}}$$

This result is a different *equation* than was obtained in part a because the center of pressure changed, altering the location of the line of action of the normal force N on the pad. This changes the moment equation. We find p_{av} to be

Answer $$p_{av} = \frac{F}{1.402} = \frac{10}{1.402} = 7.131 \text{ psi}$$

which is different from 7.207 psi found in part a. The maximum pressure occurs at $u = 1$, and, noting that C_1 is $p_{av}/2.1$, it is

$$p(1) = C_1(1.6 + 1) = (p_{av}/2.1)(1.6 + 1) = 1.238 p_{av} = 1.238(7.131)$$
$$= 8.83 \text{ psi}$$

The average pressure is $p_{av} = 7.131$ psi and the maximum pressure is $p_a = 8.83$ psi, which is approximately 24 percent higher than the average pressure. The presumption that the pressure was uniform in part a (because the pad was small, or because the arithmetic would be easier?) underestimated the peak pressure. Modeling the pad as a one-dimensional springset is better, but the pad is really a three-dimensional continuum. A theory of elasticity approach or a finite-element modeling may be overkill, given uncertainties inherent in this problem, but it represents still better modeling.

(d) To evaluate $\bar{v}$ we need to conduct two integrations:

$$\int_0^{c+w_1} p(v)v\,dv = \int_c^{c+w_1} p(v)v\,dv = \int_{1.6}^{2.6} (C_1 v)v\,dv = C_1 \left.\frac{v^3}{3}\right|_{1.6}^{2.6} = 4.49\underline{3}C_1$$

$$\int_0^{c+w_1} p(v)\,dv = \int_c^{c+w_1} p(v)\,dv = \int_{1.6}^{2.6} (C_1 v)\,dv = C_1 \left.\frac{v^2}{2}\right|_{1.6}^{2.6} = 2.1C_1$$

The distance $\bar{v}$ to the center of pressure is

$$\bar{v} = \frac{4.49\underline{3}C_1}{2.1C_1} = 2.1397$$

$$\bar{u} = \bar{v} - c = 2.1397 - 1.6 = 0.5397 \text{ in}$$

The critical coefficient of friction is given by

$$f_{cr} \geq \frac{c + \bar{u}}{a} = \frac{1.6 + 0.5397}{4} = 0.535$$

The doorstop friction pad does not have a high enough coefficient of friction to make the doorstop a self-acting brake. The pad material specification needs to be changed to sustain the function of a doorstop.

Note three lessons taught about brakes by this familiar example:

- The concept of average pressure is useful for some things and misleading for others.
- The direction of motion makes a difference.
- The location of the center of pressure depends on the pressure distribution when symmetry is absent.

Exercise caution in these areas as we address brakes and clutches, whose geometry is sometimes very complicated and therefore distracting.

The terms *fail-safe* and *dead-man* are encountered in studying the operation of brakes and clutches (as well as other mechanical elements). The terms may be considered self-explanatory. Fail-safe operating mechanisms are designed so that failure of any element to perform its function will not cause accidents to occur in the machine or to befall its operator. Dead-man, a term from the electric railway industry, refers to the train-control mechanism which stops the train should the operator suffer blackout, fail to acknowledge a signal, or die at the controls.

The analysis in this section is very useful when the dimensions of the brake or clutch are known and the characteristics of the friction material are specified. In design, however, we are interested more in synthesis than in analysis, that is, our purpose is to select a set of dimensions to obtain the best brake or clutch within the limitations of space and available friction materials.

16–2 Internal Expanding Rim Clutches and Brakes

The internal-shoe rim clutch shown in Fig. 16–3 consists essentially of three elements: the mating frictional surface, the means of transmitting the torque to and from the surfaces, and the actuating mechanism. Depending upon the operating mechanism, such clutches are further classified as *expanding-ring, centrifugal, magnetic, hydraulic,* and *pneumatic*.

The expanding-ring clutch is often used in textile machinery, excavators, and machine tools where the clutch may be located within the driving pulley. Expanding-ring clutches benefit from centrifugal effects; transmit high torque, even at low speeds; and require both positive engagement and ample release force.

The centrifugal clutch is used mostly for automatic operation. If no spring is used, the torque transmitted is proportional to the square of the speed. This is particularly useful for electric-motor drives where, during starting, the driven machine comes up to speed without shock. Springs can also be used to prevent engagement until a certain motor speed has been reached, but some shock may occur.

Magnetic clutches are particularly useful for automatic and remote-control systems. Such clutches are also useful in drives subject to complex load cycles (see Sec. 11-7).

Hydraulic and pneumatic clutches are also useful in drives having complex loading cycles and in automatic machinery, or in robots. Here the fluid flow can be controlled remotely using solenoid valves. These clutches are also available as disk, cone, and multiple-plate clutches.

Figure 16–3

An internal expanding centrifugal-acting rim clutch. (Courtesy of the Hilliard Corporation.)

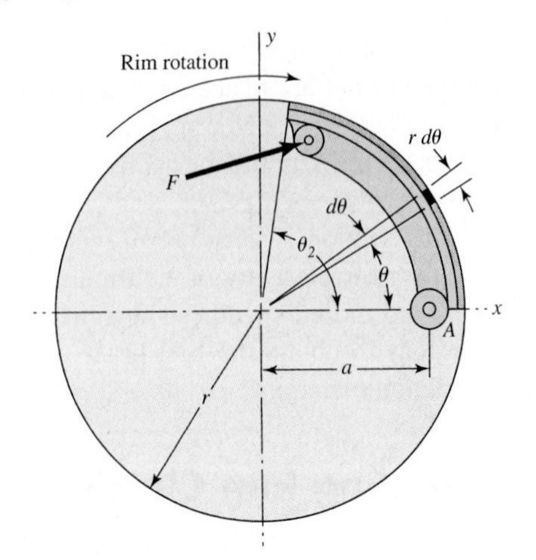

Figure 16–4

Internal friction shoe geometry.

In braking systems, the *internal-shoe* or *drum* brake is used mostly for automotive applications.

To analyze an internal-shoe device, refer to Fig. 16–4, which shows a shoe pivoted at point A, with the actuating force acting at the other end of the shoe. Since the shoe is long, we cannot make the assumption that the distribution of normal forces is uniform. The mechanical arrangement permits no pressure to be applied at the heel, and we will therefore assume the pressure at this point to be zero.

It is the usual practice to omit the friction material for a short distance away from the heel (point A). This eliminates interference, and the material would contribute little to the performance anyway, as will be shown. In some designs the hinge pin is made movable to provide additional heel pressure. This gives the effect of a floating shoe. (Floating shoes will not be treated in this book, although their design follows the same general principles.)

Let us consider the pressure p acting upon an element of area of the frictional material located at an angle θ from the hinge pin (Fig. 16–4). We designate the maximum pressure p_a located at an angle θ_a from the hinge pin. To find the pressure distribution on the periphery of the internal shoe, we place the pivot on the brake drum circle and on the abscissa, and we envision the shoe extending in a semicircle. We place a radical line at a central angle θ of approximately 135°, as shown in Fig. 16–5a, and form an isosceles triangle with a chord length of h. We note that the sum of the internal angles of this triangle sum to π radians:

$$\theta + 2\xi = \pi$$

from which

$$\xi = \frac{\pi}{2} - \frac{\theta}{2} \qquad d\xi = -\frac{d\theta}{2}$$

We also note that

$$\frac{h/2}{r} = \sin\frac{\theta}{2}$$

and it follows that

$$h = 2r\sin\frac{\theta}{2} \qquad dh = r\cos\frac{\theta}{2}d\theta$$

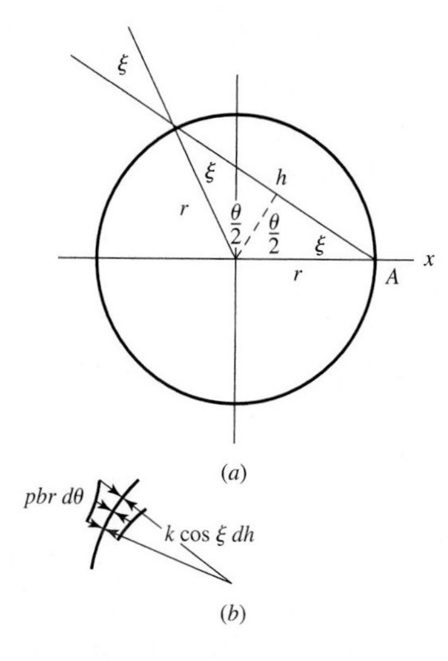

Figure 16–5

(*a*) The geometry associated with an arbitrary point on the shoe; (*b*) relating pressure on the rim to shoe lining spring rate.

Figure 16–5*b* shows, at the rim, the radical force of the rim on the shoe as $pbr\,d\theta$ where p is the pressure, and the reaction of the shoe on the rim as $k\cos\xi\,dh$, where k is the friction lining spring rate in psi per unit radial deflection. These forces are equal and opposite. We now express the shoe reaction as

$$k\cos\xi\,dh = k\cos\left(\frac{\pi}{2}-\frac{\theta}{2}\right)dh = k\left(\cos\frac{\pi}{2}\cos\frac{\theta}{2}+\sin\frac{\pi}{2}\sin\frac{\theta}{2}\right)$$

$$= k\sin\frac{\theta}{2}dh = k\sin\frac{\theta}{2}\left(r\cos\frac{\theta}{2}d\theta\right)$$

$$= \frac{kr}{2}\sin\theta\,d\theta$$

Equating the radial forces of Fig. 16–5*b* gives

$$pbr\,d\theta = \frac{kr}{2}\sin\theta\,d\theta$$

which can be expressed as the pressure p by

$$p = \frac{k}{2b}\sin\theta$$

If the largest pressure is p_a, occurring at θ_a, then

$$p_a = \frac{k}{2b}\sin\theta_a$$

defining $k/(2b)$ as

$$\frac{k}{2b} = \frac{p_a}{\sin\theta_a}$$

Backsubstituting for $k/(2b)$ gives the desired pressure distribution:

$$p = \frac{p_a}{\sin\theta_a}\sin\theta \tag{16–1}$$

This form of the pressure equation has interesting and useful attributes:

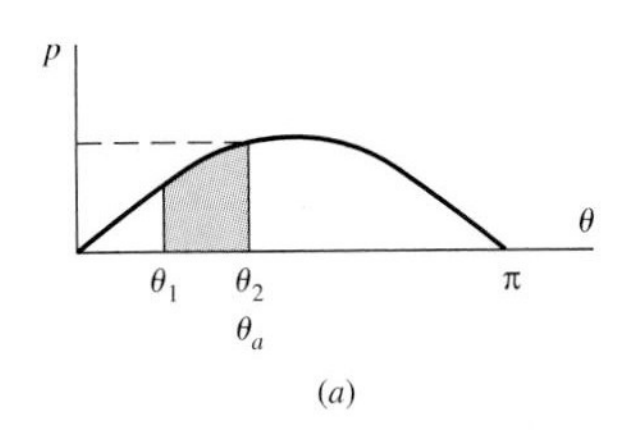

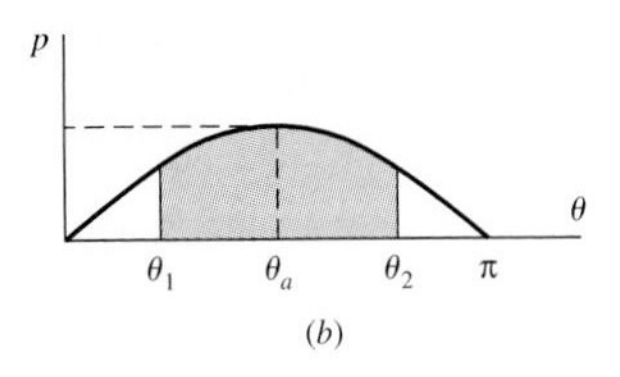

Figure 16–6

Defining the angle θ_a at which the maximum pressure p_a occurs when (a) shoe exists in zone $\theta_1 \leq \theta_2 \leq \pi/2$ and (b) shoe exists in zone $\theta_1 \leq \pi/2 \leq \theta_2$.

- The pressure distribution is sinusoidal with respect to the central angle θ.
- If the shoe is short, as shown in Fig. 16–6*a*, the largest pressure *on the shoe* is p_a occurring at the end of the shoe, θ_2.
- If the shoe is long, as shown in Fig. 16–6*b*, the largest pressure on the shoe is p_a occuring at $\theta_a = 90°$.

Since limitations on friction materials are expressed in terms of the largest allowable pressure on the lining, the designer wants to think in terms of p_a and not about the amplitude of the sinusoidal distribution that is speaking to places off the shoe. From Eq. (16–1) we see that p will be a maximum when $\theta = 90°$, or if the toe angle θ_2 is less than 90°, then p will be a maximum at the toe.

When $\theta = 0$, Eq. (16–1) shows that the pressure is zero. The frictional material located at the heel therefore contributes very little to the braking action and might as well be omitted. A good design would concentrate as much frictional material as possible in the neighborhood of the point of maximum pressure. Such a design is shown in Fig. 16–7. In this figure the frictional material begins at an angle θ_1, measured from the hinge pin A, and ends at an angle θ_2. Any arrangement such as this will give a good distribution of the frictional material.

Proceeding now (Fig. 16–7), the hinge-pin reactions are R_x and R_y. The actuating force F has components F_x and F_y and operates at distance c from the hinge pin. At any angle θ from the hinge pin there acts a differential normal force dN whose magnitude is

$$dN = pbr\,d\theta \tag{b}$$

where b is the face width (perpendicular to the paper) of the friction material. Substituting the value of the pressure from Eq. (16–1), the normal force is

$$dN = \frac{p_a br \sin\theta\,d\theta}{\sin\theta_a} \tag{c}$$

The normal force dN has horizontal and vertical components $dN\cos\theta$ and $dN\sin\theta$, as

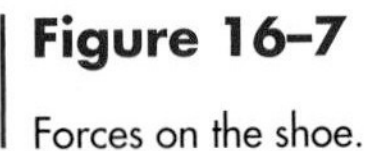

Figure 16–7

Forces on the shoe.

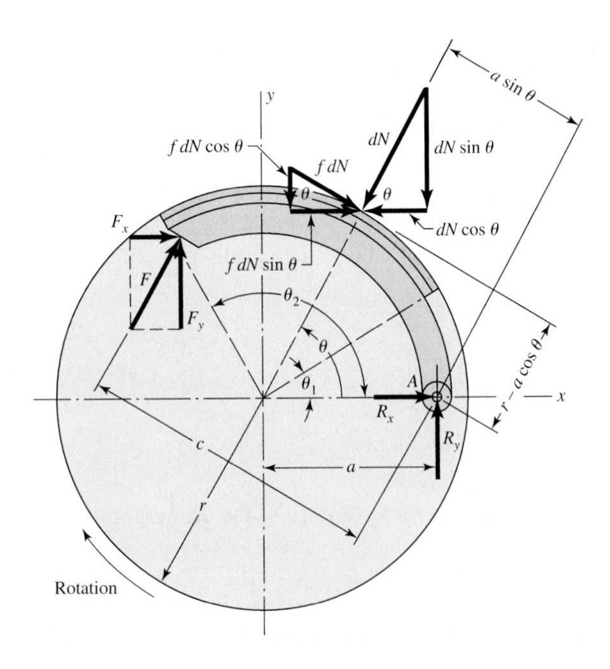

shown in the figure. The frictional force $f\,dN$ has horizontal and vertical components whose magnitudes are $f\,dN \sin\theta$ and $f\,dN \cos\theta$, respectively. By applying the conditions of static equilibrium, we may find the actuating force F, the torque T, and the pin reactions R_x and R_y.

We shall find the actuating force F, using the condition that the summation of the moments about the hinge pin is zero. The frictional forces have a moment arm about the pin of $r - a\cos\theta$. The moment M_f of these frictional forces is

$$M_f = \int f\,dN(r - a\cos\theta) = \frac{f p_a b r}{\sin\theta_a} \int_{\theta_1}^{\theta_2} \sin\theta\,(r - a\cos\theta)\,d\theta \tag{16–2}$$

which is obtained by substituting the value of dN from Eq. (c). It is convenient to integrate Eq. (16–2) for each problem, and we shall therefore retain it in this form. The moment arm of the normal force dN about the pin is $a \sin\theta$. Designating the moment of the normal forces by M_N and summing these about the hinge pin give

$$M_N = \int dN(a\sin\theta) = \frac{p_a b r a}{\sin\theta_a} \int_{\theta_1}^{\theta_2} \sin^2\theta\,d\theta \tag{16–3}$$

The actuating force F must balance these moments. Thus

$$F = \frac{M_N - M_f}{c} \tag{16–4}$$

We see here that a condition for zero actuating force exists. In other words, if we make $M_N = M_f$, self-locking is obtained, and no actuating force is required. This furnishes us with a method for obtaining the dimensions for some self-energizing action. Thus the dimension a in Fig. 16–7 must be such that

$$M_N > M_f \tag{16–5}$$

The torque T applied to the drum by the brake shoe is the sum of the frictional forces $f\,dN$ times the radius of the drum:

$$\begin{aligned} T &= \int f r\,dN = \frac{f p_a b r^2}{\sin\theta_a} \int_{\theta_1}^{\theta_2} \sin\theta\,d\theta \\ &= \frac{f p_a b r^2 (\cos\theta_1 - \cos\theta_2)}{\sin\theta_a} \end{aligned} \tag{16–6}$$

The hinge-pin reactions are found by taking a summation of the horizontal and vertical forces. Thus, for R_x, we have

$$\begin{aligned} R_x &= \int dN\cos\theta - \int f\,dN\sin\theta - F_x \\ &= \frac{p_a b r}{\sin\theta_a}\left(\int_{\theta_1}^{\theta_2} \sin\theta\cos\theta\,d\theta - f\int_{\theta_1}^{\theta_2} \sin^2\theta\,d\theta\right) - F_x \end{aligned} \tag{d}$$

The vertical reaction is found in the same way:

$$\begin{aligned} R_y &= \int dN\sin\theta + \int f\,dN\cos\theta - F_y \\ &= \frac{p_a b r}{\sin\theta_a}\left(\int_{\theta_1}^{\theta_2} \sin^2\theta\,d\theta + f\int_{\theta_1}^{\theta_2} \sin\theta\cos\theta\,d\theta\right) - F_y \end{aligned} \tag{e}$$

The direction of the frictional forces is reversed if the rotation is reversed. Thus, for counterclockwise rotation the actuating force is

$$F = \frac{M_N + M_f}{c} \tag{16–7}$$

and since both moments have the same sense, the self-energizing effect is lost. Also, for counterclockwise rotation the signs of the frictional terms in the equations for the pin reactions change, and Eqs. (d) and (e) become

$$R_x = \frac{p_a b r}{\sin\theta_a}\left(\int_{\theta_1}^{\theta_2} \sin\theta\cos\theta\, d\theta + f\int_{\theta_1}^{\theta_2} \sin^2\theta\, d\theta\right) - F_x \tag{f}$$

$$R_y = \frac{p_a b r}{\sin\theta_a}\left(\int_{\theta_1}^{\theta_2} \sin^2\theta\, d\theta - f\int_{\theta_1}^{\theta_2} \sin\theta\cos\theta\, d\theta\right) - F_y \tag{g}$$

Equations (d), (e), (f), and (g) can be simplified to ease computations. Thus, let

$$\begin{aligned} A &= \int_{\theta_1}^{\theta_2} \sin\theta\cos\theta\, d\theta = \left(\frac{1}{2}\sin^2\theta\right)_{\theta_1}^{\theta_2} \\ B &= \int_{\theta_1}^{\theta_2} \sin^2\theta\, d\theta = \left(\frac{\theta}{2} - \frac{1}{4}\sin 2\theta\right)_{\theta_1}^{\theta_2} \end{aligned} \tag{16–8}$$

Then, for clockwise rotation as shown in Fig. 16–7, the hinge-pin reactions are

$$\begin{aligned} R_x &= \frac{p_a b r}{\sin\theta_a}(A - fB) - F_x \\ R_y &= \frac{p_a b r}{\sin\theta_a}(B + fA) - F_y \end{aligned} \tag{16–9}$$

For counterclockwise rotation, Eqs. (f) and (g) become

$$\begin{aligned} R_x &= \frac{p_a b r}{\sin\theta_a}(A + fB) - F_x \\ R_y &= \frac{p_a b r}{\sin\theta_a}(B - fA) - F_y \end{aligned} \tag{16–10}$$

In using these equations, the reference system always has its origin at the center of the drum. The positive x axis is taken through the hinge pin. The positive y axis is always in the direction of the shoe, even if this should result in a left-handed system.

The following assumptions are implied by the preceding analysis:

1 The pressure at any point on the shoe is assumed to be proportional to the distance from the hinge pin, being zero at the heel. This should be considered from the standpoint that pressures specified by manufacturers are averages rather than maxima.

2 The effect of centrifugal force has been neglected. In the case of brakes, the shoes are not rotating, and no centrifugal force exists. In clutch design, the effect of this force must be considered in writing the equations of static equilibrium.

3 The shoe is assumed to be rigid. Since this cannot be true, some deflection will occur, depending upon the load, pressure, and stiffness of the shoe. The resulting pressure distribution may be different from that which has been assumed.

4 The entire analysis has been based upon a coefficient of friction which does not vary with pressure. Actually, the coefficient may vary with a number of conditions, including temperature, wear, and environment.

EXAMPLE 16–2

The brake shown in Fig. 16–8 is 300 mm in diameter and is actuated by a mechanism that exerts the same force F on each shoe. The shoes are identical and have a face width of 32 mm. The lining is a molded asbestos having a coefficient of friction of 0.32 and a pressure limitation of 1000 kPa. Estimate the

(*a*) Actuating force F.
(*b*) Braking capacity.
(*c*) Hinge-pin reactions.

Solution

(*a*) The right-hand shoe is self-energizing, and so the force F is found on the basis that the maximum pressure will occur on this shoe. Here $\theta_1 = 0°$, $\theta_2 = 126°$, $\theta_a = 90°$, and $\sin\theta_a = 1$. Also,

$$a = \sqrt{(112)^2 + (50)^2} = 123 \text{ mm}$$

Integrating Eq. (16–2) from 0 to θ_2 yields

$$M_f = \frac{f p_a b r}{\sin\theta_a}\left[\left(-r\cos\theta\right)_0^{\theta_2} - a\left(\frac{1}{2}\sin^2\theta\right)_0^{\theta_2}\right]$$
$$= \frac{f p_a b r}{\sin\theta_a}\left(r - r\cos\theta_2 - \frac{a}{2}\sin^2\theta_2\right)$$

Changing all lengths to meters, we have

$$M_f = (0.32)[1000(10)^3](0.032)(0.150)$$
$$\times\left[0.150 - 0.150\cos 126° - \left(\frac{0.123}{2}\right)\sin^2 126°\right]$$
$$= 304 \text{ N}\cdot\text{m}$$

The moment of the normal forces is obtained from Eq. (16–3). Integrating from 0 to θ_2 gives

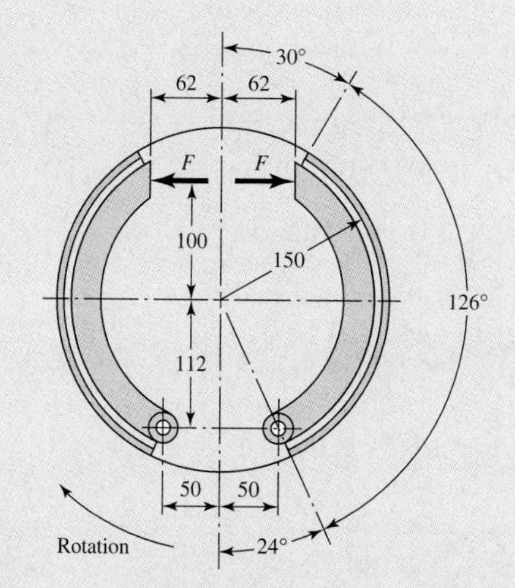

Figure 16–8

Brake with internal expanding shoes; dimensions in millimeters.

$$M_N = \frac{p_a bra}{\sin\theta_a}\left(\frac{\theta}{2} - \frac{1}{4}\sin 2\theta\right)_0^{\theta_2}$$

$$= \frac{p_a bra}{\sin\theta_a}\left(\frac{\theta_2}{2} - \frac{1}{4}\sin 2\theta_2\right)$$

$$= [1000(10)^3](0.032)(0.150)(0.123)\left[\frac{\pi}{2}\frac{126}{180} - \frac{1}{4}\sin(2)(126°)\right]$$

$$= 790 \text{ N}\cdot\text{m}$$

From Eq. (16–4), the actuating force is

Answer $$F = \frac{M_N - M_f}{c} = \frac{790 - 304}{100 + 112} = 2.29 \text{ kN}$$

(*b*) From Eq. (16–6), the torque applied by the right-hand shoe is

$$T_R = \frac{f p_a b r^2(\cos\theta_1 - \cos\theta_2)}{\sin\theta_a}$$

$$= \frac{0.32[1000(10)^3](0.032)(0.150)^2(\cos 0° - \cos 126°)}{1} = 366 \text{ N}\cdot\text{m}$$

The torque contributed by the left-hand shoe cannot be obtained until we learn its maximum operating pressure. Equations (16–2) and (16–3) indicate that the frictional and normal moments are proportional to this pressure. Thus, for the left-hand shoe,

$$M_N = \frac{790 p_a}{1000} \qquad M_f = \frac{304 p_a}{1000}$$

Then, from Eq. (16–7),

$$F = \frac{M_N + M_f}{c}$$

or

$$2.29 = \frac{(790/1000)p_a + (304/1000)p_a}{100 + 112}$$

Solving gives $p_a = 444$ kPa. Then, from Eq. (16–6), the torque on the left-hand shoe is

$$T_L = \frac{f p_a b r^2(\cos\theta_1 - \cos\theta_2)}{\sin\theta_a}$$

Since $\sin\theta_a = 1$, we have

$$T_L = 0.32[444(10)^3](0.032)(0.150)^2(\cos\ 0° - \cos\ 126°) = 162 \text{ N}\cdot\text{m}$$

The braking capacity is the total torque:

Answer $$T = T_R + T_L = 366 + 162 = 528 \text{ N}\cdot\text{m}$$

(*c*) In order to find the hinge-pin reactions, we note that $\sin\theta_a = 1$ and $\theta_1 = 0$. Then Eq. (16–8) gives

$$A = \frac{1}{2}\sin^2\theta_2 = \frac{1}{2}\sin^2 126° = 0.3273$$

$$B = \frac{\theta_2}{2} - \frac{1}{4}\sin 2\theta_2 = \frac{\pi(126)}{2(180)} - \frac{1}{4}\sin(2)(126°) = 1.3373$$

Also, let

$$D = \frac{p_a br}{\sin\theta_a} = \frac{1000(0.032)(0.150)}{1} = 4.8 \text{ kN}$$

where $p_a = 1000$ kPa for the right-hand shoe. Then, using Eq. (16–9), we have

$$R_x = D(A - fB) - F_x = 4.8[0.3273 - 0.32(1.3373)] - 2.29 \sin 24°$$
$$= -1.414 \text{ kN}$$

$$R_y = D(B + fA) - F_y = 4.8[1.3373 + 0.32(0.3273)] - 2.29 \cos 24°$$
$$= 4.830 \text{ kN}$$

The resultant on this hinge pin is

Answer $$R = \sqrt{(1.414)^2 + (4.830)^2} = 5.03 \text{ kN}$$

The reactions at the hinge pin of the left-hand shoe are found using Eqs. (16–10) for a pressure of 444 kPa. They are found to be $R_x = 0.678$ kN and $R_y = 0.535$ kN. The resultant is

Answer $$R = \sqrt{(0.678)^2 + (0.535)^2} = 0.864 \text{ kN}$$

The reactions for both hinge pins, together with their directions, are shown in Fig. 16–9.

This example dramatically shows the benefit to be gained by arranging the shoes to be self-energizing. If the left-hand shoe were turned over so as to place the hinge pin at the top, it could apply the same torque as the right-hand shoe. This would make the capacity of the brake $(2)(366) = 732$ N · m instead of the present 528 N · m, a 30 percent improvement. In addition, some of the friction material at the heel could be eliminated without seriously affecting the capacity, because of the low pressure in this area. This change might actually improve the overall design because the additional rim exposure would improve the heat-dissipation capacity.

Figure 16–9

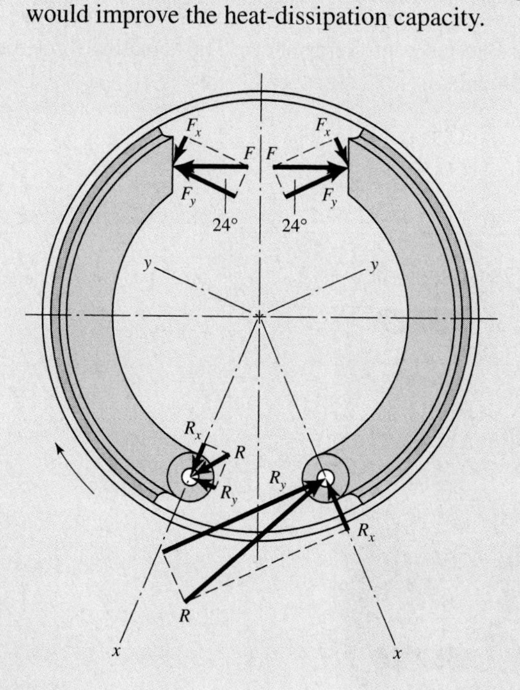

16–3 External Contracting Rim Clutches and Brakes

The patented clutch-brake of Fig. 16–10 has external contracting friction elements, but the actuating mechanism is pneumatic. Here we shall study only pivoted external shoe brakes and clutches, though the methods presented can easily be adapted to the clutch-brake of Fig. 16–10.

Operating mechanisms can be classified as:

1 Solenoids
2 Levers, linkages, or toggle devices
3 Linkages with spring loading
4 Hydraulic and pneumatic devices

The static analysis required for these devices has already been covered in Sec. 3-7. The methods there apply to any mechanism system, including all those used in brakes and clutches. It is not necessary to repeat the material in Chap. 3 that applies directly to such mechanisms. Omitting the operating mechanisms from consideration allows us to concentrate on brake and clutch performance without the extraneous influences introduced by the need to analyze the statics of the control mechanisms.

The notation for external contracting shoes is shown in Fig. 16–11. The moments of the frictional and normal forces about the hinge pin are the same as for the internal expanding shoes. Equations (16–2) and (16–3) apply and are repeated here for convenience:

$$M_f = \frac{f p_a b r}{\sin\theta_a}\int_{\theta_1}^{\theta_2} \sin\theta(r - a\cos\theta)\, d\theta \tag{16–2}$$

$$M_N = \frac{p_a b r a}{\sin\theta_a}\int_{\theta_1}^{\theta_2} \sin^2\theta\, d\theta \tag{16–3}$$

Both these equations give positive values for clockwise moments (Fig. 16–11) when used for external contracting shoes. The actuating force must be large enough to balance both moments:

$$F = \frac{M_N + M_f}{c} \tag{16–11}$$

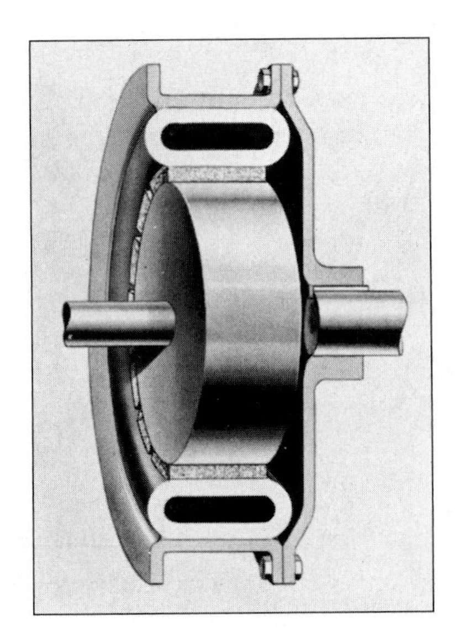

Figure 16–10

An external contracting clutch-brake that is engaged by expanding the flexible tube with compressed air. *(Courtesy of Twin Disc Clutch Company.)*

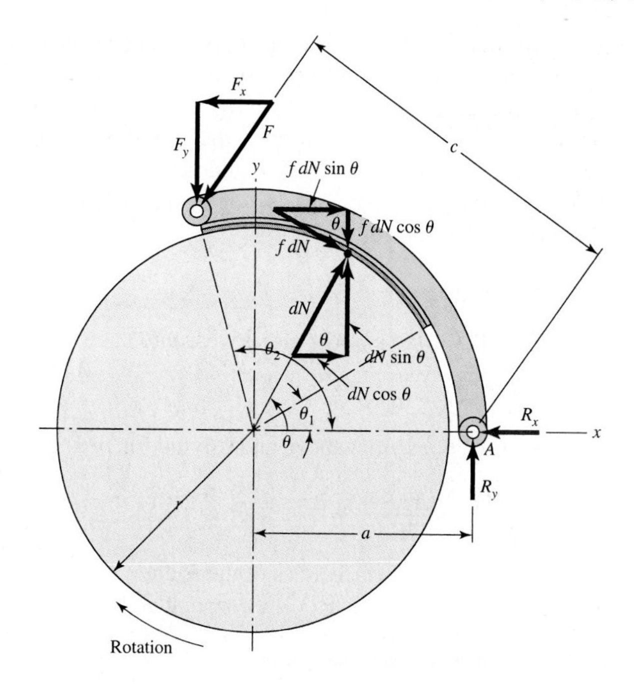

Figure 16–11

Notation of external contacting shoes.

The horizontal and vertical reactions at the hinge pin are found in the same manner as for internal expanding shoes. They are

$$R_x = \int dN \cos\theta + \int f\,dN \sin\theta - F_x \tag{a}$$

$$R_y = \int f\,dN \cos\theta - \int dN \sin\theta + F_y \tag{b}$$

By using Eq. (16–8), we have

$$\begin{aligned} R_x &= \frac{p_a br}{\sin\theta_a}(A + fB) - F_x \\ R_y &= \frac{p_a br}{\sin\theta_a}(fA - B) + F_y \end{aligned} \tag{16–12}$$

If the rotation is counterclockwise, the sign of the frictional term in each equation is reversed. Thus Eq. (16–11) for the actuating force becomes

$$F = \frac{M_N - M_f}{c} \tag{16–13}$$

and self-energization exists for counterclockwise rotation. The horizontal and vertical reactions are found, in the same manner as before, to be

$$\begin{aligned} R_x &= \frac{p_a br}{\sin\theta_a}(A - fB) - F_x \\ R_y &= \frac{p_a br}{\sin\theta_a}(-fA - B) + F_y \end{aligned} \tag{16–14}$$

It should be noted that, when external contracting designs are used as clutches, the effect of centrifugal force is to decrease the normal force. Thus, as the speed increases, a larger value of the actuating force F is required.

A special case arises when the pivot is symmetrically located and also placed so that the moment of the friction forces about the pivot is zero. The geometry of such a brake will be similar to that of Fig. 16–12*a*. To get a pressure-distribution relation, we note that lining wear is such as to retain the cylindrical shape, much as a milling machine cutter feeding in the x direction would do to the shoe held in a vise. See Fig. 16–12*b*. This means the abscissa component of wear is w_0 for all positions θ. If wear in the radial direction is expressed as $w(\theta)$, then

$$w(\theta) = w_0 \cos\theta$$

Using Eq. (12–30) to express radial wear $w(\theta)$ as

$$w(\theta) = KP'Vt$$

then denoting P' as $p(\theta)$ above and solving for $p(\theta)$ gives

$$p(\theta) = \frac{w(\theta)}{KVt} = \frac{w_0 \cos\theta}{KVt}$$

Since all elemental surface areas of the friction material see the same rubbing speed for the same duration, $w_0/(KVt)$ is a constant and

$$p(\theta) = \text{(constant)} \cos\theta = p_a \cos\theta \tag{c}$$

where p_a is the maximum value of $p(\theta)$.

Proceeding to the force analysis, we observe from Fig. 16–12*a* that

$$dN = pbr\, d\theta \tag{d}$$

or

$$dN = p_a br \cos\theta\, d\theta \tag{e}$$

The distance a to the pivot is chosen by finding where the moment of the frictional forces

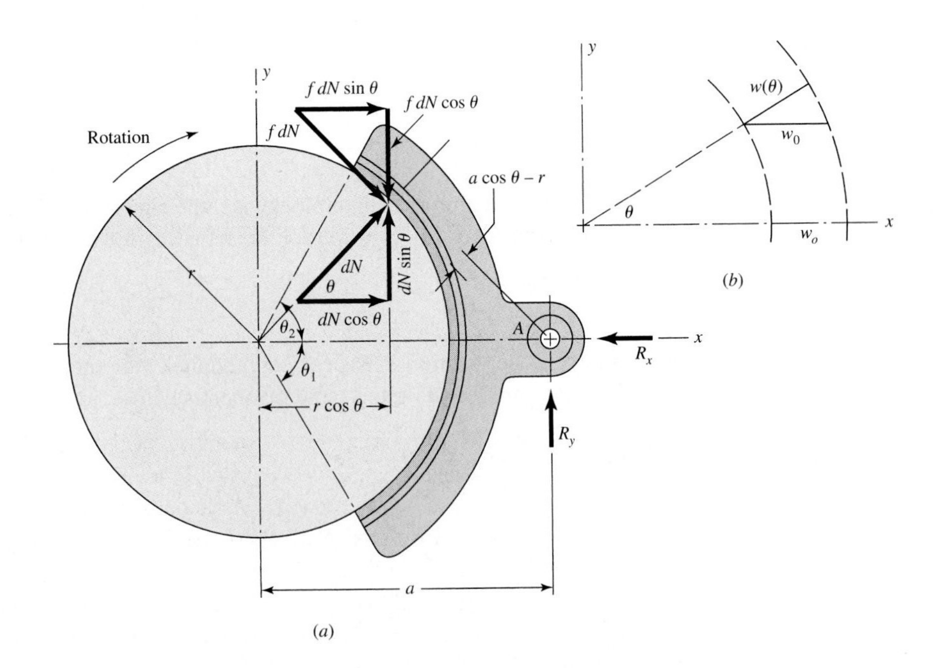

Figure 16–12

(*a*) Brake with symmetrical pivoted shoe; (*b*) wear of brake lining.

M_f is zero. First, this ensures that reaction R_y is at the correct location to establish symmetrical wear. Second, a cosinusoidal pressure distribution is sustained, preserving our predictive ability. Symmetry means $\theta_1 = \theta_2$, so

$$M_f = 2\int_0^{\theta_2} (f\,dN)(a\cos\theta - r) = 0$$

Substituting Eq. (*e*) gives

$$2fp_a br\int_0^{\theta_2} (a\cos^2\theta - r\cos\theta)d\theta = 0$$

from which

$$a = \frac{4r\sin\theta_2}{2\theta_2 + \sin 2\theta_2} \tag{16–15}$$

The distance a depends on the pressure distribution. Mislocating the pivot makes M_f zero about a different location, so the brake lining adjusts its local contact pressure, through wear, to compensate. The result is unsymmetrical wear, retiring the shoe lining, hence the shoe, sooner.

With the pivot located according to Eq. (16–15), the moment about the pin is zero, and the horizontal and vertical reactions are

$$R_x = 2\int_0^{\theta_2} dN\,\cos\theta = \frac{p_a br}{2}(2\theta_2 + \sin 2\theta_2) \tag{16–16}$$

where, because of symmetry,

$$\int f\,dN\,\sin\theta = 0$$

Also,

$$R_y = 2\int_0^{\theta_2} f\,dN\cos\theta = \frac{p_a brf}{2}(2\theta_2 + \sin 2\theta_2) \tag{16–17}$$

where

$$\int dN\,\sin\theta = 0$$

also because of symmetry. Note, too, that $R_x = -N$ and $R_y = -fN$, as might be expected for the particular choice of the dimension a. Therefore the torque is

$$T = afN \tag{16–18}$$

16–4 Band-Type Clutches and Brakes

Flexible clutch and brake bands are used in power excavators and in hoisting and other machinery. The analysis follows the notation of Fig. 16–13.

Because of friction and the rotation of the drum, the actuating force P_2 is less than the pin reaction P_1. Any element of the band, of angular length $d\theta$, will be in equilibrium under the action of the forces shown in the figure. Summing these forces in the vertical direction, we have

$$(P + dP)\sin\frac{d\theta}{2} + P\sin\frac{d\theta}{2} - dN = 0 \tag{a}$$

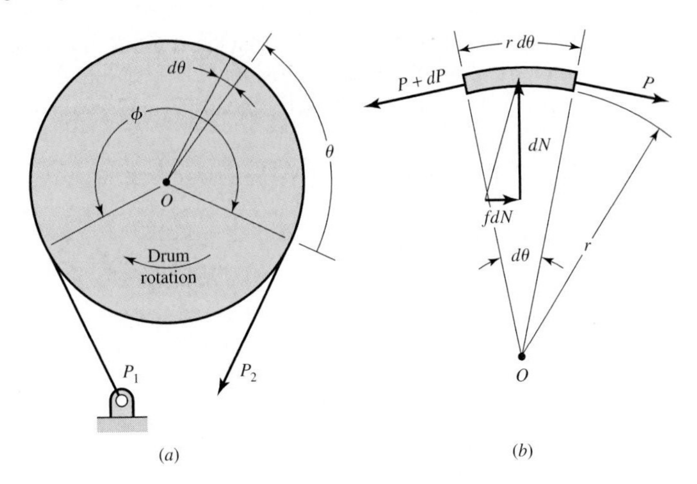

Figure 16–13

Forces on a brake band.

$$dN = P\,d\theta \tag{b}$$

since for small angles $\sin d\theta/2 = d\theta/2$. Summing the forces in the horizontal direction gives

$$(P + dP)\cos\frac{d\theta}{2} - P\cos\frac{d\theta}{2} - f\,dN = 0 \tag{c}$$

$$dP - f\,dN = 0 \tag{d}$$

Substituting the value of dN from Eq. (b) in (d) and integrating,

$$\int_{P_2}^{P_1}\frac{dP}{P} = f\int_0^{\phi} d\theta \qquad \ln\frac{P_1}{P_2} = f\phi$$

and

$$\frac{P_1}{P_2} = e^{f\phi} \tag{16–19}$$

The torque may be obtained from the equation

$$T = (P_1 - P_2)\frac{D}{2} \tag{16–20}$$

The normal force dN acting on an element of area of width b and length $r\,d\theta$ is

$$dN = pbr\,d\theta \tag{e}$$

where p is the pressure. Substitution of the value of dN from Eq. (b) gives

$$P\,d\theta = pbr\,d\theta$$

Therefore

$$p = \frac{P}{br} = \frac{2P}{bD} \tag{16–21}$$

The pressure is therefore proportional to the tension in the band. The maximum pressure p_a will occur at the toe and has the value

$$p_a = \frac{2P_1}{bD} \tag{16–22}$$

16–5 Frictional-Contact Axial Clutches

An axial clutch is one in which the mating frictional members are moved in a direction parallel to the shaft. One of the earliest of these is the cone clutch, which is simple in construction and quite powerful. However, except for relatively simple installations, it has been largely displaced by the disk clutch employing one or more disks as the operating members. Advantages of the disk clutch include the freedom from centrifugal effects, the large frictional area which can be installed in a small space, the more effective heat-dissipation surfaces, and the favorable pressure distribution. Figure 16–14 shows a single-plate disk clutch; a multiple-disk clutch-brake is shown in Fig. 16–15. Let us now determine the capacity of such a clutch or brake in terms of the material and geometry.

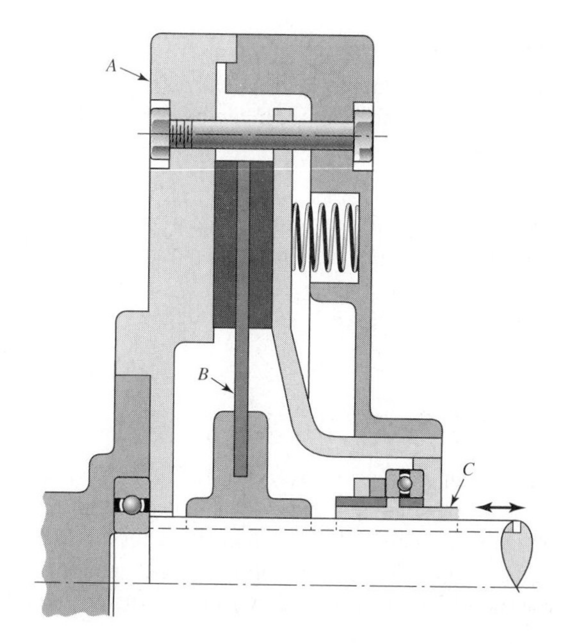

Figure 16–14

Cross-sectional view of a single-plate clutch; *A*, driver; *B*, driven plate (keyed to driven shaft); *C*, actuator.

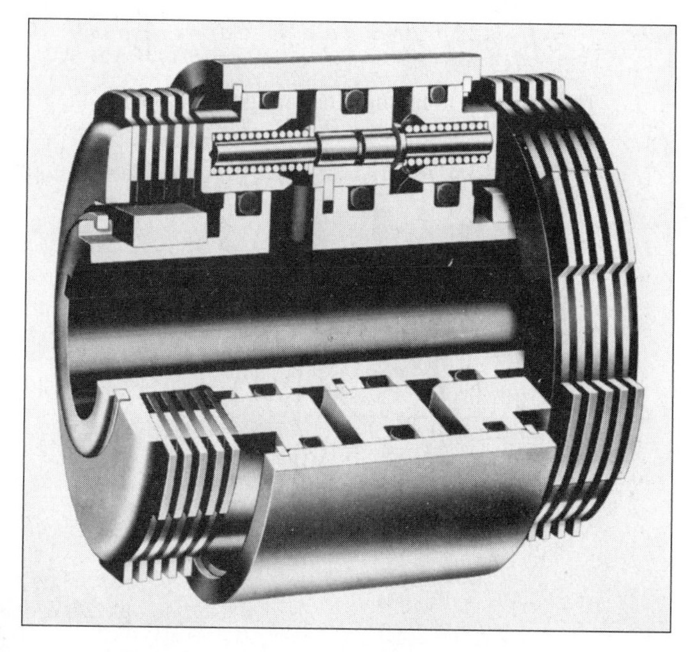

Figure 16–15

An oil-actuated multiple-disk clutch-brake for operation in an oil bath or spray. It is especially useful for rapid cycling. *(Courtesy of Twin Disc Clutch Company.)*

Figure 16–16 shows a friction disk having an outside diameter D and an inside diameter d. We are interested in obtaining the axial force F necessary to produce a certain torque T and pressure p. Two methods of solving the problem, depending upon the construction of the clutch, are in general use. If the disks are rigid, then the greatest amount of wear will at first occur in the outer areas, since the work of friction is greater in those areas. After a certain amount of wear has taken place, the pressure distribution will change so as to permit the wear to be uniform. This is the basis of the first method of solution.

Another method of construction employs springs to obtain a uniform pressure over the area. It is this assumption of uniform pressure that is used in the second method of solution.

Uniform Wear

After initial wear has taken place and the disks have worn down to a point where uniform wear is established, the axial wear can be expressed by Eq. (12–31) as

$$w = f_1 f_2 K P' V t$$

in which only P' and V vary from place to place in the rubbing surfaces. By definition uniform wear is constant from place to place; therefore,

$$P'V = (\text{constant}) = C_1$$

$$pr\omega = C_2$$

$$pr = C_3 = p_{\max} r_i = p_a r_i = p_a \frac{d}{2} \tag{a}$$

We can take an expression from Eq. (a), which is the condition for having the same amount of work done at radius r as is done at radius $d/2$. Referring to Fig. 16–16, we have an element of area of radius r and thickness dr. The area of this element is $2\pi r\, dr$, so that the normal force acting upon this element is $dF = 2\pi pr\, dr$. We can find the total normal force by letting r vary from $d/2$ to $D/2$ and integrating. Thus

$$F = \int_{d/2}^{D/2} 2\pi pr\, dr = \pi p_a d \int_{d/2}^{D/2} dr = \frac{\pi p_a d}{2}(D - d) \tag{16–23}$$

The torque is found by integrating the product of the frictional force and the radius:

$$T = \int_{d/2}^{D/2} 2\pi f p r^2\, dr = \pi f p_a d \int_{d/2}^{D/2} r\, dr = \frac{\pi p_a d}{8}(D^2 - d^2) \tag{16–24}$$

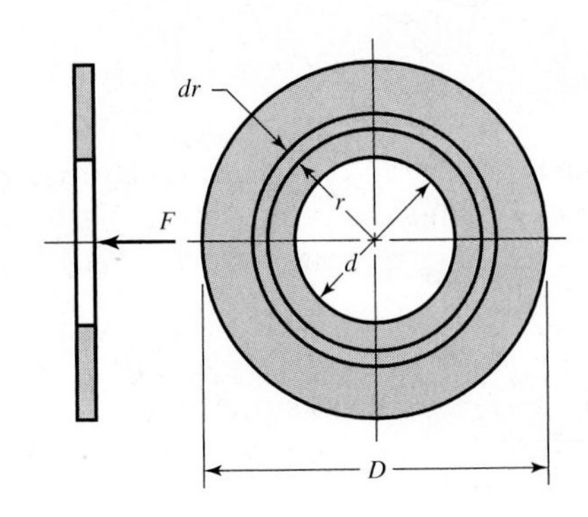

Figure 16–16

Disk friction member.

By substituting the value of F from Eq. (16–23) we may obtain a more convenient expression for the torque. Thus

$$T = \frac{Ff}{4}(D + d) \tag{16–25}$$

In use, Eq. (16–23) gives the actuating force for the selected maximum pressure p_a. This equation holds for any number of friction pairs or surfaces. Equation (16–25), however, gives the torque capacity for only a single friction surface.

Uniform Pressure

When uniform pressure can be assumed over the area of the disk, the actuating force F is simply the product of the pressure and the area. This gives

$$F = \frac{\pi p_a}{4}(D^2 - d^2) \tag{16–26}$$

As before, the torque is found by integrating the product of the frictional force and the radius:

$$T = 2\pi f p \int_{d/2}^{D/2} r^2\, dr = \frac{2\pi f p}{24}(D^3 - d^3) \tag{16–27}$$

Since $p = p_a$, we can rewrite Eq. (16–27) as

$$T = \frac{Ff}{3}\frac{D^3 - d^3}{D^2 - d^2} \tag{16–28}$$

It should be noted for both equations that the torque is for a single pair of mating surfaces. This value must therefore be multiplied by the number of pairs of surfaces in contact.

Let us express Eq. (16–25) for torque during uniform wear as

$$\frac{T}{fFD} = \frac{1 + d/D}{4} \tag{b}$$

and Eq. (16–28) for torque during uniform pressure (new clutch) as

$$\frac{T}{fFD} = \frac{1}{3}\frac{1 - (d/D)^3}{1 - (d/D)^2} \tag{c}$$

and plot these in Fig. 16–17. What we see is a dimensionless[1] presentation of Eqs. (*b*) and (*c*) which reduces the number of variables from five (T, f, F, D, and d) to three (T/FD, f, and d/D) which are dimensionless. This is the method of Buckingham. The dimensionless groups (called pi terms) are

$$\pi_1 = \frac{T}{FD} \qquad \pi_2 = f \qquad \pi_3 = \frac{d}{D}$$

This allows a five-dimensional space to be reduced to a three-dimensional space. Further, due to the "multiplicative" relation between f and T in Eqs. (*b*) and (*c*), it is possible to plot π_1/π_2 versus π_3 using a two-dimensional space (the plane of a sheet of paper)

1. Charles R. Mischke, "Minimizing Engineering Effort," Chap. 11 in Joseph E. Shigley and Charles R. Mischke (co–editors-in-chief), *Standard Handbook of Machine Design*, 2nd ed., McGraw-Hill, New York, 1996, pp. 11.1–11.9, or Charles R. Mischke, *Mathematical Model Building*, 2nd rev. ed., Iowa State University Press, Ames, 1980, pp. 139–164.

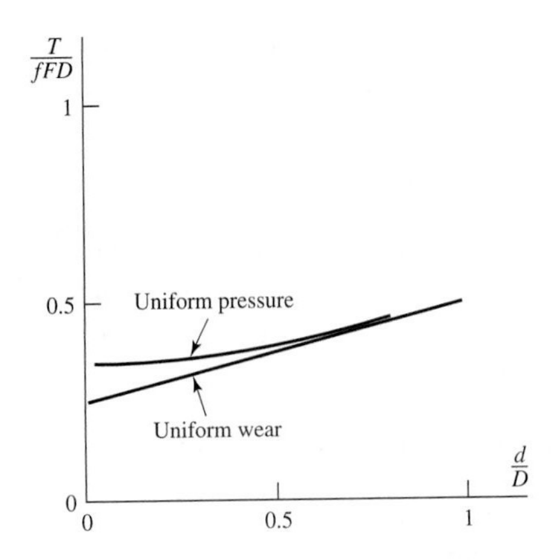

Figure 16–17

Dimensionless plot of Eqs. (*b*) and (*c*).

to view all cases over the domain of existence of Eqs. (*b*) and (*c*) and to compare, without risk of oversight! By examining Fig. 16–17 we can conclude that a new clutch, Eq. (*b*), always transmits more torque than an old clutch, Eq. (*c*). Furthermore, since clutches of this type are proportioned to make the diameter ratio d/D fall in the range $0.6 \le d/D \le 1$, the largest discrepancy between Eq. (*b*) and Eq. (*c*) will be

$$\frac{T}{fFD} = \frac{1+0.6}{4} = 0.400 \qquad \text{(old clutch, uniform wear)}$$

$$\frac{T}{fFD} = \frac{1}{3}\frac{1-0.6^3}{1-0.6^2} = 0.408\underline{3} \qquad \text{(new clutch, uniform pressure)}$$

so the proportional error is $(0.408\underline{3} - 0.400)/0.400 = 0.021$, or about 2 percent. Given the uncertainties in the actual coefficient of friction and the certainty that new clutches get old, there is little reason to use anything but Eqs. (16–23), (16–24), and (16–25).

16–6 Disk Brakes

As indicated in Fig. 16–16, there is no fundamental difference between a disk clutch and a disk brake. The analysis of the preceding section applies to disk brakes too.

We have seen that rim or drum brakes can be designed for self-energization. While this feature is important in reducing the braking effort required, it also has a disadvantage. When drum brakes are used as vehicle brakes, only a slight change in the coefficient of friction will cause a large change in the pedal force required for braking. A not unusual 30 percent reduction in the coefficient of friction due to a temperature change or moisture, for example, can result in a 50 percent change in the pedal force required to obtain the same braking torque obtainable prior to the change. The disk brake has no self-energization, and hence is not so susceptible to changes in the coefficient of friction.

Another type of disk brake is the *floating caliper brake,* shown in Fig. 16–18. The caliper supports a single floating piston actuated by hydraulic pressure. The action is much like that of a screw clamp, with the piston replacing the function of the screw. The floating action also compensates for wear and ensures a fairly constant pressure over the area of the friction pads. The seal and boot of Fig. 16–18 are designed to obtain clearance by backing off from the piston when the piston is released.

Caliper brakes (named for the nature of the actuating linkage) and disk brakes (named for the shape of the unlined surface) press friction material against the face(s) of a rotating disk. Depicted in Fig. 16–19 is the geometry of an annular-pad brake contact

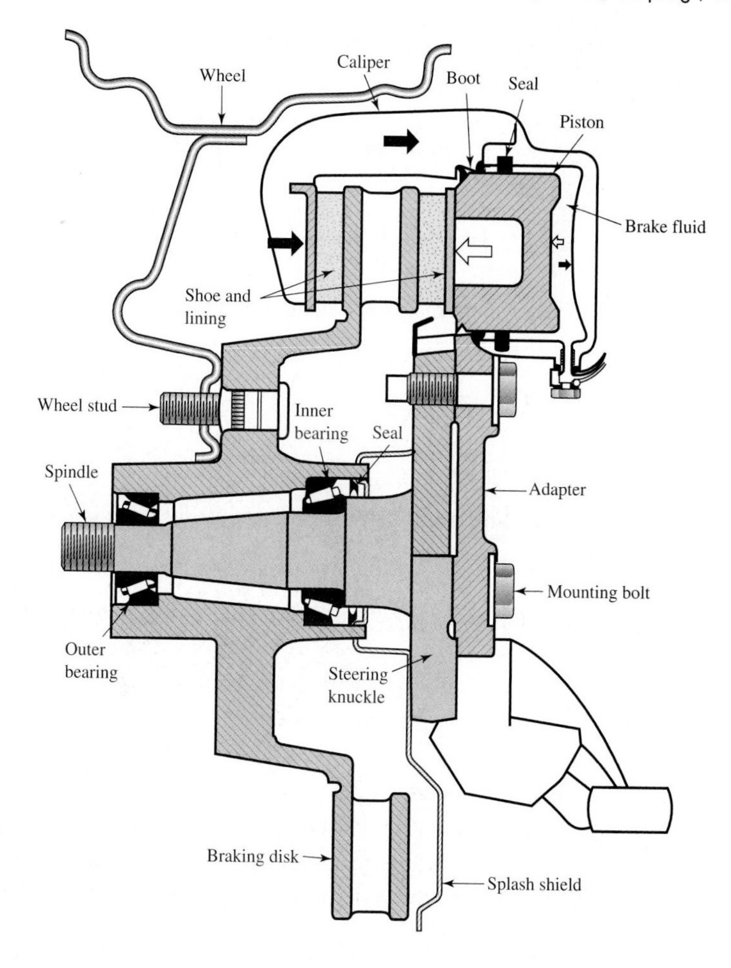

Figure 16–18

An automotive disk brake. *(Courtesy of Chrysler Corporation.)*

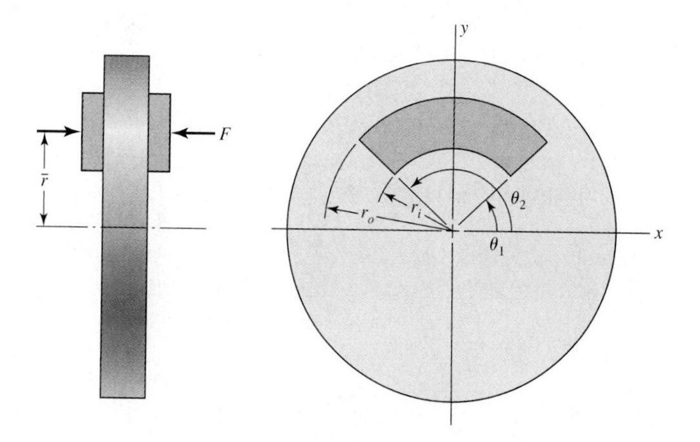

Figure 16–19

Geometry of contact area of an annular-pad segment of a caliper brake.

area. The governing axial wear equation is Eq. (12–31):

$$w = f_1 f_2 K P' V t$$

The coordinate $\bar{r}$ locates the line of action of force F that intersects the y axis. Of interest also is the effective radius r_e, which is the radius of an equivalent shoe of vanishing radial thickness. If p is the local contact pressure, the actuating force F and the friction torque T are given by

$$F = \int_{\theta_1}^{\theta_2} \int_{r_i}^{r_o} pr\,dr\,d\theta = (\theta_2 - \theta_1) \int_{r_i}^{r_o} pr\,dr \tag{16–29}$$

$$T = \int_{\theta_1}^{\theta_2} \int_{r_i}^{r_o} fpr^2\,dr\,d\theta = (\theta_2 - \theta_1) f \int_{r_i}^{r_o} pr^2\,dr \tag{16–30}$$

The equivalent radius r_e can be found from $fFr_e = T$, or

$$r_e = \frac{T}{fF} = \frac{\int_{r_i}^{r_o} pr^2 dr}{\int_{r_i}^{r_o} pr\,dr} \tag{16–31}$$

The locating coordinate $\bar{r}$ of the activating force is found by taking moments about the x axis:

$$M_x = F\bar{r} = \int_{\theta_1}^{\theta_2} pr(r \sin\theta) dr\,d\theta = (\cos\theta_1 - \cos\theta_2) \int_{r_i}^{r_o} pr^2\,dr$$

$$\bar{r} = \frac{M_x}{F} = \frac{(\cos\theta_1 - \cos\theta_2)}{\theta_2 - \theta_1} r_e \tag{16–32}$$

Uniform Wear

It is clear from Eq. (12–31) that for the axial wear to be the same everywhere, the product $P'V$ must be a constant. The pressure p can be expressed in terms of the largest allowable pressure p_a (which occurs at the inner radius r_i) as $p = p_a r_i / r$. Equation (16–29) becomes

$$F = (\theta_2 - \theta_1) p_a r_i (r_o - r_i) \tag{16–33}$$

Equation (16–30) becomes

$$T = (\theta_2 - \theta_1) f p_a r_i \int_{r_i}^{r_o} r\,dr = \frac{1}{2}(\theta_2 - \theta_1) f p_a r_i (r_o^2 - r_i^2) \tag{16–34}$$

Equation (16–31) becomes

$$r_e = \frac{p_a r_i \int_{r_i}^{r_o} r\,dr}{p_a r_i \int_{r_i}^{r_o} dr} = \frac{r_o^2 - r_i^2}{2} \frac{1}{r_o - r_i} = \frac{r_o + r_i}{2} \tag{16–35}$$

Equation (16–32) becomes

$$\bar{r} = \frac{\cos\theta_1 - \cos\theta_2}{\theta_2 - \theta_1} \frac{r_o + r_i}{2} \tag{16–36}$$

Uniform Pressure

In this situation, approximated by a new brake, $p = p_a$. Equation (16–29) becomes

$$F = (\theta_2 - \theta_1) p_a \int_{r_i}^{r_o} r\,dr = \frac{1}{2}(\theta_2 - \theta_1) p_a (r_o^2 - r_i^2) \tag{16–37}$$

Equation (16–30) becomes

$$T = (\theta_2 - \theta_1) f p_a \int_{r_i}^{r_o} r^2\, dr = \frac{1}{3}(\theta_2 - \theta_1) f p_a (r_o^3 - r_i^3) \tag{16–38}$$

Equation (16–31) becomes

$$r_e = \frac{p_a \int_{r_i}^{r_o} r^2\, dr}{p_a \int_{r_i}^{r_o} r\, dr} = \frac{r_o^3 - r_i^3}{3} \frac{2}{r_o^2 - r_i^2} = \frac{2}{3} \frac{r_o^3 - r_i^3}{r_o^2 - r_i^3} \tag{16–39}$$

Equation (16–32) becomes

$$\bar{r} = \frac{\cos\theta_1 - \cos\theta_2}{\theta_2 - \theta_1} \frac{2}{3} \frac{r_o^3 - r_i^3}{r_o^2 - r_i^2} = \frac{2}{3} \frac{r_o^3 - r_i^3}{r_o^2 - r_i^2} \frac{\cos\theta_1 - \cos\theta_2}{\theta_2 - \theta_1} \tag{16–40}$$

EXAMPLE 16–3

Two annular pads, $r_i = 3.875$ in, $r_o = 5.50$ in, subtend an angle of 108°, have a coefficient of friction of 0.37, and are actuated by a pair of hydraulic cylinders 1.5 in in diameter. The torque requirement is 13 000 in · lb. For uniform wear

(a) Find the largest normal pressure p_a.

(b) Estimate the actuating force F.

(c) Find the equivalent radius r_e and force location r.

(d) Estimate the required hydraulic pressure.

Solution

(*a*) From Eq. (16–34),

Answer

$$p_a = \frac{T}{(\theta_2 - \theta_1) f r_i (r_o^2 - r_i^2)/2}$$

$$= \frac{6500}{0.5(144° - 36°)(\pi/180)0.37(3.875)(5.5^2 - 3.875^2)} = 316 \text{ psi}$$

(*b*) From Eq. (16–33),

Answer

$$F = (\theta_2 - \theta_1) p_a r_i (r_o - r_i) = (144° - 36°)(\pi/180)316(3.875)(5.5 - 3.875)$$

$$= 3751 \text{ lbf}$$

(*c*) From Eq. (16–35),

Answer

$$r_e = \frac{r_o + r_i}{2} = \frac{5.50 + 3.875}{2} = 4.688 \text{ in}$$

From Eq. (16–36),

Answer

$$r = \frac{\cos\theta_1 - \cos\theta_2}{\theta_2 - \theta_1} \frac{r_o + r_i}{2} = \frac{\cos 36° - \cos 144°}{(144° - 36°)(\pi/180)} \frac{5.50 + 3.875}{2}$$

$$= 4.024 \text{ in}$$

(*d*) Each cylinder supplies half the actuating force, $3751/2 = 1875.5$ lb.

Answer

$$p_{\text{hydraulic}} = \frac{F}{A_P} = \frac{1875.5}{\pi(1.5^2/4)} = 1061 \text{ psi}$$

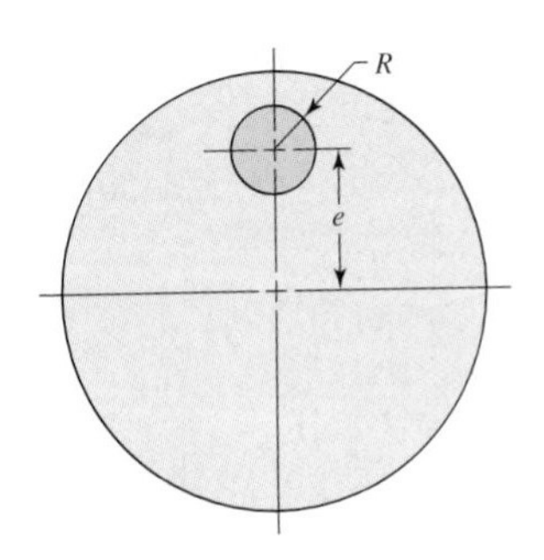

Figure 16–20

Geometry of circular pad of a caliper brake.

Circular (Button or Puck) Pad Caliper Brake

Figure 16–20 displays the pad geometry. Numerical integration is necessary to analyze this brake since the boundaries are difficult to handle in closed form. Table 16–1 gives the parameters for this brake as determined by Farekas. The effective radius is given by

$$r_e = \delta_e \tag{16–41}$$

The actuating force is given by

$$F = \pi R^2 p_{\text{av}} \tag{16–42}$$

and the torque is given by

$$T = f F r_e \tag{16–43}$$

Table 16–2 lists limits of contact pressure and rubbing speed as well as coefficients of friction for a variety of friction materials. Table 16–3 can also be useful.

Table 16–1

Parameters for a Circular-Pad Caliper Brake

Source: G. A. Fazekas, "On Circular Spot Brakes," *Trans. ASME, J. Engineering for Industry*, vol. 94, Series B, No. 3, August 1972, pp. 859–863.

$\dfrac{R}{e}$	$\delta = \dfrac{r_e}{e}$	$\dfrac{p_{\text{max}}}{p_{\text{av}}}$
0.0	1.000	1.000
0.1	0.983	1.093
0.2	0.969	1.212
0.3	0.957	1.367
0.4	0.947	1.578
0.5	0.938	1.875

EXAMPLE 16–4

A button-pad disk brake uses dry sintered metal pads. The pad radius is $\frac{1}{2}$ in, and its center is 2 in from the axis of rotation of the $3\frac{1}{2}$-in-diameter disk. Using half of the largest allowable pressure, find the actuating force and the brake torque. The coefficient of friction is 0.31.

Solution

Since the pad radius $R = 0.5$ in and eccentricity $e = 2$ in,

$$\frac{R}{e} = \frac{0.5}{2} = 0.25$$

From Table 16–1, by interpolation, $\delta = 0.963$ and $p_{\text{max}}/p_{\text{av}} = 1.290$. It follows that the effective radius e is found from Eq. (16–41):

$$r_e = \delta e = 0.963(2) = 1.926 \text{ in}$$

The average pressure, using Table 16–2 p_{max} midrange entry, is

$$p_{\text{av}} = \frac{p_{\text{max}}}{1.290} = \frac{350/2}{1.290} = 135.7 \text{ psi}$$

The actuating force F is found from Eq. (16–42) to be

Answer $$F = \pi R^2 p_{av} = \pi(0.5)^2 135.7 = 106.6 \text{ lbf} \quad \text{(one side)}$$

The brake torque T is

Answer $$T = f F r_e = 0.31(106.6)1.926 = 63.6 \text{ in} \cdot \text{lbf} \quad \text{(one side)}$$

Table 16–2

Characteristics of Friction Materials for Brakes and Clutches *Sources:* Ferodo Ltd., Chapel-en-le-frith, England; Scan-pac, Mequon, Wisc.; Raybestos, New York, N.Y. and Stratford, Conn.; Gatke Corp., Chicago, Ill.; General Metals Powder Co., Akron, Ohio; D. A. B. Industries, Troy, Mich.; Friction Products Co., Medina, Ohio.

Material	Friction Coefficient f	Maximum Pressure p_{max}, psi	Maximum Temperature		Maximum Velocity V_{max}, ft/min	Applications
			Instantaneous, °F	Continuous, °F		
Cermet	0.32	150	1500	750		Brakes and clutches
Sintered metal (dry)	0.29–0.33	300–400	930–1020	570–660	3600	Clutches and caliper disk brakes
Sintered metal (wet)	0.06–0.08	500	930	570	3600	Clutches
Rigid molded asbestos (dry)	0.35–0.41	100	660–750	350	3600	Drum brakes and clutches
Rigid molded asbestos (wet)	0.06	300	660	350	3600	Industrial clutches
Rigid molded asbestos pads	0.31–0.49	750	930–1380	440–660	4800	Disk brakes
Rigid molded nonasbestos	0.33–0.63	100–150		500–750	4800–7500	Clutches and brakes
Semirigid molded asbestos	0.37–0.41	100	660	300	3600	Clutches and brakes
Flexible molded asbestos	0.39–0.45	100	660–750	300–350	3600	Clutches and brakes
Wound asbestos yarn and wire	0.38	100	660	300	3600	Vehicle clutches
Woven asbestos yarn and wire	0.38	100	500	260	3600	Industrial clutches and brakes
Woven cotton	0.47	100	230	170	3600	Industrial clutches and brakes
Resilient paper (wet)	0.09–0.15	400	300		$PV < 500\,000$ psi · ft/min	Clutches and transmission bands

Table 16–3

Area of Friction Material Required for a Given Average Braking Power *Sources:* M. J. Neale, *The Tribology Handbook,* Butterworth, London, 1973; *Friction Materials for Engineers,* Ferodo Ltd., Chapel-en-le-frith, England, 1968.

		Ratio of Area to Average Braking Power, $in^2/(Btu/s)$		
Duty Cycle	**Typical Applications**	**Band and Drum Brakes**	**Plate Disk Brakes**	**Caliper Disk Brakes**
Infrequent	Emergency brakes	0.85	2.8	0.28
Intermittent	Elevators, cranes, and winches	2.8	7.1	0.70
Heavy-duty	Excavators, presses	5.6–6.9	13.6	1.41

16–7 Cone Clutches and Brakes

The drawing of a *cone clutch* in Fig. 16–21 shows that it consists of a *cup* keyed or splined to one of the shafts, a *cone* which must slide axially on splines or keys on the mating shaft, and a helical *spring* to hold the clutch in engagement. The clutch is disengaged by means of a fork which fits into the shifting groove on the friction cone. The *cone angle* α and the diameter and face width of the cone are the important geometric design parameters. If the cone angle is too small, say, less than about 8°, then the force required to disengage the clutch may be quite large. And the wedging effect lessens rapidly when larger cone angles are used. Depending upon the characteristics of the friction materials, a good compromise can usually be found using cone angles between 10 and 15°.

To find a relation between the operating force F and the torque transmitted, designate the dimensions of the friction cone as shown in Figure 16–22. As in the case of the axial clutch, we can obtain one set of relations for a uniform-wear and another set for a uniform-pressure assumption.

Uniform Wear

The pressure relation is the same as for the axial clutch:

$$p = p_a \frac{d}{2r} \tag{a}$$

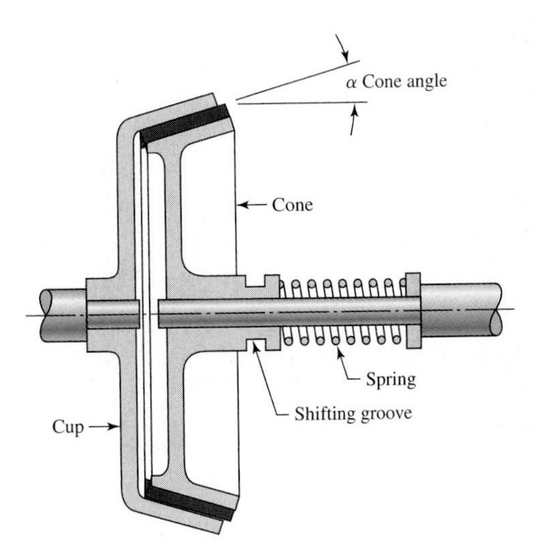

Figure 16–21

Cross section of a cone clutch.

Figure 16–22

Contact area of a cone clutch.

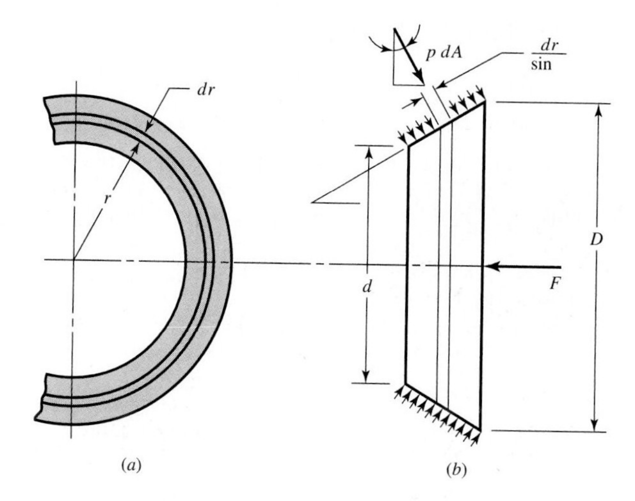

Next, referring to Fig. 16–22, we see that we have an element of area dA of radius r and width $dr/\sin\alpha$. Thus $dA = (2\pi r dr)/\sin\alpha$. As shown in Fig. 16–22, the operating force will be the integral of the axial component of the differential force $p\,dA$. Thus

$$
\begin{aligned}
F &= \int p\,dA \sin\alpha = \int_{d/2}^{D/2} \left(p_a \frac{d}{2r}\right)\left(\frac{2\pi r\,dr}{\sin\alpha}\right)(\sin\alpha) \\
&= \pi p_a d \int_{d/2}^{D/2} dr = \frac{\pi p_a d}{2}(D-d)
\end{aligned}
\tag{16–44}
$$

which is the same result as in Eq. (16–23).

The differential friction force is $fp\,dA$, and the torque is the integral of the product of this force with the radius. Thus

$$
\begin{aligned}
T &= \int rfp\,dA = \int_{d/2}^{D/2} (rf)\left(p_a \frac{d}{2r}\right)\left(\frac{2\pi r\,dr}{\sin\alpha}\right) \\
&= \frac{\pi f p_a d}{\sin\alpha}\int_{d/2}^{D/2} r\,dr = \frac{\pi f p_a d}{8\sin\alpha}(D^2-d^2)
\end{aligned}
\tag{16–45}
$$

Note that Eq. (16–24) is a special case of Eq. (16–45), with $\alpha = 90°$. Using Eq. (16–44), we find that the torque can also be written

$$
T = \frac{Ff}{4\sin\alpha}(D+d) \tag{16–46}
$$

Uniform Pressure

Using $p = p_a$, the actuating force is found to be

$$
F = \int p_a\,dA \sin\alpha = \int_{d/2}^{D/2} (p_a)\left(\frac{2\pi r\,dr}{\sin\alpha}\right)(\sin\alpha) = \frac{\pi p_a}{4}(D^2-d^2) \tag{16–47}
$$

The torque is

$$
T = \int rfp_a\,dA = \int_{d/2}^{D/2} (rfp_a)\left(\frac{2\pi r\,dr}{\sin\alpha}\right) = \frac{\pi f p_a}{12\sin\alpha}(D^3-d^3) \tag{16–48}
$$

or, using Eq. (16–47) in Eq. (16–48),

$$T = \frac{Ff}{3\sin\alpha}\frac{D^3 - d^3}{D^2 - d^2} \tag{16–49}$$

As in the case of the axial clutch, we can write Eq. (16–46) dimensionlessly as

$$\frac{T\sin\alpha}{fFd} = \frac{1 + d/D}{4} \tag{b}$$

and write Eq. (16–49) as

$$\frac{T\sin\alpha}{fFd} = \frac{1}{3}\frac{1 - (d/D)^3}{1 - (d/D)^2} \tag{c}$$

This time there are six (T, α, f, F, D, and d) parameters and four pi terms:

$$\pi_1 = \frac{T}{FD} \qquad \pi_2 = f \qquad \pi_3 = \sin\alpha \qquad \pi_4 = \frac{d}{D}$$

As in Fig. 16–17, we plot $T\ \sin\alpha/(fFD)$ as ordinate and d/D as abscissa. The plots and conclusions are the same. There is little reason for using equations other than Eqs. (16–44), (16–45), and (16–46).

16–8 Self-Locking Tapers and Torque Capacity

One way of fastening a gear to a shaft is using a conical taper on a shaft that is geometrically the frustum of a cone. If the taper is not self-locking, an axial force is necessary in order to transmit torque. A nut and thread can be supplied. Understanding self-locking tapers can begin with the slipping conical frustum of a cone clutch, and the equation for axial force F, Eq. (16–44), and the torque, Eq. (16–45). When there is no torque and no slipping, then the shearing traction is directed along the slant of the cone as shown in Fig. 16–22. The normal component is opposed by the component dF_N of the axial force dF:

$$dF_N = \left(p_a\frac{d}{2r}\right)\frac{2\pi r\,dr}{\sin\alpha}\sin\alpha$$

The frictional component dF_f of the axial force dF is

$$dF_f = f\left(\frac{p_a d}{2r}\right)\frac{2\pi r\,dr}{\sin\alpha}\cos\alpha$$

Then, summing dF_N and dF_f gives

$$dF = dF_N + dF_f = \frac{p_a d}{2r}\frac{2\pi r\,dr}{\sin\alpha}\sin\alpha + \frac{f p_a d}{2r}\frac{2\pi r\,dr}{\sin\alpha}\cos\alpha$$

$$= \pi p_a d(1 + f\cot\alpha)dr$$

Integration yields

$$F = \pi p_a d(1 + f\cot\alpha)\int_{d/2}^{D/2} dr = \frac{\pi}{2}p_a d(D - d)(1 + f\cot\alpha) \tag{16–50}$$

When the axial force F (positive when directed to the left) is removed the normal pressure tries to expel the tapered cone. The friction reverses so that

$$(1 - f \cot \alpha) = 0$$
$$\alpha_{cr} = \tan^{-1} f \tag{a}$$

There exists a critical cone angle. When $\alpha < \alpha_{cr}$ we have a self-locking taper. It takes an axial force to the right to expel the cone. Additionally, the torque capacity of the assembly is less than given by Eq. (16–45). Using Eq. (*a*) we can observe the magnitude of critical angles corresponding to coefficients of friction:

f	α_{cr}
0.0	0.0
0.1	5.7°
0.2	11.3°
0.3	16.7°
0.4	21.8°

In a cone clutch we want to avoid using an axial force to expel the cone, and we can use Eq. (16–45). This is accomplished by keeping the cone angle $\alpha > \alpha_{cr}$.

If we use a taper to hold a gear on a shaft, with no holding nut, what is the torque capacity of the mated pair? To answer this question, recognize that to expel the cone there is a component of the coefficient of friction f_a directed axially, which is "consumed" in holding. Any torque applied uses a tangential component of the coefficient of friction f_t. Each function is independent until the vector sum equals the coefficient of friction f.

When the cone angle is larger than the critical angle α_{cr} any assembly force does not result in the taper locking, and the conical cup expels the taper. When the cone angle is less than critical angle α_{cr} any assembly force sufficient to develop the full coefficient of friction results in the taper locking to its highest possible degree. Such an assembly is capable of transmitting torque.

Figure 16–23*a* shows an assembly force F_a applied. The cup resists with forces F_N and F_f. When the assembly force F_a is released the cup tries to expel the taper, but it cannot. The friction force reverses and reduces (f lowers to f_a') as indicated in Fig. 16–23*b*. No net force on the taper requires $K = K f_a' \cot \alpha$ establishing $f_a' = \tan \alpha$. A thrust F_x may be applied as shown in Fig. 16–23*c* in a direction tending to dissassemble the taper joint. No net force on the taper requires

$$f_a = (1 + F_x/K) \tan \alpha$$

The coefficient of friction f_t in tangential direction is

$$f_t = \sqrt{f^2 - f_a^2} = \sqrt{f^2 - \left(1 + \frac{F_x}{K}\right)^2 \tan^2 \alpha}$$

From Eq. (16–45),

$$T = \frac{f_t p_a d(D^2 - d^2)}{8 \sin \alpha} = \frac{f_t K (D + d)}{4 \sin \alpha}$$
$$= \frac{K(D + d)}{4 \sin \alpha} \sqrt{f^2 - \left(1 + \frac{F_x}{K}\right)^2 \tan^2 \alpha} \tag{16–51}$$

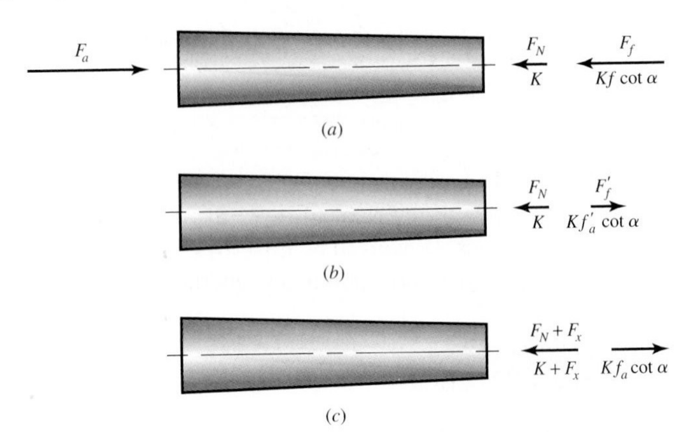

Figure 16–23

Forces on taper which sum to zero. (*a*) Assembly force F_a applied, F_N and F_f resist, f has full value. (*b*) F_a removed, F_f reversed, reduced in magnitude, $f'_a < f$. (*c*) Thrust F_x applied, $f_a > f'_a$. Below the force arrows are alternative expressions for the forces shown above the arrows.

Permissible torque is reduced by "bad" axial thrust, and by larger cone angles α, especially when they approach the critical angle α_{cr} from below.

16–9 Energy Considerations

When the rotating members of a machine are caused to stop by means of a brake, the kinetic energy of rotation must be absorbed by the brake. This energy appears in the brake in the form of heat. In the same way, when the members of a machine which are initially at rest are brought up to speed, slipping must occur in the clutch until the driven members have the same speed as the driver. Kinetic energy is absorbed during slippage of either a clutch or a brake, and this energy appears as heat.

We have seen how the torque capacity of a clutch or brake depends upon the coefficient of friction of the material and upon a safe normal pressure. However, the character of the load may be such that, if this torque value is permitted, the clutch or brake may be destroyed by its own generated heat. The capacity of a clutch is therefore limited by two factors, the characteristics of the material and the ability of the clutch to dissipate heat. In this section we shall consider the amount of heat generated by a clutching or braking operation. If the heat is generated faster than it is dissipated, we have a temperature-rise problem; that is the subject of the next section.

To get a clear picture of what happens during a simple clutching or braking operation, refer to Fig. 16–1*a*, which is a mathematical model of a two-inertia system connected by a clutch. As shown, inertias I_1 and I_2 have initial angular velocities of ω_1 and ω_2, respectively. During the clutch operation both angular velocities change and eventually become equal. We assume that the two shafts are rigid and that the clutch torque is constant.

Writing the equation of motion for inertia 1 gives

$$I_1 \ddot{\theta}_1 = -T \tag{a}$$

where $\ddot{\theta}_1$ is the angular acceleration of I_1 and T is the clutch torque. A similar equation for I_2 is

$$I_2 \ddot{\theta}_2 = T \tag{b}$$

We can determine the instantaneous angular velocities $\dot{\theta}_1$ and $\dot{\theta}_2$ of I_1 and I_2 after any period of time t has elapsed by integrating Eqs. (*a*) and (*b*). The results are

$$\dot{\theta}_1 = -\frac{T}{I_1}t + \omega_1 \tag{c}$$

$$\dot{\theta}_2 = \frac{T}{I_2}t + \omega_2 \tag{d}$$

The difference in the velocities, sometimes called the relative velocity, is

$$\begin{aligned}\dot{\theta} = \dot{\theta}_1 - \dot{\theta}_2 &= -\frac{T}{I_1}t + \omega_1 - \left(\frac{T}{I_2}t + \omega_2\right) \\ &= \omega_1 - \omega_2 - T\left(\frac{I_1 + I_2}{I_1 I_2}\right)t\end{aligned} \tag{16–52}$$

The clutching operation is completed at the instant in which the two angular velocities $\dot{\theta}_1$ and $\dot{\theta}_2$ become equal. Let the time required for the entire operation be t_1. Then $\dot{\theta} = 0$ when $\dot{\theta}_1 = \dot{\theta}_2$, and so Eq. (16–52) gives the time as

$$t_1 = \frac{I_1 I_2(\omega_1 - \omega_2)}{T(I_1 + I_2)} \tag{16–53}$$

This equation shows that the time required for the engagement operation is directly proportional to the velocity difference and inversely proportional to the torque.

We have assumed the clutch torque to be constant. Therefore, using Eq. (16–52), we find the rate of energy-dissipation during the clutching operation to be

$$u = T\dot{\theta} = T\left[\omega_1 - \omega_2 - T\left(\frac{I_1 + I_2}{I_1 I_2}\right)t\right] \tag{e}$$

This equation shows that the energy-dissipation rate is greatest at the start, when $t = 0$.

The total energy dissipated during the clutching operation or braking cycle is obtained by integrating Eq. (*e*) from $t = 0$ to $t = t_1$. The result is found to be

$$\begin{aligned}E &= \int_0^{t_1} u\,dt = T\int_0^{t_1}\left[\omega_1 - \omega_2 - T\left(\frac{I_1 + I_2}{I_1 I_2}\right)t\right]dt \\ &= \frac{I_1 I_2(\omega_1 - \omega_2)^2}{2(I_1 + I_2)}\end{aligned} \tag{16–54}$$

Note that the energy dissipated is proportional to the velocity difference squared and is independent of the clutch torque.

Note that E in Eq. (16–54) is the energy lost or dissipated; this is the energy that is absorbed by the clutch or brake. If the inertias are expressed in U.S. customary units (lb $\cdot$ s^2/in), then the energy absorbed by the clutch assembly is in lb $\cdot$ in. Using these units, the heat generated in Btu is

$$H = \frac{E}{9336} \tag{16–55}$$

In SI, the inertias are expressed in kilogram-meter units, and the energy dissipated is expressed in joules.

16–10 Temperature Rise

The temperature rise of the clutch or brake assembly can be approximated by the classic expression

$$\Delta T = \frac{H}{CW} \tag{16–56}$$

where $\Delta T =$ temperature rise, °F

$C =$ specific heat capacity, Btu/(lb$_m$ · °F); use 0.12 for steel or cast iron

$W =$ mass of clutch or brake parts, lb$_m$

A similar equation can be written using SI units. It is

$$\Delta T = \frac{E}{Cm} \tag{16–57}$$

where $\Delta T =$ temperature rise, °C

$C =$ specific heat capacity; use 500 J/kg · °C for steel or cast iron

$m =$ mass of clutch or brake parts, kg

The temperature-rise equations above can be used to explain what happens when a clutch or brake is operated. However, there are so many variables involved that it would be most unlikely that such an analysis would even approximate experimental results. For this reason such analyses are most useful, for repetitive cycling, in pinpointing those design parameters that have the greatest effect on performance.

If an object is at initial temperature T_1 in an environment of temperature T_∞, then Newton's cooling model is expressed as

$$\frac{T - T_\infty}{T_1 - T_\infty} = \exp\left(-\frac{UA}{WC}t\right) \tag{16–58}$$

$T =$ temperature at time t, °F

$T_1 =$ initial temperature, °F

$T_\infty =$ environmental temperature, °F

$U =$ overall coefficient of heat transfer, Btu/(in^2 · s · °F)

$A =$ lateral surface area, in^2

$W =$ mass of the object, lbm

$C =$ specific heat capacity of the object, Btu/(lbm · °F)

Figure 16–24 shows an application of Eq. (16–58). The curve ABC is the exponential decline of temperature given by Eq. (16–58). At time t_B a second application of the brake occurs. The temperature quickly rises to temperature T_2, and a new cooling curve is started. For repetitive brake applications, subsequent temperature peaks $T_3, T_4, \ldots,$

Figure 16–24

The effect of clutching or braking operations on temperature. T_∞ is the ambient temperature. Note that the temperature rise ΔT may be different for each operation.

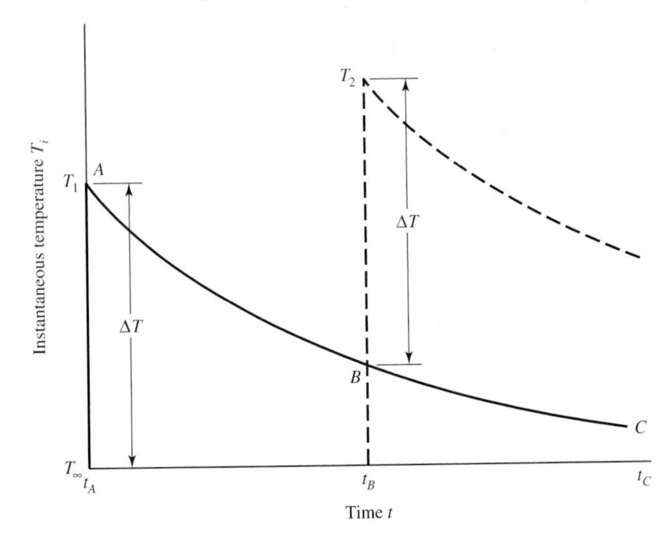

occur until the brake is able to dissipate by cooling between operations an amount of heat equal to the energy absorbed in the application. If this is a production situation with brake applications every t_1 seconds, then a steady state develops in which all the peaks T_{max} and all the valleys T_{min} are repetitive.

The heat-dissipation capacity of disk brakes has to be planned to avoid reaching the temperatures of disk and pad that are detrimental to the parts. When a disk brake has a rhythm such as discussed above, then the rate of heat transfer is described by another Newtonian equation:

$$\dot{Q} = UA(T - T_\infty) = (h_r + f_v h_c)A(T - T_\infty) \tag{16-59}$$

where $\dot{Q}$ = rate of energy loss, Btu/s

U = overall coefficient of heat transfer, Btu/(in^2 · s · °F)

h_r = radiation component of U, Btu/(in^2 · s · °F), Fig. 16–25*a*

h_c = convective component of U, Btu/(in^2 · s · °F), Fig. 16–25*a*

Figure 16–25

(*a*) Heat-transfer coefficient in still air. (*b*) Ventilation factors. *(Courtesy of Tolo-o-matic.)*

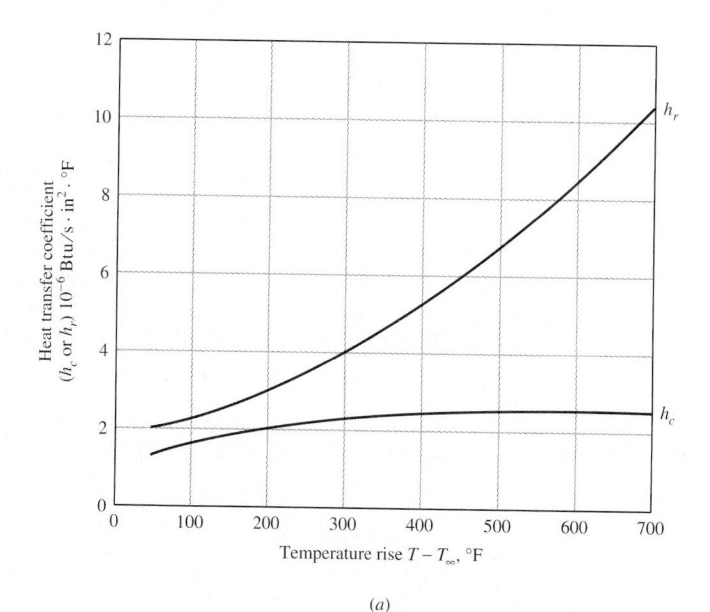

(*a*)

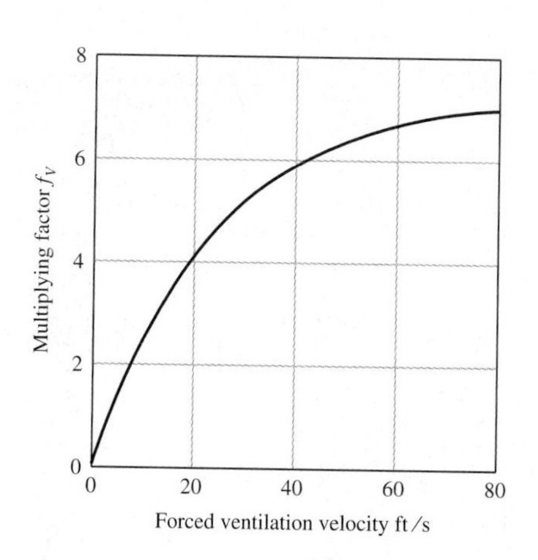

(*b*)

f_v = ventilation factor, Fig. 16–25*b*

T = disk temperature, °F

T_∞ = ambient temperature, °F

The energy E absorbed by the brake stopping an equivalent rotary inertia I in terms of original and final angular velocities ω_o and ω_f is given by

$$E = \frac{1}{2}\frac{I}{9336}(\omega_o^2 - \omega_f^2) \tag{16–60}$$

The temperature rise ΔT due to a single stop is

$$\Delta T = \frac{E}{WC} \tag{16–61}$$

T_{max} has to be high enough to transfer E Btu in t_1 seconds. We can manipulate Eq. (16–58) to the form

$$T_{max} = T_\infty + \frac{\Delta T}{1 - \exp(-\beta t_1)} \tag{16–62}$$

where β is $UA/(WC)$.

EXAMPLE 16–5 A caliper brake is used 24 times per hour to arrest a machine shaft from a speed of 250 r/min to rest. The ventilation of the brake provides a mean air speed of 25 ft/s. The equivalent rotary inertia of the machine as seen from the brake shaft is 289 lbm · in · s. The disk is steel with a density of 0.282 lb/in^3, a specific heat capacity of 0.108 Btu/(lbm · °F), a diameter of 6 in, a thickness of $\frac{1}{4}$ in. The pads are dry sintered metal. The lateral area of the brake surface is 50 in^2. Find T_{max} and T_{min} for the steady-state operation.

Solution

$$t_1 = 3600/24 = 150 \text{ s}$$

Fig. 16–24 $h_r = 3.1(10^{-6})\text{Btu/(in}^2 \cdot \text{s} \cdot \text{°F)}$

$h_c = 2.1(10^{-6})\text{ Btu/(in}^2 \cdot \text{s} \cdot \text{°F)}$

Fig. 16–25 $f_v = 4.8$

$$U = h_r + f_v h_c = 3.1(10^{-6}) + 4.8(2.1)10^{-6} = 13.1(10^{-6})\text{ Btu/(in}^2 \cdot \text{s} \cdot \text{°F)}$$

The weight of the disk is

$$W = \frac{\pi \delta D^2 h}{4} = \frac{\pi(0.282)6^2(0.25)}{4} = 1.99 \text{ lb}$$

Eq. (16–60) $$E = \frac{1}{2}\frac{I}{9336}(\omega_o^2 - \omega_f^2) = \frac{289}{2(9336)}\left(\frac{2\pi}{60}250\right)^2 = 10.6 \text{ Btu}$$

$$\beta = \frac{UA}{WC} = \frac{13.1(10^{-6})50}{1.99(0.108)} = 3.05(10^{-3})\text{ s}^{-1}$$

Eq. (16–61) $$\Delta T = \frac{E}{WC} = \frac{10.6}{1.99(0.108)} = 49.2\text{°F}$$

Answer Eq. (16–62) $$T_{max} = 70 + \frac{49.2}{1 - \exp[-3.05(10^{-3})150]} = 204\text{°F}$$

Answer $$T_{\text{min}} = 204 - 49.2 = 154.8°\text{F}$$

Table 16–2 for dry sintered metal pads gives a continuous operating maximum temperature of 570–660°F. There is no danger of overheating.

16–11 Friction Materials

A brake or friction clutch should have the following lining material characteristics to a degree which is dependent on the severity of service:

- High and reproducible coefficient of friction
- Imperviousness to environmental conditions, such as moisture
- The ability to withstand high temperatures, together with good thermal conductivity and diffusivity, as well as high specific heat capacity
- Good resiliency
- High resistance to wear, scoring, and galling
- Congenial to the environment
- Flexibility

Table 16–3 gives area of friction surface required for several braking powers.

The manufacture of friction materials is a highly specialized process, and it is advisable to consult manufacturers' catalogs and handbooks, as well as manufacturers directly, in selecting friction materials for specific applications. Selection involves a consideration of the many characteristics as well as the standard sizes available.

The *woven-cotton lining* is produced as a fabric belt which is impregnated with resins and polymerized. It is used mostly in heavy machinery and is usually supplied in rolls up to 50 ft in length. Thicknesses available range from $\frac{1}{8}$ to 1 in, in widths up to about 12 in.

A *woven-asbestos lining* is made in a similar manner to the cotton lining and may also contain metal particles. It is not quite as flexible as the cotton lining and comes in a smaller range of sizes. Along with the cotton lining, the asbestos lining was widely used as a brake material in heavy machinery.

Molded-asbestos linings contain asbestos fiber and friction modifiers; a thermoset polymer is used, with heat, to form a rigid or semirigid molding. The principal use was in drum brakes.

Molded-asbestos pads are similar to molded linings but have no flexibility; they were used for both clutches and brakes.

Sintered-metal pads are made of a mixture of copper and/or iron particles with friction modifiers, molded under high pressure and then heated to a high temperature to fuse the material. These pads are used in both brakes and clutches for heavy-duty applications.

Cermet pads are similar to the sintered-metal pads and have a substantial ceramic content.

Table 16–4 lists properties of typical brake linings. The linings may consist of a mixture of fibers to provide strength and ability to withstand high temperatures, various friction particles to obtain a degree of wear resistance as well as a higher coefficient of friction, and bonding materials.

Table 16–4

Some Properties of Brake Linings

	Woven Lining	Molded Lining	Rigid Block
Compressive strength, kpsi	10–15	10–18	10–15
Compressive strength, MPa	70–100	70–125	70–100
Tensile strength, kpsi	2.5–3	4–5	3–4
Tensile strength, MPa	17–21	27–35	21–27
Max. temperature, °F	400–500	500	750
Max. temperature, °C	200–260	260	400
Max. speed, ft/min	7500	5000	7500
Max. speed, m/s	38	25	38
Max. pressure, psi	50–100	100	150
Max. pressure, kPa	340–690	690	1000
Frictional coefficient, mean	0.45	0.47	0.40–45

Table 16–5

Friction Materials for Clutches

	Friction Coefficient		Max. Temperature		Max. Pressure	
Material	**Wet**	**Dry**	**°F**	**°C**	**psi**	**kPa**
Cast iron on cast iron	0.05	0.15–0.20	600	320	150–250	1000–1750
Powdered metal* on cast iron	0.05–0.1	0.1–0.4	1000	540	150	1000
Powdered metal* on hard steel	0.05–0.1	0.1–0.3	1000	540	300	2100
Wood on steel or cast iron	0.16	0.2–0.35	300	150	60–90	400–620
Leather on steel or cast iron	0.12	0.3–0.5	200	100	10–40	70–280
Cork on steel or cast iron	0.15–0.25	0.3–0.5	200	100	8–14	50–100
Felt on steel or cast iron	0.18	0.22	280	140	5–10	35–70
Woven asbestos* on steel or cast iron	0.1–0.2	0.3–0.6	350–500	175–260	50–100	350–700
Molded asbestos* on steel or cast iron	0.08–0.12	0.2–0.5	500	260	50–150	350–1000
Impregnated asbestos* on steel or cast iron	0.12	0.32	500–750	260–400	150	1000
Carbon graphite on steel	0.05–0.1	0.25	700–1000	370–540	300	2100

*The friction coefficient can be maintained with ±5 percent for specific materials in this group.

Table 16–5 includes a wider variety of clutch friction materials, together with some of their properties. Some of these materials may be run wet by allowing them to dip in oil or to be sprayed by oil. This reduces the coefficient of friction somewhat but carries away more heat and permits higher pressures to be used.

16–12 Miscellaneous Clutches and Couplings

The square-jaw clutch shown in Fig. 16–26*a* is one form of positive-contact clutch. These clutches have the following characteristics:

1. They do not slip.
2. No heat is generated.

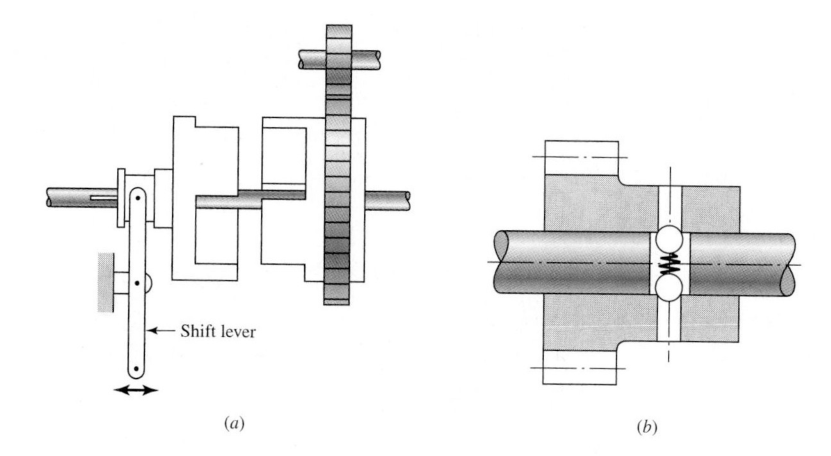

Figure 16–26

(*a*) Square-jaw clutch; (*b*) overload release clutch using a detent.

3 They cannot be engaged at high speeds.
4 Sometimes they cannot be engaged when both shafts are at rest.
5 Engagement at any speed is accompanied by shock.

The greatest differences among the various types of positive clutches are concerned with the design of the jaws. To provide a longer period of time for shift action during engagement, the jaws may be ratchet-shaped, spiral-shaped, or gear-tooth-shaped. Sometimes a great many teeth or jaws are used, and they may be cut either circumferentially, so that they engage by cylindrical mating, or on the faces of the mating elements.

Although positive clutches are not used to the extent of the frictional-contact types, they do have important applications where synchronous operation is required, as, for example, in power presses or rolling-mill screw-downs.

Devices such as linear drives or motor-operated screwdrivers must run to a definite limit and then come to a stop. An overload-release type of clutch is required for these applications. Figure 16-26*b* is a schematic drawing illustrating the principle of operation of such a clutch. These clutches are usually spring-loaded so as to release at a predetermined torque. The clicking sound which is heard when the overload point is reached is considered to be a desirable signal.

Both fatigue and shock loads must be considered in obtaining the stresses and deflections of the various portions of positive clutches. In addition, wear must generally be considered. The application of the fundamentals discussed in Parts 1 and 2 is usually sufficient for the complete design of these devices.

An overrunning clutch or coupling permits the driven member of a machine to "freewheel" or "overrun" because the driver is stopped or because another source of power increases the speed of the driven mechanism. The construction uses rollers or balls mounted between an outer sleeve and an inner member having cam flats machined around the periphery. Driving action is obtained by wedging the rollers between the sleeve and the cam flats. This clutch is therefore equivalent to a pawl and ratchet with an infinite number of teeth.

There are many varieties of overrunning clutches available, and they are built in capacities up to hundreds of horsepower. Since no slippage is involved, the only power loss is that due to bearing friction and windage.

The shaft couplings shown in Fig. 16–27 are representative of the selection available in catalogs.

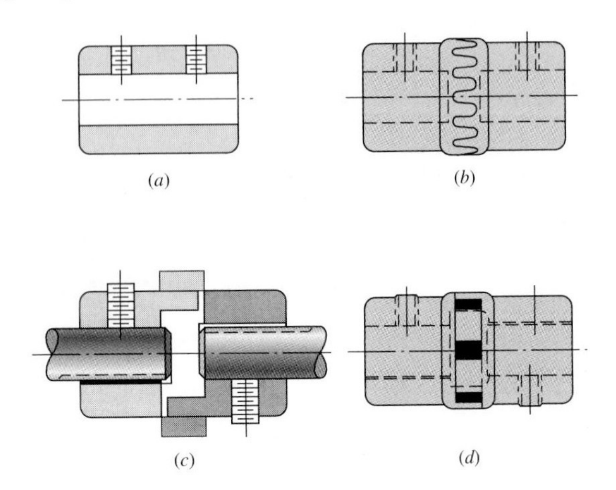

Figure 16–27

Shaft couplings. (*a*) Plain. (*b*) Light-duty toothed coupling. (*c*) BOST-FLEX® through-bore design having elastomer insert to transmit torque by compression; insert permits 1° misalignment. (*d*) Three-jaw coupling available with bronze, rubber, or polyurethane insert to minimize vibration. *(Reproduced by permission, Boston Gear Division, Incom International, Inc., Quincy, Mass.)*

16–13 Flywheels

The equation of motion for the flywheel represented in Fig. 16–1*b* is

$$\sum M = T_i(\theta_i, \dot{\theta}_i) - T_o(\theta_o, \dot{\theta}_o) - I\ddot{\theta} = 0$$

or

$$I\ddot{\theta} = T_i(\theta_i, \omega_i) - T_o(\theta_o, \omega_o) \tag{a}$$

where T_i is considered positive and T_o negative, and where $\dot{\theta}$ and $\ddot{\theta}$ are the first and second time derivatives of θ, respectively. Note that both T_i and T_o may depend for their values on the angular displacements θ_i and θ_o as well as their angular velocities ω_i and ω_o. In many cases the torque characteristic depends upon only one of these. Thus, the torque delivered by an induction motor depends upon the speed of the motor. In fact, motor manufacturers publish charts detailing the torque-speed characteristics of their various motors.

When the input and output torque functions are given, Eq. (*a*) can be solved for the motion of the flywheel using well-known techniques for solving linear and nonlinear differential equations. We can dispense with this here by assuming a rigid shaft, giving $\theta_i = \theta = \theta_o$. Thus, Eq. (*a*) becomes

$$I\ddot{\theta} = T_i(\theta, \omega) - T_o(\theta, \omega) \tag{b}$$

When the two torque functions are known and the starting values of the displacement θ and velocity ω are given, Eq. (*b*) can be solved for ω and $\ddot{\theta}$ as functions of time. However, we are not really interested in the instantaneous values of these terms at all. Primarily we want to know the overall performance of the flywheel. What should its moment of inertia be? How do we match the power source to the load? And what are the resulting performance characteristics of the system that we have selected?

To gain insight into the problem, a hypothetical situation is diagrammed in Fig. 16–28. An input power source subjects a flywheel to a constant torque T_i while the shaft rotates from θ_1 to θ_2. This is a positive torque and is plotted upward. Equation (*b*) indicates that a positive acceleration $\ddot{\theta}$ will be the result, and so the shaft velocity increases from ω_1 to ω_2. As shown, the shaft now rotates from θ_2 to θ_3 with zero torque and hence, from Eq. (*b*), with zero acceleration. Therefore $\omega_3 = \omega_2$. From θ_3 to θ_4 a load, or output torque, of constant magnitude is applied, causing the shaft to slow down from

Figure 16–28

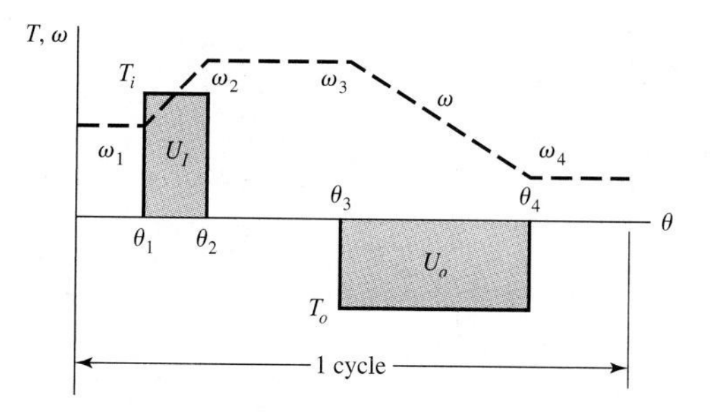

ω_3 to ω_4. Note that the output torque is plotted in the negative direction in accordance with Eq. (*b*).

The work input to the flywheel is the area of the rectangle between θ_1 and θ_2, or

$$U_i = T_i(\theta_2 - \theta_1) \tag{c}$$

The work output of the flywheel is the area of the rectangle from θ_3 to θ_4, or

$$U_o = T_o(\theta_4 - \theta_3) \tag{d}$$

If U_o is greater than U_i, the load uses more energy than has been delivered to the flywheel and so ω_4 will be less than ω_1. If $U_o = U_i$, ω_4 will be equal to ω_1 because the gains and losses are equal; we are assuming no friction losses. And finally, ω_4 will be greater than ω_1 if $U_i > U_o$.

We can also write these relations in terms of kinetic energy. At $\theta = \theta_1$ the flywheel has a velocity of ω_1 rad/s, and so its kinetic energy is

$$E_1 = \frac{1}{2} I \omega_1^2 \tag{e}$$

At $\theta = \theta_2$ the velocity is ω_2, and so

$$E_2 = \frac{1}{2} I \omega_2^2 \tag{f}$$

Thus the change in kinetic energy is

$$E_2 - E_1 = \frac{1}{2} I (\omega_2^2 - \omega_1^2) \tag{16–63}$$

Many of the torque displacement functions encountered in practical engineering situations are so complicated that they must be integrated by numerical methods. Figure 16–29, for example, is a typical plot of the engine torque for one cycle of motion of a single- cylinder internal combustion engine. Since a part of the torque curve is negative, the flywheel must return part of the energy back to the engine. Approximate integration of this curve for a cycle of 4π yields a mean torque T_m available to drive a load.

The simplest integration routine is Simpson's rule; this approximation can be handled on any computer and is short enough to use on the smallest programmable calculators. See Prob. 1–27. In fact, this routine is usually found as part of the library for most calculators and minicomputers. The equation used is

$$\int_{x_0}^{x_n} f(x)\,dx = \frac{h}{3}(f_0 + 4f_1 + 2f_2 + 4f_3 + 2f_4 + \cdots + 2f_{n-2} + 4f_{n-1} + f_n) \tag{16–64}$$

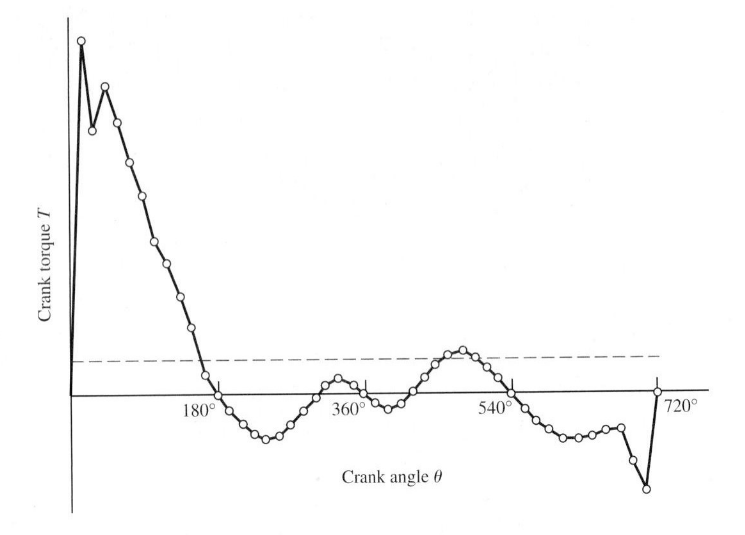

Figure 16–29

Relation between torque and crank angle for a one-cylinder, four-stroke–cycle internal combustion engine.

where

$$h = \frac{x_n - x_0}{n} \qquad x_n > x_0$$

and n is the number of subintervals used, 2, 4, 6, ... If memory is limited, solve Eq. (16–64) in two or more steps, say, from 0 to $n/2$ and then from $n/2$ to n.

It is convenient to define a *coefficient of speed fluctuation* as

$$C_s = \frac{\omega_2 - \omega_1}{\omega} \tag{16–65}$$

where ω is the nominal angular velocity, given by

$$\omega = \frac{\omega_2 + \omega_1}{2} \tag{16–66}$$

Equation (16–63) can be factored to give

$$E_2 - E_1 = \frac{I}{2}(\omega_2 - \omega_1)(\omega_2 + \omega_1)$$

Since $\omega_2 - \omega_1 = C_s\omega$ and $\omega_2 + \omega_1 = 2\omega$, we have

$$E_2 - E_1 = C_s I \omega^2 \tag{16–67}$$

Equation (16–67) can be used to obtain an appropriate flywheel inertia corresponding to the energy change $E_2 - E_1$.

EXAMPLE 16–6

Table 16–6 lists values of the torque used to plot Fig. 16–29. The nominal speed of the engine is to be 250 rad/s.

(a) Integrate the torque-displacement function for one cycle and find the energy that can be delivered to a load during the cycle.

(b) Determine the mean torque T_m (see Fig. 16–29).

Table 16–6

Plotting Data for Fig. 16–29

θ, deg	T, lb · in	θ, deg	T, lb · in	θ, deg	T, lb · in	θ, deg	T, lb · in
0	0	195	−107	375	−85	555	−107
15	2800	210	−206	390	−125	570	−206
30	2090	225	−260	405	−89	585	−292
45	2430	240	−323	420	8	600	−355
60	2160	255	−310	435	126	615	−371
75	1840	270	−242	450	242	630	−362
90	1590	285	−126	465	310	645	−312
105	1210	300	−8	480	323	660	−272
120	1066	315	89	495	280	675	−274
135	803	330	125	510	206	690	−548
150	532	345	85	525	107	705	−760
165	184	360	0	540	107	720	0
180	0						

(c) The greatest energy fluctuation is approximately between $\theta = 15°$ and $\theta = 150°$ on the torque diagram; see Fig. 16–29 and note that $T_o = -T_m$. Using a coefficient of speed fluctuation $C_s = 0.1$, find a suitable value for the flywheel inertia.

(d) Find ω_2 and ω_1.

Solution

(*a*) Using $n = 48$ and $h = 4\pi/48$, we enter the data of Table 16–6 into a computer program and get $E = 3490$ lb · in. This is the energy that can be delivered to the load.

Answer

(*b*) $T_m = \dfrac{3490}{4\pi} = 278$ lb · in

(*c*) The largest positive loop on the torque-displacement diagram occurs between $\theta = 0°$ and $\theta = 180°$. We select this loop as yielding the largest speed change. Subtracting 278 lb · in from the values in Table 16–6 for this loop gives, respectively, −278, 2522, 1812, 2152, 1882, 1562, 1312, 932, 788, 525, 254, −94, and −278 lb · in. Entering Simpson's approximation again, using $n = 12$ and $h = 4\pi/48$, gives $E_2 - E_1 = 3660$ lb · in. We now solve Eq. (16–67) for I and substitute. This gives

Answer

$$I = \frac{E_2 - E_1}{C_s\omega^2} = \frac{3660}{0.1(250)^2} = 0.586 \text{ lb} \cdot \text{s}^2 \text{ in}$$

(*d*) Equations (16–65) and (16–66) can be solved simultaneously for ω_2 and ω_1. Substituting appropriate values in these two equations yields

Answer

$$\omega_2 = \frac{\omega}{2}(2 + C_s) = \frac{250}{2}(2 + 0.1) = 262.5 \text{ rad/s}$$

Answer

$$\omega_1 = 2\omega - \omega_2 = 2(250) - 262.5 = 237.5 \text{ rad/s}$$

These two speeds occur at $\theta = 180°$ and $\theta = 0°$, respectively.

Punch-press torque demand often takes the form of a severe impulse and the running friction of the drive train. The motor overcomes the minor task of overcoming friction while attending to the major task of restoring the flywheel's angular speed. The situation can be idealized as shown in Fig. 16–30. Neglecting the running friction, Euler's equation

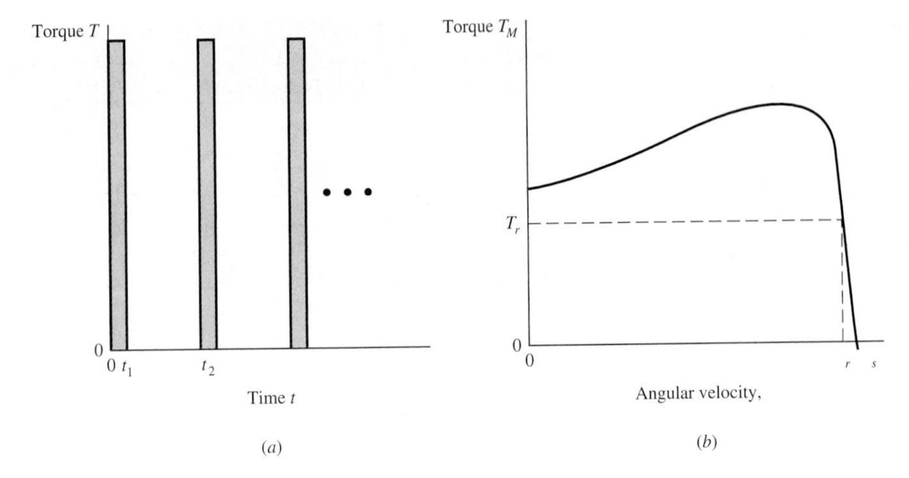

Figure 16–30

(a) Punch-press torque demand during punching. (b) Squirrel-cage electric motor torque-speed characteristic.

can be written as

$$T(t_1 - 0) = \frac{1}{2} I(\omega_1^2 - \omega_2^2) = E_2 - E_1$$

where the only significant inertia is the flywheel. Punch presses can have the motor and flywheel on one shaft then, through a gear-reduction, drive a slider-crank mechanism which carries the punching tool. The motor can be connected to the punch continuously, creating a punching rhythm, or it can be connected on command through a clutch that allows one punch and a disconnect. The motor and flywheel must be sized for the most demanding service, which is steady punching. The work done is given by

$$W = \int_{\theta_1}^{\theta_2} [T(\theta) - T]\, d\theta = \frac{1}{2} I(\omega_{\text{max}}^2 - \omega_{\text{min}}^2)$$

This equation can be arranged to include the coefficient of speed fluctuation C_s as follows:

$$W = \frac{1}{2} I(\omega_{\text{max}}^2 - \omega_{\text{min}}^2) = \frac{I}{2}(\omega_{\text{max}} - \omega_{\text{min}})(\omega_{\text{max}} + \omega_{\text{min}})$$

$$= \frac{I}{2}(C_s\bar{\omega})(2\omega_0) = IC_s\bar{\omega}\omega_0$$

When the speed fluctuation is low, $\bar{\omega} \doteq \omega_0$, $\bar{\omega}\omega_0 = \bar{\omega}^2$, and

$$I = \frac{W}{C_s\bar{\omega}^2}$$

An induction motor has a linear torque characteristic $T = a\omega + b$ in the range of operation. The constants a and b can be found from the nameplate speed ω_r and the synchronous speed ω_s:

$$a = \frac{T_r - T_s}{\omega_r - \omega_s} = \frac{T_r}{\omega_r - \omega_s} = -\frac{T_r}{\omega_s - \omega_r}$$
$$b = \frac{T_r\omega_s - T_s\omega_r}{\omega_s - \omega_r} = \frac{T_r\omega_s}{\omega_s - \omega_r} \tag{16–68}$$

For example, a 3-hp three-phase squirrel-cage a-c motor rated at 1125 r/min has a torque of 63 025(3)/1125 = 168.1 in · lb. The rated angular velocity is $\omega_r = 2\pi n_r/60 =$

$2\pi(1125)/60 = 117.81$ rad/s, and the synchronous angular velocity $\omega_s = 2\pi(1200)/60 = 125.66$ rad/s. Thus $a = -21.41$ in · lb · s/rad, and $b = 2690.9$ in · lb, and we can express $T(\omega)$ as $a\omega + b$. During the interval from t_1 to t_2 the motor accelerates the flywheel. Separating the equation $T_M = I\,d\omega/dt$ we have

$$\int_{t_1}^{t_2} dt = \int_{\omega_r}^{\omega_2} \frac{I\,d\omega}{T_M} = I \int_{\omega_r}^{\omega_2} \frac{d\omega}{a\omega + b} = \frac{I}{a} \ln \frac{a\omega_2 + b}{a\omega_r + b} = \frac{I}{a} \ln \frac{T_2}{T_r}$$

or

$$t_2 - t_1 = \frac{I}{a} \ln \frac{T_2}{T_r} \tag{16–69}$$

For the deceleration interval when the motor and flywheel feel the punch torque on the shaft as T_L, $(T_M - T_L) = I\,d\omega/dt$, or

$$\int_0^{t_1} dt = I \int_{\omega_2}^{\omega_r} \frac{d\omega}{T_M - T_L} = I \int_{\omega_2}^{\omega_r} \frac{d\omega}{a\omega + b - T_L} = \frac{I}{a} \ln \frac{a\omega_r + b - T_L}{a\omega_2 + b - T_L}$$

or

$$t_1 = \frac{I}{a} \ln \frac{T_r - T_L}{T_2 - T_L} \tag{16–70}$$

We can divide Eq. (16–69) by Eq. (16–70) to obtain

$$\frac{T_2}{T_r} = \left(\frac{T_L - T_r}{T_L - T_2}\right)^{(t_2 - t_1)/t_1} \tag{16–71}$$

Equation (16–71) can be solved for T_2 numerically. Having T_2 the flywheel inertia is, from Eq. (16–69),

$$I = \frac{a(t_2 - t_1)}{\ln(T_2/T_r)} \tag{16–72}$$

It is important that a be in units of in · lbf · s/rad so that I has proper units. The constant a should not be in in · lbf per r/min or in · lbf per r/s.

16–14 Adequacy Assessment for Clutches and Brakes

Dealing with brakes and clutches usually entails the selection of available items to be incorporated into larger projects. The design of brakes and clutches is a rarity among mechanical engineers who are not on the staff of a brake and clutch manufacturer. For this reason we will not pursue such design in this book.

The following questions are typically included in an adequacy assessment for an existing design:

- Is the torque satisfactory for intended direction(s) of operation?
- Is the pressure distribution modeled well, based on tests?
- Is the coefficient of friction stable and dependable?
- Is the largest pressure acceptable to the material?
- Is the largest temperature for intended use acceptable to the lining and drum/disk?
- If intended to be self-locking, will it always be so?
- If intended to be non–self-locking, will it always be so?

- Is the actuating force realizable from the actuator?
- Have the reactions been provided for?
- What is the factor of safety?

For selection of a brake or clutch, the engineer needs to discover if the brake or clutch

- Meets bulk, shape, and weight restrictions
- Can supply required torque at necessary speed
- Incorporates any fail-safe or dead-man features

PROBLEMS

16–1 The figure shows an internal rim-type brake having an inside rim diameter of 12 in and a dimension $R = 5$ in. The shoes have a face width of $1\frac{1}{2}$ in and are both actuated by a force of 500 lb. The mean coefficient of friction is 0.28.

(*a*) Find the maximum pressure and indicate the shoe on which it occurs.

(*b*) Estimate the braking torque effected by each shoe, and find the total braking torque.

(*c*) Estimate the resulting hinge-pin reactions.

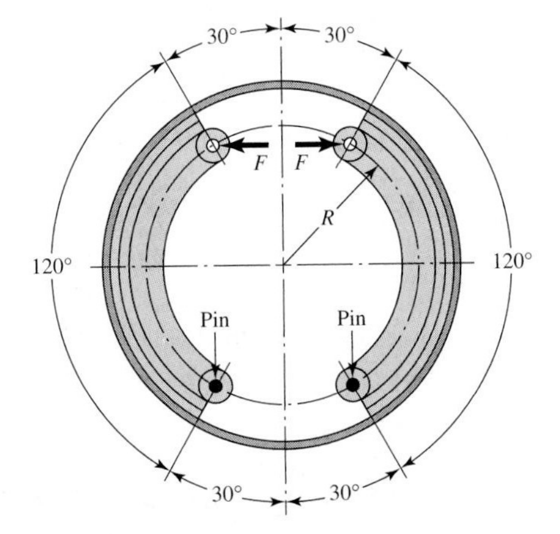

Problem 16–1

16–2 The coefficient of friction for the brake in Prob. 16–1 is a random variable $\mathbf{f} \sim N(0.28, 0.03)$. Find the $\pm 3\sigma$ bounds on the maximum pressures on each shoe, assuming a deterministic actuating force.

16–3 In the figure for Prob. 16–1, the inside rim diameter is 280 mm and the dimension R is 90 mm. The shoes have a face width of 30 mm. Find the braking torque and the maximum pressure for each shoe if the actuating force is 1000 N, the drum rotation is counterclockwise, and $f = 0.30$.

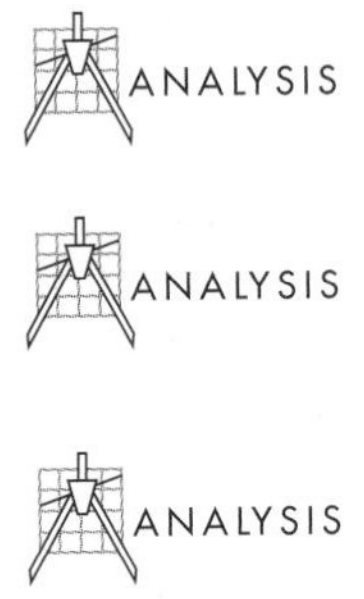

16–4 The figure shows a 400-mm-diameter brake drum with four internally expanding shoes. Each of the hinge pins A and B supports a pair of shoes. The actuating mechanism is to be arranged to produce the same force F on each shoe. The face width of the shoes is 75 mm. The material used permits a coefficient of friction of 0.24 and a maximum pressure of 1000 kPa.

(*a*) Determine the actuating force.

(*b*) Estimate the brake capacity.

(*c*) Noting that rotation may be in either direction, estimate the hinge-pin reactions.

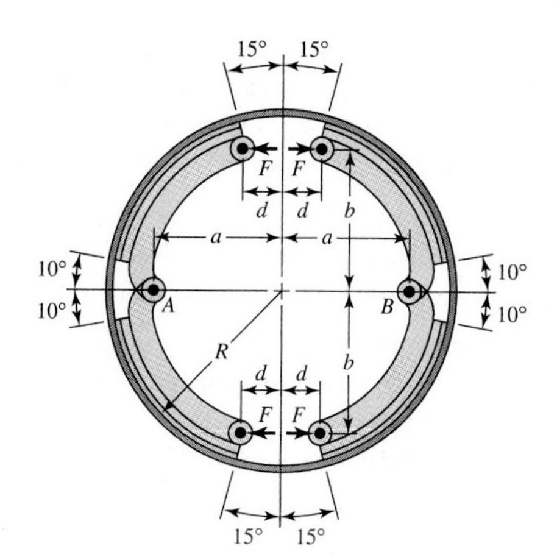

Problem 16–4
The dimensions in millimeters are $a = 150$, $b = 165$, $R = 200$, and $d = 50$.

ANALYSIS **16–5** The block-type hand brake shown in the figure has a face width of 30 mm and a mean coefficient of friction of 0.25. For an estimated actuating force of 400 N, find the maximum pressure on the shoe and find the braking torque.

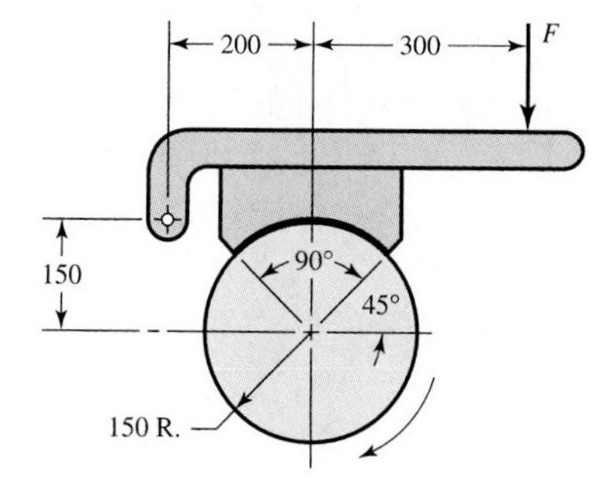

Problem 16–5
Dimensions in millimeters.

ANALYSIS **16–6** Suppose the standard deviation of the coefficient of friction in Prob. 16–5 is $\sigma_f = 0.025$, where the deviation from the mean is due entirely to environmental conditions. Find the brake torque corresponding to $\pm 3\sigma_f$.

ANALYSIS **16–7** The brake shown in the figure has a coefficient of friction of 0.30, a face width of 2 in, and a limiting shoe lining pressure of 150 psi. Find the limiting actuating force F and the torque capacity.

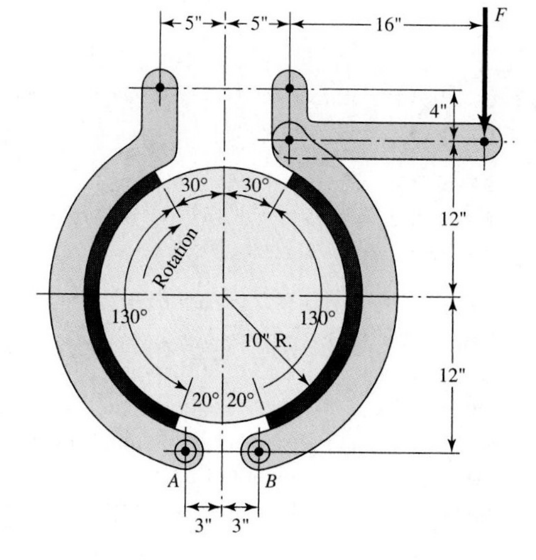

Problem 16–7

16–8 Refer to the pivoted external brake shoe and Eq. (16–15). Suppose the pressure distribution was uniform, that is, the pressure p is independent of θ. What would the pivot distance a' be? If $\theta_1 = \theta_2 = 60°$, compare a with a'.

16–9 The shoes on the brake depicted in the figure subtend a 90° arc on the drum of this external pivoted-shoe brake. The actuation force P is applied to the lever. The rotation direction of the drum is counterclockwise, and the coefficient of friction is 0.30.

(*a*) What should the dimension e be?

(*b*) Draw the free-body diagrams of the handle lever and both shoe levers, with forces expressed in terms of the actuation force P.

(*c*) Does the direction of rotation of the drum affect the braking torque?

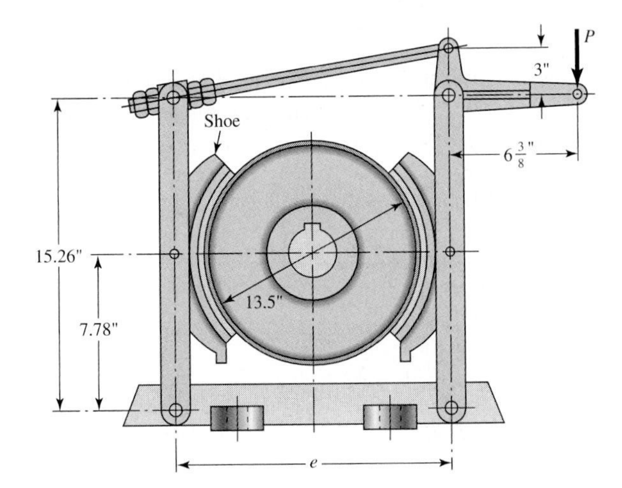

Problem 16–9

ANALYSIS

16–10 Problem 16–9 is preliminary to analyzing the brake. A molded lining is used dry in the brake of Prob. 16–9 on a cast iron drum. The shoes are 7.5 in wide and subtend a 90° arc. Find the actuation force and the braking torque.

16–11 The maximum band interface pressure on the brake shown in the figure is 90 psi. Use a 14-in-diameter drum, a band width of 4 in, a coefficient of friction of 0.25, and an angle-of-wrap of 270°. Find the band tensions and the torque capacity.

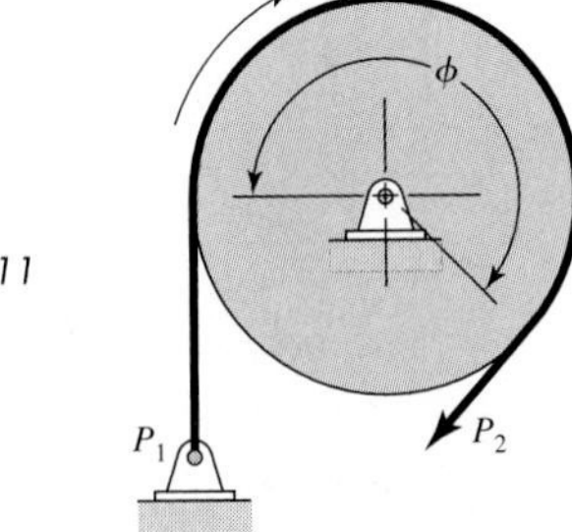

Problem 16–11

16–12 The drum for the band brake in Prob. 16–11 is 300 mm in diameter. The band selected has a mean coefficient of friction of 0.28 and a width of 80 mm. It can safely support a tension of 7.6 kN. If the angle-of-wrap is 270°, find the lining pressure and the torque capacity.

16–13 The brake shown in the figure has a coefficient of friction of 0.30 and is to operate using a maximum force F of 400 N. If the band width is 50 mm, find the band tensions and the braking torque.

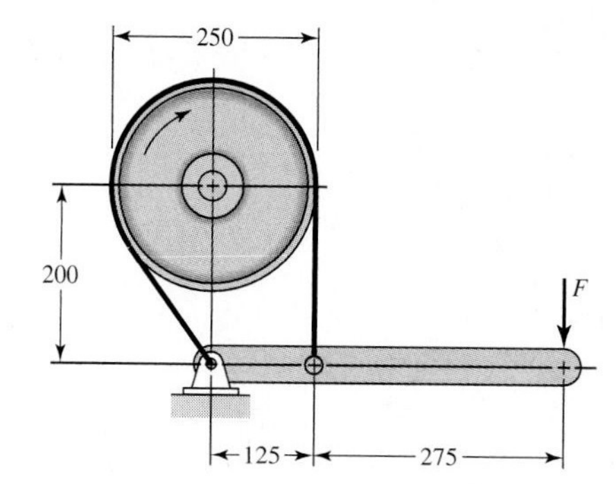

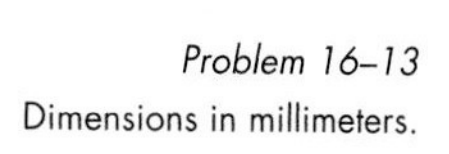

Problem 16–13
Dimensions in millimeters.

16–14 The figure depicts a band brake whose drum rotates counterclockwise at 200 r/min. The drum is 16 in in diameter supporting a band lining 3 in wide. The coefficient of friction is 0.20. The maximum lining interface pressure is 70 psi.

(*a*) Find the brake torque, necessary force P, and steady-state power.

(*b*) Complete the free-body diagram of the drum. Find the bearing radial load that a pair of straddle-mounted bearings would have to carry.

(*c*) What is the lining pressure p at both ends of the contact arc?

Problem 16–14

16–15 The figure shows a band brake designed to prevent "backward" rotation of the shaft. The angle-of-wrap is 270°, the band width is $2\frac{1}{8}$ in, and the coefficient of friction is 0.20. The torque to be resisted by the brake is 150 ft · lb.

(*a*) What dimension c_1 will just prevent backward motion?

(*b*) If the rocker was designed with $c_1 = 1$ in, what is the maximum pressure between the band and drum at 150 ft · lb back torque?

(*c*) If the back-torque demand is 100 in · lb, what is the largest pressure between the band the drum?

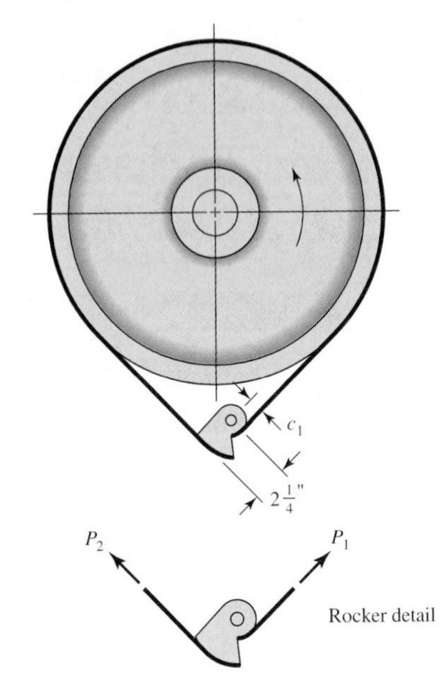

Problem 16–15

ANALYSIS

16–16 A plate clutch has a single pair of mating friction surfaces 300 mm OD by 225 mm ID. The mean value of the coefficient of friction is 0.25, and the actuating force is 5 kN.

(*a*) Find the maximum pressure and the torque capacity using the uniform-wear model.

(*b*) Find the maximum pressure and the torque capacity using the uniform-pressure model.

ANALYSIS

16–17 A hydraulically operated multidisk plate clutch has an effective disk outer diameter of 6.5 in and an inner diameter of 4 in. The coefficient of friction is 0.24, and the limiting pressure is 120 psi. There are six planes of sliding present.

(*a*) Using the uniform wear model, estimate the axial force F and the torque T.

(*b*) Let the inner diameter of the friction pairs d be a variable. Complete the following table:

d, in	2	3	4	5	6
T, in · lb					

(*c*) What does the table show?

DESIGN

16–18 Look again at Prob. 16–17.

(*a*) Show how the optimal diameter d^* is related to the outside diameter D.

(*b*) What is the optimal inner diameter?

(*c*) What does the tabulation show about maxima?

(*d*) Common proportions for such plate clutches lie in the range $0.45 \le d/D \le 0.80$. Is the result in part *a* useful?

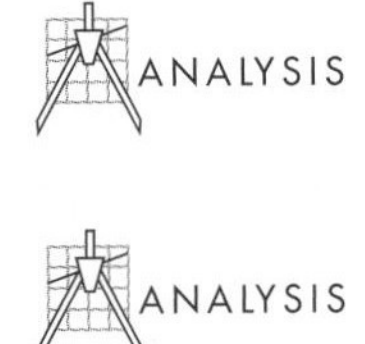

ANALYSIS

16–19 A cone clutch has $D = 330$ mm, $d = 306$ mm, a cone length of 60 mm, and a coefficient of friction of 0.26. A torque of 200 N · m is to be transmitted. For this requirement, estimate the actuating force and pressure by both models.

ANALYSIS

16–20 Consider a cone clutch with a half apex-angle of $\alpha = 30°$ and a coefficient of friction of $f = 0.30$.

(*a*) Equation (16–50) gives the sum of all the dF_N and all the dF_f. Show that the integration of the first equation of Sec. 16–8 is

$$F_N = \frac{\pi p_a d(D-d)}{2}$$

(*b*) Show that the integration of the second equation of Sec. 16–8 is

$$F_f = \frac{\pi f p_a d(D-d)\cot\alpha}{2}$$

(*c*) The results of parts *a* and *b* can be considered as partitioning of Eq. (16–50). We can view Eq. (16–50) as a leftward force applied to the cone. The partitioned values of F_N and F_f can be viewed as reactions directed rightward from the outer ring onto the cone. Set $\pi p_a d(D-d)/2 = 1$, then find F_N, F_f, and F, and show forces on a free-body diagram of the cone.

(*d*) Repeat part *c* with $\alpha = 10°$, $f = 0.30$.

(*e*) When you release (remove) F, what happens in parts *c* and *d*? What are your conclusions?

ANALYSIS

16–21 An engineer is considering holding a helical pinion on the overhang of a shaft with a Jarno (originally proposed by Oscar J. Beale of Pratt & Whitney Manufacturing Co.) #8 taper. The sizes of a Jarno #N taper are

Large end diameter:	$N/8$ in
Small end diameter:	$N/10$ in
Length of the taper:	$N/2$ in

If the coefficient of friction is 0.20 and the assembly pressure p_a is 1000 psi

(*a*) What is the cone apex half-angle of the Jarno taper?

(*b*) What are the arbor assembly forces?

(*c*) The thrust from the pinion is 200 lb. What is the torque capacity of the assembly?

ANALYSIS

16–22 For Prob. 16–21, plot Eq. (16–51) with applied thrust F_a as abscissa and permissible torque capacity as ordinate. Is this curve expected?

ANALYSIS

16–23 Show that for the caliper brake the $T/(fFD)$ versus d/D plot is the same as Eq. (*c*) of Sec. 16-5.

ANALYSIS

16–24 A two-jaw clutch has the dimensions shown in the figure and is made of ductile steel. The clutch has been designed to transmit 2 kW at 500 r/min. Find the bearing and shear stresses in the key and the jaws.

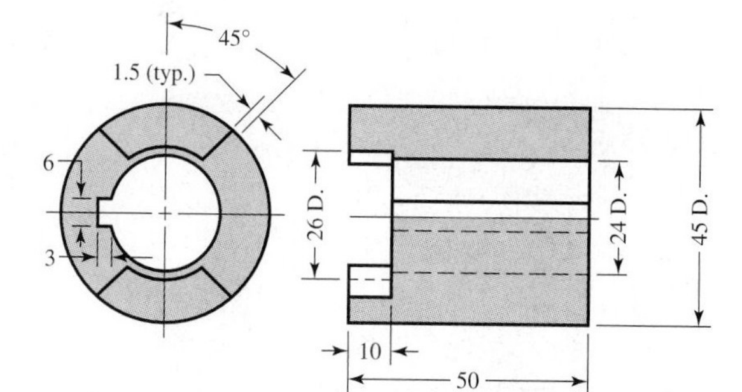

Problem 16–24
Dimensions in millimeters.

16–25 A brake has a normal braking torque of 320 N · m and heat-dissipating surfaces whose mass is 18 kg. Suppose a load is brought to rest in 8.3 s from an initial angular speed of 1800 r/min using the normal braking torque; estimate the temperature rise of the heat-dissipating surfaces.

16–26 A cast-iron flywheel has a rim whose OD is 60 in and whose ID is 56 in. The flywheel weight is to be such that an energy fluctuation of 5000 lb · ft will cause the angular speed to vary no more than 240–260 r/min. Estimate the coefficient of speed fluctuation. If the weight of the spokes is neglected, what should be the width of the rim?

16–27 A single-geared blanking press has a stroke of 8 in and a rated capacity of 35 tons. A cam-driven ram is assumed to be capable of delivering the full press load at constant force during the last 15 percent of a constant-velocity stroke. The camshaft has an average speed of 90 r/min and is geared to the flywheel shaft at a 6:1 ratio. The total work done is to include an allowance of 16 percent for friction.

(*a*) Estimate the maximum energy fluctuation.

(*b*) Find the rim weight for an effective diameter of 48 in and a coefficient of speed fluctuation of 0.10.

16–28 Using the data of Table 16–6, find the mean output torque and flywheel inertia required for a three-cylinder in-line engine corresponding to a nominal speed of 2400 r/min. Use $C_s = 0.30$.

16–29 When a motor armature inertia, a pinion inertia, and a motor torque reside on a motor shaft, and a gear inertia, a load inertia, and a load torque exist on a second shaft, it is useful to reflect all the torques and inertias to one shaft, say, the armature shaft. We need some rules to make such reflection easy. Consider the pinion and gear as disks of pitch radius.

- A torque on a second shaft is reflected to the motor shaft as the load torque divided by the negative of the stepdown ratio.
- An inertia on a second shaft is reflected to the motor shaft as its inertia divided by the stepdown ratio squared.
- The inertia of a disk gear on a second shaft in mesh with a disk pinion on the motor shaft is reflected to the pinion shaft as the *pinion* inertia multiplied by the stepdown ratio squared.

(*a*) Verify the three rules.

(*b*) Using the rules, reduce the two-shaft system in the figure to a motor-shaft shish-kebob equivalent. Correctly done, the dynamic response of the shish kebab and the real system are identical.

(*c*) For a stepdown ratio of $n = 10$ compare the shish-kebab inertias.

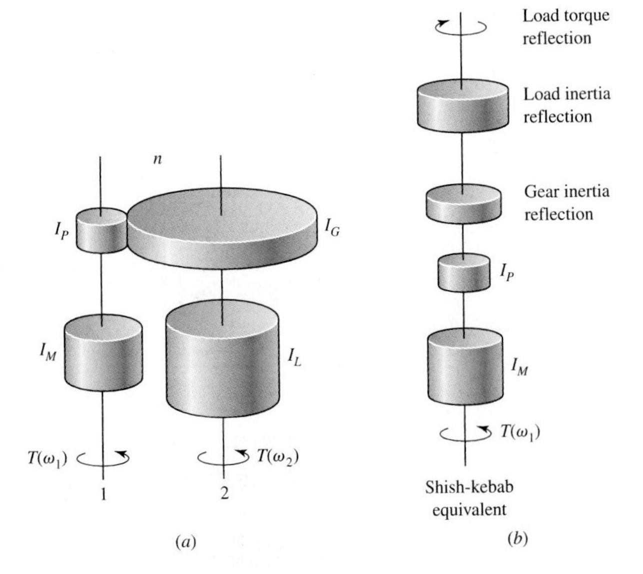

Problem 16–29
Dimensions in millimeters.

16–30 Apply the rules of Prob. 16–29 to the three-shaft system shown in the figure to create a motor shaft shish kebab.

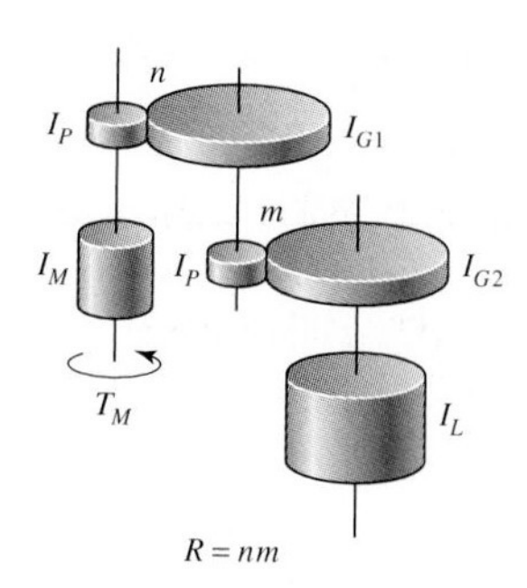

Problem 16–30

(*a*) Show that the equivalent inertia I_e is given by

$$I_e = I_M + I_P + n^2 I_P + \frac{I_P}{n^2} + \frac{m^2 I_P}{n^2} + \frac{I_L}{m^2 n^2}$$

(*b*) If the overall gear reduction R is a constant nm, show that the equivalent inertia becomes

$$I_e = I_M + I_P + n^2 I_P + \frac{I_P}{n^2} + \frac{R^2 I_P}{n^4} + \frac{I_L}{R^2}$$

(*c*) If the problem is to minimize the gear-train inertia, find the ratios n and m for the values of $I_P = 1$, $I_M = 10$, $I_L = 100$, and $R = 10$.

16–31 For the conditions of Prob. 16–30, make a plot of the equivalent inertia I_e as ordinate and the stepdown ratio n as abscissa in the range $1 \le n \le 10$. How does the minimum inertia compare to the single-step inertia?

16–32 A punch-press geared 10:1 is to make six punches per minute under circumstances where the torque on the crankshaft is 1300 ft · lb for $\frac{1}{2}$ s. The motor's nameplate reads 3 bhp at 1125 r/min for continuous duty. Design a satisfactory flywheel for use on the motor shaft to the extent of specifying material and rim inside and outside diameters as well as its width. As you prepare your specifications, note $\omega_{\max}$, $\omega_{\min}$, the coefficient of speed fluctuation C_s, energy transfer, and peak power that the flywheel transmits to the punch-press. Note power and shock conditions imposed on the gear train because the flywheel is on the motor shaft.

16–33 The punch-press of Prob. 16–32 needs a flywheel for service on the crankshaft of the punch-press. Design a satisfactory flywheel to the extent of specifying material, rim inside and outside diameters, and width. Note $\omega_{\max}$, $\omega_{\min}$, C_s, energy transfer, and peak power the flywheel transmits to the punch. What is the peak power seen in the gear train? What power and shock conditions must the gear-train transmit?

16–34 Compare the designs resulting from the tasks assigned in Probs. 16–32 and 16–33. What have you learned? What recommendations do you have?

16–35 Apply the rules from Prob. 16–29 to a triple-reduction gear train as shown in the figure. For the same parameters, $I_P = 1$, $I_M = 10$, $I_L = 100$, and $nm\lambda = R = 10$, find the stepdown ratios n^*, m^*, and λ^* which minimize the equivalent inertia I_e. How does the minimum inertia compare with those of single and double reductions?

Problem 16–35

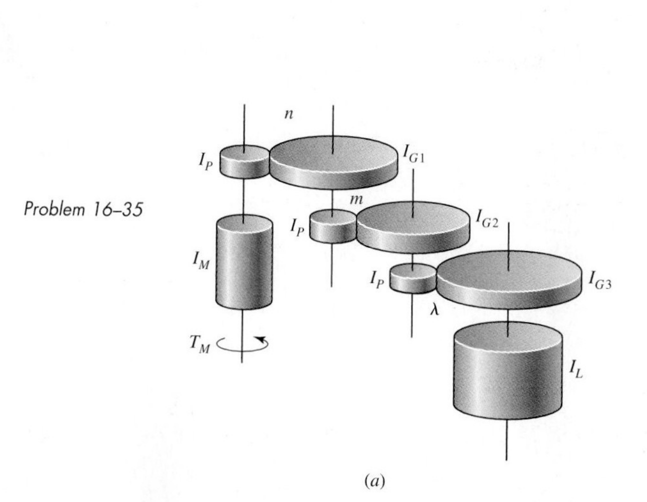

(*a*)

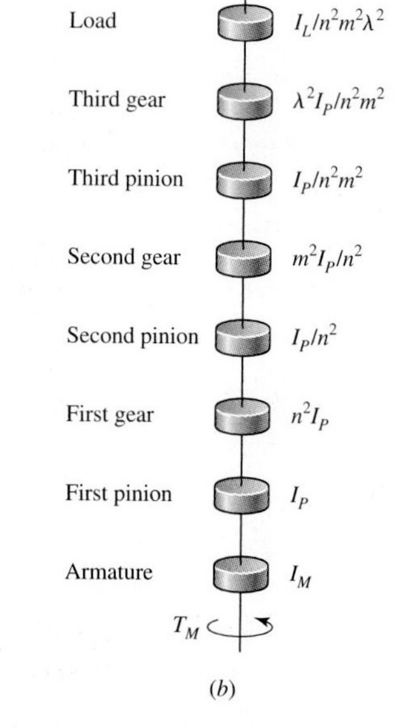

(*b*)

17

Flexible Mechanical Elements

17–1	Belts **1050**
17–2	Flat- and Round-Belt Drives **1053**
17–3	V Belts **1069**
17–4	Timing Belts **1077**
17–5	Roller Chain **1079**
17–6	Wire Rope **1088**
17–7	Flexible Shafts **1097**

Belts, ropes, chains, and other similar elastic or flexible machine elements are used in conveying systems and in the transmission of power over comparatively long distances. It often happens that these elements can be used as a replacement for gears, shafts, bearings, and other relatively rigid power-transmission devices. In many cases their use simplifies the design of a machine and substantially reduces the cost.

In addition, since these elements are elastic and usually quite long, they play an important part in absorbing shock loads and in damping out and isolating the effects of vibration. This is an important advantage as far as machine life is concerned.

Most flexible elements do not have an infinite life. When they are used, it is important to establish an inspection schedule to guard against wear, aging, and loss of elasticity. The elements should be replaced at the first sign of deterioration.

17–1 Belts

The four principal types of belts are shown, with some of their characteristics, in Table 17–1. *Crowned pulleys* are used for flat belts, and *grooved pulleys,* or *sheaves,* for round and V belts. Timing belts require *toothed wheels,* or *sprockets.* In all cases, the pulley axes must be separated by a certain minimum distance, depending upon the belt type and size, to operate properly. Other characteristics of belts are:

- They may be used for long center distances.
- Except for timing belts, there is some slip and creep, and so the angular-velocity ratio between the driving and driven shafts is neither constant nor exactly equal to the ratio of the pulley diameters.
- In some cases an idler or tension pulley can be used to avoid adjustments in center distance that are ordinarily necessitated by age or the installation of new belts.

Figure 17–1 illustrates the geometry of flat-belt drives. For a flat belt with this drive the belt tension is such that the sag or droop is visible in Fig. 7-2*a*, when the belt is running. Although the top is preferred for the loose side of the belt, for other belt types either the top or the bottom may be used, because their installed tension is usually greater.

Two types of reversing drives are shown in Fig. 17–2. Notice that both sides of the belt contact the pulleys in Figs. 17–2*b* and 17–2*c*, and so these drives cannot be used with V belts or timing belts.

Table 17–1

Characteristics of Some Common Belt Types

Belt Type	Figure	Joint	Size Range	Center Distance
Flat	t	Yes	$t = \begin{cases} 0.03 \text{ to } 0.20 \text{ in} \\ 0.75 \text{ to } 5 \text{ mm} \end{cases}$	No upper limit
Round	d	Yes	$d = \frac{1}{8}$ to $\frac{3}{4}$ in	No upper limit
V	b	None	$b = \begin{cases} 0.31 \text{ to } 0.91 \text{ in} \\ 8 \text{ to } 19 \text{ mm} \end{cases}$	Limited
Timing	p	None	$p = 2$ mm and up	Limited

Figure 17–1

Flat-belt geometry. (*a*) Open belt. (*b*) Crossed belt.

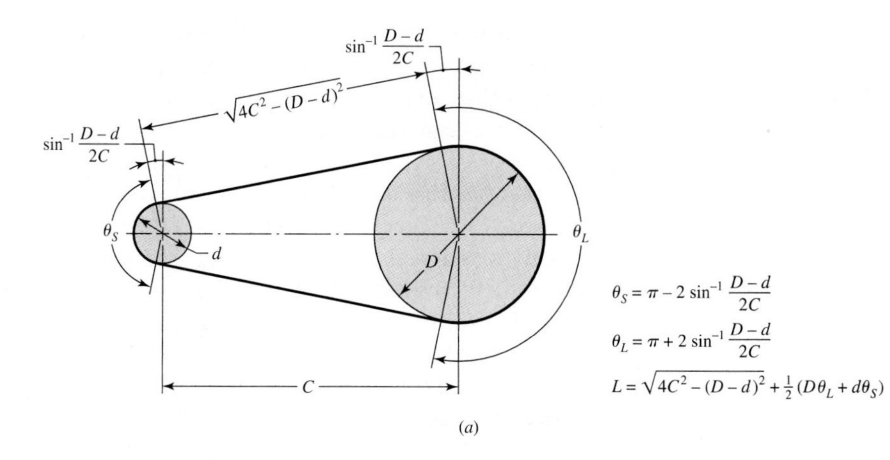

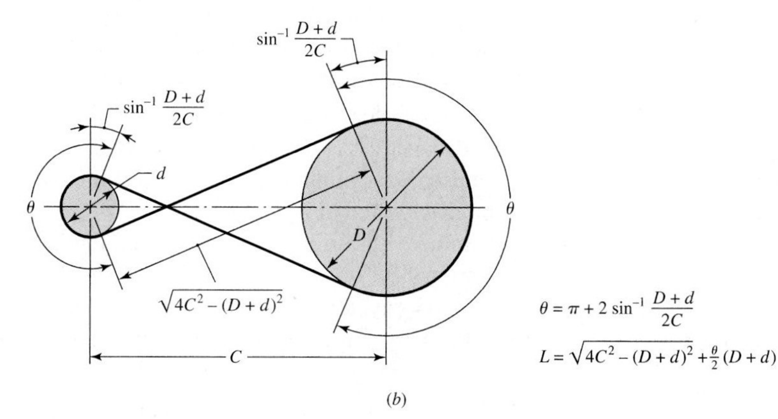

Figure 17–2

Nonreversing and reversing belt drives. (*a*) Nonreversing open belt. (*b*) Reversing crossed belt. Crossed belts must be separated to prevent rubbing if high-friction materials are used. (*c*) Reversing open-belt drive.

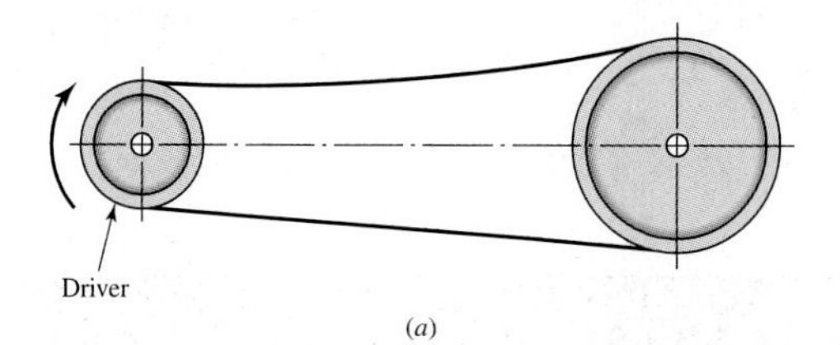

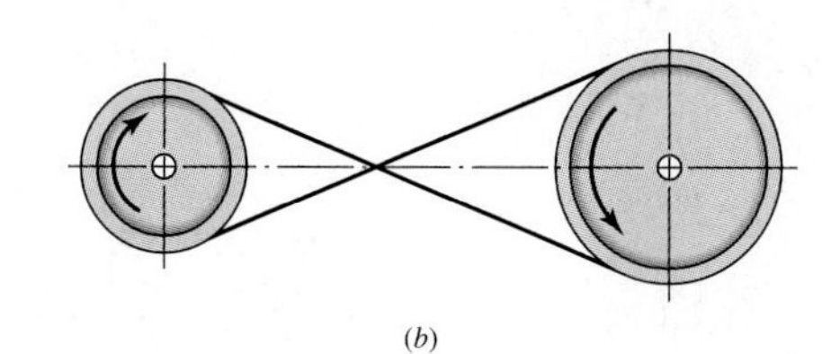

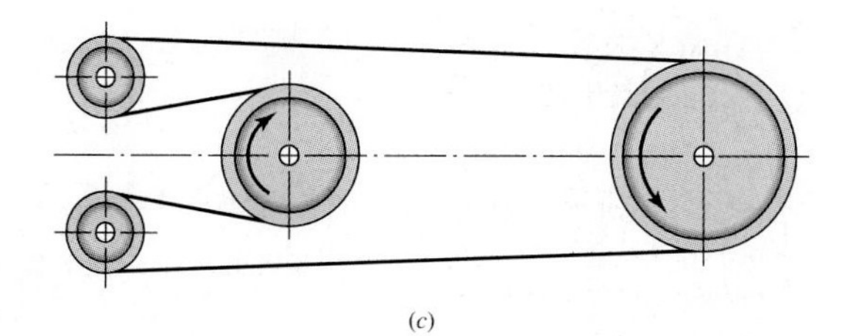

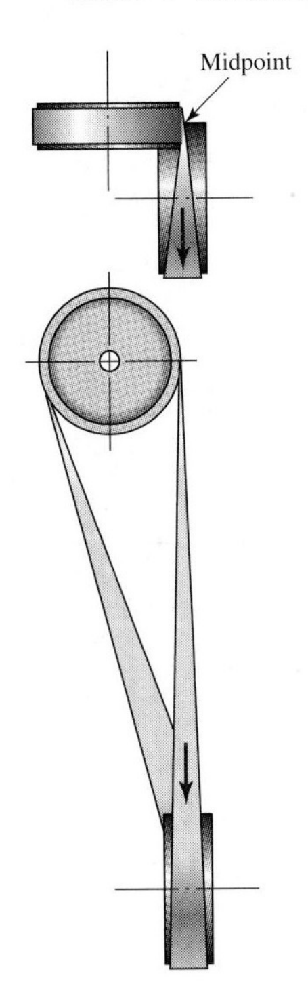

Figure 17–3

Quarter-twist belt drive; an idler guide pulley must be used if motion is to be in both directions.

Figure 17–3 shows a flat-belt drive with out-of-plane pulleys. The shafts need not be at right angles as in this case. Note the top view of the drive in Fig. 17–3. The pulleys must be positioned so that the belt leaves each pulley in the midplane of the other pulley face. Other arrangements may require guide pulleys to achieve this condition.

Another advantage of flat belts is shown in Fig. 17–4, where clutching action is obtained by shifting the belt from a loose to a tight or driven pulley.

Figure 17–5 shows two variable-speed drives. The drive in Fig. 17–5*a* is commonly used only for flat belts. The drive of Fig. 17–5*b* can also be used for V belts and round belts by using grooved sheaves.

Flat belts are made of urethane and also of rubber-impregnated fabric reinforced with steel wire or nylon cords to take the tension load. One or both surfaces may have a friction surface coating. Flat belts are quiet, they are efficient at high speeds, and they can transmit large amounts of power over long center distances. Usually, flat belting is purchased by the roll and cut and the ends are joined using special kits furnished by the manufacturer. Two or more flat belts running side by side, instead of a single wide belt, are often used to form a conveying system.

A V belt is made of fabric and cord, usually cotton, rayon, or nylon, and impregnated with rubber. In contrast with flat belts, V belts are used with similar sheaves and at shorter center distances. V belts are slightly less efficient than flat belts, but a number of them can be used on a single sheave, thus making a multiple drive. V belts are made only in certain lengths and have no joints.

Timing belts are made of rubberized fabric and steel wire and have teeth which fit into grooves cut on the periphery of the sprockets. The timing belt does not stretch or slip and consequently transmits power at a constant angular-velocity ratio. The fact that the belt is toothed provides several advantages over ordinary belting. One of these is that

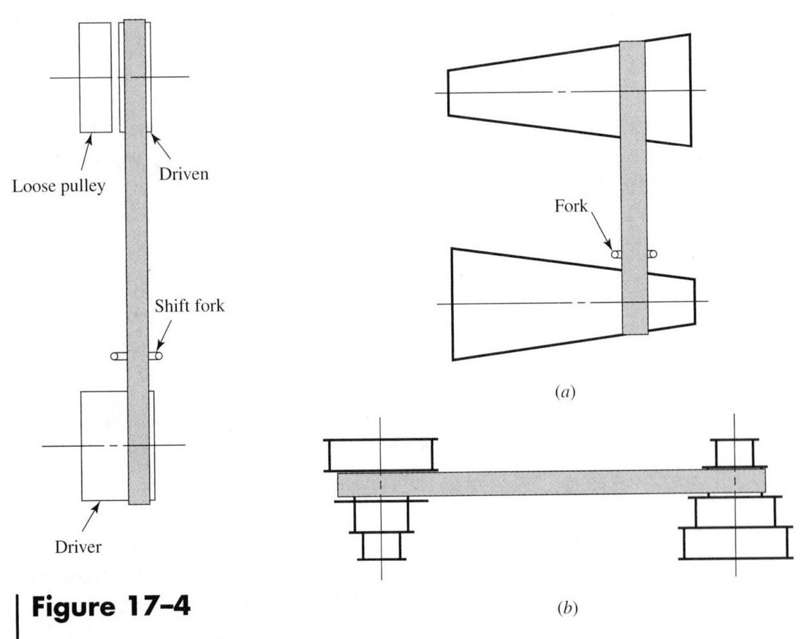

Figure 17–4

This drive eliminates the need for a clutch. Flat belt can be shifted left or right by use of a fork.

Figure 17–5

Variable-speed belt drives.

no initial tension is necessary, so that fixed-center drives may be used. Another is the elimination of the restriction on speeds; the teeth make it possible to run at nearly any speed, slow or fast. Disadvantages are the first cost of the belt, the necessity of grooving the sprockets, and the attendant dynamic fluctuations caused at the belt-tooth meshing frequency.

17–2 Flat- and Round-Belt Drives

Modern flat-belt drives consist of a strong elastic core surrounded by an elastomer; these drives have distinct advantages over gear drives or V-belt drives. A flat-belt drive has an efficiency of about 98 percent, which is about the same as for a gear drive. On the other hand, the efficiency of a V-belt drive ranges from about 70 to 96 percent.[1] Flat-belt drives produce very little noise and absorb more torsional vibration from the system than either V-belt or gear drives.

When an open-belt drive (Fig. 17–1) is used, the contact angles are found to be

$$\begin{aligned} \theta_d &= \pi - 2\sin^{-1}\frac{D-d}{2C} \\ \theta_D &= \pi + 2\sin^{-1}\frac{D-d}{2C} \end{aligned} \tag{17–1}$$

where $D =$ diameter of large pulley
$d =$ diameter of small pulley
$C =$ center distance
$\theta =$ angle of contact

The length of the belt is found by summing the two arc lengths with twice the distance between the beginning and end of contact. The result is

$$L = [4C^2 - (D-d)^2]^{1/2} + \frac{1}{2}(D\theta_D + d\theta_d) \tag{17–2}$$

A similar set of equations can be derived for the crossed belts of Fig. 17–2*b*. For these the angle of wrap is the same for both pulleys and is

$$\theta = \pi + 2\sin^{-1}\frac{D+d}{2C} \tag{17–3}$$

The belt length for crossed belts is found to be

$$L = [4C^2 - (D+d)^2]^{1/2} + \frac{\theta}{2}(D+d) \tag{17–4}$$

Firbank[2] explains flat-belt-drive theory in the following way. A change in belt tension due to friction forces between the belt and pulley will cause the belt to elongate or contract and move relative to the surface of the pulley. This motion is caused by *elastic creep* and is associated with sliding friction as opposed to static friction. The action at the driving pulley, through that portion of the angle of contact that is actually transmitting power, is such that the belt moves more slowly than the surface speed of the pulley because of the elastic creep. The angle of contact is made up of the *effective arc,* through which

1. A. W. Wallin, "Efficiency of Synchronous Belts and V-Belts," *Proc. Nat. Conf. Power Transmission,* vol. 5, Illinois Institute of Technology, Chicago, Nov. 7–9, 1978, pp. 265–271.

2. T. C. Firbank, *Mechanics of the Flat Belt Drive,* ASME paper no. 72-PTG-21.

power is transmitted, and the *idle arc*. For the driving pulley the belt first contacts the pulley with a *tight-side tension* F_1 and a velocity V_1, which is the same as the surface velocity of the pulley. The belt then passes through the idle arc with no change in F_1 or V_1. Then creep or sliding contact begins, and the belt tension changes in accordance with the friction forces. At the end of the effective arc the belt leaves the pulley with a *loose-side tension* F_2 and a reduced speed V_2.

Firbank has used this theory to express the mechanics of flat-belt drives in mathematical form and has verified the results by experiment. His observations include the finding that substantially more power is transmitted by static friction than sliding friction. He also found that the coefficient of friction for a belt having a nylon core and leather surface was typically 0.7, but that it could be raised to 0.9 by employing special surface finishes.

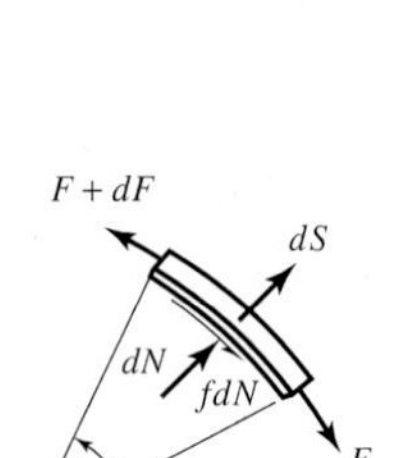

Figure 17–6

Free body of an infinitesimal element of a flat belt in contact with a pulley.

Our model will assume that the friction force on the belt is proportional to the normal pressure along the arc of contact. We seek first a relationship between the tight side tension and slack side tension, similar to that of band brakes but incorporating the consequences of movement, that is, centrifugal tension in the belt. In Fig. 17–6 we see a free body of a small segment of the belt. The differential force dS is due to centrifugal force, dN is the normal force between the belt and pulley, and $f\,dN$ is the shearing traction due to friction at the point of slip. The belt width is b and the thickness is t. The belt mass m is expressed per unit length. The centrifugal force dS can be expressed as

$$dS = (mr\,d\theta)r\omega^2 = mr^2\omega^2 d\theta = mV^2 d\theta = F_c\,d\theta \tag{a}$$

where V is the belt speed. Summing forces radially gives

$$\sum F_r = (F + dF)\frac{d\theta}{2} + F\frac{d\theta}{2} - dN - dS = 0$$

from which

$$dN = f\,d\theta - dS \tag{b}$$

Summing forces tangentially gives

$$\sum F_t = -f\,dN - F + (F + dF) = 0$$

from which, incorporating Eq. (*b*), we obtain

$$dF = f\,dN = fF\,d\theta - dS = fF\,d\theta - fmr^2\omega^2\,d\theta$$

or

$$\frac{dF}{d\theta} - fF = -fmr^2\omega^2 \tag{c}$$

The solution to this nonhomogeneous first-order linear differential equation is

$$F = A\exp(f\theta) + mr^2\omega^2 \tag{d}$$

where A is an arbitrary constant. The boundary condition that F at $\theta = 0$ equals F_2 gives $A = F_2 - mr^2\omega^2$. The solution is

$$F = (F_2 - mr^2\omega^2)\exp(f\theta) + mr^2\omega^2 \tag{17–5}$$

At the end of the angle of wrap ϕ

$$F|_{\theta=\phi} = F_1 = (F_2 - mr^2\omega^2)\exp(f\phi) - mr^2\omega^2 \tag{17–6}$$

Now we can write

$$\frac{F_1 - mr^2\omega^2}{F_2 - mr^2\omega^2} = \frac{F_1 - F_c}{F_2 - F_c} = \exp(f\phi) \tag{17–7}$$

where, from Eq. (a), $F_c = mr^2\omega^2$. It is also useful that Eq. (17–7) can be written as

$$F_1 - F_2 = (F_1 - F_c)\frac{\exp(f\phi) - 1}{\exp(f\phi)} \tag{17–8}$$

Now F_c is found as follows:

$$V = \pi\, dn/12 \text{ ft/min}$$

The weight w of a foot of belt is given in terms of the weight density γ in lbf/in^3 as $w = 12\gamma bt$ lbf/ft and F_c is written as

$$F_c = \frac{w}{g}\left(\frac{V}{60}\right)^2 = \frac{w}{32.2}\left(\frac{V}{60}\right)^2 \tag{e}$$

Figure 17–7 shows a free body of a pulley and part of the belt. The tight side tension F_1 and the loose side tension F_2 have the following additive components:

$$F_1 = F_i + F_c + \Delta F' = F_i + F_c + T/D \tag{f}$$

$$F_2 = F_i + F_c - \Delta F' = F_i + F_c - T/D \tag{g}$$

where $F_i =$ initial tension

$F_c =$ hoop tension due to centrifugal force

$\Delta F' =$ tension due to the transmitted torque T

$D =$ diameter of the pulley

The difference between F_1 and F_2 is related to the pulley torque. Subtracting Eq. (g) from Eq. (f) gives

$$\boxed{F_1 - F_2 = \frac{2T}{D} = \frac{T}{D/2}} \tag{h}$$

Adding Eqs. (f) and (g) gives

$$F_1 + F_2 = 2F_i + 2F_c$$

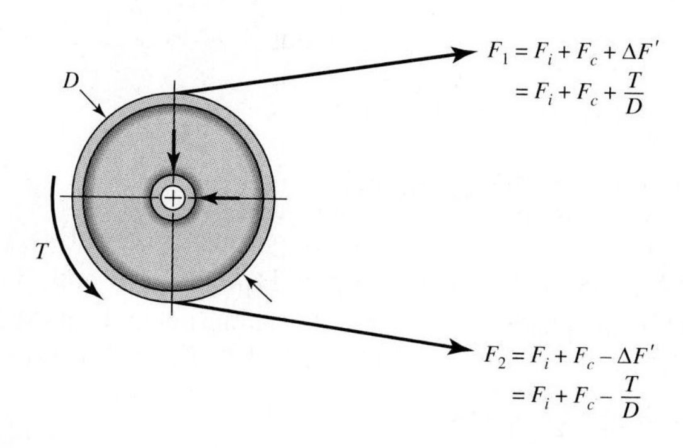

Figure 17–7

Forces and torques on a pulley.

from which

$$F_i = \frac{F_1 + F_2}{2} - F_c \tag{i}$$

Dividing Eq. (*i*) by Eq. (*h*) gives, after considerable manipulation,

$$\frac{F_i}{T/D} = \frac{(F_1 + F_2)/2 - F_c}{(F_1 - F_2)/2} = \frac{F_1 + F_2 - 2F_c}{F_1 - F_2} = \frac{(F_1 + F_c) + (F_2 - F_c)}{(F_1 - F_c) - (F_2 - F_c)}$$

$$= \frac{(F_1 - F_c)/(F_2 - F_c) + 1}{(F_1 - F_c)/(F_2 - F_c) - 1} = \frac{\exp(f\phi) + 1}{\exp(f\phi) - 1}$$

from which

$$F_i = \frac{T}{D}\frac{\exp(f\phi) + 1}{\exp(f\phi) - 1} \tag{17–9}$$

Equation (17–9) give us a fundamental insight into flat belting. If F_i equals zero, then T equals zero: no initial tension, no torque transmitted. The torque is in proportion to the initial tension. This means that if there is to be a satisfactory flat-belt drive, the initial tension must be (1) provided, (2) sustained, (3) in the proper amount, and (4) maintained by routine inspection.

From Eq. (*f*), incorporating Eq. (17–9) gives

$$F_1 = F_i + F_c + \frac{T}{D} = F_c + F_i + F_i\frac{\exp(f\phi) - 1}{\exp(f\phi) + 1}$$

$$= F_c + \frac{F_i[\exp(f\phi) + 1] + F_i[\exp(f\phi) - 1]}{\exp(f\phi) + 1}$$

$$F_1 = F_c + F_i\frac{2\exp(f\phi)}{\exp(f\phi) + 1} \tag{17–10}$$

From Eq. (*g*), incorporating Eq. (17–9) gives

$$F_2 = F_i + F_c - \frac{T}{D} = F_c + F_i - F_i\frac{\exp(f\phi) - 1}{\exp(f\phi) + 1}$$

$$= F_c + \frac{F_i[\exp(f\phi) + 1] - F_i[\exp(f\phi) - 1]}{\exp(f\phi) + 1}$$

$$F_2 = F_c + F_i\frac{2}{\exp(f\phi) + 1} \tag{17–11}$$

Equation (17–7) is called the belting equation, but Eqs. (17–9), (17–10), and (17–11) reveal how belting works. We plot Eqs. (17–10) and (17–11) as shown in Fig. 17–8 against F_i as abscissa. The initial tension needs to be sufficient so that the difference between the F_1 and F_2 curve is $2T/D$. With no torque transmitted, the least possible belt tension is $F_1 = F_2 = F_c$.

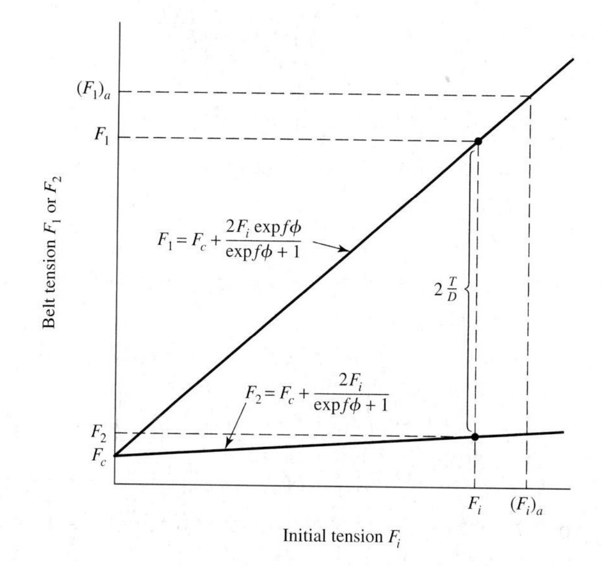

Figure 17–8

Plot of initial tension F_i against belt tension F_1 or F_2 showing the intercept F_c, the equations of the loci, and where $2T/D$ is to be found.

The transmitted horsepower is given by

$$H = \frac{(F_1 - F_2)V}{33\,000} \tag{j}$$

Manufacturers provide specifications for their belts which include allowable tension F_a (or stress σ_{all}), the tension being expressed in units of force per unit width. Belt life is usually several years. The severity of flexing at the pulley and its effect on life is reflected in a pulley correction factor C_p. Speeds in excess of 600 ft/min and their effect on life is reflected in a velocity correction factor C_v. For polyamide and urethane belts use $C_v = 1$. For leather belts see Fig. 17–9. A service factor K_s is used for excursions of load from nominal, applied to the nominal power as $H_d = H_{\text{nom}} K_s n_d$, the design factor for exigencies. These effects are incorporated as follows:

$$(F_1)_a = bF_aC_pC_v \tag{17–12}$$

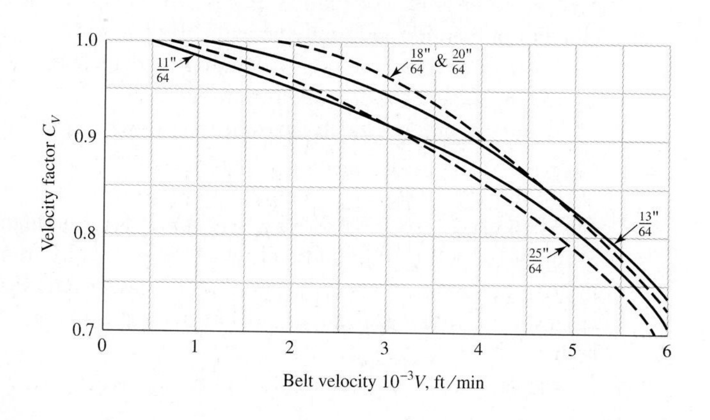

Figure 17–9

Velocity correction factor C_v for leather belts. *(Data source: Machinery's Handbook, 20th ed., Industrial Press, New York, 1976, p. 1047.)*

where $(F_1)_a$ = allowable largest tension, lbf
b = belt width, in
F_a = manufacturer's allowed tension, lbf/in
C_p = pulley correction factor
C_v = velocity correction factor

The steps in analyzing a flat-belt drive can include

1 Find $\exp(f\phi)$ from belt-drive geometry and friction
2 From belt geometry and speed find F_c
3 From $T = 63\,025 H_{\text{nom}} K_s n_d / n$ find necessary torque
4 From torque T find the necessary $(F_1)_a - F_2 = 2T/D$
5 Find F_2 from $(F_1)_a - [(F_1)_a - F_2]$
6 From Eq. (i) find the necessary initial tension F_i
7 Check the friction development, $f' < f$. Use Eq. (17–7) solved for f':

$$f' = \frac{1}{\phi} \ln \frac{(F_1)_a - F_c}{F_2 - F_c}$$

8 Find the factor of safety from $\text{fos} = H_a/(H_{\text{nom}} K_s)$

It is unfortunate that many of the available data on belting are from sources in which they are presented in a very simplistic manner. These sources use a variety of charts, nomographs, and tables to enable someone who knows nothing about belting to apply them. Little, if any, computation is needed for such a person to obtain valid results. Since a basic understanding of the process, in many cases, is lacking, there is no way this person can vary the steps in the process to obtain a better design.

Incorporating the available belt-drive data into a form which provides a good understanding of belt mechanics involved certain adjustments in the data. Because of this, the results from the analysis presented here will not correspond exactly with those of the sources from which they were obtained.

A moderate variety of belt materials, with some of their properties, are listed in Table 17–2. These are sufficient for solving a large variety of design and analysis problems. The design equation to be used is Eq. (j).

The values given in Table 17–2 for the allowable belt tension are based on a belt speed of 600 ft/min. For higher speeds, use Fig. 17–9 to obtain C_V values for leather belts. For polyamide and urethane belts, use $C_V = 1.0$.

The service factors K_S for V-belt drives, given in Table 17–15 in Sec. 17–3, are also recommended here for flat- and round-belt drives.

Minimum pulley sizes for the various belts are listed in Tables 17–2 and 17–3. The pulley correction factor accounts for the amount of bending or flexing of the belt and how this affects the life of the belt. For this reason it is dependent on the size and material of the belt used. Use Table 17–4. Use $C_P = 1.0$ for urethane belts.

Flat-belt pulleys should be crowned to keep belts from running off the pulleys. If only one pulley is crowned, it should be the larger one. Both pulleys must be crowned whenever the pulley axes are not in a horizontal position. Use Table 17–5 for the crown height.

An analysis (skill 1) example will be useful here.

Table 17–2

Properties of Some Flat- and Round-Belt Materials. (Diameter = d, thickness = t, width = w)

Material	Specification	Size, in	Minimum Pulley Diameter, in	Allowable Tension per Unit Width at 600 ft/min, lb/in	Specific Weight, lb/in³	Coefficient of Friction
Leather	1 ply	$t = \frac{11}{64}$	3	30	0.035–0.045	0.4
		$t = \frac{13}{64}$	$3\frac{1}{2}$	33	0.035–0.045	0.4
	2 ply	$t = \frac{18}{64}$	$4\frac{1}{2}$	41	0.035–0.045	0.4
		$t = \frac{20}{64}$	6[a]	50	0.035–0.045	0.4
		$t = \frac{23}{64}$	9[a]	60	0.035–0.045	0.4
Polyamide[b]	F–0[c]	$t = 0.03$	0.60	10	0.035	0.5
	F–1[c]	$t = 0.05$	1.0	35	0.035	0.5
	F–2[c]	$t = 0.07$	2.4	60	0.051	0.5
	A–2[c]	$t = 0.11$	2.4	60	0.037	0.8
	A–3[c]	$t = 0.13$	4.3	100	0.042	0.8
	A–4[c]	$t = 0.20$	9.5	175	0.039	0.8
	A–5[c]	$t = 0.25$	13.5	275	0.039	0.8
Urethane[d]	$w = 0.50$	$t = 0.062$	See	5.2[e]	0.038–0.045	0.7
	$w = 0.75$	$t = 0.078$	Table	9.8[e]	0.038–0.045	0.7
	$w = 1.25$	$t = 0.090$	17–3	18.9[e]	0.038–0.045	0.7
	Round	$d = \frac{1}{4}$	See	8.3[e]	0.038–0.045	0.7
		$d = \frac{3}{8}$	Table	18.6[e]	0.038–0.045	0.7
		$d = \frac{1}{2}$	17–3	33.0[e]	0.038–0.045	0.7
		$d = \frac{3}{4}$		74.3[e]	0.038–0.045	0.7

[a]Add 2 in to pulley size for belts 8 in wide or more.

[b] *Source: Habasit Engineering Manual,* Habasit Belting, Inc., Chamblee (Atlanta), Ga.

[c] Friction cover of acrylonitrile-butadiene rubber on both sides.

[d] *Source:* Eagle Belting Co., Des Plaines, Ill.

[e]At 6% elongation; 12% is maximum allowable value.

Table 17–3

Minimum Pulley Sizes for Flat and Round Urethane Belts. (Listed are the Pulley Diameters in Inches)

Source: Eagle Belting Co., Des Plaines, Ill.

Belt Style	Belt Size, in	Ratio of Pulley Speed to Belt Length, r/(ft · min)		
		Up to 250	250 to 499	500 to 1000
Flat	0.50 × 0.062	0.38	0.44	0.50
	0.75 × 0.078	0.50	0.63	0.75
	1.25 × 0.090	0.50	0.63	0.75
Round	$\frac{1}{4}$	1.50	1.75	2.00
	$\frac{3}{8}$	2.25	2.62	3.00
	$\frac{1}{2}$	3.00	3.50	4.00
	$\frac{3}{4}$	5.00	6.00	7.00

Table 17–4

Pulley Correction Factor C_P for Flat Belts*

Material	Small-Pulley Diameter, in					
	1.6 to 4	4.5 to 8	9 to 12.5	14, 16	18 to 31.5	Over 31.5
Leather	0.5	0.6	0.7	0.8	0.9	1.0
Polyamide, F–0	0.95	1.0	1.0	1.0	1.0	1.0
F–1	0.70	0.92	0.95	1.0	1.0	1.0
F–2	0.73	0.86	0.96	1.0	1.0	1.0
A–2	0.73	0.86	0.96	1.0	1.0	1.0
A–3	—	0.70	0.87	0.94	0.96	1.0
A–4	—	—	0.71	0.80	0.85	0.92
A–5	—	—	—	0.72	0.77	0.91

*Average values of C_P for the given ranges were approximated from curves in the *Habasit Engineering Manual*, Habasit Belting, Inc., Chamblee (Atlanta), Ga.

Table 17–5

Crown Height and ISO Pulley Diameters for Flat Belts*

ISO Pulley Diameter, in	Crown Height, in	ISO Pulley Diameter, in	Crown Height, in	
			$w \le 10$ in	$w > 10$ in
1.6, 2, 2.5	0.012	12.5, 14	0.03	0.03
2.8, 3.15	0.012	12.5, 14	0.04	0.04
3.55, 4, 4.5	0.012	22.4, 25, 28	0.05	0.05
5, 5.6	0.016	31.5, 35.5	0.05	0.06
6.3, 7.1	0.020	40	0.05	0.06
8, 9	0.024	45, 50, 56	0.06	0.08
10, 11.2	0.030	63, 71, 80	0.07	0.10

*Crown should be rounded, not angled; maximum roughness is R_a = AA 63 μin.

EXAMPLE 17–1

A polyamide A-3 flat belt 6 in wide is used to transmit 15 hp under light shock conditions where $K_s = 1.25$, and a factor of safety equal to or greater than 1.1 is appropriate. The pulley rotational axes are parallel and in the horizontal plane. The shafts are 8 ft apart. The 6-in driving pulley rotates at 1750 r/min in such a way that the loose side is on top. The driven pulley is 18 in in diameter. See Fig. 17–10. The factor of safety is for unquantifiable exigencies.

(a) Estimate the centrifugal tension F_c and the torque T.

(b) Estimate the allowable F_1, F_2 and allowable power H_a.

(c) Estimate the factor of safety. Is it satisfactory?

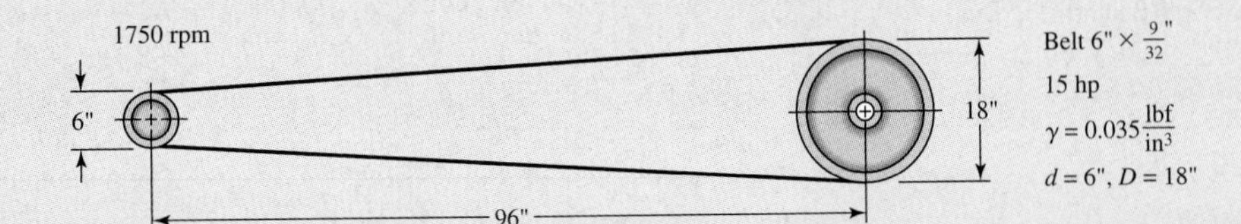

Figure 17–10

The flat-belt drive of Ex. 17–1.

Solution (*a*) Eq. (17–1) $\theta_d = \pi - 2\sin^{-1}\left[\frac{18-6}{2(8)12}\right] = 3.0165$ rad

$$\exp(f\phi) = \exp[0.8(3.0165)] = 11.17$$

$$V = \pi(6)1750/12 = 2749 \text{ ft/min}$$

Table 17–2 $w = 12\gamma bt = 12(0.042)6(0.13) = 0.393$ lb/ft

Answer Eq. (*e*) $F_c = \dfrac{wV^2}{g} = \dfrac{0.393}{32.174}\left(\dfrac{2749}{60}\right)^2 = 25.6$ lb

$$T = \frac{63\,025\,H_{\text{nom}}K_s n_d}{n} = \frac{63\,025(15)1.25(1.1)}{1750}$$

Answer $= 742.8$ in · lb

(*b*) The necessary $(F_1)_a - F_2$ to transmit the torque T is

$$(F_1)_a - F_2 = \frac{2T}{d} = \frac{2(742.8)}{6} = 247.6 \text{ lb}$$

From Eq. (17–12) the allowable largest belt tension $(F_1)_a$ is

Answer $(F_1)_a = bF_aC_pC_v = 6(100)0.70(1) = 420$ lb

then

Answer $F_2 = (F_1)_a - [(F_1)_a - F_2] = 420 - 247.6 = 172.4$ lb

and from Eq. (*i*)

$$F_i = \frac{420 + 172.4}{2} - 25.6 = 270.6 \text{ lb}$$

Answer The combination $(F_1)_a$, F_2, and F_i will transmit the design power of $15(1.25)(1.1) = 20.6$ hp and protect the belt. We check the friction development by solving Eq. (17–7) for f':

$$f' = \frac{1}{\phi}\ln\frac{(F_1)_a - F_c}{F_2 - F_c} = \frac{1}{3.0165}\ln\frac{420 - 25.6}{172.4 - 25.6} = 0.33$$

Since $f' < f$, that is, $0.33 < 0.80$, there is no danger of slipping.

(*c*)

Answer $\text{fos} = \dfrac{H}{H_{\text{nom}}K_s} = \dfrac{20.6}{15(1.25)} = 1.1$ (as expected)

Answer The belt is satisfactory and the maximum allowable belt tension exists. If the initial tension is maintained, the capacity is the design power of 20.6 hp.

An integral part of any belt design is providing a method for imposing the initial tension F_i as well as maintaining it. To do otherwise is to lose control of your design and the consequences which follow. If there is a "law of friction belting," it is *no initial tension, no power transmitted;* if there is insufficient initial tension, slipping ensues.

Initial tension is the key to the functioning of the flat belt as intended. There are ways of controlling initial tension. One way is to place the motor and drive pulley on a pivoted mounting plate so that the weight of the motor, pulley, and mounting plate and a share of the belt weight induces the correct initial tension and maintains it. A second way is use of a spring-loaded idler pulley, adjusted to the same task. Both of these methods accommodate to temporary or permanent belt stretch. See Fig. 17–11.

Because flat belts were used for long center-to-center distances, the weight of the belt itself can provide the initial tension. The static belt deflects to an approximate catenary curve, and the dip from a straight belt can be measured against a stretched music wire. This provides a way of measuring and adjusting the dip. From catenary theory the dip is related to the initial tension by

$$d = \frac{12L^2w}{8F_i} = \frac{3L^2w}{2F_i} \tag{17–13}$$

where d = dip, in
L = center-to-center distance, ft
w = weight per foot of the belt, lbf/ft
F_i = initial tension, lbf

In Ex. 17–1 the dip corresponding to a 270.6-lb initial tension is

$$d = \frac{3(8^2)0.393}{2(270.6)} = 0.14 \text{ in}$$

A decision set for a flat belt can be

- Function: power, speed, durability, reduction, service factor, C

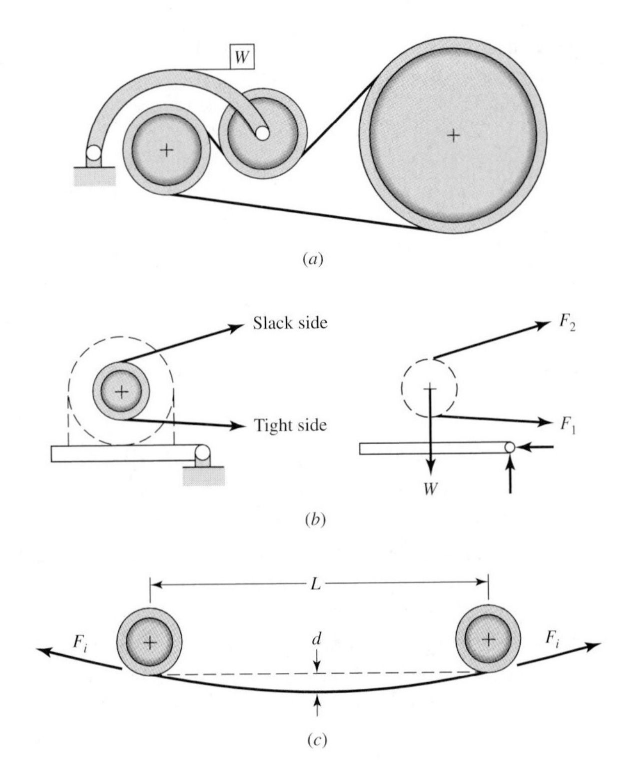

Figure 17–11

Belt-tensioning schemes.
(a) Weighted idler pulley.
(b) Pivoted motor mount.
(c) Catenary-induced tension.

- Design factor: n_d
- Initial tension maintenance
- Belt material
- Drive geometry, d, D
- Belt thickness: t
- Belt width: b

Depending on the problem, some or all of the last four could be design variables. Belt cross-sectional area is really the design decision, but available belt thicknesses and widths are discrete choices. Available dimensions are found in suppliers' catalogs.

EXAMPLE 17–2 Design a flat-belt drive to connect horizontal shafts on 16-ft centers. The velocity ratio is to be 2.25:1. The angular speed of the small driving pulley is 860 r/min, and the nominal power transmission is to be 60 hp under very light shock.

Solution

- Function: $H_{\text{nom}} = 60$ hp, 860 r/min, 2.25:1 ratio, $K_s = 1.15$, $C = 16$ ft
- Design factor: $n_d = 1.05$
- Initial tension maintenance: catenary
- Belt material: polyamide
- Drive geometry, d, D
- Belt thickness: t
- Belt width: b

The last four could be design variables. Let's make a few more a priori decisions.

Decision $d = 16$ in, $D = 2.25d = 2.25(16) = 36$ in.

Decision Use polyamide A-3 belt; therefore $t = 0.13$ in.

Now there is one design decision remaining to be made, the belt width b.

Table 17–2 $\gamma = 0.042$ lbf/in^3 $f = 0.8$ $F_a = 100$ lb/in at 600 r/min

Table 17–4 $C_p = 0.94$ $C_v = 1$

Eq. (17–12) $F_{1a} = b(100)0.94(1) = 94.0b$ lb

$$H_d = H_{\text{nom}} K_s n_d = 60(1.15)1.05 = 72.5 \text{ hp}$$

$$T = \frac{63\,025 H_d}{n} = \frac{63\,025(72.5)}{860} = 5313 \text{ in} \cdot \text{lb for } H_d \text{ power}$$

Estimate $\exp(f\phi)$ for full friction development:

$$\phi = \theta_d = \pi - 2\sin^{-1}\frac{36-16}{2(16)12} = 3.037\,399 \text{ rad}$$

$$\exp(f\phi) = \exp[0.80(3.037\,379)] = 11.38$$

Estimate centrifugal tension F_c in terms of belt width b:

$$w = 12\gamma bt = 12(0.052)b(0.13) = 0.0655b \text{ lb/ft}$$

$$V = \pi dn/12 = \pi(16)860/12 = 3602 \text{ ft/min}$$

$$F_c = \frac{wV^2}{g} = \frac{0.0655b}{32.174}\left(\frac{3602}{60}\right)^2 = 7.34b \text{ lb}$$

For design conditions, that is, at H_d power level, using Eq. (h) gives

$$(F_1)_a - F_2 = 2T/d = 2(5313)/16 = 664 \text{ lb}$$

$$F_2 = (F_1)_a - [(F_1)_a - F_2] = 94b - 664 \text{ lb}$$

Using Eq. (1),

$$F_i = \frac{(F_1)_a + F_2}{2} - F_c = \frac{94b + 94b - 664}{2} - 7.34b = 86.7b - 332 \text{ lb}$$

Place friction development at its highest level using Eq. (17–7):

$$f\phi = \ln\frac{(F_1)_a - F_c}{F_2 - F_c} = \ln\frac{94b - 7.34b}{86.7b - 664} = \ln\frac{86.7b}{86.7b - 664}$$

Solving the preceding equation for belt width b at which friction is fully developed gives

$$b = \frac{664}{86.7}\frac{\exp(f\phi)}{\exp(f\phi) - 1} = \frac{664}{86.7}\frac{11.38}{11.38 - 1} = 8.40 \text{ in}$$

A belt width greater than 8.40 in will develop friction less than $f = 0.80$. The next available larger size is 10-in width.

Decision

Use 10-in-wide belt.
It follows that for a 10-in-wide belt

$$F_c = 7.34(10) = 73.4 \text{ lb}$$

$$(F_1)_a = 94(10) = 940 \text{ lb}$$

$$F_2 = 94(10) - 664 = 276 \text{ lb}$$

$$F_i = 86.7(10) - 332 = 535 \text{ lb}$$

$$(F_1)_a - F_2 = 2T/d = 664 \text{ lb}$$

The transmitted power is

$$H_t = \frac{[(F_1)_a - F_2]V}{33\,000} = \frac{664(3602)}{33\,000} = 72.5 \text{ hp}$$

and the level of friction development f' is

$$f' = \frac{1}{\phi}\ln\frac{(F_1)_a - F_c}{F_2 - F_c} = \frac{1}{3.037}\ln\frac{940 - 73.4}{276 - 73.4} = 0.479$$

which is satisfactory. Had a 9-in belt width been available, the analysis would show $(F_1)_a = 846$ lb, $F_2 = 182$ lb, $F_i = 448$ lb, and $f' = 0.63$. With a figure of merit available reflecting cost, thicker belts (A-4 or A-5) could be examined to ascertain which of the satisfactory alternatives is best. From Eq. (17–13) the catenary dip is

$$d = \frac{3L^2w}{2F_i} = \frac{3(15^2)0.0655(10)}{2(535)} = 0.413 \text{ in}$$

Figure 17–12 illustrates the variation of flexible flat-belt tensions at some cardinal points during a belt pass.

Flat Metal Belts

Thin flat metal belts with their attendant strength and geometric stability could not be fabricated until laser welding and thin rolling technology made possible belts as thin as 0.002 in and as narrow as 0.026 in. The introduction of perforations allows no-slip applications. Thin metal belts exhibit

- High strength-to-weight ratio
- Dimensional stability
- Accurate timing
- Usefulness to temperatures up to 700°F
- Good electrical and thermal conduction properties

In addition, stainless steel alloys offer "inert," nonabsorbent belts suitable to hostile (corrosive) environments, which can be made sterile for food and pharmaceutical applications.

Thin metal belts can be classified as friction drive, timing or positioning drives, or tape drives. Among friction drives are plain, metal-coated, and perforated belts. Crowned pulleys are used to compensate for tracking errors.

Figure 17–13 shows a thin flat metal belt with the tight tension F_1 and the slack side tension F_2 revealed. The relationship between F_1 and F_2 and the driving torque T is the same as in Eq. (*h*). Equations (17–9), (17–10), and (17–11) also apply. The largest allowable tension, as in Eq. (17–12), is posed in terms of stress in metal belts. A bending stress is created by making the belt conform to the pulley, and its tensile magnitude σ_b is given by

$$\sigma_b = \frac{Et}{(1-\nu^2)D} = \frac{E}{(1-\nu^2)(D/t)} \tag{17–14}$$

Figure 17–12

Flat-belt tensions.

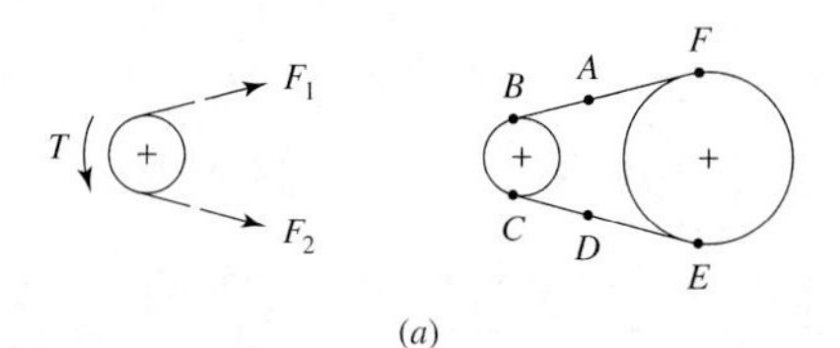

(*a*)

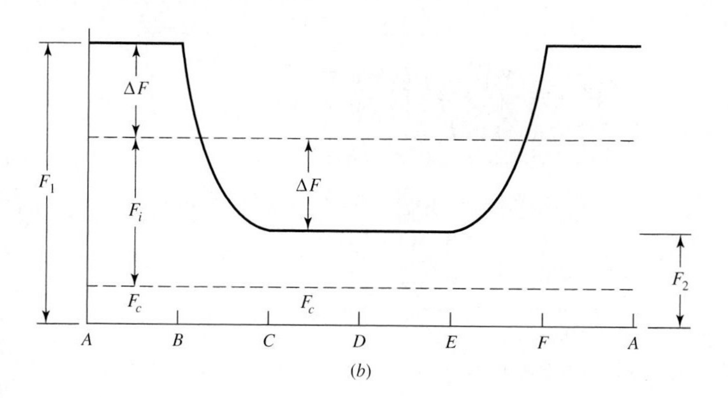

(*b*)

Figure 17–13

Metal-belt tensions and torques.

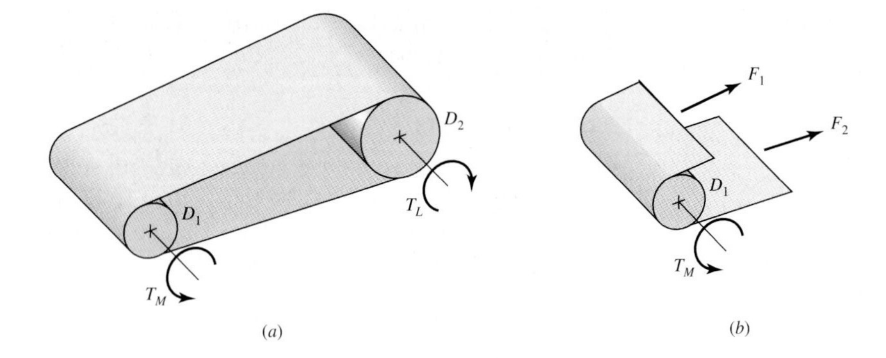

where E = Young's modulus

t = belt thickness

ν = Poisson's ratio

D = pulley diameter

The tensile stresses $(\sigma)_1$ and $(\sigma)_2$ imposed by the belt tensions F_1 and F_2 are

$$(\sigma)_1 = F_1/(bt) \qquad \text{and} \qquad (\sigma)_2 = F_2/(bt)$$

The largest tensile stress is $(\sigma_b)_1 + F_1/(bt)$ and the smallest is $(\sigma_b)_2 + F_2/(bt)$. During a belt pass both levels of stress appear.

Although the belts are of simple geometry, the method of Marin is not used because the condition of the butt weldment (to form the loop) is not accurately known, and the testing of coupons is difficult. The belts are run to failure on two equal-sized pulleys. Information concerning fatigue life, as shown in Table 17–6, is obtainable. Tables 17–7 and 17–8 give additional information.

Table 17–6 shows metal belt life expectancies for a stainless steel belt. Using Eq. (17–14) with $E = 28$ Mpsi and $\nu = 0.29$, the bending stresses corresponding to the four entries of the table are 48 914, 76 428, 91 805, and 152 855 psi. Using a natural log transformation on stress and passes, the regression line ($r = -0.96$) is

$$\sigma = 14\,169\,982N^{-0.407} = 14.17(10^6)N_p^{-0.407} \tag{17–15}$$

where N_p is the number of belt passes.

Table 17–6

Belt Life for Stainless Steel Friction Drives*

$\frac{D}{t}$	Belt Passes
625	$\geq 10^6$
400	$0.500 \cdot 10^6$
333	$0.165 \cdot 10^6$
200	$0.085 \cdot 10^6$

*Data courtesy of Belt Technologies, Agawam, Mass.

Table 17–7

Minimum Pulley Diameter*

Belt Thickness, in	Minimum Pulley Diameter, in
0.002	1.2
0.003	1.8
0.005	3.0
0.008	5.0
0.010	6.0
0.015	10
0.020	12.5
0.040	25.0

*Data courtesy of Belt Technologies, Agawam, Mass.

Table 17–8

Typical Material Properties, Metal Belts*

Alloy	Yield Strength, kpsi	Young's Modulus, Mpsi	Poisson's Ratio
301 or 302 stainless steel	175	28	0.285
BeCu	170	17	0.220
1075 or 1095 carbon steel	230	30	0.287
Titanium	150	15	—
Inconel	160	30	0.284

*Data courtesy of Belt Technologies, Agawam, Mass.

The selection of a metal flat belt can consist of the following steps:

1 Find $\exp f\phi$ from geometry and friction

2 Find endurance strength

$$S_f = 14.17(10^6)N_p^{-0.407} \qquad \text{301, 302 stainless}$$

$$S_f = S_y/3 \qquad \text{others}$$

3 Allowable tension

$$F_{1a} = \left[S_f - \frac{Et}{(1-\nu^2)D}\right]tb = ab$$

4 $\Delta F = 2T/D$

5 $F_2 = F_{1a} - \Delta F = ab - \Delta F$

6 $F_i = \dfrac{F_{1a} + F_2}{2} = \dfrac{ab + ab - \Delta F}{2} = ab - \dfrac{\Delta F}{2}$

7 $b_{\min} = \dfrac{\Delta F}{a}\,\dfrac{\exp f\phi}{\exp f\phi - 1}$

8 Choose $b > b_{\min}$, $F_1 = ab$, $F_2 = ab - \Delta F$, $F_i = ab - \Delta F/2$, $T = \Delta F D/2$

9 Check frictional development f':

$$f' = \frac{1}{\phi} \ln \frac{F_1}{F_2} \qquad f' < f$$

EXAMPLE 17–3

A friction-drive stainless steel metal belt runs over two 4-in metal pulleys ($f = 0.35$). The belt thickness is to be 0.003 in. For a life exceeding 10^6 belt passes with smooth torque ($K_s = 1$), select the belt if (*a*) the torque is to be 30 in · lb, and (*b*) find the initial tension F_i.

Solution

(*a*) $\theta_d = \pi$, therefore $\exp(0.35\pi) = 3.00$. From step 2,

$$(S_f)_{10^6} = 14.17(10^6)(10^6)^{-0.407} = 51\,212 \text{ psi}$$

From steps 3, 4, 5, and 6,

$$F_{1a} = \left[51\,212 - \frac{28(10^6)0.003}{(1 - 0.285^2)4}\right] 0.003b = 85.1b \text{ lb}$$

$$\Delta F = 2T/D = 2(30)/4 = 15 \text{ lb}$$

$$F_2 = F_{1a} - \Delta F = 85.1b - 15 \text{ lb}$$

$$F_i = \frac{F_{1a} + F_2}{2} = \frac{85.1b + 15}{2} \text{ lb}$$

From step 7,

$$b_{\min} = \frac{\Delta F}{a} \frac{\exp f\phi}{\exp f\phi - 1} = \frac{15}{85.1} \frac{3.00}{3.00 - 1} = 0.264 \text{ in}$$

Decision

Select an available 0.75-in-wide belt 0.003 in thick.

$$F_1 = 85.1(0.75) = 63.8 \text{ lb}$$

$$F_2 = 85.1(0.75) - 15 = 48.8 \text{ lb}$$

$$F_i = (63.8 + 48.8)/2 = 56.3 \text{ lb}$$

$$f' = \frac{1}{\phi} \ln \frac{F_1}{F_2} = \frac{1}{\pi} \ln \frac{63.8}{48.8} = 0.0853$$

Note $f' < f$, that is, $0.0853 < 0.35$. For the minimum-width belt,

$$\Delta F' = F b_{\min}/b = 15(0.264)/0.75 = 5.28 \text{ lb}$$

$$F_{1a} = 85.1(0.264) = 22.5 \text{ lb}$$

$$F_2 = F_{1a} - \Delta F' = 22.5 - 5.28 = 17.2 \text{ lb}$$

$$F_i = (22.5 + 17.2)/2 = 19.9 \text{ lb}$$

$$T' = \Delta F' D/2 = 5.28(4)/2 = 10.6 \text{ lb}$$

For the selected belt width multiply all results immediately preceding by $0.75/0.264 =$

2.84 to recover the 0.75-in-wide belt results:

$$\Delta F = 5.28(2.84) = 15 \text{ lb}$$

$$F_{1a} = 22.5(2.84) = 63.9 \text{ lb}$$

$$F_2 = 17.2(2.84) = 48.8 \text{ lb}$$

$$F_i = 19.9(2.84) = 56.5 \text{ lb}$$

$$T = T'(2.84) = 10.6(2.84) = 30 \text{ in} \cdot \text{lb}$$

Answer (*b*) $F_i = 56.3$ lb.

If the drive is being designed to transmit nonneglible power, the allowable power capacity H_a must be

$$H_a = H_{\text{nom}} K_s (n)_d \tag{17–16}$$

where the nominal horsepower H_{nom} is augmented by a service factor K_s and a design factor n_d as applicable. The torque to be transmitted is

$$T = \frac{63\,025 H_d}{n}$$

where n is the pulley speed in r/min. Service factors are available from manufacturers to adjust for the difference between testing conditions in the laboratory and service conditions under which your drive must perform.

17–3 V Belts

The cross-sectional dimensions of V belts have been standardized by manufacturers, with each section designated by a letter of the alphabet for sizes in inch dimensions. Metric sizes are designated in numbers. Though these have not been included here, the procedure for analyzing and designing them is the same as presented here. Dimensions, minimum sheave diameters, and the horsepower range for each of the lettered sections are listed in Table 17–9.

To specify a V belt, give the belt-section letter, followed by the inside circumference in inches (standard circumferences are listed in Table 17–10). For example, B75 is a B-section belt having an inside circumference of 75 in.

Calculations involving the belt length are usually based on the pitch length. For any given belt section, the pitch length is obtained by adding a quantity to the inside circumference (Tables 17–10 and 17–11). For example, a B75 belt has a pitch length of 76.8 in. Similarly, calculations of the velocity ratios are made using the pitch diameters of the sheaves, and for this reason the stated diameters are usually understood to be the pitch diameters even though they are not always so specified.

Additional tabular information is provided there. Table 17–12 gives power ratings of standard V belts. Table 17–13 provides angle of contact information for VV and V-flat drives. Table 17–14 displays the belt-length correction factor K_2. Table 17–15 gives values for the service factor K_s. Table 17–16 shows V-belt parameters K_b and K_c. Table 17–17 gives some durability parameters for some V-belt sections.

V-belt tensions are shown in Fig. 17–14.

Table 17–9

Standard V-Belt Sections

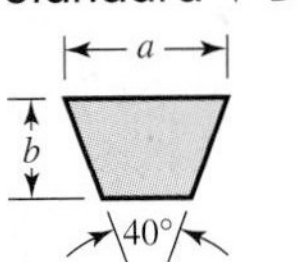

Belt Section	Width a, in	Thickness b, in	Minimum Sheave Diameter, in	hp Range, One or More Belts
A	$\frac{1}{2}$	$\frac{11}{32}$	3.0	$\frac{1}{4}$–10
B	$\frac{21}{32}$	$\frac{7}{16}$	5.4	1–25
C	$\frac{7}{8}$	$\frac{17}{32}$	9.0	15–100
D	$1\frac{1}{4}$	$\frac{3}{4}$	13.0	50–250
E	$1\frac{1}{2}$	1	21.6	100 and up

Table 17–10

Inside Circumferences of Standard V Belts

Section	Circumference, in
A	26, 31, 33, 35, 38, 42, 46, 48, 51, 53, 55, 57, 60, 62, 64, 66, 68, 71, 75, 78, 80, 85, 90, 96, 105, 112, 120, 128
B	35, 38, 42, 46, 48, 51, 53, 55, 57, 60, 62, 64, 65, 66, 68, 71, 75, 78, 79, 81, 83, 85, 90, 93, 97, 100, 103, 105, 112, 120, 128, 131, 136, 144, 158, 173, 180, 195, 210, 240, 270, 300
C	51, 60, 68, 75, 81, 85, 90, 96, 105, 112, 120, 128, 136, 144, 158, 162, 173, 180, 195, 210, 240, 270, 300, 330, 360, 390, 420
D	120, 128, 144, 158, 162, 173, 180, 195, 210, 240, 270, 300, 330, 360, 390, 420, 480, 540, 600, 660
E	180, 195, 210, 240, 270, 300, 330, 360, 390, 420, 480, 540, 600, 660

Table 17–11

Length Conversion Dimensions. (Add the Listed Quantity to the Inside Circumference to Obtain the Pitch Length in Inches)

Belt section	A	B	C	D	E
Quantity to be added	1.3	1.8	2.9	3.3	4.5

The groove angle of a sheave is made somewhat smaller than the belt-section angle. This causes the belt to wedge itself into the groove, thus increasing friction. The exact value of this angle depends on the belt section, the sheave diameter, and the angle of contact. If it is made too much smaller than the belt, the force required to pull the belt out of the groove as the belt leaves the pulley will be excessive. Optimum values are given in the commercial literature.

The minimum sheave diameters have been listed in Table 17–9. For best results, a V belt should be run quite fast: 4000 ft/min is a good speed. Trouble may be encountered if the belt runs much faster than 5000 ft/min or much slower than 1000 ft/min.

The *pitch length* L_p and the center-to-center distance C are

$$L_p = 2C + 1.57(D + d) + (D - d)^2/(4C) \tag{17–17a}$$

$$C = 0.25\left\{\left[\frac{\pi}{2}(D + d) - L_p\right] + \sqrt{\left[\frac{\pi}{2}(D + d) - L_p\right]^2 - 2(D - d)^2}\right\} \tag{17–17b}$$

Table 17–12

Horsepower Ratings of Standard V Belts

Belt Section	Sheave Pitch Diameter, in	Belt Speed, ft/min				
		1000	2000	3000	4000	5000
A	2.6	0.47	0.62	0.53	0.15	
	3.0	0.66	1.01	1.12	0.93	0.38
	3.4	0.81	1.31	1.57	1.53	1.12
	3.8	0.93	1.55	1.92	2.00	1.71
	4.2	1.03	1.74	2.20	2.38	2.19
	4.6	1.11	1.89	2.44	2.69	2.58
	5.0 and up	1.17	2.03	2.64	2.96	2.89
B	4.2	1.07	1.58	1.68	1.26	0.22
	4.6	1.27	1.99	2.29	2.08	1.24
	5.0	1.44	2.33	2.80	2.76	2.10
	5.4	1.59	2.62	3.24	3.34	2.82
	5.8	1.72	2.87	3.61	3.85	3.45
	6.2	1.82	3.09	3.94	4.28	4.00
	6.6	1.92	3.29	4.23	4.67	4.48
	7.0 and up	2.01	3.46	4.49	5.01	4.90
C	6.0	1.84	2.66	2.72	1.87	
	7.0	2.48	3.94	4.64	4.44	3.12
	8.0	2.96	4.90	6.09	6.36	5.52
	9.0	3.34	5.65	7.21	7.86	7.39
	10.0	3.64	6.25	8.11	9.06	8.89
	11.0	3.88	6.74	8.84	10.0	10.1
	12.0 and up	4.09	7.15	9.46	10.9	11.1
D	10.0	4.14	6.13	6.55	5.09	1.35
	11.0	5.00	7.83	9.11	8.50	5.62
	12.0	5.71	9.26	11.2	11.4	9.18
	13.0	6.31	10.5	13.0	13.8	12.2
	14.0	6.82	11.5	14.6	15.8	14.8
	15.0	7.27	12.4	15.9	17.6	17.0
	16.0	7.66	13.2	17.1	19.2	19.0
	17.0 and up	8.01	13.9	18.1	20.6	20.7
E	16.0	8.68	14.0	17.5	18.1	15.3
	18.0	9.92	16.7	21.2	23.0	21.5
	20.0	10.9	18.7	24.2	26.9	26.4
	22.0	11.7	20.3	26.6	30.2	30.5
	24.0	12.4	21.6	28.6	32.9	33.8
	26.0	13.0	22.8	30.3	35.1	36.7
	28.0 and up	13.4	23.7	31.8	37.1	39.1

where $D =$ pitch diameter of the large sheave

$d =$ pitch diameter of the small sheave

In the case of flat belts, there is virtually no limit to the center-to-center distance. Long center-to-center distances are not recommended for V belts because the excessive vibration of the slack side will shorten the belt life materially. In general, the center-to-center distance should be no greater than three times the sum of the sheave diameters and no less than the diameter of the larger sheave. Link-type V belts have less vibration, because of better balance, and hence may be used with longer center-to-center distances.

Table 17–13

Angle of Contact Correction Factor K_1 for VV* and V-Flat Drives

$\frac{D-d}{C}$	θ, deg	K_1 VV	K_1 V Flat
0.00	180	1.00	0.75
0.10	174.3	0.99	0.76
0.20	166.5	0.97	0.78
0.30	162.7	0.96	0.79
0.40	156.9	0.94	0.80
0.50	151.0	0.93	0.81
0.60	145.1	0.91	0.83
0.70	139.0	0.89	0.84
0.80	132.8	0.87	0.85
0.90	126.5	0.85	0.85
1.00	120.0	0.82	0.82
1.10	113.3	0.80	0.80
1.20	106.3	0.77	0.77
1.30	98.9	0.73	0.73
1.40	91.1	0.70	0.70
1.50	82.8	0.65	0.65

*A curvefit for the VV column in terms of θ is
$K_1 = 0.143\,543 + 0.007\,46\,8\,\theta - 0.000\,015\,052\,\theta^2$
in the range $90° \le \theta \le 180°$.

Table 17–14

Belt-Length Correction Factor K_2^*

Length Factor	Nominal Belt Length, in: A Belts	B Belts	C Belts	D Belts	E Belts
0.85	Up to 35	Up to 46	Up to 75	Up to 128	
0.90	38–46	48–60	81–96	144–162	Up to 195
0.95	48–55	62–75	105–120	173–210	210–240
1.00	60–75	78–97	128–158	240	270–300
1.05	78–90	105–120	162–195	270–330	330–390
1.10	96–112	128–144	210–240	360–420	420–480
1.15	120 and up	158–180	270–300	480	540–600
1.20		195 and up	330 and up	540 and up	660

*Multiply the rated horsepower per belt by this factor to obtain the corrected horsepower.

Table 17–15

Suggested Service Factors K_S for V-Belt Drives

Driven Machinery	Source of Power: Normal Torque Characteristic	High or Nonuniform Torque
Uniform	1.0 to 1.2	1.1 to 1.3
Light shock	1.1 to 1.3	1.2 to 1.4
Medium shock	1.2 to 1.4	1.4 to 1.6
Heavy shock	1.3 to 1.5	1.5 to 1.8

The basis for power ratings of V belts depends somewhat on the manufacturer; it is not often mentioned quantitatively in vendors' literature but is available from vendors. Some bases are a number of hours, 24 000, for example, or a life of 10^8 or 10^9 belt passes.

Table 17–16

Some V-Belt Parameters*

Belt Section	K_b	K_c
A	220	0.561
B	576	0.965
C	1 600	1.716
D	5 680	3.498
E	10 850	5.041
3V	230	0.425
5V	1098	1.217
8V	4830	3.288

*Data courtesy of Gates Rubber Co., Denver, Col.

Table 17–17

Durability Parameters for Some V-Belt Sections

Source: M. E. Spotts, *Design of Machine Elements*, 6th ed. Prentice Hall, Englewood Cliffs, N.J., 1985.

Belt Section	10^8 to 10^9 Force Peaks		10^9 to 10^{10} Force Peaks		Minimum Sheave Diameter, in
	K	b	K	b	
A	674	11.089			3.0
B	1193	10.926			5.0
C	2038	11.173			8.5
D	4208	11.105			13.0
E	6061	11.100			21.6
3V	728	12.464	1062	10.153	2.65
5V	1654	12.593	2394	10.283	7.1
8V	3638	12.629	5253	10.319	12.5

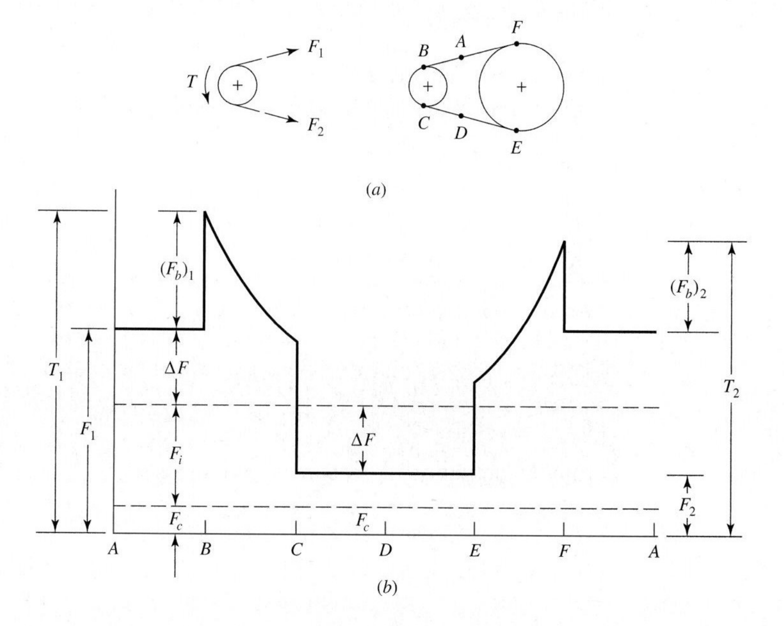

Figure 17–14

V-belt tensions.

Since the number of belts must be an integer, an undersized belt set that is augmented by one belt can be substantially oversized.

The rating, whether in terms of hours or belts passes, is for a belt running on equal-diameter sheaves (180° of wrap), of moderate length, and transmitting a steady load.

Deviations from these laboratory test conditions are acknowledged by multiplicative adjustments. If the tabulated power of a belt for a C-section belt is 9.42 hp for a 12-in-diameter sheave at a peripheral speed of 3000 ft/min (Table 17–12), then, used under other conditions, the tabulated value H_{tab} is adjusted as follows:

$$H_a = K_1 K_2 H_{\text{tab}} \tag{17–18}$$

where H_a = allowable power, per belt, Table 17–12
K_1 = angle-of-wrap correction factor, Table 17–13
K_2 = belt length correction factor, Table 17–14

The allowable power can be near to H_{tab}, depending upon circumstances.

In a V belt the effective coefficient of friction f' is $f/\sin\phi/2$, which amounts to an augmentation by a factor of about 3 due to the grooves. The effective coefficient of friction f' is sometimes tabulated against *sheave* groove angles of 30°, 34°, and 38°, the tabulated values being 0.50, 0.45, and 0.40, respectively, revealing a belt material-on-metal coefficient of friction of 0.13, 0.13, and 0.13, respectively. The Gates Rubber Company declares its effective coefficient of friction to be 0.5123 for grooves. Thus

$$\frac{F_1 - F_c}{F_2 - F_c} = \exp(0.5123\phi) \tag{17–19}$$

Since the design power is $H_d = H_{\text{nom}} K_s n_d$ and allowable power is $H_a = K_1 K_2 H_{\text{tab}}$, the number of belts is usually the next larger integer to H_d/H_a. Designers work on a per-belt basis. The centrifugal tension F_c is given by

$$F_c = K_c \left(\frac{V}{1000}\right)^2 \tag{17–20}$$

where K_c is from Table 17–12.

The power that is transmitted is based on $\Delta F = F_1 - F_2$, noting

$$H_d - H_{\text{nom}} K_s n_d \tag{17–21}$$

then

$$\Delta F = \frac{63\,025 H_d/N_b}{n(d/2)} \tag{17–22}$$

then from Eq. (17–8) the largest tension F_1 is given by

$$F_1 = F_c + \frac{\Delta F \exp f\phi}{\exp f\phi - 1} \tag{17–23}$$

From the definition of ΔF, the least tension F_2 is

$$F_2 = F_1 - \Delta F \tag{17–24}$$

From Eq. (j) in Sec. 17–2

$$F_i = \frac{F_1 + F_2}{2} - F_c \tag{17–25}$$

The factor of safety is

$$\text{fos} = \frac{H_a N_b}{H_{\text{nom}} K_s} \tag{17–26}$$

Durability (life) correlations are complicated by the fact that the bending induces flexural stresses in the belt; the corresponding belt tension that induces the same maximum tensile stress if F_{b1} at the driving sheave and F_{b2} at the driven pulley. These equivalent tensions are added to F_1 as

$$T_1 = F_1 + (F_b)_1 = F_1 + \frac{K_b}{d}$$

$$T_2 = F_1 + (F_b)_2 = F_1 + \frac{K_b}{D}$$

where K_b is given in Table 17–12. The tension–pass trade-off equation as used by the Gates Rubber Company is of the form

$$T^b N_P = K^b$$

where b is approximately 11. See Table 17–13. The Miner rule is used to sum damage incurred by the two tension peaks:

$$\frac{1}{N_P} = \left(\frac{K}{T_1}\right)^{-b} + \left(\frac{K}{T_2}\right)^{-b}$$

or

$$N_P = \left[\left(\frac{K}{T_1}\right)^{-b} + \left(\frac{K}{T_2}\right)^{-b}\right]^{-1} \quad \text{passes} \tag{17–27}$$

The lifetime in hours t is given by

$$t = \frac{N_P L_p}{720V} \tag{17–28}$$

The constants K and b have their ranges of validity. If $N_P > 10^9$, report that $N_P > 10^9$ and $t > N_P L_p/(720V)$ without placing confidence in numerical values beyond the validity interval. See the statement about N_p and t near the conclusion of Ex. 17–4.

The analysis of a V-belt drive can consist of the following steps:

- Find V, L_p, C, θ_s, and $\exp(0.5123\theta_s)$
- Find H_d, H_a, and N_b from H_d/H_a and round up
- Find F_c, ΔF, F_1, F_2, and F_i, and fos
- Find belt life in number of passes, or hours, if possible

EXAMPLE 17–4 A 10-hp split-phase motor running at 1750 r/min is used to drive a rotary pump which operates 24 hours per day. An engineer has specified a 7.4-in small sheave, an 11-in large sheave, and three B112 belts. The service factor of 1.2 was augmented by 0.1 because of the continuous-duty requirement. Analyze the drive and estimate the belt life in passes and hours.

Solution The peripheral speed V of the belt is

$$V = \pi dn/12 = \pi(7.4)1750/12 = 3390 \text{ ft/min}$$

$$L_p = L + L_c = 112 + 1.8 = 113.8 \text{ in}$$

$$\text{Eq. (17–17b)} \quad C = \frac{1}{4}\left\{ -\left[\frac{\pi}{2}(11+7.4) - 113.8\right] + \sqrt{\left[\frac{\pi}{2}(11+7.4) - 113.8\right]^2 - 2(11-7.4)^2} \right\}$$

$$= 42.4 \text{ in}$$

$$\text{Eq. (17–1)} \quad \theta_d = \pi - 2\sin^{-1}(11-7.4)/[2(42.4)] = 3.056 \text{ rad}$$

$$\exp[0.5123(3.056)] = 4.787$$

$$\text{Eq. (17–18)} \quad H_a = K_1 K_2 H_{\text{tab}} = 0.99(1.05)4.693 = 4.878 \text{ hp}$$

$$\text{Eq. (17–21)} \quad H_d \; H_{\text{nom}} K_s n_d = 10(1.3)(1) = 13 \text{ hp} \qquad N_b = 13/4.878 = 2.67 \rightarrow 3$$

$$\text{Eq. (7–20)} \quad F_c = 0.965(3390/1000)^2 = 11.1 \text{ lb}$$

$$\text{Eq.(17–22)} \quad \Delta F = \frac{63\,025(13)/3}{1750(7.4/2)} = 42.2 \text{ lb}$$

$$\text{Eq. (17–23)} \quad F_1 = 11.1 + \frac{42.2(4.787)}{4.787 - 1} = 64.4 \text{ lb}$$

$$\text{Eq. (17–24)} \quad F_2 = F_1 - \Delta F = 64.4 - 42.2 = 22.2 \text{ lb}$$

$$\text{Eq. (17–25)} \quad F_i = \frac{64.4 + 22.2}{2} - 11.1 = 32.2 \text{ lb}$$

$$\text{Eq. (17–26)} \quad \text{fos} = \frac{H_a N_b}{H_{\text{nom}} K_s} = \frac{4.878(3)}{10(1.3)} = 1.126$$

Life:

$$F_{b1} = \frac{K_b}{d} = \frac{576}{7.4} = 77.8 \text{ lb}$$

$$F_{b2} = \frac{576}{11} = 52.4 \text{ lb}$$

$$T_1 = F_1 + F_{b1} = 64.4 + 77.8 = 142.2 \text{ lb}$$

$$T_2 = F_1 + F_{b2} = 64.4 + 52.4 = 116.8 \text{ lb}$$

$$\text{Eq. (17–27)} \quad N_p = \left[\left(\frac{1193}{142.2}\right)^{-10.924} + \left(\frac{1193}{116.8}\right)^{-10.924}\right]^{-1} = 11(10^9) \text{ passes}$$

Answer Since N_P is out of the validity range of Eq. (17–27), life is reported as greater than 10^9 passes. Then

$$\text{Answer} \quad \text{Eq. (17–28)} \quad t > \frac{10^9(113.8)}{720(3390)} = 46\,624 \text{ h}$$

In Ex. 17–4 the presence of the service factor K_s, and the rounding up of the number of belts from 2.67 to 3, means that when $n_d = 1$, fos $= 1.126$, as we have seen. Consider that n_d is set to 2. Would $N_b = 6$ and would ΔF, F_1, F_2, F_c, and F_i stay the same? Would fos double? Note that F_i for the drive is $32.2(3) = 96.6$ lb.

From your experience with the analysis in Ex. 17–4, you can appreciate the following design concerns for a V-belt drive.

Specification Set

- Belt catalog number (section, inside length)
- Number of belts in drive set: N_b
- Sheave diameters D and d
- Center-to-center distance

Adequacy Assessment

- Durability goal exceeded
- Factor of safety exceeds design factor
- Center-to-center distance in range $D < C < 3(D + d)$
- Belt speed approximately 4000 ft/min
- Provision to sustain initial tension at level F_i

Decision Set

- Durability goal
- Design factor
- Small sheave diameter d
- Large sheave diameter D
- Center-to-center distance C
- Belt catalog number
- Number of belts in the drive set

Table 17–18 displays the similarities and differences among flat belts, friction metal belts, and V belts.

17–4 Timing Belts

A timing belt is made of rubberized fabric with steel wire to take the tension load. It has teeth that fit into grooves cut on the periphery of the pulleys (Fig. 17–15); these are coated with a nylon fabric. A timing belt does not stretch or slip and consequently transmits power at a constant angular-velocity ratio. No initial tension is needed. Such belts can operate over a very wide range of speeds, have efficiencies in the range of 97 to 99 percent, require no lubrication, and are quieter than chain drives. There is no chordal-speed variation, as in chain drives (see Sec. 17–5), and so they are an attractive solution for precision-drive requirements.

The steel-wire, or tension member, of a timing belt is located at the belt pitch line (Fig. 17–15). Thus the pitch length is the same regardless of the thickness of the backing.

The five standard inch-series pitches available are listed in Table 17–19 with their letter designations. Standard pitch lengths are available in sizes from 6 to 180 in. Pulleys come in sizes from 0.60 in pitch diameter up to 35.8 in and with groove numbers from 10 to 120.

Table 17–18

Similarities and Differences in Belting Equations

Flat Belts	Friction Metal Belts	V Belts
$a = F_a C_p C_v,\ F_{1a} = ab$	$a = \left[S_f - \dfrac{Et}{(1-\nu^2)D}\right] t,\ F_{1a} = ab$	$H_a = K_1 K_2 H_{\text{tab}}$ per belt
$\Delta F = F_{1a} - F_2 = 2T/D$	$\Delta F = 2T/D$	$\Delta F = \dfrac{63\,025 H_d}{n N_b (d/2)}$ per belt
$F_2 = F_{1a} - \Delta F$	$F_2 = F_{1a} - \Delta F$	$F_1 = F_c + \dfrac{\Delta F \exp f\phi}{\exp f\phi - 1}$ per belt
$F_i = \dfrac{F_{1a} + F_2}{2} - F_c$	$F_i = \dfrac{F_{1a} + F_2}{2}$	$F_2 = F_1 - \Delta F$ per belt
$f' = \dfrac{1}{\phi} \ln \dfrac{F_{1a} - F_c}{F_2 - F_c}$	$b_{\min} = \dfrac{\Delta F}{a} \dfrac{\exp f\phi}{\exp f\phi - 1}$	$F_i = \dfrac{F_1 + F_2}{2} - F_c$ per belt
Initial tension to be assured	Initial tension to be assured	Initial tension to be assured
$F_c = \dfrac{w}{g} V^2 = \dfrac{12\gamma bt}{g} \left(\dfrac{V}{60}\right)^2$	$F_c \doteq 0$	$F_c = K_c \left(\dfrac{V}{1000}\right)^2$ per belt
$\dfrac{F_1 - F_c}{F_2 - F_c} = \exp f\phi$	$\dfrac{F_1}{F_2} = \exp f\phi$	$\dfrac{F_1 - F_c}{F_2 - F_c} = \exp(0.5123\phi)$ per belt
$H_a = \dfrac{(F_{1a} - F_2)V}{33\,000} = \dfrac{(2T/D)V}{33\,000}$	$H_a = \dfrac{(F_{1a} - F_2)V}{33\,000}$	$H_a = K_1 K_2 H_{\text{tab}}$ per belt
$H_d = H_{\text{nom}} K_s n_d$	$H_d = H_{\text{nom}} K_s n_d$	$H_d = H_{\text{nom}} K_s n_d$
$\text{fos} = \dfrac{H_a}{H_{\text{nom}} K_s}$	$\text{fos} = \dfrac{H_a}{H_{\text{nom}} K_s}$	$\text{fos} = \dfrac{N_b H_a}{H_{\text{nom}} K_s}$
$\text{dip} = \dfrac{3L^2 w}{2F_i}$	301/302 stainless steel	Belt capacity rated by table, H_{tab}
	$S_f = 14.17(10^6) N_p^{-0.407}$	Durability in terms of tension peaks
	Other metal belt materials	$T_1 = F_1 + F_{b1} = F_1 + K_b/d_1$ per belt
	$(S_f)_{10^6} = \dfrac{S_y}{3}$	$T_2 = F_1 + K_b/d_2$
		$N' = \left[\left(\dfrac{K}{T_1}\right)^{-b} + \left(\dfrac{K}{T_2}\right)^{-b}\right]^{-1}$ belt passes
		$t = \dfrac{N' L_p}{720V}$ h

The design and selection process for timing belts is so similar to that for V belts that the process will not be presented here. As in the case of other belt drives, the manufacturers will provide an ample supply of information and details on sizes and strengths.

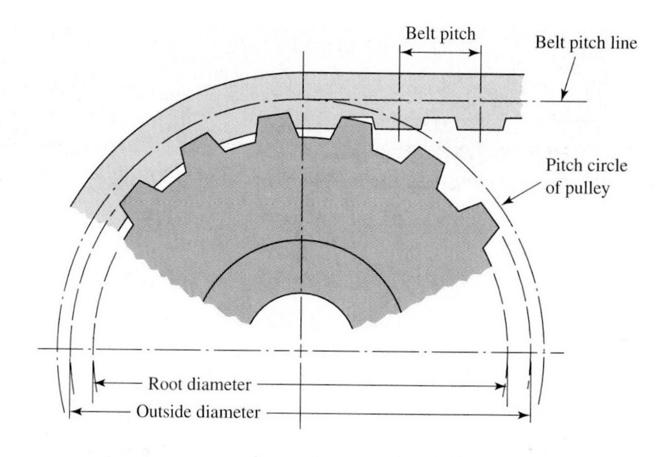

Figure 17–15

Timing-belt drive showing portions of the pulley and belt. Note that the pitch diameter of the pulley is greater than the diametral distance across the top lands of the teeth.

Table 17–19

Standard Pitches of Timing Belts

Service	Designation	Pitch p, in
Extra light	XL	$\frac{1}{5}$
Light	L	$\frac{3}{8}$
Heavy	H	$\frac{1}{2}$
Extra heavy	XH	$\frac{7}{8}$
Double extra heavy	XXH	$1\frac{1}{4}$

17–5 Roller Chains

Basic features of chain drives include a constant ratio, since no slippage or creep is involved; long life; and the ability to drive a number of shafts from a single source of power.

Roller chains have been standardized as to sizes by the ANSI. Figure 17–16 shows the nomenclature. The pitch is the linear distance between the centers of the rollers. The width is the space between the inner link plates. These chains are manufactured in single, double, triple, and quadruple strands. The dimensions of standard sizes are listed in Table 17–20.

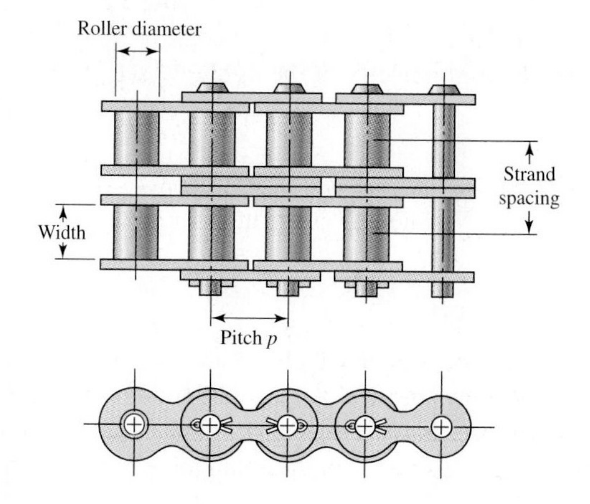

Figure 17–16

Portion of a double-strand roller chain.

Table 17–20

Dimensions of American Standard Roller Chains—Single Strand

Source: Compiled from ANSI B29.1-1975.

ANSI Chain Number	Pitch, in (mm)	Width, in (mm)	Minimum Tensile Strength, lb (N)	Average Weight, lb/ft (N/m)	Roller Diameter, in (mm)	Multiple-Strand Spacing, in (mm)
25	0.250 (6.35)	0.125 (3.18)	780 (3 470)	0.09 (1.31)	0.130 (3.30)	0.252 (6.40)
35	0.375 (9.52)	0.188 (4.76)	1 760 (7 830)	0.21 (3.06)	0.200 (5.08)	0.399 (10.13)
41	0.500 (12.70)	0.25 (6.35)	1 500 (6 670)	0.25 (3.65)	0.306 (7.77)	— —
40	0.500 (12.70)	0.312 (7.94)	3 130 (13 920)	0.42 (6.13)	0.312 (7.92)	0.566 (14.38)
50	0.625 (15.88)	0.375 (9.52)	4 880 (21 700)	0.69 (10.1)	0.400 (10.16)	0.713 (18.11)
60	0.750 (19.05)	0.500 (12.7)	7 030 (31 300)	1.00 (14.6)	0.469 (11.91)	0.897 (22.78)
80	1.000 (25.40)	0.625 (15.88)	12 500 (55 600)	1.71 (25.0)	0.625 (15.87)	1.153 (29.29)
100	1.250 (31.75)	0.750 (19.05)	19 500 (86 700)	2.58 (37.7)	0.750 (19.05)	1.409 (35.76)
120	1.500 (38.10)	1.000 (25.40)	28 000 (124 500)	3.87 (56.5)	0.875 (22.22)	1.789 (45.44)
140	1.750 (44.45)	1.000 (25.40)	38 000 (169 000)	4.95 (72.2)	1.000 (25.40)	1.924 (48.87)
160	2.000 (50.80)	1.250 (31.75)	50 000 (222 000)	6.61 (96.5)	1.125 (28.57)	2.305 (58.55)
180	2.250 (57.15)	1.406 (35.71)	63 000 (280 000)	9.06 (132.2)	1.406 (35.71)	2.592 (65.84)
200	2.500 (63.50)	1.500 (38.10)	78 000 (347 000)	10.96 (159.9)	1.562 (39.67)	2.817 (71.55)
240	3.00 (76.70)	1.875 (47.63)	112 000 (498 000)	16.4 (239)	1.875 (47.62)	3.458 (87.83)

In Fig. 17–17 is shown a sprocket driving a chain in a counterclockwise direction. Denoting the chain pitch by p, the pitch angle by γ, and the pitch diameter of the sprocket by D, from the trigonometry of the figure we see

$$\sin\frac{\gamma}{2} = \frac{p/2}{D/2} \qquad \text{or} \qquad D = \frac{p}{\sin(\gamma/2)} \tag{a}$$

Since $\gamma = 360°/N$, where N is the number of sprocket teeth, Eq. (a) can be written

$$D = \frac{p}{\sin(180°/N)} \tag{17–29}$$

The angle $\gamma/2$, through which the link swings as it enters contact, is called the *angle of articulation*. It can be seen that the magnitude of this angle is a function of the number of teeth. Rotation of the link through this angle causes impact between the rollers and the sprocket teeth and also wear in the chain joint. Since the life of a properly selected drive

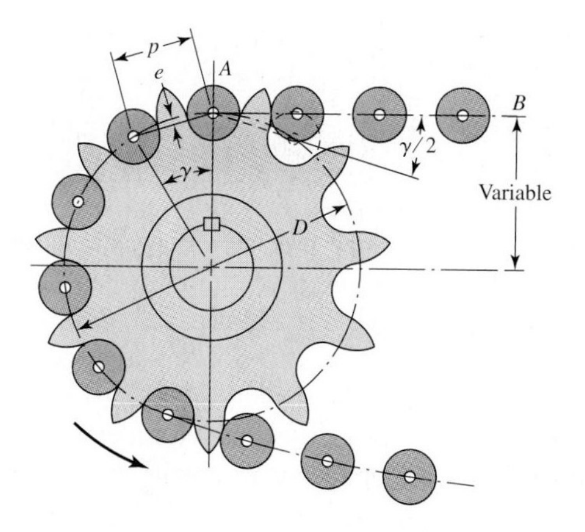

Figure 17–17

Engagement of a chain and sprocket.

is a function of the wear and the surface fatigue strength of the rollers, it is important to reduce the angle of articulation as much as possible.

The number of sprocket teeth also affects the velocity ratio during the rotation through the pitch angle γ. At the position shown in Fig. 17–17, the chain AB is tangent to the pitch circle of the sprocket. However, when the sprocket has turned an angle of $\gamma/2$, the chain line AB moves closer to the center of rotation of the sprocket. This means that the chain line AB is moving up and down, and that the lever arm varies with rotation through the pitch angle, all resulting in an uneven chain exit velocity. You can think of the sprocket as a polygon in which the exit velocity of the chain depends upon whether the exit is from a corner, or from a flat of the polygon. Of course, the same effect occurs when the chain first enters into engagement with the sprocket.

The chain velocity V is defined as the number of feet coming off the sprocket in unit time. Thus the chain velocity in feet per minute is

$$V = \frac{Npn}{12} \tag{17–30}$$

where $N =$ number of sprocket teeth
$p =$ chain pitch, in
$n =$ sprocket speed, r/min

The maximum exit velocity of the chain is

$$v_{\max} = \frac{\pi Dn}{12} = \frac{\pi np}{12\sin(\gamma/2)} \tag{b}$$

where Eq. (*a*) has been substituted for the pitch diameter D. The minimum exit velocity occurs at a diameter d, smaller than D. Using the geometry of Fig. 17–17, we find

$$d = D\cos\frac{\gamma}{2} \tag{c}$$

Thus the minimum exit velocity is

$$v_{\min} = \frac{\pi dn}{12} = \frac{\pi np}{12}\,\frac{\cos(\gamma/2)}{\sin(\gamma/2)} \tag{d}$$

Now substituting $\gamma/2 = 180°/N$ and employing Eqs. (17-30), (*b*), and (*d*), we find the speed variation to be

$$\frac{\Delta V}{V} = \frac{v_{max} - v_{min}}{V} = \frac{\pi}{N}\left[\frac{1}{\sin(180°/N)} - \frac{1}{\tan(180°/N)}\right] \tag{17–31}$$

This is called the *chordal speed variation* and is plotted in Fig. 17–18. When chain drives are used to synchronize precision components or processes, due consideration must be given to these variations. For example, if a chain drive synchronized the cutting of photographic film with the forward drive of the film, the lengths of the cut sheets of film might vary too much because of this chordal speed variation. Such variations can also cause vibrations within the system.

Although a large number of teeth is considered desirable for the driving sprocket, in the usual case it is advantageous to obtain as small a sprocket as possible, and this requires one with a small number of teeth. For smooth operation at moderate and high speeds it is considered good practice to use a driving sprocket with at least 17 teeth; 19 or 21 will, of course, give a better life expectancy with less chain noise. Where space limitations are severe or for very slow speeds, smaller tooth numbers may be used by sacrificing the life expectancy of the chain.

Driven sprockets are not made in standard sizes over 120 teeth, because the pitch elongation will eventually cause the chain to "ride" high long before the chain is worn out. The most successful drives have velocity ratios up to 6:1, but higher ratios may be used at the sacrifice of chain life.

Roller chains seldom fail because they lack tensile strength; they more often fail because they have been subjected to a great many hours of service. Actual failure may be due either to wear of the rollers on the pins or to fatigue of the surfaces of the rollers. Roller-chain manufacturers have compiled tables which give the horsepower capacity corresponding to a life expectancy of 15 kh for various sprocket speeds. These capacities are tabulated in Table 17–21 for 17-tooth sprockets. Table 17–22 displays available tooth counts on sprockets of one supplier. Table 17–23 lists the tooth correction factors for other than 17 teeth. Table 17–24 shows the multiple-strand factors K_2.

The capacities of chains are based on the following:

- 15 000-h at full load
- Single strand
- ANSI proportions
- Service factor of unity
- 100 pitches in length
- Recommended lubrication
- Elongation maximum of 3 percent
- Horizontal shafts
- Two 17-tooth sprockets

Figure 17–18

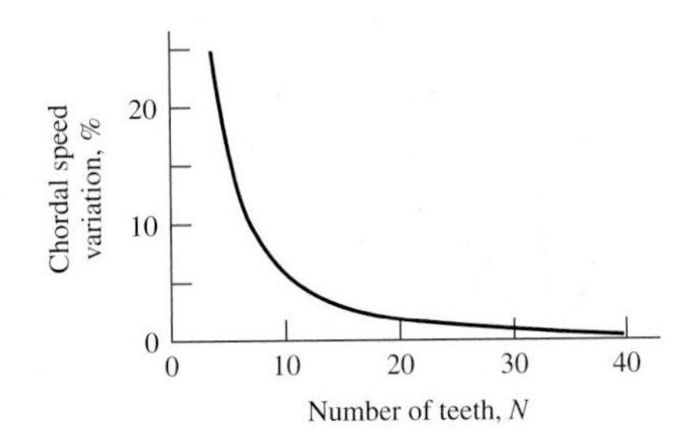

Table 17–21

Rated Horsepower Capacity of Single-Strand Single-Pitch Roller Chain for a 17-Tooth Sprocket

Source: Compiled from ANSI B29.1-1975 information only section, and from B29.9-1958.

Sprocket Speed, r/min	ANSI Chain Number 25	35	40	41	50	60
50	0.05	0.16	0.37	0.20	0.72	1.24
100	0.09	0.29	0.69	0.38	1.34	2.31
150	0.13*	0.41*	0.99*	0.55*	1.92*	3.32
200	0.16*	0.54*	1.29	0.71	2.50	4.30
300	0.23	0.78	1.85	1.02	3.61	6.20
400	0.30*	1.01*	2.40	1.32	4.67	8.03
500	0.37	1.24	2.93	1.61	5.71	9.81
600	0.44*	1.46*	3.45*	1.90*	6.72*	11.6
700	0.50	1.68	3.97	2.18	7.73	13.3
800	0.56*	1.89*	4.48*	2.46*	8.71*	15.0
900	0.62	2.10	4.98	2.74	9.69	16.7
1000	0.68*	2.31*	5.48	3.01	10.7	18.3
1200	0.81	2.73	6.45	3.29	12.6	21.6
1400	0.93*	3.13*	7.41	2.61	14.4	18.1
1600	1.05*	3.53*	8.36	2.14	12.8	14.8
1800	1.16	3.93	8.96	1.79	10.7	12.4
2000	1.27*	4.32*	7.72*	1.52*	9.23*	10.6
2500	1.56	5.28	5.51*	1.10*	6.58*	7.57
3000	1.84	5.64	4.17	0.83	4.98	5.76

Type A — Type B — Type C

*Estimated from ANSI tables by linear interpolation.

Note: Type A—manual or drip lubrication; type B—bath or disk lubrication; type C—oil-stream lubrication.

The fatigue strength of link plates governs capacity at lower speeds. The American Chain Association (ACA) publication *Chains for Power Transmission and Materials Handling* (1982) gives, for single-strand chain, the nominal power H_1, link-plate limited, as

$$H_1 = 0.004 N_1^{1.08} n_1^{0.9} p^{(3-0.07p)} \text{ hp} \tag{17–32}$$

and the nominal power H_2, roller-limited, as

$$H_2 = \frac{1000 K_r N_1^{1.5} p^{0.8}}{n_1^{1.5}} \tag{17–33}$$

where N_1 = number of teeth in the smaller sprocket

n_1 = sprocket speed, r/min

p = pitch of the chain, in

K_r = 29 for chains 25, 35; 3.4 for chain 41; and 17 for chains 40–240

The constant 0.004 becomes 0.0022 for no. 41 lightweight chain. The nominal horsepower in Table 17–21 is $H_{\text{nom}} = \min(H_1, H_2)$.

It is preferable to have an odd number of teeth on the driving sprocket (17, 19, . . .) and an even number of pitches in the chain to avoid a special link. The approximate length of the chain L in pitches is

$$\frac{L}{p} \doteq \frac{2C}{p} + \frac{N_1 + N_2}{2} + \frac{(N_2 - N_1)^2}{4\pi^2 C/p} \tag{17–34}$$

Table 17–21(continued)

Rated Horsepower Capacity of Single-Strand Single-Pitch Roller Chain for a 17-Tooth Sprocket

Source: Compiled from ANSI B29.1-1975 information only section, and from B29.9-1958.

Sprocket Speed, r/min		ANSI Chain Number 80	100	120	140	160	180	200	240
50	Type A	2.88	5.52	9.33	14.4	20.9	28.9	38.4	61.8
100		5.38	10.3	17.4	26.9	39.1	54.0	71.6	115
150		7.75	14.8	25.1	38.8	56.3	77.7	103	166
200		10.0	19.2	32.5	50.3	72.9	101	134	215
300		14.5	27.7	46.8	72.4	105	145	193	310
400		18.7	35.9	60.6	93.8	136	188	249	359
500	Type B	22.9	43.9	74.1	115	166	204	222	0
600		27.0	51.7	87.3	127	141	155	169	
700		31.0	59.4	89.0	101	112	123	0	
800		35.0	63.0	72.8	82.4	91.7	101		
900		39.9	52.8	61.0	69.1	76.8	84.4		
1000		37.7	45.0	52.1	59.0	65.6	72.1		
1200		28.7	34.3	39.6	44.9	49.9	0		
1400		22.7	27.2	31.5	35.6	0			
1600		18.6	22.3	25.8	0				
1800		15.6	18.7	21.6					
2000		13.3	15.9	0					
2500		9.56	0.40						
3000		7.25	0						
Type C				**Type C′**					

Note: Type A—manual or drip lubrication; type B—bath or disk lubrication; type C—oil-stream lubrication; type C′—type C, but this is a galling region; submit design to manufacturer for evaluation.

Table 17–22

Single-Strand Sprocket Tooth Counts Available from One Supplier*

No.	Available Sprocket Tooth Counts
25	8-30, 32, 34, 35, 36, 40, 42, 45, 48, 54, 60, 64, 65, 70, 72, 76, 80, 84, 90, 95, 96, 102, 112, 120
35	4-45, 48, 52, 54, 60, 64, 65, 68, 70, 72, 76, 80, 84, 90, 95, 96, 102, 112, 120
41	6-60, 64, 65, 68, 70, 72, 76, 80, 84, 90, 95, 96, 102, 112, 120
40	8-60, 64, 65, 68, 70, 72, 76, 80, 84, 90, 95, 96, 102, 112, 120
50	8-60, 64, 65, 68, 70, 72, 76, 80, 84, 90, 95, 96, 102, 112, 120
60	8-60, 62, 63, 64, 65, 66, 67, 68, 70, 72, 76, 80, 84, 90, 95, 96, 102, 112, 120
80	8-60, 64, 65, 68, 70, 72, 76, 78, 80, 84, 90, 95, 96, 102, 112, 120
100	8-60, 64, 65, 67, 68, 70, 72, 74, 76, 80, 84, 90, 95, 96, 102, 112, 120
120	9-45, 46, 48, 50, 52, 54, 55, 57, 60, 64, 65, 67, 68, 70, 72, 76, 80, 84, 90, 96, 102, 112, 120
140	9-28, 30, 31, 32, 33, 34, 35, 36, 37, 39, 40, 42, 43, 45, 48, 54, 60, 64, 65, 68, 70, 72, 76, 80, 84, 96
160	8-30, 32–36, 38, 40, 45, 46, 50, 52, 53, 54, 56, 57, 60, 62, 63, 64, 65, 66, 68, 70, 72, 73, 80, 84, 96
180	13-25, 28, 35, 39, 40, 45, 54, 60
200	9-30, 32, 33, 35, 36, 39, 40, 42, 44, 45, 48, 50, 51, 54, 56, 58, 59, 60, 63, 64, 65, 68, 70, 72
240	9-30, 32, 35, 36, 40, 44, 45, 48, 52, 54, 60

*Morse Chain Company, Ithaca, NY, Type B hub sprockets.

Table 17–23

Tooth Correction Factors, K_1

Number of Teeth on Driving Sprocket	K_1 Pre-extreme Horsepower	K_1 Post-extreme Horsepower
11	0.62	0.52
12	0.69	0.59
13	0.75	0.67
14	0.81	0.75
15	0.87	0.83
16	0.94	0.91
17	1.00	1.00
18	1.06	1.09
19	1.13	1.18
20	1.19	1.28
N	$(N_1/17)^{1.08}$	$(N_1/17)^{1.5}$

Table 17–24

Multiple-Strand Factors K_2

Number of Strands	K_2
1	1.0
2	1.7
3	2.5
4	3.3
5	3.9
6	4.6
8	6.0

The center-to-center distance C is given by

$$\frac{C}{p} = \frac{1}{4}\left[-A + \sqrt{A^2 - 8\left(\frac{N_2 - N_1}{2\pi}\right)^2}\right] \tag{17–35}$$

where

$$A = \frac{N_1 + N_2}{2} - \frac{L}{p} \tag{17–36}$$

The allowable power H_a is given by

$$H_a = K_1 K_2 H_{\text{tab}} \tag{17–37}$$

where $K_1 =$ correction factor for tooth number other than 17

$K_2 =$ strand correction

The horsepower that must be transmitted H_d is given by

$$H_d = H_{\text{nom}} K_s n_d \tag{17–38}$$

In Table 17–19, for $N_1 = 17$, $n_1 = 1000$ r/min, no. 40 chain with $p = 0.5$ in, from Eq. (17–32),

$$H_1 = 0.004(17)^{1.08} 1000^{0.9} 0.5^{[3-0.07(0.5)]} = 5.48 \text{ hp}$$

From Eq. (17–33),

$$H_2 = 1000(17)17^{1.5}0.5^{[3-0.07(0.5)]} = 21.64 \text{ hp}$$

The tabulated value in Table 17–21 is $H_{\text{tab}} = \min(5.48, 21.64) = 5.48$ hp.

Equation (17–32) is the basis of the pre-extreme power entries (vertical entries) of Table 17–21, and the chain power is limited by link-plate fatigue. Equation (17–33) is the basis for the post-extreme power entries of these tables, and the chain power performance is limited by impact fatigue. The entries are for chains of 100 pitch length and 17-tooth sprocket. For a deviation from this

$$H_2 = 1000\left[K_r\left(\frac{N_1}{n_1}\right)^{1.5} p^{0.8}\left(\frac{L_p}{100}\right)^{0.4}\left(\frac{15\,000}{h}\right)^{0.4}\right] \tag{17–39}$$

where L_p is the chain length in pitches and h is the chain life in hours. Viewed from a deviation viewpoint, Eq. (17–39) can be written as a trade-off equation in the following form:

$$\frac{H_2^{2.5}h}{N_1^{3.75}L_p} = \text{constant} \tag{17–40}$$

If tooth-correction factor K_1 is used, then omit the term $N_1^{3.75}$. Note that $(N_1^{1.5})^{2.5} = N_1^{3.75}$.

In Eq. (17–40) one would expect the h/L_p term because doubling the hours can require doubling the chain length, other conditions constant, for the same number of cycles. Our experience with contact stresses leads us to expect a load (tension) life relation of the form $F^a L = \text{constant}$. In the more complex circumstance of roller-bushing impact, the Diamond Chain Company has identified $a = 2.5$.

The maximum speed (r/min) for a chain drive is limited by galling between the pin and the bushing. Tests suggest

$$n_1 \leq 1000\left[\frac{82.5}{7.95^p(1.0278)^{N_1}(1.323)^{F/1000}}\right]^{1/(1.59\log p+1.873)} \text{ r/min}$$

where F is the chain tension in pounds.

The Specification Set, Adequacy Assessment, and Decision Set

For roller chain drives, the following design considerations apply.

Specification Set

- Chain number and number of strands
- Chain length in pitches
- Sprocket 1 and number of teeth
- Sprocket 2 and number of teeth
- Lubrication scheme
- Center-to-center distance (for reference)

Adequacy Assessment

- $N_1 \geq 17$, odd if possible
- $N_2 \geq 17$, even if possible
- Even number of chain pitches

- Center-to-center distance: $30p \le C \le 50p$; in any event, $C < 80p$
- Durability equals or exceeds goal
- Lubrication scheme
- Factor of safety: $K_1 K_2 H_{\text{tab}}/(K_s H_{\text{nom}})$

Decision Set

- Environment: power, speed, space, service factor
- Design factor
- Number of driving sprocket teeth
- Number of driven sprocket teeth
- Number of chain pitches
- Chain number
- Lubrication scheme
- Center-to-center distance
- Number of chain strands

EXAMPLE 17–5 Select drive components for a 2:1 reduction, 90-hp input at 300 r/min, moderate shock, an abnormally long 18-hour day, poor lubrication, cold temperatures, dirty surroundings, short drive $C/p = 25$.

Solution *Function:* $H_{\text{nom}} = 90$ hp, $n_1 = 300$ r/min, $C/p = 25$, $K_s = 1.3$
Design factor: $n_d = 1.5$
Sprocket teeth: $N_1 = 17$ teeth, $N_2 = 34$ teeth, $K_1 = 1$, $K_2 = 1, 1.7, 2.5, 3.3$
Chain number of strands:

$$H_{\text{tab}} = \frac{n_d K_s H_{\text{nom}}}{K_1 K_2} = \frac{1.5(1.3)90}{(1)K_2} = \frac{176}{K_2}$$

Form a table:

Number of Strands	$176/K_2$	Chain Number (Table 17–20)	Lubrication Type
1	$176/1 = 176$	200	C′
2	$176/1.7 = 104$	160	C
3	$176/2.5 = 70.4$	140	B
4	$176/3.3 = 53.3$	140	B

Decision 140-3 chain (H_{tab} is 72.4 hp).
Number of pitches in the chain:

$$\frac{L}{p} = \frac{2C}{p} + \frac{N_1 + N_2}{2} + \frac{(N_2 - N_1)^2}{4\pi^2 C/p}$$

$$= 2(25) + \frac{17 + 34}{2} + \frac{(34 - 17)^2}{4\pi^2(25)} = 75.79 \text{ pitches}$$

Decision Use 76 pitches. Then $L/p = 76$

Identify the center-to-center distance:

$$A = \frac{N_1 + N_2}{2} - \frac{L}{p} = \frac{17 + 34}{2} - 76 = -50.5$$

$$\frac{C}{P} = \frac{1}{4}\left[-A + \sqrt{A^2 - 8\left(\frac{N_2 - N_1}{2\pi}\right)^2}\right]$$

$$= \frac{1}{4}\left[50.5 + \sqrt{50.5^2 - 8\left(\frac{34 - 17}{2\pi}\right)^2}\right] = 25.104$$

$$C = 25.104p = 25.104(1.75) = 43.93 \text{ in}$$

Lubrication: Type B

Comment: This is operating on the pre-extreme portion of the power, so durability estimates other than 15 000 h are not available. Given the poor operating conditions life will be much shorter.

An adequacy assessment can be performed on the decision set of Ex. 17–5:

- $N_1 = 17$, which is 17 and odd OK
- $N_2 = 34$, which is even OK
- Chain pitch is 76 and even OK
- Center-to-center distance is 25.102 pitches, which is less than 30 but unavoidable, and there is bound to be some life penalty
- Durability cannot be quantitatively addressed
- Lubrication scheme is type B, plus frequent inspections
- fos $= K_1 K_2 H_{tab}/(K_s H_{nom}) = (1)(2.5)72.4/(1.3 \cdot 90) = 1.55$, which is larger than $n_d = 1.5$ OK

Lubrication of roller chains is essential in order to obtain a long and trouble-free life. Either a drip feed or a shallow bath in the lubricant is satisfactory. A medium or light mineral oil, without additives, should be used. Except for unusual conditions, heavy oils and greases are not recommended, because they are too viscous to enter the small clearances in the chain parts.

17–6 Wire Rope

Wire rope is made with two types of winding, as shown in Fig. 17-19. The *regular lay,* which is the accepted standard, has the wire twisted in one direction to form the strands, and the strands twisted in the opposite direction to form the rope. In the completed rope the visible wires are approximately parallel to the axis of the rope. Regular-lay ropes do not kink or untwist and are easy to handle.

Lang-lay ropes have the wires in the strand and the strands in the rope twisted in the same direction, and hence the outer wires run diagonally across the axis of the rope. Lang-lay ropes are more resistant to abrasive wear and failure due to fatigue than are regular-lay ropes, but they are more likely to kink and untwist.

Figure 17–19

Types of wire rope; both lays are available in either right or left hand.

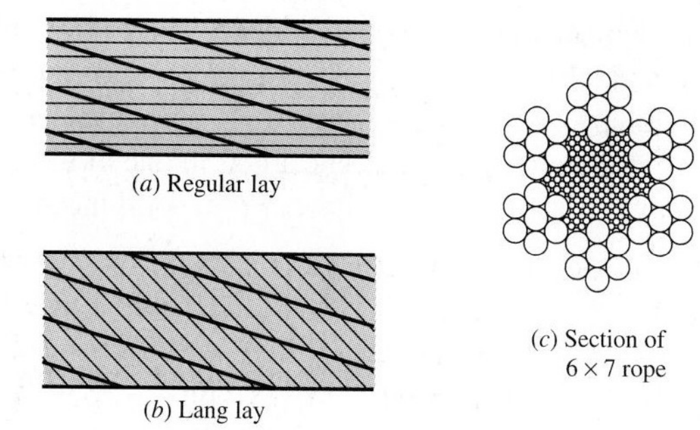

Standard-ropes are made with a hemp core which supports and lubricates the strands. When the rope is subjected to heat, either a steel center or a wire-strand center must be used.

Wire rope is designated as, for example, a $1\frac{1}{8}$-in 6 × 7 haulage rope. The first figure is the diameter of the rope (Fig. 17-19*c*). The second and third figures are the number of strands and the number of wires in each strand, respectively. Table 17-25 lists some of the various ropes which are available, together with their characteristics and properties. The area of the metal in standard hoisting and haulage rope is $A_m = 0.38d^2$.

Table 17–25

Wire-Rope Data *Source:* Compiled from *American Steel and Wire Company Handbook.*

Rope	Weight per Foot, lb	Minimum Sheave Diameter, in	Standard Sizes *d*, in	Material	Size of Outer Wires	Modulus of Elasticity,* Mpsi	Strength,† kpsi
6 × 7 haulage	$1.50d^2$	$42d$	$\frac{1}{4}-1\frac{1}{2}$	Monitor steel	$d/9$	14	100
				Plow steel	$d/9$	14	88
				Mild plow steel	$d/9$	14	76
6 × 19 standard hoisting	$1.60d^2$	$26d$–$34d$	$\frac{1}{4}-2\frac{3}{4}$	Monitor steel	$d/13$–$d/16$	12	106
				Plow steel	$d/13$–$d/16$	12	93
				Mild plow steel	$d/13$–$d/16$	12	80
6 × 37 special flexible	$1.55d^2$	$18d$	$\frac{1}{4}-3\frac{1}{2}$	Monitor steel	$d/22$	11	100
				Plow steel	$d/22$	11	88
8 × 19 extra flexible	$1.45d^2$	$21d$–$26d$	$\frac{1}{4}-1\frac{1}{2}$	Monitor steel	$d/15$–$d/19$	10	92
				Plow steel	$d/15$–$d/19$	10	80
7 × 7 aircraft	$1.70d^2$	—	$\frac{1}{16}-\frac{3}{8}$	Corrosion-resistant steel	—	—	124
				Carbon steel	—	—	124
7 × 9 aircraft	$1.75d^2$	—	$\frac{1}{8}-1\frac{3}{8}$	Corrosion-resistant steel	—	—	135
				Carbon steel	—	—	143
19-wire aircraft	$2.15d^2$	—	$\frac{1}{32}-\frac{5}{16}$	Corrosion-resistant steel	—	—	165
				Carbon steel	—	—	165

*The modulus of elasticity is only approximate; it is affected by the loads on the rope and, in general, increases with the life of the rope.

†The strength is based on the nominal area of the rope. The figures given are only approximate and are based on 1-in rope sizes and $\frac{1}{4}$-in aircraft-cable sizes.

When a wire rope passes around a sheave, there is a certain amount of readjustment of the elements. Each of the wires and strands must slide on several others, and presumably some individual bending takes place. It is probable that in this complex action there exists some stress concentration. The stress in one of the wires of a rope passing around a sheave may be calculated as follows. From solid mechanics, we have

$$M = \frac{EI}{r} \quad \text{and} \quad M = \frac{\sigma I}{c} \tag{a}$$

where the quantities have their usual meaning. Eliminating M and solving for the stress gives

$$\sigma = \frac{Ec}{r} \tag{b}$$

For the radius of curvature r, we can substitute the sheave radius $D/2$. Also, $c = d_w/2$, where d_w is the wire diameter. These substitutions give

$$\sigma = E_r \frac{d_w}{D} \tag{c}$$

To understand this equation, observe that the individual wire makes a corkscrew figure in space and if you pull on it to determine E it will stretch or give more than its native E would suggest. Therefore E is still the modulus of elasticity of the *wire,* but in its peculiar configuration as part of the rope. A value for E equal to the modulus of elasticity of the *rope* gives an approximately correct value for the stress σ. For this reason we say that E_r in Eq. (*c*) is the modulus of elasticity of the rope, not the wire, recognizing that one can quibble over the name used.

Equation (*c*) gives the tensile stress σ in the outer wires. The sheave diameter is represented by D. This equation reveals the importance of using a large-diameter sheave. The suggested minimum sheave diameters in Table 17–25 are based on a D/d_w ratio of 400. If possible, the sheaves should be designed for a larger ratio. For elevators and mine hoists, D/d_w is usually taken from 800 to 1000. If the ratio is less than 200, heavy loads will often cause a permanent set in the rope.

A wire rope tension giving the same tensile stress as the sheave bending is called the *equivalent bending load* F_b, given by

$$F_b = \sigma A_m = \frac{E_r d A_m}{D} \tag{17–41}$$

A wire rope may fail because the static load exceeds the ultimate strength of the rope. Failure of this nature is generally not the fault of the designer, but rather that of the operator in permitting the rope to be subjected to loads for which it was not designed.

The first consideration in selecting a wire rope is to determine the static load. This load is composed of the following items:

- The known or dead weight
- Additional loads caused by sudden stops or starts
- Shock loads
- Sheave-bearing friction

When these loads are summed, the total can be compared with the ultimate strength of the rope to find a factor of safety. However, the ultimate strength used in this determination must be reduced by the strength loss that occurs when the rope passes over a curved surface such as a stationary sheave or a pin; see Fig. 17–20.

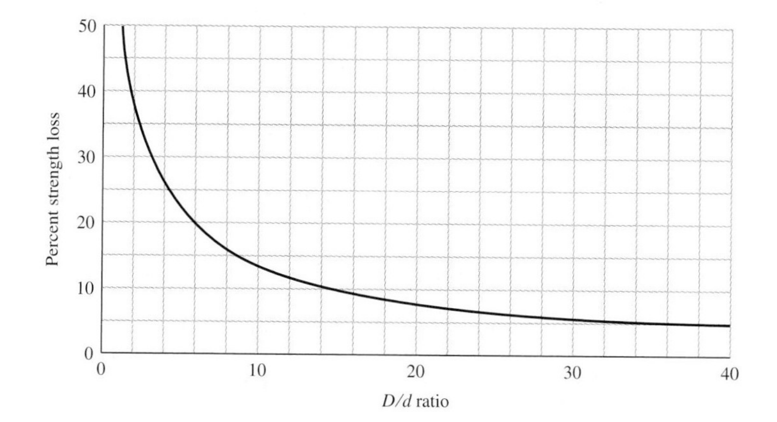

Figure 17–20

Percent strength loss due to different D/d ratios; derived from standard test data for 6×19 and 6×17 class ropes. *(From Wire Rope Users Manual; reproduced by permission from American Iron and Steel Institute.)*

For an average operation, use a factor of safety of 5. Factors of safety up to 8 or 9 are used if there is danger to human life and for very critical situations. Table 17–26 lists minimum factors of safety for a variety of design situations. Here, the factor of safety is defined as

$$n = \frac{F_u}{F_t}$$

where F_u is the ultimate wire load and F_t is the largest working tension.

Once you have made a tentative selection of a rope based upon static strength, the next consideration is to ensure that the wear life of the rope and the sheave or sheaves meets certain requirements. When a loaded rope is bent over a sheave, the rope stretches like a spring, rubs against the sheave, and causes wear of both the rope and the sheave. The amount of wear that occurs depends upon the pressure of the rope in the sheave groove. This pressure is called the *bearing pressure;* a good estimate of its magnitude is given by

$$p = \frac{2F}{dD} \tag{17–42}$$

Table 17–26

Minimum Factors of Safety for Wire Rope*

Source: Compiled from a variety of sources, including ANSI A17.1-1978.

Track cables	3.2	Passenger elevators, ft/min:	
Guys	3.5	50	7.60
Mine shafts, ft:		300	9.20
Up to 500	8.0	800	11.25
1000–2000	7.0	1200	11.80
2000–3000	6.0	1500	11.90
Over 3000	5.0	Freight elevators, ft/min:	
Hoisting	5.0	50	6.65
Haulage	6.0	300	8.20
Cranes and derricks	6.0	800	10.00
Electric hoists	7.0	1200	10.50
Hand elevators	5.0	1500	10.55
Private elevators	7.5	Powered dumbwaiters, ft/min:	
Hand dumbwaiter	4.5	50	4.8
Grain elevators	7.5	300	6.6
		500	8.0

*Use of these factors does not preclude a fatigue failure.

where F = tensile force on rope

d = rope diameter

D = sheave diameter

The allowable pressures given in Table 17–27 are to be used only as a rough guide; they may not prevent a fatigue failure or severe wear. They are presented here because they represent past practice and furnish a starting point in design.

A fatigue diagram not unlike an S-N diagram can be obtained for wire rope. Such a diagram is shown in Fig. 17–21. Here the ordinate is the pressure-strength ratio p/S_u, and S_u is the ultimate tensile strength of the *wire*. The abscissa is the number of bends which occur in the total life of the rope. The curve implies that a wire rope has a fatigue limit; but this is not true at all. A wire rope that is used over sheaves will eventually fail in fatigue or in wear. However, the graph does show that the rope will have a long life if the ratio p/S_u is less than 0.001. Substitution of this ratio in Eq. (17–42) gives

$$S_u = \frac{2000F}{dD} \tag{17–43}$$

where S_u is the ultimate strength of the *wire*, not the rope, and the units of S_u are related to the units of F. This interesting equation contains the wire strength, the load, the rope diameter, and the sheave diameter—all four variables in a single equation! Dividing both sides of Eq. (17-42) by the ultimate strength of the wires S_u and solving for F gives

$$F_f = \frac{(p/S_u)S_u dD}{2} \tag{17–44}$$

where F_f is interpreted as the fatigue tension allowable as the wire is flexed a number of times corresponding to p/S_u selected from Fig. 17–21 for a particular rope and life expectancy. The factor of safety can be defined in fatigue as

$$n = \frac{F_f - F_b}{F_t} \tag{17–45}$$

Table 17–27

Maximum Allowable Bearing Pressures of Ropes on Sheaves (in psi)

Source: Wire Rope Users Manual, AISI, 1979.

	Material				
Rope	Wood[a]	Cast Iron[b]	Cast Steel[c]	Chilled Cast Irons[d]	Manganese Steel[e]
Regular lay:					
6 × 7	150	300	550	650	1470
6 × 19	250	480	900	1100	2400
6 × 37	300	585	1075	1325	3000
8 × 19	350	680	1260	1550	3500
Lang lay:					
6 × 7	165	350	600	715	1650
6 × 19	275	550	1000	1210	2750
6 × 37	330	660	1180	1450	3300

[a] On end grain of beech, hickory, or gum.

[b] For H_B(min.) = 125.

[c] 30–40 carbon; H_B(min.) = 160.

[d] Use only with uniform surface hardness.

[e] For high speeds with balanced sheaves having ground surfaces.

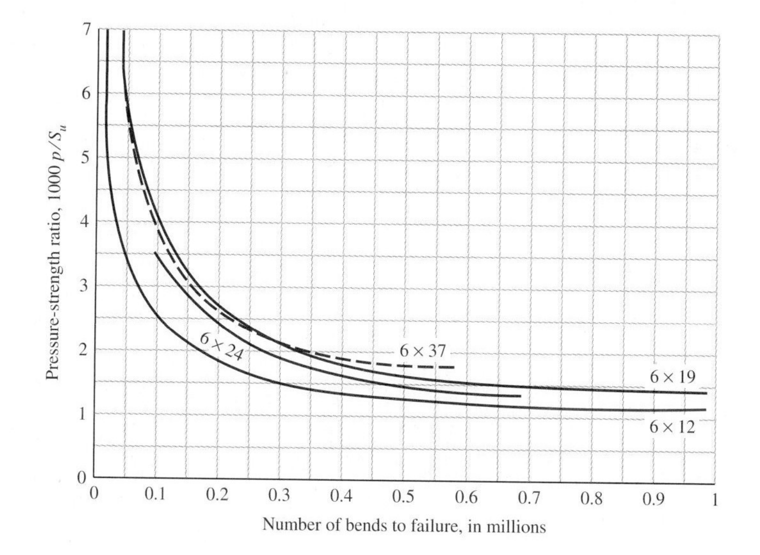

Figure 17–21

Experimentally determined relation between the fatigue life of wire rope and the sheave pressure.

where F_f is the rope tension strength under flexing and F_t is the tension at the place where the rope is flexing. Unfortunately, the designer often has vendor information which tabulates ultimate rope tension and gives no ultimate-strength S_u information concerning the wires from which the rope is made. Some guidance in strength of individual wires is

Improved Plow Steel (Monitor)	$240 < S_u < 280$ kpsi
Plow Steel	$210 < S_u < 240$ kpsi
Mild Plow Steel	$180 < S_u < 210$ kpsi

In wire-rope usage, the factor of safety has been defined in static loading as $n = F_u/F_t$ or $n = (F_u - F_b)/F_t$, where F_b is the rope tension that would induce the same outer-wire stress as that given by Eq. (c). The factor of safety in fatigue loading can be defined as in Eq. (17–45), or by using a static analysis and compensating with a large factor of safety applicable to static loading, as in Table 17–26. When using factors of safety expressed in codes, standards, corporate design manuals, or wire-rope manufacturers' recommendations or from the literature, be sure to ascertain upon which basis the factor of safety is to be evaluated, and proceed accordingly.

If the rope is made of plow steel, the wires are probably hard-drawn AISI 1070 or 1080 carbon steel. Referring to Table 10–4, we see that this lies somewhere between hard-drawn spring wire and music wire. But the constants m and A needed to solve Eq. (10–17) for S_u are lacking.

Practicing engineers who desire to solve Eq. (17–43) should determine the wire strength S_u for the rope under consideration by unraveling enough wire to test for the Brinell hardness. Then S_u can be found using Eq. (5–20). Fatigue failure in wire rope is not sudden, as in solid bodies, but progressive, and shows as the breaking of an outside wire. This means that the beginning of fatigue can be detected by periodic routine inspection.

Figure 17–22 is another graph showing the gain in life to be obtained by using large D/d ratios. In view of the fact that the life of wire rope used over sheaves is only finite, it is extremely important that the designer specify and insist that periodic inspection, lubrication, and maintenance procedures be carried out during the life of the rope. Table 17–28 gives useful properties of some wire ropes.

Figure 17–22

Service-life curve based on bending and tensile stresses only. This curve shows that the life corresponding to $D/d = 48$ is twice that of $D/d = 33$. *(From Wire Rope Users Manual; reproduced by permission from American Iron and Steel Institute.)*

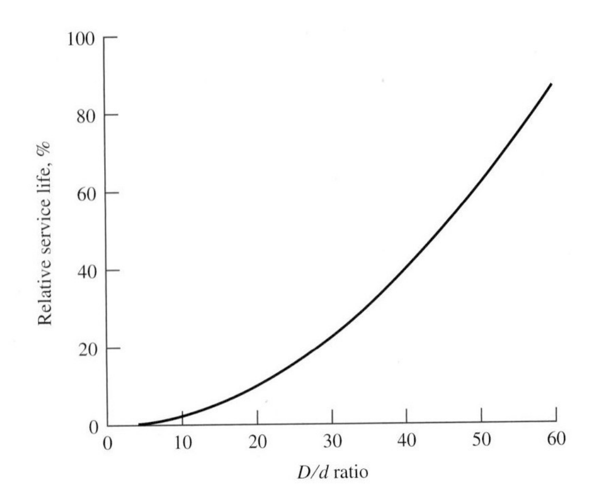

Table 17–28

Some Useful Properties of 6 × 7, 6 × 19, and 6 × 37 Wire Ropes

Wire Rope	Weight per Foot w, lbf/ft	Weight per Foot Including Core w, lbf/ft	Minimum Sheave Diameter D, in	Better Sheave Diameter D, in	Diameter of Wires d_w, in	Area of Metal A_m, in²	Rope Young's Modulus E_r, psi
6 × 7	$1.52d^2$		$42d$	$72d$	$0.111d$	$0.38d^2$	13×10^6
6 × 19	$1.60d^2$	$1.76d^2$	$30d$	$45d$	$0.067d$	$0.40d^2$	12×10^6
6 × 37	$1.55d^2$	$1.71d^2$	$18d$	$27d$	$0.048d$	$0.40d^2$	12×10^6

For a mine-hoist problem we can develop working equations from the preceding presentation. The wire rope tension F_t due to load and acceleration/deceleration is

$$F_t = \left(\frac{W}{m} + w\ell\right)\left(1 + \frac{a}{g}\right) \tag{17–46}$$

where W = weight at the end of the rope (cage and load), lbf
m = number of wire ropes supporting the load
w = weight/foot of the wire rope, lbf/ft
ℓ = pendant length of rope, ft
a = maximum acceleration/deceleration experienced, ft/s^2
g = acceleration of gravity, ft/s^2

The fatigue tensile strength in pounds for a specified life F_f is

$$F_f = \frac{(p/S_u)S_u Dd}{2} \tag{17–47}$$

where (p/S_u) = specified life, from Fig. 17–21
S_u = ultimate tensile strength of the wires, psi
D = sheave or winch drum diameter, in
d = nominal wire rope size, in

The *equivalent bending load* F_b is

$$F_b = \frac{E_r d_w A_m}{D} \tag{17–48}$$

where $E_r =$ Young's modulus for the wire rope, Table 17–28, psi
$d_w =$ diameter of the wires, in
$A_m =$ metal cross-sectional area, Table 17–28, in^2
$D =$ sheave or winch drum diameter, in

The static factor of safety n_s is

$$n_s = \frac{F_u - F_b}{F_t} \tag{17–49}$$

Be careful when comparing recommended static factors of safety to Eq. (17–49), as n_s is sometimes defined as F_u/F_t. The fatigue factor of safety n is

$$n = \frac{F_f - F_b}{F_t} \tag{17–50}$$

EXAMPLE 17–6 Given a 6 × 19 monitor steel ($S_u = 240$ kpsi) wire rope.

(a) Particularize the expressions for rope tension F_t, fatigue tension F_f, equivalent bending tensions F_b, and factor of safety n for a 531.5-ft, 1-ton cage-and-load mine hoist with a starting acceleration of 2 ft/s^2 as depicted in Fig. 17–23. The sheave diameter is 72 in.

(b) Using the expressions developed in part (*a*), examine the variation in factor of safety n for various wire rope diameters d and number of supporting ropes m.

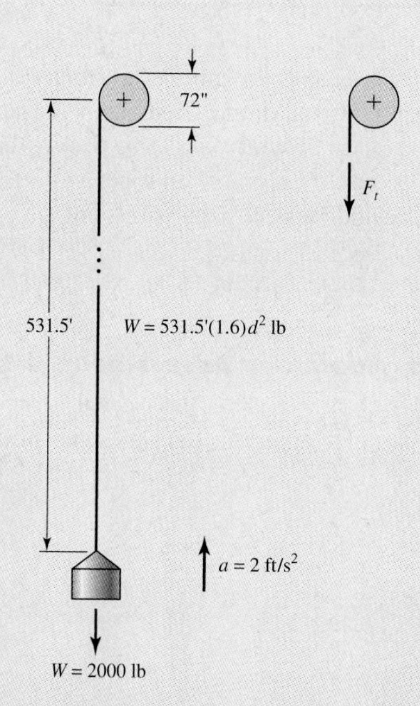

Figure 17–23

Geometry of the mine hoist of Ex. 17–6.

Solution (*a*) Rope tension F_t from Eq. (17–46) is given by

Answer
$$F_t = \left(\frac{W}{m} + wl\right)\left(1 + \frac{a}{g}\right) = \left[\frac{2000}{m} + 1.60(531.5)d^2\right]\left(1 + \frac{2}{32.2}\right)$$

$$= \frac{2124}{m} + 903d^2 \text{ lbf}$$

Fatigue tension F_f from Eq. (17–47) is given by

Answer
$$F_f = \frac{(p/S_u)S_u Dd}{2} = \frac{0.014(240\,000)72d}{2} = 12\,096d \text{ lbf}$$

Equivalent bending tension F_b from Eq. (17–48) is given by

Answer
$$F_b = \frac{E_r d_w A_m}{D} = \frac{12(10^6)0.067d(0.40d^2)}{72} = 4467d^3 \text{ lbf}$$

Factor of safety n in fatigue from Eq. (17–50) is given by

Answer
$$n = \frac{F_f - F_b}{F_t} = \frac{12\,096d - 4467d^3}{2127/m + 903d^2}$$

(*b*) Form a table as follows:

	n			
d	**m = 1**	**m = 2**	**m = 3**	**m = 4**
0.25	1.355	2.641	3.865	5.029
0.375	1.910	3.617	5.150	6.536
0.500	2.336	4.263	5.879	7.254
0.625	2.612	4.573	6.099	7.332
0.750	2.731	4.578	5.911	6.918
0.875	2.696	4.330	5.425	6.210
1.000	2.520	3.882	4.736	5.320

Wire rope sizes are discrete, as is the number of supporting wires. Note that for each m the factor of safety exhibits a maximum. Predictably the largest factor of safety increases with m. If the required factor of safety was to be 6, only three or four ropes can meet the requirement. The sizes are different: 5/8-in ropes with three wires or 3/8-in ropes with four wires. The costs include not only the wires, but the grooved winch drums.

Specification Set, Adequacy Assessment, and Decision Set for Wire Ropes

Wire rope is usually part of a larger mechanism or machine. The following factors apply for a hoisting rope.

Specification Set

- Material
- Wire size and lay

- Sheave and winch diameters
- Associated thimbles, sockets, and clips

Adequacy Assessment

- Estimate the factor of safety guarding against static failure
- Estimate the factor of safety guarding against less than 10^6 bends
- Determine whether recommended minimums of diameter for sheaves, pulleys, winch drums, and so on, are equaled or exceeded.
- Be sure that rope elongation does not compromise function of the larger machine.
- Be sure that the maintenance schedule is in accord with the wire manufacturer's recommendation and with applicable codes and standards.

Since wire rope is usually an integral part of a more complex machine, design problems, considering the brevity of this section, cannot be comprehensively treated. However, in thinking about a decision set, one can view the problem as follows.

Decision Set

- Function: load required, deflection permissible, durability goal
- Design factor: for nonquantifiable exigencies in static and fatigue loading
- Lay of rope, number of strands, number of wires per strand, core, if any
- Sheave and drum sizes
- Number of ropes supporting the load
- Wire size: rope diameter, number of strands, number of wires per strand, core, if any

A figure of merit can be complicated by the need for reliable cost information as it applies to ropes, winches, thimbles, sockets, and clips. As you get into a design problem, the first four or five groups of decisions are usually a priori, with the last one or two becoming the decision variable(s). The purpose of Ex. 17–6 was to introduce the structure of a wire rope design task.

17–7 Flexible Shafts

One of the greatest limitations of the solid shaft is that it cannot transmit motion or power around corners. It is therefore necessary to resort to belts, chains, or gears, together with bearings and the supporting framework associated with them. The flexible shaft may often be an economical solution to the problem of transmitting motion around corners. In addition to the elimination of costly parts, its use may reduce noise considerably.

There are two main types of flexible shafts: the power-drive shaft for the transmission of power in a single direction, and the remote-control or manual-control shaft for the transmission of motion in either direction.

The construction of a flexible shaft is shown in Fig. 17–24. The cable is made by winding several layers of wire around a central core. For the power-drive shaft, rotation should be in a direction such that the outer layer is wound up. Remote-control cables have a different lay of the wires forming the cable, with more wires in each layer, so that the torsional deflection is approximately the same for either direction of rotation.

Flexible shafts are rated by specifying the torque corresponding to various radii of curvature of the casing. A 15-in radius of curvature, for example, will give from 2 to 5 times more torque capacity than a 7-in radius. When flexible shafts are used in a drive

Figure 17–24

Detail of a flexible shaft.

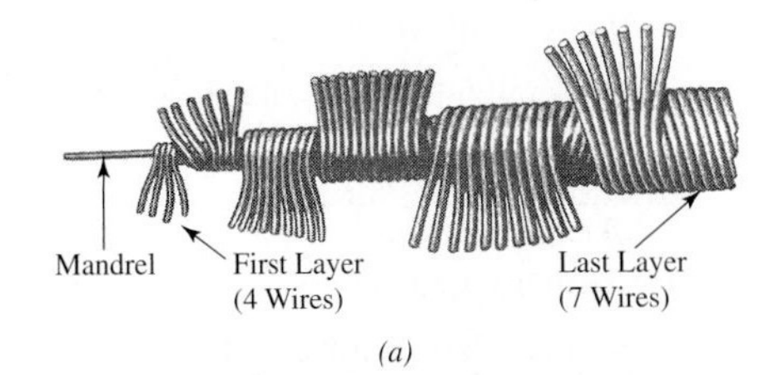

(a)

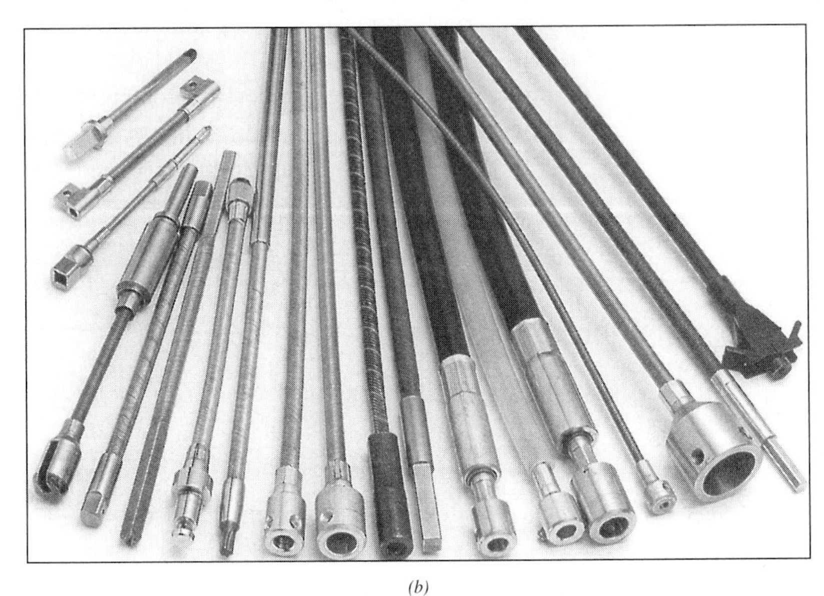

(b)

in which gears are also used, the gears should be placed so that the flexible shaft runs at as high a speed as possible. This permits the transmission of the maximum amount of horsepower.

PROBLEMS

17–1 Return to Ex. 17–1 and complete the following.

(*a*) Find the extant torque capacity, which puts the drive as built at the point of slip, as well as the initial tension F_i.

(*b*) Find the belt width b that exhibits fos $= n_d = 1.1$.

(*c*) For part *b* find the corresponding F_{1a}, F_c, F_i, F_2, power, and fos.

(*d*) What have you learned?

17–2 A 6-in-wide polyamide F-1 flat belt is used to connect a 2-in-diameter pulley to drive a larger pulley with an angular velocity ratio of 0.5. The center-to-center distance is 9 ft. The angular speed of the small pulley is 1750 r/min as it delivers 2 hp. The service is such that a service factor K_s of 1.25 is appropriate.

(*a*) Find F_c, F_i, F_{1a}, and F_2.

(*b*) Find H_a, fos, and belt length.

(*c*) Find the dip.

17–3 Take the drive of Prob. 17–1 and double the belt width. Compare F_c, F_i, F_{1a}, F_2, H_a, fos, and dip.

17–4 Perspective and insight can be gained by doubling all geometric dimensions and observing the effect on problem parameters. Take the drive of Prob. 17–2, double the dimensions, and compare.

17–5 A flat-belt drive is to consist of two 4-ft-diameter cast iron pulleys spaced 16 ft apart. Select a belt type to transmit 60 hp at a pulley speed of 380 r/min. Use a service factor of 1.1 and a design factor of 1.0.

17–6 Examine the Table 17–18 summary of friction-belt relationships in the flat nonmetal belt classification. The allowable tension is an endurance tension which should not be exceeded for long life. It is adjusted by belt width b and factors C_p and C_v in order to form $F_{1a} = C_P C_V F_a b$. The application factor K_s might have been placed in the expression in the denominator. Why was this not done?

17–7 In solving problems and examining examples, you probably have noticed some recurring forms.

$$w = 12\gamma bt = (12\gamma t)b = a_1 b,$$

$$(F_1)_a = F_a b C_p C_v = (F_a C_p C_v)b = a_0 b$$

$$F_c = \frac{wV^2}{g} = \frac{a_1 b}{32.174}\left(\frac{V}{60}\right)^2 = a_2 b$$

$$(F_1)_a - F_2 = 2T/d = 33\,000 H_d/V = 33\,000 H_{\text{nom}} K_s n_d/V$$

$$F_2 = (F_1)_a - [(F_1)_a - F_2] = a_0 b - 2T/d$$

$$f\phi = \ln \frac{(F_1)_a - F_c}{F_2 - F_c} = \ln \frac{(a_0 - a_2)b}{(a_0 - a_2)b - 2T/d}$$

Show that

$$b_{\text{min}} = \frac{1}{a_0 - a_2} \frac{33\,000 H_d}{V} \frac{\exp(f\phi)}{\exp(f\phi) - 1}$$

17–8 Belted pulleys place loads on shafts, inducing bending and loading bearings. Examine Fig. 17–7 and develop an expression for the load the belt places on the pulley, and then apply it to Ex. 17–2.

17–9 Example 17–2 resulted in a 10 in wide A-3 polyamide flat-belt selection. Show that the value of F_1 restoring f to 0.80 is

$$F_1 = \frac{(\Delta F + F_c)\exp f\phi - F_c}{\exp f\phi - 1}$$

and compare the initial tensions.

17–10 The line-shaft illustrated in the figure is used to transmit power from an electric motor by means of flat-belt drives to various machines. Pulley A is driven by a vertical belt from the motor pulley. A belt from pulley B drives a machine tool at an angle of 70° from the vertical and at a center-to-center distance of 9 ft. Another belt from pulley C drives a grinder at a center-to-center distance of 11 ft. Pulley C has a double width to permit belt shifting as shown in Fig. 17–4. The belt from pulley D drives a dust-extractor fan whose axis is located horizontally 8 ft from the axis of the lineshaft. Additional data are

Machine	Speed, r/min	Power, hp	Lineshaft Pulley	Diameter, in
Machine tool	400	12.5	B	16
Grinder	300	4.5	C	14
Dust extractor	500	8.0	D	18

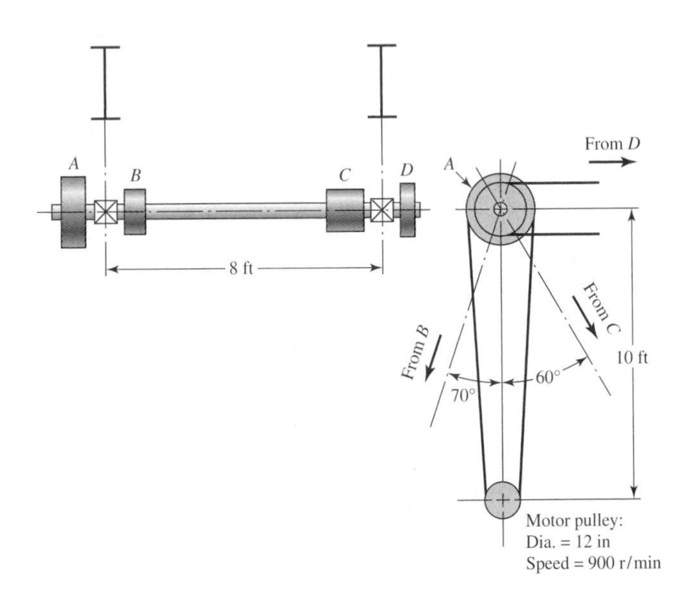

Problem 17–10
(Courtesy of Dr. Ahmed F. Abdel Azim, Zagazig University, Cairo.)

The power requirements, listed above, account for the overall efficiencies of the equipment. The two lineshaft bearings are mounted on hangers suspended from two overhead wide-flange beams. Select the belt types and sizes for each of the four drives. Make provision for replacing belts from time to time because of wear or permanent stretch.

DESIGN **17–11** Two shafts 20 ft apart, with axes in the same horizontal plane, are to be connected with a flat belt in which the driving pulley, powered by a six-pole squirrel-cage induction motor with a 100 brake hp rating at 1140 r/min, drives the second shaft at half its angular speed. The driven shaft drives light-shock machinery loads. Select a flat belt.

DESIGN **17–12** The mechanical efficiency of a flat-belt drive is approximately 98 percent. Because of its high value, the efficiency is often neglected. If a designer should choose to include it, where would he or she insert it in the flat-belt protocol?

DESIGN **17–13** In metal belts, the centrifugal tension F_c is ignored as negligible. Convince yourself that this is a reasonable problem simplification.

DESIGN **17–14** A designer has to select a metal-belt drive to transmit a power of H_{nom} under circumstances where a service factor of K_s and a design factor of n_d are appropriate. The design goal becomes $H_d = H_{\text{nom}} K_s n_d$. Use Eq. (17–8) to show that the minimum belt width is given by

$$b_{\text{min}} = \frac{1}{a}\left(\frac{33\,000 H_d}{V}\right)\frac{\exp f\theta}{\exp f\theta - 1}$$

where a is the constant from $F_{1a} = ab$.

DESIGN

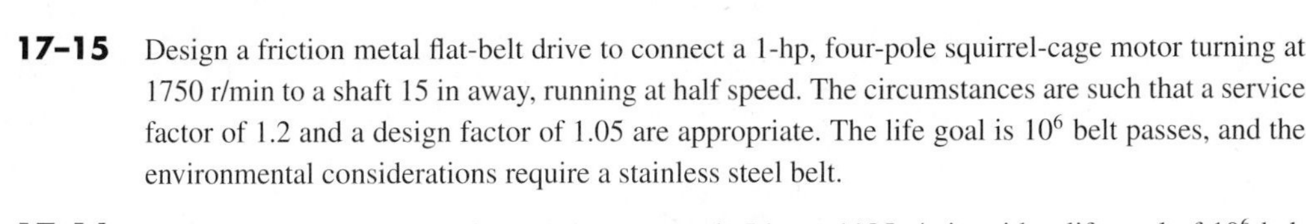

17–15 Design a friction metal flat-belt drive to connect a 1-hp, four-pole squirrel-cage motor turning at 1750 r/min to a shaft 15 in away, running at half speed. The circumstances are such that a service factor of 1.2 and a design factor of 1.05 are appropriate. The life goal is 10^6 belt passes, and the environmental considerations require a stainless steel belt.

DESIGN

17–16 A beryllium-copper metal flat belt is to transmit 5 hp at 1125 r/min with a life goal of 10^6 belt passes between two shafts 20 in apart whose centerlines are in a horizontal plane. The coefficient of friction between belt and pulley is 0.32. The conditions are such that a service factor of 1.25 and a design factor of 1.1 are appropriate. The driven shaft rotates at one-third the motor-pulley speed. Specify your belt, pulley sizes, and initial tension at installation.

DESIGN

17–17 For the conditions of Prob. 17–16 use a 1095 plain carbon-steel heat-treated belt. Conditions at the driving pulley hub require a pulley outside diameter of 3 in or more. Specify your belt, pulley sizes, and initial tension at installation.

DESIGN

17–18 Examine the definition of a in the friction metal-belt portion of Table 17–18. By setting the square brackets to zero the minimum value of d can be found. Check the minimum pulley diameter column of Table 17–7 for stainless steel belts at 10^6 belt-pass life. Any pulley diameter near the minimum diameter will produce a small value of a, and consequently a large belt width b, in order to obtain a reasonable allowable tension F_{1a}. Investigate this using Prob. 17–17 to see how much larger than $d_{\min}$ you would want to go for an initial trial of pulley size.

DESIGN

17–19 A single V belt is to be selected to deliver engine power to the wheel-drive transmission of a riding tractor. A 5-hp single-cylinder engine is used. At most, 60 percent of this power is transmitted to the belt. The driving sheave has a diameter of 6.2 in, the driven, 12.0 in. The belt selected should be as close to a 92-in pitch length as possible. The engine speed is govern-controlled to a maximum of 3100 r/min. Select a satisfactory belt and assess the factor of safety and the belt life in passes.

DESIGN

17–20 A 60-hp four-cylinder internal combustion engine is used to drive a brick-making machine under a schedule of two shifts per day. The drive consists of two 26-in sheaves spaced about 12 ft apart, with a sheave speed of 400 r/min. Select a V-belt arrangement. Find the factor of safety, and estimate the life in passes and hours.

DESIGN

17–21 A reciprocating air compressor has a 5-ft-diameter flywheel 14 in wide, and it operates at 170 r/min. An eight-pole squirrel-cage induction motor has nameplate data 50 bhp at 875 r/min.

(*a*) Design a VV drive.

(*b*) Can cutting the V-belt grooves in the flywheel be avoided by using a V-flat drive?

17–22 The geometric implications of a V-flat drive are interesting.

(*a*) If the earth's equator was an inextensible string, snug to the spherical earth, and you spliced 6 ft of string into the equatorial cord and arranged it to be concentric to the equator, how far off the ground is the string?

(*b*) Using the solution to part a, formulate the modifications to the expressions for m_G, θ_d and θ_D, L_p, and C.

(*c*) As a result of this exercise, how would you revise your solution to part b of Prob. 17–21?

17–23 A 2-hp electric motor running at 1720 r/min is to drive a blower at a speed of 240 r/min. Select a V-belt drive for this application and specify standard V belts, sheave sizes, and the resulting center-to-center distance. The motor size limits the center distance to at least 22 in.

17–24 Two B85 V belts are used in a drive composed of a 5.4-in driving sheave, rotating at 1200 r/min, and a 16-in driven sheave. Find the power capacity of the drive based on a service factor of 1.25, and find the center-to-center distance.

ANALYSIS

17–25 The standard roller-chain number indicates the chain pitch in inches, construction proportions, series, and number of strands as follows:

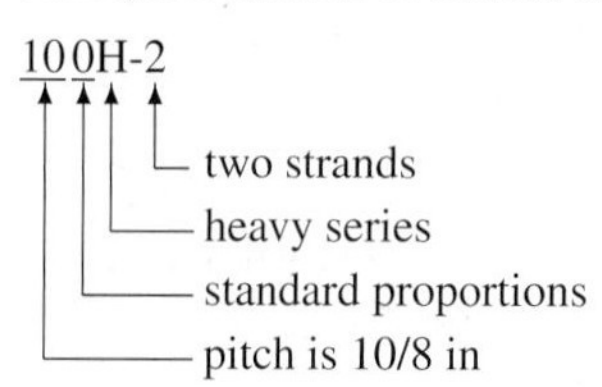

This convention makes the pitch direct-reading from the chain number. In Ex. 17–5 ascertain the pitch from the selected chain number and confirm from Table 17–20.

ANALYSIS

17–26 Equate Eqs. (17–32) and (17–33) to find the rotating speed n_1 at which the power equates and marks the division between the premaximum and the postmaximum power domains.

(*a*) Show that

$$n_1 = \left[\frac{0.25(10^6)K_r N_1^{0.42}}{p^{(2.2-0.07p)}}\right]^{1/2.4}$$

(*b*) Find the speed n_1 for a no. 60 chain, $p = 0.75$ in, $N_1 = 17$, $K_r = 17$, and confirm from Table 17–21.

(*c*) At which speeds is Eq. (17–40) applicable?

17–27 A double-strand no. 60 roller chain is used to transmit power between a 13-tooth driving sprocket rotating at 300 r/min and a 52-tooth driven sprocket.

(*a*) What is the allowable horsepower of this drive?

(*b*) Estimate the center-to-center distance if the chain length is 82 pitches.

(*c*) Estimate the torque and bending force on the driving shaft by the chain if the actual horsepower transmitted is 30 percent less than the corrected (allowable) power.

17–28 A four-strand no. 40 roller chain transmits power from a 21-tooth driving sprocket to an 84-tooth driven sprocket. The angular speed of the driving sprocket is 2000 r/min.

(*a*) Estimate the chain length if the center-to-center distance has to be about 20 in.

(*b*) Estimate the tabulated horsepower entry H'_{tab} for a 20 000-hr life goal.

(*c*) Estimate the rated (allowable) horsepower that would appear in Table 17–21 for a 20 000-hr life.

(*d*) Estimate the tension in the chain at the allowable power.

17–29 A 700 r/min 25-hp squirrel-cage induction motor is to drive a two-cylinder reciprocating pump, out-of-doors under a shed. A service factor K_s of 1.5 and a design factor of 1.1 are appropriate. The pump speed is 140 r/min. Select a suitable chain and sprocket sizes.

17–30 A centrifugal pump is driven by a 50-hp synchronous motor at a speed of 1800 r/min. The pump is to operate at 900 r/min. Despite the speed the load is smooth ($K_s = 1.2$). For a design factor of 1.1 specify a chain and sprockets that will realize a 50 000-hr life goal.

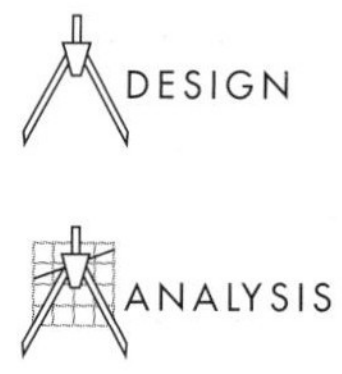

17–31 A mine hoist uses a 2-in 6 × 19 monitor-steel wire rope. The rope is used to haul loads of 4 tons from the shaft 480 ft deep. The drum has a diameter of 6 ft, the sheaves are of good-quality cast steel, and the smallest is 3 ft in diameter.

(*a*) Using a maximum hoisting speed of 1200 ft/min and a maximum acceleration of 2 ft/s^2, estimate the stresses in the rope.

(*b*) Estimate the various factors of safety.

17–32 A temporary construction elevator is to be designed to carry workers and materials to a height of 90 ft. The maximum estimated load to be hoisted is 5000 lb at a velocity not to exceed 2 ft/s. Based on minimum sheave diameters and acceleration of 4 ft/s^2, specify the number of ropes required if the 1-in plow-steel 6 × 19 hoisting strand is used.

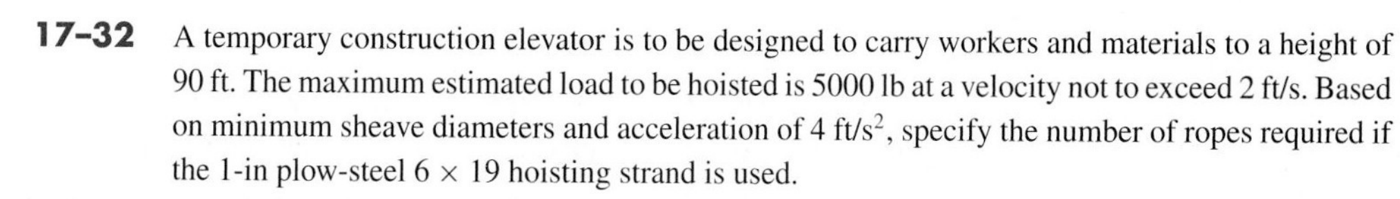

17–33 A 2000-ft mine hoist operates with a 72-in drum using 6 × 19 monitor-steel wire rope. The cage and load weigh 8000 lb, and the cage is subjected to an acceleration of 2 ft/s^2 when starting.

(*a*) For a single-strand hoist how does the factor of safety $n = F_f/F_t$ vary with the choice of rope diameter?

(*b*) For four supporting strands of wire rope attached to the cage, how does the factor of safety vary with the choice of rope diameter?

17–34 Generalize the results of Prob. 17–33 by representing the factor of safety n as

$$n = \frac{ad}{(b/m) + cd^2}$$

where m is the number of ropes supporting the cage, and a, b, and c are constants. Show that the optimal diameter is $d^* = [b/(mc)]^{1/2}$ and the corresponding maximum attainable factor of safety is $n^* = a[m/(bc)]^{1/2}/2$.

17–35 From your results in Prob. 17–34, show that to meet a fatigue factor of safety n_1 the optimal solution is

$$m = \frac{4bcn_1}{a^2} \text{ ropes}$$

having a diameter of

$$d = \frac{a}{2cn_1}$$

Solve Prob. 17–33 if a factor of safety of 2 is required. Show what to do in order to accommodate to the necessary discreteness in the rope diameter d and the number of ropes m.

17–36 For Prob. 17–31 estimate the elongation of the rope if a 9000-lb loaded mine cart is placed on the cage. The results of Prob. 4–16 may be useful.

Computer Programs

In approaching the ensuing computer problems, the following suggestions may be helpful:

- Decide whether an analysis program or a design program would be more useful. In problems as simple as these, you will find the programs similar. For maximum instructional benefit, try the design problem.
- Creating a design program without a figure of merit precludes ranking alternative designs but does not hinder the attainment of satisfactory designs. Your instructor can provide the class design library commercial catalogs, which not only have price information but define available sizes.
- Quantitative understanding and logic of interrelations are required for programming. Difficulty in programming is a signal to you and your instructor to increase your understanding. The following programs can be accomplished in 100–500 lines of code.
- Make programs interactive and user-friendly.
- Let the computer do what it can do best; the user should do what a human can do best.
- Assume the user has a copy of the text and can respond to prompts for information.

- If interpolating in a table is in order, solicit table entries in the neighborhood, and let the computer crunch the numbers.
- In decision steps, allow the user to make the necessary decision, even if it is undesirable. This allows learning of consequences and the use of the program for analysis.
- Display a lot of information in the summary. Show the decision set used up-front for user perspective.
- When a summary is complete, adequacy assessment can be accomplished with ease, so consider adding this feature.

17–37 Your experience with Probs. 17–1 through 17–12 has placed you in a position to write an interactive computer program to design/select flat-belt drive components. A possible decision set is

A Priori Decisions

- Function: H_{nom}, r/min, velocity ratio, approximate C
- Design factor: n_d
- Initial tension maintenance: catenary
- Belt material: t, d_{min}, allowable tension, density, f
- Drive geometry: d, D
- Belt thickness: t (in material decision)

Design Decisions

- Belt width: b

17–38 Problems 17–13 through 17–18 have given you some experience with flat-metal friction belts, indicating that a computer program could be helpful in the design/selection process. A possible decision set is

A Priori Decisions

- Function: H_{nom}, r/min, velocity ratio approximate C
- Design factor: n_d
- Belt material: S_y, E, ν, d_{min}
- Drive geometry: d, D
- Belt thickness: t

Design Decisions

- Belt width: b
- Length of belt (often standard loop periphery)

17–39 Problems 17–19 through 17–24 have given you enough experience with V belts to convince you that a computer program would be helpful in the design/selection of V-belt drive components. Write such a program.

17–40 Experience with Probs. 17–25 through 17–30 can suggest an interactive computer program to help in the design/selection process of roller-chain elements. A possible decision set is

A Priori Decisions

- Function: power, speed, space, K_s, life goal
- Design factor: n_d
- Sprocket tooth counts: N_1, N_2, K_1, K_2

Design Decisions

- Chain number
- Strand count
- Lubrication system
- Chain length in pitches

(center-to-center distance for reference)

18 Shafts and Axles

18–1	Introduction **1108**
18–2	Sufficing Geometric Constraints **1111**
18–3	Sufficing Strength Constraints **1120**
18–4	The Adequacy Assessment **1128**
18–5	Shaft Materials **1134**
18–6	Hollow Shafts **1135**
18–7	Critical Speeds **1135**
18–8	Shaft Design **1141**
18–9	Computer Considerations **1142**

18–1 Introduction

A *shaft* is a rotating member, usually of circular cross section, used to transmit power or motion. It provides the axis of rotation, or oscillation, of elements such as gears, pulleys, flywheels, cranks, sprockets, and the like and controls the geometry of their motion. An *axle* is a nonrotating member which carries no torque and is used to support rotating wheels, pulleys, and the like. The automotive axle is not a true axle; the term is a carryover from the horse-and-buggy era, when the wheels rotated on nonrotating members. A *spindle* is a short shaft. Terms such as *lineshaft, headshaft, stub shaft, transmission shaft, countershaft,* and *flexible shaft* are names associated with special usage.

A shaft design really begins after much preliminary work. The design of the machine itself will dictate that certain gears, pulleys, bearings, and other elements will have at least been partially analyzed and their size and spacing tentatively determined. At this stage the design must be studied from the following points of view:

1 Deflection and rigidity
- **(a)** Bending deflection
- **(b)** Torsional deflection
- **(c)** Slope at bearings and shaft-supported elements
- **(d)** Shear deflection due to transverse loading of short shafts

2 Stress and strength
- **(a)** Static strength
- **(b)** Fatigue strength
- **(c)** Reliability

The geometry of a shaft is generally that of a stepped cylinder. Gears, bearings, and pulleys must always be accurately positioned and provision made to accept thrust loads. The use of shaft shoulders is an excellent means of axially locating the shaft elements; these shoulders can be used to preload rolling bearings and to provide the necessary thrust reactions to the rotating elements. For these reasons our analysis will usually involve stepped shafts.

Figure 18–1 shows the stepped shaft supporting the gear of a worm-gear speed reducer. You should be able to determine the reason for each step of the shaft and for the sleeve. Note also how the lubricant is supplied and drained. Can this reducer be used for counterclockwise as well as clockwise rotation? Why?

In deciding on an approach to design, it is necessary to realize that a stress analysis at a specific point on a shaft can be made using only the shaft geometry in the vicinity of that point. Thus the geometry of the entire shaft is not needed. In design it is usually possible to locate the critical areas, size these to meet the strength requirements, and then size the rest of the shaft to meet the requirements of the shaft-supported elements.

Note that the deflection and slope analyses cannot be made until the geometry of the entire shaft has been defined. Thus deflection is a function of the geometry *everywhere*, whereas the stress at a section of interest is a function of *local geometry and moments.* For this reason shaft design allows a consideration of stress and strength first. Then, after tentative values for the shaft dimensions have been established, the determination of the deflection and slope can be made.

The geometric configuration of the shaft to be designed is usually determined from past experience and, most often, is simply a revision of existing models in which a limited number of changes must be made. These changes can result for a variety of reasons, such as the use of a newly designed seal or coupling, a change in the power or speed, bearings of a different size, or the use of newly designed rotating components. Such changes are easy for the designer and need no further explanation.

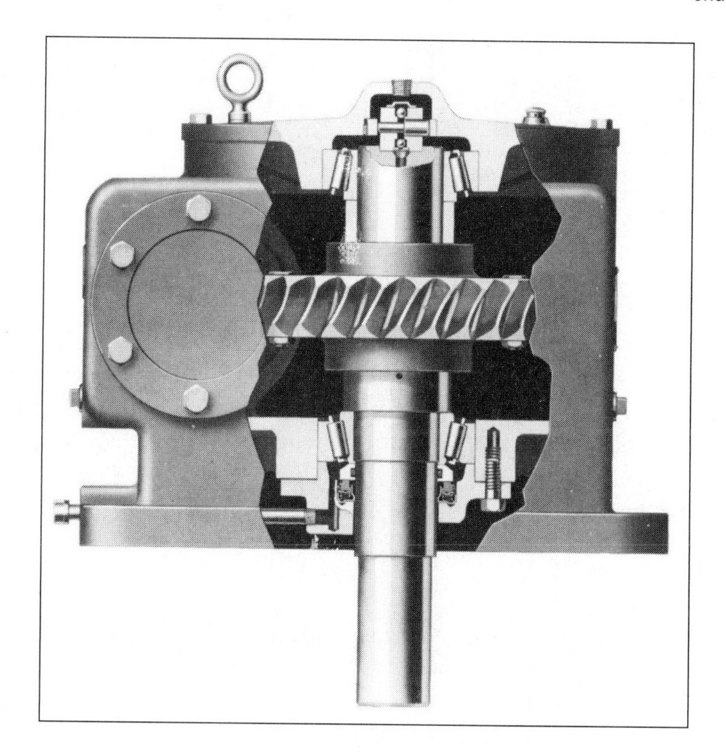

Figure 18–1

A vertical worm-gear speed reducer. *(Courtesy Cleveland Worm and Gear Company.)*

If there is no existing design to use as a starter, then the determination of the shaft geometry may have many solutions. This problem is illustrated by the two examples of Fig. 18–2. In Fig. 18–2*a* a geared countershaft is to be supported by two bearings. In Fig. 18–2*c* a fanshaft is to be configured. A variety of solutions should occur to you for each of these problems. The solutions shown are not necessarily the best ones, but they do illustrate how the shaft-mounted devices are fixed and located in the axial direction, and how provision is made for torque transfer from one element to another. Note that the shaft in Fig. 18–2*a* is subject to bending, torsional, and axial loads. The shaft in Fig. 18–2*c* is subject only to bending and torsion.

There is no magic formula to determine the shaft geometry for any given design situation. The best approach is that of studying existing designs to learn how similar problems have been solved and then of combining the best of these to solve your own problem.

Many shaft-design situations include the problem of transmitting torque from one element to another on the shaft. Common *torque-transfer elements* are:

- Keys
- Splines
- Setscrews
- Pins
- Press or shrink fits
- Tapered fits

Keys, pins, and *setscrews* have already been described in Chap. 8. Pins for this purpose include not only straight pins, and tapered pins but also a wide variety of patented pins many of which act as spring pins. Some of these are for locational purposes only—cotters, for example, should not be used to transmit very much torque—but others will

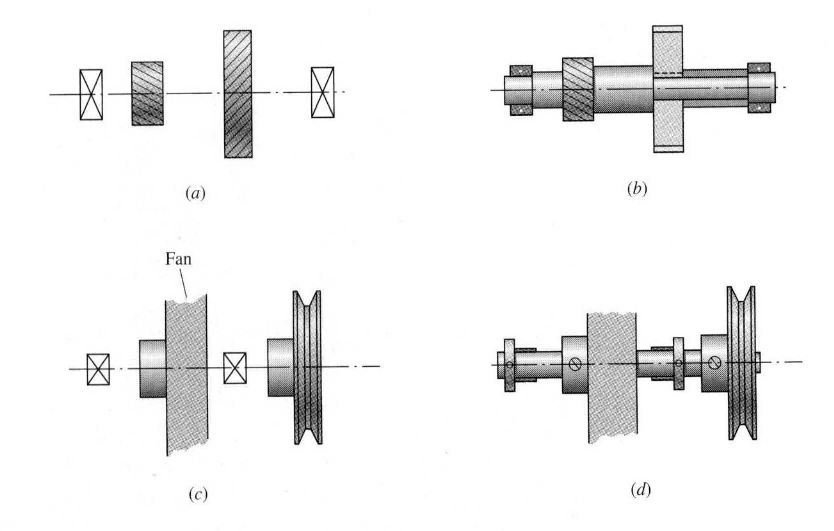

Figure 18–2

(*a*) Choose a shaft configuration to support and locate the two gears and two bearings. (*b*) Solution uses an integral pinion, three shaft shoulders, key and keyway, and sleeve. The housing locates the bearings on their outer rings and receives the thrust loads. (*c*) Choose fan-shaft configuration. (*d*) Solution uses sleeve bearings, a straight-through shaft, locating collars, and setscrews for collars, fan pulley, and fan itself. The fan housing supports the sleeve bearings.

serve as good torque transmitters. The use of these devices requires radial holes through the shaft, and hence stress concentration could be a problem, depending upon their location.

Shaft splines resemble gear teeth cut or forged into the shaft surface. They are used when large amounts of torque are to be transferred. When splines are used, stress concentration is generally quite moderate.

Press and shrink fits for securing hubs to shafts are used both for torque transfer and for preserving axial location. The resulting stress-concentration factor is usually quite small. A similar method is to use a split hub with screws to clamp the hub to the shaft. This method allows for disassembly and lateral adjustments. Another similar method uses a two-part hub consisting of a split inner member which fits into a tapered hole. The assembly is then tightened to the shaft using screws which force the inner part into the wheel and clamp the whole assembly against the shaft.

Plain tapered fits between the shaft and the shaft-mounted device are often used on the overhanging end of a shaft. Screw threads at the shaft end then permit the use of a nut to lock the wheel tightly to the shaft. This approach is useful because it can be disassembled, but it does not provide good axial location of the wheel on the shaft.

All these torque-transfer means solve the problem of securely anchoring the wheel or device to the shaft, but not all of them solve the problem of accurate axial location of the device. Some of the most-used *locational devices* include:

- Cotter and washer
- Nut and washer
- Sleeve
- Shaft shoulder
- Ring and groove
- Setscrew
- Split hub or tapered two-piece hub
- Collar and screw
- Pins

We have already discussed some of these items. The use of a ring fitted into a shaft groove is an economical solution to some problems. The grooves are quite shallow; many of the ring styles available do exert a spring force against the device to be anchored; and sometimes the grooves can be located where the effect of the stress-concentration factor is small or unimportant.

Figure 18–2*d* shows the use of a collar in which holding is obtained by tightening the setscrews against the shaft. An alternative—and better—arrangement uses a split collar and a cap screw to clamp the collar in place.

A Plan for Attacking the Shaft Design Problem

The decision sets we have encountered so far have had one or two decision variables, and the constraints upon them have often been active (tight) when we were finished. A shaft is a complicated geometric form, as you can appreciate after reading this section. A shaft may have a half-dozen different diameters and a similar number of shoulder-to-shoulder lengths. Some of the positions of elements attached to the shaft may have fixed or variable relationships to others, and the bearing locations may be decision variables too. We can quickly accumulate a dozen decision variables. This creates geometries we can no longer visualize as the figure of merit is plotted above the decision variable hyperplane. Furthermore, the many constraints present may really challenge an optimization program in finding a global optimum. When all is done, have we arrived at the correct determination?

To help reduce the dimensionality from a dozen independent variables to one or two, we can recognize that if the tight constraints are located in advance, the problem is greatly simplified. Such recognition can be based on experience or on procedures that help when experience is lacking. One method of approaching the optimum while reducing the dimensionality of the problem can be called *sufficing of constraints*. We will introduce this idea in sufficing geometric constraints, then go on to sufficing strength constraints, then we will marshall our tools.

18–2 Sufficing Geometric Constraints

Consider the geometric constraints for a transmission shaft design task. The first step is to size the gears and pulleys for the specified speed and power. The root diameter of the gear teeth or pulley groove plus the necessary radial space for a keyway fixes one constraint on shaft size. Having the gear and pulley size fixes the forces on the system. The second step is to select bearings to provide adequate life under these forces and speeds. After the bearings are selected, the distance between the bearings will be set. Also, for roller bearings, the bearing bore places a limit on shaft size. The third step is to consider shaft deflection and stress as outlined below.

In material bodies distortion is unavoidable under load. We seek to control it to avoid compromising function. The slope of a shaft centerline with respect to a rolling-bearing outer ring centerline ought to be less than 0.001 rad for cylindrical and 0.0005 rad for tapered roller bearings. Similarly, it should be less that 0.004 for deep-groove roller bearings and, typically, less than 0.0087 for spherical ball bearings. At a gear mesh the allowable relative slope of two spur gears with uncrowned teeth should be held to less than 0.0005. It is apparent that there are several stringent distortion limits on a power transmission shaft. The designer has a choice of designing for strength and checking distortion or designing for distortion and checking for strength. Most power transmission shafts have an active distortion constraint, so designing for distortion and checking for strength is appealing.

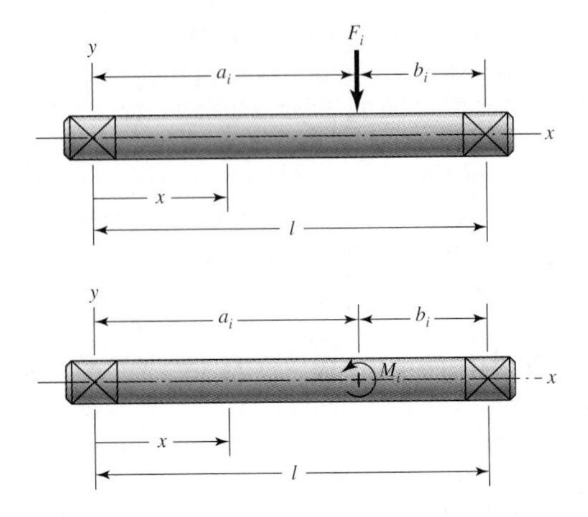

Figure 18–3

Simply supported shaft with force F_i and couple M_i applied.

For perspective, the first move is to find a uniform-diameter shaft that satisfies all the distortion constraints. Going with the odds, since bearing slope constraints are often limiting, we first establish the diameter of a uniform shaft. Consider the shafts in Fig. 18–3. We refer to Tables E–9–6 and E–9–8 to develop expressions for slope. Using superposition, for the left-bearing slope constraint active,

$$d = \left| \frac{32n}{3\pi E \ell \Sigma\theta} \left\{ \left[\sum F_i b_i \left(b_i^2 - \ell^2\right) + \sum M_i \left(3a_i^2 - 6a_i \ell + 2\ell^2\right) \right]_H^2 + \left[\sum F_i b_i \left(b_i^2 - \ell^2\right) + \sum M_i \left(3a_i^2 - 6a_i \ell + 2\ell^2\right) \right]_V^2 \right\}^{1/2} \right|^{1/4} \tag{18–1}$$

and, for the right-bearing constraint active,

$$d = \left| \frac{32n}{3\pi E \ell \Sigma\theta} \left\{ \left[\sum F_i a_i \left(\ell^2 - a_i^2\right) + \sum M_i \left(3a_i^2 - \ell^2\right) \right]_H^2 + \left[\sum F_i a_i \left(\ell^2 - a_i^2\right) + \sum M_i \left(3a_i^2 - \ell^2\right) \right]_V^2 \right\}^{1/2} \right|^{1/4} \tag{18–2}$$

where $\Sigma\theta$ is the absolute value of the allowable slope at the bearing. These equations are worth programming.

EXAMPLE 18–1

The shaft depicted in Fig. 18–4 carries two spur gears and has loadings as shown. The bearings located at A and B will be cylindrical roller bearings. The spatial centerline slope at the bearings is limited to 0.001 rad with a design factor of 1.5. Estimate the diameter of a uniform shaft that meets the slope constraints imposed by the bearings.

Figure 18–4

A contemplated shaft is to carry two spur gears between bearings A and B. Reactions and gear loadings are shown in a plane. The y axis is vertical, the other two are horizontal.

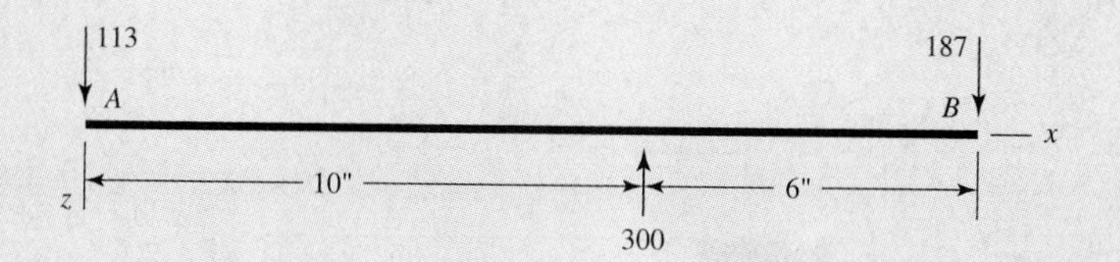

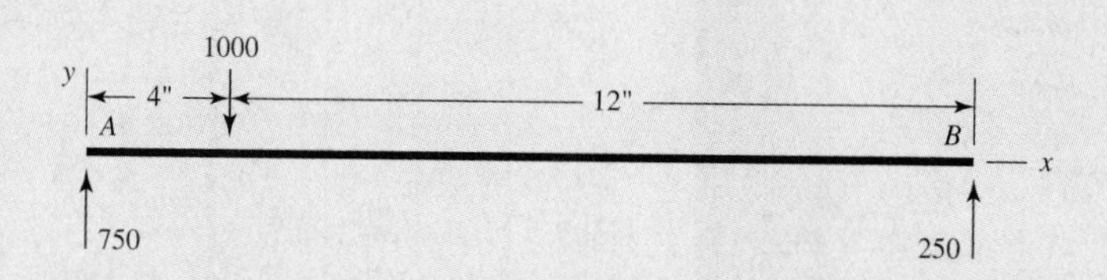

Solution From Eq. (18.1),

$$d = \left| \frac{32n_d}{3\pi E\ell\Sigma\theta} \{[F_1 b_1(b_1^2 - \ell^2)]_H^2 + [F_2 b_2(b_2^2 - \ell^2)]_V^2\}^{1/2} \right|^{1/4}$$

$$= \left| \frac{32(1.5)}{3\pi(30)10^6 16(0.001)} \{[300(6)(6^2 - 16^2)]^2 + [1000(12)(12^2 - 16^2)]^2\}^{1/2} \right|^{1/4}$$

Answer $= 1.964\text{in}$

for left-bearing constraint tight, and, using Eq. (18–2),

$$d = \left| \frac{32n_d}{3\pi E\ell\Sigma\theta} \{[F_1 a_1(\ell^2 - a_1^2)]_H^2 + [F_2 a_2(\ell^2 - a_2^2)]_V^2\}^{1/2} \right|^{1/4}$$

$$= \left| \frac{32(1.5)}{3\pi(30)(10)^6 16(0.001)} \{[300(10)(10^2 - 16^2)]^2 + [1000(4)(4^2 - 16^2)]^2\}^{1/2} \right|^{1/4}$$

Answer $= 1.835\text{ in}$

for the right-bearing constraint tight. This initial calculation tells a designer that a uniform shaft of 1.964 in in diameter will meet the bearing-slope constraints. For any decrease in diameter at the bearing journals and shoulders, diametral increase elsewhere should be made for gear seats and shoulders. The designer has an idea of the "heft" of the shaft, a useful perspective.

In Sec. 4–6 we developed a tabular method that can be used for stepped shafts. Because of your familiarity with the method, the table can take the form of that shown in Table 18–1. The deflection column entry y is formed from the prediction equation:

$$y = \int_0^x \int_0^x \frac{M}{EI} dx dx + C_1 x + C_2 \tag{18–3}$$

The slope dy/dx column is formed from the prediction equation:

$$\frac{dy}{dx} = \int_0^x \frac{M}{EI} dx + C_1 \tag{18–4}$$

Table 18–1

Form for Tabulation Method for Shaft Transverse Deflection Due to Bending Moment

	Moment M	Distance x	Diameter d	Modulus E	$\frac{M}{EI}$	$\int_0^x \frac{M}{EI}\,dx$	$\int_0^x \left(\int_0^x \frac{M}{EI}\,dx\right) dx$	Deflection y	Slope dy/dx
1						0	0		
2									
3									

where the constants C_1 and C_2 are found from

$$C_1 = \frac{\int_0^{x_a}\int_0^{x_a} M/(EI)dxdx - \int_0^{x_b}\int_0^{x_b} M/(EI)dxdx}{x_a - x_b} \tag{18–5}$$

$$C_2 = \frac{x_b\int_0^{x_a}\int_0^{x_a} M/(EI)dxdx - x_a\int_0^{x_b}\int_0^{x_b} M/(EI)dxdx}{x_a - x_b} \tag{18–6}$$

where x_a and x_b are the bearing locations. This tabular procedure can be repeated for the orthogonal deflection plane if needed, Pythagorean combinations of slopes or deflections giving the spatial values.

Given the bending-moment diagram and the shaft geometry, the deflection and slope at the station points can be found. If, in examining the deflection column, any entry is too large in absolute magnitude, a new diameter can be found from

$$d_{\text{new}} = d_{\text{old}}\left|\frac{n_d y_{\text{old}}}{y_{\text{all}}}\right|^{1/4} \tag{18–7}$$

where y_{all} is the allowable deflection at that station and n_d is the design factor. Similarly, if any slope is too large in absolute magnitude, a new diameter can be found from

$$d_{\text{new}} = d_{old}\left|\frac{n_d(dy/dx)_{\text{old}}}{(\text{slope})_{\text{all}}}\right|^{1/4} \tag{18–8}$$

where $(\text{slope})_{\text{all}}$ is the allowable slope. As a result of these calculations note the largest $d_{\text{new}}/d_{\text{old}}$ ratio, then multiply *all* diameters by this ratio. The tight constraint will be just tight, and all others will be loose. Don't be too concerned about end journal sizes, as their influence is usually negligible. The beauty of the method is that Table 18–1 needs to be completed just once and constraints can be rendered loose but for one, diameters all identified without reworking the table. Furthermore, the deflections and slopes are (usually) exact. The method lends itself to computer implementation.

EXAMPLE 18–2

The steel shaft depicted in Fig. 18–5*a* is overhung on the right-bearing side. It carries two spur gears, one at *B* and one at *D*. The radial forces of 2000 and 1100 lbf lie in the same plane. The bending moment *M*, transverse shear *V*, and torque *T* have distributions as shown in Figs. 18–5*b*, 18– 5*c*, and 18–5*d*, respectively. The gears both have a diametral pitch of 8; neither their mates-in-mesh nor detail about their sizes is shown. Because of the use of uncrowned teeth, the slopes are limited to 0.0005 rad on each gear. For a design factor $n_d = 1.5$, use the tabular method to find the diameter of a *uniform* shaft to avoid the constraints on the gears and those due to the cylindrical roller bearings expected to be used.

Solution

Using Table 18–1 and a uniform diameter of 1 in, form Table 18–2, which is abbreviated to display the results.

Bearing slopes: From Eq. (18–8), with the subscripts of *d* being the station numbers in Table 18–2,

$$d_1 = 1\left|\frac{1.5(-0.001\ 75)}{0.001}\right|^{1/4} = 1.273 \text{ in}$$

Figure 18–5

(*a*) A contemplated shaft carrying two gears and transmitting torque between the gears. The forces lie in a plane. (*b*) The bending-moment diagram. (*c*) The transverse-shear diagram. (*d*) The torque diagram.

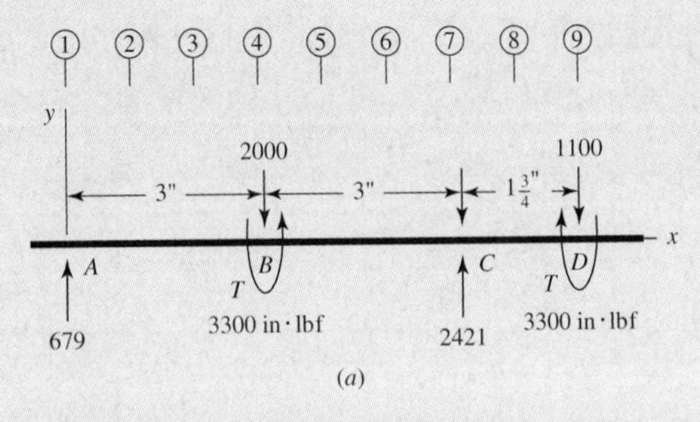

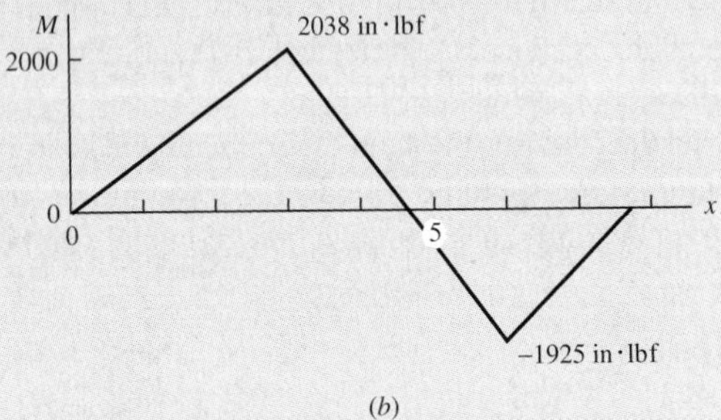

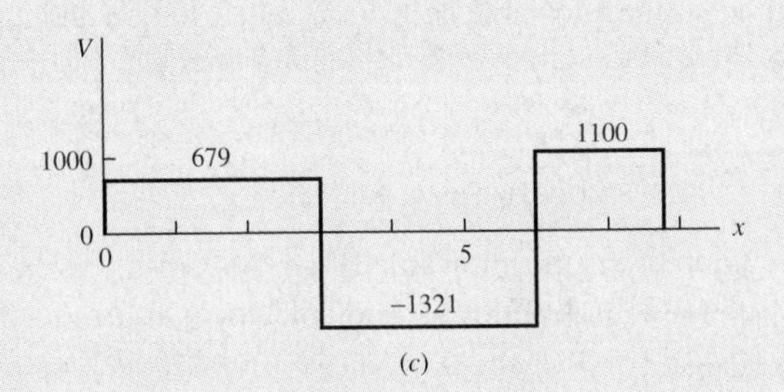

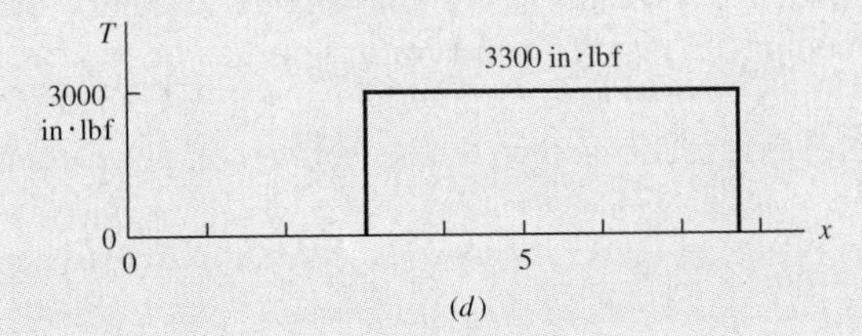

$$d_7 = 1\left|\frac{1.5(0.000\,442)}{0.001}\right|^{1/4} = 0.902\text{ in}$$

Gear-mesh slopes: From Eq. (18–8),

$$d_4 = 1\left|\frac{1.5(0.000\,327}{0.0005}\right|^{1/4} = 0.995\text{ in}$$

$$d_9 = 1\left|\frac{1.5(0.000\,279)}{0.0005}\right|^{1/4} = 0.956\text{ in}$$

Center-to-center distance: For 8-diametral pitch gears the center-to-center expansion distance on commercial-quality gears is 0.010 in. We will divide this between meshing

Table 18–2

Transverse Slope and Deflection Due to Bending Moment in Example 18–2

Station No.	M, in · lb	x, in	d, in	y, in	dy/dx
1	0	0	0	0	−0.001 75
	0		1		
2	679	1	1	−0.001 67	−0.001 52
	679		1		
3	1359	2	1	−0.002 88	−0.000 827
	1359		1		
4	2038	3	1	−0.003 17	+0.000 327
	2038		1		
5	717	4	1	−0.002 30	0.001 26
	717		1		
6	−604	5	1	−0.000 946	0.001 30
	−604		1		
7	−1925	6	1	0	0.000 442
	−1925		1		
8	−825	7	1	0.000 100	0.000 069
	−825		1		
9	0	7.75	1	0.000 256	0.000 279
	0		0		

$C_1 = -0.00175$ rad, $C_2 = 0$ in.

gears, making the allowable transverse deflection 0.005 in. From Eq. (18–7),

$$d_4 = 1\left|\frac{1.5(-0.003\ 17)}{0.005}\right|^{1/4} = 0.988 \text{ in}$$

$$d_9 = 1\left|\frac{1.5(0.0002\ 56)}{0.005}\right|^{1/4} = 0.526 \text{ in}$$

The shaft diameter, keeping all of these constraints loose but one, is 1.273 in. We also know that the tight constraint (with a margin of 1.5) is the bearing misalignment at the left (bearing A). It is left as an exercise for the student to fill in the missing columns in Table 18–2 when compared to Table 18–1.

Figures 18–6 to 18–11 are intended to generate ideas illustrating how shaft features reflect attachments.[1] These successful designs, developed and refined over a period of many years, represent very good design practice. Some of the figures in Chap. 11 will also help in the development of good shaft design. Catalog and commercial literature include illustrations and recommendations leading to good design practice.

The transverse shear V at a section of a beam in flexure imposes a shearing distortion which is superposed on the bending distortion. Usually such shearing deflection is less than 1 percent of the transverse bending deflection, and it is seldom evaluated. However, when a shaft's length-to-diameter ratio is less than 10, the shear component of transverse

1. Figures 18–6 to 18–11 were redrawn from material furnished by New Departure-Hyatt Division, General Motors Corporation.

Figure 18–6

Tapered roller bearings used in a mowing machine spindle. This design represents good practice for the situation in which one or more torque-transfer elements must be mounted outboard. *(Source: Redrawn from material furnished by The Timken Company.)*

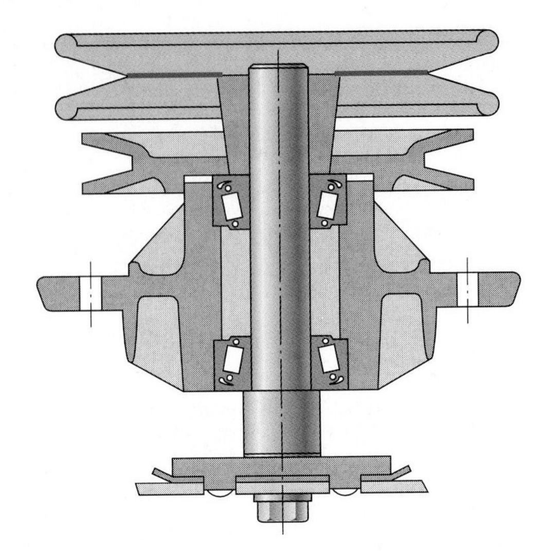

Figure 18–7

A bevel-gear drive in which both pinion and gear are straddle-mounted. *(Source: Redrawn from material furnished by Gleason Machine Division.)*

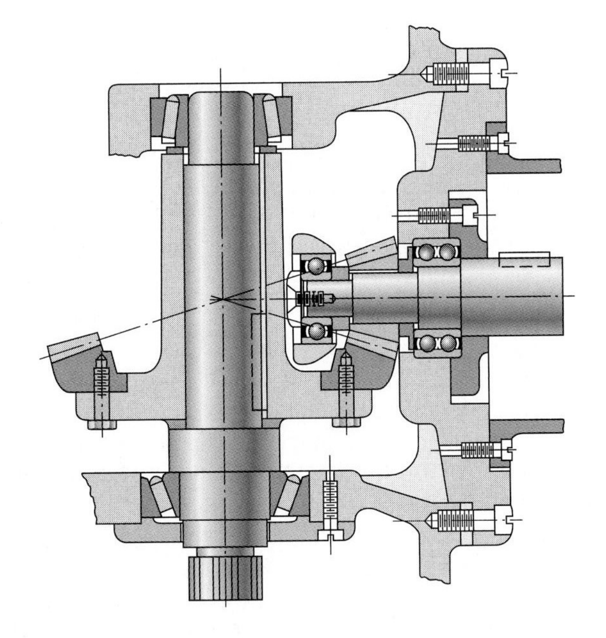

Figure 18–8

Arrangement showing bearing inner rings press-fitted to shaft while outer rings float in the housing. The axial clearance should be sufficient only to allow for machinery vibrations. Note the labyrinth seal on the right.

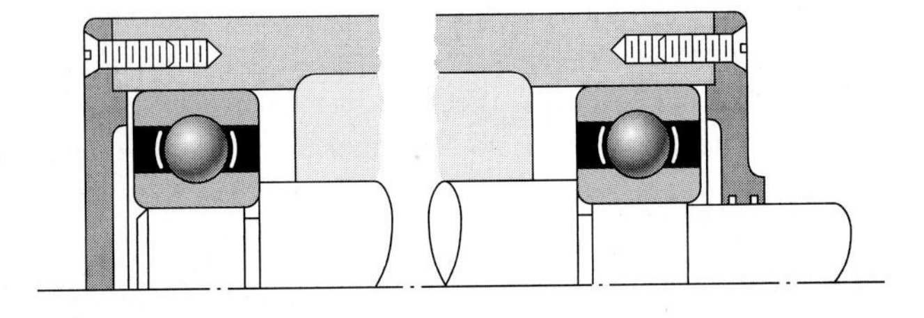

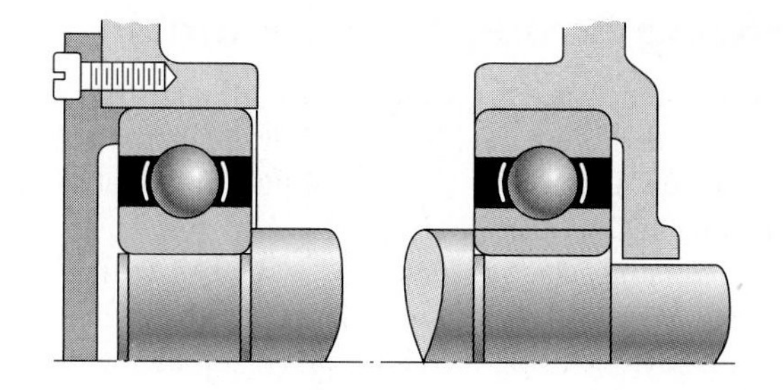

Figure 18–9

Similar to the arrangement of Fig. 18–8 except that the outer bearing rings are preloaded. Note the use of adjusting shims under the end cap.

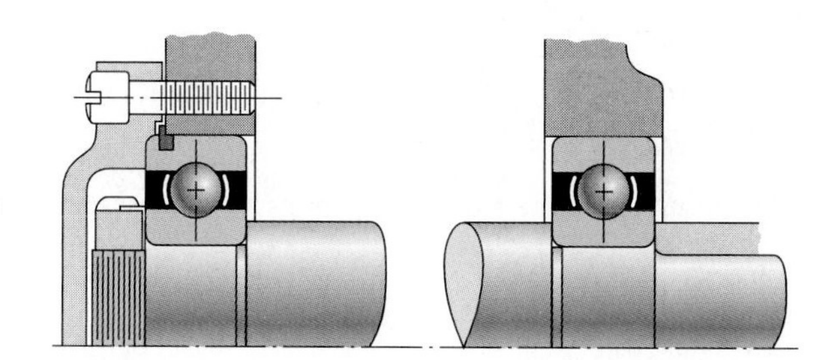

Figure 18–10

In this arrangement the inner ring of the left-hand bearing is locked to the shaft between a nut and a shaft shoulder. The locknut and washer are AFBMA standard. The snap ring in the outer race is used to positively locate the shaft assembly in the axial direction. Note the floating right-hand bearing and the grinding runout grooves in the shaft.

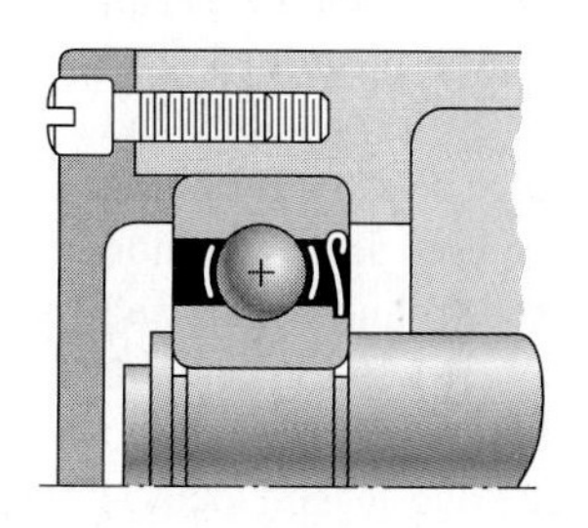

Figure 18–11

This arrangement is similar to Fig. 18–10 in that the left-hand bearing positions the entire shaft assembly. In this case the inner ring is secured to the shaft using a snap ring. Note the use of a shield to prevent dirt generated from within the machine from entering the bearing.

deflection merits attention. There are many short shafts. A tabular method [2] is explained in detail, with examples elsewhere. The method lends itself to computer implementation.

For right-circular cylindrical shafts in torsion the angular deflection θ is given in Eq. (4–5). For a stepped shaft with individual cylinder length ℓ_i and torque T_i, the angular deflection can be estimated from

$$\theta = \sum \theta_i = \sum \frac{T_i \ell_i}{G_i J_i} \tag{18–9}$$

or, for a constant torque throughout homogeneous material, from

$$\theta = \frac{T}{G} \sum \frac{\ell_i}{J_i} \tag{18–10}$$

If torsional stiffness is defined as $k_i = T_i/\theta_i$ and, since $\theta_i = T_i/k_i$ and $\theta = \sum \theta_i = \sum (T_i/k_i)$, for constant torque $\theta = T \sum (1/k_i)$, it follows that the stiffness of the shaft k in terms of segment stiffnesses is

$$\frac{1}{k} = \sum \frac{1}{k_i} \tag{18–11}$$

Note that Eq. (18–9) is not precise, since experimental evidence shows that θ is larger than given by Eq. (18–9).[3]

2. C.R. Mischke, "Tabular Method for Transverse Shear Deflection," Sec. 37.3 in Joseph E. Shigley and Charles R. Mischke (eds.-in-chief), *Standard Handbook of Machine Design*, 2nd ed., McGraw-Hill, New York, 1966.

3. R. Bruce Hopkins, *Design Analysis of Shafts and Beams*, McGraw-Hill, New York, 1970, pp. 93–99.

18–3 Sufficing Strength Constraints

We will examine sufficing shaft strength using distortion-energy–Gerber and distortion-energy–elliptic failure models under deterministic and stochastic conditions. These two failure models, DE-Gerber and DE-elliptic, fit the data. Other methods of lesser pedigree are documented in other books.

Any rotating shaft loaded by stationary bending and torsional moments will be stressed in completely reversed bending because of shaft rotation, but the torsional stress will be steady. By using the subscripts a for alternating stress amplitude σ_a and m for midrange or steady stress as in τ_{xym}, we can express the components of stress as

$$\sigma_a = \left| \frac{32M_a}{\pi d^3} \right| \tag{18–12}$$

$$\tau_{xym} = \frac{16T_m}{\pi d^3} \tag{18–13}$$

Expressing the midrange component as a von Mises normal stress gives

$$\sigma'_a = \sigma_{xa} \tag{18–14}$$

$$\sigma'_m = \sqrt{3}\tau_{xym} \tag{18–15}$$

After the curve fit to be used is determined, the fatigue-failure locus is plotted on the designer's fatigue diagram, without the data, as depicted in Fig. 18–12. The locus partitions the plane into a "safe" area (beneath the locus) and an unsafe area (above the locus).

The fatigue-failure locus of Gerber passes among fatigue test points and can be successfully used as a regression line. Statistical chances of being below the locus can be assessed from the data, and the chance of a fatigue failure can be quantified. The Gerber locus is used in conjunction with the Langer locus (first-cycle yielding) and the dog-legged function is called the Gerber-Langer failure locus. The ASME-elliptic fatigue-failure locus likewise passes among the data and can also be used as a regression line. The statistical chances of being below the line can be assessed from the data, and the chance of a fatigue can be quantified. The ANSI/ASME B106.1M-1985 standard, second printing, is the ASME standard for shafting. Even though the yield strength S_y appears in the equation, the locus crosses the Langer line twice. The dog-legged locus is

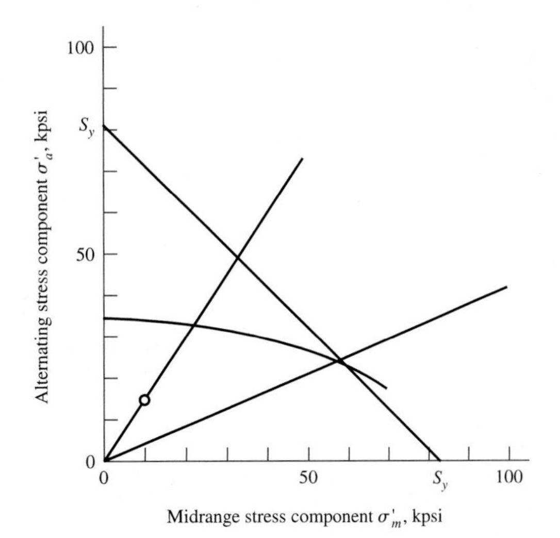

Figure 18–12

The designer's fatigue-failure diagram for Ex. 18–3. The fatigue-failure locus is DE-elliptic. The stress condition components at the shoulder are shown with the load line.

called the elliptic-Langer locus. The Gerber-Langer and ASME-elliptic–Langer curves offer the best predicators of fatigue failure.

Table 18–3 displays relationships among stress, strength, and design factor (factor of safety) for a number of models. When S_a is substituted for $n\sigma_a$ and S_m for $n\sigma_m$, the equation for the failure locus is obtained for display on the designer's fatigue diagram.

The stress condition at the critical location described by σ_a and σ_m can be plotted, along with the load line upon which it is constrained to move (if it moves). This point helps complete the visual picture.

Literature (textbooks, handbooks, technical papers, and industrial literature) concerning shafts presents a large array of equations. In an engineering analysis, a *failure locus* is used. A name is associated with each. A displayed equation uses the name in an adjectival fashion, for example, a Gerber failure locus or a Gerber-Langer failure locus. Additionally, if the stress components are von Mises components, they are identified with von Mises (or Henckey) or simply called distortion energy (easily abbreviated DE) as in DE-Gerber. Only by attention to vocabulary can we communicate essential information with precision. The short title identifies the significant stress (the failure theory), then a hyphen, followed by a failure locus name (the fatigue data model). Equations cannot usually speak for themselves.

Shaft Diameter Equation for the DE-Gerber Criterion

The von Mises stress-amplitude component σ'_a and the midrange stress component σ'_m are given by

$$\begin{aligned}\sigma'_a &= (\sigma_{xa}^2 + 3\tau_{xya}^2)^{1/2} = \frac{16}{\pi d^3}\sqrt{4(K_f M_a)^2 + 3(K_{fs} T_a)^2} = \frac{16A}{\pi d^3} \\ \sigma'_m &= (\sigma_{xm}^2 + 3\tau_{xym}^2)^{1/2} = \frac{16}{\pi d^3}\sqrt{4(K_f M_m)^2 + 3(K_{fs} T_m)^2} = \frac{16B}{\pi d^3}\end{aligned} \qquad (18\text{–}16)$$

where A and B are defined by the radicals in Eqs. (18–16). The Gerber fatigue-failure

Table 18–3

Fatigue-Strength Loci for Shafts

Locus Name	Basic Formula
Soderberg	$\frac{n\sigma_a}{S_e} + \frac{n\sigma_m}{S_y} = 1$
Goodman	$\frac{n\sigma_a}{S_e} + \frac{n\sigma_m}{S_{ut}} = 1$
Gerber	$\frac{n\sigma_a}{S_e} + \left(\frac{n\sigma_m}{S_{ut}}\right)^2 = 1$
ASME elliptic	$\left(\frac{n\sigma_a}{S_e}\right)^2 + \left(\frac{n\sigma_m}{S_y}\right)^2 = 1$
Bagci	$\frac{n\sigma_a}{S_e} + \left(\frac{n\sigma_m}{S_y}\right)^4 = 1$
Yielding (Langer)	$\frac{n}{S_y}(\sigma_a + \sigma_m) = 1$

Note: Substituting S_a for $n\sigma_a$ and S_m for $n\sigma_m$ gives a locus entirely in strength terms.

locus is defined by

$$\frac{S_a}{S_e} + \left(\frac{S_m}{S_{ut}}\right)^2 = \frac{n\sigma'_a}{S_e} + \left(\frac{n\sigma'_m}{S_{ut}}\right)^2 = \frac{16nA}{\pi d^3 S_e} + \left(\frac{16nB}{\pi d^3 S_{ut}}\right)^2 = 1$$

Solving for diameter d gives

$$d = \left(\frac{8nA}{\pi S_e}\left\{1 + \left[1 + \left(\frac{2BS_e}{AS_{ut}}\right)^2\right]^{1/2}\right\}\right)^{1/3} \tag{18–17}$$

or, solving for $1/n$,

$$\frac{1}{n} = \frac{8A}{\pi d^3 S_e}\left\{1 + \left[1 + \left(\frac{2BS_e}{AS_{ut}}\right)^2\right]^{1/2}\right\} \tag{18–18}$$

where

$$A = \sqrt{4(K_f M_a)^2 + 3(K_{fs} T_a)^2} \qquad B = \sqrt{4(K_f M_m)^2 + 3(K_{fs} T_m)^2} \tag{18–19}$$

$$r = \sigma'_a/\sigma'_m = A/B$$

Expressions for d or $1/n$ are constructed for the problem at hand by evaluating A and B and then substituting into Eqs. (18–17) and (18–18). For example, with $M_m = 0$ and $T_a = 0$,

$$A = 2K_f M_a \qquad B = \sqrt{3}K_{fs}T_m$$

and the preceding equations become

$$d = \left(\frac{16nK_f M_a}{\pi S_e}\left[1 + \sqrt{1 + 3\left(\frac{K_{fs}T_m S_e}{K_f M_a S_{ut}}\right)^2}\right]\right)^{1/3} \tag{18–20}$$

$$\frac{1}{n} = \frac{16K_f M_a}{\pi d^3 S_e}\left\{1 + \left[1 + 3\left(\frac{K_{fs}T_m S_e}{K_f M_a S_{ut}}\right)^2\right]^{1/2}\right\} \tag{18–21}$$

$$r = \frac{\sigma'_a}{\sigma'_m} = \frac{2K_f M_a}{\sqrt{3}K_{fs}T_m} \tag{18–22}$$

Shaft Diameter Equation for the DE-Elliptic Criterion

The previous definitions for σ'_a, σ'_m, A, and B apply. The elliptic fatigue-failure locus is defined by

$$\left(\frac{S_a}{S_e}\right)^2 + \left(\frac{S_m}{S_y}\right)^2 = \left(\frac{n\sigma'_a}{S_e}\right)^2 + \left(\frac{n\sigma'_m}{S_y}\right)^2 = \left(\frac{16nA}{\pi d^3 S_e}\right)^2 + \left(\frac{16nB}{\pi d^3 S_y}\right)^2 = 1$$

Solving for the diameter d gives

$$d = \left(\frac{16n}{\pi}\sqrt{\frac{A^2}{S_e^2} + \frac{B^2}{S_y^2}}\right)^{1/3}$$

Substituting for A and B gives expressions for d, $1/n$, and r:

$$d = \left\{ \frac{16n}{\pi} \left[4\left(\frac{K_f M_a}{S_e}\right)^2 + 3\left(\frac{K_{fs} T_a}{S_e}\right)^2 + 4\left(\frac{K_f M_m}{S_y}\right)^2 + 3\left(\frac{K_{fs} T_m}{S_y}\right)^2 \right]^{1/2} \right\}^{1/3} \tag{18–23}$$

$$\frac{1}{n} = \frac{16}{\pi d^3} \left[4\left(\frac{K_f M_a}{S_e}\right)^2 + 3\left(\frac{K_{fs} T_a}{S_e}\right)^2 + 4\left(\frac{K_f M_m}{S_y}\right)^2 + 3\left(\frac{K_{fs} T_m}{S_y}\right)^2 \right]^{1/2} \tag{18–24}$$

$$r = \frac{\sigma'_a}{\sigma'_m} = \frac{A}{B} = \sqrt{\frac{4(K_f M_a)^2 + 3(K_{fs} T_a)^2}{4(K_f M_m)^2 + 3(K_{fs} T_m)^2}}$$

For the case when $M_m = 0$ and $T_a = 0$, the preceding relationships become

$$d = \left\{ \frac{16n}{\pi} \left[4\left(\frac{K_f M_a}{S_e}\right)^2 + 3\left(\frac{K_{fs} T_m}{S_y}\right)^2 \right]^{1/2} \right\}^{1/3} \tag{18–25}$$

$$\frac{1}{n} = \frac{16}{\pi d^3} \left[4\left(\frac{K_f M_a}{S_e}\right)^2 + 3\left(\frac{K_{fs} T_m}{S_y}\right)^2 \right]^{1/2} \tag{18–26}$$

$$r = \frac{\sigma'_a}{\sigma'_m} = \frac{2K_f M_a}{\sqrt{3} K_{fs} T_m}$$

At a shoulder Figs. E-15-8 and E-15-9 provide information about K_t and K_{ts}. For a hole in a solid shaft, Figs. E-15-10 and E-15-11 provide K_t and K_{ts} information. For a hole in a solid shaft, use Table E-16. For grooves use Figs. E-15-14 and E-15-15.

The value of slope at which the load line intersects the failure loci junction is designated r_{crit}. It tells whether the threat is from fatigue or first-cycle yielding. If $r > r_{\text{crit}}$, the threat is from fatigue; if $r < r_{\text{crit}}$, the threat is from first-cycle yielding. For the Gerber-Langer intersection the strength components S_a and S_m are

$$\begin{aligned} S_m &= \frac{S_{ut}^2}{2S_e} \left[1 - \sqrt{1 - 4\left(\frac{S_e}{S_{ut}}\right)^2 \left(\frac{S_y}{S_e} - 1\right)} \right] \\ S_a &= S_y - S_m \\ r_{\text{crit}} &= \frac{S_a}{S_m} = \frac{S_y - S_m}{S_m} \end{aligned} \tag{18–27}$$

For the DE-elliptic–Langer intersection

$$\begin{aligned} S_a &= \frac{2S_y S_e^2}{S_e^2 + S_y^2} \\ S_m &= S_y - S_a \\ r_{\text{crit}} &= \frac{S_a}{S_m} = \frac{S_a}{S_y - S_a} \end{aligned} \tag{18–28}$$

EXAMPLE 18–3 At a shaft shoulder the small diameter d is 1.100 in, the large diameter D is 1.65 in, and the fillet radius is 0.11 in. The bending moment is 1260 in · lb and the steady torsion moment is 1100 in · lb. The heat-treated steel shaft has an ultimate strength of $S_{ut} \sim LN(105, 4.5)$ kpsi and a yield strength of $S_y \sim LN(82, 6.4)$ kpsi. The reliability goal was 0.99 under a coefficient of variation of the load C_L of 0.05.

(a) What mean design factor $\bar{n}_d$ will ensure attainment of the reliability goal against fatigue?

(b) Is the small diameter of 1.100 in adequate? Use a DE-elliptic failure locus to answer the question.

(c) Plot the designer's fatigue diagram.

Solution

(*a*) From Sec. 7–19, $z = -2.326$, $C_{kc} = 0$, $C_{Kf} = 0.11$, $C_{ka} = 0.058$, $C_{kd} = 0$, and $C_{Se'} = 0.138$ (correlation method). Then

$$C_\sigma = (C_L^2 + C_{Kf}^2)^{1/2} = (0.05^2 + 0.11^2)^{1/2} = 0.121$$

$$C_{Se} = (C_{ka}^2 + C_{kc}^2 + C_{kd}^2 + C_{Se'}^2) = (0.058^2 + 0^2 + 0^2 + 0.138^2)^{1/2}$$

$$= 0.150$$

Approximating C_{Sa} with C_{Se} we write

$$C_n \doteq [(0.150^2 + 0.121^2)/(1 + 0.121^2)]^{1/2} = 0.191$$

Answer

$$\bar{n}_d = \exp[-(-2.326)\sqrt{\ln(1 + 0.191^2)} + \ln\sqrt{1 + 0.191^2}] = 1.58$$

(*b*) $D/d = 1.65/1.100 = 1.50$, $r/d = 0.11/1.100 = 0.10$, $K_t = 1.915$ (Fig. E-15-7), $K_{fs} = 1.392$ (Fig. E-15-8). Then

$$K_f = \frac{1.915}{1 + \dfrac{2}{\sqrt{0.11}} \dfrac{1.915 - 1}{1.915} \dfrac{4}{105}} = 1.726$$

$$K_{fs} = \frac{1.392}{1 + \dfrac{2}{\sqrt{0.11}} \dfrac{1.392 - 1}{1.392} \dfrac{4}{105}} = 1.307$$

$$S_e' = 0.506(105) = 53\ 130 \text{ psi}$$

$$k_a = 2.67(105)^{-0.265} = 0.778$$

$$k_b = \left(\frac{1.100}{0.30}\right)^{-0.107} = 0.870$$

$$k_c = k_d = k_e = 1$$

$$S_e = 0.778(0.870)53\ 130 = 35\ 962 \text{ psi}$$

$$r = \frac{2}{\sqrt{3}} \frac{1.726}{1.307} \frac{1260}{1100} = 1.75$$

Eq. (18–26) $$\frac{1}{n} = \frac{16}{\pi(1.1)^3} \left\{ 4\left[\frac{1.726(1260)}{35\ 962}\right]^2 + 3\left[\frac{1.307(1100)}{82\ 000}\right]^2 \right\}^{1/2} = 0.477$$

$$n = 2.10$$

Answer Since the extant factor of safety 2.10 is greater than necessary 1.58, the fatigue strength is adequate.

$$(c)\ \sigma_a = \frac{32(1.726)1260}{\pi(1.1)^3} = 16\,643\ \text{psi}$$

$$\sigma_m = \frac{16\sqrt{3}(1.307)1100}{\pi(1.1)^3} = 9528\ \text{psi}$$

Eqs. (18–28):

$$S_a = \frac{2(82\,000)35\,962^2}{35\,962^2 + 82\,000^2} = 26\,455\ \text{psi}$$

$$S_m = 82\,000 - 26\,455 = 55\,545\ \text{psi}$$

$$r_{\text{crit}} = \frac{26\,455}{55\,545} = 0.476$$

The designer's fatigue diagram is seen in Fig. 18–12. The fatigue locus is plotted from $(S_a/36.0)^2 + (S_m/82)^2 = 1$.

Figure 18–13 shows a number of fatigue-failure loci plotted to scale for the same material. The loci labeled C and D are the DE-Gerber and DE-elliptic fatigue-failure curves, which are the best predictors of fatigue failure and are recommended for shafts. Note that the Gerber locus crosses the Langer locus once, and the elliptic locus crosses the Langer locus twice.

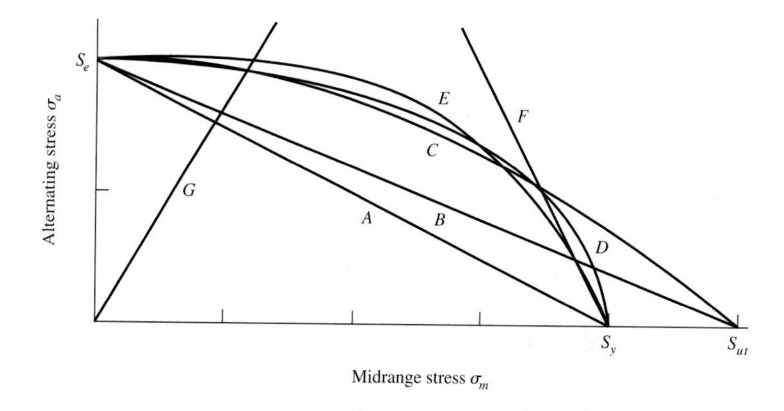

Figure 18–13

A designer's fatigue diagram plotted to scale showing the relations between some well-known fatigue-failure models. *A*, Soderberg locus: *B*, Goodman locus; *C*, Gerber locus; *D*, ASME-elliptic locus; *E*, Bagci locus; *F*, first-cycle yield locus; *G*, load line. Note that the Langer locus, curve *G*, makes a 45° angle with the abscissa *only* when the ordinate and abscissa scales are identical.

EXAMPLE 18–4 In Ex. 18–3 we found the $\bar{n}_d$ that would ensure attainment of the reliability goal of 0.99 and proceeded deterministically thereafter. We made the approximation of using C_{Se} for C_{Sa}, which we will now check.

(a) Find $\bar{S}_a$, C_{Sa}, $\bar{\sigma}_a$, z, and the extant reliability of the shoulder under fatigue.

(b) Now in possession of C_{Sa}, repeat Ex. 18–3a and find $\bar{n}_d$. Was the approximation robust?

Solution (*a*) From Ex. 18–3c $\bar{\sigma}_a = 16\ 643$ psi. The mean strength amplitude $\bar{S}_a$ is

Answer

$$\text{Eq. (7–54)} \quad \bar{S}_a = \frac{r\bar{S}_y\bar{S}_e}{\sqrt{r^2\bar{S}_y^2 + \bar{S}_e^2}} = \frac{1.75(82\ 000)35\ 962}{\sqrt{1.75^2(82\ 000)^2 + 35\ 962^2}} = 34\ 883 \text{ psi}$$

$$C_{Sy} = 6.4/82 = 0.078$$

Answer

$$\text{Eq. (7–55)} \quad C_{Sa} = (1 + 0.078)(1 + 0.150) \times \sqrt{\frac{1.75^2(82\ 000)^2 + 35\ 962^2}{1.75^2 82\ 000^2(1 + 0.078)^2 + 35\ 962^2(1 + 0.150)^2}} - 1$$

$$C_{Sa} = 0.145$$

Eq. (2–7)

Answer

$$z = -\frac{\ln\left(\dfrac{34\ 883}{16\ 662}\sqrt{\dfrac{1 + 0.121^2}{1 + 0.145^2}}\right)}{\sqrt{\ln(1 + 0.145^2)(1 + 0.121^2)}} = -3.9135, \ p_f = 0.000\ 046\ 5$$

Answer From Table E-10, $R - 1 - p_f = 1 - 0.000\ 0465 = 0.999\ 95$, and the reliability is greater than the goal of 0.99.

(*b*) $C_{Sa} = 0.145$

$$C_n = [(C_{Sa}^2 + C_{\sigma_a}^2)/(1 + C_{\sigma_a}^2)]^{1/2} = [(0.145^2 + 0.121^2)/(1 + 0.121^2]^{1/2}$$

$$= 0.187$$

$$n_f = \exp[-(-2.326)\sqrt{\ln(1 + 0.187^2)} + \ln\sqrt{1 + 0.187^2}] = 1.57$$

Answer This compares to 1.58 of Ex. 18–3*a*. The approximation is robust.

The DE–ASME-elliptic has a complication that involves the use of S_y rather than S_{ut}. A direct comparison with other methods is not possible without knowledge of S_y/S_{ut}. In the domain of most power transmission shafts, the ASME-elliptic uses a fatigue-failure locus which is practically coincident with the Gerber locus on the left portion of the curve where most power transmission shafts have their operating point on the designer's fatigue diagram. In this domain one can use either the DE-Gerber or the DE–ASME-elliptic with about the same results.

If one writes the Gerber failure locus as

$$\frac{S_a}{S_e} + \left(\frac{S_m}{S_{ut}}\right)^2 = 1$$

and the ASME-elliptic failure locus as

$$\left(\frac{S_a}{S_e}\right)^2 + \left(\frac{S_m}{S_{ut}}\frac{S_{ut}}{S_y}\right)^2 = 1$$

and then equates S_a/S_e from both expressions, the resulting equation is

$$\left(\frac{S_m}{S_{ut}}\right)^4 + \left(\frac{S_m}{S_{ut}}\right)^2 \left[\left(\frac{S_{ut}}{S_y}\right)^2 - 2\right] = 0$$

from which the S_m coordinates of the intersecting points of the two loci are

$$\frac{S_m}{S_{ut}} = 0 \quad \text{and} \quad \frac{S_m}{S_{ut}} = \sqrt{2 - \left(\frac{S_{ut}}{S_y}\right)^2}$$

The first intersection occurs at $S_a = S_e$ and $S_m = 0$. The second occurs if $S_{ut}/S_y < \sqrt{2}$. In other words, if S_{ut}/S_m is less than $\sqrt{2}$, the ASME locus is above the Gerber, then it crosses at the given nonzero abscissa. If S_{ut}/S_y is $> \sqrt{2}$, then there is no (real) crossing and the ASME-elliptic is below the Gerber for $S_m > 0$.

Combined Variable Bending and Variable Torsion

When tests are carried out with variable bending stress σ_a and variable torsional stress τ_a, in phase, the results plot as shown in Fig. 18–14*b*. For the normalized coordinates of the figure the data fall near a circular arc, which would be an elliptical arc on a $\sigma_a\tau_a$

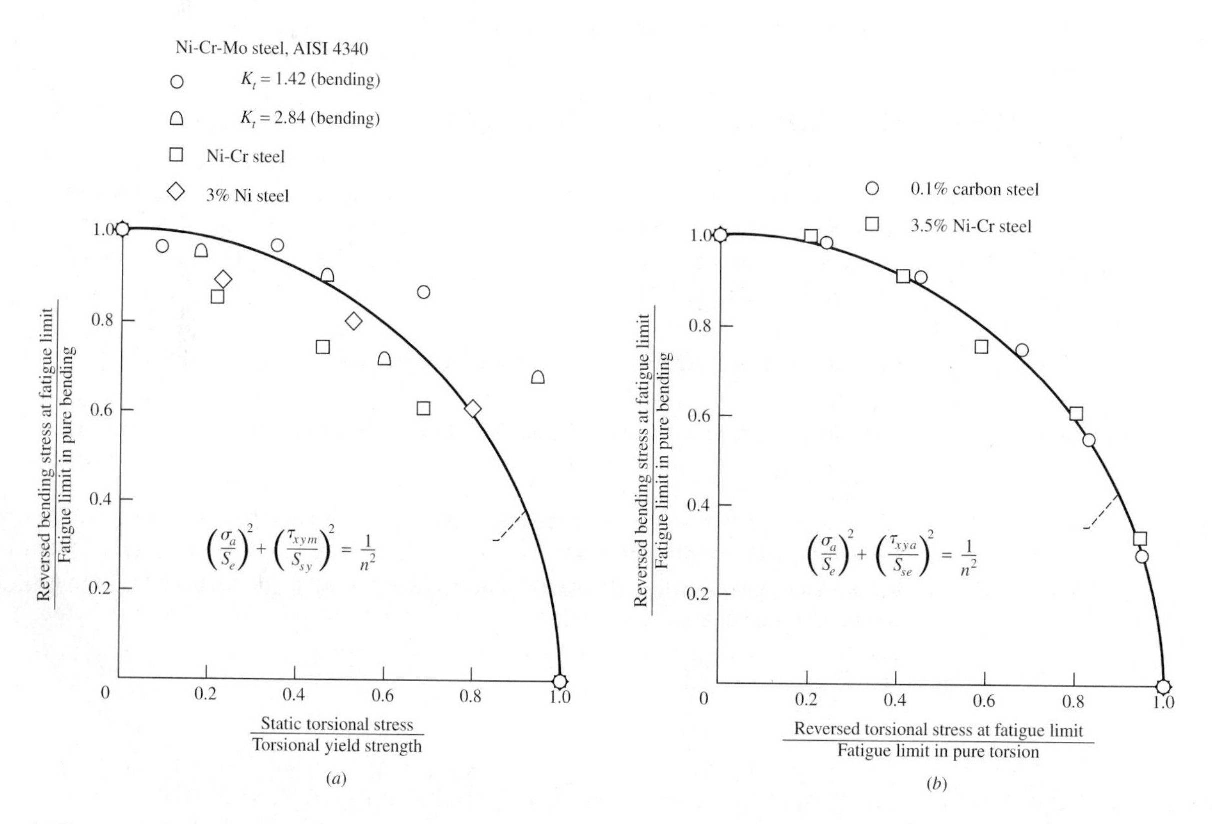

Figure 18–14

Normalized designer's fatigue-failure diagrams with test data for (*a*) reversed bending and steady torsion (DE-elliptic model) and (*b*) reversed bending and in-phase reversed torsion. *(Source: ANSI/ASME Standard B106.1.M-1985, second printing, "Design of Transmission Shafting.")*

diagram. The test failure ellipse is expressed as

$$\left(\frac{\sigma_a}{S_e}\right)^2 + \left(\frac{\tau_a}{S_{se}}\right)^2 = \frac{1}{n^2}$$

Substituting $\sigma_a = 32M_a/(\pi d^3)$, $\tau_a = 16T_a/(\pi d^3)$, $S_e = S_{se}/\sqrt{3}$ and including the fatigue stress-concentration factors K_f and K_{fs} multiplicatively associated with their nominal stresses, diameter d is found to be

$$d = \left\{ \frac{16n}{\pi} \left[4\left(\frac{K_f M_a}{S_e}\right)^2 + 3\left(\frac{K_{fs} T_a}{S_e}\right)^2 \right]^{1/2} \right\}^{1/3} \tag{18–29}$$

Also, the solution for $1/n$ gives

$$\frac{1}{n} = \frac{16}{\pi d^3} \left[4\left(\frac{K_f M_a}{S_e}\right)^2 + 3\left(\frac{K_{fs} T_a}{S_e}\right)^2 \right]^{1/2} \tag{18–30}$$

These same equations can be obtained by simplifying DE-elliptic Eqs. (18–23) and (18–24), as well as by simplifying DE-Gerber Eqs. (18–17) and (18–18) with $B = 0$. Both fatigue equations agree exactly because $r \rightarrow \infty$. Equations (18–17) and (18–18) for DE-Gerber and Eqs. (18–23) and (18–24) for DE-elliptic are general cases which can be simplified for a particular application.

18–4 The Adequacy Assessment

When any of the terms used to estimate strength, stress, and transverse deflection are random variables due to variation in strength, load, and geometry, the adequacy assessment involves stochastic methods. Shafts, being complicated geometric bodies, usually have one critical location at which failure will occur. Each discontinuity will have an associated reliability, the least reliability usually controlling.

Strength Adequacy Assessment—Stochastic

In power transmission shafts, the variability in the load transmits its coefficient of variation to the torque and bending forces, and to the stresses that are a consequence. For the common case with completely reversed bending and steady torsion one can proceed as follows:

- Identify potential critical locations.
- Estimate the mean and the coefficient of variation of loads, moment, and torques caused by the loading.
- Estimate the mean and coefficient of variation of the salient material properties: $\mathbf{S}_e$, $\mathbf{S}_{ut}$ (or $\mathbf{S}_y$), $\mathbf{E}$ (or $\mathbf{G}$).
- Estimate the Marin fatigue strength reduction factors.
- Use the correlation method with $\mathbf{S}_e = \mathbf{k}_a k_b \mathbf{k}_c \mathbf{k}_d \mathbf{k}_e \theta \bar{\mathbf{S}}_{ut}$, noting that the multiplicative association drives S_e toward a lognormal distribution.
- The expression for bending stress amplitude includes $\mathbf{K}_f$:

$$\boldsymbol{\sigma}_a = \frac{32\mathbf{K}_f \mathbf{M}_a}{\pi d^3} \qquad \boldsymbol{\sigma}_m = \frac{16\sqrt{3}\mathbf{K}_{fs}\mathbf{T}_m}{\pi d^3}$$

The distribution of $\boldsymbol{\sigma}_a$ is driven toward lognormal.

- Find the amplitude component of strength $\mathbf{S}_a$ using DE-Gerber, DE-elliptic, or DE-Smith-Dolan, as applicable.
- Find the amplitude component of stress, $\boldsymbol{\sigma}_a$.
- Interfere the $\mathbf{S}_a$ distribution with the $\boldsymbol{\sigma}_a$ distribution on a lognormal-lognormal basis using Eq. (2–55a) or Eq. (2–55b) and find the reliability estimate from Table E-10.
- Compare the estimated reliability with the reliability goal.

If you know the distribution of the load and it is not lognormal, you can conduct the interference of $\mathbf{S}_a$ with $\boldsymbol{\sigma}_a$ as a simulation to find the reliability estimate.

Strength Adequacy Assessment—Deterministic with Mean Factor of Safety

The method of Sec. 7–20 can be used to find the mean value of the factor of safety (or design factor). In light of reliability goal R and COV of the load C_L, proceed as follows:

- Identify relevant COVs: C_{ka}, C_{kc}, C_{kd}, C_{ke}, and C_{Kf}.
- $C_\sigma = [C_L^2 + C_{Kf}^2]^{1/2}$.
- If you are using the correlation method, $C_{Se'} = 0.138$; if the test method, $C_{Se'} =$ test value.
- $C_{Se} = (C_{ka}^2 + C_{kc}^2 + C_{kd}^2 + C_{Se'}^2)^{1/2}$.
- $C_n = [(C_S^2 + C_\sigma^2)/(1 + C_\sigma^2)]^{1/2}$.

If loading is completely reversed, $C_S = C_{Se}$; if not, $C_S = C_{Sa}$. If C_{Sa} is not known a priori, set $C_S = C_{Se}$, a conservative approximation.

- $n_f = \exp[-z\sqrt{\ln(1 + C_n^2)} + \ln\sqrt{1 + C_n^2}]$.
- Compare n_f with deterministic n. If $n_f < n$, the reliability goal is exceeded.

Strength Adequacy Assessment—Deterministic

A deterministic adequacy assessment is often performed with ASTM minimum properties, and results in a factor of safety. A reliability goal cannot be addressed. The adequacy assessment can also be done with mean values of properties. This changes scale for the resulting factor of safety.

- Identify potential critical locations.
- Identify loads creating bending moments and torsional moments.
- Identify the ultimate strength, yield strength, and endurance limit at the critical points.
- Estimate the stress-concentration factors K_f and K_{fs}.
- Estimate the load-line slope r and the critical slope r_{crit}.
- Estimate endurance strength S_e or S_f from Marin factors.
- Estimate stress amplitude σ_a' and strength amplitude S_a using DE-Gerber or DE-elliptic failure models.
- Estimate the factor of safety in fatigue from S_a/σ_a'.

EXAMPLE 18–5

The integral pinion-and-shaft depicted in Fig. 18–15 is one design concept for the conditions of Ex. 18–2. The shaft is to be mounted in 02-series rolling-contact bearings, and is to have a gear (not shown) keyed to the right-hand shaft overhang. The loading of Ex. 18–2 and Fig. 18–5 applies, showing transverse forces at A, B, C, and D. The bending-moment diagram shows two extremes in absolute value at B and D. The critical location

Figure 18–15

An integral pinion and shaft having an overhung gear mounted on the right, without detail. Note that the dimensioning is prepared for convenience in analysis, but it would be done differently for manufacturing and inspection.

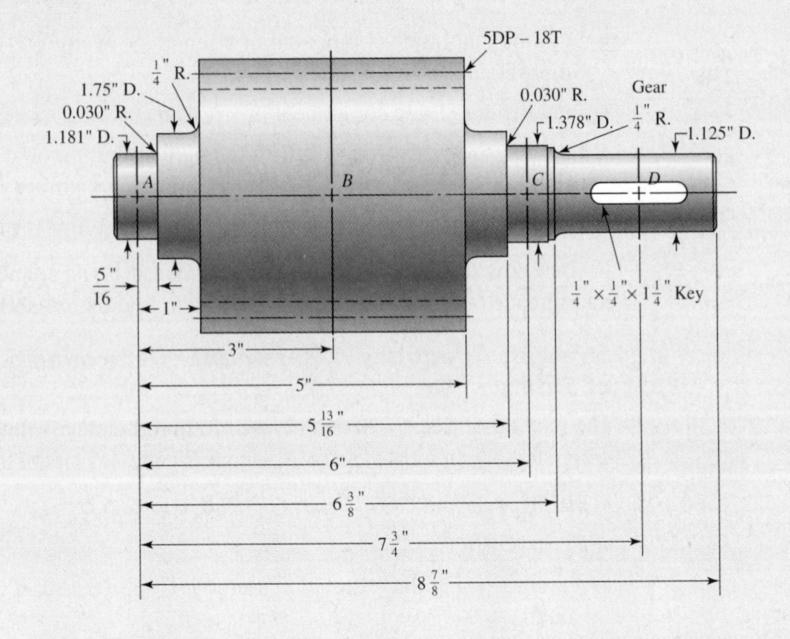

is in the neighborhood of C, where the bending moment is 1677 in · lb. Tension tests on the heat-treated 1030 steel give $S_{ut} \sim 85LN(1, 0.045)$ kpsi, and $S_y \sim 64LN(1, 0.07)$ kpsi. The load variability is expected to have a coefficient of variation of 0.05. The reliability goal was 0.999.

(a) Estimate the reliability in fatigue of the shoulder using the DE- elliptic failure locus.
(b) Estimate the reliability in first-cycle yielding.
(c) Draw the designer's fatigue diagram.

Solution (*a*) $\mathbf{M}_a = 1677\mathbf{LN}(1, 0.05)$ in · lb, $\mathbf{T}_m = 3300\mathbf{LN}(1, 0.05)$ in · lb, $D/d = 1.75/1.378 = 1.27$, $r/d = 0.030/1.378 = 0.022$. From Table E-15-9,

$$K_t = 0.622 + 0.38(1.27)^{-4.3} + 0.022^{-0.5}\sqrt{\frac{-0.322 - 0.277(1.27)^2 + 0.599(1.27)^4}{1 - 2.55(1.27)^2 + 5.27(1.27)^4}}$$

$$K_t = 2.598$$

From Table E-15-8,

$$K_{ts} = 0.78 + 0.2(1.27)^{-10} + 0.022^{-0.46}\sqrt{\frac{0.002 - 0.125(1.27)^2 + 0.123(1.27)^4}{1 - 2.75(1.27)^2 + 2.55(1.27)^4}}$$

$$K_{ts} = 1.921$$

$$\bar{K}_f = \frac{2.598}{1 + \dfrac{2}{\sqrt{0.030}}\dfrac{2.598 - 1}{2.598}\dfrac{4}{85}} = 1.947$$

$$\bar{K}_{fs} = \frac{1.921}{1 + \dfrac{2}{\sqrt{0.030}}\dfrac{1.921 - 1}{1.921}\dfrac{4}{85}} = 1.524$$

$$\bar{\sigma}_a = \frac{32K_f M_a}{\pi d^3} = \frac{32(1.947)1677}{\pi(1.378)^3} = 12\,712$$

$$\bar{\sigma}_m = \frac{16\sqrt{3}K_{fs}T_m}{\pi d^3} = \frac{16\sqrt{3}\,1.524(3300)}{\pi(1.378)^3} = 16\,954 \text{ psi}$$

Eq. (18–26) $r = \frac{2}{\sqrt{3}} \frac{1.947}{1.524} \frac{1677}{3300} = 0.750$

$$\mathbf{k}_a = 2.67(85)^{-0.265}\mathbf{LN}(1, 0.058) = 0.823\,\mathbf{LN}(1, 0.058)$$

$$k_b = \left(\frac{1.378}{0.30}\right)^{-0.107} = 0.849$$

$$\mathbf{S}_e = 0.823(1, 0.058)0.849(1, 0)(1, 0)(1, 0)0.506(1, 0.138)85 \text{ kpsi}$$

$$\bar{S}_e = 0.823(0.849)0.506(85) = 30.06 \text{ kpsi}$$

$$C_{Se} = (0.058^2 + 0.138^2)^{1/2} = 0.150$$

Eqs. (18–28) $S_a = \frac{2S_y S_e^2}{S_e^2 + S_y^2} = \frac{2(64)30.06^2}{30.06^2 + 64^2} = 23.13 \text{ kpsi}$

$$S_m = S_y - S_a = 64 - 23.13 = 40.87 \text{ kpsi}$$

$$r_{\text{crit}} = \frac{S_a}{S_m} = \frac{23.13}{40.87} = 0.566$$

Eqs. (7–50) $S_a = \frac{r S_y S_e}{\sqrt{r^2 S_y^2 + S_e^2}} = \frac{0.750(64)30.06}{\sqrt{0.750^2(64)^2 + 30.06^2}} = 25.48 \text{ kpsi}$

Eq. (7–51) $C_{Sa} = (1 + 0.07)(1 + 0.150)$

$$\cdot \sqrt{\frac{0.750^2(64)^2 + 30.06^2}{0.750^2(64)^2(1 + 0.07)^2 + 30.06^2(1 + 0.150)^2}} - 1$$

$$= 0.126$$

$$C_\sigma = (C_L^2 + C_{Kf}^2)^{1/2} = (0.05^2 + 0.11^2)^{1/2} = 0.121$$

Eq. (2–7) $z = \frac{-\ln\left(\frac{25.48}{12.712}\sqrt{\frac{1 + 0.121^2}{1 + 0.126^2}}\right)}{\sqrt{\ln(1 + 0.121^2)(1 + 0.126^2)}} = -3.992 \doteq -4$

Answer From Table E-10, $p_f = 0.000\,031\,7$, $R = 0.999\,97$. Since $0.999\,97 > 0.999$, shoulder fatigue strength is satisfactory.

(*b*) The yield for $\mathbf{n}_y$, from Eq. (18–34), is

$$\mathbf{n}_y = \frac{\mathbf{S}_y}{\boldsymbol{\sigma}_a + \boldsymbol{\sigma}_m} \qquad \bar{n}_y = \frac{\bar{S}_y}{\bar{\sigma}_a + \bar{\sigma}_m} = \frac{64}{12.712 + 16.954} = 2.16$$

The sum $\boldsymbol{\sigma}_a + \boldsymbol{\sigma}_m$ is correlated $\rho = 1$. The mean of the sum is $\bar{\sigma}_a + \bar{\sigma}_m$ and the COV can be shown to be $C_\sigma = 0.121$. The numerator of $\mathbf{n}_y$ is $\mathbf{S}_y(1, 0.07)$ and the denominator is $(\boldsymbol{\sigma}_a + \boldsymbol{\sigma}_m)(1, C_\sigma)$, so from Eq. (2–7),

$$z = -\frac{\ln\left(\frac{\bar{S}_y}{\bar{\sigma}_a + \bar{\sigma}_m}\sqrt{\frac{1 + C_\sigma^2}{1 + C_{Sy}^2}}\right)}{\sqrt{\ln(1 + C_\sigma^2)(1 + C_{Sy}^2)}} = -\frac{\ln\left(\frac{64}{29.7}\sqrt{\frac{1 + 0.121^2}{1 + 0.07^2}}\right)}{\sqrt{\ln(1 + 0.121^2)(1 + 0.07^2)}} \doteq -5.5$$

Answer From Table E-10, $p_f = 0.0^7190$, $R = 0.9^781$. That is approximately one failure by yielding in 200 million. The approximation occurs because the denominator is the sum

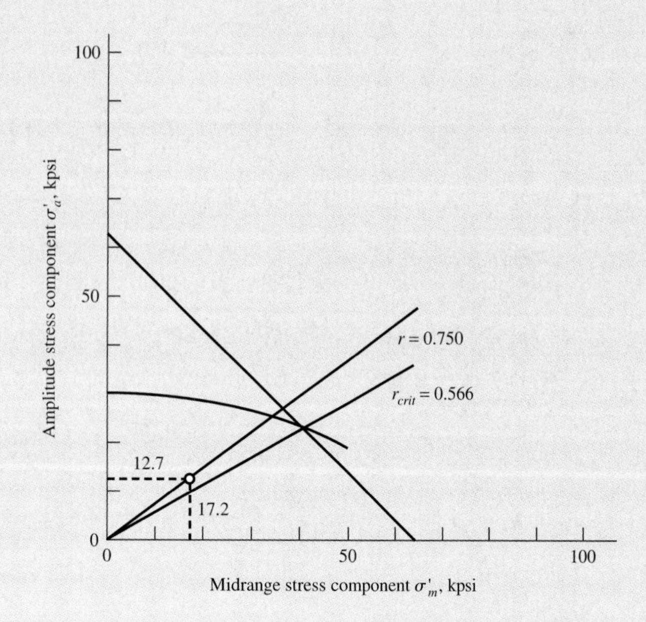

Figure 18–16

The designer's fatigue diagram of the shoulder in Ex. 18–5, showing the DE-elliptic fatigue-failure locus, the stress components at the shoulder, and the load line.

of two lognormal variates, and that starts to deviate from lognormality. The value of z is based on the quotient of two lognormal variates.

The probability of failure is much lower than in fatigue because the COV of the strength $\mathbf{S}_y$ is smaller than the COV of $\mathbf{S}_a$. Now you can appreciate that when we say "The threat is from fatigue," this is a deterministic statement. The probability of a fatigue failure is $0.0^4314/0.0^7190$, or 1653, times more likely from fatigue than from yielding.
(c) The elliptic failure locus is plotted from $(S_a/30)^2 + (S_m/85)^2 = 1$ and is shown in Fig. 18–16.

EXAMPLE 18–6

Repeat Ex. 18–5 with a deterministic adequacy assessment using DE-elliptic fatigue locus. Find the factor of safety in fatigue and in first-cycle yielding.

Solution

We will use mean values. The values that we will need are $M_a = 1677$ in · lb, $T_m = 3300$ in · lb, $K_f = 1.947$, $K_{fs} = 1.524$, $\sigma_a = 12\,712$ psi, $\sigma_m = 16\,954$ psi, $k_a = 0.823$, $k_b = 0.849$, $S_e = 30.06$ kpsi, $S_a = 25.48$ kpsi. From Eq. (18–26),

$$\frac{1}{n} = \frac{16}{\pi(1.378)^3}\left\{4\left[\frac{1.947(1677)}{30\,060}\right]^2 + 3\left[\frac{1.524(3300)}{64\,000}\right]^2\right\}^{1/2} = 0.499$$

$$n = 2.004$$

which agrees with $S_a/\sigma_a = 25.48/12.712 = 2.004$. Now n_y is given by

$$n_y = \frac{S_y}{\sigma_a + \sigma_m} = \frac{64}{12.712 + 16.954} = 2.16$$

We cannot answer the question about the reliability, although we can say "The threat is from fatigue." We could use the method of Ex. 7–20, noting that $C_\sigma = 0.121$, $C_{Se} =$

0.150, and write

$$C_n = [(0.150^2 + 0.121^2)/(1 + 0.121^2)]^{1/2} = 0.191$$

$$n_f = \exp[-(-3.09)\sqrt{\ln(1 + 0.191^2)} + \ln\sqrt{1 + 0.191^2}] = 1.83$$

and since $2.004 > 1.83$, reliability exceeds 0.999.

Distortion Adequacy Assessment

Adequacy assessment in power transmission shafting includes checking distortional constraints for looseness.

- Identify deflection at gear meshes.
- Identify shaft slope at gear meshes.
- Identify shaft slope at bearings.
- Identify center-to-center growth at gear meshes when second shaft information is available. See Table 18–4.
- Compare the above with allowable deflections and slopes.

The tabular method exemplified by Table 18–1 is a useful tool.

Table 18–4

Upper Bound on Center-to-Center Expansion ΔC in Commercial-Quality Spur Gearing

Diametral Pitch, teeth/in	ΔC, in
Up to 10	+0.010
11 to 19	+0.005
20 to 50	+0.003

If stochastic information is necessary, notice that the fundamental calculation kernel is $M/(EI)$, which introduces load variability through M and material variability through E. The geometric variability is usually an order of magnitude smaller and can be neglected. If the coefficient of variation of the load C_L is transmitted to the bending moment $\mathbf{M}$ as C_M and the coefficient of variation of the Young modulus $\mathbf{E}$ is C_E, then the coefficient of variation of the deflection and slope is $(C_M^2 + C_E^2)^{1/2}$. The coefficient of variation of the load usually dominates as the coefficient of variation of the Young modulus is small $(0.02 \le C_E \le 0.03)$, depending on the methodology used, which is rarely reported. Reflect on how you would experimentally measure the Young modulus in a tension test.

If the shaft is short (diameter of a uniform shaft greater than 1/10 of the span between bearings), transverse shear deflection and slope should be added to the analysis.[4]

EXAMPLE 18–7

Continue the adequacy assessment of Ex. 18–5 applied to the design concept of Fig. 18–15, finding the slope and deflection at bearings and gears using the tabular method.

4. "Tabular Method for Transverse Shear Deflection," Sec. 37.3 in Joseph E. Shigley and Charles R. Mischke (eds.-in-chief), *Standard Handbook of Machine Design*, 2nd ed., McGraw-Hill, New York, 1996.

Solution The tabular method requires stations wherever diameter changes and where transverse loads (and couples) are applied. In this case a minimum of nine stations are needed. We will use nine stations. From Figs. 18–5 and 18–15 we form the input side of the table, the first four columns.

Station	Moment M, in · lbf	Location x, in	Diameter d, in	Deflection y, in	Slope dy/dx
1	0	0	1.181	0	−0.000 043 6
	0		1.181		
2	212.3	0.3125	1.181	−0.000 012 4	−0.000 032 0
	212.3		1.750		
3	679.3	1.00	1.75	−0.000 028 1	−0.000 009 84
	679.3		3.00		
4	2038	3.00	3.00	−0.000 028 8	0.000 012 9
	2038		3.00		
5	−604	5.00	3.00	0.000 016 4	0.000 025 0
	−604		1.75		
6	−1677.3	5.8125	1.75	0.000 013 7	−0.000 042 1
	−1677.3		1.378		
7	−1925	6.00	1.378	0.00	−0.000 106
	−1925		1.378		
8	−1512.5	6.375	1.378	−0.000 063 3	−0.000 227
	−1512.5		1.125		
9	0	7.75	1.125	−0.000 780	−0.000 668
	0		1.125		

Completing the table gives $C_1 = -0.436(10^{-4})$ and $C_2 = 0$, from which the prediction equations allow the right-hand pair of columns to be generated.

Bearing slopes: The bearings occur at stations 1 and 7, and both slopes are less than the allowable slope for roller bearings.

Gear-mesh slopes: The central planes of the gears are at stations 4 and 9. The slope at station 4 is 0.000 012 9, which is less than 0.0005. The slope at station 9 is 0.000 668, which is greater than 0.0005. (As a matter of perspective, had the shear slope been evaluated and added to the slope at station 9, the slope would increase to 0.000 725.) The use of crowned teeth will improve the situation.

Center-to-center distance: At stations 4 and 9 the allowable transverse center-to-center distance increase is 0.010 in for a five-pitch spur-gear pair (0.005 in each) and the deflections at stations 4 and 9 are less than the limit.

As it stands, the slope at station 9 exceeds the allowable slope and the design concept is not satisfactory. The violated constraint is the slope of the gear teeth at station 9.

The designer's task now is to coax the violated constraint to appropriate tightness (or looseness) without violating any other constraint.

18–5 Shaft Materials

Various steels have comparable Young moduli. For that reason, rigidity cannot be controlled by material decisions, but only by geometric decisions. Necessary strength to resist loading stresses affects the choice of materials and their treatments. ANSI 1020–1050 steels are a common choice, as is a 11xx free-machining steel. Heat-treating increases expense; if used, include 1340-50, 3140-50, 4140, 4340, 5140, and 8650. It should be used only if there is not other way to obtain necessary strengths. Carburizing grades of 1020, 4320, 4820, and 8620 are chosen when surface hardness is important.

Cold-rolled sections are available to about $3\frac{1}{2}$ in in diameter and hot-rolled sections up to 6 in can be obtained. Larger sizes require forging before machining.

In approaching material selection, the amount to be produced is a salient factor. For low production, turning is the usual primary shaping process. An economic viewpoint may require removing the least material. High production may permit a volume-conservative shaping method (hot or cold forming, casting) and minimum material in the shaft can become a design goal. Properties of the shaft locally depend on its history: cold work, cold forming, rolling of fillet features, heat treatment, including quenching medium, agitation, and tempering regimen.[5]

18–6 Hollow Shafts

A hollow shaft can reduce weight and allow fluids to circulate to moving parts for cooling or lubrication. Thick-walled seamless tubing is available for simpler, smaller shafts.

- Thick-walled tubing may not have sufficient wall material for shouldering. Split-ring shaft collars, tapered hubs, and other less common methods of locating and fastening may have to be used.
- In hollow shafting, whether the geometry of the tube was created by a drawing process, drilling, or boring, it will have to be checked for balance and corrected if necessary.

The equations developed in this chapter can be conveniently modified to apply to the hollow shaft. If the outer diameter is d_o and the inner diameter is d_i, define $K = d_i/d_o$. For torsional and bending loading the quantity $d_o(1 - K^4)^{1/3}$ may be substituted for the diameter d in Eqs. (18–16) through (18–25), for example, as applicable. The algebra of the equations allows explicit solutions for outside diameter d_o when K is known. However, when d_i is fixed, then K is not known, and iterative procedures are necessary to find the outside diameter d_o.

18–7 Critical Speeds

When a shaft is turning, eccentricity causes a centrifugal force deflection which is resisted by the shaft's flexural rigidity EI. As long as deflections are small, no harm is done. The culprit, however, is called *critical speeds:* at certain speeds the shaft is unstable, with deflections increasing without upper bound. It is fortunate that although the dynamic deflection shape is unknown, using a static deflection curve gives an excellent estimate of the critical speed. Such a curve meets the boundary condition of the differential equation (zero moment and deflection at both bearings) and the shaft energy is not particularly sensitive to the anatomy of the deflection curve. Designers seek first critical speeds at least twice the operating speed.

The shaft, due to its own mass, has a critical speed. The ensemble of attachments to a shaft likewise has a critical speed which is much lower than the shaft's intrinsic critical speed. Estimating these critical speeds (and harmonics) is a task of the designer. When

5. See Joseph E. Shigley and Charles R. Mischke (eds.-in- chief), *Standard Handbook of Machine Design*, 2nd ed., McGraw-Hill, New York, 1996. For cold-worked property prediction see Chap. 8, and for heat-treated property prediction see Chaps. 8 and 13.

geometry is simple, as in a shaft of uniform diameter, simply supported, the task is easy. It can be expressed[6] as

$$\omega_1 = \left(\frac{\pi}{l}\right)^2 \sqrt{\frac{EI}{m}} = \left(\frac{\pi}{l}\right)^2 \sqrt{\frac{gEI}{A\gamma}} \tag{18–31}$$

where m is the mass per unit length, A the cross-sectional area, and γ the specific weight. For an ensemble of attachments, Rayleigh gave

$$\omega_1 = \sqrt{\frac{g\Sigma w_i y_i}{\Sigma w_i y_i^2}} \tag{18–32}$$

where w_i is the weight of the ith attachment and y_i is the deflection at the ith body location. It is possible to use Eq. (18–32) for the case of Eq. (18–31) by partitioning the shaft into segments and placing its weight force at the segment centroid as seen in Fig. 18–17. Computer assistance is often used to lessen the difficulty in finding transverse deflections of a stepped shaft. Rayleigh's equation overestimates the critical speed.

To counter the increasing complexity of detail, we adopt a useful viewpoint. Inasmuch as the shaft is an elastic body, we can use *influence coefficients*. An influence coefficient is the transverse deflection at location i on a shaft due to a unit load at location j on the shaft. From Table E-9-6 we obtain, for a simply supported beam with a single unit load as shown in Fig. 18–18,

$$\delta_{ij} = \begin{cases} \dfrac{b_j x_i}{6EIl}(l^2 - b_j^2 - x_i^2) & x_i \le a_i \\[2ex] \dfrac{a_j(l - x_i)}{6EIl}(2lx_i - a_j^2 - x_i^2) & x_i > a_i \end{cases} \tag{18–33}$$

For three loads the influence coefficients may be displayed as

	j		
i	**1**	**2**	**3**
1	δ_{11}	δ_{12}	δ_{13}
2	δ_{21}	δ_{22}	δ_{23}
3	δ_{31}	δ_{32}	δ_{33}

Maxwell's reciprocal theorem states that there is a symmetry about the main diagonal, composed of δ_{11}, δ_{22}, and δ_{33}, of the form $\delta_{ij} = \delta_{ji}$. This relation reduces the work of finding the influence coefficients. From the influence coefficients above, one can find the deflections y_1, y_2, and y_3 of Eq. (18–32) as follows:

$$\begin{aligned} y_1 &= F_1\delta_{11} + F_2\delta_{12} + F_3\delta_{13} \\ y_2 &= F_1\delta_{21} + F_2\delta_{22} + F_3\delta_{23} \\ y_3 &= F_1\delta_{31} + F_2\delta_{32} + F_3\delta_{33} \end{aligned} \tag{18–34}$$

The forces F_i can arise from weight attached w_i or centrifugal forces $m_i\omega^2 y_i$. The

6. Charles R. Mischke, *Elements of Mechanical Analysis,* Addison-Wesley, Reading, Mass., 1963, p. 216.

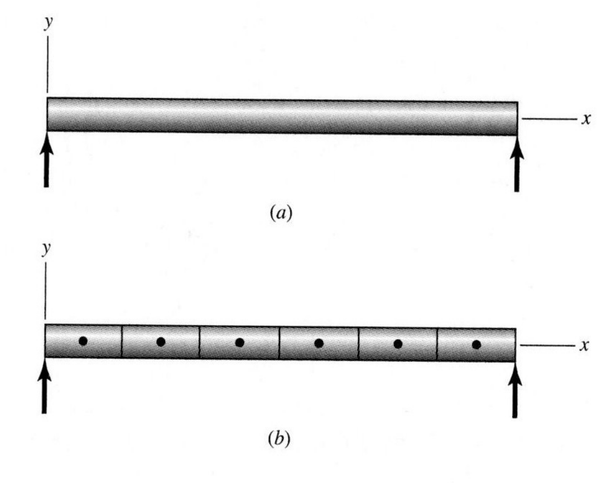

Figure 18–17

(*a*) A uniform-diameter shaft for Eq. (18–31). (*b*) A segmented uniform-diameter shaft for Eq. (18–32).

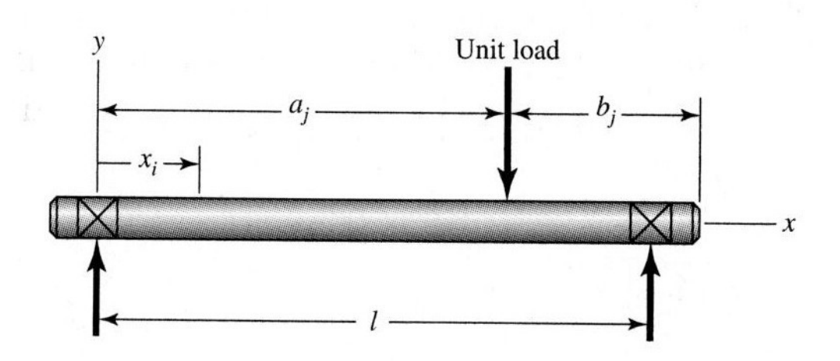

Figure 18–18

The influence coefficient δ_{ij} is the deflection at i due to a unit load at j.

equation set (18–34) written with inertial forces can be displayed as

$$\begin{vmatrix} (m_1\delta_{11} - 1/\omega^2) & m_2\delta_{12} & m_3\delta_{13} \\ m_1\delta_{21} & (m_2\delta_{22} - 1/\omega^2) & m_3\delta_{23} \\ m_1\delta_{31} & m_2\delta_{32} & (m_3\delta_{33} - 1/\omega^2) \end{vmatrix} = 0 \tag{18–35}$$

which says that a deflection other than zero exists only at the definite values of w, at the critical speeds. If there is a single attached body, $i = 1$, then

$$m_1\delta_{11} - \frac{1}{\omega^2} = 0 \qquad \frac{1}{\omega^2} = m_1\delta_{11} = \frac{w_1}{g}\delta_{11} \tag{18–36}$$

where ω is the first critical speed with a single attached body. If there are two attached bodies, the 2×2 determinant vanishes and the resulting quadratic in $(1/\omega^2)$ is

$$\left(\frac{1}{\omega^2}\right)^2 - (m_1\delta_{11} + m_2\delta_{22})\frac{1}{\omega^2} + m_1 m_2(\delta_{11}\delta_{22} - \delta_{12}\delta_{21}) = 0 \tag{18–37}$$

Equation (18–37) has two roots, $1/\omega_1^2$ and $1/\omega_2^2$, which have the cardinal property

$$\left(\frac{1}{\omega^2} - \frac{1}{\omega_1^2}\right)\left(\frac{1}{\omega^2} - \frac{1}{\omega_2^2}\right) = 0$$

The multiplication process produces

$$\left(\frac{1}{\omega^2}\right)^2 - \left(\frac{1}{\omega_1^2} + \frac{1}{\omega_2^2}\right)\left(\frac{1}{\omega^2}\right) + \frac{1}{\omega_1^2}\frac{1}{\omega_2^2} = 0 \tag{18–38}$$

Corresponding terms of Eq. (18–37) and (18–38) must have equal coefficients. Examine the coefficients of $1/\omega^2$:

$$\frac{1}{\omega_1^2} + \frac{1}{\omega_2^2} = m_1\delta_{11} + m_2\delta_{22} \tag{18–39}$$

If only body one was present $m_1\delta_{11} = 1/\omega_{11}^2$. If only body two was present, $m_2\delta_{22} = 1/\omega_{22}^2$, where ω_{11} and ω_{22} are the first critical speeds for the shaft body one assembly and the shaft body two assembly. It follows, from Eq. (18–39),

$$\frac{1}{\omega_1^2} + \frac{1}{\omega_2^2} = \frac{1}{\omega_{11}^2} + \frac{1}{\omega_{22}^2} \tag{18–40}$$

If ω_2 is the second (higher) critical speed, $1/\omega_2^2$ can be neglected and the first critical speed can be approximated from

$$\frac{1}{\omega_1^2} \doteq \frac{1}{\omega_{11}^2} + \frac{1}{\omega_{22}^2} \tag{18–41}$$

This idea can be extended to an n-body shaft:

$$\frac{1}{\omega_1^2} \doteq \sum_{1=1}^{n} \frac{1}{\omega_{ii}^2} \tag{18–42}$$

This is Dunkerley's approximation. By ignoring the higher mode term(s), the first critical speed estimate is *lower* than actually is the case.

The simple additive character of Dunkerley's equation, Eq. (18–42) suggests a useful superposition. Consider three loads P, Q, and R, at positions 1, 2, and 3 on the shaft. Noting that

$$\omega_{11}^2 = \frac{g}{P\delta_{11}} \qquad \omega_{22}^2 = \frac{g}{Q\delta_{11}} \qquad \omega_{33}^2 = \frac{g}{R\delta_{11}}$$

we can find the first critical speed of loads at station 1 as

$$\frac{1}{\omega_1^2} = \frac{1}{\omega_{11}^2} + \frac{1}{\omega_{22}^2} + \frac{1}{\omega_{33}^2} = \frac{P+Q+R}{g}\delta_{11}$$

In the determination of critical speeds the loads at the same position add arithmetically. Since Eq. (18–42) has no loads appearing in the equation, it follows that if each load could be placed at some convenient location transformed into an equivalent load, then the critical speed of an array of loads could be found by summing the equivalent loads, all placed at a single convenient location. For the load at station 1, placed at the center of span, denoted with the subscript cc, the equivalent load is found from

$$\omega_{11}^2 = \frac{g}{P_1\delta_{11}} = \frac{g}{P_{1c}\delta_{cc}}$$

or

$$P_{1c} = P_1\frac{\delta_{11}}{\delta_{cc}} \tag{18–43}$$

EXAMPLE 18–8 Consider a simply supported steel shaft as depicted in Fig. 18–19, with 1 in diameter and a 31-in span between bearings, carrying two gears weighing 35 and 55 lb.

(a) Find the influence coefficients.

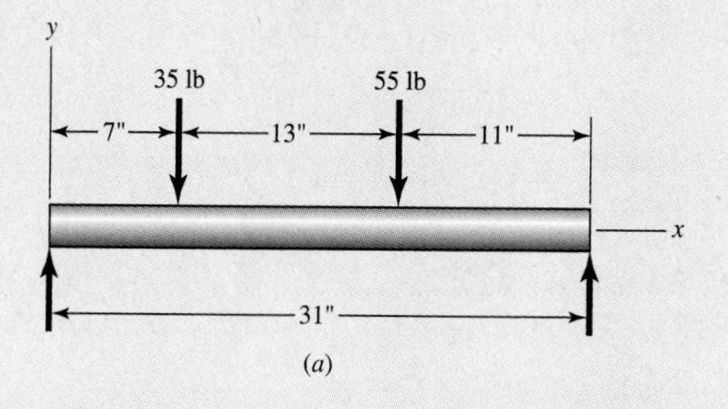

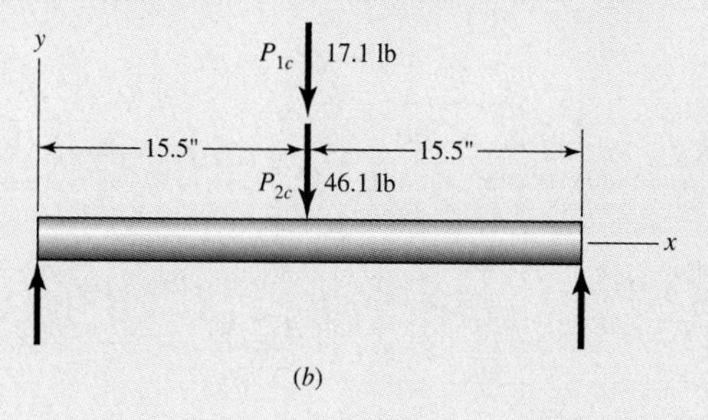

Figure 18–19

(*a*) A 1-in uniform-diameter shaft for Ex. 18–8. (*b*) Superposing of equivalent loads at the center of the shaft for the purpose of finding the first critical speed.

(b) Find ΣPy and ΣPy^2 and the first critical speed using Rayleigh's equation, Eq. (18–32).

(c) From the influence coefficients, find ω_{11} and ω_{22}.

(d) Using Dunkerley's equation, Eq. (18–42), estimate the first critical speed.

(e) Use superposition to estimate the first critical speed.

(f) Estimate the shaft's intrinsic critical speed. Suggest a modification to Dunkerley's equation to include the effect of the shaft's mass on the first critical speed of the attachments.

Solution (*a*)

$$I = \frac{\pi d^4}{64} = \frac{\pi(1)^4}{64} = 0.049\,09 \text{ in}^4$$

$$6EIl = 6(30)10^6(0.049\,09)31 = 0.274(10^9)$$

$$\delta_{11} = \frac{24(7)(31^2 - 24^2 - 7^2)}{0.274(10^9)} = 2.060(10^{-4}) \text{ in/lb}$$

$$\delta_{22} = \frac{11(20)(31^2 - 11^2 - 20^2)}{0.274(10^9)} = 3.533(10^{-4}) \text{ in/lb}$$

$$\delta_{12} = \delta_{21} = \frac{11(7)(31^2 - 11^2 - 7^2)}{0.274(10^9)} = 2.223(10^{-4}) \text{ in/lb}$$

Answer

i	*j* = 1	*j* = 2
1	2.060(10⁻⁴)	2.233(10⁻⁴)
2	2.233(10⁻⁴)	3.533(10⁻⁴)

$$y_1 = P_1\delta_{11} + P_2\delta_{12} = 35(2.060)10^{-4} + 55(2.233)10^{-4} = 0.0195 \text{ in}$$

$$y_2 = P_1\delta_{21} + P_2\delta_{22} = 35(2.233)10^{-4} + 55(3.533)10^{-4} = 0.0272 \text{ in}$$

(*b*) $\Sigma P_i y_i = 35(0.0195) + 55(0.0272) = 2.179 \text{ in} \cdot \text{lb}$

Answer $\Sigma P_i y_i^2 = 35(0.0194)^2 + 55(0.0272)^2 = 0.0540 \text{ in}^2 \cdot \text{lb}$

Answer $$\omega = \sqrt{\frac{386(2.179)}{0.0540}} = 124.7 \text{ rad/s, or } 1192 \text{ r/min}$$

(*c*)

Answer $$\frac{1}{\omega_{11}^2} = \frac{w_1}{g}\delta_{11} \qquad \omega_{11} = \sqrt{\frac{g}{w_1\delta_{11}}} = \sqrt{\frac{386}{35(2.060)10^{-4}}} = 231.4 \text{ rad/s, or } 2210 \text{ r/min}$$

Answer $$\omega_{22} = \sqrt{\frac{g}{w_2\delta_{22}}} = \sqrt{\frac{386}{55(3.533)10^{-4}}} = 140.9 \text{ rad/s, or } 1346 \text{ r/min}$$

$$(d)\ \frac{1}{\omega_1^2} = \sum \frac{1}{\omega_{ii}^2} = \frac{1}{231.4^2} + \frac{1}{140.9^2} = 6.905(10^{-5})$$

Answer $$\omega_1 = \sqrt{\frac{1}{6.905(10^{-5})}} = 120.3 \text{ rad/s, or } 1149 \text{ r/min}$$

which is less than part *b*, as expected.

$$(e)\ \delta_{cc} = \frac{b_{cc}x_{cc}(l^2 - b_{cc}^2 - x_{cc}^2)}{6EIl} = \frac{15.5(15.5)(31^2 - 15.5^2 - 15.5^2)}{0.274(10^9)}$$

$$= 4.213(10^{-4}) \text{ in/lb}$$

$$P_{1c} = P_1\frac{\delta_{11}}{\delta_{cc}} = 35\frac{2.060(10^{-4})}{4.213(10^{-4})} = 17.1 \text{ lb}$$

$$P_{2c} = P_2\frac{\delta_{22}}{\delta_{cc}} = 55\frac{3.533(10^{-4})}{4.213(10^{-4})} = 46.1 \text{ lb}$$

Answer $$\omega = \sqrt{\frac{g}{\delta_{cc}\Sigma P_{ic}}} = \sqrt{\frac{386}{4.213(10^{-4})(17.1 + 46.1)}} = 120.4 \text{ rad/s, or } 1149.8 \text{ r/min}$$

which agrees with part *d*, as expected.

(*f*) $A = \pi(1^2)/4 = 0.7854 \text{ in}^2$. From Eq. (18–31),

Answer $$\omega_s = \left(\frac{\pi}{31}\right)^2 \sqrt{\frac{386(30)10^6(0.049\,09)}{0.7854(0.282)}} = 520.3 \text{ rad/s, or } 4968.5 \text{ r/min}$$

An augmentation term for Dunkerley's equation would be $1/520.3^2$, so

$$\frac{1}{\omega^2} = \frac{1}{520.3^2} + 6.905(10^{-5}) = 7.274(10^{-5})$$

$$= 117.2 \text{ rad/s, or } 1119.6 \text{ r/min}$$

which is slightly less than part *d*, as expected.

The shaft's intrinsic first critical speed ω_s is just one more single effect to add to Dunkerley's equation. Since it does not fit into the summation, it is usually written up front.

Answer

$$\frac{1}{\omega_1^2} \doteq \frac{1}{\omega_s^2} + \sum_{i=1}^{n} \frac{1}{\omega^2} \tag{18–44}$$

Common shafts are complicated by the stepped-cylinder geometry which makes the influence-coefficient determination part of a numerical solution.

18–8 Shaft Design

Many shaft designs are modifications to existing shafts. The modification results from minor changes in bulk geometry of speed reducers, small changes in gear ratio, substitution of helical gears for spur gears or herringbone gears for helical gears. The point is that the designer has a head start. He or she knows the previous shaft's geometry and has a feel for the general size of the new design. The shaft to be designed is nebulous from the beginning. Moreover, other shaft designs are first-time efforts and no approximation is available to help in the visualization.

A shaft embodies a large number of decisions. To view all these decisions as "open" is to create a design space (really a hyperspace) that cannot be visualized. As a result, a designer's most valuable tool—geometric thinking—is severely hampered.

An approach for the designer is to come closer to the position of the engineer who has to modify an existing design and has an approximate picture of the result at the outset. Since the active constraint is likely to be distortional, consider distortion first.

- Find the diameter of the uniform-diameter shaft which meets the deflection and slopes at the bearings and at the power transmission elements.
  Equations (18–1) and (18–2) will be helpful. Finding that diameter will give you an average size for the shaft. Bearing seats and shoulders will create geometric features at the bearing locations. If the bearing locations are "outboard" of the power transmission features, the bearing journal diameter will have little influence on deflection and slope; thus either it can be ignored, or the central diameter can be increased slightly.
- Consider the power transmission features, shoulders, and hub bores, and make some tentative decisions on step geometry (diameters and length). Now the geometry is becoming less opaque.
- Take your approximate idea of the shaft geometry and perform a tabular deflection and slope analysis resembling Table 18–1. Use Eqs. (18–7) and (18–8) to find the largest $d_{\text{new}}/d_{\text{old}}$ ratio, then multiply *all* diameters by this ratio. See Ex. 18–2. You will have a stepped shaft that meets all the deflection and slope constraints. Check the new shoulders (especially with rolling-contact bearings) to see if shoulder height still is within manufacturers' recommended range. If adjustments are necessary, construct a new table.
- Begin a strength analysis using DE-Gerber or DE-elliptic theory. Use a desirable shaft material that avoids heat treatment. Heat treatment may increase strength. The cost of heat treatment will increase the cost of the shaft severalfold. Examine, feature by

feature, left bearing shoulder, left gear shoulder, gear keyway, right gear shoulder, right gear keyway, right bearing shoulder, shaft collar locations, snap-ring locations, for example. Examine these features for adequate diameter to see if material strength or diameter needs improvement. This feature-by-feature examination will tell the designer where the critical location is. If the critical feature is strength-sufficed, the designer has a clearer picture of the final design. The engineer may now have a satisfactory shaft. If some material can be pared off the shaft at other features (still keeping relevant constraints loose), consider this. In smaller production runs where volume-conservative initial forming methods are not used, the cost of additional turning to remove material (make chips) may mitigate against any such size reduction.

18–9 Computer Considerations

The previous section suggested how to proceed in the design of a shaft to slowly build the proportions of a satisfactory machine element. The intricacies of the computational chores—especially if the chores need to be repeated—suggest that computer assistance is attractive.

PROGRAMMING TASK NO. 1

Uniform Shaft to Meet Geometric Constraints

To estimate a uniform-diameter shaft, use Eqs. (18–1) and (18–2). Program them to handle bending moments and couples in space, using orthogonal planes (say horizontal and vertical). Plan the program to be interactive and solicit the Young modulus E, bearing span l, design factor n_d, allowable slope $\Sigma\theta$, and number of stations ns.

1. Enter E, l, n_d, $\Sigma\theta$, and ns.
2. From a DO-loop solicit the x coordinate of each station.
3. From a DO-loop, for the horizontal plane, enter the loads at all stations other than bearing locations, then all the loads for the vertical plane at all stations other than the bearing locations.
4. Repeat Step 4 for all couples causing transverse bending in the horizontal and vertical planes.
5. Give the user a chance to view a table of all input information.
6. Give the user a chance to correct any errors made in entry.
7. Allow the programmed equations to perform the calculations necessary to produce the uniform diameter.
8. Test such a program on problems with known answers (benchmarking).

With the uniform-diameter shaft in mind, the next step is to consider shoulders, hub bores, and the like, and make tentative decisions on the between-the-bearings shaft geometry, putting more than uniform-diameter material in the middle and, perhaps, less near the bearings. This step is better done manually so that the designer sees what he or she is doing.

PROGRAMMING TASK NO. 2

Implementation of Table 18–1 for Stepped Shafts

This second useful program is to build a table such as Table 18–1, perhaps with fewer columns to reflect the monitor's screen width. It is possible to produce tables for both the horizontal and vertical planes and let the computer make the Pythagorean combinations of deflection and slope and complete a combined table, but the designer might be better served by seeing the intermediate tables. Viewing two tables and making the Pythagorean combinations by hand will help the designer keep in touch with important numbers and their relative sizes. Also, plotting an end view of the y deflection from one table and the z direction from the other is useful in visualizing the lay in space of the shaft centerline. In planning such a program, the following features are useful.

❶ Give the user the option of writing to a file.

❷ Enter the number of stations, station number of the left bearing, station number of the right bearing, and the Young modulus.

From within a DO-loop, solicit

❶ Moment to the left of station (number called out on screen).

❷ Moment to the right of station (number called out on screen).

❸ x Coordinate of station (number called out on screen).

❹ Diameter to left of station (number called out on screen).

❺ Diameter to right of station (number called out on screen).

Then

❶ Show a table of entries.

❷ Give the user the option of correcting entries containing errors without starting over.

❸ Print the horizontal and vertical tables with C_1, C_2, and E showing in the heading (so prediction equations can be constructed outside the computer, if needed). Follow with the body of Table 18–1. If the combined table is displayed, the user can identify any violated constraints and multiply *all* the diameters [using Eqs. (18–7) and (18–8)] by $(d_{new}/d_{old})_{max}$. Alternatively, if the input data initially solicited by the program asked the user to supply all constraints on deflection and slope, calculated all d_{new}/d_{old}, identified $(d_{new}/d_{old})_{max}$, applied it to all diameters and redid the tables, the final table would be effortless, at the price of understanding.

Be careful not to give the computer so much to do behind the scenes that it prevents the user from seeing and understanding what is going on. You can program the computer, by using options, to show a little or show a lot.

The user now checks to see if shoulder heights are within the bearing manufacturer's recommended range. This final adjustment, if necessary, may not change the table much, as the near-the-bearing geometry has little effect on deflection and slope.

Strength sufficing has two common forms:

- Finding the required strength for extant geometry.
- Finding required geometry at potential critical locations.

The deterministic adequacy assessment of Ex. 18–6 is a straightforward sequence of calculations which are mildly nonlinear and cannot be easily inverted. For example, if a factor of safety of 2.5 was required, and the ultimate strength of the material to meet it needs to be identified, the sequence of calculations is

$$\bar{k}_a = 2.67\bar{S}_{ut}^{-0.265}$$

$$k_b = (1.378/0.30)^{-0.107} \qquad \text{(unvarying)}$$

$$\bar{S}_e = \bar{k}_a k_b 0.506\bar{S}_{ut}$$

$$\bar{K}_t = 2.598 \qquad \bar{K}_{ts} = 1.921 \qquad \text{(both unvarying)}$$

$$\bar{K}_f = \frac{2.598}{1 + \dfrac{2}{\sqrt{0.030}}\dfrac{2.598 - 1}{2.598}\dfrac{4}{\bar{S}_{ut}}} \qquad \bar{K}_{fs} \doteq \frac{1.921}{1 + \dfrac{2}{\sqrt{0.030}}\dfrac{1.921 - 1}{1.921}\dfrac{4}{\bar{S}_{ut}}}$$

$$r = \frac{2\bar{K}_f 1677}{\sqrt{3}\bar{K}_{fs}3300}$$

$$\bar{\sigma}'_a = \frac{32\bar{K}_f 1677}{\pi(1.378)^3}$$

$$\bar{S}_a = \frac{(rS_{ut})^2}{2S_e}\left[-1 + \sqrt{1 + \left(\frac{2S_e}{rS_{ut}}\right)^2}\right] \qquad \text{or} \qquad \bar{S}_a = \frac{r\bar{S}_y\bar{S}_e}{\sqrt{r^2\bar{S}_y^2 + \bar{S}_e^2}}$$

$$\bar{n} = \bar{S}_a/\bar{\sigma}'_a$$

This sequence of calculations can be programmed on a hand-held programmable calculator giving, for a DE-Gerber locus, similar to Ex. 18–5:

$\bar{S}_{ut}$, kpsi	$\bar{k}_a$	k_b	$\bar{S}_e$, kpsi	$\bar{K}_f$	$\bar{K}_{fs}$	r	$\bar{\sigma}'_a$, kpsi	$\bar{S}_a$, kpsi	$\bar{n}$
85	0.823	0.849	30.06	1.947	1.544	0.740	12.71	25.22	1.984
95	0.799	0.849	32.62	2.000	1.579	0.743	13.06	27.63	2.116
105	0.778	0.849	35.11	2.045	1.628	0.746	13.35	27.97	2.245
115	0.759	0.849	37.53	2.083	1.633	0.749	13.60	32.26	2.372
125	0.743	0.849	39.91	2.117	1.655	0.751	13.82	34.51	2.497

The table above shows that the Marin surface factor $\bar{k}_a$ is a decreasing monotonic function, whereas the stress-concentration factors in bending fatigue and torsional fatigue, $\bar{K}_f$ and $\bar{K}_{fs}$, as well as load-line slope r, stress amplitude mean $\bar{\sigma}'_a$, and strength amplitude component $\bar{S}_a$ are all increasing monotonic functions of $\bar{S}_{ut}$. Furthermore, we learn that doing just two lines in the table and interpolating/extrapolating is effective. A formal root-finding procedure to search for the zero place of $(\bar{S}_{ut}/\bar{\sigma}'_a - 2.5)$ takes longer to set up than the computation to prepare the table.

For stochastic problems, the calculations are only a little more involved with lognormal load and lognormal strength. When the load distribution is not lognormal, it is still possible to avoid a computer simulation. If the load distribution is normal or robustly so, then an interference can be performed using the idea of Table 6–3. Equation (6–29) is the basis for interference of any two distributions.

PROGRAMMING TASK NO. 3

Mean Design Factor and Reliability Goal

Refresh your memory by reviewing Sec. 7–20 on finding the mean value of the design factor for attaining a reliability goal when the COV of the loading is known. Then reread Sec. 18–4 on the same subject. Now consider programming a protocol which will allow you to do this.

❶ Note the reliability goal R and z of $(1 - R)$, and COV of loading C_L. At the critical section if loading is

- axial, $C_{kc} = 0.125$
- bending, $C_{kc} = 0.0$
- torsion, $C_{kc} = 0.125$

If critical feature is a

- hole, $C_{Kf} = 0.10$
- shoulder, $C_{Kf} = 0.11$
- groove, $C_{Kf} = 0.15$

❷ Find COV of stress from $C_\sigma = (C_L^2 + C_{Kf}^2)^{1/2}$. If surface condition is

- ground, $C_{ka} = 0.120$
- machined or cold-rolled, $C_{ka} = 0.058$
- hot-rolled, $C_{ka} = 0.110$
- as-forged, $C_{ka} = 0.145$

If temperature is 70°F, $C_{kd} = 0.0$; otherwise, $C_{kd} = 0.11$.
If the correlation method is used, $C'_{Se} = 0.138$; if from test, $C_{Se'} =$ test value.

❸ Estimate the COV of the endurance strength from

$$C_{Se} = (C_{ka}^2 + C_{kc}^2 + C_{kd}^2 + C_{Se'}^2)^{1/2}$$

For fully reversed loading the COV of the design factor is

$$C_n = [(C_{Se}^2 + C_\sigma^2)/(1 + C_\sigma^2)]^{1/2}$$

❹ For fatigue loading with amplitude and midrange components, when S_a is not known a priori, substitute C_{Se} for C_{Sa} (this is conservative). The mean value of the design factor is

$$\bar{n}_f = \exp\left[-z\sqrt{\ln(1 + C_n^2)} + \ln\sqrt{1 + C_n^2}\right]$$

The above procedure is independent of the mean values of stress amplitude $\bar{\sigma}'_a$ and the mean value of the strength $\mathbf{S}_e$. Therefore, this is known before any geometric or strength decisions are made. The beauty of the program is that it can easily be modified and extended as engineering looks further into the dark places.

With just a little stochastic attention, we have tamed a problem that has plagued engineering since the beginning. Reflect also on the role *viewpoint* contributed to our understanding.

If you complete programming task 3, understand it, and use it, you have journeyed through the book to reach a valuable reward. Without the statistical attention and practice you would not be in a position to understand variability and its quantification as it relates to shafting.

PROBLEMS

18–1 The 16-tooth pinion in the figure drives a double-reduction gear train as shown. All gears have 25° pressure angle. The pinion rotates counterclockwise at 1200 r/min and transmits 50 hp to the gear train. Our focus is on the pinion shaft. None of the shafts have been designed.

(*a*) For the pinion shaft develop the moment diagram.

(*b*) For cylindrical roller bearings the shaft slope at the journal should be less than 0.001 rad. For a design factor of $n_d = 2$ estimate the uniform shaft diameter that would meet the deflection constraints.

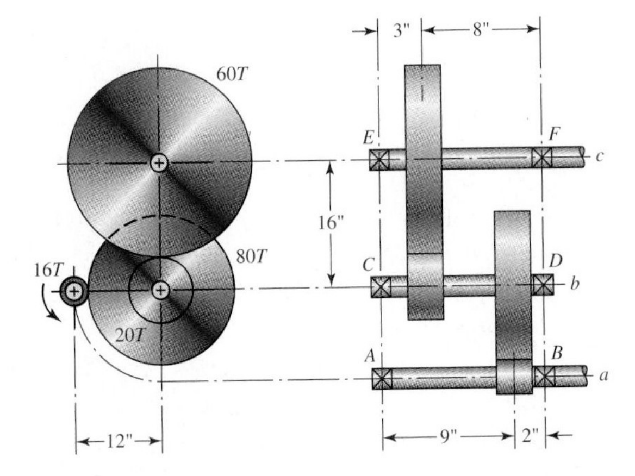

Problem 18–1

18–2 Investigate the result of Prob. 18–1, for fatigue strength sufficiency in a preliminary way, whether strength or deflection controls. Use a 1030 hot-rolled steel for this purpose. The likely reliability goal for this shaft is 0.999, with a coefficient of load variation of $C_L = 0.10$.

(*a*) Show, using the discussion at the end of Sec. 18–8, that the design factor should be about 2.19.

(*b*) Use the DE-elliptic fatigue-failure locus for your factor of safety estimate.

(*c*) Use the DE-Gerber fatigue-failure locus in your estimate.

18–3 Having found the pinion shaft of Probs. 18–1 and 18–2 has a tight deflection constraint, design as much of the shaft as you can. The overhang to mount a coupling upon can presume a 6-in extension of the shaft beyond the bearing. Draw your shaft, showing all dimensional decisions.

18–4 All the designs completed in response to Prob. 18–3 will differ. For the sake of discussion, consider the design shown in the figure. The designer specified 02–40-mm ball bearing on the left and an 03–40-mm cylindrical roller bearing on the right. Check this design for adequacy with respect to distortion using the tabular method of Table 18–1 (and details found in Chap. 4).

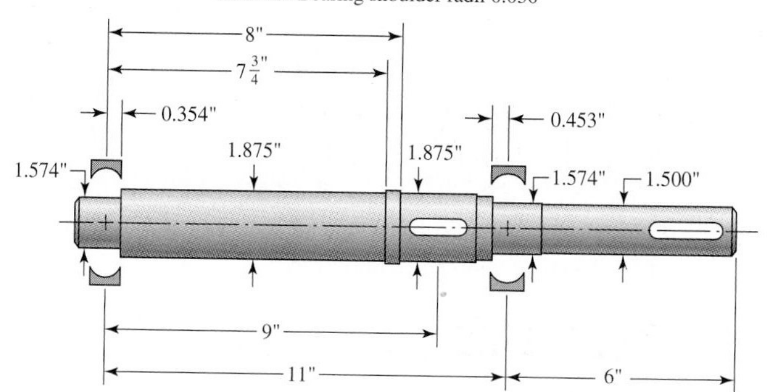

Problem 18–4
Shoulder fillets at bearing seat 0.030-in radius, others $\frac{1}{8}$-in radius, except right-hand bearing seat transition, $\frac{1}{4}$ in. The material is 1030 HR. Keyways $\frac{3}{8}$ in wide by $\frac{3}{16}$ in deep.

DESIGN

18–5 Check the design in Prob. 18–3 for adequacy in fatigue using the DE-elliptic fatigue-failure locus.

18–6 Now you have the opportunity to see how your design (Prob. 18–3) compares. Using your design completed in response to Prob. 18–3, perform the adequacy checks suggested in Probs. 18–4 and 18–5.

DESIGN

18–7 The experience you have gained with Probs. 18–1 through 18–6 will make the design of shaft c of Prob. 18–1 easier. Design shaft c of Prob. 18–1 for a reliability goal of 0.995 with a coefficient of variation of the load $C_L = 0.06$.

DESIGN

18–8 The design of the intermediate shaft b of the speed reducer of Prob. 18–1 is more difficult. The moment diagram does not lie in a plane. It is necessary to use moment diagrams in orthogonal planes. Design shaft b of Prob. 18–1 for $R = 0.995$ and $C_L = 0.06$.

ANALYSIS

18–9 Whether you have had the time to accomplish the design of the shafts a, b, and c of Prob. 18–1 or not, it is useful to consider the interrelationship to the geometry of the shafts, which the rolling-contact bearings will have. Avoid engaging in an iterative design situation unnecessarily. There will be a bearing reliability goal for the six bearings of the reducer. Plan a strategy for identifying the individual bearing reliabilities. The design work you will save will be your own.

DESIGN

18–10 A geared industrial roll shown in the figure is driven at 300 r/min by a force F acting on a 3-in-diameter pitch circle as shown. The roll exerts a normal force of 30 lb/in of roll length on the material being pulled through. The material passes under the roll. The coefficient of friction is 0.40. Develop the moment and shear diagrams for the shaft modeling the roll force as (a) a concentrated force at the center of the roll, and (b) a uniformly distributed force along the roll. These diagrams will appear on two orthogonal planes.

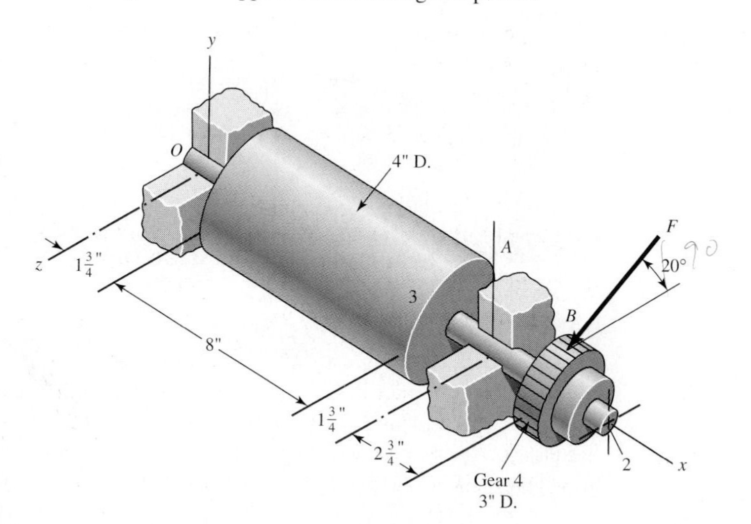

Problem 18–10
Material moves under the roll.

DESIGN

18–11 Using a 1035 hot-rolled steel, estimate the necessary diameters at the locations of peak bending moment using a design factor of 2. There are likely to be fillets at both ends of the right-hand bearing seat, where the bending moment is slightly less than the local extreme. Estimating the fatigue stress-concentration factor as 2, and using a design factor of 2, what is the approximate necessary diameter of the bearing seat using the DE-elliptic fatigue-failure locus in Prob. 18–10?

DESIGN

18–12 For the situation in Prob. 18–10, what diameter uniform shaft will meet a deflection bearing slope of (*a*) 0.001 rad and (*b*) 0.0005 rad?

DESIGN

18–13 Use the tabulation method (Table 18–1) to find the diameter of a uniform shaft to meet the slope limitation of 0.0005 rad at the gear mesh. What is the diameter to meet the slope with a factor of safety of 2?

DESIGN

18–14 Design a shaft for the situation of the industrial roll of Prob. 18–10 with a reliability goal of 0.999 against fatigue failure, with a coefficient of variation of the load C_L of 0.2. Plan for a ball bearing on the left and a cylindrical roller on the right. For distortion use a factor of safety of 2.

ANALYSIS

18–15 The figure shows a proposed design for the industrial roll shaft of Prob. 18–10. Hydrodynamic film bearings were to be used. All surfaces are machined except the journals, which are ground and polished. The material is 1035 HR steel (Table E-25). Perform an adequacy assessment. Is the design satisfactory?

Problem 18–15
Bearing shoulder fillets 0.030 in, others 1/16 in. Sled-runner keyway is $3\frac{1}{2}$ in long.

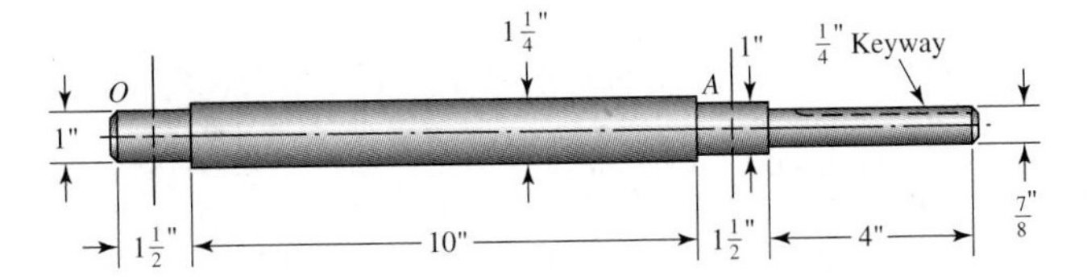

ANALYSIS

18–16 As shown in the figure, the axle of a railroad freight car is tapered, with the least diameter between the rails. This problem will give some insight as to why this is so. A freight-car axle is force-fitted to its wheels with the journals outboard of the wheels, and with journal centers about 80 in apart. The track gauge is $56\frac{1}{2}$ in between the rails, and so the span between the rail centers is about $59\frac{1}{2}$ in. Freight-car wheels are usually of diameter 33 in for cars up to 70-ton capacity. Consider the worst-case location of the center of mass of the car and load to be 72 in above the rails. The worst-case vertical load is to be 42 750 lb per axle. The worst-case horizontal load due to crosswinds and track curvature is 17 100 lb per axle, through the center of mass. Construct the bending-moment diagram for the axle.

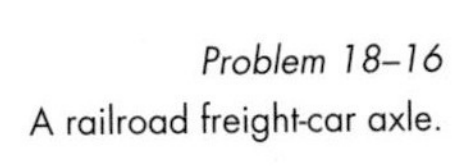
Problem 18–16
A railroad freight-car axle.

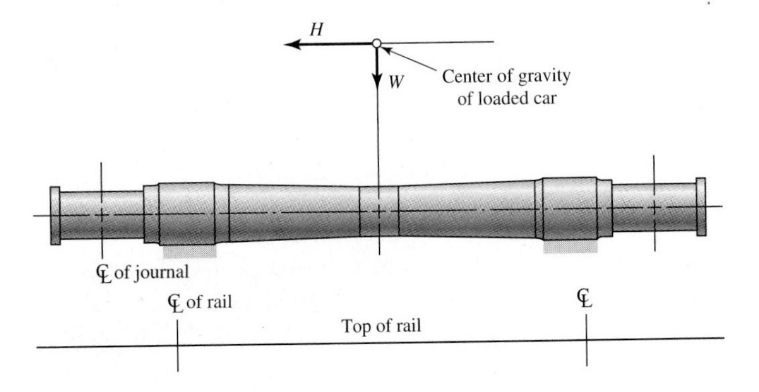

ANALYSIS

18–17 In a class C American Association of Railroads standard axle, the diameter of the wheel seat is 7 in, and the diameter of the axle at midspan is $5\frac{3}{8}$ in. Using the results of Prob. 18–16, compare the bending stress level at the center of the wheel seat and at the axle midspan. Is this difference to be expected?

18–18 The section of shaft shown in the figure is to be designed to approximate relative sizes of $d = 0.75D$ and $r = D/20$ with diameter d conforming to that of standard metric rolling-bearing bore sizes. The shaft is to be made of SAE 2340 steel, heat-treated to obtain minimum strengths in the shoulder area of 1226-MPa ultimate tensile strength and 1130-MPa yield strength with a Brinell hardness not less than 368. At the shoulder the shaft is subjected to a completely reversed bending moment of 70 N · m, accompanied by a steady torsion of 45 N · m. Use a design factor of 2.5 and size the shaft for an infinite life. The results should be based on DE-elliptic fatigue-failure locus.

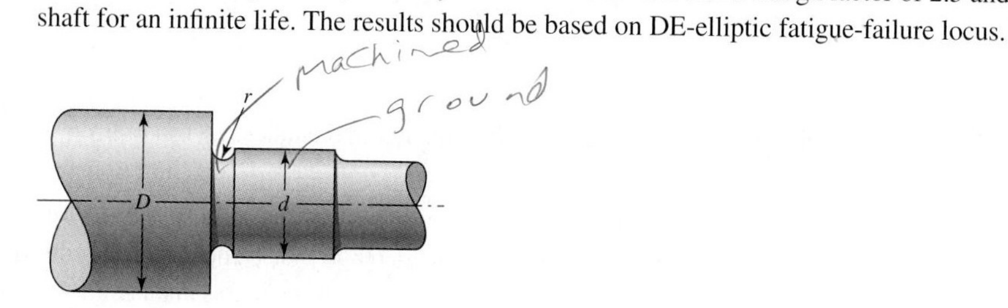

Problem 18–18
Section of a shaft containing a grinding-relief groove. Unless otherwise specified, the diameter at the root of the groove $d_R = d - 2r$, and though the section of diameter d is ground, the root of the groove is still a machined surface.

18–19 A transverse drilled and reamed hole can be used in a solid shaft to hold a pin which locates and holds a mechanical element, such as the hub of a gear, in axial position, and allows for the transmission of torque. Since a small-diameter hole introduces high stress concentration, and a larger diameter hole erodes the area resisting bending and torsion, investigate the existance of a pin diameter with minimum adverse affect on the shaft. Then formulate a design rule. (*Hint:* Use Table E–16.)

DESIGN

18–20 A slow-speed spur gear with a 1.75-in bore and a $1\frac{1}{2}$-in-long hub has a pitch diameter of 8 in and involute teeth of 20° pressure angle, and it transmits 4.5 hp at 112 r/min. The shaft is to have a bearing span of 10 in, with gear placed 3 in from the right-hand bearing. Deep-groove ball bearings are to be used. A 2-in overhang for a coupling having a 1-in diameter seat is to be provided at the left of the left-hand bearing. The bearing life is to be 10 kh at a reliability of 0.995 for the bearing pair. Some a priori decisions are shown in the figure. Select appropriate bearings, then dimension the shaft in the neighborhood of the bearings.

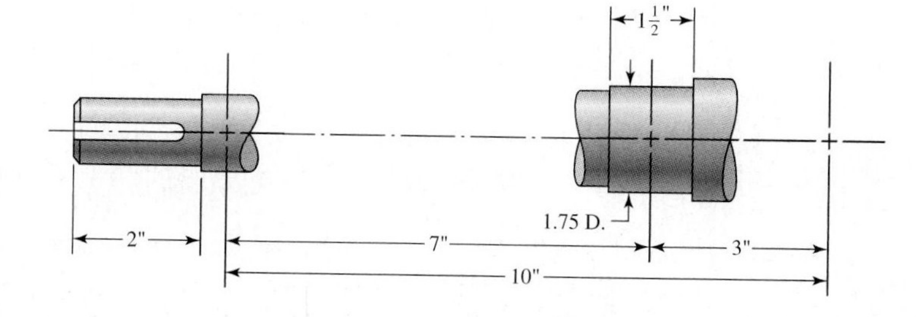

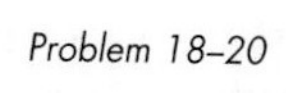
Problem 18–20

DESIGN

18–21 The design of Prob. 18–20 has been completed, resulting in a shaft which is 13 in long overall. It is important to have assurance that the deflection is within limits necessary to the functioning of the gear mesh and bearing life. A testing machine which can accurately apply a compressive load is available, but the throat area is such that a specimen and fixture only up to 8 in long can be handled. For an iconic scale model made half-size, what load should be applied? How should a measured deflection or slope be interpreted if the model is made of the same material as the prototype?

Problem 18–21

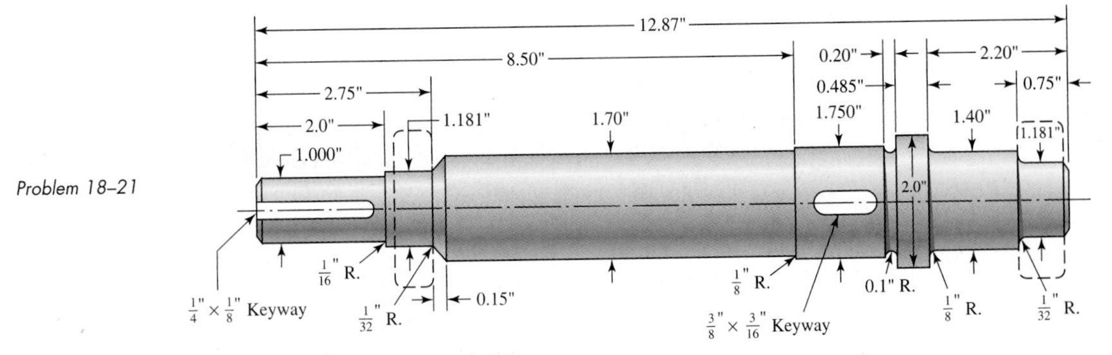

ADEQUACY ASSESSMENT

18–22 An AISI 1020 cold-drawn steel shaft with the geometry shown in the figure carries a transverse load of 7 kN and a torque of 107 N · m. Examine the shaft for strength and deflection. If the largest allowable slope at the bearings is 0.001 rad and at the gear mesh is 0.0005 rad, what is the factor of safety guarding against damaging distortion? What is the factor of safety guarding against a fatigue failure? If the shaft turns out to be unsatisfactory, what would you recommend to correct the problem?

Problem 18–22

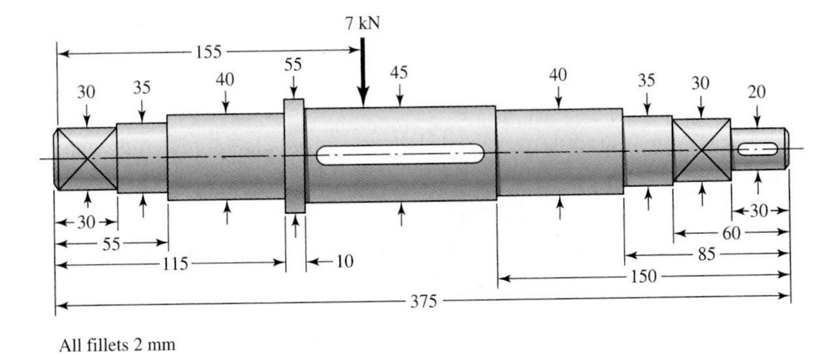

ANALYSIS

18–23 A 1-in-diameter uniform shaft is 24 in long between bearings.

(*a*) Find its intrinsic critical speed.

(*b*) If the goal is to double the critical speed, find the new diameter.

(*c*) A half-size model of the original shaft has what critical speed?

ANALYSIS

18–24 Demonstrate how rapidly Rayleigh's method converges for a uniform-diameter solid shaft, by partitioning the shaft into first one, then two, and finally three elements.

ANALYSIS

18–25 Compare Eq. (18–37) for the angular frequency of a two-disk shaft with Eq. (18–38), and note that the constants in the two equations are equal.

(*a*) Develop an expression for the *second* critical speed.

(*b*) Estimate the second critical speed of the shaft addressed in Ex. 18–8, parts *a* and *b*.

ANALYSIS

18–26 For a uniform-diameter shaft, does hollowing the shaft increase or decrease the critical speed?

18–27 The shaft shown in the figure carries a 20-lb gear on the left and a 35-lb gear on the right. Estimate the first critical speed due to the loads, the shaft's intrinsic critical speed, and the critical speed of the combination.

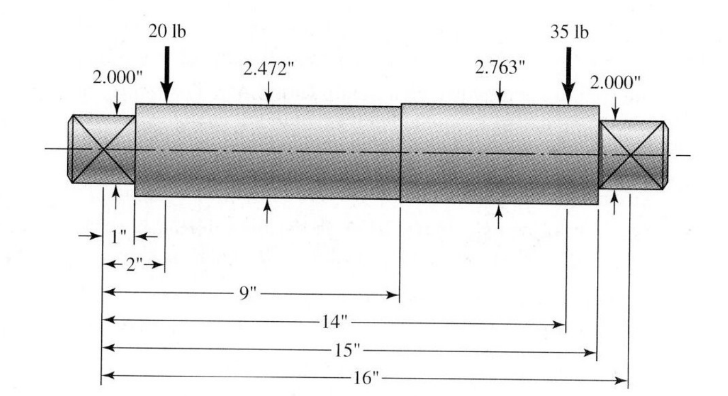

Problem 18–27

DESIGN

18–28 A shaft is to be designed to support the spur pinion and helical gear shown in the figure on two bearings spaced 28 in center-to-center. Bearing *A* is a cylindrical roller and is to take only radial load; bearing *B* is to take the thrust load of 220 lb produced by the helical gear and its share of the radial load. The bearing at *B* can be a ball bearing. The radial loads of both gears are in the same plane, and are 660 lb for the pinion and 220 lb for the gear. The bearings are to have a life of 2000 h at a combined reliability of 0.995. The shaft speed is 1150 r/min. Select the bearings and design the shaft. Make a sketch to scale of the shaft showing all fillet sizes, keyways, shoulders, and diameters. Specify the material and its heat treatment.

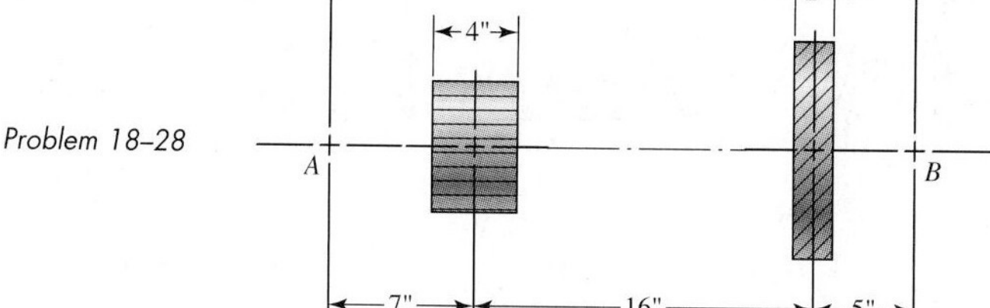

Problem 18–28

18–29 A heat-treated steel shaft is to be designed to support the spur gear and the overhanging worm shown in the figure. A bearing at *A* takes pure radial load. The bearing at *B* takes the worm-thrust load for either direction of rotation. The dimensions and the loading are shown in the figure; note that the radial loads are in the same plane. Make a complete design of the shaft, including a sketch of the shaft showing all dimensions. Identify the material and its heat treatment (if necessary). Provide an adequacy assessment of your final design. The shaft speed 310 r/min.

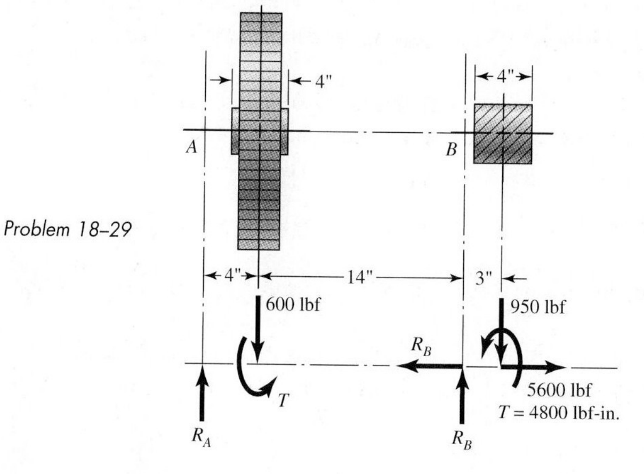

Problem 18–29

REDESIGN

18–30 A bevel-gear shaft mounted on two 40-mm 02-series ball bearings is driven at 1720 r/min by a motor connected through a flexible coupling. The figure shows the shaft, the gear, and the bearings. The shaft has been giving trouble—in fact, two of them have already failed—and the down time on the machine is so expensive that you have decided to redesign the shaft yourself rather than order repacements, which would, it is more than likely, also fail in a few months. You have not been able to assemble much information. A hardness check of the two shafts in the vicinity of the fracture of the two shafts showed an average of 198 Bhn for one and 204 Bhn of the other. As closely as you can estimate the two shafts failed at a life measure between 600 000 and 1 200 000 cycles of operation. The surfaces of the shaft were machined, but not ground. The fillet sizes were not measured, but they correspond with the recommendations for the ball bearings used. You know that the load is a pulsating or shock-type load, but you have no idea of the magnitude, because the shaft drives an indexing mechanism, and the forces are inertial. The keyways are $\frac{3}{8}$ in wide by $\frac{3}{16}$ in deep. The straight-toothed bevel pinion drives a 48-tooth bevel gear. Specify a new shaft in sufficient detail to ensure a long and trouble-free life.

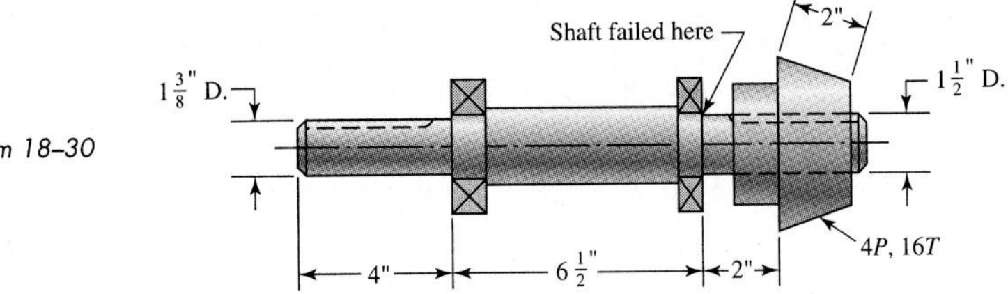

Problem 18–30

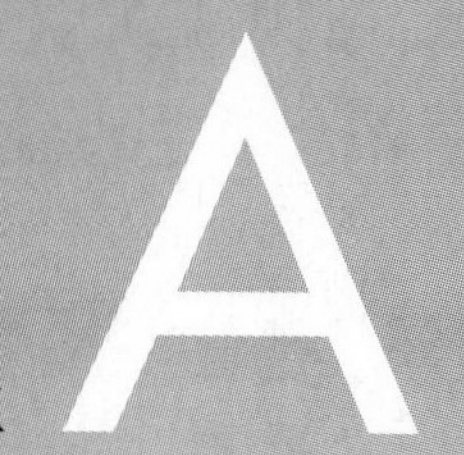

Statistical Relations **Appendix** A

A–1 Real Numbers

Engineering calculations involve a blend of numbers:

- Mathematical constants, such as π or e
- Toleranced dimensions
- Measurement numbers
- Mathematical functions, themselves approximate
- Unit-conversion constants
- Mechanically generated digits from calculators and computers
- Rule-of-thumb numbers

Scientific notation is a display consisting of a lead zero, followed by a decimal point, then a display of digits, followed by 10 raised to an integer power. For example, the number 987 654 321 is displayed in scientific notation as 0.987 654 321(10^9). Thus economy of expression is achieved with no ambiguity in referring to digits within the number. *All* are to the right of the decimal point.

The set of all *integers* is the set of the counting numbers 1, 2, . . . , augmented by negative numbers and zero. The set of all *rational numbers* m/n is constructed from the integers, dividing by zero excepted. The set of all *real numbers* is constructed by adding limits of all bounded monotone increasing sequences (*irrational numbers*) to the set of rational numbers. Each set of numbers contains the previous set. Each point in a *number line* corresponds to a real number.

The display of a real number often raises the problem of economy of notation. If the true number is 3.141 592 654. . . , then an *approximate number* 3.14 is useful to someone for some purpose, and it is given without qualification. A *significant number* is a number that does not differ from the true number by more than one-half in the last recorded digit. Thus 3.1416 is a significant number corresponding to π and is bounded by (implied) *range numbers* computable from the $\pm$ 0.000 05 implied, for certain.

A significant number is a special form of an *approximation-error number*. For instance,

$$\pi = 3.141^{+0.0006}_{-0.0001}$$

The true value lies within the range numbers (explicitly displayed), *for certain*.

Another form of number is the *incomplete number*. It can be formed from the true number by simply truncating after a prescribed number of digits. Every incomplete number encountered is treated as exact in computation. Computational precision is ensured by selecting the number of computational digits appropriately larger, say double-precision rather than single-precision, than the results require.

A *rounded number* is a number that has been shortened for display, essentially by the rules for significant numbers. No significance is implied, and none is to be inferred. Only the approximation-error number can be recognized from the display because of the presence of the range numbers.

For numbers resulting from measurement, significant, approximate, range, and approximation-error numbers are not useful.

A–2 Random Variables

A *random variate*[1] is characterized by a distribution and statistical parameters; it is denoted by boldface as **x**. For example, a variate exhibiting a *normal* distribution is expressed as $N(\mu, \sigma)$ where the parameter μ is an incomplete number representing the distribution *mean* and σ is the *standard deviation*. Any qualification on the mean is addressed by the standard deviation and the distribution identity. Range numbers do not exist, but *confidence bounds* (which are incomplete numbers) can be expressed at some stated probability level.

A–3 Estimators

A *best estimator* of a statistical parameter must be consistent, unbiased, and efficient.[2] Denote θ as the parameter being estimated from the data, and $\hat{\theta}$ as its estimator.

- The estimator $\hat{\theta}$ must be a *consistent estimator* of θ. As the sample size N grows larger, the difference $|\theta - \hat{\theta}|$ becomes smaller.
- The estimator $\hat{\theta}$ must be an *unbiased estimator* of θ, that is, the expected value of $\hat{\theta}$ must equal θ, symbolically $E[\hat{\theta}] = \theta$.
- The estimator $\hat{\theta}$ must be an *efficient estimator* of θ. The variance of an efficient estimator θ must be smaller than the variance of any other estimator θ', symbolically $V[\theta] \leq V[\theta']$.

An efficient estimator must be consistent and unbiased. Common notation for the estimator of the population mean μ is the sample mean $\bar{x}$. Similarly, the estimator of the population variance σ^2 is the sample variance s^2.

A–4 Sampling Statistics

The *coefficient of variation,* abbreviated c.o.v. or cov or COV, of **x** is C_x, which is the ration of the standard deviation to the mean,

$$C_x = \sigma_x/\mu_x \tag{A–1}$$

$$\mathbf{x} = \mathbf{N}(\mu_x, \sigma_x) = \mu_x \mathbf{N}(1, C_x) \tag{A–2}$$

read as "x is a normally distributed variate with a mean of μ_x and a standard deviation of σ_x." This is equal to the mean μ_x multiplied by a normal variate with a mean of 1 and a standard deviation of C_x. The standard deviation of $\mathbf{N}(1, C_x)$ happens to be the COV of $\mathbf{N}(\mu_x, \sigma_x)$. A general form of data display is the histographic form. See Fig. 2–1. If

1. *Stochastic variable, random variable,* and *variate* are all synonyms.

2. Leland Blank, *Statistical Procedures for Engineering, Management, and Science,* McGraw-Hill, New York, 1980.

N is the total number of observations, f_i the census of the ith class of observations, and k the number of classes, then

$$N = \sum_{i=1}^{k} f_i$$

If data are drawn from a population with a mean of μ_x, ungrouped, the best estimate of μ_x is $\bar{x}$, given by

$$\bar{x} = \frac{1}{N}\sum x_i \tag{A-3}$$

The sample standard deviation s, the best estimate of σ, is given by

$$s = \sqrt{\frac{\sum (x_i - \bar{x})^2}{N-1}} = \sqrt{\frac{\sum x_i^2 - \left(\sum x_i\right)^2/N}{N-1}} \tag{A-4}$$

If the data are grouped as in Fig. 2–1*a* and f_i is the class frequency,

$$\bar{x} = \frac{\sum_{i=1}^{k} f_i x_i}{\sum_{i_n=1}^{k} \frac{f_i x_i}{f_i}} = \frac{1}{N}\sum_{i=1}^{k} f_i x_i \tag{A-5}$$

and x_i is the class midpoint. The sample standard deviation is

$$s = \sqrt{\frac{\sum_{i=1}^{k} f_i (x_i - \bar{x})^2}{N-1}} = \sqrt{\frac{\sum f_i x_i^2 - \left[\left(\sum f_i x_i\right)^2/N\right]}{N-1}} \tag{A-6}$$

The histographic display indicates the distribition's shape. The number of classes k can be too small or too large. As a guide, the Sturges[3] rule for the desirable number of classes gives

$$k = 1 + 3.3 \log_{10} N \tag{A-7}$$

For example, with 30 observations, the Sturges rule suggests

$$k = 1 + 3.3 \log_{10}(30) = 5.9 \doteq 6$$

A–5 Cumulative Distribution Function

If a histogram had the number of observation increase without upper bound, and the class interval approach zero, the limiting shape and function are valuable. The probability that an observation of $\mathbf{x}$ lies between x and $x + \Delta x$, per unit of x, is given by $f_i/(N\Delta x)$. If the limit of $f_i/(N\Delta x)$ exists and is denoted as dF, then

$$F(x) = \int_{-\infty}^{x} f(x)\, dx \tag{A-8}$$

3. H. A. Sturges, "The Choice of Class Interval," *J. Am. Statistical Association*, vol. 21 (1926), pp. 65–66.

where F is the *cumulative distribution function* of x, abbreviated CDF. When $x \to \infty$ then

$$\int_{-\infty}^{\infty} f(x)\,dx = 1 \tag{A-9}$$

The function $f(x)$ is called the *probability density function* of x, or, simply, the *density function* of x; it is denoted in abbreviated form as PDF. Differentiation of the CDF of x with respect to x gives

$$\frac{dF(x)}{dx} = f(x) \tag{A-10}$$

showing that the density function $f(x)$ is the slope of the CDF at x. The histographic counterpart is

$$F_i = \frac{f_i w_i}{2} + \sum_{j=1}^{i-1} f_j w_j \tag{A-11}$$

where w_i is the width of the ith class. Plotting F_i versus x_i with finite data gives a fuzzy snapshot of the CDF, and plotting $f_i(Nw_i)$ versus x_i gives an approximation to the PDF. The first move of a statistician is to plot the data, and you should do the same.

A–6 The Gaussian (Normal) Distribution

The most likely parent distribution from which data come is the distribution of Gauss, which bears his name. It minimizes the set of errors. The probability density function of the normal distribution is

$$f(x) = \frac{1}{\sigma_x\sqrt{2\pi}} \exp\left[-\frac{1}{2}\left(\frac{x-\mu_x}{\sigma_x}\right)^2\right] \tag{A-12}$$

Integration of Eq. (A–12) requires numerical methods. Table E-10 shows the results for a unit normal $N(0, 1)$. Entries in the body of the table are denoted as $\Phi(z)$. For any normal distribution other than $\mathbf{N}(0, 1)$ the value of z of $(x - \mu_x)/\sigma_x$ and the integral of Eq. (A–12) is expressed as

$$F(z_\alpha) \int_{-\infty}^{z_\alpha} \frac{1}{\sqrt{2\pi}} \exp\left(-\frac{u^2}{2}\right) du = \Phi(z_\alpha) \tag{A-13}$$

where u is a dummy variable of integration. The cumulative distribution function is rectified when plotted with the data as ordinate and the abscissa as the z value corresponding to the ith CDF, namely, $\bar{F}_i = i/(n+1)$ or $\tilde{F}_i = (i - 0.3)/(n + 0.4)$, as desired, to seek mean or median loci, respectively.

A–7 Lognormal Distribution

A lognormal distribution is denoted as $\mathbf{x} \sim LN(\mu_x, \sigma_x)$. The logarithm of $\mathbf{x}$ is distributed normally, hence the name. If

$$y = \ln \mathbf{x} \tag{A-14}$$

then

$$\mathbf{y} \sim N(\mu_y, \sigma_y) \tag{A-15}$$

The distribution of **y** is a companion or subsidiary distribution to $\mathbf{x} \sim LN(\mu_x, \sigma_x)$. The PDF of **x** is

$$f(x) = \frac{1}{x\sigma_y\sqrt{2\pi}} \exp\left[-\frac{1}{2}\left(\frac{\ln x - \mu_y}{\sigma_y}\right)^2\right] \tag{A–16}$$

where

$$\mu_y = \ln \mu_x - \ln\sqrt{1 + C_x^2} \doteq \ln \mu_x - \frac{C_x^2}{2} \tag{A–17}$$

$$\sigma_y = \sqrt{\ln(1 + C_x^2)} \doteq C_x \tag{A–18}$$

These equations make it possible to use Table E-10 for statistical computations involving a lognormal distribution. Also useful are the transformations from **y** to **x**:

$$\mu_x = \exp(\mu_y + \sigma_y^2/2) \quad \text{(mean)} \tag{A–19}$$

$$\sigma_x = \exp(2\mu_y + 2\sigma_y^2) - \exp(2\mu_y + \sigma_y^2) \quad \text{(standard deviation)} \tag{A–20}$$

$$\tilde{x} = \exp(\mu_y) \quad \text{(median)} \tag{A–21}$$

$$\hat{x} = \exp(\mu_y - \sigma_y^2) \quad \text{(mode)} \tag{A–22}$$

The CDF is rectified when plotting the logarithm of the data as ordinate and the z variable corresponding to the CDF, namely, $\bar{F}_i = i/(n+1)$ or $\tilde{F}_i = (i - 0.3)/(n + 0.4)$, as abscissa, as desired, to seek mean or median loci, respectively.

A–8 The Uniform Distribution

The uniform distribution is a closed-interval distribution which arises when the chance of an observation is the same as the chance for any other observation. If a is the lower bound and b the upper bound, then the PDF is

$$f(x) = \begin{cases} 1/(b-a) & a \le x \le b \\ 0 & \text{otherwise} \end{cases} \tag{A–23}$$

The CDF is given by

$$F(x) = \begin{cases} 0 & x < a \\ (x-a)/(b-a) & a \le x \le b \\ 1 & x > b \end{cases} \tag{A–24}$$

The mean and standard deviation are given by

$$\mu_x = \frac{a+b}{2} \tag{A–25}$$

$$\sigma_x = \frac{b-a}{2\sqrt{3}} \tag{A–26}$$

The uniform distribution is a two-parameter distribution. Equations (A–23) and (A–24) use the range numbers a and b. Equations (A–25) and (A–26), the PDF and the CDF, can be expressed in terms of the mean and standard deviaion as

$$f(x) = \begin{cases} 1/(2\sqrt{3}\sigma_x) & \mu_x - \sqrt{3}\sigma_x \le x \le \mu_x + \sqrt{3}\sigma_x \\ 0 & \text{elsewhere} \end{cases} \tag{A–27}$$

$$F(x) = \begin{cases} 0 & x < \mu_x - \sqrt{3}\sigma_x \\ \dfrac{x - (\mu_x - \sqrt{3}\sigma_x)}{2\sqrt{3}\sigma_x} & \mu_x - \sqrt{3}\sigma_x \le x \le \mu_x + \sqrt{3}\sigma_x \\ 1 & x \ge \mu_x + \sqrt{3}\sigma_x \end{cases} \tag{A–28}$$

The CDF is rectified by plotting the data as ordinate and the CDF as $\bar{F}_i = i/(n+1)$ or $\tilde{F}_i = (i - 0.3)/(n + 0.4)$, as desired, to seek mean or median loci, respectively.

A–9 The Weibull Distribution

The three-parameter Weibull, $\mathbf{x} \sim W[x_0, \theta, b]$, has an explicit survival fraction equation:

$$R(x) = \exp\left[-\left(\frac{x - x_0}{\theta - x_0}\right)^b\right] \qquad x \ge x_0 \tag{A–29}$$

The two-parameter Weibull, $\mathbf{x} \sim W[\theta, b]$, has the survival equation

$$R(x) = \exp\left[-\left(\frac{x}{\theta}\right)^b\right] \qquad x \ge 0 \tag{A–30}$$

where $x_0 =$ guaranteed value of x, $x \ge x_0$

$\theta =$ characteristic or scale value, $\theta \ge x_0$

$b =$ shape parameter, $b > 0$

The mean value μ_x is

$$\mu_x = x_0 + (\theta - x_0)\Gamma(1 + 1/b) \tag{A–31}$$

The standard deviaion is

$$\sigma_x = (\theta - x_0)[\Gamma(1 + 2/b) - \Gamma^2(1 + 1/b)]^{1/2} \tag{A–32}$$

where Γ is the gamma function. See Table E–34. The median $\tilde{x}$ is

$$\tilde{x} = x_0 + (\theta - x_0)(\ln 2)^{1/b} \tag{A–33}$$

The mode $\hat{x}$ is

$$\hat{x} = x_0 + (\theta - x_0)\left(\frac{b - 1}{b}\right)^{1/b} \tag{A–34}$$

The Weibull distribution contains within it an excellent approximation to the normal, and it displays the exponential distribution exactly. The CDF is rectified by plotting $\ln(x - x_0)$ as ordinate and $\ln\ln\,[1/(1 - F_i)]$ as abscissa using $\bar{F}_i = 1/(n + 1)$ or $\tilde{F}_i = (i - 0.3)/(n + 0.4)$, as desired, to seek mean or median loci, respectively.

Professor Waloddi Weibull originated what came to be known as the Weibull distribution in 1937. In his American paper in 1951 he claimed that it applied to a wide range of problems, citing seven examples from the strength of steel to the height of adult men in the British Isles. He also claimed the function "may sometimes render good service." He did not claim that it always worked. Time has shown him correct in both statements. Since the Weibull distribution did not arise for classical reasons, the statistical and technical communities were skeptical of a distribution that both fit the data and selected the distribution.

Leonard Johnson and Dorian Shainin applied and improved the technique. The U.S. Air Force funded Weibull's research until 1975. The Weibull distribution has become the leading method worldwide for fitting life data.[4]

4. See Robert B. Abernethy, *The New Weibull Handbook*, 2nd ed., 1996, R. B. Abernethy, Publisher, North Palm Beach, Fla.; distributed by Gulf Publishing Co., Houston, Tex.

Linear Regression

Appendix B

Placing straight lines among data and being able to make probabilistic statements about line parameters, ordinate intercept a, and slope b is part of the subject of regression. If $\mathbf{y}$ is a random variable which occurs as the result of an observation at a deterministic level of x, then

$$\mathbf{y} = a + bx + \boldsymbol{\epsilon} \tag{B–1}$$

The variability in $\mathbf{y}$ is the result of the variability in $\boldsymbol{\epsilon}$, which is normally distributed, $\boldsymbol{\epsilon} \sim N(0, s_{y \cdot x})$. The mean value is unaffected by $\boldsymbol{\epsilon}$ since its mean is zero. The standard deviation $s_{y \cdot x}$ (read as the standard deviation of y given x) gives the variability to $\mathbf{y}$. In this way the mean of $\mathbf{y}$ comes from $a + bx$ and the variability from $\boldsymbol{\epsilon}$. For each data point x_i, y_i one can write

$$y_i = a + bx_i + \epsilon_i \tag{B–2}$$

Solving for ϵ_i and squaring and summing over the number of data points n gives

$$\sum \epsilon_i^2 = P = \sum (y_i - a - bx_i)^2 \tag{B–3}$$

Minimizing P, which is minimizing the sum of the squared errors, expecting a stationary point minimum, requires $\partial P / \partial a = 0$ and $\partial P / \partial b = 0$. This results in *normal equations,* which, solved simultaneously for $\hat{a}$ and $\hat{b}$, give

$$\hat{a} = \frac{\sum x_i^2 \sum y_i - \sum x_i \sum x_i y_i}{n \sum x_i^2 - \left(\sum x_i\right)^2} = \bar{y} - \hat{b}\bar{x} \tag{B–4}$$

$$\hat{b} = \frac{n \sum x_i y_i - \sum x_i \sum y_i}{n \sum x_i^2 - \left(\sum x_i\right)^2} \tag{B–5}$$

The straightness of the data string on xy coordinates and the tightness of the line to the points are quantitatively measured by the *correlation coefficient* $\hat{r}$. Using the notation $X = x - \bar{x}$ and $Y = y - \bar{y}$, one can write

$$\hat{r} = \frac{n \sum xy - \sum x \sum y}{\sqrt{\left[n \sum x^2 - \left(\sum x\right)^2\right]\left[n \sum y^2 - \left(\sum y^2\right)\right]}} = \frac{\sum XY}{\sqrt{\sum X^2 \cdot \sum Y^2}} = \frac{\hat{b} s_x}{s_y} \tag{B–6}$$

where s_x and s_y are the standard deviations of the x coordinates and y coordinates of the data, respectively. The subscripts i have been dropped for economy of notation. The standard deviation $s_{y \cdot x}$ of the variate ϵ is found from

$$s_{y \cdot x} = \sqrt{\frac{\sum \epsilon_i^2}{n-2}} = \sqrt{\frac{\sum y^2 - \hat{a}\sum y - \hat{b}\sum xy}{n-2}} \tag{B–7}$$

The estimate of the standard deviation of the regression line slope b is given by

$$s_b = \sqrt{\frac{s_{y \cdot x}}{\sum (x - \bar{x})^2}} = \frac{s_{y \cdot x}}{\sqrt{\sum X^2}} \tag{B–8}$$

and the confidence bounds on b at confidence level β are

$$b = \hat{b} \pm t(\beta, n-2)s_b \tag{B–9}$$

where t is Student's statistic.[1] See Table E-36.

The estimate of the standard deviation on $\hat{a}$ is given by

$$s_a = s_{y \cdot x}\left[\frac{1}{n} + \frac{(x - \bar{x})^2}{\sum (x - \bar{x})^2}\right]_{x=0}^{1/2} = s_{y \cdot x}\left[\frac{1}{n} + \frac{\bar{x}^2}{\sum (x - \bar{x})^2}\right]^{1/2}$$

$$= s_{y \cdot x}\left[\frac{1}{n} + \frac{\bar{x}^2}{\sum X^2}\right]^{1/2} \tag{B–10}$$

and the confidence bounds on a, at confidence level β, are

$$a = \hat{a} + t(\beta, n-2)s_a \tag{B–11}$$

1. "Student" was the pen name used by William Gosset.

Propagation of Error Relations **Appendix** C

When random variables are algebraically related by addition, subtraction, multiplication, and division, there are simple ways in which the means and standard deviations are propagated through such relationships. If random variables **x** and **y** are related to create another random variable **z**, by (say) addition or subtraction, as

$$\mathbf{z} = \mathbf{x} \pm \mathbf{y}$$

the means are related by

$$\mu_z = \mu_x \pm \mu_y \tag{C–1}$$

For **x** and **y** independent, for *both* addition and subtraction, the standard deviation is given by

$$\sigma_z = (\sigma_x^2 + \sigma_y^2)^{1/2} \tag{C–2}$$

For the product **xy**, for **x** and **y** independent, the mean and standard deviation are

$$\mu_{xy} = \mu_x \mu_y \tag{C–3}$$

$$\sigma_{xy} = \mu_{xy}(C_x^2 + C_y^2 + C_x^2 C_y^2)^{1/2} \tag{C–4}$$

For the division **x/y**, **x** and **y** independent,

$$\mu_{x/y} \doteq \mu_x/\mu_y \tag{C–5}$$

$$\sigma_{x/y} \doteq \mu_{x/y}[(C_x^2 + C_y^2)/(1 + C_y^2)]^{1/2} \tag{C–6}$$

For more complex relations, say a function $\boldsymbol{\phi}$ of $\mathbf{x}_1, \mathbf{x}_2, \ldots, \mathbf{x}_n$, $\boldsymbol{\phi}$ can be expanded by Taylor series, and using the first terms

$$\mu_\phi = \phi(\mathbf{x}_1, \mathbf{x}_2, \ldots, \mathbf{x}_n)_\mu \tag{C–7}$$

$$\sigma_\phi = \left[\sum_{i=1}^{n} \left(\frac{\partial \phi}{\partial x_i}\right)_\mu^2 \sigma_{xi}^2\right]^{1/2} \tag{C–8}$$

The first two terms of the Taylor series are

$$\mu_\phi = \phi(\mathbf{x}_1, \mathbf{x}_2, \ldots, \mathbf{x}_n)_\mu + \frac{1}{2}\sum_{i=1}^{n} \left(\frac{\partial^2 \phi}{\partial x_i^2}\right)_\mu \sigma_{xi}^2 \tag{C–9}$$

$$\sigma_\phi = \left[\sum_{i=1}^{n} \left(\frac{\partial \phi}{\partial x_i}\right)_\mu^2 \sigma_{xi}^2 + \frac{1}{2}\sum_{i=1}^{n} \left(\frac{\partial^2 \phi'}{\partial x_i^2}\right)_\mu^2 \sigma_{xi}^4\right]^{1/2} \tag{C–10}$$

Table 2–4 shows a convenient tabulation of common algebraic relationships.

Simulation — Appendix D

Monte Carlo simulations are used to find the probability of success or failure using the following protocol:

- Generate instances of all random variables in the problem by drawing instances from the appropriate distributions.
- Perform the appropriate calculations using these values.
- Note the number of successes (or failures) and the total number of trials to find a reliability (or probability of failure) estimate.
- As the number of trials increases, the reliability estimate improves.

D–1 Generating Instances of Random Variables

The fundamental random number generator of a computer is supplied by the computing machinery manufacturer. The routine is *machine-specific*. Such a generator develops a pseudo-random number list from a starting seed integer (or a pair). The list is

- Approximately uniformly random, passing various statistical tests.
- Finite, and it may be of the order of 1 million before repeating. This information is supplied by the manufacturer.
- Capable of regeneration, because of the seed(s) trigger.

Moreover, the random number program may be called in a repetitive fashion. Consult the operator's manual provided by the computer manufacturer.

Other distributions are software-generated, including user-written programs.

D–2 Normal-Distributed Variate with a Mean of $\bar{x}$ and Standard Deviation of σ_x

- Name your routine suggestively, perhaps *normal* or *gauss*, with arguments consisting of the seed integer(s), $\bar{x}$, σ_x, returning the random number x.
- Set a running summation to zero.
- Set up a repetitive loop with 12 passes:
  Call randu (ix, iy, u)
  Set sum $=$ sum $+ u$
  End of loop
  Set $x = \bar{x} + (\text{sum} - 6.)\sigma_x$
- Return to the calling program.

This subprogram generates an instance $\mathbf{x} \sim N(\bar{x}, \sigma_x)$ each time it is called.

D–3 Lognormal-Distributed Variate with a Mean of $\bar{x}$ and a Standard Deviation of σ_x

- Name your routine suggestively, perhaps *lognorm* with arguments consisting of (the seed integer(s), $\bar{x}$, σ_x, x).
- Set $C_x = \sigma_x/\bar{x}$
  Set $\bar{y} = \ln(\bar{x}) - 0.5\ln(1 + C_x^2)$
  Set $\sigma_y = \sqrt{\ln(1 + C_x^2)}$
  Call normal (seed(s), $\bar{y}$, σ_y, y)
  Set $x = \exp(y)$
- Return to the calling program.

The subprogram lognorm has generated an instance $\mathbf{x} \sim LN(\bar{x}, \sigma_x)$.

D–4 Weibull-Distributed Variate Given x_0, θ, and b

To generate a Weibull-distributed variate given the guaranteed value x_0, the characteristic value θ, and the shape factor b, do the following:

- Name your routine suggestively, perhaps *weibull* with the arguments (seed(s), x_0, θ, b, x).
- Call randu (ix, iy, u).
- Set $x = x_0 + (\theta - x_0)\ln(1/u)^{1/b}$.
- Return to calling program.

The subprogram has generated one instance of $\mathbf{x} \sim W[x_0, \theta, b]$.
You are free to add other random number generators as the need arises.

D–5 Questions Concerning Confidence in Monte Carlo Simulation Results

A common procedure in Monte Carlo simulations is to generate random numbers from appropriate distributions and substitute them in evaluating algebraic equations, eventually finding an instance w of $\mathbf{w}$. The probability or chance of observing $w > w_{\text{critical}}$ (or, alternatively, $w < w_{\text{critical}}$) is found by counting successes as 1 and failures as 0, and estimating the probability of success as the number of successes divided by the number of trials. Two questions are ever-present: "How many trials?" and "How accurate is the estimate of the probability?"

Consider a column vector, say $\{x\}$, composed of zeros (failures) and ones (successes). The sum of the elements of $\{x\}$ is Np where N is the number of entries (trials) and p is the probability of success. The mean of the elements in $\{x\}$ is

$$\bar{x} = \frac{\sum x}{N} = \frac{Np}{N} = \hat{p} \tag{D–1}$$

where $\hat{p}$ is an unbiased estimator of the probability of success. The column vector of the squares of the entries in $\{x\}$ has a sum of N_p also, *because the column vectors are identical*. The standard deviation of x from Eq. (A–4) is

$$\sigma_x = \sqrt{\frac{\sum x^2 - (\sum x)^2/N}{N-1}} \doteq \sqrt{\frac{N_p - N^2p^2/N}{N}} = \sqrt{p(1-p)} \tag{D–2}$$

The standard deviation of the mean of **x** itself is

$$\sigma_{\bar{x}} = \frac{\sum x}{\sqrt{N}} = \sqrt{\frac{p(1-p)}{N}} \tag{D–3}$$

In a simulation involving N_1 trials, the bilateral tolerance at the β confidence level (two-tailed) $z_\beta \sigma_{\bar{x}}$ is denoted as the bilateral error in p, that is, e_1, so

$$e_1 = z_\beta(\sigma_{\bar{x}})_1 = z_\beta \sqrt{\frac{p_1(1-p_1)}{N_1}} \tag{D–4}$$

The number of trials N_2 necessary to attain an error e_2 associated with m significant figures to the right of the decimal point is $(10^{-m})/2$, so e_2 is given by

$$e_2 = (10^{-m})/2 = z_\beta \sqrt{\frac{p_2(1-p_2)}{N_2}} \tag{D–5}$$

Arguing that $p_1(1-p_1) \doteq p_2(1-p_2)$ allows us to write

$$e_1\sqrt{N_1} = e_2\sqrt{N_2} \tag{D–6}$$

showing that the error is reduced proportionally to the square root of the number of trials. Solving Eq. (D–5) for N in terms of m gives

$$N = 4z_\beta^2 10^{2m} p(1-p) \tag{D–7}$$

When p is 0 or 1 in Eq. (D–7), N equals zero, so N exhibits a stationary-point maximum when $p = 1/2$. The maximum number of trials N_{max} is

$$N_{\text{max}} = 4z_\beta^2 10^{2m}(1/2)(1-1/2) = z_\beta^2 10^{2m} \tag{D–8}$$

Equation (D–8) estimates the largest number of trials to realize m significant figures at the β confidence level. Since p is usually small, it is advantageous to run the simulation for (say) 1000 trials, estimate p, then, from Eq. (D–7), estimate the number of additional trials from $N_2 = N - 1000$. Then extend the simulation for N_2 more trials and check again. This reduces the number of trials to that needed.

While m is couched in significant-digit ideas, remember that $m - 1$ digits to the right of the decimal point in a p-by-simulation will be correct, and the next *may* be off by half the time.

Useful Tables

Appendix E

E–1 Standard SI Prefixes **1171**

E–2 Conversion Factors **1172**

E–3 Optional SI Units for Bending, Torsion, Axial, and Direct Shear Stresses **1173**

E–4 Optional SI Units for Bending and Torsional Deflection **1173**

E–5 Physical Constants of Materials **1173**

E–6 Properties of Structural-Steel Angles **1174**

E–7 Properties of Structural-Steel Channels **1176**

E–8 Properties of Round Tubing **1178**

E–9 Shear, Moment, and Deflection of Beams **1179**

E–10 Cumulative Distribution Function of Normal (Gaussian) Distribution **1187**

E–11 A Selection of International Tolerance Grades—Metric Series **1188**

E–12 Fundamental Deviations for Shafts—Metric Series **1188**

E–13 A Selection of International Tolerance Grades—Inch Series **1189**

E–14 Fundamental Deviations for Shafts—Inch Series **1190**

E–15 Charts for Theoretical Stress-Concentration Factors K_t **1191**

E–16 Approximate Stress-Concentration Factors K_t and K_{ts} for Bending a Round Bar or Tube with a Transverse Round Hole **1198**

E–17 Preferred Sizes and Renard (R-series) Numbers **1200**

E–18 Geometric Properties **1201**

E–19 American Standard Pipe **1204**

E–20 Deterministic ASTM Minimum Tensile and Yield Strengths for HR and CD Steels **1205**

E–21 Mean Mechanical Properties of Some Heat-Treated Steels **1206**

E–22 Results of Tensile Tests of Some Metals* **1208**

E–23 Mean Monotonic and Cyclic Stress-Strain Properties of Selected Steels **1209**

E–24 Mechanical Properties of Two Common Non-Steels **1211**

E–25 Stochastic Yield and Ultimate Strengths for Selected Materials **1213**

E–26 Stochastic Parameters from Finite Life Fatigue Tests on Selected Metals **1214**

E–27 Finite Life Fatigue Strengths of Selected Plain Carbon Steels **1215**

E–28 Decimal Equivalents of Wire and Sheet-Metal Gauges **1216**

E–29 Dimensions of Square and Hexagonal Bolts **1218**

E–30 Dimensions of Hexagonal Cap Screws and Heavy Hexagonal Screws **1219**

E–31 Dimensions of Hexagonal Nuts **1220**

E–32 Basic Dimensions of American Standard Plain Washers **1221**

E–33 Dimensions of Metric Plain Washers **1222**

E–34 Gamma Function **1223**

E–35 Values of Correlation Coefficient *r* **1224**

E–36 Distribution of Student's *t* **1226**

E–37 Helical Spring Specification Checklists **1227**

E–38 Greek Alphabet **1229**

Table E–1

Standard SI Prefixes*†

Name	Symbol	Factor
exa	E	1 000 000 000 000 000 000 = 10^{18}
peta	P	1 000 000 000 000 000 = 10^{15}
tera	T	1 000 000 000 000 = 10^{12}
giga	G	1 000 000 000 = 10^{9}
mega	M	1 000 000 = 10^{6}
kilo	k	1 000 = 10^{3}
hecto‡	h	100 = 10^{2}
deka‡	da	10 = 10^{1}
deci‡	d	0.1 = 10^{-1}
centi‡	c	0.01 = 10^{-2}
milli	m	0.001 = 10^{-3}
micro	μ	0.000 001 = 10^{-6}
nano	n	0.000 000 001 = 10^{-9}
pico	p	0.000 000 000 001 = 10^{-12}
femto	f	0.000 000 000 000 001 = 10^{-15}
atto	a	0.000 000 000 000 000 001 = 10^{-18}

*If possible use multiple and submultiple prefixes in steps of 1000.

†Spaces are used in SI instead of commas to group numbers to avoid confusion with the practice in some European countries of using commas for decimal points.

‡Not recommended but sometimes encountered.

Table E–2

Conversion Factors A to Convert Input X to Output Y Using the Formula $Y = AX$*

Multiply Input X	By Factor A	To Get Output Y	Multiply Input X	By Factor A	To Get Output Y
British thermal unit, Btu	1055	joule, J	moment of inertia, lbm · ft^2	0.0421	kilogram-meter^2, kg · m^2
Btu/second, Btu/s	1.05	kilowatt, kW			
calorie	4.19	joule, J	moment of inertia, lbm · in^2	293	kilogram-millimeter^2 kg · mm^2
centimeter of mercury (0°C)	1.333	kilopascal, kPa			
centipoise, cP	0.001	pascal-second, Pa · s	moment of section (second moment of area), in^4	41.6	centimeter^4, cm^4
degree (angle)	0.0174	radian, rad			
foot, ft	0.305	meter, m			
foot^2, ft^2	0.0929	meter^2, m^2	ounce-force, oz	0.278	newton, N
foot/minute, ft/min	0.0051	meter/second, m/s	ounce-mass	0.0311	kilogram, kg
			pound, lb†	4.45	newton, N
foot-pound, ft · lb	1.35	joule, J	pound-foot, lb · ft	1.36	newton-meter, N · m
foot-pound/second, ft · lb/s	1.35	watt, W	pound/foot^2, lb/ft^2	47.9	pascal, Pa
			pound-inch, lb · in	0.113	joule, J
			pound-inch, lb · in	0.113	newton-meter N · m
foot/second, ft/s	0.305	meter/second, m/s			
gallon (U.S.), gal	3.785	liter, l	pound/inch, lb/in	175	newton/meter, N/m
horsepower, hp	0.746	kilowatt, kW			
inch, in	0.0254	meter, m	pound/inch^2, psi (lb/in^2)	6.89	kilopascal, kPa
inch, in	25.4	millimeter, mm			
inch^2, in^2	645	millimeter^2, mm^2	pound-mass, lbm	0.454	kilogram, kg
inch of mercury (32°F)	3.386	kilopascal, kPa	pound-mass/second, lbm/s	0.454	kilogram/second, kg/s
kilopound, kip	4.45	kilonewton, kN	quart (U.S. liquid), qt	946	milliliter, ml
kilopound/inch^2, kpsi (ksi)	6.89	megapascal, MPa (N/mm^2)	section modulus, in^3	16.4	centimeter^3, cm^3
mass, lb · s^2/in	175	kilogram, kg			
mile, mi	1.610	kilometer, km	slug	14.6	kilogram, kg
mile/hour, mi/h	1.61	kilometer/hour, km/h	ton (short 2000 lbm)	907	kilogram, kg
mile/hour, mi/h	0.447	meter/second, m/s	yard, yd	0.914	meter, m

*Approximate.

†The U.S. Customary System unit of the pound-force is often abbreviated as lbf to distinguish it from the pound-mass, which is abbreviated as lbm. In most places in this book the pound-force is usually written simply as the pound and abbreviated as lb.

Table E–3

Optional SI Units for Bending Stress $\sigma = Mc/I$, Torsion Stress $\tau = Tr/J$, Axial Stress $\sigma = F/A$, and Direct Shear Stress $\tau = F/A$

Bending and Torsion				Axial and Direct Shear		
M, T	I, J	c, r	σ, τ	F	A	σ, τ
N · m*	m^4	m	Pa	N*	m^2	Pa
N · m	cm^4	cm	MPa (N/mm^2)	N†	mm^2	MPa (N/mm^2)
M · m†	mm^4	mm	GPa	kN	m^2	kPa
kN · m	cm^4	cm	GPa	kN†	mm^2	GPa
N · mm†	mm^4	mm	MPa (N/mm^2)			

*Basic relation.

†Often preferred.

Table E–4

Optional SI Units for Bending Deflection $y = f(F\ell^3/EI)$ or $y = f(w\ell^4/EI)$ and Torsional Deflection $\theta = T\ell/GJ$

Bending Deflection					Torsional Deflection				
F, $w\ell$	ℓ	I	E	y	T	ℓ	J	G	θ
N*	m	m^4	Pa	m	N · m*	m	m^4	Pa	rad
kN†	mm	mm^4	GPa	mm	N · m†	mm	mm^4	GPa	rad
kN	m	m^4	GPa	μm	N · mm	mm	mm^4	MPa (N/mm^2)	rad
N	mm	mm^4	kPa	m	N · m	cm	cm^4	MPa (N/mm^2)	rad

*Basic relation.

†Often preferred.

Table E–5

Physical Constants of Materials

Material	Modulus of Elasticity E		Modulus of Rigidity G		Poisson's Ratio ν	Unit Weight w		
	Mpsi	GPa	Mpsi	GPa		lb/in³	lb/ft³	kN/m³
Aluminum (all alloys)	10.3	71.0	3.80	26.2	0.334	0.098	169	26.6
Beryllium copper	18.0	124.0	7.0	48.3	0.285	0.297	513	80.6
Brass	15.4	106.0	5.82	40.1	0.324	0.309	534	83.8
Carbon steel	30.0	207.0	11.5	79.3	0.292	0.282	487	76.5
Cast iron (gray)	14.5	100.0	6.0	41.4	0.211	0.260	450	70.6
Copper	17.2	119.0	6.49	44.7	0.326	0.322	556	87.3
Douglas fir	1.6	11.0	0.6	4.1	0.33	0.016	28	4.3
Glass	6.7	46.2	2.7	18.6	0.245	0.094	162	25.4
Inconel	31.0	214.0	11.0	75.8	0.290	0.307	530	83.3
Lead	5.3	36.5	1.9	13.1	0.425	0.411	710	111.5
Magnesium	6.5	44.8	2.4	16.5	0.350	0.065	112	17.6
Molybdenum	48.0	331.0	17.0	117.0	0.307	0.368	636	100.0
Monel metal	26.0	179.0	9.5	65.5	0.320	0.319	551	86.6
Nickel silver	18.5	127.0	7.0	48.3	0.322	0.316	546	85.8
Nickel steel	30.0	207.0	11.5	79.3	0.291	0.280	484	76.0
Phosphor bronze	16.1	111.0	6.0	41.4	0.349	0.295	510	80.1
Stainless steel (18-8)	27.6	190.0	10.6	73.1	0.305	0.280	484	76.0

Table E–6

Properties of Structural-Steel Angles*†

w = weight per foot, lb/ft
m = mass per meter, kg/m
A = area, in^2 (cm^2)
I = second moment of area, in^4 (cm^4)
k = radius of gyration, in (cm)
y = centroidal distance, in (cm)
Z = section modulus, in^3, (cm^3)

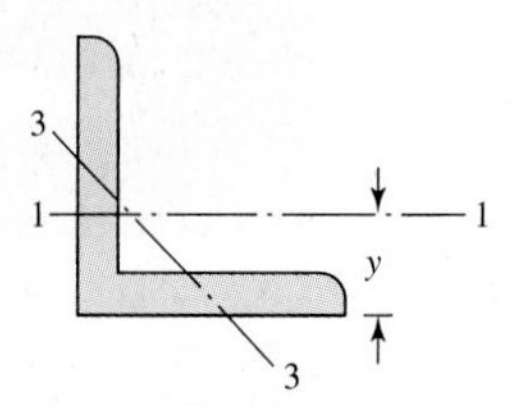

Size, in	w	A	I_{1-1}	k_{1-1}	Z_{1-1}	y	k_{3-3}
$1 \times 1 \times \frac{1}{8}$	0.80	0.234	0.021	0.298	0.029	0.290	0.191
$\times \frac{1}{4}$	1.49	0.437	0.036	0.287	0.054	0.336	0.193
$1\frac{1}{2} \times 1\frac{1}{2} \times \frac{1}{8}$	1.23	0.36	0.074	0.45	0.068	0.41	0.29
$\times \frac{1}{4}$	2.34	0.69	0.135	0.44	0.130	0.46	0.29
$2 \times 2 \times \frac{1}{8}$	1.65	0.484	0.190	0.626	0.131	0.546	0.398
$\times \frac{1}{4}$	3.19	0.938	0.348	0.609	0.247	0.592	0.391
$\times \frac{3}{8}$	4.7	1.36	0.479	0.594	0.351	0.636	0.389
$2\frac{1}{2} \times 2\frac{1}{2} \times \frac{1}{4}$	4.1	1.19	0.703	0.769	0.394	0.717	0.491
$\times \frac{3}{8}$	5.9	1.73	0.984	0.753	0.566	0.762	0.487
$3 \times 3 \times \frac{1}{4}$	4.9	1.44	1.24	0.930	0.577	0.842	0.592
$\times \frac{3}{8}$	7.2	2.11	1.76	0.913	0.833	0.888	0.587
$\times \frac{1}{2}$	9.4	2.75	2.22	0.898	1.07	0.932	0.584
$3\frac{1}{2} \times 3\frac{1}{2} \times \frac{1}{4}$	5.8	1.69	2.01	1.09	0.794	0.968	0.694
$\times \frac{3}{8}$	8.5	2.48	2.87	1.07	1.15	1.01	0.687
$\times \frac{1}{2}$	11.1	3.25	3.64	1.06	1.49	1.06	0.683
$4 \times 4 \times \frac{1}{4}$	6.6	1.94	3.04	1.25	1.05	1.09	0.795
$\times \frac{3}{8}$	9.8	2.86	4.36	1.23	1.52	1.14	0.788
$\times \frac{1}{2}$	12.8	3.75	5.56	1.22	1.97	1.18	0.782
$\times \frac{5}{8}$	15.7	4.61	6.66	1.20	2.40	1.23	0.779
$6 \times 6 \times \frac{3}{8}$	14.9	4.36	15.4	1.88	3.53	1.64	1.19
$\times \frac{1}{2}$	19.6	5.75	19.9	1.86	4.61	1.68	1.18
$\times \frac{5}{8}$	24.2	7.11	24.2	1.84	5.66	1.73	1.18
$\times \frac{3}{4}$	28.7	8.44	28.2	1.83	6.66	1.78	1.17

Table E–6

Properties of Structural-Steel Angles *(Continued)*

Size, mm	m	A	I_{1-1}	k_{1-1}	Z_{1-1}	y	k_{3-3}
25 × 25 × 3	1.11	1.42	0.80	0.75	0.45	0.72	0.48
× 4	1.45	1.85	1.01	0.74	0.58	0.76	0.48
× 5	1.77	2.26	1.20	0.73	0.71	0.80	0.48
40 × 40 × 4	2.42	3.08	4.47	1.21	1.55	1.12	0.78
× 5	2.97	3.79	5.43	1.20	1.91	1.16	0.77
× 6	3.52	4.48	6.31	1.19	2.26	1.20	0.77
50 × 50 × 5	3.77	4.80	11.0	1.51	3.05	1.40	0.97
× 6	4.47	5.59	12.8	1.50	3.61	1.45	0.97
× 8	5.82	7.41	16.3	1.48	4.68	1.52	0.96
60 × 60 × 5	4.57	5.82	19.4	1.82	4.45	1.64	1.17
× 6	5.42	6.91	22.8	1.82	5.29	1.69	1.17
× 8	7.09	9.03	29.2	1.80	6.89	1.77	1.16
× 10	8.69	11.1	34.9	1.78	8.41	1.85	1.16
80 × 80 × 6	7.34	9.35	55.8	2.44	9.57	2.17	1.57
× 8	9.63	12.3	72.2	2.43	12.6	2.26	1.56
× 10	11.9	15.1	87.5	2.41	15.4	2.34	1.55
100 × 100 × 8	12.2	15.5	145	3.06	19.9	2.74	1.96
× 12	17.8	22.7	207	3.02	29.1	2.90	1.94
× 15	21.9	27.9	249	2.98	35.6	3.02	1.93
150 × 150 × 10	23.0	29.3	624	4.62	56.9	4.03	2.97
× 12	27.3	34.8	737	4.60	67.7	4.12	2.95
× 15	33.8	43.0	898	4.57	83.5	4.25	2.93
× 18	40.1	51.0	1050	4.54	98.7	4.37	2.92

*Metric sizes also available in sizes of 45, 70, 90, 120, and 200 mm.

†These sizes are also available in aluminum alloy.

Table E–7

Properties of Structural-Steel Channels*

a, b = size, in (mm)
w = weight per foot, lb/ft
m = mass per meter, kg/m
t = web thickness, in (mm)
A = area, in^2 (cm^2)
I = second moment of area, in^4 (cm^4)
k = radius of gyration, in (cm)
x = centroidal distance, in (cm)
Z = section modulus, in^3 (cm^3)

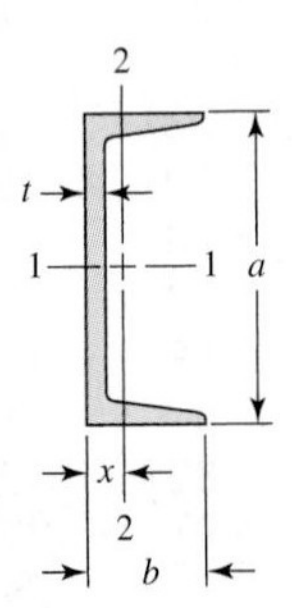

a, in	b, in	t	A	w	I_{1-1}	k_{1-1}	Z_{1-1}	I_{2-2}	k_{2-2}	Z_{2-2}	x
3	1.410	0.170	1.21	4.1	1.66	1.17	1.10	0.197	0.404	0.202	0.436
3	1.498	0.258	1.47	5.0	1.85	1.12	1.24	0.247	0.410	0.233	0.438
3	1.596	0.356	1.76	6.0	2.07	1.08	1.38	0.305	0.416	0.268	0.455
4	1.580	0.180	1.57	5.4	3.85	1.56	1.93	0.319	0.449	0.283	0.457
4	1.720	0.321	2.13	7.25	4.59	1.47	2.29	0.433	0.450	0.343	0.459
5	1.750	0.190	1.97	6.7	7.49	1.95	3.00	0.479	0.493	0.378	0.484
5	1.885	0.325	2.64	9.0	8.90	1.83	3.56	0.632	0.489	0.450	0.478
6	1.920	0.200	2.40	8.2	13.1	2.34	4.38	0.693	0.537	0.492	0.511
6	2.034	0.314	3.09	10.5	15.2	2.22	5.06	0.866	0.529	0.564	0.499
6	2.157	0.437	3.83	13.0	17.4	2.13	5.80	1.05	0.525	0.642	0.514
7	2.090	0.210	2.87	9.8	21.3	2.72	6.08	0.968	0.581	0.625	0.540
7	2.194	0.314	3.60	12.25	24.2	2.60	6.93	1.17	0.571	0.703	0.525
7	2.299	0.419	4.33	14.75	27.2	2.51	7.78	1.38	0.564	0.779	0.532
8	2.260	0.220	3.36	11.5	32.3	3.10	8.10	1.30	0.625	0.781	0.571
8	2.343	0.303	4.04	13.75	36.2	2.99	9.03	1.53	0.615	0.854	0.553
8	2.527	0.487	5.51	18.75	44.0	2.82	11.0	1.98	0.599	1.01	0.565
9	2.430	0.230	3.91	13.4	47.7	3.49	10.6	1.75	0.669	0.962	0.601
9	2.485	0.285	4.41	15.0	51.0	3.40	11.3	1.93	0.661	1.01	0.586
9	2.648	0.448	5.88	20.0	60.9	3.22	13.5	2.42	0.647	1.17	0.583
10	2.600	0.240	4.49	15.3	67.4	3.87	13.5	2.28	0.713	1.16	0.634
10	2.739	0.379	5.88	20.0	78.9	3.66	15.8	2.81	0.693	1.32	0.606
10	2.886	0.526	7.35	25.0	91.2	3.52	18.2	3.36	0.676	1.48	0.617
10	3.033	0.673	8.82	30.0	103	3.43	20.7	3.95	0.669	1.66	0.649
12	3.047	0.387	7.35	25.0	144	4.43	24.1	4.47	0.780	1.89	0.674
12	3.170	0.510	8.82	30.0	162	4.29	27.0	5.14	0.763	2.06	0.674

Table E–7

Properties of Structural-Steel Channels *(Continued)*

$a \times b$, mm	m	t	A	I_{1-1}	k_{1-1}	Z_{1-1}	I_{2-2}	k_{2-2}	Z_{2-2}	x
76 × 38	6.70	5.1	8.53	74.14	2.95	19.46	10.66	1.12	4.07	1.19
102 × 51	10.42	6.1	13.28	207.7	3.95	40.89	29.10	1.48	8.16	1.51
127 × 64	14.90	6.4	18.98	482.5	5.04	75.99	67.23	1.88	15.25	1.94
152 × 76	17.88	6.4	22.77	851.5	6.12	111.8	113.8	2.24	21.05	2.21
152 × 89	23.84	7.1	30.36	1166	6.20	153.0	215.1	2.66	35.70	2.86
178 × 76	20.84	6.6	26.54	1337	7.10	150.4	134.0	2.25	24.72	2.20
178 × 89	26.81	7.6	34.15	1753	7.16	197.2	241.0	2.66	39.29	2.76
203 × 76	23.82	7.1	30.34	1950	8.02	192.0	151.3	2.23	27.59	2.13
203 × 89	29.78	8.1	37.94	2491	8.10	245.2	264.4	2.64	42.34	2.65
229 × 76	26.06	7.6	33.20	2610	8.87	228.3	158.7	2.19	28.22	2.00
229 × 89	32.76	8.6	41.73	3387	9.01	296.4	285.0	2.61	44.82	2.53
254 × 76	28.29	8.1	36.03	3367	9.67	265.1	162.6	2.12	28.21	1.86
254 × 89	35.74	9.1	45.42	4448	9.88	350.2	302.4	2.58	46.70	2.42
305 × 89	41.69	10.2	53.11	7061	11.5	463.3	325.4	2.48	48.49	2.18
305 × 102	46.18	10.2	58.83	8214	11.8	539.0	499.5	2.91	66.59	2.66

*These sizes are also available in aluminum alloy.

Table E–8

Properties of Round Tubing

w_a = unit weight of aluminum tubing, lb/ft
w_s = unit weight of steel tubing, lb/ft
m = unit mass, kg/m
A = area, in^2 (cm^2)
I = second moment of area, in^4 (cm^4)
J = second polar moment of area, in^4 (cm^4)
k = radius of gyration, in (cm)
Z = section modulus, in^3 (cm^3)
d, t = size (OD) and thickness, in (mm)

Size, in	w_a	w_s	A	I	k	Z	J
$1 \times \frac{1}{8}$	0.416	1.128	0.344	0.034	0.313	0.067	0.067
$1 \times \frac{1}{4}$	0.713	2.003	0.589	0.046	0.280	0.092	0.092
$1\frac{1}{2} \times \frac{1}{8}$	0.653	1.769	0.540	0.129	0.488	0.172	0.257
$1\frac{1}{2} \times \frac{1}{4}$	1.188	3.338	0.982	0.199	0.451	0.266	0.399
$2 \times \frac{1}{8}$	0.891	2.670	0.736	0.325	0.664	0.325	0.650
$2 \times \frac{1}{4}$	1.663	4.673	1.374	0.537	0.625	0.537	1.074
$2\frac{1}{2} \times \frac{1}{8}$	1.129	3.050	0.933	0.660	0.841	0.528	1.319
$2\frac{1}{2} \times \frac{1}{4}$	2.138	6.008	1.767	1.132	0.800	0.906	2.276
$3 \times \frac{1}{4}$	2.614	7.343	2.160	2.059	0.976	1.373	4.117
$3 \times \frac{3}{8}$	3.742	10.51	3.093	2.718	0.938	1.812	5.436
$4 \times \frac{3}{16}$	2.717	7.654	2.246	4.090	1.350	2.045	8.180
$4 \times \frac{3}{8}$	5.167	14.52	4.271	7.090	1.289	3.544	14.180

Size, mm	m	A	I	k	Z	J
12 × 2	0.490	0.628	0.082	0.361	0.136	0.163
16 × 2	0.687	0.879	0.220	0.500	0.275	0.440
16 × 3	0.956	1.225	0.273	0.472	0.341	0.545
20 × 4	1.569	2.010	0.684	0.583	0.684	1.367
25 × 4	2.060	2.638	1.508	0.756	1.206	3.015
25 × 5	2.452	3.140	1.669	0.729	1.336	3.338
30 × 4	2.550	3.266	2.827	0.930	1.885	5.652
30 × 5	3.065	3.925	3.192	0.901	2.128	6.381
42 × 4	3.727	4.773	8.717	1.351	4.151	17.430
42 × 5	4.536	5.809	10.130	1.320	4.825	20.255
50 × 4	4.512	5.778	15.409	1.632	6.164	30.810
50 × 5	5.517	7.065	18.118	1.601	7.247	36.226

Table E–9

Shear, Moment, and Deflection of Beams

1 Cantilever—end load

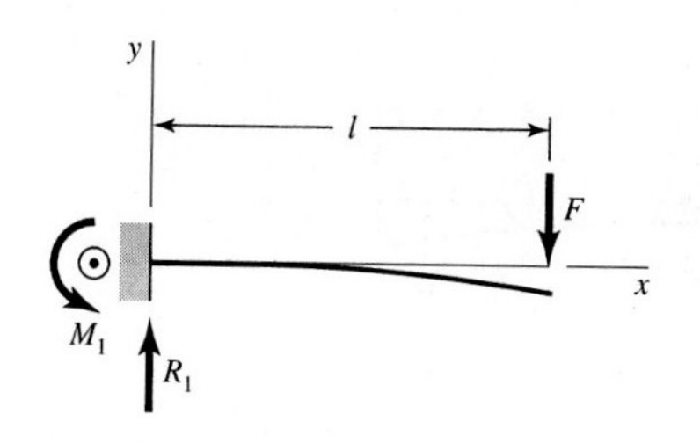

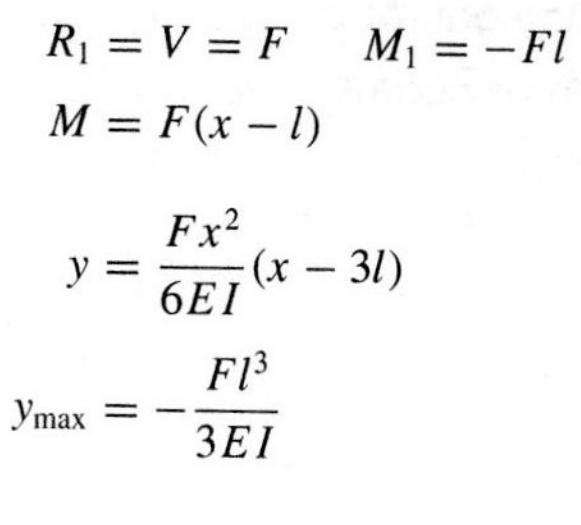

$$R_1 = V = F \qquad M_1 = -Fl$$

$$M = F(x - l)$$

$$y = \frac{Fx^2}{6EI}(x - 3l)$$

$$y_{max} = -\frac{Fl^3}{3EI}$$

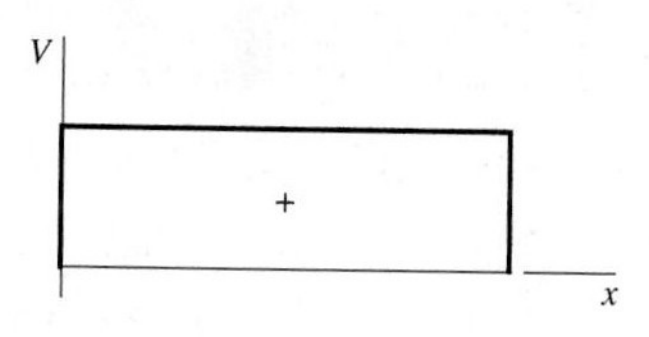

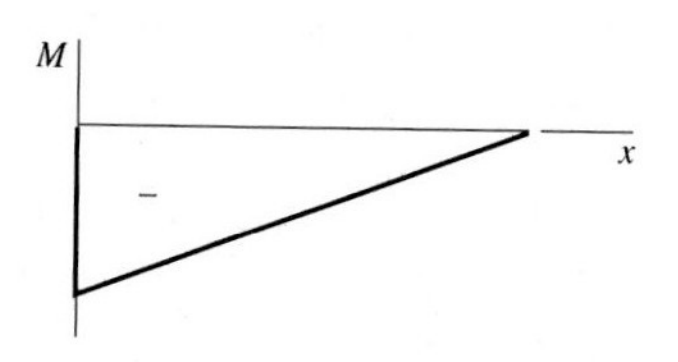

2 Cantilever—intermediate load

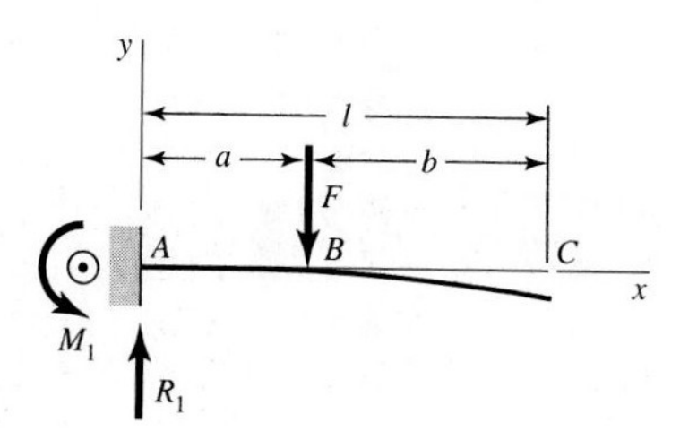

$$R_1 = V = F \qquad M_1 = -Fa$$

$$M_{AB} = F(x - a) \qquad M_{BC} = 0$$

$$y_{AB} = \frac{Fx^2}{6EI}(x - 3a)$$

$$y_{BC} = \frac{Fa^2}{6EI}(a - 3x)$$

$$y_{max} = \frac{Fa^2}{6EI}(a - 3l)$$

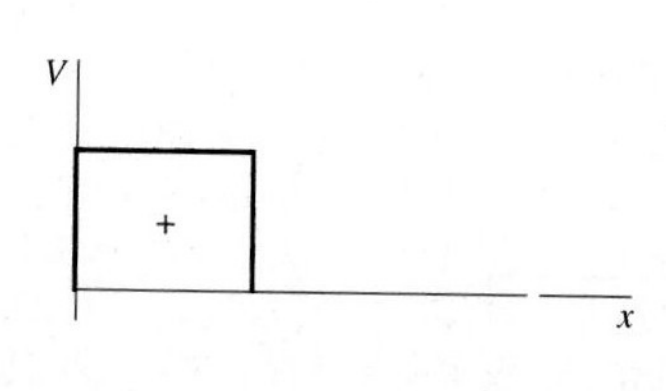

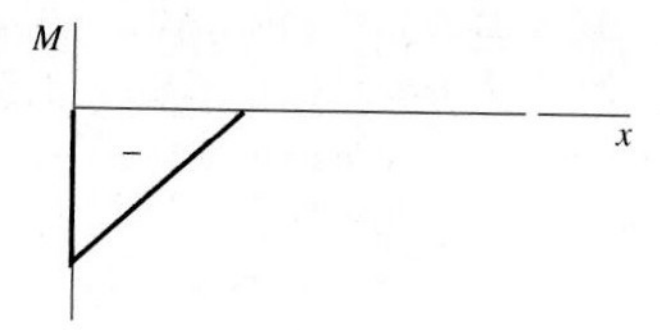

Table E–9

Shear, Moment, and Deflection of Beams *(Continued)*

3 Cantilever—uniform load

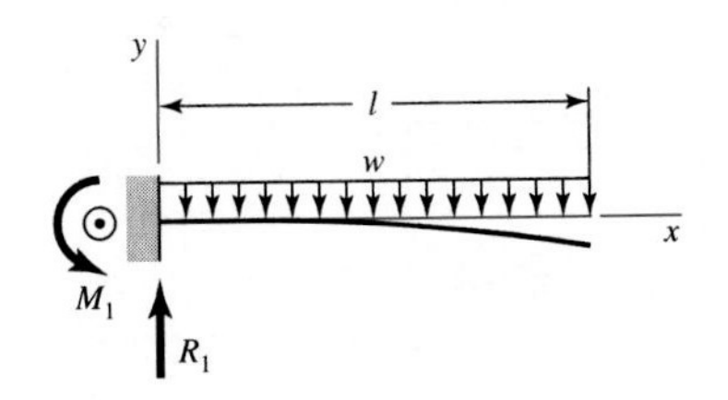

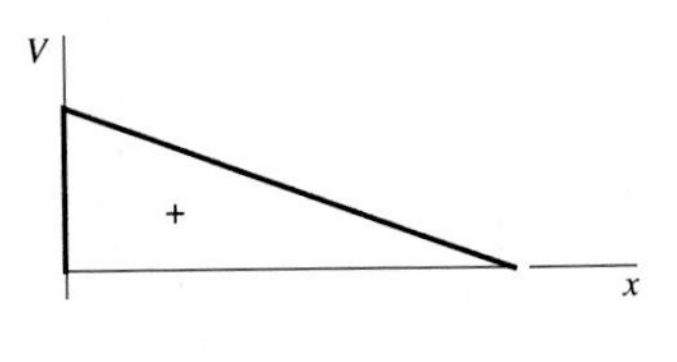

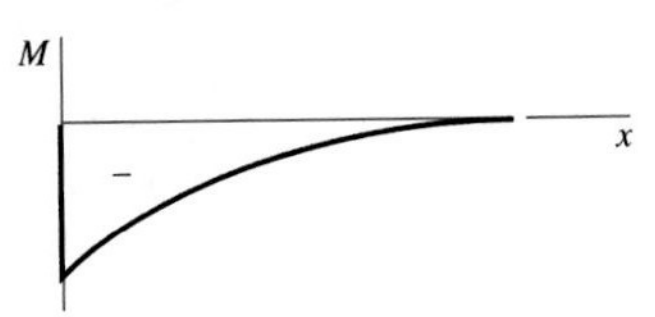

$$R_1 = wl \qquad M_1 = -\frac{wl^2}{2}$$

$$V = w(l - x) \qquad M = -\frac{w}{2}(l - x)^2$$

$$y = \frac{wx^2}{24EI}(4lx - x^2 - 6l^2)$$

$$y_{\max} = -\frac{wl^4}{8EI}$$

4 Cantilever—moment load

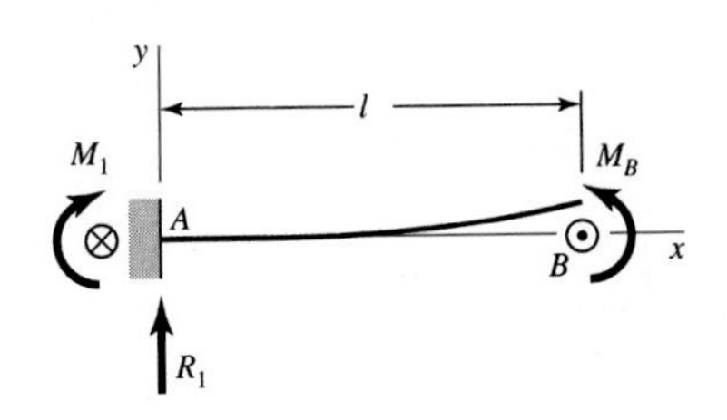

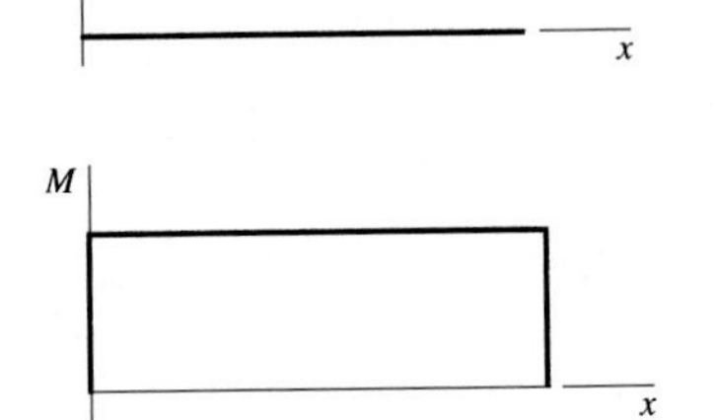

$$R_1 = 0 \qquad M_1 = M_B \qquad M = M_B$$

$$y = \frac{M_B x^2}{2EI} \qquad y_{\max} = \frac{M_B l^2}{2EI}$$

Table E–9

Shear, Moment, and Deflection of Beams *(Continued)*

5 Simple supports—center load

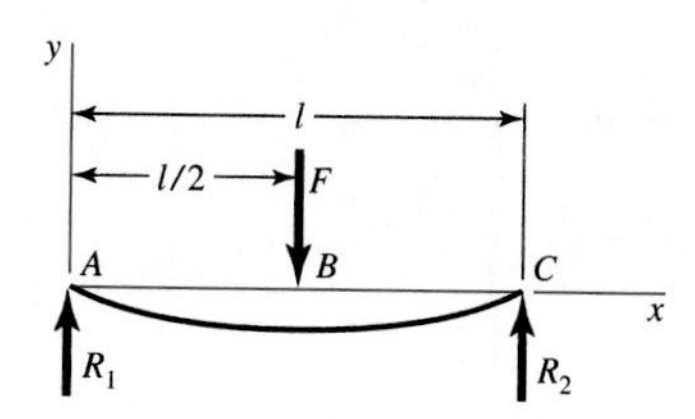

$$R_1 = R_2 = \frac{F}{2} \qquad V_{AB} = R_1$$

$$V_{AB} = R_1 \qquad V_{BC} = -R_2$$

$$M_{AB} = \frac{Fx}{2} \qquad M_{BC} = \frac{F}{2}(l - x)$$

$$y_{AB} = \frac{Fx}{48EI}(4x^2 - 3l^2)$$

$$y_{\max} = -\frac{Fl^3}{48EI}$$

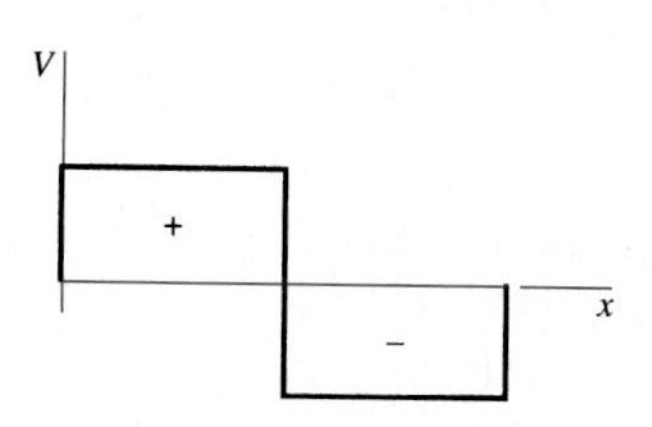

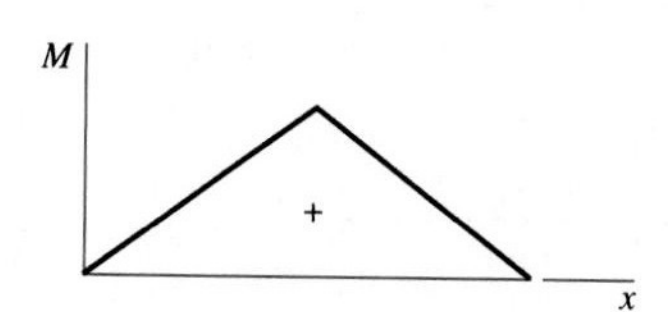

6 Simple supports—intermediate load, $a < b$

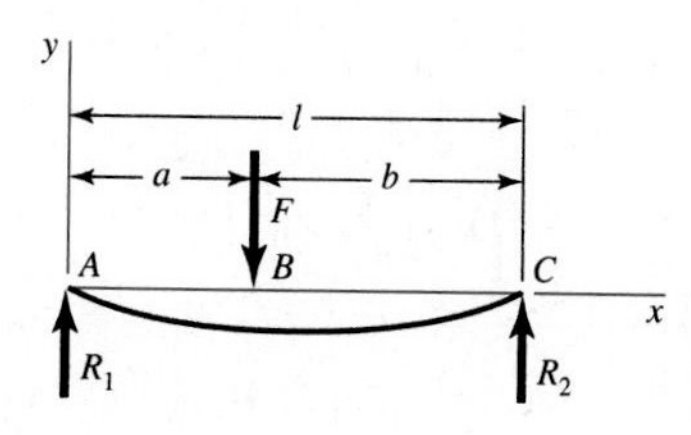

$$R_1 = \frac{Fb}{l} \qquad R_2 = \frac{Fa}{l}$$

$$V_{AB} = R_1 \qquad V_{BC} = -R_2$$

$$M_{AB} = \frac{Fbx}{l} \qquad M_{BC} = \frac{Fa}{l}(l - x)$$

$$y_{AB} = \frac{Fbx}{6EIl}(x^2 + b^2 - l^2)$$

$$y_{BC} = \frac{Fa(l - x)}{6EIl}(x^2 + a^2 - 2lx)$$

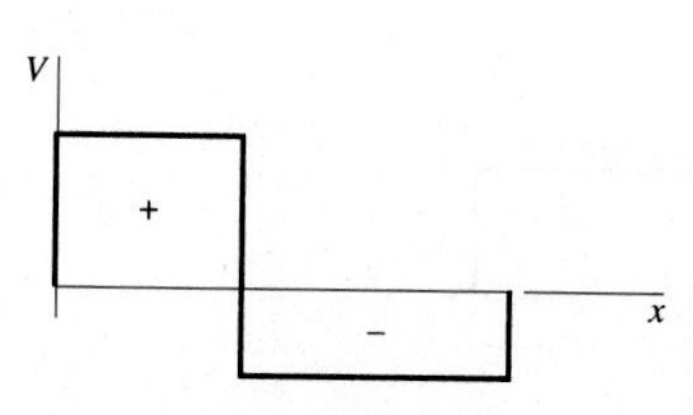

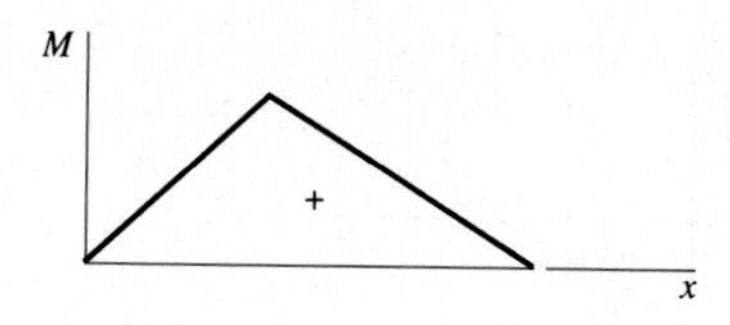

Table E–9

Shear, Moment, and Deflection of Beams *(Continued)*

7 Simple supports—uniform load

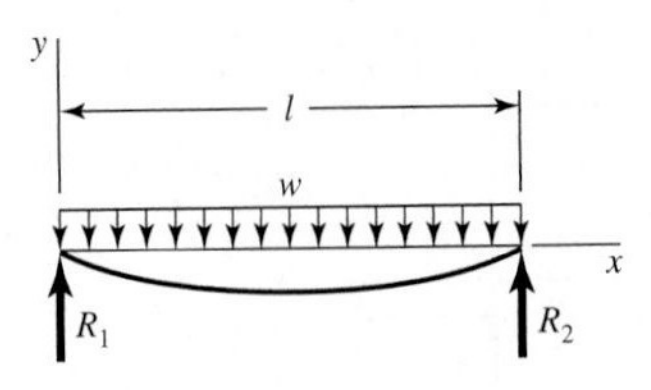

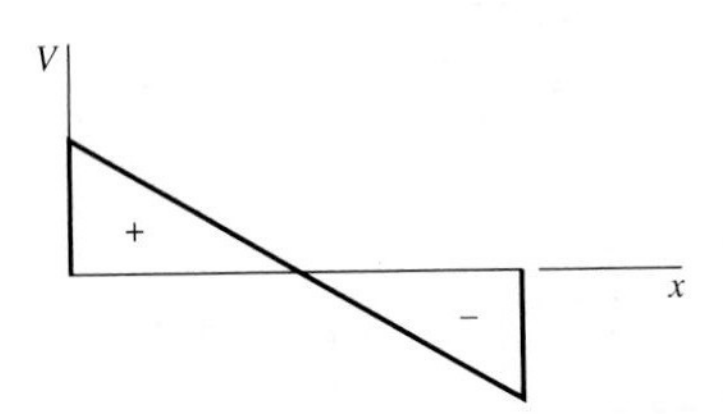

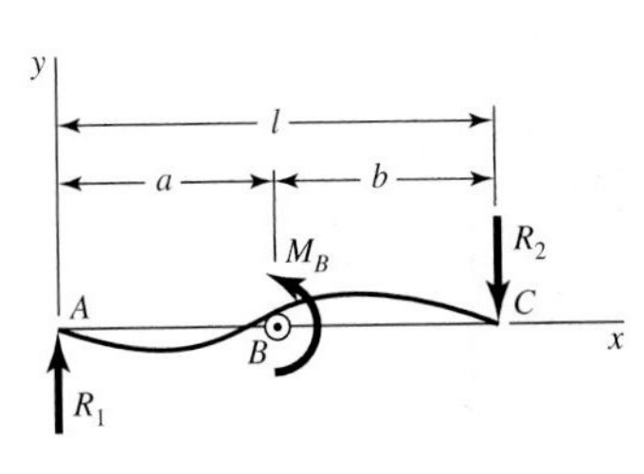

$$R_1 = R_2 = \frac{wl}{2} \qquad V = \frac{wl}{2} - wx$$

$$M = \frac{wx}{2}(l - x)$$

$$y = \frac{wx}{24EI}(2lx^2 - x^3 - l^3)$$

$$y_{\max} = -\frac{5wl^4}{384EI}$$

8 Simple supports—moment load

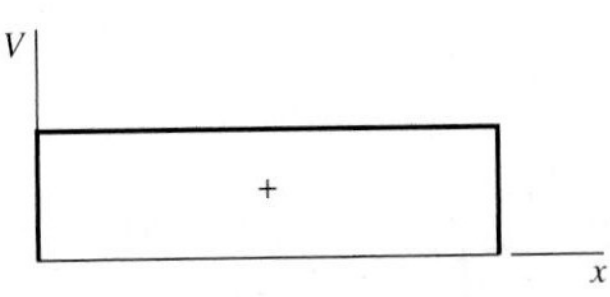

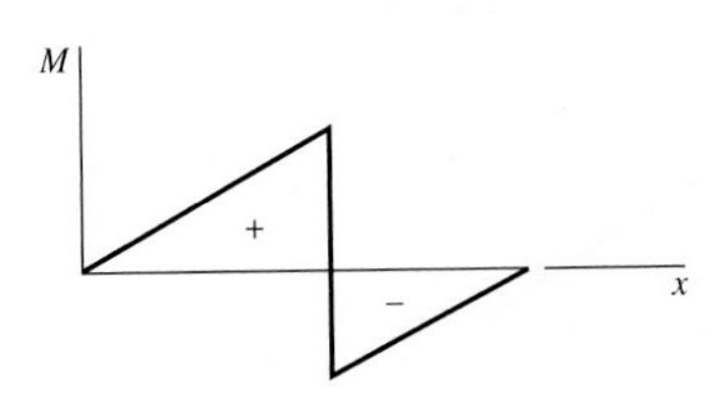

$$R_1 = -R_2 = \frac{M_B}{l} \qquad V = \frac{M_B}{l}$$

$$M_{AB} = \frac{M_B x}{l} \qquad M_{BC} = \frac{M_B}{l}(x - l)$$

$$y_{AB} = \frac{M_B x}{6EIl}(x^2 + 3a^2 - 6al + 2l^2)$$

$$y_{BC} = \frac{M_B}{6EIl}[x^3 - 3lx^2 + x(2l^2 + 3a^2) - 3a^2 l]$$

Table E–9

Shear, Moment, and Deflection of Beams *(Continued)*

9 Simple supports—twin loads

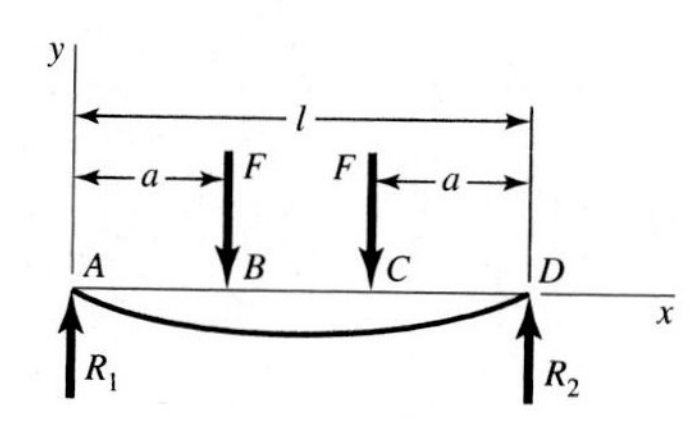

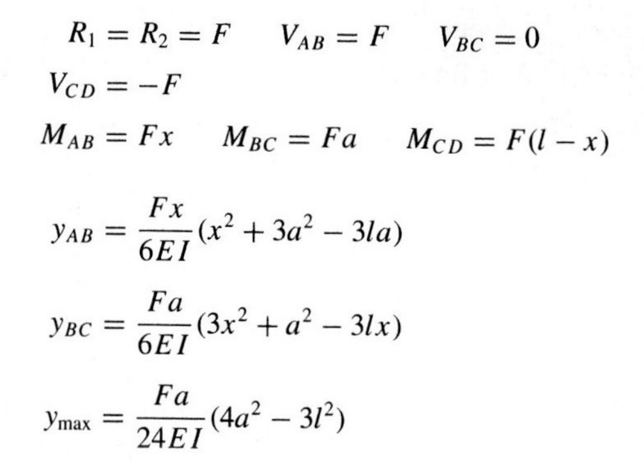

$$R_1 = R_2 = F \qquad V_{AB} = F \qquad V_{BC} = 0$$

$$V_{CD} = -F$$

$$M_{AB} = Fx \qquad M_{BC} = Fa \qquad M_{CD} = F(l - x)$$

$$y_{AB} = \frac{Fx}{6EI}(x^2 + 3a^2 - 3la)$$

$$y_{BC} = \frac{Fa}{6EI}(3x^2 + a^2 - 3lx)$$

$$y_{max} = \frac{Fa}{24EI}(4a^2 - 3l^2)$$

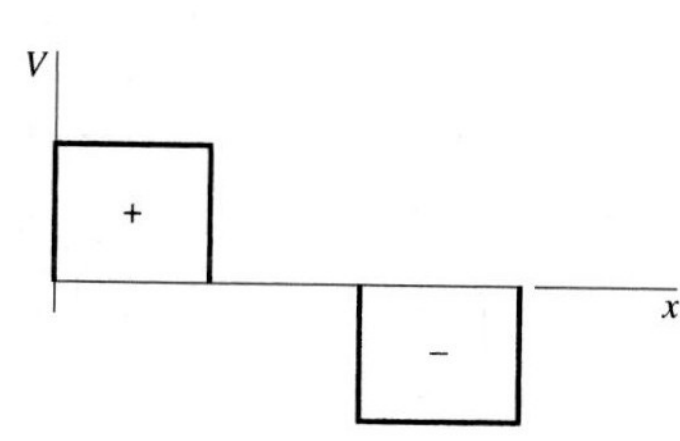

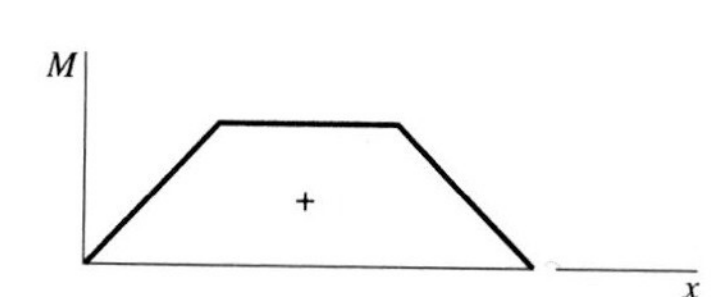

10 Simple supports—overhanging load

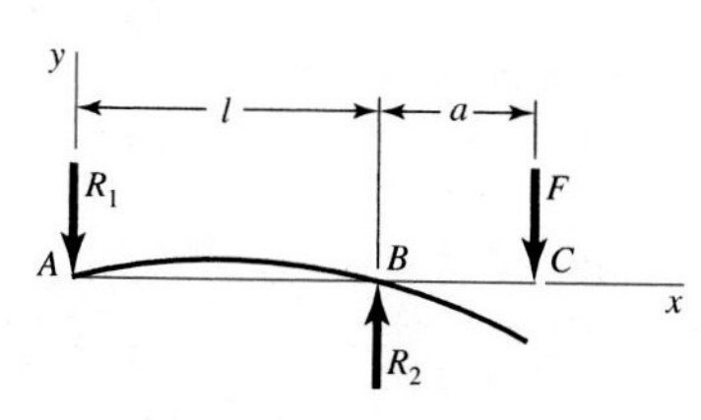

$$R_1 = -\frac{Fa}{l} \qquad R_2 = \frac{F}{l}(l + a)$$

$$V_{AB} = -\frac{Fa}{l} \qquad V_{BC} = F$$

$$M_{AB} = -\frac{Fax}{l} \qquad M_{BC} = F(x - l - a)$$

$$y_{AB} = \frac{Fax}{6EIl}(l^2 - x^2)$$

$$y_{BC} = \frac{F(x - l)}{6El}[(x - l)^2 - a(3x - l)]$$

$$y_c = -\frac{Fa^2}{3El}(l + a)$$

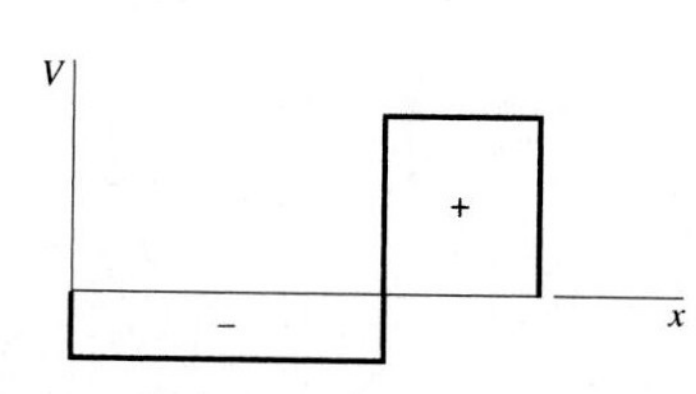

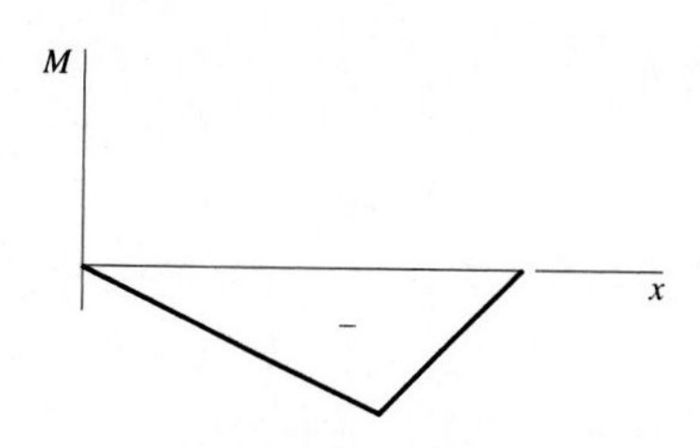

Table E–9

Shear, Moment, and Deflection of Beams *(Continued)*

11 One fixed and one simple support—center load

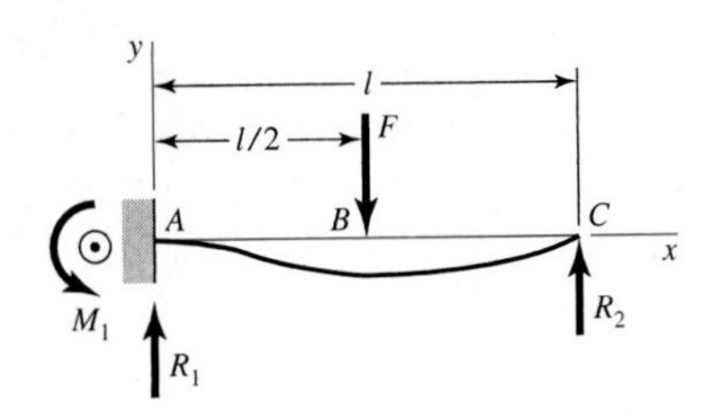

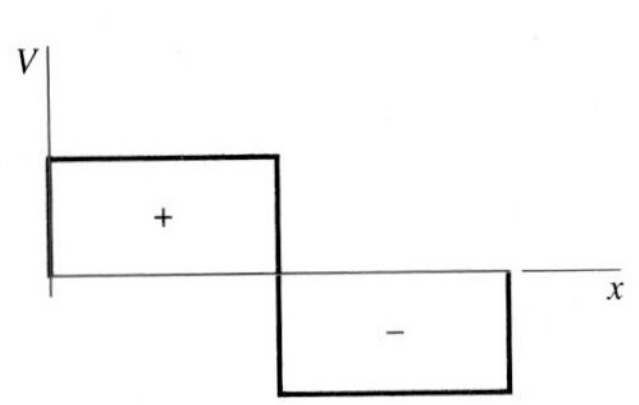

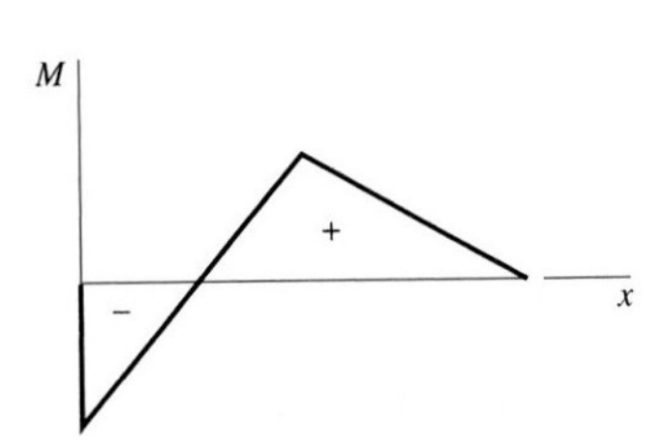

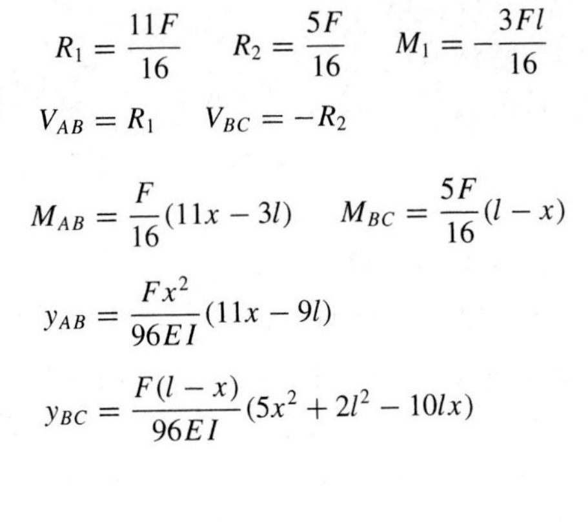

$$R_1 = \frac{11F}{16} \qquad R_2 = \frac{5F}{16} \qquad M_1 = -\frac{3Fl}{16}$$

$$V_{AB} = R_1 \qquad V_{BC} = -R_2$$

$$M_{AB} = \frac{F}{16}(11x - 3l) \qquad M_{BC} = \frac{5F}{16}(l - x)$$

$$y_{AB} = \frac{Fx^2}{96EI}(11x - 9l)$$

$$y_{BC} = \frac{F(l - x)}{96EI}(5x^2 + 2l^2 - 10lx)$$

12 One fixed and one simple support—intermediate load

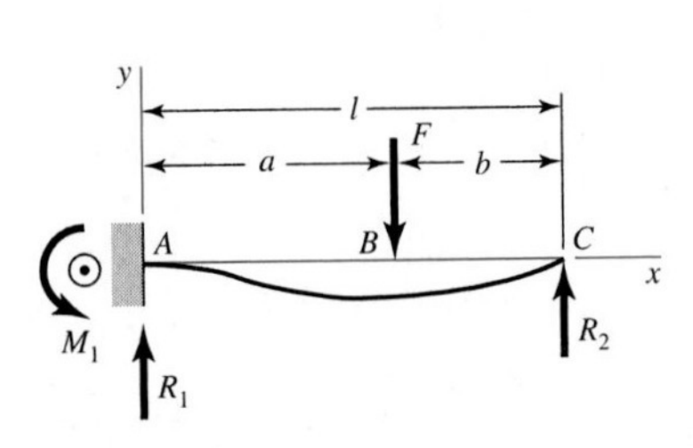

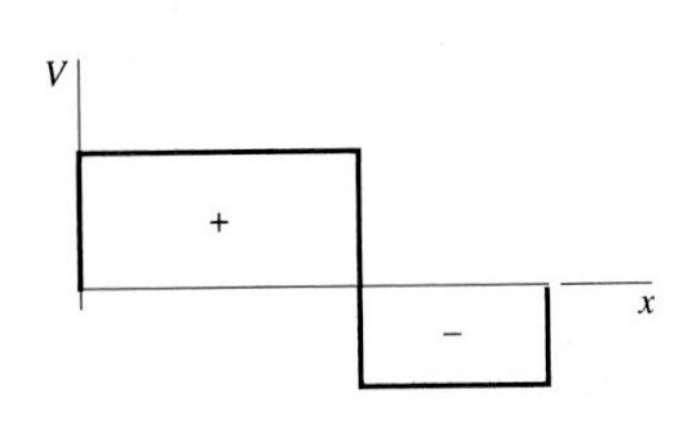

$$R_1 = \frac{Fb}{2l^3}(3l^2 - b^2) \qquad R_2 = \frac{Fa^2}{2l^3}(3l - a)$$

$$M_1 = \frac{Fb}{2l^2}(b^2 - l^2) \qquad V_{AB} = R_1$$

$$V_{AB} = R_1 \qquad V_{BC} = -R_2$$

$$M_{AB} = \frac{Fb}{2l^3}[b^2 l - l^3 + x(3l^2 - b^2)]$$

$$M_{BC} = \frac{Fa^2}{2l^3}(3l^2 - 3lx - al + ax)$$

$$y_{AB} = \frac{Fbx^2}{12EIl^3}[3l(b^2 - l^2) + x(3l^2 - b^2)]$$

$$y_{BC} = y_{AB} - \frac{F(x - a)^3}{6EI}$$

Table E–9

Shear, Moment, and Deflection of Beams *(Continued)*

13 One fixed and one simple support—uniform load

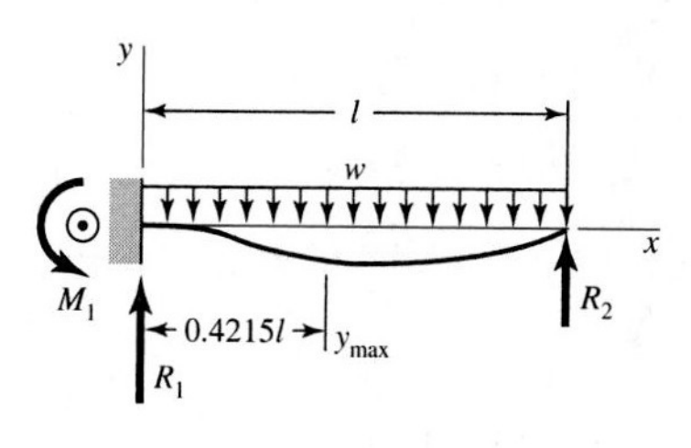

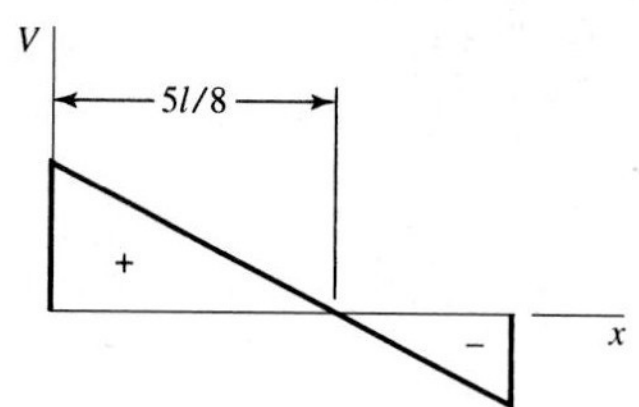

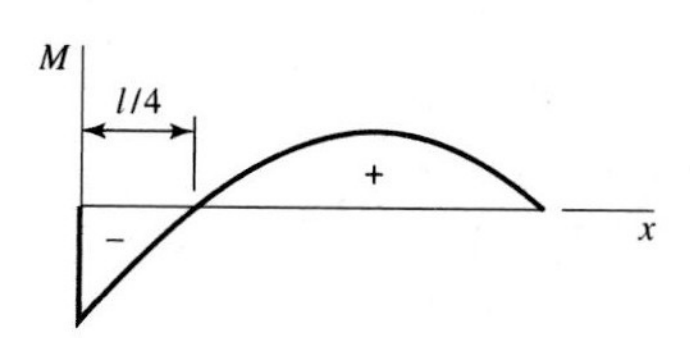

$$R_1 = \frac{5wl}{8} \qquad R_2 = \frac{3wl}{8} \qquad M_1 = -\frac{wl^2}{8}$$

$$V = \frac{5wl}{8} - wx$$

$$M = \frac{w}{8}(4x^2 + 5lx - l^2)$$

$$y = \frac{wx^2}{48EI}(l - x)(2x - 3l)$$

$$y_{max} = -\frac{wl^4}{185EI}$$

14 Fixed supports—center load

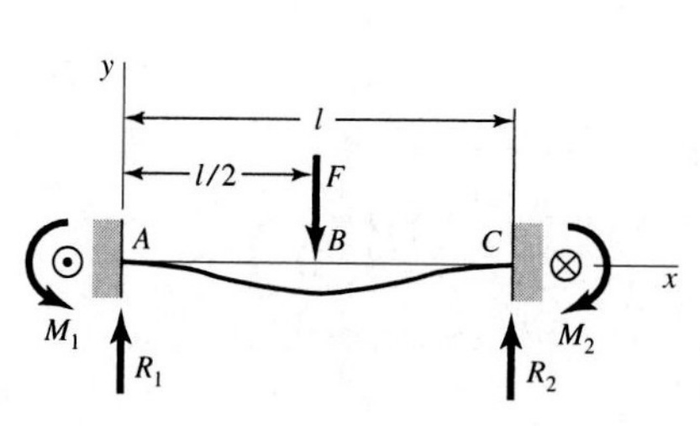

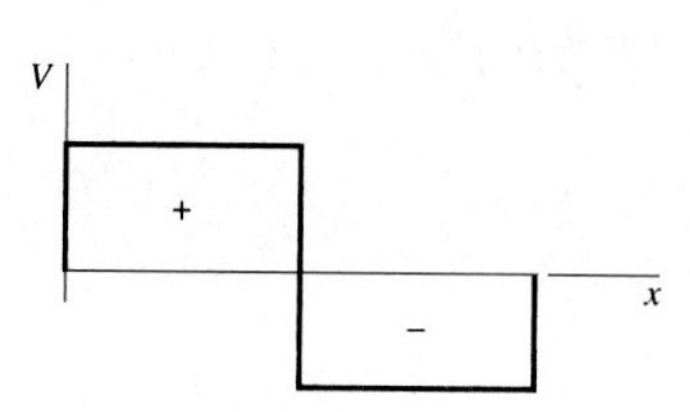

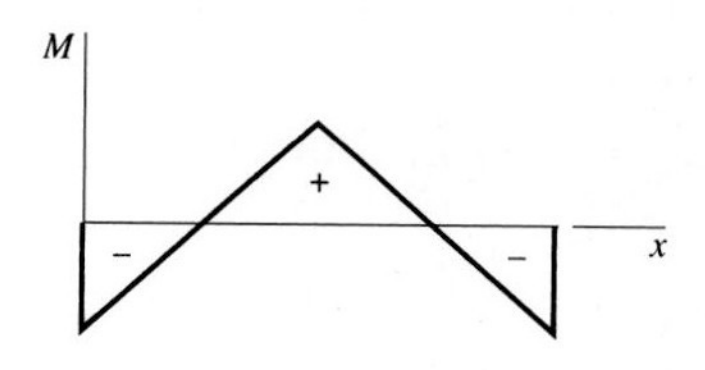

$$R_1 = R_2 = \frac{F}{2} \qquad M_1 = M_2 = -\frac{Fl}{8}$$

$$V_{AB} = -V_{BC} = \frac{F}{2}$$

$$M_{AB} = \frac{F}{8}(4x - l) \qquad M_{BC} = \frac{F}{8}(3l - 4x)$$

$$y_{AB} = \frac{Fx^2}{48EI}(4x - 3l)$$

$$y_{max} = -\frac{Fl^3}{192EI}$$

Table E–9

Shear, Moment, and Deflection of Beams *(Continued)*

15 Fixed supports—intermediate load

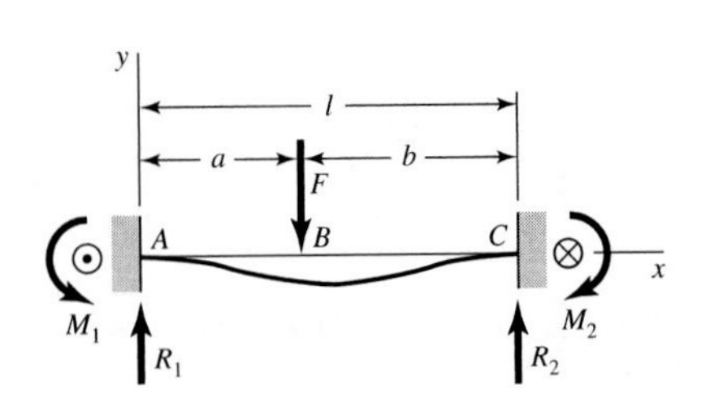

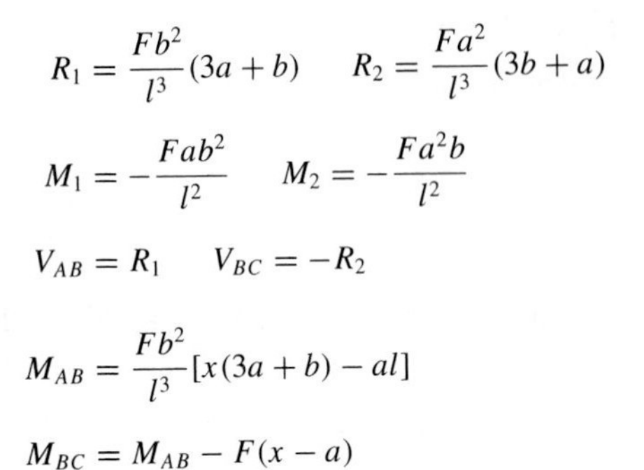

$$R_1 = \frac{Fb^2}{l^3}(3a + b) \qquad R_2 = \frac{Fa^2}{l^3}(3b + a)$$

$$M_1 = -\frac{Fab^2}{l^2} \qquad M_2 = -\frac{Fa^2b}{l^2}$$

$$V_{AB} = R_1 \qquad V_{BC} = -R_2$$

$$M_{AB} = \frac{Fb^2}{l^3}[x(3a + b) - al]$$

$$M_{BC} = M_{AB} - F(x - a)$$

$$y_{AB} = \frac{Fb^2x^2}{6EIl^3}[x(3a + b) - 3al]$$

$$y_{BC} = \frac{Fa^2(l - x)^2}{6EIl^3}[(l - x)(3b + a) - 3bl]$$

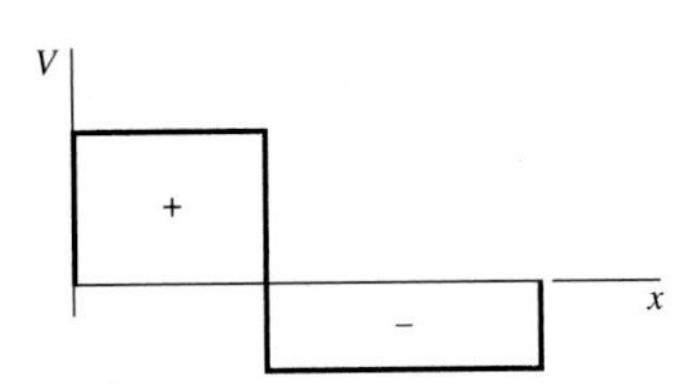

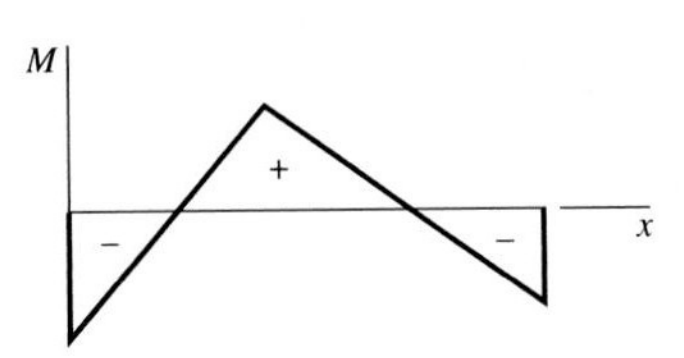

16 Fixed supports—uniform load

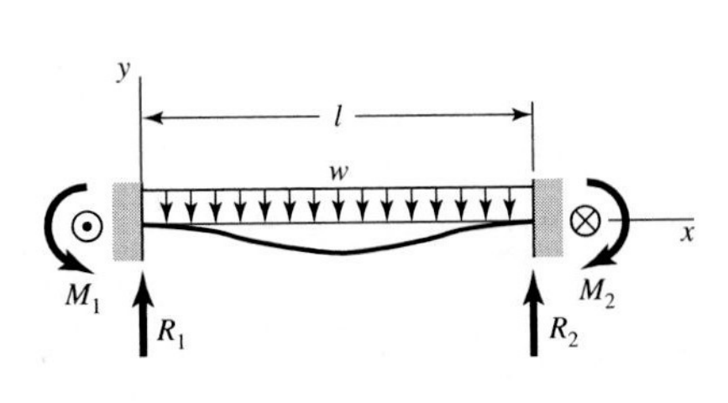

$$R_1 = R_2 = \frac{wl}{2} \qquad M_1 = M_2 = -\frac{wl^2}{12}$$

$$V = \frac{w}{2}(l - 2x)$$

$$M = \frac{w}{12}(6lx - 6x^2 - l^2)$$

$$y = -\frac{wx^2}{24EI}(l - x)^2$$

$$y_{\max} = -\frac{wl^4}{384EI}$$

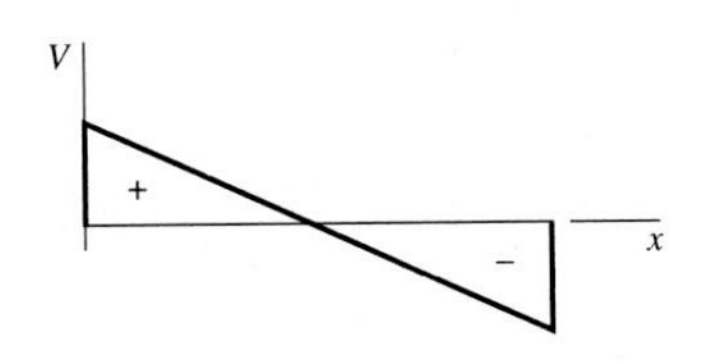

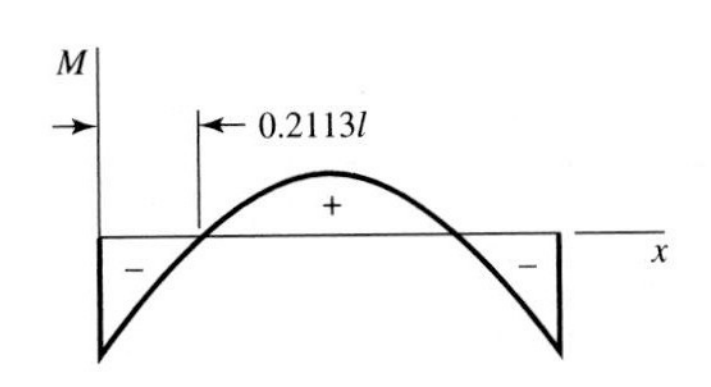

Table E–10

Cumulative Distribution Function of Normal (Gaussian) Distribution

$$\Phi(z_\alpha) = \int_{-\infty}^{z_\alpha} \frac{1}{\sqrt{2\pi}} \exp\left(-\frac{u^2}{2}\right) du$$

$$= \begin{cases} \alpha & z_\alpha \le 0 \\ 1-\alpha & z_\alpha > 0 \end{cases}$$

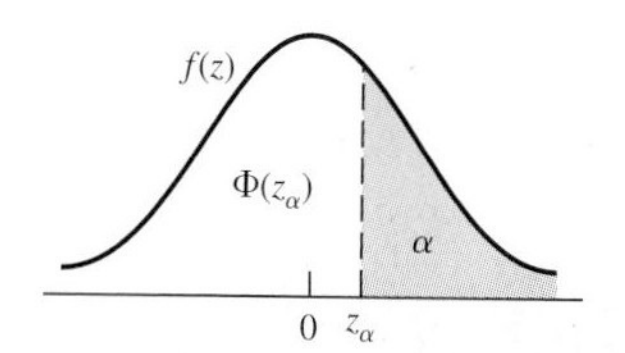

Z_α	0.00	0.01	0.02	0.03	0.04	0.05	0.06	0.07	0.08	0.09
0.0	0.5000	0.4960	0.4920	0.4880	0.4840	0.4801	0.4761	0.4721	0.4681	0.4641
0.1	0.4602	0.4562	0.4522	0.4483	0.4443	0.4404	0.4364	0.4325	0.4286	0.4247
0.2	0.4207	0.4168	0.4129	0.4090	0.4052	0.4013	0.3974	0.3936	0.3897	0.3859
0.3	0.3821	0.3783	0.3745	0.3707	0.3669	0.3632	0.3594	0.3557	0.3520	0.3483
0.4	0.3446	0.3409	0.3372	0.3336	0.3300	0.3264	0.3238	0.3192	0.3156	0.3121
0.5	0.3085	0.3050	0.3015	0.2981	0.2946	0.2912	0.2877	0.2843	0.2810	0.2776
0.6	0.2743	0.2709	0.2676	0.2643	0.2611	0.2578	0.2546	0.2514	0.2483	0.2451
0.7	0.2420	0.2389	0.2358	0.2327	0.2296	0.2266	0.2236	0.2206	0.2177	0.2148
0.8	0.2119	0.2090	0.2061	0.2033	0.2005	0.1977	0.1949	0.1922	0.1894	0.1867
0.9	0.1841	0.1814	0.1788	0.1762	0.1736	0.1711	0.1685	0.1660	0.1635	0.1611
1.0	0.1587	0.1562	0.1539	0.1515	0.1492	0.1469	0.1446	0.1423	0.1401	0.1379
1.1	0.1357	0.1335	0.1314	0.1292	0.1271	0.1251	0.1230	0.1210	0.1190	0.1170
1.2	0.1151	0.1131	0.1112	0.1093	0.1075	0.1056	0.1038	0.1020	0.1003	0.0985
1.3	0.0968	0.0951	0.0934	0.0918	0.0901	0.0885	0.0869	0.0853	0.0838	0.0823
1.4	0.0808	0.0793	0.0778	0.0764	0.0749	0.0735	0.0721	0.0708	0.0694	0.0681
1.5	0.0668	0.0655	0.0643	0.0630	0.0618	0.0606	0.0594	0.0582	0.0571	0.0559
1.6	0.0548	0.0537	0.0526	0.0516	0.0505	0.0495	0.0485	0.0475	0.0465	0.0455
1.7	0.0446	0.0436	0.0427	0.0418	0.0409	0.0401	0.0392	0.0384	0.0375	0.0367
1.8	0.0359	0.0351	0.0344	0.0336	0.0329	0.0322	0.0314	0.0307	0.0301	0.0294
1.9	0.0287	0.0281	0.0274	0.0268	0.0262	0.0256	0.0250	0.0244	0.0239	0.0233
2.0	0.0228	0.0222	0.0217	0.0212	0.0207	0.0202	0.0197	0.0192	0.0188	0.0183
2.1	0.0179	0.0174	0.0170	0.0166	0.0162	0.0158	0.0154	0.0150	0.0146	0.0143
2.2	0.0139	0.0136	0.0132	0.0129	0.0125	0.0122	0.0119	0.0116	0.0113	0.0110
2.3	0.0107	0.0104	0.0102	0.00990	0.00964	0.00939	0.00914	0.00889	0.00866	0.00842
2.4	0.00820	0.00798	0.00776	0.00755	0.00734	0.00714	0.00695	0.00676	0.00657	0.00639
2.5	0.00621	0.00604	0.00587	0.00570	0.00554	0.00539	0.00523	0.00508	0.00494	0.00480
2.6	0.00466	0.00453	0.00440	0.00427	0.00415	0.00402	0.00391	0.00379	0.00368	0.00357
2.7	0.00347	0.00336	0.00326	0.00317	0.00307	0.00298	0.00289	0.00280	0.00272	0.00264
2.8	0.00256	0.00248	0.00240	0.00233	0.00226	0.00219	0.00212	0.00205	0.00199	0.00193
2.9	0.00187	0.00181	0.00175	0.00169	0.00164	0.00159	0.00154	0.00149	0.00144	0.00139

Z_α	0.0	0.1	0.2	0.3	0.4	0.5	0.6	0.7	0.8	0.9
3	0.00135	$0.0^{3}968$	$0.0^{3}687$	$0.0^{3}483$	$0.0^{3}337$	$0.0^{3}233$	$0.0^{3}159$	$0.0^{3}108$	$0.0^{4}723$	$0.0^{4}481$
4	$0.0^{4}317$	$0.0^{4}207$	$0.0^{4}133$	$0.0^{5}854$	$0.0^{5}541$	$0.0^{5}340$	$0.0^{5}211$	$0.0^{5}130$	$0.0^{6}793$	$0.0^{6}479$
5	$0.0^{6}287$	$0.0^{6}170$	$0.0^{7}996$	$0.0^{7}579$	$0.0^{7}333$	$0.0^{7}190$	$0.0^{7}107$	$0.0^{8}599$	$0.0^{8}332$	$0.0^{8}182$
6	$0.0^{9}987$	$0.0^{9}530$	$0.0^{9}282$	$0.0^{9}149$	$0.0^{10}777$	$0.0^{10}402$	$0.0^{10}206$	$0.0^{10}104$	$0.0^{11}523$	$0.0^{11}260$

z_α	−1.282	−1.643	−1.960	−2.326	−2.576	−3.090	−3.291	−3.891	−4.417
$F(z_\alpha)$	0.10	0.05	0.025	0.010	0.005	0.001	0.005	0.00005	0.000005
$R(z_\alpha)$	0.90	0.95	0.975	0.999	0.995	0.999	0.9995	0.9999	0.999995

Table E–11

A Selection of International Tolerance Grades—Metric Series (Size Ranges Are for *Over* the Lower Limit and *Including* the Upper Limit. All Values Are in Millimeters)

Source: Preferred Metric Limits and Fits, ANSI B4.2-1978. See also BSI 4500.

Basic Sizes	Tolerance Grades					
	IT6	IT7	IT8	IT9	IT10	IT11
0–3	0.006	0.010	0.014	0.025	0.040	0.060
3–6	0.008	0.012	0.018	0.030	0.048	0.075
6–10	0.009	0.015	0.022	0.036	0.058	0.090
10–18	0.011	0.018	0.027	0.043	0.070	0.110
18–30	0.013	0.021	0.033	0.052	0.084	0.130
30–50	0.016	0.025	0.039	0.062	0.100	0.160
50–80	0.019	0.030	0.046	0.074	0.120	0.190
80–120	0.022	0.035	0.054	0.087	0.140	0.220
120–180	0.025	0.040	0.063	0.100	0.160	0.250
180–250	0.029	0.046	0.072	0.115	0.185	0.290
250–315	0.032	0.052	0.081	0.130	0.210	0.320
315–400	0.036	0.057	0.089	0.140	0.230	0.360

Table E–12

Fundamental Deviations for Shafts—Metric Series
(Size Ranges Are for *Over* the Lower Limit and *Including* the Upper Limit. All Values Are in Millimeters)

Source: Preferred Metric Limits and Fits, ANSI B4.2-1978. See also BSI 4500.

Basic Sizes	Upper-Deviation Letter					Lower-Deviation Letter				
	c	d	f	g	h	k	n	p	s	u
0–3	−0.060	−0.020	−0.006	−0.002	0	0	+0.004	+0.006	+0.014	+0.018
3–6	−0.070	−0.030	−0.010	−0.004	0	+0.001	+0.008	+0.012	+0.019	+0.023
6–10	−0.080	−0.040	−0.013	−0.005	0	+0.001	+0.010	+0.015	+0.023	+0.028
10–14	−0.095	−0.050	−0.016	−0.006	0	+0.001	+0.012	+0.018	+0.028	+0.033
14–18	−0.095	−0.050	−0.016	−0.006	0	+0.001	+0.012	+0.018	+0.028	+0.033
18–24	−0.110	−0.065	−0.020	−0.007	0	+0.002	+0.015	+0.022	+0.035	+0.041
24–30	−0.110	−0.065	−0.020	−0.007	0	+0.002	+0.015	+0.022	+0.035	+0.048
30–40	−0.120	−0.080	−0.025	−0.009	0	+0.002	+0.017	+0.026	+0.043	+0.060
40–50	−0.130	−0.080	−0.025	−0.009	0	+0.002	+0.017	+0.026	+0.043	+0.070
50–65	−0.140	−0.100	−0.030	−0.010	0	+0.002	+0.020	+0.032	+0.053	+0.087
65–80	−0.150	−0.100	−0.030	−0.010	0	+0.002	+0.020	+0.032	+0.059	+0.102
80–100	−0.170	−0.120	−0.036	−0.012	0	+0.003	+0.023	+0.037	+0.071	+0.124
100–120	−0.180	−0.120	−0.036	−0.012	0	+0.003	+0.023	+0.037	+0.079	+0.144
120–140	−0.200	−0.145	−0.043	−0.014	0	+0.003	+0.027	+0.043	+0.092	+0.170
140–160	−0.210	−0.145	−0.043	−0.014	0	+0.003	+0.027	+0.043	+0.100	+0.190
160–180	−0.230	−0.145	−0.043	−0.014	0	+0.003	+0.027	+0.043	+0.108	+0.210
180–200	−0.240	−0.170	−0.050	−0.015	0	+0.004	+0.031	+0.050	+0.122	+0.236
200–225	−0.260	−0.170	−0.050	−0.015	0	+0.004	+0.031	+0.050	+0.130	+0.258
225–250	−0.280	−0.170	−0.050	−0.015	0	+0.004	+0.031	+0.050	+0.140	+0.284
250–280	−0.300	−0.190	−0.056	−0.017	0	+0.004	+0.034	+0.056	+0.158	+0.315
280–315	−0.330	−0.190	−0.056	−0.017	0	+0.004	+0.034	+0.056	+0.170	+0.350
315–355	−0.360	−0.210	−0.062	−0.018	0	+0.004	+0.037	+0.062	+0.190	+0.390
355–400	−0.400	−0.210	−0.062	−0.018	0	+0.004	+0.037	+0.062	+0.208	+0.435

Table E–13

A Selection of International Tolerance Grades—Inch Series (Size Ranges Are for *Over* the Lower Limit and *Including* the Upper Limit. All Values Are in Inches, Converted from Table E–11)

Basic Sizes	Tolerance Grades					
	IT6	IT7	IT8	IT9	IT10	IT11
0–0.12	0.0002	0.0004	0.0006	0.0010	0.0016	0.0024
0.12–0.24	0.0003	0.0005	0.0007	0.0012	0.0019	0.0030
0.24–0.40	0.0004	0.0006	0.0009	0.0014	0.0023	0.0035
0.40–0.72	0.0004	0.0007	0.0011	0.0017	0.0028	0.0043
0.72–1.20	0.0005	0.0008	0.0013	0.0020	0.0033	0.0051
1.20–2.00	0.0006	0.0010	0.0015	0.0024	0.0039	0.0063
2.00–3.20	0.0007	0.0012	0.0018	0.0029	0.0047	0.0075
3.20–4.80	0.0009	0.0014	0.0021	0.0034	0.0055	0.0087
4.80–7.20	0.0010	0.0016	0.0025	0.0039	0.0063	0.0098
7.20–10.00	0.0011	0.0018	0.0028	0.0045	0.0073	0.0114
10.00–12.60	0.0013	0.0020	0.0032	0.0051	0.0083	0.0126
12.60–16.00	0.0014	0.0022	0.0035	0.0055	0.0091	0.0142

Table E–14

Fundamental Deviations for Shafts—Inch Series (Size Ranges Are for *Over* the Lower Limit and *Including* the Upper Limit. All Values Are in Inches, Converted from Table E–12)

Basic Sizes	Upper-Deviation Letter					Lower-Deviation Letter				
	c	d	f	g	h	k	n	p	s	u
0–0.12	−0.0024	−0.0008	−0.0002	−0.0001	0	0	+0.0002	+0.0002	+0.0006	+0.0007
0.12–0.24	−0.0028	−0.0012	−0.0004	−0.0002	0	0	+0.0003	+0.0005	+0.0007	+0.0009
0.24–0.40	−0.0031	−0.0016	−0.0005	−0.0002	0	0	+0.0004	+0.0006	+0.0009	+0.0011
0.40–0.72	−0.0037	−0.0020	−0.0006	−0.0002	0	0	+0.0005	+0.0007	+0.0011	+0.0013
0.72–0.96	−0.0043	−0.0026	−0.0008	−0.0003	0	+0.0001	+0.0006	+0.0009	+0.0014	+0.0016
0.96–1.20	−0.0043	−0.0026	−0.0008	−0.0003	0	+0.0001	+0.0006	+0.0009	+0.0014	+0.0019
1.20–1.60	−0.0047	−0.0031	−0.0010	−0.0004	0	+0.0001	+0.0007	+0.0010	+0.0017	+0.0024
1.60–2.00	−0.0051	−0.0031	−0.0010	−0.0004	0	+0.0001	+0.0007	+0.0010	+0.0017	+0.0028
2.00–2.60	−0.0055	−0.0039	−0.0012	−0.0004	0	+0.0001	+0.0008	+0.0013	+0.0021	+0.0034
2.60–3.20	−0.0059	−0.0039	−0.0012	−0.0004	0	+0.0001	+0.0008	+0.0013	+0.0023	+0.0040
3.20–4.00	−0.0067	−0.0047	−0.0014	−0.0005	0	+0.0001	+0.0009	+0.0015	+0.0028	+0.0049
4.00–4.80	−0.0071	−0.0047	−0.0014	−0.0005	0	+0.0001	+0.0009	+0.0015	+0.0031	+0.0057
4.80–5.60	−0.0079	−0.0057	−0.0017	−0.0006	0	+0.0001	+0.0011	+0.0017	+0.0036	+0.0067
5.60–6.40	−0.0083	−0.0057	−0.0017	−0.0006	0	+0.0001	+0.0011	+0.0017	+0.0039	+0.0075
6.40–7.20	−0.0091	−0.0057	−0.0017	−0.0006	0	+0.0001	+0.0011	+0.0017	+0.0043	+0.0083
7.20–8.00	−0.0094	−0.0067	−0.0020	−0.0006	0	+0.0002	+0.0012	+0.0020	+0.0048	+0.0093
8.00–9.00	−0.0102	−0.0067	−0.0020	−0.0006	0	+0.0002	+0.0012	+0.0020	+0.0051	+0.0102
9.00–10.00	−0.0110	−0.0067	−0.0020	−0.0006	0	+0.0002	+0.0012	+0.0020	+0.0055	+0.0112
10.00–11.20	−0.0118	−0.0075	−0.0022	−0.0007	0	+0.0002	+0.0013	+0.0022	+0.0062	+0.0124
11.20–12.60	−0.0130	−0.0075	−0.0022	−0.0007	0	+0.0002	+0.0013	+0.0022	+0.0067	+0.0130
12.60–14.20	−0.0142	−0.0083	−0.0024	−0.0007	0	+0.0002	+0.0015	+0.0024	+0.0075	+0.0154
14.20–16.00	−0.0157	−0.0083	−0.0024	−0.0007	0	+0.0002	+0.0015	+0.0024	+0.0082	+0.0171

Table E–15

Charts of Theoretical Stress-Concentration Factors K_t^*

Figure E–15–1

Bar in tension or simple compression with a transverse hole. $\sigma_0 = F/A$, where $A = (w - d)t$ and t is the thickness.

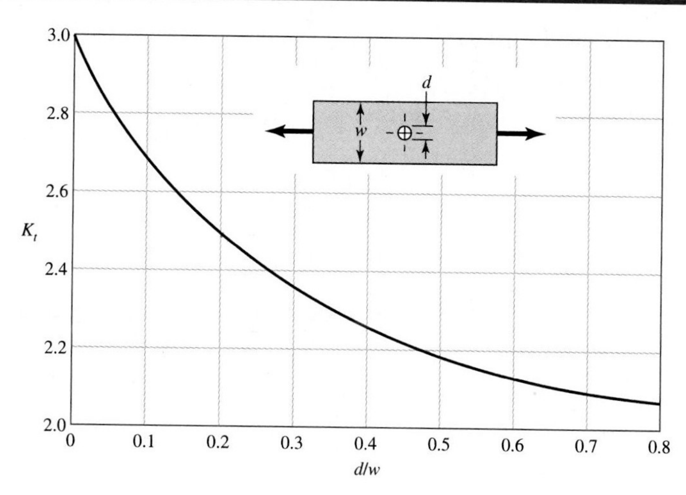

Figure E–15–2

Rectangular bar with a transverse hole in bending. $\sigma_0 = Mc/I$, where $I = (w - d)h^3/12$.

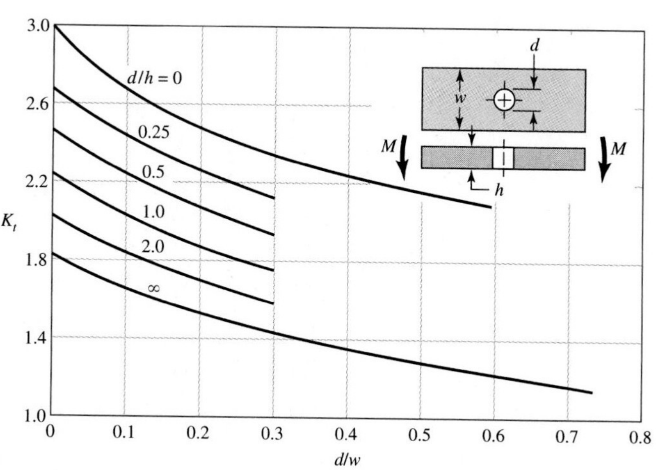

Figure E–15–3

Notched rectangular bar in tension or simple compression. $\sigma_0 = F/A$, where $A = dt$ and t is the thickness.

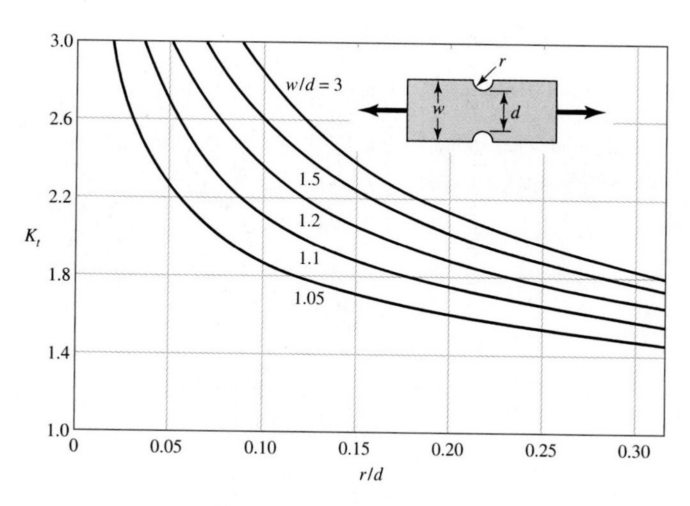

Table E-15

Charts of Theoretical Stress-Concentration Factors K_t *(Continued)*

Figure E-15-4

Notched rectangular bar in bending. $\sigma_0 = Mc/I$, where $c = d/2$, $I = td^3/12$, and t is the thickness.

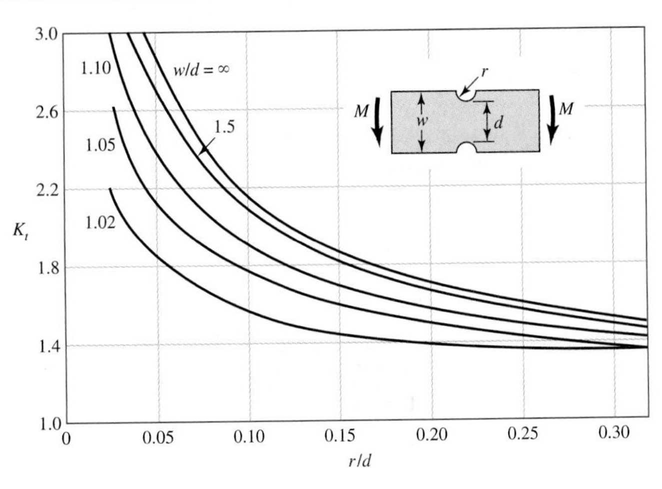

Figure E-15-5

Rectangular filleted bar in tension or simple compression. $\sigma_0 = F/A$, where $A = dt$ and t is the thickness.

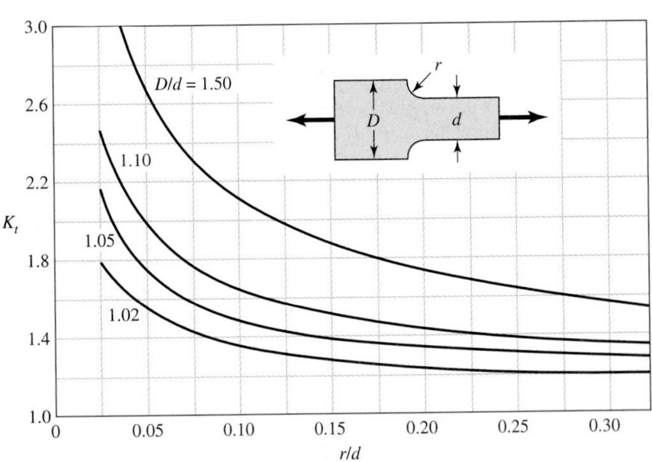

Figure E-15-6

Rectangular filleted bar in bending. $\sigma_0 = Mc/I$, where $c = d/2$, $I = td^3/12$, t is the thickness.

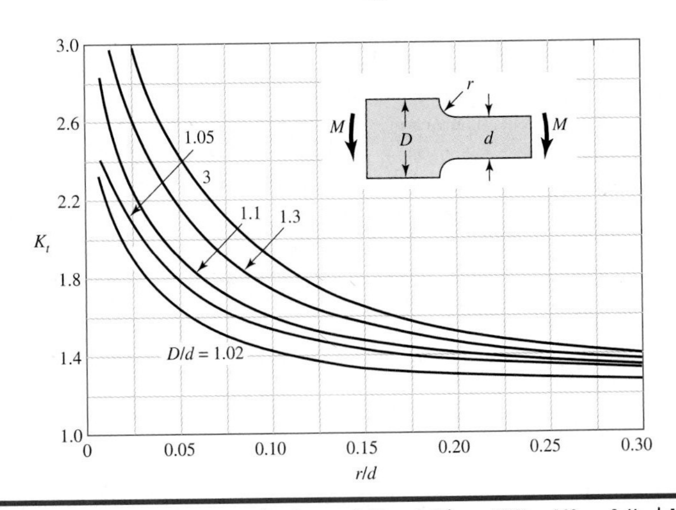

*Unless otherwise stated, these factors are from R. E. Peterson, "Design Factors for Stress Concentration," *Machine Design*, vol. 23, no. 2, February 1951, p. 169; no. 3, March 1951, p. 161; no. 5, May 1951, p. 159; no. 6, June 1951, p. 173; no. 7, July 1951, p. 155; reproduced with the permission of the author and publisher.

Figure E–15–7

Round shaft with shoulder fillet in tension $\sigma_0 = F/A$ where $A = \pi d^2/4$.*

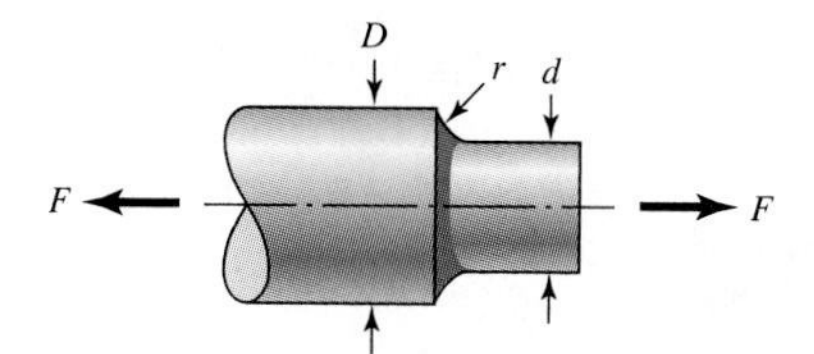

Stress-concentration factor K_t related to maximum ordered principal stress $(\sigma_1)_{max} = K_t\sigma_0$

$$K_t = 0.493 + 0.48\left(\frac{D}{d}\right)^{-2.43} + \left(\frac{r}{d}\right)^{-0.48}\sqrt{\frac{3.43 - 3.41(D/d)^2 - 0.0232(D/d)^4}{1 - 8.85(D/d)^2 - 0.078(D/d)^4}}$$

	r/d						
D/d	**0.01**	**0.05**	**0.10**	**0.15**	**0.20**	**0.25**	**0.30**
1.01	1.825	1.360	1.247	1.197	1.167	1.146	1.130
1.05	2.753	1.766	1.526	1.419	1.355	1.310	1.278
1.10	3.360	2.022	1.697	1.551	1.464	1.404	1.360
1.50	5.006	1.674	2.107	1.853	1.701	1.597	1.519
2.00	5.551	2.871	2.227	1.936	1.762	1.642	1.553
6.00	5.947	3.105	2.302	1.984	1.793	1.661	1.564

Stress-concentration factor K_t related to maximum von Mises stress $(\sigma')_{max} = K_t\sigma_0$

$$K_t = 0.496 + 0.472\left(\frac{D}{d}\right)^{-2.85} + \left(\frac{r}{d}\right)^{-0.48}\sqrt{\frac{2.921 - 2.945(D/d)^2 + 0.0217(D/d)^4}{1 - 9.59(D/d)^2 + 0.053(D/d)^4}}$$

	r/d						
D/d	**0.01**	**0.05**	**0.10**	**0.15**	**0.20**	**0.25**	**0.30**
1.01	1.715	1.306	1.206	1.162	1.135	1.117	1.103
1.05	2.525	1.654	1.443	1.348	1.291	1.252	1.223
1.10	3.052	1.870	1.583	1.454	1.377	1.324	1.285
1.50	4.482	2.417	1.915	1.691	1.556	1.463	1.395
2.00	4.965	2.595	2.020	1.762	1.607	1.501	1.422
6.00	5.256	2.696	2.074	1.796	1.628	1.514	1.429

*S. M. Tipton, J. R. Sorem, and R. D. Rolovic, "Updated Stress-Concentration Factors for Filleted Shafts in Bending and Tension," *Trans. ASME Journal of Mechanical Design*, vol. 118, No. 3, (September 1996), pp. 321–327.

Figure E–15–8

Round shaft with shoulder fillet in torsion, $\tau_0 = Tc/J$ where $c = d/2$ and $J = \pi d^4/32$.*

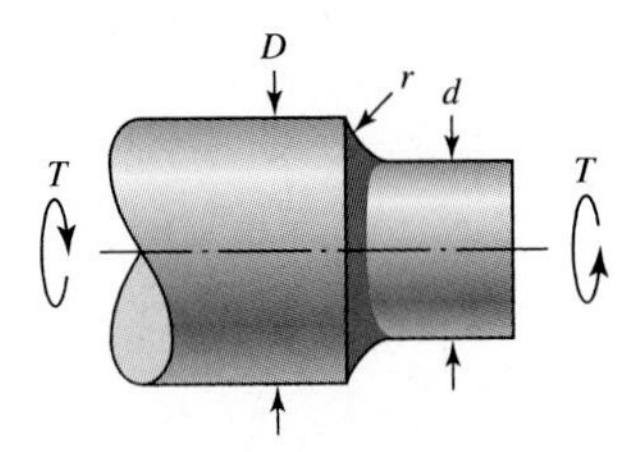

The stress-concentration factor K_{ts} is related to either the maximum ordered principal stress $(\sigma_1)_{max} = K_{ts}\tau_0$ or the von Mises stress $(\sigma')_{max} = K_{ts}\sigma_0 = \sqrt{3}K_{ts}\tau_0$,

$$K_{ts} = 0.78 + 0.2\left(\frac{D}{d}\right)^{-10} + \left(\frac{r}{d}\right)^{-0.46}\sqrt{\frac{0.002 - 0.125(D/d)^2 + 0.123(D/d)^4}{1 - 2.75(D/d)^2 + 2.55(D/d)^4}}$$

	r/d						
D/d	**0.02**	**0.05**	**0.10**	**0.15**	**0.20**	**0.25**	**0.30**
1.01	1.288	1.176	1.117	1.091	1.074	1.063	1.055
1.05	1.588	1.352	1.229	1.174	1.140	1.117	1.100
1.10	1.753	1.445	1.284	1.212	1.168	1.137	1.115
1.50	2.059	1.620	1.392	1.288	1.226	1.183	1.150
2.00	2.101	1.647	1.410	1.303	1.238	1.194	1.160
6.00	2.109	1.652	1.414	1.306	1.241	1.196	1.162

*R. D. Rolovic, S. M. Tipton, and J. R. Sorem, Jr., "Multiaxial Stress Concentration in Filleted Shafts," submitted to ASME Journal of Mechanical Design, March 2000 (In review).

Figure E–15–9

Round shaft with shoulder fillet in bending $\sigma_0 = Mc/I$ where $c = d/2$ and $I = \pi d^4/64$.*

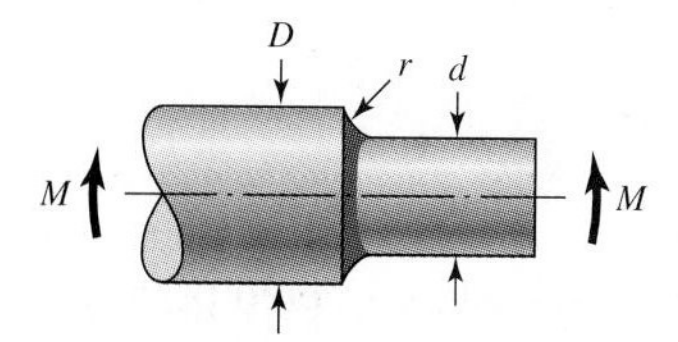

The stress-concentration factor K_t is related to maximum ordered principal stress $(\sigma_1)_{\max} = K_t \sigma_0$,

$$K_t = 0.632 + 0.377\left(\frac{D}{d}\right)^{-4.4} + \left(\frac{r}{d}\right)^{-0.5}\sqrt{\frac{-0.14 - 0.363(D/d)^2 + 0.503(D/d)^4}{1 - 2.39(D/d)^2 + 3.368(D/d)^4}}$$

	r/d						
D/d	**0.01**	**0.05**	**0.10**	**0.15**	**0.20**	**0.25**	**0.30**
1.01	1.790	1.349	1.245	1.199	1.171	1.152	1.138
1.05	2.638	1.697	1.474	1.376	1.317	1.276	1.247
1.10	3.154	1.897	1.599	1.467	1.388	1.355	1.295
1.50	4.237	2.279	1.815	1.610	1.487	1.404	1.342
2.00	4.424	2.388	1.843	1.624	1.494	1.405	1.339
6.00	4.495	2.360	1.854	1.630	1.496	1.405	1.337

Stress-concentration factor K_t related to maximum von Mises stress $(\sigma')_{\max} = K_t \sigma_0$,

$$K_t = 0.622 + 0.38\left(\frac{D}{d}\right)^{-4.3} + \left(\frac{r}{d}\right)^{-0.5}\sqrt{\frac{-0.322 - 0.277(D/d)^2 + 0.599(D/d)^4}{1 - 2.55(D/d)^2 + 5.27(D/d)^4}}$$

	r/d						
D/d	**0.01**	**0.05**	**0.10**	**0.15**	**0.20**	**0.25**	**0.30**
1.01	1.681	1.297	1.206	1.166	1.141	1.125	1.113
1.05	2.411	1.592	1.398	1.313	1.261	1.226	1.200
1.10	2.850	1.758	1.499	1.384	1.316	1.269	1.235
1.50	3.773	2.068	1.664	1.485	1.378	1.305	1.252
2.00	3.936	2.115	1.683	1.492	1.378	1.300	1.243
6.00	3.994	2.130	1.688	1.493	1.376	1.296	1.238

*S. M. Tipton, J. R. Sorem, and R. D. Rolovic, "Updated Stress-Concentration Factors for Filleted Shafts in Bending and Tension," *Trans. ASME Journal of Mechanical Design*, vol. 118, No. 3, (September 1996), pp. 321–327.

Table E–15

Charts of Theoretical Stress-Concentration Factors K_t *(Continued)*

Figure E–15–10

Round shaft in torsion with transverse hole.

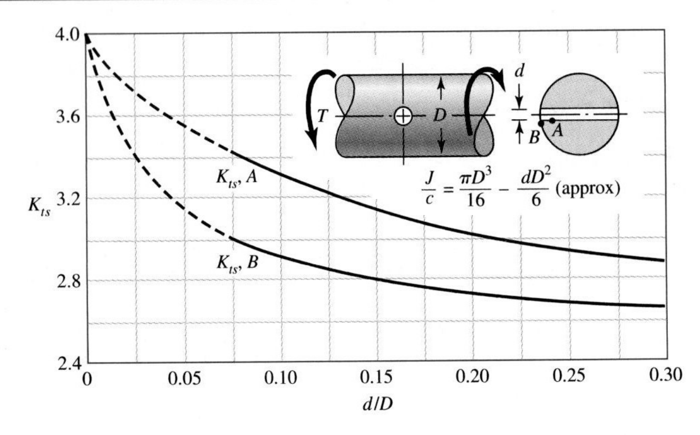

Figure E–15–11

Round shaft in bending with a transverse hole. $\sigma_0 = M/[(\pi D^3/32) - (dD^2/6)]$, approximately.

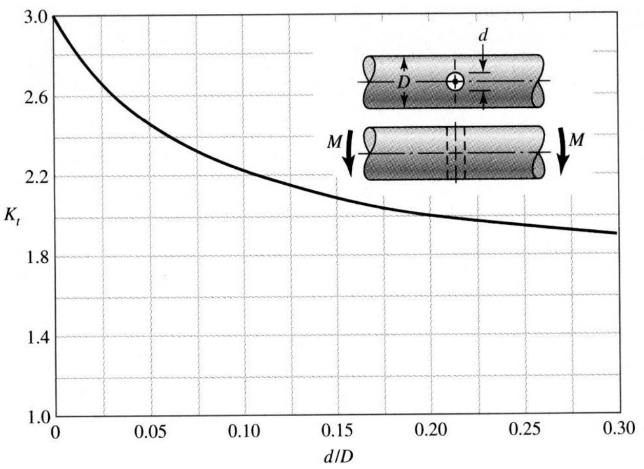

Figure E–15–12

Plate loaded in tension by a pin through a hole. $\sigma_0 = F/A$, where $A = (w - d)t$. When clearance exists, increase K_t 35 to 50 percent. *(M. M. Frocht and H. N. Hill, "Stress Concentration Factors around a Central Circular Hole in a Plate Loaded through a Pin in Hole," J. Appl. Mechanics, vol. 7, no. 1, March 1940, p. A-5.)*

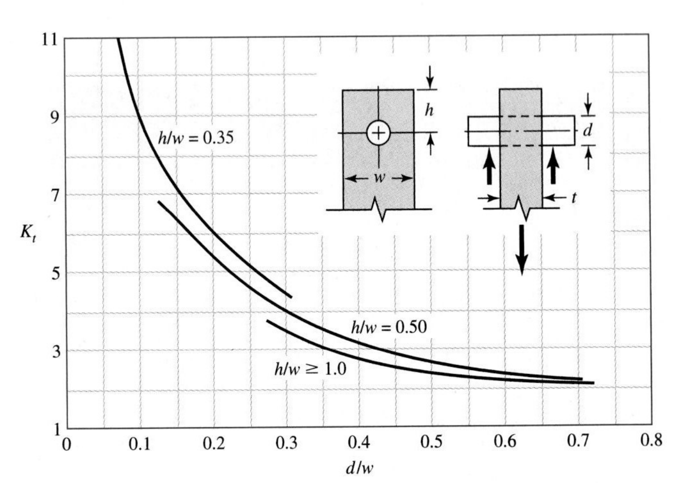

Table E–15

Charts of Theoretical Stress-Concentration Factors K_t *(Continued)*

Figure E–15–13

Grooved round bar in tension. $\sigma_0 = F/A$, where $A = \pi d^2/4$.

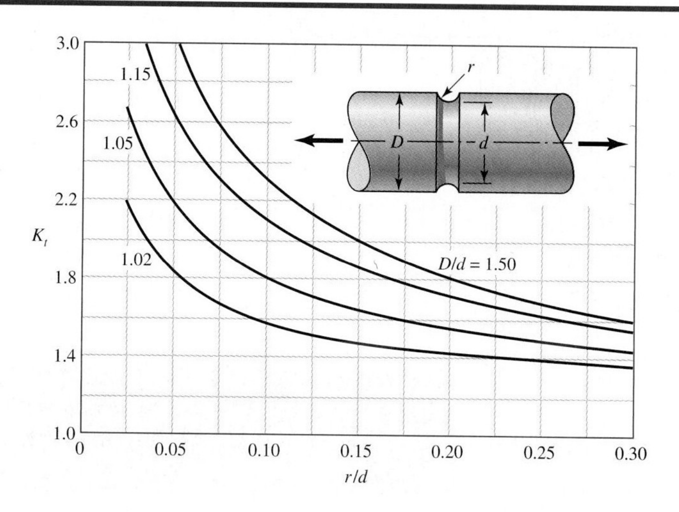

Figure E–15–14

Grooved round bar in bending. $\sigma_0 = Mc/I$, where $c = d/2$ and $I = \pi d^4/64$.

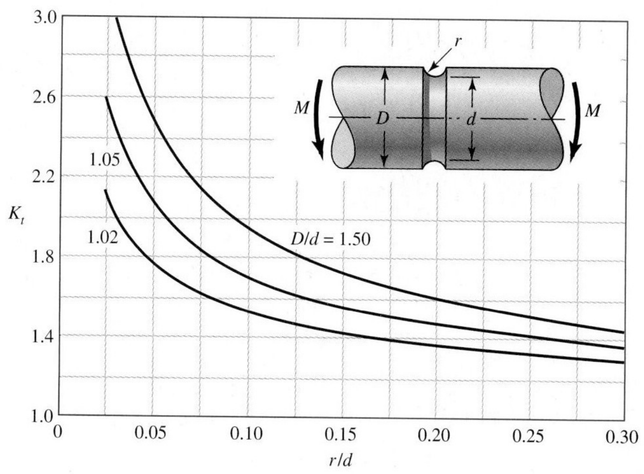

Figure E–15–15

Grooved round bar in torsion. $\tau_0 = Tc/J$, where $c = d/2$ and $J = \pi d^4/32$.

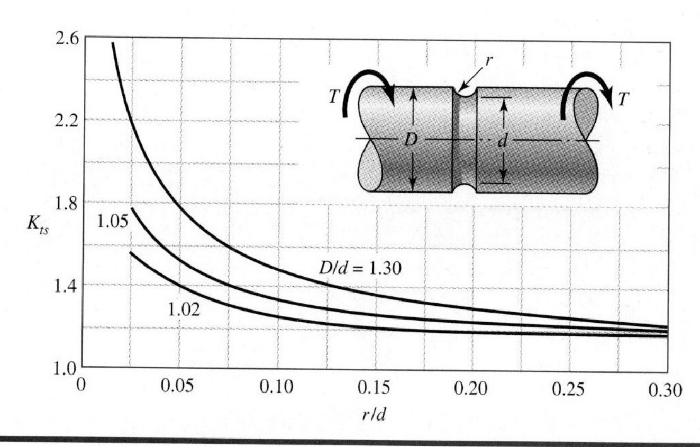

Table E–16

Approximate Stress-Concentration Factors K_t for Bending of a Round Bar or Tube with a Transverse Round Hole *Source:* R. E. Peterson, *Stress Concentration Factors,* Wiley, New York, 1974, pp. 146, 235.

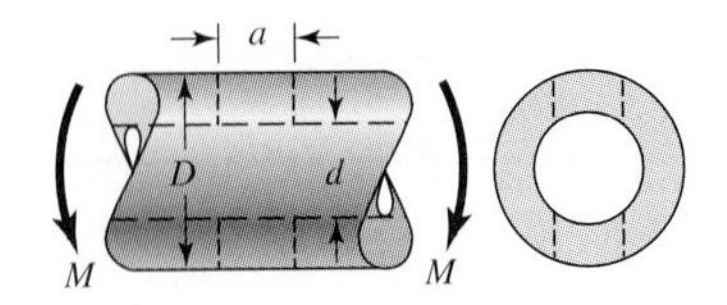

The nominal bending stress is $\sigma_0 = M/Z_{net}$ where Z_{net} is a reduced value of the section modulus and is defined by

$$Z_{net} = \frac{\pi A}{32D}(D^4 - d^4)$$

Values of A are listed in the table. Use $d = 0$ for a solid bar

	d/D					
	0.9		0.6		0	
a/D	A	K_t	A	K_t	A	K_t
0.050	0.92	2.63	0.91	2.55	0.88	2.42
0.075	0.89	2.55	0.88	2.43	0.86	2.35
0.10	0.86	2.49	0.85	2.36	0.83	2.27
0.125	0.82	2.41	0.82	2.32	0.80	2.20
0.15	0.79	2.39	0.79	2.29	0.76	2.15
0.175	0.76	2.38	0.75	2.26	0.72	2.10
0.20	0.73	2.39	0.72	2.23	0.68	2.07
0.225	0.69	2.40	0.68	2.21	0.65	2.04
0.25	0.67	2.42	0.64	2.18	0.61	2.00
0.275	0.66	2.48	0.61	2.16	0.58	1.97
0.30	0.64	2.52	0.58	2.14	0.54	1.94

Table E–16

Approximate Stress-Concentration Factors K_{ts} for a Round Bar or Tube Having a Transverse Round Hole and Loaded in Torsion *(Continued)* *Source:* R. E. Peterson, *Stress Concentration Factors,* Wiley, New York, 1974, pp. 148, 244.

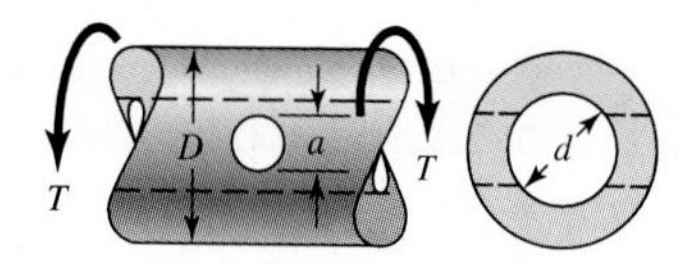

The maximum stress occurs on the inside of the hole, slightly below the shaft surface. The nominal shear stress is $\tau_0 = TD/2J_{net}$, where J_{net} is a reduced value of the second polar moment of area and is defined by

$$J_{net} = \frac{\pi A(D^4 - d^4)}{32}$$

Values of A are listed in the table. Use $d = 0$ for a solid bar

	d/D									
	0.9		0.8		0.6		0.4		0	
a/D	A	K_{ts}	A	K_{ts}	A	K_{ts}	A	K_{ts}	A	K_{ts}
0.05	0.96	1.78							0.95	1.77
0.075	0.95	1.82							0.93	1.71
0.10	0.94	1.76	0.93	1.74	0.92	1.72	0.92	1.70	0.92	1.68
0.125	0.91	1.76	0.91	1.74	0.90	1.70	0.90	1.67	0.89	1.64
0.15	0.90	1.77	0.89	1.75	0.87	1.69	0.87	1.65	0.87	1.62
0.175	0.89	1.81	0.88	1.76	0.87	1.69	0.86	1.64	0.85	1.60
0.20	0.88	1.96	0.86	1.79	0.85	1.70	0.84	1.63	0.83	1.58
0.25	0.87	2.00	0.82	1.86	0.81	1.72	0.80	1.63	0.79	1.54
0.30	0.80	2.18	0.78	1.97	0.77	1.76	0.75	1.63	0.74	1.51
0.35	0.77	2.41	0.75	2.09	0.72	1.81	0.69	1.63	0.68	1.47
0.40	0.72	2.67	0.71	2.25	0.68	1.89	0.64	1.63	0.63	1.44

Table E–17

Preferred Sizes and Renard (R-Series) Numbers (When a Choice Can Be Made, Use One of These Sizes. However, Not All Parts or Items Are Available in All the Sizes Shown in the Table)

Fraction of Inches

$\frac{1}{64}$, $\frac{1}{32}$, $\frac{1}{16}$, $\frac{3}{32}$, $\frac{1}{8}$, $\frac{5}{32}$, $\frac{3}{16}$, $\frac{1}{4}$, $\frac{5}{16}$, $\frac{3}{8}$, $\frac{7}{16}$, $\frac{1}{2}$, $\frac{9}{16}$, $\frac{5}{8}$, $\frac{11}{16}$, $\frac{3}{4}$, $\frac{7}{8}$, 1, $1\frac{1}{4}$, $1\frac{1}{2}$, $1\frac{3}{4}$, 2, $2\frac{1}{4}$, $2\frac{1}{2}$, $2\frac{3}{4}$, 3, $3\frac{1}{4}$, $3\frac{1}{2}$, $3\frac{3}{4}$, 4, $4\frac{1}{4}$, $4\frac{1}{2}$, $4\frac{3}{4}$, 5, $5\frac{1}{4}$, $5\frac{1}{2}$, $5\frac{3}{4}$, 6, $6\frac{1}{2}$, 7, $7\frac{1}{2}$, 8, $8\frac{1}{2}$, 9, $9\frac{1}{2}$, 10, $10\frac{1}{2}$, 11, $11\frac{1}{2}$, 12, $12\frac{1}{2}$, 13, $13\frac{1}{2}$, 14, $14\frac{1}{2}$, 15, $15\frac{1}{2}$, 16, $16\frac{1}{2}$, 17, $17\frac{1}{2}$, 18, $18\frac{1}{2}$, 19, $19\frac{1}{2}$, 20

Decimal Inches

0.010, 0.012, 0.016, 0.020, 0.025, 0.032, 0.040, 0.05, 0.06, 0.08, 0.10, 0.12, 0.16, 0.20, 0.24, 0.30, 0.40, 0.50, 0.60, 0.80, 1.00, 1.20, 1.40, 1.60, 1.80, 2.0, 2.4, 2.6, 2.8, 3.0, 3.2, 3.4, 3.6, 3.8, 4.0, 4.2, 4.4, 4.6, 4.8, 5.0, 5.2, 5.4, 5.6, 5.8, 6.0, 7.0, 7.5, 8.0, 8.5, 9.0, 9.5, 10.0, 10.5, 11.0, 11.5, 12.0, 12.5, 13.0, 13.5, 14.0, 14.5, 15.0, 15.5, 16.0, 16.5, 17.0, 17.5, 18.0, 18.5, 19.0, 19.5, 20

Millimeters

0.05, 0.06, 0.08, 0.10, 0.12, 0.16, 0.20, 0.25, 0.30, 0.40, 0.50, 0.60, 0.70, 0.80, 0.90, 1.0, 1.1, 1.2, 1.4, 1.5, 1.6, 1.8, 2.0, 2.2, 2.5, 2.8, 3.0, 3.5, 4.0, 4.5, 5.0, 5.5, 6.0, 6.5, 7.0, 8.0, 9.0, 10, 11, 12, 14, 16, 18, 20, 22, 25, 28, 30, 32, 35, 40, 45, 50, 60, 80, 100, 120, 140, 160, 180, 200, 250, 300

Renard Numbers*

1st choice, R5: 1, 1.6, 2.5, 4, 6.3, 10
2d choice, R10: 1.25, 2, 3.15, 5, 8
3d choice, R20: 1.12, 1.4, 1.8, 2.24, 2.8, 3.55, 4.5, 5.6, 7.1, 9
4th choice, R40: 1.06, 1.18, 1.32, 1.5, 1.7, 1.9, 2.12, 2.36, 2.65, 3, 3.35, 3.75, 4.25, 4.75, 5.3, 6, 6.7, 7.5, 8.5, 9.5

*May be multiplied or divided by powers of 10.

Table E–18

Geometric Properties

Part 1 Properties of Sections

A = area

G = location of centroid

$I_x = \int x^2\, dA$ = second moment of area about x axis

$I_{xy} = \int xy\, dA$ = mixed moment of area about x and y axes

$J_G = \int r^2\, dA = \int (x^2 + y^2)\, dA = I_x + I_y$

= second polar moment of area about axis through G

$k_x^2 = I_x/A$ = squared radius of gyration about x axis

Rectangle

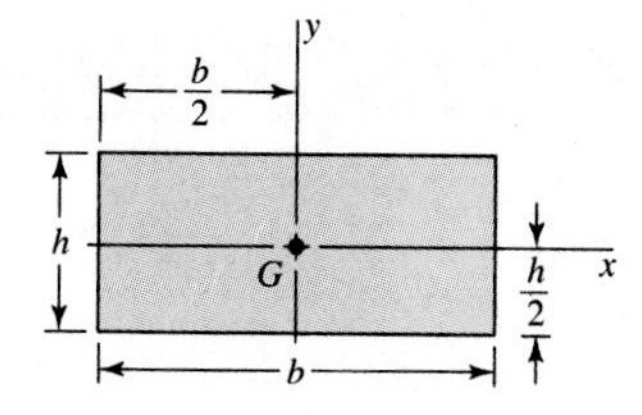

$A = bh \qquad I_x = \dfrac{bh^3}{12} \qquad I_y = \dfrac{b^3h}{12} \qquad I_{xy} = 0$

Circle

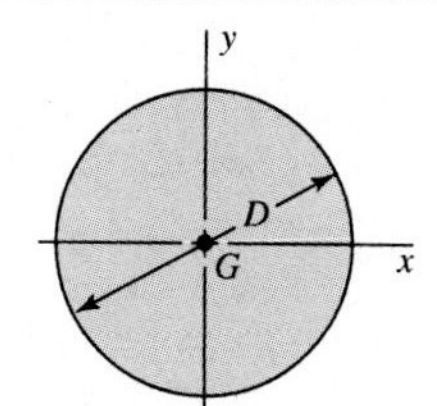

$A = \dfrac{\pi D^2}{4} \qquad I_x = I_y = \dfrac{\pi D^4}{64} \qquad I_{xy} = 0$

Hollow circle

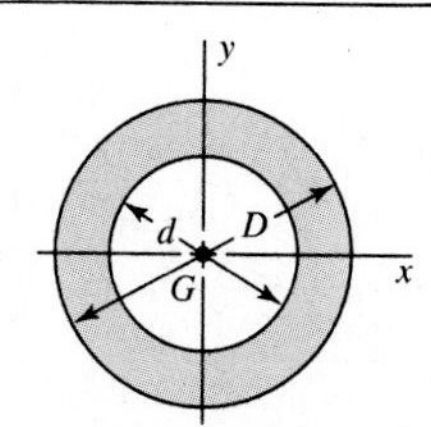

$A = \dfrac{\pi}{4}(D^2 - d^2) \qquad I_x = I_y = \dfrac{\pi}{64}(D^4 - d^4) \qquad I_{xy} = 0$

Table E–18

Geometric Properties *(Continued)*

Right triangles

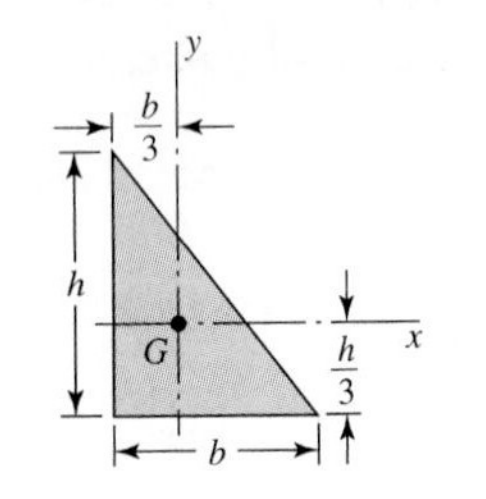

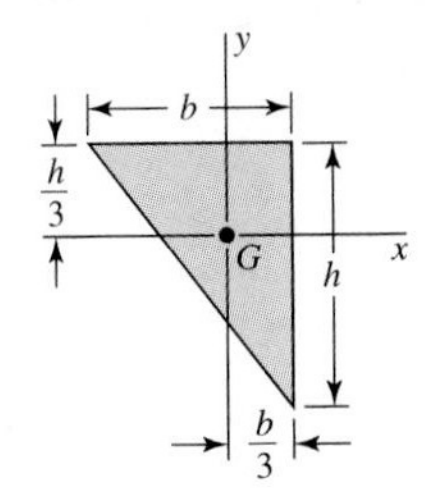

$$A = \frac{bh}{2} \qquad I_x = \frac{bh^3}{36} \qquad I_y = \frac{b^3 h}{72} \qquad I_{zy} = \frac{-b^2 h^2}{72}$$

Right triangles

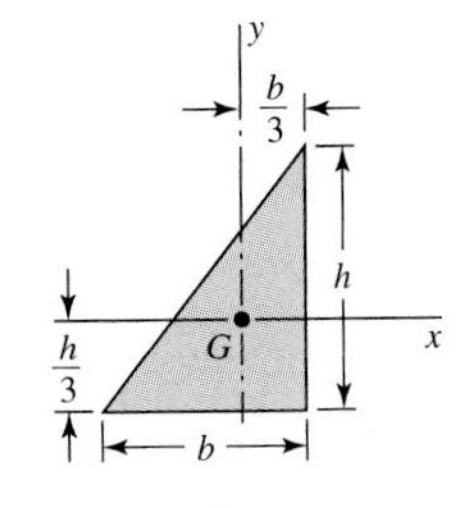

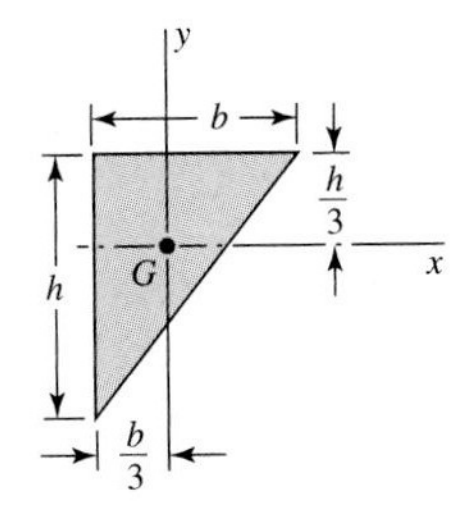

$$A = \frac{bh}{2} \qquad I_x = \frac{bh^3}{36} \qquad I_y = \frac{b^3 h}{36} \qquad I_{xy} = \frac{b^2 h^2}{72}$$

Quarter-circles

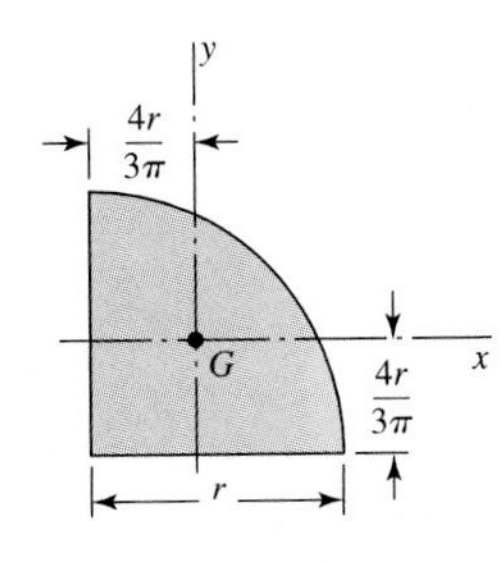

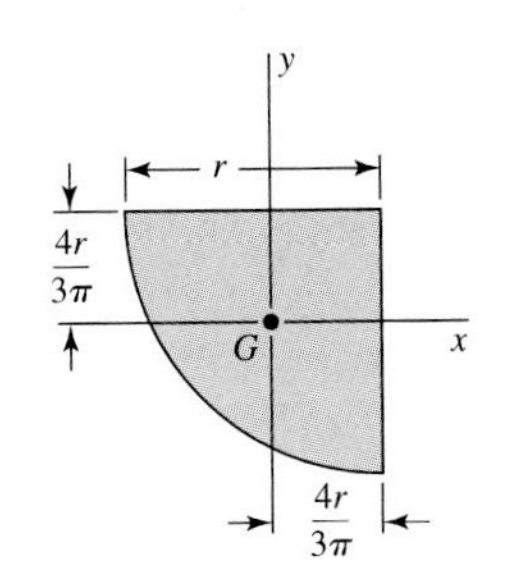

$$A = \frac{\pi r^2}{4} \qquad I_x = I_y = r^4\left(\frac{\pi}{16} - \frac{4}{9\pi}\right) \qquad I_{xy} = r^4\left(\frac{1}{8} - \frac{4}{9\pi}\right)$$

Quarter-circles

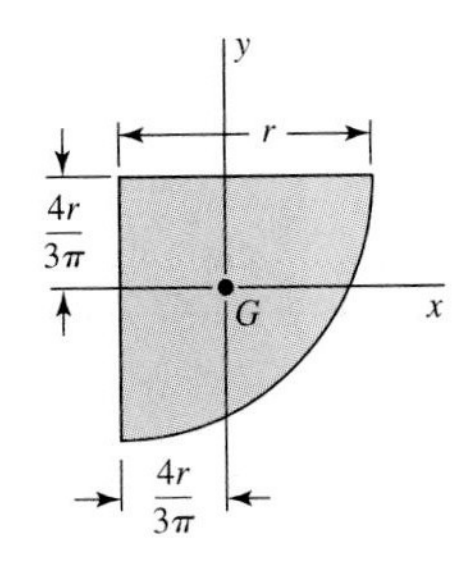

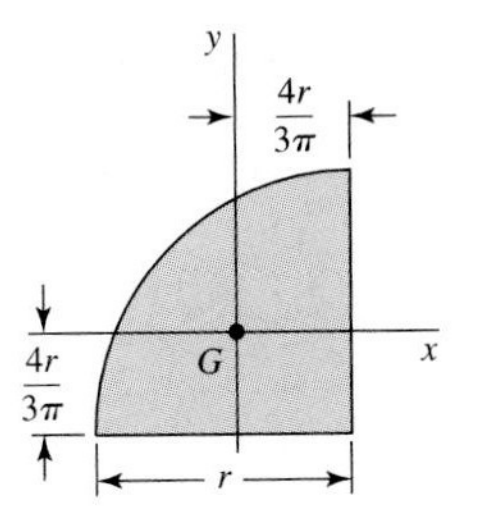

$$A = \frac{\pi r^2}{4} \qquad I_x = I_y = r^4\left(\frac{\pi}{16} - \frac{4}{9\pi}\right) \qquad I_{xy} = r^4\left(\frac{4}{9\pi} - \frac{1}{8}\right)$$

Table E–18

Geometric Properties *(Continued)*

Part 2 Properties of Solids (ρ = Density, Weight Unit Volume)

Rods

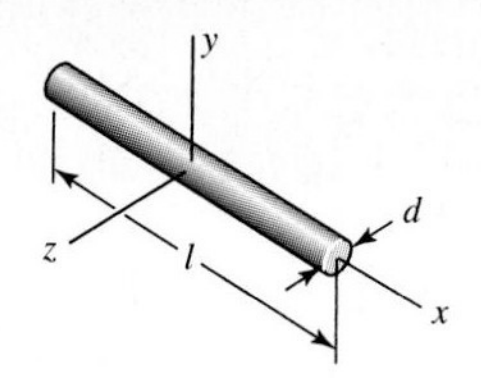

$$m = \frac{\pi d^2 \ell \rho}{4g} \qquad I_y = I_z = \frac{m\ell^2}{12}$$

Round disks

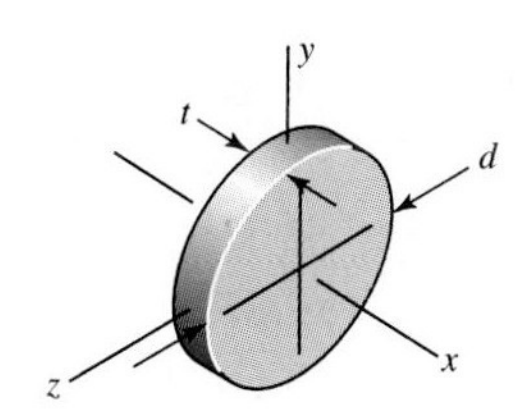

$$m = \frac{\pi d^2 t \rho}{4g} \qquad I_x = \frac{md^2}{8} \qquad I_y = I_z = \frac{md^2}{16}$$

Rectangular prisms

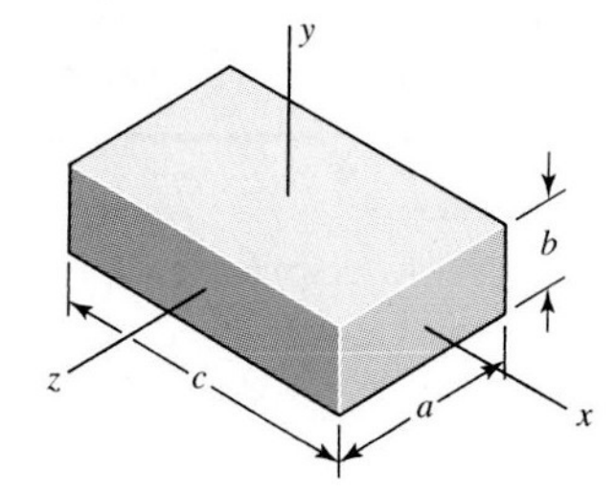

$$m = \frac{abc\rho}{g} \qquad I_x = \frac{m}{12}(a^2 + b^2) \qquad I_y = \frac{m}{12}(a^2 + c^2) \qquad I_z = \frac{m}{12}(b^2 + c^2)$$

Cylinders

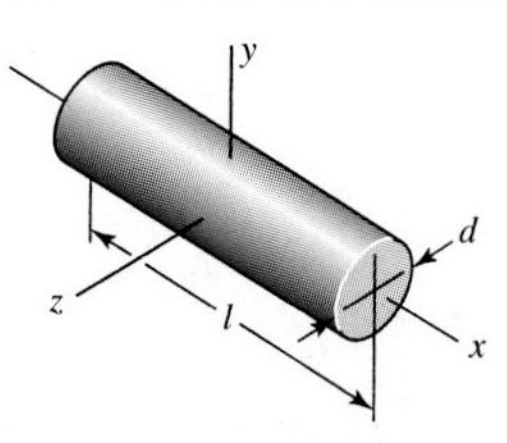

$$m = \frac{\pi d^2 \ell \rho}{4g} \qquad I_x = \frac{md^2}{8} \qquad I_y = I_z = \frac{m}{48}(3d^2 + 4\ell^2)$$

Hollow cylinders

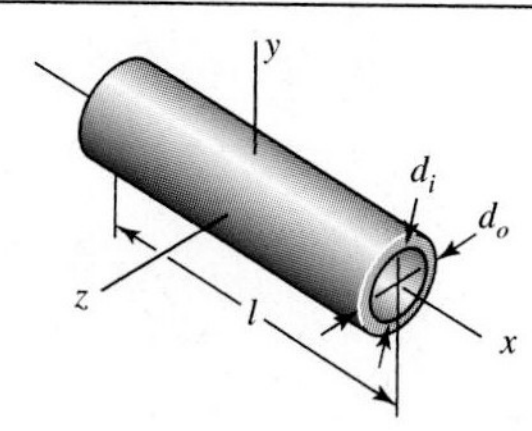

$$m = \frac{\pi (d_o^2 - d_i^2)\ell \rho}{4g} \qquad I_x = \frac{m}{8}(d_o^2 + d_i^2) \qquad I_y = I_z = \frac{m}{48}(3d_o^2 + 3d_i^2 + 4\ell^2)$$

Table E–19

American Standard Pipe

Nominal Size, in	Outside Diameter, in	Threads per inch	Wall Thickness, in		
			Standard No. 40	Extra Strong No. 80	Double Extra Strong
$\frac{1}{8}$	0.405	27	0.070	0.098	
$\frac{1}{4}$	0.540	18	0.090	0.122	
$\frac{3}{8}$	0.675	18	0.093	0.129	
$\frac{1}{2}$	0.840	14	0.111	0.151	0.307
$\frac{3}{4}$	1.050	14	0.115	0.157	0.318
1	1.315	$11\frac{1}{2}$	0.136	0.183	0.369
$1\frac{1}{4}$	1.660	$11\frac{1}{2}$	0.143	0.195	0.393
$1\frac{1}{2}$	1.900	$11\frac{1}{2}$	0.148	0.204	0.411
2	2.375	$11\frac{1}{2}$	0.158	0.223	0.447
$2\frac{1}{2}$	2.875	8	0.208	0.282	0.565
3	3.500	8	0.221	0.306	0.615
$3\frac{1}{2}$	4.000	8	0.231	0.325	
4	4.500	8	0.242	0.344	0.690
5	5.563	8	0.263	0.383	0.768
6	6.625	8	0.286	0.441	0.884
8	8.625	8	0.329	0.510	0.895

Table E–20

Deterministic ASTM Minimum Tensile and Yield Strengths for Some Hot-Rolled (HR) and Cold-Drawn (CD) Steels
[The Strengths Listed Are Estimated ASTM Minimum Values in the Size Range 18 to 32 mm ($\frac{3}{4}$ to $1\frac{1}{4}$ in). These Strengths Are Suitable for Use with the Design Factor Defined in Sec. 1–9, Provided the Materials Conform to ASTM A6 or A568 Requirements or Are Required in the Purchase Specifications. Remember that a Numbering System is Not a Specification. See Table 1–1 for Certain ASTM Steels]
Source: 1986 SAE Handbook, p. 2.15.

1 UNS No.	2 SAE and/or AISI No.	3 Proces-sing	4 Tensile Strength, MPa (kpsi)	5 Yield Strength, MPa (kpsi)	6 Elongation in 2 in, %	7 Reduction in Area, %	8 Brinell Hardness
G10060	1006	HR	300 (43)	170 (24)	30	55	86
		CD	330 (48)	280 (41)	20	45	95
G10100	1010	HR	320 (47)	180 (26)	28	50	95
		CD	370 (53)	300 (44)	20	40	105
G10150	1015	HR	340 (50)	190 (27.5)	28	50	101
		CD	390 (56)	320 (47)	18	40	111
G10180	1018	HR	400 (58)	220 (32)	25	50	116
		CD	440 (64)	370 (54)	15	40	126
G10200	1020	HR	380 (55)	210 (30)	25	50	111
		CD	470 (68)	390 (57)	15	40	131
G10300	1030	HR	470 (68)	260 (37.5)	20	42	137
		CD	520 (76)	440 (64)	12	35	149
G10350	1035	HR	500 (72)	270 (39.5)	18	40	143
		CD	550 (80)	460 (67)	12	35	163
G10400	1040	HR	520 (76)	290 (42)	18	40	149
		CD	590 (85)	490 (71)	12	35	170
G10450	1045	HR	570 (82)	310 (45)	16	40	163
		CD	630 (91)	530 (77)	12	35	179
G10500	1050	HR	620 (90)	340 (49.5)	15	35	179
		CD	690 (100)	580 (84)	10	30	197
G10600	1060	HR	680 (98)	370 (54)	12	30	201
G10800	1080	HR	770 (112)	420 (61.5)	10	25	229
G10950	1095	HR	830 (120)	460 (66)	10	25	248

Table E–21

Mean Mechanical Properties of Some Heat-Treated Steels
[These are Typical Properties for Materials Normalized and Annealed. The Properties for Quenched and Tempered (Q&T) Steels Are from a Single Heat. Because of the Many Variables, the Properties Listed are Global Averages. In All Cases, Data Were Obtained from Specimens of Diameter 0.505 in, Machined from 1-in Rounds, and of Gauge Length 2 in. Unless Noted, All Specimens Were Oil-Quenched]

Source: ASM Metals Reference Book, 2d ed., American Society for Metals, Metals Park, Ohio, 1983.

1	2	3	4	5	6	7	8
AISI No.	Treatment	Temperature °C (°F)	Tensile Strength MPa (kpsi)	Yield Strength, MPa (kpsi)	Elongation, %	Reduction in Area, %	Brinell Hardness
1030	Q&T*	205 (400)	848 (123)	648 (94)	17	47	495
	Q&T*	315 (600)	800 (116)	621 (90)	19	53	401
	Q&T*	425 (800)	731 (106)	579 (84)	23	60	302
	Q&T*	540 (1000)	669 (97)	517 (75)	28	65	255
	Q&T*	650 (1200)	586 (85)	441 (64)	32	70	207
	Normalized	925 (1700)	521 (75)	345 (50)	32	61	149
	Annealed	870 (1600)	430 (62)	317 (46)	35	64	137
1040	Q&T	205 (400)	779 (113)	593 (86)	19	48	262
	Q&T	425 (800)	758 (110)	552 (80)	21	54	241
	Q&T	650 (1200)	634 (92)	434 (63)	29	65	192
	Normalized	900 (1650)	590 (86)	374 (54)	28	55	170
	Annealed	790 (1450)	519 (75)	353 (51)	30	57	149
1050	Q&T*	205 (400)	1120 (163)	807 (117)	9	27	514
	Q&T*	425 (800)	1090 (158)	793 (115)	13	36	444
	Q&T*	650 (1200)	717 (104)	538 (78)	28	65	235
	Normalized	900 (1650)	748 (108)	427 (62)	20	39	217
	Annealed	790 (1450)	636 (92)	365 (53)	24	40	187
1060	Q&T	425 (800)	1080 (156)	765 (111)	14	41	311
	Q&T	540 (1000)	965 (140)	669 (97)	17	45	277
	Q&T	650 (1200)	800 (116)	524 (76)	23	54	229
	Normalized	900 (1650)	776 (112)	421 (61)	18	37	229
	Annealed	790 (1450)	626 (91)	372 (54)	22	38	179
1095	Q&T	315 (600)	1260 (183)	813 (118)	10	30	375
	Q&T	425 (800)	1210 (176)	772 (112)	12	32	363
	Q&T	540 (1000)	1090 (158)	676 (98)	15	37	321
	Q&T	650 (1200)	896 (130)	552 (80)	21	47	269
	Normalized	900 (1650)	1010 (147)	500 (72)	9	13	293
	Annealed	790 (1450)	658 (95)	380 (55)	13	21	192
1141	Q&T	315 (600)	1460 (212)	1280 (186)	9	32	415
	Q&T	540 (1000)	896 (130)	765 (111)	18	57	262
4130	Q&T*	205 (400)	1630 (236)	1460 (212)	10	41	467
	Q&T*	315 (600)	1500 (217)	1380 (200)	11	43	435
	Q&T*	425 (800)	1280 (186)	1190 (173)	13	49	380
	Q&T*	540 (1000)	1030 (150)	910 (132)	17	57	315
	Q&T*	650 (1200)	814 (118)	703 (102)	22	64	245
	Normalized	870 (1600)	670 (97)	436 (63)	25	59	197
	Annealed	865 (1585)	560 (81)	361 (52)	28	56	156
4140	Q&T	205 (400)	1770 (257)	1640 (238)	8	38	510
	Q&T	315 (600)	1550 (225)	1430 (208)	9	43	445

Table E–21

(Continued)

[These are Typical Properties for Materials Normalized and Annealed. The Properties for Quenched and Tempered (Q&T) Steels Are from a Single Heat. Because of the Many Variables, the Properties Listed Are Global Averages. In All Cases, Data Were Obtained from Specimens of Diameter 0.505 in, Machined from 1-in Rounds, and of Gauge Length 2 in. Unless Noted, All Specimens Were Oil-Quenched]

Source: ASM Metals Reference Book, 2d ed., American Society for Metals, Metals Park, Ohio, 1983.

1	2	3	4	5	6	7	8
AISI No.	Treatment	Temperature °C (°F)	Tensile Strength MPa (kpsi)	Yield Strength, MPa (kpsi)	Elongation, %	Reduction in Area, %	Brinell Hardness
4140	Q&T	425 (800)	1250 (181)	1140 (165)	13	49	370
	Q&T	540 (1000)	951 (138)	834 (121)	18	58	285
	Q&T	650 (1200)	758 (110)	655 (95)	22	63	230
	Normalized	870 (1600)	1020 (148)	655 (95)	18	47	302
	Annealed	815 (1500)	655 (95)	417 (61)	26	57	197
4340	Q&T	315 (600)	1720 (250)	1590 (230)	10	40	486
	Q&T	425 (800)	1470 (213)	1360 (198)	10	44	430
	Q&T	540 (1000)	1170 (170)	1080 (156)	13	51	360
	Q&T	650 (1200)	965 (140)	855 (124)	19	60	280

*Water-quenched

Table E–22

Results of Tensile Tests of Some Metals*

Source: J. Datsko, "Solid Materials," chap. 7 in Joseph E. Shigley and Charles R. Mischke (eds.-in-chief), *Standard Handbook of Machine Design*, 2nd ed., McGraw-Hill, New York, 1996, pp. 7.47–7.50.

			Strength (Tensile)					
Number	**Material**	**Condition**	**Yield S_y, MPa (kpsi)**	**Ultimate S_u, MPa (kpsi)**	**Fracture, σ_f, MPa (kpsi)**	**Coefficient σ_0, MPa (kpsi)**	**Strain Strength, Exponent m**	**Fracture Strain ε_f**
1018	Steel	Annealed	220 (32.0)	341 (49.5)	628 (91.1)†	620 (90.0)	0.25	1.05
1144	Steel	Annealed	358 (52.0)	646 (93.7)	898 (130)†	992 (144)	0.14	0.49
1212	Steel	HR	193 (28.0)	424 (61.5)	729 (106)†	758 (110)	0.24	0.85
1045	Steel	Q&T 600°F	1520 (220)	1580 (230)	2380 (345)	1880 (273)†	0.041	0.81
4142	Steel	Q&T 600°F	1720 (250)	1930 (210)	2340 (340)	1760 (255)†	0.048	0.43
303	Stainless steel	Annealed	241 (35.0)	601 (87.3)	1520 (221)†	1410 (205)	0.51	1.16
304	Stainless steel	Annealed	276 (40.0)	568 (82.4)	1600 (233)†	1270 (185)	0.45	1.67
2011	Aluminum alloy	T6	169 (24.5)	324 (47.0)	325 (47.2)†	620 (90)	0.28	0.10
2024	Aluminum alloy	T4	296 (43.0)	446 (64.8)	533 (77.3)†	689 (100)	0.15	0.18
7075	Aluminum alloy	T6	542 (78.6)	593 (86.0)	706 (102)†	882 (128)	0.13	0.18

*Values from one or two heats and believed to be attainable using proper purchase specifications. The fracture strain may vary as much as 100 percent.

†Derived value.

Table E–23

Mean Monotonic and Cyclic Stress-Strain Properties of Selected Steels

Source: ASM Metals Reference Book, 2nd ed., American Society for Metals, Metals Park, Ohio, 1983, p. 217.

Grade (a)	Orientation (e)	Description (f)	Hardness HB	Tensile Strength S_{ut} MPa	Tensile Strength S_{ut} ksi	Reduction in Area %	True Strain at Fracture ε_f	Modulus of Elasticity E GPa	Modulus of Elasticity E 10^4 psi	Fatigue Strength Coefficient σ'_f MPa	Fatigue Strength Coefficient σ'_f ksi	Fatigue Strength Exponent b	Fatigue Ductility Coefficient ε'_F	Fatigue Ductility Exponent c
A538A (b)	L	STA	405	1515	220	67	1.10	185	27	1655	240	−0.065	0.30	−0.62
A538B (b)	L	STA	460	1860	270	56	0.82	185	27	2135	310	−0.071	0.80	−0.71
A538C (b)	L	STA	480	2000	290	55	0.81	180	26	2240	325	−0.07	0.60	−0.75
AM-350 (c)	L	HR, A		1315	191	52	0.74	195	28	2800	406	−0.14	0.33	−0.84
AM-350 (c)	L	CD	496	1905	276	20	0.23	180	26	2690	390	−0.102	0.10	−0.42
Gainex (c)	LT	HR sheet		530	77	58	0.86	200	29.2	805	117	−0.07	0.86	−0.65
Gainex (c)	L	HR sheet		510	74	64	1.02	200	29.2	805	117	−0.071	0.86	−0.68
H-11	L	Ausformed	660	2585	375	33	0.40	205	30	3170	460	−0.077	0.08	−0.74
RQC-100 (c)	LT	HR plate	290	940	136	43	0.56	205	30	1240	180	−0.07	0.66	−0.69
RQC-100 (c)	L	HR plate	290	930	135	67	1.02	205	30	1240	180	−0.07	0.66	−0.69
10B62	L	Q&T	430	1640	238	38	0.89	195	28	1780	258	−0.067	0.32	−0.56
1005-1009	LT	HR sheet	90	360	52	73	1.3	205	30	580	84	−0.09	0.15	−0.43
1005-1009	LT	CD sheet	125	470	68	66	1.09	205	30	515	75	−0.059	0.30	−0.51
1005-1009	L	CD sheet	125	415	60	64	1.02	200	29	540	78	−0.073	0.11	−0.41
1005-1009	L	HR sheet	90	345	50	80	1.6	200	29	640	93	−0.109	0.10	−0.39
1015	L	Normalized	80	415	60	68	1.14	205	30	825	120	−0.11	0.95	−0.64
1020	L	HR plate	108	440	64	62	0.96	205	29.5	895	130	−0.12	0.41	−0.51
1040	L	As forged	225	620	90	60	0.93	200	29	1540	223	−0.14	0.61	−0.57
1045	L	Q&T	225	725	105	65	1.04	200	29	1225	178	−0.095	1.00	−0.66
1045	L	Q&T	410	1450	210	51	0.72	200	29	1860	270	−0.073	0.60	−0.70
1045	L	Q&T	390	1345	195	59	0.89	205	30	1585	230	−0.074	0.45	−0.68
1045	L	Q& T	450	1585	230	55	0.81	205	30	1795	260	−0.07	0.35	−0.69
1045	L	Q&T	500	1825	265	51	0.71	205	30	2275	330	−0.08	0.25	−0.68
1045	L	Q&T	595	2240	325	41	0.52	205	30	2725	395	−0.081	0.07	−0.60
1144	L	CDSR	265	930	135	33	0.51	195	28.5	1000	145	−0.08	0.32	−0.58
1144	L	DAT	305	1035	150	25	0.29	200	28.8	1585	230	−0.09	0.27	−0.53
1541F	L	Q&T forging	290	950	138	49	0.68	205	29.9	1275	185	−0.076	0.68	−0.65
1541F	L	Q&T forging	260	890	129	60	0.93	205	29.9	1275	185	−0.071	0.93	−0.65
4130	L	Q&T	258	895	130	67	1.12	220	32	1275	185	−0.083	0.92	−0.63
4130	L	Q&T	365	1425	207	55	0.79	200	29	1695	246	−0.081	0.89	−0.69

Table E–23

Mean Monotonic and Cyclic Stress-Strain Properties of Selected Steels *Source: ASM Metals Reference Book*, 2nd ed., American Society for Metals, Metals Park, Ohio, 1983, p. 217. *(Continued)*

Grade (a)	Orientation (e)	Description (f)	Hardness HB	Tensile Strength S_{ut} MPa	Tensile Strength S_{ut} ksi	Reduction in Area %	True Strain at Fracture ε_f	Modulus of Elasticity E GPa	Modulus of Elasticity E 10^4 psi	Fatigue Strength Coefficient σ'_f MPa	Fatigue Strength Coefficient σ'_f ksi	Fatigue Strength Exponent b	Fatigue Ductility Coefficient ε'_F	Fatigue Ductility Exponent c
4140	L	Q&T, DAT	310	1075	156	60	0.69	200	29.2	1825	265	−0.08	1.2	−0.59
4142	L	DAT	310	1060	154	29	0.35	200	29	1450	210	−0.10	0.22	−0.51
4142	L	DAT	335	1250	181	28	0.34	200	28.9	1250	181	−0.08	0.06	−0.62
4142	L	Q&T	380	1415	205	48	0.66	205	30	1825	265	−0.08	0.45	−0.75
4142	L	Q&T and deformed	400	1550	225	47	0.63	200	29	1895	275	−0.09	0.50	−0.75
4142	L	Q&T	450	1760	255	42	0.54	205	30	2000	290	−0.08	0.40	−0.73
4142	L	Q&T and deformed	475	2035	295	20	0.22	200	29	2070	300	−0.082	0.20	−0.77
4142	L	Q&T and deformed	450	1930	280	37	0.46	200	29	2105	305	−0.09	0.60	−0.76
4142	L	Q&T	475	1930	280	35	0.43	205	30	2170	315	−0.081	0.09	−0.61
4142	L	Q&T	560	2240	325	27	0.31	205	30	2655	385	−0.089	0.07	−0.76
4340	L	HR, A	243	825	120	43	0.57	195	28	1200	174	−0.095	0.45	−0.54
4340	L	Q&T	409	1470	213	38	0.48	200	29	2000	290	−0.091	0.48	−0.60
4340	L	Q&T	350	1240	180	57	0.84	195	28	1655	240	−0.076	0.73	−0.62
5160	L	Q&T	430	1670	242	42	0.87	195	28	1930	280	−0.071	0.40	−0.57
52100	L	SH, Q& T	518	2015	292	11	0.12	205	30	2585	375	−0.09	0.18	−0.56
9262	L	A	260	925	134	14	0.16	205	30	1040	151	−0.071	0.16	−0.47
9262	L	Q&T	280	1000	145	33	0.41	195	28	1220	177	−0.073	0.41	−0.60
9262	L	Q&T	410	565	227	32	0.38	200	29	1855	269	−0.057	0.38	−0.65
950C (d)	LT	HR plate	159	565	82	64	1.03	205	29.6	1170	170	−0.12	0.95	−0.61
950C (d)	L	HR bar	150	565	82	69	1.19	205	30	970	141	−0.11	0.85	−0.59
950X (d)	L	Plate channel	150	440	64	65	1.06	205	30	625	91	−0.075	0.35	−0.54
950X (d)	L	HR plate	156	530	77	72	1.24	205	29.5	1005	146	−0.10	0.85	−0.61
950X (d)	L	Plate channel	225	695	101	68	1.15	195	28.2	1055	153	−0.08	0.21	−0.53

Notes: (a) AISI/SAE grade, unless otherwise indicated. (b) ASTM designation. (c) Proprietary designation. (d) SAE HSLA grade. (e) Orientation of axis of specimen, relative to rolling direction; L is longitudinal (parallel to rolling direction); LT is long transverse (perpendicular to rolling direction). (f) STA, solution treated and aged; HR, hot rolled; CD, cold drawn; Q&T, quenched and tempered; CDSR, cold drawn strain relieved; DAT, drawn at temperature; A, annealed.

Table E–24

Mechanical Properties of Two Common Materials (Non-Steels)
(a) Typical Properties of Gray Cast Iron
[The American Society for Testing Materials (ASTM) Numbering System for Gray Cast Iron Is Such that the Numbers Correspond to the *Minimum Tensile Strength* in kpsi. Thus an ASTM No. 20 Cast Iron Has a Minimum Tensile Strength of 20 kpsi. Note Particularly that the Tabulations Are *Typical* of Several Heats]

ASTM Number	Tensile Strength S_{ut}, kpsi	Compressive Strength S_{uc}, kpsi	Shear Modulus of Rupture S_{su}, kpsi	Modulus of Elasticity, Mpsi		Endurance Limit* S_e, kpsi	Brinell Hardness H_B	Fatigue Stress-Concentration Factor K_f
				Tension†	Torsion			
20	22	83	26	9.6–14	3.9–5.6	10	156	1.00
25	26	97	32	11.5–14.8	4.6–6.0	11.5	174	1.05
30	31	109	40	13–16.4	5.2–6.6	14	201	1.10
35	36.5	124	48.5	14.5–17.2	5.8–6.9	16	212	1.15
40	42.5	140	57	16–20	6.4–7.8	18.5	235	1.25
50	52.5	164	73	18.8–22.8	7.2–8.0	21.5	262	1.35
60	62.5	187.5	88.5	20.4–23.5	7.8–8.5	24.5	302	1.50

*Polished or machined specimens.

†The modulus of elasticity of cast iron in compression corresponds closely to the upper value in the range given for tension and is a more constant value than that for tension.

Table E–24

Mechanical Properties of Two Common Materials (Non-Steels) *(Continued)*
(b) Mechanical Properties of Some Aluminum Alloys
[These Are *Typical* Properties for Sizes of About $\frac{1}{2}$ in; Similar Properties Can Be Obtained Using Proper Purchase Specifications. The Values Given for Fatigue Strength Correspond to $50(10^7)$ Cycles of Completely Reversed Stress. Alluminum Alloys Do Not Have an Endurance Limit. Yield Strengths Were Obtained by the 0.2 Percent Offset Method]

Aluminum Association Number	Temper	*Strength* Yield, S_y, MPa (kpsi)	Tensile, S_u, MPa (kpsi)	Fatigue, S_f, MPa (kpsi)	Elongation in 2 in, %	Brinell Hardness H_B
Wrought:						
2017	O	70 (10)	179 (26)	90 (13)	22	45
2024	O	76 (11)	186 (27)	90 (13)	22	47
	T3	345 (50)	482 (70)	138 (20)	16	120
3003	H12	117 (17)	131 (19)	55 (8)	20	35
	H16	165 (24)	179 (26)	65 (9.5)	14	47
3004	H34	186 (27)	234 (34)	103 (15)	12	63
	H38	234 (34)	276 (40)	110 (16)	6	77
5052	H32	186 (27)	234 (34)	117 (17)	18	62
	H36	234 (34)	269 (39)	124 (18)	10	74
Cast:						
319.0*	T6	165 (24)	248 (36)	69 (10)	2.0	80
333.0†	T5	172 (25)	234 (34)	83 (12)	1.0	100
	T6	207 (30)	289 (42)	103 (15)	1.5	105
335.0*	T6	172 (25)	241 (35)	62 (9)	3.0	80
	T7	248 (36)	262 (38)	62 (9)	0.5	85

*Sand casting.
†Permanent-mold casting.

Table E–25

Stochastic Yield and Ultimate Strengths for Selected Materials *Source:* Data compiled from "Some Property Data and Corresponding Weibull Parameters for Stochastic Mechanical Design," *Trans. ASME Journal of Mechanical Design*, vol. 114 (March 1992), pp. 29–34.

Material		μ_{Sut}	σ_{Sut}	x_0	θ	b	μ_{Sy}	σ_{Sy}	x_0	θ	b	C_{Sut}	C_{Sy}
1018	CD	87.6	5.74	30.8	90.1	12	78.4	5.90	56	80.6	4.29	0.0655	0.0753
1035	HR	86.2	3.92	72.6	87.5	3.86	49.6	3.81	39.5	50.8	2.88	0.0455	0.0768
1045	CD	117.7	7.13	90.2	120.5	4.38	95.5	6.59	82.1	97.2	2.14	0.0606	0.0690
1117	CD	83.1	5.25	73.0	84.4	2.01	81.4	4.71	72.4	82.6	2.00	0.0632	0.0579
1137	CD	106.5	6.15	96.2	107.7	1.72	98.1	4.24	92.2	98.7	1.41	0.0577	0.0432
12L14	CD	79.6	6.92	70.3	80.4	1.36	78.1	8.27	64.3	78.8	1.72	0.0869	0.1059
1038	HT bolts	133.4	3.38	122.3	134.6	3.64						0.0253	
ASTM40		44.5	4.34	27.7	46.2	4.38						0.0975	
35018	Malleable	53.3	1.59	48.7	53.8	3.18	38.5	1.42	34.7	39.0	2.93	0.0298	0.0369
32510	Malleable	53.4	2.68	44.7	54.3	3.61	34.9	1.47	30.1	35.5	3.67	0.0502	0.0421
Malleable	Pearlitic	93.9	3.83	80.1	95.3	4.04	60.2	2.78	50.2	61.2	4.02	0.0408	0.0462
604515	Nodular	64.8	3.77	53.7	66.1	3.23	49.0	4.20	33.8	50.5	4.06	0.0582	0.0857
100-70-04	Nodular	122.2	7.65	47.6	125.6	11.84	79.3	4.51	64.1	81.0	3.77	0.0626	0.0569
201SS	CD	195.9	7.76	180.7	197.9	2.06						0.0396	
301SS	CD	191.2	5.82	151.9	193.6	8.00	166.8	9.37	139.7	170.0	3.17	0.0304	0.0562
	A	105.0	5.68	92.3	106.6	2.38	46.8	4.70	26.3	48.7	4.99	0.0541	0.1004
304SS	A	85.0	4.14	66.6	86.6	5.11	37.9	3.76	30.2	38.9	2.17	0.0487	0.0992
310SS	A	84.8	4.23	71.6	86.3	3.45						0.0499	
403SS		105.3	3.09	95.7	106.4	3.44	78.5	3.91	64.8	79.9	3.93	0.0293	0.0498
17-7PSS		198.8	9.51	163.3	202.3	4.21	189.4	11.49	144.0	193.8	4.48	0.0478	0.0607
AM350SS	A	149.1	8.29	101.8	152.4	6.68	63.0	5.05	38.0	65.0	5.73	0.0556	0.0802
Ti-6AL-4V		175.4	7.91	141.8	178.5	4.85	163.7	9.03	101.5	167.4	8.18	0.0451	0.0552
2024	0	28.1	1.73	24.2	28.7	2.43						0.0616	
2024	T4	64.9	1.64	60.2	65.5	3.16	40.8	1.83	38.4	41.0	1.32	0.0253	0.0449
	T6	67.5	1.50	55.9	68.1	9.26	53.4	1.17	51.2	53.6	1.91	0.0222	0.0219
7075	T6 .025"	75.5	2.10	68.8	76.2	3.53	63.7	1.98	58.9	64.3	2.63	0.0278	0.0311

Table E–26

Stochastic Parameters for Finite Life Fatigue Tests in Selected Metals *Source:* E. B. Haugen, *Probabilistic Mechanical Design*, Wiley, New York, 1980, Appendix 10–B.

1	2	3	4	5	6	7	8	9
		TS	*YS*	Distri-	Stress Cycles to Failure			
Number	Condition	MPa (kpsi)	MPa (kpsi)	bution	10^4	10^5	10^6	10^7
1046	WQ&T, 1210°F	723 (105)	565 (82)	W x_0	544 (79)	462 (67)	391 (56.7)	
				θ	594 (86.2)	503 (73.0)	425 (61.7)	
				b	2.60	2.75	2.85	
2340	OQ&T 1200°F	799 (116)	661 (96)	W x_0	579 (84)	510 (74)	420 (61)	
				θ	699 (101.5)	588 (85.4)	496 (72.0)	
				b	4.3	3.4	4.1	
3140	OQ&T, 1300°F	744 (108)	599 (87)	W x_0	510 (74)	455 (66)	393 (57)	
				θ	604 (87.7)	528 (76.7)	463 (67.2)	
				b	5.2	5.0	5.5	
2024	T-4	489 (71)	365 (53)	N σ	26.3 (3.82)	21.4 (3.11)	17.4 (2.53)	14.0 (2.03)
Aluminum				μ	143 (20.7)	116 (16.9)	95 (13.8)	77 (11.2)
Ti-6A1-4V	HT-46	1040 (151)	992 (144)	N σ	39.6 (5.75)	38.1 (5.53)	36.6 (5.31)	35.1 (5.10)
				μ	712 (108)	684 (99.3)	657 (95.4)	493 (71.6)

Statistical parameters from a large number of fatigue tests are listed. Weibull distribution is denoted *W* and the parameters are x_0, "guaranteed" fatigue strength; θ, characteristic fatigue strength; and *b*, shape factor. Normal distribution is denoted *N* and the parameters are μ, mean fatigue strength; and σ, standard deviation of the fatigue strength. The life is in stress-cycles-to-failure. TS = tensile strength, YS = yield strength. All testing by rotating-beam specimen.

Table E–27

Finite Life Fatigue Strengths of Selected Plain Carbon Steels *Source:* Compiled from Table 4 in H. J. Grover, S. A. Gordon, and L. R. Jackson, *Fatigue of Metals and Structures*, Bureau of Naval Weapons Document NAVWEPS 00-25-534, 1960.

						Stress Cycles to Failure							
Material	Condition	BHN*	Tensile Strength	Yield Strength	RA*	10^4	$4(10^4)$	10^5	$4(10^5)$	10^6	$4(10^6)$	10^7	10^8
1020	Furnace cooled		58	30	0.63			37	34	30	28	25	
1030	Air-cooled	135	80	45	0.62		51	47	42	38	38	38	
1035	Normal	132	72	35	0.54			44	40	37	34	33	33
	WQT	209	103	87	0.65		80	72	65	60	57	57	57
1040	Forged	195	92	53	0.23				40	47	33	33	
1045	HR, N		107	63	0.49	80	70	56	47	47	47	47	
1050	N, AC	164	92	47	0.40	50	48	46	40	38	34	34	
	WQT 1200	196	97	70	0.58		60	57	52	50	50	50	50
.56 MN	N	193	98	47	0.42	61	55	51	47	43	41	41	41
	WQT 1200	277	111	84	0.57	94	81	73	62	57	55	55	55
1060	As Rec.	67 Rb	134	65	0.20	65	60	55	50	48	48	48	
1095		162	84	33	0.37	50	43	40	34	31	30	30	30
	OQT 1200	227	115	65	0.40	77	68	64	57	56	56	56	56
10120		224	117	59	0.12		60	56	51	50	50	50	
	OQT 860	369	180	130	0.15		102	95	91	91	91	91	

BHN = Brinell Hardness Number; RA = Fractional Reduction in Area, strengths in kpsi.

Table E–28

Decimal Equivalents of Wire and Sheet-Metal Gauges* (All Sizes Are Given in Inches)

Name of Gauge:	American or Brown & Sharpe	Birmingham or Stubs Iron Wire	United States Standard†	Manufacturers Standard	Steel Wire or Washburn & Moen	Music Wire	Stubs Steel Wire	Twist Drill
Principal Use:	**Nonferrous Sheet, Wire, and Rod**	**Tubing, Ferrous Strip, Flat Wire, and Spring Steel**	**Ferrous Sheet and Plate, 480 lb/ft³**	**Ferrous Sheet**	**Ferrous Wire Except Music Wire**	**Music Wire**	**Steel Drill Rod**	**Twist Drills and Drill Steel**
7/0			0.500		0.490			
6/0	0.580 0		0.468 75		0.461 5	0.004		
5/0	0.516 5		0.437 5		0.430 5	0.005		
4/0	0.460 0	0.454	0.406 25		0.393 8	0.006		
3/0	0.409 6	0.425	0.375		0.362 5	0.007		
2/0	0.364 8	0.380	0.343 75		0.331 0	0.008		
0	0.324 9	0.340	0.312 5		0.306 5	0.009		
1	0.289 3	0.300	0.281 25		0.283 0	0.010	0.227	0.228 0
2	0.257 6	0.284	0.265 625		0.262 5	0.011	0.219	0.221 0
3	0.229 4	0.259	0.25	0.239 1	0.243 7	0.012	0.212	0.213 0
4	0.204 3	0.238	0.234 375	0.224 2	0.225 3	0.013	0.207	0.209 0
5	0.181 9	0.220	0.218 75	0.209 2	0.207 0	0.014	0.204	0.205 5
6	0.162 0	0.203	0.203 125	0.194 3	0.192 0	0.016	0.201	0.204 0
7	0.144 3	0.180	0.187 5	0.179 3	0.177 0	0.018	0.199	0.201 0
8	0.128 5	0.165	0.171 875	0.164 4	0.162 0	0.020	0.197	0.199 0
9	0.114 4	0.148	0.156 25	0.149 5	0.148 3	0.022	0.194	0.196 0
10	0.101 9	0.134	0.140 625	0.134 5	0.135 0	0.024	0.191	0.193 5
11	0.090 74	0.120	0.125	0.119 6	0.120 5	0.026	0.188	0.191 0
12	0.080 81	0.109	0.109 357	0.104 6	0.105 5	0.029	0.185	0.189 0
13	0.071 96	0.095	0.093 75	0.089 7	0.091 5	0.031	0.182	0.185 0
14	0.064 08	0.083	0.078 125	0.074 7	0.080 0	0.033	0.180	0.182 0
15	0.057 07	0.072	0.070 312 5	0.067 3	0.072 0	0.035	0.178	0.180 0

Table E–28

Decimal Equivalents of Wire and Sheet-Metal Gauges* (All Sizes Are Given in Inches) *(Continued)*

Name of Gauge:	American or Brown & Sharpe	Birmingham or Stubs Iron Wire	United States Standard†	Manufacturers Standard	Steel Wire or Washburn & Moen	Music Wire	Stubs Steel Wire	Twist Drill
Principal Use:	Nonferrous Sheet, Wire and Rod	Tubing Ferrous Strip, Flat Wire, and Spring Steel	Ferrous Sheet and Plate, 480 lb/ft^3	Ferrous Sheet	Ferrous Wire Except Music Wire	Music Wire	Steel Drill Rod	Twist Drills and Drill Steel
16	0.050 82	0.065	0.062 5	0.059 8	0.062 5	0.037	0.175	0.177 0
17	0.045 26	0.058	0.056 25	0.053 8	0.054 0	0.039	0.172	0.173 0
18	0.040 30	0.049	0.05	0.047 8	0.047 5	0.041	0.168	0.169 5
19	0.035 89	0.042	0.043 75	0.041 8	0.041 0	0.043	0.164	0.166 0
20	0.031 96	0.035	0.037 5	0.035 9	0.034 8	0.045	0.161	0.161 0
21	0.028 46	0.032	0.034 375	0.032 9	0.031 7	0.047	0.157	0.159 0
22	0.025 35	0.028	0.031 25	0.029 9	0.028 6	0.049	0.155	0.157 0
23	0.022 57	0.025	0.028 125	0.026 9	0.025 8	0.051	0.153	0.154 0
24	0.020 10	0.022	0.025	0.023 9	0.023 0	0.055	0.151	0.152 0
25	0.017 90	0.020	0.021 875	0.020 9	0.020 4	0.059	0.148	0.149 5
26	0.015 94	0.018	0.018 75	0.017 9	0.018 1	0.063	0.146	0.147 0
27	0.014 20	0.016	0.017 187 5	0.016 4	0.017 3	0.067	0.143	0.144 0
28	0.012 64	0.014	0.015 625	0.014 9	0.016 2	0.071	0.139	0.140 5
29	0.011 26	0.013	0.014 062 5	0.013 5	0.015 0	0.075	0.134	0.136 0
30	0.010 03	0.012	0.012 5	0.012 0	0.014 0	0.080	0.127	0.128 5
31	0.008 928	0.010	0.010 937 5	0.010 5	0.013 2	0.085	0.120	0.120 0
32	0.007 950	0.009	0.010 156 25	0.009 7	0.012 8	0.090	0.115	0.116 0
33	0.007 080	0.008	0.009 375	0.009 0	0.011 8	0.095	0.112	0.113 0
34	0.006 305	0.007	0.008 593 75	0.008 2	0.010 4		0.110	0.111 0
35	0.005 615	0.005	0.007 812 5	0.007 5	0.009 5		0.108	0.110 0
36	0.005 000	0.004	0.007 031 25	0.006 7	0.009 0		0.106	0.106 5
37	0.004 453		0.006 640 625	0.006 4	0.008 5		0.103	0.104 0
38	0.003 965		0.006 25	0.006 0	0.008 0		0.101	0.101 5
39	0.003 531				0.007 5		0.099	0.099 5
40	0.003 145				0.007 0		0.097	0.098 0

*Reproduced by courtesy of the Reynolds Metal Company. Specify sheet, wire, and plate by stating the gauge number, the gauge name, and the decimal equivalent in parentheses.

†Reflects present average and weights of sheet steel.

Table E–29

Dimensions of Square and Hexagonal Bolts

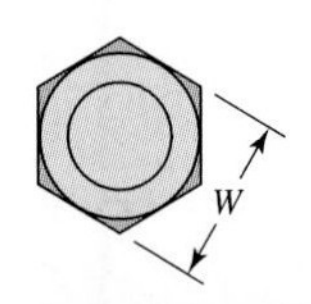

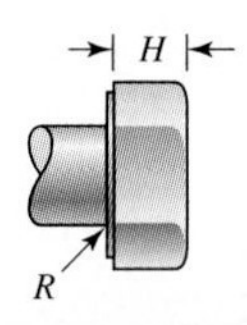

Nominal Size, in	Head Type: Square W	Square H	Regular Hexagonal W	Regular Hexagonal H	Regular Hexagonal R_{min}	Heavy Hexagonal W	Heavy Hexagonal H	Heavy Hexagonal R_{min}	Structural Hexagonal W	Structural Hexagonal H	Structural Hexagonal R_{min}
$\frac{1}{4}$	$\frac{3}{8}$	$\frac{11}{64}$	$\frac{7}{16}$	$\frac{11}{64}$	0.01						
$\frac{5}{16}$	$\frac{1}{2}$	$\frac{13}{64}$	$\frac{1}{2}$	$\frac{7}{32}$	0.01						
$\frac{3}{8}$	$\frac{9}{16}$	$\frac{1}{4}$	$\frac{9}{16}$	$\frac{1}{4}$	0.01						
$\frac{7}{16}$	$\frac{5}{8}$	$\frac{19}{64}$	$\frac{5}{8}$	$\frac{19}{64}$	0.01						
$\frac{1}{2}$	$\frac{3}{4}$	$\frac{21}{64}$	$\frac{3}{4}$	$\frac{11}{32}$	0.01	$\frac{7}{8}$	$\frac{11}{32}$	0.01	$\frac{7}{8}$	$\frac{5}{16}$	0.009
$\frac{5}{8}$	$\frac{15}{16}$	$\frac{27}{64}$	$\frac{15}{16}$	$\frac{27}{64}$	0.02	$1\frac{1}{16}$	$\frac{27}{64}$	0.02	$1\frac{1}{16}$	$\frac{25}{64}$	0.021
$\frac{3}{4}$	$1\frac{1}{8}$	$\frac{1}{2}$	$1\frac{1}{8}$	$\frac{1}{2}$	0.02	$1\frac{1}{4}$	$\frac{1}{2}$	0.02	$1\frac{1}{4}$	$\frac{15}{32}$	0.021
1	$1\frac{1}{2}$	$\frac{21}{32}$	$1\frac{1}{2}$	$\frac{43}{64}$	0.03	$1\frac{5}{8}$	$\frac{43}{64}$	0.03	$1\frac{5}{8}$	$\frac{39}{64}$	0.062
$1\frac{1}{8}$	$1\frac{11}{16}$	$\frac{3}{4}$	$1\frac{11}{16}$	$\frac{3}{4}$	0.03	$1\frac{13}{16}$	$\frac{3}{4}$	0.03	$1\frac{13}{16}$	$\frac{11}{16}$	0.062
$1\frac{1}{4}$	$1\frac{7}{8}$	$\frac{27}{32}$	$1\frac{7}{8}$	$\frac{27}{32}$	0.03	2	$\frac{27}{32}$	0.03	2	$\frac{25}{32}$	0.062
$1\frac{3}{8}$	$2\frac{1}{16}$	$\frac{29}{32}$	$2\frac{1}{16}$	$\frac{29}{32}$	0.03	$2\frac{3}{16}$	$\frac{29}{32}$	0.03	$2\frac{3}{16}$	$\frac{27}{32}$	0.062
$1\frac{1}{2}$	$2\frac{1}{4}$	1	$2\frac{1}{4}$	1	0.03	$2\frac{3}{8}$	1	0.03	$2\frac{3}{8}$	$\frac{15}{16}$	0.062
Nominal Size, mm											
M5	8	3.58	8	3.58	0.2						
M6			10	4.38	0.3						
M8			13	5.68	0.4						
M10			16	6.85	0.4						
M12			18	7.95	0.6	21	7.95	0.6			
M14			21	9.25	0.6	24	9.25	0.6			
M16			24	10.75	0.6	27	10.75	0.6	27	10.75	0.6
M20			30	13.40	0.8	34	13.40	0.8	34	13.40	0.8
M24			36	15.90	0.8	41	15.90	0.8	41	15.90	1.0
M30			46	19.75	1.0	50	19.75	1.0	50	19.75	1.2
M36			55	23.55	1.0	60	23.55	1.0	60	23.55	1.5

Table E–30

Dimensions of Hexagonal Cap Screws and Heavy Hexagonal Screws (W = Width Across Flats; H = Height of Head; See Figure in Table E–29)

Nominal Size, in	Minimum Fillet Radius	Type of Screw: Cap W	Type of Screw: Heavy W	Height H
$\frac{1}{4}$	0.015	$\frac{7}{16}$		$\frac{5}{32}$
$\frac{5}{16}$	0.015	$\frac{1}{2}$		$\frac{13}{64}$
$\frac{3}{8}$	0.015	$\frac{9}{16}$		$\frac{15}{64}$
$\frac{7}{16}$	0.015	$\frac{5}{8}$		$\frac{9}{32}$
$\frac{1}{2}$	0.015	$\frac{3}{4}$	$\frac{7}{8}$	$\frac{5}{16}$
$\frac{5}{8}$	0.020	$\frac{15}{16}$	$1\frac{1}{16}$	$\frac{25}{64}$
$\frac{3}{4}$	0.020	$1\frac{1}{8}$	$1\frac{1}{4}$	$\frac{15}{32}$
$\frac{7}{8}$	0.040	$1\frac{5}{16}$	$1\frac{7}{16}$	$\frac{35}{64}$
1	0.060	$1\frac{1}{2}$	$1\frac{5}{8}$	$\frac{39}{64}$
$1\frac{1}{4}$	0.060	$1\frac{7}{8}$	2	$\frac{25}{32}$
$1\frac{3}{8}$	0.060	$2\frac{1}{16}$	$2\frac{3}{16}$	$\frac{27}{32}$
$1\frac{1}{2}$	0.060	$2\frac{1}{4}$	$2\frac{3}{8}$	$\frac{15}{16}$
Nominal Size, mm				
M5	0.2	8		3.65
M6	0.3	10		4.15
M8	0.4	13		5.50
M10	0.4	16		6.63
M12	0.6	18	21	7.76
M14	0.6	21	24	9.09
M16	0.6	24	27	10.32
M20	0.8	30	34	12.88
M24	0.8	36	41	15.44
M30	1.0	46	50	19.48
M36	1.0	55	60	23.38

Table E–31

Dimensions of Hexagonal Nuts

Nominal Size, in	Width W	Height H Regular Hexagonal	Thick or Slotted	JAM
$\frac{1}{4}$	$\frac{7}{16}$	$\frac{7}{32}$	$\frac{9}{32}$	$\frac{5}{32}$
$\frac{5}{16}$	$\frac{1}{2}$	$\frac{17}{64}$	$\frac{21}{64}$	$\frac{3}{16}$
$\frac{3}{8}$	$\frac{9}{16}$	$\frac{21}{64}$	$\frac{13}{32}$	$\frac{7}{32}$
$\frac{7}{16}$	$\frac{11}{16}$	$\frac{3}{8}$	$\frac{29}{64}$	$\frac{1}{4}$
$\frac{1}{2}$	$\frac{3}{4}$	$\frac{7}{16}$	$\frac{9}{16}$	$\frac{5}{16}$
$\frac{9}{16}$	$\frac{7}{8}$	$\frac{31}{64}$	$\frac{39}{64}$	$\frac{5}{16}$
$\frac{5}{8}$	$\frac{15}{16}$	$\frac{35}{64}$	$\frac{23}{32}$	$\frac{3}{8}$
$\frac{3}{4}$	$1\frac{1}{8}$	$\frac{41}{64}$	$\frac{13}{16}$	$\frac{27}{64}$
$\frac{7}{8}$	$1\frac{5}{16}$	$\frac{3}{4}$	$\frac{29}{32}$	$\frac{31}{64}$
1	$1\frac{1}{2}$	$\frac{55}{64}$	1	$\frac{35}{64}$
$1\frac{1}{8}$	$1\frac{11}{16}$	$\frac{31}{32}$	$1\frac{5}{32}$	$\frac{39}{64}$
$1\frac{1}{4}$	$1\frac{7}{8}$	$1\frac{1}{16}$	$1\frac{1}{4}$	$\frac{23}{32}$
$1\frac{3}{8}$	$2\frac{1}{16}$	$1\frac{11}{64}$	$1\frac{3}{8}$	$\frac{25}{32}$
$1\frac{1}{2}$	$2\frac{1}{4}$	$1\frac{9}{32}$	$1\frac{1}{2}$	$\frac{27}{32}$
Nominal Size, mm				
M5	8	4.7	5.1	2.7
M6	10	5.2	5.7	3.2
M8	13	6.8	7.5	4.0
M10	16	8.4	9.3	5.0
M12	18	10.8	12.0	6.0
M14	21	12.8	14.1	7.0
M16	24	14.8	16.4	8.0
M20	30	18.0	20.3	10.0
M24	36	21.5	23.9	12.0
M30	46	25.6	28.6	15.0
M36	55	31.0	34.7	18.0

Table E–32

Basic Dimensions of American Standard Plain Washers (All Dimensions in Inches)

Fastener Size	Washer Size	Diameter		Thickness
		ID	OD	
# 6	0.138	0.156	0.375	0.049
# 8	0.164	0.188	0.438	0.049
# 10	0.190	0.219	0.500	0.049
$\frac{3}{16}$	0.188	0.250	0.562	0.049
#12	0.216	0.250	0.562	0.065
$\frac{1}{4}$ N	0.250	0.281	0.625	0.065
$\frac{1}{4}$ W	0.250	0.312	0.734	0.065
$\frac{5}{16}$ N	0.312	0.344	0.688	0.065
$\frac{5}{16}$ W	0.312	0.375	0.875	0.083
$\frac{3}{8}$ N	0.375	0.406	0.812	0.065
$\frac{3}{8}$ W	0.375	0.438	1.000	0.083
$\frac{7}{16}$ N	0.438	0.469	0.922	0.065
$\frac{7}{16}$ W	0.438	0.500	1.250	0.083
$\frac{1}{2}$ N	0.500	0.531	1.062	0.095
$\frac{1}{2}$ W	0.500	0.562	1.375	0.109
$\frac{9}{16}$ N	0.562	0.594	1.156	0.095
$\frac{9}{16}$ W	0.562	0.625	1.469	0.109
$\frac{5}{8}$ N	0.625	0.656	1.312	0.095
$\frac{5}{8}$ W	0.625	0.688	1.750	0.134
$\frac{3}{4}$ N	0.750	0.812	1.469	0.134
$\frac{3}{4}$ W	0.750	0.812	2.000	0.148
$\frac{7}{8}$ N	0.875	0.938	1.750	0.134
$\frac{7}{8}$ W	0.875	0.938	2.250	0.165
1 N	1.000	1.062	2.000	0.134
1 W	1.000	1.062	2.500	0.165
$1\frac{1}{8}$ N	1.125	1.250	2.250	0.134
$1\frac{1}{8}$ W	1.125	1.250	2.750	0.165
$1\frac{1}{4}$ N	1.250	1.375	2.500	0.165
$1\frac{1}{4}$ W	1.250	1.375	3.000	0.165
$1\frac{3}{8}$ N	1.375	1.500	2.750	0.165
$1\frac{3}{8}$ W	1.375	1.500	3.250	0.180
$1\frac{1}{2}$ N	1.500	1.625	3.000	0.165
$1\frac{1}{2}$ W	1.500	1.625	3.500	0.180
$1\frac{5}{8}$	1.625	1.750	3.750	0.180
$1\frac{3}{4}$	1.750	1.875	4.000	0.180
$1\frac{7}{8}$	1.875	2.000	4.250	0.180
2	2.000	2.125	4.500	0.180
$2\frac{1}{4}$	2.250	2.375	4.750	0.220
$2\frac{1}{2}$	2.500	2.625	5.000	0.238
$2\frac{3}{4}$	2.750	2.875	5.250	0.259
3	3.000	3.125	5.500	0.284

N = narrow; W = wide; use W when not specified.

Table E–33

Dimensions of Metric Plain Washers (All Dimensions in Millimeters)

Washer Size*	Minimum ID	Maximum OD	Maximum Thickness	Washer Size*	Minimum ID	Maximum OD	Maximum Thickness
1.6 N	1.95	4.00	0.70	10 N	10.85	20.00	2.30
1.6 R	1.95	5.00	0.70	10 R	10.85	28.00	2.80
1.6 W	1.95	6.00	0.90	10 W	10.85	39.00	3.50
2 N	2.50	5.00	0.90	12 N	13.30	25.40	2.80
2 R	2.50	6.00	0.90	12 R	13.30	34.00	3.50
2 W	2.50	8.00	0.90	12 W	13.30	44.00	3.50
2.5 N	3.00	6.00	0.90	14 N	15.25	28.00	2.80
2.5 R	3.00	8.00	0.90	14 R	15.25	39.00	3.50
2.5 W	3.00	10.00	1.20	14 W	15.25	50.00	4.00
3 N	3.50	7.00	0.90	16 N	17.25	32.00	3.50
3 R	3.50	10.00	1.20	16 R	17.25	44.00	4.00
3 W	3.50	12.00	1.40	16 W	17.25	56.00	4.60
3.5 N	4.00	9.00	1.20	20 N	21.80	39.00	4.00
3.5 R	4.00	10.00	1.40	20 R	21.80	50.00	4.60
3.5 W	4.00	15.00	1.75	20 W	21.80	66.00	5.10
4 N	4.70	10.00	1.20	24 N	25.60	44.00	4.60
4 R	4.70	12.00	1.40	24 R	25.60	56.00	5.10
4 W	4.70	16.00	2.30	24 W	25.60	72.00	5.60
5 N	5.50	11.00	1.40	30 N	32.40	56.00	5.10
5 R	5.50	15.00	1.75	30 R	32.40	72.00	5.60
5 W	5.50	20.00	2.30	30 W	32.40	90.00	6.40
6 N	6.65	13.00	1.75	36 N	38.30	66.00	5.60
6 R	6.65	18.80	1.75	36 R	38.30	90.00	6.40
6 W	6.65	25.40	2.30	36 W	38.30	110.00	8.50
8 N	8.90	18.80	2.30				
8 R	8.90	25.40	2.30				
8 W	8.90	32.00	2.80				

N = narrow; R = regular; W = wide.

*Same as screw or bolt size.

Table E–34

Gamma Function*

Source: Reproduced from William H. Beyer (ed.), *Handbook of Tables for Probability and Statistics,* 2nd ed., The Chemical Rubber Company, Cleveland, 1966.

Values of $\Gamma(n) = \int_0^\infty e^{-x} x^{n-1} dx;\ \Gamma(n+1) = n\Gamma(n)$

n	Γ(n)	n	Γ(n)	n	Γ(n)	n	Γ(n)
1.00	1.000 00	1.25	.906 40	1.50	.886 23	1.75	.919 06
1.01	.994 33	1.26	.904 40	1.51	.886 59	1.76	.921 37
1.02	.988 84	1.27	.902 50	1.52	.887 04	1.77	.923 76
1.03	.983 55	1.28	.900 72	1.53	.887 57	1.78	.926 23
1.04	.978 44	1.29	.899 04	1.54	.888 18	1.79	.928 77
1.05	.973 50	1.30	.897 47	1.55	.888 87	1.80	.931 38
1.06	.968 74	1.31	.896 00	1.56	.889 64	1.81	.934 08
1.07	.964 15	1.32	.894 64	1.57	.890 49	1.82	.936 85
1.08	.959 73	1.33	.893 38	1.58	.891 42	1.83	.939 69
1.09	.955 46	1.34	.892 22	1.59	.892 43	1.84	.942 61
1.10	.951 35	1.35	.891 15	1.60	.893 52	1.85	.945 61
1.11	.947 39	1.36	.890 18	1.61	.894 68	1.86	.948 69
1.12	.943 59	1.37	.889 31	1.62	.895 92	1.87	.951 84
1.13	.939 93	1.38	.888 54	1.63	.897 24	1.88	.955 07
1.14	.936 42	1.39	.887 85	1.64	.898 64	1.89	.958 38
1.15	.933 04	1.40	.887 26	1.65	.900 12	1.90	.961 77
1.16	.929 80	1.41	.886 76	1.66	.901 67	1.91	.965 23
1.17	.936 70	1.42	.886 36	1.67	.903 30	1.92	.968 78
1.18	.923 73	1.43	.886 04	1.68	.905 00	1.93	.972 40
1.19	.920 88	1.44	.885 80	1.69	.906 78	1.94	.976 10
1.20	.918 17	1.45	.885 65	1.70	.908 64	1.95	.979 88
1.21	.915 58	1.46	.885 60	1.71	.910 57	1.96	.983 74
1.22	.913 11	1.47	.885 63	1.72	.912 58	1.97	.987 68
1.23	.910 75	1.48	.885 75	1.73	.914 66	1.98	.991 71
1.24	.908 52	1.49	.885 95	1.74	.916 83	1.99	.995 81
						2.00	1.000 00

*For large positive values of x, $\Gamma(x)$ approximates the asymptotic series

$$x^x e^{-x}\sqrt{\frac{2x}{x}}\left[1+\frac{1}{12x}+\frac{1}{288x^2}-\frac{139}{51\,840x^3}-\frac{571}{2\,488\,320x^4}+\cdots\right]$$

Table E–35

Values of Correlation Coefficient r

Source: Table reproduced with the permission of the authors and publisher from E. L. Crow, F. A. Davis, and M. W. Maxfield, *Statistical Manual* (New York: Dover Publications, 1960).

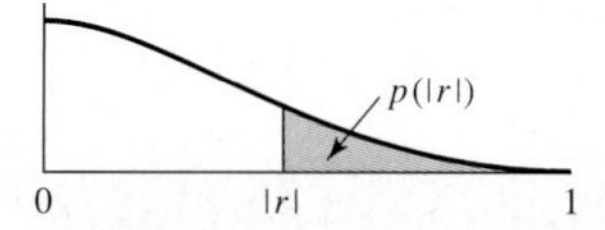

	5 Percent Level of Significance				1 Percent Level of Significance				
	Total Number of Variables				Total Number of Variables				
ν	2	3	4	5	2	3	4	5	ν
1	.997	.999	.999	.999	1.000	1.000	1.000	1.000	1
2	.950	.975	.983	.987	.990	.995	.997	.998	2
3	.878	.930	.950	.961	.959	.976	.983	.987	3
4	.811	.881	.912	.930	.917	.949	.962	.970	4
5	.754	.836	.874	.898	.874	.917	.937	.949	5
6	.707	.795	.839	.867	.834	.886	.911	.927	6
7	.666	.758	.807	.838	.798	.855	.885	.904	7
8	.632	.726	.777	.811	.765	.827	.860	.882	8
9	.602	.697	.750	.786	.735	.800	.836	.861	9
10	.576	.671	.726	.763	.708	.776	.814	.840	10
11	.553	.648	.703	.741	.684	.753	.793	.821	11
12	.532	.627	.683	.722	.661	.732	.773	.802	12
13	.514	.608	.664	.703	.641	.712	.755	.785	13
14	.497	.590	.646	.686	.623	.694	.737	.768	.14
15	.482	.574	.630	.670	.606	.677	.721	.752	15
16	.468	.559	.615	.655	.590	.662	.706	.738	16
17	.456	.545	.601	.641	.575	.647	.691	.724	17
18	.444	.532	.587	.628	.561	.633	.678	.710	18
19	.433	.520	.575	.615	.549	.620	.665	.698	19
20	.423	.509	.563	.604	.537	.608	.652	.685	20
21	.413	.498	.552	.592	.526	.596	.641	.674	21
22	.404	.488	.542	.582	.515	.585	.630	.663	22
23	.396	.479	.532	.572	.505	.574	.619	.652	23
24	.388	.470	.523	.562	.496	.565	.609	.642	24
25	.381	.462	.514	.553	.487	.555	.600	.633	25
26	.374	.454	.506	.545	.478	.546	.590	.624	26
27	.367	.446	.498	.536	.470	.538	.582	.615	27
28	.361	.439	.490	.529	.463	.530	.573	.606	28
29	.355	.432	.482	.521	.456	.522	.565	.598	29
30	.349	.426	.476	.514	.449	.514	.558	.591	30
35	.325	.397	.445	.482	.418	.481	.523	.556	35
40	.304	.373	.419	.455	.391	.454	.494	.526	40
45	.288	.353	.397	.432	.372	.430	.470	.501	.45
50	.273	.336	.379	.412	.354	.410	.449	.479	50

Table E–35

Values of Correlation Coefficient *r (Continued)*
Source: Table reproduced with the permission of the authors and publisher from E. L. Crow, F. A. Davis, and M. W. Maxfield, *Statistical Manual* (New York: Dover Publications, 1960).

	5 Percent Level of Significance				1 Percent Level of Significance				
	Total Number of Variables				Total Number of Variables				
ν	2	3	4	5	2	3	4	5	ν
60	.250	.308	.348	.380	.325	.377	.414	.442	60
70	.232	.286	.324	.354	.302	.351	.386	.413	70
80	.217	.269	.304	.332	.283	.330	.362	.389	80
90	.205	.254	.288	.315	.267	.312	.343	.368	90
100	.195	.241	.274	.300	.254	.297	.327	.351	100
125	.174	.216	.246	.269	.228	.266	.294	.316	125
150	.159	.198	.225	.247	.208	.244	.270	.290	150
200	.138	.172	.196	.215	.181	.212	.234	.253	200
300	.113	.141	.160	.176	.148	.174	.192	.208	300
400	.098	.122	.139	.153	.128	.151	.167	.180	400
500	.088	.109	.124	.137	.115	.135	.150	.162	500
1000	.062	.077	.088	.097	.081	.096	.106	.116	1000

The critical value of r at a given level of significance, total number of variables, and degrees of freedom ν, is read from the table. If the computed $|r|$ exceeds the critical value, then the null hypothesis that there is no association between the variables is rejected at the given level. The test is an equal-tails test, since we are usually interested in either positive or negative correlation. The shaded portion of the figure is the stipulated probability as a level of significance.

Table E–36

Distribution of Student's t
Source: This table is taken from Table III of R. A. Fisher and F. Yates, *Statistical Tables for Biological, Agricultural, and Medical Research*, published by Oliver & Boyd Ltd., Edinburgh, Scotland, by permission of the authors and publishers.

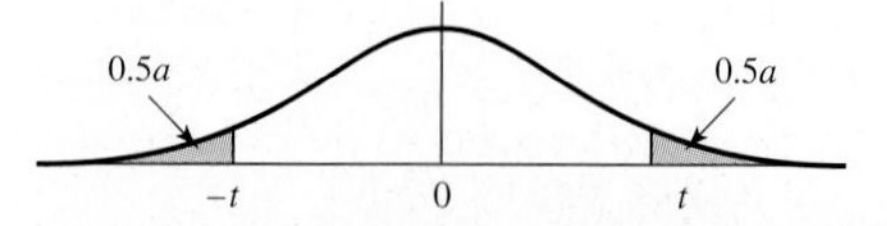

Degrees of freedom, ν	Probability, α 0.10	0.05	0.01	0.001
1	6.314	12.706	63.657	636.619
2	2.920	4.303	9.925	31.598
3	2.353	3.182	5.841	12.941
4	2.132	2.776	4.604	8.610
5	2.015	2.571	4.032	6.859
6	1.943	2.447	3.707	5.959
7	1.895	2.365	3.499	5.405
8	1.860	2.306	3.355	5.041
9	1.833	2.262	3.250	4.781
10	1.812	2.228	3.169	4.587
11	1.796	2.201	3.106	4.437
12	1.782	2.179	3.055	4.318
13	1.771	2.160	3.012	4.221
14	1.761	2.145	2.977	4.140
15	1.753	2.131	2.947	4.073
16	1.746	2.120	2.921	4.015
17	1.740	2.110	2.898	3.965
18	1.734	2.101	2.878	3.922
19	1.729	2.093	2.861	3.883
20	1.725	2.086	2.845	3.850
21	1.721	2.080	2.831	3.819
22	1.717	2.074	2.819	3.792
23	1.714	2.069	2.807	3.767
24	1.711	2.064	2.797	3.745
25	1.708	2.060	2.787	3.725
26	1.706	2.056	2.779	3.707
27	1.703	2.052	2.771	3.690
28	1.701	2.048	2.763	3.674
29	1.699	2.045	2.756	3.659
30	1.697	2.042	2.750	3.646
40	1.684	2.021	2.704	3.551
60	1.671	2.000	2.660	3.460
120	1.658	1.980	2.617	3.373
∞	1.645	1.960	2.576	3.291

This table gives the values of t corresponding to various values of the probability α (level of significance) of a random variable falling inside the shaded areas in the figure, for a given number of degrees of freedom ν available for the estimation of error. For a one-sided test the confidence limits are obtained for 2α.

Table E–37

Helical Spring Specification Checklists

COMPRESSION SPRING SPECIFICATION CHECKLIST
(Fill in required data only)

Material: ______

Working Conditions:

To work in ______ mm (in) diameter hole
To work over ______ mm (in) diameter shaft
Load ______ N (lbf), ± ______ N (lbf)
at ______ mm (in)
Load ______ N (lbf), ± ______ N (lbf)
at ______ mm (in)
Rate ______ N/mm (lbf/in), ± ______ N/mm (lbf/in)
between ______ mm (in) ______ and ______ mm (in)
Maximum solid height ______ mm (in)
Direction of coil (right-hand, left-hand or optional) ______
Type of ends ______
Allowable relaxation ______ % Hours/days ______
Impact loading ______ mm/s (in/s)

Frequency of loading ______ Hz
Required life ______ cycles
Required reliability ______

Special Information:

Squareness ______ Parallelism ______
Finish ______
Maximum operating temperature ______ °C(°F)
Operating environment ______
Electrical/magnetic ______

Design Data (Reference):

Wire diameter ______ mm (in)
Outside diameter ______ mm (in)
Inside diameter ______ mm (in)
Free length ______ mm (in)
Total number of coils ______

HELICAL EXTENSION SPRINGS SPECIFICATION CHECKLIST
(Fill in required data only)

Material ______

Working Conditions:

Maximum outside diameter ______ mm (in)
Initial tension ______ N (lbf)
Load ______ N (lbf), ± ______
at ______ length mm (in)
Load ______ N (lbf), ± ______
at ______ length mm (in)
Impact Loading ______ m/s (in/s)
Rate ______ N/mm (lbf/in)
Maximum extended length in service ______ mm (in)
______ mm (in) during installation
Direction of coil: right ______ left ______
optional ______
Type of ends ______
Position of ends and tolerance ______
Gap opening and tolerance ______ mm (in)

Suggested Design Data:

Wire diameter ______ mm (in)
Outside diameter ______ mm (in)
Total number of coils ______
Free length inside ends ______ mm (in)

Special Information:

Finish ______
Maximum operating temperature ______ °C(°F)
Operating environment ______
Frequency of Loading ______ hertz
Required life ______ cycles
Required Reliability ______

Table E–37

Helical Spring Specification Checklists *(Continued)*

HELICAL TORSION SPRING SPECIFICATION CHECKLIST

(Fill in required data only)

Material ____________________

Working Conditions:

To work in __________ mm (in) diameter hole
To work over __________ mm (in) diameter shaft
Torque ______ N·mm (lb-in), ± ______ N·mm (lb-in)
when angle between ends is ______ degrees
Torque ______ N·mm (lb-in), ± ______ N·mm (lb-in)
when angle between ends is ______ degrees
Axial space required __________ mm (in)
Direction of coil (right or left-hand) __________
Maximum wound position ______ revolutions or degrees

Reference Data:

Wire diameter __________ mm (in)
Mean diameter __________ mm (in)
Number of coils __________
Rate ______ N·mm/revolution (lb-in/revolution)

Special Information:

Finish __________
Loading (cyclic, impact, static, other) __________
Required life __________ cycles
Required reliability __________
Operating deflection range __________ revolutions
Maximum operating temperature __________ °C(°F)
Operating environment __________

Table E–38

Greek Alphabet

Alpha	A	α		Nu	N	ν	
Beta	B	β		Xi	Ξ	ξ	
Gamma	Γ	γ		Omicron	O	o	
Delta	Δ	δ	∂	Pi	Π	π	
Epsilon	E	ϵ		Rho	P	ρ	
Zeta	Z	ζ		Sigma	Σ	σ	ς
Eta	H	η		Tau	T	τ	
Theta	Θ	θ	ϑ	Upsilon	Υ	υ	
Iota	I	ι		Phi	Φ	ϕ	φ
Kappa	K	κ	$\varkappa$	Chi	X	χ	
Lambda	Λ	λ		Psi	Ψ	ψ	
Mu	M	μ		Omega	Ω	ω	

Solutions to Selected Problems

Appendix F

F–1 Chapter 1

1–10 A 4 in × 4 in × 160 in simply-supported fir beam can carry a 372 lb load anywhere along its span.

1–12 (a) $e_1 = 006\ 067\ 977$, $e_2 = 0.009\ 489\ 743$, $e = 0.015\ 557\ 720$
(b) $e_1 = -0.003\ 932\ 023$, $e_2 = 0.000\ 510\ 257$, $e = -0.004\ 442\ 280$

1–14 $e_1/X_1 = 0.002\ 988\ 342$, $e_2/X_2 = 0.001\ 185\ 438$, $(e_1/X_1)(e_2/X_2) = 0.000\ 003\ 542$

1–16 (a) 0, $0.01\underline{9}$

1–18 One digit is correct

1–33 $\theta_{crit} = \tan^{-1} f$

1–36 $F = 7 + 448y$, y in inches, F in lb

F–2 Chapter 2

2–1 $\bar{x} = 122.9$ kilocycles, $s = 30.3$ kilocycles

2–3 $f(x) = 1/[x(0.2423)\sqrt{2\pi}]\exp\{-[\ln x - 4.7779)/0.2423]^2/2\}$

2–5 $\bar{x} = 198.55$ kpsi, $s = 9.55$ kpsi

2–36 (a) $\bar{w} = 0.020$ in, $t_w = 0.015$ in, $w = \bar{w} \pm t_w = 0.020 \pm 0.015$ in
(b) $\bar{w} = 0.010 \pm 0.005$ in uniform

2–38 (a) $\bar{w} = 0.009 \pm 0.005$ in, $a = 1.051 \pm 0.001$ in
(b) $\bar{w} = 0.009$, $\sigma_w = 0.005/\sqrt{3}$ in, $a = 1.051$, $\sigma_a = 0.002\ 38$ in (uniform)

2–42 $\bar{D}_0 = 4.012$ in, $t_{D_0} = 0.036$ in; $D_0 = 4.012 \pm 0.036$ in

2–51 $\bar{x} = 98.26$ kpsi, $s = 4.30$ kpsi

2–59 $\mu_n = 122.6$ kcycles, $\sigma_n = 34.5$ kcycles

F–3 Chapter 3

3–1 (a) $\sigma_1 = 14$, $\sigma_2 = 4$, $\sigma_3 = 0$, $2\theta = 53.1°$ cw
(b) $\sigma_1 = 18.6$, $\sigma_2 = 6.4$, $\sigma_3 = 0$, $2\theta = 55°$ ccw
(c) $\sigma_1 = 26.2$, $\sigma_2 = 7.78$, $\sigma_3 = 0$, $2\theta = 139.7°$ ccw
(d) $\sigma_1 = 23.4$, $\sigma_2 = 4.57$, $\sigma_3 = 0$, $2\theta = 122°$ cw

3–6 $\sigma = 10.2$ kpsi, $\delta = 0.0245$ in, $\epsilon_1 = 0.000\ 340$, $\nu = 0.292$, $\epsilon_2 = -0.000\ 099$
$\Delta d = 0.5\epsilon_2 = -0.000\ 049$ in

3–13 (a) $M_{max} = 253$ in·lb, (b) $(a/\ell)^* = 0.207$, $M^* = 214$ in·lb

3–15 The same

3–19 From Ex. 3–9, $T = 29.94$ in lb. Two strips, same torque on each, $\theta = 0.192$ rad, $k_t = 155.9$ in lb/rad. For 1/8-in thick strip, $T = 59.9$ in·lb, $\theta = 0.0960$ rad, $k_t = 62.4$ in lb/rad

3–32 $M_{max} = 219$ in·lb, $\sigma = 17.8$ kpsi, $\tau_{max} = 3.4$ kpsi, both models

3–41 $p_i = 640$ psi

3–45 $\sigma_{r\,max} = 3656$ psi

3–58 $\sigma_i = 26\ 300$ psi, $\sigma_0 = -15\ 800$ psi

3–59 At AA, $\sigma_i = 42\ 100$ psi, $\sigma_0 = -25\ 200$ psi; at BB, $\sigma_i = 21\ 039$ psi, $\sigma_0 = -12\ 623$ psi

3–63 $\sigma_i = 17.1$ kpsi, $\sigma_0 = -5.7$ kpsi

3–67 $p_{max} = 399F^{1/3}$ MPa, $\sigma_1 = -0.206p_{max}$, $\sigma_2 = \sigma_3 = -p_{max}$, $\tau_{xy} = -0.104\ p_{max}$, $\tau_{oct} = 0.069\underline{3}p_{max}$

F–4 Chapter 4

4–1 (a) $k = (1/k_1 + 1/k_2 + 1/k_3)$,
(b) $k = k_1 + k_2 + k_3$, $\tau_{max} = 197.3$ N/mm^2, $z = 0.034$ mm

4–5 $\sigma_{max} = 20.4$ kpsi, $y = -0.908$ in

4–8 $y_{\text{left}} = -0.0508$, $y_{\text{right}} = -0.0508$, $y_{\text{midspan}} = 0.0191$ in

4–11 $y_{\max} = -0.0130$ in

4–17 $(a/\ell) = 0.293$ in

4–20 $d_L = 1.31$ in, $d_R = 1.35$ in, specify $d = 1\ 3/8$ in

4–21 $d = 36.4$ mm

4–32 $y_B = 0.0419$ in

4–33 $k = 10.60$ kN/m

4–37 $R_0 = 3.57$ kip, $R_c = 1.43$ kip, both in same direction

4–45 $\delta = 66.4F$ mm when F is in kN

4–52 $\sigma_c = -9218$, $\sigma_b = \pm 6409$ psi, $y = -0.877$ in

F–5 Chapter 5

5–1 $E = 30$ Mpsi, $S_y = 45.5$ kpsi, $S_{ut} = 85.5$ kpsi, area reduction = 45.8 percent

5–3 Perform regression ala Prob. 2–14; $E = 21(10^6)$ psi at origin, ${}_{0.001}S_y = 43.8$ kpsi

5–5 $\bar{m} = 0.735$, $C_m = 0.0741$, $\bar{\sigma}_o = 3.765$ kpsi, $C_{\sigma_o} = 0.0337$, $\bar{\sigma}_0 = 3.765\varepsilon^{0.735}$

5–7 (a) $|\varepsilon_i| = |\varepsilon_0| = 0.314$, (b) $S_u = 67.4$ kpsi, (c) $S_y = 66.3$ kpsi

5–8 Linear regression: $a = 0.274\ 106$, $b = -0.002\ 903$, $\bar{x} = 13$, $\bar{y} = 0.23\underline{6}$, $s_{y \cdot x} = 0.016\ 449$, $s_a = 0.007\ 528$, $s_b = 0.000\ 436$, $r = -0.912$, fit significant at 0.01 level

5–9 Plot suggests yes, $a = 1.266\ 364$, $b = -1.980\ 015(10^{-3})$, $\bar{x} = 399.2857$, $\bar{y} = 0.475\ 714$

5–14 No

F–6 Chapter 6

6–1 (a) Expect $S_{yt} = S_{yc}$, Fig. 6–27 use distortion energy theory, $\sigma' = 12.29$ kpsi, $n = 3.42$

6–2 (a) Use ductile Coulomb-Mohr; $\sigma_1 = 30$, $\sigma_2 = 0$, $\sigma_3 = -50$, $r = -0.56$, $n = 1.77$

6–4 (a) Use brittle maximum normal stress hypothesis; $n = S_{ut}/\sigma_A = 36/9 = 4.0$

6–5 (a) $\sigma_A = 20$ kpsi, $\sigma_B = 20$ kpsi, $r = 1$, $n = 1.5$

6–11 $\sigma_{t\ \max} = 13\ 212$ psi, $\sigma_\ell = 6418$ psi, $\sigma' = 9431$ psi, $n = 4.88$

6–14 Solve as a static problem using Mod. II-Mohr hypothesis

6–21 $\mathbf{M}_{\max} = 328\mathbf{N}(1, 0.1)$ in·lb, $\boldsymbol{\sigma}_x = 26.8\mathbf{N}(1, 0.1)$ kpsi, $z = -3.5$, $R = 0.9998$

6–26 (a) $\delta = 0.0005$ in, $p = 3517$ psi, $\sigma_{it} = -5864$ psi, $\sigma_{ir} = -3717$ psi, $\sigma_{ot} = -9140$ psi, $\sigma_{or} = -3517$ psi

F–7 Chapter 7

7–1 $S'_e = 33.5$ kpsi, $\sigma'_F = 116.2$ kpsi, $b = -0.0857$, $f = 0.915$, $a = 109.5$ kpsi, $S_f = 48.8$ kpsi, $N = 433\ 754$ cycles

7–3 $(S_f)_{ax} = 161.65N^{-0.0864}$ kpsi, $10^3 \le N \le 10^6$ cycles

7–5 $S_e = 94.1$ kpsi

7–7 $S_e = 288.5$ MPa

7–9 $S'_e = 222.6$ MPa, $k_a = 0.857$, $k_b = 1$, $k_c = 0.890$, $S_e = 175.7$ MPa, $K_t = 2.5$, $K_f = 2.09$, $F_a = 22.4$ kN, $F_y = 98.7$ kN

7–12 Rotation presumed. $\mathbf{S}'_e = 55.7\mathbf{LN}(1, 0.138)$ kpsi, $\mathbf{k}_a = 0.768\ \mathbf{LN}(1.\ 0.058)$, $k_b = 0.858$, $\mathbf{S}_e = 36.7\ \mathbf{LN}(1, 0.150)$ kpsi, $\mathbf{K}_f = 1.567\ \mathbf{LN}(1, 0.11)$, $\boldsymbol{\sigma} = 22.3\ \mathbf{LN}(1, 0.11)$ kpsi, $z = -1.163$, $R = 0.887$

7–21 At the fillet $n_f = 1.71$

7–31 $N_T = 10\ 770$ cycles

F–8 Chapter 8

8–1 (a) Thread depth 2.5 mm, thread width 2.5 mm, $d_m = 22.5$ mm, $d_r = 20$ mm, $\ell = p = 5$ mm

8–4 $T_{\text{raise}} = 16.23$ N·m, $T_{\text{lower}} = 6.62$ N·m, $e = 0.294$

8–8 $T = 16.5$ lb in, $d_m = 0.547$ in, $\ell = 0.16\underline{6}$ in, $\sec\alpha = 1.143$, Eq. (8–1) $T = 0.7421F$, $T_c = 0.0328F$, total torque $= 0.107F$, $F = 154$ lb

8–11 $L_T = 1.25$ in, $L = 1.109$ in, $H = 0.4375$ in, $L_G + H = 1.5465$ in, round to 1.75 in, $\ell_d = 0.500$ in, $\ell_T = 0.609$ in

8–13 $L_T = 1.25$ in, $L > h + 1.5d = 1.625 \rightarrow 1.75$ in, $\ell_d = 0.500$ in, $\ell_T = 0.625$ in

8–15 (a) $A_d = 0.442$ in^2, $A_{\text{tube}} = 0.552$ in^2, $k_b = 1.02(10^6)$ lb/in, $k_m = 1.27(10^6)$ lb/in, $C = 0.445$; (b) $F_i = 11\ 808$ lb

8–18 Frusta to Wileman, in ratio, is 1.11/1.08

8–27 $C = 0.238$

8–28 $C = 0.263$

8–39 (a) $L = 2.5$ in; (b) $k_b = 6.78$ Mlb/in, $k_m = 14.41$ Mlb/in, $C = 0.320$

8–55 Bearing on bolt, $n = 8.85$

F–9 Chapter 9

9–1 $F = 17.7$ kip

9–3 $F = 11.3$ kip

9–5 (a) $\tau' = 1.13F$ kpsi, $\tau''_x = \tau''_y = 5.93F$ kpsi, $\tau_{\max} = 9.22F$ kpsi, $F = 2.17$ kip; (b) $\tau_{\text{all}} = 11$ kpsi, $F_{\text{all}} = 1.19$ kip

9–9 $\tau' = 0$ (why?), $F = 49.2$ kN

9–10 A two-way tie for first, vertical parallel beads, and square beads

9–12 First; horizontal parallel beads, second, square beads.

9–13 Decisions: Pattern; all-around square
Electrode: E60XX
Type: two parallel fillets, two transverse fillets
Length of beads: 12 in
Leg: 1/4-in

9–24 $\tau_{\max} = 18$ kpsi

9–36 $\rho' = 9.48$ in concave

F–10 Chapter 10

10–3 (a) $L_0 = 5.17$; (b) $F_{Sy} = 45.2$ lb; (c) $k = 11.55$ lb/in; (d) $(L_0)_{cr} = 5.89$, guide spring

10–5 (a) $L_s = 17$ mm; (b) $p = 6.33$ mm; (c) $F_s = 97.3$ N; (d) $k = 2643$ N/m; (e) $L_0 = 53.8$ mm, $(L_0)_{cr} = 105.2$ mm, needs guidance

10–9 Not solid safe

10–15 Not solid safe

10–19 (a) 9.72 mm, $L_s = 42.5$ mm, $N_a = 11.5$; (b) $k = 1148$ N/m; (c) $F_i = 88.97$ N; (d) $\tau_s = 294$ MPa

10–29 (a) $k' = 24.69$ in·lb/turn each; (b) 307 kpsi

F–11 Chapter 11

11–1 $x_D = 540$, $F_D = 2.278$ kN, $C_{10} = 18.55$ kN, 02–30 mm deep-groove ball bearing, $R = 0.919$

11–4 $R = R_1 R_2 = 0.927(0.941) = 0.892$, goal not met

11–10 $x_D = 180$, $C_{10} = 57.0$ kN

11–13 $C_{10} = 8.88$ kN

F–12 Chapter 12

12–1 $c_{\min} = 0.000\,75$ in, $r = 0.500$ in, $r/c = 667$, $N_j = 18.3$ r/s, $\mathcal{S} = S = 0.261$, $h_0/c = 0.595$, $rf/c = 5.8$, $Q_s/Q = 0.5$, $h_0 = 0.000\,446$ in, $H = 0.0134$ Btu/s, $Q = 0.0273$ in^3/s, $Q_s = 0.0137$ in^3/s

12–3 $h_0 = 0.000\,275$ in, $p_{\max} = 848$ psi, $C_{\min} = 0.0025$ in

12–7 $h_0 = 0.022$ mm, $f = 0.006\,83$, $Q = 1055$ mm^3/s

12–9 $h_0 = 0.0099$ mm, $H = 34.3$ W, $Q = 1072$ mm^3/s, $Q_s = 973$ mm^3/s

12–18 Take bearing specification of Ex. 12–7, $D = 1.996$ in, $d = 0.0006$ in, $B = 2.000$ in, $b = 0.003$ in and construct the design window. $r_{\max} = 0.998$ in, $\Delta r = 0.0003$ in, $r_{\min} = 0.9977$ in, $\bar{r} = 0.997\,85$ in, $\Delta c = 0.0018$ in, $c_{\min} = 0.002$ in, $c_{\max} = 0.0038$ in, $\bar{c} = 0.0029$ in. Analyze for $W = 900$ lb, $n_d W = 1800$ lb $\bar{r} = 0.997\,85$ in, and three clearances, 0.002 in, 0.0029 in and 0.0038 in

12–21 From $B = 2.000$ in, $b = 0.003$ in, $D = 1.996$ in and $d = 0.0006$ in, find $r_{\max} = 0.998$ in, $c_{\min} = 0.002$ in, $c_{\max} = 0.0038$ in, $\Delta r = 0.0003$ in. Make $c_{\min} = 0.001$ in, and it follows that the corresponding bearing is $D = 1.996$ in, $d = 0.0006$ in, $B = 1.998$ in, $b = 0.0036$ in

12–22 Next available bushing $B = 1.875$ in, assume $b = 0.003$ in, $r_{\max} = 0.998$ in, $c_{\min} = 0.002$ in, $\Delta c = 0.0018$ in, $\Delta r = 0.0003$ in, move window to the left, b still 0.003 in, $\Delta c = 0.0018$ in $D = 1.871$ in, $d = 0.0006$ in, $r_{\max} = 0.9355$ in. The corresponding new bearing is $D = 1.871$ in, $d = 0.0006$ in, $B = 1.875$ in, and $b = 0.003$ in.

12–23 Bore $= 6.0045^{+0.0015}_{-0}$ in, journal $= 6.0005^{+0}_{-0.0005}$ in

12–38 15.2 mPa·s

12–40 $V = 0.5$ in, $L = 0.875$ in, ΔT tight constraint

F–13 Chapter 13

13–1 35 teeth, 3.25 in

13–2 400 r/min, $p = 3\pi$ mm, $C = 112.5$ mm

13–4 $a = 0.333$ in, $b = 0.417$ in, $c = 0.0837$ in, $p = 1.047$ in, $t = 0.523$ in, $d_1 = 7$ in, $d_{1b} = 6.578$ in, $d_2 = 9.333$ in, $d_{2b} = 8.77$ in, $p_b = 0.984$ in

13–5 $d_P = 5.333$ in, $\gamma = 23.63°$, $\Gamma = 66.37°$, $A_0 = 2.911$ in, $F = 0.970$ in

13–8 (a) 13, (b) 15, (c) 18, (d) 101

13–10 10:20 and higher

13–13 (a) $p_n = 3\pi$ mm, $p_t = 10.44$ mm, $p_x = 22.30$ mm (b) $m_t = 3.310$ mm, $\emptyset_t = 21.88°$ (c) $d_P = 59.58$ mm, $d_G = 105.92$ mm

13–15 $e = 4/51$, $n_d = 47.1$ r/min cw

13–21 $n_A = -68.57$ r/min cw

13–26 $n_9 = 11/36$ same sense

F–14 Chapter 14

14–1 $W^t = 1349$ lb, $H = 31.96$ hp (pinion bending) $W^t = 1729$ lb, $H = 41.0$ hp (gear bending) $W^t = 255$ lb, $H = 6$ hp (pinion and gear wear)

14–4 $W^t = 5869$ N, $H = 23.3$ kW (pinion bending) $W^t = 7698$ N, $H = 30.6$ kW (gear bending) $W^t = 1135$ N, $H = 4.5$ kW (pinion and gear wear)

14–7 $W^t = 16\,725$ N, $H = 96.3$ kW (pinion bending) $W^t = 3433$ N, $H = 19.75$ kW (pinion and gear wear)

14–9 $W^t = 726$ lb, $H = 17.2$ hp (pinion bending) $W^t = 284$ lb, $H = 6.7$ hp (pinion wear) AGMA method accounts for more conditions

14–11 Rating power = min(152.6, 190.5, 53.0, 59.0) = 53 hp

14–15 Rating power = min(270, 335, 240, 267) = 240 hp

14–11 Prob. 14–11 half-sized: power is 1/6.98 instead of 1/8, W^t is 1/3.49 instead of 1/4

F–15 Chapter 15

15–1 $W_P^t = W_1^t = 690$ lb, $H_1 = 16.4$ hp, $W_2^t = 618.2$ lb, $H_2 = 14.7$ hp

15–2 $W_3^t = 480$ lb, $H_3 = 11.4$ hp; $W_4^t = 548$ lb, $H_4 = 13.0$ hp

15–8 Pinion core 300 Bhn, case, 373 Bhn; gear core 339 Bhn, case, 345 Bhn

15–9 All four $W^t \doteq 690$ lb

15–11 Pinion core 180 Bhn, case, 205 Bhn; Gear core, 180 Bhn, case, 205 Bhn

F–16 Chapter 16

16–1 (a) Right shoe: $p_a = 111.6$ psi cw rotation (b) Right shoe: $T = 2530$ in·lb; left shoe: 13.0 in·lb; total $T = 3840$ in·lb (c) RH shoe: $R^x = -229$ lb, $R^y = 942$ lb, $R = 969$ lb LH shoe: $R^x = 130$ lb, $R^y = 171$ lb, $R = 215$ lb

16–3 LH shoe: $M_f = 95.5$ N·m, $p_a = 361$ kPa; RH shoe: $T = 78.8$ N·m, $p_a = 298$ kPa. $T_{\text{total}} = 174.3$ N·m

16–5 $p_a = 203$ kN, $T = 38.76$ N·m

16–8 $a' = 1.209r$, $a = 1.170r$

16–10 $P = 1038$ lb, $T = 13\,428$ in·lb

16–14 (a) $t = 8200$ in·lb, $P = 197$ lb, $H = 26$ hp; (b) $R = 901.5$ lb; (c) $p|_{\theta=0} = 27.3$ psi, $p|_{\theta=270°} = 70$ psi

16–17 (a) $F = 1885$ lb, $T = 7125$ in·lb; (c) torque capacity exhibits a stationary point maximum

16–18 (a) $d^* = D/\sqrt{3}$; (b) $d^* = 3.75$ in, $T^* = 7173$ in·lb (c) $(d/D)^* = 1/\sqrt{3} = 0.577$

16–19 (a) Uniform wear: $p_a = 82.3$ kPa, $F = 949$ N (b) Uniform pressure: $p_a = 79.1$ kPa, $F = 948$ N

16–21 (a) $\alpha = 1.432°$ for all Jarno tapers; (b) $F_{\text{assembly}} = 2262$ lb; (c) $T = 882$ in·lb

16–26 $C_s = 0.08$, $t = 1931/364$ in

16–29 (b) $I_e = I_M + I_P + n^2 I_P + I_1/n^2$ (c) $I_e = 10 + 1 + 10^2(1) + 100/10^2 = 10 + 1 + 100 + 1 = 112$

16–30 (c) $n^* = 2.43$, $m^* = 4.12$ which are independent of I_L

F–17 Chapter 17

17–1 (a) $T = 850$ in lb, $F_i = 169.5$ lb; (b) $b = 5.24$ in; (c) $F_{1a} = 293.4$ lb, $F_c = 22.3$ lb, $F_i = 147.6$ lb, $F_2 = 46.6$ lb, $H = 20.6$ hp, fos = 1.1

17–2 (a) $F_c = 0.913$ lb, $F_i = 88.4$ lb, $F_{1a} = 147$ lb, $F_2 = 31.6$ lb; (b) $H_a = 3.20$ hp, fos = 1.28; (c) 0.173 in

17–5 A–3 polyamide belt, $b = 7$ in, $F_c = 90.3$ lb, $T = 10\,946$ in·lb, $F_{1a} = 700$ lb, $F_2 = 243.9$ lb, $F_i = 381.7$ lb, dip = 0.461 in

17–8 $R^x = (F_1 + F_2)[1 - 0.5((D-d)/(2C))^2]$, $R^y = (F_1 - F_2)(D-d)/(2C)$ From Ex. 17–2, $R^y = 1214.4$ lb, $R^x = 34.6$ lb

17–15 With $d = 2$ in, $D = 4$ in, life of 10^6 passes, $b = 4.5$ in, fos = 1.1118

17–19 Select one B90 belt

17–21 Select eight D240 belts, life > 10^9 passes, life > 121261 h

17–26 (b) $n_1 = 1227$ r/min. Table 17–21 confirms this point occurs in the range 1200 ± 200 rpm.
(c) Eq. (17–40) applicable at speeds exceeding 1227 r/min for No. 60 chain

17–27 (a) $H_a = 7.91$ hp; (b) $C \doteq 18$ in;
(c) $T = 1163$ in·lb

17–29 Four-strand No. 60 chain, $N_1 = 17$ teeth, $N_2 = 84$ teeth, rounded $L/p = 134$, fos = 1.17, life 15 000 h (pre-extreme)

F–18 Chapter 18

18–1 (a) Maximum bending moment 2371 in·lb
(b) $d_A = 1.625$ in, $d_B = 1.870$ in, average diameter ≥ 1.870 in

18–2 (a) $C_\sigma = 0.149$, $C_{Se} = 0.150$, $C_n = 0.219$, $n_f = 2.19$; (b) $n_f = 2.93$; (c) $n = 3.11$, deflection likely constraint

18–19 From Table E–16, set $K_{fs}(32)T/(\pi A D^3) = K'_{ts}(32)T/(\pi D^3)$, from which $K'_{ts} = K_{ts}/A$. Tabulate a/D and K_{ts}. Note stationary point minimum at $a/D = 0.10$. Design rule: make pin 1/10 of shaft diameter. A similar result is obtainable for bending.

18–21 Slope: $\xi_m = \xi$, deflection $y_m = sy = y/2$, $M_m = s^3 M = M/8$ Force: $F_m = s^2 F = F/4$, same material, same stress levels

18–23 (a) $\omega = 868.1$ rad/s; (b) $d = 2$ in;
(c) $\omega = 1736$ rad/s (doubles)

18–25 (b) $\omega_2 = 567$ rad/s $= 5414$ r/min

Index

Page references in *italic* refer to information in the form of tables and/or figures; *n*. cites a footnote.

AA (Aluminum Association), 10
Absolute system of units, 32
Absolute viscosity, 742
Acme threads, 448
Action, conjugate, 837
Action, line of, 837
Addendum, 834
Adequacy assessment, 10, 20
  bolted tension joint, 492
  clutches and brakes, 1039–1040
  gear mesh, 940, 941
  helical compression spring, static service, 611, 612
  journal bearing, 776-777, 780
  roller-chain drive, 1086–1087, 1088
  rolling-contact bearings, 724
  shafts, distortion, 1133–1134
  shafts, strength, 1128–1129
  welded joint, 544
  wire rope, hoisting, 1097
Adhesive bonding, 562
  beam on an elastic foundation, 576
  cautions on adhesive use, 574
  design with adhesives, 573
  fracture mechanics, 572
  mechanisms of adhesion, 563
  references, 576
  residual stresses, 572
  shear lag, 567
  shear concentration, 571
  two basic stress distributions, 566
  types of adhesives, 564
AFBMA (Anti-Friction Bearing Manufacturers Association), 18
AGMA (American Gear Manufacturers Association), 18, 898, 952
AGMA approach, 848
AGMA bending strength, 909, 957
AGMA contact strength, 913, 957
AGMA equations factors:
  bevel gearing, 957–967
  spur and helical gearing, 915–929
  worm gearing, 974–978
AGMA equations, spur and helical gearing, 909
  bending strength, 911
  bending stress, 909
  contact strength, 913
  contact stress, 910
AGMA pitting strength, 913, 957
AGMA safety factors, spur and helical gearing, 929
AGMA symbols:
  bevel gearing, 955–956
  spur and helical gearing, 899–900
AISC (American Institute of Steel Construction), 18, 27, 28
AISC code for welds, *543*, 544
AISC specification for strength, 28
AISI (American Iron and Steel Institute), 18
AISI system of numbering, 272
Alignment limits in bearings, 732
Allowable stress, 24
Allowable stress numbers used by AGMA, 909
Allowable stresses:
  AISC values, 28
  in welds, *543*
Allowables, 28
Allowance, 72
Alloy steels, 279
  chromium, 279
  manganese, 280
  molybdenum, 280
  nickel, 279
  silicon, 280
  tungsten, 280
  vanadium, 280
Alternative:
  acceptable, 8–9
  feasible, 8–9
  satisfactory, 9
  suitable, 8–9
Aluminum, 283
  alloys of, *273*, 283
Aluminum Association (AA), 18
American Gear Manufacturers Association (AGMA), 18, 895, 952
American Institute of Steel Construction (AISC), 18
American Iron and Steel Institute (AISI), 18
American National Standards Institute (ANSI), 18
American Society of Mechanical Engineers (ASME), 18
American Society for Metals (ASM), 18
American Society of Testing and Materials (ASTM), 18
American Welding Society (AWS), 18
Angles:
  of action, 842
  of approach, 841
  helix, 852
  lead, 856
  of recess, 842
  of trist, 123
Angular alignment, 732
Angular deflection, 177
Angular gear, 843

Annealing, 277
ANSI (American National Standards Institute), 18
Anti-Friction Bearing Manufacturers Association (AFBMA), 18
Application factors:
  in gearing, 930, 957
  in rolling-contact bearings, *703, 708*
Approach action, 844
Arc:
  of action, 844
  of approach, 844
  of recess, 844
Area-moment method, 187
Areas of geometric shapes, *188*
ASA (American Standards Association), 18*n*.
Asbestos lining, *1031*
ASM (American Society for Metals), 18
ASME (American Society of Mechanical Engineers), 18
ASME elliptic failure locus, 1122–1123, *1125*
Assessment, adequacy, 10
Associations, 18
ASTM (American Society of Testing and Materials), 18
ASTM minimum strength, 27
ASTM numbering system for cast irons, 274
ASTM specification for bolts, 468
ASTM steels, *27*
ASTM system for steels, *273*
Asymmetrical beams, 118
Atkins, A. G., 292*n*.
Average life of bearings, 694
AWS (American Welding Society), 18
Austenite, 278
Automotive brake, 1017
Axial clutches, 1013
Axial pitch, 852
Axle, 1108

B10 life of bearings, 694
Babbitt alloys, *804*
Back cone, 851
Backlash, 834
Bagci failure locus, *1121, 1125*
Bainite, 278
Bairstow, I. L., 361*n*.
Ball bearing, 690–693
Band brakes, 1011
Bannantine, J. A., 364*n*.
Barson, J. M., 299*n*.
Barth equations, velocity factor, 902
Base circle, 838
Base pitch, 843
Base units, 31
  SI, 33
Basic load rating, 694
  catalog loading rating, 694
Basic size, 69
Bauschinger's theory, 361
Beams with asymmetrical sections, 118
Bearing alignment, 732
Bearing alloys, *804*
Bearing bore, 691, 783, 790
Bearing characteristic number, 745, 748, Case Study 12a
Bearing clearance, 781–787
Bearing dimensions, *701, 702*
Bearing enclosures, 729, 732
Bearing failure, 696–699
Bearing life, 694, *703*
Bearing materials, 808
Bearing mountings, 729–732
Bearing oilgrooves, *806*
Bearing performance, *781*
Bearing rating, 695
Bearing types, 691–692
Bearings:
  alignment limits in, 732
  average life of, 694
  clearance in, 781
  corner radius of, 701
  deep groove, 699
  diameter series, 701
  double row, 691
  duplex, 731
  filling notch, 691
  film formation in, 691
  film thickness in, 751
  fitted, 752
  friction in, 744–745
  grooves in, 806
  heating of, 772
  inner ring, 701
  L10 life of, 694–696
  life of, 693
  load center of, 715
  load ratings of, 701
  lubrication of, 728–729
  minimum life of, 693–694
  nomenclature of, 691, 752
  outer ring, 701
  partial, 752
  pressure in, 769
  pressure-fed, 792–798
  ratings of, 695
  reliability of, 725–726
  ring-oiled, 773
  roller, 691
  sealed, 691
  self-aligning, 115, 691
  separation of, 691
  shielding, 692
  shoulder sizes of, 701
  silver, *604*
  spherical, 692
  tapered roller, 692
  temperature in, 770–771
Belleville springs, 678
Belts:
  flat- and round-belt drives, 1053–1062
  flat-belt geometry, 1050–1053
  flat metal belts, 1065–1069
  summary, *1078*
  timing belts, 1077–1079
  types, 1050
Bending, normal stresses, 111
Bending of gear teeth, 898–901
Bending moment, 106–108
  sign of, 107
Bending strength in gears, charts, *911, 913*
Bending stress, 113–114
Bending in welded joints, 540–542
Bergsträsser factor, 591
Beryllium bronze, 285
Bevel gear forces, *864*
Bevel gearing:
  bending stress, 957
  contact stress, 954
  general, 952
  permissible bending stress, 957
  permissible contact stress, 957

Bevel gearing, variations of, 952
Bevel gears:
  geometry factors, 960
  straight, 850
  strength, 954, 957
  stresses, 954, 957
  symbols, *955–956*
Biaxial stress, 97
BIPM (International Bureau of Weights and Measures), 18
BIPM rules, 33
Blake, J. C., 472*n.*
Blob, 32
Blodgett, O. W., 546
Boller, C., 364*n.*
Bolt:
  drawing, *458*
  elongation, 471
  failure, photo, *316*
  length, 462
  preload, 479–480
  specifications for, 467–469
  stiffness, 459–462
  strength, 466
  stresses, 484–485
  torque factor, *473*
Bolt torque—bolt tension relation:
  deterministic, 471, 473
  stochastic, 473–476
Booser, E. R., 772*n.*
Boresi, A. P., 144*n.*
Boundary-lubricated bearings, 807
  design data, 816–817
  design procedure, 819–821
  oscillating journal motion, 815–816
  rotating-pin wear, 812–814
  sliding wear, 808–809
  temperature rise, 819–821
  thrust washers, 809–811
Boyd, John, 769*n.*
Brake analysis, general, 993–995
  band type, 1011–1012
  cone, 1022–1024
  disk, 1016–1022
  external contacting, 1008–1011
  internal expanding, 999–1005
  temperature rise, 1027–1030
Brandes, E. A., 292*n.*
Brass, 284–285
Breakeven point, 19
Bridgman, P. W., 259*n.*
Bridgman's correction, 259
British Standards Institute (BSI), 18
Brinell hardness, 268
Broghammer, E. I., 906*n.*
Buchingham wear load, 985–986
Buckingham, Earle, 424*n.*
Buckingham, Edgar, 740, 1015
Buckling, 102, 206–208
Butt and fillet welds, 530
  stress distribution, *532*
  throat shear model, 533–535

C clamp, 514
Caliper brake, 1016–1020
Cams, gear teeth as, 837
Carburizing, 278
Carlson, H., 609
Carnahan, B., 143*n.*
Cartwright, D. J., 292*n.*
Case hardening, 278
Case Study 1, 35
Case Study 2a, 48
Case Study 2b, 62
Case Study 2c, 76
Case Study 3a, 111
Case Study 3b, 131
Case Study 4, 203
Case Study 5, 271
Case Study 6, 350
Case Study 7, 433
Case Study 8, 512
Case Study 10a, 680
Case Study 10b, 682
Case Study 12a, 749
Case Study 12b, 785
Case Study 13a, 881
Castigliano Theorem, 195
Castleberry, G. A., 148*n.*
Casting materials:
  alloy cast iron, 282
  ductile and modular cast iron, 281
  gray cast iron, 281
  malleable cast iron, 282
  steel, cast, 282
  white cast iron, 282
CDF (cumulative distribution function), 53
Centroids:
  of areas, 188
  of bolt groups, 498
  of weld groups, 537, 541–542
Chains, roller, 1079–1088
Chariot, south-pointing, 881–883
Characteristic parameter of Weibull, 1158
Choudry, M., 484*n.*
Chromium, 279
Church, Irving P., 26
Circle, Mohr, 96, *97*
Circles:
  base, 838
  clearance, 840
  dedendum, 840
  pitch, 833
Circular pitch, 833
  normal, 852
  transverse, 852
Clamp, roundbar, 216
Claussen, G. E., 532*n.*
Clearance, 72, 834
  diametral, 72
  radial, 72
Clutch analysis:
  cone, 1022–1024
  frictional contact, axial, 1013–1016
  miscellaneous, 1032–1034
  temperature rise, 1017–1030
Codes and standards, 17
Coefficient(s):
  correlation, 116
  elastic (*see* Elastic coefficient)
  fatigue ductility, 362
  fatigue strength, 363
  of friction
    belts, *1059*
    chart, *767*
  of thermal expansion, 137, *138*
  of worm gearing, 871, 976
  of variation in fatigue, 371, 375, 378, 380, 383
  (*see also specific coefficients*)
Coffin, Jr., L. F., 363*n.*
Cold work, 265
  factor, 266
Cold-working processes, 276

Collins, J. A., 366*n*.
Column:
  critical load, 208
  ends, *208*
  stability, 207
  stress, 211
Combined loading modes, 408
Companion (subsidiary) distribution of a lognormal, 67–68, 1157
Competence, 12
Compression members
  critical load, 207
  eccentricially loaded, 210
  general, 204
  intermediate length, 210
  long length, 206
  short length (strut), 214
Computer consideration, shaft design, 1142–1146
Computer tools, 13
Conjugate action, 837
Contact ratio, 834
Contact stresses, 144
Copper alloys, 214, 215
Corner, J. J., 354*n*.
Corner radius of bearing, 701
Correlation coefficient, 1161
Corrosion-resistant steels, 280
Coulomb-Mohr hypothesis, 333
  modifications:
    mod. I-Mohr, 337
    mod. II-Mohr, 337, 338, *339*
Coupling equations, 64, 65
Crafts, W., 307*n*.
Critical speeds of shafts, 1135–1141
Cumulative damage, 414–417
Cumulative Distribution Function (CDF), 53
Curvature:
  of beams, 178
  of springs, 591–592
Curved beams, 138
  formulas, 140–142
Curved members, 200–202
Cycle counting in fatigue, 416
Cyclic properties, 164

Dandage, S., 762*n*.
Datsko, Joseph, 261*n*.
Datsko method, *304*
Decision-making:
  acceptability, 8, 9
  feasibility, 8, 9
  suitability, 8, 9
Decision set, 8
  bolted tension joint, 492
  boundary-lubricated bearing, 819–820
  design plane, 800
  flat-belt drive, 1062–1063
  helical compression spring, dynamic, 626, 629
  helical compression spring, static, 612
  helical torsion spring, 673
  pressure-fed journal bearing, 799
  self-contained full journal bearing, 791
  spur and helical gear mesh, 942, 943
  straight bevel-gear mesh, 972
  V-belt drive, 1077
  welded joint, 549
  wire hoisting rope, 1097
  worm-gear mesh, 980–981
Decisions:
  a priori, 10
  design, 10
Dedendum, 834
Deflection, bending, 178
  by adjusting tight constraints, 186
  by area-moment method, 187
  by integration, 180
  by singularity functions, 190
  by tabular worksheet, numerical, *182*
Deflection of curved members, 200
Deflection of viscous dampers, 220
Deformation, plastic, 261
Design algorithm, bolted tension joint, 493
Design considerations, 16
Design factor, 29
  in fatigue, 431–433
  in helical compression spring soliding, 618–620
  in shaft design, 1145
Design imperative, 5
Design of helical-coil extension springs, 637–639
  coil body, 640
  endhook bending, 641
  endhook torsion, 642
  free-length tolerances, 643
Design process elements, 14
Designer's fatigue diagram, 426
  brittle material, 429
  ductile material, 427
Designer's notebook, 13
Deviation, 68
  fundamental, 68
  lower, 68
  upper, 68
Diagram, stress-strain, 256
Diameter:
  major, 446
  minor, 446
Diameter, pitch, 833
Dillard, D. A., 562*n*.
Dimensional analysis, 1015, 1024
Distinctions between science and engineering, *8*
Distortion-energy theory, 326–329
Distributions, statistical:
  Gaussian, 1156
  lognormal, 1156–1157
  normal, 1156
  triangular, 88
  uniform, 1157
  Weibull, 1158–1159
Dolan, T. J., 906*n*.
Dowling, N. E., 397*n*.
Dreyfus, H. L., 12*n*.
Ductility, 266

Eccentric loading, 498
Effective diameter, 376
Efficiency, 876
  spur gearing, 876
  worm gearing, 870, 982
Elastic coefficient, spur and helical gearing, 908
Elastic limit, 256
Elastic strain range, 363
Electric motor, frame sizes, 892
Electrode properties, 543
Elliptic failure locus, 405, 1125
Elongation, 103, 177
End-condition constants, 208
Endurance limit, 369
Endurance limit modification factors, 374
Energy, strain, 193
Engineering societies, 18
Engineering stress, 258, 262
Engle, R. M., 299*n*.
EP lubricants, 807

Epicyclic gear trains, 858
Error, propagation of, 58, *59*
Euler column equation, 207
Eutectoid steel, 277
Extension spring, 593
Extrusion, 275

Face contact ratio, 915
Face width, 834, 837
Factor of safety, *25*
Factor of safety method, 23
Factor(s):
- Bergsträsser, 591
- design, 23, 29
- of safety, 29
  - for bolted joints, 478, 479, 487
  - for wire rope, *1091*, 1092, 1095
- service, 29
- shear-stress augmentation factor, 591
- strain energy, shear, 112
- Wahl, 415

Failure hypotheses, 322–333
- criticism, 234–242

Failure models:
- Philon, 342
- stochastic design factor, 342

Failures resulting from:
- static loading, 314
- variable loading, 360

Faires, V. M., 635*n*.
Farahmand, B., 300*n*.
Fasteners, threaded, 457
Fatigue ductility coefficient, 362
Fatigue ductility exponent, 363
Fatigue equation summary for steels, *388*, *390*
Fatigue failure loci, 398
- distortion-energy elliptic, 401
- Gerber parabolic, 399, 401
- modified Goodman, 398, 402
- Soderberg, 401
- superposed, 400, 1125

Fatigue ratio ϕ_b of Gough, 370
Fatigue ratio $\phi_{0.30}$ for some classes of metals, *371*
Fatigue strength, 372
Fatigue strength, wire rope, 1093
Fatigue strength coefficient, 363
Fatigue strength exponent, 363
Fatigue strength in metals, 360
Fatigue stress-concentration factor K_f, 383
Felbeck, D. K., 292*n*.
Field, J., 308*n*.
Figure of merit, 10
Filling notch bearing, 691
Film formation in bearings, 757
Film thickness in bearings, 752
Finite life in fatigue, 368
Firbank, T. C., 1053*n*.
Fit, classes, 70
Fits, press and shring, 135
Flat belts, 1050
- crossed, 1057
- equations, 1055–1056
- free body, 1054
- geometry, 1051
- initial tension maintenance, 1062
- life, metal belts, 1066
- power, 1057
- quarter twist, 1052
- uncrossed, 1057
- variable speed, 1052

Flexible shafts, 1097–1098
Flexure, curved members, 138
Flexure, normal stresses, 111
Fluctuating stress characterization, 396
Flywheels, 1034–1038
Foot-pound-second system (fps), 32
Force, definition, 105
Force analysis:
- bevel gearing, 863
- helical gearing, 866
- spur gearing, 860
- worm gearing, 869

Force vector, 104–105
Form factor, Lewis, 901
Forman, R. G., 299*n*.
Forys, E., 595*n*.
Fracture mechanics, 288
- crack growth, 297
- fatigue, 297
- FNK equation, 300
- fracture toughness, 289, 290
- life prediction, 299
- linear elastic, 288
- stress-intensity factor, 289
  - information, 292, *293–295*

Fragile technology, 12
Free-body diagram, 26, 105
Freche, J. C., 420*n*.
Friction materials, 1031–*1032*
Fuchs, H. O., 303*n*.
Fundamental deviation, 68

Gamma function, 1158
- table, 1223

Gasketed joints, 483
Gauge length, 256
Gear descriptors:
- addendum, 835
- backlash, 834
- bottom land, 834
- circular pitch, 833
- clearance, 834
- dedendum, 834
- module, 833
- pitch circle, 833
- pitch diameter, 833
- pitch, diametral, 834
- top land, 834
- tooth thickness, 833
- width of space, 833

Gear
- external, 843
- failure, photo, *319*
- internal, 843

Gear materials:
- bevel gears, 968, 969
- spur and bevel gears, 941

Gear ratios and numbers of teeth, 874
Gear teeth, forming of, 848
- by finishing, 850
- by hobbing, 849
- by milling, 848
- by shaping, 848

Gear trains, 856
Gearing:
- AGMA factors for, 915–929, 957–964
- allowable stresses for, 911, 913
- crossed helical, 975
- Fillet radius in, 834
- helical tooth proportions, 836
- Hertzian stresses in, 907
- nomenclature, 899–900, 955–956
- reliability factors for, 927–928, 963
- size factor for, 923, 959
- surface condition factor for, 922
- temperature factors for, 928, 963

Gears, families of:
- bevel, 832

Gears, families of: (*cont.*)
helical, 832
hypoid, 832
spur, 832
worm, 833, 855
Gears, helical:
herringbone, 852
parallel, 851
Generating line, 840
Gerber failure line, 400
for bolts, 489
illustrated, *1125*
Gravitational system, 31
Gravity, 32
Grease lubrication, 729
Green, I., 464*n*.
Gordon, S. A., 370*n*.
Grooves in bearings, 793, 806
Grossman, M. A., 308*n*.
Grover, H. J., 370*n*.
Guaranteed variate, 1158

Handrock, J. H., 364*n*.
Hagen-Poiseuille law, 743
Hand relations for worm and helical gears, 857
Hard-drawn wire, 599
Hardness, 268
Brinell, 268
Rockwell, 268
Hardness protocol, spur and helical gearing, 937–940
Haringx, J. A., 597*n*.
Haugen, E. B., 1214
Heating:
of bearings, 772–774, 816–818
of brakes, 1027–1030
Heat-treatment of steel, 277
annealing, 277
case-hardening, 278
quenching, 277
tempering, 278
Helical coil torsion springs, 664
bending stress, 669
describing end location, 664
dynamic strength, 668
static strength, 667
Helical compression spring shear stress components, 625
Helical compression springs, 590
Bergstrasser factor, 591
curvature correction factor, 592
dead turns corrections Q and Q', 595
deflection, 592
free length, 595
shear stress, 590
shear stress augmentation factor, 590
solid length, 595
spring rate, 592
stability, 596, 597
Wahl curvature factor, 591
Helical compression springs for static service, 609
Helical extension springs, 592
initial tension, 594
stress augmentation factor, 592, 593
Helical gear force, illustrated, *867*
Helical gearing, tooth proportions, 836
Helical gears, parallel, 851
Helical spring fatigue locus:
Gerber, 622
Wahl, 623
Zimmerli, 623
Helical springs, first critical frequency, 620–622
Helix angle, 456, 852
Herringbone gears, 852
Hertzian stresses, 144
Higdon, A., 102*n*.
Helicoid, 852
Hip prosthesis, 169–170
Herringbone gears, 852
Histogram, *50*
Hobbing, 849
Holmes, O. W., 940*n*.
Hopkins, R. B., 1119
Hooke's law, 103, 256
Horger, O. J., 623*n*.
Horsepower formula:
customary U. S., 124
SI power, 124
Hot-rolled shapes, 276
Hot rolling, 275
Hot-working processes, 295
Hydrodynamic lubrication, 274
Hydrostatic lubrication, 274
Hydrostatic stresses, 326
Hypotheses of failure, 322
distortion energy, 327
internal friction, 332
maximum normal stress, 323
maximum shear stress, 324
strain energy, 326
Hysteresis loops, 362

I. Mech. E. (Institution of Mechanical Engineers), 18
IFI (Industrial Fasteners Institute), 18
Impact, shock, 229
Impact properties, 270
Incomplete number, 51, 1153
Industrial Fasteners Institute (IFI), 18
Integration:
meaning, 180
numerical:
Simpson's rule, 182
trapezoidal rule, 181
Interference, 72, 343
general equation, 344
Interference in gears, 845
Interference of several kinds:
of gear teeth, 845–848, 854–855
general, 343–347
stresses, 135–137
types of, 63–68
Institution of Mechanical Engineers (I. Mech. E.), 18
International Bureau of Weights and Measures (BIPM), 18
International Standards Organization (ISO), 18
International tolerance grade (ITO), 68
Interpolation equation:
Raimondi and Boyd charts, 772
spur gear mesh reliability, 927–928
straight bevel-gear mesh reliability, 963–967
Investment casting, 275
Involute curve, generation of, 838, 839
Involute profile, 837
Involute profile properties, 838, 843
Irwin, G. R., 292*n*.
ISO (International Standards, Association), 18
IT numbers, 68
Ito, Y., 463*n*.
Izod test, 271

J:
in gearing, 916–919
in welds, 536–537

Jack, automotive, 249
Jackson, L. R., 370*n*
Jansen, J. K., 946*n*.
Joerres, R. E., 408*n*.
Johnson, J. B., column formula, 210
Joints:
  fastener stiffness, 458–461
  member stiffness, 461–466
  separation, 478–479
Joints loaded in shear:
  bolted, 560
  riveted, 558
Journal, 470
  in decision set, 791–792
  in design window protocol, 785–787
  diameter in clearance equations, 783
  in wear equations, 810–814
Journal bearing loads and materials, 803
  bearing types, 805
  characteristics of bearing metals, *804*
  range of unit loads, *804*
Journal bearings:
  design considerations, 757
  hydrodynamic theory, 752
  Petroff's equation, 744
  Raimondi and Boyd charts, 759
    interpolation equation, 772
  Reynold's equation, 756
  Sommerfeld number, 748
  stable lubrication, 750
  thick-film lubrication, 751
  Trumpler's criteria, 757
  types of lubrication, 741
  viscosity, 741
Juvinall, R. C., 397*n*.

Kamm, Lawrence J., 20*n*.
Karelitz, G. G., 772*n*.
Kearney, V. E., 299*n*.
Keith, T. G., 827*n*.
Keys, 506
Krause, D. E., 268*n*.
Kuguel, R., 376*n*.
Kurtz, H. J., 472*n*.

L10 life in bearings, 693–694
Land:
  bottom, 834
  top, 834
Lamont, J. L., 307*n*.
Landgraf, R. W., 81–82, 361*n*.
Langer line, 401–402
Lapedes, D. W., 29*n*.
Lead, 447
Lead angle, 856
Lemmon, D. C., 772*n*.
Length of springs, 594–596, 638
Lewis bending equation, 898
Lewis form factors, 901
Liability, 20
Life:
  in bearings, 693–694, 695, 698
  in springs, 633, 634, 644, 669
Life factor (stress cycle factor):
  in spur and helical gears, 926–927
  in straight bevel gears, 961–962
Limit:
  elastic, 256
  proportional, 256
Limits, 71
Line, pressure, 840
  generating, 840
Line of action, 837
Linear elastic fracture mechanics, 421
Linear regression, 55, 1161–1162
Lipp, Robert, 854*n*.
Lipson, C., 375*n*.
Little, R. E., 463
Liu, C. K., 148*n*.
Loading, suddenly applied, 230
Loading factor, 377
Logarithmic (true) strain, 259
Lognormal distribution, 54, 64–65, 1156–1157
Lognormal-lognormal interference, 64
Low-cycle fatigue, 368
Lower deviation, 68
Lubrication, 741
Luther, H. A., 143*n*.

M threads, 447
Machine screws, 458–459
McKee, S. A., 484*n*.
McKee, T. R., 484*n*.
Mabie, H. H., 706*n*.
Magnaflux inspection, 360
Magnesium, 283–284
Major diameter, 446
Malleable cast iron, 282
Manganese, 280
Manson, S. S., 420*n*.
Manson-Coffin relation, 363
Manual of Steel Construction, 27
Marin, Joseph, 374*n*.
Marin fatigue modification factors, 374
Marshek, K. M., 397*n*.
Mathematical models, 15
Maximum-shear-stress hypothesis, 324–326
Maximum-normal-stress hypothesis, 323–324
Mean design factor, 65
Mean value, A-3, 1154
Means of algebraic operations, table, *59*
Mechanical design, 7
Metric threads, 447
Merit, figure of, 10
Method, offset, 256
Miner, M. A., 416*n*.
Miner's rule, 416–421
Minimum film thickness variable, 764–765
Minimum life in bearings, 696
Minimum (ASTM) strength, definition, 27–28
Minimum tooth numbers for bevel gears, 836
Minor diameter, 446
Miscellaneous effects factor k_e, 381
Mischke, C. R., 8*n*.
Mitchiner, R. G., 906*n*.
Miter gears, 860
MJ threads, 447
Modified Goodman fatigue diagram, 398, 400
Mod. I-Mohr (M1M) hypothesis, 337–338
Mod. II-Mohr (M2M) hypothesis, 338–339
Module(s), 833–834
  standard values, 835
Modulus:
  of elasticity (definition), 103
  of rigidity, 103
  of rupture, 260
  of section, 114
Mohr circle, 96–97
  analysis, 98–100
  for triaxial stress, 101
Mohr hypothesis, 335

Molybdenum, 280
Moment:
external, 105
sign convention, 107
Moment diagram, 116
Moment load, 499, 1112
Monte Carlo simulation, 74
accuracy, 75, 1165
Moore, R. R., specimen, 367
Motor frame, 892
Mounting tapered roller bearings, direct or indirect mounting, 714, 718
Music wire, 598–600

Nachtgull, R. C., 420*n*.
Nagata, S., 463*n*.
National Bureau of Standards (NBS), 18*n*.
National Institute of Standards and Technology, (NIST), 18
Natural tolerance, 71
Necking, 259, 260
Needle bearings, 692
Neutral plane, 112
Newton, 32
Newton's law, 31, 32, 104, 754
Nickel, 279
Nitriding, 278
Noll, C. G., 375*n*.
Nominal size, 71
Normal distribution, 53–54
cumulative distribution, 1156
probability density function, 1156
table, 1187
Normal diametral pitch, 852
Normal stresses, 94
maximum, 95
Normalizing, 277
Norris, C. H., 532
Notation of journal bearings, 766
Notch sensitivity, 287
charts, 287–288
Notebook designers, 13
Nonferrous metals:
aluminum, 283
brass, 284
bronze, 285
copper-base alloys, 284
magnesium, 283
Normal circular pitch, 852
Normal-normal interference, 63
Norris, C. H., 532*n*.
Numbers, 1153–1154
errors in computation, 41
Number of teeth:
for avoiding interference, 848
for analog computation, 848–849
for minimum center-to-center distance, 874–878

O ring, 86, 461
Octahedral planes, 101
Octahedral stress, 101–102
normal, 102
shear, 102, 147
Offset method, 256
Offset yield strength, 259
Oil flow, 768, 795–796
Oil-tempered wire, 599
Oiliness, 807
Olsen, E. H., 102*n*.
Optimization strategy:
helical compression spring, dynamic service, 626, 632
helical compression spring, static service, *613*
Osgood, C. C., 463*n*.
Outer ring of bearing, 691
Over-yielding, 262

Palmgren, A., 416*n*.
Palmgren-Miner rule, 416
applied, 418
Parabolic curve for columns, 209–210
Parabolic relation in fatigue, 400–410
Parallel helical gears, 851
Paris, P. C., 292*n*.
Partial bearing, 752
Pearlite, 277
Permissible-stress method, 23
Peterson, R. E., 131*n*.
Petroff's equation, 744–746
Phosphor bronze, 285
Piecewise equations, 231
Pinion, 833
Pins, 504
loose pins, 511
Piotrowski, George, 538*n*.
Pitch, 446
axial, 852–853
base, 843
circular, 834, 841
diametral, 834, 839
normal, 852
preferred values, 835
of timing belts, 1079
transverse, 853
Pitch circle, 833
diameter, 833
Pitch point, 837
Pitting resistance equation (AGMA), 913, 957
Pitting resistance table:
spur, helical gears, 915, 916
straight bevel gears, 964–965
Plane stress, 95
Planet:
carrier, 858
gear, 858
Planetary gear trains, 858
Plastic deformation, 261
Plastic strain:
strain line, 363
strain range, 362
Plasticity, 261
Plastics, 285, *286*
Plasting, 382
Poisson ratio, 103
Pocius, A. V., 563*n*.
Pope, J. A., 370*n*.
Powder-metallurgy process, 275
Power equation:
customary U. S., 124
SI, 124
Power transmission of belts:
comparison table, 1078
of flat, 1057–1058
of metal, 1069
of V belts:
equation, 1074
table, *1070–1071*
Prediction equations, 181
Preferred sizes, 16, 1200
Preferred units, 31
Prefixes, use of, 34
Preload:
in bolts, 470, 477
recommendations for, 480
Presetting of springs, 596
Press, 514

Press-fit stresses, 135–137
Pressure:
  angle, 840
  line, 840
  point, 840
Pressure angle, 835–836
  normal, 852–853
  standard values, 835–836
  transverse, 852–853
  worm gearing, 836
Pressure in bearings, 769
Pressure-fed bearings, 792
  decision set, 799
  design plane, 800
  theory, 793–798
Pressure line, 845
Pressure vessels, 132–135
Pretension in springs, 594
Principal shear stresses, 101
Principal strains, table, *104*
Principal stresses, *104*
Probability density function (PDF), 53, 1156
Product design attributes, 4, 5
Profile, involute, 837
Proof load, 466
Proof stress, 259
Proof strength, 466
Propagation of error, 58–60
  in calculation, 41
  in distortion, 233–237
  in stresses, 149–154
  table, 59
Properties of:
  cold-worked metals, 303
  heat-treated metals, 307
Proportional limit, 257
Pulleys, 1051–1052
Pure tension, 102
PV values, tables *808*, *812*, *817*

Quadratic function, 182
Quenching, 277–278

Rack, 842
Rack cutter, 849
Radial factors X_1, Y_1, X_2, Y_2 in bearings:
  equation, 699
  table, *700*
Radial load, 961
Radius:
  of curvature, 112, 178
  of gyration, 207
Raimondi, A. A., 759*n*.
Ramp function, 108
Random variable:
  definition, 1154
  mathematical operations, 59, 1163
Range number, 73
Rat trap, 687, 688
Rating life, 693
Ratings of bearings, 694
  catalog rating, 695
Ratio, angular velocity, 839
Ratio, contact, 844
Ratio, slenderness, 207
Recess action, 842
Rectification of data strings, *78*
Reduction of area, 266
Reece, C. K., 946*n*.
Reemsnyder, H. H., 421*n*.
Regression, linear, 55
  variation in parameters, 78, 82
Regression, nonlinear, 81, 82
Reliability, 30
Reliability analysis, 343–345
Reliability of bearings, 696, 699
Reliability factors, 927, 928, 963–964
Reliability method, 30
Residual stresses, 381–382
Resistance welding, 557
Resonance of springs, 570
Retaining rings, 570
Reversing belt drives, 1051
Reynolds, Sir Osborne, 442, 753–754
Reynolds' assumptions, 454
Richardson's error estimate, 143
Right-hand rule, 457
Rigidity, assumption of, 176
Riley, W. F., 102*n*.
Rim clutches, 999
Ring-oiled bearings, 773
Rings:
  analysis of, 200–204
  stresses in, 135
Rings, rotation, 135
Rippel, H. C., 806*n*.
Roadmaps, 930, 931, 956, 968
Rockwell hardness, 768
Rolfe, S. T., 299*n*.
Roll threading, 277
Roller bearing:
  application factor, 708
  bearing life, 693
  combined radial and thrust loading, 699
  life adjustment factors, 712
  load-life-reliability tradeoff, 697
  load-life tradeoff, 694
  matters of fit, lubrication, mounting, and enclosure, 728, 729
  oscillatory motion, 709
  reliability-life tradeoff, 696
  selecting ball and cylindrical roller bearing, 709
  types, 690–693
Roller chain, 1079
  power ratings, 1083–1086
Rolovic, R. D., 321*n*.
Rooke, D. P., 292*n*.
Roman method, 23
Root diameter, *449*
Rope, wire, 1088–1097
Rotating beam fatigue specimen, 367
Rotational directions, helical gears, 887
Round belts, 1053, 1059
Roundbar clamp, 216
Rounded number, 51, 1154

SAE (Society of Automotive Engineers), 18
SAE specifications for bolts, *467*
SAE system, 272
Safety, 5, 16, 20, 29–30
Salakian, A. G., 532*n*.
Samonov, C., 591*n*.
Sample variance, 51, 1155
Sand casting, 274
Satisfactory alternative, 9
Saybolt viscosity, 743
Screw jack, 450
Screw threads, 446–450
  gauge size, 449
  terminology, 446
Screws, power, 450
  equations, 452, 453
Sealed bearing, 691
Sealing methods, 432
Seals, 732
Secant column equation, 211

Second moment of area:
 definition, 113
 polar, 123–124
 tables, *537*, *541–542*
 in welds, 535–536
Secondary shear, 535
Section modulus, 114
Seeger, T., 364*n*.
Seely, F. B., 144*n*.
Seireg, A. S., 762*n*.
Self-aligning bearings, 115, 691
Separator of a bearing, 691
Service factor, *29*
Set removal in springs, 596
Setscrews, 504
 holding power, 505
 seating torque, 505
 types, 504
Shaft coupling, 1032–1034
Shaft geometry, 1108–1111
Shaft materials, 1134–1135
Shafts, flexible, 1097–1098
Shafts, introduction, 1108–1111
 critical speeds, 1135–1141
 computer considerations, 1142–1146
 hollow, 1135
Shape parameter, Weibull, 78, 696, 1158
Shear:
 of bolts, 487–504
 in torsion, picture, *123*
Shear and moment in beams, 106
Shear energy, correction factors, 194
Shear energy hypothesis, 328
Shear force, 107
 diagram, 116
 relations in bending, 107
 sign of, 107
Shear joints:
 bolts or rivets carrying load, 503
 centroid of pattern, 498
 primary and secondary shear, 499
Shear loading of bolts, 458
Shear modulus of elasticity, 103
Shear stress, principal, 101
Shell molding, 274
Shielded bearing, 691
Shigley, J. E., 138*n*.
Shock, 229
Shoulder sizes in bearings, 701
Shrink-fit pressure, 137
Shrink fit stresses, 133
SI (International System of Units), 32
SI units, 33
 preferred, 35
Side leakage, 768
Sidebottom, D. M., 144*n*.
Sih, C. M., 292*n*.
Silicon, 280
Simulation, 61
Simpson's rule integration, 142, 143
Singularity functions, *108*
Size, basic, 68, 71
 nominal, 71
Size factor k_b, 376
Skill 1, 8, *9*
Skill 2, 8, *9*
Slenderness ratio, 207
Slope of beams, 179
Slope limits of bearing-supported shafts, 732
Slope of shaft, 242
Slug, 32
Smith, J. O., 144*n*., 408
S-N:
 diagram, 368
 line, 372–373
Snug-tight condition, 471, 516–517
Society of Automotive Engineers (SAE), 18
Solid-film lubrication, 741
Sommerfeld, number, 743, Case Study 12a
Sorem, J. R., 321*n*.
South-pointing chariot, 881–883
Space, width of, 833
Specification set, 8
 bolted tension joint, 492
 helical compression spring, dynamic service, 605
 helical compression spring, static service, 625
 helical torsion spring, 669
 roller-chain drive, 1086
 V-belt drive, 1077
 wire hoisting rope, 1096–1097
Speed fluctuation, coefficient of, 1036
Speed ratio, 858
Spherical bearings, 691
Spiral angle, 952
Spiral bevel gears, 832
Splines, 1109, 1110
Spot welding, 557
Spotts, M. F., 876*n*.
Spring life in dynamic service, 633–635
 cycles-strength tradeoff, *644*
 maximum allowable stresses, *643*
Spring rates, 176
Spring wire materials, 598
 coil diameter tolerances, *604*
 constants A and m, *600*
 free length tolerances, *603*
 load tolerances, *605*
 mechanical properties, *606*
 preferred wire diameters, *601*
 relative costs, *602*
 ultimate strength, 598
 usage table, *599*
 wire tolerances, *603*
Square threads, 448
Static shaft loading, 347
Statically indeterminate problems, 198
Statistical parameters, estimating, 50
Steady-state in self-contained bearings, 772
 clearance, 781
 computer considerations, 790–791
 design plane, 785
 design window, 784
 tolerances and clearance, 783
Steel numbering systems, 272
 AISI, 272
 SAE, 272
 unified, 272
Stephens, R. I., 303*n*.
Stiffness constant:
 bolt, 461
 members, 464–465
Stiles, W. B., 102*n*.
Stochastic design factor, 24, 64, 65
Stochastic design factor method, 24, 431–433, 1145–1146
Stochastic failure loci, 411
 ASME elliptic, 412
 Gerber, 411
 Smith-Dolan, 412
Stochastics of spring rate, 608–609
Straight bevel gear, 850, 952
Straight bevel gear analysis, 969
Strain, elastic, 103
Strain energy, 193

Strain life fatigue failure model, *366*
Strain strengthening coefficient, 261
Strength:
  compression, 260
  definition, 26, 314
  fracture, 257
  minimum, meaning of, 27
  notation, 27
  static, 316–319
  stochastic values, *1213*, *1214*
  tensile, 357
  ultimate, 257
  of wire, 598, 600
Strength, static, 256
Strength-hardness relation, 268
Strength of welded joints, 542
  welding code method, fatigue loading, 548
Stress, shear in beams, 118
Stress, shear in noncircular sections, 124, 125
Stress, thermal, 137
Stress, torsion, 123
Stress, triaxial, 100
Stress, true, 259
Stress, von Mises antecedents:
  octahedral, 101
  principal, 97
Stress components, 94
Stress concentration, 319
Stress-concentration factor:
  geometric, 130
  model summary, 374
  theoretical, 130
Stress concentration and notch sensitivity, 383
Stress-corrosion cracking, 303
Stress distribution in welded joints, 532
Stress intensity factor, 290
Stress-life relationship, 367
Stress margin, 63
Stress range, 362, 398
Stress-strain diagram, 257
Stress-strain relations, *104*
Stress-strength interference, 343–345
Stress symbols, 27, 94–95
Stresses, contact, 144
  Hertzian, 144
  Smith-Liu, 148
Stresses, uniformly distributed, 102
Stresses in cylinders, 132
  thick-walled, 133
  thin-walled, 134
Structural shapes, 162, *1174–1178*
Struts, 214–216
Suitability-feasibility-acceptability test, 8
Suddenly-applied loading, 230
Sun gear, 858
Surface condition factor for gearing, 922
Surface durability, spur and helical gearing, 907
Surface endurance strength, 423–426
Surface factor, 375
Surface fatigue strength, 423
Surface fatigue strength, AGMA, 913
Surface stress in gears, 910
Survival equation, 696
Symbols for gears, *899–900*, *955–956*
Symmetry in beams, 112
Synthesis, 14–15
Systems of units, 31–32
Systems, tooth, *835*

T section, 114
Tada, H., 292*n*.
Talent:
  in engineering, 7
  in science, 7
Taper pins, sizes, 507
Tapered roller bearings:
  catalog sheets, *716–717*
  form, 714
  load-life-reliability tradeoff, 718
  location of reactions, 715
  notation, 715
Tapers, 1024–1027
Tavernelli, J. F., 363*n*.
Teeth, number of:
  largest
    helical, 855
    spur, 897
  smallest
    helical, 854–855
    spur, 846–847
Temperature in bearings:
  boundary-lubricated, 816–818
  natural circulation, 770–771
  pressure-fed, 797–798
Temperature effects, 271
Temperature factor k_d, 379
Temperature limits for oil, 782
Trumpler criterion, 758
Temperature rise:
  boundary-lubricated bearing, 820
  natural circulation bearing, 771
  pressure-fed bearing, 797–798
Temperature stresses, 137–138
Tempering, 278
Tensile strength, 257
  histogram for bolts, 466
  of spring materials, 598, *600*
Tensile strength, correlation method, 370
  correlation equations, 371
Tension joints, external load, 470
Tension joints under dynamic loading, 484
  stress component equations, 485
Tension test, 256
Tension test specimen, *256*
Terminology:
  of bevel gears, 850–851
  of gear teeth, 834
  of helical gears, 851–854
  of worm gears, 855–856
Test, suitability-feasibility-acceptability, 8
Thermal expansion, coefficient of, *138*
Thermoplastics, 285–287
Thickness, tooth, 833
Thin-film lubrication, 751, 807
Thread terminology, 446–450
Thread torque, 450–453
Threaded fasteners, 457
Threads:
  metric coarse pitch, *448*
  metric fine pitch, *448*
  unified coarse pitch, *449*
  unified fine pitch, *449*
Throat area of weld-bead patterns, 537, 541–542
Thrust bearing, 691, 806–807, 809–811
Thrust collar, *453*
Thrust factor, 699, *700*
Timing belts, 1050, *1079*
Timoshenko, S., 124*n*.
Tipton, S. M., 321*n*.
Toggle press, *250*
Tolerance, 58
  bilateral, 71
  natural, 71

Tolerance (*cont.*)
- stacking of, 73
- statistical, 73
- unilateral, 73

Tolerance letter, 69
Tolerance systems, 73
Tolerances, 19
- of springs, *603*, *604*
- of steel wire, *603*

Tooth numbers, *848*
Tooth proportions, *835*, *836*
Tooth thickness, 833
Top land, 834
Torque factors for bolts, *473*
Torque transfer elements, 1109, 1110
Torque-twist diagram, 260
Torsional fatigue strength, 408
Torsional properties of fillet welds, 537
Torsional strength, allowable values, *643*, *644*
Torsional yield strength, 260
Tower, Beauchamo, 753*n*.
Tower's experiment, 753
Train value, 857
Trains, gear, 856
- epicyclic, 858
- planetary, 858

Transition temperature, ductile/brittle, 270
Transmitted load, 861
Transverse circular pitch, 852
Triaxial stress, 100
True strain, 259
True stress, 259
True stress-true strain diagram, 261
Tubes, thin-walled:
- closed, 127
- open, 128

Tungsten, 280
Twist angle, 123

U weld, *530*
Uicker, J. L., 919*n*
Ultimate strength, 257
Uncertainty, 22, 48
Undercutting, 846
Under-yielding, 262
Unified numbering system, 272
Unified threads, 447
Unit load, critical, 207
United States of America Standards Institute (USAS), 18*n*.
U.S. Customary units, 32
Units, 31, 32
Unstable equilibrium, 207
Unstable lubrication, *751*
Upper deviation, 69
USAS (United States of America Standards Institute), 18*n*.

V belts, 1069
- rating, *1071*
- sizes, *1070*

V-belt lengths, equations, 1070
Van Gerpen, H. W., 946*n*.
Variable(s), design, 10
Vanadium, 280
Variance (standard deviation squared), 1154
Variates:
- definition of, 1154
- designation of, 1154
- manipulation of, *59*

Variation, coefficient of, 1154
Viewpoint, 8
Virtual number of teeth, 854

Wahl factor, 591, 597*n*.
Wallin, A. W., 1053*n*.
Walton, C. F., 268*n*.
Washerface, *458*
Wear factor:
- Buckingham's, 424
- sliding, 809

Weibull, Waladdi, 1158–1159
Weibull distribution, 1158–1159
Weight, 32
Weld-metal properties, 513
Weld symbols, 528–529
Weld types, 529, 530
Welded joints:
- fatigue loading, 554
- static loading, 549

Welded joints in bending:
- primary shear, 540
- second moment of area, 540
- table of weld bead pattern properties, *541–542*

Welded joints in torsion:
- line bead model, 536–537
- primary shear, 535
- polar second area moment, 536
- secondary shear, 535
- table of weld bead pattern properties, *537*
- throat area, 535

Welding symbols, 528
Welds in fatigue, *546*, *547*, 548
Whole depth of gear teeth, 834
Width of face, spur gear, *836*
Width of space, 833
Wileman, J., 464*n*.
Wilkes, J. O., 143*n*.
Wire rope, 1088–1097
- sizes, 1089
- strength, 1092

Wire tolerances, *603*
Wolford, J. C., 622*n*.
Woodruff key, 507
- sizes, *508*, *509*

Working depth, spur gear, *836*
Worm-gear speed reducer, picture, 1109
Worm gearing:
- AGMA transmitted load, 975
- analysis, 973–980
- Buckingham transmitted load, 986
- design of, 980–985

Worm gears, 855
Worm gears, AGMA, 974
Wrenching methods, 471

Yield line, *400*
Yield point, 256
Yield strength, 256
Yield stress, 259
Young, W. C., 125*n*.

z-variable of the normal distribution, 64, *1187*
Zinc alloys, 284–285
Zimmerli, F. P., 623*n*.

Table 2–6

Means and Standard Deviations for Simple Algebraic Operations on Independent (Uncorrelated) Random Variables

Function	Mean	Standard Deviation
a	a	0
$\mathbf{x}$	μ_x	σ_x
$\mathbf{x}+a$	μ_x+a	σ_x
$a\mathbf{x}$	$a\mu_x$	$a\sigma_x$
$\mathbf{x}+\mathbf{y}$	$\mu_x+\mu_y$	$(\sigma_x^2+\sigma_y^2)^{1/2}$
$\mathbf{x}-\mathbf{y}$	$\mu_x-\mu_y$	$(\sigma_x^2+\sigma_y^2)^{1/2}$
$\mathbf{xy}$	$\mu_x\mu_y$	$\mu_{xy}\left(C_x^2+C_y^2+C_x^2C_y^2\right)^{1/2}$
$\mathbf{x}/\mathbf{y}$	μ_x/μ_y	$\mu_{x/y}\left[(C_x^2+C_y^2)/(1+C_y^2)\right]^{1/2}$
$\mathbf{x}^n$	$\mu_x^n\left[1+\frac{n(n-1)}{2}C_x^2\right]$	$\lvert n\rvert\mu_x^nC_x\left[1+\frac{(n-1)^2}{4}C_x^2\right]$
$1/\mathbf{x}$	$\frac{1}{\mu_x}(1+C_x^2)$	$\frac{C_x}{\mu_x}(1+C_x^2)$
$1/\mathbf{x}^2$	$\frac{1}{\mu_x^2}(1+3C_x^2)$	$\frac{2C_x}{\mu_x^2}\left(1+\frac{9}{4}C_x^2\right)$
$1/\mathbf{x}^3$	$\frac{1}{\mu_x^3}(1+6C_x^2)$	$\frac{3C_x}{\mu_x^3}(1+4C_x^2)$
$1/\mathbf{x}^4$	$\frac{1}{\mu_x^4}(1+10C_x^2)$	$\frac{4C_x}{\mu_x^4}\left(1+\frac{25}{4}C_x^2\right)$
$\sqrt{\mathbf{x}}$	$\sqrt{\mu_x}\left(1-\frac{1}{8}C_x^2\right)$	$\frac{\sqrt{\mu_x}}{2}C_x\left(1+\frac{1}{16}C_x^2\right)$
$\mathbf{x}^2$	$\mu_x^2(1+C_x^2)$	$2\mu_x^2C_x\left(1+\frac{1}{4}C_x^2\right)$
$\mathbf{x}^3$	$\mu_x^3(1+3C_x^2)$	$3\mu_x^3C_x(1+C_x^2)$
$\mathbf{x}^4$	$\mu_x^4(1+6C_x^2)$	$4\mu_x^4C_x\left(1+\frac{9}{4}C_x^2\right)$

Note: The coefficient of variation of variate $\mathbf{x}$ is $C_x=\sigma_x/\mu_x$. For small COVs their square is small compared to unity, so the first term in the powers of $\mathbf{x}$ expressions are excellent approximations, and these are used in the partial derivative estimation method of Sec. 2-5. For correlated products and quotients see Charles R. Mischke, *Mathematical Model Building*, 2nd rev. ed., Iowa State University Press, Ames, 1980, app. C.